Behavioural Neuroscience

INTERNATIONAL NINTH EDITION

S. Marc Breedlove
Michigan State University

Neil V. Watson
Simon Fraser University

This version of the text has been adapted and customized. Not for sale in the USA or Canada.

SINAUER ASSOCIATES

NEW YORK OXFORD
OXFORD UNIVERSITY PRESS

Oxford University Press is a department of the University of Oxford. It furthers the University's objective of excellence in research, scholarship, and education by publishing worldwide.

Sinauer Associates is an imprint of Oxford University Press.

Published in the United States of America by Oxford University Press
198 Madison Avenue, New York, NY 10016, United States of America

Oxford is a registered trade mark of Oxford University Press.

ISBN 9780197613351
Printing number: 9 8 7 6 5 4 3 2 1

Printed by LSC Communications, United States of America

For Stacey, Collin, and John
S.M.B.

For Scott and Sherry
N.V.W.

Brief Contents

Contents

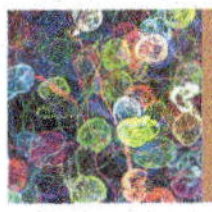

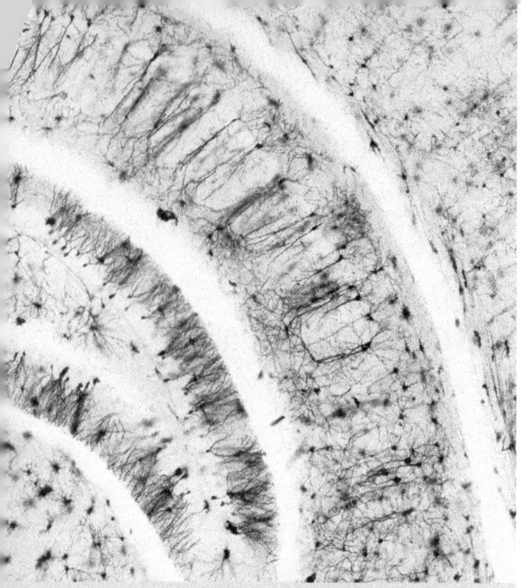

Courtesy of Dr. Sarah Moghadam, VA Palo Alto Health Care System, Palo Alto, CA and Dr. Ahmad Salehi, Dept. of Psychiatry & Behavioural Sciences, Stanford Medical School

Preface

Twenty-four years ago, a new kind of textbook was published for University courses that were often called "Brain and Behaviour." As the field evolved, the book's title metamorphosed from *Biological Psychology* to *Behavioural Neuroscience*, but the same drive to provide a definitive and comprehensive survey of the neurosciences lies at the heart of all our efforts. We strive to keep the book up-to-date while keeping a conversational tone to make this wealth of information not just accessible, but fascinating. The biggest change in this new edition is the development of **Learning Objectives** for each segment of the book, with the idea that telegraphing what's to come will focus readers' attention and facilitate learning. As you finish each section of text, it would be a good idea to go back and read the associated Learning Objectives to see whether in fact you incorporated the material. If not, a quick review of that text may be in order.

As always, there have been plenty of new findings to add to this edition. In fact, the problem we face is which of the many, many new findings to *leave out*—those that are not quite essential for a survey of the field. We are pretty picky about what we add, and still it seems like a deluge of new information and ideas. Hundreds of new papers are cited in this edition. If that sounds like a lot, let us give you a perspective on how many new papers were omitted. On our newsfeed site (www.biopsychology.com/news/), 1,299 new links were added in 2018 alone. Those are just the findings that were important enough to get the attention of mass media reporters. As we note in Chapter 0, over 40,000 new articles indexed under "neuroscience" appeared that year in PubMed. It would take a thick tome just to list the *titles* of the papers from 2018!

While being very, very selective in sampling this flood of findings, we have made substantial changes in every chapter. For example, in Chapter 2 we have a new figure comparing "kiss and run" synapses with more traditional models of synaptic transmission. Chapter 4 has new material about a hormone secreted from bone that acts on the hypothalamus to reduce appetite. We totally reorganised Chapter 6 for a more streamlined approach and discuss the growing doubts about whether amyloid deposits cause Alzheimer's. Chapter 9 needed a new figure comparing transduction in the five taste receptors. Chapter 14 talks about yet another factor affecting appetite, glucagon-like peptide 1. Chapter 12 now discusses the logic of pharmacogenomics to treat depression, while Chapter 17 was thoroughly reorganised and includes more about executive function. Honestly, we could go on like this for every chapter. Clearly this is an exciting era in the neurosciences. As Lewis Carrol put it, "We must run as fast as we can just to stay in place!"

We've also kept several very popular features from previous editions: **The Cutting Edge** appears in each chapter, where we explore some of the most exciting examples of recent research, and each chapter ends with a **Visual Summary**, where you can see graphic reminders as you review the principal findings that we just presented. These Visual Summaries really shine online, where with just a click you can review figures, animations, and quizzes to help integrate the material. We also continue to open each chapter with a gripping **vignette**, relating someone's real-life experiences that will be better understood as the content of the chapter unfolds, and we again replaced several of these vignettes as more recent events bring to the surface many of the important issues in behavioural neuroscience. Likewise we've retained the marginal glossary that makes it easy to find the definitions that unlock the material, as well as two features to let you burrow in on a particular subject: the online supplements called **A Step Further** cited throughout the text, and the **Recommended Reading** at the close of each chapter.

You might think that approaching the quarter-century mark we'd be jaded about improving and revising our presentations, but we still love it, perhaps because the dynamic and exciting pace of neuroscience research shows no sign of abating soon. As always, we welcome all feedback, praise or criticism, cuts or additions, from our readers. You can email us directly at behavneuro@gmail.com.

Acknowledgments

We continue to feel so lucky to work with the inestimable team at Sinauer Associates, now a part of Oxford University Press, whose deep skills and generous guidance transform our hundreds of files, thousands of email attachments, and sometimes scrambled emails into yet another beautiful book. Again, we feel so grateful to benefit from the experience and exquisite taste of others. In particular, the book could not exist without the contributions of Senior Acquisition Editor Syd Carroll, Production Editor Alison Hornbeck, Production Manager Joan Gemme, Book Designer and Production Specialist Annette Rapier, and Media and Supplements Editor Zan Carter and her crew. We also fondly bid adieu to the recently retired Chris Small, Production Manager for all our previous editions. We hope you're enjoying yourself, Chris, but how could you abandon us!? A cadre of commandos delved deep in the archives to deal with copyrights and permissions, so we salute you Michele Beckta, Mark Siddall, and Tracy Marton. We'd also like to thank our copy editor Lou Doucette, and our longtime art studio, Dragonfly Media, who bring amazing skill and commitment to make us look good.

We must also thank the founder of Sinauer Associates, Andy Sinauer, for his unwavering support over the years, with a touch of sadness upon his retirement. We are so proud to be a part of Andy's tremendous legacy, begun all those years ago with *From Neuron to Brain*, creating gorgeous books that make even the most complex topics accessible and enjoyable.

By this point in the evolution of the book, we have benefitted from the wisdom and advice of hundreds of colleagues who have generously served as reviewers of past editions. Although we don't have the space to list them all, we want to acknowledge that in many ways the book you are holding is the product of a whole community of neuroscientists. In this, the Ninth Edition, the following colleagues have provided invaluable critique and commentary:

Susan Bachus, *University of Maryland, Baltimore County*
Susan Barron, *University of Kentucky*
Christopher Beeman, *Central Washington University*
Jin Bo, *Eastern Michigan University*
David Brodbeck, *Algoma University*
Elizabeth Caldwell, *University of New Hampshire*
James Cherry, *Boston University*
Michael Cohen, *Loyola University Chicago*
Paul J. Currie, *Reed College*
Patrick Cushen, *Murray State University*
Deana Davalos, *Colourado State University*
Darragh P. Devine, *University of Florida*
Christopher W. Drapeau, *Valparaiso University*
Kelli A. Duncan, *Vassar College*
Raymond H. Dye, Jr., *Loyola University Chicago*
Taffeta Elliott, *New Mexico Institute of Mining and Technology*
Alison A. Fedio, *Argosy University, Northern Virginia*
Sara B. Festini, *University of Tampa*

Cynthia Michelle Finley, *College of Marin*
Jonathan Franz, *SUNY Empire State College*
Koren Ganas, *University of Illinois*
Sophie George, *Dixie State University*
Aaron Godlaski, *Centre College*
Brian J. Hock, *Austin Peay State University*
Jennifer Ingemi, *Northeastern University*
Mary Ellen Kelly, *Haverford College*
Susan Kennedy, *Denison University*
Michael Kerchner, *Washington College*
Sarita Lagalwar, *Skidmore College*
Stephen Lippi, *George Mason University*
Mario L. Mata, *California State University, Los Angeles*
Alexandra Roach, *University of South Carolina, Aiken*
Russell Romeo, *Barnard College of Columbia University*
Timothy Roth, *Franklin and Marshall College*
Emma Sarro, *Dominican College*
Peter A. Serrano, *Hunter College, City University of New York*
Fredric Shaffer, *Truman State University*
KatieAnn Skogsberg, *Centre College*
Lucy J. Troup, *University of the West of Scotland*
Adriana Uruena-Agnes, *St. Petersburg College*
Jennifer Wilhelm, *College of Charleston*
Jan R. Wessel, *University of Iowa*
Susan Zup, *University of Massachusetts Boston*

Finally, we thank all those tireless colleagues trying to understand the neural basis of Behaviour, with techniques that would have seemed like sorcery only a few years ago, and who share their hard-won findings with us all.

S. Marc Breedlove

Neil V. Watson

Media and Supplements

to accompany *Behavioural Neuroscience*, Ninth Edition

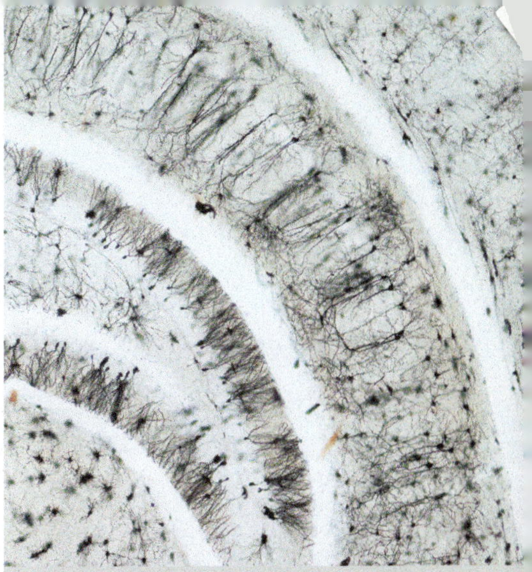

Courtesy of Dr. Sarah Moghadam, VA Palo Alto Health Care System, Palo Alto, CA and Dr. Ahmad Salehi, Dept. of Psychiatry & Behavioural Sciences, Stanford Medical School

For the Student

BioPsychology NewsLink (bn9e.com/news)

This invaluable online resource helps students make connections between the science of behavioural neuroscience and their daily lives and keeps them appraised of the latest developments in the field. The site includes links to thousands of news stories, all organised both by keyword and by textbook chapter. The site is updated 3–4 times per week, so it includes up-to-the-minute information. NewsLink updates are also available on Facebook (facebook.com/Behaviouralneuroscience).

For the Instructor

learninglink.oup.com/Oxford Learning Link

The Oxford Learning Link provides instructors using *Behavioural Neuroscience* 9e with a wide variety of resources to aid in course planning, lecture development, and student assessment. Content includes:

- *Figures & Tables*: All the figures, photos, and tables from the textbook are provided as JPEGs, all optimised for use in presentations.
- *PowerPoint Presentations*: Two PowerPoint presentations are provided for each chapter of the textbook:
  - *Figures*: All the chapter's figures, photos, and tables, with titles and complete captions
  - *Lectures*: Complete lecture outlines, including selected figures
- *Instructor's Manual*: The Instructor's Manual includes useful resources for planning your course, lectures, and exams. For each chapter of the textbook, the IM includes a chapter overview, a chapter outline, the chapter's key concepts, additional references for course and lecture development, and a list of the chapter's key terms.
- *Videos*: A robust collection of video segments from the BBC and other sources bring to life many important concepts discussed in the textbook. These videos can be used as excellent lecture-starters and/or discussion topics.

- *Animations*: These detailed animations from the Companion Website help enliven lectures and illustrate dynamic processes.
- *Animation Quizzes*: These quizzes test the student's understanding of the topic (NEW for this edition).
- *Chapter Quizzes*: Quiz questions for each chapter in two formats: Available in Blackboard, Canvas, D2L platform, or as MS Word files.
  - Multiple choice tests student comprehension of the material covered in each chapter.
  - Essays challenge students to synthesise and apply what they have learned.
- *Test Bank*: The Test Bank consists of a broad range of questions covering key facts and concepts in each chapter. Multiple choice, essay, and paragraph development questions are included. Questions are ranked according to Bloom's Taxonomy and referenced to specific textbook sections. NEW for this edition, questions are also aligned to the textbook Learning Objectives. (Available in Blackboard, Canvas, D2L platform, or as MS Word files.)

Introduction

Scope and Outlook

0

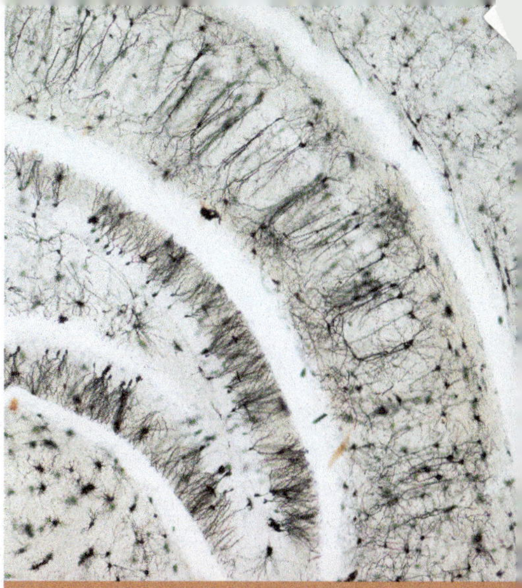

Courtesy of Dr. Sarah Moghadam, VA Palo Alto Health Care System, Palo Alto, CA and Dr. Ahmad Salehi, Dept. of Psychiatry & Behavioural Sciences, Stanford Medical School

Machine or Human?

In the near future depicted in the HBO series *Westworld*, people visit a theme park set in the Old West, with steam locomotives, saloons, and brothels, populated with androids, called "hosts," to entertain humans. The mechanical hosts provide their guests with anything, from casual banter to gunfights, harmless flirting to kinky sex, the only restriction being that the robots are never to harm the humans. The android hosts are so lifelike in appearance and behaviour that visitors may have a hard time distinguishing whether someone is a fellow guest or a robot. To make the androids' simulation of humans complete, they are given backstories, false memories of a life before their appearance for each new batch of guests. Importantly, none of the androids know that they are mechanical beings rather than humans. It's probably not much of a spoiler to say that several plot lines in the series hinge on androids slowly discovering their true nature, moving from shock and shame that they are mere machines, to openly rebelling from the notion that they are to be used, and abused, as mere playthings for the humans.

We aren't told too much about how the android "brains" in *Westworld* work, because, of course, such technology remains far outside our grasp, so the writers, reduced to mere speculation, remain rather vague. But apparently the knowledge and personality for any particular android lies in a "control unit," a golf-ball-size device that can be extracted from the head of one host and implanted into the head of another, interchangeable body. Presumably, if we had enough knowledge and surgical skill, we could remove your brain from your head and connect it up to the head of some other body. Would you still be you? Even if we put your brain into a body of the opposite sex? Come to think of it, are you entirely sure there is a brain in your head, and not one of those control units?

Our aim in this book is to help you learn what is known so far about how brains work, and about how much more we have yet to learn. We will explore the many ways in which the structures and actions of the brain produce mind and behaviour. But that is only half of our task. We are also interested in the ways in which behaviour and experience modify the structures and actions of the brain. One of the most important lessons we want to convey is that interactions between brain and behaviour are reciprocal. The brain controls behaviour and, in turn, behaviour and experience alter the brain.

We hope to give an interesting account of the main ideas and research in behavioural neuroscience, which is of great popular as well as scientific interest. Most important, we try to communicate our own interest and excitement about the mysteries of mind and body.

0.1 The Brain Is Full of Surprises

Learning Objectives

After reading this section, you should be able to:

0.1.1 **Name the main type of cells found in the brain, and name the connections between them.**

0.1.2 **List the names of some of the many fields of study related to behavioural neuroscience.**

0.1.3 **Describe five different perspectives taken in understanding the biology of behaviour.**

> *I used to think that the brain was the most wonderful organ in my body. Then I realized who was telling me this.*
>
> —Emo Philips
> (American comedian)

Of course we should always consider the source when evaluating an idea, but even so, the brain indeed seems like a pretty wonderful organ. For one thing, brains produced the entire extent of human knowledge, everything we understand about the universe, however limited that may be. Brains also produced every written description of that hard-won knowledge (including this book you hold in your hands), as well as every work of visual art, from doodles to the sweeping frescos on the ceiling of the Sistine Chapel.

Most of us have a hard time grasping the idea of a billion of anything, but your head contains an estimated 86 billion nerve cells, or **neurons** (from the Greek word for "nerve" or "cord") (Herculano-Houzel, 2012). Each neuron contacts many other cells at points called *synapses*, so there are trillions of those between your ears. A specialized extension of neurons, called an *axon*, is microscopically slender, yet it may be several feet long. We'll learn that axons produce electrical impulses that travel hundreds of miles per hour. **FIGURE 0.1** offers a list of just a few of the things we will

neuron Also called *nerve cell*. The basic unit of the nervous system.

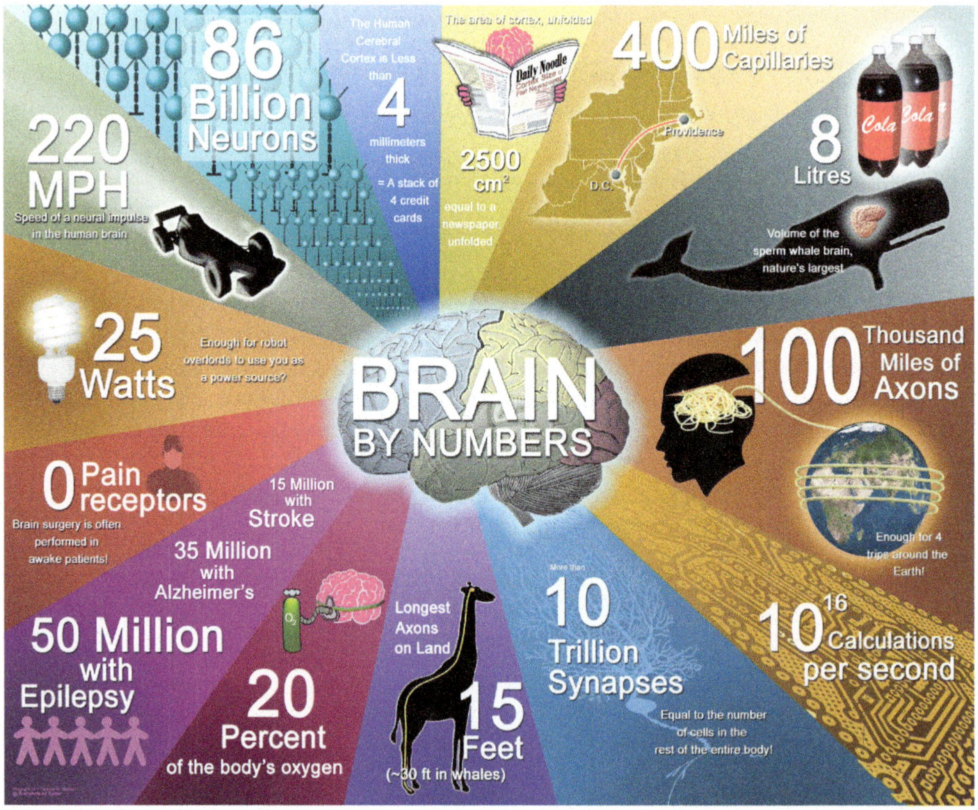

0.1 Your Brain by the Numbers The cerebral cortex is the outermost portion of the brain.

learn about the human brain in the course of this book. All this hardware isn't just for show—it allows you to take in all the information in that figure in less than a minute.

neuroscience The study of the nervous system.

behavioural neuroscience Also called *biological psychology*. The study of the neural bases of behaviour and mental processes.

What is behavioural neuroscience?

No treaty or trade union agreement defines the boundaries of behavioural neuroscience. The first people to study the relationships between brain and behaviour regarded themselves as philosophers, and their findings contributed to the births of biology and psychology. Those disciplines merged in the twentieth century to form *biological psychology*, the field that relates behaviour to bodily processes. With the modern explosion of **neuroscience**, the study of the brain, this research has evolved to the point that **behavioural neuroscience** offers a more accurate description. Whichever name is used, the main goal of this field is to understand the neuroscience underlying behaviour and experience.

Behavioural neuroscience is a field that includes many players who come from quite different backgrounds: psychologists, biologists, physiologists, engineers, neurologists, psychiatrists, and many others. Thus, there are many career opportunities, in both universities and private industry, for people with interests in this field (Hitt, 2007). **FIGURE 0.2** maps the relations of behavioural neuroscience to these many other disciplines. Clearly, the behavioural neuroscience umbrella opens very wide.

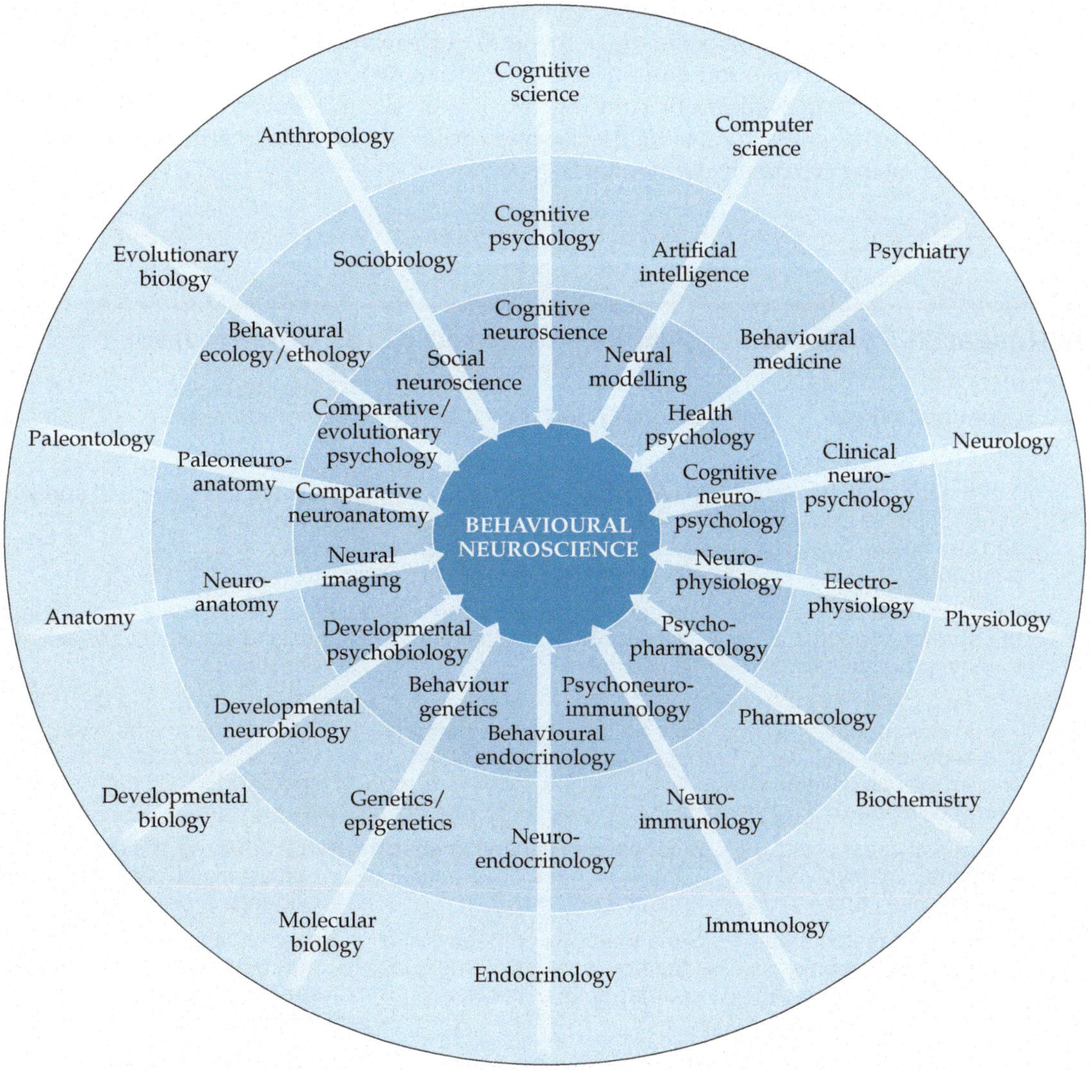

0.2 What's in a Name? In this graphical representation of the relationships among behavioural neuroscience and other scientific disciplines, fields towards the centre of the map are closest to behavioural neuroscience in their history, outlook, aims, and/or methods.

Five viewpoints explore the biology of behaviour

In our effort to understand the neuroscience bases of behaviour, we use several different perspectives. Because each one yields information that complements the others, the combination of perspectives is especially powerful. We will discuss five major perspectives:

1. *Describing* behaviour
2. Observing the *development* of behaviour and its biological characteristics over the lifespan
3. Studying the biological *mechanisms* of behaviour
4. Studying *applications* of behavioural neuroscience—for example, its application to dysfunctions of human behaviour
5. Studying the *evolution* of behaviour

These perspectives are discussed in the sections that follow, and **TABLE 0.1** illustrates how each perspective can be applied to three kinds of behaviour.

Behaviour can be described according to different criteria

Until we describe what we want to study, we cannot accomplish much. Depending on our goals, we may describe behaviour in terms of detailed acts or processes, or in terms of results or functions. An analytical description of arm movements might record the successive positions of the limb or the contraction of different muscles. A functional behavioural description, by contrast, would state whether the limb was being used in walking or running, texting or sexting. To be useful for scientific study, a description must be precise and reveal the essential features of the behaviour, using accurately defined terms and units.

TABLE 0.1 Five Research Perspectives Applied to Three Kinds of Behaviour

Research perspective	Sexual behaviour	Learning and memory	Language and communication
DESCRIPTION			
Structural	What are the main patterns of reproductive behaviour and sex differences in behaviour?	In what main ways does behaviour change as a consequence of experience—for example, conditioning?	How are the sounds of speech patterned?
Functional	How do specialized patterns of behaviour contribute to mating and to care of young?	How do certain behaviours lead to rewards or avoidance of punishment?	What behaviour is involved in making statements or asking questions?
ONTOGENY (development)	How do reproductive and secondary sex characteristics develop over the lifespan?	How do learning and memory change as we grow older?	What changes in the brain when a child learns to speak?
MECHANISMS	What neural circuits and hormones are involved in reproductive behaviour?	What anatomical and chemical changes in the brain hold memories?	What brain regions are particularly involved in language?
APPLICATIONS	Low doses of testosterone restore libido in some postmenopausal women.	Gene therapy and behavioural therapy improve memory in some senile patients.	Speech therapy, in conjunction with amphetamine treatment, speeds language recovery following stroke.
EVOLUTION	How does mating depend on hormones in different species?	How do different species compare in kinds and speed of learning?	How did the human speech apparatus evolve?

The body and behaviour develop over the life-span

Ontogeny is the process by which an individual changes in the course of its lifetime—that is, grows up and grows old. Observing the way in which a particular behaviour changes during ontogeny may give us clues to its functions and mechanisms. For example, we know that learning ability in monkeys increases over the first years of life. Therefore, we can speculate that prolonged maturation of brain circuits is required for complex learning tasks. In rodents, the ability to form long-term memories lags somewhat behind the maturation of learning ability. So, young rodents learn well but forget more quickly than older ones, suggesting that learning and memory involve different processes. Studying the development of reproductive capacity and of differences in behaviour between the sexes, along with changes in body structures and processes, throws light on body mechanisms underlying sexual behaviours.

conserved In the context of evolution, referring to a trait that is passed on from a common ancestor to two or more descendant species.

ontogeny The process by which an individual changes in the course of its lifetime—that is, grows up and grows old.

Biological mechanisms underlie all behaviour

To learn about the mechanisms of an individual's behaviour, we study how his or her *present* body works. To understand the underlying mechanisms of behaviour, we must regard the organism (with all due respect) as a "machine," made up of billions of neurons. We must ask, How is this thing constructed to be able to do all that? These are sometimes described as *proximate* questions—questions about the physical interactions that control a particular behaviour. How cells in your eye respond differently to light of different wavelengths is a proximate question. On the other hand, why colour vision, once it arose, benefitted our ancestors is an evolutionary question.

Our major aim in behavioural neuroscience is to examine body mechanisms that make particular behaviours possible. In the case of learning and memory, for example, we would like to know the sequence of electrical and biochemical processes that occur when we learn something and retrieve it from memory. What parts of the nervous system are involved in that process? In the case of reproductive behaviour, we also want to understand the neuronal and hormonal processes that underlie mating behaviours.

Research can be translated to address human problems

Like other sciences, behavioural neuroscience is also dedicated to improving the human condition. Numerous human diseases involve malfunctioning of the brain. Many of these are already being alleviated as a result of research in the neurosciences, and the prospects for continuing advances are good. Attempts to apply knowledge also benefit basic research. For example, the study of memory disorders in humans has pushed investigators to extend our knowledge of the brain regions involved in different kinds of memory (see Chapter 16).

We compare species to learn how the brain and behaviour have evolved

Nature is conservative. Once particular features of the body or behaviour evolve, they may be maintained for millions of years and may be seen in animals that otherwise appear very different. For example, the electrical messages used by nerve cells (see Chapter 2) are essentially the same in a jellyfish, a cockroach, and a human being. Some of the chemical compounds that transmit messages through the bloodstream (hormones) are also the same in diverse animals (see Chapter 4). Species share these **conserved** characteristics because the features first arose in a shared ancestor (**BOX 0.1** on the next page). But mere similarity of a feature between species does not guarantee that the feature came from a common ancestral species. Similar solutions to a problem may have evolved independently in different classes of animals.

Charles Darwin's theory of evolution through natural selection is central to all modern biology. From this perspective emerge two rather different emphases: (1) the *continuity* of behaviour and biological processes among species that reflects shared ancestry and (2) the species-specific *differences* in behaviour and biology that have evolved as adaptations to different environments.

BOX 0.1 We Are All Alike, and We Are All Different

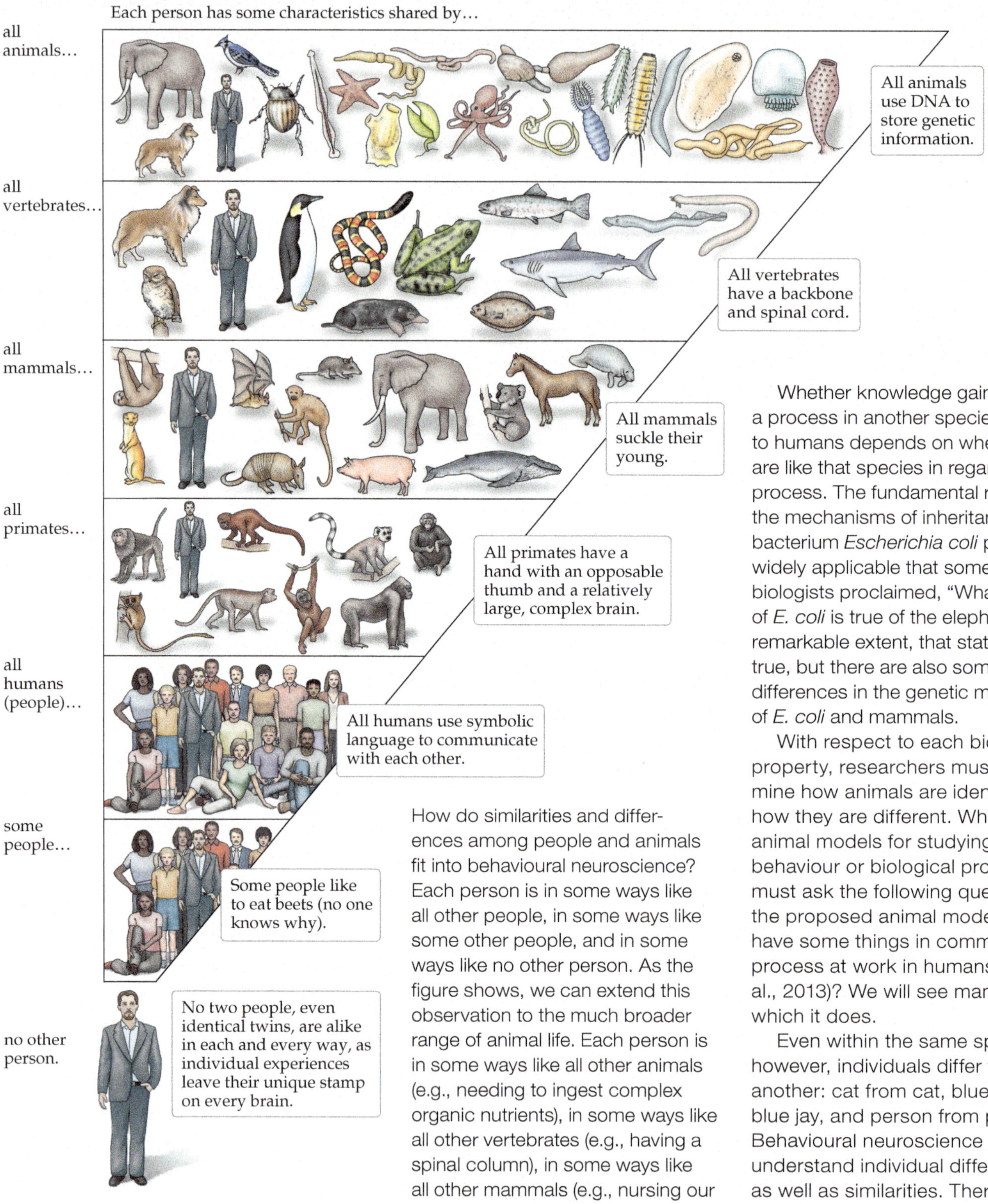

How do similarities and differences among people and animals fit into behavioural neuroscience? Each person is in some ways like all other people, in some ways like some other people, and in some ways like no other person. As the figure shows, we can extend this observation to the much broader range of animal life. Each person is in some ways like all other animals (e.g., needing to ingest complex organic nutrients), in some ways like all other vertebrates (e.g., having a spinal column), in some ways like all other mammals (e.g., nursing our young), and in some ways like all other primates (e.g., having a hand with an opposable thumb and a relatively large, complex brain).

Whether knowledge gained about a process in another species applies to humans depends on whether we are like that species in regard to that process. The fundamental research on the mechanisms of inheritance in the bacterium *Escherichia coli* proved so widely applicable that some molecular biologists proclaimed, "What is true of *E. coli* is true of the elephant." To a remarkable extent, that statement is true, but there are also some important differences in the genetic mechanisms of *E. coli* and mammals.

With respect to each biological property, researchers must determine how animals are identical and how they are different. When we seek animal models for studying human behaviour or biological processes, we must ask the following question: Does the proposed animal model really have some things in common with the process at work in humans (Seok et al., 2013)? We will see many cases in which it does.

Even within the same species, however, individuals differ from one another: cat from cat, blue jay from blue jay, and person from person. Behavioural neuroscience seeks to understand individual differences as well as similarities. Therefore, the way in which each person is able to process information and store the memories of these experiences is another part of our story.

0.2 Three Approaches Relate Brain and Behaviour

Learning Objectives

After reading this section, you should be able to:

0.2.1 **Differentiate between the independent and dependent variables in scientific experiments.**

0.2.2 **Name the type of research in which a part of the brain is manipulated to observe effects on behaviour, and offer examples.**

0.2.3 **Name the type of research in which behaviour or experience is manipulated to observe effects on the brain, and offer examples.**

0.2.4 **Describe correlational research about the brain and behaviour, and offer examples.**

0.2.5 **Explain why the brain must be capable of changing its structure, and name the term to describe that changeability.**

Behavioural neuroscientists use three approaches to understand the relationship between brain and behaviour: somatic intervention, behavioural intervention, and correlation. In the most common approach, **somatic intervention** (**FIGURE 0.3A**), we alter a structure or function of the brain or body to see how this alteration changes behaviour. Here, somatic intervention is the **independent variable**, and the behavioural effect is the **dependent variable**; that is, the resulting behaviour depends on how the brain has been altered. For example, in response to mild electrical stimulation of one part of her brain, not only did one patient laugh, but she found whatever she happened to be looking at amusing (Fried et al., 1998).

In later chapters we describe many kinds of somatic intervention with both humans and other animals, as in the following examples:

- A hormone is administered to some animals but not to others; various behaviours of the two groups are later compared.
- A part of the brain is stimulated electrically, or by use of light to stimulate only a particular class of neurons, and behavioural effects are observed.
- A connection between two parts of the nervous system is cut, and changes in behaviour are measured.

somatic intervention An approach to finding relations between body variables and behavioural variables that involves manipulating body structure or function and looking for resultant changes in behaviour.

independent variable The factor that is manipulated by an experimenter.

dependent variable The factor that an experimenter measures to monitor a change in response to manipulation of an independent variable.

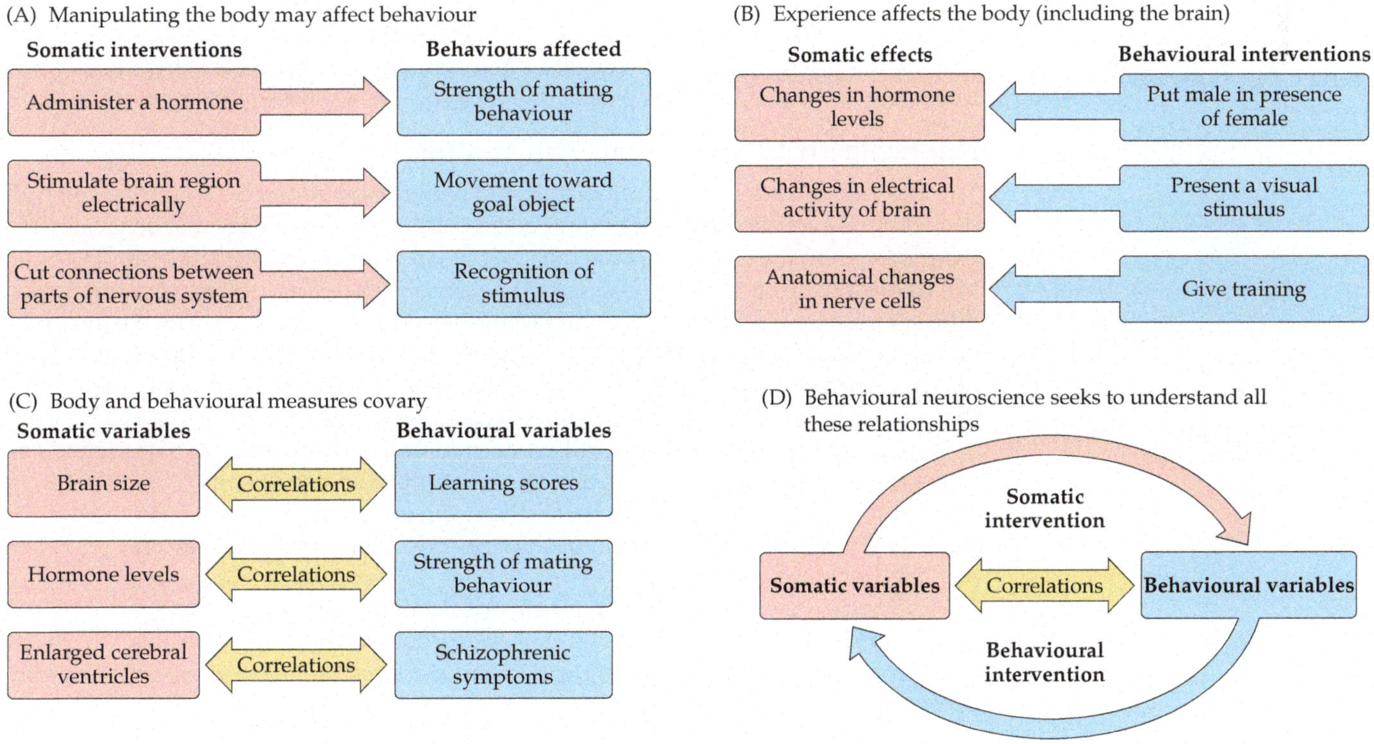

0.3 Three Main Approaches to Studying the Neuroscience of Behaviour (A) In *somatic intervention*, investigators change the body structure or chemistry of an animal in some way and observe and measure any resulting behavioural effects. (B) Conversely, in *behavioural intervention*, researchers change an animal's behaviour or its environment and try to ascertain whether the change results in physiological or anatomical changes. (C) Measurements of both kinds of variables allow researchers to arrive at *correlations* between somatic changes and behavioural changes. (D) Each approach enriches and informs the others.

behavioural intervention An approach to finding relations between body variables and behavioural variables that involves intervening in the behaviour of an organism and looking for resultant changes in body structure or function.

correlation The covariation of two measures.

neuroplasticity Also called *neural plasticity*. The ability of the nervous system to change in response to experience or the environment.

The approach opposite to somatic intervention is psychological or **behavioural intervention** (**FIGURE 0.3B**). In this approach, the scientist intervenes in the behaviour or experience of an organism and looks for resulting changes in body structure or function. Here, behaviour is the independent variable, and change in the body is the dependent variable. Among the examples that we will consider in later chapters are the following:

- Putting two adults of opposite sex together may lead to increased secretion of certain hormones.
- Exposing a person or animal to a visual stimulus provokes changes in electrical activity and blood flow in parts of the brain.
- Training of animals in a maze is accompanied by electrical, biochemical, and anatomical changes in parts of their brains.

The third approach to brain-behaviour relations, **correlation** (**FIGURE 0.3C**), consists of finding the extent to which a given body measure varies with a given behavioural measure. Later we will examine the following questions, among others:

- Are people with large brains more intelligent than people with smaller brains (a topic we'll take up later in this chapter)?
- Are individual differences in sexual behaviour correlated with levels of certain hormones in the individuals?
- Is the severity of schizophrenia correlated with the magnitude of changes in brain structure?

Such correlations should not be taken as proof of causal relationship. For one thing, even if a causal relation exists, the correlation does not reveal its direction—that is, which variable is independent and which is dependent. For another, two factors might be correlated only because a third, unknown factor affects the two factors measured. If you and your study partner get similar scores on an exam, that's not because your performance *caused* her to get the score she did, or vice versa. What a correlation does suggest is that the two variables are linked in some way—directly or indirectly. Such a correlation often stimulates investigators to formulate hypotheses and to test them by somatic or behavioural intervention. Only by moving on to such intervention approaches can we establish whether one variable is *causing* changes in the other.

Combining these three approaches yields the circle diagram of **FIGURE 0.3D**, incorporating the basic approaches to studying relationships between bodily processes and behaviour. It also emphasizes the theme that the relations between brain and behaviour are reciprocal: each affects the other in an ongoing cycle of bodily and behavioural interactions. We will see examples of this reciprocal relationship throughout the book.

Neuroplasticity: behaviour can change the brain

The idea that there is a reciprocal relationship between brain and behaviour has embedded within it a concept that is, for most people, startling. When we say that behaviour and experience affect the brain, we mean that they, literally, physically alter the brain. The brain of a child growing up in a French-speaking household assembles itself into a configuration different from that of the brain of a child who hears only English. That's why the first child, as an adult, understands French effortlessly while the second does not. In this case we cannot tell you what the structural differences are exactly, but we do know one part of the brain that is being altered by these different experiences (see Chapter 18).

Numerous examples, almost all in animal subjects, show that experience can affect the number or size of neurons, or the number or size of connections between neurons. This ability of the brain, both in development and in adulthood, to be changed by the environment and by experience is called **neuroplasticity** (or *neural plasticity*, or simply *plasticity*).

Today when we hear the word *plastic*, we think of the class of materials found in so many modern products. But originally, *plastic* meant "flexible, malleable" (from the Greek *plassein*, "to mould or form"), and the modern materials were named *plastics* because they can be moulded into nearly any shape. In 1890, William James (1842–1910) described plasticity as the possession of a structure weak enough to yield to an influence but strong enough not to yield all at once.

In the ensuing years, research has shown that the brain is even more plastic, more yielding, than James suspected. For example, parts of neurons known as *dendritic spines* (see Chapter 1) appear to be in constant motion, changing shape in the course of seconds. We will see many examples in which experience alters the structure and/or function of the brain: In Chapter 4, you'll read that hearing a baby cry causes the mother's brain to secrete a hormone. In Chapter 6, we'll see that visual experience in kittens directs the formation of connections in the brain. In Chapter 13, we'll discuss how a mother rat's grooming of her pups affects the survival of spinal cord neurons. And Chapter 16 talks about how a sea slug learning a task changes the connections between two particular neurons.

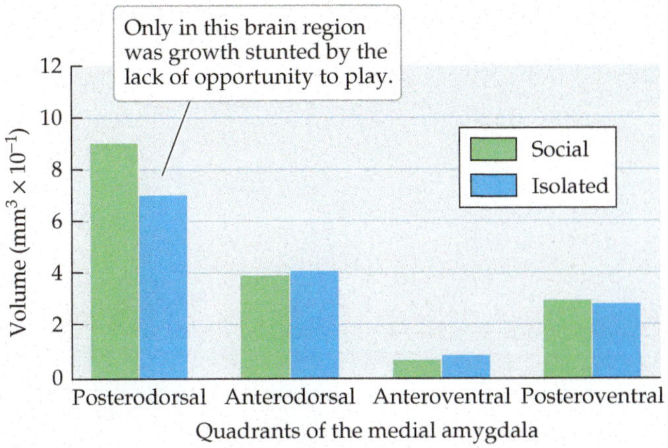

0.4 The Role of Play in Brain Development A brain region involved in processing odours (the posterodoursal portion of the medial amygdala) was smaller in male rats housed individually than in males housed together and allowed to play. Other nearby regions were identical in the two groups. (After B. M. Cooke et al., 2000. *Behav Brain Res* 117: 107–113.)

Behavioural neuroscience and social psychology are related

The plasticity of the human brain has a remarkable consequence: other individuals can affect the physical structure of your brain! Indeed, the whole point of coming to a lecture hall is to have the instructor use words and figures to alter your brain so that you can retrieve that information in the future (in other words, teach you something). Many of these alterations in your brain last only until you take an exam, but every once in a while the instructor may tell you something that you'll remember for the rest of your life. Most aspects of our social behaviour are learned—from the language we speak to the clothes we wear and the kinds of food we eat—so the mechanisms of learning and memory (see Chapter 16) are important for understanding social behaviour.

For an example from an animal model, consider the fact that rats spend a lot of time investigating the smells around them, including those coming from other rats. Cooke et al. (2000) took young male rats, just weaned from their mother, and raised them in two different ways: either alone in separate cages, or with other males in group cages so they could engage in play (including a lot of sniffing of each other's butts). Examination of these animals as adults found only one brain difference between the groups: a region of the brain known to process odours was smaller in the isolated males than in the males raised with playmates (**FIGURE 0.4**). Was it the lack of play (N. S. Gordon et al., 2003), the lack of odours to investigate, or the stress of isolation that made the region smaller? Whatever the mechanism, social experience affects this brain structure. In Chapter 16 we'll see more examples of social experience altering the brain.

Here's an example of how social influences can affect human brain function. When people were asked to put a hand into moderately hot water (47°C), part of the brain became active, presumably because of the discomfort involved (Rainville et al., 1997). But people who were led to believe the water would be *very* hot had a more activated brain than did those led to believe the discomfort would be minimal (**FIGURE 0.5**), even though the water was the same temperature for everyone. The socially induced psychological expectation affected the magnitude of the brain response,

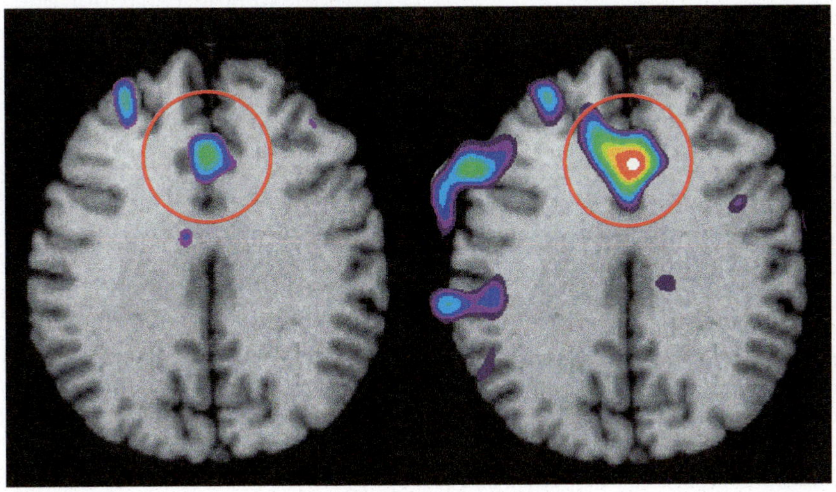

0.5 Pictures of Pain People told to expect only mild discomfort from putting a hand into 47°C water (left) showed less activation in a particular brain region (the anteriour cingulate cortex) than did people expecting more discomfort (right) from water of the very same temperature. Areas of high activation are indicated by orange, red, and white.

From P. Rainville, 1998. Presented at INABIS '98, 5th Internet World Congress on Biomedical Sciences at McMaster University, Canada, Dec 7–16th. Available at http://www.mcmaster.ca/inabis98/woody/rainville0419/index.html

even though the physical stimulus was exactly the same. (By the way, the people with the more activated brains also reported feeling more pain.)

In most cases, biological and social factors continually interact and affect each other in an ongoing series of events as behaviour unfolds. For example, the level of the hormone testosterone in circulation can affect dominance behaviour and aggression (see Chapter 11). The dominance may be exhibited in a great variety of social settings, ranging from playing chess to physical aggression. In humans and other primates, the level of testosterone correlates positively with the degree of dominance and with the amount of aggression exhibited. Winning a contest, whether a game of chess or a boxing match, raises the level of testosterone; losing a contest lowers the level. Thus, at any moment the level of testosterone is determined, in part, by recent dominant-submissive social experience, and the level of testosterone determines, in part, the degree of dominance and aggression in the future. Of course, social and cultural factors also help determine the frequency of aggression; cross-cultural differences in rates of aggression exist that cannot be correlated with hormone levels, and ways of expressing aggression and dominance are influenced by sociocultural factors.

Perhaps nothing distinguishes neuroscience from the other sciences more clearly than this fascination with neuroplasticity and the role of experience. Neuroscientists have a pervasive interest in how experience physically alters the brain and therefore affects future behaviour. We will touch on this theme in every chapter of this book.

0.3 Behavioural Neuroscientists Use Several Levels of Analysis

Learning Objectives

After reading this section, you should be able to:

0.3.1 **Name and describe the scientific approach of explaining mechanisms at simpler and simpler levels.**

0.3.2 **Give a survey of important ongoing questions about the relationship between the brain and behaviour.**

0.3.3 **Offer estimates of the extent of neurological and psychiatric disorders.**

0.3.4 **Explain the importance of research with animals for neuroscience, and discuss the ethics of such research.**

Scientific explanations of systems or structures or functions usually involve breaking them down into smaller parts, as a way of understanding them. This approach is known as **reductionism**. In principle, it is possible to reduce each explanatory series down to the molecular or atomic level, though for practical reasons this extent of reductionism is rare. For example, most chemists deal with large, complex molecules and the laws that govern them; seldom do they seek explanations in terms of subatomic quarks and bosons.

Understanding behaviour often requires several levels of biological analysis. The units of each level of analysis are simpler in structure and organisation than those of the level above. The **levels of analysis** range from social interactions to the brain, continuing to successively less complex units until we arrive at single nerve cells and their even simpler, molecular constituents.

Naturally, in all fields different problems are carried to different levels of analysis, and fruitful work is often being done simultaneously by different workers at several levels (**FIGURE 0.6**). Thus, in their research on visual perception, cognitive neuroscientists advance analytical descriptions of behaviour. They try to determine how the eyes move while looking at a visual pattern, or how the contrast among parts of the pattern determines its visibility. Meanwhile, other behavioural neuroscientists study the differences in visual abilities among species and try to determine the adaptive significance of these differences. For example, how is the presence (or absence) of colour vision related to the life of a species? At the same time, other investigators trace out brain structures and networks involved in different kinds of visual discrimination. Still other scientists try to ascertain the electrical and chemical events that occur in the brain during vision.

reductionism The scientific strategy of breaking a system down into increasingly smaller parts in order to understand it.

levels of analysis The scope of experimental approaches. A scientist may try to understand behaviour by monitoring molecules, nerve cells, brain regions, or social environments, or some combination of these levels of analysis.

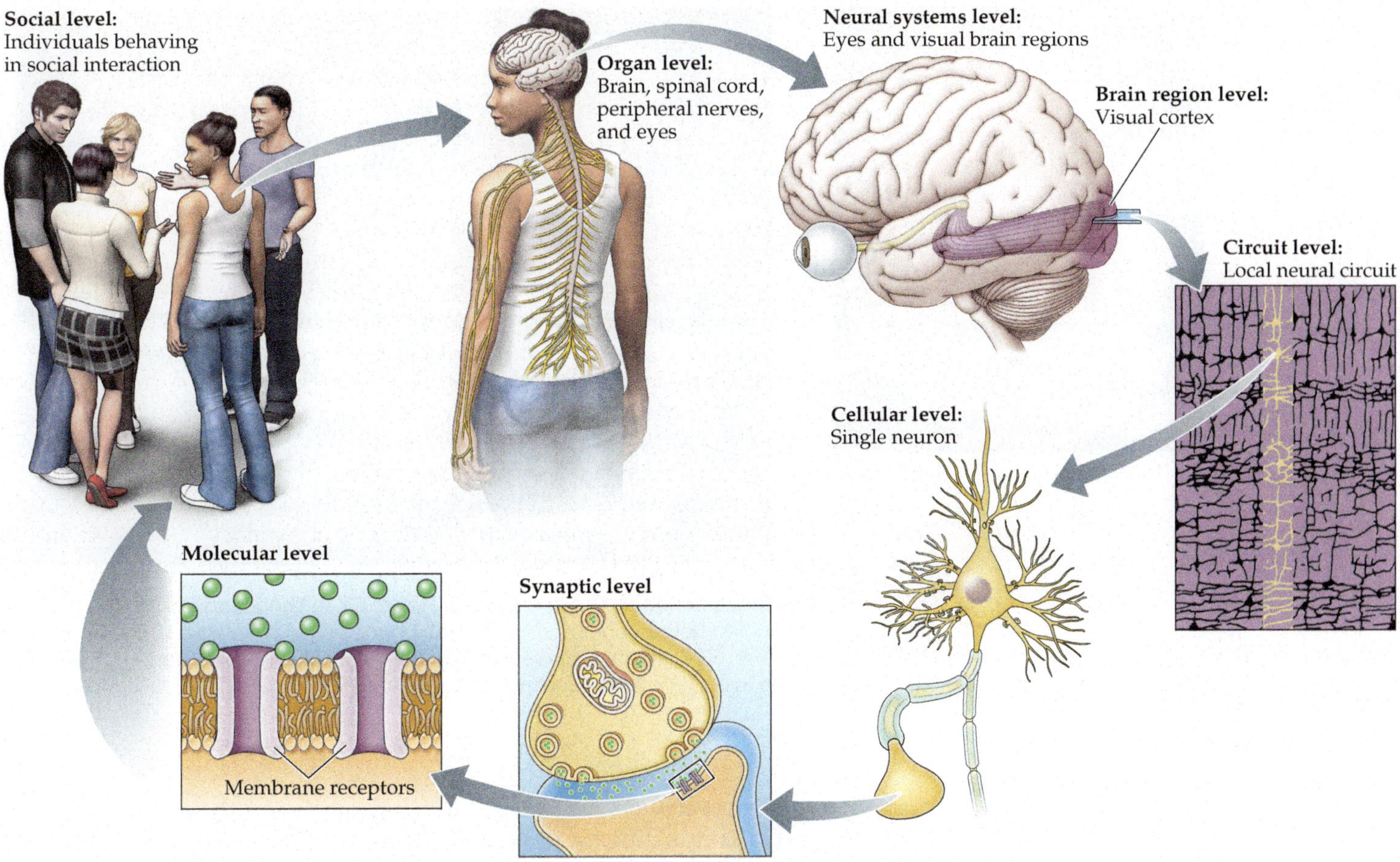

0.6 Levels of Analysis in Behavioural Neuroscience The scope of behavioural neuroscience ranges from the level of the individual interacting with others, to the level of the molecule. Depending on the question at hand, investigators use different techniques to focus on these many levels, but always with an eye toward how their findings apply to behaviour.

We will encounter many diverse brain and behaviour topics

Here are some examples of research topics considered in this book:

- How does the brain grow, maintain, and repair itself over the lifespan (see Chapter 6), and how are these capacities related to the growth and development of the mind and behaviour from the womb to the tomb?
- How does the nervous system capture, process, and represent information about the environment? For example, sometimes brain damage causes a person to lose the ability to identify other people's faces (see Chapter 17); what does that tell us about how the brain recognizes faces?
- How does sexual orientation develop? Some brain regions are different in straight versus gay men (see Figure 13.26); what, if anything, do those differences tell us about the development of human sexual orientation?
- Some people suffer damage to the brain and afterward seem alarmingly unafraid in dangerous situations and unable to judge the fearfulness of other people; what parts of the brain are damaged to cause such changes (see Figure 11.16)?
- How does the brain manage to change during learning (see Chapter 16), and how are memories retrieved?
- What brain sites and activities underlie feelings and emotional expression? Are particular parts of the brain active in romantic love, for example (**FIGURE 0.7A**)?
- Why are different brain regions active during different language tasks (**FIGURE 0.7B**)?

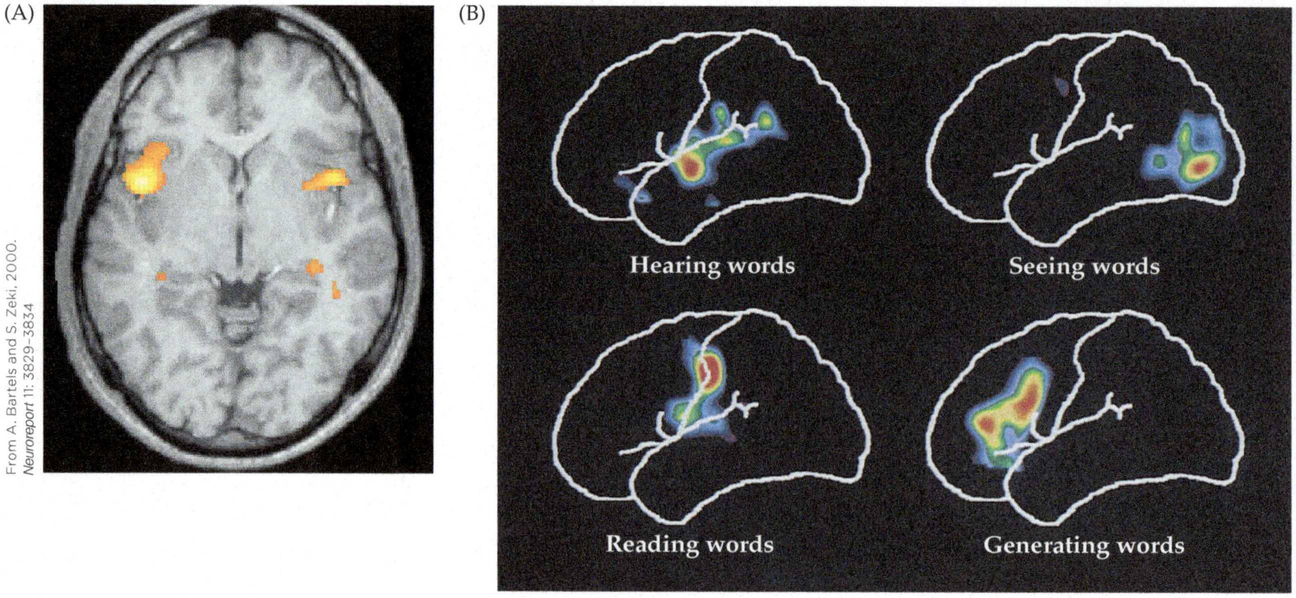

From A. Bartels and S. Zeki, 2000. *Neuroreport* 11: 3829–3834

Courtesy of Marcus Raichle

0.7 "Tell Me Where Is Fancy Bred, Or in the Heart Or in the Head?" (A) The parts of the brain highlighted here become especially active when a person thinks about his or her romantic partner. (B) Different brain regions are activated when people perform four different language tasks. The techniques used to generate such images are described in Chapter 1.

The relationship between the brain and behaviour is, on the one hand, very mysterious because it is difficult to understand how a physical device, the brain, could be responsible for our subjective experiences of fear, love, and awe. Yet despite this mystery, we all use our brains every day. Perhaps it is the "everyday miracle" aspect of the topic that has generated so much folk wisdom about the brain. Think of it as "neuromythology."

Sometimes these popular ideas about the brain are in line with our current knowledge, but in many cases we know they are false. For example, the notion that we normally use only 10% of our brain is commonplace—a survey of teachers found that nearly half of them agreed with this notion (Howard-Jones, 2014)—but it is patent nonsense. Brain scans make it clear that the entire brain is activated by even fairly mundane tasks. Indeed, although the areas of activation shown in Figure 0.7 appear rather small and discrete, we will show in Box 1.3 that experimenters must work very hard to create images that separate activation related to a particular task from the background of widespread, ongoing brain activity.

We offer a list of other commonly held beliefs about the brain and behaviour in **A STEP FURTHER 0.1 : NEUROMYTHOLOGY: FACTS OR FABLES?** Throughout the book we offer such opportunities for you to explore a given topic in more detail.

Behavioural neuroscience contributes to our understanding of human disorders

One of the great promises of neuroscience is that it can help us understand brain disorders and devise treatment strategies. Like any other complex mechanism, the brain is subject to a variety of malfunctions and breakdowns. People afflicted by disorders of the brain are not an exotic few—a European survey estimated that at least 38% of the population would suffer from a mental disorder at some point in a typical year (Wittchen et al., 2011). At least one person in five around the world currently has neurological and/or psychiatric disorders that vary in severity from mild challenges to complete disability. **FIGURE 0.8A** shows the estimated numbers of U.S. residents afflicted by some of the main neurological disorders. **FIGURE**

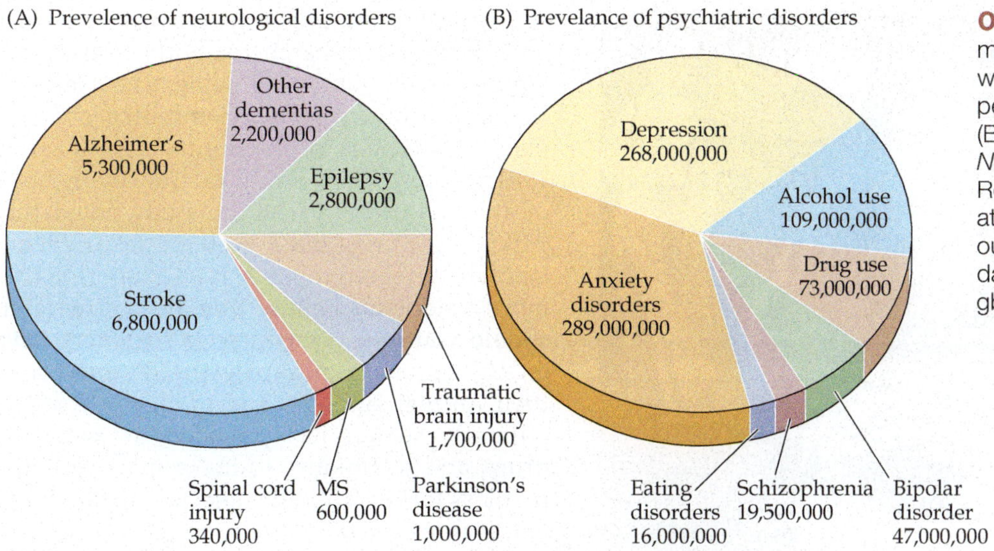

0.8 The Toll of Brain Disorders Estimated numbers of people in the United States with neurological disorders (A) and number of people worldwide with psychiatric disorders (B). (Part A after C. L. Gooch et al., 2017. *Ann Neurol* 81: 479–484; B after H. Ritchie and M. Roser, 2019. "Mental Health". Published online at OurWorldInData.org. Retrieved from: https://ourworldindata.org/mental-health. Underlying data available from http://ghdx.healthdata.org/gbd-results-tool.)

0.8B gives estimates of the numbers of adults worldwide with certain psychiatric disorders. The percentage of U.S. adults suffering from mental illness may be increasing (Twenge, 2015; Twenge et al., 2010).

The toll of these disorders is enormous, in terms of both individual suffering and social costs (Demyttenaere et al., 2004). The National Advisory Mental Health Council has estimated that direct and indirect costs of behavioural and brain disorders amount to $400 billion a year in the United States alone. For example, the cost for treatment of dementia (severely disordered thinking) exceeds the costs of treating cancer and heart disease combined. The World Health Organisation (2004) estimates that over 15% of all disease burden, in terms of lost productivity, is due to mental disorders. The high cost in suffering and expense has compelled researchers to try to understand the mechanisms involved in these disorders and to try to alleviate or even prevent them.

In this quest, the distinction between clinical and laboratory approaches begins to fade away. For example, when clinicians encounter a pair of twins, one of whom has schizophrenia while the other seems healthy, the discovery of structural differences in their brains (**FIGURE 0.9**) immediately raises questions for laboratory scientists: Did the structural differences arise before the symptoms of schizophrenia, or the other way around? Were the brain differences present at birth, or did they arise during puberty? Does medication that reduces symptoms affect brain structure? When genes associated with schizophrenia in people are introduced into mice, will that change the mouse brains (see Figure 12.7)? This last question is just one instance of when working with animals is essential, an issue we address next.

0.9 Identical Twins but Nonidentical Brains and Behaviour In these images of the brains of identical twins, the fluid-filled cerebral ventricles are prominent as dark "butterfly" shapes. The brain of the twin with schizophrenia (A) has the enlarged cerebral ventricles that some researchers believe are characteristic of this disorder. The other twin does not have schizophrenia; his brain (B) clearly has smaller ventricles.

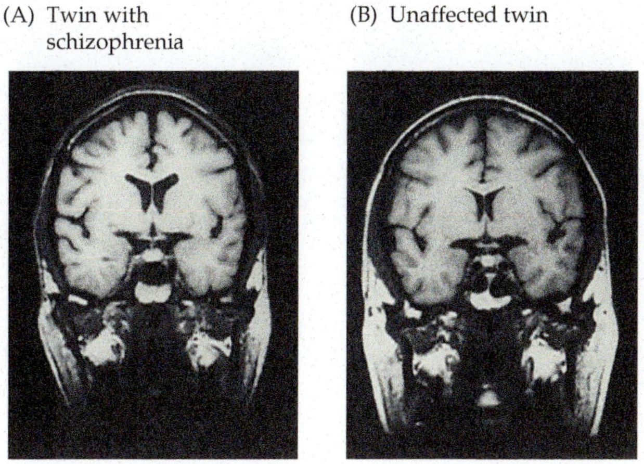

MRIs courtesy of E. Fuller Torrey and Daniel Weinberger

Animal research makes vital contributions

Because we will draw on animal research throughout this book, we want to comment on some of the ethical issues of experimentation on animals. Human beings' involvement and concern with other species predates recorded history. Early humans had to study animal behaviour and physiology in order to escape some species and hunt others. To study biological bases of behaviour inevitably requires research on animals of other species as well as on human beings. Psychology students usually underestimate the contributions of animal research because the most widely used introductory psychology textbooks often present major findings from animal research as if they were obtained with human participants (Domjan and Purdy, 1995).

© Santa Cruz Sentinel/ZUMAPRESS.com

0.10 Car Firebombed by Animal Rights Activists The extremists targeted the cars and homes of two scientists who work with animals at the University of California in Santa Cruz in 2008. The next year, the car of a researcher at UCLA was torched.

Because of the importance of carefully regulated animal research for both human and animal health and well-being, the National Research Council (NRC Committee on Animals as Monitors of Environmental Hazards, 1991) undertook a study on the many uses of animals in research. The study noted that 93% of the mammals used in research are laboratory-reared rodents. It also reported that most Americans believe that animal research should continue. Of course, researchers have an obligation to minimize the discomfort of their animal subjects, and ironically enough, animal research has provided us with the drugs and techniques to make most research painless for the animal subjects (Sunstein and Nussbaum, 2004).

Nevertheless, a very active minority of people believe that research with animals, even if it does lead to lasting benefits, is unethical. For example, in his 1975 book *Animal Liberation,* Peter Singer asserts that research with animals can be justified only if it actually produces benefits. The trick, of course, is how to predict which experiment will lead to a breakthrough. Thus Singer refuses to say that animal experimentation is never justified (Neale, 2006). In the meantime, animal rights groups have vandalized labs, burned down buildings, and exploded bombs in laboratories (Conn and Parker, 2008). In 2008, animal rights extremists set off firebombs at the homes of two scientists in Santa Cruz, California. One scientist's family, including two young children, had to flee their home through a second-story window (**FIGURE 0.10**) (Paddock and La Ganga, 2008). These personal attacks on individuals appear to be intended to intimidate and frighten scientists (D. Grimm, 2014), and they have already hounded at least one researcher out of the field (Nature Neuroscience, 2015).

Perhaps in a future where robots can be made that look and act like humans, methods will be available to clearly see all the processes at work in a living, working human brain. In the meantime, there's no substitute for research with animal subjects. Every chapter in this book is teeming with information that was gathered from humane experiments with animals.

0.4 The History of Research on the Brain and Behaviour Begins in Antiquity

Learning Objectives

After reading this section, you should be able to:

0.4.1 **Trace the historical point at which the brain was recognized as the control unit for behaviour.**

0.4.2 **Discuss the importance of the Renaissance in better understanding human anatomy.**

0.4.3 **Explain Descartes's contribution to early neuroscience and his now-discredited ideas of dualism.**

0.4.4 **Trace the history of phrenology and the relationship to modern thinking about brain and behaviour.**

0.4.5 **Discuss the difficult question of consciousness and the explosion of neuroscience research.**

Only recently have scientists recognized the central role of the brain in controlling behaviour. When Egyptian pharaoh Tutankhamen was mummified (about 1300 BCE), five important organs were preserved in his tomb: liver, lungs, stomach, intestines, and heart. All these organs were considered necessary to ensure the pharaoh's continued existence in the afterlife. The brain, however, was picked out through the nostrils (**FIGURE 0.11**) and thrown away. Although the Egyptian version of the afterlife entailed considerable struggle, the brain was not considered an asset.

Neither the Hebrew Bible (written from the twelfth to the second century BCE) nor the New Testament ever mentions the brain. However, the Bible mentions the heart hundreds of times and makes several references each to the liver, the stomach, and the bowels as the seats of passion, courage, and pity, respectively. "Teach us … that we may gain a heart of wisdom" (Psalms 90:12).

The heart is also where Aristotle (about 350 BCE), the most prominent scientist of ancient Greece, located mental capacities. We still reflect this ancient notion when we call people *kindhearted, openhearted, fainthearted, hardhearted,* or *heartless* and when we speak of learning *by heart*. Aristotle considered the brain to be only a cooling unit to lower the temperature of the hot blood from the heart.

Also about 350 BCE, the Greek physician Herophilus (the "Father of Anatomy") advanced our knowledge of the nervous system by dissecting bodies of both people and animals. He traced nerves from muscles and skin into the spinal cord and noted that each region of the body is connected to separate nerves.

A second-century Greco-Roman physician, Galen, treated the injuries of gladiators. His reports of behavioural changes caused by injuries to the heads of gladiators drew attention to the brain as the controller of behaviour. Galen advanced the idea that animal spirits—a mysterious fluid—passed along nerves to all regions of the body. But Galen's ideas about the anatomy of the human brain were very inaccurate because dissecting humans was illegal in Rome at that time.

Renaissance scientists began to understand brain anatomy

The eminent Renaissance painter and scientist Leonardo da Vinci (1452–1519) studied the workings of the human body and laid the foundations of anatomical drawing. He especially pioneered in providing views from different angles and cross-sectional representations. His artistic renditions of the body included portraits of the nerves in the arm and the fluid-filled ventricles in the brain (**FIGURE 0.12**).

Renaissance anatomists emphasized the shape and appearance of the external surfaces of the brain because these were the parts that were easiest to see when the skull was removed. It was immediately apparent to anyone who looked that the brain

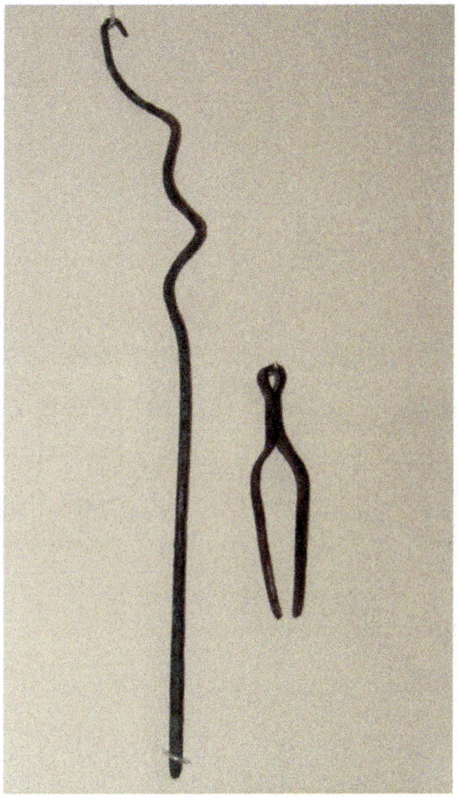

0.11 Brain Removal Kit Ancient Egyptians had little regard for the brain. During mummification, they would use tools like these to reach through the nostrils to pick out brain pieces and throw them away.

(A) Early drawing

(B) Later drawing based on observation

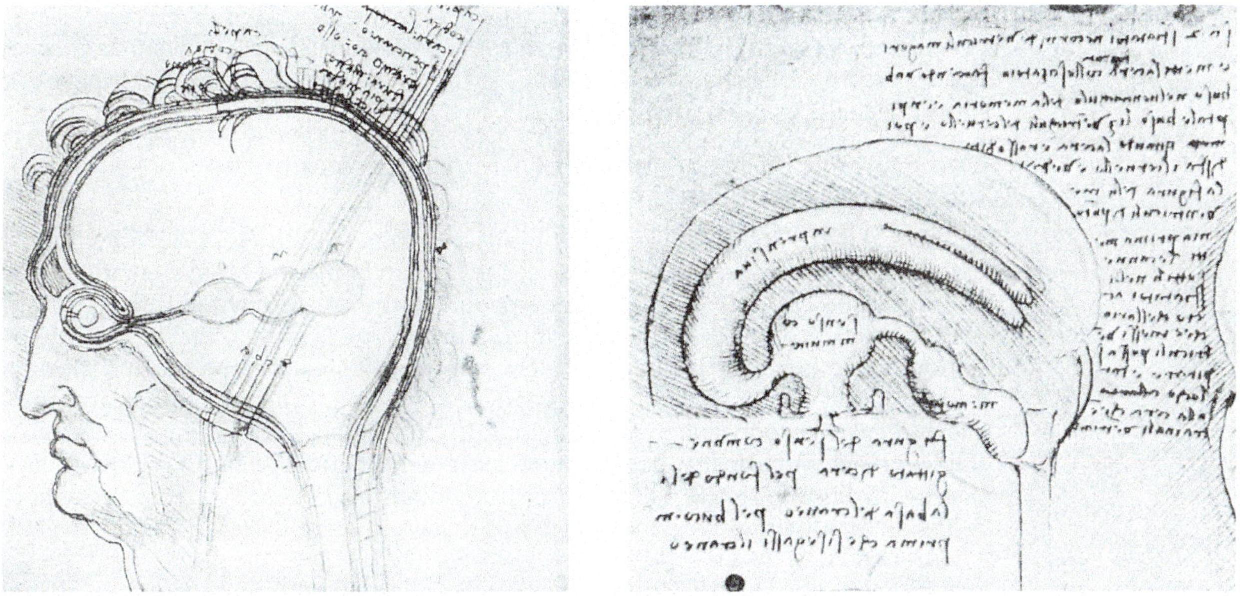

0.12 Leonardo da Vinci's Changing View of the Brain (A) In an early representation, Leonardo simply copied old schematic drawings that represented the fluid-filled cerebral ventricles as a linear series of chambers. (B) Later he made a drawing based on direct observation: after making a cast of the ventricles of an ox brain by pouring melted wax into the brain and letting it set, he cut away the tissue to reveal the true shape of the ventricles.

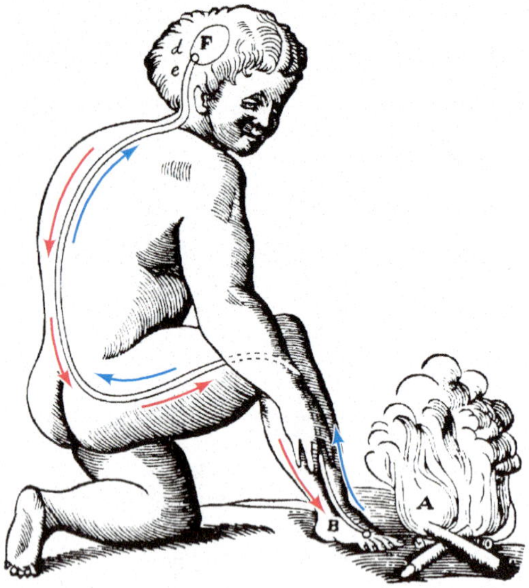

0.13 An Early Account of Reflexes In this depiction of an explanation by Descartes, when a person's toe touches fire, the heat causes nervous activity to flow up the nerve to the brain (blue arrows). From there the nervous activity is "reflected" back down to the leg muscles (red arrows), which contract, pulling the foot away from the fire; the idea of activity being reflected back is what gave rise to the word *reflex*. In Descartes's time, the difference between sensory and motor nerves had not yet been discovered, nor was it known that nerve fibres normally conduct in only one direction. Nevertheless, Descartes promoted thinking about bodily processes in scientific terms, and this focus led to steadily more accurate knowledge and concepts.

has an extraordinarily complex shape. To Renaissance artists like Michelangelo (1475–1564), this marvelous structure was God's greatest gift to humankind. So, in Michelangelo's painting on the ceiling of the Sistine Chapel, God seems to ride the form of the human brain when bestowing life to Adam (Meshberger, 1990), while in another scene God's throat resembles the base of the brain (Suk and Tamargo, 2010).

In 1633, René Descartes (1596–1650) wrote an influential book (*De Homine* [*On Man*]) in which he tried to explain how the behaviour of animals, and to some extent the behaviour of humans, could be like the workings of a machine.

Descartes proposed the concept of spinal reflexes and a neural pathway for them (**FIGURE 0.13**). Attempting to relate the mind to the body, Descartes suggested that the two come into contact in the pineal gland, located within the brain. He suggested the pineal gland for this role because (1) whereas most brain structures are double, located symmetrically in the two hemispheres, the pineal gland is single, like consciousness, and (2) he believed, erroneously, that the pineal gland exists only in humans and not in animals.

As Descartes was preparing to publish his book, he learned that the Catholic Church had forced Galileo to renounce his teaching that Earth revolves around the sun, threatening to execute him if he did not recant. Fearful that his own speculations about mind and body could also incur the wrath of the Church, Descartes withheld his book from publication. It did not appear in print until 1662, after his death. Descartes believed that if people were nothing more than intricate machines, they could have about as much free will as a pocket watch, with no opportunity to make the moral choices that were so important to the Church. He asserted that humans, at least, had a nonmaterial soul as well as a material body. This notion of **dualism** spread widely and left other philosophers with the task of determining how a nonmaterial soul could exert influence over a material body and brain. Mainstream neuroscientists reject dualism and insist that all the workings of the mind can also, in theory, be understood as purely physical processes in the material world, specifically in the brain.

The concept of localisation of function arose in the nineteenth century

By the end of the 1600s, the English physician Thomas Willis (1621–1675), with his detailed descriptions of the structure of the human brain and his systematic study of brain disorders, convinced educated people in the Western world that the brain is the organ that coordinates and controls behaviour (Zimmer, 2004). A popular notion of the nineteenth century, called **phrenology**, elaborated on this idea by asserting that the cerebral cortex consisted of separate functional areas and that each area was responsible for a behavioural faculty such as love of family, perception of colour, or curiosity. Investigators assigned functions to brain regions anecdotally, by observing the behaviour of individuals and inferring, from the shape of the skull, which underlying regions of the brain were more or less developed (**FIGURE 0.14A**).

Opponents rejected the entire concept of localisation of brain function, insisting that the brain, like the mind, functions as a whole. Today we know that the whole brain is indeed active when we are doing almost any task. When we are performing particular tasks, however (as we saw earlier in this chapter), certain brain regions become even more activated. Different tasks activate different brain regions. Modern

dualism Here, the notion promoted by René Descartes that the mind is subject only to spiritual interactions while the body is subject only to material interactions.

phrenology The outmoded belief that bumps on the skull reflect enlargements of brain regions responsible for certain behavioural faculties.

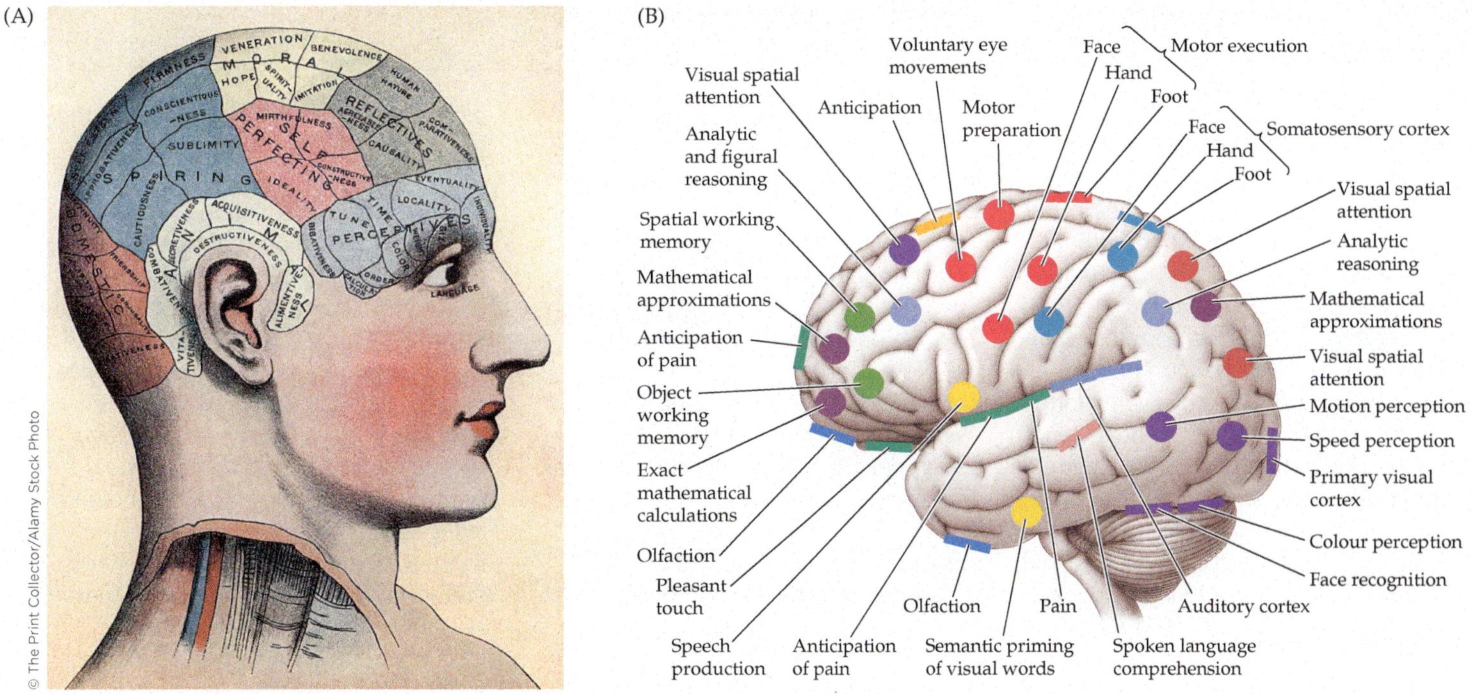

0.14 Old and New Phrenology (A) In the early nineteenth century, certain "faculties," such as skill at mathematics or a tendency toward aggression, were believed to be directly associated with particular brain regions. Phrenologists used diagrams like this one to measure bumps on the skull, which they took as an indication of how fully developed each brain region was in an individual, and hence how fully that person should display particular qualities. (B) Today, technology enables us to roughly gauge how active different parts of the brain are when a person is performing various tasks (see Chapter 1). But virtually the entire brain is active during any task, so the localisation of function that such studies provide is really a measure of where *peak* activity occurs, rather than a suggestion of a single region involved in a particular task. (B after M. J. Nichols and W. T. Newsome, 1999. *Nature* 402: C35–C38.)

brain maps of these places where *peaks* of activation occur (**FIGURE 0.14B**) bear a passing resemblance to their phrenological predecessors, differing only in the specific locations of functions. But unlike the phrenologists, we confirm these modern maps by other methods, such as examining what happens after brain damage.

Even as far back as the 1860s, the French surgeon Paul Broca (1824–1880) argued that language ability was not a property of the entire brain but rather was localized in a restricted brain region. Broca presented a postmortem analysis of a patient who had been unable to talk for several years. The only portion of the patient's brain that appeared damaged was a small region within the frontal portions of the brain on the left side—a region now known as *Broca's area* (labelled "Speech production" in Figure 0.14B). The study of additional patients further convinced Broca that language expression is mediated by this specific brain region rather than reflecting activities of the entire brain.

In 1890, William James's book *The Principles of Psychology* signaled the beginnings of a modern approach to behavioural neuroscience. The strength of the ideas described in this book is evident from the continuing frequent citation of the work. In James's work, psychological ideas such as consciousness and other aspects of human experience came to be seen as properties of the nervous system. A true behavioural neuroscience began to emerge from this approach.

These nineteenth-century observations form the background for a continuing theme of research in behavioural neuroscience—notably, the search for distinguishing differences among brain regions on the basis of their structure, and the effort to relate different kinds of behaviour to different brain regions (M. Kemp, 2001). An additional theme emerging from these studies is the relation of brain size to ability across species (see Figure 5.9), and even across various people (**BOX 0.2** on the next page).

BOX 0.2 Bigger Better? The Case of the Brain and Intelligence

Does a bigger brain indicate greater intelligence? Brain size does seem to explain many species differences in complex behaviour, and the human brain has expanded remarkably over the past few million years (see Chapter 5). But do variations in brain size within our species correlate with intelligence? This question has been the subject of lively controversy for at least two centuries. Sir Francis Galton (1822–1911), who invented the correlation coefficient, stated that the greatest disappointment in his life was his failure to find a significant relationship between head size and intelligence. But Galton had to use head size, when he really wanted to measure brain size. In addition, he had to rely on teachers' estimates of their students' intelligence, and every student knows that teachers can be quite wrong. Other investigators in the nineteenth century measured the volumes of skulls of various groups and estimated intelligence on the basis of people's occupations or other doubtful criteria (S. J. Gould, 1981).

The development and standardisation of intelligence quotient (IQ) tests in the twentieth century provided invaluable help for one side of the question, and these scores indeed correlate, with ranges from +0.08 to +0.22, with estimates of brain size based on known head size (Van Valen, 1974).

Newer, noninvasive techniques (discussed in detail in Chapter 1) to visualise the brains of healthy people now make it possible to directly measure brain size in living humans. One study found a significant correlation coefficient of about +0.26 between brain size and IQ (Posthuma et al., 2002). In another study, brain scans were used for measuring the sizes of different brain regions (Figure A). After correction for body size, the correlation between brain size and IQ scores was +0.38 (Andreasen et al., 1993). IQ seems to correlate better with the volume of the front of the brain than that of the back (Colom et al., 2013). When the brains of children were measured at age 6 and again at 11, those with the highest IQs displayed the greatest thickening of the outer layer of the brain, especially in the front (P. Shaw et al., 2006). Other brain-imaging studies report correlations between IQ scores and the extent of *connectivity* between brain regions of about +0.50 (Figure B) (Haász et al., 2013; Malpas et al., 2016).

Thus, on the basis of modern techniques, the long-standing controversy appears to have been settled in favor of a significant correlation between brain size and intelligence (as shown in Figure B). Note, however, that the modest size of the correlations, while statistically significant, indicates that only about 10–20% of variability in IQ is accounted for by brain size. So, there is plenty of room for other factors to contribute to overall IQ. There is abundant evidence that early experience can affect IQ (Chapter 16). In addition, many people dispute whether IQ tests really measure a general property of intelligence (Stanovich, 2009). (Figure B after C. B. Malpas et al., 2016. *J Clin Neurosci* 24: 128–134.)

(A)

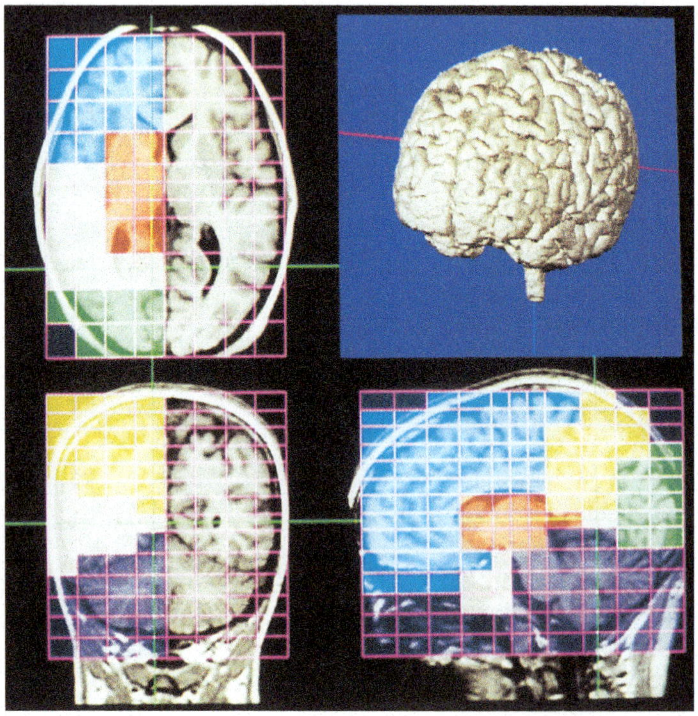

Courtesy of Nancy Andreasen

(A) Examples of measuring specific features of brain size from scans.

(B)

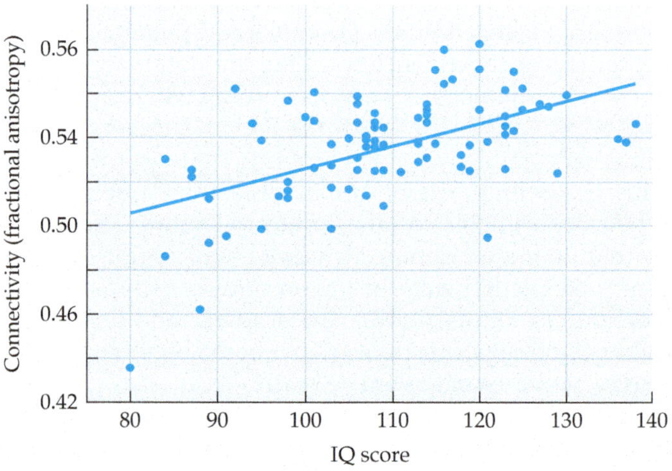

(B) Correlation between IQ and brain connectivity.

Modern behavioural neuroscience arose in the twentieth century

The end of the nineteenth century brought many important developments for behavioural neuroscience. German psychologist Hermann Ebbinghaus showed in 1885 how to measure learning and memory in humans. In 1898, American psychologist Edward L. Thorndike demonstrated how to measure learning and memory in animals. Early in the twentieth century, Russian physiologist Ivan P. Pavlov announced research in his laboratory on conditioning in animals.

American psychologist Shepard I. Franz (1902) sought the site of learning and memory in the brain by removing different brain regions in animal subjects. This work started a search for the traces of experience in the brain—a quest that Karl S. Lashley (1890–1958) referred to as the "search for the engram."

Behavioural neuroscience bears the strong imprint of Canadian psychologist Donald O. Hebb (1904–1985), a student of Lashley (P. M. Milner, 1993). In his book *The Organisation of Behaviour* (1949), Hebb showed, in principle, how complex cognitive behaviour could be accomplished by networks of active neurons. He suggested how brain cell connections that are initially more or less random could become organised, by sensory input and stimulation, into strongly interconnected groups that he called *cell assemblies*. His hypothesis about how neurons strengthen their connections through use gave rise to the concept of the *Hebbian synapse*, a topic much studied by current neuroscientists (see Chapters 6 and 16).

Consciousness is a thorny problem

Almost anyone using this book has at some time wondered about **consciousness**: the personal, private awareness of our emotions, intentions, thoughts, and movements and of the sensations that impinge upon us. How is it possible that you are aware of the words on this page, the room you are occupying, the goals you have in life?

In his review of theories of consciousness, Adam Zeman (2002) notes that almost all scientists agree on some aspects of consciousness:

- Consciousness matters; it permits us to do certain important things, like planning and mentally "simulating" what might happen in the future.
- Consciousness is bound up somehow with the activity of the brain.
- We are not aware of all of our brain's activities. Some brain activity, and therefore some of our behaviour, is unconscious.
- The deepest parts of our brain are important for arousal.
- The topmost parts of the brain are responsible for whatever we experience from moment to moment.

consciousness The state of awareness of one's own existence and experience.

In the chapters to come, we will see many examples of experiments that demonstrate these properties of consciousness. However consciousness is brought about, any satisfying understanding would be able, for example, to explain why a certain pattern of activity in your brain causes you to experience the sensation of blue when looking at the sky (**FIGURE 0.15**), or the smell of cinnamon when entering a bakery. A good theory would let us predict that after messing about with your brain, changing particular connections or activating particular neurons, you would experience yellow when seeing the sky.

Unfortunately, we are nowhere near understanding consciousness this clearly. We describe some intriguing (and disturbing) experiments explicitly directed at human consciousness in Chapter 17. In the rest of the book, we rarely use the words *conscious* or *consciousness*. Normally we cannot say anything about the particulars of what humans or animal subjects are *experiencing*, but only whether their behaviour suggests that the brain detected a signal or event. Thus, we are in no position to know whether complicated machines like androids are, or might one day be, conscious.

0.15 How Blue IS the Sky? We would all agree that this sky is the colour everyone calls "blue." But in Chapter 17 we will ask whether everyone who sees that sky has the same experience of the colour.

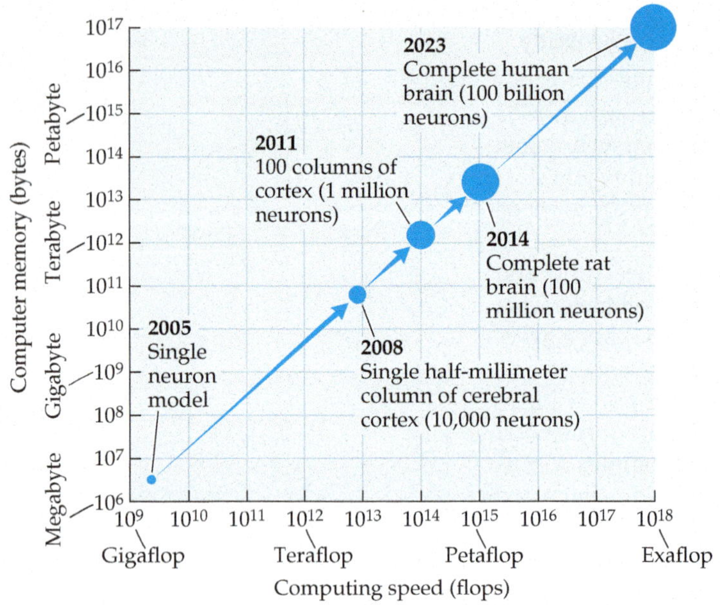

0.16 The Human Brain Project The goal of massive projects in both the United States and Europe is to have a digital re-creation of the neurons and connections found in a human brain. As this projection from the European project (bluebrain.epfl.ch) shows, this would require a computer that is vastly faster than any made to date, as well as a truly staggering amount of computer memory. (After M. M. Waldrop, 2012. *Nature* 482: 456–458.)

Both the United States and Europe have begun projects hoping to map an entire human brain, a truly formidable task (Waldrop, 2012), which we discuss in Chapter 1. It will require a computer with vastly more memory and faster processing than any yet devised (**FIGURE 0.16**). Some people even doubt whether our "merely human" brains will ever be able to understand something as complicated as consciousness (Yong, 2019). Nevertheless, any gains we make in understanding how the brain works, which is the subject of this book, will bring us closer to that goal.

The Cutting Edge

Behavioural Neuroscience Is Advancing at a Tremendous Rate

It is difficult to convey the rate at which we are learning new things about the brain. Each year, over 25,000 neuroscientists meet at the annual meeting of the Society for Neuroscience (sfn.org). On the website the NewsLink tab directs you to news stories about behavioural neuroscience for a general audience, with more than 20 articles added every week, over a thousand per year. The predominant index for biomedical research, PubMed (pubmed.gov), classified over 40,000 articles under *neuroscience* in 2018 alone (**FIGURE 0.17**). Somehow, we didn't quite get around to reading them all. This incredible pace of research is driven, in part, by talented young scientists who are more excited by neuroscience than competing fields. Excitement about

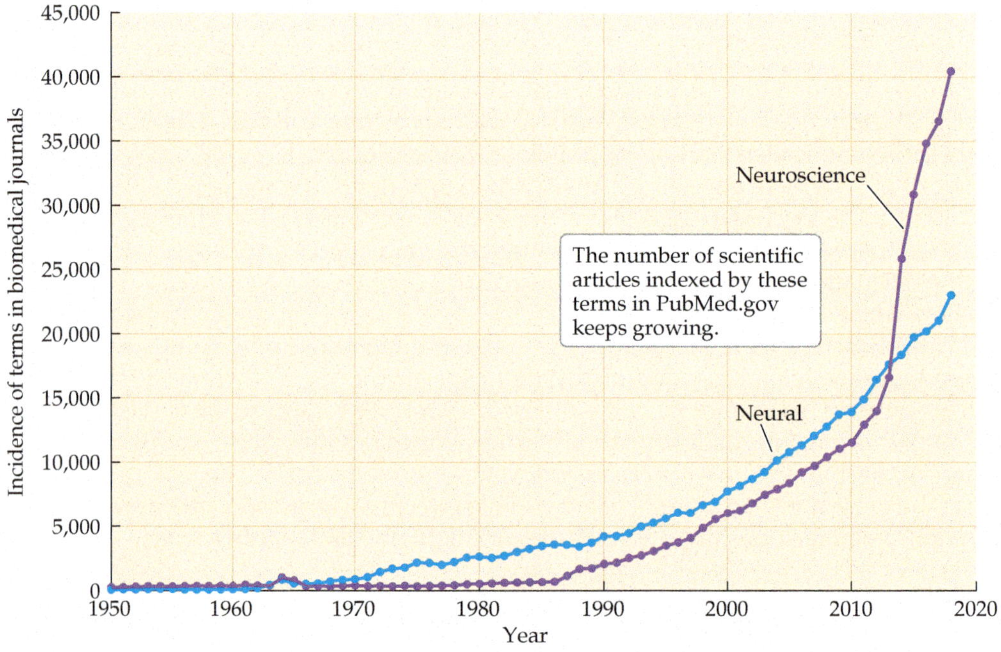

0.17 Neuroscience on the Rise In the major index of biomedical journals, PubMed (pubmed.gov), occurrence of the term *neuroscience* was rather rare in the first 20 years covered by the index. However, use of the term has increased steadily since 1990 and risen *very* sharply in the past decade. It seems safe to say that no one has read, or ever will read, the 40,451 such articles published in 2018 alone. (www.pubmed.gov. Accessed July 2019.)

understanding the brain is also the reason that undergraduate majors in neuroscience are now being offered in dozens of colleges and universities around the United States and the world.

Given this explosion of information in behavioural neuroscience, it is difficult for us to keep this book up to date. We've done our best in every chapter to convey those exciting new concepts that seem to be holding up to the scrutiny of the field, citing over 500 new articles to keep the text current. Each chapter includes a special feature: **The Cutting Edge**. Here we present exciting new findings about the area under discussion. These are the types of findings that have scientists in the field buzzing among themselves. In addition to highlighting new and exciting ideas, we use The Cutting Edge to describe experimental approaches in more detail, to give you a better feel for the process of scientific reasoning and hypothesis testing. We really enjoy writing these breaking news stories about behavioural neuroscience, and we hope that they excite you too. ■

Recommended Reading

Blackmore, S., and Troscianko, E. T. (2018). *Consciousness: An Introduction* (3rd ed.). London: Routledge.

Carter, R. (2009). *The Human Brain Book.* London: Dorling Kindersley.

Dehaene, S. (2014). *Consciousness and the Brain: Deciphering How the Brain Codes Our Thoughts.* New York: Penguin Books.

Godfrey-Smith, P. (2017). *Other Minds: The Octopus, the Sea, and the Deep Origins of Consciousness.* New York: Farrar, Straus and Giroux.

Kaku, M. (2015). *The Future of the Mind: The Scientific Quest to Understand, Enhance, and Empower the Mind.* New York: Random House.

Koch, C. (2012). *Consciousness: Confessions of a Romantic Reductionist.* Cambridge, MA: MIT Press.

Wickens, A. P. (2014). *A History of the Brain: From Stone Age Surgery to Modern Neuroscience.* New York: Psychology Press.

Zimmer, C. (2004). *The Soul Made Flesh: The Discovery of the Brain—and How It Changed the World.* New York: Basic Books.

0 Visual Summary

You should be able to relate each summary to the adjacent illustration, including structures and processes. You can also refer back to the links to figures, animations, and activities that will help you consolidate the material.

1 **Behavioural neuroscience** is a branch of **neuroscience** that focuses on the biological bases of behaviour. It is closely related to many other neuroscience disciplines. Review Figure 0.2

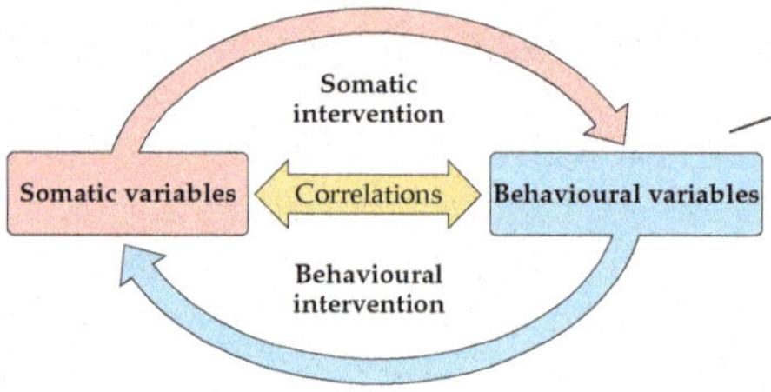

2 Behavioural neuroscientists balance three general research perspectives—**correlation**, **somatic intervention**, and **behavioural intervention**—in designing their research. Review Figure 0.3

3 Research in behavioural neuroscience is conducted at **levels of analysis** ranging from molecular events to the functioning of the entire brain and complex social situations. Review Figure 0.6

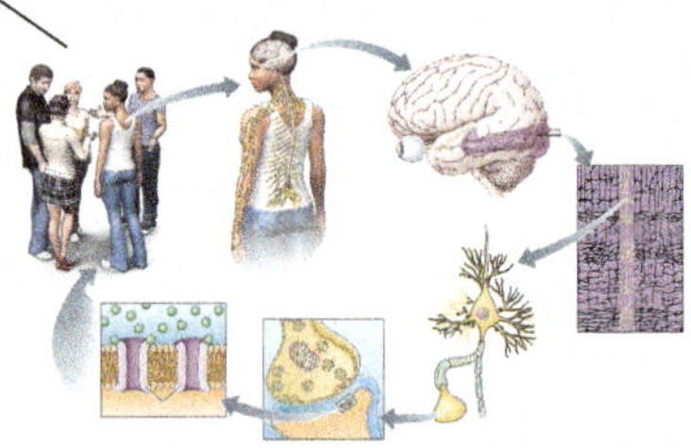

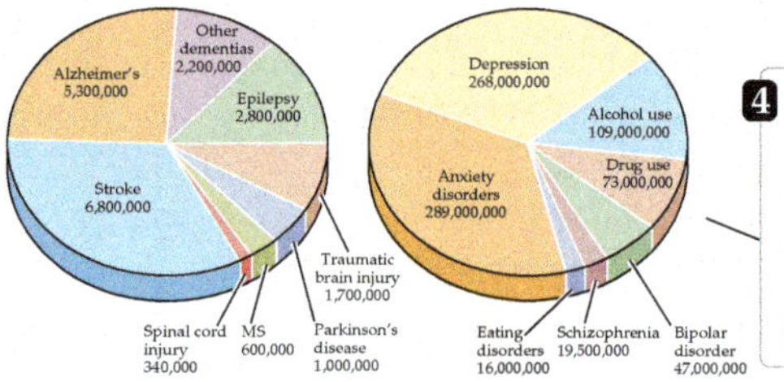

4 The prevalence of neurological and psychiatric disorders exacts a very high emotional and economic toll. Review Figure 0.8

5 Although genes can have a major impact on brain function, it is clear that experience physically alters the brain and that genetically identical people will not necessarily have the same brain disorders. Review Figure 0.9

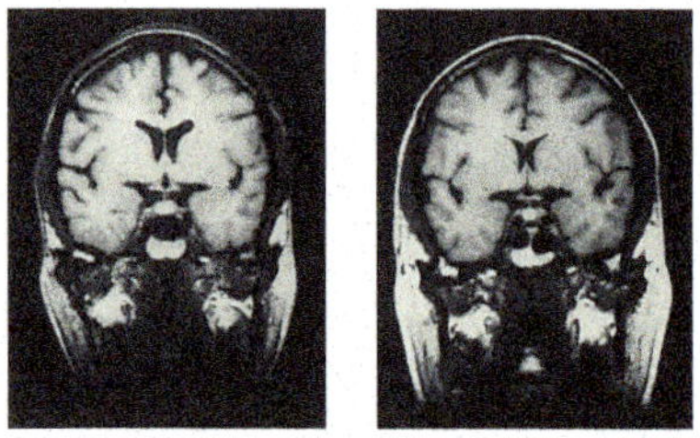

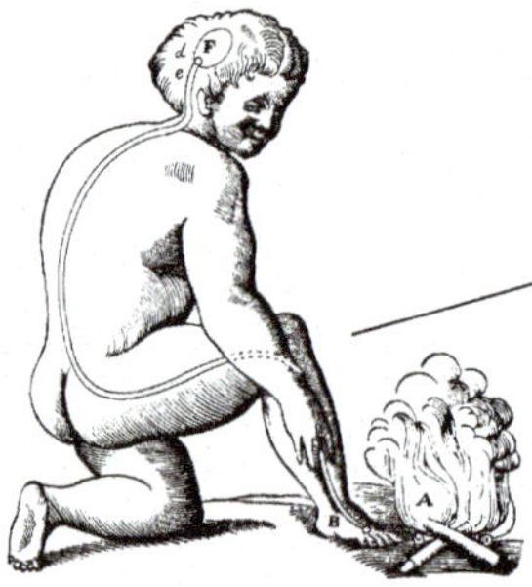

6 Although humans have wondered about the control of behaviour for thousands of years, only comparatively recently has a mechanistic view of the brain taken hold. Review Figure 0.13

7 The concept of localisation of function, which originated in **phrenology**—despite obvious flaws with the phrenologists' methodology—was an important milestone for behavioural neuroscience. Today we know that the part of the brain that shows a peak of activity varies in a predictable way depending on what task we're doing. Review Figure 0.14

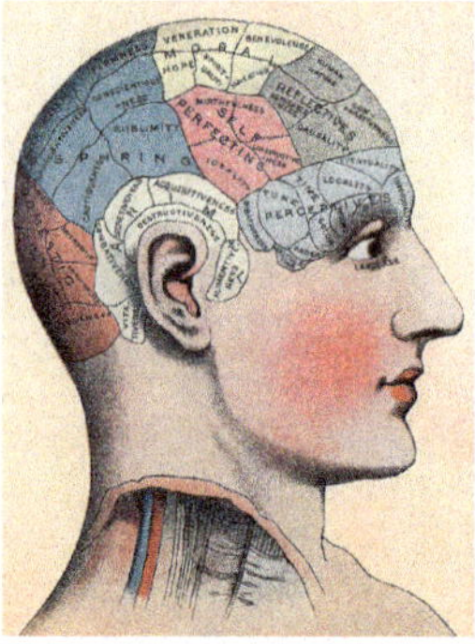

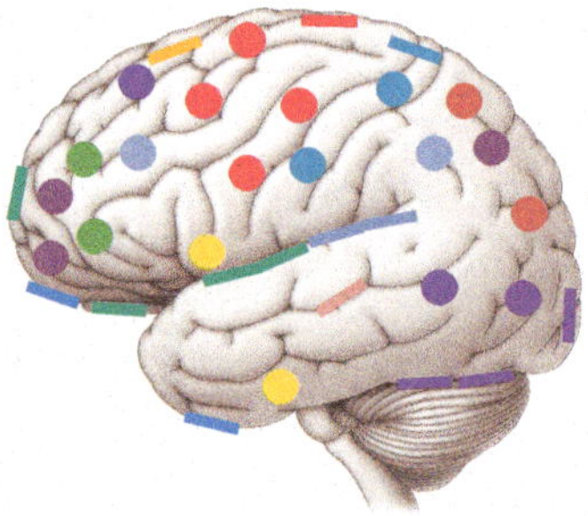

8 Localisation of cognitive functions remains a major focus of behavioural neuroscience. With modern imaging technology and a more carefully validated understanding of cognitive abilities, a detailed view of the organisation of the brain is emerging. Review Figure 0.14

Biological Foundations of Behaviour

PART I

Courtesy of Dr. Ryo Egawa

Functional Neuroanatomy

The Cells and Structures of the Nervous System

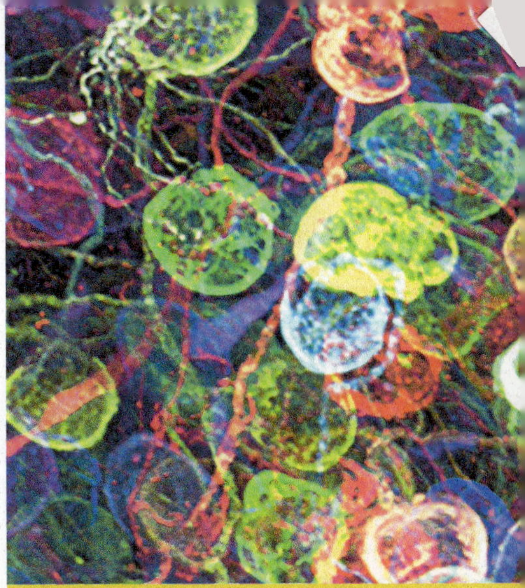

Courtesy of Dr. Ryo Egawa

1

A Stimulating Experience

It was like a scene from a science fiction film. While she remained conscious, comfortable, and aware of her surroundings, Bev's skull was opened and the surface of her brain was exposed. Using a tiny electrode, the surgeon stimulated her brain in precise locations, while sensory experiences and behavioural responses were carefully noted. But this was not make-believe; like thousands of people before her, Bev was undergoing a procedure called *cortical electrical-stimulation mapping*, developed in the mid-twentieth century by neurosurgeon Wilder Penfield (1891–1976). Penfield had learned that mapping the locations of specific functions in someone's brain made it possible to remove diseased brain tissue without harming neighbouring regions involved in crucial behaviours like speech or movement. In Bev's case, the target was a patch of diseased tissue that was causing her to experience frequent *seizures*, uncontrollable convulsions of her body.

But beyond its utility as a surgical tool, brain stimulation offered a way to ask more-profound questions about the organisation of the human brain. Across many individuals, researchers found that stimulation of some brain regions reliably provoked specific movements; stimulating other regions produced specific sensations, like a tingling hand or flashes of blue light. Elsewhere, stimulations could evoke clear and nuanced vignettes of past experiences, such as the smell of a childhood haunt or a fragment of a favorite song. Furthermore, although some regions are organised much the same from person to person, other regions seem to be organised in ways that are unique to each individual, and at the highest levels the brain's control of complex cognition remains mostly a tantalising mystery. New brain-imaging technology is improving our ability to map the brain and start answering fundamental questions about brain organisation: Does each brain region control a specific behaviour, or is the pattern of connections within the brain more important? Do some brain regions act as general-purpose processors? Is everybody's brain organised in the same way?

Thoughts, feelings, behaviours—almost everything that defines you as a person is produced by the three-pound organ inside your head that, despite its unremarkable appearance, is the most complicated object in the known universe. Neuroscientists adopt several different perspectives in describing the physical properties of the brain. Accordingly, in this chapter we look at the structure of the brain from several different perspectives: we open with its cellular composition, go on to its major anatomical divisions, and finally discuss its appearance in computerised brain imaging. In later chapters we'll build on this information as we learn how cells within the brain communicate through electrical (Chapter 2), chemical (Chapter 3), and hormonal (Chapter 4) signals.

1.1 Specialised Cells Make Up the Nervous System

Learning Objectives

After reading this section, you should be able to:

1.1.1 **Give a brief historical account of the study of cellular neuroanatomy.**

1.1.2 **Name and describe the general functions of the four main parts of a neuron.**

1.1.3 **Classify neurons according to both structure and function.**

1.1.4 **Outline the key components of a synapse and the major steps in neurotransmission.**

1.1.5 **Describe the four principal types of glial cells and their functions.**

neuron Also called *nerve cell*. The basic unit of the nervous system, each composed of a cell body, receptive extension(s) (dendrites), and a transmitting extension (axon).

glial cells Also called *glia* or *neuroglia*. Nonneuronal brain cells that provide structural, nutritional, and other types of support to the brain.

neuron doctrine The hypothesis that the brain is composed of cells that are distinct structurally, metabolically, and functionally.

synapse The tiny gap between neurons where information is passed from one to the other.

cell body Also called *soma*. The region of a neuron that is defined by the presence of the cell nucleus.

mitochondrion A cellular organelle that provides metabolic energy for the cell's processes.

cell nucleus The spherical central structure of a cell that contains the chromosomes.

ribosomes Structures in the cell body where genetic information is translated to produce proteins.

Every organ and muscle in your body is in communication with your nervous system. Like all other living tissue, the nervous system is made up of cells, the most important of which are the **neurons** (or *nerve cells*). Each of the 80–90 billion neurons in the adult human brain (Herculano-Houzel, 2012) is a tiny, discrete information-processing unit, receiving inputs from other cells, integrating those inputs, and then distributing the processed information to other neurons. Arranged in networks that vary enormously in complexity, circuits of neurons define and control all of our abilities and behaviours, from the simplest reflexes to the most complex intellectual processes. An even greater number of **glial cells** (or just *glia*) provide various forms of support and also contribute to information processing. But because neurons produce readily measured electrical signals and do most of the work of the brain, we know much more about them than about glial cells.

The interconnection of nerve cells was a hotly debated question in the early days of modern neuroscience. Rapid advances in microscopy allowed nineteenth-century neuroanatomists to visualise the cells of the brain with high resolution, revealing an astonishing variety of shapes and sizes of neurons. However, the manner in which neurons interacted remained mysterious. Some neuroscientists, like Italian Camillo Golgi (1843–1926), thought that neurons were *continuous* with one another, forming a nearly endless network of connected tubes through which information flowed. But the great Spanish neuroscientist Santiago Ramón y Cajal (1852–1934) developed a convincing alternative. Exploiting Golgi's revolutionary staining techniques (see Box 1.1) to create pen-and-ink studies of neurons so precise that they remain accurate and useful to the present day (**FIGURE 1.1**), Cajal proposed that although neurons come very close to one another (i.e., they are *contiguous*), they are not quite continuous with one another. He argued that at each point of contact between neurons, a tiny gap keeps the cells separate.

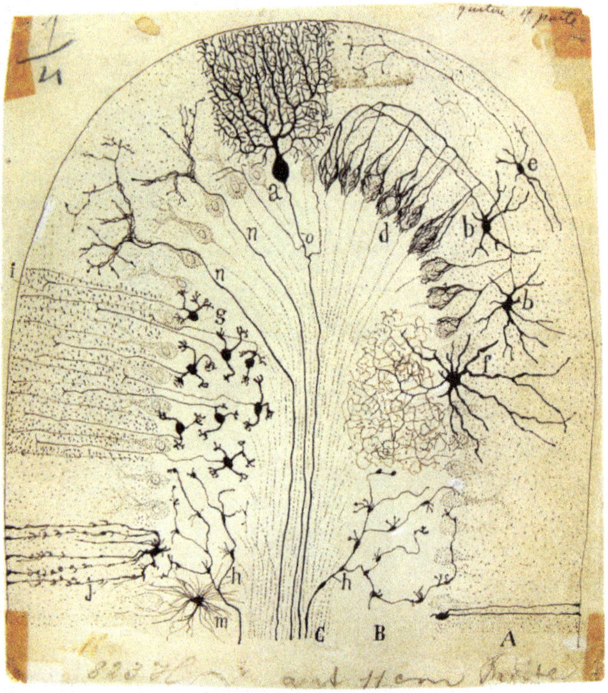

1.1 Nineteenth-Century Drawing of Neurons The great Spanish neuroanatomist Santiago Ramón y Cajal created detailed renderings of the cells of the nervous system, such as this drawing of mammalian brain neurons (labelled by Cajal with lowercase letters).

From these studies emerged a new perspective—the **neuron doctrine**—which proposed (1) that the cells of the brain are independent from one another structurally, metabolically, and functionally and (2) that information is transmitted from one neuron to the next across tiny gaps. The existence and function of these gaps were later demonstrated by Sir Charles Sherrington (1857–1952), who named them **synapses**.

Although estimates vary somewhat, it is thought that the human brain may have as many as 10^{15} synapses. This number, called a *quadrillion*, is so large that it is hard to comprehend: if you gathered that many grains of sand, each a millimeter in diameter, they would fill a cube that is longer on each side than an American football field—over a million cubic yards, or 750,000 cubic meters (in more familiar units, that's about 260 million gallons of sand, or 750 million liters)! These vast networks of connections are responsible for all of humanity's achievements.

The neuron has four structural divisions specialised for information processing

Although neurons come in hundreds of different shapes and sizes (**FIGURE 1.2**), they all share certain features. Like any of the other cells of the body, a neuron has a **cell body** (or *soma*, plural *somata*) that

1.2 **The Great Diversity of Neurons** Neurons come in a bewildering variety of shapes and sizes. These examples, drawn to scale, are taken from the nervous systems of various species.

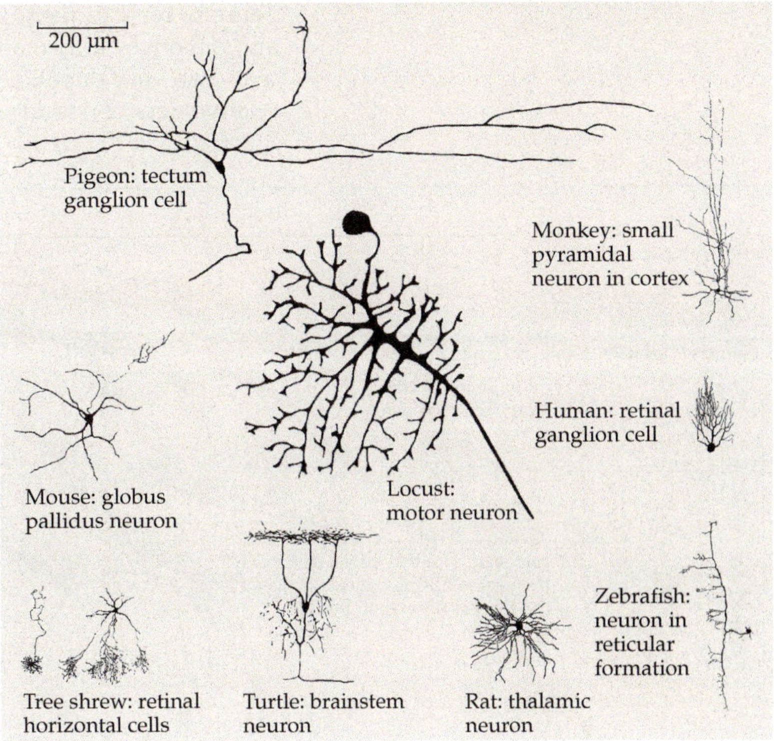

contains a variety of organelles such as the **mitochondria** (singular *mitochondrion*) that produce energy, the **cell nucleus** that contains genes encoded in DNA, and the **ribosomes** and related machinery that translate genetic instructions from the cell nucleus into the proteins needed for the construction and operation of the neuron (consult the Appendix if you need a refresher on cellular actions). But the neuron also has some unique components that allow it to collect input signals from other sources, process and combine this information, and distribute the result of this processing to other cells by means of its own electrochemical output signals. Therefore, neurons all share some distinctive structures that are directly related to information processing. These structures, illustrated in **FIGURE 1.3**, can be viewed as belonging to four distinct functional "zones":

1. **Input zone** At cellular extensions called **dendrites** (from the Greek *dendron*, "tree"), neurons receive information via synapses from other neurons. Dendrites may be elaborately branched to accommodate synaptic contacts from many other neurons.
2. **Integration zone** In addition to receiving additional synaptic inputs, the cell body serves as an integration zone, combining (*integrating*) the information that the neuron has received to determine whether or not to send a signal of its own.
3. **Conduction zone** A single extension, the **axon**, carries (*conducts*) the neuron's own electrical signals away from the cell body. Before its end, the axon may split into multiple branches called **axon collaterals**.
4. **Output zone** At specialised swellings at the ends of the axon, called **axon terminals** (or *synaptic boutons*), the neuron's activity is transmitted across synapses to other cells.

Neurons can be classified by size, shape, or function

The tremendous diversity of sizes, shapes, and functions of neurons reflects their varying roles, so neuroscientists rely on all three attributes in order to classify neurons. First, vertebrate nerve cell bodies vary tremendously in *size*, ranging from as small as 10 micrometers (μm) to as large as 100 μm or more in diameter; this variability in neuronal sizes is evident in Figure 1.2. In general, larger

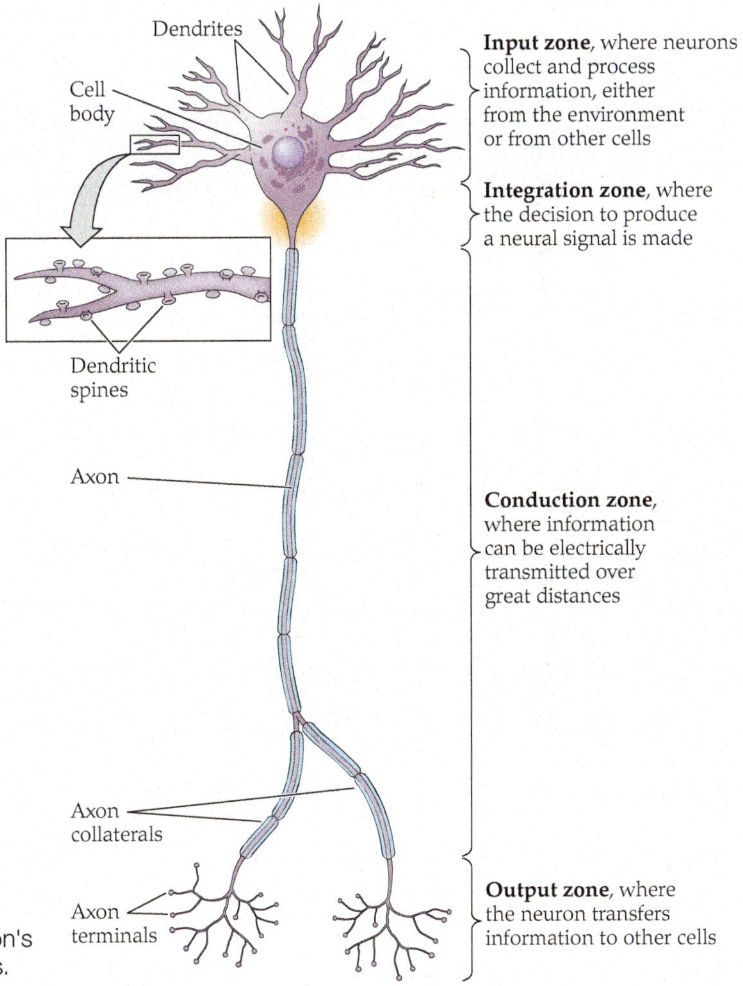

1.3 **The Major Parts of the Neuron** The dendrites and cell body of the neuron may be covered with thousands of synaptic contacts providing inputs from other neurons. Likewise, the neuron's axon terminals pass information on to other neurons via synapses.

neurons tend to have more-complex inputs and outputs, cover greater distances, and/or convey information more rapidly than smaller neurons. The size of neurons may also vary depending on their activity level and rate of protein synthesis. The relative sizes of neural structures that we will be discussing throughout the book are illustrated in **FIGURE 1.4**.

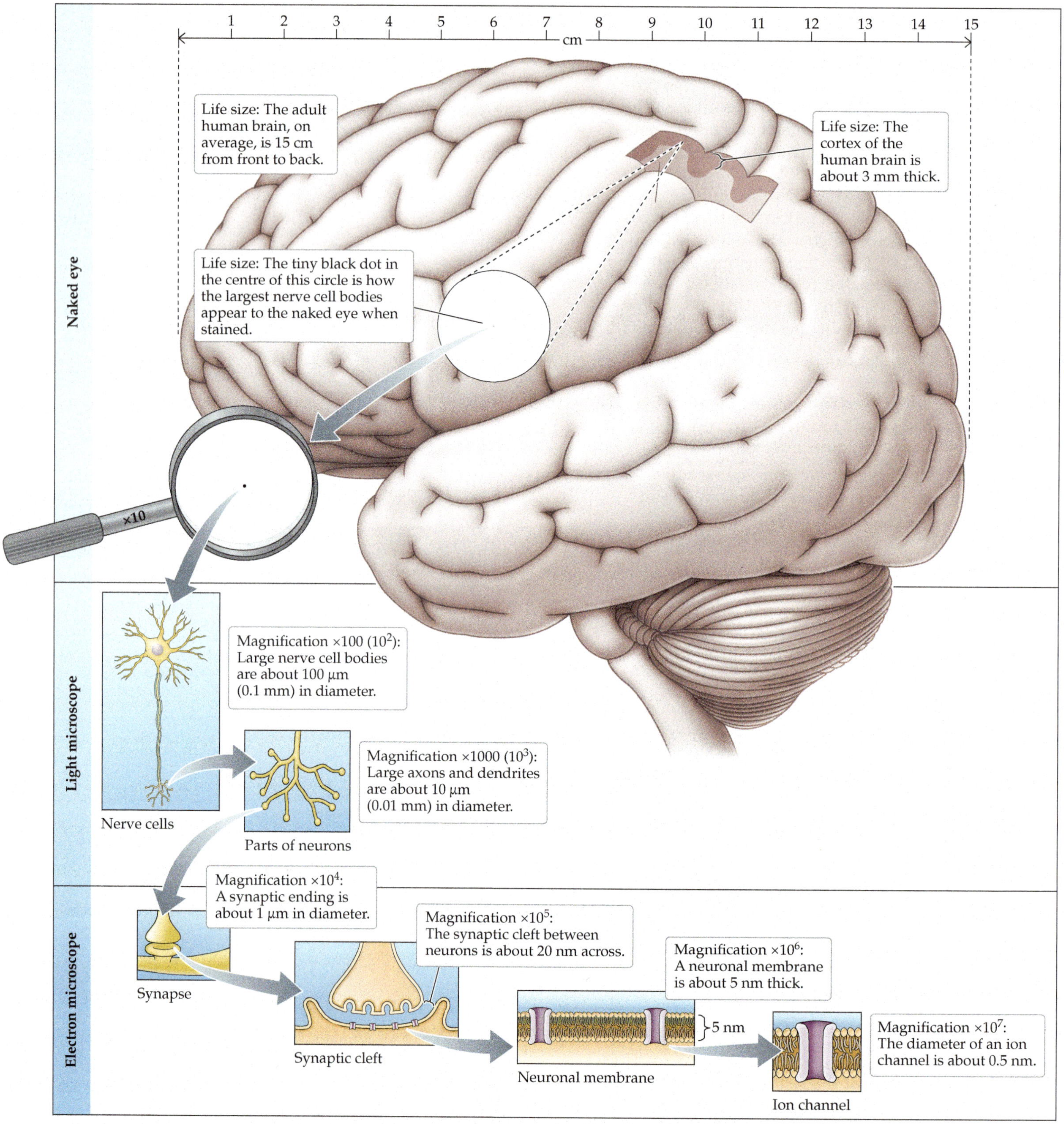

1.4 Sizes of Some Neural Structures and the Units of Measure and Magnification Used in Studying Them

Furthermore, we can use the *shapes* of cell bodies, dendrites, and axons to classify the many varieties of nerve cells into three principal forms (**FIGURE 1.5**), each specialised for a particular kind of information processing:

1. **Multipolar neurons** have many dendrites and a single axon, and they are the most common type of neuron (**FIGURE 1.5A**).
2. **Bipolar neurons** have a single dendrite at one end of the cell and a single axon at the other end (**FIGURE 1.5B**). This type of neuron is especially common in sensory systems, such as vision.
3. **Unipolar neurons** (also called *monopolar*) have a single extension (or *process*), usually thought of as an axon, that branches in two directions after leaving the cell body (**FIGURE 1.5C**). One end is the input zone with branches like dendrites; the other, the output zone. Such cells transmit touch information from the body into the spinal cord.

In all three types of neurons, the dendrites are in the input zone. In multipolar and bipolar cells, the cell body also receives synapses and so is also part of the input zone.

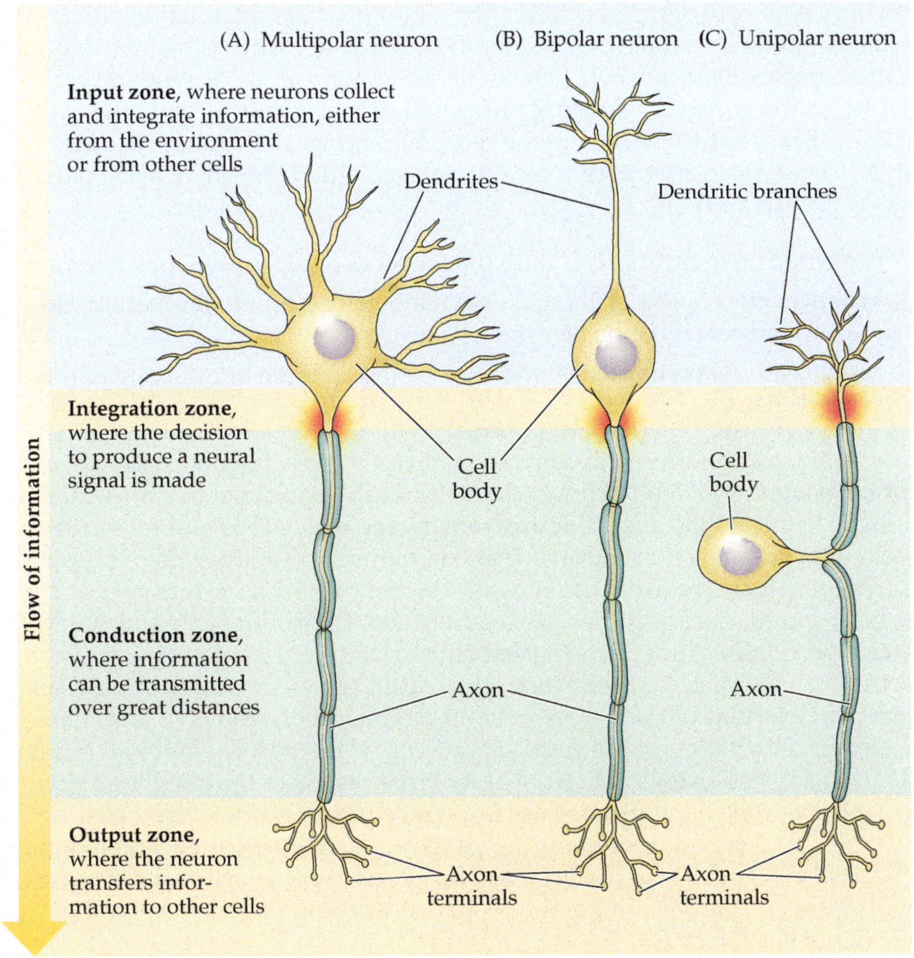

1.5 Three Principal Forms of Neurons Despite their variation in shape, neurons all feature input, integration, conduction, and output zones. Three important forms of neurons are (A) multipolar neurons with multiple dendrites and a single axon, (B) bipolar neurons with a single dendrite and a single axon, and (C) unipolar neurons with a single process that emerges from the cell body and extends in two directions; in this case the integration zone is not in the cell body but at the base of the dendritic branches.

input zone The part of a neuron that receives information from other neurons or from specialised sensory structures. Usually corresponds to the cell's dendrites.

dendrite One of the extensions of the cell body through which synaptic inputs are received.

integration zone The part of the neuron that initiates nerve electrical activity, described in detail in Chapter 3. Usually corresponds to the neuron's axon hillock.

conduction zone The part of the neuron over which the nerve's electrical signal may be actively propagated. Usually corresponds to the cell's axon.

axon A single extension from the nerve cell that carries action potentials from the cell body to other neurons.

axon collateral A branch of an axon from a single neuron.

output zone The part of a neuron, usually corresponding to the axon terminals, at which the cell sends information to another cell.

axon terminal Also called *synaptic bouton*. The end of an axon or axon collateral, which forms a synapse on a neuron or other target cell.

multipolar neuron A nerve cell that has many dendrites and a single axon.

bipolar neuron A nerve cell that has a single dendrite at one end and a single axon at the other end.

unipolar neuron Also called *monopolar neuron*. A nerve cell with a single branch that leaves the cell body and then extends in two directions; one end is the receptive pole, the other end the output zone.

Go to Activity 1.1
Major Components and Classification of Neurons

interneuron A neuron that is neither a sensory neuron nor a motor neuron; it receives input from and sends output to other neurons.

motor neuron Also called *motoneuron*. A nerve cell that transmits motor messages, stimulating a muscle or gland.

sensory neuron A neuron that is directly affected by changes in the environment, such as light, odour, or touch.

arborisation The elaborate branching of the dendrites of some neurons.

presynaptic Referring to the region of a synapse that releases neurotransmitter.

postsynaptic Referring to the region of a synapse that receives and responds to neurotransmitter.

presynaptic membrane The specialised membrane of the axon terminal of the neuron that transmits information by releasing neurotransmitter.

synaptic cleft The space between the presynaptic and postsynaptic membranes.

postsynaptic membrane The specialised membrane on the surface of the cell that receives information by responding to neurotransmitter from a presynaptic neuron.

synaptic vesicle A small, spherical structure that contains molecules of neurotransmitter.

neurotransmitter Also called *synaptic transmitter*, *chemical transmitter*, or simply *transmitter*. The chemical released from the presynaptic axon terminal that serves as the basis of communication between neurons.

receptor Also called *receptor molecule*. A protein that binds and reacts to molecules of a neurotransmitter or hormone.

Neurons also vary in their basic *function*. The great majority of neurons in the brain, often featuring complex dendrites and relatively short axons (because they convey information only short distances) are classified as **interneurons**. These are the neurons that receive information from other neurons, process it, and pass the integrated information to other neurons. Interneurons make up the hugely intricate networks and circuits that perform the complex functions of the brain. In contrast, the neurons that connect the brain and spinal cord to the rest of the body have axons of up to a meter in length (in fact, the giraffe has axons that are *several* meters in length). In order for you to wiggle your toes (or for the giraffe to wiggle hers), individual axons must carry the instructions from the spinal cord to muscles of the foot. **Motor neurons** (or *motoneurons*)—the neurons that govern movements—have long axons reaching out to synapse on muscles, causing them to contract in response to commands from the brain. Other motor neurons contact and control organs and glands. The long axons of **sensory neurons** carry messages from peripheral tissue (sensors in the skin, for example) back to the spinal cord and brain. Sensory neurons take many different shapes, depending on whether they detect light or sound or touch and so on. So, neurons are remarkably diverse, their forms reflecting their highly specialised functions.

Information is transmitted across synapses

The arrangement of a neuron's dendrites—its tree branch–like **arborisation**—reflects the complexity of the neuron's information-processing function. Some simple neurons have just a couple of short dendritic branches, while other neurons have huge and complex dendritic trees receiving many thousands of synaptic contacts from other neurons. At each synapse, information is transmitted from the axon terminal of the **presynaptic** neuron to the receptive surface of the **postsynaptic** neuron (**FIGURE 1.6A**), so a synapse typically can be divided into three principal components (**FIGURE 1.6B** and **C**):

1. The **presynaptic membrane** of the axon terminal of the presynaptic neuron
2. The **synaptic cleft**, a gap of about 20–40 nanometers (nm) that separates the presynaptic and postsynaptic neurons
3. The specialised **postsynaptic membrane** on the dendrite or cell body of the postsynaptic neuron

Presynaptic axon terminals contain numerous tiny hollow spheres, called **synaptic vesicles**, each 30–140 nm in diameter. Each vesicle contains molecules of a specialised chemical substance, a **neurotransmitter**, which the neuron uses to communicate with postsynaptic neurons. This communication starts when, in response to electrical activity in the axon, the synaptic vesicles fuse with the presynaptic membrane and rupture, releasing their payload of neurotransmitter molecules into the cleft (see Figure 1.6B). After diffusing across the cleft, the released neurotransmitter interacts with postsynaptic **receptors**: specialised protein molecules that capture and react to molecules of the neurotransmitter. This action results in local changes in the electrical excitation of the postsynaptic cell. If the postsynaptic cell is a neuron (as most are), the events occurring at its synapses affect the likelihood that the postsynaptic neuron will in turn release its own neurotransmitter. Molecules of neurotransmitter generally do not enter the postsynaptic neuron; they simply bind to the receptors momentarily to induce a response, and then detach and diffuse away. Many different substances are known to act as neurotransmitters; we will discuss them in depth in Chapter 3.

The postsynaptic membrane contains a high density of receptors. And because each synapse occupies a very small patch of the postsynaptic neuron—less than 1 μm^2—a large number of synapses can cover the surfaces of the dendrites and cell body. In fact, some neurons receive as many as 100,000 synaptic contacts, although a more common number is about 5,000–10,000. As you might expect, neurons with elaborate dendrites tend to have more synaptic inputs.

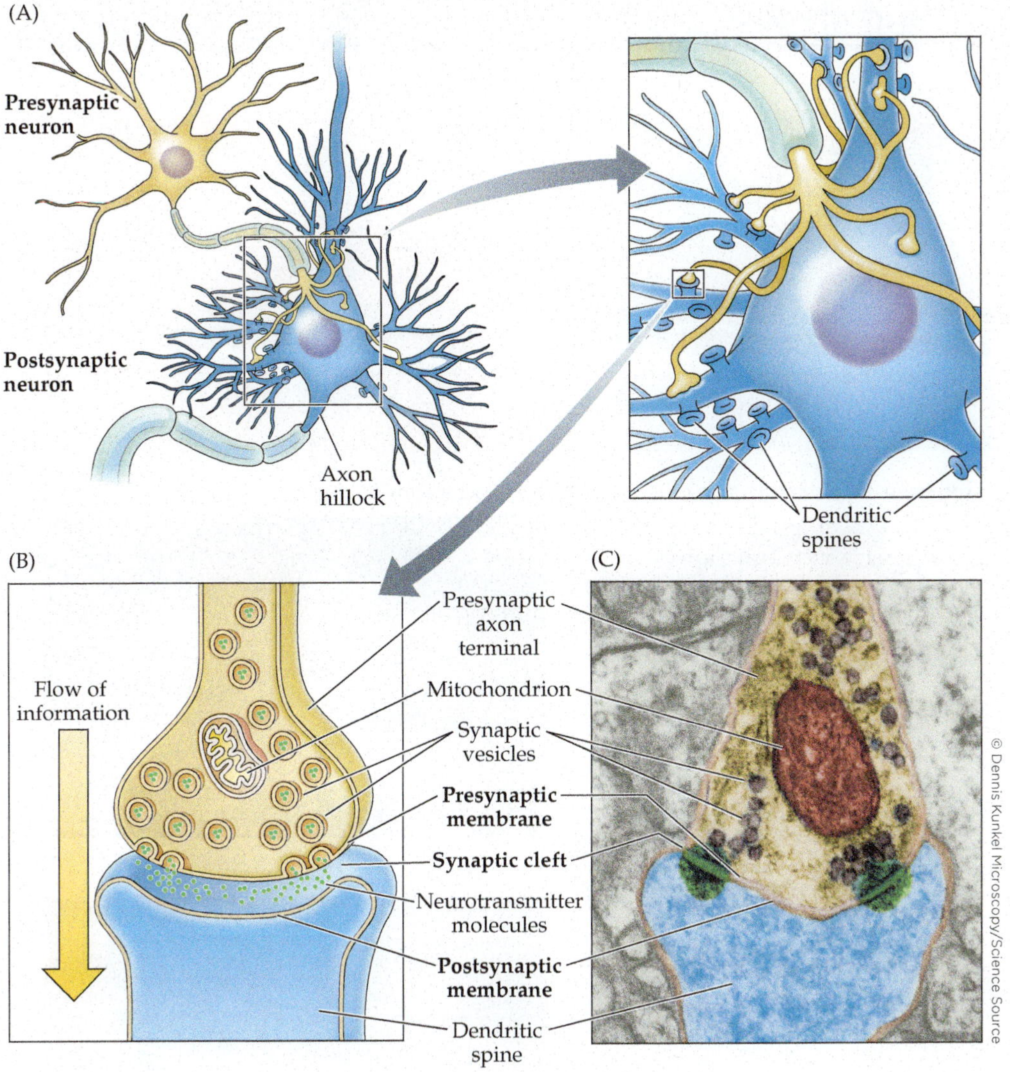

1.6 Synapses (A) Axon terminals typically form synapses on the cell body or dendrites of a neuron. On dendrites, synapses may form on dendritic spines or on the shaft of a dendrite. (B) Information flows through a synapse from the presynaptic membrane across a gap called the *synaptic cleft* to the postsynaptic membrane. (C) This photomicrograph shows a synapse with some structures colour coded.

The configuration of synapses on a neuron's dendrites and cell body is constantly changing—synapses come and go, and dendrites change their shape—in response to new patterns of synaptic activity and the formation of new neural circuits. We use the general term **neural plasticity** to refer to this continual remodeling of the connections between neurons. Studding the dendrites of many neurons are outgrowths called *dendritic spines* (see Figure 1.3) that, by effectively increasing the surface area of the dendrites, allow for extra synaptic contacts. Both the number and structure of dendritic spines may be rapidly altered by experience, such as training or exposure to sensory stimuli (see Chapter 17). This plastic property of dendritic spines has made them the focus of intensive research efforts. Some dendritic spines change from minute to minute, while others may be stable for a lifetime (Moyer and Zuo, 2018; Stein and Zito, 2018).

The axon integrates and then transmits information

A typical axon has several distinct regions. The axon arises from the **axon hillock** ("little hill"), a cone-shaped projection of the cell body (see Figure 1.6A). The axon hillock is the neuron's integration zone, gathering and integrating information from all the synapses on the dendrites and cell body, then converting the processed information into a code of electrical impulses that carries the neuron's message down the axon toward its targets (see Chapter 3). The axon is a hollow tube, with a diameter ranging from 0.2 to 20 μm in mammals, but up to 500 μm or more in the "giant" axons of some invertebrates (Debanne et al., 2011).

neural plasticity Also called *neuroplasticity*. The ability of the nervous system to change in response to experience or the environment.

axon hillock A cone-shaped area from which the axon originates out of the cell body. Functionally, the integration zone of the neuron.

BOX 1.1

Visualising the Cells of the Brain

(A) Nissl stain

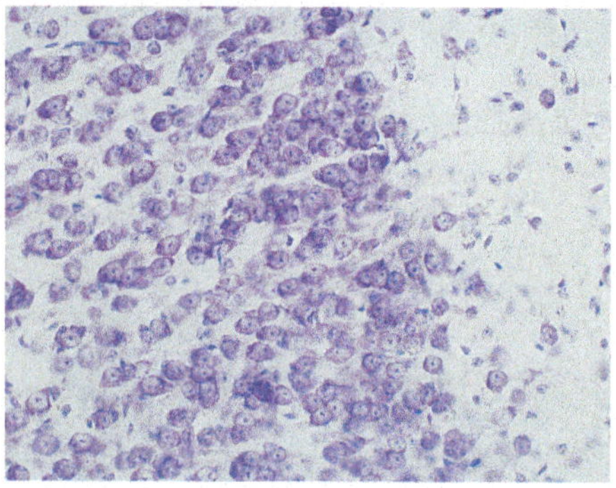

Courtesy of Cynthia L. Jordan (Michigan State University)

(B) Golgi stain

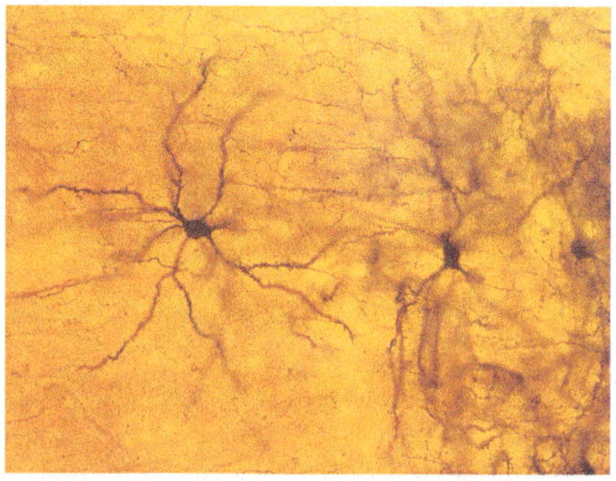

Courtesy of Dr. Timothy DeVoogd

(C) Expression of *c-fos* in activated cells

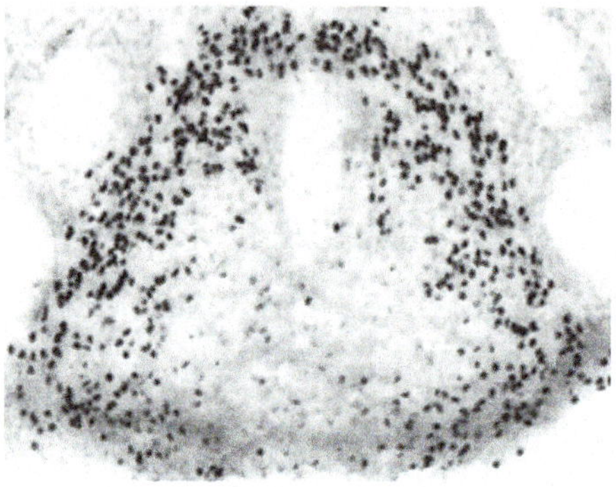

From N. Sunn et al., 2001. *Proc Natl Acad Sci* 99: 1701–1706. © 2002 National Academy of Sciences, U.S.A.

(D) Tract tracing

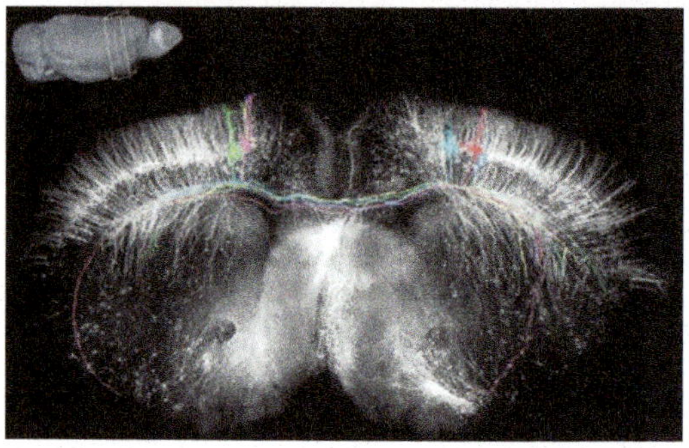

J. Yuan et al., 2015. *Front Neuroanat* 9: 70 CC BY 4.0

(E) Brainbow

Courtesy of Ryo Egawa

Histology—the scientific study of the composition of tissue—underwent a revolution beginning in the mid-1800s, when derivatives of fabric dyes were found to vividly stain cells in ways that revealed previously hidden microscopic structure. In the nervous system, it became possible to selectively stain different parts of neurons and glia, such as cell membranes, the cell body, or the sheaths surrounding axons. Nowadays, scientists use specialised staining procedures to study the numbers, shapes, distribution, and interconnections of neurons within targeted regions of the brain.

Counting Cells in Brain Regions

Some traditional cell stains, collectively known as **Nissl stains**, outline *all* cell bodies because the dyes are attracted to RNA distributed within the cell. Nissl stains allow us to measure cell body size and the density of cells in particular regions (Figure A). Many types of Nissl stains and other traditional general-purpose cell stains are available.

Examining the Forms of Individual Neurons

Mysteriously, and in contrast to Nissl stains, **Golgi stains** label only a small minority of neurons in a sample, but the affected cells are stained very deeply and completely, revealing fine details of cell structure such as the branches of dendrites and axons (Figure B). Golgi-stained neurons stand out in sharp contrast to their unstained neighbours, so Golgi staining is useful for quantifying the types and precise shapes of neurons in a region. There are several variants on this strategy, such as filling cells with fluorescent molecules.

Mapping the Expression of Cellular Products

Often, neuroscientists would like to know the distribution of neurons that exhibit a specific property, such as sensitivity to a hormone or drug, production of particular proteins, or synthesis of the RNA message that indicates that a specific gene has been activated. Numerous clever techniques have been developed to trick such neurons into revealing themselves. In **autoradiography**, for example, animals are treated with radioactive versions of experimental drugs, and then thin slices of the brain are placed alongside photographic film. Radioactivity emitted by the labelled compound in the tissue "exposes" the film's emulsion in the same way that light normally would. Thus, the brain essentially takes a picture of itself, highlighting the specific brain regions where the drug has become selectively concentrated. An alternative way to visualise cells that have an attribute in common—termed **immunohistochemistry (IHC)**—involves creating antibodies against a protein of interest (we can create antibodies to almost any protein). Equipped with colourful labels, these antibodies can selectively seek out and attach themselves to their target proteins within neurons in a brain slice, thereby revealing the distribution of only those neurons that make the target protein. In the example in Figure C, antibodies have labelled only those cells containing a protein expressed by *c-fos*, which is an immediate early gene (IEG) that is expressed in cells that have been recently active. Localising IEG proteins allows researchers to identify brain regions that were active during particular behaviours performed by an animal shortly before it was euthanised.

A related procedure called **in situ hybridisation** goes a step further and, using radioactively labelled strands of nucleic acid (RNA or DNA, see the Appendix), labels only those neurons in which a gene of interest has been turned on.

Tracing Interconnections between Neurons

Many research questions are more concerned with the pattern of connections between neurons than with their cellular structure (Figure D). But tracing the interconnections between regions is a technical challenge because axons are profuse and tiny, follow intricate routes, and are difficult to disentangle from one another. To accomplish this goal, scientists have developed many sorts of **tract tracers**, substances that are taken up by neurons and transported over the routes of their axons. In *anterograde labeling* the tract tracer is injected near the dendrites and cell bodies of a region of interest, where it is taken up and transported to the tips of the axons, thus revealing the *targets* of the neurons in the region under study. Conversely, in *retrograde labeling*, when a different kind of tract tracer is injected into a region of interest, it is taken up by axon terminals and then transported back to their originating cell bodies, thus revealing the *sources* of innervation of the region. A few specialised retrograde tract tracers (such as labelled pseudorabies virus) can even work *transsynaptically*: they jump backward across synapses and work their way "upstream," back towards higher levels of the nervous system, leaving visible molecules of label all along the way. More recently, genetic manipulations have been devised to insert and then activate the genes for multiple colourful fluorescent proteins within neurons; the resultant Brainbow imagery, in which the neurons express hundreds of possible hues (Figure E), has revolutionised the study of interconnections of neurons (Lichtman et al., 2008; Weissman and Pan, 2015).

histology The scientific study of the composition of tissues.

Nissl stain A cell stain that reveals all cell bodies by staining RNA.

Golgi stain A cell stain that fills a small proportion of neurons with a dense dark product.

autoradiography A histological technique that shows the distribution of radioactive chemicals in tissues.

immunohistochemistry (IHC) A technique in which labelled antibodies are used to visualise the histological distribution of specific proteins.

in situ hybridisation A technique in which labelled complementary nucleic probes are used to identify cells expressing specific messenger RNA transcripts, reflecting the activation of specific genes of interest.

tract tracer A compound used to identify the routes and interconnections of neuronal projections.

With very few exceptions, neurons have only one axon. But as noted earlier, this solitary axon often divides into several axon collaterals, allowing the neuron to **innervate** (influence) numerous postsynaptic cells. **TABLE 1.1** compares the main structural features of axons and dendrites. Some of the wide variety of technologies used to visualise cell bodies, axons, and dendrites are described in **BOX 1.1**.

innervate To provide neural input.

The cell body manufactures various materials, such as enzymes and structural proteins, under the guidance of the DNA (deoxyribonucleic acid) contained in the cell nucleus (see the Appendix). Important substances needed at the axon terminals are loaded into *transport vesicles*—hollow spheres with specialised leglike *motor proteins* on their outer surface. When activated, the motor proteins literally "walk" the vesicles

TABLE 1.1 Distinctions between Axons and Dendrites

Property	Axons	Dendrites
Number	Usually one per neuron, with many terminal branches	Usually many per neuron
Diameter	Uniform until start of terminal branching	Tapering progressively toward ending
Axon hillock	Present	No hillock-like region
Sheathing	Usually covered with myelin	No myelin sheath
Length	Ranging from practically nonexistent to several meters long	Often much shorter than axons

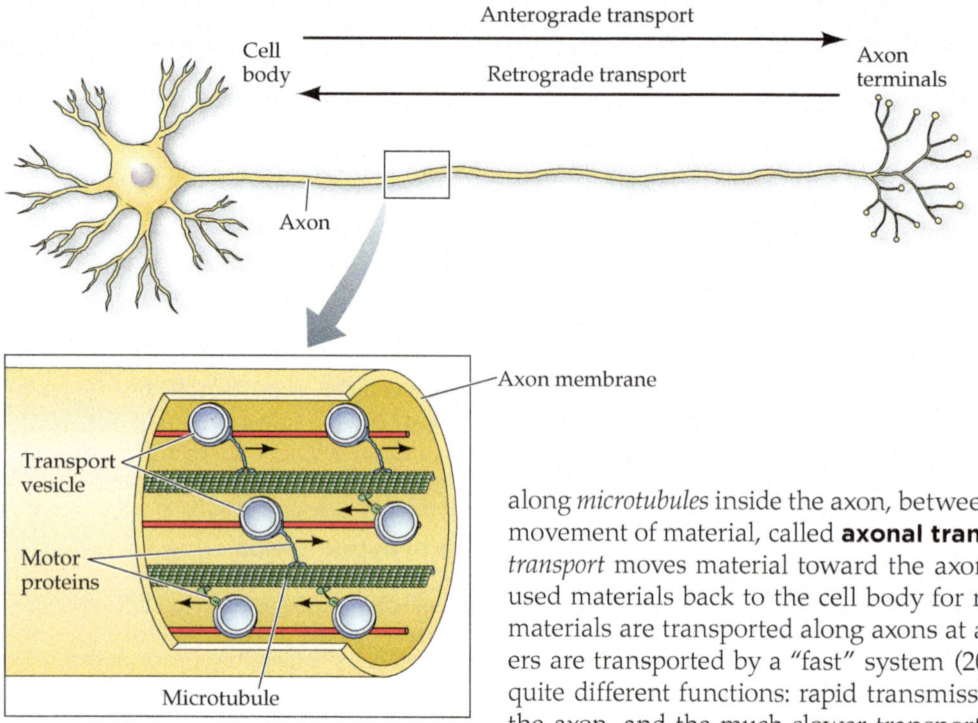

1.7 Axonal Transport Motor proteins literally "walk" transport vesicles, laden with important molecules, along the length of the axon in both directions.

along *microtubules* inside the axon, between the cell body and the axon terminals. This movement of material, called **axonal transport**, works in both directions: *anterograde transport* moves material toward the axon terminals, and *retrograde transport* moves used materials back to the cell body for recycling (**FIGURE 1.7**). Furthermore, some materials are transported along axons at a "slow" rate (less than 8 mm per day); others are transported by a "fast" system (200–400 mm per day). So, the axon has two quite different functions: rapid transmission of electrical signals along the *outside* of the axon, and the much slower transportation of substances *inside* the axon, to and from the axon terminals. You can find links to videos of both types of axonal activity on the website.

Glial cells support and enhance neural activity

Glial cells were originally believed to hold the nervous system together (the Greek *glia* means "glue"). But glial cells can also communicate with each other and with neurons, and they directly affect neuronal functioning by providing neurons with raw materials and chemical signals that alter neuronal structure and excitability (**FIGURE 1.8**). In its exclusion of glial cells, the neuron doctrine was perhaps an oversimplification (Bullock et al., 2005).

There are at least as many glial cells as neurons in the brain—even more than this 1 to 1 ratio in some regions (Azevedo et al., 2009; Herculano-Houzel, 2014)—and yet in contrast to the hundreds of types of neurons, glial cells come in only four basic forms, three of which are found in the brain. One type, called an **astrocyte** (from the Greek *astron*, "star"), is a star-shaped cell with numerous processes extending in all directions (**FIGURE 1.8A**), weaving among neurons. Some astrocytes form sucker-like *end feet* on blood vessels, regulating local blood flow to provide more supplies to neurons when they are active (Schummers et al., 2008). Astrocytes receive synapses directly from neurons and surround and monitor the activity of nearby neuronal synapses. There is evidence that cross talk among astrocytes and neighbouring neurons constitutes a "tripartite synapse" (*tripartite* means "three-part"), with astrocytes directly participating in the transmission of information between neurons (Eroglu and Barres, 2010; Perea et al., 2009), although the extent and significance of this sort of "gliotransmission" remains controversial (Agulhon et al., 2010). Astrocytes are also involved in the formation of new synapses, as well as the pruning of surplus synapses that is a normal part of brain development (W. S. Chung et al., 2013).

A second type of glial cell is the **microglial cell** (**FIGURE 1.8B**). As the name suggests, microglial cells are very small. They are also remarkably active, continually extending and withdrawing very fine processes that, when they contact a site of damage, form a spherical containment zone around the injury (Wolf et al., 2017). The brain's cleanup crew, microglial cells migrate to sites of injury or disease in the nervous system to remove debris from injured or dead cells. But we are beginning to realise that microglial cells are involved in more than just damage control: for example, microglial cells appear to be involved in aspects of pain perception (S. Beggs et al., 2012) and neuronal-modelling (Schafer et al., 2012).

axonal transport The transportation of materials between the neuronal cell body and axon terminals.

astrocyte A star-shaped glial cell with numerous processes (extensions) that run in all directions.

microglial cells Also called *microglia*. Extremely small glial cells that remove cellular debris from injured or dead cells.

oligodendrocyte A type of glial cell that forms myelin in the central nervous system.

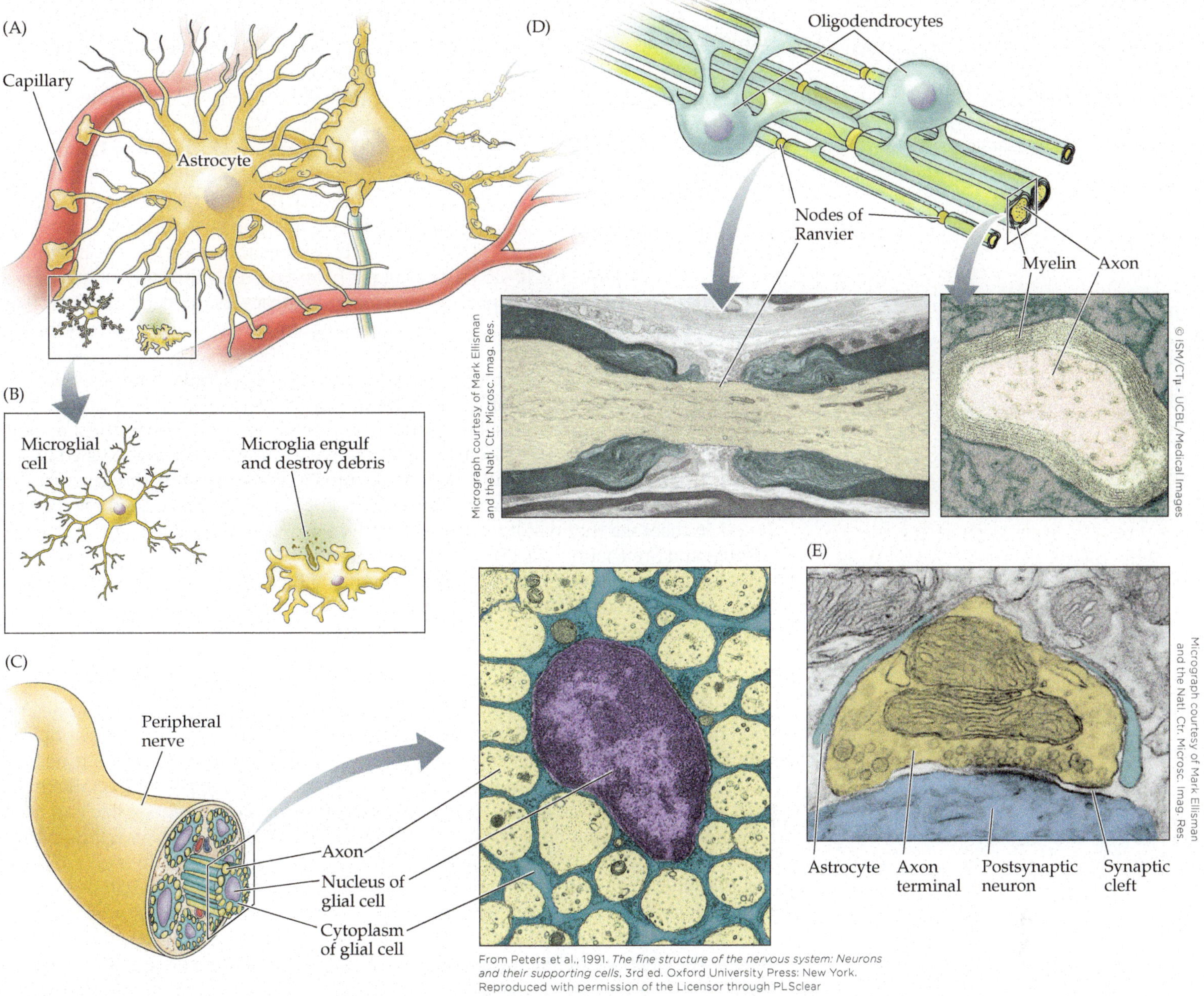

1.8 Glial Cells (A) Star-shaped astrocytes detect neural activity and regulate adjacent capillaries to control blood flow, supplying neurons with more energy when they are active. (B) Tiny microglial cells surround and break down any debris that forms, especially after damage to the brain. (C) Unmyelinated axons are embedded in the troughs of glial cells. The light-coloured circular shapes in the photograph are unmyelinated axons surrounded by the cytoplasm (blue) of a glial cell (the large dark area is the glial nucleus). (D) Extensions of oligodendrocytes form myelin wrapping (blue) on axons (yellow). The colourised electron micrograph of a myelinated axon (lower right) shows the many layers of the myelin sheath. The longitudinal micrograph of an axon (lower left) shows a node of Ranvier, the gap between adjacent myelinated segments. (E) Processes from astrocytes (blue) surround and insulate synapses and directly modify synaptic activity.

The remaining types of glial cells—**oligodendrocytes** and **Schwann cells**—perform a very different yet vital function called **myelination**. All along the axons of many neurons, these glial cells wrap sections of the axon in multiple layers of **myelin**, a fatty insulating substance, giving the axon the appearance of a string of slender beads. Between adjacent beads, small uninsulated patches of axonal membrane, called **nodes of Ranvier**, remain exposed (**FIGURE 1.8D**).

Within the brain and spinal cord, myelination is provided by the oligodendrocytes, with each cell typically providing myelin beads to several nearby axons (see Figure 1.8D). In the rest of the body, Schwann cells do the ensheathing; a single Schwann cell ensheathes a limited length of a single axon. But whether it is provided

Schwann cell A glial cell that forms myelin in the peripheral nervous system.

myelination The process of ensheathing axons in myelin.

myelin The fatty insulation around an axon, formed by glial cells, that speeds the conduction of action potentials.

node of Ranvier A gap between successive segments of the myelin sheath where the axon membrane is exposed.

edema The swelling of tissue, such as in the brain, in response to injury.

multiple sclerosis Literally, "many scars." A disorder characterised by widespread degeneration of myelin.

by oligodendrocytes or Schwann cells, myelination's result is the same—a large increase in the speed with which electrical signals pass down the axon, jumping from one node to the next (see Chapter 3). Many thin, short axons lack myelin but still are surrounded by oligodendrocytes or Schwann cells, which segregate the unmyelinated axons (**FIGURE 1.8C**). Furthermore, the manner in which glial cells surround some synaptic contacts suggests that one of their roles is to insulate and isolate synapses to prevent one from affecting the other (**FIGURE 1.8E**). Oligodendrocytes also provide chemical signals (*trophic signals*) that enhance the structural integrity of axons (Nave, 2010). The process of myelination continues for a long time in humans—in some brain regions for 10–15 years after birth, and possibly throughout life. In fact, more than 75% of the brain's glial cells are oligodendrocytes, reflecting the complexity and importance of myelination. In contrast, about 17% of glia are astrocytes, and microglia make up just 7% of the total (Pelvig et al., 2008).

Glial cells are of clinical interest for a variety of reasons. Unlike neurons, glial cells continue to divide throughout life, and consequently they form many of the types of tumours that may arise in the brain. Some glial cells, especially astrocytes, respond to brain injury by changing in size—that is, by swelling. **Edema**—the swelling of tissue—damages neurons and is responsible for many symptoms of brain injuries. Astrocytes also directly influence local brain chemistry, so they have been implicated in diseases that result from changes in neuronal excitability, such as epilepsy (Robel and Sontheimer, 2015). Astrocytes and microglia are also increasingly implicated in degenerative processes like Alzheimer's disease, Parkinson's disease, and amyotrophic lateral sclerosis (Liddelow et al., 2017). Normally, microglia remove debris and also help maintain synapses, but pathological changes in their microenvironment can cause microglia to show a damaging inflammation response, making them a possible target in the search for Alzheimer's treatments (W. S. Chung et al., 2015; Perry and Holmes, 2014). Disease processes that interfere with the myelination provided by oligodendrocytes can have a wide variety of effects. For example, in **multiple sclerosis** the loss of the insulating myelin sheath from axons in various regions of the brain can have severe consequences that vary depending on the region that is affected, just as losing the outer insulation from a computer cable can cause short circuits and the loss of vital information. Changes in all three kinds of glial cells, especially the loss of oligodendrocytes and their associated myelin, are implicated in the onset and symptoms of schizophrenia (Bernstein et al., 2015).

Supported and influenced by glial cells, and sharing information through synapses, neurons form the vast ensembles of circuits that intricately process information, giving the brain its visible form. Powerful anatomical and genetic initiatives, such as the Allen Institute's brain-mapping project (www.portal.brain-map.org) and database of individual cell characteristics (celltypes.brain-map.org), are now creating detailed maps of the cellular compositions of the major divisions of the brain. These major divisions are our next topic.

1.2 The Nervous System Consists of Central and Peripheral Divisions

Learning Objectives

After reading this section, you should be able to:

1.2.1 **Describe the composition of nerves, and distinguish between somatic and autonomic nerves.**

1.2.2 **Give a general description of the naming conventions and functions of the cranial and spinal nerves.**

1.2.3 **Name and describe the key structures, innervation patterns, and general functions of the sympathetic, parasympathetic, and enteric divisions of the autonomic nervous system.**

1.2.4 **Provide an inventory of the main anatomical structures that make up the two cerebral hemispheres.**

1.2.5 **Explain the difference between gray matter and white matter, with examples.**

1.2.6 **Summarize the sequence of events in the development of the fetal nervous system.**

Neuronal cell bodies, dendrites, axons, and glia mass together to form the tissues that define the **gross neuroanatomy** of the nervous system—the neural structures that are visible to the unaided eye. **FIGURE 1.9A** presents a view of the entire human nervous system. This viewpoint reveals a natural subdivision into a **peripheral nervous system,** consisting of all nervous system parts that are outside the bony skull and spinal column, and a **central nervous system** (**CNS**), consisting of the brain and spinal cord (**FIGURE 1.9B**).

The peripheral nervous system has two divisions

The peripheral nervous system consists of **nerves**—collections of axons bundled together—that extend throughout the body. Some nerves, called **motor nerves**, transmit information from the spinal cord and brain to the muscles, organs, and glands. Others, called **sensory nerves**, arise from sensory surfaces and convey information from the body to the spinal cord and brain. The various nerves of the body are divided into two distinct systems: (1) the **somatic nervous system**, which consists of nerves that interconnect the brain and the major muscles and sensory systems of the body, and (2) the **autonomic nervous system**, the nerves that primarily control the viscera (internal organs).

THE SOMATIC NERVOUS SYSTEM Taking its name from the Latin word for "body"—soma—the somatic nervous system is the main pathway through which the brain controls movement and receives sensory information from the body and from the sensory organs of the head. The nerves that make up the somatic nervous system form two anatomical groups: the **cranial nerves** and the **spinal nerves**.

We each have 12 pairs of cranial nerves—one left-sided and one right-sided nerve in each pair—that serve the sensory and motor systems of the head and neck. These nerves pass through small openings in the skull, directly entering or leaving the brain without ever joining the spinal cord. Each cranial nerve is known both by name and by Roman numeral.

gross neuroanatomy Anatomical features of the nervous system that are apparent to the naked eye.

peripheral nervous system The portion of the nervous system that includes all the nerves and neurons outside the brain and spinal cord.

central nervous system (CNS) The portion of the nervous system that includes the brain and the spinal cord.

nerve A collection of axons bundled together outside the central nervous system.

motor nerve A nerve that conveys neural activity to muscle tissue and causes it to contract.

sensory nerve A nerve that conveys sensory information from the periphery into the central nervous system.

somatic nervous system The part of the peripheral nervous system that provides neural connections to the skeletal musculature.

autonomic nervous system The part of the peripheral nervous system that supplies neural connections to glands and to smooth muscles of internal organs.

cranial nerve A nerve that is connected directly to the brain.

spinal nerve Also called *somatic nerve*. A nerve that emerges from the spinal cord.

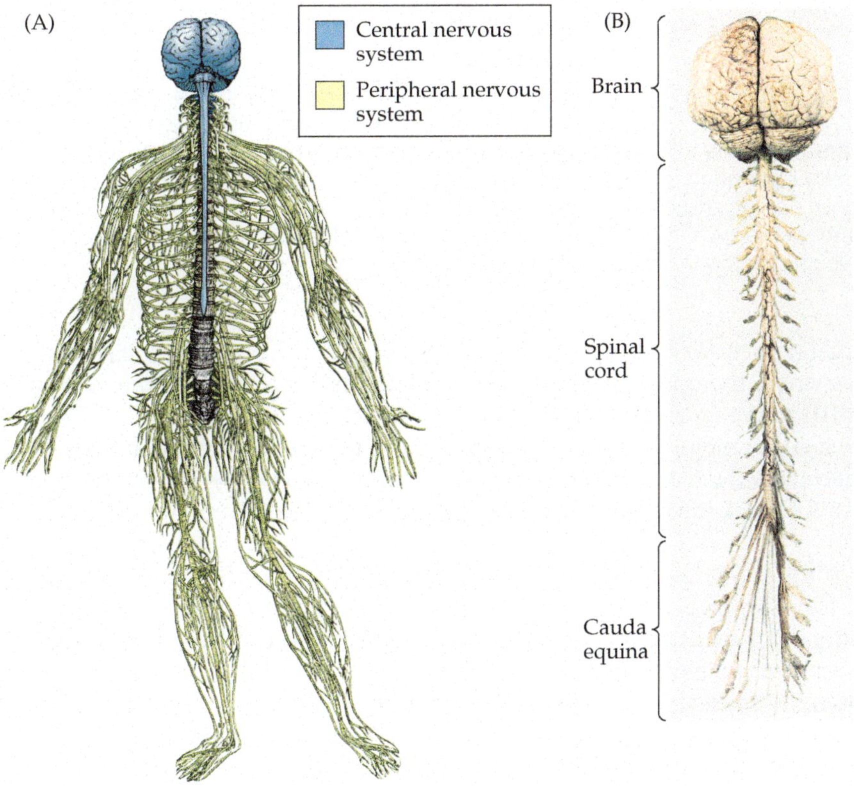

From *Gray's Anatomy*, 35th ed., Figure 1.9, page 807. © Elsevier, 1973. Dissection by M. C. E. Hutchinson, photograph by Kevin Fitzpatrick, Guy's Hospital Medical School, London

1.9 The Central and Peripheral Nervous Systems (A) This view of the nervous system is a composite of two drawings. A modern view of the central nervous system (the brain and spinal cord), shown in blue, is superimposed on a rendering of the peripheral nervous system by the great sixteenth-century anatomist Andreas Vesalius (1514–1564). The peripheral nervous system, shown in yellow, courses through the body and connects all body organs and systems to the central nervous system (CNS). (B) The brain and spinal cord together form the CNS. The solid part of the spinal cord ends in the middle of the lower back. Below this point a spray of fibres called the cauda equina (Latin for "horse's tail") continues downward inside the spinal column.

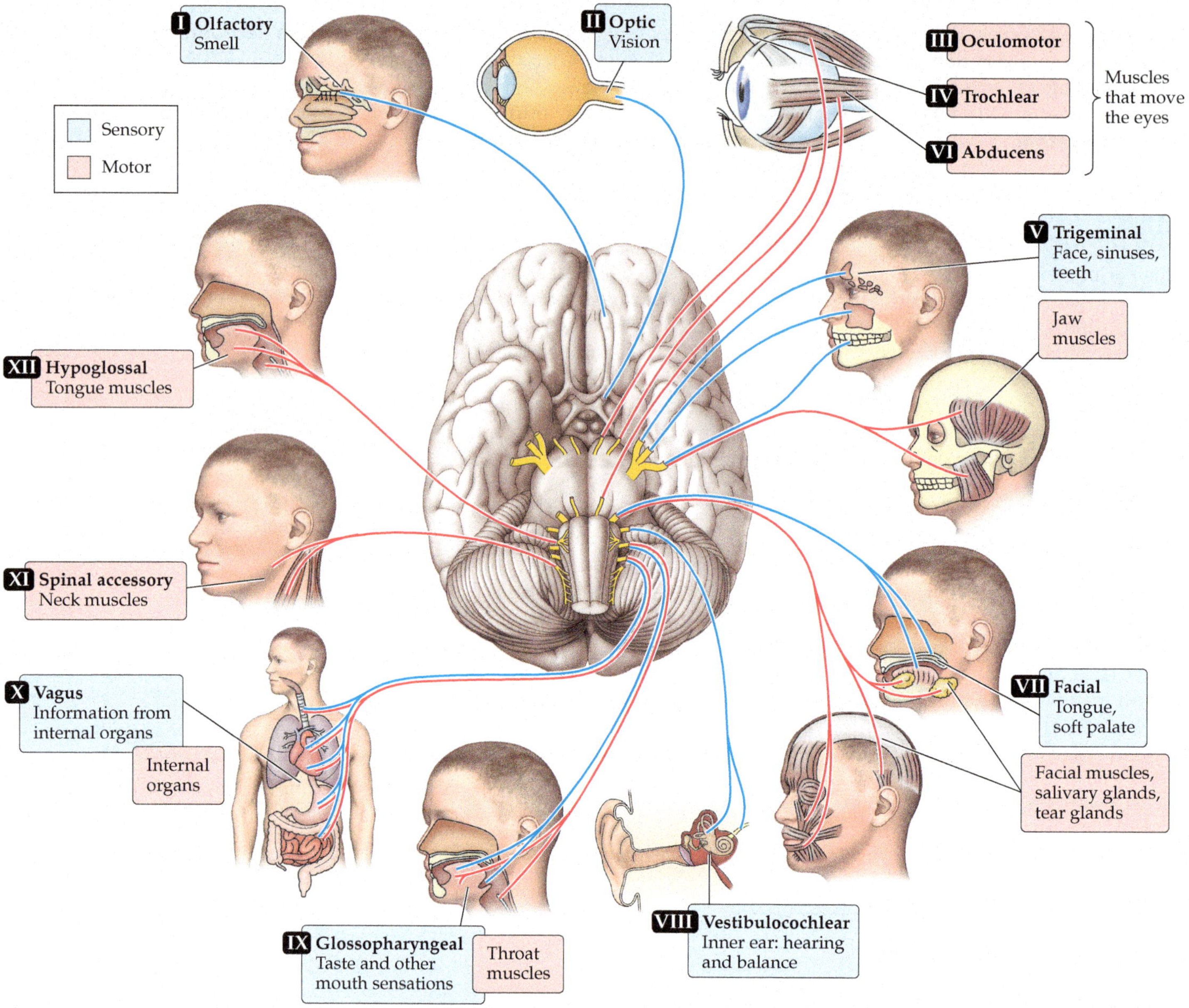

1.10 The Cranial Nerves Bundles of axons form the 12 pairs of cranial nerves, which are conventionally referred to by the Roman numerals I–XII. This ventral view of the brain shows the cranial nerves and their primary functions. Blue represents sensory nerves; red represents motor nerves. For cranial nerves with both types of functions, separate axons within the nerve carry either motor or sensory information.

Go to Activity 1.2
The Cranial Nerves

dorsal root The branch of a spinal nerve, entering the dorsal horn of the spinal cord, that carries sensory information from the peripheral nervous system to the spinal cord.

ventral root The branch of a spinal nerve, arising from the ventral horn of the spinal cord, that carries motor messages from the spinal cord to the peripheral nervous system.

Three cranial nerves are exclusively sensory pathways to the brain: the olfactory (I) nerve conveys smell, the optic (II) nerve carries visual information, and the vestibulocochlear (VIII) nerve is concerned with hearing and balance. Five pairs of cranial nerves are exclusively motor pathways from the brain: the oculomotor (III), trochlear (IV), and abducens (VI) nerves innervate muscles to move the eye; the spinal accessory (XI) nerves control neck muscles; and the hypoglossal (XII) nerves control the tongue (**FIGURE 1.10**).

The remaining cranial nerves have both sensory and motor functions. The trigeminal (V), for example, carries facial sensation through some axons, and it controls chewing movements through other axons. The facial (VII) nerves control facial muscles and receive some taste sensation, and the glossopharyngeal (IX) nerves receive additional taste sensations and sensations from the throat and also control the muscles there. The vagus (X) nerve extends far from the head, innervating the heart, liver, and intestines. Its long, convoluted route is the reason for its name, which is

Latin for "wandering." The vagus is the primary route by which the brain both controls and receives information from many visceral organs, and it participates in such varied functions as sweating, digestion, and heart rate (Shaffer et al., 2014).

Along the length of the spinal cord, an additional 31 pairs of spinal nerves emerge at regularly spaced intervals through openings in the backbone (**FIGURE 1.11**). As with the cranial nerves, one member of each pair of spinal nerves serves each side of the body. Furthermore, each spinal nerve consists of the fusion of two distinct branches, called *roots*, which are functionally different. The **dorsal** (back) **root** of each spinal nerve consists of sensory projections from the body to the spinal cord. The **ventral** (front) **root** consists of motor projections from the spinal cord to the muscles.

Each spinal nerve is named according to the segment of spinal cord to which it is connected: there are 8 **cervical** (neck), 12 **thoracic** (trunk), 5 **lumbar** (lower back), 5 **sacral** (pelvic), and 1 **coccygeal** (bottom) spinal segments. So, we refer to the spinal nerve that is connected to the twelfth segment of the thoracic portion of the spinal cord as *T12*, the nerve connected to the third segment of the sacral portion as *S3*, and so on. Axons from different spinal nerves merge to form peripheral nerves.

THE AUTONOMIC NERVOUS SYSTEM Although it is "autonomous" in the sense that we don't have very much conscious, voluntary control over its actions, the autonomic nervous system is the brain's main system for controlling the organs of the body. Supporting these functions are aggregates of neurons called **autonomic ganglia**, found in various locations in the body outside of the CNS. Autonomic neurons within the brain and spinal cord send their axons to innervate neurons in the ganglia, which in turn send their axons to innervate all the major organs. The central neurons that

cervical Referring to the topmost eight segments of the spinal cord, in the neck region.

thoracic Referring to the 12 spinal segments below the cervical (neck) portion of the spinal cord, corresponding to the chest.

lumbar Referring to the five spinal segments that make up the upper part of the lower back.

sacral Referring to the five spinal segments that make up the lower part of the lower back.

coccygeal Referring to the lowest spinal vertebra (which is also called the *tailbone*).

autonomic ganglia Collections of nerve cell bodies, belonging to the autonomic division of the peripheral nervous system, that are found in various locations and innervate the major organs.

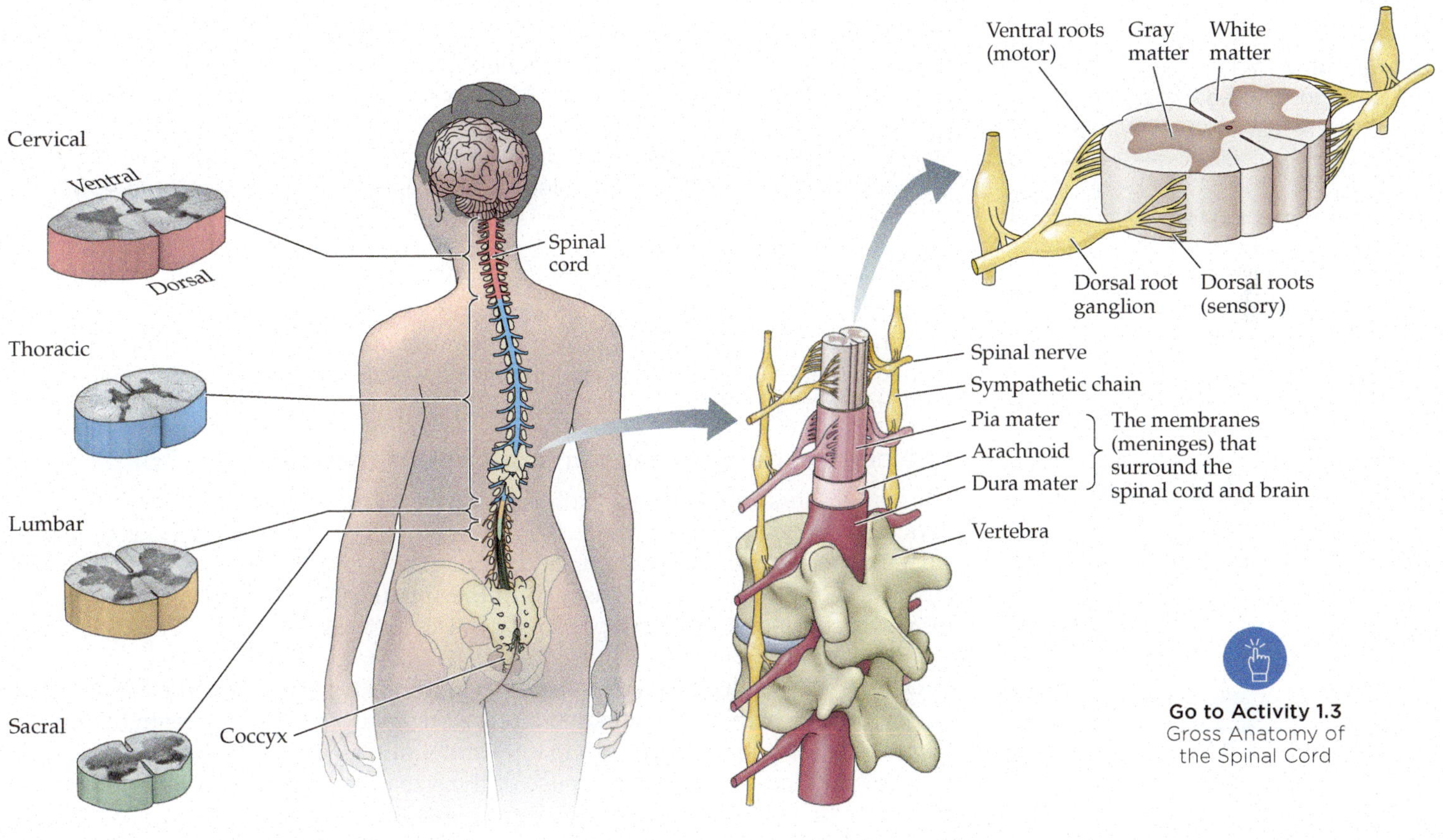

1.11 The Spinal Cord and Spinal Nerves (Middle) The spinal column runs from the base of the brain to the coccyx (tailbone); a pair of nerves emerges from each level (see Figure 1.9B). (Bottom right) The spinal cord is surrounded by bony vertebrae and is enclosed in three membrane layers (the meninges). Each vertebra has an opening on each side through which the spinal nerves pass. (Top right) The spinal cord gray matter is located in the centre of the cord and is surrounded by white matter. In the gray matter are interneurons and the motor neurons that send axons to the muscles. The white matter consists of myelinated axons that run up and down the spinal column. (Left) These stained cross sections show the spinal cord at the cervical, thoracic, lumbar, and sacral levels.

innervate the ganglia are known as **preganglionic** autonomic neurons; the ganglionic neurons that innervate the body are known as **postganglionic** neurons.

The autonomic nervous system has three major divisions: the sympathetic nervous system, the parasympathetic nervous system, and the enteric nervous system. The sympathetic and parasympathetic nervous systems act more or less in opposition. The preganglionic cells of the **sympathetic nervous system** are found in the middle parts of the spinal cord—the thoracic and lumbar regions. They send their axons only a short distance, innervating the **sympathetic chain** of autonomic ganglia that runs along each side of the spinal column (**FIGURE 1.12, LEFT**). Postganglionic cells in the sympathetic chain send axons throughout the body, innervating all of the major organ systems. In general, sympathetic activation prepares the body for action: blood pressure increases, the pupils of the eyes dilate, and the heart rate quickens. This set of reactions is sometimes called simply the *fight-or-flight* response.

In contrast to the sympathetic system, the **parasympathetic nervous system** generally helps the body to relax, recuperate, and prepare for future action, sometimes called the *rest-and-digest* response. The parasympathetic system gets its name from its anatomical points of origin in the brainstem and sacral spinal cord, thus arising above and below the sympathetic nerves (the Greek *para* means "around"). Compared with sympathetic nerves, parasympathetic axons travel a longer distance before terminating in parasympathetic ganglia (**FIGURE 1.12, RIGHT**), because parasympathetic ganglia are not collected in a chain as sympathetic ganglia are. Instead, parasympathetic ganglia are dispersed throughout the body, usually positioned near the organs they affect.

The sympathetic and parasympathetic systems have very different effects on individual target organs because the organs receive different neurotransmitters from the two opposing systems. The sympathetic system uses **norepinephrine** (also known as *noradrenaline*), which tends to accelerate activity, while the parasympathetic system uses **acetylcholine**, which tends to slow down activity. The balance between these two opposing systems determines the state of the internal organs at any given moment. So, for example, when parasympathetic activity predominates, heart rate slows, blood pressure drops, and digestive processes are activated. As the brain causes the balance of autonomic activity to become predominantly sympathetic, opposite effects are seen: increased heart rate and blood pressure, inhibited digestion, and so on. This tension between parasympathetic and sympathetic activity ensures that the individual is appropriately prepared for current circumstances.

Resembling a mesh embedded within the walls of the digestive organs, the **enteric nervous system** is a local network of sensory and motor neurons that regulates the functioning of the gut, under the control of the CNS. Because it regulates digestive activities of the gut, the enteric nervous system plays a key role in maintaining fluid and nutrient balances in the body (discussed in Chapter 13).

The central nervous system consists of the brain and spinal cord

The spinal cord funnels sensory information from the body up to the brain and conveys the brain's motor commands out to the body. The spinal cord also contains circuits that perform local processing and that control simple units of behaviour, such as reflexes. We will discuss other aspects of the spinal cord in later chapters, so for now we will focus on the executive portion of the CNS: the brain.

BRAIN FEATURES THAT ARE VISIBLE TO THE NAKED EYE Given the importance of the adult human brain, it is surprising that it only weighs about 1400 grams (about 3 pounds), accounting for just 2% of the average body weight. Put your two fists together and you get a sense of the size of the two **cerebral hemispheres**: smaller than most people expect. However, even casual inspection reveals that what the brain lacks in weight it makes up for in intricacy.

The lumpy convolutions of the paired cerebral hemispheres are the result of elaborate folding together of a thick sheet of brain tissue called the **cerebral cortex** (or sometimes just *cortex*), which is made up mostly of neuronal cell bodies, dendrites, and axons. The resulting ridges of tissue, called **gyri** (singular *gyrus*), are separated from each other by furrows called **sulci** (singular *sulcus*). Folding up the tissue in this

preganglionic Literally, "before the ganglion." Referring to neurons in the autonomic nervous system that run from the central nervous system to the autonomic ganglia.

postganglionic Literally, "after the ganglion." Referring to neurons in the autonomic nervous system that run from the autonomic ganglia to various targets in the body.

sympathetic nervous system A component of the autonomic nervous system that arises from the thoracic and lumbar spinal cord.

sympathetic chain A chain of ganglia that runs along each side of the spinal column; part of the sympathetic nervous system.

parasympathetic nervous system A component of the autonomic nervous system that arises from both the cranial nerves and the sacral spinal cord.

norepinephrine Also called *noradrenaline*. A neurotransmitter produced and released by sympathetic postganglionic neurons to accelerate organ activity. Also produced in the brainstem and found in projections throughout the brain.

acetylcholine A neurotransmitter produced and released by parasympathetic postganglionic neurons, by motor neurons, and by neurons throughout the brain.

enteric nervous system An extensive mesh-like system of neurons that governs the functioning of the gut.

cerebral hemispheres The right and left halves of the forebrain.

cerebral cortex Also called simply *cortex*. The outer covering of the cerebral hemispheres that consists largely of neuronal cell bodies and their branches.

gyrus A ridged or raised portion of a convoluted brain surface.

sulcus A furrow of a convoluted brain surface.

Go to Activity 1.4
Concept Matching:
Sympathetic vs. Parasympathetic

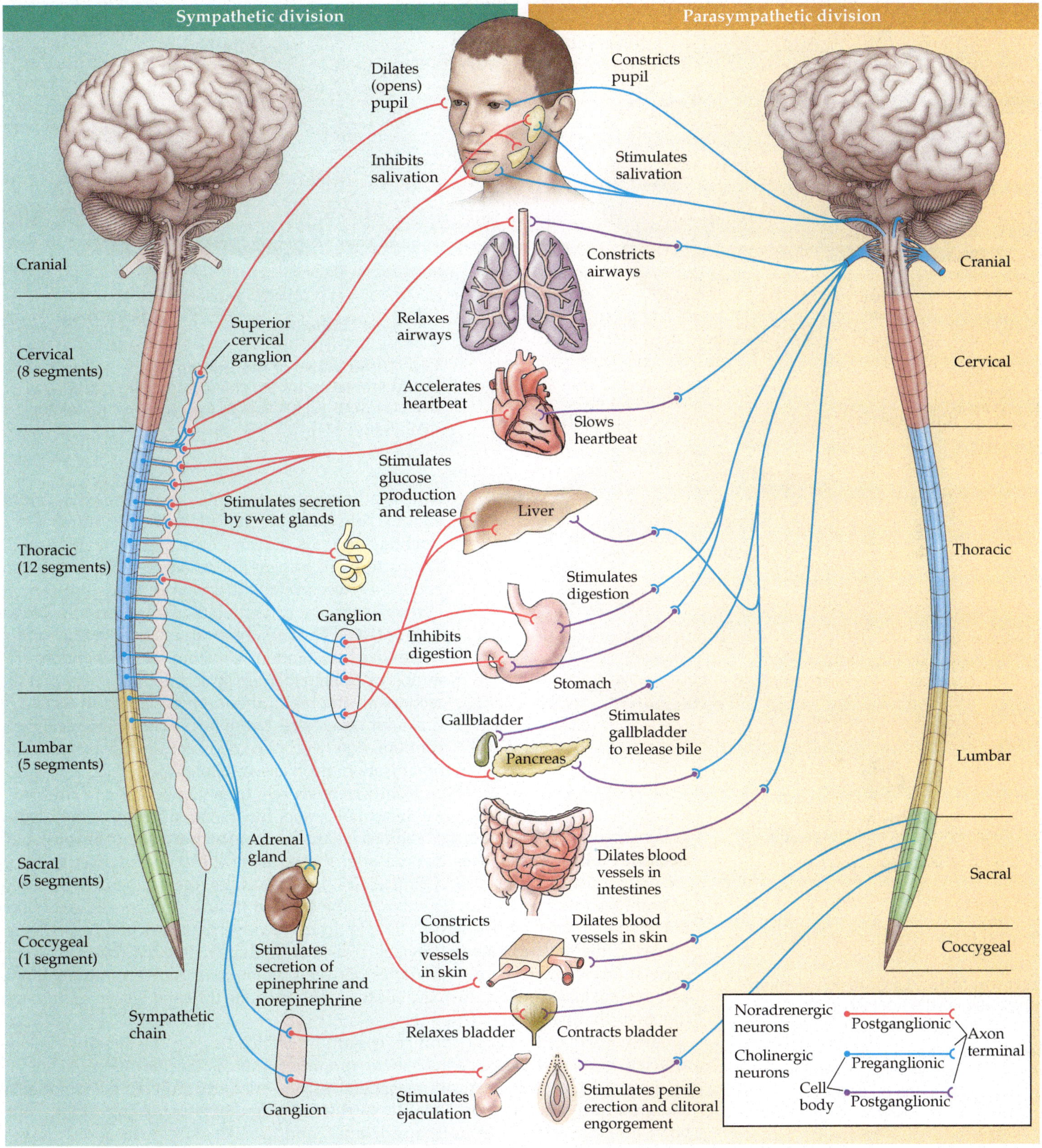

1.12 The Autonomic Nervous System (Left) The sympathetic division of the autonomic nervous system consists of the sympathetic chains and the nerve fibres that flow from them. (Right) The parasympathetic division arises from both the brain and the sacral parts of the spinal cord. Parasympathetic postganglionic cells and all preganglionic axons, whether sympathetic or parasympathetic, produce and release acetylcholine (from which we get the adjective *cholinergic*) as a neurotransmitter. Sympathetic postganglionic cells produce and use norepinephrine (also known as *noradrenaline*; hence the adjective *noradrenergic*) as a neurotransmitter. The different postganglionic transmitters have opposing effects on target organs, allowing precise control. Neurotransmitters are discussed in detail in Chapter 4. A third division of the autonomic nervous system, the enteric nervous system, is discussed in Chapter 13.

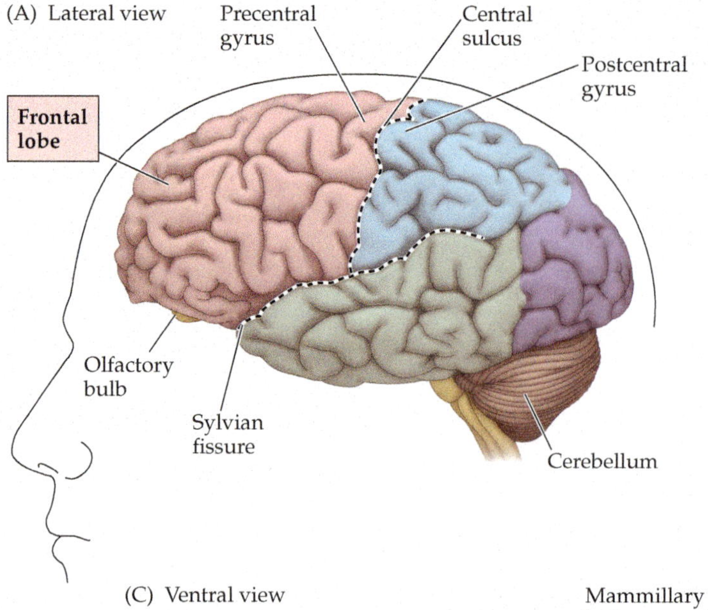

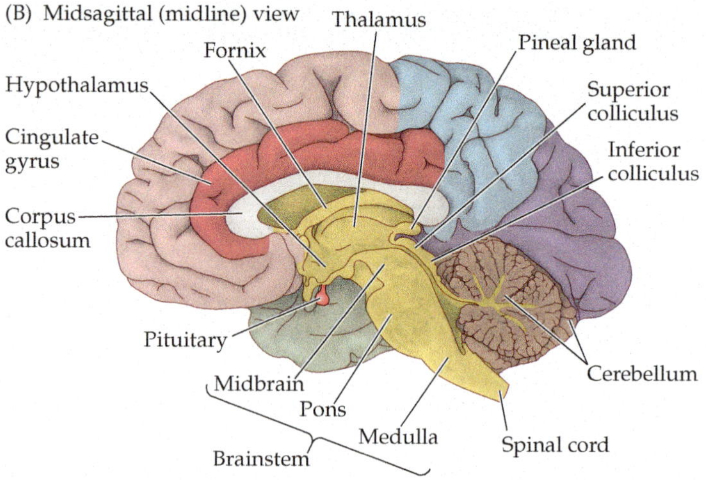

1.13 Three Views of the Human Brain The four lobes of the cerebral cortex are colour coded. (A) Lateral view (from the side). (B) Midsagittal (midline) view. (C) Ventral view (from below).

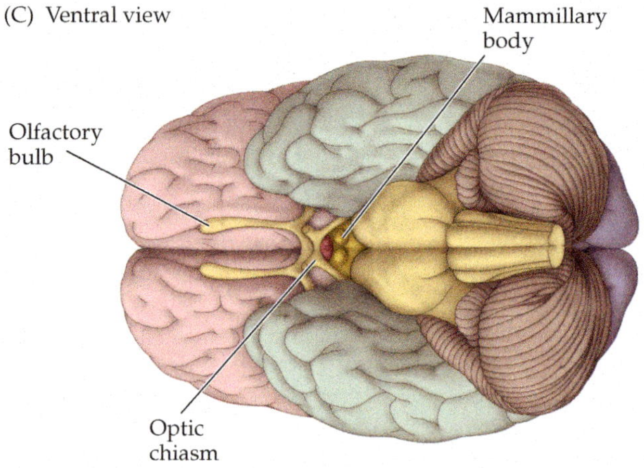

Go to Activities 1.5–1.7
Gross Anatomy of the Human Brain
Lateral View
Midsaggital View
Ventral View

way greatly increases the amount of cortex that can be jammed into the skull; indeed, about two-thirds of the cerebral surface is hidden in the depths of these folds. The pattern of folding is not random; in fact, it is similar enough between brains that we can name the various gyri and sulci and can group them together into *lobes*.

FIGURE 1.13 offers three views of the human brain in standard orientations. Anatomists use standard terminology to help identify structures, locations, and directions in the brain; these are described in **BOX 1.1**. (It's a bit of a chore, but learning the anatomical conventions now will make our later discussions of brain organisation much easier to follow.)

Neuroscientists rely on a combination of anatomical landmarks and functional differences to distinguish among four major cortical regions called the **frontal**, **parietal**, **temporal**, and **occipital lobes**. These lobes, named after the bones of the skull that overlie them, are distinguished by colours in Figure 1.13. In some cases, the boundaries between adjacent lobes are very clear; for example, the **Sylvian fissure** (a deep sulcus) divides the temporal lobe (just beside your temple) from the other regions of the hemisphere. Likewise, the **central sulcus** provides a distinct landmark dividing the frontal and parietal lobes. The physical boundaries between the occipital lobe and the temporal and parietal lobes are less obvious, but the lobes are quite different with regard to the functions they perform.

The cortex is the seat of complex cognition. Depending on the specific regions affected, cortical damage can cause far-ranging symptoms including impairments of movement or body sensation, speech errors, memory problems, personality changes, or many kinds of visual impairments. Some life-sustaining functions—heart rate and respiration, reflexes, balance, and the like—are governed by lower, subcortical brain regions.

We can identify certain general categories of processing that are particularly associated with specific cortical lobes. For example, the occipital lobes receive and process information from the eyes, giving rise to the sense of vision. Auditory information is directed to the temporal lobes, and damage there can impair hearing (the temporal lobes are also particularly associated with the sense of smell and with aspects of learning and memory). The parietal lobes receive sensory information from the body and participate in spatial cognition. The sense of touch is mediated by a strip of parietal

frontal lobe The most anteriour portion of the cerebral cortex.

parietal lobe Large region of cortex lying between the frontal and occipital lobes of each cerebral hemisphere.

temporal lobe Large lateral cortical region of each cerebral hemisphere, continuous with the parietal lobe posteriourly and separated from the frontal lobe by the Sylvian fissure.

occipital lobe Large region of cortex covering much of the posteriour part of each cerebral hemisphere.

Sylvian fissure Also called *lateral sulcus*. A deep fissure that demarcates the temporal lobe.

central sulcus A fissure that divides the frontal lobe from the parietal lobe.

BOX 1.2 Three Customary Orientations for Viewing the Brain and Body

Because the nervous system is a three-dimensional structure, two-dimensional illustrations and diagrams cannot represent it completely. The brain is usually cut in one of three main planes to obtain a two-dimensional section from this three-dimensional object. Although it takes time and practice to master, it is useful to know the terminology that applies to these sections, as shown in the figure.

The plane that bisects the body into right and left halves is called the **sagittal plane** (from the Latin *sagitta,* "arrow"). The plane that divides the body into a front (anteriour) and a back (posteriour) part is called by several names: **coronal plane** (from the Latin *corona*, "crown"), *frontal plane, or transverse plane*. The third main plane, which divides the brain into upper and lower parts, is called the **horizontal plane**.

In addition, several directional terms are used. **Medial** means "toward the middle" and is contrasted with **lateral**, "toward the side." Relative to one location, a second location is **ipsilateral** if it is on the same side of the body, and **contralateral** if on the opposite side of the body. These terms are all relative, as are the terms **superiour** ("above") and **inferiour** ("below"); for example, the eye is lateral to the nose but medial to the ear, and the mouth is inferiour to the nose but superiour to the chin.

The head end is referred to as **anteriour** or **rostral**, and the tail end is called **posteriour** or **caudal** (from the Latin *cauda*, "tail"). **Proximal** means "near the centre," and **distal** means "toward the periphery" or "toward the end of a limb." We call an axon, tract, or nerve **afferent** if it carries information into a region that we are interested in, and **efferent** if it carries information away from the region of interest (a handy way to remember this is that ***e***fferents ***e***xit but ***a***fferents ***a***rrive, relative to the region of interest).

Dorsal means "towards or at the back," and **ventral** means "towards the belly." In four-legged animals, such as the cat or the rat, *dorsal* refers to both the back of the body and the top of the head and brain. For consistency in comparing brains among species, this term is also used to refer to the top of the brain of a human or of a chimpanzee, even though in such two-legged animals the top of the brain is not at the back of the body. Similarly, *ventral* is understood to designate the bottom of the brain of a two-legged as well as of a four-legged animal.

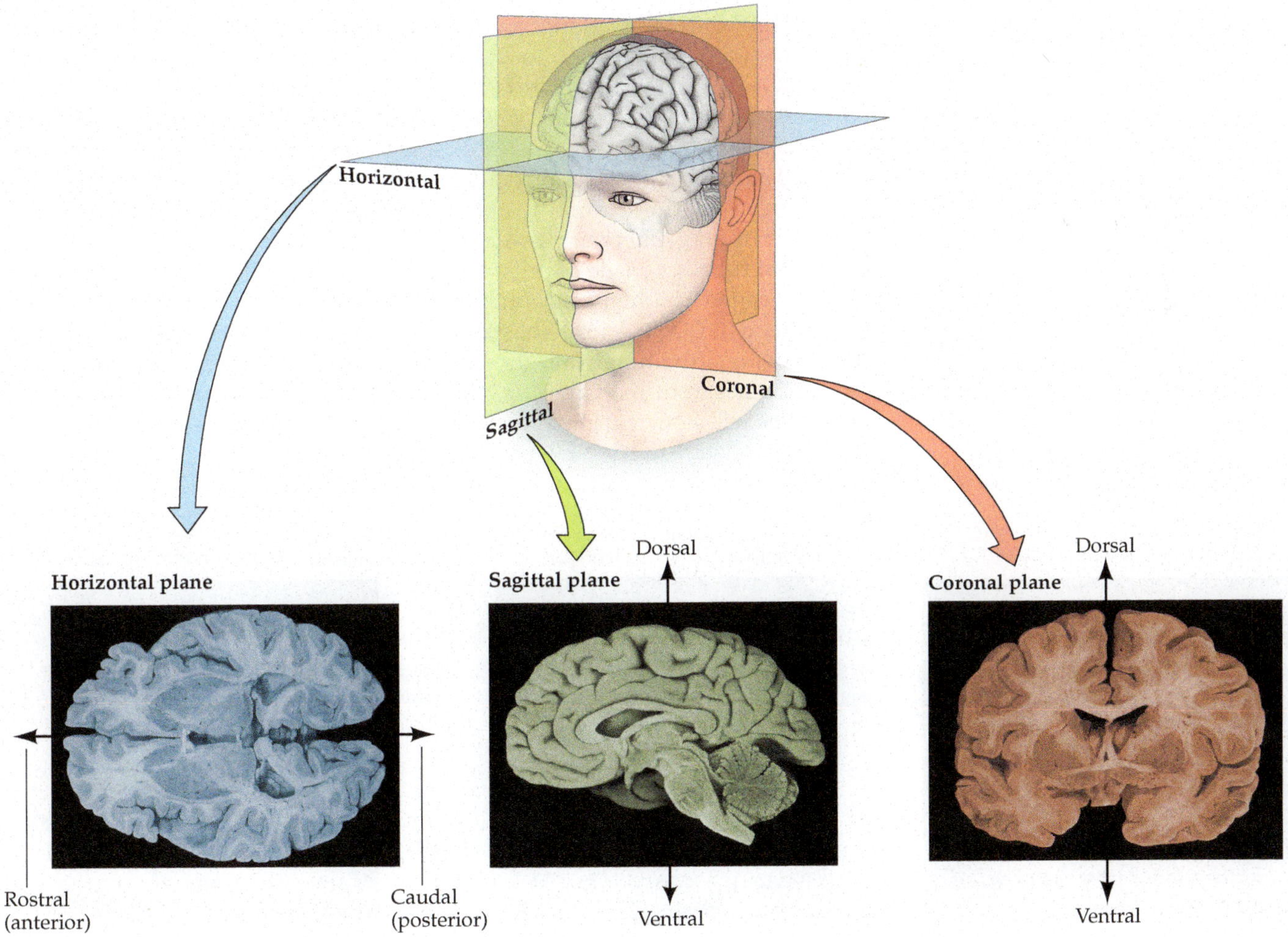

All photos courtesy of Drs. S. Mark Williams and Dale Purves, Duke University Medical Centre

postcentral gyrus The strip of parietal cortex, just behind the central sulcus, that receives somatosensory information from the entire body.

precentral gyrus The strip of frontal cortex, just in front of the central sulcus, that is crucial for motor control.

corpus callosum The main band of axons that connects the two cerebral hemispheres.

gray matter Areas of the brain that are dominated by cell bodies and are devoid of myelin.

white matter A pale-coloured layer underneath the cortex that consists largely of axons with white myelin sheaths.

neural tube An embryonic structure with subdivisions that correspond to the future forebrain, midbrain, and hindbrain.

forebrain Also called *prosencephalon*. The anteriour division of the brain, containing the cerebral hemispheres, the thalamus, and the hypothalamus.

midbrain Also called *mesencephalon*. The middle division of the brain.

hindbrain Also called *rhombencephalon*. The rear division of the brain, which in the mature vertebrate contains the cerebellum, pons, and medulla.

cortex that is located just posteriour to the central sulcus, thus cleverly called the **postcentral gyrus**.

In general, the frontal lobes are important for movement and high-level cognition. Immediately anteriour to the central sulcus, the **precentral gyrus** of the frontal lobe is crucial for motor control. In fact, Wilder Penfield's experiments with stimulation mapping of the brain, which we discussed at the beginning of the chapter, revealed that the movement-controlling neurons within the precentral gyrus are organised into an orderly map of the muscles of the other side of the body (see Figure 11.13); in general, the left hemisphere controls the right side of the body, and vice versa. Similarly, the postcentral gyrus contains a sensory map of the other side of the body (Penfield and Rasmussen, 1950).

Beyond the generalities presented here, each of the lobes also performs a wide variety of other high-level functions; these will be major topics in later chapters. In people with undamaged brains, the four cortical lobes of each hemisphere are continually communicating among themselves and with their counterparts in the other hemisphere, their collaboration producing the seamless experiences and complex behaviours that distinguish us as individuals. In fact, hundreds of millions of axons cross the midline in a large C-shaped bundle called the **corpus callosum** (see Figure 1.13B), enabling communication between the right and left cerebral hemispheres and allowing the brain to act as a single entity during complex processing.

When you slice into the brain, two distinct shades of colour are evident. The darker-coloured **gray matter** of the exteriour contains a preponderance of neuronal cell bodies and dendrites, which are devoid of myelin (**FIGURE 1.14**). Beneath the outer surface is the lighter-coloured **white matter**, which consists mostly of axonal fibre tracts. It gains its appearance from the whitish fatty myelin that ensheathes and insulates the axons of many neurons. A simple view is that gray matter primarily receives and processes information, while white matter mostly transmits information to other locations.

DEVELOPMENTAL SUBDIVISIONS OF THE BRAIN It can be difficult to understand some of the anatomical distinctions applied to the adult human brain. For example, the part of the cortex closest to the back of the head is anatomically identified as being part of the *fore*brain. Why? The key to understanding this confusing terminology is to consider how the brain develops early in life.

In a very young embryo of any vertebrate, the CNS looks like a tube. The walls of this **neural tube** are made of cells, and the interiour is filled with fluid. A few weeks after conception, the human neural tube begins to show three separate swellings at the head end (**FIGURE 1.15A**): the **forebrain** (or *prosencephalon*), the **midbrain** (or *mesencephalon*), and the **hindbrain** (or *rhombencephalon*). (The term *encephalon*, meaning "brain," comes from the Greek *en*, "in," and *kephale*, "head.") The remainder of the neural tube will form the spinal cord.

1.14 Inside the Brain (A) The coloured lines here indicate the planes of section shown in (B) and (C). The lighter-coloured interiour is white matter, packed with fatty myelin that surrounds axons sending information in and out of the cortex. Gray matter consists of cell bodies that form the outer layers of the cortex and nuclei within the brain.

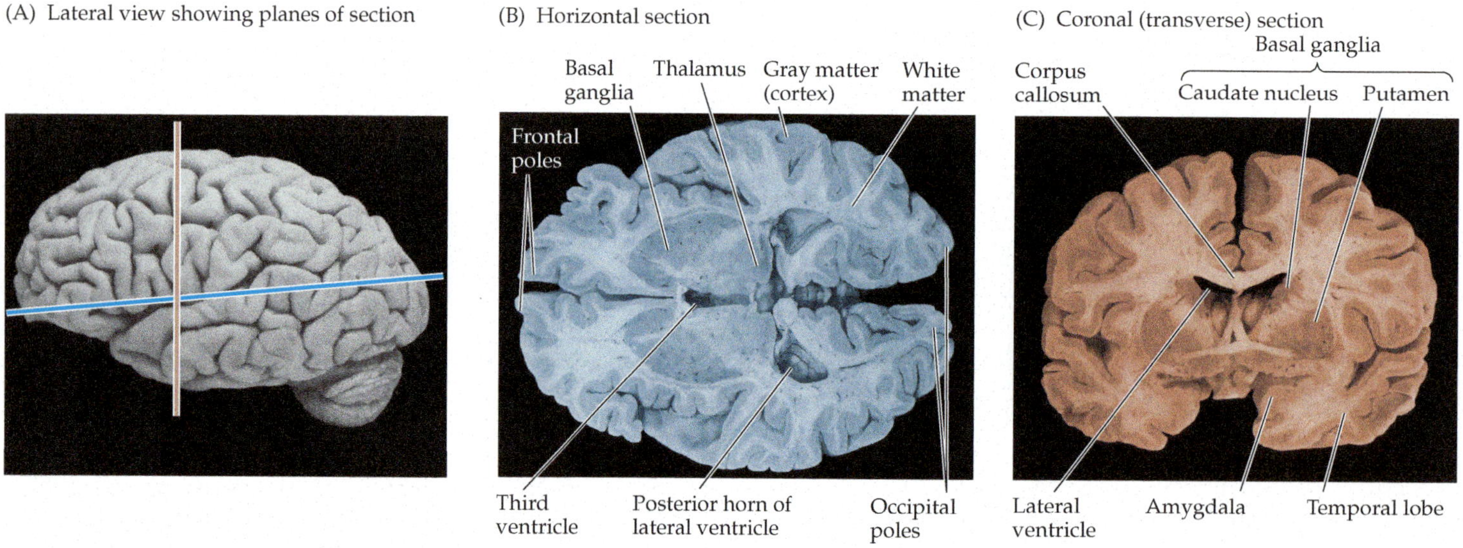

All photos courtesy of Drs. S. Mark Williams and Dale Purves, Duke University Medical Centre

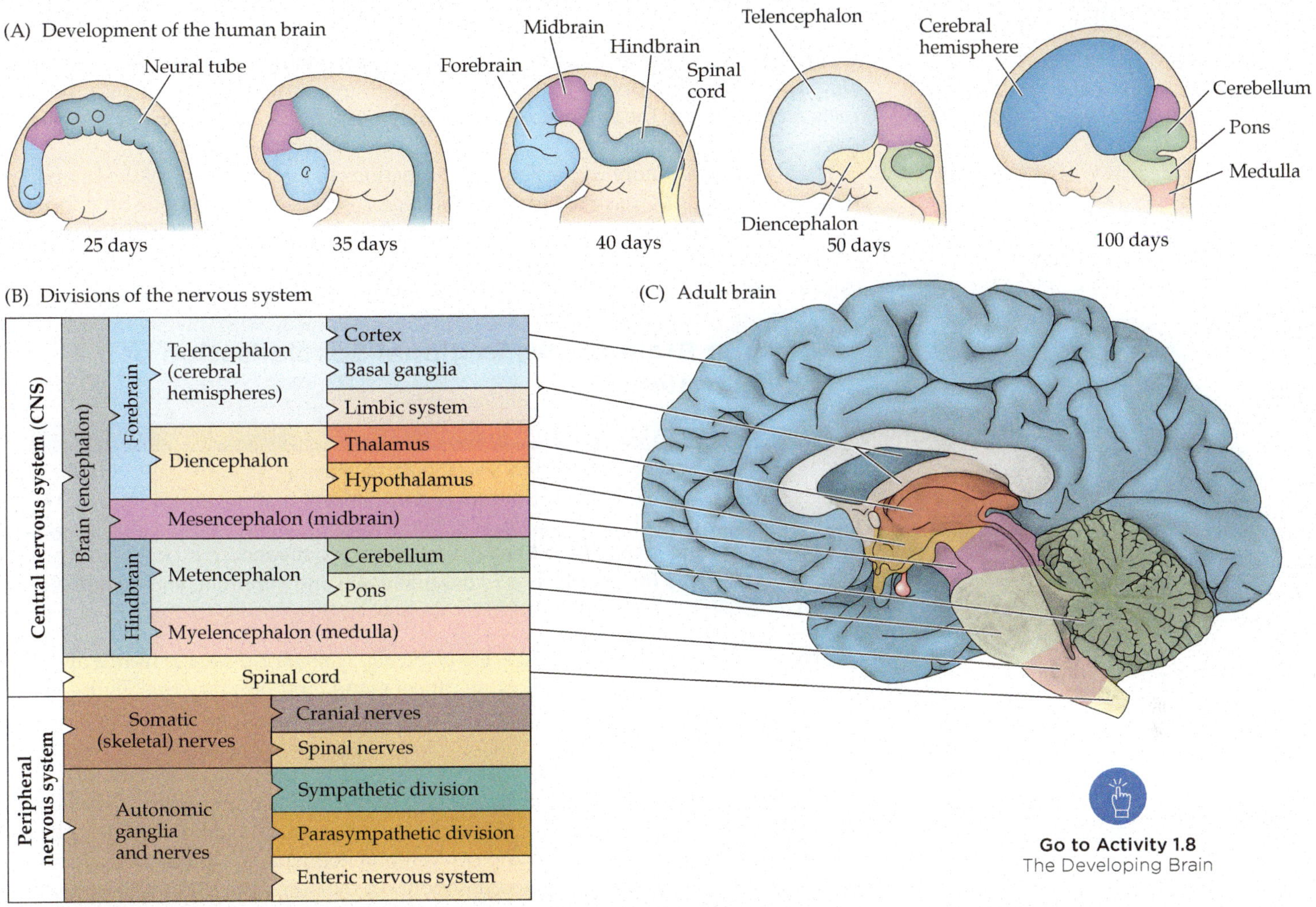

Go to Activity 1.8
The Developing Brain

1.15 Divisions of the Human Nervous System in the Embryo and the Adult
(A) A few weeks after conception, the head end of the neural tube shows three main divisions. About 50 days after conception, five main divisions of the brain are visible. (B) The organisation of these divisions schematically. (C) Their positions in the adult brain.

About 50 days after conception, the forebrain and hindbrain have already developed clear subdivisions. At the very front of the developing brain is the **telencephalon**, which will become the cerebral hemispheres (consisting of cortex plus some deeper structures belonging to two functionally related groups, the basal ganglia and the limbic system). The other part of the forebrain is the **diencephalon** (or "between brain"), which will go on to become the regions called the *thalamus* and the *hypothalamus*. We'll discuss these various forebrain systems later in the chapter.

Behind the midbrain (mesencephalon), the hindbrain further develops into several principle structures: the **metencephalon**, made up of the **cerebellum** ("little brain") and **pons** ("bridge"), and the **medulla**, also called the *myelencephalon*. The term **brainstem** usually refers to the midbrain, pons, and medulla combined (some scientists also include the diencephalon). **FIGURE 1.15C** shows the positions of these structures and their relative sizes in the adult human brain. Even when the brain achieves its adult form, it is still a fluid-filled tube, but a tube of very complicated shape.

Each of the five main sections (telencephalon, diencephalon, mesencephalon, metencephalon, and myelencephalon) can be subdivided in turn. We can work our way from the largest, most general divisions of the nervous system on the left of the schematic in **FIGURE 1.15B** to more-specific ones on the right.

Within and between the major brain regions are aggregations of neurons called **nuclei** (singular *nucleus*) and bundles of axons called **tracts**. (Recall that in the periphery, aggregations of neurons are called *ganglia*, and bundles of axons are called *nerves*.) Unfortunately, the word *nucleus* can mean either "a collection of neurons

telencephalon The frontal subdivision of the forebrain that includes the cerebral hemispheres when fully developed.

diencephalon The posteriour part of the forebrain, including the thalamus and hypothalamus.

metencephalon A subdivision of the hindbrain that includes the cerebellum and the pons.

cerebellum A structure located at the back of the brain, dorsal to the pons, that is involved in the central regulation of movement.

pons A portion of the metencephalon; part of the brainstem connecting midbrain to medulla.

medulla Also called *myelencephalon*. The posteriour part of the hindbrain, continuous with the spinal cord.

brainstem The region of the brain that consists of the midbrain, the pons, and the medulla.

nucleus Here, a collection of neurons within the central nervous system (e.g., the caudate nucleus).

tract A bundle of axons found within the central nervous system.

within the CNS" or "the spherical DNA-containing organelle within a single cell," so you must rely on the context to understand which meaning is intended. Because brain tracts and nuclei are the same from individual to individual, and often from species to species, they have names too.

You are probably more interested in the functions of all these parts of the brain than in their names, but as we noted earlier, each region serves more than one function, and our knowledge of the functional organisation of the brain is continually being updated with new research findings. With that caution in mind, we'll briefly survey functions of specific brain structures next, leaving the detailed discussion for later chapters.

1.3 The Brain Shows Regional Specialisation of Functions

Learning Objectives

After reading this section, you should be able to:

1.3.1 **Describe the cellular structure and organisation of the cortex.**

1.3.2 **Identify the major components of the basal ganglia and the limbic system, and state some of the behavioural functions of each.**

1.3.3 **Name the major divisions of the brainstem and midbrain, and identify key functions performed by each.**

Like the rest of our bodies, our brains are *bilaterally symmetrical*—aside from a few midline structures like the corpus callosum, pineal gland, and pituitary (see Figure 1.13B), most structures of the brain are paired, in left- and right-sided mirror-image versions. One important principle of the vertebrate brain is that each side of the brain generally controls the opposite (or *contralateral;* see Box 1.2) side of the body. The right side of the brain thus controls movements of the left side of the body and receives left-sided sensory information. Likewise, the left side of the brain monitors and controls the right side of the body. We'll learn about how the two cerebral hemispheres interact in Chapter 19, but for now let's review the various components of the brain and their functions.

The cerebral cortex performs complex cognitive processing

Neuroscientists agree that understanding human cognition depends on unravelling the structure and fundamental functions of the cerebral cortex. If the cerebral cortex

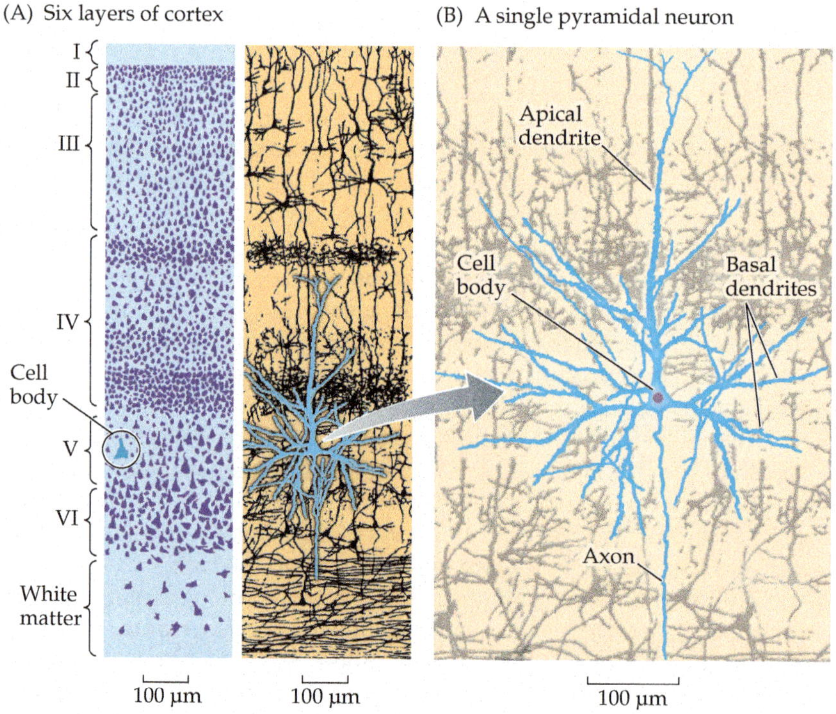

1.16 Layers of the Cerebral Cortex (A) The six layers of cortex can be distinguished with stains that reveal all cell bodies (left) or with stains that reveal a few neurons in their entirety (right). (B) This pyramidal cell has been enlarged about 100 times. (After P. Rakic, 1979. In *The neurosciences: Fourth study program*, F. O. Schmitt and F. G. Worden [Eds.]. MIT Press: Cambridge, MA.)

were unfolded, it would occupy an area of about 2000 square centimeters (cm^2), or 315 square inches, more than 3 times the area of this book's cover. How are those millions of cells arranged? And how do the arrangements perform particular feats of information processing?

The neurons of the cerebral cortex are arranged in six distinct layers (**FIGURE 1.16A**) (in mammals this tissue is sometimes referred to as *neocortex*). Each cortical layer has a unique appearance because it consists of either a band of similar neurons or a particular pattern of dendrites or axons. For example, the outermost layer, layer I, is distinct because it has few cell bodies, and layers V and VI stand out because of their many neurons with large cell bodies. While the great majority of the cortex consists of this six-layered tissue, a few telencephalic structures are instead made up of *allocortex* (from the Greek *allos*, "other"), tissue with three layers or unlayered organisation (previously known as *archi-* or *paleocortex*).

The most prominent kind of neuron in the cerebral cortex—the **pyramidal cell** (**FIGURE 1.16B**)—usually has its pyramid-shaped cell body in layer III or V. One dendrite of each pyramidal cell, called the *apical dendrite*, extends from the top of the cell body (its *apex*) to the outermost layer of the cortex. Each pyramidal cell also has several dendrites, called *basal dendrites*, that spread out horizontally from the base of the cell body.

In classic research, Vernon Mountcastle (1918–2015) found that in many regions of the cerebral cortex, neurons are organised into regular columns, perpendicular to the layers, that seem to serve as information-processing units (Mountcastle, 1979). These **cortical columns** extend through the entire thickness of the cortex, from the white matter to the surface. Within each column, most of the synaptic interconnections of neurons are vertical, although there are some horizontal connections as well (Sakmann, 2017).

Subcortical structures are involved in movement and the regulation of emotions

Buried within the cerebral hemispheres are several large gray matter structures, richly interconnected with each other and with other brain regions and contributing to a wide variety of behaviours. One prominent cluster—the **basal ganglia**—includes the **caudate**

pyramidal cell A type of large nerve cell that has a roughly pyramid-shaped cell body; found in the cerebral cortex.

cortical column One of the vertical columns that constitute the basic organisation of the neocortex.

basal ganglia A group of forebrain nuclei, including caudate nucleus, globus pallidus, and putamen, found deep within the cerebral hemispheres.

caudate nucleus One of the basal ganglia; it has a long extension or tail.

Go to Activities 1.9–1.10
The Basal Ganglia
The Limbic System

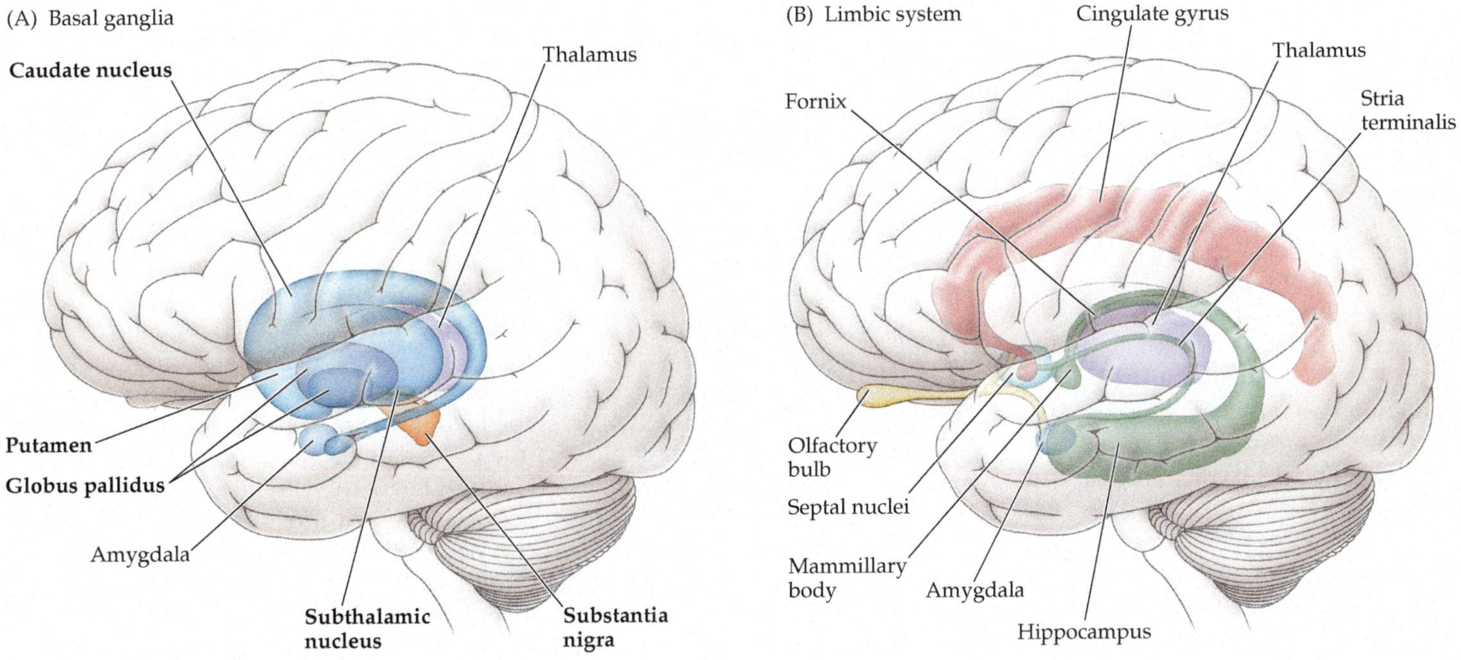

1.17 Two Important Brain Systems (A) The basal ganglia—caudate nucleus, putamen, globus pallidus, subthalamic nucleus, and substantia nigra—are important in movement. (B) The limbic system—including hippocampus, cingulate gyrus, fornix, septal nuclei, stria terminalis, olfactory bulb, amygdala, and mammillary bodies—is important for emotion, learning, and memory.

putamen One of the basal ganglia.

globus pallidus One of the basal ganglia.

substantia nigra A brainstem structure in humans that innervates the basal ganglia and is named for its dark pigmentation.

limbic system A loosely defined, widespread network of brain nuclei that are implicated in emotional processes.

amygdala A group of nuclei in the medial anteriour part of the temporal lobe.

mammillary body One of a pair of nuclei at the base of the brain.

hippocampus A medial temporal lobe structure that is important for learning and memory.

fornix A fibre tract that extends from the hippocampus to the mammillary body.

septal nuclei A collection of gray matter structures lying medially below the corpus callosum, implicated in the perception of reward.

cingulate gyrus A cortical portion of the limbic system, found in the frontal and parietal midline.

olfactory bulb An anteriour projection of the brain that terminates above the upper nasal passages and, through small openings in the skull, provides receptors for smell.

stria terminalis A limbic axonal pathway connecting the amygdala and hypothalamus.

thalamus The brain regions that surround the third ventricle.

hypothalamus Part of the diencephalon, lying ventral to the thalamus.

tectum The dorsal portion of the midbrain, including the inferiour and superiour colliculi.

superiour colliculi Paired gray matter structures of the dorsal midbrain that receive visual information.

inferiour colliculi Paired gray matter structures of the dorsal midbrain that receive auditory information.

red nucleus A brainstem structure related to motor control.

reticular formation An extensive region of the brainstem (extending from the midbrain to the medulla).

Purkinje cell A type of large nerve cell in the cerebellar cortex.

granule cell A type of small nerve cell.

parallel fibre One of the axons of the granule cells that form the outermost layer of the cerebellar cortex.

nucleus, the **putamen**, the **globus pallidus** in the telencephalon under the cerebral cortex, and the **substantia nigra** in the midbrain (**FIGURE 1.17A**; see also Figure 1.14B and C). These nuclei (not really ganglia, despite the unfortunate name *basal ganglia*) are reciprocally connected with the cerebral cortex, forming a looping neural system. The basal ganglia are very important in motor control, as we will see in Chapter 10.

Curving through each hemisphere, alongside the basal ganglia, lies a loose network of structures called the **limbic system** (**FIGURE 1.17B**) that is critical for emotion and learning. Limbic components include the **amygdala** (Latin for "almond," because it has that shape), which has several subdivisions with diverse functions such as emotional regulation (Chapter 11) and odour perception (Chapter 8), as well as the **mammillary bodies**, **hippocampus** (from the Greek *hippokampos*, "sea horse," which it resembles), and **fornix**, all of which contribute to learning and memory (Chapter 16). The **septal nuclei** play a role in reward and reinforcement in learning (Chapter 3). Other components of the limbic system include a strip of cortex atop the corpus callosum in each hemisphere called the **cingulate gyrus**, which is implicated in the direction of attention and many other cognitive functions, as well as the **olfactory bulb**, which is involved in the sense of smell. The **stria terminalis** is a fibre pathway that connects the amygdala to limbic structures near the base of the brain, especially the hypothalamus, participating in highly motivated behaviours such as sex and threat responses, as well as the integration of hormonal signals.

The diencephalon directs sensory information and controls basic physiological functions

Toward the medial (middle) and basal (bottom) aspects of the forebrain are found the two main components of the diencephalon: the **thalamus** and the **hypothalamus** (which simply means "under thalamus"). You can see these structures most clearly in Figures 1.13B, 1.15C, and 1.17B. The thalamus is a complex cluster of nuclei that acts as a switch box, directing almost all incoming sensory information to the appropriate regions of the cortex, to be processed further, and receiving instructions back from the cortex to control which sensory information is transmitted.

The small-but-mighty hypothalamus has a much different role: it is packed with discrete nuclei involved in many vital functions, such as hunger, thirst, temperature regulation, sex, and many more. Furthermore, because the hypothalamus also controls the pituitary gland, it serves as the brain's main interface with the hormonal systems of the body (Chapter 5). We'll encounter the hypothalamus again in several later chapters.

The midbrain has sensory and motor components

In comparison to the forebrain and hindbrain, the midbrain is quite small and contains only a few obvious landmarks. The top part of the midbrain, which is called the **tectum** (from the Latin for "roof," because it's atop the midbrain), features two pairs of bumps—one pair in each hemisphere—with specific roles in sensory processing. The more rostral bumps are called the **superiour colliculi** (singular *colliculus*), and they have specific roles in visual processing. The more caudal bumps, called the **inferiour colliculi** (see Figure 1.13B), process information about sound.

Two important motor centres are embedded in the midbrain. One, the substantia nigra that we mentioned in discussing the basal ganglia (see Figure 1.17A), contains neurons that release the transmitter dopamine. Loss of this system leads to Parkinson's disease (Chapter 10). The other motor centre is the **red nucleus** (named for its reddish tint), which communicates with motor neurons in the spinal cord. Several cranial nerves, including those controlling eye movements, originate in the midbrain.

Also found in the midbrain is a distributed network of neurons collectively referred to as the **reticular formation** (from the Latin *reticulum*, "network"). The reticular formation stretches from the midbrain down to the medulla and is implicated in a variety of behaviours, including sleep and arousal (Chapter 15), temperature regulation (Chapter 14), and motor control.

The cerebellum is attached to the pons and is crucial for motor coordination

The lateral, midsagittal, and ventral views of the brain in Figure 1.13 reveal the cerebellum, the smaller hemispheres tucked up under the posteriour cortex and attached to the dorsal brainstem. Like the cortex, the surface of the cerebellum is elaborately convoluted, giving it more surface area. The arrangement of cells within this folded sheet is relatively simple, consisting of three layers (**FIGURE 1.18**). A middle layer is composed of a single sheet of enormous neurons called **Purkinje cells** after the anatomist who first described their elaborate, fan-shaped dendritic patterns. Axons from the small neurons of the **granule cell** layer, lying below the Purkinje cells, rise to the surface of the cerebellum to form the **parallel fibres** of the outermost layer (called the *molecular layer*). The cerebellum has long been known to be crucial for motor coordination and control, but we now know it also participates in certain aspects of cognition, including learning.

Immediately below (ventral to) the cerebellum lies the pons (see Figure 1.13B and C), a part of the brainstem. Within the pons are important motor control and sensory nuclei, including several nuclei from which cranial nerves arise. Information from the ears first enters the brain in the pons, via the nuclei of the vestibulocochlear (VIII) cranial nerves.

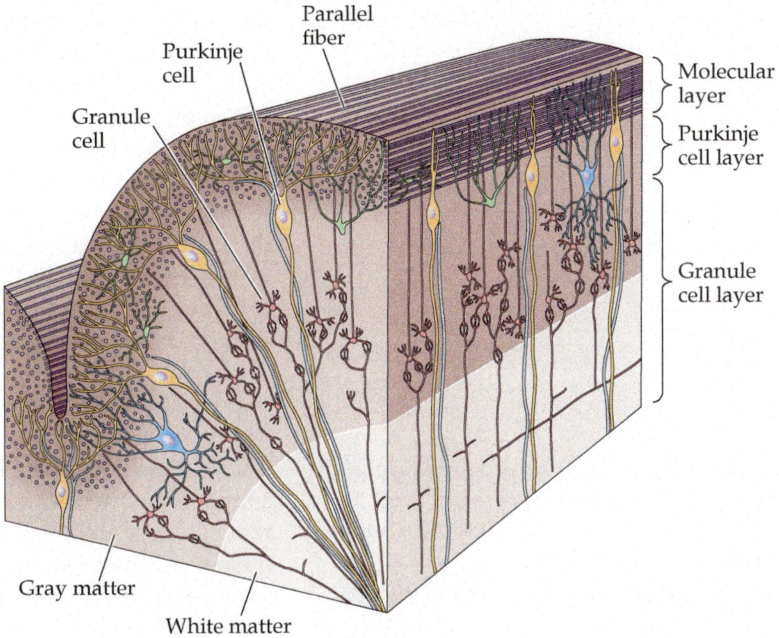

1.18 The Arrangement of Cells within the Cerebellum Large Purkinje cells dominate the cerebellum, dividing it into three layers. Innervation between the various types of cells in the cerebellum forms a very consistent pattern. A variety of scattered cells, depicted here in black, inhibit the activity of other cells.

The medulla maintains vital basic body functions

The medulla is the most caudal portion of the brainstem, and it marks the transition from brain to spinal cord (see Figure 1.13B). Within the medulla are the nuclei of cranial nerves XI and XII, containing the cell bodies of the neurons that control the neck and tongue muscles, respectively. The reticular formation, which we first saw in the midbrain, stretches through the pons and ends in the medulla. Because the medulla contains nuclei that regulate breathing and heart rate, damage there is often fatal. All axons passing between the brain and spinal cord necessarily course through the medulla, and several medullary nuclei add their own axons to the descending fibre tracts.

Go to Activity 1.11
The Cerebellum

Behaviours and cognitive abilities are determined by functional connections between brain regions

In order to understand the neural origins of our most complex behaviours and experiences—thought, language, music—it will be necessary to understand how different brain regions with distinct functions collaborate in larger-scale networks. This applies to functional units as small as the individual cortical columns we mentioned earlier and to much larger assemblages of millions of cells making up substantial parts of cortical lobes.

Cortical regions communicate with one another via tracts of axons looping through the underlying white matter (**FIGURE 1.19**). Some of these connections are short pathways to nearby cortical regions; others travel longer distances through the cerebral hemispheres. Some pathways link corresponding areas in the two hemispheres, traversing the corpus

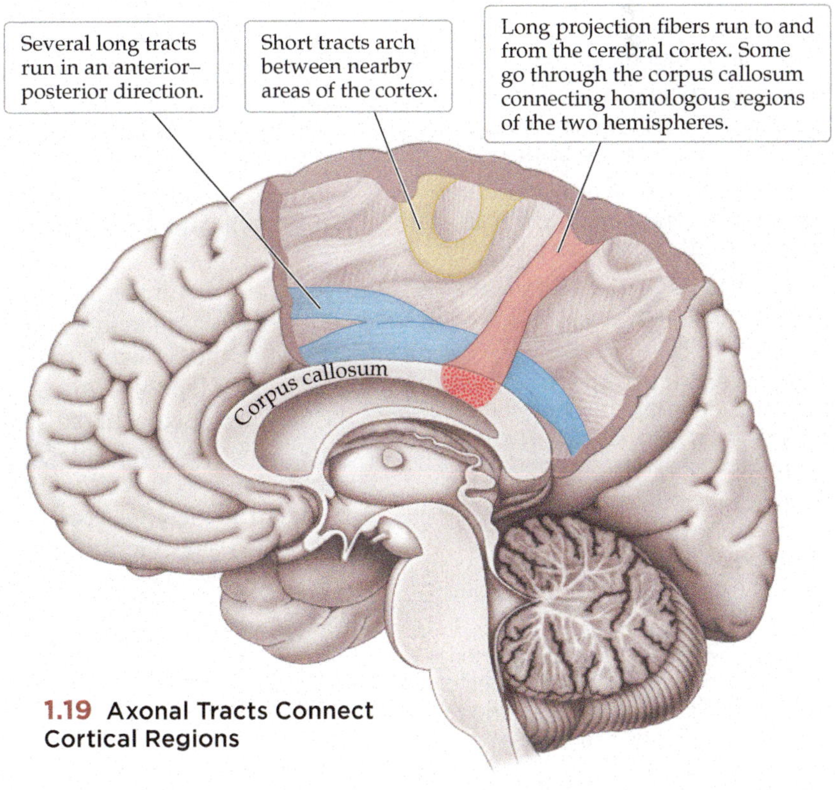

1.19 Axonal Tracts Connect Cortical Regions

callosum to go from one hemisphere to the other. Some longer links between cortical regions involve multisynaptic chains of neurons that loop through subcortical regions such as the thalamus and the basal ganglia, allowing processing and integration of information at several levels.

Research at the frontier of neuroscience aims to describe the "connectome" of the human brain: the network map that completely describes the functional connections within and between brain regions (Glasser, Smith et al., 2016). Due to the complexity of the problem, large, international collaborations have been established with two goals: (1) New technologies for probing brain pathways are being developed by the BRAIN Initiative (Yuste and Church, 2014) and the Human Connectome Project (www.humanconnectomeproject.org). (2) Available technology is being used to map brain network activity (e.g., the Brain Activity Map [BAM] project), to develop computer simulations of human brain activity (e.g., the Human Brain Project [https:/www.humanbrainproject.eu]), and to develop other "Big Data" approaches to the problem (Huang and Luo, 2015). Dealing with the immense volume of data generated by such approaches, where connections are tracked from the level of individual synapses to large fibre pathways in the brain, presents a substantial technological challenge (Lichtman et al., 2014). Nevertheless, advances in our understanding of brain organisation—for example, the parcellation of cerebral cortex into almost 200 distinct regions, using data from the Human Connectome Project (Glasser, Coalson et al., 2016) combined with anatomical data from nonhuman animals (van Essen and Glasser, 2018)—demonstrate that Big Data research is rapidly transforming the field of behavioural neuroscience.

1.4 Specialised Support Systems Protect and Nourish the Brain

Learning Objectives

After reading this section, you should be able to:

1.4.1 **Name and describe the meninges and components of the ventricular system, and explain their clinical significance.**

1.4.2 **Briefly describe the production and circulation of cerebrospinal fluid.**

1.4.3 **Give an overview of the vascular supply of the brain, and note the signs and symptoms of stroke.**

Within the bony skull and vertebrae, the brain and spinal cord are enveloped by three protective membranes called the **meninges** (see Figure 1.11). Between a tough outer sheet called the **dura mater** (in Latin, literally "tough mother") and the delicate **pia mater** ("tender mother") that adheres tightly to the surface of the brain, a webby substance called the **arachnoid** ("spiderweb-like") creates a reservoir called the subarachnoid space that suspends the brain in a bath of **cerebrospinal fluid (CSF)**. **Meningitis**, an inflammation of the meninges usually caused by viral infection, is a potentially lethal medical emergency characterised in early stages by headache, fever, and stiff neck as the inflamed meninges press on the brain. Similarly, haemorrhages within the layers of the meninges can be caused by trauma or disease, and they have varying, often life-threatening symptoms. For example, in epidural hematoma, blood accumulates between the skull and dura, with rapid effects on the brain directly below the expanding pool of blood, whereas in subarachnoid haemorrhage, leaking blood seeps into the CSF-filled subarachnoid space, squeezing the brain and producing intense headache and other neurological signs. Any haemorrhage in the skull is a medical emergency and can often be successfully treated if medical attention is obtained promptly. Sometimes, the meninges can form large tumours called **meningiomas**, which are usually classified as benign in the sense that they are noncancerous, but of course in an enclosed space like the cranium, any mass that takes up space is far from harmless.

The cerebral ventricles are chambers filled with fluid

Inside the brain is a series of chambers—the **ventricular system**—filled with CSF (**FIGURE 1.20**). The CSF circulating through the ventricular system has at least two main functions. First, it acts mechanically as a shock absorber for the brain: floating in

meninges The three protective sheets of tissue—dura mater, pia mater, and arachnoid—that surround the brain and spinal cord.

dura mater The outermost of the three meninges that surround the brain and spinal cord.

pia mater The innermost of the three meninges that surround the brain and spinal cord.

arachnoid The thin covering (one of the three meninges) of the brain that lies between the dura mater and pia mater.

cerebrospinal fluid (CSF) The fluid that fills the cerebral ventricles.

meningitis An acute inflammation of the meninges, usually caused by a viral or bacterial infection.

meningiomas Several classes of noncancerous tumours arising from the meninges.

ventricular system A system of fluid-filled cavities inside the brain.

lateral ventricle A complexly shaped lateral portion of the ventricular system within each hemisphere of the brain.

choroid plexus A highly vascular portion of the lining of the ventricles that secretes cerebrospinal fluid.

third ventricle The midline ventricle that conducts cerebrospinal fluid from the lateral ventricle to the fourth ventricle.

fourth ventricle The passageway within the pons that receives cerebrospinal fluid from the third ventricle and releases it to surround the brain and spinal cord.

hydrocephalus A ballooning of the ventricles, at the expense of the surrounding brain, which may occur when the circulation of cerebrospinal fluid is blocked.

carotid arteries The major arteries that ascend the left and right sides of the neck to the brain, supplying blood to the anteriour and middle cerebral arteries.

1.20 The Cerebral Ventricles These views of an adult human brain show the position of the cerebral ventricles within it. Cerebrospinal fluid (CSF) is made by the choroid plexus in the lateral ventricles and exits from the fourth ventricle to surround the brain and spinal cord.

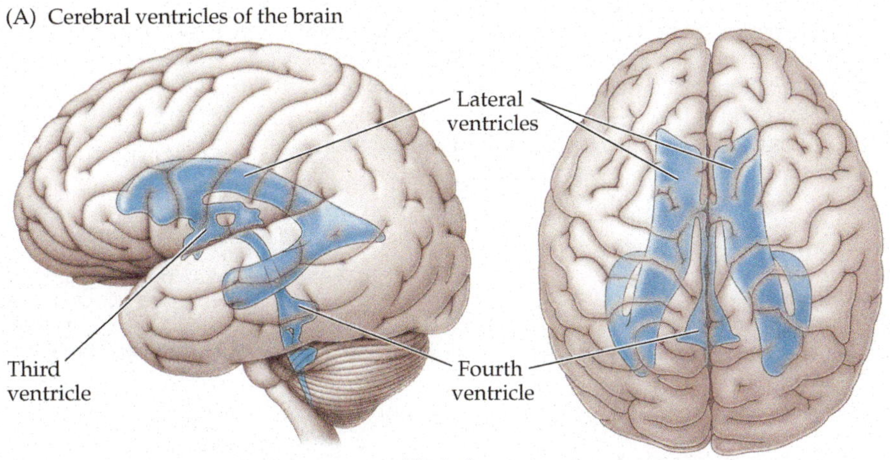

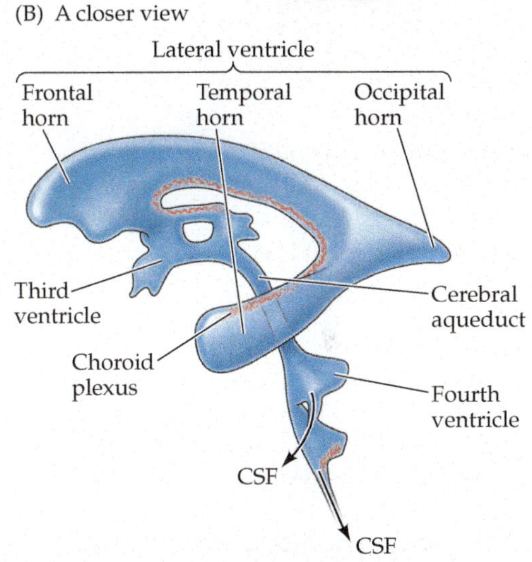

CSF within the subarachnoid layer, the brain is protected from sudden movements of the head that would smash it against the inside of the skull. Second, CSF provides a medium for the exchange of materials, including nutrients, between blood vessels and brain tissue.

Each hemisphere of the brain contains a **lateral ventricle**, extending into all four lobes of the hemisphere. The lateral ventricles are lined with a specialised membrane called the **choroid plexus**, which produces CSF by filtering blood. CSF flows from the lateral ventricles into the **third ventricle** (located in the midline) and continues down a narrow passage to the **fourth ventricle**, which lies between the cerebellum and the pons. Just below the cerebellum, three small openings allow CSF to exit the ventricular system and circulate over the outer surface of the brain and spinal cord. CSF is absorbed back into the circulatory system through large veins beneath the top of the skull. Any problem that blocks the flow of CSF through the ventricular system may result in **hydrocephalus**, a ballooning of the ventricles as they accumulate CSF, resulting in symptoms that vary depending on which structures are compromised by the expanding ventricles.

The brain has an elaborate vascular system

Robbed of the oxygen and nutrients that the blood supplies, the brain will swiftly die. That's because the brain is so needy: it accounts for only 2% of the weight of the average human body, but it consumes more than 20% of the body's energy at rest. Supplying these life-giving metabolic fuels to the brain is the job of a complex set of large blood vessels (**FIGURE 1.21**). The **carotid arteries** ascend the left and right sides of the neck and branch into external and internal carotid arteries;

Go to Activity 1.12
The Cerebral Ventricles

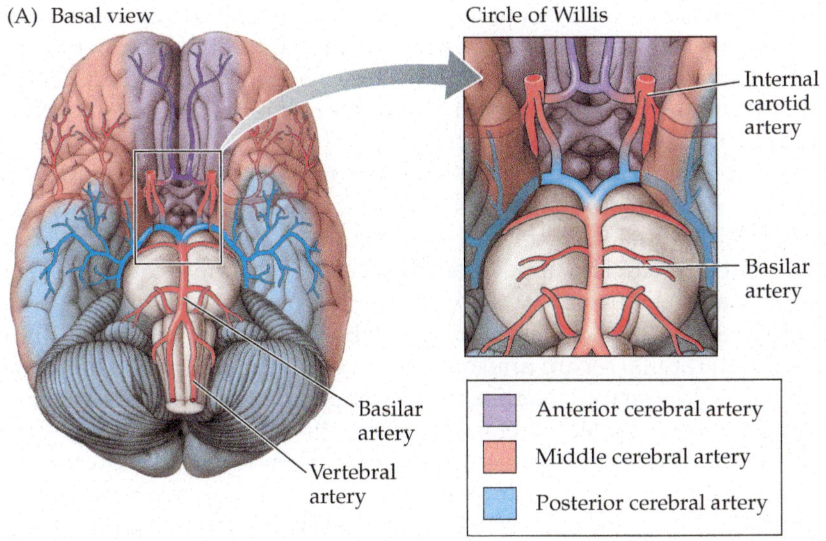

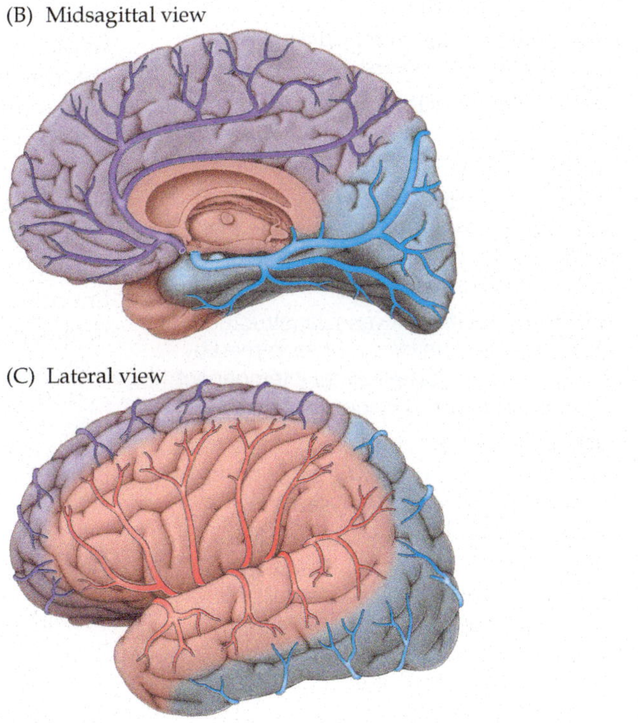

1.21 The Blood Supply of the Human Brain The anteriour, middle, and posteriour cerebral arteries—the three principal arteries that provide blood to the cerebral hemispheres—are depicted here in basal (A), midsagittal (B), and lateral (C) views of the brain. The basilar and internal carotid arteries form a circle at the base of the brain known as the *circle of Willis*.

1.22 Stroke A stroke is the result of any situation that cuts off blood supply to the brain. The effect of a stroke on cognition depends on which artery—and associated brain tissue—is compromised. The impact on the brain can be greatly reduced in some cases if medical attention is obtained promptly.

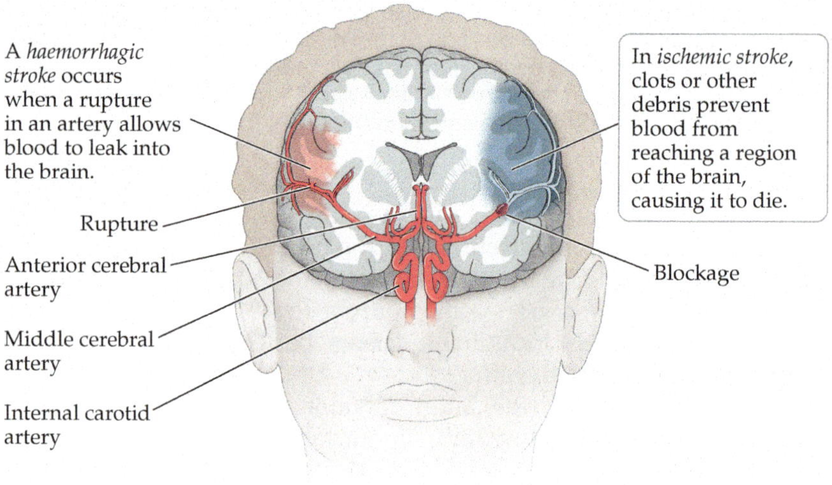

these are the major arteries that you can feel pulsing in each side of your neck after exertion. The internal carotid artery enters the skull and branches into **anteriour** and **middle cerebral arteries**, which supply blood to about two-thirds of the cerebral hemispheres, indicated in purple and pink in Figure 1.21. The rest of the cortex (indicated in blue in Figure 1.21) is supplied by the **posteriour cerebral arteries**. Blood is provided to these arteries via the left and right **vertebral arteries**, which enter at the base of the neck and rise, to either side of the spinal cord, through the cervical vertebrae and into the base of the skull, where they fuse together to form the **basilar artery**. In addition to feeding the posteriour cerebral arteries, the basilar artery has branches supplying blood to the hindbrain.

At the base of the brain, the major cerebral arteries are joined via communicating arteries to form a structure called the **circle of Willis** (see Figure 1.21A). This joining of arterial paths may provide an alternate route for blood flow if any of the main arteries to the brain should be damaged or blocked by disease. The general term **stroke** applies to a situation in which a clot, a narrowing, or a rupture interrupts the supply of blood to a discrete brain region, causing the affected region to stop functioning or die (**FIGURE 1.22**). Although the exact effects of a stroke depend on the region of the brain that is affected, the five most common warning signs are sudden numbness or weakness, altered vision, dizziness, severe headache, and confusion or difficulty speaking or understanding speech. Effective treatments are available to help restore blood flow and minimize the long-term effects of a stroke, but only if the person receives emergency medical treatment within a few hours (Albers et al., 2018).

Fine vessels and capillaries branch off from the main arteries and deliver nutrients and other substances to brain cells and remove waste products. Thanks to regulation by CNS cells that surround the vessels, especially astrocytes and pericytes (small cells that encircle, support, and regulate the brain's capillaries), the endothelial cells that make up the walls of the capillaries in the brain have much tighter junctions than is the case elsewhere in the body (Daneman et al., 2010). As a consequence, brain capillaries are unusually resistant to the passage of large molecules across their walls and into neighbouring neurons. This **blood-brain barrier** may have evolved to help protect the brain from infections and blood-borne toxins, but it also makes the delivery of drugs to the brain more difficult. However, the blood-brain barrier is not completely impenetrable.

Most of the body contains elements of the lymphatic system, which drains waste-bearing fluids from body tissues and from the newly discovered *interstitium*, an organ made up of interconnected reservoirs lining many organs, ducts, and vessels (Benias et al., 2018). Although the brain had long been thought to lack the lymphatic system found in other tissues, recent detailed anatomical research in

anteriour cerebral arteries Two large arteries, arising from the internal carotid arteries, that provide blood to the anteriour poles and medial surfaces of the cerebral hemispheres.

middle cerebral arteries Two large arteries, arising from the internal carotid arteries, that provide blood to most of the lateral surfaces of the cerebral hemispheres.

posteriour cerebral arteries Two large arteries, arising from the basilar artery, that provide blood to posteriour aspects of the cerebral hemispheres, cerebellum, and brainstem.

vertebral arteries Arteries that ascend the vertebrae, enter the base of the skull, and join together to form the basilar artery.

basilar artery An artery, formed by the fusion of the vertebral arteries, that supplies blood to the brainstem and to the posteriour cerebral arteries.

circle of Willis A vascular structure at the base of the brain that is formed by communicating arteries that interconnect the major cerebral arteries.

stroke Damage to a region of brain tissue that results from blockage or rupture of vessels that supply blood to that region.

blood-brain barrier The mechanisms that make the movement of substances from blood vessels into brain cells more difficult than exchanges in other body organs.

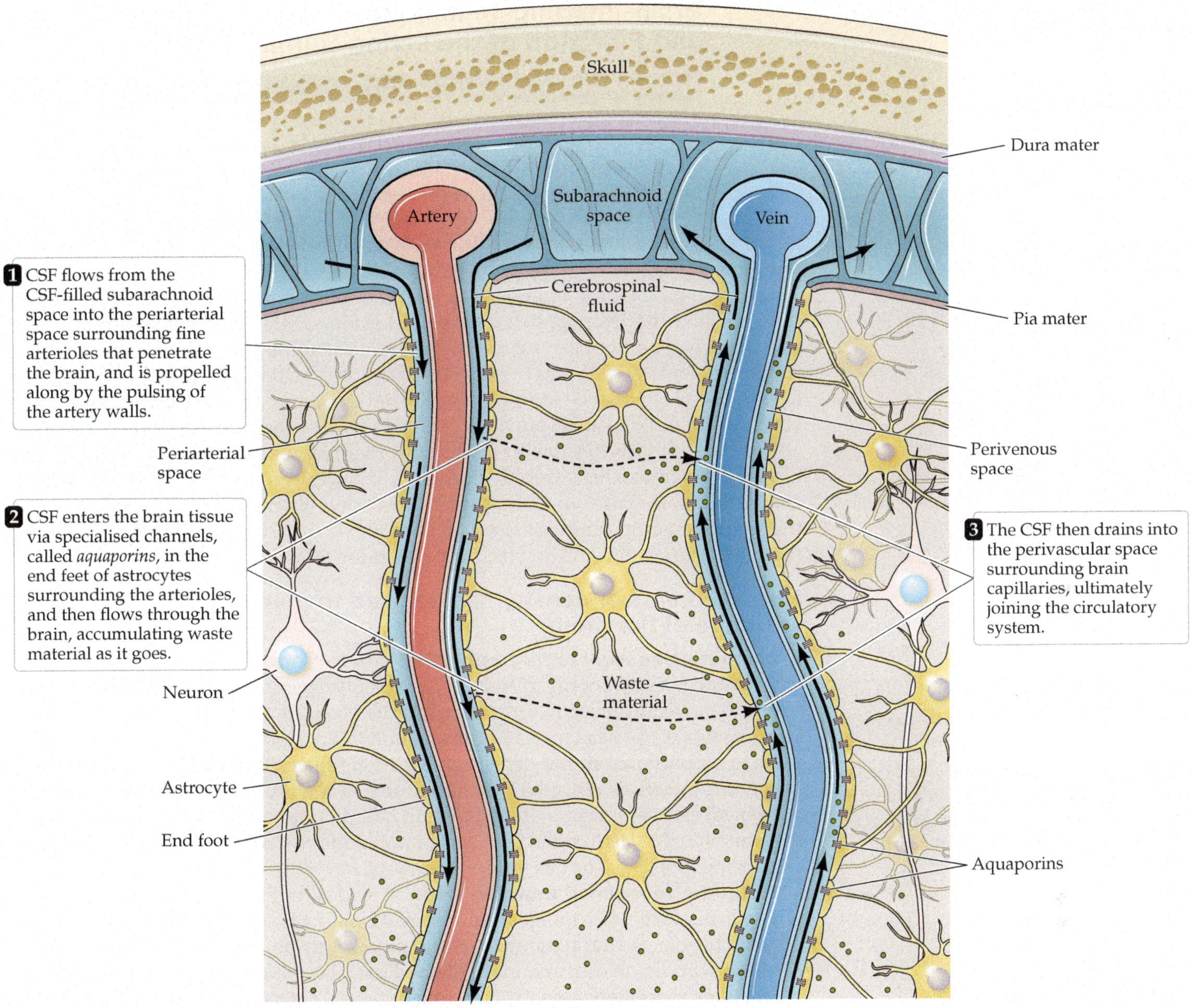

1.23 The Glymphatic System This recently discovered system provides a flow of CSF through the interiour of the brain that helps to clear cellular debris, proteins, and other wastes. By removing harmful substances such as the amyloid proteins implicated in Alzheimer's disease, glymphatic flow may protect us from various neurological diseases. (After M. Nedergaard and S. A. Goldman, 2016. *Sci Am* 314: 44–49.)

mice revealed a **glymphatic system** (**FIGURE 1.23**; the name reflects the involvement of glial cells) that similarly provides drainage of wastes in the brain and may provide a route for the distribution of various nutrients and signalling substances (Aspelund et al., 2015; Jessen et al., 2015). Researchers are working to understand possible roles of the glymphatic system in various neurological problems, including Alzheimer's disease and cancer. Because it allows the immune system to access the brain, the glymphatic system also may be important in diseases with an immune component, like multiple sclerosis. Researchers also hope to learn how to exploit glymphatic mechanisms to deliver drugs to the brain, perhaps bypassing the blood-brain barrier.

glymphatic system A lymphatic system in the brain that participates in removal of wastes and the movement of nutrients and signalling compounds.

1.5 Brain-Imaging Techniques Reveal the Structure and Function of the Living Human Brain

Learning Objectives

After reading this section, you should be able to:

1.5.1 **Identify and briefly describe the basis of anatomical imaging technologies like X-ray, CT, and MRI.**

1.5.2 **Describe and compare functional-imaging technologies, including PET, fMRI, and MEG, and explain how each is useful for research.**

1.5.3 **Discuss important caveats and potential pitfalls in interpreting the results of imaging studies.**

Researchers have long sought ways to peer into the living human brain to see structures and how they work during different behaviours. X-rays of the head are of limited usefulness because they cannot resolve the brain's small variations in density. Attempts to improve contrast in X-rays through the injection of radiopaque (X-ray-blocking) dye led to a useful technique—**angiography** (from the Greek *angeion*, "blood vessel," and *graphein*, "to write")—that provides detailed views of the cerebral blood vessels and aids in the diagnosis of vascular disease. But more recent technological developments have allowed researchers to study the brains of healthy humans too. These techniques are used in some cases to identify brain structure; other approaches focus more on tracking changes in brain *activity* as behaviour occurs. Here we briefly describe the most important brain-imaging technologies in use today.

Measurements of density can be used to map the structure of the brain

In **computerised axial tomography** (**CAT** or **CT**) (from the Greek *tomos*, "crosscut" or "section"), X-ray energy is used to generate images. In a CT scanner, an X-ray source is moved by steps in an arc around the head. At each point, detectors on the opposite side of the head measure the amount of X-ray radiation that is absorbed; this value is proportional to the density of the tissue through which the X-rays passed. When this process is repeated from many angles and the results are mathematically combined, an anatomical map of the brain based on tissue density can be generated by computer (**FIGURE 1.24A**). CT scans are medium-resolution images, useful for visualising problems such as strokes, tumours, or cortical atrophy.

Magnetic resonance imaging (**MRI**) provides higher-resolution images than CT, and because MRI images are derived from radio frequency energy, an additional benefit is that patients are not exposed to potentially damaging X-rays. An MRI scan involves three main steps. First, the patient's head is placed in the centre of an extremely powerful circular magnet that causes all the protons in the brain's tissues to line up in parallel, instead of in their usual random orientations. Protons are found in the nuclei of atoms, notably the hydrogen atoms making up water within body tissues. Next, the protons are knocked over by a strong pulse of radio waves. When this pulse is turned off, the protons relax back to their original configuration, emitting radio waves as they go. The emitted radio frequency energy, measured by detectors ringing the head, varies depending on the density of various tissues. Finally, a powerful computer uses this density-based information to generate a detailed cross-sectional map of the brain that scientists and physicians use to evaluate the size and shape of distinct brain regions (**FIGURE 1.24B**) (Elster and Burdette, 2001). With their higher resolution, MRI images can reveal subtle changes in the brain, such as the loss of myelin that is characteristic of multiple sclerosis.

In **diffusion tensor imaging** (**DTI**), MRI technology is used in a different way to specifically study white matter tracts—axon bundles—within the living brain. If the water molecules (i.e., their protons) affected by an MRI scanner's magnets are in a relatively unconfined volume, such as a cerebral ventricle, then as they relax, they can diffuse in any direction. This is known as *isotropy*. But water molecules contained within the narrow tubes of axons are constrained, and they diffuse more readily in some directions (lengthwise, in particular) than others. This

angiography A brain-imaging technique in which a specialised X-ray image of the head is taken shortly after the cerebral blood vessels have been filled with a radiopaque dye by means of a catheter.

computerised axial tomography (CAT or CT) A noninvasive technique for examining brain structure in humans through computer analysis of X-ray absorption at several positions around the head.

magnetic resonance imaging (MRI) A noninvasive technique that uses magnetic energy to generate images that reveal some structural details in the living brain.

diffusion tensor imaging (DTI) A modified form of MRI in which the diffusion of water in a confined space is exploited to produce images of axonal fibre tracts.

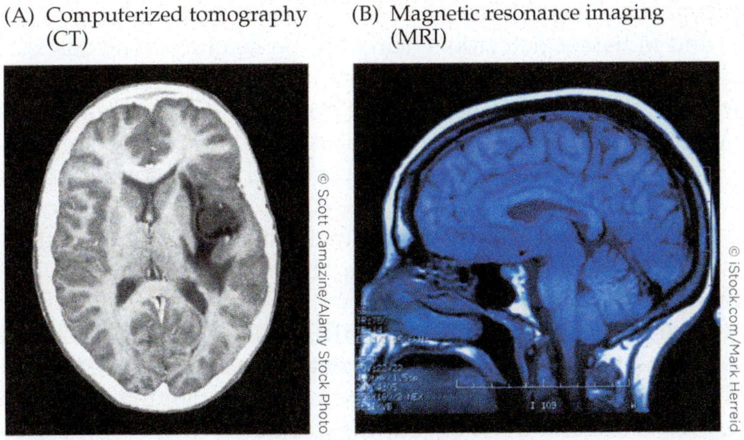

1.24 Visualising the Living Human Brain (A) CT scan from a patient with a brain tumour, visible as the dark area in the right hemisphere. (B) Midsagittal MRI image of a healthy brain. (C) Diffusion tensor imaging (DTI) is an alternative application of MRI that exploits fractional anisotropy—the diffusion of water in axons—to visualise axonal connections between regions. Data from multiple scans can be combined to create three-dimensional models of the interconnections of brain regions, known as *DTI tractography.* (D) PET scans from a healthy human brain and the brain of a patient with Alzheimer's, showing levels of metabolic activity. Note the greater level of activity in the healthy brain. (E) Functional-MRI images showing changes in regional brain metabolism recorded during the presentation of visual or auditory stimuli (images of a romantic partner; see Figure 1.7).

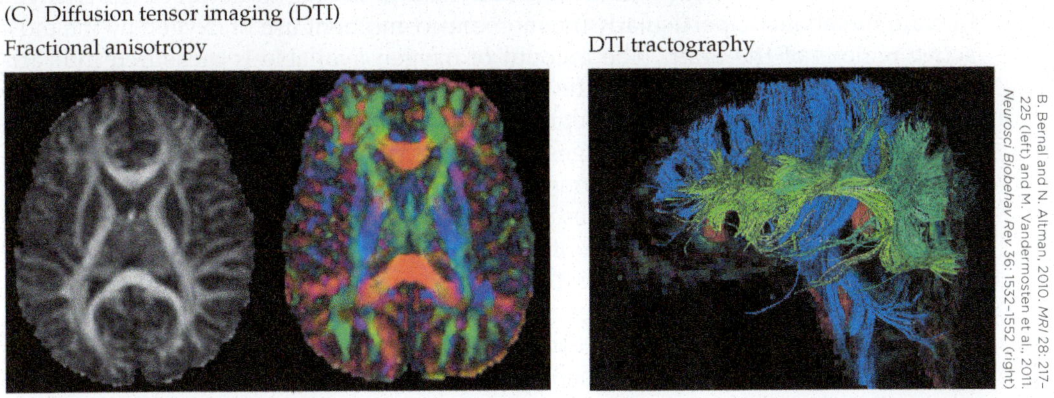

Go to Animation 1.1
Visualising the Living Human Brain

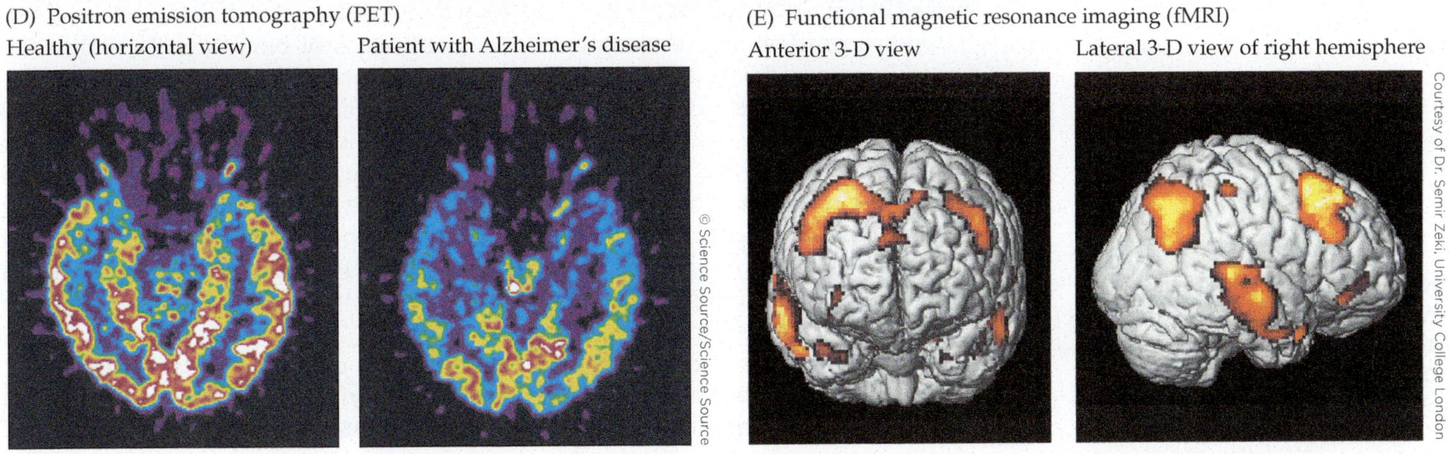

property, known as **fractional anisotropy** (**FA**) (**FIGURE 1.24C, LEFT**), reflects connections between brain regions and can be mathematically exploited in DTI to produce *structural* images of axonal fibre pathways (**FIGURE 1.24C, RIGHT**), a procedure called **DTI tractography** or *fibre tracking.* Although the technology doesn't have the resolution to portray individual axons, the origin, orientation, course, and termination of axonal bundles can be effectively visualised using tractographic analysis. DTI has thus become one of the most important tools in efforts to map the human connectome.

Functional-imaging techniques map regional brain activity during behaviours

In **positron emission tomography** (**PET**) the objective is to depict the brain's *activity* rather than details of its structure, and it has proven to be very valuable for both

fractional anisotropy (FA) The tendency of water to diffuse more readily along the long axis of an enclosed space, such as an axon. FA is the basis of diffusion tensor imaging.

DTI tractography Also called fibre tracking. Visualisation of the orientation and terminations of white matter tracts in the living brain via diffusion tensor imaging.

positron emission tomography (PET) A technique for examining brain function by combining tomography with injections of radioactive substances used by the brain.

functional MRI (fMRI) Magnetic resonance imaging that detects changes in blood flow and therefore identifies regions of the brain that are particularly active during a given task.

optical imaging A method for visualising brain activity in which near-infrared light is passed through the scalp and skull.

transcranial magnetic stimulation (TMS) Localised, noninvasive stimulation of cortical neurons through the application of strong magnetic fields.

magnetoencephalography (MEG) A noninvasive functional brain-imaging technique that measures the tiny magnetic fields produced by active neurons, in order to identify regions of the brain that are particularly active during a given task.

lesions Regions of damage within the brain.

experimental and medical purposes. Short-lived radioactive chemicals are injected into the bloodstream, and radiation detectors encircling the head map the destination of these chemicals in the brain. A particularly effective strategy is to administer radioactive glucose while a participant is performing a specific cognitive task. Because the radioactive glucose is preferentially taken up and used by the brain regions that are most active from moment to moment, a computer-generated, colour-coded portrait of brain activity can be created (**FIGURE 1.24D**) (Roland, 1993). Through precise experimental control and the use of special mathematical techniques (described in **BOX 1.3**), we can generate metabolic maps of the brain that identify the regions that contribute to specific functions.

Since its introduction in the 1990s, **functional MRI** (**fMRI**) has revolutionised cognitive neuroscience research, producing images with reasonable speed (temporal resolution) and excellent sharpness (spatial resolution). Exploiting the same basic technology used in the structural MRI scanning described earlier, fMRI uses high-powered, rapidly oscillating magnetic-field gradients to detect small changes in brain metabolism, particularly the moment-to-moment use of oxygen by the most active regions of the brain. The amount of oxygen available is measured indirectly, on the basis of blood flow or the state of hemoglobin in blood (called the *blood-oxygen-level-dependent*, or *BOLD*, signal). As with PET, scientists can use fMRI data to create computer-generated images that reflect the *activity* of different parts of the brain while people engage in various experimental tasks (see Box 1.3), trading PET's quicker response for the higher resolution offered by MRI. Although fMRI cannot resolve the fine cellular structure of the brain, and is too slow to track rapid, moment-by-moment changes in the activity of networks of neurons, the detailed activity maps provided by fMRI have revealed many important clues about how networks of brain structures collaborate on complex cognitive processes (**FIGURE 1.24E**). The fMRI image generally reflects synaptic inputs and local processing, rather than the production of neural impulses (Logothetis, 2008) and is altered by aging and disease states (Rypma et al., 2005).

Other investigators are using light to make images of brain activity within the head (Gratton and Fabiani, 2001; Villringer and Chance, 1997). In **optical imaging**, researchers capitalise on the observation that near-infrared light (having wavelengths of 700–1000 nm) passes easily through skin, scalp, and skull and penetrates a short distance into the cortex. When such light is transmitted into the brain and detectors pick up the reflections through the scalp, the responses reveal the activity of cortical regions. Some components of the optical responses represent the electrical signals of neurons, and other components represent blood flow. The relatively low expense and small size of the optical imaging apparatus may allow many more laboratories to use brain imaging in their research.

As with light, it is a relatively simple matter to pass magnetic fields into the brain, but it is technically much more challenging to do so in a highly focused and precise manner. In **transcranial magnetic stimulation** (**TMS**) (**FIGURE 1.25**), strong, focal magnetic currents are used to briefly stimulate the cortex of an alert participant directly, without needing a hole in the scalp or skull. This approach enables experimenters to activate (or interfere with) discrete areas of the cortex through a process of electromagnetic induction. The regions stimulated and the resultant behavioural effects can then be carefully tracked and mapped. In *repetitive TMS* (*rTMS*), this focal magnetic stimulation of the brain is cycled several times per second, producing transient but measurable changes in behaviour that are useful in clinical settings, as well as in research.

Magnetism and brain function are linked in another way too. Like any electrical system, active circuits of the brain produce their own magnetic fields. Although they are minuscule, the magnetic fields created by activity in local circuits of neurons can be detected by ultrasensitive detectors called *SQUIDs* (superconducting quantum interference devices). In **magnetoencephalography** (**MEG**), a large array of SQUIDs is used to create real-time maps of cortical activity during cognitive processing (**FIGURE 1.26**). Because the temporal resolution of MEG is so good—it responds very quickly to moment-by-moment changes in brain activity—it is

1.25 Transcranial Magnetic Stimulation Magnetic fields induced by electromagnetic coils stimulate neurons of the underlying cortical surface.

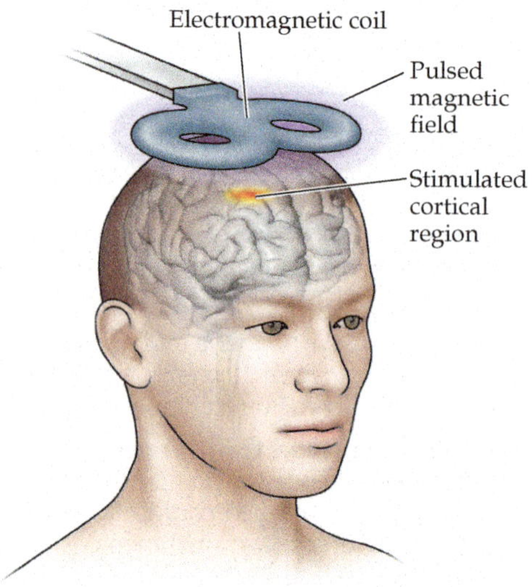

BOX 1.3 Isolating Specific Brain Activity

The advent of modern brain imaging has enabled dramatic images of the brain showing the particular brain regions that are activated during specific cognitive processes; there are many such images in this book. But normally, as you might expect, almost all of the brain is active at any given moment (showing that the old notion that "we use only 10% of our brain" is nonsense). So how do we get these highly specific images of brain activity?

The PET scan shown beneath the box labelled "Visual stimulus" in the figure was made while a person looked at a fixation point surrounded by a flickering checkerboard ring. The scan next to it (beneath the box labelled "Control") was made while a person looked at a fixation point alone. In a comparison of the two, it is hard to see differences, but subtracting the control values from the stimulation values yields an image such as that shown at the upper right ("Difference image"); in this scan it is easy to see that the main difference in activity is in the posteriour part of the brain (the visual cortex).

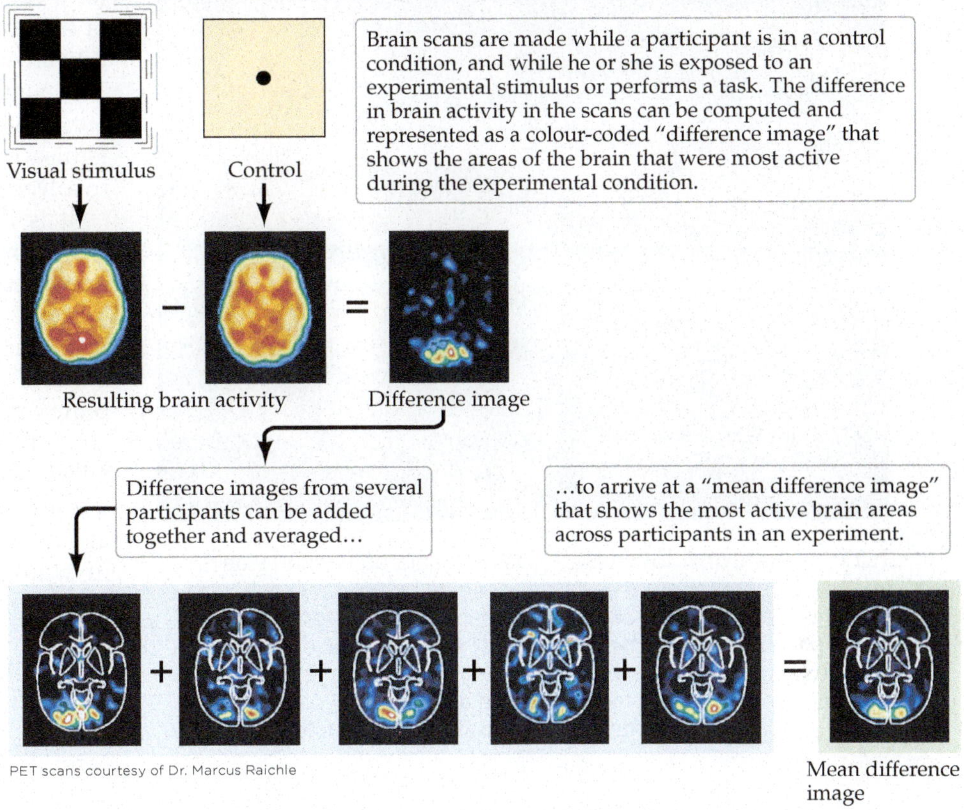

PET scans courtesy of Dr. Marcus Raichle

The PET scans shown in the bottom row are difference images for five individuals who performed the same two stimulation and control tasks. Averaging these five scans yields the mean difference image for all five individuals that is shown at far right. Averaged images yield more-reliable results than individual images, but they lack some of the specificity of the individual images.

excellent for studying the rapidly shifting patterns of brain activity in cortical circuits that fMRI is too slow to track (Baillet, 2017), particularly when used in conjunction with MRI (F. H. Lin et al., 2004).

Sophisticated imaging techniques are powerful tools requiring cautious interpretation

In many areas of behavioural neuroscience, technological innovations in brain imaging have provided researchers with exceptional, noninvasive tools that supplement traditional studies of brain **lesions** (regions of damage). They also give us insights into processes that previously were completely hidden. For example, through

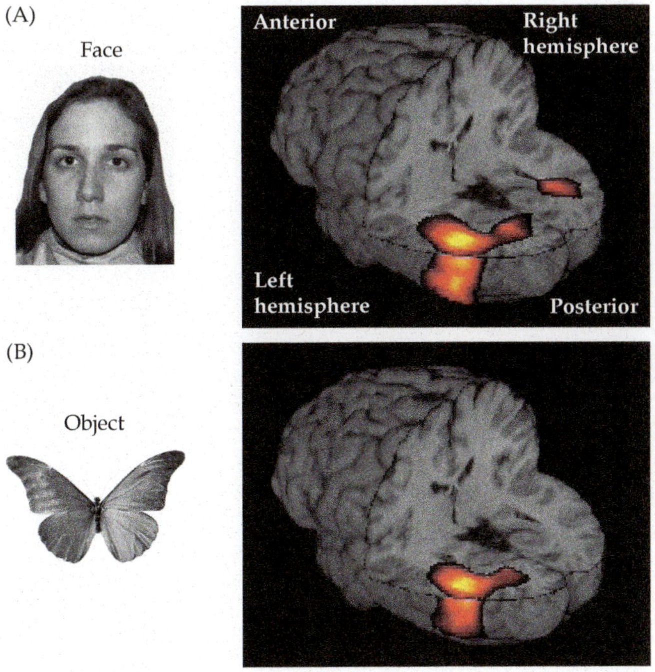

Courtesy of Drs. Mario Liotti and Anthony Herdman, Simon Fraser University and Down Syndrome Research Foundation

1.26 Animal Magnetism Using measurements of the minuscule magnetic fields given off by ensembles of cortical cells during specific behavioural functions, magnetoencephalography (MEG) provides a real-time map of brain activity. In these images, MEG data have been superimposed on structural MRIs of a participant's brain, creating maps of brain activity associated with viewing faces (A) versus nonface objects (B).

the use of functional brain imaging, evidence has emerged that someone who appears to be in a coma (or a vegetative state) may be somewhat conscious of her condition and surroundings (**FIGURE 1.27**) (Owen et al., 2006), a topic we return to in Chapter 18. Images made with PET, MEG, fMRI, and DTI are providing us with new appreciation of the organisation and dynamics of massive neural circuits involved in all forms of mental life.

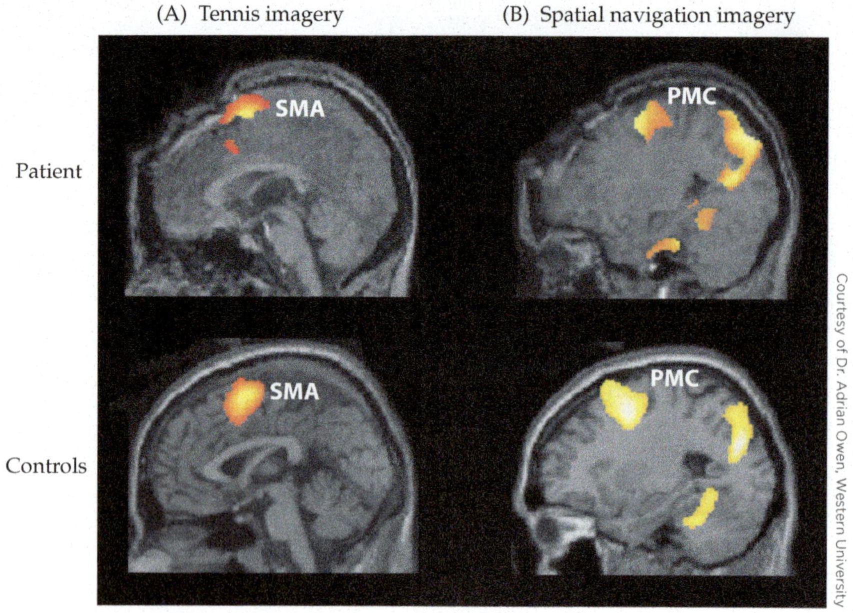

1.27 Coma Consciousness This brain-injured woman was apparently unresponsive and diagnosed as being in a coma, but when researchers asked her to imagine playing tennis (A) or walking through her home (B), fMRI of her brain showed the same patterns of activation as in controls during the same tasks. Note the damage to the skull caused by the accident that caused the coma. SMA, supplementary motor area; PMC, premotor cortex.

It is important to keep in mind that although the colourful images of brain activity that we see everywhere in the popular media seem unambiguous and easy to label, they suffer from a variety of procedural and experimental limitations (Racine et al., 2005). For one thing, technology for functional brain imaging is subject to what the engineers call a "speed-accuracy trade-off": some technologies, like fMRI, provide very fine structural detail (high *spatial resolution*) but are relatively slow to create images and thus are poor at tracking dynamic changes in brain activity (low *temporal resolution*). Electrophysiological measures like MEG and electroencephalography (EEG), conversely, are fast enough to provide excellent temporal resolution but lack the spatial resolution needed to provide highly detailed images. Researchers must weigh these trade-offs carefully in designing experiments (**FIGURE 1.28**).

It is likewise important to realise that these technologies do not produce actual images of the brain, but rather computer-generated composites based on measurements such as density, blood flow, or electromagnetic activity—these measurements are presumed to reflect the structure, metabolism, and electrical activity of the brain. The colours assigned to patterns of activation are arbitrary—obviously, your frontal lobes don't actually glow red hot when you are working on a tricky logic problem—and are based on statistical estimations. The computer algorithms underlying functional brain imaging are the subject of considerable debate and controversy, and there is growing concern that they may sometimes be misleading (Eklund et al., 2016; Poldrack et al., 2017). Further, it is all too easy to interpret the results of brain imaging as if many behaviours and cognitive processes can be associated with specific locations in the brain; some authors have gone so far as to warn of a modern-day version of phrenology, which we discussed in Chapter 1 (G. A. Miller, 2010). But there is no doubt that modern

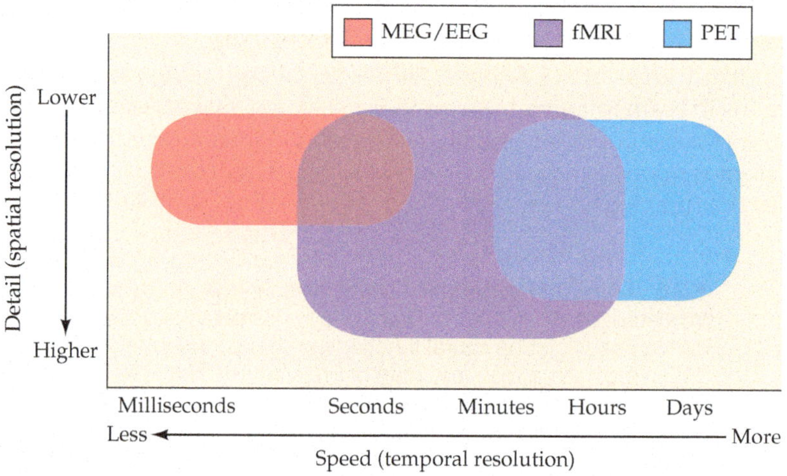

1.28 Spatial versus Temporal Resolution in Functional Brain Imaging No current technique offers both highly detailed imagery and rapid tracking of changes in brain activity. Development of ever-faster and more-accurate technology is a major research focus in neuroscience.

functional brain imaging technologies, interpreted with caution, are helping us to understand the complexities of neural activity. The origins of these neural signals are the topic of Chapter 3.

The Cutting Edge

Two Heads Are Better Than One

Conventional functional-imaging technology does an amazing job of visualising brain activity during behaviour, but the behaviour itself is necessarily quite unnatural. For one thing, participants must lie completely motionless in a very narrow, very noisy tunnel. Material for cognitive processing is generally delivered by computer monitor and headphones, and behavioural responses are often limited to button presses. In real life, our behaviour is more complex and generally takes place in a social setting, directed towards and shaped by other people.

The burgeoning field of **social neuroscience** aims to understand brain activity as it relates to our interactions with others (McEwen and Akil, 2011). It is relatively easy to study the social interactions of people in the field, where researchers can observe natural behaviour at a distance. We can also take measurements of peripheral physiological responses to social interactions, such as salivary measurements of hormonal changes (e.g., van Anders and Watson, 2006). But how can we introduce the social dimension to brain-imaging research?

To approach this question, researchers would like to obtain simultaneous brain scans from both partners in a social dyad (pair) while they interact with each other. One initial approach, known as *fMRI hyperscanning*, involved the use of separate—but linked—fMRI machines to record from subjects' brains simultaneously during social interactions (Montague et al., 2002). The problem, of course, is that separating the participants into separate MRI machines makes for rather unnatural social interactions, even if they are linked by video (Redcay et al., 2010). But what if we could cram both participants into the same scanner? In **dyadic functional MRI** (**dfMRI**) (R. F. Lee et al., 2012), an MRI scanner is fitted with specially designed dual head coils—a head coil is the part of the apparatus that encircles the head and deals with the radio frequency energy that is used to create brain images. With the use of dfMRI, two people's brains can be scanned while they lie side by side in the machine, interacting through a window linking the two coils (**FIGURE 1.29**). The dfMRI technique thus adds an additional level of validity to the interactions—aside from the more intangible aspects of the three-dimensional "realness" of the situation, participants can simultaneously observe small movements, touch each other, and so on.

Initial research with dfMRI has started to reveal some of the subtleties of collaboration between two brains involved in the same behavioural situation. For example, as a pair of friends blinked and gazed at each other's faces, enhanced BOLD signal was observed in the region of the temporoparietal junction (TPJ, a region previously

social neuroscience A field of study devoted to uncovering brain mechanisms and patterns of neural activity that are both cause and consequence of social behaviours such as affiliation, competition, cooperation, etc.

dyadic functional MRI (dfMRI) A form of fMRI in which the brains of two socially-interacting people are scanned simultaneously in a single scanner.

(A) (B)

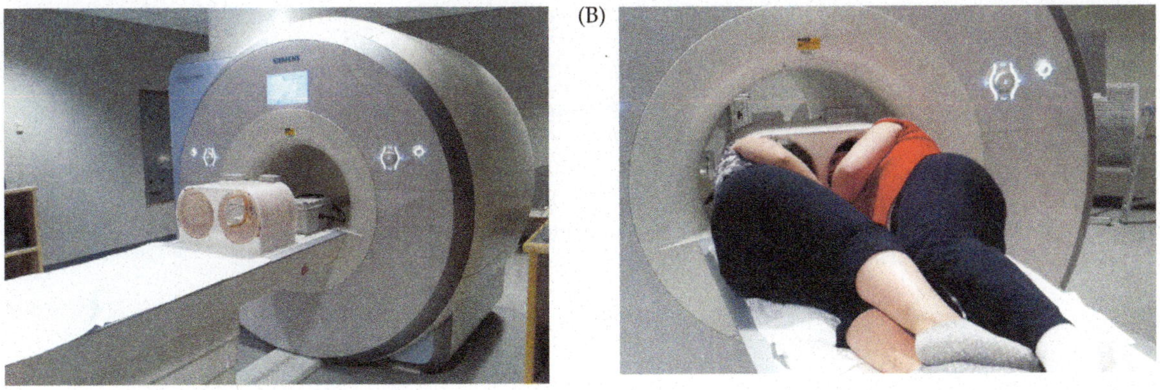

1.29 Dyadic Functional MRI (A) This unique MRI apparatus allows two people to interact while their brains are simultaneously scanned, side by side. (B) Given the narrow confines of the MRI machine, close friendship and a couple of breath mints will be helpful.

From R. F. Lee et al., 2011. *Magn Reson Med* 68: 1087–1096

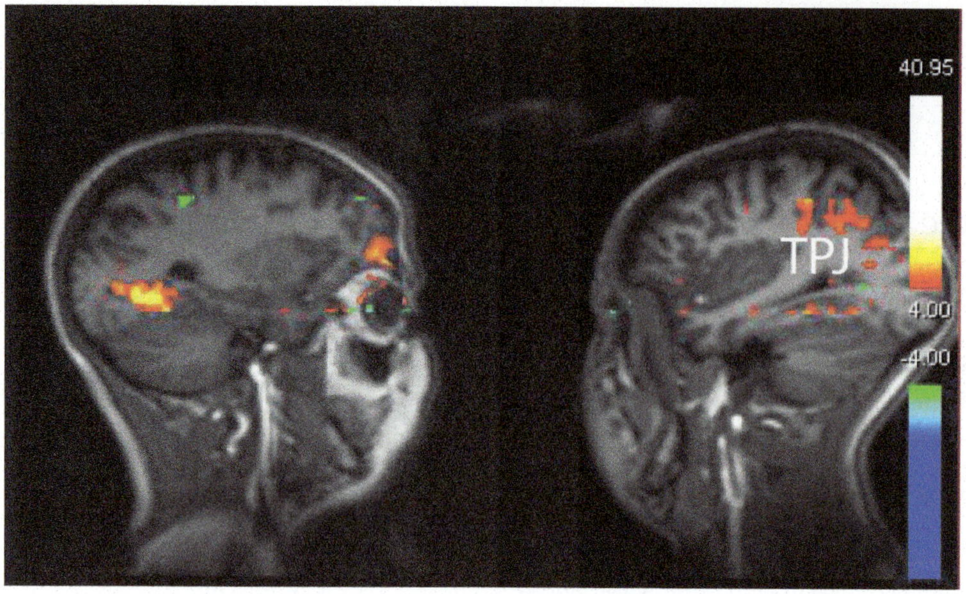

From R. F. Lee et al. 2011. *Magn Reson Med* 68: 1087–1096

1.30 Dyadic Functional MRI Images during Social Interaction Here, two friends are observing each other's faces while opening and closing their eyes. This is the first time that researchers have imaged the activity of two brains as they are directly interacting. TPJ, temporoparietal junction.

implicated in social cognition), in the fusiform face area (see Chapter 19), and in frontal cortex (**FIGURE 1.30**; Lee et al., 2012). What is additionally interesting to researchers is how the BOLD responses differ between the participants; as you can see, the left and right individuals in the scans show their own unique patterns of activation, suggesting that individual responses to the social situation may reflect personality and other intrinsic differences and perhaps also differences in the individuals' roles within the pair. Although this approach to social neuroscience is still in its infancy, dfMRI and other emerging technologies hold the promise of unlocking the brain's activity during much richer social phenomena, such as the emergence of reciprocity in social exchanges (Shaw et al., 2018), and the possibility that brain-to-brain alignment explains the role of empathetic social touch, like handholding, in pain reduction (Goldstein et al., 2018). ■

Recommended Reading

Allen Brain Map, http://portal.brain-map.org/.

Allen Brain Atlas, Cell Types, neural cell features database, celltypes.brain-map.org.

EU Human Brain Project, https://www.humanbrainproject.eu/en/.

NIH Brain Research through Advancing Innovative Neurotechnologies (BRAIN) Initiative, https://www.braininitiative.nih.gov/.

Blumenfeld, H. (2010). *Neuroanatomy through Clinical Cases* (2nd ed.). Sunderland, MA: Oxford University Press/Sinauer.

Catani, M., and Thiebaut de Schotten, M. (2012). *Atlas of Human Brain Connections.* Oxford, UK: Oxford University Press.

Huettel, S. A., Song, A. W., and McCarthy, G. (2014). *Functional Magnetic Resonance Imaging* (3rd ed.). Sunderland, MA: Oxford University Press/Sinauer.

Mai, J. K., Majtanik, M., and Paxinos, G. (2016). *The Human Brain* (4th ed.). San Diego, CA: Academic Press.

Mendoza, J., and Foundas, A. L. (2010). *Clinical Neuroanatomy: A Neurobehavioural Approach.* Heidelberg, Germany: Springer.

Miller, G. A. (2010). Mistreating psychology in the decades of the brain. *Perspectives on Psychological Science, 5,* 716–743.

Papanicolau, A. C. (Ed.). (2017). *The Oxford Handbook of Functional Brain Imaging in Neuropsychology and Cognitive Neurosciences.* Oxford, UK: Oxford University Press.

Schoonover, C. (2010). *Portraits of the Mind: Visualising the Brain from Antiquity to the 21st Century.* New York: Abrams.

Swanson, L. W., Newman, E., Araque, A., and Dubinsky, J. M. (2017). *The Beautiful Brain: The Drawings of Santiago Ramón y Cajal.* New York: Abrams.

Vanderah, T. W., and Gould, D. J. (2015). *Nolte's the Human Brain: An Introduction to Its Functional Neuroanatomy* (7th ed.). Philadelphia: Elsevier.

Woolsey, T. A., Hanaway, J., and Gado, M. H. (2007). *The Brain Atlas: A Visual Guide to the Human Central Nervous System.* New York: Wiley-Liss.

1 Visual Summary

You should be able to relate each summary to the adjacent illustration, including structures and processes. You can also refer back to the links to figures, animations, and activities that will help you consolidate the material.

1 The nervous system is extensive—monitoring, regulating, and modulating the activities of all parts and organs of the body. Review **Figure 1.9**

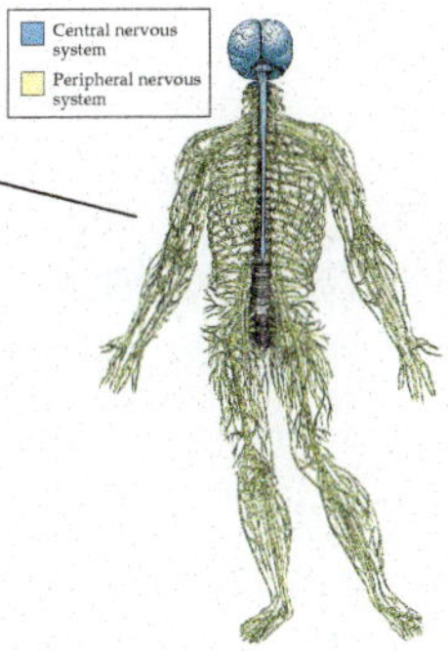

2 At the microscopic level, **neurons** are the basic information-processing units of the nervous system. The typical neuron has four main parts: (1) the **cell body**, which contains the **nucleus**; (2) **dendrites**, which receive information; (3) an **axon**, which conducts information, in the form of electrical potentials, away from the cell body; and (4) **axon terminals**, which transmit the neuron's activity to other cells. Because of the variety of functions they serve, neurons are extremely varied in size, shape, and chemical activity. Review **Figures 1.2**, **1.3**, **1.5**, **Table 1.1**, **Activity 1.1**

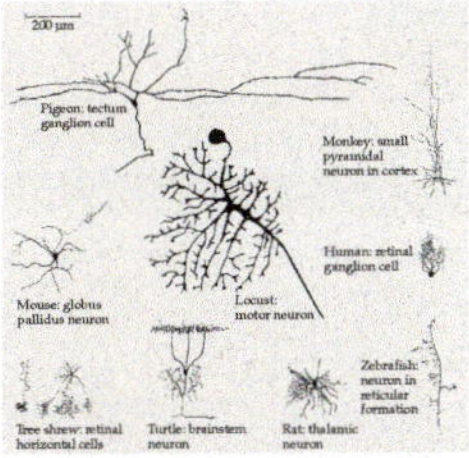

3 Neurons make functional contacts with other neurons, or with muscles or glands, at specialised junctions called **synapses**. Synapses may be on dendritic spines, which exhibit **neural plasticity**, changing shape in response to experience. At most synapses a chemical transmitter liberated by the **presynaptic** terminal diffuses across the **synaptic cleft** and binds to special receptor molecules in the **postsynaptic membrane**. Review **Figure 1.6**

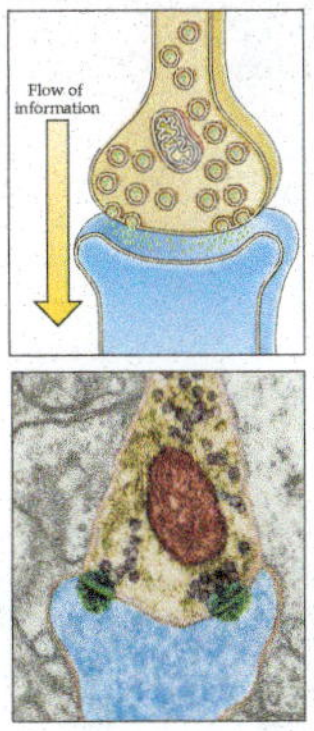

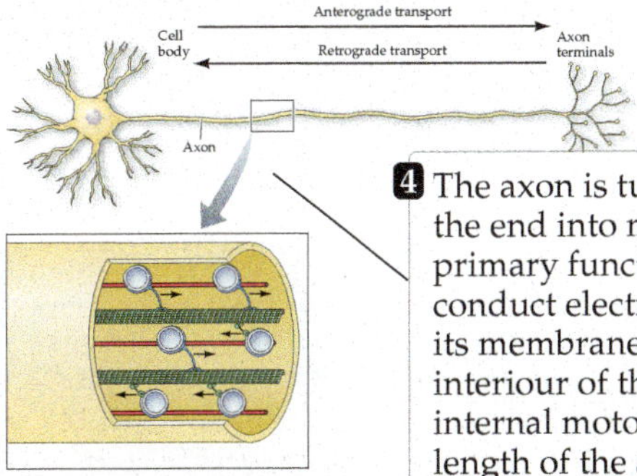

4 The axon is tubular, branching at the end into many collaterals. The primary function of the axon is to conduct electrical signals along its membrane. In addition, in the interiour of the axon, specialised internal motor proteins "walk" the length of the axon in both directions, carrying vesicles filled with important substances. Review **Figure 1.7**

5 **Glial cells** serve many functions, including the breakdown of transmitters, the production of **myelin** sheaths around axons, the exchange of nutrients and other materials with neurons, the direct regulation of the interconnections and activity of neurons, and the removal of cellular debris. Review **Figure 1.8**

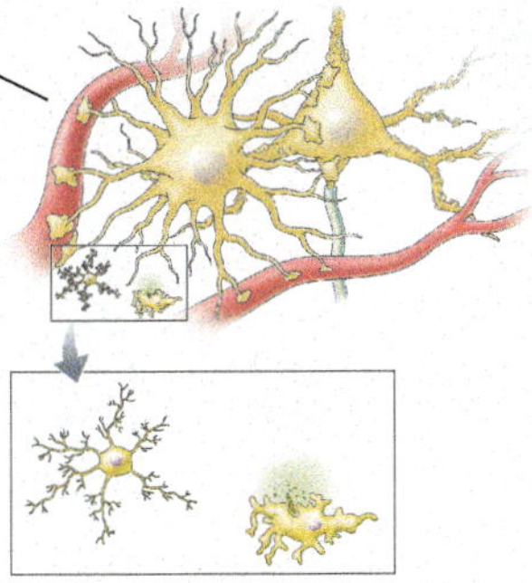

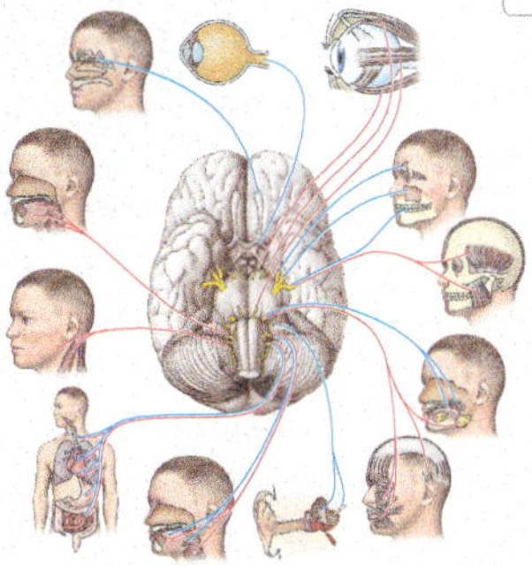

6 At the gross anatomical level (i.e., to the naked eye), the nervous system of vertebrates is divided into peripheral and central nervous systems. The **peripheral nervous system** includes the **cranial nerves**, **spinal nerves**, and **autonomic nervous system**. Review **Figures 1.10**, **1.11**, **Activities 1.2**, **1.3**

7 The autonomic nervous system consists of the **sympathetic nervous system**, which tends to ready the body for action; the **parasympathetic nervous system**, which tends to have an effect opposite to that of the sympathetic system; and the **enteric nervous system**, which innervates the gut. Review **Figure 1.12**, **Activity 1.4**

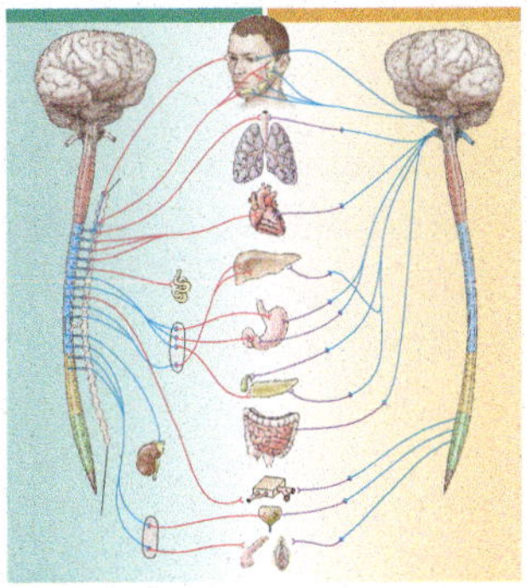

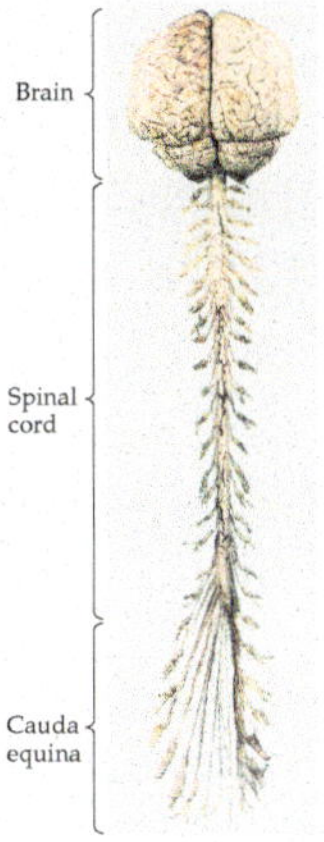

8 The **central nervous system (CNS)** consists of the brain and spinal cord. Review **Figure 1.9**

(Continued)

9 The main divisions of the brain are the **forebrain** (**telencephalon** and **diencephalon**), the **midbrain** (mesencephalon), and the hindbrain (metencephalon and myelencephalon). Review **Figure 1.15**, **Activity 1.8**

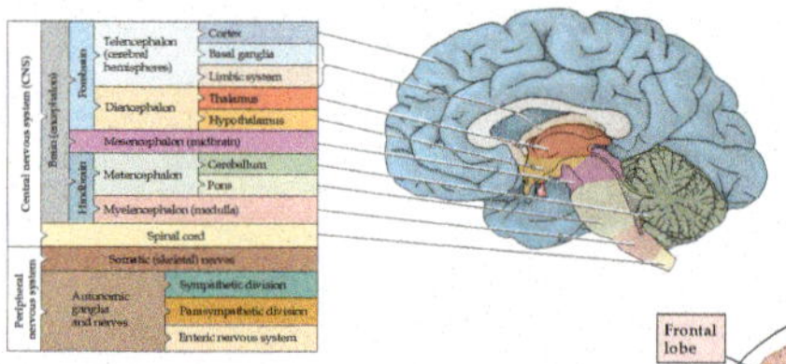

10 The human brain is dominated by the cerebral hemispheres, which include the **cerebral cortex**, an extensive sheet of folded tissue. The six-layered cerebral cortex is responsible for higher-order functions such as vision, language, and memory. Other neural systems include the **basal ganglia**, which regulate movement; the **limbic system**, which controls emotional behaviours; and the **cerebellum**, which aids motor control. Review **Figures 1.13**, **1.17**, **1.18**, **Activities 1.5–1.7**, **1.9–1.11**

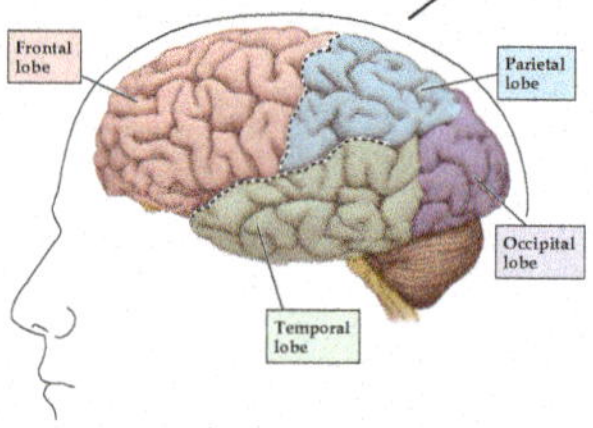

11 The brain and spinal cord, surrounded and protected by the three **meninges**, float in **cerebrospinal fluid** (**CSF**), which surrounds and infiltrates the brain (via cerebral ventricles). Review **Figure 1.20**, **Activity 1.12**

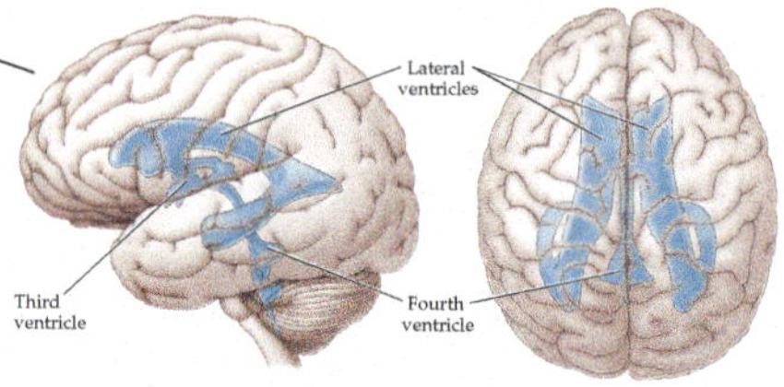

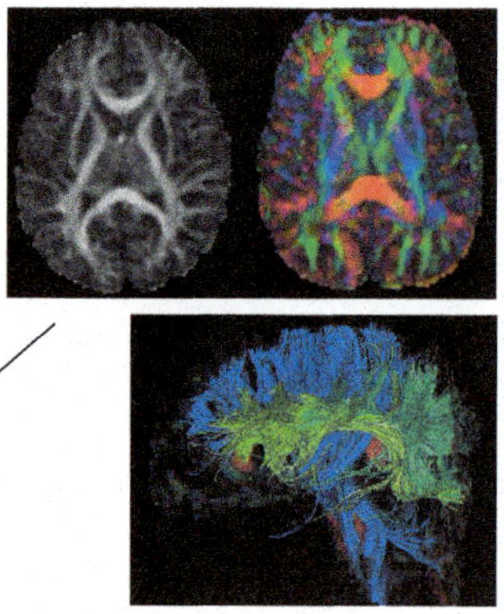

12 The vascular system of the brain is an elaborate array of blood vessels that deliver nutrients and other substances to the brain. The walls of the blood vessels in the brain form the **blood-brain barrier**, restricting the flow of large, potentially harmful molecules into the brain, while a flow of CSF into, through, and out of brain tissues via the perivascular **glymphatic system** drains wastes and provides some immune access to the brain. Disruption of the blood supply to the brain results in a **stroke**. Review **Figures 1.21–1.23**

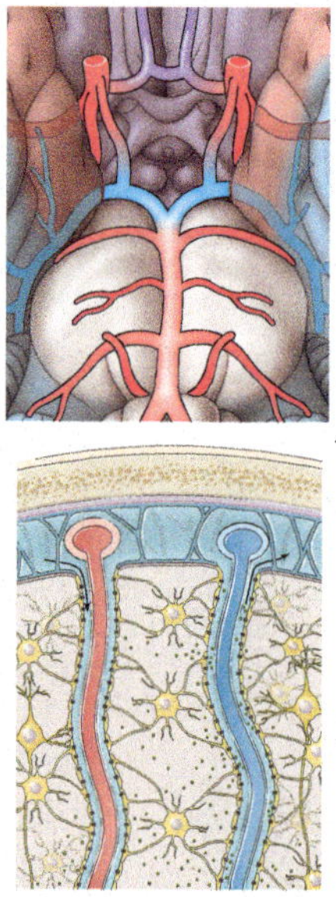

13 Modern imaging techniques make it possible to visualise the anatomy of the living human brain and regional metabolic differences. These techniques include **computerised axial tomography (CT)**, **positron emission tomography** (PET), **magnetic resonance imaging** (MRI), **functional MRI** (fMRI), **diffusion tensor imaging** (DTI), near-infrared optical imaging, and **magnetoencephalography** (MEG). Review **Figures 1.24**, **1.26**, **1.28**, **Box 1.3**, **Animation 1.1**

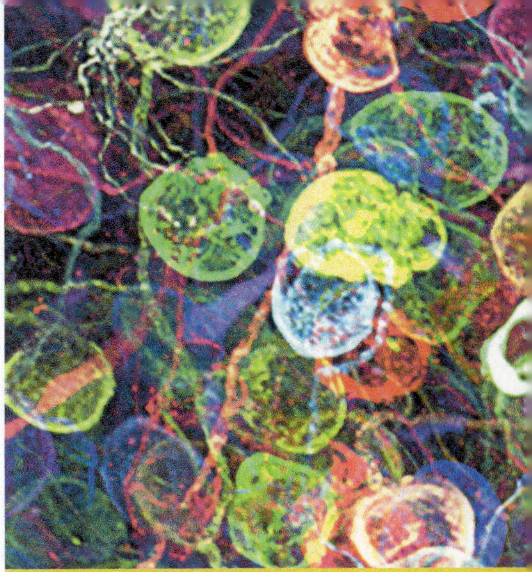
Courtesy of Dr. Ryo Egawa

2

Neurophysiology

The Generation, Transmission, and Integration of Neural Signals

The Laughing Brain

Deidre never knew when she might suddenly suffer a seizure, that is, a loss of consciousness accompanied by convulsions, the uncontrollable, rhythmic movements of her whole body. The seizures resulted from electrical malfunctions in her brain and were a symptom of a disorder called *epilepsy*. Medications that usually control epileptic seizures in other people didn't work for Deidre. The seizures weren't just embarrassing—they left her vulnerable to accidents. So Deidre agreed to try something drastic. She let doctors implant tiny wires through her skull to pinpoint exactly where in the brain the electrical problems began. From this information the doctors could decide whether to remove that part of the brain, which should stop the seizures.

When the doctors carefully passed a tiny electrical current through each wire in turn, they found that stimulating the wire to a particular part of the brain elicited a reliable change in Deidre's behaviour—she laughed. You might think that Deidre would be puzzled by her own sudden laughter, but she wasn't told when the wires were stimulated and she never seemed to guess they were causing her to laugh. When the doctors asked why she was laughing, Deidre always offered a reason. "You guys standing there, you're so funny!" Stimulating that part of the brain didn't just cause Deidre to laugh, it also affected her mind, causing her to be amused by whatever happened to be going on. If two doctors were standing there in lab coats, she interpreted their behaviour as humourous. Presumably this same part of her brain is normally active when Deidre hears a funny joke.

Deidre's behaviour demonstrates that even a mental process as elusive as a sense of humour is a product of a machine: give the machine a tiny zap of electricity, and she'll be amused without even hearing a joke. Deidre's epilepsy is also a result of electrical activity—uncontrolled electrical activity—in that marvelously complex machine between her ears. In this chapter we'll find out why electrical activity in the brain affects behaviour so profoundly.

In Chapter 1 we learned that the brain consists of billions of neurons that make trillions of connections with one another. In this chapter we see that each of those neurons is an information-processing device, taking in lots of information, analysing that information, then passing along the results of that analysis to other cells. To understand how a neuron takes in, analyses, and transmits information, we delve into **neurophysiology**, the study of electrical and chemical processes in neurons.

We'll learn that information flows *within* a neuron via electrical signals, while information passes *between* neurons through chemical signals. This alternating series of electrical and chemical signalling underlies all neuronal function, including our ability to shoot a basket, or get a joke and laugh. We trace that sequence of electrical and chemical signalling in this chapter. First we explain how neurons produce electrical signals, called *action potentials,* and send them along their axons. Then we describe how such an electrical signal causes axon terminals to release a chemical messenger, called a *neurotransmitter,* into the synapse. Next we discuss how the neurotransmitter affects the electrical state of a neuron on the other side of the synapse. Finally, we talk about how we can eavesdrop on all that electrical activity in the amasing human brain.

2.1 Electrical Signals Are the Vocabulary of the Nervous System

Learning Objectives

After reading this section, you should be able to:

2.1.1 **Explain how the distribution of ions results in a negative charge inside neurons.**

2.1.2 **List the sequence of ion flows that underlies an important neuronal signal called an action potential.**

2.1.3 **Describe how specialised ion channels propagate the action potential from the start of an axon to the tips of its every branch.**

2.1.4 **Contrast the way the action potentials spread down myelinated versus unmyelinated axons.**

All living cells possess an electrical charge—they are more negative on the inside than on the outside—that is a legacy of their evolutionary origins. Early single-celled organisms living in the primordial sea contained many proteins, which are negatively charged. Long ago, neurons began to exploit this electrical property to keep track of information, resulting in a cellular signalling system that is much the same in jellyfish, insects, and human beings. These electrical signals underlie the whole range of thought and action, from composing music to feeling an itch on the skin and swatting a mosquito. To understand this electrical signalling system, we'll first review the physical forces at work and then discuss some details of why neurons are electrically polarised, how neuronal polarisation is influenced by other neurons, and how a change of polarity in one part of a neuron can spread throughout the cell.

neurophysiology The study of electrical and chemical processes in neurons.

ion An atom or molecule that has acquired an electrical charge by gaining or losing one or more electrons.

anion A negatively charged ion, such as a protein or chloride ion.

cation A positively charged ion, such as a potassium or sodium ion.

intracellular fluid Also called *cytoplasm*. The watery solution found within cells.

extracellular fluid The fluid in the spaces between cells (interstitial fluid) and in the vascular system.

cell membrane The lipid bilayer that ensheathes a cell.

lipid bilayer The structure of the neuronal cell membrane, which consists of two layers of lipid molecules.

microelectrode An especially small electrode used to record electrical potentials from living cells.

resting membrane potential A difference in electrical potential across the membrane of a nerve cell during an inactive period.

millivolt (mV) A thousandth of a volt.

negative polarity A negative electrical-potential difference relative to a reference electrode.

diffusion The spontaneous spread of molecules of one substance among molecules of another substance until a uniform concentration is achieved.

A balance of electrochemical influences produces the resting membrane potential of neurons

Let's start by considering a neuron at rest, neither perturbed by other neurons nor producing its own signals. Of the many **ions** (electrically charged molecules) that a neuron contains, a majority are **anions** (negatively charged ions), especially large protein anions that cannot exit the cell; the rest are **cations** (positively charged ions; to help remember that cations are positive, note the word contains a *t*, which resembles a plus sign). All of these ions are dissolved in an **intracellular fluid**, which is separated from the **extracellular fluid** by the **cell membrane**, which is made up of a **lipid bilayer**—two layers of linked fatty molecules.

There are more negatively charged ions inside neurons relative to the outside. We can measure that difference in charge across the membrane by inserting a fine **microelectrode** inside a neuron and using a voltmeter to compare the cell's interiour with the extracellular fluid surrounding it (as illustrated in **FIGURE 2.1**). Such measures show that a neuron at rest exhibits a characteristic **resting membrane potential** of about –50 to –80 thousandths of a volt, or **millivolts** (**mV**). The negative sign indicates the **negative polarity** of the cell's interiour.

To understand why neurons have a resting membrane potential of about –65 mV or so, we need to consider two influences at work. One is electrostatic pressure, the force that causes oppositely charged particles to come together and causes particles with the same electrical charge to repel one another (**FIGURE 2.2A**). The other influence to consider is **diffusion** (**FIGURE 2.2B**), the tendency of molecules of a substance to spread away from regions of high concentration to regions of low concentration. For example, if a drop of ink is placed in a glass

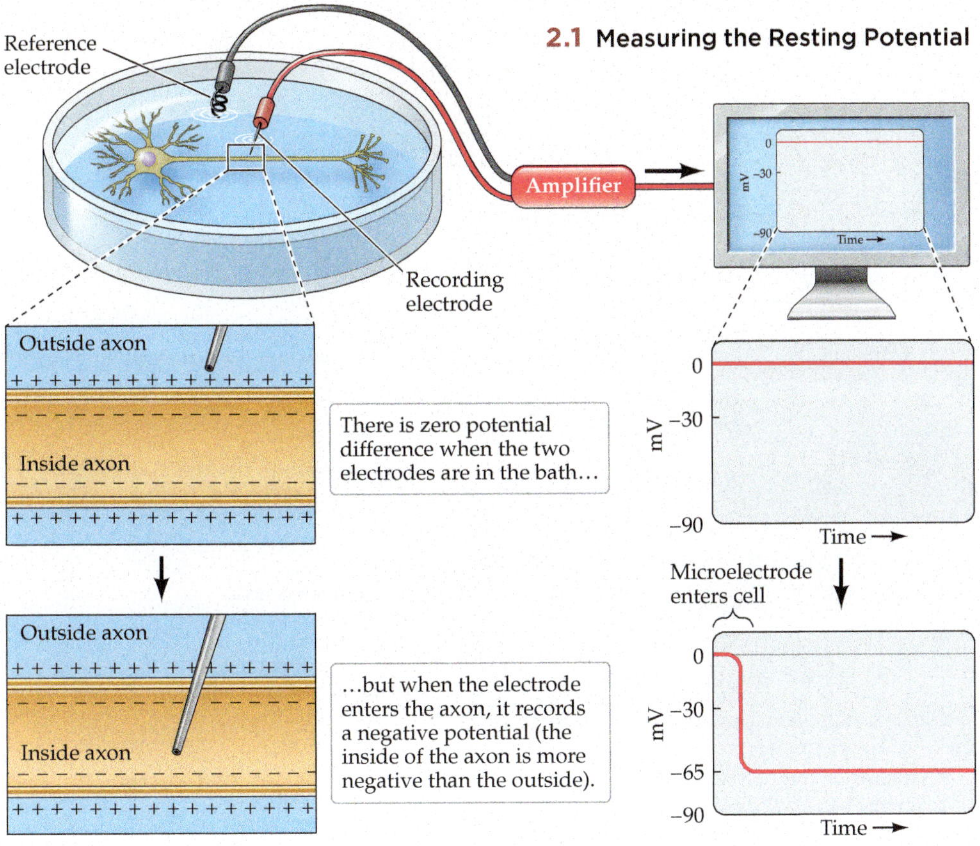

of water, the ink molecules tend to move from the drop, where they are highly concentrated, into the rest of the glass, where they are less concentrated. In other words, molecules tend to move down their **concentration gradient** until they are evenly distributed.

These two influences are at work across the neuronal membrane because it contains membrane-spanning proteins called **ion channels**, tubelike pores that allow ions of a specific type to pass through the membrane. Some ion channels stay open all the time, and the cell membrane of a neuron contains many such channels that selectively allow **potassium ions** (**K^+**) to cross the membrane. Because it is studded with these K^+ channels, we say that the cell membrane of a neuron exhibits **selective permeability** to potassium; that is, K^+ ions can enter or exit the cell fairly freely, while other ions are impeded by the cell membrane (**FIGURE 2.2C**).

The resting potential of the neuron reflects a balancing act between two opposing influences that drive K^+ ions in and out of the neuron. If K^+ ions were equally concentrated on both sides of a selectively permeable membrane, our voltmeter would show a reading of 0 volts difference between the two sides (**FIGURE 2.2D**). On the other hand, if we purposely put 10 times more K^+ ions on one side than the other, diffusion would push many of them across the selectively permeable membrane (**FIGURE 2.2E**). However, this movement of positive ions from one side of the membrane to the other would cause a difference in charge that our voltmeter could measure. Eventually the diffusion pushing the K^+ ions in one direction would be counteracted by the electrostatic charge pulling them back again. In a case where there were 10 times more K^+ ions on one side than the other, once a voltage of about –58 mV was reached, the system would be at **equilibrium**, meaning that every K^+ ion pushed in one direction by diffusion would be matched by one pulled back by electrostatic charge (**FIGURE 2.2F**). Importantly, electrostatic pressure and diffusion would be acting on ions in the immediate vicinity of the membrane, and only a tiny fraction of the K^+ ions would need to move to reach this equilibrium point of –58 mV, so the concentration of ions would not significantly change on either side of the membrane.

concentration gradient Variation of the concentration of a substance within a region.

ion channel A pore in the cell membrane that permits the passage of certain ions through the membrane when the channel is open.

potassium ion (K^+) A potassium atom that carries a positive charge because it has lost one electron.

selective permeability The property of a membrane that allows some substances to pass through, but not others.

equilibrium Here, the state in which the number of ions crossing a membrane in one direction is matched by the number crossing in the opposite direction.

2.2 Influences Underlying Electrical Signalling in Neurons

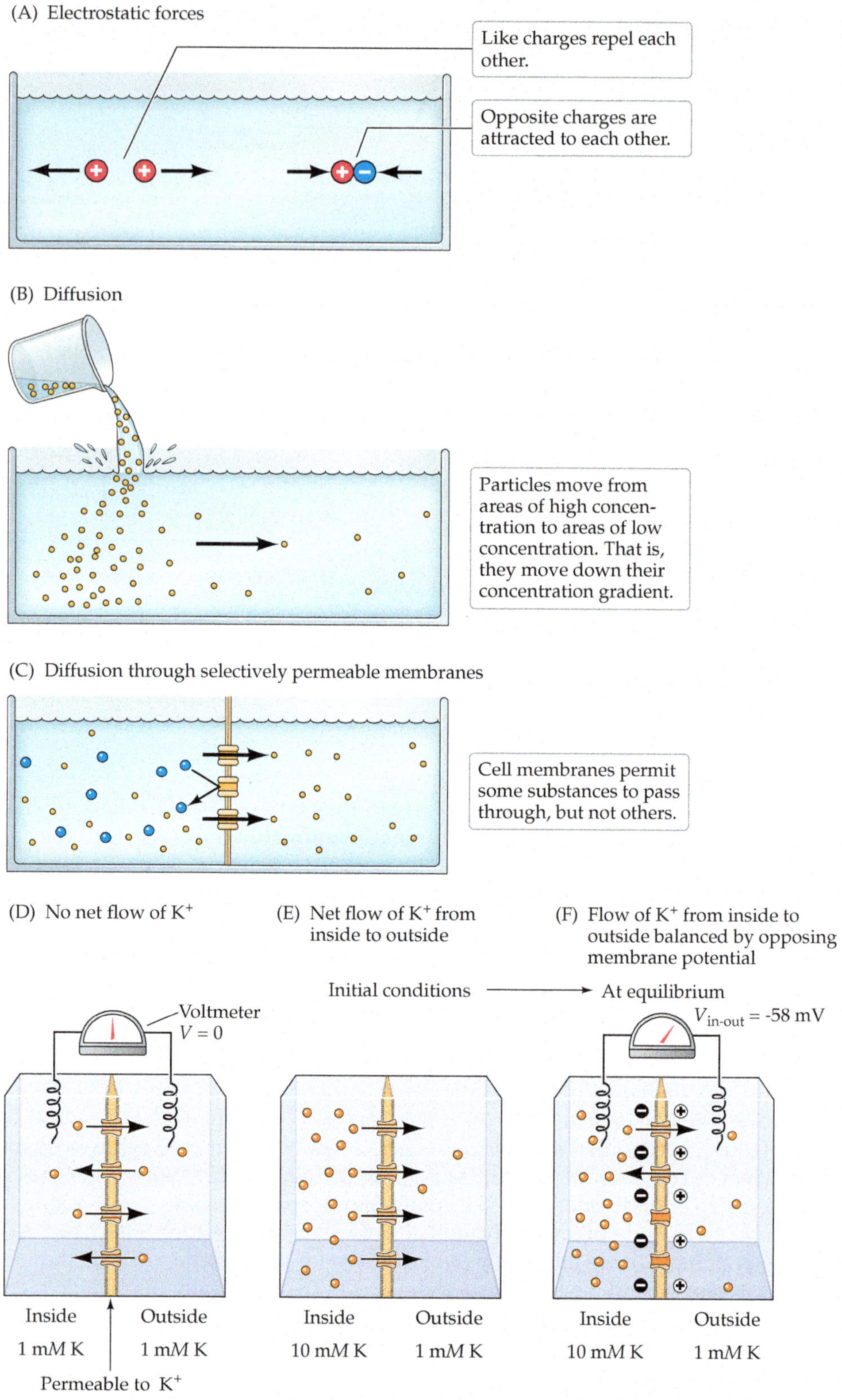

sodium ion (Na^+) A sodium atom that carries a positive charge because it has lost one electron.

sodium-potassium pump The energetically expensive mechanism that pushes sodium ions out of a cell, and potassium ions in.

K^+ ions are more concentrated inside neurons than outside. In contrast, **sodium ions** (**Na^+**) are much more concentrated outside neurons than inside because neurons use a mechanism, the **sodium-potassium pump**, that pumps three Na^+ ions out of the cell for every two K^+ ions pumped in. This action consumes energy (**FIGURE 2.3A**). In fact, a large fraction of the energy consumed by the brain is used to maintain these ionic differences across neuronal membranes.

2.3 The Ionic Basis of the Resting Potential

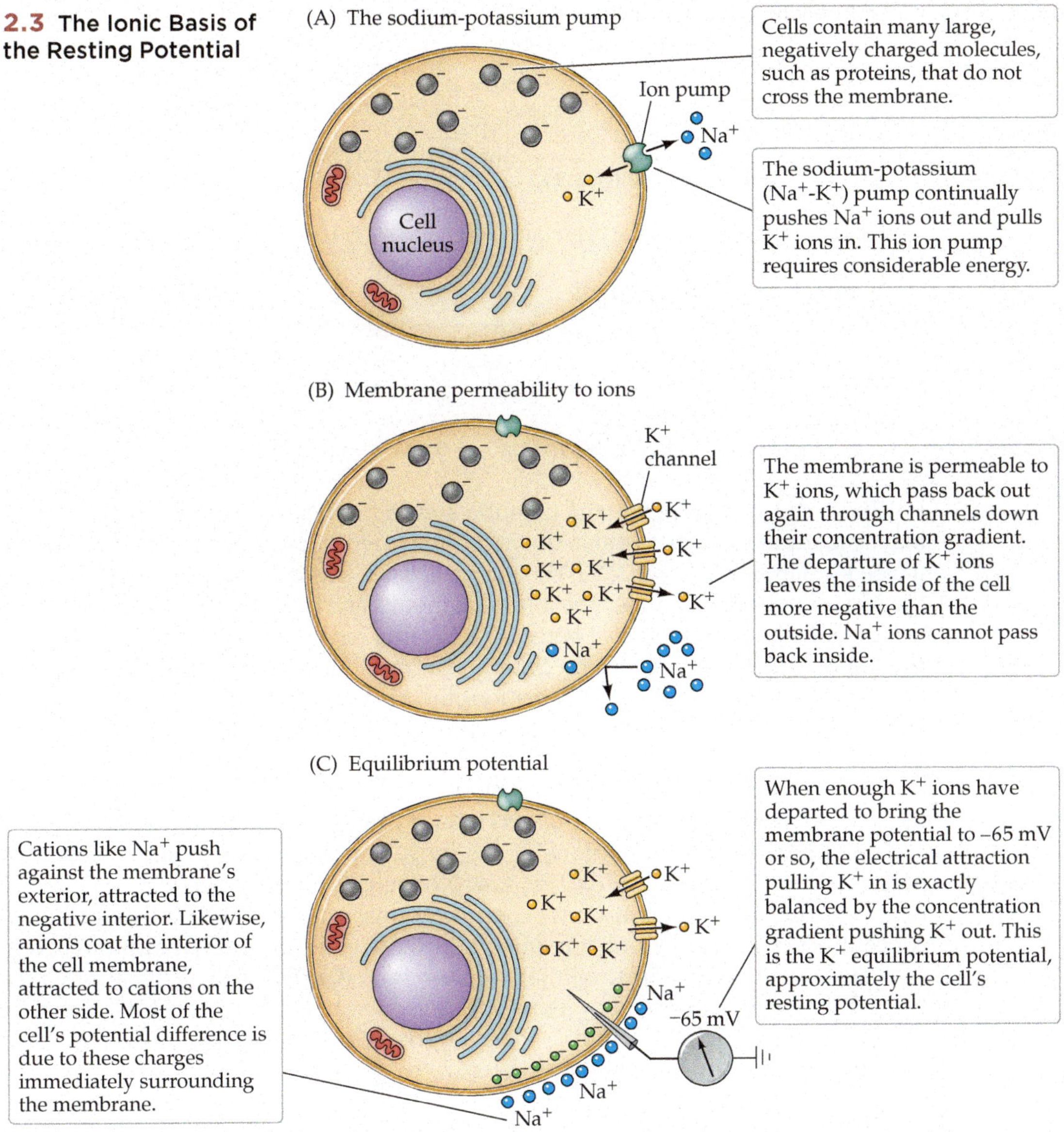

Go to Animation 2.1
The Resting Membrane Potential

The sodium-potassium pump causes a buildup of K^+ ions inside the cell, but recall that at rest the membrane is much more permeable to K^+ ions than other ions like Na^+. That means K^+ ions will tend to leave the interiour, down their concentration gradient, causing a net buildup of negative charges inside the cell (**FIGURE 2.3B**). As negative charge builds up inside the cell, it begins to exert **electrostatic pressure** to pull positively charged K^+ ions back inside. When these opposing influences—the concentration gradient pushing K^+ ions out and the electrostatic pressure pulling them in—cause the cell to reach K^+ equilibrium, any further movement of K^+ ions into the cell (drawn by electrostatic pressure) is matched by the flow of K^+ ions out of the cell (moving down the concentration gradient), as **FIGURE 2.3C** depicts.

How negative does the neuron's interiour need to be to reach this equilibrium? We've seen that when K^+ ions are 10 times more concentrated on one side of a membrane, they will reach equilibrium once a voltage difference of –58 mV is established (see Figure 2.2F). In real neurons, K^+ ions are in fact about 24 times more concentrated inside than outside, so at what potential would they reach equilibrium?

The **Nernst equation** is a mathematical function predicting the **equilibrium potential**, the voltage difference across a permeable membrane needed to exactly counterbalance diffusion pushing an ion from the side of the membrane with a high concentration to the side with a low concentration. The full equation includes several physical constants that can be filled in to simplify the equation at room temperature:

electrostatic pressure The propensity of charged molecules or ions to move toward areas with the opposite charge.

Nernst equation An equation predicting the equilibrium potential for a given ion based on the concentrations of the ion on opposite sides of a permeable membrane.

equilibrium potential The voltage across a permeable membrane that exactly counteracts the movement of ions from the side with a high concentration to the side with a low concentration.

Goldman equation An equation predicting the potential difference across a membrane based on the concentrations of multiple ions on opposite sides of the membrane, as well as its relative permeability to each ion.

calcium ion (Ca^{2+}) A calcium atom that carries a double positive charge because it has lost two electrons.

action potential The propagated electrical message of a neuron that travels along the axon to the presynaptic axon terminals.

hyperpolarisation An increase in membrane potential (the interiour of the neuron becomes even more negative).

depolarisation A reduction in membrane potential (the interiour of the neuron becomes less negative).

Go to Activity 2.1
Distribution of Ions

Equilibrium potential for ion *X*:

$$E_x = \frac{58}{\text{(electrical charge ion } x)} \text{ mV} \times \log \frac{\text{(concentration of X outside)}}{\text{(concentration of X inside)}} = \frac{58}{\text{(electrical charge ion } x)} \text{ mV} \times \log \frac{[X]_{outside}}{[X]_{inside}}$$

Because the K^+ ion has an electrical charge of $^+1$, the Nernst equation for K^+ can be further simplified:

Equilibrium potential for K^+:

$$E_{K^+} = 58\text{mV} \times \log \frac{[K^+]_{outside}}{[K^+]_{inside}}$$

Because K^+ is about 24 times more concentrated inside neurons than out, the equation becomes:

$$E_{K^+} = 58\text{mV} \times \log {}^1\!/_{24} = 58\text{mV} \times -1.38 = -80\text{mV}$$

When the resting membrane potentials of neurons are actually measured, they may sometimes reach this value of –80 mV but are typically closer to –65 mV as indicated in Figures 2.1 and 2.3C. What accounts for this discrepancy? For one thing, the Nernst equation assumes the neuronal membrane is permeable *only* to K^+, but in fact the membrane is *somewhat* permeable to other ions. Another equation, called the **Goldman equation**, takes into account the intracellular and extracellular concentrations of K^+ and Na^+ and several other ions but also accounts for the *degree* of permeability to each. Taking those additional factors into account, the Goldman equation predicts a voltage potential that is quite close to the resting potentials observed in neurons. You can learn more about the Goldman equation in A STEP FURTHER 2.1: CALCULATING THE RESTING POTENTIAL on the website.

The approximate distributions of the most important ions inside and outside neurons are illustrated in **FIGURE 2.4**. Notice the high intracellular concentration, reported as millimolar (m*M*) concentrations, of K^+ and the high extracellular concentration of Na^+, which are enforced by the sodium-potassium pump. Neurons also keep levels of intracellular **calcium ions (Ca^{2+})** low by using another ion pump to eject Ca^{2+} ions and by using specialised proteins that store Ca^{2+} to use for intracellular signalling, including the synaptic release of neurotransmitter, which we'll take up later in this chapter.

The resting potential of a neuron provides a baseline level of polarisation. But neurons routinely undergo a brief but radical *change* in polarisation, sending an electrical signal from one end of the axon to the other, as we'll discuss next.

A threshold amount of depolarisation triggers an action potential

Action potentials are very brief but large changes in neuronal polarisation that arise in the initial segment of the axon and are propagated at high speed along the axon's length. The information that a neuron sends to its postsynaptic targets is encoded in patterns of these action potentials, so we need to understand their properties—where they come from, how they race down the axon, and how they communicate their information across synapses to other cells. Let's turn first to the creation of the action potential.

Two concepts are central to understanding how action potentials are triggered. **Hyperpolarisation** is an increasing negativity of the membrane potential (i.e., the neuron becomes even more negative on the inside, relative to the outside). So if the neuron already has a resting membrane potential of, say, –65 mV, hyperpolarisation makes it even *farther from zero*, maybe –70 mV. **Depolarisation** is the

2.4 The Distribution of Ions Inside and Outside of a Neuron Most potassium ions (K^+) are found inside the neuron. Most sodium ions (Na^+), chloride ions (Cl^-), and calcium ions (Ca^{2+}) are in the extracellular fluid. These ions are exchanged through specialised channels in the cell membrane. The large, negatively charged protein molecules stay inside the neuron. The measures are the standard millimoles per liter, reported as millimolar (mM) concentrations.

	Na^+	K^+	Cl^-	Ca^{2+}	Proteins
Concentration outside cell (m*M*)	145	5	110	1–2	low
Concentration inside cell (m*M*)	5–15	140	4–30	0.0001	high

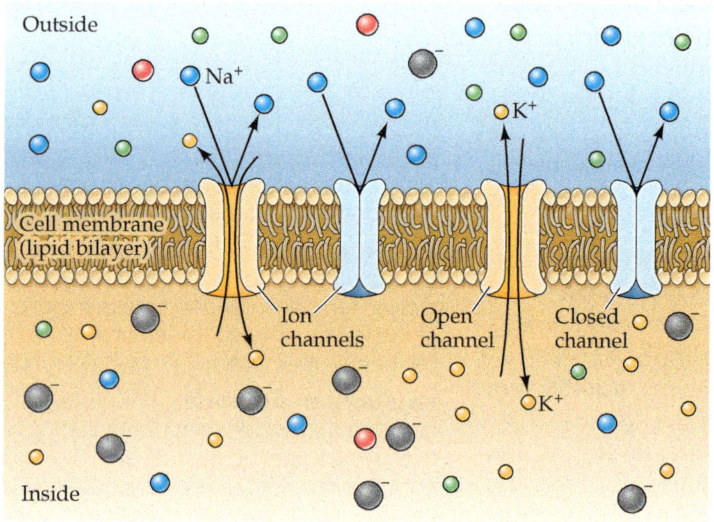

reverse, referring to a decreased polarisation of the cell membrane. The depolarisation of a neuron from a resting potential of –65 mV to, say, –60 mV makes the inside of the neuron more like the outside. In other words, depolarisation of a neuron brings its membrane potential *closer to zero*.

FIGURE 2.5A illustrates an apparatus for experimentally hyperpolarising and depolarising a neuron with an electrode. (Later we'll talk about how synapses from other neurons produce similar hyperpolarisations and depolarisations.) Applying a *hyperpolarising* stimulus to the membrane produces an immediate response that passively follows the stimulus pulse (**FIGURE 2.5B**; the distortions at the beginning and end of the neuron's response are caused by the membrane's ability to store electrical charge, known as *capacitance*). The greater the stimulus, the greater the response, so the neuron's change in potential is called a *graded response*.

If we measured the membrane response at locations successively farther and farther away from the stimulus location, we would see another way in which the membrane response seems passive and graded. Like ripples spreading from a pebble dropped in a pond, the potentials produced by stimulation of the membrane diminish as they spread away from the point of stimulation (see Figure 2.5B, bottom). A simple law of physics describes this phenomenon: as the potential spreads

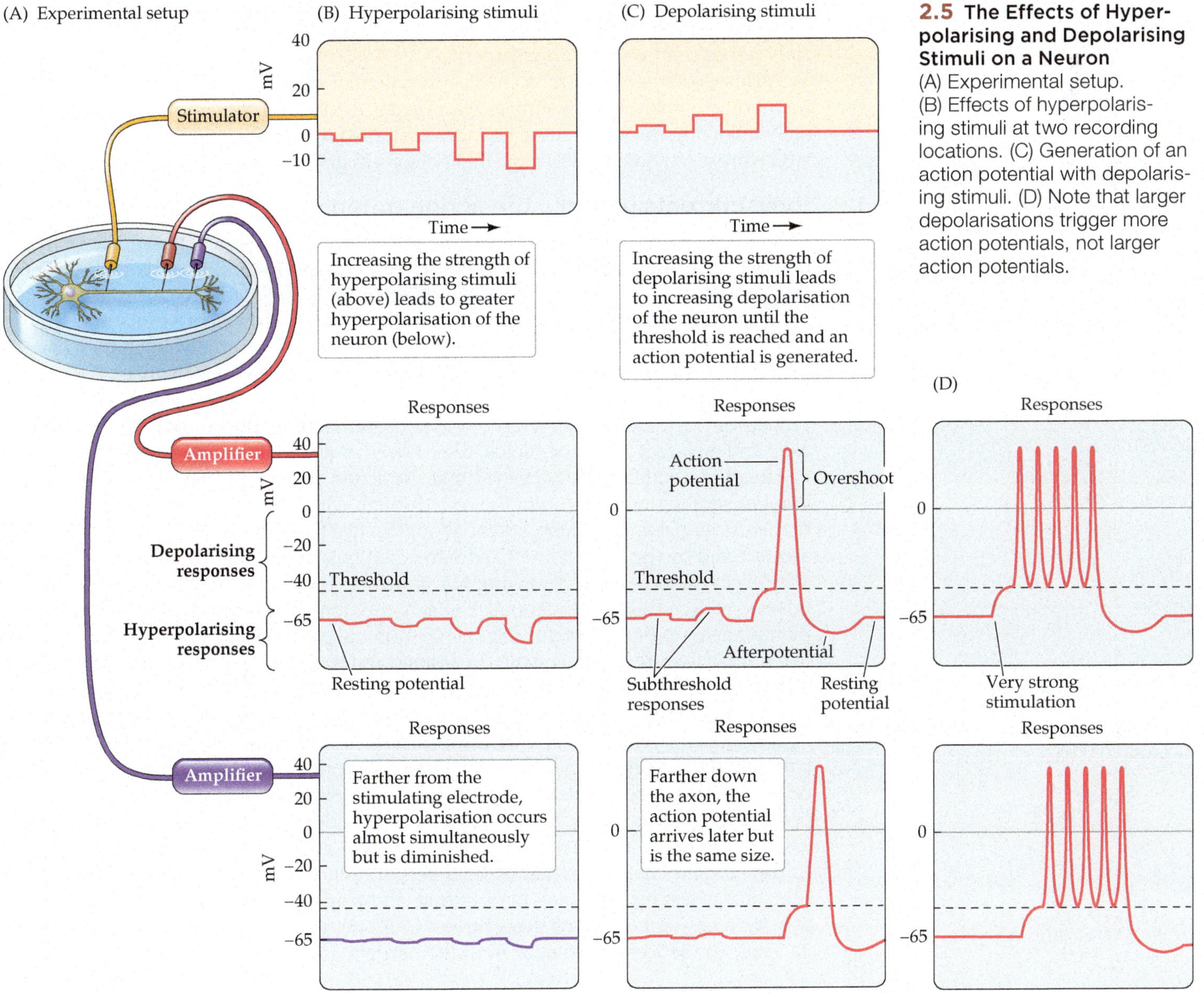

2.5 The Effects of Hyperpolarising and Depolarising Stimuli on a Neuron
(A) Experimental setup. (B) Effects of hyperpolarising stimuli at two recording locations. (C) Generation of an action potential with depolarising stimuli. (D) Note that larger depolarisations trigger more action potentials, not larger action potentials.

local potential An electrical potential that is initiated by stimulation at a specific site, which is a graded response that spreads passively across the cell membrane, decreasing in strength with time and distance.

threshold The stimulus intensity that is just adequate to trigger an action potential.

all-or-none property The fact that the amplitude of the action potential is independent of the magnitude of the stimulus.

afterpotential The positive or negative change in membrane potential that may follow an action potential.

voltage-gated Na^+ channel A Na^+-selective channel that opens or closes in response to changes in the voltage of the local membrane potential; it mediates the action potential.

across the membrane, its sise decays as a function of the square of the distance. Such **local potentials**, which are graded and diminish over time and distance, also arise at synapses in response to other neurons, as we will see later in this chapter.

Up to a point, the application of *depolarising* pulses to the membrane follows the same pattern as for hyperpolarising stimuli, producing local, graded responses. However, the situation changes suddenly if the stimulus depolarises the cell to −40 mV or so (the exact value varies slightly among neurons). At this point, known as the **threshold**, a sudden and brief (0.5–2.0 millisecond [ms]) response—the action potential—is provoked (**FIGURE 2.5C**). The action potential (sometimes referred to as a *spike* because of its shape) is a rapid reversal of the membrane potential that momentarily makes the inside of the membrane *positive* with respect to the outside. That brief period when the neuron interiour is positive is referred to as an *overshoot* (see Figure 2.5C). Unlike the passive spread of the graded potentials that we have been discussing, the action potential is actively propagated (or regenerated) down the axon, through ionic mechanisms that we'll discuss shortly.

Applying strong stimuli to produce depolarisations that far exceed the neuron's threshold reveals another important property of action potentials: larger depolarisations produce *more* action potentials, not larger action potentials (**FIGURE 2.5D**). In other words, the sise (or *amplitude*) of the action potential is independent of stimulus magnitude. This characteristic is referred to as the **all-or-none property** of the action potential: either it fires at its full amplitude, or it doesn't fire at all. It turns out that information is encoded by changes in the *frequency* of action potentials rather than in their amplitude.

A closer look at the form of the action potential shows that the return to baseline membrane potential is not simple. Many axons exhibit electrical oscillations immediately following the spike; these changes are called **afterpotentials** (see Figure 2.5C), which are also related to the movement of ions in and out of the cell, as we'll see next.

Ionic currents underlie the action potential

What events explain the action potential? To answer this question, English neurophysiologists Alan Hodgkin (1914–1998) and Andrew Huxley (1917–2012) took advantage of the giant axon of the squid, from a neuron involved in the animal's emergency escape behaviour. More than half a millimeter in diameter, the squid's giant axon is readily apparent to the naked eye and therefore much better suited to experimentation than mammalian axons, which range in sise from 0.1 to 20 micrometers (μm) in diameter (Debanne et al., 2011). Microelectrodes can be inserted into a giant axon without greatly altering the properties of the axon; it is even possible to push the intracellular fluid out of the squid axon and replace it with other fluids to see how that affects the action potential.

Hodgkin and Huxley established that the action potential is created by the movement of sodium ions (Na^+) into the cell, through channels that open up in the membrane. At its peak, the action potential approaches the equilibrium potential for Na^+ as predicted by the Nernst and Goldman equations: about +65 mV. At that potential, the concentration gradient pushing Na^+ ions into the cell is exactly balanced by the positive charge pushing them out. The action potential thus involves a rapid shift in membrane properties, switching suddenly from the potassium-dependent resting state to a primarily sodium-dependent active state, and then swiftly returning to the resting state. This shift is accomplished through the actions of a very special ion channel: the voltage-gated Na^+ channel.

Like other ion channels, the **voltage-gated Na^+ channel** includes a tubular, membrane-spanning protein, but this Na^+-selective pore is ordinarily closed. When the cell membrane becomes depolarised to threshold levels, the channel's shape changes, opening the pore to allow Na^+ ions through. Consider what happens when a patch of axonal membrane depolarises (**FIGURE 2.6**). As long as the depolarisation is below threshold, Na^+ channels remain closed (step 1). But when the depolarisation reaches threshold, a few Na^+ channels open at first, allowing ions to start entering the neuron, depolarising the membrane even further and opening still more Na^+ channels (step 3). Thus, the process accelerates until the barriers are removed and Na^+ ions rush in,

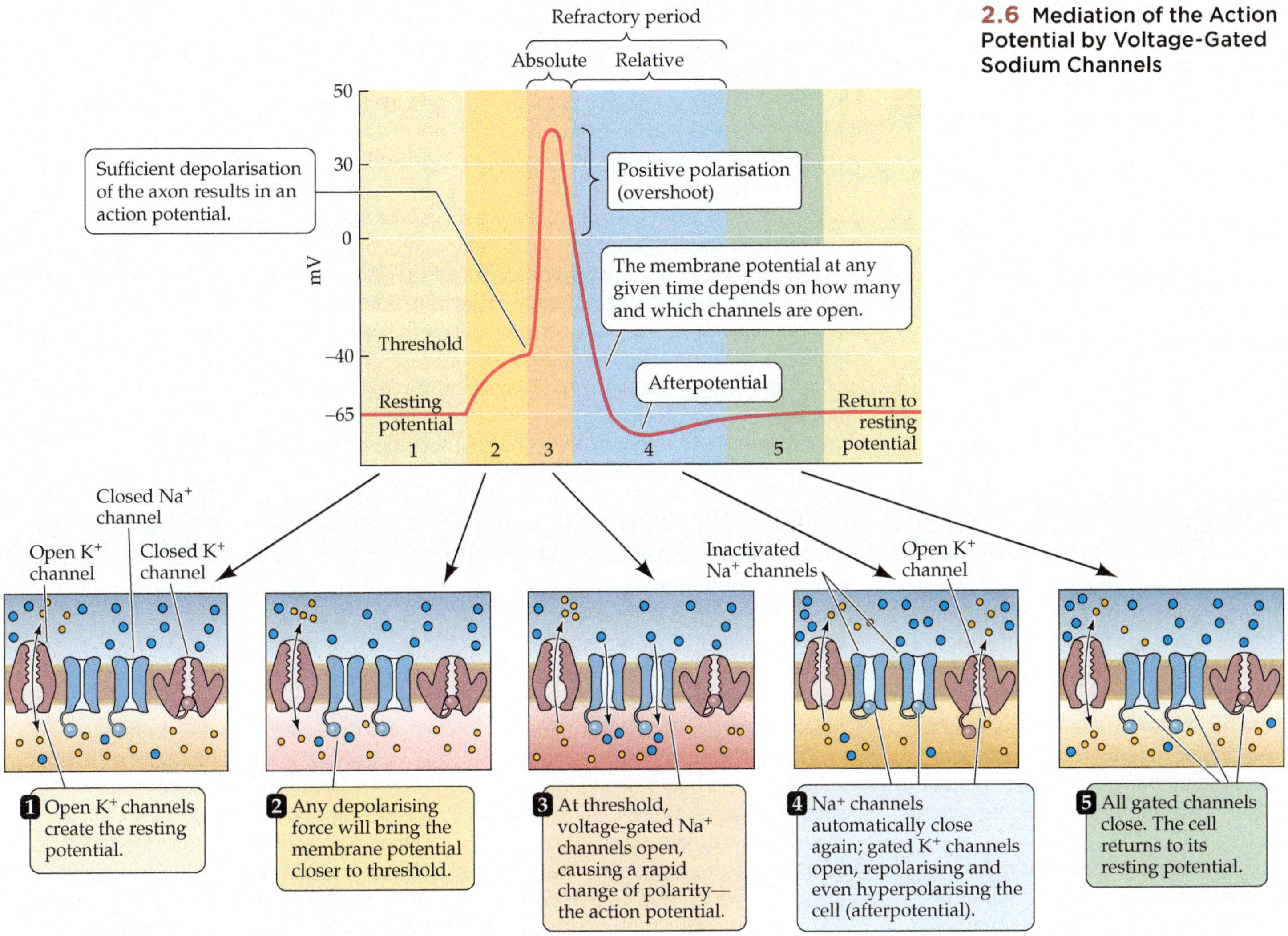

2.6 Mediation of the Action Potential by Voltage-Gated Sodium Channels

both because they are attracted to the negatively charged interiour of the neuron and because they are flowing down their concentration gradient.

But the voltage-gated Na^+ channels stay open for only about a millisecond; then they close again before the neuron can reach the Na^+ equilibrium potential of about +65 mV. Now, the positive charge inside the nerve cell, the overshoot, pushes K^+ ions out the channels that are always open, plus voltage-gated K^+ channels open as well, making the membrane even more permeable to K^+, so the resting potential is quickly restored.

Applying very strong stimuli reveals another important property of axonal membranes. As we bombard the beleaguered axon with ever-stronger stimuli, an upper limit to the frequency of action potentials becomes apparent at about 1200 spikes per second. (Many neurons have even slower maximum rates of response.) Similarly, applying pairs of stimuli that are spaced closer and closer together reveals a related phenomenon: beyond a certain point, only the first stimulus is able to elicit an action potential. The axonal membrane is said to be **refractory** (unresponsive) to the second stimulus.

Refractoriness has two phases: During the **absolute refractory phase**, a brief period immediately following the production of an action potential, no amount of stimulation can induce another action potential, because the voltage-gated Na^+ channels are unresponsive (see Figure 2.6, step 3). The absolute phase is followed by a period of reduced sensitivity, the **relative refractory phase**, during which only a very strong stimulation can produce another action potential, because the flow of K^+ ions out has temporarily hyperpolarised the neuron, so a stronger stimulus would be needed to reach threshold (see Figure 2.6, step 4).

refractory Referring to transiently inactivated or exhausted axonal membrane.

absolute refractory phase A brief period of complete insensitivity to stimuli.

relative refractory phase A period of reduced sensitivity during which only strong stimulation produces an action potential.

Go to Animation 2.2
The Action Potential

This tiny protein molecule, the voltage-gated Na^+ channel, is really quite a complicated machine. It monitors the axon's polarity, and at threshold it changes its shape to open the channel, shutting down again just a millisecond later. The channel then "remembers" that it was recently open and refuses to open again for a short time. These properties produce and enforce the characteristics of the action potential. As you might expect, anything that alters the functioning of neuronal ion channels will affect action potentials and therefore behaviour.

You might wonder if the repeated inrush of Na^+ ions would allow them to build up, affecting the cell's resting potential. In fact, relatively few Na^+ ions need to enter to change the membrane potential (Alle et al., 2009). Plus, most of the change in ion concentration taking place during an action potential is happening right next to the membrane, leaving the ionic concentrations in the interiour of the axon relatively unaffected. In the long run, the sodium-potassium pump enforces the low concentrations of Na^+ ions inside the neuron. Hodgkin and Huxley detected these ion movements by manipulating the concentration of ions in and out of the axon and then monitoring current across the membrane, using methods described in **BOX 2.1**.

BOX 2.1 Voltage Clamping and Patch Clamping

One important tool for Hodgkin and Huxley's study of the action potential in the squid giant axon was the preparation known as **voltage clamping**, in which a stimulator is used to force the axon membrane to remain at a particular potential (voltage) (Figure A). One electrode is placed inside the axon, and another is used as a reference electrode in the bath surrounding the axon. A voltmeter measures the potential difference between the two electrodes and sends that information to a voltage clamp amplifier, which compares the present membrane potential to a command voltage that the experimenter has set. If the membrane potential matches the command potential, the machinery does nothing. But if there's a difference between the two voltages, the clamp amplifier injects electrical current into another electrode inside the axon to force the membrane potential to match the command potential. Importantly, another device measures how much current is being injected into the axon to keep it at the command potential. Initially, injection of a constant amount of current will clamp the axon at the desired voltage. After that, the machine should not have to change the amount of injected current unless current flows across the axon's membrane, as when another electrode depolarises the axon to open Na^+ channels. When that happens, then the apparatus must change the amount of injected current to maintain the voltage clamp. Keeping track of these changes in current from the apparatus tells you how much current passes through the Na^+ channels.

This tool was originally developed to voltage clamp squid axons, but scientists learned various tricks to "pop" the tip of a microelectrode into a neuron such that the neuron's membrane forms a seal, known as *whole-cell recording* (Figure B, middle). Thus you can either measure various synaptic potentials and action potentials, or use the voltage clamp apparatus to measure synaptic currents across the neuron's membrane.

To get even more detailed information, researchers learned how to pull away a tiny patch of membrane covering the microelectrode tip, a technique called **patch clamping**. In some cases, you might end up with a single ion channel, such as a voltage-gated Na^+ channel, on that tiny bit of membrane (see Figure B). Now you voltage clamp this tiny fragment and monitor any current flowing across the channel. When the ion channel opens, you can detect a tiny current (measured here in picoamperes [pA]) passing across the membrane caused by the flow of Na^+ ions, which stops when the channel closes again (Figure C, top). You can then ask how many channel openings there are during a particular time period, say in one second. Then you can purposely vary the membrane potential by varying the command potential in the voltage clamping apparatus and note how the probability of an opening decreases if the membrane is hyperpolarised, and increases if it is depolarised (Figure C, bottom).

Several other ingenious variations of patch clamping are possible. Depending on how you manipulate your electrode (and with luck and repeated trials), you can end up with an "inside-out" patch of membrane, where the part of the membrane that had been facing the cytoplasm is now facing the bath (see Figure B, middle). If there is a neurotransmitter receptor on this bit of membrane, you can now expose its cytoplasmic surface to drugs, for example, to see how manipulating second-messenger systems inside the neuron might affect receptor function, again measuring any currents that cross the membrane in response to manipulations. Removing the electrode another way can result in an "outside-out" patch of membrane where the extracellular side is facing the bath (Figure B, bottom). Now you can apply neurotransmitters or drugs to the extracellular portion of the receptor to see how they affect it. (C [left] after F. Bezanilla and A. M. Correa, 1995. In *Cephalopod neurobiology*, N. J. Abbott et al. [Eds.], pp. 131–151. Oxford University Press: New York; [right] after A. M. Correa et al., 1992. *Biophys J* 61: 1332–1352.)

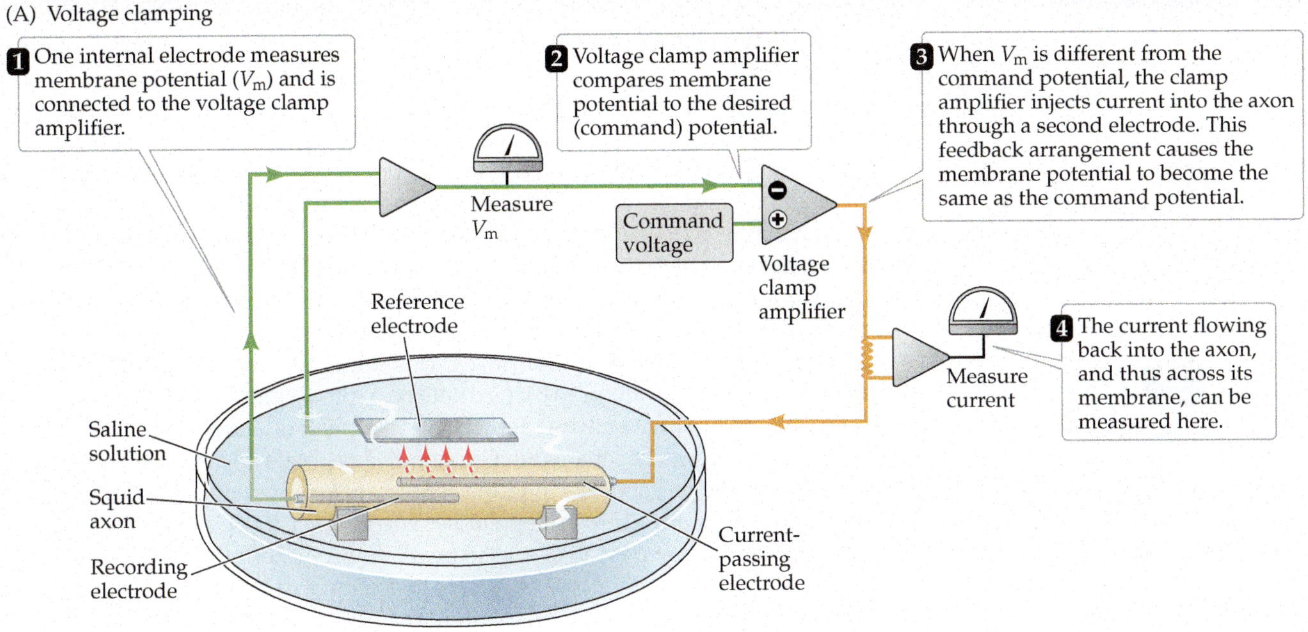

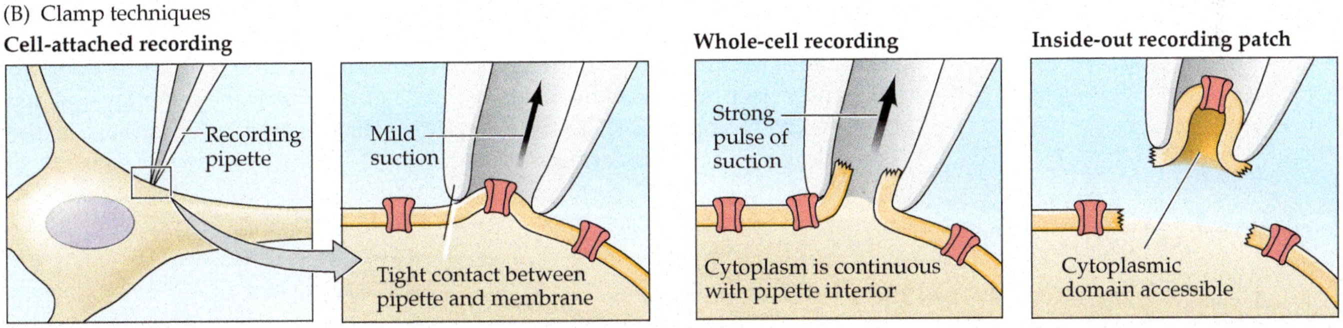

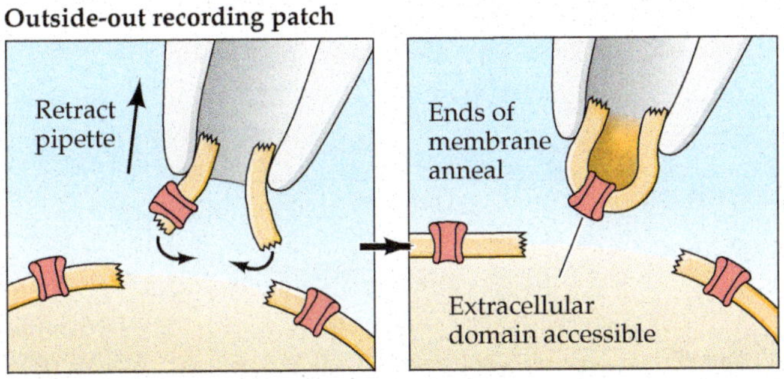

voltage clamping The use of electrodes to inject current into an axon or neuron to keep the membrane potential at a set value. The apparatus measures how much current must be injected to counteract any ion channel openings.

patch clamping The use of voltage clamping to monitor current flow across a tiny patch of membrane taken from a cell.

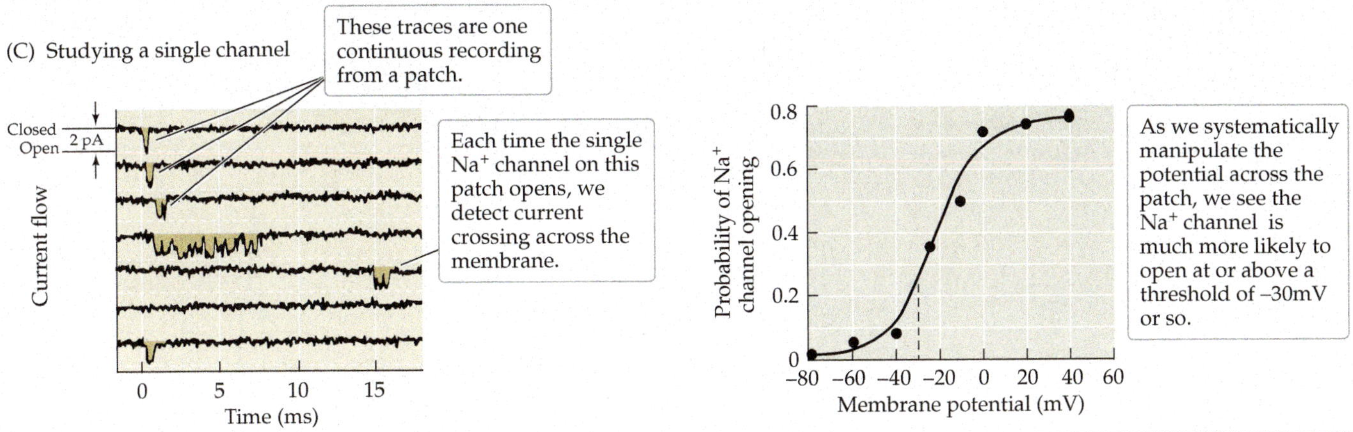

axon hillock A cone-shaped area from which the axon originates out of the cell body; functionally, the integration zone of the neuron.

Action potentials are actively propagated along the axon

In general, the transmission of action potentials is limited to axons. Cell bodies and dendrites usually have few voltage-gated Na^+ channels, so they do not conduct action potentials. The cell body and dendrites have very different ion channels that are stimulated chemically at synapses, as we'll discuss later in this chapter. Because the axon has many voltage-gated Na^+ channels, once an action potential starts just past the **axon hillock** (the slight swelling of the axon where it emerges from the cell body; see Figure 1.6A), it regenerates itself down the length of the axon—another function for which voltage-gated channels are crucial. If we use several different recording electrodes to record an action potential as it races toward the axon terminals, we see that an action potential initiated near the cell body spreads in a sort of chain reaction along the length of the axon, traveling at speeds that range from less than 1 meter per second (m/s) in some axons to more than 100 m/s in others (**FIGURE 2.7**).

Go to Animation 2.3
Conduction along Unmyelinated vs. Myelinated Axons

How does the action potential travel? It is important to understand that the action potential is *regenerated* along the length of the axon. Remember, the action potential is a spike of depolarising electrical activity (with a peak of about +40 mV), so it strongly depolarises the next adjacent axon segment. Because this adjacent segment is similarly covered with voltage-gated Na^+ channels, the depolarisation immediately creates a new action potential, which in turn depolarises the next patch of membrane, which generates yet another action potential, and so on all down the length of the axon. An analogy is the spread of fire along a row of closely spaced match heads in a matchbook. When one match is lit, its heat is enough to ignite the next match and so on along the row.

The axon normally conducts action potentials in only one direction—from the axon hillock toward the axon terminals—because as it progresses along the axon, the action potential leaves in its wake a stretch of refractory membrane (**FIGURE 2.8A**). Propagated activity does not spread from the axon hillock back over the cell body and dendrites, because the membranes there have very few voltage-gated Na^+ channels, so they cannot produce an action potential.

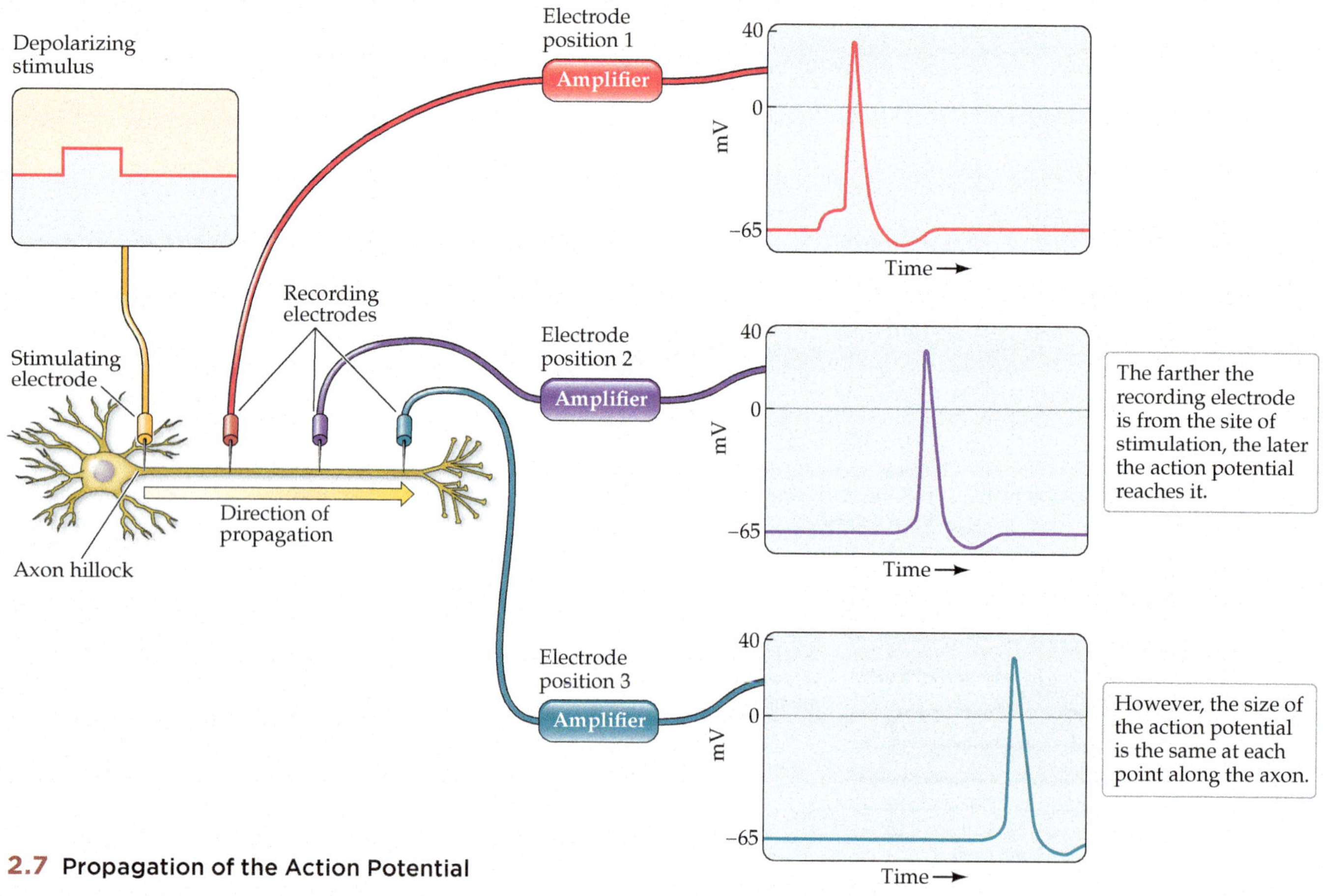

2.7 Propagation of the Action Potential

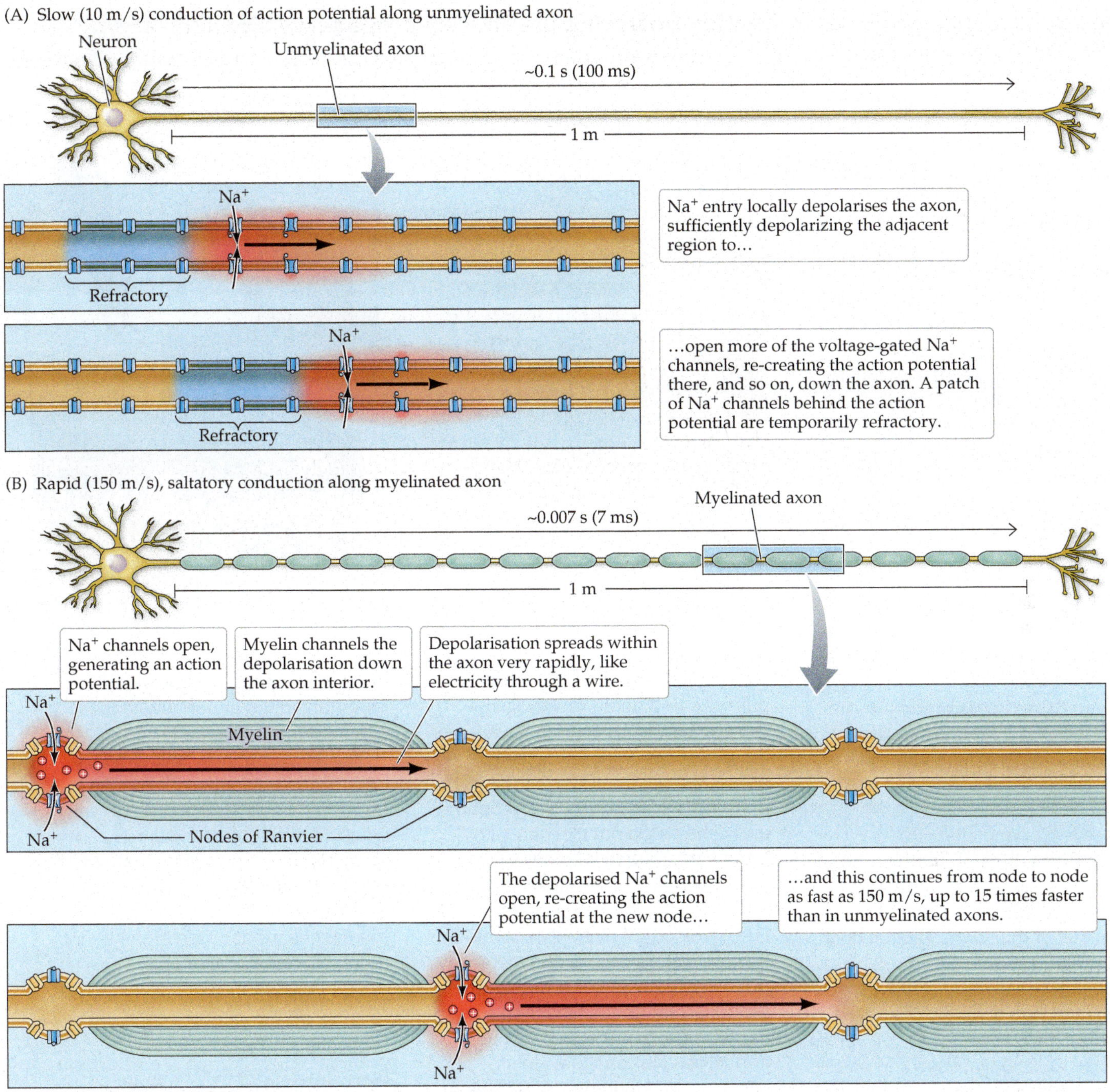

2.8 **Conduction along Unmyelinated Axons and Saltatory Conduction along Myelinated Axons**

If we record the speed of action potentials along axons that differ in diameter, we see that **conduction velocity** varies with the diameter of the axon. Larger axons allow the depolarisation to spread faster through the interiour. In mammals, the conduction velocity in large fibres may be as fast as 120 m/s, about one-third the speed of sound in air. We'll discuss axon diameter and conduction velocity again when we describe touch and pain sensation (see Table 8.2).

The highest conduction velocities require more than just large axons. Myelin sheathing also greatly speeds conduction. As we described in Chapter 2, the myelin sheath that encases some axons is interrupted by **nodes of Ranvier**, small gaps spaced about every millimeter along the axon (see Figure 1.7D). Because the myelin insulation offers considerable resistance to the flow of ionic currents across the membrane, the action potential jumps from node to node. This process is called **saltatory conduction** (from the Latin *saltare*, "to leap or jump") (**FIGURE 2.8B**).

conduction velocity The speed at which an action potential is propagated along the length of an axon (or section of peripheral nerve).

node of Ranvier A gap between successive segments of the myelin sheath where the axon membrane is exposed.

saltatory conduction The form of conduction that is characteristic of myelinated axons, in which the action potential jumps from one node of Ranvier to the next.

BOX 2.2 Changing the Channel

(A) Potassium channel selectivity filter

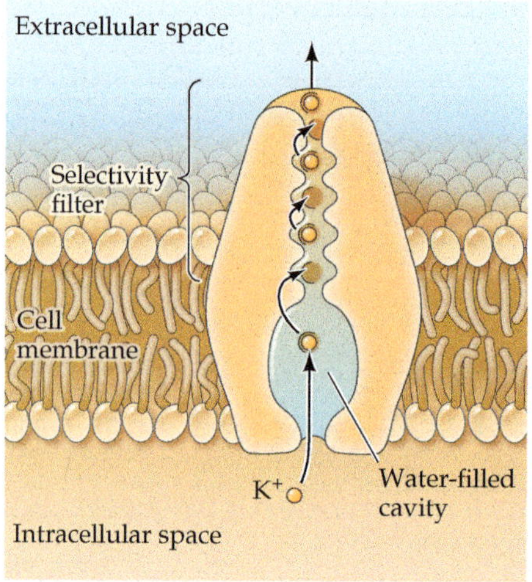

(B) Some sources of channel toxins

© iStock.com/John Arnold

© iStock.com/Julie de Leseleuc

© iStock.com/Kerry Werry

The cell membrane is made up of fatty molecules, so it tends to repel water. Because ions in body fluids are usually surrounded by clusters of water molecules, they cannot easily pass directly through neuronal membranes. Instead, they must pass through membrane-spanning ion channels, which are highly selective for particular types of ions. Research has begun to reveal some of the functional details of these channels, such as the K^+ channel shown in Figure A (Berneche and Roux, 2001). The inner surfaces of the K^+ channel are lined with oxygen atoms that mimic water molecules. With the oxygen atoms substituting for their usual escort of water molecules, K^+ ions fit exactly into this *selectivity filter*. Other ions, such as Na^+ ions, do not fit as comfortably and thus remain outside. The end result is a 10,000-fold selectivity for K^+ ions!

Given the extreme precision with which these ion channels must operate, it's no surprise that even minor alteration of channel functioning can cause serious health problems (Spillane et al., 2016). **Channelopathies** are medical conditions that arise from abnormalities in the form and function of ion channels, as a result of mutation. Sodium channelopathy—a problem with sodium channels—is associated with a variety of seizure disorders (Erro et al., 2017), as well as heritable muscle diseases and certain types of cardiac ailments. Chloride channel disorders can result in deafness, kidney problems, and movement disorders (Duran et al., 2010), as well as seizures. In fact, evidence is mounting that mutations in ion channel genes of all sorts may be a major cause of epilepsy (Ptáček, 2015).

The critical importance of channels is also exploited by a variety of potent animal toxins (Figure B). For example, **tetrodotoxin** (**TTX**) and **saxitoxin** (**STX**) selectively block voltage-gated sodium channels, thereby preventing the production of action potentials; paralysis and death rapidly follow. Tetrodotoxin is found in the ovaries of the puffer fish, which is esteemed as a delicacy in Japan. If the ovaries of the puffer fish are not removed properly and if the fish is not cleaned with great care, people who eat it may be poisoned by TTX. Saxitoxin is likewise a seafood threat: it derives from "red tide," a bloom of algae that sometimes affects shellfish. The extremely potent neurotoxin **batrachotoxin**, produced by South American poison arrow frogs, has the reverse effect and forces Na^+ channels to stay open, with equally lethal results. Scorpions are a rich source of channel-specific toxins, as we explore in Chapter 8 (see Figure 8.23).

Tiny quantities of any of these toxins can dramatically impede the ability of neurons to function. However, the same specificity that makes channel toxins so deadly also makes them useful tools in the laboratory. For example, the venom of the tarantula spider contains toxins that very specifically target the voltage sensor of certain voltage-gated ion channels; in the lab, these toxins provided important clues about how those channels work (Stevens et al., 2011).

Blocking channels isn't always a bad thing. Local anaesthetics like lidocaine can be injected into nerves to block voltage-gated sodium channels, stopping action potentials that would otherwise signal pain to the brain. Some visits to the dentist are much more pleasant because of the temporary blockage of *those* channels.

channelopathy A genetic abnormality of ion channels, causing a variety of symptoms.

tetrodotoxin (TTX) A toxin from puffer fish ovaries that blocks the voltage-gated sodium channel, preventing action potential conduction.

saxitoxin (STX) An animal toxin that blocks sodium channels when applied to the outer surface of the cell membrane.

batrachotoxin A toxin, secreted by poison arrow frogs, that selectively interferes with Na^+ channels.

The evolution of rapid saltatory conduction in vertebrates gives them a major behavioural advantage over invertebrates, in which axons are unmyelinated and mostly small in diameter and thus slower in conduction. To address this problem, many invertebrates have a few giant axons that mediate essential motor responses, such as escape behaviour. The squid's giant axon, which we mentioned earlier, has an unusually high conduction rate for an invertebrate, but that rate is only about 20 m/s, slower than for even small myelinated axons of mammals.

To conduct action potentials as swiftly as a myelinated vertebrate axon does, an unmyelinated invertebrate axon would have to be 100 times larger in volume. It has been estimated that at least 10% of the volume of the human brain is occupied by myelinated axons. To maintain the conduction velocity of our cerebral neurons without the help of myelin, our brains would have to be 10 times as large as they are! This fact helps explain why myelination is an important index of maturation of the developing nervous system (see Chapter 6).

There are times when we would rather axons did *not* conduct action potentials, as when a dentist works on an ailing tooth. Then we can turn to drugs that block Na^+ channels to stop the action potential from propagating pain signals to the brain. Scientists have discovered a host of such drugs and toxins that block specific channels, which have helped us understand neural transmission, as we discuss in **BOX 2.2**.

2.2 Synapses Transmit Information from One Neuron to Another

Learning Objectives

After reading this section, you should be able to:

- **2.2.1** **Explain why some synapses make the postsynaptic cell more likely to fire.**
- **2.2.2** **Understand how a neuron integrates information from all its synaptic inputs.**
- **2.2.3** **Contrast spatial summation and temporal summation in postsynaptic cells.**

The point of a neuron generating an action potential is to transmit information to other cells. The action potential is re-created down every branch of the axon until it reaches the terminals of each. There, the electrical signal of the action potential is converted into a chemical signal as the axon terminal releases a **neurotransmitter**, a chemical released from a presynaptic terminal that serves to communicate with the postsynaptic cell. Now let's see how neurotransmitters affect the electrical properties of the postsynaptic neuron.

Synapses can be excitatory or inhibitory

Neurotransmitters released into synapses briefly alter the membrane potential of the postsynaptic cell. We call these brief changes **postsynaptic potentials**. A given neuron, receiving synapses from hundreds of other cells, is subject to hundreds or thousands of postsynaptic potentials. When integrated, this massive array of local potentials determines whether the neuron will reach threshold and therefore generate an action potential of its own.

We can study postsynaptic potentials with a setup like that shown in **FIGURE 2.9**. This setup enables us to compare the effects of excitatory versus inhibitory synapses on the local membrane potential of a postsynaptic cell. The responses of the presynaptic and postsynaptic cells are shown on the same graphs in Figure 2.9 for easy comparison of their timing. It is important to remember that excitatory and inhibitory neurons get their names from their actions on *postsynaptic neurons*, not from their effects on behaviour.

Stimulation of a presynaptic neuron (red) causes it to produce an all-or-none action potential that spreads to the end of the axon, releasing transmitter. After a brief delay, the postsynaptic cell (yellow) displays a small local depolarisation, as channels open to let cations in. This local postsynaptic membrane depolarisation is known as an **excitatory postsynaptic potential** (**EPSP**) because it pushes the postsynaptic

neurotransmitter Also called *synaptic transmitter*, *chemical transmitter*, or simply *transmitter*. The chemical released from the presynaptic axon terminal that serves as the basis of communication between neurons.

postsynaptic potential A local potential that is initiated by stimulation at a synapse, can vary in amplitude, and spreads passively across the cell membrane, decreasing in strength with time and distance.

excitatory postsynaptic potential (EPSP) A depolarising potential in the postsynaptic neuron that is caused by excitatory connections. EPSPs increase the probability that the postsynaptic neuron will fire an action potential.

2.9 Recording Postsynaptic Potentials

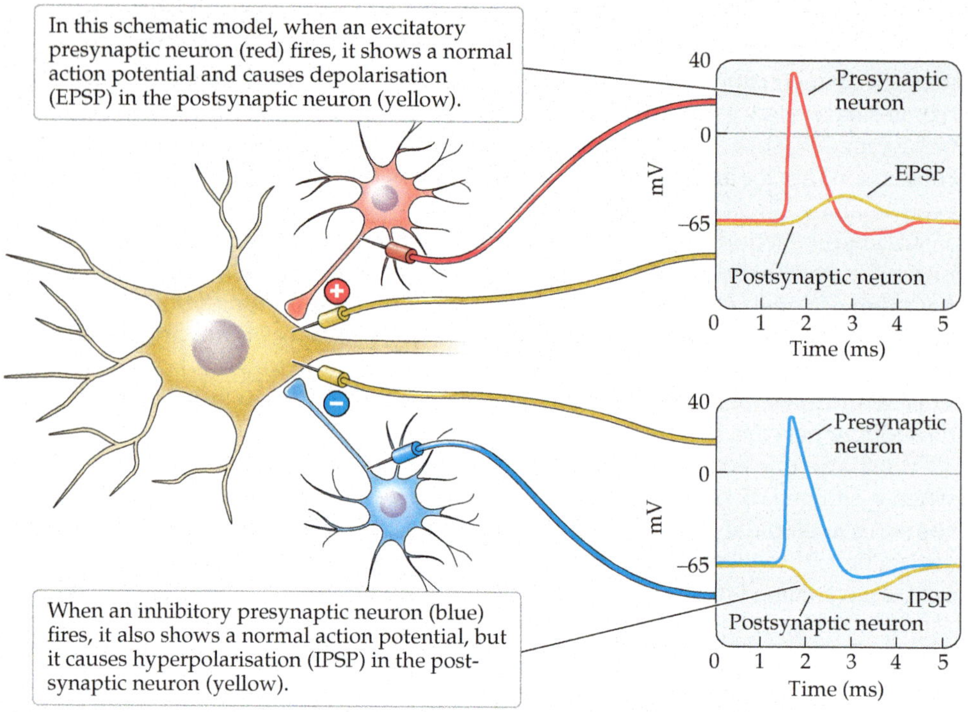

cell a little closer to the threshold for an action potential. Thus the red neuron forms an excitatory synapse upon the yellow target neuron.

Generally, the combined effect of many excitatory synapses is needed to elicit an action potential in a postsynaptic neuron. If EPSPs are elicited by many neurons that converge on the postsynaptic cell, these potentials can produce a depolarisation large enough to reach threshold and trigger an action potential. Note that there is a delay: in the fastest cases, the postsynaptic depolarisation begins about half a millisecond after the action potential arrives at the presynaptic terminal. This **synaptic delay** reflects the time needed for the neurotransmitter to be released and diffuse across the synapse.

The action potential of another presynaptic neuron (the blue cell in Figure 2.9) looks exactly like that of the excitatory presynaptic neuron; all neurons use the same kind of propagated signal. But the effect of this neuron on the *post*synaptic cell is quite different. When the blue neuron fires, the postsynaptic effect is an *increase* of the resting membrane potential. This hyperpolarisation moves the cell membrane potential away from threshold—*decreasing* the probability that the neuron will fire an action potential—so it is called an **inhibitory postsynaptic potential** (**IPSP**).

Usually IPSPs result from the opening of channels that permit **chloride ions** (**Cl^-**) to enter the cell. Because Cl^- ions are much more concentrated outside the cell than inside (see Figure 2.4), they rush into the cell, making it even more negative. Although in this discussion we have been paying more attention to excitation, inhibition also plays a vital role in the neural processing of information. Just as driving a car requires brakes as well as an accelerator, neural switches must be turned off as well as on. The nervous system treads a narrow path between overexcitation, which leads to seizures such as those that plagued Deidre, and underexcitation, which leads to coma and death.

So, excitatory and inhibitory presynaptic neurons both work in the same way, with only one exception: they have opposite effects on the postsynaptic cell. What determines whether a synapse excites or inhibits the postsynaptic cell? One factor is the particular neurotransmitter released by the presynaptic cell. Some transmitters generate an EPSP in the postsynaptic cell; others generate an IPSP. Plus, as we'll see in Chapter 4, the same neurotransmitter may be excitatory at one synapse and inhibitory at another, depending on the receptors present in the postsynaptic cell.

Whether a neuron fires an action potential at any given moment is decided by the balance between the number of excitatory and the number of inhibitory signals that it is receiving, so let's talk about that next.

synaptic delay The brief delay between the arrival of an action potential at the axon terminal and the creation of a postsynaptic potential.

inhibitory postsynaptic potential (IPSP) A hyperpolarising potential in the postsynaptic neuron that is caused by inhibitory connections. IPSPs decrease the probability that the postsynaptic neuron will fire an action potential.

chloride ion (Cl^-) A chlorine atom that carries a negative charge because it has gained one electron.

Spatial summation and temporal summation integrate synaptic inputs

Complex behaviour requires neurons to integrate and transform the messages they receive. In other words, they perform *information processing*: using a sort of neural algebra, each neuron adds and subtracts the myriad inputs it receives from other neurons. These operations are possible because of the characteristics of synaptic inputs, the way in which the neuron integrates the postsynaptic potentials, and the trigger mechanism that determines whether a neuron will fire an action potential.

Postsynaptic potentials are caused by transmitter chemicals that can be either depolarising (excitatory) or hyperpolarising (inhibitory). From their points of origin on the dendrites and cell body, these graded EPSPs and IPSPs spread passively over the neuron, decreasing in strength over time and distance. Whether the postsynaptic neuron will fire an action potential is determined by whether a depolarisation exceeding threshold reaches the portion of the axon just beyond the hillock, where action potentials begin.

The presynaptic terminals provide excitatory (depolarising) or inhibitory (hyperpolarising) stimulation to the postsynaptic cell membrane. If the membrane potential rises (depolarises) above a threshold level, an action potential is fired.

Suppose two excitatory terminals are activated, as shown in **FIGURE 2.10A**, causing local depolarisations (in red) of the cell body. These depolarisations spread out over the neuron, dissipating as they spread, so only a small proportion of the original depolarisation reaches the axon hillock. Taken alone, neither would be sufficient to cause threshold depolarisation, but when they both arrive at about the same time, the two depolarisations sum to push the membrane potential of the hillock region to threshold.

Go to Animation 2.4
Spatial Summation

FIGURE 2.10B shows what happens when inhibitory synapses also are active, creating postsynaptic hyperpolarisations. These hyperpolarisations also spread passively, dissipating as they travel. Because some potentials excite and others inhibit the hillock, these effects partially cancel each other. Thus, the net effect is the difference between the two: the neuron subtracts the IPSPs from the EPSPs and no action potential arises. Simple arithmetic, right?

When summed, EPSPs and IPSPs do tend to cancel each other out. But because postsynaptic potentials spread passively and dissipate as they cross the cell membrane, the resulting sum is also influenced by *distance*. For example, simultaneous EPSPs from two synapses close to the hillock will produce a larger sum there than will two EPSPs from farther away. Only if the overall sum of *all* the potentials—both EPSPs and IPSPs—is sufficient to depolarise the cell to threshold at the axon hillock is an action potential triggered (**FIGURE 2.10C**). Usually the convergence of excitatory messages from many presynaptic neurons is required for a neuron to fire an action potential. This summation of potentials from different physical locations across the cell body is called **spatial summation** (**FIGURE 2.11A**).

spatial summation The summation at the axon hillock of postsynaptic potentials from across the cell body. If this summation reaches threshold, an action potential is triggered.

2.10 Integration of Excitatory and Inhibitory Inputs

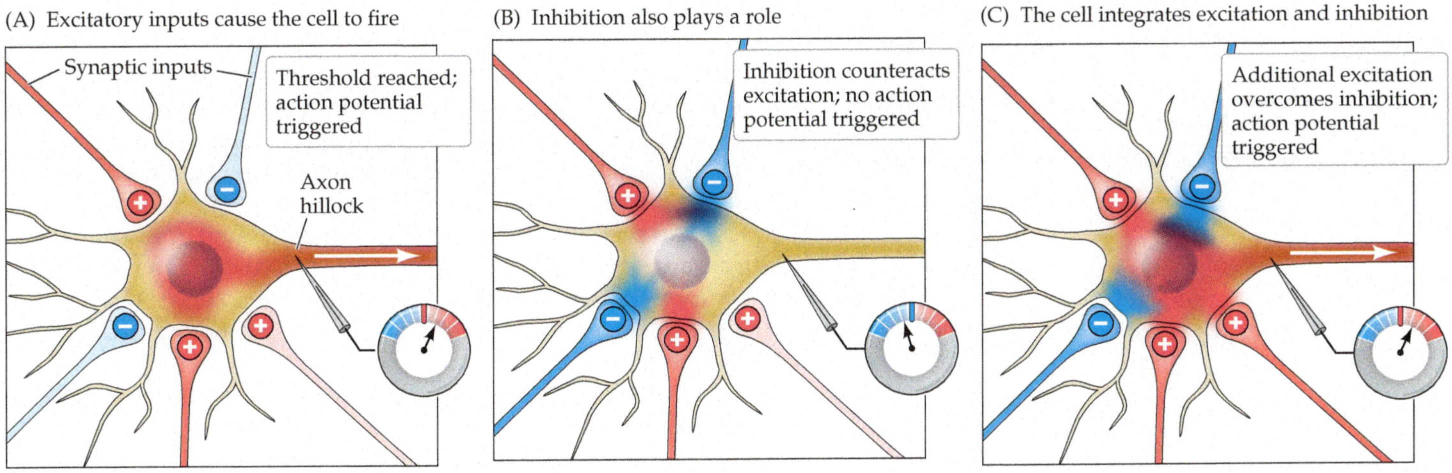

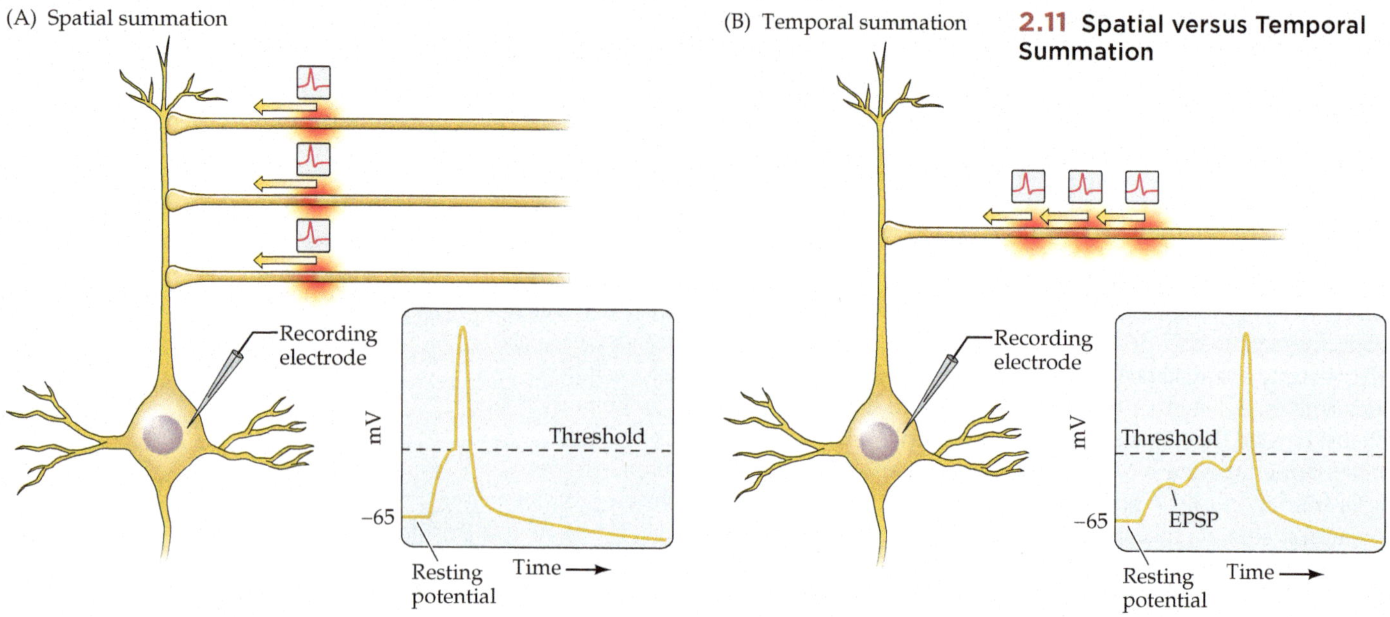

2.11 Spatial versus Temporal Summation

temporal summation The summation of postsynaptic potentials that reach the axon hillock at different times. The closer in time the potentials occur, the more complete the summation.

Postsynaptic effects that are not absolutely simultaneous can also be summed, because the postsynaptic potentials last a few milliseconds before fading away. The closer they are in time, the greater is the overlap and the more complete is the summation, which in this case is called **temporal summation**. Temporal summation is easily understood if you imagine a neuron with only one input. If EPSPs arrive one right after the other, they sum and the postsynaptic cell eventually reaches threshold and produces an action potential (**FIGURE 2.11B**). If too much time passes between EPSPs, each will fade away before the next occurs, and the neuron will never fire. Likewise, EPSPs from two different synapses may push the target to fire if they arrive at nearly the same time (they sum temporally) but will not if the second arrives after the first EPSP has faded away.

It should now be clear that, although action potentials are all-or-none phenomena, the overall postsynaptic effect is graded in sise. The membrane potential at the axon hillock thus reflects the millisecond-to-millisecond integration of all the neuron's inputs, which the axon encodes into an ongoing pattern of action potentials.

Dendrites add to the story of neuronal integration. A vast number of synaptic inputs, arrayed across the dendrites and cell body, can induce postsynaptic potentials. Dendrites therefore augment the receptive surface of the neuron and increase the amount of information that the neuron can take in. All other things being equal, the farther out on a dendrite a potential is produced, the less effect the potential should have at the axon hillock, because the potential decreases in amplitude (i.e., strength) as it passively spreads. When the potential arises at a dendritic spine, its effect is further reduced because it has to spread down the shaft of the spine. Thus, information arriving at various parts of the neuron is weighted in terms of the distance and path resistance to the axon hillock.

Interestingly, in some types of neurons, synapses that are distant from the axon hillock compensate by producing larger postsynaptic potentials. These large local potentials boost the effectiveness of the synapse on the integration process occurring at the axon, making distant synapses more comparable to nearer synapses. Furthermore, some neurons have *dendritic* integration zones, featuring voltage-gated ion channels, which sum and amplify local postsynaptic potentials, increasing their eventual impact at the axon hillock (Harnett et al., 2013). And finally, glial cells also play a role in synaptic transmission: they increase the strength of the postsynaptic potential (Eroglu and Barres, 2010), overlying the presynaptic terminal and thereby preventing neurotransmitter from leaking out of the synaptic cleft.

TABLE 2.1 summarises the many properties of action potentials, EPSPs, and IPSPs, noting the principal similarities and differences among the three kinds of neural

TABLE 2.1 Characteristics of Electrical Signals of Nerve Cells

	Type of Signal		
Characteristic	**Action potential**	**Excitatory postsynaptic potential (EPSP)**	**Inhibitory postsynaptic potential (IPSP)**
Location	Axon	Dendrites and soma	Dendrites and soma
Signalling role	Conduction along an axon	Transmission between neurons	Transmission between neurons
Typical duration (ms)	1–2	10–100	10–100
Amplitude	Overshooting, 100 mV	Depolarising, from less than 1 to more than 20 mV	Hyperpolarising, from less than 1 to about 15 mV
Character	All-or-none, digital	Graded, analog	Graded, analog
Mode of propagation	Actively propagated, regenerative	Local, passive spread	Local, passive spread
Ion channel opening	First Na^+, then K^+, in different channels	Na^+–K^+	Cl^-–K^+
Channel sensitive to:	Voltage (depolarisation)	Chemical (neurotransmitter)	Chemical (neurotransmitter)

potentials. Now let's look in detail at how the electrical signal arriving at the presynaptic terminal sends a chemical signal to the postsynaptic cell.

2.3 Action Potentials Cause the Release of Transmitter Molecules into the Synaptic Cleft

Learning Objectives

After reading this section, you should be able to:

2.3.1 **List the sequence of events underlying the release of neurotransmitter and its effects on a target neuron.**

2.3.2 **Describe the specialised proteins that cause the release of neurotransmitter into the synapse.**

2.3.3 **Define the different mechanisms of action that drugs may have on a neurotransmitter receptor.**

2.3.4 **Contrast the two fundamentally distinct types of neurotransmitter receptors.**

2.3.5 **Describe the different classes of synapses based on the part of a target neuron that is innervated.**

The transfer of information across a synapse is called *synaptic transmission*:

1. The action potential traveling down the axon arrives at the axon terminal.
2. This depolarisation opens voltage-gated calcium channels in the membrane of the axon terminal, allowing calcium ions (Ca^{2+}) to enter the terminal.
3. The Ca^{2+} causes synaptic vesicles filled with neurotransmitter to fuse with the presynaptic membrane and rupture, releasing the transmitter molecules into the synaptic cleft.
4. Transmitter molecules cross the cleft to bind to special receptor molecules in the postsynaptic membrane, leading to the opening of ion channels in the postsynaptic membrane.
5. This ion flow creates a local EPSP or IPSP in the postsynaptic neuron.
6. Synaptic transmitter is either (a) inactivated (degraded) by enzymes or (b) removed from the synaptic cleft by *transporters*, so the transmission is brief and accurately reflects the activity of the presynaptic cell.
7. Synaptic transmitter may also activate presynaptic *autoreceptors*, regulating future transmitter release.

These steps are illustrated in **FIGURE 2.12** on the next page.

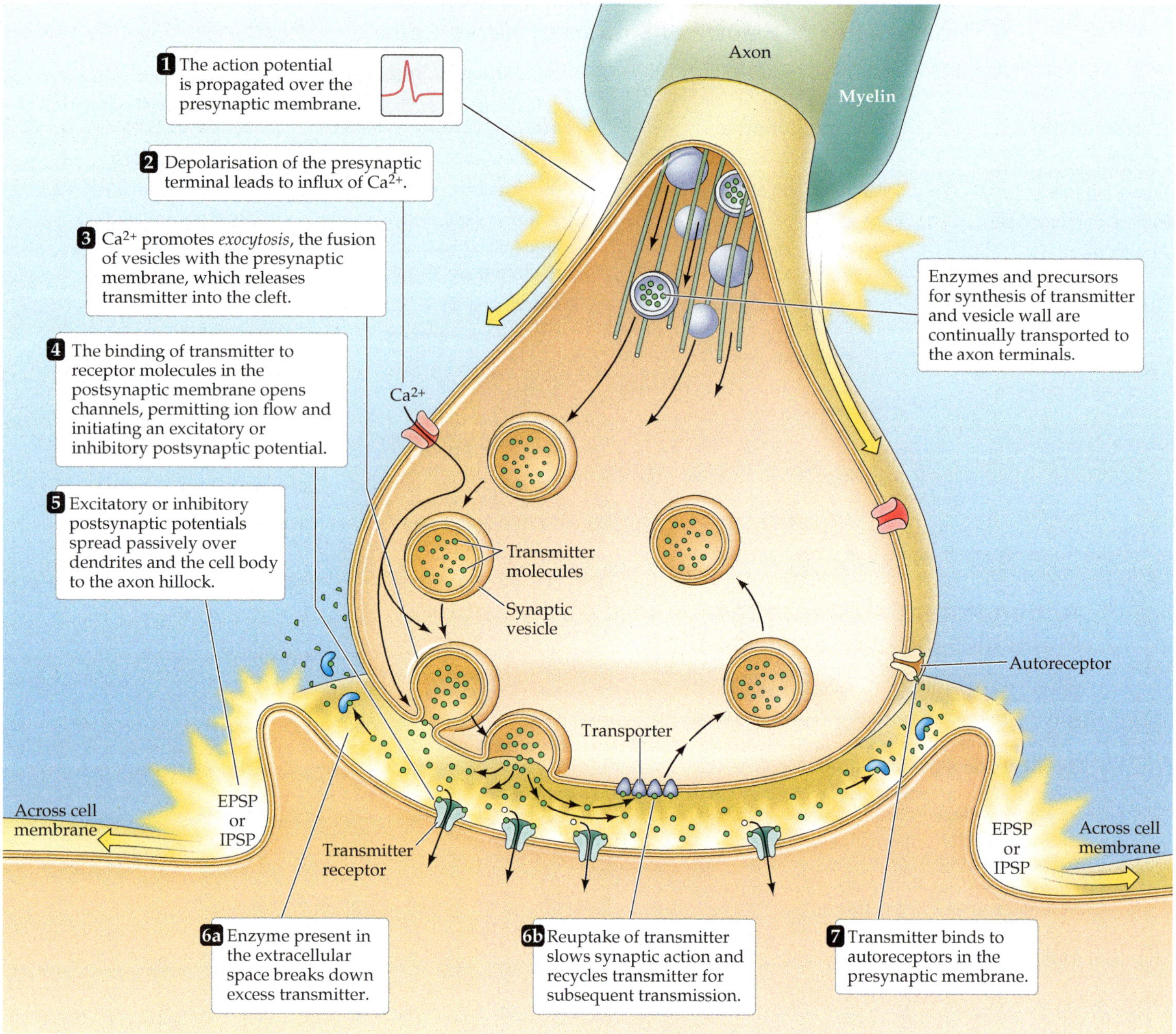

2.12 Steps in Transmission at a Chemical Synapse

exocytosis The process by which a synaptic vesicle fuses with the presynaptic terminal membrane to release neurotransmitter into the synaptic cleft.

v-SNARE A specialised protein anchored to vesicles to aid their fusing to the presynaptic membrane to release neurotransmitter.

t-SNARE A specialised protein anchored to the presynaptic "target" membrane to bind v-SNAREs to dock vesicles, making them ready for release.

synaptotagmin A specialised protein that responds to calcium ions to trigger vesicular exocytosis.

The IPSPs and EPSPs in the postsynaptic cell spread throughout its interiour. If the integration of all the EPSPs and IPSPs depolarises the axon hillock enough, the postsynaptic neuron will fire an action potential of its own. Now let's examine these steps in detail.

When an action potential reaches a presynaptic terminal, it opens voltage-gated *calcium* channels that allow an influx of calcium ions (Ca^{2+}), rather than K^+ or Na^+, into the axon terminal. These Ca^{2+} ions activate enzymes that cause vesicles near the presynaptic membrane to fuse with the membrane and discharge their contents into the synaptic cleft (see Figure 2.12, steps 1–3). The higher the frequency of action potentials arriving at the terminal, the greater the influx of Ca^{2+}, and the more vesicles that dump transmitter into the synapse. Most synaptic delay is caused by the time needed for Ca^{2+} to enter the terminal and the vesicles to fuse. Both the diffusion of the transmitter across the cleft and the interaction of transmitter molecules with their receptors also take some time.

Synaptic vesicles, which are about 50 nanometers (nm) in diameter, release their neurotransmitter contents by fusing with the presynaptic membrane, in a process called **exocytosis**. Several specialised long-chain proteins mediate exocytosis. One family of proteins, called SNAREs, serve as tethers: those attached to vesicles are called **v-SNAREs**, while those attached to the presynaptic membrane are called **t-SNAREs** (the *t* stands for "target"; **FIGURE 2.13A**). When the v-SNARES on the vesicle attach to the t-SNARES, the vesicle is said to be *docked*, ready to be released (**FIGURE 2.13B**). Another protein attached to the vesicle, called **synaptotagmin**, serves as a Ca^{2+} sensor (Q. Zhou et al., 2015). When the action potential arrives at the axon terminal, the incoming Ca^{2+} ions bind and activate synaptotagmin (**FIGURE 2.13C**), which then triggers the final fusion of the vesicular and presynaptic membranes, allowing the neurotransmitter molecules to enter the synaptic cleft (**FIGURE 2.13D**). **Botulinum toxin** (Botox) and **tetanus toxin** both silence synapses by cutting up SNARE proteins, effectively disabling exocytosis and therefore synaptic transmission.

Because all the synaptic vesicles in an axon terminal contain about the same number of molecules of transmitter, estimated to be tens of thousands, each vesicle produces about the same change in postsynaptic potential when it ruptures and releases its contents. Typically, an action potential causes the exocytosis of several hundred vesicles at a time.

The presynaptic terminal normally produces and stores enough transmitter to ensure that it is ready for activity. The rate of production of transmitter is governed by enzymes that are manufactured in the neuronal cell body and transported down the axons to the terminals. Intense activity of the neuron reduces the number of available vesicles, but soon more vesicles are produced to replace those that were discharged. In a variation of regular exocytosis, sometimes a vesicle fuses with the postsynaptic membrane just long enough to spill neurotransmitter into the cleft, then pinches off again to return to the presynaptic terminal, a process termed *kiss and run* (**FIGURE 2.14**), and there is evidence for a special, ultrafast process of vesicle formation (Gan and Watanabe, 2018).

Receptor molecules recognise transmitters

The action of a key in a lock is a good analogy for the action of a transmitter on a receptor protein. Just as a particular key can open a door, a molecule of the correct shape, called a **ligand** (see Chapter 4), can fit into a receptor protein and activate or block it. So, for example, at synapses where **acetylcholine** (**ACh**) is the transmitter, it fits into *ligand-binding sites* in **receptor molecules** located in the postsynaptic membrane.

The nature of the postsynaptic receptors at a given synapse determines the action of the transmitter (see Chapter 4). For example, ACh can function as either an inhibitory or an excitatory neurotransmitter, at different synapses. At excitatory synapses, binding of ACh opens channels for Na^+ and K^+ ions

2.13 SNAREs and Synaptotagmin Mediate Exocytosis

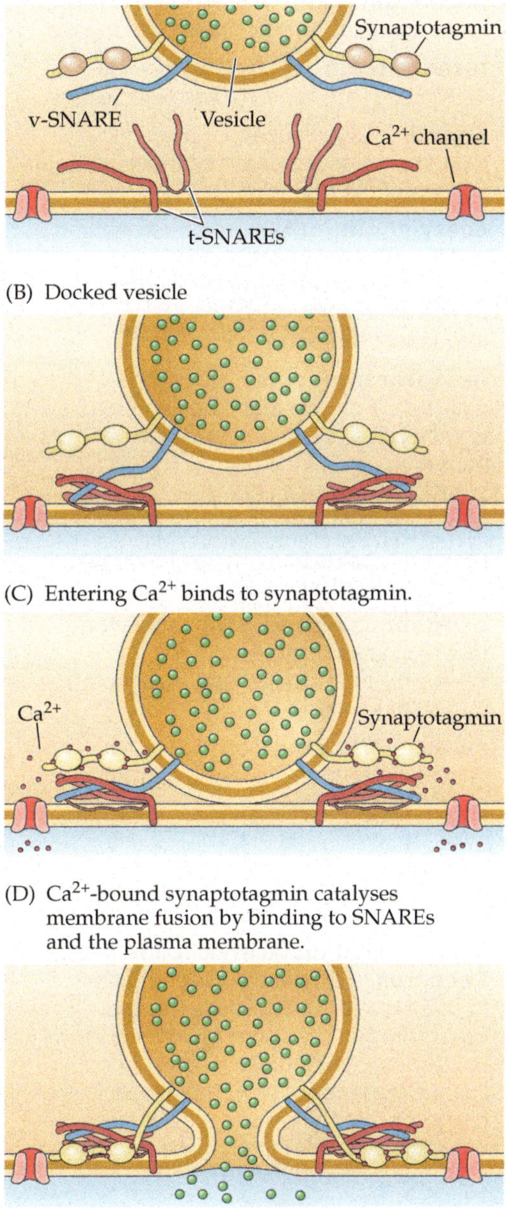

Go to Animation 2.5
Synaptic Transmission

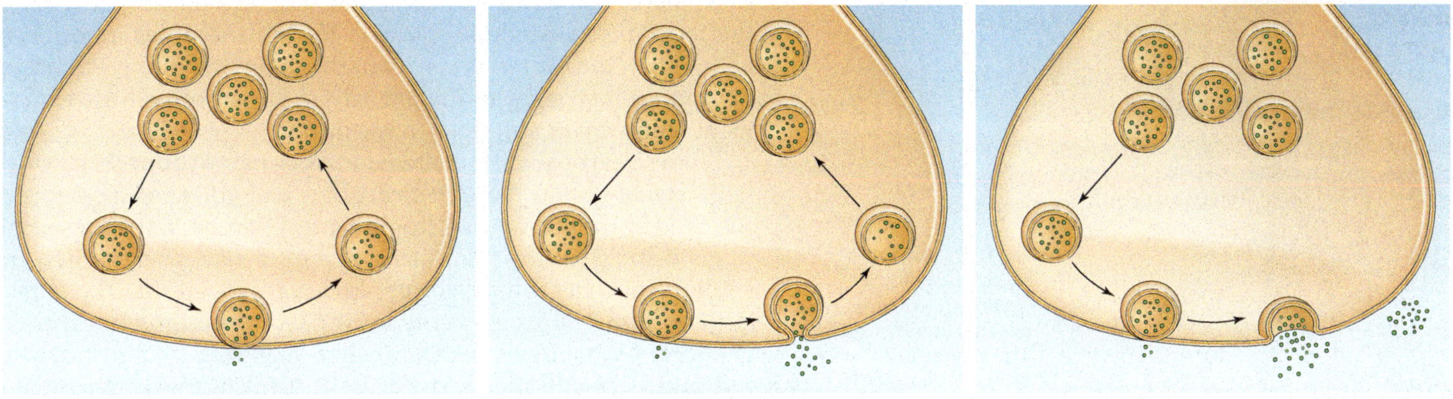

2.14 Kiss-and-Run Synapses (Adapted from N. T. N. Phan et al., 2017. *Nat Rev Chem* 1: 48.)

botulinum toxin A toxin that cleaves SNAREs, disabling neurotransmitter release.

tetanus toxin A toxin that cleaves SNAREs, disabling neurotransmitter release.

ligand A substance that binds to receptor molecules, such as those at the surface of the cell.

acetylcholine (ACh) A neurotransmitter produced and released by parasympathetic postganglionic neurons, by motor neurons, and by neurons throughout the brain.

receptor molecule Also called *receptor*. A protein that binds and reacts to molecules of a neurotransmitter or hormone.

endogenous ligand Any substance that is produced within the body and selectively binds to the type of receptor that is under study.

exogenous ligand Any substance that originates outside the body and selectively binds to the type of receptor that is under study.

curare An alkaloid neurotoxin that causes paralysis by blocking acetylcholine receptors in muscle.

(**FIGURE 2.15**). In this way, ACh released onto muscle fibres excites them to contract. At inhibitory synapses, ACh typically acts on a different type of receptor to open channels that allow chloride ions (Cl^-) to enter, thereby hyperpolarising the membrane (i.e., making it more negative and so less likely to create an action potential).

The lock-and-key analogy is strengthened by the observation that various chemicals can fit onto receptor proteins and block the entrance of the key. Neurotransmitters and hormones made inside the body are examples of **endogenous ligands**; drugs and toxins from outside the body are **exogenous ligands**. Some exogenous ligands resemble the ingredients of a witches' brew. As an example, consider a couple of potent poisons that block ACh receptors: curare and bungarotoxin. **Curare** is an arrowhead poison, extracted from a plant, that is used by indigenous South Americans. If the hunter hits any part of the prey, the arrow's poison soon blocks ACh receptors on muscles, paralysing the animal. **Bungarotoxin**, another blocker of ACh receptors, is found in the venom of the banded krait (*Bungarus multicinctus*), a snake native to Asia. This toxin has proven very useful in research because a fluorescent or radioactive label can be attached to molecules of bungarotoxin without causing any change in their function. The labelled bungarotoxin can then be used to permanently bind and mark ACh receptors for monitoring their number, distribution, and function.

Another poison, muscarine (extracted from the toxic mushroom *Amanita muscaria*), mimics the action of ACh on some receptors. Molecules such as muscarine and nicotine that act like a transmitter at a receptor are called **agonists** (from the Greek *agon*, "contest" or "struggle") of that transmitter. Conversely, molecules that interfere with or prevent the action of a transmitter—in the manner that curare or bungarotoxin blocks the action of ACh—are called **antagonists**. We'll learn more about agonists and antagonists in Chapter 4.

Just as there are master keys that fit many different locks, there are submaster keys that fit a certain subgroup of locks, and keys that fit only a single type of lock. Similarly, each chemical transmitter may bind to different subtypes of receptor molecules that vary in their anatomical distribution and functions. Receptors that normally respond to ACh, for example, are called **cholinergic** receptors; the two main subtypes are nicotinic and muscarinic.

Most ACh receptors in the brain are muscarinic. Muscarinic cholinergic receptors are also found on organs innervated by the parasympathetic division of the autonomic system (e.g., the intestines, the salivary glands, and heart muscle). These receptors were key to a famous experiment. In the early twentieth century, debate raged over the basic nature of neural communication. The discovery that individual nerve cells contact each other at thousands of points was fresh knowledge. What happened at these synapses to convey information from one cell to the next? How could we find out for sure?

One night in 1921, Otto Loewi (1873–1961) dreamed of a simple experiment that would definitively discriminate between the two candidate modes of transmission—chemical versus electrical. In excitement, Loewi sat up in bed and scribbled a few notes, but in the morning he was disappointed to find the notes indecipherable. When the dream came again the following night, Loewi got up and went straight to the lab, where he performed the experiment while it was still fresh in his mind.

Loewi electrically stimulated the vagus nerve of one frog, which he knew would decrease its heart rate, and collected a sample of the fluid surrounding that frog's heart. Then he bathed a second frog's heart with the fluid sample from the first frog. When the second frog's heart also slowed, Loewi reasoned that the stimulation of the first frog's nerve must have caused the release of a chemical—what Loewi initially called *Vagusstoff* ("substance from the vagus")—into the fluid (**FIGURE 2.16**). Vagusstoff was later found to be ACh. This is how scientists learned that the nervous

2.15 A Nicotinic Acetylcholine (ACh) Receptor Each nicotinic ACh receptor consists of five subunits. The two ligand-binding sites normally bind ACh molecules, but they also bind exogenous ligands like nicotine and other nicotinic drugs. The ACh molecule and Na^+ ions are enlarged here for diagrammatic purposes.

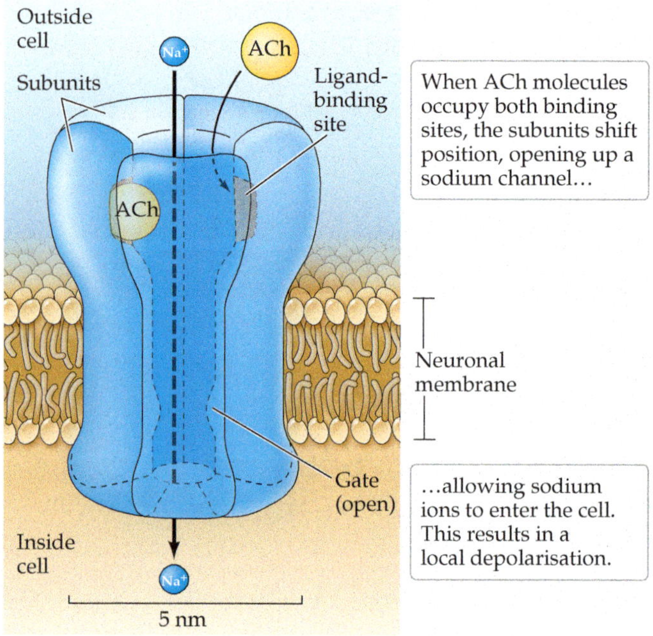

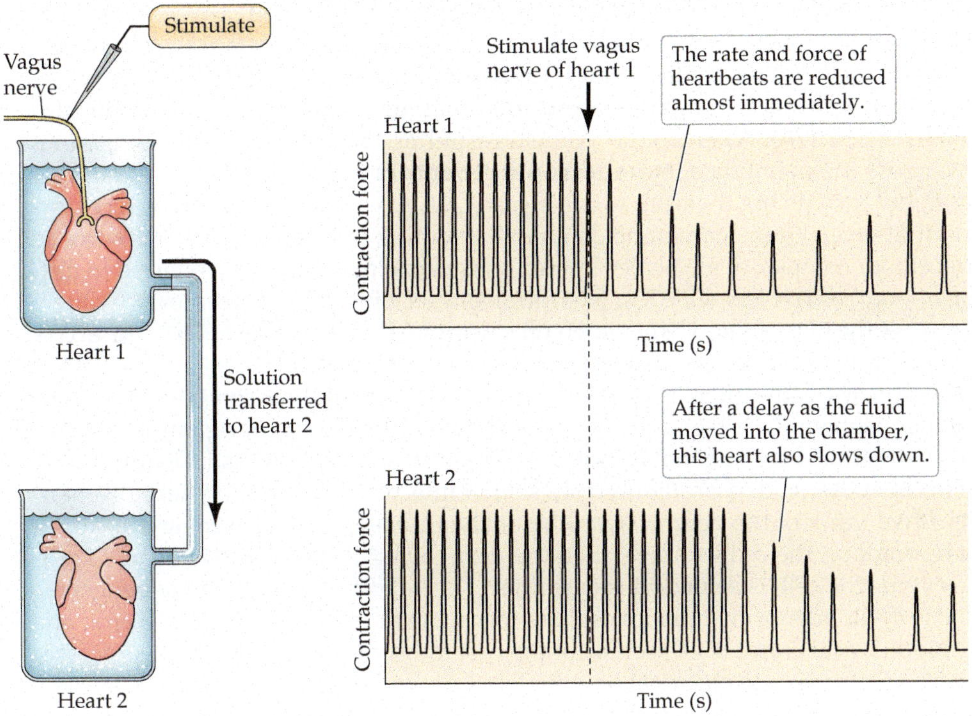

2.16 Loewi's Demonstration of a Chemical Messenger Loewi reasoned that some chemical released into the bath when he stimulated the nerve to the first heart must have slowed down the second heart. We now know that chemical is the neurotransmitter acetylcholine (ACh) acting on inhibitory muscarinic receptors in heart muscle.

system, long known to use electrical signals, also uses chemical signals. This breakthrough won Loewi a Nobel Prise in 1936.

Nicotinic cholinergic receptors are found at synapses on muscles and in autonomic ganglia; it is the blockade of these receptors that is responsible for the paralysis caused by curare and bungarotoxin Most nicotinic cholinergic synapses are excitatory, but there are also inhibitory nicotinic synapses (Wersinger and Fuchs, 2011), and there are both excitatory and inhibitory muscarinic synapses, making at least four subtypes of cholinergic receptors. The evolution of many subtypes of receptors for each transmitter enables a given transmitter to have subtly different effects in different parts of the brain.

The nicotinic ACh receptor resembles a lopsided dumbbell with a tube running down its central axis (see Figure 2.15). The handle of the dumbbell spans the cell membrane (which is about 6 nm thick). The sides of the ion channel that runs through the handle of the receptor consist of five protein subunits arranged like staves in a barrel. Two subunits are alike and, in conjunction with neighbouring subunits, provide two recognition sites for ACh (Changeux, 2018); the other three subunits are all different. For the channel to open, both of the ACh-binding sites must be occupied.

After the structure of the nicotinic ACh receptor was determined, similar analyses were carried out for other receptors, including receptors for some other synaptic transmitter molecules such as GABA (gamma-aminobutyric acid), glycine, and glutamate. These receptors also consist of multiple subunits, which resemble each other, suggesting that they have a common evolutionary origin.

The coordination of different transmitter systems of the brain is incredibly complex. Each subtype of neurotransmitter receptor has a unique pattern of distribution within the brain. Different receptor systems become active at different times in fetal life. The number of any given type of receptor remains plastic in adulthood: not only are there seasonal variations, but many kinds of receptors show a regular daily variation of 50% or more in number, affecting the sensitivity of cells to the related variety of transmitter. Similarly, the numbers of some receptors have been found to vary with the use of drugs. An increase in receptor numbers is generally referred to as **up-regulation**, and a process that decreases receptor density is called **down-regulation** of that receptor type (see Chapter 3).

bungarotoxin A neurotoxin from the venom of the banded krait that selectively blocks acetylcholine receptors.

agonist A molecule, usually a drug, that binds a receptor molecule and initiates a response like that of another molecule, usually a neurotransmitter.

antagonist A molecule, usually a drug, that interferes with or prevents the action of a transmitter.

cholinergic Referring to cells that use acetylcholine as their synaptic transmitter.

up-regulation A compensatory increase in receptor availability at the synapses of a neuron.

down-regulation A compensatory reduction in receptor availability at the synapses of a neuron.

ionotropic receptor Also called *ligand-gated ion channel* and *chemically gated ion channel*. A receptor protein that includes an ion channel that is opened when the receptor is bound by an agonist.

ligand-gated ion channel Also called *chemically gated ion channel*. An ion channel that opens or closes in response to the presence of a particular chemical.

metabotropic receptor A receptor protein that does not contain an ion channel but may, when activated, use a G protein system to alter the functioning of the postsynaptic cell.

G proteins A class of proteins that are next to the intracellular portion of a receptor and that are activated when the receptor binds an appropriate ligand on the extracellular surface.

second messenger A slow-acting substance in the postsynaptic cell that amplifies the effects of synaptic activity and signals synaptic activity within the postsynaptic cell.

Go to Animation 2.6
Ionotropic and Metabotropic Receptors

Transmitters bind to receptors, gating ion channels

The recognition of transmitter molecules by receptor molecules controls the opening of ion channels in two different ways. **Ionotropic receptors**, also known as *chemically gated ion channels* or **ligand-gated ion channels**, directly control an ion channel (**FIGURE 2.17A**). When bound by the transmitter, the ion channel opens and ions flow across the membrane. **Metabotropic receptors** recognise the synaptic transmitter, but they do not themselves contain any ion channels. Instead, these receptors consist of seven "transmembrane domains" spanning the membrane and control an intracellular complex of molecules known as **G proteins**, so they are referred to as *G protein-coupled receptors* (*GPCRs*). Neurotransmitter binding the receptor activates a subunit of the G protein to either open up a nearby ion channel, as in **FIGURE 2.17B**, or trigger other biochemical reactions in the postsynaptic cell.

G protein is a convenient designation for proteins that bind the compounds guanosine diphosphate (GDP), guanosine triphosphate (GTP), and other guanine nucleotides. Sometimes the G protein itself acts to open ion channels, as in Figure 2.17B. But in other cases the G protein activates other internal chemicals to affect ion channels. If we think of the neurotransmitter as the first, external messenger arriving at the receptor on the cell's surface, then the next chemical signal, activated by G proteins *inside* the cell, is a **second messenger**. Several different second messengers—such as cyclic adenosine monophosphate (cyclic AMP or cAMP), diacylglycerol, and arachidonic acid—amplify the effect of the first messenger and can initiate processes that lead to changes in electrical potential at the membrane. An important feature of second-messenger systems is their ability to amplify and prolong the synaptic signals that a neuron receives.

About 80% of the known neurotransmitters and hormones activate G protein-coupled receptors, so these are very important signalling systems. The protein is located on the inner side of the neuronal membrane. When a transmitter molecule binds to the extracellular part of a GPCR, intracellular subunits of the G protein complex separate from each other. The *alpha subunit* splits away from the combined *beta* and *gamma subunits*, and now both can migrate to modulate the activity of their target molecules. Depending on the type of cell and receptor, the target may be a second-messenger system, an enzyme that works on an ion channel, or an ion pump (Calebiro and Jobin, 2018).

2.17 Two Types of Chemical Synapses

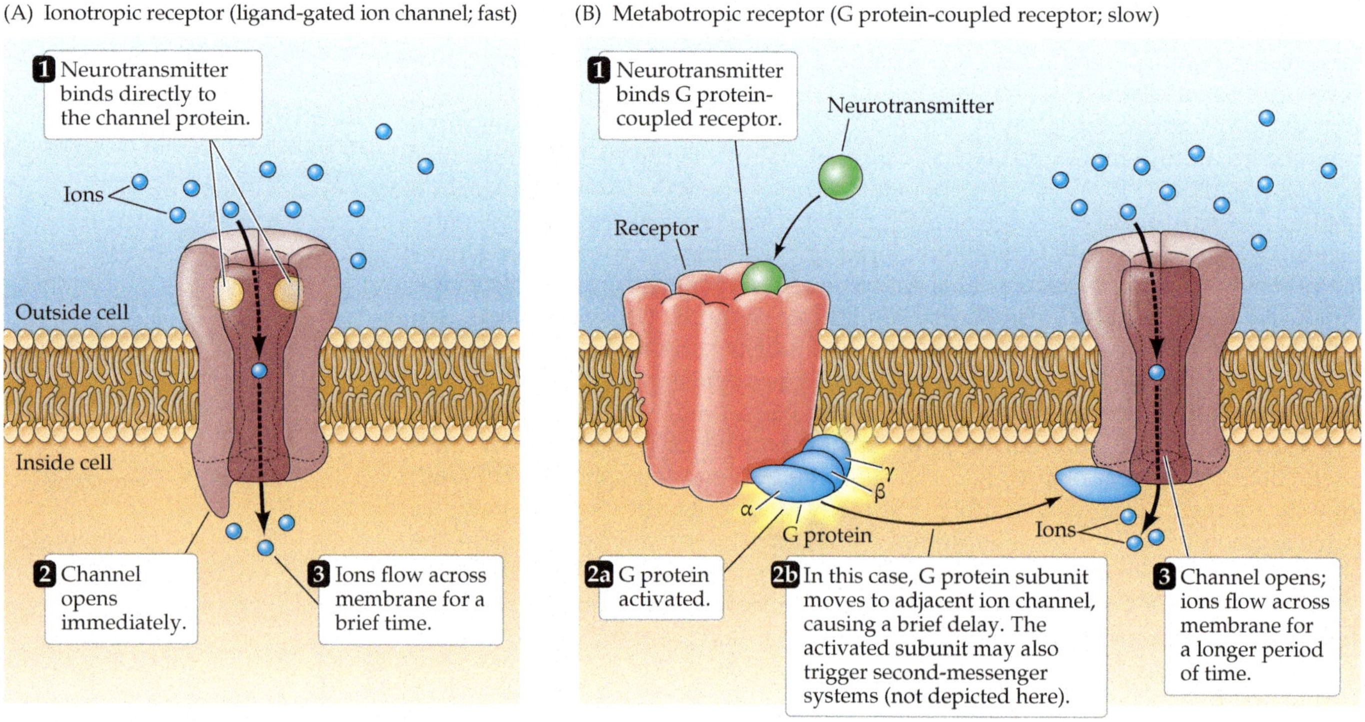

The action of synaptic transmitters is stopped rapidly

When a chemical transmitter such as ACh is released into the synaptic cleft, its postsynaptic action is not only prompt but usually very brief as well. This brevity is important for maximising the transmission of information at the synapse—by clearing away the effects of the previous release of neurotransmitter, the postsynaptic machinery is ready to respond to the next release.

The prompt cessation of transmitter effects is achieved in one of two ways:

1. *Degradation* Transmitter can be rapidly broken down and thus inactivated by a special enzyme—a process known as **degradation** (see step 6a in Figure 2.12). For example, the enzyme that inactivates ACh is acetylcholinesterase (AChE). AChE breaks down ACh very rapidly into choline and acetic acid, and these products are recycled (at least in part) to make more ACh in the axon terminal. AChE is especially concentrated at synapses but also found elsewhere in the brain. Thus, if any ACh escapes from a synapse where it is released, it is unlikely to reach other synapses intact.
2. *Reuptake* Alternatively, transmitter molecules may be rapidly cleared from the synaptic cleft by being taken up into the presynaptic terminal—a process known as **reuptake** (see step 6b in Figure 2.12). Norepinephrine, dopamine, and serotonin are examples of transmitters whose activity is terminated mainly by reuptake. In these cases, special receptors for the transmitter, called **transporters**, are located on the presynaptic axon terminal and bring the transmitter back inside. Once taken up into the presynaptic terminal, transmitter molecules may be repackaged into newly formed synaptic vesicles. Certain drugs that interfere with reuptake mechanisms are effective in the treatment of depression (see Chapter 12).

Several factors regulate neurotransmitter release

Some neurotransmitter molecules never make it to the postsynaptic membrane. They may bind to receptors on the presynaptic membrane, so-called **autoreceptors** (see Figure 2.12). Through autoreceptors, the presynaptic cell is informed about the net concentration of transmitter in the synaptic cleft and may regulate future neurotransmitter release to adjust that concentration.

For simplicity, we have been focussing on the classic **axo-dendritic** and **axo-somatic** synapses. But many nonclassic forms of chemical synapses are found in the nervous system (**FIGURE 2.18A,B**). As the name implies, **axo-axonic** synapses form on axons, often near the axon terminal, allowing the presynaptic neuron to regulate how much neurotransmitter will be released from the targeted terminal.

degradation The chemical breakdown of a neurotransmitter into inactive metabolites.

reuptake The process by which released synaptic transmitter molecules are taken up and reused by the presynaptic neuron, thus stopping synaptic activity.

transporter A specialised receptor in the presynaptic membrane that recognises transmitter molecules and returns them to the presynaptic neuron for reuse.

autoreceptor A receptor for a synaptic transmitter that is located in the presynaptic membrane, telling the axon terminal how much transmitter has been released.

axo-dendritic Referring to a synapse in which a presynaptic axon terminal synapses onto a dendrite of the postsynaptic neuron, either via a dendritic spine or directly onto the dendrite itself.

axo-somatic Referring to a synapse in which a presynaptic axon terminal synapses onto the cell body (soma) of the postsynaptic neuron.

axo-axonic Referring to a synapse in which a presynaptic axon terminal synapses onto another axon's terminal.

2.18 Different Types of Synaptic Connections Most synapses are formed by an axon stimulating a dendrite (A), but axons also sometimes synapse upon cell bodies (B) or even other axons (C). And in some instances, specialised dendrites synapse upon other dendrites (D).

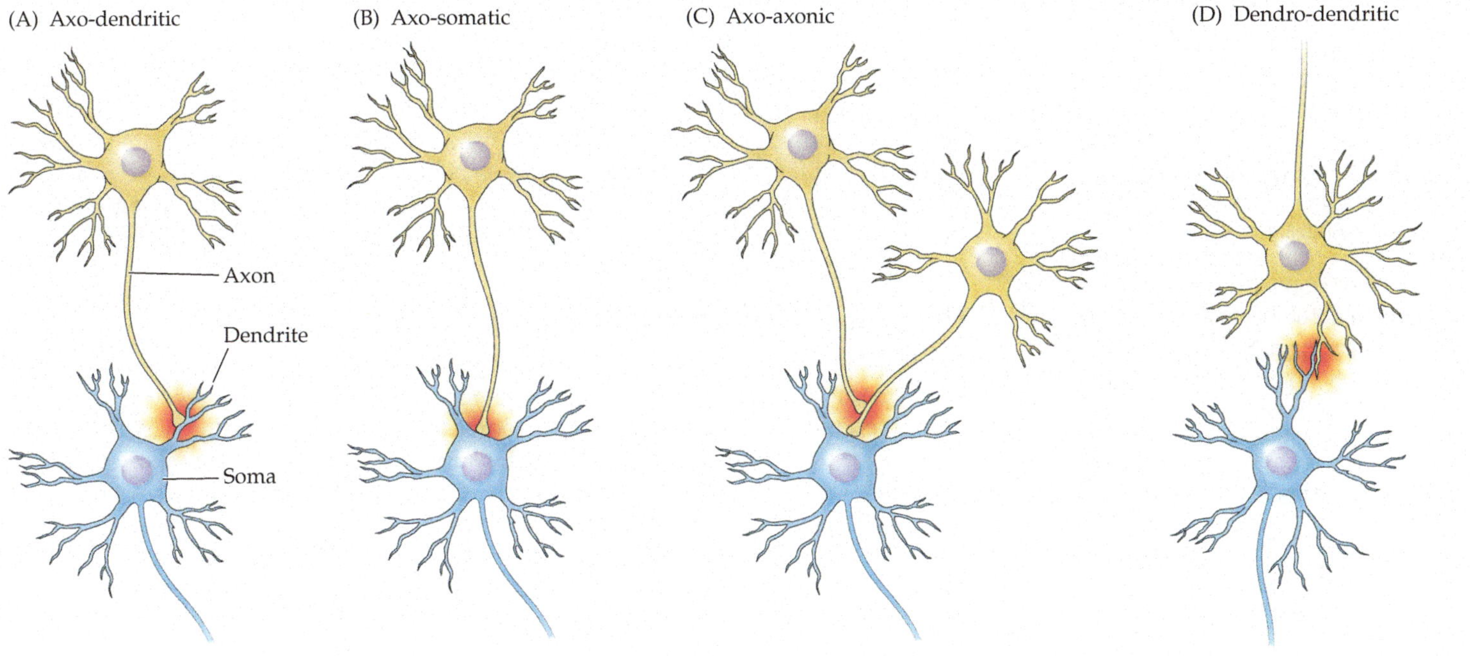

(A) Electron micrograph of an electrical synapse

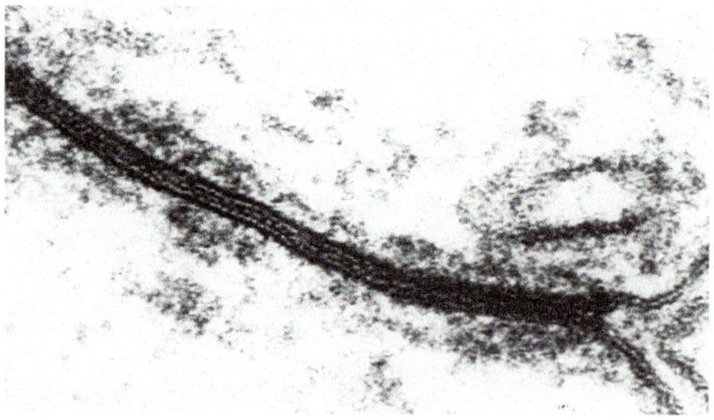

(B) Diagram of an electrical synapse

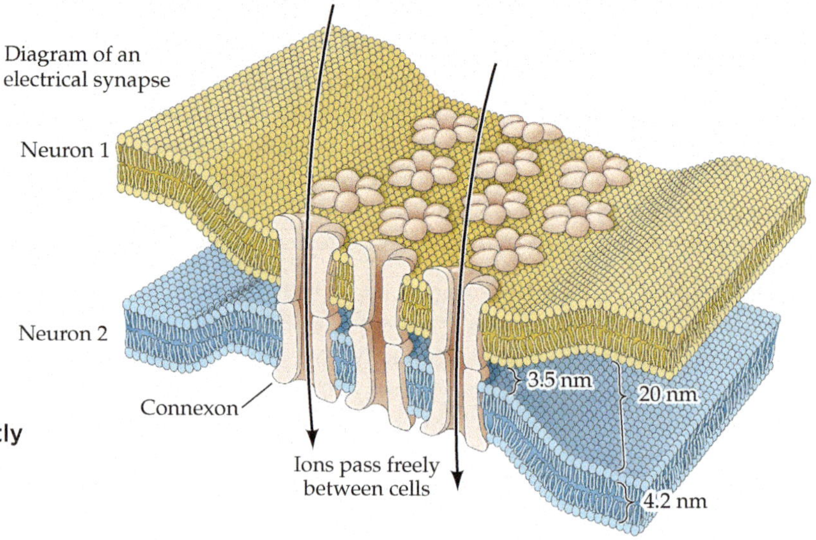

2.19 Electrical Synapses Allow Ions to Flow Directly between Neurons

Neurons can also form **dendro-dendritic** contacts, allowing coordination of their activities. Evidence is also mounting that **ectopic transmission** occurs between many neurons; in this mode of transmission, the location of transmitter release and the sites at which the transmitter acts are both well outside the conventional boundaries of nearby synapses (Coggan et al., 2005). And throughout the brain are found axons with regular swellings, called **varicosities**, along their length; like a drip irrigation system, these **nondirected synapses** steadily release neurotransmitter to broadly affect surrounding areas.

At a **retrograde synapse**, transmission starts with classic axo-dendritic synaptic activity, but the postsynaptic cell subsequently releases a gas neurotransmitter, such as carbon monoxide or nitric oxide (see Chapter 4), which signals the presynaptic cell to release more transmitter.

Finally, although we will emphasise synapses that use chemical neurotransmitters to communicate between cells, the brain also has widespread *electrical* synapses (Harris, 2018). At an **electrical synapse** (or *gap junction*) the presynaptic membrane comes even closer to the postsynaptic membrane than it does at a chemical synapse; the gap at an electrical synapse (**FIGURE 2.19A**) measures only 2–4 nm. In contrast, the synaptic cleft of a chemical synapse is 20–40 nm. At electrical synapses, the facing membranes of the two cells have relatively large channels, called **connexons**, arranged to allow ions to flow from one neuron directly into the other (**FIGURE 2.19B**). As a consequence, the electrical current that is associated with neural activity in one neuron can flow directly across the gap junction to affect the other neuron.

Transmission at these synapses closely resembles action potential conduction along the axon. Electrical synapses therefore work with practically no time delay, in contrast to chemical synapses, where the delay is on the order of a millisecond—slow in terms of neurons. Because of the speed of their transmission, electrical synapses are frequently found in neural circuits that mediate escape behaviours in invertebrates. They are also found where many fibres must be activated synchronously, as in the system for moving our eyes. Electrical synapses may contribute to the spread of synchronised seizure discharges in epilepsy (Q. Li et al., 2019).

Neurons and synapses combine to make circuits

Use of the term *circuit* for a group of neurons and their synaptic interconnections is an analogy to electrical circuits, in which an arrangement of components (e.g., resistors, capacitors, transistors, and their connecting wires) accomplishes a particular function. Electrical circuits can represent signals in either analog or digital ways—that is, in terms of continuously varying values or in terms of integers. Neurons similarly feature two kinds of processes: analog-like signals that vary in strength (such as graded potentials at synapses) and digital-like, all-or-none signals (such as action potentials) that vary in frequency. The nervous system comprises many different types of neural circuits to accomplish basic functions in cognition, emotion, and action—all the categories of behaviour and experience.

dendro-dendritic Referring to a synapse in which a synaptic connection forms between the dendrites of two neurons.

ectopic transmission Cell-cell communication based on release of neurotransmitter in regions outside traditional synapses.

varicosity The axonal swelling from which neurotransmitter diffuses in a nondirected synapse.

nondirected synapse A type of synapse in which the presynaptic and postsynaptic cells are not in close apposition; instead, neurotransmitter is released by axonal varicosities and diffuses away to affect wide regions of tissue.

retrograde synapse A synapse in which a signal (usually a gas neurotransmitter) flows from the postsynaptic neuron to the presynaptic neuron, thus counter to the usual direction of synaptic communication.

electrical synapse Also called *gap junction*. The region between neurons where the membranes are so close that changes in potential can flow from one to the other without being translated into a chemical message.

connexon A protein assembly that provides an open ion channel between two neurons, forming an electrical synapse between them.

2.20 The Knee-Jerk Reflex

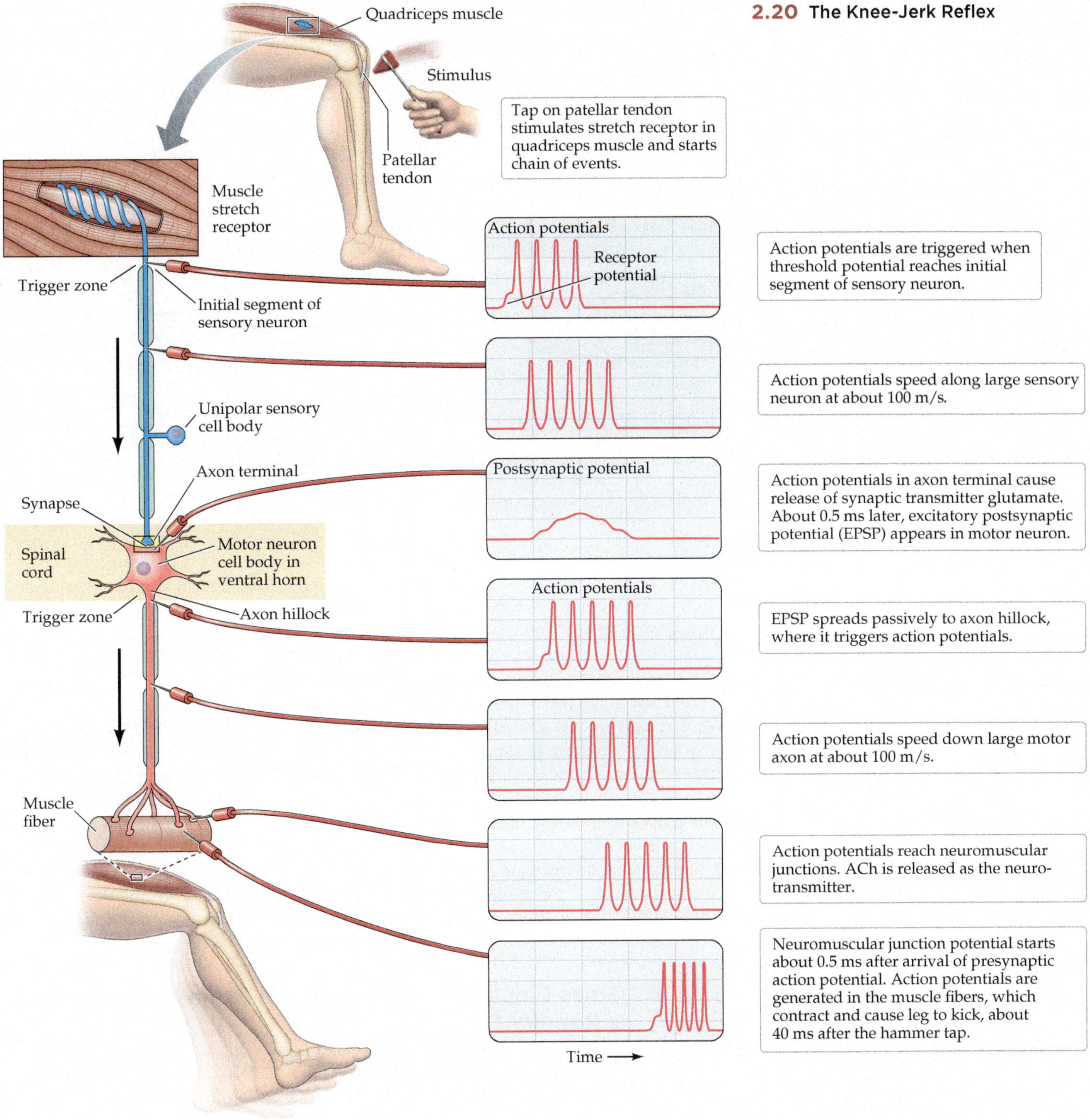

The simplest neural circuit that is routinely encountered in the nervous system is the **neural chain**, a straightforward linking of a series of neurons. (A look at some other simple neural circuits is presented on the website in **A STEP FURTHER 2.2: CIRCUITS OF NEURONS PROCESS INFORMATION**.) From the seventeenth century until well into the twentieth century, most attempts to understand behaviour in neural terms were based on chains of neurons, which do indeed account for some behaviours.

For example, the basic circuit for the stretch reflex, such as the **knee-jerk reflex**, consists of a sensory neuron, a motor neuron, and a single synapse where the sensory neuron communicates with the motor neuron. **FIGURE 2.20** details the sequence and

neural chain A simple kind of neural circuit in which neurons are attached linearly, end to end.

knee-jerk reflex A variant of the stretch reflex in which stretching of the tendon below the knee leads to an upward kick of the leg.

2.21 Two Representations of Neural Circuitry (A) This simple representation shows the input part of the visual system. (B) This more realistic representation illustrates convergence and divergence.

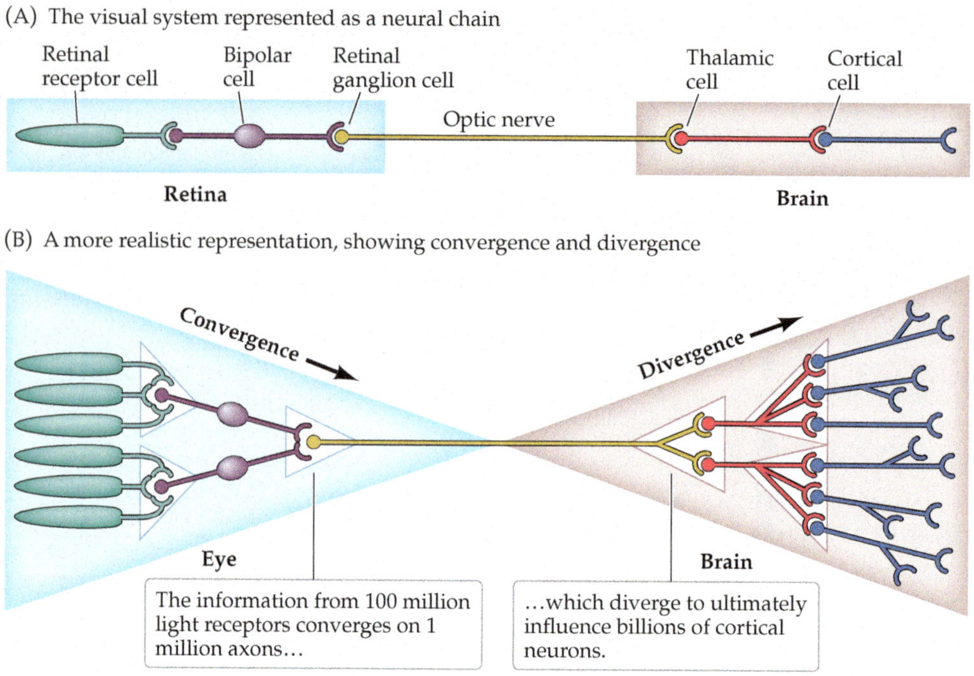

timing of events in the knee-jerk reflex. Note that this reflex is extremely rapid: only about 40 ms elapse between the stimulus and the initiation of the response. Three factors account for this rapidity: (1) both the sensory and the motor axons involved are myelinated and of large diameter, so they conduct rapidly; (2) the sensory cells synapse directly on the motor neurons; and (3) both the central synapse and the neuromuscular junction are fast, ionotropic synapses.

For some purposes the afferent (input) parts of the visual system can be represented as a neural chain (**FIGURE 2.21A**) (in reality, however, the retina contains many kinds of neural circuits, which we will discuss in Chapter 10). A more accurate schematic diagram of the visual system (**FIGURE 2.21B**) highlights two other features that are common to many kinds of neural circuits: **convergence** and **divergence**. In many parts of the nervous system, the axons from large numbers of neurons converge on a small number of cells; in each human eye, about 100 million receptor cells concentrate their information on about 1 million axons that carry the information from the eye to the brain (see Figure 2.21B). Higher in the visual system there is much divergence; the 1 million axons of the optic nerve communicate to billions of neurons in several different specialised regions of the cerebral cortex.

2.4 Gross Electrical Activity of the Brain Is Readily Detected

Learning Objectives

After reading this section, you should be able to:

2.4.1 **Describe the apparatus used to record widespread electrical signals from the brain.**

2.4.2 **Understand the basis of seizures and list the major categories of epilepsy.**

2.4.3 **Explain how electrical recordings of the brain in response to stimuli can reveal brain function.**

2.4.4 **Describe exciting new technology to control brain function using light.**

The electrical activity of millions of cells working together combines to produce electrical potentials large enough that we can detect them at the surface of the skull. Such recordings of electrical activity in the brain, made with large electrodes either on the scalp or within the brain, can provide useful glimpses of the simultaneous workings of large populations of neurons (A. F. Jackson and Bolger, 2014). Investigators divide

convergence The phenomenon of neural connections in which many cells send signals to a single cell.

divergence The phenomenon of neural connections in which one cell sends signals to many other cells.

electroencephalogram (EEG) A recording of gross electrical activity of the brain recorded from electrodes placed on the scalp.

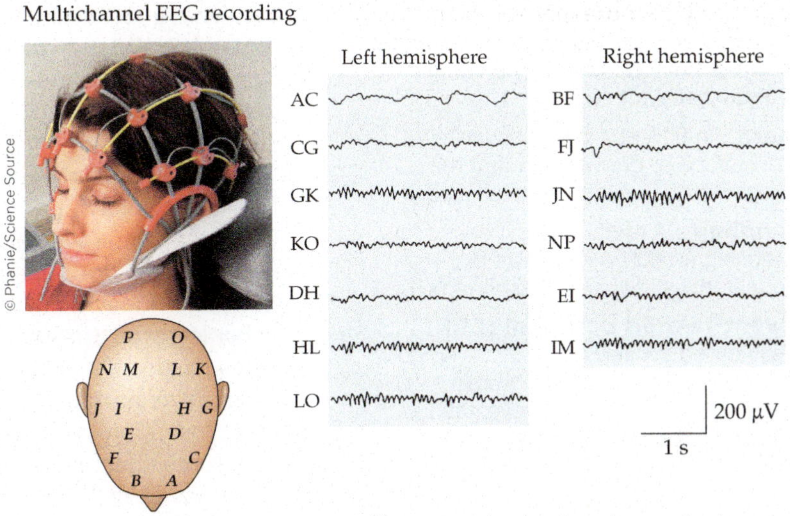

2.22 Gross Potentials of the Human Nervous System (Top left) Electrode array for EEG recording. (Bottom left) Each electrode can be assigned a letter on a map of the scalp. (Right) Typical EEG recordings showing potential measured between various points on the scalp.

these gross brain potentials into two principal classes: those that appear spontaneously without specific stimulation, and those that are evoked by particular stimuli.

Electroencephalograms offer a window into brain function

A recording of spontaneous brain potentials, or *brain waves*, is called an **electroencephalogram (EEG)** (**FIGURE 2.22**). As we will see in Chapter 14, EEG recordings of a sleeping person allow investigators to detect different kinds and stages of sleep. Brain potentials also provide significant diagnostic data—for example, they may offer predictions about the functional effects of brain injury (Conley et al., 2018). EEGs also help distinguish types of seizure disorders, as we discuss next.

epilepsy A brain disorder marked by major sudden changes in the electrophysiological state of the brain that are referred to as *seizures*.

seizure An epileptic episode.

tonic-clonic seizure Formerly called *grand mal seizure*. A type of generalised epileptic seizure in which nerve cells fire in high-frequency bursts.

simple partial seizure Also called *absence attack*. A seizure that is characterised by a spike-and-wave EEG and often involves a loss of awareness and inability to recall events surrounding the seizure.

Seizure disorders result from electrical storms in the brain

Epilepsy (from the Greek *epilepsia*, a form of the verb meaning "to seise") has provoked wonder and worry since the dawn of civilisation. Through the ages the seizures that accompany this disease have spawned much speculation about the cause—from demons to gods. At least 50 million people, worldwide, have epilepsy (Behr et al., 2016).

Seizures are an unfortunate manifestation of the electrical character of the nervous system. Because of the extensive connections among nerve cells, the brain can generate massive waves of intense nerve cell activity that seem to involve almost the entire brain. In the normal, active brain, electrical activity tends to be desynchronised; that is, different brain regions carry on their functions more or less independently. In contrast, a **seizure** features widespread synchronisation of electrical activity: broad swaths of the brain start firing in simultaneous waves of excitation, which are evident in the EEGs as an abnormal "spike-and-wave" pattern of brain activity. Many abnormalities of the brain, such as trauma, injury, or metabolic problems, can predispose brain tissue to produce such synchronised activity, which can easily spread.

There are several major categories of seizure disorders. Generalised seizures are characterised by loss of consciousness and symmetrical involvement of body musculature. In **tonic-clonic seizures** (previously known as *grand mal seizures*), abnormal EEG activity is evident all over the brain (**FIGURE 2.23A**). The person loses consciousness and makes characteristic movements: an enduring *tonic* contraction of the muscles for 1 or 2 minutes, followed by jerky, rhythmic *clonic* contractions and relaxations. Minutes or hours of confusion and sleep follow the seizure.

Simple partial seizures (also known as *absence attacks*) are a more subtle variant of generalised seizures,

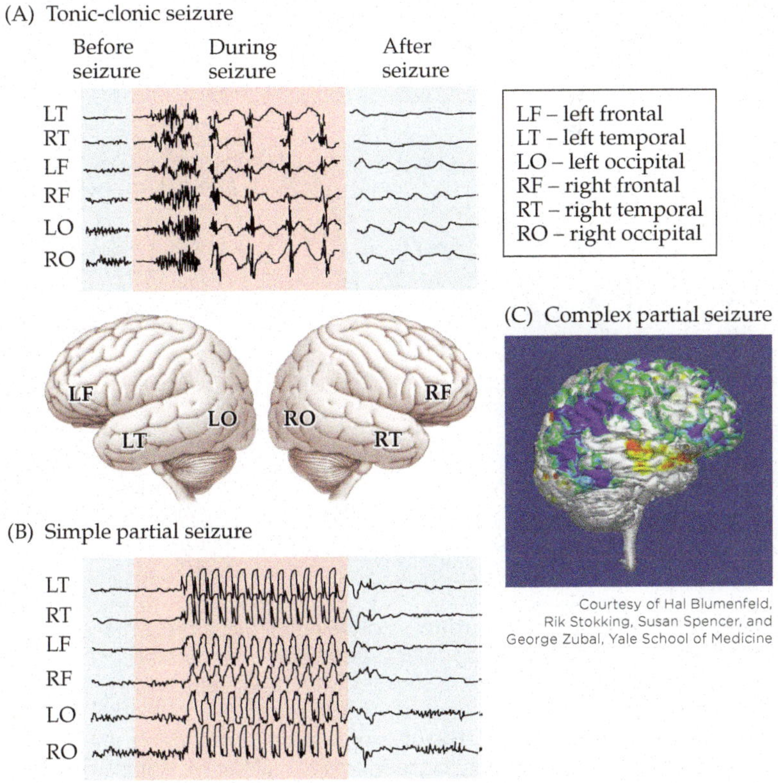

2.23 Discharge Patterns during Seizures

complex partial seizure In epilepsy, a type of seizure that doesn't involve the entire brain and therefore can cause a wide variety of symptoms.

aura In epilepsy, the unusual sensations or premonition that may precede the beginning of a seizure.

kindling A method of experimentally inducing an epileptic seizure by repeatedly stimulating a brain region.

event-related potential (ERP) Also called *evoked potential*. Averaged EEG recordings measuring brain responses to repeated presentations of a stimulus.

in which the characteristic spike-and-wave EEG activity is evident for 5–15 seconds at a time (**FIGURE 2.23B**), sometimes occurring many times per day. The person is unaware of the environment during these periods and later cannot recall events that occurred during the simple partial episode. Behaviourally, the person does not show unusual muscle activity, except for a cessation of ongoing activity and sustained staring.

Complex partial seizures do not involve the entire brain and thus can produce a wide variety of symptoms, often preceded by an unusual sensation, or **aura**. In one example, a woman felt an unusual sensation in the abdomen, a sense of foreboding, and tingling in both hands before the seizure spread. At the height of it, she was unresponsive and rocked her body back and forth while speaking nonsensically, twisting her left arm, and looking toward the right. **FIGURE 2.23C** is a three-dimensional reconstruction showing where the seizures occurred in her brain. In some individuals, complex partial seizures may be provoked by environmental stimuli and may produce strikingly abnormal behaviour.

Seizures affect nonhuman animals too, and such cases are studied as a model of human epilepsy. In **kindling** (Gorter et al., 2016), animals receive repeated electrical stimulation that is too weak to cause a seizure on its own. Although the individual stimuli are small, eventually their effects accumulate to cause spontaneous seizures. In other words, the kindling stimulations somehow change the tissue and make it more epilepsy prone. Interestingly, after years of epilepsy some human patients develop multiple foci for the initiation of seizures, perhaps because of a kindling process (Avansini et al., 2013).

Many seizure disorders can be effectively controlled with the aid of antiepileptic drugs. Although these drugs have a wide variety of different targets, they have in common a tendency to selectively modulate the excitability of neurons, either by counteracting problems with ionic balance (see Box 2.1) or by promoting inhibitory processes (Sankaraneni and Lachhwani, 2015). As we learned in Chapter 1, in serious cases of epilepsy that don't respond to medication, the patient may choose to determine the part of the brain where the seizures begin and have it surgically removed. This was the case with Deidre, whom we met at the start of the chapter.

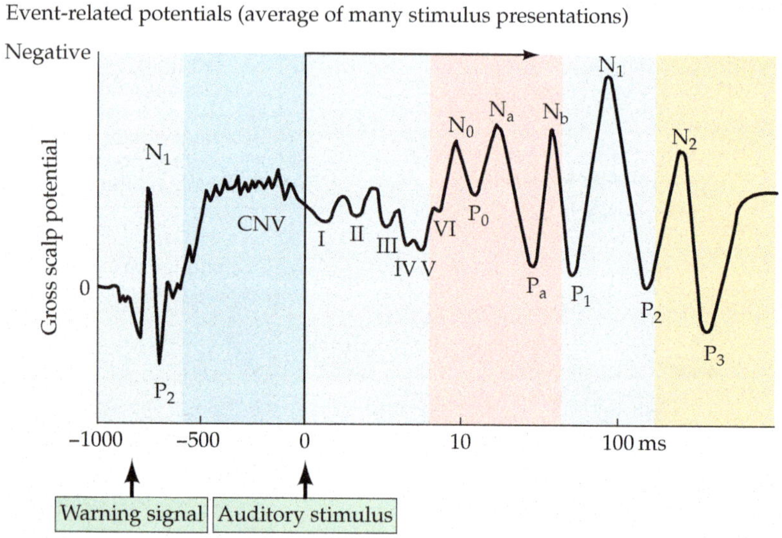

2.24 Event-Related Potentials Following stimulus presentation, a fixed sequence of processing-related potentials is generated. Early components (labelled I–VI) are associated with brainstem activity, followed by large-amplitude negative- and positive-voltage events (labelled N0–N2 and P0–P3). The later-appearing components are associated with cognitive processing in the cortex. (After M. Kutas and S. A. Hillyard, 1984. *Handbook of Cognitive Neuroscience*. Plenum Press: New York.)

Event-related potentials measure changes resulting from discrete stimuli

Gross potential changes evoked by discrete sensory stimuli, such as light flashes or clicks, are called **event-related potentials** (**ERPs**) (**FIGURE 2.24**). Typically, many ERPs are averaged to obtain a reliable estimate of stimulus-elicited brain activity. Sensory-evoked potentials have very distinctive characteristics of wave shape and latency (time delay) that reflect the type of stimulus, the state of the individual reacting to the stimulus, and the site of recording (Luck, 2014).

Computer techniques enable researchers to record brain potentials using electrodes that are located far from the sites at which the potentials are generated. For example, auditory-evoked potentials from the brainstem (see Figure 2.24) can be recorded through electrodes on the scalp. Decreases in the amplitude of certain waves or increases in their latency have been valuable for the detection of hearing impairments in very young children and noncommunicative persons (**FIGURE 2.25**). Infants with impaired hearing produce reduced auditory ERPs or no ERP at all in response to sounds (Musacchia et al., 2015). ERPs are also used in the assessment of brainstem injury or damage.

The long-latency components of scalp-recorded ERPs reflect the operation of information-processing mechanisms of the brain, such as attention, decision making, and other complex cognitive processes, as we'll discuss in Chapter 17. In contrast, short-latency responses are determined more by exogenous factors, such as the physical characteristics of the stimulus. For example, dimensions like stimulus intensity have a bigger effect on early components of ERPs than on longer-latency components.

Although it is usually difficult to localise which brain region has produced a given component of the ERP, such changes are detected quickly, within a fraction of a second. In contrast, computer-coordinated imaging of brain activity, such as functional MRI (fMRI) (see Chapter 1), indicates which brain region is active, but because such imaging techniques must average activity over seconds or minutes, they are slower than ERPs. Perhaps the greatest promise for future research is a melding of the two techniques—combining techniques that measure rapid changes in electrical activity, such as ERPs and magnetoencephalography (MEG) (see Chapter 1), with slower but higher-resolution techniques such as fMRI or PET scans to identify probable sites of origin of the evoked activity (Vitali et al., 2015). These techniques have given us a better glimpse of brain functioning than Otto Loewi ever dreamed of.

Courtesy of Brett Martin, Jen Gerometta, and Christine Rota-Donahue

2.25 Did You Hear That? When hearing is normal, presentation of sounds should evoke an ERP from auditory centres in an infant's brain.

The Cutting Edge

Optogenetics: Using Light to Probe Brain-Behaviour Relationships

A remarkable new tool makes it possible to precisely stimulate or inhibit electrical activity in the brain. **Optogenetics** uses genetic tools to insert light-sensitive ion channels into neurons so that stimulating the brain with light, delivered by fibre-optic cables, can excite or inhibit those targeted neurons (Tye and Deisseroth, 2012). This breakthrough was made possible by the discovery that various microbial organisms, such as some algae and bacteria, produce light-sensitive proteins, called *opsins*, that resemble mammalian opsins, such as those found in light-receptor cells in our eyes. Unlike those mammalian opsins underlying vision, which rely on other signalling proteins that we'll describe in Chapter 10, these microbial opsins can themselves open an ion channel in response to light. The first opsin that was studied, **channelrhodopsin**, responds to blue light by allowing Na+ ions to enter the cell, depolarising it (Kato et al., 2012). Another example is **halorhodopsin**, which when stimulated by yellow light, pumps Cl– ions into the cell, hyperpolarising it (Wiegert et al., 2017) (**FIGURE 2.26**).

optogenetics The introduction of gene(s) into neurons that makes them electrically respond to light.

channelrhodopsin An ion channel protein that, when stimulated by light, depolarises neurons.

halorhodopsin An ion channel protein that, when stimulated by light, hyperpolarises neurons.

In the microbes that first made them, these opsins guide light-directed growth. But when neurons are induced to make these opsins, it becomes possible to excite or inhibit the neuron's electrical activity simply by illuminating the cell with a blue or yellow light, respectively. For example, a scientist might inject into one part of an animal's brain a virus that has been engineered to infect neurons and cause them to make channelrhodopsin. Then a tiny fibre-optic probe can be surgically implanted to provide light stimulation to that same region. When the animal recovers, it can move freely about, and whenever the investigator chooses to direct blue light to the end of the probe, that will stimulate the neurons to fire (**FIGURE 2.27A**).

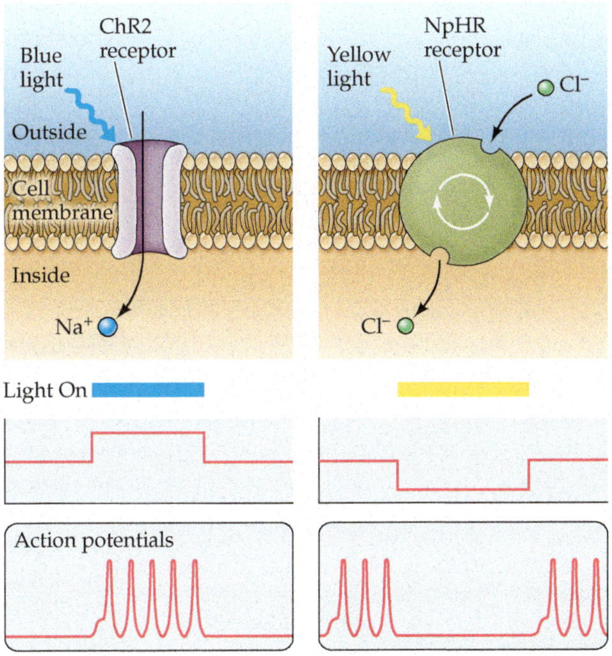

2.26 Optogenetic Proteins The protein called *channelrhodopsin* (ChR2), when stimulated by blue light, opens up a channel to let Na^+ ions into the neuron, depolarising it to cause it to fire. In contrast, halorhodopsin (NpHR), in response to yellow light, pumps Cl– ions into the neuron, hyperpolarising it and preventing it from reaching threshold. (After F. Zhang et al., 2007. *Nat Rev Neurosci* 8: 577–581; based on M. Häusser and S. L. Smith, 2007. *Nature* 446: 617–619; F. Zhang, 2007. *Nature* 446: 633–639.)

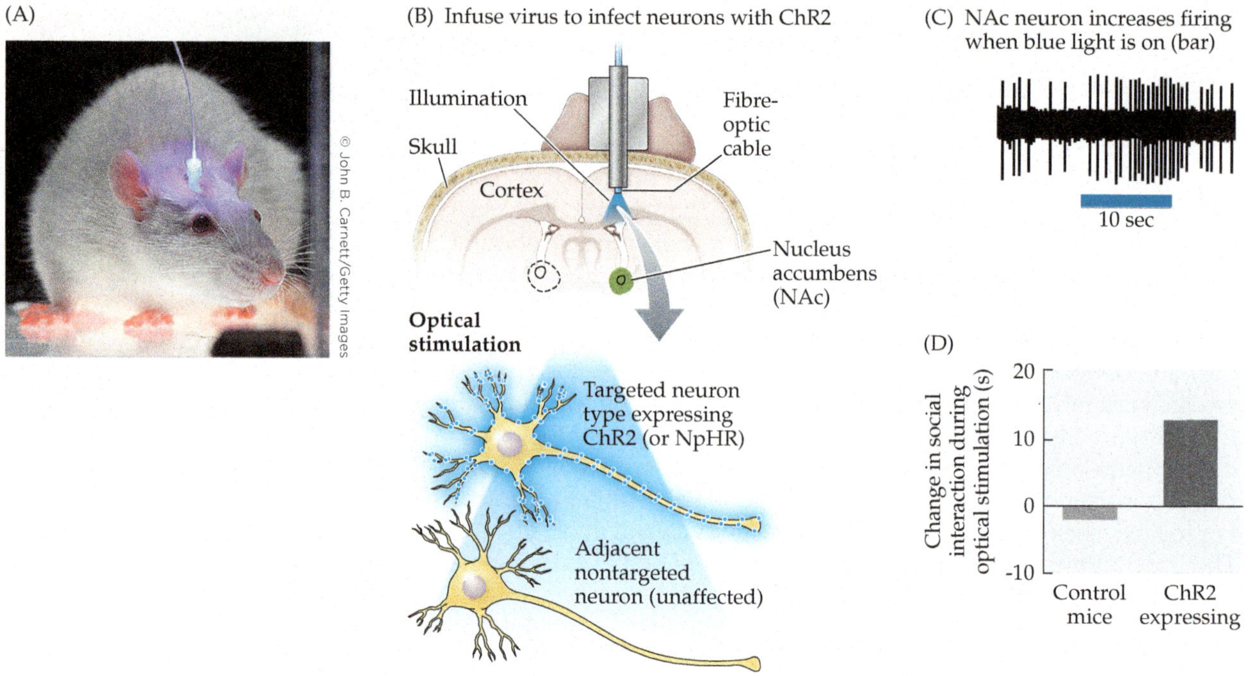

2.27 Optogenetic Tools (A) Once the animal recovers from implant surgery, it can move freely about. The researcher can stimulate the particular class of cells making the opsin during spontaneous behaviour. (B) Using genetic methods, scientists can arrange for only certain types of neurons to make the opsin so that only those cells will respond to the light stimulation. (C) In this example, the neurons in the nucleus accumbens reliably increase their firing when illuminated with blue light for 10 seconds (bar). (D) Prolonged light stimulation of the nucleus accumbens increased the social behaviour in mice expressing channelrhodopsin (ChR2) there, but not in control mice. (B, C, and D after L. A. Gunaydin et al., 2014. *Cell* 157: 1535–1551; B also after F. Zhang et al., 2007. *Nat Rev Neurosci* 8: 577–581.)

Thus far, optogenetics might seem no different from using electrical wires to stimulate the brain, as with Deidre at the start of the chapter. The difference is that in optogenetic approaches, one can use genetic tricks so that, for example, only neurons that use GABA as a neurotransmitter will make the opsin. Then shining the light will excite only those neurons, not their thousands of neighbors that use other transmitters (**FIGURE 2.27B**). Or the investigator can arrange it so that only those neurons in brain region *X* that send axons to brain region *Y* will make the opsin, and therefore only those will respond to the light. One can even use channelrhodopsin and halorhodopsin in the same animal so that sending blue light into the fibre-optic cable into the brain excites the neurons, while shining yellow light shuts them down. Or one can have one class of neurons respond to the blue light while another responds to yellow (C. K. Kim et al., 2017). In one example, shining light into the nucleus accumbens increased the firing of the targeted neurons there (**FIGURE 2.27C**), and prolonged light stimulation increased social behaviour in the mice (**FIGURE 2.27D**) (Gunaydin et al., 2014). These powerful techniques allow unprecedented control of neural circuits and promise to cast new light on the relationship between brain and behaviour. ■

Recommended Reading

Kandel, E. R., Schwartz, J. H., Jessell, T. M, Siegelbaum, S.A., et al. (2013). *Principles of Neural Science* (5th ed.). New York: McGraw-Hill.

Nicholls, J. G., Martin, A. R., Fuchs, P. A., Brown, D. A., et al. (2012). *From Neuron to Brain* (5th ed.). Sunderland, MA: Oxford University Press/Sinauer.

Purves, D., Augustine, G. J., Fitzpatrick, D., LaMantia, A.-S., et al. (2017). *Neuroscience* (6th ed.). Sunderland, MA: Oxford University Press/Sinauer.

Valenstein, E. S. (2005). *The War of the Soups and the Sparks: The Discovery of Neurotransmitters and the Dispute over How Neurons Communicate.* New York: Columbia University Press.

Zheng, J., and Trudeau, M. C. (Eds.). (2017). *Handbook of Ion Channels.* Boca Raton, FL: CRC Press.

2 Visual Summary

You should be able to relate each summary to the adjacent illustration, including structures and processes. You can also refer back to the links to figures, animations, and activities that will help you consolidate the material.

1 Chemical signals transmit information between neurons; electrical signals transmit information within a neuron. Neurons exhibit a small **electrical potential** difference across the cell membrane; neural signals are changes in this **resting membrane potential**. Review Figure 2.1, Animation 2.1

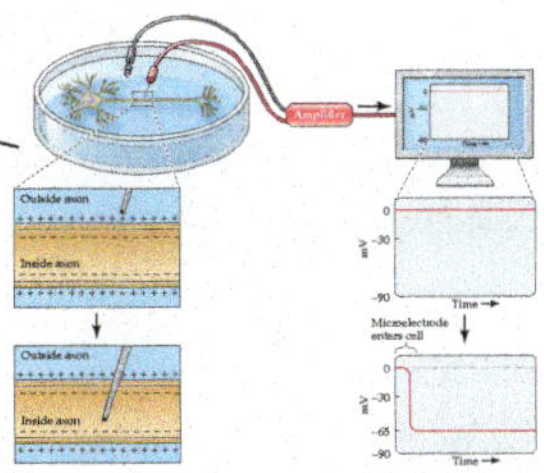

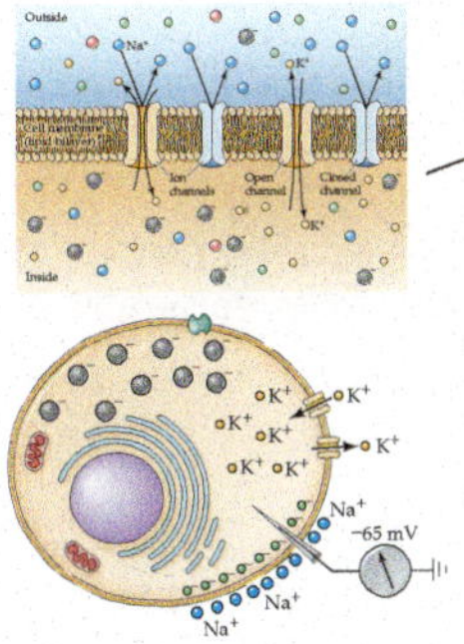

2 Different concentrations of **ions** inside and outside the neuron—especially **potassium ions (K^+)**, to which the resting membrane is **selectively permeable**—account for the resting membrane potential. At the **K^+ equilibrium potential**, the **electrostatic pressure** pulling K^+ ions into the neuron is balanced by the **concentration gradient** pushing them out; at this point, the resting membrane potential is about –65 mV. Review Figures 2.2–2.4, Activity 2.1

3 **Depolarisation** (reduction of the resting potential) of an axon until it reaches a **threshold** value opens **voltage-gated sodium (Na^+) channels**, making the membrane completely permeable to **sodium ions (Na^+)**. The Na^+ ions rush in, and the axon becomes briefly more positive inside than outside. This event is called an **action potential**. Review Figure 2.5

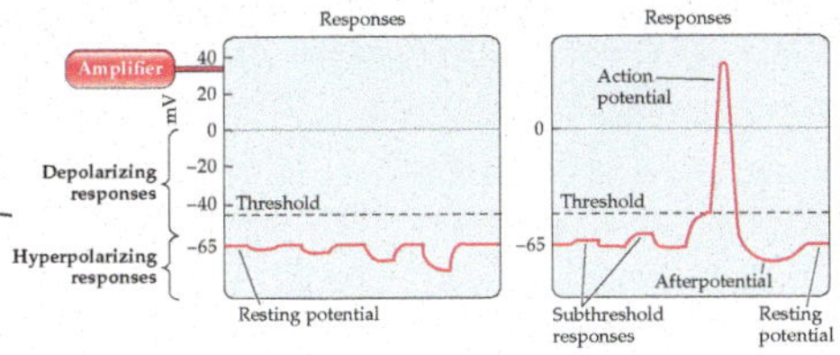

4 Following the action potential, the resting potential is quickly restored by the outflow of K^+ ions. The **sodium-potassium pumps** maintain the resting potential in the long run, counteracting the influx of Na^+ ions during action potentials. Review Figure 2.6, Animation 2.2

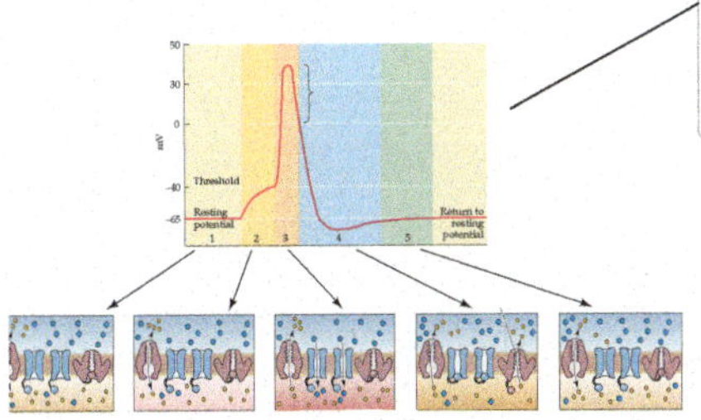

5 The action potential strongly depolarises the adjacent patch of axonal membrane, causing it to generate its own action potential. In this regenerative manner, the action potential spreads down the axon. **Saltatory conduction** of the action potential along the **nodes of Ranvier** between **myelin** sheaths speeds propagation down the axon. Review Figure 2.8, Box 2.2, Animation 2.3

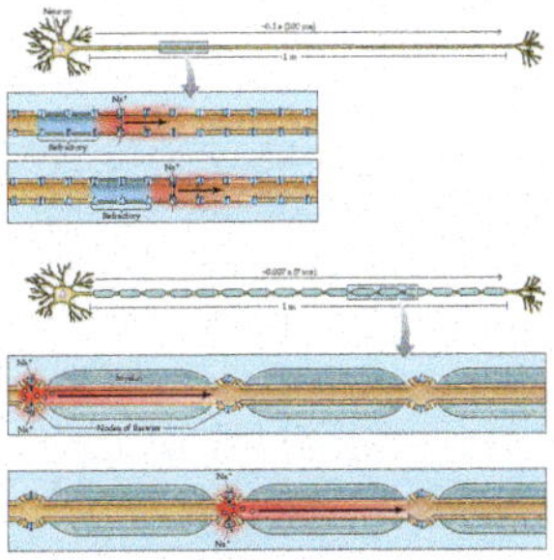

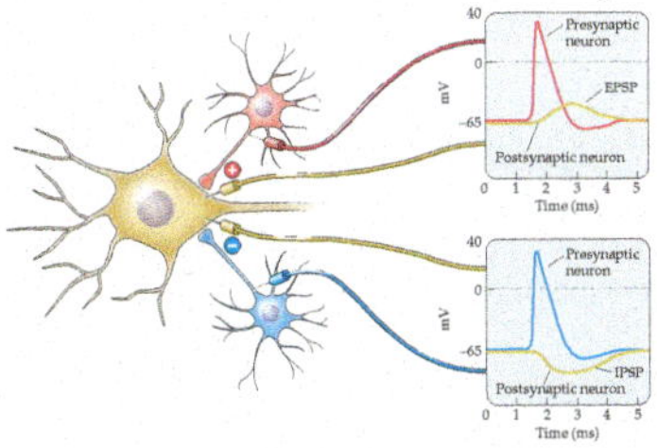

6 Like all other **local potentials, postsynaptic potentials** spread very rapidly but are not regenerated, so they diminish as they spread passively along dendrites and the cell body. **Excitatory postsynaptic potentials (EPSPs)** are depolarising (they decrease the resting potential), increasing the likelihood that the neuron will fire an action potential. **Inhibitory postsynaptic potentials (IPSPs)** are **hyperpolarising** (they make the cell more polarised), decreasing the likelihood that the neuron will fire. Review Figure 2.9

7 Neurons process information by integrating (summing algebraically) the postsynaptic potentials through both **spatial summation** (summing potentials from different locations) and **temporal summation** (summing potentials across time). Review Figure 2.11, Animation 2.4

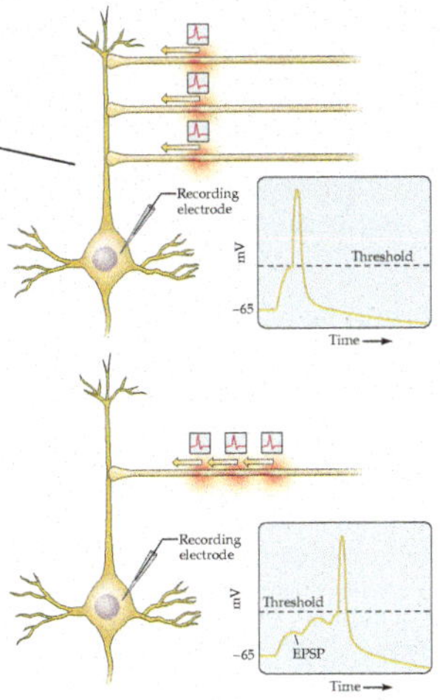

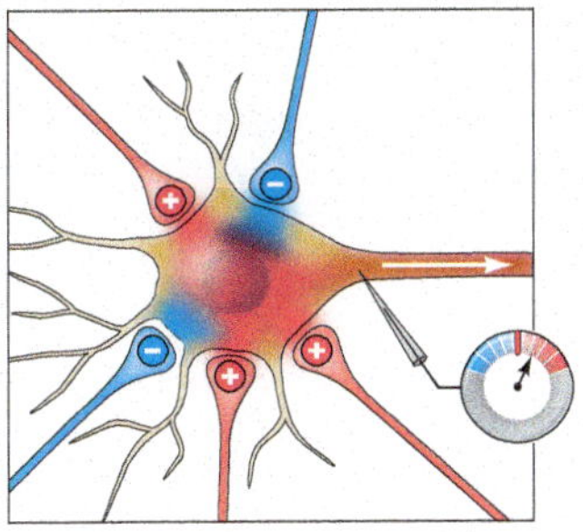

8 Action potentials are initiated just past the **axon hillock** when the excess of EPSPs over IPSPs reaches threshold. During the action potential, the neuron cannot be excited by a second stimulus; it is **absolutely refractory**. For a few milliseconds afterward, the hyperpolarised neuron is **relatively refractory**, requiring a stronger stimulation than usual in order to fire. Review Figure 2.10

(Continued)

9 Synaptic transmission occurs when a chemical neurotransmitter diffuses across the **synaptic cleft** and binds to **neurotransmitter receptors** in the postsynaptic membrane. **Ionotropic receptors** contain an ion channel; **metabotropic receptors** use **G proteins** to either open an ion channel or activate **second messengers** to affect the target cell. Review **Figures 2.12**, **2.13**, **2.17**, **Animations 2.5, 2.6**

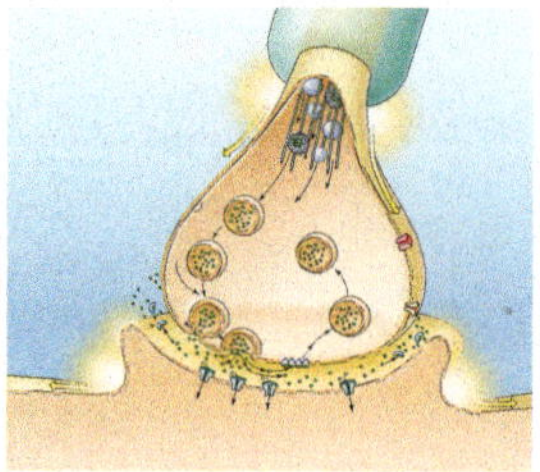

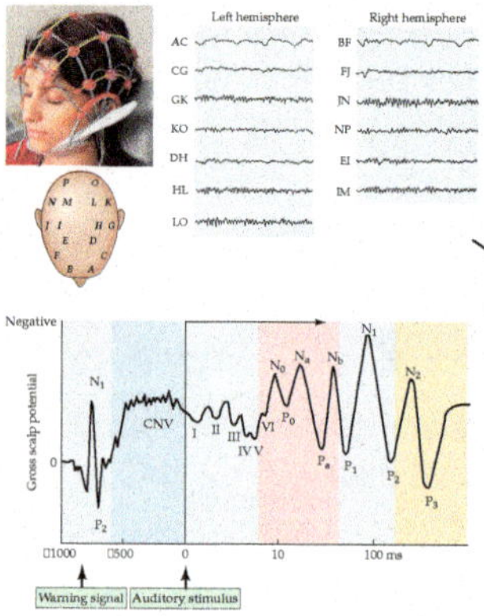

10 Summing electrical activity over millions of nerve cells as detected by electrodes on the scalp, **electroencephalograms (EEGs)** can reveal rapid changes in brain function—for example, in response to a brief, controlled stimulus that evokes an **event-related potential (ERP)**. They can also reveal the highly synchronised electrical outbursts of a **seizure** in people with **epilepsy**. Review **Figures 2.22–2.24**

The Chemistry of Behaviour

Neurotransmitters and Neuropharmacology

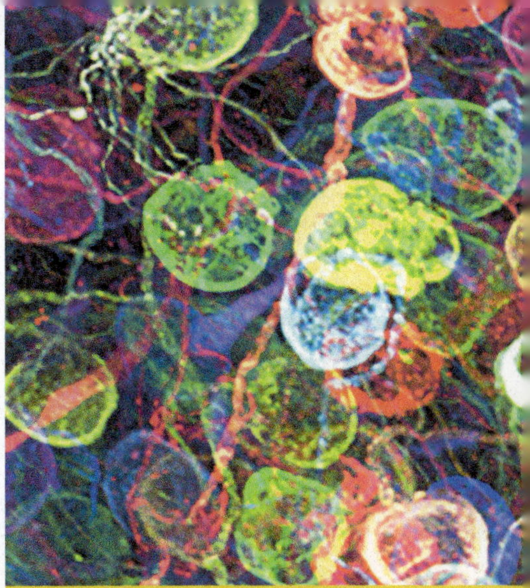

Courtesy of Dr. Ryo Egawa

3

The Birth of a Pharmaceutical Problem Child

Swiss pharmacologist Albert Hofmann was working for Sandoz Pharmaceuticals and studying compounds derived from ergot, a fungus that grows on grain, in the hope of synthesising new and useful drugs from it. One day in April of 1943, Dr. Hofmann began to feel unwell at work and, believing he was coming down with a cold, left for home. Soon, however, he began experiencing bizarre visual phenomena: "an uninterrupted stream of fantastic pictures, extraordinary shapes with intense, kaleidoscopic play of colours" (A. Hofmann, 1981). When he closed his eyes, the luridly coloured, oddly shifting forms seemed to surge toward him. The state lasted about 2 hours. Suspecting that he had accidentally ingested a small amount of an experimental compound, Hofmann began to research its properties, intentionally taking larger doses and noting the details of its effects.

We can debate the wisdom of testing unknown drugs on yourself, but Hofmann soon realized he had a revolutionary drug on his hands. Even a tiny dose of this stuff caused incapacitating changes in brain function and astonishing, sometimes frightening, visual experiences. In addition to the visual phenomena, he reported sometimes feeling that his sense of self was "loosened." Careful analyses confirmed the new drug to be amazingly potent: a few millionths of a gram was enough to induce substantial effects. What was this substance? How can such a tiny amount of a drug exert so large an effect on the brain?

As far back as we can trace human history, people have experimented with all kinds of **exogenous** substances—animal, vegetable, and mineral compounds from *external* sources—in order to change the functioning of their bodies and brains. From these experiences, people have learned to consume some substances and shun others. Social customs and dietary codes evolved to help people benefit from helpful substances and to protect them from consuming toxins. Our forebears sipped, swallowed, and smoked their way to euphoria, calmness, pain relief, and hallucination. They discovered deadly poisons in frogs, miraculous antibiotics in mold, powerful painkillers in poppies, and all the rest of a vast catalogue of helpful and harmful substances.

The brain is an electrochemical system, so it's no surprise that most drugs that affect the nervous system do so by altering brain chemistry and synaptic transmission. We begin our tour of **neurochemistry** and **neuropharmacology** by briefly reviewing the structure and function of the synapse, before delving more deeply into the neurotransmitter mechanisms that we introduced in Chapter 2. Then we discuss many of the major classes of drugs that affect the nervous system and behaviour, before we finally turn to mechanisms of drug abuse and dependency.

exogenous Arising from outside the body.

neurochemistry The branch of neuroscience concerned with the fundamental chemical composition and processes of the nervous system.

neuropharmacology Also called *psychopharmacology*. The scientific field concerned with the discovery and study of compounds that selectively affect the functioning of the nervous system.

3.1 Synaptic Transmission Involves a Complex Electrochemical Process

Learning Objectives

After reading this section, you should be able to:

3.1.1 **Recap the processes leading up to the production of an action potential.**

3.1.2 **Describe the two main families of neurotransmitter receptors.**

3.1.3 **Explain how the existence of subtypes of receptors that vary in anatomical distribution adds complexity to neural processes.**

As we learned in Chapter 3, the typical neuron integrates a variety of inputs, and if sufficiently excited (i.e., depolarised), it fires a distinctive brief electrical impulse, called an *action potential*, that rapidly sweeps down the length of the axon toward the axon terminals. **FIGURE 3.1** recaps the events that follow the arrival of the action potential. First, because the action potential strongly depolarises the axon terminal, voltage-gated calcium (Ca^{2+}) channels in the terminal membrane are induced

3.1 Synapses Convert Electrical Signals into Chemical Signals

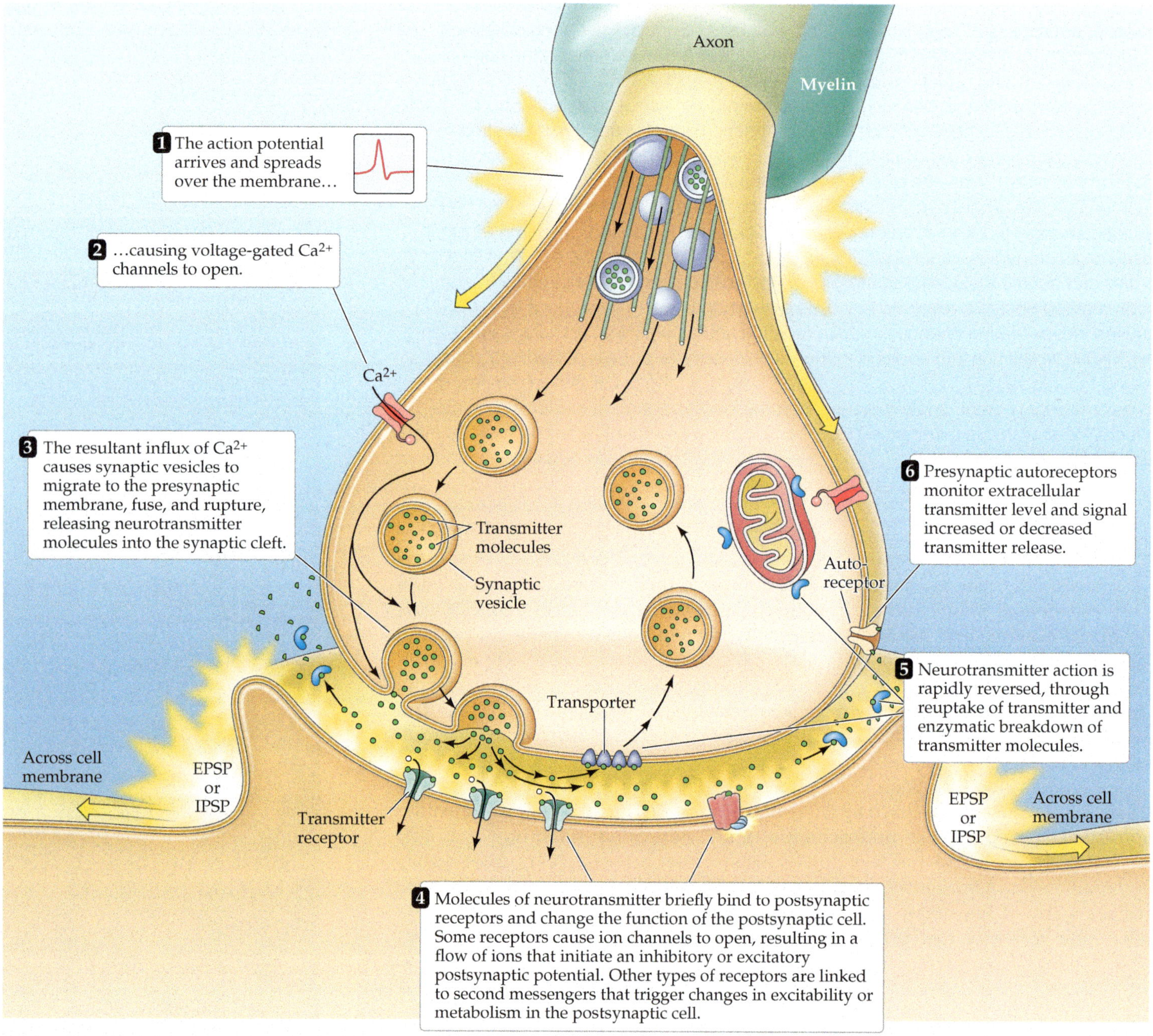

to open. The resultant influx of Ca^{2+} ions triggers the SNAREs, large proteins on the presynaptic membrane and on the synaptic vesicles that cause the vesicles to fuse to the presynaptic membrane (see Figure 2.13) and release their cargo of *neurotransmitter* molecules into the synaptic cleft (a process called *exocytosis*). In Chapter 2 we also saw that following their diffusion across the cleft, neurotransmitter molecules briefly bind to their corresponding **receptors**, which then mediate a response on the postsynaptic side. Receptors are protein molecules embedded in the postsynaptic membrane that recognize a specific transmitter. The transmitter molecule binds to the receptor, either changing its shape to open an ion channel, as with fast, **ionotropic receptors**, or altering chemical reactions within the target cell, as with slow, **metabotropic receptors** (see Figure 3.15).

Receptors add an important layer of complexity in neural signalling, because any given transmitter may affect various kinds of receptors that differ from one another in structure. This diversity of **receptor subtypes** is true for both metabotropic receptors and ionotropic receptors. In mammals, a gene superfamily encodes hundreds of different **G protein-coupled receptors (GPCRs)**, many of which are metabotropic neurotransmitter receptors. In the case of ionotropic receptors, families of genes encode a variety of protein subunits that combine to make up the ion channels at the core of the receptors. The characteristics of each subtype of ionotropic receptor—the specific neurotransmitter it recognizes and the type of ions that it selectively conducts—are determined by the unique combination of subunits that make up the receptor. Furthermore, neurons may vary these combinations over time, thereby altering the functioning of synapses—a type of neuroplasticity (Bar-Shira et al., 2015). To get a sense of this complexity, consider that there are at least 14 different subtypes of receptors for the neurotransmitter serotonin, 5 for dopamine, and 5 for norepinephrine.

A neurotransmitter's different receptor subtypes may trigger very different responses in target cells (**FIGURE 3.2**), and in many cases they are also distributed differently within the nervous system (e.g., Beliveau et al., 2017). As we will see shortly, drug development capitalises on the existence and varying neuroanatomical distribution of receptor subtypes. Although a given neurotransmitter will interact with *all* the subtypes of its receptors, it is possible to design drugs that primarily affect a specific subtype, thereby producing the specific effects associated with that receptor subtype.

receptor Also called *receptor molecule*. A protein that binds and reacts to molecules of a neurotransmitter or hormone.

ionotropic receptor A receptor protein that includes an ion channel that is gated by the binding of neurotransmitter molecules.

metabotropic receptor A receptor protein that does not contain an ion channel but may, when activated, use a second messenger system to alter the functioning of the postsynaptic cell.

receptor subtype Any type of receptor having functional characteristics that distinguish it from other types of receptors for the same neurotransmitter.

G protein-coupled receptor (GPCR) A cell surface receptor that, when activated extracellularly, initiates G protein signalling mechanisms inside the cell.

3.2 The Versatility of Neurotransmitters A single neurotransmitter may interact with many different receptor subtypes in different parts of the brain—binding to fast, ionotropic receptors on some target cells and to slow, metabotropic receptors on other cells. Both types of receptors may either excite or inhibit the target cell.

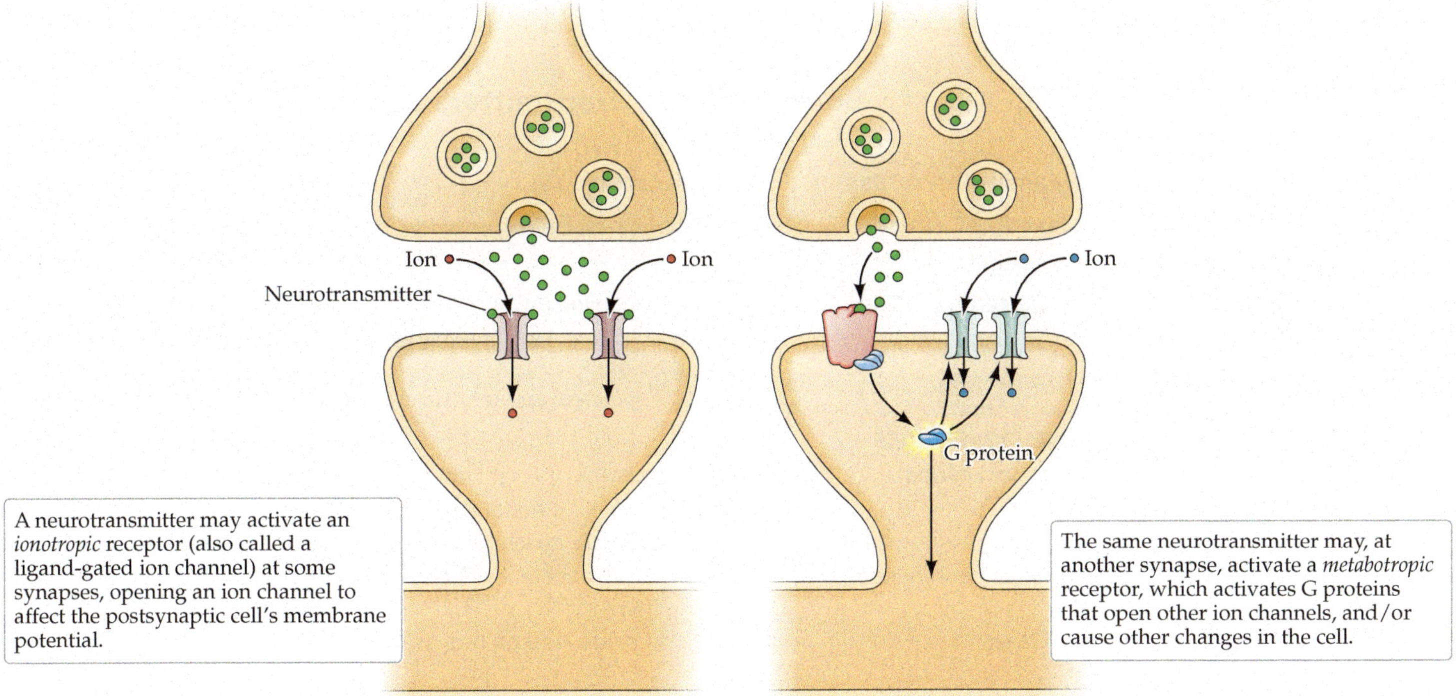

3.2 Many Neurotransmitters Have Been Identified

Learning Objectives

After reading this section, you should be able to:

3.2.1 **Summarize the criteria for establishing that a substance acts as a neurotransmitter.**

3.2.2 **Name the major chemical families of transmitters, and give examples of each.**

We learned in Chapter 2 that each neuron integrates electrical information from many synapses and may then release a chemical—an **endogenous** substance (a substance from an internal source) called a *neurotransmitter*—that communicates the result of that information processing to the postsynaptic cell. As you might have guessed, most drugs that affect behaviour do so by altering this chemical communication process at millions, or even billions, of synapses. Identification of neurotransmitters, their receptors, and the effects of drugs that modify neurotransmission is one of the most active areas of research in all of neuroscience.

To be considered a classic neurotransmitter, a substance should meet the following criteria:

- The substance exists in presynaptic axon terminals.
- The presynaptic cell contains appropriate enzymes for synthesising the substance.
- The substance is released in significant quantities when action potentials reach the terminals.
- Specific receptors that recognize the released substance exist on the postsynaptic membrane.
- Experimental application of the substance produces changes in postsynaptic cells.
- Blocking release of the substance prevents presynaptic activity from affecting the postsynaptic cell.

endogenous Produced inside the body.

amine neurotransmitter A neurotransmitter based on modifications of a single amino acid.

amino acid neurotransmitter A neurotransmitter that is itself an amino acid.

peptide neurotransmitter Also called *neuropeptide*. A neurotransmitter consisting of a short chain of amino acids.

TABLE 3.1 summarizes the major categories of some of the many neurotransmitters presently known. Substances that meet these criteria and are thus considered classic transmitters include various **amine neurotransmitters**, such as acetylcholine, dopamine, and serotonin; **amino acid neurotransmitters**, like GABA and glutamate; and a wide variety of **peptide neurotransmitters** (or *neuropeptides*), made up of short chains of amino acids. As the search goes on, the number of probable

TABLE 3.1 Some Synaptic Transmitters and Families of Transmitters

Family and subfamily	Transmitter(s)
AMINO ACIDS	Gamma-aminobutyric acid (GABA), glutamate, glycine, histamine
AMINES	
Quaternary amines	Acetylcholine (ACh)
Monoamines	*Catecholamines*: norepinephrine (NE), epinephrine (adrenaline), dopamine (DA)
	Indoleamines: serotonin (5-hydroxytryptamine; 5-HT), melatonin
NEUROPEPTIDES	
Opioid peptides	*Enkephalins*: met-enkephalin, leu-enkephalin
	Endorphins: β-endorphin
	Dynorphins: dynorphin A
Other neuropeptides	Oxytocin, substance P, cholecystokinin (CCK), vasopressin, neuropeptide Y (NPY), hypothalamic releasing hormones
Gases	Nitric oxide, carbon monoxide

Go to Activity 3.1
Families of Transmitters

synaptic transmitters continues to grow, and the effort occasionally yields surprises like the **gas neurotransmitters**: soluble gases that diffuse between neurons to alter ongoing processes.

Even if a substance is known to be a transmitter in one location, proving that it acts as a transmitter at another location may be difficult. For example, acetylcholine was long known to be a transmitting agent in the peripheral nervous system, but it was harder to prove that it serves as a transmitter in the central nervous system as well. Now it is recognized that acetylcholine is widely distributed in the brain, and many scientists study its possible relationship to the cognitive deficits seen in Alzheimer's disease. Considering the rate at which these substances are being discovered and characterised, it would not be surprising if there turned out to be several hundred different neurotransmitters conveying information at synapses in different subsets of neurons.

3.3 Neurotransmitter Systems Form a Complex Array in the Brain

Learning Objectives

After reading this section, you should be able to:

3.3.1 **Review the structure, distribution, and functions of the peptide neurotransmitters.**

3.3.2 **Categorize and describe the synthesis and distribution of acetylcholine, the catecholamines, and the indoleamines.**

3.3.3 **Discuss atypical aspects of the synthesis and function of the diffusible gas transmitters.**

3.3.4 **Review major behavioural functions associated with each type of transmitter.**

Powerful methods for probing the composition of neural tissue (see Box 2.1) reveal that brain activity depends on a remarkably diverse assortment of neurotransmitters, distributed through intricate anatomical networks that overlap and interact in highly complex ways. Although at one time it was thought that each neuron contained only one transmitter, we now know that some neurons contain more than one—a phenomenon known as neurotransmitter **co-localisation** or *co-release*. In this section we discuss the distribution of just a few of the major neurotransmitters, and their receptors.

The most abundant excitatory and inhibitory neurotransmitters in the brain are amino acids

The most plentiful excitatory neurotransmitters in the brain are **glutamate** and **aspartate**. **Glutamatergic** transmission employs what are called *AMPA*, *kainate*, and *NMDA* receptors (the names of these receptor subtypes refer to drugs that activate them), which are ionotropic. Because NMDA-type glutamate receptors are active in a fascinating model of learning and memory, they have been studied very closely and will be discussed more fully in Chapter 16. There are also several metabotropic glutamate receptors (mGluR's), which act more slowly because they work through second messengers. Glutamate is associated with **excitotoxicity**, a phenomenon in which neural injury, such as a stroke or trauma, provokes an excessive release of glutamate that overexcites neurons, eventually killing them.

In contrast to glutamate and aspartate, the amino acid transmitters **gamma-aminobutyric acid** (**GABA**) and **glycine** typically have an inhibitory effect. GABA receptors are divided into several large classes: the $GABA_A$, $GABA_B$, and $GABA_C$ receptors. Although they all normally respond to GABA, the subtypes of GABA receptors exhibit quite different properties.

$GABA_A$ receptors are ionotropic (i.e., they are ligand-gated ion channels; see Figure 3.2, left). Activation of $GABA_A$ receptors results in rapid-onset inhibitory postsynaptic potentials due to a localized influx of Cl^- ions. Each $GABA_A$ receptor is made up of five protein subunits surrounding a Cl^- ion channel that can be widened or narrowed depending on the state of the surrounding complex. By mixing

gas neurotransmitter A soluble gas, such as nitric oxide or carbon monoxide, that is produced and released by a neuron to alter the functioning of another neuron.

co-localisation Also called *co-release*. Here, the appearance of more than one neurotransmitter in a given presynaptic terminal.

glutamate An amino acid transmitter, the most common excitatory transmitter.

aspartate An amino acid transmitter that is excitatory at many synapses.

glutamatergic Referring to cells that use glutamate as their synaptic transmitter.

excitotoxicity The property by which neurons die when overstimulated, as with large amounts of glutamate.

gamma-aminobutyric acid (GABA) A widely distributed inhibitatory amino acid transmitter.

glycine An amino acid transmitter, often inhibitory.

acetylcholine (ACh) A neurotransmitter produced and released by parasympathetic postganglionic neurons, by motor neurons, and by neurons throughout the brain.

cholinergic Referring to cells that use acetylcholine as their synaptic transmitter.

nicotinic Referring to cholinergic receptors that respond to nicotine as well as to acetylcholine.

muscarinic Referring to cholinergic receptors that respond to the chemical muscarine as well as to acetylcholine.

catecholamine A class of monoamines that serve as neurotransmitters, including dopamine and norepinephrine.

indoleamine A class of monoamines, including serotonin and melatonin, that serve as neurotransmitters.

dopamine (DA) A monoamine transmitter found in the midbrain—especially the substantia nigra—and basal forebrain.

and matching of the various protein subunits that make up the $GABA_A$ receptors, the brain may produce dozens of different kinds (Chua and Chebib, 2017).

$GABA_B$ receptors are metabotropic receptors, typically producing a slow-occurring inhibitory postsynaptic potential due to changes in K^+ ion conductance (Gassman and Bettler, 2012). Drugs that selectively alter $GABA_B$ activity may be helpful in domains ranging from pain relief to mood disorders (Pin and Bettler, 2016). $GABA_C$ receptors are ionotropic with a chloride channel, but they differ from other GABA receptors in certain details of their subunit structure.

Given GABA's inhibitory actions, it is not surprising that some drugs that activate $GABA_A$ receptors—for example, benzodiazepines like Xanax (alprazolam) and Ativan (lorazepam)—are potent sedatives that also act to reduce anxiety and panic attacks, block epileptic seizures, and aid relaxation and sleep. Drugs that interfere with the actions of GABA receptors can provoke seizures by blocking the important inhibitory influence of GABA. Interestingly, some GABA-ergic neurons appear to co-release the excitatory transmitter glutamate; perhaps these cells continually rebalance their inhibitory versus excitatory influence on postsynaptic targets (El Mestikawy et al., 2011).

Acetylcholine was the first neurotransmitter to be identified

As we saw in Chapter 2, Otto Loewi's dream-inspired experiments of the 1920s famously established that neurons use chemical messages to communicate (see Figure 2.14). The anatomical distribution of **acetylcholine** (**ACh**), the chemical eventually identified as the transmitter at work in Loewi's experiments, was subsequently mapped by staining cells containing the enzymes needed to synthesise ACh. **FIGURE 3.3** shows the distribution of **cholinergic** (ACh-using) neurons and their projections in the brain. Important clusters of cholinergic cells are found in the basal forebrain, including the medial septal nucleus, the nucleus of the diagonal band, and the nucleus basalis. These cholinergic cells project to the hippocampus and amygdala, as well as throughout the cerebral cortex. Widespread loss of cholinergic neurons is evident in Alzheimer's disease, suggesting that cholinergic systems are crucial for learning and memory. Similarly, the anticholinergic drug scopolamine interferes with learning and memory in experimental settings.

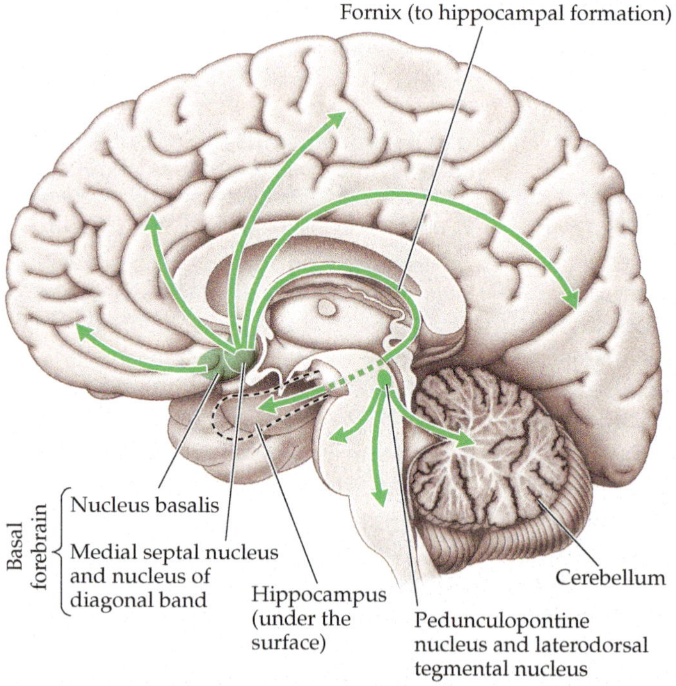

3.3 Cholinergic Pathways in the Brain In this midsagittal view, the brain nuclei containing cell bodies of neurons that release ACh are shown in green; the projections of axons from these neurons are indicated by green arrows. Because they use ACh as a transmitter, these neurons are said to be *cholinergic*.

In Chapter 2 we noted that there are two families of ACh receptors in the peripheral and central nervous systems: **nicotinic** (nACh) and **muscarinic** (mACh) receptors. Each family contains subtypes of receptors. Most nicotinic receptors are ionotropic, responding rapidly and usually having an excitatory effect (see Figure 3.2, left). Muscles use nACh receptors, so drugs that block nACh receptors, like curare, cause widespread paralysis. The mACh receptors are (metabotropic) GPCRs (see Figure 3.2, right), so they have slower responses when activated, and they can be either excitatory or inhibitory (see Figure 3.15B). Muscarinic receptors can be blocked by atropine or scopolamine; either drug produces pronounced changes in cognition, including drowsiness, confusion, memory problems, and blurred vision.

Five monoamines act as neurotransmitters

There are two principal classes of neurotransmitters that, because they are modified amino acids, are called *monoamines*: catecholamines and indoleamines. The **catecholamine** neurotransmitters—each derived from the amino acid tyrosine and featuring a six-sided *catechol* ring within its molecular structure—are dopamine, epinephrine, and norepinephrine. The **indoleamine** neurotransmitters—each derived from the amino acid tryptophan and containing a five-sided *indole* ring—are melatonin and serotonin. Let's take a closer look at three especially important monoamines: dopamine, serotonin, and norepinephrine. (The neuronal synthesis of the monoamines and acetylcholine is summarized in **BOX 3.1**.)

BOX 3.1 Pathways for Neurotransmitter Synthesis

Here we provide reference information on the chemical pathways by which the classic neurotransmitters are synthesised. By understanding these pathways, researchers can target drug discoveries toward affecting specific transmitter systems. Furthermore, because enzyme action is crucial for transmitter synthesis, neuroanatomists can use the anatomical distribution of these enzymes to determine which transmitters are used by different brain regions. For example, the enzyme **choline acetyltransferase** (**ChAT**) catalyses the synthesis of ACh from its precursor, choline:

Acetyl CoA + choline
↓ ChAT
ACh + coenzyme A

The enzyme **acetylcholinesterase** (**AChE**) breaks down the ACh, leaving choline and acetic acid:

ACh
↓ AChE
Choline + acetic acid

As it turns out, AChE is very widely distributed, but ChAT is found primarily in the nuclei shown in Figure 3.3.

All of the catecholamine transmitters (norepinephrine, epinephrine, and dopamine) are synthesised from the amino acid tyrosine in a succession of metabolic steps:

Tyrosine
↓ Tyrosine hydroxylase
L-dopa
↓ Aromatic L-amino acid decarboxylase
Dopamine
↓ Dopamine β-hydroxylase
Norepinephrine
↓ Phenylethanolamine *N*-methyltransferase
Epinephrine

Note that only neurons that possess the enzyme tyrosine hydroxylase have the capacity to produce any catecholamine transmitter and that l-dopa is a precursor for all three.

The indoleamine serotonin is produced from the amino acid tryptophan in two chemical steps:

Tryptophan
↓ Tryptophan hydroxylase
5-hydroxytryptophan (5-HTP)
↓ Aromatic L-amino acid decarboxylase
5-hydroxytryptamine (5-HT; serotonin)

The monoamines (dopamine, norepinephrine, epinephrine, serotonin, and melatonin) are inactivated through a combination of presynaptic reuptake and enzymatic breakdown. Most of this enzymatic action is performed by a class of enzymes called **monoamine oxidases** (**MAOs**).

Neuropeptides are synthesised like any other peptide or protein—through transcription of a gene and translation of messenger RNA (mRNA)—so we can find neurons making those transmitters by looking for the appropriate mRNA transcript (see the Appendix).

choline acetyltransferase (ChAT) An enzyme involved in the synthesis of the neurotransmitter acetylcholine.

acetylcholinesterase (AChE) An enzyme that inactivates the transmitter acetylcholine both at synaptic sites and elsewhere in the nervous system.

monoamine oxidase (MAO) An enzyme that breaks down and thereby inactivates monoamine transmitters.

DOPAMINE Of the 80 to 90 billion neurons in the human brain, only about a million synthesise **dopamine** (**DA**), but they have an outsized influence on behaviour. Several subtypes of DA receptors have been discovered and have been numbered D_1, D_2, D_3, D_4, and D_5, in the order of their discovery; these receptors vary widely in their actions, anatomical distributions, and behavioural functions, as we discuss later in the chapter. **FIGURE 3.4** shows the locations of dopaminergic neurons and their projections in the brain, focussing on the **mesostriatal pathway** and the **mesolimbocortical pathway**, two of the main groups of these neurons.

The mesostriatal pathway, as the name indicates, originates from the mesencephalon (midbrain)—specifically the **substantia nigra** and nearby areas—and ascends as part of the *medial forebrain bundle* to innervate the **striatum**: the caudate nucleus and putamen (see Figure 3.4, left). All these structures are part of the *basal ganglia* described in Chapter 1 (see Figure 1.17A). The mesostriatal DA pathway plays a crucial role in motor control, and significant loss of these neurons produces the movement problems of Parkinson's disease (described in Chapter 10).

mesostriatal pathway A set of dopaminergic axons arising from the midbrain and innervating the basal ganglia, including those from the substantia nigra to the striatum.

mesolimbocortical pathway A set of dopaminergic axons arising in the midbrain and innervating the limbic system and cortex.

substantia nigra Literally, "black substance." A group of pigmented neurons in the midbrain that provides dopaminergic projections to areas of the forebrain, especially the basal ganglia.

striatum The caudate nucleus and putamen together.

ventral tegmental area (VTA) A portion of the midbrain that projects dopaminergic fibres to the nucleus accumbens.

serotonin (5-HT) A synaptic transmitter that is produced in the raphe nuclei and is active in structures throughout the cerebral hemispheres.

serotonergic Referring to neurons that use serotonin as their synaptic transmitter.

raphe nuclei A string of nuclei in the midline of the midbrain and brainstem that contain most of the serotonergic neurons of the brain.

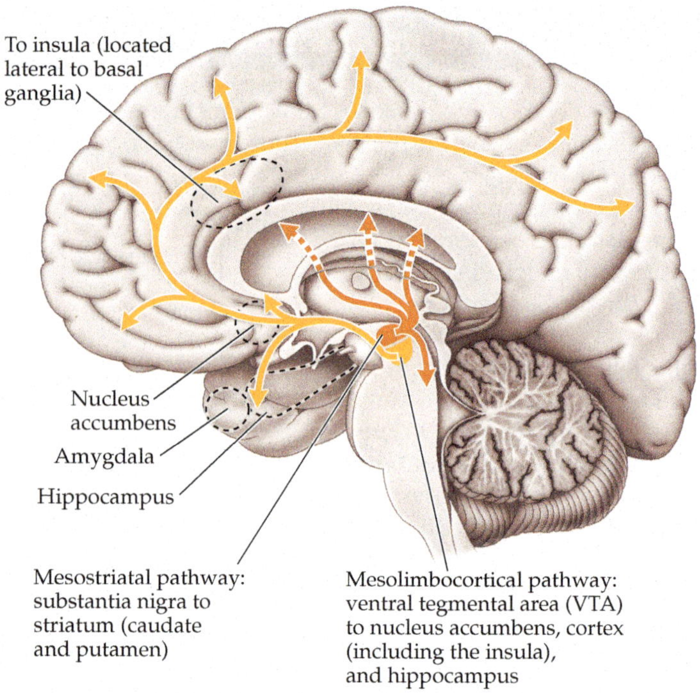

Go to Animation 3.1
Neurotransmitter Pathways in the Brain

3.4 Dopaminergic Pathways in the Brain Neurons in the pathways represented in this midsagittal view release dopamine and thus are called *dopaminergic*.

The mesolimbocortical pathway also originates in the midbrain, in the **ventral tegmental area** (**VTA**) (see Figure 3.4, right), and projects to the limbic system (amygdala, nucleus accumbens, hippocampus) and the cortex. This system is important in reward and reinforcement, especially via the dopamine D_2 receptor subtype (Glimcher, 2011); we'll revisit this topic at the end of the chapter. Abnormalities in the mesolimbocortical pathway are associated with some of the symptoms of schizophrenia, as we'll discuss in Chapter 12.

SEROTONIN Because its chemical name is *5 hydroxytryptamine*, **serotonin** is abbreviated **5-HT**. Large areas of the brain are innervated by **serotonergic** fibres, although 5-HT cell bodies are relatively few and are concentrated along the midline in the **raphe nuclei** (pronounced "rafay"; Latin for "seam") of the midbrain and brainstem. **FIGURE 3.5** shows the distribution of serotonergic cell bodies and their axonal projections. Only about 200,000 neurons of the human brain are serotonergic, but like dopaminergic neurons, they exert widespread influence through the rest of the brain.

Serotonin has been implicated in the control of sleep states (see Chapter 15), mood, sexual behaviour, anxiety, and many other functions. Several families of drugs that are used as antidepressants share an action on serotonin: they act to increase its synaptic availability. Prozac, for example (see Chapter 12), slows the clearance of serotonin from synapses by inhibiting its reuptake into axon terminals. At least 14 types of 5-HT receptors (5-HT_1, 5-HT_2, and so on) have been described, differing widely in their neuroanatomical distribution, and all but one type are metabotropic receptors. The effects of serotonergic drugs on behaviour depend on which subtypes of 5-HT receptors are affected (Carr and Lucki, 2011; Gorzalka et al., 1990).

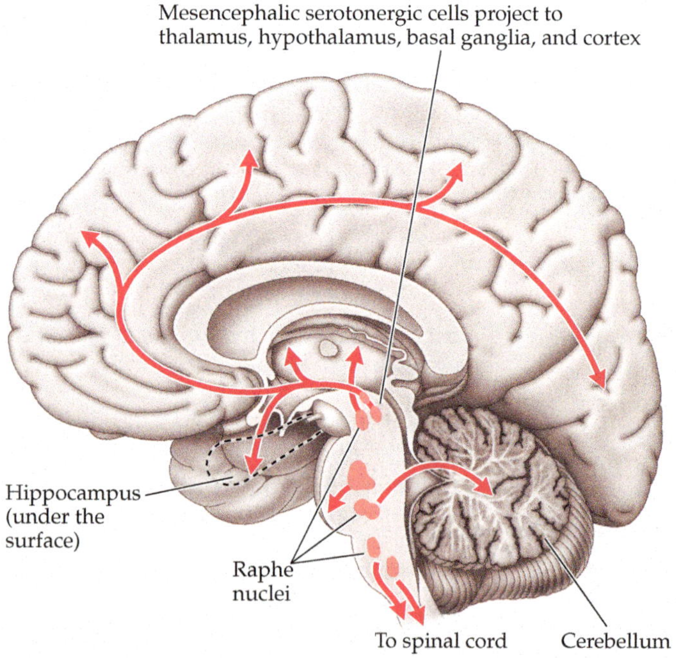

3.5 Serotonergic Pathways in the Brain The neurons in the nuclei shown in this midsagittal view release serotonin (5-HT) and thus are said to be *serotonergic*.

NOREPINEPHRINE The two main clusters of neurons in the brainstem releasing **norepinephrine** (**NE**) are the **locus coeruleus**, ("blue spot") in the pons, and the lateral tegmental area of the midbrain (**FIGURE 3.6**). Because norepinephrine is also known as *noradrenaline*, NE-producing cells are said to be **noradrenergic**. Recall from Chapter 1 that sympathetic fibres innervating the body are also noradrenergic (see Figure 1.12).

Fibres from the noradrenergic cells of the locus coeruleus project broadly throughout the cerebrum, including the cerebral cortex, limbic system, and thalamus. The cerebellum and spinal cord also receive noradrenergic innervation. The CNS contains four subtypes of NE receptors—α_1-, α_2-, β_1-, and β_2-adrenoceptors—all of which are metabotropic receptors. Given the brain's wide noradrenergic projections, it's no surprise that there are noradrenergic contributions to diverse behavioural and physiological processes, including mood, overall arousal, and sexual behaviour.

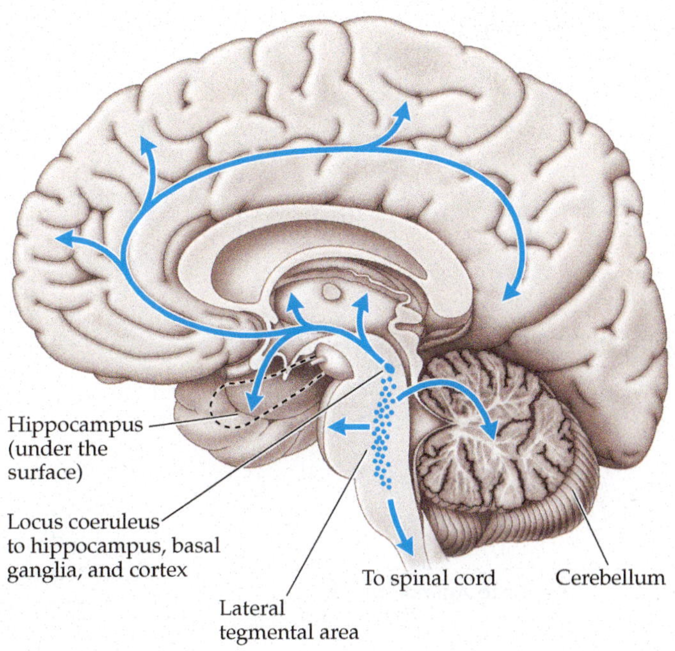

3.6 Noradrenergic Pathways in the Brain The neurons in the pathways shown in this midsagittal view release norepinephrine (noradrenaline) as a transmitter and thus are said to be *noradrenergic*.

Many peptides function as neurotransmitters

We can't catalog all the peptide neurotransmitters here, but these are some that we'll see again in later chapters:

- The **opioid peptides** are a group of endogenous substances with actions that resemble those of opiate drugs like morphine. Some key opioids are met-enkephalin, leu-enkephalin, β-endorphin, and dynorphin (see Chapter 7).
- A large group of peptides discovered in the periphery—especially in the organs of the gut (which explains some of their names)—are also made by neurons in the spinal cord or brain, where they participate in synaptic transmission. Examples include substance P, cholecystokinin (CCK), vasoactive intestinal polypeptide (VIP), neurotensin, and neuropeptide Y (NPY) (see Chapter 14).
- Various peptide hormones, such as oxytocin and vasopressin, originate in the brain or pituitary (see Chapters 4 and 13). Peptides are involved in an astonishing variety of functions, ranging from basic housekeeping like urine production through to higher-level functions like memory, pair-bonding (see Chapter 4), and social processes.

Some neurotransmitters are gases

It may surprise you to learn that neurons use certain gas molecules that dissolve in water (and so are called *soluble gases*) to communicate information. The best studied of these is **nitric oxide**, or **NO** (distinct from *nitrous* oxide, or laughing gas, which is N_2O).

The actions of NO and other gas neurotransmitters are different from those of the classic transmitters in several important ways. First, the gas transmitter molecules are produced in cellular locations other than the axon terminals, especially the dendrites, and molecules of gas transmitters are not held in or released from vesicles; the substance simply diffuses out of the neuron as soon as it is produced. Second, the released transmitter doesn't interact with membrane-bound receptors on the surface of the target cell, but rather it diffuses *into* the target cell and stimulates the production of second messengers. And third, gas transmitters can serve as **retrograde transmitters**, diffusing from the postsynaptic neuron back to the presynaptic neuron, where they stimulate changes in synaptic efficacy that may be involved in learning and memory (which will be covered in more detail in Chapter 16). Found widely throughout the body, gas transmitters have been implicated in processes as diverse as hair growth and penile erection (Meldrum et al., 2014), in addition to their role in the brain.

norepinephrine (NE) Also called *noradrenaline*. A neurotransmitter produced and released by sympathetic postganglionic neurons to accelerate organ activity. Also produced in the brainstem and found in projections throughout the brain.

locus coeruleus Literally, "blue spot." A small nucleus in the brainstem whose neurons produce norepinephrine.

noradrenergic Referring to systems using norepinephrine (noradrenaline) as a transmitter.

opioid peptide A type of endogenous peptide that mimics the effects of morphine in binding to opioid receptors and producing marked analgesia and reward.

nitric oxide (NO) A soluble gas that serves as a retrograde gas neurotransmitter in the nervous system.

retrograde transmitter A neurotransmitter that diffuses from the postsynaptic neuron back to the presynaptic neuron.

3.4 The Effects of a Drug Depend on Its Site of Action and Dose

Learning Objectives

After reading this section, you should be able to:

3.4.1 **Distinguish between agonist, antagonist, inverse agonist, and partial agonist drug actions.**

3.4.2 **Explain the distinction between competitive and noncompetitive ligands.**

3.4.3 **Summarise the key events that occur when a ligand interacts with a receptor.**

3.4.4 **Define and distinguish between binding affinity and efficacy.**

3.4.5 **Define the dose-response curve, and explain ways in which it is useful in studies of pharmacodynamics.**

3.4.6 **Review the processes by which drug tolerance develops.**

Go to Animation 3.2
Agonists and Antagonists

In everyday English, we use the term *drug* in different ways. One common meaning is "a medicine used in the treatment of a disease" (as in *prescription drug* or *over-the-counter drug*). Many *psychoactive drugs*—compounds that alter the function of the brain and thereby affect conscious experiences—fall into this category, and they are useful in psychiatric settings. Other psychoactive drugs are used recreationally with varying degrees of risk to health; these are *drugs of abuse,* although such substances may also have therapeutic value. Some drugs affect the brain by altering enzyme action or modifying other internal cellular processes, but as you may have guessed, most drugs of interest to behavioural neuroscience are the ones that bind to specific receptors. Any substance that binds to a receptor is termed a **ligand**. This includes hormones and neurotransmitters, which we would classify as *endogenous ligands,* and also receptor-selective drugs, which are *exogenous ligands.* Receptor ligands typically have one of several kinds of effects (**FIGURE 3.7**):

1. A ligand that is classified as an **agonist** initiates the normal effects of the transmitter on that receptor.
2. A receptor **antagonist** is a ligand that binds to a receptor and does not activate it, thereby blocking it from being activated by other ligands (including the native neurotransmitter).
3. An **inverse agonist**—a less common type of ligand—binds to the receptor and initiates an effect that is the *reverse* of the normal function of the receptor.

ligand A substance that binds to receptor molecules, such as those at the surface of the cell.

agonist A molecule, usually a drug, that binds a receptor molecule and initiates a response like that of another molecule, usually a neurotransmitter.

antagonist A molecule, usually a drug, that interferes with or prevents the action of a transmitter.

inverse agonist A substance that binds to a receptor and causes it to do the opposite of what the naturally occurring transmitter does.

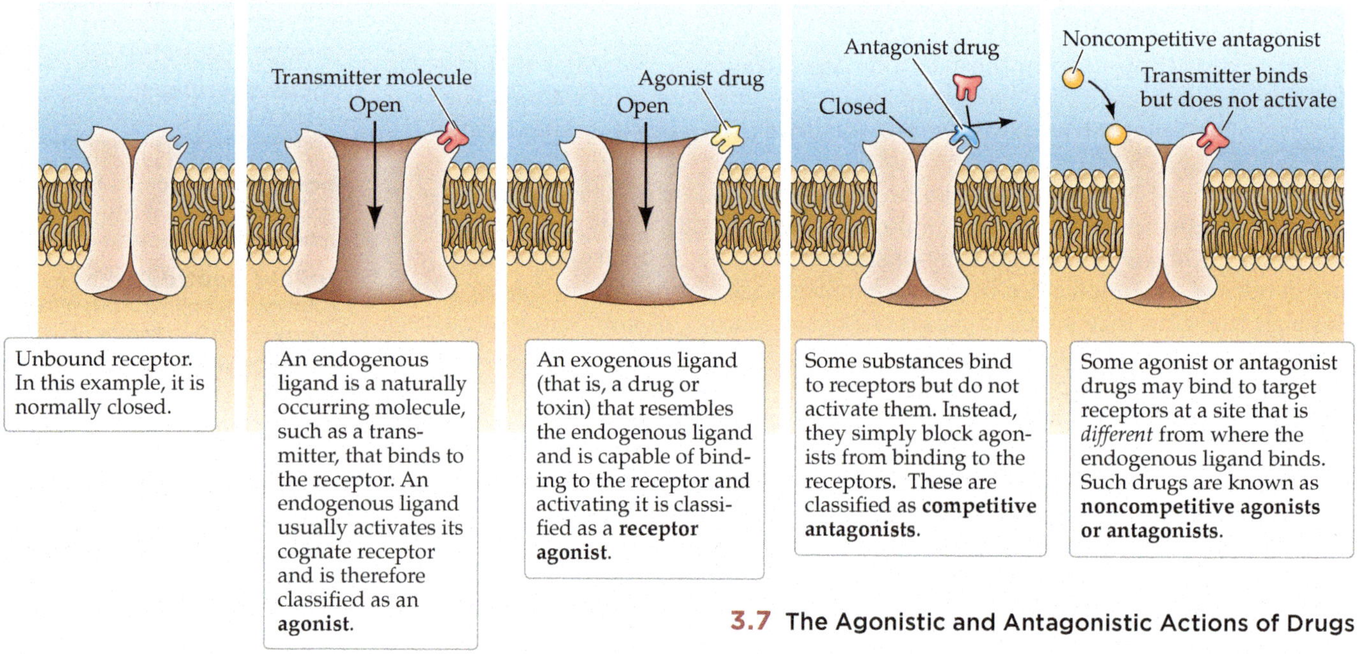

3.7 The Agonistic and Antagonistic Actions of Drugs

But there is one more layer of complexity. Drugs with any of the three actions just described are properly called **competitive ligands** because they compete with the endogenous transmitter for binding to the same part of the receptor complex. But some drugs bind to a part of the receptor complex that does *not* normally bind the transmitter (see Figure 3.7, far right). Because this sort of drug does not directly compete with the transmitter for its binding site, we say that the drug is a **noncompetitive ligand** (or *neuromodulator*) binding to a **modulatory site** on the receptor. Noncompetitive ligands may either activate the receptor, thereby acting as noncompetitive agonists, or prevent the receptor from being activated by the transmitter, thus acting as noncompetitive antagonists. We'll see several examples of these actions when we discuss modulatory sites on the $GABA_A$ receptor complex, later in this chapter.

competitive ligand A substance that directly competes with the endogenous ligand for the same binding site on a receptor molecule.

noncompetitive ligand Also called *neuromodulator.* A substance that alters the response to an endogenous ligand without interacting with the endogenous ligand's recognition site.

modulatory site A portion of a receptor that, when bound by a compound, alters the receptor's response to its transmitter.

Drugs fit like keys into molecular locks

Because a given neurotransmitter interacts with a variety of different subtypes of receptors—a few are discussed in **TABLE 3.2**—it is possible to develop drugs that primarily interact with just one or a few receptor subtypes. The various subtypes of receptors generally differ in their distribution within the brain, and they also serve very different cellular functions, so selectively activating or blocking specific subtypes of receptors can have widely varying effects. For example, treating someone with serotonin would activate all of that person's serotonin receptors, regardless of subtype, causing a debilitating constellation of physiological and behavioural changes. But drugs that are selective antagonists of 5-HT_3 receptors, showing little activity

TABLE 3.2 The Bewildering Multiplicity of Transmitter Receptor Subtypes

Transmitter	Known receptor subtypes	Function
Glutamate	AMPA, kainate, and NMDA receptors (ionotropic)	Glutamate is the most abundant of all neurotransmitters and the most important excitatory transmitter.
	mGluR's (***m**etabotropic **glu**tamate **r**eceptors*)	Glutamate receptors are crucial for excitatory signals, and NMDA receptors are especially implicated in learning and memory.
Gamma-aminobutyric acid (GABA)	$GABA_A$ (ionotropic)	GABA receptors mediate most of the brain's inhibitory activity, balancing the excitatory actions of glutamate. $GABA_A$ receptors are inhibitory in many brain regions, reducing excitability and preventing seizure activity.
	$GABA_B$ (metabotropic)	$GABA_B$ receptors are also inhibitory, by a different mechanism.
Acetylcholine (ACh)	Muscarinic receptors (metabotropic)	Both mediate cholinergic transmission in the cortex.
	Nicotinic receptors (ionotropic)	Nicotinic receptors are crucial for muscle contraction.
Dopamine (DA)	D_1 through D_5 receptors (all metabotropic)	Found throughout the forebrain
	D_6 and D_7 probable	Involved in complex behaviours, including motor function, reward, higher cognition
Norepinephrine (NE)	α_1-, α_2-, β_1-, and β_2-adrenergic receptors (all metabotropic)	NE has multiple effects in visceral organs and is important in sympathetic nervous system and "fight or flight" responses. In the brain, NE transmission provides an alerting and arousing function.
Serotonin	5-HT_1 receptor family (5 members)	Different subtypes differ in their distribution in the brain.
	5-HT_2 receptor family (3 members)	May be involved in mood, sleep, and higher cognition
	5-HT_3 through 5-HT_7 receptors	5-HT_3 receptors are particularly involved in nausea.
	All but one subtype (5-HT_3) are metabotropic.	
Miscellaneous peptides	Many specific receptors for peptides such as opiates (delta, kappa, and mu receptors), cholecystokinin (CCK), neurotensin, neuropeptide Y (NPY), and dozens more (metabotropic)	Peptide transmitters have many different functions, depending on their anatomical localisation. Some important examples include the control of feeding, sexual behaviours, and social functions.

binding affinity Also called simply *affinity.* The propensity of molecules of a drug (or other ligand) to bind to receptors.

efficacy Also called *intrinsic activity.* The extent to which a drug activates a response when it binds to a receptor.

partial agonist A drug that, when bound to a receptor, has less effect than the endogenous ligand would.

at other 5-HT receptor subtypes, produce a powerful and specific anti-nausea effect. We have also discussed how a single transmitter, GABA, may have a huge variety of different receptor subtypes, depending on variations in the proteins making up the GABA receptors. Apparently, evolution tinkers with the structure of *receptors* more than with the transmitters themselves.

Drug molecules do not actively seek out particular receptor molecules; rather, drug molecules spread widely throughout the body, and when they happen to come in contact with a receptor molecule possessing the correct shape, the two molecules bind together briefly and begin a chain of events. The lock-and-key analogy is often used for this binding action, as mentioned in Chapter 3. In the case of receptor-selective drugs, though, we have to think of keys (drug molecules) trying to insert themselves into all the locks (receptor molecules) in the neighborhood; each such key fits into only a particular subset of the locks. Once the drug (the key) binds to the receptor (the lock), it alters the activity of the receptor, activating it or blocking it. But the binding is usually temporary, and when the drug or transmitter breaks away from the receptor, the receptor returns to its unbound shape and function.

Drug-receptor interactions vary in specificity and activity

The tuning of drug molecules to receptors is not absolutely specific. That is, a particular drug molecule will generally bind strongly to one kind of receptor, more weakly to some other types, and not at all to many others. A drug molecule that has more than one kind of action in the body exhibits this flexibility because it affects more than one kind of receptor molecule. For example, some drugs combat anxiety at low doses without producing sedation (relaxation, drowsiness), but at higher doses they cause sedation, probably because at those doses they activate additional types of receptors.

The degree of chemical attraction between a ligand and a receptor is termed **binding affinity** (or simply *affinity*). A drug with high affinity for a particular type of receptor will selectively bind to that type of receptor even at low doses, and it will stay bound for a relatively long time. The same dose of a lower-affinity drug will bind fewer receptor molecules. **FIGURE 3.8** illustrates how binding affinity is measured. It is interesting to note that the neurotransmitters themselves are low-affinity ligands—their weak binding lets them rapidly dissociate from receptors, allowing the synapse to reset for the next signal's arrival.

After binding, the propensity of a ligand to *activate* the receptor to which it is bound is termed its **efficacy** (or *intrinsic activity*): agonists have high efficacy and antagonists have low efficacy. **Partial agonists**, unsurprisingly, are drugs that

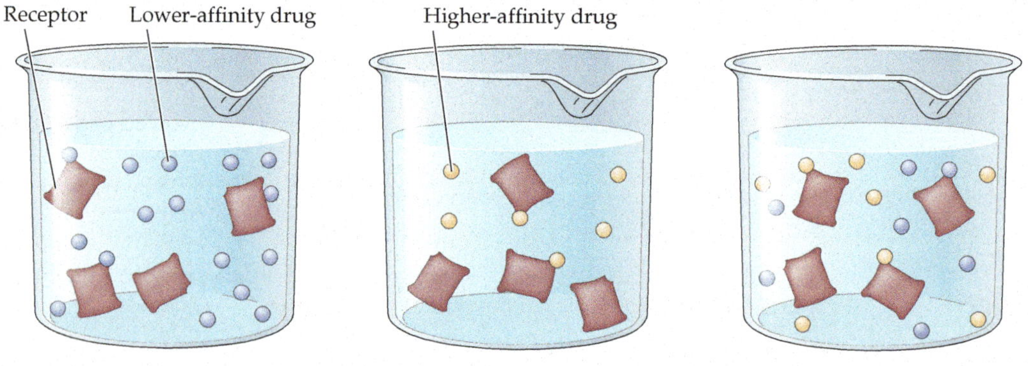

If a particular drug has a low affinity for a receptor, then it will quickly uncouple from the receptor. To bind half the receptors at any given time, a higher concentration of the drug is needed.

If a drug has a high affinity for a receptor, the two will stay together for a longer time, and a lower concentration of drug will be sufficient to bind half the receptors.

If equal concentrations of the two drugs are present, the high-affinity drug will be bound to more receptors at any given time. If the drugs have an equivalent effect on the receptors, then the higher-affinity drug will be more potent.

3.8 Using Binding Affinity to Compare Drug Effectiveness

have appreciable but submaximal efficacy; they thus produce a middling response. So, it is a combination of affinity and efficacy—where it binds and what it does—that determines the overall action of a drug. To some extent we can compare the effectiveness of different drugs by comparing their affinity for a receptor of interest (see Figure 12.13 for an example).

Dose-response relationships reflect the potency and safety of drugs

As you would probably predict, administering larger doses of a drug increases the proportion of receptors that are bound by the drug. Within certain limits, this increase in receptor binding also increases the response to the drug; in other words, greater doses tend to produce greater effects. When plotted as a graph, the relationship between drug doses and observed effects is called a **dose-response curve** (**DRC**). Careful analysis of DRCs reveals many aspects of drug activity and is one of the main tools for understanding **pharmacodynamics** (the functional relationships between drugs and their targets).

FIGURE 3.9 illustrates how DRCs are used to measure key characteristics of drugs. For example, DRCs reveal the effective dose range of a drug and allow comparison of the potencies and efficacies of different drugs. Sometimes a DRC gives us hints that the drug may be binding more than one type of receptor. By directly comparing the DRCs for two drugs, we can determine which drug would be safer to use.

dose-response curve (DRC) A formal plot of a drug's effects (on the *y*-axis) versus the dose given (on the *x*-axis).

pharmacodynamics Collective name for the factors that affect the relationship between a drug and its target receptors, such as affinity and efficacy.

3.9 The Dose-Response Curve (DRC)

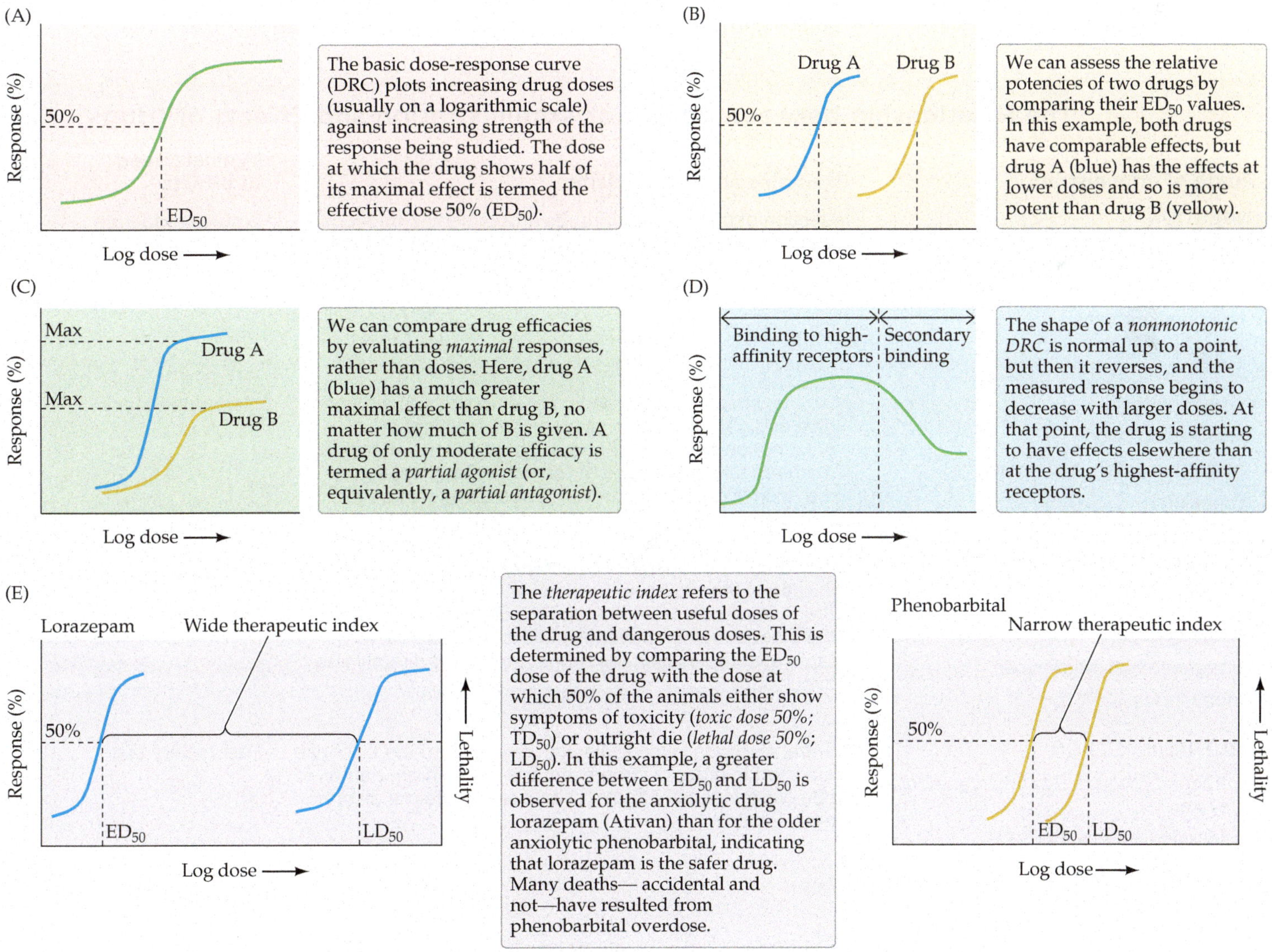

bioavailable Referring to a substance, usually a drug, that is present in the body in a form that is able to interact with physiological mechanisms.

biotransformation The process in which enzymes convert a drug into a metabolite that is itself active, possibly in ways that are substantially different from the actions of the original substance.

pharmacokinetics Collective name for all the factors that affect the movement of a drug into, through, and out of the body.

blood-brain barrier The mechanisms that make the movement of substances from blood vessels into brain cells more difficult than exchanges in other body organs, thus affording the brain greater protection from exposure to some substances found in the blood.

Drugs are administered and eliminated in many different ways

The amount of drug that reaches the brain and the speed with which it starts acting are determined in part by the drug's route of administration. Some routes, such as smoking or intravenous injection, rapidly increase the concentration of drug in the body that is **bioavailable**: free to act on the target tissue and therefore not bound to other proteins or in the process of being metabolised or excreted. With other routes, such as oral ingestion, drug concentration builds up more slowly over longer periods of time.

The duration of a drug effect is largely determined by the manner in which the drug is metabolised and excreted from the body—via the kidneys, liver, lungs, and other routes. In some cases, the metabolites of drugs are themselves active; this **biotransformation** of drugs can be a source of unwanted side effects. Some drugs may be stored in *depots* (collecting in fat or bone, for example), only to reemerge and have physiological effects after long periods of time. The factors that affect the movement of a drug into, through, and out of the body are collectively referred to as **pharmacokinetics**.

Humans have devised a variety of ingenious techniques for introducing substances into the body; these are summarized in **TABLE 3.3**. In Chapter 1 we discussed the **blood-brain barrier**: the tight junctions between the endothelial cells of blood vessels within the CNS that inhibit the movement of larger molecules out of the bloodstream and into the brain. This barrier poses a major challenge for neuropharmacology because many drugs that might be clinically or experimentally useful are too large to pass the blood-brain barrier to enter the brain. To a limited extent this problem can be circumvented by delivering drugs directly into the brain, but that is a drastic step. Alternatively, some drugs can take advantage of active transport systems that normally move nutrients out of the bloodstream and into the brain.

TABLE 3.3 The Relationship between Routes of Administration and Effects of Drugs

Route of administration	Examples and mechanisms	Typical speed of effects
INGESTION Tablets and capsules Syrups Infusions and teas Suppositories	Ingestion of many sorts of drugs and remedies depends on absorption by the gut, which is somewhat slower than most other routes, and is affected by digestive processes and the presence of food	Slow to moderate
INHALATION Nasal absorption Inhaled powders and sprays Smoking	Nicotine, cocaine, and organic solvents such as airplane glue and gasoline are inhaled, as are a variety of prescription drugs and hormone treatments. Inhalation methods take advantage of the rich vascularisation of the nose and lungs to convey drugs directly into the bloodstream.	Moderate to fast
PERIPHERAL INJECTION Subcutaneous Intramuscular Intraperitoneal (abdominal) Intravenous	Many drugs are injected. Subcutaneous (under the skin) injections tend to have the slowest effects because drugs must diffuse into nearby tissue in order to reach the bloodstream; intravenous injections have very rapid effects because the drugs are placed directly into circulation.	Moderate to fast
CENTRAL INJECTION Intracerebroventricular (into ventricular system) Intrathecal (into spinal CSF) Epidural (under the dura mater) Intracerebral (directly into a brain region)	Central methods involve injection directly into the CNS and are used in order to circumvent the blood-brain barrier, to rule out peripheral effects, or to directly affect a discrete brain location.	Fast to very fast

Repeated treatments may reduce the effectiveness of drugs

Our bodies are impressively adaptable. Many body systems change their functioning in order to accommodate environmental challenges, and in most ways drug treatments can be viewed as changes in the body's chemical environment. This adaptability is evident in the development of drug **tolerance**, where a drug's effectiveness diminishes with repeated treatments. Consequently, successively larger and larger doses of drug are needed to get the same effect.

Drug tolerance can develop in several different ways. Some drugs provoke **metabolic tolerance**, in which the body's metabolic systems and organs (such as the liver and its specialised enzymes) become increasingly effective at eliminating the drug from the bloodstream before it has a chance to affect the brain or another target.

Alternatively, the target tissue itself may show altered sensitivity to the drug, or **functional tolerance**. Although we tend to think of receptors in postsynaptic membranes as being fairly static, in fact the reverse is true—receptor densities are continually waxing and waning as new receptors are inserted into the synapse or as old ones are internalised by the cell for recycling (Anggono and Huganir, 2012; Choquet and Triller, 2013). In neuropharmacology, this dynamic regulation of receptor proteins—changing the number of receptors present in the cell membrane—is an important source of functional tolerance. Receptor regulation tends to alter neuronal sensitivity in the direction opposite to the drug's effect. Thus, over the course of repeated exposures to an *agonist* drug, neurons may **down-regulate** (decrease) the number of available receptors to which the drug can bind, thereby becoming less sensitive and countering the drug effect. If the drug is an *antagonist*, target neurons may instead **up-regulate** (increase) the number of receptors and thereby become more sensitive (**FIGURE 3.10**).

Tolerance to a drug often generalises to other drugs belonging to the same chemical class; this effect is termed **cross-tolerance**. For example, people who have developed tolerance to heroin tend to exhibit a degree of tolerance to all the

tolerance A condition in which, with repeated exposure to a drug, an individual becomes less responsive to a constant dose.

metabolic tolerance The form of drug tolerance that arises when repeated exposure to the drug causes the metabolic machinery of the body to become more efficient at clearing the drug.

functional tolerance Decreased responding to a drug after repeated exposures, generally as a consequence of up- or down -regulation of receptors.

down-regulation A compensatory reduction in receptor availability at the synapses of a neuron.

up-regulation A compensatory increase in receptor availability at the synapses of a neuron.

cross-tolerance A condition in which the development of tolerance for one drug causes an individual to develop tolerance for another drug.

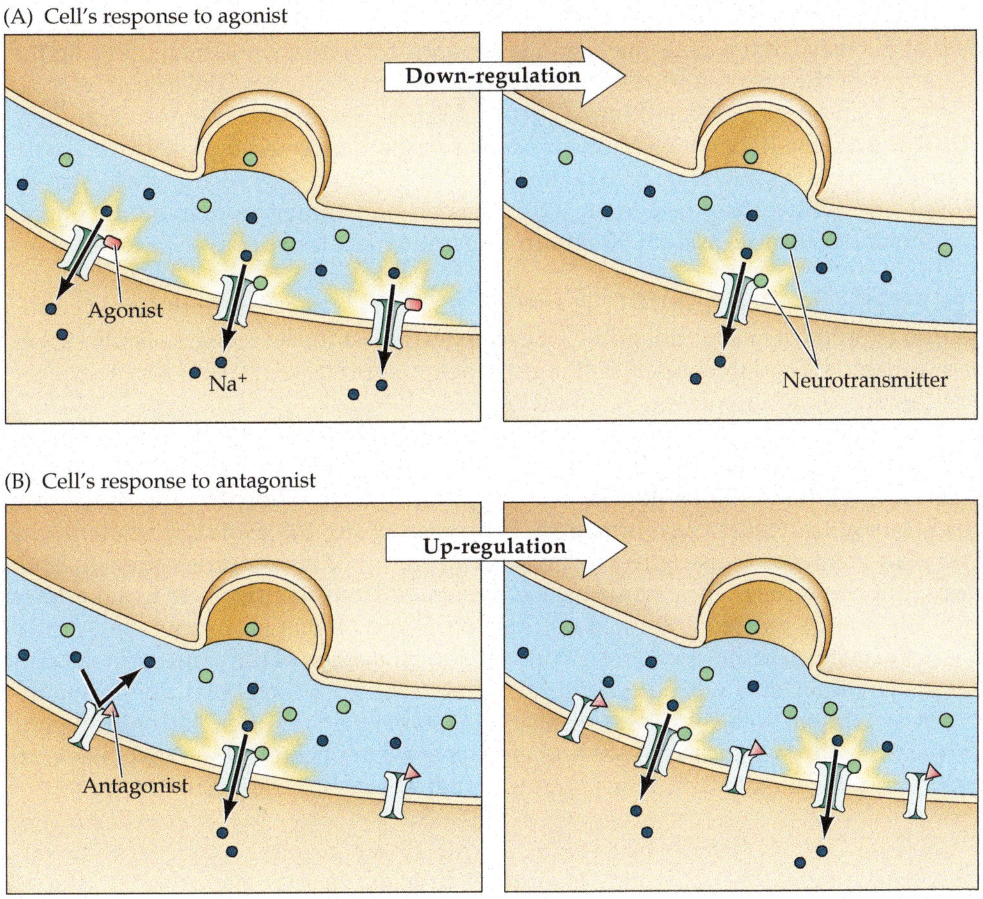

3.10 Receptor Regulation Over time, cells respond to the actions of receptor-selective drugs by altering their own sensitivity, through changes in the number of receptors they make available. In (A), the effects of an agonist drug augment the effects of the endogenous ligand, enhancing the effect on the postsynaptic cell. The postsynaptic cell may respond by down-regulating (decreasing) the number of receptors it places into the synapse, in order to become less sensitive and more like the predrug state. In (B), a competitive antagonist instead blocks an endogenous substance from having its usual effect, to which the postsynaptic cell may respond by up-regulating (increasing) the number of its receptors in order to become more sensitive and compensate for the lessened effect of the endogenous ligand.

withdrawal symptom An uncomfortable symptom that arises when a person stops taking a drug that he or she has used frequently, especially at high doses.

sensitisation A process in which the body shows an enhanced response to a given drug after repeated doses.

local anaesthetic A drug, such as procaine or lidocaine, that blocks sodium channels to stop neural transmission in pain fibres.

other drugs in the opiate category, including codeine, morphine, and methadone. For drugs that have multiple effects in the body, tolerance to the various effects may develop at different rates. Once established, drug tolerance is believed to be a major cause of **withdrawal symptoms**, the unpleasant sensations that occur when one stops using a drug. As we will discuss later, one factor that maintains drug addiction is the avoidance of the physical discomfort of withdrawal symptoms.

Furthermore, some drug responses can become *stronger* with repeated treatments, rather than weaker. Termed **sensitisation**, this effect is thought to contribute to the drug craving that people with addiction may experience. This heightened sensitivity may last for a prolonged period and is associated with long-term brain changes (Peris et al., 1990), altering the function of the mesolimbocortical DA reward pathway (Vialou et al., 2012).

3.5 Drugs Affect Each Stage of Neural Conduction and Synaptic Transmission

Learning Objectives

After reading this section, you should be able to:

3.5.1 **Summarise the ways in which drugs alter presynaptic processes, with examples.**

3.5.2 **Summarise drug effects on postsynaptic processes, with examples.**

3.5.3 **Define autoreceptors and explain their function, using caffeine as an example.**

3.5.4 **Describe the processes that terminate transmitter action at synapses.**

Local anaesthetics such as procaine (trade name Novocain) block sodium channels and therefore action potentials (see Chapter 3) in the axons reporting pain from, say, that tooth your dentist is working on. But the great majority of drugs that act on the nervous system to produce changes in behaviour do so by affecting synaptic transmission.

Some drugs alter presynaptic events

One of the ways that a drug may change synaptic transmission is by modifying the behaviour of the presynaptic neuron, changing the system that converts an electrical signal (an action potential) into a chemical signal (secretion of neurotransmitter). **FIGURE 3.11** illustrates presynaptic processes that are targeted by CNS drugs, with examples of each kind of drug. The most common presynaptic drug effects can be grouped into three main categories: (1) effects on transmitter production, (2) effects on transmitter release, and (3) effects on transmitter clearance.

EFFECTS ON TRANSMITTER PRODUCTION In order for the presynaptic neuron to produce neurotransmitter, the axon terminals must receive a steady supply of raw materials and the enzymes that use them to synthesise transmitter molecules, vesicles, and other important components.

Some drugs alter this process in various ways (**FIGURE 3.11A**). For example, a drug may inhibit the enzymes that neurons use to synthesise a particular neurotransmitter, resulting in depletion of that transmitter. Alternatively, drugs that block axonal transport prevent materials from reaching the axon terminals in the first place, likewise causing the terminals to run out of neurotransmitter. In both cases, affected presynaptic neurons are prevented from having their usual effects on postsynaptic neurons, with sometimes profound effects on behaviour. A third class of drug (e.g., reserpine) doesn't prevent the *production* of transmitter molecules but instead interferes with the ability of neurons to *store* catecholamine transmitters in synaptic vesicles, for later release. This has the effect of depleting transmitter at catecholaminergic synapses and may have a complicated effect on behaviour, depending on whether and how much transmitter reaches the postsynaptic cell.

EFFECTS ON TRANSMITTER RELEASE Even if a presynaptic terminal has an adequate supply of transmitter stored in vesicles, various agents or conditions

A Effects on Transmitter Production

1 Inhibition of transmitter synthesis
Example: *Para*-chlorophenylalanine inhibits tryptophan hydroxylase, preventing synthesis of serotonin from its metabolic precursor.

2 Blockade of axonal transport
Example: Colchicine impairs maintenance of microtubules and blocks axonal transport.

3 Interference with the storage of transmitters
Example: Reserpine blocks the packaging of transmitter molecules within vesicles, thereby allowing the transmitter to be broken down by enzymes.

B Effects on Transmitter Release

4 Prevention of synaptic transmission
Example: Tetrodotoxin, found in puffer fish, blocks voltage-gated Na^+ channels and prevents nerve conduction.

5 Alteration of synaptic transmitter release
Example: Calcium channel blockers (e.g., verapamil) inhibit release of transmitters. Amphetamine stimulates release of catecholamine transmitters. Black widow causes overrelease, and thus depletion, of ACh.

6 Alteration of transmitter release through modulation of presynaptic activity
Example: Caffeine competes with adenosine for presynaptic receptors, thus preventing its inhibitory effects.

C Effects on Transmitter Clearance

7 Inactivation of transmitter reuptake
Example: Cocaine and amphetamine inhibit reuptake mechanisms, thus prolonging synaptic activity. Certain antidepressants inhibit serotonin or norepinephrine reuptake.

8 Blockade of transmitter degradation
Example: Some drugs (e.g., monoamine oxidase [MAO] inhibitors) inhibit enzymes that normally break down neurotransmitter molecules in the axon terminal or in the synaptic cleft. As a result, transmitter remains active longer and to greater effect.

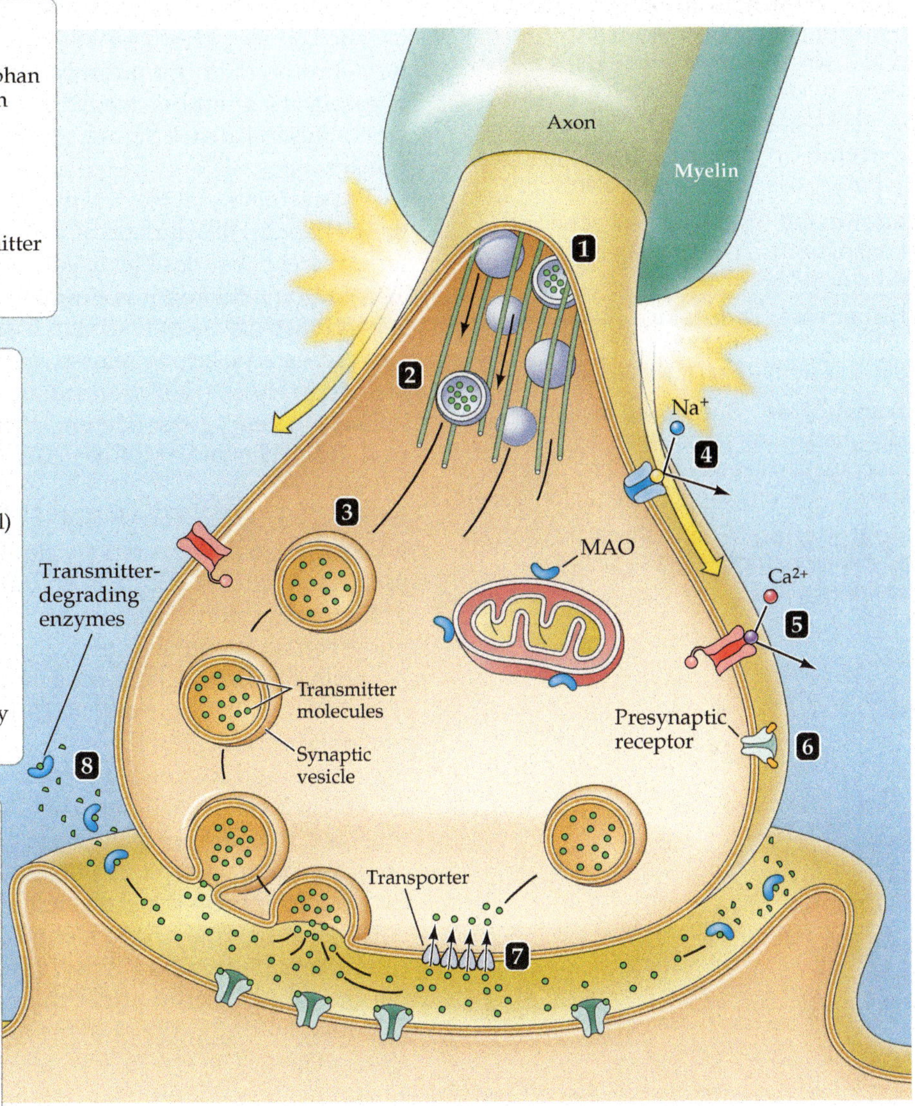

3.11 Drug Effects on Presynaptic Mechanisms

can prevent the *release* of transmitter when an action potential reaches the terminal (**FIGURE 3.11B**). For example, compounds that block sodium channels (like *tetrodotoxin*, the toxin that makes puffer fish a dangerous delicacy) prevent axons from firing action potentials, shutting down synaptic transmission with deadly results. And drugs called *calcium channel blockers* do exactly as their name suggests, blocking the calcium influx that normally drives the release of transmitter into the synapse.

Some toxins prevent the release of *specific* kinds of transmitters (de Paiva et al., 1993). For instance, the active ingredient in Botox—botulinum toxin, which is formed by a bacterium that multiplies in improperly canned food—binds to specialized receptors in nicotinic cholinergic membranes and is transported into the cell, where it cuts up SNARE proteins, preventing the Ca^{2+}-dependent release of acetylcholine (McMahon et al., 1992), resulting in muscle paralysis. The widespread paralysis that occurs after eating contaminated food can be lethal, but when the dilute toxin is selectively injected into facial muscles, the resulting local paralysis reduces wrinkling of the overlying skin. Tetanus (lockjaw) bacteria produce a toxin, called *tetanospasmin*, that also interferes with the SNAREs. By blocking exocytosis, particularly at inhibitory synapses in the CNS, tetanospasmin causes the loss of normal inhibitory influences on motor neurons, resulting in strong involuntary contractions of the muscles.

autoreceptor A receptor for a synaptic transmitter that is located in the presynaptic membrane, telling the axon terminal how much transmitter has been released.

caffeine A stimulant compound found in coffee, cacao, and other plants.

adenosine In the context of neural transmission, a neuromodulator that alters synaptic activity.

transmitter reuptake The reabsorption of synaptic transmitter by the axon terminal from which it was released.

transporter Specialized receptor in the presynaptic membrane that recognizes neurotransmitter molecules and returns them to the presynaptic neuron for reuse.

degradation The chemical breakdown of a neurotransmitter into inactive metabolites.

A different approach involves modifying the systems that the neuron normally uses to monitor and regulate its own transmitter release. For example, presynaptic neurons often use **autoreceptors** to monitor how much transmitter they have released; it's a kind of feedback system. Drugs that alter autoreceptor signals provide a false feedback signal, prompting the presynaptic cell to release more or less transmitter.

The **caffeine** that we obtain from coffee and other beverages (worldwide, we drink more than 2.2 *billion* cups of coffee *every day*) acts by blocking the autoreceptor effect of an endogenous ligand called **adenosine** (Lindskog et al., 2002). Adenosine acts as a neuromodulator: it is normally co-released with primary transmitters to control synaptic activity by inhibiting transmitter release. So, by blocking adenosine, caffeine *increases* catecholamine release, resulting in the arousal for which coffee is renowned (McLellan et al., 2016). Interestingly, consuming caffeine after a period of studying may improve memory consolidation in humans for some, but not all, learning tasks (Borota et al., 2014; Hussain and Cole, 2015).

EFFECTS ON TRANSMITTER CLEARANCE Immediately after being released, transmitter is rapidly cleared from the synapse. Obviously, getting rid of the used transmitter is normally an important step because until it is gone, new releases of transmitter from the presynaptic side won't be able to have much extra effect. But under certain circumstances, a significant reduction of synaptic transmitter availability may contribute to disorders such as depression. There are several pharmacological strategies for altering the clearance process. Some drugs interfere with **transmitter reuptake** by blocking the specialized proteins, called **transporters**, that normally remove neurotransmitter from the cleft for reuse (**FIGURE 3.11C**) (see also Chapter 2). For example, we'll see that some very common antidepressants inhibit the reuptake of serotonin, allowing the transmitter to accumulate and have a larger effect on postsynaptic receptors. Such drugs are said to work presynaptically, because the transporters are on the presynaptic terminal.

Other drugs may have a similar effect by inhibiting **degradation**, the chemical process of breaking down neurotransmitter into inactive metabolites, allowing the transmitter to affect the postsynaptic cell longer (see Figure 3.11, step 8). Agents that inhibit the enzyme acetylcholinesterase (AChE), called *acetylcholinesterase inhibitors*, allow ACh to remain active at the synapse and alter the timing of synaptic transmission. They include certain pesticides and chemical weapons, and they produce prolonged contraction of muscles and resultant paralysis, as well as overactivity of the parasympathetic nervous system.

Some drugs may alter postsynaptic processes

Another powerful way for drugs to affect synaptic transmission is to modify the activity of the postsynaptic neuron. As illustrated in **FIGURE 3.12**, there are two major classes of postsynaptic drug actions: (1) direct effects on transmitter receptors and (2) effects on cellular processes within the postsynaptic neuron.

EFFECTS ON TRANSMITTER RECEPTORS As we discussed earlier in the chapter, receptor antagonists bind directly to postsynaptic receptors and block them from being activated by their neurotransmitters (**FIGURE 3.12A**). This can have immediate and dramatic effects. Curare, for example, blocks the nicotinic ACh receptors found on muscles, resulting in immediate paralysis of all skeletal muscles, including those used for breathing (which is why curare is an effective arrow poison). Selective receptor agonists bind to specific receptors and activate them, mimicking the natural neurotransmitter at those receptors. These drugs are often very potent, with effects that vary depending on the particular types of receptors activated. LSD is an example, producing bizarre visual experiences through strong stimulation of a subtype of serotonin receptor (5-HT_{2A} receptors) found in visual cortex.

EFFECTS ON CELLULAR PROCESSES When they bind to their matching receptors on postsynaptic membranes, neurotransmitters can stimulate a variety

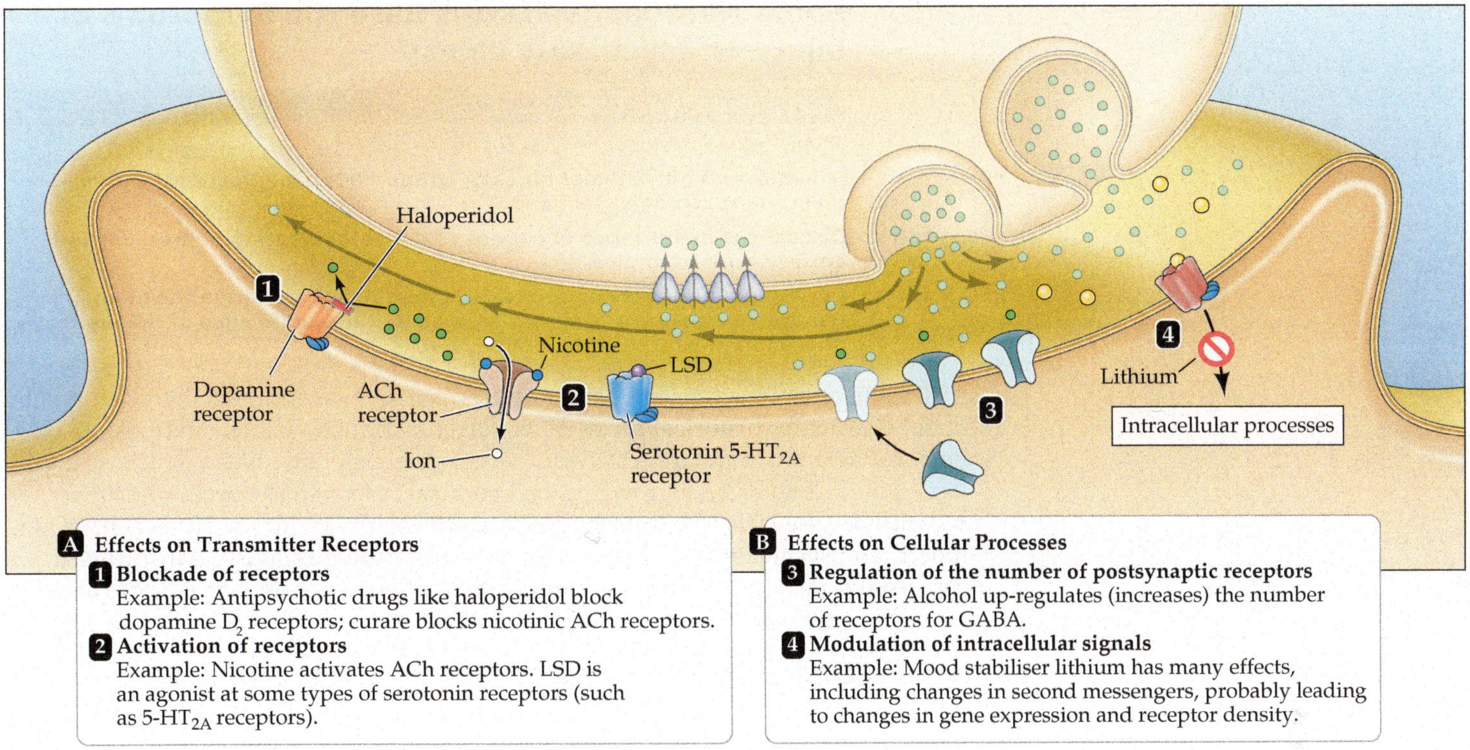

3.12 Drug Effects on Postsynaptic Mechanisms

of changes within the postsynaptic cell, such as activation of second messengers, modulation of gene expression and the production of various proteins (**FIGURE 3.12B**). As we saw earlier, some drugs induce the postsynaptic cell to up-regulate (increase) its receptors, thus changing the sensitivity of the synapse, while other drugs cause a down-regulation of receptor density. Some drugs directly alter second-messenger systems, with widespread effects in the brain; this is one of the ways in which the mood-stabilising drug lithium chloride is believed to act. Future research will probably focus on drugs to selectively activate, alter, insert, or block targeted genes within the DNA of neurons. These *genomic* effects could produce profound long-term changes in the structure and function of the targeted neurons.

DREADDs (designer receptors exclusively activated by designer drugs) Artificially modified ACh receptors that respond solely to matching exogenous ligands.

Exploiting the selective binding of drugs to postsynaptic receptors, researchers have learned how to force cells to make custom-designed receptors that don't normally occur in nature—and which respond only to custom-made drugs (B. L. Roth, 2016; Urban and Roth, 2015). By arranging to have these **DREADDs**, that is, **designer receptors exclusively activated by designer drugs**, expressed *only* by neurons of interest—say, within a particular region, or making a particular kind of transmitter or hormone—researchers can then use the designer drugs to selectively activate or inhibit only the targeted cells (**FIGURE 3.13**). This provides a powerful way to assess the function of the targeted neurons in an otherwise unaltered animal.

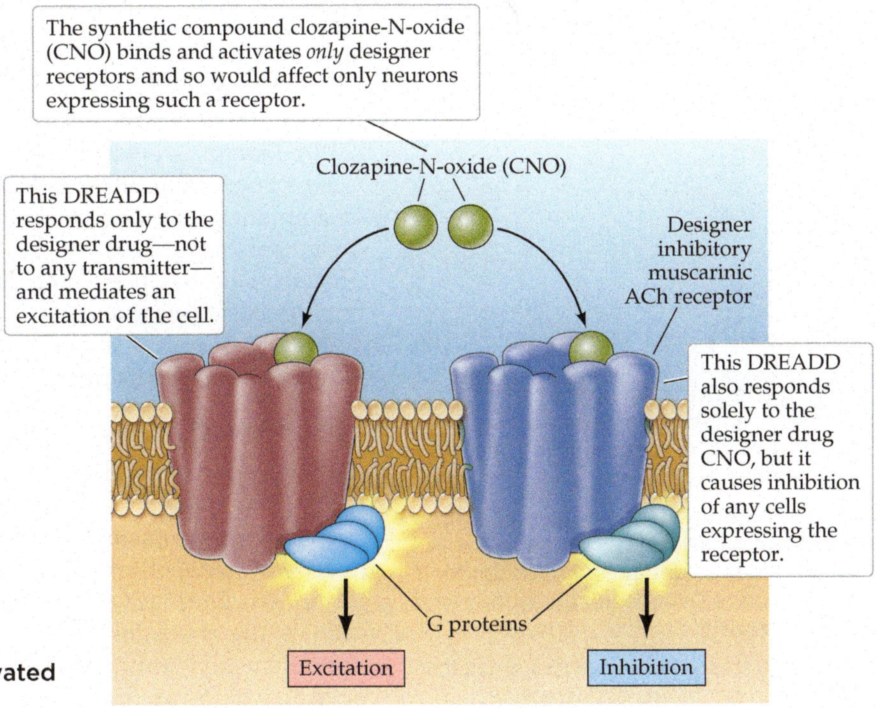

3.13 Designer Receptors Exclusively Activated by Designer Drugs (DREADDs)

3.6 Some Neuroactive Drugs Ease the Symptoms of Injury or Psychiatric Illness

Learning Objectives

After reading this section, you should be able to:

3.6.1 **Summarise the two major types of antipsychotic medications, and review their pharmacological actions.**

3.6.2 **Discuss the major types of actions of drugs for treating depression and anxiety, with examples.**

3.6.3 **Review the discovery of opiates and their major actions in the brain. Using the opiate receptors as examples, discuss the significance of the discovery of orphan receptors in the brain.**

Mental disorders have tormented humanity across the centuries: there can be little doubt that many historical accounts of sorcery, strange visions, and possession by demons had their foundation in symptoms of severe mental illness, which we discuss in detail in Chapter 12. But where the historical response to psychiatric illness was to lock away the afflicted, neuroscience breakthroughs of the last 70 years have revolutionised psychiatry and liberated millions from the purgatory of institutionalised care. In the sections that follow, we will briefly review some of the major categories of psychoactive drugs, based on how they affect behaviour.

Antipsychotics relieve symptoms of schizophrenia

It's hard to believe, but prior to the 1950s about half of all hospital beds in the United States were taken up by psychiatric patients (Menninger, 1948), and a majority of institutionalised psychiatric patients were experiencing the delusions and hallucinations of schizophrenia. This awful situation was suddenly dramatically improved by the development of a family of drugs called *antipsychotics* (or *neuroleptics*). The first of these drugs, chlorpromazine (Thorazine), and successors like haloperidol (Haldol) and loxapine (Loxitane) all share one crucial feature: they act as selective antagonists of dopamine D_2 receptors in the brain. These **first-generation antipsychotics** are so effective in relieving emergent *positive symptoms* of schizophrenia, such as hallucinations and delusions, that a dopaminergic model of the disease became dominant.

The 1990s saw the advent of **second-generation antipsychotics** that have both dopaminergic activity and additional, nondopaminergic actions, especially the blockade of certain serotonin receptors. These drugs may be helpful in relieving symptoms that are resistant to the first-generation antipsychotics—especially the *negative symptoms* that involve impairment or loss of a behaviour, such as social withdrawal and blunted emotional responses—but early hopes that these drugs would generally outperform first-generation antipsychotics have not been borne out (Kane and Correll, 2010). Emerging evidence that schizophrenia also involves transmitters beyond the classic targets has since prompted an intense research effort to develop third-generation antipsychotics, with novel targets like glutamate and oxytocin, but effective treatments remain elusive (M. C. Davis et al., 2014; O'Tuathaigh et al., 2017). So, although there has been progress in its treatment, schizophrenia remains a difficult, multifaceted disease and a major health problem, with no single type of recommended treatment strategy yet agreed upon (Correll et al., 2017). We will return to the topic of schizophrenia and its treatment in Chapter 12.

Antidepressants reduce chronic mood problems

Disturbances of mood, or *affective disorders*, are among the most common of all psychiatric complaints (World Health Organisation, 2001). The first generation of effective **antidepressant** medications, developed in the 1950s, were the monoamine oxidase (MAO) inhibitors, such as tranylcypromine (Parnate) and isocarboxazid (Marplan). Normally, MAOs break down monoamine neurotransmitters at axon terminals, thereby reducing transmitter activity. By blocking this process, MAO inhibitors allow monoamine neurotransmitters to accumulate at synapses (see Figure 3.11, step 8), with an associated improvement in mood.

first-generation antipsychotics A class of antischizophrenic drugs whose principal mode of action is antagonism of dopamine D_2 receptors.

second-generation antipsychotics A class of antischizophrenic drugs that have actions other than or in addition to the dopamine D2 receptor antagonism that characterises the typical antipsychotics.

antidepressants A class of drugs that relieve the symptoms of depression.

Increasing synaptic monoamine availability appears to be a key activity of all antidepressants. The second generation of antidepressants—the **tricyclics**—combat depression by increasing the synaptic content of the monoamines norepinephrine and serotonin. Named for their three-ringed molecular structure, tricyclics such as imipramine (Tofranil) block the reuptake of neurotransmitters into presynaptic axon terminals (see Figure 3.11, step 7), thereby allowing the transmitter to accumulate in the synapse. More recently developed antidepressants, such as fluoxetine (Prozac), sertraline (Zoloft), and citalopram (Celexa), are **selective serotonin reuptake inhibitors** (**SSRIs**); as their names indicate, these drugs alleviate depression by selectively causing serotonin to accumulate in synapses. Related compounds classified as **serotonin-norepinephrine reuptake inhibitors** (**SNRIs**) such as venlafaxine (Effexor), additionally inhibit the reuptake of norepinephrine and have therapeutic applications that extend beyond their antidepressant applications to include anxiety disorders. These newer antidepressants lack some of the undesirable side effects of older drugs, although they have side effects of their own (such as disturbances of sexual function) and can take as long as 6–8 weeks to have full effect. We will discuss the causes and treatment of affective disorders in more detail in Chapter 12.

Anxiolytics combat anxiety

Most of us occasionally suffer feelings of vague dissatisfaction or apprehension that we call anxiety, but some people are stricken by disabling emotional distress that resembles abject fear and terror. These clinical states of anxiety include panic attacks, phobias (such as the fear of taking an airplane or even of leaving the house), and generalized anxiety.

Humans have long sought relief from anxiety through the ingestion of **anxiolytics** (from the word *anxiety* and the Greek *lytikos*, "able to loosen"). Sometimes also called *tranquilizers*, anxiolytics belong to the general category of **depressants**: drugs that depress or reduce nervous system activity. Alcohol and opiates are perhaps the original anxiolytics, but their anxiety-fighting properties come at the cost of intoxication, addiction potential, and neuropsychological impairment with long-term abuse, so they are not suitable for therapeutic use.

Barbiturate drugs were originally developed to reduce anxiety, promote sleep, and prevent epileptic seizures. They are still used for those purposes but are also addictive and easily overdosed, often fatally, as illustrated in Figure 3.9E. Since the 1970s the most widely prescribed anxiolytics have been the **benzodiazepine agonists**, which are both more specific and safer than the barbiturates. Members of this class of drug, like alprazolam (trade name Xanax) and lorazepam (Ativan), bind to specific sites on $GABA_A$ receptors and enhance the activity of GABA (R. W. Olsen, 2018). Because $GABA_A$ receptors are inhibitory, benzodiazepines help GABA to produce larger inhibitory postsynaptic potentials than would be caused by GABA alone. This has the end result of reducing the excitability of neurons.

$GABA_A$ receptors have several different binding sites—some that facilitate and some that inhibit the effect of GABA (**FIGURE 3.14**)—therefore many different drugs can interact with this receptor complex to alter its shape and function (Masiulis et al., 2019). For example, benzodiazepines bind to a unique modulatory site on the receptor complex that is distant from where GABA itself binds.

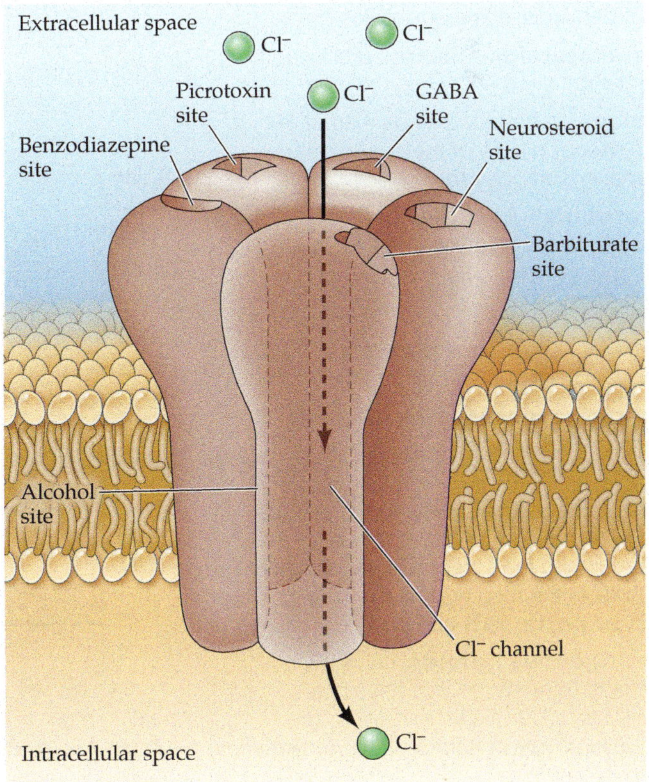

3.14 The $GABA_A$ Receptor Has Many Different Binding Sites The $GABA_A$ complex is made up of five protein subunits that penetrate the cell membrane and surround a Cl– ion channel at the core. These subunits contain many different recognition and receptor sites; some of the most important ones are labelled here. $GABA_A$ receptors are widespread in the brain and are crucial for normal inhibitory processes. Many depressant drugs work by increasing the responsiveness of $GABA_A$ receptors to GABA.

tricyclics A class of drugs that act by increasing the synaptic accumulation of serotonin and norepinephrine.

selective serotonin reuptake inhibitor (SSRI) A drug that blocks the reuptake of transmitter at serotonergic synapses.

serotonin-norepinephrine reuptake inhibitor (SNRI) A drug that blocks the reuptake of transmitter at both serotonergic and noradrenergic synapses.

anxiolytics A class of substances that are used to combat anxiety.

depressants A class of drugs that act to reduce neural activity.

barbiturate A powerful sedative anxiolytic derived from barbituric acid, with dangerous addiction and overdose potential.

benzodiazepine agonists A class of antianxiety drugs that bind to sites on $GABA_A$ receptors.

3.15 The Source of Opium and Morphine The opium poppy has a distinctive flower and seedpod. The bitter flavour and CNS actions of opium may provide the poppy plant a defense against being eaten.

The benzodiazepine-binding site is thus an **orphan receptor**—a receptor for which an endogenous ligand has not been conclusively identified—and the hunt for its endogenous ligand has been intense. **Allopregnanolone**, a steroid derived from the hormone progesterone, acts on yet another site on the $GABA_A$ receptor. Allopregnanolone is elevated during stress and has a calming effect (Melcangi and Panzica, 2014). Alcohol ingestion also increases brain concentrations of allopregnanolone, so this steroid may mediate some of the calming influence of alcohol. Several other progesterone-like **neurosteroids** (steroids produced in the brain) may act on $GABA_A$ receptors to produce anxiolytic, analgesic, and anticonvulsant effects (Belelli et al., 2018; Gunn et al., 2015).

Other categories of anxiolytics under active investigation affect other transmitter systems, such as serotonergic and neuropeptide mechanisms (Griebel and Holmes, 2013). The serotonergic agonist buspirone (BuSpar) is an effective anxiolytic that lacks the sedative effects of benzodiazepines. Buspirone is a $5\text{-}HT_{1A}$ agonist and a partial agonist of dopamine D_2 receptors, but the exact mechanism of its anxiolytic action remains unknown.

Opiates potently relieve pain

Opium, extracted from poppy flower seedpods (**FIGURE 3.15**), has been used by humans since at least the Stone Age. **Morphine**, the major active substance in opium, is a very effective **analgesic** (painkiller) that has brought relief from severe pain to many millions of people (see Chapter 7). Unfortunately, morphine also has a strong potential for addiction, as does its close relative **heroin** (diacetylmorphine) and powerful opiate painkillers like oxycodone (OxyContin) and fentanyl, a synthetic opiate that is 30–40 times stronger than heroin. Accidental fentanyl overdose is a growing epidemic: the fatal fentanyl overdose of the musician Prince, in 2016, was just one of thousands of similar tragedies each year (**FIGURE 3.16**).

The opiates morphine, heroin, and codeine bind to specific receptors—**opioid receptors**—that are concentrated in certain regions of the brain. Opioid receptors are found in the limbic and hypothalamic areas of the brain, and they are particularly rich in the locus coeruleus and in the **periaqueductal gray**—the gray matter that surrounds the cerebral aqueduct in the brainstem (**FIGURE 3.17**). Injection of morphine directly into the periaqueductal gray produces strong analgesia, indicating that this is a region where morphine acts to reduce pain perception (see Chapter 7). The discovery of orphan receptors for opiates came as a surprise, and because the presence of these receptors implied that there must be an endogenous ligand produced within the body, it prompted an intense scientific effort to identify **endogenous opioids**.

orphan receptor Any receptor for which no endogenous ligand has yet been discovered.

allopregnanolone A naturally occurring steroid that modulates GABA receptor activity in much the same way that benzodiazepine anxiolytics do.

neurosteroids Steroids produced in the brain.

opium A heterogeneous extract of the seedpod juice of the opium poppy, *Papaver somniferum*.

morphine An opiate compound derived from the poppy flower.

analgesic Referring to painkilling properties.

heroin Diacetylmorphine; an artificially modified, very potent form of morphine.

opioid receptor A receptor that responds to endogenous and/or exogenous opiates.

periaqueductal gray The neuronal body–rich region of the midbrain surrounding the cerebral aqueduct that connects the third and fourth ventricles. It is involved in pain perception.

endogenous opioids A family of peptide transmitters that have been called the body's own narcotics. The three kinds are enkephalins, endorphins, and dynorphins.

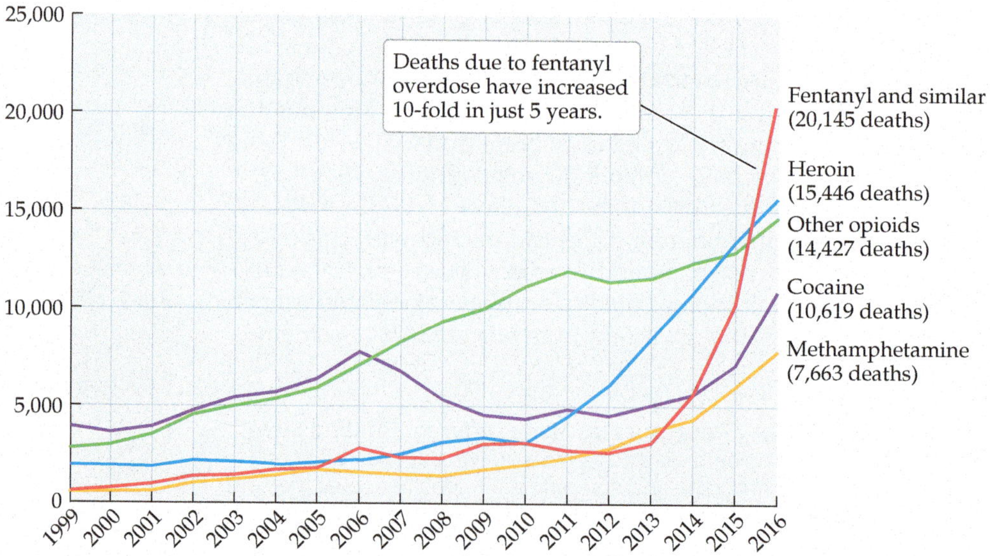

3.16 An Epidemic of Overdose Deaths (After NIDA, 2017, based on data from National Centre for Health Statistics, CDC Wonder Database.)

3.17 The Distribution of Opioid Receptors in the Rat Brain This horizontal section (rostral is at the top) shows opioid receptors widely distributed in the brain (the areas of highest binding are shown in yellow, orange, and red). They are concentrated in the medial thalamus and in some brainstem areas: the periaqueductal gray and the inferior colliculus.

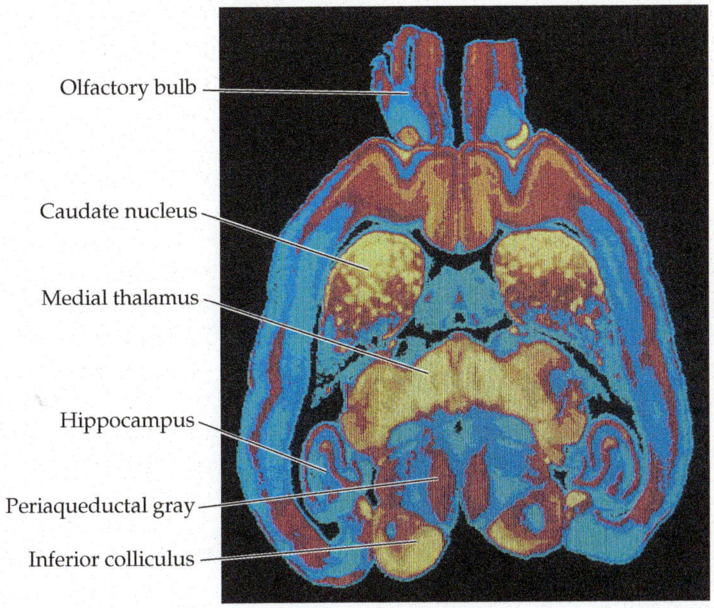

Courtesy of Miles Herkenham, National Institute of Mental Health

Researchers have now identified three major families of these potent peptides (see Table 3.1): the **enkephalins** (from the Greek *en*, "in," and *kephale*, "head"), the **endorphins** (a contraction of *endogenous morphine*), and the **dynorphins** (short for *dynamic endorphins*, in recognition of their potency and speed of action). Like morphine, enkephalins relieve pain and are addictive. Only a small part of the enkephalin molecule is the same as the morphine molecule, but this common part is what binds to the opioid receptor.

There are three main kinds of opioid receptors—the delta (δ), kappa (κ), and mu (μ) opioid receptors—all of which are metabotropic receptors. Powerful drugs that block opiate receptors—naloxone and naltrexone are examples—can rapidly reverse the effects of opiates and rescue people from overdose. Opiate antagonists also block the rewarding aspects of drugs like heroin, so they can be helpful for treating addiction, as we'll discuss at the end of this chapter. For some alcoholics, naltrexone treatment blocks the euphoria that normally results from alcohol, suggesting that in these individuals alcohol causes the release of endogenous opioids that brings pleasure. Interestingly, only that minority of alcoholics who carry a gene for a particular variant of the mu opioid receptor benefit from naltrexone treatment (Oroszi et al., 2009).

enkephalins One of three kinds of endogenous opioids.

endorphins One of three kinds of endogenous opioids.

dynorphins One of three kinds of endogenous opioids.

cannabis A dried preparation of the *Cannabis sativa* plant, typically smoked to obtain THC and CBD.

3.7 Some Neuroactive Drugs Are Used to Alter Conscious Experiences

Learning Objectives

After reading this section, you should be able to:

3.7.1 **Identify the main active ingredients in cannabis, their sites of action in the brain, and their roles in behavioural processes.**

3.7.2 **Compare and contrast the main families of stimulant drugs, and explain how they interact with neurons, their major effects, and risks they pose to human health.**

3.7.3 **Review the modes of action of alcohol in the brain.**

3.7.4 **Discuss the prevalence of alcohol use disorders, negative effects on the brain, and scope for recovery of the brain in recovering alcoholics.**

3.7.5 **Define hallucinogenic drugs, and highlight the ways in which the major hallucinogens have shared or different actions in the brain.**

Humans have a long history of tinkering with their conscious experience of the world. Some of the major classes of neuroactive drugs that modify consciousness include cannabinoids, stimulants, alcohol, and hallucinogenic and dissociative drugs.

Cannabinoids have a wide array of effects

Cannabis and related preparations, such as hashish, are derived from the *Cannabis sativa* plant (**FIGURE 3.18**) and have been used in various cultures for thousands of years

© David Buzzard/Alamy Stock Photo

3.18 Relaxation Cannabis laws are being relaxed in many jurisdictions. Legal access to cannabis, via licensed shops like this one in Canada, acknowledges existing widespread use of cannabis for recreational and medicinal purposes and is expected to reduce criminal activity. However, concerns about health risks in adolescents and heavy users remain.

Δ9-tetrahydrocannabinol (THC) The major active ingredient in cannabis thought to be responsible for the "high" feeling.

cannabidiol (CBD) An active ingredient in cannabis thought to be responsible inducing a feeling of relaxation.

cannabinoid receptor A category of cell membrane receptor proteins that selectively recognize and respond to cannabinoid transmitters and drugs.

endocannabinoid An endogenous ligand of cannabinoid receptors, thus an analog of cannabis that is produced by the brain.

anandamide An endogenous substance that binds the cannabinoid receptor molecule.

(E. B. Russo et al., 2008). The common alternative name *marijuana* is considered by some to have dubious, possibly racist origins (Halperin, 2018). Typically administered by smoking, by vaping, or via edible products such as cookies and candies, cannabis contains dozens of active ingredients, especially the compounds **Δ9-tetrahydrocannabinol** (**THC**), which is thought to produce the psychoactive effect of "feeling high" (Boggs et al., 2018), and **cannabidiol** (**CBD**). Cannabis use usually produces pleasant relaxation and mood alteration, although the drug can occasionally cause stimulation and paranoia instead.

Occasional use of cannabis seems to be mostly harmless, but as with other substances, heavy use can be harmful. For example, persistent heavy use (i.e., ongoing use of cannabis four or more times each week) may be associated with respiratory problems, addiction, psychiatric disorders, and cognitive decline (Meier et al., 2012). Adolescents who use cannabis are more likely to develop schizophrenia in adulthood (H. J. Jones et al., 2018; Marconi et al., 2016). However, it remains to be determined whether the cannabis use causes the illness or, conversely, whether adolescents who are already experiencing symptoms of mental illness may be more drawn to cannabis use (J. Bourque et al., 2017).

As was the case with opiates and benzodiazepines, researchers found that the brain contains **cannabinoid receptors** that mediate the effects of compounds like THC. Cannabinoid receptors are found in the substantia nigra, the hippocampus, the cerebellar cortex, and the cerebral cortex (**FIGURE 3.19**) (Devane et al., 1988); other regions, such as the brainstem, show few of these receptors. There are at least two subtypes of cannabinoid receptors—CB_1 and CB_2 (Gerard et al., 1991; Pertwee, 1997)—both of which are GPCRs. Genetic disruption of CB_1 receptors is sufficient to make mice unresponsive to the rewarding properties of cannabinoid drugs (Ledent et al., 1999). Only the CB_1 receptor is found in the nervous system; CB_2 receptors are especially prominent in the immune system.

The discovery of cannabinoid receptors touched off an intensive search for an endogenous ligand, and several such compounds—termed **endocannabinoids**—were identified. Interestingly, endocannabinoids can function as retrograde messengers, conveying messages from the postsynaptic cell to the presynaptic cell. This retrograde signal is thought to modulate the release of neurotransmitter by the presynaptic nerve terminal (Murray et al., 2007). The most studied endocannabinoid is **anandamide** (from the Sanskrit *ananda*, "bliss") (Devane et al., 1992), which has diverse functional effects, including alterations of memory formation, appetite stimulation, reduced sensitivity to pain, and protection from excitotoxic brain damage (Marsicano et al., 2003; P. B. Smith et al., 1994). Most CB_1 receptors are found on GABA-ergic axon terminals, and cannabinoids acting on these receptors modulate GABA release. In the case of hippocampal neurons, this action has the result—familiar to chronic cannabis users—of transient amnesia secondary to changes in protein synthesis in the postsynaptic cells (Puighermanal et al., 2012).

Cannabinoids are thus targets of a large-scale research effort aimed at developing drugs with some of the specific beneficial effects of cannabis: improving mood, relieving pain, lowering blood pressure, combating nausea, lowering eye pressure in glaucoma, and so on. Furthermore, recreational sale and use of cannabis is being legalised increasingly. It is legal in various U.S. states, nationwide in Canada, and to varying extents in other countries, and possession of cannabis for recreational or medical purposes has been "decriminalised" (i.e., tolerated while technically illegal) in many additional jurisdictions. It seems likely that the relaxation of cannabis laws will continue, so a fuller understanding of the beneficial and adverse effects of cannabis, and potential for cannabis addiction, is a high priority for researchers (Curran et al., 2016).

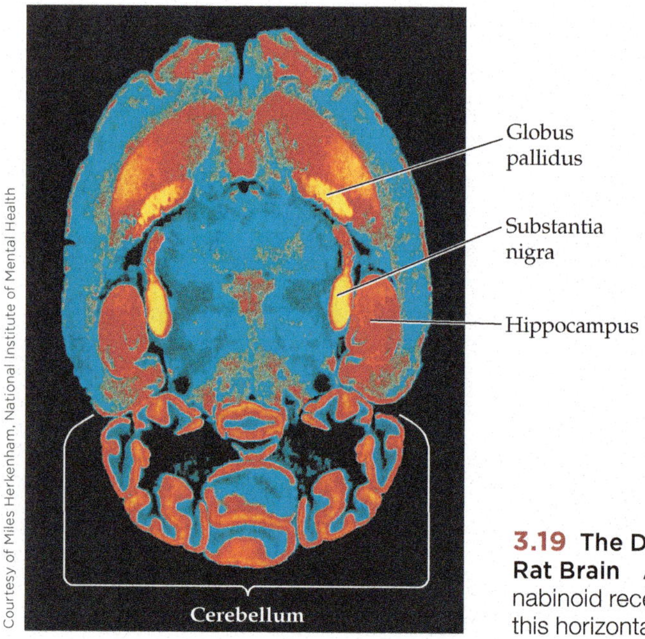

3.19 The Distribution of Cannabinoid Receptors in the Rat Brain Areas of the brain with high concentrations of cannabinoid receptors are indicated by yellow, orange, and red in this horizontal section.

Stimulants increase the activity of the nervous system

Proper functioning of the nervous system involves a fine balance between excitatory and inhibitory influences. Stimulants are drugs that tip the balance towards ard the excitatory side; they therefore have an alerting, activating effect. Many naturally occurring and artificial stimulants are widely used, including nicotine, caffeine, amphetamine, and cocaine. Some stimulants act directly by increasing excitatory synaptic potentials. Others act by blocking normal inhibitory processes; we've already seen that caffeine blocks adenosine receptors. Paradoxically, stimulants such as methylphenidate (Ritalin) have a calming effect in people with attention deficit hyperactivity disorder (ADHD); this activity may be mediated by changes in activity at specific dopamine and serotonin receptors and transporters, especially in the frontal lobes (Berridge and Devilbiss, 2011; Faraone, 2018).

nicotine A compound found in plants, including tobacco, that acts as an agonist on a large class of cholinergic receptors.

cocaine A drug of abuse, derived from the coca plant, that acts by potentiating catecholamine stimulation.

NICOTINE Tobacco is native to the Americas, where European explorers first encountered smoking; these explorers brought tobacco back to Europe with them. Tobacco use became much more widespread following technological innovations that made it easier to smoke, in the form of cigarettes (W. Bennett, 1983). Exposed to the large surface of the lungs, the **nicotine** from cigarettes enters the blood and brain much more rapidly than does nicotine from other tobacco products. Nicotine increases heart rate, blood pressure, secretion of hydrochloric acid in the stomach, and intestinal activity. In the short run, these effects make tobacco use pleasurable. But these neural effects on body function, quite apart from the effects of tobacco tar on the lungs, make prolonged tobacco use very unhealthful. Smoking and nicotine exposure in adolescence have an adverse effect on attention and cognitive development, likely as a consequence of impairments of glutamatergic synapses in the prefrontal cortex (Counotte et al., 2011), as well as lasting epigenetic modifications of neural function (M. Yuan et al., 2015).

The *nicotinic* ACh receptors didn't get their name by coincidence—it is through these receptors that the nicotine from tobacco exerts most of its effects in the body. Nicotinic receptors drive the contraction of skeletal muscles and the activation of various visceral organs, but they are also found in high concentrations in the brain, including the cortex. This is one way in which nicotine enhances some aspects of cognitive performance. Astonishingly, the nicotine from one cigarette can occupy 88% of the brain's nicotinic receptors (A. L. Brody et al., 2006). Using a sophisticated optogenetic model that allowed activation or inactivation of nicotinic ACh receptors in discrete brain regions at will, researchers found that nicotine acts directly on receptors within the ventral tegmental area (see Figure 3.4) to exert its rewarding/addicting effects (Durand-de Cuttoli et al., 2018).

COCAINE For hundreds of years, people in Bolivia, Colombia, and Peru have used the leaves of the coca shrub—either chewed or brewed as a tea—to increase endurance, alleviate hunger, and promote a sense of well-being. This use of coca leaves does not seem to cause problems. But processing and purifying an extract from this plant produces a much more potent and dangerous compound: **cocaine**.

First isolated in 1859, cocaine was added to beverages (such as *Coca-Cola*) and tonics for its stimulant qualities, and subsequently it was used as a local anesthetic and as an antidepressant. But people soon discovered that the rapid hit of cocaine resulting from snorting or smoking (see Table 3.3) has a stimulant effect that is powerful and pleasurable. *Crack*, a smokable form of cocaine that appeared in the mid-1980s, enters the blood and the brain even more rapidly and thus is even more addictive than cocaine powder. However it is consumed, cocaine is highly addictive. Moreover, heavy cocaine use raises the risk of serious side effects such as stroke, psychosis, loss of gray matter in the frontal lobes, and severe mood disturbances (Crunelle et al., 2014), and it causes many changes in the structure and function of the brain, leading to high rates of relapse in users attempting to quit cocaine use (Hanlon et al., 2013). Like other psychostimulants, cocaine acts by blocking monoamine transporters, especially those for dopamine (**FIGURE 3.20**), slowing reuptake of the transmitters and therefore boosting their effects.

3.20 Cocaine-Binding Sites in the Monkey Brain This autoradiograph of a coronal section shows the distribution of cocaine-binding sites. The areas of highest binding, shown by orange and yellow, include many regions that receive dopaminergic innervation.

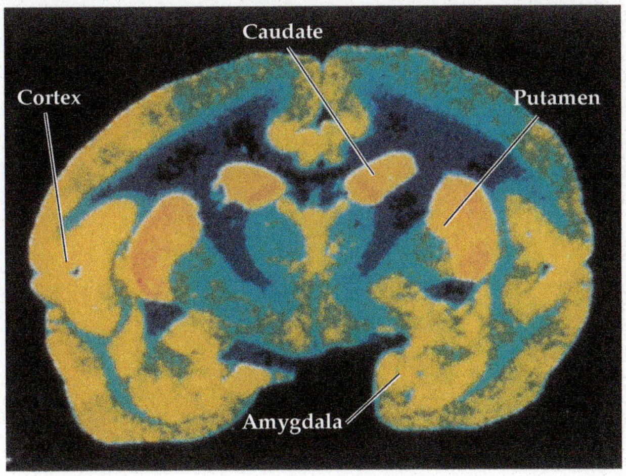

Courtesy of Bertha K. Madras and Marc J. Kaufman

dual dependence Dependence for emergent drug effects that occur only when two drugs are taken simultaneously.

amphetamine A molecule that resembles the structure of the catecholamine transmitters and enhances their activity.

methamphetamine A stimulant amphetamine derivative with strongly addictive qualities.

khat Also spelled *qat*. An African shrub that, when chewed, acts as a stimulant.

As a consequence of sensitisation, which we discussed earlier, chronic cocaine use can provoke symptoms similar to psychosis. Cessation of cocaine use often produces very uncomfortable withdrawal symptoms: initial agitation and powerful drug cravings, followed by depression and an inability to enjoy anything else in life. Cerebral glucose metabolism is decreased for months after cocaine use is discontinued and may contribute to that depression. People who use cocaine along with other substances run the added risk of **dual dependence**, in which the interaction of two (or more) drugs produces another addictive state. For example, cocaine metabolized in the presence of ethanol (alcohol) yields an active metabolite called *cocaethylene*, to which the user may develop an additional addiction (Y. Liu et al., 2018).

AMPHETAMINE The molecular structure of the synthetic psychostimulant **amphetamine** closely resembles that of several monoamine transmitters (norepinephrine, epinephrine, and dopamine). Amphetamine and its more potent relatives, especially **methamphetamine** ("meth" or "speed"), amplify the effects of monoamine transmitters in four distinct ways. First, amphetamines induce the release of these transmitters from presynaptic terminals even in the absence of action potentials. Second, when action potentials *do* reach the axon terminals, these drugs also potentiate the subsequent release of transmitter. Third, after the transmitter has been released, amphetamine analogs block the reuptake of transmitter molecules into the presynaptic terminals. And fourth, by providing an alternative target for the enzyme (monoamine oxidase) that normally inactivates transmitter molecules, amphetamines inhibit the metabolic breakdown of the released transmitter, with the result that it is present at the synapse longer.

The result of meth's various actions is thus that monoaminergic synapses become unnaturally potent, strongly affecting behaviour. Over the short term, amphetamine drugs cause increased vigour and stamina, wakefulness, decreased appetite, and feelings of euphoria. For these reasons amphetamine has historically been used in military applications and other settings where intense sustained effort is required. However, the quality of the work being performed may suffer, and the costs of amphetamine use soon outweigh the benefits. Addiction and tolerance to amphetamine and methamphetamine develops rapidly, requiring ever larger doses and leading to sleeplessness, severe weight loss, and general deterioration of mental and physical condition (**FIGURE 3.21**).

Prolonged use of amphetamine may lead to symptoms that closely resemble those of paranoid schizophrenia: compulsive, agitated behaviour and irrational suspiciousness. In fact, some amphetamine users have been misdiagnosed as having schizophrenia (see Chapter 12). Amphetamine also acts on the autonomic nervous system to produce high blood pressure, tremor, dizziness, sweating, rapid breathing, and nausea. Worst of all, people who chronically abuse meth display symptoms of brain damage long after they quit using the drug (Q. Wu et al., 2018), probably resulting from neurotoxic damage to monoaminergic neurons (Moratalla et al., 2017).

Amphetamine-like stimulants called *cathinones* are released when the African shrub **khat**

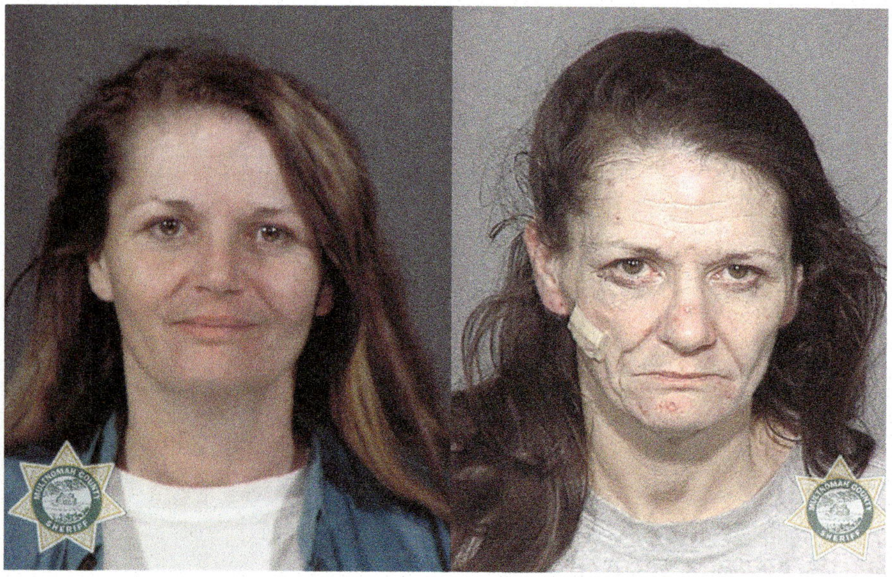

Mug shots courtesy of the Multnomah County Sheriff's Office, Oregon

3.21 Faces of Meth These before and after photos, taken just two-and-a-half years apart, are a stark testament to the heavy toll taken by chronic methamphetamine abuse. Meth causes multiple severe problems such as motor disorders, cognitive impairment, psychosis, and rapid changes in appearance due to accelerated tooth decay ("meth mouth"), skin pathology, and excessive weight loss.

(or *qat*, pronounced "cot") is chewed. Many types of synthetic cathinones—known collectively as "bath salts"—have been developed and marketed in recent years (Baumann et al., 2018). These new designer drugs, especially *mephedrone* ("plant food" or "meow-meow"), have shown a significant growth in popularity in the early twenty-first century, despite multiple serious health consequences, including psychosis, muscle disease, and kidney damage (C. M. White, 2016; Zawilska, 2014).

fetal alcohol syndrome A disorder, including intellectual disability and characteristic facial anomalies, that affects children exposed to too much alcohol (through maternal ingestion) during fetal development.

Alcohol acts as both a stimulant and a depressant

Alcohol has traveled the full route of human history, no doubt because it is so easily produced by the fermentation of fruit or grains and thus is an ingredient of many types of beverages. Taken in moderation, alcohol is harmless or even beneficial to the health of adults. For example, moderate consumption—one drink per day for women, up to two per day for men—may be associated with reduced risk of cardiovascular and Alzheimer's diseases and with improved control of blood sugar levels (Katsiki et al., 2014; Scarmeas et al., 2018). *Heavy* alcohol consumption, however, is very damaging and linked to a multitude of serious diseases (Bell et al., 2017; J. H. O'Keefe et al., 2014). About 7% of adults in the United States (approximately 17 million people) meet the diagnostic criteria for alcohol use disorder (Substance Abuse and Mental Health Services Administration, 2018).

The psychoactive effect of alcohol in the nervous system is biphasic: an initial stimulant phase is followed by a more prolonged depressant phase. This complex action reflects the fact that alcohol alters the function of multiple neurotransmitter systems, such as glutamate and GABA (Roberto and Varodayan, 2017). In particular, by acting in a manner similar to the benzodiazepines but via a different recognition site, alcohol activates the $GABA_A$ receptor–coupled chloride channel (see Figure 3.14), thereby altering postsynaptic excitability in many brain regions. This action contributes to the social disinhibition, impaired motor coordination, and sensory disturbances that we call *drunkenness* (Hanchar et al., 2005; Harrison et al., 2017). Alcohol also affects other transmitters. For example, low doses of alcohol stimulate dopamine pathways, and the resulting increase in dopamine may be related to the slightly euphoric feelings that many people experience when having a drink.

Chronic abuse of alcohol damages neurons. Cells of the superior frontal cortex, Purkinje cells of the cerebellum, and hippocampal pyramidal cells show particularly prominent pathological changes. Some of these degenerative effects of chronic alcohol use may be due to a secondary consequence of alcoholism: poor diet. For example, chronic alcoholism is accompanied by severe thiamine deficiency, which can lead to neural degeneration and Korsakoff's syndrome (see Chapter 16). Heavy drinking by expectant mothers can cause grievous permanent brain damage to the developing fetus termed **fetal alcohol syndrome**, a topic to which we will return in Chapter 6 (see Figure 6.17). Furthermore, although it was previously thought that low levels of alcohol consumption were safe during pregnancy, as few as one or two drinks per week may be associated with lower fetal weight and preterm birth (Mamluk et al., 2017), changes in craniofacial development (Muggli et al., 2017), and later behavioural problems (J. Murray et al., 2016). Such findings, combined with a paucity of information on the impact of light drinking on other aspects of fetal health, such as cognitive function, and clear evidence of fetal risks from alcohol in studies with lab animals (Valenzuela et al., 2012), encourage an abundance of caution: the CDC currently advises that there is *no* known safe level of alcohol use in pregnancy (Centres for Disease Control, 2016).

Chronic heavy alcohol abuse is associated with a spectrum of pathological changes in the brain. Especially prominent is widespread degeneration of cerebral white matter pathways, along with extensive atrophy of cortical regions, including the frontal and cingulate cortices, hippocampus, and medial temporal lobes, and damage to the diencephalon and cerebellum (de la Monte and Kril, 2014). However, some of the anatomical changes associated with chronic alcoholism may be reversible with abstinence. In humans recovering from alcoholism, MRI studies show an increase in the volume of cortical gray matter and an associated reduction in ventricular volume within days of giving up alcohol (**FIGURE 3.22**). Improved volumes are evident

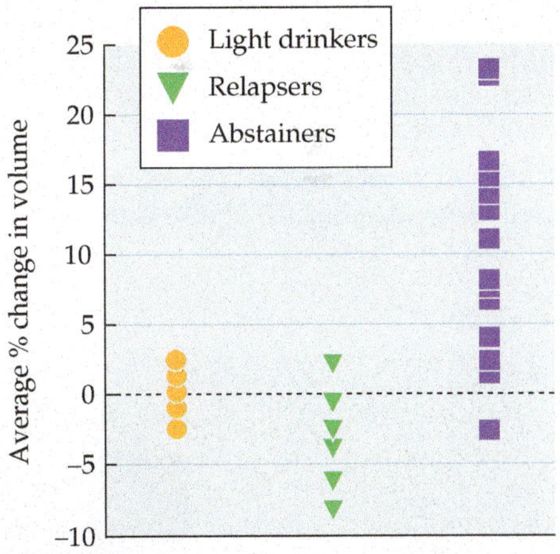

3.22 Changes in Brain Volume after Just One Week of Abstinence in Alcoholics Compared with a nonalcoholic control group and a relapsed alcoholic group, the average volume of ten selected brain regions was notably increased in a group of alcoholics after just one week of abstinence. While the extent of cognitive improvement that accompanies these changes is uncertain, such results illustrate the potential benefit of reduced intake in heavy drinkers. (After V. A. Cardenas et al., 2007. *NeuroImage* 34: 879–887.)

throughout the cerebral cortex, especially frontal, temporal, cingulate, and insular regions, along with increases in the amygdala, thalamus, and cerebellum (Cardenas et al., 2007). Likewise, atrophic changes in white matter associated with alcoholism appear to at least partly recover in abstinence (Pfefferbaum et al., 2014; Zahr and Pfefferbaum, 2017). While these results are encouraging, it remains unclear to what extent cognitive function improves along with the recovery of brain volume.

There is a strong hereditary component to alcoholism, as indicated by studies of alcohol use and abuse within human families and by selective breeding experiments in rats (Crabbe, 2014; Reilly et al., 2017). The transition from intermittent overconsumption to a more stable and treatment-resistant pattern of alcoholism may involve up-regulation of a specific form of NMDA receptor in the nucleus accumbens, part of the brain's reward circuitry (Seif et al., 2013) that we'll discuss shortly.

Even in the absence of clear-cut alcoholism, periodic overconsumption of alcohol, or *bingeing*—defined as four (female) or five (male) drinks on a single occasion (Substance Abuse and Mental Health Services Administration, 2018)—may cause brain damage. After only 4 days of bingeing on alcohol, rats exhibit neural degeneration in several brain regions. Damage is especially evident in the olfactory bulbs and in limbic structures connected with the hippocampus, and it is associated with impairments of cognitive ability (Obernier et al., 2002). Evidence suggests that some of the damage caused by bingeing may be reversible (Zahr et al., 2009), although the extent of recovery remains to be determined. Alcohol bingeing also significantly reduces the rate of neurogenesis and carries the highest risk of fetal alcohol syndrome (Gupta et al., 2016). The damaging effects of alcohol on the brain are particularly evident in adolescents and may predispose the future development of more severe alcohol use disorders (Crews et al., 2016; S. Wilson et al., 2015). Alcohol bingeing can depress breathing enough to kill, as happens every year to a few college students. Many additional college students die from alcohol-related accidents (Hingson et al., 2005).

Hallucinogenic and dissociative drugs alter sensory perception

hallucinogens A class of drugs that alter sensory perception and produce peculiar experiences.

LSD Also called *acid*. Lysergic acid diethylamide, a hallucinogenic drug.

Drugs classified as **hallucinogens** (also called *psychedelics* or, in recognition of their more spiritual qualities, *entheogens*) have long been sought out by humans as a way of profoundly altering consciousness. The effects of **LSD**, or *acid* (lysergic acid diethylamide), and related substances like mescaline (from the peyote cactus) and psilocybin ("magic mushrooms") are predominantly visual (**FIGURE 3.23**), producing bizarre and mysterious sensory experiences that

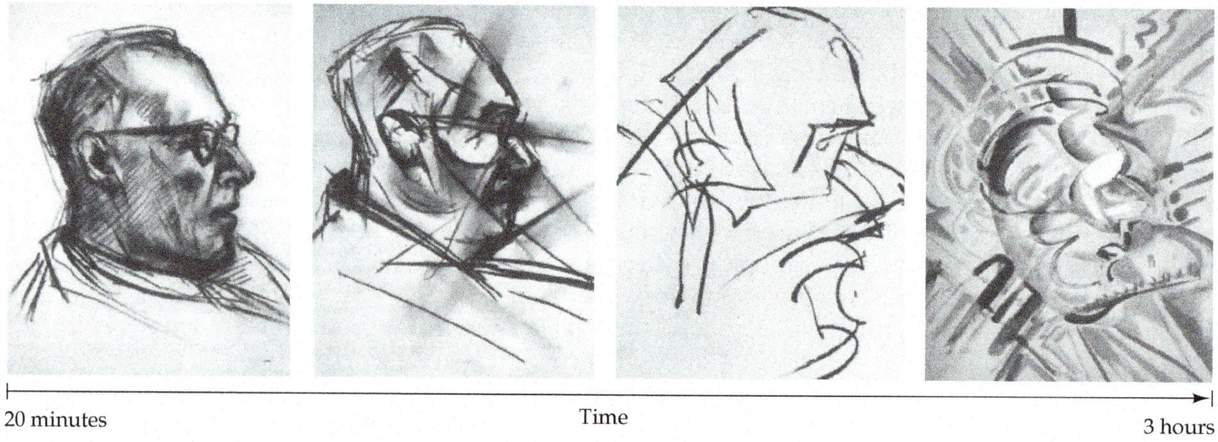

3.23 Perceptual Alteration with LSD These images are portraits produced by a professional artist just after taking LSD (leftmost drawing) and then at three successive time points while the drug was active. The model for all four drawings is the same (the researcher, in fact).

users may report as having deep spiritual or psychological meaning. But the term *hallucinogen* is a misnomer because, whereas a hallucination is a novel perception that takes place in the absence of sensory stimulation (hearing voices, or seeing something that isn't there), the drugs in this category tend to alter or distort *existing* perceptions. Users often see fantastic images with intense colours, but they are often aware that these altered perceptions are not real events.

Hallucinogenic agents are diverse in their neural actions. For example, the Mexican herb *Salvia divinorum* is unusual among hallucinogens because it acts on the opioid kappa receptor. Other hallucinogens, such as muscarine (another mushroom compound), affect the ACh system. Mescaline, the drug extracted from the peyote plant, affects noradrenergic and serotonergic systems. Many hallucinogens, including LSD, mescaline, and psilocybin, act as serotonin receptor agonists or partial agonists, especially at 5-HT_{2A} receptors. These receptors are found in high concentration in the visual cortex of the brain, which may account for the fantastical images that users experience. People treated with psilocybin also show reduced activity in medial prefrontal cortex and anterior cingulate cortex (Carhart-Harris et al., 2012). These structures inhibit limbic emotion-processing regions, so the drug's emotional, mystical qualities may be the product of an uninhibited limbic system.

You may have guessed that Albert Hofmann (**FIGURE 3.24**), whose story opened this chapter, was the discoverer of LSD. In fact, the study of LSD's activities became the focus of his professional career, summarized in his 1981 book, *LSD: My Problem Child*. Following its discovery, LSD was intensively studied as a possible psychiatric treatment, but public concern about possible harm and abuse potential led to strict legal controls on availability and research use of these compounds for much of the twentieth century. However, in addition to their impressive perceptual effects, LSD, psilocybin, and other hallucinogens can produce beneficial mood changes, introspective states, and feelings of creativity that have led to renewed interest in the possibility of using hallucinogens to treat specific psychiatric disorders, including depression, anxiety, and obsessive-compulsive disorder (Kyzar et al., 2017).

Unlike the hallucinogens we've described so far, **ketamine** (known as *Special K*) is a drug that is already in widespread use in medical settings. Developed in the 1960s as a potent analgesic and anesthetic agent, ketamine is classified as a **dissociative** because at moderate doses it produces feelings of depersonalisation and detachment from reality. Ketamine has several actions in the brain, especially blockade of NMDA receptors. PET studies indicate that ketamine increases metabolic activity in the prefrontal cortex (Breier et al., 1997; Zorumski et al., 2016), and while high doses produce transient hallucinogenic effects and occasional psychotic symptoms in volunteers, low doses have a potent and rapid antidepressant effect that may help ease symptoms in resistant cases (Williams and Schatzberg, 2016).

Ecstasy is the street name for the hallucinogenic amphetamine derivative **MDMA** (3,4-methylenedioxymethamphetamine). Major actions of MDMA in the brain include an increase in the release of serotonin, stimulation of 5-HT_{2A} receptors, and changes in the levels of dopamine and certain hormones, notably the peptides prolactin and oxytocin, which have been implicated in prosocial feelings and behaviours. But exactly how these drug activities account for the subjective effects of MDMA—positive emotions, empathy, euphoria, a sense of well-being, and colourful visual phenomena—remains to be established. In lab animals, chronic treatment with MDMA alters the structure and

ketamine A dissociative anesthetic drug that acts as an NMDA receptor antagonist.

dissociative A type of drug that produces a dreamlike state in which consciousness is partly separated from sensory inputs.

MDMA Also called *Ecstasy*. A drug of abuse, 3,4-methylenedioxymethamphetamine.

3.24 The Father of LSD Albert Hofmann discovered LSD in 1943 and devoted the remainder of his career to studying it. A prohibited drug in most jurisdictions, LSD is distributed on colourful blotter paper. This example, picturing Hofmann and the LSD molecule, is made up of 1036 individual doses (or "hits").

function of serotonergic neurons (T. J. Monks et al., 2004), although it's not clear whether there is lasting damage, at least at the lower doses typically used by humans. Several possible psychiatric and cognitive consequences of chronic MDMA use have been described, including memory disturbances and depression (Parrott, 2013; Sumnall and Cole, 2005). However, shorter-term MDMA treatment is also being studied in clinical settings, as a possible therapy for treatment-resistant posttraumatic stress disorder (Amoroso, 2015; Mithoefer et al., 2013). The neural actions, recreational properties, and possible psychiatric uses of some of the major hallucinogens are summarized in **TABLE 3.4**.

TABLE 3.4 Possible Clinical Applications for Hallucinogens

Drug name and date of discovery	Action in brain	Recreational use	Possible clinical application
Psilocybin / Psilocin (*Psilocybe* mushroom) (Archaeological evidence indicates use in prehistory)	Partial agonist of 5-HT receptors, especially 5-HT_{2A} receptors that occur in high density in visual cortex. Modifies activity of frontal and occipital cortex.	Users of "shrooms" often report spiritual experiences and feelings of transcendence, along with intense visual experiences and alterations in the perception of time. The exact effects are strongly influenced by the expectations and surroundings of the user.	Recent studies suggest that psilocybin—administered in controlled settings—can offer substantial and enduring improvements in the symptoms of obsessive-compulsive disorder (OCD), cluster headache (a type of migraine), and debilitating anxiety and anguish (as in a sample of terminal cancer patients) (Grob et al., 2011; E. A. Schindler et al., 2015).
Lysergic acid diethylamide (LSD) (1938)	Activates many subtypes of monoamine receptors, especially DA and 5-HT, resulting in heightened activity in many cortical regions, especially frontal, cingulate, and occipital cortex.	"Acid" produces pronounced perceptual changes that resemble hallucinations. Intense colours in geometric patterns, novel visual objects, and an altered sense of time are common.	LSD may be an effective treatment for alcoholism and other addictions and may also be an effective treatment for some types of debilitating anxiety (Bogenschutz and Johnson, 2016; Gasser et al., 2014).
Ketamine (1962)	Widespread effects in the brain, especially blockade of NMDA receptors, and stimulation of opioid and ACh receptors.	"Special K" creates a detached, trancelike state, in keeping with its routine medical use as an anaesthetic. It may also produce hallucinogenic perceptual alterations.	Recent experiments have revealed a potent antidepressant effect of ketamine at lower doses, even in cases that resist other types of treatments (Williams and Schatzberg, 2016).
3,4-Methylenedioxymethamphetamine (MDMA) (1912/1970s)	Stimulates release of monoamine transmitters and the pro-social hormone oxytocin	Users of "Ecstasy" experience intense visual phenomena, empathy, strongly pro-social feelings, and euphoria.	MDMA treatment may effectively reduce symptoms of posttraumatic stress disorder (PTSD), especially in combination with conventional psychotherapy, but concerns remain regarding drug safety (Amoroso, 2015; Parrott, 2014).

3.8 Substance Abuse and Addiction Are Worldwide Social Problems

Learning Objectives

After reading this section, you should be able to:

3.8.1 **Provide a formal definition of substance abuse, and distinguish between mild, moderate, and severe forms.**

3.8.2 **Summarise the major theoretical models of substance abuse.**

3.8.3 **Describe some individual differences that affect susceptibility to addiction.**

3.8.4 **Review leading categories of treatments for addiction, briefly explaining the logic of each.**

Substance abuse and addiction afflict many millions of people and disrupt the lives of their families, friends, and associates. Just one example reveals the extent of the problem: in the United States each year, more men and women die of smoking-related lung cancer than of colon, breast, and prostate cancers *combined*. In addition to the personal impact of so much illness and early death, there are dire social costs: huge expenses for medical and social services, millions of hours lost in the workplace, elevated rates of crime associated with illicit drugs, and scores of children who are damaged by their parents' substance abuse behaviour, in the uterine environment as well as in the childhood home.

For medical purposes, addiction is defined as "substance use disorder" (SUD) in the *Diagnostic and Statistical Manual of Mental Disorders*, fifth edition (*DSM-5*; American Psychiatric Association, 2013). SUD can take multiple forms, and it varies in severity from mild to severe. The *DSM-5* criteria for a diagnosis of alcohol use disorder appear in **TABLE 3.5**; identical criteria are used for diagnosis of all other types of substance use disorders, including use of opioids like heroin, stimulants like cocaine and meth, tobacco, cannabis, hallucinogens, and so on. Almost everyone will meet some of the criteria some of the time; diagnosis of substance use disorder requires more-sustained problems and a pattern of use that interferes with normal daily functioning. A mild disorder is diagnosed if two or three of the listed criteria are met. People meeting four or five of the criteria are classified as having a moderate substance use disorder, and a severe substance use disorder is diagnosed if six or more of the criteria are met.

TABLE 3.5 *DSM-5* **Diagnostic Criteria for Alcohol Use Disorder**

A problematic pattern of alcohol use leading to clinically significant impairment or distress, as manifested by at least two of the following, occurring within a 12-month period:

1. Alcohol is often taken in larger amounts or over a longer period than was intended.
2. There is a persistent desire or unsuccessful efforts to cut down or control alcohol use.
3. A great deal of time is spent in activities necessary to obtain alcohol, use alcohol, or recover from its effects.
4. Craving, or a strong desire or urge to use alcohol.
5. Recurrent alcohol use resulting in a failure to fulfill major role obligations at work, school, or home.
6. Continued alcohol use despite having persistent or recurrent social or interpersonal problems caused or exacerbated by the effects of alcohol.
7. Important social, occupational, or recreational activities are given up or reduced because of alcohol use.
8. Recurrent alcohol use in situations in which it is physically hazardous.
9. Alcohol use is continued despite knowledge of having a persistent or recurrent physical or psychological problem that is likely to have been caused or exacerbated by alcohol.
10. Tolerance, as defined by either of the following:
    a. A need for markedly increased amounts of alcohol to achieve intoxication or desired effect.
    b. A markedly diminished effect with continued use of the same amount of alcohol.
11. Withdrawal, as manifested by either of the following:
    a. The characteristic withdrawal syndrome for alcohol [listed elsewhere in *DSM-5* as an alcohol-induced disorder].
    b. Alcohol (or a closely related substance, such as a benzodiazepine) is taken to relieve or avoid withdrawal symptoms.

Source: Reprinted with permission from the *Diagnostic and Statistical Manual of Mental Disorders*, Fifth Edition, © 2013. American Psychiatric Association.

According to the 2016 National Survey on Drug Use and Health, some 20.1 million people in the United States alone suffer from substance-related disorders (Substance Abuse and Mental Health Services Administration, 2017). Worldwide, the number is probably in the hundreds of millions. Sex differences are evident in multiple aspects of drug abuse and dependence (Becker et al., 2017), but although historically men have been more likely than women to engage in substance abuse, rates in young people are now roughly equivalent between the sexes (Substance Abuse and Mental Health Services Administration, 2017).

Researchers have proposed numerous models of substance abuse and addiction that vary in their emphasis on physiological, cognitive, behavioural, and environmental factors (Sloboda et al., 2012; Thombs and Osborn, 2013). Some of these models stem from social forces; others are more deeply rooted in scientific observations and theories. But for any model of drug abuse, the challenge is to come up with a single account that can explain the addicting power of substances as diverse as, for example, cocaine (a stimulant), heroin (an analgesic and euphoriant), and alcohol (largely a sedative).

Addictive substances are no exception to the general rule that any given drug may produce multiple effects, so it is difficult to determine exactly which drug actions are most important in producing dependence. For example, we have already noted that cocaine (1) is a local anaesthetic, (2) produces intensely pleasurable feelings, and (3) is a psychomotor stimulant. The opiate drugs, like morphine and heroin, also produce intensely rewarding sensations, but they are depressants, not stimulants. Like cocaine, the opiates produce a strong physical dependence and powerful withdrawal symptoms on cessation of use. But opiates tend to produce only tolerance, whereas at least some of the actions of cocaine induce sensitisation. So a comprehensive theory of drug abuse and addiction must be able to account for dependence across this wide and confusing range of addictive compounds and effects.

In the discussion that follows, we will focus primarily on addiction to cocaine, the opiate drugs (such as morphine and heroin), nicotine, and alcohol because these substances have been studied the most thoroughly, but these perspectives apply to most abused substances.

Competing models of substance abuse have been proposed

Any comprehensive model of substance abuse has to answer several difficult questions: What social and environmental factors in people's lives cause them to start abusing substances? What factors cause them to continue? What physiological mechanisms make a substance rewarding? What is addiction, physiologically and behaviourally, and why is it so hard to quit?

THE MORAL MODEL The earliest approach to explaining drug abuse was to simply blame the substance abuser for a failure of moral character or a lack of self-control. Explanations of this sort often have a religious aspect and hold that only divine help will free a person from addiction. Applications based on the moral model can occasionally be effective. The "gin epidemic" of eighteenth-century England was eventually curbed through a combination of regulatory changes and public health education (Warner et al., 2001), and in the United States, beginning around the 1830s, the temperance movement is estimated to have cut per capita consumption of alcohol to about one-third its prior level (Rorabaugh, 1976). However, despite good intentions, high hopes, and multi-billion-dollar budgets, there is little evidence that modern morality-based campaigns—Project D.A.R.E., for example—have any appreciable effect on rates of drug abuse (Vincus et al., 2010; S. L. West and O'Neil, 2004). The minimal success of these approaches probably stems from their lack of a scientific framework that addresses the neurobiological roots of addiction.

THE DISEASE MODEL According to the disease model, the person who abuses drugs requires medical treatment rather than moral exhortation or punishment. This view also justifies spending money to research drug abuse in the same way that money is spent to research other diseases. However, the term *disease* is usually reserved for a state in which we can identify an abnormal physical or biochemical condition that initiates the problem. No abnormal physical or biochemical condition has been found in the case of drug addiction, and the disease model is mute about how addiction initially arises, although evidence indicates that some people are genetically more susceptible to addiction than others. Nevertheless, this model continues to appeal to many, and an intensive effort is under way to identify the physiological "switch" that establishes addiction after exposure to a drug.

A related formulation of the disease model views drug abuse as a type of self-medication, in which the addict is drawn to specific drugs in an effort to compensate for a deficiency of a corresponding endogenous substance: taking opiates to compensate for a lack of endorphins, for example. The causes of the endogenous shortages are generally unspecified, however.

dysphoria Unpleasant feelings; the opposite of euphoria.

nucleus accumbens A region of the forebrain that receives dopaminergic innervation from the ventral tegmental area.

THE PHYSICAL DEPENDENCE MODEL The physical dependence model, sometimes called the *withdrawal avoidance model*, argues that people keep taking drugs in order to avoid unpleasant withdrawal symptoms. The specific withdrawal symptoms depend on the drug, but they are often the opposite of the effects produced by the drug itself. For example, withdrawal from morphine causes irritability, a racing heart, and waves of goose bumps (that's where the term *cold turkey* comes from—the skin looks like the skin of a plucked turkey).

Whereas most drugs of abuse produce pleasurable feelings, withdrawal usually induces the opposite: **dysphoria**. Withdrawal symptoms can develop rapidly; some teenagers can experience withdrawal symptoms within 2 days of starting smoking (DiFranza et al., 2007). And withdrawal symptoms can be suppressed quickly (within a few minutes in the case of morphine) by administration of the withdrawn drug or a chemically similar compound. So the model does a good job of explaining why addicts will go to great lengths to obtain their addicted drug, but it has an important shortcoming: it can't explain how the addiction gets established in the first place. Why do some people, but not all, start to abuse a drug before physical dependence (tolerance) has ever developed? And how is it that some people can become addicted to some drugs even in the absence of clear physical withdrawal symptoms? For example, cocaine seems to be at least as addictive as heroin and produces intense drug craving in users, and yet cocaine withdrawal is not accompanied by the shaking, vomiting, and other physical symptoms seen during heroin withdrawal.

THE POSITIVE REWARD MODEL The positive reward model of addictive behaviour proposes that people get started with drug abuse, and become addicted, because the abused drug provides powerful reinforcement. Using a self-administration apparatus that allows animals to administer drugs to themselves (**FIGURE 3.25**), it is possible to quantify the motivation of animals to consume drugs (McKim, 1991).

Morphine-dependent rats or monkeys quickly learn to repeatedly press a lever in order to receive a small morphine injection (e.g., T. Thompson and Schuster, 1964). The drug infusion therefore acts like any other experimental reward, such as food or water. Furthermore, even animals that are not already morphine-dependent learn to press a lever for morphine, and they will happily self-administer doses of morphine that are so low that no physical dependence ever develops (Schuster, 1970). Animals will also furiously press a lever to self-administer cocaine and other stimulants that do not produce withdrawal symptoms as marked as those that opiates produce (Pickens and Thompson, 1968; Tanda et al., 2000). In fact, cocaine supports some of the highest rates of lever pressing ever recorded.

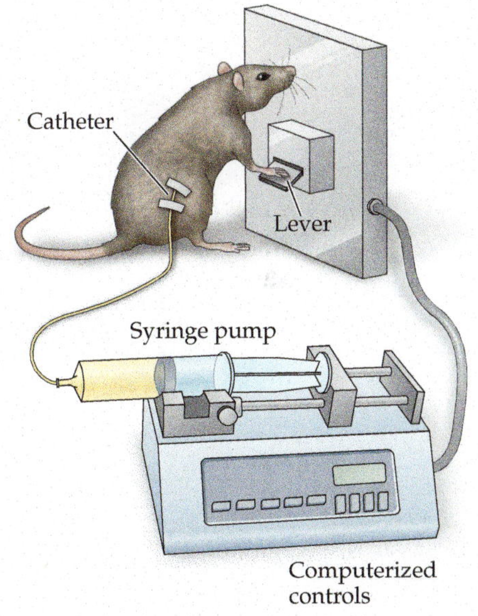

3.25 Experimental Setup for Self-Administration of a Drug by an Animal Here a computer is programmed to administer a small dose of an experimental drug after a certain number of lever presses. The number of lever presses that the animal will perform to receive the drug is a measure of its rewarding properties and addictive potential. Lab animals will press the lever many thousands of times to receive a single small dose of a highly addictive compound like cocaine or methamphetamine.

Experiments using drug self-administration thus suggest that by itself, the physical dependence model is inadequate to explain drug addiction, although physical dependence and tolerance may contribute to drug hunger. The more comprehensive view of drug self-administration interprets it as a behaviour controlled by a powerful pattern of positive and negative rewards (a variant of operant conditioning; see Chapter 12), without the need to implicate a disease process.

Many—but not all—addictive drugs cause the release of dopamine in the **nucleus accumbens**, just like more-conventional rewards such as food, sex, or winning money (Nutt et al., 2015; Volkow et al., 2017); interestingly, dopamine release is also linked to pathological gambling (Clark, 2018; Pettorruso et al., 2018). As we mentioned previously, dopamine released from axons originating from the VTA, part of the mesolimbocortical dopaminergic pathway illustrated in Figure 3.4, has been widely implicated in the perception of reward (**FIGURE 3.26**). Endocannabinoid actions within the reward circuitry may also contribute to the rewarding aspects of drug use (Parsons and Hurd, 2015).

If the dopaminergic pathway from the VTA to the nucleus accumbens serves as a reward system for a wide variety of experiences, then the addictive power of drugs may come from their artificial stimulation of this pathway. When the drug hijacks this system, providing an unnaturally powerful reward, the user learns to associate the drug-taking behaviour with that pleasure and begins seeking out drugs more and more until life's other pleasures fade into the background. These higher-order cognitive aspects of addiction depend on glutamatergic inputs from the prefrontal cortex, integrating aspects of memory, attention, and self-control to regulate the functioning of the dopamine reward system (e.g., R. Z. Goldstein and Volkow, 2011).

Cocaine, methamphetamine, and other stimulants all appear to cause long-lasting changes in dopaminergic circuitry, down-regulating dopamine D_2 and D_3

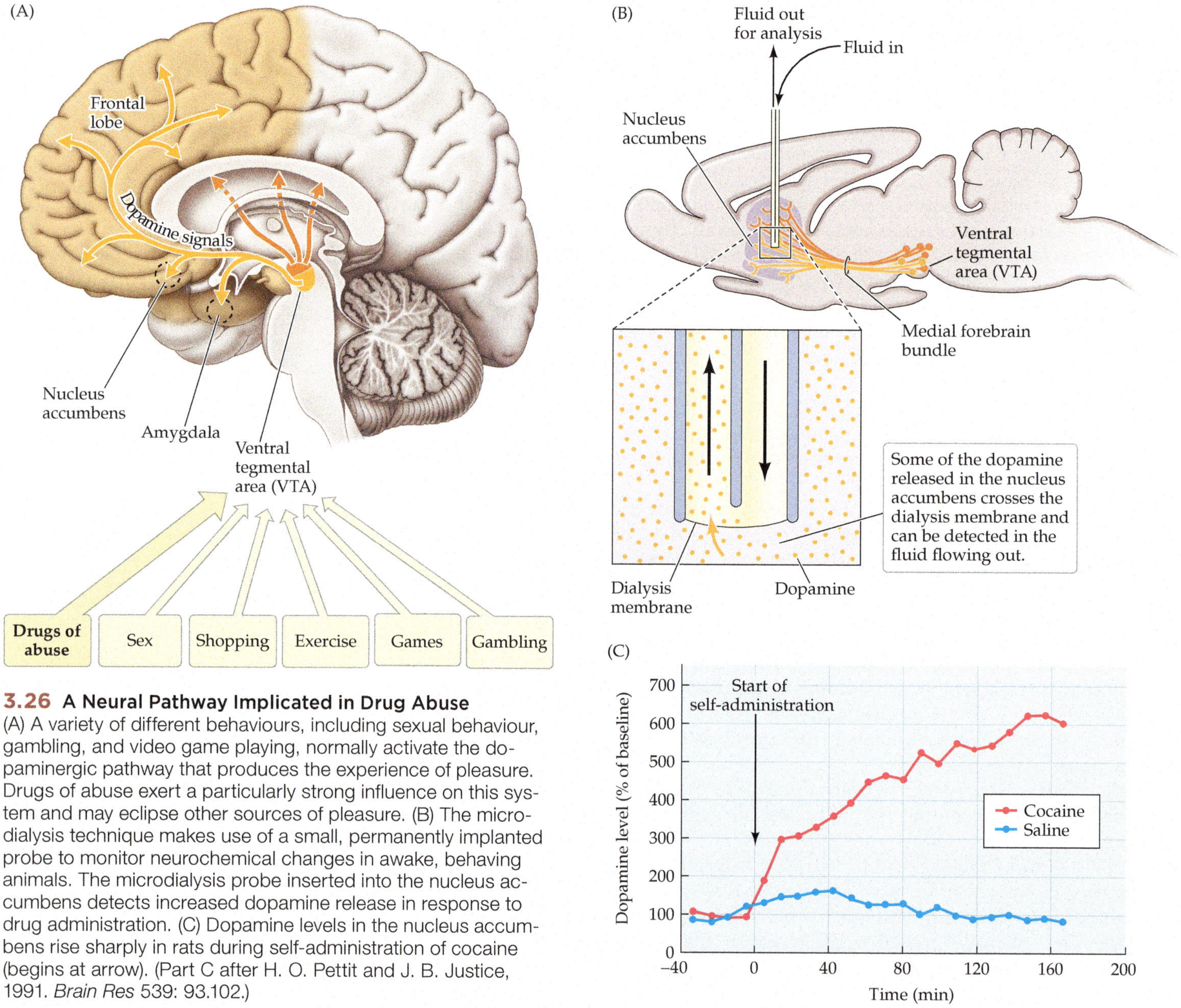

3.26 A Neural Pathway Implicated in Drug Abuse
(A) A variety of different behaviours, including sexual behaviour, gambling, and video game playing, normally activate the dopaminergic pathway that produces the experience of pleasure. Drugs of abuse exert a particularly strong influence on this system and may eclipse other sources of pleasure. (B) The microdialysis technique makes use of a small, permanently implanted probe to monitor neurochemical changes in awake, behaving animals. The microdialysis probe inserted into the nucleus accumbens detects increased dopamine release in response to drug administration. (C) Dopamine levels in the nucleus accumbens rise sharply in rats during self-administration of cocaine (begins at arrow). (Part C after H. O. Pettit and J. B. Justice, 1991. *Brain Res* 539: 93.102.)

receptors, inhibiting dopamine transporters, and eventually decreasing dopaminergic activity (Ashok et al., 2017; Volkow et al., 2006), as well as altering other neurotransmitter systems within the nucleus accumbens (B. R. Lee et al., 2013; X. Li et al., 2015)—changes that collectively seem to augment drug hunger and the pleasure associated with drug use, while simultaneously impairing the pleasure experienced from other behaviours. The drug's "pathological" reinforcement of drug-associated behaviours leads to exclusive, compulsive drug seeking. If natural activities like conversation, eating, and even sex no longer provide appreciable pleasure, addicts may seek drugs as the only source of pleasure available to them.

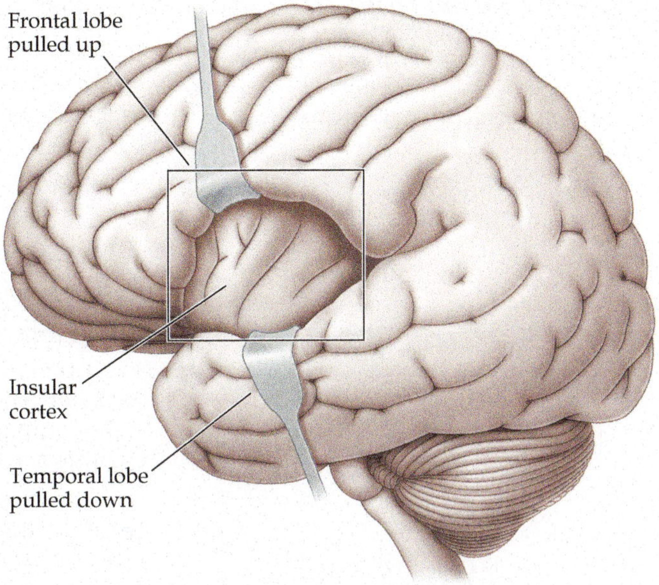

3.27 The Insula and Addiction Buried deep within the frontal lobes, changes in the insula are associated with the development of drug craving. Reduced volume of the left insula is observed in individuals addicted to various substances.

The Cutting Edge

Uncovering the Insula

Tucked deep within the folds of the frontal cortex, the **insula** (Latin for "island"; **FIGURE 3.27**) appears to be an important—yet understudied—participant in addiction, craving, and pleasure. For example, people who suffer damage to the left insula reportedly lose their urge to smoke tobacco and are able to effortlessly quit smoking (Naqvi et al., 2007). Addiction to numerous substances is associated with abnormalities of the insula and shrinkage to less than 80% of insula volumes in healthy controls (S. Mackey et al., 2018). It is thought that via its rich interconnections with prefrontal and sensory cortex and the VTA, the insula interacts with several large brain networks to mediate multiple features of addiction (Droutman et al., 2015), including motivational processes, physical sensory states associated with drug use, and external factors that promote substance use (Naqvi et al., 2014). In particular, the insula may be important for mediating the effects of social phenomena and pressures, such as social exclusion, that feature in the development and repeated triggering of drug hunger (Heilig et al., 2016). Future research aimed at better understanding the unique contributions of the insula in consciousness, subjective feelings, and social processes (Namkung et al., 2017) may also lead toward a more thorough understanding of the origins of addictive behaviours. ■

People differ in their vulnerability to drug abuse

Not everyone who uses an addictive drug becomes addicted. For example, most hospitalized patients treated with opiates for pain relief do not go on to abuse opiates after their pain has resolved. Of Vietnam War veterans who had used heroin overseas, only a minority relapsed to dependence within 3 years after their return (Robins and Slobodyan, 2003). Nevertheless, modern prescription painkillers are highly effective at activating the dopamine reward system, so the use of these drugs outside of medical contexts carries a high risk of addiction. The individual and environmental factors that account for differential susceptibility to addiction are the subject of active investigation and fall into several general categories:

- *Biological factors* Sex is an important variable: sex differences are evident in every aspect of drug addiction, ranging from the immediate reinforcing aspects of drug use to the longer-term consequences of addiction, withdrawal, and relapse (J. B. Becker and Chartoff, 2019). There is also evidence for genetic predisposition. For example, having a biological parent with alcoholism makes drug abuse more likely, even for children adopted away soon after birth (Cadoret et al., 1986). A tendency to use opiates and cocaine also appears to be heritable (Kendler et al., 2000), and adolescents carrying a specific version of the gene for one serotonin transporter are more likely to abuse drugs (Caspi et al.,

insula A region of cortex lying below the surface, within the lateral sulcus, of the frontal, temporal, and parietal lobes.

cue-induced drug use An increased likelihood to use a drug (especially an addictive drug) because of the presence of environmental stimuli that were present during previous use of the same drug.

2003), although this risk can be mitigated by involved and attentive parenting (G. H. Brody et al., 2009).

- *Family situation* Family breakup, a poor relationship with parents, or the presence of an antisocial sibling are associated with drug abuse.
- *Personal characteristics* Certain traits, such as aggressiveness and poor emotional control, are especially associated with drug abuse. Strong educational goals and maturity are associated with lower likelihood of drug abuse.
- *Environmental factors* A high prevalence of drug use in the community, and especially in the peer group, predisposes an individual toward drug abuse.

It is now clear that environmental stimuli—locations, social settings, sensory stimuli, and so on—can rapidly become strongly associated with the subjective effects of abused substances. These environmental stimuli can then become risk factors for deepening addiction or relapse: simply being in a setting where a person previously used drugs can trigger drug craving in that person (Ciccocioppo et al., 2004). This phenomenon, termed **cue-induced drug use**, is thought to be a product of long-lasting associations due to neuroplastic changes in the mesolimbic dopamine reward pathway (J. Wang et al., 2018; M. E. Wolf, 2016): neurons that encode learned associations between drugs and drug cues are found in the nucleus accumbens, a major component of the reward pathway of the brain (Koya et al., 2009). Emotional aspects of cue-induced craving and drug use may be controlled by the central nucleus of the amygdala (Y. Q. Li et al., 2008; Warlow et al., 2017).

Drug abuse and dependence can be prevented or treated in multiple ways

Annual economic losses due to substance abuse—the costs of law enforcement, medical care, and lost productivity—are estimated to exceed $700 billion in the United States alone (National Institute on Drug Abuse). The social costs of abuse and addiction, furthermore, are incalculable. So, the development of effective treatment programs is a pressing concern.

Many of those who become dependent on substances are able to overcome their addiction without outside help: more than 90% of ex-smokers and about half of those who recover from alcoholism appear to have quit on their own (S. Cohen et al., 1989; Institute of Medicine, 1990). Many others gain great benefit from counselling or social interventions such as the 12-step program developed by Alcoholics Anonymous in the 1930s. However, some people are unable to overcome powerful addictions without medical intervention. Medical strategies for treating drug dependence can be grouped into several categories:

- *Lessening the discomfort of withdrawal and drug craving* Benzodiazepines and drugs that suppress central adrenergic activity (e.g., clonidine) help reduce withdrawal symptoms during the early drug-free period, and anti-nausea medication and sleep aids may also be of benefit. Other medications help reduce uncomfortable cravings for the abused substance; for example, acamprosate (trade name Campral) eases alcohol-associated withdrawal symptoms. Similarly, noninvasive stimulation of prefrontal cortex using rTMS (repetitive transcranial magnetic stimulation; see Chapter 1) reportedly reduces drug craving and relapse in addicts (M. Diana et al., 2017).
- *Providing alternatives to the addictive drug* Agonist or partial agonist analogs of the addictive drug weakly activate the same mechanisms that the addictive drug activates, allowing the individual to gradually wean off the addictive drug. For example, the opioid receptor agonist methadone reduces heroin appetite and lessens withdrawal symptoms; similarly, nicotine patches or vaping (e-cigarettes, which are actually vaporisers) provide doses of nicotine without many of the other harmful components of cigarette smoke. While safer than smoking, vaping nonetheless exposes the user to multiple toxins or substances of unknown safety.

- *Directly blocking the actions of the addictive drug* Specific receptor antagonists can prevent an abused drug from interacting with its receptors. For example, the opiate receptor antagonist naloxone (trade name Narcan) blocks heroin's actions, but it also may produce harsh withdrawal symptoms.
- *Altering the metabolism of the abused drug* Changing the breakdown of a drug can change, reduce, or reverse its rewarding properties. Disulfiram (Antabuse) changes alcohol metabolism such that a nausea-inducing metabolite (acetaldehyde) accumulates, counteracting the rewarding aspects of alcohol abuse.
- *Blocking the brain's reward system* Treatment with dopamine receptor blockers reduces the activity of the mesolimbocortical dopamine reward system, causing drugs of abuse to lose much of their pleasurable quality. One problem with this approach is a generalized loss of pleasurable feelings, termed *anhedonia*.
- *Immunisation to render the drug ineffective* What if we could get the user's body to reject the abused substance? We all know **vaccination** as a way to prime the immune system to produce selective antibodies that combat infectious diseases, but researchers have succeeded in creating vaccines against drugs like cocaine, heroin, methamphetamine, and nicotine too (Nguyen et al., 2017; Skolnick, 2015). Here, the strategy is to prompt the individual's immune system to produce antibodies in sufficient quantities to remove the targeted drugs from circulation before they ever reach the brain.

vaccination Injection of a foreign substance in order to provoke an immune response and the formation of antibodies against that substance.

One such approach is illustrated in **FIGURE 3.28A**, using cocaine conjugated with protein from an inactivated virus (this adenovirus provokes a strong immune reaction) to create a vaccine. Just as with other forms of vaccination, the immune system has a "memory" for the cocaine vaccine and continues to make the anti-cocaine antibodies. Now, when cocaine enters the bloodstream, the antibodies immediately bind to the drug molecules, forming large conglomerates that cannot reach the brain and are subsequently cleared from the body. In mice, the vaccine prevents the hyperactivity normally associated with cocaine exposure (**FIGURE 3.28B**).

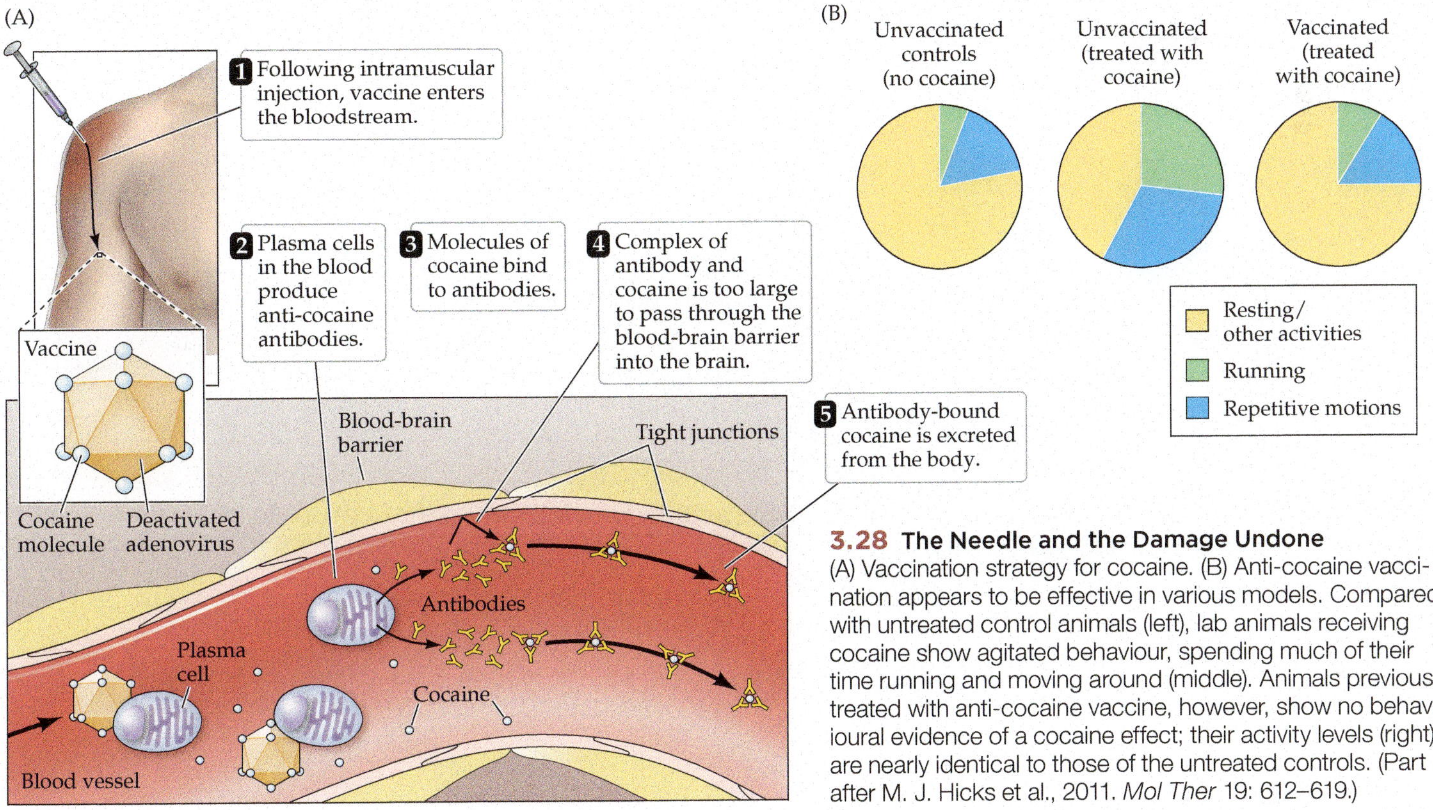

3.28 The Needle and the Damage Undone
(A) Vaccination strategy for cocaine. (B) Anti-cocaine vaccination appears to be effective in various models. Compared with untreated control animals (left), lab animals receiving cocaine show agitated behaviour, spending much of their time running and moving around (middle). Animals previously treated with anti-cocaine vaccine, however, show no behavioural evidence of a cocaine effect; their activity levels (right) are nearly identical to those of the untreated controls. (Part B after M. J. Hicks et al., 2011. *Mol Ther* 19: 612–619.)

Preliminary studies indicate that vaccination is effective for blunting the rewarding properties of cocaine and reducing addiction (Haney et al., 2010; Orson et al., 2014). Other vaccines under study target heroin, fentanyl, nicotine, and methamphetamine, with varying degrees of success so far (Nguyen et al., 2017; Skolnick, 2015).

Recommended Reading

Advokat, C. D., Comaty, J. E., and Julien, R. M. (2018). *Julien's Primer of Drug Action* (14th ed.). New York: Worth.

Erickson, C. K. (2018). *The Science of Addiction: From Neurobiology to Treatment* (2nd ed.). New York: Norton.

Grilly, D. M., and Salamone, J. (2011). Drugs, Brain, and Behaviour (6th ed.). Boston: Allyn & Bacon.

Karch, S. B., and Drummer, O. (2015). *Karch's Pathology of Drug Abuse* (5th ed.). Boca Raton, FL: CRC Press.

Meyer, J. S., and Quenzer, L. F. (2018). *Psychopharmacology: Drugs, the Brain, and Behaviour* (3rd ed.). Sunderland, MA: Oxford University Press/Sinauer.

Nestler, E., Hyman, S., and Malenka, R. (2014). *Molecular Neuropharmacology* (3rd ed.). New York: McGraw-Hill.

Nutt, D. (2012). *Drugs Without the Hot Air.* Cambridge, UK: UIT Cambridge Press.

Pollan, M. (2018). *How to Change Your Mind: What the New Science of Psychedelics Teaches Us about Consciousness, Dying, Addiction, Depression, and Transcendence.* New York: Penguin.

Schatzberg, A. F., and Nemeroff, C. B. (Eds.). (2013). *Essentials of Clinical Psychopharmacology* (3rd ed.). Arlington, VA: American Psychiatric Publishing.

Thombs, D. L., and Osborn, C. J. (2013). *Introduction to Addictive Behaviours* (4th ed.). New York: Guilford.

3 Visual Summary

You should be able to relate each summary to the adjacent illustration, including structures and processes. You can also refer back to the links to figures, animations, and activities that will help you consolidate the material.

1 The major categories of **neurotransmitters** are **amine, amino acid, peptide**, and soluble **gas** neurotransmitters. Because many drugs work by acting on receptor molecules, investigators search for the receptor molecules and for the **endogenous** substances that work on the receptors. A given neurotransmitter may normally bind several different subtypes of receptors. Review **Figures 3.1, 3.2, Table 3.1, Activity 3.1**

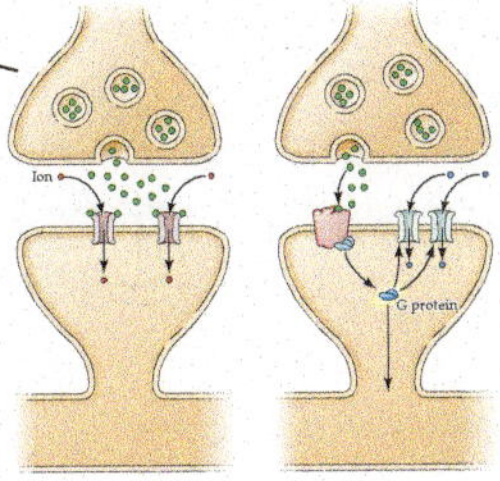

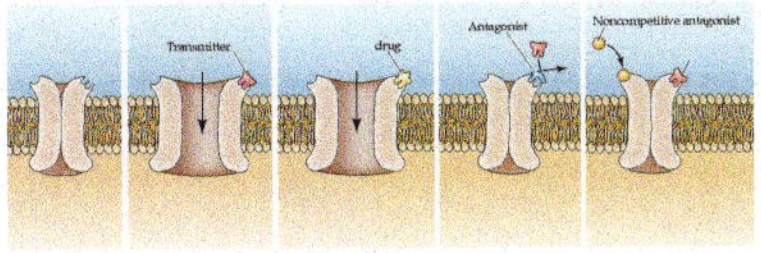

2 A **ligand** is any substance that binds to a receptor. **Agonists** activate transmitter pathways, **antagonists** block transmitter pathways, and **inverse agonists** have effects opposite to a transmitter's normal effects. The classic neurotransmitters are found in segregated regions that project widely throughout the brain. Review **Figures 3.3–3.7, Animations 3.1, 3.2**

3 Drugs vary in their binding affinity for different types of receptors, as well as in efficacy—ability to produce effects—once they are bound. The relationship between concentrations of a drug and its physiological effects is studied by use of a dose-response curve, which reveals its activity, specificity, potency, and safety. Review **Figures 3.8, 3.9**

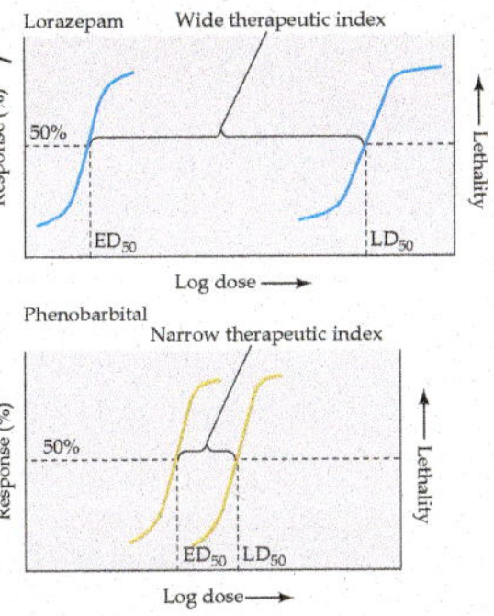

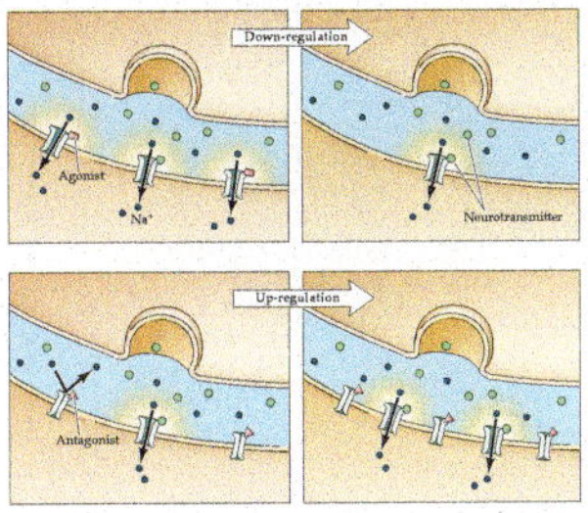

4 Repeated treatments with a drug can produce **tolerance** to its effects, often through the **up-** or **down-regulation** of receptors. Most CNS drugs alter neural transmission. Many drugs selectively affect presynaptic mechanisms, such as by inhibiting axonal transport, affecting **transmitter reuptake**, or acting on presynaptic receptors. Neuromodulators such as **caffeine** may affect the release of the transmitter or the receptor's response to the transmitter. Review **Figures 3.10, 3.11**

5 Other CNS drugs selectively modify postsynaptic mechanisms, directly blocking or activating receptors, or modifying processes within the postsynaptic cell. **First-generation antipsychotics** (antischizophrenics) generally block dopamine D2 receptors; **second-generation antipsychotics** add actions at other types of transmitter receptors to D2 antagonism. **Antidepressants** improve synaptic availability of **serotonin** and **norepinephrine**. Review **Figure 3.12**

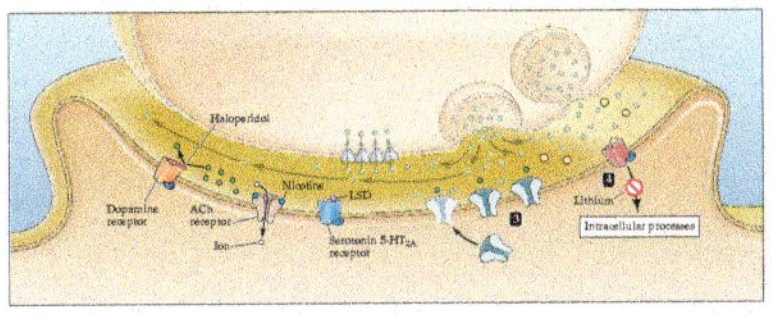

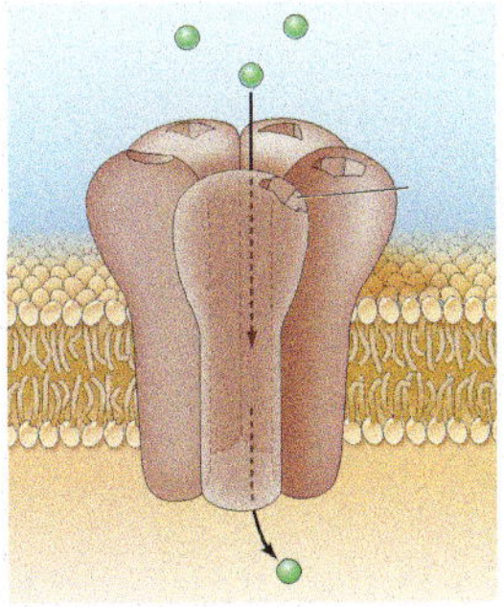

6 The benzodiazepines are potent **anxiolytic** drugs. The benzodiazepines synergize the activity of the inhibitory transmitter **GABA** at some of its receptors. Alcohol acts on GABA receptors to produce some of its effects. Alcohol in moderation has beneficial effects, but in higher doses it is very harmful, damaging neurons in many areas of the brain. Review **Figure 3.14**

7 Opiates are potent painkillers; **endogenous opioids** include the **endorphins**, and **exogenous** opiates include **morphine** and **heroin.** Review **Figures 3.15–3.17**

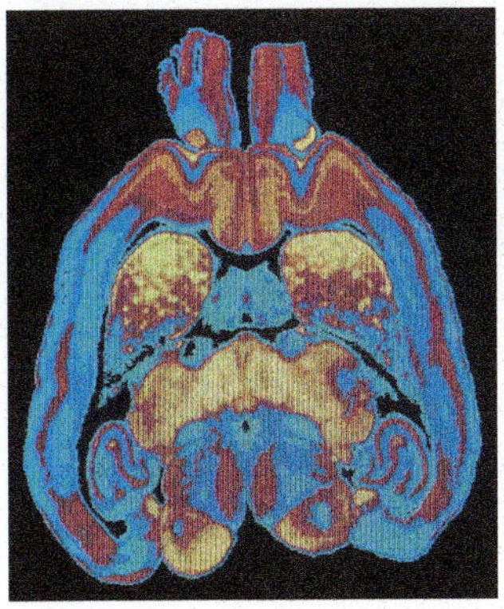

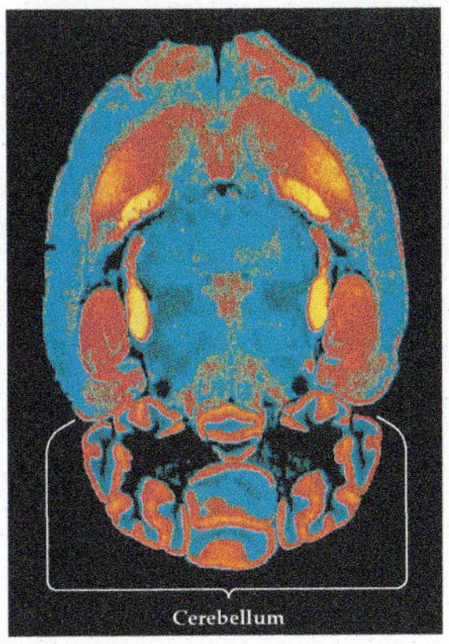

8 The active ingredient in **cannabis, THC**, acts on cannabinoid receptors to produce its effects. An endogenous cannabinoid, **anandamide**, serves as a **retrograde transmitter** in some synapses. Review **Figures 3.18, 3.19**

(Continued)

9 Some stimulants, such as **nicotine**, imitate an excitatory synaptic transmitter. Others, such as **amphetamine** and **cocaine**, cause the release of excitatory synaptic transmitters and block the reuptake of transmitters. Still others, such as caffeine, block the activity of an inhibitory neuromodulator. Review Figure 3.20

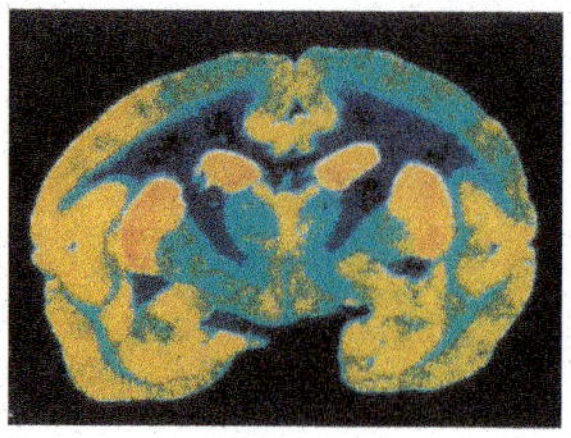

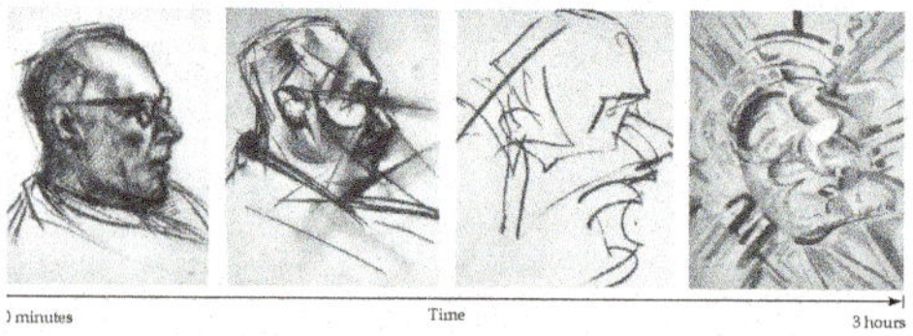

10 Some drugs are called **hallucinogens** because they alter sensory perception and produce extraordinary visual and emotional experiences. Different hallucinogens act on different kinds of synaptic receptors, and it is not yet clear what causes the hallucinogenic effects. Review Figure 3.23, Table 3.4

11 Substance abuse and addiction are being studied intensively, and several models have been proposed: the moral model, the disease model, the physical dependence model, and the positive reward model. Review Figures 3.25, 3.26

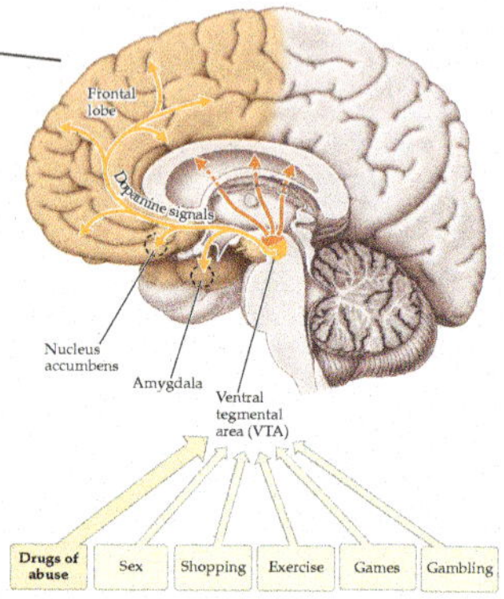

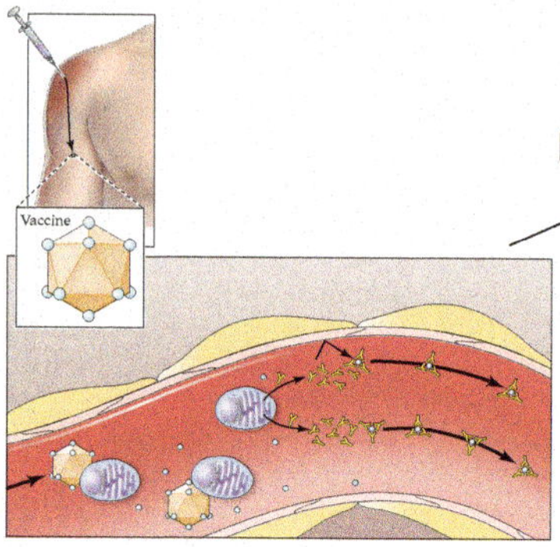

12 People differ in their vulnerability to drug abuse according to several factors: genetic predisposition, personality characteristics, and family and social context. There are several medicinal approaches to treating drug addiction, including antiwithdrawal and anticraving medication and vaccination. Review Figure 3.28

4

Hormones and the Brain

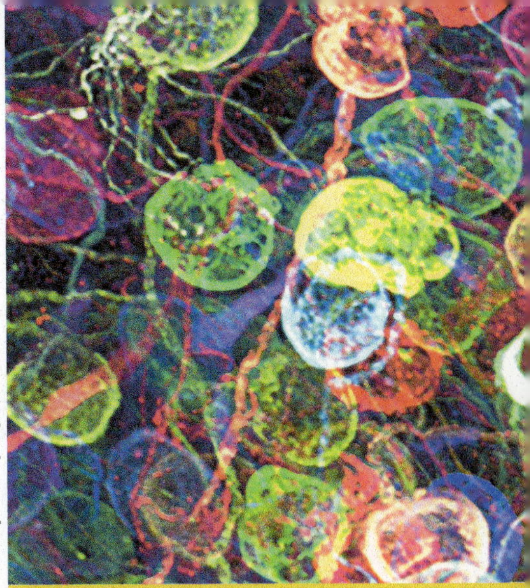

Courtesy of Dr. Ryo Egawa

Crafting a Personality through Hormones

When she was 13, Mary Lou spent a whole summer in the intensive care ward of a hospital with a grave illness that doctors could not diagnose. When she finally got out of the hospital, she assumed that her life would be short, so Mary Lou pursued it with gusto, working as an artist, playing in a band, working on a PhD in electrical engineering at an Ivy League school. But by her early 20s, Mary Lou's medical condition was deteriorating. She slept 20 hours per day, coped with constant headaches and vomiting. But the worst part for Mary Lou was that, for the first time in her life, she felt stupid.

Then doctors finally detected a slow-growing brain tumour that had disrupted function of a pea-sized structure called the *pituitary gland*. The tumour was removed, but the pituitary was damaged beyond repair. As we'll learn in this chapter, the lack of a functioning pituitary (panhypopituitarism) means Mary Lou fails to make a host of hormones, some of which are crucial for survival. Fortunately, pharmacies can provide many of these missing hormones, so in the 20 years after surgery, a recovered Mary Lou has been able to found several technology start-ups and has headed projects, for Google, Oculus, and Facebook, before starting a new initiative to build an affordable infrared 3D camera to provide high-resolution images inside the body (https://www.maryloujepsen.com/). Would such imaging have revealed Mary Lou's pituitary tumour before her doctors found it?

Taking hormones saved Mary Lou's life. But in the process of tinkering to get the right doses and combinations of hormones, Mary Lou learned that hormones influence not only her physical health and intelligence, but also her moods and personality. As she says, "It took me years to craft a better 'me' after my personality was essentially killed by the effects of the tumour and surgery" (Jepsen, 2013). What are these hormones that Mary Lou must tinker with, and how can they have such a large effect on her brain and behaviour?

The cells in your body use chemicals, including an extensive array of hormones, to communicate, so alterations in hormones can produce striking changes in brain function. Cognitive abilities, emotions, our appetite for food or drink or sex, our aggressiveness or submissiveness, our care for children—the scope of hormonal influences on behaviour is vast. Furthermore, hormones do more than influence adult behaviour. Early in life, thyroid and sex hormones regulate brain development. Later, the changing outputs of endocrine glands and the body's changing sensitivity to hormones are prominent aspects of adolescence and aging.

In this chapter we consider the major hormones, their anatomical sources, their physiological actions, and their effects on behaviour. This discussion sets the stage for topics in later chapters, such as hormonal effects in reproductive behaviour (Chapter 13); feeding, drinking, and body maintenance (Chapter 14); and stress and emotion (Chapter 11).

4.1 Hormones Have Many Actions in the Body

Learning Objectives

After reading this section, you should be able to:

4.1.1 **Contrast the different chemical signalling systems within the body and between individuals.**

4.1.2 **Classify the major classes of hormones, and describe how they affect target cells.**

4.1.3 **Describe the structure and function of neuroendocrine cells.**

hormone A chemical secreted by cells that is conveyed by the bloodstream and regulates target organs or tissues.

endocrine gland A gland that secretes hormones into the bloodstream to act on distant targets.

exocrine gland A gland whose secretions exit the body via ducts.

The ancient Greeks emphasized body *humours*, or fluids, as an explanation of temperament and emotions. It was believed that these fluids—phlegm, blood, black bile, and yellow bile (also known as *choler*)—all interacted to produce health or disease. The notion that these fluids jointly determine personality has survived to the present day in our use of certain unusual adjectives, such as *phlegmatic* ("impassive"), *sanguine* ("cheerful"; *sanguis* is Latin for "blood"), *bilious* ("irritable"), and *choleric* ("hot-tempered").

We now know there are a lot more than four hormones. Most **hormones** (from the Greek *horman*, "to excite") are chemicals secreted by a group of cells in one part of the body and carried through the bloodstream to other parts of the body, where they act on specific target tissues to produce specific physiological effects. Many hormones are produced by **endocrine glands** (from the Greek *endon*, "within," and *krinein*, "to secrete"), so called because they release their hormones *inside* the body. Endocrine glands are sometimes contrasted with **exocrine glands** (tear glands, salivary glands, sweat glands), which use ducts to secrete fluid outside the body (the Greek *exo* means "out").

Endocrine glands come in a variety of sizes, shapes, and locations in the body (**FIGURE 4.1**). The definition of *hormone* is more inclusive than it used to be,

4.1 Major Endocrine Glands and Their Functions Table 4.2 lists the hormones secreted by the glands shown here and the effects of those hormones in the body.

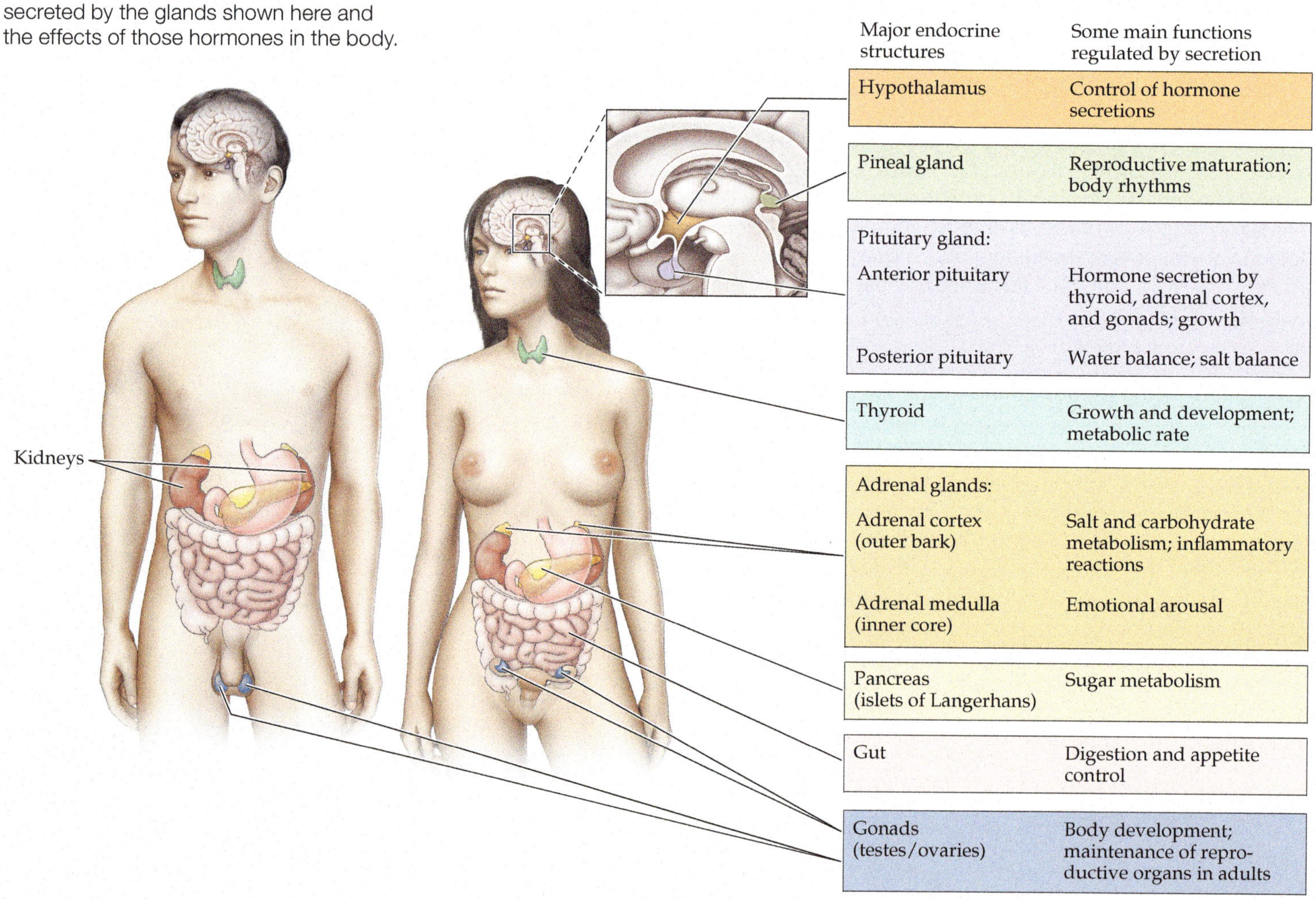

recognising that not just traditional endocrine glands, but other tissues, such as the heart and kidneys, secrete hormones too. Even plants, which have no endocrine glands, use chemical signals that are considered to be hormones.

castration Removal of the gonads, usually the testes.

The scientific method established the importance of testicular hormones

The importance of hormones was anticipated in ancient civilisation. In the fourth century BCE, Aristotle accurately described the effects of **castration** (removal of the testes) in birds, and he compared the behavioural and bodily effects with those seen in *eunuchs* (castrated men). Perhaps this practice built on prehistoric agricultural knowledge of the effects of castration on cattle. Although no one knew what mechanism was involved, clearly the testes were important for the reproductive capacity and sexual characteristics of males.

The first published experiment in behavioural endocrinology showed that the effects of castration were due to a loss of hormones from the testes. In 1849 German physician Arnold Adolph Berthold (1803–1861) tried replicating some studies that were rumored to have been done by English physician John Hunter (Sawin, 1996). When roosters are castrated as juveniles, they fail to develop normal reproductive behaviour and secondary sexual characteristics, such as the rooster's comb, in adulthood (**FIGURE 4.2**). Berthold observed, however, that placing one testis (either from the same chick or even from another chick) into the body cavity of a young castrate could preserve normal development of adult anatomy and behaviour. Animals so treated began crowing and showed the usual male sexual behaviours. Because the nerve supply to the testis had not been reestablished, Berthold concluded that the testes release a chemical into the blood that affects both male

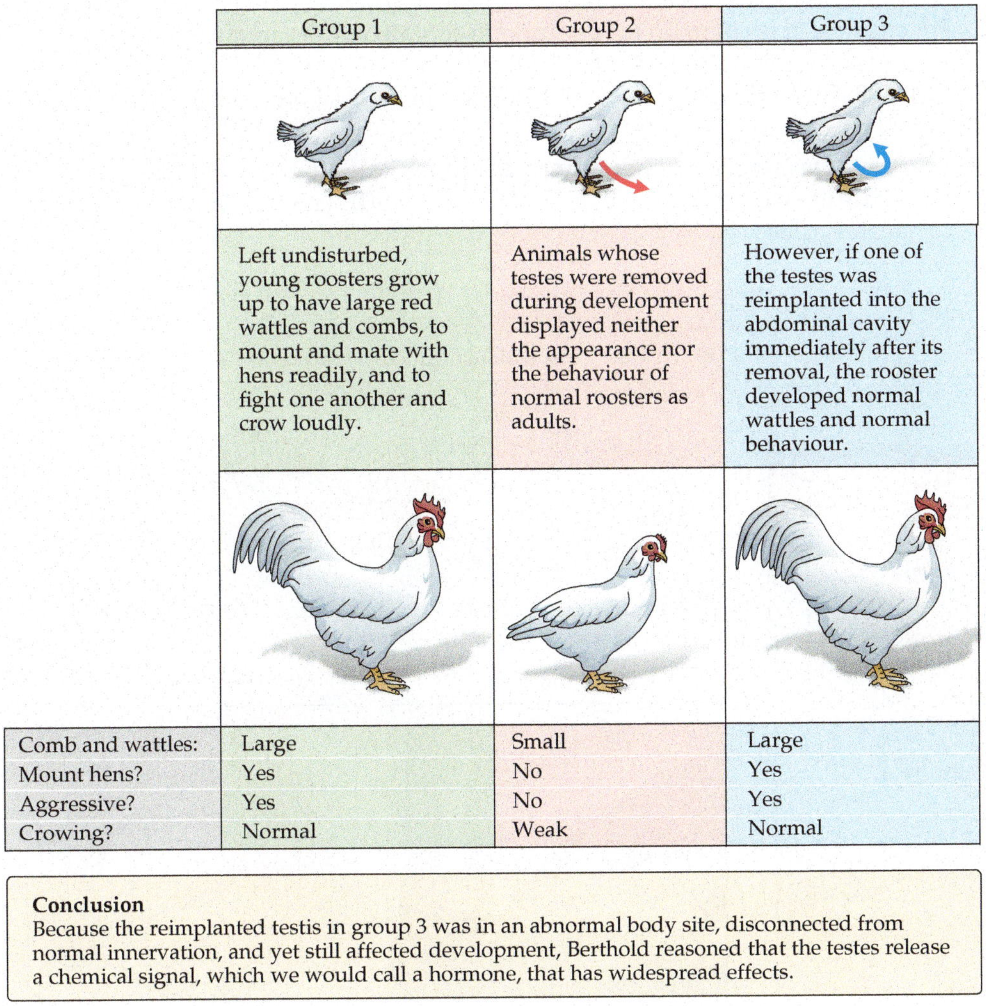

4.2 The First Experiment in Behavioural Endocrinology Berthold's nineteenth-century experiment demonstrated the importance of hormones for behaviour. The hormone responsible for these changes, testosterone, must be present early in life to have these effects in roosters.

endocrine Referring to glands that release chemicals to the interior of the body. These glands secrete the principal hormones.

autocrine Referring to a signal that is secreted by a cell into its environment and that feeds back to the same cell.

paracrine Referring to cellular communication in which a chemical signal diffuses to nearby target cells through the extracellular space.

pheromone A chemical signal that is released outside the body of an animal and affects other members of the same species.

allomone A chemical signal that is released outside the body by one species and affects the behaviour of other species.

Go to Animation 4.1
Chemical Communication Systems

behaviour and male body structures. Today we know that hormone from the testes is *testosterone*.

Although Berthold didn't know it, experiments like this also illustrate another distinction in hormonal action. If he had waited until the castrated chicks were adults before transplanting the testes, Berthold would have seen little effect. The testosterone must be present *early* in life to have such dramatic effects on the adult body and behaviour. We say that the brain and body are "organized" by exposure to hormones early in life, and sometimes these changes can be dramatic and long lasting. If you wait until adulthood to provide hormones, they still affect the body and behaviour, but the changes are less dramatic and tend to be short-lived. In that case, the hormones are said to "activate" behaviour. We'll discuss organisational and activational effects of hormones in more detail in Chapter 13.

Organisms use several types of chemical communication

By reviewing the several categories of chemical signals used by the body, we can see how hormonal communication by endocrine glands compares with other methods of communication:

- *Endocrine communication* In **endocrine** communication, our topic for this chapter, the chemical signal is a hormone released into the bloodstream to selectively affect distant target organs (**FIGURE 4.3A**).
- *Synaptic communication* As we saw in Chapters 2 and 3, synaptic transmission (sometimes called *neurocrine* function) is the release of a chemical signal—a neurotransmitter—into the synaptic cleft to affect the postsynaptic membrane (**FIGURE 4.3B**). Typically, synaptic transmitter function is highly localised.
- *Autocrine communication* In **autocrine** communication, a released chemical acts on the releasing cell itself and thereby affects its own activity (**FIGURE 4.3C**). For example, it is common for a neuron to contain autoreceptors that detect neurotransmitter molecules released by that neuron; the cell can thus monitor its own activity. In this case, the neurotransmitter serves both an autocrine and a synaptic function.
- *Paracrine communication* In **paracrine** communication, the released chemical signal diffuses to nearby target cells (**FIGURE 4.3D**). The strongest impact is on the nearest cells.

4.3 Chemical Communication Systems

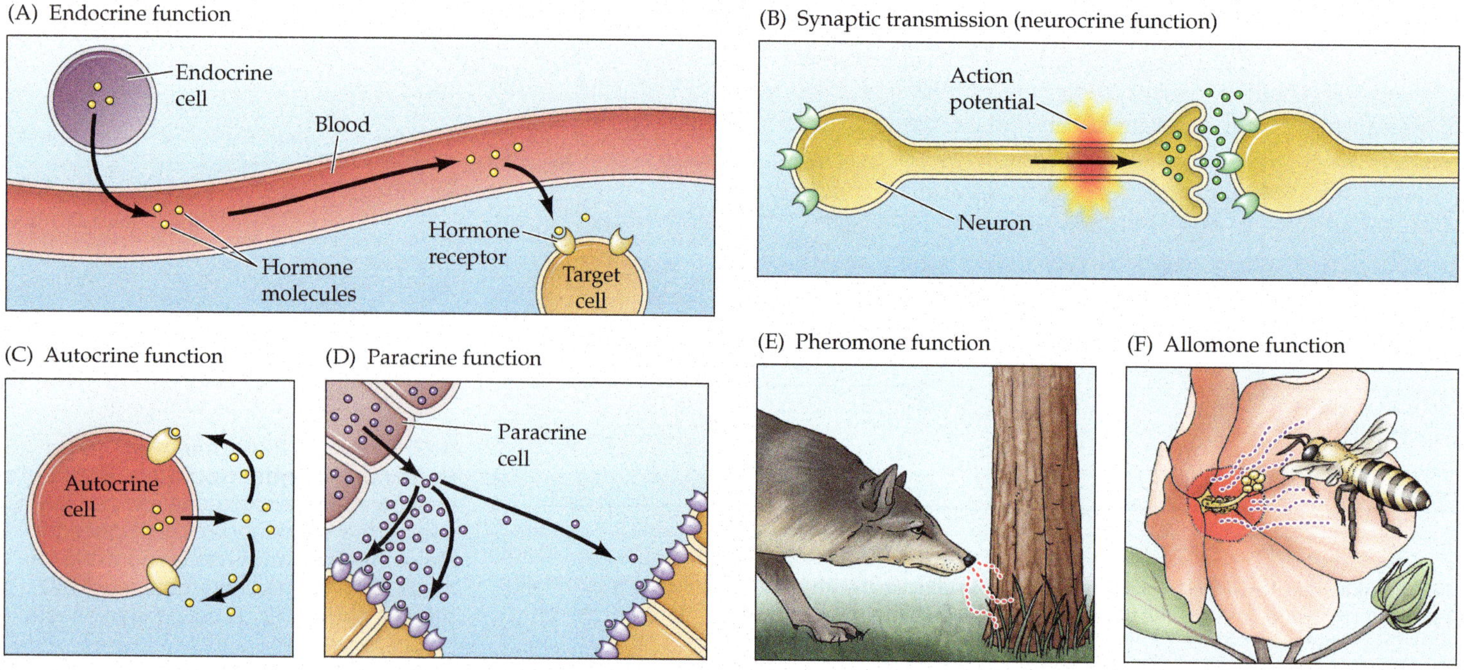

- *Pheromone communication* Chemicals can be used for communication not only within an individual, but also between individuals. **Pheromones** (from the Greek *pherein*, "to carry") are released into the outside environment to affect other individuals of the same species (**FIGURE 4.3E**). For example, ants produce pheromones that communicate the presence of intruders in the nest or that identify the route to a rich food source (to the annoyance of picnickers). Dogs and wolves urinate on landmarks to designate their territory; other members of the species smell the pheromones in the urine and either respect or challenge the territorial claim. In Chapters 9 and 13 we'll discuss pheromones in more detail.
- *Allomone communication* Some chemical signals are released by members of one species to affect the behaviour of individuals of *another* species. These substances are called **allomones** (from the Greek *allos*, "other") (**FIGURES 4.3F** and **4.4**). Flowers exude scented allomones to attract insects and birds in order to distribute pollen. And the bolas spider—nature's femme fatale—releases a moth sex pheromone to attract male moths to their doom (Haynes et al., 2002).

4.4 Scents and Sensibility Skunks produce a very effective allomone.

Hormonal actions can be organized according to general principles

Although there are some exceptions, the following rules are general principles of hormonal action:

1. Hormones frequently act in a *gradual* fashion, activating behavioural and physiological responses hours or weeks after entering the bloodstream. The changes may persist for days, weeks, or years after hormone release is over.
2. When hormones alter behaviour, they tend to act by changing the *intensity* or *probability* of evoked behaviours, rather than acting as a switch to turn behaviours on or off regardless of context.
3. Both the quantities and the types of hormones released are influenced by environmental factors. Therefore, *the relationship between behaviour and hormones is reciprocal*; that is, hormones change behaviours and behaviours change hormone levels. For example, high levels of testosterone are related to aggression, and in some species, males who lose in aggressive encounters show a reduction in testosterone levels, while the winners may show little change in testosterone levels. This example illustrates the reciprocal relation between behavioural and somatic (body) events that we discussed in Chapter 1 (see Figure 1.2).
4. A hormone may have multiple effects on different cells, organs, and behaviours; conversely, a single type of behaviour or physiological change can be affected by many different hormones (**FIGURE 4.5**).
5. Hormones are produced in small amounts and often are secreted in bursts. This *pulsatile* secretion pattern is sometimes crucial for the small amount of hormone to be effective.
6. The levels of most hormones vary rhythmically throughout the day, and many hormonal systems are controlled by circadian "clocks" in the brain, as we'll see in Chapter 15. A shortcoming of taking hormones for therapy, as Mary Lou must do, is that we can't really replicate the pulsatile and rhythmic changes in levels seen in naturally released hormones.
7. Hormones interact; the effects of one hormone can be markedly changed by the actions of another hormone.
8. Hormones can affect only those cells that possess corresponding receptor proteins to recognize the hormones and alter cell function.

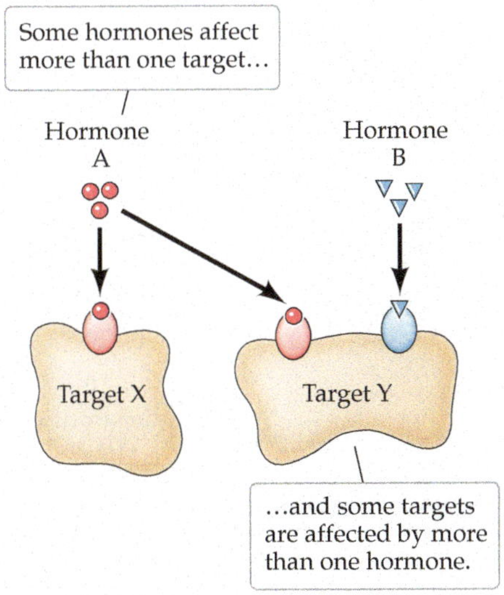

4.5 The Multiplicity of Hormone Action A single hormone (hormone A in this illustration) may affect multiple target tissues in various locations throughout the body. Similarly, a single process or body organ (target Y here) may be sensitive to several hormones.

Neuroendocrine cells blend neuronal and endocrine functions

Synaptic transmission—chemical communication between neurons—and hormonal communication are both secretory events, and they seem similar in several ways. For example, the neuron produces particular transmitter chemicals and releases them, just as an endocrine gland produces and releases hormones. Similarities between

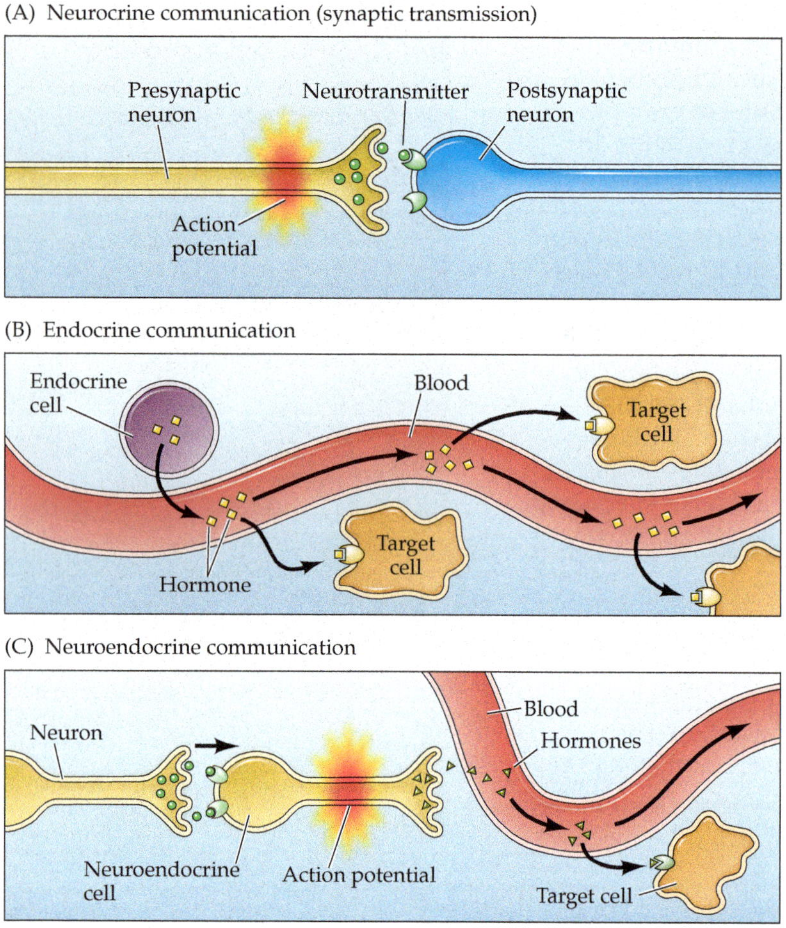

4.6 Neuroendocrine Cells Blend Neuronal and Endocrine Mechanisms (A) Neurons can communicate only with the particular neurons, muscle cells, or glands on which they synapse. The target cell is determined by the synaptic anatomy. (B) Endocrine signals are transmitted through the bloodstream and are recognized by appropriate receptors wherever they occur in the body. (C) Neuroendocrine (neurosecretory) cells are the interface between neurons and endocrine glands. They receive synaptic signals from other neurons, yet they secrete hormones into the bloodstream. In this way electrical signals are converted into hormonal signals.

neuronal and hormonal communication are exemplified by the **neuroendocrine cells** (or *neurosecretory cells*) of the hypothalamus (**FIGURE 4.6**). These cells are neurons in almost every way—they receive synaptic input, reach threshold, and produce action potentials—except that at their axon terminals they release hormone into the bloodstream rather than neurotransmitter into a synapse. Later we'll learn that neuroendocrine cells are crucial for brain control of endocrine glands.

Some peptides are used both as hormones by endocrine glands and as **neuropeptides** (peptides used by neurons) in the brain. Because these same peptides are found in single-celled organisms, the nervous system and the endocrine system may share an evolutionary origin from chemical communication systems in our remote single-celled ancestors (Elphick et al., 2018).

Whereas neural messages are rapid and are measured in milliseconds, hormonal messages are slower and are measured in seconds and minutes. But this distinction is blurred sometimes: **neuromodulators** alter the reactivity of cells to specific transmitters (see Chapter 3), acting more slowly and having longer-lasting effects than neurotransmitters have (Marder et al., 2014). So neuromodulators are something of a blend of neurotransmitter (because they're released into synapses) and hormone (because they act gradually).

Hormones can be classified by chemical structure

Most hormones fall into one of three categories: peptide hormones, amine hormones, or steroid hormones. Like any other protein, a **peptide hormone** is composed of a string of amino acids (**FIGURE 4.7A**). (Recall that a peptide is simply a small protein—a short string of amino acids. Thus these may sometimes be referred to as *protein hormones*.) Different peptide

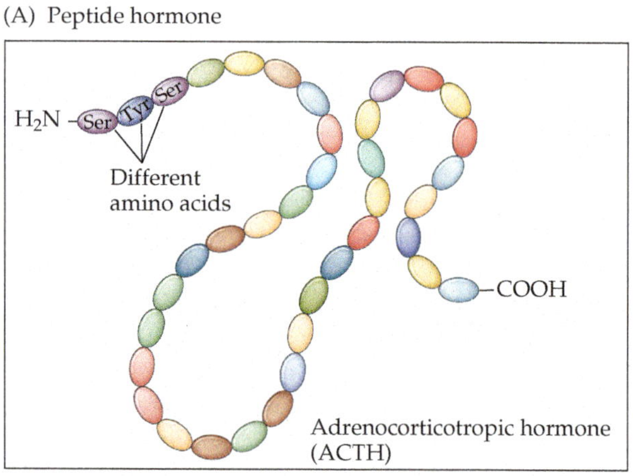

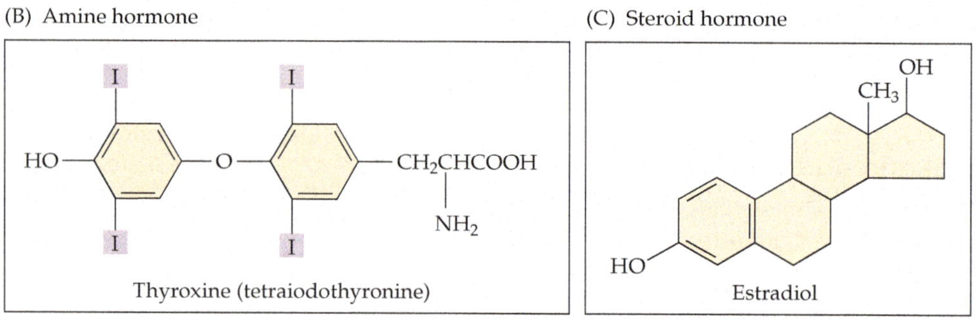

4.7 Chemical Structures of the Three Main Hormone Types (A) Peptide hormones consist of strings of amino acids, as adrenocorticotropic hormone (ACTH) here. If the string is longer, it may be referred to as a *protein hormone*. (B) Amine hormones, such as thyroxine, are modified single amino acids. (C) Steroid hormones, such as estradiol, are derived from cholesterol and consist of four interconnected rings of carbon atoms, to which are attached different numbers and types of atoms.

TABLE 4.1 Examples of Major Classes of Hormones

Class	Hormone
Peptide hormones	Adrenocorticotropic hormone (ACTH)
	Follicle-stimulating hormone (FSH)
	Luteinizing hormone (LH)
	Thyroid-stimulating hormone (TSH)
	Growth hormone (GH)
	Prolactin
	Insulin
	Glucagon
	Oxytocin
	Vasopressin (arginine vasopressin, AVP; antidiuretic hormone, ADH)
	Releasing hormones, such as:
	Corticotropin-releasing hormone (CRH)
	Gonadotropin-releasing hormone (GnRH)
Amine hormones	Epinephrine (adrenaline)
	Norepinephrine (NE)
	Thyroid hormones
	Melatonin
Steroid hormones	Estrogens (e.g., estradiol)
	Progestins (e.g., progesterone)
	Androgens (e.g., testosterone, dihydrotestosterone)
	Glucocorticoids (e.g., cortisol)
	Mineralocorticoids (e.g., aldosterone)

hormones consist of different combinations of amino acids. **Amine hormones** are smaller and simpler, consisting of a modified version of a single amino acid (hence their alias, *monoamine* hormones) (**FIGURE 4.7B**). **Steroid hormones** are derivatives of cholesterol, sharing its structure of four rings of carbon atoms (**FIGURE 4.7C**). Different steroid hormones vary in the number and kinds of atoms attached to the rings. Steroids dissolve readily in lipids, so they easily pass through membranes (recall from Chapter 1 that the cell membrane is a lipid bilayer). **TABLE 4.1** gives examples of each class of hormones.

The different classes of hormones act through very different mechanisms, as we'll see next.

4.2 Hormones Have a Variety of Cellular Actions

Learning Objectives

After reading this section, you should be able to:

4.2.1 **Contrast the receptors and mechanisms of action of peptide and amine hormones versus steroid hormones.**

4.2.2 **Describe how G protein-coupled receptors (GPCRs) activate second messenger systems.**

4.2.3 **Compare the genomic versus non-genomic actions of steroid hormones.**

4.2.4 **Understand how the brain regulates hormone production and secretion.**

In later chapters we will be considering the effects of specific hormones on behaviour. In preparation for that discussion, let's examine the mechanisms by which hormones affect neuronal targets, and the regulation of hormone secretion.

neuroendocrine cell Also called *neurosecretory cell.* A neuron that releases hormones into local or systemic circulation.

neuropeptide Also called *peptide neurotransmitter.* A peptide that is used by neurons for signalling.

neuromodulator A substance that influences the activity of synaptic transmitters.

peptide hormones A class of hormones, each of which consists of a string of amino acids. If the string of amino acids is long enough, it may be called a *protein hormone.*

amine hormones Also called *monoamine hormones.* A class of hormones, each composed of a single amino acid that has been modified into a related molecule, such as melatonin or epinephrine.

steroid hormones A class of hormones, each of which is composed of four interconnected rings of carbon atoms.

G protein-coupled receptors (GPCRs) A large family of membrane-spanning receptors that use G proteins on their intracellular surface to affect the receiving cell.

second messenger A slow-acting substance in a target cell that amplifies the effects of synaptic or hormonal activity and regulates activity within the target cell.

cyclic adenosine monophosphate (cyclic AMP, or cAMP) A second messenger activated in many target cells in response to synaptic or hormonal stimulation.

inositol triphosphate (IP_3) A member of a class of second-messenger compounds (*phosphoinositides*) common in postsynaptic cells.

diacylglycerol (DAG) A common second-messenger compound in postsynaptic cells.

Hormones initiate actions by binding to receptor molecules

The three classes of hormones exert their influences on target organs in two different ways. Peptide and amine hormones bind to specific receptors (proteins that recognize only one hormone or class of hormones) that are usually found *on the surface* of target cell membranes. In contrast, steroid hormones easily pass through cell membranes, so they generally bind to specific receptor proteins located *inside* the cell.

Let's look at these two main modes of action in a little more detail and examine the ways in which hormones affect cells.

PEPTIDE AND AMINE HORMONES What determines whether a cell responds to a particular peptide hormone? Only those cells that produce the appropriate receptor proteins for a hormone and insert them into the membrane can respond to that hormone. As we saw with neurotransmitter receptors in Chapters 2 and 3, the receptor protein spans the cellular membrane. When a hormone binds to the extracellular portion of the receptor, the receptor molecule changes its overall shape. The alteration in the intracellular portion of the receptor then changes the internal chemistry of the cell. Most of these hormone receptors on the surface of cells are **G protein-coupled receptors** (**GPCRs**) (Kleinau et al., 2016), related to the metabotropic neurotransmitter receptors we discussed earlier (see Figures 2.17 and 3.2). The activation of the G protein complex causes its subunits to separate, and both the alpha subunit and the combined beta and gamma subunits can then activate a **second messenger** (**FIGURE 4.8A**). Second-messenger systems within the target cell can bring about changes in metabolism, membrane potentials, and other cellular functions.

One second-messenger compound in particular—**cyclic adenosine monophosphate** (**cyclic AMP** or **cAMP**)—transmits the messages of many of the peptide and amine hormones. It may seem surprising that the same second messenger can mediate the effects of many different hormones, but a change in cAMP levels can cause many different outcomes, depending on which cells are affected, on which part of a cell is affected, and on the prior biochemical activity inside the cell. Other widespread second-messenger compounds include **inositol triphosphate** (**IP_3**) and **diacylglycerol** (**DAG**), both part of the phospholipase C signalling pathway.

Go to Animation 4.2
Mechanisms of Hormone Action

The specificity of hormonal effects is determined in large part by the selectivity of receptors. Only a minority of cells produce the receptor that recognises and reacts to a particular hormone, and thus only those cells can respond to it. For

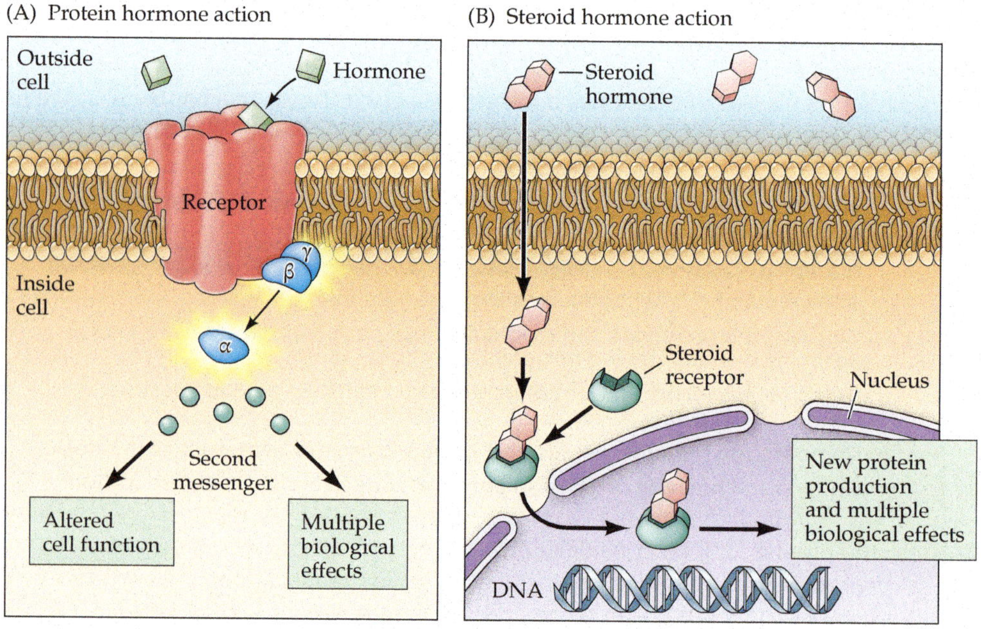

4.8 Two Main Mechanisms of Hormone Action (A) Protein hormone receptors are found in the cell membrane. When the hormone binds to the receptor, a second-messenger system is activated, which affects various cellular processes. (B) Steroid hormones diffuse passively into cells. Inside the target cells are large receptor molecules that bind to the steroid hormones. The steroid-receptor complexes then bind to DNA, causing an increase in the production of some gene products and a decrease in the production of others. This is the mechanism by which steroids exert a genomic effect, which is distinct from the nongenomic effects mentioned in the text.

example, the peptide adrenocorticotropic hormone (ACTH) interacts with receptors on the membranes of cells in the adrenal gland, and in these cells an increase in cAMP leads to the synthesis and release of other hormones.

Peptide hormones usually act relatively rapidly, within seconds to minutes. (Although rapid for a hormone, this action is much slower than neural activity.) There can also be prolonged effects. For example, ACTH promotes the proliferation and growth of some adrenal cells, thereby increasing their long-term capacity to produce hormones. A cell may increase or decrease the number of hormone receptors it makes, and these changes are sometimes referred to as *up-regulation* and *down-regulation*, respectively.

STEROID HORMONES Steroid hormones typically act more slowly than protein or amine hormones, usually requiring hours to take effect. The specific actions of steroid hormones are determined by the receptors that reside *inside* target cells. These receptors are truly ancient, in an evolutionary sense, having arisen in primordial invertebrate species a billion years ago (Eick and Thornton, 2011). Steroid hormones pass in and out of many cells in which they have no effect. If appropriate receptor proteins are inside a cell, however, these receptors bind to the hormone (**FIGURE 4.8B**). The resulting steroid-receptor complex then binds to specific regions of the DNA in the nucleus of the cell, where it acts as a **transcription factor**, controlling the expression of specific genes. This action results in increases or decreases in the production of the proteins encoded by the regulated genes (see the Appendix). By regulating gene expression in this manner, steroid hormones can affect a vast array of cellular processes.

Simple possession of appropriate steroid receptors is not enough to ensure that a given neuron will respond to that particular steroid hormone. Cells make a wide variety of **steroid receptor cofactors** that may be required, along with the bound steroid receptors, in order for the cell to respond. Furthermore, the nature of the target cell's response may be determined by the type of cofactor present: two different cells containing the same steroid receptors may respond quite differently to the steroid hormone if they are producing different steroid receptor cofactors (Molenda-Figueira et al., 2006; Charlier, 2009).

By altering protein production, steroids have slow but long-lasting effects on the development or adult function of cells. We will discuss such effects of steroids further in Chapter 13. There is a large "superfamily" of steroid receptor genes (Sever and Glass, 2013), and some steroids act on more than one receptor. For example, a second estrogen receptor is named estrogen receptor β (Kuiper et al., 1996) to distinguish it from the previously discovered estrogen receptor α. The brain contains both types of estrogen receptors, which differ in their anatomical distributions within the brain and in their effects on behaviour (Borrow and Handa, 2017).

Because steroid-receptor complexes become concentrated in the nuclei of target cells, we can study where a steroid hormone is active by injecting radioactively tagged molecules of the steroid and observing where they accumulate. For example, tagged estrogens accumulate not only in the reproductive tract (as you might expect), but also in the nuclei of some neurons throughout the hypothalamus. Because neurons producing hormone receptors are found in only a limited number of brain regions, we can begin to learn how hormones affect behaviour by finding those brain sites and asking what happens when the hormones arrive there. This strategy for learning about hormones and behaviour is discussed in **BOX 4.1**.

Steroids can also affect cells through mechanisms other than the classic nuclear steroid receptor. For example, estradiol, in addition to its slow, long-lasting action on gene expression, can have a rapid, brief effect on some neurons without affecting gene expression. This rapid **nongenomic effect** of steroids involves a separate receptor located in the neuronal *membrane* (Mani et al., 2012), a GPCR (see Figure 3.17B; Barton et al., 2018) that modulates neural excitability. Similarly, testosterone can have effects that are too rapid to involve the transcription of genes (Foradori et al., 2008).

Sometimes the brain makes its own steroid hormone. Steroids made in the brain are called **neurosteroids**. Progesterone-like neurosteroids act as noncompetitive

transcription factor A substance that binds to recognition sites on DNA and alters the rate of expression of particular genes.

steroid receptor cofactors Proteins that affect the cell's response when a steroid hormone binds its receptor.

nongenomic effect An effect of a steroid hormone that is not mediated by direct changes in gene expression.

neurosteroids Steroid molecules produced within the brain that affect neurons.

BOX 4.1 Techniques of Modern Behavioural Endocrinology

To establish that a particular hormone affects behaviour, investigators usually begin with the type of experiment that Berthold performed in the nineteenth century: observing the behaviour of the intact animal and then removing the endocrine gland and looking for a change in behaviour (see Figure 4.2). Berthold was limited to this type of experiment, but modern scientists have many additional options available. Let's imagine that we're investigating a particular effect of hormones on behaviour to see how we might proceed.

Which Hormones Affect Which Behaviours?

First we must carefully observe the behaviour of several individuals, seeking ways to classify and quantify the different types of behaviour and to place them in the context of the behaviour of other individuals. For example, most adult male rats will try to mount and copulate with a receptive female placed in their cage. If the testes are removed from a male rat, he will eventually stop copulating with females. We know that one of the hormones produced by the testes is testosterone. Is it the loss of testosterone that causes the loss of male copulatory behaviour?

To explore this question, we inject some synthetic testosterone into castrated males and observe whether the copulatory behaviour returns. (It does.) Another way to ask whether a steroid hormone is affecting a particular behaviour is to examine the behaviour of animals that lack the receptors for that steroid. We can delete the gene for a given hormone receptor, making a **knockout organism** (so named because the gene for the receptor has been "knocked out"), and ask which behaviours are different in the knockouts versus normal animals (see Box 6.3).

Next we might examine individual male rats and ask whether the ones that copulate a lot have more testosterone circulating in their blood than those that copulate only a little. To investigate this question, we measure individual differences in the amount of copulatory behaviour, take a sample of blood from each individual, and measure levels of testosterone with **radioimmunoassay** (**RIA**), a technique using an antibody that binds to a particular hormone. By adding many such antibodies to each blood sample and measuring how many of the antibodies find a hormone molecule to bind, we can estimate the total number of molecules of the hormone per unit volume of blood.

It turns out that individual differences in the sexual behaviour of normal male rats (and male humans) do *not* correlate with differences in testosterone levels in the blood. In both rats and humans, a drastic loss of testosterone, as occurs after castration, results in a gradual decline in sexual behaviour. All healthy males, however, appear to make more than enough testosterone to maintain sexual behaviour, so something else must modulate this behaviour. In other words, the hormone acts in a permissive manner: it permits the display of the behaviour, but something else determines how much of the behaviour each individual exhibits.

Where Are the Target Cells?

What does testosterone do to permit sexual behaviour? One step toward answering this question is to ask another question: Which parts of the brain are normally affected by this hormone? We have several methods at our disposal for investigating this question.

In one method, we can inject a castrated animal with radioactively labelled testosterone and wait for the hormone to accumulate in the brain regions that have receptors for the hormone (Figure A). Then we can sacrifice the animal, remove the brain, freeze it, cut thin sections from it, and place the thin sections on photographic film. Radioactive emissions from the tissue will expose the film, revealing which brain regions have accumulated the most labelled testosterone. This method is known as **autoradiography** because the tissue "takes its own picture" with radioactivity.

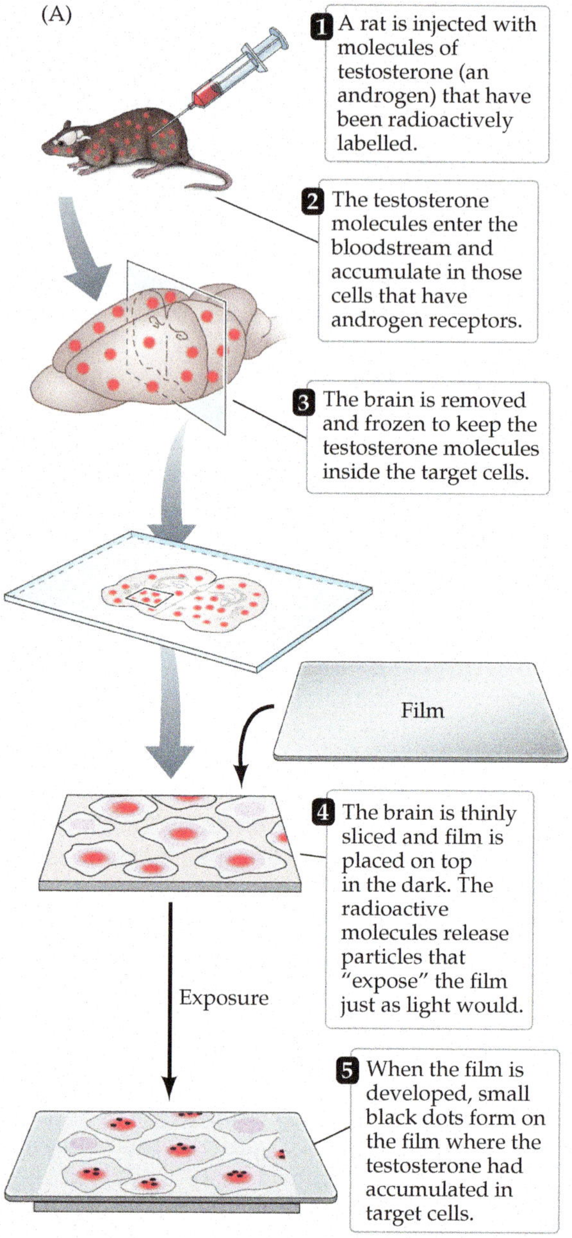

(A) Steps in steroid autoradiography.

(B) Autoradiogram

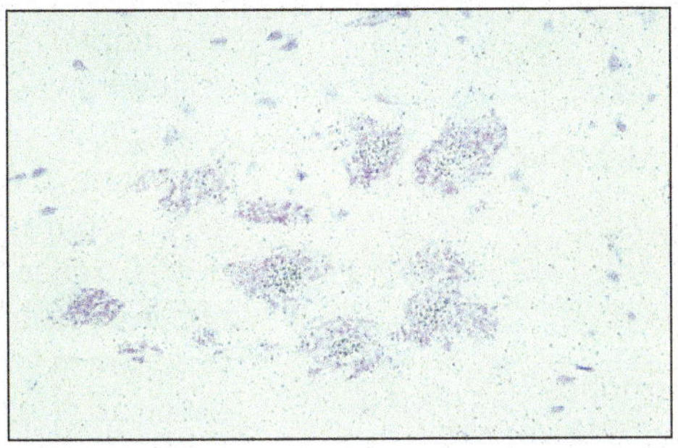

(C) Autoradiogram

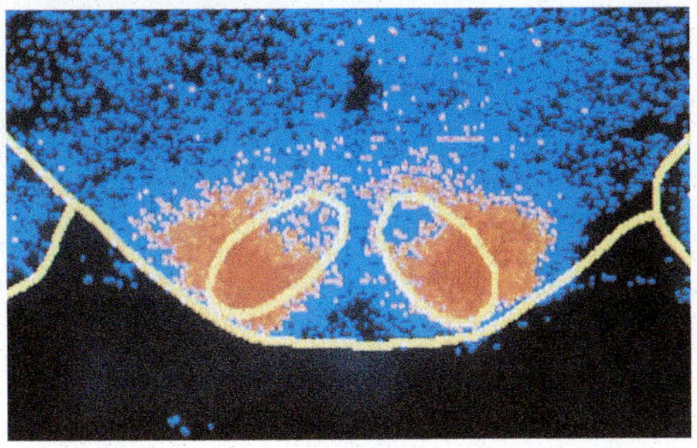

Courtesy of Bruce McEwen

(D) Immunocytochemistry

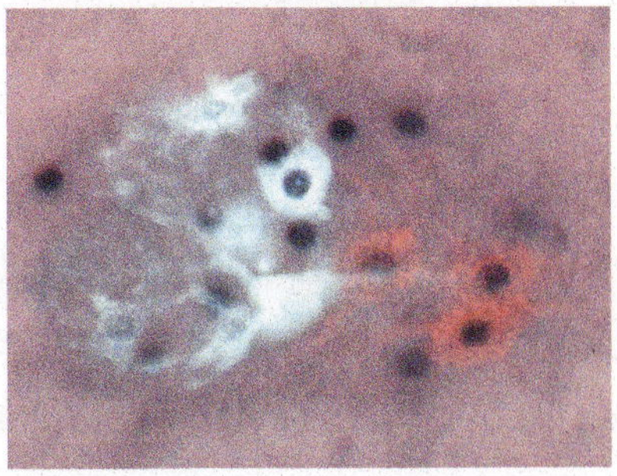

Courtesy of Cynthia Jordan

(B) An autoradiogram showing that spinal motor neurons (purple cell profiles) accumulate radioactive testosterone (small dots). (C) An autoradiogram showing the concentration of oxytocin receptors in the ventromedial hypothalamus (oval outlines). (D) Immunocytochemistry revealing cells with nuclei that contain androgen receptors (dark circles), to which testosterone can bind. The somata of these neurons have been labelled with the tracers Fluoro-Gold (white) and Fluoro-Ruby (red).

When the labelled hormone is a steroid such as testosterone, the radioactivity accumulates in the nuclei of neurons and leaves small black specks on the film (Figure B).

When the radiolabelled hormone is a peptide hormone such as oxytocin, the radioactivity accumulates in the membranes of cells and appears in particular layers of the brain. Computers can generate colour maps that highlight regions with high densities of receptors (Figure C).

Another method for detecting hormone receptors is **immunocytochemistry** (**ICC**). In this method (described in more detail in Box 1.1), we use antibodies that recognise the hormone receptor (Figure D). This method allows us to map the distribution of hormone receptors in the brain. We put the antibodies on slices of brain tissue, wait for them to bind to the receptors, wash off the unbound antibodies, and use chemical methods to visualise the antibodies by creating a tiny dark spot in the nuclei of target brain cells. We can also use **in situ hybridisation** (see Box 1.1) to look for the neurons that make the mRNA for the steroid receptor. Because these cells make the transcript for the receptor, they are likely to possess the receptor protein itself.

What Happens at the Target Cells?

Once we have used autoradiography, immunocytochemistry, or in situ hybridisation (or, better yet, all three) to identify brain regions that have receptors for the hormone, those regions become candidates for the places at which the hormone works to change behaviour. Now we can take castrated males and implant tiny pellets of testosterone into one of those brain regions. We use RIA to ensure that the pellets are small enough that they have no effect on hormone levels in the blood. Then we ask whether the small implant in that brain region restores the behaviour. If not, then in other animals we can implant pellets in a different region or try placing implants in a combination of brain sites.

It turns out that such implants can restore male sexual behaviour in rats only if they are placed in the medial preoptic area (mPOA) of the hypothalamus. Thus, we have found so far that testosterone does something to the mPOA to permit individual males to display sexual behaviour. Now we can examine the mPOA in detail to learn which changes in the anatomy, physiology, or protein production of this region are caused by testosterone. We have more or less caught up to modern-day scientists who work on this very question. Some of the preliminary answers suggested by their research will be discussed in Chapter 13 (see Figure 13.6).

knockout organism An individual in which a particular gene has been disabled by an experimenter.

radioimmunoassay (RIA) A technique that uses antibodies to measure the concentration of a substance, such as a hormone, in blood.

autoradiography A histological technique that shows the distribution of radioactive chemicals in tissues.

immunocytochemistry (ICC) A method for detecting a particular protein in tissues, in which an antibody recognizes and binds to the protein and then chemical methods are used to leave a visible reaction product around each antibody.

in situ hybridisation A method for detecting particular RNA transcripts in tissue sections by providing a nucleotide probe that is complementary to, and will therefore hybridize with, the transcript of interest.

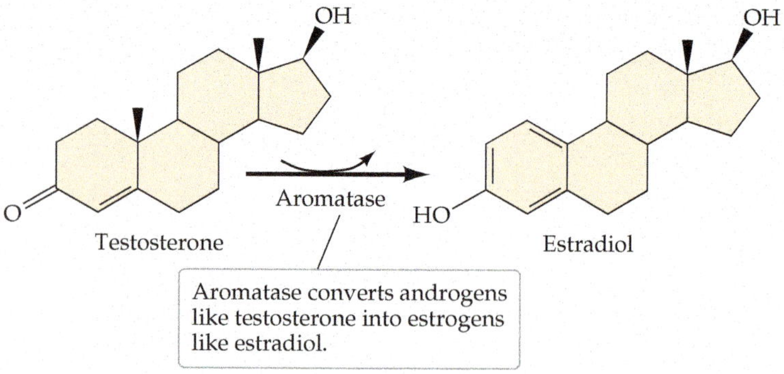

4.9 Enzymatic Conversion of Steroid Hormones

agonists at $GABA_A$ receptors to reduce anxiety (Belelli et al., 2018; see Figure 3.12). The brain also makes "gonadal" steroids like testosterone and estrogens. Furthermore, the brain may transform one steroid into another. For example, in Chapter 13 we'll see that testosterone released from the testes may, upon entering the brain, be converted into estrogens! An enzyme called **aromatase**, which converts testosterone to estradiol in a single chemical reaction (**FIGURE 4.9**), is abundant in the hypothalamus. In fact, testosterone is the major precursor for making estrogens. The reason ovaries release so much estrogen is because they are loaded with aromatase, rapidly converting most testosterone and other androgens into estrogens. Testes, on the other hand, have far less aromatase on hand and so release smaller amounts of estrogens. That's one reason why *no steroid hormone is found exclusively in one sex.*

Feedback control mechanisms regulate the secretion of hormones

One of the major features of almost all hormonal systems is that they don't just manufacture a hormone; they also detect and evaluate the effects of the hormone. Thus, secretion is usually monitored and regulated so that the rate is appropriate to ongoing activities and needs of the body. The basic control used is a **negative feedback** system: output of the hormone *feeds back* to inhibit the drive for more of that same hormone. This negative feedback action of a hormonal system is like that of a thermostat, and just as the thermostat can be set to different temperatures at different times, the set points of a person's endocrine feedback systems can be changed to meet varying

aromatase An enzyme that converts some androgens into estrogens.

negative feedback The property by which some of the output of a system feeds back to counteract input signals.

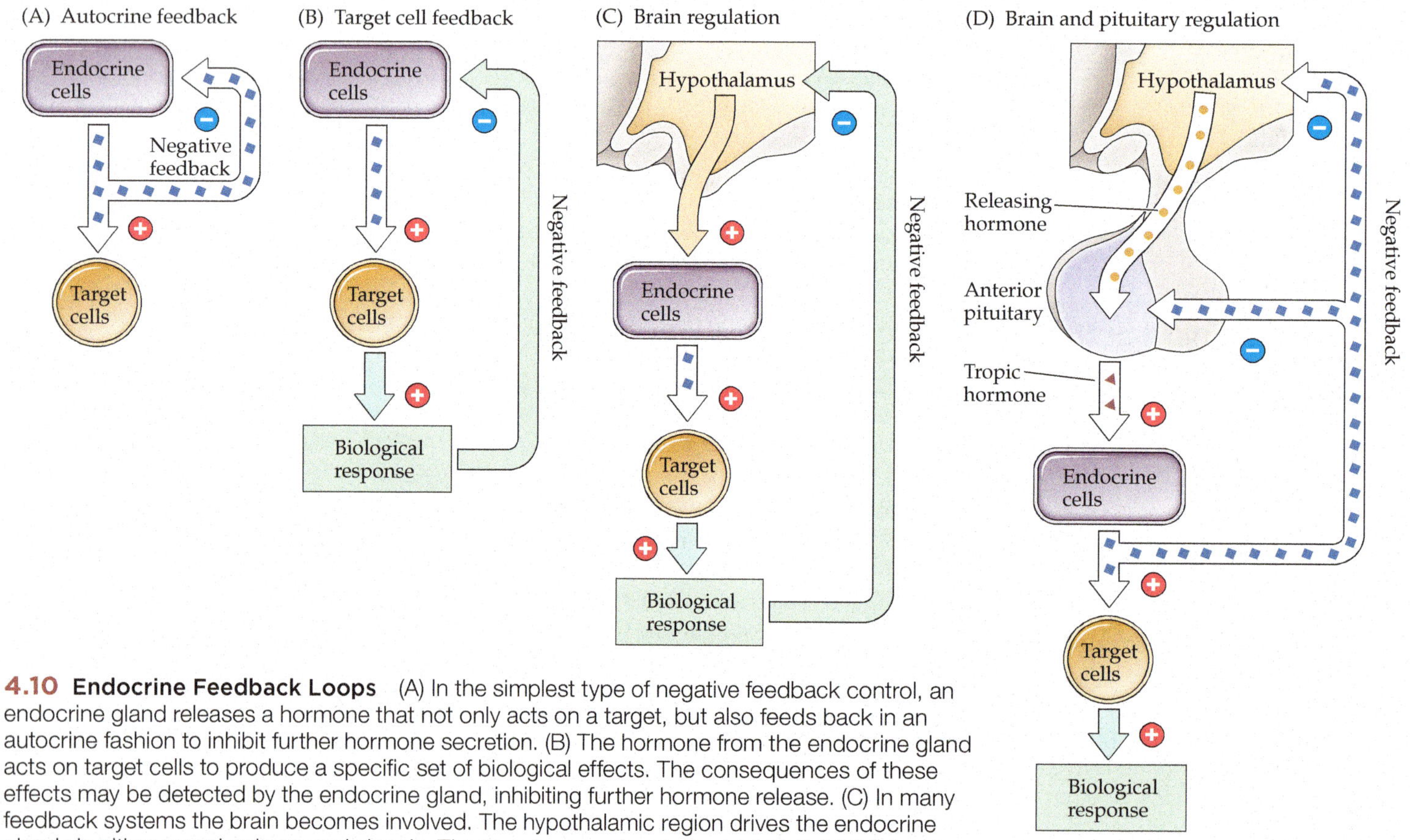

4.10 Endocrine Feedback Loops (A) In the simplest type of negative feedback control, an endocrine gland releases a hormone that not only acts on a target, but also feeds back in an autocrine fashion to inhibit further hormone secretion. (B) The hormone from the endocrine gland acts on target cells to produce a specific set of biological effects. The consequences of these effects may be detected by the endocrine gland, inhibiting further hormone release. (C) In many feedback systems the brain becomes involved. The hypothalamic region drives the endocrine gland via either neural or hormonal signals. The target organ signals the brain to inhibit this drive. (D) More-complex feedback mechanisms involve the hypothalamus and the anterior pituitary, as well as the endocrine gland. Feedback is regulated by a variety of hormones via multiple routes.

4.11 An Example of Complex Endocrine Regulation The brain funnels information to the hypothalamus, which then controls the anterior pituitary, which in turn stimulates the thyroid gland. Note that three hormones and at least four cell groups are interacting in this instance.

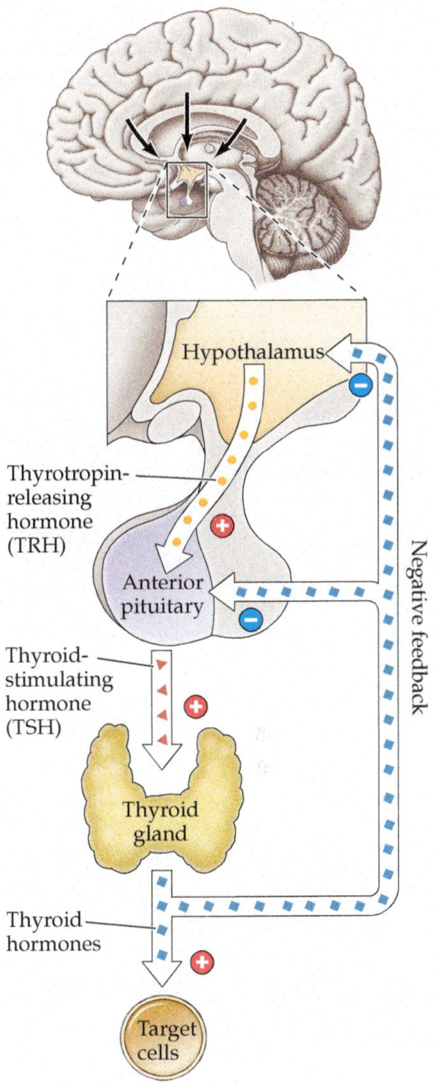

circumstances. We'll discuss negative feedback regulation of other processes in Chapter 14 (see Figure 14.1).

In the simplest kind of hormone regulation system, diagrammed in **FIGURE 4.10A**, an endocrine cell releases a hormone that acts on target cells, but the same hormone also feeds back to inhibit the gland that released it. This is an autocrine response.

In other cases, the endocrine cell reacts not to its own hormone, but to the biological response that its hormone elicits from the target cells (**FIGURE 4.10B**). If the initial effect is too small, additional hormone is released; if the effect is sufficient, no further hormone is released. For example, the hormone insulin is released to control the level of glucose circulating in our blood. After a meal, glucose from the food enters the bloodstream, causing insulin to be released from the pancreas. The insulin causes glucose to enter muscle and fat cells. As the level of glucose in the blood falls, the pancreas secretes less insulin, so a balance tends to be maintained (Chapter 13).

A more complex endocrine system includes the brain, usually the hypothalamus, as part of the circuit that controls an endocrine gland (**FIGURE 4.10C**). When we are alarmed, for example, the hypothalamus directs the adrenal medulla to secrete the hormone epinephrine (also called *adrenaline*), which affects many target cells. The brain detects these effects and exerts negative feedback on the hypothalamus to reduce further hormone output.

An even greater degree of complexity is encountered when the anterior pituitary becomes involved (**FIGURE 4.10D**). As we'll see in the next section, several anterior pituitary hormones regulate hormone secretion by other endocrine glands; all of these pituitary hormones are called **tropic hormones**. (*Tropic*, pronounced with a long *o* as in *toe*, means "directed toward." There is nothing "tropical" about tropic hormones.) The hypothalamus uses another set of hormones, called **releasing hormones**, to control the pituitary release of tropic hormones. Thus, the brain's releasing hormones affect the pituitary's tropic hormones, which affect the release of hormones from endocrine glands. Negative feedback in this case goes from the hormone of the endocrine gland to both the hypothalamus and the anterior pituitary (**FIGURE 4.11**). We'll explore the role of the pituitary in more detail, next.

4.3 Each Endocrine Gland Secretes Specific Hormones

Learning Objectives

After reading this section, you should be able to:

4.3.1 **Distinguish the mechanisms of hormone secretion in the posterior versus anterior pituitary gland.**

4.3.2 **Survey the many endocrine hormones regulated by the anterior pituitary and their effects.**

4.3.3 **List the six anterior pituitary tropic hormones that affect endocrine glands.**

4.3.4 **Describe the neurobehavioural functions of hormones secreted from the adrenal, thyroid, and pineal glands.**

Go to Animation 4.3
The Hypothalamus and Endocrine Function

We restrict our account of endocrine glands in this chapter to some of the main glands because a thorough treatment would fill an entire book (e.g., Melmed et al., 2016). We will discuss hormones from the pancreas and stomach in Chapter 14 when we consider hunger. Keep in mind that most hormones have more functions than are mentioned here.

The pituitary gland releases many important hormones

Resting in a socket in the base of the skull is the **pituitary gland**, or *hypophysis* (see Figure 4.1), occupying a volume of about 1 cubic centimeter and weighing about 1 gram. The hypothalamus sits just above it. The term *pituitary* comes from the Latin *pituita*, "mucus," reflecting the erroneous historical belief that waste products

tropic hormones A class of anterior pituitary hormones that affect the secretion of other endocrine glands.

releasing hormones A class of hormones, produced in the hypothalamus, that traverse the hypothalamic-pituitary portal system to control the pituitary's release of tropic hormones.

pituitary gland Also called *hypophysis*. A small, complex endocrine gland located in a socket at the base of the skull.

TABLE 4.2 Main Endocrine Glands, Their Hormone Products, and Principal Effects of the Hormones

Gland	Hormones	Principal effects
POSTERIOR PITUITARY (storage organ for certain hormones produced by hypothalamus)	Oxytocin	Stimulates contraction of uterine muscles; stimulates release of milk by mammary glands
	Vasopressin (AVP), or antidiuretic hormone (ADH)	Stimulates increased water reabsorption by kidneys; stimulates constriction of blood vessels
ANTERIOR PITUITARY	Growth hormone (GH)	Stimulates growth
	Thyroid-stimulating hormone (TSH)	Stimulates the thyroid
	Adrenocorticotropic hormone (ACTH)	Stimulates the adrenal cortex
	Follicle-stimulating hormone (FSH)	Stimulates growth of ovarian follicles and of seminiferous tubules of the testes
	Luteinizing hormone (LH)	Stimulates conversion of follicles into corpora lutea; stimulates secretion of sex hormones by gonads
	Prolactin	Stimulates milk production by mammary glands
HYPOTHALAMUS	Releasing hormones	Regulate hormone secretion by anterior pituitary
	Oxytocin; vasopressin	*See* "Posterior pituitary" above
PINEAL	Melatonin	Regulates seasonal changes; regulates puberty
ADRENAL CORTEX	Glucocorticoids (corticosterone, cortisol, hydrocortisone, etc.)	Inhibit incorporation of amino acids into protein in muscle; stimulate formation and storage of glycogen; help maintain normal blood sugar level
	Mineralocorticoids (aldosterone, deoxycorticosterone, etc.)	Regulate metabolism of sodium and potassium
	Sex hormones (especially androstenedione)	Regulate facial and body hair
ADRENAL MEDULLA	Catecholamines (epinephrine, norepinephrine)	Prepare body for action
GONADS		
Testes	Androgens (testosterone, dihydrotestosterone, etc.)	Stimulate development and maintenance of male primary and secondary sexual characteristics and behaviour
Ovaries	Estrogens (estradiol, estrone, etc.)	Stimulate development and maintenance of female secondary sexual characteristics and behaviour
	Progestins (progesterone)	Stimulate female secondary sexual characteristics and behaviour; maintain pregnancy
THYROID	Thyroxine (tetraiodothyronine); triiodothyronine	Stimulate oxidative metabolism
	Calcitonin	Prevents excessive rise in blood calcium
PANCREAS	Insulin	Stimulates glycogen formation and storage
	Glucagon	Stimulates conversion of glycogen into glucose
STOMACH	Secretin	Stimulates secretion of pancreatic juice
	Cholecystokinin (CCK)	Stimulates release of bile by gallbladder
	Gastrin	Stimulates secretion of gastric juice
	Ghrelin	Provides appetite signal to the hypothalamus
HEART	Atrial natriuretic peptide	Promotes salt loss in urine

Go to Activity 4.1
Major Endocrine Glands

dripped down from the brain into the pituitary, which secreted them out through the nose. (The ancients may have thought you could literally sneeze your brains out!) The pituitary used to be referred to as the *master gland* because it controls hormone release from several other endocrine glands. That's why loss of her pituitary left Mary Lou, whom we met at the start of this chapter, with deficits in many hormones. But this gland is itself enslaved by the hypothalamus above it, as we'll see. **TABLE 4.2** gives a fuller but far from complete listing of hormones and their functions.

The pituitary gland consists of two main parts: the **anterior pituitary** (or *adenohypophysis*) and the **posterior pituitary** (or *neurohypophysis*). The anterior and posterior pituitary develop from different embryonic tissues and are completely separate in function. The pituitary is connected to the hypothalamus by a thin piece of tissue called the **pituitary stalk** (or *infundibulum*). The stalk contains many axons and is richly supplied with blood vessels. The axons extend only to the posterior pituitary, which we will consider next. The blood vessels, as we will see later, carry information exclusively to the anterior pituitary.

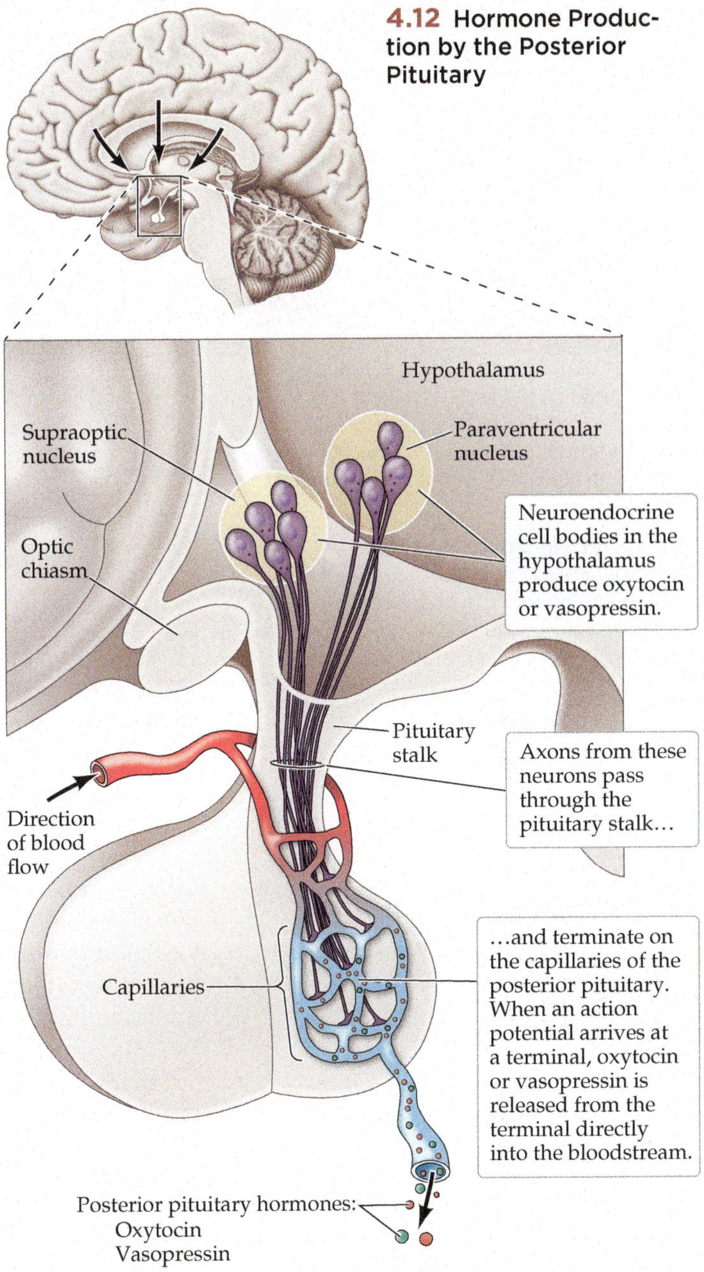

4.12 Hormone Production by the Posterior Pituitary

THE POSTERIOR PITUITARY The posterior pituitary gland secretes two principal hormones: **oxytocin** and **arginine vasopressin (AVP)**, often called just **vasopressin**. Neurons in the **supraoptic nuclei** and the **paraventricular nuclei** of the hypothalamus synthesise these two hormones and transport them along their axons to the axon terminals (**FIGURE 4.12**). Action potentials in these hypothalamic neuroendocrine cells travel down the axons in the pituitary stalk and reach the axon terminals in the posterior pituitary, causing release of the hormones from the terminals into the rich vascular bed of the neurohypophysis. The axon terminals abut capillaries (small blood vessels), allowing the hormones to enter circulation immediately.

Some of the signals that activate the nerve cells of the supraoptic and paraventricular nuclei are related to thirst and water regulation, which we will discuss in Chapter 14. Secretion of vasopressin increases blood pressure by causing blood vessels to contract. Vasopressin also inhibits the formation of urine, so it is sometimes called *antidiuretic hormone* (ADH) (a diuretic is a food or drug that promotes urination). This action of vasopressin helps conserve water. In fact, the major physiological role of vasopressin is its potent antidiuretic activity; it exerts this effect with less than one-thousandth of the dose needed to alter blood pressure.

Oxytocin is involved in many aspects of reproductive and parental behaviour. One of its functions is to stimulate contractions of the uterus in childbirth (the word *oxytocin* is derived from the Greek *oxys*, "rapid," and *tokos*, "childbirth"). Injections of oxytocin (or the synthetic version, Pitocin) are frequently used in medical settings to induce or accelerate labour and delivery.

Oxytocin also triggers the **milk letdown reflex**, the contraction of mammary gland cells that ejects milk into the breast ducts. This phenomenon exemplifies the reciprocal relationship between behaviour and hormone release. When an infant or young animal first begins to suckle, the arrival of milk at the nipple is delayed by 30–60 seconds. This delay is caused by the sequence of steps that precedes letdown. Stimulation of the nipple activates receptors in the skin, which transmit this information through a chain of neurons and synapses to hypothalamic cells that contain oxytocin. Once these cells have been sufficiently stimulated, the oxytocin is released from the posterior pituitary and travels via the bloodstream to the mammary glands, where it produces a contraction of the tissues storing milk, making the milk available at the nipple (**FIGURE 4.13**).

For mothers, this reflex response to suckling frequently becomes conditioned to baby cries, so milk appears promptly at the start of nursing. Because the mother

anterior pituitary Also called *adenohypophysis*. The front division of the pituitary gland. It secretes tropic hormones.

posterior pituitary Also called *neurohypophysis*. The rear division of the pituitary gland.

pituitary stalk Also called *infundibulum*. A thin piece of tissue that connects the pituitary gland to the hypothalamus.

oxytocin A hormone, released from the posterior pituitary, that triggers milk letdown in the nursing female.

arginine vasopressin (AVP) Also called *vasopressin* or *antidiuretic hormone (ADH)*. A peptide hormone from the posterior pituitary that promotes water conservation.

supraoptic nucleus A hypothalamic nucleus containing neuroendocrine cells that send axons to the posterior pituitary to release oxytocin or vasopressin.

paraventricular nucleus A nucleus of the hypothalamus implicated in the release of oxytocin and vasopressin and in the control of feeding and other behaviours.

milk letdown reflex The reflexive release of milk in response to suckling or to stimuli associated with suckling.

median eminence Midline feature on the base of the brain marking the point at which the pituitary stalk exits the hypothalamus to connect to the pituitary. It contains elements of the hypophyseal portal system.

hypophyseal portal system Also called *pituitary portal system.* A system of capillaries spanning between the neurosecretory cells of the hypothalamus and the secretory tissue of the anterior pituitary.

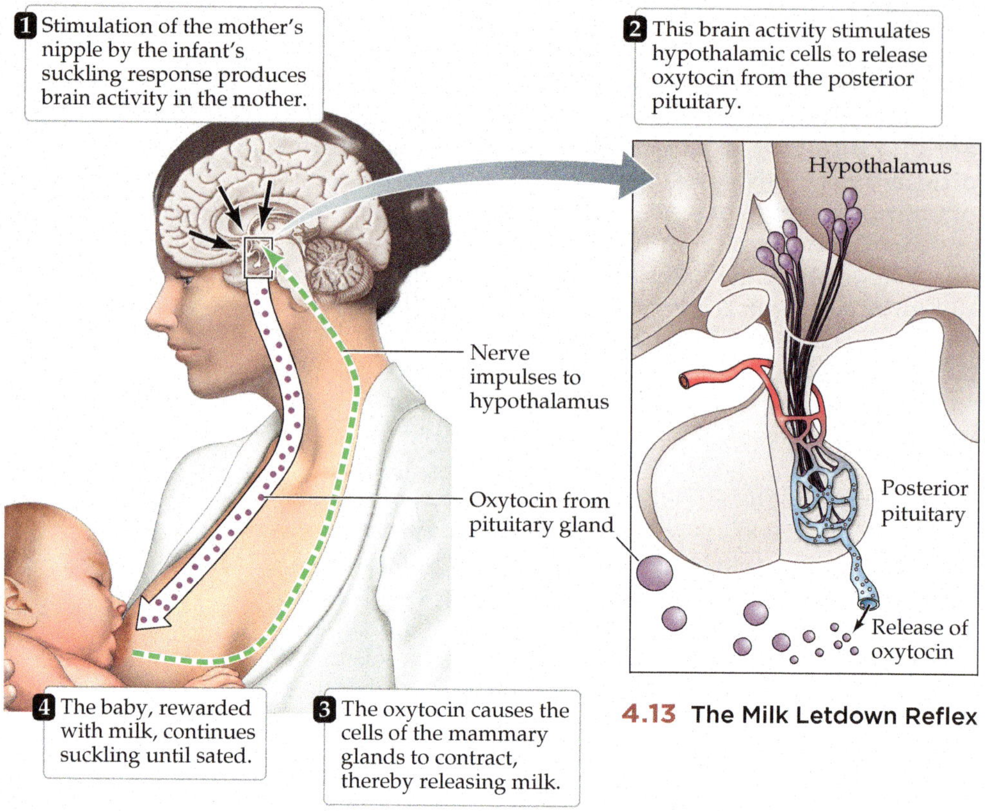

4.13 The Milk Letdown Reflex

learns to release oxytocin *before* the suckling begins, sometimes the cries of someone else's baby in public may trigger an inconvenient release of milk. Oxytocin and vasopressin also serve as neurotransmitters from hypothalamic cells (**FIGURE 4.14**), projecting widely through the nervous system. Oxytocin and vasopressin have been implicated in other social behaviours, as we'll discuss at the end of this chapter.

THE ANTERIOR PITUITARY Different cells of the anterior lobe of the pituitary synthesise and release different tropic hormones, which we'll discuss in the next section. Secretion of these tropic hormones, however, is under the control of releasing hormones, as mentioned earlier. Let's explore how this regulation works.

Hypothalamic releasing hormones govern the anterior pituitary

The neurons that synthesise the different releasing hormones are neuroendocrine cells residing in various regions of the hypothalamus. The axons of these neuroendocrine cells converge on the **median eminence**, just above the pituitary stalk. This region contains an elaborate profusion of blood vessels that form the **hypophyseal portal system** (or *pituitary portal system*). Here, in response to inputs from the rest of the brain, the axon terminals of the hypothalamic neuroendocrine cells secrete their releasing hormones into the local bloodstream (**FIGURE 4.15**).

Blood carries the various releasing hormones only a few millimeters, into the anterior pituitary. The rate at which releasing hormones arrive at their target cells in the anterior pituitary controls the rate at which the anterior pituitary cells, in turn, release their tropic hormones into the general circulation. These tropic hormones then regulate the activity of major endocrine organs throughout the body.

The hypothalamic neuroendocrine cells that synthesise the releasing hormones are themselves subject to two kinds of influences:

(A) Intact (B) Castrated

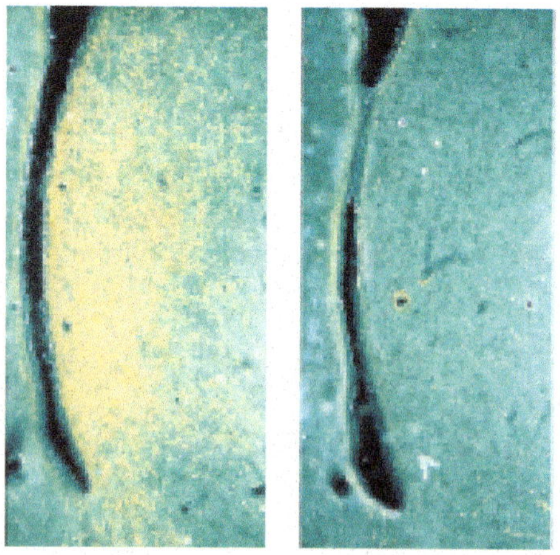

4.14 Vasopressin Can Serve as a Neurotransmitter Revealed here by immunocytochemistry are vasopressin-filled axonal fibres (yellow) in the septum of intact (A) and castrated (B) male rats.

1. They are directly affected by *circulating messages*, such as other hormones (especially hormones that have themselves been secreted in response to tropic hormones), and by blood sugar and products of the immune system. The hypothalamus is not shielded by the blood-brain barrier (see Chapter 1) to the same extent that other brain regions are, which makes it possible for a wide variety of blood-borne material to access the hypothalamic neuroendocrine cells.
2. They receive *synaptic inputs* (either excitatory or inhibitory) from many other brain regions. A wide range of neural signals, reflecting both internal and external events, can thereby influence the endocrine system. As a result, hormonal actions can be coordinated with ongoing events, and conditioning (learning) can alter endocrine status. We saw an example of such influence in our discussion of the milk letdown reflex, and we'll see other examples throughout the book.

The hypothalamic releasing hormone system therefore exerts high-level control over endocrine organs throughout the body and provides a route by which brain activity is translated into hormonal action. Cutting the pituitary stalk interrupts the portal blood vessels and the flow of releasing hormones, leading to profound atrophy of the pituitary and major hormonal disruptions.

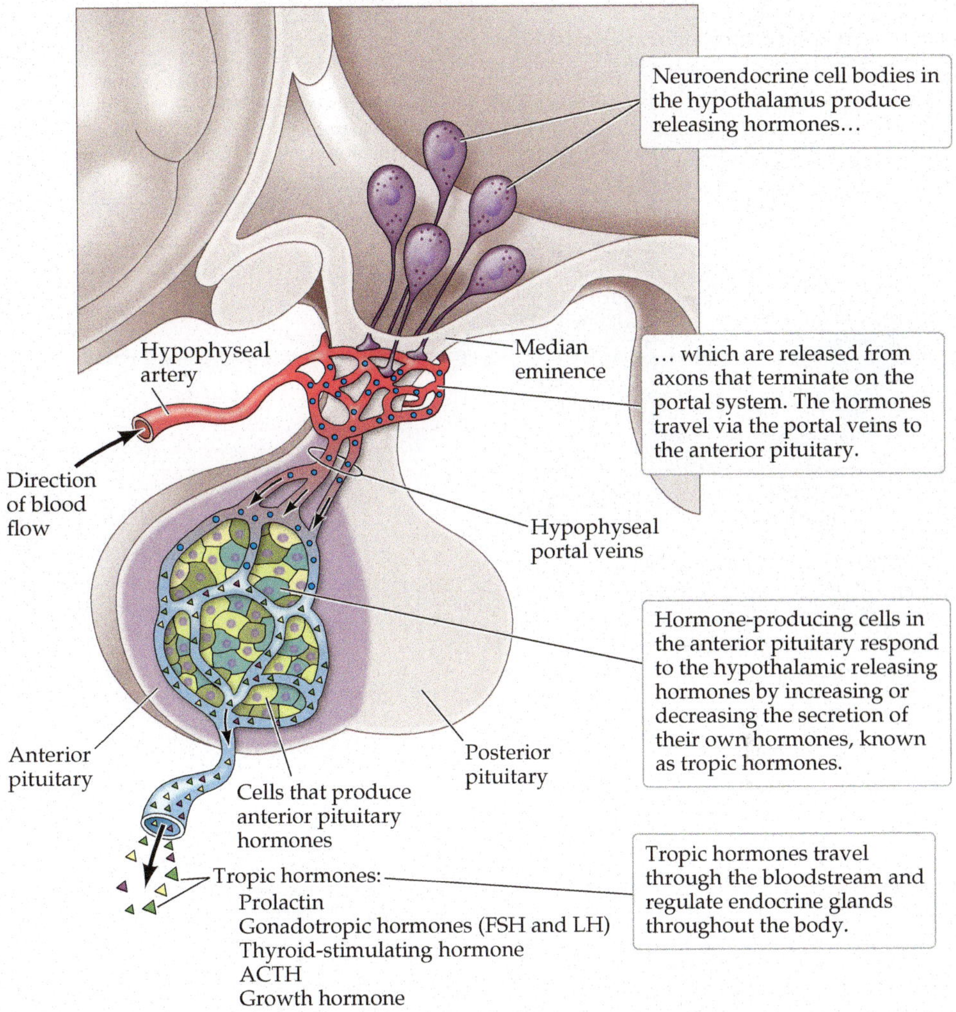

4.15 Hormone Release by the Anterior Pituitary

TROPIC HORMONES OF THE ANTERIOR PITUITARY Driven by various releasing hormones from the hypothalamus, the anterior pituitary gland secretes six main tropic hormones (**FIGURE 4.16**; see also Table 4.2). Two of these regulate the function of the adrenal cortex and the thyroid gland:

1. **Adrenocorticotropic hormone** (**ACTH**) controls the production and release of hormones of the adrenal cortex. The adrenal cortex, in turn, releases steroid hormones. The levels of ACTH and adrenal steroids show a marked daily rhythm (see Chapter 15) and respond to stress (see Chapter 11).
2. **Thyroid-stimulating hormone** (**TSH**) increases the release of thyroid hormones from the thyroid gland and markedly affects thyroid gland size.

Two other tropic hormones of the anterior pituitary influence the gonads and consequently are termed **gonadotropins**:

3. **Follicle-stimulating hormone** (**FSH**) gets its name from its actions in the ovary, where it stimulates the growth and maturation of egg-containing **follicles** and the secretion of estrogens from the follicles. In males, FSH governs sperm production.
4. **Luteinizing hormone** (**LH**) stimulates the follicles of the ovary to rupture, release their eggs, and form into structures called **corpora lutea** (singular *corpus luteum*) that secrete the sex steroid hormone progesterone. In males, LH

adrenocorticotropic hormone (ACTH) A tropic hormone secreted by the anterior pituitary gland that controls the production and release of hormones of the adrenal cortex.

thyroid-stimulating hormone (TSH) A tropic hormone, released by the anterior pituitary gland, that signals the thyroid gland to secrete its hormones.

gonadotropin Any anterior pituitary hormone that selectively stimulates the gonads to produce steroids and gametes.

follicle-stimulating hormone (FSH) A gonadotropin, named for its actions on ovarian follicles.

follicles Ovarian structures containing immature ova.

luteinizing hormone (LH) A gonadotropin, named for its stimulatory effects on the ovarian corpora lutea.

corpora lutea The structures formed from collapsed ovarian follicles subsequent to ovulation. The corpora lutea are a major source of progesterone.

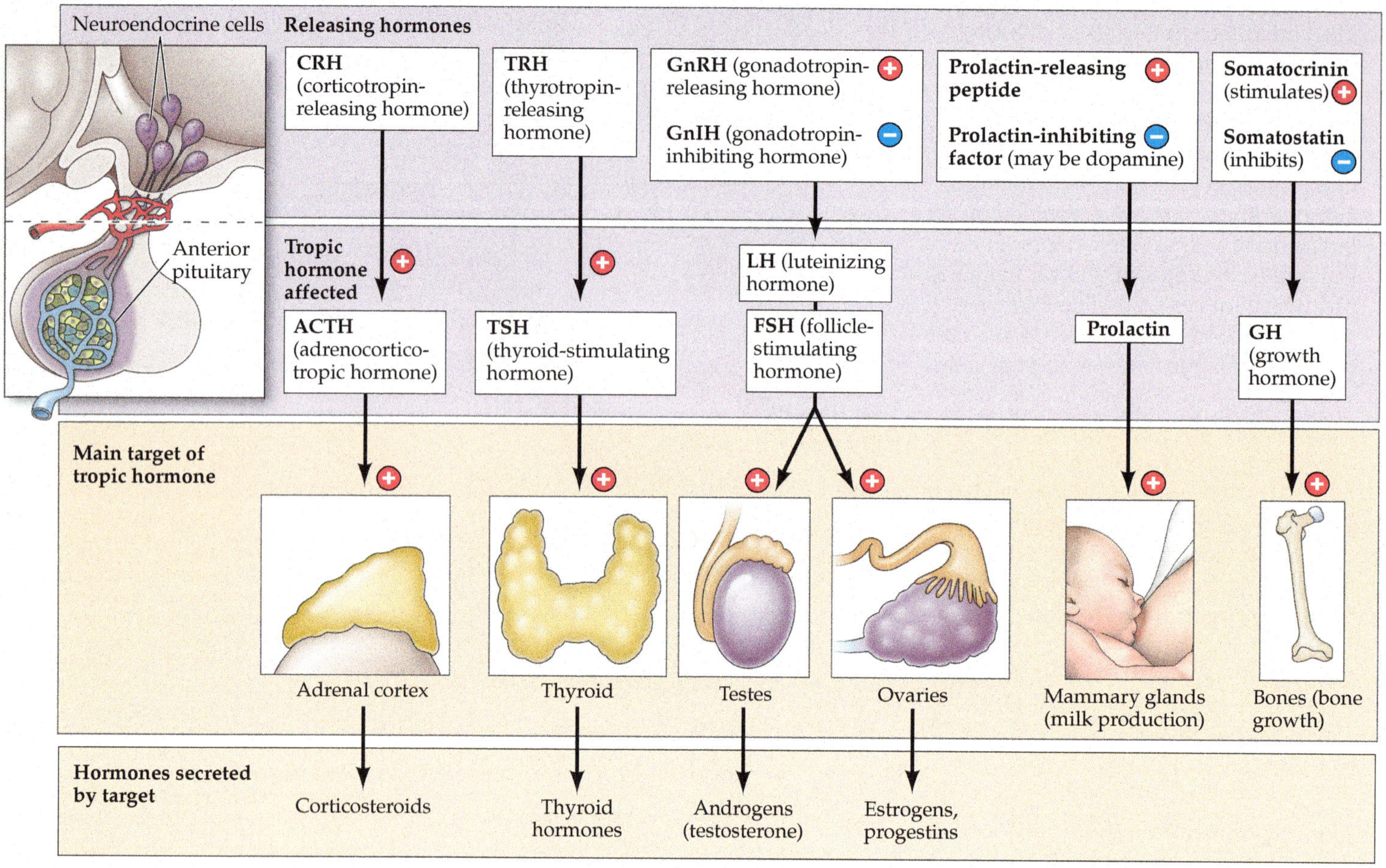

4.16 Secretions of the Anterior Pituitary Hormones produced in the anterior pituitary include tropic hormones, which control endocrine glands and directly affect other structures, such as bones.

stimulates the testes to produce testosterone. We will discuss the gonadal steroid hormones in more detail in the next section. We'll discuss gonadotropins again when we consider social behaviour like mating, at the end of this chapter.

The two remaining tropic hormones control milk production and body growth:

5. **Prolactin** is so named because it promotes lactation in female mammals. But prolactin has a number of roles in addition to its actions on breast tissue. For example, it is closely involved in the parental behaviour of a wide variety of vertebrate species.
6. **Growth hormone** (**GH**), also known as *somatotropin* or *somatotropic hormone*, acts throughout the body to influence the growth of cells and tissues by affecting protein metabolism. GH is released almost exclusively during sleep. Starvation, vigourous exercise, and intense stress can all profoundly inhibit GH release. Other hormones also affect GH secretion. The stomach secretes a hormone called *ghrelin* (from the Proto-Indo-European root for "grow"), which evokes GH release from the anterior pituitary (Müller et al., 2015). Ghrelin is discussed in more detail in Chapter 14.

Now let's consider three endocrine glands: the adrenal gland, the thyroid gland, and the pineal gland.

Two divisions of the adrenal gland produce hormones

Resting on top of each kidney is an **adrenal gland** (see Figure 4.1), which secretes a large variety of hormones. In mammals, the adrenal structure is divided into two major portions. The outer 80% of the gland, the **adrenal cortex**, is composed of distinct layers of cells, each producing different steroid hormones. The core of the gland is the **adrenal medulla**, which is richly innervated by autonomic nerves.

prolactin A peptide hormone, produced by the anterior pituitary, that promotes mammary development for lactation in female mammals.

growth hormone (GH) Also called *somatotropin* or *somatotropic hormone*. A tropic hormone, secreted by the anterior pituitary, that influences the growth of cells and tissues.

adrenal gland An endocrine gland atop the kidney.

adrenal cortex The outer rind of the adrenal gland.

adrenal medulla The inner core of the adrenal gland, which secretes epinephrine and norepinephrine.

epinephrine Also called *adrenaline*. A compound that acts both as a hormone, secreted by the adrenal medulla under the control of the sympathetic nervous system, and as a synaptic transmitter.

As part of the "fight or flight" reaction to threat, the adrenal medulla secretes hormones—the catecholamines **epinephrine** (adrenaline) and **norepinephrine (NE)** (noradrenaline)—that prepare the body for action, raising heart rate and respiration, among other things. Because emergencies demand quick action, secretion of these hormones is under direct control of the brain, via sympathetic nerve terminals that release acetylcholine in the adrenal medulla. In Chapter 3 we saw that epinephrine and norepinephrine are also synaptic transmitters at certain sites in the nervous system.

The adrenal cortex produces and secretes a variety of steroid hormones, collectively called the **adrenocorticoids** (or *adrenal steroids*). One subgroup consists of the **glucocorticoids**, so named because of their effects on the metabolism of glucose. Hormones of this type, such as **cortisol**, increase the level of blood glucose and accelerate the breakdown of proteins. It was probably the lack of glucocorticoids that caused Mary Lou to feel weak and nauseous from the lack of circulating glucose. She must take glucocorticoids every day to stay healthy.

In high concentrations, glucocorticoids have a marked anti-inflammatory effect; that is, they inhibit the swelling around injuries or infections. This action normally results in the temporary decrease of bodily responses to tissue injury, which is why synthetic glucocorticoids (such as prednisone) are important and useful drugs. However, sustained high levels of circulating glucocorticoids are harmful to the brain (see Chapter 11), so Mary Lou must adjust her dosage to avoid harmful effects of too much hormone.

A second subgroup of adrenal steroids consists of the **mineralocorticoids**, so named because of their effects on minerals such as sodium and potassium. The primary mineralocorticoid hormone is **aldosterone**, which acts on the kidneys to retain sodium and thus reduces the amount of urine produced, conserving water. This action helps maintain a healthful concentration of ions in blood and extracellular fluids (see Chapter 14). The adrenal cortex also produces **sex steroids**, notably **androstenedione**. Androstenedione contributes to the adult pattern of body hair in men and women (see Chapter 13).

Levels of circulating adrenal cortical hormones are regulated in several steps (**FIGURE 4.17**). The pituitary hormone ACTH promotes steroid synthesis in the adrenal gland. Adrenal steroids in turn exert a negative feedback effect on ACTH release. As the level of adrenal cortical hormones increases, the secretion of ACTH is suppressed, so the output of hormones from the adrenal cortex diminishes. When the levels of adrenal steroids fall, the pituitary ACTH-secreting cells are released from suppression, and the concentration of ACTH in the blood rises again.

Thyroid hormones regulate growth and metabolism

Situated in the throat, around the location of the Adam's apple, is the **thyroid gland** (see Figure 5.1). This gland produces and secretes several hormones. Two of these—*thyroxine* (or *tetraiodothyronine*) and *triiodothyronine*—are usually referred to as **thyroid hormones**; a third—*calcitonin*—promotes calcium deposition in bones and will not be discussed further in this chapter.

norepinephrine (NE) Also called *noradrenaline*. A compound that acts both as a neurotransmitter, produced and released by sympathetic postganglionic neurons to accelerate organ activity, and as a hormone released from the adrenal medulla.

adrenocorticoids Also called *adrenal steroids*. A class of steroid hormones that are secreted by the adrenal cortex.

glucocorticoids A class of steroid hormones, released by the adrenal cortex, that affect carbohydrate metabolism and inflammation.

cortisol A glucocorticoid stress hormone of the adrenal cortex.

mineralocorticoids A class of steroid hormones, released by the adrenal cortex, that affect ion concentrations in body tissues.

aldosterone A mineralocorticoid hormone, secreted by the adrenal cortex, that induces the kidneys to conserve sodium ions.

sex steroids Steroid hormones secreted by the gonads: androgens, estrogens, and progestins.

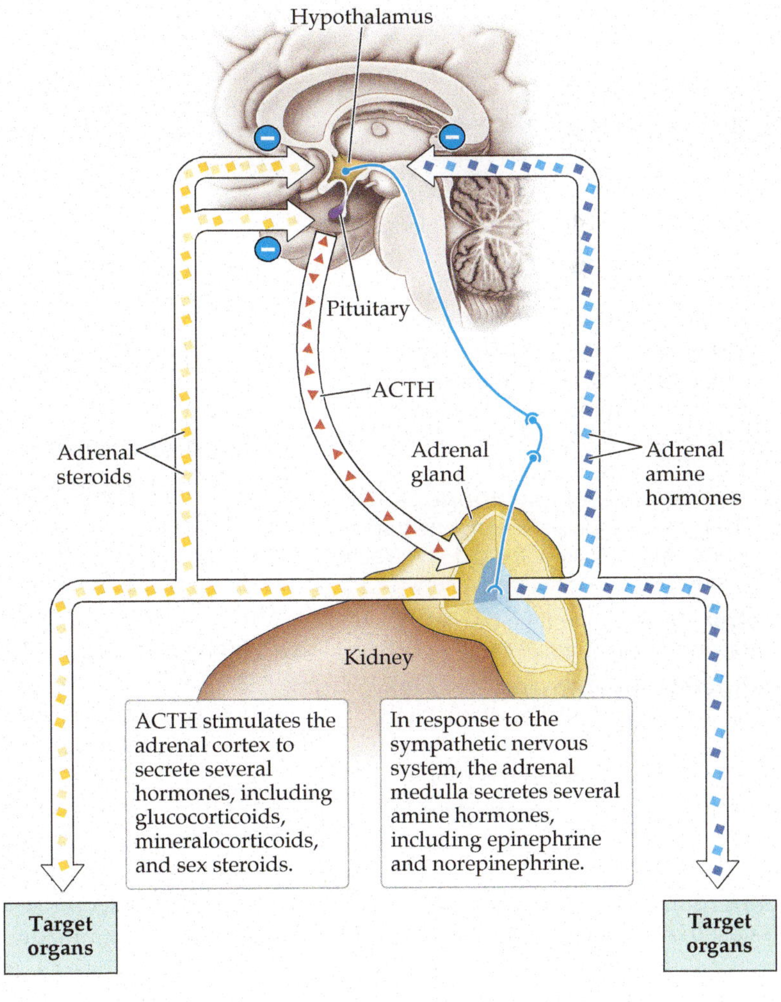

4.17 Regulation of Hormones Produced by the Adrenal Glands Situated above each kidney, each adrenal gland consists of an outer cortex (in yellow) and an inner medulla (blue).

Painting by Johann Friedrich Dietler

4.18 Isn't That a Stylish Collar? Because there is little iodine in the soil in Switzerland, vegetables grown there provide insufficient iodine even for prosperous people like the novelist Jeremias Gotthelf (1797–1854), who suffered from a large goiter that he routinely concealed behind elaborate collars.

The thyroid is unique among endocrine glands because it stores a large amount of hormone—at least a 100-day supply—which it slowly releases. Although thyroid hormones are amines, they behave like steroids. They bind to specialised receptors (part of the steroid receptor superfamily) found inside cells. The thyroid hormone-receptor complex then binds to DNA and regulates gene expression.

Figure 4.11 shows the control network for regulating thyroxine levels in blood. The major control is exerted by TSH from the anterior pituitary gland. The secretion of TSH by the pituitary is controlled by the hypothalamic release of **thyrotropin-releasing hormone** (**TRH**), which stimulates the pituitary to secrete TSH. When the level of circulating thyroid hormone falls, both TRH and TSH are secreted. When TSH reaches the thyroid gland, it stimulates the production and release of thyroid hormones. The thyroid hormones then have a negative feedback effect, inhibiting further TRH and TSH release.

Thyroid hormones are the only substances produced by the body that contain iodine, and their manufacture is critically dependent on the supply of iodine. In parts of the world where foods contain little iodine, many people have hypothyroidism. Driven by higher and higher TSH levels, the thyroid gland swells in its attempt to produce more thyroid hormones, causing a **goiter** to form. Because the soil in Switzerland has little iodine, through the nineteenth century even well-fed citizens there often had goiters (**FIGURE 4.18**). Today the addition of a small amount of iodine to table salt—producing *iodized salt*—ensures that we won't develop goiters even if we get insufficient iodine from our vegetables.

Thyroid hormones have a general effect on the nervous system, maintaining alertness and reflexes. People with hypothyroidism may be depressed (Loh et al., 2019), so Mary Lou must also adjust her dosages of thyroid hormones to keep her mind sharp and avoid feeling fatigue.

When thyroid deficiency starts early in life, body growth is stunted and the face malformed. Thyroid deficiency also produces a marked reduction in brain size and in the branching of axons and dendrites (Stenzel and Huttner, 2013). This state, called **congenital hypothyrodism** or *cretinism*, is accompanied by intellectual disability.

The pineal gland secretes melatonin

The **pineal gland** sits atop the brainstem and in mammals is overlain by the cerebral hemispheres (**FIGURE 4.19A**; see also Figure 4.1). Most brain structures are paired (with symmetrical left and right sides), but the pineal gland is a single structure. It is this unusual aspect of the pineal that led the seventeenth-century philosopher René Descartes to propose it to be "the seat of the soul"; religious dogma of his day held that the soul, like the pineal, is indivisible.

Today we know that the pineal plays a crucial role in biological rhythms. Governed by the superior cervical ganglion—part of the sympathetic nervous system—the pineal releases an amine hormone called **melatonin**. The melatonin receptor is a G protein-coupled receptor residing in cell membranes and is similar to receptors for peptide hormones. Because melatonin is released almost exclusively at night (**FIGURE 4.19B**), it provides a signal that tracks day length and, by extension, the seasons (see Chapter 15).

Melatonin secretion controls breeding condition in many seasonally breeding mammals. In hamsters, for example, the lengthening of nights in autumn causes

androstenedione The chief sex hormone secreted by the human adrenal cortex.

thyroid gland An endocrine gland, located in the throat, that regulates cellular metabolism throughout the body.

thyroid hormones Two hormones, triiodothyronine and thyroxine (also called *tetraiodothyronine*), that are released from the thyroid gland and have widespread effects, including growth and maintenance of the brain.

thyrotropin-releasing hormone (TRH) A hypothalamic hormone that regulates the release of thyroid-stimulating hormone from the anterior pituitary.

goiter A swelling of the thyroid gland resulting from iodine deficiency.

congenital hypothyroidism Also called *cretinism*. Reduced stature and intellectual disability caused by thyroid deficiency during early development.

pineal gland A secretory gland in the brain midline that is the source of melatonin release.

melatonin An amine hormone that is released by the pineal gland.

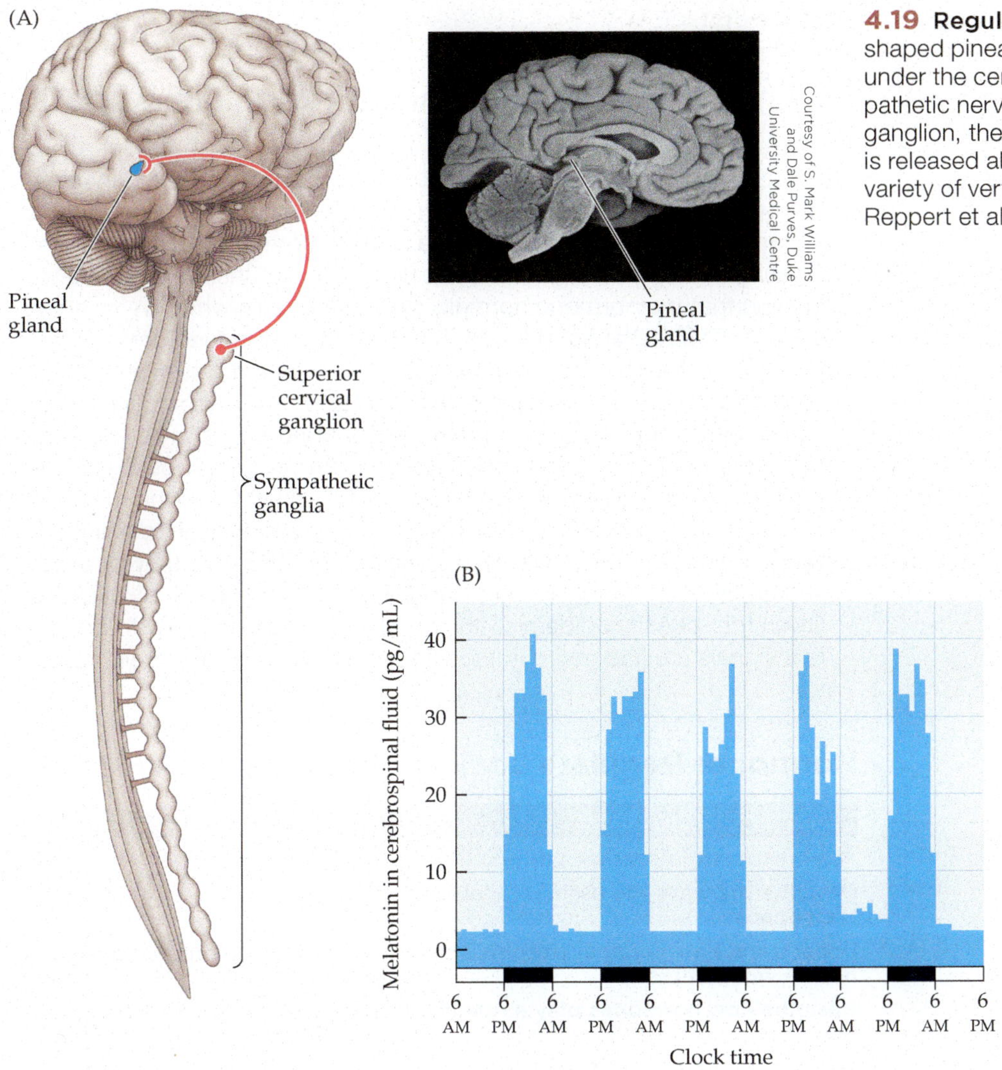

4.19 Regulation of the Pineal Gland (A) The pea-shaped pineal gland sits atop the brainstem, tucked under the cerebral hemispheres. Innervated by the sympathetic nervous system, specifically the superior cervical ganglion, the pineal releases melatonin. (B) Melatonin is released almost exclusively during the night in a wide variety of vertebrates, including humans. (Part B after Reppert et al. 1979. *Endocrinol* 104: 294.301.)

the pineal to prolong its nocturnal release of melatonin. The hypothalamus responds to the prolonged exposure to melatonin by releasing less and less gonadotropin-releasing hormone (see next section), resulting in atrophy of the gonads (Revel et al., 2009). In the spring, as days lengthen and the breeding season approaches, the process reverses and the animal prepares to breed.

In birds, light from the environment penetrates the thin skull and reaches the pineal gland directly. Photosensitive cells in the bird pineal gland monitor daily light durations (Cassone, 2014). In several reptile species the pineal is close to the skull and even has an extension of photoreceptors providing a "third eye" in the back of the head. The pineal photoreceptors do not form images but simply monitor day length to regulate seasonal functions.

Humans are not, strictly speaking, seasonal breeders, but melatonin plays a role in our biological rhythms, especially the timing of sleep onset. Like other vertebrates, we release melatonin at night, and administering exogenous melatonin reportedly induces sleep sooner. This is why melatonin has been used to treat jet lag (Herxheimer, 2014). Interestingly, Mary Lou's loss of the pituitary should not affect her pineal function, yet she reports she no longer suffers from jet lag during her frequent international travels (Jepsen, 2013). So perhaps jet lag is caused by shifts in the daily rhythms of hormones other than melatonin (see Chapter 15).

Dialing Up a Personality Mary Lou Jepsen has learned to fine-tune her hormone treatments in order to feel like her true self.

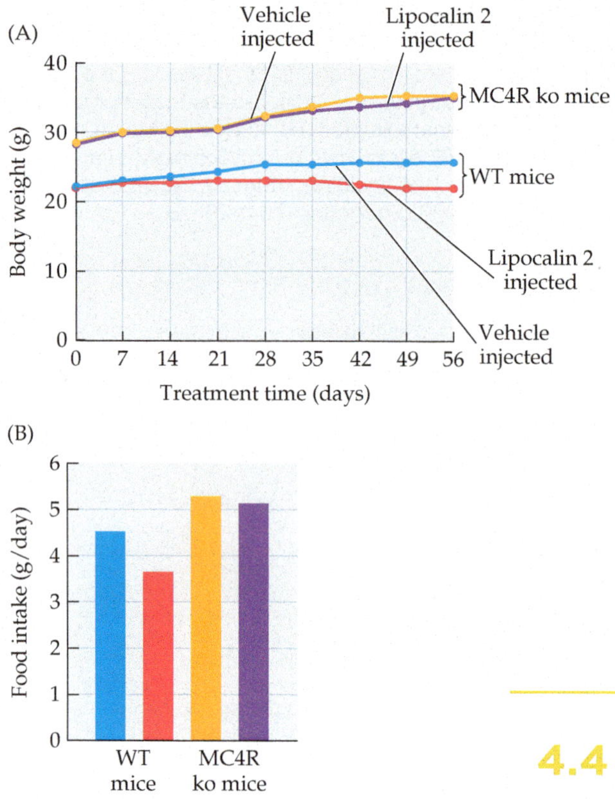

4.20 Lipocalin 2 Affects Appetite by Acting upon the MC4R Receptor (A) MC4R knockout (ko) mice gain weight faster than WT mice, but lipocalin 2 treatment retards WT growth even more. (B) Lipocalin 2 treatment reduces food intake in wild-type (WT) mice but not in MC4R knockout (ko) mice, which eat more than WT mice whether they receive lipocalin 2 or not. (After I. Mosialou et al. 2017. *Nature* 4.3: 384.390.)

The Cutting Edge

Bones Secrete Hormones to Regulate Appetite

It's been known for some time that bones also release hormones to act upon other parts of the body, including the kidneys and the pancreas, but bone hormones affect the brain too. After mice finish a meal, bone-forming cells (osteoblasts) release a peptide hormone called *lipocalin 2*. Treating mice with exogenous lipocalin 2 reduced their food intake, suggesting that the hormone normally signals the brain when we've eaten enough. The researchers found that the hormone indeed crosses the blood-brain barrier and binds to cells in the hypothalamus (Mosialou et al., 2017). They then tested the idea that lipocalin 2 binds to the melanocortin 4 receptor (MC4R) by giving the hormone to wild-type mice and mice with the MC4R gene knocked out. Sure enough, lipocalin 2 treatment reduced food intake and body weight gain in wild-type mice but had no effect upon MC4R knockouts, who ate and weighed more than wild-type mice (**FIGURE 4.20**). In Chapter 14 we'll learn that fat cells secrete a different hormone to signal the hypothalamus to stop eating. Given the "obesity epidemic," there's a lot of interest in how these various signals are integrated in the brain to control our body weight. ■

4.4 Hormones Regulate Social Behaviours and Vice Versa

Learning Objectives

After reading this section, you should be able to:

4.4.1 **Describe the ways the brain regulates secretion of hormones from the gonads.**

4.4.2 **Understand how gonadal steroids activate mating behaviour in male and female vertebrates.**

4.4.3 **Describe how hormones play a role in pair-bonding in monogamous species of voles.**

4.4.4 **Appreciate the reciprocal nature of hormonal effects on behaviour—that behaviours affect hormone release too.**

Sometimes hormonal conditions produce symptoms that physicians may mistake for psychiatric disorders, as **A STEP FURTHER 4.1: ENDOCRINE PATHOLOGY CAN PRODUCE EXTREME EFFECTS ON HUMAN BEHAVIOUR** details on the website. For now, to get an idea of how hormones affect behaviour, let's consider their role in social interactions, including mating behaviour.

The gonads produce steroid hormones, regulating reproduction

Almost all aspects of reproductive behaviour, including mating and parenting, depend on hormones. Since Chapter 13 is devoted to reproductive behaviour and physiology, at this point we will only briefly note relevant hormones and some pertinent aspects of anatomy and physiology. Female and male **gonads** (ovaries and testes, respectively; see Figure 4.1) contain two different subcompartments—one to produce hormones (the sex steroids we mentioned earlier) and another to produce gametes (eggs or sperm). The gonadal hormones are critical for triggering both reproductive behaviour, controlled by the brain, and gamete production.

The hypothalamus controls gonadal hormone production by releasing **gonadotropin-releasing hormone** (**GnRH**), which drives the anterior pituitary to release the gonadotropins FSH or LH, which we mentioned earlier. Although named for their effects on ovaries, FSH and LH drive development and steroid production in both testes and ovaries. The GnRH neurons in turn are stimulated by a hypothalamic peptide, **kisspeptin** (Kriegsfeld, 2006; J. T. Smith et al., 2006), which appears to play an important role in governing the onset of puberty (Han et al., 2005).

The hypothalamus uses GnRH to *stimulate* pituitary gonadotropin secretion, but it also *inhibits* gonadotropin secretion with a hormone named, sensibly enough, **gonadotropin-inhibiting hormone** (**GnIH**) (Tsutsui et al., 2006). GnRH and GnIH thus work in opposition, translating inputs from the brain into controls on the pituitary, and therefore the gonads, like an accelerator and a brake, respectively.

THE TESTES Within the **testes** are Sertoli cells, which produce sperm, and Leydig cells, which produce and secrete the sex steroid **testosterone**. Testosterone and other male hormones are called **androgens** (from the Greek *andro-*, "man," and *gennan*, "to produce").

Testosterone controls a wide range of bodily changes that become visible at puberty, including changes in voice, hair growth, and genital size (**FIGURE 4.21A**). In species that breed only in certain seasons of the year, testosterone has especially marked effects on appearance and behaviour—for example, the antlers and fighting between males that are displayed by many species of deer. As men age, testosterone levels tend to decline. Although elderly men who happen to maintain high levels of circulating testosterone perform better on tests of memory and attention than do those with low levels (Yaffe et al., 2002), there have been too few studies to tell whether taking supplemental testosterone actually helps aging men (Hua et al., 2016; P. J. Snyder et al., 2018). Furthermore, treatment with exogenous testosterone reportedly increases aggressive behaviours in men, and it decreases cooperative behaviours (Carré et al., 2017; B. M. Bird et al., 2019) and may increase prostate cancer risk.

THE OVARIES The paired female gonads, the **ovaries**, also produce both the mature gametes—called *ova* (singular *ovum*) or *eggs*—and sex steroid hormones. However, hormone secretion is more complicated in ovaries than in testes. Ovarian hormones are produced in cycles, the duration of which varies with the species. Human ovarian cycles last about 4 weeks; rat cycles last only 4 days.

Normally, the ovary produces two major classes of steroid hormones: **progestins** (from the Latin *pro*, "favoring," and *gestare*, "to bear," because these hormones help to maintain pregnancy) and **estrogens** (from the Latin *oestrus*, "gadfly" or "frenzy"—*estrus* is the scientific term for the periodic sexual receptivity of females of many species—and the Greek *gennan*, "to produce"). Estrogens drive the development of female bodily changes at puberty, including the growth of breasts. The most important naturally occurring estrogen is **17β-estradiol** (or just *estradiol*). The primary progestin is **progesterone** (**FIGURE 4.21B**). Interestingly, estrogens make the brain sensitive to progesterone by promoting the production of progestin receptors there.

Oral contraceptives contain small doses of synthetic estrogen and/or progestin, which exert a negative feedback effect on the hypothalamus, inhibiting the release of GnRH. The lack of GnRH prevents the release of FSH and LH from the pituitary, and therefore the ovary fails to release an egg for fertilisation.

Estrogens may improve aspects of cognitive functioning (Ycaza Herrera and Mather, 2015), although this topic is still debated (e.g., Korol and Pisani, 2015). Estrogens may also protect the brain from some of the effects of stress and stroke (S. Suzuki et al., 2009). For these reasons and others, estrogen replacement therapy has been a popular postmenopausal treatment, but evidence that these treatments increase the risk of serious diseases like cancer and heart disease has raised questions about their safety (M. Hickey et al., 2012; Lisabeth and Bushnell, 2012). Many synthetic estrogens are produced and tested in search of drugs that have only the beneficial effects of the hormone, without its harmful side effects.

RELATIONS AMONG GONADAL HORMONES As we mentioned earlier, the steroid hormones—androgens, estrogens, progestins, and the adrenal steroids—are all based on the chemical structure of cholesterol, featuring a backbone of four interconnected carbon rings (see Figure 4.7C). Furthermore, progestins can be converted to androgens, and androgens in turn can be converted into estrogens.

gonads The sexual glands (ovaries in females, testes in males), which produce gametes for reproduction.

gonadotropin-releasing hormone (GnRH) A hypothalamic hormone that controls the release of luteinizing hormone and follicle-stimulating hormone from the pituitary.

kisspeptin A hypothalamic peptide hormone that increases gonadotropin secretion by facilitating the release of gonadotropin-releasing hormone.

gonadotropin-inhibiting hormone (GnIH) A hypothalamic peptide hormone that reduces gonadotropin secretion from the pituitary.

testes The male gonads, which produce sperm and androgenic steroid hormones.

testosterone A hormone, produced by male gonads, that controls a variety of bodily changes that become visible at puberty.

androgens A class of hormones that includes testosterone and other male hormones.

ovaries The female gonads, which produce eggs for reproduction.

progestins A major class of steroid hormones, including progesterone, that are produced by the ovary.

estrogens A class of steroid hormones produced by female gonads.

17β-estradiol Also called simply *estradiol*. The primary type of estrogen that is secreted by the ovary.

progesterone The primary type of progestin secreted by the ovary.

oral contraceptive A birth control pill, typically consisting of steroid hormones to prevent ovulation.

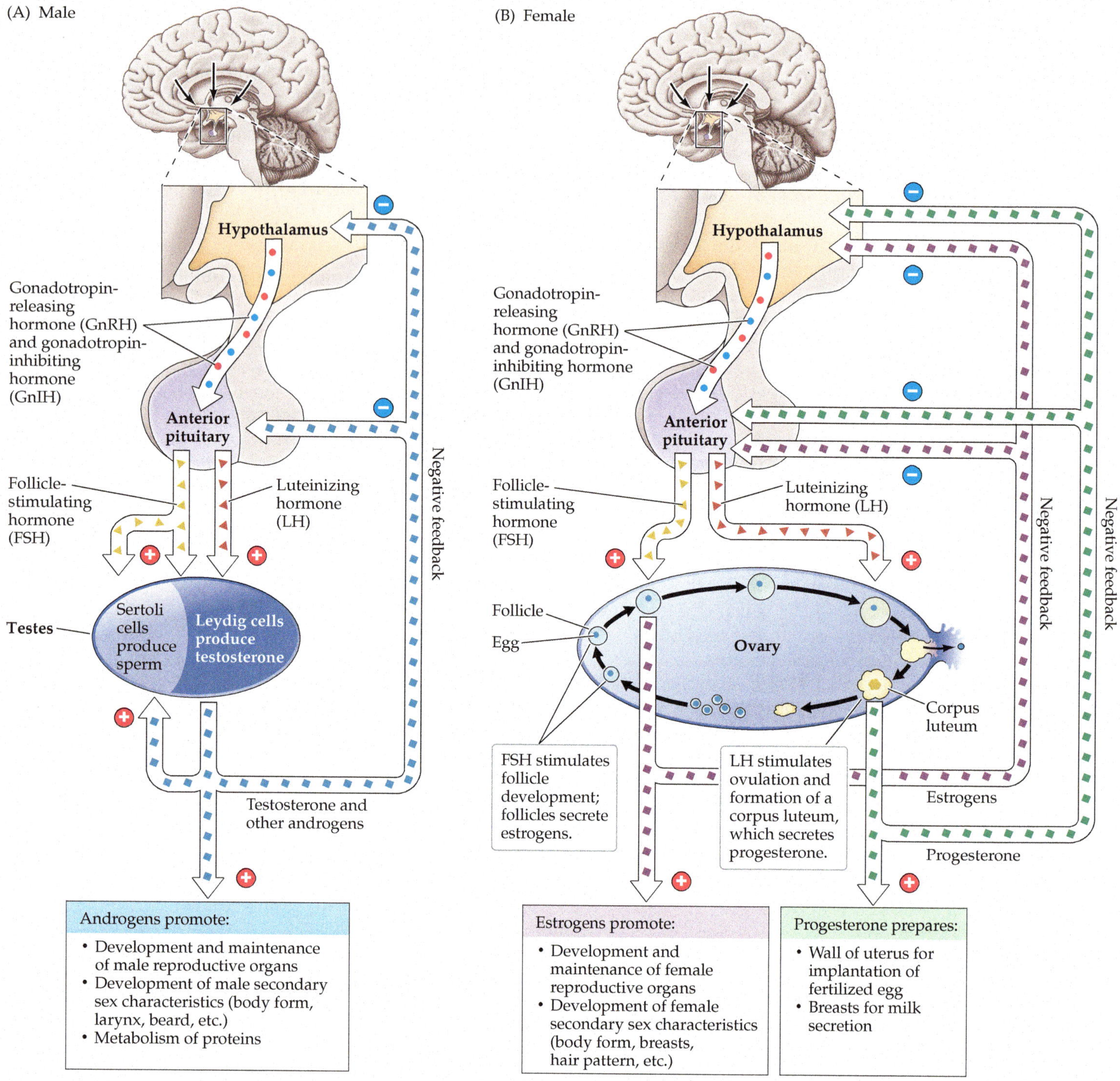

4.21 Regulation of Gonadal Steroid Hormones

Each of these conversions is controlled by specific enzymes. The structural similarity among steroids reflects their evolutionary history: as enzymes evolved to modify old steroids, new steroids became available for signalling.

Different organs—and the two sexes—differ in the relative amounts of gonadal hormones that they produce. For example, whereas the testis converts only a relatively small proportion of testosterone into estradiol, the ovary converts most of the testosterone it makes into estradiol. But it is important to appreciate that *no steroid is found exclusively in either males or females*; rather, the two sexes differ in the proportion of these steroids.

Hormones can affect social behaviour

We've already seen the role that the hormone oxytocin plays in the interaction of nursing babies and their mothers (see Figure 4.13). It turns out that this hormone is involved in several other social behaviours too. For one thing, a pulse of oxytocin is released during orgasm in both men and women (Carmichael et al., 1994), adding to the pleasurable feelings accompanying sexual encounters.

In nonhuman animals, oxytocin and vasopressin modulate many social processes (Lim and Young, 2006). Rodents given supplementary doses of oxytocin spend more time in physical contact with each other (Carter, 1992). Male mice with the oxytocin gene knocked out are unable to produce the hormone, and they display social amnesia: they seem unable to recognize the scents of female mice that they have met before (Ferguson et al., 2000). These oxytocin knockout males can be cured of their social amnesia with brain infusions of oxytocin (Winslow and Insel, 2002). In mice, the oxytocin released during delivery appears to protect fetal neurons from injury (Tyzio et al., 2006) and to improve the mother's ability to navigate a maze (Tomizawa et al., 2003).

In another rodent, the prairie vole (*Microtus ochrogaster*), in which couples form stable monogamous pair-bonds, oxytocin infusions in the brains of females help them bond to their mates. In male prairie voles, it is vasopressin rather than oxytocin that facilitates the formation of a preference for a specific female partner. In fact, the distribution of vasopressin receptors in the brains of male prairie voles may be what makes them monogamous.

Supporting this idea is the finding that in the closely related meadow voles (*M. pennsylvanicus*), which do not form pair-bonds and instead have multiple mating partners, the males have far fewer vasopressin receptors in certain brain regions than do prairie voles (**FIGURE 4.22**) (Lim et al., 2004). Furthermore, laboratory mice that have been genetically engineered to produce vasopressin receptors in their brains in the same pattern that is seen in prairie vole males are much more interested in associating with females, almost as if they were trying to form pair-bonds.

Thus, it appears that oxytocin and vasopressin regulate a range of social behaviours and that natural selection sometimes alters the social behaviours of a species through changes in the brain distribution of receptors for these two peptides (Donaldson and Young, 2008).

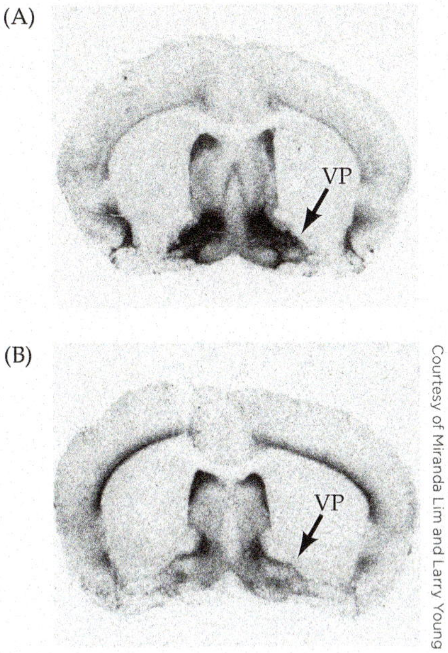

4.22 Vasopressin and the Monogamous Brain (A) Monogamy in male prairie voles seems to be due to the dense concentration of vasopressin receptors in the ventral pallidum (VP). (B) Males of the closely related meadow vole species have fewer vasopressin receptors in the VP (notice lighter area compared with A), which may explain why they are not monogamous.

Social behaviour can affect hormones

There have been several dramatic cases of children raised with little or no family attention, which devastatingly hinders development of the body and the brain (Rymer, 1993). Even less drastic family deprivation can result in failure of growth. This syndrome is referred to as **psychosocial dwarfism** to emphasize that the growth failure arises from psychological and social factors mediated through the CNS and its control over endocrine functions (W. H. Green et al., 1984). When children with psychosocial dwarfism are removed from stressful circumstances, many begin to grow rapidly. The growth rates of five such children, before and after periods of emotional deprivation, are shown in **FIGURE 4.23**. These children seem to have compensated for much of the growth deficit that occurred during prolonged stress periods (Sirotnak et al., 2004; Kumaran and Kershaw, 2014).

psychosocial dwarfism Reduced stature caused by stress early in life that inhibits growth.

How do stress and emotional deprivation impair growth? Growth impairments appear to be mediated by changed outputs of several hormones, including cortisol, GH, and other hormones known as *somatomedins* (normally released by the liver in response to GH).

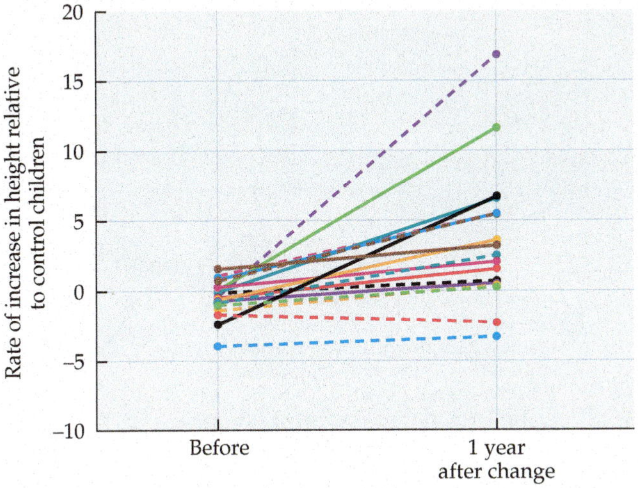

4.23 Psychosocial Dwarfism While the children were in abusive situations, their growth rate was slower than age-matched controls. A year following their removal from abuse, their height increased much faster than controls. By adulthood, they nearly matched their parents in height (data not shown). (After B. C. Gohlke and R. Stanhope, 2002. *Acta Paediatr* 91: 961–965.)

Some children with psychosocial dwarfism show almost a complete lack of GH release, which may be caused by an absence of the releasing hormone somatocrinin from the hypothalamus (Albanese et al., 1994). Disturbed sleep has also been suggested as a cause of this failure, because GH is typically released during certain stages of sleep and children under stress show disturbed sleep patterns (L. I. Gardner, 1972). Other children who exhibit psychosocial dwarfism show normal levels of GH but low levels of somatomedins, and these hormones, along with GH, appear to be necessary for normal growth. Still other children with this condition show elevated levels of cortisol, probably as a result of stress, that inhibit growth. Some affected children show none of these hormonal disturbances, so there must also be other routes through which emotional experiences affect growth.

Psychosocial dwarfism is just one example of the reciprocal relationship between hormones and behaviour, which we explore to conclude the chapter.

Hormonal and neural systems interact to produce integrated responses

In many ways, the endocrine system and the nervous system can be viewed simply as divisions of a single master control mechanism. The two divisions work together, in an intimate and reciprocal manner, to seamlessly integrate the systems of the body and produce adaptive responses to environmental challenges.

Incoming environmental stimuli elicit activity in sensory pathways that project to a wide variety of brain regions, including the cerebral cortex, cerebellum, and

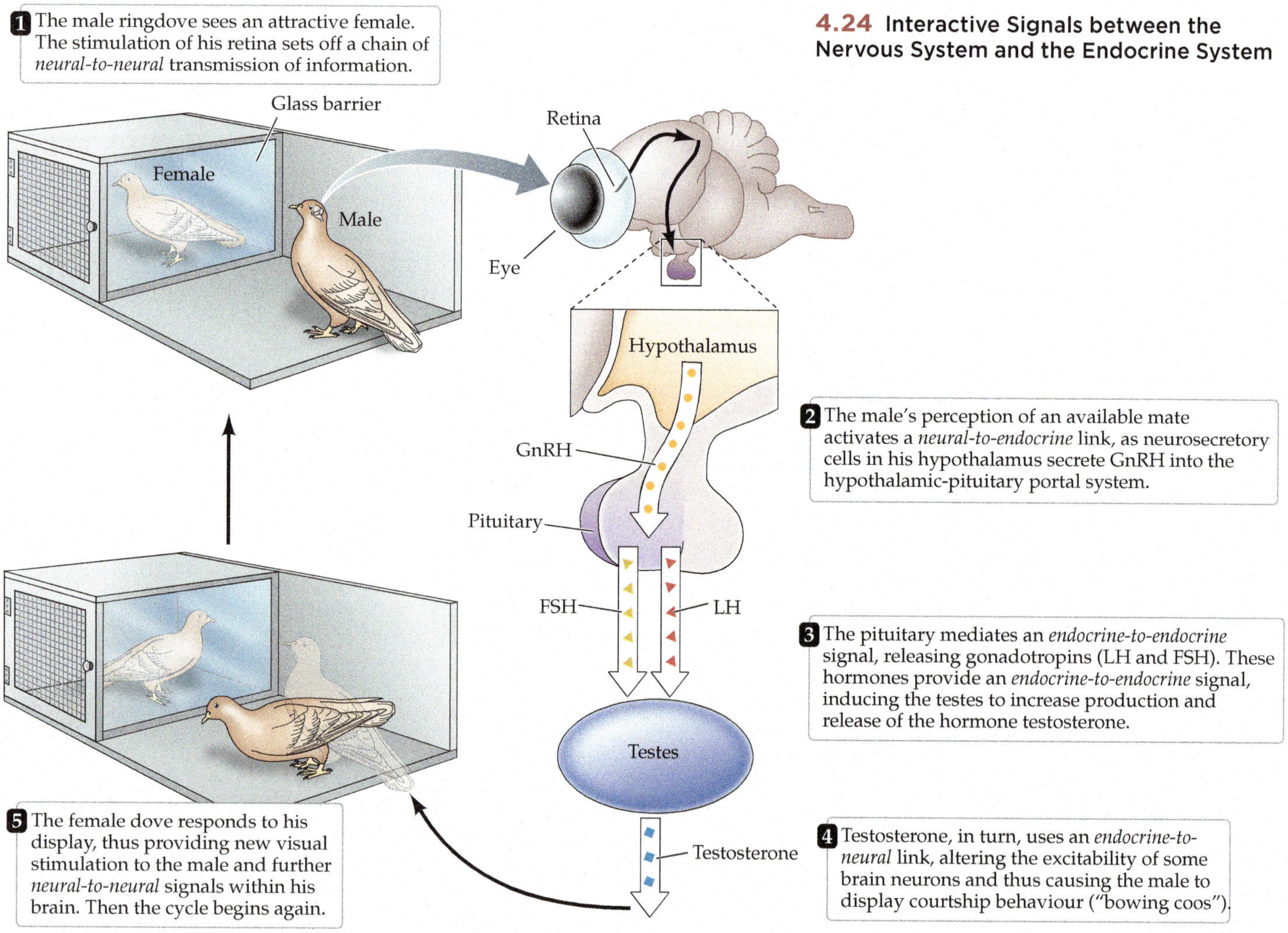

4.24 Interactive Signals between the Nervous System and the Endocrine System

hypothalamus. Our own behavioural responses to environmental circumstances bring further changes in stimulation. For example, if we approach an object or a sound source, we cause the visual image to become larger or the sound to become louder. Meanwhile the endocrine system is tuning our response characteristics to be consistent with the nature of the stimulus. If the stimulus calls for action—that faint buzzing sound turns out to be coming from a nest of angry wasps, for example—energy is mobilized through hormonal routes to prepare for appropriate behaviours (namely, sprinting and swatting and maybe some judicious swearing). The behaviours themselves, of course, are executed under the control of the nervous system. Sensory receptor organs are also subject to continual adjustment, thus modifying further processing of stimuli. Another example of neural and hormonal coordination is the milk letdown reflex (see Figure 4.13).

Four kinds of signals are possible between neurons and endocrine cells: neural-to-neural, neural-to-endocrine, endocrine-to-endocrine, and endocrine-to-neural. All four types are illustrated in one particular social behaviour: the courtship of ring-doves (**FIGURE 4.24**). Experience activates the brain, which alters hormone secretion. The release of one hormone often affects release of other hormones. These hormones in turn can affect brain functioning and therefore future behaviour. That behaviour will affect the animal's future experience, completing the circle of influences.

Thus the interactions between endocrine activity and behaviour are cyclical, as depicted by the circle schema in **FIGURE 4.25**. The level of circulating hormones can be altered by experience, which in turn can affect future behaviour and future experience. For example, starting to exercise or stepping out in the cold increases the release of thyroid hormones. Men rooting for a sports team will produce more testosterone if their team wins (Bernhardt et al., 1998). Physical stresses, pain, and unpleasant emotional situations decrease thyroid output and trigger the release of adrenal glucocorticoids (see Chapter 15).

Each of these hormonal events will affect the brain, shaping behaviour, which will once more affect the person's future hormone production. Any thorough understanding of the relationship between hormones and behaviour must include these reciprocal interactions.

In later chapters about mating (Chapter 13), eating and drinking (Chapter 14), biological rhythms (Chapter 15), and stress (Chapter 11), this reciprocal relationship between hormones and behaviour will recur, reinforcing Mary Lou Jepson's observation that hormones affect not just our health, but our moods and personality as well.

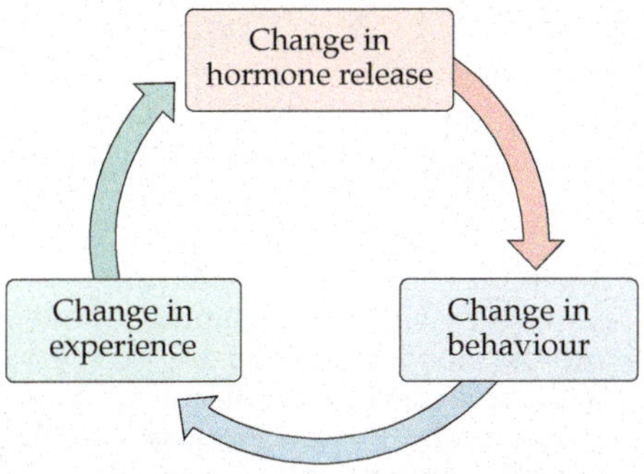

4.25 The Reciprocal Relations between Hormones and Behaviour

Recommended Reading

Becker, J. B., Breedlove, S. M., Crews, D., and McCarthy, M. M. (Eds.). (2002). *Behavioural Endocrinology* (2nd ed.). Cambridge, MA: MIT Press.

Fink, G., Pfaff, D. W., and Levine, J. (2011). *Handbook of Neuroendocrinology*. San Diego, CA: Academic Press.

Garcia-Segura, L. M. (2009). *Hormones and Brain Plasticity*. New York: Oxford University Press.

Jameson, J. L. (Ed.). (2017). *Harrison's Endocrinology* (4th ed.). New York: McGraw-Hill.

Melmed, S., Polonsky, K. S., Larsen, P. R., and Kronenberg, H. M. (2016). *Williams Textbook of Endocrinology* (13th ed.). Philadelphia: Elsevier.

Nelson, R. J., and Kriegsfeld, L. J. (2017). *An Introduction to Behavioural Endocrinology* (5th ed.). Sunderland, MA: Oxford University Press/Sinauer.

Patisaul, H. B., and Belcher, S. M. (2017). *Endocrine Disruptors, Brain, and Behaviour*. New York: Oxford University Press.

4 Visual Summary

You should be able to relate each summary to the adjacent illustration, including structures and processes. You can also refer back to the links to figures, animations, and activities that will help you consolidate the material.

1 **Hormones** are chemicals that are secreted by **endocrine glands** into the bloodstream and are taken up by receptor molecules in target cells. Review **Figures 4.1**, **4.2**, **Table 4.2**

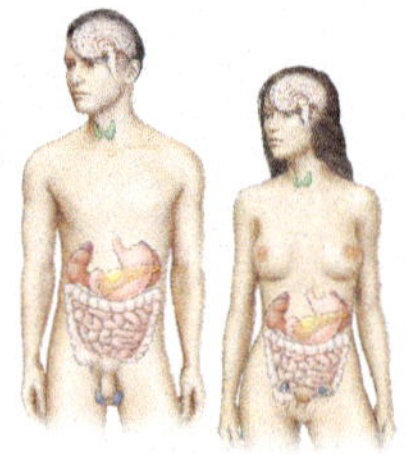

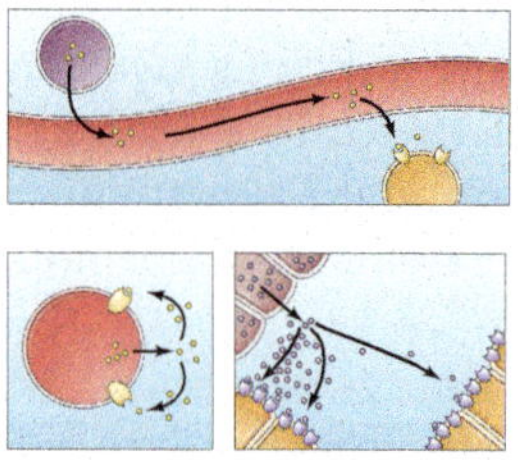

2 Hormones are just one of a variety of modes of chemical communication between cells. Neurotransmitters cross a tiny gap during synaptic transmission. An **autocrine** signal acts upon the cell that released it, while **paracrine** signals act on nearby cells. **Pheromones** are chemical signals to individuals of the same species, while **allomones** communicate with individuals of other species. Review **Figures 4.3**, **4.4**, **Animation 4.1**

3 Unlike neuronal signalling, hormones spread more slowly and act throughout the body. Some hormones act on receptors in a wide variety of cells and can therefore coordinate their influences on the activities of most cells in the body. Other hormones have receptors in only certain special cells or organs. Review **Figures 4.5**, **4.6**

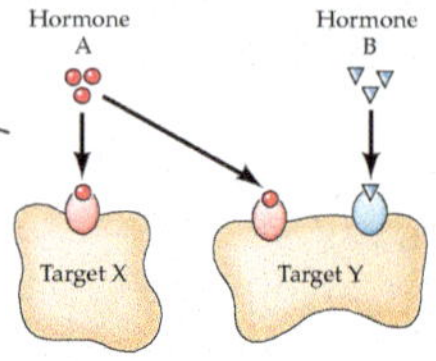

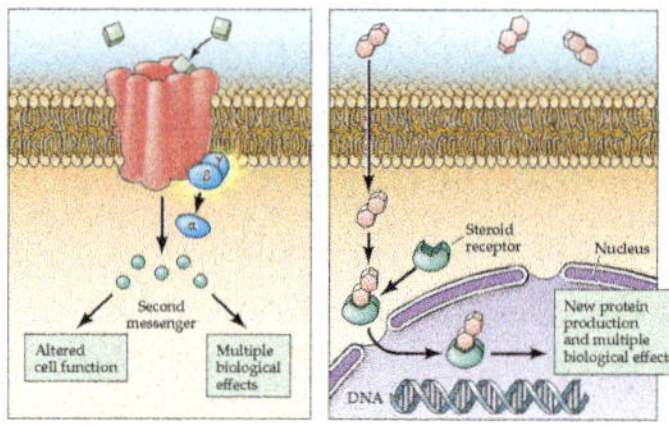

4 **Peptide** and **amine hormones** bind to receptor molecules at the surface of the target cell membrane and activate second-messenger molecules inside the cell. **Steroid hormones** pass through the membrane and bind to receptor molecules inside the cell, ultimately regulating gene expression. Review **Figures 4.7**, **4.8**, **Animation 4.2**

5 A **negative feedback** system monitors and controls the rate of secretion of each hormone. The hormone acts on target cells, leading them to change the amount of a substance they release. In the simplest case the hormone also acts on the endocrine cells, and this regulates further output of the endocrine gland. Review **Figure 4.10**

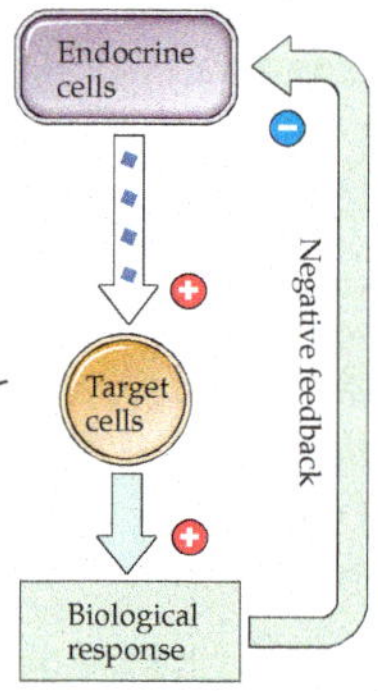

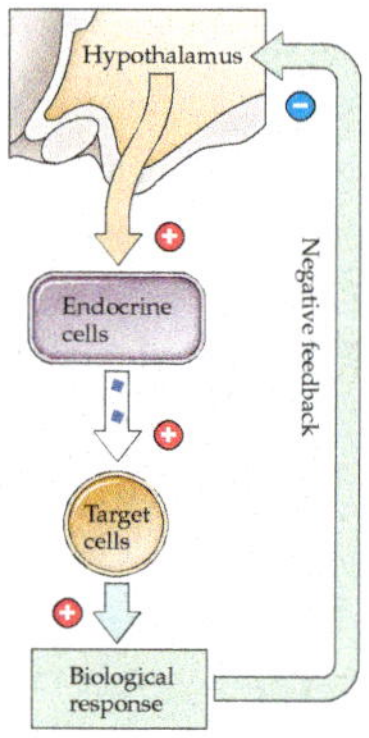

6 Other hormones are controlled by a releasing hormone from the hypothalamus that regulates the release of an anterior pituitary **tropic hormone**, which in turn controls secretion by an endocrine gland. The endocrine gland hormone then provides negative feedback to the **hypothalamus** and pituitary. Review **Figures 4.10**, **4.11** **Animation 4.3**

7 **Posterior pituitary** hormones are manufactured by **neuroendocrine cells** in the **supraoptic** and **paraventricular nuclei** of the hypothalamus, which send their axons down the **pituitary stalk** to terminate on capillaries there. When these neuroendocrine cells are stimulated to produce an action potential, they release **oxytocin** or **arginine vasopressin** into circulation. Review **Figures 4.12–4.14**, **Activity 4.1**, **Animation 4.3**

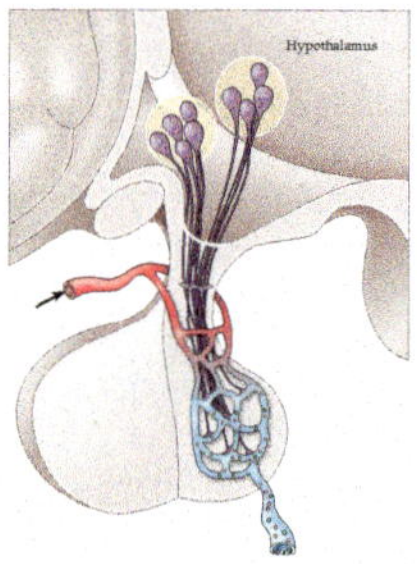

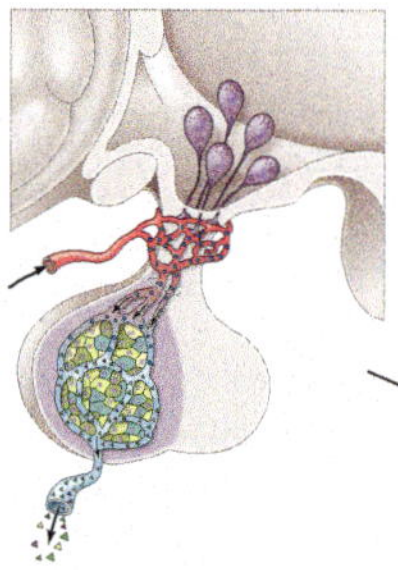

8 **Anterior pituitary** hormones are controlled by the overlying hypothalamus. Hypothalamic neuroendocrine cells send axons to the **median eminence** to secrete releasing hormones into the **hypophyseal portal system**, which transports the releasing hormones to the pituitary. Different hypothalamic releasing hormones either stimulate or inhibit anterior pituitary cells that secrete tropic hormones. Review **Figures 4.15–4.17**

9 Hypothalamic cells secrete **gonadotropin-releasing hormone (GnRH)** into the hypophyseal portal system to stimulate anterior pituitary cells to release **follicle-stimulating hormone (FSH)** and **luteinizing hormone (LH)**, which stimulate the gonads to release steroid hormones. The principle gonadal steroids in males are **androgens** such as **testosterone**, while ovaries release **estrogens** such as **17β-estradiol** and **progestins** such as **progesterone**. Review **Figure 4.21**

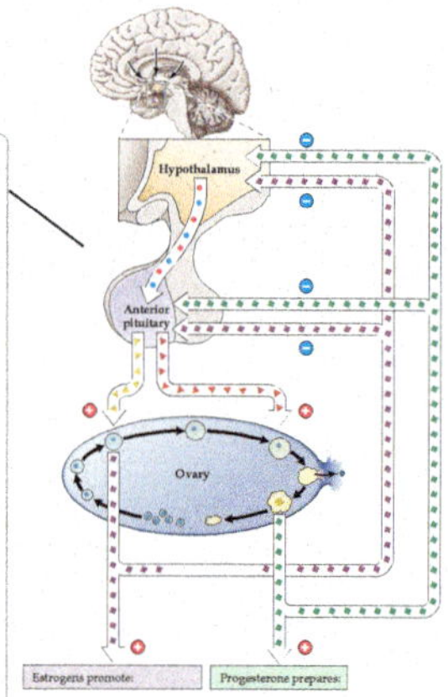

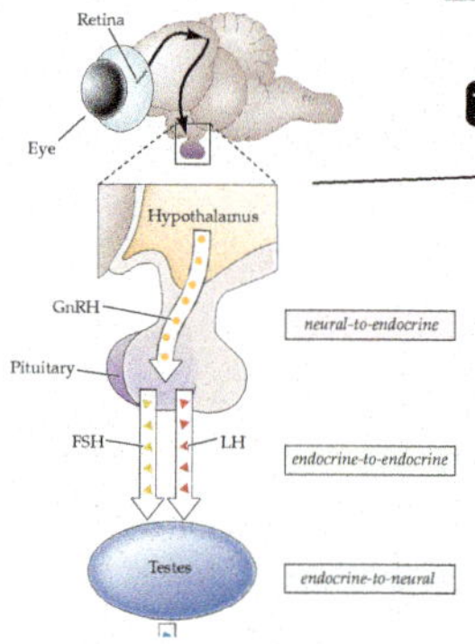

10 Many behaviours require the coordination of neural and hormonal components. Messages may be transmitted in the body via neural-to-neural, neural-to-endocrine, endocrine-to-endocrine, or endocrine-to-neural links. Experience affects hormone secretion, and hormones affect behaviour and therefore future experiences. Review **Figures 4.22–4.25**

Evolution and Development of the Nervous System

PART II

Evolution of the Brain and Behaviour

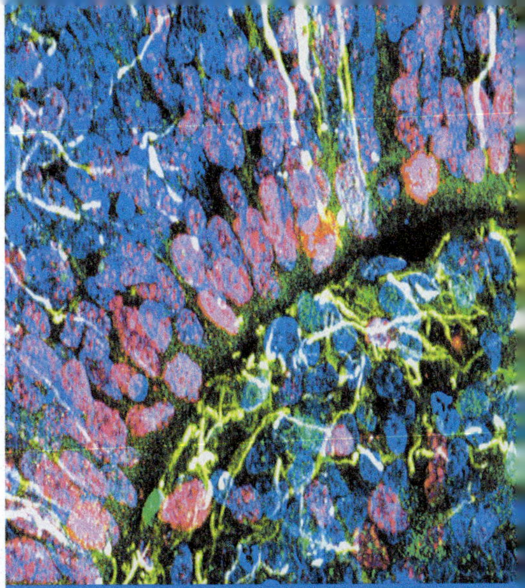

Courtesy of Muotri Lab/UC San Diego

5

We Are Not So Different, Are We?

It probably comes as no surprise that the species that are our closest relatives are the chimpanzees: the common chimpanzee (*Pan troglodytes*) and the bonobo (*Pan paniscus*). So many of their expressions and behaviours seem oddly, well, *human*. But despite the apparent similarities, humans, chimps, and bonobos are also strikingly dissimilar in many fundamental ways. Whereas humans have complex languages, chimps and bonobos make only a small variety of vocal sounds. And whereas humans walk erect and have long legs, chimps and bonobos travel mostly on all fours and have relatively long arms. The human brain is about twice as large. Humans have spread wide from their origins in Africa, populating (or overpopulating) the globe, but chimps and bonobos have remained in Africa, their numbers now dwindling at an alarming rate.

Given the many differences between humans and our closest ape cousins, it is remarkable that the genetic material of the species differs by so little: the chimpanzee and bonobo genomes each differ from the human genome by about 1.3%, and chimps and bonobos differ from one another by only 0.4% (Prüfer et al., 2012). One prominent scientist suggested that, since the genes of chimps and humans differ so little, the social context provided during the rearing of human children must be what causes them to come out so different from chimpanzees. If that suggestion strikes you as unlikely, you are probably right. Several people have tried to rear chimpanzees as you would human children (see Chapter 18), and none of the chimps ever won a spelling bee or got a driver's license (not even a learner's permit).

Of course social rearing is crucial for human development, but it cannot explain the vast differences between humans and other great apes. So the problem remains: given their tiny genetic differences, how can we explain the striking differences in behaviour, anatomy, and neurobiology between humans versus chimpanzees and bonobos? In other words, what makes humans human (and chimpanzees chimpanzee)? Progress in neuroscience is suggesting answers to this puzzle, as we will see.

Our major objective in this chapter is to explore the intriguing story of how brains and behaviour have evolved. We will see that brain size in primates, especially in humans, increased rapidly in our recent evolution. This enlarged brain doubtless increased our capacity for higher cognitive abilities, yet it is difficult to determine which expanded brain regions brought us which additional abilities. Thus, scientists study the nervous system in a wide variety of animals to understand how the evolution of a particular brain feature affects particular behaviours. Describing the relationships between the nervous system and behaviour in even a small fraction of Earth's inhabitants would be an immense (and dull) task unless we had a rationale beyond mere completeness. Choosing the right species to compare, however, reveals principles of nervous system organisation.

5.1 How Did the Enormous Variety of Species Arise on Earth?

Learning Objectives

After reading this section, you should be able to:

5.1.1 **Review the historical origins of evolutionary theory.**

5.1.2 **Describe how mutation and selection pressures combine to give rise to new characteristics.**

5.1.3 **Discuss speciation, methods for studying the relatedness of species, and the significance of taxonomy.**

Until about 200 years ago, the dominant view among Western scholars was that each species had been created separately. Then, some **naturalists**—students of the forms and classification of organisms—began to have doubts. For example, researchers observed that the limb bones of all mammals, no matter what the animal's way of life, are remarkably similar in many details. If these species had been specifically created for different ways of locomotion, the naturalists reasoned, they should have been built on different plans rather than all being modifications of the same single plan.

Nineteenth-century geologists showed that Earth has been changing for millions of years, and studies of the fossils of extinct species provided evidence of **evolution**—the process of "descent with modification." It thus became clear that, far from being static categories, animal species are continually changing across the generations, gradually gaining some features and losing others, and sometimes spinning off new species (while others become extinct). But a plausible *mechanism* for evolution was lacking.

Evolution is the result of individual differences in reproduction

In 1858, Charles Darwin (1809–1892) and Alfred Russel Wallace (1823–1913) described the process of **evolution by natural selection**. Darwin and Wallace had each hit upon the idea independently. The idea came to Wallace out of the blue, while he was suffering from a fever. In contrast, Darwin had been accumulating and considering evidence for over 20 years, ever since he had voyaged on the HMS *Beagle* to South America and the Galápagos Islands. The governor of the Galápagos had pointed out to Darwin that the giant tortoises differed in their shell patterns from island to island. Later, examining the specimens of finches that he had collected on different Galápagos islands, Darwin observed that the birds also differed from island to island. Although the birds resembled finches on the mainland, those on the Galápagos appeared to represent several different species, suited by their body sizes and beaks to obtaining different kinds of food, such as nuts or seeds or insects. Darwin speculated that the different birds had all descended from a single ancestral species long ago but that, isolated on the various islands, they had gradually diverged from their ancestors and from one another.

In 1859, Darwin published his revolutionary book *On the Origin of Species by Means of Natural Selection*. The hypothesis he stated was based on four main observations and one important inference. The observations were these:

1. Reproduction will tend to increase a population rapidly unless factors limit it.
2. Individuals of a given species are not identical.
3. Some of the variation among individuals is inherited.
4. Not all the offspring of a given generation survive to reproduce.

The inference was that the variations among individuals affect the probability of their surviving long enough to reproduce, thereby passing on their individual characteristics to their offspring. Individuals better suited to the prevailing conditions enjoy more success in reproduction, so their descendants will make up a greater proportion of successive generations. Through this mechanism, new **adaptations**—traits that increase the probability of successful reproduction—will

naturalist A student of natural history, including the forms and classification of organisms.

evolution In biology, the processes by which successive generations of individuals change over very long periods of time.

evolution by natural selection The Darwinian theory that evolution proceeds by differential success in reproduction.

adaptation Here, a trait that increases the probability that an individual will leave offspring in subsequent generations.

eventually predominate in the population. Over long spans of time and many generations, this process of selection acting at the level of individuals (and their reproductive output) can substantially change a species.

The concept of evolution by natural selection has become the most important organizing principle in the life sciences, directing the study of behaviour and its mechanisms, as well as the study of morphology (form and structure). Darwin (1859) felt confident that psychological functions are as much products of evolution as are the organs of the body.

Darwin later proposed an additional mechanism of natural selection: **sexual selection** (1871). This principle holds that members of each sex exert selective pressures on the other in terms of both anatomical and behavioural features that favor reproductive success. For example, choosiness on the part of peahens has led to the ornamental but costly tails of peacocks. We will discuss this principle later in this chapter and again in Chapter 13.

At first, Darwin's theory suffered from uncertainty about two important processes: (1) the mechanism by which an individual inherits its characteristics from its parents, and (2) the source of individual variation upon which natural selection acts. Working with pea plants, an Austrian monk and botanist named Gregor Johann Mendel (1822–1884) provided the needed breakthrough. Mendel's observations on the inheritance of traits across generations resulted in his publication, in 1866, of a set of formal laws of heredity that would eventually form the foundation of the modern science of **genetics**: the study of the mechanisms of inheritance (Mendel, 1967). It wasn't until 1900, however, that Mendel's work was linked to evolution—by the Dutch biologist Hugo de Vries (1848–1935), who was conducting experiments with primroses. De Vries went beyond Mendel in one very important respect. Noting that occasionally a new feature arose spontaneously and was then passed on to successive generations, de Vries reasoned that evolution could proceed by sudden jumps, or **mutations**, as he called these changes (**FIGURE 5.1**).

Mutations occur spontaneously and randomly in plants and animals, and because they are the result of changes in the organism's genes, mutations are heritable. (Nowadays, scientists also deliberately induce mutations in plants and animals, as we will discuss in Box 6.1) Depending on how it modifies the individual, a given mutation may be harmful, neutral, or beneficial. A beneficial mutation will generally give the individual at least a slight reproductive advantage over other individuals of the same species, called *conspecifics*, with the result that the mutation will become more and more widespread as it is passed down through subsequent generations. This gradual, incremental process of selectively transmitting new features and modifications can account for the evolution of even fantastically complex structures, such as eyes or brains—it just takes a lot of generations (and of course the evolutionary timescale runs to millions of generations).

Progress in genetics research was accelerated by focussing on organisms that reproduce rapidly. An example is the fruit fly *Drosophila melanogaster*, whose generation time is 10 days and whose salivary glands produce giant chromosomes that are visible with a light microscope. **Chromosomes** (from the Greek *chroma*, "colour," and *soma*, "body") are the supercoiled lengths of DNA, found within the cell nucleus, that contain the **genes** that encode the tens of thousands of proteins that make up the body (see the Appendix). An additional complexity is the discovery that

sexual selection A form of evolution through natural selection in which members of one sex favour specific heritable traits in the other sex when choosing a reproductive partner.

genetics The study of inheritance, including the genes encoded in DNA.

mutation A change in the nucleotide sequence of a gene as a result of unfaithful replication.

chromosome A complex of condensed strands of DNA and associated protein molecules. It is found in the cell nucleus.

gene A length of DNA that encodes the information for constructing a particular protein.

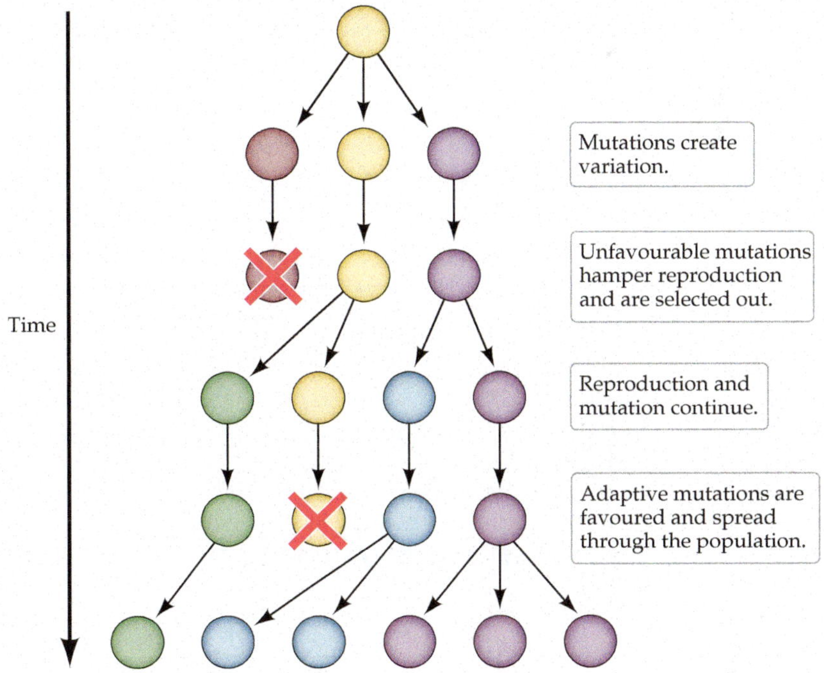

5.1 Natural Selection at the Genetic Level A gene mutation that affects behaviour may be selected for if the alteration in behaviour is adaptive. Thus, not just physical traits, but also behaviours evolve in a species over time, regardless of the details of the reproductive biology of the species. (After László Szalai/Wikimedia/CC BY-SA 3.0.)

directional selection Selection that favours a shift in a characteristic within a population.

an individual's experiences and environment can modify the expression of certain genes in a way that can be transmitted to offspring without changing the sequence of nucleotides in the affected genes. We'll have more to say about this field of study, called *epigenetics*, in Chapter 6.

Many of the findings about genetic mechanisms based on work with *Drosophila* and even with bacteria hold true for larger organisms, such as humans, that reproduce much more slowly. Gradual changes within a species, as well as the formation of new species, can now be understood in light of modern evolutionary theory, which combines Darwin's hypothesis of natural selection with modern genetics and molecular biology.

Natural selection stabioses some characteristics but drives change in others

Evolution through natural selection is a process of continually trying on new features, driven by differential reproductive success, modelling and remodelling with neither a goal nor an endpoint. In the usual evolutionary scenario, random mutation endows some lucky individual with a new characteristic that gives them a slight reproductive advantage, and the mutation spreads through subsequent generations, eventually changing most or all of the population. This form of natural selection, called **directional selection** is exemplified by the classic study of the peppered moth (*Biston betularia*; Kettlewell, 1955), as recently confirmed in a large-scale field experiment (Cook et al., 2012). Typical peppered moths are gray (actually, a mix of black and white spots), which camouflages them from their natural predators—birds—while they rest on the bark of similar-coloured trees. During the industrial revolution of the nineteenth and twentieth centuries, when large quantities of smoke from coal and other sources were being released into the air of Northern England, tree bark became darkened by soot, and the pale peppered moths became easy for birds to spot (and eat). Moth colouration therefore came under strong directional selection pressure, and sure enough, darker-coloured moths soon came to predominate (**FIGURE 5.2A**).

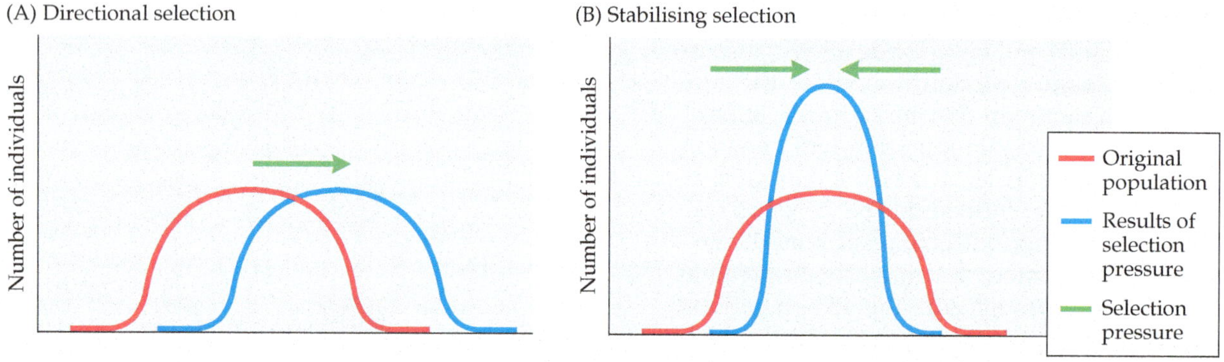

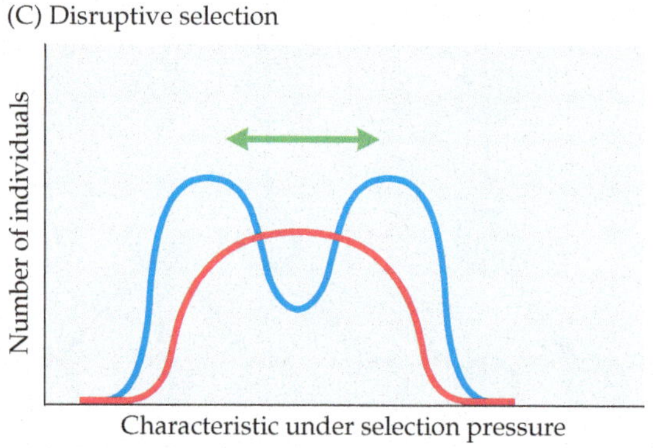

(D)

©Ian Redding / Shutterstock.com

5.2 Forms of Natural Selection (A) Directional selection pressure produces a population-level change in a trait, whereas stabilising selection (B) reduces variation in a trait. In disruptive selection (C), subgroups of individuals diverge in the expression of a trait. (D) Light and dark forms of the peppered moth. (After Ealbert17/Wikimedia/CC BY-SA 4.0.)

In modern postindustrial England the air is now much cleaner: tree trunks have reverted to their original lighter colour, and so have the moths. Further, we would expect that moth colouration would start to show evidence of **stabilising selection** (**FIGURE 5.2B**)—namely, we would expect to see variation in moth colouration decrease over time because being lighter or darker than the optimal camouflage colour would incur reproductive disadvantage to the moth (by its becoming a meal).

But now imagine a scenario in which the local trees with gray-coloured bark are gradually infiltrated by a different species of trees that all have white bark; what would happen to moth colouration? Evolutionary theory suggests that the moths could now undergo **disruptive selection** (**FIGURE 5.2C**), whereby a subpopulation of white-coloured moths (suited to resting on white-barked trees) might emerge, splitting off from the grey-coloured moths that continue to enjoy camouflage when on grey-barked trees. In fact, scientists think this is one of the ways in which new species arise.

stabilising selection Selection that favours reduced variation in a characteristic within a population.

disruptive selection Selection that favours a divergence in a characteristic within a population.

convergent evolution The evolutionary process by which responses to similar ecological features bring about similarities in behaviour or structure among animals that are only distantly related (i.e., that differ in genetic heritage).

homoplasy A physical resemblance between physical or behavioural characteristics due to convergent evolution, such as the similar body forms of tuna and dolphins.

homology A physical resemblance that is based on common ancestry, such as the similarity in forelimb structures of different mammals.

analogy Similarity of function, although the structures of interest may look different. The human hand and an elephant's trunk are analogous features.

genus A group of species that resemble each other because of shared inheritance.

species A group of individuals that can readily interbreed to produce fertile offspring.

Evolution may converge upon similar solutions in different species

Adaptation to similar ecological features may bring about similarities in behaviour or structure among animals that are only distantly related. These similarities are referred to as examples of **convergent evolution**. For example, the body forms of a tuna and a dolphin resemble each other because they each evolved for efficient swimming, even though the tuna is a fish and the dolphin is a mammal descended from terrestrial ancestors. Such a resemblance is an example of **homoplasy**, a resemblance between physical or behavioural characteristics that is due to convergent evolution. By contrast, a **homology** is a resemblance based on common *ancestry*, such as the similarities in forelimb structures of mammals (**FIGURE 5.3**). **Analogy** refers to similarities in *function*, despite differences in structure and evolution. For example, the legs of insects, starfish, and humans are analogous features in that they all serve locomotion.

How closely related are two species?

People have probably always classified the animals around them and realised that some forms resemble each other more closely than others. The Swedish biologist Carolus Linnaeus (1707–1778) proposed the basic classification system that we use today. In Linnaeus's system, each species is assigned two names—the first name identifying the **genus** (plural *genera*), the second name indicating the **species**. Both names are always italicized, and the genus name is capitalized. According to this system, the modern human species is *Homo sapiens*.

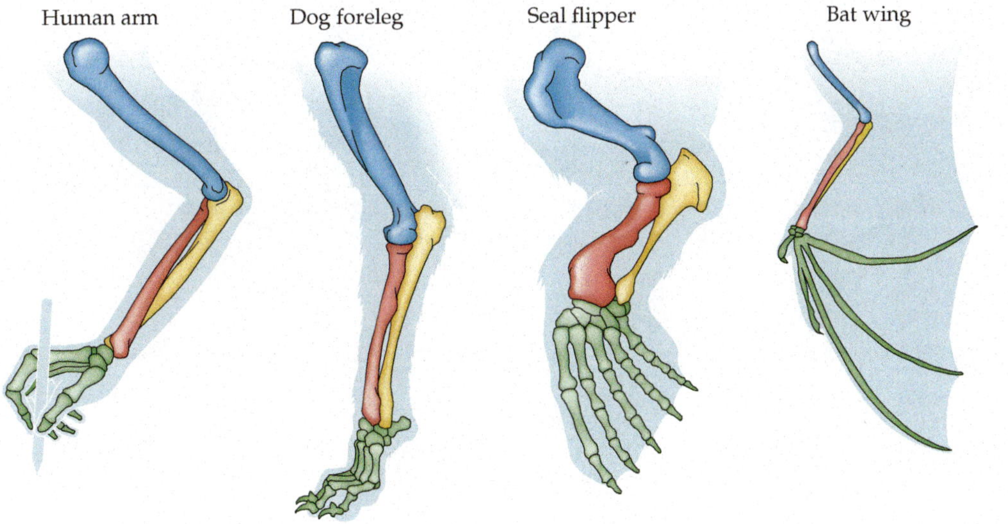

5.3 Homology of Forelimb Structures Bones of the same sort are shown here in the same colour in all species. The sizes and shapes of the bones of the forelimb have evolved so that they are adapted to widely different functions: skilled manipulation in humans, locomotion in dogs, swimming in seals, flying in bats. The similarities among the sets of bones reflect descent from a common ancestor.

The different levels of classification are illustrated and defined in **FIGURE 5.4**. The main trunk of the "tree," the animal kingdom, includes all animal species. As branches divide and subdivide toward the outer reaches of the tree, each successive category includes fewer species, and the species are more closely related. The order of categories, from most broad to most narrow, is this: kingdom, phylum, class, order, family, genus, species. (Here's a handy mnemonic to help you remember them: ***k**indly **p**ut **c**lothes **o**n, **f**or **g**oodness' **s**ake.*)

5.4 Linnaean Classification of the Domestic Dog

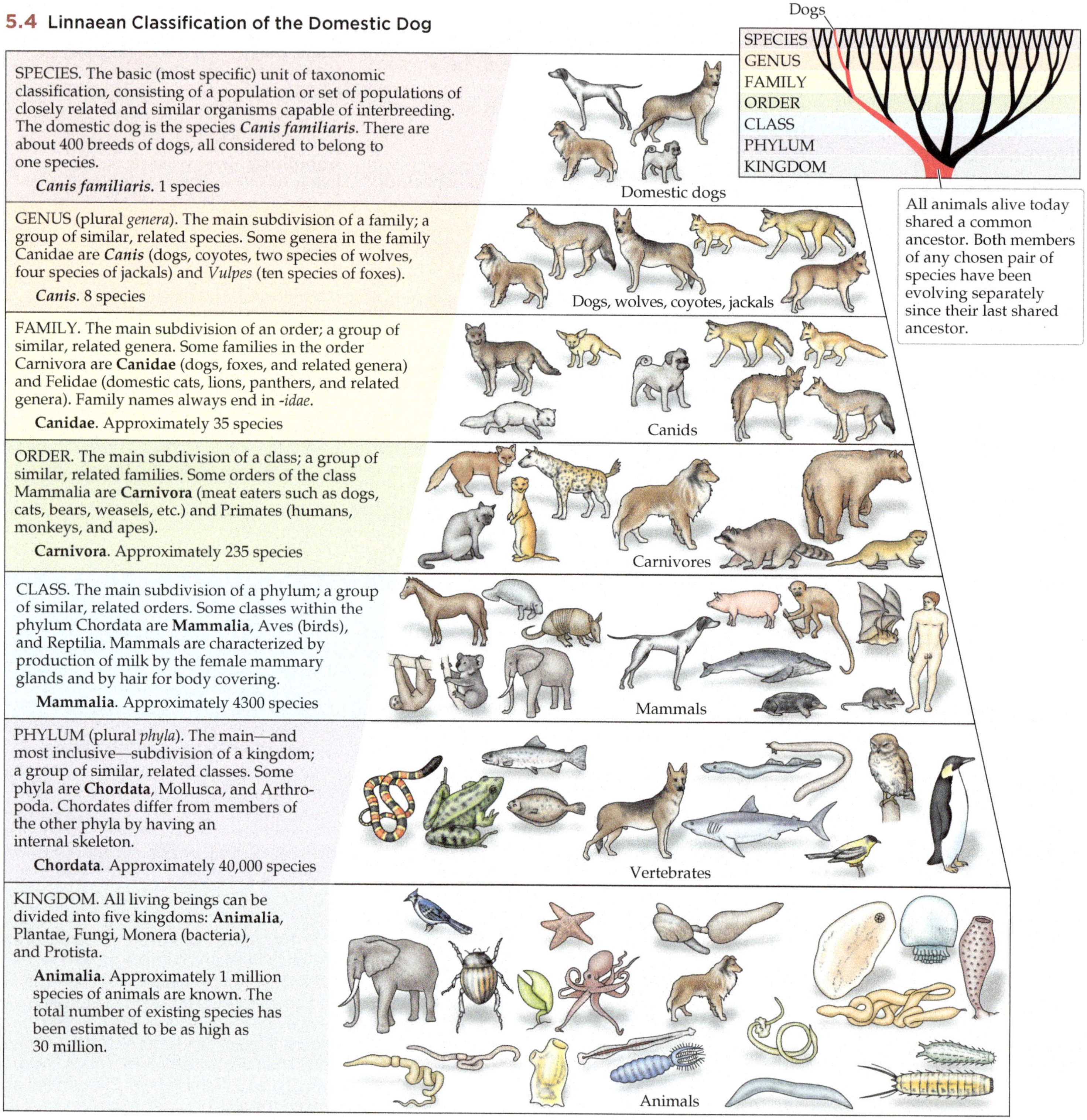

Today we understand that the similarities between some species of organisms reflect **phylogeny** (from the Greek *phylon*, "tribe" or "kind," and *genes*, "born"), the evolutionary history of a particular group of organisms. Phylogeny is often represented as a *family tree* that shows which species may have given rise to others. Comparisons among extant animals, coupled with fragmentary but illuminating data from fossils, allow us to hypothesise about the history of the body and brain, and the forces that shaped them through countless generations (Pennisi, 2003). The phylogenetic approach—looking at patterns among related species—also allows scientists to make inferences about the evolution of behaviours (e.g., Delbarco-Trillo et al., 2011).

phylogeny The evolutionary history of a particular group of organisms.

taxonomy The classification of organisms.

Newer methods aid in classifying animals and inferring evolution

Today the field of **taxonomy** (from the Greek *taxis*, "arrangement," and *nomos*, "law"), or *classification*, makes use of our understanding of genetics to reconstruct phylogeny. DNA appears to change at a relatively steady average rate in all lineages within a given order of animals (Luo and Ho, 2018). Thus, the proportion of differences between DNA samples from two species can be used as a "molecular clock" to estimate how long ago they diverged from a common ancestor. For example, **FIGURE 5.5** shows an attempt to reconstruct the family tree of apes and humans according to the genetic similarity of the species (Wildman et al., 2003). The diagram depicts humans and chimpanzees as more closely related to each other than either species is to gorillas. The timelines in Figure 5.5 should be viewed as approximations, however, because scientists are still discussing and testing ideas about calibration of the molecular clocks. For example, some estimates based on mutation rate suggest that humans and chimpanzees actually may have split from a common ancestor earlier than previously believed, in the range of 7–13 million years ago (Langergraber et al., 2012). If this is true, the evolutionary tree describing the relatedness of the great apes—including us—will need another revision.

Estimates from fossil and DNA evidence do not always agree completely in dating branches of the evolutionary tree. Fossil dates tend to be too recent, because we can never find the first specimen of a given species. Molecular dates have tended to be too old, because of problems with calibrating rates of change of DNA over time. As technology improves, however, these discrepancies are diminishing (Ho and Duchêne, 2014).

Go to Activity 5.1
Family Tree of Apes and Humans

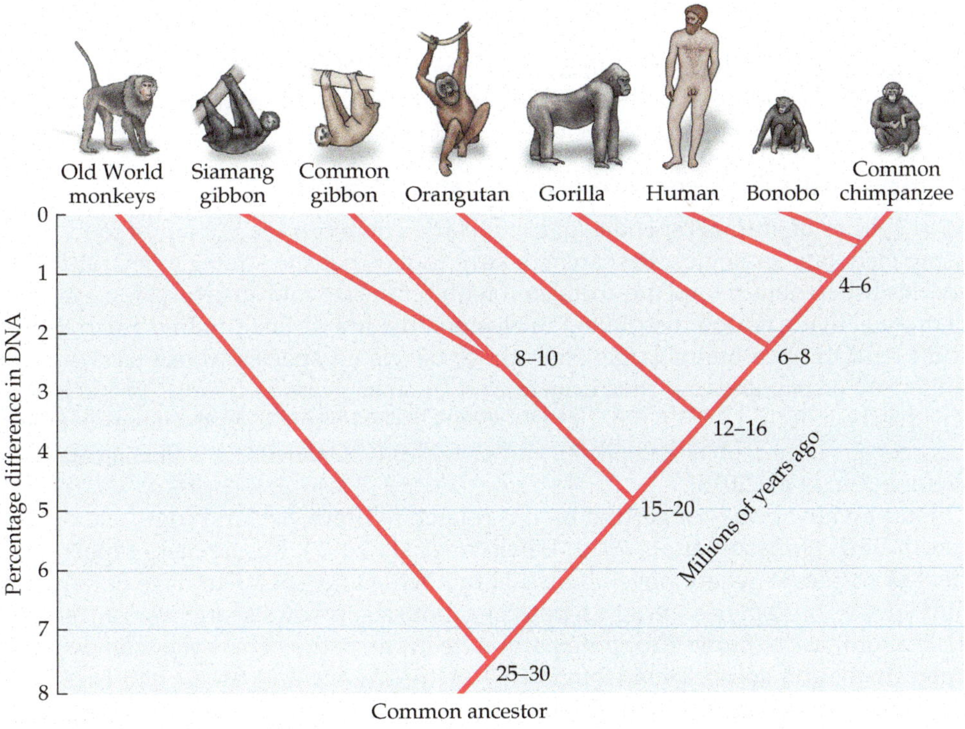

5.5 Family Tree of Apes and Humans This tree was derived from measurements of differences between pairs of species in samples of their genetic material, molecules of deoxyribonucleic acid (DNA). To see how different two species are in their genetic makeup, trace the lines from the two members of a pair to the point that connects them and then match the point with the scale on the left. For example, the line from humans and the line from chimpanzees converge at a point indicating that human DNA differs from chimpanzee DNA by just over 1%. The DNA of humans and of chimpanzees differs from that of gorillas, in turn, by about 2.3%. The scale on the right gives the estimated amount of time, in millions of years, since any pair of species shared a common ancestor. For example, humans and chimpanzees diverged from a common ancestor about 4–6 million years ago. (After D. E. Wildman et al., 2003. *Proc Nat Acad Sci USA* 100: 7181–7188.)

5.2 Why Should We Study Other Species?

Learning Objectives

After reading this section, you should be able to:

5.2.1 **Explain the importance of ecological niches.**

5.2.2 **Relate brain evolution to key environmental circumstances.**

5.2.3 **Describe the logic of comparing brain evolution across species, in order to uncover general principles.**

One old-fashioned reason for comparing species was based on the unfounded assumption that humans were the pinnacle achievement of evolution, as though evolution had an objective (and we were it). This human-centred perspective was properly criticized because it implicitly pictured other animals as incomplete "little humans" or "subhumans." It also embraced an old notion referred to as *linear descent*: the idea that evolution had proceeded along a single trajectory from simple to complex, culminating in humans. Today, scientists instead understand evolution as a multibranching set of radiations, and we can use comparisons of different species to gather clues about this evolutionary history.

Different kinds of animals have evolved specific behaviours and neural mechanisms that allow them to exploit specific sets of environmental opportunities, or **ecological niches**. The anteater's long face and tongue, for example, allow it to take advantage of the presence of huge ant colonies far more effectively than its precursor species could do. Every species has its own evolutionary history of modification to exploit the ecological niches arising in the local environment. Examples and comparisons of animals in different ecological niches appear in every chapter of this book. Comparing two or more carefully chosen species leads to a much deeper understanding because the evolutionary framework provides additional explanatory power. Species with varying biological histories show different solutions to the challenges of survival and reproduction. In many cases, these pressures to adapt have led to changes in brain structure. **BOX 5.1** explains some of the factors that scientists consider in choosing particular species to study.

One important adaptation is the ability to learn and remember, in order to successfully predict where, when, and how to obtain food and mates, and to avoid danger. This capability to store information must have arisen early in evolution, because even simple animals show lasting changes in behaviour following important experiences. Understanding how simpler animals form and store memories has provided many insights into memory mechanisms in more-complex animals, including human beings, as we will see in Chapter 16. For now, let's look at a few examples of how comparing species can inform us about brain function.

Complicated lives require complicated brains

Most species of animals spend a lot of time and energy in the pursuit of food, often using elaborate strategies. Researchers have found that the strategies that different species use to obtain food are correlated with brain size and structure. For example, mammals that eat food distributed in clusters that are difficult to find (such as ripe fruit) tend to have brains larger than those of related species whose food is more uniformly distributed and thus easy to find (such as grass or leaves). This relationship has been found within families of rodents, insectivores (e.g., shrews and moles), lagomorphs (e.g., rabbits and pikas), and primates (Clutton-Brock and Harvey, 1980; Steinhausen et al., 2016).

Finding novel ways of getting food is related to brain size and cognitive capacities in birds (Jønsson et al., 2012; Lefebvre et al., 1997). Researchers report a variety of instances where novel food-seeking behaviour seems to have emerged in bird species: magpies digging up potatoes, house sparrows searching car radiator grilles for insects, crows dropping palm nuts in the paths of cars that run over and open them, and so on. Data from both North America and the British Isles indicate that the more innovative species have relatively larger forebrains. These results thus suggest that selection pressures favored increased size of the forebrain,

ecological niche The unique assortment of environmental opportunities and challenges to which each organism is adapted.

BOX 5.1

Why Should We Study Particular Species?

With all the species that are available, why should we choose certain ones for study? In selecting species for their research, scientists apply several criteria, including the following:

1. *Outstanding features* Some species are champions at specific behaviours and abilities, such as sensory discrimination (like the incredible auditory localisation abilities of owls) or control of movement (such as the flight behaviour of the housefly). These abilities are often linked to highly specialised neuronal structures that incorporate and optimise particular designs that may be less conspicuous in other organisms (Bullock, 1986). Study of such species may yield general principles that apply to other species.
2. *Convenience* Some species, such as the laboratory rat, are particularly convenient for study because they breed readily, are inexpensive to maintain, are not endangered, have relatively short life-spans, and have been studied extensively already, so there is a good base of knowledge about them at the outset. In addition, some (not all) aspects of their brains and behaviour are similar enough to other species to allow for generalisations. Some species are convenient for methodological reasons. For example, some mollusks have relatively simple nervous systems that aid in tracing neural circuits, and fruitflies (*Drosophila melanogaster*) are a classic model species in genetics because they have a simple genome, short generation time, and numerous genes with mammalian homologues? (R. Lewis, 1998).
3. *Comparison* Close relationships between species that behave very differently enable the testing of hypotheses. For example, species of voles that are closely related, and otherwise very similar, show large differences in the size of the hippocampus that appear to reflect differences in the sizes of their home ranges (the hippocampus is implicated in spatial navigation, which is of course more challenging in larger ranges). A similar difference is found in birds, with species that cache food in dispersed spatial locations showing larger hippocampi (see Chapter 17). In New World monkeys, like those in the figure, the structure of the retina differs markedly between species that are active at night and species that are active in daylight (Finlay et al., 2008).
4. *Preservation* Studies of rare and/or endangered species, conducted in the field or in zoos, can help set priorities and assess options for the conservation of biodiversity (Mace et al., 2003).
5. *Economic importance* Studying animals that are important for the economy—agricultural animals, food species like fishes, predators, and crop-damaging species—can provide information that helps increase production and/or decrease losses.
6. *Treatment of disease* Some species are subject to the same diseases as other species and therefore are valuable models for investigation. For example, in neuroscience alone there are animal models of Alzheimer's disease, Down syndrome, epilepsy, mood disorders, amyotrophic lateral sclerosis, narcolepsy, stroke, and many other disorders.
7. *Curiosity and the advancement of knowledge* The organisms with which we share the planet are intrinsically fascinating and often mysterious. Studying them out of pure curiosity can yield unexpected benefits: who could've guessed that studying the light sensitivity of microorganisms would lead to the development of a powerful tool—optogenetics (see Figure 2.26 and 2.27)?

Owl monkeys

© Kevin Schafer / Getty Images

Howler monkey

© Amazon-Images/Alamy Stock Photo

The retina of the nocturnal owl monkey features high densities of rod photoreceptors for excellent performance in low light. In contrast, the retina of the howler monkey contains a high concentration of cone photoreceptors, giving it excellent acuity and colour sensitivity, consistent with its diurnal (daytime) lifestyle.

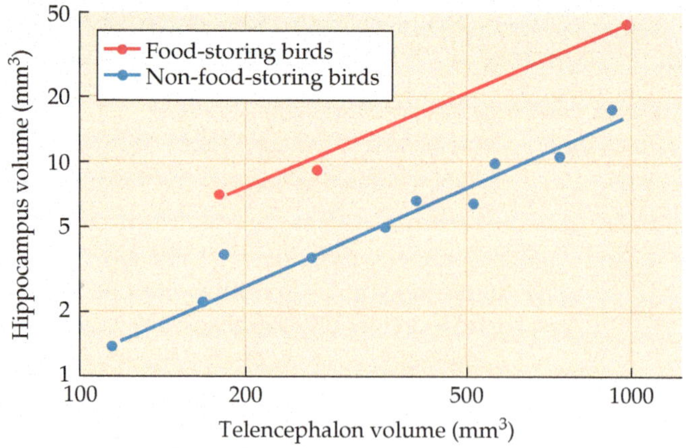

5.6 Food Storing in Birds as Related to Hippocampal Size Food-storing species of birds have twice as large a hippocampus in relation to their forebrain (telencephalon) as do species that do not store food. Note that both axes on this graph are logarithmic. (After D. F. Sherry et al., 1989. *Brain Behav Evol* 34: 308–317.)

allowing these species to cope with environmental challenges and opportunities in new, flexible ways. Later in the chapter we will see an extension of this kind of study to species of primates, showing that increased size of the forebrain seems to be related to innovation and sociality.

Other behavioural adaptations have also been related to differences in relative sizes of certain brain structures. For example, some species of bats find their way and locate prey by hearing; others rely almost entirely on vision. In the midbrain, the auditory centre (inferior colliculus) is much larger in bats that depend on hearing; bats that depend on sight have a larger visual centre (superior colliculus).

Birds in families that store bits of food for later use (e.g., the acorn woodpecker, Clark's nutcracker, and the black-capped chickadee) have a larger hippocampus relative to the forebrain and to body weight than do birds in families that do not store food (Sherry, 1992), which may relate to differences in gene expression, neuron number, and neural connectivity (T. C. Roth et al., 2010). This difference has been found among both North American species (**FIGURE 5.6**) and European species. For more on hippocampal size and memory for food storage, see **A STEP FURTHER 5.1: FOOD STORING DEPENDS ON HIPPOCAMPAL SIZE** on the website.

A different evolutionary pressure affected the brains of songbirds—species in which males sing to attract mates. In some species, each male may sing only a single song, while in other species males have repertoires of ten or more different tunes. The females prefer to mate with males with larger repertoires, so large repertoires are definitely adaptive in those species. Among the closely related species of European warblers, repertoire size correlates with the volume of a brain region known as the *HVC* (**FIGURE 5.7**). This correlation strongly suggests that the HVC is important for song production in birds, an inference that has been confirmed in lab studies (in fact, the HVC is sometimes referred to as the *higher vocal centre*). It is the females that are exerting the evolutionary pressure, by choosing males with large repertoires and thereby also selecting for larger HVCs. This is therefore an example of sexual selection: large song repertoires are adaptive only because the opposite sex prefers them.

BOX 5.2 provides other examples of solutions that different species employ to solve the dilemmas of adaptation. As a general rule, the relative size of a brain region is a rough guide to the importance of the function of that region for the adaptations of the species.

Invertebrate nervous systems provide models of neural function

Most of the animals on Earth are invertebrates—animals without backbones—and they far exceed vertebrates in many ways, including diversity of appearance, variety of habitat, and overall numbers. For each person on Earth, there are at least 1 billion insects, which are just one type of invertebrate!

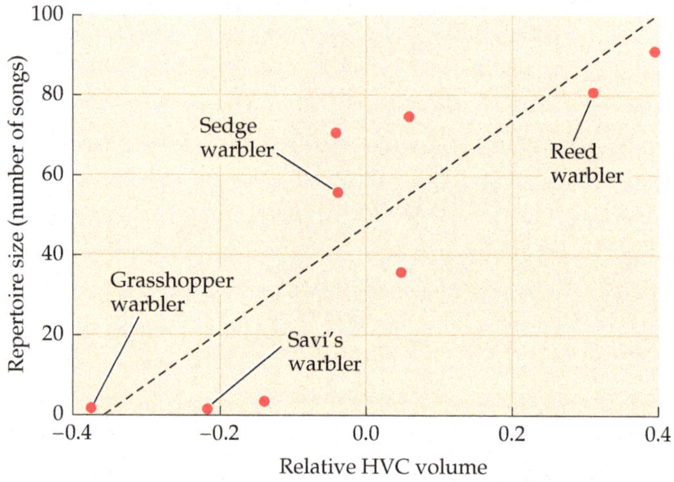

5.7 Brainy Warblers Sing More Songs (After T. Székely et al., 1996. *Proc R Soc Lond B Biol Sci* 263: 5.7–5.0.)

BOX 5.2 To Each Its Own Sensory World

Differences in the lifestyles of various mammals, reflecting the varying ecological niches they occupy, are evident in the organisation of the cerebral cortex. The common rat (*Rattus norvegicus*) is nocturnal and uses its whiskers to find its way in the dark. About 28% of the cortical representation of the rat's body surface is devoted to the whiskers (vibrissae), compared with only about 9% in the gray squirrel (*Sciurus carolinensis*) (Figure A). In addition, the nocturnal rat makes little use of vision, and its primary visual cortex (V1) is relatively small compared with that of the squirrel, which is diurnal. (Figure A) (K. L. Campi and L. Krubitzer, 2010. *J Comp Neurol* 518: 491–512.)

The remarkable platypus (*Ornithorhynchus anatinus*) is an egg-laying mammal that lives in and around streams in eastern Australia and Tasmania. Because of its duck-like bill and webbed feet, some scientists thought it might be a hoax when preserved animals were brought to Europe in the nineteenth century. The platypus is largely nocturnal and dives into murky waters, closing its eyes, ears, and nostrils as it hunts for insects, shrimp, and crayfish. How it senses its prey remained a mystery until the 1980s, when investigators found that the main sensory organ of the platypus is its bill, which is about 7 centimeters (cm) long in an adult of 160 cm.

The bill has about 16 longitudinal stripes of receptors: stripes of touch (mechanosensory) receptors alternate with touch-electrical (electrosensory) receptors (Figure B) (P. R. Manger et al. 1998. *Phil Trans R Soc Lond B* 353: 1171–1186). As the platypus moves its bill underwater, it can detect prey by both the mechanical ripples and the changes in electrical fields that they cause. In keeping with the importance of the bill in locating prey, almost all somatosensory cortex (S1 and S2) in the platypus is devoted to the bill (see Figure B), and the primary visual (V1) and auditory (A1) areas of the cortex are small (L. Krubitzer et al., 1995. *J Comp Neurol* 351: 261–306).

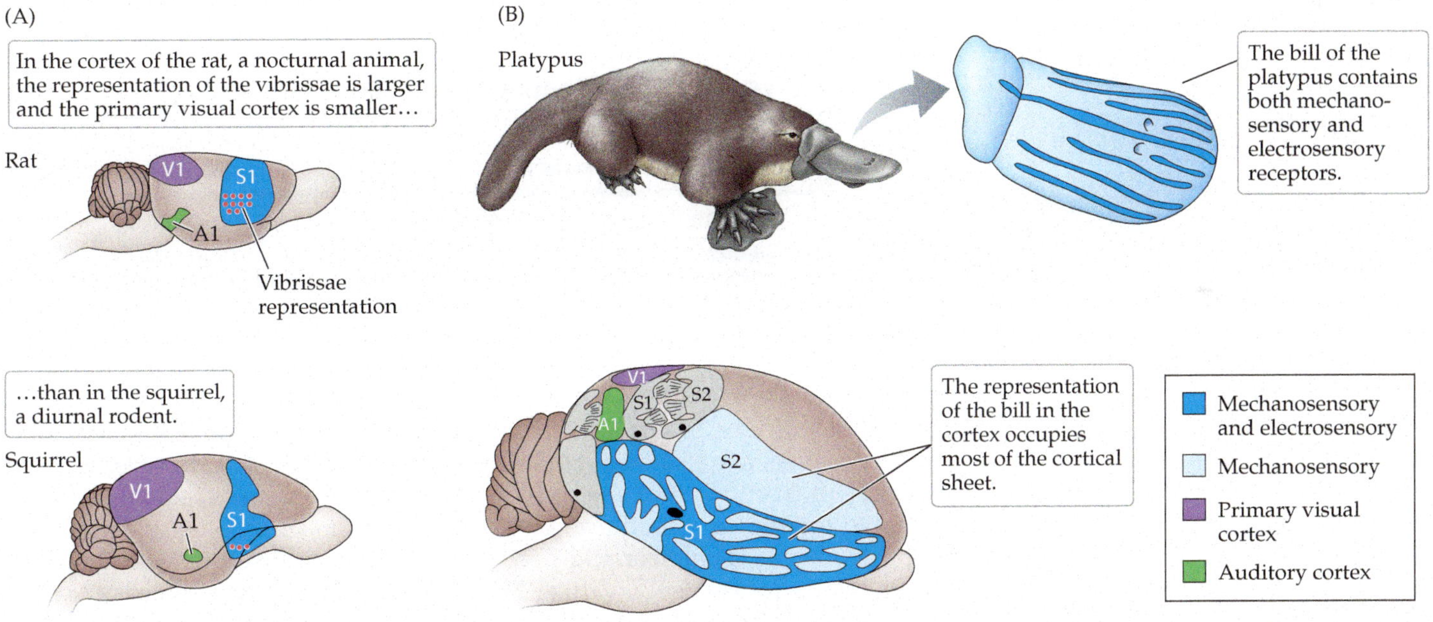

The gross anatomy of the nervous systems of some representative animals is illustrated in **FIGURE 5.8**. The enormous complexity of the vertebrate brain, with its billions of nerve cells, presents major difficulties for understanding the fundamental neuronal processes underlying behaviours. To address this problem, many researchers have turned instead to the nervous systems of invertebrates that have only hundreds or thousands of neurons. If we can understand the cellular processes for memory formation in a model system like the worm *Caenorhabditis elegans*, which has only 302 neurons (see Figure 6.2), perhaps we can uncover basic cellular mechanisms of learning and memory that are shared across species (Byrne and

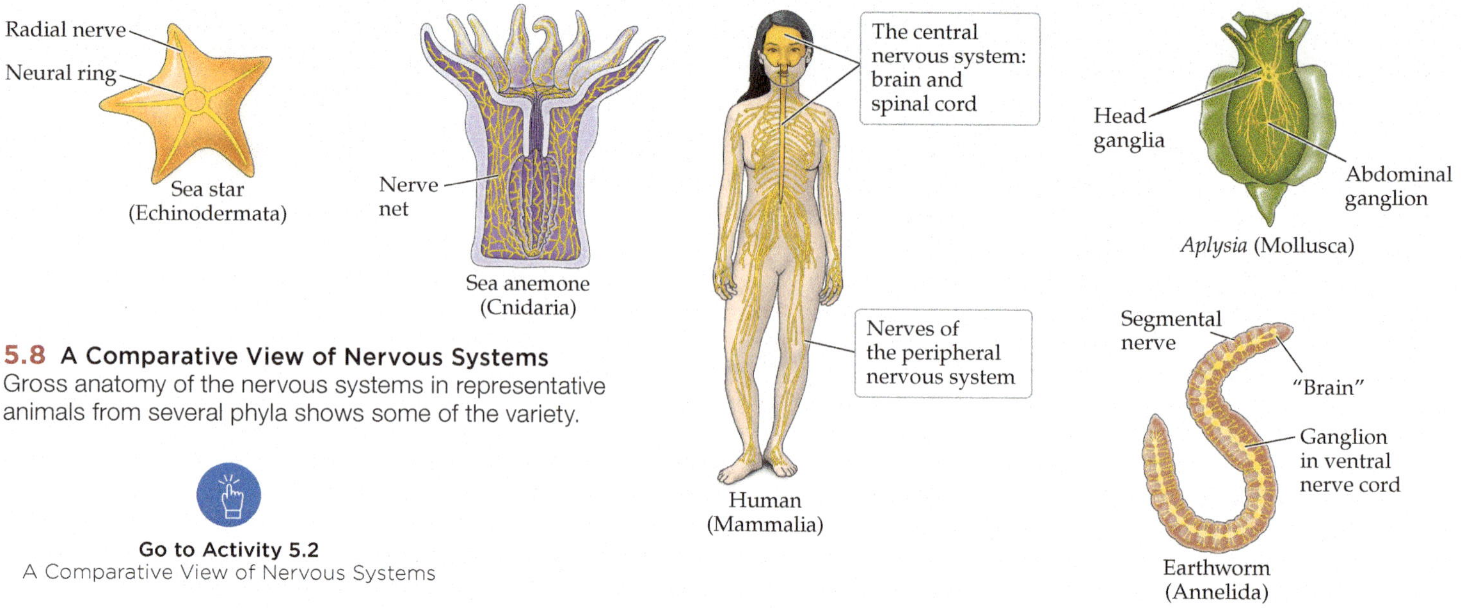

5.8 A Comparative View of Nervous Systems
Gross anatomy of the nervous systems in representative animals from several phyla shows some of the variety.

Go to Activity 5.2
A Comparative View of Nervous Systems

Hawkins, 2015; McDiarmid et al., 2018). We compare invertebrate and vertebrate nervous systems in **A STEP FURTHER 5.2: INSECT NERVOUS SYSTEMS** on the website. Since we are primarily interested in understanding human behaviour, let's focus here on brains more similar to our own.

5.3 All Vertebrate Brains Share the Same Basic Structures

Learning Objectives

After reading this section, you should be able to:

5.3.1 **Discuss how selection pressures have affected the proportions of brain regions in different species of mammals, using humans and rats as examples.**

5.3.2 **Catalogue key features of the brain shared by all vertebrates.**

As you can see in **FIGURE 5.9**, the various orders of mammals all share the same basic set of brain regions devoted to visual, auditory, and somatosensory processing. The regions are also arranged in the same basic pattern. However, the relative sizes, proportions, and anatomical locations of these brain regions have been subject to evolutionary modification as the species have adapted to their unique ecological niches (Krubitzer and Seelke, 2012). For example, species with a high degree of olfactory sensitivity, like the rat, have relatively large olfactory regions compared with humans, whereas the expansion of some regions in the human temporal lobes seems to be related to language function (see Chapter 18). The diversity of mammalian brains can be seen online at brainmuseum.org.

A comparison of the human brain with the brain of the lab rat, perhaps the most completely studied brain, reveals basic similarities and differences (**FIGURE 5.10**). Each of the main structures in the human brain has a counterpart in the rat brain. This comparison can be extended in great detail to include nuclei, fibre tracts, and types of cells and even the many molecules specific to brain function. These similarities in structure and organisation of mammalian brains reflect the heritage of our evolution from a common ancestor long ago.

There are differences between the brains of humans and the brains of other mammals, of course, but they are mainly quantitative. Whereas the brain of an adult human being weighs about 1400 grams (g), that of an adult rat weighs a little less than 2 g, but in each case the brain represents about 2% of total body weight. The cerebral hemispheres occupy a much greater proportion of the brain in the

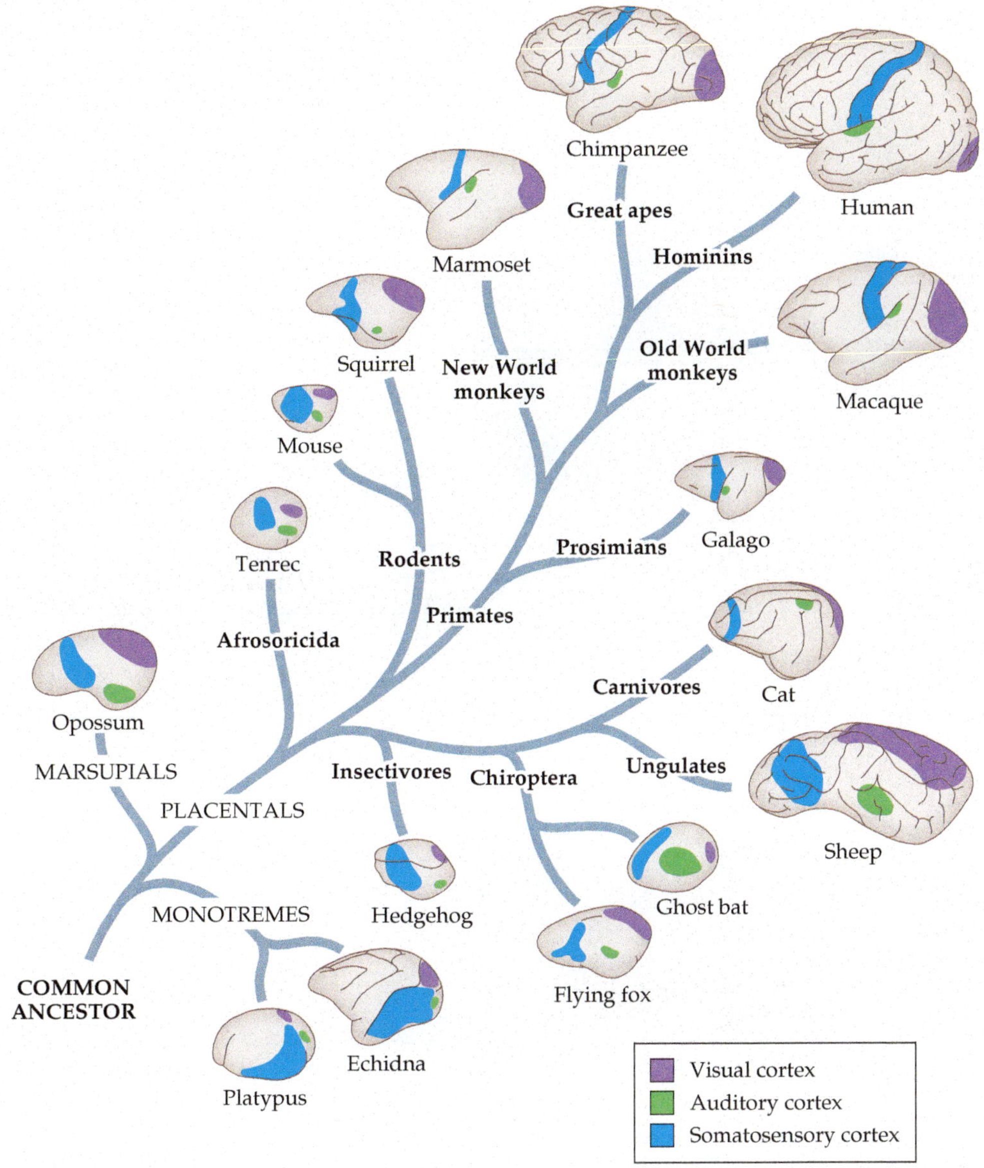

5.9 Functions and Relations This family tree of mammalian brains shows that while we mammals all share the same basic sensory cortical regions, their size and placement are determined by how each species makes its living. Brains are not drawn to scale. (Courtesy of Leah Krubitzer, U.C. Davis.)

human than in the rat, and the human cerebral cortex is highly convoluted (i.e., covered in gyri and sulci) (see Chapter 1), whereas the rat cerebral cortex is smooth. The olfactory bulb is relatively larger in rats than in humans, probably because of the rat's much greater use of the sense of smell.

All vertebrate nervous systems share certain main features but differ in others

Let's extend our view to the basic features of the nervous systems of vertebrates more generally, not just mammals. Vertebrate nervous systems share many characteristics:

- *Development from a hollow dorsal neural tube* The head of the embryonic neural tube goes on to form the major subdivisions of the brain, but a fluid series of hollow spaces within the brain—the ventricular system—persists into adulthood (see Chapters 1 and 6).
- *Bilateral symmetry* The cerebral hemispheres are almost mirror images. (We'll see some interesting exceptions in Chapter 18.)

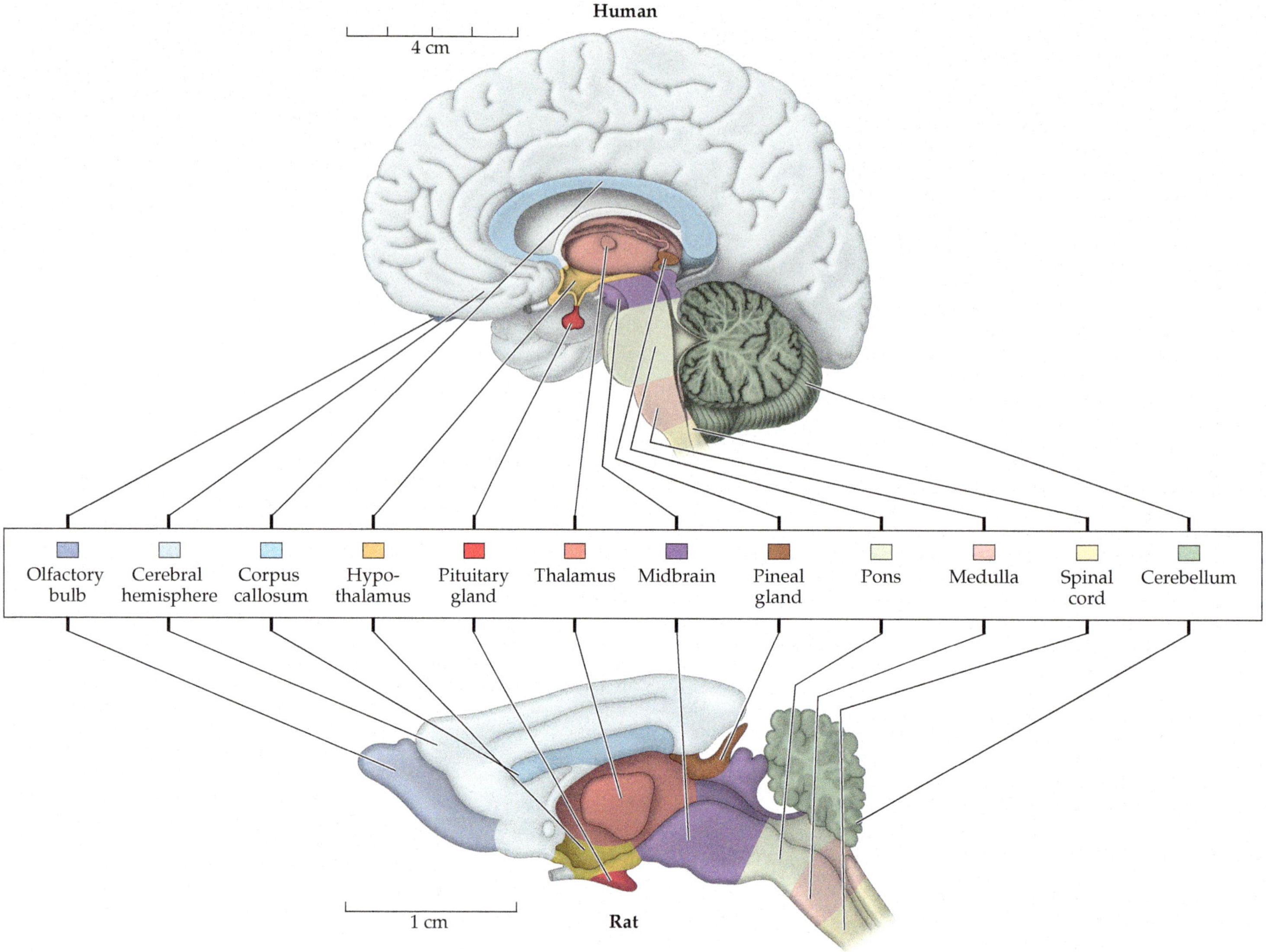

5.10 Human and Rat Brains Compared Midsagittal views of the right hemisphere in human and rat brains show that the main structures are the same in both, and they have the same topological relations to each other. Note, however, that the cerebral hemispheres are relatively much larger in the human brain, whereas the rat has a relatively larger midbrain and olfactory bulb. (The rat brain has been enlarged here about 6 times in linear dimensions relative to the human brain.)

- *Segmentation* Pairs of spinal nerves extend from each level of the spinal cord.
- *Hierarchical control* Higher levels of the nervous system tend to control or modulate lower levels. Within each level of the hierarchy, however, multiple systems may act together in a parallel manner. So, for example, networks of cortical regions modulate midbrain and brainstem mechanisms.
- *Separate systems* The central nervous system (brain and spinal cord) is clearly separate from the peripheral nervous system, as shown for humans in Figure 5.8.
- *Functional specialisation* Certain physiological and behavioural processes are controlled by specific neural mechanisms, often consisting of specific networks of brain regions operating together.

Vertebrates descended from a common ancestor that possessed all these features, so all vertebrates have all of these features in common. In general, vertebrate species with larger bodies tend to possess larger brains, with larger neurons and larger dendritic trees, as **FIGURE 5.11** shows for some mammals. However, some classes of vertebrates have proportionately larger brains than others, as we'll see next.

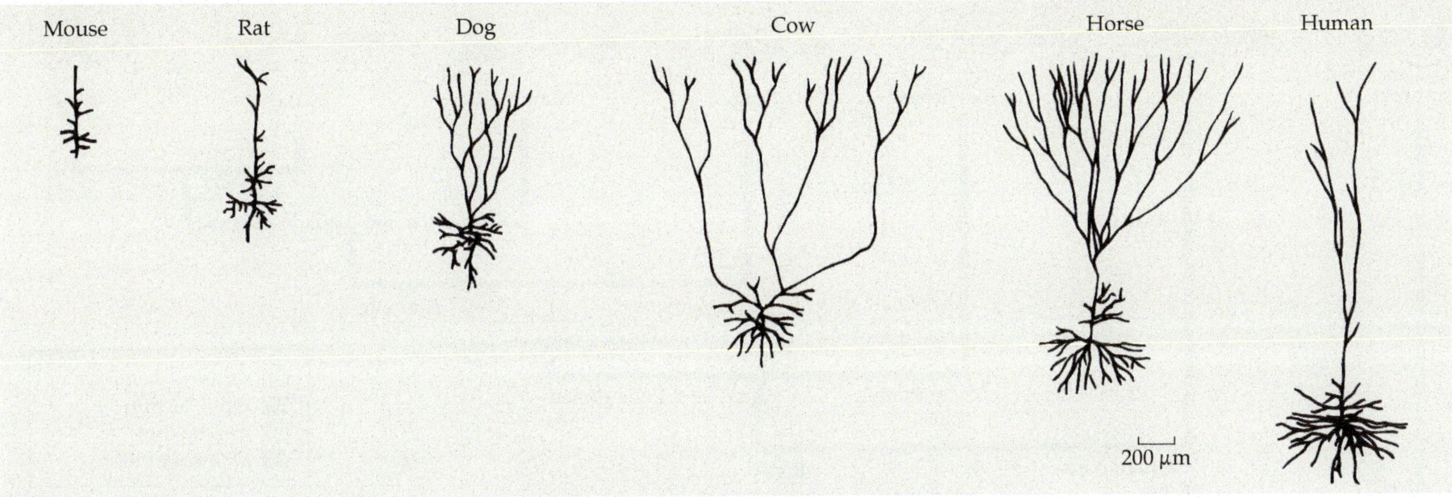

5.11 The Same Kind of Neuron in Different Species These pyramidal neurons from the motor cortices of different mammals are all drawn to the same scale. (After A. Barasa, 1960. *Z Zellforsch* 53: 69–89.)

5.4 The Evolution of Vertebrate Brains Reflects Changes in Behaviour

Learning Objectives

After reading this section, you should be able to:

5.4.1 **Discuss methods for inferring brain size in ancient species.**

5.4.2 **Summarize major anatomical developments in the evolution of vertebrate brains.**

5.4.3 **Review theoretical and technical considerations in comparing relative brain sizes between species.**

During the course of evolution, the characteristics of the nervous system changed progressively. One especially prominent change in the last 100 million years has been a general tendency for the brain size of vertebrates to increase, and the brains of our human ancestors have shown a particularly striking increase in size over the last 2 million years. How does evolution of the brain relate to changes in behaviour?

Present-day animals and fossils reveal evolution of the brain

Theoretically, we could learn more about the evolution of the brain by studying the brains of fossil animals, but brains themselves do not fossilise—at least, not literally. Two methods of analysis have proven helpful. One is to use the cranial cavity of a fossil skull to make a cast of the brain that once occupied that space. These **endocasts** (the Greek *endon* means "within") give a reasonable indication of the size and shape of the brain, but no fine detail.

The other method is to study present-day animals, choosing species that show various degrees of similarity to (or difference from) the ancestral species. Although no modern animal is an ancestor of any other living species—because all species are constantly evolving—some present-day species resemble ancestral forms more closely than others do. For example, present-day salamanders are much more similar to the fossils of vertebrates of 300 million years ago than are any mammals. Among mammals, some species, such as the opossum, resemble fossil mammals of 50 million years ago more than do other species, such as the dog. Thus, a species such as the opossum is said to retain primitive or ancestral states of particular anatomical features. In studying the brains of living species, anatomists can obtain far more detailed information than they get from endocasts, because they can investigate the internal structure of the brain: its nuclei, its fibre tracts, and the circuitry formed by connections of its neurons.

endocast A cast of the cranial cavity of a skull, especially useful for studying fossils of extinct species.

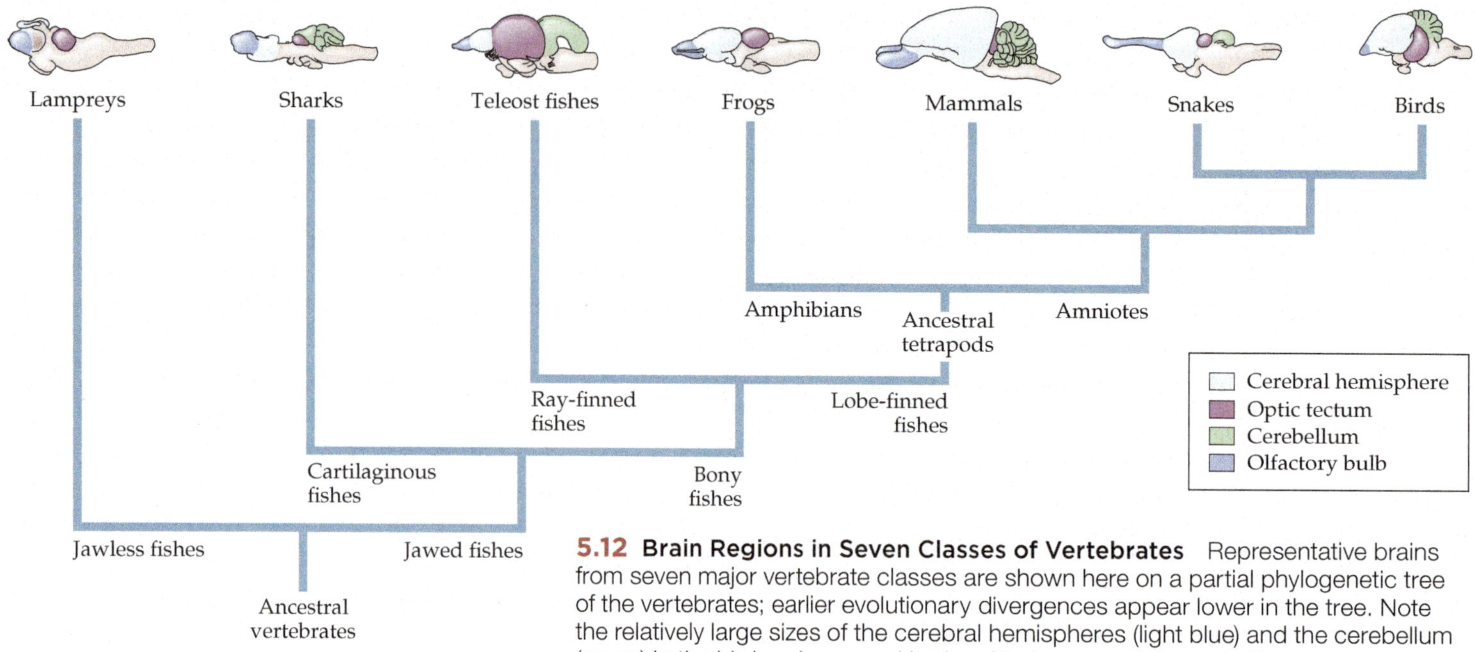

5.12 Brain Regions in Seven Classes of Vertebrates Representative brains from seven major vertebrate classes are shown here on a partial phylogenetic tree of the vertebrates; earlier evolutionary divergences appear lower in the tree. Note the relatively large sizes of the cerebral hemispheres (light blue) and the cerebellum (green) in the bird and mammal brains. (Brains are not drawn to the same scale.)

As we noted earlier, we must be careful not to interpret change over time, including the change in brain size, as if it were a linear evolutionary sequence. The main classes of vertebrates in **FIGURE 5.12** represent different lines or radiations of evolution that have been proceeding separately and simultaneously for at least 200 million years. For example, today's sharks have much larger brains than primitive sharks had, but the evolution of large-brained sharks had nothing to do with the development of large brains in mammals. The line of descent that eventually led to mammals had separated from the shark line long before the large-brained sharks evolved.

Through evolution, vertebrate brains have changed in both size and organisation

Let's consider some examples of changes in the size and organisation of vertebrate brains. Even the living vertebrate that has the most primitive features—the lamprey (a jawless fish)—has a fairly complex brain. The lamprey has not only the basic neural chassis of spinal cord, hindbrain, and midbrain, but also a diencephalon and a telencephalon. Its telencephalon has cerebral hemispheres and other subdivisions that are also found in the mammalian brain. So all vertebrate brains appear to have these regions.

One difference in basic brain structure between the lamprey and other vertebrates is that the cerebellum in the lamprey is very small (too small to be depicted in Figure 5.12). The evolution of large cerebellar hemispheres in birds and mammals appears to be a case of independent evolution from the small cerebellum in their common reptilian ancestor; the increased size of the cerebellum may be responsible for increased complexity of sensory processing and increased motor agility.

The differences among the brains of vertebrate species, then, lie not in the existence of basic subdivisions, but in their *relative size*. At what stages of vertebrate evolution do various brain regions first become important? Large, paired optic lobes in its midbrain probably represent the lamprey's highest level of visual integration. In bony fishes, amphibians, and reptiles, the relatively large optic tectum in the midbrain (see Figure 5.12) is the main brain centre for vision. In birds and mammals, however, complex visual perception requires an enlarged telencephalon.

All mammals have a six-layered **cortex**, sometimes called *neocortex* (from the Greek *neo*, "new," and the Latin *cortex*, "bark of a tree"). In more-recent mammals the cortex accounts for more than half the volume of the brain and is responsible for higher-order functions, such as the conscious perception of visual objects.

cortex Also sometimes called *neocortex*. The outer covering of the cerebral hemispheres that consists largely of nerve cell bodies and their branches.

Brain regions that serve as higher-order centres in more primitive species—the optic tectum of lampreys or frogs, for example—are also present in the mammalian nervous system, but they tend to serve lower-order processing functions. So in the example of vision, the mammalian superior colliculus is a midbrain structure that serves in reflexively orienting gaze toward visual targets.

Reptiles were the first vertebrates to exhibit relatively large cerebral hemispheres. Reptiles were also the first vertebrates to have a cerebral cortex, but they have only three cortical layers, unlike the six cortical layers of mammals. (Three-layered cortex is sometimes referred to as *archicortex,* from the Greek *arche,* "ancient.") Part of the cortex in reptiles may be homologous to the three-layered hippocampus in mammals.

Brain size evolved independently in multiple lineages

If we compare animals of similar body size, we see considerable variation in brain size within each line of evolution. For example, within the ancient class of jawless fishes, more-recent members called *hagfishes* have forebrains that are 4 times as large as those of lampreys of comparable body size. Such increases in brain size have been related to behavioural capacity most thoroughly in mammals, as we'll see next.

THE ENCEPHALISATION FACTOR The study of brain size across species is complicated by the wide variation in body sizes, raising the question, How are body size and brain size related? A general relationship was found first for present-day species and then applied successfully to fossil species. This function turns out to be a useful starting point in understanding relationships between brain and behaviour.

We humans long believed our own brains to be the largest, but this belief was upset in the seventeenth century when the elephant brain was found to weigh 3 times as much as our own. Later, whale brains were found to be even larger. Scientists of the day proposed that brain weight should be considered as a fraction of body weight, with the reassuring outcome that humans then outranked elephants, whales, and all other animals of large or moderate body size. But a mouse has about the same ratio of brain weight to body weight as a human, and the tiny shrew outranks a human on this measure. So, from a comparative point of view, what is the general relation between brain size and body size?

When we plot brain weights and body weights for many species of mammals, we see some generalities (**FIGURE 5.13**). There is a distinct overall correlation between body and brain weight across many species, so all of the points on the plot fall within a narrow polygon. Using a statistical procedure called *linear regression,* a "line of best fit" can be created that passes as closely as possible through the collection of points representing mammalian species. This line has a slope of about 0.69 (Harvey and Krebs, 1990), representing the general mathematical relationship between body weight and brain weight across all mammals.

Having created this general model of braininess, relative to body weight, we can then consider how much any individual species deviates from the expectation for mammals as a group. This deviation (known as a *residual* in linear regression procedures) corresponds to the vertical distance for that species above or below the diagonal line on the graph. This distance, *k,* is known as the **encephalisation factor**. The greater the encephalisation factor is for a species—that is, the higher its value is above the diagonal line—the more exaggerated the brain size is for that species relative to the mathematical prediction for a mammal of its

encephalisation factor A measure of brain size relative to body size.

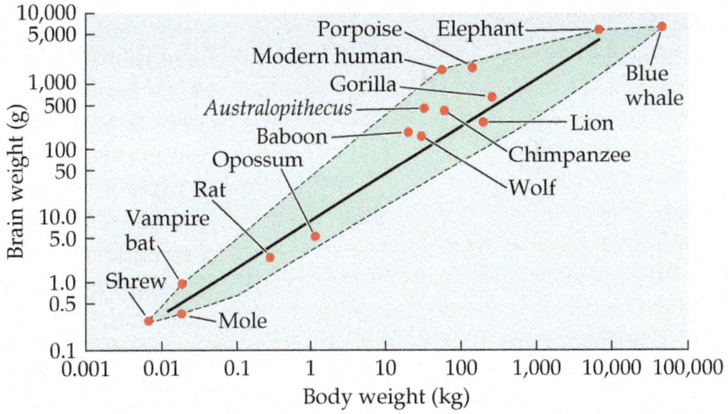

5.13 The Relation between Brain Weight and Body Weight Brain weight is related here to body weight in several mammalian species. Note that both axes are logarithmic, so the graph includes a wide range of brain weights and body weights. A polygon has been drawn to connect the extreme cases and include the whole sample. The black diagonal line is a line of best fit (also known as the *regression line*), indicating a prediction line for brain size among mammals as a group. For each species, the encephalisation factor, *k,* corresponds to the distance (the *residual*) between the line and the brain weight value for that species. Although not shown, bird species are about as "brainy" as mammals, but reptile species are substantially less so. Although the encephalisation factor is not a good basis for ranking species, large deviations from predicted encephalisation may signal strong selection pressures (see text for explanation). (After H. J. Jerison, 1969. *Am Nat* 103: 575–588; data from G. Crile and D. P. Quiring, 1940. *Ohio J Sci* 40: 219–259.)

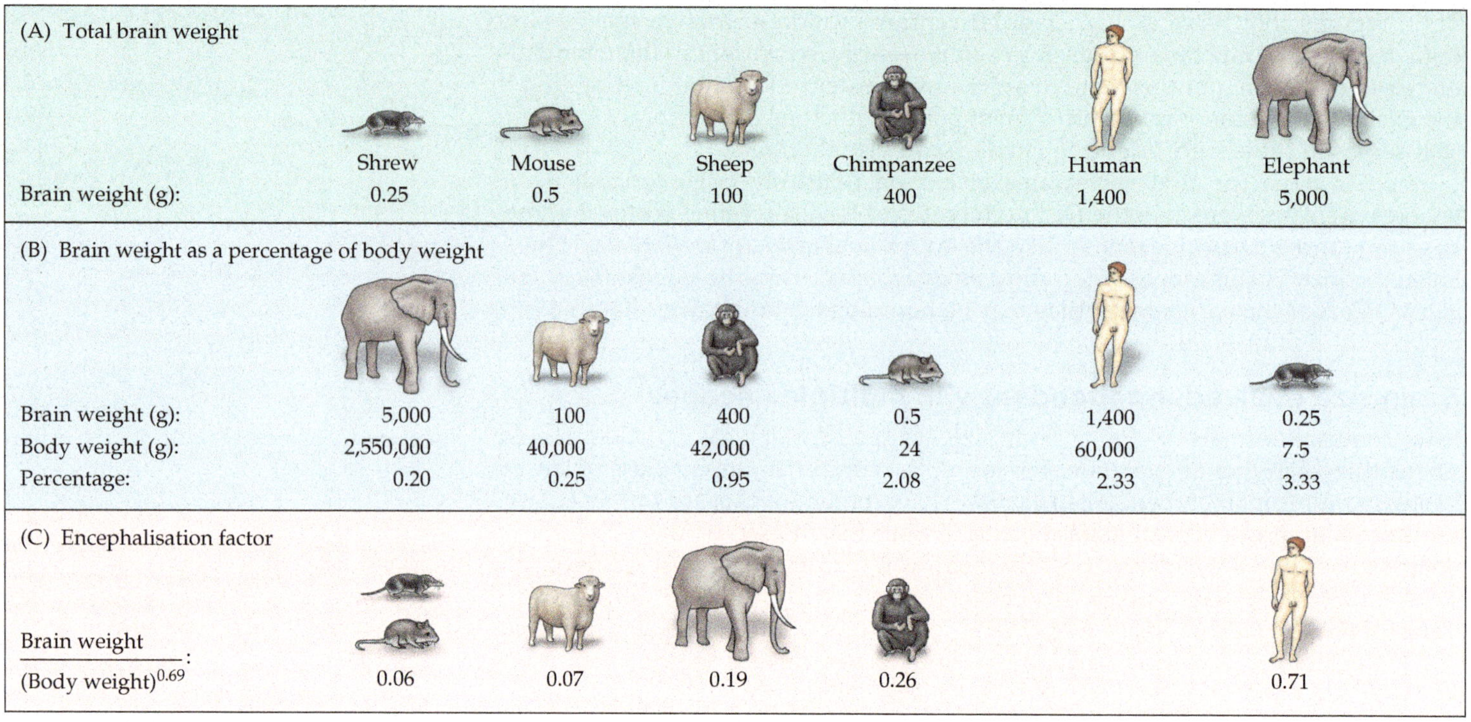

5.14 Who Is the Brainiest? For this sample of small to large mammals, the answer depends on what measure is used: total brain weight (A), brain weight as a percentage of body weight (B), or the encephalisation factor, *k* (C). For each measure, the animals are ranked here from lowest value to highest. (Data from G. Crile and D. P. Quiring, 1940. *Ohio J Sci* 40: 219–259.)

size. As summarised in **FIGURE 5.14**, among mammals the encephalisation factor is greatest for humans and quite a bit less for chimpanzees, despite our evolutionary closeness (see Figure 5.14C). The encephalisation factor for opossums indicates that they have *less* brain for their size than would be typical of mammals as a group. Nevertheless, it is important to guard against a tendency to selectively emphasize data that confirm our biased expectation that we are brainier than other species: our brains are substantially larger than predicted for other animals of the same size, but dolphins have comparable proportions, as do some birds, and we have historically overestimated the number of neurons in the human brain by as much as 15% (Scudellari, 2015). Furthermore, the simple application of the encephalisation factor across species fails to take into consideration the great variation in other selection pressures that shaped the bodies of those species (Willemet, 2013). For one example, there is reduced selection pressure constraining body size in whales (because when they are floating in the open ocean, the effects of gravity are reduced, dissipation of heat is efficient, and locomotion can be lower in intensity than in animals that must run or fly), with the result that by the standard of the encephalisation factor, they seem to be less brainy than many other mammals. The unique niches occupied by other species likewise influence their encephalisation twice scores; perhaps the encephalisation factor is best viewed as potential indicator of strong selection pressures, rather than as a way of ranking species.

You may have heard that most species of dinosaurs became extinct because of their inadequate ("walnut-sized") brains. But is this notion correct? No. The endocasts of dinosaurs indicate that their brain weights are typical of the encephalisation of reptiles. The brain of *Tyrannosaurus rex* probably weighed about 700 g: roughly half the size of a human brain, but still a lot larger than a walnut, and appropriate for a reptile of its size (Jerison, 1991). So it seems unlikely that dinosaurs perished because of a lack of brains. In fact, some species of dinosaurs survived and were the ancestors of today's birds, some of whom can outperform primates on certain types of learning tasks (A. A. Wright et al., 2017) and, as we noted earlier, exhibit encephalisation metrics comparable to those of mammals.

BRAIN SIZE AS ADAPTATION As we saw earlier in the chapter, certain capabilities—navigating by sound, foraging for food, performing a large repertoire of songs—are linked to sizes of particular brain regions. But some capabilities, because they are related to overall cortical volume rather than the volume of any particular region, may have propelled a more general increase in brain size. Most major parts of the brain increase roughly in proportion to total increases in brain size, but even at the gross anatomical level, the rates of increase do show subtle differences (Finlay and Darlington, 1995). For example, the olfactory bulb has a larger relative size in some species, presumably because natural selection pressures have favored a highly developed sense of smell in those species.

As we compare brains across different species of primates, from small to large, it is evident that the medulla becomes *proportionally smaller* relative to brain weight, the cerebellum keeps pace with overall brain weight, and the cortex becomes proportionally larger than any other part (**FIGURE 5.15**). This pattern suggests that the cortex has grown disproportionally over the course of human evolution. It seems that the brain regions that have most expanded over primate evolution are the ones that develop later in life and serve more complex functions (Finlay and Darlington, 1995); for example, the human medulla is fully developed at birth, but the cortex continues adding neurons throughout childhood. This has led theorists to propose that larger brains evolved by prolonging the later stages of development. A mutation that prolonged the last stages of brain development, when neurons are being added almost exclusively in cortex, would result in a larger cortex relative to the rest of the brain. And to the extent that the larger cortex gave individuals advantages over conspecifics, the altered gene that caused it would be favoured by natural selection. With 16 billion cortical neurons—including many large von Economo neurons (or spindle neurons) that are found only in primates and a few other large and brainy mammals (Allman et al., 2010; Butti et al., 2013)—humans have far more cortical neurons than our closest relatives; we have twice as many as great apes like gorillas and orangutans, for example (Herculano-Houzel, 2018).

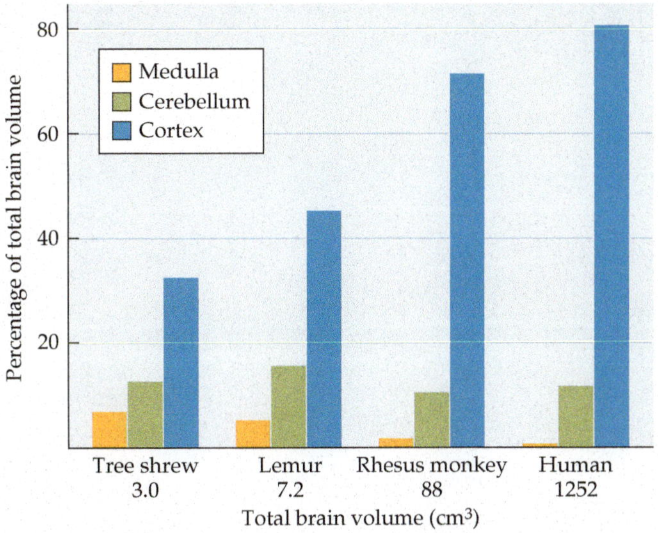

5.15 Changes in the Apportionment of Brain Regions among Primates This graph shows the percentage of brain volume occupied by three different parts of the brain in four different primates. As the overall size of the brain increases, the sizes of its different parts increase at different rates. The size of the cortex increases steadily as a proportion of total brain size, while that of the cerebellum stays about the same and the relative size of the medulla *decreases*. (Data from H. Stephan et al., 1981. *Folia Primatologica* 35: 1–29.)

Such a "later becomes larger" pattern of evolution may even explain changes in the fine structure of the cortex. During fetal development, the outermost layer (I) forms first with a few neurons. Then new neurons populate the innermost layer (VI), and new neurons are added to form each subsequent outer layer. A comparison across mammals suggests that the later-added, outer layers of cortex (e.g., layers II and III) have enlarged more in primates than the innermost layers (V and VI; **FIGURE 5.16**). Let's consider primate evolution further to see if we can understand what selective pressures favored the expansion of cortex in humans.

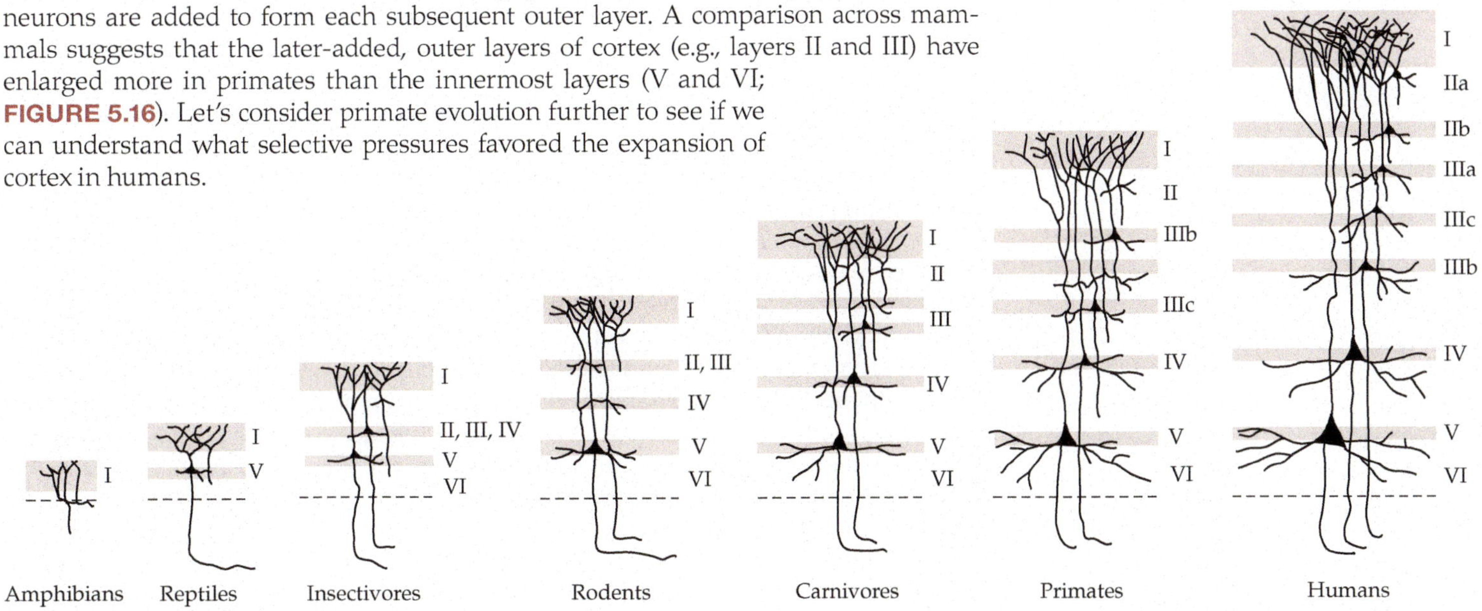

5.16 Evolution Allows Later-Developing Brain Regions to Grow Larger The expansion of cortex that took place in mammals, especially among primates, is due primarily to greater growth of the outermost layers of cortex, which are the last to arise during development. Note that reptiles have a three-layered cortex, compared with the basic six-layered cortex in the insectivores and other mammals. (After R. S. Hill and C. A. Walsh, 2005. *Nature* 437: 64–67; based on M. Marín-Padilla, 1992. *J Comp Neurol* 321: 223–240.)

5.5 Many Factors Led to the Rapid Evolution of a Large Cortex in Primates

Learning Objectives

After reading this section, you should be able to:

5.5.1 **Summarize the taxonomy and evolution of hominin species.**

5.5.2 **Describe factors both promoting and limiting brain expansion in early hominins.**

5.5.3 **Discuss ways in which natural selection for key cognitive abilities, sexual selection, and differential gene expression may all contribute to primate brain evolution.**

We've seen that the comparative approach—studying general principles across different species—reveals that humans are exceptionally brainy for their size, compared with other vertebrates. This additional brain power confers advantages in cognitive abilities, but how did we come to evolve such a large cortex? The study of **hominins**—primates within the family Hominidae (which includes four genera of great apes)—and particularly those hominin species that are ancestral to modern *Homo sapiens*, can help us understand how our body and brain adapted to the environment through natural selection.

Hominin brains enlarged rapidly in our recent evolution

The structural and behavioural features that we consider characteristic of humans did not develop simultaneously (Falk, 1993). Our large brain is a relatively late development. The trunk and arms of hominins reached their present form about 10 million years ago, and it appears that hominins became bipedal more than 4 million years ago. The fossil record indicates that by at least 2.5 million years ago, hominins were manufacturing and using stone tools in their daily lives (Semaw, 2000).

Yet these early bipedal, tool-making hominins, called **australopithecines**, still had relatively small brains of 350–400 cubic centimeters (cm^3), about the size of the modern chimpanzee brain. Wild chimpanzees do not appear to make tools from flaked stone, although they collect stones to use as hammering tools and may inadvertently produce tool-like modifications to stones as they use them to open nuts—perhaps this was a precursor to toolmaking in our own lineage (Bril et al., 2012). But the australopithecines also clearly made and used sharpened stone tools for hunting and butchering animals (Domínguez-Rodrigo et al., 2005). The ability to use tools reduced the selection pressure to maintain large jaws and teeth, and hominin jaws and teeth thus became steadily smaller than those of other apes and more like those of modern human beings. Our australopithecine ancestors were successful, lasting some 2 million years relatively unchanged, living in nomadic social groups of 20–50 individuals, cooperatively hunting and gathering plant foods—a new lifestyle that was continued by later hominins.

By about 2 million years ago, the last of the *Australopithecus* species had developed modern features, such as dexterous hands and smaller teeth, and a trend toward increasing cranial capacity (Berger et al., 2010; Kivell et al., 2011). These features became much more developed as the genus *Homo* appeared. One early representative of the genus, *Homo erectus*, had a cranial capacity of about 700 cm^3 and a smaller face than the australopithecines. As *Homo erectus* evolved, the brain became steadily larger, reaching the present-day volume of about 1400 cm^3, and the face continued to become smaller. *Homo erectus* made elaborate stone tools, used fire, and killed large animals. Fossils and tools of *Homo erectus* are found throughout three continents, whereas australopithecine remains are found only in Africa. *Homo erectus* may have represented a level of capacity and of cultural adaptation that enabled the hominins to expand into new ecological niches and to overcome barriers that kept earlier hominins in a narrower range. Although we don't yet have a clear picture of the entire *Homo* family tree (**FIGURE 5.17A**), it is clear that *Homo* species have undergone rapid evolution of the brain and behaviour as our ancestors have diversified. Thus, after remaining little changed in size during about 2 million years of tool use by the australopithecines, the brain of descendant *Homo* species almost tripled in volume during the next 1.5 million years and reached

hominin The subgroup of Hominidae that contains modern humans and their ancestral species.

australopithecine Referring to *Australopithecus*, a primate genus, known only from the fossil record and thought to be an ancestor to humans.

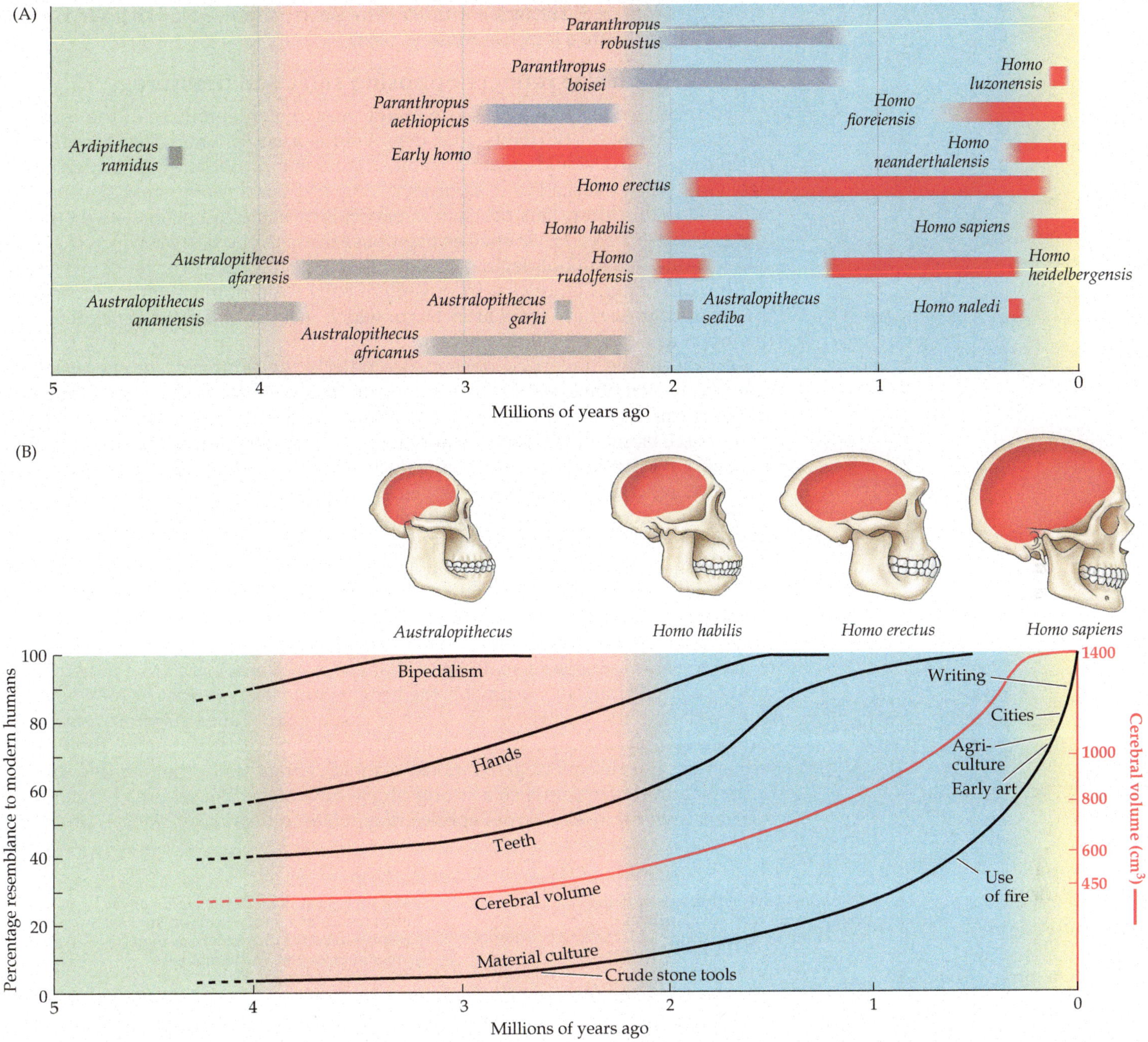

5.17 Hominin Evolution (A) The *Homo* family tree is not yet complete, but considerable diversity of species and features is already evident within the main expansion period, ranging from 2 million years ago to the present. *H. naledi*, for example, shares features more closely associated with early *Homo* species than with *H. sapiens*, despite evidence that *H. naledi* was alive as recently as 250,000 years ago. (B) The bipedal (two-footed) gait was similar to that of modern humans even in *Australopithecus*, but cerebral volume reached its current size only in *Homo sapiens*. High culture (art, agriculture, cities, writing) emerged only relatively recently and was not associated with any additional change in brain size. (Part A after Smithsonian Institution. National Museum of Natural History. 2019. http://humanorigins.si.edu/evidence/human-fossils/species; B after P. V. Tobias, 1980. *La Recherche* 11: 282–292; updated with the assistance of Tim White.)

its current form when *Homo sapiens* appeared, currently believed to be about 300,000 years ago (Hublin et al., 2017; Schlebusch et al., 2017).

The size of the human brain now appears to be at a plateau. The recent changes in human lifestyle indicated in **FIGURE 5.17B**—such as the appearance of written language, the introduction of agriculture and animal husbandry (about 10,000 years ago), and urban living (the last few thousand years)—have all been accomplished and assimilated by a brain that does not seem to have altered in size since *Homo* *sapiens* first appeared. The lack of further increase in brain size may be related to the costs of a large brain, a topic we consider next.

Hominin brain size reflects opposing selection pressures

While directional selection pressure might favor the enlargement of the brain—and the behavioural flexibility that it provides—a larger brain also entails significant costs that would create a stabilising selection pressure restricting brain size. Growth of a large brain involves a long and burdensome gestation period and difficult birthing. This problem is particularly acute in us bipeds, where the babies of big-brained *Homo* species must pass through the narrow pelvis that our small-brained australopithecine ancestors evolved in order to balance over two feet instead of four. Once that inconvenient baby is born, much of the growth of the brain continues for years after birth, which means prolonged dependence of the infant and prolonged parental care.

Of course, this prolonged dependence reflects the protracted development of humans we mentioned earlier. The extended growth period of humans compared with other primates is obvious in brain weight after birth. While the growth of the chimpanzee brain levels off shortly after birth, the human brain keeps growing after birth, at a rate rivalling that of the fetus (**FIGURE 5.18**). Indeed, the newborn human is, in a sense, continuing fetal development outside the womb. This is also another example of "later becomes larger" that we discussed earlier (see Figure 5.17), as most of the postnatal brain growth is due to expansion of the cortex.

The human brain makes up only about 2% of our total adult body weight, but it is exceedingly expensive to build and maintain, consuming disproportionately more of our metabolic budget, even when at rest. The large number of genes that encode the features of the brain are vulnerable to accidents; mutations of any of them are likely to lead to behaviour disorders. Despite all these evolutionary costs, we nonetheless evolved a large brain, so presumably the benefits were even greater.

Any change that evolves in an organ is assumed to confer an improvement in the fitness of individuals: an increase in the likelihood that the genes they carry will appear in subsequent generations. A rapid change, as in the increase in size of the hominin brain, implies a strong fitness advantage. We've already seen that differences in brain size across species may relate to one key ecological problem:

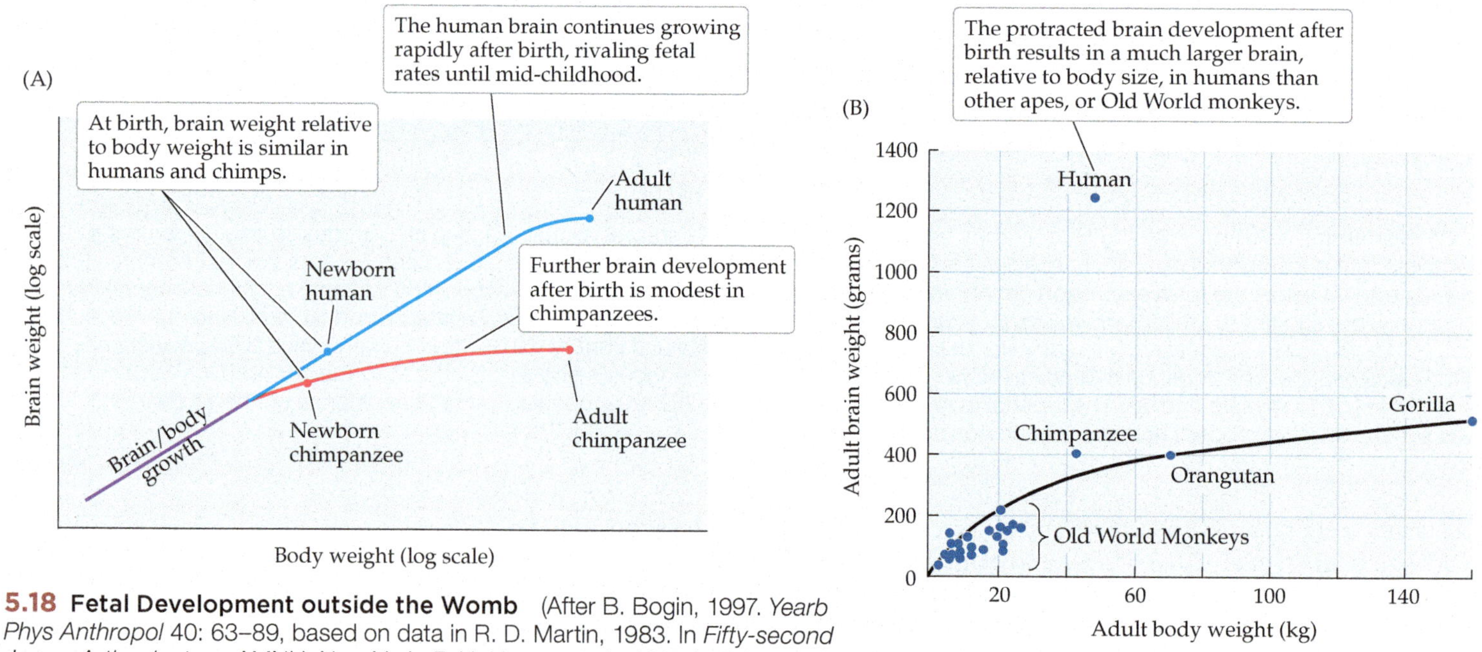

5.18 Fetal Development outside the Womb (After B. Bogin, 1997. *Yearb Phys Anthropol* 40: 63–89, based on data in R. D. Martin, 1983. In *Fifty-second James Arthur lecture*. AMNH: New York; P. H. Harvey et al., 1987. In *Primate societies*, B. B. Smuts et al. [Eds.], pp. 181–196. University of Chicago Press: Chicago.)

obtaining and consuming suitable food. An alternative perspective on the positive selection pressures driving expansion of the hominin brain emphasizes skill development. A major study correlated brain size in 116 species of primates with three different factors that have been proposed to account for the enhanced size of primate brains: (1) innovations in behaviour, (2) use of tools, and (3) social learning—that is, learning by observing others (Reader and Laland, 2002). Rather than testing animals for these traits or observing them directly, the scholars surveyed about 1000 articles in primate journals and other relevant literature for instances of innovation, tool use, and social learning (**FIGURE 5.19**). In addition to using total brain weights, they used the ratio of what has been called the *executive brain* (the forebrain) to the brainstem. Both total brain size and relative forebrain size correlated positively with the frequency of each of the three behavioural indices, indicating that each is related to expansion of the primate brain. Thus, the results indicate that these multiple, interrelated sources of selection favored evolution of the large primate brain. Similar relationships between forebrain size and innovation may apply to birds as well, because members of the crow family have been observed to use tools (C. D. Bird and Emery, 2009), and they have relatively larger forebrains than other birds have (Cnotka et al., 2008).

5.19 Transmitting Culture Culture has been observed in nonhuman primates. For example, a population of Japanese macaques developed a set of behaviours that included washing food, playing in water, and eating marine food items, and they transmitted this culture of water-related behaviours from generation to generation.

Other researchers have argued that the recent expansion of primate brains may instead reflect specific selection pressures in the social domain. The *social brain hypothesis*, for example, suggests that a larger cortex is needed to handle the complex cognitive task of maintaining social relationships with other large-brained individuals (Dunbar, 2009; Dunbar and Shultz, 2017). Based on correlations between social group size and brain size across primate species, researchers have estimated the maximum size of various species' social groups from their cortex sizes (Dunbar, 2009). For humans, this value—an average social circle of about 150, predicted from the size of our cortex—may indicate the maximum number of individuals with whom we can have a meaningful social relationship. Perhaps our brains are too small to let us really keep tabs on more than about 150 "friends" on Facebook or Instagram. (If you haven't reached your limit yet, you can "like" us at facebook.com/Behaviouralneuroscience to learn about new developments in behavioural neuroscience.) However, contradictory findings from large-scale research across many primate species indicate that primate brain size today may correlate better with ecological factors like diet than with social factors (DeCasien et al., 2017), highlighting the difficulty of inferring the complex cognitive factors driving differences in brain expansion between species in the ancient past (Dunbar and Shultz, 2017; L. E. Powell et al., 2017).

Brain size predicts success in adapting to a novel environment

We have seen that brain size correlates with the complexity of feeding behaviour, and with social group size in primates, but is it possible to test the hypothesis that enlarged brains evolved as a more general adaptation to allow us to cope with sudden changes in environmental conditions? One way would be to introduce species with relatively large or small brains into novel environments and see whether they were able to establish themselves and thrive. However, experimental introductions of species into novel environments would be unethical because of the unpredictable ecological problems that often occur when nonnative species are released into a new area. An alternative is to study the rich record of past human-mediated introductions. A comprehensive study that used a global database to examine more than 5.0 examples of introduction involving 195 bird species found that the species with larger brains, relative to body size, tended to be more successful in establishing themselves in novel environments (Sol et al., 2005).

Sexual selection may have contributed to hominin brain expansion

Using a different approach, we can evaluate the rapid expansion of the human brain over the last 1.5 million years in terms of Darwin's second evolutionary principle: sexual selection. Based on a reanalysis of the fossil record, Geoffrey F. Miller (2000) suggested that natural selection to obtain food and shelter is not likely to account completely for the large brain and complex intelligence of *Homo sapiens*. For example, he notes, brain size tripled in our ancestors between 2.5 million years ago and 200,000 years ago, yet during this period our ancestors continued to make the same kind of stone ax (see Figure 5.18). Only well *after* the human brain stopped expanding did technological progress develop, so brain growth did not correlate well with this supposed fitness benefit of enlarged brains.

Rather, Miller proposes, an additional factor accounts for large human brains: in humans, he postulates, much creativity, along with related brain growth, is due to sexual selection for abilities to attract attention, stimulate, and surprise a potential mate. This hypothesis, Miller claims, has the further value of presenting an evolutionary theory for such characteristic human traits as humor, art, music, language, and creativity (**BOX 5.3**).

The hypothesis that sexual selection for artistry and creativity may lead to increased brain size is supported by findings regarding the bowerbird family (Ptilonorhynchidae). To attract and impress females, male bowerbirds construct elaborate structures of twigs, decorated with colourful objects such as shiny beetles, shells, and petals (**FIGURE 5.20**). Bowerbirds have large brains, compared with

BOX 5.3 Evolutionary Psychology

This brightly coloured peacock must impress the plain-coloured peahen before she will accept his sperm.

Zoologists have long viewed the behaviour of animals through an evolutionary prism: just like any other adaptation, the specific behaviours of animals can be seen as a set of adaptations that evolved in response to specific pressures within the ecological niches of the species, such as the need to find food, attract mates, and avoid predators. Application of this perspective to the study of behaviour—especially the behaviour of humans—is the focus of a lively and controversial field called **evolutionary psychology** (Buss, 2019; Tooby and Cosmides, 2015). Applying evolutionary principles to the human mind, researchers in evolutionary psychology view the mind as a large collection of cognitive "modules," each shaped through natural selection to solve a specific adaptive problem that confronted our distant hominin ancestors. This makes for a rich source of theories about how evolution might have shaped our behaviour—the hard part is to come up with experiments that put these ideas to the test, because they depend on conditions unique to a long-gone ancestral environment. Ethical considerations prevent direct experimentation in humans on many of the variables of interest, so studies in evolutionary psychology tend to rely more often on indirect surveys and correlational designs (see Chapter 0) (Confer et al., 2010).

There's no doubt that humans looking for a mate find some traits more attractive than others, and just as sexual selection pressures have generated sex differences in size and ornamentation in many species (as in the figure), sexual selection has also shaped behaviours such as vocalisations, territoriality, courtship behaviours, and so on. This leads inevitably to questions about the extent to which our own behaviours have been affected by the difference

5.20 A Bowerbird Bower To attract mates, male bowerbirds build elaborate bowers of twigs, such as this structure, and decorate them with colourful objects. The architectural complexity and ornate decoration of the bowers may be the reason for the relatively large brains of bowerbirds. If a human moves one item while no birds are around, the male bowerbird will promptly put it back in place when he returns.

other birds, and within the bowerbirds, species that build more-elaborate bowers have relatively larger brains (Madden, 2001).

Primate species differ in gene expression

At the start of this chapter, we asked why humans and chimpanzees, which are identical in almost 99% of their genomic DNA sequences, differ in many morphological, behavioural, and cognitive aspects. It is important to understand that humans and their closest relatives can differ on a genetic basis in two principle ways: (1) the DNA *sequences* of specific genes may vary in important ways between the species, and (2) how those genes are *expressed* to

between male and female reproductive strategies. For example: *Are women inherently more selective than men about choosing mating partners?* Such a difference could be a result of the tremendous investment of time, energy, and resources that a female mammal must make in each offspring. Conversely, men might be more promiscuous than women because of the low cost of producing sperm, zero gestational costs, and potential for a man to expend *no* energy on child rearing. *Is there an "ideal" waist-to-hip ratio that indicates maximal fertility in women?* Correlational studies suggest that a particular ratio is especially attractive to men, across different cultures and historical eras (Singh, 2002). *Are women attracted to power, and men to youth, because natural selection favored these preferences?* Such mate preferences are hypothesised to confer specific advantages for producing numerous strong offspring (youth) or for providing resources and protection to offspring (power).

Sexual selection may have been crucial for evolution of the human brain. If early hominins had come to favor mates who sang, made jokes, or produced artistic works, such high-order functioning would have evolved rapidly in an "arms race," as the ever more discriminating brains of one sex demanded ever more impressive performances from the brains of the other sex and, reciprocally, may have created pressures that favored mate guarding and sensitivity to infidelity (Andrews et al., 2008; G. F. Miller, 2000). Did humor, song, and art originate from the drive to be sexually attractive? And does sexual selection account for the large size of the human brain?

Of course, other aspects of natural selection also shape human behaviour. For example, the particular forms of learning and memory that we exhibit are presumed to reflect our evolutionary history. Our most striking adaptation is the use of language to communicate concepts between individuals and across generations, so it is perhaps no surprise that one of the greatest feats of human memory is the seemingly effortless learning of language in childhood. Our memory for the symbols and grammar of language is *domain-specific*; that is, we possess evolved mental mechanisms that seem specialised for learning to talk, and those mechanisms cannot easily be pressed into the service of behaviours in other domains. Researchers similarly seek adaptation-centred explanations for behaviours as diverse as altruism, religion, sensory perception, emotional expression, innovation, and cultural behaviours (Muthukrishna and Henrich, 2016; Norenzayan et al., 2016). Further, researchers are now rethinking the notion that cognitive modules are *exclusively* highly specialised, and they are broadening the evolutionary view of psychology to include the evolution of at least a few *domain-general* mechanisms, such as memory processes that can function as general-purpose "scratch pads," thus allowing flexible responses to unpredictable circumstances (Heyes, 2012).

evolutionary psychology A research field that asks how evolutionary selection pressures have shaped the behaviour of humans and other animals.

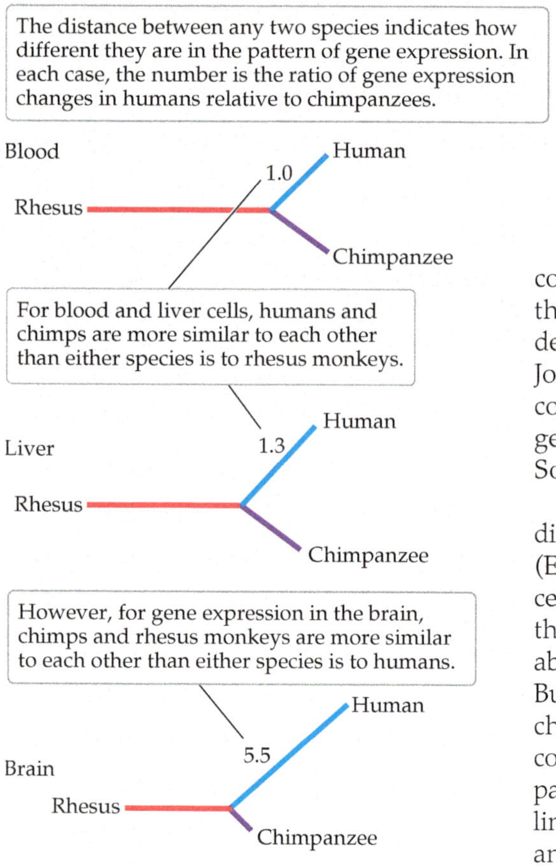

5.21 Genetic Similarities of Primate Species Vary across Different Tissues Although the chimpanzee and human genomes are 98.7% identical, the *expression* of these genes differs between species and between organ systems, which may help explain some of the striking differences between the two species. In particular, there has been a 5.5-fold increase in the rate of change in expression levels of genes in the brains of humans, compared with chimpanzees. (After W. Enard et al., 2002. *Science* 296: 340–343.)

construct a complex brain may also vary. For example, the gene *ASPM* influences the size of the cerebral cortex; interruptions in the function of *ASPM* result in the development of abnormally tiny brains (*microcephaly*) and mental disability (M. B. Johnson et al., 2018; Passemard et al., 2016). The protein encoded by *ASPM* differs considerably between humans and chimps, suggesting that the sequence of this gene evolved rapidly in the ancestral line leading to humans (Evans et al., 2004). So, even moderate changes in just a few crucial genes may make a big difference.

In addition to differences in the structure of genes, researchers report that humans differ considerably from other primates in patterns of gene *expression* in the brain (Enard et al., 2002; Somel et al., 2013). When gene expression in blood cells and liver cells is compared, humans and chimpanzees are more similar to each other than either species is to rhesus monkeys, as **FIGURE 5.21** shows. These relationships probably reflect the well-documented evolutionary relationships among the three species. But when gene expression is studied in the *brain,* we humans are so different from chimpanzees that, mathematically speaking, their expression pattern has more in common with that of the monkeys than with us. These observations suggest that the pattern of gene expression in the brain has changed and accelerated in the human lineage, presumably under selection pressure, since the time we shared a common ancestor with chimpanzees. No such acceleration is evident in the liver or blood.

Even a small change in gene expression can cause a dramatic difference in brain development (Chenn and Walsh, 2002). For example, overexpression of just one gene in mice causes so much more growth in the cortex that the normally smooth surface develops gyri and sulci (**FIGURE 5.22**). So, to answer the question we raised at the start of this chapter, the *pattern* of gene expression has a tremendous effect on the developing brain and may be what makes humans so different from chimps, despite our nearly 99% identical genomes.

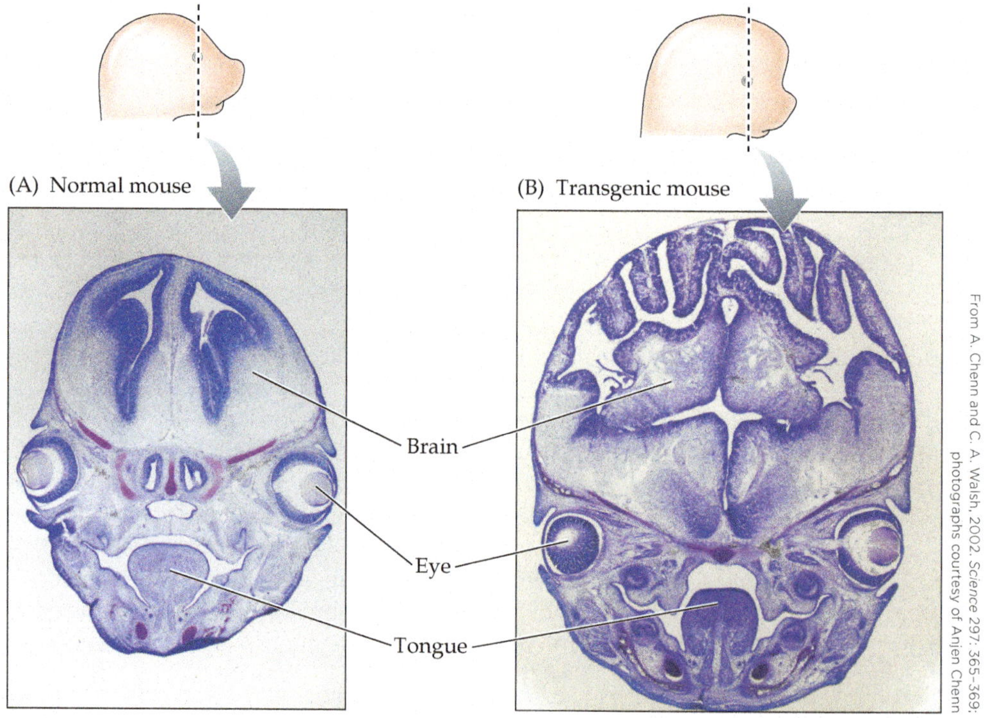

5.22 Over Your Head Normally, mice at birth have a fairly simple cortex with no sulci or gyri (A), but increasing the expression of just one gene (that for β-catenin) in transgenic mice results in a monstrously complex, highly folded cortex (B). These animals die shortly after birth.

5.6 Evolution Continues Today

Learning Objectives

After reading this section, you should be able to:

5.6.1 Discuss processes of rapid and recent evolution, with examples.

5.6.2 Evaluate evidence that humans continue to evolve, despite recent cultural and technological developments.

5.6.3 Explain how recent changes in the human genome may be detected.

Although some examples of evolution have occurred slowly—over millions of years, as shown by the aspects of hominin evolution illustrated in Figure 5.18—other examples occur in a matter of years or decades. These rapid instances of evolution would have surprised Darwin, who thought that natural selection required vast periods of time to be effective. In some cases, selection is driven by human behaviour; this is sometimes called evolution by *artificial* selection. One obvious example is the selective breeding that farmers have employed for millennia: by deciding which individuals will reproduce, the farmer can encourage specific traits to manifest in the offspring (giving us oranges without seeds, for example, and the small and companionable wolves that we call *dogs*). Humans can also induce evolution accidentally; for example, inappropriate use of antibiotic medicines eliminates all but a very few resistant bacteria, leaving the survivors to reproduce and rapidly spread antibiotic resistance through subsequent generations (Andersson et al., 2019).

Another striking example of human-generated evolution is the case of bighorn rams in the Rocky Mountains of Alberta, Canada. The massive, curling horns of these rams make them a prized hunting trophy (**FIGURE 5.23**). Under strictly enforced laws, only mature rams with horns reaching an almost 360° curl may be hunted. In some areas of Alberta, most rams are shot within a year or two after reaching this status. As a result, selection has worked in favor of rams whose horns never reach trophy status. In fact, the average horn size has dropped by about 25% over the past 30 years and has shown only minor rebound now that intensive hunting of this species has ceased (Coltman et al., 2003; Pigeon et al., 2016).

© iStock.com / Paul Tessier

5.23 Bigger Is No Longer Better Because human hunters selectively shoot rams with large horns, the population of bighorn rams has changed. Males these days have smaller horns than males just a few decades ago, providing a clear example of evolution in action.

Darwin's finches continue to evolve in the Galápagos Islands. In 1973, the medium ground finch (*Geospiza fortis*) predominated on one small island, eating both large and small seeds. Then, in 1982, a breeding population of the large ground finch (*G. magnirostris*) became established on that island. These larger birds eat large seeds more easily than the medium ground finches can, even those with relatively large beaks. For two decades there were not enough individuals of the larger species to make much difference. But then in 2004, a drought sharply reduced the food supply. Competition for larger seeds became severe, and many of the medium ground finches with larger beaks died off. The smaller-beaked medium ground finches, however, survived and passed along the small-beaked trait to their offspring (Grant and Grant, 2006).

Economically important cases of recent evolution concern commercial fishing. In many regions, fishermen are allowed to keep only fish that are larger than a particular size. Atlantic cod off the coast of Newfoundland began maturing at smaller sizes over several decades, probably because the largest fish were the ones being captured (E. M. Olsen et al., 2004; Zimmerman and Jørgensen, 2017). The remaining small fish produce fewer eggs than large fish, and this reduced egg production could help explain the collapse in the cod population. Similar fishing-induced evolutionary changes have been described in various commercial species, such as the whitefish of the Great Lakes (Morbey and Mema, 2018).

New developments in radiocarbon dating have shown that the colonisation of Europe and Asia by *Homo sapiens* was more rapid than previously believed (Mellars, 2006), occurring less than 50,000 years ago

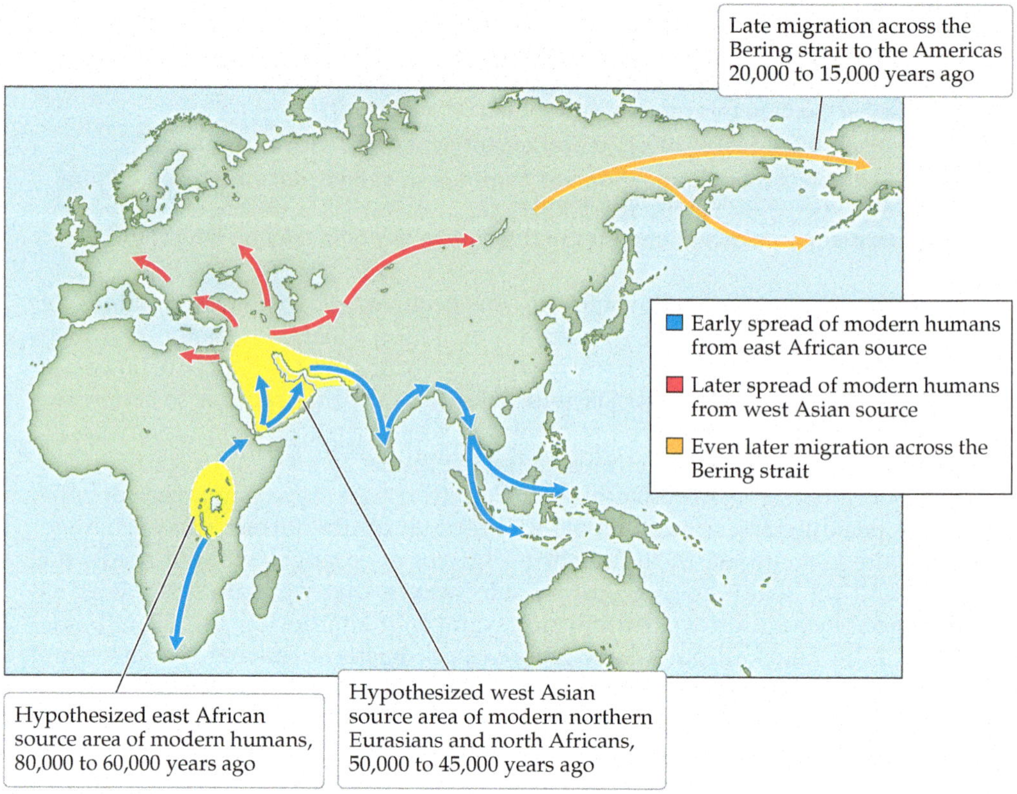

5.24 The Colonisation of Europe and Asia by *Homo sapiens* (After T. Goebel, 2007. *Science* 315: 194–196.)

(**FIGURE 5.24**). Thus, the differences in skin colour, stature, and facial traits that characterize Asian, African, and European populations evolved in less than 50,000 years in response to different climatic conditions. Human variation, then, is a reminder of both how quickly evolutionary processes can work and how very closely related we all are.

The Cutting Edge

Are Humans Still Evolving?

When we study the evolution of the brain and behaviour, we are looking backward through time. The human brain is the product of selection forces encountered over an extended period of time in an ancestral environment thousands of years ago. You can use your brain and muscles to drive a car, but they were optimised by natural selection for life in the African bush. Each of us is doing modern things with an ancient brain, because there hasn't been enough time for evolution to change it very much.

So what does the future hold for human evolution? Because big brains have already proven advantageous, science fiction writers sometimes imagine our distant descendants as huge-brained superhumans—but that won't happen as long as the route into the world is through Mother's narrow bipedal pelvis. And anyway, there is a more immediate issue. We are living in a world in which cultural and medical developments have radically altered two main ingredients of evolution: life-span and fertility. Historical data suggest a role of natural and sexual selection in our very recent past (Courtiol et al., 2012; M. L. Wilson et al., 2017), but medical science has now largely replaced natural selection in determining who dies and who gets to reproduce. So if rates of reproduction no longer reflect the genetic fitness of the parents, and all the beneficial mutations they have accumulated, are human beings evolving at all?

Many writers argue that recent changes in human brains and behaviour are due to cultural evolution, the ability of humans to pass hard-won knowledge along to the next generation. Are humans also going through genetic changes as a result of recent

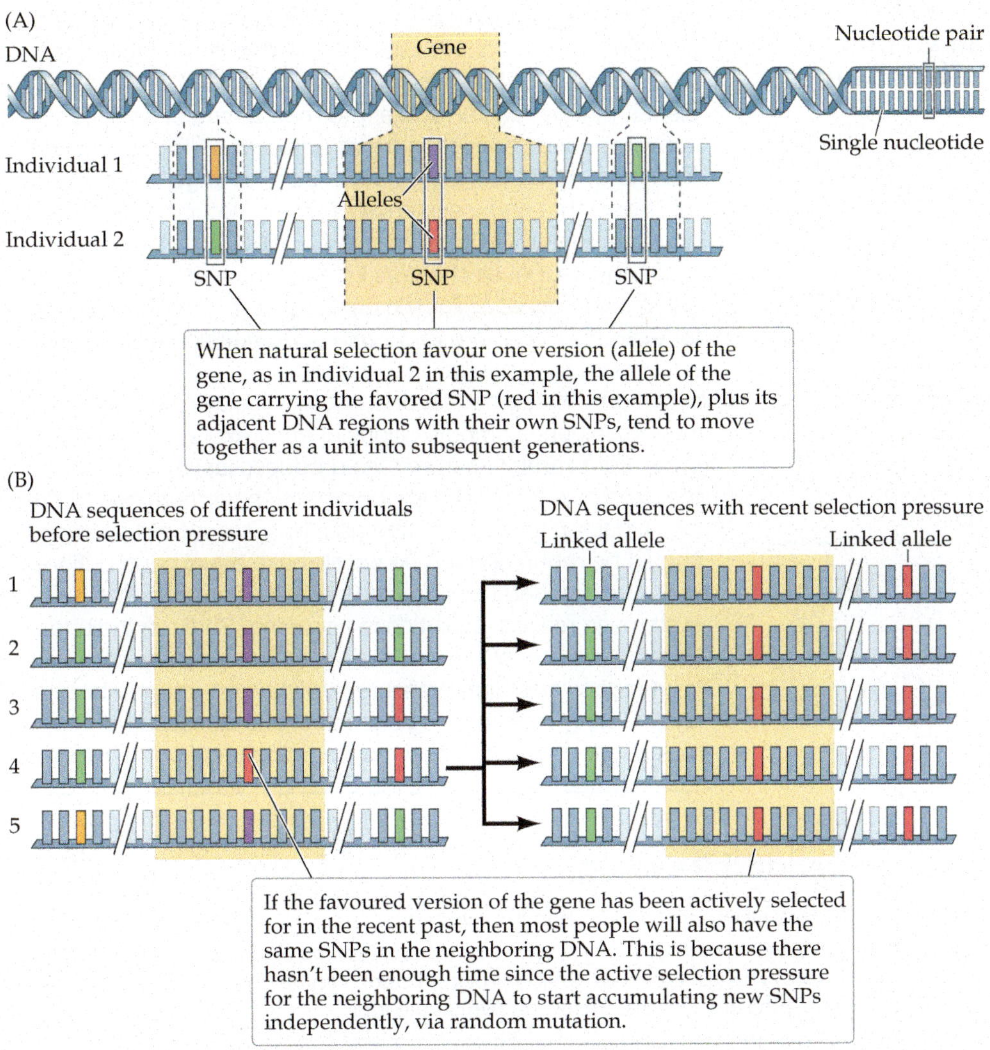

5.25 SNP Variation and Natural Selection (A) Stretches of DNA that contain genes are bracketed by noncoding regions (lengths of DNA that are not part of a gene) that tend to be passed to offspring along with the gene. If the gene contains an allele that enhances reproductive success (as in Individual 2), that version of the gene (plus the bracketing DNA) will be favoured in subsequent generations. (B) In the absence of selection pressure, DNA from different individuals shows a random mix of SNP alleles in the gene and surrounding DNA (left). But recent active selection pressure that favors a specific allele of the gene results in the pattern at right: the SNPs in the gene and its surrounding regions are invariant because they have recently spread from an originating individual (in this case, Individual 2 in part A).

natural selection? The answer, now that we know how to sequence the genome, appears to be a qualified yes (Okada et al., 2018; Stearns et al., 2010).

Recall that a gene is a stretch of DNA, made up of nucleotides, that encodes a particular protein (see the Appendix). To determine whether a candidate gene has been under strong natural selection pressure in the recent past, scientists look at the gene bundled together with its neighbouring stretches of DNA; collectively, this is known as an individual's *haplotype*. As shown in **FIGURE 5.25A**, minor variations where one nucleotide substitutes for another can occur at certain locations within a gene (or neighbouring stretch of DNA); these are called **single-nucleotide polymorphisms (SNPs)** (*polymorphism* means "many shapes"). Most SNPs (not all) have just two different versions, or **alleles**. Because SNPs result in slightly differing versions of genes, they cause variation within traits: eye colour variants, susceptibility to disease, levels of enzyme function, and so on.

If a mutation in a gene confers a slight reproductive advantage on those who possess it, it will come under natural selection pressure—it will be "selected for" as we described earlier in the chapter—and will gradually spread throughout subsequent generations, until most people possess the same allele. But simply sequencing a gene in a sample of people doesn't reveal whether the gene has been subject to *recent* evolutionary pressure. That additional information comes from looking at SNPs in the DNA that brackets the gene under study. Because these neighbouring stretches of DNA tend to "tag along" with genes during the recombination processes of reproduction,

single-nucleotide polymorphism (SNP) Variation in a single base pair within a longer DNA sequence, such as a gene

allele Any particular version of a gene.

TABLE 5.1 Examples of Human Genes Subject to Recent Selection Pressure

Gene	Benefit or selection pressure
AGT, CYP3A	Protection against hypertension
ASPM, NPAS3	Brain development
DRD4, MAOA	Cognition and behaviour
FOXP2	Language
TAS2R38	Bitter taste perception
CCR5	Protection against smallpox and AIDS
G6PD, Duffy blood group, *HBC* (hemoglobin C), *TNFSF5*	Protection against malaria
Lactase	Improved nutrition from milk

geneticists can ask whether the adjacent SNPs show independent variation or whether they appear to be locked to the SNP allele in the gene being studied (**FIGURE 5.25B**). Genes that have been subject to recent natural selection show reduced variation in these adjacent SNPs, as shown in the figure. **TABLE 5.1** provides a sample of genes (and their functions) that show evidence of recent natural selection in humans (Balter, 2005; Kamm et al., 2013).

Of course, not all changes in recent humans are evolutionary, or even genetic. The increased stature of modern humans, for example, has more to do with good nutrition and medical advances than changes in any genes—another form of cultural evolution. And some responses to environmental pressures involve epigenetic modifications of gene expression (see Figure 6.20). For example, early stress produces lasting changes in the expression of genes encoding stress hormone receptors (see Chapter 11). Perhaps the question, Are we still evolving? should be rephrased as, Do we all make equal contributions to the next generations? As long as we don't, and there is a systematic reason, evolution is occurring. Sadly, in the underdeveloped parts of the world, disease and poverty still exert tremendous selection pressure, and new environmental challenges like global warming and infectious diseases like HIV, malaria, and Zika virus may continue to mould our genome. ■

Recommended Reading

Understanding Evolution, http://evolution.berkeley.edu.

Darwin, C. R. (1859). *On the Origin of Species by Means of Natural Selection.* London: John Murray. (Note: This classic book, long out of copyright, can be obtained as a free e-book from various internet sources, including Amazon.com.)

De Waal, F. (2016). *Are We Smart Enough to Know How Smart Animals Are?* New York: Norton.

Dunbar, R. (2016). *Human Evolution: Our Brains and Behaviour.* New York: Oxford University Press.

Futuyma, D. J., and Kirkpatrick, M. (2017). *Evolution* (4th ed.). Sunderland, MA: Oxford University Press/Sinauer.

Gray, P. B., and Garcia, J. R. (2013). *Evolution and Human Sexual Behaviour.* Cambridge, MA: Harvard University Press.

Shubin, N. J. (2009). *Your Inner Fish: A Journey into the 3.5-Billion Year History of the Human Body.* New York: Vintage Books.

5 Visual Summary

You should be able to relate each summary to the adjacent illustration, including structures and processes. You can also refer back to the links to figures, animations, and activities that will help you consolidate the material.

1 Darwin's theory of **evolution by natural selection** posits that individuals with adaptive traits produce more offspring, so species evolve over time. This process of natural selection favors newly arisen **genes** (**mutations**) that confer adaptive traits, including behavioural traits. By these gradual changes, all animal species arose from a common ancestor. Review **Figure 5.1**

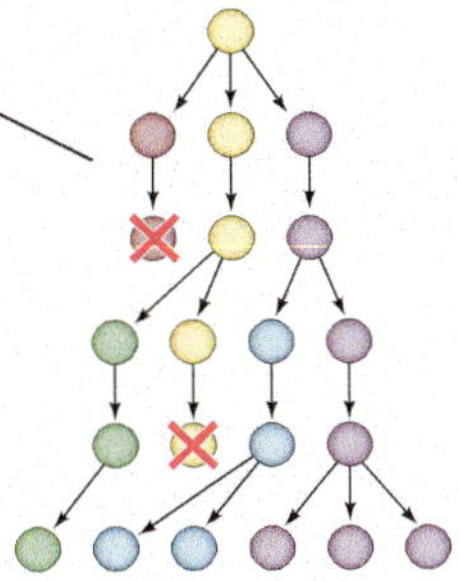

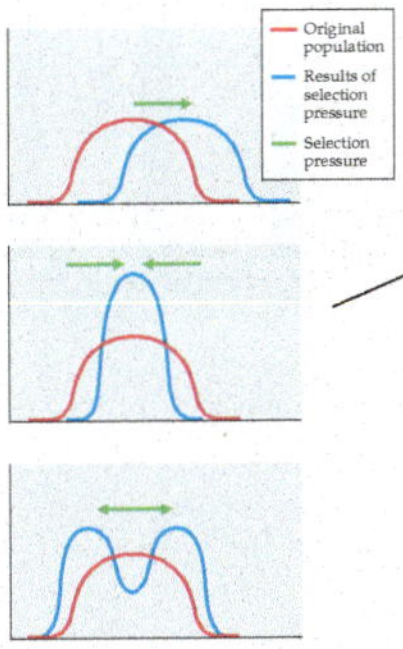

2 Selection pressures may act to shift, stabilise, or split the expression of a trait in a population. Review **Figure 5.2**

3 Some species feature similar structures and functions due to shared ancestry; in other cases the similarity is the result of **convergent evolution**. Review **Figure 5.3**

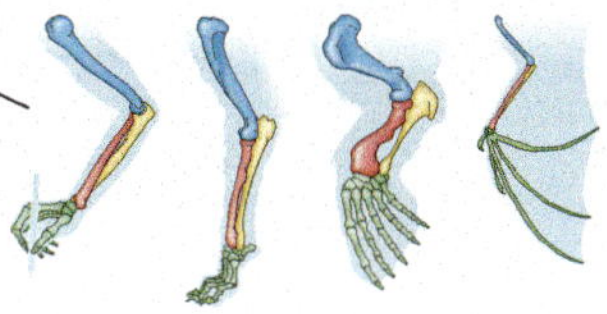

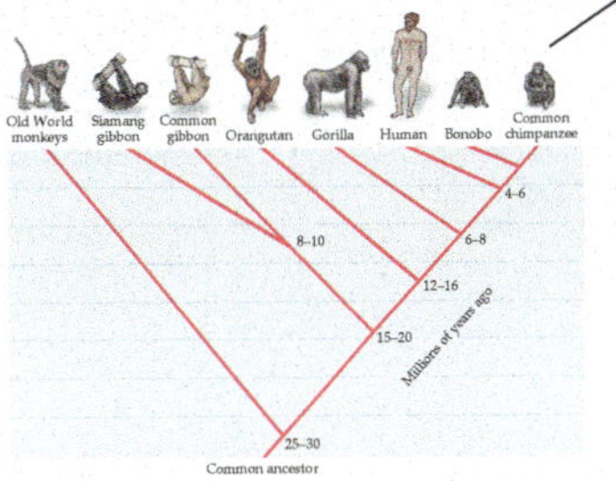

4 Studies of the classification of animals help determine how closely related different **species** are. Knowing this relationship, in turn, helps us interpret similarities and differences in behaviour and structure among different species. Review **Figures 5.4**, **5.5**, **Activity 5.1**

5 Comparative studies help us understand the evolution of the nervous system, including the human brain. They also provide a perspective for understanding species-typical behavioural **adaptations**. Review **Box 5.2**

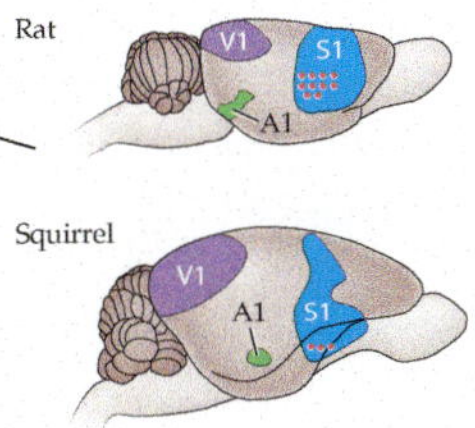

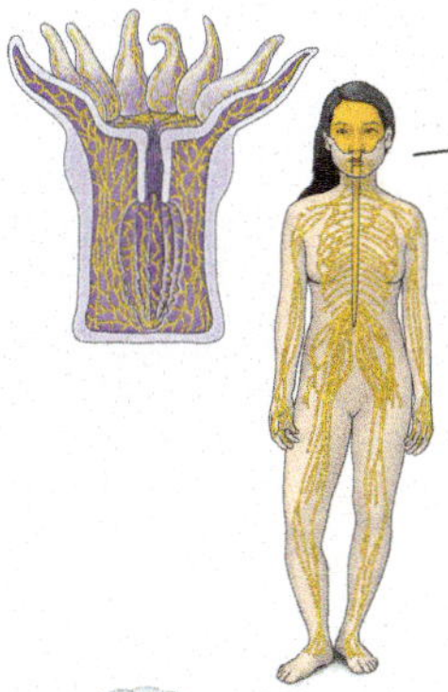

6 The nervous systems of invertebrates range in complexity from a simple nerve net to the complex structures of mollusks. The nervous systems of certain invertebrates may provide a simplified model for understanding some aspects of vertebrate nervous systems. Review **Figure 5.8**, **Activity 5.2**

7 Differences in the size and organisation of specific brain regions are sometimes related to distinctive forms of behavioural adaptation in different species. Review **Figure 5.9**

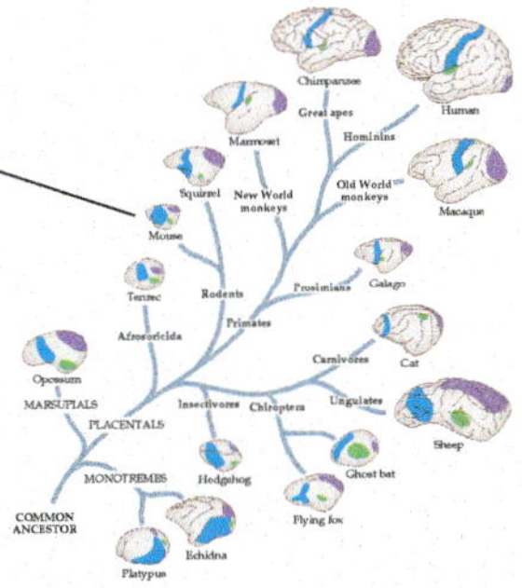

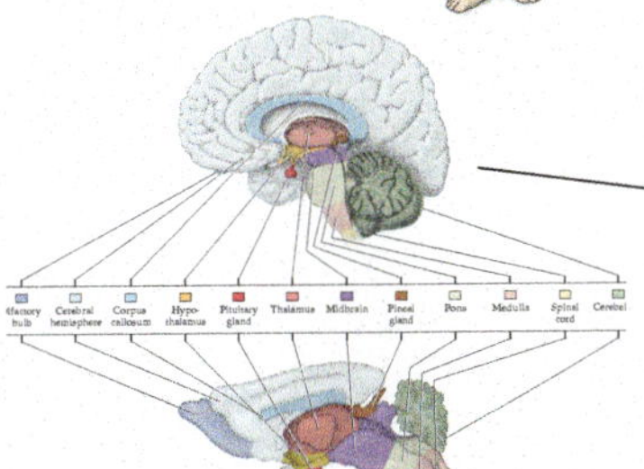

8 The main divisions of the brain are the same in all vertebrates. Differences among these animals are largely quantitative, as reflected in the relative sizes of nerve cells and brain regions and in the amount of dendritic branching in neurons. Review **Figures 5.9**, **5.10**

9 Fossil **endocasts** of brains from extinct species indicate that the main result of mammalian evolution has been larger overall brain size. The brain size of a species must be interpreted in terms of body size. As a rough rule of thumb, vertebrate brain weight is proportional to body weight to the power of 0.69. Review **Figures 5.13**, **5.14**

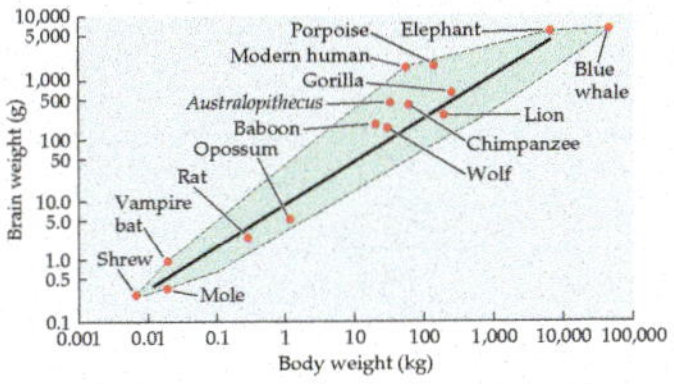

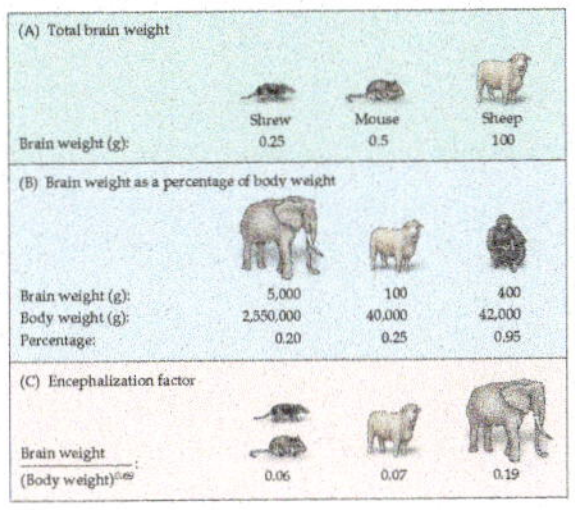

10 Some animals have larger brains and some have smaller brains than the general relation between brain and body weights predicts; that is, they differ in **encephalisation factor**. Large deviations from predicted encephalisation are presumed to reflect unusual selection pressures. Review **Figures 5.13**, **5.14**

(Continued)

11 Primates have an especially large **cortex** relative to overall brain size. This relative enlargement of the cortex appears to have evolved because the later stages of brain development are prolonged, resulting in a disproportionally large cortex. Review **Figures 5.15**, **5.16**

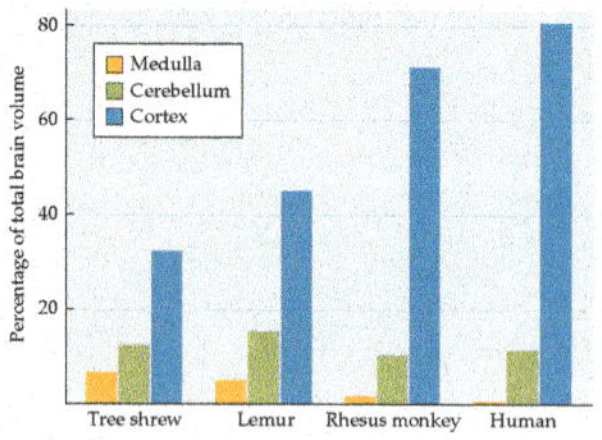

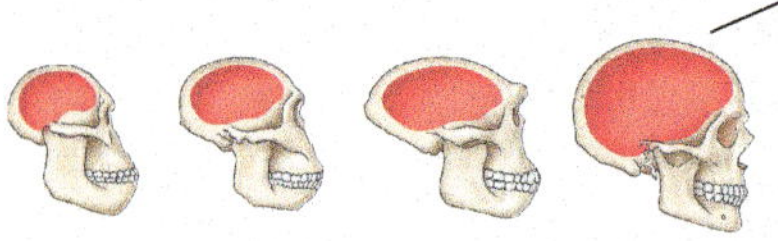

12 Several factors, including tool use, innovation, and social relationships, are thought to have driven enlargement of the primate cortex. Review **Figures 5.17**, **5.18**

13 Humans more closely resemble their nearest relatives, the chimpanzees, in the blood and liver than in the brain. Review **Figure 5.21**

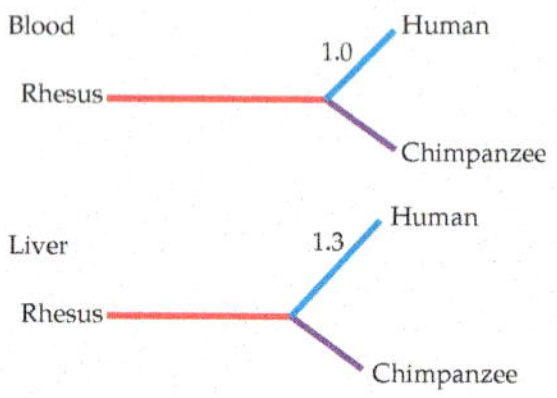

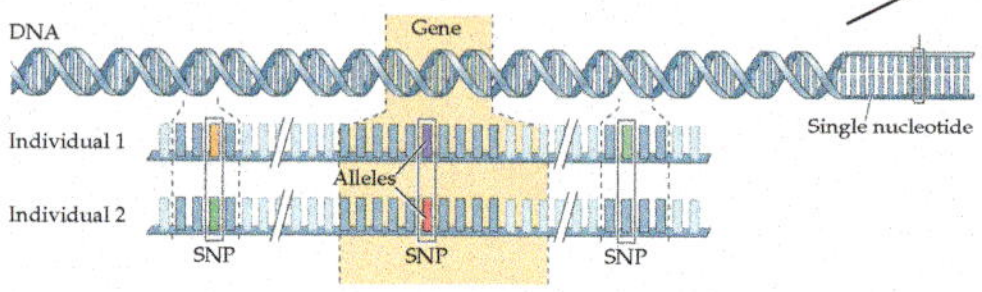

14 Analysis of patterns of reproduction as well as genetic studies of the frequencies of **single-nucleotide polymorphisms (SNPs)** in specific chromosomal loci indicate that, while cultural influences have an important impact, evolution through natural selection continues in humans to this day. Review **Figure 5.25**, **Table 5.1**

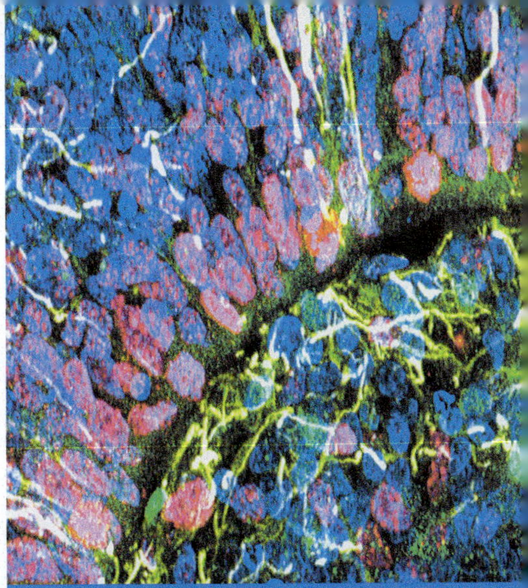

6

Life-Span Development of the Brain and Behaviour

Overcoming Blindness

As a 3-year-old, Michael May was injured by a chemical explosion that destroyed his left eye and damaged the surface of his right eye so badly that he was blind. He could tell whether it was day or night, but otherwise he couldn't see anything. An early attempt to restore his sight with corneal transplants failed, but Mike seemed undaunted. He learned to play Ping-Pong using his hearing alone (but only on the table at his parents' house, where he learned to interpret the sound cues). Mike also enjoyed riding a bicycle, but his parents made him stop after he crashed first his brother's bike and then his sister's.

As an adult, Mike became a champion skier, marrying his instructor and raising two sons. He also started his own company, making equipment to help blind people navigate on their own. Then, when Mike was 46, technical advances made it possible to restore vision in his right eye. As soon as the bandages were removed, he could see his wife's blue eyes and blond hair. But even years later, he could not recognize her face unless she spoke to him, or recognize three-dimensional objects like cubes or spheres unless they were moving. Mike could still ski, but he found that he had to close his eyes to avoid falling over. On the slopes, seeing was more distracting than helpful.

The doctors could tell that images were focussing properly on Mike's retina, so why was his vision so poor?

In this chapter we describe brains in terms of their progress through life from the womb to the tomb. Picture, if you can, the number of neurons in the mature human brain—over 80 billion (Herculano-Houzel, 2012). There are many types of neurons, and most are connected to many other neurons, forming vast networks. Indeed, there are an estimated quadrillion synaptic connections in the brain. Yet each of us began as a single microscopic cell, the fertilized egg. How can one cell divide and grow to form one of the most complicated machines on Earth, perhaps in the universe?

We'll see that the emergence and migration of nerve cells, their initial connections and even whether they live or die, are crucially dependent on genes early in development. But as development proceeds, experience gained through behavioural interactions with the environment decides which synapses wither and which persist, meaning both genes and the environment guide the construction of this fabulous gadget between our ears.

6.1 Neural Development Is Guided by the Interaction of Genetic and Environmental Factors

Learning Objectives

After reading this section, you should be able to:

6.1.1 **Describe the early stages of neural development, including the first three brain subdivisions.**

6.1.2 **Understand why genotype is fixed but phenotype constantly changes.**

6.1.3 **Explain how gene expression determines a developing cell's fate, including whether it will be a neuron.**

6.1.4 **Contrast control of cellular differentiation in C. elegans versus that in other animals.**

6.1.5 **Describe how several environmental factors can lead to intellectual disability.**

The fertilized egg, or **zygote**, has 46 chromosomes—23 from each parent—which contain genetic recipes for the development of a new individual. (A summary of the life cycle of cells, including a discussion of the basic genetic materials and how they direct cell activities, is provided in the Appendix.) Within 12 hours after fertilisation, the single cell begins dividing, and after 3 days it has become a small mass of homogeneous cells, like a cluster of grapes, a mere 200 micrometers (μm) in diameter.

Within a week the emerging embryo shows three distinct cell layers (**FIGURE 6.1A**) that are the beginnings of all tissues. The outer layer, called the **ectoderm** (from the Greek *ektos*, "out," and *derma*, "skin") will form the nervous system. As the ectoderm thickens, it grows into a flat neural plate. Uneven rates of cell division form the **neural groove**, which will become the midline (**FIGURE 6.1B**). The ridges of the groove join together to form the **neural tube** (**FIGURE 6.1C**), which will form the central nervous system (brain and spinal cord). The peripheral nervous system will be formed by cells from columns of ectoderm that are pinched off from each side of the neural tube, the *neural crest*. At the head end of the neural tube, three subdivisions become apparent, corresponding to the future **forebrain** (*prosencephalon*, consisting of the *telencephalon* and *diencephalon*), **midbrain** (*mesencephalon*), and **hindbrain** (*rhombencephalon*, consisting of the *metencephalon* and *myelencephalon*) (**FIGURE 6.1D**), which were discussed in Chapter 2 (see Figure 2.15). The interior of the neural tube becomes the fluid-filled cerebral ventricles of the brain, the central canal of the spinal cord, and the passages that connect them. By the end of the eighth week, the human embryo shows the rudimentary beginnings of most body organs. The rapid development of the brain is reflected in the fact that by this time the head is half the total size of the embryo. (Note that the developing human is called an **embryo** during the first 10 weeks after fertilisation; thereafter it is called a **fetus**.) **FIGURE 6.1E** shows the gross development of the human brain from prenatal weeks 10–41. While these dramatic developments are happening at the macroscopic level, even more elaborate changes are happening among the cells that make up the brain. Before we consider those changes at the cellular level, we need a framework for considering two different influences guiding development.

Neural development relies on signals that guide gene expression

Throughout this chapter, we will be discussing what is sometimes called the **nature/nurture debate**: how much of our brain development is due to the genes we inherit (nature) versus our experiences growing up (nurture). Two terms help us to understand this debate: The sum of all the genetic information that an individual has inherited is its **genotype**. The sum of all the physical characteristics that make up an individual is its **phenotype**. Your genotype, some 20,000 genes (Pennisi, 2012), was determined at the moment of fertilisation and remains the same throughout your life. But your *phenotype* changes constantly, as you grow up and grow old and even, in a tiny way, as you take each breath. In other words, phenotype is determined by the interaction of genotype and other factors, including experience. Thus, individuals who have identical genotypes do not have identical phenotypes, because they have not received identical

zygote The fertilized egg.

ectoderm The outer cellular layer of the developing embryo, giving rise to the skin and the nervous system.

neural groove In the developing embryo, the groove between the neural folds.

neural tube An embryonic structure with subdivisions that correspond to the future forebrain, midbrain, and hindbrain.

forebrain Also called *prosencephalon*. The anterior division of the brain, containing the cerebral hemispheres, the thalamus, and the hypothalamus.

midbrain Also called *mesencephalon*. The middle division of the brain.

hindbrain Also called *rhombencephalon*. The rear division of the brain, which, in the mature vertebrate, contains the cerebellum, pons, and medulla.

embryo The earliest stage in a developing animal.

fetus A developing individual after the embryo stage.

nature/nurture debate The question of how much of brain development is due to inherited genes (nature) versus experiences growing up (nurture).

genotype All the genetic information that one specific individual has inherited.

phenotype The sum of an individual's physical characteristics at one particular time.

Go to Videos 6.1–6.2
Early Nervous System Development
Brain Development

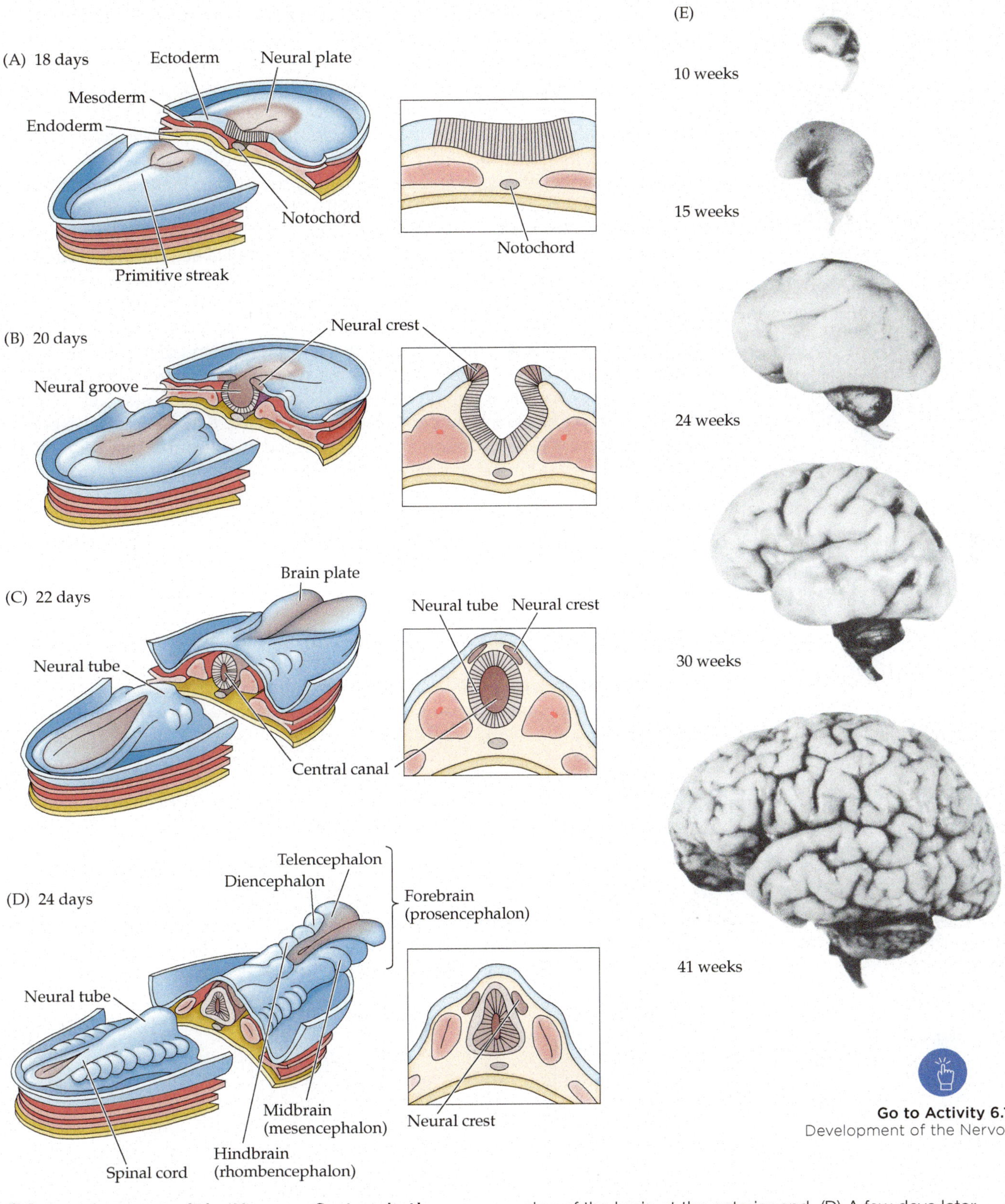

Part E from J.-C. Larroche, 1977. *Developmental pathology of the neonate.* Excerpta Medica: Amsterdam

Go to Activity 6.1
Development of the Nervous System

6.1 Development of the Nervous System in the Human Embryo and Fetus (A) At 18 days the embryo has begun to implant in the uterine wall and consists of three layers of cells: endoderm, mesoderm, and ectoderm. A thickening of the ectoderm leads to development of the neural plate (insets). (B) At 20 days the neural groove begins to develop. (C) At 22 days the neural groove has closed to form the neural tube, with the rudimentary beginning of the brain at the anterior end. (D) A few days later, three major divisions of the brain—forebrain (prosencephalon, consisting of the telencephalon and diencephalon), midbrain (mesencephalon), and hindbrain (rhombencephalon)—are discernible. (E) In these lateral views of the human brain (shown at one-third size) during fetal development, note the gradual emergence of gyri and sulci.

gene expression The process by which a cell transcribes a particular gene and makes the protein it encodes.

cell differentiation The developmental stage in which cells acquire distinctive characteristics, such as those of neurons, as the result of expressing particular genes.

cell-cell interactions The general process during development in which one cell affects the differentiation of other, usually neighbouring, cells.

experiences. And since their nervous system phenotypes are somewhat different, they will not behave exactly the same in the future.

It is easy enough to say, as many authors have, that the nature/nurture debate is settled, that in fact both factors are crucial for normal development. But to understand *why* both factors are crucial, you need to know that no matter what genes you inherit, *whether and when cells actually use those genes during neural development is controlled by environmental influences, including experience*. We say a cell expresses a gene when it actually transcribes the gene and starts making the protein it encodes. Such **gene expression** determines whether a particular cell will become a neuron or glia, a heart cell or muscle cell, and so on. This process by which a given cell expresses a particular subset of genes and so becomes a very particular type of cell is called **cell differentiation**.

To understand the control of cell differentiation, consider a microscopic worm called *Caenorhabditis elegans*, so tiny that it consists of only about a thousand cells, precisely 302 of which are neurons. Because the body of *C. elegans* is almost transparent (**FIGURE 6.2A**), scientists were able to watch the fertilized egg divide and divide again, to trace the origin of every cell (Pines, 1992). In *C. elegans*, we can exactly predict the fate of each cell in the adult—whether it will be a sensory neuron, muscle cell, skin cell, or other type of cell—on the basis of its mitotic "ancestors" (**FIGURE 6.2B**). Every cell expresses a particular subset of genes to become a particular kind of cell, and in these worms that gene expression is determined entirely by the cell's mitotic lineage.

Whereas gene expression in *C. elegans* is a highly predetermined and unvarying result of mitotic lineage, in vertebrates the paths of cellular differentiation are more complex. Unlike in *C. elegans*, gene expression is totally independent of mitotic lineage in most other species, including all vertebrates. Instead, gene expression and the eventual fate of cells is regulated locally via **cell-cell interactions**, where cells communicate with their neighbours to coordinate differentiation. Thus, development is less predetermined and more susceptible to environmental signals in most species, compared with *C. elegans*. We'll see that gene expression in neural cells depends upon environmental factors early in development and upon experience later in development. Let's consider some environmental factors that can affect brain development

(A)

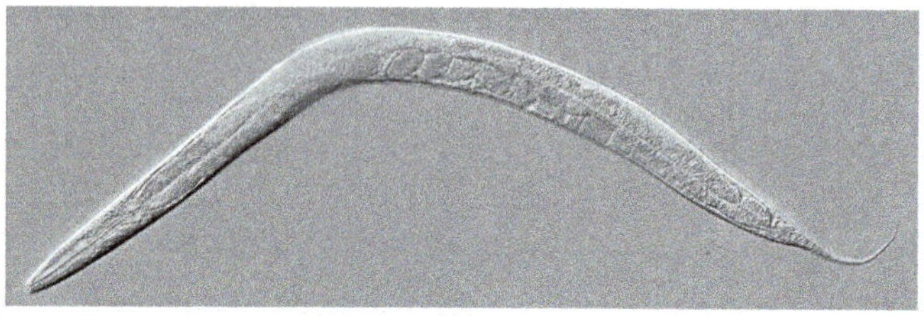

(B)

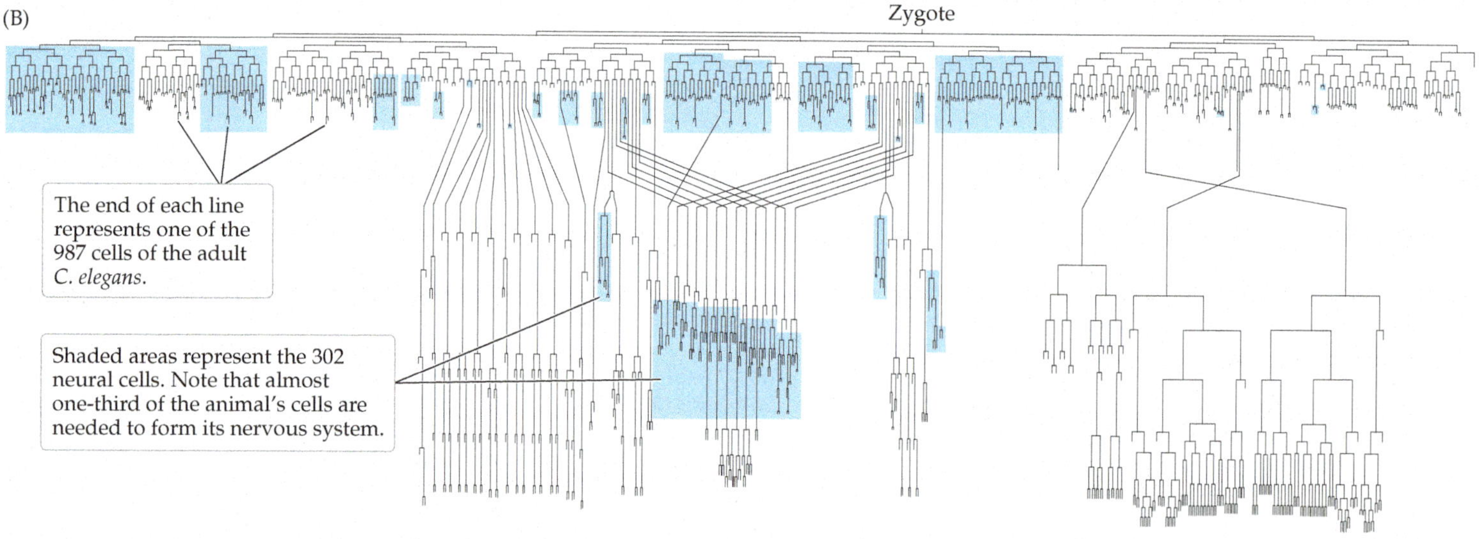

6.2 Cell Fate in a Simple Organism (A) This photomicrograph shows the transparent body of *Caenorhabditis elegans*. (B) In this mitotic lineage of cells that give rise to the body of the adult *C. elegans*, nervous system cells are highlighted in blue. The structure and function of every cell can be predicted from its mitotic lineage. Such mitotic determination of cell differentiation does not seem important to the development of vertebrates or more complex invertebrates like fruit flies. (Part B after J. E. Sulston et al., 1983. *Dev Biol* 100: 64–119; J. Pines, 1992. *Curr Opin Cell Biol* 4: 144–148.)

in unfortunate ways, no matter what genes you inherited. Environmental factors may limit brain development.

Many psychological disorders, like depression and schizophrenia, affect mostly adults. But other disorders are seen primarily in children. **Intellectual disability** refers to a variety of conditions that impede mental growth. Some advocates for the intellectually disabled regard the older term *mental retardation* as demeaning, so it has fallen from use.

Certain basic environmental factors are required for proper brain development. For example, if children experience complicated delivery at birth, when transient **hypoxia** (lack of oxygen) may affect the brain, they are at greater risk for intellectual disability than are children who have a problem-free birth. Similarly, if the mother does not get enough to eat, the fetal brain may have insufficient energy and nutrients to develop properly: Fetuses carried by malnourished Dutch women during the "hunger winter" of 1944 were underweight at birth, as would be expected, and more likely to be intellectually disabled or develop schizophrenia in adulthood (Chapter 12). A similar outcome followed a famine in China in the 1950s (A. S. Brown and Susser, 2008).

Even when nutrition is not a problem, the embryo and fetus are not immune to outside influence; what is taking place in the mother's body directly affects them. Maternal conditions such as viral infection and exposure to drugs are especially likely to result in developmental disorders in the unborn child. Concern with the maternal environment affecting development spawned the field of **teratology** (from the Greek *teras*, "monster"), the study of prenatal malformations. One powerful **teratogen**, called Accutane, is found in certain acne medications, so it's important that women who are pregnant—or could possibly become pregnant—avoid using it.

There is a long history of concern about alcohol and pregnancy. About 40% of children born to alcoholic mothers show a distinctive profile of anatomical, physiological, and behavioural impairments known as **fetal alcohol syndrome** (**FAS**) (Murawski et al., 2015). Prominent anatomical effects of fetal exposure to alcohol include distinctive changes in facial features (e.g., a sunken nasal bridge and altered shape of the nose and eyelids) and stunted growth. In severe cases, children with FAS may lack a corpus callosum (**FIGURE 6.3**). Few FAS children catch up in the years following birth. The most common problem associated with FAS is intellectual disability, which varies in severity. No threshold has yet been established for this syndrome, but it can occur with relatively moderate alcohol intake during pregnancy. Even when FAS is not diagnosed, prenatal exposure to alcohol is correlated with behavioural impairments. Thus, the CDC asserts, "there is no known safe amount of alcohol to drink while pregnant." The spreading legalisation of cannabis use has raised concerns about the possible effects of active compounds in cannabis on developing fetuses (Metz and Strickrath, 2015).

The effects of genetic disorders may depend on environmental factors

Autism spectrum disorder (**ASD**) is a developmental disorder characterized by impaired social interactions and language and a narrow range of interests and activities. As the term *spectrum* indicates, the severity of ASD varies considerably. Severe cases of autism are usually discovered when apparently normal toddlers begin regressing, losing language skills, and withdrawing from family interaction. Children with severe ASD tend to **perseverate** (such as by continually nodding the head or making stereotyped finger movements), actively avoid making eye contact with other people, and have a difficult time judging other people's thoughts or feelings (C. E. Robertson

intellectual disability A disability characterized by significant limitations in intellectual functioning and adaptive behaviour.

hypoxia A lack of oxygen.

teratology The study abnormal development.

teratogen Substances that induce malformations in development.

fetal alcohol syndrome (FAS) A disorder, including intellectual disability and characteristic facial anomalies, that affects children exposed to too much alcohol (through maternal ingestion) during fetal development.

autism spectrum disorder (ASD) A range of conditions characterized by social, communication and behavioural challenges.

perseverate Extended repetition of a particular behaviour.

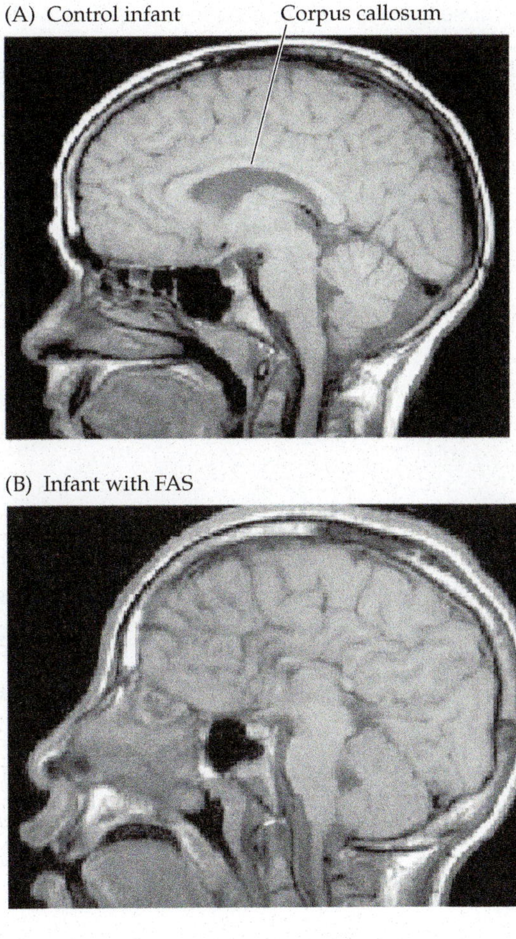

Courtesy of E. Riley

6.3 Abnormal Brain Development Associated with Fetal Alcohol Syndrome (A) The brain of a control infant. (B) The brain of an infant of the same age with FAS. This brain shows microcephaly (abnormal smallness), fewer cerebral cortical gyri, and the absence of a corpus callosum connecting the two hemispheres.

© Noah Goodrich/Caters News/Zuma Press

Be Careful What You Eat Millie Lonergan, who has phenylketonuria, eats fruit and protein-free rice and pasta (without cheese) for a diet low in phenylalanine.

phenylketonuria (PKU) An inherited disorder of protein metabolism in which the absence of an enzyme leads to a toxic buildup of certain compounds, causing intellectual disability.

Go to Animation 6.3
Stages of Neuronal Development

and Baron-Cohen, 2017). Children with the disorder are born with larger brains than controls (Hazlett et al., 2017).

There is clearly a genetic component to ASD (Krishnan et al., 2016), yet many children with ASD are helped by highly structured training in language and behaviour (Weitlauf et al., 2014). The number of children diagnosed with ASD is increasing steadily, and no one knows why. There are concerns that environmental conditions, including air pollution, may increase the likelihood of ASD (K. Lyall et al., 2017). However, the notion that childhood vaccines may cause autism was triggered by a fraudulent publication. This purported cause has been thoroughly and repeatedly debunked (Aschner and Ceccatelli, 2010; Hviid et al., 2019), yet some celebrities and trial lawyers keep the discredited idea in the public eye.

Several hundred different genetic disorders affect the metabolism of proteins, carbohydrates, or lipids, having a profound impact on the developing brain (Najmabadi et al., 2011). Characteristically, the genetic defect is the absence of a particular enzyme that controls a critical biochemical step in the synthesis or breakdown of some vital body product.

One of the most common of these is **phenylketonuria** (**PKU**), a recessive hereditary disorder of protein metabolism that at one time resulted in many people with intellectual disability. About one out of 100 persons is a carrier; one in 10,000 births produces an affected individual. The basic defect is the absence of an enzyme that metabolizes phenylalanine, an amino acid that is present in many foods. As a result, the brain is damaged by an enormous buildup of phenylalanine, which becomes toxic.

The discovery of PKU marked the first time that an inborn error of metabolism was associated with intellectual disability. Screening for PKU looks for excess levels of phenylalanine in children a few days after birth. Early detection is important because brain impairment can be prevented simply by reducing phenylalanine in the diet. Such dietary control of PKU is critical during the early years of life, especially before age 2; after that, diet can be relaxed somewhat. Note this important example of the interaction of genes and the environment in PKU: the dysfunctional genotype causes a phenotype of intellectual disability *only* in the presence of phenylalanine. Reducing phenylalanine consumption reduces or prevents this effect of the gene.

Phenylketonuria illustrates one reason why, despite the importance of genes for nervous system development, understanding the genome alone could never enable an understanding of the developing brain. Knowing that a baby inherits PKU doesn't tell you anything about how that child's brain will develop *unless* you also know something about the child's diet.

Next, we'll examine brain development at the cellular level, where initial stages are controlled by genes, to prepare us for understanding how subsequent brain development is guided by sensory experience.

6.2 Development of the Nervous System Can Be Divided into Six Distinct Stages

Learning Objectives

After reading this section, you should be able to:

6.2.1 **Describe the main cellular stages of neural development.**

6.2.2 **List some of the factors that guide migrating neurons and growing axons.**

6.2.3 **Describe the role of trophic factors like nerve growth factor (NGF) in neural development.**

6.2.4 **Explain why myelination is important for brain function.**

It is useful to consider brain development as a sequence of distinct cellular stages. The six stages, which we will take up in order, are depicted in **FIGURE 6.4**:

1. *Neurogenesis*, the mitotic division of nonneuronal cells to produce neurons
2. *Cell migration*, the movements of cells to establish distinct nerve cell populations (brain nuclei, layers of the cerebral cortex, and so on)

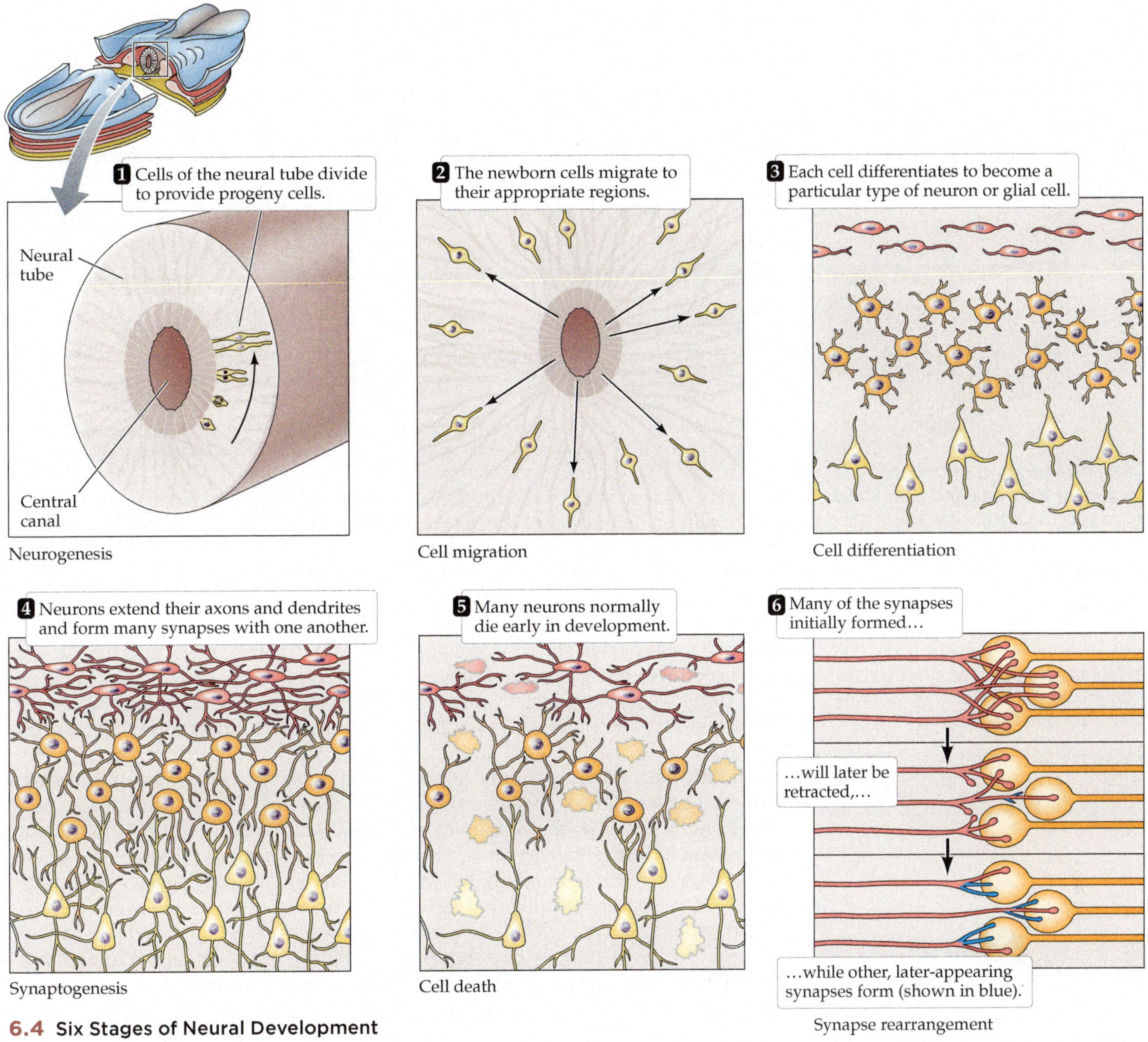

6.4 Six Stages of Neural Development

3. *Differentiation,* the transformation of precursor cells into distinctive types of neurons and glial cells
4. *Synaptogenesis,* the establishment of synaptic connections, as axons and dendrites grow
5. *Neuronal cell death,* the selective death of many nerve cells
6. *Synapse rearrangement,* the loss of some synapses and development of others, to refine synaptic connections

Cell proliferation produces cells that become neurons or glial cells

The production of nerve cells is called **neurogenesis**. Nerve cells themselves do not divide, but the cells that will give rise to neurons begin as a single layer of cells along the inner surface of the neural tube. These cells divide, in a process called **mitosis**, and gradually form a closely packed layer of cells called the **ventricular zone**. All neurons and glial cells are derived from cells that originate from this ventricular mitosis.

neurogenesis The mitotic division of nonneuronal cells to produce neurons.

mitosis The process of division of somatic cells that involves duplication of DNA.

ventricular zone Also called *ependymal layer*. A region lining the cerebral ventricles that displays mitosis, providing neurons early in development and glial cells throughout life.

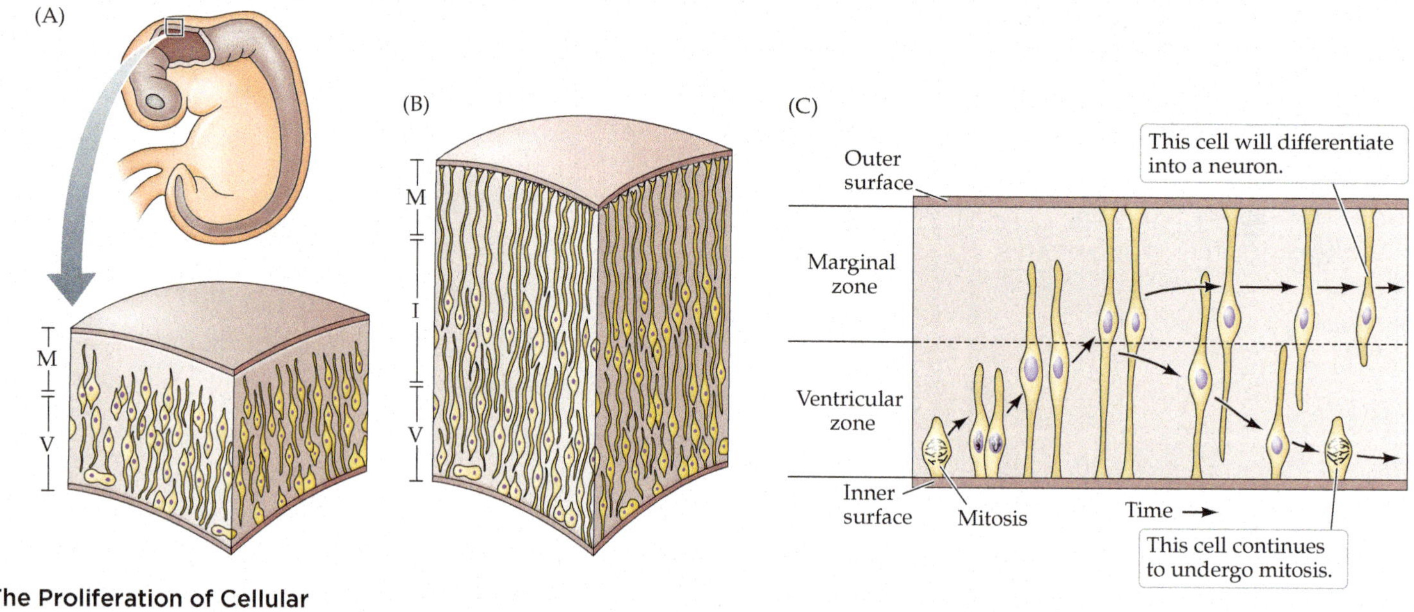

6.5 The Proliferation of Cellular Precursors of Neurons and Glial Cells (A) In this small section of the wall of the neural tube at an early stage of embryonic development, only ventricular (V) and marginal (M) layers are visible. (B) Later an intermediate (I) layer develops as the wall thickens. (C) Nuclei (within their cells) migrate from the ventricular layer to the outer layers. Some cells then become neurons while others send their nuclei back to the ventricular zone to divide again.

Eventually, some cells leave the ventricular zone and begin transforming into either neurons or glial cells.

Each part of an animal's brain has a species-characteristic "birth date." That is, there is an orderly chronological program for brain development, and we know on approximately which days of embryonic development the precursor cells of each group of neurons will stop dividing (**FIGURE 6.5**). At birth, mammals have already produced most of the neurons they will ever have. The postnatal increase of human brain weight is primarily due to growth in the size of neurons, branching of dendrites, elaboration of synapses, increase in myelin, and addition of glial cells.

But early reports that new neurons are added just after birth in some brain regions (Altman, 1969) have been supplemented with findings of **adult neurogenesis**, the generation of new neurons in adulthood, in humans (Bergmann et al., 2015) and other animals (Shingo et al., 2003). Adult neurogenesis is particularly prominent in the dentate gyrus of the hippocampal formation (Moreno-Jiménez et al., 2019), where it's been estimated we add 1400 new neurons per day (Spalding et al., 2013), replacing neurons that have died. Likewise, nerve cells of the olfactory organ (which we use to detect odours) are replaced throughout life (D. A. Lim and Alvarez-Buylla, 2016) (see Chapter 9). The addition of new neurons in adulthood is especially welcome because after age 40, our brain mass declines about 5% each decade (R. Peters, 2006), which likely reflects the loss of thousands of neurons every day. Good thing we start with lots!

Go to Video 6.4
Cell Migration

While the new neurons acquired in adulthood represent a tiny minority of neurons, there's reason to think they are important (J. S. Snyder, 2019). Enriched experience, such as learning, increases the rate of neurogenesis in adult mammals (Opendak and Gould, 2015). So by studying this chapter, you may be giving your brain a few more neurons to use on exam day! Physical exercise also boosts neurogenesis, at least in rats. This effect can be blocked by stresses such as social isolation (Stranahan et al., 2006), so invest in exercise and a network of friends too.

adult neurogenesis The creation of new neurons in the brain of an adult.

cell migration The movement of cells from site of origin to final location.

radial glial cells Glial cells that form early in development, spanning the width of the emerging cerebral hemispheres, and guide migrating neurons.

cell adhesion molecule (CAM) A protein found on the surface of a cell that guides cell migration and/or axonal pathfinding.

New nerve cells migrate

Newly born neurons are on the move. At some stage the cells that form in the ventricular layer through mitotic division move away, in a process known as **cell migration**. In primates, by the time of birth almost all future neurons have completed their migration, but in rats, cells that will become neurons continue to migrate in some regions for several weeks after birth.

Cells do not move in an aimless, haphazard manner. Cells in the developing brain move along the surface of a particular type of glial cell (Rakic, 1985). Like spokes (radii) of a wheel, these **radial glial cells** extend from the inner to the outer

6.6 Glial Spokes Guide Migrating Cells Early in development, radial glial cells span the width of the emerging cerebral hemispheres. (A) This enlargement shows how radial glial cells act as guide wires for the migration of neurons. New cells shinny past established neurons to become neurons in successively higher (outer) layers of the cortex. (B) Further enlargement shows a single neuron migrating out along a radial glial fibre. (After W. M. Cowan, 1979. *Sci Am* 241: 112–133, based on P. Rakic, 1971. *Brain Res* 33: 471–476.)

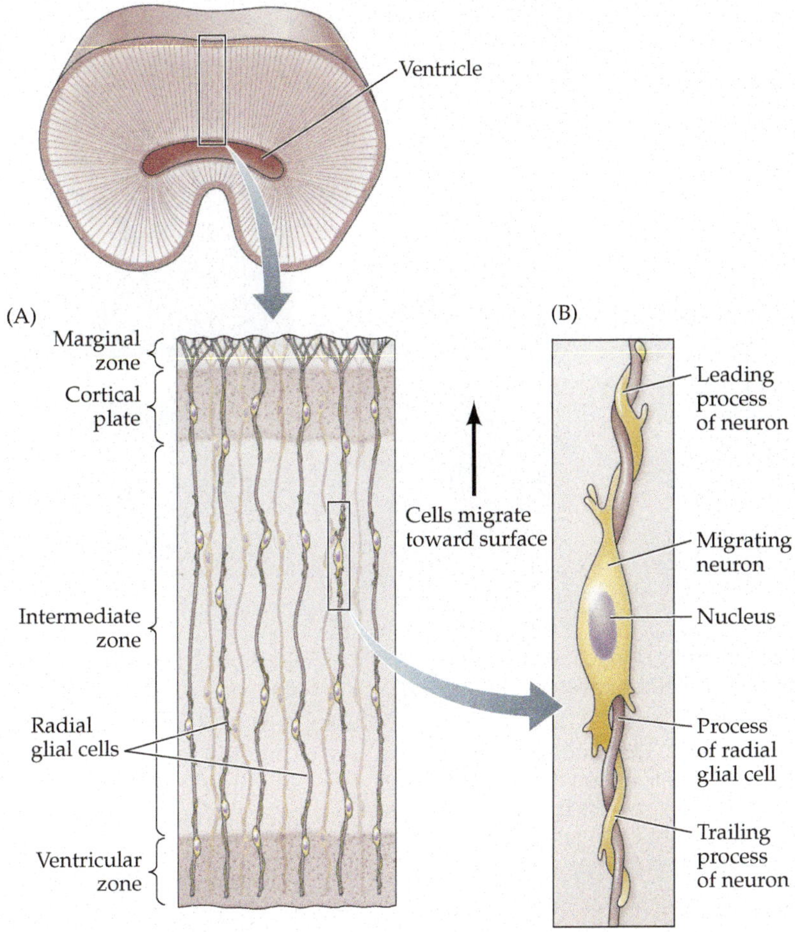

surfaces of the emerging nervous system (**FIGURE 6.6**). The radial glial cells act as a series of guides, and the newly formed cells mostly creep along them, as if "riding the glial monorail." Some migrating cells move in a direction perpendicular to the radial glial cells (Parnavelas et al., 2000), like Tarzan swinging from vine to vine; others move in a rostral stream to settle in the olfactory bulbs (Eom et al., 2010).

The migration of cells and the outgrowth of nerve cell extensions (dendrites and axons) involve various chemicals that promote the adhesion of developing elements of the nervous system. These **cell adhesion molecules** (**CAMs**) guide migrating cells and growing axons (McKeown et al., 2013). The CAMs laid down by cells along the path is another example of cell-cell interaction guiding neural development.

The single-file migration of nerve cell precursors (see Figure 6.6A) is followed by the aggregation of cells into the brain nuclei we discussed in Chapter 1. For example, cells of the cerebral cortex arrive in waves during fetal development, each successive wave forming a new outer layer, until the six layers of the adult cortex are formed, with the latest arrivals near the top.

Many mutations in mice affect the nervous system by disrupting migration. A group of mouse mutants includes those with different single-gene mutations that affect migration of neurons in the cerebellum (Tissir and Goffinet, 2003). The names of these mutant mice—*reeler, staggerer,* and *weaver*—reflect their locomotor impairment (**FIGURE 6.7**). Today, scientists deliberately delete or introduce a particular gene in

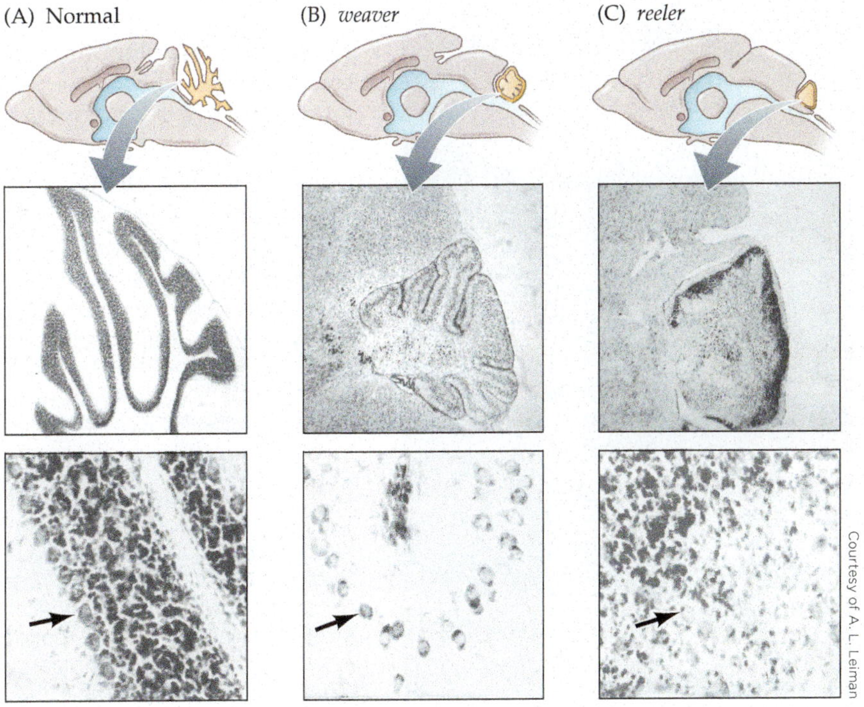

6.7 Consequences of Disordered Migration in the Cerebellum of Mice The cerebellum in a normal mouse (A) and two mutants with abnormal neuronal migration (B, C) at two levels of magnification (top: ×25; bottom: ×250). In the mutant *weaver* (B), note the almost complete absence of the tiny granule cells (bottom), while the alignment of the large Purkinje cells (arrow) is normal. The mutant *reeler* (C) shows marked derangement of the customary layering of cells. Both mutants show overall shrinkage of the cerebellum (top).

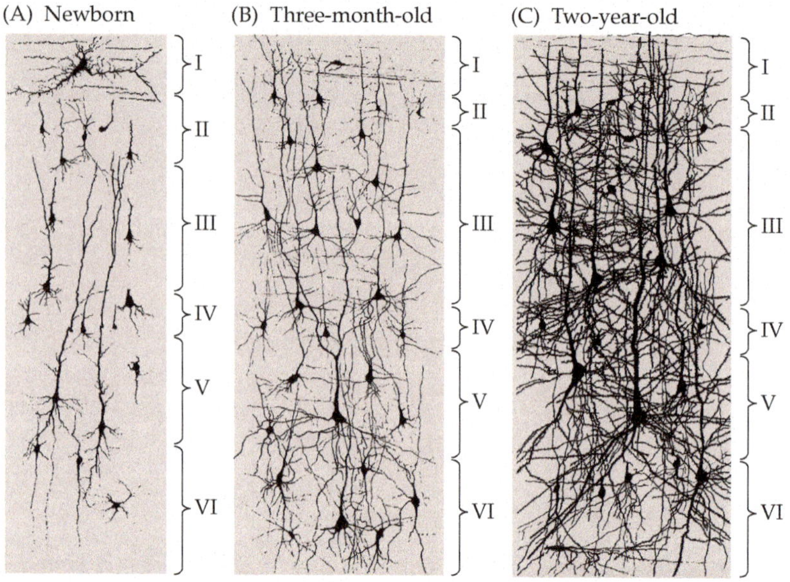

6.8 Cerebral Cortex Tissue in the Early Development of Humans These representations of cerebral cortex show the extent of neural connections and neuronal differentiation at birth (A), at 3 months of age (B), and at 2 years of age (C). Numerals refer to the six cortical layers. (From J. L. Conel, 1939, 1947, 1959. *The postnatal development of the human cerebral cortex*: *Vols. 1, 3, 6.* Harvard University Press: Cambridge, MA.)

mice in order to study the role of that gene on the nervous system (**BOX 6.1**).

Cells in newly formed brain regions differentiate into neurons

Newly arrived cells in the brain bear no more resemblance to mature nerve cells than to the cells of other organs. Once they reach their destinations, however, the cells begin to express particular genes. This means that the cell transcribes a particular subset of genes to make the specific proteins it needs. This process of cell differentiation shapes the cell into the distinctive forms and functions of neurons found in that particular region (**FIGURE 6.8**). This differentiation reflects how each neuron is expressing a particular subset of genes (Bakken et al., 2016) and therefore taking a particular structure and playing a particular function.

What controls differentiation is not completely understood, but we know the local environment greatly influences the process. In other words, neighbouring cells are a major influence on the differentiation of neurons. In vertebrates (unlike the nematode *C. elegans*), young neural cells seem to have the capacity to become many varieties of neurons, and the particular type of neuron that a cell becomes depends on where it happens to be and what its neighbouring cells are. For example, consider spinal motor neurons—cells in the spinal cord that send their axons out to control muscles. These large, multipolar cells are found in the left and right sides of the spinal cord in the ventral horns of gray matter. Motor neurons are among the first recognizable neurons in the spinal cord, and they send their axons out early in fetal development. How do these cells "know" they should express motor neuron–specific genes and differentiate into motor neurons?

Examination of the late divisions giving rise to motor neurons makes it clear that the cells are not attending to mitotic lineage (Leber et al., 1990). Instead, some spinal cells are directed to become motor neurons under the influence of other cells lying just ventral to the developing spinal cord—in the **notochord**, a rodlike structure that forms along the midline (see Figure 6.1A) (Miri et al., 2013). The notochord releases a protein (playfully named Sonic hedgehog) that diffuses to the spinal cord and directs some (but not all) cells to become motor neurons (**FIGURE 6.9**).

Such an influence of one set of cells on the fate of neighbouring cells is known as **induction**; the notochord releases a protein that induces some spinal cord cells to differentiate into motor neurons. Induction of this sort has been demonstrated many times in the developing vertebrate body and brain. This is an example of the extensive cell-cell interaction in developing vertebrates, each cell taking cues from its neighbours.

notochord A midline structure arising early in the embryonic development of vertebrates.

induction The process by which one set of cells influences the fate of neighbouring cells, usually by secreting a chemical factor that changes gene expression in the target cells.

Courtesy of Thomas Jessell

6.9 The Induction of Spinal Motor Neurons In this cross section of embryonic chick spinal cord, the notochord (green circle at bottom) lies just beneath the spinal cord and secretes the protein Sonic hedgehog. A moderate concentration of this protein in the ventral spinal cord induces the cells there to develop as motor neurons (gold), forming columns of motor neurons on the left and right sides. Another protein (blue) expressed only in the dorsal spinal cord inhibits differentiation of motor neurons.

BOX 6.1 Transgenic and Knockout Mice

Animals with mutations in specific genes offer clues about the role of genes in development and brain function. Among the many new tools brought about by the revolution in molecular biology is *site-directed mutagenesis*, the ability to cause a mutation in a particular gene. Researchers using such techniques, including CRISPR, must know the sequence of nucleotides in the gene of interest (see Appendix). Then they can use the tendency of complementary nucleotides to hybridize with that part of the gene to induce changes.

The easiest change to understand is total disruption of the gene, making it nonfunctional. If this is done in embryonic mouse cells, these cells may form the testes or ovaries of a developing mouse. That mouse can then produce offspring that are missing the gene. We call the resulting animal a **knockout organism** because the gene of interest has been *knocked out*.

By following the development of knockout mice, we can obtain clues about the roles of particular genes in normal animals. For example, the motor neurons of mice whose genes for brain-derived neurotrophic factor (BDNF) have been knocked out survive despite the absence of BDNF (Sendtner et al., 1996), so we know that BDNF is not crucial for motor neuron survival. On the other hand, some parasympathetic ganglia fail to form in BDNF knockout mice (J. T. Erickson et al., 1996), suggesting that these neurons depend on BDNF for survival. Use of Cre-lox technology (see Appendix) makes it possible to knock out the gene only in the brain, or perhaps even only a portion of the brain. As we'll see in Chapter 17, several genes suspected of playing a role in learning have been knocked out in mice, and the resulting animals indeed show deficits in learning.

There are some problems in interpreting results in knockout mice, because the missing gene may have contributed only very indirectly to the learning process, or the animal's poor performance may have been due to a distraction caused by the knockout. For that matter, even normal behaviour by animals missing the gene does not prove that the gene is unimportant for behaviour. Perhaps the developing animal, in the absence of that gene, somehow has compensated for the loss and found a new way to solve the problem. This would be another example of the embryonic regulation that is so common in vertebrate development.

In other cases, a functional, manipulated copy of a gene can be introduced into the mouse. This animal is described as **transgenic** because a gene has been *transferred* into its *genome*. Sometimes the introduction of just a single new gene can have a dramatic effect on brain development; for example, compare the brains of newborn mice that are normal with those of transgenic mice carrying a modified gene for β-catenin (see Figure 6.23). Modifying this one gene caused the mouse to make far too many neurons, so extra gyri and sulci developed (Chenn and Walsh, 2002).

The transgenic approach is often used as a method for improving our understanding of genetic disorders. For example, in Chapter 10 we'll learn that when a gene mutation that causes severe motor impairments in people is transferred into mice, the mice develop symptoms similar to those that appear in humans. It may be possible to study the disease more closely in these mice and test possible therapies.

Knockout and transgenic animals have another limitation: they possess the genetic manipulation from the moment of conception and in every cell in the body. However, neuroscientists have begun knocking out or replacing genes in adult animals—by injecting the animals with a triggering substance such as tetracycline or by replacing or knocking out a gene in only one region of the brain. These manipulations allow the animals to develop with a normal genotype, thereby making it easier to interpret the result of the gene manipulation in adulthood. It may even be possible to knock out and then restore a gene in the same individual mouse, tracking its behaviour as the gene is lost and regained.

knockout organism An individual in which a particular gene has been disabled by an experimenter.

transgenic Referring to an animal in which a new or altered gene has been deliberately introduced into the genome.

Because each cell influences the differentiation of others, vertebrate neural development is very complex, but also very flexible. For example, cells differentiate into the type of neuron that is appropriate for wherever they happen to be in the brain; thus, cell-cell interaction coordinates development—directing differentiation to provide the right type of neuron for each part of the brain. Another benefit of cell-cell interactions in development is that if a few cells are injured or lost, other cells will "answer the call"—that is, respond to the inducing factors—and fill in for the missing cells. Embryologists refer to such adaptive responses to early injury as **regulation** (or *self-regulation*): the developing animal compensates for missing or injured cells.

This phenomenon can be observed in embryos from which some cells have been removed. For example, if cells are removed early enough from a developing limb bud

regulation An adaptive response to early injury, as when developing individual cells compensate for missing or injured cells.

(A)

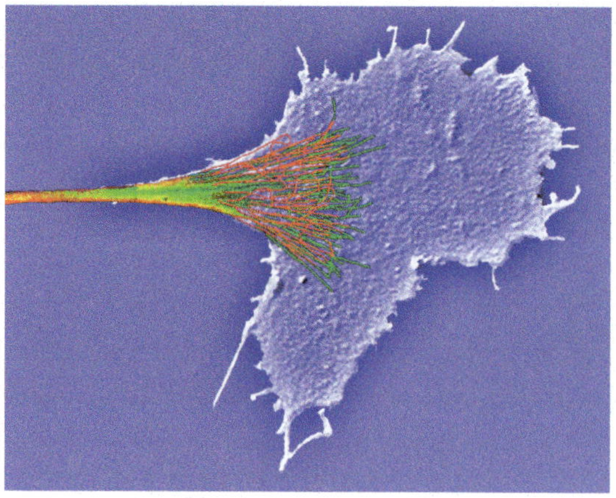

Courtesy of Paul Bridgman

(B)

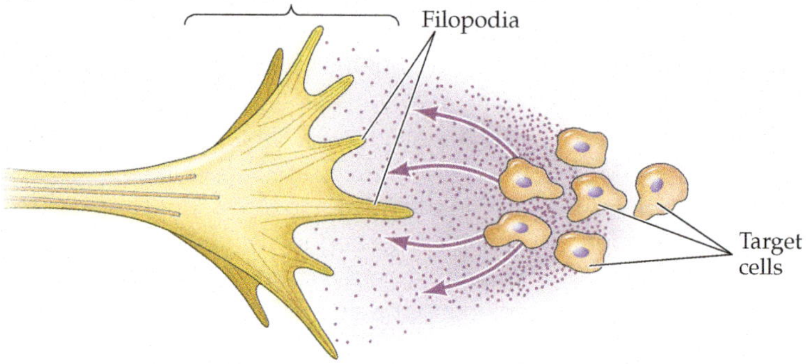

6.10 The Growth Cones of Growing Axons and Dendrites (A) The fine, threadlike extensions pictured here are filopodia, which find adhesive surfaces and pull the growth cone, and therefore the growing axon, to the right. (B) Target cells release a chemical that creates a gradient (dots) around them. Growth cones orient to and follow the gradient to the cells. (Part B after M. Tessier-Lavigne et al., 1988. *Nature* 336: 775–778.)

in a chick embryo, other cells pitch in, and by the time the chick hatches, the limb looks normal, with no parts missing. Because cell fate is so tightly coupled with mitotic lineage in *C. elegans*, this organism shows little or no regulation. If a cell in *C. elegans* is killed, no other cells take its place; the worm must do without that cell.

The more complicated vertebrate system, in which cells take cues from their neighbours to guide their gene expression and functional fate, has another consequence: if cells that have not yet differentiated extensively can be obtained and placed into a particular brain region, they will differentiate in an appropriate way and become properly integrated. Such undifferentiated cells, called **stem cells**, are present throughout embryonic tissues, so they can be gathered from umbilical cord blood, miscarried embryos, or unused embryos produced during in vitro fertilisation.

It may be possible to take cells from adult tissue and, by treating them with various factors in a dish, induce them to become "adult" stem cells (Scudellari, 2016). Researchers hope that, someday, placing stem cells in areas of brain degeneration—loss of myelination in multiple sclerosis or loss of dopaminergic neurons in Parkinson's disease (see Chapter 10)—might reverse such degeneration as the implanted cells differentiate to fill in for the missing components.

stem cell A cell that is undifferentiated and therefore can take on the fate of any cell that a donor organism can produce.

synaptogenesis The establishment of synaptic connections as axons and dendrites grow.

growth cone The growing tip of an axon or a dendrite.

filopodia Very fine, tubular outgrowths from the growth cone.

chemoattractants Compounds that attract particular classes of axonal growth cones.

The axons and dendrites of young neurons grow extensively and form synapses

The biggest change in brain cells early in life is the extensive growth of axons and dendrites (termed *process outgrowth*) and the proliferation of synapses, or **synaptogenesis**. At the tips of axons and dendrites alike, specialized swellings called **growth cones** are found. Very fine extensions, called **filopodia** (singular *filopodium*, from the Latin *filum*, "thread," and the Greek *pod*, "foot"), extend from the growth cone (**FIGURE 6.10**). Just as migrating cells are guided by CAMs, the filopodia of growth cones adhere to CAMs laid down by other cells and then contract to pull the growth cone in a particular direction, the growing axon or dendrite trailing behind. Dendritic growth cones are found in adults, mediating the continued elongation and change in dendrites that occurs throughout life in response to experience (see Chapter 16).

What guides axons along the paths they take? The CAMs guiding growth cones are released by the target nerve cells or other tissues, such as muscles. The axon growth cone responds to the concentration gradients of these chemicals that provide directional guidance (**FIGURE 6.11A**). CAMs that attract certain

(A)

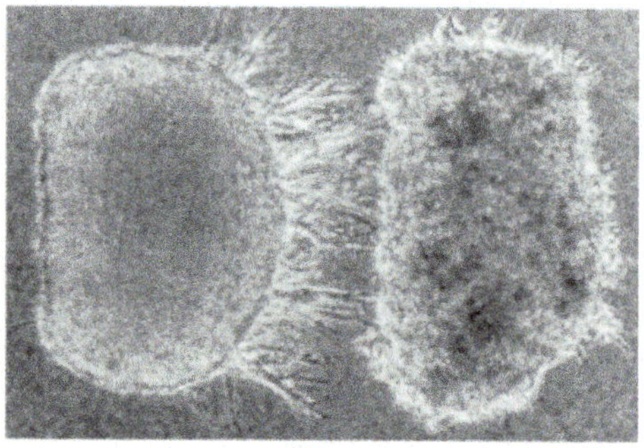

Courtesy of Marc Tessier-Lavigne

Go to Video 6.5
Growth Cones

(B)

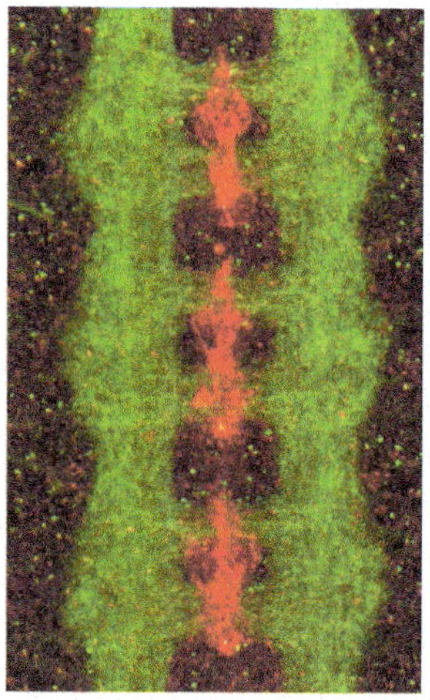

Courtesy of Julie Simpson and Corey S. Goodman

6.11 Cues for Axonal Growth Cones (A) The extensions visible here are growing out of a sensory ganglion (left) toward their normal target tissue. (B) The chemorepellent protein Slit (red), shown here in an embryo of the fruit fly *Drosophila*, repels most axons (green), preventing them from crossing the midline.

growth cones are called **chemoattractants**; chemicals that repel growth cones are **chemorepellents** (McKeown et al., 2013). For example, because it is important for some axons to remain on one side of the body and for others to cross over, the protein Slit repels some axons to prevent them from crossing the midline (**FIGURE 6.11B**) (Dickson and Gilestro, 2006). The same secreted protein may act as a chemoattractant to some growth cones and a chemorepellent to others (Polleux et al., 2000). These various signals also guide axons to regenerate when they are cut in adulthood (**BOX 6.2**).

chemorepellents Compounds that repel particular classes of axonal growth cones.

BOX 6.2

Degeneration and Regeneration of Nervous Tissue

When a mature nerve cell is injured, it can regrow in several ways. Complete replacement of injured nerve cells is rare in mammals, but Figures A and B illustrate characteristic forms of degeneration and regeneration in the mammalian peripheral and central nervous systems. Injury close to the cell body of a neuron produces a series of changes resulting in the eventual destruction of the cell; this process is called **retrograde degeneration** (Figure A, 2). If the injured neuron dies, the target cells formerly innervated by that neuron may show signs of *transneuronal degeneration* (Figure A, 3).

Cutting through the axon also produces loss of the distal part of the axon (the part that is separated from the cell body). This process is called **anterograde degeneration** (Figure B, 2). The part of the axon that remains connected to the cell body may regrow. Severed axons in the peripheral nervous system regrow readily. Sprouts emerge from the part of the axon that is still connected to the nerve cell body and advance slowly toward the periphery (Figure B, 3). Cell adhesion molecules (CAMs) help guide the regenerating peripheral axons. No one knows why mammalian axons in the CNS seem unable to regenerate in this manner. Some fishes and amphibians have an enviable advantage over humans: after an injury to the brain, they can regenerate many of the lost connections.

retrograde degeneration Destruction of the nerve cell body following injury to its axon.

anterograde degeneration Also called *Wallerian degeneration*. The loss of the distal portion of an axon resulting from injury to the axon.

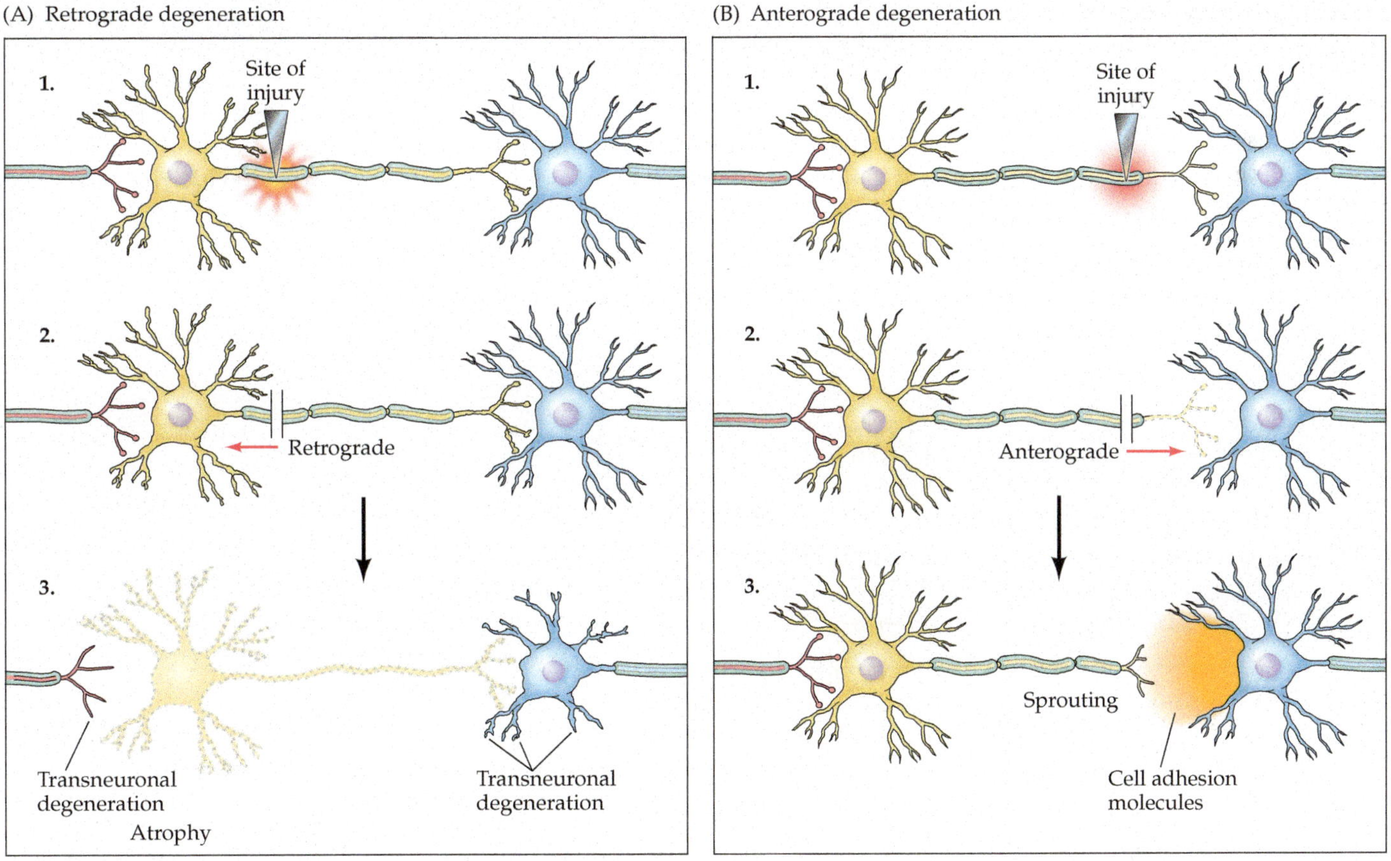

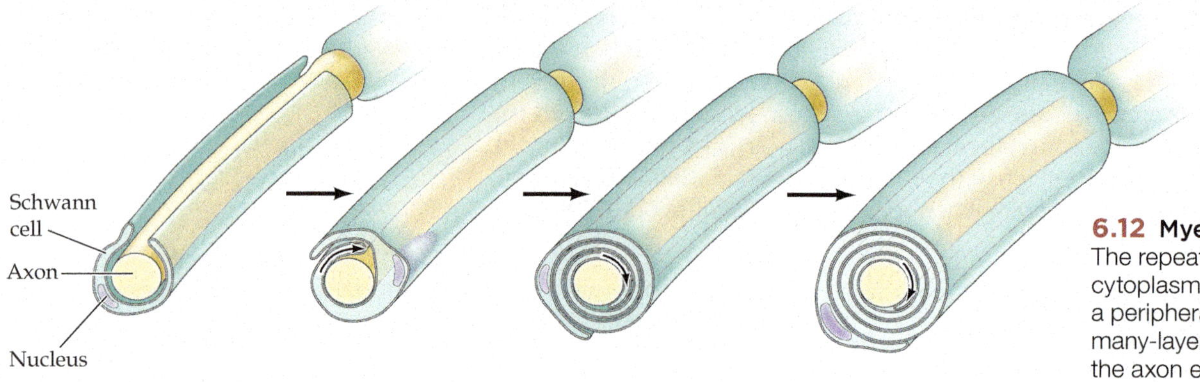

6.12 Myelin Formation The repeated wrapping of the cytoplasm of a Schwann cell around a peripheral axon results in a many-layered sheath that insulates the axon electrically, speeding the conduction of electrical signals along its length.

myelination The process of myelin formation.

cell death Also called *apoptosis*. The developmental process during which "surplus" cells die.

death gene A gene that is expressed only when a cell becomes committed to natural cell death (apoptosis).

caspases A family of proteins that regulate cell death (apoptosis).

Diablo A protein released by mitochondria, in response to high calcium levels, that activates apoptosis.

inhibitors of apoptosis proteins (IAPs) A family of proteins that inhibit caspases and thereby stave off apoptosis.

Once axons have reached their final destination, they induce nearby glia to ensheath them in myelin, which will greatly speed conduction (Chapter 2). The process of **myelination** (**FIGURE 6.12**) has a big impact on behaviour by allowing neuronal networks to communicate speedily. In humans, the earliest myelination is evident in cranial and spinal nerves about 24 weeks after conception. But the most *intense* phase of myelination occurs shortly after birth, extending into young adulthood and even, to some extent, throughout life (R. A. Hill et al., 2018).

The death of many neurons is a normal part of development

As strange as it may seem, cell death is a crucial phase of brain development, especially during embryonic stages. This developmental stage is not unique to the nervous system. Naturally occurring **cell death**, also called *apoptosis* (from the Greek *apo*, "away from," and *ptosis*, "act of falling"), is a kind of sculpting process in the emergence of tissues in both animals and plants.

The number of neurons that die during early development is quite large. In some regions of the brain and spinal cord, *most* of the young nerve cells die during prenatal development. In 1958, Viktor Hamburger (1900–2001) first described naturally occurring neuronal cell death in chicks, in which nearly half the originally produced spinal motor neurons die before the chick hatches (**FIGURE 6.13**). Genetically interfering with neural apoptosis in fetal mice causes them to grow brains that are too large to fit in the skull (Depaepe et al., 2005), so we can see how vital it is that some cells die.

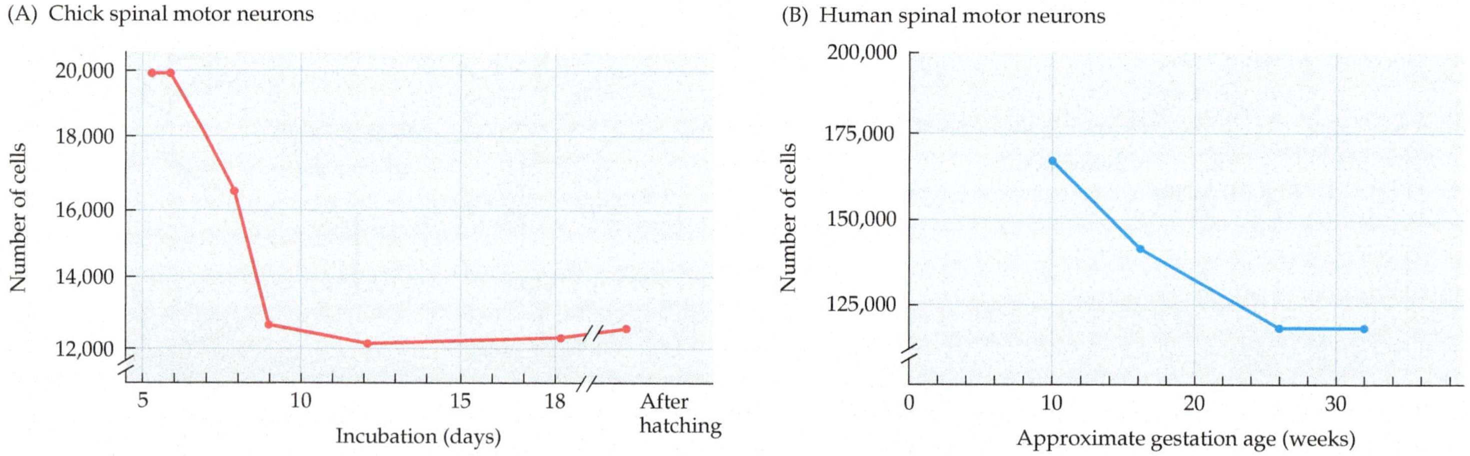

6.13 Many Neurons Die during Normal Early Development The pattern of neuronal cell death in spinal motor neurons of chicks (A) and humans (B). Many neuronal populations show a similar pattern of apoptosis. (Part A from V. Hamburger, 1975. *J Comp Neurol* 160: 535–546; B after N. G. Forger and S. M. Breedlove, 1987. *J Comp Neurol* 264: 118–122.)

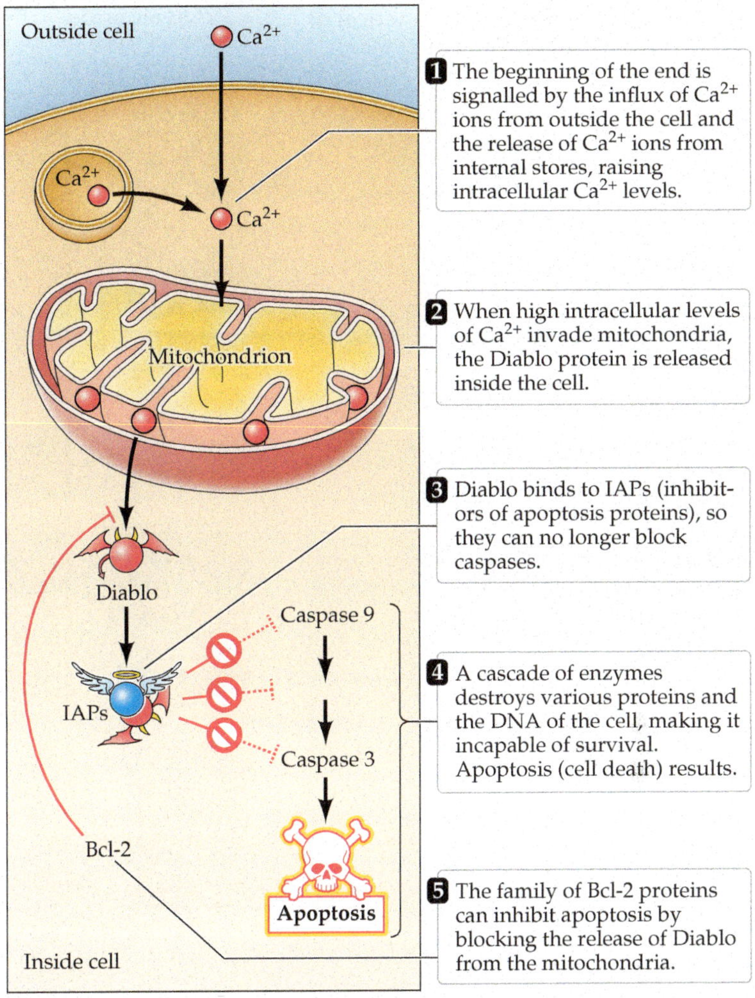

6.14 Death Genes Regulate Apoptosis

These cells are not dying because of a defect. Rather, it appears that these cells have "decided" to die and are actively committing suicide. Your chromosomes carry **death genes**—genes that are expressed only when a cell undergoes apoptosis (H. Park and Poo, 2013). For example, the **caspases** are a family of proteases (protein-dissolving enzymes) that cut up proteins and nuclear DNA. Apoptosis appears to begin with a sudden influx and release of calcium (Ca2+) ions that cause the mitochondria inside the cell to release a protein called, devilishly enough, **Diablo** (Tait and Green, 2010).

Diablo binds to a family of proteins, the well-named **inhibitors of apoptosis proteins (IAPs)** (Silke and Meier, 2013). The IAPs normally inhibit the caspases, so when Diablo binds the IAPs, the caspases are freed to dismantle the cell. **Bcl-2** proteins block apoptosis by preventing Diablo release from the mitochondria. This intricate system of checks and balances, which determines whether a cell gives up the ghost (**FIGURE 6.14**), must have been established long ago in evolution, because homologues of the genes that produce these proteins function similarly in *C. elegans*. In the worm, mitotic lineage determines which cells are fated to die, but what determines which cells will die in vertebrates?

As you may have guessed, apoptosis in vertebrates is regulated by cell-cell interactions, such as the availability of synaptic targets. Reducing the size of the synaptic target invariably reduces the number of surviving nerve cells. If the leg of a tadpole is removed in development, for instance, many more developing spinal motor neurons die than would die if the leg remained. Conversely, grafting an extra leg onto one side of the body—a technique that is possible with chicken embryos and tadpoles—reduces the usual loss of cells; in such cases the mature spinal cord has more than the usual number of motor neurons on that side.

Thus neurons compete for connections to target structures (other nerve cells or end organs, such as muscle). Neurons that make adequate synapses survive and grow; those that fail to form synaptic connections die. The neurons compete not just for synaptic sites, but also for a chemical that the target structure makes and releases. Neurons that receive enough of the chemical survive; those that do not, die. Such target-derived chemicals are called **neurotrophic factors** (or simply *trophic factors*) because they act as if they "feed" the neurons to help them survive (in Greek, *trophe* means "nourishment"). The neurotrophic factor that was the first to be identified prevents the death of developing sympathetic neurons, as we'll discuss next.

Neurotrophic factors allow neurons to survive and grow

In the 1950s, investigators discovered a substance—called **nerve growth factor (NGF)**—that markedly affects the growth of neurons in spinal ganglia and in the ganglia of the sympathetic nervous system (Levi-Montalcini, 1982). Administered to a chick embryo, NGF resulted in the survival of many more sympathetic neurons than usual. These cells were also larger and had more extensive processes (**FIGURE 6.15**).

Bcl-2 A family of proteins that regulate apoptosis.

neurotrophic factor Also called *trophic factor*. A target-derived chemical that acts as if it "feeds" certain neurons to help them survive.

nerve growth factor (NGF) A substance that markedly affects the growth of neurons in spinal ganglia and in the ganglia of the sympathetic nervous system.

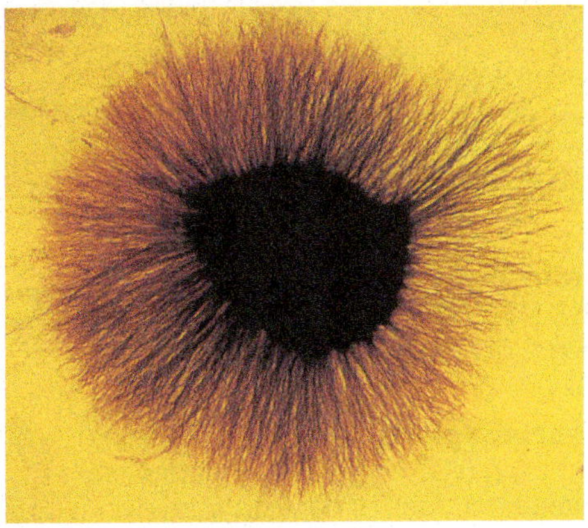

From D. Purves and J. W. Lichtman, 1985. *Principles of neural development*. Sinauer Associates: Sunderland, MA. Courtesy of R. Levi-Montalcini

6.15 The Effects of Nerve Growth Factor If NGF is added to the solution bathing a spinal ganglion grown in vitro, neuronal processes grow outward in an exuberant, radiating fashion.

6.16 A Model for the Action of Neurotrophic Factors

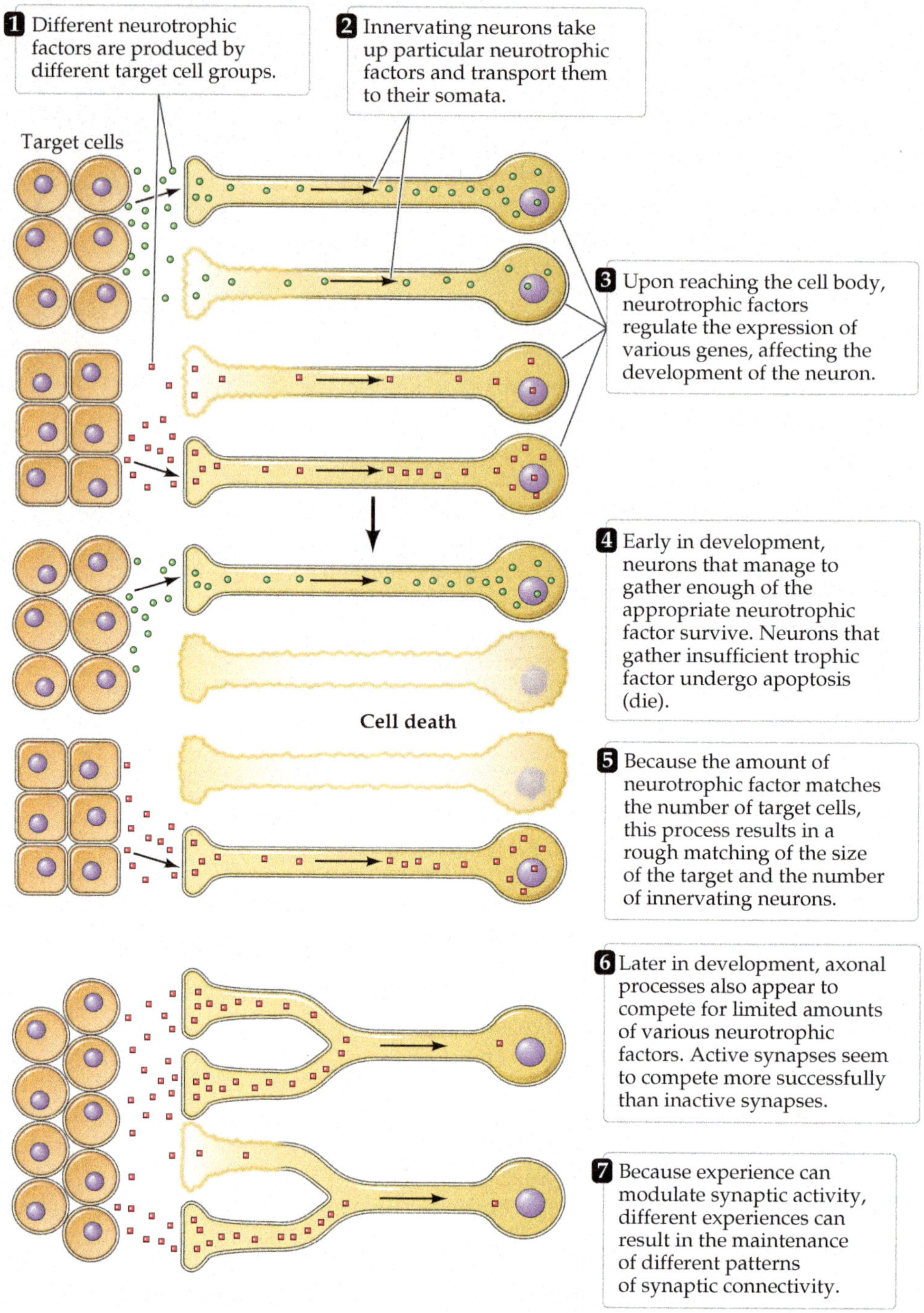

Various target organs normally produce NGF during development. It is taken up by the axons of sympathetic neurons that innervate those organs and is transported back to the cell body, where NGF prevents the sympathetic neurons from dying. The amount of NGF produced by targets during development is roughly correlated with the amount of sympathetic innervation that the targets maintain into adulthood. Thus, cell death, controlled by access to NGF, provides each target with an appropriate amount of sympathetic innervation.

There are additional neurotrophic factors, each one affecting the survival of a particular cell type during a specific developmental period (**FIGURE 6.16**). One such factor, named **brain-derived neurotrophic factor** (**BDNF**), is very similar to NGF. Investigators used molecular techniques to search for other NGF-related molecules and found several more. This family of NGF-like molecules—now named the **neurotrophin** family—includes neurotrophin-1 (i.e., NGF), neurotrophin-2 (BDNF), neurotrophin-3, and neurotrophin-4/5 (the fifth neurotrophin discovered turned out to be identical to the fourth—oops). Neurotrophic factors that

brain-derived neurotrophic factor (BDNF) A protein, purified from the brains of animals, that can keep some classes of neurons alive.

neurotrophins A family of proteins, including NGF and BDNF, that prevent different classes of neurons from dying.

are unrelated to NGF also have been found, including ciliary neurotrophic factor (named after its ability to keep neurons from ciliary ganglia alive in vitro).

The exact roles of these various factors (and other neurotrophic factors yet to be discovered) are under intense scientific scrutiny as potential ways to prevent neuronal death in various diseases like Parkinson's (Chapter 10). One role of neurotrophic factors seems to be guiding the final stage of brain development at the cellular level, the rearrangement of synaptic connections, which we discuss next.

synapse rearrangement Also called *synaptic remodelling*. The loss of some synapses and the development of others; a refinement of synaptic connections that is often seen in development.

6.3 Lifelong Synapse Rearrangement Is Guided by Experience

Learning Objectives

After reading this section, you should be able to:

6.3.1 **Describe the process of synapse rearrangement and offer evidence that a net loss of synapses may be adaptive.**

6.3.2 **Explain how regenerating visual systems in amphibians were eventually found to depend upon experience.**

6.3.3 **Describe studies showing that visual experience early in life is required to develop normal vision.**

The sixth stage of neural development begins late, but it never quite stops, going on to some extent until we die. It turns out that many of those synapses formed early in development are later retracted or reorganised, a process termed **synapse rearrangement**, or *synaptic pruning* (**FIGURE 6.17**). For example, as we learned already, about half of the

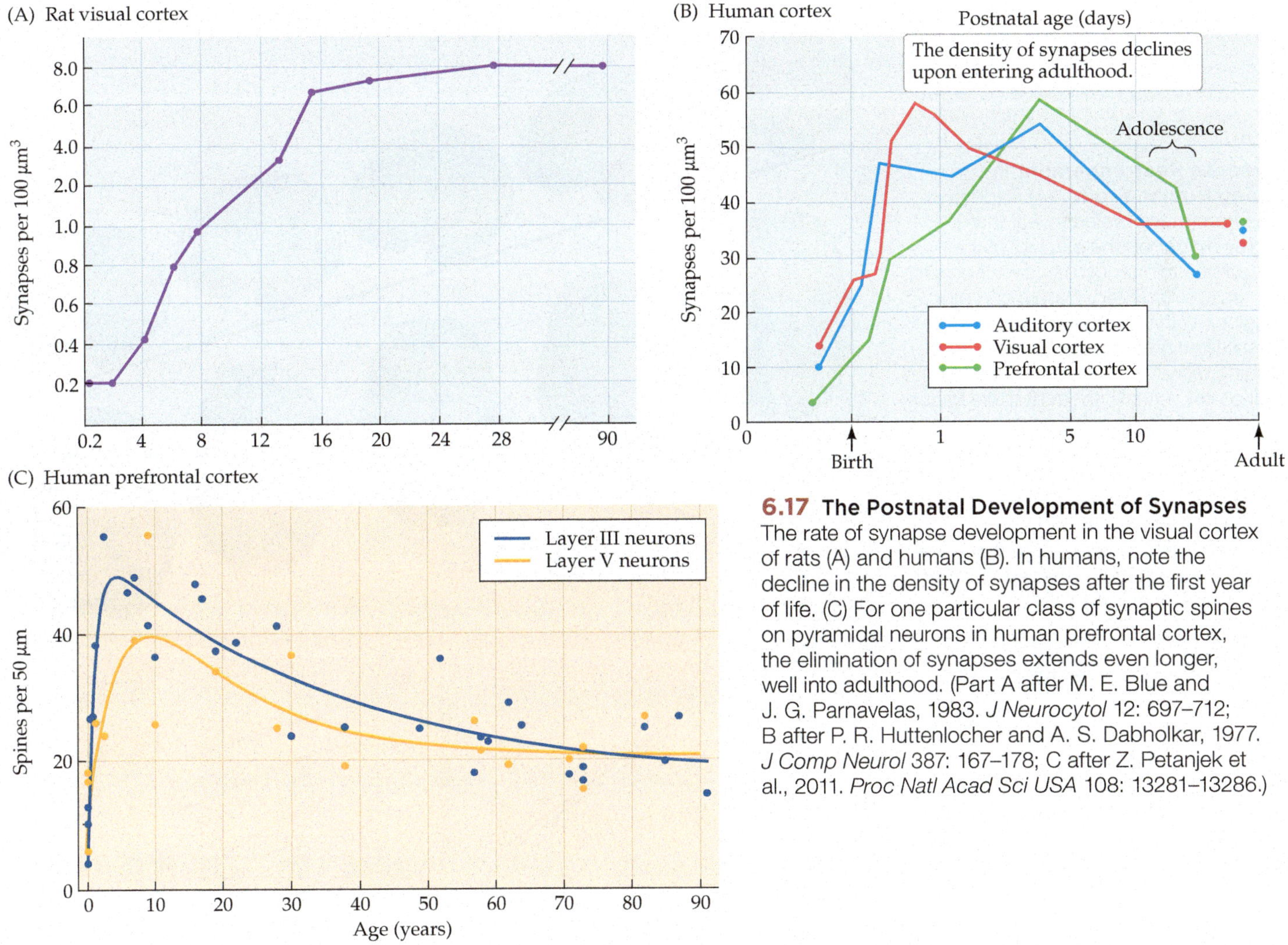

6.17 The Postnatal Development of Synapses The rate of synapse development in the visual cortex of rats (A) and humans (B). In humans, note the decline in the density of synapses after the first year of life. (C) For one particular class of synaptic spines on pyramidal neurons in human prefrontal cortex, the elimination of synapses extends even longer, well into adulthood. (Part A after M. E. Blue and J. G. Parnavelas, 1983. *J Neurocytol* 12: 697–712; B after P. R. Huttenlocher and A. S. Dabholkar, 1977. *J Comp Neurol* 387: 167–178; C after Z. Petanjek et al., 2011. *Proc Natl Acad Sci USA* 108: 13281–13286.)

spinal motor neurons that form die later (see Figure 6.13). By the end of the cell death period, each surviving motor neuron innervates many muscle fibres, and every muscle fibre is innervated by several motor neurons. But later the surviving motor neurons retract many of their axon collaterals, until each muscle fibre comes to be innervated by only one motor neuron.

Similar events occur in other neural regions, including the cerebellum (Mariani and Changeaux, 1981), the brainstem (H. Jackson and Parks, 1982), the visual cortex (Hubel et al., 1977), and the autonomic ganglia (Lichtman and Purves, 1980). In human cerebral cortex there seems to be a net loss of synapses from late childhood through adolescence (see Figure 6.17B and C). This synaptic remodelling causes thinning of the gray matter in the cortex as pruning of dendrites and axon terminals progresses. The thinning process continues in a caudal–rostral direction during maturation (**FIGURE 6.18A**), so prefrontal cortex matures last (Gogtay et al., 2004). Indeed, some synapse elimination there continues well into adulthood (Mills et al., 2014) (**FIGURE 6.18B**). During this process, connections between prefrontal cortex and some brain regions become more numerous, while others thin out (Dosenbach et al., 2013). Since prefrontal cortex is important for inhibiting behaviour (see Chapter 18), this delayed brain maturation may contribute to teenagers' impulsivity and lack of control. Furthermore, because the synaptic pruning going on at this stage is critical for future functioning of the brain, the tendency of psychiatric disorders, such as schizophrenia and mood disorders, to emerge in adolescence may reflect the vulnerability of this developmental stage (Paus et al., 2008).

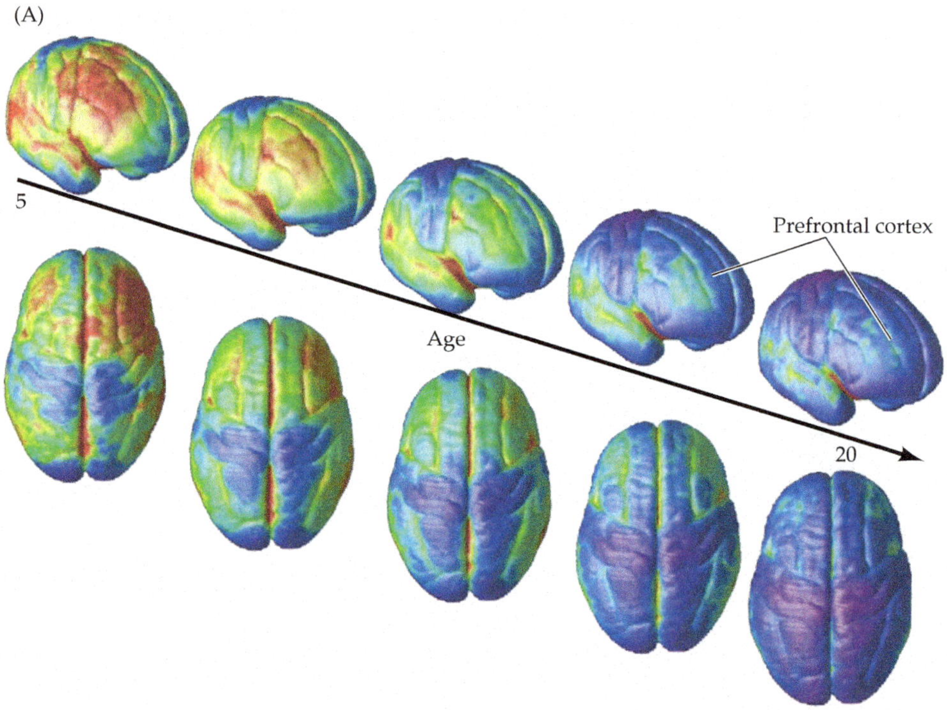

From N. Gogtay et al., 2004. *Proc Natl Acad Sci USA* 101: 8174–8179

6.18 Synapse Rearrangement in the Developing Human Brain (A) Repeated measures from many brains reveal that the layer of gray matter on the exterior of the cortex becomes thinner across development, as synapses are retracted. Purple and blue depict regions with little change in cortical thickness; yellow and red depict areas that are changing rapidly at that age. Note that the prefrontal cortex, usually thought to be important in inhibiting behaviour, does not finish maturation until late adolescence. (B) A sample from the prefrontal cortex shows how rapidly cortical thickness changes during puberty and well into adulthood. Because the y axis does not begin at zero, note that these sex differences are in fact rather subtle. (Part B after K. L. Mills et al., 2014. *Soc Cogn Affect Neurosci* 9: 123–131.)

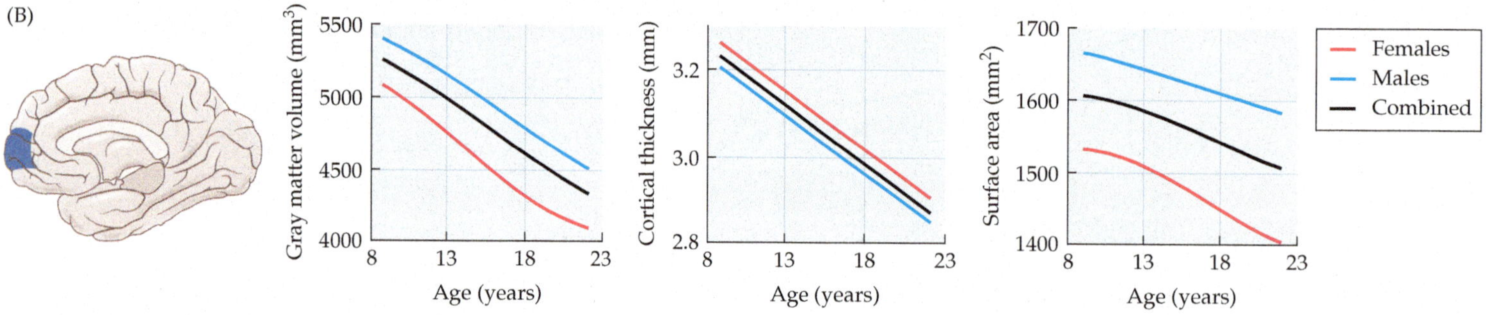

Retaining too many synapses can impair intellectual development

Now that most cases of PKU are identified at birth to avoid its harmful effects, the most frequent remaining inherited cause of intellectual disability is **fragile X syndrome** (**FIGURE 6.19**), which is more common in males than in females. At the end of the long arm of the X chromosome is a site that seems fragile—prone to breaking because the DNA there is unstable (Lyons et al., 2015). A person with this condition has a modified facial appearance, including elongation of the face, large prominent ears, and a prominent chin. A wide range of cognitive effects—from mild to severe impairment—are associated with the syndrome. Cortical neurons from the brains of people with fragile X syndrome, as well as mice genetically engineered to have this syndrome, possess an *excess* of small, immature dendritic spines (Bagni and Greenough, 2005). These findings suggest that the syndrome affects mental development by blocking the normal elimination of synapses after birth (see Figure 6.18).

The molecular basis of fragile X syndrome provided a surprise for geneticists because it demonstrated that we don't always pass on faithful copies of our DNA to our offspring. The fragile site in the DNA consists of three nucleotides (CGG; see the Appendix for a review of nucleotides) repeated over and over. Most people have only 6–50 of these **trinucleotide repeats** at this site. But during the production of sperm or eggs, the number of repeats sometimes changes, so a mother who has only 50 trinucleotide repeats may provide 100 repeats to her son. Any children who receive more than 200 repeats will display fragile X syndrome (Lozano et al., 2014). Trinucleotide repeats in a different gene cause another behaviour disorder: Huntington's disease (see Chapter 10).

What determines which synapses are kept and which are lost? Although we don't know all the factors, one important influence is neural activity. One theory is that active synapses take up some neurotrophic factor that maintains the synapse, while inactive synapses get too little trophic factor to remain stable (see Figure 6.16, bottom). Intellectual stimulation probably contributes, as suggested by the fact that teenagers with the highest IQs show an especially prolonged period of cortical thinning (P. Shaw et al., 2006).

We'll run into lifelong synaptic rearrangement repeatedly in this book. In Chapter 7 we'll see that regions of cerebral cortex that process touch information undergo synapse rearrangement throughout life, and in Chapter 16 we'll find that learning and memory happen through the strengthening and weakening of existing synapses or through synapse rearrangement. Next we'll consider another example of adult synaptic rearrangement, the visual system of fishes and amphibians, which sets the stage for understanding why experience is crucial for the development of vision.

The visual systems of some vertebrates can regenerate in adulthood

If the optic nerve that connects an eye to the brain is cut in an adult mammal, the animal is blinded in that eye and never recovers. In fishes and amphibians such as frogs, however, the animal is only temporarily blinded; in a few months the axons from the eye (specifically, from the ganglion cells of the retina) reinnervate the brain (specifically the dorsal portion of the midbrain, called the *tectum*) and the animal recovers its eyesight. When food is presented on the left or right, above or below, the animal flicks its tongue accurately to retrieve it. Thus, either (1) the retina reestablishes the same pattern of connections to the tectum that was there before surgery, and the brain interprets visual information as before, or (2) the retina reinnervates the tectum at random, but the rest of the brain learns to interpret the information presented in this new pattern.

Several lines of evidence established that the first hypothesis is correct. One such piece of evidence is that the first-arriving retinal axons sometimes pass over uninnervated tectum to reach their original targets. In the classic case illustrating this phenomenon, the optic nerve was cut and the eye was rotated 180°; when the animal recovered eyesight, it behaved as if the visual image had been rotated 180°, moving to the left when trying to get food that was presented on the right, and flicking its tongue up when food was presented below. The only explanation for this behaviour

Courtesy of Kelley Devine and Patrick Devine

6.19 Too Many Synapses? A man with fragile X syndrome.

fragile X syndrome A condition that is a frequent cause of inherited intellectual disability and is produced by a fragile site on the X chromosome that seems prone to breaking because the DNA there is unstable.

trinucleotide repeat Repetition of the same three nucleotides within a gene, which can lead to dysfunction, as in the cases of Huntington's disease and fragile X syndrome.

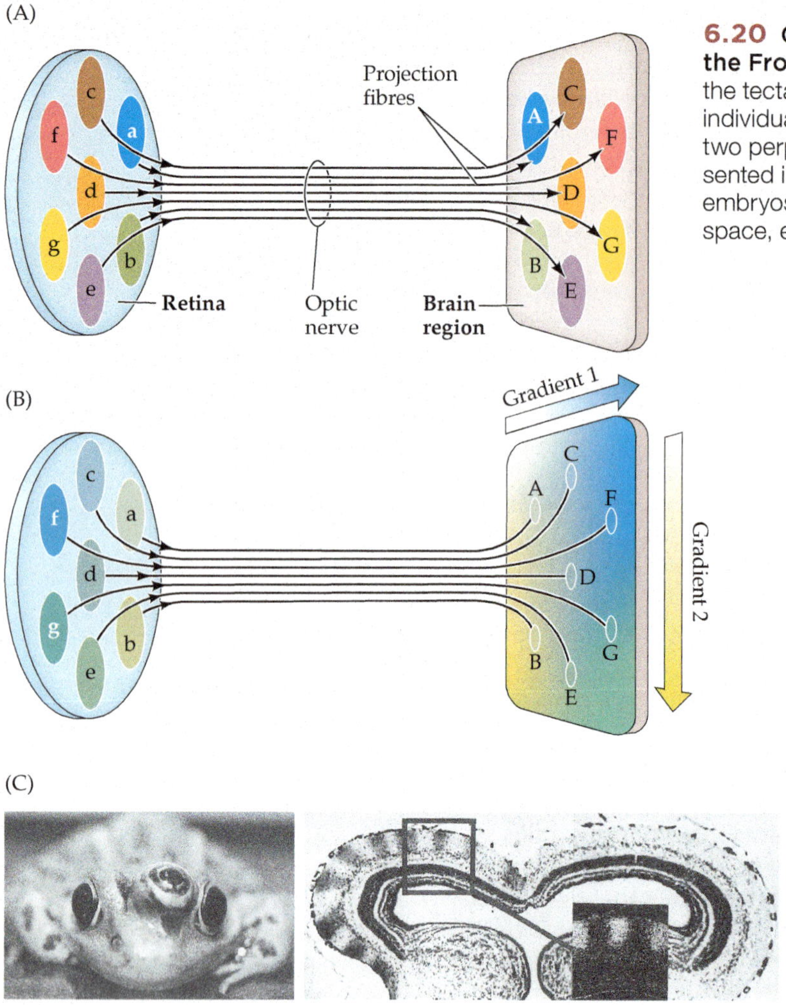

Courtesy of Martha Constantine-Paton

6.20 Chemoaffinity Guides the Initial Growth of Axons to the Frog Brain Chemical signatures between retinal neurons and the tectal neurons they innervate either could be very elaborate and individualistic, as in (A), or could be relatively simple, consisting of two perpendicular gradients of just two different cues, as represented in (B). (C) When frogs were implanted with a third eye as embryos, the two eyes innervating the same tectum competed for space, each innervating a separate patch of the brain.

is that the retinal axons had grown back to their *original* positions on the tectum, regardless of the rotation of the eye. Furthermore, once the original connections had been reestablished, the brain interpreted the information as if the eye were in its original position. Even months later, animals that underwent this treatment had not learned to make sense of information from the rotated eye.

Roger Sperry (1913-1994) proposed the **chemoaffinity hypothesis** to explain how retinal axons know which part of the tectum to innervate. Suppose each retinal cell and each tectal cell had a specific chemical identity—an address of sorts. Then each retinal cell would need only to seek out the proper address in the tectum, and the entire pattern would be reestablished; many chemical cues (represented by many colours in **FIGURE 6.20A**, top) or only a few (two colour gradients in Figure 6.20A, bottom) may be involved.

But it turned out that after arriving at the roughly appropriate region of tectum, retinal connections are fine-tuned by neural activity and visual experience. Normally, each retina innervates only the tectum on the opposite side. When implantation of a third eye forces two retinas to innervate a single tectum (**FIGURE 6.20B**, left), they each do so in the same rough pattern, but they segregate; axons from one retina predominate in one area, and axons from the other retina predominate in neighbouring tectum, so there are alternating stripes of innervation from the two eyes (Figure 6.20B, right).

This segregation depends on activity (Constantine-Paton et al., 1990). If neural activity in one eye is silenced (by injection of drugs), the eye loses its connections to the tectum and the other eye takes over, innervating the entire tectum. If both eyes are silenced (by keeping the animal in the dark), neither eye predominates, their axons fail to segregate in the tectum, and the detailed pattern of innervation fails to appear. Presumably the two eyes are competing for limited supplies of neurotrophic factor from the tectum, and active synapses take up more of the factor(s).

Thus, the retinotectal system appears to reestablish the original pattern of innervation in two steps: (1) chemical cues bring retinal axons to the approximately correct region of tectum, and (2) the neural activity of the retinal cells, normally driven by visual experience, directs these axons to innervate or maintain innervation of the precise tectal region. A similar competition goes on in young mammals as information from the two eyes competes to form synapses in visual cortex, as we'll see next.

Visual deprivation can lead to blindness

Some people do not see forms clearly with one of their eyes, even though the eye is intact and a sharp image is focused on the retina. Such impairments of vision are known as **amblyopia** (from the Greek *amblys*, "dull," and *ops*, "eye"). Some people with this disorder have an eye that is turned inward (are cross-eyed) or outward. Children born with such a misalignment see a double image rather than a single fused image. By the time an untreated person reaches the age of 7 or 8, pattern vision in the deviated eye is

chemoaffinity hypothesis The notion that each cell has a chemical identity that directs it to synapse on the proper target cell during development.

amblyopia Reduced visual acuity of one eye that is not caused by optical or retinal impairments.

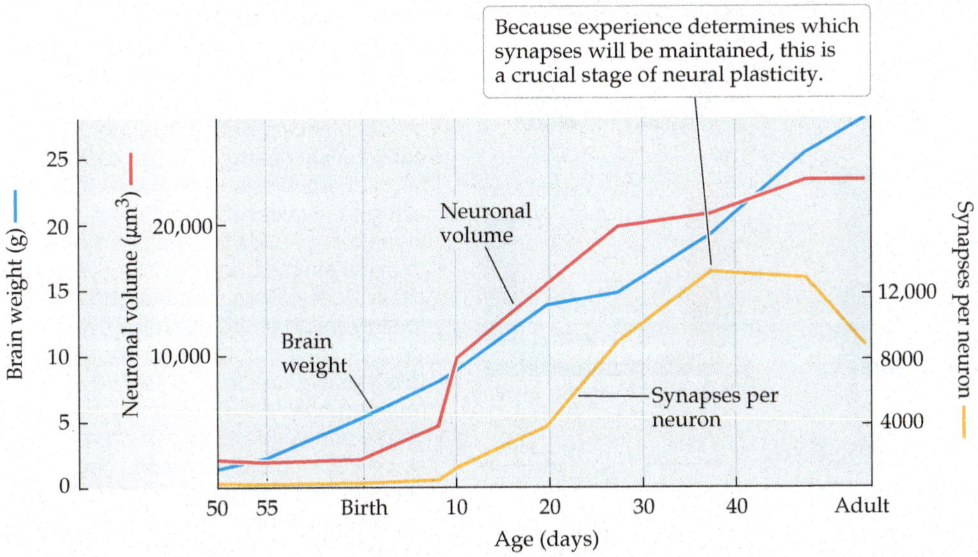

6.21 Brain Development in the Visual Cortex of Cats (After B. G. Cragg, 1975. *J Comp Neurol* 160: 147–166.)

almost completely suppressed. If the eyes are realigned during childhood, the person learns to fuse the two images and has good depth perception. But if realignment is done in adulthood, it's too late to restore acute vision to the turned eye.

Much of what we know about the causes of amblyopia comes from visual deprivation experiments with lab animals, in which the eyelids are reversibly sutured shut or the animal is fitted with frosted contact lenses, thereby preventing focused images on the retina. These experiments revealed startling changes related to disuse of the visual system in early life. **Binocular deprivation**, depriving animals of sight in both eyes, produces structural changes in visual cortical neurons: a loss of dendritic spines and a reduction in synapses. If such deprivation is maintained for several weeks during development, when the animal's eyes are opened, it will be blind. Although light enters its eyes, and the cells of the eyes send messages to the brain, the brain seems to ignore the messages, and the animal is unable to detect visual stimuli. If the deprivation lasts long enough, the animal is *never* able to recover eyesight. Thus, early visual experience is crucial for the proper development of vision, and there is a **sensitive period** during which these manipulations of experience can exert long-lasting effects on the system (**FIGURE 6.21**). These effects are most extensive during the early period of synaptic development in the visual cortex. After the sensitive period, the manipulations have little or no effect.

Depriving only one eye of light, that is, **monocular deprivation**, during development produces profound structural and functional changes in the thalamus and visual cortex and impairs vision in the deprived eye when the animal reaches adulthood. The effect of visual deprivation can be illustrated graphically by an **ocular dominance histogram**, which portrays the strength of response of a brain neuron to stimuli presented to either the left or the right eye. Normally, most cortical neurons (except those in layer IV) are excited equally by light presented to either eye (**FIGURE 6.22A**). Keeping one eye closed or covered in development results in a striking shift from the normal graph; most cortical neurons respond only to input from the non-deprived eye (**FIGURE 6.22B**). In cats the critical period for this effect is the first 4 months of life. In rhesus monkeys the sensitive period extends to age 6 months. After these ages, visual deprivation has little effect.

During early development, synapses are rearranged in the visual cortex, and axons representing input from each eye "compete" for synaptic places. Active, effective synapses predominate over inactive synapses. Thus, if one eye is "silenced," synapses carrying information from that eye are retracted while synapses driven by the other eye are maintained. Donald O. Hebb (1949) proposed that effective synapses (those that successfully drive the postsynaptic cell) might grow stronger at the expense of ineffective synapses. Thus, synapses that grow stronger or weaker depending on their effectiveness in driving their target cell are known as

binocular deprivation Depriving both eyes of form vision, as by sealing the eyelids.

sensitive period The period during development in which an organism can be permanently altered by a particular experience or treatment.

monocular deprivation Depriving one eye of light.

ocular dominance histogram A graph that portrays the strength of response of a brain neuron to stimuli presented to either the left eye or the right eye.

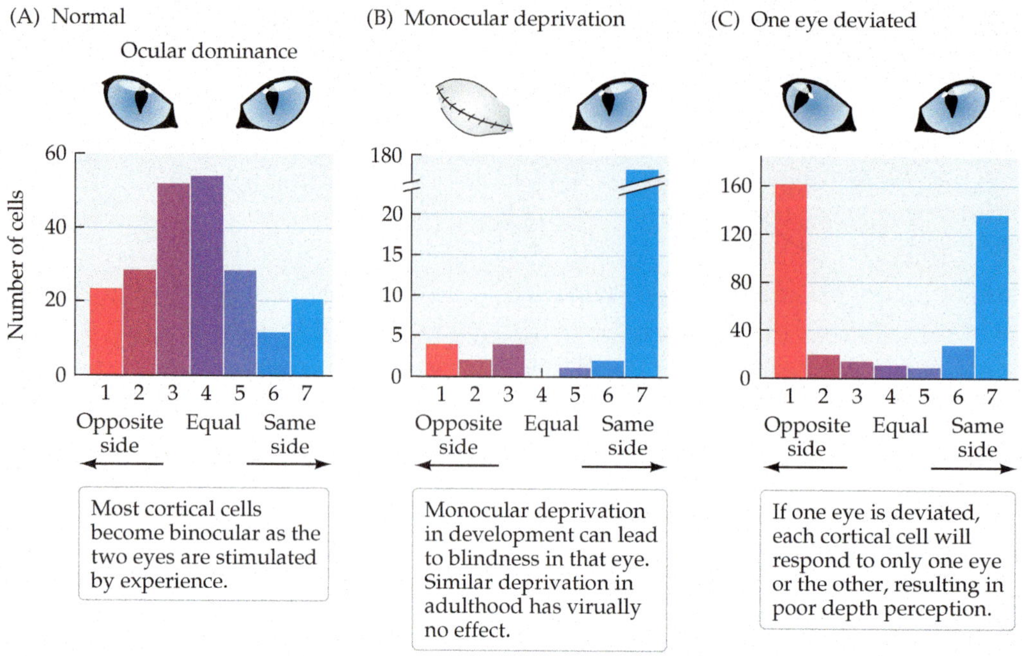

6.22 Ocular Dominance Histograms These histograms show responses of cells in the visual cortex of cats: (A) normal adults; (B) after monocular deprivation through the early critical period; (C) after early deviation of one eye (squint). The numbers along the x-axis represent a gradation in response: Cells that respond *only* to stimulation of the opposite eye are class 1 cells. Cells that respond *mainly* to stimulation of the opposite eye are class 2. Cells that respond equally to either eye are class 4. Cells that respond only to stimulation of the eye on the same side are class 7, and so on. (D) Hebbian synapses can account for changes after monocular deprivation. (Parts A and C after D. H. Hubel and T. N. Wiesel, 1965. *J Neurophysiol* 28: 1041–1059; B after T. N. Wiesel and D. H. Hubel, 1965. *J Neurophysiol* 28: 1029–1040.)

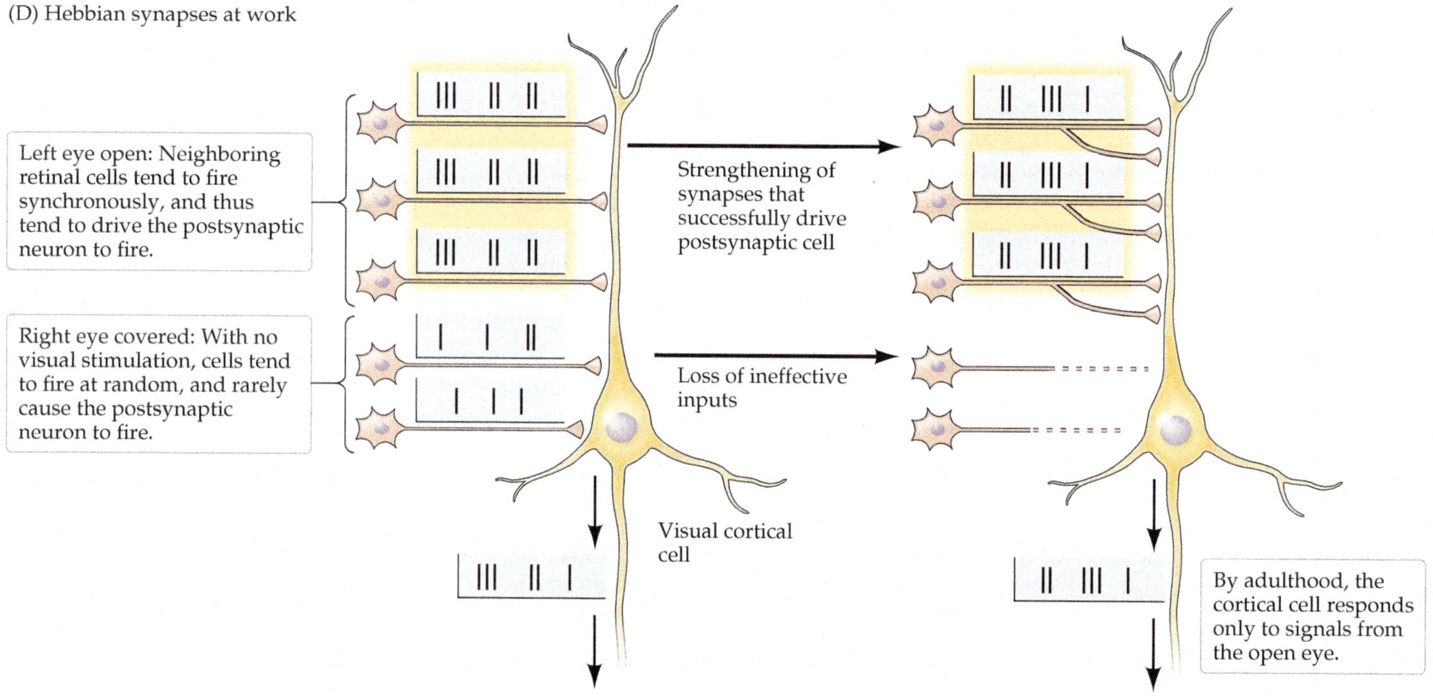

Hebbian synapse A synapse that is strengthened when it successfully drives the postsynaptic cell.

Hebbian synapses (**FIGURE 6.22D**). In Chapter 16 we will see that Hebbian synapses likely play a role in learning and memory.

Researchers offer a similar explanation for amblyopia produced by misalignment of the eyes. Hubel and Wiesel (1965) produced an animal replica of this human condition by surgically causing the eyes to diverge in kittens. The ocular dominance histogram of these animals reveals that the normal binocular sensitivity of visual cortical cells is greatly reduced (**FIGURE 6.22C**). A much larger proportion of visual cortical cells is excited by stimulation of either the right or the left eye in these animals than in control animals. The reason for this effect is that after surgery, visual stimuli falling on the misaligned eyes no longer provide simultaneous, convergent input to the cells of the visual cortex.

Neurotrophic factors may be playing a role in experience-driven synapse rearrangement. For example, if the postsynaptic cells are making a limited supply of a neurotrophic factor, and if active synapses take up more of the factor than inactive synapses do, then perhaps the inactive axons retract for lack of neurotrophic factor. BDNF has been implicated as the neurotrophic factor being competed for in the mammalian visual cortex (McAllister et al., 1997; Sansevero et al., 2019) and in the frog retinotectal system (Du and Poo, 2004). So perhaps ineffective synapses wither for lack of neurotrophic support. In **A STEP FURTHER 6.1: EXPERIENCES IN NONVISUAL SENSES ALSO AFFECT NEURAL DEVELOPMENT** on the website, you can learn how mouse whiskers compete for synapses in the cortex.

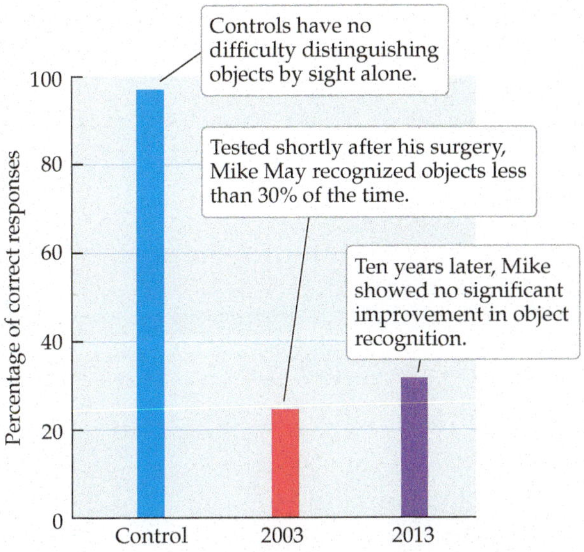

6.23 What Do I See? Despite having the damage to his eye surgically corrected, Mike May was rather poor at distinguishing between basic objects, like spheres versus cubes. When retested 10 years later, there was no significant improvement in his ability to distinguish objects by sight alone. (After E. Huber et al., 2015. *Psychol Sci* 26: 393–401.)

Early exposure to visual patterns helps fine-tune connections in the visual system

Human disorders have also proven that early experience is crucial for vision. Babies born with cataracts (cloudy lenses) in industrialized countries usually have them removed a few months after birth and will have good vision. But if such a child grows up with the cataracts in place, removing them in adulthood is much less effective; the adults acquire the use of vision slowly (Ostrovsky et al., 2009) and to only a limited extent (Bower, 2003). Early visual experience is known to be especially crucial for learning to perceive faces, because infants with cataracts that occlude vision for just the first 6 months of life are impaired at recognizing faces even 9 years later (Le Grand et al., 2001). These experience-dependent effects are probably mediated by synapse rearrangement within the visual cortex (Ruthazer et al., 2003) like that seen in kittens.

Why does Mike May, whom we met at the start of the chapter, have such poor vision despite the clear images entering his eye? Ten years after his vision was restored, he still could not reliably distinguish objects by sight alone (**FIGURE 6.23**). Had the accident happened to him as an adult, the surgery to let light back into his eye would have restored normal vision. But, like a kitten growing up with opaque contact lenses, Mike was deprived of form vision—in his case, for over 40 years. Because this deprivation began when he was a child, synaptic connections within his visual cortex were not strengthened by the patterns of light moving across the retina. In the absence of patterned stimulation, synapses between the eye and the brain languished and disappeared.

In one sense, Mike was lucky that his blindness came as late as it did. He had normal form vision for the first 3½ years of his life, and that stimulation may have been sufficient to maintain some synapses that would otherwise have been lost. These residual synapses are probably what allow him to make any sense whatsoever of his vision. Yet, despite more than a decade of visual experience as an adult, Mike still has problems distinguishing three-dimensional objects or faces (Huber et al., 2015). Other people who gain vision for the first time as adults have similar difficulties recognizing objects and faces (Gregory and Wallace, 1963; Sikl et al., 2013).

One demonstration shows how visual experience in everyday life may affect our perception. In **FIGURE 6.24**, the numbers and letters along the bottom line appear more slanted than those above, but in fact the slant is the same. One theory of why we see a difference here that doesn't exist is that our experience reading digital clock readouts and italic fonts may tune synapses in the brain to perceive them as more upright than they really are—an effect lost if the figures are backward (Whitaker and McGraw, 2000).

© Florence Low

Learning to See Despite over a decade of vision in one eye, Mike May still has a hard time identifying faces.

6.24 Which Line Is More Slanted? The numbers and letters on the lower line look more slanted than those on the upper line, but in fact the characters on the two lines are equally slanted. If you look at them in a mirror, the upper line will look more slanted. Does this optical illusion result from modification of synapses caused by looking at digital clocks and italic font?

6.4 Experience Can Alter Gene Expression to Affect Brain Development

Learning Objectives

After reading this section, you should be able to:

6.4.1 **Define epigenetic changes and explain how they might play a role in the interaction of genes and the environment.**

6.4.2 **Describe the mechanism by which rodent dams can alter their offspring's lifelong stress response.**

6.4.3 **Offer evidence that glia play a central role in Rett syndrome.**

We saw that in the case of PKU, the genotype alone does not determine whether there will be disability. Only if sufficient phenylalanine is in the diet will brain development be impaired. A much more prominent mechanism by which the environment influences gene action concerns gene expression. It turns out that no matter which version of a gene you inherit, whether and how much that gene will be expressed may depend on experience. In other words, experience can control how genes are used in the developing brain. Thus, every behaviour is affected not only by genes, but also by experience.

Experience regulates gene expression in the developing and mature brain

We've already seen how neural activity and experience can affect gene expression. There is growing excitement about how experience can also exert long-lasting changes in gene expression, and how that can affect neural development.

EPIGENETICS Recall that although nearly all of the cells in your body have a complete copy of your genome, each cell uses only a small subset of those genes at any one time. We told you earlier that when a cell transcribes a particular gene and makes the encoded protein, we say the cell has *expressed* that gene. **Epigenetics** is the study of factors that affect gene *expression* without making any changes in the nucleotide sequence of the genes themselves. One important epigenetic factor affecting the developing brain in mice is the mothering they receive. If genetically identical embryos of one mouse strain are implanted into the wombs of two foster mothers, of either their own strain or another strain, their behaviour is affected (Francis et al., 2003). Strain B6 males carried and raised by mothers from another strain (Balb) show significant differences in several behaviours, including maze running and measures of anxiety (**FIGURE 6.25**). Since the various B6 males are *genetically identical* to one another, their different behaviours must be due to the effects of both different prenatal environments and different postnatal experiences on how those genes are expressed.

One particular influence of mothering on gene expression has been well documented. **Methylation** is a chemical modification of DNA that does not affect the nucleotide sequence of a gene but makes that gene less likely to be expressed. There are patterns of gene methylation in the developing cortex that are consistent from one individual to another (R. Lister et al., 2013), indicating that this regulation of gene expression is important for proper development. Neuronal activity can affect the methylation of genes, and therefore the likelihood those genes will be expressed, even in adulthood (Guo et al., 2011). Thus experience at one time in life may affect gene expression later. One well-characterized example is when rodent pups are provided with inattentive mothers, which leads to methylation of the gene for the glucocorticoid receptor (discussed in Chapters 4 and 11) in the pups' brains. This methylation reduces expression of the gene in the pups, so they secrete more glucocorticoids in response to stress (T. Y. Zhang and Meaney, 2010), making them hyperresponsive to stress as adults (**FIGURE 6.26**).

A similar mechanism may apply to humans, because this same gene is also more likely to be methylated in the postmortem brains of suicide victims than of controls, *but only if the victim was subjected to childhood abuse*. Suicide victims who did not suffer childhood abuse were no more likely to have the gene methylated than were controls (McGowan et al., 2009). These results suggest that methylation of the gene in abused

epigenetics The study of factors that affect gene expression without making any changes in the nucleotide sequence of the genes themselves.

methylation A chemical modification of DNA that does not affect the nucleotide sequence of a gene but makes that gene less likely to be expressed.

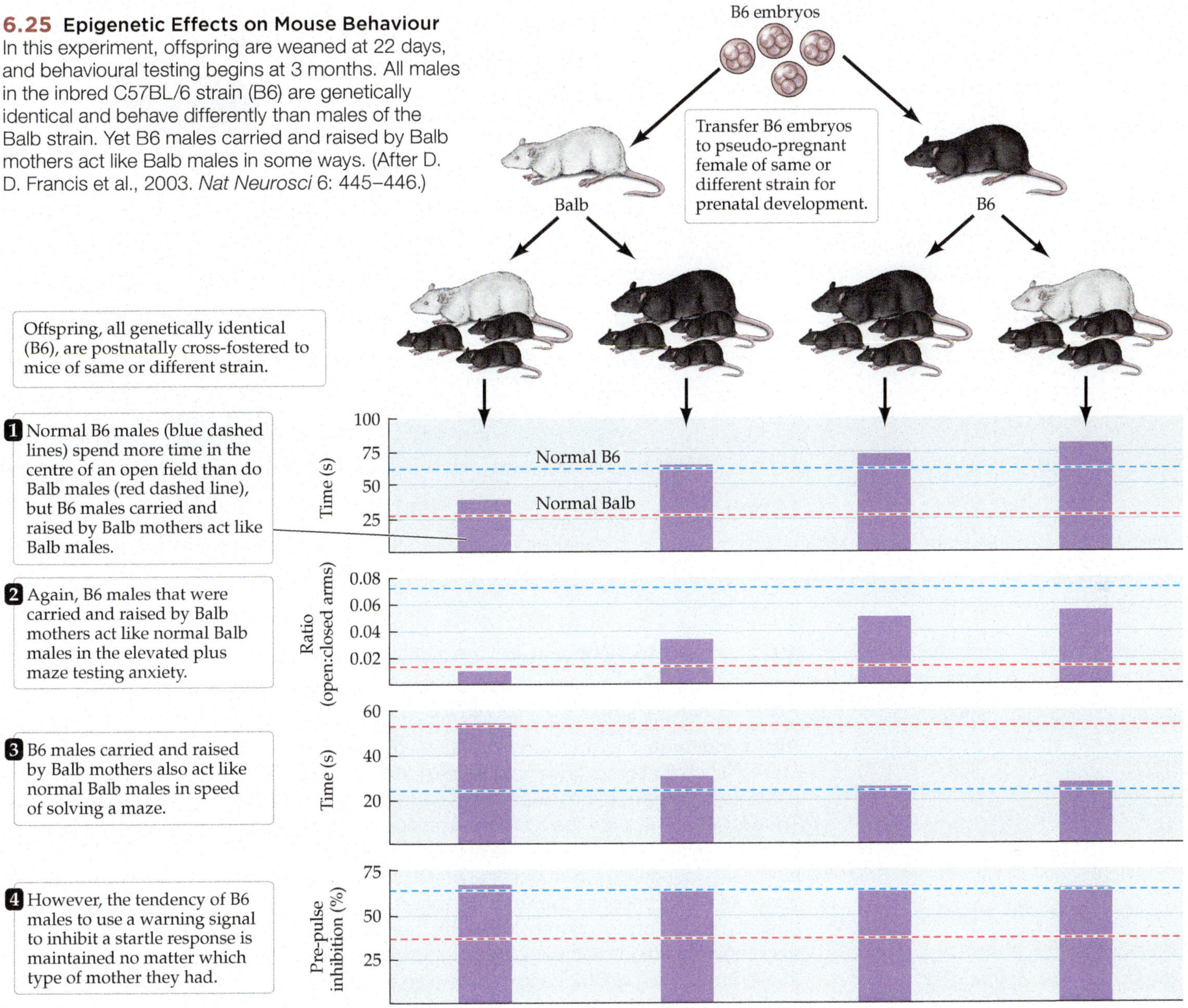

6.25 Epigenetic Effects on Mouse Behaviour In this experiment, offspring are weaned at 22 days, and behavioural testing begins at 3 months. All males in the inbred C57BL/6 strain (B6) are genetically identical and behave differently than males of the Balb strain. Yet B6 males carried and raised by Balb mothers act like Balb males in some ways. (After D. D. Francis et al., 2003. *Nat Neurosci* 6: 445–446.)

children may have made them hyperresponsive to stress as adults—a condition that may have led them to take their own lives. This is a powerful demonstration of epigenetic influences on our behaviour.

EPIGENETIC REGULATION IN ADULTS Neurons in adults also alter gene expression in response to synaptic stimulation. Some genes, called *immediate early genes*, are briefly expressed by almost any neuron that has been stimulated (see Box 1.1). Neuroscientists exploit this process by exposing an animal to, say, a sound of a particular frequency and then examining the brain to see which neurons altered gene expression in response to different frequencies. Likewise, lights, odours, or touches will all affect neuronal expression of immediate early genes in particular regions of the brain and spinal cord that receive information about those sensations. Experience also affects the expression of many other genes (Mayfield et al., 2002), not just immediate early genes. Thus, one reason why genetically identical individuals do not have identical brains or behaviour is that they are inevitably exposed to different experiences, so they grow up expressing their identical genes in very nonidentical ways (Ridley, 2003).

Before we conclude the chapter by examining old age, let's look at exciting work that suggests it may be possible to reverse one cause of severe intellectual disability.

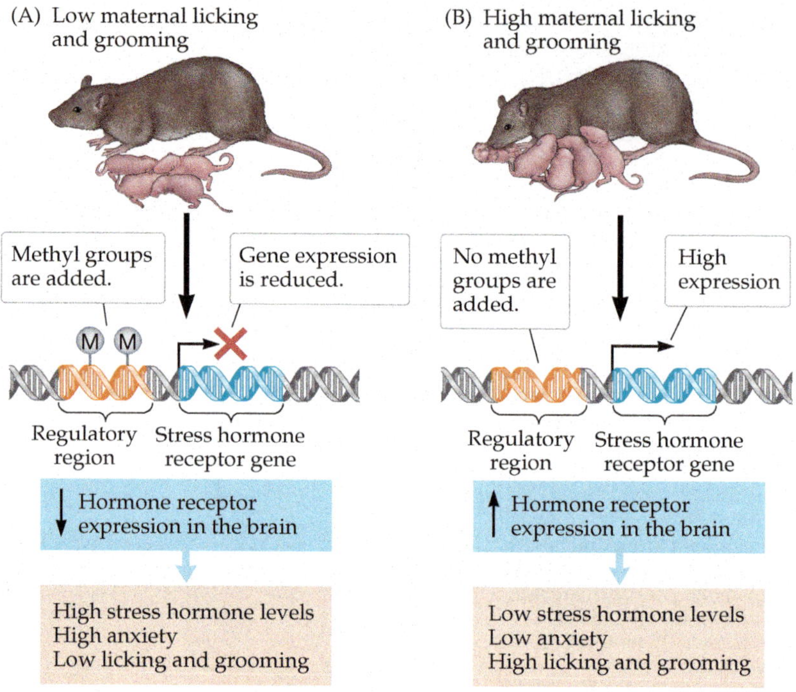

6.26 Early Experience Imprints Genes to Affect the Stress Response in Adulthood Attentive rodent mothers prevent methylation of the stress hormone receptor gene in their pups, so their daughters grow up to be attentive mothers themselves. In this way, maternal care can have epigenetic effects that can be transmitted across generations. (After D. A. Hackman et al., 2010. *Nat Rev Neurosci* 11: 651–659.)

The Cutting Edge

Harnessing Glia to Reverse an Inherited Brain Disorder

Rett syndrome is a rare disorder caused by a mutation that disables a gene called *MeCP2* such that it no longer produces a functional protein (Ip et al., 2018). Absence of the protein disrupts development in many parts of the body, including the brain. The *MeCP2* gene is on the X chromosome, so males who inherit a dysfunctional copy on their only X cannot make any functional protein and almost always die before birth. Girls inheriting a dysfunctional *MeCP2* on one X chromosome are not as severely affected, because they can still make some functioning protein from the gene on their *other* X chromosome. However, their body and brain growth is impaired, and they show severe intellectual disability, typically never learning to speak.

Disrupting the *MeCP2* gene in mice also severely affects body and brain growth, as well as behaviour. You might expect that the impaired brain growth would be due to the lack of *MeCP2* protein in neurons. However, recent experiments suggest that the problem in brain growth in these knockout mice (see Box 6.1) may be caused by the absence of the protein in *glial cells*. Recall from Chapter 1 that microglia in the brain remain tiny (as their name implies) unless there is damage nearby. In that case, the microglia divide and grow to engulf and clean up any debris.

The strongest evidence that the symptoms of Rett syndrome in these mice are caused by microglia comes from dramatic experiments where animals lacking a functional *MeCP2* gene are rescued by transplants of normal microglia (Derecki et al., 2012). This is possible because microglia are part of the immune system, originating from bone marrow cells that enter the brain, where inducers transform the cells into microglia. The researchers subjected the knockout mice to radiation that is known to destroy the entire immune system, including microglia in the brain. Then they infused the animals with bone marrow cells, which went throughout the body, replacing the entire immune system, including microglia (**FIGURE 6.27A**). When 28-day-old *MeCP2* knockout animals were treated this way and given bone marrow from *normal* mice, their brain and body growth was restored (**FIGURE 6.27B**). What's more, the knockout animals given normal microglia also developed more normal behaviour (**FIGURE 6.27C**). Notice that when the knockout mice were provided bone marrow from other

Rett syndrome A rare genetic disorder, occurring almost exclusively in girls, of slowing development resulting in intellectual disability, stereotyped movements, and loss of language.

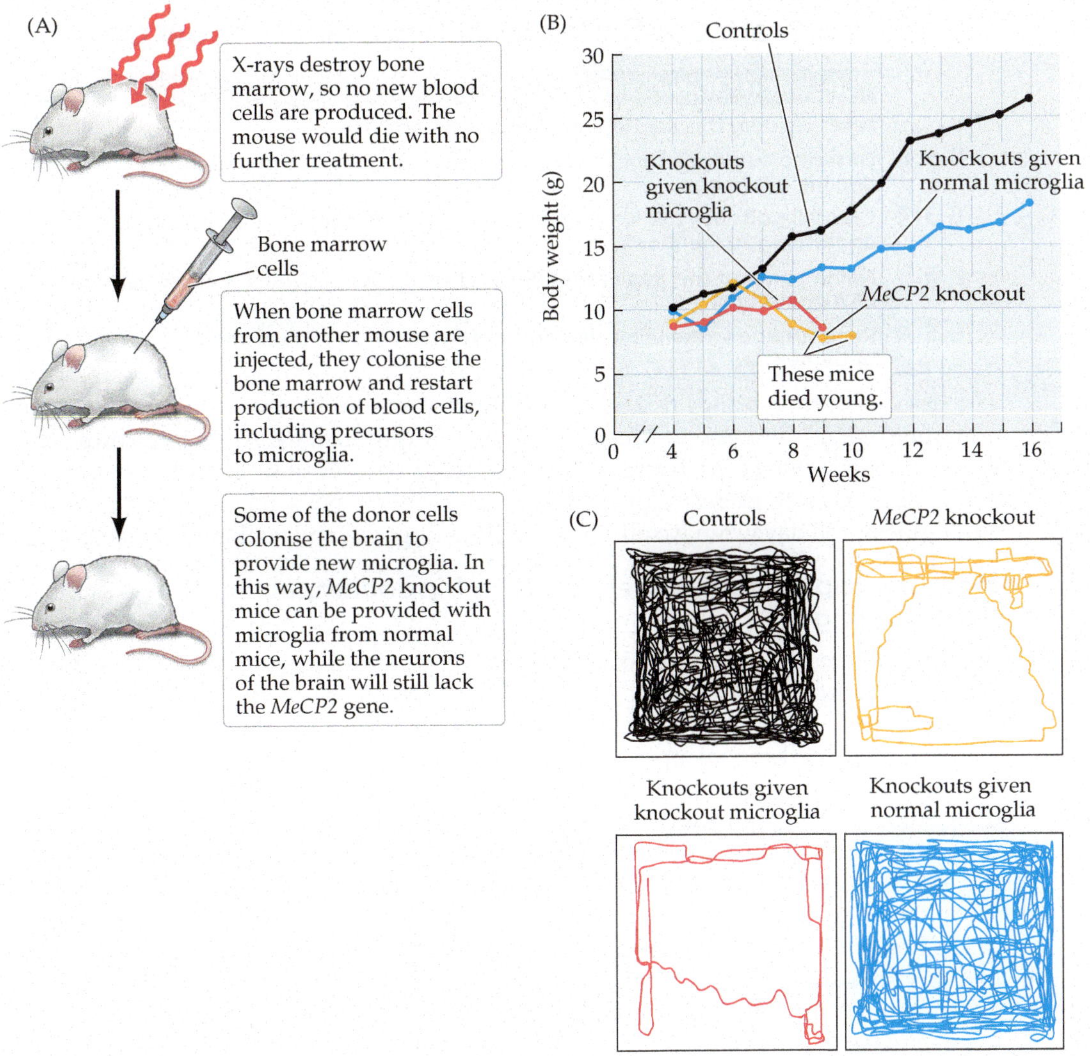

6.27 Rescuing Mice from Rett Syndrome (A) Rett syndrome is caused by a dysfunctional *MeCP2* gene, so an animal model was created by knocking out that gene in mice. Researchers destroyed the mutant microglia by exposing the knockout mice to radiation, then infused bone marrow cells from normal animals. Some bone marrow cells entered the brain and differentiated into microglia. (B) The normal microglia also allowed the animals to gain weight and avoid early death. Note that providing knockout mice with microglia from other knockouts was ineffective. (C) Providing the knockout mice with normal microglia also rescued their behaviour. The lines represent the paths that a representative mouse from each group took while exploring an open field. (After N. Derecki et al., 2012. *Nature* 484: 105–109.)

knockouts, there was no benefit. If researchers waited until the animals were older before replacing the microglia, it was too late. The mice had to get normal microglia in place early in development to be rescued.

This rescue is impressive, but you might wonder whether the benefit really came from replacing microglia. Maybe what helped the knockout mice was replacing the immune system *outside* the brain. But the researchers showed that the benefit came from the brain. When they put a lead shield over the animals' heads during the irradiation, protecting microglia in the brain, then the normal bone marrow cells did not enter the brain (because the abnormal microglia were already there). In that case, infusing normal bone marrow did no good—the animals died young. Only by killing off the microglia lacking the gene, so that invading bone marrow cells would replace them, could they rescue the animals.

These results suggest that the symptoms of Rett syndrome are caused by a failure of microglia to clear away debris in the developing brain. Perhaps the buildup of waste products interferes with synaptic rearrangement or leads to loss of neurons (Petrelli et al., 2016). If so, then it might be possible to treat newborns who inherit the mutant gene by harvesting their bone marrow, introducing a functional copy of the *MeCP2* gene into the harvested cells, then implanting them in the brain to provide functional microglia. These results also raise the question of whether other mental disorders might be the result of more subtle dysfunction in microglia. Long ignored by scientists, glia are turning out to be a lot more important to neural development than just as "glue" for neurons. ■

6.5 The Brain Continues to Change as We Grow Older

Learning Objectives

After reading this section, you should be able to:

6.5.1 Describe evidence that the hippocampus plays a role in memory decline in old age.

6.5.2 Critically discuss the leading theory about the sequence of events underlying Alzheimer's.

6.5.3 Name some of the genes that have been associated with greater risk of Alzheimer's.

6.5.4 Recommend some lifestyle choices that seem to reduce the risk of dementia.

The passage of time brings us an accumulation of joys and sorrows—perhaps riches and fame—and a progressive decline in many of our abilities. Although slower responses seem inevitable with aging, many cognitive abilities show little change during the adult years, until we reach an advanced age. What happens to brain structure from adulthood to the day when we all become a little forgetful and walk more hesitantly?

Memory impairment correlates with hippocampal shrinkage during aging

In a study of healthy people age 55–87, investigators asked whether mild impairment in memory is specifically related to reduction in size of the hippocampal formation (HF) or is better explained by generalised shrinkage of brain tissue. (In Chapter 16 we'll see that the hippocampus is implicated in memory.) Volunteers took a series of memory tests and were scored for both immediate recall and delayed recall. MRI images for each person were measured for three variables (**FIGURE 6.28**): (1) volume of the HF; (2) volume of the supratemporal gyrus, a region that is close to the HF and is known to shrink with age but has not been implicated in memory; and (3) volume of the fluid-filled space between the interior of the skull and the surface of the brain (which yields a measure of overall shrinkage of the brain). Immediate memory showed very little decline with age, but delayed memory did decline. When effects of sex, age, IQ, and overall brain atrophy were eliminated statistically, HF volume was the only brain measure that correlated significantly with the delayed-memory score (Golomb et al., 1994). Later studies confirmed this correlation between volume of the medial temporal lobe, which encases the hippocampus, and memory in the elderly (H. R. Bailey et al., 2013).

Two other brain regions show how different the effects of aging can be. In the motor cortex, a type of large neuron—called the *Betz cell*—starts to decline in number by about age 50, and by the time a person reaches age 80, many of these cells have shriveled away (M. E. Scheibel et al., 1977). In contrast, other cells involved in motor circuitry—for example, those in an area of the brainstem called the *inferior olive*—remain about the same in number through at least eight decades of life (Sjöbeck et al., 1999).

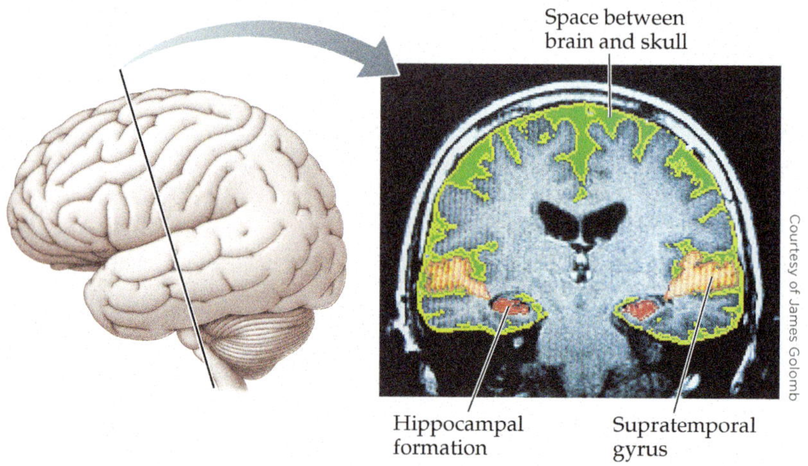

6.28 Hippocampal Shrinkage Is Correlated with Age-Related Memory Decline MRI images like the one on the right, taken from the sectional plane shown on the left, illustrated that shrinkage of the hippocampal formation was correlated with memory decline in cognitively healthy elderly people. Overall brain shrinkage and shrinkage of the supratemporal gyrus were not. (Based on J. Golomb et al., 1994. *J Neurol Neurosurg Psychiatry* 57: 590–593.)

PET scans of elderly people add a new perspective to aging-related changes. Studies of healthy people reveal that cerebral metabolism remains almost constant. This stability is in marked contrast to the decline of cerebral metabolism in Alzheimer's disease (see Figure 1.24D), which we will consider next.

Is Alzheimer's disease caused by amyloid plaques?

The population of elderly people in the United States is increasing dramatically. Most people reaching an advanced age lead happy, productive lives, although at a slower pace than they did in their earlier years. In some elderly people, however, age has brought a particular agony: the disorder called **Alzheimer's**

disease, named after Alois Alzheimer (1864–1915), the neurologist who first described a type of **dementia** (drastic failure of cognitive ability, including memory failure and loss of orientation) appearing before the age of 65.

Over 35 million people worldwide have some type of senile dementia (Abbott, 2011), and the progressive aging of our population means that these ranks will continue to swell. The frequency of Alzheimer's increases with aging up to age 85–90, but several people have lived over 110 years without showing signs of mental impairment (Z. Yang et al., 2013). This last finding indicates that Alzheimer's is in fact a disease, and not simply the result of wear and tear in the brain. The fact that remaining physically and mentally active reduces the risk of developing Alzheimer's disease (Smyth et al., 2004) also refutes the notion that brains simply "wear out" with age. Extensive use of the brain makes Alzheimer's *less* likely.

Alzheimer's disease begins as a loss of memory of recent events. Eventually this memory impairment becomes all-encompassing, so extensive that Alzheimer's patients cannot maintain any form of conversation because both the context and prior information are rapidly lost. They cannot answer simple questions such as, What year is it? Who is the president of the United States? or Where are you now? Cognitive decline is progressive and relentless, until the patient needs constant supervision.

The cerebral cortex of a patient with Alzheimer's shows striking atrophy, especially in the frontal, temporal, and parietal areas. PET scans show marked reduction of metabolism in posterior parietal cortex and some portions of the temporal lobe (C. Marcus et al., 2014) (see Figure 2.24D). The brains of individuals with Alzheimer's also reveal progressive changes at the cellular level (**FIGURE 6.29**):

- Patches termed **senile plaques** appear in the cortex, the hippocampus, and associated limbic system sites. The plaques, formed by the buildup of a substance called **β-amyloid**, impair synaptic function (Wei et al., 2010).
- Many cells show abnormalities called **neurofibrillary tangles**, which are abnormal whorls of neurofilaments, including a protein called **Tau**, that form a tangled array inside the cell. The number of neurofibrillary tangles is directly related to the magnitude of cognitive impairment (Y. Wang and Mandelkow, 2016), and they may be a secondary response to amyloid plaques.
- These degenerative events may cause the cholinergic neurons in the basal forebrain to disappear in Alzheimer's patients, either because the cells die or because they stop producing acetylcholine. The latter possibility is more likely, because providing these neurons with NGF restores their cholinergic characteristics in aged monkeys (D. E. Smith et al., 1999).

The only surefire diagnosis for Alzheimer's at present is postmortem examination of the brain for senile plaques and neurofibrillary tangles. But one approach

Alzheimer's disease A form of dementia that may appear in middle age but is more frequent among the aged.

dementia Drastic failure of cognitive ability, including memory failure and loss of orientation.

senile plaques Also called *amyloid plaques*. Small areas of the brain that have abnormal cellular and chemical patterns. Senile plaques correlate with senile dementia.

β-amyloid A protein that accumulates in senile plaques in Alzheimer's disease.

neurofibrillary tangle An abnormal whorl of neurofilaments within nerve cells.

Tau A protein associated with neurofibrillary tangles in Alzheimer's disease.

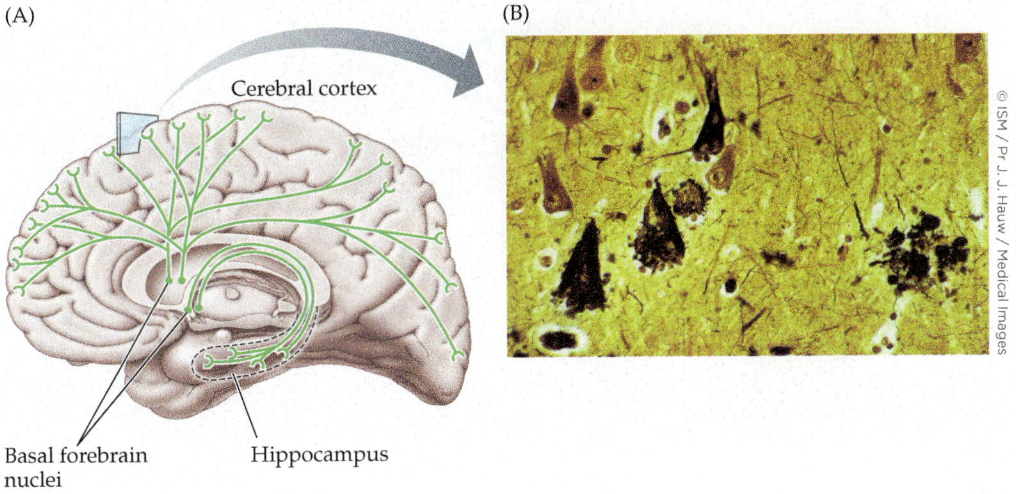

6.29 People with Alzheimer's Show Structural Changes in the Brain
(A) This representation of the brain shows the location of the basal forebrain nuclei and the distribution of their axons, which use acetylcholine as a neurotransmitter. These cells seem to disappear in Alzheimer's patients. (B) Neurofibrillary tangles (the flame-shaped objects) and senile plaques (the darkly stained clusters) are visible in this micrograph of the cerebral cortex of an aged patient with Alzheimer's.

amyloid precursor protein (APP) A protein that, when cleaved by several enzymes, produces β-amyloid.

β-secretase An enzyme that cleaves amyloid precursor protein, forming β-amyloid, which is suspected of causing Alzheimer's disease.

presenilin An enzyme that cleaves amyloid precursor protein, forming β-amyloid.

apolipoprotein E (ApoE) A protein that may help break down amyloid.

is to inject a dye, called *Pittsburgh Blue* (*PiB*), that has an affinity for β-amyloid. Then, a PET scan determines whether the dye has accumulated in the brain (Wolk et al., 2009). The brain of virtually every patient diagnosed with Alzheimer's accumulates the dye, as do the brains of many elderly people showing mild cognitive impairment. Scientists are working on similar scans to monitor Tau tangles, especially in the hippocampus (M. Maruyama et al., 2013), which might offer even better predictions about the onset of memory problems in people with Alzheimer's (**FIGURE 6.30**) and provide a way to watch the progress of drugs being tested to treat Alzheimer's (Sevigny et al., 2016). Other groups hope to detect Alzheimer's by measuring β-amyloid in blood (Nakamura et al., 2018).

If amyloid plaques are the primary cause of Alzheimer's disease, what causes the buildup of β-amyloid? **Amyloid precursor protein** (**APP**) is cleaved by two enzymes—**β-secretase** and **presenilin**—to form extracellular β-amyloid that builds up. This buildup may be accelerated by a breakdown in the lymphatic system of waste disposal (Da Mesquita et al., 2018) that we discussed in Chapter 1 (see Figure 1.23). Once a buildup of β-amyloid forms, it seems to attract more β-amyloid molecules to join, reminiscent of infectious prion proteins (Jaunmuktane et al., 2015) that we'll discuss in Chapter 10. Perhaps in response to the β-amyloid, some neurons form neurofibrillary tangles of Tau. Another enzyme, **apolipoprotein E** (**ApoE**), normally breaks down β-amyloid (Bu, 2009) and Tau (Shi et al., 2017). Mutations in each of the genes that produce these proteins can increase the risk of Alzheimer's disease (Bertram and Tanzi, 2008), but most patients have normal copies of the genes (**FIGURE 6.31**). There is growing evidence that β-amyloid normally encases invading microbes (Kumar et al., 2016), so Alzheimer's might be the result of an overly vigourous response to infection.

Several treatment strategies are being explored. For example, injection of antibodies that bind β-amyloid should slow the formation of plaques, but they do not seem to delay symptoms (C. Holmes et al., 2008). Another strategy is to develop drugs that interfere with enzymes that produce β-amyloid or that boost enzymes like ApoE that break down the amyloid (Cramer et al., 2012). Unfortunately, all the drug trials to date have been disappointing, which is leading at least some scientists to question whether amyloid actually causes the dementia or is a brain response to stave off whatever changes actually cause the symptoms (Makin, 2018). In the meantime, and in keeping with the repeated theme of this chapter that genes and

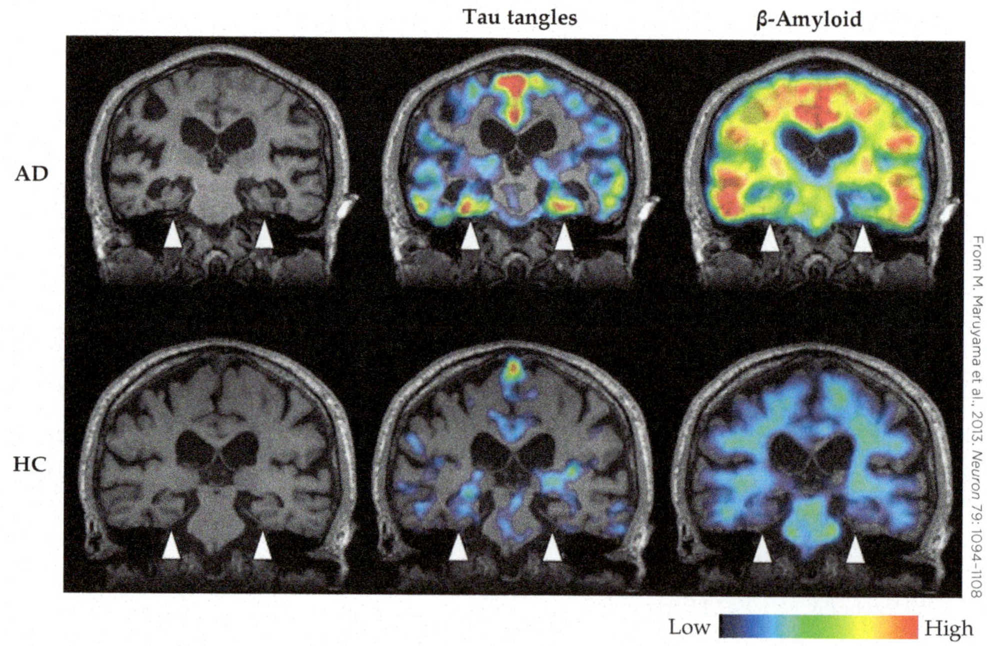

6.30 Imaging Tau Tangles and Amyloid Plaques in the Brain
By injecting radiolabelled markers, scientists can image Tau tangles, which are concentrated in the hippocampus (indicated by arrowheads) of an Alzheimer's disease (AD) patient compared with a healthy control (HC). Similarly, Pittsburgh Blue can reveal β-amyloid, which is abundant throughout the brain of the AD patient.

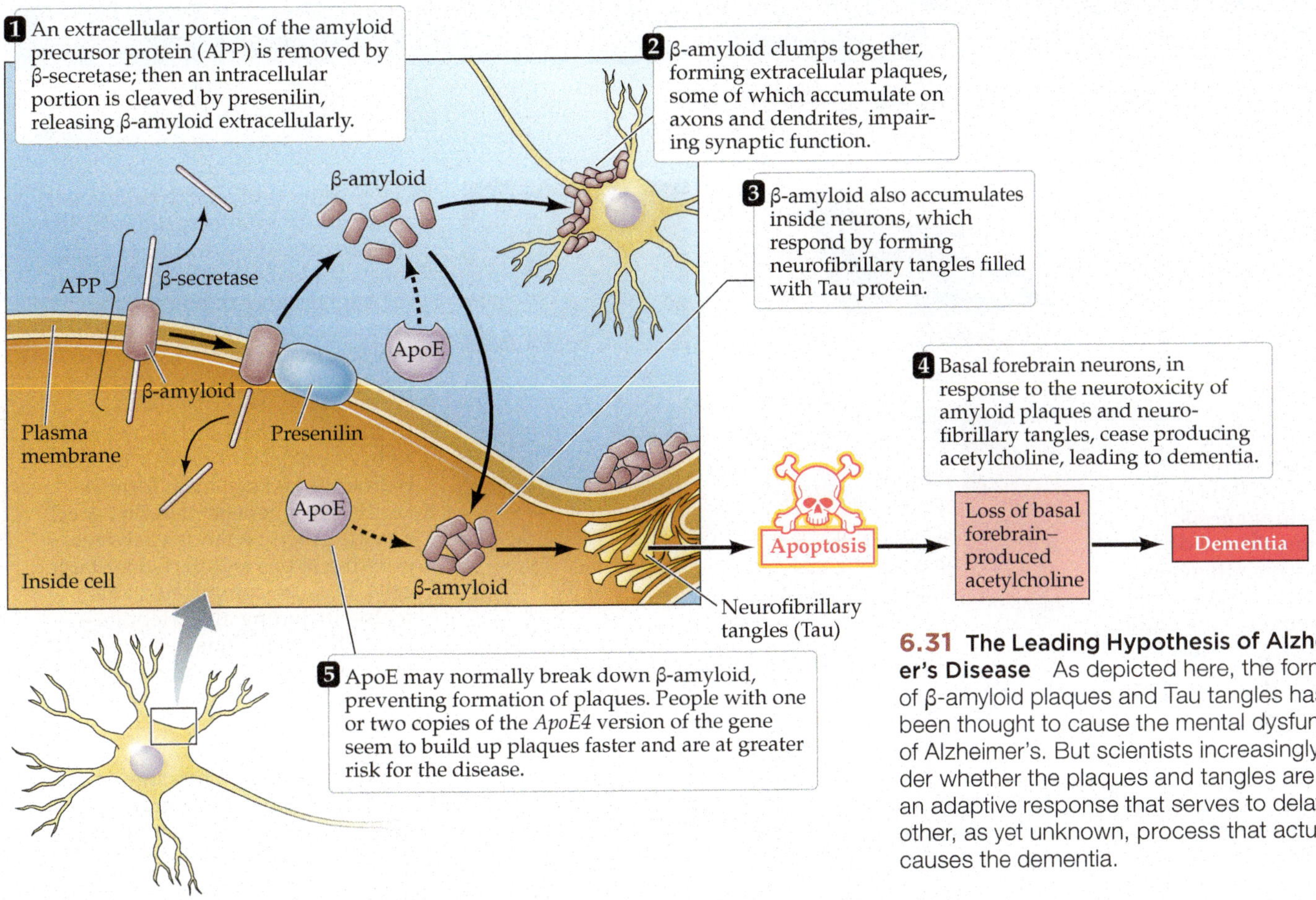

6.31 The Leading Hypothesis of Alzheimer's Disease As depicted here, the formation of β-amyloid plaques and Tau tangles has long been thought to cause the mental dysfunction of Alzheimer's. But scientists increasingly wonder whether the plaques and tangles are in fact an adaptive response that serves to delay some other, as yet unknown, process that actually causes the dementia.

experience interact, there is good evidence that physical activity, mental activity (Winblad et al., 2016), and adequate sleep (Ju et al., 2014) can postpone the appearance of Alzheimer's disease, no matter what genes you inherited.

Recommended Reading

Bazzett, T. J. (2008). *An Introduction to Behaviour Genetics.* Sunderland, MA: Oxford University Press/Sinauer.

Breedlove, S. M. (2017). *Foundations of Neural Development.* Sunderland, MA: Oxford University Press/Sinauer.

Gilbert, S. F., and Barresi, M. J. F. (2016). *Developmental Biology* (11th ed.). Sunderland, MA: Oxford University Press/Sinauer.

Sanes, D. H., Reh, T. A., and Harris, W. A. (2019). *Development of the Nervous System* (4th ed.). San Diego, CA: Academic Press.

Steinberg, L. (2014). *Age of Opportunity: Lessons from the New Science of Adolescence.* New York: Houghton Mifflin Harcourt.

Wolfe, M. S. (2016). *Developing Therapeutics for Alzheimer's Disease: Progress and Challenges.* San Diego, CA: Academic Press.

6 Visual Summary

You should be able to relate each summary to the adjacent illustration, including structures and processes. You can also refer back to the links to figures, animations, and activities that will help you consolidate the material.

1 The vertebrate brain develops from a **neural tube** with three subdivisions that will become the **forebrain**, **midbrain**, and **hindbrain**. Review Figure 6.1, Activity 6.1, Videos 6.1, 6.2

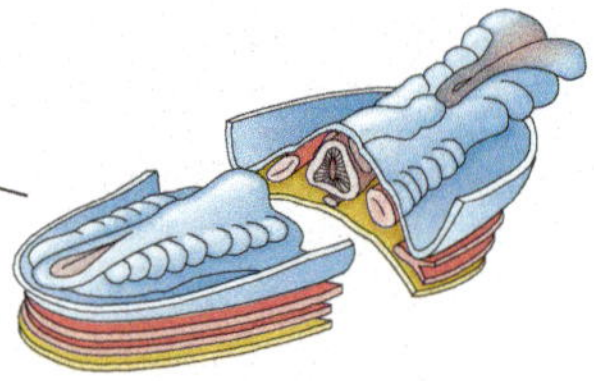

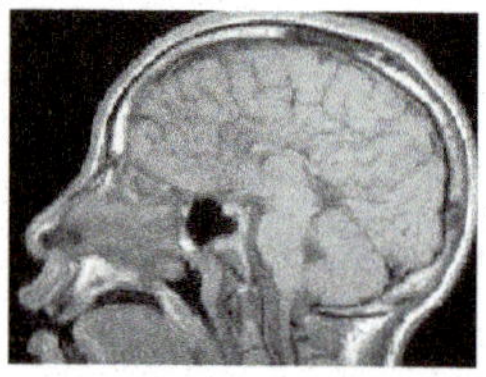

2 Impairments of fetal development that lead to **intellectual disability** can be caused by the use of drugs such as alcohol during pregnancy. The inheritance of many different genes can lead to intellectual disability. Review Figure 6.3

3 Early embryological events in the formation of the nervous system include a sequence of six cellular processes: (1) **neurogenesis**, (2) **cell migration**, (3) **cell differentiation**, (4) **synaptogenesis**, (5) **cell death**, and (6) **synapse rearrangement**. Review Figure 6.4, Animation 6.3, Video 6.4

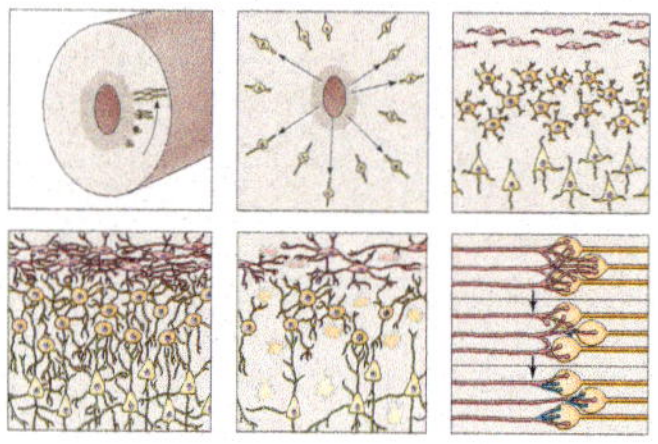

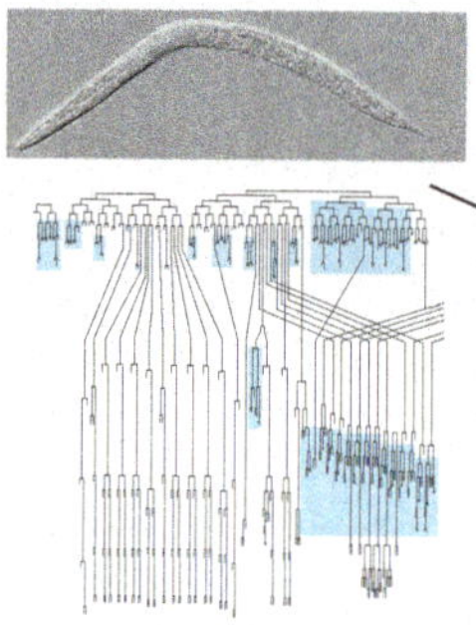

4 In simple animals such as the nematode *Caenorhabditis elegans*, neural pathways and synapses form according to a genetic plan that specifies the precise relations between growing axons and particular target cells. In more-complicated animals—including all vertebrates—**cell-cell interactions** determine the fates of individual neurons and glia. Review Figures 6.2–6.9

5 Although in humans most neurons are present at birth, most synapses develop *after* birth and continue developing into adulthood. Fetal and postnatal changes in the brain include the **myelination** of axons by glial cells and the development of dendrites and synapses by neurons. Review Figures 6.7–6.12, Video 6.5

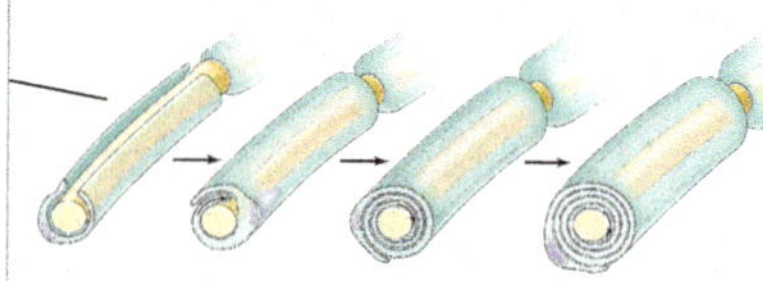

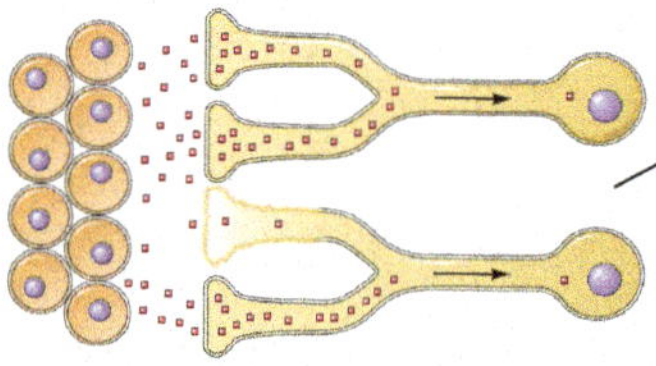

6 Among the many influences on brain development are genetic information and a multitude of environmental factors, such as **neurotrophic factors**, nutrition, and experience. Review Figures 6.13–6.16

7 Maldevelopment of the brain can occur as a result of mutations or other genetic disorders, including **fragile X syndrome**, and metabolic disorders, such as **phenylketonuria (PKU)**. However, the severity of PKU depends on an environmental factor—the amount of phenylalanine in the diet. Review Figure 6.19

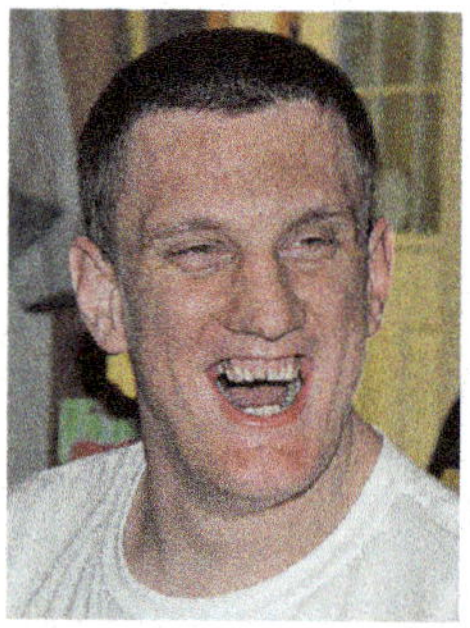

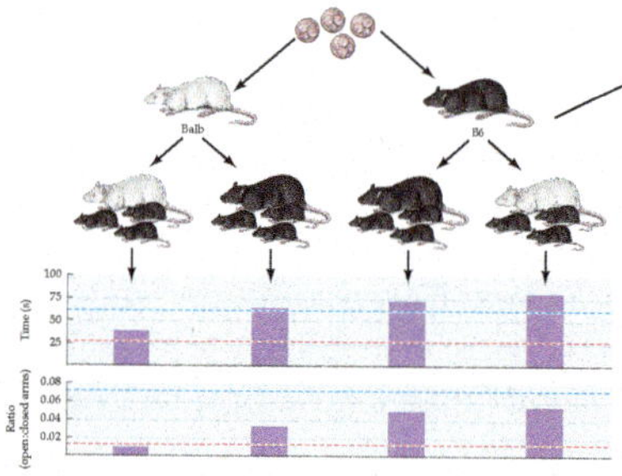

8 Gene expression is affected by environmental factors and experience, so **epigenetic** influences can profoundly affect brain development without altering the sequence of nucleotides in any genes. Similarly, genetically identical individuals do not display identical behaviours. Review Figures 6.25, 6.26

9 Experience affects the growth and development of the brain, inducing and modulating the formation of synapses and determining which synapses will survive and which will be eliminated. Such visually guided synaptic rearrangement is crucial for proper development of the visual system. Review Figures 6.21–6.24

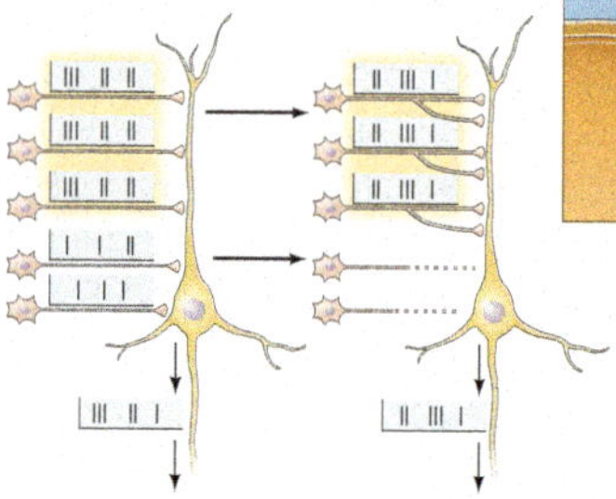

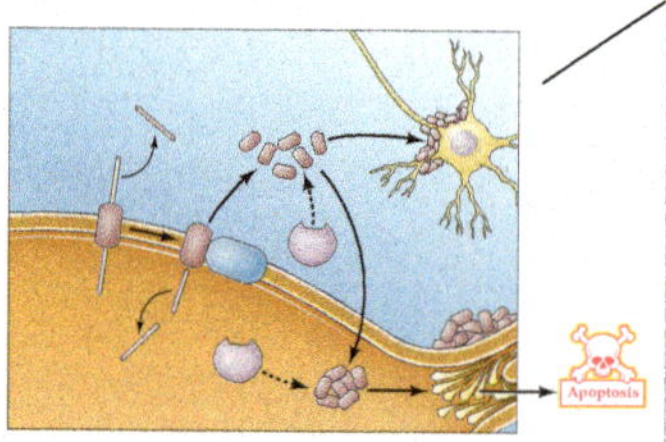

10 **Alzheimer's disease** seems to be caused by a buildup of **β-amyloid**, causing degenerative extracellular **senile plaques** and intracellular **neurofibrillary tangles** made up of **Tau**, through much of the cortex. Several genes, including those that encode the enzymes presenilin and **apolipoprotein E (ApoE)**, influence the rate of amyloid accumulation and the risk of Alzheimer's. Mental activity, physical activity, and adequate sleep seem to postpone the onset of Alzheimer's. Review Figures 6.28–6.31

Perception and Action

PART III

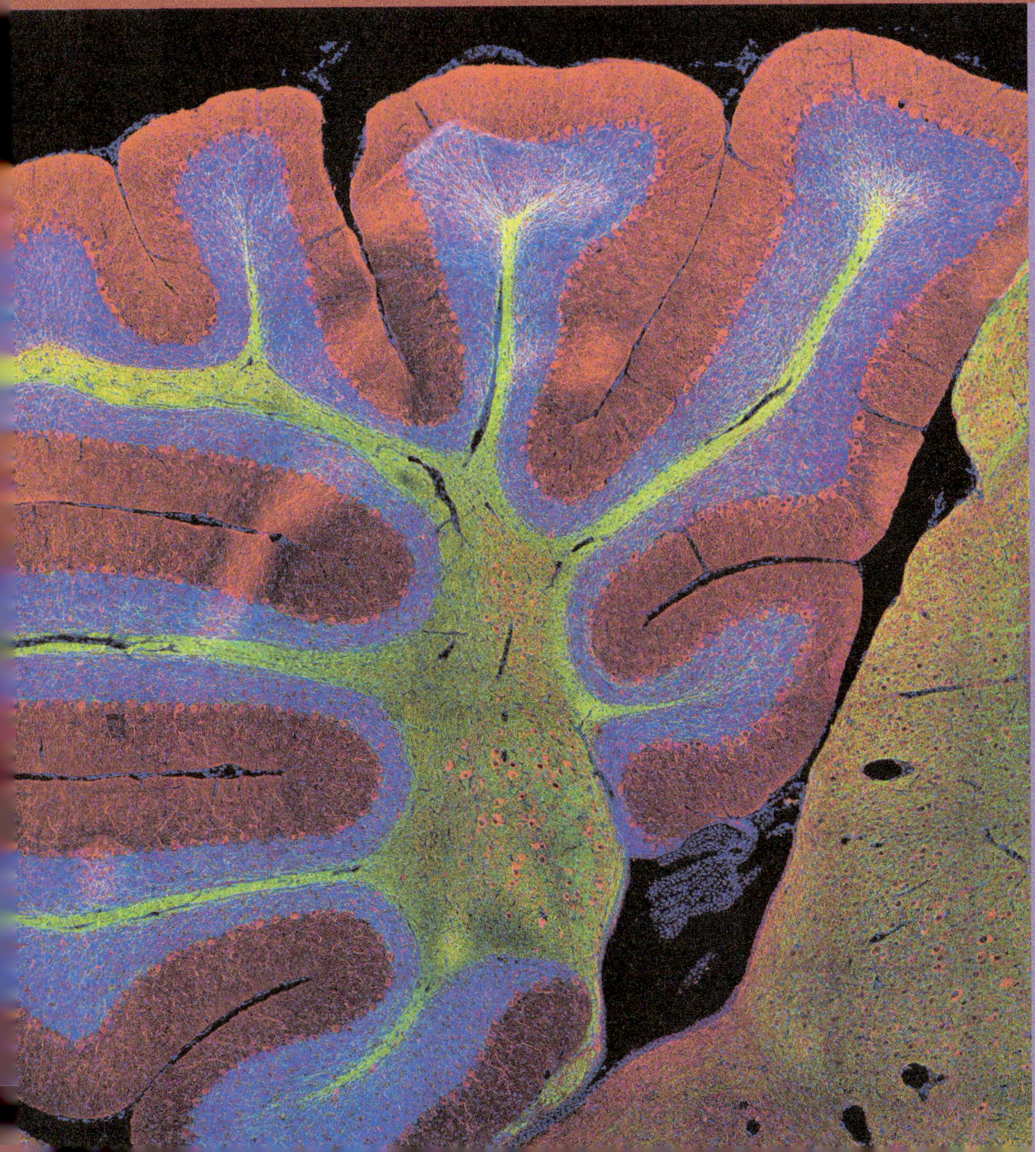

General Principles of Sensory Processing, Touch, and Pain

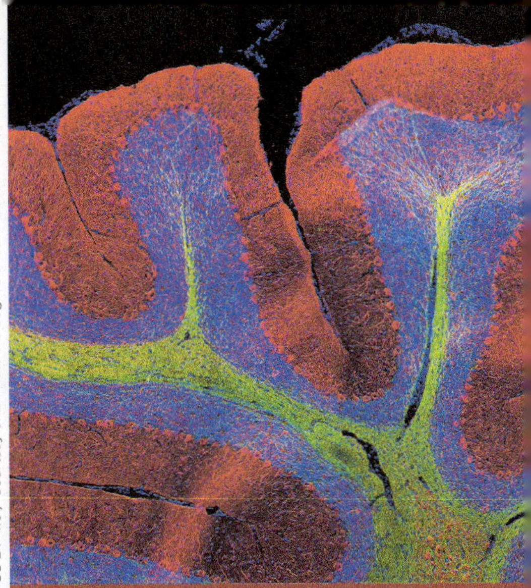

Photo from M. E. Terzenidou et al. 2017. *PLOS Genet* 13: e1006656 / CC BY 4.0 / Courtesy of Dr. Katerina Segklia and Dr. Rebecca Matsas

7

What's Hot? What's Not?

Ashlyn B. has never felt pain. While most babies cry after the arduous birth process, newborn Ashlyn calmly stared out from her blankets. Later, she developed a terrible diaper rash, and it didn't seem to bother her at all. That seemed strange to her mother, although the doctors dismissed it. But when Ashlyn's teeth came in, she nearly chewed off part of her tongue! When she reached up to her eye and scratched the cornea deeply, that should have been excruciating, but her parents only found out about the injury when the eye swelled and grew bloodshot. Soon her mother had to wrap Ashlyn's hands to keep her from biting them and rubbing her face raw (Heckert, 2012). Despite feeling no pain—in fact *because* she feels no pain—Ashlyn's daily life is full of peril. For instance, as a teenager, she was stirring noodles in boiling water when the spoon slipped in, and Ashlyn reflexively reached in to retrieve it. With wonderful support from her family, Ashlyn has learned to think carefully about what she does to avoid injury because, although she doesn't feel *pain*, she can be *damaged*, just like everyone else, and that could lead to disability or death.

How did Ashlyn come to have this dangerous "gift" of feeling no pain? What's going on inside her so that experiences that would bring us agony cause her no discomfort at all? What can she teach us about the neuroscience of pain, and about the importance of pain for survival?

All around us, many different types of energy affect us in various ways. Some molecules travelling through the air cause us to note particular odours. We detect waves of compressed air molecules as sounds. Our abilities to detect, recognise, and appreciate these varied energies depend on the characteristics of our sensory systems.

For each species, however, certain environmental features have become especially significant for adaptive success. For example, the bat darting through the evening sky is specially equipped to navigate using the echoes of its own ultrasonic cries, which we humans are unable to hear. Some snakes have infrared-sensing organs in their faces that allow them to "see" heat sources, enabling them to locate warm-blooded prey in the dark. How do animals, including humans, detect changes in the world around them?

sensory receptor organ An organ (such as the eye or ear) specialized to receive particular stimuli.

stimulus A physical event that triggers a sensory response.

receptor cell A specialized cell that responds to a particular energy or substance in the internal or external environment and converts energy into a change in the electrical potential across its membrane.

adequate stimulus The type of stimulus for which a given sensory organ is particularly adapted.

SENSORY PROCESSING

Each species has distinctive windows on the world, based on which energies it detects and how its nervous system processes that information. We open this chapter by considering some of the basic principles of sensory processing. Then we look at how those principles apply first to touch and then to pain.

7.1 Sensory Receptor Organs Detect Energy or Substances

Learning Objectives

After reading this section, you should be able to:

7.1.1 **Understand the concept of adequate stimulus.**

7.1.2 **Explain the concept of labelled lines.**

7.1.3 **Describe sensory transduction and sensory threshold.**

All animals have specialized body parts that are particularly sensitive to some forms of energy. These **sensory receptor organs** act as filters of the environment: they detect and respond to some events but not others. We call an event that affects the sensory organ a **stimulus** (plural stimuli). Stimuli may be sound waves reaching the ear, light entering the eye, or food touching the tongue. **Receptor cells** within the organs detect particular kinds of stimuli and convert them into the language of the nervous system: electrical signals. Eventually, information from sensory receptor organs enters the brain as a series of action potentials travelling along millions of axons, and our brains must make sense of it all.

Across the animal kingdom, receptor organs offer enormous diversity. For example, some snakes use detectors of infrared radiation, and several species of fishes detect electrical fields (Bellono et al., 2018). These specialised sensors evolved to detect signals that are crucial for survival in particular environmental niches. Thus, receptor organs reflect

7.1 The Variety of Eyes (A) Scanning electron micrograph of blackfly (*Simulium damnosum*) showing the compound eye magnified ×13. (B) The panther chameleon can move its two eyes independently. (C) The eyes of the Philippine tarsier are specialised for nighttime foraging. (D) The vision of the American bald eagle is particularly sharp.

TABLE 7.1 Classification of Sensory Systems

Type of sensory system	Modality	Adequate stimuli
Mechanical	Touch	Contact with or deformation of body surface
	Pain	Tissue damage
	Hearing	Sound vibrations in air or water
	Vestibular	Head movement and orientation
	Joint	Position and movement
	Muscle	Tension
Visual	Seeing	Visible radiant energy
Thermal	Cold	Decrease in skin temperature
	Warmth	Increase in skin temperature
Chemical	Smell	Odourous substances dissolved in air or water
	Taste	Substances in contact with the tongue or palate
	Common chemical	Changes in CO_2, pH, osmotic pressure
	Vomeronasal	Pheromones in air or water
Electrical	Electroreception	Differences in density of electrical currents
Magnetic	Magnetoreception	Orientation of Earth's magnetic field

strategies for success in particular worlds. Some animals detect Earth's magnetic field (Mouritsen, 2018), and maybe people do too (Wang et al., 2019). It's easy to see how such a sense would help navigation, but scientists have yet to pin down the sensory organ(s) that detect this force.

Even after we identify a single receptor organ, such as the eye, a wide array of sizes, shapes, and forms reflects the varying survival needs of different animals (**FIGURE 7.1**). Different kinds of energy, such as light and sound, need different receptor organs to convert them into neural activity, just as taking a photograph requires a camera, not a microphone.

TABLE 7.1 classifies sensory systems, identifying the kinds of stimuli detected by sensory receptor organs in each system. An **adequate stimulus** is the type of stimulus for which a given sensory organ is particularly adapted. The adequate stimulus for the eye is photic (light) energy; an electrical shock or pressure on your eye can create an illusory sensation of light (called a phosphene), but neither electricity nor mechanical pressure is considered an adequate stimulus for the eye.

Sensory systems of particular animals have restricted ranges of responsiveness

For any single form of physical energy, the sensory systems of a particular animal are quite selective. For example, humans do not hear sounds with frequencies exceeding 20,000 cycles per second (i.e., hertz, or Hz)—a range we call ultrasonic. Many bats, however, detect vibrations in air of 50,000 Hz or more. The range of hearing of some larger mammals includes infrasound: frequencies lower than humans can hear. **FIGURE 7.2** compares the auditory ranges of some animals. In the visual realm, too, some animals can detect stimuli that humans cannot. For example, birds and bees can see light in the ultraviolet range.

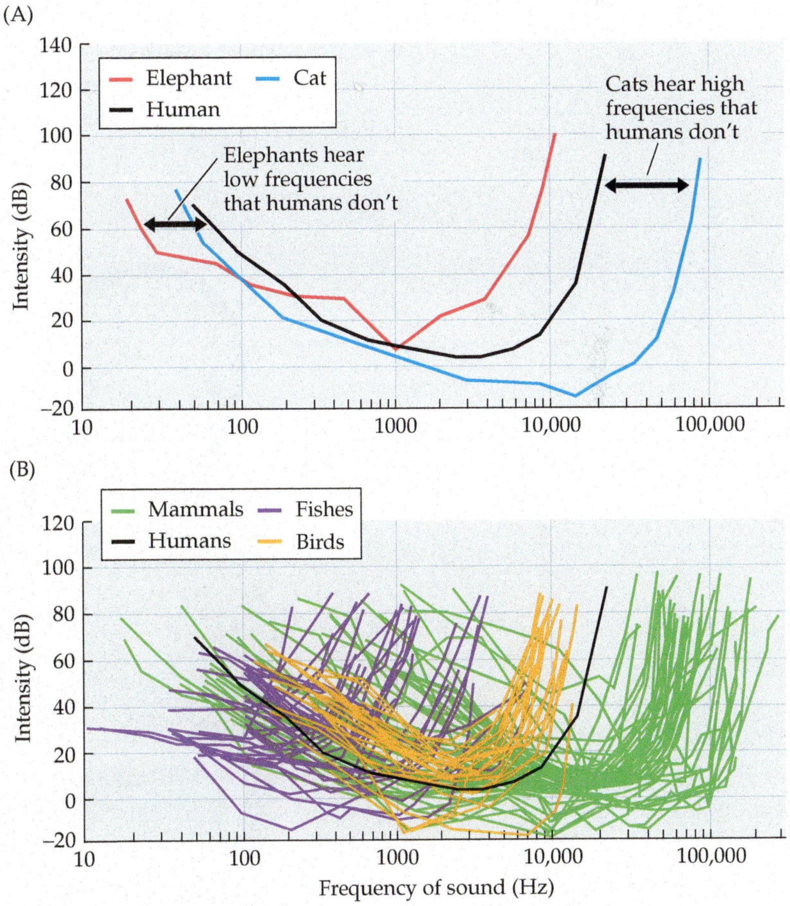

7.2 Do You Hear What I Hear? For comparison, the auditory sensitivity ranges of three mammals (A) and of many species of fishes, birds, and mammals (B) are plotted here together. Note that the species within a class detect a similar range of frequencies. dB, decibel; Hz, hertz. For a discussion of the measurement of sound, see Box 9.1. (After R. R. Fay, 1987. *Hearing in vertebrates: A psychophysics databook*. Hill-Fay Associates: Winnetka, IL. Courtesy of Dr. Richard Fay and the Fay Foundation.)

specific nerve energies The doctrine that the receptors and neural channels for the different senses are independent and operate in their own special ways and can produce only one particular sensation each.

labelled lines The concept that each nerve input to the brain reports only a particular type of information.

What type of stimulus was that?

How do we know whether a sudden event was a noise, a flash, or a slap? The physiologist Johannes Müller (1801–1858) proposed the doctrine of **specific nerve energies**, suggesting that neural channels for the different senses each use a different nerve "energy." For example, no matter how the eye is stimulated—by light or mechanical pressure or electrical shock—the resulting sensation is always visual. Müller formulated his hypothesis before anyone knew about action potentials. He imagined that different receptor organs might each use a different type of energy to communicate with the brain and that the brain knew which type of stimulus had happened by which type of energy was received.

Today we know that the messages for the different senses—such as seeing, hearing, touching, sensing pain, and sensing temperature—all use the same type of "energy": action potentials. But the brain recognises the different kinds of sensation (modalities) as separate and distinct because each modality sends its action potentials along separate nerve tracts. This is the concept of **labelled lines**: particular neurons are, at the outset, labelled for distinctive sensory experiences. Neural activity in one line signals a sound, activity in another line signals a smell, and activity in other lines signals touch. We can even distinguish different types of touch because some lines signal light touch, others indicate vibration, and yet other lines report stretching of the skin (**FIGURE 7.3**).

You can demonstrate this effect right now. If you take your finger and gently press on your eyelid, you'll see a dark blob appear on the edge of your field of view (it helps to look at a blank white wall). Of course, your skin also feels the touch of your finger, but why do you see a blob with your eye? The energy you applied, pressure, affected action potentials coming from your eye. Because your brain that line as always carrying visual information, what you experienced was a change in vision.

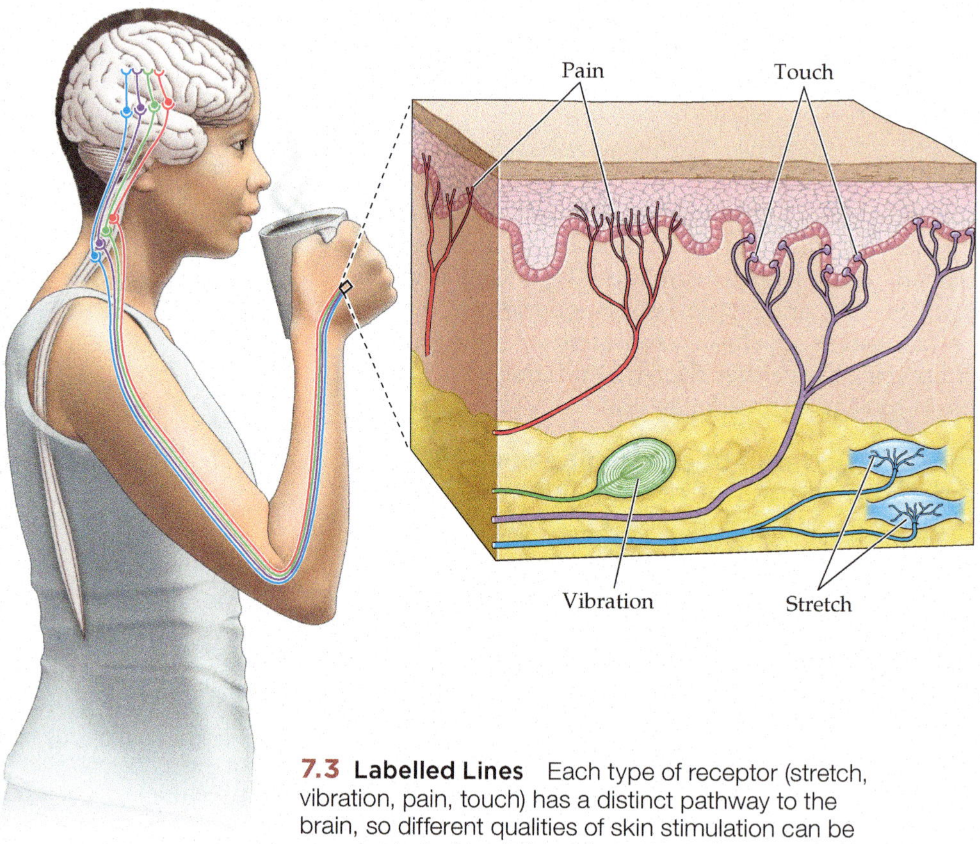

7.3 Labelled Lines Each type of receptor (stretch, vibration, pain, touch) has a distinct pathway to the brain, so different qualities of skin stimulation can be communicated to distinct places in the brain.

Sensory processing begins in receptor cells

Detection of sensory stimuli starts with receptor cells. A given receptor cell is specialised to detect particular energies or chemicals. Upon exposure to a suitable stimulus, a receptor cell converts that energy into a change in the electrical potential across its membrane. Changing the signal in this way is called **sensory transduction** (devices that convert energy from one form to another are known as transducers, and the process is called transduction). Receptor cells are transducers that convert energy around us into neural activity that leads to sensory perception. **FIGURE 7.4** shows some different receptor cells in skin. We will look at these types in more detail later in the chapter.

Some receptor cells have axons to transmit information. Other receptor cells have no axons of their own but stimulate associated nerve endings, either mechanically or chemically. For example, various kinds of energy-detecting corpuscles are associated with nerve endings in the skin. The eye has specialised receptor cells that convert light energy into electrical changes that cause neurotransmitter to be released onto nearby neurons. The inner ear has specialised hair cells that transduce mechanical energy into electrical signals that stimulate the fibres of the auditory nerve.

The initial stage of sensory processing is a change in electrical potential in receptor cells

The structure of a receptor determines the forms of energy to which it will respond. The steps between the arrival of energy at a receptor cell and the initiation of action potentials in a nerve fibre involve local changes of membrane potential called **receptor potentials** (or generator potentials). In most instances, the receptor potential resembles the excitatory postsynaptic potentials discussed in Chapter 2.

One example of the generator potential can be studied in a receptor called the **Pacinian corpuscle** (or lamellated corpuscle) (Loewenstein, 1971). This receptor, which detects vibration, is found throughout the body in skin and muscle. It consists of an axon surrounded by a structure that resembles a tiny onion because it has concentric layers of tissue (see Figure 7.4).

Mechanical stimuli (in this case vibration) delivered to the corpuscle produce a graded electrical potential with an amplitude that is directly proportional to the

sensory transduction The process in which a receptor cell converts the energy in a stimulus into a change in the electrical potential across its membrane.

receptor potential Also called *generator potential*. A local change in the resting potential of a receptor cell that mediates between the impact of stimuli and the initiation of action potentials.

Pacinian corpuscle Also called *lamellated corpuscle*. A skin receptor cell type that detects vibration.

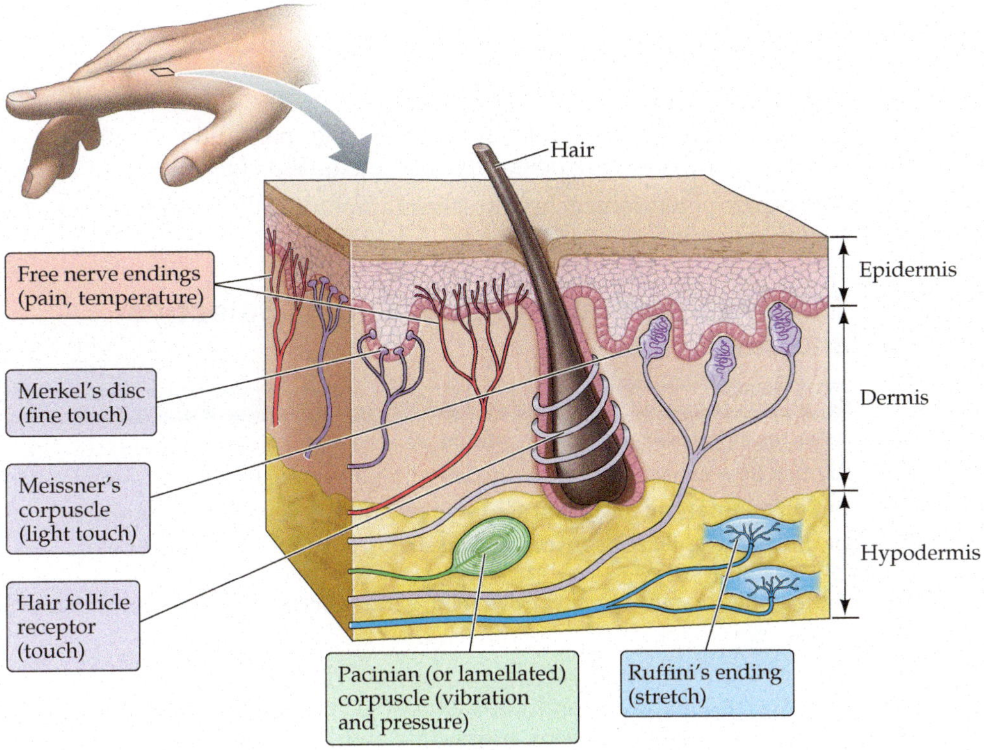

Go to Activity 7.1
Skin Receptors

7.4 Receptors in Skin The main receptors found in human skin are Pacinian corpuscles, Meissner's corpuscles, Merkel's discs, Ruffini's endings, and free nerve endings. The different functions of several of these receptors are compared in Figure 7.14.

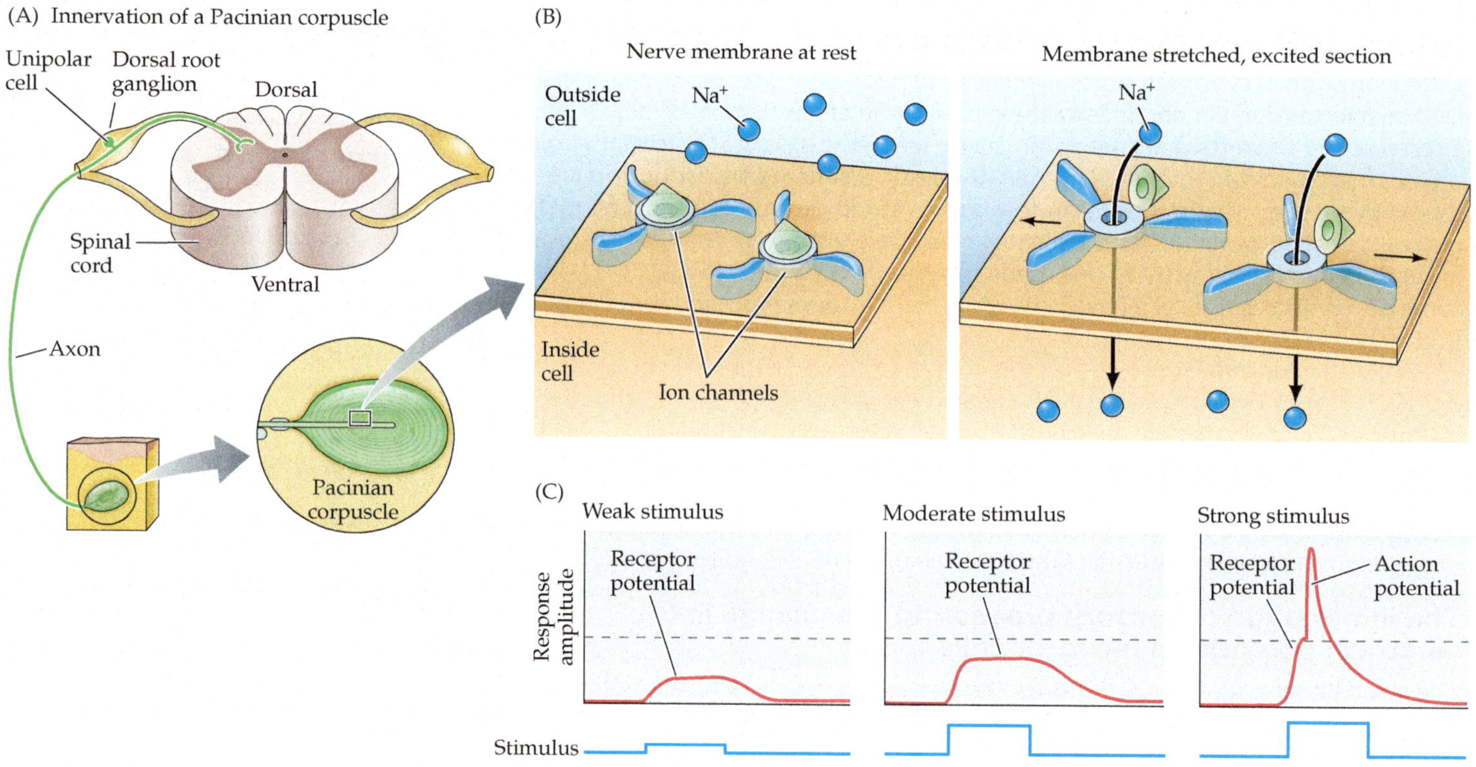

7.5 The Structure and Function of the Pacinian Corpuscle (A) The Pacinian corpuscle surrounds an afferent nerve fibre ending. (B) When the nerve membrane is at rest (left), the ion channels are closed to sodium ions (Na^+). Vibration applied to the corpuscle (right) stretches part of the neuronal membrane, opening the ion channels and permitting the entry of Na^+, which initiates an action potential (Lumpkin and Caterina, 2007). (C) The neuron shows increasing response to stimuli of increasing intensity until it reaches threshold, triggering an action potential. (Part B after Y. N. Jan and L. Y. Jan, 2017. *Nature* 554: 469–470.).

threshold The stimulus intensity that is just adequate to trigger an action potential.

strength of the stimulus. When this receptor potential gets big enough, an action potential is generated and we say that the receptor has reached **threshold** (**FIGURE 7.5A**). The sequence of excitatory events is as follows:

1. Mechanical stimulation deforms the corpuscle.
2. Deformation of the corpuscle stretches the tip of the axon.
3. Stretching the axon opens mechanically gated ion channels in the membrane (Brohawn et al., 2014), allowing positively charged ions to enter (**FIGURE 7.5B**).
4. When the receptor potential reaches threshold amplitude, the axon produces one or more action potentials (**FIGURE 7.5C**).

7.2 Sensory Information Processing Is Selective and Analytical

Learning Objectives

After reading this section, you should be able to:

7.2.1 **Explain how sensory systems encode stimulus intensity and location.**

7.2.2 **Define adaptation, and contrast tonic and phasic sensory receptors.**

7.2.3 **Describe the levels of sensory processing, from sensory receptor to cortex, in mammals.**

7.2.4 **Explain the concept of a receptive field, with examples from the sense of touch.**

Thinkers in ancient Greece believed that nerves were tubes through which tiny bits of stimulus objects traveled to the brain, to be analysed and recognized there. (Imagine the nerves in your tongue sending minuscule chunks of garlic to your brain for analysis.) Even after learning about neural conduction in the twentieth century, many investigators thought that the sensory nerves simply transmitted accurate information about stimuli to the brain centres. Now, however, it is clear that the sensory organs and peripheral sensory pathways convey only limited—even distorted—information to the brain. A good deal of selection and analysis takes place along sensory pathways. In the discussion that follows, we will examine six aspects of sensory processing: coding, adaptation, suppression, pathways, receptive fields, and attention.

Coding: Sensory events are represented by action potentials

Information about the world is represented in the nervous system by electrical potentials in cells. We have already considered the first steps in this process—the transduction of energy at receptors, the receptor potential, and the creation of action potentials in sensory neurons. But how does this neural activity "stand for" (or represent) the stimuli impinging on the organism? Through some form of **coding**, the pattern of electrical activity in the sensory system must convey information about the original stimulus. Neural codes are limited in that each action potential is always the same size and duration, so sensory information is encoded by other features of neural activity, such as the number and frequency of the action potentials, the rhythm in which clusters of action potentials occur, and so on. Let's examine neural representations of the intensity and location of stimuli.

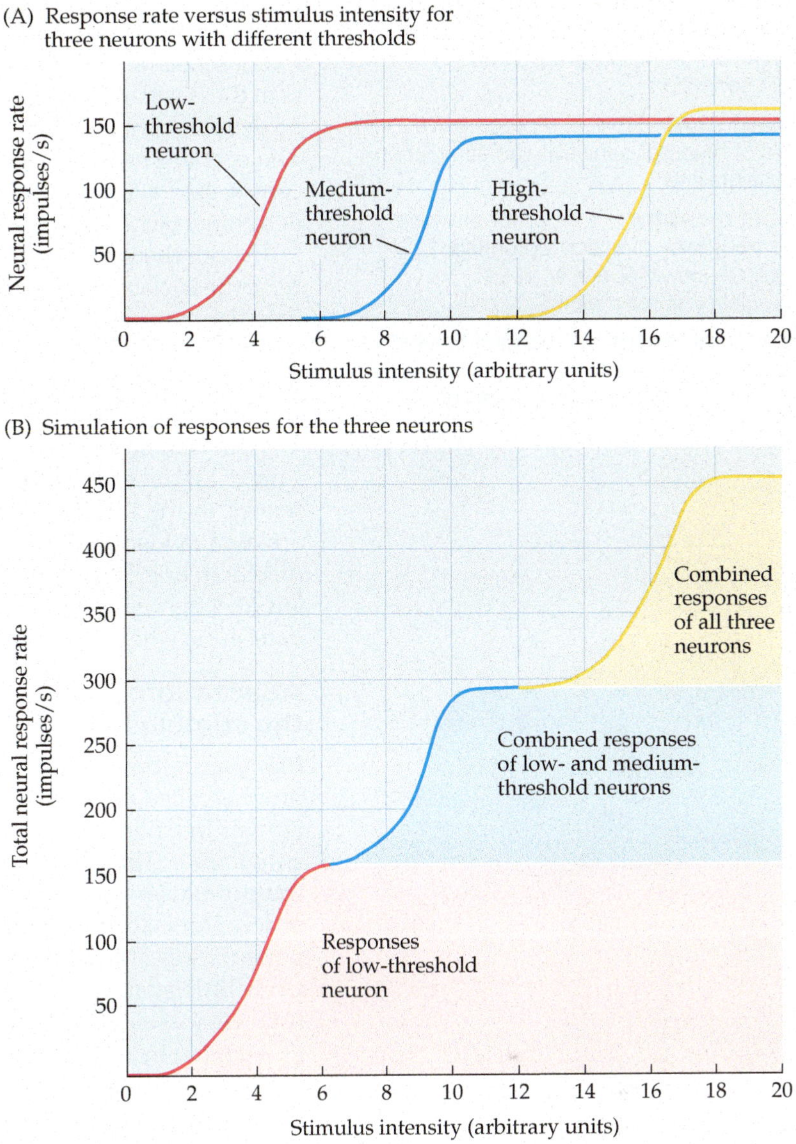

7.6 Intensity Coding (A) Each of the three nerve cells represented here has a different threshold—low, medium, or high—and thus a different response to stimuli. Each cell varies its response over a *fraction* of the total *range* of stimulus intensities. (B) Although none of these nerve cells can respond faster than 150 times per second, the *sum* of all three can vary in response rate from 0 to 450 action potentials per second, accurately indicating the intensity of the stimulus—an example of range fractionation.

STIMULUS INTENSITY We respond to sensory stimuli over a wide range of intensities. Furthermore, within this range we can detect small differences of intensity. How are different intensities of a stimulus represented in the nervous system? In general, more intense stimuli generate more rapid action potentials, but only up to some limit because a given neuron can only fire so fast (**FIGURE 7.6A**). Thus only a limited range of different sensory intensities can be represented in this manner.

As we noted in Chapter 2 the maximal rate of firing for any single neuron is about 1200 action potentials per second, and most sensory fibres do not fire that fast. Multiple receptor cells acting in a parallel manner provide a broader range for coding the intensity of a stimulus. As the strength of a stimulus increases, new neurons are "recruited"; thus, intensity can be represented by the number of active cells. **Range fractionation** takes place when different receptor cells are "specialists" in particular segments, or fractions, of an intensity scale (**FIGURE 7.6B**). This mode of stimulus coding requires an array of receptors and neurons that differ in the stimulus intensity needed to reach threshold. Some sensory neurons have a very low threshold (so they are highly sensitive); others have a much higher threshold (so they are less sensitive). Thus, one clue to the intensity of a stimulus is whether it activates only low-threshold receptors, or both low- and high-threshold receptors.

STIMULUS LOCATION The position of an object or event, either outside or inside the body, is an important piece of information. Did something just poke my foot, or my hand? Some sensory systems reveal this information by the position of excited

coding The rules by which action potentials in a sensory system reflect a physical stimulus.

range fractionation A hypothesis of stimulus intensity perception stating that a wide range of intensity values can be encoded by a group of cells, each of which is a specialist for a particular range of stimulus intensities.

somatosensory Referring to body sensation, particularly touch and pain sensation.

adaptation Here, the progressive loss of receptor sensitivity as stimulation is maintained.

tonic receptor A receptor in which the frequency of action potentials declines slowly or not at all as stimulation is maintained.

phasic receptor A receptor in which the frequency of action potentials drops rapidly as stimulation is maintained.

receptors on the sensory surface. This feature is most evident in the **somatosensory** (body sensation) system. You know that an object is on your back if a receptor in the skin there is stimulated. If a receptor on your palm is stimulated, then the object must be there. Each receptor activates pathways that convey specific positional information. The spatial properties of a stimulus are represented by labelled lines that uniquely convey spatial information. Similarly, in the visual system an object's spatial location determines which receptors in the eye are stimulated.

In both the visual and the touch systems, cells at all levels of the nervous system—from the surface sheet of receptors to the cerebral cortex—are arranged in an orderly, maplike manner. The map at each level is not exact but reflects both position and receptor density. More cells are allocated to the spatial representation of sensitive, densely innervated sites, like the skin of the lips or the centre of the retina, than to sites that are less sensitive, such as the skin of the back or the periphery of the retina.

With bilateral receptor systems—the two ears or the two nostrils—the relative time of arrival of the stimulus at the two receptors, or the relative intensity, is directly related to the location of the stimulus. For example, the only time when both ears are excited identically is when the sound source is equidistant from the ears, in the median plane of the head. As the stimulus moves to the left or right, receptors of the left and right sides are excited asymmetrically. The use of inputs from both ears to determine where sounds come from is illustrated in Figure 8.10.

Adaptation: Receptor response can decline even if the stimulus is maintained

Many receptors show progressive loss of response when stimulation is maintained, a process called **adaptation**. We can demonstrate adaptation by recording action potentials in a fibre leading from a sensory receptor that is receiving a constant level of stimulation. The frequency of action potentials progressively declines, even though the stimulus is continued (**FIGURE 7.7**). In terms of adaptation, there are two kinds of receptors: **Tonic receptors** show little or no decrease in the frequency of action potentials as stimulation is maintained; in other words, these receptors show relatively little adaptation. **Phasic receptors** display adaptation, rapidly decreasing the frequency of action potentials when the stimulus is maintained.

Adaptation means that there is a progressive shift in neural activity away from accurate portrayal of physical events. Thus, the sensory system may fail to register neural activity even though the stimulus continues. Such a discrepancy is no accident; sensory systems emphasize *change* in stimuli because changes are more likely to be significant for survival. Sensory adaptation prevents the nervous system from becoming overwhelmed by stimuli that offer very little "news" about the world. For example, your pants may press a hair on your leg continually, but you're saved from a constant neural barrage from this stimulus by several mechanisms, including adaptation.

The basis of adaptation includes both neural and nonneural events. For example, in some mechanical receptors, adaptation develops from the elasticity of the receptor cell itself. This situation is especially evident in the Pacinian corpuscle, which detects vibration (see Figure 7.5). Maintained vibration on the receptor results in an initial burst of neural activity and a rapid decrease to almost nothing. But when the corpuscle (which is a separate, accessory cell) is removed, the same constant stimulus applied to the uncovered sensory nerve fibre produces a continuing discharge of action potentials. So for this receptor, at least some adaptation is

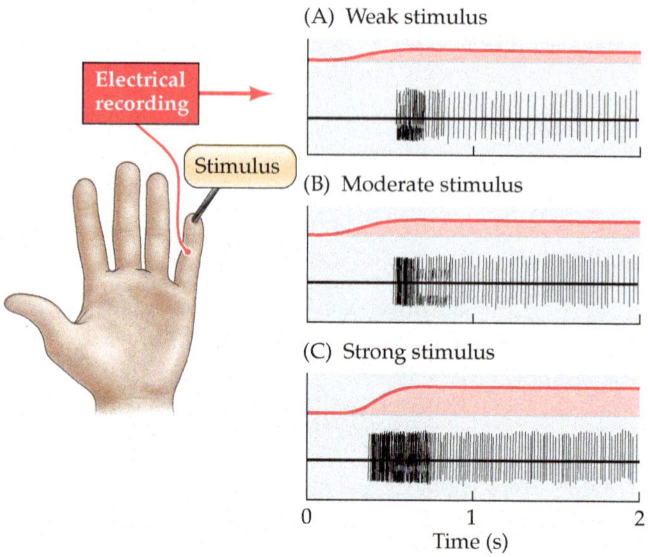

7.7 Sensory Adaptation The neuron represented here responds to a touch on the fifth finger. It fires rapidly when the stimulus—whether weak (A), moderate (B), or strong (C)—is first applied, but then it adapts, slowing to a steady rate. (After M. Knibestol and A. B. Valbo, 1970. *Acta Physiol Scand* 80: 178–195.)

due to mechanical properties of the nonneural component, the corpuscle rather than the axon.

Suppression: Sometimes we need receptors to be quiet

We have noted that successful survival does not depend on exact reporting of stimuli. Rather, our success as a species demands that our sensory systems accentuate, from among the many things happening about us, the important changes of stimuli. We just discussed how sensory receptor adaptation can suppress a constant stimulus, but two other suppression strategies are also available.

In many sensory systems, accessory structures can reduce the level of input in the sensory pathway. For example, closing the eyelids reduces the amount of light that enters the eye. In the auditory system, contraction of the middle-ear muscles reduces the intensity of sounds that reach the inner ear. In this form of sensory control, the relevant mechanisms change the intensity of the stimulus before it reaches the receptors.

In a second form of information control, neural connections descend from the brain to lower stations in the sensory pathway, in some cases as far as the receptor surface. This is sometimes referred to as a **top-down process**—the upper brain regions modulate the activity of lower centres. For example, higher centres in the pain system (discussed later in this chapter) send axons down the spinal cord, where they can inhibit incoming pain signals. Top-down processing is also evident in the auditory system, where cells in the brainstem send axons along the auditory nerve to connect with the base of the receptor cells to adjust their response to certain sounds.

top-down process A process in which higher-order cognitive processes control lower-order systems, often reflecting conscious control.

sensory pathway The chain of neural connections from sensory receptor cells to the cortex.

thalamus The brain regions at the top of the brainstem that trade information with the cortex.

receptive field The stimulus region and features that affect the activity of a cell in a sensory system.

Pathways: Successive levels of the nervous system process sensory information

Sensory information travels from the sensory surface to the highest levels of the brain, and each sensory system has its own distinctive pathway. Pathways from receptors lead into the spinal cord or brainstem, where they connect to distinct clusters of neurons. These cells, in turn, have axons that connect to other groups of neurons. Each sensory modality—such as touch, vision, or hearing—has a distinct hierarchy of tracts and stations in the brain that are collectively known as the **sensory pathway** for that modality.

Each station in the pathway accomplishes a basic aspect of information processing. For example, painful stimulation of the finger leads to reflex withdrawal of the hand, which is mediated by spinal circuits. At the brainstem level, other circuits can turn the head towards the source of pain. Eventually, sensory pathways reach the cerebral cortex, where the most complex aspects of sensory processing take place. For most senses, information reaches the **thalamus** before being relayed to the cortex (**FIGURE 7.8**). Information about each sensory modality is sent to a separate division of the thalamus. One way for the brain to suppress particular stimuli is for the cortex to direct the thalamus to emphasize some sensory information and suppress other information (Briggs and Usrey, 2008), another example of top-down processing.

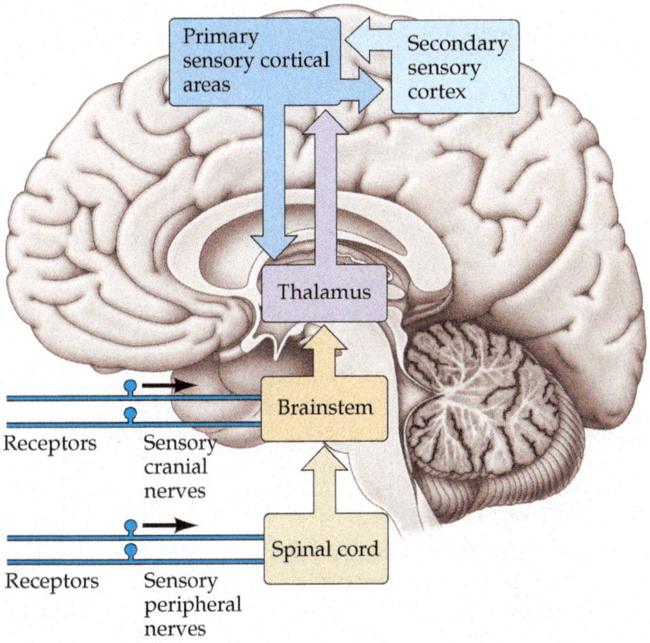

7.8 Levels of Sensory Processing Sensory information enters the CNS through the brainstem or spinal cord and then reaches the thalamus. The thalamus transmits the information to the cerebral cortex; the cortex directs the thalamus to suppress some sensations. Primary sensory cortex swaps information with nonprimary sensory cortex. This organisation is present in all sensory systems except smell, which bypasses the thalamus, going directly to cortex (see Figure 9.26).

Receptive fields: What turns on this particular receptor cell?

The **receptive field** of a sensory neuron consists of a region of space in which a stimulus will alter that neuron's firing rate. To determine this receptive field, investigators record the neuron's electrical responses to a variety of stimuli to see what makes the activity of the cell change from its resting rate. For example, which patch of skin must we vibrate to change the activity of a particular Pacinian corpuscle? Such experiments show that somatosensory receptive fields have either an excitatory centre and an inhibitory surround,

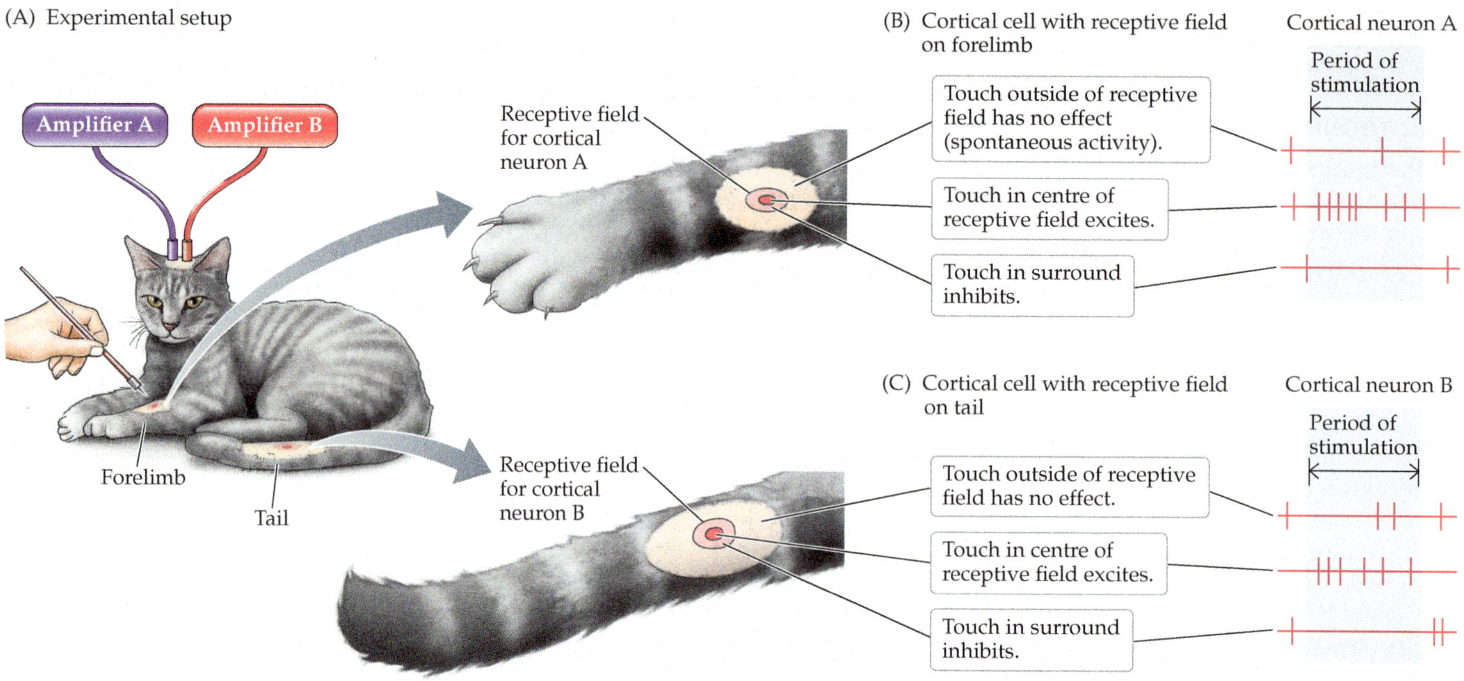

7.9 Identifying Somatosensory Receptive Fields The procedures illustrated here are used to record from somatosensory neurons of the cerebral cortex. Changes in the position of the stimulus affect the rate of action potentials. Neuron A responds to touch on a region of the forepaw; neuron B, only a few centimetres away in the somatosensory cortex, responds to stimulation of the tail. The receptive fields of these neurons include an excitatory centre and an inhibitory surround, but other neurons have receptive fields with the reverse organisation: inhibitory centres and excitatory surrounds.

Go to Animation 7.1
Somatosensory Receptive Fields

or an inhibitory centre and an excitatory surround (**FIGURE 7.9**). Such receptive fields make it easier to detect edges and discontinuities on the objects we feel.

Receptive fields differ also in size and shape and in the quality of stimulation that activates them. For example, some neurons respond preferentially to light touch, while others fire most rapidly in response to painful stimuli, and still others respond to cooling.

Following sensory information from the receptor cell in the periphery into the brain shows that neurons all along the pathway will respond to particular stimuli, so each of these cells has a receptive field too. But as each successive level combines information from prior cells in the pathway, the receptive fields change considerably. Receptive fields have been studied for cells at all levels, from the periphery to the brain, and we will see many examples of receptive fields later in this chapter and in Chapters 8 and 9.

RECEPTIVE FIELDS IN THE CEREBRAL CORTEX For a given sensory modality we can find several different regions (fields) of cortex that receive information about that sense. Each of these cortical regions has a separate map of the same receptive surface, but the different cortical regions process the information differently and make different contributions to perceptual experiences.

By convention, one of the cortical maps is designated as **primary sensory cortex** for that particular modality. Thus, there is primary somatosensory cortex, primary auditory cortex, and so on. The other cortical maps for a given modality are said to be **secondary sensory cortex**, or nonprimary sensory cortex (see Figure 7.8). The primary cortical area is the main source of input to the other fields for the same modality, even though these other fields also have direct thalamic inputs. Information is sent back and forth between the primary and nonprimary sensory cortex through subcortical loops.

Primary somatosensory cortex (somatosensory 1, or **S1**) of each hemisphere lies in the postcentral gyrus of the parietal cortex, just behind the central sulcus

primary sensory cortex For a given sensory modality, the region of cortex that receives most of the information about that modality from the thalamus or, in the case of olfaction, directly from the secondary sensory neurons.

secondary sensory cortex Also called *nonprimary sensory cortex*. For a given sensory modality, the cortical regions receiving direct projections from primary sensory cortex for that modality.

primary somatosensory cortex (S1) Also called *somatosensory 1*. The gyrus just posterior to the central sulcus, in the parietal lobe, where sensory receptors on the body surface are mapped; primary cortex for receiving touch and pain information.

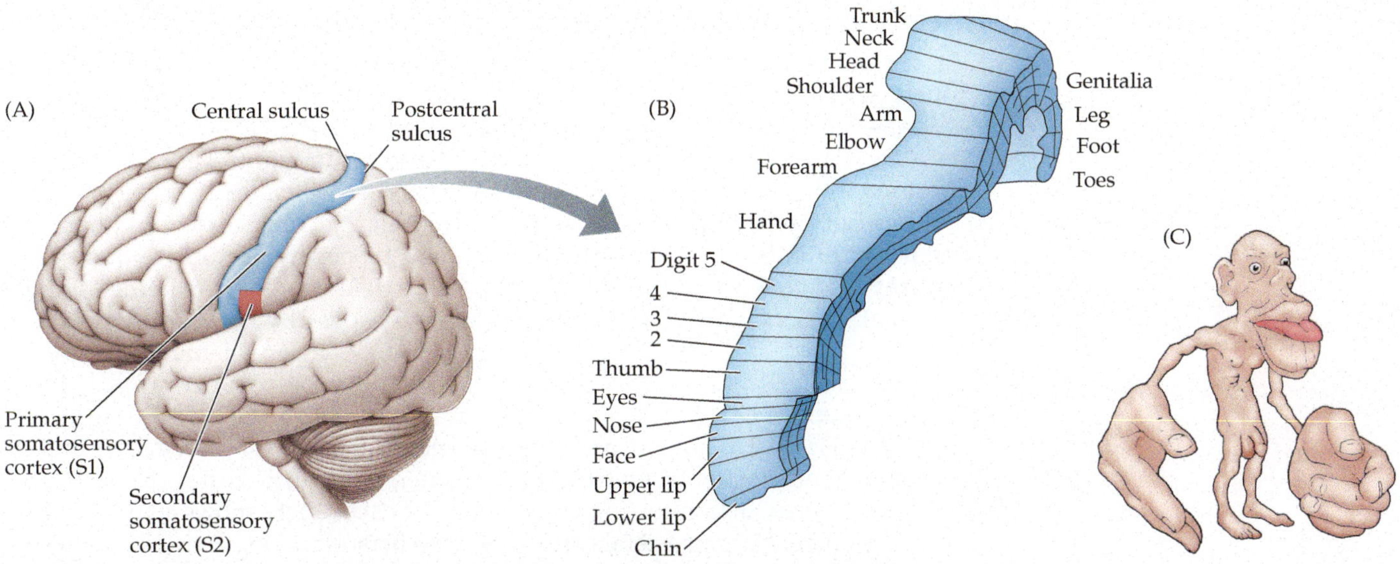

7.10 Representation of the Body Surface in Somatosensory Cortex (A) The locations of primary somatosensory cortex (S1) on the lateral surface of the parietal cortex. Secondary somatosensory cortex (S2) is much smaller than S1. (B) The order and size of cortical representations of different regions of skin. Note that information from the various parts of the hand and fingers takes up much more room than does information from the shoulder. (C) The *homunculus* (literally, "little man") depicts the body surface with each area drawn in proportion to the size of its representation in the primary somatosensory cortex.

dividing the parietal lobe from the frontal lobe (**FIGURE 7.10**). Each S1 receives touch information from the opposite side of the body. The cells in S1 are arranged according to the plan of the body surface (Kell et al., 2005). Each region is a map of the body in which the relative areas devoted to body regions reflect the density of body innervation. Thus, parts of the body where we are especially sensitive to touch (like the hand and fingers) send information to a larger area of S1 than do less sensitive body regions (like the shoulder). **Secondary somatosensory cortex** (somatosensory 2, or **S2**) maps both sides of the body in registered overlay; that is, the left-arm and right-arm representations occupy the same part of the map, and so forth.

Some mammals have a very different pattern of representation in somatosensory cortex. For example, the nose of the star-nosed mole is a very sensitive organ for touch, and a considerable portion of its somatosensory cortex is devoted to responding to each of the rays of the "star" (**FIGURE 7.11**) (Catania, 2019).

secondary somatosensory cortex (S2) Also called *somatosensory 2*. The region of cortex that receives direct projections from primary somatosensory cortex.

(A)

(B)

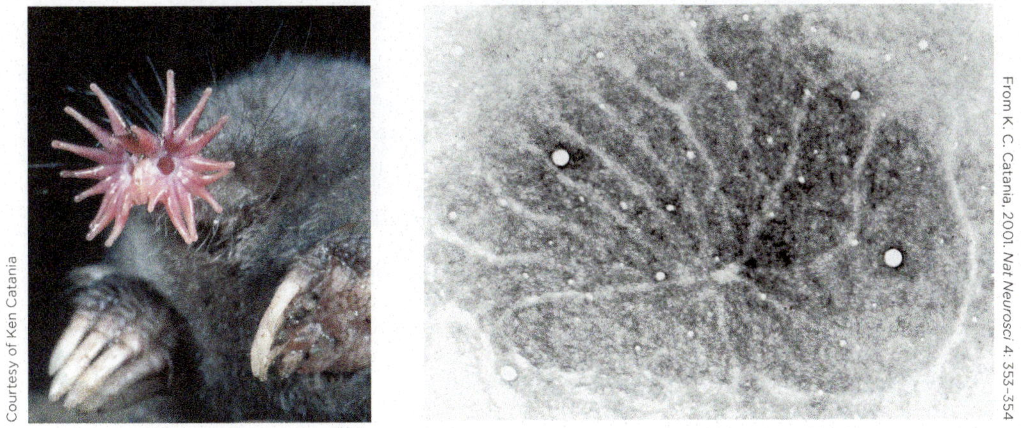

7.11 Hey There, You with the Star on Your Nose (A) The tip of the star-nosed mole's nose is a very delicate organ for touch. (B) In this section from the somatosensory cortex of a star-nosed mole, we can see how each of the 11 rays from one-half of the star-shaped nose projects to its own patch of cortex. The two bottommost rays are the most sensitive; each innervates a larger piece of cortex than do the other rays.

attention A state or condition of selective awareness or perceptual receptivity, by which specific stimuli are selected for enhanced processing.

cingulate cortex Also called *cingulum*. A region of medial cerebral cortex that lies dorsal to the corpus callosum.

polymodal Involving several sensory modalities.

Attention: How do we notice some stimuli but not others?

Attention is the process by which we select or focus on one or more specific stimuli for enhanced processing and analysis. Sometimes attention is a top-down process, when we decide to concentrate our attention on a particular task. In other cases, a sudden loud noise may pull our attention to some dramatic event. We'll discuss these and other aspects of attention at some length in Chapter 17.

In sensory processes, a cortical region that plays a special role in attention is the posterior parietal lobe. Many cells here are especially responsive in a trained monkey that is expecting a stimulus (Mountcastle et al., 1981), whether auditory or visual. Lesions of this area in monkeys, and especially the right parietal lobe in humans (see Chapter 17), result in inattention, or neglect of stimuli, on the opposite side.

The **cingulate cortex** (the portion of cortex along and just above the corpus callosum; see Figure 1.17) has also been implicated in attention. We'll see an example of that later in this chapter when we learn that anterior cingulate cortex seems to mediate the emotional, discomforting aspect of pain. **FIGURE 7.12** shows activation in the cingulate and posterior parietal cortex when the person was expecting visual stimuli in a particular location (Gitelman et al., 1996; Nobre et al., 1997).

Sensory systems influence one another

Often the use of one sensory system influences perception derived from another sensory system. For example, cats may not respond to birds unless they can both see and hear the birds; neither sense alone may be sufficient to elicit a response (B. Stein and Meredith, 1993). Similarly, humans detect a visual signal more accurately if it is accompanied by a sound from the same part of space (Feng et al., 2017).

Many sensory areas in the brain—so-called association areas—do not represent exclusively a single modality but show a mixture of inputs from different modalities. Some "visual" cells, for instance, also respond to auditory or touch stimuli. Perhaps loss of input from one modality allows these cells to analyse input from the remaining senses better—as happens, for example, in cases of people who become blind early in life and are better than sighted people at localising auditory stimuli (Gougoux et al., 2005). The normal stimulus convergence on such **polymodal** cells provides a mechanism to analyse an object that affects two modalities (B. E. Stein and Stanford, 2008). For a few people, a stimulus in one modality may evoke an additional perception in another modality, as when seeing a letter evokes a colour, or hearing different tones evokes different flavours—a situation that is described further in **BOX 7.1**.

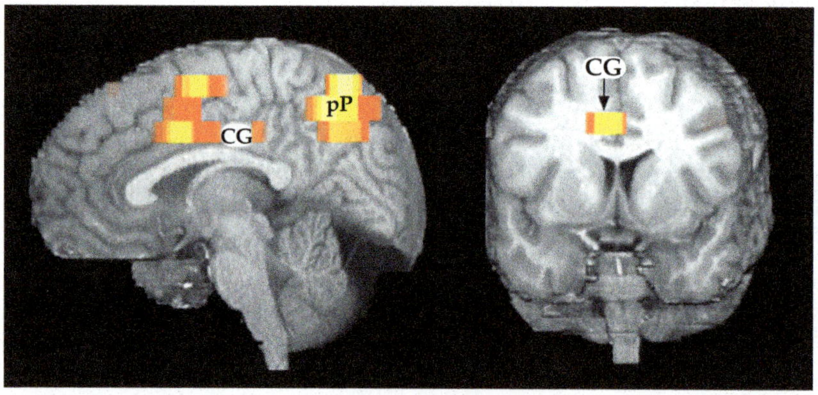

Courtesy of Darren Gitelman

7.12 Brain Regions Activated When We Are Attending Functional MRI images of a person cued to expect a stimulus in a particular portion of the visual field show right-hemisphere activation in posterior parietal cortex (pP) and cingulate cortex (CG) in midsagittal (left) and frontal (right) views. Areas of highest activation are shown in yellow.

BOX 7.1 Synesthesia

For a few people, stimuli in one modality evoke the involuntary experience of an additional sensation in another modality—a condition known as **synesthesia** (from the Greek *syn*, "union," and *aesthesis*, "sensation"). For example, a person with synesthesia (a "synesthete") may perceive different colours when seeing different letters of the alphabet ("*D* looks green, but *E* is blue") or words for days of the week ("*Tuesday* is a yellow word"). In one documented example, a musician experienced a particular taste whenever she heard a specific musical tone interval (Beeli et al., 2005). Discordant tones evoked unpleasant tastes.

How common synesthesia is depends to some extent on how it's defined, but it is estimated that as much as 2–4% of the population displays some form of synesthesia, and as much as 1% reports experiencing a colour along with particular days of the week or numbers (Simner et al., 2006). In a sample of 11 people reporting the common synesthesia of colour associated with letters, many associations were identical across the individuals. For example, most reported experiencing red with the letters *A*, *M*, and *S*; yellow with the letters *C*, *O*, and *U*; and green with *D*, *P*, and *V* (Witthoft and Winawer, 2013). The authors noted that these colour associations match those in a popular toy set of magnetic letters (Figure A), which at least some of the synesthetes remembered having as a child. So, did the early exposure to the toy inspire these particular associations of letters with colours? (If you look at Figure A, you'll see the toy letters are coloured in the order of the rainbow—red, orange, yellow, green, blue, violet [ROYGBV]—over and over.) Brain imaging shows that such synesthetes have more axonal connections across cortex, especially in the temporal lobe (Rouw and Scholte, 2007), suggesting that the experience of colour is due to extensive connections between brain regions.

Since you probably do not have this experience, you may doubt whether other people really do, if the basis for knowing that they do is self-report alone. For some forms of synesthesia, however, a clever test may confirm whether, for example, "2 is red but 5 is green." Ramachandran and Hubbard (2001) showed a page like that in Figure B to someone who reported having this experience. The person was then asked to quickly point to all of the 2s. Because the shapes of 2 and 5 are so similar, it takes a certain amount of time for most people (nonsynesthetes) to pick out each 2 among all those 5s. But this person found them much faster than that. For such a synesthete, the numbers look coloured, as in Figure C; with the addition of colours, the red 2s seem to pop out from the field of green 5s.

You can test whether *you* have synesthesia online at http://synesthete.org (Eagleman et al., 2007).

(A) Magnetic letter set

Courtesy of R. Berry / My Toys R Your Toys / Etsy

(B) Difficult search

(C) Easy search

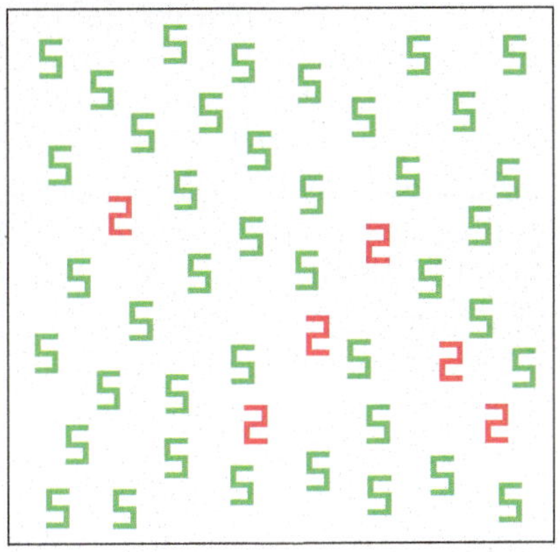

synesthesia A condition in which stimuli in one modality evoke the involuntary experience of an additional sensation in another modality.

TOUCH: MANY SENSATIONS BLENDED TOGETHER

The skin that envelops our bodies is a delicate yet durable boundary that separates us from our surroundings. It also presents to the world a massive array of sensory receptors monitoring many types of stimuli. But touch is not just touch. Careful studies of skin sensations reveal qualitatively different sensory experiences: pressure, vibration, tickle, "pins and needles," and more-complex dimensions, such as smoothness or wetness—all recorded by the receptors in the skin.

7.3 Skin Is a Complex Organ That Contains a Variety of Sensory Receptors

Learning Objectives

After reading this section, you should be able to:

7.3.1 **Name the four main types of touch receptors, and compare their receptive fields.**

7.3.2 **Compare the four categories of sensory axons by size, conduction velocity, and which sensory information is conveyed by each.**

7.3.3 **Trace the neural pathway for touch from receptors in the skin to the brainstem, thalamus, and cortex.**

7.3.4 **Describe evidence of plasticity in representation in somatosensory cortex, in monkeys and people.**

Because the average person has about 1–2 square metres (10–20 square feet) of skin, it is sometimes considered the largest human organ. Skin is made up of three separate layers; the relative thickness of each varies over the body surface. The outermost layer—the **epidermis**—is the thinnest. The middle layer—the **dermis**—contains a rich web of nerve fibres in a network of connective tissue and blood vessels. The innermost layer—the **hypodermis** (or subcutaneous tissue)—provides an anchor for muscles, contains Pacinian corpuscles, and helps shape the body (see Figure 7.4).

Pain, heat, and cold at the skin are detected by free nerve endings (see Figure 7.4), which are described later in this chapter. In contrast, light touch is detected by four highly sensitive **tactile** (touch) receptors (**FIGURE 7.13**). We mentioned earlier the Pacinian corpuscles, which are found in the hypodermis. The onion-like outer portion of the corpuscle acts as a filter, shielding the underlying nerve fibre from most stimulation. Only vibrating stimuli of more than 200 Hz will pass through the corpuscle and stretch the nerve fibre to reach threshold. Normally, the skin receives this sort of rapid vibration when it is moving across the texture of an object's surface. The ridges provided by fingerprints mechanically filter out vibrations of some frequencies and amplify others, apparently optimising the stimulation of Pacinian corpuscles (Scheibert et al., 2009). Pacinian corpuscles are fast-responding and fast-adapting receptors (**FIGURE 7.13A**).

Most of our ability to perceive the form of an object we touch comes from the fast-adapting **Meissner's corpuscles** (**FIGURE 7.13B**) and the slow-adapting, oval **Merkel's discs** (**FIGURE 7.13C**). These receptors are densely distributed in skin regions where we can discriminate fine details by touch (fingertips, tongue, and lips).

The receptive fields of Merkel's discs usually have an inhibitory surround, which increases their spatial resolution. This field also makes them especially responsive to isolated points on a surface (such as the dots for Braille characters). Genetically modified mice that lack Merkel's discs no longer respond to light touch (Maricich et al., 2009).

Meissner's corpuscles are more numerous than Merkel's discs but offer less spatial resolution. Meissner's corpuscles seem specialized to respond to changes in stimuli (as one would expect from rapidly adapting receptors) to detect localized movement between the skin and a surface (Heidenreich et al., 2012). This sensitivity to change in stimuli provides detailed information about texture (K. O. Johnson and Hsiao, 1992).

Both Meissner's corpuscles and Merkel's discs preferentially respond to edges on a surface (Pruszynski and Johansson, 2014). These receptors respond to touch because they make a specialized ion channel, called **Piezo**, which opens when

epidermis The outermost layer of skin, over the dermis.

dermis The middle layer of skin, between the epidermis and the hypodermis.

hypodermis Also called *subcutaneous tissue*. The innermost layer of skin, under the dermis.

tactile Of or relating to touch.

Meissner's corpuscle A skin receptor cell type that detects light touch.

Merkel's disc A skin receptor cell type that detects fine touch.

Piezo A family of two proteins that respond to mechanical stretch by opening channels to let cations in to depolarise the cell.

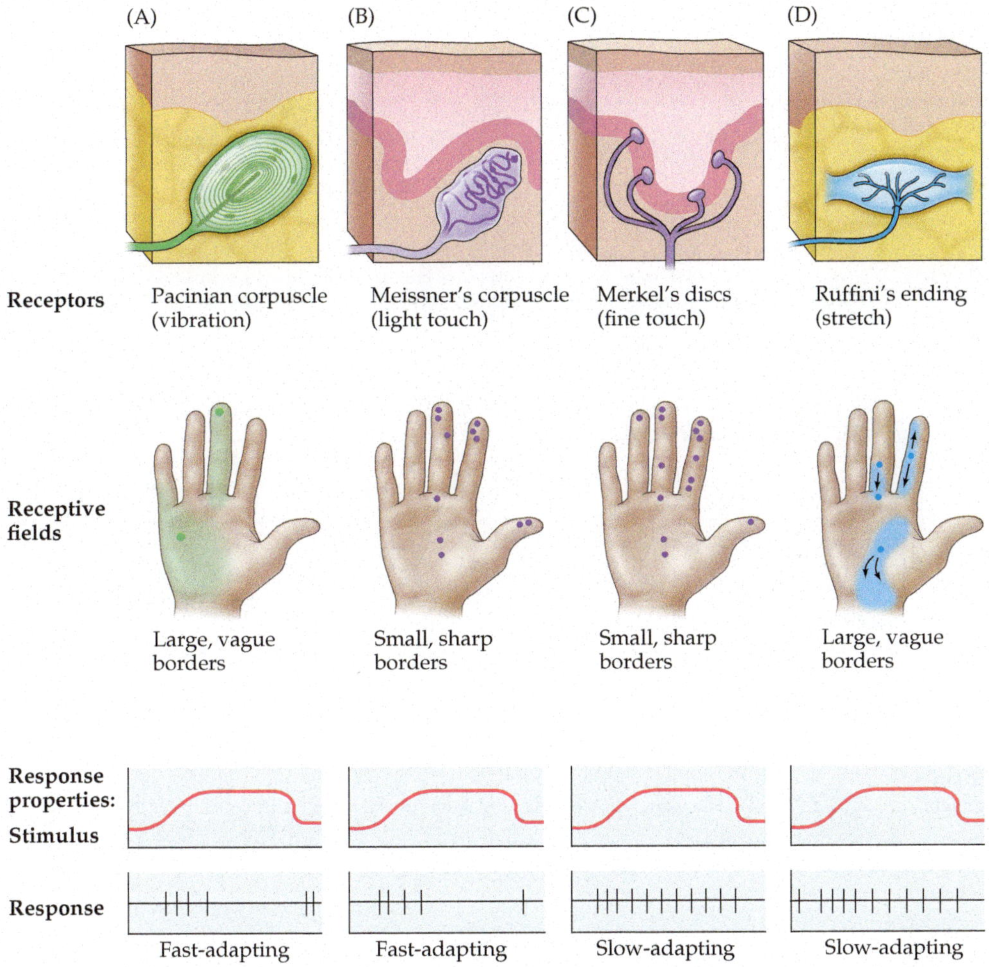

7.13 Properties of Skin Receptors Related to Touch Shown here for each skin receptor is the type of receptor (top), the size and type of the receptive field (middle), and the electrophysiological response (bottom). (A) Pacinian corpuscles activate fast-adapting fibres with large receptive fields. (B) Meissner's corpuscles are fast-adapting mechanoreceptors with small receptive fields. (C) Merkel's discs are slow-adapting receptors with small receptive fields. (D) Ruffini's endings are slow-adapting receptors with large receptive fields. The locations of these receptors in the skin are illustrated in Figure 7.4. (After R. S. Johansson and Å. B. Vallbo, 1983. *Trends Neurosci* 6: 27–31; Å. B. Vallbo and R. S. Johansson, 1984. *Hum Neurobiol* 3: 3–14.)

Go to Activity 7.2
Properties of Skin Receptors Related to Touch

mechanically stretched and so depolarizes the cell (see Figure 7.5B for one model of how Piezo proteins work). Disabling the Piezo2 gene in mice severely disrupts their response to light touch, without affecting response to painful stimuli or temperatures (Ranade et al., 2014).

The fourth type of touch receptors are the slow-adapting **Ruffini's endings**, which detect stretching of the skin when we move fingers or limbs. The very few Ruffini's endings (Pare et al., 2003) have large receptive fields (Johansson and Flanagan, 2009) (**FIGURE 7.13D**).

Recall from Chapter 2 that large axons conduct action potentials faster than small axons do and that myelination speeds conduction even more. All four of these light-touch receptors utilize moderately large (so-called Aβ, or A beta) myelinated fibres (**TABLE 7.2**). So the light-touch receptors send information very rapidly to the CNS. Later in this chapter we'll learn that some pain fibres are large and conduct rapidly, while others are small and conduct slowly. When a finger is moved across the raised dots of Braille, the Merkel's discs seem to convey the most information about the position of the dots as **FIGURE 7.14** shows on the next page.

Ruffini's ending A skin receptor cell type that detects stretching of the skin.

7.14 Various Touch Receptors Responding to Braille
(After J. R. Phillips et al., 1990. *Exp Brain Res* 81: 589–592.)

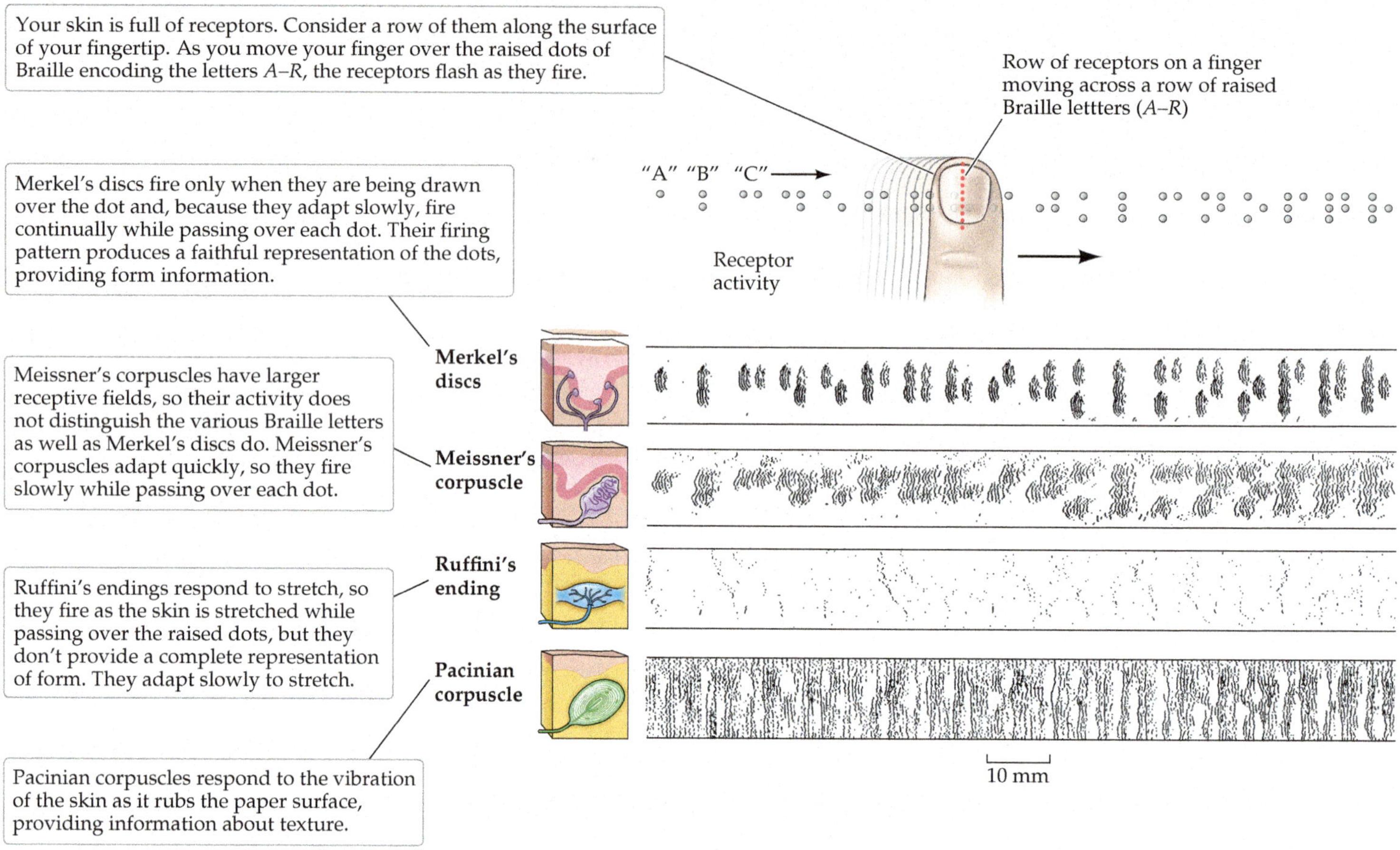

TABLE 7.2 Fibres That Link Receptors to the CNS

Sensory function(s)	Receptor type(s)	Axon type	Diameter (μm)	Conduction speed (m/s)
Proprioception (body sense)	Muscle spindle (see Chapter 11)	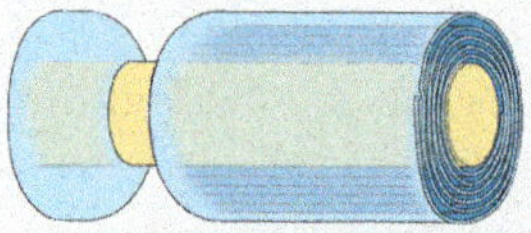 Aα (A alpha)	13–20	80–120
Touch (see Figures 7.13 and 7.14)	Pacinian corpuscles, Ruffini's endings, Merkel's discs, Meissner's corpuscles	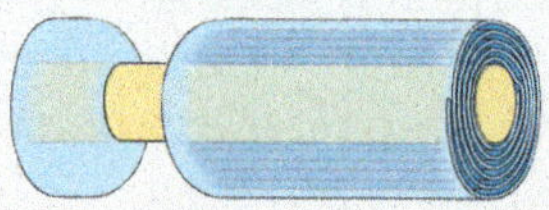 Aβ (A beta)	6–12	35–75
Pain, temperature	Free nerve endings	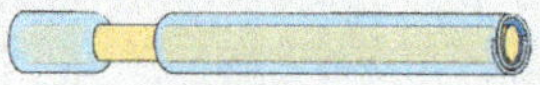 Aδ (A delta)	1–5	5–30
Temperature, pain, itch	Free nerve endings	 C	0.2–1.5	< 1

The dorsal column system carries somatosensory information from the skin to the brain

The touch receptors that we have described (Pacinian corpuscles, Merkel's discs, Meissner's corpuscles, and Ruffini's endings) send their axons to the spinal cord, where they enter the dorsal horn and turn upward, travelling to the brain along the spinal cord's dorsal column of white matter, which is why this is called the **dorsal column system**. These axons go all the way up to the brainstem, where they synapse on neurons of the **dorsal column nuclei** in the medulla (**FIGURE 7.15**). The axons of these medullary neurons then cross the midline to the opposite side and ascend to a group of nuclei of the thalamus. Outputs of the thalamus are directed to primary somatosensory cortex (S1). Note this means that touch receptors in your toe that are too small to see with the naked eye nevertheless extend their axons all the way to your medulla, a distance of 6 feet or more for some of you.

The skin surface can be divided into bands corresponding to the spinal nerves that carry the axons from the different regions. A **dermatome** (from the Greek derma, "skin," and tome, "part" or "segment") is the region of skin innervated by a particular spinal nerve. The pattern of dermatomes is hard to understand in an upright human, but remember that our erect posture is a recent evolutionary development. The mammalian dermatomal pattern evolved among our quadrupedal (four-legged) ancestors. Thus, the dermatomal pattern makes more sense when depicted on a person in a

dorsal column system A somatosensory system that delivers most touch stimuli via the dorsal columns of spinal white matter to the brain.

dorsal column nuclei Collection of neurons in the medulla that receive somatosensory information via the dorsal columns of the spinal cord. These neurons send their axons across the midline and to the thalamus.

dermatome A strip of skin innervated by a particular spinal nerve.

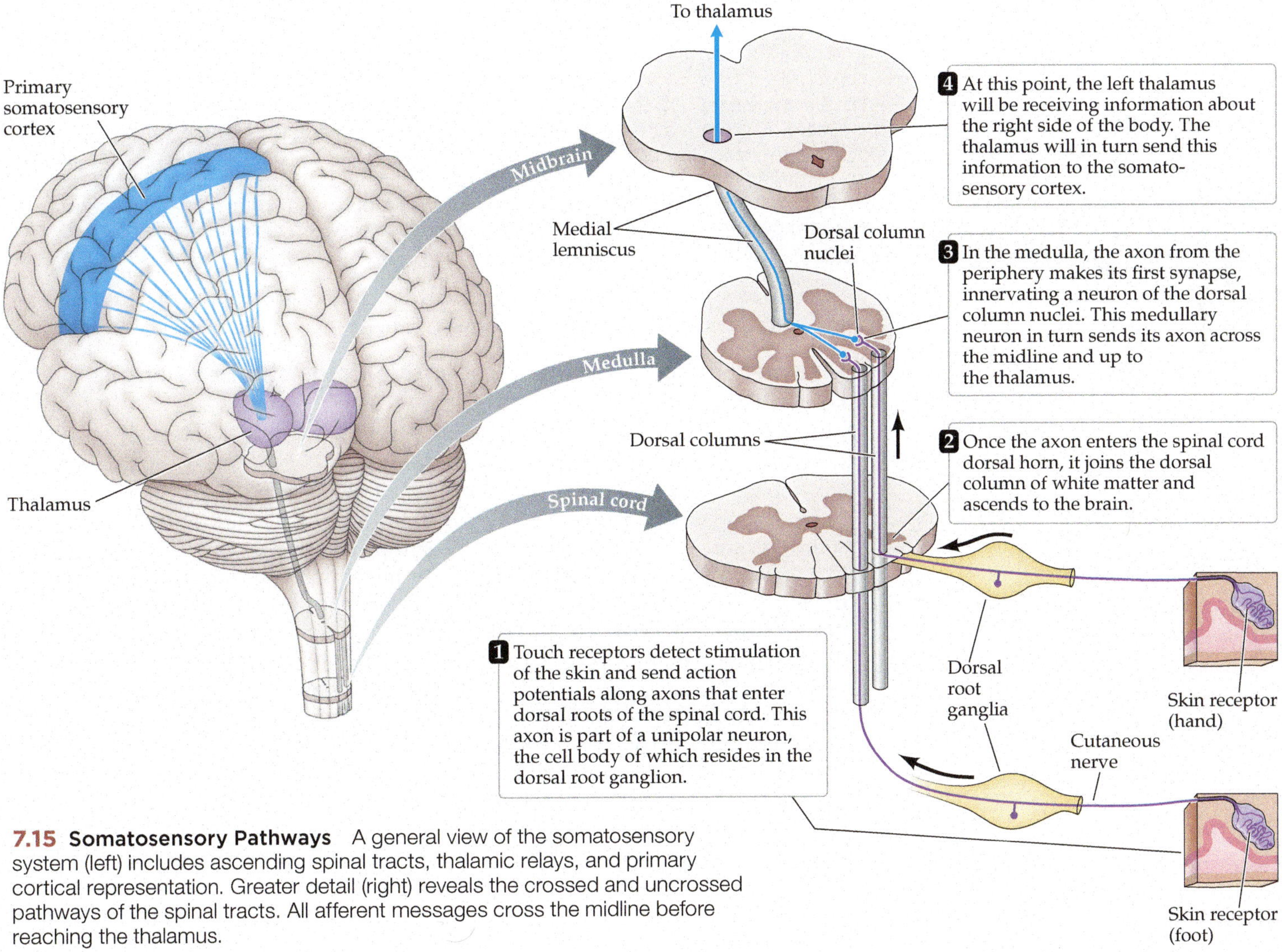

7.15 Somatosensory Pathways A general view of the somatosensory system (left) includes ascending spinal tracts, thalamic relays, and primary cortical representation. Greater detail (right) reveals the crossed and uncrossed pathways of the spinal tracts. All afferent messages cross the midline before reaching the thalamus.

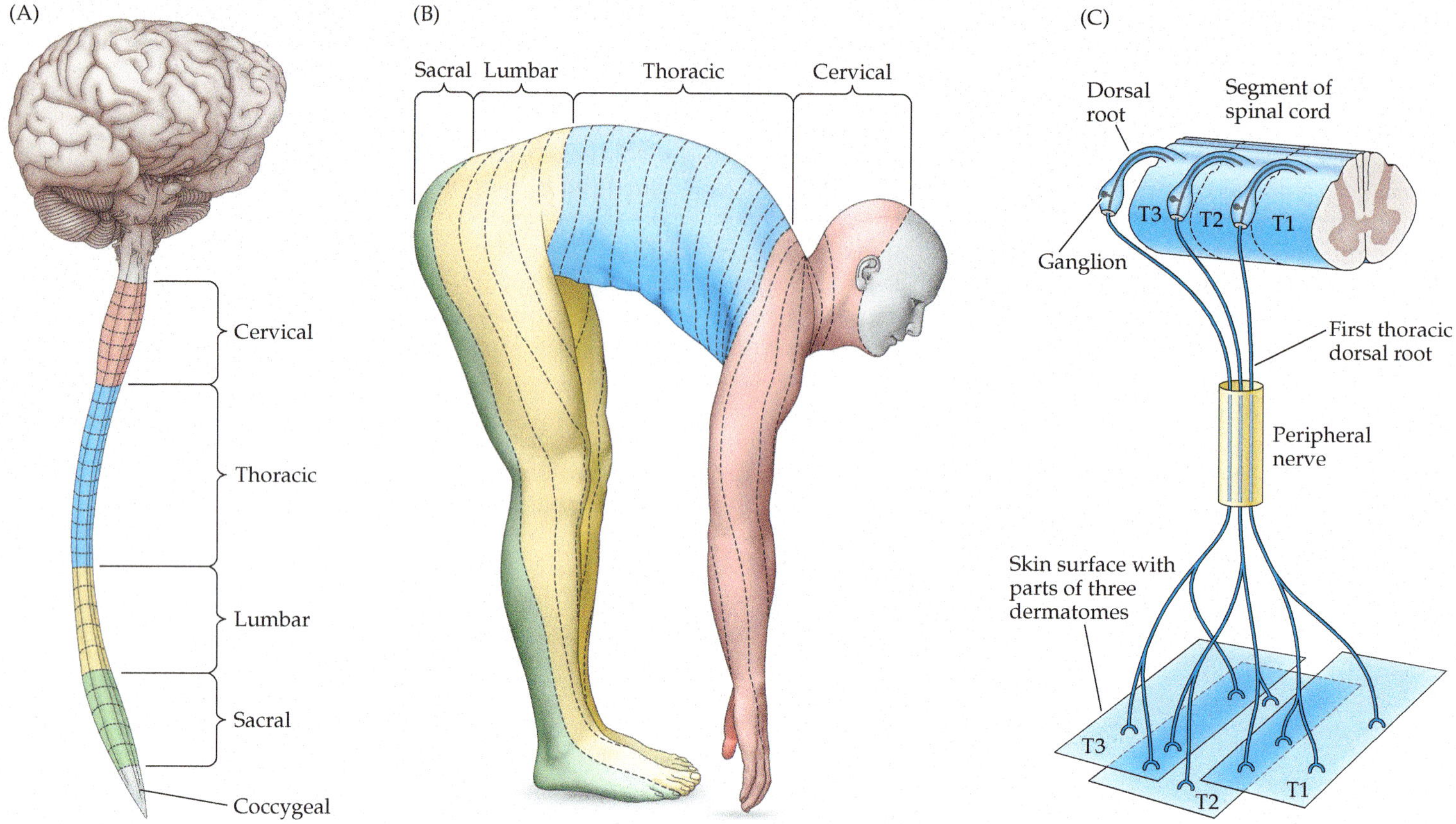

7.16 Dermatomes (A) Bands of skin send their sensory inputs to different dorsal roots of the spinal cord. Each dermatome is the section of skin that is innervated primarily by a given dorsal root. (B) In this side view of the human body in quadrupedal position, the pattern of dermatomes is colour coded to correspond to the spinal regions in A, and it appears more straightforward than it would in the erect posture. (C) Adjacent dorsal roots of the spinal cord collect sensory fibres from overlapping strips of skin, so the boundaries between the dermatomes overlap.

quadrupedal posture. The dermatomes overlap a modest amount (**FIGURE 7.16**) and may vary somewhat from person to person (Challoumas et al., 2018).

Plasticity in cortical maps: Receptive fields can be changed by experience

At one time, most researchers thought that cortical maps were fixed early in life and were invariant among all members of the same species. Now, however, we know that cortical maps can change with experience (D. T. Blake et al., 2002).

In one experiment, the receptive field of a monkey's hand was mapped in detail in the somatosensory cortex (**FIGURE 7.17A**). Then the middle finger was surgically removed. The region of cortex that responded to each remaining, adjacent finger expanded, so the region that had formerly responded to the removed finger now responded to the neighbouring fingers (**FIGURE 7.17B**). In another experiment, a monkey was trained to rest two fingers on a rotating disk in order to obtain food rewards. After several weeks of training, the hand area was mapped again, and the stimulated fingers were found to have enlarged representations compared with their previous areas (**FIGURE 7.17C**).

Similar findings were noted in rats exposed to differential tactile experiences (Xerri et al., 1996). Professional musicians who play stringed instruments have expanded cortical representations of their left fingers, presumably because they have been using these fingers to depress the strings for precisely the right note (Münte et al., 2002). Brain imaging also reveals cortical reorganisation in people who lose a hand in adulthood (**FIGURE 7.18**). One man received a transplanted hand (from an accident victim) 35 years after losing his own. Despite the length of time that had passed, his brain reorganized in just a few months to receive sensation from the new hand in the appropriate part of S1 (S. H. Frey et al., 2008).

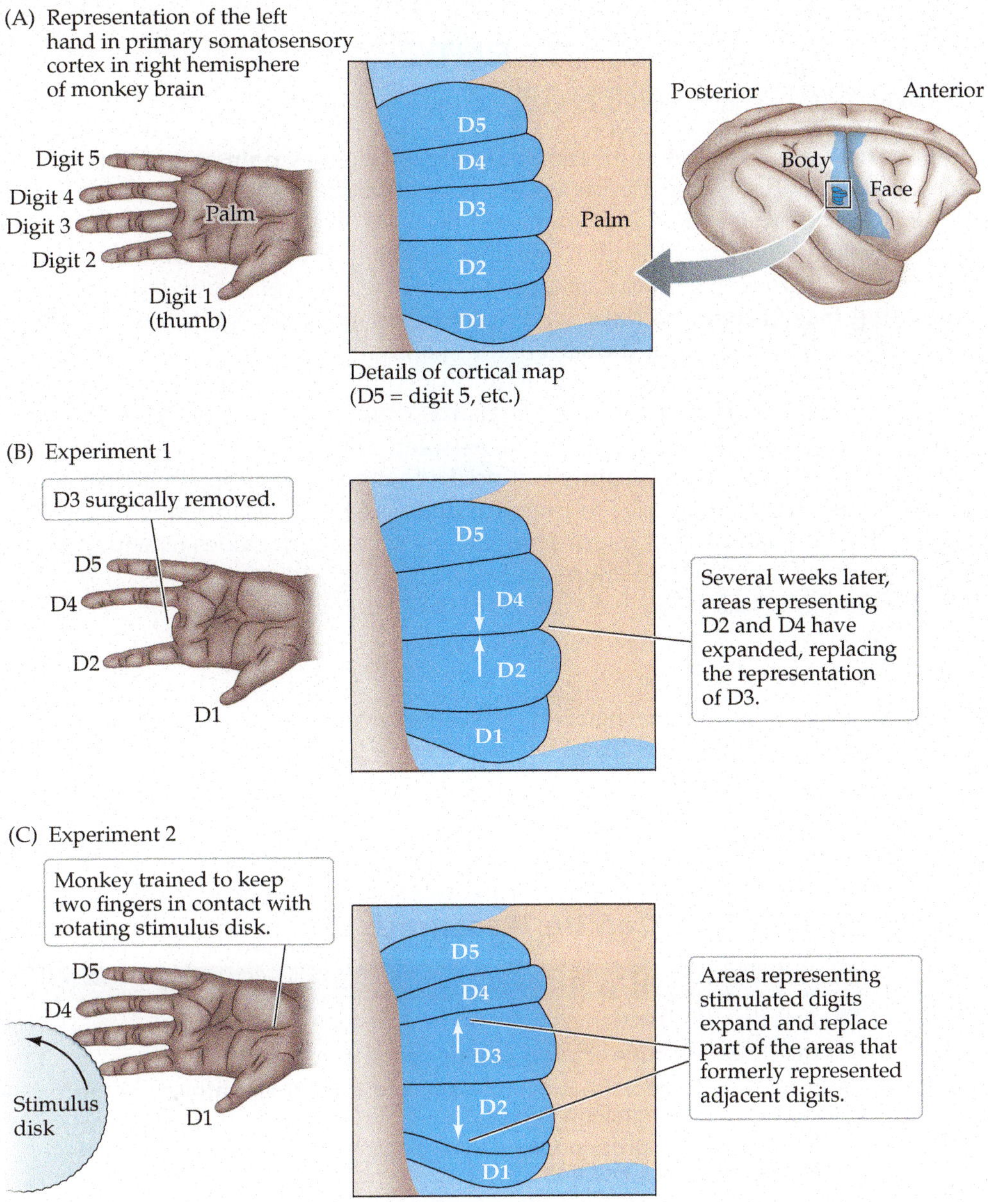

7.17 The Plasticity of Somatosensory Representations These experiments demonstrate that the adult brains of monkeys can be altered by experience. (After M. M. Merzenich and W. M. Jenkins, 1993. *J Hand Ther* 6: 89–104.)

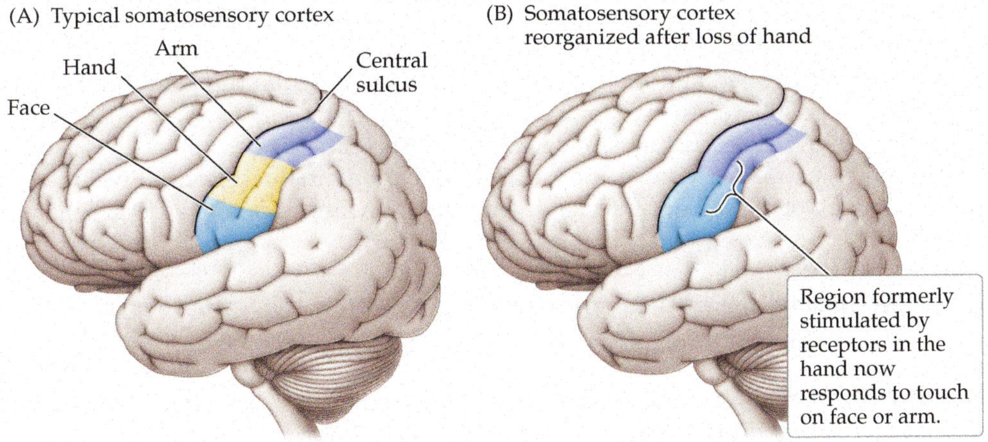

7.18 Typical and Reorganized Somatosensory Cortex In humans we can map S1 using functional brain imaging to determine which parts of cortex are activated by touch on different parts of the body. (A) Normally, the region of S1 receiving information from the hand is interposed between the regions representing the upper arm and the face. (B) In a person who, as an adult, loses one hand, the cortical regions representing the upper arm and face expand, taking over the cortical region previously representing the missing hand. Presumably the loss of sensory input from the lost hand allows those cortical neurons to become innervated by neighbouring inputs. (B after T. T. Yang et al., 1994. *NeuroReport* 5: 701–704.)

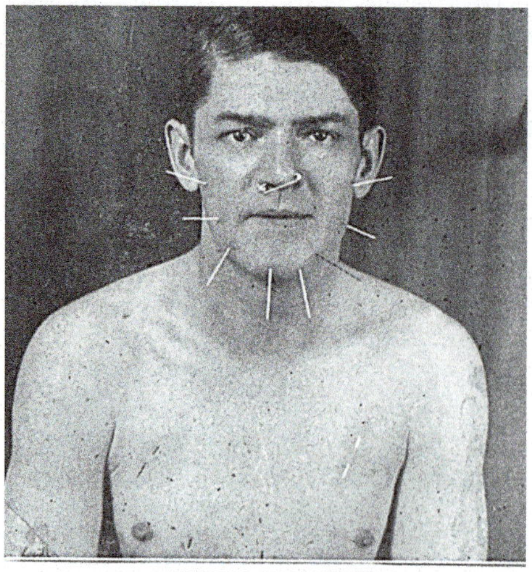

© Chronicle/Alamy Stock Photo

7.19 Doesn't That Hurt? The earliest scientific report of a person with congenital insensitivity to pain was of a man working in the theater, like the man shown here, as a "human pincushion."

PAIN: AN UNPLEASANT BUT ADAPTIVE EXPERIENCE

The International Association for the Study of Pain defines **pain** as "an unpleasant sensory and emotional experience associated with actual or potential tissue damage, or described in terms of such damage." Because pain is unpleasant and causes great suffering, it may be difficult to imagine a biological role for it. But clues to the adaptive significance of pain can be gleaned from the study of rare individuals with **congenital insensitivity to pain**, who never experience pain (Hirsch et al., 1995), like Ashlyn, whom we met at the start of the chapter.

The first clinically reported person with congenital insensitivity to pain worked on the stage as a "human pincushion" (**FIGURE 7.19**) (Dearborn, 1932). People who display pain insensitivity can discriminate between the touch of the point or head of a pin, but they experience no pain when pricked with the point. Such people show extensive scarring from injuries to fingers, hands, and legs (Manfredi et al., 1981), and they tend to die young of injuries, suggesting that pain provides us with powerful lessons about how to avoid injury.

Dennis and Melzack (1983) list three ways that pain helps us:

1. Short-lasting pain causes us to withdraw from the source, often reflexively, thus preventing further damage.
2. Long-lasting pain promotes behaviours, such as sleep, inactivity, grooming, feeding, and drinking, that speed recuperation.
3. The expression of pain serves as a social signal to other animals. For example, screeching after a painful stimulus signals the potential harm to genetically related individuals and elicits caregiving behaviour from them.

7.4 Human Pain Can Be Measured

Learning Objectives

After reading this section, you should be able to:

7.4.1 List the three principle aspects of pain perception.

7.4.2 Describe the receptors that detect pain, and trace their pathway to the spinal cord, brainstem, and cortex.

7.4.3 Name the receptor protein family that allows free nerve endings to detect temperature change and itch.

7.4.4 Describe neuropathic pain, including phantom limb pain and migraine.

In some parts of the world, people endure, with stoic indifference, rituals (including body mutilation) that would cause most other humans to cry out in pain. Incisions of the face, hands, arms, legs, or chest; walking on hot coals; and other treatments clearly harmful to the body can be part of the ritual. Comparable experiences are also seen in more ordinary circumstances, such as when a highly excited athlete continues to play a game with a broken arm or leg. Learning, experience, emotion, and culture all affect the perception of pain in striking ways.

The mere terms mild and intense are inadequate to describe the pain that is distinctive to a particular disease or injury. The physician's simple query "Is the pain still there?" has been replaced by a more detailed analysis that provides better clues about how to control pain. For example, the McGill Pain Questionnaire (Melzack, 1984) consists of a list of words arranged into classes that describe three different aspects of pain:

1. The *sensory-discriminative* quality (e.g., throbbing, gnawing, shooting)
2. The *motivational-affective* (emotional) quality (e.g., tiring, sickening, fearful)
3. An overall *cognitive* evaluative quality (e.g., no pain, mild, excruciating)

Patients are asked to select the set of words that best describes their pain. These three components reflect different aspects of pain perception (**FIGURE 7.20**).

pain The discomfort normally associated with tissue damage.

congenital insensitivity to pain The condition of being born without the ability to perceive pain.

One of the interesting aspects of the McGill scale is that it can distinguish among pain syndromes, meaning that patients use a distinctive constellation of words to describe a particular pain experience. For example, tooth pain is described differently from arthritic pain, which in turn is described differently from menstrual pain.

Contemporary studies of pain mechanisms have revealed receptors in the skin that transmit pain information to the central nervous system. In this section we will discuss some features of peripheral and CNS pathways that mediate pain.

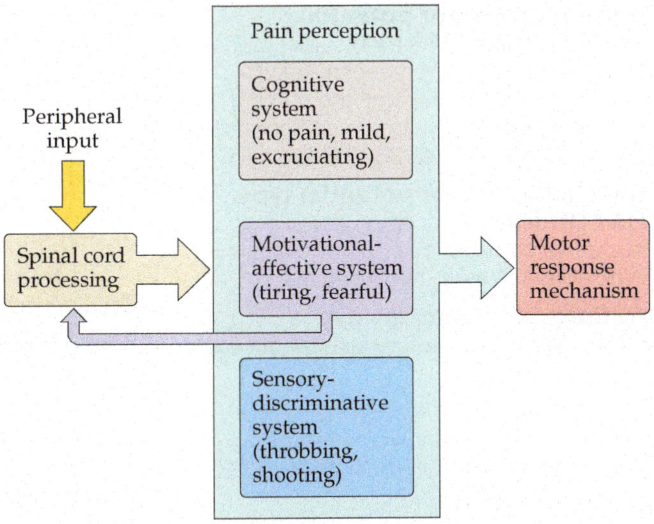

7.20 The Multifaceted Character of Pain

Peripheral receptors get the initial message

In most cases the initial stimulus for pain is the destruction or injury of tissue adjacent to certain nerve fibres. The damaged tissue releases chemicals that activate pain fibres in the skin. Various substances have been suggested as the chemical mediators of pain, including neuropeptides, serotonin, histamine, various proteolytic (protein-metabolizing) enzymes, and prostaglandins (a group of widespread hormones) (**FIGURE 7.21**).

Peripheral receptors and nerve fibres that are specialized to respond to pain are called **nociceptors**. **Free nerve endings** in the dermis display no specialized structures (they look like naked nerve endings), but they have specialized receptor proteins on the cell membrane that respond to various signals. Different free nerve endings produce different receptor proteins, so they report different stimuli, such as pain and/or changes in temperature.

HOT, COLD, OR COOL? The best evidence for specialized pain fibres in the periphery came from studies of **capsaicin**, the chemical that makes chili peppers spicy hot. Investigators isolated a receptor found in some free nerve endings that binds capsaicin, and they found that the receptor also responds to sudden increases in temperature (Caterina et al., 1997). The receptor was cloned and found to be a member of a family of proteins called transient receptor potential (TRP) ion channels.

nociceptor A receptor that responds to stimuli that produce tissue damage or pose the threat of damage.

free nerve ending An axon that terminates in the skin without any specialised cell associated with it and that detects pain and/or changes in temperature.

capsaicin A compound synthesised by various plants to deter predators by mimicking the experience of burning.

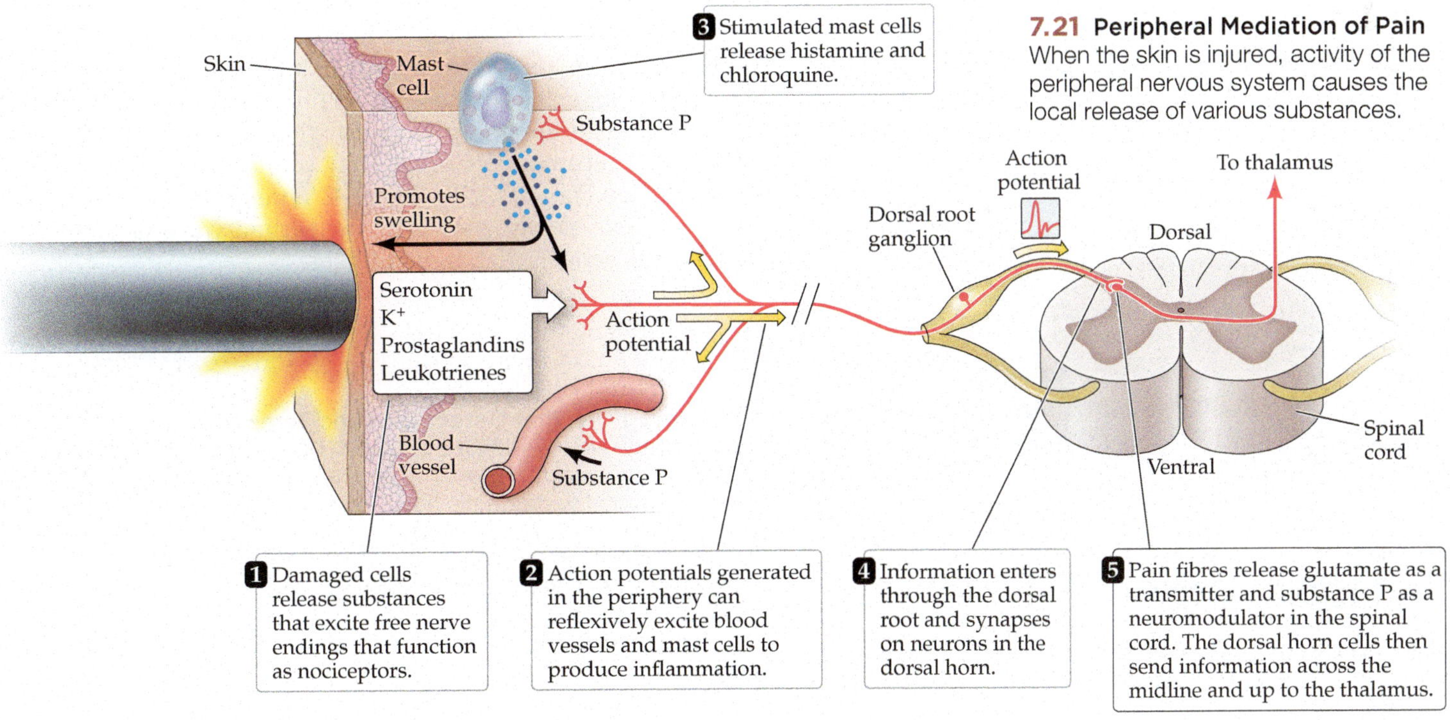

7.21 Peripheral Mediation of Pain When the skin is injured, activity of the peripheral nervous system causes the local release of various substances.

transient receptor potential vanilloid type 1 (TRPV1) Also called *vanilloid receptor 1*. A receptor that binds capsaicin to transmit the burning sensation from chili peppers and normally detects sudden increases in temperature.

transient receptor potential type M3 (TRPM3) A receptor, found in some free nerve endings, that opens its channel in response to rising temperatures.

Aδ fibre A moderately large, myelinated, and therefore fast-conducting axon, usually transmitting acute pain information.

C fibre A small, unmyelinated axon that conducts pain information slowly and adapts slowly.

They named the capsaicin receptor **transient receptor potential vanilloid type 1** (**TRPV1**), or vanilloid receptor 1, because the crucial component of the capsaicin molecule is a chemical known as vanilloid. Mice lacking the gene for this receptor still responded to mechanosensory pain, but not to mild heat or capsaicin (Caterina et al., 2000). So the reason that chili peppers taste "hot" is that the capsaicin in the peppers activates TRPV1 receptors in the body that normally detect noxious heat. The venom of a Caribbean tarantula contains peptides that activate this same receptor, causing an intense burning sensation (Siemens et al., 2006).

TRPV1's normal job is to report a drastic rise in temperature to signal burning. Chili peppers evolved the chemical capsaicin to activate TRPV1 to ward off mammalian predators. Capsaicin molecules have the perfect shape to bind to TRPV1 and open its ion channel, which is normally opened by heat. Because the brain interprets action potentials from that nerve as signalling painful heat (another example of a labelled line), we mammals experience painful heat.

Interestingly, the bird version of TRPV1 still responds to heat but is not affected by capsaicin (Tewksbury and Nabhan, 2001). In this case both the animal and the plant benefit from capsaicin: the birds are able to get food from the plants—food that might be gone if mammals were able to eat it—and then fly off to spread the chili seeds far and wide. Clever, clever chili plants.

Why do our lips swell when we have overindulged in capsaicin? As Figure 7.21 illustrates, the action potentials generated by the chemical travel back out other branches of the axon to affect blood vessels and mast cells, triggering the swelling. Paradoxically, rubbing capsaicin into the skin overlying arthritic joints brings some pain relief, perhaps because overactivated pain fibres temporarily run out of transmitter.

A related receptor, **transient receptor potential type M3** (**TRPM3**), detects even higher temperatures than does TRPV1 (Vendewauw et al., 2018). This is an example of range fractionation we discussed earlier (see Figure 7.6), as different receptors respond to different ranges of temperature. TRPM3 differs from TRPV1 in two other ways as well: TRPM3 does not respond to capsaicin, and TRPM3 receptors are found on nerve fibres larger than those carrying TRPV1. Remember that large axons conduct action potentials more rapidly than do small axons. TRPM3 receptors are found on relatively large axons known as **Aδ** (or A delta) **fibres**: large-diameter, myelinated axons (**FIGURE 7.22**). Because of the relatively large axon diameter and myelination, these fibres report to the spinal cord very quickly. When you burn your finger on a hot skillet, the first, sharp pain you feel is conducted by these fat Aδ fibres that detected the heat with TRPM3 receptors. In contrast, the nerve fibres that possess TRPV1 receptors consist of thin, unmyelinated fibres called **C fibres** (see Figure 7.22B). These C fibres conduct slowly, and TRPV1 adapts slowly, providing the second wave of pain—the dull, lasting ache in that darned finger.

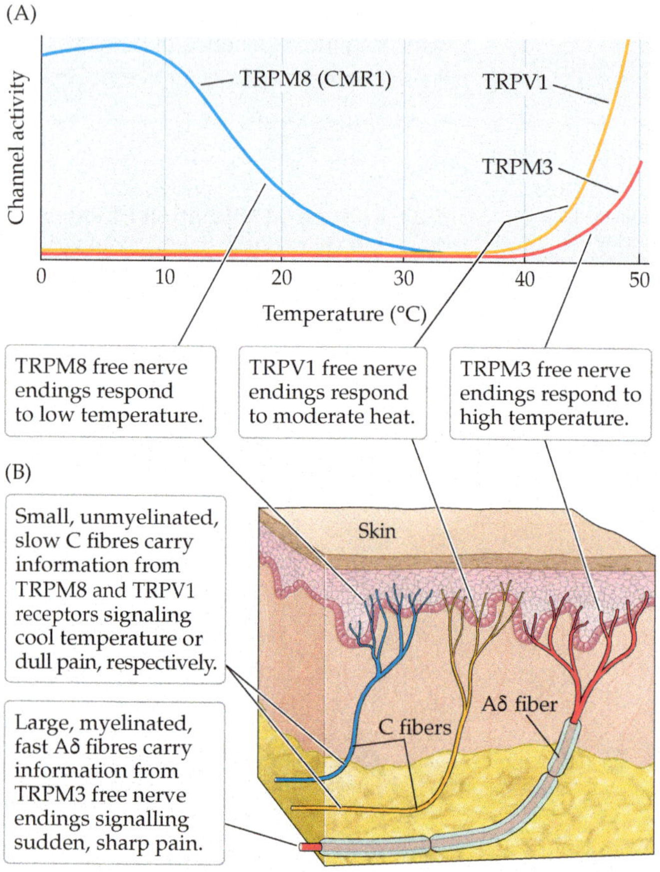

7.22 Receptors That Detect Pain and Temperature (A) Free nerve endings with TRPM8 are activated by temperatures below normal body temperature. Free nerve endings with the capsaicin receptor (TRPV1) respond to moderate heat and the capsaicin found in chili peppers. Other free nerve endings, with the related receptor protein TRPM3, detect high temperatures. (B) The TRPM3 free nerve endings transmit a fast action potential along large, myelinated Aδ fibres to the spinal cord. TRPM8 and TRPV1 receptors transmit along slower, unmyelinated C fibres. (Part A after J. Vriens et al., 2014. *Nat Rev Neurosci* 15: 573–589.)

This is the reason for the slight delay between licking the cut surface of a chili pepper and feeling the burn. Only the TRPV1 fibres, with their slowly conducting C fibres, have been activated, not the fast Aδ fibres using TRPM3. Table 7.2 compares these different fibres.

Taking a cue from the success with capsaicin, another group of investigators tried looking for the "coolness" receptor, reasoning that it should respond to the chemical menthol. They found another member of the TRP family, **TRPM8** (or *cool-menthol receptor 1, CMR1*). TRPM8 responds to cool temperatures, and it is found on small C fibres (Bautista et al., 2007), so it transmits information about cool temperatures rather slowly (see Figure 7.22). Another member of the TRP family is expressed by other free nerve endings to make them responsive to the spices oregano, thyme, and clove (H. Xu et al., 2006), while another receptor responds to garlic and onion (Salazar et al., 2008), and yet another receptor from this family responds to the mustard oils that are found in wasabi (Jordt et al., 2004). Spices add so much to our experience of food because they stimulate these various receptors in our mouth (in addition to the olfactory receptors we'll discuss in Chapter 9).

TRPM8 Also called *cool-menthol receptor 1* (*CMR1*). A sensory receptor, found in some free nerve endings, that opens an ion channel in response to a mild temperature drop or exposure to menthol.

Na_v1.7 Also called *SCN9A*. A voltage-gated sodium channel used almost exclusively by nociceptors to initiate action potentials.

PAIN-SPECIFIC SODIUM CHANNELS The search for the protein receptor used by free nerve endings to detect mechanical damage has been hindered by the fact that so many chemicals are released by damaged tissue. It remains unclear which chemical is crucial for activating nociceptors. However they are initially activated, nociceptors then use a very particular type of voltage-gated sodium channel to produce action potentials to report pain to the brain. We learned this fascinating aspect of nociception from a family in northern Pakistan. A 10-year-old boy there was giving street performances of piercing his arms with knives and walking on hot coals without pain. Before scientists could locate him, he died after jumping off a roof to impress his friends. The boy got up, saying he felt fine, unaware of the internal bleeding that killed him.

This boy's apparent fearlessness, which cost him his life, reminds us that pain, and fear derived from pain, really can be our friend. Six of the boy's relatives also felt no pain. They could detect light touch, coolness, warmth, and other sensations just fine. Scientists isolated the gene, called SCN9A, responsible for pain insensitivity in this family and found that it encodes a voltage-gated sodium channel **Na_v1.7** that is expressed in some free nerve endings (Cox et al., 2006). For some reason, pain receptors rely primarily on this particular voltage-gated sodium channel to initiate action potentials (Waxman and Zamponi, 2014). A mutation of the SCN9A gene in this family prevents the channel from opening, and so the pain receptors do not start action potentials. Ashlyn, the teenager we described at the start of the chapter, has a similar mutation in this gene. She can feel light touch, vibrations, coolness, and warmth, but not mechanical damage or the high temperatures that cause burns. Axons that would normally signal pain don't produce action potentials to send the message to her brain. Other mutations of the same channel may make it open more readily, causing a life of chronic pain (Catterall and Yu, 2006). Interestingly, this same channel is also used by odour-detecting sensory cells to produce action potentials (J. Weiss et al., 2011), as we'll discuss further in Chapter 9. That is why Ashlyn and others with this mutation cannot detect odours (Heckert, 2012). Why these particular sensory neurons use this particular sodium channel for producing action potentials, while the vast majority of neurons use other sodium channels, remains a mystery. But it's exciting to think that we may be able to synthesise drugs that specifically block that channel, and therefore pain, without silencing the entire brain. Work with grasshopper mice suggested yet another pain-specific sodium channel, as we'll see next.

A Painless Life? Unable to feel pain, Ashlyn must be more vigilant than other teenagers to avoid injury.

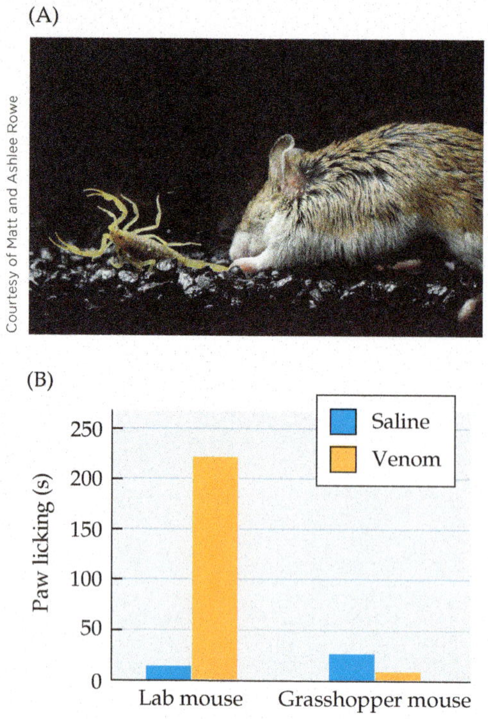

7.23 An Indifference to Scorpion Toxin (After A. H. Rowe et al., 2013. *Science* 342: 441–446.)

The Cutting Edge

Evolving an Indifference to Toxins

Because sensory neurons signalling pain use very specific versions of the voltage-gated sodium channel that triggers action potentials, we have the exciting opportunity to produce novel compounds that might reduce pain by interfering with only those sodium channels. But a long-running arms race in the U.S. Southwestern desert between grasshopper mice and the bark scorpions they love to eat beautifully illustrates how natural selection can also produce novel compounds, in this case for *causing pain*. Bark scorpions evolved a toxin that selectively activates that same Na_V1.7 that drives pain fibres to fire, so we humans find their sting quite painful (but not fatal for adults). Laboratory mice also seem to find the sting painful—if a tiny bit is injected into a paw, they lick it vigourously. But grasshopper mice that normally live in that same desert don't seem to mind the scorpion toxin (**FIGURE 7.23**). When scientists recorded from sensory cells in the two species of mice, they found that the toxin strongly activates sensory cells in lab mice but *inhibits* them in grasshopper mice (A. H. Rowe et al., 2013).

Did the grasshopper mice evolve a new Na_V1.7 that ignores the toxin? Apparently not, because the current opened by the toxin in grasshopper mice was unaffected by tetrodotoxin which blocks the Na_V1.7 channel. Instead, the grasshopper mice seem to have evolved a mutation in *another* voltage-gated sodium channel that is found exclusively in pain receptors—Na_V1.7. The Na_V1.7 channel seems to *initiate* firing of pain sensory neurons, while Na_V1.8 mediates the *sustained*, continuous firing of the nociceptive neurons, and the toxin activates both channels in humans. But when the scientists sequenced the gene for Na_V1.8 from grasshopper mice, they found it differed from that of lab mice in several amino acids. By systematically inducing different mutations in the gene, they identified two amino acid changes that are necessary for the difference in response to scorpion toxin. While the toxin opens the Na_V1.7 channel in both species, and also opens the Na_V1.8 channel in lab mice, that channel in grasshopper mice is *blocked* by the toxin. So the grasshopper mice probably feel the first sting of the scorpion toxin, when Na_V1.7 responds, but without the sustained firing of nociceptors normally mediated by Na_V1.8, they probably never feel the long, slow burn that we (and lab mice) feel. Unless the scorpions evolve a new toxin that can activate rather than block the grasshopper mouse Na_V1.8 channel, they're going to have to be careful out there. In the meantime, the discovery that Na_V1.8 mediates sustained firing of nociceptors offers yet another target for relieving pain if scientists can find a compound that blocks the human version of the channel. ■

Spinal pathways transmit pain information

In the central nervous system, special pathways mediate pain and temperature information. Earlier we discussed the dorsal column system that carries touch information to the brain (see Figure 7.15). The sensations of pain and temperature are transmitted separately by the **anterolateral**, or spinothalamic system. Free nerve endings in the skin send their axons to synapse on neurons in the dorsal horn of the spinal cord. These spinal cord neurons send their axons across the midline to the opposite side and up the anterolateral column of the spinal cord to the thalamus (hence the term spinothalamic). So, in the spinothalamic system, pain information crosses the midline in the spinal cord before ascending to the brain (**FIGURE 7.24**); recall that touch information, in contrast, first ascends to the brainstem and then crosses the midline (see Figure 7.15). Either way, sensory information from one side of the body ends up in the opposite side of the brain.

The afferent fibres from the periphery that carry nociceptive information use **glutamate** as a neurotransmitter to excite spinal cells in the dorsal horn (Dubin and Patapoutian, 2010), but they also release the neuromodulator **substance P**, which is a neuropeptide (the P stands for peptide). Injection of capsaicin into the skin provides a specific painful stimulus that leads to the release of substance P in the dorsal horn.

anterolateral system Also called *spinothalamic system*. A somatosensory system that carries most of the pain and temperature information from the body to the brain.

glutamate An amino acid transmitter; the most common excitatory transmitter.

substance P A peptide transmitter implicated in pain transmission.

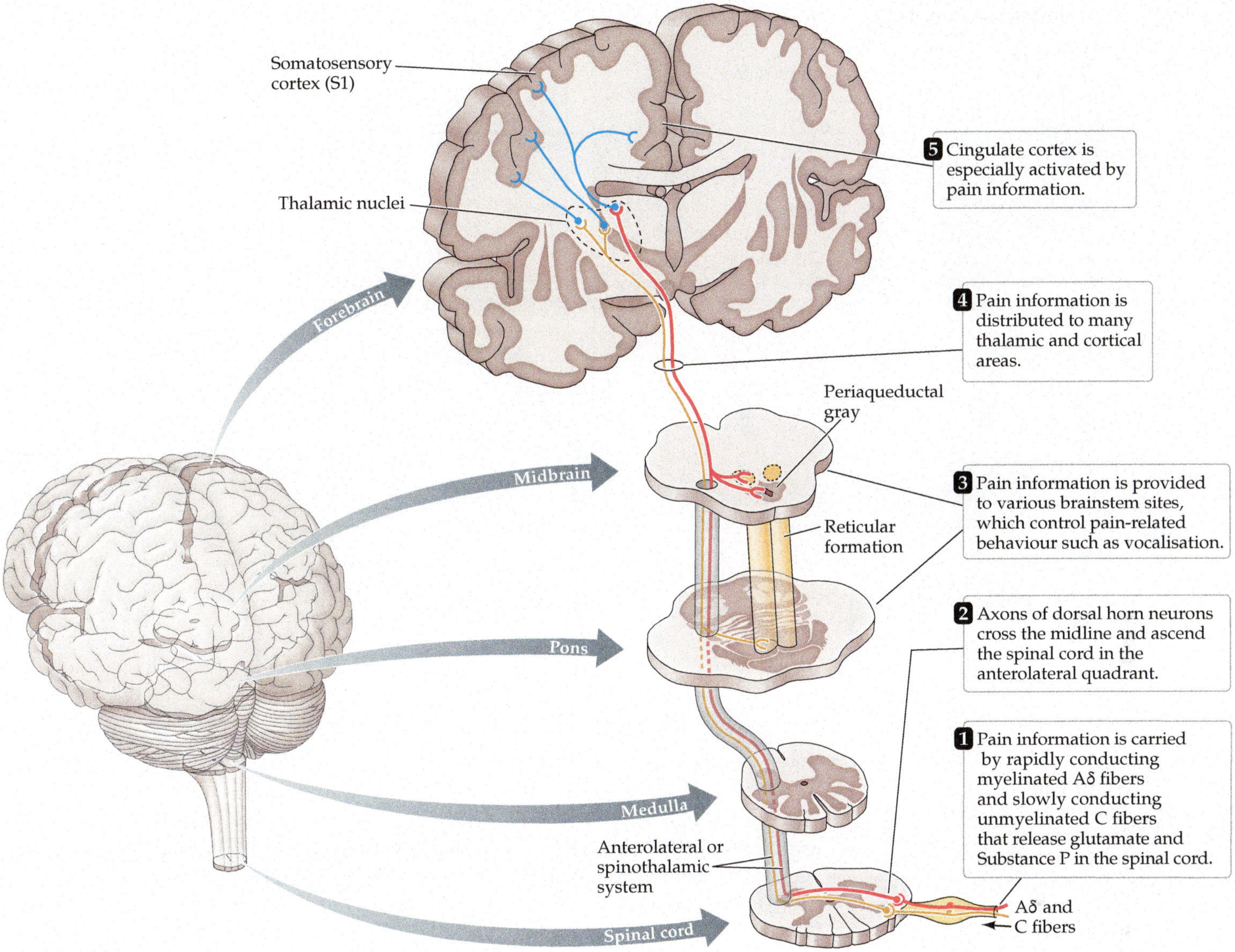

7.24 Pain Ascends the Spinothalamic System to Reach the Brain Sensation of pain travels from its origin to the brain via the spinothalamic system, crossing the midline in the spinal cord before ascending to the thalamus. Shortly we'll see that fibres signalling itch also cross the midline in the spinal cord, but those fibres report to the pons rather than the thalamus.

There, the postsynaptic neurons take up the substance P and begin remodelling their dendrites; investigators have speculated that this neural plasticity later affects pain perception (Sahbaie et al., 2010), as we'll discuss shortly.

Further evidence that substance P plays a role in pain comes from experiments with knockout mice that lack either a gene for substance P (Y. Q. Cao et al., 1998) or the substance P receptor (De Felipe et al., 1998). These mice are unresponsive to certain kinds of intense pain. Interestingly, they still respond to mildly painful stimuli, suggesting that other signals, perhaps the glutamate neurotransmitter, can carry that information.

Itch uses privileged pathways

Yet other free nerve endings provide us with the sensation of itch (gee, thanks!). Some of these fibres respond to histamine that is released from mast cells in the skin (see Figure 7.21). Other causes of itch—like certain medications and plants such as cowhage—do not respond to antihistamines (Gawande, 2008), indicating that a second signal released from mast cells, chloroquine, can also provoke the sensation of itch (Ru et al., 2017). These two itch systems both use members of the TRP receptor family to respond to histamine and chloroquine from mast cells, and both use slowly conducting C fibres to send action potentials to the spinal cord. But unlike heat and mechanical damage–triggered nociceptors that release glutamate, the itch-specific fibres

Go to Activity 7.3
Ascending Pain Pathways in the CNS

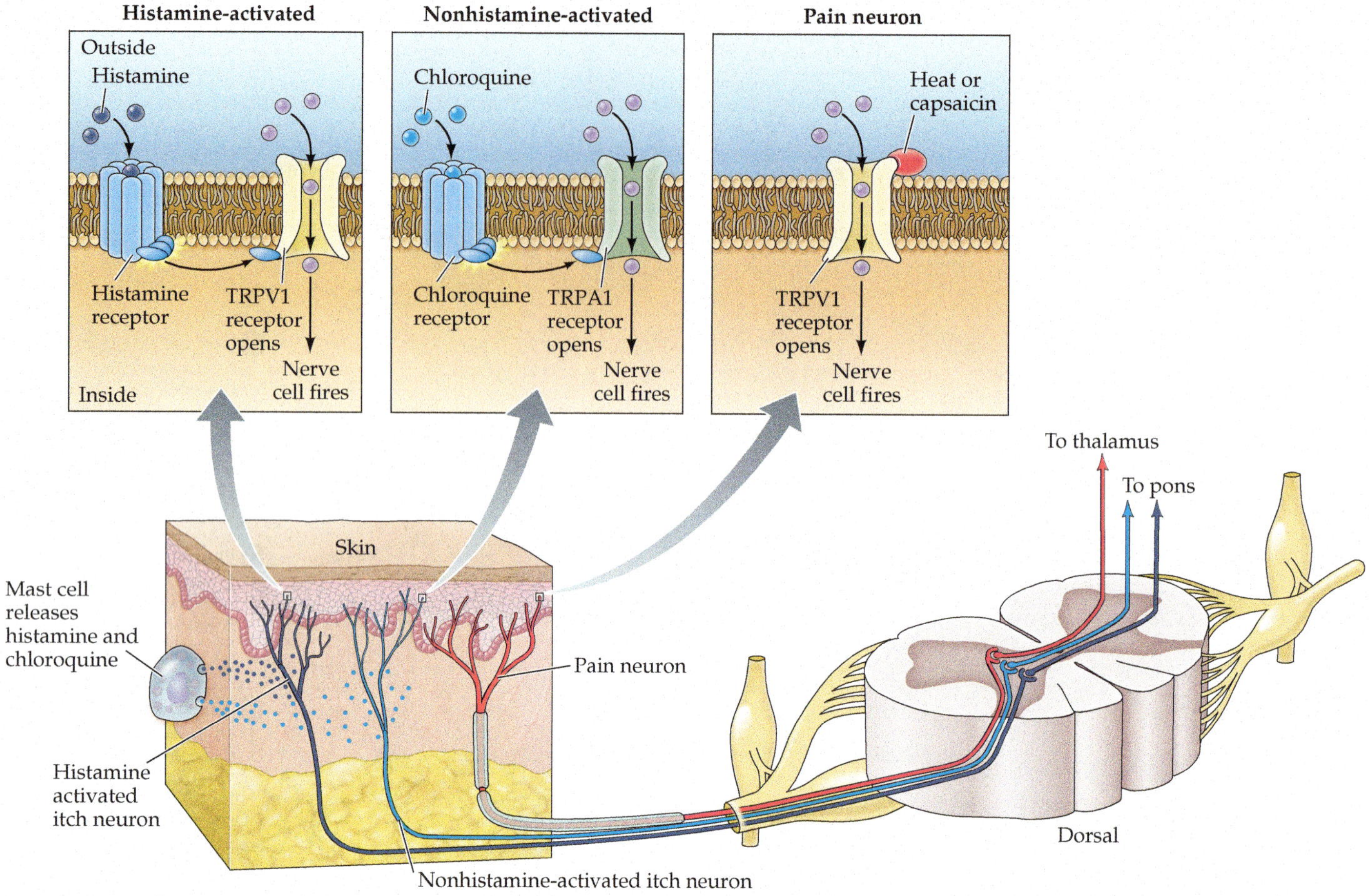

7.25 Itch versus Pain Stimuli cause itch by triggering mast cells to release histamine and other chemicals. These act upon separate cohorts of free nerve endings with the receptors to respond to mast cell signals. These itch receptors synapse upon neurons in the dorsal horn, which in turn send their axons across the midline and ascend to terminate in the pons (rather than the thalamus that receives pain information). (After S. Sutherland, 2015. *Sci Am* 314: 38–43.)

use another neurotransmitter, called **natriuretic polypeptide B (Nppb)**, to stimulate neurons in the dorsal horn (LaMotte et al., 2014). These neurons in turn send their axons across the midline in the spinal cord, but instead of projecting to the thalamus, they terminate in the pons (Mu et al., 2017) (**FIGURE 7.25**). Mice lacking the gene for Nppb fail to show an itch response (S. K. Mishra and Hoon, 2013). Scratching the affected area of skin provides temporary relief by activating touch receptors that silence the spinal neurons that normally respond to the Nppb fibres to report itch sensation to the brain. This spinal mechanism to suppress itch signalling uses neuropeptide Y (NPY) (see Chapter 14) as a transmitter (Bourane et al., 2015), so perhaps it will be possible to develop drugs to activate these NPY-sensitive circuits to suppress itch. One peculiarity of itch is that it is socially contagious—seeing someone else scratching makes us more likely to feel an itch—in both humans and mice (Y. Q. Yu et al., 2017).

The reign of pain is mainly in the brain

Sometimes pain persists long after the injury that gave rise to it has healed. The most dramatic example is a person's continued perception of chronic pain coming from a missing limb after loss of an arm or leg. Called phantom limb pain, this sensation is an example of **neuropathic pain**, so called because the pain seems to be due to inappropriate signalling of pain by neurons (rather than to tissue damage). The pain experienced may also be called neuralgia. Such cases also can be seen as a disagreeable example of neural plasticity, because the nervous system seems to have amplified its response to the pain signal (S. P. Cohen and Mao, 2014). Some of this amplification is taking place in the cortex (Flor et al., 2006), but changes in the spinal cord also contribute (Kuner and Flor, 2017). After peripheral nerve injury (such as a damaged back or lost limb), microglial cells surround the synapses between pain fibres and neurons in the dorsal horn of the spinal cord. These microglial cells release chemicals that make the dorsal horn neurons hyperexcitable (Inoue and Tsuda, 2018). Thus, the dorsal horn

natriuretic polypeptide B (Nppb) A peptide neurotransmitter used by neurons reporting itch to the spinal cord.

neuropathic pain Pain caused by damage to peripheral nerves. It is often difficult to treat.

neurons become chronically active, flooding the thalamus with action potentials signalling pain. Other immune system responses, including inflammation, are also suspected of a role in neuropathic pain (R. R. Ji et al., 2016).

Chronic pain can have dramatic effects on the brain. One study found that the gray matter in the dorsolateral prefrontal cortex of people with chronic back pain shrinks faster than in normal aging, averaging 1.3 cubic millimetres more per year of pain (Apkarian et al., 2004). This shrinkage is equivalent to 10–20 years of normal aging.

Cases of phantom limb pain are notoriously difficult to treat. In one type, the person perceives that a missing limb is twisted and therefore hurts. Vilayanur Ramachandran has pioneered a treatment for this condition: the person looks at an image in a mirror that is positioned in such a way that the reflection of the intact limb seems to have filled in for the missing limb (**FIGURE 7.26**). The person then repeatedly moves "both limbs" (by moving the remaining one), watching closely the whole while, and sometimes reports that the phantom limb feels as though it has straightened out and no longer hurts (Ramachandran and Rogers-Ramachandran, 2000). This result suggests that the brain interprets the signals coming from the limb stump as painful, but visual stimuli may lead to a reinterpretation.

Pain information is eventually integrated in the cingulate cortex. The cingulate cortex is much more activated by a stimulus if people are led to believe that the stimulus will be painful (see Figure 0.5) (Rainville et al., 1997). The extent of activation in the cingulate (as well as in somatosensory cortex) also correlates with how much discomfort different people report in response to the same mildly painful stimulus (Coghill et al., 2003; Vogt, 2005). In people experiencing pain, both anterior and posterior cingulate were activated; but when they were empathizing with a loved one who was experiencing pain, experiencing the emotional components of pain, only the anterior cingulate was activated (T. Singer et al., 2004).

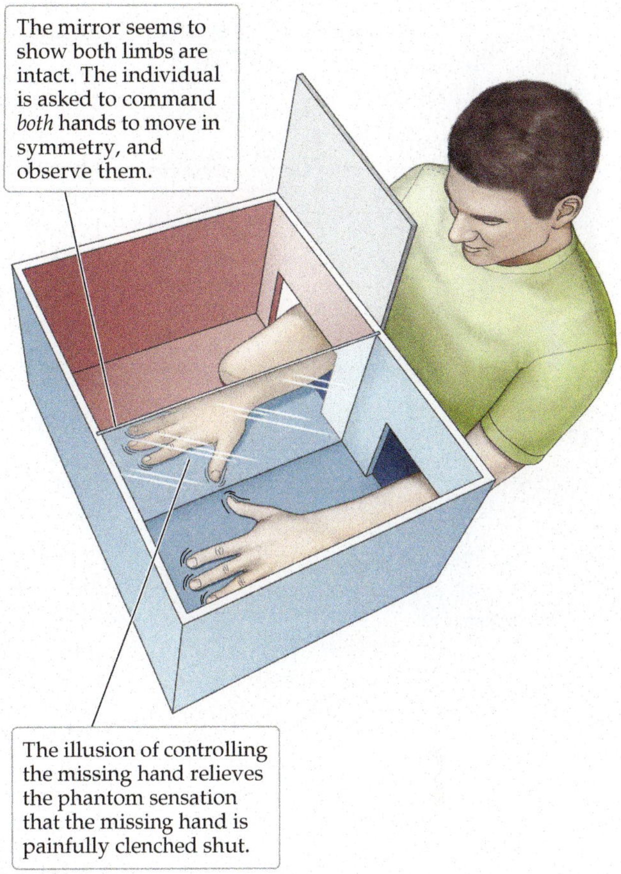

7.26 Using a Visual Illusion to Relieve Phantom Limb Pain (After V. S. Ramachandran and D. Rogers-Ramachandran, 2000. *Arch Neurol* 57: 317–320.)

The cingulate cortex is also activated in people experiencing illusory pain (Craig et al., 1996): placing your hand over alternating pipes of cool and warm (not hot) water produces the sensation of pain, as though the pipes were hot. Presumably this illusory pain sensation is produced by the unusual circumstance of neighbouring TRPM8 and TRPV1 receptors being stimulated at the same time. Since no tissue is actually being damaged, this is a wonderful illustration that "pain is in the brain."

Migraines, intense headaches that typically affect one side of the head and recur at regular intervals, also illustrate the central role of the brain in pain. Some migraines are associated with a wave of neuronal hyperexcitation that spreads across the cortex, leaving in its wake a wave of inhibition or cortical spreading depression. The wave of activity stimulates trigeminal nerves that normally signal damage in blood vessels, causing the experience of pain (Pietrobon and Moskowitz, 2014). Another theory is that the trigeminal nerves release a neuropeptide called CGRP, which actually causes the pain. Intravenous infusion of antibodies to CGRP halted migraines in people who got no relief from other medications (Tepper, 2019). Genetic susceptibility to migraine has been traced to genes encoding ion channels and pumps (Chasman et al., 2011) that trigger the hyperexcitation.

Social rejection hurts too

We've noted that while it is very unpleasant, pain is our friend. The neural machinery for pain was favoured by natural selection to shape our behaviour, causing us to avoid doing things that are bad for us (which might leave us with no descendants). Perhaps natural selection has also exploited that same machinery underlying pain to shape our social behaviour (Panksepp, 1998). We are born small and defenseless, utterly dependent on others for protection. In evolutionary times, our dependence on others continued in adulthood—only groups of humans could benefit from cooperative

migraines Intense headaches, typically perceived from one half of the head, that recur regularly and can be difficult to treat.

7.27 Social Rejection There is evidence that social rejection activates the same emotional brain circuits as physical pain.

hunting and ensure safety from predators whose claws and teeth far outmatch our own. Does being excluded from the group, which could be fatal in those days, activate the emotional component of the pain system (**FIGURE 7.27**)?

Indeed, several findings indicate that social rejection also activates the anterior cingulate cortex and that the extent to which a person is upset by rejection correlates with activation of this region (Eisenberger, 2012). In one study, participants were brought to the lab to play a virtual game of catch with other people through a computer. The other two players were supposed to be on other computers, but in fact the researchers determined whether the participant would get to participate in throwing the ball (social inclusion). For some participants, the other two "players" would toss the ball back and forth exclusively with each other, excluding the participant. This game took place while the participant's head was in an fMRI scanner monitoring brain activity.

As you might expect, being excluded from a virtual ball toss bothered some participants more than others (Eisenberger and Lieberman, 2004). Remarkably, the more distress or feelings of rejection the participant felt, the greater the activity in anterior cingulate cortex (**FIGURE 7.28A**). If in fact social rejection activates the same emotional pain as physical damage, then can the same drugs that relieve the affective discomfort of physical ills also relieve the pain of rejection? Yes. When participants were given either placebos or a painkiller (acetaminophen, often marketed as Tylenol) before playing virtual catch, the ones given placebo reported more hurt feelings from social rejection than those given Tylenol (DeWall et al., 2010). What's more, there was less activation of the anterior cingulate in the participants taking Tylenol before being socially rejected (**FIGURE 7.28B**). Participants taking the painkiller also reported fewer hurt feelings in their everyday lives than those taking placebo.

So, does Ashlyn, the girl who's never felt physical pain, suffer hurt feelings if she is socially rejected? Do the higher brain centres, never receiving signals of physical pain, fail to develop enough for her to have hurt feelings? Ashlyn cried when her dog ran away, so she can feel sad (Heckert, 2012). And surely she can tell if people reject her, but does that hurt?

These findings that social hurt activates the same brain centres as physical pain refute the old saying "Sticks and stones may break my bones, but words will never hurt me." That outmoded notion may explain why historically our society has been more concerned about physical bullying—pushing, biting, hitting—than verbal bullying such as yelling, name-calling, and social rejection. Playgrounds often display a sex difference in bullying styles. Boys tend to resort to physical bullying, while girls tend to use social rejection to hurt each other (Simmons, 2003). Adults who would quickly intervene if boys were hitting each other might ignore a group of girls rejecting a playmate. But if social rejection and physical bullying activate the same brain regions mediating pain, and if the discomfort experienced is the same, the distinction seems unfounded.

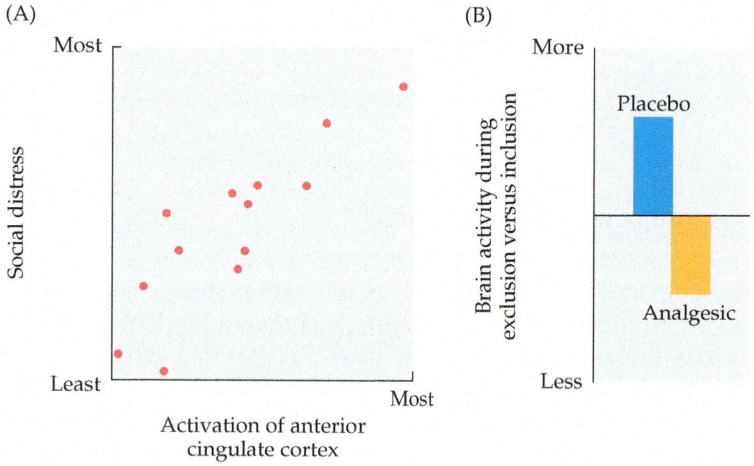

7.28 Social Rejection Activates Brain Regions for Affective Pain (A) Participants' emotional response to social rejection when playing a virtual game of catch correlates with activation in anterior cingulate cortex, the region that mediates the emotional distress of pain from physical damage. (B) People who had been taking Tylenol before experiencing social rejection reported less distress than those taking placebo and also displayed less activation of this brain region. (Part A after N. I. Eisenberger et al., 2003. *Science* 302: 290Y2; B after C. N. DeWall et al., 2010. *Psychol Sci* 21: 931–937.)

7.5 Pain Can Be Difficult to Control

Learning Objectives

After reading this section, you should be able to:

7.5.1 **Differentiate between opiates and opioids, and name several endogenous opioids.**

7.5.2 **Describe an array of approaches that have been tried to control chronic pain, listing the advantages and disadvantages of each.**

7.5.3 **Discuss the effects of placebos and the factors that make them more effective.**

Having learned that there are specific pathways that transmit pain signals to the brain, you might think that simply cutting the axons signalling pain would take care of the problem. But one frustrating, puzzling aspect of neuropathic pain is that cutting the pathway may reduce pain perception only temporarily. After a pathway in the spinal cord is cut, pain is initially diminished, but it returns in a few weeks or months. The usual way that these data are interpreted is that nociceptive input from the remaining intact pathways becomes abnormally effective.

Different strategies can alleviate pain

The oldest approaches to contol pain are pharmacological, so we'll consider those first.

CANNABIS **Cannabis** (also called marijuana) has been used for centuries for **analgesia** (loss of pain sensation; from the Greek an-, "not," and algesis, "feeling of pain"). The dried leaves and flowers of the plant Cannabis sativa reduce pain by stimulating endogenous cannabinoid receptors (CB1 receptors), both in the spinal cord (Pernía-Andrade et al., 2009) and, surprisingly, in the free nerve endings of the nociceptors themselves (Clapper et al., 2010). This latter discovery suggests it may be possible to devise topical creams that provide cannabinoid stimulation through the skin to relieve pain. The potential for cannabinoid analgesia was emphasized recently by the discovery of a woman with mutations disabling an enzyme that normally metabolizes cannabinoids—she had high levels of circulating endocannabinoids and was totally indifferent to pain (Habib et al., 2019). Intriguingly, her frequent cuts and burns seemed to heal very quickly. She felt no discomfort at giving birth and has been described as relentlessly upbeat (Sample, 2019).

OPIATES Opium has also been used for analgesia for centuries. In attempts to determine how **opiates** (drugs that are derived from or related to opium) control pain, modern researchers found that the brain contains natural opiate-like substances, or **endogenous opioids**. In effect, the brain modulates pain by releasing substances that work the same way as opiates (exogenous opioids) such as morphine. Three classes of endogenous opioids—**endorphins**, **enkephalins**, and **dynorphins**—have been discovered (see Chapter 3), and several classes of **opioid receptors** have been identified and designated by Greek letters. Of these, the mu (μ) receptor seems to be most affected by morphine. There are interesting individual differences in opioid receptor function. For example, analgesics that act on μ receptors are more effective in men than in women, while drugs that act on the kappa (κ) receptors are more effective in women than in men (Mogil and Chanda, 2005). Furthermore, the κ receptor–specific drugs are especially effective in redheaded women (Mogil et al., 2003).

In an enormously influential paper, Melzack and Wall (1965) suggested that pain is subject to many modulating influences, including some that can close spinal "gates" controlling the flow of pain information from the spinal cord to the brain. Indeed, a brainstem system projects to the spinal cord to close the pain gate. Electrical stimulation of the **periaqueductal gray** area (see Figure 7.24) of the brainstem produces potent analgesia. Periaqueductal gray neurons send endorphin-containing axons to stimulate neurons in the medulla. These medullary neurons send axons to the spinal cord, eventually stimulating neurons to release endogenous opioids there. In this way, pain information is blocked by a direct gating action

cannabis Also called *marijuana*. Dried leaves and flowers of the plant Cannabis sativa, typically smoked to obtain THC for a psychoactive effect.

analgesia Absence of or reduction in pain.

opiates A class of compounds that exert an effect like that of opium, including reduced pain sensitivity.

endogenous opioids A class of peptides produced in various regions of the brain that bind to opioid receptors and act like opiates.

endorphins One of three kinds of endogenous opioids, substances that reduce pain perception.

enkephalins One of three kinds of endogenous opioids, substances that reduce pain perception.

dynorphins One of three kinds of endogenous opioids, substances that reduce pain perception.

opioid receptor A receptor that responds to endogenous and/or exogenous opioids.

periaqueductal gray The neuronal body–rich region of the midbrain surrounding the cerebral aqueduct that connects the third and fourth ventricles. It is involved in pain perception.

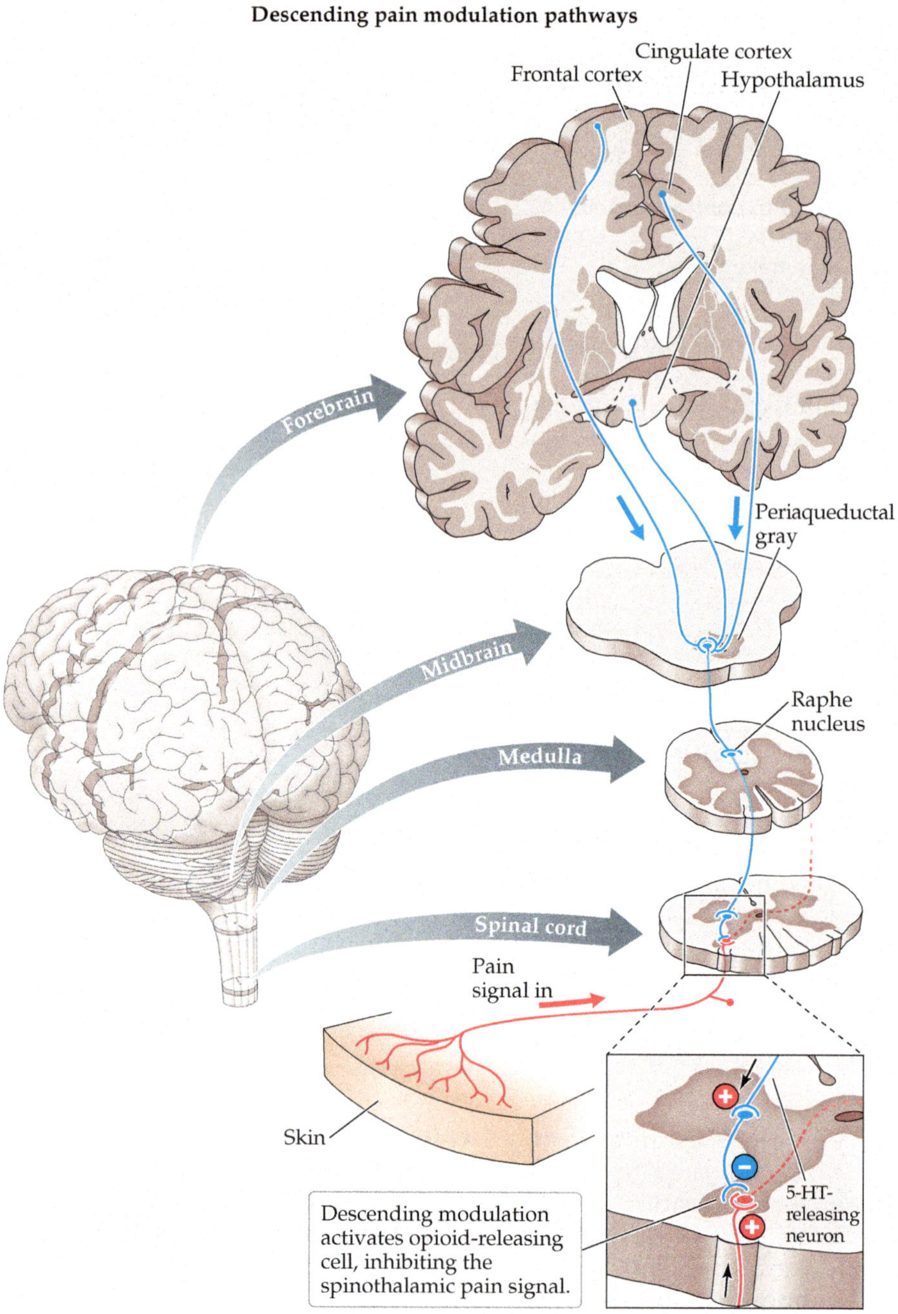

7.29 Descending Pain Control Pathways Pain sensation can be suppressed by descending brain pathways. (After A. Basbaum and H. L. Fields, 1984. *Ann Rev Neurosci* 7: 309–339.)

in the spinal cord (**FIGURE 7.29**). Morphine provides analgesia by stimulating the opioid receptors in this descending pain control system, in both the brainstem and the spinal cord. Frontal cortex, cingulate cortex, and the hypothalamus provide top-down input to the periaqueductal gray to suppress pain perception (T. V. Bliss et al., 2016).

In addition to their beneficial pain-relieving effects, opiates and other analgesics (painkillers) often produce side effects such as confusion, drowsiness, vomiting, constipation, and depression of the respiratory system. Now that we know the circuitry of the pain relief system, why give large doses of the drug systemically (i.e., throughout the body)? Instead, physicians can administer very small doses of opiates directly to the spinal cord to relieve pain, thus avoiding many of the side effects. The drugs can be administered epidurally (just outside the spinal cord's dura mater) or intrathecally (between the dura mater and the spinal cord). Both routes are somewhat invasive and therefore are restricted to surgical anaesthesia, childbirth, or the management of severe chronic pain (Landau and Levy, 1993).

Because of long-standing concerns about the addictive potential of morphine and other opiates, the traditional standard for using them to relieve pain was to prescribe them only infrequently and in low doses. However, more modern medical standards emphasize swift use of painkillers after surgery. Once chronic pain develops, it is extremely difficult to overcome, so the best approach is to prevent the onset of chronic pain by early, aggressive treatment. As we noted in Chapter 4, use of opiates for chronic pain comes with the risk of addiction and accidental overdose, so the search for alternative means of controlling pain, such as those summarized in **TABLE 7.3**, goes on.

Go to Activity 7.4
Ascending and Descending Pain Pathways

PLACEBOS The search for pain relief has led people to consume many unusual substances; even chemically inert pills alleviate pain in many patients. The term **placebo** (Latin for "I shall please") refers to an inert substance (such as a sugar pill) or other treatment that has no obvious direct physiological effect. In one example, volunteers who had just had their wisdom teeth extracted were told that they were being given an analgesic but were not told what kind (J. D. Levine et al., 1978). Some of these patients received morphine-based drugs, and some were given saline solutions—the placebo. One out of three patients given the placebo experienced pain relief. (Morphine produced relief in most, but not all, patients.)

placebo A substance that is known to be ineffective or inert but that, when administered like a drug, can sometimes bring relief.

naloxone A potent antagonist of opiates that is often administered to people who have taken drug overdoses. It blocks receptors for endogenous opioids.

TABLE 7.3 Types of Pain Relief Intervention

Approach	Mechanism	Limitations/comments
PSYCHOGENIC		
Placebo	May activate endorphin-mediated pain control system	Ethical concerns of deceiving patient
Hypnosis	Alters brain's perception of pain	Unaffected by opiate antagonists
Stress	Both opioid and non-opioid mechanisms	Clinically impractical and inappropriate
Cognitive (learning, coping strategies)	May activate endorphin-mediated pain control system	Limited usefulness for severe pain
PHARMACOLOGICAL		
Opiates	Bind to opioid receptors in periaqueductal gray and spinal cord	Severe side effects due to binding in other brain regions
Spinal block	Drugs block pain signals in spinal cord	Avoids side effects of systemic administration
Anti-inflammatory drugs	Block prostaglandin and/or leukotriene synthesis at site of injury (see Figure 7.21)	May have side effects
Cannabinoids	Act in spinal cord and on nociceptor endings	Illegal in some regions; smoke damages lungs
STIMULATION		
Acupuncture	Seems similar to placebo	Sometimes affected by opiate antagonists
Central gray	Electrical stimulation activates endorphin-mediated pain control systems, blocking pain signal in spinal cord	Inhibited by opiate antagonists; invasive surgery to implant electrodes
SURGICAL		
Cut peripheral nerve cord Rhizotomy (cutting dorsal root) Cord hemisection	Create physical break in pain pathway	Considerable risk of failure or return of pain
Frontal lobotomy	Disrupts affective response to pain	Irreversible; risky; severe effects on behaviour

The researchers gave the opioid antagonist **naloxone** to other patients, who were also administered the placebo. Patients given the placebo and naloxone did not experience pain relief; this result implies that placebo relieves pain by causing the release of endogenous opioids. Functional brain imaging indicates that opioids and placebos activate the same brain regions (Petrovic et al., 2002) and that both treatments reduce the activity of neural regions responding to pain, including cingulate cortex (Wager et al., 2004) and spinal cord (Eippert et al., 2009). A consistent finding is that some people experience relief from a placebo and others do not. People who respond to placebo show a greater activation of brain regions with opioid receptors than do nonresponders (**FIGURE 7.30**), further implicating endogenous opioids in the placebo effect.

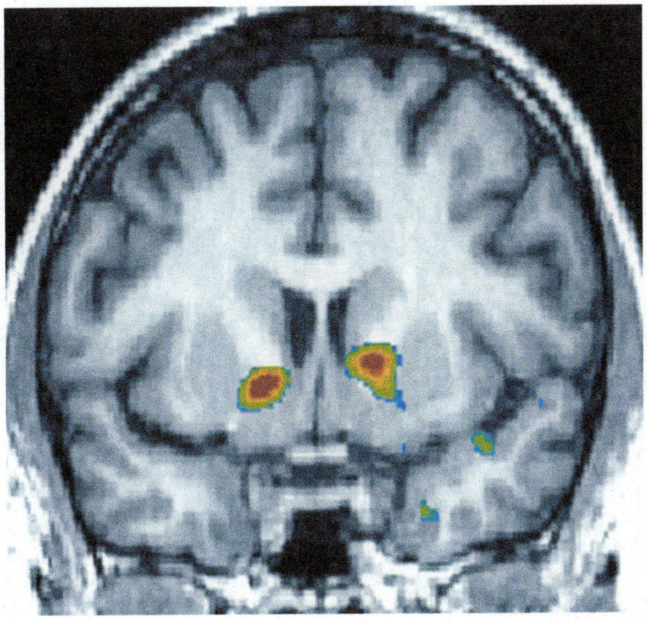

From D. J. Scott et al., 2007. *Arch Gen Psychiatry* 65: 220–231. Courtesy of Jon-Kar Zubieta

7.30 Placebo Affects Opioid Systems in the Brain Volunteers were subjected to a painful procedure and given a placebo for the pain. Some people seemed to respond to the placebo, but others did not. This difference image shows brain regions that were more activated in responders than in nonresponders. These regions are rich in receptors for endogenous opioids.

TABLE 7.4 Environmental Cues That Affect the Placebo Effect

Type of Cue	Examples
EXTERNAL	
Verbal suggestion	"This treatment will make you feel better."
Place	Doctor's office
Social	Body language, eye gaze, voice cues, "White coat"
INTERNAL	
Outcome expectancies	"My pain will go away."
Emotions	Reduced anxiety
Meaning schema	"I am being cared for."
Explicit memories	"This brought relief before."

Source: T. D. Wager and L. Y. Atlas, 2015. *Nat Rev Neurosci* 16: 403–417.

Careful experimentation has revealed many environmental cues that can enhance the placebo effect (Wager and Atlas, 2015) (**TABLE 7.4**). For example, large pills are more effective than small pills, and medicines are more effective if they are thought to be expensive. Having a doctor or nurse in a white coat assert that the treatment will work also boosts the effectiveness of placebos (**FIGURE 7.31A**). In other words, the expectation that the treatment will be effective actually contributes to that effectiveness. The expectation of relief may also explain why hypnosis effectively controls pain in many people (Jensen et al., 2014).

At least part of the pain-blocking character of acupuncture appears to be mediated by the release of endorphins, because administering opioid antagonists such as naloxone prior to acupuncture blocks or reduces its pain control effects (Z. Q. Zhao, 2008). But pain relief from acupuncture is also at least partly due to placebo effect (**FIGURE 7.31B**). One report that acupuncture relieves headaches also found that where the needles were inserted made no difference (Linde et al., 2009). Again, the expectation that the needles will reduce pain helps bring relief.

(A)

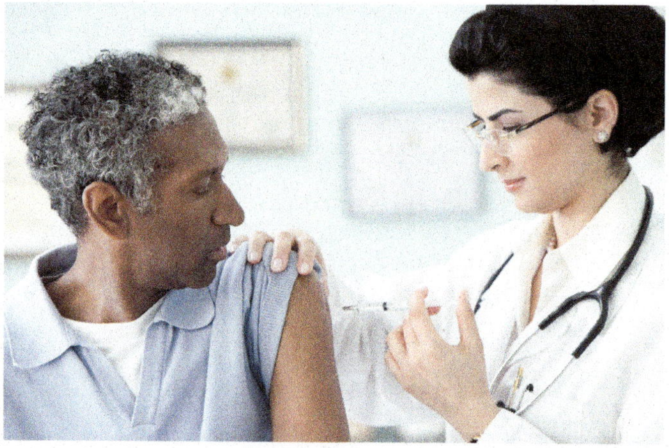

(B)

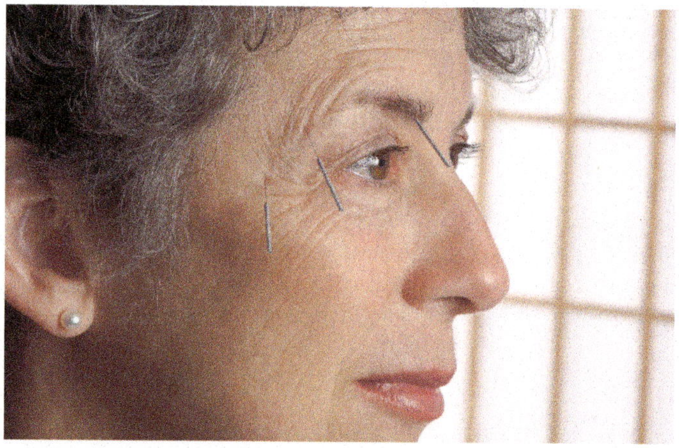

7.31 Placebos Can Be Effective (A) Having the doctor express confidence that the treatment will work can boost the placebo effect. (B) In some cases, acupuncture controls pain. Opioid antagonists block this analgesia, indicating that acupuncture, like some placebos, activates endogenous opioids.

Sometimes people in clinical trials who are unknowingly receiving a placebo control drug nevertheless report unpleasant side effects. Such results are sometimes called **nocebo** effects—experiencing harm from an inert treatment. Here, too, treatments that are perceived as expensive are more likely to elicit nocebo responses from participants (Tinnermann et al., 2017).

nocebo An inert substance that causes discomfort due to the patient's expectations.

STRESS A deer fleeing an encounter with a mountain lion would do well to ignore any pain from injuries for the moment. Similarly, sometimes people badly hurt in traumatic circumstances report little or no immediate pain (Beecher, 1956). Laboratory studies demonstrate the existence of pain control circuitry, but they don't tell us how the inhibitory systems are normally activated.

To answer this question, researchers have examined pain inhibition that arises in stressful circumstances. It appears that brain systems produce analgesia when pain threatens to overwhelm effective coping strategies. For example, exposing rats to several different kinds of inescapable footshock induces analgesia (Terman et al., 1984). Stress analgesia may be blocked by the opiate antagonists, depending on the type of stress, indicating that stress activates both an opioid-sensitive analgesic system and a pain control system that does not involve opioids but may involve release of endogenous cannabinoids (A. G. Hohmann et al., 2005).

The many types of pain relief strategies, including surgical and pharmacological strategies, psychological treatments, brain and spinal cord stimulation, and sensory stimulation, reflect how powerfully pain affects us. Only by gaining a better understanding of this mixed blessing can we avoid those nightmarish times when the agony of pain serves no purpose.

Recommended Reading

Ballantyne, J. C., Fishman, S. M., and Rathmell, J. P. (Eds.). (2019). *Bonica's Management of Pain* (5th ed.). Philadelphia: Lippincott.

Cytowic, R. E., and Eagleman, D. M. (2009). *Wednesday Is Indigo Blue*. Cambridge, MA: MIT Press.

Delmas, P., Hao, J., and Rodat-Despoix, L. (2011). Molecular mechanisms of mechanotransduction in mammalian sensory neurons. *Nature Reviews Neuroscience*, 12(3),139–153.

Greenberg, G. (2018, January 11). Why nothing works. *The New York Times, Sunday Magazine*, pp. 50–56.

McMahon, S., Koltzenburg, M., Tracey, I., and Turk, D. C. (2013). *Wall & Melzack's Textbook of Pain* (6th ed.). Philadelphia: Saunders.

Wolfe, J. M., Kluender, K. R., Levi, D. M., Bartoshuk, L. M., et al. (2018). *Sensation & Perception* (5th ed.). Sunderland, MA: Oxford University Press/Sinauer.

7 Visual Summary

You should be able to relate each summary to the adjacent illustration, including structures and processes. You can also refer back to the links to figures, animations, and activities that will help you consolidate the material.

1 A sensory system furnishes selected information to the brain about internal and external events. It captures and processes only information that is significant for the particular organism. Some **stimuli** are detected readily by some species but have no effect on others. Review Figure 7.2, Table 7.1

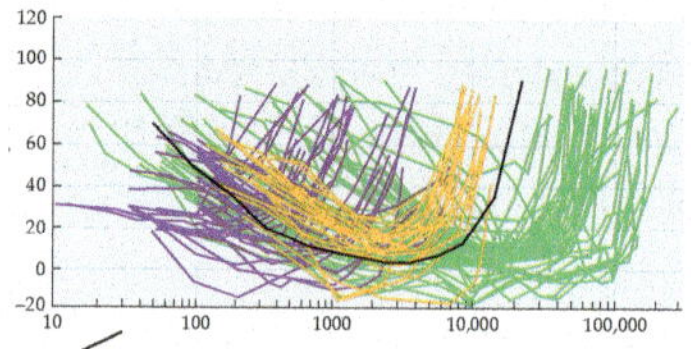

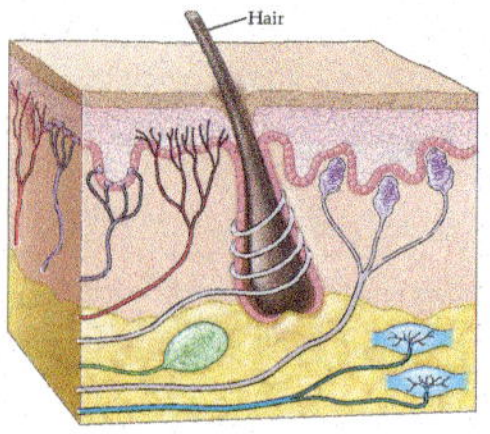

2 **Sensory receptor organs** furnish only selected information to the brain. For example, our skin contains four kinds of touch receptors, each specialized for detecting a specific stimulus. The brain decodes sensory information by using **labelled lines**, nerves that normally transmit only certain types of sensory information. Review Figures 7.3, 7.4, Activity 7.1

3 Energy is transduced at sensory receptors by the production of a **receptor potential** that triggers action potentials. **Coding** translates receptor information into patterns of neural activity. The frequency and pattern of action potentials signal the intensity and type of stimulus encountered. Review Figures 7.5, 7.6

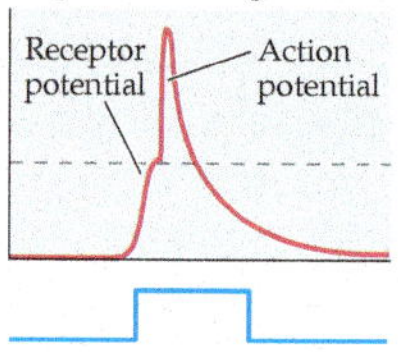

4 In **adaptation**, the rate of action potentials progressively decreases as the same stimulation is maintained. This decline is slow in the case of **tonic receptors** but rapid for **phasic receptors**. Adaptation protects the nervous system from redundant stimulation. Review Figure 7.7

5 The succession of levels in a **sensory pathway** allows for increasingly elaborate kinds of processing. Information enters the spinal cord or brainstem and reaches the **thalamus**, which communicates to the cortex. **Attention**, the temporary enhancement of certain sensory messages, is modulated at higher levels of the sensory pathway. Review Figures 7.8, 7.12

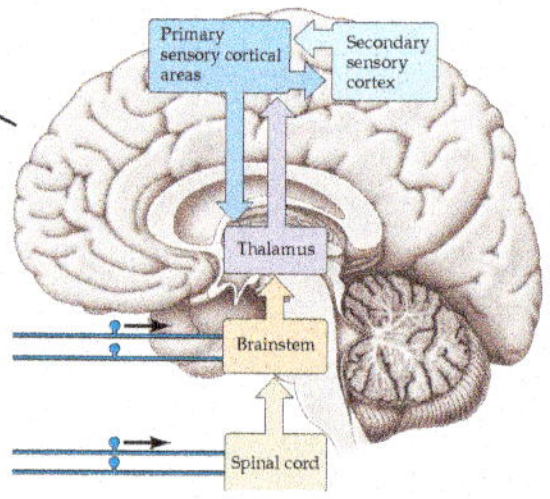

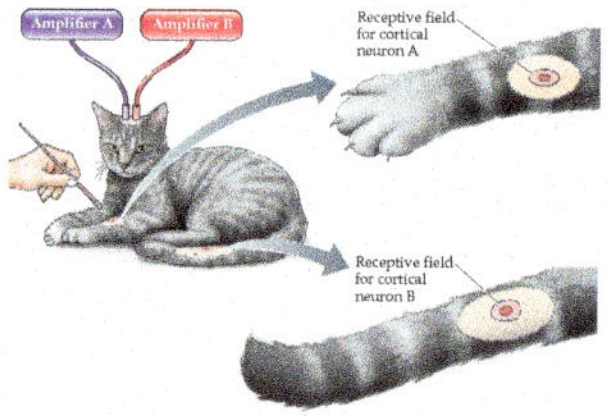

6 The **receptive field** of a neuron is the region in space where a stimulus will change the firing of that cell. The receptive fields of neurons may be very different at successive levels of the sensory pathway. Review Figures 7.9, 7.13, Animation 7.1

7 At the level of the cerebral cortex, there are multiple maps of the body surface. **Primary somatosensory cortex**, or **S1**, located in the postcentral gyrus, contains a map of the contralateral body, which overrepresents highly sensitive body regions. Review Figures 7.10, 7.11, 7.17, 7.18

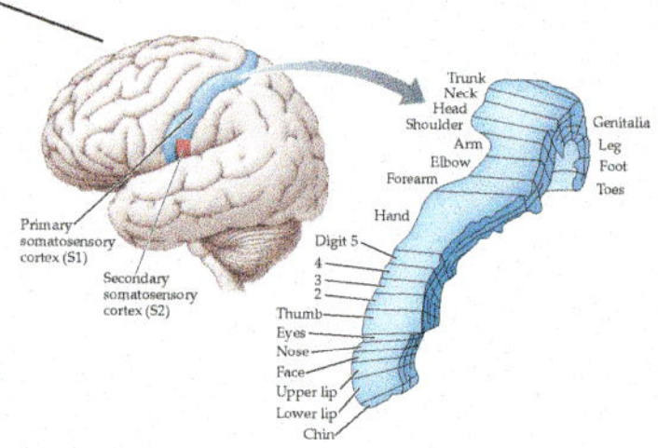

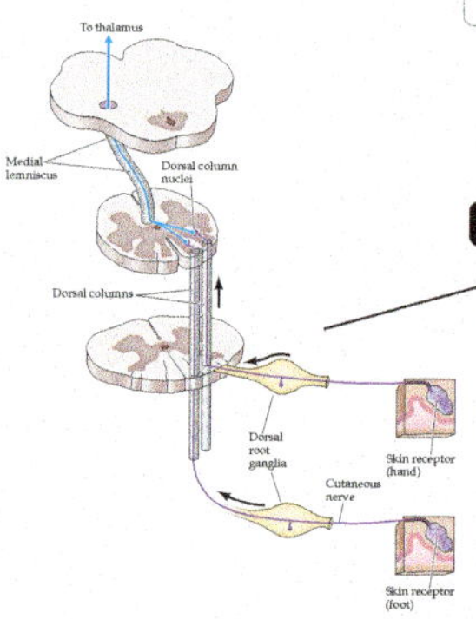

8 Touch information from the body courses through a distinct spinal pathway, the **dorsal column system**. Review Figures 7.15, 7.16

9 Some sensory **receptor cells** are specialised to transduce particular kinds of energy. **Merkel's discs** and **Meissner's corpuscles** detect fine touch, while **Pacinian corpuscles** and **Ruffini's endings** respond to vibration and stretch, providing information about textures. Review Figures 7.13, 7.14, Table 7.2, Activity 7.2

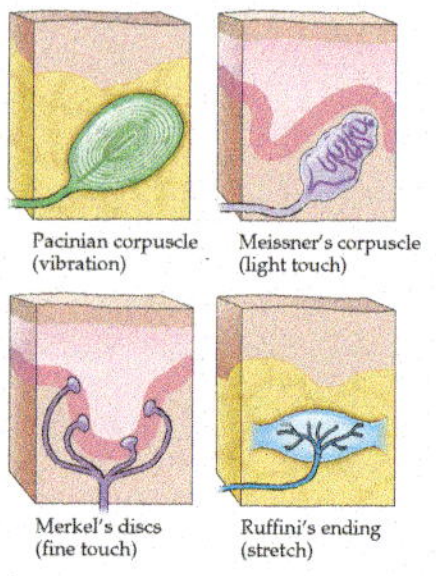

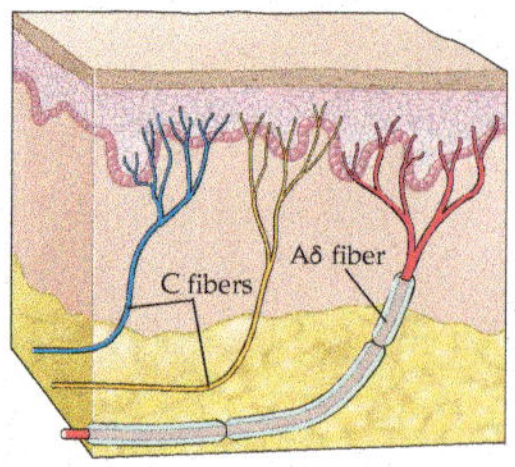

10 **Free nerve endings** act as **nociceptors**. They detect mechanical damage or temperature changes, using their specialised receptor proteins that detect these conditions, and then open up ion channels to trigger action potentials. For example, **transient receptor potential vanilloid 1 (TRPV1)** detects heat and is activated by chili peppers. Review Figures 7.20–7.22

11 Some peripheral pain fibres have fairly large, myelinated **Aδ fibres** to transmit sharp pain rapidly; others use small, unmyelinated **C fibres** to slowly transmit dull, aching pain after injury. Review Figure 7.22, Table 7.2

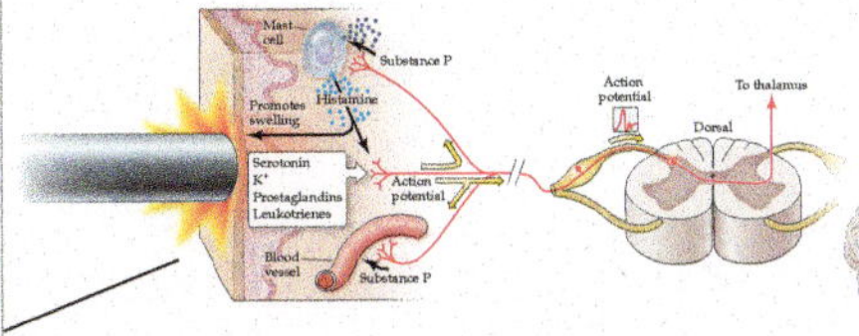

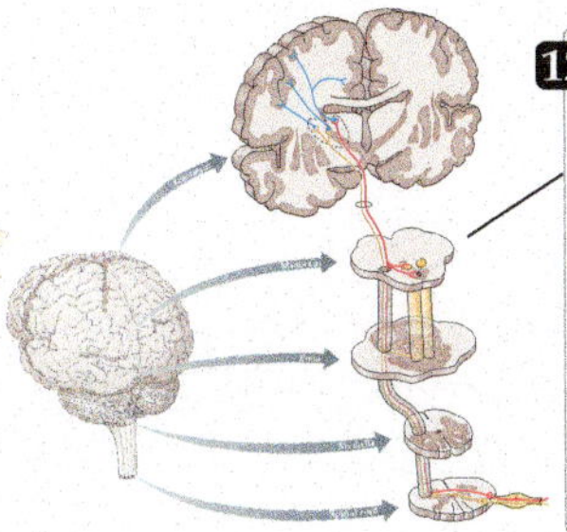

12 Pain and temperature information enters the spinal cord, crosses the midline, and ascends through the **anterolateral (spinothalamic) system** to the brain. Itch information crosses the midline in the spinal cord and ascends to the pons. Review Figures 7.24, 7.25, Activity 7.3

13 Pain sensation is subject to many controlling or modulating conditions, including circuitry within the brain and spinal cord that employs **endogenous opioid** synapses. One way to modulate pain is to affect the descending pathways arising in the brain that inhibit incoming neural activity at synapses in the **periaqueductal gray** and even within the spinal cord. Review Figure 7.29, Activity 7.4

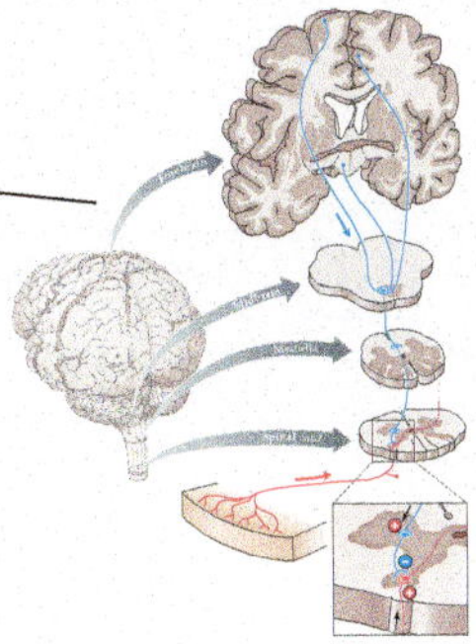

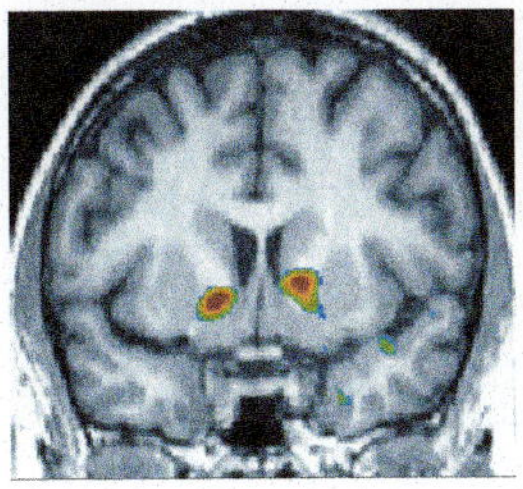

14 Pain control has been achieved by the administration of drugs (including **placebos**), electrical and mechanical stimulation of the skin, **acupuncture**, and surgery, among other methods. Review Figures 7.26, 7.30, 7.31, Table 7.3

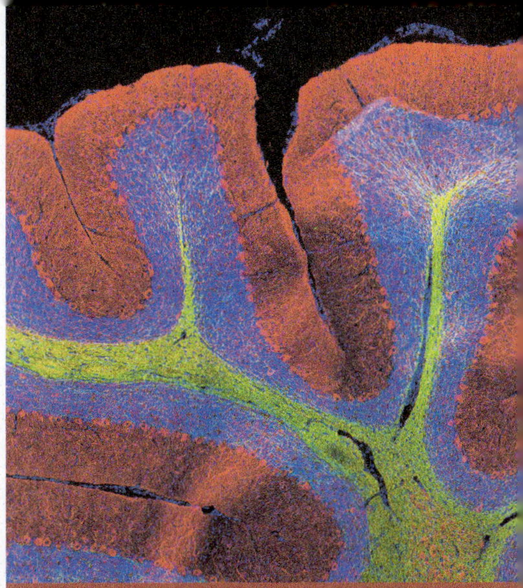

Photo from M. E. Terzenidou et al. 2017. *PLOS Genet* 13: e1006656 / CC BY 4.0 / Courtesy of Dr. Katerina Segklia and Dr. Rebecca Matsas

8

Vision

From Eye to Brain

When Seeing Isn't Seeing

The bathroom was cold, so the young woman turned on a small heater before getting in the shower. She didn't know that the heater was malfunctioning, filling the room with deadly, odourless carbon monoxide. Her husband found her unconscious on the floor and called for an ambulance to rush her to the emergency room. When she regained consciousness, "D.F." seemed to have gotten off lightly, avoiding what could have been a fatal accident. She could understand the doctors' questions and reply sensibly, move all her limbs, and perceive touch on her skin. But something was wrong with her sight.

D.F. couldn't recognise faces, even her husband's, nor could she name any objects presented to her view. D.F. still cannot recognise objects, decades after her accident. Yet she is not entirely blind. Show her a flashlight and she can tell you that it's made of shiny aluminum with some red plastic, but she doesn't recognise it ("Is it a kitchen utensil?"). Without telling her what it is, if you ask her to pick it up, D.F.'s hand goes directly to the flashlight and holds it exactly as one normally holds a flashlight. Show D.F. a slot in a piece of plastic, and she cannot tell you whether the slot is oriented vertically, horizontally, or diagonally; but if you hand her a disk and ask her to put it through the hole, D.F. invariably turns the disk so that it goes smoothly through the slot (Goodale et al., 1991).

Can D.F. see or not?

Vision offers tremendous benefits for vital behaviours such as finding food, avoiding predators, recognising a mate, and locating shelter, so a variety of visual systems have evolved. Beyond serving basic needs, vision affords most of us the pleasures of nature and art, reading and writing, and watching videos. Because vision is so important, a lot of effort is exerted to prevent its deterioration, and even to restore vision to the blind.

However, the sheer volume of visual information poses a serious problem. Seeing the surrounding world has been compared to drinking from a waterfall. How does the visual system avoid being overwhelmed by the flood of information entering the eyes? The answer seems to be that the visual perceptions of each species depend on how their eyes and brains evolved to process information about light and attend to the aspects likely to be important for survival in that particular species.

Research on vision is one of the most active fields of neuroscience, as it should be, since about one-third of the human cerebral cortex is devoted to visual analysis and perception.

8.1 The Retina Transduces Light into Neuronal Activity

Learning Objectives

After reading this section, you should be able to:

8.1.1 **Explain the basic physics of light entering the eye, and name the various structures of the eye.**

8.1.2 **List the six basic types of retinal neurons, and explain their patterns of innervation with one another.**

8.1.3 **Review the steps by which light affects the membrane potential of photoreceptors.**

8.1.4 **Contrast the two different classes of photoreceptors and the scotopic and photopic visual systems they provide.**

wavelength The length between two peaks in a repeated stimulus such as a wave, light, or sound.

photon A quantum of electromagnetic energy in the range of wavelengths we call light.

cornea The transparent outer layer of the eye, whose curvature is fixed. It bends light rays and is primarily responsible for forming the image on the retina.

lens A structure in the eye that helps focuss an image on the retina.

refraction The bending of light rays by a change in the density of a medium, such as the cornea and the lens of the eye.

ciliary muscle One of the muscles that controls the shape of the lens inside the eye, focussing an image on the retina.

accommodation The process of focussing by the ciliary muscles and the lens to form a sharp image on the retina.

pupil The aperture, formed by the iris, that allows light to enter the eye.

iris The circular structure of the eye that provides an opening to form the pupil.

extraocular muscle One of the muscles attached to the eyeball that controls its position and movements.

To understand how the visual system constructs our vision, let's follow the path of information triggered by light, starting with the optical properties of the eye. Next, we'll look at the cells and processes intrinsic to the back of the eye that encode light as neural activity.

Visible light is a narrow band of electromagnetic radiation

The physical energy to which our visual system responds is a band of electromagnetic radiation. This radiation comes in very small packets of energy called *quanta* (singular *quantum*). Each quantum can be described by a single number, representing its **wavelength** (the distance between two adjacent crests of vibratory activity).

The human visual system responds only to quanta whose wavelengths lie within a very narrow section of the total electromagnetic range, from about 400 to 700 nanometers (nm), as **FIGURE 8.1** shows. Quanta in this range are called **photons** (from the Greek *phos*, "light"). This band of radiant energy visible to animals may be narrow, but it provides for accurate reflection from the surface of objects in the size range that matters for survival. Among the other wavelengths of electromagnetic radiation, radio waves are good for imaging objects of astronomical size, and X-rays penetrate below the surfaces of objects. Each photon has a tiny amount of energy; the exact amount depends on the wavelength. A typical lightbulb gives off about 8 quintillion (8×10^{18}) photons per second.

When quanta within the visible spectrum enter the eye, they can evoke visual sensations. The exact nature of such sensations depends both on the wavelengths of the photons and on the number of photons per second, as we'll see.

The vertebrate eye acts in some ways like a camera

The eye is an elaborate structure with optical functions (capturing light and forming detailed spatial images) and neural functions (transducing light into neural signals and processing those signals). Accurate optical images are crucial for discerning the shapes of objects; that is, light from a point on a target object must end up as a point—rather than a blur—in the retinal image. Without optical images, light-sensitive cells would be able to detect only the presence or absence of light and would not be able to see forms.

To produce optical images, the eye has many of the features of a camera, starting with the **cornea** and **lens** to focuss light (**FIGURE 8.2**). Light travels through a medium, such as air, in a straight line until it encounters a change in the density of the medium, which causes light rays to bend. This bending of light rays, called **refraction**, is the basis of such instruments as eyeglasses, telescopes, and microscopes. Like those instruments, the cornea of

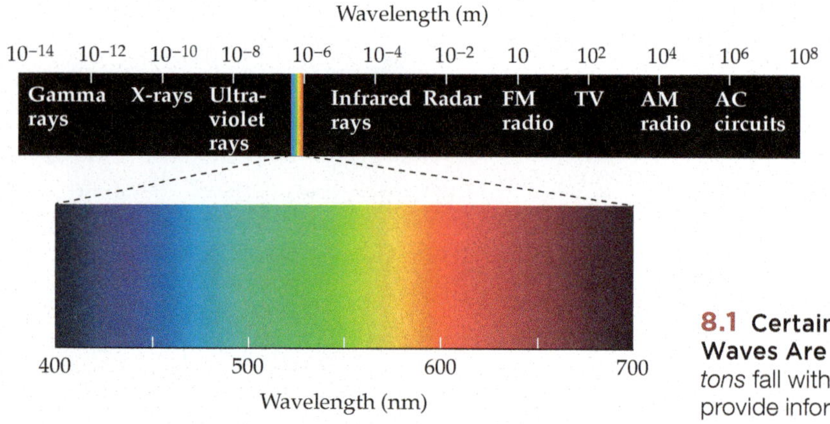

8.1 Certain Frequencies of Electromagnetic Waves Are Called Light Light particles called *photons* fall within a particular range of frequencies that can provide information about physical objects around us.

8.2 Structures of the Human Eye Here the right eye is viewed in cross section from above. The visual image focussed on the retina is inverted top to bottom and reversed right to left. The gap in the retina where the optic nerve leaves the eyeball is called the *optic disc*. The optic disc is also shown in Figure 8.6A.

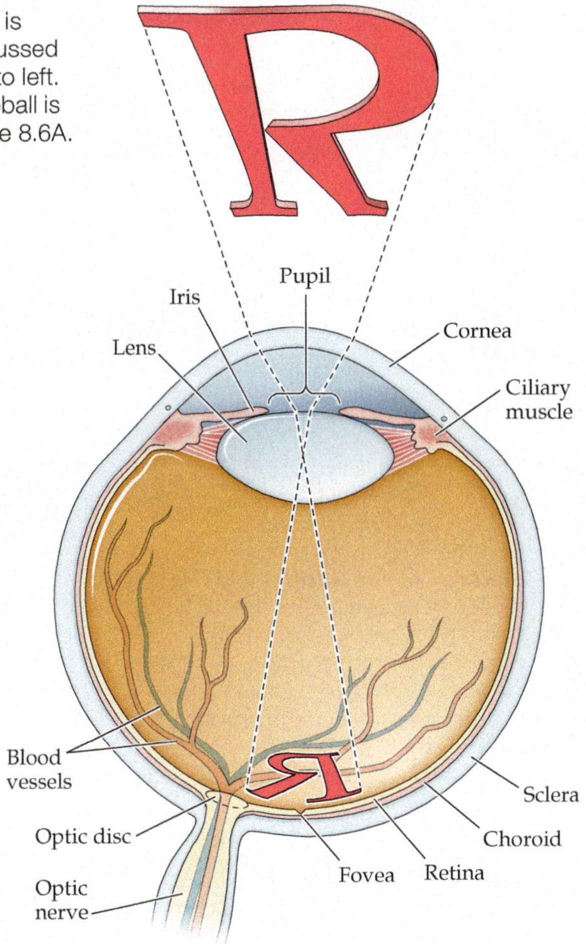

the eye—the curvature of which is fixed—bends light rays and is primarily responsible for forming the image on the retina.

In a camera, the lens moves in or out to adjust focuss, and the same system is used in the eyes of fishes, amphibians, and reptiles. In mammals and birds, however, focuss is adjusted by changes in the *shape* of the lens, which is controlled by the **ciliary muscles** inside the eye. As the degree of contraction of the ciliary muscles varies, the lens focusses images of nearer or farther objects so that they form sharp images on the retina; this process of focussing is called **accommodation**. As mammals age, their lenses become less elastic and therefore less able to change curvature to bring nearby objects into focuss. Aging humans correct this problem by wearing reading glasses (or holding the menu farther away).

The amount of light that enters the eye is controlled by the size of the **pupil**, which is an opening in the colourful disc called the **iris** (see Figure 8.2). In Chapter 2 we mentioned that dilation of the pupils is controlled by the sympathetic division of the autonomic nervous system, and constriction is controlled by the parasympathetic division. Because usually both divisions are active, pupil size reflects a balance of influences.

During an eye examination, the doctor may use a drug to block acetylcholine transmission in the parasympathetic synapses of your iris; this drug relaxes the sphincter muscle fibres and permits the pupil to open widely. One drug that has this effect got its name—*belladonna* (Italian for "beautiful lady")—because it was thought to make a woman more beautiful by giving her the wide-open pupils of an attentive person. Other drugs, such as morphine, constrict the pupils.

The movement of the eyes is controlled by the **extraocular muscles**, three pairs of muscles that extend from the outside of the eyeball to the bony socket of the eye, known as the *orbit*.

Visual processing begins in the retina

The first stages of visual-information processing occur in the **retina**, the receptive surface inside the back of the eye (**FIGURE 8.3A**). The retina is only 200–300 micrometers (μm) thick—as thick as 2–3 sheets of paper—but it contains several types of cells in distinct layers (**FIGURE 8.3B**). The receptor cells that detect light are called **photoreceptors**. There are two kinds of photoreceptors: some are called **rods** because of their relatively long, narrow form; others are called **cones** (**FIGURE 8.3C**). There are several different types of cones, which respond differently to light of varying wavelengths, providing us with colour vision, as described later in the chapter. Both rod and cone photoreceptors release neurotransmitter molecules (**FIGURE 8.3D**) that control the activity of the **bipolar cells** they synapse upon (see Figure 8.3B). The bipolar cells, in turn, synapse upon **ganglion cells**. The axons of the ganglion cells form the **optic nerve**, which carries information from the eyes to the brain.

Horizontal cells and **amacrine cells** are especially significant in lateral interactions within the retina, which we'll discuss later. The horizontal cells make contacts among the receptor cells and bipolar cells; the amacrine cells contact both the bipolar and the ganglion cells.

Interestingly, the rods, cones, bipolar cells, and horizontal cells generate only graded local potentials; they do not produce action potentials. These neurons affect each other through release of neurotransmitters proportional to graded changes in electrical potentials. The ganglion cells, on the other hand, do conduct action potentials. Because the ganglion cells have action potentials and are relatively large,

retina The receptive surface inside the eye that contains photoreceptors and other neurons.

photoreceptors Neural cells in the retina that respond to light.

rods A class of light-sensitive receptor cells (photoreceptors) in the retina that are most active at low levels of light.

cones A class of photoreceptor cells in the retina that are responsible for colour vision.

bipolar cells A class of interneurons of the retina that receive information from rods and cones and pass the information to retinal ganglion cells.

ganglion cells A class of cells in the retina whose axons form the optic nerve.

optic nerve Cranial nerve II; the collection of ganglion cell axons that extend from the retina to the optic chiasm.

horizontal cells Specialised retinal neurons that contact both the receptor cells and the bipolar cells.

amacrine cells Specialised retinal neurons that contact both the bipolar cells and the ganglion cells and are especially significant in inhibitory interactions within the retina.

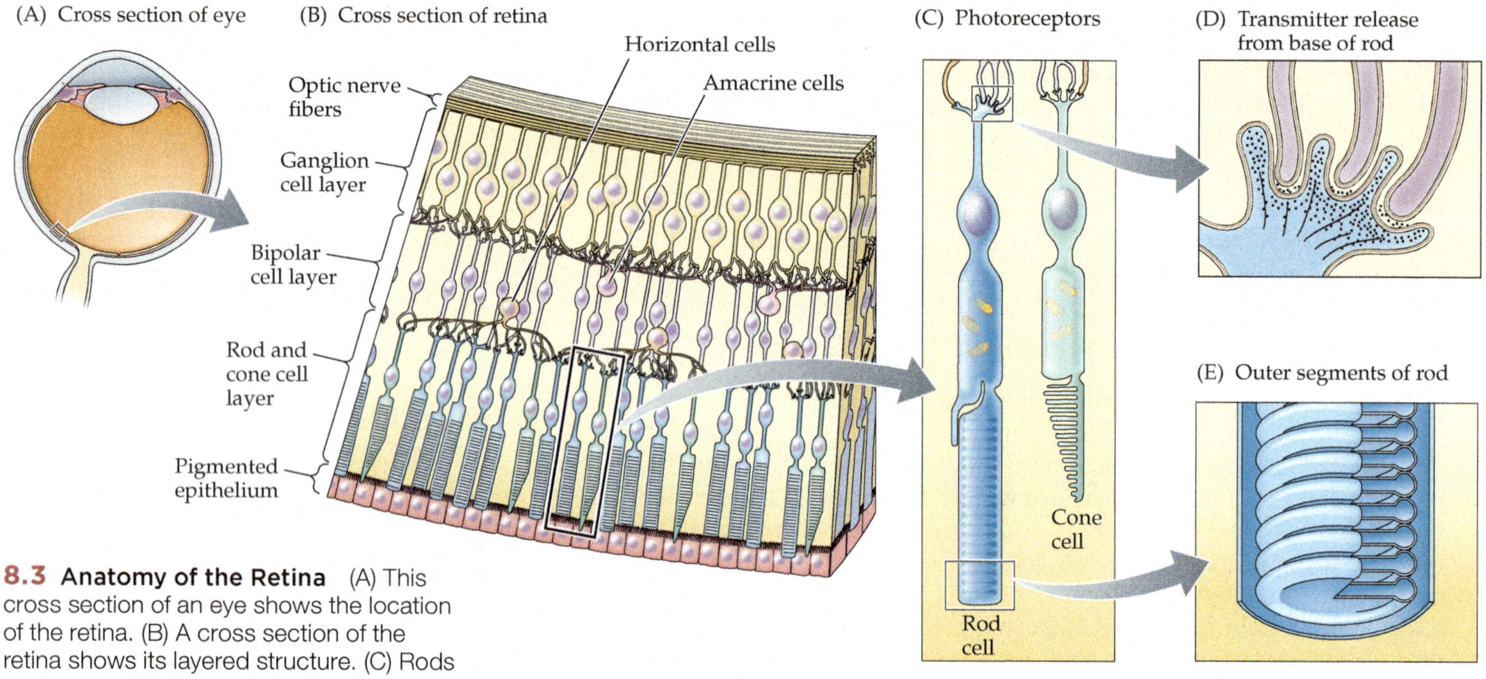

8.3 Anatomy of the Retina (A) This cross section of an eye shows the location of the retina. (B) A cross section of the retina shows its layered structure. (C) Rods and cones differ from each other in structure, but both release neurotransmitter (D) into bipolar neurons. (E) Both photoreceptors contain stacks of discs that catch and react to light.

they were the first retinal neurons to have their electrical activity recorded. From the receptor cells to the ganglion cells, enormous amounts of data converge and are compressed; the human eye contains about 100 million rods and 4 million cones, but there are only 1 million ganglion cells to transmit that information to the brain. Thus, a great deal of information processing is done in the eye.

Photoreceptors transduce light into chemical reactions

The extraordinary sensitivity of rods and cones is the result of their unusual structure and biochemistry. A portion of these cells, when magnified, looks like a large stack of pancakes (**FIGURE 8.3E**) and is where photons are detected. Because light is reflected in many directions by the lens and the fluid inside the eye, only a fraction of the light that passes through the cornea actually reaches the retina. The stacking of discs increases the likelihood that one will capture those photons that make it through to the retina.

Go to Activity 8.1
The Structure of the Eye

The photons that strike the discs are captured by special photopigment receptor molecules. In the rods this photopigment is **rhodopsin** (from the Greek *rhodon*, "rose," and *opsis*, "vision"). Cones use similar photopigments, as we will see later. The rod and cone photopigments in the eye consist of two parts: **RETINAL** (an abbreviated name for *retinaldehyde*) and **opsin**. (In this book the noun *RETINAL*, standing for the molecule, is printed in small capital letters to distinguish it from the adjective *retinal*, meaning "pertaining to the retina.") The visual receptor molecules span the membranes of receptor discs and have structures that are similar to those of the G protein-coupled receptors (GPCRs) for neurotransmitters that we discussed in Chapter 3.

When rhodopsin is hit by light (**FIGURE 8.4A**), RETINAL dissociates rapidly from the opsin molecule to reveal an enzymatic site. This altered opsin molecule rapidly activates many molecules of the G protein transducin (**FIGURE 8.4B**). Transducin, in turn, acts through an enzyme, phosphodiesterase (PDE), to break apart cyclic GMP (cyclic guanosine monophosphate, or cGMP) to 5′-GMP. Cyclic GMP holds channels for sodium ions (Na^+) open; stimulation by light initiates a cascade of events that *closes* these channels. Capture of a *single quantum* of light can lead to the closing of *hundreds* of Na^+ channels in the photoreceptor membrane, thereby blocking the entry of *more than a million* Na^+ ions (Schnapf and Baylor, 1987). Closing the Na^+ channels creates a hyperpolarising potential (**FIGURE 8.4C**).

rhodopsin The photopigment in rods that responds to light.

RETINAL Abbreviation for *retinaldehyde*, also called *vitamin A aldehyde*. One of the two components of photopigments in the retina.

opsin One of the two components of photopigments in the retina.

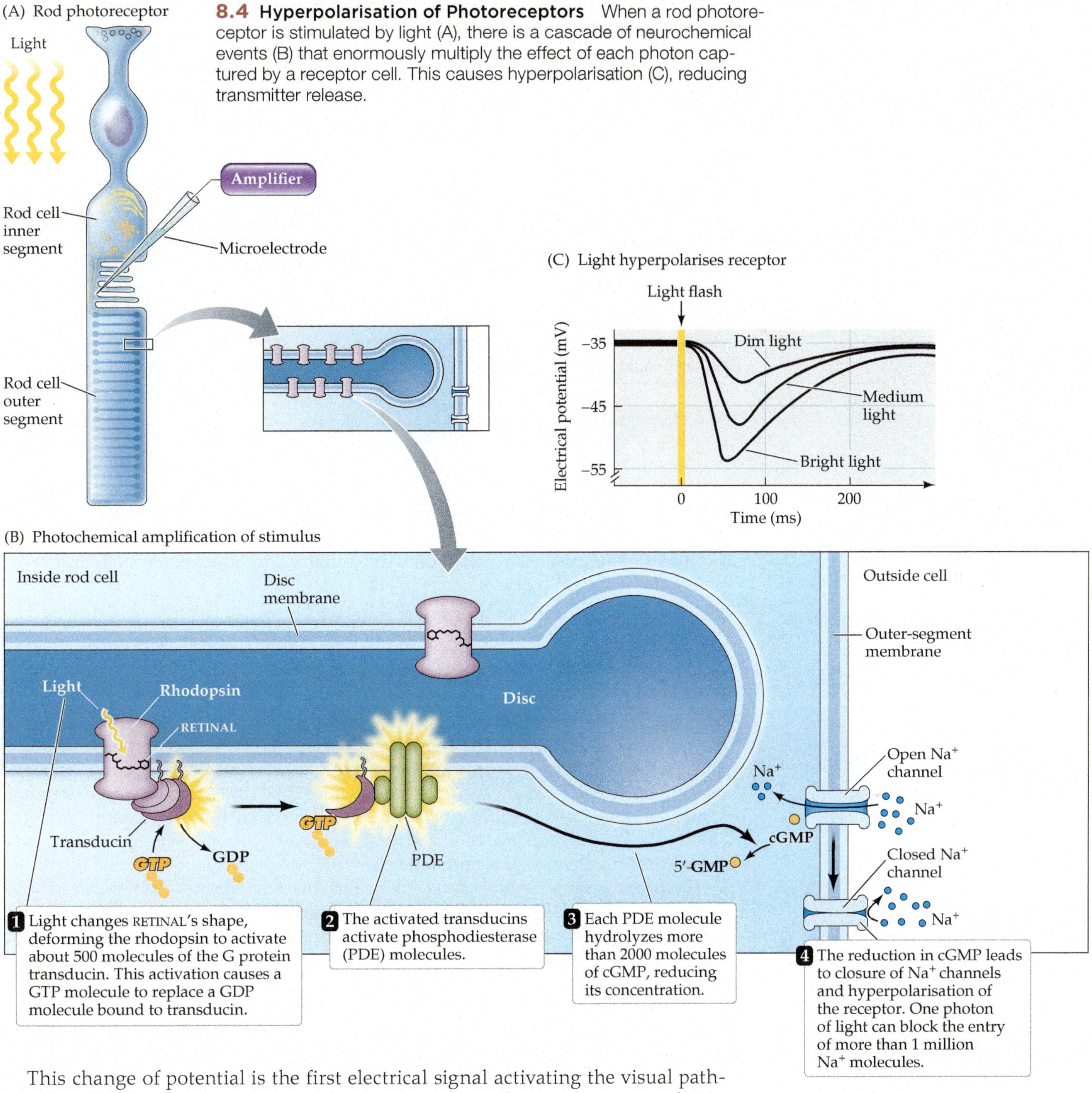

8.4 Hyperpolarisation of Photoreceptors When a rod photoreceptor is stimulated by light (A), there is a cascade of neurochemical events (B) that enormously multiply the effect of each photon captured by a receptor cell. This causes hyperpolarisation (C), reducing transmitter release.

This change of potential is the first electrical signal activating the visual pathway: light hitting rhodopsin hyperpolarises the rods. In the same way, stimulation of the cone pigments by light hyperpolarises the cones. For both rods and cones, the size of the hyperpolarising photoreceptor potential determines how much *less* synaptic transmitter will be released (see Figure 8.4C).

It may seem puzzling at first that stimulation by light *hyper*polarises vertebrate retinal photoreceptors and causes them to release *less* neurotransmitter, since sensory stimulation depolarises most other receptor cells. But remember that the visual system responds to *changes* in light. Either an increase or a decrease in the intensity of light can stimulate the visual system, and hyperpolarisation is just as much a neural signal as depolarisation.

TABLE 8.1 Properties of the Human Photopic and Scotopic Visual Systems

Property	Photopic system	Scotopic system
Receptors[a]	Cones	Rods
Approximate number of receptors per eye	4 million	100 million
Photopigments[b]	Three classes of cone opsins; the basis of colour vision	Rhodopsin
Sensitivity	Low; needs relatively strong stimulation; used for day vision	High; can be stimulated by weak light intensity; used for night vision
Location in retina[c]	Concentrated in and near fovea; present less densely throughout retina	Outside fovea
Receptive field size and visual acuity	Small in fovea, so acuity is high; larger outside fovea	Larger, so acuity is lower
Temporal responses	Relatively rapid	Slow

[a]Cones and rods are illustrated in Figure 8.3.
[b]Figure 8.25 shows the spectral sensitivities of the photopigments.
[c]See Figure 8.6.

scotopic system A system in the retina that operates at low levels of light and involves the rods.

photopic system A system in the retina that operates at high levels of light, shows sensitivity to colour, and involves the cones.

The cascade of processes required to stimulate the visual receptors helps account for three characteristics of the visual system:

1. Its *sensitivity*, because weak stimuli are amplified to produce physiological effects (one photon blocks *millions* of Na^+ ions)
2. The *integration* of the stimulus over time, which makes vision relatively slow (compared, for example, with audition) but increases its sensitivity
3. The *adaptation* of the visual system to a wide range of light intensities, as we will discuss shortly.

Two different functional systems arise from the two different populations of receptors (rods and cones) in the retina. One system uses the rods and works in dim light, so it is called the **scotopic system** (from the Greek *skotos*, "darkness," and *ops*, "eye"). The scotopic system has only one receptor type (rods) and therefore does not respond differentially to different colours, which is the basis for the saying "at night, all cats are gray." There is a lot of convergence in the scotopic system, as many rods provide information to each ganglion cell.

The other system requires more light and, in some species, shows differential sensitivity to wavelengths, enabling colour vision. This system uses the cones and is called the **photopic system** (from the Greek *phos*, "light"). Compared with the scotopic system, the photopic system has less convergence, as some ganglion cells report information from only a single cone.

At moderate levels of illumination, both rods and cones function, providing us the benefit of both the scotopic and photopic systems, which are compared in **TABLE 8.1**. These two systems also help us see over a wide range of illumination, as we will discuss next.

8.2 Properties of the Retina Shape Many Aspects of Our Vision

Learning Objectives

After reading this section, you should be able to:

8.2.1 **Contrast the rapid versus gradual ways the visual system adapts to a wide range of light levels.**

8.2.2 **Provide three reasons why our vision is sharpest at the centre of the visual field.**

8.2.3 **Offer visual illusions to demonstrate why our sense of brightness is not a simple function of the amount of light striking the retina.**

Different mechanisms enable the eyes to work over a wide range of light intensities

Many sensory systems have to work over wide ranges of stimulus intensity, as we learned in Chapter 8. This is certainly true of the visual system: a very bright light is about 10 billion times as intense as the weakest lights we can see. How do we do it?

One way the visual system deals with a large range of intensities is by adjusting the size of the pupil. In bright light the pupil constricts quickly to admit less light. Interestingly, the pupillary response to light is controlled not by rods or cones, but by specialised retinal ganglion cells that possess their own photopigment, which makes them sensitive to light (Xue et al., 2011). These ganglion cells project to the superior colliculus to control pupil diameter, matching the level of ambient light (Lucas et al., 2003). We'll see in Chapter 15 that these same ganglion cells also help control daily cycles of behaviour called *circadian rhythms*. Although rapid, this pupillary response reduces light only about 16-fold, so it cannot account for the billionfold range of visual sensitivity (**FIGURE 8.5**).

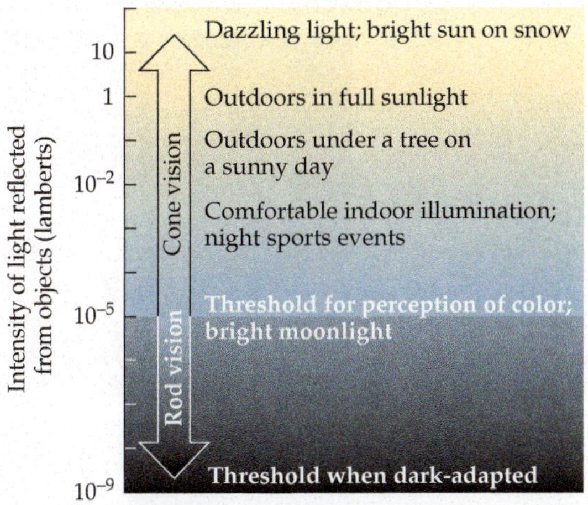

8.5 The Wide Range of Sensitivity to Light Intensity

Another mechanism for dealing with different light intensities is **range fractionation**, the handling of different intensities by different receptors—some with high sensitivity (rods) and others with low sensitivity (cones) (see Table 8.1). Figure 7.6 illustrated range fractionation for the somatosensory system. Additional range fractionation would carry an unacceptable cost: If, at a particular light level, several sets of receptors were not responding, sharpness of vision would be impaired. If only a fraction of the receptors responded to the small changes in the intensity of light around a given level, the active receptors would be spaced apart from each other in the retina, and the image would be "grainy" or "pixelated."

Natural selection solved this problem by giving every photoreceptor an enormous range of adaptation; that is, each photoreceptor adjusts its sensitivity to match the ambient illumination. At any given time, a photoreceptor operates over a range of intensities of about a hundredfold; that is, it is completely depolarised by a stimulus about one-tenth the ambient level of illumination, and a light 10 times brighter than the ambient level will completely hyperpolarise it. But the receptors constantly shifpolarise their whole range of response, to work around the prevailing level of illumination. This tremendous range of **photoreceptor adaptation** is the main reason we can see over such wide ranges of light. Thus, the visual system is concerned with *differences*, or changes, in brightness—not with the absolute level of illumination.

Two main factors account for most of the adaptation of photoreceptors. Probably the most important factor is one shared by other sensory modalities: varying the concentration of calcium (Ca^{2+}) ions (E. N. Pugh and Lamb, 1990). The photoreceptors regulate the release and storage of intracellular Ca^{2+} ions to control their sensitivity to light. Also, when the photoreceptor pigment is split apart by light, its two components—RETINAL and opsin—recombine slowly. This recombination takes time, affecting how much photopigment is available to respond to light. If you go from bright daylight into a dark theater, it takes several minutes until enough rhodopsin recombines to restore your dark vision (E. N. Pugh and Lamb, 1993). Further adaptation occurs in the ganglion cells and the thalamus and probably at higher levels too.

Acuity is best in foveal vision

The **visual field** is the whole area you can see without moving your head or eyes. You may think that your vision is equally good across the visual field, but this is an illusion. **Visual acuity**, the sharpness of vision, is especially fine in the centre of the visual field and falls off rapidly toward the periphery. We can understand this difference in visual acuity by learning more about the retina and successive levels of the visual system.

The central region of the retina, called the **fovea** (Latin for "pit"), has a dense concentration of cones. The **optic disc**, to the nasal side of the fovea, is where blood vessels and ganglion cell axons leave the eye. There are no photoreceptors at the optic disc, so there is a **blind spot** here that we normally do not notice. To locate your blind spot

range fractionation A hypothesis of stimulus intensity perception stating that a wide range of intensity values can be encoded by a group of cells, each of which is a specialist for a particular range of stimulus intensities.

photoreceptor adaptation The tendency of rods and cones to adjust their light sensitivity to match ambient levels of illumination.

visual field The whole area that you can see without moving your head or eyes.

visual acuity Sharpness of vision.

fovea The central portion of the retina, packed with the most photoreceptors and therefore the centre of our gaze.

optic disc The region of the retina devoid of receptor cells because ganglion cell axons and blood vessels exit the eyeball there.

blind spot The portion of the visual field from which light falls on the optic disc. Because there are no receptors in this region, light striking it cannot be seen.

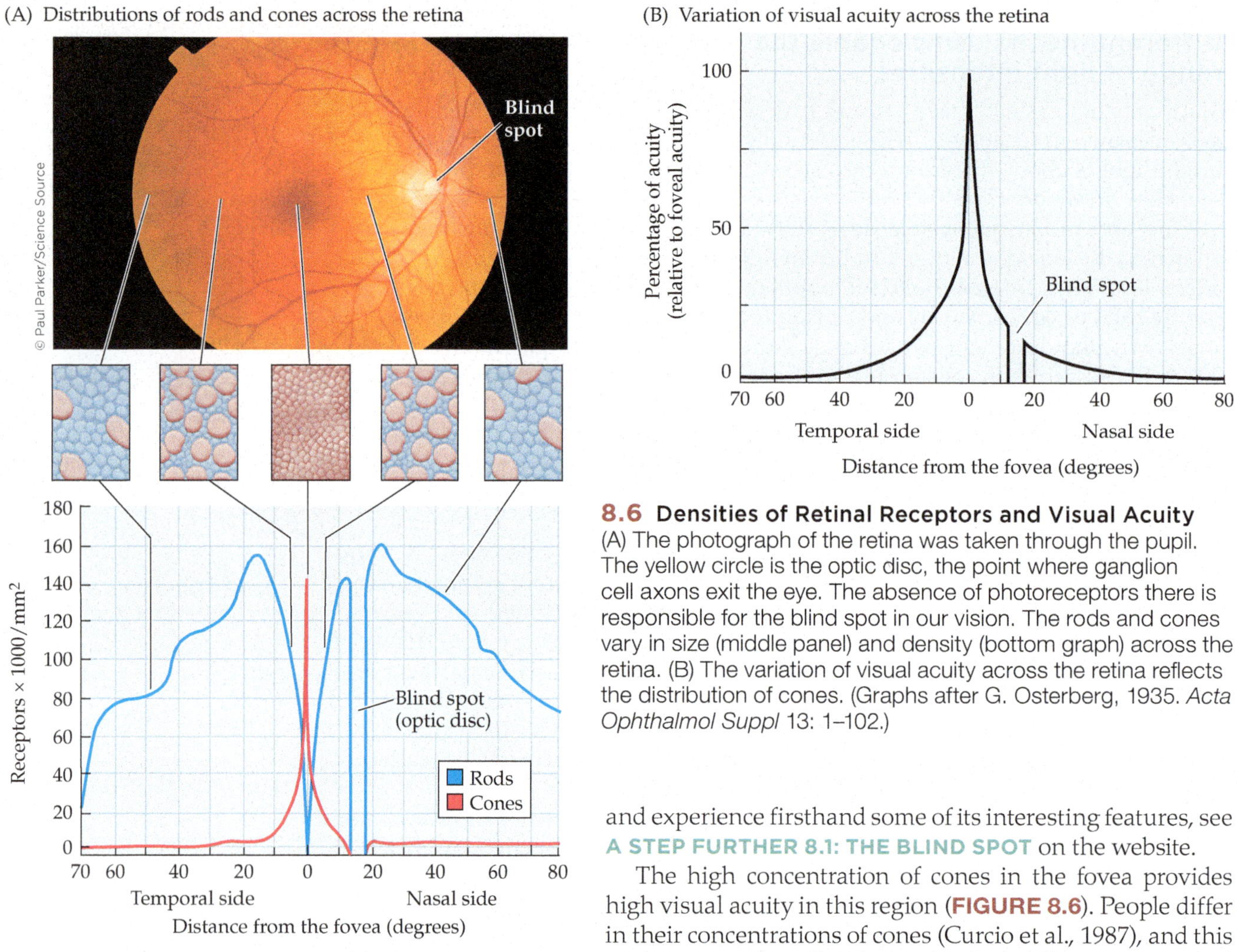

8.6 Densities of Retinal Receptors and Visual Acuity (A) The photograph of the retina was taken through the pupil. The yellow circle is the optic disc, the point where ganglion cell axons exit the eye. The absence of photoreceptors there is responsible for the blind spot in our vision. The rods and cones vary in size (middle panel) and density (bottom graph) across the retina. (B) The variation of visual acuity across the retina reflects the distribution of cones. (Graphs after G. Osterberg, 1935. *Acta Ophthalmol Suppl* 13: 1–102.)

and experience firsthand some of its interesting features, see **A STEP FURTHER 8.1: THE BLIND SPOT** on the website.

The high concentration of cones in the fovea provides high visual acuity in this region (**FIGURE 8.6**). People differ in their concentrations of cones (Curcio et al., 1987), and this variation may be related to individual differences in visual acuity. Species differences in visual acuity also reflect the density of cones in the fovea. For example, hawks, whose acuity is much greater than that of humans, have much narrower and more densely packed cones in the fovea than we do. In the human retina, both cones and rods are larger toward the periphery. Therefore, our acuity falls off about as rapidly in the horizontal direction as in the vertical direction. But species that live in open, flat environments (such as the cheetah and antelopes) have fields of acute vision that extend farther horizontally than vertically. In all vertebrates, there are fewer blood vessels overlying the foveal region of the retina

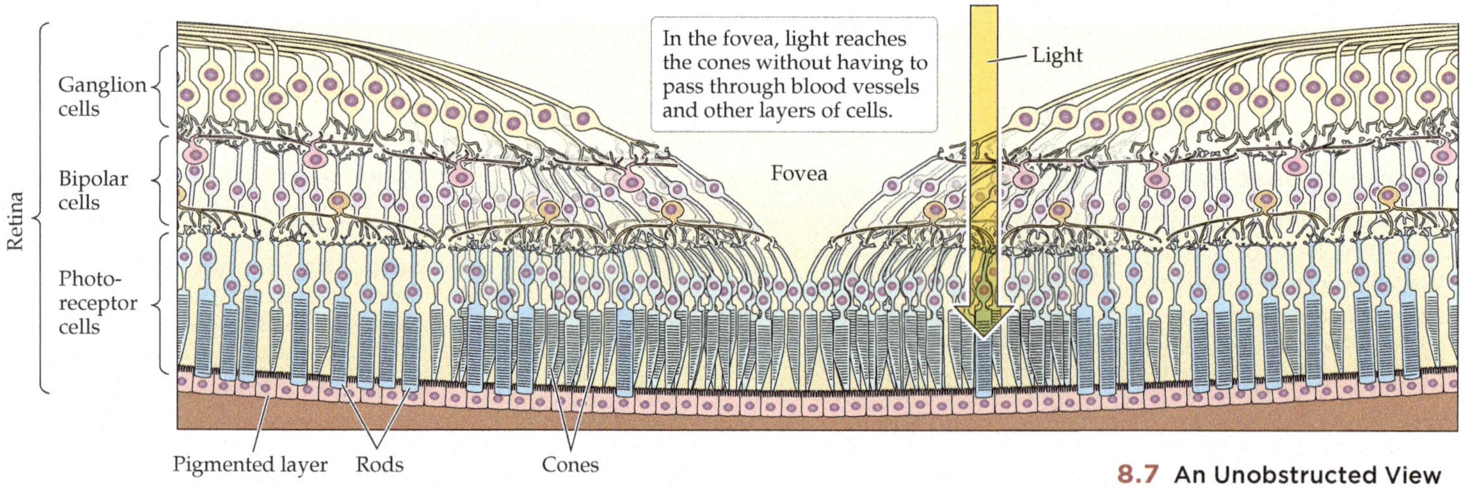

8.7 An Unobstructed View

(**FIGURE 8.7**), so more light reaches the photoreceptors there.

Rods are distributed differently than cones: rods are absent in the fovea but more numerous than cones in the periphery of the retina (see Figure 8.6A). They are the most concentrated in a ring about 20° away from the centre of the retina. This is why if you want to see a dim star, you do best to search for it a little off to the side of your centre of gaze. This directs the dim light to the periphery of your visual field, onto rods driving the scotopic system. Not only are the rods more sensitive to dim light than the cones, but as we mentioned earlier, input from many rods converges on ganglion cells in the scotopic system, further increasing the system's sensitivity to weak stimuli. On the other hand, that greater convergence of information from photoreceptors to ganglion cells in rods than in cones is another reason why acuity is greater in the cone-rich fovea. A cone here is more likely to have exclusive control over a single ganglion cell. Rods provide high sensitivity with limited acuity; cones provide high acuity with limited sensitivity.

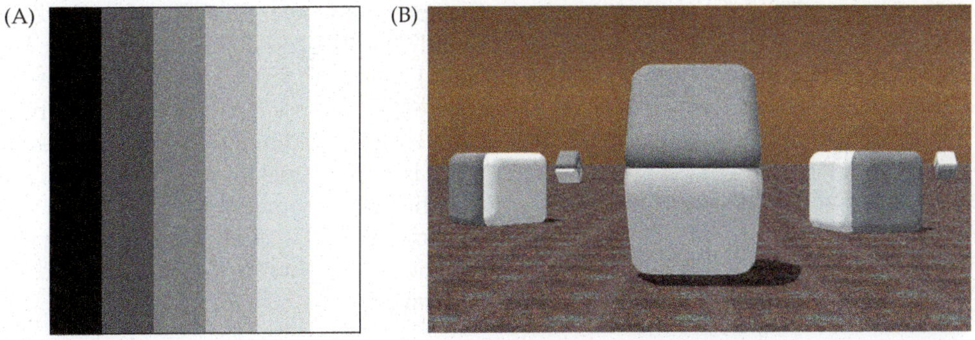

8.8 The Effect of Context on the Perception of Brightness (A) Each bar shown here is uniform in colour, but each of the gray bars appears ligher on its left edge and darker on its right. (B) Shading around a patch of light affects the perception of brightness. For example, even though the two tiles in the middle are the same shade of gray, one appears darker than the other. If you don't believe this, place your pinky finger over the place where the two tiles meet. See? (Part B from D. Purves et al., 1999. *J Neurosci* 19: 8543–8551. © 1998 Society for Neuroscience.)

saccades Fast movements of the eyes that present various parts of the visual scene to the fovea. When we fix our gaze, small saccades that we are unaware of avert photoreceptor adaptation.

lateral inhibition The phenomenon by which interconnected neurons inhibit their neighbours, producing contrast at the edges of regions.

Quick movements of the eyes, called **saccades**, bring various parts of the visual scene to the fovea to take advantage of the sharp visual acuity there. Even when we think we are holding our eyes still to fixate on a scene, they are in fact making tiny saccades (Martinez-Conde et al., 2013), constantly shifting the scene to different parts of the retina to be detected by fresh photoreceptors. Otherwise, the photoreceptors would adapt and stop responding to the light, causing the scene to disappear.

Brightness is created by the visual system

The brightness dimension of visual perception is created in part by the visual system, not simply by the amount of light reflected. **FIGURE 8.8** presents two examples. The enhancement of the boundaries of the bars in **FIGURE 8.8A**, each of which is uniformly gray but looks as though it varies in brightness, is based on a neural process called **lateral inhibition**.

Lateral inhibition occurs where the neurons in a region—in this case, retinal cells—are interconnected, either through their own axons or by means of interneurons, and each neuron tends to inhibit its neighbours (**FIGURE 8.9**). The photoreceptors stimulated by the right edge of each gray bar in Figure 8.8A are inhibited by the neighbouring photoreceptors stimulated by the lighter bar next door. Thus, photoreceptors on the right edge report receiving less light than they actually do (i.e., that edge looks darker to us).

In **FIGURE 8.8B**, two tiles that clearly differ in brightness *reflect the same amount of light*. If you use your pinkie to cover the boundary between

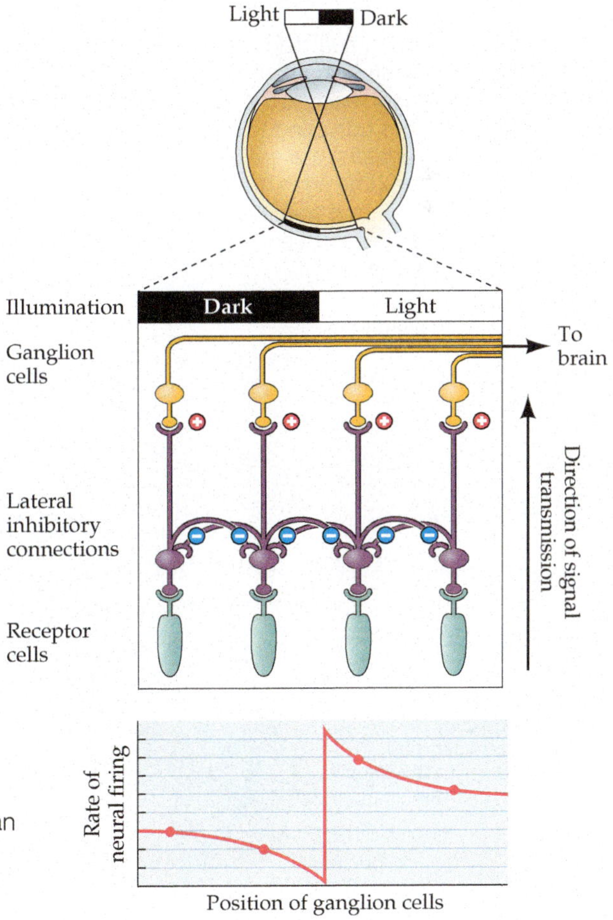

8.9 Lateral Inhibition in the Retina Because of lateral inhibitory connections, the two central receptor cells differ more in their rates of firing than do either the two left-hand cells or the two right-hand cells, as the graph shows. This is a simplified view; in fact, each purple cell inhibits many of its neighbours and not just those directly adjacent.

optic chiasm The point at which the two optic nerves meet.

optic tract The axons of retinal ganglion cells after they have passed the optic chiasm. Most of the axons terminate in the lateral geniculate nucleus.

lateral geniculate nucleus (LGN) The part of the thalamus that receives information from the optic tract and sends it to visual areas in the occipital cortex.

the upper and lower tiles, they appear equally dark. How are such puzzling effects produced? Although the contrast effect in Figure 8.8A is determined, at least in part, by lateral inhibition among adjacent retinal cells, the two indicated patches of the tiles in Figure 8.8B are not immediately adjacent (they are separated by shading), so the effect must be produced higher in the visual system (Purves and Lotto, 2010). The important point is that our visual experience is not a simple reporting of the physical properties of light. Rather, our experience of light versus dark is created by the brain in response to many factors, including surrounding stimuli. Later in the chapter we'll find that our experience of colour is also created by the visual system, and not a simple reporting of the wavelengths of light.

8.3 Neural Signals Travel from the Retina to Several Brain Regions

Learning Objectives

After reading this section, you should be able to:

8.3.1 **Trace the neuronal pathway carrying information from the eye to the visual cortex.**

8.3.2 **Describe the topographic projection of the visual field onto the two retinas and to primary visual cortex.**

8.3.3 **Outline the steps by which visual systems evolved to provide highly accurate portraits of the world.**

The ganglion cells in each eye produce action potentials that are conducted along their axons to send visual information to the brain. These axons make up the optic nerve (also known as *cranial nerve II*) that brings visual information into the brain on each side, eventually reaching visual cortex in the occipital lobe at the back of the brain. Primary visual cortex passes along the information to other visual cortical regions, as we will discuss shortly.

In all vertebrates, some or all of the axons of each optic nerve cross to the opposite cerebral hemisphere. The optic nerves cross the midline at the **optic chiasm** (named for the Greek letter χ [chi] because of its crossover shape). In humans, axons from the half of the retina toward the nose (called the *nasal hemiretina*) cross over to the opposite side of the brain (**FIGURE 8.10A**). The half of the retina toward the temple (the *temporal hemiretina*) projects its axons to its own side of the head. This distribution of retinal fibres means the right side of the brain receives input

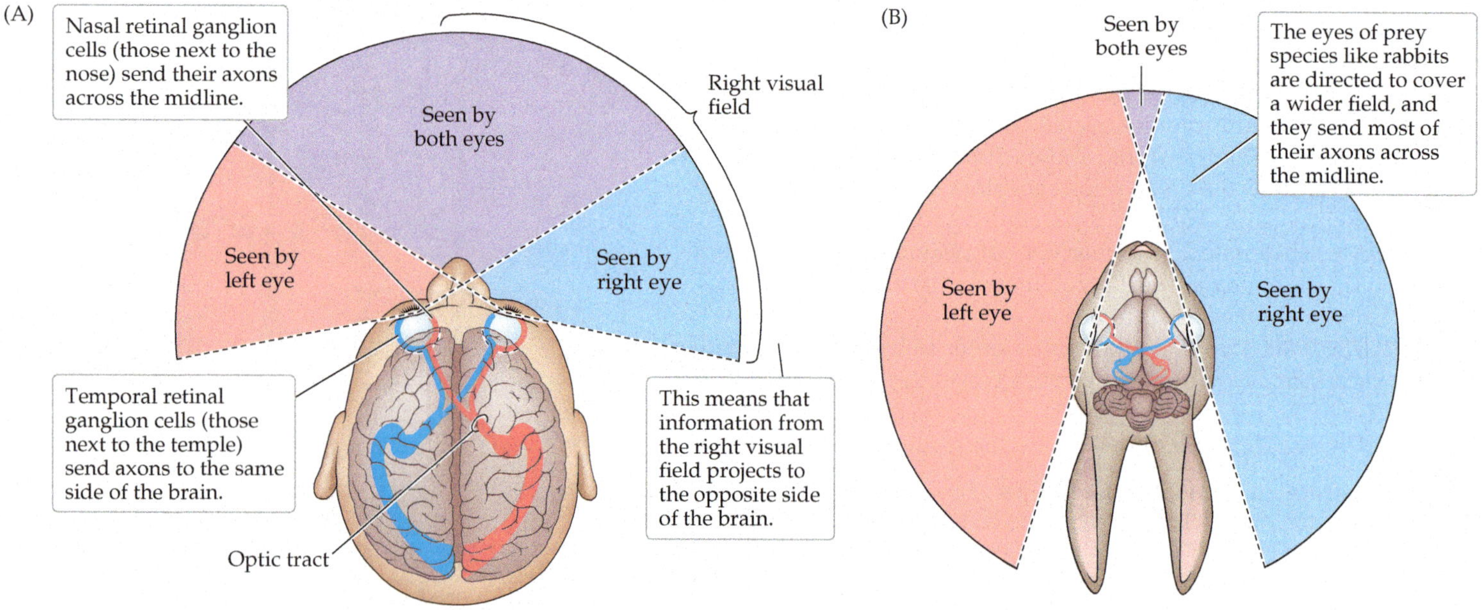

8.10 The Visual Fields of the Two Eyes Overlap to Some Extent (A) In humans, both eyes receive information from both the left and right visual field. This means that the central portion of the visual field projects to both eyes. However, since axons from the portion of the retina next to the nose (nasal retina) cross the midline on the way to the brain while the portion of the retina next to the temple (temporal retina) projects to the same side of the brain, information from the left visual field reaches the right side of the brain, and information from the right visual field reaches the left side. (B) In prey species that need to be on the lookout for predators, the eyes may be farther apart so they can scan more of the space around them, and consequently almost all retinal ganglion cells send their axons across the midline.

from the left half of the visual field, while the left side of the brain receives input from the right half of the visual field.

Having our eyes cover overlapping sections of the visual field helps us judge the distance of objects, by (unconsciously) comparing the inputs from the two eyes. In prey species such as rabbits, natural selection has sacrificed some of this binocular depth perception, favoring instead the broader visual field that results from having the eyes positioned farther apart (**FIGURE 8.10B**); this broader visual field helps the rabbit detect approaching predators. In such species, most of the optic nerve fibres cross the midline.

After they pass the optic chiasm, the axons of the retinal ganglion cells are known collectively as the **optic tract**. As **FIGURE 8.11** shows, most axons of the optic tract in mammals terminate on cells in the visual portion of the thalamus, the **lateral geniculate nucleus** (**LGN**; step 4a in Figure 8.11). The axons of neurons in the LGN form the **optic radiations** (step 5), which terminate in **primary visual cortex** (**V1**) of the **occipital lobe** at the back of the brain (step 6). The primary visual cortex is often called *striate cortex* because a broad stripe, or *striation*, is visible in anatomical sections through this region; the stripe represents layer IV of the cortex, where the optic-radiation fibres arrive. Information from the two eyes converges on cells beyond layer IV of the primary visual cortex, making binocular, three-dimensional vision possible, as we discussed in Chapter 6.

Some retinal ganglion cells send their optic-tract axons to the superior colliculus in the midbrain (see Figure 8.11, step 4b) to help coordinate rapid movements of the eyes toward a target and to control the size of the pupil, as we mentioned earlier.

optic radiation Axons from the lateral geniculate nucleus that terminate in the primary visual areas of the occipital cortex.

primary visual cortex (V1) Also called *striate cortex* or *area 17*. The region of the occipital lobe where most visual information first arrives in the cortex.

occipital lobe Large region of cortex covering much of the posterior part of each cerebral hemisphere, specialised for visual processing.

Go to Animation 8.1
Visual Pathways in the Human Brain

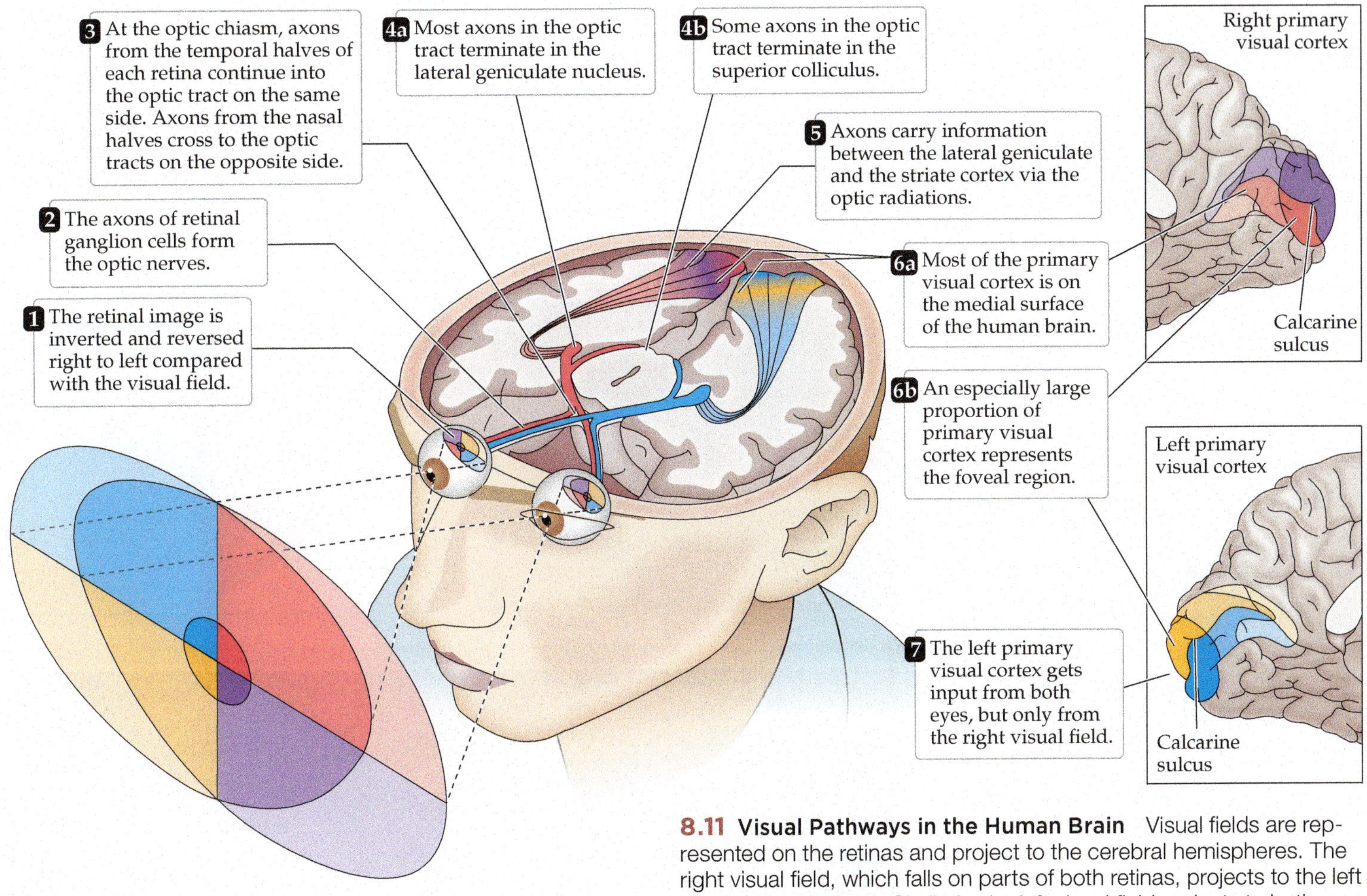

8.11 Visual Pathways in the Human Brain Visual fields are represented on the retinas and project to the cerebral hemispheres. The right visual field, which falls on parts of both retinas, projects to the left cerebral hemisphere. Similarly, the left visual field projects to both eyes and then to the right cerebral hemisphere.

extrastriate cortex Visual cortex outside of the primary visual (striate) cortex.

scotoma A region of blindness caused by injury to the visual pathway or brain.

In addition to V1 shown in Figure 8.11, numerous surrounding regions of the cortex are also largely visual in function. These visual cortical areas outside the striate cortex are sometimes called **extrastriate cortex**. These different cortical regions work in parallel to process different aspects of visual perception, such as form, colour, location, and movement, as we will discuss later in this chapter. In striate cortex, as well as most extrastriate regions, there is a topographic projection of the retina, which means there's a topographic projection of the visual field, discussed next.

The retina projects to the brain in a topographic fashion

The retina represents a two-dimensional map of visual space. As this information courses through the brain, the point-to-point correspondence between neighbouring parts of visual space is maintained, forming a maplike projection (see Figure 8.11). Much of this topographic projection of visual space is devoted to the foveal region (**FIGURE 8.12A**) (Tootell et al., 1982). Although the monkey V1 is located on the lateral surface of the occipital area, human V1 is located mainly on the medial surface of the cortex (**FIGURE 8.12B**; see also Figure 8.20D). The fact that about half of the human V1 is devoted to the fovea and the retinal region just around the fovea does not mean that our spatial perception is distorted. Rather, this representation makes possible the great acuity of spatial discrimination in the central part of the visual field. In other words, another reason why we see better in the fovea is because we devote more brain circuits to analyzing it. Apparently those brain circuits are somewhat plastic, benefiting from practice, because people who intensively play "point and shoot" video action games display improved visual acuity (R. Li et al., 2009).

Because of the orderly mapping of the visual field (known as *retinotopic mapping*) at the various levels of the visual system, damage to parts of the visual system can be diagnosed from defects in perception of the visual field. If we know the site of injury in the visual pathway, we can predict the location of such a perceptual gap, or **scotoma** (plural *scotomas* or *scotomata*), in the visual field. Although the word *scotoma* comes from the Greek *skotos*, meaning "darkness," a scotoma is not perceived as a dark patch in the visual field; rather, it is a spot where nothing can be perceived, and usually rigorous testing is required to demonstrate its existence.

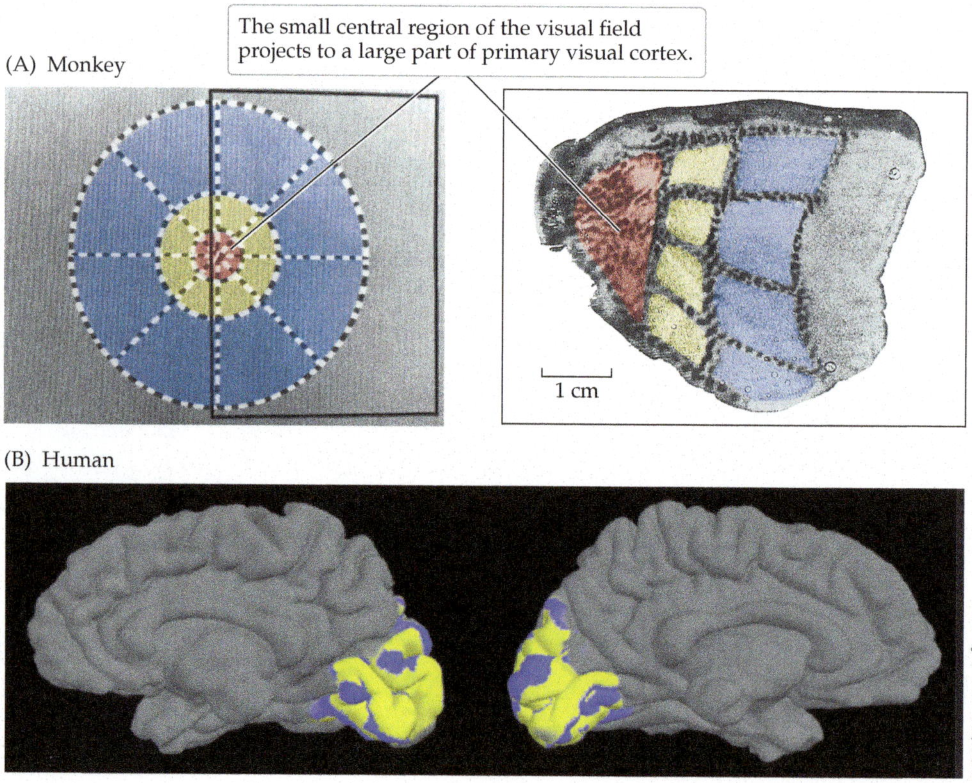

8.12 Location of the Primary Visual Cortex (A) A pattern of flickering lights (left) was shown in a monkey's visual field, and a map of the visual field (right) was revealed by autoradiography in a flattened portion of the primary visual cortex. (B) Maps of human visual cortex derived from fMRI measurements show primary visual cortex as the innermost yellow region on each of these medial views.

Within a scotoma, a person cannot consciously perceive visual cues, but some visual discrimination in this region may still be possible; this paradoxical phenomenon has been called *blindsight*. In other cases, stimuli that cannot be seen within a scotoma affect judgments of stimuli outside it (Stoerig and Cowey, 1997). Blindsight may also be related to the phenomenon known as *hemispatial neglect*—neglect of the side opposite to an injured cerebral hemisphere—which we will discuss in Chapter 17.

When we look at a complex organ like the eye of a mammal, an octopus, or a fly, it is hard at first to understand how it could have evolved. But inspection of different living species reveals a gradation from very simple light-sensitive cells to increasingly complex organs with focussing devices (**BOX 8.1**), and each gradation of a visual system helps the animal survive.

BOX 8.1 Eyes with Lenses Have Evolved in Several Phyla

Phylogenetic studies indicate that the evolution of eyes with lenses, like those of a mammal or an octopus, included the following steps (Fernald, 2000), as illustrated in the figure:

1. *Concentrating light-sensitive cells* into localised groups that serve as photoreceptor organs. Animals with such photoreceptor organs have better chances of surviving and reproducing, because photoreceptor organs make it easier to respond differently to stimuli that strike different parts of the body surface.
2. *Clustering light receptors* at the bottom of pitlike or cuplike depressions in the skin. Animals with this adaptation can discriminate better among stimuli that come from different directions.
3. *Narrowing the top of the cup* into a small aperture so that, like a pinhole camera, the eye can focuss well.
4. *Closing the opening with transparent skin or filling the cup with a transparent substance.* This covering protects the eye against the entry of foreign substances that might injure the receptor cells or block vision.
5. *Forming a lens* either by thickening the transparent skin or by modifying other tissue in the eye. This adaptation improves the focussing of the eye while allowing the aperture to be relatively large to let in more light.

The only requirement for the evolution of eyes to begin appears to be the existence of light-sensitive cells. Natural selection then favors the development of auxiliary mechanisms to improve vision.

Eyes evolved independently in many different phyla (Salvini-Plawen and Mayr, 1977)—an example of convergent evolution. The fact that the cephalopods (such as squid and octopuses) evolved a visual system that in many ways resembles that of vertebrates (from fishes to humans) suggests that major constraints limit the development of a visual system for a large, rapidly moving animal. In both cephalopods and vertebrates, the eyes are relatively large, allowing for many receptors and the ability to gather large amounts of light. The incoming light is regulated by a pupil and focussed by a lens. In both cephalopods and vertebrates, three sets of extraocular muscles move the eyeballs.

However, it is still uncertain whether eyes of all species have evolved from a single progenitor or have arisen more than once during evolution (Fernald, 2000). The eyes of all seeing animals share at least two important genetic features: First, they all use opsin pigments. Second, from the compound eye of the fruit fly *Drosophila*, to the vertebrate eye, to the cephalopod eye, they all develop through genes of the Pax family. The finding that homologous molecules are key regulators of eye development in different phyla suggests that eyes in all phyla share a common origin.

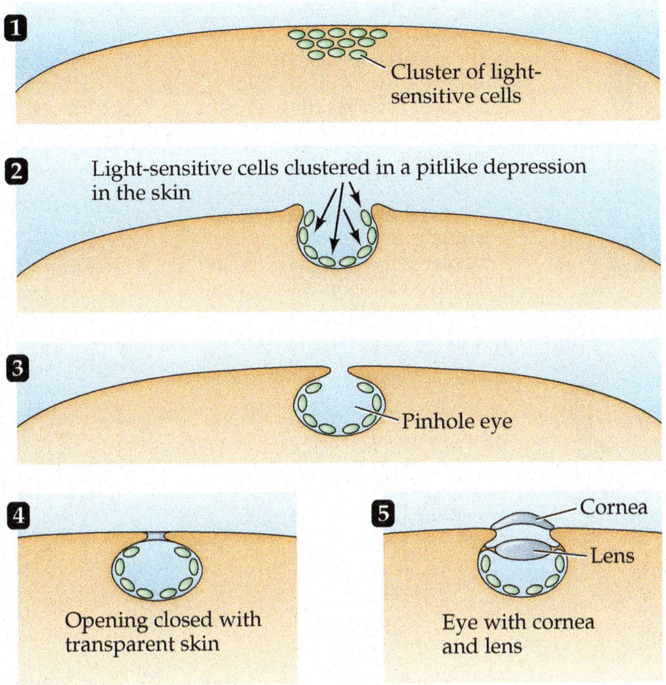

8.4 Neurons at Different Levels of the Visual System Have Very Different Receptive Fields

Learning Objectives

After reading this section, you should be able to:

8.4.1 **Define what is meant by a visual receptive field.**

8.4.2 **Describe the two basic types of receptive fields of retinal bipolar cells and ganglion cells.**

8.4.3 **Compare the receptive fields of simple versus complex visual cortical neurons, and explain how integrating information from several ganglion cells could produce them.**

8.4.4 **Critically discuss the limitations of a visual system based only on feature detectors, and describe the alternative, spatial-frequency filter model.**

8.4.5 **Contrast receptive fields of neurons in primary versus nonprimary visual cortex.**

8.4.6 **Describe the columnar organisation of primary visual cortex.**

receptive field The stimulus region and features that affect the activity of a cell in a sensory system.

on-centre bipolar cell A retinal bipolar cell that is excited by light in the centre of its receptive field.

off-centre bipolar cell A retinal bipolar cell that is inhibited by light in the centre of its receptive field.

on-centre ganglion cell A retinal ganglion cell that is activated when light is presented to the centre, rather than the periphery, of the cell's receptive field.

off-centre ganglion cell A retinal ganglion cell that is activated when light is presented to the periphery, rather than the centre, of the cell's receptive field.

As we noted in Chapter 7, the **receptive field** of a sensory cell consists of the stimulus region and the features that excite or inhibit the cell. The nature of the receptive field of a cell gives us good clues about its function(s) in perception. In the next sections we will see that neurons at lower levels in the visual system seem to account for some perceptual phenomena while neurons at higher levels account for others.

Photoreceptors excite some retinal neurons and inhibit others

At their resting potentials, both rod and cone photoreceptors steadily release the synaptic neurotransmitter glutamate. Light always hyperpolarises the photoreceptors, causing them to release *less* glutamate. But the responses of the bipolar cells that receive this glutamate differ, depending on what type of glutamate receptor they possess. Glutamate depolarises one group of bipolar cells but hyperpolarises another group. These groups differ in their receptive fields.

One group of bipolar cells are called **on-centre bipolar cells**: turning *on* a light directed at the centre of an on-centre bipolar cell's receptive field excites the cell because it receives *less* glutamate, which inhibits on-centre bipolar cells (**FIGURE 8.13, LEFT**). A second group of bipolar cells are **off-centre bipolar cells**: turning *off* light in the centre of an off-centre bipolar cell's receptive field excites the cell because it receives *more* glutamate, which depolarises off-centre bipolar cells (**FIGURE 8.13, RIGHT**).

Bipolar cells also *release* glutamate, and glutamate always depolarises the ganglion cells. Therefore, when light is turned on, on-centre bipolar cells depolarise (excite) **on-centre ganglion cells**; and when light is turned off, off-centre bipolar cells depolarise (excite) **off-centre ganglion cells** (see Figure 8.13). The stimulated on-centre and off-centre ganglion cells then fire nerve impulses and report "light" or "dark" to higher visual centres.

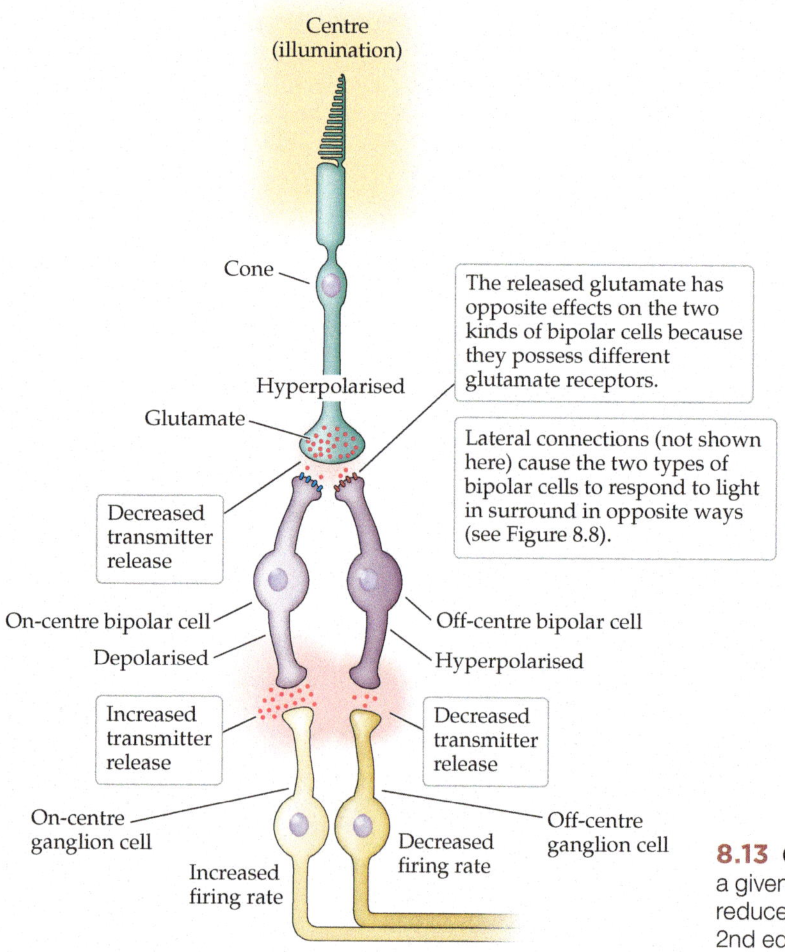

8.13 Connections of Cones to Bipolar Cells Light falling on a given photoreceptor may increase the firing of one ganglion cell and reduce the firing of another. (After D. Purves et al., 2001. *Neuroscience*, 2nd ed. Sinauer: Sunderland, MA.)

Neurons in the retina and the LGN have concentric receptive fields

The pattern of connections of photoreceptors to bipolar cells shapes the receptive fields of the ganglion cells. Scientists can record from single ganglion cells while moving a small spot of light across the visual field, keeping the animal's eye still. These studies show that the receptive fields of retinal ganglion cells are concentric, consisting of a roughly circular central area and a ring around it. Through various retinal connections, including the lateral inhibition we discussed earlier (see Figure 8.9), the photoreceptors in the central area and those in the ring around it tend to have the opposite effects on the next cells in the circuit. Thus, both bipolar cells and ganglion cells have two basic types of retinal receptive fields: **on-centre/off-surround** (**FIGURE 8.14A**) and **off-centre/on-surround** (**FIGURE 8.14B**). These antagonistic effects of the centre and its surround explain why uniform illumination of the visual field is less effective in activating a ganglion cell than is a well-placed small spot or a line or edge passing through the centre of the cell's receptive field.

We told you earlier that most ganglion cell axons synapse in the LGN of the thalamus in mammals. The primate LGN has six main layers and some smaller layers in between. The structure is called *geniculate* because the layers are bent like a knee, which in Latin is *genu*. The four dorsal, or outer, layers of the primate LGN are called **parvocellular** (from the Latin *parvus*, "small") because their cells are relatively small. The two ventral, or inner, layers are called **magnocellular** (from the Latin *magnus*, "large") because the neurons there are large. LGN neurons of all six layers have concentric receptive fields. For most neurons in the magnocellular layers, these concentric receptive fields are relatively large, receiving input from large ganglion cells, which receive their input from bipolar cells contacting many neighbouring photoreceptors. Most magnocellular LGN neurons do not show differential wavelength responses; that is, they cannot be involved in colour discrimination. In contrast, the neurons of the parvocellular layers have relatively small

on-centre/off-surround Referring to a concentric receptive field in which the centre excites the cell of interest while the surround inhibits it.

off-centre/on-surround Referring to a concentric receptive field in which the centre inhibits the cell of interest while the surround excites it.

parvocellular Referring to relatively small cells.

magnocellular Referring to relatively large cells.

Go to Animation 8.2
Receptive Fields in the Retina

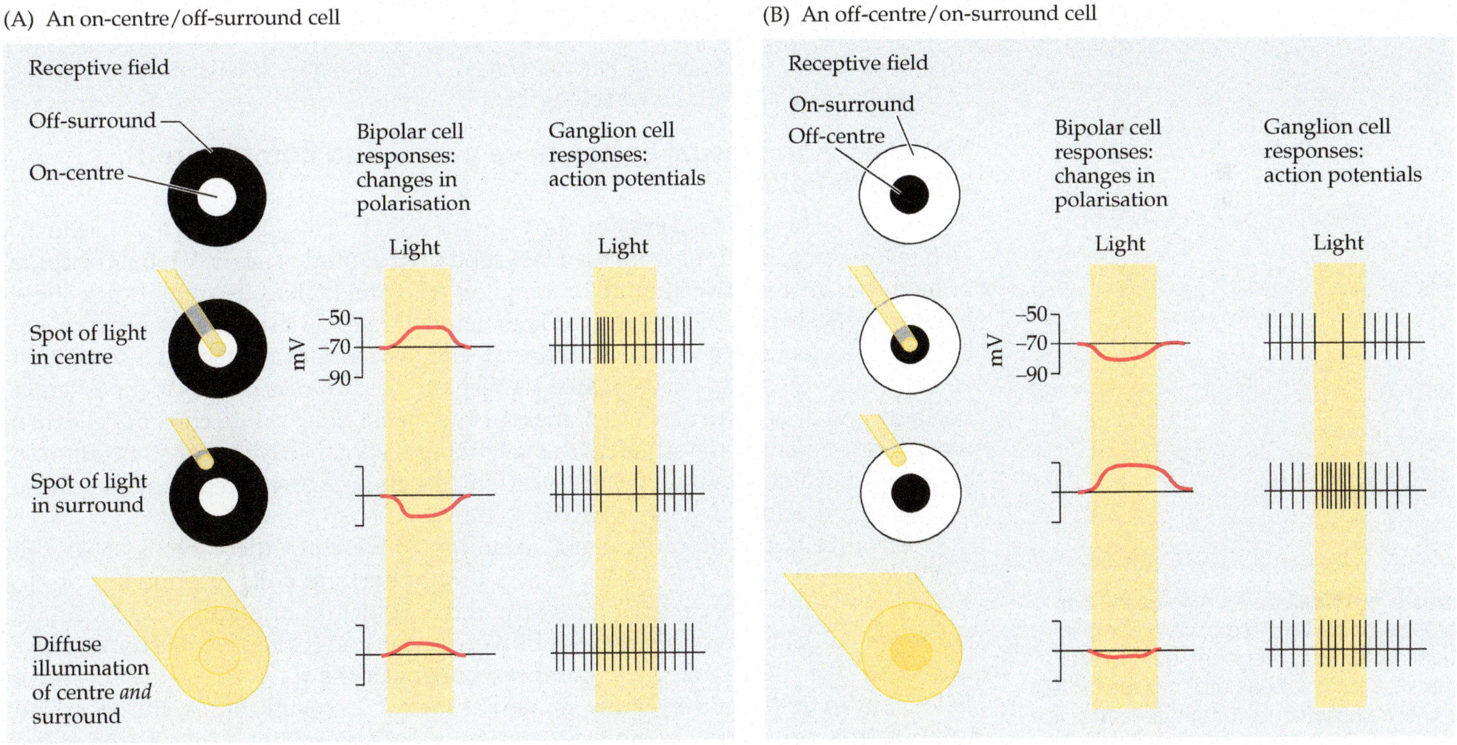

8.14 Receptive Fields of Retinal Cells In primates, each retinal bipolar cell, as well as each retinal ganglion cell, has a concentric receptive field, with antagonistic centre and surround. Bipolar cells respond by changes in local membrane potentials; ganglion cells respond with action potentials. (A) An on-centre/off-surround cell is excited by an increase of illumination in the centre of its receptive field and inhibited by illumination in the surround. (B) Changes in illumination have the opposite effects on an off-centre/on-surround cell. Note that uniform light across the entire receptive field (bottom) has little effect on the firing of either type of ganglion cell.

8.15 The Lateral Geniculate Nucleus
This cross section shows the layered structure of the LGN in primates.

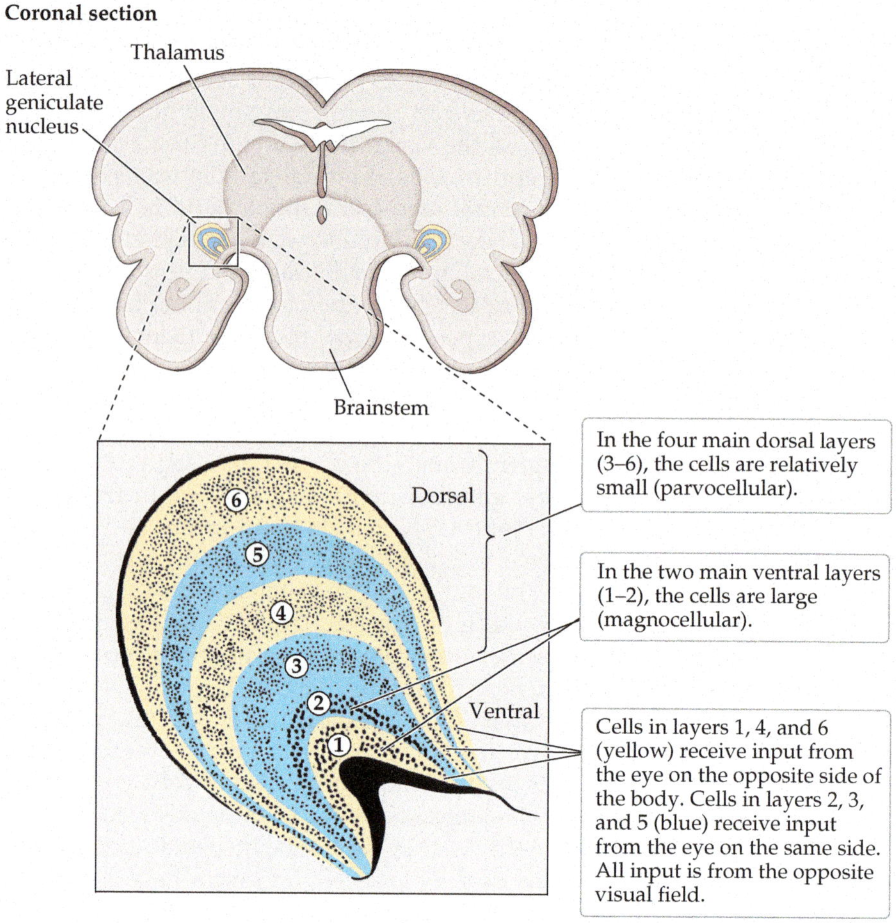

receptive fields, whose input can be traced back to cones (**FIGURE 8.15**). These LGN neurons discriminate wavelengths.

Neurons in the visual cortex have varied and complicated receptive fields

The next level of the visual system, the primary visual cortex, provided a puzzle to vision researchers. Neurons from the LGN send their axons to cells in V1, but the spots of light that are effective stimuli for ganglion cells and LGN cells are not very effective for cortical cells. In 1959, David Hubel and Torsten Wiesel reported that visual cortical cells require more-specific stimuli that are more elongated than those that activate LGN cells. Most cells in area V1 respond best to lines or bars in a particular position and at a particular orientation in the visual field. Some cortical cells also require movement of the stimulus to make them respond actively. For some of these cells, any movement in their field is sufficient; others are more demanding, requiring motion in a specific direction (**FIGURE 8.16**).

Cortical cells can be categorised according to which of these types of stimuli produce maximum responses. So-called **simple cortical cells** respond best to an edge or a bar that has a particular width and a particular orientation and location in the visual field. Therefore these cells were sometimes called *bar detectors* or *edge detectors*. Like the simple cells, **complex cortical cells** have elongated receptive fields, but in addition they show some latitude for location; that is, they respond to a bar of a particular size and orientation anywhere within a larger area of the visual field.

This categorisation of receptive fields can be described as hierarchical; that is, more-complex events are built up from inputs of simpler ones. For example, a simple cortical cell can be thought of as receiving input from a row of LGN cells, and a complex cortical cell can be thought of as receiving input from a row of

simple cortical cell Also called *bar detector* or *edge detector*. A cell in the visual cortex that responds best to an edge or a bar that has a particular width, as well as a particular orientation and location in the visual field.

complex cortical cell A cell in the visual cortex that responds best to a bar of a particular size and orientation anywhere within a particular area of the visual field.

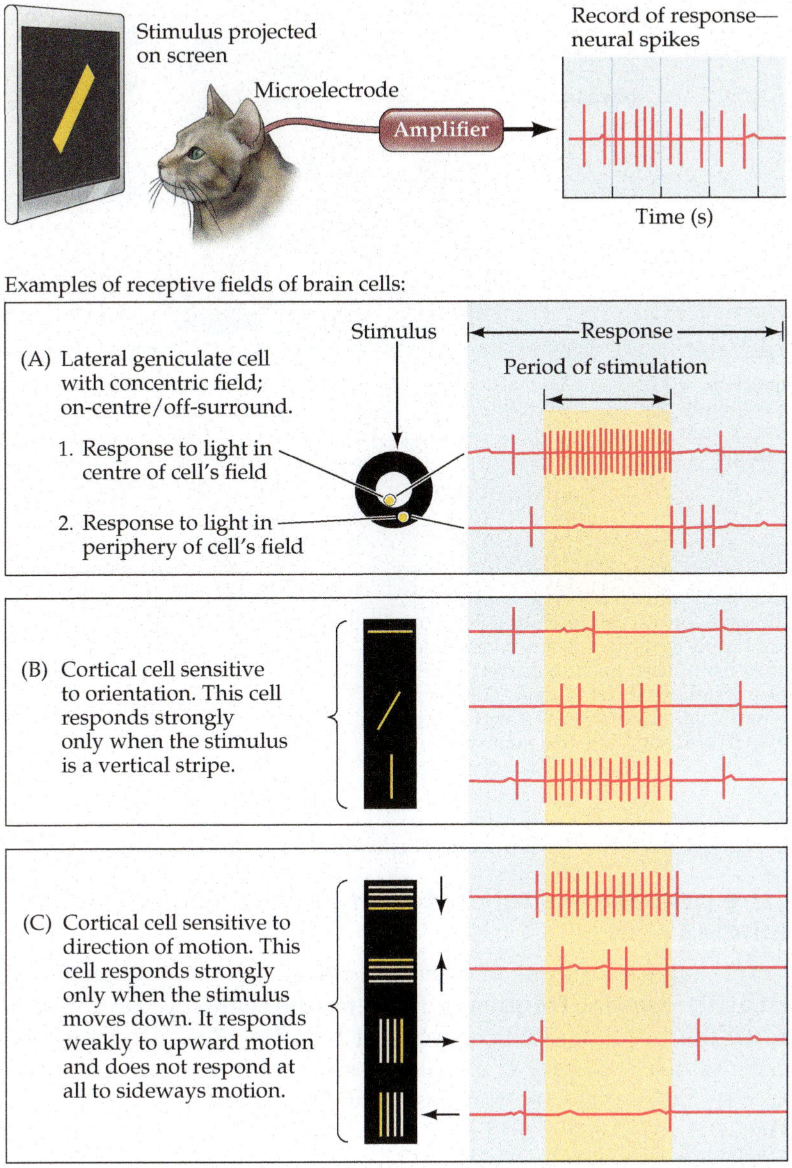

8.16 Receptive Fields of Cells at Various Levels in the Cat Visual System Microelectrode recordings reveal that cells differ greatly in their receptive fields. (A) Visual cells in the lateral geniculate nucleus (LGN) have concentric receptive fields, like those of retinal bipolar cells and ganglion cells (see Figure 8.14). (B) Visual cells in the cerebral cortex are more responsive to bars of light and may show orientation specificity or respond only to motion, or (C) they may respond only to motion in a particular direction.

simple cortical cells (**FIGURE 8.17**). Other theorists extrapolated from this model, suggesting that higher-order circuits of cells could detect any possible form. Thus it was suggested that, by integration of enough successive levels of analysis, a neuron could be constructed that would enable a person to recognise his or her grandmother, and such hypothetical "grandmother cells" were frequently mentioned in the literature. According to this view, whenever such a cell was excited, up would pop a mental picture of one's grandmother. This hypothesis was given as a possible explanation for facial recognition.

Critics soon pointed out both theoretical and empirical problems with this hierarchical model of vision. For one thing, a hierarchical system like this would require a vast number of cells—perhaps more neurons than the cortex possesses—in order to account for all the visual objects that we may encounter. Although some neurons are activated by the sight of very specific faces (e.g., "Halle Berry neurons" are some neurons that were activated by photos of that actress) in both humans (Pedreira et al., 2010) and monkeys (Freiwald et al., 2009), these neurons do not respond to specific features of the face, as you would expect if they were feature detectors. Rather, they respond only when the whole face or most of the face is presented (Freiwald et al., 2009). Thus, we need an alternative model of vision, which we describe next.

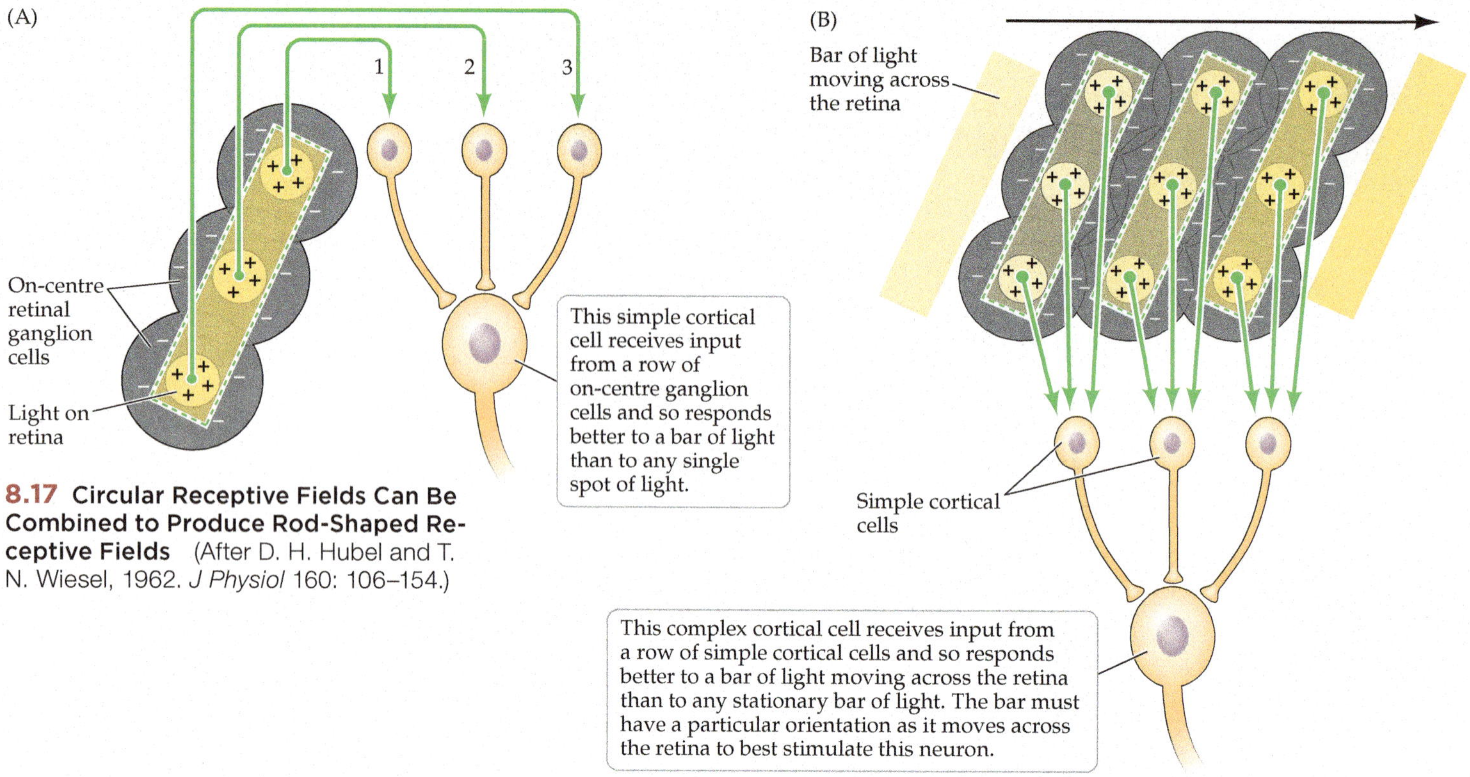

8.17 Circular Receptive Fields Can Be Combined to Produce Rod-Shaped Receptive Fields (After D. H. Hubel and T. N. Wiesel, 1962. *J Physiol* 160: 106–154.)

Most cells in the primary visual cortex are tuned to particular spatial frequencies

Concepts of pattern analysis in terms of lines and edges have largely given way to what is known as the **spatial-frequency filter model**, which is more powerful but less intuitive than the hierarchical model it replaces (R. L. De Valois and De Valois, 1988). By *spatial frequency of a visual stimulus* we mean the number of light-dark (or colour) cycles that the stimulus shows per degree of visual space. For example, **FIGURE 8.18A AND B** differ in the spacing of the bars: Figure 8.18A has twice as many bars in the same horizontal space and is therefore said to have double the spatial frequency of Figure 8.18B. The spatial-frequency technique applies Fourier analysis (see Box 9.1) or linear systems theory, rather than breaking down visual patterns into bars and angles.

In Box 9.1 we saw that we can use Fourier analysis to decompose any complex sound into a sum of simple sine waves. Fourier analysis can similarly be applied to *visual* patterns. If a span of the visual field varies between dark and light bands according to a simple sine wave function with a fixed frequency, visual patterns like the ones in **FIGURE 8.18C AND D** result. Any series of dark and light stripes, like those in Figure 8.18A and B, can be analysed into the sum of a visual sine wave and its odd harmonics (multiples of the basic frequency).

A complex visual pattern or scene can also be analysed by the Fourier technique; in this case, frequency components at different angles of orientation are also used. A given spatial frequency can exist at any level of contrast; Figure 8.18C and D show examples of high and low contrast, respectively. To reproduce or perceive a complex pattern or scene accurately, the system has to handle all the spatial frequencies present. If the high frequencies are filtered out, the small details and sharp contrasts are lost; if the low frequencies are filtered out, the large uniform areas and gradual transitions are lost. **FIGURE 8.18E–G** shows how the filtering of spatial frequencies affects a photograph. The photograph is still recognisable after either the high visual frequencies (Figure 8.18F) or low frequencies (Figure 8.18G) are filtered out.

spatial-frequency filter model
A model of pattern analysis that emphasises Fourier analysis of visual stimuli.

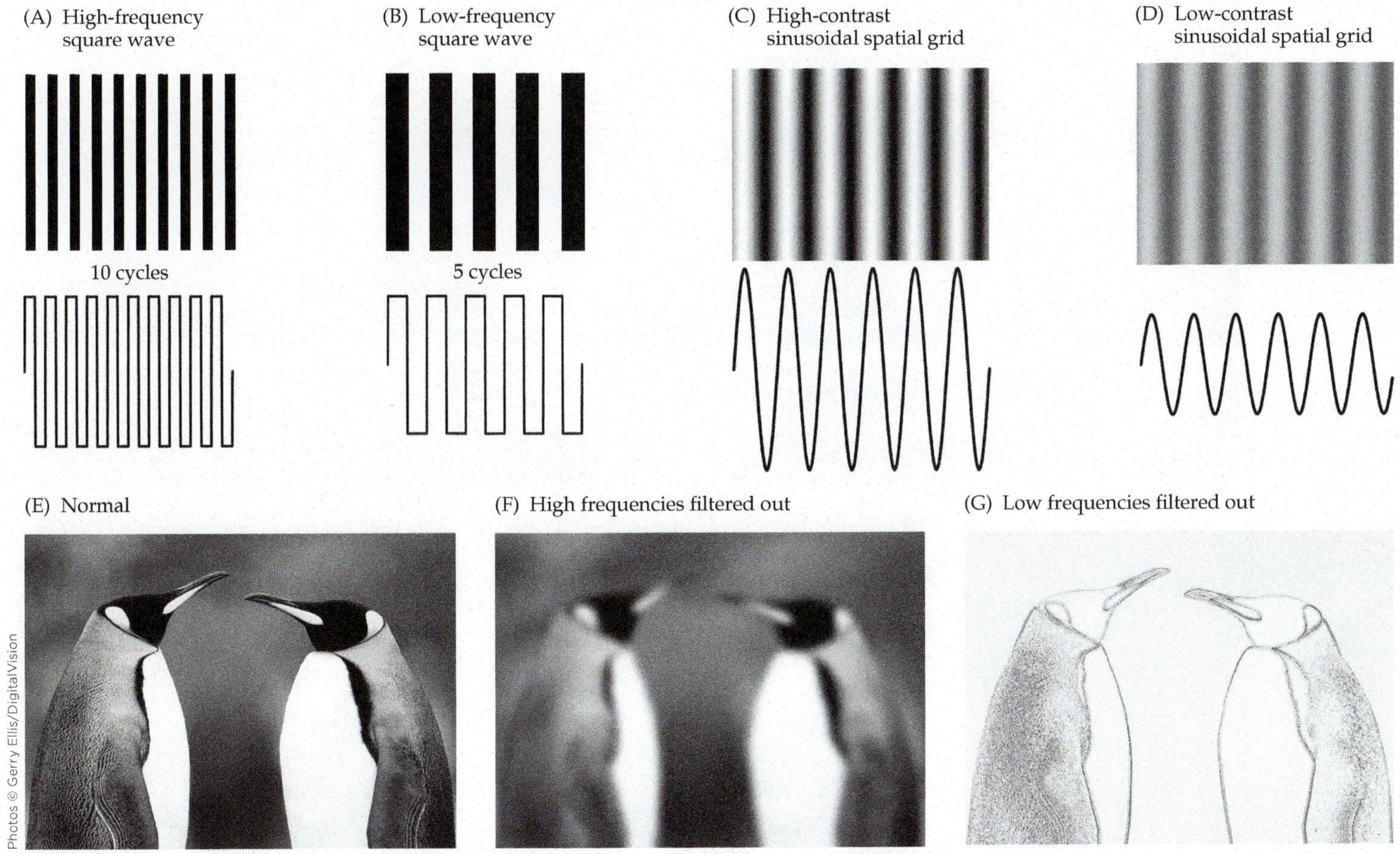

8.18 Spatial Frequencies (A, B) The spacing between dark and light stripes shows that the grating in part A has double the spatial frequency of the grating in part B. (C, D) These visual grids show sinusoidal modulation of intensity: (C) high contrast; (D) low contrast. (E) A photograph of two penguins. (F) Filtering out high spatial frequencies fuzzes out edges. (G) Conversely, when low spatial frequencies are filtered out, contrasting edges are emphasized.

The suggestion that the nervous system detects different spatial frequencies (F. W. Campbell and Robson, 1968) was soon supported by experiments on selective adaptation to spatial patterns. In these experiments a person spent a minute or more inspecting a visual grating with a given spacing (spatial frequency), such as those in Figure 8.18A and B. Looking at the grating made the cells that are tuned to that frequency adapt (become less sensitive). Then the person's sensitivity to gratings of different spacings was determined.

The results showed that sensitivity to the subsequent gratings was reduced briefly at the particular frequency to which the person had adapted. The suggestion of multiple spatial-frequency channels had revolutionary impact because it led to entirely different conceptions of how the visual system might go about dealing with spatial stimuli. The idea suggests that, rather than specifically detecting such seminaturalistic features as bars and edges, the system is breaking down complex stimuli into their individual spatial-frequency components in a kind of crude Fourier analysis (R. L. De Valois and De Valois, 1988, p. 320). In such a system, we might require a view of the whole face, which includes the low-frequency components, for recognition. This could explain why "Halle Berry neurons" do not respond to small portions of a face, because those have only high-frequency components.

Similarly, Leonardo da Vinci's *Mona Lisa* is famous because sometimes the model seems to be smiling but other times she doesn't (**FIGURE 8.19**). That ambiguity may be due to differences in spatial frequency (Livingstone, 2000). The low-spatial-frequency components of the picture (left two panels in the figure) make it look as if she is smiling, but the high-spatial-frequency components (right) give her a rueful, almost sad expression. As we run our eyes over the original, views from the fovea report the sad, high-frequency components; but views from the peripheral vision, with large receptor fields that can detect only low-frequency components, emphasize the smile (Bohrn et al., 2010). The spatial-frequency approach has proven useful in the analysis of many aspects of human pattern vision, and it provides the basis of high-definition television.

Go to Animation 8.3
Spatial Frequencies

8.19 Mona Lisa's Ambiguous Smile Leonardo da Vinci's Mona Lisa sometimes seems to be smiling, but other times she doesn't. Filtering that reveals very low spatial frequencies (left panel) or moderately low spatial frequencies (middle) makes it look as if she is smiling, but the high-spatial-frequency components (right) make her look almost sad. Is this why when we see her from the corner of our eye, she seems to be smiling, but when we look directly at her mouth, it conveys sadness?

Area V1 is involved in the formation of mental images

Neurons of V1 appear to be involved not only in perceiving objects and events, but also in forming mental images. For example, imagined objects activate regions that correspond to the retinotopic mapping of V1; when people imagined small letters, PET recording showed activation of the foveal representation; when they imagined large letters, the parafoveal representation was also activated (Kosslyn et al., 1993).

Impairing the function of V1 through repetitive transcranial magnetic stimulation (rTMS) also affected mental imagery. Before each set of trials in which participants formed and inspected mental images, rTMS was administered for 10 minutes. This stimulation did not prevent the formation of images, but it significantly impaired the process (Kosslyn et al., 1999), thus adding further evidence of the necessity of V1 for the formation of images.

Neurons in the visual cortex beyond area V1 have complex receptive fields and contribute to the identification of forms

Area V1 is only a small part of the portion of cortex that is devoted to vision. Area V1 sends axons to other visual cortical areas, including areas that appear to be involved in the perception of form: V2, V4, and the inferior temporal area (see Figure 8.20C). Some of these extrastriate areas also receive direct input from the LGN. The receptive fields of the cells in many of these extrastriate visual areas are even more complex than those of cells in area V1. Anatomical, physiological, and behavioural investigations with macaque monkeys reveal at least 32 distinct cortical areas that are directly involved in visual function (Van Essen and Drury, 1997) (**FIGURE 8.20A–C**).

The visual areas of the human brain (**FIGURE 8.20D**) have been less thoroughly mapped than those of the monkey brain, and mainly by neuroimaging (the spatial resolution of which is not as fine as that of the electrophysiological recording used in the monkey brain), but the general layout appears to be similar in the two species, especially for V1 (Tootell et al., 2003).

An astonishing proportion of primate cortex analyses visual information. The areas that are largely or entirely visual in function occupy about 55% of the surface of the macaque cortex, and about 30% of human cortex (Tootell et al., 2003). We will discuss only a few of the main visual cortical areas and their functions.

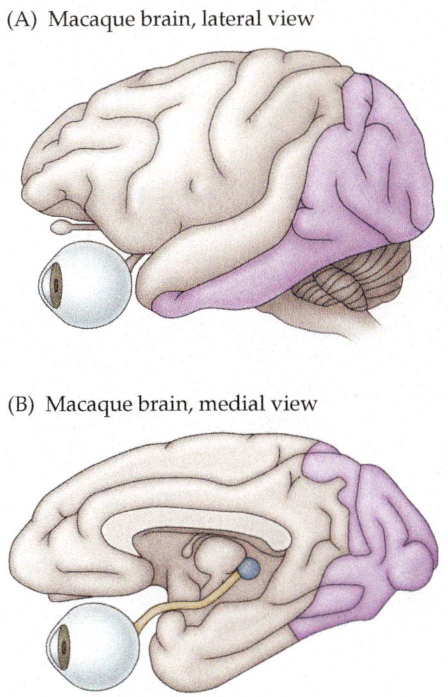

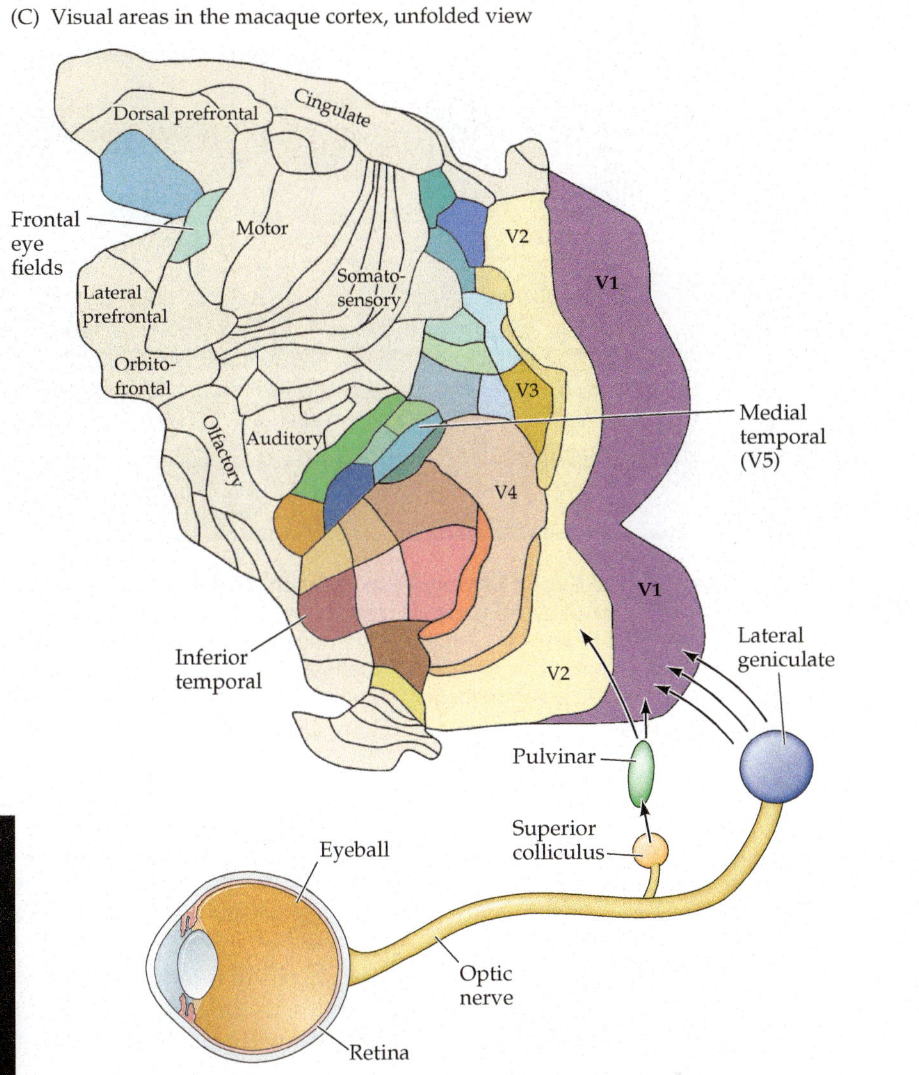

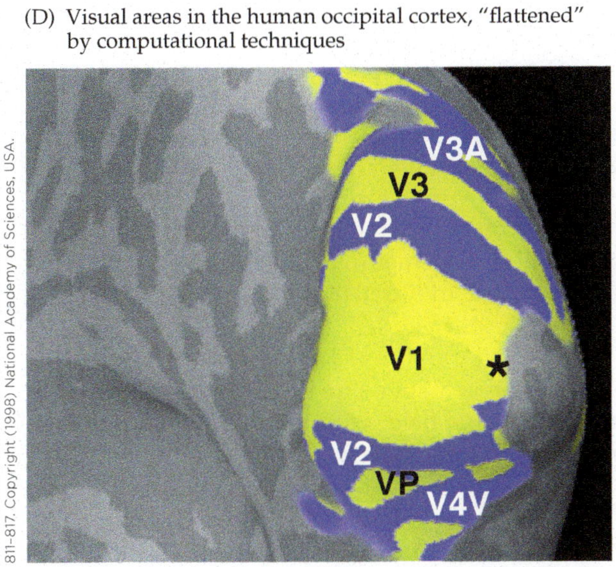

8.20 Main Visual Areas in Monkey and Human Brains (A, B) Macaque visual areas in occipital and temporal cortex are shown in purple. (C) All the known visual areas of the macaque on a flattened cortex in colour. (D) Through computational techniques, the occipital regions of human brain shown in Figure 8.12B were "inflated," flattening the brain and bringing sulci in the cortex to the surface, thus revealing the relative size and extent of various cortical visual areas. The asterisk identifies the representation of the centre of the fovea. (Parts A–C after D. J. Felleman and D. C. Van Essen, 1991. *Cereb Cortex* 1: 1–47)

Area V2 is adjacent to V1, and many of its cells respond to *textures* in naturalistic scenes (Freeman et al., 2013). Many V2 cells can respond to illusory contours, which may help explain how we perceive contours such as the boundaries of the upright triangle in **FIGURE 8.21** (Peterhans and von der Heydt, 1989). Clearly, such cells respond to complex relations among the parts of their receptive fields.

Area V4 receives axons from V2 and has cells that give their strongest responses to the sinusoidal frequency gratings that we discussed earlier (see Figure 8.18C

8.21 A Geometric Figure with "Illusory" or "Subjective" Contours Cells have been found in visual cortical areas that respond to illusory contours such as those of the upright triangle shown here. These contours thus have neurophysiological meaning.

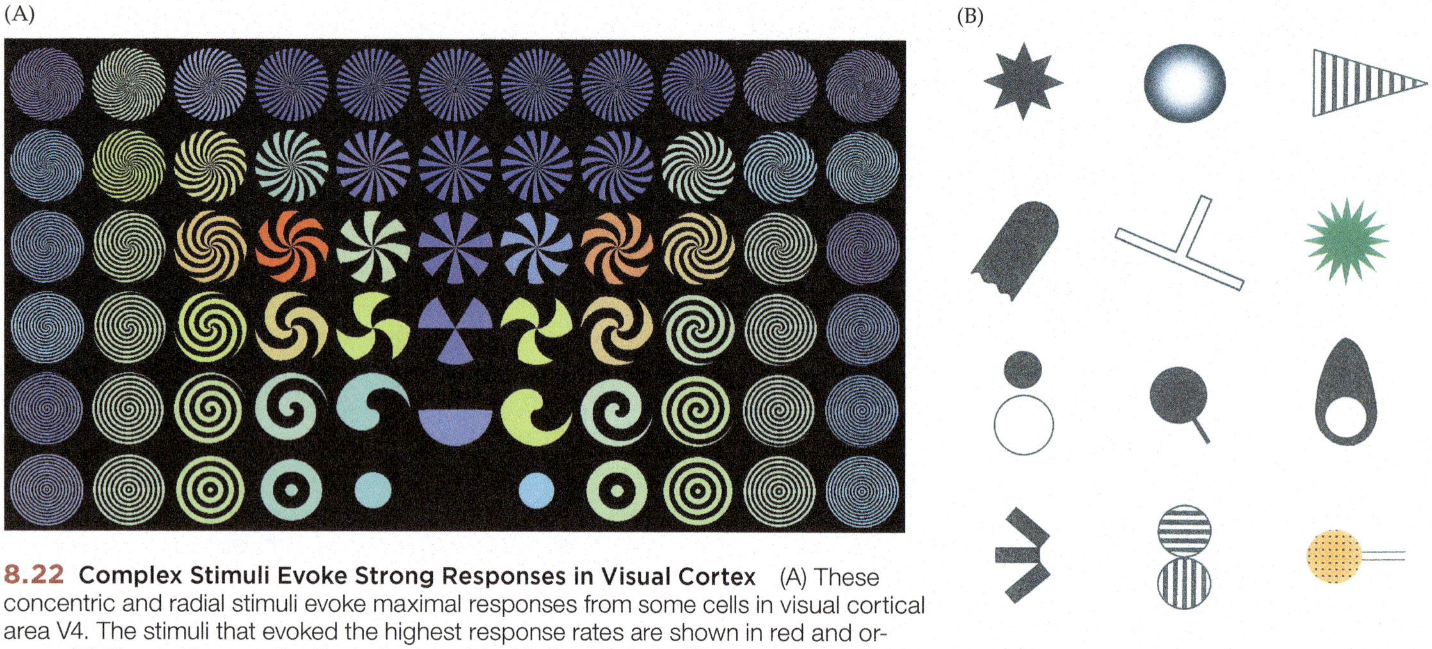

From J. L. Gallant et al., 1993. *Science* 259: 100–103

8.22 Complex Stimuli Evoke Strong Responses in Visual Cortex (A) These concentric and radial stimuli evoke maximal responses from some cells in visual cortical area V4. The stimuli that evoked the highest response rates are shown in red and orange. (B) These 12 examples illustrate the critical features of stimuli that evoke maximal responses from cells in the anterior inferior temporal area. (Part B from K. Tanaka, 1993. *Science* 262: 685–688.)

and D). However, many V4 cells give even stronger responses to concentric and radial stimuli, such as those in **FIGURE 8.22A** (Gallant et al., 1993). Area V4 also has many cells that respond most strongly to wavelength differences, as we will see later when we discuss colour vision. Area V5, also called the *medial temporal area* or *MT*, appears to be specialised for the perception of motion, as we will also discuss later in the chapter.

Most cells in inferior temporal visual cortex do not require a natural object such as a face to activate them; instead, they require moderately complex shapes, sometimes combined with colour or texture, such as those in **FIGURE 8.22B**.

Area V1 is organised in columns

The primary visual cortex has separate representations for at least four dimensions of the visual stimulus: (1) location in the visual field, with larger, finer mapping of the central region of the visual field than of the periphery; (2) ocular dominance; (3) orientation; and (4) colour.

Ocular dominance columns were first discovered by electrophysiological recording. Although the receptive field of an individual V1 neuron is the same for vision through either eye, some cells are equally activated by the two eyes while other cells respond preferentially (i.e., more strongly) to stimulation of one eye. However, all the cells in a vertical column of cells have the same ocular dominance. The vertical columns are arranged into **ocular dominance slabs** about 0.5 millimeters wide, all cells of which respond preferentially to stimulation of one eye. A given point in the visual field elicits responses in cells in adjacent left-eye-preferring and right-eye-preferring ocular dominance slabs. Ocular dominance is especially clear in layer IV, where each cell is monocular, responding to only one eye. Above and below the (monocular) ocular dominance stripes in layer IV, most of the cells respond to stimulation of both eyes but still prefer one eye over the other.

Optical imaging of cortical activity allows us to see the ocular dominance stripes in the primary visual cortex of an awake monkey (**FIGURE 8.23A AND B**). Ocular dominance columns develop during the first months of life in cats and monkeys. As we saw in Chapter 6, both eyes must be exposed to the visual environment if each eye is to obtain its own cortical representation (see Figure 6.22). Up to the age of 3 or 4 months, human infants are unimpressed by stereograms (pairs of pictures showing somewhat different left-eye and right-eye views that most adult observers perceive

ocular dominance column A region of cortex in which one eye or the other provides a greater degree of synaptic input.

ocular dominance slab A slab of visual cortex, about 0.5 mm wide, in which the neurons of all layers respond preferentially to stimulation of one eye.

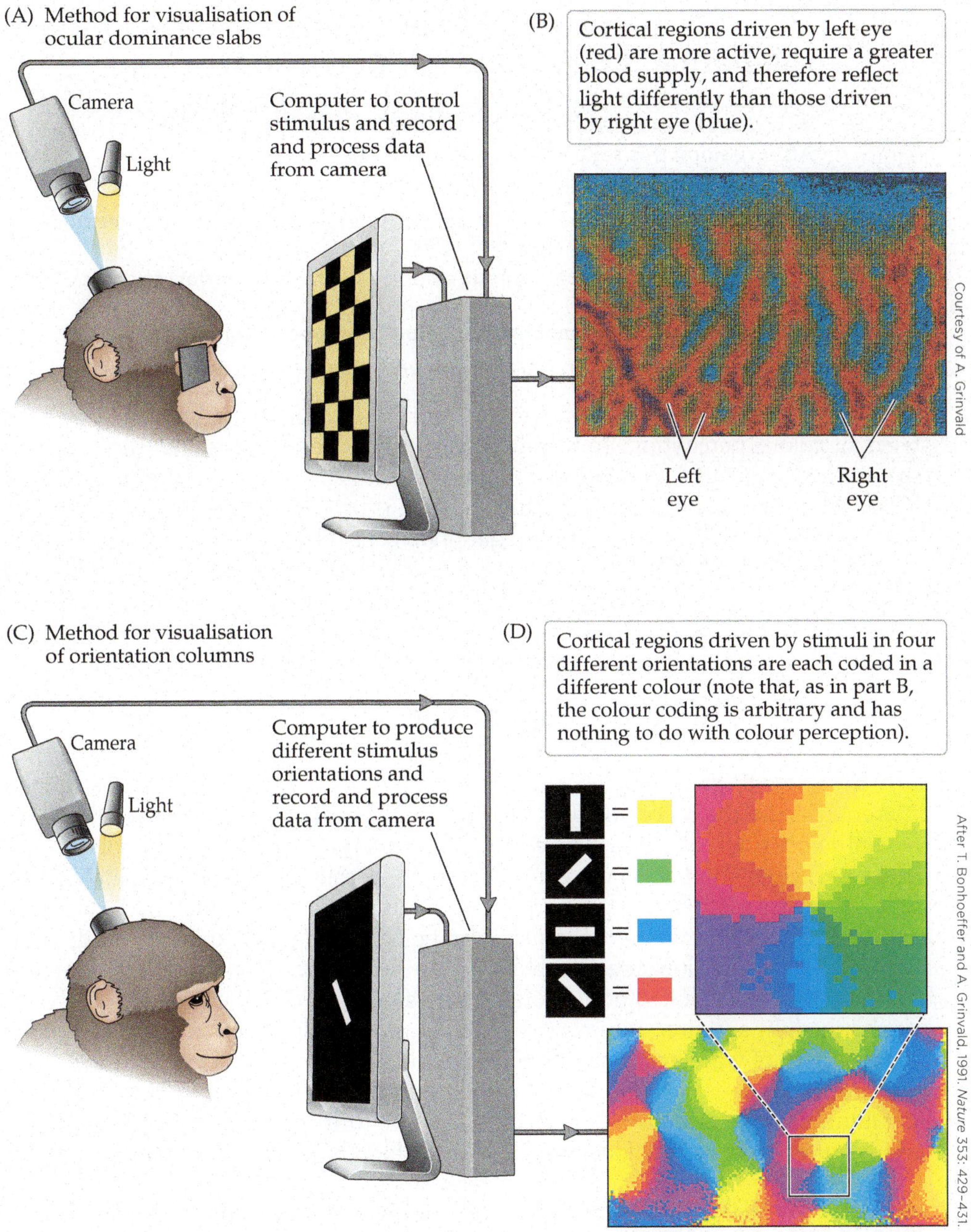

8.23 Visualisation of Ocular Dominance Columns and Orientation Columns by Optical Imaging (A) In this method for visualisation of ocular dominance, a camera records changes in light reflected from the cortex when the monkey views a twinkling checkerboard with one eye. Small differences in reflected light are amplified, and intensity is coded by colour (red for strong intensity, blue for weak). (B) After the recording is processed, regions activated by the active eye are seen as red stripes. (C) In this method for visualisation of orientation preference, stimuli at different orientations (vertical, horizontal, diagonal) are presented to reveal groups of neurons that respond most strongly to a particular orientation. The stimuli are usually black or white, but here they are colour coded to correspond to colour-coded responses to four different orientations combined into a single pattern. (D) Although the pattern at first seems disorderly, closer inspection reveals several regions where four orientations converge in a pinwheel pattern (inset). The foci of pinwheels occur at regular intervals, each orientation is represented only once within a pinwheel, and the sequence of orientations is consistent across pinwheels. (Parts A and C after A. Grinvald et al., 1991. *Proc Natl Acad Sci USA* 88: 11559–11563.)

as a three-dimensional view). Beginning at the age of 3 or 4 months, however, most infants are captivated by stereograms (Held, 1993). Presumably, before that age the cortex is unable to separate the information from the two eyes because the information reaches the same cortical neurons.

Primary visual cortex has a columnar organisation for stimulus orientation as well (**FIGURE 8.23C AND D**). That is, in one **orientation column** all the cells may be "tuned" to upright stimuli (at an orientation of 0°); in an adjacent column, all cells may respond best to another orientation, perhaps at 10° from the vertical; in the next column, perhaps at 25°; and so forth. There are also columns of cells in V1 that respond to particular spatial frequencies (Nauhaus et al., 2012), which we discussed earlier.

These columns are Organised parallel to the surface of the cortex. In Figure 8.23D, optical recordings of the surface show the regions of primary visual cortex that respond best to stimuli of four different orientations. Adjacent orientation columns shift by 90°, creating regularly spaced "pinwheels" in which responses to the different stimulus orientations pivot around a centre (Paik and Ringach, 2011).

orientation column A column of visual cortex that responds to rod-shaped stimuli of a particular orientation.

8.5 Colour Vision Depends on Special Channels from the Retinal Cones through Cortical Area V4

Learning Objectives

After reading this section, you should be able to:

8.5.1 **Describe the various dimensions of colour.**

8.5.2 **Offer illusions that demonstrate that our perception of colour does not simply reflect the wavelength of light.**

8.5.3 **Contrast the trichromatic versus opponent-process models for colour discrimination.**

8.5.4 **Explain why there are no "red" or "blue" or "green" cones.**

8.5.5 **List the four types of spectral opponency seen in LGN neurons.**

For most people, colour is a fundamental aspect of vision. Here we discuss several stages of colour perception. In the first stage, the cones—the retinal receptor cells that are specialised to respond to certain wavelengths of light—receive visual information. In the second stage, this information is processed by neurons in the local circuits of the retina, leading to retinal ganglion cells that are excited by light of some wavelengths and inhibited by light of other wavelengths. The ganglion cells send the wavelength information via their axons to the LGN, mainly in the parvocellular layers. From there this information goes to area V1, from which it is relayed to other visual cortical areas, where additional stages of colour perception take place.

Colour is created by the visual system

What most people mean when they use the term *colour* is **hue**, which varies continuously from blue to green, yellow, and red and their intermediates. These hues appear different because the reflected light that reaches our eyes can vary in wavelength, and we can detect some of these differences. Perception of hue does not simply depend upon a particular stimulus (a wavelength of light), because we perceive a patch illuminated by a particular wavelength as various different hues, depending on factors such as the intensity of illumination, the surrounding field, and prior exposure to a different stimulus. As illumination fades, the blues in a painting or a rug appear more prominent and the reds appear duller, even though the wavelength distribution in the light has not changed. In addition, the hue perceived at a particular point is strongly affected by the pattern of wavelengths and intensities in other parts of the visual field. For example, our interpretation of a particular hue depends on what we perceive about the light illuminating the scene (**FIGURE 8.24A–C**). To understand how the visual system creates our experience of colour, we must understand how cone photoreceptors work.

Colour perception requires receptor cells that differ in their sensitivities to different wavelengths

Artists have long known that all the hues can be obtained from a small number of primary colours. On the basis of observations of mixing pigments and lights, nineteenth-century scientists hypothesised that three separate kinds of receptors in the retina provide the basis for colour vision. Endorsed in 1852 by the great physiologist-physicist-psychologist Hermann von Helmholtz, this **trichromatic hypothesis** (from the Greek *tri-*, "three," and *chroma*, "colour") became the dominant position.

Helmholtz predicted that blue-sensitive, green-sensitive, and red-sensitive receptors would be found; that each would be sharply tuned to its part of the spectrum; and that each type would have a separate path to the brain. The colour of an object would be recognised, then, on the basis of which colour receptors were activated. This system would be like the mechanisms for discriminating touch and temperature on the basis of which skin receptors and labelled neural lines are activated (see Chapter 7).

Later in the nineteenth century, physiologist Ewald Hering proposed a different explanation. He argued, on the basis of visual experience, that there are four unique hues and three opposed pairs of colours—blue versus yellow, green versus red, and black versus white—and that three physiological processes with opposed positive and negative values must therefore be the basis of colour vision. As we will see, both this

hue A dimension of light perception, varying around the colour circle through blue, green, yellow, orange, and red.

trichromatic hypothesis A hypothesis of colour perception stating that there are three different types of cones, each excited by a different region of the spectrum and each having a separate pathway to the brain.

8.24 Colour Perception Our perception of colour is not a simple reflection of the wavelength of light reaching the retina, as the following illusions show. (A) The young woman's two eyes are the same shade of gray in both pictures. But when a red screen is placed over one eye, we perceive the gray eye as blue, while superimposing a blue screen causes us to perceive the gray eye as red. (B) Here the dotted squares in the figures reflect the exact same wavelength of light, but most of us perceive one as green and the other as yellow. (C) The dress that broke the internet. People perceive this dress as either white and gold or as blue and black, depending on how they interpret the lighting of the scene. If you cut a rectangle 2 cm × 1 cm out of a sheet of paper and place the opening over the dress, you'll see it's the same colour in the two panels.

opponent-process hypothesis and the trichromatic hypothesis are encompassed in current colour vision theory, but neither old hypothesis is sufficient by itself.

Measurements of photopigments in cones have borne out the trichromatic hypothesis, in part. Each cone of the human retina has one of three types of pigments. These pigments do not, however, have the narrow spectral distributions that Helmholtz predicted. The colour system that Helmholtz postulated would have given rather poor colour vision and poor visual acuity. Colour vision would be poor because only a few different hues would be discriminable; within the long-wavelength region of the spectrum, there would be only red, not the whole range of hues that we see. Acuity would be poor because the grain of the retinal mosaic would be coarse; a red stimulus would be able to affect only one-third of the receptors. (In reality, though, acuity is as good in red light as it is in white light.)

The human visual system does not have receptors that are sensitive to only a narrow part of the visible spectrum. Two of the three retinal cone pigments show some response to light of almost *any* wavelength. The pigments have different *peaks* of sensitivity, but even the peaks are not as far apart as Helmholtz predicted. As **FIGURE 8.25** shows, the cone pigment peaks occur at about 420 nm (in the part of the spectrum where we usually see violet under photopic conditions), about 530 nm (where most of us see green), and about 560 nm (where most of us see yellow-green). Despite Helmholtz's prediction, none of the curves peak in the long-wavelength part of the spectrum, where most of us see red (about 630 nm).

Under ordinary conditions, almost any visual object stimulates at least two kinds of cones, thus ensuring high visual acuity and good

opponent-process hypothesis The idea that colour vision depends on systems that produce opposite responses to light of different wavelengths.

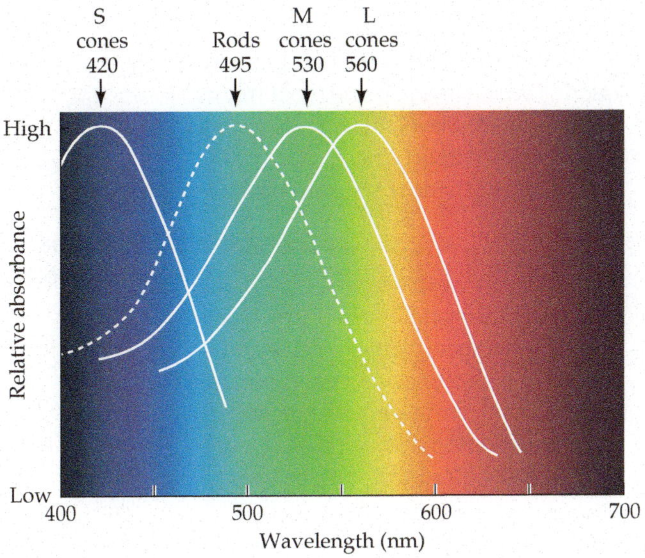

8.25 Spectral Sensitivities of Human Photopigments Each pigment has a different peak sensitivity (S, short-wavelength; M, medium-wavelength; L, long-wavelength) but responds to a wide range of wavelengths. Knowing only that an M cone is active, you cannot tell whether it was stimulated by weak light at 530 nm ("green") or by strong light at anywhere from 450 nm ("blue") to 620 nm ("red"). Only by comparing responses of different cones can the brain extract colour information.

perception of form. The spectral sensitivities of the three cone types differ from each other, and the nervous system detects and processes these *differences between cones* to extract the colour information. Thus, certain ganglion cells and certain cells at higher stations in the visual system are colour-specific, even though the receptor cells are not. Similarly, photoreceptors are not form-specific (they respond to single points of light), but form is detected later in the system from the joint activity of multiple different receptors.

Because the cones are not colour detectors, the most appropriate brief names for them can be taken from their peak areas of wavelength sensitivity: *short* (*S*) for the receptor with peak sensitivity at about 420 nm, *medium* (*M*) for peak sensitivity at about 530 nm, and *long* (*L*) for 560 nm (see Figure 8.25). There are typically twice as many L as M receptors, but far fewer S receptors (Brainard et al., 2000; Carroll et al., 2000); this difference explains why acuity is much lower with short-wavelength illumination (blue light) than in the other parts of the visible spectrum. Inheriting genes that cannot produce functional opsins (sometimes called *photopigments*) can affect colour discrimination, as **BOX 8.2** explains.

BOX 8.2 Most Mammalian Species Have Some Colour Vision

Even though the term *colour blindness* is commonly used, most people with impaired colour vision, about 8% of human males and about 0.5% of females, are able to distinguish some hues. Complete colour blindness can be caused by brain lesions or by the congenital absence of specialised receptors, but in humans it is extremely rare.

Animals exhibit different degrees of colour vision. Many species of birds, fishes, and insects have excellent colour vision. Humans and Old World monkeys also have an excellent ability to discriminate wavelengths, but many other mammals (e.g., cats) cannot discriminate wavelengths very well. Still, most mammals have at least some degree of colour vision (G. H. Jacobs, 1993). Although only certain primates have good *tri*chromatic colour vision based on *three* classes of cone photopigments (each of which is an opsin), most mammalian species have at least *di*chromatic colour vision (based on *two* classes of cone pigments). Most so-called colour-blind (actually colour-*deficient*) people have dichromatic vision and can distinguish short-wavelength stimuli (blue) from long-wavelength stimuli (not blue). When a gene encoding a third photopigment was introduced into photoreceptors of adult male squirrel monkeys with such dichromatic vision, they soon displayed excellent trichromatic vision (Mancuso et al., 2009). Likewise, introducing photopigment genes in adult mice (G. H. Jacobs et al., 2007) enabled them to discriminate colours they normally could not see, so it may be possible to correct dichromatic vision in people one day (Neitz and Neitz, 2014).

There is a continuum of colour vision capabilities, including at least four categories among mammalian species:

1. *Excellent trichromatic colour vision* is found in diurnal primates such as humans and the rhesus monkey.
2. *Robust dichromatic colour vision* is found in species that have two kinds of cone photopigments and a reasonably large population of cones. Examples of such species include the dog, the pig, and males of the South American squirrel monkey.
3. *Feeble dichromatic colour vision* occurs in species that have two kinds of cone pigments but very few cones. An example is the domestic cat.
4. *Minimal colour vision* is possessed by species that have only a single kind of cone pigment and that must rely on interactions between rods and cones to discriminate wavelength. Examples include the owl monkey and the raccoon.

Among those species of South American monkeys that are generally dichromats, some females are actually trichromatic. Why? Because the gene encoding one photopigment is on the X chromosome. Since females have two X chromosomes, if the two chromosomes carry *different* genes for the photopigment, the female will possess a total of three different kinds of cones and will therefore have trichromatic vision.

Unlike South American monkeys that carry a single gene on the X chromosome for photopigment, many other primates, like humans, have *two different* photopigment genes on the X chromosome: one for medium- and one for long-wavelength cones. The short-wavelength photopigment gene is on an autosome rather than on the X. Women are much less likely than men to have colour deficiency, because if one of their X chromosomes has a defective gene for either photopigment, the copy on the other X is likely to be fine. Because men have only one X, if they have a defective copy of the gene for the medium- or long-wavelength pigment, there's no backup copy to compensate. They will see blue (light stimulating the short-wavelength cones) and not blue (light stimulating the other functioning cones), as the figure illustrates. City University of London offers a video that will test your ability to discriminate colours; see Activity 8.2

Interestingly, a woman may carry slightly different genes for the long-wavelength photopigment on her

Some retinal ganglion cells and parvocellular LGN cells show spectral opponency

Recordings made from retinal ganglion cells in Old World monkeys reveal a second stage of processing of colour vision. Most ganglion cells and neurons in the parvocellular layers of the LGN fire in response to some wavelengths and are inhibited by other wavelengths. In other words, they show the sort of opponent process that Hering predicted.

FIGURE 8.26A shows the response of an LGN cell as a large spot of light centred on its receptive field changes from one wavelength to another. The neuron responds to wavelengths above 600 nm, where the L cones are most sensitive, and is inhibited at shorter wavelengths, where the L cones are less sensitive than the M cones. A cell exhibiting this response pattern is therefore called a *plus L/minus M cell* (*+L/–M*). This is an example of a **spectrally opponent cell** because two regions of the spectrum have opposite effects on the rate at which the neuron fires action potentials.

spectrally opponent cell A visual receptor cell that has opposite firing responses to different regions of the spectrum.

Each spectrally opponent ganglion cell receives input from two or three different kinds of cones through bipolar cells. The connections from at least one type of cone are

two X chromosomes (Neitz et al., 1998) and therefore have *four* different kinds of cones. Such so-called *tetra*chromats tend to be very good at discriminating colours and very sensitive to clashing colours (G. Jordan et al., 2010). For a real mind-bender, consider what colour perception might be like for mantis shrimp, which have 12 different photoreceptors, each responding to a different range of wavelengths (Thoen et al., 2014)!

Go to Activity 8.2
Colour Vision Test

(A)

Photo by David McIntyre; simulation created using software from Vischeck (www.vischeck.com)

(B)

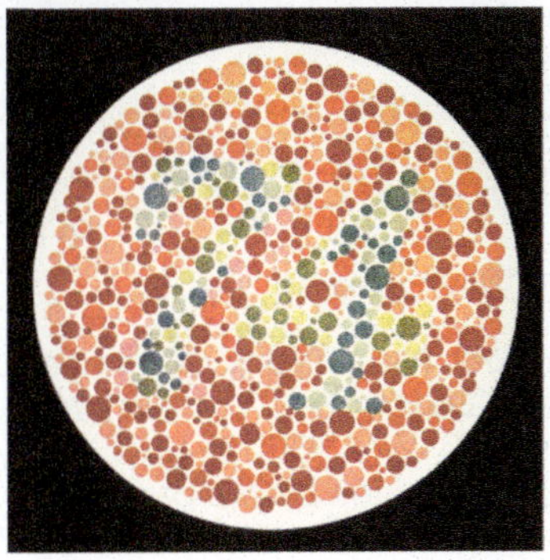

© Brand X Pictures/Alamy Stock Photo

Simulating Colour Blindness The photograph on the right has been adjusted to simulate the experience of the most common form of colour blindness in humans, which is the absence of cones sensitive to medium-wavelength light (M cones). For such individuals, the world's colours consist of blue (detected by short-wavelength photopigment encoded on chromosome 7) and not blue (detected by the long-wavelength photopigment encoded on the X chromosome). (B) In a typical test for colour vision, dichromats may have a difficult time detecting the numerals displayed in a figure like this. (A [right] was produced by software available from Vischeck.)

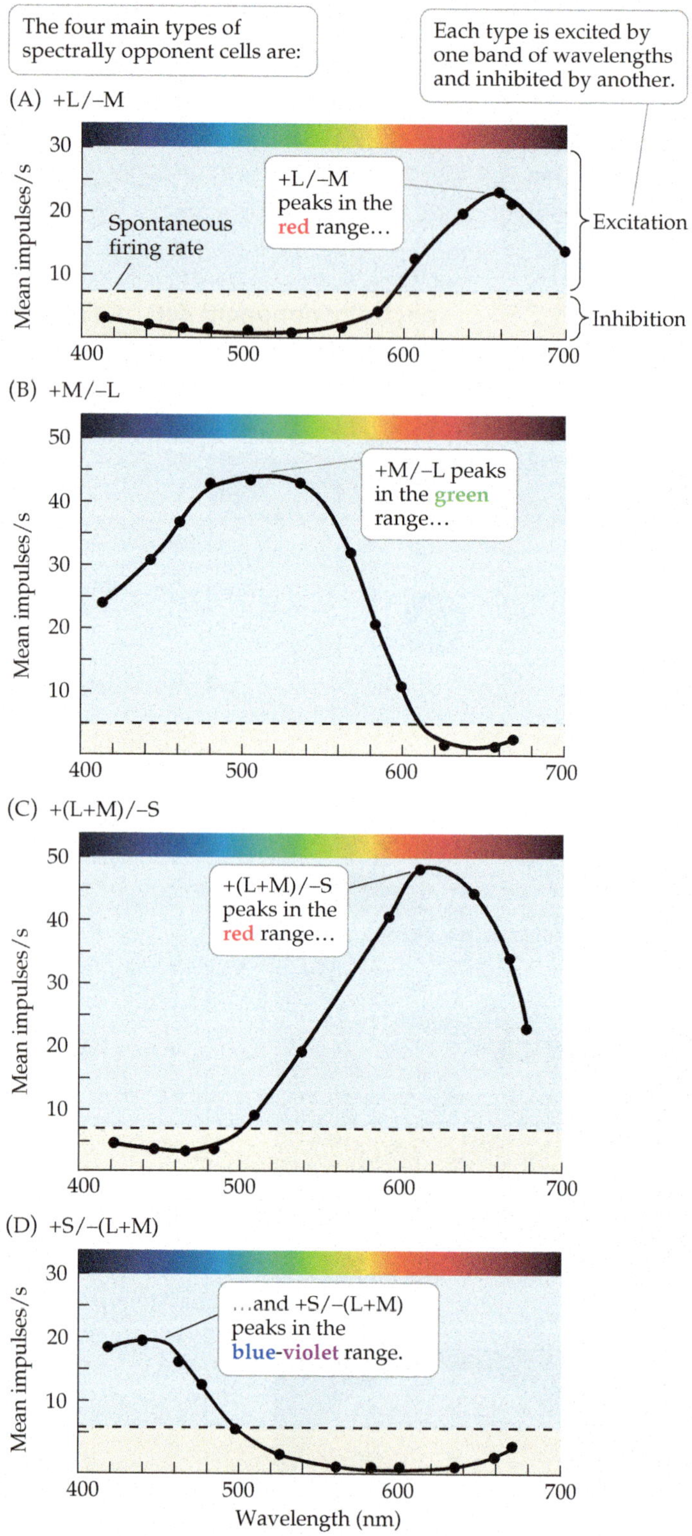

8.26 Responses of the Four Main Types of Spectrally Opponent Cells in Monkey LGN The four main types of spectrally opponent cells are (A) +L/–M, (B) +M/–L, (C) +(L+M)/–S, and (D) +S/–(L+M). Each type is excited by one band of wavelengths and inhibited by another. (After R. L. De Valois and K. K. De Valois, 1993. *Vision Res* 33: 1053–1065.)

excitatory, and those from at least one other type are inhibitory. The spectrally opponent ganglion cells thus record the *difference* in stimulation of different populations of cones. For example, a plus M/minus L (+M/–L) cell responds to the difference in the excitation of M and L cones.

The peaks of the sensitivity curves of the M and L cones are not very different (see Figure 8.25). However, whereas the M-minus-L difference curve (**FIGURE 8.26B**) shows a clear peak at about 500 nm (in the green part of the spectrum), the L-minus-M difference function (see Figure 8.26A) shows a peak at about 650 nm (in the red part of the spectrum). Thus, +M/–L and +L/–M cells yield distinctly different neural response curves. LGN cells excited by the L and M cells but inhibited by S cells—that is, +(L+M)/–S cells—peak in the red range (**FIGURE 8.26C**); while cells excited by S but inhibited by L and M—that is, +S/–(L+M) cells—peak in the blue-violet range (**FIGURE 8.26D**). These spectrally opponent neurons are the second stage in the system for colour perception, but they still cannot be called *colour cells,* because their peak wavelength sensitivities do not correspond precisely to the wavelengths that we see as the principal hues.

FIGURE 8.27 diagrams the presumed inputs to not only the four kinds of spectrally opponent ganglion cells, but also the ganglion cells that detect brightness and darkness. The brightness detectors receive stimulation from both M and L cones (+M/+L); the darkness detectors are inhibited by both M and L cones (–M/–L).

In the monkey LGN, 70–80% of the cells are spectrally opponent; in the cat, very few spectrally opponent cells are found—only about 1%. This difference explains why it is easy to train monkeys to discriminate between stimuli of different colours, and why it is so hard to train cats to notice even large differences in colour.

Some visual cortical cells and regions appear to be specialised for colour perception

In the cortex, spectral information appears to be used for various kinds of information processing. Forms are segregated from their background by differences in colour or intensity (or both). The most important role that colour plays in our perception is to denote which parts of a complex image belong to one object and which belong to another. Some animals use displays of brightly coloured body parts to call attention to themselves, but colour can also be used as camouflage.

Some *spectrally opponent cortical cells* contribute to the perception of colour, providing a third stage of the colour vision system. Adding and subtracting the outputs of spectrally opponent ganglion cells can yield cortical cells that are perceptually opponent: red versus green, blue versus yellow, and black versus white (R. L. De Valois and De Valois, 1993). The spectral responses of these cells correspond to the wavelengths of the principal hues specified by human observers. Visual cortical region V4 is particularly rich in colour-sensitive cells (Schein and Desimone, 1990). These cells may be important for colour constancy and for discrimination between a figure and background (Zeki et al., 1991), but cells in V4 are also tuned in the spatial domain, for orientation and for spatial frequency (Tanigawa et al., 2010).

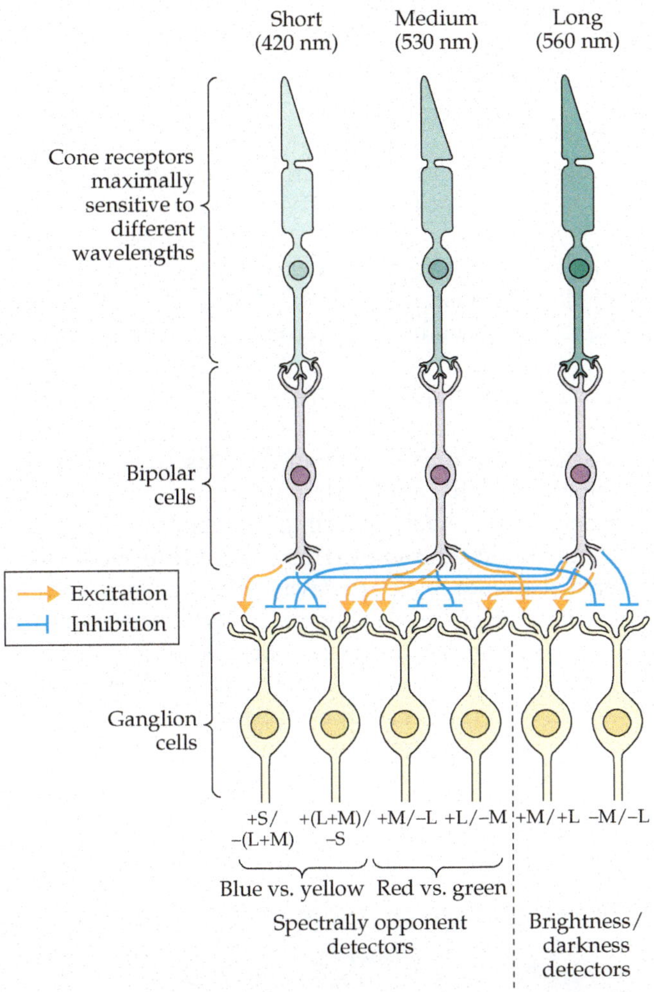

8.27 A Model of the Connections of Wavelength Discrimination Systems in the Primate Retina The connections from the cones yield four kinds of spectrally opponent ganglion cells, as well as ganglion cells that detect brightness or darkness. (After R. L. De Valois and K. K. De Valois, 1980. *Ann Rev Psychol* 31: 309–341.)

8.6 The Many Cortical Visual Areas Are Organised into Two Major Streams

Learning Objectives

After reading this section, you should be able to:

8.6.1 **Describe evidence for separable systems analyzing what we see and where it is, and relate this evidence to the case of D.F. from the opening of the chapter.**

8.6.2 **Name the cortical region that detects motion in the visual field, and describe the results of damage to this area.**

8.6.3 **Describe mirror neurons and how they might relate to behaviours like empathy and autism.**

Many investigators have wondered why primate visual systems contain so many distinct regions. Certain regions specialise in processing different attributes of visual experience (such as shape, location, colour, motion, and orientation). But the number of visual fields—over 30—is larger than the number of basic attributes. Perhaps the reason that so many separate visual regions have been found is simply that investigators, being visually oriented primates themselves, have lavished special attention on the visual system.

Separate systems analyse what we see and where we see it

Earlier work with hamsters led to the hypothesis that there are two visual systems: one, for *identification* of objects, involves especially the visual cortex; the other, for *location* of objects, involves especially the superior colliculus (Kravitz et al., 2011). Mortimer Mishkin and Leslie Ungerleider (1982) proposed that primates also have

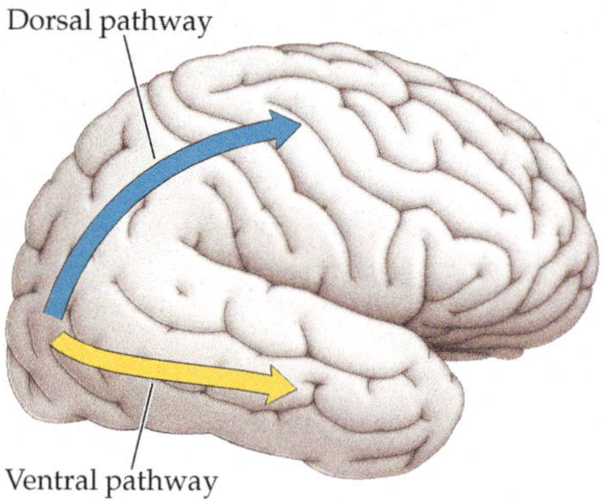

8.28 Parallel Processing Pathways in the Visual System The dorsal *where* pathway (shown in blue) and the ventral *what* pathway (shown in yellow) serve different functions.

two main cortical processing streams, both originating in primary visual cortex: a ventral processing stream responsible for visually identifying objects, and a dorsal stream responsible for perceiving the location of objects and for guiding movement toward them (**FIGURE 8.28**). These processing streams were called, respectively, the *what* and *where* streams (Ungerleider and Pasternak, 2004). The two streams are not completely separate, because there are normally many cross-connections between them. In the ventral stream, including regions of the occipitotemporal and inferior temporal areas, information about faces is analysed (Keller et al., 2017), as we'll discuss in further detail in Chapter 17.

These separate visual cortical streams help us understand the case of patient D.F., described at the start of this chapter. Recall that after carbon monoxide poisoning, she lost the ability to perceive faces and objects while retaining the ability to reach and grasp under visual control. The investigators who studied her (A. D. Milner et al., 1991) hypothesised that D.F.'s ventral visual stream had been devastated but that her dorsal stream was unimpaired. An opposite kind of dissociation had already been reported: damage to the posterior parietal cortex often results in **optic ataxia** in which patients have difficulty using vision to reach for and grasp objects, yet these patients may retain the ability to correctly identify objects (Perenin and Vighetto, 1988).

optic ataxia Difficulty using vision to reach and manipulate objects.

Investigation with MRI supports the idea that D.F.'s ventral stream was damaged while her dorsal stream was relatively unaffected (T. W. James et al., 2003).

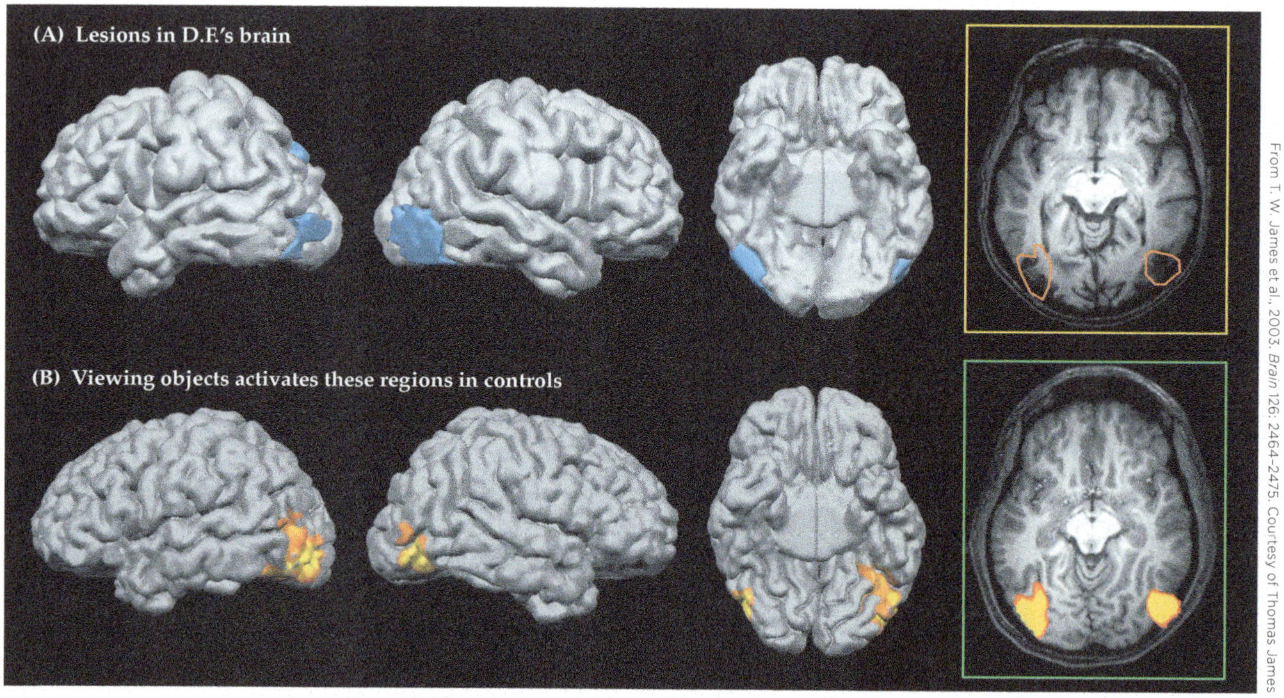

8.29 Object Recognition Centres in the Brain (A) In this reconstruction from MRI images, the brain region that was damaged in patient D.F. (blue) is seen from lateral views and from below (outlined in orange on the right). (B) In neurologically intact people, this same brain region is activated (yellow) when they are looking at recognisably intact pictures of various objects rather than scrambled pictures. D.F.'s inability to recognise the objects she sees appears to be due to the damage to this region on the border of the occipital and temporal cortices.

High-resolution MRI of D.F.'s brain (**FIGURE 8.29A**) reveals damage concentrating in the ventrolateral occipital cortex. Throughout her brain, there is evidence of atrophy, indicated by shrunken gyri and enlarged sulci. **FIGURE 8.29B** shows the area activated in fMRI recordings when controls viewed pictures of objects; it corresponds to D.F.'s lateral occipital lesion. When D.F. reached for and grasped objects, her fMRI activation in the parietal lobe was similar to that of controls, indicating that her dorsal stream is largely intact. D.F.'s intact dorsal pathway not only tells her where objects are but also guides her movements to use these objects properly.

It is still puzzling that one part of D.F. knows exactly how to grasp a pencil put before her, while another part of her—the part that talks to you—has no idea whether it's a pencil, a ruler, or a bouquet of flowers. Imagining what this disjointed visual experience must be like for D.F. allows us to appreciate how effortlessly our brains usually bind together information to give us the marvelous sense of sight.

Perception of visual motion is analysed by a special system that includes cortical area V5

Some retinal ganglion cells respond preferentially to a certain direction of motion of objects; for example, they may respond to stimuli that move to the left but not to stimuli that move to the right. Motion is analysed by the cortex, partly in regions close to those that control eye movements. In monkeys, all of the neurons in area V5 (MT) (see Figure 8.20C), an early portion of the dorsal *where* stream, are specialised for the perception of motion and its direction. PET studies show that moving stimuli also evoke responses in human area V5 (Zeki, 2015).

Experimental lesions of area V5 in monkeys trained to report the direction of perceived motion impaired their performance, at least temporarily (M. R. Cohen and Newsome, 2008). Altering the activity of V5 neurons can affect a monkey's perception of movement. Electrically stimulating an area of V5 that normally responds to stimuli moving upward caused the monkeys to report that dots on the screen were moving upward even when they were actually moving in another direction.

One striking case study described a woman who had lost the ability to perceive motion after a stroke damaged her area V5 (Zihl et al., 1983). The woman was unable to perceive continuous motion and saw only separate, successive still images. This impairment led to many problems in her daily life. She had difficulty crossing streets, because she could not follow the positions of automobiles in motion: "When I'm looking at the car at first, it seems far away. But then when I want to cross the road, suddenly the car is very near." She also complained of difficulties in following conversations because she could not see the movements of speakers' lips.

The anterior part of the dorsal stream includes mirror neurons

The anterior part of the dorsal visual stream merges with the motor cortex and includes neurons that have both visual and motor functions. These neurons were discovered by investigators who were recording the activity of neurons in the motor cortex of monkeys. To their surprise, the investigators found cells that fired not only when the monkey grasped and moved an object but also when the monkey observed *another monkey*, or even a person, grasping and moving an object.

It took the investigators several years to believe and publish their finding (Di Pelligrino et al., 1992). They proposed that these **mirror neurons** mediate the understanding of actions performed by others: when an individual sees an action performed by another, neurons that represent that action are activated in the observer's premotor cortex (see Figure 8.18) (Rizzolatti and Craighero, 2004).

Several studies provide data for the existence of a mirror neuron system in humans (e.g., Buccino et al., 2004). Researchers have speculated that mirror neurons could be involved in domains as varied as empathy, learning by imitation, and the evolution of language; damage to the mirror neuron system has been proposed as a cause of autism, which is characterized by impairments in understanding other people's behaviour. We will consider additional aspects of mirror neurons in Chapter 10 in the context of the motor system.

mirror neuron A neuron that is active both when an individual makes a particular movement and when that individual sees another individual make that same movement.

myopia Nearsightedness; the inability to focuss the retinal image of objects that are far away.

8.7 Visual Neuroscience Can Be Applied to Alleviate Some Visual Deficiencies

Learning Objectives

After reading this section, you should be able to:

8.7.1 **Describe how improper focussing of images on the retina causes myopia.**

8.7.2 **Discuss theories of why myopia is more common than it used to be.**

8.7.3 **Describe amblyopia and how it is treated.**

8.7.4 **Describe macular degeneration and ways it might be treated in the future.**

We humans are a very vision-oriented species, so it's no surprise that a great deal of research effort goes into finding ways to prevent impairment, to improve inadequate vision, and to restore sight to the blind. In the United States, half a million people are blind. Recent medical advances have reduced some causes of blindness but have increased blindness from other causes. For example, medical advances permit people with diabetes to live longer, but because we don't know how to prevent blindness associated with diabetes, there are more people alive today with diabetes-induced blindness. We'll first consider ways of avoiding impairment of vision. Then we'll take up ways of training that are designed to improve an impaired visual system.

Impairment of vision often can be prevented or reduced

Studies of the development of vision in children and nonhuman animals show that the incidence of **myopia** (nearsightedness, from the Greek *myein*, "to be closed," and *ops*, "eye") can be reduced. Myopia develops if the eyeball is too long, causing the eye to focuss objects in front of the retina rather than on the retina (**FIGURE 8.30**). As a result, distant objects appear blurred. Considerable evidence suggests that the reason some children develop myopia is that certain environmental factors cause the eyeball to grow excessively. Previously it was thought that the modern habit of looking closely at nearby objects (books, computer screens, and so on) might be responsible for myopia (Marzani and Wallman, 1997), but mounting evidence suggests that indoor lighting may be to blame.

Before civilisation, most people spent the bulk of their time outdoors, looking at objects illuminated by sunlight. But with the advent of indoor lighting, we've come to spend a lot of time looking at things with light that, while containing many

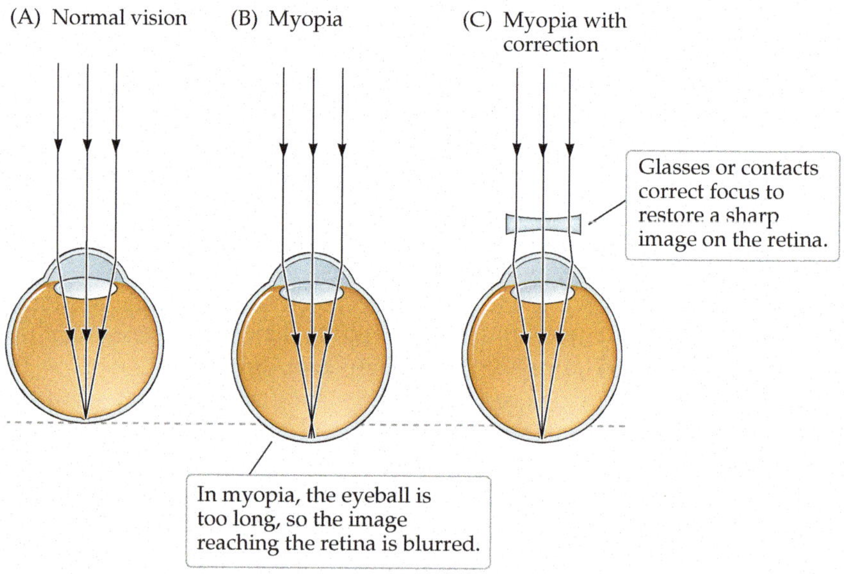

8.30 Focussing Images on the Retina
(A) Normally, the cornea and lens refract light to focuss a sharp image of the outside world on the retina. (B) In myopia, the eyeball is too long, so the image is in focuss in front of the retina. In this case, the image that actually reaches the retina is blurred. (C) Eyeglasses refract the light before it reaches the cornea, to bring the image into sharp focuss on the retina.

wavelengths, does not exactly match the composition of sunlight and is much less bright than outdoor light. Several studies found that children with myopia spend less time outdoors than do other children, but that correlation could be caused by genes that favor both myopia and indoor activities, like reading. Indeed, the advent of public schools in various nations is accompanied by increased rates of myopia. However, one of these studies focussed on people of Chinese origin who lived in either Singapore, where crowded conditions mean that people spend little time outdoors, or Sydney, Australia. Even though these populations should be genetically very similar, 30% of the Chinese children living in Singapore, who average only 30 minutes a day outdoors, were myopic, versus only 3% of those living in Sydney, who average 2 hours a day outdoors (Rose et al., 2008). What's more, in these populations myopia correlates much more strongly with time spent indoors than with time spent reading. In another study, requiring Taiwanese children to spend their entire 80 minutes of recess outdoors significantly reduced the incidence of myopia compared with a nearby control school (P. C. Wu et al., 2013).

Of course, too much sunlight can be a bad thing, especially for our skin. But researchers note that almost all children in Australia wear hats to shield their faces when outdoors, yet they still benefit from being outdoors in terms of avoiding myopia. Likewise, there's no evidence that wearing glasses blocks the benefit of light from the sun. The next challenge will be determining what it is about indoor lighting, as opposed to sunlight, that encourages the eyeball in children to grow excessively, leading to myopia. One possibility is simply the higher illumination levels encountered outdoors, since greater light exposure can slow experimentally induced myopia in animals (Norton and Siegwart, 2013).

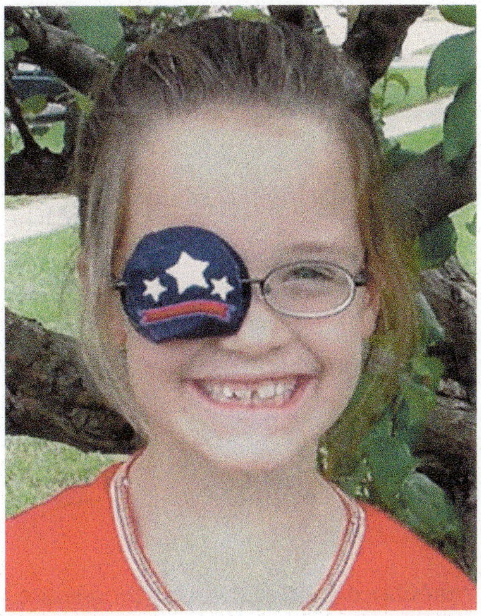
Courtesy of Patch Pals, www.PatchPals.com

8.31 Hey There, You with the Stars in Your Eye As a treatment for amblyopia, this girl is wearing a patch over her "good" eye—the one she has been relying on while ignoring information from her other, "weak" eye. Visual experience through the weak eye will strengthen its influence on the cortex.

Increased exercise can restore function to a previously deprived or neglected eye

In Chapter 6 we considered the misalignment of the two eyes (*lazy eye*), which can lead to the condition called **amblyopia** (from the Greek *amblys*, "dull" or "blunt," and *ops*, "eye"), in which acuity is poor in one eye even though the eye and retina are normal. If the two eyes are not aligned properly during the first few years of life, the primary visual cortex of the child tends to suppress the information that arrives in the cortex from one eye, and that eye becomes functionally blind. Studies of the development of vision in children and other animals also show that most cases of amblyopia are avoidable.

The balance of the eye muscles can be surgically adjusted to bring the two eyes into better alignment. Alternatively, if the weak eye is given regular practice, with the good eye covered, then vision can be preserved in both eyes. Attempts to alleviate amblyopia by training, however, have produced mixed results and a great deal of controversy. In a study sponsored by the American Academy of Ophthalmology, 507 patients, age 7–17, were separated at random into two groups. The participants in one group were given eyeglasses. Those in the other group were given not only eyeglasses but also a patch covering the good eye (**FIGURE 8.31**), and they participated in visual exercises for 26 weeks. One-fourth of those who received only eyeglasses showed improvement, but among those who also had eye patches and engaged in visual exercises, half improved (Pediatric Eye Disease Investigator Group, 2005).

Other studies have reported considerable improvement, even with adults, if the eye is exercised sufficiently and if the amblyopia is not too severe. In Chapter 6 we saw that Michael May, who was blind from the age of 3 to 46, regained some aspects of vision quickly following surgery to let light back into one of his eyes. Other aspects, such as depth perception and the perception of faces, remain severely impaired.

Frequent causes of visual impairment, especially with age, include diseases that damage the rods and cones. A variety of strategies are being followed to reverse these disorders, representing a cutting edge of visual science, as we'll discuss next.

amblyopia Reduced visual acuity of one eye that is not caused by optical or retinal impairments.

The Cutting Edge

Seeing the Light

Macular degeneration is a visual impairment caused by damage to the retina, specifically the photoreceptors. The most common type, "dry" macular degeneration, is caused by atrophy of the retinal pigmented epithelium (see Figure 8.3B), resulting in death of the overlying photoreceptors. In the more severe, "wet" macular degeneration, abnormal growth of retinal capillaries leads to detachment of the retina and/or death of photoreceptors. The damage is mostly restricted to the fovea, and because visual acuity is poor for the rest of the retina, visual impairment is quite severe (**FIGURE 8.32**). An NIH-conducted trial (Age-Related Eye Disease Study [AREDS]) found that supplemental vitamins and antioxidants could slow the disease only slightly (Evans and Lawrenson, 2017).

It might be possible to restore vision in these cases by transplanting new photoreceptors to replace those that have been lost. In one dramatic "proof of principle," scientists were able to provide vision to mice with a mutation that had disabled their rod photoreceptors (Pearson et al., 2012). The scientists gathered retinas from normal mice that were 3 days old, then injected some of their young rod cells into the retinas of the mutant adults. Some of the transplanted rod cells integrated into the retinal circuitry (**FIGURE 8.33A**). To test whether the transplanted rods actually allowed the mice to see, each animal was surrounded by computer monitors showing black vertical bars moving either clockwise or counterclockwise, to determine whether the mouse would move its head to follow the bars (**FIGURE 8.33B**). The researchers could vary the contrast of the bars to ask how much contrast was needed before the mouse could see them. They could also vary the width of the bars to measure the animal's visual acuity—how wide the bars must be for the mouse to see them (**FIGURE 8.33C**). These measures of each individual mouse's contrast sensitivity and acuity correlated with how many rods had integrated into its retinas (**FIGURE 8.33D**). Because the extent of vision following the procedure actually reflects the number of transplanted rods, the current challenge is how to maximise the number of rods that get incorporated into the retina.

macular degeneration A progressive loss of central vision due to death or obstruction of photoreceptors in the retina.

(A) Normal vision

(B) Vision with macular degeneration

Photos by M. H. Siddall

8.32 Simulating Visual Experience with Macular Degeneration Macular degeneration occurs in the fovea, deleting vision in the centre of gaze. The low density of photoreceptors outside the fovea means poor visual acuity in the rest of the visual field, as if it were out of focuss. (After D. J. Marmor and M. F. Marmor, 208. *Arch Ophthalmol* 128: 117–125.)

Thus it may be possible someday to restore vision in people who have lost their photoreceptors to macular degeneration, but the big problem is where to get young human photoreceptors (S. D. Schwartz et al., 2012). It might be possible to culture cells from a blind person needing a transplant and get the cells to differentiate into photoreceptors. Then these transformed photoreceptors could be injected back into the blind person's eyes. Using a virus to infect retinal glial cells to express β-catenin (see Figure 6.22) turned them into photoreceptors and restored vision in a mouse model of blindness (Yao et al., 2018). Perhaps such manipulations can restore vision in humans one day.

An alternative method for restoring vision does not require introducing more cells at all. One study used viruses to induce retinal bipolar cells to express a light-sensitive ion channel (channelrhodopsin, see Figure 2.26). Then the bipolar cells themselves fired in response to light, without need of photoreceptor cells to stimulate them. This procedure restored vision in several mouse models of blindness, including mice that had no photoreceptor cells (Doroudchi et al., 2011). ■

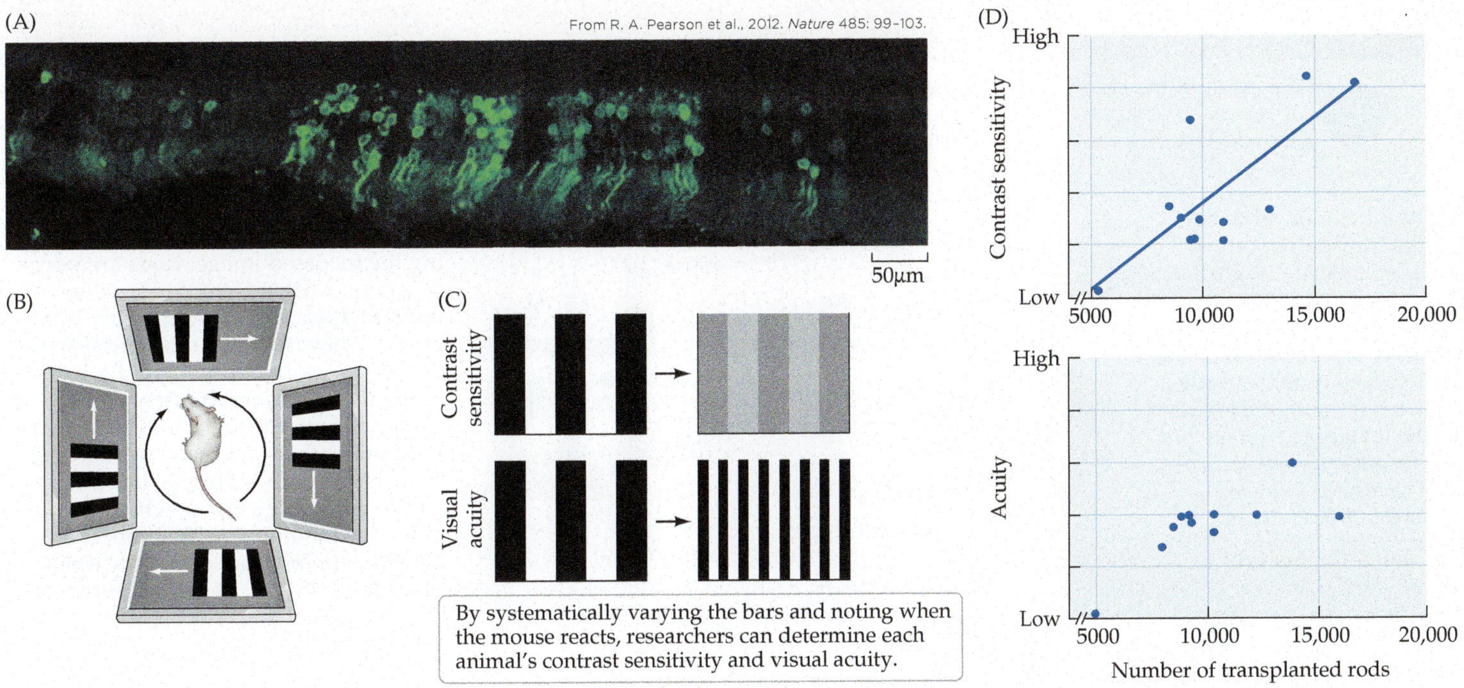

8.33 Transplanted Photoreceptors Allow Blind Mice to See (A) Pearson et al. (2012) harvested photoreceptors from 3-day-old mice and transplanted them into adult mice with a mutation that eliminates photoreceptors. After a few weeks, the transplanted rods (green cells) were integrated into the retina, making synaptic contact with bipolar cells and horizontal cells. (B) To see if the transplanted photoreceptors provided the blind mice with vision, the animals were placed in a chamber surrounded by computer monitors that displayed bars moving around the mouse, either clockwise or counterclockwise. If the mouse moved its head to follow the bars, that demonstrated that it could see them. (C) The bars could be systematically altered to vary either contrast (top) or spatial frequency (bottom) to measure each animal's contrast sensitivity and acuity. (D) There was a correlation between the number of integrated rod cells found in an individual mouse's retina and the animal's contrast sensitivity and acuity. Efforts are under way to optimise the procedures to provide as many rods as possible. (B–D after R. A. Pearson et al., 2012. *Nature* 485: 99–103.)

Recommended Reading

De Valois, R. L., and De Valois, K. K. (1988). *Spatial Vision*. New York: Oxford University Press.

Dowling, J. E., and Dowling, J. L. (2016). *Vision: How It Works and What Can Go Wrong*. Cambridge, MA: MIT Press.

Purves, D., and Lotto, R. B. (2011). *Why We See What We Do Redux: A Wholly Empirical Theory of Vision*. Sunderland, MA: Oxford University Press/Sinauer.

Sacks, O. (2010). *The Mind's Eye*. New York: Knopf.

Schiller, P. H., and Tehovnik, E. J. (2015). *Vision and the Visual System*. New York: Oxford University Press.

Werner, J. S., and Chalupa, L. M. (Eds.). (2014). *The New Visual Neurosciences*. Cambridge, MA: MIT Press.

Wolfe, J. M., Kluender, K. R., Levi, D. M., Bartoshuk, L. M., et al. (2017). *Sensation & Perception* (5th ed.). Sunderland, MA: Oxford University Press/Sinauer.

8 Visual Summary

You should be able to relate each summary to the adjacent illustration, including structures and processes. You can also refer back to the links to figures, animations, and activities that will help you consolidate the material.

1 Light is bent, or **refracted**, by the transparent outer layer of the eye, the **cornea**, focussing an image on the **retina** in the back of the eye. We vary the thickness of the **lens** to fine-tune that refraction, sharpening the focus of the image. The image on the retina is upside-down and reversed. Review Figure 8.2, Activity 8.1

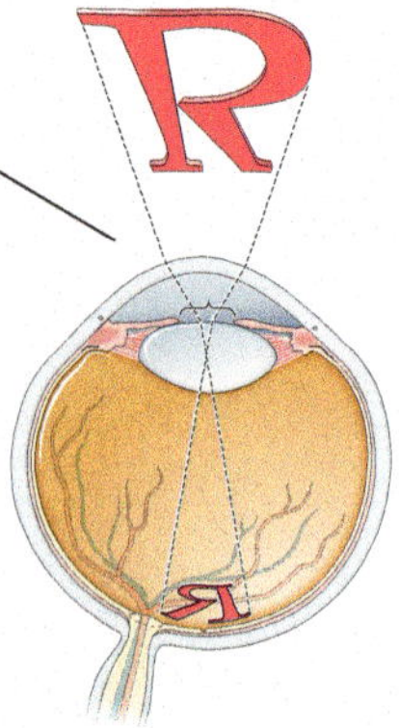

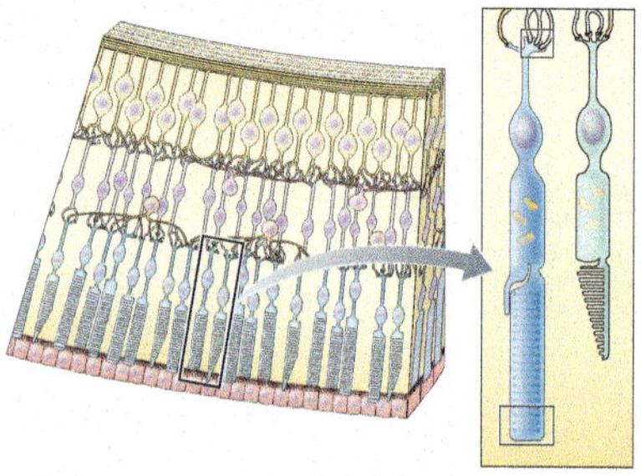

2 The retina contains two different types of **photoreceptors** to detect light forming the focussed image. **Rods** are very sensitive, working even in very low light, and respond to light of any wavelength. They drive the **scotopic system**, which can work in dim light. Each of the three different types of **cones** responds better to some wavelengths of light than others, allowing us to detect colours. The cones provide information for the **photopic system**, which needs more light to function. Photoreceptors **adapt** to function across a wide range of light intensities. Review Figures 8.3–8.7, Table 8.1

3 The retina consists of layers of neurons, with the photoreceptors in the very back stimulating **bipolar cells**, which stimulate **ganglion cells**. The ganglion cells of the retina project their axons to the brain via the **optic nerve**. **Amacrine cells** and **horizontal cells** communicate across the retina, using processes such as lateral inhibition to analyse brightness. Review Figures 8.3, 8.7–8.9

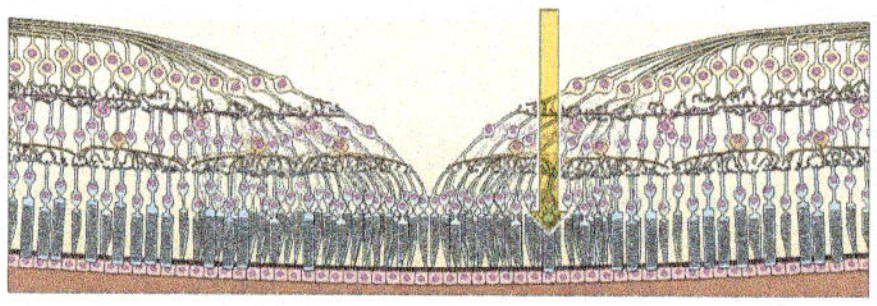

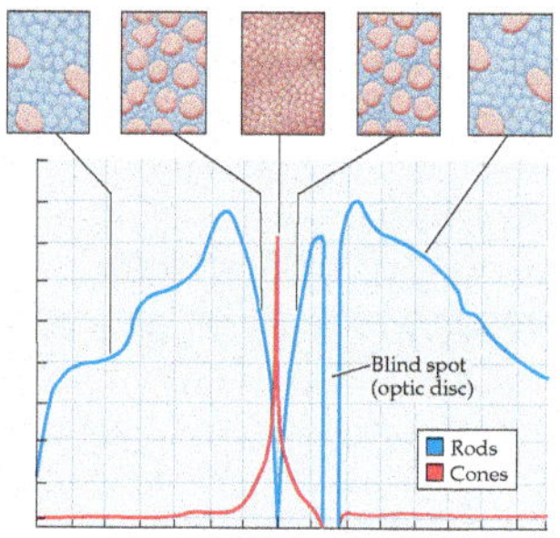

4 The centre of our **visual field** lands on the **fovea**, the portion of the retina with the greatest density of photoreceptors, an absence of overlying cell layers, and more direct synaptic connections to ganglion cells, providing us with our greatest **visual acuity** (sharpness of vision). Cones are concentrated in the fovea, and rods in the peripheral retina, so our peripheral vision is best for seeing dim objects but provides no colour information. Review Figures 8.6, 8.10–8.12

5 The left visual field falls upon the nasal retina of the left eye and the temporal retina of the right eye. Only nasal retinal ganglion cells of each eye send their axons across the midline, forming the **optic chiasm**, so the left visual field projects to the right hemisphere, and the right visual field projects to the left hemisphere. Review Figures 8.10, 8.11, Animation 8.1

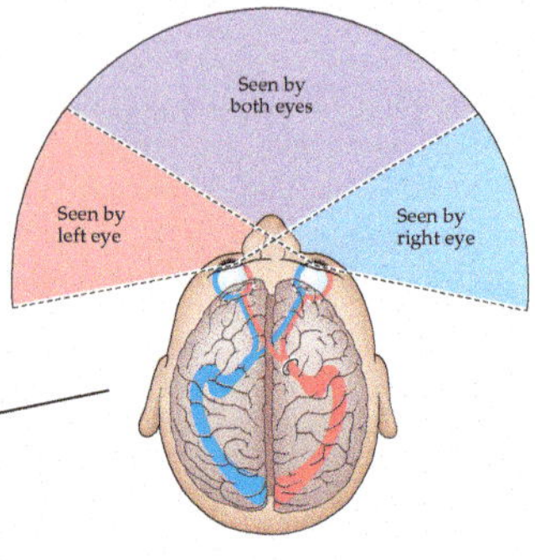

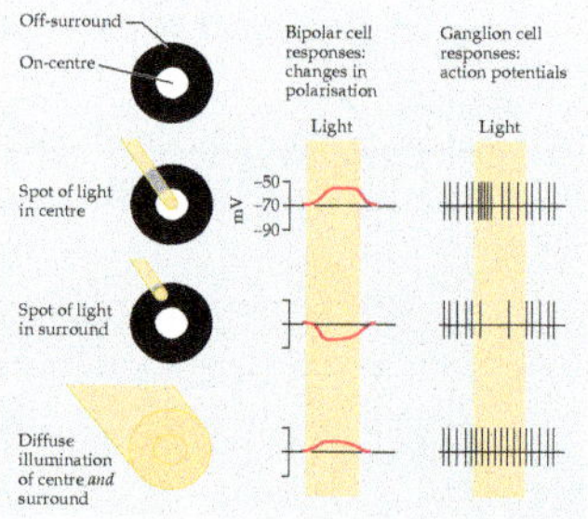

6 The **receptive fields** of bipolar cells and ganglion cells consist of a circular centre and a surround that have opposing effects: either **on-centre/off-surround** or **off-centre/on-surround**. Review Figures 8.13, 8.14, Animation 8.2

7 Ganglion cells of the retina synapse upon neurons in the **lateral geniculate nucleus** (**LGN**) of the thalamus. The LGN neurons send axons to synapse on neurons in layer IV of the **primary visual cortex** (also called **V1** or *striate cortex*) in the **occipital lobe**. V1 sends information to many different cortical areas, called **extrastriate cortex** (nearly one-third of the human cortex), to further analyse visual information. Review Figures 8.15, 8.20

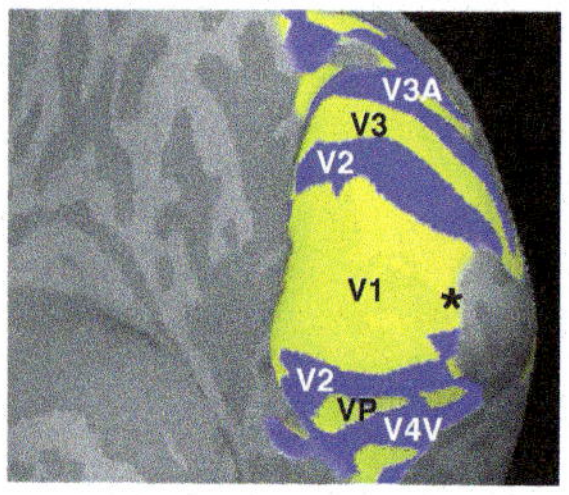

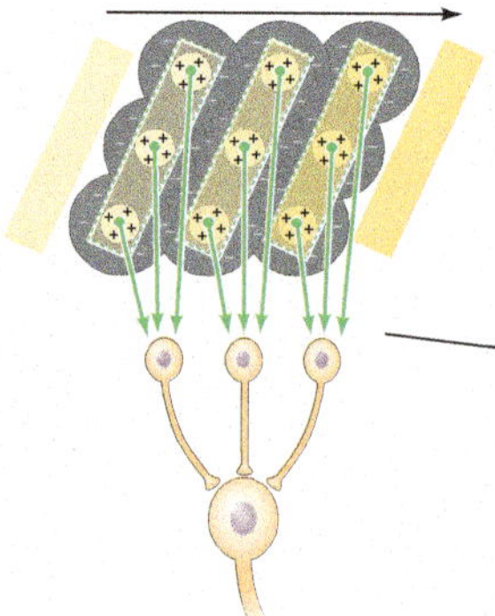

8 Receptive fields of cells at successively higher levels in the visual system change in two main ways: (1) they become larger (occupy larger parts of the visual field), and (2) they require increasingly specific stimuli to evoke responses. For example, they best respond to a bar of light at a particular angle, or to bars that move in a particular direction. Review Figures 8.16–8.23, Animation 8.3

9 Our detection of **hue** (colour) depends on the three different cone **opsins**, or photopigments. Each cone responds to a wide range of wavelengths, not just a single colour. Our perception of hue results from the *relative activity* of each type of cone. One way of assessing this relative activity is by retinal connections that yield **spectrally opponent** retinal ganglion cells. Review **Figures 8.24–8.27**, **Activity 8.2**

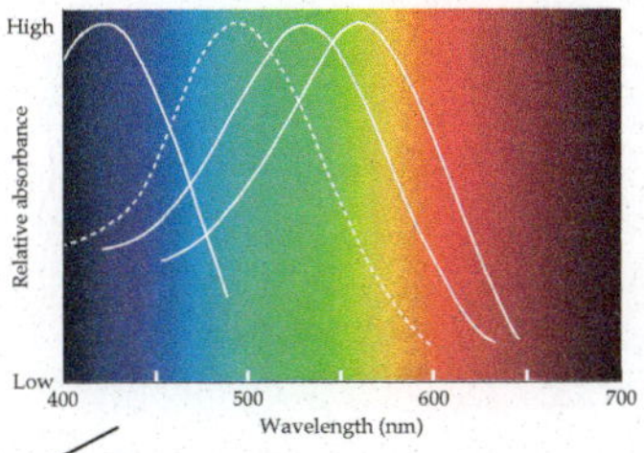

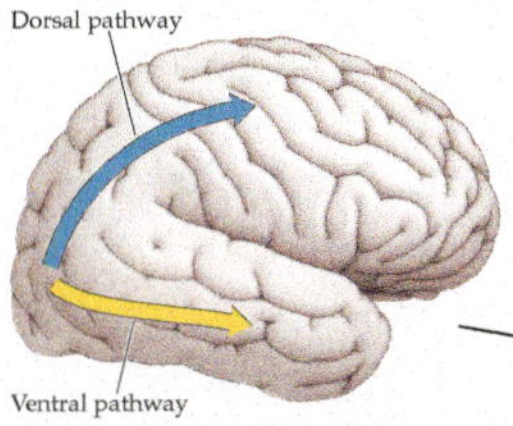

10 Visual cortical areas are organised into two main streams: a dorsal *where* stream that serves in location and visuomotor skills, and a ventral *what* stream that serves in the recognition of faces and objects. Review **Figures 8.28**, **8.29**

11 Myopia (nearsightedness) results when the eyeball is too long to focuss the visual image onto the retina properly. Prolonged exposure to indoor lighting during childhood may favor such growth. Review **Figure 8.30**

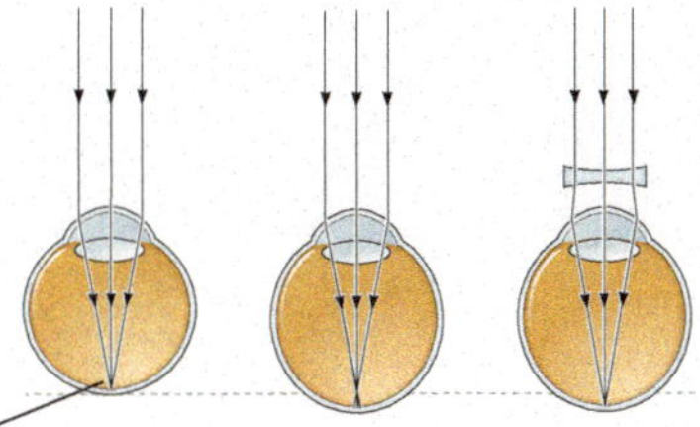

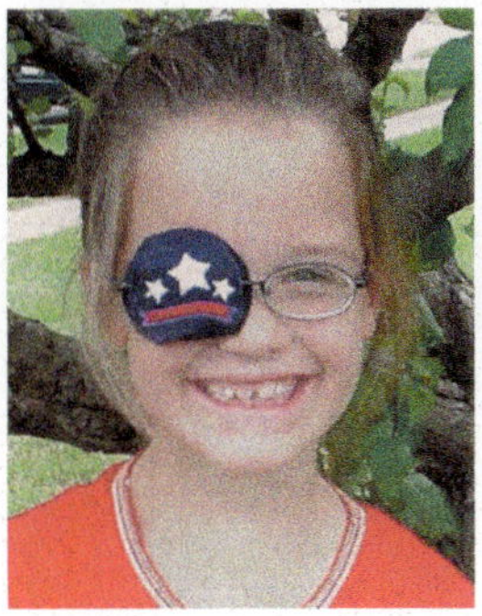

12 Treating **amblyopia** works best when the retraining starts early in life, but the visual nervous system retains some plasticity even in adults. Review **Figures 8.31–8.32**

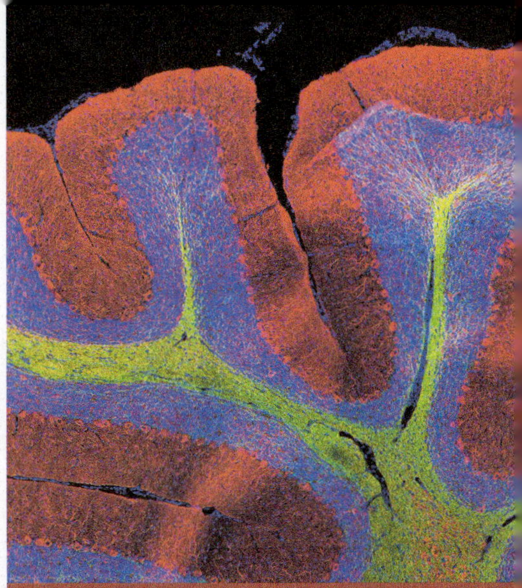

Photo from M. E. Terzenidou et al., 2017. *PLOS Genet* 13: e1006656 / CC BY 4.0 / Courtesy of Dr. Katerina Segklia and Dr. Rebecca Matsas

9

Hearing, Balance, Taste, and Smell

No Ear for Music

Given the category "wedding music," you will probably immediately think of a tune or two that you can hear in your head—perhaps "The Wedding March" or (alas?) "Uptown Funk." The same goes for Beethoven's Symphony No. 5 or the theme from *Star Wars*. These pieces of music, and others like them, are tunes that most people can identify right away, after hearing just a few bars. But Tony is different.

Tony is completely baffled by music. He is much less sensitive than most people to the difference between musical tones. Tony's ability to understand pitch, and the relationships between chords, is nearly at chance level. Although he is happy to try to belt out a song, he is unable to sing in tune and doesn't even recognise that he is singing off-key unless he is told so by a (grimacing) friend. He cannot identify the tune to "Happy Birthday" unless he also hears the lyrics, in which case he can identify it immediately. Yet he has no problem with nonmusical uses of pitch, such as the rising tone at the end of a question.

Tony is a man of better-than-average intelligence, with perfectly normal cognitive functioning, except for his lifelong inability to appreciate music despite years of childhood music lessons. This highly specific difficulty with music is an affliction that Tony shares with many other people; notable examples include Pope Francis, revolutionary Che Guevara, U.S. President Theodoure Roosevelt, and Nobel Prize–winning economist Milton Friedman.

What is the nature of this mysterious problem? Is it a kind of hearing problem? A learning disability?

Your existence is a testament to the keen senses of your ancestors—senses that helped them to find food and mates and to avoid predators and other dangers. In this chapter we consider some of the amazing sensory systems that allow us to monitor important signals from distant sources, especially sounds (audition) and smells (olfaction). We also discuss related systems for detecting position and movement of the body (the vestibular system, related to the auditory system) and tastes of foods (the gustatory or taste sense, which, like olfaction, is a chemical sense). We begin with hearing because audition evolved from special mechanical receptors related to the somatosensory elements that we discussed in Chapter 7.

HEARING

9.1 Pressure Waves in the Air Are Perceived as Sound

Learning Objectives

After reading this section, you should be able to:

9.1.1 **Explain how the external ear and middle ear capture and concentrate sound energy and convey it to the inner ear.**

9.1.2 **Sketch the anatomy of the middle and inner ears, highlighting the locations of sensorineural components.**

9.1.3 **Describe the propagation of vibrations in the cochlea and the process by which they are transduced into neural activity.**

9.1.4 **Explain how frequencies are encoded by the organ of Corti and how its tuning is sharpened by active processes in the cochlea.**

Hearing is vital for the survival of many animals. Humans employ an impressive variety of vocalisations—from barely audible murmurs to soaring flights of song—but we especially rely on speech sounds to provide a basis for our social relations and for the transmission of knowledge between individuals. Helen Keller, who was both blind and deaf, said, "Blindness deprives you of contact with things; deafness deprives you of contact with people."

The sounds produced by other animals—from insects to whales—also have a wide range of complexity, in keeping with their adaptive significance. Birds sing and crickets chirp in order to attract mates; monkeys grunt and screech and burble to signal comfort, danger, and pleasure. Elephants employ vocalisations too deep for us to hear (see Figure 8.2) and can recognise individuals by their calls (Herbst et al., 2012). Owls and

---

BOX 9.1 The Basics of Sound

We perceive a repetitive pattern of local increases and decreases in air pressure as sound. Usually this oscillation is caused by a vibrating object, such as a loudspeaker or a person's larynx during speaking. A single alternation of compression and expansion of air is called one *cycle*.

The figure illustrates the oscillations in pressure produced by loudspeakers, each vibrating at a single frequency. Such sounds, called **pure tones**, are represented by sine waves that plot the oscillation between the compressions (peaks) and expansions (troughs) of air in front of the vibrating cone of the speaker. A pure tone is described physically in terms of two measures:

1. **Amplitude**, or *intensity*—usually measured as sound pressure level, in dynes per square centimetre (dyne/cm^2). Our perception of amplitude is termed **loudness**, expressed in **decibels** (**dB**). The

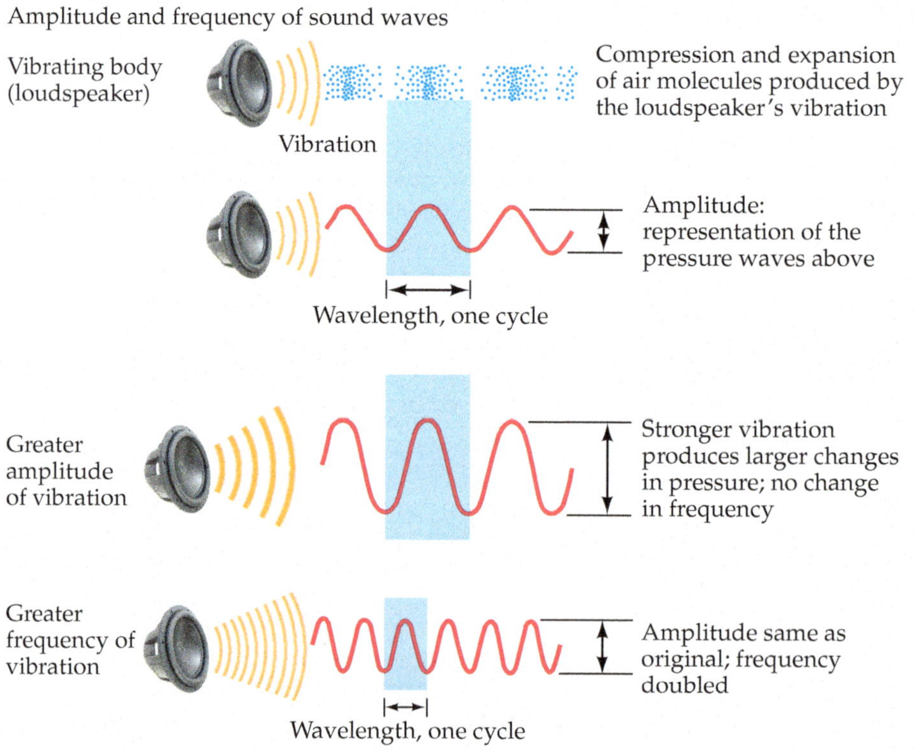

bats exploit the directional property of sound to locate prey and avoid obstacles in the dark, and whales employ sounds that can travel hundreds of miles in the ocean. Unlike light, sounds can go around obstacles and can be detected in the darkest night.

How does energy transmitted through air become the speech, music, and other sounds we hear? Your auditory system detects changes in the vibration of air molecules that are caused by sound sources, encoding rapid changes in the *intensity* of sounds (measured in decibels [dB] and perceived as *loudness*) along with variations in the *frequency* of sounds (measured in cycles per second, or hertz [Hz] and perceived as *pitch*). **BOX 9.1** describes some of the basic properties of sound relevant to our discussion of the auditory sense.

The outer parts of the auditory system have been shaped through evolution to capture biologically important sound vibrations and direct them into the inner parts of the ear, which accomplishes the **transduction** of the mechanical energy of sound into the form of electrical energy—action potentials—that the brain can understand. Your ears are incredibly sensitive organs; in fact, one of the main jobs of your powers of attention is to filter out the constant barrage of unimportant little noises that your ears are able to detect (see Chapter 17).

The external ear captures, focuses, and filters sound

The oddly shaped fleshy objects that most people call *ears* are properly known as **pinnae** (singular *pinna*, Latin for "wing"). Aside from their occasional utility as handles and jewelry hangers, the pinnae funnel sound waves into the second part of the **external ear**: the **ear canal**. The pinna is a distinctly mammalian characteristic, and mammals show a wide array of ear shapes and sizes. Furthermore, although only a minority of humans can move their ears—and even then, only enough to entertain children—many

transduction The conversion of one form of energy to another.

pinna The external part of the ear.

external ear The part of the ear that we readily see (the pinna) and the canal that leads to the eardrum.

ear canal A tube leading from the pinna to the middle ear.

decibel scale is logarithmic: 1 dB is the threshold for human hearing, a faint whisper is about 20 dB, and a jet airliner 500 feet overhead (140 dB) is about a million times as intense. Normal conversation is about 60 dB.

2. **Frequency**, or the number of cycles per second, measured in **hertz** (**Hz**). For example, middle A on a piano has a frequency of 440 Hz. Our perception of frequency is termed **pitch**.

Most sounds are more complicated than a pure tone. For example, a sound made by a musical instrument contains a **fundamental** frequency plus one or more integer multiples of the fundamental, called **harmonics**. So, middle A on the piano has a fundamental frequency of 440 Hz plus some combination of harmonics at 880 Hz, 1320 Hz, 1760 Hz, and so on. When different instruments play the same note, the notes differ in the relative intensities of the various harmonics, and there are subtle qualitative differences between instruments in the way they commence, shape, and sustain the sound; these differences are what give each instrument its characteristic voice, or **timbre**. As a result, most "real-world" sounds have a wave form that is much more complex than the simple sine wave of a pure tone. However, using a mathematical process called **Fourier analysis**, a complex wave form can be decomposed into an algebraic sum of many simple sine waves. (We will see in Chapter 10 that Fourier analysis can also be applied to visual patterns.)

pure tone A tone with a single frequency of vibration.

amplitude The force that sound exerts per unit area, usually measured as dynes per square centimetre.

loudness The subjective experience of the pressure level of a sound.

decibel (dB) A measure of sound intensity.

frequency The number of cycles per second in a sound wave; measured in hertz (Hz).

hertz (Hz) Cycles per second, as of an auditory stimulus.

pitch A dimension of auditory experience in which sounds vary from low to high.

fundamental The predominant frequency of an auditory tone or a visual scene.

harmonics Multiples of a particular frequency called the fundamental.

timbre The characteristic sound quality of a musical instrument, as determined by the relative intensities of its various harmonics.

Fourier analysis The mathematical decomposition of a complex pattern into a sum of sine waves of various frequencies and amplitudes.

9.1 The Ears Have It The external ears, or pinnae, of mammals come in a variety of shapes, each adapted to a particular niche. Many mammals can move their ears to direct them towards a particular sound. In such cases, the brain must account for the position of the ear to judge where a particular sound came from.

other mammals deftly shape and swivel their pinnae to help determine the source of a sound (**FIGURE 9.1**). Animals with exceptional auditory localisation abilities, such as bats, may have especially mobile ears. In these animals, proprioceptors (position sensors; see Chapter 10) within the pinnae provide information to the auditory system about the positions of the external ears as an aid to localising sound sources.

The "ridges and valleys" of the pinna modify the character of sound that reaches the middle ear. Some frequencies of sound are enhanced; others are suppressed. For example, the shape of the human ear especially increases the reception of sounds between 2000 and 5000 Hz—a frequency range that is important for speech perception. The shape of the external ear is also important in identifying the direction and distance of the source of a sound (discussed later in this chapter).

The middle ear concentrates sound energies

A collection of tiny structures made of membrane, muscle, and bone—essentially a tiny biological microphone—links the ear canal to the neural receptor cells of the inner ear (**FIGURE 9.2A**). This **middle ear** (**FIGURE 9.2B**) consists of the taut **tympanic membrane** (eardrum), sealing the end of the ear canal, and a chain of tiny bones, the **ossicles**, that mechanically couple the tympanic membrane to the inner ear at a special location called the **oval window**. These ossicles, the smallest bones in the body, are called the **malleus** (Latin for "hammer"), the **incus** (Latin for "anvil"), and the **stapes** (Latin for "stirrup").

Sound waves in the air strike the tympanic membrane and cause it to vibrate with the same frequency as the sound. The movement of the tympanic membrane moves the chain of ossicles, concentrating the tiny mechanical forces of vibrating air particles, captured from the relatively large tympanic membrane, onto the small oval window. This arrangement vastly amplifies sound pressure so that it can produce movement in the fluid of the inner ear, as we'll see.

Two muscles vary the mechanical linkage between the ossicles to improve auditory perception and protect the delicate inner ear from loud, potentially damaging sounds. One of these muscles, the **tensor tympani** (see Figure 9.2B), is attached to the malleus, which is connected to the tympanic membrane. The other muscle of the middle ear is attached to the stapes and thus is called the **stapedius**. Within 200 milliseconds of the arrival of a loud sound, the brain signals the muscles via a reflex arc in the brainstem, causing the muscles to contract and stiffen the chain of ossicles (Mukerji

middle ear The cavity between the tympanic membrane and the cochlea.

tympanic membrane Also called *eardrum*. The partition between the external ear and the middle ear.

ossicles Three small bones (incus, malleus, and stapes) that transmit sound across the middle ear, from the tympanic membrane to the oval window.

oval window The location on the surface of the cochlea at which vibrations are received from the ossicles.

malleus Latin for "hammer." A middle-ear bone that is connected to the tympanic membrane.

incus Latin for "anvil." A middle-ear bone situated between the malleus and the stapes.

stapes Latin for "stirrup." A middle-ear bone that is connected to the oval window.

tensor tympani The muscle attached to the malleus that modulates mechanical linkage to protect the delicate receptor cells of the inner ear from damaging sounds.

stapedius A middle-ear muscle that is attached to the stapes.

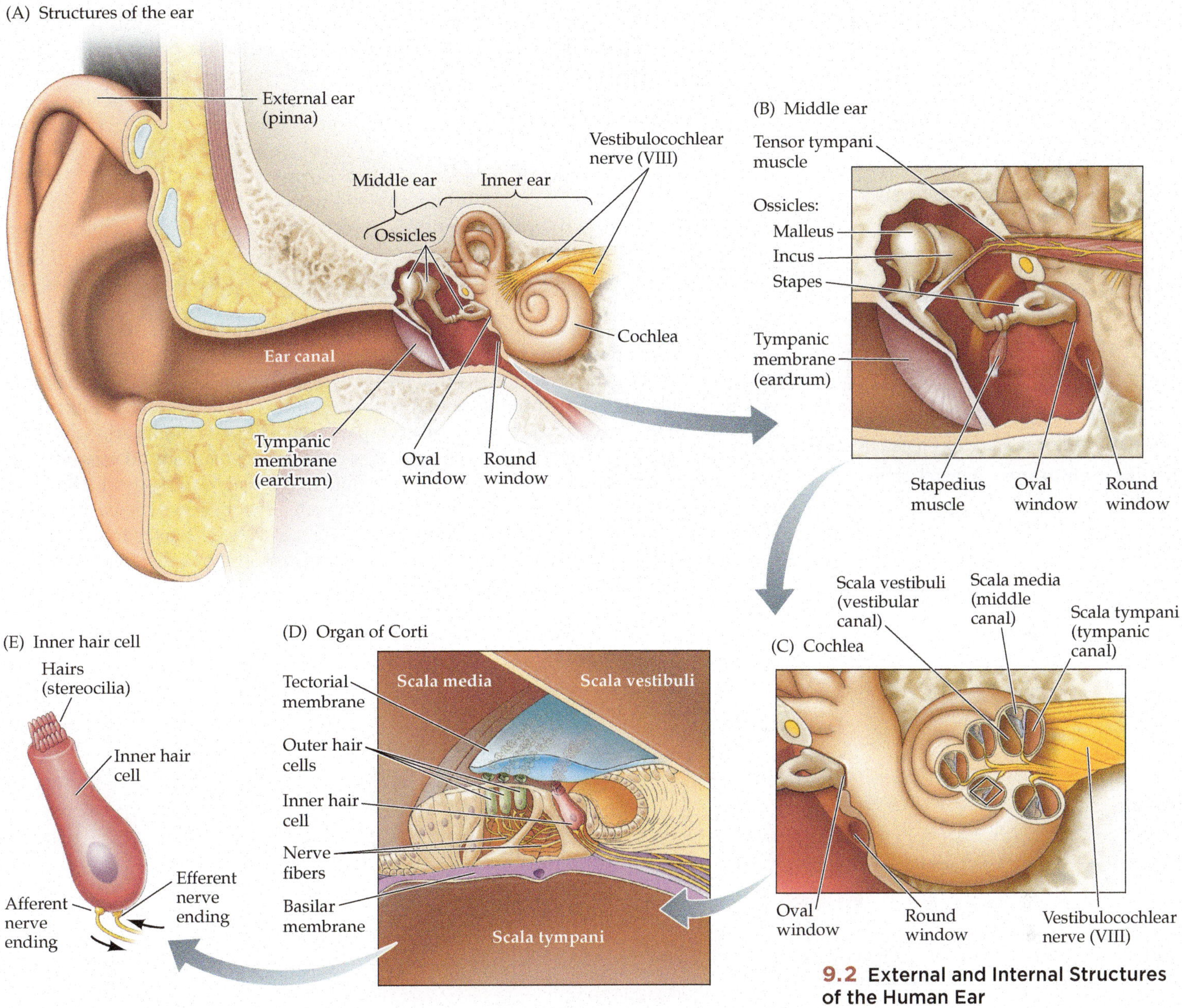

9.2 External and Internal Structures of the Human Ear

et al., 2010). Because the stiffened chain of ossicles is less able to transmit energy, the power of the vibrations reaching the inner ear is significantly reduced. The middle-ear muscles also attenuate self-made sounds; without this system, body movement, swallowing, vocalisations, and other internally produced sounds would be painfully loud. Species that produce especially loud calls, such as bats, rely on this system to protect their auditory receptors from physical damage. Importantly, the middle-ear muscles activate just *before* we produce self-made sounds like speech or coughing, which is why we don't perceive our own sounds as distractingly loud.

Go to Animation 9.1
Sound Transduction

The cochlea converts vibrational energy into waves of fluid

The complex structures of the **inner ear** ultimately convert sound into neural activity. In mammals the auditory portion of the inner ear is a coiled, fluid-filled structure called the **cochlea** (from the Greek *kochlos*, "snail") (**FIGURE 9.2C AND D**). Embedded in the temporal bone of the skull, the human cochlea is a marvel of miniaturisation. In an adult, the cochlea measures only about 9 millimetres (mm) in diametre at its widest point —roughly the size of a pea. Unrolled, the cochlea would measure about 35–40 mm in length.

inner ear The cochlea and vestibular apparatus.

cochlea A snail-shaped structure in the inner ear that contains the primary receptor cells for hearing.

scala vestibuli Also called *vestibular canal*. One of the three canals running the length of the cochlea.

scala media Also called *middle canal*. The central of the three canals running the length of the cochlea, situated between the scala vestibuli and the scala tympani.

scala tympani Also called *tympanic canal*. One of three canals running the length of the cochlea.

round window A membrane separating the cochlear duct from the middle-ear cavity.

organ of Corti A structure in the inner ear that lies on the basilar membrane of the cochlea and contains the hair cells and terminations of the auditory nerve.

hair cell A cochlear auditory receptor cell.

basilar membrane A membrane in the cochlea that contains the principal structures involved in auditory transduction.

The cochlea is a spiral of three parallel canals: (1) the **scala vestibuli** (*vestibular canal*), (2) the **scala media** (*middle canal*), and (3) the **scala tympani** (*tympanic canal*). In Figures 9.2C and 9.3 you can see that these canals—and the membranes that divide them—are tapered along their length. The oval window is adjacent to the *base* of the spiral, where the canals and membranes are narrow; the distant wide end is referred to as the *apex*. Because these canals are filled with noncompressible fluid, movement inside the cochlea in response to a push on the oval window requires a second membrane-covered window that can bulge outward a bit. This membrane is the **round window**, which separates the scala tympani from the middle ear (see Figure 9.2B).

The scala media contains the receptor system, called the **organ of Corti**, that converts vibration (from sound) into neural activity (see Figure 9.2D). It consists of three main structures: (1) the auditory sensory cells, called **hair cells** (**FIGURE 9.2E**), which are embedded in the **basilar membrane**; (2) an elaborate framework of supporting cells; and (3) the auditory nerve terminals that transmit neural signals to and from the brain.

The flexible basilar membrane, which is about 5 times wider at the apex of the cochlea than at the base, separates the scala tympani from the scala media and vibrates in response to sound. When the stapes moves in and out as a result of sound waves hitting the eardrum, it sets up waves or ripples in the fluid of the scala vestibuli, which in turn cause the basilar membrane to ripple, like shaking out a rug. Because the basilar membrane is tapered—again, it's wider at the apex than at the base—each successive location along the basilar membrane shows its strongest response to a different frequency of sound (**FIGURE 9.3**). High frequencies have their greatest effect near the base, where the basilar membrane is narrow and relatively stiff, whereas low-frequency sounds produce a larger

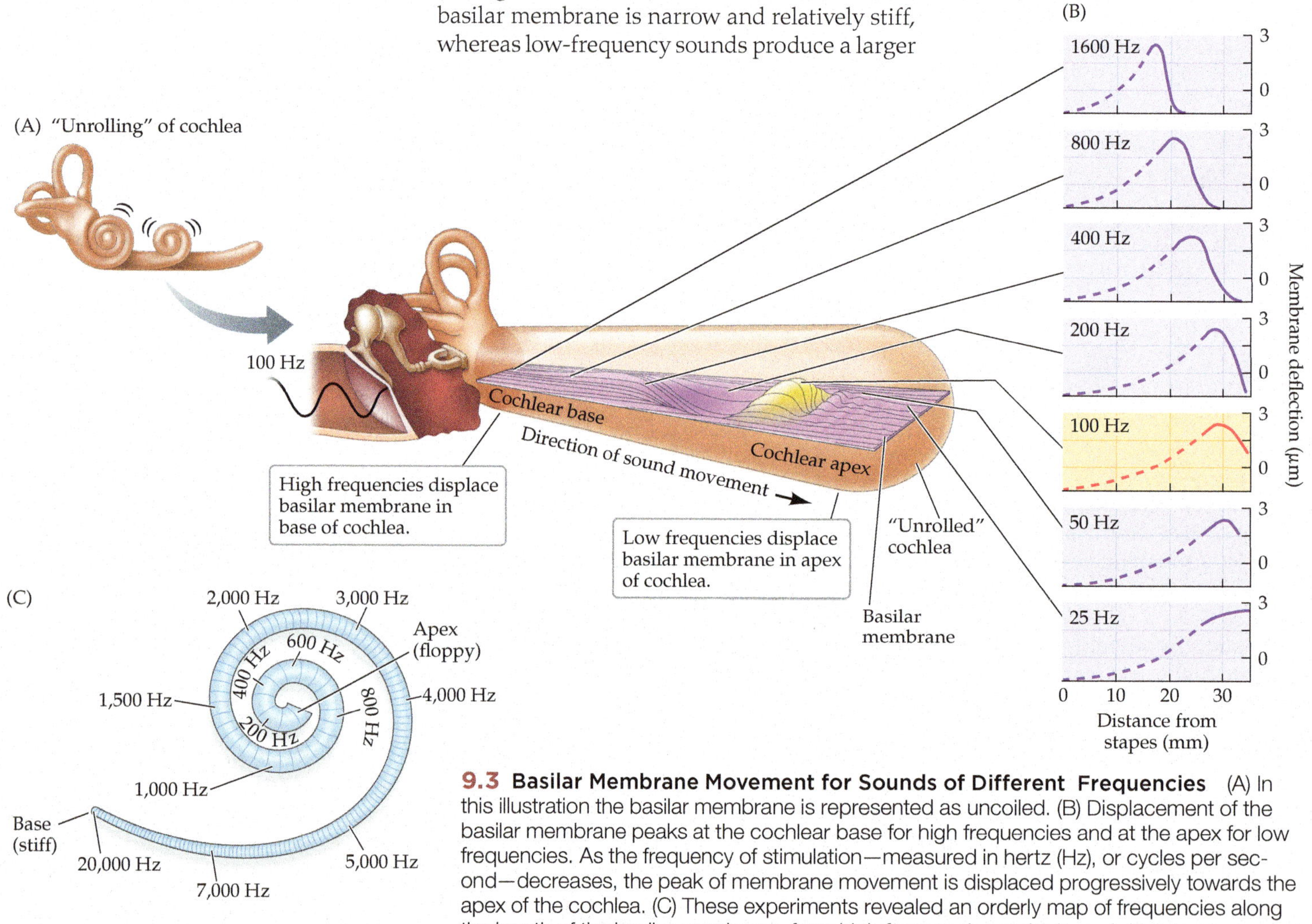

9.3 Basilar Membrane Movement for Sounds of Different Frequencies (A) In this illustration the basilar membrane is represented as uncoiled. (B) Displacement of the basilar membrane peaks at the cochlear base for high frequencies and at the apex for low frequencies. As the frequency of stimulation—measured in hertz (Hz), or cycles per second—decreases, the peak of membrane movement is displaced progressively towards the apex of the cochlea. (C) These experiments revealed an orderly map of frequencies along the length of the basilar membrane, from high frequencies at the base (which is narrow and stiff) to low frequencies at the apex (which is broad and floppy).

response near the apex, where the membrane is wider and floppier (Yoon et al., 2011). The hair cells transduce movements of the basilar membrane into electrical signals.

There are two sets of hair cells within the organ of Corti: a single row of about 3500 **inner hair cells** (**IHCs**; called *inner* because they are closer to the central axis of the coiled cochlea) and about 12,000 **outer hair cells** (**OHCs**) in three rows (see Figure 9.2D). From the upper end of each hair cell protrude tiny hairs that range from 2 to 6 micrometres in length (see Figure 9.2E). Each hair cell has 50–200 of these relatively stiff hairs, called **stereocilia** (singular *stereocilium*; from the Greek *stereos*, "solid," and the Latin *cilium*, "eyelid/eyelash") or simply *cilia*. The heights of the stereocilia increase progressively across the hair cell, so the tops form a slope. Atop the organ of Corti is the **tectorial membrane** (see Figure 9.2D). The stereocilia of the hair cells extend into indentations in the bottom of this membrane, thereby forming a mechanical bridge between the basilar and tectorial membranes.

inner hair cell (IHC) One of the two types of cochlear receptor cells for hearing.

outer hair cell (OHC) One of the two types of cochlear receptor cells for hearing.

stereocilium A relatively stiff hair that protrudes from a hair cell in the auditory or vestibular system.

tectorial membrane A membrane that sits atop the organ of Corti in the cochlear duct.

tip link A fine, threadlike fibre that runs along and connects the tips of stereocilia.

Auditory nerve fibres contact the hair cells at the base (see Figure 9.2E). The organ of Corti has four kinds of synapses and nerve fibres. Two of these (1 and 3 in **FIGURE 9.4A**) are *afferents* that convey messages from the hair cells to the brain; the other two (2 and 4 in Figure 9.4A) are *efferents* that convey messages from the brain to the hair cells. Several different synaptic transmitters—especially glutamate and acetylcholine (ACh), but also GABA and dopamine—appear to be involved in activity at the various hair cell synapses (**FIGURE 9.4B**) (Goutman et al., 2015). Each IHC is associated with 16–20 auditory nerve fibres; relatively few nerve fibres contact the many OHCs. Importantly, it is the afferent nerve fibres running from the IHCs, accounting for 90–95% of all afferent auditory fibres, that give rise to our perception of sound. The importance of the IHC afferents is illustrated by experiments in mice with a type of glutamate knockout: these mice have normal OHCs, but their IHCs are nonfunctional, and they are deaf. Restoration of the knocked out IHC neurotransmission via gene therapy restores their hearing (Akil et al., 2012). The IHCs also receive efferent messages (see Figure 9.4B), probably to inhibit afferent responses to loud sounds.

Go to Activity 9.1
Organ of Corti

So how do IHCs transduce sound into neural activity? As sounds induce vibrations of the basilar membrane, the vibrations bend the hair cell stereocilia that nestle under the tectorial membrane (see Figure 9.2D). Very small displacements of hair bundles cause rapid changes in ion channels of the stereocilia. Each hair cell has only about 100 such ion channels, about one or two per stereocilium, located towards the tops of the cilia. Fine, rodlike fibres called **tip links** run along the tips of the stereocilia. These tip links play a key role in the generation of hair cell potentials. Sounds that cause the stereocilia to sway, even only very slightly, increase the tension on the tip links and pop open the ion channels to which they are attached (Hudspeth, 2014) (**FIGURE 9.5A**). The channels snap shut again in a fraction of a millisecond as the hair cell sways back. The ion channel of a stereocilium thus resembles a trapdoor or porthole, and it appears that a springlike mechanism in the tip link's anchor in the neighbouring stereocilium provides the force to rapidly shut the channel (Powers et al., 2012)—so in a real sense, a stereocilium ion channel is spring-loaded with a hair trigger.

The mechanical opening of the stereocilium ion channels allows an inrush of potassium (K^+) and calcium (Ca^{2+})

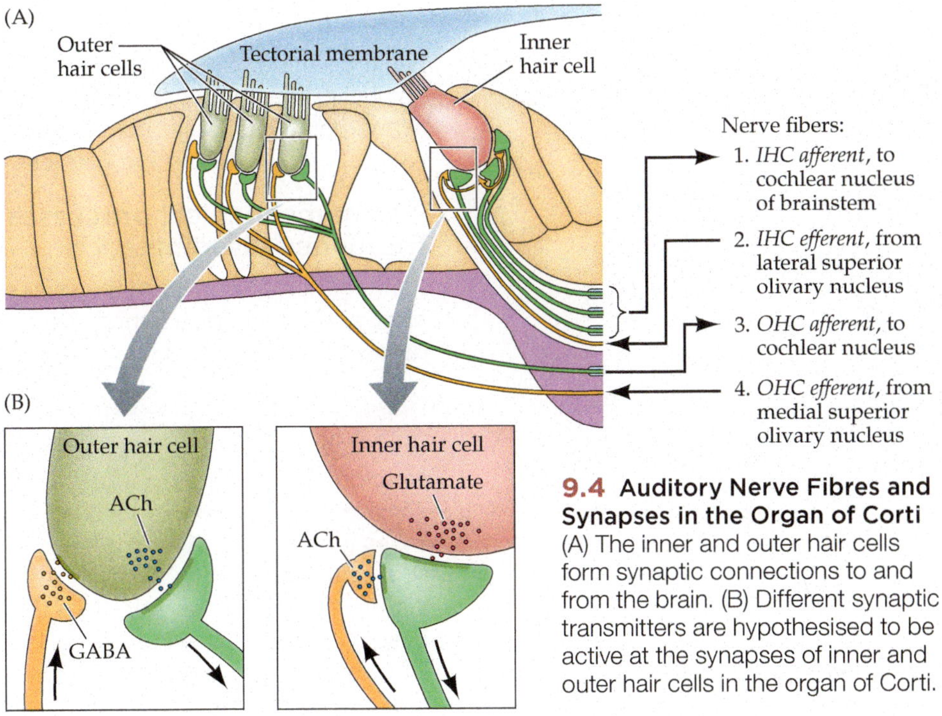

9.4 Auditory Nerve Fibres and Synapses in the Organ of Corti
(A) The inner and outer hair cells form synaptic connections to and from the brain. (B) Different synaptic transmitters are hypothesised to be active at the synapses of inner and outer hair cells in the organ of Corti.

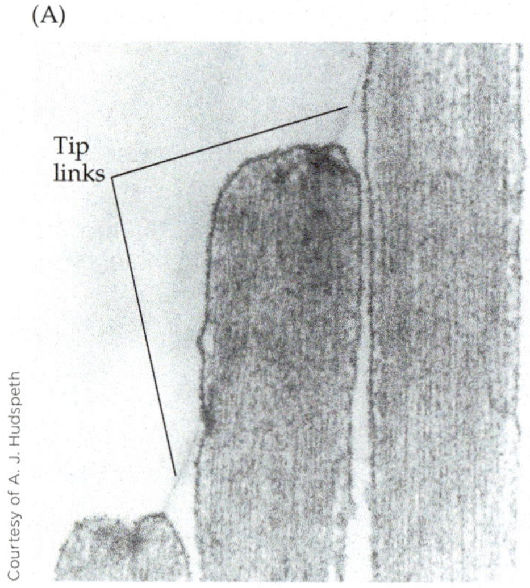

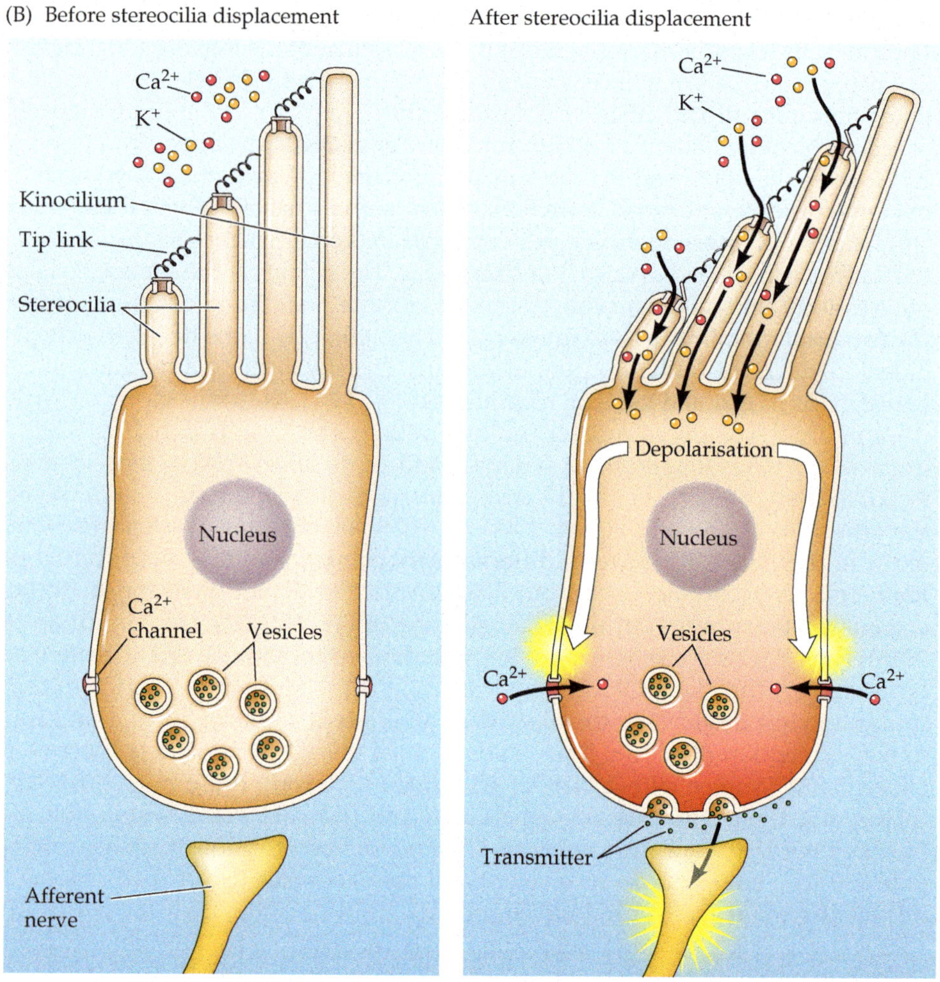

9.5 How Stereocilia Sense Auditory Stimulation (A) This micrograph of stereocilia shows the threadlike tip links. (B) Bending of the stereocilia (right) opens large, nonselective ion channels, allowing K^+ and Ca^{2+} to enter the stereocilia. The resulting depolarisation opens Ca^{2+} channels in the cell's base, causing the release of neurotransmitter to excite afferent nerves (Hudspeth, 1992).

ions—they are in relatively higher concentration outside than inside the cell—resulting in rapid depolarisation of the entire hair cell (**FIGURE 9.5B**). This initial depolarisation leads to a rapid influx of Ca^{2+} at the *base* of the IHC, which causes synaptic vesicles there to fuse with the presynaptic membrane and release their transmitter contents—probably glutamate—from the base of the hair cell, stimulating the afferent nerve fibre to trigger action potentials in these afferent axons (see Figure 9.5). The action potentials reach the brain via the vestibulocochlear nerve (cranial nerve VIII), as we'll discuss shortly, but first let's take a closer look at the actions of OHCs.

Active mechanical processes in the cochlea enhance frequency discrimination

Humans can discriminate between sounds that differ in frequency by just 2 Hz. The basic physical characteristics of the basilar membrane cannot account for such sharp discrimination of sounds, so an additional tuning process must be at work. It turns out that the OHCs fine-tune the cochlea to help discriminate frequencies.

Rather than conveying sound information, the afferents from the OHCs report the mechanical state of the basilar membrane to the brain. OHC *efferents* returning from the brain activate a most remarkable property of OHCs: a near instantaneous change in the length of OHCs! Hyperpolarisation causes the OHCs to lengthen by as much as 5%, using a specialised motor protein called **prestin**; depolarisation causes the OHCs to shorten (He et al., 2014). Through this electromechanical action, the brain uses OHCs to continually modify the stiffness of regions of the basilar membrane, resulting in both sharpened tuning and pronounced amplification (Hudspeth, 2014). A curious consequence of this mechanical manipulation of the basilar membrane is that the cochlea itself produces sounds—called

prestin A motor protein that allows outer hair cells to change length.

otoacoustic emissions—by pushing back on the eardrum. (Otoacoustic emissions are discussed in more detail on the website in **A STEP FURTHER 9.1: THE EARS EMIT SOUNDS AS PART OF THE HEARING PROCESS.**)

Now that the inner ear has transduced sound waves into trains of action potentials, the auditory signals must leave the cochlea and enter the brain.

otoacoustic emission A sound produced by the cochlea itself, either spontaneously or in response to an environmental noise.

vestibulocochlear nerve Cranial nerve VIII, which runs from the cochlea to the brainstem auditory nuclei.

tuning curve A graph of the responses of a single auditory nerve fibre or neuron to sounds that vary in frequency and intensity.

cochlear nuclei Brainstem nuclei that receive input from auditory hair cells and send output to the superiour olivary complex.

superiour olivary nuclei Brainstem nuclei that receive input from both right and left cochlear nuclei and provide the first binaural analysis of auditory information.

inferiour colliculi Paired gray matter structures of the dorsal midbrain that receive auditory information.

medial geniculate nuclei Nuclei in the thalamus that receive input from the inferiour colliculi and send output to the auditory cortex.

9.2 Auditory Signals Run from Cochlea to Cortex

Learning Objectives

After reading this section, you should be able to:

9.2.1 **Describe the auditory pathways, from cochlea to cortex.**

9.2.2 **Summarise the integration of signals from the left and right ears, at various levels of the auditory pathway.**

9.2.3 **Describe the orderly map of frequencies found at each level of the auditory system.**

On each side of your head, about 30,000–50,000 auditory axons from the cochlea make up the auditory part of the **vestibulocochlear nerve** (cranial nerve VIII). Recall that most of these afferent fibres are carrying messages from the IHCs, each of which stimulates several nerve fibres. Each auditory neuron responds to a very precise frequency at its threshold, but for more-intense stimuli the neuron responds to a broader range of frequencies. For example, the fibre whose responses are shown in red in **FIGURE 9.6** has its *best frequency* at 1200 Hz; that is, it responds to a very weak tone at 1200 Hz. When sounds are 20 dB louder, however, the fibre responds to any frequency from 500 to 1800 Hz. We call this the **tuning curve** of that cell's sensitivity to sounds of various frequencies. Thus, although an auditory nerve fibre transmits exclusively auditory information, it does not respond to just one frequency of stimulation. If the brain received a signal from only one such fibre, it would not be able to tell whether the stimulus was a weak tone of 1200 Hz or a stronger tone of 500 or 1800 Hz, or any frequency in between. Instead, the brain analyses signals from thousands of such units to calculate the intensity and frequency of each sound.

Input from the auditory nerves is distributed to both sides of the brain via the ascending network depicted in **FIGURE 9.7**. Each auditory nerve divides into two main branches as it enters the brainstem, with one branch going to the dorsal **cochlear nucleus** and the other terminating in the ventral cochlear nucleus. The output of neurons in the cochlear nuclei also travels via multiple paths. One path from each cochlear nucleus projects bilaterally to the **superiour olivary nuclei**, so each superiour olivary nucleus receives input from both the right and left cochlear nuclei. This bilateral input is the first level in the auditory system at which *binaural* (two-ear) effects are processed; as you might expect, this mechanism plays a key role in localising sounds by comparing the two ears, as we'll discuss shortly. Several other parallel paths converge on the **inferiour colliculi**, which are the primary auditory centres of the midbrain. Outputs of the inferiour colliculi go to the **medial geniculate nuclei** of the thalamus. Neurons of the medial geniculate project axons to several auditory cortical areas.

Throughout the auditory pathways, neuronal responses are frequency-sensitive, as with the vestibulocochlear nerve fibres that we discussed earlier (see Figure 9.6). The ability to discriminate frequencies becomes sharper at higher stations of the auditory nervous system. At

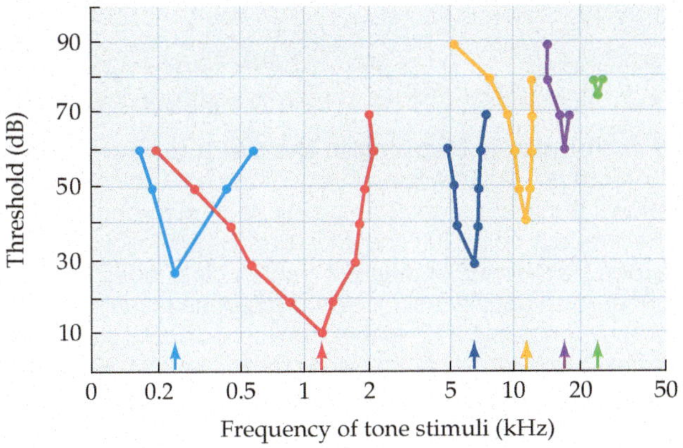

9.6 Examples of Tuning Curves of Auditory Neurons These curves depict the sensitivity of six different neurons (represented by six different colours) to sounds of different intensities and frequencies. Because they represent threshold measurements, the *lowest* point on each curve (indicated by arrows on the *x*-axis for six cells) corresponds to that neuron's preferred frequency. (After N. Y. S. Kiang, 1965. *Discharge patterns of single fibres in the cat's auditory nerve.* MIT Press: Cambridge, MA.)

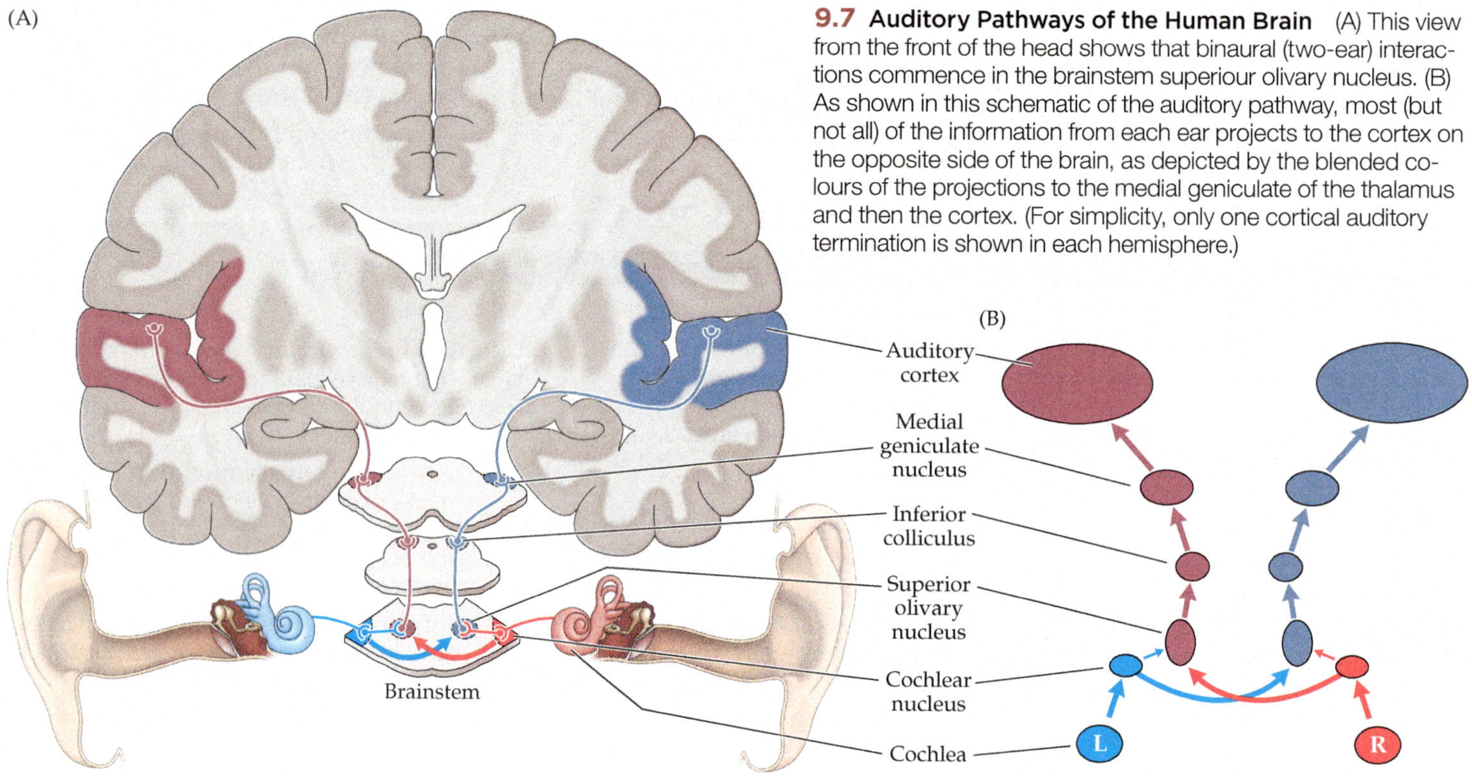

9.7 Auditory Pathways of the Human Brain (A) This view from the front of the head shows that binaural (two-ear) interactions commence in the brainstem superiour olivary nucleus. (B) As shown in this schematic of the auditory pathway, most (but not all) of the information from each ear projects to the cortex on the opposite side of the brain, as depicted by the blended colours of the projections to the medial geniculate of the thalamus and then the cortex. (For simplicity, only one cortical auditory termination is shown in each hemisphere.)

the medial geniculate nucleus and the auditory cortex, not only are neurons excited by certain frequencies, but they are also *inhibited* by neighbouring frequencies. This interplay of excitation and inhibition further sharpens the frequency responses, allowing us to discriminate very small frequency differences.

All levels of the auditory pathway display *tonotopic organisation*; that is, they are spatially arranged in an orderly map according to the auditory frequencies to which they respond. This organisation begins with the cochlea—remember, frequency is ordered along the length of the basilar membrane (see Figure 9.3), and because the IHC fibres exiting the cochlea are organised according to their points of origin along the basilar membrane, analysis of the auditory projection reveals an orderly map from low frequencies (sounds that we perceive as lower pitched, or "bass") to high frequencies (perceived as higher pitched, or "treble"). This can be seen when the responses of auditory brain regions to tones of different frequencies are mapped, as shown in **FIGURE 9.8**. (The organisation of auditory cortical areas in other species

Go to Animation 9.2
Mapping Auditory Frequencies

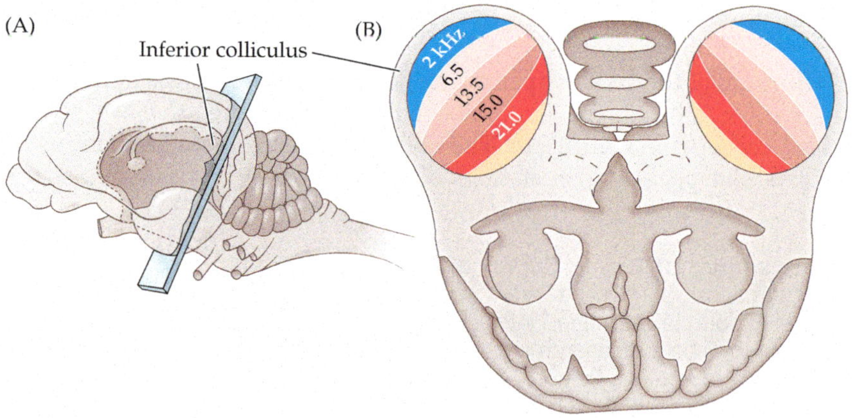

9.8 Tonotopic Mapping in the Cat Inferiour Colliculus (A) This lateral view of the cat brain shows the plane of the transverse section through the inferiour colliculus shown in part B. (B) Following exposure to tones of different frequencies, labeling of neuronal activity using 2-deoxyglucose reveals an orderly tonotopic map within the inferiour colliculus. (After J. Serviere et al., 1984. *J Comp Neurol* 228: 463–477.)

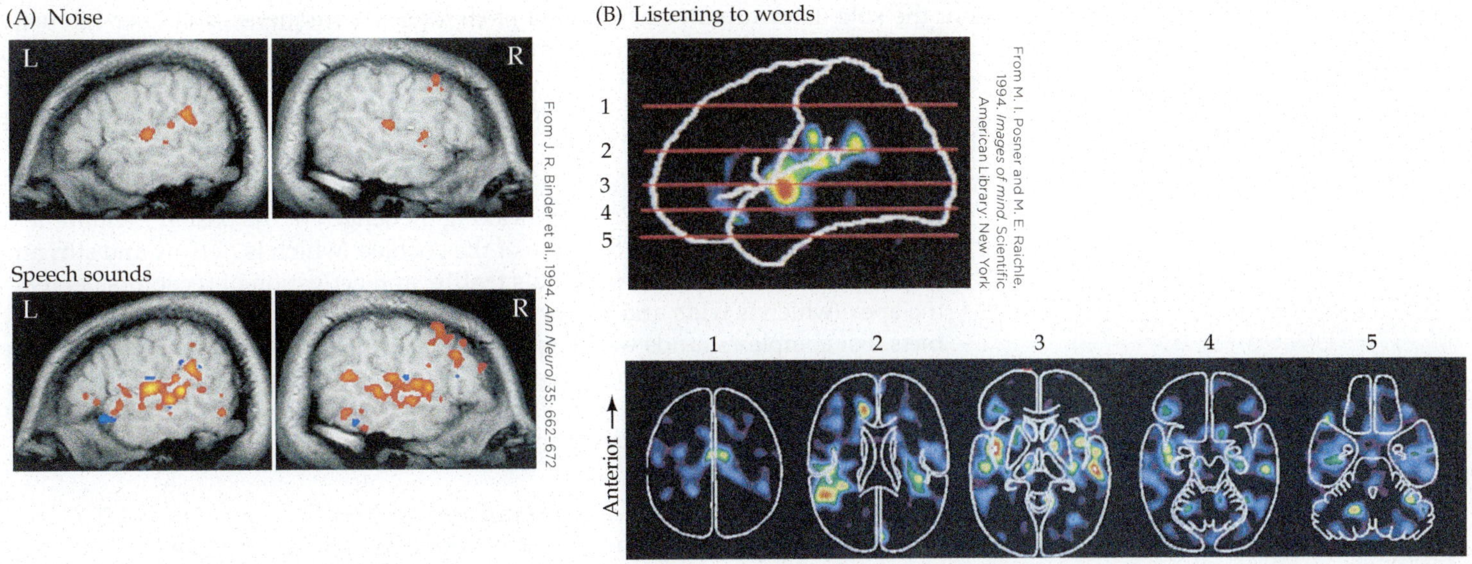

9.9 Responses of the Human Auditory Cortex to Random Sounds versus Speech A) Functional MRI scans of the cerebral hemispheres show that pure tones or noise (top) activate chiefly the primary auditory area on the superiour aspect of the temporal lobe, while speech sounds (bottom) activate other auditory cortical regions, as well as the primary auditory area. (B) Lateral (top) and horizontal (bottom) PET scans show that listening to words activates not only several regions of the cerebral cortex but also regions of the thalamus and the cerebellum. The numbered horizontal lines in the top panel correspond to the levels of the horizontal sections in the bottom panel.

is described on the website in **A STEP FURTHER 9.2: THE AUDITORY CORTICAL REGIONS OF MANY SPECIES SHOW TONOTOPIC ORGANISATION**.)

Brain-imaging studies in humans confirm that many sounds (tones, noises, and so on) result in activation of **primary auditory cortex (A1)**, located on the upper surface of the temporal lobes (**FIGURE 9.9A**). Speech sounds produce similar activation but additionally activate other, more specialised auditory areas (**FIGURE 9.9B**). Interestingly, at least some of these regions are activated when hearing people try to lip-read—that is, to understand someone by watching that person's lips without auditory cues (L. E. Bernstein et al., 2002); this result suggests that the auditory cortex integrates other, nonauditory, information with sounds.

9.3 Pitch Information Is Encoded in Two Complementary Ways

Learning Objectives

After reading this section, you should be able to:

9.3.1 **Review the frequency discrimination ability of the human auditory system, and distinguish between frequency and pitch.**

9.3.2 **Describe the two major accounts of encoding of frequency information by the cochlea.**

9.3.3 **Contrast human frequency sensitivity with that of other animals.**

Most of us can discriminate very small differences in frequency of sound over the entire audible range—from 20 Hz to 15,000 or even 20,000 Hz. The ability to detect a change in frequency is usually measured as the **minimal discriminable frequency difference** between two tones. The detectable difference is about 2 Hz for sounds up to 2000 Hz; at higher frequencies, larger differences are required.

Note that *frequency* and *pitch* are not synonymous terms. *Frequency* describes a physical property of sounds; *pitch* relates solely to the subjective sensory experience of sounds (see Box 9.1). This is an important distinction because frequency is

primary auditory cortex (A1) The region of superiour temporal cortex in which auditory processing occurs.

minimal discriminable frequency difference The smallest change in frequency that can be detected reliably between two tones.

not the sole determinant of perceived pitch (at some frequencies, higher-intensity sounds may seem higher pitched), and changes in pitch do not precisely parallel changes in frequency.

How do we distinguish pitches? Two signals from the cochlea appear to inform the brain about pitch:

1. According to **place coding** theory, the pitch of a sound is encoded in the physical location of the activated receptors along the length of the basilar membrane: activation of receptors near the base of the cochlea (which is narrow and stiff and responds to high frequencies) signals **treble**, and activation of receptors nearer the apex (which is wide and floppy and responds to low frequencies) signals **bass**. For complex sounds with components at several different frequencies, the cochlea maps simultaneous peaks of vibration at different places along the basilar membrane (something like Fourier analysis; see Box 9.1). This is another version of **labelled lines** (see Chapter 7)—each axon fires in response to one particular frequency.
2. A complementary account called **temporal coding** theory proposes that the frequency of an auditory stimulus is encoded in the rate of firing of auditory neurons. In such cases the rate of firing of action potentials by auditory neurons is *phase-locked* to the stimulus; that is, the spacing of action potentials is some integer multiple of the period (the duration) of each sound cycle. The frequency of action potentials produced by the neuron is thus directly mathematically related to the number of cycles per second (i.e., hertz) of the sound. Because individual cells tend not to encode every sound cycle, especially as frequencies increase, an accurate phase-locked representation may be encoded within the combined output of several cells rather than in the output of a single cell. Encoding sound frequency within volleys of action potentials in this manner allows the auditory system to overcome the maximal neuronal firing rate of about 1000 action potentials per second. In general, temporal coding is most evident at the lower end of the hearing range, but it functions up to about 4000 Hz.

Because each mechanism makes up for certain deficiencies in the other, the contemporary view of pitch perception incorporates elements of both models. Thus, the frequency properties of a sound are coded in two ways: (1) according to the distribution of excitation among cells, that is, place coding or tonotopic representation; and (2) according to the temporal pattern of discharge in cells projecting to the auditory cortex.

Mammalian species employ an astonishing range of frequencies in their vocalisations, from **infrasound** (less than 10 Hz) in elephants and whales to **ultrasound** (greater than 20,000 Hz) in bats and porpoises and many other species (the ghost-faced bat emits vocalisations at an incredible 160,000 Hz). These sounds have been shaped by evolution to suit each species' particular ecological specialisations (see Figure 9.1). For example, many species of bats analyse the echoes of their ultrasonic vocalisations, in order to navigate and hunt in the dark; porpoises likewise employ powerful ultrasonic sonar to scan the ocean around them. Their extraordinary sensitivity and accuracy are conferred by elaborate adaptations of their bodies and brains. At the other end of the spectrum, homing-pigeon races are severely disrupted if the birds encounter the sonic-boom shock wave of a supersonic jet (an infrasound pulse that can travel hundreds of miles), suggesting that the pigeons use infrasound cues to establish a navigational map (Hagstrum, 2000). Elephants' voices include infrasonic components so powerful that they can be detected *seismically* by other elephants (Herbst et al., 2012; O'Connell-Rodwell, 2007) while conveying enough detail to distinguish between human-related threats and bee-related threats (Soltis et al., 2014). Tigers may use infrasound to add impact to their roars (E. J. Walsh et al., 2003), and infrasound experimentally inserted into concerts heightens the music's emotional effect on human listeners, but it's not yet known how we detect infrasound.

place coding The encoding of sound frequency as a function of the location on the basilar membrane that is most stimulated by that sound.

treble An aspect of pitch corresponding to the subjective experience of high-frequency sounds (especially musical sounds, such as cymbals).

bass An aspect of pitch corresponding to the subjective experience of low-frequency sounds (especially musical sounds, such as bass guitar).

labelled lines The concept that each nerve input to the brain reports only a particular type of information.

temporal coding The encoding of sound frequency in terms of the number of action potentials per second produced by an auditory nerve.

infrasound Very low-frequency sound; in general, below the threshold for human hearing, at about 20 Hz.

ultrasound High-frequency sound; in general, above the threshold for human hearing, at about 20,000 Hz.

9.4 Brainstem Auditory Systems Are Specialised for Localising Sounds

Learning Objectives

After reading this section, you should be able to:

9.4.1 **Explain the two principal features of sound that the nervous system uses for sound localisation.**

9.4.2 **Compare and contrast the brainstem sound localisation mechanisms of birds and mammals, highlighting key similarities and differences.**

9.4.3 **Describe how sound positions are determined in elevation, rather than azimuth.**

Being able to quickly identify where a sound is coming from—be it the crack of a twig under a predator's foot or the alluring murmur of a would-be mate—is a matter of great evolutionary significance. So it's no surprise that we are remarkably good at locating a sound source (our accuracy is within about 1° horizontally around the head, and many animals are even better). The auditory system accomplishes this feat by analyzing two kinds of **binaural** (two-ear) cues that signal the location of a sound source in the horizontal plane (this is called *azimuth*):

1. **Interaural intensity differences** (IIDs) result from comparison of the *intensity* of the sound—the physical property that we consciously perceive as loudness—between the left versus right ears (*interaural* means "between the two ears"). Depending on the species—and the placement and characteristics of their pinnae—intensity differences occur because one ear is pointed more directly towards the sound source or because the head casts what is called a *sound shadow* (**FIGURE 9.10A**), preventing sounds located to one side (called *off-axis sounds*) from reaching both ears with equal loudness. The sound shadow effect is most pronounced for higher-frequency sounds. Low-frequency sounds have longer sound waves that reach around the head more effectively, reducing the head's sound shadow (**FIGURE 9.10B**).
2. **Interaural temporal differences** (ITDs) are differences between the two ears in the *time of arrival* of sounds. They arise because one ear is always a little closer to an off-axis sound than is the other ear. Two kinds of temporal (time) differences are present in a sound: *onset disparity*, which is the difference between the two ears in hearing the beginning of the sound, and *ongoing phase disparity*, which is the continuing mismatch between the two ears in the time of arrival of all the peaks and troughs that make up the sound wave (these cues are illustrated in **FIGURE 9.10C**).

binaural Referring to two ears.

interaural intensity differences Perceived differences in loudness between the two ears, which can be used to localise a sound source.

interaural temporal differences Differences between the two ears in the time of arrival of a sound, which can be employed by the nervous system to localise sound sources.

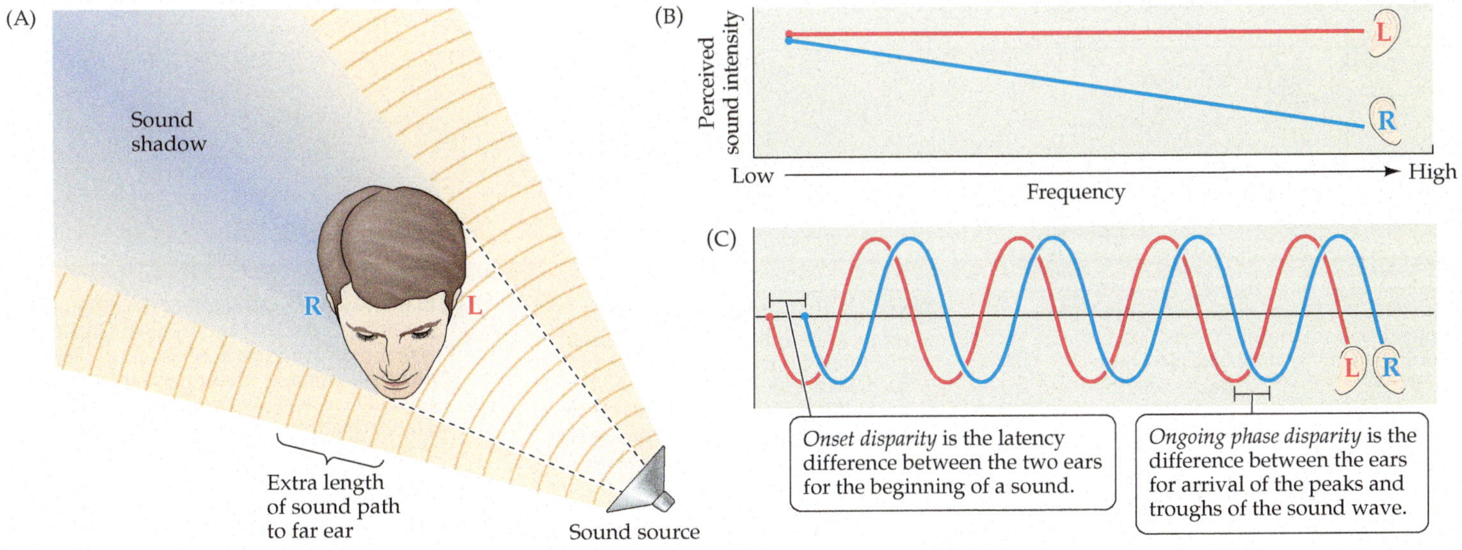

9.10 Cues for Binaural Hearing The two ears receive somewhat different information from sound sources located to one side or the other of the observer's midline. (A) The head casts a sound shadow, producing binaural differences in sound intensity. (B) The resulting differences in perceived intensity are greater at higher frequencies. (C) Sounds also take longer to reach the more distant ear, resulting in binaural differences in time of arrival.

duplex theory A theory that we localise sound by combining information about intensity differences and latency differences between the two ears.

coincidence detector A device that senses the co-occurrence of two events.

Researchers now know that sound localisation involves processing of *both* intensity differences and temporal differences; this is known as the **duplex theory**. Low-frequency sounds, no matter where they originate in azimuth, reach the two ears with comparable intensity. For these frequencies, differences in times of arrival are the principal cues for sound localisation (and at very low frequencies, neither cue is much help; this is why you can place the subwoofer of an audio system anywhere you want). At higher frequencies, however, the sound shadow cast by the head produces significant binaural intensity differences. Of course, you can't perceive which types of processing you're relying on for any given sound; in general, we are aware of the *results* of neural processing but not the processing itself.

What brain systems analyse binaural cues? Birds and mammals both have highly specialised brainstem mechanisms that receive information from the two ears, using arrays of neurons that compare the inputs from the left and right ears with extreme precision (Seidl et al., 2014). But they do it in different ways. In birds, the organisation of neurons within the primary sound localisation nucleus (called the *nucleus laminaris*) constitutes an auditory map of potential sound sources in the external world. In this model, originally proposed by Lloyd Jeffress (1948), each binaural neuron of the nucleus laminaris functions as a **coincidence detector** and is maximally excited by a specific latency difference between inputs from the two ears, corresponding to a particular place in space (Joris et al., 1998) (**FIGURE 9.11**). This map is further refined at higher levels of the auditory system, especially the *tectum* (equivalent to the mammalian inferiour colliculus), which contains a complete map of space.

Mammals apparently do things quite differently (Franken et al., 2015). The superiour olivary nucleus is the primary sound localisation nucleus in the mammalian brain, and its two main divisions serve different functions. The *lateral superiour olive* (LSO) processes intensity differences. The *medial superiour olive* (MSO) processes temporal differences between the two ears, but in contrast to the nucleus laminaris in birds, the MSO does not seem to contain a map of auditory space. Instead, sound location is encoded by the overall relative activity of the neurons of the left MSO compared with the right MSO, through a combination of excitatory and inhibitory outputs from cells that respond to coincidence of sound arriving at the left and right ears (Grothe et al., 2010). So, for example, a sound on the midline would activate the left and right MSOs equally, and the two signals would effectively cancel each other out. But a sound from the right side of space would produce more excitation of the left MSO than the right MSO, and the converse

9.11 The Classic (Jeffress) Model of Sound Localisation in the Auditory Brainstem of Birds

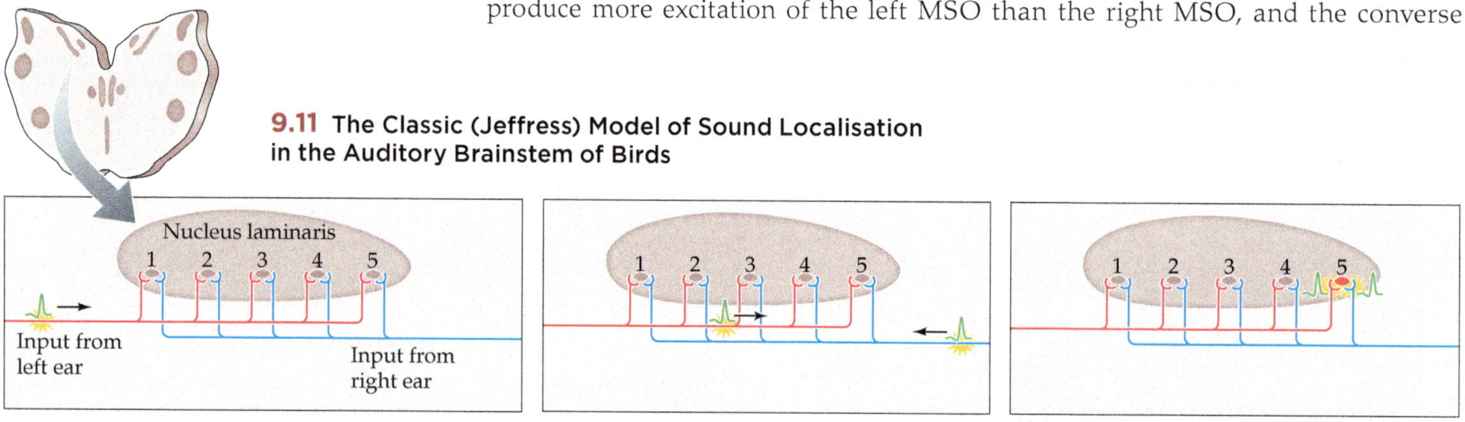

1 Sound to the left midline causes input action potentials from the left cochlea/cochlear nucleus to arrive at the nucleus laminaris.

2 Moments later, the same sound reaches the right ear, causing action potentials from the right ear/right cochlear nucleus to arrive at the nucleus laminaris slightly later than the action potentials from the left ear.

3 In this example, the result of the difference in timing of the left- and right-ear signals is that they converge at neuron 5 in the nucleus laminaris. Neuron 5 is thus a coincidence detector that localises sounds originating from a specific point to the left of midline. In this simple example only 5 neurons are modelled; in reality, the operation of thousands of such coincidence-detecting neurons provides a detailed spatial map of sound sources.

would be true for sounds on the left, proportionate to degree the sound is off axis. This disparity is passed along for further processing at other levels of the auditory system. The avian and mammalian sound localisation systems are so different that they probably arose independently in these classes of animals: an example of convergent evolution that we discussed in Chapter 6 (Grothe and Pecka, 2014).

The structure of the external ear provides yet another sort of localisation cue. As we mentioned earlier, the hills and valleys of the external ear selectively reinforce some frequencies in a complex sound and diminish others (**FIGURE 9.12**). This process is known as **spectral filtering**, and the frequencies that are affected depend on where the sound originates (Zonooz et al., 2019).

The relationship between spectral cues and location is learned and calibrated during development. Unlike the binaural intensity and latency cues that localise a sound in azimuth (horizontal location), spectral cues provide critical information about *elevation* (vertical location). Researchers can place a speaker directly in the ear canal and, by varying the binaural cues and the spectral filtering of natural sounds, fool a person or animal into believing that the sound came from a particular point in space (L. Xu et al., 1999).

Orchestra conductors are especially good at identifying the spatial location of sound sources (Münte et al., 2001), presumably because of extensive practice ("Ms. Hornbeck, could you *please* play in tune?"). Single-cell recordings indicate that the multiple binaural and spectral cues involved in sound localisation converge in the auditory system at the level of the inferiour colliculus (Slee and Young, 2014).

9.12 All Ears As in other mammals, including humans, the ridges and valleys of this bat's outer ear produce precise changes, called *spectral filtering*, in the sounds being funnelled into the ear. This process provides additional cues about the location of a sound source. Some moths produce ultrasonic clicks that interfere with these signals (Corcoran et al., 2009), confusing the bat about where to find that tasty moth.

9.5 The Auditory Cortex Processes Complex Sounds

Learning Objectives

After reading this section, you should be able to:

9.5.1 **Discuss the functions of auditory cortex from an ecological perspective.**

9.5.2 **Evaluate the importance of experience in the development and tuning of the auditory system throughout the life-span.**

9.5.3 **Describe the relationship between musical experience and the development of auditory competencies in music and other domains.**

The historical view of auditory function was that the various subcortical auditory areas performed only basic processing, serving mostly as stepping-stones in a pathway to the auditory cortex (Masterton, 1993). The cortex, it was believed, was where auditory sensation and discrimination really arose. But behavioural testing suggested otherwise; for example, although auditory cortical neurons do respond to pure tones, cats can still discriminate different tones following surgical removal of primary auditory cortex (M. R. Rosenzweig, 1946). So what does auditory cortex do, specifically?

Most of the sounds in nature—vocalisations, footsteps, snaps, crackles, and pops—contain many frequencies and change rapidly. The auditory cortex seems to be specialised for the detection of these more complex "biologically relevant" sounds (Theunissen and Elie, 2014). In other words, the auditory cortex evolved to process the sounds of everyday life. Indeed, most cortical auditory neurons habituate rapidly to continuous tones, ceasing to respond after only a few milliseconds, but brief or abruptly changing sounds usually evoke responses from many neurons, from the cochlear nuclei to the auditory cortex.

Other cortical regions have rich interconnections with auditory cortex. Two streams of auditory processing are found in the cortex (Hackett, 2011; Kaas and Hackett, 1999): a dorsal stream, involving the parietal lobe, is concerned with spatial *location* of sounds; a ventral stream through the temporal lobe may analyse the various *components* of sounds (Romanski et al., 1999), which in the human left hemisphere might include

spectral filtering Alteration of the amplitude of some, but not all, frequencies in a sound.

processing of speech sounds (see Chapter 19). This organisation suggests the existence of separate auditory processing streams for *what* versus *where*, neatly parallelling a similar scheme that has been proposed for visual processing that we will discuss in Chapter 8 (see Figures 8.27 and 8.28). Furthermore, during locomotion the activated motor cortex selectively modifies the responsiveness of auditory cortex, helping it to focus on specific stimuli that are of immediate importance (D. M. Schneider et al., 2014).

Experience during early development and throughout life affects auditory perception and the auditory pathways

The unique capabilities of the auditory cortex result from hardwired sensitivity that is fine-tuned by experience as we grow (Kandler et al., 2009). Human infants have diverse hearing capabilities at birth, but hearing for complex speech sounds in particular becomes more precise and rapid through exposure to the speech of their family and other people. Similarly, early experience with binaural hearing (via hearing aids), compared with equivalent monaural hearing or no hearing, has a significant effect on the ability of children to localise sound sources later in life (W. D. Beggs and Foreman, 1980). Studies with lab animals confirm that experience with sounds of a particular frequency can cause a rapid retuning of auditory neurons (**FIGURE 9.13**) (Fritz et al., 2003; N. M. Weinberger, 1998). (You can learn about the role of experience during auditory development in owls on the website in **A STEP FURTHER 9.3: AUDITORY SYSTEMS ARE CALIBRATED THROUGH POLYMODAL INTEGRATION**.) As we grow older, age gradually impairs the responsiveness of auditory cortex neurons and reduces our ability to distinguish between simultaneously occurring auditory stimuli (Overton and Recanzone, 2016; Recanzone, 2018); there's a reason why grandparents complain that they can't understand a conversation in a noisy restaurant.

Music is also effective in shaping the responses of auditory cortex. It might not surprise you to learn that the auditory cortex of trained musicians shows a bigger response to musical sounds than does the same cortex in nonmusicians. After all, when two people differ in any skill, their brains *must* be different in some way, and maybe people born with brains that are more responsive to complex sounds are also more likely to become musicians. The surprising part is the discovery that the extent to which a musician's brain is extra sensitive to musical notes is correlated with the age at which she began her serious training in music—the earlier the training began, the bigger the difference in auditory cortex in adulthood (Pantev et al., 1998). Furthermore, musical training and experience are associated with sharply improved perception and discrimination of basic speech sounds later in life (Intartaglia et al., 2017; M. W. Weiss and Bidelman, 2015). These and similar findings show that intensive musical experience in development potently alters the functioning of auditory pathways in the brain.

Certainly there are big differences by adulthood: the portion of primary auditory cortex where music is first processed, called *Heschl's gyrus*, is more than twice as large in professional musicians as in nonmusicians, and more than twice as strongly activated by music (P. Schneider et al., 2002). Debate continues about whether musical ability is a specific and hardwired human trait or just a happy by-product of systems that evolved for other purposes (Balter, 2004), but the existence of disorders in which people cannot accurately discern tunes (called **amusia**, from the Greek *amousia*, "want of harmony") suggests that at least the rudiments of a musical sense are inborn (Münte, 2002). Further,

amusia A disorder characterised by the inability to discern tunes accurately.

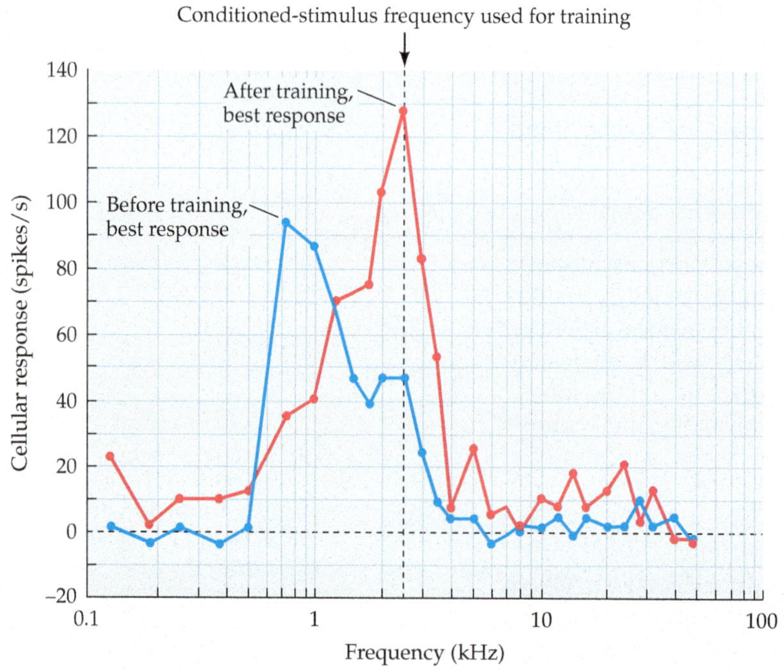

9.13 Plasticity in the Tuning of an Auditory Receptive Field This graph plots the response of a single guinea pig auditory cortex neuron to varying frequencies. Before training, the "best" frequency of the cell was about 0.7 kHz (blue line). After training with a 1.1 kHz tone, the best frequency had shifted to 1.1 kHz (red line), and it stayed there for at least 4 weeks. (After N. M. Weinberger, 1998. *Neurobiol Learn Mem* 70: 226–251.)

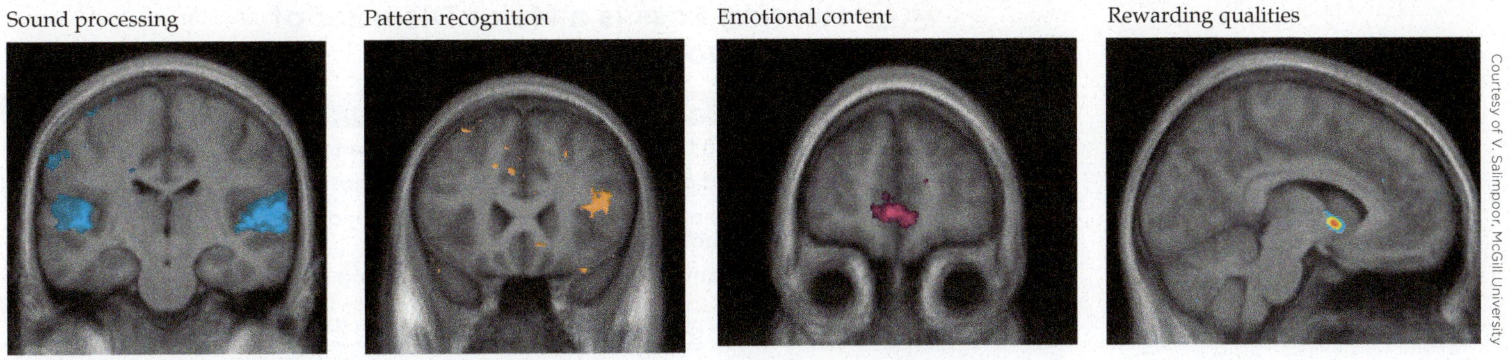

9.14 Making Sweet Music These fMRI images show 4 distinct brain mechanisms that are activated when we concentrate on new music. Separable systems analyse complex sounds (blue) and process temporal patterns of the music (orange). Frontal cortical regions (magenta) are associated with processing the emotional content of the music. The overall pleasurableness of a new song is associated with activity in the nucleus accumbens (rainbow), part of the brain's reward system, and the strength of this signal predicts the amount that people will spend in order to buy it.

auditory cortex that is involved in perceptual analysis of music interacts directly with the mesolimbic dopamine-based reward circuitry of the brain (see Chapter 3) to attach a reward value to music that is new to us (Gold et al., 2019). The degree of activation of this system predicts an individual's willingness to pay for a music recording (Salimpoor et al., 2015; **FIGURE 9.14**).

diffusion tensor imaging (DTI) A modified form of MRI in which the diffusion of water in a confined space is exploited to produce images of axonal fibre tracts.

Tone-deaf Tony, whom we described at the beginning of the chapter, exemplifies the dyslexia-like problem of amusia. Whereas most infants clearly understand the basics of musical relationships almost from birth, people with congenital amusia never develop that ability (Peretz and Hyde, 2003). This seems to be a problem in recognizing pitch, not the timing aspects of music (*rhythm*), since that facet of music sensation may be unaffected in amusia (K. L. Hyde and Peretz, 2004). Brain imaging suggests that the right inferiour frontal gyrus, which is normally active in pitch perception, contains less white matter in people with amusia than in people with normal music perception (K. L. Hyde et al., 2006). Images of fibre tracts between brain regions, created using **diffusion tensor imaging** (**DTI**) (see Chapters 2 and 19), reveal that tone-deaf people have fewer connections between frontal cortex and the temporal auditory cortex (Loui et al., 2009; **FIGURE 9.15**). The result is an inability to consciously access pitch information, even though cortical pitch processing systems are intact (Zendel et al., 2015). However, music is a complex stimulus, and many additional regions of the brain are active in people listening to or producing music; dysfunction in some of these other regions may result in other types of musical defects. If you're worried about your own ability to carry a tune, the National Institutes of Health (NIH) provides an online test of pitch perception at nidcd.nih.gov/tunestest/test-your-sense-pitch.

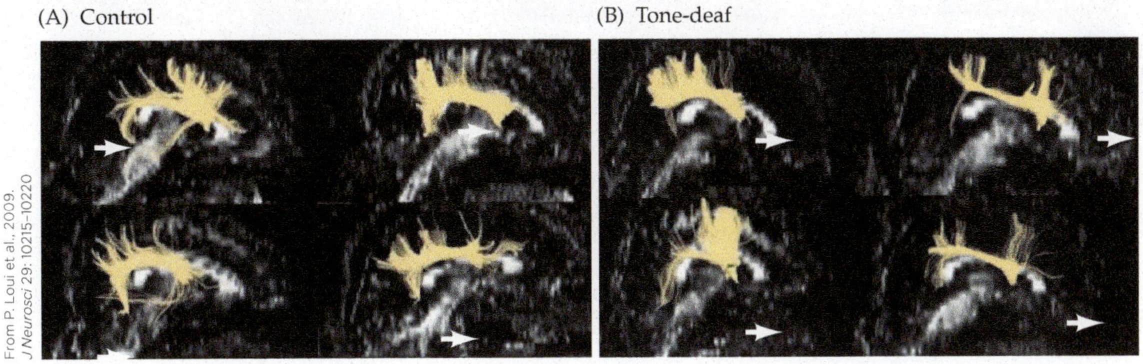

9.15 Brain Connections in Tone-Deaf People Diffusion tensor imaging (DTI) reveals myelinated tracts in these midsagittal images of people facing to the right. The group of fibres connecting the frontal cortex, which is active during pitch discrimination, to the temporal lobe, where auditory processing begins, is called the *arcuate fasciculus* because of its arc shape (arrow). The arcuate fasciculus is much more prominent in four controls (A) than in four people with amusia (B).

9.6 Hearing Loss Is a Major Disorder of the Nervous System

Learning Objectives

After reading this section, you should be able to:

9.6.1 **Describe and contrast the three major categories of hearing loss.**

9.6.2 **Identify potential harmful noise intensities and the ways in which noise harms the auditory system.**

9.6.3 **Summarise and evaluate methods for treating each form of hearing loss.**

Disorders of hearing include **hearing loss** (defined as decreased sensitivity to sound, ranging from moderate to severe) and **deafness** (a loss of hearing so profound that speech cannot be perceived even with the use of hearing aids). Hearing loss affects as much as 15% of the population—about 37.5 million people in the United States alone (Blackwell et al., 2014)—with more than 3% categorized as experiencing serious hearing loss or deafness (Mitchell, 2006). The prevalence of hearing loss is probably similar in other developed countries. Auditory problems can come about in several different ways.

There are three main causes of hearing loss and deafness

Many severe hearing impairments arise early in life and impair language acquisition. Others occur later in life as a consequence of environmental factors, such as exposure to loud sounds, infections, or side effects of certain drugs. **FIGURE 9.16** depicts the three kinds of problems that can prevent sound waves in the air from being transformed into conscious auditory perception: problems with vibrations reaching the cochlea, trouble within the cochlea in converting sound waves into action potentials, and dysfunction of brain regions that process sound information:

hearing loss Decreased sensitivity to sound, in varying degrees.

deafness Hearing loss so profound that speech perception is lost.

conduction deafness A hearing impairment that is associated with pathology of the external-ear or middle-ear cavities.

1. Before anything even happens in the nervous system, the ear must convert the sound vibrations in air into waves of fluid within the cochlea. **Conduction deafness** or hearing loss arises when disorders of the outer or middle ear prevent vibrations produced by auditory stimuli from reaching the cochlea

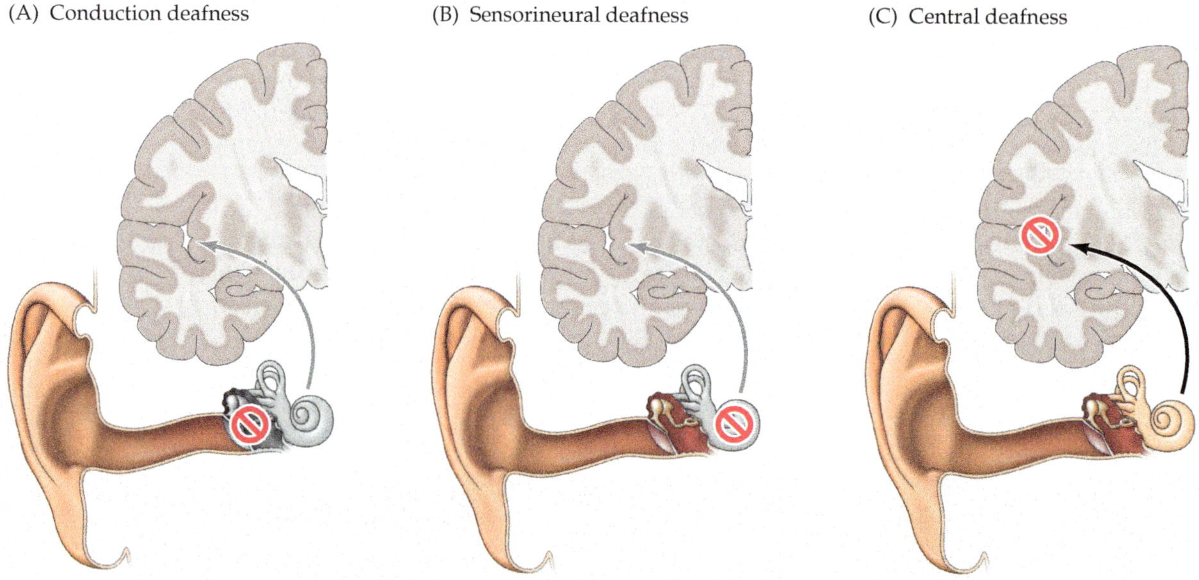

9.16 Three Major Categories of Hearing Loss and Deafness Most deafness is the result of either (A) a problem with the outer or middle ear that prevents the transmission of sound energy to the cochlea, called *conduction deafness*, or (B) a problem with the cochlea or its neural projections that interferes with the transduction of mechanical sound energy into action potentials that the brain can understand, termed *sensorineural deafness*. (C) Less commonly, people may also experience hearing difficulties as a result of lesions of auditory areas in the brain.

(**FIGURE 9.16A**). In one common form of conduction deafness, the ossicles become fused and can no longer transmit sound vibrations effectively. Surgery to free up the ossicles is helpful in some cases. The nervous system is generally not involved in this form of hearing loss.

2. Even if vibrations are successfully conducted to the cochlea, the sensory apparatus of the cochlea—the organ of Corti—may fail to encode them in the form of action potentials, a problem known as **sensorineural deafness** or hearing loss. This type of hearing loss can be the result of loud noise exposure, metabolic problems, infections, exposure to toxins, trauma, or hundreds of different hereditary disorders (**FIGURE 9.16B**). Sensorineural hearing loss is usually permanent because damage to the structure of the organ of Corti, particularly the loss of hair cells, is generally irreversible. Defects in certain genes affecting hair cell structure and function are prominent causes of sensorineural deafness that is present from birth (Petit and Richardson, 2009). For example, a mutations in the gene *GJB2* is a major cause of congenital or early-onset hearing loss (Azadegan-Dehkordi et al., 2019); this gene encodes the protein connexin-26, which is involved in the formation of electrical synapses (known as *gap junctions*; see Figure 3.19).

   Drug-induced hearing loss sometimes results from the toxic properties of a group of antibiotics that includes streptomycin and gentamicin. The **ototoxic** (ear-damaging) properties of streptomycin were discovered when many patients who received it as treatment for tuberculosis subsequently developed cochlear damage. In severe cases the hair cells of the cochlea were completely destroyed, producing total, irreversible loss of hearing.

   Noise pollution and loud sounds—industrial noise, loud engines, gunfire—can severely damage the cochlea in a short period of time. Once again, the hair cells suffer the brunt of the damage: in the affected part of the cochlea, the stereocilia appear shattered and broken, like a flattened forest (**FIGURE 9.17**). There is growing concern about hearing loss caused by listening to personal music players over headphones, which can produce sounds above 100 dB. Those who listen for more than 5 hours per week at 89 dB or louder are exceeding workplace limits, so they are exposing themselves to levels known to induce permanent hearing loss (SCENIHR, 2008). Loud sounds coupled with the use of some over-the-counter drugs, such as aspirin, can also have profound cumulative effects on hearing (McFadden and Champlin, 1990), reducing sensitivity to certain tones by up to 40 dB and/or leading to the development of **tinnitus**, a sensation of noises or ringing in the ears (Zenner et al., 2017).

3. **Central deafness** (hearing loss caused by brain lesions such as stroke) (**FIGURE 9.16C**) is seldom a simple loss of auditory sensitivity. An example illustrating the complexity of changes in auditory perception following cerebral cortical damage is **word deafness**, a disorder in which people show normal speech and hearing for simple sounds but cannot recognise spoken words. Word deafness may be due to an abnormally slow analysis of auditory inputs. Another example of central

sensorineural deafness A hearing impairment that originates from cochlear or auditory nerve lesions.

ototoxic Toxic to the ears, especially the middle or inner ear.

tinnitus A sensation of noises or ringing in the ears.

central deafness A hearing impairment that is related to lesions in auditory pathways or centres, including sites in the brainstem, thalamus, or cortex.

word deafness The specific inability to hear words, although other sounds can be detected.

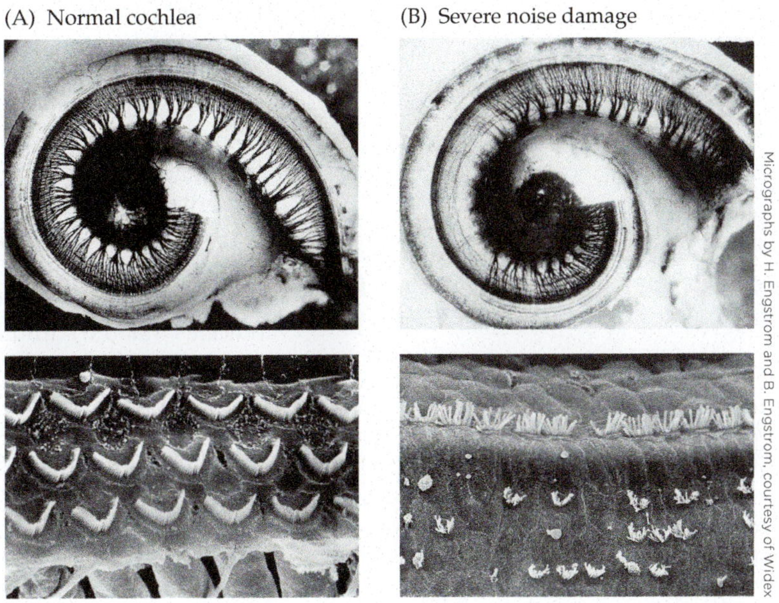

9.17 Cochlear Damage Due to Loud Noise (A) In a normal cochlea, hair cells line the organ of Corti throughout its length (top). With scanning electron microscopy, the stereocilia are easily visualised as orderly rows of bristles (bottom) protruding into the tectorial membrane. (B) Exposure to excessively loud sounds can have rapid destructive effects in the cochlea. Here, the hair cells are completely missing from a wide extent of the cochlea (top) corresponding to the frequencies of the intense sounds that the person was exposed to. The stereocilia (bottom) have been crushed and flattened by the noise exposure.

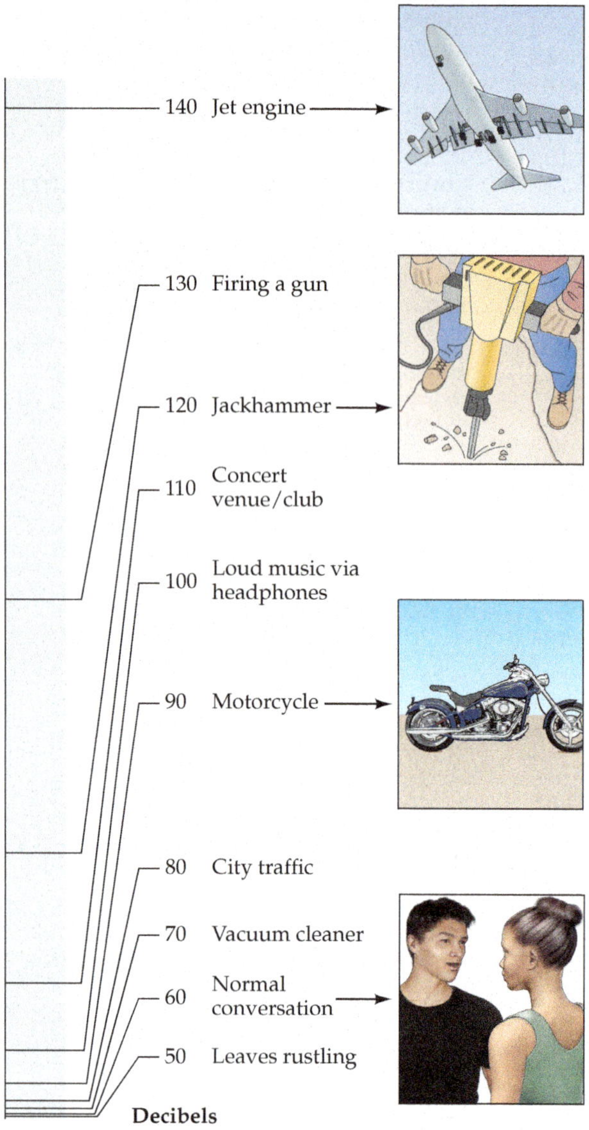

9.18 How Loud Is Too Loud? Because the decibel scale is logarithmic, each increment of 10 dB represents a 100-fold increase in loudness. Listening to sounds over 85–90 dB for extended periods of time can cause hearing loss, and by 110 dB or so, exposure for as little as 30 minutes can be harmful. Earplugs are available that attenuate all frequencies equally, making concerts a little quieter without muffling the music.

deafness is **cortical deafness**, in which patients have difficulty recognizing both verbal and nonverbal auditory stimuli. Cortical deafness is a rare syndrome because it requires bilateral damage to the auditory cortex.

Strokes that interrupt all of the projection fibres from the medial geniculate nucleus to the various auditory cortical regions also cause deafness. People with stroke-induced deafness still show various acoustic reflexes mediated by the brainstem—such as a startle response to a sudden, loud noise—although they deny hearing the sounds to which they are reacting. In contrast, patients with bilateral destruction of only the *primary* auditory cortex often have less-severe hearing loss, presumably because other auditory cortical regions contribute to hearing, as discussed earlier. And central hearing impairment doesn't necessarily involve gross tissue damage: for example, long-term exposure to noise that isn't loud enough to hurt the cochlea may degrade hearing by provoking plastic changes in auditory cortex (Gourévitch et al., 2014). A comparison of sound sources is depicted in **FIGURE 9.18**; if you are concerned about your own exposure, excellent sound level metre apps for smartphones are available at little or no cost (including one from the National Institute for Occupational Safety and Health (NIOSH) at www.cdc.gov/niosh/topics/noise/app.html).

Treatments for deafness focus on replacing missing stimulation

Generations ago, people experiencing hearing loss due to conduction problems used ear horns (basically reversed megaphones) to capture sounds and direct them into the ear more loudly than usual; cupping one hand over your ear has a similar but more modest effect. Similarly, today's hearing aids use electronic amplification to deliver louder sounds to the impaired—but still functional—auditory system, and surgery may be used to free up the fused ossicles or replace them with a Teflon prosthetic, restoring the transmission of vibrations to the cochlea.

Unfortunately, amplifying sounds is less useful in sensorineural hearing loss, where neural elements have been irreversibly damaged or were absent from birth. But while the hair cells may be lost, the electrical excitability of the auditory nerve often remains unchanged. So one approach in sensorineural deafness involves the use of **cochlear implants** to directly stimulate the auditory nerve, thereby bypassing the ossicles and hair cells altogether (**FIGURE 9.19**). Cochlear implants provide only a limited range of frequencies and loudness, so they do not restore normal hearing, but they nonetheless provide stimulation that the brain can interpret as sound. Functional brain imaging indicates that the stimulation activates auditory cortex in a tonotopic manner, and it appears to be processed by auditory cortex in similar fashion to "natural" inputs (Lazeyras et al., 2002). For example, modern cochlear implants provide enough information to allow recipients to distinguish between voiced and unvoiced speech sounds (e.g., "v" and "f"), which cannot be distinguished in lipreading. They likewise allow reasonably good (but not perfect) perception of tonal aspects of speech, which are especially important in languages like Mandarin (S. Wang et al., 2013). The best linguistic outcomes seem to occur when implants are received early in life (Geers et al., 2017), capitalising on the plasticity of the young brain.

Go to Video 9.3
A Baby Hears His Mother's Voice

cortical deafness A hearing impairment that is caused by a fault or defect in the cortex.

cochlear implant An electromechanical device that detects sounds and selectively stimulates nerves in different regions of the cochlea via surgically implanted electrodes.

(A)

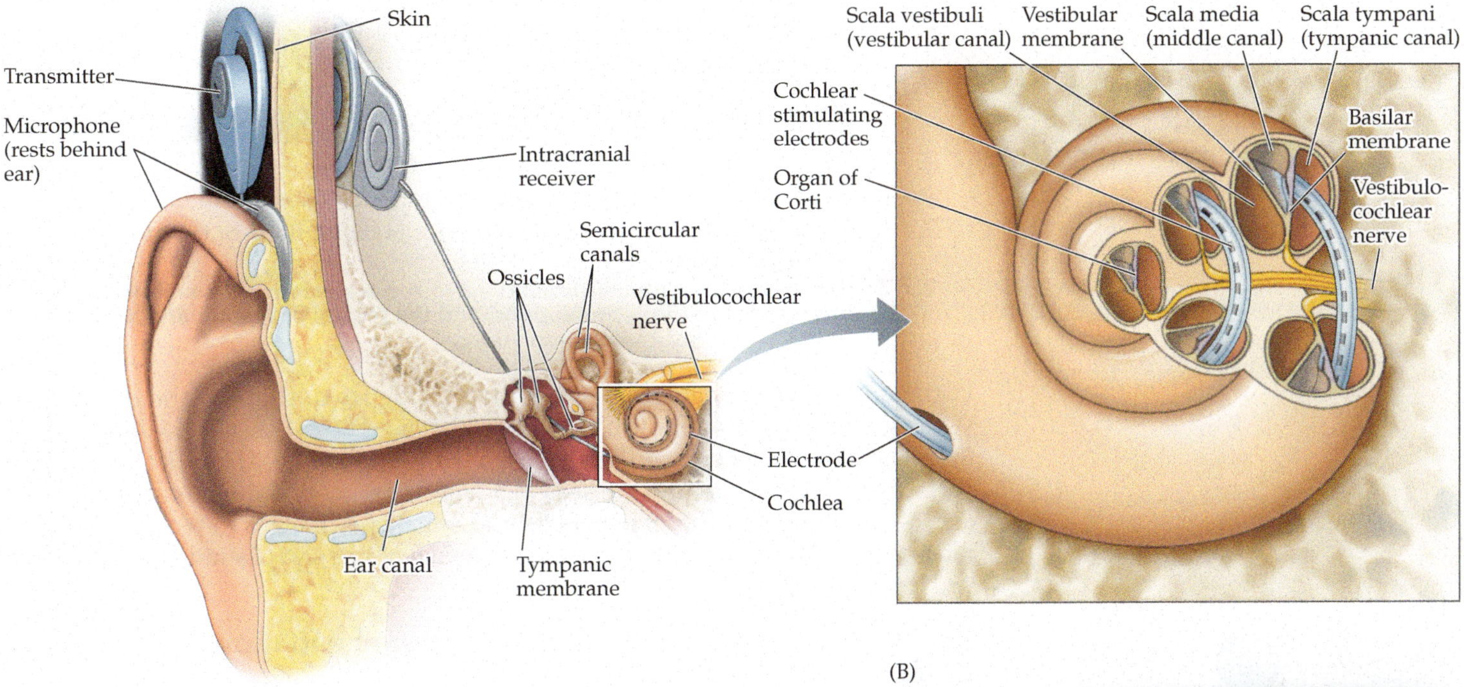

(B)

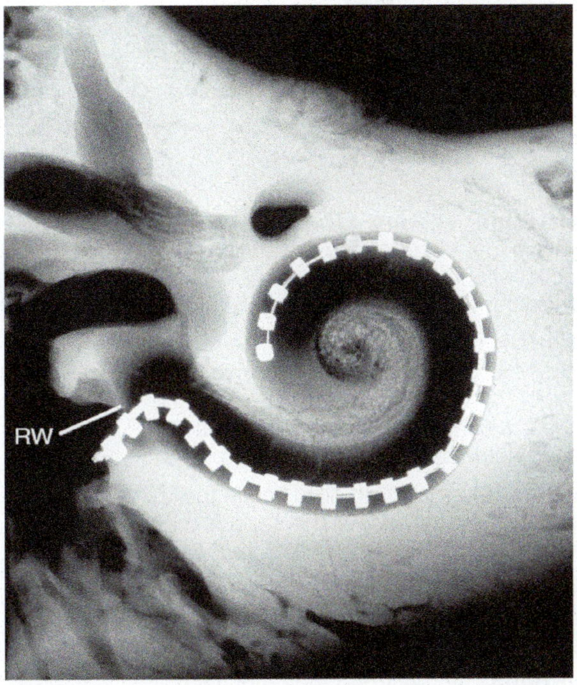

From G. Clark, 2003. *Cochlear implants: Fundamentals and applications.* Springer Nature: New York. Reprinted by permission.

9.19 Cochlear Implants Provide Hearing in Some Deaf People
A microphone detects sound and directs the cochlear implant circuitry to stimulate the auditory nerve. Although this apparatus provides only a crude simulation of ordinary auditory nerve activity, the brain can learn to use the information to decipher speech. (B) This X-ray reveals a cochlear implant in place within the cochlea, inserted via the round window (RW). Each of the 29 small beads is a stimulating electrode that excites the auditory nerve at positions along the organ of Corti that correspond to successively higher frequencies

Within the Deaf community, the use of cochlear implants is the subject of a lively controversy. While many hearing people might assume that a deaf person would automatically want to have his hearing restored, some deaf advocates are concerned that cochlear implants threaten a vibrant Deaf culture. The solution may lie in ensuring through education that a child with cochlear implants retains a sense of belonging in both the hearing and Deaf communities, rather than being forced to identify with one or the other.

Can damaged hair cells be regrown? Although fishes and amphibians produce new hair cells throughout life, mammals have long been viewed as incapable of regenerating hair cells. This conclusion may have been too hasty, however (Géléoc and Holt, 2014). As in fishes and birds, the supporting cells packed around the hair cells of the mammalian organ of Corti do remain capable of dividing and differentiating into hair cells in adulthood (Rubel et al., 2013). Deletion of the gene *Rb1* or blockade of a gene in the Notch signalling system in mice enables hair cells themselves to divide, providing new, functional hair cells in a deafened cochlea (W. Li et al., 2015; Mizutari et al., 2013). Alternatively, a gene therapy that uses a virus to insert the gene *Atoh1* into the cochlea of deafened mice results in the growth of new hair cells and significant restoration of hearing within 1–2 months (Kraft et al., 2013). Human clinical trials of *Atoh1* and other drugs and gene therapies are currently under way, raising hopes of an effective treatment for sensorineural hearing loss in the near future.

Now let's consider another sensory system that relies on hair cells, this time to track the position and motion of the head.

VESTIBULAR PERCEPTION

9.7 An Inner Ear System Senses Gravity and Acceleration

Learning Objectives

After reading this section, you should be able to:

9.7.1 **Describe the anatomical features of the vestibular system.**

9.7.2 **Explain how accelerations and changes in the position of the head are transduced into sequences of action potentials.**

9.7.3 **Describe the vestibular projections to the brainstem, the functional importance of these projections, and the results of vestibular dysfunction or abnormal vestibular stimulation.**

Without our sense of balance, it would be a challenge to simply stand on two feet. When you use an elevator, you clearly sense that your body is rising or falling, despite the sameness of your surroundings. When you turn your head, take a tight curve in your car, or bounce through the seas in a boat, your continual awareness of motion allows you to plan further movements and anticipate changes in perception due to movement of your head. And of course, too much of this sort of stimulation can make you lose your lunch.

Go to Animation 9.4
The Vestibular System

Specialised vestibular structures capture head movements in three-dimensional space

The position and movement of the head are detected by the **vestibular system**, which is part of the inner ear, attached to the adjacent cochlea within a hollow in the temporal bone (the Latin term *vestibulum* means "entrance hall"). The most obvious components of the vestibular system are the three **semicircular canals**, fluid-filled tubes oriented at right angles to each other (**FIGURE 9.20A**). The canals are oriented in the three different planes in which the head can rotate (**FIGURE 9.20B**)—nodding up and down (technically known as *pitch*), shaking side to side (*yaw*), and tilting left or right (*roll*). The three canals are connected at their ends to a saclike structure called the **utricle** (literally, "little uterus"). Lying below the utricle is another small fluid-filled sac, the **saccule** ("little sac").

Receptors in these structures, just like those of the auditory system, are groups of hair cells, whose stereocilia bend, leading to the excitation of vestibular neurons. Each semicircular canal terminates near the utricle in an enlarged region—the **ampulla** (plural *ampullae*)—that contains the hair cells that signal movement in the corresponding plane (see Figure 9.20B). Here the stereocilia of the hair cells are embedded in a gelatinous structure, the **cupula** (plural *cupulae*). The orientation of the stereocilia is quite precise and determines the direction of mechanical force to which they are especially sensitive. Movement of the head in any one axis sets up a relative flow of the fluid in one of the three semicircular canals, deflecting the stereocilia in the ampulla and signalling the brain that the head has moved. Working together, the three semicircular canals provide accurate tracking of the rotation of the head.

The utricle and saccule each contain an *otolithic membrane*—a gelatinous sheet studded with tiny crystals called **otoliths**, which literally means "ear stones"—which, thanks to its mass, lags slightly when the head moves. This bends nearby hair cells, stimulating them to track straight-line acceleration and deceleration—the final signals that the brain needs in order to calculate the position and movement of the body in three-dimensional space. At the base of the hair cells are nerve fibres connected in much the same way as those that connect auditory hair cells.

It is generally accepted that the auditory organ evolved from the vestibular system, although the ossicles probably evolved from parts of the jaw. Before that, the vestibular system evolved from the **lateral-line system**, a sensory system found in many kinds of fishes and some amphibians. The lateral-line system consists of an array of receptors along the side of the body. Tiny hairs that emerge from sensory cells in the skin are embedded in cupulae, like those in the mammalian ampulla (**FIGURE 9.20C**).

vestibular system The inner ear system that encodes the orientation and acceleration of the head in three axes, crucial for the sense of balance.

semicircular canal One of the three fluid-filled tubes in the inner ear that are part of the vestibular system. Each of the tubes, which are at right angles to each other, detects angular acceleration.

utricle A small, fluid-filled sac in the vestibular system above the saccule that responds to static positions of the head.

saccule A small, fluid-filled sac under the utricle in the vestibular system that responds to static positions of the head.

ampulla An enlarged region of each semicircular canal that contains the receptor cells (hair cells) of the vestibular system.

cupula A small gelatinous structure, containing hair cells that detect fluid movement within the semicircular canals of the vestibular system.

otolith A small crystal on the gelatinous membrane in the vestibular system.

lateral-line system A sensory system, found in many kinds of fishes and some amphibians, that informs the animal of water motion in relation to the body surface.

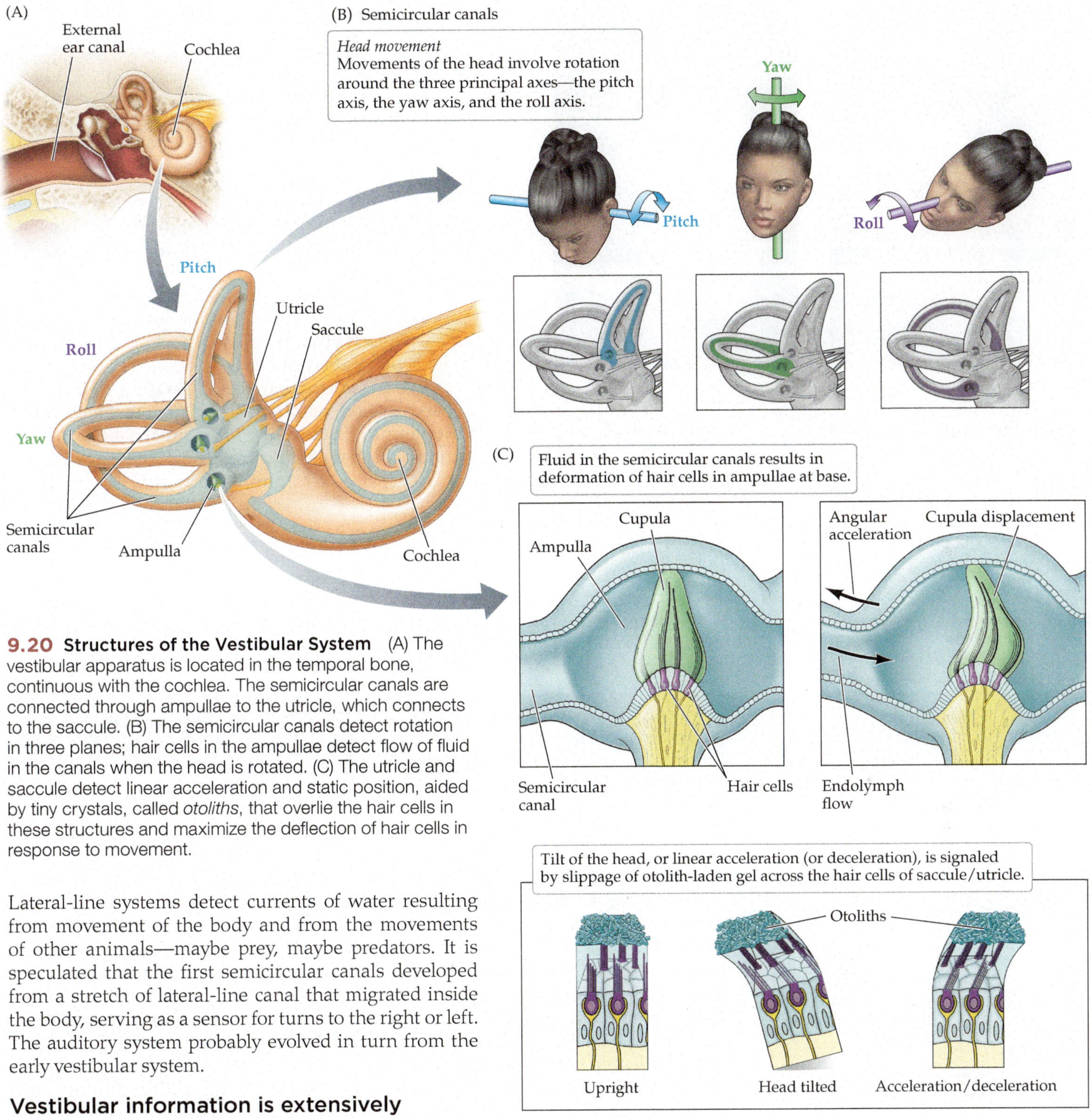

9.20 Structures of the Vestibular System (A) The vestibular apparatus is located in the temporal bone, continuous with the cochlea. The semicircular canals are connected through ampullae to the utricle, which connects to the saccule. (B) The semicircular canals detect rotation in three planes; hair cells in the ampullae detect flow of fluid in the canals when the head is rotated. (C) The utricle and saccule detect linear acceleration and static position, aided by tiny crystals, called *otoliths*, that overlie the hair cells in these structures and maximize the deflection of hair cells in response to movement.

Lateral-line systems detect currents of water resulting from movement of the body and from the movements of other animals—maybe prey, maybe predators. It is speculated that the first semicircular canals developed from a stretch of lateral-line canal that migrated inside the body, serving as a sensor for turns to the right or left. The auditory system probably evolved in turn from the early vestibular system.

Vestibular information is extensively processed in the brainstem

The brain pathways carrying vestibular system information reflect its importance to motor control and posture. Nerve fibres from the vestibular receptors are part of the vestibulocochlear cranial nerve (VIII) entering the brainstem to synapse in the **vestibular nuclei**. Some of the fibres bypass this structure and go directly to the cerebellum, contributing to its motor functions. The outputs of the vestibular nuclei are complex, as is appropriate, considering their influences on the motor system. These outputs go to the motor nuclei of the eye muscles, the thalamus, and the cerebral cortex, among others.

vestibular nuclei Brainstem nuclei that receive information from the vestibular organs through cranial nerve VIII (the vestibulocochlear nerve).

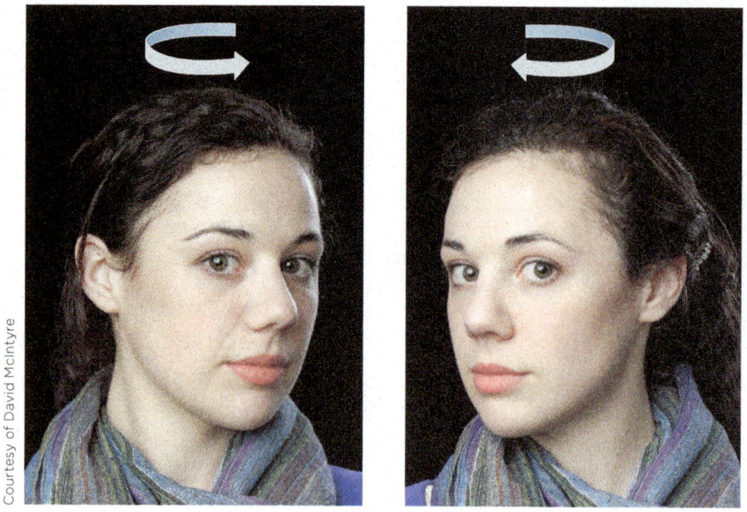

9.21 The Vestibulo-Ocular Reflex Maintaining fixed gaze while the head moves requires ongoing rapid integration of vestibular information into the motor program controlling the six muscles that move each eye. The vestibulo-ocular reflex (VOR) employs a high-speed brainstem network, with inputs from the nearby vestibular nuclei providing near-instantaneous updates about head movements so that the eyes can move in compensation.

vestibulo-ocular reflex (VOR) The brainstem mechanism that maintains gaze on a visual object despite movements of the head.

motion sickness The experience of nausea brought on by unnatural passive movement, as in a car or boat.

One of the many important functions of the vestibular system is the precise control of eye movements. Try moving your head around while staring at a particular spot on the wall. It seems almost effortless to maintain your gaze on the fixation point, but in reality it is a very complex processing problem; the six muscles that control the movement of each eye must rapidly and precisely counter the movements of the head as they occur. This function is called the **vestibulo-ocular reflex** (**VOR**) and involves a high-speed network within the brainstem that uses vestibular information about head rotations to move the eyes in compensation (**FIGURE 9.21**). The VOR functions accurately even when the eyes are closed, showing that vestibular inputs—rather than visual ones—control the accuracy of gaze as we move through the world.

Some forms of vestibular excitation produce motion sickness

There is one aspect of vestibular activation that many of us would gladly do without. Too much strong vestibular stimulation—as when riding in a boat, car, plane, or roller coaster—can produce the misery of **motion sickness**. Motion sickness is caused especially by movements of the body that we cannot control. For example, passengers in a car are more likely to suffer from motion sickness than is the driver.

Why do we experience motion sickness? The *sensory conflict theory* argues that we feel bad when we receive contradictory sensory messages, especially a discrepancy between vestibular and visual information. When an airplane bounces around in turbulence, for instance, the vestibular system signals that various changes in direction and accelerations are occurring, but as far as the visual system is concerned, nothing is happening; the plane's interiour is a constant. One hypothesis is that the stimulation is activating a system that originally evolved to rid the body of swallowed poison (M. Treisman, 1977). According to this hypothesis, discrepancies in sensory information might normally signal a dangerous neurological impairment, triggering dizziness and vomiting to get rid of potentially toxic food. However, data to support the "poison hypothesis" are scanty at best, and overall the evolutionary origins of motion sickness remain a mystery (Oman, 2012). The observation that virtual reality devices frequently induce motion sickness, and that susceptibility to this sickness is associated with individual differences in pretest body sway, has been interpreted as evidence that motion sickness results from postural instability rather than sensory conflict (Munafo et al., 2017).

A different type of vestibular disorientation can strike airplane pilots. Note from Figure 9.20C that either linear acceleration or an upward head tilt can produce the same stimulation of the otolith organs (the saccule and utricle). So in conditions of very low visibility, an acceleration of the plane may be misinterpreted as a climb (an upward tilt of the plane) (P. R. MacNeilage et al., 2007; Sánchez-Tena et al., 2018). This *false-climb illusion* (also called the *somatogravic illusion*) can be very compelling, leading pilots to force their planes into a dive because they mistakenly believe they are climbing. Because the dive itself produces further acceleration, heightening the illusion of climbing, an even steeper dive may be adopted, with tragic consequences. For this reason, among others, pilots of modern aircraft making landings in low-visibility conditions learn to rely on a variety of instruments rather than on their vestibular systems.

THE CHEMICAL SENSES: TASTE AND SMELL

9.8 Chemicals in Foods Are Perceived as Tastes

Learning Objectives

After reading this section, you should be able to:

9.8.1 **Describe the structure, function, and distribution of the papillae on the tongue.**

9.8.2 **Identify the five basic tastes, and explain how each is transduced by specialised mechanisms of the tongue.**

9.8.3 **Trace the neural projection of gustatory information to the brainstem and higher-order systems.**

9.8.4 **Discuss the discovery and significance of taste receptors in tissues other than the tongue.**

Delicious foods, poisons, dangerous adversaries, and fertile mates—these are just a few of the sources of chemical signals i n the environment. Sensitive detection of these signals is vital for survival throughout the animal kingdom. The sense of taste provides an immediate assessment of foods (Lindemann, 1995): sweet indicates high-calorie foods; savoury tastes signal a protein source; salty and sour relate to important aspects of homeostasis; bitter warns of toxic constituents. The sense of smell is critical for appreciating the rich and complex flavours of individual foods but has additional important functions as well, such as signalling the presence of prey, predators, or potential mates.

Most people derive great pleasure from eating delicious food, and because we recognise many substances by their distinct flavours, we tend to think that we can discriminate many different tastes. In reality, though, humans detect only a small number of basic tastes: the huge variety of sensations aroused by different foods are actually **flavours** rather than simple tastes, and they rely on the sense of smell as well as taste. (To appreciate the importance of smell to flavour, block your nose while eating first some raw potato and then some apple: without the sense of smell, you can't tell much difference between them!) Sensitivity to a large number of different odours is responsible for the complexities and subtleties of your favourite dishes, whether your tastes run to BBQ, sushi, pasta, or, inexplicably, surströmming (Google it, we dare you).

flavour The sense of taste combined with the sense of smell.

taste bud A cluster of 50–150 cells that detects tastes. Taste buds are found in papillae.

papilla A small bump that projects from the surface of the tongue. Papillae contain most of the taste receptor cells.

circumvallate papillae One of three types of small structures on the tongue that contain taste receptors, located in the back.

foliate papillae One of three types of small structures on the tongue that contain taste receptors, located along the sides of the tongue.

fungiform papillae One of three types of small structures on the tongue that contain taste receptors, located in the front of the tongue.

Specialised receptor cells on the tongue encode five distinct tastes

Scientists are in broad agreement that we possess five basic tastes: salty, sour, sweet, bitter, and umami. (*Umami*, Japanese for "delicious taste," is the term for the savoury, meaty taste that is characteristic of gravy or soy sauce.) These tastes are determined genetically, as we will see shortly, but there is considerable genetic variation across the globe in both the strength and pleasurable qualities of the basic tastes (Pirastu et al., 2016; Robino et al., 2014). Further, the hunt continues for additional basic tastes. For example, studies suggest that humans may possess a primary fat taste (Mattes, 2011); another candidate, called *kokumi*, is described as the full-bodied, thick, mouth-filling quality of some foods (Brennan et al., 2014).

It is a common misconception that the myriad little bumps on the tongue are the **taste buds** that give us our perceptions of the basic tastes, but they aren't. They are actually **papillae** (singular *papilla*, Latin for "nipple") (**FIGURE 9.22**), tiny lumps of tissue that serve to increase the surface area of the tongue. There are three kinds of papillae, distributed on the tongue as shown in **FIGURE 9.23**. Each of the relatively few **circumvallate papillae** and **foliate papillae** contains many taste buds in its sides. **Fungiform papillae**, which contain only about six taste buds each, resemble button mushrooms in shape (*fungus* is Latin for "mushroom"). The tongue contains hundreds of fungiform papillae, but we'll see that the numbers vary greatly among individuals.

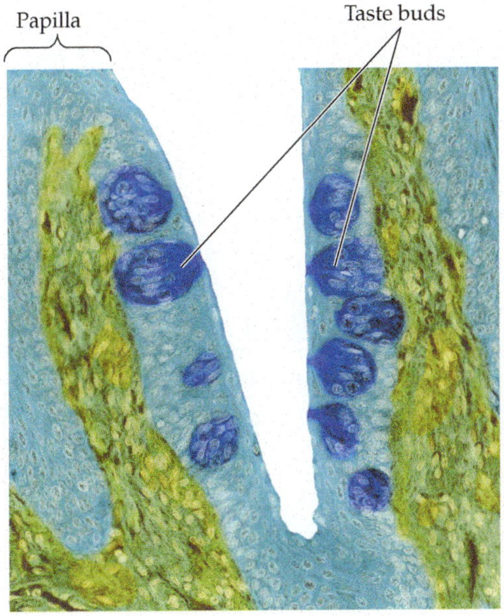

9.22 A Cross Section of the Tongue

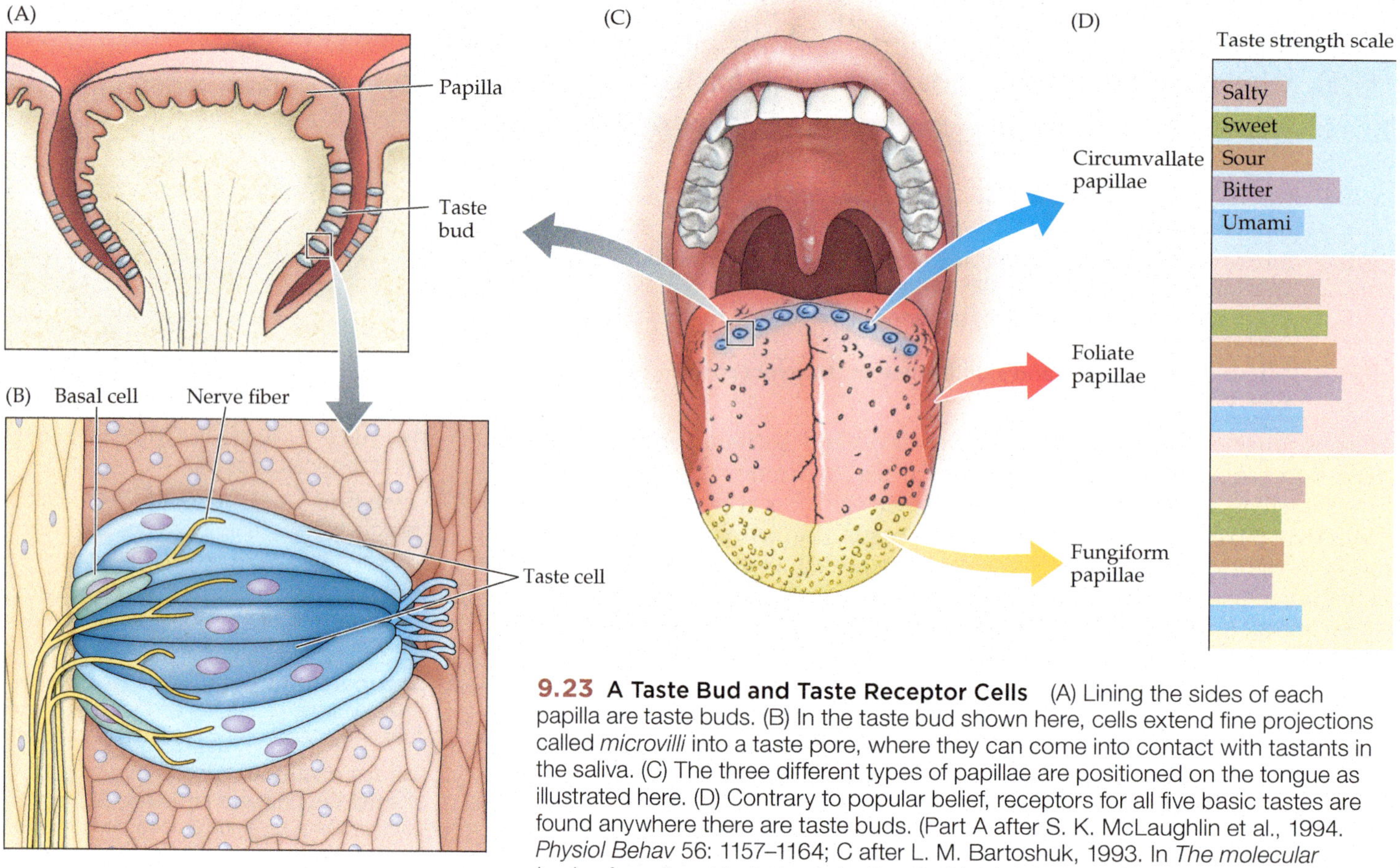

9.23 A Taste Bud and Taste Receptor Cells (A) Lining the sides of each papilla are taste buds. (B) In the taste bud shown here, cells extend fine projections called *microvilli* into a taste pore, where they can come into contact with tastants in the saliva. (C) The three different types of papillae are positioned on the tongue as illustrated here. (D) Contrary to popular belief, receptors for all five basic tastes are found anywhere there are taste buds. (Part A after S. K. McLaughlin et al., 1994. *Physiol Behav* 56: 1157–1164; C after L. M. Bartoshuk, 1993. In *The molecular basis of smell and taste transduction*, D. Chadwick, J. Marsh, and J. Goode [Eds.], pp. 251–257. Wiley: New York; J. Chandrashekar et al., 2006. *Nature* 444: 288–294.)

taste pore The small aperture through which tastant molecules are able to access the sensory receptors of the taste bud.

tastant A substance that can be tasted.

Go to Activity 9.2
Taste

Each papilla holds one or more taste buds, and each taste bud consists of a cluster of 50–150 taste receptor cells (**FIGURE 9.23A**). At the surface end of the taste bud is an opening called the **taste pore**. The taste cells extend fine cilia into the taste pore, where they come into contact with **tastants** (substances that can be tasted). Each taste cell is sensitive to just one of the five basic tastes, and with a life span of only 10–14 days, taste cells are constantly being replaced. Taste buds also contain immature taste cells that eventually become new taste cells (**FIGURE 9.23B**); it's not clear whether this turnover occurs spontaneously or is prompted by the loss of taste cells through daily use. And as our varied experience with hot drinks, frozen flagpoles, or spicy foods informs us, taste is not the only sensory capability of the tongue. Some of the sensory cells within taste buds signal temperature, pain, and touch.

Many books show a map of the tongue indicating that each taste is perceived mainly in one region (sweet at the tip of the tongue, bitter at the back, and so on), but this notion is just an enduring myth. All five basic tastes can be perceived *anywhere* on the tongue where there are taste receptors (Collings, 1974; Yanagisawa et al., 1992). The areas do not differ greatly in the strength of taste sensations that they mediate (**FIGURE 9.23D**).

The ability to taste many substances is already well developed in humans at birth. Even premature infants seem to show a clear preference for sweet substances, as well as an aversion to bitter substances (Steiner, 1974). Newborns seem to be relatively insensitive to salty tastes, but a preference for mildly salty substances develops in the first few months. This preference does not seem to be related to experience with salty tastes; rather it probably indicates maturation of the mechanisms of salt perception (Beauchamp et al., 1994).

Different cellular processes transduce the basic tastes

The tastes salty and sour are evoked when taste cells are stimulated by simple ions acting on ion channels in the membranes of the taste cells. Sweet, bitter, and umami tastes are perceived by specialised G protein-coupled receptors (GPCRs). The five taste transduction pathways are illustrated in **FIGURE 9.24**.

SALTY The transduction of the taste of salt (NaCl) is perhaps the easiest to understand, because it mostly relies on ion channels of the sort we have seen in previous chapters. Sodium ions (Na^+) from salty food enter taste cells via sodium channels in the cell membrane, causing a depolarisation of the taste cell and release of neurotransmitter that stimulates the afferent neurons that relay the information to the brain. We know that this is a crucial mechanism for perceiving saltiness, because blocking the sodium channels with a drug greatly reduces our ability to taste salt. Blockade of sodium channels does not completely eradicate salt taste, however. A second salt receptor, a variant of the receptor TRPV1 (transient receptor potential vanilloid type 1), may provide this additional sensitivity to Na^+, along with sensitivity to cations from other salts in foods, such as potassium (K^+) and calcium ($Ca2^+$) (V. Lyall et al., 2004). Transient receptor potential (TRP) proteins form a large family of receptors that are sensitive to external stimuli of many kinds, including temperature detectors in skin, as we discussed in Chapter 8 (see Figure 8.25). Mice in which TRPV1 has been genetically knocked out show little change in salt sensitivity, however (Treesukosol et al., 2007), so the search for additional, possibly redundant salt receptors continues. One strong candidate for this role is TMC-1, an ionotropic sodium receptor that acts as a salt sensor in invertebrates and is also found in mammals, including humans (Chatzigeorgiou et al., 2013). Interestingly, while the other tastes are generally *either* appetizing (sweet and umami) *or* unpalatable (sour and bitter), salt is *both*: it is appetizing at lower concentrations but powerfully aversive at high concentrations. It turns out that the aversive quality of strong salt taste is accomplished by temporarily recruiting the aversive taste pathways through activation of sour- and bitter-sensing cells (Oka et al., 2013).

SOUR Acids in food taste sour—the more acidic the food, the sourer it tastes—but no one knows exactly how sour tastants are detected. The property that all acids share is that they release protons (H^+; also called hydrogen ions). A number of mechanisms have been proposed for sour detection, especially acid-sensing ion channels (thus resembling the ionotropic receptors described in Chapters 2 and 3). For example, one type of potassium channel is blocked by H^+, preventing the release of K^+ ions from taste cells and causing depolarisation and neurotransmitter release. Although the exact mechanisms of sour detection are not completely understood, it seems that all sour-sensitive taste cells contain a particular type of ion channel protein (called PKD2L1) and share an inward flow of protons that depolarises the cell (Bushman et al., 2015). Interestingly, this same receptor detects the taste of carbonation in drinks (Chandrashekar et al., 2009) and cues water consumption in thirsty animals (Zocchi et al., 2017).

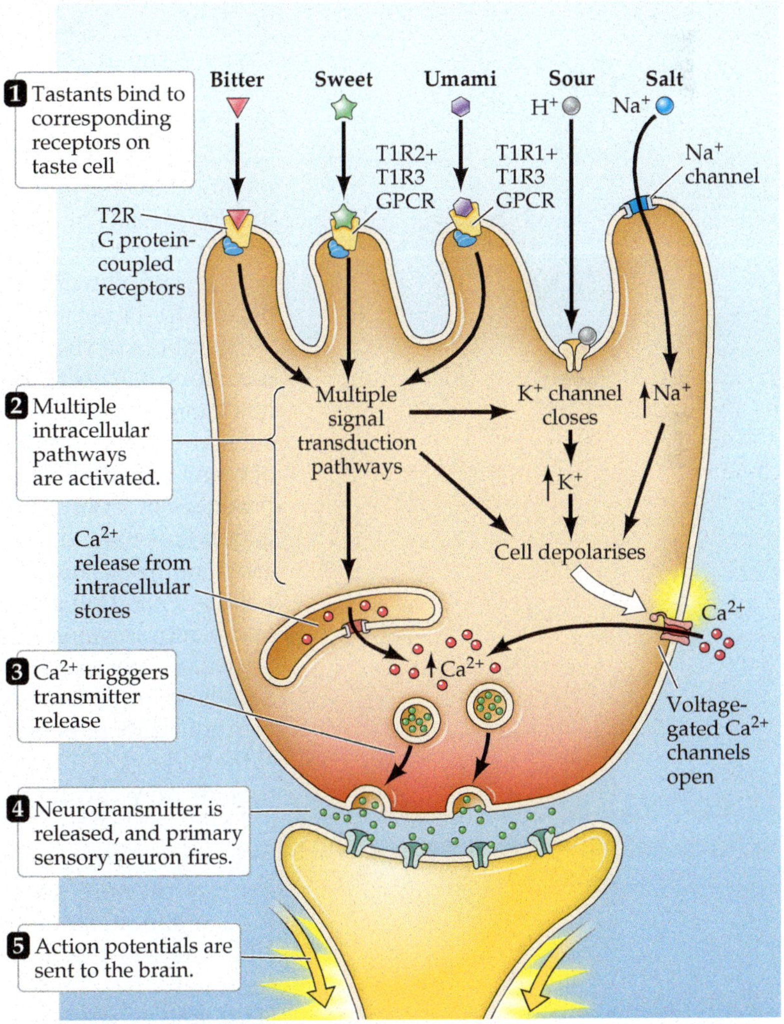

9.24 Five Kinds of Taste Transduction Here all five taste pathways are illustrated, but in reality each taste cell transduces only one of the five tastes. (After D. U. Silverthorn, 2007. *Human physiology: An integrated approach*, 4th ed. Pearson/Benjamin Cummings: San Francisco.)

T1R A family of taste receptor proteins that, when particular members heterodimerise, form taste receptors for sweet flavours and umami flavours.

T2R A family of bitter taste receptors.

umami One of the five basic tastes (along with salty, sour, sweet, and bitter), probably mediated by amino acids in foods.

SWEET The molecular mechanisms in the transduction of sweet, bitter, and umami tastes are more complex than those responsible for salty and sour tastes. Sweet, bitter, and umami tastants appear to stimulate specialised GPCR molecules embedded in the membranes of the taste cells, resulting in intracellular activation of second messengers. These receptors are made up of simpler protein subunits belonging to two families—designated **T1R** and **T2R**—that are combined in various ways, as we'll see shortly. The receptors function much like the slow metabotropic receptors that we considered in Chapter 3, although they may employ a different G protein alpha subunit, called *gustducin* (Kinnamon, 2012).

When two members of the T1R family—T1R2 and T1R3—combine (heterodimerise), they make a receptor that selectively detects sweet tastants (G. Nelson et al., 2001). Mice lacking T1R2 or T1R3—and particularly knockout mice that lack both components—are insensitive to sweet tastes (G. Q. Zhao et al., 2003). Dozens of substances taste sweet, so how can a single receptor be sensitive to sweet molecules that are as different as, for example, sugar, aspartame, and saccharin? Different sweet tastants appear to interact with different recognition sites within the T1R2+T1R3 receptor complex (M. Cui et al., 2006). Furthermore, certain sweet tastes may be sensed independently of the T1R2+T1R3 system, via a different receptor that has yet to be characterised (Treesukosol and Spector, 2012).

If you've spent any time around cats, you may be aware that most of them seem to be indifferent to sweets. It turns out that in all cats, from tabbies to tigers, the gene that encodes T1R2 is disabled, so their sweet receptors don't work (X. Li et al., 2009). Your cat may like ice cream, but that's more likely because of the fat in it than the sugar.

BITTER Bitter sensations are evoked by many different tastants. The association of bitter tastes with many toxic substances—such as nicotine, caffeine, strychnine, and morphine—provided strong evolutionary pressure to develop a high sensitivity to bitterness. The fact that different bitter substances can be discriminated—as shown by psychophysical work with human tasters (McBurney et al., 1972) and with animal subjects (Lush, 1989)—suggested that there is more than one receptor protein for bitterness.

Members of the T2R family of G protein-coupled receptors appear to be bitter receptors (Chandrashekar et al., 2000; Matsunami et al., 2000). The T2R family has about 30 members, and this large number may reflect the wide variety of bitter substances encountered in the environment, as well as the adaptive importance of being able to detect and avoid them. Interestingly, each bitter-sensing taste cell produces most or all of the different types of T2R bitter receptors (E. Adler et al., 2000). This means that these taste cells serve as broadly tuned bitter sensors: they are maximally sensitive to any of a wide variety of bitter substances but not very good at encoding qualitative distinctions *between* them. Nonetheless, behavioural and electrophysiological data indicate that at least certain broad classes of subjectively bitter substances *are* discriminable (Travers and Geran, 2009). The full details of bitter sensitivity remain to be established.

About 25% of people in the United States cannot taste the chemical phenylthiocarbamide (PTC) and the related compound 6-*n*-propylthiouracil (PROP), even though they can taste other bitter substances. Family studies indicate that tasters and nontasters are genetically different. Furthermore, some people, referred to as *supertasters*, exhibit heightened sensitivity to some bitter tastes, suggesting that they are genetically different as well (Bartoshuk and Beauchamp, 1994). In fact, supertasters also enjoy stronger sweet sensations from some substances. Nontasters of PROP have the fewest fungiform papillae, averaging 96 per square centimetre (cm^2) on the tongue tip, medium tasters have an intermediate number (184), and supertasters have the most (425) (**FIGURE 9.25**).

UMAMI The fifth basic taste, **umami**, is described as a meaty, savoury flavour. For most of the twentieth century, researchers argued about whether a distinct umami taste existed, but at least two types of receptors appear to be specialised to respond to exactly this sort of tastant. One of these, a variant of the metabotropic glutamate receptor, is expressed in certain taste buds (Chaudhari et al., 2000; Y. Maruyama et al., 2006) and most likely responds to the amino acid glutamate, which is found in high concentrations in meats, cheeses, kombu, and other savoury foodstuffs (that's why MSG—monosodium glutamate—is used as a "flavour enhancer").

A second probable umami receptor is a heterodimer of T1R1 and T1R3 proteins. Despite the similarity to the T1R2+T1R3 sweet receptor described already, the T1R1+T1R3 receptor selectively responds to most of the 20 standard amino acids that might be encountered in the diet (G. Nelson et al., 2002).

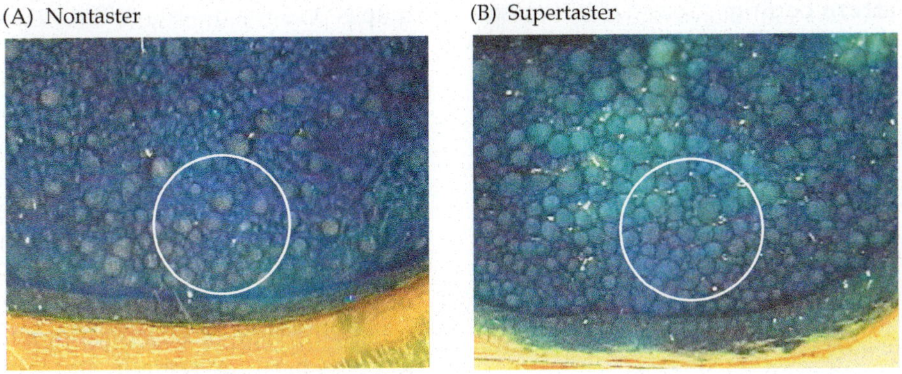

Courtesy of Linda Bartoshuk and the Bartoshuk Lab

9.25 It's All a Matter of Taste Buds We can count the number of fungiform papillae (larger blue-green circles) in a unit area (the white circle is about the size of a hole made by a paper punch). (A) The tongue of this person has only about 22 fungiform papillae in the defined area. (B) This supertaster's tongue has about twice as many. Some individuals have as few as 5 fungiform papillae in this space, while others pack in as many as 60.

In mice lacking the gene that encodes the T1R3 protein, sensitivity to sweet and umami tastants is greatly reduced but not abolished (Damak et al., 2003), implying that there must be additional receptor systems for these tastes that are T1R3-independent. It also suggests that the receptors for things that taste especially good—sweet and umami—may have shared evolutionary origins. In giant pandas, the *T1R1* gene became a nonfunctional "pseudogene" around the time in the animals' evolutionary history when they switched from a carnivorous diet to an exclusive diet of bamboo (H. Zhao et al., 2010). This is another illustration of the ecological molding of the senses. Similarly, birds (like their house cat enemies) lack the *T1R2* gene and thus ordinarily can't taste sweet, so evolution repurposed the hummingbirds' T1R1+T1R3 umami receptor into a new class of sweet receptor (Baldwin et al., 2014), allowing them to sense the nectar they need for survival.

gustatory system The taste system.

Taste information is transmitted to several parts of the brain

The **gustatory system** (from the Latin *gustare*, "to taste") extends from the taste receptor cells through brainstem nuclei and the thalamus to the cerebral cortex (**FIGURE 9.26**). Each taste cell transmits information to several afferent fibres, and each afferent fibre receives information from several taste cells. The afferent fibres run along three different cranial nerves—the facial (VII), glossopharyngeal (IX), and vagus (X) nerves (see Figure 2.9). The gustatory fibres in each of these nerves run to the brainstem. Here they synapse with second-order gustatory neurons that project to the ventral posteriour medial nucleus of the thalamus. Those thalamic neurons extend their third-order gustatory fibres to the cortical taste area located in the insula, just anteriour to the central sulcus. Even within the cortical taste area, a "gustatopic"

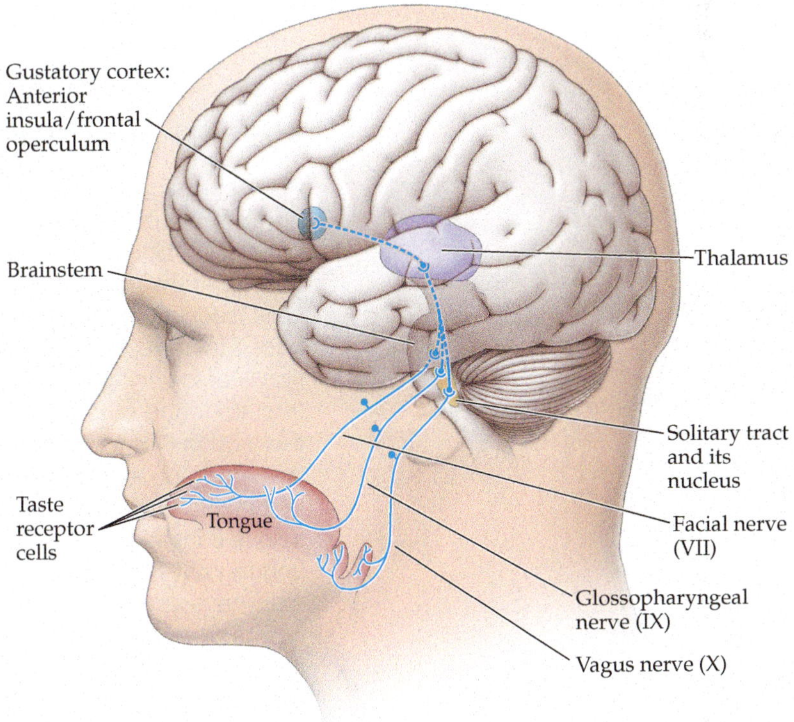

9.26 Anatomy and Main Pathways of the Human Gustatory System Taste information is conveyed via several cranial nerves to nuclei in the brainstem, and then to the thalamus. From the thalamus, a taste-specific pathway projects to cortical taste areas that give us our conscious perception of tastes.

pattern coding Coding of information in sensory systems based on the temporal pattern of action potentials.

map is preserved—the different tastes are represented in distinct areas within gustatory cortex (X. Chen et al., 2011). Researchers have used optogenetic techniques (see Figures 3.26 and 3.27) to control the behaviour of mice by selectively stimulating different cortical taste representations: activation of neurons in the sweet area elicits behavioural signs of attraction, whereas activation of neurons in the bitter area provokes strong behavioural avoidance (Peng et al., 2015).

A long-standing debate among sensory neuroscientists concerns the encoding of incoming information in the nerves leading to the brain. Some gustation researchers argue that the brain extrapolates taste information from **pattern coding**, the relative activity across ensembles of axons from different taste receptor cells. This type of encoding provides great capacity for distinguishing subtle differences between stimuli, and it provides a rich, parallel set of inputs upon which the brain can act to tune the stimulation it receives (as we will see again in describing colour perception in Chapter 8). Other researchers believe that taste sensation involves a simple system of labelled lines (which we touched on in Chapter 7), in which there are five distinct types of axon "lines" corresponding to five classes of taste receptor cells—one line for each taste (plus the sensors for heat, pain, and so on that we discussed earlier). Imagine five separate telephone lines, one for each taste. In this case, taste perception is believed to be determined simply by which line is active, rather than by the pattern of activity across groups of axons (Chandrashekar et al., 2006).

It has been reported that selectively inactivating taste cells that express receptors for just one of the five tastes tends to eradicate sensitivity to that one taste while leaving the other four tastes unaffected (A. L. Huang et al., 2006). This result would seem to support the position that taste is a labelled-line system, where the line for the knocked-out taste has been lost. However, the results are not inconsistent with the pattern-coding account, either, because when one of the five taste categories is knocked out, *patterns* of activity must also be affected. In other words, how could pattern coding work properly if one-fifth of the pattern had been inactivated? Furthermore, it remains to be established how a purely labelled-line system could account for the ability to discriminate different forms of salty, or different forms of sweet. In fact, it is difficult to imagine an experiment that would allow us to unambiguously establish whether labelled lines or pattern coding accounts for taste perception. The resolution of the debate, if there is to be one, will require new technical developments in sensory neuroscience.

The Cutting Edge

More Than a Matter of Taste

For most people, the concept of "taste" doesn't extend beyond the tongue; after all, our conscious experience of the gustatory sense is limited to the things we have in our mouth. Scientists have historically shared this bias until fairly recently. But now that we know how to search for some of the molecular components of taste-sensing cells, such as the taste-associated G protein gustducin and the T1R- and T2R-based taste receptors, objective analyses are revealing that taste-like chemosensation is astonishingly widespread within the body. Within the oral cavity, taste receptors are found outside the tongue, such as in the soft palate (the posteriour roof of the mouth) (Tomonari et al., 2012). Perhaps that's not too surprising, considering that food contacts all parts of the mouth. But how about taste cells in the stomach? In the colon? In the lungs, brain, and gonads? Although they don't give rise to conscious sensations of taste, "taste" cells have now been found in all of those locations and more (**FIGURE 9.27**). If flies can taste the surfaces they land on, do we mammals have similar subconscious capabilities?

The discovery of cells in the stomach and intestine that express gustducin and T1R3 sweet receptors (S. V. Wu et al., 2002) suggested that the gut actively samples the composition of food. These cells help organise appropriate digestive responses, such as the mobilisation of insulin from the pancreas in response to sweet substances (Kokrashvili et al., 2009). Other cells in the stomach and colon express T2R bitter receptors that may alter the transit time of foods or cause diarrhea, possibly in response to the presence of bitter toxins (Avau et al., 2015; Kaji et al., 2014). Further, by altering the release of gut hormones like ghrelin, glucagon-like peptide-1 (GLP-1), and cholecystokinin (CCK) (see Chapter 13), and altering stomach activity (Janssen et al., 2011), taste-transducing cells of the gut may protectively adjust appetite and regulate the absorption of ingested material (Behrens and Meyerhof, 2019).

Taste-transducing cells are also found in components of the airways, such as the nasal cavity and the bronchioles, and in the cilia that line the airways. These taste-sensitive cells, which primarily express T2R bitter receptors, appear to regulate protective reflexes like sneezing and coughing, probably to protect from noxious substances and infectious agents (Tizzano and Finger, 2013). Bitter and sweet receptors throughout the upper respiratory tract help to activate and direct innate immune responses against microbial infections (R. J. Lee and Cohen, 2015). Because the taste-transducing cells of the lungs likewise mediate bronchial inflammation, drugs that interact with these receptors may provide a new generation of asthma drugs (Avau and Depoortere, 2016; Deshpande et al., 2010).

The role of bitter taste receptors in the testes remains somewhat mysterious, but as elsewhere, they may serve to sense toxins (J. Xu et al., 2013). In any case, these taste receptors appear to play a crucial role—genetic knockout of testis bitter receptors in mice results in small testes and a great reduction in spermatogenesis (F. Li and Zhou, 2012). Likewise, bitter taste receptors are expressed in the skin (L. Shaw et al., 2018) and in the mammalian brain (Dehkordi et al., 2012)—discoveries that raise more questions than they answer, seeing as skin is arguably the organ most exposed to environmental chemicals, and the brain is the least (see Chapter 1). Perhaps we will eventually find that the taste transduction mechanism plays a more fundamental role in cellular signalling apart from gustation—and we can finally begin accounting for taste. ■

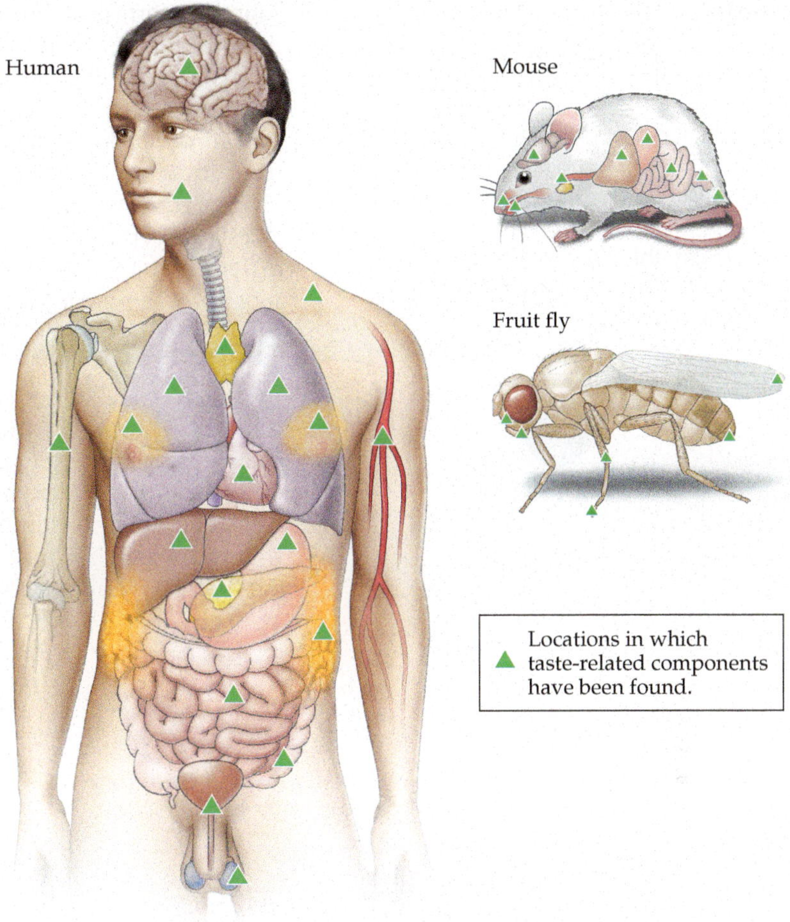

9.27 Locations Expressing Taste Receptors The search for molecular components of taste-transducing cells has revealed taste-like sensors in a surprising variety of tissues. (After an illustration by Nicole Rager Fuller in R. Ehrenberg, 2010. *Science News* 177: 22–25.)

9.9 Chemicals in the Air Elicit Odour Sensations

Learning Objectives

After reading this section, you should be able to:

9.9.1 **Describe the main structures of the olfactory system, with a focus on the cells and projections of the olfactory epithelium.**

9.9.2 **Explain the process of olfactory transduction, and discuss the function and variety of olfactory receptors that have been discovered.**

9.9.3 **Compare and contrast human olfactory capabilities with those of other species.**

9.9.4 **Describe the structure and function of the vomeronasal system, and weigh the evidence for and against the idea that humans respond to pheromones.**

Many aspects of an animal's world are determined by chemicals carried in the air. Olfactory sensitivity varies widely across species of mammals, reflecting differences between species in the evolutionary importance of smell for survival and

anosmia The inability to smell.

olfactory epithelium A sheet of cells, including olfactory receptors, that lines the dorsal portion of the nasal cavities and adjacent regions, including the septum that separates the left and right nasal cavities.

olfactory receptor neuron A type of neuron, found in the olfactory epithelium, that senses airborne odourants via specialised receptor proteins.

Go to Animation 9.5
The Human Olfactory System

reproduction (Bear et al., 2016; Freitag et al., 1998). Cats and mice, dogs and rabbits—all have a sharper sense of smell than humans. Most birds have only basic olfactory abilities, and dolphins don't have functional olfactory receptors at all. Surveys of olfaction in large populations of healthy people reveal surprising variation in odour sensitivity, ranging from fairly widespread **anosmia** (odour blindness, in varying degrees; Hoffman et al., 2016) to olfactory supersensitivity, with olfactory performance slightly better among women than men (Sorokowski et al., 2019) and typically worse in older populations (Kern et al., 2014). But despite these differences, the various mammals, including humans, all accomplish olfaction in much the same way, as we'll see next.

The sense of smell starts with receptor neurons in the nose

In humans, a sheet of cells called the **olfactory epithelium** (**FIGURE 9.28**) lines the dorsal portion of the nasal cavities and adjacent regions, including the septum that separates the left and right nasal cavities. Within the 5–10 cm^2 of olfactory epithelium that we possess, three types of cells are found: supporting cells, basal cells, and about 6 million **olfactory receptor neurons**. In many other mammals this number is an order of magnitude greater; for example, popular dog breeds generally have about 100–200 million receptor neurons in the olfactory epithelium,

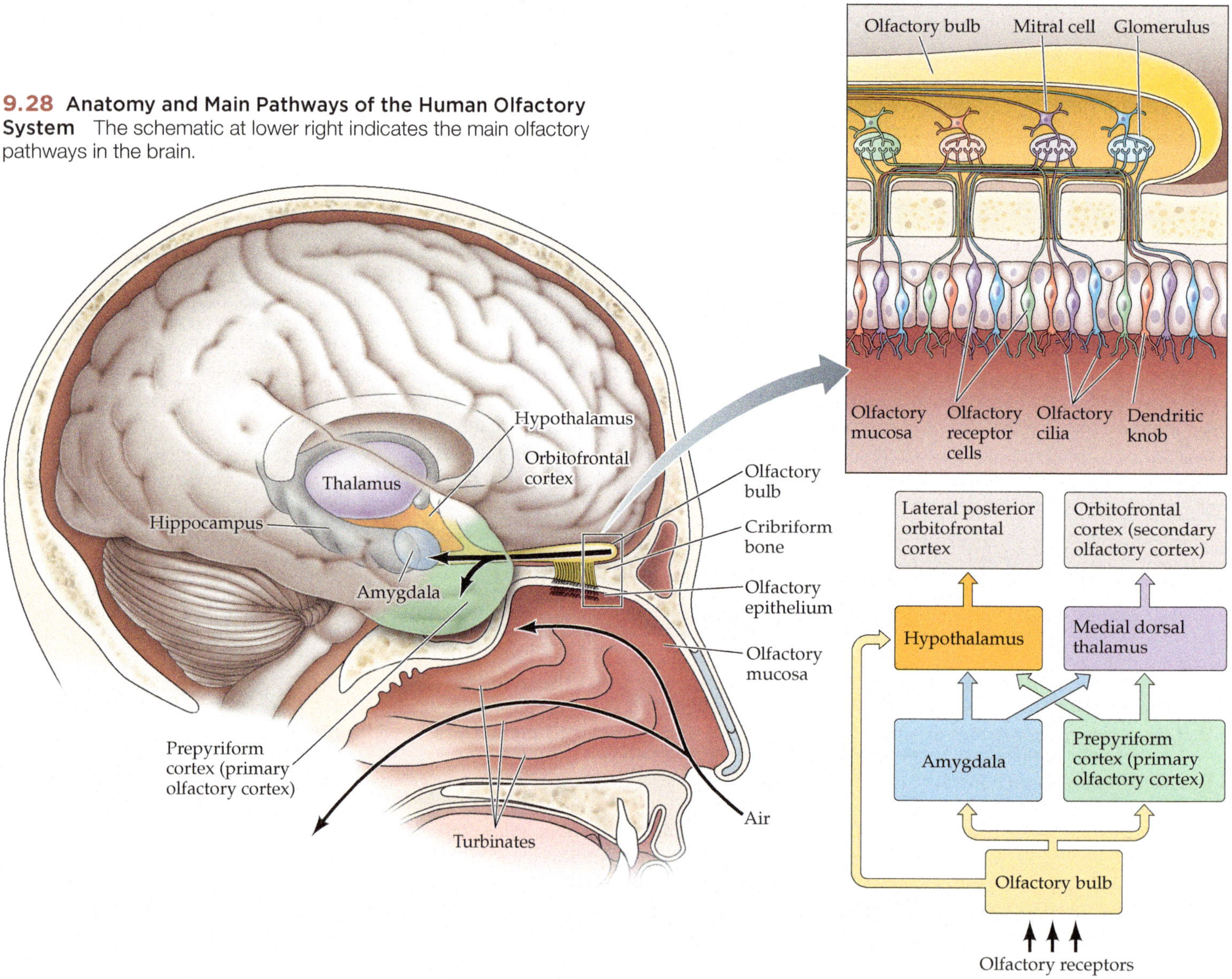

9.28 Anatomy and Main Pathways of the Human Olfactory System The schematic at lower right indicates the main olfactory pathways in the brain.

and in bloodhounds the number is closer to 300 million. This is one reason why they can follow an odour trail so much better than we can, although, interestingly, people can learn to do this pretty well too (J. Porter et al., 2007). In fact, dogs possess a sense of smell that is so sensitive that they can detect the banana-like scent of *n*-amyl acetate at less than 2 parts per *trillion*—like tasting a small pinch of sugar dissolved in a billion cups of tea (A. King, 2013).

cilium A hairlike cellular extension.

dendritic knob A portion of an olfactory receptor cell present in the olfactory epithelium.

Each olfactory receptor cell has a long, slender apical dendrite that extends to the outermost layer of the epithelium, the mucosal surface. There, numerous **cilia** (singular *cilium*) emerge from the **dendritic knob** and extend along the mucosal surface. At the opposite end of each bipolar olfactory receptor cell, a fine, unmyelinated axon, which is among the smallest-diametre axons in the nervous system, runs through tiny holes in the skull directly into the olfactory bulb (to be discussed shortly). These myriad clumps of axons, referred to as the olfactory nerve (cranial nerve I), are really quite short—not much longer than the thickness of the bone they pass through.

In contrast with many other receptor neurons in the body, olfactory receptor neurons can be replaced in adulthood (Costanzo, 1991). One theory is that these receptor neurons normally degenerate after a few weeks because they are in direct contact with external irritants, such as chemicals and viruses, so they must constantly be replaced. It's clear that, if destroyed, an olfactory receptor cell will be replaced as an adjacent basal cell differentiates into a neuron and extends dendrites to the mucosal surface and an axon into the brain that ultimately forms new synapses in the correct portion of the olfactory bulb (C. T. Leung et al., 2007). What's not clear is whether receptor cells are *normally* "disposable," subject to constant turnover even in the absence of a particular trauma.

If the olfactory epithelium is damaged, it can be regenerated and will properly reconnect to the olfactory bulb. The functional capability of these new connections has been clearly demonstrated in both behavioural and electrophysiological studies of animals with completely regenerated olfactory epithelium. Investigators are trying to determine how these neurons can regenerate while those in most other parts of the CNS cannot.

Odourants excite specialised receptor molecules on olfactory receptor neurons

Odourants enter the nasal cavity during inhalation and especially during periods of sniffing; they also rise to the nasal cavity from the mouth when we chew food. The direction of airflow in the nose is determined by complex curved surfaces called *turbinates* that form the nasal cavity (see Figure 9.28). Airborne molecules initially encounter the fluids of the mucosal layer, which contain binding proteins that transport odourants to receptor surfaces (Farbman, 1994).

Once at a receptor cell, the odourant stimulus interacts with receptor proteins located on the surface of the olfactory cilia and the dendritic knob. These receptor proteins are members of a superfamily of GPCRs (Buck and Axel, 1991). Interactions of odourants with their receptors trigger the synthesis of second messengers, such as cyclic AMP (cAMP). Cyclic AMP opens cation channels (L. J. Brunet et al., 1996), resulting in depolarisation of the olfactory receptor neuron, which in turn leads to the generation of action potentials.

This sensory transduction process, portrayed in **FIGURE 9.29**, is similar to the activation of other sensory systems, such as those for sweet and bitter tastes and those in the eye (see Chapter 8). A specific G protein, named G_{olf} in recognition of its importance in olfaction, must be used by all the olfactory receptors, because mice in which the gene that encodes G_{olf} is knocked out are generally anosmic (Belluscio et al., 1998).

Mice have about 2 million olfactory receptor neurons, each of which expresses only one of about 1000 different receptor proteins. These receptor proteins can be divided into four different subfamilies of about 250 receptors each (Mori et al., 1999). Within each subfamily, members have a very similar structure and presumably recognise

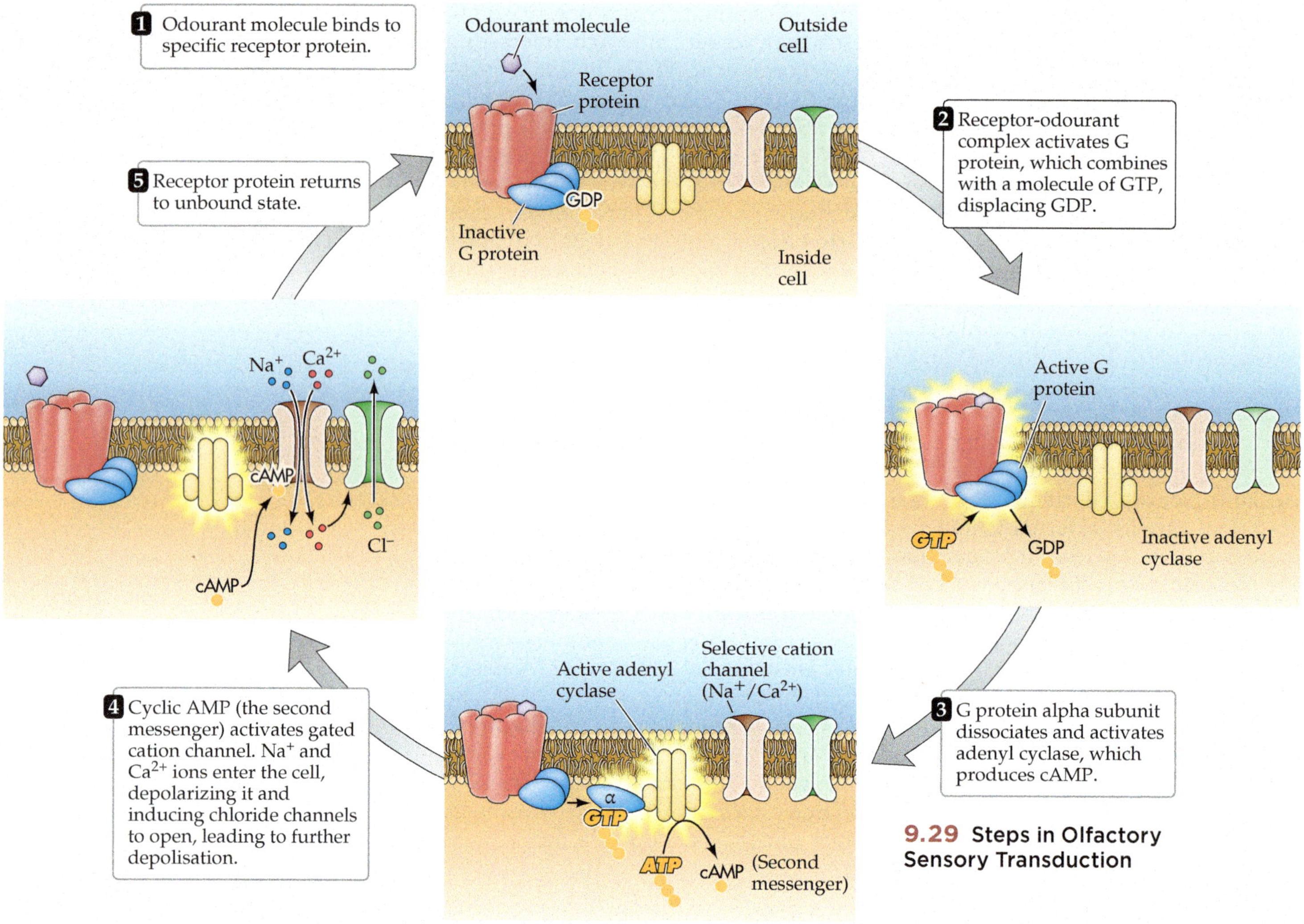

9.29 Steps in Olfactory Sensory Transduction

similar odourants. In rodents, each subfamily of receptors is synthesised in a separate band of the epithelium, as shown in **FIGURE 9.30** (Vassar et al., 1993).

Now that it has been fully mapped, we know that the human genome also contains about 1000 apparent olfactory receptor genes, but only about 400 appear to be fully functional (Gilad et al., 2003). The rest have apparently accumulated mutations and become nonfunctional during the course of evolution (Olender et al., 2008), implying that whatever they detected ceased to be important to our ancestors' survival and reproduction.

For a long time the orthodox view was that humans could distinguish 5,000–10,000 different odours at most, but subsequent psychophysical experiments suggest this is too low a figure. Although controversy swirls around the actual number of odours that we can distinguish—a recent estimate places the number as high as 1 trillion different odours (Bushdid et al., 2014), but this may be a significant overestimation (Gerkin and Satro, 2015)—the widely held belief that we humans have a poor sense of smell relative to other animals has been overstated historically (McGann, 2017). Whatever turns out to be the actual number of odours humans can distinguish, our ability to discriminate thousands, millions, or perhaps billions of odours using just 400 kinds of receptors indicates that each type of functional odourant receptor must interact with many different odourants in order to make up the many distinct patterns that encode the spectrum of scents (Duchamp-Viret et al., 1999). Furthermore, any two people differ by about 30% in the makeup of their

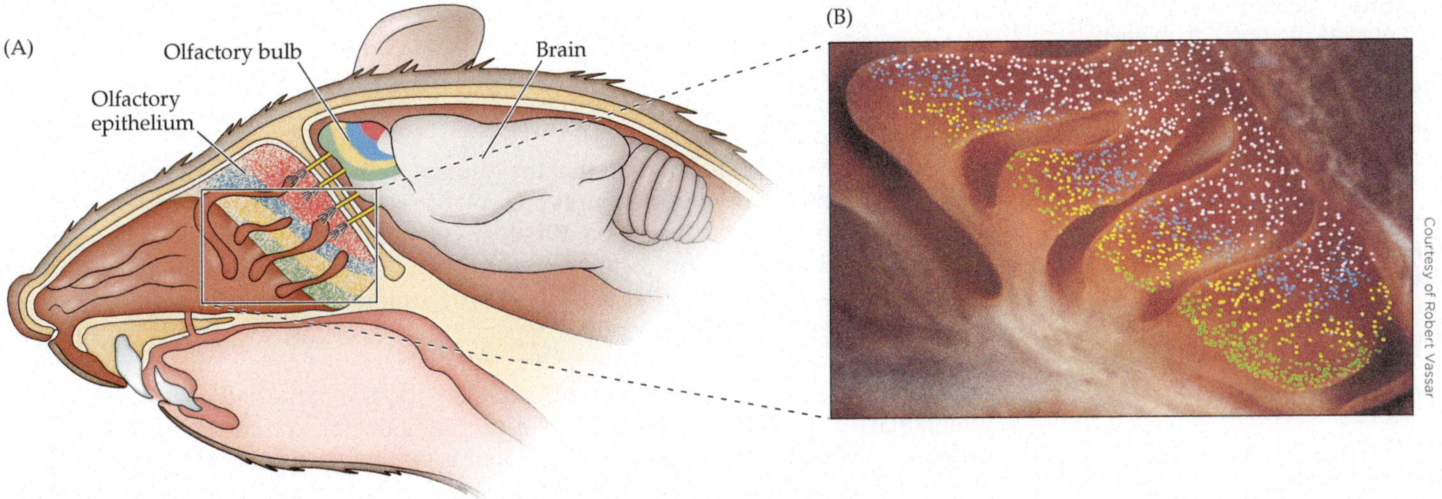

9.30 Different Kinds of Olfactory Receptor Molecules on the Olfactory Epithelium (A) In this diagram showing the anatomy of the rat olfactory organ, colour swathes correspond to the distribution of four receptor subfamilies. (B) The different colours in this photograph of rat olfactory epithelium show receptor locations for the four receptor subfamilies and also correspond to the topographic representation in the olfactory bulb. The different receptor types have distinct but overlapping spatial distributions. (After J. Vassar et al., 1993. *Cell* 74: 309–318.)

olfactory receptors (Frumin et al., 2014), suggesting that to some extent we each inhabit a personalised olfactory environment (Trimmer et al., 2019).

Olfactory axons connect with the olfactory bulb, which sends its output to several brain regions

The numerous axons of the olfactory receptor neurons pass through the cribriform plate (see Figure 9.28) to terminate in a complex structure at the anteriour end of the brain called the **olfactory bulb**. The olfactory bulb, in relation to the rest of the brain, is much smaller in humans than in animals, such as rats, that depend extensively on olfaction (compare Figures 9.28 and 9.30). The olfactory bulb is organised into many roughly spherical neural circuits called **glomeruli** (singular *glomerulus*, from the Latin *glomus*, "ball"), within which the axon terminals of olfactory receptor neurons synapse on the dendrites of the specialised **mitral cells** of the olfactory bulb (see Figure 9.28). The intrinsic circuitry within and between these glomeruli contributes to an elaborate system for modulating, tuning, and sharpening olfactory bulb activity (Aungst et al., 2003).

Mice have about 1800 glomeruli. Each one receives inputs exclusively from olfactory neurons that are expressing the same type of olfactory receptor, and the glomeruli are organised in functional zones according to the four receptor protein subfamilies described in the previous section (Mori et al., 1999). So there appears to be a topographic distribution of smells in the bulb, and within each functional zone neighbouring glomeruli tend to receive inputs from receptors that are closely related. This glomerular map of the olfactory bulb is established during an early developmental critical period, becoming essentially immutable in early postnatal life (Ma et al., 2014; Tsai and Barnea, 2014).

Perhaps surprisingly, olfactory receptors are found in the axon terminals of olfactory neurons, as well as in their dendrites (Barnea et al., 2004). Experiments with knockout mice (see Box 7.3) suggest that the olfactory receptor proteins help guide the innervating axons to their corresponding glomeruli (Imai et al., 2009). Disruptions of the receptor protein prevent newly generated olfactory receptor axons from reaching their normal targets (F. Wang et al., 1998).

Output from the olfactory bulb consists of the axons from mitral cells, which extend to a variety of brain regions. These include the prepyriform and entorhinal

olfactory bulb An anteriour projection of the brain that terminates in the upper nasal passages and, through small openings in the skull, receives axons from olfactory receptor neurons.

glomerulus A complex arbor of dendrites from a group of olfactory cells.

mitral cell A type of cell in the olfactory bulb that conducts smell information from the glomeruli to the rest of the brain.

pheromone A chemical signal that is released outside the body of an animal and affects other members of the same species.

vomeronasal system A specialised chemical detection system that detects pheromones and transmits information to the brain.

vomeronasal organ (VNO) A collection of specialised receptor cells, near to but separate from the olfactory epithelium, that detect pheromones and send electrical signals to the accessory olfactory bulb in the brain.

cortex (note that smell is the only sensory modality that can synapse directly in the cortex rather than having to pass through the thalamus), the amygdala, and the hypothalamus. This close relation between olfactory inputs and limbic system structures involved in memory and emotion may help explain the potent ability of odours to evoke strong nostalgic memories of childhood (M. Larsson and Willander, 2009).

Functional MRI studies suggest that the human prepyriform cortex is activated during sniffing, whether or not an odour is present, because the airflow induced by sniffing provides somatosensory stimulation. When an odour is present during a sniff, primary olfactory cortex (prepyriform cortex) and secondary olfactory cortex (orbitofrontal cortex) are both activated (Sobel et al., 2000). Furthermore, the same chemical mix may produce a different odour perception, depending on how fast the air enters during a sniff (Sobel et al., 1999). So the brain gauges the airflow rate during a sniff in order to interpret olfactory information properly. In addition, the size of the sniff appears to reflect attentional processes: larger sniffs occur during sampling of *pleasant* odours even when they are only imagined, that is, when no odour is present (Bensafi et al., 2003).

Many vertebrates possess a vomeronasal system

Many vertebrates have a second chemical detection system that appears to specialise in detecting **pheromones**, the odour signals or trails that many animals secrete (see Chapter 5). This **vomeronasal system** (**FIGURE 9.31**), as it is called, is present in most terrestrial mammals, amphibians, and reptiles. The receptors for the system are found in a **vomeronasal organ** (**VNO**) of epithelial cells near the olfactory epithelium.

Rodents express two major families of vomeronasal receptors—V1R and V2R—that encode hundreds of different types of receptors (Dulac and Torello, 2003). Although both are families of GPCRs, they are quite different from each other; in fact, the V1Rs are more similar to the T2R bitter taste receptors (discussed earlier) than they are to the receptors of the main olfactory system. Interestingly, the distribution of certain V2Rs differs between male and female rats, in keeping with the critical role of pheromones in organizing rodent reproductive behaviour (Herrada and Dulac, 1997). V2Rs are also sensitive to certain major histocompatibility complex (MHC) molecules (Loconto et al., 2003); detection of MHCs is thought to be one of the main ways in which animals can assess their degree of relatedness to other animals, which has implications for mating strategies (see Chapter 5). The VNO-specific ion channel TRPC2 (transient receptor potential, subfamily C, 2)—a member of the TRP family of sensory receptors that we mentioned earlier—is also essential for VNO function (Stowers et al., 2002).

VNO receptors are remarkably sensitive, detecting extremely low concentrations of pheromone molecules (Leinders-Zufall et al., 2000). The receptors send their information to the accessory olfactory bulb (adjacent to the main olfactory bulb), which projects to the medial amygdala, which in turn projects to the hypothalamus. Relying in part on MHCs, hamsters (Mateo and Johnston, 2000) and mice (Isles et al., 2001) can distinguish relatives from nonrelatives just by smell, even if they were raised by foster parents. Presumably these animals compare whether other animals smell like themselves, perhaps to avoid mating with kin. In parallel, dedicated mechanisms in olfactory cortex activate fear responses and stress hormone secretion in response to predator odour signals, helping the animal to avoid their toothy source (Kondoh et al., 2016).

9.31 The Vomeronasal System

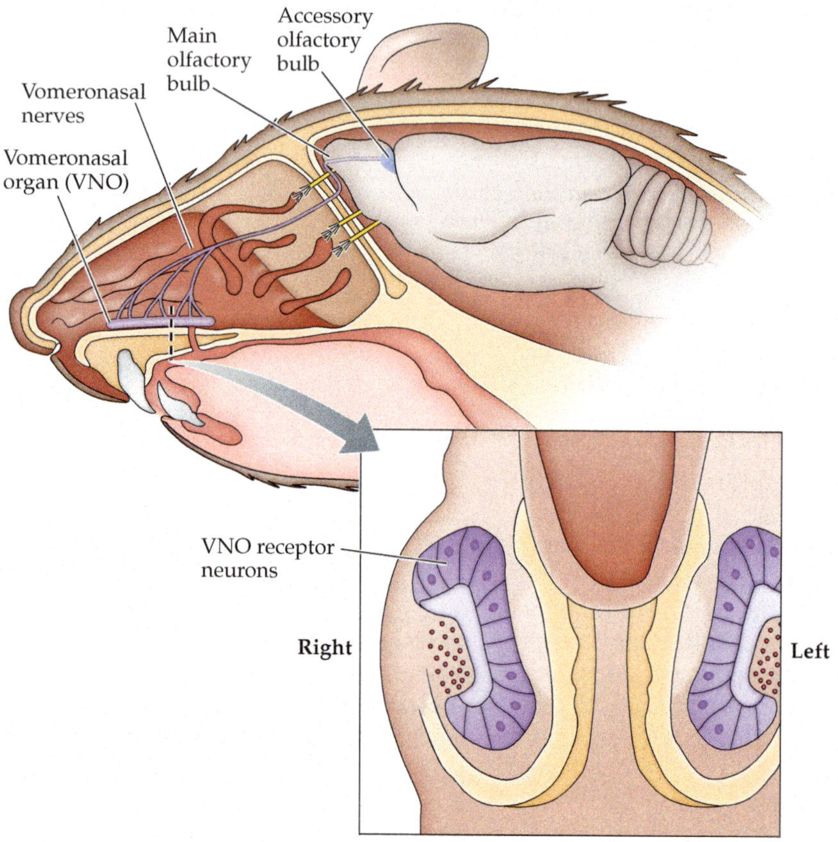

In humans, molecular and anatomical evidence for a functional VNO is scanty at best. Humans have a structure that resembles a VNO and show behavioural evidence of some pheromone-like capabilities: for example, early work found that simple exposure to bodily odours of other women can shift women's menstrual cycles (Stern and McClintock, 1998), and exposure to female tears reportedly can cause reductions in testosterone and sexual arousal in men (Gelstein et al., 2011). However, analysis of the genome indicates that almost all of the human variants of the genes that encode the V1R and V2R vomeronasal receptors, and the TRPC2 channel, have become nonfunctional pseudogenes over evolutionary time, and the question of whether humans and other mammals really employ pheromonal cues that significantly affect behaviour remains controversial (Doty, 2010; Lübke and Pause, 2015).

The discovery of a new class of olfactory receptors within the main olfactory epithelium (Liberles and Buck, 2006) added a new dimension to the question of pheromone sensitivity. Called **TAARs**, for **trace amine-associated receptors**, these receptors are expressed in specific olfactory neurons within the main olfactory epithelium, just like the regular olfactory receptors, but at least in mice they seem to respond to volatile sex-specific urinary compounds that may be pheromones. Mice in which the TAAR genes have been knocked out lose their normal aversion to certain urinary volatiles, even including the urine of predators (Dewan et al., 2013). Thus it seems that the old dichotomy that the olfactory epithelium detects odours while the VNO detects pheromones is an oversimplification, and the priour evidence suggesting that humans respond to pheromones, despite a vestigial or nonfunctional VNO, isn't necessarily paradoxical. If rodents can detect pheromones through the main olfactory epithelium, perhaps we can too. Whatever the details of the mechanism may be, evidence is rapidly accumulating that odour is an ecologically important channel for human social communication (J. H. de Groot et al., 2017).

trace amine-associated receptors (TAARs) A family of probable pheromone receptors produced by neurons in the main olfactory epithelium.

Recommended Reading

Bartoshuk, L. M., and Beauchamp, G. K. (2005). *Tasting and Smelling* (2nd ed.). New York: Academic Press.

Doty, R. L. (2015). *Handbook of Olfaction and Gustation* (3rd ed.). New York: Wiley-Blackwell.

Finger, T. E., Silver, W. L., and Restrepo, D. (Eds.). (2000). *The Neurobiology of Taste and Smell* (2nd ed.). New York: Wiley-Liss.

Horowitz, S. S. (2012). *The Universal Sense: How Hearing Shapes the Mind*. London: Bloomsbury.

Menini, A. (2009). *The Neurobiology of Olfaction*. Boca Raton, FL: CRC Press.

Musiek, F. E., and Baran, J. A. (2018). *The Auditory System: Anatomy, Physiology, and Clinical Correlates* (2nd ed.). Boston: Pearson.

Palmer, A., and Rees, A. (2010). *Oxford Handbook of Auditory Science*. Oxford, UK: Oxford University Press.

Wolfe, J. M., Kluender, K. R., Levi, D. M., Bartoshuk, L. M., et al. (2017). *Sensation & Perception* (5th ed.). New York: Oxford University Press/Sinauer.

Wyatt, T. D. (2014). *Pheromones and Animal Behaviour: Chemical Signals and Signatures* (2nd ed.). Cambridge, UK: Cambridge University Press.

Yost, W. A. (2013). *Fundamentals of Hearing* (5th ed.). San Diego, CA: Academic Press.

9 Visual Summary

You should be able to relate each summary to the adjacent illustration, including structures and processes. You can also refer back to the links to figures, animations, and activities that will help you consolidate the material.

1 The **external ear** captures, focuses, and filters sound. Vibrations of the **tympanic membrane** (eardrum) are focused by the three ossicles of the **middle ear** onto the **oval window** to stimulate the **cochlea**. Review Figure 9.2, Animation 9.1

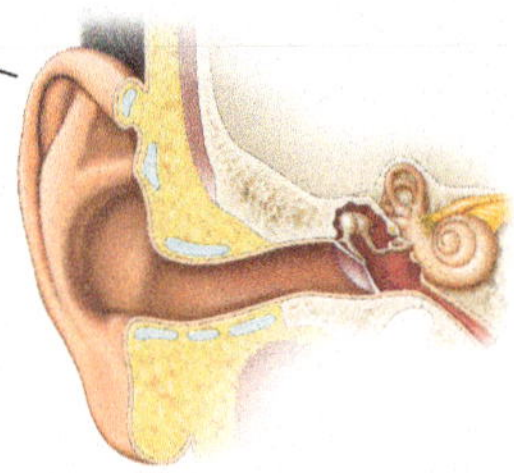

2 Sound arriving at the oval window causes ripples on the **basilar membrane** of the cochlea. High-frequency sounds affect the basilar membrane near the base of the cochlea; low-frequency sounds affect the apex. Review Figure 9.3

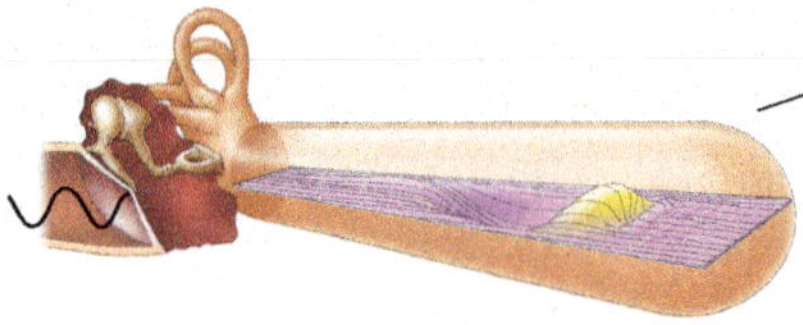

3 The **organ of Corti** has both **inner hair cells** and **outer hair cells**. The inner hair cells convey most of the information about sounds. The outer hair cells change their length with the help of a specialised motor protein called prestin, under the control of the brain, tuning the basilar membrane. Review Figure 9.4, Activity 9.1

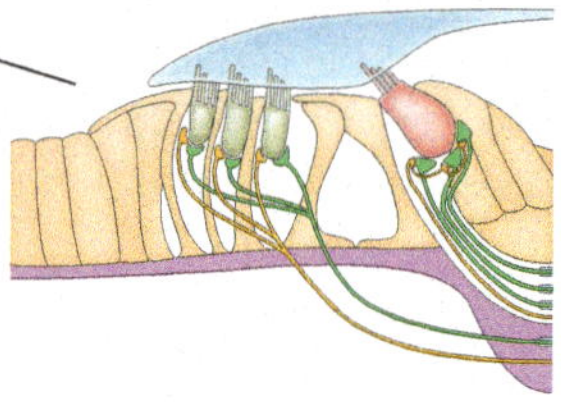

4 Movement of the **stereocilia** of the hair cells causes the opening and closing of ion channels, thereby **transducing** mechanical movement into changes in electrical potential. These changes in potential stimulate the nerve cell endings that contact the hair cells. Review Figure 9.5

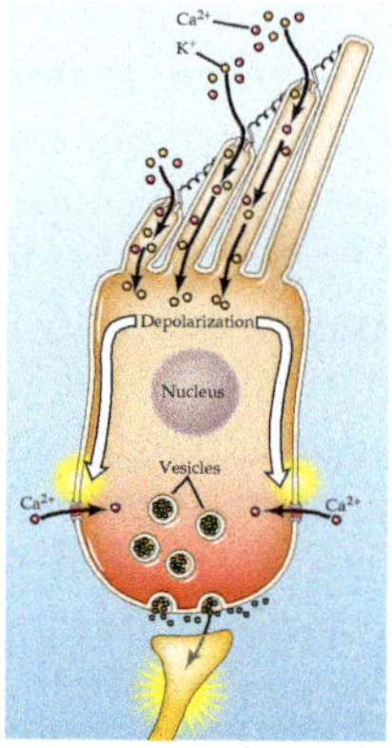

5 Afferents from the inner hair cells transmit auditory information first to the **cochlear nuclei** of the brainstem and then bilaterally to the **superiour olivary nuclei**, which in turn innervate the **inferiour colliculi**. From there auditory information is relayed to the **medial geniculate nuclei** and then **primary auditory cortex** in the temporal lobe. Review Figure 9.7

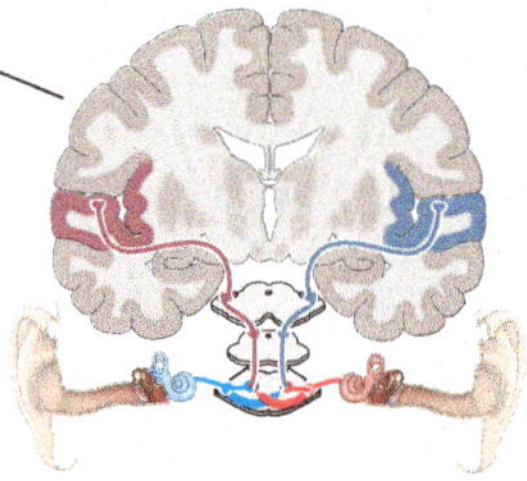

6 At each level of the auditory system, sound frequencies are mapped in an orderly succession called **tonotopic organisation**. Pitch is encoded two ways: as the place on the basilar membrane most strongly stimulated, in **place coding**; and in the rate of firing in the auditory nerve, in **temporal coding**. Review Figure 9.8, Animation 9.2

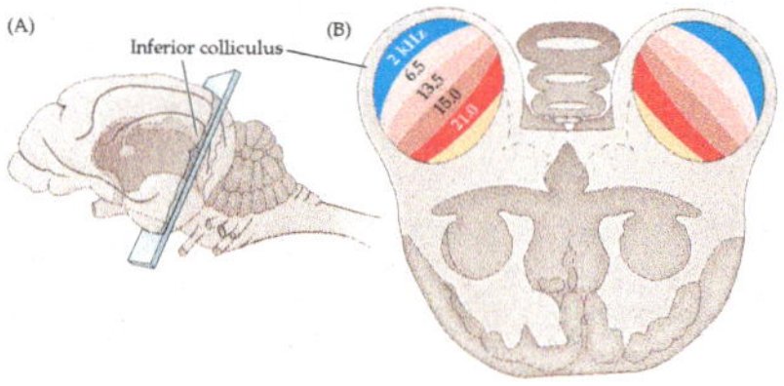

7 Auditory localisation depends on differences in the sounds arriving at the two ears. For low-frequency sounds, differences in time of arrival at the two ears, that is, **interaural temporal differences**, are especially important. For high-frequency sounds, **interaural intensity differences** are especially important. Review Figure 9.10

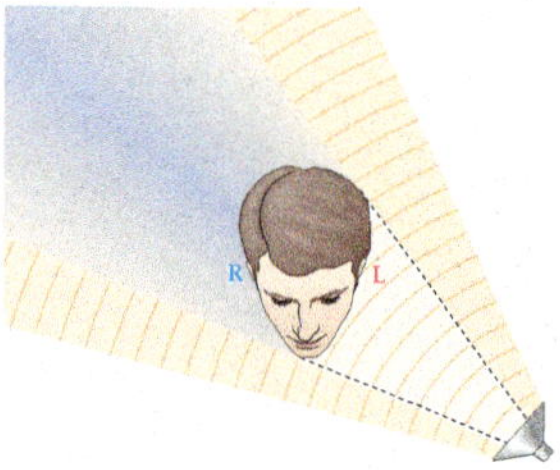

8 Birds and mammals evolved different neural mechanisms for localising sounds; in mammals the lateral superiour olive of the brainstem processes differences in intensity, and the medial superiour olive processes differences in time of arrival. Review Figure 9.11

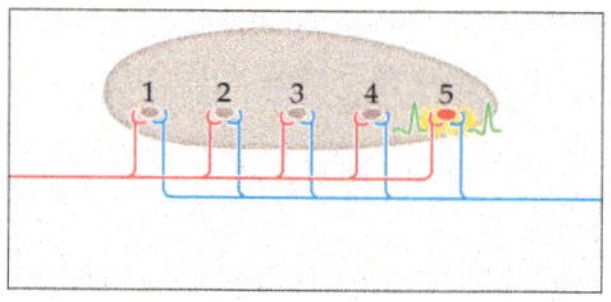

9 **Conduction deafness** consists of impairments in the transmission of sound through the external or middle ear to the cochlea. Sensorineural deafness arises in the cochlea, often because of the destruction of hair cells, or in the auditory nerve. **Central deafness** stems from brain damage. Some forms of deafness may be alleviated by direct electrical stimulation of the auditory nerve by a **cochlear implant**. Review Figures 9.16–9.19, Video 9.3

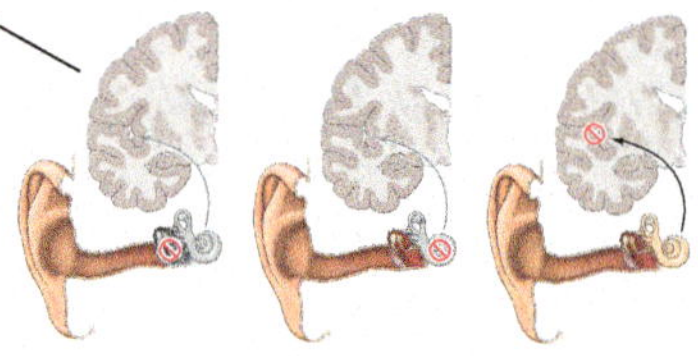

10 The receptors of the **vestibular system** lie within the inner ear next to the cochlea. In mammals the vestibular system consists of three **semicircular canals**, plus the **utricle** and the **saccule**. Within these structures hair cells detect rotation of the body in three planes, and the utricle and saccule sense static positions and linear accelerations. Review Figure 9.20, Animation 9.4

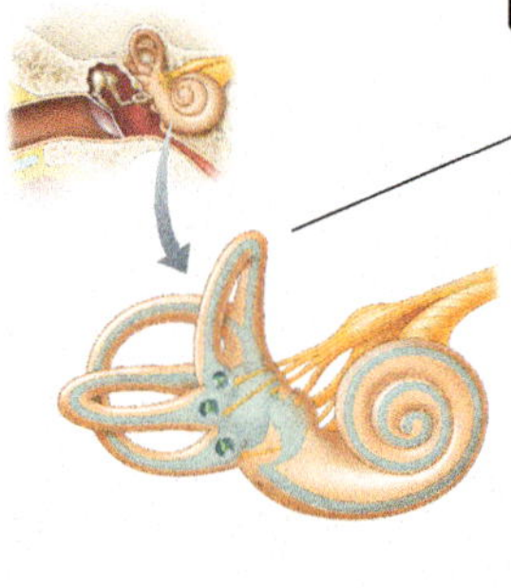

11 Nerve fibres from the vestibular receptors enter the brainstem and synapse in the **vestibular nuclei,** which send their outputs to the motor nuclei of numerous structures. One important function of the vestibular system is the **vestibulo-ocular reflex (VOR)**, which uses vestibular information about head rotations to precisely move the eyes to keep our gaze fixed. Review **Figure 9.21**

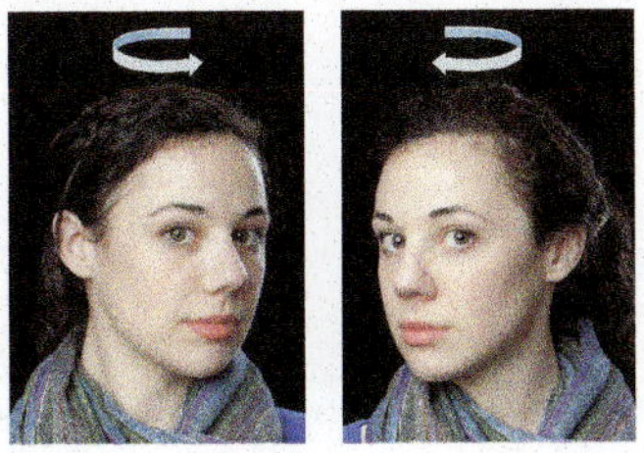

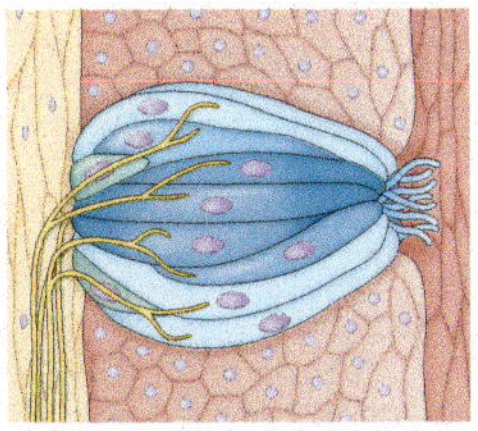

12 Humans detect only five main tastes: salty, sour, sweet, bitter, and **umami**. In mammals, most taste receptor cells are located in clusters of cells called **taste buds**, situated within small projections from the surface of the tongue called **papillae**. Review **Figures 9.22**, **9.23**, **Activity 9.2**

13 The tastes of salty and sour are evoked primarily by the action of simple ions on ion channels in the membranes of taste cells. Sweet and bitter tastes are perceived by specialised receptor molecules belonging to the **T1R** and **T2R** families, which are coupled to G proteins. Heterodimers of T1R proteins function as sweet and umami receptors. The **gustatory system** extends from the taste receptor cells through brainstem nuclei to the thalamus and then to the cerebral cortex. Review **Figures 9.24**, **9.26**

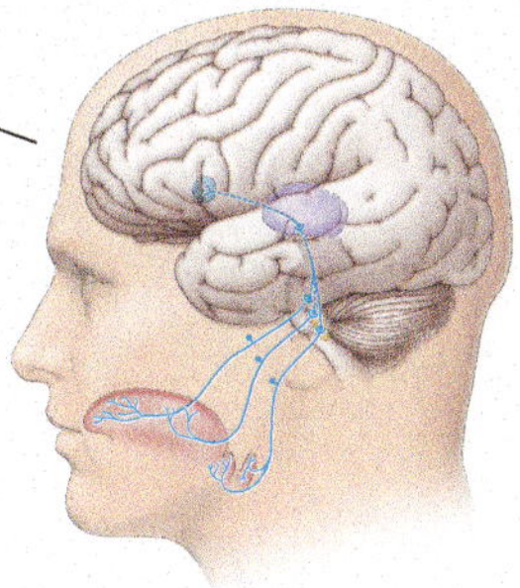

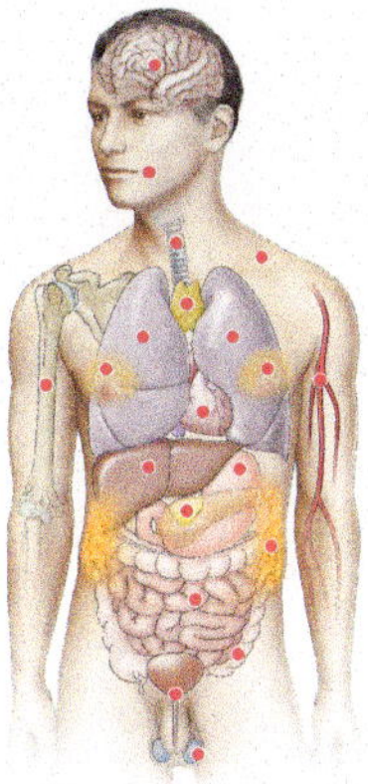

14 Taste-transducing cells have been located throughout the body. These cells may serve a variety of chemosensory and other functions. Review **Figure 9.27**

15 Humans can discriminate a huge variety of different odours, but some other species have more **olfactory receptor neurons** and larger **olfactory bulbs**. Each olfactory receptor neuron has dendrites in the **olfactory epithelium** in the nose and a fine, unmyelinated axon that runs to a **glomerulus** within an olfactory bulb. If an olfactory receptor cell dies, an adjacent cell will replace it. Review **Figure 9.28**, **Animation 9.5**

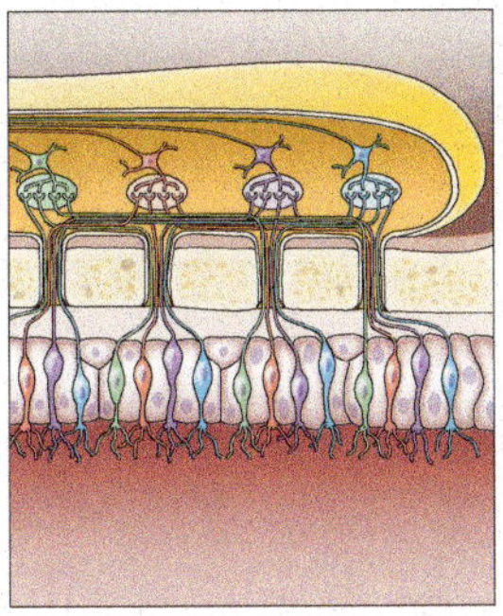

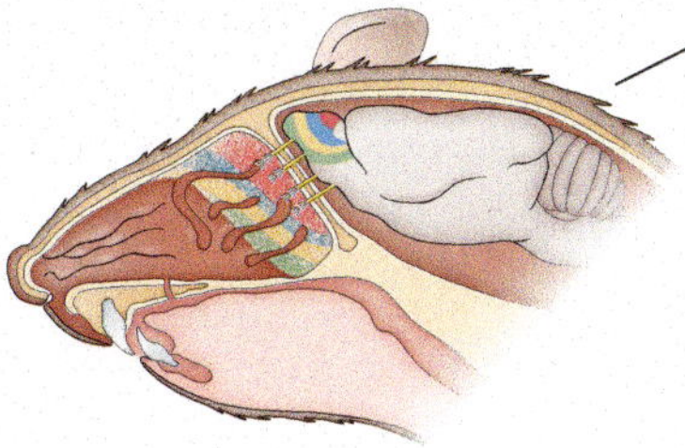

16 All olfactory neurons expressing a particular receptor synapse in the same glomerulus in the olfactory bulb. The projection from the epithelium to the bulb and then to the brain maintains a zonal distribution for different kinds of receptor molecules. Review **Figures 9.29**, **9.30**

17 The **vomeronasal organ** contains receptors to detect **pheromones** released from other individuals of the species. These receptors transmit signals to the accessory olfactory bulb, which in turn communicates with the amygdala. Pheromones can also be detected by specialised receptors in the main olfactory epithelium. Review **Figure 9.31**

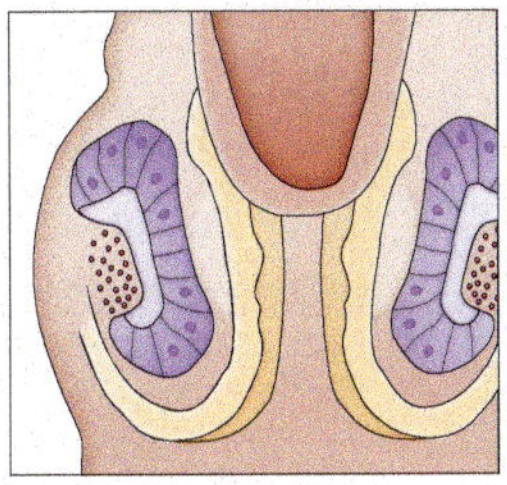

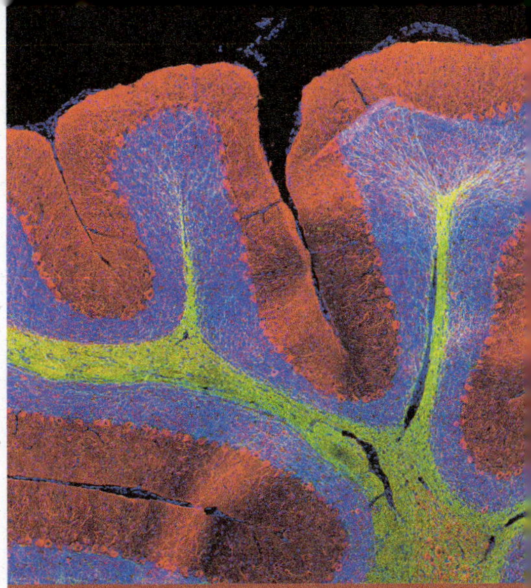

10

Motor Control and Plasticity

What You See Is What You Get

Ian Waterman had a perfectly ordinary life until he caught a viral infection at age 19. For reasons that no one understands, the infection targeted a very specific set of nerves sending information from his body to his brain. Ian can still feel pain or deep pressure, as well as warm and cool surfaces on his skin, but he has no sensation of light touch below his neck. What's more, although Ian can still move all of his muscles, he receives no information about muscle activity or body position (Cole, 1995). You might think this deficiency wouldn't cause any problem, because you've probably never thought much about your "body sense"; it's not even one of the five senses that people talk about, is it?

In fact, however, the loss of this information was devastating. Ian couldn't walk across a room without falling down, and he couldn't walk up or down stairs. The few other people suffering a loss like this spent the rest of their lives in wheelchairs. But Ian was young and determined, so he started teaching himself how to walk using another source of feedback about his body: his vision. Now, as long as the lights are on, Ian can carefully watch his moving body and gauge what motor commands to send out to keep walking. If the lights go out, however, he collapses; and he has learned that he just has to lie where he is until the lights come on again. He has so finely honed this ability to guide movements with vision that if asked to point repeatedly to the same location in the air, he does so more accurately than other people do. Still, it's a mental drain to have to watch and attend constantly to his body just to do everyday tasks.

Ian has a good job and an active, independent life, but he is always vigilant. Lying in bed, he has to be very careful to remain calm, tethering his limbs with the covers to prevent them from flailing about.

And the lights are always on at Ian's house.

Our emphasis in this chapter shifts to the motor system as we complete the circuit from sensory input to behaviour. It is important to consider sensory and motor functions together. Just as we saw in Chapters 7–9 that motor activities are important for sensory and perceptual functions—movements of the fingers in active touch perception, sniffing in smell, and movements of the eyes and head in vision—so we will see in this chapter that sensory and perceptual processes guide and correct our actions.

In addition, just as our apparently effortless perception turns out to depend on intricate sensory mechanisms and perceptual processes, so, too, our apparently effortless adult motor abilities—such as reaching out and picking up an object, walking across the room, sipping a cup of coffee—require complex muscular systems with constant feedback from the body, as Ian's case shows. After examining movements and their coordination from the behavioural view, we'll take up the neuroscience hierarchical view of spinal cord mechanisms and brain control, concluding with a survey of the major motor disorders.

10.1 The Behavioural View Considers Reflexes versus Plans

Learning Objectives

After reading this section, you should be able to:

10.1.1 Explain how we know that some reflexes are mediated by the spinal cord.

10.1.2 Offer evidence of motor planning versus connecting reflexes.

10.1.3 Contrast closed-loop versus open-loop motor control, and the trade-off of accuracy for speed.

By the early nineteenth century, scientists knew that the dorsal roots of the spinal cord transmit sensory information and that the ventral roots contain motor fibres; connections between the two seemed to provide the basis for simple movements. Research with **spinal animals** (animals in which the spinal cord has been disconnected from the brain) led British physiologist Sir Charles Sherrington (1857–1952) to argue that the basic units of behaviour are **reflexes**: simple, unvarying, and unlearned responses to sensory stimuli such as touch, pressure, and pain. We analysed the famous knee-jerk reflex in Figure 2.20.

Sherrington's work inspired a rush to catalogue the various reflexes and their neural circuitry, particularly in the spinal cord. These studies showed that some reflexes involve only short pathways in the spinal cord, directly linking dorsal and ventral roots; others involve longer loops, connecting spinal cord segments to each other or to brain regions.

Is normal behaviour simply the connecting together of different reflexes, the sensation from one reflex triggering the next? No. The limitations of this perspective are apparent when we think about complex sequences of behaviour, such as speech. A speaker has a plan in which several units (speech sounds and words) are placed in a larger pattern (the intended complete statement). Sometimes the units are misplaced, although the pattern is preserved: "Our queer old dean," said history professor William Spooner when he meant "Our dear old queen." Or, "You hissed all my mystery lectures." (This type of speech error is called a spoonerism.) Such mistakes reveal the overall plan: the speaker is anticipating a later sound and executing it too soon. A chain of reflexes, each one triggering the next, could not generate such an error.

The **motor plan**, also called the *motor program*, is a complex set of commands to muscles that is established *before* the behaviour starts. Feedback from movements informs and fine-tunes the motor program as it unfolds, but the basic sequence of movements is planned. Examples of behaviours that exhibit this kind of internal plan range from highly skilled acts, such as piano playing, to the simple escape behaviours of animals such as crayfish.

Motor behaviour can be analysed and measured in a variety of ways

We can readily analyse movements using high-speed video, which provides an intimate frame-by-frame portrait of even the most rapid events. Computer programmes process images and sensors attached at joints to quantify performance, enabling detailed measurement of the positions of different body parts in successive instants. Other devices record the direction, strength, and speed of motions. **FIGURE 10.1** illustrates the paths of normal and impaired reaching movements, which we will consider at several points in this chapter.

Another approach to the fine-grained analysis of movements is to record the electrical activity of muscles—**electromyography** (**EMG**). Like neurons, muscles produce action potentials when they contract, as we'll see later in this chapter. Therefore, fine needle electrodes placed in a muscle, or electrodes placed on the skin over a muscle, can detect electrical indications of muscle activity. If electrodes are placed over several different muscles, we get a record of the timing and strength of contraction of the muscles involved in a movement (Hanakawa et al., 2003). The EMGs in **FIGURE 10.2** show that a person pulling a knob will adjust his legs just before moving his arm—another example of motor planning.

spinal animal An animal whose spinal cord has been surgically disconnected from the brain to enable the study of behaviours that do not require brain control.

reflex A simple, highly stereotyped, and unlearned response to a particular stimulus (e.g., an eye blink in response to a puff of air).

motor plan Also called *motor program*. A plan for action in the nervous system.

electromyography (EMG) The electrical recording of muscle activity.

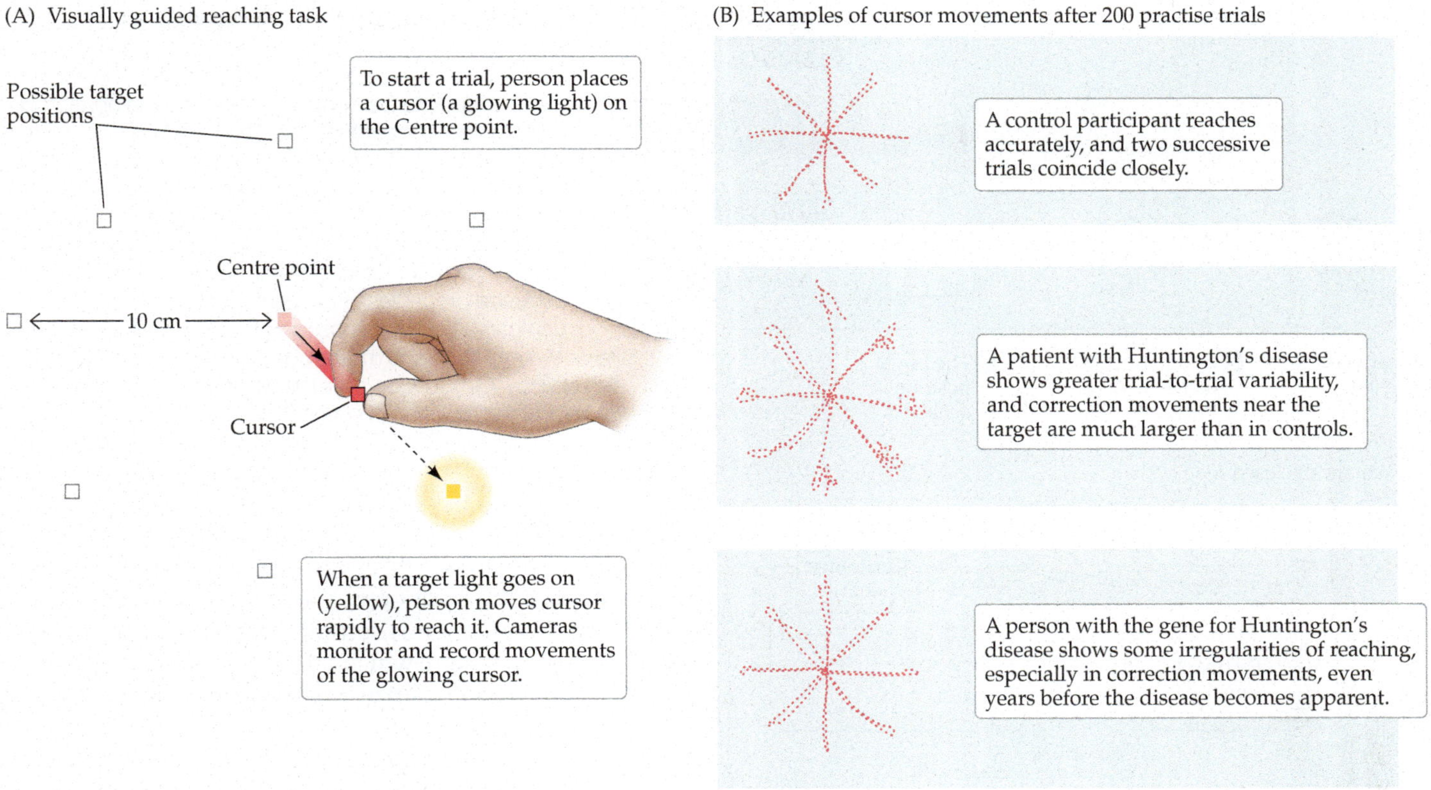

10.1 Measurement of Reaching Movements (A) An experimental setup to study reaching movements. (B) Recorded movement trajectories of a control (top), a patient with Huntington's disease (middle), and a carrier of the gene for Huntington's disease (bottom). We'll discuss this disorder in detail later in this chapter. (Part B courtesy of Maurice A. Smith.)

closed-loop control mechanism A control mechanism that provides a flow of information from whatever is being controlled to the device that controls it.

Motor control contrasts accuracy versus speed

One way to look at the mechanisms that regulate and control our movements employs the language of engineering. In designing machines, engineers commonly have two goals: (1) accuracy, to prevent or minimize error, and (2) speed, to complete a task quickly and efficiently. It is difficult to accomplish both these goals at once; usually there is a trade-off between the two. Two forms of control mechanisms—closed-loop and open-loop—are used to optimize one at the expense of the other.

Closed-loop control mechanisms maximise *accuracy*: information from whatever is being controlled flows back to the device that controls it. Driving a car is an example of a closed-loop

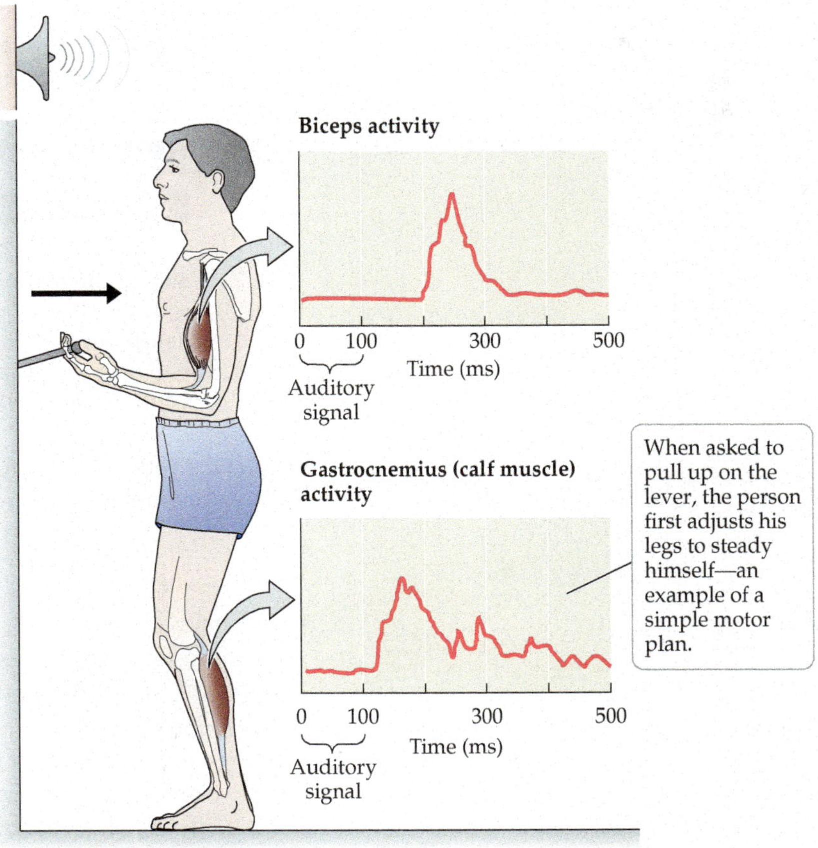

10.2 Electromyography For these recordings made from biceps and gastrocnemius (calf) muscles, the study participant was instructed to pull the handle as soon as a tone sounded. He sets his legs before pulling the knob, evidence of a motor plan. (After D. Purves et al., 2001. *Neuroscience*, 2nd ed. Oxford University Press/Sinauer: Sunderland, MA.)

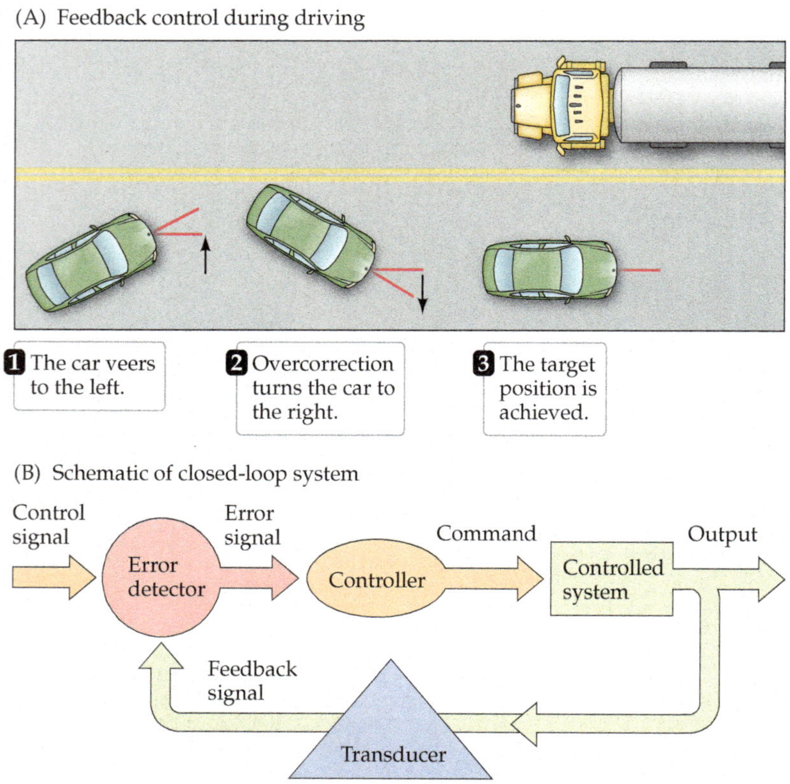

10.3 A Closed-Loop System (A) Automobile driving provides an example of feedback control. (B) In the example in part A, the *controlled system* is the automobile. The *control signal* (i.e., the input) to the controlled system is the position of the steering wheel; the *output* is the position of the car on the road. In any closed-loop system, the transducer is an element that measures output, and the error detector measures differences between actual output and desired output (the control signal). In this example the *transducer* (the visual system), the *error detector* (the perceptual system), and the *controller* (the muscles) are all properties of the person driving the car. The driver compares the actual position of the car on the road with the desired position and makes corrections to minimize the discrepancy. Closed-loop systems emphasise accuracy and flexibility at the expense of speed.

open-loop control mechanism A control mechanism in which feedback from the output of the system is not provided to the input control.

ballistic movement A rapid muscular movement that is generally preprogrammed.

motor system; in this case the variable being controlled is the position of the car on the road (**FIGURE 10.3**). Information is continually provided by the driver's visual system, which guides corrections.

Open-loop control mechanisms maximise *speed;* there are no external forms of feedback, and the activity is preprogrammed. Open-loop controls are needed in systems that must respond so rapidly that there is no time to wait for a feedback signal. For example, once a baseball pitcher begins throwing a fastball, the pitch will be completed as programmed no matter what sensory feedback is received. Such open-loop movements are called **ballistic movements**. Because there is no feedback, open-loop systems need other ways to reduce error and variability. They must anticipate potential error. As living systems, we learn to anticipate and avoid error, so (for example) we can learn to rapidly type our names with our eyes shut, or deliver a smoking fastball across the plate. Of course, open-loop movements are fast, but they're no way to drive a car. Instead, most of our movements in daily life depend on a blend of open-loop and closed-loop components, depending on the need for speed versus accuracy at that moment.

10.2 Neuroscience Reveals Hierarchical Systems

Learning Objectives

After reading this section, you should be able to:

10.2.1 **Sketch the hierarchy of motor control from muscle to forebrain.**

10.2.2 **Describe pathologies that act upon muscle and upon neuromuscular junctions.**

10.2.3 **Describe the structure of muscle spindles and Golgi tendon organs, and contrast the proprioception they provide.**

We can distinguish several different levels of hierarchically organized motor control systems:

- The *skeletal system* and the muscles attached to it determine which movements are possible.
- The *spinal cord* controls skeletal muscles in response to sensory information. In the simplest case, the response may be a reflex. The spinal cord also implements motor commands from the brain.
- The *brainstem* integrates motor commands from higher levels of the brain and transmits them to the spinal cord. It also relays sensory information about the body from the spinal cord to the forebrain.
- Some of the main commands for action are initiated in the *primary motor cortex*.

10.4 The Hierarchy of Movement Control The motor cortex receives information from other cortical areas and sends commands to the thalamus and brainstem, which pass commands to the spinal cord. Both the cerebellum and the basal ganglia adjust these commands.

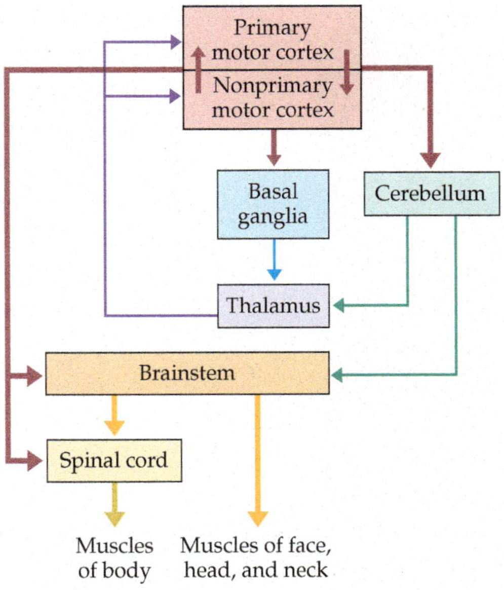

- Areas adjacent to the primary motor cortex, *nonprimary motor cortices*, provide an additional source of motor commands, acting indirectly via primary motor cortex and through direct connections to lower levels of the motor hierarchy.
- Other brain regions—the *cerebellum* and *basal ganglia*—modulate the activities of these hierarchically organized control systems. Some of their contributions are routed via the *thalamus* in a loop back to the cortex.

We will examine each of these levels of control, outlined in **FIGURE 10.4**, in the following discussions, beginning with the skeletal system.

The skeletal system enables particular movements and precludes others

Some properties of behaviour arise from physical characteristics of the skeleton. For example, the length, form, and weight of the limbs shape an animal's stride. The primary sites for bending are the joints, where bones meet. Some joints, such as the hip, are almost "universal" joints, permitting movement in many planes. Others, like the elbow or knee, are more limited and swing mostly in one direction. **FIGURE 10.5** illustrates the human skeleton and shows examples of joints and their possible movements. Interestingly, the octopus has no bones or joints, yet when it grabs food with a tentacle, the octopus will place "bends" in the tentacle, as if it had an elbow and wrist, to bring the food to its mouth (Sumbre et al., 2005), suggesting that joints may be the optimal solution for precise movements.

smooth muscle A type of muscle fibre, as in the heart, that is controlled by the autonomic nervous system rather than by voluntary control.

tendon Strong tissue that connects muscle to bone.

antagonist Here, a muscle that counteracts the effect of another muscle.

synergist Here, a muscle that acts together with another muscle.

motor neuron Also called *motoneuron*. A neuron in the brain or spinal cord that transmits motor messages to a muscle.

Muscles control the actions of the skeletal system

Our bare skeleton must now be clothed with muscles. Muscles work solely through contraction (shortening), so the skeletal connections of a muscle give us clues about the particular movements it causes. Muscles have springlike properties that influence the timing of behaviour and the forces that can be generated; the rate and force of muscular contractions limit some responses. A variety of visceral organs and blood vessels also employ a type of muscle, called **smooth muscle** (because of its appearance), but since its functions and control are very different from those of the skeletal muscles and since smooth muscle is not generally involved in voluntary behaviour, we will not concern ourselves with smooth muscle in this chapter.

Muscles are connected to bone by **tendons**. Around a joint, different muscles are arranged in a reciprocal fashion: when one muscle group contracts, it stretches the other group; that is, the muscles are **antagonists**. Muscles that act together are called **synergists**. For example, several synergistic muscles act together to extend the arm at the elbow. Another set of muscles that flex the arm are antagonists to those that extend it. Movement around a joint requires one set of **motor neurons** (the neurons that send axons to activate muscle contraction) to be excited while the antagonistic set of motor neurons is

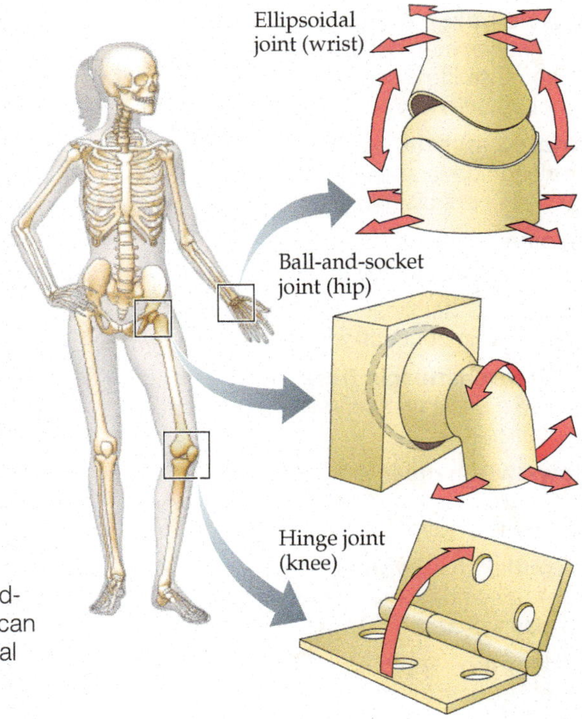

10.5 Joints and Movements Enlarged mechanical models indicate the kinds of movements that each type of joint can perform. The wrist joint moves in two principal planes: lateral and vertical. The hip joint is a "universal" joint, moving in all three planes. The knee joint has a single plane of motion.

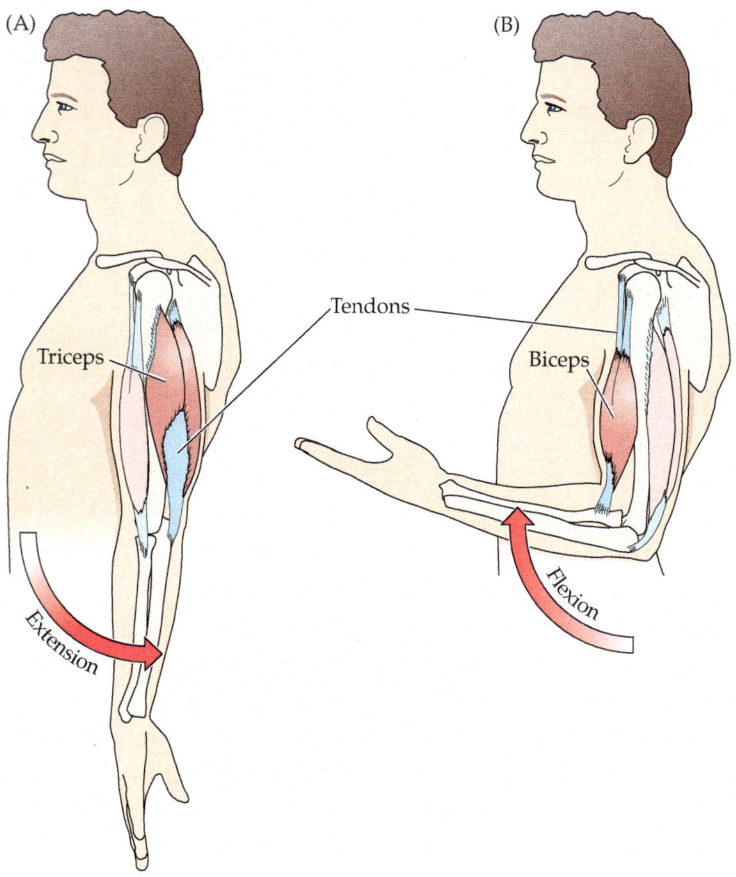

10.6 The Arrangement of Muscles around the Elbow
Because muscles exert force only by contracting, muscle attachments determine the resulting movement. (A) The triceps contracts to extend the arm. (B) When the biceps contracts, it flexes the arm. Because these two muscles mediate opposite movements, they are known as *antagonists*.

inhibited (**FIGURE 10.6**). We can lock a limb in position by contracting antagonistic muscles simultaneously.

THE MOLECULAR MACHINERY OF MUSCLES A skeletal muscle is composed of thousands of individual **muscle fibres** working together under voluntary control. Each muscle fibre contains many filaments of two kinds arranged in a regular manner (**FIGURE 10.7**), giving the fibres a striped, or "striated," appearance. In this **striated muscle** the thick and thin filaments (made up of the complex proteins **myosin** and **actin**, respectively) overlap. Contraction of the muscle increases the overlap: the filaments slide past each other, shortening the overall length of the muscle fibre.

Because of the varying tasks they perform, different muscles require different speeds, precision, strength, and endurance. So there are two main types of striated muscle fibres: fast-twitch and slow-twitch. Eye movements, for example, must be quick and accurate so that we can follow moving objects and shift our gaze from one target to another. But fibres in the extraocular muscles, which control eye movements, do not have to maintain tension for long periods of time; accordingly, they are mostly **fast-twitch muscle fibres**. In leg muscles, fast-twitch fibres react promptly and strongly but tire rapidly; they are used mainly for activities in which muscle tension changes frequently, as in walking or running. Mixed in with the fast-twitch muscle fibres are **slow-twitch muscle fibres**, which are not as fast but have greater resistance to fatigue; they are used chiefly to maintain posture. These properties of muscle fibres are determined, in part, by use. Athletes who train for endurance boost the proportion of their muscle fibres that display slow-twitch properties (Putman et al., 2004).

If you eat chicken or turkey, you've already contrasted fast-twitch and slow-twitch muscles. The "white meat" of the breast is fast-twitch muscle for the rapid wing beats needed for flight. These birds fly for only short periods, so there's no need for these muscles to resist fatigue. In contrast, the "dark meat" of the leg is slow-twitch muscle, which continually supports the animals as they walk about. These muscles are darker because they have higher concentrations of blood vessels, mitochondria, and the oxygen-binding protein myoglobin than fast-twitch muscles. They contract slowly but have lots of endurance. Most muscles consist of a mixture of slow-twitch and fast-twitch muscle fibres.

MUSCULAR DYSTROPHY Numerous muscle diseases are characterised by biochemical abnormalities that lead to structural changes in muscle; collectively, these disorders are referred to as **muscular dystrophy** (**MD**) (from the Greek *dys*, "bad," and *trophe*, "nourishment"). As the name implies, in various muscular dystrophies the muscles waste away (D. J. Blake et al., 2002).

Duchenne muscular dystrophy, the most prevalent form of MD, strikes almost exclusively boys, beginning at the age of about 4 to 6 years and usually leading to death in early adulthood. Studies of family pedigrees show that the disorder is a simple Mendelian trait—caused by a single gene—carried on the X chromosome. When the gene was identified, it was named *dystrophin*. In some ways the name is unfortunate because the dystrophin protein, when normal, does *not* lead to dystrophy. **Dystrophin** is normally produced in muscle cells and is part of a vital structural

muscle fibre A large cylindrical cell that can contract in response to neurotransmitter released from a motor neuron.

striated muscle A type of muscle with a striped appearance, generally under voluntary control.

myosin A protein that, along with actin, mediates the contraction of muscle fibres.

actin A protein that, along with myosin, mediates the contraction of muscle fibres.

fast-twitch muscle fibre A type of striated muscle that contracts rapidly but fatigues readily.

slow-twitch muscle fibre A type of striated muscle fibre that contracts slowly but does not fatigue readily.

muscular dystrophy (MD) A disease that leads to degeneration of and functional changes in muscles.

dystrophin A protein that is needed for normal muscle function.

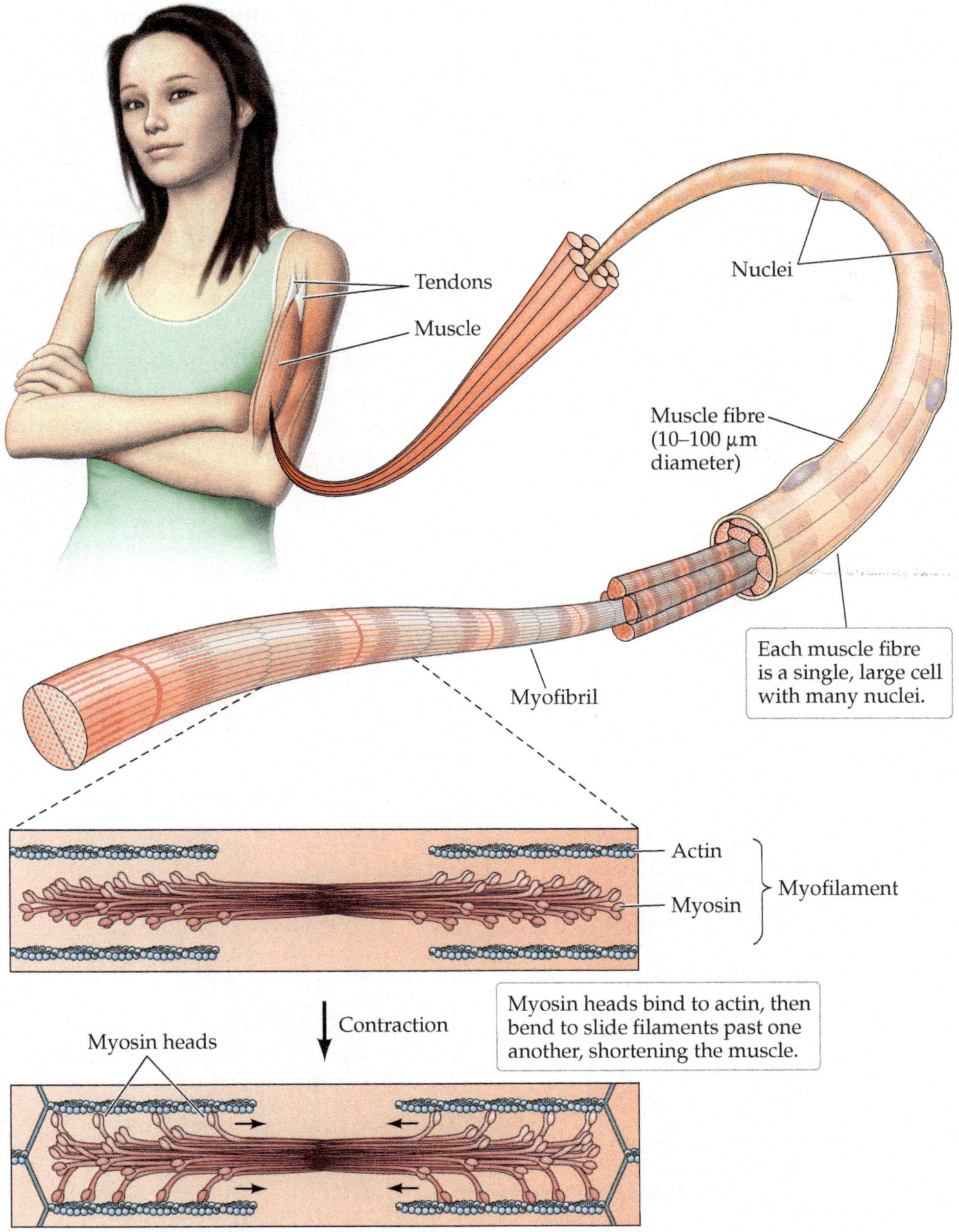

10.7 The Composition of Muscles and the Mechanism of Muscle Contraction Muscle fibres are shown here at progressively greater magnifications, with the bottom image 2 million times life size. The actions of myosin and actin cause muscle contraction.

Go to Animation 10.1
Muscle Contraction

component of muscle fibres. Because a woman has two X chromosomes, even if one carries the defective copy of the *dystrophin* gene, the other X chromosome can still direct production of sufficient normal dystrophin. But about 50% of her sons will receive the defective gene and, because they have only the one X chromosome, will be afflicted with the disease. In an effort to halt the loss of muscle fibres in Duchenne, scientists are trying to use gene therapy to induce the muscles of boys with this disease to produce normal dystrophin. One approach, using CRISPR (see Appendix Figure A.5) to insert the *dystrophin* gene into muscle, worked in a dog model of Duchenne muscular dystrophy (Amoasii et al., 2018), so hopes are high.

Neural messages reach muscle fibres at the neuromuscular junction

As each motor neuron integrates the information bombarding it through hundreds or thousands of synapses, it may produce an action potential. The action potential travels down the axon, which splits into many branches near the target muscle

acetylcholine (ACh) A neurotransmitter produced and released by parasympathetic postganglionic neurons, by motor neurons, and by neurons throughout the brain.

neuromuscular junction (NMJ) The region where the motor neuron terminal and the adjoining muscle fibre meet; the point where the nerve transmits its message to the muscle fibre.

motor unit A single motor axon and all the muscle fibres that it innervates.

innervation ratio The ratio expressing the number of muscle fibres innervated by a single motor axon.

(**FIGURE 10.8A AND B**). Each axonal branch carries an action potential to its axon terminal, which then (in vertebrates) releases the neurotransmitter **acetylcholine** (**ACh**). Muscle fibres respond to the ACh by producing action potentials of their own. The action potentials travel along the muscle fibres, permitting sodium (Na^+) and calcium (Ca^{2+}) ions to enter and then trigger the molecular changes in actin and myosin that produce contraction.

The region where a motor neuron terminal and adjoining muscle fibre meet is called the **neuromuscular junction** (**NMJ**) (**FIGURE 10.8C–E**). The NMJ is large and very effective: normally every action potential that reaches an axon terminal releases enough ACh to trigger an action potential and contraction in the innervated muscle fibre.

The **motor unit** consists of a single motor neuron and all the muscle fibres innervated by its various axonal branches (see Figure 10.8C). When the motor neuron fires, each of the muscle fibres that it innervates is stimulated. The **innervation ratio** is the number of muscle fibres innervated by one motor neuron. Low innervation

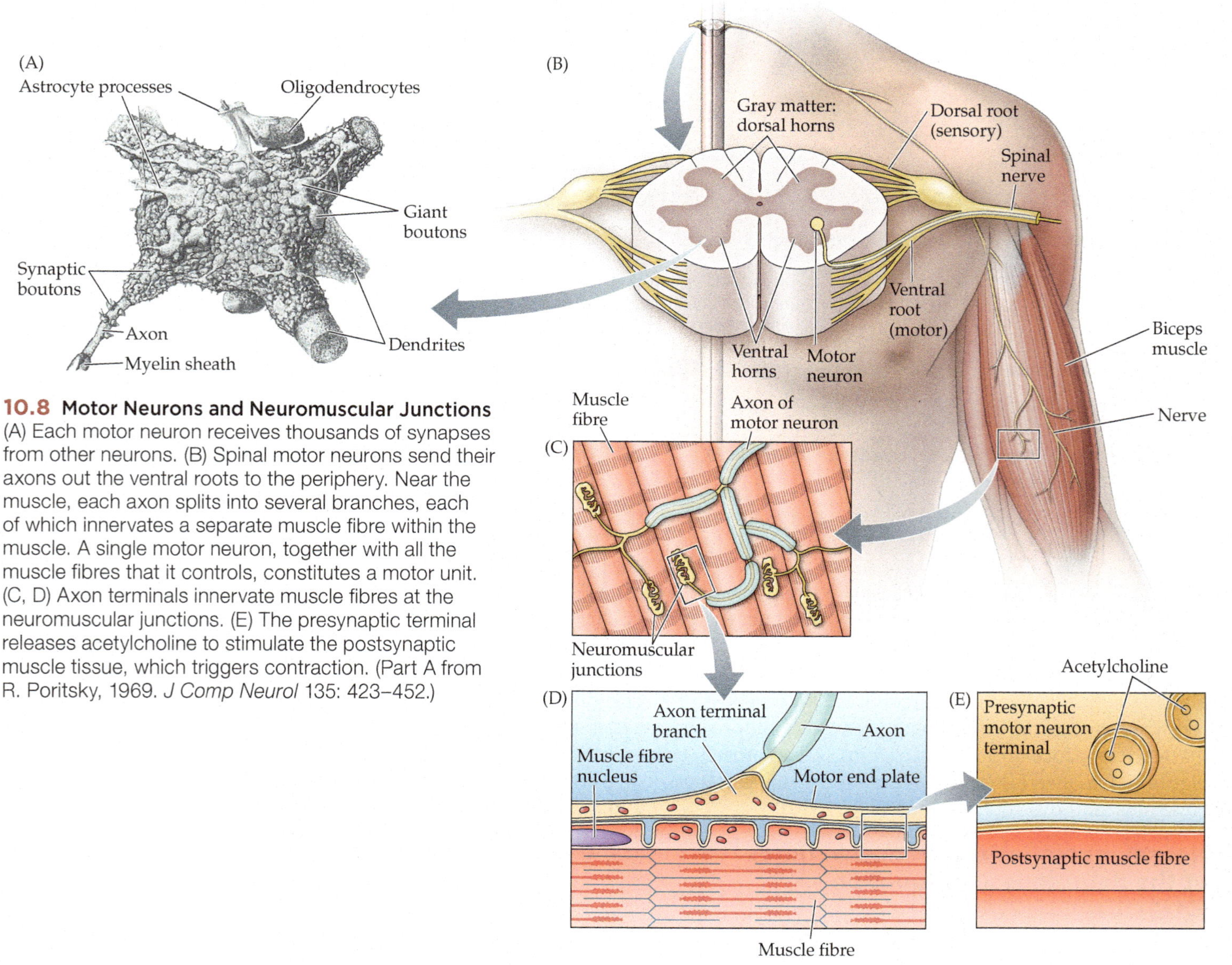

10.8 Motor Neurons and Neuromuscular Junctions
(A) Each motor neuron receives thousands of synapses from other neurons. (B) Spinal motor neurons send their axons out the ventral roots to the periphery. Near the muscle, each axon splits into several branches, each of which innervates a separate muscle fibre within the muscle. A single motor neuron, together with all the muscle fibres that it controls, constitutes a motor unit. (C, D) Axon terminals innervate muscle fibres at the neuromuscular junctions. (E) The presynaptic terminal releases acetylcholine to stimulate the postsynaptic muscle tissue, which triggers contraction. (Part A from R. Poritsky, 1969. *J Comp Neurol* 135: 423–452.)

ratios characterise delicate muscles involved in fine movements, like those that move the eye—which have one motor neuron for every three fibres (a 1:3 ratio), allowing exquisitely fine gradations of contraction. Motor units of massive muscles such as those of the leg have high innervation ratios, with each motor neuron innervating hundreds of muscle fibres, so a single motor neuron contracts many muscle fibres at once and produces a lot of force.

Motor neurons are sometimes called the **final common pathway**: the sole route through which the spinal cord and brain can control our many muscles. Because they respond to inputs from so many sources, motor neurons often have very widespread dendritic fields, and they are the largest cells in the spinal cord. Furthermore, while motor neurons *release* only ACh, they respond to a tremendous variety of synaptic transmitters, both excitatory and inhibitory, released by the diverse inputs that each motor neuron receives. Virtually all motor neuron axons are myelinated, so their action potentials are conducted quickly. Once the motor plan is formed and the motor neurons stimulated to enact it, you want the muscles to respond right away.

ATTACKS ON THE NEUROMUSCULAR JUNCTION Several poisons affect neuromuscular junctions. For example, snake bites can cause neuromuscular blocks because the venom of some highly poisonous snakes contains substances (such as bungarotoxin) that block postsynaptic ACh receptors. If the neuromuscular junctions of the muscles for breathing are blocked, the victim suffocates.

Studies of bungarotoxin led to an understanding of a debilitating neuromuscular disorder, **myasthenia gravis** (from the Greek *mys*, "muscle," and *asthenes*, "weak," and the Latin *gravis*, "grave" or "serious"). This disorder is characterised by a profound weakness of skeletal muscles. The disease often first affects the muscles of the head, producing symptoms such as drooping of the eyelids, double vision, and slowing of speech. In later stages, paralysis of the muscles that control swallowing and respiration becomes life-threatening. The weakness happens because the neuromuscular junctions are not working—the muscles aren't getting the message to contract.

Myasthenia gravis is an **autoimmune disorder**: most cases result when antibodies develop and attack a patient's own ACh receptors, disrupting neuromuscular junctions. In other cases, the antibodies are directed toward other proteins that are associated with the ACh receptor. Treatment often consists of suppressing the immune system with drugs (Dalakas, 2019) or of surgically removing the thymus (Cataneo et al., 2018), one source of antibodies.

Sensory feedback from muscles, tendons, and joints monitors movements

To produce rapid coordinated movements of the body, the brain and spinal cord must continually monitor the state of the muscles and the positions of the limbs. This perception of body movements and positions is called **proprioception** (from the Latin *proprius*, "own," and *recipere*, "to receive"). Because it is crucial for coordinating movement, this information is transmitted to the brain by large, myelinated axons, which conduct action potentials swiftly (see Chapter 2). For some reason, the virus that attacked Ian, whom we met at the start of this chapter, killed only these large axons, leaving him without proprioception below the neck. His predicament illustrates how important this "sixth sense" is for movement.

Two major kinds of proprioceptive receptors that normally report the state of muscles and joints to the brain are (1) muscle spindles, which monitor muscle *length*, and (2) Golgi tendon organs, which monitor muscle *tension*. We'll discuss each in turn.

THE MUSCLE SPINDLE The **muscle spindle** is a complicated structure consisting of both afferent and efferent elements. Each spindle (a tapered cylinder) contains small muscle fibres called **intrafusal fibres** (from the Latin *intra*, "within,"

final common pathway The information-processing pathway consisting of all the motor neurons in the body. Motor neurons are known by this collective term because they receive and integrate all motor signals from the brain and then direct movement accordingly.

myasthenia gravis A disorder characterised by a profound weakness of skeletal muscles. It is caused by a loss of acetylcholine receptors.

autoimmune disorder A disorder caused when the immune system mistakenly attacks a person's own body, thereby interfering with normal functioning.

proprioception Body sense; information about the position and movement of the body that is sent to the brain.

muscle spindle A muscle receptor that lies parallel to a muscle and sends action potentials to the central nervous system when the muscle is stretched.

intrafusal fibre One of the small muscle fibres that lie within each muscle spindle, controlling its length.

extrafusal fibre One of the ordinary muscle fibres that lie outside the spindles and provide most of the force for muscle contraction.

primary sensory ending Also called annulospiral ending. The axon that transmits information from the central portion of a muscle spindle.

secondary sensory ending Also called flower spray ending. The axon that transmits information from the ends of a muscle spindle.

gamma motor neuron Also called gamma efferent. A motor neuron that innervates the contractile tissue (the intrafusal fibre) in a muscle spindle.

alpha motor neuron A motor neuron that controls the main contractile fibres (extrafusal fibres) of a muscle.

and *fusus*, "spindle"); the ordinary muscle fibres that lie outside the spindles are called **extrafusal fibres** (**FIGURE 10.9**).

The muscle spindle contains two kinds of receptor endings: **primary sensory endings** (also called *annulospiral endings*) and **secondary sensory endings** (also called *flower spray endings*). These endings are related to different parts of the spindle (see Figure 10.9C). The primary ending wraps in a spiral fashion around the central region of the spindle. The secondary endings terminate toward the thin ends of the spindle. These different receptor elements of the muscle spindle respond to two different aspects of changes in muscle length. The primary (annulospiral) endings show a maximum discharge early in stretch and then adapt and show a lower discharge rate, a phasic response (see Chapter 7). They therefore communicate *velocity of muscle stretch* to the nervous system. In contrast, the secondary (flower spray) endings are slow to change their rate during the early phase of stretch and are maximally sensitive to maintained, or *static*, muscle length, a tonic response. In this way, they communicate *muscle length* to the nervous system. The nervous system thus has information about the movement and length of every muscle.

For these receptors to work properly, the overall muscle spindle needs to be about the same length as the muscle. But muscles change their length as they contract or relax. To keep the muscle spindle about the same length as the muscle, a special motor neuron stimulates the intrafusal fibre portion of the spindle, altering the tension, thus maintaining the sensitivity of its receptors. These motor neurons are called **gamma motor neurons**, or *gamma efferents*, to distinguish them from the larger **alpha motor neurons**, which go to the extrafusal muscle fibres that do all the work (see Figure 10.9C).

How do the muscle spindle afferents become excited? Suppose a muscle is stretched, as when a load is placed on it. For example, imagine that you hold out your hand and someone hands you a heavy book. The additional load would move your arm downward, stretching the biceps muscle and the spindles it contains. The resulting deformation of the endings on the spindle would trigger action potentials in the afferent fibres. These afferents would inform the spinal cord, and the spinal cord would tell the brain about the muscle stretch and therefore the load imposed. Muscle spindle afferents also trigger changes in the motor neurons, as part of the stretch reflex we'll discuss shortly. Deprived of this proprioceptive information, Ian can estimate the weight of objects only by seeing how fast and far they move when he lifts them (Proske and Gandevia, 2016).

THE GOLGI TENDON ORGAN While muscle spindles respond primarily to stretch, the other proprioceptive receptors in muscle—Golgi tendon organs—are

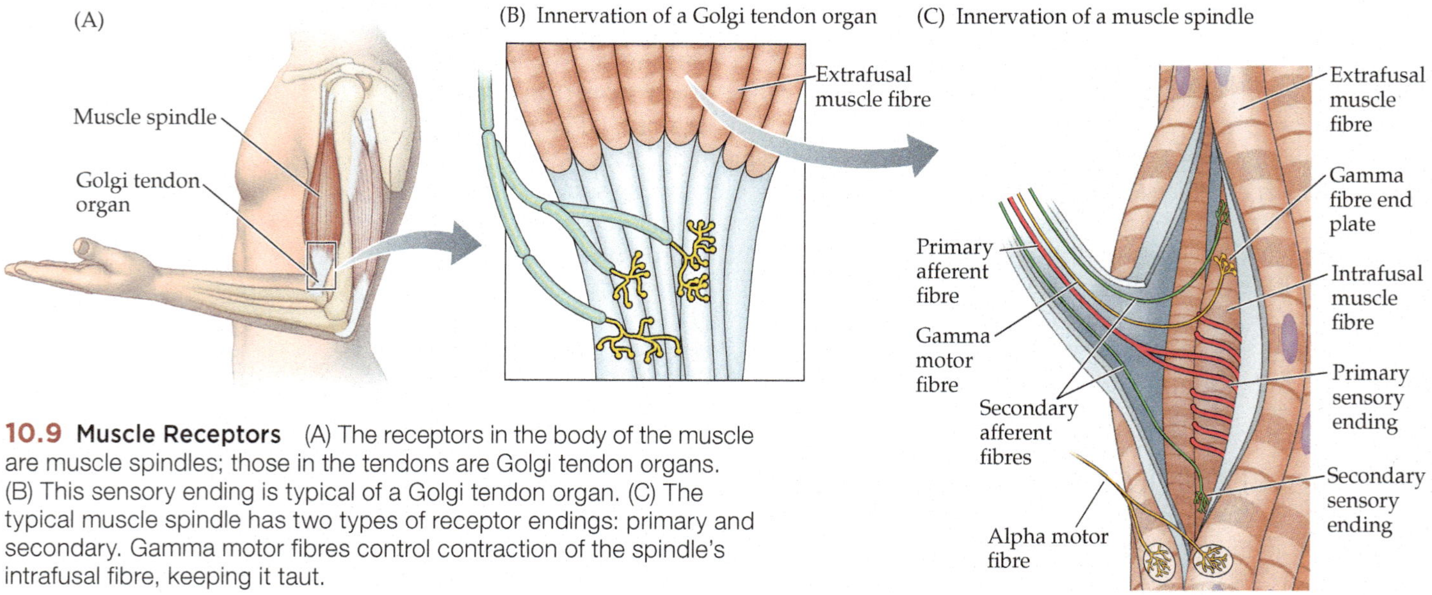

10.9 Muscle Receptors (A) The receptors in the body of the muscle are muscle spindles; those in the tendons are Golgi tendon organs. (B) This sensory ending is typical of a Golgi tendon organ. (C) The typical muscle spindle has two types of receptor endings: primary and secondary. Gamma motor fibres control contraction of the spindle's intrafusal fibre, keeping it taut.

especially sensitive to muscle *tension* (**FIGURE 10.10**). Structurally, Golgi tendon organs consist of sensory nerve endings interwoven through the tough proteins that give tendons their strength and elasticity. Because they are part of a robust structure, **Golgi tendon organs** are relatively insensitive to passive muscle lengthening (see Figures 10.9 and 10.10), such as simply extending an arm. But large loads (requiring strong contractions) will stretch the tendon organ and stimulate the nerve endings to fire. Thus, the Golgi tendon organs monitor the force of muscle contractions, providing a second source of sensory information about the muscles to help control movement. This arrangement makes the tendon organs useful in another important way: they detect overloads that threaten to tear muscles and tendons. Strong stimulation of the Golgi tendon organs inhibits the motor neurons supplying the muscles that pull on the tendon and thus, by relaxing the tension, prevents mechanical damage. The afferent neurons in both muscle spindles and Golgi tendon organs use the same stretch-sensitive ion channel protein (Piezo2) as the touch receptor (see Figure 7.5) to trigger action potentials (Woo et al., 2015).

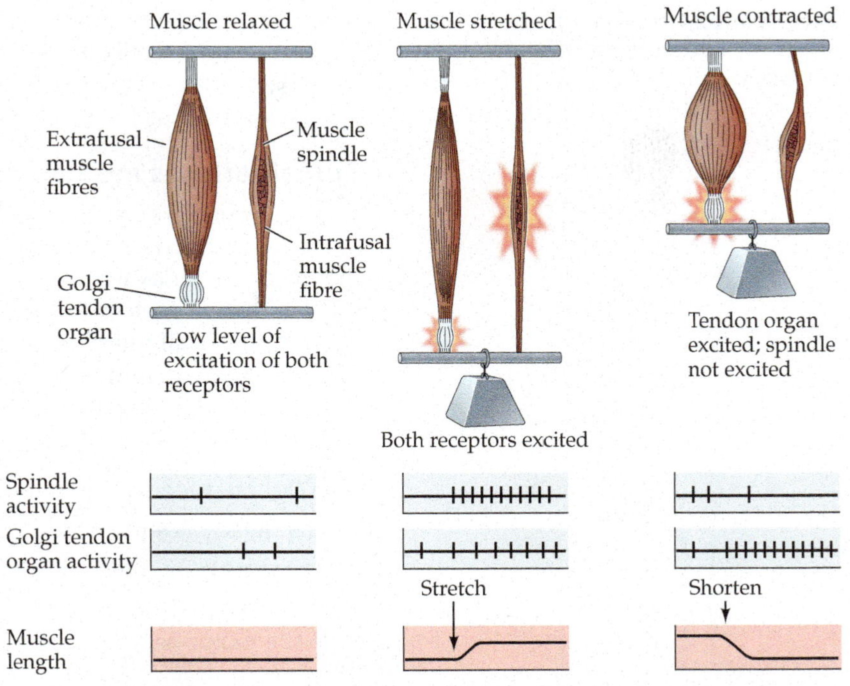

10.10 Spindles Measure Stretch; Tendon Organs Measure Tension

Classic studies in physiology emphasised the importance of information from muscle spindles and Golgi tendon organs for controlling movement. Cutting the afferent fibres from muscles in monkeys caused them to stop using the deafferented limb, even if the efferent connections from motor neurons to muscles were preserved (Mott, 1895; Sherrington, 1898). The deafferented limb was not paralysed, since it could be activated (the motor neurons still innervated the muscles), but lack of information from the muscle led to relative disuse. The arm dangled, apparently useless. However, if the good arm was restrained (by a ball placed around the hand), the animal soon learned to use the deafferented arm and could become quite dexterous (Taub, 1976). Monkeys managed to do this the way Ian does, guiding movements by using vision for feedback about their arm movements.

Probably all of us use visual information to guide the motor system without being aware that we're doing so. Remember the young woman D.F. from Chapter 8? She could not report whether a slot was vertical or horizontal; yet when she was asked to insert a disk in the slot, she consistently rotated her hand to put it in smoothly. This finding suggests that even neurologically intact people unknowingly use visual cues to guide their movements (Goodale and Haffenden, 1998). These visual cues are the only ones available for Ian.

Golgi tendon organ One of the receptors located in tendons that send action potentials to the central nervous system reporting muscle tension.

10.3 The Spinal Cord Is a Crucial Link in Controlling Body Movement

Learning Objectives

After reading this section, you should be able to:

10.3.1 **List the sequence of events that underlie the stretch reflex.**

10.3.2 **Describe the role of central pattern generators in locomotion.**

10.3.3 **List four strategies being explored to restore movement control following spinal cord damage.**

10.3.4 **Describe the pathology of amyotrophic lateral sclerosis and theories of its origins.**

Motor neurons are the final common pathway for behaviour, because all action depends on activating these cells to determine the onset, coordination, and termination

stretch reflex The contraction of a muscle in response to stretch of that muscle.

of muscle activity. But to really understand the physiology of movement, we need to know the source of the inputs to motor neurons. First we'll consider the organisation of the spinal cord, and the disorders that disrupt that organisation, before considering the descending brain pathways that activate this final common pathway.

Spinal reflexes mediate "automatic" responses

A good example of automatic control at the spinal level is the **stretch reflex**—the contraction that results when a muscle stretches. One version of the stretch reflex is the knee-jerk reflex we discussed in Chapter 2 (see Figure 2.17). This reflex isn't just present to give doctors (or comedians) something to do—it evolved to help us with load control. That might be more obvious if we consider the stretch reflex in an arm rather than a leg. Adding a weight to the hand in **FIGURE 10.11** imposes a sudden stretch on muscle A (labelled "M_A"). The circuit that keeps us from dropping the load is one that links muscle spindles and the relevant muscles:

1. Weight is added, pulling the hand down.
2. The muscle is stretched.
3. Afferents from the muscle spindle are excited ("SN_A" in Figure 10.11).
4. These spindle afferents connect directly—that is, monosynaptically—to the motor neurons that control the stretched muscle, exciting them.
5. The motor neurons stimulate the muscle to oppose muscle stretch, keeping the hand from dropping farther.

This sequence describes a simple negative feedback system that tends to restore the limb to its "desired" position. Muscle spindle afferents also inhibit the motor neurons that supply the antagonistic muscle ("M_B" in Figure 10.11). Two synapses are required for this inhibition of the antagonistic motor neuron. Spindle information terminates on a spinal interneuron, which inhibits the motor neuron that supplies M_B. The relaxation of antagonistic muscles prevents them from being injured by the sudden movement. In these and other movements, spinal circuits modulate the inputs from muscle spindles to produce smooth movements (A. J. Fink et al., 2014).

Other spinal circuits can generate more-complex movements. Most locomotion is rhythmic, consisting of repetitive cycles of the same movement, whether it be the beating of wings or the swinging of legs. Many such rhythmic movements are generated by mechanisms within the spinal cord (Kiehn, 2016). These endogenous rhythms are normally modulated by sensory feedback, but they can function independently

Go to Animation 10.2
The Stretch Reflex Circuit

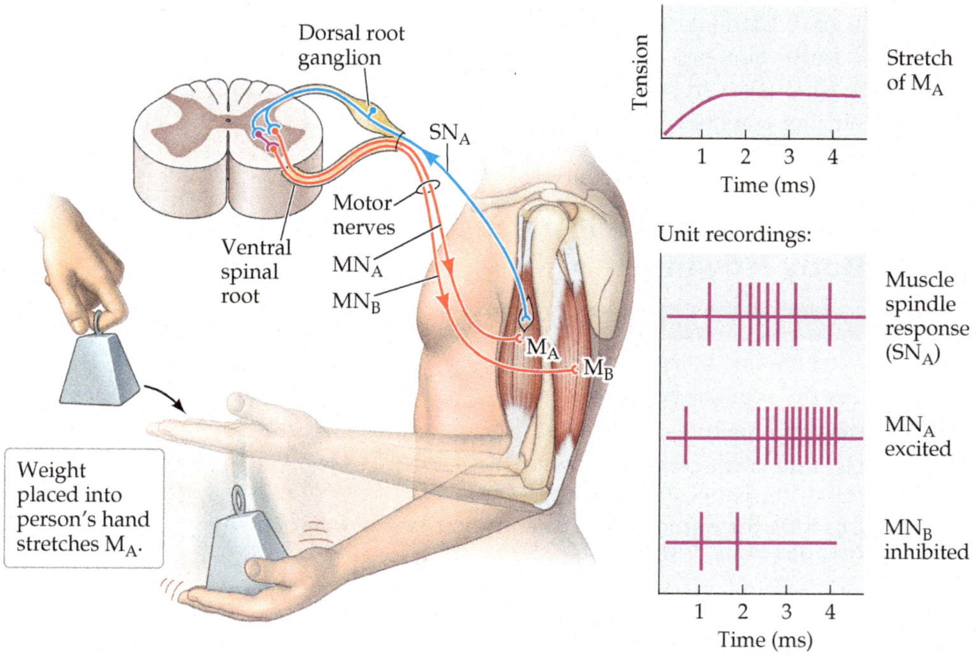

10.11 The Stretch Reflex Circuit MN_A is the motor nerve to muscle A (M_A); MN_B is the motor nerve to muscle B (M_B), an antagonist to M_A. SN_A is the sensory nerve from the muscle spindle of M_A. Characteristic responses at different stages in the circuit are shown at right.

of brain influences or afferents. The term **central pattern generator** refers to the neural circuitry responsible for generating rhythmic patterns of behaviour, as seen in walking. The central pattern generators for walking seem much the same in cats, birds, and humans (Dominici et al., 2011). This essential rhythm of walking generated by spinal cord mechanisms is activated and modulated by the brain.

central pattern generator Neural circuitry that is responsible for generating the rhythmic pattern of a behaviour such as walking.

flaccid paralysis A loss of reflexes below the level of transection of the spinal cord.

Spinal cord injuries cause severe motor impairments

Vehicular accidents, falls, violence, and sports injuries cause many human spinal injuries that result in motor impairment. Injuries to the human spinal cord commonly develop from force to the neck or back, breaking a bone that then compresses the spinal cord. If the spinal cord is severed completely, immediate paralysis results, and reflexes below the level of injury may be lost—a condition known as **flaccid paralysis**. Flaccid paralysis generally results only when a considerable length of the spinal cord has been destroyed. When the injury severs the spinal cord without causing extensive destruction of spinal cord tissue, reflexes below the level of injury may become abnormally strong because the intact tissue has been disconnected from the brain's descending inhibitory projections.

Over 250,000 individuals in the United States have spinal cord injuries, and about 12,000 new traumatic spinal injuries occur each year. Although much remains to be discovered, the hope of reconnecting the injured spinal cord no longer seems far-fetched. Four main strategies for reconnecting the brain and spinal cord in humans are being investigated (**FIGURE 10.12**):

1. Inserted stem cells might differentiate into new neurons to send new axons across the break (Assinck et al., 2017; Cyranoski, 2019).

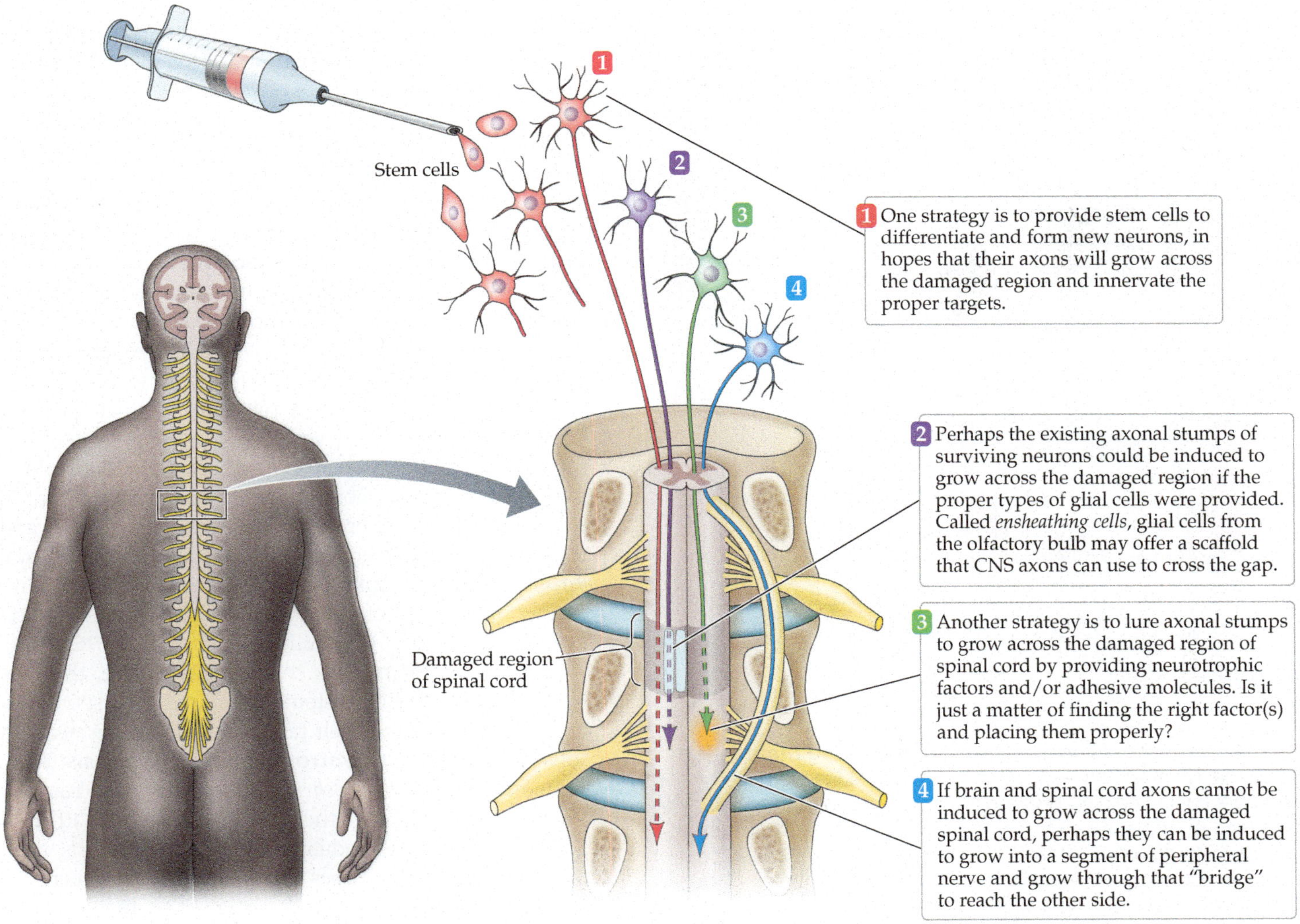

10.12 Research Strategies for Reconnecting the Brain and Spinal Cord

polioviruses A class of viruses that destroy motor neurons of the spinal cord and brainstem.

acute flaccid myelitis A sudden weakness or paralysis in limb(s) with decreased or absent reflexes.

amyotrophic lateral sclerosis (ALS) Also called Lou Gehrig's disease. A disease in which motor neurons and their target muscles waste away.

2. Transplanted glial cells can promote regeneration in the CNS. Recall from Chapter 9 that olfactory receptor neurons are continually produced, and somehow the axons of the new neurons find the way to their proper targets. Specialized glial cells called olfactory *ensheathing cells* appear to play a central role in guiding this axon growth, so researchers have tried transplanting ensheathing cells from the olfactory bulbs into spinal cord cuts in rats. In most cases, and for various forms of spinal cord damage, at least some function is restored (Khankan et al., 2016). Unlike mammals, zebrafish can regenerate spinal cord connections after injury, and glia seem to direct this regeneration (Mokalled et al., 2016).
3. Neurotrophic factors and/or cell adhesion molecules (see Chapter 6) can be used to entice the axons of surviving neurons to grow across the damaged region of spinal cord (Anderson et al., 2015) and reconnect to their targets (Hodgetts and Harvey, 2017). In dogs, a nonspecific polymer, polyethylene glycol, seems to repair broken membranes when injected into spinal cord within a few days of injury (X. Lu et al., 2018). Dogs given this treatment were more than twice as likely to walk again (Laverty et al., 2004).
4. Implanting special "regeneration-friendly" materials at the injury site may allow nerves to regenerate past the break (Marchesan et al., 2017). A variant of this is to transplant peripheral nerves to connect the brain and lower spinal cord by forming a "bridge" around the injured spinal cord (Estrada and Müller, 2016). This approach centres on the observation that axons in peripheral nerves, when cut by injury, will regrow and reconnect to their targets; yet cut axons in the CNS almost never accomplish this feat. No one really knows why peripheral and central axons differ in this regard, but it might be possible to exploit regeneration-friendly peripheral nerves to reconnect the spinal cord.

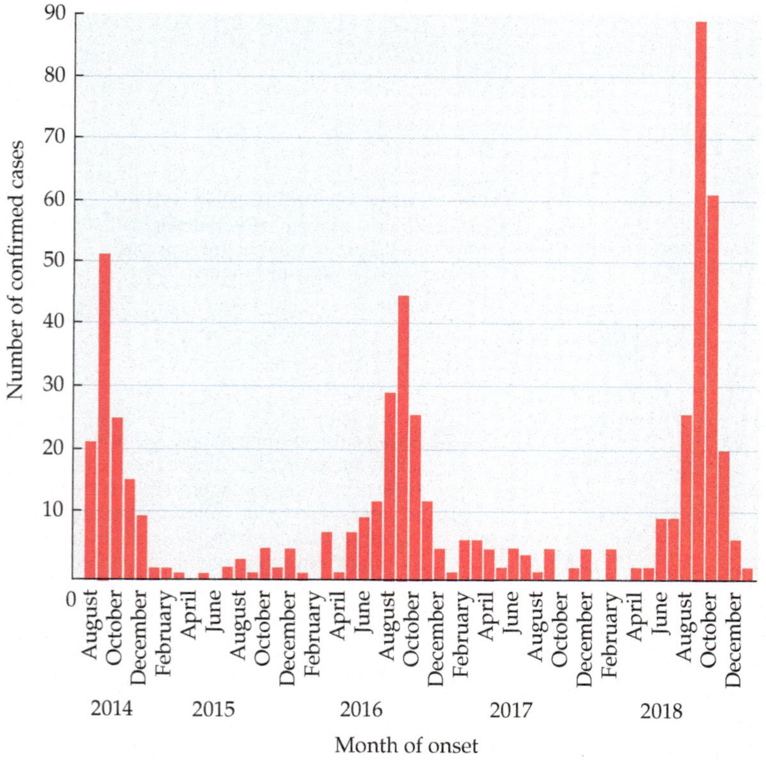

10.13 Acute Flaccid Myelitis Cases in the United States The periodic peaks in new cases every 2 years resembles the patterns seen in viral infections. The condition mostly strikes young children, many of whom recover spontaneously in a year or two. (After D. M. Morens et al., 2019. *mBio* 10: e00521–19/CC BY 4.0; Centres for Disease Control and Prevention, 2019. *AFM confirmed U.S. cases*. Centres for Disease Control and Prevention: Atlanta, GA. https://www.cdc.gov/acute-flaccid-myelitis/afm-cases.html.)

While spinal injuries severely impair movement, another movement disorder amounts to a death sentence, as we'll see next.

Motor neuron pathology leads to motor impairments and death

Virus-induced destruction of motor neurons—for example, by the disease known as *polio*—was once a frightening prospect around the world. **Polioviruses** destroy motor neurons of the spinal cord and, in more severe types of the disease, cranial motor neurons of the brainstem. Because the muscles are no longer called on to contract, they atrophy. If the muscles controlling breathing deteriourate sufficiently, the person must rely on a ventilator to stay alive. In 2014, a formerly very rare condition began striking young children worldwide—**acute flaccid myelitis**, which leaves one or more limbs paralysed. Most children recover within a year or two (J. A. Martin et al., 2017), and there is growing evidence that the condition is caused by a virus attacking the spinal cord (Morens et al., 2019). Puzzlingly, new cases of acute myelitis have been peaking every 2 years, typically in mid-September (**FIGURE 10.13**).

Sometimes, for reasons that remain elusive, the motor neurons of the brainstem and spinal cord spontaneously start to die and their target muscles waste away. In this disease, called **amyotrophic lateral sclerosis** (**ALS**; sometimes called *Lou Gehrig's disease* after the baseball player who lost his life to the disorder in 1941), the afflicted person experiences gradually worsening paralysis until most skeletal muscle ceases to function. There is currently no treatment for ALS (S. Perrin, 2014), so premature death is the usual outcome, although a few people with ALS can survive for long periods; celebrated British physicist Stephen Hawking (1942–2018), who was diagnosed with ALS in the early 1960s, may hold the record for survival after such a diagnosis (T. McCoy, 2015). A wide range of possible causal factors are under investigation, including premature aging, neurotoxins, viruses, immune responses, and endocrine dysfunction (Petrov et al., 2017).

About 10% of ALS cases are hereditary (Cirulli et al., 2015). The pedigrees of several afflicted families indicate that mutation in a single gene, called *SOD1*, accounts for about 20% of the heritable cases of ALS. This gene was isolated (P. M. Andersen et al., 1995) and found to encode an enzyme: copper-zinc superoxide dismutase. When scientists produced transgenic mice that, in addition to their own normal copies of the enzyme, carried a copy of the defective human *SOD1* gene, these animals displayed an ALS-like syndrome (Gurney et al., 1994). Their muscles wasted away and their motor neurons died, leading to an early death. Because the mice still had their own *SOD1* genes to produce the normal enzyme, these findings suggest that the familial ALS resulted not from the loss of enzyme action but rather from what is called a *toxic gain of function*—a harmful new action of the mutant protein. The abnormal SOD1 protein does not act within the motor neurons themselves (Philips and Rothstein, 2015), but rather acts through nearby astrocytes (Nagai et al., 2007) or through the muscle targets (Wong and Martin, 2010) to kill the motor neurons.

Another protein that has been implicated in ALS is TDP43 (Majumder et al., 2018), and again the mutant protein seems to have a toxic gain of function when expressed in mice (E. B. Lee et al., 2012). Understanding how the mutant TDP43 and SOD1 proteins cause ALS in that minority of cases that are inherited may shed light on what goes wrong in the more common, uninherited forms of ALS.

10.4 Pathways from the Brain Control Different Aspects of Movements

Learning Objectives

After reading this section, you should be able to:

10.4.1 Trace the pathways of the pyramidal motor system from cortex to spinal cord.

10.4.2 Distinguish the contributions of primary versus nonprimary motor cortex.

10.4.3 Contrast the types of motor activity influenced by the supplementary motor cortex versus premotor cortex.

10.4.4 Describe the activity of mirror neurons, and explain their potential role in empathy and autism.

Some muscles are controlled directly by the brain. The cranial motor nuclei of the brainstem send their axons to innervate muscles of the head and neck (**FIGURE 10.14**;

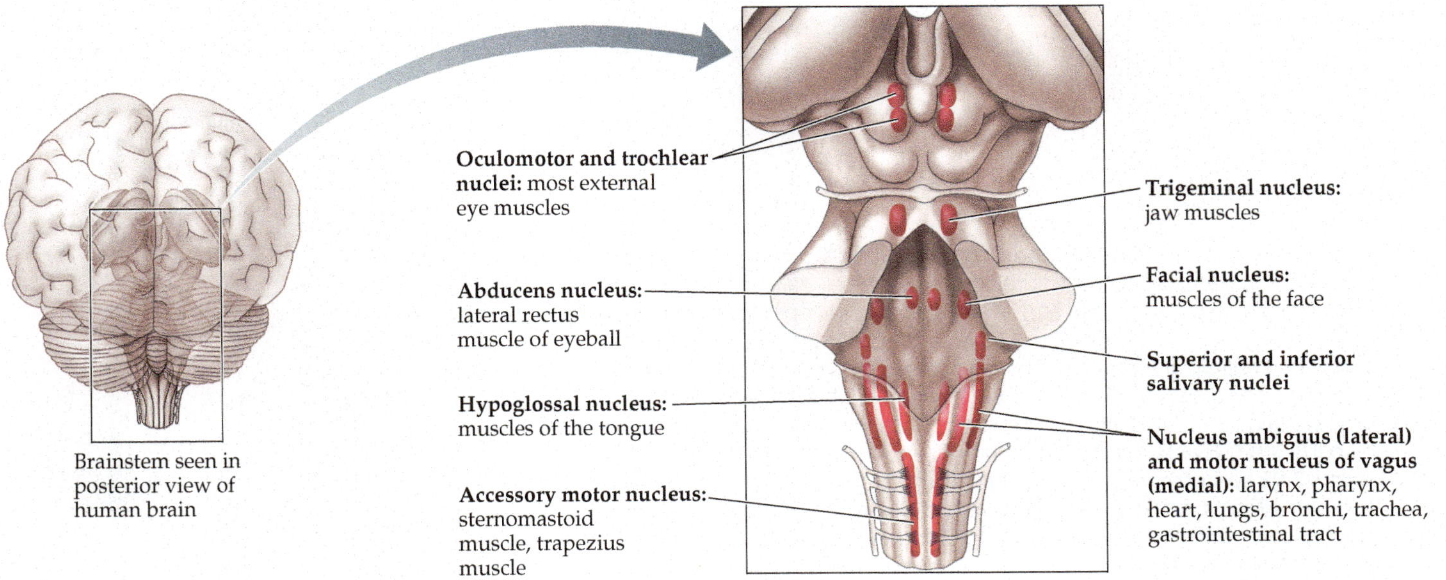

10.14 Motor Nuclei in the Brainstem Shown here from the rear, motor nuclei in the brainstem control muscles of the head and neck. The cranial nerves corresponding to these muscles are shown in Figure 2.9.

pyramidal system Also called corticospinal system. The motor system that includes neurons within the cerebral cortex that send axons to form the pyramidal tract.

primary motor cortex (M1) The apparent executive region for the initiation of movement; primarily the precentral gyrus.

see also Figure 1.9). But for all the other muscles, the brain has to send commands to the spinal cord, and then the spinal cord motor neurons control the muscles.

The **pyramidal system** (or *corticospinal system*) consists of neuronal cell bodies within the cerebral cortex and their axons, which pass through the brainstem, forming the pyramidal tract to the spinal cord (**FIGURE 10.15A**). The pyramidal tract is seen most clearly where it passes through the floor of the medulla. In a cross section of the medulla, the tract is a wedge-shaped ventral protuberance (pyramid) on each side of the midline. In the medulla the pyramidal tract from the right hemisphere crosses the midline to innervate the left spinal cord, and vice versa. Because the pyramidal tract crosses the midline (in what is technically known as a *decussation*), the right cortex controls the left side of the body while the left cortex controls the right.

Many of the axons of the pyramidal tract originate from neurons in the **primary motor cortex (M1)**, which consists mainly of the precentral gyrus, just anteriour to the central sulcus (**FIGURE 10.15B**; see also Figure 1.12B). The cell bodies of many of these large neurons are found in layer V of the primary motor cortex.

Primary motor cortex is an executive motor control mechanism—and more

In humans, brain damage to M1 produces partial paralysis on the side of the body opposite the brain lesion (i.e., the contralateral side of the body). The disturbance is greatest in distal muscles, such as those of the hand. Humans with these lesions are generally disinclined to use the affected limb.

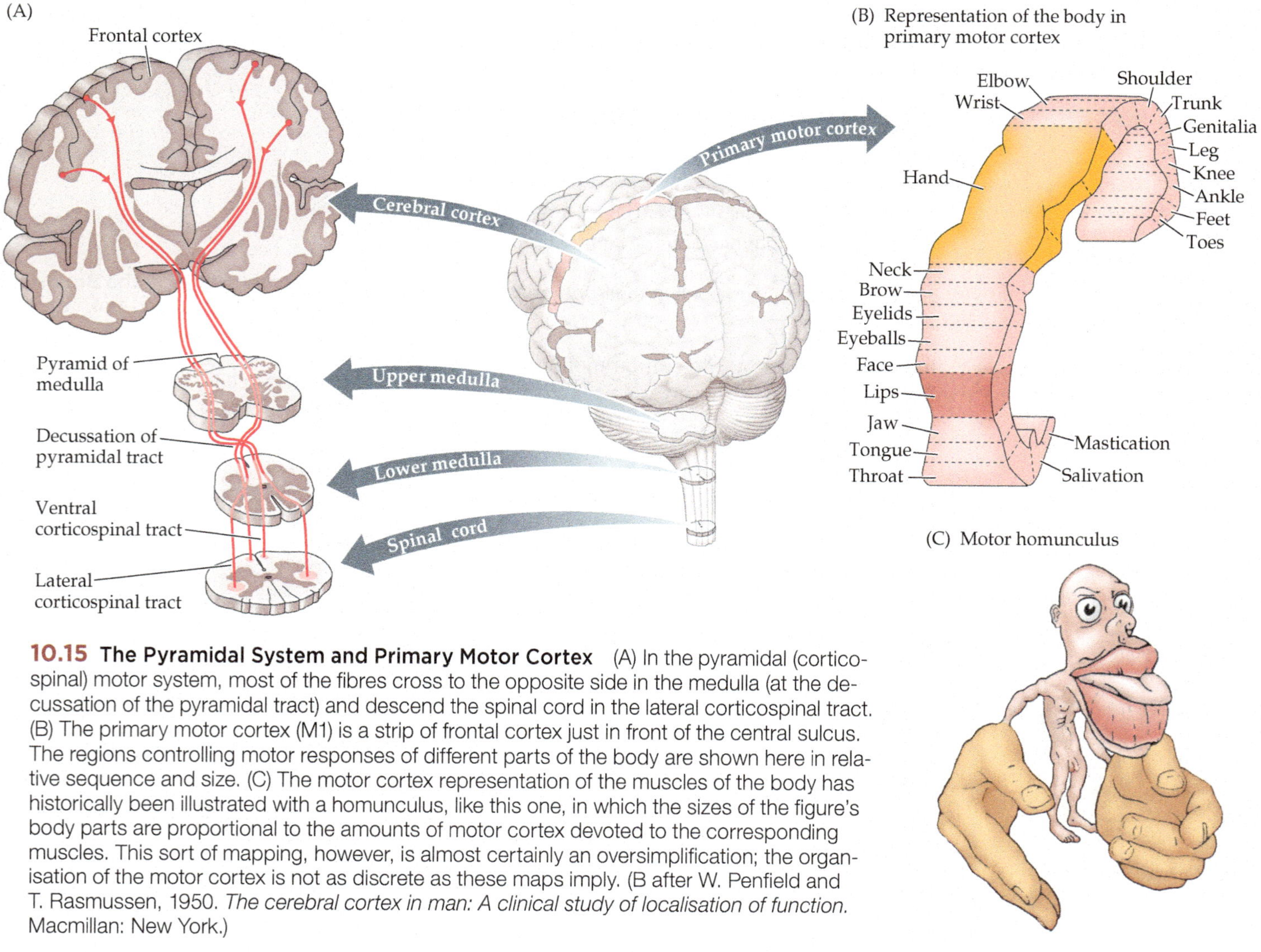

10.15 The Pyramidal System and Primary Motor Cortex (A) In the pyramidal (corticospinal) motor system, most of the fibres cross to the opposite side in the medulla (at the decussation of the pyramidal tract) and descend the spinal cord in the lateral corticospinal tract. (B) The primary motor cortex (M1) is a strip of frontal cortex just in front of the central sulcus. The regions controlling motor responses of different parts of the body are shown here in relative sequence and size. (C) The motor cortex representation of the muscles of the body has historically been illustrated with a homunculus, like this one, in which the sizes of the figure's body parts are proportional to the amounts of motor cortex devoted to the corresponding muscles. This sort of mapping, however, is almost certainly an oversimplification; the organisation of the motor cortex is not as discrete as these maps imply. (B after W. Penfield and T. Rasmussen, 1950. *The cerebral cortex in man: A clinical study of localisation of function.* Macmillan: New York.)

Early-twentieth-century researchers like Wilder Penfield, whom we discussed in Chapter 1, used electrical stimulation to develop functional maps of human primary motor cortex. Disproportionately large regions in the maps of M1 are devoted to the body parts involved in the most elaborate and complex movements. For example, humans and other primates have extremely large cortical fields concerned with hand movements (**FIGURE 10.15C**). In general, the layout of the motor homunculus in the precentral gyrus is aligned with the somatosensory homunculus in the postcentral gyrus (see Figure 7.10C).

Using nonhuman primates, it is possible to trace motor pathways in a retrograde manner from specific muscles back to corresponding regions of motor cortex, using a type of virus that can jump backward across synapses. This research shows that the organisation of M1 is far less discrete than was originally thought; for example, the cortical neurons controlling the individual fingers are extensively intermingled and arise from a variety of locations (Rathelot and Strick, 2006). This arrangement likely aids the coordination of movements that involve multiple muscles. In primates the pyramidal tract has some monosynaptic connections with spinal motor neurons (D. M. Griffin et al., 2015), but most pyramidal-tract neurons influence spinal motor neurons through polysynaptic routes and share control with other descending influences.

By recording from M1 neurons of monkeys that were trained to make free arm movements in eight possible target directions (Georgopoulos and Carpenter, 2015), scientists can eavesdrop on the commands originating there. Many M1 cells change their firing rates according to the direction of the movement, but for any one cell, discharge rates are highest in one particular direction. **FIGURE 10.16** shows a cell with a preference for upper leftward movement, but other cells may have a preference for upward, downward, rightward, or diagonal movement or any angle in between.

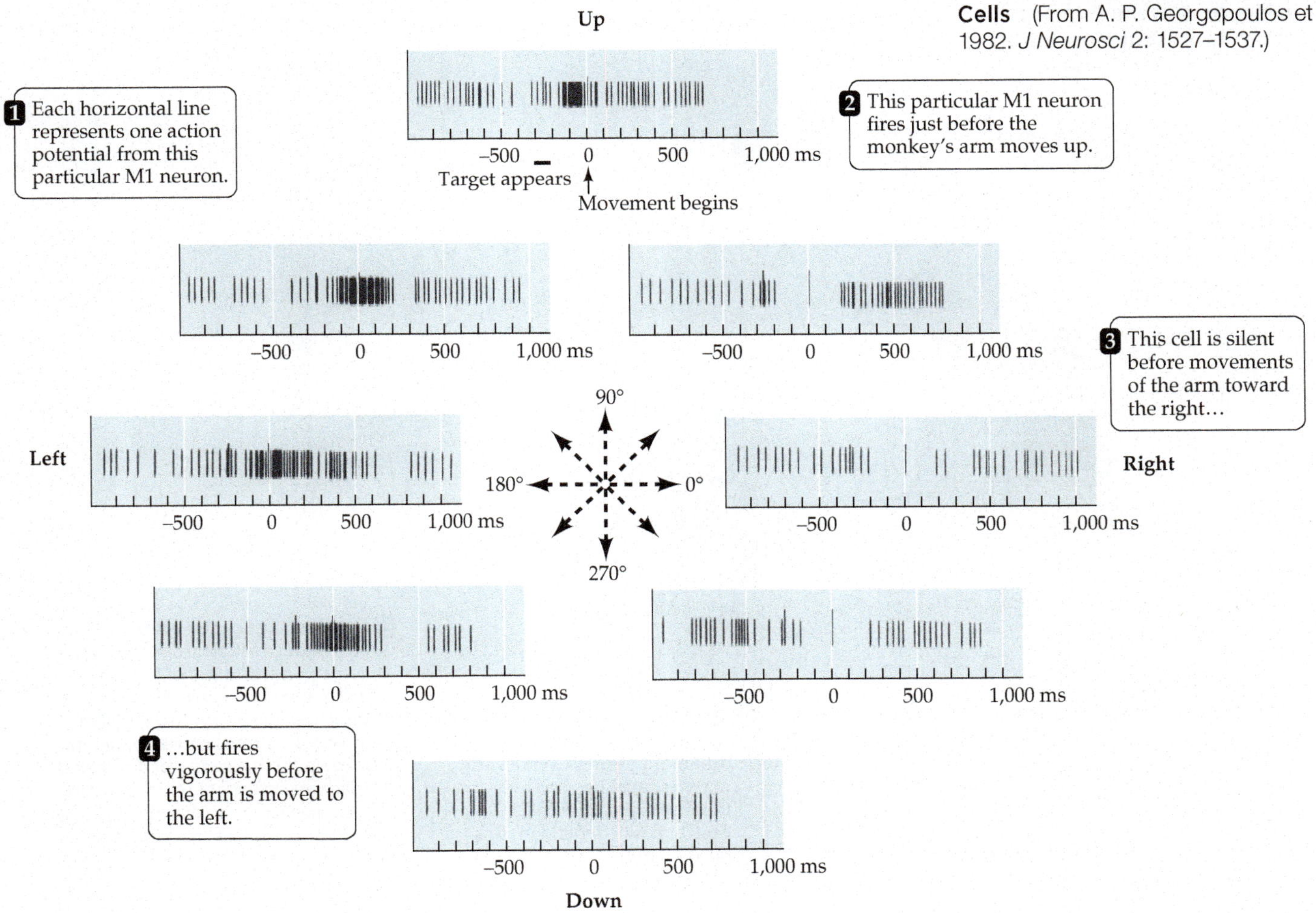

10.16 Directionality of Motor Cortex Cells (From A. P. Georgopoulos et al., 1982. *J Neurosci* 2: 1527–1537.)

Therefore, each cell carries only partial information about the direction of reaching, but collectively they represent any direction of reaching. When we average the activity of several hundred M1 neurons at once, we can predict fairly well the direction toward which the animal will reach with its arm. Given the millions of neurons in this region, a larger sampling would presumably provide an even more accurate prediction.

A long-standing controversy raged over whether *muscles* or *movements* are represented in M1 (Shenoy et al., 2013). That is, does activity of a cortical motor neuron encode a relatively simple parameter, such as contraction of a particular muscle, or does it encode a more abstract parameter, such as a particular movement of the hand through space? To address this controversy, experimenters trained a monkey to perform rapid tracking movements using the hand and wrist (Kakei et al., 1999). As **FIGURE 10.17** shows, the animal grasped a handle that could be rotated along the two axes of wrist motion. Movement of the handle controlled the position of a cursor on a computer screen. After much training (over 8 years!), the monkey performed well.

The experimenters then recorded the activity of single cells in M1 as the monkey performed from different starting positions. A substantial group of M1 neurons displayed changes in activity that corresponded to *muscle contractions*, but an even larger group of neurons showed activity that corresponded to *particular movements*, regardless of which muscles had to be used to accomplish those movements (which could vary depending on the position of the hand at the start). In other words, these cells were active whenever the monkey moved its hand in a particular direction, no matter which muscles were needed to accomplish that movement. Thus, *both* movements (Kaufman et al., 2014) and, to a lesser extent, muscles (Griffin et al., 2015) are represented in M1. Recordings of the activity of motor cortex, to predict what movement the person is calling for, can guide a robotic arm—the subject of **BOX 10.1**.

10.17 Do Neurons in the Primary Motor Cortex Represent Muscles or Movements? (After S. Kakei et al., 1999. *Science* 285: 2136–2139.)

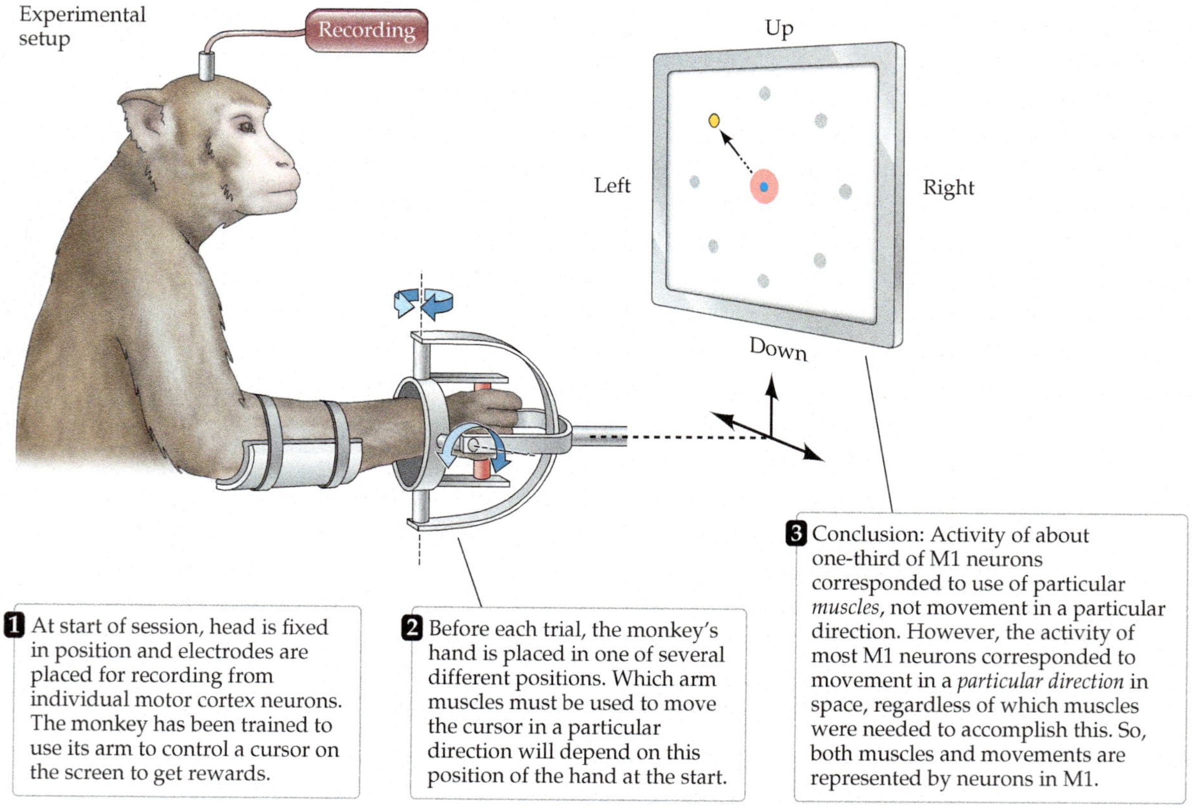

PRIMARY MOTOR CORTEX AND LEARNING Studies with both experimental animals and humans demonstrated that motor representations in M1 change as a result of training. Maps prepared from electrical stimulation in M1 show changes as an animal learns new skills—for example, a visually guided tracking movement with the arm (J. N. Sanes and Donoghue, 2000) or a precision grasping task (Nudo et al., 1996). As monkeys learn a particular skill, the total metabolic activity in M1 declines (Picard et al., 2013), as if it has become more efficient at producing the required movements.

In humans, the width of the precentral gyrus as seen with MRI offers an estimate of the size of M1. The gyrus is significantly wider in piano players, especially in the hand representation area (see Figure 10.15B and C), than in nonmusicians. The younger the musician was at the start of musical training, the larger the gyrus is in adulthood (J. A. Bailey et al., 2014), so this expansion of M1 seems to be in

BOX 10.1 Cortical Neurons Can Guide a Robotic Arm

With the help of recordings such as those in Figure 10.16, signals from neurons in the motor cortex of an owl monkey (Figure A) can be used to control a robotic arm in three dimensions, reproducing movements to reach for pieces of food placed in different positions, pick them up, carry them to the mouth, and return to the start position (Wessberg et al., 2000). The results of such studies suggested that paralysed patients might learn to operate a robotic arm by cortical activity, even though they couldn't move their own limbs.

One system, BrainGate, uses many microwires to record the activity of neurons in M1. Two people who had lost control of all four limbs volunteered to have the wires implanted into their brains. They then underwent training where they would watch a robot arm make regular, repetitive movements and *imagine* that the arm was *their* arm and that they were commanding the movements. The researchers carefully monitored the activity of the many M1 neurons during this phase, to "decode" the pattern of activity that signalled when the person was sending commands for movements up or down, left or right, and so on. The next phase of training was to monitor the person's M1 activity and decode those virtual commands to actually move the robot arm. Watching the arm as it moved, the volunteers could then learn to fine-tune their mental commands to move the robot arm the way they wanted. One person was able to use this system for relatively fine movements, using the robot arm to drink coffee from a bottle (Figure B).

Animal work indicates that recording more neurons should give an even better idea of the intended direction of movement, as well as providing the data needed to control grip strength and other more subtle aspects of reaching and grasping (Velliste et al.,

(A) Owl monkey perched atop a robotic arm

Courtesy of Miguel Nicolelis

(B)

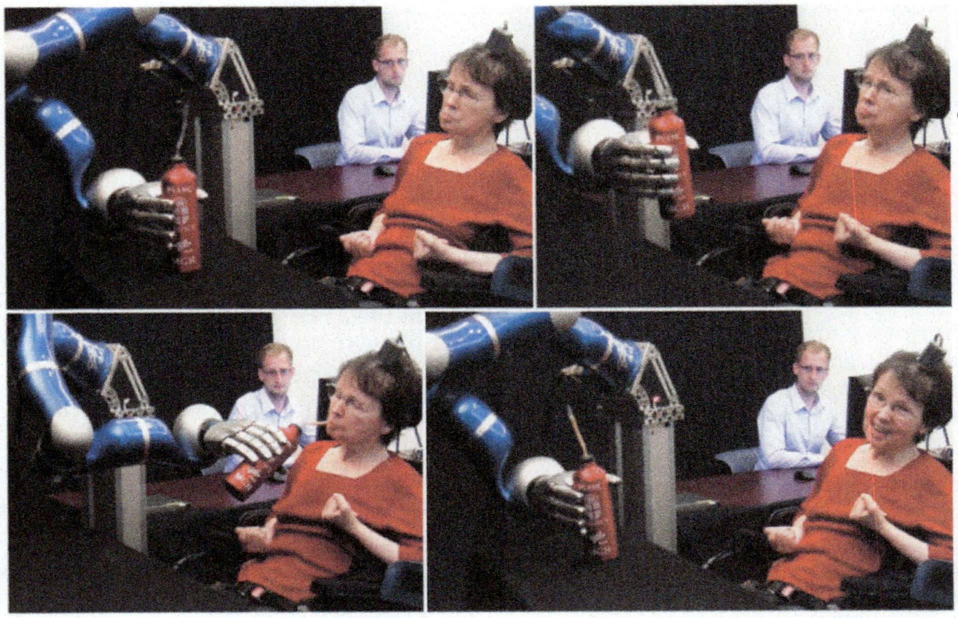

From Hochberg et al., 2012. *Nature* 485: 372–375, courtesy of BrainGate, braingate.org

(Continued)

BOX 10.1 *(continued)*

2008). Another possibility is to implant the electrodes in premotor cortex, with the idea that these are the neurons that call on primary motor cortex, so the neuronal signals for the intent to move the arm may be detected sooner. Perhaps recording from neurons in even more posteriour parietal cortex will reveal even more high-level motor planning—the person's intent (Aflalo et al., 2015). Again, the neuronal activity would be detected by a computer and used to move the computer cursor or a robotic limb. Monkeys given such an implant became very proficient at moving a cursor on a computer screen (Santhanam et al., 2006). A more difficult approach, still in the early stages, is to use the cortical activity to control electrical stimulation of muscles in the paralysed arm itself (Bouton et al., 2016), rather than in a robot arm. Or have the decoder stimulate the spinal cord to move leg muscles so the person can walk (F. B. Wagner et al., 2018).

Plasticity in motor cortices is gaining new appreciation in the field of neural prosthetics. It takes months for a human to learn to move a robotic arm (Hochberg et al., 2012). During this training, it is vital that the person be able to see how the robotic device is moving, so motor cortex can reorganise to work with the device (Koyama et al., 2010). Monkey research on brain-computer interfaces has shown a remapping of the direction preferences of cortical neurons to work with specific computational decoders; in other words, closed-loop control (seeing the arm move) is vital for motor plasticity to allow individuals to operate an artificial limb. Optimised computational solutions can improve performance, but equal or better performance can be reached with simple algorithms if the individuals have visual feedback (i.e., they are in a closed-loop control situation). The next challenge is to provide closed-loop control by providing the client with robotically generated sensory feedback from the artificial limb (Bensmaia and Miller, 2014). Otherwise, people using the robotic arms or legs will be like Ian, whom we met at the start of the chapter, dependent on vision to regulate movement.

Given these successes, it is conceivable that people may learn to use devices such as these to control a prosthetic "exoskeleton" (Figure C) (J. Zhang et al., 2017). No longer an idea limited to science fiction, it is now possible that one day people who have lost the ability to command their own bodies to move may be able to move robotic bodies so they can take care of themselves.

(C)

response to the experience of musical training. In rats, cortical remodelling due to motor learning has been directly observed by means of sophisticated mapping of the motor cortex before and after training of a new skill (Monfils et al., 2005) (**FIGURE 10.18**).

Nonprimary motor cortex contributes to complex behaviours

Just anteriour to M1 are cortical regions that are also important for motor control. Because they are not primary motor cortex, they are called **nonprimary motor cortex**. Despite the name, nonprimary motor systems can contribute to behaviour directly, through communication with lower levels of the motor hierarchy in the brainstem and spinal cord systems, as well as indirectly through M1. The traditional account of nonprimary motor cortex emphasises two main regions: the **supplementary motor area** (**SMA**), which lies mainly on the medial aspect of the hemisphere, and the **premotor cortex**, which is anteriour to the primary motor cortex (**FIGURE 10.19**). A rough idea of the different functions of these two nonprimary motor regions is that the SMA is important for planning movements that are internally generated, while the premotor cortex directs movement in response to external cues.

Patients with bilateral damage to the SMA are unable to move voluntarily (only some automatic and reflex movements remain), suggesting that this region is

nonprimary motor cortex Frontal lobe regions adjacent to the primary motor cortex that contribute to motor control and modulate the activity of the primary motor cortex.

supplementary motor area (SMA) A region of nonprimary motor cortex that receives input from the basal ganglia and modulates the activity of the primary motor cortex.

premotor cortex A region of nonprimary motor cortex just anteriour to the primary motor cortex.

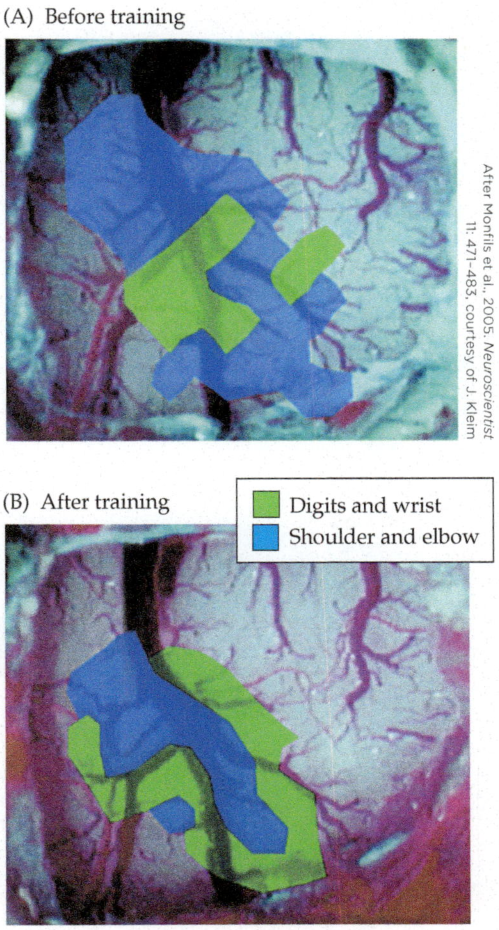

10.18 Motor Learning Causes Remapping of Motor Cortex (A) A map of forelimb control in rat motor cortex, priour to training (green areas control digits and wrist; blue areas, shoulder and elbow). (B) The same rat's motor cortex, following 10 days of training on a task requiring precise reaching and grasping with the forepaw. The representation of the digits and wrist (green) has expanded into areas previously associated with the shoulder and elbow.

important for planning movement sequences (Tanji, 2001). In one example, cerebral blood flow was measured during simple tasks, such as keeping a spring compressed between two fingers of one hand (Roland, 1984). As you might expect, this task produced a marked increase in blood flow to the contralateral M1, but increasing the complexity of the tasks to form a sequence of behaviours extended the area of increased blood flow to include the SMA. When people simply mentally *rehearse* the complex movement sequence without actually doing it, the enhanced blood flow is seen only in the SMA, not in M1.

In contrast to the observations from the SMA, the premotor cortex is activated when motor sequences are guided externally by stimuli (Ohbayashi et al., 2016) rather than generated internally. The premotor cortex is not a single system, but really a mosaic of different units serving separate but overlapping functions in organising complex motor behaviour (Graziano, 2006; Graziano and Aflalo, 2007). These behaviours cluster together in major categories: defensive movements, feeding behaviour, and so on. For example, a subset of premotor neurons is activated when objects are brought close to a monkey's face or hand. If the lights are then turned off, some of these neurons continue to fire even if the object is silently moved away, suggesting that the cells are coding for where the monkey *thinks* the object is (Graziano et al., 1997). When the lights are turned back on and the monkey sees that the object is gone, the neurons cease firing. Such neurons may help us reach out for objects that are no longer visible.

In general, evidence is mounting that we should think of the organisation of motor and premotor areas in terms of the mapping of behaviours, rather than the mapping of specific movements (Graziano and Aflalo, 2007). It's an idea that is nicely illustrated by recent discoveries about a unique population of premotor neurons, called *mirror neurons*, which we'll discuss next.

Mirror neurons in premotor cortex track movements in others

One of the subregions making up the mosaic of premotor cortex, an area known as *F5*, contains a population of remarkable neurons that seem to fulfill two functions. These neurons fire shortly before a monkey makes a very particular movement of the hand and arm to reach for an object; different neurons fire during different reaching

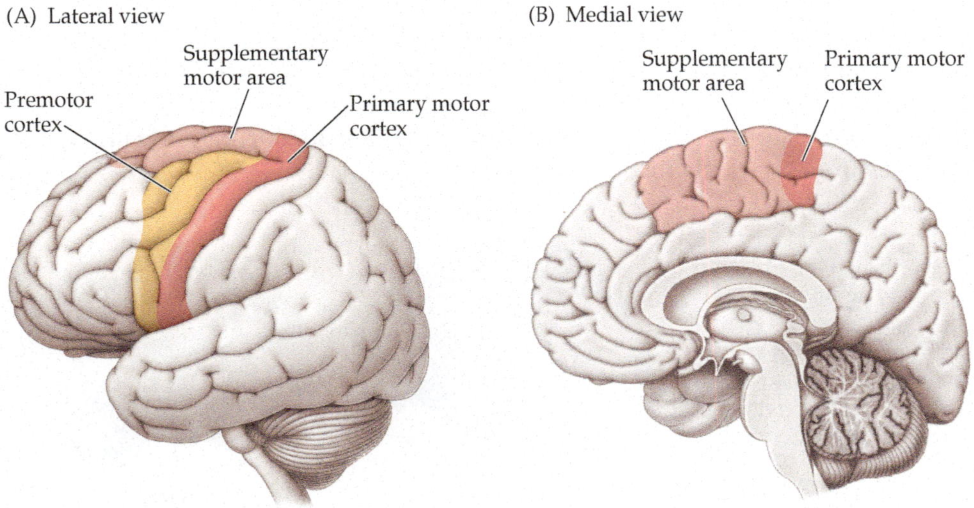

10.19 Human Motor Cortical Areas (A) The primary motor cortex (M1) lies just anteriour to the central sulcus. Anteriour to the primary motor cortex are the premotor cortex and the supplementary motor area (SMA), which together make up the nonprimary motor cortex. Although we show the premotor cortex here as a unitary structure for the sake of clarity, recent evidence from nonhuman primates indicates that it is a mosaic of subareas with distinct properties. (B) The SMA lies mainly on the medial surface of the cerebral hemispheres.

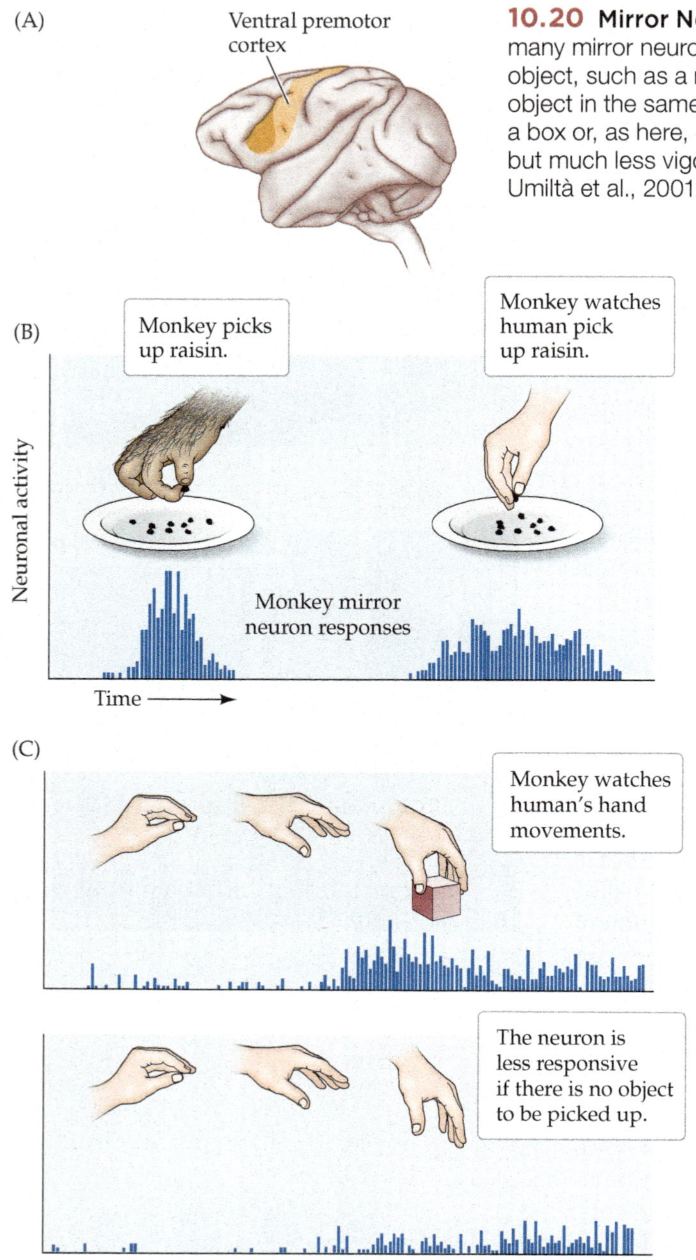

10.20 Mirror Neurons (A) In the ventral portion of premotor cortex known as *area F5*, many mirror neurons are found. (B) This neuron fires just before a monkey reaches for an object, such as a raisin (left), or when the monkey sees a human experimenter reach for that object in the same manner (right). (C) This mirror neuron fires when the monkey reaches for a box or, as here, it observes a human reaching for a box (top). The mirror neuron also fires, but much less vigourously, if there is no box for the human to grasp (bottom). (After M. A. Umiltà et al., 2001. *Neuron* 31: 155–165; G. Rizzolatti et al., 2006. *Sci Am* 295: 54–61.)

movements. These results suggest that these neurons trigger specific movements. But these neurons also fire whenever the monkey sees *another* monkey (or a human) make that same movement. These cells are called **mirror neurons** because they fire as though the monkey were imitating the movements of the other individual (**FIGURE 10.20**). They are matching the observed behaviour with an internal motor representation of that behaviour, as if the monkey were imagining doing the same thing as the individual it is watching.

MRI and EEG studies indicate that mirror neurons are also found in adult humans (Buccino et al., 2004) and children (J. F. Lepage and Theoret, 2006). In the classic "mirror game," in which one person moves slowly and the other tries to imitate the first person as closely as possible, these neurons are probably at work. They have also been seen in other areas of frontal cortex and parietal cortex (Rizzolatti and Sinigaglia, 2016). The activity of these neurons suggests that they are important in the understanding of other individuals' actions and in attempts to imitate those actions. Indeed, using transcranial magnetic stimulation to disrupt premotor cortex made it more difficult for participants to guess what sort of actions other people were miming (J. Michael et. al., 2014). It's possible that this function of mirror neurons is the basis for acting jointly with others to accomplish a cooperative task.

It is intriguing to think that mirror neurons may help us understand what other people are doing, because the idea suggests a neural basis for empathy. Thus, there has been a great deal of speculation about the function of mirror neurons in the imitating behaviour of human infants, the evolution of language, and other behaviour (Caramazza et al., 2014). Some have speculated that people with autism, which is characterised by a failure to anticipate other people's thinking and actions, may have a deficit in mirror neuron activity (J. H. Williams et al., 2006). When people with autism make "copycat" movements of the fingers or body, which they find difficult, a brain region that contains mirror neurons (the inferiour frontal cortex) is less activated than in people without autism (Villalobos et al., 2005). The same region is also underactivated when people with autism try to mimic emotional facial expressions of others (**FIGURE 10.21**) (Dapretto et al., 2006). Perhaps a deficit in mirror neurons in such a brain region renders other people's behaviours so bewildering that some children with autism withdraw from social relations and language.

mirror neuron A neuron that is active both when an individual makes a particular movement and when that individual sees another individual make that same movement.

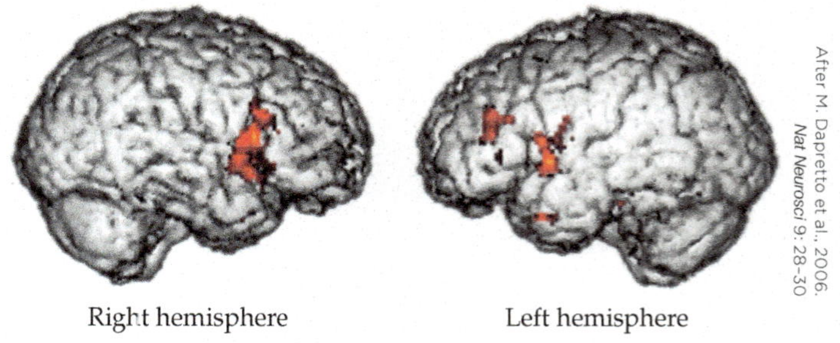

10.21 Underactivation of Mirror Cells in Autism When control children and children with autism are asked to imitate the emotional facial expressions displayed in photographs of other people, many brain regions show activation in both groups. But the indicated region, the pars opercularis region, is less activated in children with autism than in controls. This is the region that contains "mirror neurons." Perhaps deficits in activating brain regions underlying imitation and empathy contribute to the social impairments of autism.

10.5 Extrapyramidal Systems Also Modulate Motor Commands

Learning Objectives

After reading this section, you should be able to:

10.5.1 **Describe the two major extrapyramidal pathways to the spinal cord.**

10.5.2 **Compare the contributions of the basal ganglia versus the cerebellum to motor behaviour.**

10.5.3 **List the three major components of the cerebellar cortex and the results of damage to each.**

In addition to the corticospinal outflow from primary and nonprimary motor cortices through the pyramidal tract, many other motor tracts run from the forebrain to the brainstem and spinal cord. Because these tracts are outside the pyramids of the medulla, they and their connections are called the **extrapyramidal system**. The two most important components of the extrapyramidal system are the basal ganglia and the cerebellum.

How do these components of the extrapyramidal system communicate with the spinal cord? Their messages are transmitted via two brainstem pathways: the **reticulospinal tract**, which originates in the **reticular formation** of the brainstem, and the **rubrospinal tract**, which originates from the midbrain's **red nucleus** (the Latin *ruber* means "red"). Both tracts send axons down the spinal cord to synapse on spinal interneurons.

Using these brainstem pathways, the basal ganglia and cerebellum play slightly different roles in controlling motor output, as we'll see next.

The basal ganglia modulate movements

The **basal ganglia** include a group of interconnected forebrain nuclei: the caudate nucleus, putamen, and globus pallidus. Closely associated with these structures are two nuclei in the midbrain: the **substantia nigra** and the **subthalamic nucleus**. **FIGURE 10.22** shows the locations of these structures. The caudate nucleus and putamen together are referred to as the **striatum**.

Each of these structures receives input from wide areas of the cerebral cortex and sends much of its output back to the cortex via the thalamus, forming a loop from the cortex through the basal ganglia and thalamus and back to the cortex. Lesions of basal ganglia in humans produce movement impairments that seem quite different from those that follow interruption of the pyramidal system. Two disorders described at the end of this chapter—Parkinson's disease and Huntington's disease—are caused by degeneration of the basal ganglia.

The basal ganglia play a role in determining the amplitude and direction of movement and the initiation of movement (Yttri and Dudman, 2016). Much of the motor function of the basal ganglia appears to be accomplished by modulating other brain circuits, such as the motor pathways of the cortex (see Figure 10.4). The basal ganglia are especially important in the performance of movements influenced by memories, in contrast to those guided by sensory control (Graybiel and Grafton, 2015). The basal ganglia are also important in skill learning (Rueda-Orozco and Robbe, 2015), as we'll discuss in Chapter 16 (see Figure 16.17).

The cerebellum affects programmes, coordination, and learning of acts

Across vertebrate groups, the size of the cerebellum varies according to the range and complexity of movements. For example, the cerebellum is much larger in fishes with

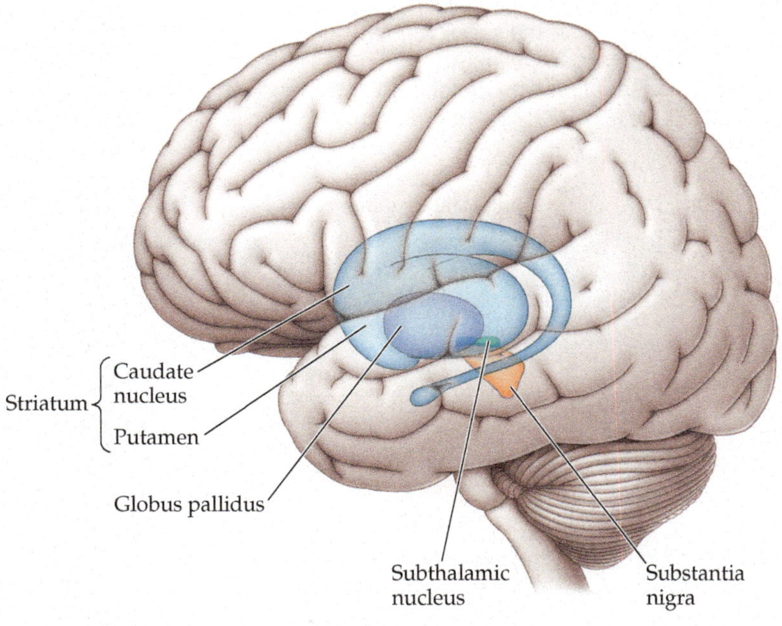

10.22 Basal Ganglia Involved in Movement Several structures within the basal ganglia are involved in modulating and tuning the movements programmed by other systems.

extrapyramidal system A motor system that includes the basal ganglia and some closely related brainstem structures.

reticulospinal tract A tract of axons arising from the brainstem reticular formation and descending to the spinal cord to modulate movement.

reticular formation An extensive region of the brainstem (extending from the medulla through the thalamus) that is involved in arousal and motor control.

rubrospinal tract A tract of axons arising from the red nucleus in the midbrain and innervating neurons of the spinal cord.

red nucleus A brainstem structure related to motor control.

basal ganglia A group of forebrain nuclei, including caudate nucleus, globus pallidus, and putamen, found deep within the cerebral hemispheres.

substantia nigra A brainstem structure in humans that innervates the basal ganglia and is named for its dark pigmentation.

subthalamic nucleus A nucleus just ventral to the thalamus that interacts with the basal ganglia. It is a favoured site for deep brain stimulation to treat Parkinson's disease.

striatum The caudate nucleus and putamen together.

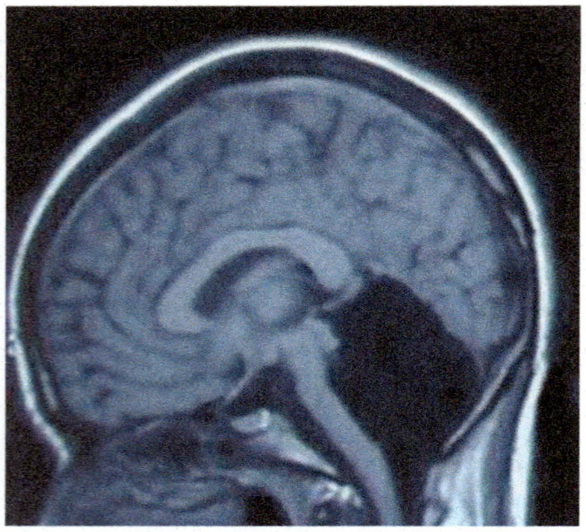

10.23 A Woman without a Cerebellum

extensive locomotor behaviour than it is in less active fishes; it is also larger in flying birds than in bird species that do not fly.

The outer layers of the cerebellum are called the *cerebellar cortex* and are dominated by a sheet of large multipolar cells called *Purkinje cells* (see Figure 2.18). All output of the cerebellar cortex travels via the axons of Purkinje cells, all of which synapse with the deep cerebellar nuclei. At these synapses, Purkinje cells produce only inhibitory postsynaptic potentials. Hence, all the circuitry of the extensive cortical portion of the cerebellum, which also includes 10–20 billion granule cells in humans, guides movement by *inhibiting* neurons.

Inputs to the cerebellar cortex come both from sensory sources and from other brain motor systems. Sensory inputs include the muscle and joint receptors and the vestibular, somatosensory, visual, and auditory systems. Both pyramidal and nonpyramidal pathways contribute inputs to the cerebellum and in turn receive outputs from the deep nuclei of the cerebellum (Gao et al., 2018). It has been suggested that the cerebellum elaborates neural "programmes" for the control of skilled movements, particularly rapid, repeated movements that become automatic. In fact, we now know that the cerebellum is crucial for a wide variety of motor (Brooks et al., 2015) and nonmotor learning (Katz and Steinmetz, 2002), as discussed in more detail in Chapter 16 (see Figure 16.17).

Despite the importance of the cerebellum in movement and learning, sometimes people are born without a cerebellum and the plasticity of the developing brain compensates so well that the missing structure may be discovered only by accident, as at routine autopsy (Boyd, 2010) or from a brain scan (F. Yu et al., 2015) (**FIGURE 10.23**).

The cerebellum and the basal ganglia contribute differently to the modulation of motor functions

The foregoing discussion and Figure 10.4 indicate that the cerebellum and the basal ganglia occupy rather similar positions in modulating motor functions. However, differences in the activity of these brain regions appeared in fMRI studies of people performing a tactile discrimination task (Y. Liu et al., 1999). The people were given two similarly shaped objects—one in each hand—and they had to decide by active touching whether the objects were the same or different.

Go to Activity 10.1
Basal Ganglia Involved in Movement

Changing patterns of activity during this task were found in M1, the SMA, the cerebellum, and the basal ganglia. Cerebellar activity correlated significantly with activity of the SMA but not with activity of M1, whereas basal ganglia activity correlated more strongly with activity of M1 than of the SMA. The difference in times of activation of the two regions reinforces the notion that cerebellum and basal ganglia play different roles in behaviour. The basal ganglia work with motor cortex to initiate and terminate movements (G. Cui et al., 2013), while the cerebellum and SMA (Bonini et al., 2014) monitor ongoing activity to produce smooth movements (**FIGURE 10.24**).

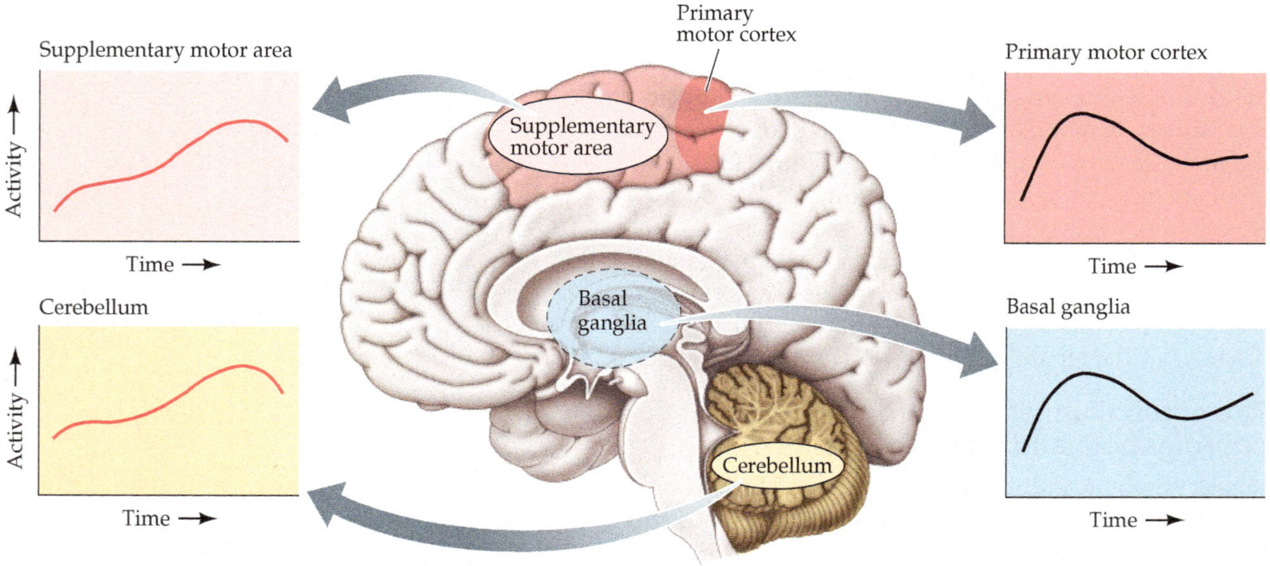

10.24 Contributions of the Cerebellum and Basal Ganglia to the Modulation of Movements Cerebellar activity correlates with that of the supplementary motor area; basal ganglia activity correlates with that of primary motor cortex. (After Y. Liu et al., 1999. *Nature* 400: 364–367.)

The basal ganglia and cerebellum are extensively connected to coordinate their contributions (Bostan and Strick, 2018).

We'll see these different contributions of the cerebellum and basal ganglia in the task of reaching, when we consider brain disorders of movement at the end of the chapter. Patients with Huntington's disease, who sustain damage to the basal ganglia, show impairment especially *terminating* a reach (see Figure 10.1B), because their movements become ballistic and sensorimotor feedback is delayed. Parkinson's disease, which damages the basal ganglia in a different way, is characterised by difficulty *initiating* a reach but smooth movement once the reach has been started (because, once the movement begins, the intact cerebellum takes over ongoing coordination). In contrast, people with cerebellar damage are able to initiate movements without difficulty and typically stop at the final target, but they take serpentine paths to the target, suggesting difficulty in *ongoing* motor control. Let's explore effects of cerebellar damage in more detail.

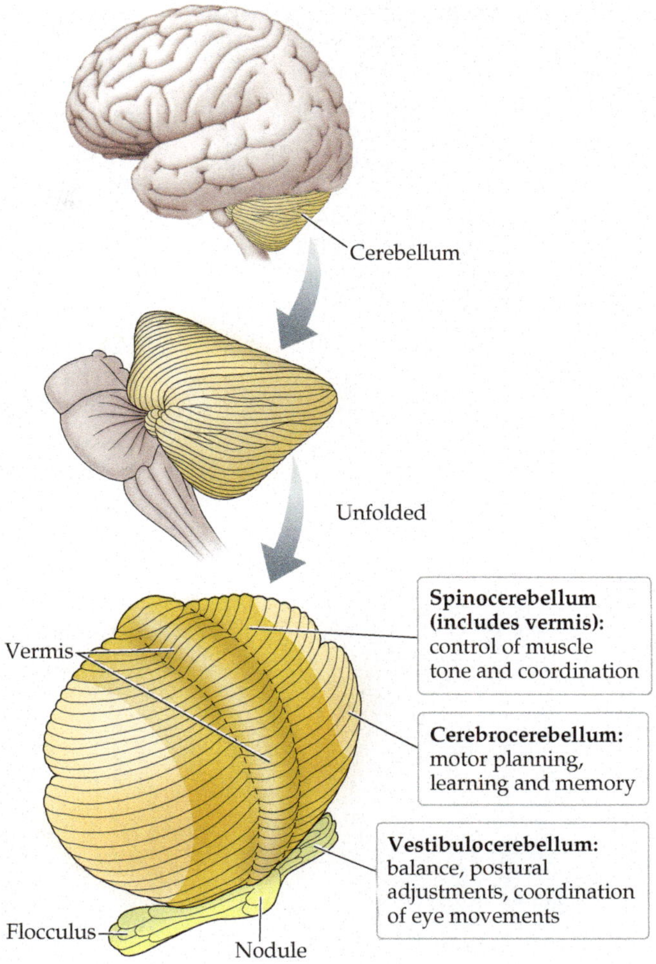

10.25 Functional Organisation of the Cerebellum

Cerebellar damage causes many types of impairment

Because the cerebellum modulates many aspects of motor performance, it is not surprising that its impairment leads to many abnormalities of behaviour. These symptoms permit an examiner to identify with considerable accuracy which part of the cerebellum is damaged (Haaland et al., 2017).

The cerebellum has three major functional divisions, and a unique pattern of impairment results from damage to each. The physical layout of these three regions is most easily seen if we unfold the cerebellar cortex and lay it flat, as we might flatten out the intact peel of an orange (**FIGURE 10.25**). The centremost part, the **spinocerebellum**, including a midline ridge called the *vermis* (for its shape; *vermis* is Latin for "worm"), receives sensory information about the current spatial locations of the parts of the body and anticipates subsequent movement. The spinocerebellum has rich connections with descending motor pathways, which it modulates. Damage here produces characteristic abnormalities of gait and posture, especially **ataxia** (loss of coordination) of the legs. Long-term alcoholism can cause degeneration of the anteriour lobe of the spinocerebellum, resulting in characteristic weaving and erratic gait.

The lateral portions of the cerebellum combined are called the **cerebrocerebellum** in recognition of its close relationship with the cerebral cortex. The cerebrocerebellum is implicated in planning complex movements, so damage here can cause diverse motor problems, such as **decomposition of movement**, in which gestures are broken up into individual segments instead of being executed smoothly. Because the cerebrocerebellum also functions in higher-level cognition, such as motor learning, damage here can also cause cognitive deficits.

Normally tucked beneath the other two major divisions of the cerebellum is the **vestibulocerebellum**, made up of small and somewhat primitive structures called the *nodule* and *flocculus*. As its name suggests, the vestibulocerebellum has close connections with the vestibular nuclei of the brainstem, through which it receives information about body orientation. Its outputs help the motor systems to maintain posture and guide eye movements (Herzfeld et al., 2015); for example, damage to this system produces errors in gaze and difficulty with tracking visual objects as the head moves.

With more research, we are coming to realize that the cerebellum plays a crucial role in an astonishingly wide variety of behaviours. So maybe we shouldn't be too surprised by the report that this compact structure contains more than half of all the neurons in the human nervous system (B. B. Andersen et al., 1992). There is also emerging evidence that glia play an important role in cerebellar function, as we explore next.

spinocerebellum The central part of the cerebellum, consisting mostly of the vermis and anteriour lobe.

ataxia An impairment in the direction, extent, and rate of muscular movement. It is often caused by cerebellar pathology.

cerebrocerebellum The lateral portions of each cerebellar hemisphere.

decomposition of movement Difficulty of movement in which gestures are broken up into individual segments instead of being executed smoothly. It is a symptom of cerebellar lesions.

vestibulocerebellum The portion of the cerebellum tucked next to the brainstem, consisting of the nodule and the flocculus.

Cre-recombinase An enzyme normally made by bacteria that removes a segment of DNA flanked by two lox sites.

loxP Also called *lox*. A specific sequence of nucleotides recognized by the enzyme Cre-recombinase. If the enzyme encounters a pair of loxP sites in a gene, it will remove the DNA between the two sites and recombine the gene, usually rendering the gene product dysfunctional.

The Cutting Edge

Cerebellar Glia Play a Role in Fine Motor Coordination

As we noted above, damage to the cerebellum often results in disruption of locomotion—the person may have difficulty walking, or navigating stairs. Historically, neuroscientists focussed on the activity of neurons to understand such behaviours. However, recent years have brought a greater appreciation for the importance of glial cells in neural function, including those in the cerebellum. For example, we now know that many glial cells possess receptors that respond to neurotransmitter released from nearby synapses.

In the cerebellum, special astrocytes called *Bergmann glia* send processes along Purkinje cell dendrites, surrounding synapses between parallel fibres from granule cells and the Purkinje cell dendrites (**FIGURE 10.26A**). These Bergmann glia possess AMPA-type receptors that normally recognize the neurotransmitter glutamate (see Chapter 3). Do those AMPA receptors on the glia have any effect on cerebellar function? To address this issue, Saab et al. (2012) produced mice in which they could disable genes for the AMPA receptor in Bergmann glia but leave AMPA receptors functioning normally in other cells, including the two neurons making the synapse.

To accomplish this neat trick, the researchers exploited an enzyme, normally found only in bacteria, called **Cre-recombinase**. This enzyme recognizes certain very specific sequences of nucleotides in DNA called **loxP** or simply *lox*. When the enzyme encounters a pair of lox sites in DNA, it snips the DNA at those points, throws out the intervening bit of DNA, then joins the cut ends of DNA back together (see the Appendix). If the lox sites are

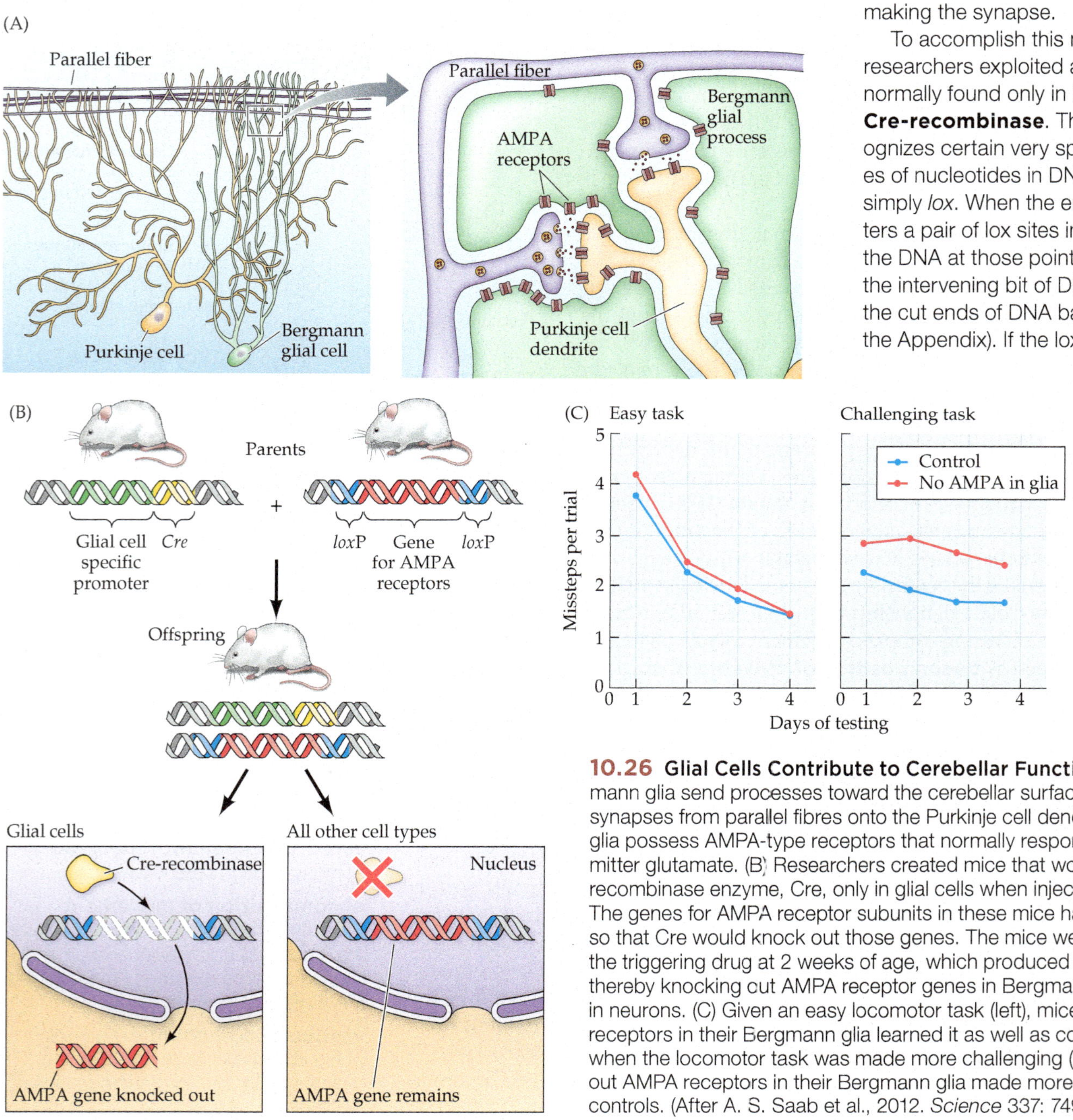

10.26 Glial Cells Contribute to Cerebellar Function (A) Bergmann glia send processes toward the cerebellar surface to "cap" over synapses from parallel fibres onto the Purkinje cell dendrites. These glia possess AMPA-type receptors that normally respond to the transmitter glutamate. (B) Researchers created mice that would produce a recombinase enzyme, Cre, only in glial cells when injected with a drug. The genes for AMPA receptor subunits in these mice had been altered so that Cre would knock out those genes. The mice were injected with the triggering drug at 2 weeks of age, which produced Cre only in glia, thereby knocking out AMPA receptor genes in Bergmann glia, but not in neurons. (C) Given an easy locomotor task (left), mice missing AMPA receptors in their Bergmann glia learned it as well as control mice. But when the locomotor task was made more challenging (right), mice without AMPA receptors in their Bergmann glia made more mistakes than controls. (After A. S. Saab et al., 2012. *Science* 337: 749–753.)

placed properly, the recombined gene will be disabled, unable to make a functional protein. This Cre-lox system can be used to "knock out" particular genes in some cells while leaving the genes alone in other cells in the same mouse.

In this case, the researchers first made mice that express the bacterial recombinase gene, called *Cre*, only in glial cells. Then they bred them with mice in which lox sites had been inserted into the genes for AMPA receptors. In offspring that received all these genes, *Cre* was expressed only in glia, which meant that AMPA receptors were knocked out of Bergmann glia but were undisturbed in cerebellar neurons (**FIGURE 10.26B**).

Did the absence of AMPA receptors in the Bergmann glia make any difference to cerebellar function? Yes. First, the researchers found that synapses between parallel fibres and Purkinje cells were weaker in animals without AMPA receptors on Bergmann glia. And while the genetically manipulated animals' behaviour was relatively normal, when they were given a challenging locomotor task (walking along a special ladder), they made more mistakes than control mice (**FIGURE 10.26C**). Thus the AMPA receptors in Bergmann glia somehow strengthen synaptic connections between parallel fibres and Purkinje cells, and this appears to be important for fine motor coordination. Now researchers are eager to find out how AMPA receptors in Bergmann glia accomplish all this. As usual with cutting-edge science, answering one question brings up many more. ■

10.6 Brain Disorders Can Disrupt Movement

Learning Objectives

After reading this section, you should be able to:

10.6.1 **Define common terms describing motor dysfunction, including plegia, paresis, spasticity, and apraxia.**

10.6.2 **Describe the symptoms of Parkinson's disease, its underlying brain cause, and treatment approaches.**

10.6.3 **Describe the genetic cause, symptoms, and outlook for Huntington's disease.**

We've already described movement impairment caused by disorders of muscle, motor neuron, and neuromuscular junctions, as well as spinal cord injury. Other movement disorders are caused by disorders of the brain, including the cortex and basal ganglia, which we'll discuss in order.

Cerebral cortex pathology causes some motor impairments

The most common motor impairments that follow strokes or injury to the human cerebral cortex are **plegia** (paralysis) or **paresis** (weakness) of voluntary movements, usually on one side of the body (known as *hemiplegia* and *hemiparesis*). Generally the paralysis appears on the side of the body opposite the injured hemisphere. In addition, affected patients show some **spasticity**: increased rigidity in response to forced movement of the limbs. This rigid spasticity after cortical damage contrasts with the flaccid paralysis seen when the spinal cord is damaged.

Spasticity reflects the exaggeration of stretch reflexes that have been released from the inhibition they usually receive from the cortex. Abnormal reflexes occur, such as the flaring and extension of the toes elicited by stroking the sole of the foot (the Babinski reflex). In the months following cerebral cortical injury, the clinical picture changes. The initial paralysis slowly diminishes, and some voluntary movements of the limbs return, although fine motor control of fingers is seldom regained. In humans the symptoms are more severe than in many other mammals.

Damage to nonmotor zones of the cerebral cortex, such as some regions of parietal or frontal association cortex, produces more-complicated changes in motor control. One condition characterised by such damage is **apraxia** (from the Greek *a-*, "not," and *praxis*, "action"), the inability to carry out complex movements even though paralysis or weakness is not evident and language comprehension and motivation are intact. Apraxia is illustrated in the following example: When asked to smile, a patient is unable to do so, although he certainly attempts to. If asked to use a comb placed in front of him, he seems unable to figure out what to do. But things aren't as simple as they

plegia Paralysis; the loss of the ability to move.

paresis Partial paralysis.

spasticity Markedly increased rigidity in response to forced movement of the limbs.

apraxia An impairment in the ability to begin and execute skilled voluntary movements, even though there is no muscle paralysis.

ideomotor apraxia The inability to carry out a simple motor activity in response to a verbal command, even though this same activity is readily performed spontaneously.

ideational apraxia An impairment in the ability to carry out a sequence of actions, even though each element or step can be done correctly.

Parkinson's disease A degenerative neurological disorder, characterised by tremors at rest, muscular rigidity, and reduction in voluntary movement, that involves dopaminergic neurons of the substantia nigra.

α-synuclein A protein that has been implicated in Parkinson's disease.

parkin A protein that has been implicated in Parkinson's disease.

might appear. At one point in the discussion, the patient spontaneously smiles, and at another point he retrieves a comb from his pocket and combs his hair with ease and accuracy. Apraxia is a symptom of a variety of disorders, including stroke, Alzheimer's disease, and developmental disorders of children.

Neurologists studying patients who have experienced strokes have discovered several different types of apraxia. **Ideomotor apraxia** is characterised by the inability to carry out a *simple* motor activity, either in response to a verbal command ("Smile" or "Use this comb") or by copying someone else's gesture, even though this same activity is readily performed spontaneously. **Ideational apraxia** is an impairment in carrying out a *sequence* of actions, although each step can be done correctly (Zadikoff and Lang, 2005). Patients with ideational apraxia have difficulty carrying out instructions for a sequence of acts—"Push the button, then pull the handle, then depress the switch"—but they can do each of these tasks in isolation. Apraxia may be somewhat independent of language deficits; the patients may be perfectly good at naming objects (Rosci et al., 2003) yet unable to perform the requested sequence of actions with those objects.

In Parkinson's disease the death of dopaminergic neurons alters activity of the basal ganglia

About 200 years ago, physician James Parkinson noted people in London who moved quite slowly, showed regular tremors of the hands and face while at rest, and walked with a rigid bearing (Przedborski, 2017). Another feature of what is now known as **Parkinson's disease** is a loss of facial muscle tone, giving the face a masklike appearance. People who have Parkinson's also show few spontaneous actions and have great difficulty in all motor efforts, no matter how routine. The hands may display tremors while at rest but move smoothly while performing a task. Parkinson's disease afflicts almost 1% of the U.S. population age 65 and older, but it sometimes unaccountably occurs in younger people, such as actor Michael J. Fox (**FIGURE 10.27**).

People with Parkinson's show progressive degeneration of dopamine-containing cells in the substantia nigra that project to the basal ganglia, particularly the caudate nucleus and putamen. The loss of cells in this area is continual, but symptoms begin to appear only after extensive cell death (Coelho and Ferreira, 2012). The discovery of a form of the disorder induced by illicit drugs (Hamadjida et al., 2018) suggested that exposure to toxins over a prolonged period underlies the development of the disorder.

Most cases of Parkinson's disease are not inherited (Trinh and Farrer, 2013), but in one large Italian family, Parkinson's disease develops in members who inherit a defective copy of the gene that encodes **α-synuclein** (S. A. Schneider and Alcalay, 2017), a protein normally expressed in the basal ganglia. Another family with inherited Parkinson's disease turned out to have a defective copy of another gene, named *parkin*, which encodes the protein **parkin** (Corti et al., 2011).

Mutation of the gene for α-synuclein, or having an overabundance of the protein due to a duplication of the normal gene (Soldner et al., 2016), leads to abnormal accumulation of α-synuclein into clumps called *Lewy bodies* in the substantia nigra and other brain regions (Surmeier et al., 2017). The misfolding of α-synuclein to form these clumps is reminiscent of prion diseases, where misfolded proteins can transmit disease from one individual to another (**BOX 10.2**).

In the noninherited cases of Parkinson's, other factors, such as toxins or brain injury, may

10.27 Leaders in the Campaign against Parkinson's Disease Parkinson's disease can strike at any point in adulthood. Actor Michael J. Fox was just 30 years old when he was diagnosed with Parkinson's, in 1991. Shown here testifying before a U.S. Senate health committee with Muhammad Ali (1942–2016), who also developed Parkinson's, Fox heads a research foundation (see https://www.michaeljfox.org/) that provides funding and advocacy for Parkinson's disease research.

BOX 10.2

Prion-Like Neurodegeneration May Be at Work in Parkinson's

Over two centuries ago, shepherds in Europe recognized a fatal disease in sheep that was called *scrapie* because the animals "scraped" their skin, presumably in an attempt to relieve itching. The shepherds learned that the only way they could stop an outbreak was to kill all the sheep, burn the carcasses and the fields they had used, and keep new sheep away from those fields for years. They didn't know it, but these drastic measures were needed because the disease is not transmitted by a virus or bacterium, which would rely on relatively fragile DNA or RNA for reproduction. Rather, scrapie arises when a particular endogenous protein that normally takes one shape takes on a new, abnormal shape. Once one protein molecule is misfolded, it induces the other molecules of that protein to fold abnormally around it. The accumulation of abnormally folded proteins forms toxic clumps that lead to brain degeneration. These infectious protein particles were named **prions** (pronounced "PREE-ons").

At some point, feed that contained protein derived from sheep with scrapie was fed to some cows in England and caused the cow version of the prion protein to fold abnormally. The result was a bovine version of scrapie called **bovine spongiform encephalopathy** (**BSE**, or mad cow disease) because the massive brain degeneration leaves the brain "spongy" (figure). Unfortunately, before BSE was detected, infected cows provided beef for Britons and caused a similar disorder called **Creutzfeldt-Jakob disease** (**CJD**), infecting over 100,000 people in the United Kingdom. CJD is fatal, causing widespread brain degeneration and therefore dementia, sleep disorders (see Chapter 15), schizophrenia-like symptoms, and death. Although a few cases of BSE have now been detected in North American cattle, currently they are believed to pose little risk to human health, thanks to changes in screening and feed production procedures.

There is increasing speculation that similar misfolding of other proteins, leading to clumping and toxicity, might underlie other neurodegenerative diseases, including Alzheimer's disease (D. M. Walsh and Selkoe, 2016), which we discussed in Chapter 6, and Parkinson's (Theillet et al., 2016). As we noted in the main text, mutations in α-synuclein can cause Parkinson's disease, and these mutations make the protein more likely to misfold and form toxic fibrils that clump to form Lewy bodies. Infusing misfolded α-synuclein fibrils into the brain of healthy mice or rats recapitulates the Lewy bodies and motor defects seen in mutant mice (Luna and Luk, 2015). Thus progressive misfolding of more and more α-synuclein molecules may spread degeneration, in which case therapies to prevent or reverse misfolding may one day slow or reverse Parkinson's. Inspired by these results, other scientists are exploring the possibility that spontaneous ALS may arise from misfolded SOD (a protein) (Furukawa et al., 2016) and that misfolding of amyloid (see Figure 6.31) may lead to Alzheimer's (Chiti and Dobson, 2017).

prion A protein that can become improperly folded and thereby become an infectious agent, spreading diseases such as bovine spongiform encephalopathy.

bovine spongiform encephalopathy (BSE) Also called *mad cow disease*. A disorder caused by improperly formed prion proteins, leading to altered behaviour and death.

Creutzfeldt-Jakob disease (CJD) A brain disorder in humans, leading to abnormal behaviour and death, that is caused by improperly folded prion proteins; the human equivalent of mad cow disease.

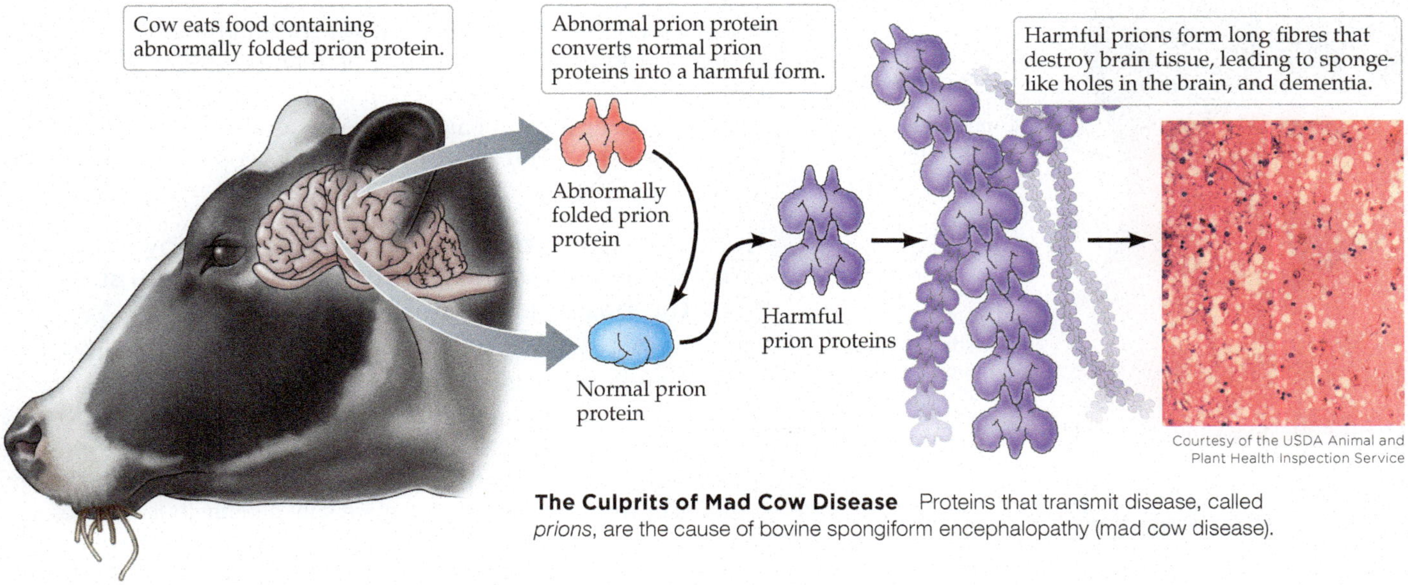

The Culprits of Mad Cow Disease Proteins that transmit disease, called *prions*, are the cause of bovine spongiform encephalopathy (mad cow disease).

L-dopa The immediate precursor of the transmitter dopamine.

deep brain stimulation (DBS) Mild electrical stimulation through an electrode that is surgically implanted deep in the brain.

Huntington's disease Also called *Huntington's chorea.* A progressive genetic disorder characterised by abrupt, involuntary movements and profound changes in mental functioning.

accelerate the formation of Lewy bodies that are associated with the death of the dopaminergic cells. Thus, it might be possible someday to prevent Parkinson's by developing drugs to stop or reverse the accumulation of α-synuclein (Menzies et al., 2015).

There was no treatment for Parkinson's disease until the late 1960s, when administration of a precursor to dopamine was found to enhance the dopamine levels of surviving cells. The precursor, called **L-dopa**, markedly reduces symptoms in patients with Parkinson's; notably, it decreases tremors and increases the speed of movements. Although L-dopa can reverse some symptoms of Parkinson's disease, neuron loss in the substantia nigra is relentless. Because the cell bodies in the brainstem degenerate, dopamine-containing terminals in the caudate nucleus and putamen also disappear. Eventually, too few dopamine-containing neurons remain in the substantia nigra, and L-dopa stops being effective.

Electrical stimulation of sites within the basal ganglia, often the subthalamic nucleus, can reduce the symptoms of Parkinson's disease (de Hemptinne et al., 2015). This **deep brain stimulation (DBS)** requires the surgical implantation of electrodes into the brain (**FIGURE 10.28**), and it carries the risk of cognitive side effects such as impulsivity and difficulty making decisions (M. J. Frank et al., 2007).

Another way to compensate for the loss of dopaminergic neurons might be to replace them with new ones. In several experimental treatments, fetal neurons or stem cells have been transplanted into the brains of people with severe Parkinson's in the hope that enough of the cells would establish dopaminergic synaptic connections in the basal ganglia to alleviate the symptoms of the disease. Although significant symptom relief has been reported in some people with Parkinson's (Barker et al., 2015), it's not clear whether these transplants will become a viable treatment option (Cyranoski, 2018). For one thing, evidence suggests that the transplanted cells may somehow go on to develop Parkinson's disease themselves (Kordower et al., 2008). And although the graft of fetal cells clearly improves the condition of some patients, other patients with grafts become afflicted with severe involuntary movements (Olanow et al., 2003). Debate about this procedure will continue for the foreseeable future.

Figure 10.28 Deep Brain Stimulation to Treat Parkinson's Disease For many patients, mild electrical stimulation of the brain brings relief of symptoms such as tremors and slow movement. The most common brain targets are the globus pallidus and the subthalamic nucleus (see Figure 10.22). (After NIH/National Institute of Neurological Disorders and Stroke (NINDS). 2010. *Deep Brain Stimulation at Two Different Targets Produces Similar Motor Improvements in Parkinson's Disease.* https://www.nih.gov/news-events/news-releases/.)

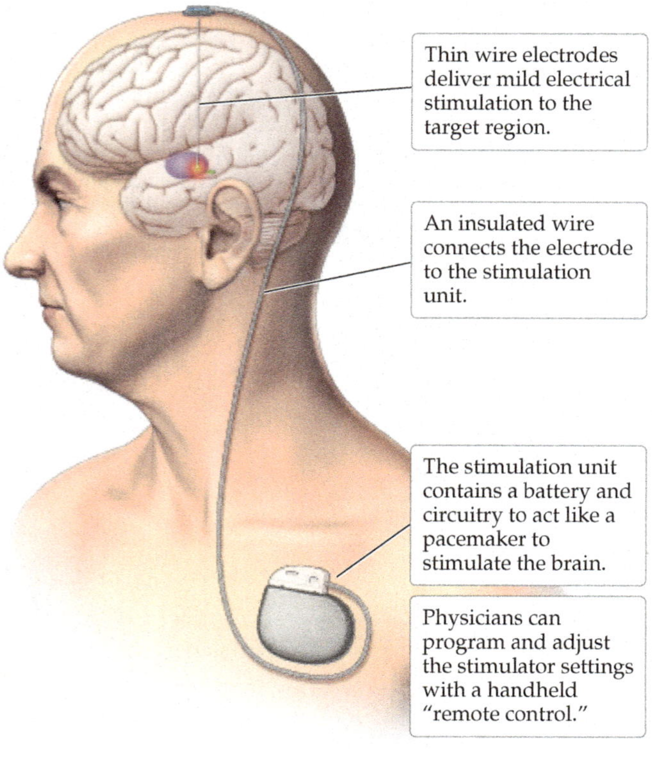

Parkinson's disease is often accompanied by nonmotor symptoms, including sleep disorders (Chapter 15), depression, and cognitive decline (Schapira et al., 2017). Some researchers have attributed the depression to the consequences of reduced mobility and the general stress of such incapacity. However, L-dopa treatment for Parkinson's disease sometimes produces impulsive behaviours such as gambling (Szarfman et al., 2006), probably by enhancing dopamine signalling in the pathway from the ventral tegmental area (VTA) to the nucleus accumbens that we discussed in Chapter 3 (see Figure 3.22). So depression in Parkinson's may be a direct effect of depleted dopamine in that pathway.

A woman who noticed that her husband developed a distinctive musky odour years before he developed Parkinson's disease inspired researchers to develop a new screening test for the disease, looking for four particular compounds in swabs from the skin (Trivedi et al., 2019). No one knows why people developing Parkinson's disease would produce this odour.

Huntington's disease is characterised by excessive movement caused by deteriouration of the basal ganglia

Whereas damage to the basal ganglia in Parkinson's disease *slows* movement, other kinds of basal ganglia disorders cause *excessive* movement. An example of the latter type was reported by George Huntington (1850–1916), a young American physician whose publication (1872) described a strange motor affliction. The first symptoms of **Huntington's disease** are subtle behaviour changes: clumsiness, and twitches in the fingers and face. Subtlety is rapidly lost as the illness progresses; a continuing stream of involuntary jerks engulfs the entire body. Aimless movements of the eyes, jerky leg movements, and writhing of the body turn the routine activities of the day into insurmountable obstacles. Worse yet, as the disease progresses, marked behaviour changes include intellectual deterioura-tion, depression, and, in a minority of patients, a psychotic state that resembles schizophrenia. In some patients, cognitive and emotional changes may appear many years before obvious motor impairments do (Wexler et al., 2016). Huntington's disease usually develops over a period of 15–20 years. The neuroanatomical basis of this disorder is the profound, progressive destruction of the basal ganglia, especially the caudate nucleus and the putamen (**FIGURE 10.29**).

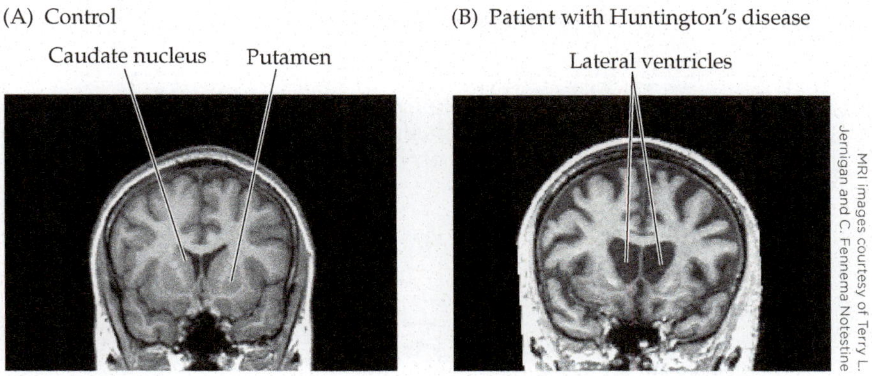

10.29 Neuropathology in Huntington's Disease Compared with the control (A), a coronal MRI section through the brain of a patient with Huntington's disease (B) shows marked enlargement of the lateral ventricles, caused by atrophy of the neighbouring caudate nucleus and putamen. Note also the shrunken cortical gyri and enlarged sulci of the patient compared with those of the brain of a healthy person.

George Huntington correctly deduced that the disorder is inherited, passed from generation to generation. Careful analysis of family pedigrees eventually revealed that Huntington's disease is transmitted by a single dominant gene on chromosome 4 (Gusella et al., 2014). Because we have two copies of every gene but pass only one of them on to any one child, each child of a person with Huntington's has a 50% chance of inheriting the bad gene and eventually developing the disease. The affected gene, *HTT*, normally encodes a protein called **huntingtin**. In Huntington's disease, the huntingtin protein that is produced is abnormally lengthened because of a series of three nucleotides, CAG (see the Appendix), that is repeated over and over in the *HTT* gene. If the gene contains fewer than 30 of these **trinucleotide repeats**, no symptoms appear, but if there are 35 or more CAG trinucleotide repeats in the *HTT* gene, the person will develop Huntington's disease (Squitieri, 2013).

We don't yet know what the function of the normal huntingtin protein is, or how the abnormal version of the protein causes the symptoms of Huntington's disease (G. P. Bates et al., 2015). Another mystery concerns the tissue specificity of Huntington's disease: why does the mutant huntingtin protein selectively damage the basal ganglia, when it is also being expressed in neurons and glial cells throughout the brain as well as in cells in muscle, liver, and testes? There's evidence that the abnormal protein, after killing one neuron, then kills the neurons that had innervated the dying cell (Pecho-Vrieseling et al., 2014), spreading the damage within the basal ganglia. There is currently no treatment for Huntington's, but this may change. When antisense RNA for huntingtin was injected into the cerebrospinal fluid of the spinal cord, it arrived in the basal ganglia and appeared to bind the mRNA for the gene, slowing down production of this destructive protein. Small trials suggest this therapy may also slow down progression of the disease (Wadman, 2018), illustrating the value of ongoing research and signalling a new hope for many thousands of people.

huntingtin A protein produced by a gene called *HTT* that may contain too many trinucleotide repeats. When it does, the protein causes Huntington's disease in a carrier.

trinucleotide repeat Repetition of the same three nucleotides within a gene, which can lead to dysfunction, as in the cases of Huntington's disease and fragile X syndrome (see Chapter 6).

Recommended Reading

Bates, G., Tabrizi, S., and Jones, L. (2014). *Huntington's Disease* (4th ed.). Oxford, UK: Oxford University Press.

Georgopoulos, A. P., and Carter, A. F. (2015). Coding of movements in the motor cortex. *Current Opinion in Neurobiology, 33*, 34–39.

Graziano, M. (2006). The organisation of behavioural repertoire in motor cortex. *Annual Review of Neuroscience, 29*, 105–134.

Palfreman, J. (2015). *Brain Storms: The Race to Unlock the Mysteries of Parkinson's Disease*. New York: Farrar, Straus and Giroux.

Purves, D., Augustine, G. J., Fitzpatrick, D., Hall, W., et al. (Eds.). (2017). *Neuroscience* (6th ed.). Sunderland, MA: Oxford University Press/Sinauer. (See Unit III: Movement and Its Central Control, Chapters 16–21.)

Shenoy, K. V., Sahani, M., and Churchland, M. M. (2013). Cortical control of arm movements: A dynamical systems perspective. *Annual Review of Neuroscience, 36*, 337–359.

10 Visual Summary

You should be able to relate each summary to the adjacent illustration, including structures and processes. You can also refer back to the links to figures, animations, and activities that will help you consolidate the material.

1 **Reflexes** are patterns of relatively simple and stereotyped movements elicited by the stimulation of sensory receptors. More-complex motor behaviours indicate the existence of a **motor plan**. Review **Figures 10.1**, **10.2**

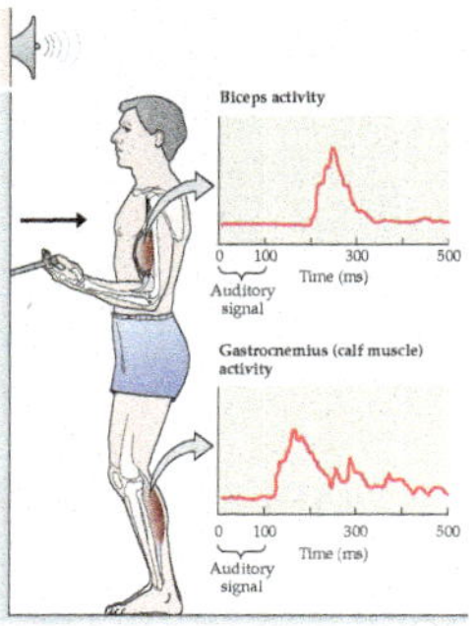

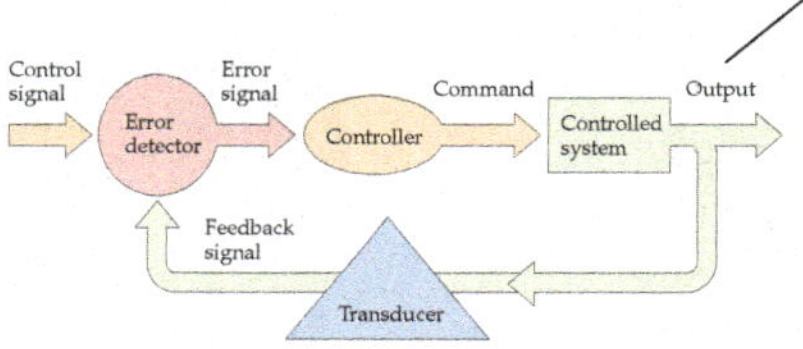

2 In **closed-loop control**, movements may be corrected while they are being produced on the basis of feedback from sensory systems. Some behaviours are so rapid, however, that they are controlled by **open-loop** systems; that is, the pattern is preset and not modified by feedback once it has started. Review **Figure 10.3**

3 Muscles around a joint work in pairs. **Antagonists** work in opposition; **synergists** work together. **Smooth muscles**, such as those in the intestines, are under involuntary control; **striated muscles** are under voluntary control. Action potentials travel down motor nerve fibres (axons from motor neurons) and reach **muscle fibres** at the **neuromuscular junction**, releasing acetylcholine to trigger muscle contraction. Review **Figures 10.5–10.7**, **Animation 10.1**

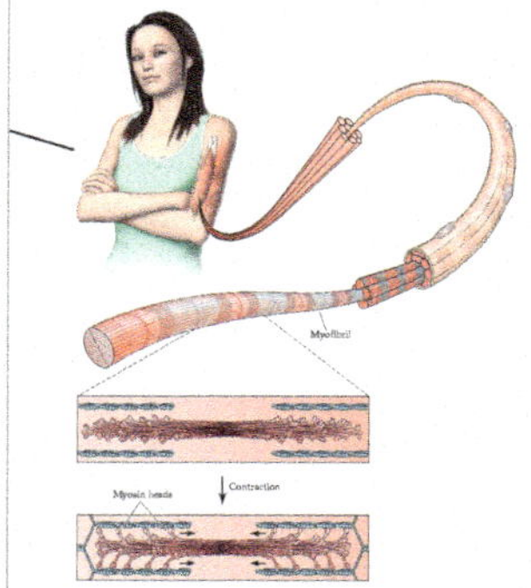

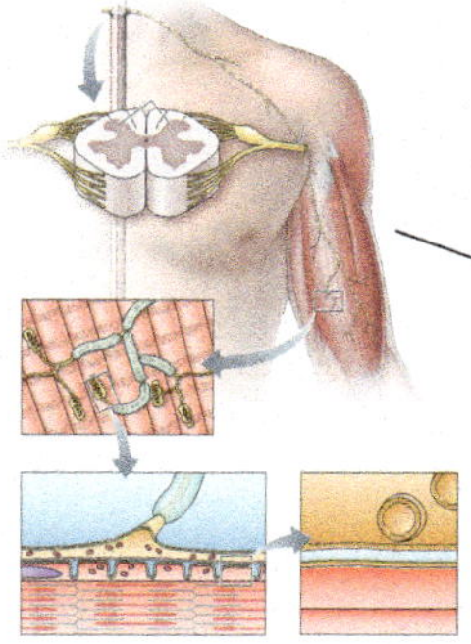

4 The final common pathway for action potentials to skeletal muscles consists of **motor neurons** whose cell bodies in vertebrates are located in the ventral horn of the spinal cord and within the brainstem. The motor neurons receive information from a variety of sources, including sensory input from the dorsal spinal roots, other spinal cord neurons, and descending fibres from the brain. Review **Figures 10.8**, **10.14**

5 **Muscle spindles** and **Golgi tendon organs** transmit crucial information about muscle activities to the CNS. The sensitivity of the muscle spindle can be adjusted by efferent impulses that control the length of the spindle. When a muscle is stretched, a reflex circuit triggers contraction, which works to restore the muscle to its original length; this response is called the **stretch reflex**. Review **Figures 10.9–10.11**, **Animation 10.2**

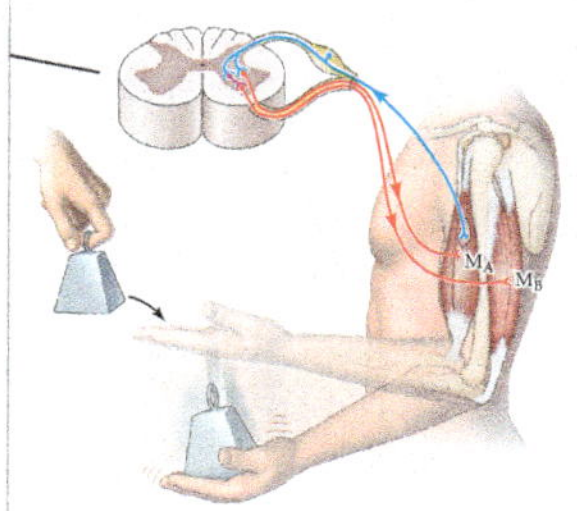

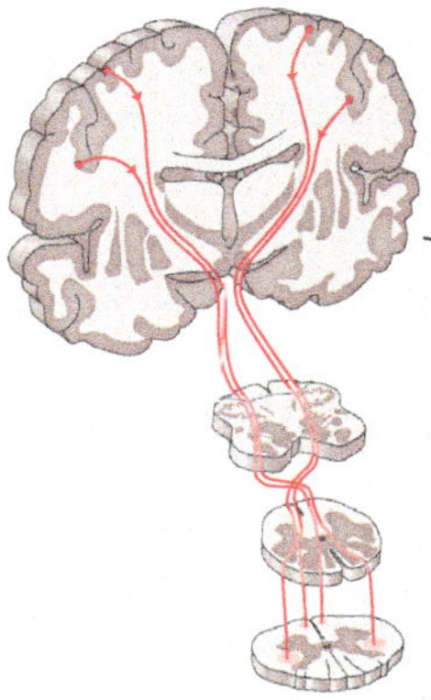

6 Motor control systems are organized into a hierarchy that includes primary and nonprimary motor cortex, the cerebellum, and the basal ganglia. The fibres of the **pyramidal** (corticospinal) **system** originate mainly in the **primary motor cortex (M1)** and run directly to spinal motor neurons or to interneurons in the spinal cord. Review **Figures 10.4**, **10.15–10.17**

7 **Nonprimary motor cortex** includes the **supplementary motor area (SMA)**, which helps control the sequence of internally generated movements, and the **premotor cortex**, which guides movement in response to external cues. Premotor cortex and other frontal lobe regions contain **mirror neurons**, which are active when an individual is moving an object in a particular fashion or when the individual sees someone else moving an object in that manner. Review **Figures 10.19–10.21**

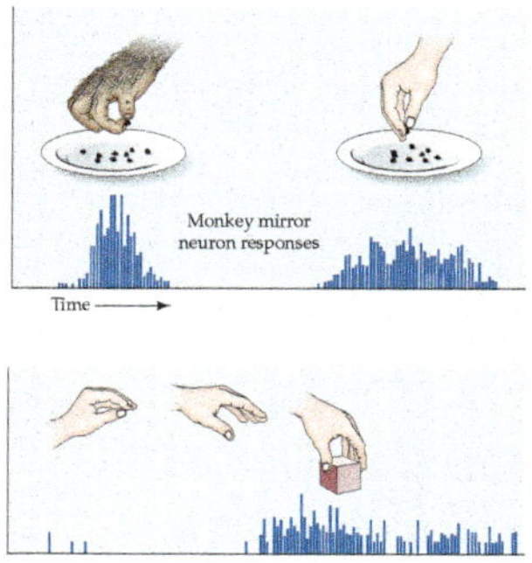

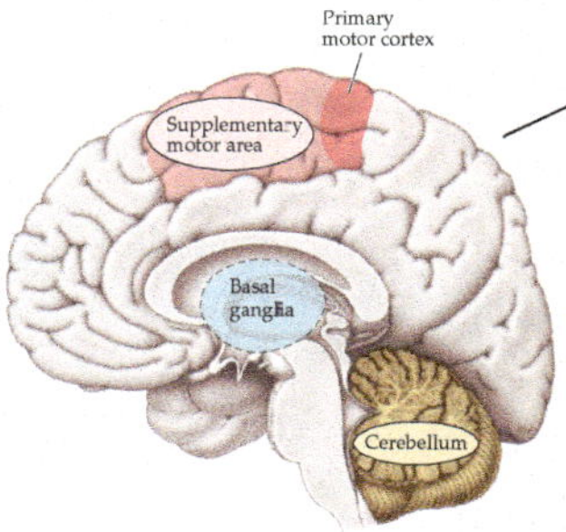

8 **Extrapyramidal** brain regions that modulate movement include the **basal ganglia** (caudate nucleus, putamen, globus pallidus, **substantia nigra**, and **subthalamic nucleus**), some major brainstem nuclei (thalamic nuclei, **reticular formation**, and **red nucleus**), and the cerebellum. Review **Figures 10.22–10.24**, **Activity 10.1**

9 The cerebellum is made up of three major functional divisions: **spinocerebellum, cerebrocerebellum, and vestibulocerebellum**. Damage in each division is associated with specific motor impairments. Cerebellar glia also contribute to fine motor coordination. Review **Figures 10.25–10.26**

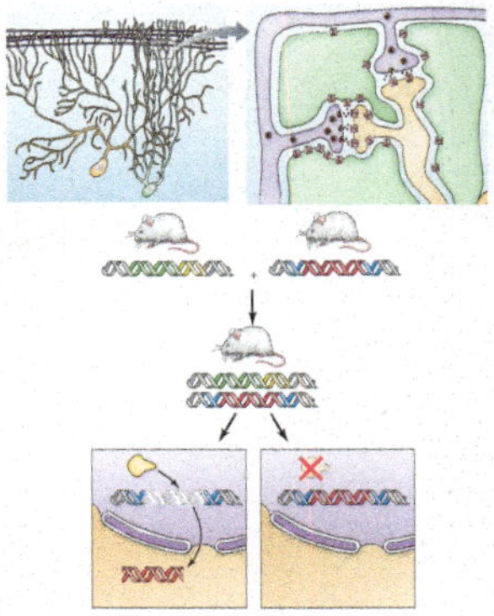

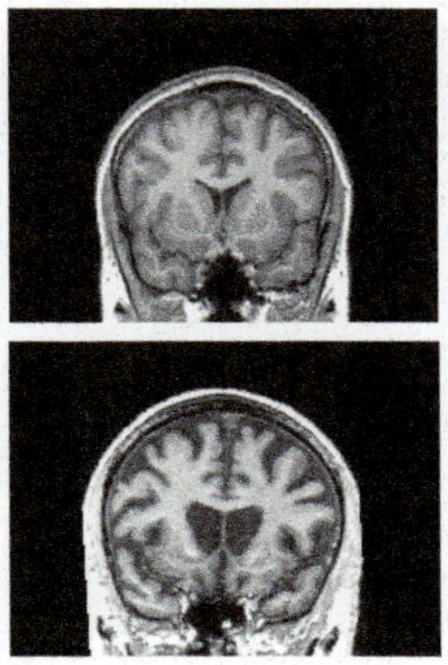

10 Movement disorders, such as **muscular dystrophy**, **amyotrophic lateral sclerosis (ALS)**, and **Parkinson's** and **Huntington's diseases**, can result from impairment at any of several levels of the motor system: muscles, neuromuscular junctions, motor neurons, spinal cord, brainstem, cerebral cortex, basal ganglia, or cerebellum. Review **Figures 10.12, 10.27–10.29**

Emotions and Mental Disorders

PART IV

© Eye of Science/Science Source

© Eye of Science/Science Source

11

Emotions, Aggression, and Stress

The Hazards of Fearlessness

No one enjoys being afraid, so you might think it would be great to never know fear. When we say that heroes are "fearless," what we really mean is that they manage to function effectively *despite* the fear they experience, not that they never feel afraid. However, there really are people who *literally* do not experience fear. One such woman, a patient known as S.M., lost her fear in late childhood because of a genetic disorder so rare, fewer than 300 cases have been reported (Feinstein et al., 2011).

In Chapter 7 we saw how the absence of an unpleasant experience, pain, can be hazardous to your health. S.M. similarly shows us the survival value of fear. Not only is she unafraid of snakes or spiders, but she once walked right up to a knife-wielding drug addict who was trying to rob her and basically dared him to stab her. He was so disquieted by her strange response that *he* ran away. Another time she was nearly killed in an act of domestic violence. While her behavioural responses and self-report appear typical for other emotions, S.M. shows very little of the physiological response, organised by the sympathetic nervous system, that the rest of us experience in response to frightening situations. Similarly, S.M. produces almost no startle response to a sudden, loud noise (Aschwanden, 2013), and she is insensitive to facial expressions of fear in others, especially the wide-eyed gasp that universally communicates alarm (L. F. Barrett, 2018).

S.M. is not deliberately reckless; she has learned to follow simple safety rules like looking both ways before crossing the street. But there are other consequences of S.M.'s fearlessness that you might not predict. When talking to someone, she tends to get much closer than other people do, sometimes just a foot away (Kennedy et al., 2009), suggesting she has a distorted sense of personal space. When strangers talk to her in public, like the mugger she encountered, she tends to stroll right up to them. S.M. also fails to perceive risk in more mundane social situations, so she's an easy target for internet scams. Although she's very outgoing, and might fondly address a waiter she'd only met once before, she has few long-term friendships, perhaps because she speaks without caution. Maybe being fearless isn't all it's cracked up to be.

What happened to S.M. to make her this way, and is there really nothing she is afraid of?

The sound of unexpected footsteps in the eerie quiet of the night brings fear to many of us. But the sound of music we enjoy and the voice of someone we love summon feelings of warmth. For some of us, feelings and emotions can become vastly exaggerated; fears, for example, can become paralysing attacks of anxiety and panic. No story about our behaviour is complete without considering the everyday events that involve feelings.

Our chapter begins with emotions, the bodily responses and brain mechanisms underlying happiness and joy, fear and loathing. We'll discuss both the development of emotions in individuals and how evolution shaped emotions to organise adaptive responses to daily threats and opportunities. The first sections of the chapter examine brain mechanisms related to emotional states, emphasising fear and aggression because both are important for survival and they are readily studied in animals. That discussion leads us to the last section, where we consider stress. Stress is associated with a variety of health problems because it involves not only the nervous and endocrine systems but also the immune system, and the interactions between all three.

11.1 Broad Theories of Emotion Emphasize Bodily Responses

Learning Objectives

After reading this section, you should be able to:

11.1.1 **Describe and compare the three dominant theories of the relationship of emotion to physiological changes.**

11.1.2 **Discuss the integration of autonomic responses with the perception of specific emotions.**

11.1.3 **Evaluate the ways in which individuals differ in their responses to emotional stimuli.**

11.1.4 **Discuss the pros and cons of trying to measure emotional states using techniques like polygraph testing.**

The topic of emotions is complicated by the fact that we apply the word *emotion* to several different things. Emotion is a private, subjective *feeling* that we may have without anyone else being aware of it. But the word *emotional* is also used to describe many *behaviours* that people show, such as fearful facial expressions, frantic arm movements, or angry shouting. Furthermore, during strong emotion we often experience *physiological* changes, such as a rapidly beating heart, shortness of breath, or excessive sweating. To encompass all three of these aspects, we will define **emotion** as a subjective mental state that is usually accompanied by distinctive behaviours as well as involuntary physiological changes.

In many emotional states the heart races, the hands and face become warm, the palms sweat, and the stomach feels queasy. Common expressions capture this emotional-physical association: "my hair stood on end" or "a sinking feeling in my stomach." These sensations are the result of activation of the autonomic nervous system—either the **sympathetic nervous system** (the fight-or-flight system that generally activates the body for action) or the **parasympathetic nervous system** (which generally prepares the body to relax and recuperate) (see Figure 1.12).

Several theories have tried to explain the close ties between the subjective feelings of emotions and the autonomic activity. Folk wisdom suggests that the autonomic reactions are caused by the emotions—"I was so angry, my stomach was churning," as though the anger produces the churning sensation (**FIGURE 11.1A**). Yet research indicates that the relationship between emotion and physiological arousal is more subtle.

Do emotions cause bodily changes, or vice versa?

William James (1842–1910) and Carl G. Lange (1834–1900) turned the folk notion on its head, suggesting that the emotions we experience are *caused by* the bodily changes. From this perspective, we experience fear because we perceive the activity that dangerous conditions trigger in our body (**FIGURE 11.1B**). Different emotions thus feel different because they are generated by different constellations of physiological responses.

The **James-Lange theory** inspired many attempts to link specific emotions to specific bodily responses. These attempts mostly failed because it turns out that there is no distinctive autonomic pattern for each emotion. Fear, surprise, and anger, for example, all tend to be accompanied by sympathetic activation, while parasympathetic activation tends to accompany both joy and sadness.

emotion A subjective mental state that is usually accompanied by distinctive behaviours, feelings, and involuntary physiological changes.

sympathetic nervous system A component of the autonomic nervous system that arises from the thoracic and lumbar spinal cord.

parasympathetic nervous system A component of the autonomic nervous system that arises from both the cranial nerves and the sacral spinal cord.

James-Lange theory The theory that our experience of emotion is a response to the physiological changes that accompany it.

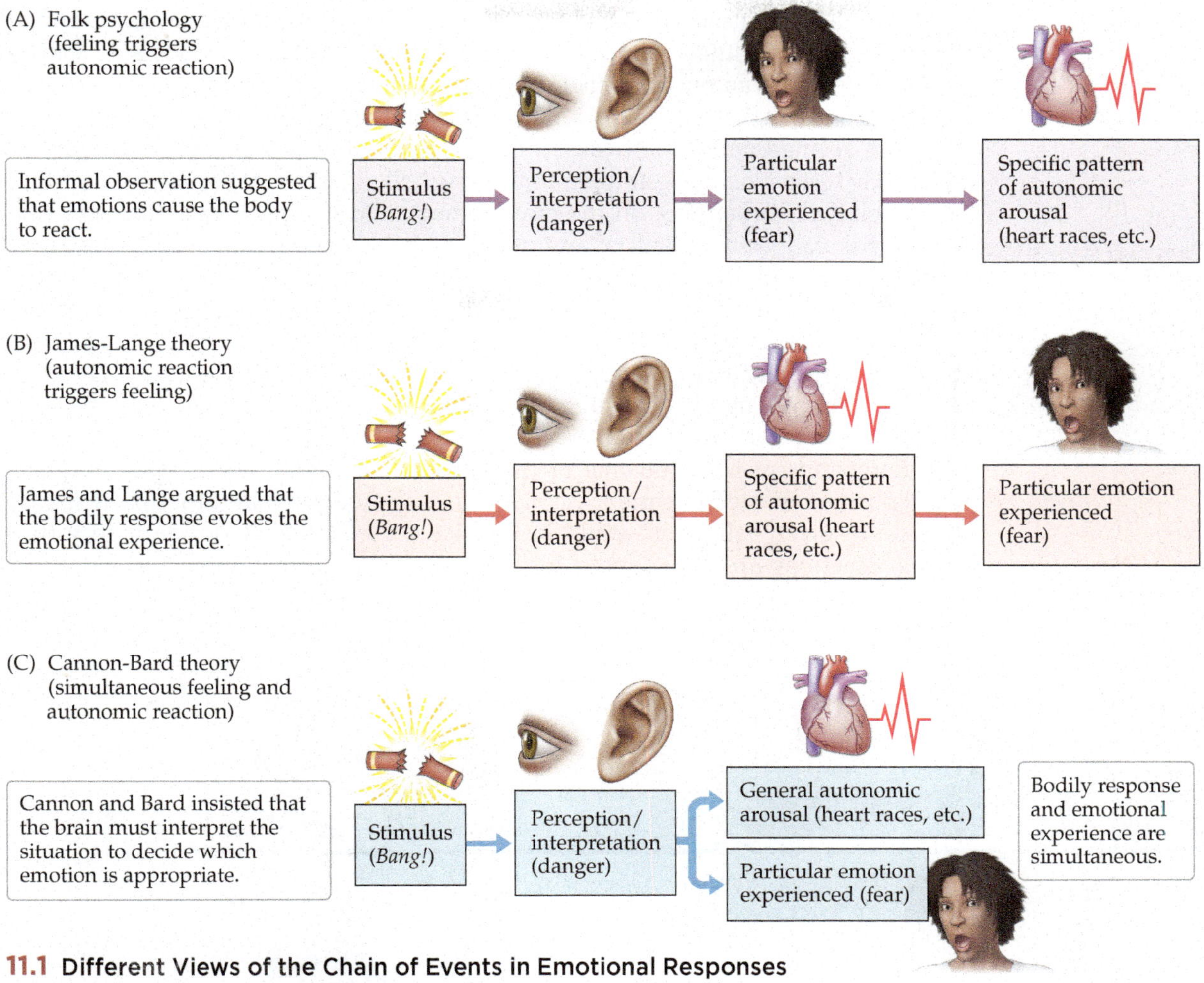

11.1 Different Views of the Chain of Events in Emotional Responses

In addition, the physiological reactions are rather slow, as physiologists Walter B. Cannon (1871–1945) and Philip Bard (1898–1977) pointed out (Cannon, 1929). In the **Cannon-Bard theory**, the brain decides which particular emotion is an appropriate response to the stimuli. According to this model, the cerebral cortex *simultaneously* decides on the appropriate emotional experience (fear, surprise, joy) and activates the autonomic nervous system to prepare the body, using either the sympathetic system, to ready the body for action, or the parasympathetic system to help the body relax (**FIGURE 11.1C**).

Stanley Schachter proposed a cognitive interpretation of stimuli and visceral states

Like Cannon and Bard, Stanley Schachter (1975) emphasised cognitive mechanisms in emotion. Under Schachter's **cognitive attribution theory**, however, emotional labels (e.g., *anger, fear, joy*) are attributed to relatively nonspecific feelings of physiological arousal. *Which* emotion we experience depends on *cognitive* systems that assess the context—our current social, physical, and psychological situation.

In a famous test of this idea, people were injected with epinephrine (adrenaline) and told either that there would be no effect or that their hearts would race (Schachter and Singer, 1962). Participants who were warned of the reaction reported no emotional experience, but some participants who were not forewarned experienced emotions when their bodies responded to the drug.

However, *which* emotion was experienced could be affected by whether a confederate in the room acted angry or happy. The unsuspecting people injected with epinephrine were much more likely to report feeling angry when in the presence of an "angry" confederate, and more likely to report feeling elated when with a

Cannon-Bard theory The theory that our experience of emotion is independent of the simultaneous physiological changes that accompany it.

cognitive attribution theory The theory that our emotional experience results from cognitive analysis of the context around us, such that physiological changes may accentuate emotions but not specify which emotion we experience.

"happy" confederate (**FIGURE 11.2A**). These findings contradict the James-Lange prediction that feelings of anger or happiness should each be associated with a unique profile of autonomic reactions. Schachter and Singer concluded that the misinformed participants experienced their physiological arousal as whichever emotion seemed appropriate, based on their cognitive assessment of the situation: "My heart's really pounding; I'm so angry!" or "My heart's really pounding; I'm so elated!" depending on the environment. Thus, Schachter and Singer saw emotional state as the result of an interaction between physiological arousal and cognitive interpretation of that arousal. This cognitive theory of emotions proposes that our interpretation of the context, including social factors such as other people's emotions, determines which emotion we'll experience. This may be why being in a theater full of laughing people helps us find the movie funny. The cognitive attribution model also suggests that our emotional experience at one time may affect how we interpret later events (**FIGURE 11.2B**).

Another interesting outcome of the experiment is that the people receiving epinephrine reported experiencing *more intense* emotions than other participants who were given saline, as would be predicted by the James-Lange theory that bodily reactions are experienced as emotion. The autonomic responses did not

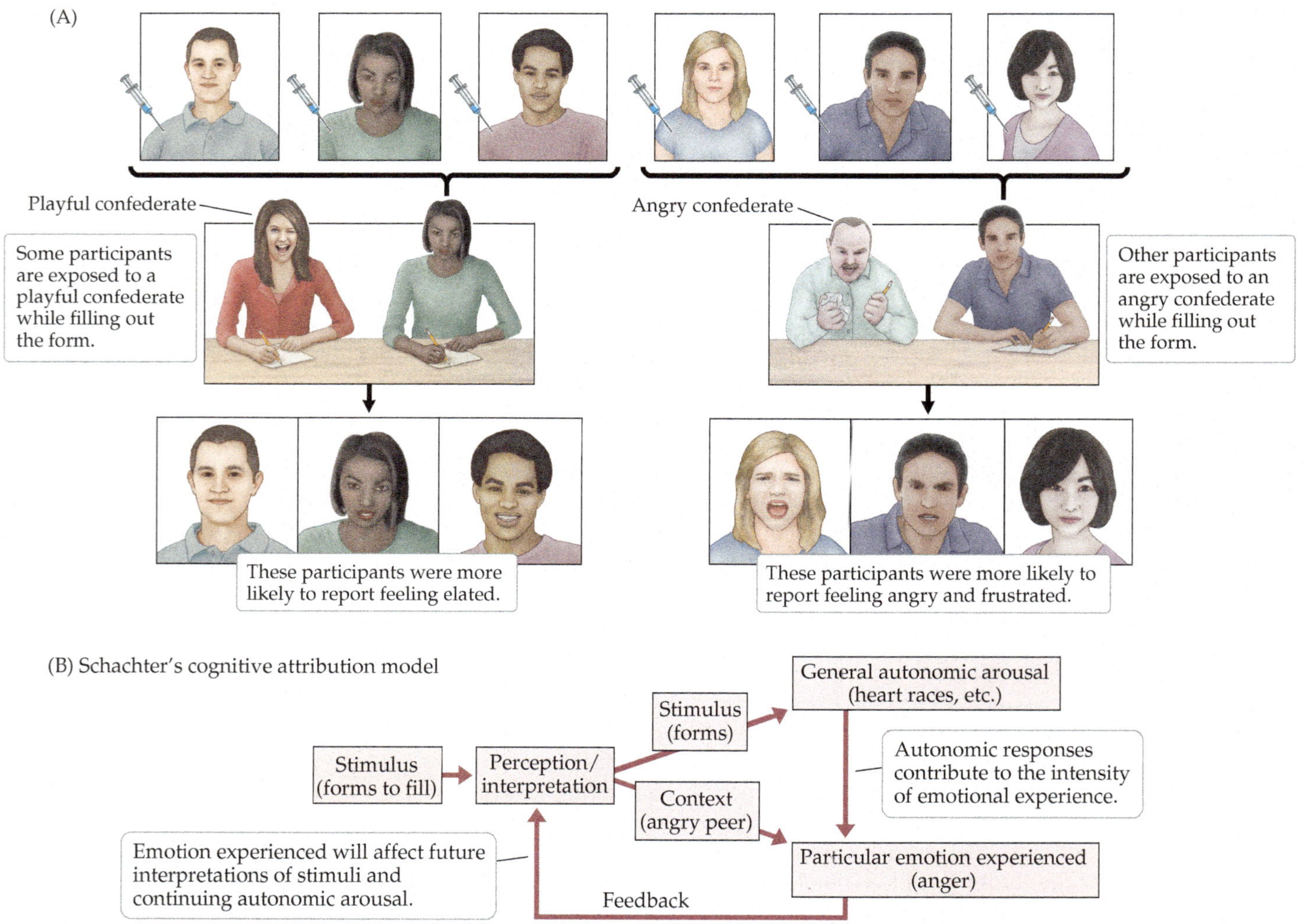

11.2 Cognition Plays a Role in the Experience of Emotion (A) Schachter and Singer found that study participants' emotional experiences after having been (unknowingly) injected with epinephrine could vary depending on how people around them were acting. (B) These results suggest that cognitive analysis of the situation affects what emotion we experience. Thus, sympathetic activation may increase the *intensity* of emotion we experience, but it does not completely determine *which* emotion we experience.

specify *which* emotion was being experienced, but awareness of the body's autonomic responses intensified that experience (G. W. Hohmann, 1966).

Schachter's theory has its critics (B. H. Friedman, 2010). For example, the theory asserts that physiological arousal is nonspecific, affecting only the intensity of a perceived emotion but not its quality. Yet evidence subsequently emerged that shows that patterns of autonomic activity seem to differ between several emotions, such as fear versus sadness versus anger (Levenson et al., 1990; Stemmler et al., 2001). Indeed, positive emotions more generally elicit a different array of autonomic responses than do negative emotions (Cacioppo et al., 2000), although, within those categories, different emotions elicit approximately the same autonomic profile. The fact that all negative emotions involve the same physiological responses is one reason why the **polygraph**—which measures various aspects of autonomic arousal, such as heart rate, blood pressure, sweating, and so on—is so poor at distinguishing liars from, say, anxious innocents (**BOX 11.1**).

polygraph Also called a *lie detector*. A device that measures several bodily responses, such as heart rate and blood pressure.

BOX 11.1 Lie Detector?

One of the most controversial attempts to apply biomedical science in legal settings is the so-called lie detector test. In this procedure, properly called a *polygraph* test (from the Greek *polys*, "many," and *graphein*, "to write"), multiple physiological measures are recorded in an attempt to detect lying during a carefully structured interview. The test is based on the assumption that people have emotional responses when lying because they fear detection and/or feel guilty about lying. Emotions are usually accompanied by bodily responses that are difficult to control, such as changes in respiratory rate, heart rate, blood pressure, and skin conductance (a measure of sweating). In polygraph recordings like the one in the figure, each wiggly line, called a *trace*, provides a measurement of one of these physiological variables, and taken together, the measures are assumed to track the participant's physiological arousal over time. When participants lie in response to direct questions (arrows), momentary changes in several of the measured variables may occur.

People who administer polygraph examinations for a living claim that polygraphs are accurate in 95% of tests, but the estimate from impartial research is an overall accuracy of about 65% (A. Eriksson and Lacerda, 2007; Nietzel, 2000). Even if the higher figure were correct, the fact that these tests are widely used would mean that thousands of truthful people could be branded as liars and fired, disciplined, or not hired. On the other hand, many criminals and spies have been able to pass the tests without detection. For example, longtime CIA agent Aldrich Ames, who was sentenced in 1995 to life in prison for espionage, successfully passed polygraph tests after becoming a spy. His Russian spymaster's advice for passing? Make small talk with the person conducting the test. In fact, how-to guides for beating polygraph tests, written by former polygraph operators, are readily available on the internet (e.g., www.polygraph.com). In the wake of the terrorist attacks of 2001, a federally appointed panel of scientists noted that even if polygraphs were correct 80% of the time (which is much higher than impartial research suggests), then giving the test to 10,000 people that included 10 spies would condemn 1,600 innocent people and let 2 spies go free (National Academy of Sciences, 2003)! Some scientists believe that modern neuroscience may provide new methods of lie detection someday, perhaps using functional brain-imaging technology such as PET or fMRI (Abe et al., 2007), but results to date have not been reliable enough for most purposes (Rusconi and Mitchener-Nissen, 2013), and even if such lie detectors are someday validated, they will be more costly and less widely available than polygraphs.

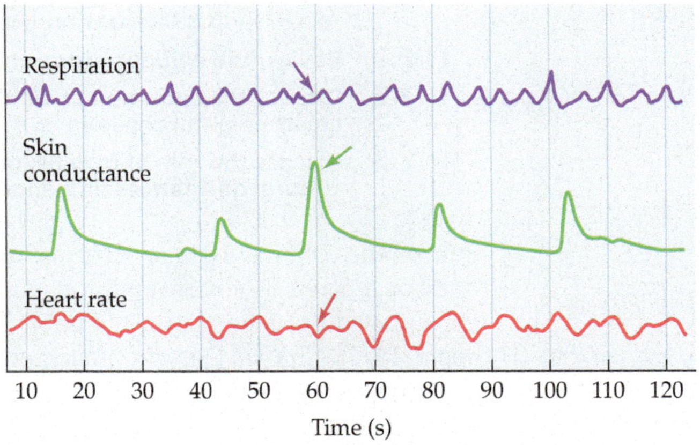

The polygraph measures signs of arousal.

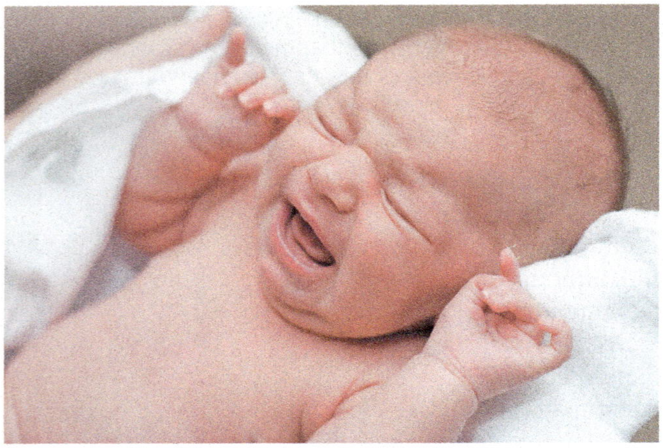

© iStock.com/NuStock

11.3 Individual Differences in the Emotional Responses of Infants Some infants, when swabbed with rubbing alcohol, react more strongly than others to the unpleasant cold sensation. As adults, these people are more likely to be shy and have phobias. They also show greater activation of the amygdala in response to photographs of strangers.

Individuals differ in their emotional responsiveness

Even as newborns, people differ in their emotional reactivity and physiological responses to emotional situations—a characteristic known as **individual response stereotypy** (Lacey and Lacey, 1970). In longitudinal studies (extending over many years), researchers have tracked the emotional reactivity of people from early childhood through adulthood, using stimuli that provoke autonomic responses, such as immersion of the hand in ice-cold water, performance of rapidly paced arithmetic calculations, and intense stimulation of the skin. Across these conditions, investigators have observed individual profiles of response that are evident even in newborns. For example, some newborns respond vigourously with heart rate changes, others with gastric contractions, and still others with blood pressure responses.

individual response stereotypy The tendency of individuals to show the same response pattern to particular situations throughout their life span.

The response patterns remain remarkably consistent throughout life. When newborns were classified on the basis of their behavioural responses to being swabbed with rubbing alcohol (the sudden cold sensation surprises and sometimes upsets babies), about 20% of the infants were classified as "high reactives" because they gave especially strong reactions to the stimuli (**FIGURE 11.3**) (Kagan, 1997). Many of these high reactives went on to develop anxiety disorders, such as social anxiety (severe shyness), and to become extremely shy children, so by the time they were old enough for school, about a third of them displayed diagnosable extreme phobias (compared with fewer than 10% of the other children). Furthermore, these individual differences in temperament create a self-reinforcing bias in the way that attention is paid to relevant emotional cues; for example, socially anxious high reactives may process facial cues more deeply and show exaggerated responses, such as strong activation of the amygdala, to photographs of strangers' faces (Morales et al., 2016; C. E. Schwartz et al., 2003). Because the amygdala has been implicated in fear (as we'll see later in this chapter), high reactives may have a lifelong aversion to new acquaintances, which would certainly affect many aspects of life. These infants may have inherited greater vigilance for danger, which might have been more adaptive in the past than it is today.

11.2 Did a Core Set of Emotions Evolve in Humans and Other Animals?

Learning Objectives

After reading this section, you should be able to:

11.2.1 **Review the evidence for a core set of emotions.**

11.2.2 **Discuss the evolution of emotions as preprogramming for responding to environmental challenges.**

11.2.3 **Discuss the role of facial expressions of emotion and the ways in which cultural differences influence facial displays.**

Just as the colours of the spectrum combine into subtle hues, researchers think there may be a core set of basic emotions underlying the more varied and delicate nuances of our world of feelings. In his book *The Expression of the Emotions in Man and Animals* (1872), Charles Darwin presented evidence that certain expressions of emotions are universal among people of all regions of the world. Furthermore, Darwin asked whether nonhuman animals show comparable expressions of some emotions and argued that aspects of emotional expression may have originated in a common ancestor.

11.4 Emotional Expression in Animals (A) Crested black macaque monkey "in a placid condition." (B) "The same when pleased by being caressed." (From C. Darwin, 1872. *The expression of the emotions in man and animals.* J. Murray: London.)

Darwin's perspective encompassed behavioural phenomena—the apparent emotional expressions of various mammals (**FIGURE 11.4**)—as well as the physiological mechanisms of emotional display, such as facial muscles and their innervation. Earlier scholars had believed the facial muscles were given uniquely to humans so that they could express their feelings, but Darwin emphasised that nonhuman primates have the same facial muscles that humans have. A century later, Redican (1982) noted distinct facial expressions in nonhuman primates that may represent similar emotions across primate species (**FIGURE 11.5**), including *grimace*, perhaps analogous to human expressions of fear or surprise; *tense mouth*, akin to human expressions of anger; and *play face*, which may be homologous to the human laugh. This connection may even extend beyond primates: for example, tickling and playing with rats can elicit ultrasonic vocalisations that resemble laughter in some ways, and that may facilitate social contact and learning (Burgdorf et al., 2008; Panksepp, 2007).

Emotions organise our responses to environmental challenges

How do emotions and their expression help individuals survive and reproduce? Darwin offered several suggestions:

> *The movements of expression in the face and body ... are ... of much importance for our welfare. They serve as the first means of communication between the mother and her infant; she smiles approval, and thus encourages her child on the right path, or frowns disapproval. We readily perceive sympathy in others by their expression; our sufferings are thus mitigated and our pleasure increased; and mutual good feeling is thus strengthened. The movements of expression give vividness and energy to our spoken words. They reveal the thoughts and intentions of others more truly than do words, which may be falsified. (Darwin, 1872, p. 365)*

Proponents of **evolutionary psychology** point to additional ways in which emotions are adaptive and could have developed through natural selection (Cosmides and Tooby, 2000). They suggest that emotions are broad motivational programs that coordinate various responses to solve specific adaptive problems, including

evolutionary psychology A field devoted to asking how natural selection has shaped behaviour in humans and other animals.

11.5 Facial Expression of Emotions in Nonhuman Primates (A) A juvenile bonobo (pygmy chimpanzee) shows a play face while being tickled. According to Frans de Waal (2003), bonobo laughter reveals both upper and lower teeth, making it closer in appearance to human laughter than that of chimpanzees, that tend to reveal only the lower teeth while laughing. (B) The female chacma baboon on the left bares her teeth, grinning to signal submission to a dominant male. In some primates, including humans, teeth baring has gained a different, friendlier meaning.

11.6 Basic Emotions According to one popular scheme, the eight basic emotions are arrayed as four pairs of opposite emotions. Lower- and higher-intensity forms of each basic emotion appear at the bottom and top levels, respectively. (After R. Plutchik, 1994. *The psychology and biology of emotion.* HarperCollins: New York.)

maintaining cooperative relations with other members of your group, choosing a mate, avoiding predators, and finding food sources.

For example, most of us have experienced the frightening nighttime perception of being stalked by a predator, real or imagined, human or nonhuman. The intensity of the experience may be a legacy of the ancestral environment in which our behavioural systems evolved, where the presence of an unseen predator would be a situation of utmost urgency. Individuals would have differed in their responses to this life-threatening situation. Some individuals' poor choices would have resulted in reduced reproductive success. Other individuals would have made more-effective choices, and to the extent that this behaviour was heritable, their descendants would also be more likely to survive in similar situations. Thus, through natural selection, an effective program for dealing with danger could have evolved: fear. The emotion of fear calls forth shifts in perception, attention, cognition, and action that focus on avoiding danger and seeking safety, as well as physiologically preparing for fighting or flight, while other activities, such as seeking food, sleep, or mates, are suppressed. From an evolutionary perspective, in the face of an imminent threat to life, it is better to be afraid, calling on this recipe for action, developed and tested over the ages, than to ad-lib something.

Viewed in this way, emotions can be seen as evolved preprogramming that helps us deal quickly and effectively with a wide variety of situations. To give you another example, feelings of disgust for body fluids may help us avoid exposure to germs (Curtis et al., 2004), so it would be adaptive to recognise disgust in others. Our human tendency to make snap judgments about other people, based on their appearance and facial expressions, may be an unfortunate overgeneralisation of mechanisms that evolved to help us recognise signs of threat or danger in others (Todorov et al., 2008).

There may be a basic core set of emotions underlying the more varied and delicate nuances of our world of feelings. One formulation (Plutchik, 1994) suggests there are eight basic emotions, grouped in four pairs of opposites—joy/sadness, affection/disgust, anger/fear, and expectation/surprise—with all other emotions arising from combinations of this basic array (**FIGURE 11.6**). But researchers do not yet agree about the number of basic emotions (six, seven, eight?). Although there may be no way to determine once and for all the number of basic emotions, one clue comes from examining the number of different kinds of facial expressions that we produce and can recognise in others.

Facial expressions have complex functions in communication

How many different emotions can be detected in facial expressions? According to Paul Ekman and collaborators, there are distinctive expressions for anger, sadness, happiness, fear, disgust, surprise, contempt, and embarrassment (**FIGURE 11.7**)

11.7 Universal Facial Expressions of Emotion According to Paul Ekman and colleagues, the eight basic emotional facial expressions shown here are displayed in all cultures.

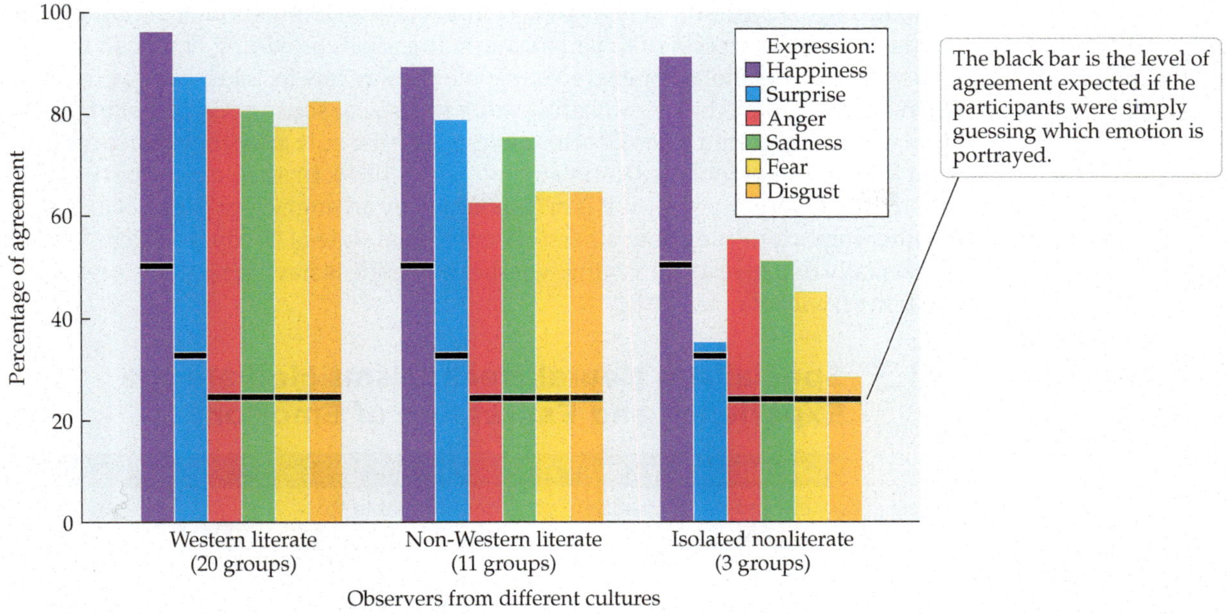

11.8 Cultural Differences in Recognising Facial Expressions of Emotion Within Western and non-Western literate groups (left and middle), there is widespread agreement about the emotions represented by photographs of basic facial expressions. But people from isolated nonliterate groups (right) are much less likely to agree with literate people's judgments of some facial expressions, especially those of surprise and disgust. The black horizontal bars indicate the percentage of agreement that would be expected by chance alone. (After J. A. Russell, 1994. *Psychol Bull* 115: 102–141.)

(Keltner and Ekman, 2000). Facial expressions of these emotions are interpreted similarly across many cultures without explicit training. (In case you're keeping track, whereas Plutchik included affection and expectation in his eight basic emotions, Keltner and Ekman include, instead, facial expressions of contempt and embarrassment. The other six emotions—anger, sadness, happiness, fear, disgust, and surprise—are recognised in both schemes.)

Cross-cultural similarity is also noted in the *production* of expressions specific to particular emotions. For example, people in a nonliterate New Guinea society show emotional facial expressions like those of people in industrialised societies. However, facial expressions apparently are not unfailingly universal. Although some degree of agreement is generally evident across cultures, researchers have repeatedly found isolated groups whose identifications of emotions from facial expressions did not fully agree with those of Westerners, such as for surprise and disgust (**FIGURE 11.8**). These cultural differences suggest that different cultures have adopted different rules for the facial expression of emotions, and that they control and enforce those rules by cultural conditioning (**FIGURE 11.9**). Everyone agrees that cultures affect the facial display of emotion; the remaining controversy is over the form and extent of the cultural influence (C. Chen et al., 2018; Crivelli et al., 2017). In everyday life we do not see emotional faces in isolation, so context normally colours our judgment about what emotion a person is experiencing (L. F. Barrett et al., 2011).

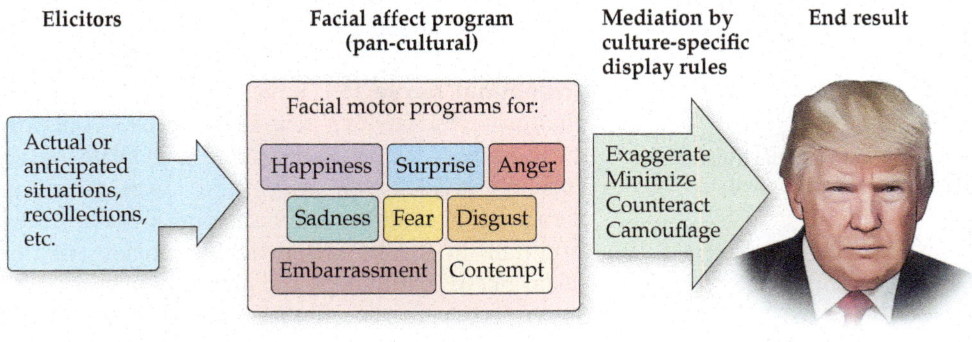

Based on photo © Evan El-Amin/Shutterstock.com

11.9 A Model for Emotional Facial Expressions across Cultures

According to Fridlund (1994), a major role of facial expression is paralinguistic; that is, the face is accessory to verbal communication, providing emphasis and direction in conversation. For example, people display few facial responses to odour when smelling alone but significantly more in a social setting (Gilbert et al., 1986). Similarly, bowlers tend to smile after making a strike only after they turn around to meet the faces of onlookers (Kraut and Johnston, 1979). In simulations, participants take a threat more seriously if it is accompanied by an angry face (Reed et al., 2014). Another important cue about a person's emotional state is "body language," which is especially effective at conveying whether someone is having negative or positive emotions (Aviezer et al., 2012).

11.3 Specialised Neural Mechanisms Mediate the Experience and Expression of Emotions

Learning Objectives

After reading this section, you should be able to:

11.3.1 **Review the neuromotor bases of facial expression.**

11.3.2 **Sketch and describe the major brain pathways involved in emotional behaviours, noting the behavioural manifestations of activity in the major pathways.**

11.3.3 **Describe the process of fear conditioning, the neural mechanisms responsible for fear reactions, and the role of this system in pathological states.**

What brain mechanisms give us our feelings of emotion, and how do these feelings get translated into the pattern of facial gestures that communicates our feelings to others? A rich field of study has addressed the anatomical and neurophysiological bases of emotion, and taken together, the evidence indicates that specialised brain systems are involved in emotions but also that the same mechanisms seem to be involved in many different emotions.

facial nerve Cranial nerve VII, which receives information from the face and controls the superficial muscles there.

trigeminal nerve Cranial nerve V, which receives information from the face and controls jaw musculature.

Facial expressions are mediated by muscles, cranial nerves, and CNS pathways

Within the human face is an elaborate network of finely innervated muscles whose functional roles, in addition to facial expression, include speech production, eating, and respiration, among others. Facial muscles can be divided into two categories:

1. *Superficial facial muscles* attach to facial skin (**FIGURE 11.10**). On contraction they change the shape of the mouth, eyes, or nose, for example; or they pull on their attachment to the skin. One such muscle, the *frontalis*, wrinkles the forehead and raises the eyebrow.
2. *Deep facial muscles* attach to bone. These muscles enable movements such as chewing and large movements of the face. An example of a deep muscle is the *temporalis*, a powerful jaw muscle.

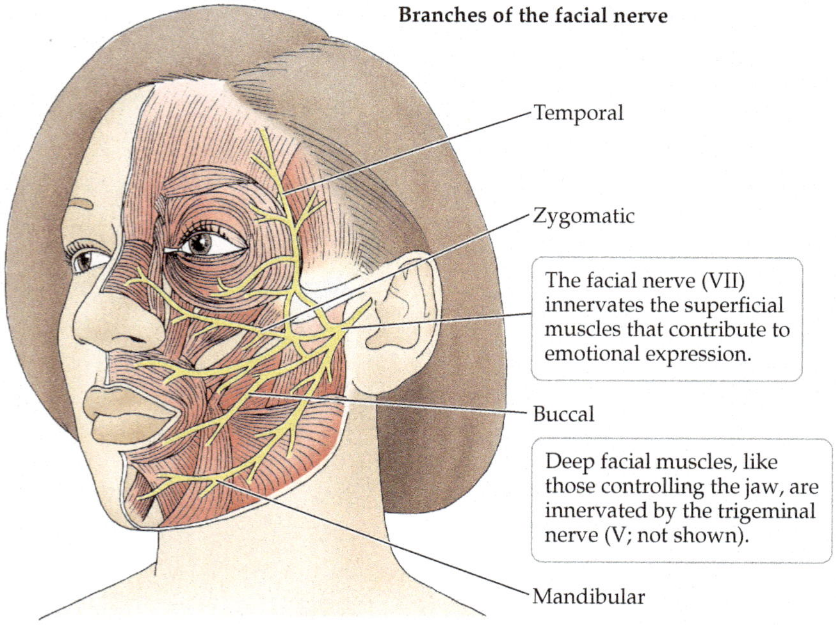

11.10 Facial Muscles and Their Neural Control

Human facial muscles are innervated by two cranial nerves: (1) the **facial nerve** (VII), which innervates the superficial muscles of facial expression (see Figure 1.10); and (2) the motor branch of the **trigeminal nerve** (V), which innervates muscles that move the jaw (e.g., the temporalis). As Figure 11.10 shows, the main trunk of the facial nerve (yellow) divides into upper and lower divisions shortly after entering the face. These nerve fibres originate in the brainstem in the nucleus of the facial nerve. Ultimately, the activity of the cranial nerves is governed by the face area of motor cortex:

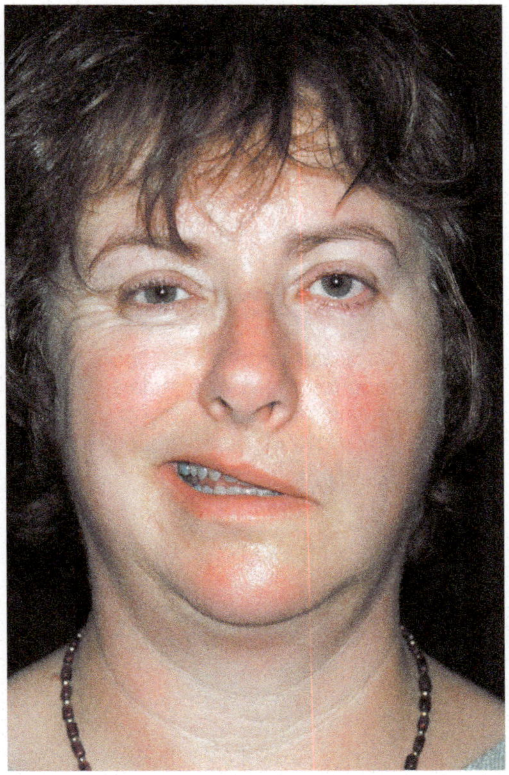

11.11 Bell's Palsy Leaves Half of the Face Paralysed This woman is smiling, but only the muscles on the right side of her face (left half of photograph) respond to her commands.

this is a disproportionately large brain region in humans (see Figure 10.14), probably reflecting the importance of emotional expression in our species.

An intriguing but controversial proposal—the *facial feedback hypothesis*—builds on the James-Lange notion that sensations from our body inform us about our emotions, suggesting that sensory feedback from facial expressions has effects on mood. In one test of this idea, participants were given a task such as using the upper lip to hold a pencil under the nose, versus clenching it between the teeth, creating expressions resembling a sad or happy face, respectively. When asked how happy or sad they *felt*, or how funny they found a cartoon, participants who had been simulating a smile reported more positive feelings than those who had been simulating a frown (J. I. Davis et al., 2009). Due to difficulties in replicating the original research findings, there is an important controversy about the strength of this phenomenon (Wagenmakers et al., 2016), but if it proves to be true that simply forcing a smile can actually make you happier, then the old song that advises us to "just put on a happy face" may have its basis in fact. People who *can't* put on a convincing happy face, because they've received Botox injections that have paralysed their facial muscles to reduce wrinkles, reportedly experience emotions less intensely after the treatment than before (J. I. Davis et al., 2010). Being forced to display false emotional expressions in stressful situations, however, may have negative consequences for well-being and job happiness (Hülsheger and Schewe, 2011).

Numerous medical conditions can impair facial expression. Chronic selective inhibition of the facial musculature is one symptom of Parkinson's disease (which is discussed in Chapter 10) and of schizophrenia (Kring, 1999). Sometimes viruses infect the facial nerve and damage it enough to cause paralysis of facial muscles. This condition, known as **Bell's palsy**, usually affects just one side, resulting in a variety of symptoms, including drooping eyelid and mouth (**FIGURE 11.11**). There is no standard treatment, but happily most people recover on their own within a few weeks, and almost everyone recovers within 6 months. Regardless of the cause, impaired ability to use facial expression to communicate emotional information can impoverish social interactions.

Electrical stimulation of the brain can produce emotional effects

One way to study the neuroanatomy of emotion is to electrically stimulate sites in the brains of awake, freely moving animals and then observe the effects on behaviour. Such stimulation can have either rewarding or aversive effects, or it may elicit sequences of emotional behaviour. Such experiments in the 1950s produced an intriguing finding: rats will readily press a lever in order to receive a brief burst of electrical stimulation in a brain region called the *septum* (Olds and Milner, 1954) (**FIGURE 11.12**). This phenomenon, called **brain self-stimulation**, can also happen in humans. People receiving electrical

Bell's palsy A disorder, usually caused by viral infection, in which the facial nerve on one side stops conducting action potentials, resulting in paralysis of one side of the face.

brain self-stimulation The process in which animals will work to provide electrical stimulation to particular brain sites, presumably because the experience is very rewarding.

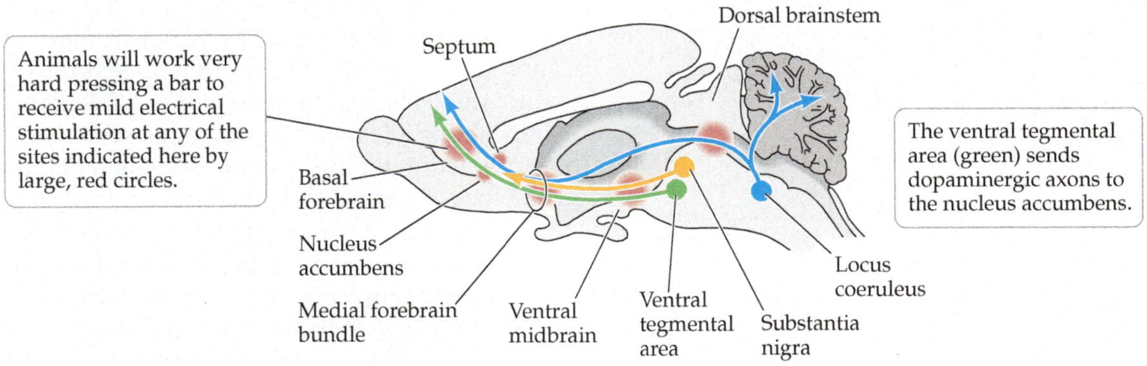

11.12 Brain Self-Stimulation Sites in Rodents

medial forebrain bundle A collection of axons travelling in the midline region of the forebrain.

decorticate rage Also called *sham rage*. Sudden intense rage characterised by actions (such as snarling and biting in dogs) that lack clear direction.

limbic system A loosely defined, widespread group of brain nuclei that innervate each other to form a network. These nuclei are implicated in emotions.

Klüver-Bucy syndrome A condition, brought about by bilateral amygdala damage, that is characterised by dramatic emotional changes including reduction in fear and anxiety.

stimulation in this region feel a sense of pleasure or warmth, and in some instances stimulation in this region provokes sexual excitation (Heath, 1972).

Building on the discovery of self-stimulation, a rush of experimentation subsequently mapped the distribution of brain sites that support self-stimulation responses. Animals will work to receive electrical stimulation of many different subcortical sites, but stimulation of the cortex does not seem to have such rewarding effects. Effective stimulation sites for self-stimulation are especially concentrated in a large axon tract that ascends from the midbrain, projecting through the hypothalamus before terminating in multiple forebrain sites: the **medial forebrain bundle**. One important target for the axons is the nucleus accumbens, a major component of the brain's reward circuitry (see Chapter 3). Dopaminergic stimulation of this site appears to be very pleasurable.

One theory is that electrical stimulation taps into the circuits mediating more-customary rewards, such as the arrival of food to a hungry animal or water to a thirsty animal (N. M. White and Milner, 1992). As we discussed in Chapter 3, researchers have proposed that drugs of abuse are addictive because they activate these same neural circuits with an artificial intensity (Pierce and Kumaresan, 2006).

Brain lesions affect emotions

Early in the twentieth century, decorticate dogs (dogs from which the cortex had been removed) were found to respond to routine handling with sudden intense **decorticate rage**—snarling, biting, and so on—sometimes referred to as *sham rage* because it seemed undirected. Clearly, then, emotional behaviours of this type must be organised at a subcortical level, with the cerebral cortex normally providing *inhibition* of emotional responsiveness. On the basis of studies of the spread of rabies virus in the brains of cats, combined with brain autopsies of humans with emotional disorders, James W. Papez (1937) proposed a subcortical circuit of emotion. Papez (which rhymes with "capes") noted associations between emotional changes and specific sites of brain damage. These interconnected regions, known as the *Papez circuit*, include the mammillary bodies of the hypothalamus, the anterior thalamus, the cingulate cortex, the hippocampus, and the fornix. The arrows in **FIGURE 11.13** schematically depict this circuit. Later, Paul MacLean (1949) suggested that the amygdala and several other regions also interacted with the components of this circuit, and he proposed that the entire system be called the **limbic system**.

Early support for the limbic model of emotion came from studies of monkeys following removal of large portions of their temporal lobes (Klüver and Bucy, 1938). The animals' behaviour changed dramatically after surgery; the highlight of this behavioural change was an extraordinary taming effect, part of what became known as the **Klüver-Bucy syndrome**. Animals that had been wild and fearful of humans prior to surgery became tame and showed neither fear nor aggression afterward. They also showed strong oral tendencies, eating a variety of objects, including rocks! Frequent and often inappropriate sexual behaviour was also observed and was described as *hypersexuality*.

Because lesions restricted to the cerebral cortex did not produce these results, deeper regions of the temporal lobe, including sites within the limbic system (see Figure 11.13), were implicated, and more-detailed investigation focused on the amygdala. Because this type of behavioural change is seen in monkeys in which *only* the left and right amygdalas have been destroyed—without damaging any adjacent tissue (Emery et al., 2001)—it appears that the amygdala is a key structure in the behavioural changes in Klüver-Bucy syndrome, specifically the loss of fear.

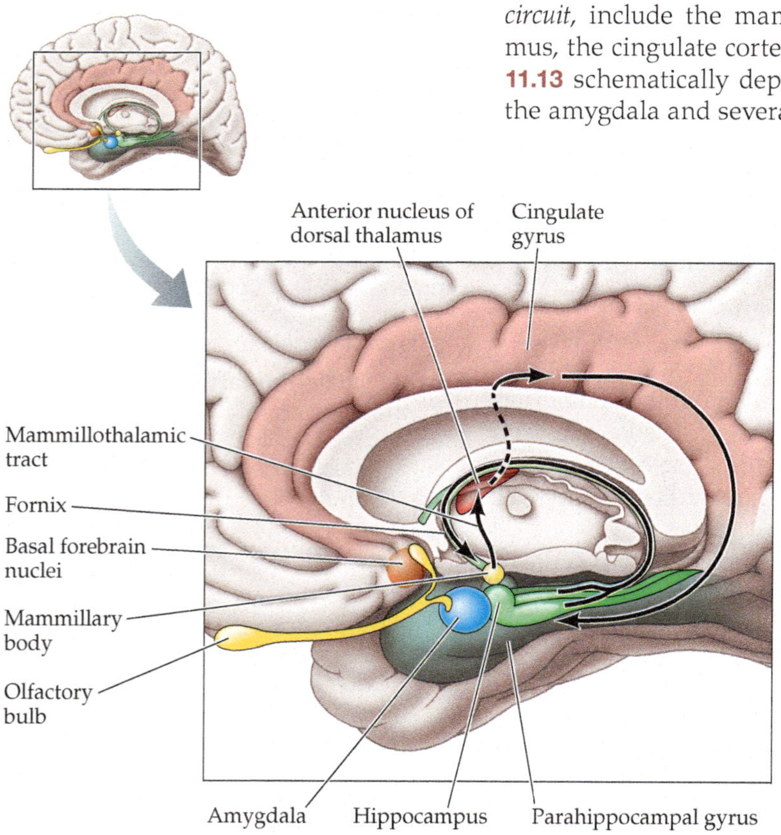

11.13 Medial Regions of the Brain Involved in Emotions Brain regions included in the Papez circuit are overlaid with arrows and are shown with structures included in MacLean's later conception of the limbic system.

Fear is mediated by circuitry that includes the amygdala

There is nothing subtle about fear. Fear-provoking situations elicit similar behaviours from individuals of many different species. For example, it is very easy to reliably elicit fear by using a type of classical conditioning called **fear conditioning** (see Chapter 16), in which the person or animal is presented with a stimulus such as light or sound that is paired with a brief aversive stimulus such as mild electrical shock. After several such pairings, the sound or light by itself effectively elicits typical fear-associated behaviours, such as freezing in position, and autonomic signs like rapid heart rate and heavy breathing (**FIGURE 11.14A**).

fear conditioning A type of classical conditioning where a previously neutral stimulus is repeatedly paired with shock or some other unpleasant experience, causing the individual to act fearful in response to the stimulus.

Go to Activity 11.1
Conditioned Fear Response

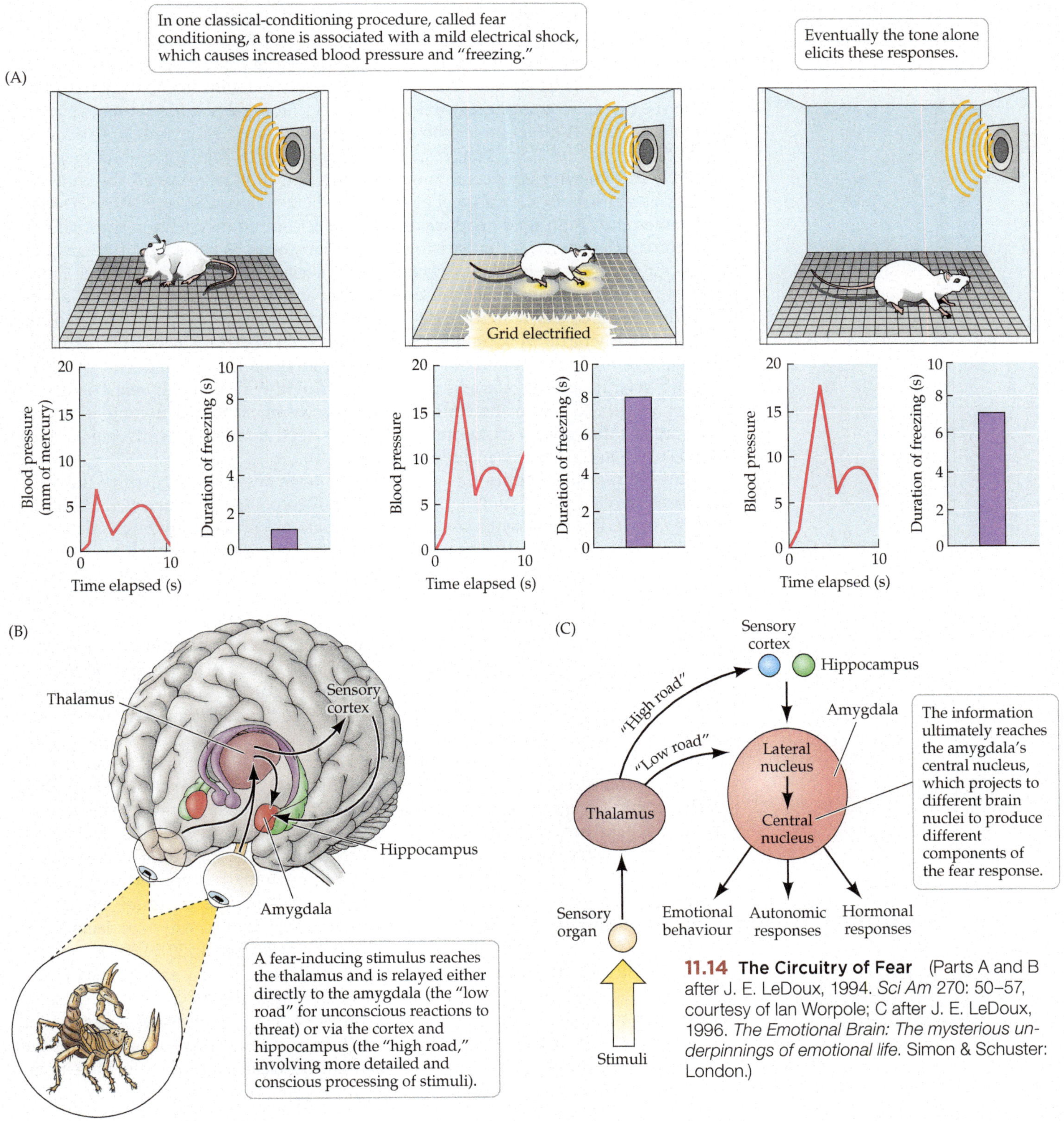

11.14 The Circuitry of Fear (Parts A and B after J. E. LeDoux, 1994. *Sci Am* 270: 50–57, courtesy of Ian Worpole; C after J. E. LeDoux, 1996. *The Emotional Brain: The mysterious underpinnings of emotional life.* Simon & Schuster: London.)

amygdala A group of nuclei in the medial anterior part of the temporal lobe.

Studies of such fear conditioning have allowed researchers to develop a map of the neural circuitry of emotional learning, which confirmed that the **amygdala** is a key structure in the mediation of fear (**FIGURE 11.14B**). Located at the anterior medial portion of the temporal lobe, the amygdala is composed of about a dozen different nuclei, each with a distinct set of connections. Recall that lesions of the entire amygdala seemed to abolish fear in monkeys. Lesioning just the central nucleus of the amygdala has the same effect, preventing blood pressure increases and freezing behaviour in response to a conditioned fear stimulus. Subsequent research has confirmed that the amygdala is crucial not only for aversive conditioning but also *appetitive learning*: conditioned positive emotional reactions to attractive stimuli, such as to sex-related stimuli, or other pleasurable signals. In both cases, the amygdala is thought to help form associations between emotional responses and specific memories of stimuli that are stored elsewhere in the brain (Janak and Tye, 2015; Paton et al., 2006).

On its way to the amygdala, sensory information about emotion-provoking stimuli reaches a fork in the road at the level of the thalamus (recall from Chapter 1 that the thalamus acts like a switchboard, directing sensory information to specific brain regions). A direct projection from the thalamus to the amygdala, nicknamed the "low road" for emotional responses in the original fear-conditioning studies (LeDoux, 1996), bypasses conscious processing and allows for immediate emotional reactions to stimuli (Celeghin et al., 2015; de Gelder et al., 2012) (**FIGURE 11.14C**). An alternate "high road" pathway routes the incoming information through sensory cortex, allowing for processing that, while slower, is conscious, fine-grained, and integrated with higher-level cognitive processes, such as memory (see Figure 11.14B). Contributions from prefrontal cortex and anterior cingulate offer an additional level of fear conditioning called *observational fear learning*, in which fear of potentially harmful stimuli is learned through social transmission (Olsson and Phelps, 2007). So, to extend our example, an individual can learn to fear scorpions by observing signs of fear and pain in others, without personally experiencing a scorpion attack. Given the considerable adaptive benefits that it confers, it's not surprising that observational fear learning is seen in species as diverse as mice, cats, cows, and primates, including humans.

Interconnections within the amygdala also form an important part of the story. Information about the stimulus (the sound in Figure 11.14A) from several brain regions, including sensory cortex, reaches the lateral portion of the amygdala first, and evidence suggests that neurons here encode the association between specific stimuli and aversive events, like electrical shocks. Further, the basolateral amygdala plays a key role in selecting between active escape behaviours versus passive freezing (Maren and Quirk, 2004; Terburg et al., 2018).

Activation of the lateral amygdala triggers a wider network within the amygdala, ultimately activating the central nucleus (Namburi et al., 2015). The central nucleus then transmits information to various brainstem centres to evoke three different aspects of emotional responses (see Figure 11.14C): pathways through the central gray (or periaqueductal gray) evoke emotional behaviours, those through the lateral hypothalamus evoke autonomic responses, and those through the bed nucleus of the stria terminalis evoke hormonal responses. Different types of fearful stimuli, such as odours from an aggressive male versus odours from a predator, may activate slightly different pathways to and from the amygdala (C. T. Gross and Canteras, 2012).

Learned fears are notoriously slow to extinguish; in our example (see Figure 11.14A), once the shock and the sound have been paired, the auditory tone must be presented without shock many times before animals stop freezing in response. During such extinction training, activity in the prefrontal cortex, which projects to the amygdala, is necessary for the animal to benefit (Adhikari et al., 2015). If the prefrontal cortex is silenced during extinction training, the next day the animal responds to the sound as if no extinction trials took place (Do-Monte et al., 2015). Mice missing one of the two types of cannabinoid receptors (see Chapter 3) also have a harder time unlearning their fearful reaction to a tone (Marsicano et al., 2002), suggesting that stimulation of these receptors normally extinguishes learned fears. If so, then it may be possible to develop cannabinoid drugs to treat phobias in humans.

Data from rats and mice on the role of the amygdala in fear mesh well with observations in humans. When humans are shown visual stimuli associated with pain or fear, blood flow to the amygdala increases (LaBar et al., 1998), even in the absence of conscious perception of the stimuli (Pegna et al., 2005). Similarly, when people view facial expressions of fear, electrophysiological responses occur much more quickly in the amygdala than in visual cortex, reflecting the privileged "low road" access of the fear-inducing stimulus (Méndez-Bértolo et al., 2016). People who have temporal lobe seizures that include the amygdala commonly report that intense fear heralds the start of a seizure (Engel, 1992), and electrical stimulation of temporal lobe sites during brain surgery likewise may elicit feelings of fear (Bancaud et al., 1994). The neural mechanisms of fear conditioning are also thought to play a central role in posttraumatic stress disorder (PTSD; see Chapter 12), in which memories of horrible events repeatedly intrude into consciousness, reawakening all the autonomic and psychological symptoms of fear.

But perhaps the most compelling evidence that the amygdala is important for fear in our species comes from people like patient S.M., the woman we met at the start of the chapter, who is literally fearless. The fearlessness that she and other people with the disorder display seems almost certainly due to the loss of the amygdala. Her very rare genetic condition causes the accumulation of calcium deposits in the amygdala, starting in late childhood, which eventually destroys the amygdala in both cerebral hemispheres (**FIGURE 11.15**). When S.M. is shown movie clips that other people find frightening, she reports being unmoved (**FIGURE 11.16**). S.M. is also very poor at recognising the facial expressions of fear in other people, but she recognises other emotional expressions—

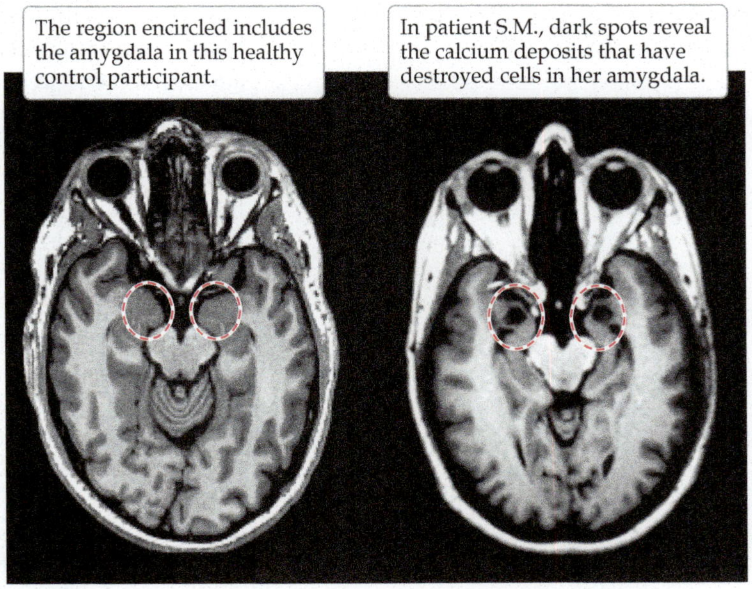

From J. S. Feinstein et al. 2013. *Nat Neuro* 16: 270–272

11.15 The Locus of Fear Unlike a control participant (left), patient S.M. (right) displays dark deposits of calcium in the amygdalas of both hemispheres. This remarkably specific lesion seems to have made her fearless without affecting other behaviours.

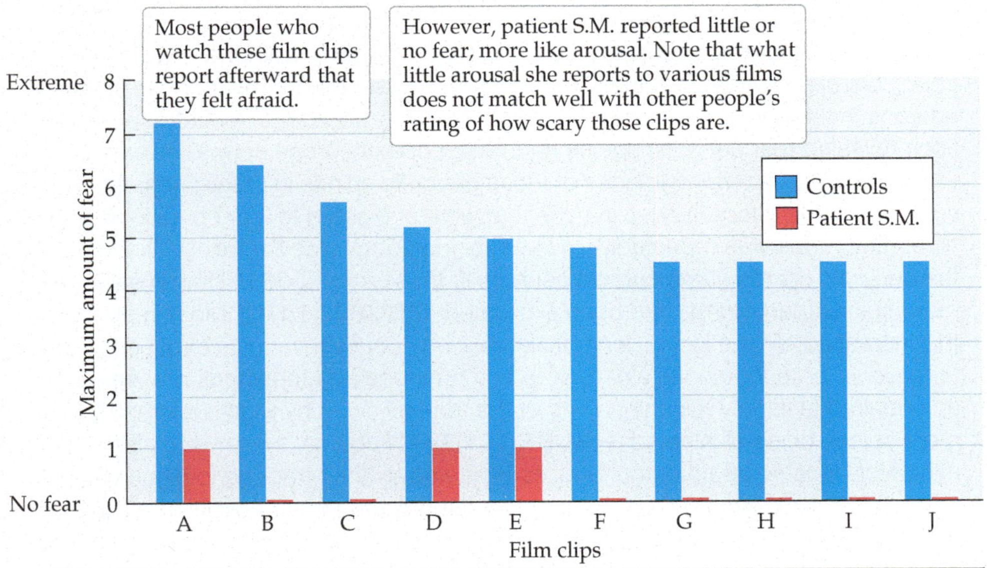

A – *The Ring* (2002) The ghost of a murdered child infiltrates the lives of her soon-to-be victims.
B – *Blair Witch Project* (1999) Campers are attacked by an unknown apparition during the middle of the night.
C – *CSI* (2009) A man struggles to survive after being buried alive.
D – *The English Patient* (1996) A man is tortured during World War II.
E – *Se7en* (1995) A mutilated man awakes from the dead.
F – *Cry Freedom* (1987) Armed trespassers attack a woman who is home alone during the night.
G – *Arachnophobia* (1990) A large poisonous spider attacks a girl in the shower.
H – *Halloween* (1978) A woman is being chased by a murderer.
I – *The Shining* (1980) A young boy hears voices in the hallway of a haunted hotel.
J – *The Silence of the Lambs* (1991) A female FBI agent tries to capture a twisted serial killer who is hiding in a dark basement.

11.16 What Scary Movie? (After J. S. Feinstein et al., 2011. *Curr Biol* 21: 34–38.)

a pattern seen in other people with damaged amygdalas (Adolphs et al., 2005). Her inability to recognise fear may be caused by her tendency not to look at people's eyes, so she doesn't see fear cues like wide eyes (L. F. Barrett, 2018; Kennedy and Adolphs, 2011).

What's fascinating about people like S.M. is that they have never known fear since the disease destroyed their amygdalas in late childhood. But they *can* be made afraid—by being given only air with a high concentration of carbon dioxide. Within 10 seconds, S.M. was flailing her hands and feeling a strong, panicky fear (Feinstein et al., 2013), crying out, "Help me" (Aschwanden, 2013). Similarly, a pair of monozygotic twins with the same genetic disorder and fearless behaviour as S.M. both showed evidence of panic, including a full-blown panic attack in one twin, when injected with an epinephrine-like drug (Khalsa et al., 2016). So despite the central role of the amygdala in most fear, the brain has alternative ways of making us very afraid and inducing panic. One possibility is that the amygdala is important for recognising and responding to *external* threats, but the brain relies on other systems, perhaps in the brainstem, to detect and respond to signs of *internal* threats, like insufficient oxygen in the blood.

The Cutting Edge

Synaptic Changes during Fear Conditioning

When microscopes use infrared light, it penetrates deep into tissue, and when the light is finally absorbed, it is unlikely to do any damage. Normally such light is also too weak to bounce back to our eye to let us see anything. If a fluorescent molecule is in the tissue, an infrared photon is also too weak to stimulate it to fluorescence so we can see it. To get around this problem, **two-photon excitation microscopy** uses lasers to shine a *lot* of infrared light onto tissue so that when two infrared photons hit a fluorescent molecule at the same time, they excite it enough to cause it to emit a higher-powered photon that we can see (**FIGURE 11.17A**). If such a microscope is centreed over the cortex of a living mouse (**FIGURE 11.17B**), fluorescent molecules (either fluorescent proteins, produced by a transgenic mouse, or fluorescent neuroanatomical tracers implanted by the scientist) can be excited and seen deep under the cortical surface (Helmchen and Denk, 2005).

Using this method, scientists visualised changes in the brain that may underlie fear conditioning. They directed their two-photon microscope on the dendritic spines of neurons in the frontal cortex of mice (C. S.-W. Lai et al., 2012). Repeated observations revealed that dendritic spines on frontal cortex neurons normally wax and wane over time. After recording images of these dendritic spines in many mice, the researchers looked for changes in mice that were subjected to fear conditioning, when an auditory tone was repeatedly followed by footshock (see Figure 11.14). Examining the dendritic spines of the cortical neurons in these mice 2 days later revealed that more spines than normal had been eliminated (**FIGURE 11.17C**). Most important, this increased elimination of dendritic spines was not seen in mice exposed to the tone alone, or to footshock alone, or to both the tone and footshock presented in an unpaired fashion. Only the *pairing* of the tone followed by footshock resulted in more spines being eliminated (**FIGURE 11.17D**). Of course, these were also the only mice that became afraid of the tone, as indicated by their freezing in place when it occurred. In fact, among the mice that were subjected to fear conditioning, those that showed the greatest fear response to the tone (freezing) also showed the most dendritic spine elimination (**FIGURE 11.17E**).

Thus the elimination of these dendritic spines, along with the loss of synaptic input on the spines, may play a role in learning to freeze in response to the fearful signal. Later, the researchers repeatedly exposed fear-conditioned animals to the tone *without* shock, and they learned to stop freezing in response to the sound; this process is called *extinction*. Fear extinction was associated with the formation of more dendritic spines in frontal cortex (Lai et al., 2012) *and* with the elimination of synaptic spines in auditory cortex (Lai et al., 2018); just as we might expect as the auditory-specific association was being "unlearned." With powerful techniques like these, scientists hope to someday fully understand the entire circuitry of fear. ■

two-photon excitation microscopy Method of providing many low-energy photons that can penetrate deep into tissues, such that the simultaneous arrival of two photons at a fluorescent molecule is sufficient to elicit a visible photon in response.

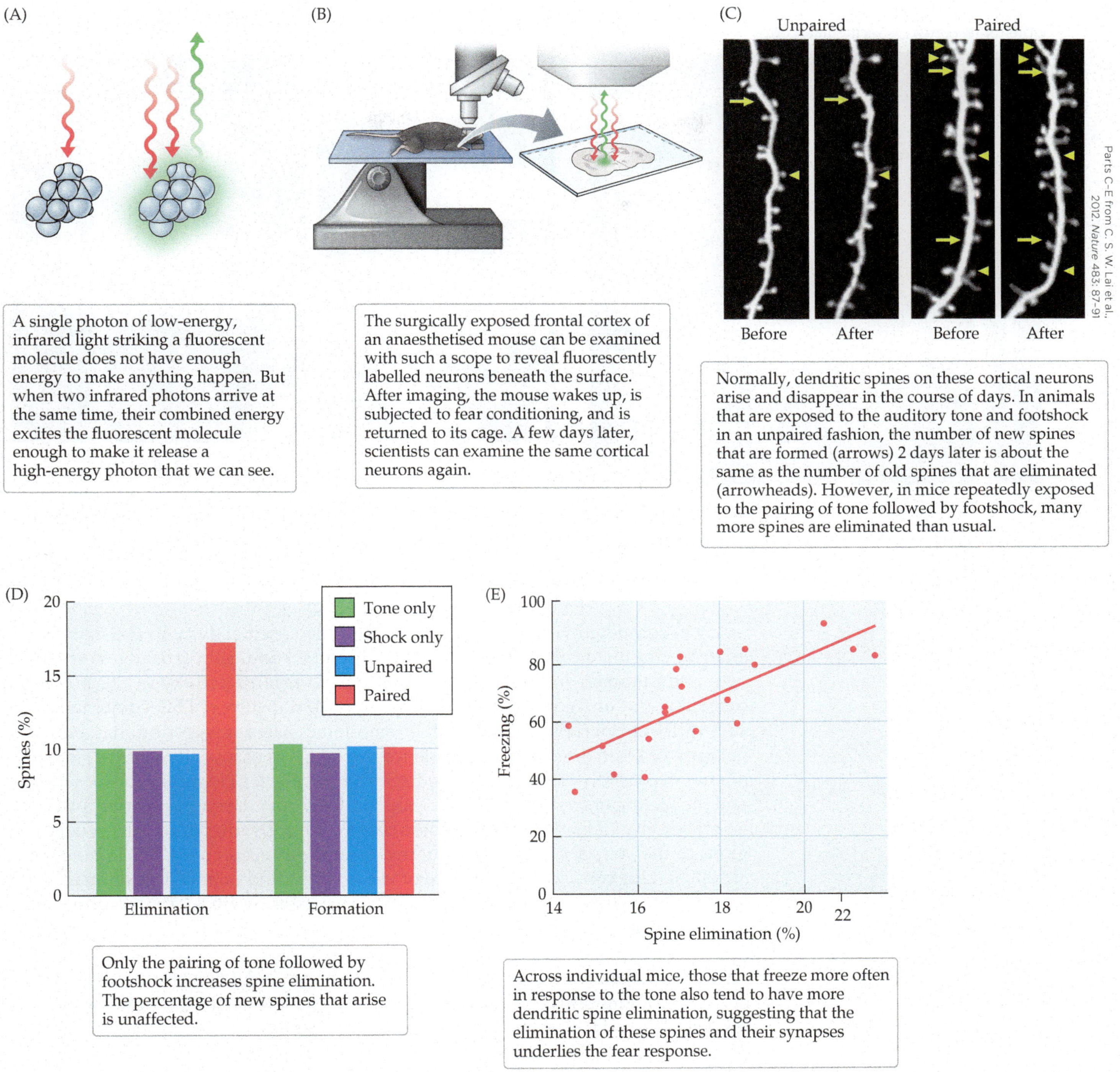

11.17 Changes in Dendritic Spines during Fear Conditioning

Different emotions activate different regions of the human brain

Several brain regions, and their networks, are consistently implicated in varying emotions (**FIGURE 11.18**) (for examples, see A. R. Damasio et al., 2000, and Panksepp, 2011). Andreas Bartels and Semir Zeki (2000) recruited volunteers who professed to be "truly, deeply, and madly in love." Each participant furnished four colour photographs: one of their romantic partner and three of nonromantic friends who shared the same age, gender, and length of friendship as the loved partner. Functional MRI brain scans were taken while each participant was shown counterbalanced sequences of the four photographs. Brain activity elicited by viewing the loved person was compared with that elicited by viewing friends.

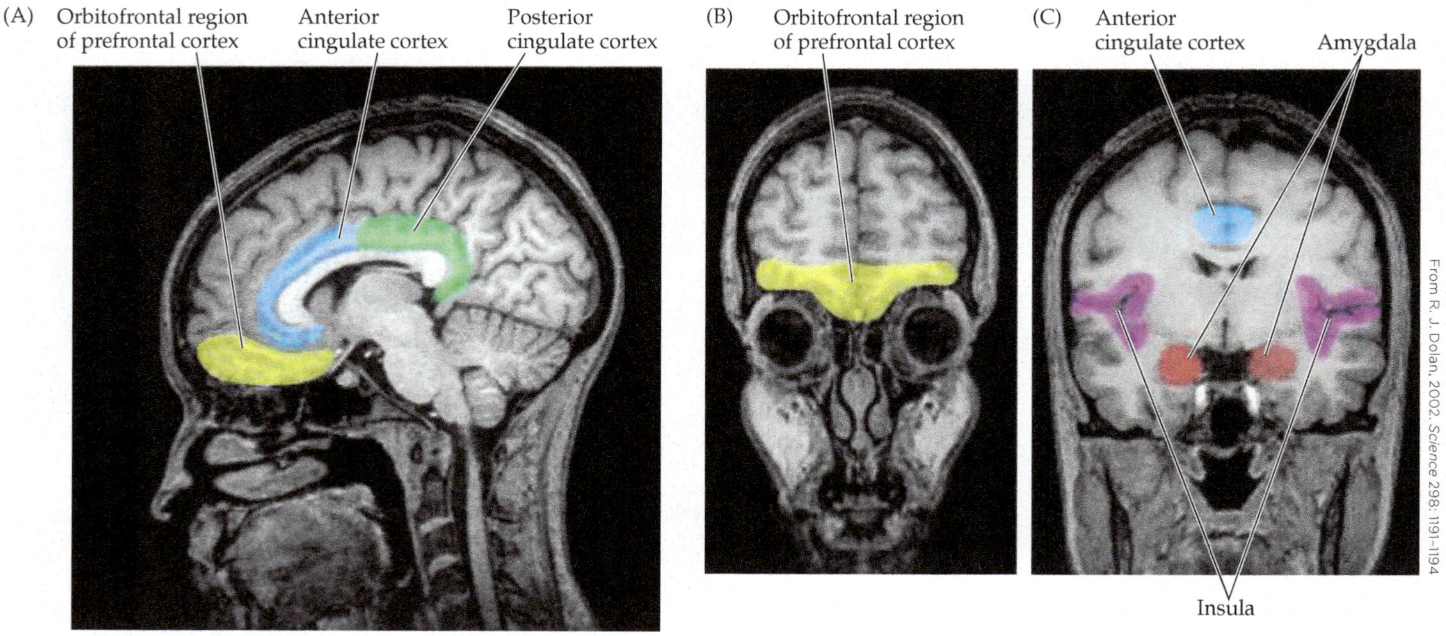

11.18 The Emotional Brain Brain regions implicated in emotions are depicted here in midsagittal (A), anterior coronal (B), and posterior coronal (C) sections.

Love, compared with friendship, involved increased activity in the insula and anterior cingulate cortex (see Figure 11.18A and C) and, subcortically, in the caudate and putamen—all bilaterally. It also led to *reduced* activity in the posterior cingulate and amygdala and in the right prefrontal cortex. This combination of sites differs from those found in other emotional states, suggesting that a unique network of brain areas is responsible for the emotion of love. In contrast, feelings of envy reportedly involve increased activity of anterior cingulate cortex activity; and the feeling of *schadenfreude* (literally "dark joy" in German), experienced when an envied rival falls from grace, is associated with activation of reward-related regions of the ventral forebrain, including the nucleus accumbens (H. Takahashi et al., 2009). Our conscious experience of these varied and nuanced emotional states is believed to involve the anterior insular cortex (see Figure 4.27); impairments in emotional awareness, termed *alexithymia*, are associated with dysfunction of the insula (Gu et al., 2013).

Antonio Damasio et al. (2000) compared brain activation during four different kinds of emotion, and again, the insula, cingulate cortex, and prefrontal cortex (see Figure 11.18) were among the regions implicated. In a screening session, adults were asked to recall and attempt to reexperience episodes involving sadness, happiness, fear, or anger, as well as an equally specific but emotionally neutral episode. During the experimental session, the participant was asked to signal as soon as the desired emotion was experienced, and PET images of brain activity were made. Interestingly, physiological responses (skin conductance response and change in heart rate) *preceded* the signal, supporting the idea that at least some physiological responses precede the feeling of emotion (as the James-Lange theory would predict). The PET images were averaged for all participants experiencing a given emotion, and activity during the neutral state was subtracted from activity during the emotion (see Box 2.3). Results showed that activity was altered in many brain regions during emotional experience, and even though we don't understand what role each region plays, it does appear that the four emotions were accompanied by significantly different patterns of brain activity (**FIGURE 11.19**).

These studies confirm that there is no simple, one-to-one relation between a specific emotion and changed activity of a brain region. There is no "happy centre"

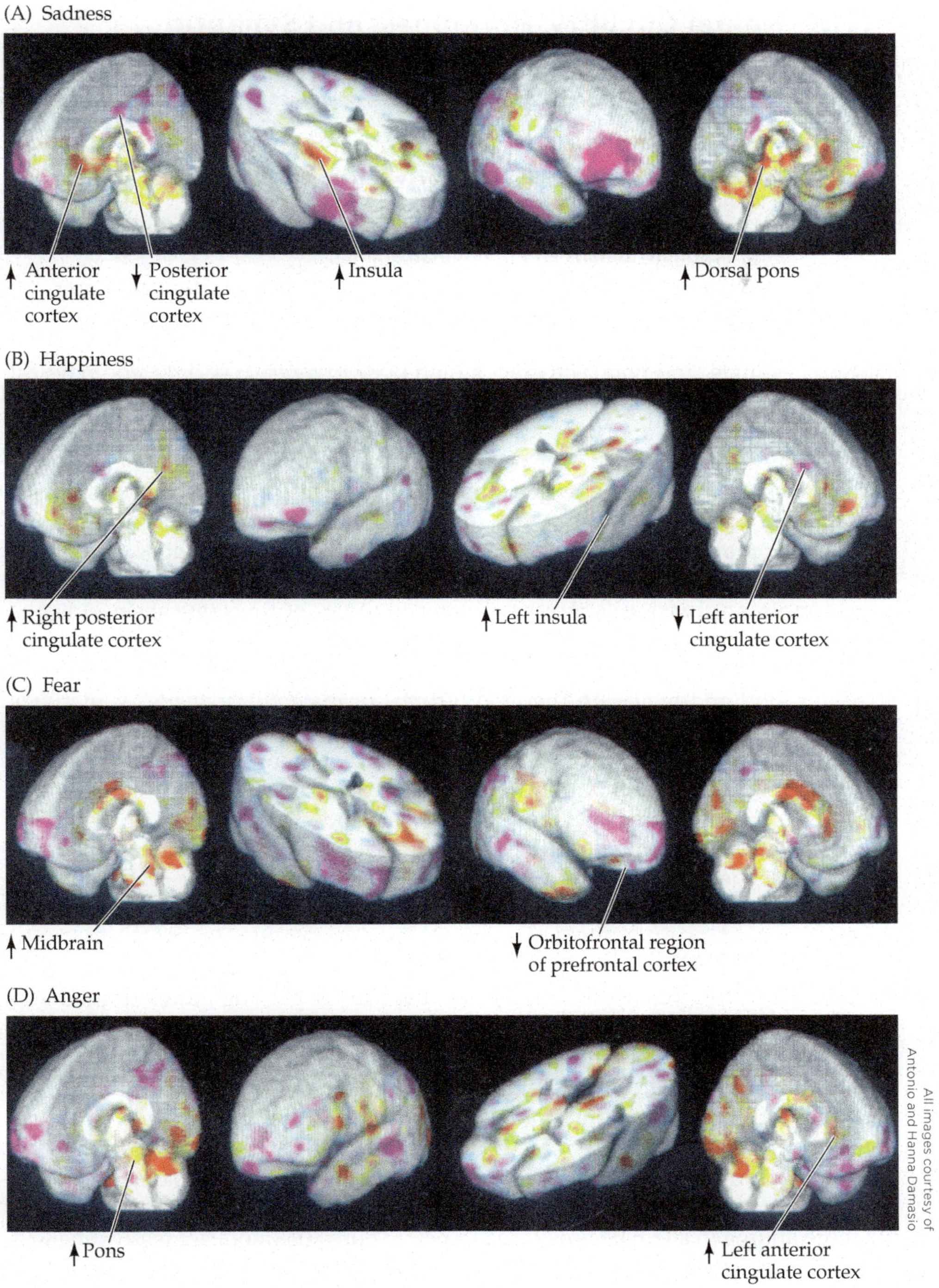

11.19 Brain Regions Active in Four Emotions Red and yellow indicate areas of increased activity; purple indicates areas of decreased activity. For the identified sites, an upward arrow indicates increased activity; a downward arrow, decreased activity.

or "sad centre." Instead, each emotion involves differential patterns of activation across a network of brain regions associated with emotion. For example, activity of the cingulate cortex is altered in sadness, happiness, and anger; the left somatosensory cortex is deactivated in both anger and fear. Feelings of regret over costly decisions (a topic we'll return to in Chapter 17) apparently involve activation of the amygdala and orbitofrontal cortex (see Figure 17.27). Although different emotions are associated with different patterns of activation, there is a lot of overlap among patterns for different emotions.

11.4 Neural Circuitry, Hormones, and Synaptic Transmitters Mediate Violence and Aggression

Learning Objectives

After reading this section, you should be able to:

11.4.1 **Define aggression, distinguishing between proactive and reactive forms, and discuss the adaptive significance of aggression.**

11.4.2 **Evaluate the role of testosterone, and other hormones and transmitters, in aggression in humans and nonhuman animals.**

11.4.3 **Summarise brain mechanisms believed to control aggressive behaviours.**

11.4.4 **Describe the behavioural and physiological correlates of psychopathy.**

aggression Behaviour that is intended to cause pain or harm (whether physical or emotional) to others, either individually or in groups.

intermale aggression Aggression between males of the same species.

Violence, assaults, and homicide exact a high price in modern society, and physical assault is not the only form of aggression. Verbal and symbolic aggression—name-calling, horn honking, angry glares—also take their toll. We can define **aggression** as behaviour that is intended to cause pain or harm (whether physical or emotional) to others, either individually or in groups. We will focus primarily on physical aggression and violence between individuals, excluding the aggression of predators toward their prey, which is perhaps better viewed as feeding behaviour (Glickman, 1977).

Intermale aggression (aggression between males of the same species) is observed in most vertebrates, and whatever we may think about aggression, it seems clear that in many species aggressive behaviour in males is adaptive for gaining access to food and mates. In the wild, groups of male chimpanzees sometimes band together to kill a rival male (Wilson et al., 2014), increasing the attackers' chances of mating in the future. Human males are 5 times as likely as females to be arrested on charges of murder in the United States. Aggressive behaviour between boys, but not girls, is evident early in childhood, in the form of vigourous and destructive play behaviour (J. Archer, 2006). These and similar observations suggest that the hormone that prepares males for reproduction—testosterone—also plays a role in their aggressive behaviour.

Androgens seem to increase aggression

Because aggressive behaviour in males is adaptive for gaining access to food and mates in many species, it makes sense that the same testosterone that prepares males for reproduction would also make them more aggressive. For example, at sexual maturity, when levels of circulating androgens such as testosterone rise, intermale aggression markedly increases in many species (Svare, 2013). In seasonally breeding species as diverse as birds and primates, male aggression waxes and wanes in concert with changes in testosterone across the seasons (Bronsard and Bartolomei, 2013). In spotted hyenas, the females are more aggressive than the males (see Box 13.1), and the more androgen a female is exposed to before birth, the more aggressive she will be growing up (Dloniak et al., 2006). In more-typical mammals, such as mice or rats, decreasing circulating androgens by castration usually reduces intermale aggressive behaviour profoundly (**FIGURE 11.20**). Treating castrated males with testosterone restores fighting behaviour. Fruit flies don't make testosterone, but the same genes that trigger male courtship and copulatory behaviour also promote intermale aggression (Vrontou et al., 2006), so the link between reproduction and intermale aggression seems to extend throughout the animal kingdom.

The relationship between testosterone and aggression in humans is less clear-cut. Treating adult volunteers with extra testosterone does not increase their aggression (O'Connor et al., 2004). Similarly, young men going through puberty experience a sudden large increase in circulating testosterone and yet do not show a correlated increase in aggressive behaviour (J. Archer, 2006). Nevertheless, some human studies have shown a positive correlation between testosterone levels and the magnitude of hostility, as measured by behaviour rating scales. Comprehensive studies of military veterans suggest that testosterone *is* related to antisocial behaviour (Dabbs and Morris, 1990). Nonaggressive tendencies in

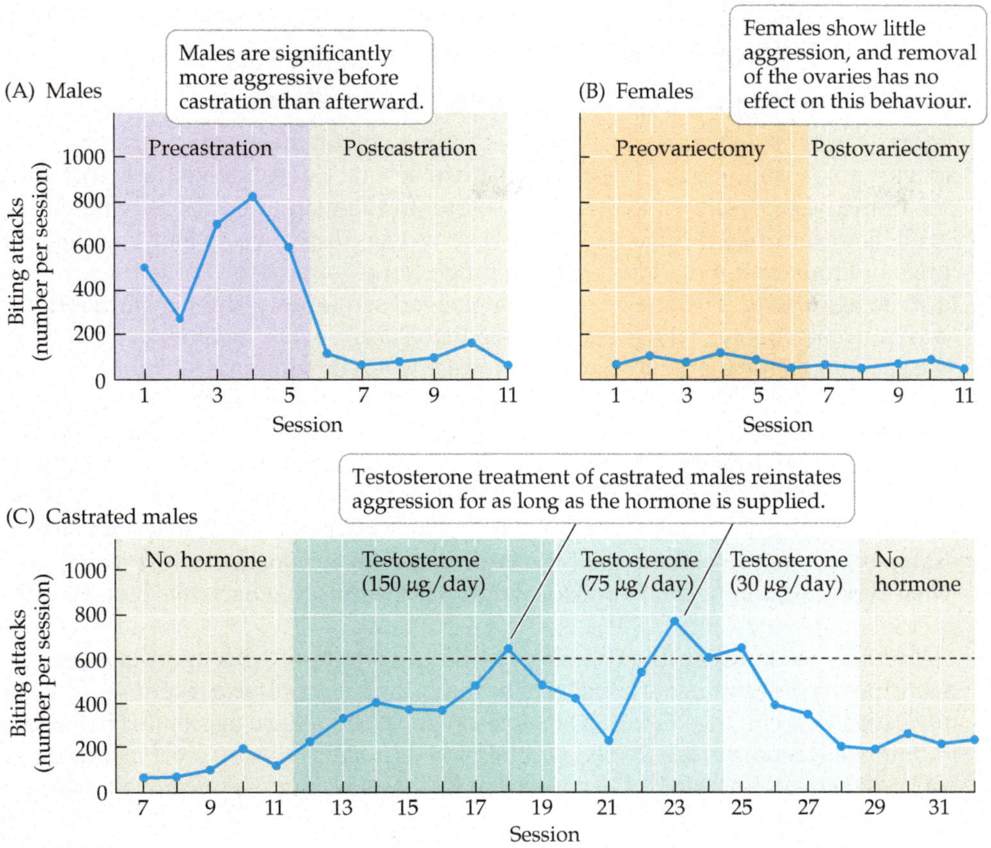

11.20 The Effects of Androgens on the Aggressive Behaviour of Mice Counts of the number of biting attacks initiated by males before and after castration (A) and by females before and after removal of the ovaries (B) reveal significantly higher aggression in males before castration. When castrated males are treated with testosterone (C), aggressive behaviour is reinstated. (After G. C. Wagner et al., 1980. *Aggress Behav* 6: 1–7.)

males are associated with satisfaction in family functioning and with lower levels of serum testosterone (Julian and McKenry, 1979). Among female convicts, testosterone concentrations are highest in women convicted of unprovoked violence and lowest among women convicted of defensive violent crimes (Dabbs et al., 1988).

At least two variables complicate the correlations between testosterone and aggression. First is the observation that experience can affect testosterone levels. In mice and monkeys, the loser in aggressive encounters shows reduced androgen levels (I. S. Bernstein and Gordon, 1974; Lloyd, 1971), so measured levels of testosterone sometimes may be a *result*, rather than a *cause*, of behaviour. In men, testosterone levels rise in the winners and fall in the losers after competitions ranging from wrestling to chess. Male sports fans even show a vicarious competition effect in response to simply watching "their" team win or lose a sporting event (Bernhardt, 1997), and as poll results rolled in during the 2008 U.S. presidential election, male McCain voters experienced a sharp drop in circulating testosterone, compared with backers of the winner, Obama (Stanton et al., 2009).

These observations suggest that a second confounding variable between testosterone and aggression is *dominance* (Mazur and Booth, 1998), since most chess players could hardly be said to be physically aggressive. For example, women treated with testosterone reportedly spend more time staring into the eyes of threatening faces (Terburg et al., 2012), which is a socially dominant behaviour. According to this model, testosterone levels should be associated with behaviours that confer or protect the individual's social status (and thus reproductive fitness). This type of aggression is said to be *proactive*—part of an offense strategy to improve the individual's standing in comparison to others, or achieve a desired social outcome—in contrast to *reactive* aggression that encompasses the many sorts of defensive behaviour, ranging from freezing to preemptively attacking, that help protect against external threats (LeDoux and Daw, 2018; Wrangham, 2018). While surgical or "chemical" castration may reduce violence in some sex offenders (L. E. Weinberger et al., 2005), the main effect is a reduction in sexual motivation more than a direct effect on aggression. Many ethical issues raised by this approach to the rehabilitation of sex offenders and the intricacies of such intervention have yet to be worked out.

Aggression has several neurochemical correlates

Aggressive behaviour in various animals, including humans, is modulated by brain activity associated with the systems that use several neurotransmitters, including dopamine, vasopressin, and especially serotonin (Numan, 2015; Rosell and Siever, 2015). For example, Higley et al. (1992) observed aggressive behaviour and fight injuries in monkeys from a large, free-ranging colony, to rank the animals from least to most aggressive. The researchers then gauged serotonin activity by measuring a serotonin metabolite, 5-HIAA (5-hydroxyindoleacetic acid), in cerebrospinal fluid. The most aggressive monkeys had the lowest levels of serotonin metabolites, suggesting that they had the least serotonin being released at synapses in the brain. In agreement with this finding, mice with one of the serotonin receptor genes knocked out are hyperaggressive (Bouwknecht et al., 2001), as one would expect if serotonin normally inhibits aggression (Siever, 2008). This inhibitory role of serotonin in aggression is probably evolutionarily ancient, as it is evident even in invertebrates: for example, enhanced serotonergic activity prompts solitary locusts to overcome their usual aversion to one another and form the huge swarms that decimate broad swaths of cropland (Anstey et al., 2009).

The balance between the inhibitory neurotransmitter GABA and the excitatory neurotransmitter glutamate appears to be important in aggressive responses to stimuli. Drugs that enhance GABA transmission generally reduce aggressive behaviour in humans (Lieving et al., 2008), although, paradoxically, GABA-ergic agonists like benzodiazepines and alcohol can occasionally have an aggression-provoking effect in a minority of users (Albrecht et al., 2016; Guina and Merrill, 2018). A variety of peptide hormones, including vasopressin, oxytocin, and the endogenous opioids, likewise have all been implicated in the control of aggression (Siever, 2008). Increased aggression is often seen in knockout mice (see Box 7.3), no matter which of several genes is deleted (Bedrosian and Nelson, 2014). So it's clear that aggression is regulated by many systems. Development of antiaggression treatments based on these mechanisms is an active area of research.

The **medial amygdala** analyses olfactory and pheromonal information, allowing male rats and mice to distinguish between male rivals to be attacked and females to be courted. That information is relayed to the ventrolateral portion of the **ventromedial hypothalamus** (**VMH**), which serves as a switch to trigger aggression. When experimenters optogenetically stimulated neurons in the VMH in male mice, the animals would stop mounting a receptive female and begin attacking her. Conversely, optogenetically inhibiting VMH activity reduced the likelihood that the subject would attack a male (Falkner et al., 2016; H. Lee et al., 2014). A direct input to the VMH from the suprachiasmatic nucleus—the brain's circadian clock—appears to regulate the daily variation in aggression seen in many species, including our own (Todd et al., 2018).

So far we've discussed aggression in males, but females are also aggressive at times, particularly when they are caring for their young. This **maternal aggression** is typically studied by introducing an intruder mouse, usually a male, into the cage of a mother nursing a litter. In such conditions, she may immediately attack the intruder. Neural circuits in the VMH are again implicated in maternal aggression, along with other hypothalamic regions, including the preoptic area (POA) and the premammillary nucleus (Motta et al., 2013), and serotonergic mechanisms originating from the midbrain raphe nuclei (Holschbach et al., 2018).

medial amygdala A region of the amygdala involved in processing olfactory stimuli.

ventromedial hypothalamus (VMH) A hypothalamic region involved in eating, sexual, and aggressive behaviours.

maternal aggression The fierce defensive aggression of females defending their offspring.

The biopsychology of human violence is a topic of controversy

Some forms of human violence are characterised by sudden, intense physical assaults. A long-standing controversy surrounds the idea that some forms of intense human violence are derived from temporal lobe disorders (Mark and Ervin, 1970). Aggression is sometimes a prominent symptom in people with temporal lobe seizures, and a significant percentage of people arrested for violent crimes have abnormal EEGs or

11.21 Psychopathic Impulsivity Serial killer Ted Bundy displayed many characteristics of a psychopath. He was superficially charming and, as shown here acting out in the courtroom when the judge was away, impulsive in nature. This scene also hints that, like other psychopaths, Bundy felt little or no remorse for his actions.

© Bill Frakes/Time and Life Pictures/Getty Images

other forms of neuropathology, often associated with temporal lobe function (L. M. Cope et al., 2014; D. O. Lewis, 1990).

Psychopathy is not a psychiatric disorder with formal diagnostic criteria in the *DSM-5*—instead it describes a well-studied cluster of personality traits that may be associated with antisocial behaviours. Psychopaths are often intelligent people with superficial charm who may have poor self-control, a grandiose sense of self-worth, and little or no feeling of remorse (Hare et al., 1990). And while most people who score high on psychopathic tendencies lead normal, often highly successful lives, psychopaths have sometimes committed horribly violent acts without compunction (**FIGURE 11.21**). Compared with controls, psychopaths do not react as negatively to words about violence (Gray et al., 2003), and they show blunted responses to aversive cues associated with fear conditioning that typically cause strong reactions in other people (Glenn and Raine, 2014). Imaging studies suggest that psychopaths have both reduced size and reduced activity of the prefrontal cortex (**FIGURE 11.22**), which may impair their ability to control impulsive behaviour (Y. Yang and Raine, 2009; Y. Yang et al., 2012).

Undoubtedly, human violence and aggression stem from many sources. Biological studies of aggression have been vigourously criticised by both politicians and social scientists. These critics argue that, as a result of emphasising biological factors such as genetics or brain mechanisms, the most evident origins of human violence and aggression (poverty and lack of education) might be overlooked, and odious forms of biological controls of social dysfunction might be instituted. However, the quality of life of some violent persons might be significantly improved if biological factors could be identified and addressed. For example, treatments that enhance serotonin activity in the brain might be an important addition to a social-environmental or psychotherapeutic intervention (George et al., 2011).

Next, let's look at one of the principal consequences of aggression and other difficult situations: stress.

psychopathy A cluster of personality traits associated with superficial charm, lying, and diminished remorse.

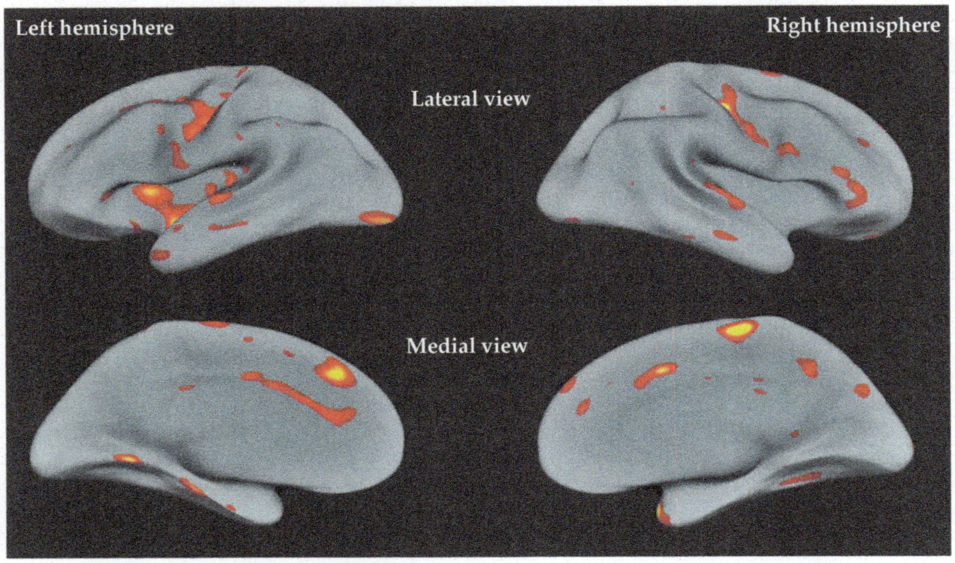

After M. Ly et al., 2012. *Am J Psychiatry* 169: 743–749

11.22 Brain Abnormalities in Psychopaths Hot colours (red/yellow) indicate areas of cortical thinning in incarcerated psychopaths, relative to nonpsychopathic inmates. Thinning is especially evident in left frontal cortex and bilateral temporal and cingulate cortex.

11.5 Stress Activates Many Bodily Responses

Learning Objectives

After reading this section, you should be able to:

11.5.1 **Describe the physiological correlates of stress, and contrast acute and chronic stress responses.**

11.5.2 **Discuss individual differences in stress vulnerability and responses to stressful situations.**

11.5.3 **Evaluate the role of early life experiences in lifelong stress responding.**

11.5.4 **Describe the form and function of communication between the nervous system and the immune system.**

11.5.5 **Summarize the impact of chronic stress on health, and explain some possible ways to mitigate these effects.**

We all experience stress, but what is it? Attempts to define *stress* tend to be somewhat vague. Hans Selye (1907–1982), whose work launched the modern field of stress research, broadly defined stress as "the rate of all the wear and tear caused by life" (Selye, 1956). Nowadays, researchers try to sharpen their focus by treating **stress** as a multidimensional concept that encompasses stressful stimuli, the stress-processing system (including cognitive assessment of the stimuli), and the responses to stress. While many different parts of the body respond to stress, it's clear that the brain carefully monitors and controls those responses (McEwen et al., 2015).

stress Any circumstance that upsets homeostatic balance.

alarm reaction The initial response to stress.

adrenal medulla The inner core of the adrenal gland, which secretes epinephrine and norepinephrine.

epinephrine Also called *adrenaline*. Here, a hormone secreted by the adrenal medulla under the control of the sympathetic nervous system, which prepares the body for action.

norepinephrine Also called *noradrenaline*. Here, a hormone secreted by the adrenal medulla under the control of the sympathetic nervous system, which prepares the body for action.

adrenal cortex The outer rind of the adrenal gland, which secretes steroid hormones, including cortisol.

adrenal steroids Steroid hormones secreted by the adrenal cortex, including glucocorticoids such as cortisol and mineralocorticoids such as aldosterone.

cortisol A glucocorticoid stress hormone of the adrenal cortex.

The stress response has multiple stages

Selye emphasised a close connection between stress and disease that he termed "general adaptation syndrome." He called the initial response to stress the **alarm reaction**. As one part of the alarm reaction, the hypothalamus activates the sympathetic nervous system to ready the body for action; this is the fight-or-flight system we mentioned at the start of the chapter. The sympathetic system stimulates the core of the adrenal gland, the **adrenal medulla**, to release the hormones **epinephrine** (also known as *adrenaline*) and **norepinephrine** (*noradrenaline*). These hormones act on many parts of the body to boost heart rate, breathing, and other physiological processes that prepare the body for action. As another part of the alarm reaction, the hypothalamus stimulates the anterior pituitary to release a hormone that drives the outer layer of the adrenal gland, the **adrenal cortex**, to release **adrenal steroids** such as **cortisol** (FIGURE 11.23). These hormones act more slowly than

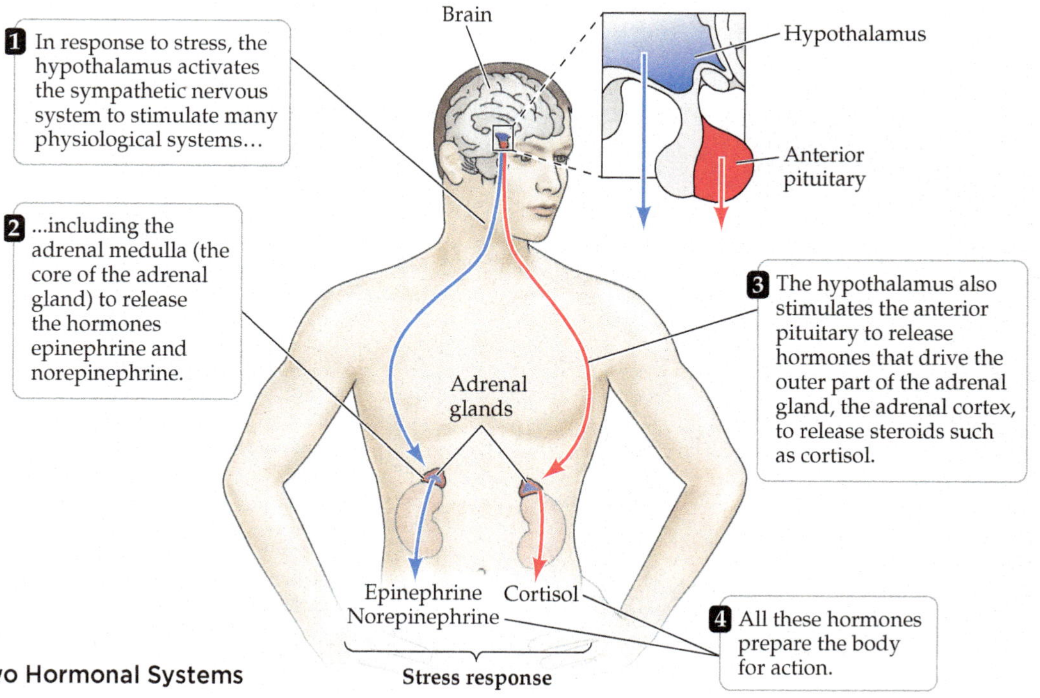

11.23 Stress Activates Two Hormonal Systems

epinephrine, but they also ready the body for action, including releasing body stores of energy. Glucocorticoid receptors—the receptors that respond to cortisol—are found in many locations in the brain, where they are thought to mediate the formation of memories associated with stress and fear (de Quervain et al., 2017).

A classic study examined these hormonal responses to stress in a group of young recruits in the Norwegian military both before and during scary parachute training (Ursin et al., 1978). On each jump day, the anterior pituitary released enhanced levels of hormones, and both the sympathetic and parasympathetic systems were activated (**FIGURE 11.24A**). Initially, cortisol levels were elevated in the blood before each jump, but with more and more successful jumps over successive days, the pituitary-adrenal response soon declined. Epinephrine and norepinephrine were also elevated before the first jumps, but eventually they returned to normal before jumps. Testosterone showed the reverse pattern, falling far below control levels on the first day of training but returning to normal with subsequent jumps (**FIGURE 11.24B**). Once the soldiers mastered the jumps, they no longer showed hormonal responses, having reached the adaptation stage that Selye had described.

Less-dramatic real-life situations also evoke clear endocrine responses (Frankenhaeuser, 1978). Factory work leads to the release of epinephrine; the shorter the work cycle—that is, the more frequently the person has to repeat the same operations—the higher the levels of epinephrine. Riding in a commuter train provokes the release of epinephrine; the longer the ride and the more crowded the train, the

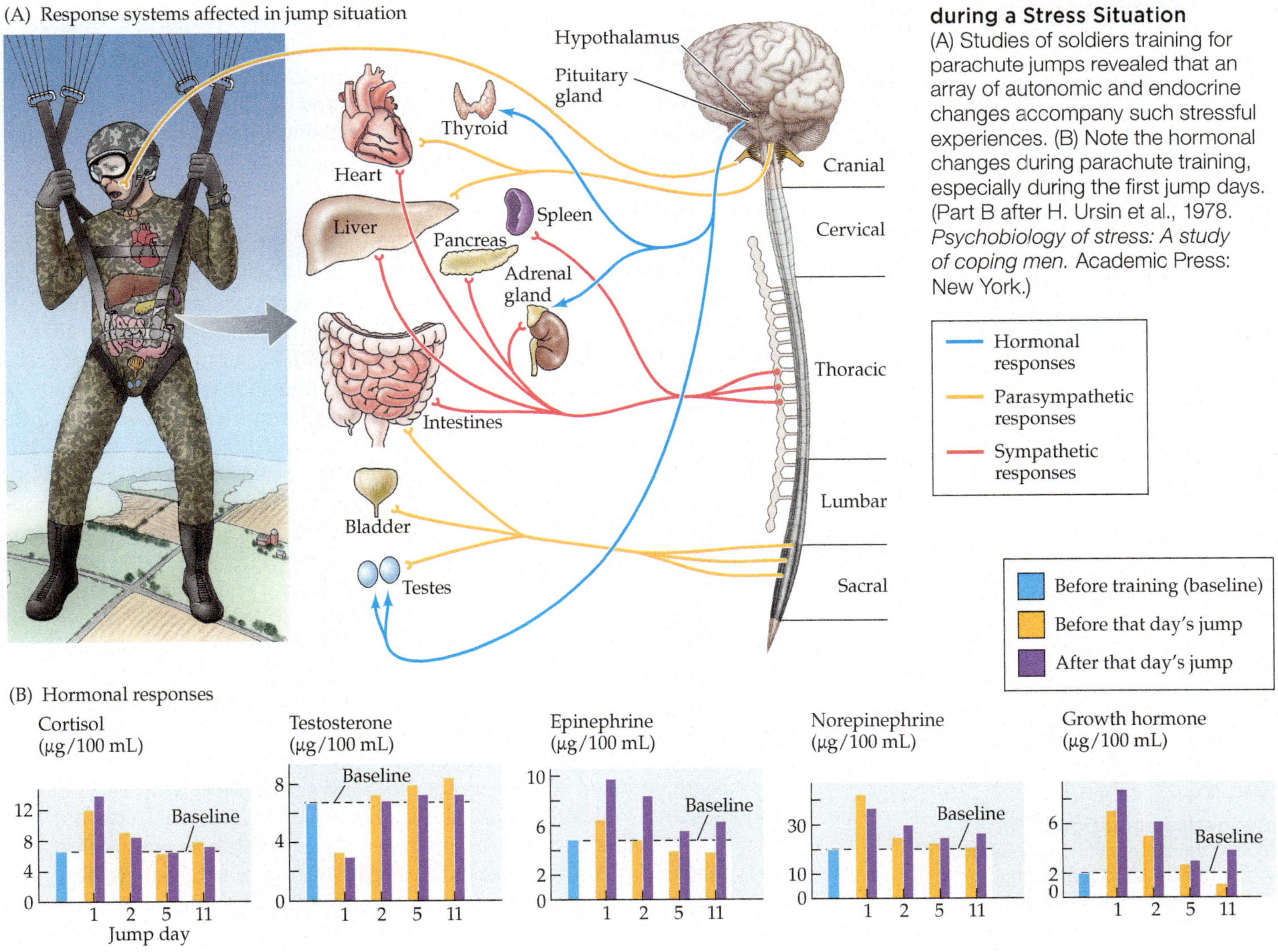

11.24 Autonomic Activation during a Stress Situation
(A) Studies of soldiers training for parachute jumps revealed that an array of autonomic and endocrine changes accompany such stressful experiences. (B) Note the hormonal changes during parachute training, especially during the first jump days. (Part B after H. Ursin et al., 1978. *Psychobiology of stress: A study of coping men.* Academic Press: New York.)

11.25 Hormonal Changes in Humans in Response to Social Stresses (A) Small changes in crowding on a morning commuter train ride affect hormone levels in humans. A 10% increase in the number of passengers during a period of gasoline rationing (right) resulted in a much higher level of epinephrine secretion. (B) Levels of epinephrine and norepinephrine in a graduate student in the weeks before, during, and after a thesis exam reflect levels of stress. (Part A after U. Lundberg, 1976. *J Human Stress* 2: 26–32; B after M. Frankenhaeuser, 1978. *Nebr Symp Motiv* 26: 123–162.)

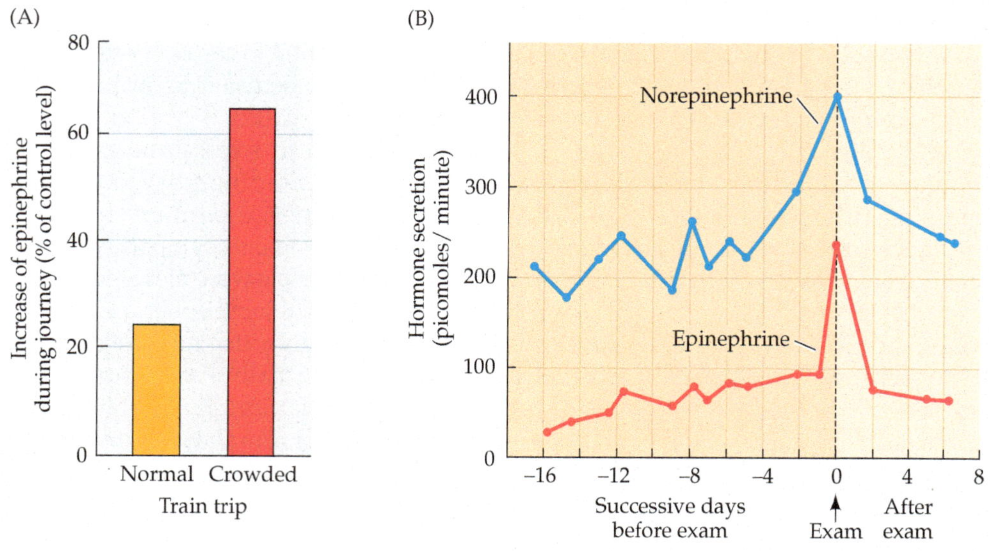

greater the hormonal response (**FIGURE 11.25A**). Likewise, the stress of a PhD oral exam leads to a dramatic increase in both epinephrine and norepinephrine (**FIGURE 11.25B**), and medical students stressed by preparing for their licensing exams show fMRI evidence of impairments in the brain mechanisms controlling attention; following the exam, their fMRI scans return to normal (Liston et al., 2009). Sustained social stressors can also exact lasting medical costs: for example, young people with asthma who experience stress due to peer rejection or a negative emotional climate at home have a more responsive adrenal system, more severe asthma symptoms, and impaired expression of anti-inflammatory genes (Farrell et al., 2018; M. L. Murphy et al., 2015). In general, adverse childhood experiences have an enduring effect on multiple aspects of health in later life, including neural and cognitive development, emotional regulation, and measures of lifetime achievement, as we discussed in Chapter 6 (Cameron et al., 2017; Felitti et al., 1998).

Individuals differ in the stress response

Why do individuals differ in their responses to stress (Infurna and Luthar, 2016)? One hypothesis focuses on early experience. Rat pups clearly find it stressful to have a human pick them up and handle them. Yet several studies found that rats that had been briefly handled as pups were *less* susceptible to adult stress than were rats that had been left alone as pups (S. Levine et al., 1967). For example, the previously handled rats secreted lower adrenal steroid amounts in response to a wide variety of adult stressors. Researchers dubbed this effect **stress immunisation** because a little stress early in life seemed to make the animals more resilient to later stress.

Follow-up research showed there was more to the story. The pups did not benefit because they were stressed; they benefitted because their mothers *comforted* them *after* the stress. When pups are returned to their mother after a separation, she spends considerable time licking and grooming them. And she will lick the pups much longer if they were handled by humans during the separation. Michael Meaney and colleagues suggest that this gentle tactile stimulation from Mom is crucial for the stress immunisation effect. They found that, even among undisturbed litters, the offspring of mother rats that exhibited more licking and grooming behaviour were more resilient in their responses to adult stress than other rats were (D. Liu et al., 1997). So the "immunising" benefit of early stressful experience happens only if the pups are promptly comforted after each stressful event.

If the pups are deprived of their mother for long periods, receiving very little of her attention, then as adults they exhibit a greater stress response, have difficulty learning mazes, and show reduced neurogenesis in the hippocampus (Mirescu et al., 2004). Maternal deprivation exerts this negative effect on adult stress responses by causing long-lasting changes in the expression of adrenal steroid receptors in the brain. This change is termed **epigenetic regulation** because it represents a change

stress immunisation The concept that mild stress early in life makes an individual better able to handle stress later in life.

epigenetic regulation Process affecting the expression of a particular gene or genes without affecting the sequence of nucleotides making up the gene itself.

in the *expression* of the gene, rather than a change in the encoding region of the gene (see Figure 7.21).

Dramatic evidence for the same phenomenon has been seen in humans. For example, examination of the brains of people who commit suicide has revealed the same epigenetic change in expression of the adrenal steroid receptor, but only in those who had a history of being abused or neglected as children (McGowan et al., 2009). The implication is that the early abuse epigenetically modified expression of the gene, making the person less able to handle stress and thus more likely to develop significant psychiatric disturbances—especially mood and anxiety disorders—that heighten the risk of suicide (Nock et al., 2010). People who commit suicide who had no history of early neglect did not show the epigenetic change, so their suicidality may have been the result of mental health issues of other origin.

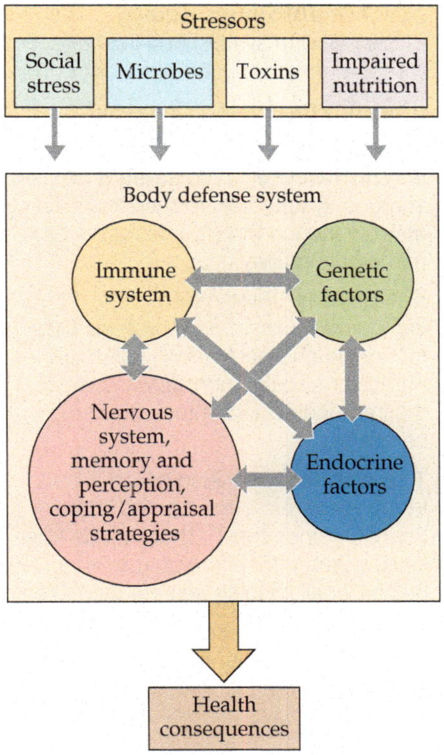

11.26 Factors That Interact during the Development and Progression of Disease

Stress and emotions affect our health

During the past 50 years, researchers and clinicians have begun to understand some of the ways in which psychological, behavioural, and social factors play a role in health and disease, forming a field known as **psychosomatic medicine**. The related field called **health psychology** (or *behavioural medicine*) emphasizes the role of psychological factors in the cause, progression, and consequences of health and illness (Ogden, 2012). **FIGURE 11.26** shows how several factors interact to affect human health and disease. Some consistent correlations between stressful events and illness have been found (N. Adler and Matthews, 1994). For example, men who reported frequent and severe stress during a period of 1–5 years prior to interviews were more likely to experience heart disease during the next 12 years than were those who reported little stress (Rosengren et al., 1991).

People have long understood that there's a link between strong emotions and heart attacks. An important development in understanding this relationship was the identification of two general behaviour patterns and their correlation to the development and maintenance of heart disease (M. Friedman and Rosenman, 1974). Type A behaviour is characterised by excessive competitive drive, impatience, hostility, and accelerated speech and movements, while type B behaviour patterns are more relaxed. Initial work indicated that type A people were more likely to develop heart disease than type B individuals. Most of these associations between type A/B personalities were later accounted for by a single factor, *hostility*: angry, mistrustful, antagonistic behaviour. So studies asking why people who display hostility are more likely to suffer heart disease (Almada et al., 1991) have largely replaced investigations of the type A/B concept (Matthews, 2005). Other risk factors for dying of heart disease include poor sleep, depression, stressful jobs and/or primary relationships (Rozanski, 2014), and social isolation (Holt-Lunstad et al., 2010). Another psychological factor that appears to affect heart disease is a sense of life purpose (Sone et al., 2008). Among people diagnosed with heart disease, those who had a low sense of purpose in life were more likely to die within a few years than people with a strong sense of purpose (**FIGURE 11.27**).

psychosomatic medicine A field of study that emphasizes the role of psychological factors in disease.

health psychology Also called *behavioural medicine*. A field that studies psychological influences on health-related processes, such as why people become ill or how they remain healthy.

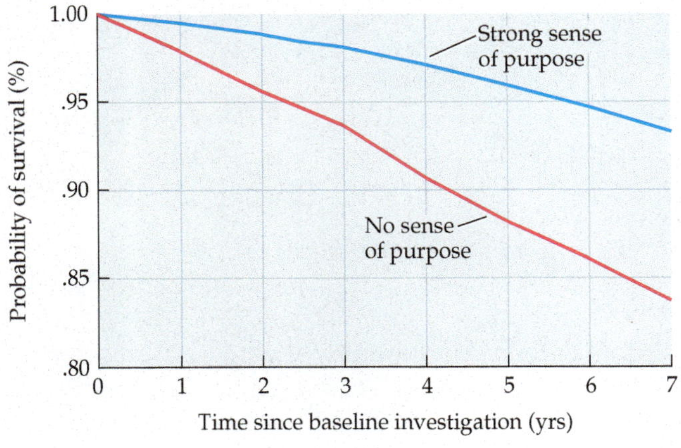

11.27 A Sense of Purpose People who felt they had little or no sense of purpose in their lives were more likely to die within a few years after diagnosis of heart disease than people with a strong sense of life purpose. (From T. Sone et al., 2008. *Psychosom Med* 70: 709–711.)

psychoneuroimmunology The study of the immune system and its interaction with the nervous system and behaviour.

phagocyte An immune system cell that engulfs invading molecules or microbes.

B lymphocyte Also called *B cell*. An immune system cell, formed in the bone marrow (hence the *B*), that mediates humoral immunity.

antibodies Also called *immunoglobulins*. Large proteins that recognise and permanently bind to particular shapes, normally as part of the immune system attack on foreign particles.

T lymphocytes Also called *T cells*. Immune system cells, formed in the thymus (hence the *T*), that include killer T cells, which attack foreign microbes, and helper T cells, which secrete cytokines.

cytokines Proteins that induce the proliferation of other cells, as in the immune system.

In the 1980s a new field, **psychoneuroimmunology**, emphasised that the immune system—with its collection of cells that recognise and attack intruders—interacts with other systems, especially hormone systems and the nervous system (Ader, 2001). Studies of both human and nonhuman subjects now clearly show psychological and neurological influences on the immune system. For example, people with happy social lives are less likely to develop a cold when exposed to the virus (S. Cohen et al., 2006). In fact, social support seems to buffer against social stress. People exposed to a cold virus have more severe symptoms if they are experiencing conflict with others. But individuals who feel they have more social support, and who *receive more hugs from others*, are protected from that effect of conflict (S. Cohen et al., 2015). Likewise, people who tend to feel positive emotions will also produce more antibodies in response to a flu vaccination (Rosenkranz et al., 2003), which should help them fight off sickness. These interactions go in both directions: the brain influences responses of the immune system, and immune cells and their products affect brain activities.

THE IMMUNE SYSTEM To understand this intriguing story, we need to consider some of the main features of the immune system. In your blood are different classes of white blood cells (leukocytes). The **phagocytes** ("eating" cells) are specialised to engulf and destroy invading germs (**FIGURE 11.28**). But phagocytes rely on other white blood cells (the lymphocytes) to tell them what to attack. **B lymphocytes** (or *B cells*, because they form in the ***b***one marrow), produce proteins called **antibodies** (or *immunoglobulins*). Antibodies latch onto foreign molecules such as viruses or bacteria and summon phagocytes and circulating proteins to destroy the invaders. **T lymphocytes** (*T cells*), so called because they form in the ***t***hymus gland, include some that act as *killer T cells*, forming a strong part of the body's attack against foreign substances. Other T lymphocytes called *helper T cells* secrete **cytokines**, cell-signalling proteins that regulate the activity of B lymphocytes and phagocytes. These immune system cells form in the thymus gland, bone marrow, spleen, and lymph nodes (**FIGURE 11.29**), which release the cells into the bloodstream.

COMMUNICATION AMONG THE NERVOUS, IMMUNE, AND ENDOCRINE SYSTEMS The brain can affect the immune system through autonomic nerve fibres that innervate immune system organs such as the spleen and thymus gland. These fibres are usually noradrenergic, sympathetic postganglionic axons that affect antibody production and immune cell proliferation (Bellinger et al., 1992). This communication may explain why simply seeing photographs of people displaying infectious disease symptoms, such as a runny nose and cough, can quickly trigger an increased release of cytokines (Schaller et al., 2010). Likewise, people who are lonely show an exaggerated release of cytokines in response to stress (Jaremka et al., 2013).

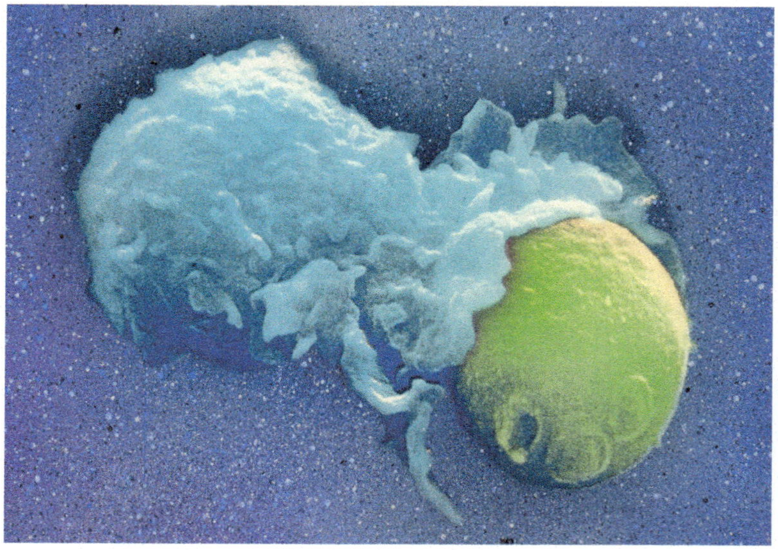

11.28 A Phagocyte at Work Phagocytes engulf and consume invaders, such as this yeast cell (right).

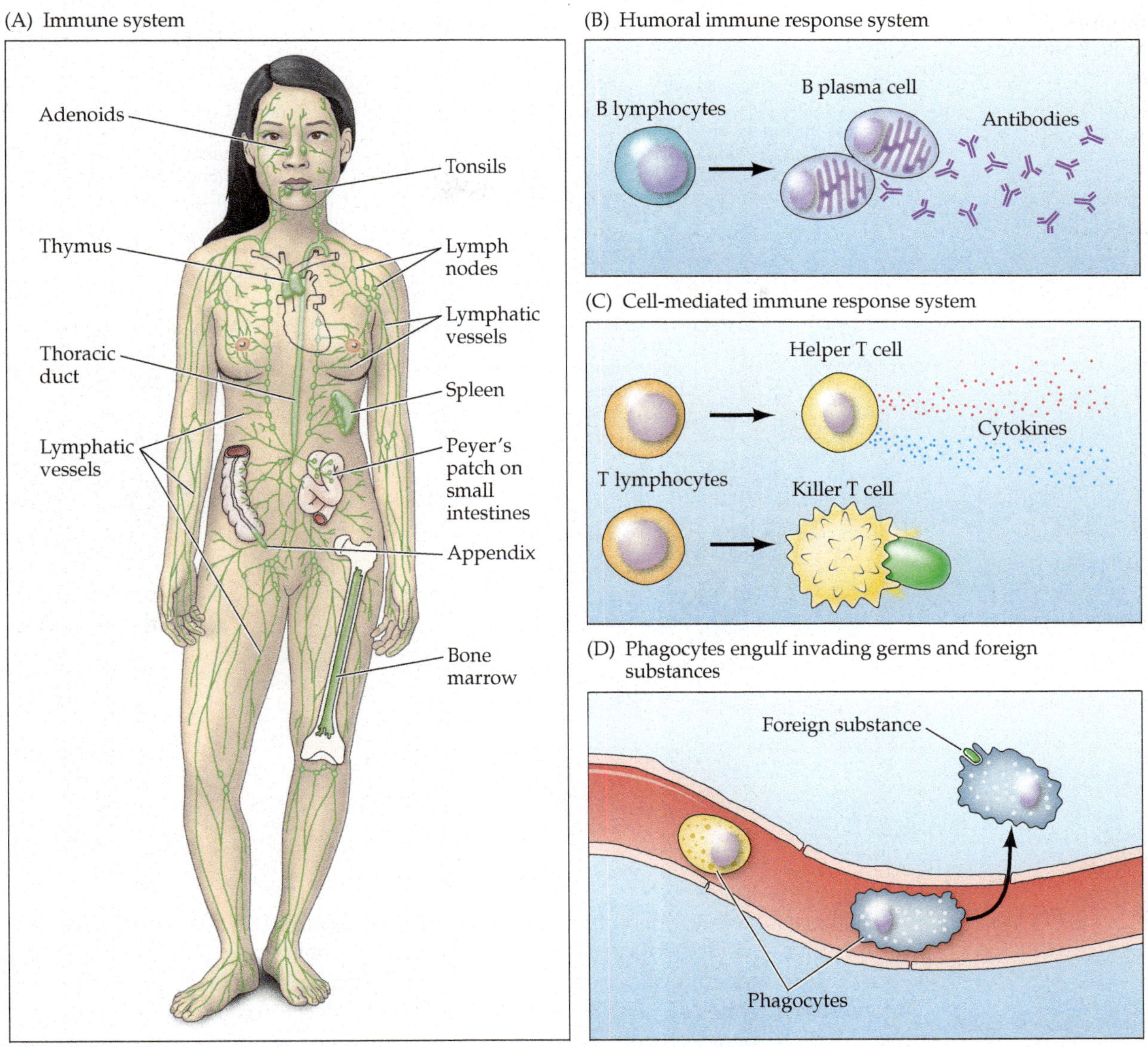

11.29 Main Components of the Human Immune System (A) The various components of the immune system protect us by means of three classes of white blood cells: B lymphocytes (B) produce antibodies to attack invading microbes. T lymphocytes (C) form helper cells that release cytokines to regulate B cells to divide or die. T cells also form killer cells that, together with phagocytes ("eating" cells) (D), directly attack foreign tissues or microbes.

The brain also carefully monitors immune reactions to make sure they are not too extreme and ultimately harmful to the body. For example, peripheral axons of the vagus nerve have receptors to detect high levels of cytokines and relay the information to the brain. Then, brainstem neurons with axons that lead back out the vagus nerve release acetylcholine, which inhibits cytokine release from immune organs (H. Wang et al., 2003). Hypothalamic neurons and neurons located in the walls of cerebral ventricles also monitor cytokines in circulation (Dantzer et al., 2008; Samad et al., 2001). Thus, the brain is directly informed about the actions of the immune system, which serves as an early warning system to alert the brain when microbes invade the body.

There is an interesting theory about why our brains monitor the immune system so closely. Although that achy, lethargic feeling that we have with the flu is unpleasant, it is also adaptive because it forces us to rest and keep out of trouble until we recover (Hart, 1988). Perhaps high levels of cytokines are what cause the brain to enforce that sick feeling. This suggestion has given rise to the idea that some people's depression may be due to a broad set of physiological changes in the brain, including

11.30 Examples of Reciprocal Relations of the Nervous, Endocrine, and Immune Systems

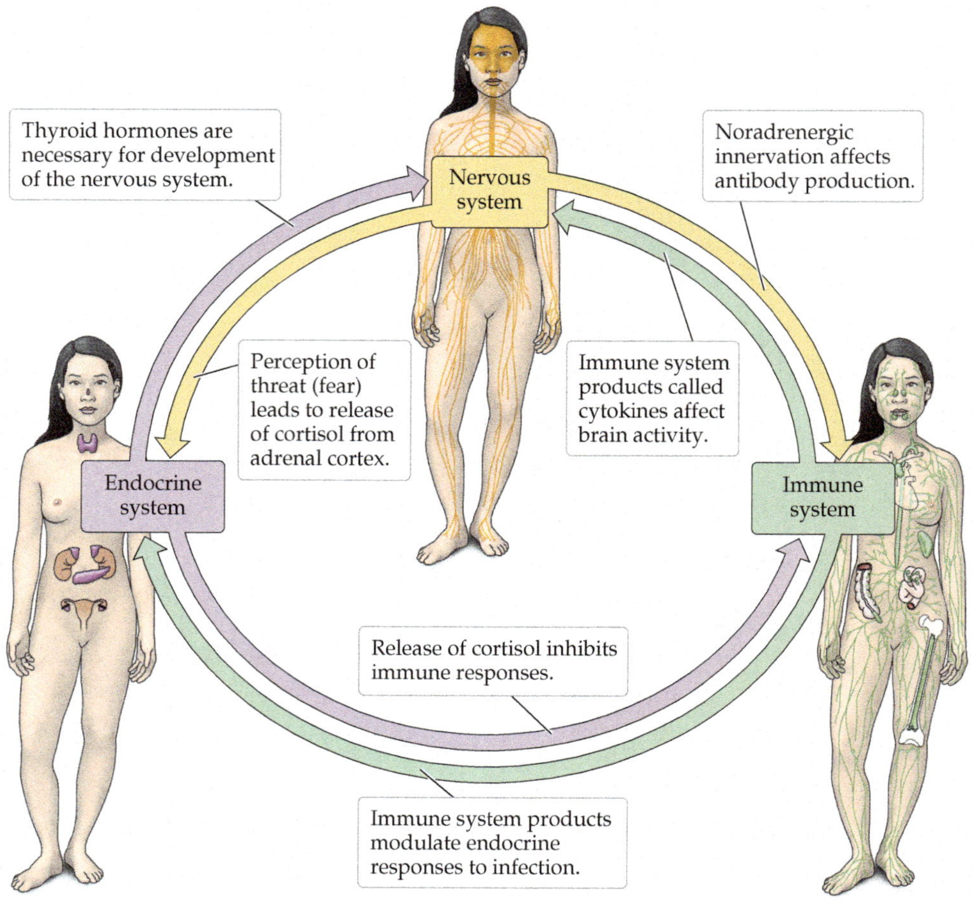

large alterations in availability of several neurotransmitters, that are brought about by excessive quantities of cytokines in circulation and penetrating the brain (Hodes et al., 2015). Indeed, one action of antidepressant drugs is to reduce cytokine production (Dantzer et al., 2008). Immune factors are able to access the brain via the glymphatic system—the perivascular astrocyte-mediated lymphatic system that we described in Figure 1.23—with effects ranging from supporting damaged tissue to roles in learning and social behaviours (Da Mesquita, Fu et al., 2018; Kipnis, 2018).

The immune system and nervous system also interact extensively with the endocrine system (**FIGURE 11.30**). All three systems interact reciprocally, so there is a constant state of flux, carefully tuning the immune system so that it vigorously attacks foreign cells but leaves the body's own cells alone.

PSYCHOLOGICAL STRESS AND IMMUNITY The anatomical and physiological systems described in the previous sections give us some hints about ways in which psychological factors can alter immune system responses. For example, several lines of evidence indicate that the immune system is compromised during depression (M. Stein et al., 1991)—a situation that, if sustained, could increase susceptibility to infectious diseases, cancer, and autoimmune disorders. Altered immune function is also observed in people who are grieving the death of a relative, especially a spouse (M. Stein and Miller, 1993).

Stressful exam periods usually produce a decline in the number of immune cells and in levels of cytokines (Glaser et al., 1986). Most important, some studies have noted that the student's *perception* of the stress of the academic program is a predictor of the level of circulating antibodies: those who perceive the program as stressful show the lowest levels. One experiment considered the effects of university examinations on wound healing in dental students (Marucha et al., 1998). Two small wounds were placed on the roof of the mouth of 11 dental students (sounds like revenge, doesn't it?). The first wound was timed during summer vacation; the second was inflicted 3 days before the first major examination of the term. Two independent daily measures showed that no student healed as rapidly during the exam period, when healing took 40% longer. A measure of immunological response declined 68% during the exam period. The experimenters concluded that even something as transient, predictable, and relatively benign (do students agree with this description?) as examination stress can significantly impede wound healing.

Another connection between the nervous and immune systems was described in Chapter 15, where we learned that sleep deprivation impairs the responsiveness of the immune system (see Box 15.1).

Why does chronic stress suppress the immune system?

Although brief stress doesn't impair immune function, and may even enhance it (Dhabhar, 2018), longer-lasting stress has a pronounced suppressive effect on the immune system. We discussed earlier how, in response to stressful conditions, the brain causes adrenal steroid hormones such as cortisol to be released from the adrenal cortex. These adrenal steroids directly suppress the immune system. But doesn't it seem like a bad idea to suppress immunity just when you are more likely to sustain an injury, and maybe an infection? Modern evolutionary theory offers some possible explanations for this seemingly maladaptive situation (for a very readable account, see Sapolsky, 2004).

To the extent that stress might be a sudden emergency, the *temporary* suppression of immune responses makes some sense because the stress response demands a rapid Mobilisation of energy. Slow and long-lasting immune responses consume energy that otherwise could be used for dealing with the emergency at hand. A zebra wounded by a lion must first escape and hide, and only then does infection of the wound pose a threat. So the stress of the encounter first suppresses the immune system, conserving resources until a safe haven is found. Later the animal can afford to mobilize the immune system to heal the wound. The adrenal steroids also suppress the swelling (inflammation) of injuries, especially of joints, to help the animal remain mobile long enough to find refuge. It is precisely this action that makes glucocorticoids like prednisone such useful medicines for treating inflammation.

In the wild, animals are under stress for only a short while; any animal stressed for a *prolonged* period dies. So natural selection favoured stress reactions as a drastic effort to deal with a short-term problem. What makes humans "special" is that, with our highly social lives and keen analytical minds, we are capable of experiencing stress for prolonged periods—months or even years. The bodily reactions to stress, which evolved to deal with short-term problems, become a handicap when extended too long (Sapolsky, 2004). Chronic stress leads to degeneration in the hippocampus (Egeland et al., 2015) and prefrontal cortex, while the amygdala may expand (Radley and Sawchenko, 2011). **TABLE 11.1** lists a variety of stress responses that are beneficial in the short term but detrimental in the long term.

Go to Activity 11.2
The Stress Response and Consequences of Prolonged Stress

TABLE 11.1 The Stress Response and Consequences of Prolonged Stress

Principal components of the stress response	Common pathological consequences of prolonged stress
Mobilisation of energy at the cost of energy storage	Fatigue, muscle wasting, steroid diabetes
Increased cardiovascular and cardiopulmonary tone	Hypertension (high blood pressure)
Suppression of digestion	Ulcers
Suppression of growth	Psychogenic dwarfism, bone decalcification
Suppression of reproduction	Suppression of ovulation, impotence, loss of libido
Suppression of immunity and of inflammatory response	Impaired disease resistance
Analgesia	Apathy
Neural responses, including altered cognition and sensory thresholds	Neural degeneration in hippocampus and prefrontal cortex

Source: R. M. Sapolsky, 1992. In J. B. Becker et al. (Eds.), *Behavioural endocrinology* (pp. 287–324). Cambridge, MA: MIT Press.

mindfulness-based stress reduction (MBSR) A therapy to reduce stress that pairs relaxation with efforts to focus attention on the present moment, rather than past or future problems.

What we have described so far is a really depressing picture. If you are stressed for long periods of time, your health suffers, which brings another source of stress to your life. But don't give up hope. Even if there are some sources of stress you cannot avoid altogether, there are things you can do to reduce the impact of stress. *Relaxation training* involves focussing your attention on something calming while becoming more aware of your body, trying to relax muscles as much as you can (McGuigan and Lehrer, 2007). A program of therapy to deal with stress, partially inspired by various practices of meditation, is **mindfulness-based stress reduction** (**MBSR**). MBSR pairs relaxation with efforts to focus attention on the present moment, including current sensations, thoughts, and bodily states, in an open, nonjudgmental way. MBSR is focused on results and does not require practitioners to adopt any particular religious or spiritual views. It has been shown to reduce activity in the amygdala (Goldin and Gross, 2010) and prevent relapses of anxiety disorders or depression (S. G. Hofmann et al., 2010). You can learn more about MBSR at umassmed.edu/cfm.

Recommended Reading

Davidson, R. J., and Begley, S. (2012). *The Emotional Life of Your Brain.* New York: Hudson Street Press.

Davis, K. L., and Panksepp, J. (2018). *The Emotional Foundations of Personality: A Neurobiological and Evolutionary Approach.* New York: W. W. Norton.

Ekman, P. (2007). *Emotions Revealed: Recognising Faces and Feelings to Improve Communication and Emotional Life.* New York: Owl Books.

Fields, D. R. (2016). *Why We Snap: Understanding the Rage Circuits in Your Brain.* New York: Dutton.

Gilbert, D. (2007). *Stumbling on Happiness.* New York: Vintage.

Hodgins, S., Viding, E., and Plodowski, A. (2009). *The Neurobiological Basis of Violence: Science and Rehabilitation.* New York: Oxford University Press.

LeDoux, J. (2015). *Anxious: Using the Brain to Understand and Treat Fear and Anxiety.* New York: Viking.

Oatley, K., Keltner, D., and Jenkins, J. M. (2013). *Understanding Emotions* (2nd ed.). New York: Wiley.

Raine, A. (2013). *The Anatomy of Violence: The Biological Roots of Crime.* New York: Pantheon.

Sapolsky, R. (2004). *Why Zebras Don't Get Ulcers* (3rd ed.). New York: Holt.

11 Visual Summary

You should be able to relate each summary to the adjacent illustration, including structures and processes. You can also refer back to the links to figures, animations, and activities that will help you consolidate the material.

1 **Emotions** are feelings, actions, physiological arousal, and motivational programs. The **James-Lange theory** considered emotions to be the perceptions of stimulus-induced autonomic activity, while the **Cannon-Bard theory** emphasised *simultaneous* emotional experience and bodily response. Review **Figure 11.1**

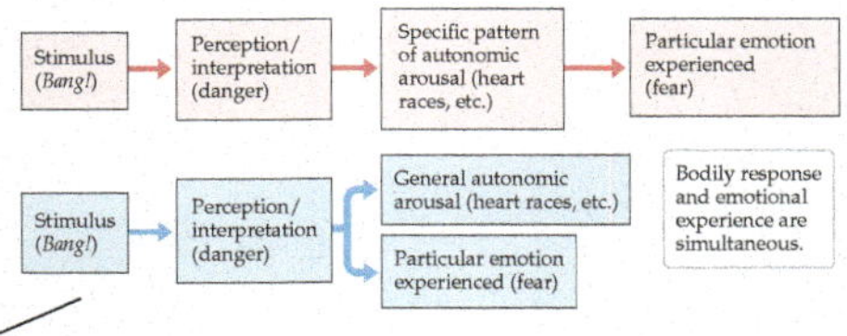

2 According to Schachter's **cognitive attribution model**, autonomic activity may intensify the emotion we feel, but not *which* emotion we experience. We attribute sympathetic arousal to specific emotions by analysing the physical and social context, including the emotions of those around us. Emotions evolved to organise adaptive responses to varying situations. Individuals differ, even as babies, in how much they react to unpleasant stimuli. Review **Figures 11.2, 11.3**

3 Distinctive facial expressions represent anger, sadness, happiness, fear, disgust, surprise, contempt, and embarrassment, which are interpreted similarly across many cultures. **Polygraphs** actually measure activation of the sympathetic nervous system and therefore reflect stress, not lying. Review **Figures 11.6–11.9, Box 11.1**

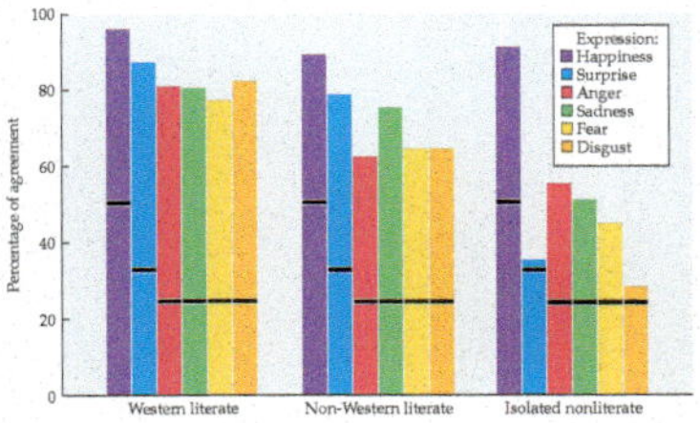

4 Facial expressions are controlled by distinct sets of facial muscles controlled by the **facial** and **trigeminal nerves**. Review **Figures 11.10, 11.11**

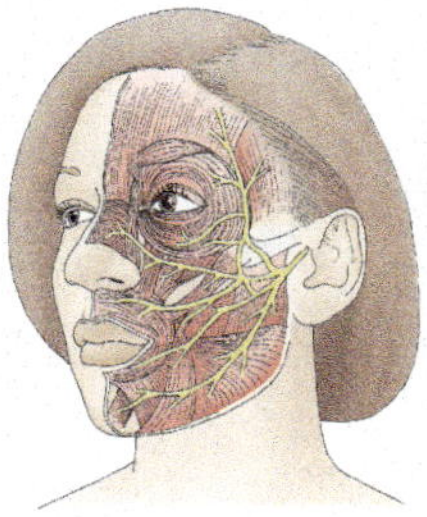

5 Lesions revealed an interconnected brain circuit, the **limbic system** (which includes the amygdala) that mediates and controls emotions. Electrical **self-stimulation** of some brain regions is rewarding. Review **Figures 11.12, 11.13**

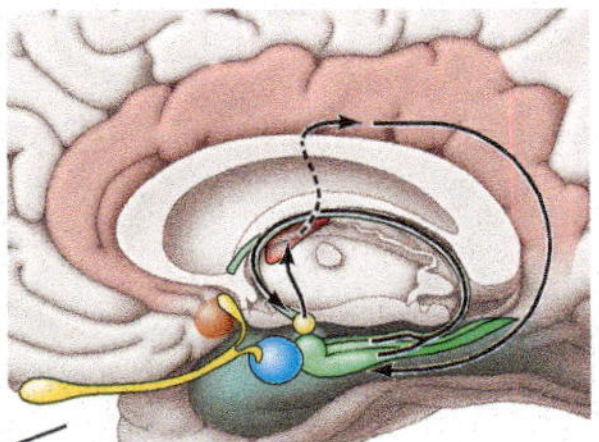

6 Fear is mediated by circuitry involving the **amygdala**, which receives information both through a rapid direct route and via cortical sensory regions, allowing for both immediate responses and cognitive processing. The prefrontal cortex is important for the extinction of fear responses to a stimulus. Review **Figures 11.14–11.19**, **Activity 11.1**

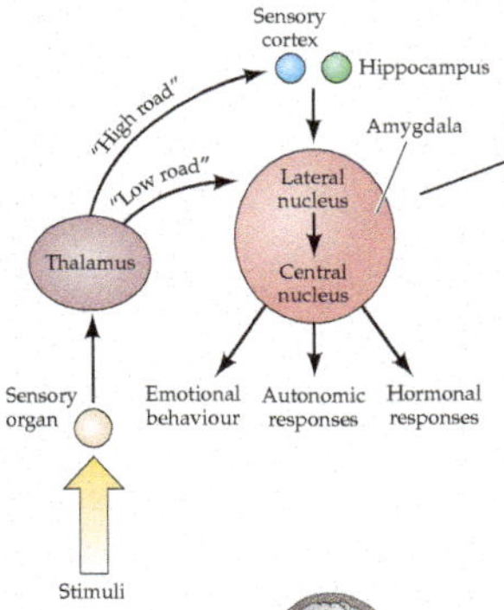

7 Aggression may be proactive or reactive. **Intermale aggression** is increased by androgens such as testosterone. Stimulation of some limbic system regions elicits a species-typical pattern of aggression. Serotonin levels are negatively correlated with aggression. Review **Figures 11.20, 11.22**

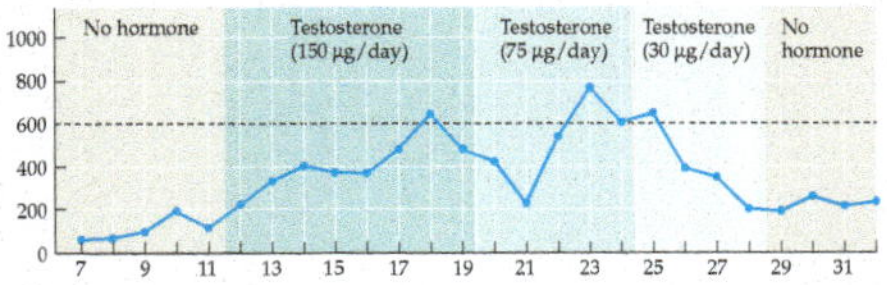

8 **Stress** elevates levels of some hormones—**cortisol** from the **adrenal cortex** and **epinephrine** and **norepinephrine** from the **adrenal medulla**—while suppressing other hormones (e.g., testosterone). Review **Figures 11.23–11.25**

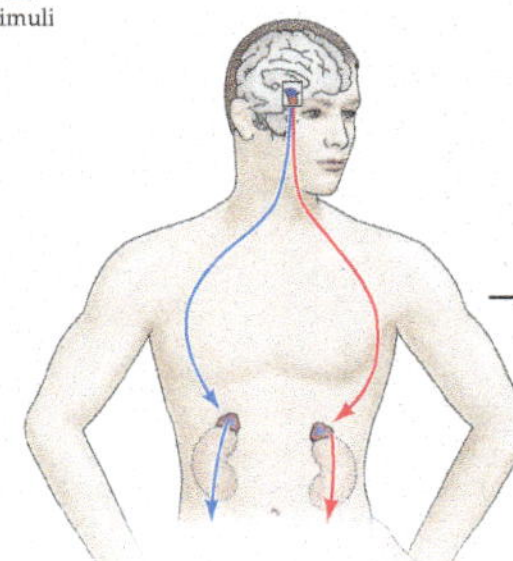

9 Physiological responses to stress are adaptive in the short run. However, in socially complex species that can experience stress for long periods, including humans, these hormonal responses decrease immune system competence, damaging our health. Review **Figure 11.26**, **Table 11.1**, **Activity 11.2**

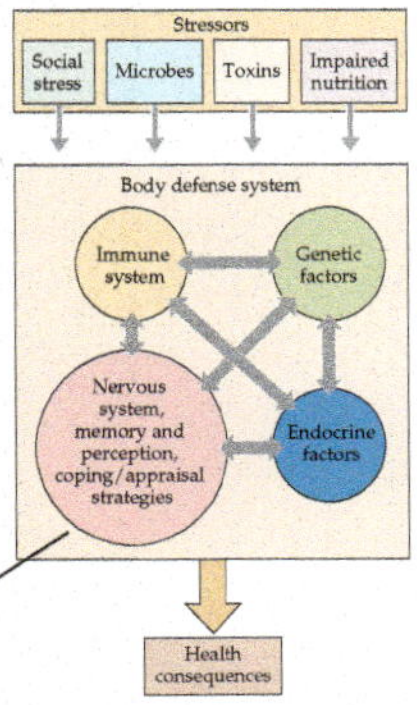

10 The nervous, endocrine, and immune systems interact reciprocally to monitor and maintain health. Emotional traits such as hostility and depression can damage our health by, for example, increasing the risk of heart attack. Review **Figures 11.27–11.30**

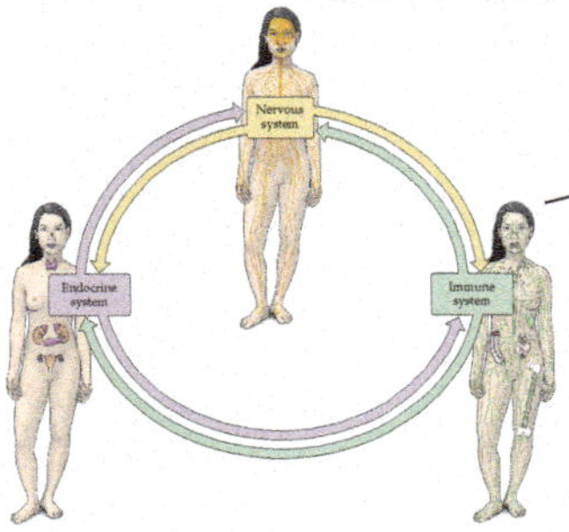

© Eye of Science/Science Source

12

Psychopathology

Biological Basis of Behavioural Disorders

"The Voice"

Eleanor was off to a great start in college, eager to partake of a life of lectures and parties. The first crack in her optimism came when she was packing up to leave a classroom and heard a voice calmly say, "She is leaving the building." No one was in the room with her, yet the voice was so distinct. Spooked, Eleanor hurried home only to hear, as she arrived, the same voice say, "She is opening the door." For weeks, this disembodied voice dispassionately narrated everything Eleanor did, in the third person. When she finally told a friend about "the Voice," Eleanor was met with suspicion and fear. She went to a physician, getting his full attention when she started talking about the Voice. Thus began Eleanor's experience of schizophrenia, a disorder that afflicts about 1% of individuals in every human society.

Eleanor came to hate the Voice that brought her so much attention and suspicion. It seemed as if her hostility to the Voice was soon reciprocated. She began hearing not just one, but several voices, and they became progressively more menacing and demanding. They began goading Eleanor to injure herself, so her parents started hiding the cutlery. When a voice told Eleanor to pick up the glass of water on her instructor's desk and throw it in his face, he was not amused. Eleanor became a target, someone marked as different and disturbed, suffering ostracism and verbal taunts from her peers. The life she had hoped for seemed lost forever. One psychiatrist said, "Eleanor, you'd be better off with cancer because cancer is easier to cure than schizophrenia" (Longden, 2013, p. 37).

Was Eleanor's life really over?

Debilitating mental problems have plagued humankind throughout history, afflicting their victims with disordered thought and emotional chaos. Psychopathology affects hundreds of millions of people worldwide, not just an exotic few.

The seeds for a biological perspective in psychiatry were sown at the start of the twentieth century, when thousands of people in mental hospitals had a disorder called *paralytic dementia*. It was characterised by the sudden onset of **delusions** (false beliefs strongly held in spite of contrary evidence), grandiosity (boastful self-importance), euphoria, poor judgment, and impulsive and capricious behaviour. One sign of the disease was the "Argyll-Robertson pupil": the pupil in the eye did not constrict in response to light (as pupils normally do) but could still constrict when the person tried to look at something close up (Argyll-Robertson, 1869).

Recognised for centuries, this disorder was often attributed to "weak character." In 1911, however, microbiologist Hideyo Noguchi (1876–1928) discovered that the brains of people with this disorder had been extensively damaged by syphilis, a sexually transmitted bacterial infection. The mental disorder got a new name, *syphilitic psychosis*, and the subsequent discovery of antibiotic drugs soon made syphilitic psychosis a thing of the past, prompting researchers to consider whether other psychiatric disorders result from biological processes.

TABLE 12.1 Standardized 12-Month and Lifetime Prevalence of Disorders in the United States

Disorders	Rate (%)[a] Previous 12 months	Lifetime
ANY PSYCHIATRIC DISORDER	26.2	46.4
ANY SUBSTANCE USE DISORDER	3.8	14.6
Alcohol abuse	3.1	13.2
Drug abuse	1.4	7.9
ANY MOOD DISORDER	9.5	20.8
Major depressive episode	6.7	12.6
Minor depression (dysthymia)	1.5	2.5
Manic episode	2.6	3.9
ANY ANXIETY DISORDER	18.1	28.8
Phobia	15.5	24.6
Panic	2.7	4.7
OBSESSIVE-COMPULSIVE DISORDER (OCD)	1.0	1.6

[a]The rates are from surveys of the noninstitutionalised adult population of the United States.
Source: Data from R. C. Kessler et al., 2005. *Arch Gen Psychiatry* 62: 593–602; R. C. Kessler et al., 2005. *Arch Gen Psychiatry* 62: 617–627.

delusion A false belief strongly held in spite of contrary evidence.

epidemiology The scientific study of patterns of disease in a population.

Because they are mostly diagnosed on the basis of behavioural symptoms rather than physical laboratory tests, the precise classification of psychiatric disorders is subject to periodic revision, but we know from **epidemiology** (the scientific study of the incidence and distribution of diseases) that psychiatric disorders are startlingly prevalent in modern society. More than one-third of the U.S. population *at some point in life* reports symptoms that match the defining features of a major psychiatric disorder (**TABLE 12.1**). At any one time, men and women show comparable prevalence of mental disorders, although depression is somewhat more prevalent in females, and drug dependency and alcoholism are more prevalent in males. Certain psychiatric disorders—for example, schizophrenia—tend to appear in adolescence and young adulthood. Peaks for depression and antisocial personality disorders appear in 25- to 44-year-olds (**FIGURE 12.1**), whereas cognitive impairment occurs especially in people older than 65. Clearly, mental disorders exact an enormous toll on us throughout our lives.

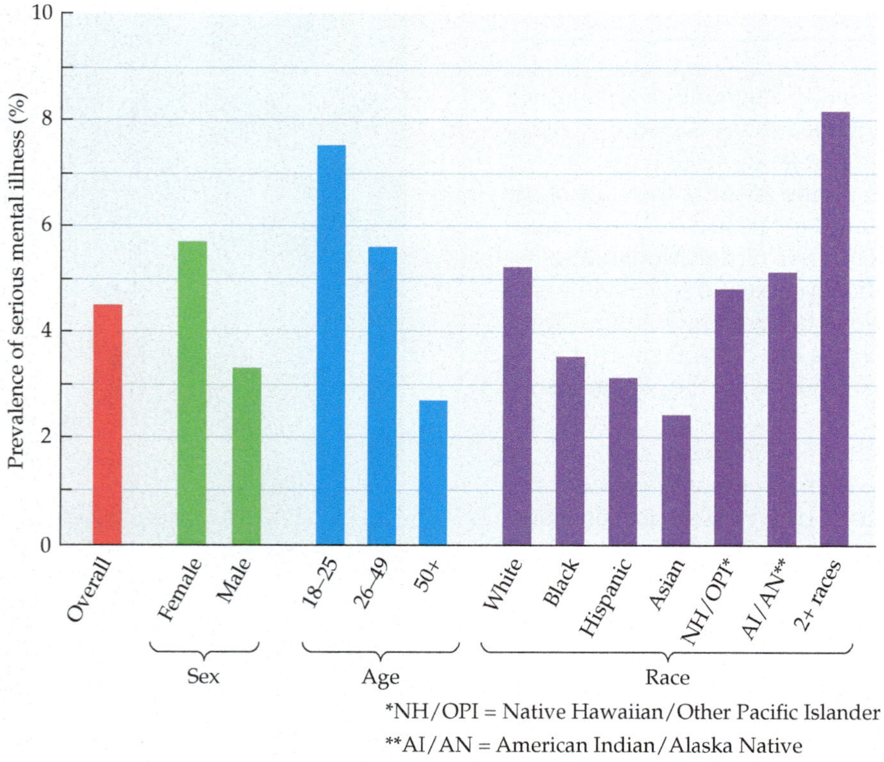

12.1 Prevalence of Serious Mental Illness among U.S. Adults Prevalence is the percentage of people suffering from a disorder at any one point in time. The greater prevalence of mental illness in females is mostly due to their greater susceptibility to depression. Males, on the other hand, are more susceptible to drug addiction. Episodes of schizophrenia are most common in young adults. Note that mental illness occurs in every racial group. (After National Institute of Mental Health, 2018. Past year prevalence of serious mental illness among U.S. adults (2017). Bethesda, MD. Data courtesy of SAMHSA. Last updated February 2019. www.nimh.nih.gov/health/statistics/prevalence/serious-mental-illness-smi-among-us-adults.shtml.)

12.1 Schizophrenia Is the Major Neurobiological Challenge in Psychiatry

Learning Objectives

After reading this section, you should be able to:

12.1.1 **Identify the most common symptoms of schizophrenia.**

12.1.2 **Discuss the strong influence of both genes and the environment on the chances of developing schizophrenia.**

12.1.3 **Describe several of the major brain differences of people with schizophrenia versus controls.**

12.1.4 **Review the possible neurochemical bases of schizophrenia, and discuss several classes of antipsychotic drugs and their mechanisms of action.**

Medical records tell us that around the world and across the centuries, certain people's lives have been shattered by bizarre and debilitating symptoms—hearing voices that aren't there, feeling intense fear, paranoia, and acting strangely (Heinrichs, 2003). For some people with schizophrenia, this state lasts a lifetime, but some 30–40% will recover (Harrow et al., 2012). Schizophrenia is also a "public" disorder because many people who have it become homeless on our streets. Epidemiological surveys of schizophrenia reveal a prevalence of about 1% of the population—about 3 million people in the United States alone. This disorder consumes a disproportionate share of community health resources because of its chronic and overwhelming character.

Schizophrenia is characterised by an unusual array of symptoms

Eugen Bleuler (1857–1939) introduced the term *schizophrenia* (from the Greek *schizein*, "to split," and *phren*, "mind") in 1911 (Bleuler, 1950 translation). Despite the name he chose, Bleuler was not thinking of a "split personality." Rather, he identified the key symptom as **dissociative thinking**, a major impairment in the logical structure of thought, as well as emotional disturbance, delusions, and hallucinations. German psychiatrist Emil Kraepelin (1856–1926) described numerous clinical features common to the varied forms of schizophrenia: paranoia, grandiose delusions, abnormal emotional regulation, bizarre disturbances of thought, and auditory hallucinations (Kraepelin, 1919). Although movies often portray people with schizophrenia experiencing elaborate visual hallucinations (**FIGURE 12.2**), in fact the hallucinations are almost always purely auditory (Javitt and Sweet, 2015), like the Voice heard by Eleanor, whom we met at the start of the chapter.

Today, the major symptom categories associated with **schizophrenia** include (1) auditory hallucinations, (2) highly personalised delusions (false beliefs), (3) changes in affect (emotion), and (4) cognitive impairments. An additional major division of schizophrenic symptoms distinguishes positive and negative symptoms (Crow, 1980; Loch, 2019). The term **positive symptoms** refers to behavioural states that have been *gained*; examples include hallucinations, delusions, and excited motor behaviour. The term **negative symptoms** refers to behavioural functions that have been *lost*—for example, slow and impoverished thought and speech, emotional and social withdrawal (M. F. Green et al., 2015). Since people with schizophrenia report that they feel very strong emotions, the blunted emotions in schizophrenia may be limited to emotional *expression*—facial and body signals (Kring and Caponigro, 2010). The fact that positive and negative symptoms respond differently to drug treatments suggests that they arise from different neural disruptions. Likewise, there are probably several different kinds of schizophrenia, which vary in the relative degree of paranoia, blunted affect, or cognitive impairment (**TABLE 12.2** on the next page).

dissociative thinking A condition, seen in schizophrenia, that is characterised by disturbances of thought and difficulty in relating events properly.

schizophrenia A severe psychopathology characterised by negative symptoms such as emotional withdrawal and impoverished thought and also by positive symptoms such as hallucinations and delusions.

positive symptom In psychiatry, a behaviour that is gained in a disorder. Examples include hallucinations, delusions, and excited motor behaviour in schizophrenia.

negative symptom In psychiatry, a symptom that reflects insufficient functioning. Examples include emotional and social withdrawal, blunted affect, and slowness and impoverishment of thought and speech.

12.2 Not So Beautiful Voices Mathematician John Nash's struggle with schizophrenia is depicted in the Academy Award–winning movie *A Beautiful Mind.* The movie portrays him as having elaborate visual hallucinations, but in fact his hallucinations were exclusively auditory, and he suffered from delusions that fueled his paranoid thinking. Auditory hallucinations are common in schizophrenia; visual hallucinations are quite rare.

TABLE 12.2 Symptoms of Schizophrenia

Symptom dimension	Symptom category
POSITIVE SYMPTOMS Refers to symptoms that are present but should not be	PSYCHOSIS Hallucinations Delusions Disorganised thought and speech Bizarre behaviours
NEGATIVE SYMPTOMS Refers to characteristics of the individual that are absent but should be present	EMOTIONAL DYSREGULATION Lack of emotional expression Reduced facial expression (flat affect) Inability to experience pleasure in everyday activities (anhedonia)
	IMPAIRED MOTIVATION Reduced conversation (alogia) Diminished ability to begin or sustain activities Social withdrawal
COGNITIVE SYMPTOMS Refers to problems with processing and acting on external information	NEUROCOGNITIVE IMPAIRMENT Memory problems Poor attention span Difficulty making plans Reduced decision-making capacity Poor social cognition Abnormal movement patterns

Schizophrenia has a heritable component

A variety of approaches make it clear that there is a strong genetic component to schizophrenia, but later we'll see that there's also strong evidence that stress increases the likelihood of the disorder. Thus you should consider the studies we'll take up next as evidence of a genetic *susceptibility* to stress that can lead to schizophrenia.

FAMILY STUDIES If schizophrenia is inherited, relatives of people with schizophrenia should show a higher incidence of the disorder than is found in the general population. In addition, the risk of schizophrenia among relatives should increase with the closeness of the relationship, because closer relatives share a greater number of genes. Indeed, parents and siblings of people with schizophrenia have a higher risk of developing schizophrenia than do individuals in the general population (**FIGURE 12.3**). However, the mode of inheritance of schizophrenia is not simple; that is, it does not involve a single recessive or dominant gene (Hyman, 2018). Rather, many genes play a role in the emergence of schizophrenia.

ADOPTION STUDIES It is easy to find fault with family studies. They confuse hereditary and environmental factors because members of a family share both. But what about children who are not raised with their biological parents? In fact, studies of adopted persons confirm the significance of genetic factors in schizophrenia. The biological parents of adoptees with schizophrenia are far more likely than the adopting parents to have also had this disorder (Kety et al., 1994).

TWIN STUDIES In twins, nature provides researchers with what seem to be the perfect conditions for a genetic experiment. Human twins from the same fertilised egg—called **monozygotic** (*identical*) twins—share a nearly identical set of genes (Bruder et al., 2008). Twins from two different eggs—**dizygotic** (*fraternal*) twins—have only half of their genes in common, like other full siblings. When both individuals of a twin pair have schizophrenia, they are described as being **concordant** for this trait. If only one member of the pair exhibits the disorder, the pair is described as **discordant**. Whereas about half of the monozygotic twins of people with schizophrenia are concordant for the disorder, the rate of concordance for dizygotic twins is only about 17% (see Figure 12.3) (Cardno and Gottesman, 2000; Gottesman, 1991). The significantly higher concordance rate among monozygotic twins (who are twice as closely related genetically as dizygotic twins are) is strong evidence of a genetic factor. After all, environmental variables like family structure and socioeconomic stress would presumably be comparable for the two kinds of twins.

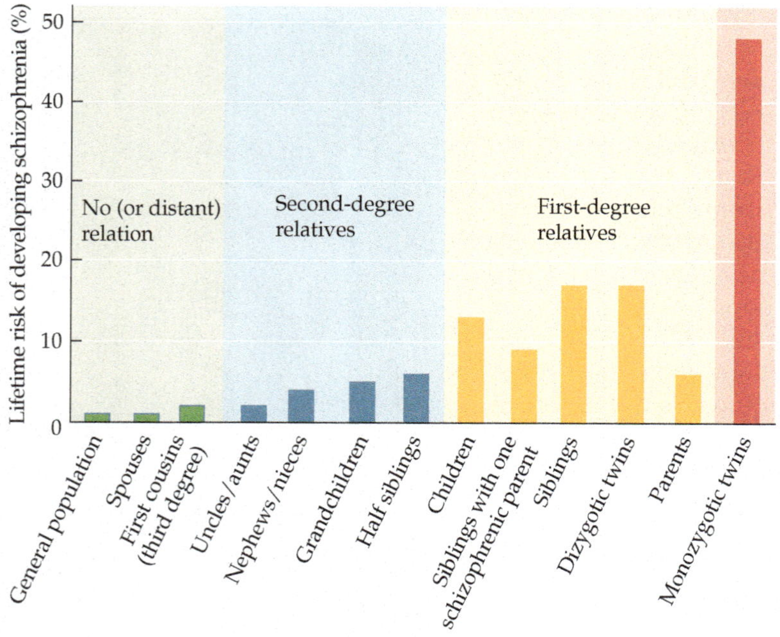

12.3 The Heritability of Schizophrenia The more closely related a person is to someone with schizophrenia, the greater are that person's chances of also developing schizophrenia. (After I. I. Gottesman, 1991. *Schizophrenia genesis: The origins of madness*. Freeman: New York.)

Even with identical twins, however, the concordance rate for schizophrenia is only about 50% (see Figure 12.3), so genes alone cannot fully explain whether a person will develop schizophrenia. What accounts for the discordance of the other 50% of monozygotic twins? E. Fuller Torrey noted that the twin who went on to develop schizophrenia tended to be the one who was more abnormal throughout life. The symptomatic twin frequently weighed less at birth and had an early developmental history that included more instances of physiological distress (Torrey et al., 1994). During development, this twin was more submissive, tearful, and sensitive than the identical sibling and often was viewed by the parents as being more vulnerable.

The childhood difficulties of twins who later develop schizophrenia are reflected in behavioural, cognitive, and other neurological signs, such as impairments in motor coordination (Torrey et al., 1994). These early signs are sufficiently evident that observers watching home videos of children can, with uncanny accuracy, pick out children who went on to develop schizophrenia in adulthood (J. Schiffman et al., 2004; Walker, 1991).

These sorts of behavioural distinctions can be objectively measured by various neuropsychological tests. For example, eye-tracking measurements, recording eye movements while the person's gaze follows a moving target on a computer screen, are abnormal in people with schizophrenia (Stuve et al., 1997). These people tend to be unable to use smooth movements of the eyes to follow the moving target and instead show an intrusion of rapid, jerky eye movements (**FIGURE 12.4A**). These

monozygotic Referring to twins derived from a single fertilised egg (*identical* twins). Such individuals have the same genotype.

dizygotic Referring to twins derived from separate eggs (*fraternal* twins). Such twins are no more closely related genetically than are other full siblings.

concordant Referring to any trait that is seen in both individuals of a pair of twins.

discordant Referring to any trait that is seen in only one individual of a pair of twins.

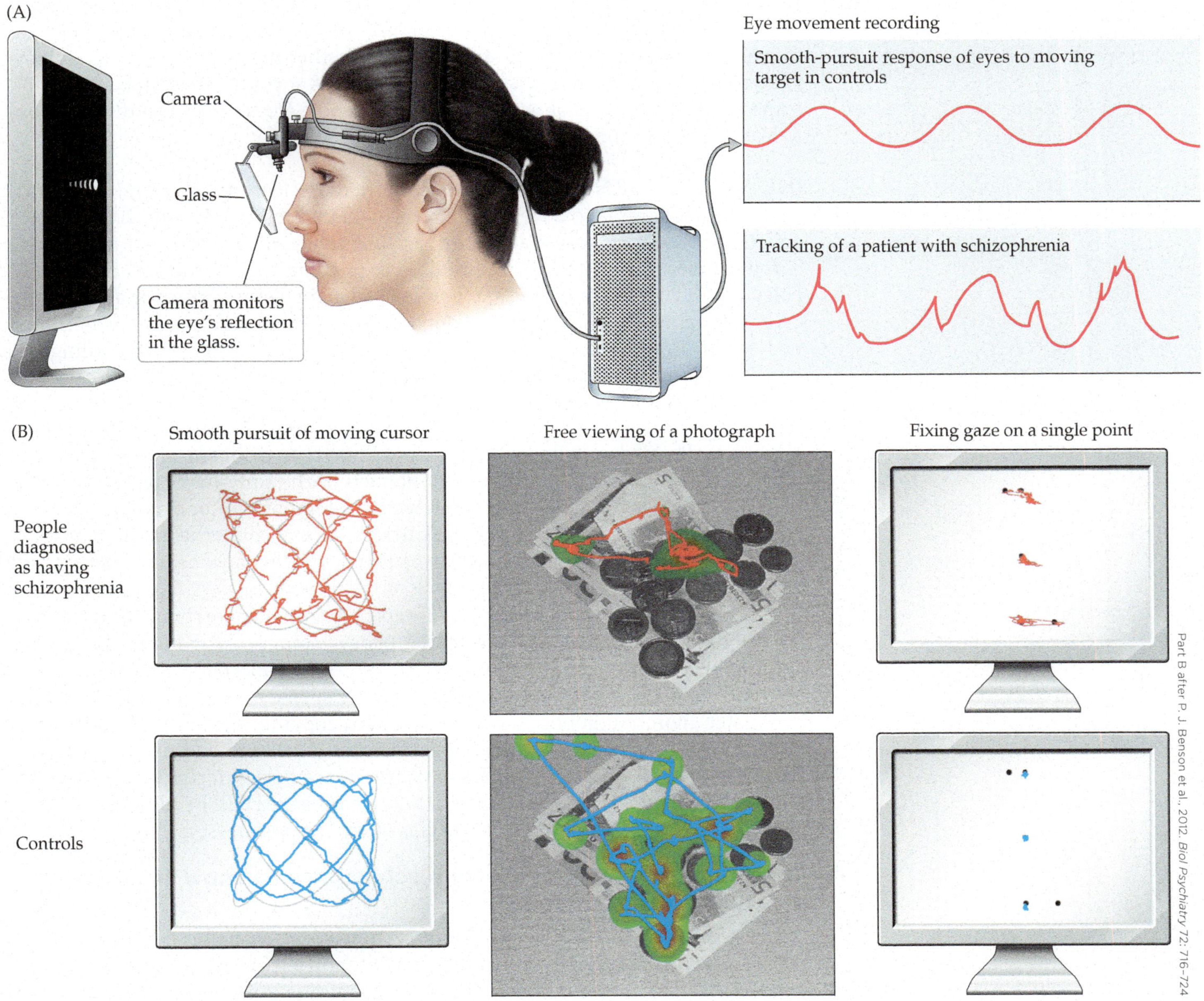

12.4 Eye Tracking in People with Schizophrenia versus Controls (A) People who have been diagnosed with schizophrenia have a more difficult time with "smooth-pursuit" visual tracking of moving objects than do controls. (B) Examples of disordered smooth-pursuit movements. Can this characteristic be used to identify people at risk of developing schizophrenia?

endophenotype Behavioural or physical characteristics accompanying susceptibility to a particular disorder, which may be used to identify those at risk.

findings suggest that schizophrenia may have an associated **endophenotype**: a group of behavioural or physical characteristics that accompany an inherited susceptibility to a particular disorder.

In a test of this idea, experimenters monitored the eye movements of people with schizophrenia and a control group of healthy people, while they performed three tasks: (1) watching a cursor on a computer screen travel a complex weaving pathway (smooth pursuit), (2) freely viewing a series of photographs, and (3) trying to fix their gaze on one point while distracting stimuli appeared elsewhere on the screen (**FIGURE 12.4B**). As in previous studies, the people with schizophrenia had a hard time controlling eye movements. When the researchers built mathematical models based on the performance of the participants, they found they could correctly discriminate between people with and people without schizophrenia in another set of participants with 88% accuracy (P. J. Benson et al., 2012). When the researchers fine-tuned their mathematical model further, it discriminated 298 total cases with 98% accuracy! No people with schizophrenia were misclassified by the model. Five controls showed abnormalities, but you have to wonder whether they were, in fact, at risk for schizophrenia. Mathematical analysis showed that the differences in eye movement were seen in both sexes and were not due to differences in whether the people were in remission, or whether they were receiving antipsychotics, or any other factor examined. Tests of this sort may thus make it possible some day to screen lots of people to identify people at risk and start interventions to avert the condition.

INDIVIDUAL GENES It has proven difficult to identify any single gene that causes schizophrenia to develop or increases susceptibility (Hyman, 2018). In fact, genetic analyses suggest that more than 100 genes, distributed among multiple chromosomes, influence schizophrenia risk in some way (Birnbaum et al., 2017; Foley et al., 2017). Nonetheless, a few genes have been identified that appear to be abnormal in some cases of schizophrenia. These include the genes encoding neuregulin 1 (Mei and Xiong, 2008), which participates in synaptic signalling and myelination; dysbindin, which is implicated in synaptic plasticity; and catechol-O-methyltransferase (COMT), which is involved in metabolizing dopamine. In one large Scottish family, the several members who had schizophrenia shared a mutant, disabled version of a gene now known as *disrupted in schizophrenia 1* (*DISC1*). The DISC1 protein normally interacts with many other proteins during brain development, including some known to be important for synaptic plasticity (Tomoda et al., 2017; Tsuboi et al., 2015). In fact, several genes that seem to increase the risk of schizophrenia are known to be involved in synapse rearrangement (see Figure 6.18) (A. Sekar et al., 2016).

An interesting *epigenetic* factor (see Chapter 6) in schizophrenia is paternal age: the children of older men are at greater risk of developing schizophrenia than are the children of younger men (de Kluiver et al., 2017). The sperm of older men, which are the product of more cell divisions than the sperm of younger men, have accumulated more mutations caused by errors in copying the chromosomes (Kong et al., 2012); these new mutations may increase the likelihood of schizophrenia (Fromer et al., 2014). Another epigenetic difference between people with schizophrenia and controls is the likelihood that certain genes will be methylated (Jaffe et al., 2016), which would make them less likely to be expressed.

The brains of some people with schizophrenia show structural and functional changes

Because the symptoms of schizophrenia can be so marked and persistent, investigators hypothesised early in the twentieth century that people with this illness would show distinctive and measureable structural anomalies in the brain. Only with the advent of CT and MRI scans, however, did it become possible to study brain anatomy in living people at all stages of their illness. Such studies reveal the presence of significant, consistent anatomical differences in the brains of many people with schizophrenia.

12.5 Ventricular Enlargement in Schizophrenia The volume of the cerebral ventricles, relative to overall brain volume, is greater in people with schizophrenia than in controls. (After N. C. Andreasen et al., 1990. *Arch Gen Psychiatry* 47: 1008–1015.)

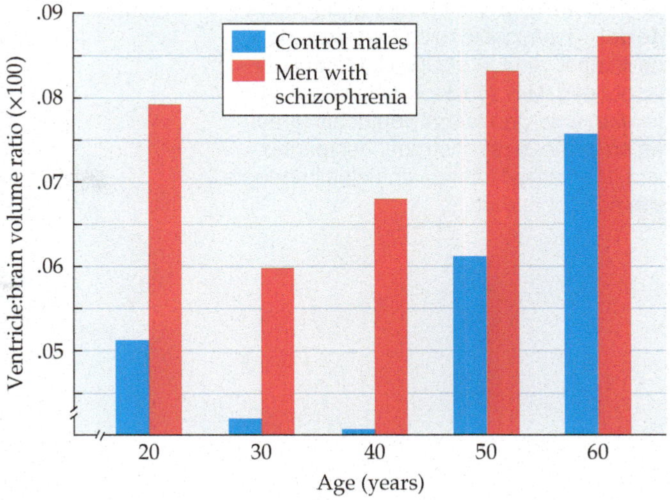

VENTRICULAR ABNORMALITIES Many people with schizophrenia have enlarged cerebral ventricles, especially the lateral ventricles (**FIGURE 12.5**) (T. M. Hyde and Weinberger, 1990), and ventricular enlargement predicts later development of schizophrenia in adolescents (Y. Chung et al., 2017). The extent of enlargement is also related to responsiveness to antipsychotic drugs: people with more-enlarged ventricles tend to show poorer response to these drugs (Garver et al., 2000).

Enlargement of the ventricles appears to be a stable trait, remaining for many years after the initial onset of the disease (Andreasen, 1994). Studies of identical twins discordant for schizophrenia yielded startlingly clear results: twins with schizophrenia have decidedly enlarged lateral ventricles compared with their well counterparts, whose ventricles are of normal size (**FIGURE 12.6**; also see Figure 0.9) (Torrey et al., 1994).

Recall that a disabled version of the gene *DISC1* is closely associated with schizophrenia (Weng et al., 2018). It affects neuroplasticity through multiple molecular interactions in both neurons and astrocytes (Terrillion et al., 2018; Tropea et al., 2018). Producing transgenic mice that express the mutant version of *DISC1* that is associated with schizophrenia in humans has a dramatic effect: the mice develop enlarged lateral ventricles (**FIGURE 12.7**) (Pletnikov et al., 2008) that are very reminiscent of the enlarged ventricles in people with schizophrenia.

What is the significance of enlarged ventricles? Because overall brain size does not seem to be affected in people with schizophrenia or mice expressing mutant *DISC1*, the enlarged ventricles must come at the expense of brain tissue. Therefore, interest has centred on brain structures that run alongside the lateral ventricles, as we discuss next.

LIMBIC SYSTEM ABNORMALITIES Because the hippocampus and amygdala help form some of the walls of the lateral ventricles, atrophy in these regions could cause the ventricular enlargement in people with schizophrenia. Indeed, among twins who are discordant for schizophrenia, the hippocampus and the amygdala are smaller in the twin who is ill.

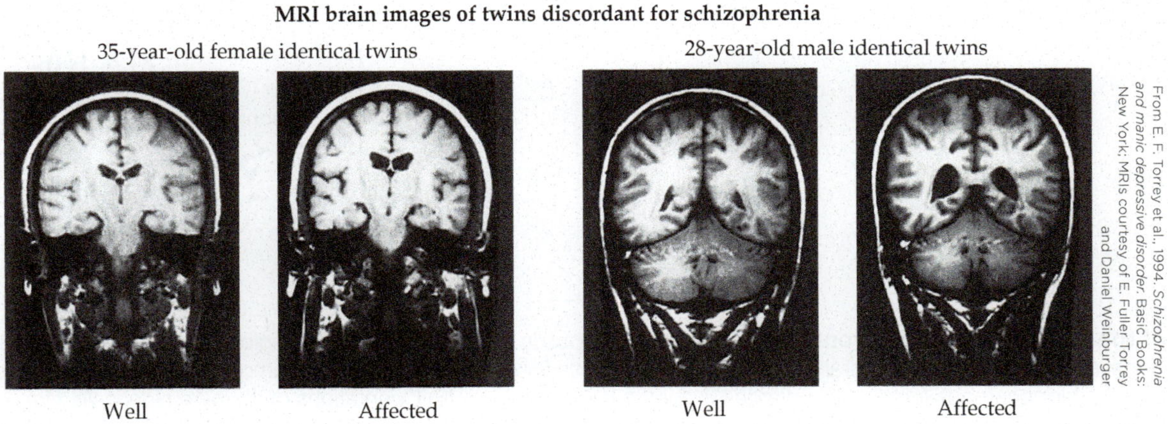

12.6 Identical Genes, Different Fates Although the two members of each set of monozygotic twins shown here have the same genes, only one of the twins (the one with larger ventricles) developed schizophrenia.

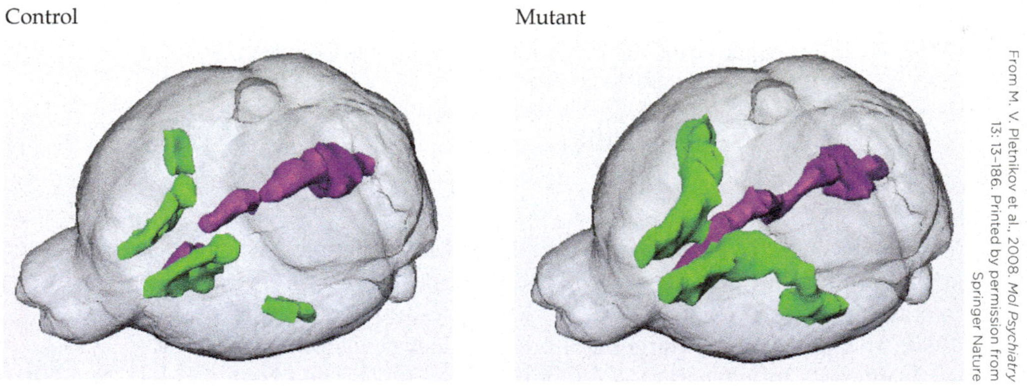

12.7 Enlarged Ventricles in a Mouse Model Transgenic mice that express the mutant version of *DISC1*, the gene associated with schizophrenia in humans, develop enlarged lateral ventricles (green region in these reconstructions) reminiscent of enlarged lateral ventricles in people with schizophrenia.

Postmortem and brain-imaging studies of people with schizophrenia reveal abnormalities in several parts of the limbic system, including the hippocampus, amygdala, and parahippocampal regions. The hippocampal pyramidal cells of people with schizophrenia appear Disorganised (**FIGURE 12.8**), possibly resulting from abnormal synaptic arrangements of both the inputs and outputs of these cells (Heckers and Konradi, 2002). Evidence suggests that there may be widespread abnormalities in the activity of limbic networks extending from the hippocampus to other limbic structures, including the entorhinal cortex, parahippocampal cortex, and cingulate cortex, as well as connectivity between the hippocampus and cortical regions (Avery et al., 2018; Bubb et al., 2018).

The cellular disorganisation of schizophrenia probably arises during early cell development. But because we now know that humans make new neurons throughout life, especially in the hippocampus (see Chapter 6), we can reason that abnormal neurogenesis or disordered integration of newly born cells could contribute to the development of schizophrenia.

CORTICAL ABNORMALITIES People with schizophrenia differ from controls in the structure and functional activity of the corpus callosum (Olabi et al., 2011) that connects the two cortical hemispheres. Several studies have reported

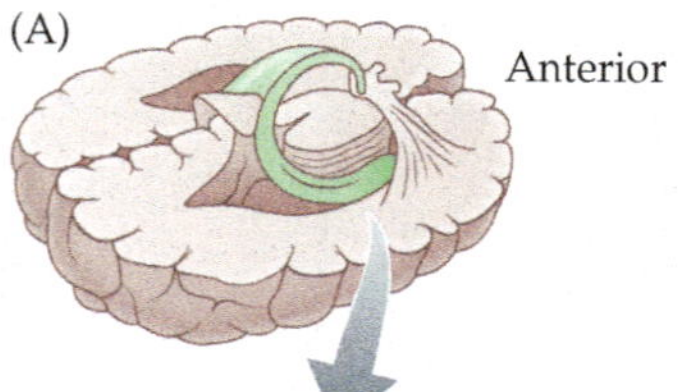

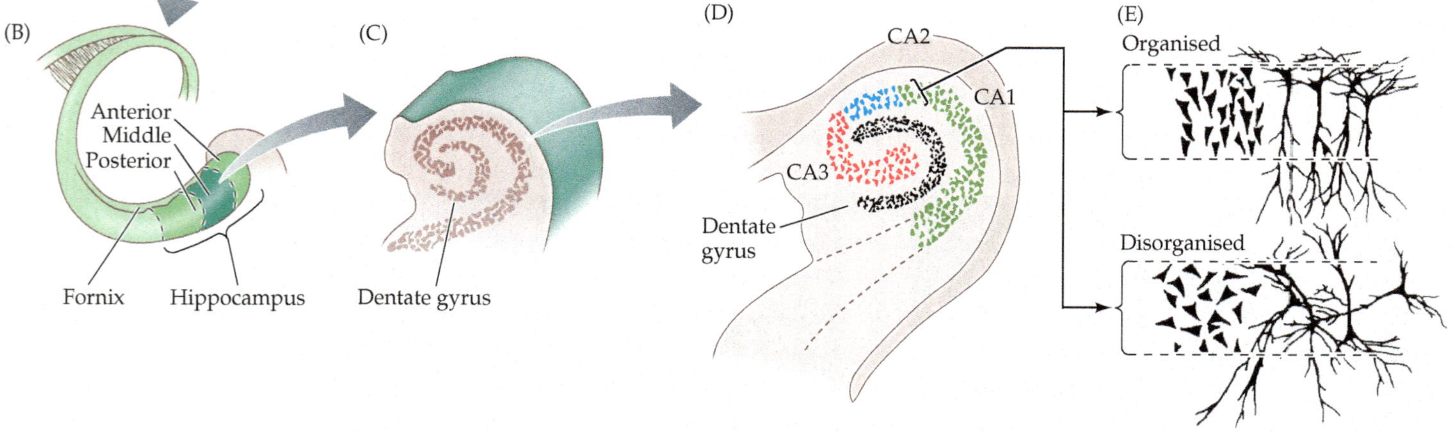

12.8 Cellular Disarray of the Hippocampus in Chronic Schizophrenia (A) This horizontal section of the cerebral hemispheres shows the location of the hippocampus (green). (B) Enlargements of the hippocampus and fornix show the location of anteriour, middle, and posteriour hippocampal segments. (C) The hippocampus and dentate gyrus are enlarged in this cross section. (D) The hippocampus is subdivided into three regions: CA1, CA2, and CA3 ("CA" stands for *cornu ammonis*, or "Ammon's horn," another name for the hippocampus). (E) Orientations of the pyramidal cells of a control (top) and a person with schizophrenia (bottom) are compared in these hippocampal cross sections. (After J. A. Kovelman and A. B. Scheibel, 1984. *Biol Psychiatry* 19: 1601–1621.)

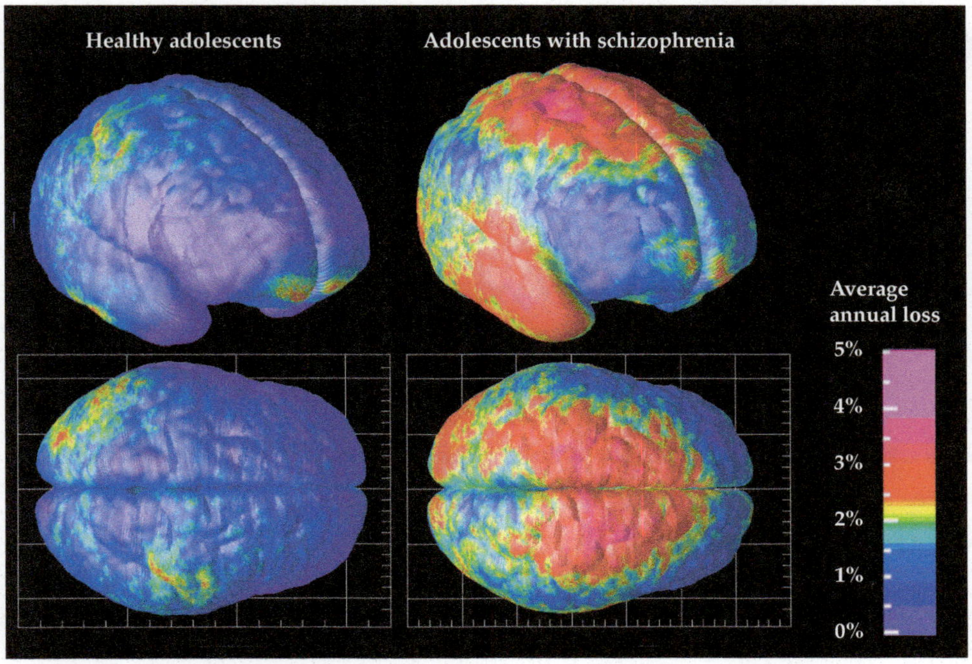

12.9 Accelerated Loss of Gray Matter in Adolescents with Schizophrenia Although neuron loss is a normal part of development, adolescents with schizophrenia (right) lose gray matter over wide regions at a faster rate than do unaffected adolescents (left).

a more accelerated loss of gray matter at adolescence in people with schizophrenia than in controls (Y. Chung et al., 2015; Karlsgodt et al., 2010) (**FIGURE 12.9**). This thinning of gray matter is thought to reflect the net loss of synapses during development (see Figures 6.17 and 6.18), which is intriguing when you recall that several genes associated with schizophrenia are known to be involved in synapse rearrangement. In addition to structural changes in the cortex, what about the *activity* of cortex in schizophrenia?

DIFFERENCES IN BRAIN ACTIVATION People with schizophrenia tend to be impaired on neuropsychological tests that are sensitive to frontal cortical lesions. These findings raised the possibility that frontal cortex was impaired in schizophrenia. Early observations using PET indicated that people with schizophrenia show relatively less metabolic activity in the frontal lobes (compared with their posteriour lobes), while controls have more-equal activation of frontal and posteriour cortex (M. S. Buchsbaum et al., 1984). This observation, referred to as the **hypofrontality hypothesis**, fueled interest in the role of the frontal lobes in schizophrenia (Minzenberg et al., 2009; Penadés et al., 2017). In discordant identical twins, reduced frontal blood flow and associated gray matter changes are evident only in the twin with schizophrenia (Andreasen et al., 1986; T. D. Cannon et al., 2002).

Treatment with drugs that alleviate symptoms of schizophrenia, discussed in the next section, is associated with increased activation of frontal cortex (Honey et al., 1999). Neurons in the frontal cortex of people with schizophrenia have dendrites with a reduced density of synaptic spines compared with controls (L. A. Glantz and Lewis, 2000), which may contribute to a less active frontal cortex. Some experiments show the hypofrontality effect only during difficult cognitive tasks that particularly depend on the frontal lobes for accurate performance, such as the Wisconsin Card Sorting Task (see Figure 17.28). Unlike controls, people with schizophrenia

hypofrontality hypothesis The hypothesis that schizophrenia may result from underactivation of the frontal lobes.

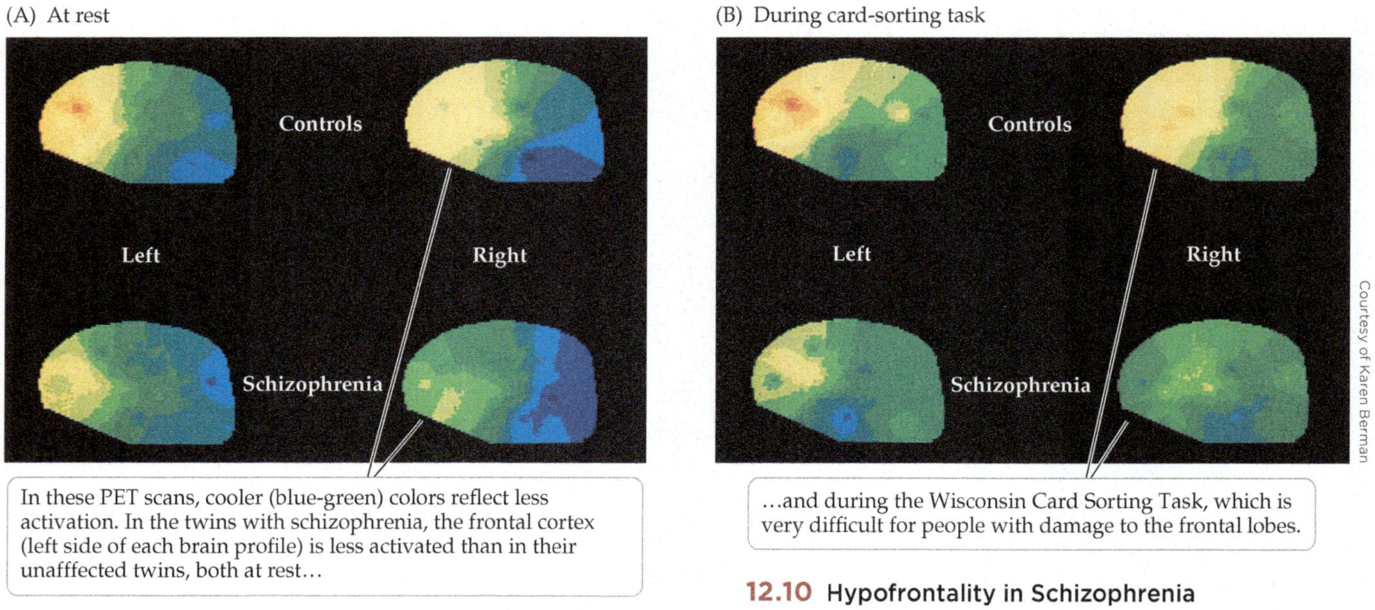

12.10 Hypofrontality in Schizophrenia

lobotomy The surgical detachment of a portion of the frontal lobe from the rest of the brain, once used as a treatment for schizophrenia and many other ailments.

psychosurgery Surgery in which brain lesions are produced to alleviate severe psychiatric disorders.

chlorpromazine An antipsychotic drug, one of the class of phenothiazines.

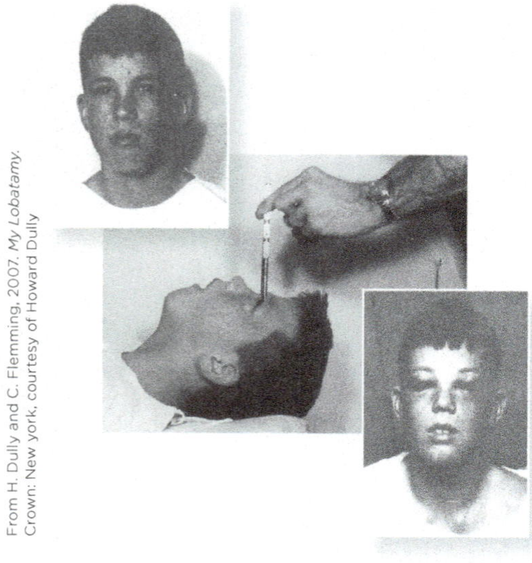

12.11 The Teenaged Howard Dully before, during, and after His Lobotomy These photographs appear in Mr. Dully's memoir, *My Lobotomy.*

show no increase in their prefrontal activation above resting levels during the task (D. R. Weinberger et al., 1994) (**FIGURE 12.10**).

Perhaps the most revolutionary insights into schizophrenia came not from studies of structural or functional differences, but from the discovery of antipsychotic drugs, as we discuss next.

Antipsychotic medications revolutionized the treatment of schizophrenia

In the 1930s, there were no effective treatments for schizophrenia. Because people were often unable to take care of themselves, they were placed in caregiving institutions. In many cases, the health and welfare of people in these (poorly funded) institutions were badly neglected, leading to recurrent scandals. So perhaps it was in desperation that psychiatrists turned to **lobotomy**, the surgical separation of a portion of the frontal lobes from the rest of the brain, as a treatment for schizophrenia. Certainly there was little scientific evidence to think the surgery would be effective. But early practitioners reported nearly miraculous recoveries that, in retrospect, must be regarded as wishful thinking on the part of the physicians. The surgery may well have made the people easier to handle, but they were rarely able to return to independent living. Used for almost any mental disorder, not just schizophrenia, lobotomies were performed on some 40,000 people in the United States alone (Kopell et al., 2005). One of the last people to be given a lobotomy was Howard Dully, whose stepmother took the rebellious 12-year-old boy to several psychiatrists until finding one who was willing to say he had schizophrenia. That was used to justify giving the boy a lobotomy in 1960 (**FIGURE 12.11**). Not until he was 50 was Howard able to find out what had happened to him as a child, a journey he recounts movingly in his memoir, *My Lobotomy* (Dully and Fleming, 2007).

Today **psychosurgery**, surgical modification of the brain to treat psychiatric disorders, involves much smaller lesions and is rarely performed, and then only as a last resort. For example, for some people with severe epilepsy, surgical removal of the damaged brain tissue can help, as we discussed in Chapter 2. Psychosurgery is also being explored for other disorders, as we'll note later in this chapter.

By the 1950s, more and more physicians were sceptical that lobotomy was effective for any disorder, and a drug discovered in 1954—**chlorpromazine** (trade name Thorazine)—quickly replaced lobotomy as a treatment for schizophrenia. Although chlorpromazine was originally developed as an anesthetic (Ban, 2007), a lucky observation revealed that it could powerfully reduce the "positive" symptoms of schizophrenia, including auditory hallucinations, delusions, and disordered thinking. Ironically, the drug was first promoted as "a lobotomy in a bottle" because at the time, that was viewed as a good thing! But chlorpromazine was much better than a lobotomy, not only because drug treatment is reversible while lobotomy is not, but because the drug actually worked. In fact, the symptoms of schizophrenia that responded to chlorpromazine were exactly those that kept people in mental institutions. The introduction of chlorpromazine truly revolutionized psychiatry, relieving symptoms for millions of people and freeing them from a lifetime in psychiatric hospitals.

THE DOPAMINE HYPOTHESIS Chlorpromazine and other antipsychotic drugs, known as **antipsychotics**, that came along a little later were eventually found to share a specific action: they block postsynaptic dopamine receptors, particularly dopamine D_2 receptors. Because all of the earliest antipsychotic drugs blocked dopamine D_2 receptors, researchers proposed the **dopamine hypothesis**: that people with schizophrenia have an excess of either dopamine release or dopamine receptors.

At about the same time, support for the dopamine hypothesis came from another source: people who were abusing drugs. As a consequence of drug tolerance (see Chapter 3), some daily users of amphetamine reach the point of taking astonishingly high doses—as much as 3000 milligrams (mg) per day—in order to experience the drug's "high." (Compare this with the normal 5 mg dose used to combat sleepiness.) Many individuals taking these large doses of amphetamine develop symptoms of paranoia, often involving delusions of persecution with auditory hallucinations, and they exhibit suspiciousness and bizarre motor behaviour. These symptoms strikingly resemble those of schizophrenia and are referred to as **amphetamine psychosis**. What's more, amphetamine exacerbates the symptoms of schizophrenia. It turns out that amphetamine promotes the release of dopamine and prolongs the action of the released transmitter by blocking reuptake (see Chapter 3). Antipsychotic drugs rapidly reverse amphetamine psychosis. In contrast, drugs such as LSD and mescaline, despite being called "hallucinogens," act principally on serotonin receptors and produce effects that bear only superficial resemblance to psychosis. People with schizophrenia who are given LSD report that the experience produced by the drug is very different from their experiences of the disorder, and in lab animals LSD actually *enhances* measures of cognitive function; in short, the hallucinogens make poor models of schizophrenia (Ham et al., 2017).

All of the various antipsychotic drugs that are now classified as **first-generation antipsychotics** (or *typical antipsychotics*) are D_2 receptor antagonists. In fact, the clinically effective dose of the first-generation antipsychotic can be predicted from its affinity for D_2 receptors (**FIGURE 12.12**), as the dopamine hypothesis would predict. For example, haloperidol, discovered a few years after chlorpromazine, has a greater affinity for D_2 receptors and quickly became the more widely used drug. Other clinical findings bolster the dopamine hypothesis; for example, treating people who have

antipsychotics A class of drugs that alleviate symptoms of schizophrenia, typically by blocking dopamine receptors.

dopamine hypothesis The hypothesis that schizophrenia results from either excessive levels of synaptic dopamine or excessive postsynaptic sensitivity to dopamine.

amphetamine psychosis A delusional and psychotic state, closely resembling acute schizophrenia, that is brought on by repeated use of high doses of amphetamine.

first-generation antipsychotics A major class of antischizophrenic drugs that share an antagonist activity at dopamine D_2 receptors.

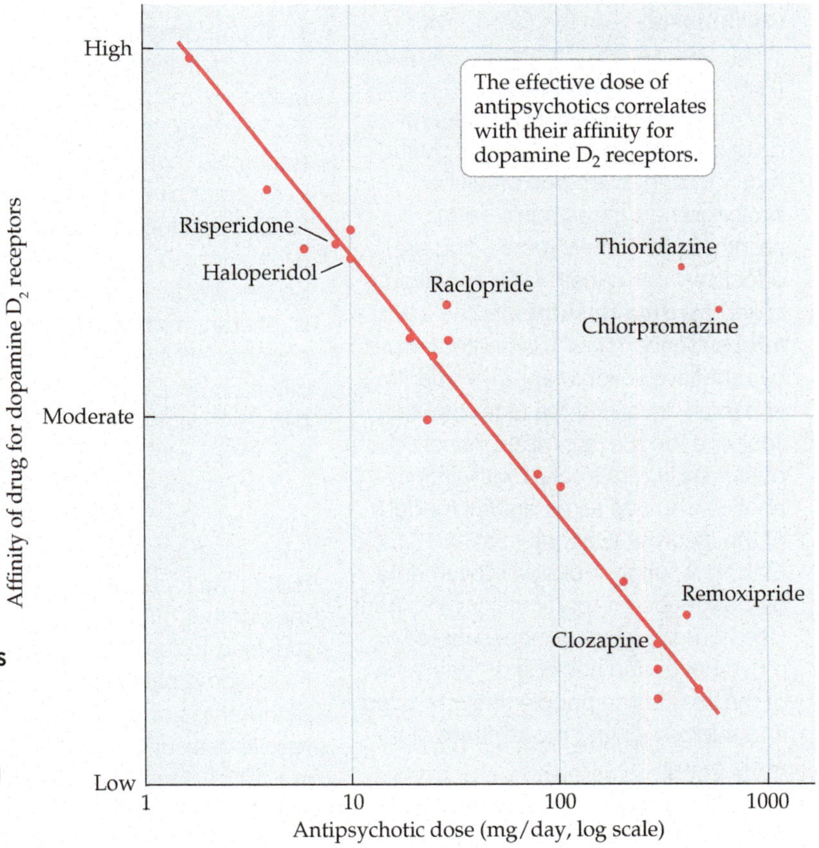

12.12 Antipsychotic Drugs Affect Dopamine Receptors
An early observation was that for each of the major antipsychotic medications, the recommended dose for reduction of the symptoms of schizophrenia and avoidance of major side effects was correlated with the dopamine D_2 receptor binding affinity of that drug. (After P. Seeman and T. Tallerico, 1998. *Mol Psychiatry* 3: 123–134.)

Parkinson's disease with L-dopa (the metabolic precursor of dopamine) may induce schizophrenia-like symptoms, presumably by boosting the synaptic availability of dopamine.

Although some evidence supports the dopamine hypothesis of schizophrenia, there are also several problems with the hypothesis. For example, there is no correspondence between the speed with which drugs block dopamine receptors (quite rapidly—within hours) and how long it takes for the symptoms to diminish (usually on the order of weeks). Thus, the relation of dopamine to schizophrenia is more complex than just hyperactive dopamine synapses. Another weakness of the dopamine model of schizophrenia is that some people with schizophrenia don't respond to dopamine antagonists at all (Alpert and Friedhoff, 1980).

The search for new antipsychotics to avoid motor side effects, which are debilitating in some people (**BOX 12.1**), led to the development of drugs called

BOX 12.1 Long-Term Effects of Antipsychotic Drugs

Clearly, antipsychotic drugs have revolutionized the treatment of schizophrenia. With these drugs, many people who might otherwise have been in mental hospitals their whole lives can take care of themselves in nonhospital settings. The drugs can justly be regarded as "miracle drugs."

Unfortunately, traditional antipsychotic drugs can have other, undesirable effects. After delays ranging from days to years, a minority of people receiving the drugs start to show maladaptive motor symptoms (**dyskinesia**, from the Greek *dys*, "bad," and *kinesis*, "motion"). Although many of these symptoms are transient and disappear when the dosage of drug is reduced, some drug-induced motor changes emerge only after prolonged drug treatment—after months, sometimes years—and are effectively permanent. The condition, called **tardive dyskinesia** (the Latin *tardus* means "slow"), is characterised by repetitive, involuntary movements, especially involving the face, mouth, lips, and tongue; a similar phenomenon observed in rats treated with antipsychotics is in use as an animal model of the disorder (Bachus et al., 2012). Elaborate, uncontrollable movements of the tongue are particularly prominent (see figure), including incessant rolling movements and sucking or smacking of the lips. Some people show twisting and sudden jerking movements of the arms or legs.

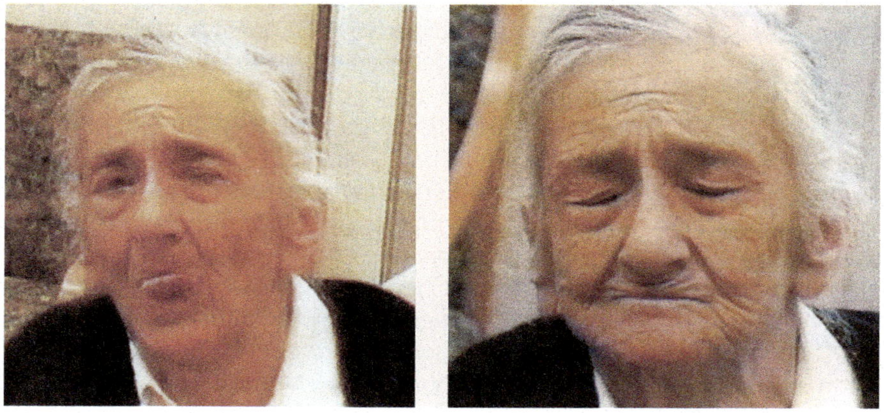

Photographs courtesy of Steven J. Frucht

The underlying mechanism for tardive dyskinesia continues to be a puzzle. Some researchers claim that it arises from the chronic blocking of dopamine receptors, which results in receptor site supersensitivity. But tardive dyskinesia frequently takes a long time to develop and tends to be irreversible—a time course that is different from dopamine receptor supersensitivity. In addition, there is no difference in D_1 or D_2 receptor binding between people with tardive dyskinesia and those without these symptoms.

Long-term treatment with traditional antipsychotic drugs has another unusual effect: Prolonged blockage of dopamine receptors seems to increase the number of dopamine receptors and lead to receptor supersensitivity. In some patients, discontinuation of the drugs or a lowering of dosage results in a sudden, marked increase in positive symptoms of schizophrenia, such as delusions or hallucinations. This **supersensitivity psychosis** can often be reversed by administration of increased dosages of dopamine receptor–blocking agents. The second-generation antipsychotics discussed in the text have fewer dyskinesia side effects than traditional antipsychotics have, but unfortunately they are more likely to lead to weight gain.

dyskinesia Difficulty or distortion in voluntary movement.

tardive dyskinesia A disorder characterised by involuntary movements, especially involving the face, mouth, lips, and tongue. It can be caused by prolonged use of antipsychotic drugs, such as chlorpromazine.

supersensitivity psychosis An exaggerated psychosis that may emerge when doses of antipsychotic medication are reduced, probably as a consequence of the up-regulation of receptors that occurred during drug treatment.

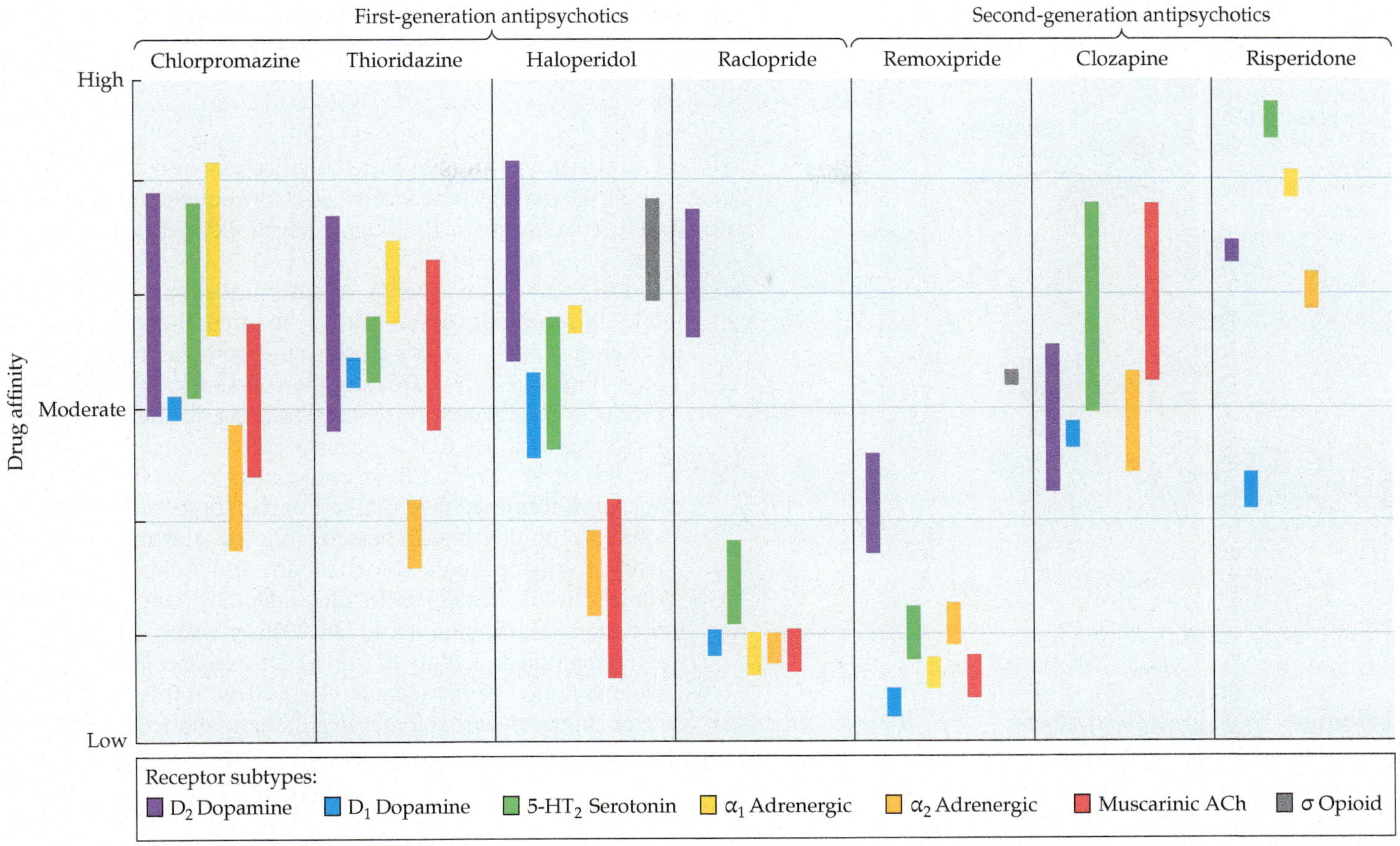

12.13 Antipsychotic Drugs That Affect Dopamine Receptors Drugs vary widely in the affinity with which they bind to various neurotransmitter receptors. Drugs that block dopamine receptors, specifically the D_2 variety (purple), are more effective at combatting symptoms of schizophrenia. Second-generation antipsychotics such as clozapine tend to block 5-HT_2 receptors (green) more effectively than they block D_2 receptors. (After P. Seeman, 1990. *Acta Psychiatr Scand Suppl* 358: 14–20.)

second-generation antipsychotics (or *atypical antipsychotics*). These drugs generally have affinity for dopamine receptors, like the first-generation antipsychotics, but they also feature high affinity for other types of receptors. For example, the second-generation antipsychotic **clozapine** selectively blocks serotonin receptors (especially 5-HT_{2A} receptors), as well as other receptor types (**FIGURE 12.13**).

Second-generation antipsychotics are just as effective as the older generation of drugs for relieving the symptoms of schizophrenia. So, if the problem is as simple as an overstimulation of dopamine receptors, why are the second-generation antipsychotics effective? For example, clozapine can *increase* dopamine release in frontal cortex (Hertel et al., 1999)—hardly what we would expect if excess dopaminergic activity lies at the root of schizophrenia. In fact, it seems that supplementing antipsychotic treatments with L-dopa (thereby increasing dopaminergic activity) actually helps reduce symptoms of schizophrenia (Jaskiw and Popli, 2004).

Until recently, almost all clinicians believed that second-generation antipsychotics were more effective than first-generation antipsychotics for treating schizophrenia, especially for relieving negative symptoms. But several studies comparing the outcome for people with schizophrenia who had been given the two types of drugs found no difference in overall effectiveness (Crossley et al., 2010; Saha et al., 2016). Although the second-generation antipsychotics are less likely than first-generation antipsychotics to cause side effects in motor function (see Box 12.1), they are more likely to cause weight gain (Sikich et al., 2008), which causes other health problems.

THE GLUTAMATE HYPOTHESIS Another drug that, like chlorpromazine, was initially developed as an anesthetic has a much different relationship to schizophrenia. **Phencyclidine** (**PCP**) was soon found to be a potent **psychotomimetic**; that is, PCP produces phenomena strongly resembling both the positive and negative symptoms of schizophrenia. Users of PCP ("angel dust") often experience auditory

second-generation antipsychotics A class of antischizophrenic drugs that act at other receptors in addition to blocking dopamine receptors.

clozapine A second-generation antipsychotic.

phencyclidine (PCP) Also called *angel dust*. An anesthetic agent that is also a psychedelic drug. PCP makes many people feel dissociated from themselves and their environment.

psychotomimetic A drug that induces a state resembling schizophrenia.

A model of PCP action on the NMDA receptor

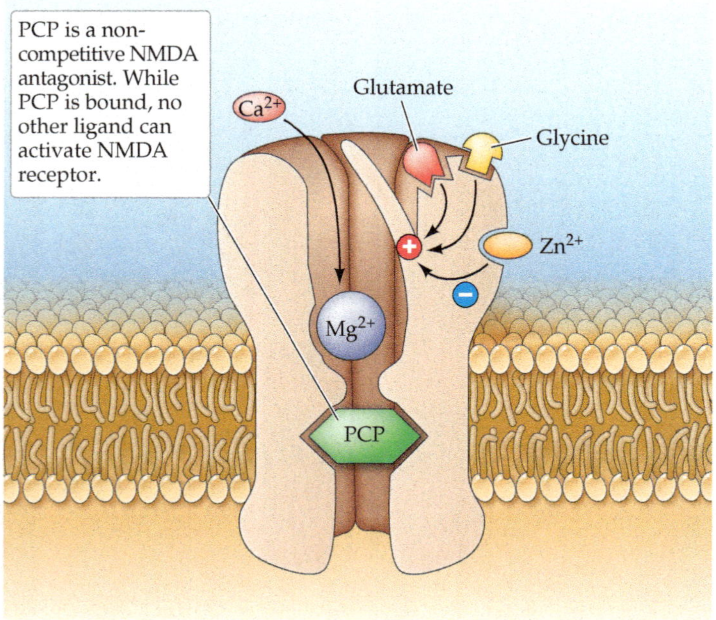

12.14 The Effects of PCP on the NMDA Receptor The resemblance of PCP-induced psychosis to schizophrenia prompted the development of a glutamate hypothesis of schizophrenia.

hallucinations, strange depersonalisation, and disorientation; and they may become violent as a consequence of their drug-induced delusions. Prolonged psychotic states can develop with chronic use of PCP.

PCP acts as an NMDA receptor antagonist, blocking the NMDA receptor's central calcium channel, thereby preventing the endogenous ligand—glutamate—from having its usual effects (**FIGURE 12.14**). Treating monkeys with PCP for 2 weeks produces a schizophrenia-like syndrome, including poor performance on a test that is sensitive to prefrontal damage (J. D. Jentsch et al., 1997). Other antagonists of NMDA receptors, such as **ketamine**, have similar effects. These and other observations prompted researchers to propose a **glutamate hypothesis** of schizophrenia (Moghaddam and Adams, 1998), that schizophrenia results from an underactivation of glutamate receptors (Hardingham and Do, 2016), possibly accounting for the reduction in activity of frontal cortex—*hypofrontality*—that we mentioned earlier (Marek et al., 2010). If this hypothesis is true, you might ask whether compounds that increase glutamatergic activity would be effective antischizophrenic drugs. Unfortunately, selective NMDA receptor agonists tend to produce seizures, so glutamatergic-boosting compounds are not an option.

ketamine A dissociative anesthetic drug, similar to PCP, that acts as an NMDA receptor antagonist.

glutamate hypothesis The hypothesis that schizophrenia may be caused, in part, by understimulation of glutamate receptors.

The glutamate hypothesis has been expanded to suggest that underactivation of *all* glutamate receptors, not just the NMDA subtype, contributes to schizophrenia (González-Maeso et al., 2008). There are at least eight different subtypes of metabotropic glutamate receptors (mGluR's) (see Chapter 3), so drugs that modify the activity of the proper class of mGluR's might someday provide a third generation of antipsychotics (Stansley and Conn, 2019).

An integrative psychobiological model of schizophrenia emphasises the interaction of multiple factors

We've established that there is genetic influence on schizophrenia, but also that genes alone cannot account for the disorder. What environmental factors contribute to the probability of developing schizophrenia? Research suggests that a variety of stressful events significantly increase the risk. For example, schizophrenia usually appears during a time in life that many people find stressful—the transition from childhood to adulthood, when people deal with physical, emotional, and lifestyle changes (e.g., going away to college).

Prenatal stress—for example, if the mother has an infection while carrying the child—is also a risk factor. The baby of a pregnant woman who contracts influenza in the first trimester of pregnancy is 7 times more likely to develop schizophrenia (P. H. Patterson, 2007). Several other maternal infections also increase chances that the fetus will develop schizophrenia one day (A. S. Brown, 2011). This correlation may be why people born in late winter and early spring are more likely to develop schizophrenia (Messias et al., 2004) (**FIGURE 12.15**): their mothers may be more likely to have gotten sick during the winter before, perhaps at some fetal stage that is particularly vulnerable. Likewise, if the mother and baby have incompatible blood types, or if the mother develops diabetes during pregnancy, or if for some other reason there is a low birth weight, the baby is more likely to develop schizophrenia (S. King et al., 2010). Birth complications that deprive the baby of oxygen also increase the probability of schizophrenia (M. C. Clarke et al., 2011).

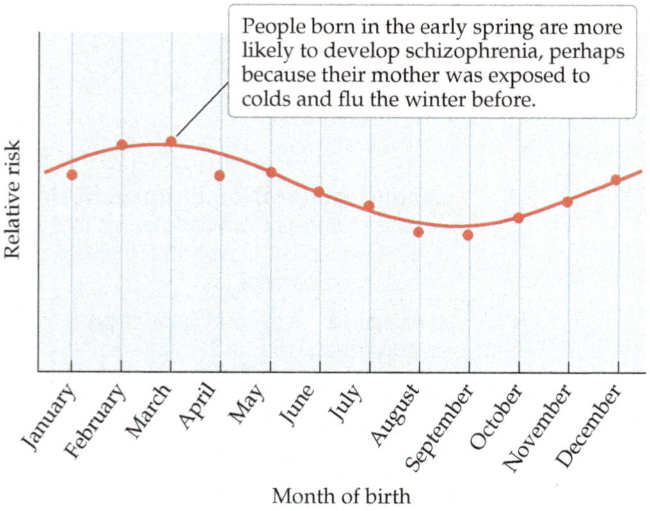

12.15 Seasonal Birth Effect on Schizophrenia (After P. B. Mortensen et al., 1999. *N Engl J Med* 340: 603–608.)

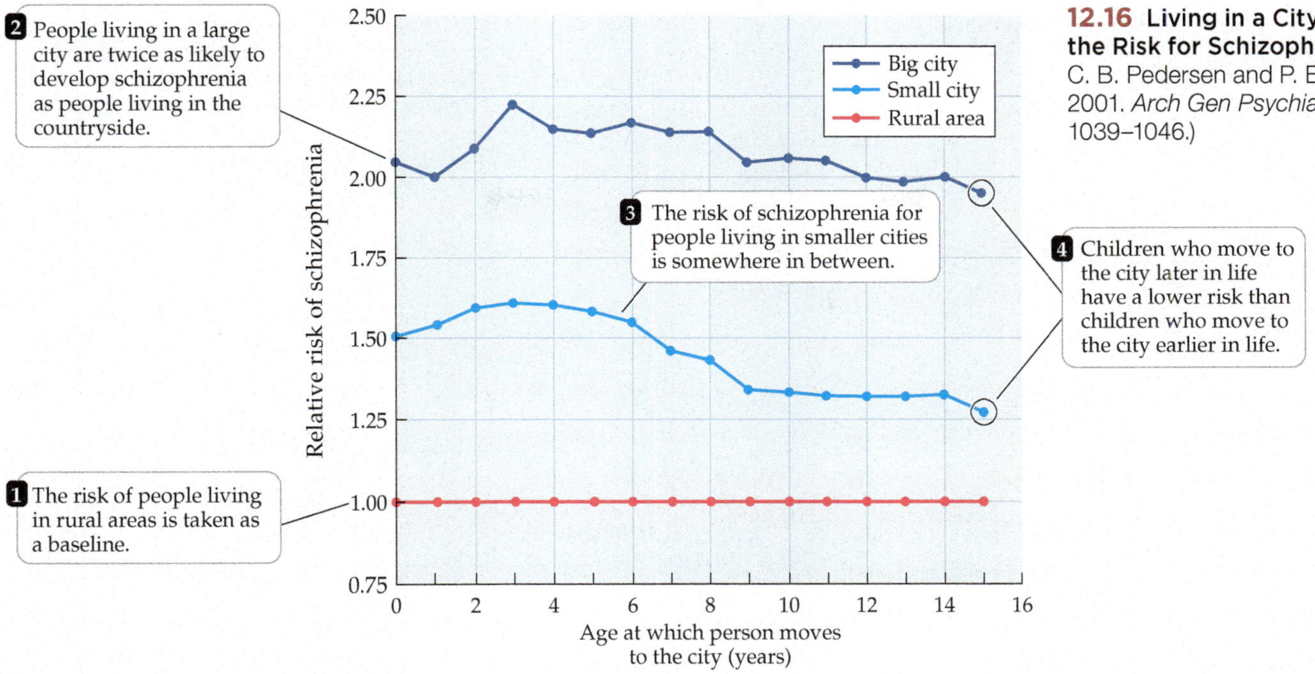

12.16 Living in a City Increases the Risk for Schizophrenia (After C. B. Pedersen and P. B. Mortensen, 2001. *Arch Gen Psychiatry* 58: 1039–1046.)

Because we know that some people are genetically more susceptible to schizophrenia than others, these findings suggest that relatively minor stress during development can make the difference in whether schizophrenia emerges. It is fascinating, and frightening, to think that events in the womb can affect the outcome 16 or 20 years later, when the schizophrenia appears.

Another risk factor seen in multiple studies is the stress of city living. As **FIGURE 12.16** shows, people born and raised in a medium-size city are about 1½ times more likely to develop schizophrenia than people living in the country. What's more, the earlier in life a person begins living in the city, the greater the risk. People moving into a big city are even more likely to develop the disorder (Pedersen and Mortensen, 2001). Conversely, children who move from the city to the country have a *reduced* risk of developing schizophrenia (van Os et al., 2010). We don't know what it is about living in a city that makes schizophrenia more likely. Pollutants, greater exposure to minor diseases, crowded conditions, tense social interactions—all of these could be considered stressful. That may be why people who live in cities show increased activity of the amygdala compared with people living in rural areas, and why adults who grew up in cities show greater activity in the anteriour cingulate (Lederbogen et al., 2011), which communicates extensively with the amygdala.

Thus, the modern view is that schizophrenia is caused by the interaction of genetic factors and stress. Each life stage has its own specific features that increase vulnerability to schizophrenia: infections before birth, complications at delivery, urban living in childhood, social stress in adolescence and adulthood (Powell, 2010). From this perspective, the emergence of schizophrenia and related disorders depends on whether a person who is genetically susceptible to schizophrenia is subjected to environmental stressors (**FIGURE 12.17**). This model of genes interacting with environmental stressors also suggests that we may someday be able to reduce the likelihood that a child at risk will develop schizophrenia. New biological aids, such as functional brain imaging and genetic tools, might help us identify and understand the at-risk child early in life, when interventions to reduce stress might prevent schizophrenia later in life.

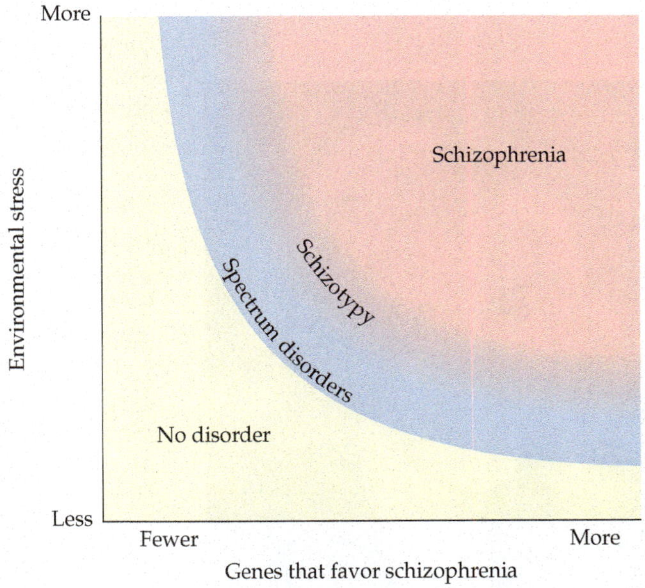

12.17 A Model of the Interaction between Stress and Genetic Influences in Schizophrenia Environmental stress and genetic susceptibility may combine to produce a schizophrenic disorder. Disorders ranging from more mild to more severe are called, respectively, *spectrum disorders*, *schizotypy*, and *schizophrenia*. (After A. F. Mirsky and C. C. Duncan, 1986. *Annu Rev Psychol* 37: 291–321.)

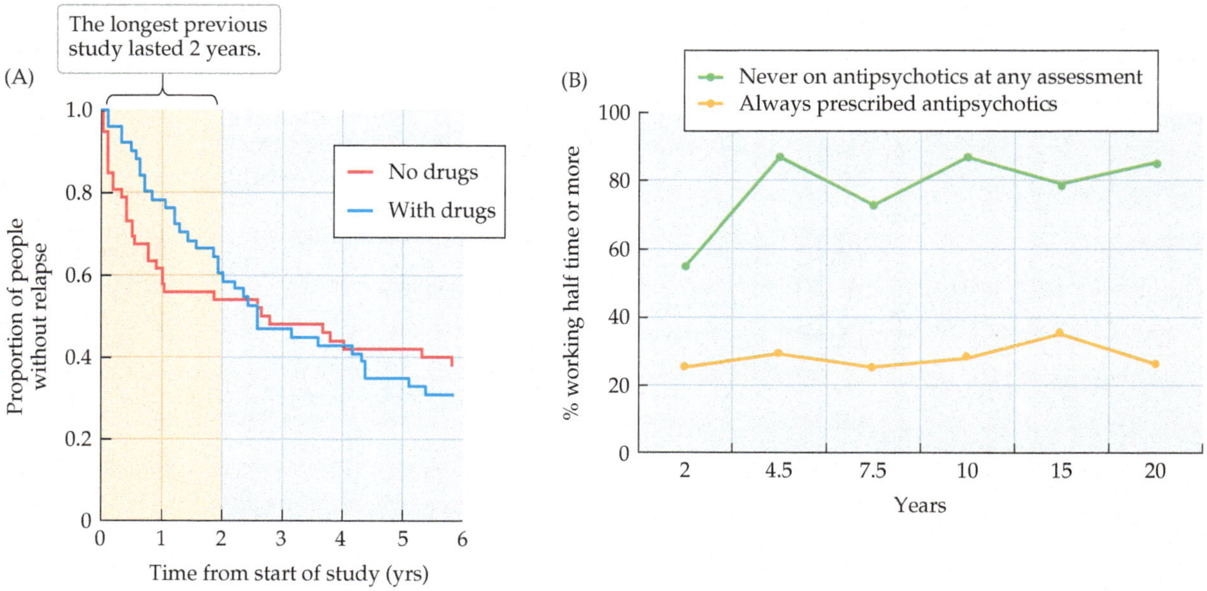

12.18 Long-Term Outcomes with and without Antipsychotics (A) Although shorter-term relapse rates seemed to be improved by treatment with antipsychotic drugs, this beneficial effect is lost, and eventually reversed, over the longer term. (B) In a group of people with schizophrenia followed for 20 years, chronic treatment with antipsychotics was associated with impaired ability to work. Although it is possible that the two groups differed in the severity of their symptoms at outset, the results are at odds with one of the arguments for using antipsychotic medication: that they will help the person to return to work. (Part A after L. Wunderink, 2019. *Ther Adv Psychopharmacol* 9: 1–14; B after M. Harrow et al., 2017. *Psychiatry Res* 256: 267–274.)

There is a controversial question of whether continued use of antipsychotics is the best pathway to recover from schizophrenia. For one thing, people who develop schizophrenia in developing countries are twice as likely to recover as people in industrialized nations like the United States (Hopper and Wanderling, 2000; Jablensky et al., 1992; Myers, 2010; Sartorius et al., 1972), even though the latter are more likely to receive medication. Plus programs that only provide medication are less successful than those that wean people with schizophrenia from medication and help them become reintegrated in society (DeSisto et al., 1995). Long-term studies indicate that, no matter how useful antipsychotics are for controlling symptoms at the onset of schizophrenia, relapses are less common in people who have weaned themselves from the drugs (Wunderink, 2019) , and people with schizophrenia who have been on antipsychotic medication for long periods of time are less likely to return to work than people with schizophrenia who did not receive antipsychotics (Harrow et al., 2017) (**FIGURE 12.18**). There is also growing support for cognitive behavioural therapy to help people with schizophrenia reduce stress (and therefore reduce outbreaks of symptoms) and to help them feel less disturbed by the voices they hear (Morrison et al., 2014). This allows them to disengage from the voices and realize the voices have no power over them.

That was the approach that worked for Eleanor, whom we met at the start of the chapter. She eventually decided that the voices were a result of childhood traumas she had experienced. She began to regard the voices as ways to understand what had happened and how she could cope with the consequences. Eleanor weaned herself from medication and returned to school, finishing her bachelor's degree. As she began regarding the voices more positively, they became less hostile, even helpful (Longden, 2013). During one of her exams, a voice dictated the answers to Eleanor, prompting her to wonder whether that was cheating! She went on to earn a master's degree and doctorate in psychology and has published a dozen scientific articles already in her young career. As you may have guessed, the main topic of Eleanor's research is the role of early life adversity in schizophrenia (Longden and Read, 2016a) and efforts to reduce the stigma of psychosis (Longden and Read, 2017). She also serves on the board of the international Hearing Voices Movement (intervoiceonline.org), working to generate acceptance and understanding of people like herself, who hear voices.

Courtesy of Eleanor Longden

Hearing Voices? Once Eleanor learned the voices she heard had no control over her life, she no longer feared them.

12.2 Depression Is the Most Prevalent Mood Disorder

Learning Objectives

After reading this section, you should be able to:

12.2.1 **Define depression and discuss its relationship to suicide risk.**

12.2.2 **Discuss the evidence that depression can be inherited.**

12.2.3 **Identify known neurophysiological correlates of depression.**

12.2.4 **Summarise the major clinical interventions used to treat depression, and catalog the main pharmacological modes of action of antidepressant medications.**

12.2.5 **Discuss ways in which sex differences, sleep, and hormone levels may all contribute to mood disorders.**

Disturbances of mood are a fact of life for humans; most of us experience periods of unhappiness that we commonly describe as depression. But for some people, an unhappy mood state is more than a passing malaise. This condition is most common in people over 40 years of age, especially women, but it can affect people of any age, race, or ethnicity (CDC, 2010). Clinically significant **depression** is characterised by an unhappy mood; loss of interests, energy, and appetite; difficulty in concentration; and restless agitation. Pessimism seems to seep into every act (Solomon, 2001). Periods of such *major depressive disorder* (sometimes called *unipolar depression*) can occur with no readily apparent stress, and without treatment the depression often lasts several months (Kupfer et al., 2012). U.S. government surveys show that each year, more than 7% of American adults experience at least one episode of clinically significant depression, with sharply increasing incidence among people age 18–24 (Substance Abuse and Mental Health Services Administration, 2018).

depression A psychiatric condition characterised by such symptoms as an unhappy mood; loss of interests, energy, and appetite; and difficulty concentrating.

Along with other mental illnesses, depression can be lethal, as it may lead to suicide. But whether or not the person is depressed, many suicides appear to be impulsive acts or are prompted by time-limited crises that would have eventually resolved themselves (Kleiman et al., 2017). For example, one classic study found that of more than 500 people who were prevented from jumping off the Golden Gate Bridge in San Francisco, only 6% later went on to commit suicide (Seiden, 1978). Similarly, suicide rates went down by a third in Britain when the switch was made in the 1960s and 70s from coal gas, which contains lots of deadly carbon monoxide, to natural gas for heating and cooking. The suicide rate has remained at that reduced level since. Apparently those thousands of Britons who would have found it easy to follow a suicidal impulse by turning on the kitchen oven did not kill themselves when more planning was required. Thus, it is important for society to erect barriers, either literally (e.g., on bridges) or metaphorically, to make it difficult for people to kill themselves regardless of whether they are experiencing depression or severe anxiety or, as in about 40% of cases of suicide, do not have an established mental disorder (Chesney et al., 2014; Leavey et al., 2016). If suicide is averted the first time it is seriously attempted, the person is very unlikely to ever try it again; similarly, reducing access to means of suicide, such as through firearm legislation that mandates waiting periods, could help reduce suicide rates in some regions (Anestis et al., 2019). But despite awareness campaigns and various preventative measures, suicide remains a significant and growing problem (CDC 2018). Warning signs that indicate a person may be contemplating suicide are listed in **TABLE 12.3**.

TABLE 12.3 Warning Signs of Suicide[a]

Threatening to hurt or kill oneself or talking about wanting to hurt or kill oneself
Looking for ways to kill oneself by seeking access to firearms, pills, or other means
Talking or writing about death, dying, or suicide when these actions are out of the ordinary for the person
Feeling hopeless
Feeling rage or uncontrolled anger or seeking revenge
Acting reckless or engaging in risky activities—seemingly without thinking
Feeling trapped—like there's no way out
Increasing alcohol or drug use
Withdrawing from friends, family, and society
Feeling anxious, agitated, or unable to sleep, or sleeping all the time
Experiencing dramatic mood changes
Seeing no reason for living or having no sense of purpose in life

Adapted from: https://www.nimh.nih.gov/health/topics/suicide-prevention/index.shtm

[a]Developed by the U.S. Department of Health and Human Services, these warning signs offer guidance about how to recognise someone at risk for suicide. If you or someone you know exhibits even a few of these signs, you can call the National Suicide Prevention Lifeline at 1-800-273-TALK (1-800-273-8255) at any time of day, any day of the year.

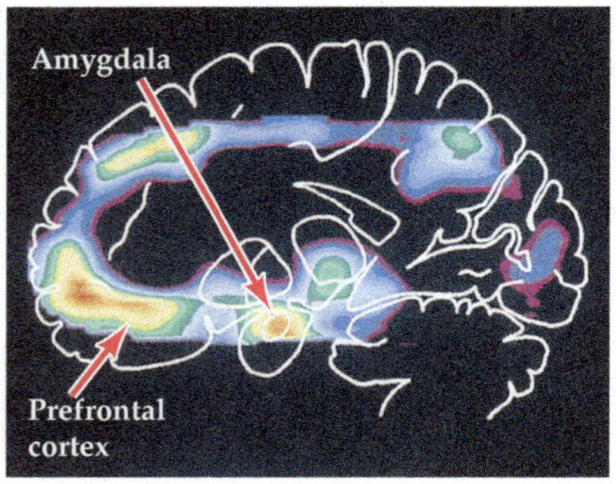

From W. C. Drevets, 1998. *Annu Rev Med* 49: 341–361

12.19 Brain Activity Patterns in Depression This PET scan reveals increased activity in the prefrontal cortex and the amygdala of depressed patients. This combined image is the result of the subtraction of brain scans of controls from those of depressed individuals. Areas of highest activation are shown in red and orange.

Inheritance is an important determinant of depression

Genetic studies of depressive disorders reveal strong hereditary contributions, with a much higher concordance rate for monozygotic twins than for dizygotic twins, from early adulthood through late middle age and regardless of whether the twins are reared apart or together (Kendler et al., 1999; Torvik et al., 2018). Adoption studies show higher rates of depression in the *biological* parents of a depressed person than in the foster parents. Although several early studies implicated specific chromosomes in susceptibility for depression, subsequent linkage studies revealed that, as in schizophrenia, there is no single gene for depression (Risch et al., 2009). Rather, many genes contribute to making a person more or less susceptible, and environmental factors determine whether depression results (Gotlib et al., 2014).

The brain changes with depression

Most reports of differences in the brains of depressed people focus on functional changes. PET scans of depressed people show increases in blood flow, suggesting greater activity, in the prefrontal cortex and the amygdala compared with controls (**FIGURE 12.19**) (Drevets, 1998). In addition to increasing in the prefrontal cortex, blood flow decreases in the parietal and posteriour temporal cortex and in the anteriour cingulate, systems that have been implicated in attention (see Chapter 17). The increase in blood flow in the amygdala—a structure involved in mediating fear (see Chapter 15)—persists even after the alleviation of depression over time.

Descendants of people with severe depression also have a thinner cortex across large swaths of the right hemisphere than do controls (B. S. Peterson et al., 2009), which might make them vulnerable to depression. Many studies report hippocampal volume is reduced in people with depression (Sexton et al., 2013), and there is reduced activation of the hippocampal region in depressed people during memory tasks (K. D. Young et al., 2011). But whether these changes in the hippocampus are present before the depression, and therefore may be a contributing cause of the disorder, or are a result of the depression remains unknown. In any case, there are effective treatments for depression, as we discuss next.

A wide variety of treatments are available for depression

Electroconvulsive therapy (ECT)—the intentional induction of a large-scale seizure (Weiner, 1994)—was originally deployed during the 1930s in a desperate and unsuccessful attempt to relieve the symptoms of schizophrenia. Despite ECT's failure at helping people with schizophrenia, however, clinical observations soon revealed that it *could* rapidly reverse severe depression. Demonizing portrayals of ECT in popular entertainment, and the advent of antidepressant drugs, have made ECT less common, but it remains an important tool for treating severe, drug-resistant depression (M. Fink and Taylor, 2007). ECT may work by reducing functional connectivity between an overactive prefrontal cortex and other cortical regions (J. S. Perrin et al., 2012).

A more modern technique for altering cortical electrical activity, *repetitive transcranial magnetic stimulation* (rTMS; see Chapter 2), is being explored as a treatment for depression (D. R. Kim et al., 2009), but it appears to be less effective than ECT (Health Quality Ontario, 2016). The milder treatment of direct current stimulation through the skull appears to be effective for relieving depression in some studies (Brunoni et al., 2016), but not in others (Murphy et al., 2009). Another electrophysiological approach involves *vagal nerve stimulation,* in which electrodes are surgically wrapped around the vagus nerve (see Chapter 1) in the neck, and a pacemaker provides mild electrical stimulation at intervals. The treatment is offered for people who have not experienced relief from drugs or ECT, but the effectiveness of vagal nerve stimulation for depression, versus possible placebo effects, has not been well established (Blumberger et al., 2015; Vonck et al., 2014). In **deep brain stimulation (DBS)**, mild electrical stimulation is applied to brain sites through a surgically implanted electrode, directed at several different target regions (Crowell et al., 2014), as a last-ditch effort to relieve depression that resists other treatments (Kringelbach et al., 2007). But despite over 10 years of reports

electroconvulsive therapy (ECT) A last-resort treatment for intractable depression, in which a strong electrical current is passed through the brain, causing a seizure.

deep brain stimulation (DBS) Mild electrical stimulation through an electrode that is surgically implanted deep in the brain.

using DBS for depression (Mayberg et al., 2005), double-blind controlled studies are only recently being published, and they call into question whether it is more effective than placebo at reducing depression (Bergfeld et al., 2016; Kisely et al., 2018).

Today, the most common treatment for depression is the use of drugs that affect monoamine transmitters. The **monoamine hypothesis** (Schildkraut and Kety, 1967) was suggested by the first antidepressants, which were inhibitors of **monoamine oxidase** (**MAO**), the enzyme that normally inactivates the monoamines: norepinephrine, dopamine, and serotonin. The fact that MAO inhibitors raise the level of monoamines present in synapses suggests that depressed people do not get enough stimulation at those synapses. This would also explain why the drug **reserpine**, which reduces norepinephrine and serotonin release in the brain, can cause profound depression. Inducing the release of monoamines may be the way ECT helps depression. Second-generation antidepressants called **tricyclics** conform to the monoamine hypothesis because they inhibit the reuptake of monoamines, boosting their synaptic activity.

Among the monoamines, serotonin may play the most important role in depression. A major class of modern antidepressants is the **selective serotonin reuptake inhibitors** (**SSRIs**), such as Prozac (**TABLE 12.4**) (see Chapter 3). These drugs are more effective than MAO inhibitors and tricyclics and have fewer side effects. In rats and mice, SSRIs increase neurogenesis in the hippocampus (Sahay and Hen, 2007; Samuels et al., 2015), which may mediate some of the mood effects of the drugs. SSRI treatment also increases the production of brain steroids (L. D. Griffin and Mellon, 1999), such as allopregnanolone, which may contribute to the effectiveness of SSRIs by stimulating GABA receptors and reducing anxiety.

However, there are problems with the theory that reduced serotonin stimulation causes depression. We know that SSRI drugs increase synaptic serotonin within hours of administration. Yet it typically takes several weeks of SSRI treatment before people feel better. This paradox suggests that it is the brain's *response* to increased synaptic serotonin that relieves the symptoms, and that this response takes time. One suggestion is that SSRIs may immediately improve a person's emotional outlook (Capitão et al., 2015) but that several weeks of experiencing this rosier outlook may be required to overturn the depression. Thus, even though boosting serotonin helps some people, their depression may originally have been caused by other factors in the brain.

Second-generation monoamine reuptake inhibitors, called **serotonin-norepinephrine reuptake inhibitors** (**SNRIs**), affect the synaptic availability of norepinephrine in addition to blocking reuptake of serotonin (Hillhouse and Porter, 2015). It is not yet clear whether these drugs, such as Effexor (venlafaxine), Cymbalta (duloxetine), and Pristiq (desvenlafaxine), offer significantly greater benefit in depression than the SSRIs, and they are similarly slow to have effects on mood (e.g., Magni et al., 2013). Drugs affecting other signalling pathways are also under investigation as possible antidepressants; for example, the glutamate receptor antagonist ketamine (see Chapter 4) relieves depression almost instantly (Sanacora et al., 2017), in contrast to SSRIs and SNRIs, which typically must be taken for a number of weeks before mood effects are observed. There are also reports that hallucinogens like psilocybin can rapidly relieve depression (Baumeister et al., 2014; Carhart-Harris, Erritzoe et al., 2012) (see Table 3.4).

Clinical research has confirmed that when averaged across groups of people with

monoamine hypothesis The hypothesis that depression is caused by reduced activity of one or more monoamine transmitters, such as serotonin.

monoamine oxidase (MAO) An enzyme that breaks down and thereby inactivates monoamine transmitters.

reserpine A drug that causes the depletion of monoamines and can lead to depression.

tricyclics A class of drugs that act by increasing the synaptic accumulation of serotonin and norepinephrine.

selective serotonin reuptake inhibitors (SSRIs) A class of drugs that block the reuptake of transmitter at serotonergic synapses. They are commonly used to treat depression.

serotonin-norepinephrine reuptake inhibitor (SNRI) A drug that blocks the reuptake of transmitter at both serotonergic and noradrenergic synapses.

TABLE 12.4 Drugs Used to Treat Depression

Drug class	Mechanism of action	Examples[a]
Monoamine oxidase (MAO) inhibitors	Inhibit the enzyme monoamine oxidase, which breaks down serotonin, norepinephrine, and dopamine	Marplan, Nardil, Parnate
Tricyclics and heterocyclics	Inhibit the reuptake of norepinephrine, serotonin, and/or dopamine	Wellbutrin, Elavil, Aventyl, Ludiomil, Norpramin
Selective serotonin reuptake inhibitors (SSRIs)	Block the reuptake of serotonin, having little effect on norepinephrine or dopamine synapses	Prozac, Paxil, Zoloft
Second-generation antidepressants and investigational drugs	Norepinephrine and dopamine reuptake inhibitors (NDRIs), serotonin-norepinephrine reuptake inhibitors (SNRIs), noradrenergic and specific serotonergic antidepressants (NaSSAs), serotonin antagonist and reuptake inhibitors (SARIs), opioid receptor modulators, ketamine	Wellbutrin/Zyban (NDRI), Effexor (SNRI), Remeron (NaSSA), Oleptro (SARI), Buprenex (opioid receptor modulator)

[a]The names given are the more commonly used trade names rather than chemical names.

12.20 **When Are Antidepressants More Effective than Placebo?** (After J. C. Fournier et al., 2010. *JAMA* 303: 47–53.)

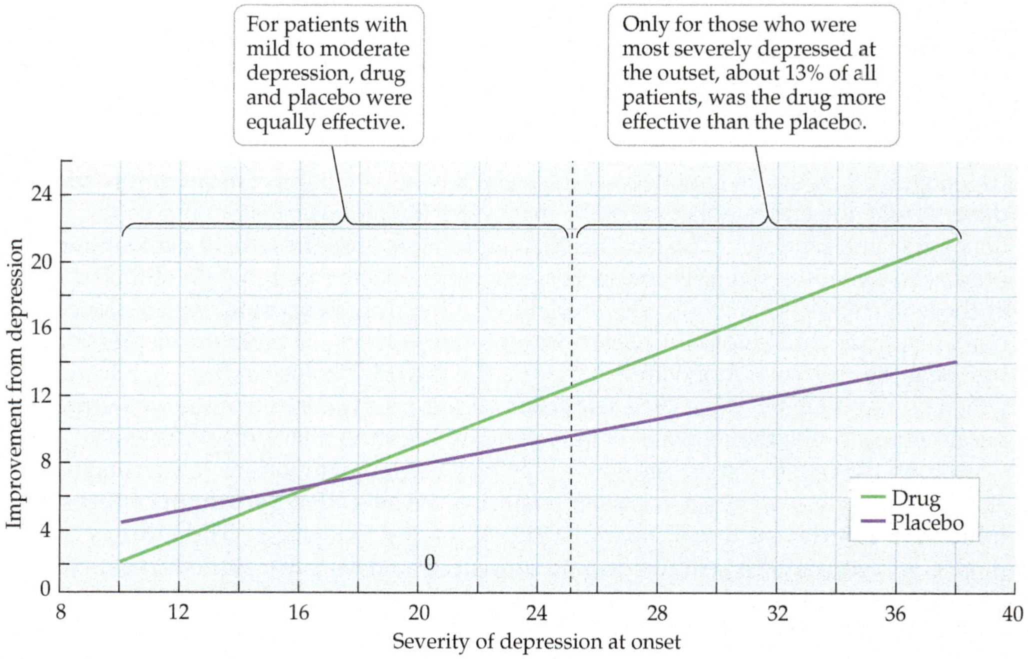

serotonin syndrome Syndrome of confusion, muscle spasms, and fever that may occur when brain levels of serotonin are too high and is a risk of taking SSRIs.

cognitive behavioural therapy (CBT) Psychotherapy aimed at correcting negative thinking and improving interpersonal relationships.

depression, the first- and second-generation antidepressants are indeed effective, but the overall beneficial effects appear to be modest in size (Cipriani et al., 2018). However, the group data may underestimate the benefit of antidepressants for many people, because the drugs don't help everyone with depression, and antidepressants can be accompanied by undesirable side effects that prompt people to stop taking their medication, leading to lower *group* averages on measures of effectiveness. In some placebo-controlled trials, a significant proportion of people taking the *placebo* also report feeling better, suggesting that at least some people helped by SSRI treatment are actually benefitting from a placebo effect (Berton and Nestler, 2006), and other analysis suggests that SSRIs show their greatest benefit in people with more severe cases of depression (Fournier et al., 2010) (**FIGURE 12.20**).

Millions of children and adolescents have been given prescriptions for antidepressants, but there is little or no evidence that any of these drugs, with the possible exception of fluoxetine (Prozac), have beneficial effects in pediatric populations (Cipriani et al., 2016). This is worrisome, because evidence also suggests that reuptake inhibitors—including fluoxetine—increase the risk of suicidal thoughts and actions in children and adolescents (Schneeweiss et al., 2010). Another risk of SSRIs, for people of all ages, is that a variety of over-the-counter drugs may synergize with the drugs to push synaptic serotonin levels too high, triggering **serotonin syndrome**, which includes confusion, muscle spasms, and fever. Thousands of cases are reported each year, causing over 100 deaths (Dvir and Smallwood, 2008).

Despite the huge popularity of drug treatments for depression, a course of 20 or so sessions of **cognitive behavioural therapy (CBT)**—psychotherapy aimed at correcting negative thinking, reducing stress, and improving interpersonal relationships—is about as effective as SSRI treatment (Butler et al., 2006). Furthermore, the rate of relapse is lower for CBT than for SSRI treatment (DeRubeis et al., 2008; Kuyken et al., 2016). Interestingly, a combination of CBT and SSRIs can be more effective in combatting depression than either treatment by itself, especially in more resistant forms of the disorder (Nakagawa et al., 2017; Schramm et al., 2008). Typically, CBT helps the client to recognise self-defeating modes of thinking and encourages breaking out of a cycle of self-fulfilling depression (**FIGURE 12.21**).

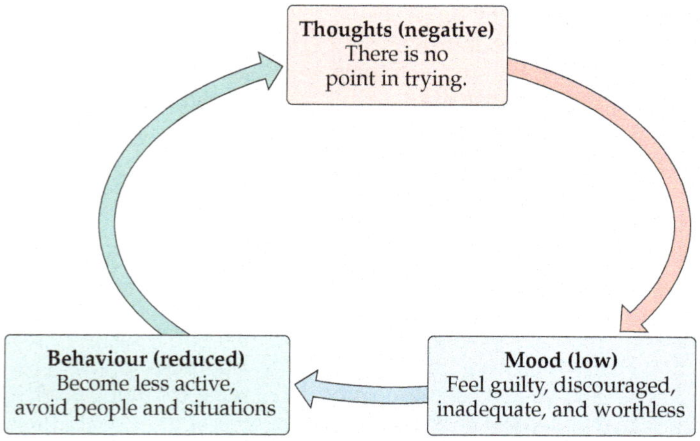

12.21 **The Treadmill of Depression** Cognitive behavioural therapy (CBT) aims to break the cycle at one or more points. Thus even a modest program of physical exercise may help because it interrupts the "actions" or "behaviour" part of the cycle.

The Cutting Edge

Can Our Genes Tell Us Which Drugs to Use?

Despite decades of dedicated research efforts, swift and effective treatments to relieve the crushing symptoms of major depression remain an unrealized dream. Many patients don't respond to antidepressants at all, and many more experience only partial relief, or difficult side effects, leading to a trial-and-error approach to treatment in which people with depression may be cycled through many different drugs, hoping to find one that will provide that elusive cure. A large part of the problem is that the major psychiatric illnesses are rarely caused by deletion, mutation, or variation of a single gene. As we've discussed, hundreds or thousands of genes are implicated in each of the major psychiatric disorders, which makes it unlikely that we will soon have a simple genetic screen to tell us who will, and who will not, develop major depression. So, researchers are instead trying to use genetic screening a different way—to make personalised predictions about which of the many available drugs will be most effective.

Among people who share the diagnosis of major depression, subgroups will vary in whether they will benefit from a particular antidepressant and whether they will experience adverse side effects. Some of the variability in patients' responses to particular drugs can be explained by a collection of **genetic polymorphisms**—individual small variations within certain genes—so perhaps screening for specific combinations of polymorphisms could guide individualized selection and dosing of antidepressant medications (Fabbri and Serretti, 2018; Tansey et al., 2013). In **pharmacogenomics**, each person's genome is scanned for polymorphisms in genes belonging to two general classes: *pharmacokinetic genes* that determine how the drug is metabolised and moves through the body, and *pharmacodynamic genes* related to the brain mechanisms through which the drug has its intended effects on neural function (**FIGURE 12.22**). The

genetic polymorphisms The phenomenon in which there exist multiple (sometimes many) different alleles at an individual locus within a gene.

pharmacogenomics The identification of combinations of alleles affecting the metabolism and actions of drugs, in order to personalise drug treatments.

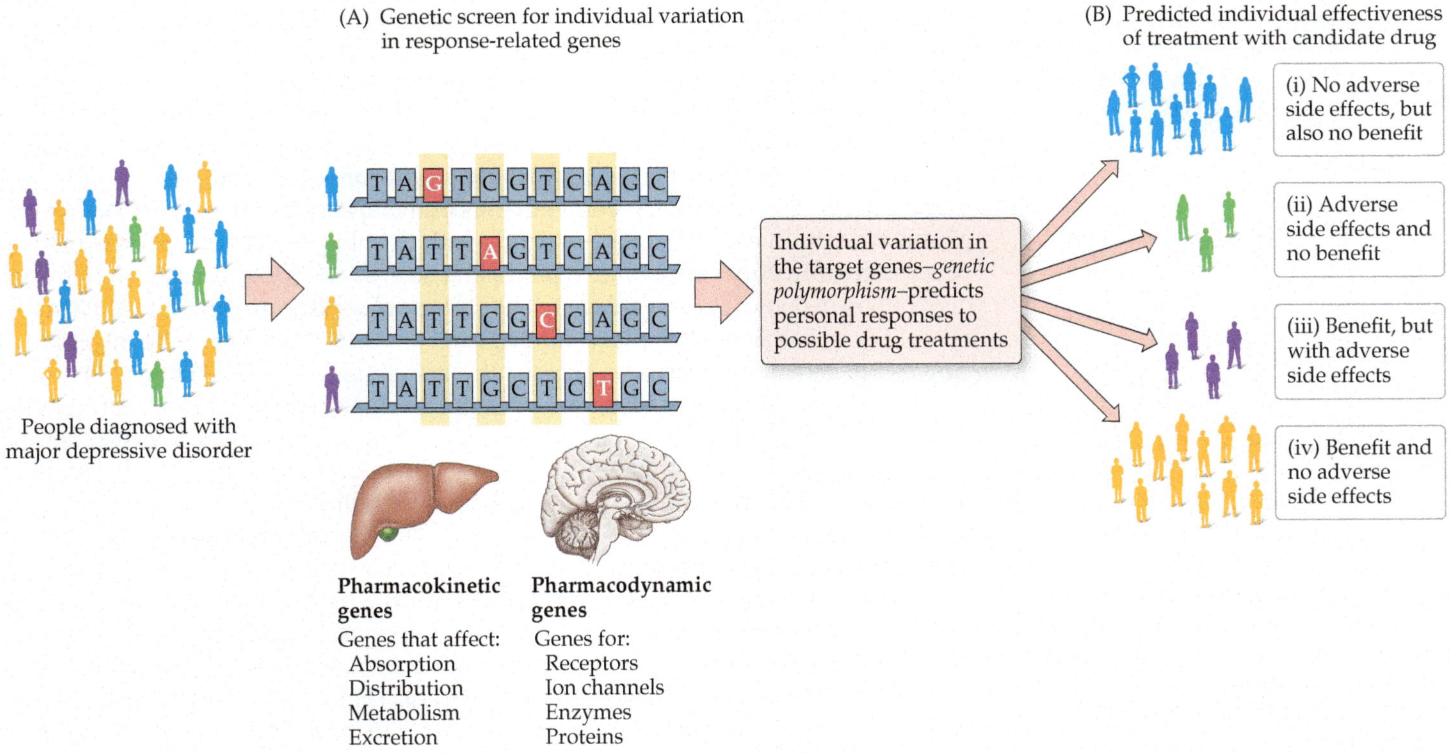

12.22 Pharmacogenomics-Guided Treatment A group of people all sharing the same diagnosis—in this case, major depressive disorder—differ in the extent to which they will benefit from any particular drug treatment, as well as the extent to which they will experience adverse side effects. (A) Individuals' DNA is screened for specific variations—*genetic polymorphisms*—in genes related to the drugs' effects in the brain, and in genes related to metabolism and clearance of the drug. (B) Researchers hope that the combination of polymorphisms in each individual can be used to predict which drugs are most likely to be effective for them, as well as least likely to produce serious side effects.

various combinations of polymorphisms that individuals possess are then used to make predictions about how they will respond to any specific antidepressant drug and whether they will experience toxicity or other adverse effects from the drug.

Does it work? Maybe. At this early stage, data are somewhat scanty, but preliminary meta-analyses suggest that pharmacogenomics-guided treatment leads to a more than 50% increase in remission in people with major depression (Bousman et al., 2019; Rosenblat et al., 2018). In the largest clinical trial yet, more than 1100 people with major depressive disorder that had previously shown poor response to antidepressants were assigned to receive either treatment as usual or pharmacogenomics-guided treatment. Participants and raters were blind to experimental condition. At week 8, significant improvements in response were observed in the pharmacogenomics-guided group (Greden et al., 2019).

However, one of the primary goals of the study—a general improvement in symptoms independent of whether remission occurred—was not observed more often in the pharmacogenomics-guided participants. Another study (with a small sample) found that the effect of pharmacogenomic assays could be predicted based on knowledge of one or two of the pharmacological properties of the drugs under study—no genetic testing required (Macaluso and Preskorn, 2018). In general, authorities warn that there is not yet sufficient basis to incorporate pharmacogenomic screening into clinical practice (Zeier et al., 2018), with the FDA noting that "the relationship between DNA variations and the effectiveness of antidepressant medication has never been established" (FDA, 2018).

While the effectiveness of pharmacogenomics-guided treatment for depression has yet to be conclusively determined, it appears to hold great promise. Part of the process will involve accurately identifying genes within which variation leads to disease. For example, a polymorphic region called 5-HTTLPR within the serotonin transporter gene, previously believed to be involved in stress-related depression, has recently been shown to have little relationship to that process (Culverhouse et al., 2018). Future developments may also involve measuring *epigenetic* modifications of key genes (Hack et al., 2019) (see Chapter 7). ■

The hypothalamic-pituitary-adrenal axis is involved in depression

People who have very high levels of circulating glucocorticoids such as cortisol are prone to depression. This condition, called **Cushing's syndrome**, may have several different causes, including hormone-secreting tumours or therapeutic treatments with synthetic glucocorticoids. In more than 85% of people with Cushing's syndrome, depression appears quite early, even before other typical signs, such as obesity or unusual growth and distribution of body hair (Krystal et al., 1990). These observations suggest that dysfunction of the hypothalamic-pituitary-adrenal axis (**FIGURE 12.23A**) may be involved in depression, perhaps as part of a depression-inducing stress reaction (Herbert, 2013).

People who commit suicide show very high levels of circulating cortisol (Pompili et al., 2010), and people hospitalized with depression show elevated cortisol levels (**FIGURE 12.23B**). These findings suggest that adrenocorticotropic hormone (ACTH) is released in excessive amounts by the anteriour pituitary. A standard method for assessing hypothalamic-pituitary-adrenal function—the **dexamethasone suppression test**—can reveal a tendency to release excess cortisol.

Cortisol is normally secreted in a daily rhythm, with higher levels in the morning; in depressed people the rhythm is blunted and the overall levels of cortisol are higher (**FIGURE 12.23C**). Dexamethasone is a potent synthetic glucocorticoid that ordinarily suppresses the early-morning rise in ACTH that is typical in most people. When given late at night, dexamethasone seems to "fool" the hypothalamus into believing that there is a high level of circulating cortisol. In healthy individuals, the dexamethasone suppresses cortisol release the next day, but in many individuals with depression it fails to have this effect (**FIGURE 12.23D**) (Belvederi et al., 2014). As depression is relieved, dexamethasone again suppresses cortisol normally, no matter what caused the relief—passage of time, psychotherapy, pharmacotherapy, or electroconvulsive therapy.

Cushing's syndrome A condition in which levels of adrenal glucocorticoids are abnormally high.

dexamethasone suppression test A test of pituitary-adrenal function in which the participant is given dexamethasone, a synthetic glucocorticoid hormone, which should cause a decline in the production of adrenal corticosteroids.

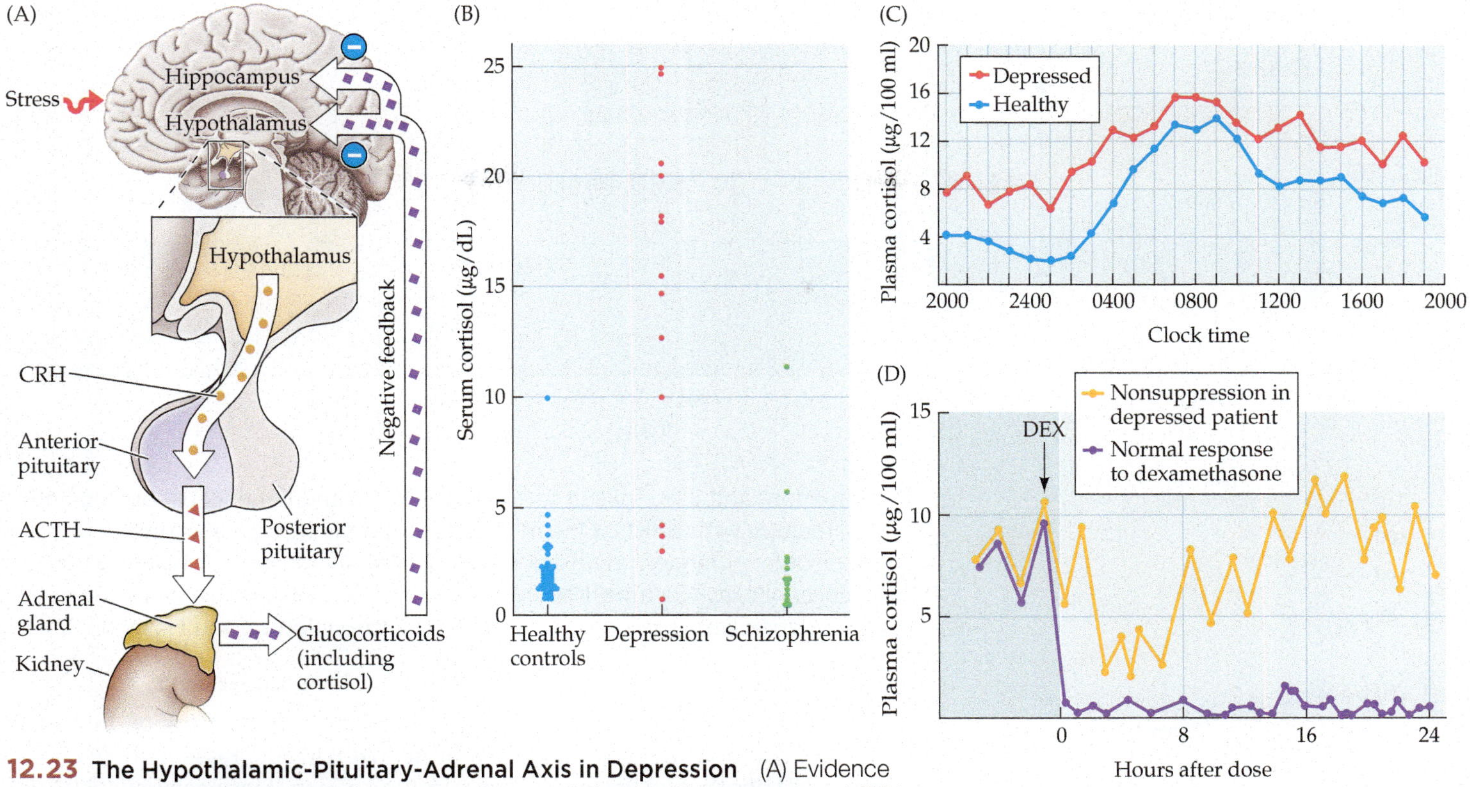

12.23 The Hypothalamic-Pituitary-Adrenal Axis in Depression (A) Evidence shows that the hypothalamic-pituitary-adrenal system is involved in depression. ACTH, adrenocorticotropic hormone; CRH, corticotropin-releasing hormone. (B) Circulating cortisol levels are usually higher in people with depression than in psychiatric or healthy controls. In this plot, each dot represents an individual case. (C) People with depression may display increased levels of cortisol across the circadian rhythm of release. (D) The same dose of dexamethasone is far less effective at suppressing the next day's cortisol release in people with depression. (Part B after A. J. Rosthschild et al., 1982. *Brit J Psychiat* 141: 471–474; C after E. R. Kandel, 2000. In *Principles of neural science*, 4th ed., E. R. Kandel et al. [Eds.], pp. 1209–1226. McGraw-Hill: New York; D after S. B. Klein, 2000. *Biological psychology*. Prentice-Hall: Upper Saddle River, NJ.)

Why is depression more prevalent in women than in men?

Studies all over the world show that more women than men have major depression. In the United States, women are twice as likely as men to have major depression (D. J. Brody et al., 2018). Some researchers suggest that the apparent sex difference reflects patterns of help-seeking by males and females—notably, that women use health facilities more than men do. But sex differences in the incidence of depression also are evident in door-to-door surveys (Robins and Regier, 1991), which would appear to rule out the simple explanation that women seek treatment more often than men do.

Some researchers have emphasised gender differences in endocrine physiology. The occurrence of clinical depression often is related to events in the female reproductive cycle—for example, before menstruation, during use of contraceptive pills, following childbirth, and during menopause. Although there is little relation between circulating levels of individual hormones and measures of depression, the phenomenon of **postpartum depression**, a bout of depression either immediately preceding or following childbirth, suggests that some combination of hormones can precipitate depression. About one out of every seven pregnant women will show symptoms of depression (Dietz et al., 2007).

Because postpartum depression may affect the mother's relationship with her child, causing long-lasting deleterious effects on the child's behaviour (Tronick and Reck, 2009), there is alarm about this problem. Unfortunately, there is also evidence that prenatal exposure to SSRIs taken by the mother may affect the child's later behaviour (Brandlistuen et al., 2015; Lupattelli et al., 2018), and no one knows whether SSRIs in breast milk affect the child. Thus cognitive behavioural therapy offers the safest treatment for postpartum

postpartum depression A bout of depression that afflicts a woman around the time she gives birth.

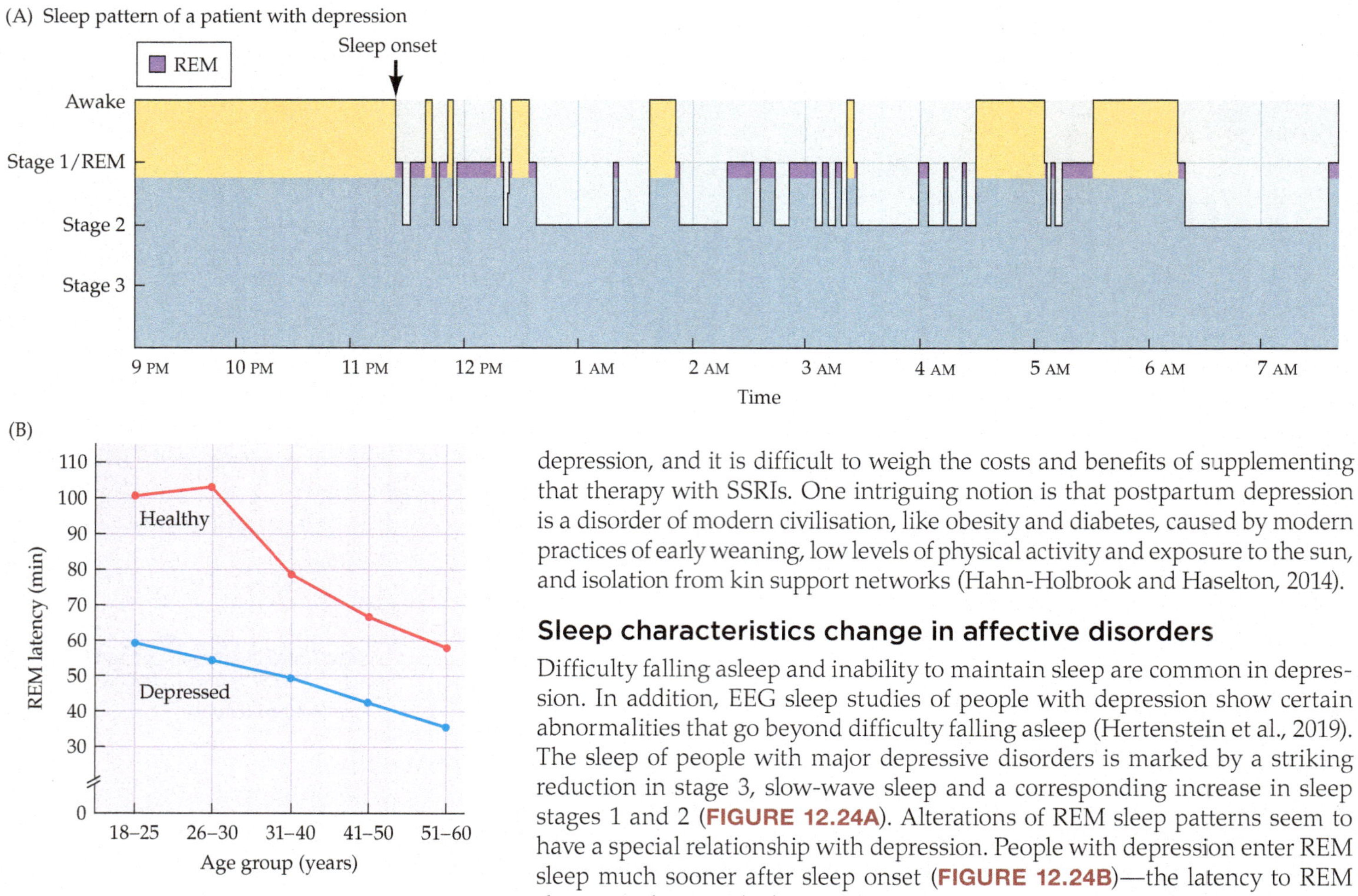

12.24 Sleep and Depression
(A) Depressed people spend little or no time in stage 3 sleep, also known as *slow-wave sleep*. (Compare with Figure 14.12.) (B) People with depression also enter their first REM period earlier in the night. Thus, REM sleep seems to be distributed differently in people with depression. (Part B after J. C. Gillin et al., 1981. *Psychiatry Res* 4: 73–78; D. J. Kupfer et al., 1982. *Neurobiol Aging* 3: 351–360.)

depression, and it is difficult to weigh the costs and benefits of supplementing that therapy with SSRIs. One intriguing notion is that postpartum depression is a disorder of modern civilisation, like obesity and diabetes, caused by modern practices of early weaning, low levels of physical activity and exposure to the sun, and isolation from kin support networks (Hahn-Holbrook and Haselton, 2014).

Sleep characteristics change in affective disorders

Difficulty falling asleep and inability to maintain sleep are common in depression. In addition, EEG sleep studies of people with depression show certain abnormalities that go beyond difficulty falling asleep (Hertenstein et al., 2019). The sleep of people with major depressive disorders is marked by a striking reduction in stage 3, slow-wave sleep and a corresponding increase in sleep stages 1 and 2 (**FIGURE 12.24A**). Alterations of REM sleep patterns seem to have a special relationship with depression. People with depression enter REM sleep much sooner after sleep onset (**FIGURE 12.24B**)—the latency to REM sleep correlates with the severity of depression—and their REM sleep is unusually vigourous. Furthermore, the temporal distribution of REM sleep is altered, with an increased amount of REM sleep occurring during the first half of sleep, as though REM sleep were displaced toward an earlier period in the night (Palagini et al., 2013). In addition to these links between the daily sleep rhythm and depression, seasonal rhythms have been implicated in a particular depressive condition known as *seasonal affective disorder* (*SAD*), which is described in **BOX 12.2**. Effectively treating insomnia is associated with positive changes in mood (Gebara et al., 2018).

Scientists are still searching for animal models of depression

Because a monkey or a cat or a rat can't tell us if it has delusions of persecution or if it hears voices telling it what to do, animal models of schizophrenia are limited. But many of the signs of depression—such as decreased social contact, problems with eating, and changes in activity—are behaviourally overt (Nestler and Hyman, 2010).

In one type of stress model—**learned helplessness**—an animal is exposed to a repetitive stressful stimulus, such as an electrical shock, that it cannot escape (Seligman and Beagley, 1975). Like depression, learned helplessness has been linked to a decrease in serotonin function (Petty et al., 1994) and also dopamine (B. Li et al., 2011). Removing the olfactory bulb from rodents also creates a model of depression: the animals display irritability, preferences for alcohol, and elevated levels of corticosteroids—all of which are reversed by many antidepressants. A strain of rats created through selective breeding—the Flinders-sensitive line—has been proposed as a model of depression because these animals show reduced locomotor activity, reduced body weight, increased REM sleep, learning difficulties, and exaggerated immobility in response to chronic stress (Overstreet, 1993). So, while these varied animal models may be useful in finding the essential mechanisms that cause and maintain depression in humans, there remains no consensus as to how depression is best modelled in animals (Harro, 2019).

learned helplessness A learning paradigm in which individuals are subjected to inescapable, unpleasant conditions.

BOX 12.2 The Season to Be Depressed?

Seasonal rhythms characterize the behaviour and physiology of many animals, including humans. For some unfortunate people, winter seems to bring a low period that may become a profound depression. Sometimes the winter depression alternates with summertime mania (Blehar and Rosenthal, 1989). In wintertime, affected people feel depressed, slow down, generally sleep a lot, and overeat. Come summer, they are elated, energetic, and active, and they become thinner. This syndrome—called **seasonal affective disorder (SAD)**—appears predominantly in women and generally starts in early adulthood.

Seasonal rhythms in animals are controlled by the length of the day, so researchers asked whether seasonal changes in exposure to sunlight might cause SAD. Some people with SAD were treated with doses of bright light, to see whether it acts as an antidepressant. The efficacy of light therapy, also called *phototherapy*, in SAD is fairly well established (Golden et al., 2005), and in many ways light therapy resembles treatment with traditional antidepressant drugs (A. Moscovitch et al., 2004). Light therapy may be most effective when administered immediately upon awakening in the morning (Lewy et al., 1998). A person receiving phototherapy is shown in the figure.

However, given the known susceptibility of depression to placebo effects (see Figure 12.20), it is possible that light is only effective because people diagnosed with SAD expect it to be. In fact, we have known for over a century that the seasonal peak in suicide is not in winter (and certainly not around the December holidays, as a persistent myth indicates), but during the lengthening days of spring and summer (Durkheim, 1897; Galvão et al., 2018). That well-replicated finding seems difficult to reconcile with the idea that short days can cause depression in any significant fraction of the population. In fact, several studies found no relationship between daylength and depression, casting doubt on whether SAD is a valid concept (Brancaleoni et al., 2009; Mersch et al., 1999; Traffanstedt et al., 2016).

Light therapy for SAD

seasonal affective disorder (SAD) A depression thought to be caused by the short days of winter.

12.3 Extreme Mood Cycles Define Bipolar Disorder

Learning Objectives

After reading this section, you should be able to:

12.3.1 **Define bipolar disorder in terms of its cardinal symptoms and relation to the other major psychiatric illnesses.**

12.3.2 **Summarise treatment options for bipolar disorder and their probable mode of action.**

12.3.3 **Discuss the genetic and physiological components of bipolar disorder, including differences between the sexes.**

bipolar disorder Formerly called *manic-depressive illness*. A psychiatric disorder characterised by periods of depression that alternate with excessive, expansive moods.

Bipolar disorder is characterised by periods of depression alternating with periods of excessively expansive mood (or *mania*) that includes sustained overactivity, talkativeness, strange grandiosity, and increased energy. Imaging confirms that

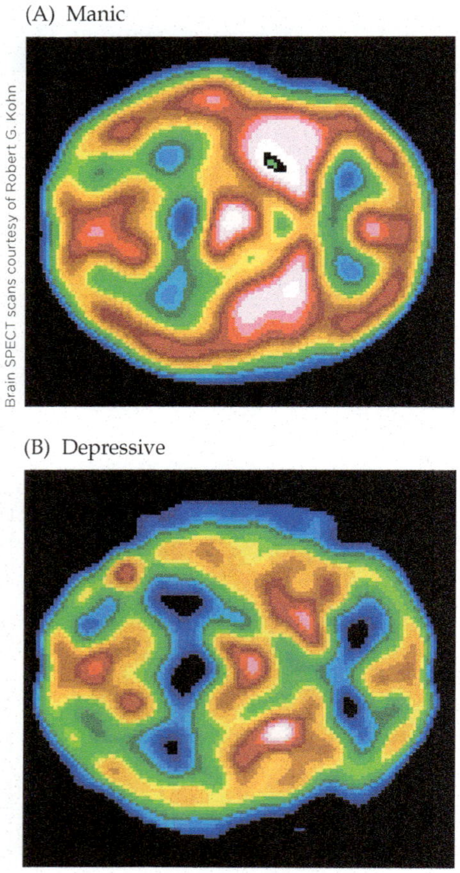

12.25 Functional Images of Bipolar Disorder Dramatic differences in brain activity are evident between the manic (A) and depressive (B) phases of this patient's bipolar illness. Areas of highest activation are shown in red and orange.

the brain is more active during these bouts too (**FIGURE 12.25**). The rate at which mood alternates varies between individuals: Some people exhibit *rapid-cycling* bipolar disorder, defined as consisting of four or more distinct cycles in one year (and some individuals have many more cycles than that; some may even show several cycles per *day*). Other people experience a milder, subclinical, related state called *cyclothymia*, in which the person experiences less-extreme moods, cycling between *dysthymia* (poor mood or mild depression) and *hypomania* (a state of increased energy and positive mood that lacks some of the bizarre aspects of frank mania).

Many researchers believe that bipolar disorder has more in common with schizophrenia than with "ordinary" depression. This is one reason why old terms for bipolar disorder—*bipolar depression* and *manic depression*—have fallen out of favour. For example, the self-aggrandizing ideas and extreme talkativeness of people in the manic phase of bipolar disorder (e.g., "The president called me this morning to thank me for my efforts") resemble the delusions seen in schizophrenia. In addition, families in which some individuals have been diagnosed with bipolar disorder are more likely than other families to have individuals with a diagnosis of schizophrenia (Lichtenstein et al., 2009; van Os and Kapur, 2009).

The neural basis of bipolar disorder is not fully understood, but people with bipolar disorder exhibit enlarged ventricles (Arnone et al., 2009), which is also reminiscent of changes seen in schizophrenia (see Figures 12.5 and 12.6). The more manic episodes the person has experienced, the greater the ventricular enlargement, suggesting a worsening of brain loss over time. The changes probably include subcortical limbic structures, such as the amygdala (DelBello et al., 2004) and hippocampus (Moorhead et al., 2007). Although first-generation antipsychotics do not seem to help people with bipolar disorder, the newer, second-generation antipsychotics may dampen the manic phase.

However, most people with bipolar disorder benefit from treatment with the element **lithium** (Kingsbury and Garver, 1998). The benefits of lithium for bipolar disorder were discovered purely by accident when it was intended as an inert control for another drug, and the mechanism of action remains mysterious. Lithium has wide-ranging effects on the brain, including interacting with a protein that is part of the circadian molecular clock (L. Yin et al., 2006) (see Chapter 15), boosting activity of BDNF (brain-derived neurotrophic factor) (M. K. Rowe and Chuang, 2004), and reducing neuronal activity (Mertens et al., 2015). Lithium occurs naturally in low levels in drinking water, and there's evidence that communities with more lithium in their water have a lower incidence of depression and suicide (Sugawara et al., 2013; Vita et al., 2015). For people taking lithium medications, care must be taken to avoid overdose, because it has a narrow therapeutic index (the range of safe doses; see Figure 3.9E). Nevertheless, well-managed lithium treatment produces marked relief for many people and even has been reported to increase the volume of gray matter in the brain (G. J. Moore et al., 2000).

Perhaps the fact that the manic phases blocked by lithium are so exhilarating is the reason that some people with bipolar disorder stop taking the medication. Unfortunately, doing so means that the depressive episodes return as well. But other drug treatments are available, and repetitive transcranial magnetic stimulation may provide a nonpharmacological treatment alternative in difficult cases of bipolar disorder (N. Michael and Erfurth, 2004). Furthermore, evidence suggests that CBT can be as effective as drug treatments (Hollon et al., 2002) and perhaps can be beneficially combined with other forms of treatment.

Men and women are equally affected by bipolar disorder, and the age of onset is usually much earlier than that of unipolar major depression. Bipolar disorder has a complex heritability: many different genes affect the probability of the disorder (Faraone et al., 2004; Smoller and Finn, 2003). One specific gene implicated in bipolar disorder is the gene that encodes BDNF (E. Green and Craddock, 2003; Neves-Pereira et al., 2002), which we discussed in Chapter 6. Many of the genes that increase the risk of bipolar disorder also increase the risk of schizophrenia, another indication that the disorders may be related. Finally, it seems that people who are susceptible to schizophrenia and bipolar disorder are more likely to be creative, artistic types.

lithium An element that, when administered as a drug, often relieves the symptoms of bipolar disorder.

12.26 A Connection between Schizophrenia, Bipolar Disorder, and Creativity? (After M. C. Keller and P. M. Visscher, 2015. *Nat Neurosci* 18: 928–929.)

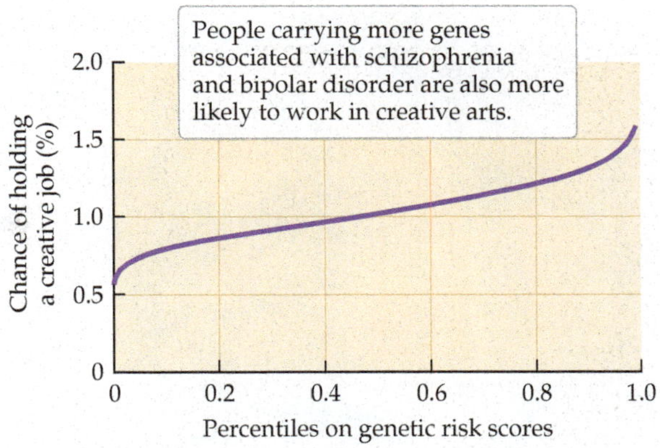

Based on two mega-analyses of genes associated with increased risk for schizophrenia and bipolar disorder, researchers assigned risk scores to 100,000 people in Iceland and Sweden (who were not part of the previous analyses). Then they ascertained which of those people held creative positions, based on whether they were members of societies of actors, dancers, musicians, writers, or visual artists. In this population, the more genes associated with schizophrenia or bipolar disorder people carried, the more likely they were to be associated with one of those creative enterprises (**FIGURE 12.26**). There was no such relationship to creative occupations when the population was ranked for genetic risk for 20 other (nonpsychiatric) disorders (Power et al., 2015).

12.4 There Are Several Types of Anxiety Disorders

Learning Objectives

After reading this section, you should be able to:

12.4.1 **Define and distinguish the main types of anxiety disorders, as described in the *DSM-5*.**

12.4.2 **Summarise the most common anxiolytic medications, and explain their modes of action.**

12.4.3 **Discuss the symptoms, causes, and possible physiological bases of posttraumatic stress disorder (PTSD).**

12.4.4 **Describe the symptoms, causes, and various treatments of obsessive-compulsive disorder (OCD).**

All of us have at times felt apprehensive and fearful. Some people experience this state with an intensity that is overwhelming and includes irrational fears, a sense of terror, unusual body sensations such as dizziness, difficulty breathing, trembling, shaking, and a feeling of loss of control. Anxiety can be lethal: a follow-up of people with panic disorder revealed an increased mortality in men with this disorder, resulting from cardiovascular disease and suicide (Kleiman et al., 2017).

The *DSM-5* distinguishes several major types of **anxiety disorders**. **Phobic disorders** are intense, irrational fears that become centred on a specific object, activity, or situation that the person feels compelled to avoid. Anxiety disorders also include *generalised anxiety disorder*, in which persistent and excessive anxiety and worry are experienced for months, and *panic disorder*, characterised by recurrent transient attacks of intense fearfulness. Some people who have recurrent panic attacks have temporal lobe abnormalities, and overall temporal lobe volumes tend to be lower in people with panic disorder (van Tol et al., 2010), while hippocampal volumes are normal. Given the special role of the amygdala in mediating fear (see Chapter 11), changes in the amygdala and associated circuitry may underlie the symptoms (Rauch et al., 2003).

Other anxiety-related disorders that we'll consider shortly are posttraumatic stress disorder and obsessive-compulsive disorder. There is a strong genetic contribution to each of these disorders (Oler et al., 2010; Shih et al., 2004).

anxiety disorder Any of a class of psychological disorders that include recurrent panic states and generalised anxiety disorder.

phobic disorder An intense, irrational fear that becomes centred on a specific object, activity, or situation that a person feels compelled to avoid.

benzodiazepines A class of antianxiety drugs that bind with high affinity to receptor molecules in the central nervous system. One example is diazepam (Valium).

Drug treatment of anxiety provides clues to the mechanisms of this disorder

Throughout history, people have consumed all sorts of substances in the hopes of controlling anxiety; examples include alcohol, bromides, scopolamine, opiates, and barbiturates. In the 1960s the tranquilising agent meprobamate (Miltown) was introduced and became an instant best seller, ushering in the modern age of anxiety pharmacotherapy. Soon researchers discovered a new class of drugs called **benzodiazepines**, which quickly became the favoured drugs for treating anxiety.

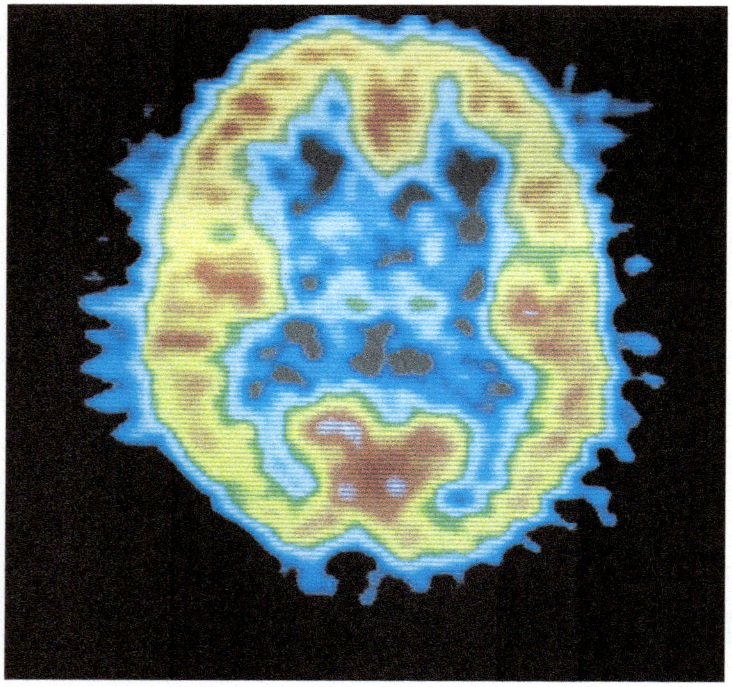

Courtesy of Goran Sedvall

12.27 The Distribution of Benzodiazepine Receptors in the Human Brain This PET scan of benzodiazepine receptors shows their wide distribution in the brain, especially the cortex. Highest concentrations are in orange and red.

One type of benzodiazepine—*diazepam* (Valium)—is one of the most prescribed drugs in history. Other commonly prescribed benzodiazepines include Xanax, Halcion, and Ativan. These drugs are so widely used that they can be detected in lakes and streams—having first passed through human consumers—at concentrations that have been demonstrated to inhibit social behaviours in fish (Brodin et al., 2013). Such drugs that combat anxiety are termed **anxiolytics** ("anxiety dissolving"), although at high doses they also have anticonvulsant and sleep-inducing properties. The anxiolytic drugs were also discussed in Chapter 3.

Anxiolytic benzodiazepines bind to GABA receptors, where they act as noncompetitive agonists (recall from Chapter 4 that GABA is the most common inhibitory transmitter in the brain). This interaction with GABA receptors results in the enhancement of GABA's action at inhibitory synapses in the brain. In other words, GABA-mediated postsynaptic inhibition is facilitated by benzodiazepines. Benzodiazepines preferentially bind to the many $GABA_A$ receptors that are widely distributed throughout the brain, especially in the cerebral cortex and some subcortical areas, such as the hippocampus and the amygdala (**FIGURE 12.27**).

The ultimate function of the benzodiazepine–$GABA_A$ receptor complex is to regulate the permeability of neuronal membranes to chloride ions (Cl^-). When GABA is released from a presynaptic terminal and activates postsynaptic receptors, chloride channels within the receptors open, allowing more negatively charged chloride ions to enter the postsynaptic neuron. This creates a local hyperpolarisation (an inhibitory postsynaptic potential, or IPSP) that inhibits the neuron from firing. Benzodiazepines alone do little to change chloride conductance, but in the presence of GABA, they markedly enhance GABA-provoked increases in chloride permeability and thus potentiate the inhibitory effect of GABA. Interestingly, the brain probably makes its own anxiety-relieving substances that interact with the benzodiazepine-binding site; the neurosteroid allopregnanolone is a prime candidate for this function (see Chapter 3). Drugs developed to act at this site are effective anxiolytics in both rats and humans (Rupprecht et al., 2009).

anxiolytics A class of substances that are used to combat anxiety. Examples include alcohol, opiates, barbiturates, and the benzodiazepines.

Although the benzodiazepines remain an important category of anxiolytics, other types of anxiety-relieving drugs have been developed. A notable example is the drug buspirone (BuSpar), an agonist at serotonin $5\text{-}HT_{1A}$ receptors that has been shown to provide relief from anxiety. This observation is consistent with findings from functional-imaging research confirming that $5\text{-}HT_{1A}$ receptor density is abnormal in anxiety disorders (Neumeister et al., 2004). SSRI antidepressants, such as paroxetine (Paxil) and fluoxetine (Prozac), which increase stimulation of serotonin receptors, are also sometimes effective treatments for anxiety disorders.

In posttraumatic stress disorder, horrible memories won't go away

Some people experience especially awful moments in life that seem indelible, resulting in vivid impressions that persist the rest of their lives. The kind of event that seems particularly likely to produce subsequent stress disorders is one that is intense and usually associated with witnessing abusive violence and/or death. Examples include the sudden loss of a close friend, rape, torture, kidnapping, or profound social dislocation, such as in forced migration. In these cases, memories of horrible events intrude into consciousness and produce the same intense visceral arousal—the fear and trembling and general autonomic activation—that the original event caused. These traumatic memories are easily reawakened by stressful circumstances and even by seemingly benign stimuli that somehow prompt recollection of the original event. An ever-watchful and fearful stance becomes the portrait of individuals afflicted with what is called **posttraumatic stress disorder** (**PTSD**, formerly called *combat fatigue, war neurosis,* or *shell shock*).

Analysis of a random sample of Vietnam War veterans indicates that 19% had PTSD at some point after service. This was the rate for all Vietnam veterans; of those exposed to high levels of war zone stressors, more than 35% developed PTSD at some point, and most of them still had the disorder decades later (Dohrenwend et al., 2006). Familial factors affect vulnerability, as shown in twin studies of Vietnam veterans who had seen combat. Monozygotic twins were more similar than dizygotic twins. People who display combat-related PTSD show (1) memory changes such as amnesia for some war experiences, (2) flashbacks, and (3) deficits in short-term memory. These memory disturbances suggest involvement of the hippocampus, and indeed the volume of the right hippocampus is smaller in combat veterans with PTSD than in controls, with no differences in other brain regions (Bremner et al., 1995). It was once widely assumed that the stressful episode(s) caused the hippocampus to shrink, but some veterans with PTSD left a monozygotic twin at home, and it turned out that the nonstressed twin without PTSD also tended to have a smaller hippocampus (Gilbertson et al., 2002). So an inherited tendency to have a small hippocampus, and perhaps lower rates of adult neurogenesis (J. S. Snyder et al., 2011), may be what makes a person more susceptible to PTSD if exposed to stress.

A comprehensive psychobiological model of the development of PTSD draws connections among the symptoms and the neural mechanisms of fear conditioning, behavioural sensitisation, and extinction (Charney et al., 1993). Perhaps people learn to avoid a large range of stimuli associated with the original trauma. Work in animals has revealed that this type of memory—**fear conditioning**—is very persistent and involves the amygdala (see Chapter 11).

Finally, the persistence of memory and fear in PTSD may depend on the failure to *forget*. In experimental animals, NMDA antagonists delivered to the amygdala prevent the extinction of fear-mediated startle. Sites projecting to the amygdala, such as the hippocampus and prefrontal cortex, may also lose their effectiveness in suppressing learned fear responses (**FIGURE 12.28**). There is also a hormonal link, because people with PTSD exhibit abnormalities in their cortisol response to stressful stimuli (Wichmann et al., 2017). One possibility is that people with PTSD have persistent alterations in *sensitivity* to cortisol. If, as a result, they feel the effect of stress hormones more strongly than other people do, it might be harder for them to forget stressful events. In Chapter 16 we will discuss research-based methods that have been proposed to help people forget traumatic life events. For now, the most effective treatment for PTSD is CBT involving prolonged exposure, gradually increasing

posttraumatic stress disorder (PTSD) Formerly called *combat fatigue*, *war neurosis*, or *shell shock*. A disorder in which memories of an unpleasant episode repeatedly plague the person.

fear conditioning A form of learning in which fear comes to be associated with a previously neutral stimulus.

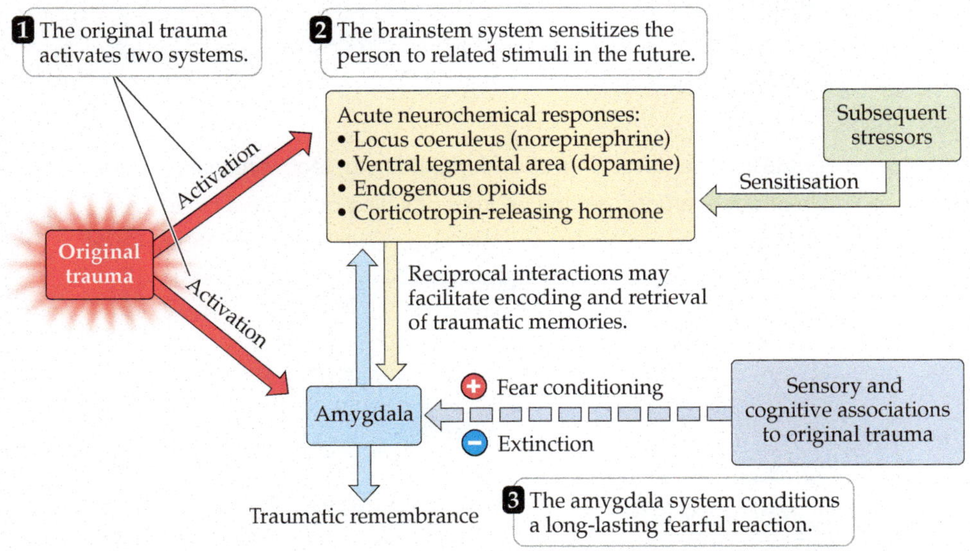

12.28 **A Neural Model of Posttraumatic Stress Disorder**

the client's (imagined or real) exposure to those stimuli and conditions that trigger flashbacks, until they no longer trigger responses (Foa et al., 2013).

In obsessive-compulsive disorder, thoughts and acts keep repeating

Neatness, orderliness, and similar traits are attributes we tend to admire, especially during those chaotic moments when we realize we have created another tottering pile of papers, bills, or the like. But when does orderliness and routine cross the line into pathology? People with **obsessive-compulsive disorder** (**OCD**) lead lives riddled with repetitive rituals and persistent thoughts that they feel powerless to control or stop, despite recognising that the behaviours are abnormal. In OCD patients, routine acts that we all engage in, such as checking whether the door is locked when we leave our home, become *compulsions*, acts that are repeated over and over. Recurrent thoughts, or *obsessions*, such as fears of germs or other potential harms in the world, invade the consciousness. These symptoms progressively isolate a person from ordinary social engagement with the world. For many patients, hours of each day are consumed by compulsive acts such as repetitive handwashing. The symptoms of OCD are summarised in **TABLE 12.5**.

obsessive-compulsive disorder (OCD) A syndrome in which the affected individual engages in recurring, repetitive acts that are carried out without rhyme, reason, or the ability to stop.

Determining the number of persons afflicted with OCD is difficult, especially because many people with this disorder tend to hide their symptoms. It is estimated that nearly 1% of adults will be affected by "severe" OCD in the United States in any given year (Kessler, Chiu et al., 2005). In many cases, the initial symptoms of this disorder appear in childhood; the peak age group for onset of OCD, however, is 25–44 years.

TABLE 12.5 Symptoms of Obsessive-Compulsive Disorders

Symptoms (most to least common by type)
OBSESSIONS (THOUGHTS)
Dirt, germs, or environmental toxins
Something terrible happening (fire, death, or illness of self or loved one)
Symmetry, order, or exactness
Religious obsessions
Body wastes or secretions (urine, stool, saliva)
Lucky or unlucky numbers
Forbidden, aggressive, or perverse sexual thoughts, images, or impulses
Fear of harming self or others
Household items
Intrusive nonsense sounds, words, or music
COMPULSIONS (ACTS)
Performing excessive or ritualised handwashing, showering, bathing, teeth brushing, or grooming
Repeating rituals (going in or out of a door, getting up from or sitting down on a chair)
Checking (doors, locks, stove, appliances, emergency brake on car, paper route, homework)
Engaging in miscellaneous rituals (writing, moving, speaking)
Removing contaminants from contacts
Touching
Counting
Ordering or arranging
Preventing harm to self or others
Cleaning household or inanimate objects

Source: After S. E. Swedo et al., 1989. *Arch Gen Psychiatry* 46: 335–341.

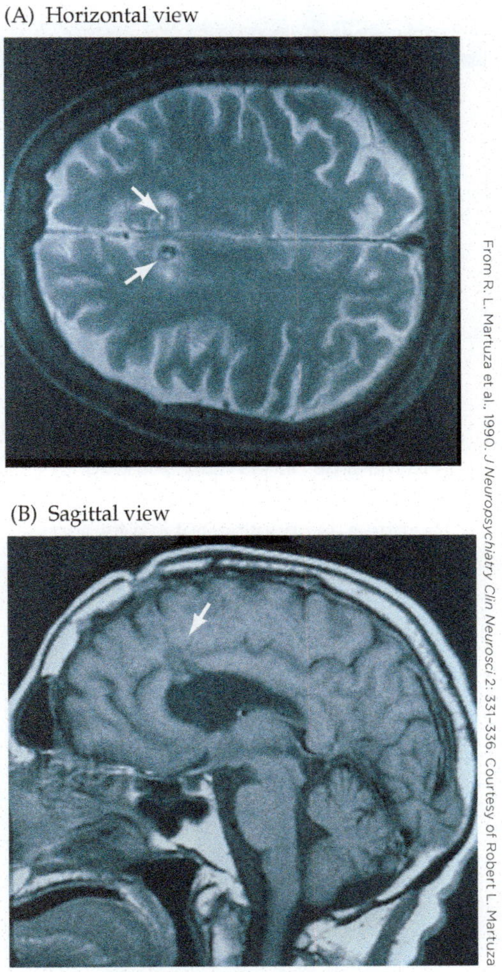

12.29 Neurosurgery to Treat Obsessive-Compulsive Disorder These horizontal (A) and sagittal (B) MRIs show the brain of a person who underwent a cingulotomy—the disruption of cingulate cortex connections (arrows)—in an attempt to treat OCD.

People with OCD display increased metabolic rates in the orbitofrontal cortex, cingulate cortex, and caudate nuclei (Chamberlain et al., 2008).

Happily, OCD responds to treatment in most cases. It shows an excellent response to CBT (Abramowitz et al., 2018; Öst et al., 2016) and also to several drugs. What do effective OCD drugs—like fluoxetine (Prozac), fluvoxamine (Luvox), and clomipramine (Anafranil)—tend to have in common? They share the ability to inhibit the reuptake of serotonin at synapses, thereby increasing the synaptic availability of serotonin. This observation suggests that dysfunction of serotonergic neurotransmission plays a central role in OCD. Recall that we already discussed SSRIs like Prozac that inhibit the reuptake of serotonin when we discussed treatments for depression. How can the same drug help two disorders that seem so different? For one thing, depression often accompanies OCD, so the two disorders may be related. Furthermore, functional brain imaging suggests that the same SSRI drugs alter the activity of the orbitofrontal prefrontal cortex in people with OCD (Saxena et al., 2001) and also affect primarily ventrolateral prefrontal cortex in people with depression (see Figure 12.19).

There is growing evidence that overactivity of a circuit from prefrontal cortex to striatum to thalamus and back to cortex may underlie OCD. Induced overactivity of this circuit in mice resulted in prolonged stereotyped grooming behaviour that was relieved by SSRI treatment (Ahmari et al., 2013; Burguière et al., 2013). Deep brain stimulation (DBS) of the nucleus accumbens in people with OCD disrupted this same circuit and reportedly relieved symptoms (van Westen et al., 2015), but such studies tend to have small sample sizes and lack proper control groups, making it difficult to conclusively establish the efficacy of DBS for OCD (Naesström et al., 2016). Occasionally, people with treatment-resistant OCD may undergo psychosurgery as a last-ditch measure. A significant fraction of people with severe and disabling OCD who undergo *cingulotomy*, surgical disruption of circuits in the cingulate cortex (**FIGURE 12.29**), appear to benefit from this intervention (L. T. Brown et al., 2016). A similar surgery, making small, discrete lesions in the anteriour part of the internal capsule (the white matter projections underlying the cortex), is likewise beneficial in some patients (Brown et al., 2016) but can cause apathy, a common symptom of lobotomy (Rück et al., 2008).

There is a heritable component to OCD; as with schizophrenia and depression, numerous genes contribute to OCD susceptibility, notably including genes affecting serotonergic, dopaminergic, and glutamatergic systems (Pauls et al., 2014; Sinopoli et al., 2017). There is also evidence that perinatal events and infections can trigger OCD (Orlovska et al., 2017). Upon observing that numerous children exhibiting OCD symptoms had recently been treated for strep throat, R. C. Dale et al. (2005) found that many children with OCD were producing antibodies to brain proteins. They theorised that, in mounting an immune response to the streptococcal bacteria, these children also made antibodies that attacked their own brains. The genetic link may be that some people are more likely than others to produce antibodies to the brain proteins.

Many researchers also believe that OCD and another disease involving repetitive behaviours, known as *Tourette's syndrome*, are part of a spectrum of related disorders (Olson, 2004). Tourette's and OCD are often **comorbid** (occurring together), and both disorders involve abnormalities of the basal ganglia. However, drug therapy in Tourette's syndrome has typically focused on modifying the actions of dopamine rather than of serotonin (**BOX 12.3**).

comorbid Referring to the tendency of certain diseases or disorders to occur together in individuals.

BOX 12.3 Tics, Twitches, and Snorts: The Unusual Character of Tourette's Syndrome

Their faces twitch in an insistent way, and every now and then, out of nowhere, they blurt out odd sounds. At times they fling their arms, kick their legs, or make violent shoulder movements. People with **Tourette's syndrome** are also supersensitive to tactile, auditory, and visual stimuli (A. J. Cohen and Leckman, 1992). Many people report that an urge to emit verbal or phonic tics builds up and that giving in to the urge brings relief. Although popular media often portray people with Tourette's as shouting out insults and curse words (coprolalia), verbal tics of that sort are rare.

Tourette's syndrome begins early in life; the mean age of diagnosis is 6–7 years (de Groot et al., 1995), and the syndrome is 3 to 4 times more common in males than in females. Figure A draws a portrait of the chronology of symptoms. Often people with Tourette's also exhibit attention deficit hyperactivity disorder (ADHD) or OCD (S. Park et al., 1993). Children with Tourette's display a thinning of primary somatosensory and motor cortex representing facial, oral, and laryngeal structures (Sowell et al., 2008), suggesting that the tics mediated by these regions may be underinhibited by cortex.

Family studies indicate that genetics plays an important role in this disorder. Twin studies of the disorder reveal a concordance rate among monozygotic twins of 53–77%, contrasted with a concordance rate among dizygotic twins of 8–23% (T. M. Hyde et al., 1992). Among discordant monozygotic twin pairs, the twin with Tourette's has a greater density of dopamine D2 receptors in the caudate nucleus of the basal ganglia than the unaffected twin has (S. S. Wolf et al., 1996). This observation suggests that differences in the dopaminergic system, especially in the basal ganglia, may be important (D2 receptor binding in an affected twin is illustrated in Figure B). The contemporary view is that Tourette's syndrome is mediated in a complex manner by more than one gene, but the precise genes that are involved remain unidentified (Abelson et al., 2005; Pauls, 2003).

Treatment with haloperidol, a dopamine D_2 receptor antagonist that is an effective antischizophrenic drug, significantly reduces tic frequency and is a primary treatment for Tourette's syndrome. Unfortunately, this treatment has the side effects we mentioned when discussing schizophrenia, and as with people with schizophrenia, some people with Tourette's respond well to the second-generation antipsychotics, which bring fewer side effects. Behaviour modification techniques that aim at reducing the frequency of some symptoms, especially tics, help some people learn to substitute more subtle or more socially acceptable behaviours in place of the more obvious tics (Himle et al., 2006). Some people with Tourette's that has not responded to medication have gained relief from deep brain stimulation (DBS) (Baldermann et al., 2016; Porta et al., 2009), but there has yet to be a consensus on which brain targets are most effective for combatting symptoms (Fraint and Pal, 2015).

Tourette's syndrome A heightened sensitivity to tactile, auditory, and visual stimuli that may be accompanied by the buildup of an urge to emit verbal or phonic tics.

(A) The chronology of Tourette's symptoms

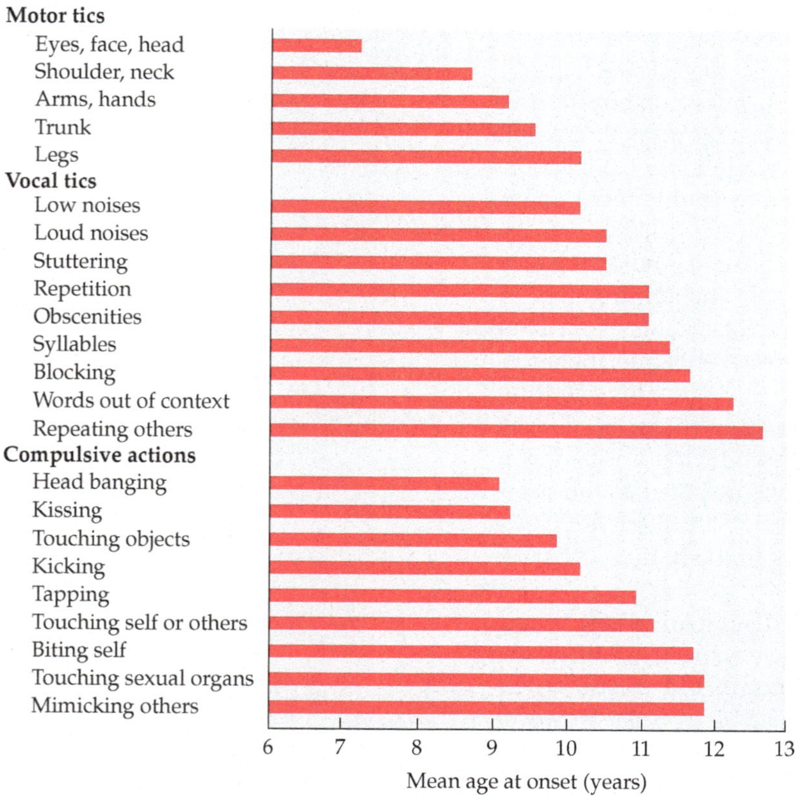

(B) D_2 receptor binding in Tourette's syndrome. (Left) PET scan of D_2 binding. (Right) MRI scan illustrating the location of the caudate nuclei.

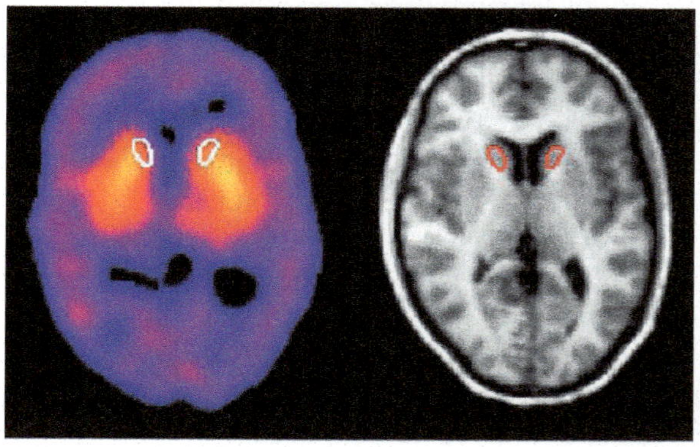

From S. S. Wolf et al., 1996. *Science* 273: 1225–1227

Recommended Reading

Charney, D., Nestler, E. J., Sklar, P., and Buxbaum, J. D. (Eds.). (2018). *Charney & Nestler's Neurobiology of Mental Illness* (5th ed.). New York: Oxford University Press.

Denys, D., Feenstra, M., and Schuurman, R. (Eds.). (2012). *Deep Brain Stimulation: A New Frontier in Psychiatry.* New York: Springer.

Hersen, M., and Beidel, D. C. (2014). *Adult Psychopathology and Diagnosis* (7th ed.). New York: Wiley.

Huettel, S. A., Song, A. W., and McCarthy, G. (2014). *Functional Magnetic Resonance Imaging* (3rd ed.). Sunderland, MA: Oxford University Press/Sinauer.

Kessler, D. A. (2016). *Capture: Unraveling the Mystery of Mental Suffering.* New York: HarperCollins.

Martino, D., and Leckman, J. F. (Eds.). (2013). *Tourette Syndrome.* Oxford, UK: Oxford University Press.

Meyer, J. S., and Quenzer, L. F. (2018). *Psychopharmacology: Drugs, the Brain, and Behaviour* (3rd ed.). Sunderland, MA: Oxford University Press/Sinauer.

Solomon, A. (2001). *The Noonday Demon: An Atlas of Depression.* New York: Scribner.

Steketee, G. (Ed.). (2011). *The Oxford Handbook of Obsessive Compulsive and Spectrum Disorders.* New York: Oxford University Press.

Go to Activity 12.1
Concept Matching: Psychopathology

12 Visual Summary

You should be able to relate each summary to the adjacent illustration, including structures and processes. You can also refer back to the links to figures, animations, and activities that will help you consolidate the material.

1 Psychiatric disorders are prevalent in modern society. Studies of families, twins, and adoptees demonstrate a strong role of genetic factors in **schizophrenia**. Rather than a single gene determining whether a person will develop schizophrenia, many genes contribute to the risk. Review **Figures 12.1–12.3**, **Tables 12.1**, **12.2**

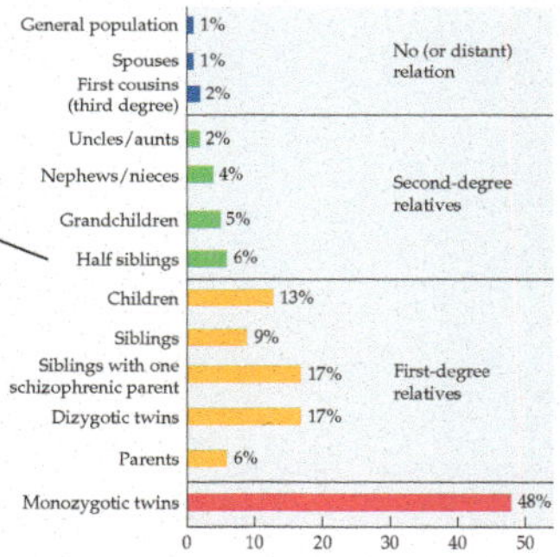

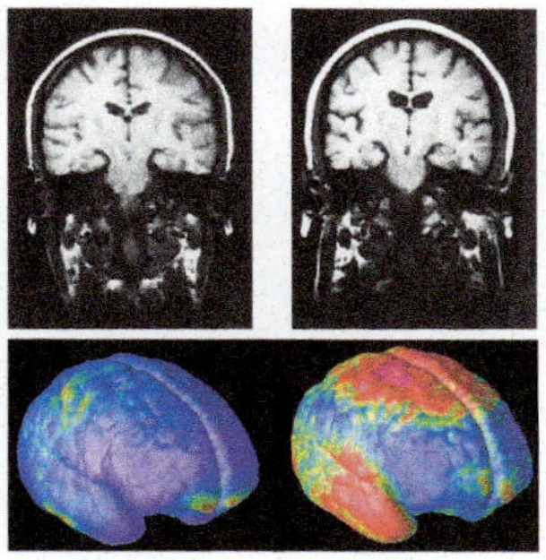

2 Structural and functional changes in the brains of people with schizophrenia—including enlarged ventricles and eye-tracking abnormalities—may arise from early developmental problems. The emergence of schizophrenia depends on the interaction of genes that make a person vulnerable to environmental stressors. Review **Figures 12.4–12.9**, **12.15–12.17**

3 The frontal lobes are less active in people with schizophrenia than in controls. Biochemical theories of schizophrenia emphasise the importance of the dopamine, glutamate, and serotonin receptors. **First-generation antipsychotics** block dopamine D_2 receptors, while **second-generation antipsychotics** block 5-HT_{2A} receptors. Review **Figures 12.10–12.14**, **12.18**, **Box 12.1**

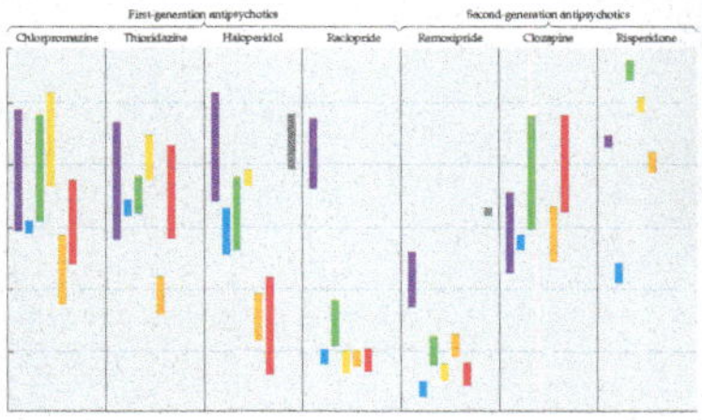

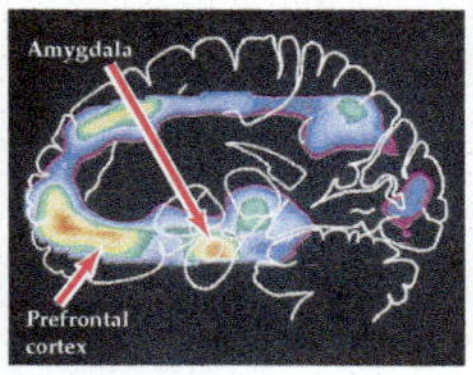

4 **Depression** also has a strong genetic factor. In general, females are more likely than males to have depression. People with depression show increased activity in the frontal cortex and the amygdala, as well as disruption of sleep patterns and adrenal hormone regulation. Review **Figures 12.19**, **12.23**, **12.24**, **Box 12.2**, **Table 12.3**

5 The most effective treatment for most cases of depression is a combination of **cognitive behavioural therapy** (**CBT**) and a **selective serotonin reuptake inhibitor (SSRI)**. Review **Figures 12.20–12.22**, **Table 12.4**

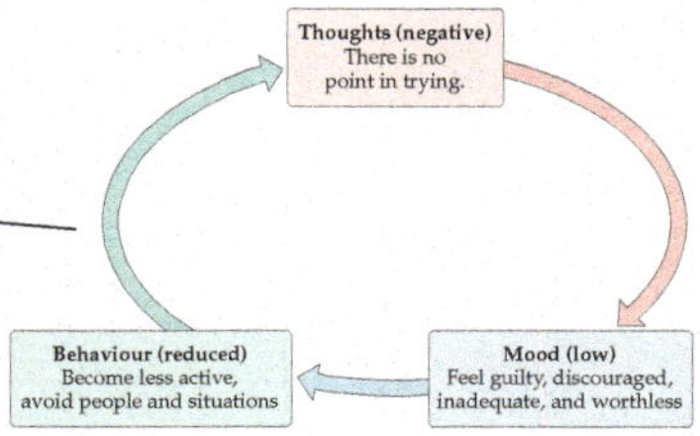

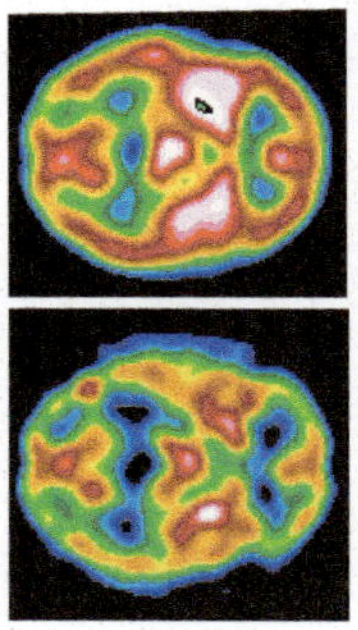

6 **Bipolar disorder** is characterised by extreme mood swings and subtle changes in the brain and is commonly treated with **lithium**. As in schizophrenia, many genes contribute to the risk of bipolar disorder, and genetic risk for both disorders seems to be associated with creativity. Review **Figures 12.25**, **12.26**

(Continued)

7 Anxiety states are characterised by functional changes in the temporal lobes, particularly the amygdala. **Benzodiazepine** antianxiety drugs, known as **anxiolytics**, enhance the inhibitory effects of receptors for the transmitter GABA. Drugs affecting serotonergic synapses may also reduce anxiety. Review Figure 12.27

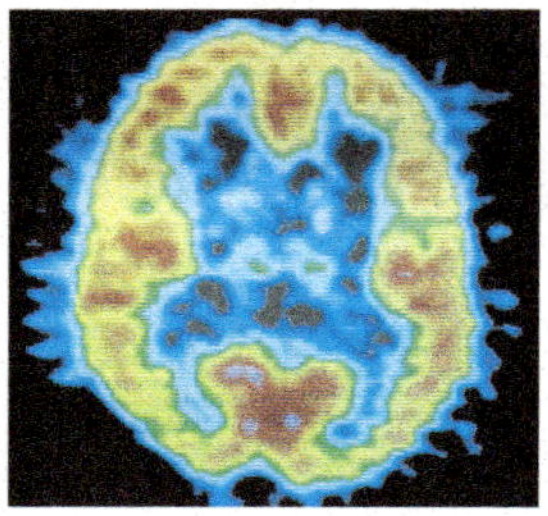

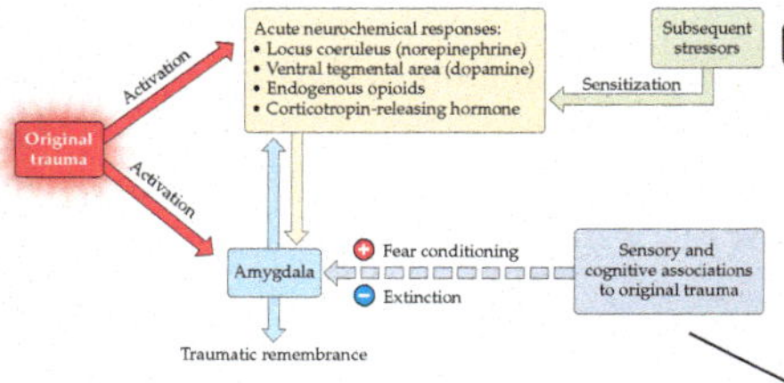

8 **Posttraumatic stress disorder (PTSD)** is characterised by an inability to forget horrible experiences. Temporal lobe atrophy in this disorder may be caused by exposure to stress hormones, such as cortisol, and prolonged sensitivity to those hormones. Review Figure 12.28

9 **Obsessive-compulsive disorder (OCD)** is characterised by changes in basal ganglia and frontal cortex and linked to serotonin. In **Tourette's syndrome,** overstimulation of dopamine receptors induces motor and verbal tics and compulsions. A restricted type of **psychosurgery** is sometimes used to treat the most severe cases of anxiety disorders. Review Figure 12.29, Table 12.5, Box 12.3, Activity 12.1

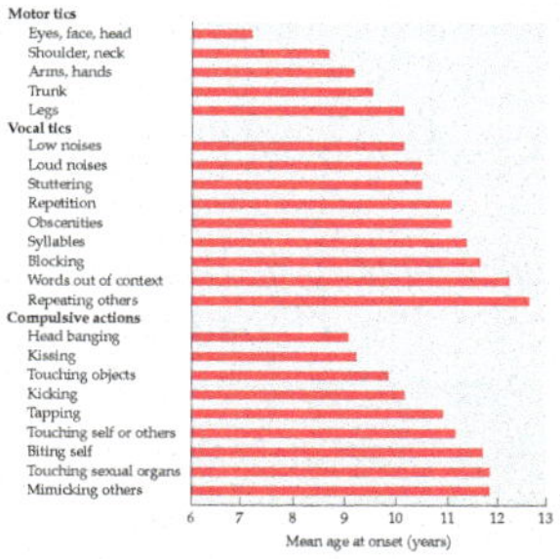

Regulation and Behaviour

PART V

© Michael W. Davidson/Science Source

13

Sex

Evolutionary, Hormonal, and Neural Bases

Genitals and Gender: What Makes Us Male and Female?

The night before "Bella" was to start 6th grade, her parents finally told her a secret. She had been born a boy with testes, but with a life-threatening condition called *cloacal exstrophy*, a developmental defect that affects about one in 400,000 live births. Basically, the skin covering the lower abdomen and forming the genitalia was missing. Surgery was needed immediately to cover the exposed intestines to avoid injury and infection, but like other genetic males born with this condition, Bella had been born without a penis. The parents had faced a terrible dilemma: Would it be better to raise the child as a boy without a penis, despite the emotional costs of the deformity? Or would it be better for the child to be unambiguously assigned to the female gender, to have early surgery to remove the testes and fashion female-looking genitals, and then to be raised as a girl? In the end, Bella's parents had decided on the surgery to remove the testes and make her genitalia appear like that of a girl, and to raise her as their daughter.

You might think Bella would have found this story hard to believe. But in fact, she believed them immediately. "I knew that it was true. Even if it had been another person that had told me I would have believed them because it made sense to me" (Reiner and Gearhart, 2004). Too confused to start school the next morning, "Benjamin" began school a week later as a boy. How did Benjamin make the transition from girl to boy? "Cut my hair. That's all I did."

Sexual behaviours are almost as diverse as the species that employ them, and they are the products of millions of years of evolution that has favoured continual mixing of genes in an attempt to outrun parasites and other threats (Morran et al., 2011). But in every case, males and females must produce a specific set of behaviours, in a precise and intricately coordinated sequence, in order to reproduce successfully. In this chapter we review sexual behaviours, which include the sex act itself—copulation—as well as the parental behaviours required for the newborns of many species, including our own, to survive. Then we consider sexual differentiation, the process by which an individual's body and brain develop in a male or female fashion. For humans, an important aspect of sexual differentiation is the emergence of sexual orientation. Do we *decide* whether to be attracted to men or to women, are we *taught* whom we should find attractive, or do prenatal events have an influence?

SEXUAL BEHAVIOUR

Two barriers have limited our understanding of sexual behaviour: (1) cultural barriers to the dissemination of knowledge about sexual behaviour and (2) the remarkable variety of sexual behaviours in which various species engage. Nevertheless, the basic biology of reproduction and studies of other animals offer a starting framework for the rest of the chapter.

13.1 Reproductive Behaviour Can Be Divided into Four Stages

sexual attraction The first step in the mating behaviour of many animals, in which animals produce stimuli to attract sexual partners.

pair-bond A durable and exclusive relationship between two individuals, such as mates.

Learning Objectives

After reading this section, you should be able to:

13.1.1 **Categorise and describe the four stages of mating behaviour.**

13.1.2 **Compare the various methods of sexual and asexual reproduction.**

13.1.3 **Describe the role of gonadal steroids in regulating sexual behaviour.**

There are four major stages of reproductive behaviour: (1) sexual attraction, (2) appetitive behaviour, (3) copulation, and (4) postcopulatory behaviour.

Sexual attraction is the first stage in bringing mates together. In many species, females are attracted to males that display the best assortment of species-typical ornaments or traits, such as the feathers of the peacock in **FIGURE 13.1**. Such traits are the products of sexual selection pressures exerted by the opposite sex across many generations (see Chapter 5). Sexually selected traits may signal superiour genetic fitness or other characteristics that make the individual desirable as a mate, and thus attractive. The process of attraction and mate selection can have high stakes: for example, in some species (such as voles) mated males and females will form lasting **pair-bonds**, living together before and long after copulation (see Chapter 4). Interestingly, there is a distinction between social pair-bonds and sexual (or genetic) pair-bonds. In the case of voles, the formation of exclusive social pair-bonds leads to the greatest reproductive success, even though the partners may occasionally engage in extrapair copulation (Ophir et al., 2008). They are *socially* monogamous (having one partner) but not quite *sexually* monogamous.

In many species, sexual attraction is closely synchronised with physiological readiness and appetite for sex. That's not to say that we are perfectly accurate in gauging sexual interest—in fact, it seems that men overestimate women's interest, and women underestimate men's interest (Perilloux et al., 2012). Attraction and sexual response are also strongly shaped by learned associations, varying from one individual to the next on the basis of life experience (Pfaus et al., 2001). This process, by which a previously neutral item becomes sexually attractive, is the foundation of the fashion industry.

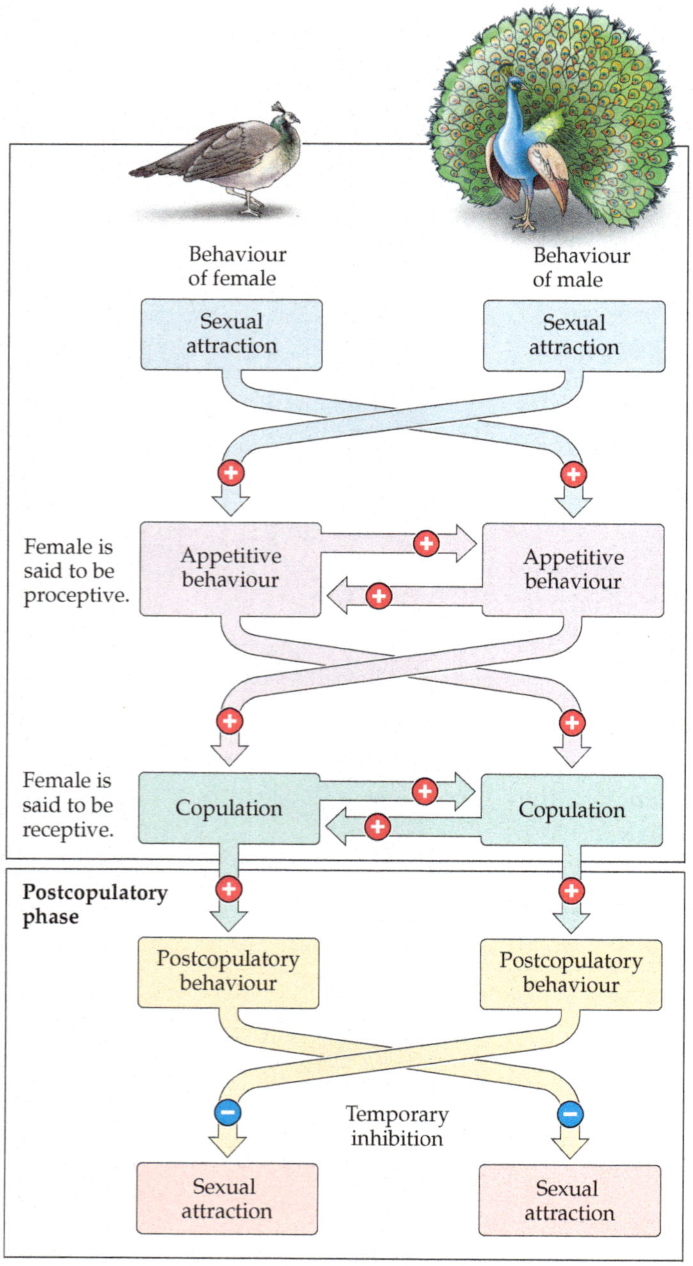

13.1 Stages of Reproductive Behaviour Interaction between male and female partners in sexual reproduction is extensive, progressing in four stages: sexual attraction, appetitive behaviour, copulation, and postcopulatory behaviour. The postcopulatory phase includes a temporary decrease in the sexual attractiveness of the partner. (After F. A. Beach, 1976. *Horm Behav* 7: 105.)

In experiments we gauge an individual's attractiveness by observing the responses of potential mates: how rapidly they approach, how hard they work to gain access, and so on. By manipulating the appearance of individuals, we can deduce which special features are most attractive. For example, males of many primate species are strongly attracted to the "sex skin" that swells on a female's rump when her ovaries are secreting estrogens. Most male mammals are attracted by particular female odours, which also tend to reflect estrogen levels. Because estrogen secretion is associated with the release of eggs, these mechanisms tend to synchronise female sexual attractiveness with peak fertility. Of course, the female may find a particular male to be unattractive—perhaps his feathers are underwhelming—and refuse to mate with him. Although apparent rape has been described in some nonhuman species, including such close relatives of ours as orangutans (Knott et al., 2010), for most species copulation is controlled by the female and requires her active cooperation.

If the animals are mutually attracted, they may progress to the next stage: **appetitive behaviours**—species-specific behaviours that establish, maintain, or promote sexual interaction (see Figure 13.1). A female displaying these behaviours is said to be **proceptive**: she may approach males, remain close to them, or show alternating approach and retreat behaviour. Proceptive female rats typically exhibit "ear wiggling" and a hopping and darting gait that induce a male to mount. Male appetitive behaviours usually consist of staying near the female. In many mammals the male may sniff around the female's face and vagina. Male birds may engage in elaborate songs or nest-building behaviours.

If both animals display appetitive behaviours, they may progress to the *consummatory* stage of reproduction: **copulation**, also known as *coitus*. In many vertebrates, including all mammals, copulation involves one or more **intromissions**, in which the male inserts his penis into the female's vagina, followed by a variable amount of genital stimulation, usually through pelvic thrusting. When stimulation reaches a threshold level, the male **ejaculates** sperm-bearing **semen** into the female; the length of time and quantity of stimulation that are required vary greatly between species and between individuals.

After one bout of copulation, the animals will not mate again until a **refractory period** has elapsed. The refractory period varies from minutes to months, depending on the species and circumstances. Many animals will resume mating sooner if they are provided with a new partner—a phenomenon known as the **Coolidge effect** (**FIGURE 13.2**).

appetitive behaviour The second stage of mating behaviour. Helps establish or maintain sexual interaction.

proceptive Referring to a state in which an animal advertises its readiness to mate through species-typical behaviours, such as ear wiggling in the female rat.

copulation Also called *coitus*. The sexual act.

intromission Insertion of the erect penis into the vagina during copulation.

ejaculation The forceful expulsion of semen from the penis.

semen A mixture of fluids and sperm that is released during ejaculation.

refractory period A period following copulation during which an individual cannot recommence copulation.

Coolidge effect The propensity of an animal that has appeared sexually satiated with a present partner to resume sexual activity when provided with a novel partner.

Sooty enjoyed two nights of passion among 24 females.

Guinea pig Don Juan sires 43 offspring in 2 nights

PONTYPRIDD, WALES, 1 DECEMBER 2000

HAVING ESCAPED from captivity at Little Friend's Farm earlier this year, a male guinea pig named Sooty chose to re-enter captivity immediately—in the nearby cage housing 24 females. Two months later he is now the father of 43 offspring.

According to his owner, Carol Feehan, Sooty was missing for two whole days before the staff checked the females' pen. "We did a head count and found 25 guinea pigs," she told the press. "Sooty was fast asleep in the corner.

"He was absolutely shattered. We put him back in his cage and he slept for two days."

13.2 The Coolidge Effect By reducing the refractory period when a sexually exhausted male encounters an unfamiliar female, the Coolidge effect permits him to take advantage of a new reproductive opportunity and sire more offspring. (Of course, encountering 24 lovelorn females at once is a situation few males—guinea pig or otherwise—could even dream of.)

sexually receptive Referring to the state in which an individual (in mammals, typically the female) is willing to copulate.

estrus The period during which female animals are sexually receptive.

postcopulatory behaviour The final stage in mating behaviour. Species-specific postcopulatory behaviours include rolling (in the cat) and grooming (in the rat).

copulatory lock Reproductive behaviour in which the male's penis swells after ejaculation so that the male and female are forced to remain joined for 5–15 minutes. Occurs in dogs and some rodents, but not in humans.

internal fertilisation The process by which sperm fertilise eggs inside the female's body, as in all mammals, birds, and reptiles.

gamete A sex cell (sperm or ovum) that contains only unpaired chromosomes and therefore has only half the usual number of chromosomes in other cells.

sperm The gamete produced by males for fertilisation of eggs (ova).

ovum An egg, the female gamete.

zygote The fertilised egg.

external fertilisation The process by which eggs are fertilised outside of the female's body, as in many fishes and amphibians.

parthenogenesis "Virgin birth," a process of reproduction by which a female produces live offspring without need of a male.

The female often appears to be the one to choose whether copulation will take place; when she is willing to copulate, she is said to be **sexually receptive**, in heat, or in **estrus**. In some species the female may show proceptive behaviours days before she will participate in copulation itself. In most (but not all) species, females are receptive only when mating is likely to produce offspring. Most species are seasonal breeders, with females that are receptive only during the breeding season, and some—such as Pacific salmon, octopuses, and cicadas—reproduce only once, at the end of life.

The fourth stage of reproductive behaviour consists of **postcopulatory behaviours**. These behaviours are especially varied across species. In some mammals—dogs and southern grasshopper mice, for example—the male's penis swells so much after ejaculation that he can't remove it from the female for a while (10–15 minutes in dogs), and the animals are said to be in a **copulatory lock** (Dewsbury, 1972)—just one of the many strategies employed by males of different species to try and ensure their paternity. (Despite wild stories you may have heard or read, humans never experience copulatory lock; that urban myth started in 1884 when a physician submitted a fake report as a practical joke on a journal editor [Nation, 1973].) For most birds and some mammals, postcopulatory behaviours include extensive parental behaviours to nurture the offspring, which we'll discuss later in this chapter.

Copulation brings gametes together

All mammals, birds, and reptiles employ **internal fertilisation**: the fusion of their **gametes**—**sperm** and **ovum** (plural *ova*)—within the female's body to form a **zygote**. Although there is room for only one or a few zygotes to grow, the safe and stable internal environment maximises the probability that each embryo will eventually develop into a new individual. Other vertebrates—aquatic species including most fishes and amphibians—employ **external fertilisation**, releasing their gametes into the outside world in a reproductive gamble that pits sheer numbers against a high rate of loss. Of course, the reproductive behaviour of each species is shaped by its particular ecological niche, accounting for the impressive diversity that we observe in the sex lives of different species. **TABLE 13.1** summarises the different types of sexual reproduction in animals.

You may be surprised to learn that a few vertebrate species, including turkeys, Komodo dragons, and pit vipers (Booth et al., 2012), can reproduce without sex—the female lays an unfertilised egg that hatches a clone of herself. This asexual mode of reproduction is **parthenogenesis** ("virgin birth"). Even more surprising, there are about 90 vertebrate species, including fish called Amazon mollies as well as some species of whiptail lizards and mole salamanders (Lampert and Schartl, 2010), that rely *exclusively* on parthenogenesis, so there are no males of these species. Interestingly, no such "unisexual" species of birds or mammals have been found.

TABLE 13.1 Types of Sexual Reproduction in Animals

Type	Strategy	Examples
EXTERNAL FERTILISATION		
Aquatic species	Many gametes are released into the environment	Many invertebrates, fishes, amphibians
INTERNAL FERTILISATION		
Oviparous species	Female lays a few eggs	Insects, all birds, many reptiles
Viviparous species	Female nourishes embryos in her body, giving birth to a few, relatively mature young	Some fishes, some reptiles, all mammals except monotremes (echidna and platypus)
Ovoviviparous species	Female gives birth to live young that developed from eggs carried internally	Some fishes, some reptiles

Most of what we know about the copulatory behaviour of mammals derives from studies of lab animals, especially rats and mice. Like most other rodents, rats do not engage in lengthy courtship, nor do the partners tend to remain together after copulation. Rodents are attracted to each other largely through odours. Females are spontaneous ovulators; that is, even when left alone, they **ovulate** (release eggs from the ovary) every 4–5 days. For those few hours around the time of ovulation, the female seeks out a male and displays proceptive behaviours, and both animals produce *ultrasonic vocalisations*: sounds in the range of 22–50 kilohertz that are too high-pitched for humans to detect but audible to each other (Egnor and Seagraves, 2016).

13.3 Copulation in Rats The raised rump and deflected tail of the female (the lordosis posture) make intromission possible in rats.

These behaviours prompt the male to mount the female from the rear, grasp her flanks with his forelegs, and rhythmically thrust his hips against her rump. If she is receptive, the female adopts a stereotyped posture called **lordosis** (**FIGURE 13.3**), elevating her rump and moving her tail to one side, allowing intromission. Once intromission has been achieved, the male rat makes a single deep thrust and then springs back off the female. During the next 6–7 minutes the male and female orchestrate seven to nine such intromissions; then, instead of springing away, the male raises the front half of his body up for a second or two while he ejaculates. Finally, he falls backward off the female.

After copulation, the male and female separately engage in grooming their genitalia, and the male pays little attention to the female for the next 5 minutes or so, until, often in response to the female's proceptive behaviours, the two engage in another bout of intromissions and ejaculation. This pattern of multiple intromissions before ejaculation is an obligatory part of rat fertility: only after repeated intromissions will the female's brain cause the release of hormones to support pregnancy. In this instance, then, the behaviour of the male rat directly affects the hormonal secretions of his mate.

ovulation The production and release of an egg (ovum).

lordosis A female receptive posture in quadrupeds in which the hindquarters are raised and the tail is turned to one side, facilitating intromission by the male.

castration Removal of the gonads, usually the testes.

activational effect A temporary change in behaviour resulting from the administration of a hormone to an adult animal.

Gonadal steroids activate sexual behaviour

Hormones play an important role in almost all aspects of rodent mating behaviours. Testosterone mediates the male's interest in copulation: if he is **castrated** (his testes removed), he will eventually stop mounting receptive females. Although testosterone disappears from the bloodstream within a few hours after castration, the hormone's effects on the nervous system take days or weeks to dissipate. Treating a castrated male with testosterone eventually restores mating behaviour; if testosterone treatment is stopped, the mating behaviour fades away again. This is an example of a hormone exerting an **activational effect**: the hormone transiently promotes certain behaviours. In normal development, the rise of androgen secretion at puberty activates masculine behaviour in males.

Although individual male rodents differ considerably in how eagerly they will mate, blood levels of testosterone clearly are *not* responsible for these differences. For one thing, animals displaying different levels of sexual vigour do not show reliable differences in blood levels of testosterone. In a classic experiment, when male guinea pigs were castrated and subsequently all treated with exactly the same doses of testosterone, their precastration differences in sexual activity persisted (**FIGURE 13.4**). Not only that, it also turns

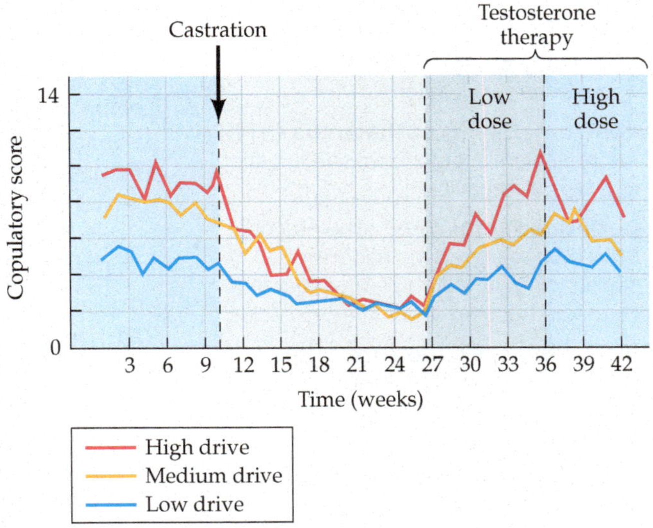

13.4 Androgens Permit Male Copulatory Behaviour Although androgens—especially testosterone—are important for normal male sexual function, individual differences in sexual activity are *not* determined by differences in androgen levels. Even after testosterone levels were made uniform in all the male guinea pigs in a study group, they returned to the individual levels of sexual activity that they had exhibited priour to castration. Even doubling the amount of hormone at week 36 did not increase the mating activity of any group. Presumably Sooty from Figure 13.2 would have fallen into the "high drive" group. (After J. A. Grunt and W. C. Young, 1953. *J Comp Physiol Psychol* 46: 138–144.)

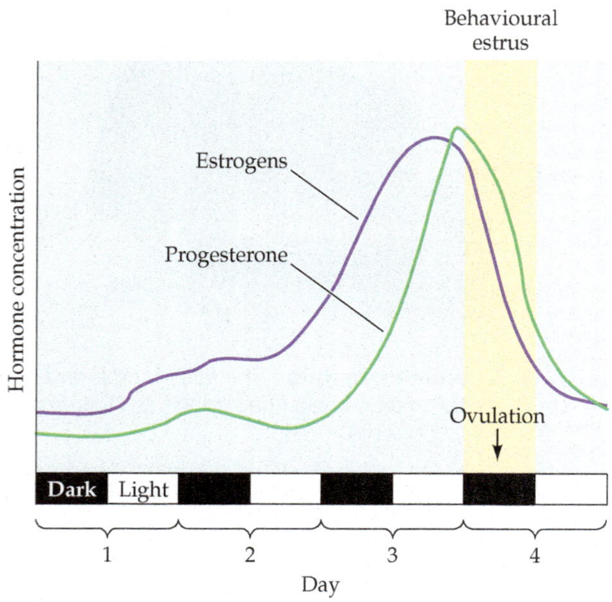

13.5 The Ovulatory Cycle of Rats

out that a very small amount of testosterone—one-tenth the amount normally produced by the animals—is enough to fully maintain the mating behaviour of male rats. Thus, since all male rats make more testosterone than is required to maintain their copulatory behaviour, some other factor, which we can call *drive,* must differ across individual males.

Estrogens secreted at the beginning of the 4- to 5-day **ovulatory cycle** facilitate the proceptive behaviour of the female rat, and the subsequent production of progesterone increases proceptive behaviour and activates receptivity (**FIGURE 13.5**). An adult female whose ovaries have been removed will show neither proceptive nor receptive behaviours. However, 2 days of estrogen treatment followed by a single injection of progesterone will, about 6 hours later, make the female rat proceptive and receptive for a few hours. Only the correct combination of estrogens and progesterone will fully activate copulatory behaviours in female rats—another example of activational effects of gonadal steroids.

13.2 The Neural Circuitry of the Brain Regulates Reproductive Behaviour

Learning Objectives

After reading this section, you should be able to:

13.2.1 **Describe the basic neural circuitry controlling sexual behaviour in rodent males versus females.**

13.2.2 **List the brain regions within those circuits where gonadal hormones regulate mating behaviour.**

13.2.3 **Describe evidence that experience plays a role in finalizing that circuitry.**

13.2.4 **Delineate the neural pathway reporting pheromonal stimuli, and describe how they affect mating.**

Although most of what we know about the neural circuitry of sexual behaviour comes from studies of rats, steroid receptors are found in the same specific brain regions across a wide variety of vertebrate species. Steroid-sensitive regions include the cortex, brainstem nuclei, medial amygdala, hippocampus, and much of the hypothalamus, which plays a particularly important role in regulating copulatory behaviour.

Estrogens and progesterone regulate a lordosis circuit that spans from brain to muscle

Scientists have exploited the steroid sensitivity of the rat lordosis response to develop a map of the neural circuitry that controls this behaviour. In particular, the **ventromedial hypothalamus** (**VMH**) is crucial for lordosis, because lesions there abolish the response. Furthermore, tiny quantities of estradiol implanted directly into the brain can induce receptivity in females, but only when placed in the VMH (Lisk, 1962).

Estrogens potently affect neurons of the VMH, reorganising parts of their dendritic trees (G. D. Griffin and Flanagan-Cato, 2011). Later, when both estrogens and progesterone are in circulation, the effect of estrogens on the length and connectivity of VMH dendrites is reversed. The VMH sends axons to the **periaqueductal gray** region of the midbrain, where again, lesions greatly diminish lordosis. The periaqueductal gray neurons project to the **medullary reticular formation**, which in turn projects to the spinal cord via the **reticulospinal tract**. In the spinal cord, this descending signal from the brain is integrated with sensory inputs from the flanks and rump, which are stimulated by the grasping and mounting behaviour of the male. The result of this processing is a pattern of motor activity

ovulatory cycle The periodic occurrence of ovulation.

ventromedial hypothalamus (VMH) A hypothalamic region involved in eating and sexual behaviours.

periaqueductal gray The neuronal body–rich region of the midbrain surrounding the cerebral aqueduct that connects the third and fourth ventricles.

medullary reticular formation The hindmost portion of the brainstem reticular formation, implicated in motor control and copulatory behaviour.

reticulospinal tract A tract of axons arising from the brainstem reticular formation and descending to the spinal cord to modulate movement.

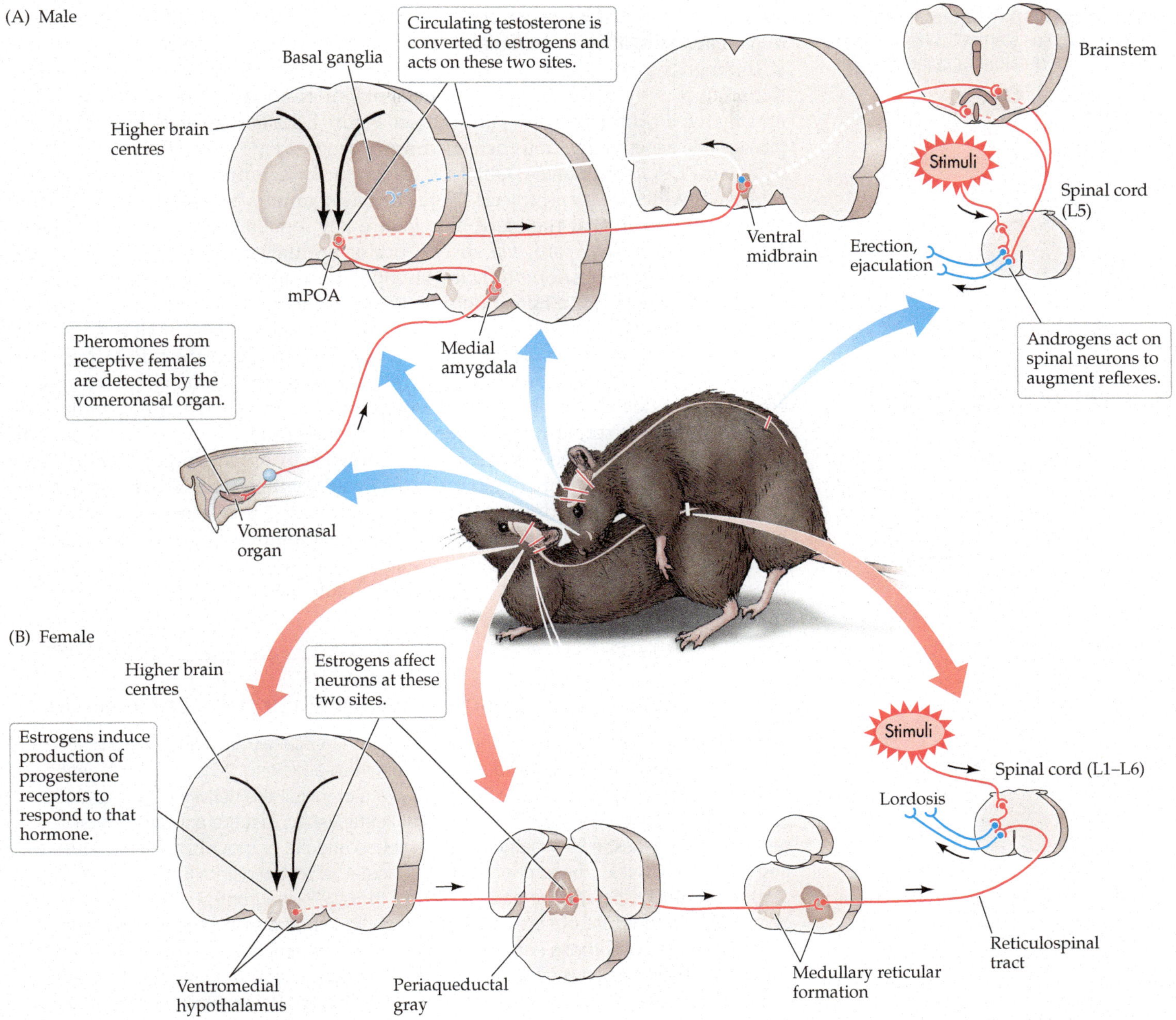

13.6 Neural Circuits for Reproduction in Rodents (Part B after D. W. Pfaff, 1980. *Estrogens and brain function: Neural analysis of a hormone-controlled mammalian reproductive behaviour.* Springer, New York.)

that produces lordosis. So the role of the VMH is to monitor steroid hormone concentrations and, at the right time in the ovulatory cycle, activate a multisynaptic pathway that induces spinal motor neurons to contract back muscles, producing lordosis in response to male mounting (Pfaff et al., 2018). **FIGURE 13.6B** schematically represents this neural pathway and its steroid-responsive components.

Androgens act on a neural system for male reproductive behaviour

As with the lordosis circuit, mapping the sites of steroid action has provided important clues about the neural circuitry controlling male copulatory behaviour in rodents (**FIGURE 13.6A**). The hypothalamic **medial preoptic area (mPOA)** is chock-full of steroid-sensitive neurons, and lesions of the mPOA abolish male copulatory behaviour in a wide variety of vertebrate species (Dominguez, 2009). Furthermore, mating can be reinstated in castrated males by small implants of testosterone placed directly in the mPOA, but not in other brain regions. Note that lesions of the mPOA do not

medial preoptic area (mPOA)
A region of the anteriour hypothalamus implicated in the control of many behaviours, including thermoregulation, sexual behaviour, and gonadotropin secretion.

paragigantocellular nucleus (PGN) A region of the brainstem reticular formation implicated in modulation of spinal reflexes.

interfere with males' *motivation* for copulating with females; they will still press a bar to gain access to a receptive female (Everitt and Stacey, 1987), but they seem unable to commence mounting. Thus, the mPOA seems to be a final common pathway for the production of male copulatory behaviours; motivation relies on other systems, notably the dopamine reward pathway (see Figure 3.22) that appears to mediate the pleasurable aspects of various behaviours and drugs (Pfaus et al., 2012).

The mPOA coordinates copulatory behaviour by sending axons to the ventral midbrain, which in turn projects (1) to the basal ganglia to coordinate mounting behaviours (recall that the basal ganglia are key regulators of motor behaviour) and (2) to the spinal cord, via several brainstem nuclei (Hamson and Watson, 2004) that regulate various reflexes of copulation. One of these brainstem nuclei, the **paragigantocellular nucleus (PGN)** in the pons, sends serotonergic fibres down into the spinal cord, where they inhibit a circuit responsible for penile erection (McKenna, 1999). So, in order for erections to occur, the mPOA must inhibit the inhibitory PGN projection, thereby releasing the spinal circuit to permit erection. Antidepressant drugs that boost serotonergic activity in the brain—for example, selective serotonin reuptake inhibitors like Prozac (see Chapter 12)—can cause problems with erection and orgasm, probably by enhancing serotonergic inhibition of the spinal cord circuitry. (In case you're wondering, Viagra [sildenafil] promotes erection by acting directly on the penis, not the spinal cord or brain.) Because the reflex circuits and ejaculation-generating neurons are in the lumbar spinal cord (Truitt and Coolen, 2002), men with damage at higher levels of the spinal cord often remain capable of copulation and ejaculation.

The Cutting Edge

Sexual Experience Solidifies Neural Circuits for Mating

As neuroscience technology advances, we gain a finer-detail view of how the brain steers a mouse from one behaviour to another, from attacking to mating. Optogenetic methods (see Figures 2.26 and 2.27) to stimulate or inhibit the activity of only estrogen receptor (ER)–containing neurons in the VMH established that the activity of these neurons is responsible for whether a male mouse attacks or mounts another mouse (H. Lee et al., 2014). Next, researchers surgically installed an optic fibre into the VMH of a mouse to monitor the spontaneous activity of those ER+ neurons. Those neurons, and only those neurons, are observed with this technique because a transgene the mouse inherited causes only those neurons to glow when they fire. The ER+ neurons rarely fire when a male mouse is alone, but if another mouse is introduced to the cage, the neurons increase firing dramatically. Importantly, different ensembles of neurons fire, depending on whether the intruder is male or female (Remedios et al., 2017; **FIGURE 13.7**). Of course, the male mouse also displays very different behaviours towards intruder males, which they attack, versus females, which they investigate and eventually attempt to mount.

These are the responses you see in *adult* male mice that have had social and sexual experience. But an inexperienced young adult male mouse responds differently. When it encounters another male for the first time, it does not attack the intruder. If the male intruder is replaced by a female, the ensemble of neurons activated in the young male's VMH is the same as was activated by the male intruder. However, once our male subject has sexual experience with a female, its behaviour and VMH responses diverge, and thereafter distinctly different ensembles of neurons are activated in response to male versus female intruders, and the resident male will vigourously attack another male. Thus it appears that sexual experience is crucial for distinct ensembles of VMH neurons to be activated in response to male versus female intruders and, therefore, for the resident to display the appropriate behaviour. One implication of this research calls into question the idea that either mating or aggression (see Chapter 11) are "instinctive" or "hardwired," since social experience is required for the fine-tuning of circuits directing those behaviours. ■

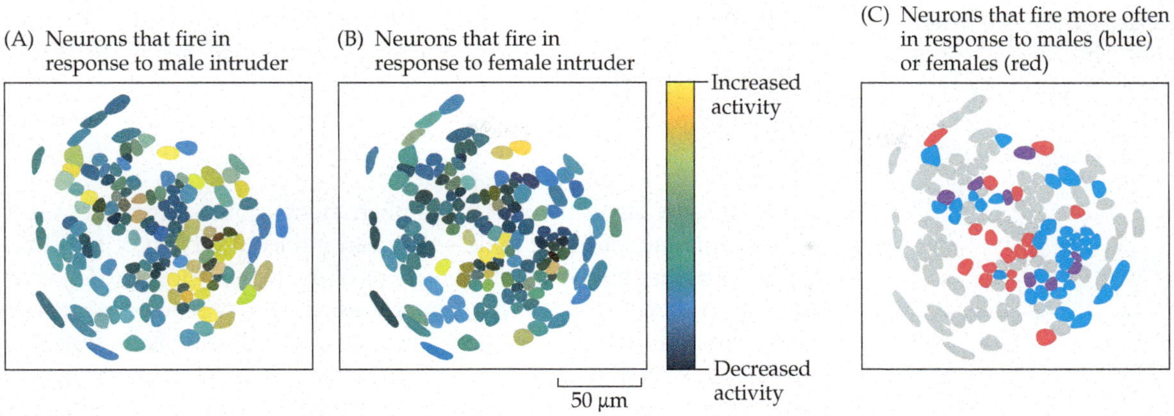

13.7 Distinct Neuronal Ensembles Respond to Male or Female Intruders (A) Using fibre optics to observe neurons in the VMH that emit light when they fire, researchers mapped which cells were activated in response to a male intruder. (B) A different ensemble of VMH neurons fire in response to a female intruder. (C) Here the neurons are colour coded to display those that were activated in response to a female intruder (red) or a male intruder (blue) or the few that were activated by intruders of either sex (purple). (After R. Remedios et al., 2017. *Nature* 550: 388–392.)

Pheromones guide reproductive behaviour in many species

When steroid hormones from the gonads affect the brain to activate mating behaviour, the activation is not absolute; individuals are simply *more likely* to engage in mating behaviours when steroid levels are adequate. This activation can be thought of as communication between the gonads and the brain: by producing steroids to make gametes, the gonads also inform the brain that the body is ready to mate. This signalling takes place inside the individual, while **pheromones** (see Chapters 4 and 9) are chemical signals that communicate information *between* animals to help coordinate their reproductive activities.

An additional branch of male copulatory circuitry in rodents involves the sensory system detecting pheromones to activate male arousal. The **vomeronasal organ** (**VNO**) (see Chapter 9) consists of specialized receptor cells near to but separate from the olfactory epithelium. These sensory cells send axons to the accessory olfactory bulb in the brain. Neurons in the accessory olfactory bulb project to the **medial amygdala**, which depends on circulating sex steroids to maintain masculine form and function (Cooke et al., 2003). Lesions here will abolish the penile erections that normally occur in response to odours from receptive females (Y. Kondo et al., 1997). The medial amygdala, in turn, sends axons to the mPOA. So the mPOA appears to integrate hormonal and sensory information, such as pheromones, and to coordinate the motor patterns of copulation. Figure 13.6A summarises the neural circuitry for male rat copulatory behaviour.

In many species, chemicals that are part of the basic reproductive physiology have been co-opted to also serve as pheromones. In female goldfish, for example, a hormone called *F prostaglandin* that is required for normal ovulation also escapes the female's body and passes into her watery surroundings, where it acts as a pheromone to stimulate mating behaviours in male goldfish (K. Li et al., 2018). The most likely scenario for the evolution of this relationship is that long ago, females released F prostaglandin only as a by-product of ovulation, but because the presence of the hormone conveyed important information about the female's reproductive status, natural selection favoured males that detected the hormone and began courting in response to the signal. Even very simple unicellular organisms, such as yeasts, prepare each other for mating by releasing and detecting pheromones (S. Fields, 1990).

There are multiple examples of pheromonal signalling in rodents. Pheromones in the urine of male mice can accelerate puberty in young females (Stowers and Kuo, 2015). Female mice can also uniquely identify an individual male by the particular mix

pheromone A chemical signal that is released outside the body of an animal and affects other members of the same species.

vomeronasal organ (VNO) A collection of specialized receptor cells, near to but separate from the olfactory epithelium, that detect pheromones and send electrical signals to the accessory olfactory bulb in the brain.

medial amygdala A portion of the amygdala that receives olfactory and pheromonal information.

© FLPA/Andrew Forsyth/AGE Fotostock

13.8 Musth During the reproductive season, bull elephants secrete a pheromone-laden fluid from glands behind the eyes. You can see the darkened skin where the fluid is trickling down.

of pheromones in his urine, and if the pheromones come from a dominant male—but not a subordinate male—they can induce the birth of new neurons in the olfactory bulbs and hippocampus of the female mouse (Mak et al., 2007). When prairie voles mate, the female is exposed to pheromones from her mate's mouth and urine. If she is then isolated and has urine from that male or any other male applied to her snout, pregnancy will be blocked; the fetuses will be resorbed by the female, and she will soon be ready to mate again. So when the female remains with her original mate, she is careful not to let her mate's urine touch her vomeronasal organ; otherwise she will lose their offspring (Smale, 1988). Terminating pregnancy and absorbing the fetuses in the presence of a new male, a phenomenon known as the *Bruce effect*, may be an attempt to make the best of a bad situation: if her original mate is gone, a female may be better off beginning a new litter with a different male. The Bruce effect has even been observed in primates, specifically the gelada, an African monkey (E. K. Roberts et al., 2012).

Pheromones can also convey important information between individuals of the same sex. During **musth**, the mating season of elephants, males secrete a pheromone-laden liquid from glands on their temples, just behind their eyes (**FIGURE 13.8**). In pubescent males, this substance has a distinctive honey-like odour that may attract bees and birds, but mature bulls signal their rank and attract females with a stronger-smelling secretion that contains a pheromone called *frontalin* (curiously, frontalin is also an important pheromone in insects). By broadcasting their low rank and avoiding mature-smelling males, the honey-scented juveniles thus avoid aggressive encounters (J. L. Brown, 2014).

musth An annual period of heightened aggressiveness and sexual activity in male elephants.

13.3 The Hallmark of Human Sexual Behaviour Is Diversity

Learning Objectives

After reading this section, you should be able to:

13.3.1 **Describe Masters and Johnson's model of four phases of human copulation.**
13.3.2 **Describe in broad strokes brain activation changes that accompany orgasm.**
13.3.3 **Critically discuss the role of gonadal hormones in human sexual behaviour.**

How much of what we've described so far about sexual behaviour in animals is relevant to human sexuality? Until the 1940s, when biology professor Alfred Kinsey began to ask friends and colleagues about their sexual histories, there was virtually no scientific study of human sexual behaviour. Kinsey constructed a standardised set of questions and procedures to obtain information for samples of the U.S. population categorised by sex, age, religion, and education. Eventually he and his collaborators published extensive surveys (based on thousands of respondents) of the sexual behaviour of American males (Kinsey et al., 1948) and females (Kinsey et al., 1953).

Controversial in their time, these eye-opening surveys indicated that nearly all men masturbated, that college-educated people were more likely to engage in oral sex than were non-college-educated people, that many people had at one time or another engaged in sex with people of the same sex, and that a stable proportion of the population preferred such sex over heterosexual sex.

Another way to investigate human sexual behaviour is to make behavioural and physiological observations of people engaged in sexual intercourse or masturbation, but the squeamishness of the general public impeded such research for many years.

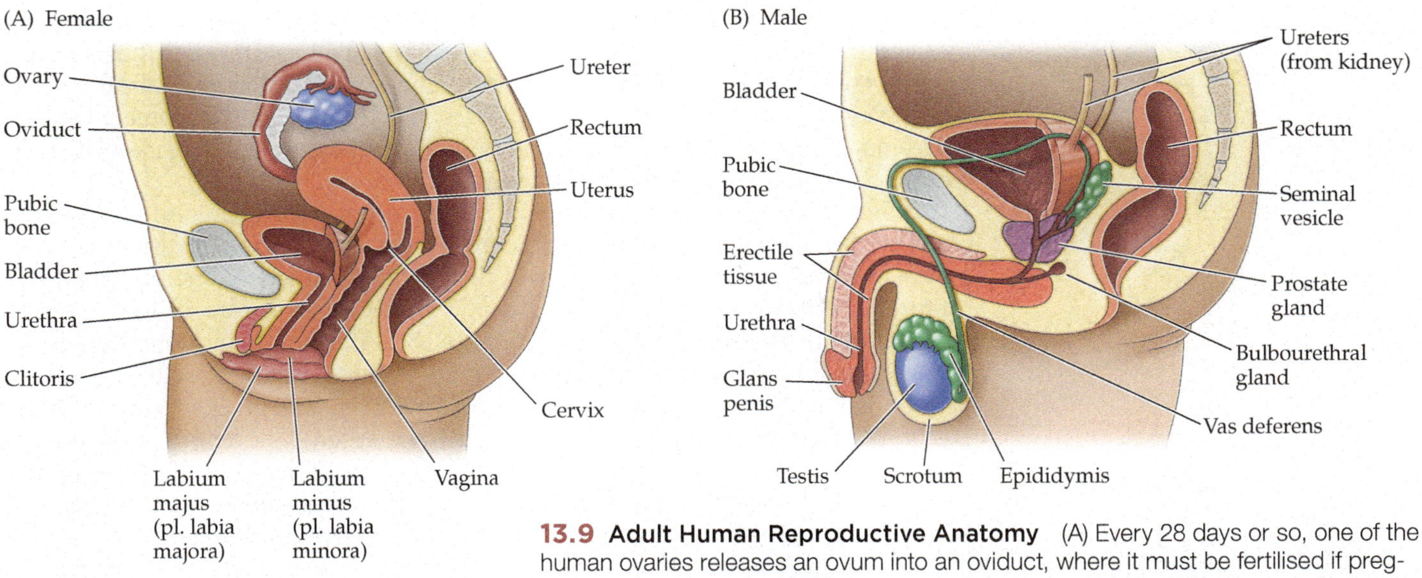

13.9 Adult Human Reproductive Anatomy (A) Every 28 days or so, one of the human ovaries releases an ovum into an oviduct, where it must be fertilised if pregnancy is to occur. The fertilised zygote implants in the wall of the uterus, and a placenta develops. If no pregnancy occurs, the uterine wall (endometrium) sloughs off during menstruation. Most mammalian species lose less endometrial tissue than humans do and thus do not menstruate. (B) In males, sperm originate in the testes, mature in the adjacent epididymis, and are expelled during ejaculation via the muscular vas deferens. Along the way, structures such as the seminal vesicles and prostate gland add their secretions, forming semen.

Go to Activities 13.1–13.2
Female Reproductive Anatomy
Male Reproductive Anatomy

Finally, after Kinsey's surveys were published, physician William Masters and psychologist Virginia Johnson began a large, famous project of this kind (Masters et al., 1994), documenting the impressively diverse sexuality of humans.

Among most mammalian species, including most nonhuman primates, the male mounts the female from the rear; but among humans, face-to-face postures are most common. A great variety of coital positions have been described, and many couples vary their positions from session to session or even within a session. It is this variety in reproductive behaviours, rather than differences in reproductive anatomy (**FIGURE 13.9**), that distinguishes human sexuality from that of most other species.

Another difference between species is that, unlike other animals, humans can report their subjective experiences of sexual behaviour—specifically **orgasm**, the (all-too) brief, extremely pleasurable sensations experienced by most men during ejaculation and by most women during copulation or masturbation. In the original conceptual model of human sexuality, Masters and Johnson (1966) summarised the typical response patterns of both men and women as consisting of four phases: increasing excitement, plateau, orgasm, and resolution (**FIGURE 13.10**). During the excitement phase, the **phallus** (the penis in men, the clitoris in women) becomes engorged with blood, making it erect.

orgasm The climax of sexual experience, marked by extremely pleasurable sensations.

phallus The clitoris or penis.

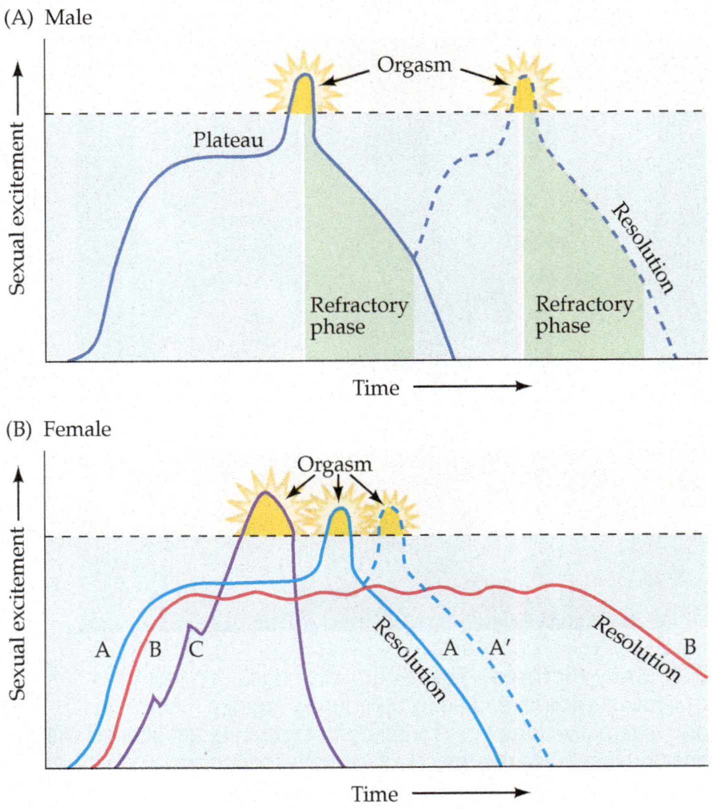

13.10 Human Sexual Response Patterns (A) The typical male pattern includes an absolute refractory period after orgasm, which may be followed by renewed arousal (dotted line). (B) These three patterns (A, B, C) are often observed in women. These diagrams are schematic and do not represent a particular physiological measure, although heart rate varies in roughly this manner. The patterns vary considerably from one individual to another. (After W. H. Masters and V. E. Johnson, 1966. *Human sexual response.* Boston: Little, Brown.)

In women, parasympathetic activity during the excitement phase causes changes in vaginal blood vessels, resulting in the production of lubricating fluids that facilitate intromission. Stimulation of the penis, clitoris, and vagina during rhythmic thrusting accompanying intromission may lead to orgasm. In both men and women, orgasm is accompanied by waves of contractions of genital muscles (mediating ejaculation in men and contractions of the uterus and vagina in women).

In spite of some basic similarities, the sexual responses of men and women differ in important ways. For one thing, women show a much greater variety of commonly observed copulatory sequences. Whereas men have only one basic pattern, captured by the linear model of Masters and Johnson (see Figure 13.10A), women have at least three typical patterns (see Figure 13.10B). Another important aspect of human sexuality is that most men, but not most women, have an absolute refractory period following orgasm (see Figure 13.10A). That is, most men cannot achieve full erection and another orgasm until some time has elapsed—the length of time varying from minutes to hours, depending on individual differences and other factors. Many women, on the other hand, can have multiple orgasms in rapid succession. Functional imaging of the brains of men and women during sexual activity suggests that substantially different networks are active in men's and women's brains during sexual activity *priour* to orgasm (Georgiadis et al., 2009).

Perhaps the most striking observation so far from functional imaging studies of the brain during orgasm is how widespread the changes in activation are. Dozens of brain sites are active at this time; nonetheless, the findings map onto some of what we know about sexual circuitry in lab rats. In men, sexual arousal, and especially ejaculation, activates subcortical sites that include the ventral tegmental area and right basal forebrain (Georgiadis et al., 2010; Holstege et al., 2003), part of the dopamine-based reward system discussed in Chapter 3, where we talked about the role of this reward system in drug abuse (see Figure 3.24). Penile stimulation activates the right insula and secondary somatosensory cortex (Georgiadis and Holstege, 2005). Subcortically, penile stimulation first activates a part of the basal ganglia, followed by activation in the lateral hypothalamus and anteriour middle cingulate cortex (Georgiadis et al., 2010).

In women, orgasms change activity in a variety of brain regions, as shown in **FIGURE 13.11**. Subcortically, clitoral stimulation and orgasm are associated with alterations in activity in the hypothalamus, amygdala, cerebellum, cingulate, and brainstem. Notably, in parallel with men, several sites in the basal forebrain are active during female orgasm, including the nucleus accumbens, likewise implicating the brain's reward system in sexual responses (Komisaruk and Whipple, 2005). Interestingly, sexual activity also results in pronounced increases in activity of orbitofrontal cortex and the anteriour cingulate (Wise et al., 2017), suggesting alteration of consciousness during orgasm.

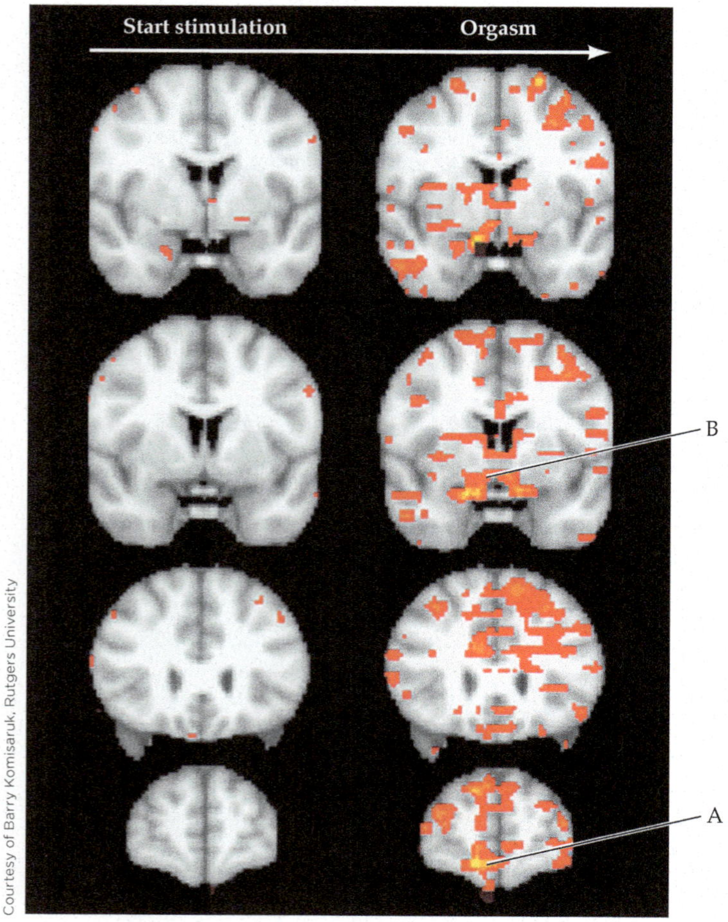

13.11 Sexy Pictures These coronal images of a woman's brain reveal widespread activation during orgasm. Notice, in particular, activation of prefrontal cortex (point A), associated with conscious, self-aware aspects of sexual experiences, and of the basal forebrain (point B), associated with pleasure and reward.

Taking a broader perspective on sexuality reveals additional distinctions between men and women (Peplau, 2003). Research has generally found that basic sex drive is greater in men, on average, reflected in more frequent masturbation, sexual fantasies, and pursuit of sexual contacts. Emotional components and cognitive factors play a stronger role in the sex lives of women than of men. In addition to a somewhat flexible sexuality that adapts to new experiences and situations over time, most women place more emphasis on sexual intimacy within the context of committed relationships. These observations and others have led sex researchers to adopt a more nuanced view of female sexuality. While Masters and Johnson simply adapted the linear model of male sexuality to women, the more modern perspective views women's sexuality as a cycle, governed in large measure by

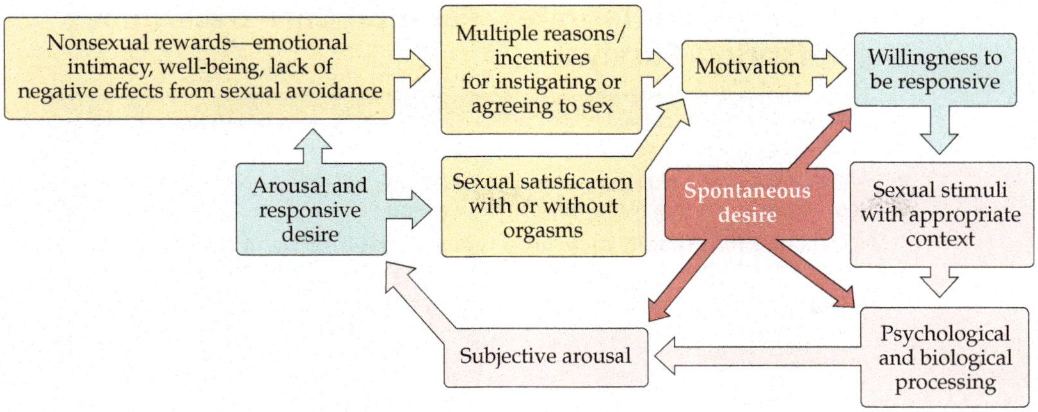

13.12 Women's Sexual/Emotional Response According to one model of female sexual responses (Basson, 2008), a willingness to become receptive (top left) interacts with sexual drive (red arrows). A combination of sexual stimuli (top) and biological factors result in feelings of arousal and desire (bottom right). A variety of factors, including rewarding nonsexual intimacy and/or sexual gratification from this encounter, provide feedback that affects levels of motivation (left) for subsequent sexual behaviour. (After R. Basson, 2008. *Int J Impot Res* 20: 466–478; R. Basson, 2001. *Obstet Gynecol* 98: 350–353.)

emotional factors (Basson, 2008). According to this model, emotional intimacy and desire (more than physiological arousal) are crucial in the initiation of sexual responses, and following a sexual encounter, a combination of both emotional and physical satisfaction affects the likelihood of further sexual activity (**FIGURE 13.12**).

Although male and female sexuality may bear the imprint of our evolutionary history, on an individual basis it is also shaped by sociocultural pressures and experience. Sexual therapy, for example, usually consists of helping the person to relax, to recognise the sensations associated with coitus, and to learn the behaviours that produce the desired effects in both partners. Masturbation during adolescence, rather than being harmful as suggested in previous times, may help prevent sexual problems in adulthood. As with other behaviours, practice, practice, practice helps.

Sexual behaviour may also aid overall health; epidemiological studies indicate that men who have frequent sex tend to live longer than men who do not (Davey-Smith et al., 1997), and sexual experiences effectively induce adult neurogenesis in the hippocampus in lab animals, suggesting that sexual behaviours may be directly beneficial to the brain (Leuner et al., 2010). Of course, it is also important to one's health to take precautions (such as using condoms) to avoid contracting sexually transmitted infections.

Hormones play only a permissive role in human sexual behaviour

We saw that a little bit of testosterone must be in circulation to activate male-typical mating behaviour in rodents. The same relation seems to hold for human males. For example, boys who fail to produce testosterone at puberty show little interest in sex unless they receive androgen supplements (Stamou and Georgopoulos, 2018). These males, as well as men who have lost their testes as a result of cancer or accident, made it possible to conduct experiments showing that testosterone indeed stimulates sexual interest and activity, as well as a sense of heightened energy, in men (J. M. Davidson et al., 1979).

Recall that, in rats, additional testosterone has no effect on the vigour of mating. Consequently, there is no correlation between the amount of androgens produced by an individual male rat and his tendency to copulate. In humans, too, just a little testosterone is sufficient to restore behaviour, and despite popular misconceptions, there is no correlation between systemic androgen levels and sexual activity among men, provided they have at least *some* androgen. Some women experience sexual dysfunction after menopause, reporting decreased sexual desire and difficulty achieving comfortable coitus. There are many possible reasons for such a change, including several hormonal changes. Providing postmenopausal women with low doses of both estrogens and androgens can have beneficial effects on the genital experience of sex and a modest increase in women's sexual interest (Basson, 2008; Sherwin, 2002). Attempts by pharmaceutical companies to find a "female Viagra," a drug that would more effectively boost sexual desire in women, have had only modest effects so far (Cha and McGinley, 2019).

There have been several attempts to determine whether women's interest or participation in sexual behaviour varies with the menstrual cycle. For example, one study found women were more likely to wear red when at the peak of their fertility (Beall and Tracy, 2013). But in general, such effects are small, and several studies have failed to see any significant change in interest in sex across the menstrual cycle.

13.4 For Many Vertebrates, Parental Care Determines Offspring Survival

Learning Objectives

After reading this section, you should be able to:

13.4.1 **Offer evidence that rats normally find pups aversive, and describe the role of experience in overcoming that aversion.**

13.4.2 **Describe four readily measured types of maternal behaviour and the evidence that the hormones of pregnancy regulate them.**

13.4.3 **Outline the neural pathways that seem to mediate parental behaviour in rodents.**

parental behaviour Behaviour of adult animals with the goal of enhancing the well-being of their own offspring, often at some cost to the parents.

altricial Referring to animals that are born in an undeveloped state and depend on maternal care, as human infants do.

precocial Referring to animals that are born in a relatively developed state and that are able to survive with relatively little maternal care.

parabiotic Referring to a surgical preparation that joins two animals to share a single blood supply.

In many vertebrate species, copulation is not enough to ensure reproduction. Extensive **parental behaviour** is necessary for the survival of many young vertebrates, and all newborn mammals. This is especially true of species with **altricial** young, which start life with poorly developed motor or sensory systems (e.g., songbirds, humans) and are so generally helpless that a very substantial parental investment, perhaps by both parents, is required just for them to survive. Although **precocial** young (e.g., ducks, horses) likewise require parental investments in the form of food and protection, at least they can move around, see, and keep themselves warm. In these species the female alone may care for the offspring.

In Chapter 4 we discussed the milk letdown reflex, when the infant's suckling on the nipple triggers the secretion of oxytocin to promote the release of milk (see Figure 4.12). In rats, the pregnant female prepares for her pups by licking all of her nipples. Doing so probably helps clean the nipples before the pups arrive, but it also makes them more sensitive to touch. This self-grooming actually expands the amount of sensory cortex that responds to skin surrounding the nipples (Xerri et al., 1994), which probably sets the stage for the letdown reflex. This is a wonderful example of an animal's behaviour altering its own brain and therefore changing its future behaviour.

Rat mothers (called *dams*) show four easily measured maternal behaviours: nest building, crouching over pups, retrieving pups, and nursing (**FIGURE 13.13**). A virgin female or male rat won't normally show these behaviours towards rat pups. In fact, a virgin female finds the smell of newborn pups aversive. Information about the odour from pups projects via the olfactory bulb to the medial amygdala and on to the VMH. Lesions anywhere along that path will cause a virgin rat to show maternal behaviour right away (Lonstein et al., 2015), because she no longer detects the smell.

However, if a virgin female is exposed to newborn pups a few hours a day for several days in a row, she (or almost any adult rat, male or female) will start building a nest, crouching over pups, and retrieving them. As the rat gradually habituates to the smell of the pups, it starts taking care of them. But the rat dam that gives birth to her first litter will instantly show these behaviours. It turns out that the rather complicated pattern of hormones during pregnancy shapes her brain to display maternal behaviours before she is exposed to the pups.

The effect of hormones on a rat's maternal behaviours is demonstrated by a **parabiotic** preparation in which two female rats are surgically joined, sharing a single blood supply, such that each is exposed to any hormones secreted by the other (**FIGURE 13.14**). If one of those females is pregnant, then at the end of her pregnancy, the other female, who was never pregnant but was exposed to the pregnant rat's hormones, will also immediately show maternal behaviour (Terkel and Rosenblatt, 1972). Which hormone is responsible for promoting maternal behaviour? No single hormone alone can do it; the combination of several hormones, including estrogens, progesterone, and prolactin, is required. There is ample evidence that the hormones of pregnancy also prepare human mothers to nurture their newborns (A. S. Fleming et al., 2002).

13.13 Parental Behaviour in Rats

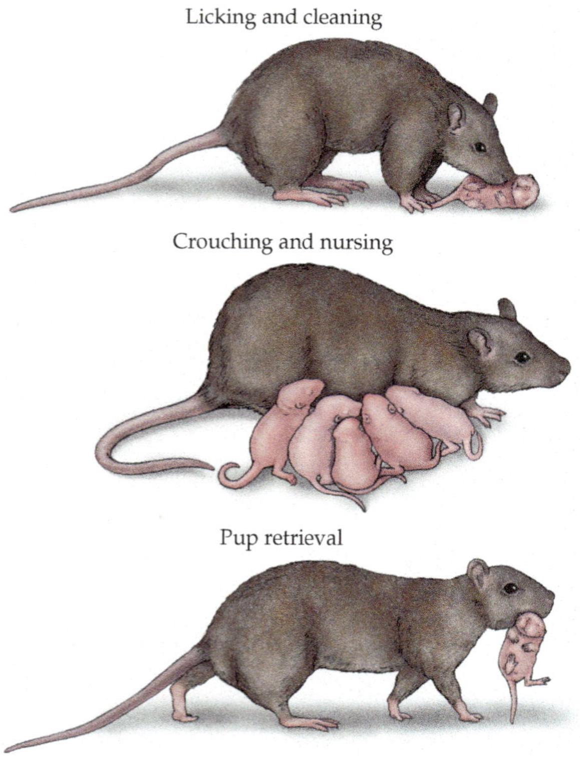

The network of brain regions that controls maternal behaviour shows considerable overlap with the circuitry for sexual behaviour. It's not too surprising that the same brain regions are involved in mating and maternal behaviour, when you consider that these behaviours are simply successive stages in the overall process of reproduction. One brain region that regulates maternal behaviour is the anteroventral periventricular nucleus (AVPV), which is larger in females than males and which grows even larger after pregnancy (N. Scott et al., 2015). As we noted earlier, the mPOA is sensitive to many steroid hormones, and lesions there severely reduce or eliminate oral maternal behaviours such as licking and pup retrieval, but they have relatively little effect on crouching or nursing. Lesioning the periaqueductal gray, conversely, has no effect on the oral maternal behaviours, but it virtually eliminates the crouching position females take to nurse pups (Lonstein and Stern, 1997). Some of the mPOA neurons that play a role in regulating pup care, in mice of either sex, express the neuropeptide galanin (J. Kohl et al., 2018). Interestingly, pregnancy leaves its mark on the human brain. Several cortical regions shrink during pregnancy, and this shrinkage correlates with the mother's attachment to her baby, a structural change still evident 2 years later (Hoekzema et al., 2017). Recall from Chapter 6 that such cortical shrinkage is associated with brain maturation and probably represents more effective functioning.

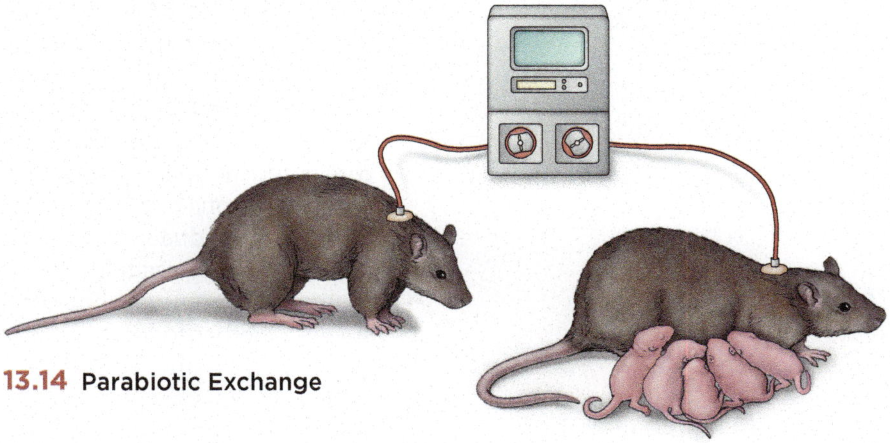

13.14 Parabiotic Exchange

Now that we've described the hormonal and neural interactions that influence male and female reproductive behaviours, let's discuss how males and females come to be different in the first place.

SEXUAL DIFFERENTIATION

Sexual differentiation is the process by which individuals develop either male or female bodies and behaviours. In mammals this process begins before birth and continues into adulthood. As we'll see, some people may be very male-like (masculine) in some parts of the body and very female-like (feminine) in others, so sometimes a person is neither male nor female but may be a blend of the two sexes.

13.5 Sex Determination and Sexual Differentiation Occur Early in Development

Learning Objectives

After reading this section, you should be able to:

- **13.5.1** **Outline the mechanisms directing sexual differentiation of the mammalian body, including hormonal factors.**
- **13.5.2** **Describe how three different genetic conditions can result in an intersex phenotype.**
- **13.5.3** **Explain how human sex may differ in various parts of the same body.**

In mammals, every egg carries an X chromosome from the mother; fusion with an X- or Y-bearing sperm is the key event in **sex determination**, the developmentally early event that typically establishes the course of subsequent sexual differentiation of the body. With only occasional exceptions, mammals that receive an X chromosome from the father will become females with XX sex chromosomes; those that receive the father's Y chromosome will become XY males.

In vertebrates the first major consequence of sex determination is in the gonads. Very early in development each individual has a pair of **indifferent gonads**,

sexual differentiation The process by which individuals develop male-like or female-like bodies and behaviour.

sex determination The process by which the initial decision is made for a fetus to develop as a male or a female.

indifferent gonads The undifferentiated gonads of the early vertebrate fetus, which will eventually develop into either testes or ovaries.

glands that vaguely resemble both testes and ovaries. During the first month of gestation in humans, differential genetic instructions determine whether the indifferent gonads begin changing into ovaries or testes. In mammals with a Y chromosome, a gene called ***SRY*** (for ***s***ex-determining ***r***egion on the ***Y*** chromosome) is responsible for the development of testes. The Sry protein causes the cells in the core of the indifferent gonad to proliferate at the expense of the outer layers, and the indifferent gonad develops into a testis.

In XX individuals (or if the *SRY* gene is defective) no Sry protein is produced, and the indifferent gonad takes a different course: cells of the outer layers of the gonad proliferate more than those of the inner core, and an ovary forms. This early decision of whether to form testes or ovaries has a domino effect, setting off a chain of events that usually results in either a male or a female.

A persistent bias in biomedical science has resulted in males being studied exclusively in most cases, on the assumption that female ovulatory cycles would introduce more variability and that whatever was found in males would hold true for both sexes. However, analysis of the animal literature revealed that in fact there is more variability in males than females (Prendergast et al., 2014). This research bias has had a real impact. For example, heart attacks in women tend to produce symptoms that are very different from those symptoms, derived from observations of men, that were the only ones publicized until recently. In the United States, the National Institutes of Health (NIH) has now instituted a policy that future research grant applications must either include both sexes or provide a sound rationale for excluding one sex (Sandberg et al., 2015).

Gonadal hormones direct sexual differentiation of the body

For all mammals, including humans, sexual differentiation of the bodies of males and females is the product of differential hormone secretion. Whereas developing testes produce several hormones, fetal ovaries produce very little hormone. When exposed to testicular hormones, many tissues of the body begin developing masculine characters; if the cells are not exposed to testicular hormones, they develop feminine characters. As a result of their differential exposure to steroid hormones, thousands of genes in tissues throughout the body are expressed differently in developing males and females (van Nas et al., 2009).

We can chart masculine or feminine development by examining the structures that connect the gonads to the outside of the body. The internal reproductive tracts are quite different in adult males and females (see Figure 13.9), but at the embryonic stage all individuals have the precursor tissues of both systems. The early fetus has a genital tubercle that can form either a clitoris or a penis, as well as two sets of ducts—the **wolffian ducts** and the **müllerian ducts** (**FIGURE 13.15A**)—that connect the indifferent gonads to the outer body wall. In females, the müllerian ducts develop into the oviducts (also called *fallopian tubes*), uterus, and inner vagina (**FIGURE 13.15B AND C, RIGHT**), and most of the wolffian system degenerates. In males, hormones secreted by the testes orchestrate the converse outcome: the wolffian ducts develop into epididymis, vas deferens, and seminal vesicles (**FIGURE 13.15B AND C, LEFT**), while the müllerian ducts shrink to mere remnants.

Two testicular secretions drive masculinisation: testosterone, which promotes the development of the wolffian system, and **anti-müllerian hormone** (**AMH**), which causes regression of the müllerian system. In the absence of testes and their secretions, the genital tract develops in a feminine pattern, as the wolffian ducts regress and the müllerian ducts develop into components of the female internal reproductive tract.

Testosterone also masculinises other, non-wolffian-derived structures such as the fetal genitalia. Local conversion of testosterone into a more potent androgen, **dihydrotestosterone** (**DHT**), accomplished by the enzyme **5α-reductase** in the genital skin, induces the bipotential genitals to form a scrotum and penis (see Figure 13.15). We'll see later that without local conversion into DHT, testosterone is able to masculinise the genitalia only partially. If androgens are absent altogether, the genital tissues develop into the female labia and clitoris.

SRY* gene** A gene on the Y chromosome that directs the developing gonads to become testes. The name *SRY* stands for ***sex-determining ***r***egion on the ***Y*** chromosome.

wolffian ducts A duct system in the embryo that will develop into male structures (the epididymis, vas deferens, and seminal vesicles) if testes are present.

müllerian ducts A duct system in the embryo that will develop into female reproductive structures (oviduct, uterus, and inner vagina) if testes are not present.

anti-müllerian hormone (AMH) A protein hormone secreted by the fetal testis that inhibits müllerian duct development.

dihydrotestosterone (DHT) The 5α-reduced metabolite of testosterone; a potent androgen that is principally responsible for the masculinisation of the external genitalia in mammalian sexual differentiation.

5α-reductase An enzyme that converts testosterone into dihydrotestosterone (DHT).

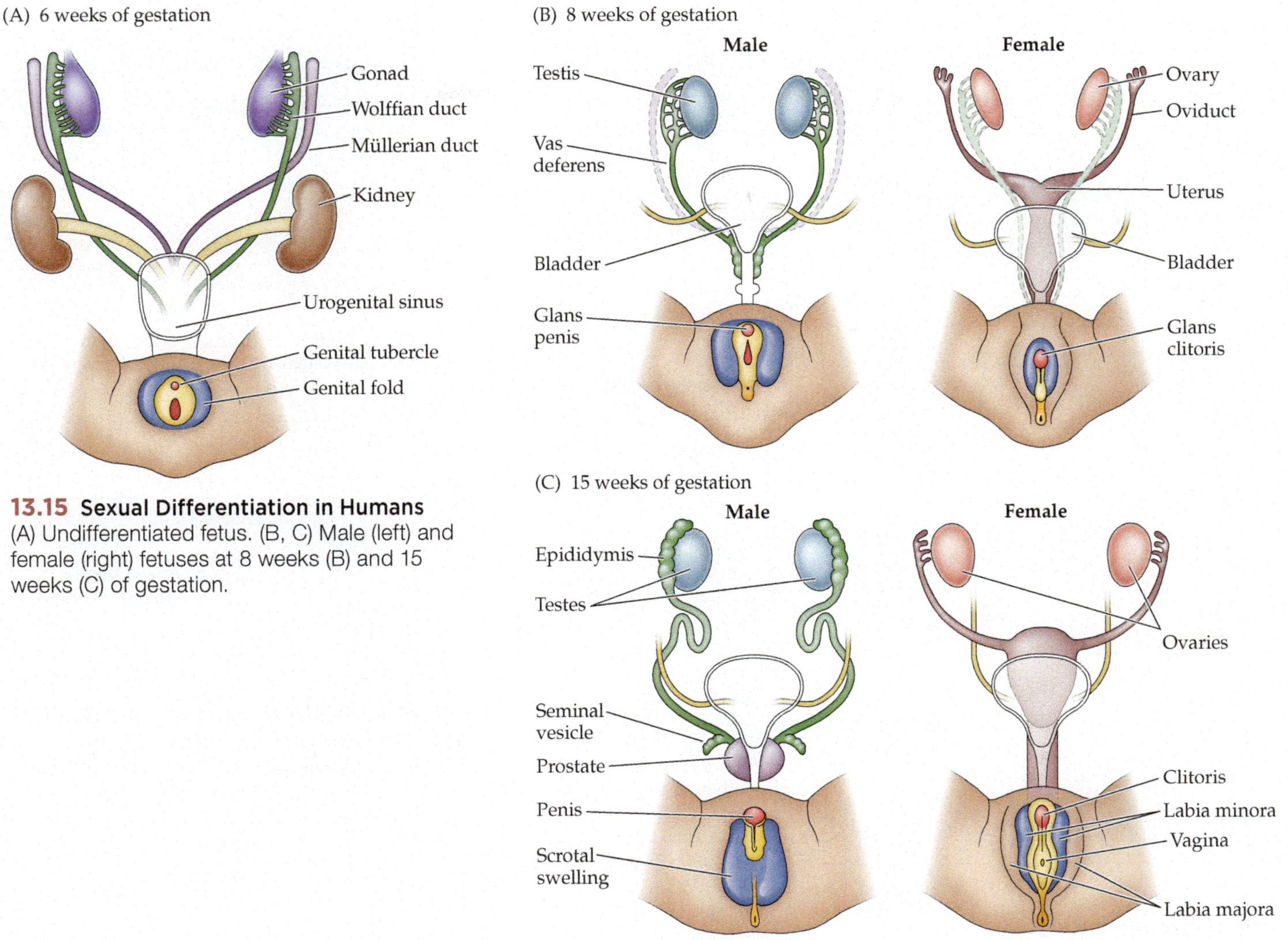

13.15 Sexual Differentiation in Humans
(A) Undifferentiated fetus. (B, C) Male (left) and female (right) fetuses at 8 weeks (B) and 15 weeks (C) of gestation.

In general, unless the indifferent gonad becomes a testis and begins secreting hormones, an immature mammal develops as a female in most respects. As noted earlier, immature ovaries produce few hormones. So the sex chromosomes determine the sex of the gonad, and differential exposure to gonadal hormones controls subsequent sexual differentiation of the rest of the body (**FIGURE 13.16A**). Later in life, hormones and experience act in concert to guide further sexual differentiation of **gender**, the behaviours and attitudes that a given culture considers to be masculine or feminine, and *gender identity*, whether one identifies as male, female, or an alternative gender (**FIGURE 13.16B**).

Differences in the sequence of sexual differentiation result in predictable changes in development

Some people have only one sex chromosome: a single X (embryos without at least one X chromosome do not survive). This genetic makeup results in **Turner's syndrome**, in which the person has recognisable ovaries, as you might expect because no *SRY* gene is present, but they are underdeveloped.

Sometimes XX individuals with well-formed ovaries are exposed to androgens in utero, and depending on the degree of exposure, they may be masculinised. For example, most fetal rats develop in the uterus sandwiched between two siblings. If a female is bracketed by brothers, some of the androgen from the siblings must reach the female, because although her gross appearance will be feminine at

gender The behaviours and attitudes that a given culture considers to be masculine or feminine, or an alternative.

Turner's syndrome A condition seen in individuals carrying a single X chromosome but no other sex chromosome.

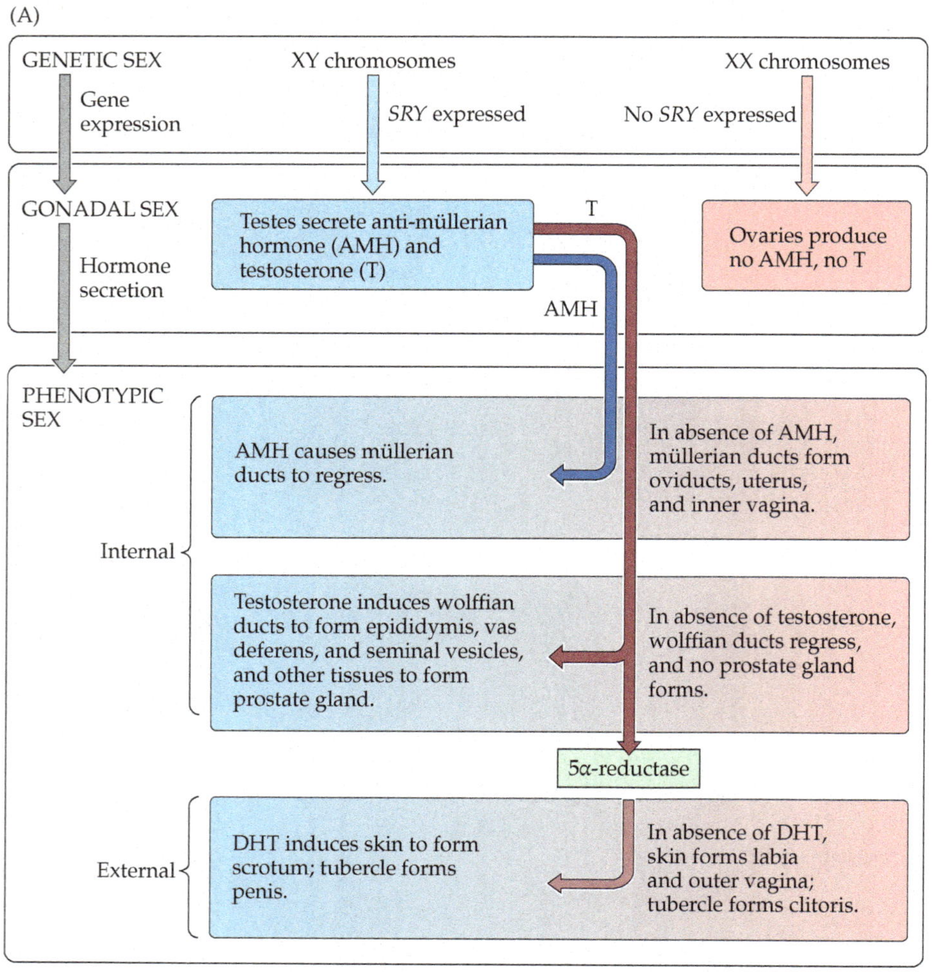

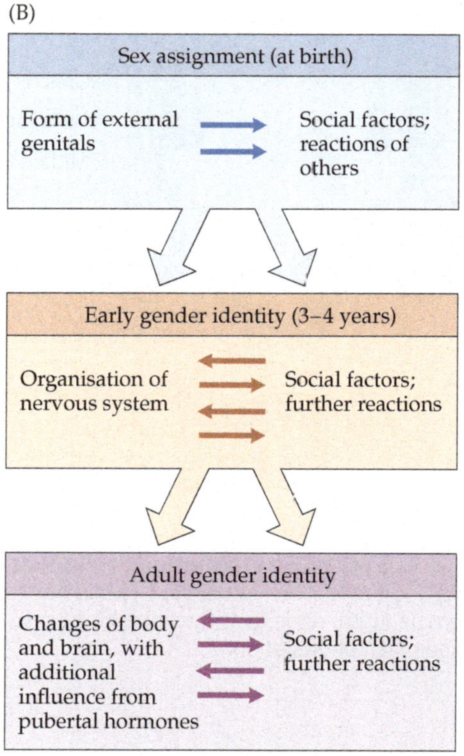

13.16 Sexual Differentiation and Gender Identity (A) Genetic and hormonal mechanisms of embryonic sexual differentiation. (B) Steps toward adult gender identity in humans.

congenital adrenal hyperplasia (CAH) Any of several genetic conditions that result in exposure of a fetus to excessive adrenal androgens.

intersex Referring to an individual with atypical genital development and sexual differentiation that generally resembles a form intermediate between typical male and typical female genitals.

birth, her anogenital distance (the distance from the base of the clitoris [or penis in a male] to the anus) will be slightly greater (i.e., more male-like) than that of a female developing between two sisters (Clemens et al., 1978).

In humans, **congenital adrenal hyperplasia** (**CAH**) can cause a developing female to be exposed to excess androgens before birth. In CAH, the adrenal glands fail to produce sufficient corticosteroids, producing instead considerable amounts of androgens. In XX individuals with this condition, the androgen levels produced are usually intermediate between those of typical females and males, and the newborn often has an **intersex** appearance: a phallus that is intermediate in size between a typical clitoris and a typical penis, and skin folds that resemble both labia and scrotum (**FIGURE 13.17**). In severe cases the individual may appear to have a well-formed penis and scrotum, but no testes are present in the "scrotum"; instead, these individuals have abdominal ovaries. Once children with CAH are born, the excess adrenal androgen production can be corrected, but by then substantial sexual differentiation has already occurred.

The result is a vigourous controversy over the best course of action for the parents of girls with CAH: to opt for immediate surgical modification of the genitalia, or to wait until adulthood, when the CAH-affected individuals can decide for

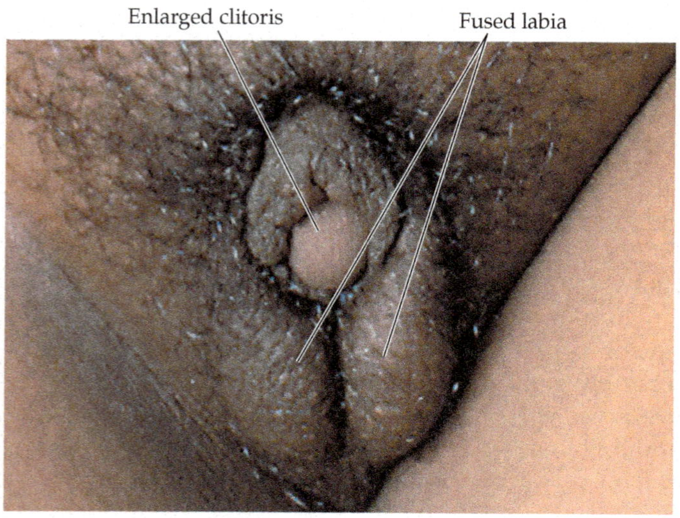

13.17 An Intersex Phenotype The partially masculinised genitalia of this woman with CAH are the result of excessive androgen production by the adrenals before birth.

themselves whether to have surgery, which is, after all, purely cosmetic. Females with CAH are much more likely to be described by their parents (and themselves) as tomboys than are other girls, and they exhibit enhanced spatial abilities on cognitive tests that usually favour males (Puts et al., 2008). In adulthood, although a majority of females with CAH describe themselves as heterosexual, they are more likely to report a lesbian orientation than the general population of women.

13.18 Women with Androgen Insensitivity Syndrome These women are part of a national AIS-DSD Support Group (aisdsd.org).

Dysfunctional androgen receptors can block male masculinisation

The importance of androgens for masculine sexual differentiation is illustrated by the condition known as **androgen insensitivity syndrome (AIS)**. The gene for the androgen receptor is found on the X chromosome. If this gene is dysfunctional in XY individuals, their functional *SRY* gene will lead to the development of testes that secrete AMH and testosterone. But in the absence of functional androgen receptors, their tissues will not respond to the masculinising effects of testosterone.

In the absence of working androgen receptors, the wolffian ducts fail to sense testosterone and thus regress; the external genital tissue forms labia and a clitoris. At puberty women with AIS develop breasts; breast development depends on the ratio of estrogenic to androgenic stimulation at puberty, and since androgen-insensitive individuals receive little androgenic *stimulation*, the functional estrogen-to-androgen ratio is high. However, lacking ovaries or uterus, women with AIS do not start menstruating and are infertile with a shallow vagina (due to the missing müllerian contribution). But in almost all other respects, women with AIS look like other women (**FIGURE 13.18**) and, as we'll see shortly, behave like other women.

androgen insensitivity syndrome (AIS) A syndrome caused by a mutation of the androgen receptor gene that renders tissues insensitive to androgenic hormones like testosterone. Affected XY individuals are phenotypic females, but they have internal testes and regressed internal genital structures.

Some people seem to change sex at puberty

Babies are occasionally born with a rare genetic mutation that disables the enzyme (5α-reductase) that converts testosterone to DHT. An XY individual with this condition will develop testes and normal male internal reproductive structures (because testosterone and AMH function normally), but the external genitalia will not masculinise fully (**FIGURE 13.19**). The reason for this is that the genital epithelium, which normally possesses 5α-reductase, is unable to amplify the androgenic signal by converting the testosterone to the more active DHT. Consequently, the phallus is only slightly masculinised and resembles a large clitoris, and the genital folds resemble labia, although they contain the testes. Usually there is no vaginal opening. The prostate and seminal vesicles are also affected, resulting in abnormalities of semen production, but paternity can be achieved with medical assistance (H. J. Kang et al., 2011).

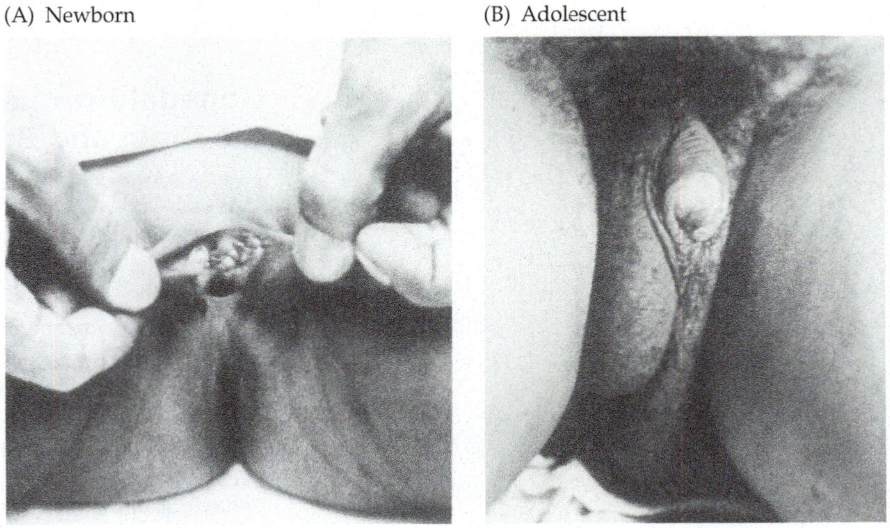

13.19 *Guevedoces* In the Dominican Republic, some individuals, called *guevedoces*, are born with ambiguous genitalia (A) and are raised as girls. At puberty, however, the phallus grows into a recognisable penis (B), and the individuals begin acting like young men.

guevedoces Literally, "eggs at 12" (in Spanish). A nickname for individuals who are raised as girls but at puberty change appearance and begin behaving as boys do.

A particular village in the Dominican Republic is home to several families that carry the mutation causing 5α-reductase deficiency. Children born with this appearance seem to be regarded as girls in the way they are dressed and raised (Imperato-McGinley et al., 1974). At puberty, however, the testes increase androgen production, and the external genitalia become more fully masculinised. The phallus grows into a small but recognisable penis; the body develops narrow hips and a muscular build, without breasts; and the individuals begin acting like young men. The villagers have nicknamed such individuals ***guevedoces***, meaning "eggs (testes) at 12 (years)." These men never develop facial beards (or balding), but they usually have girlfriends, indicating that they are sexually interested in women. We will discuss the sexual identities of *guevedoces*, as well as women with CAH or AIS, a little later.

How should we define gender—by genes, gonads, genitals, or the brain?

Most people are either male or female, and whether we examine their chromosomes, gonads, external genitalia, or internal structures, we see a consistent pattern: they are all either feminine or masculine in character. Compared with physical features, behaviour is much more difficult to define as feminine or masculine. The only behaviour displayed *exclusively* by one sex is childbirth. Even behaviours that are very rarely displayed by members of one sex or the other (e.g., sexual assault by women or breast-feeding by men) occur sometimes in the unexpected sex. For behaviours that can be measured and studied experimentally in humans or other animals, we have to resort to group means and statistical tests to see the differences. Almost all individuals display some behaviours that are more common in the opposite sex.

The various syndromes that we have been discussing show us that even physical features can be confusing criteria by which to judge sex (see Figure 13.18), and they indicate that sex, gender, and sexual orientation are distinct and independent from one another. Androgen-insensitive humans have male XY sex chromosomes and internal testes, and like most males they do not have oviducts or a uterus. But they do have a vagina and breasts, and in most respects their behaviour is typical of females: they dress like females, they are attracted to and marry males, and perhaps most important, even after they learn the details of their condition, they strongly identify themselves as women (Bakker, 2019).

As we will see next, there is evidence that the same hormones that masculinise the genitalia of male animals in utero also masculinise the brain to affect later behaviour. We'll also see that there are structural sex differences in the brains of humans and other animals. In androgen-insensitive rats, some brain regions are masculine and others are feminine. Thus, from a scientific standpoint we cannot regard an animal, especially a human, as simply masculine or feminine. Rather we must specify which structure or behaviour we are talking about when we say it is typical of females or of males. As we turn our attention to the effects of hormones on the brain, it is especially important to maintain the perspective that individuals may be relatively masculine in some traits and more feminine in others.

13.6 Gonadal Hormones Direct Sexual Differentiation of the Brain and Behaviour

Learning Objectives

After reading this section, you should be able to:

13.6.1 **Describe the organisational effect of hormones on sexual differentiation of the brain.**

13.6.2 **Explain the importance of sensitive periods for hormonal effects on the brain.**

13.6.3 **Describe the aromatisation hypothesis for masculinisation of the rodent brain, and offer evidence that this mechanism does not apply to primates.**

13.6.4 **Detail the cellular actions of hormones in the differentiation of two sexual dimorphisms in the brain and one in the spinal cord.**

13.6.5 **Explain how experience also plays a role in sexual differentiation of behaviour in animals and humans.**

As scientists began discovering that testicular hormones direct masculine development of the body, behavioural researchers found evidence for a similar influence on the brain. A female guinea pig, like most other rodents, normally displays the lordosis posture in response to a mounting male for only a short period around the time of ovulation, when her fertility is highest. If a male mounts her at other times, she does not show lordosis. Experimenters can induce female rodents to display lordosis by injecting ovarian steroids in the sequence that normally occurs during ovulation—giving her estrogens for a few days and then progesterone. A few hours after the progesterone injection, the female will display lordosis in response to male mounting.

Phoenix et al. (1959) exposed female guinea pigs to testosterone in utero. As adults, these females did *not* show lordosis. Even if their ovaries were removed and they were given the steroidal regimen that reliably activated lordosis in control females, these fetally androgenised females did not show lordosis. On the basis of these data, the researchers inferred that the same testicular steroids that masculinise the genitalia during early development also masculinise the developing brain. In other words, they proposed that the brain was just one more target tissue that is masculinised by androgens acting early in life (**FIGURE 13.20**). This type of lasting change due to steroid exposure is known as an **organisational effect**.

A steroid has an organisational effect only when present during a specific **sensitive period**, generally in early development. Unlike the transient nature of activational effects of hormones, which we discussed earlier, the organisational effects of hormones tend to be permanent. The exact boundaries of the sensitive period of development depend on which behaviour and which species are being

organisational effect A permanent alteration of the nervous system, and thus permanent change in behaviour, resulting from the action of a steroid hormone on an animal early in its development.

sensitive period The period during development in which an organism can be permanently altered by a particular experience or treatment.

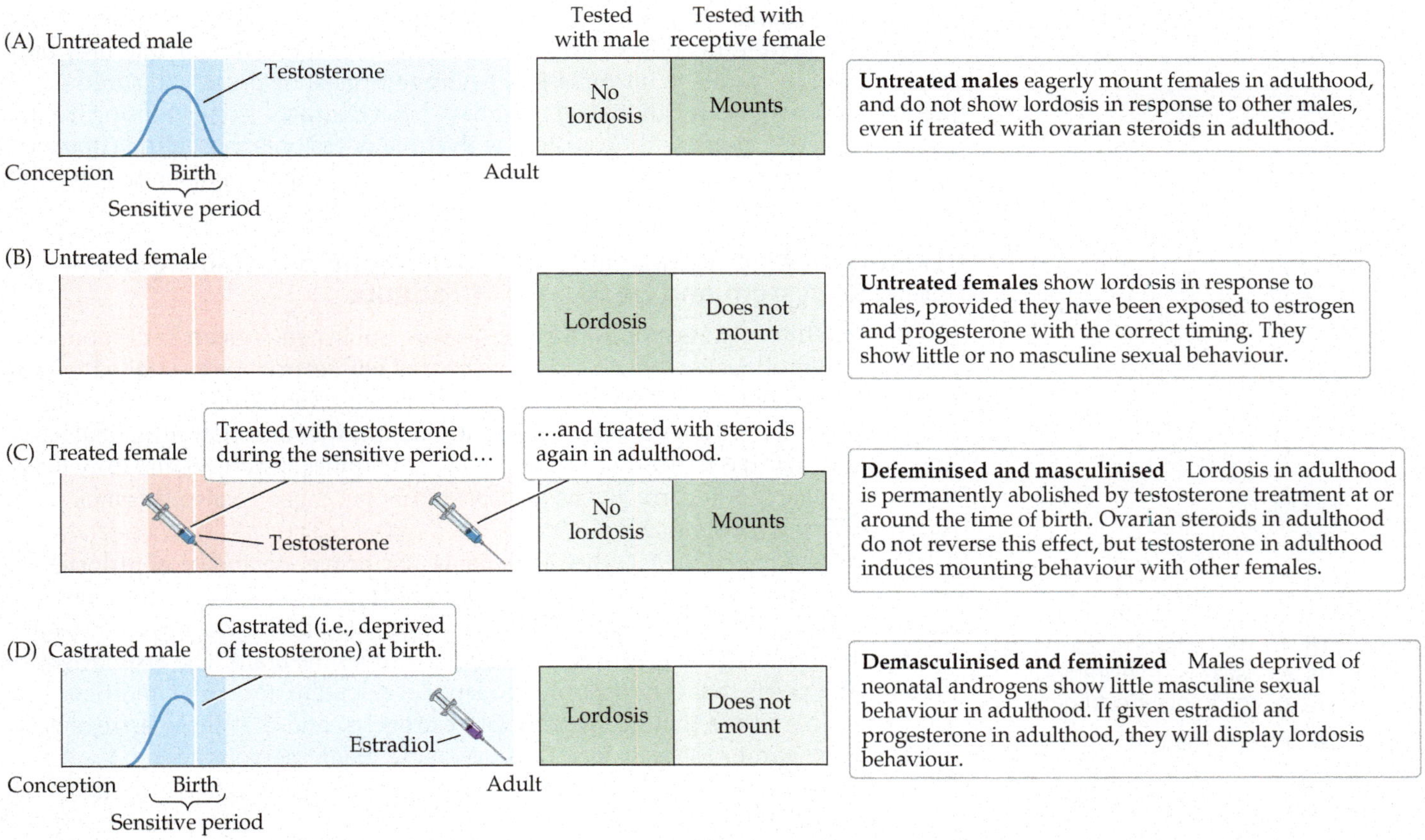

13.20 Organisational Effects of Sex Hormones on Rodent Behaviour Untreated males (A) produce a testosterone surge around birth and go on to develop normal adult sexual behaviour. Untreated females (B) that are not exposed to gonadal hormones during the perinatal sensitive period develop normal adult female sexual behaviour. However, (C) providing steroids to females during the sensitive period or (D) preventing neonatal steroid exposure in males can cause abnormal organisation of sexual control mechanisms and the expression of cross-sex sexual behaviour in adulthood.

studied. For rats, androgens given just after birth (the **neonatal** period) can permanently affect later behaviour. Guinea pigs, however, must be exposed to androgens *before* birth for adult lordosis behaviour to be affected. In mammals, puberty can be viewed as a second sensitive period; for example, steroid exposure during puberty causes the addition of new cells (an organisational effect) to sex-related brain regions of rats (Ahmed et al., 2008).

Early testicular secretions result in masculine behaviour in adulthood

The *organisational hypothesis* provides a unitary explanation for sexual differentiation: that a single steroid signal (androgen) diffuses through all tissues, masculinising the body, the brain, and behaviour. From this point of view the nervous system is simply another type of tissue listening for the androgenic signal that will instruct it to organise itself in a masculine fashion. If the nervous system does not detect androgens, it will organise itself in a mostly feminine fashion.

What was demonstrated originally for the lordosis behaviour of guinea pigs has been observed in a variety of vertebrate species and for many behaviours. Exposing female rat pups to testosterone either just before birth or during the first 10 days after birth greatly reduces their lordosis responsiveness as adults. This explains the observation that adult male rats show very little lordosis even when given estrogens and progesterone. However, male rats that are castrated on the day of birth readily display lordosis responses in adulthood when injected with estrogens and progesterone.

In most cases, full masculine behaviour requires androgens both in development (to organise the nervous system to enable the later behaviour) and in adulthood (to activate that behaviour). For example, the copulatory behaviour of male rats can be quantified in terms of how often they mount a receptive female and how often such mounting results in intromission. Androgens must be present in adulthood to activate this behaviour: adult males that have been castrated stop mounting in a few weeks; injecting them with testosterone eventually restores masculine copulatory behaviour. Only animals exposed to androgen both in development and in adulthood show fully masculine behaviour (see Figure 13.4).

The estrogenic metabolites of testosterone masculinise the nervous system and behaviour of rodents

Soon after the organisational hypothesis was published, researchers reported a paradoxical finding: newborn female rats treated with estrogens also failed to show lordosis behaviour in adulthood (Feder and Whalen, 1965). In fact, if such females were also treated with testosterone in adulthood, they would show *mounting* behaviour, indicating that the early estrogens had masculinised their brains. Why would estradiol, regarded (incorrectly) as a *female* hormone, cause a *masculinisation* of the brain's sexual circuitry in females?

The answer lies in the fact that estradiol is a principle metabolite of testosterone. In a single chemical reaction, called **aromatisation**, the enzyme **aromatase** converts testosterone to estradiol (and other androgens to other estrogens). The ovaries normally contain a great deal of aromatase, but so does the brain. The **aromatisation hypothesis** posits that during normal rodent development, testicular androgens enter the brain and are there converted into estrogens, and that these estrogens are what act on neurons to *masculinise* the developing rodent nervous system.

All rat fetuses are exposed to high levels of estrogens that originate in the mother and cross the placenta, so why aren't the brains of females masculinised by the maternal estrogens? A blood protein called **α-fetoprotein** binds estrogens and prevents them from entering the brain (Bakker et al., 2006), but it does not bind testosterone. So the male rat's brain is masculinised when testosterone from his testes bypasses the α-fetoprotein and enters his brain, where it is aromatised to estradiol within individual neurons. This newly synthesised estrogen binds to local estrogen receptors (not androgen receptors), and the steroid-receptor complex regulates gene expression to cause the brain to develop in a masculine

neonatal Referring to newborns.

aromatisation The chemical reaction that converts testosterone to estradiol, and other androgens to other estrogens.

aromatase An enzyme that converts some androgens into estrogens.

aromatisation hypothesis The hypothesis that testicular androgens enter the brain and are converted there into estrogens to masculinise the developing nervous system in some rodents.

α-fetoprotein A protein found in the plasma of fetuses. In rodents, α-fetoprotein binds estrogens and prevents them from entering the brain.

fashion (**FIGURE 13.21**). If no androgens are present, there can be no estrogenic action in the brain (because α-fetoprotein has blocked estrogens of peripheral origin), so the fetus develops in a feminine fashion. Injections of estrogens in rat pups affected lordosis in the original research because the injection flooded the bloodstream with so many estradiol molecules that they overwhelmed the circulating α-fetoprotein, allowing estrogen to enter the brain. A lack of aromatase seems to play a role in the unusual sexual differentiation of the spotted hyena (**BOX 13.1**, next page).

The aromatisation hypothesis is fully applicable to masculine copulatory behaviour in rats. If a male rat is castrated at birth, it grows up to have a small penis and show few intromissions, even when given replacement testosterone in adulthood. Castration plus the androgen DHT, which cannot be aromatised into an estrogen, results in a male with normal genitalia but little or no masculine sexual behaviour, because of the lack of organisational estrogenic stimulation of the brain. Conversely, males castrated and treated with estrogens as newborns mount and show intromission behaviour when treated with testosterone in adulthood, despite having very small penises (due to the lack of early DHT). The important point in all this is that it is hormonal masculinisation of the *brain*, not the genitalia, that organises male rat copulatory behaviour.

In primates, including humans, aromatisation cannot play an important role in masculinisation of the nervous system. The human brain produces significant quantities of aromatase, but men who have mutations in the aromatase gene—and are thus unable to produce aromatase and therefore estrogens—nonetheless have masculine gender identities and sexual development (Grumbach and Auchus, 1999). Likewise, men with defective estrogen receptors nevertheless develop masculine gender attributes, and people with AIS display feminine behaviour and gender identity, even though they produce lots of testosterone and have functional estrogen receptors. The details of the masculinisation of the primate nervous system remain to be worked out, but any hormonal effects must be accomplished through androgen receptors rather than estrogen receptors. And whichever specific steroid receptor is involved, many vertebrate species display distinct sex differences in the brain, as we'll see next.

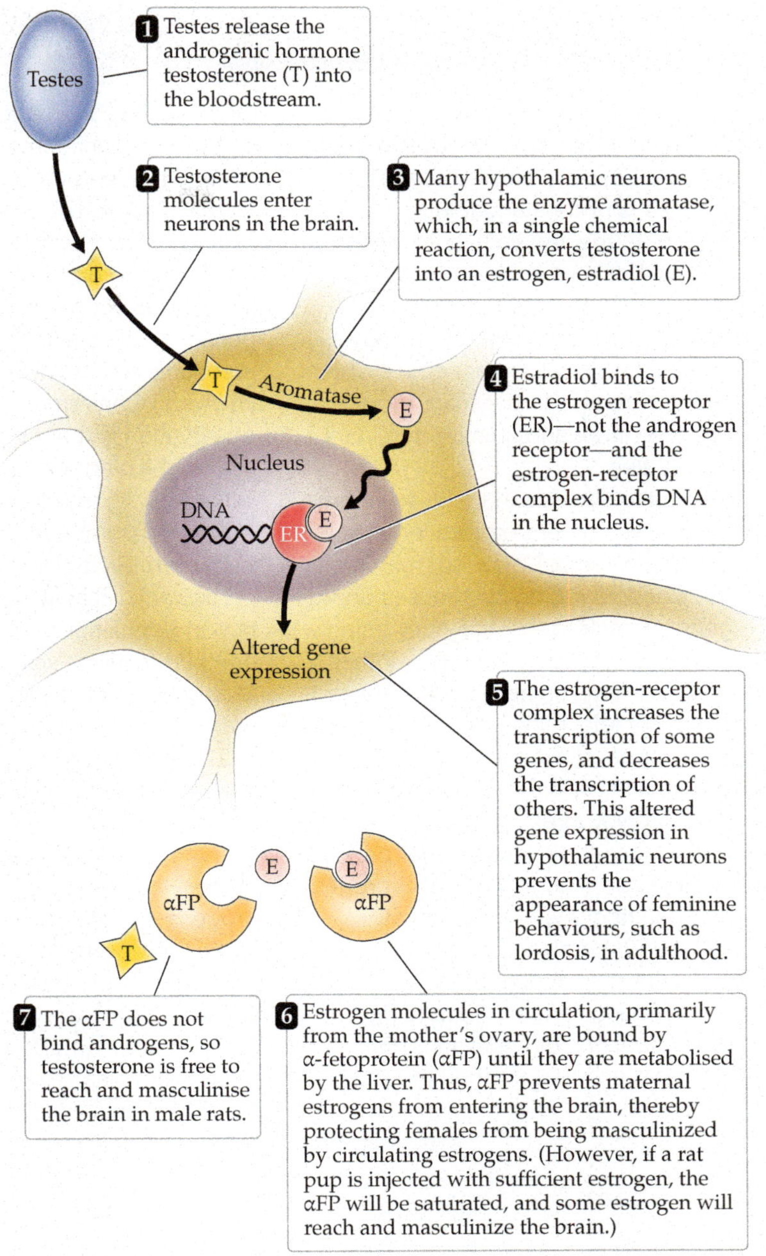

13.21 The Aromatisation Hypothesis

Several regions of the nervous system differ between males and females

The fact that male and female animals behave differently means that their brains must be different in some way, and according to the organisational hypothesis, this difference results primarily from androgenic masculinisation of the developing brain. The exact form of these neural sex differences can be very subtle; the same basic circuit of neurons will produce very different behaviour if the pattern of synapses varies. Sex differences in the number of synapses were identified in the preoptic area (POA) of the hypothalamus as early as 1971 (Raisman and Field, 1971). But scientists soon found that there are much more obvious sex differences in the brain, including differences in neuronal numbers, sizes, and shapes. Darwin coined a term, **sexual dimorphism**, to describe the condition in which males and females show pronounced sex differences in appearance. Let's discuss a few well-known models of sexual dimorphism in the brain.

Go to Animation 13.1
The Aromatisation Hypothesis

sexual dimorphism The condition in which males and females show pronounced sex differences in appearance.

BOX 13.1

The Paradoxical Sexual Differentiation of the Spotted Hyena

Scholars of antiquity believed that spotted hyenas were **hermaphrodites** (both male and female). The mistake is understandable because from birth the female hyena has a clitoris that is as large as the penis of the male (as the photograph shows), through which she urinates, receives semen, and gives birth. Female hyenas are also larger than, more aggressive than, and socially dominant over males.

What difference in the hyena's prenatal development causes females to develop a "penis"? In other mammals the placenta rapidly aromatises androgens into estrogens; this conversion may be a way to protect the mother and fetal females from androgens produced by fetal males. But the placenta of the spotted hyena is remarkably deficient in the aromatase enzyme (A. J. Conley et al., 2007).

Because the hyena mother produces large amounts of the androgen androstenedione (Hammond et al., 2012) and the placenta fails to convert the androstenedione to estrogens, all the fetuses are exposed to high concentrations of androgens, which may help account for their masculine appearance (Cunha et al., 2014). (However, female hyenas treated prenatally with androgen-blocking drugs still develop masculinised exteriours [Place et al., 2011], suggesting that an additional mechanism must be involved.) One remarkable observation is that female pups, which are born with teeth, fight viciously with siblings immediately from birth. Sisters fight particularly violently, and in wild populations it is common for one pup to kill its sibling, at least when food is scarce (Smale et al., 1999). It remains to be established whether this extreme aggression is due to prenatal stimulation of the brain with androgens.

Even if the extreme aggressiveness of the female hyena is due to fetal androgens, females do mate with males, so their brains have not been made permanently unreceptive (as would happen in prenatally androgenised rats). Indeed, female hyenas seem to have a typically feminine sexually dimorphic nucleus of the preoptic area (SDN-POA) (Fenstemaker et al., 1999) and spinal nucleus of the bulbocavernosus (SNB) (Forger et al., 1996). Just the same, mating in the spotted hyena is a tense affair: the female seems to just barely tolerate the male's proximity, and the male alternates between approaching and retreating from his alluring but dangerous mate.

hermaphrodite An individual possessing the reproductive organs of both sexes, either simultaneously or at different points in time.

SONG CONTROL REGIONS IN MALE SONGBIRDS Male canaries and zebra finches produce elaborate songs to attract mates, but females do not, and the brain nuclei controlling song are 5 to 6 times larger in males than in females (Nottebohm and Arnold, 1976). One of these nuclei, called the *robustus archistriatum* (RA), controls motor neurons innervating the muscular singing organ, called the **syrinx**. The RA is controlled, in turn, by the *high vocal centre* (HVC), so named because it exerts top-level control over the song production hierarchy. As we would expect, lesions of RA or HVC disrupt singing, and electrical stimulation can elicit song snippets. (Details of the song system can be seen in Figure 18.3.)

Exposing a hatchling female zebra finch to either testosterone or estradiol causes the development of a larger, male-like HVC and RA, in keeping with the organisational hypothesis, and if she receives additional testosterone in adulthood, she will sing like a male (Gurney and Konishi, 1979). Female zebra finches treated with androgens *only* in adulthood do not sing, so we know that early hormone organises a masculine song system and adult hormone activates it. Singing in canaries, in contrast, requires only *adult* androgen treatment; previously untreated female canaries start singing after a few weeks of androgen treatment in adulthood. The androgens cause both HVC and RA to become larger, and their neurons grow and form new synaptic connections.

syrinx The vocal organ in birds.

13.22 A Sex Difference in the Hypothalamus The sexually dimorphic nucleus of the preoptic area (SDN-POA) is much larger in male rats than in females.

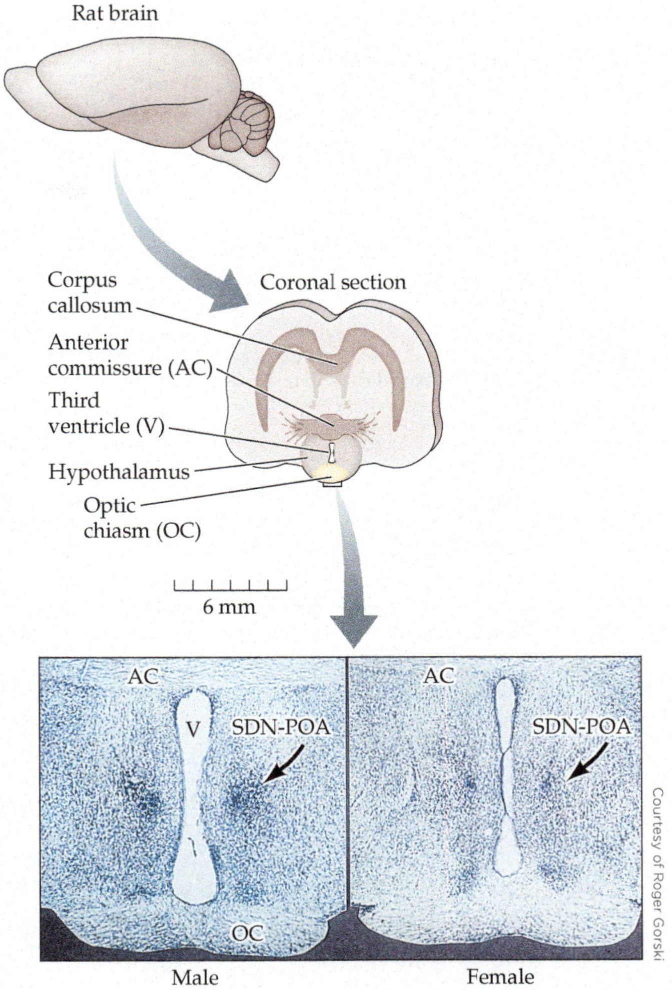

The difference in the hormonal controls of song in zebra finches and canaries may be related to the ecological niche occupied by each. Zebra finches are ready to breed at any time of the year, but canaries are seasonal breeders whose reproductive tracts shut down in the fall. The male canary song system tracks this ebb and flow: HVC and RA grow large in the spring, when the males are singing and their testosterone levels are high, and shrink again in the fall, as testosterone levels and singing decline.

THE PREOPTIC AREA OF RATS Where else might there be neural sex differences? Roger Gorski et al. (1978) examined the POA of the hypothalamus in rats because of the earlier reports that the numbers of synapses in this region were different in males and females and because lesions of the POA disrupt ovulatory cycles in female rats and reduce copulatory behaviour in males. Sure enough, the investigators found a nucleus within the POA that has a much larger volume in males than in females.

This nucleus, dubbed the **sexually dimorphic nucleus of the POA** (**SDN-POA**), is much more evident in male rats than in females (**FIGURE 13.22**). Like the song control nuclei in zebra finches, the SDN-POA conformed beautifully to the organisational hypothesis: males castrated at birth had much smaller SDN-POAs in adulthood, while females androgenised at birth had large, male-like SDN-POAs as adults. Castrating male rats in adulthood, however, did not alter the size of the SDN-POA. Thus, testicular androgens somehow alter the development of the SDN-POA, resulting in a nucleus permanently larger in males than in females (**FIGURE 13.23**). There is evidence that androgens exert epigenetic effects of the sort described in Chapter 6, controlling the methylation of genes during development and thus affecting their expression in adulthood (M. M. McCarthy and Nugent, 2015).

The rat SDN-POA system also conforms to the aromatisation hypothesis: testosterone is converted to an estrogen in the brain and binds estrogen receptors to masculinise the nucleus. For example, XY rats that are androgen-insensitive (like the people with AIS discussed earlier) have testes but a feminine exteriour. These rats have a masculine SDN-POA because their estrogen receptors are normal. Androgen-insensitive rats also do not display lordosis in response to estrogens and progesterone, because the testosterone that they secreted early in life was converted to an *estrogen* in the brain and masculinised their behaviour (K. L. Olsen, 1979). Instead, the androgen-insensitive rats show normal male attraction to receptive females, with whom they may attempt to mate despite the lack of a penis (Hamson et al., 2009).

THE SPINAL CORD IN MAMMALS In rats, the bulbocavernosus (BC) muscles that surround the base of the penis are innervated by motor neurons in the **spinal nucleus of the bulbocavernosus** (**SNB**). Male rats have about 200 SNB cells, but females have far fewer motor neurons in this region of the spinal cord.

On the day before birth, female rats have BC muscles attached to the base of the clitoris that are nearly as large as those of males and that are innervated by motor neurons in the SNB region (Rand and Breedlove, 1987). In the days just before and after birth, however, many SNB cells die, especially in females (Nordeen et al., 1985), and the BC muscles of females die.

sexually dimorphic nucleus of the preoptic area (SDN-POA) A region of the preoptic area that is 5 to 6 times larger in volume in male rats than in females.

spinal nucleus of the bulbocavernosus (SNB) A group of motor neurons in the spinal cord of rats that innervate striated muscles controlling the penis.

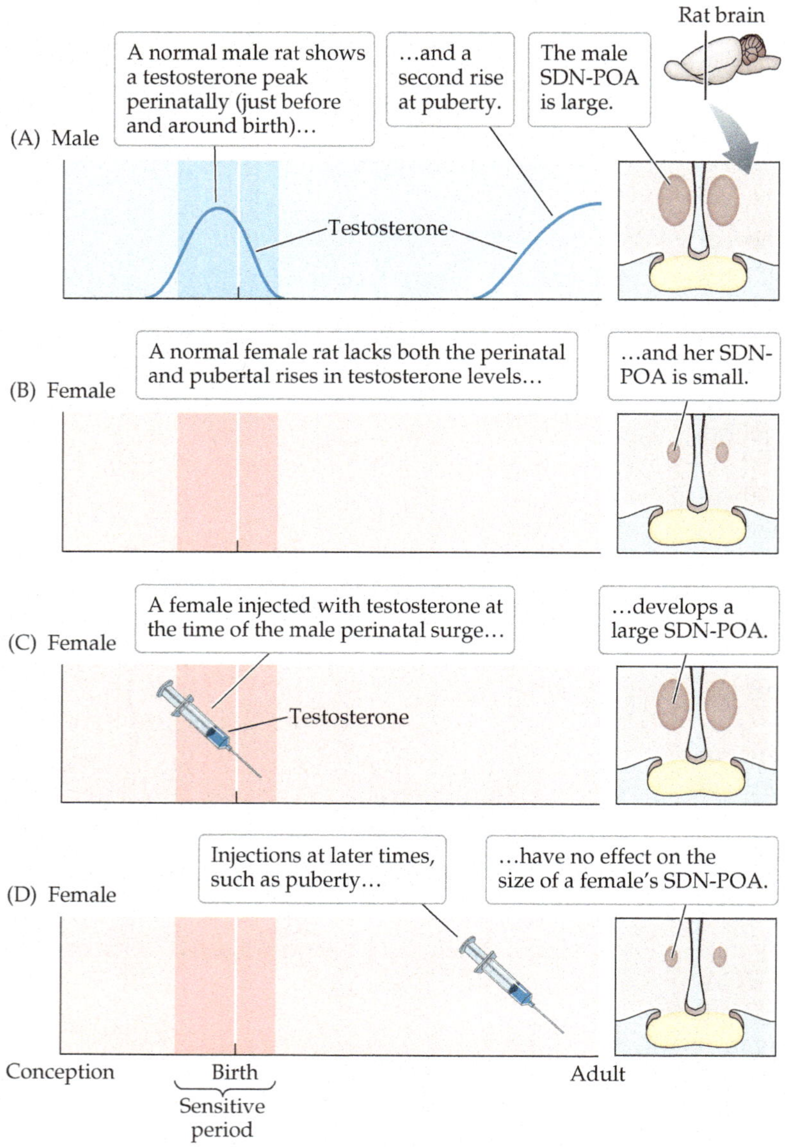

13.23 Sexually Dimorphic Nucleus of the Preoptic Area (SDN-POA) Testosterone can permanently enlarge the SDN-POA in rats, but only if given during a "sensitive period" early in life.

A single injection of androgens given to a newborn female rat permanently spares some SNB motor neurons and their muscles: a powerful organisational effect. Castration of newborn males, accompanied by prenatal blockade of androgen receptors, causes the BC muscles and SNB motor neurons to die as in females. The system also dies in newborn androgen-insensitive rats, so aromatisation seems to be unimportant for masculine development of the SNB.

Androgens act on the BC muscles to prevent their demise, and this sparing of the muscles likewise causes the innervating SNB motor neurons to survive (Fishman et al., 1990; C. L. Jordan et al., 1991). The exact site within the BC muscle where androgen has its effects is still unknown (Niel et al., 2009). But whatever the site of action is, the likely result is that the muscles are induced by androgen to provide a signal, such as a neurotrophic factor, that prevents SNB motor neurons from dying (**FIGURE 13.24**) (Forger et al., 1997).

The developmental rescue of SNB motor neurons is accomplished indirectly as a consequence of actions on muscle, but androgens can also directly affect the neurons themselves. SNB neurons contain androgen receptors and retain androgen sensitivity throughout life. In adulthood, androgen acts directly on the neurons to cause them to grow (Watson et al., 2001) and start producing substances to aid in the formation of new connections. For example, androgens directly stimulate SNB neurons to produce N-cadherin, a cell adhesion molecule that mediates the formation of new contacts between cells (D. A. Monks and Watson, 2001). These are examples of activational effects of androgens.

All male mammals have BC muscles, but in nonrodents the BC motor neurons are found in a slightly different spinal location and are known as **Onuf's nucleus**. You might be surprised to learn that most female mammals retain a BC muscle into adulthood; in women, for example, the BC (also known as *constrictor vestibule*) helps constrict the vaginal opening. But, as in rodents, the system is profoundly sexually dimorphic. Men have larger BC muscles and more numerous Onuf's motor neurons, most likely determined by testosterone exposure during fetal development (Forger and Breedlove, 1986, 1987).

These model systems, and others like them, demonstrate the power of androgens to mold the brain in sex-typical ways. By controlling the amount and timing of testosterone exposure, researchers can make the various brain structures as masculine or feminine as they like. However, despite the crucial role of testosterone in masculinising sexually dimorphic nuclei in the nervous system, there's evidence that experience affects these systems too.

Environmental influences also affect sexual differentiation of the nervous system

Environmental factors of many sorts, ranging from the chemical composition of the environment to the behaviours of other individuals, can potently modulate the development of sexual dimorphism. Neonatal exposure to the ubiquitous plastics component bisphenol

Onuf's nucleus The human homolog of the spinal nucleus of the bulbocavernosus (SNB) in rats.

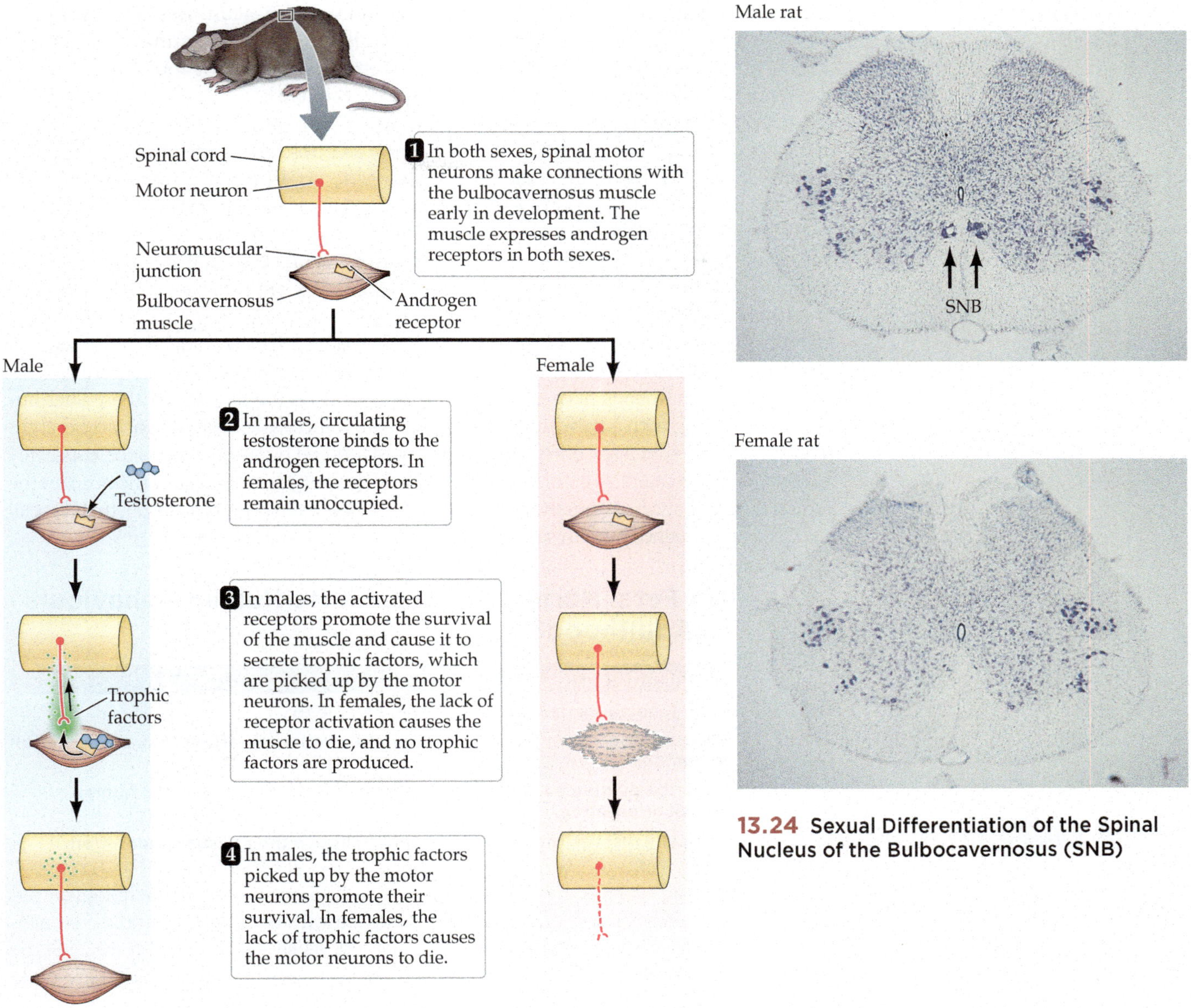

13.24 Sexual Differentiation of the Spinal Nucleus of the Bulbocavernosus (SNB)

A (BPA), for example, interferes with the development of adult sexual behaviours and demasculinises male rats in tests of emotional behaviour (B. A. Jones and Watson, 2012; B. A. Jones et al., 2011). BPA, which acts as a steroid mimic, is found in manufactured products ranging from drinking bottles and baby toys to food can liners.

Purely social inputs can likewise alter the sexual differentiation of the nervous system. The development of the SNB system offers a clear example: newborn rat pups can neither urinate nor defecate on their own; the mother (dam) must lick the anogenital region of each pup to elicit a spinal reflex to empty the bladder and colon. (Incidentally, the dam ingests at least some of the wastes to monitor pheromones from the pups, adjusting the composition of her milk as the pups mature. Another reason not to be a rat!)

Celia Moore et al. (1992) noticed that dams spend more time licking the anogenital regions of male pups than of females. If the dam is anosmic (unable to smell), she licks all the pups less and does not distinguish between males and females. Males raised by anosmic mothers thus receive less anogenital licking, and remarkably, fewer of their SNB cells survive the period around birth. The dam's stimulation of a male's anogenital region helps to masculinise his spinal cord.

On the one hand, this masculinisation is still an effect of androgens because the dam identifies male pups by detecting androgen metabolites in their urine. On the other hand, this effect is clearly the result of a social influence: the dam treats a pup differently because he's a male, and thereby she masculinises his developing nervous system. This example illustrates the futility of trying to distinguish "biological" from "social" influences. Any social influence that alters the individual's future behaviour *must* have exerted a physical effect on the brain.

Attention from the dam has a different organising effect on female rat pups. In adulthood, females who were licked frequently as pups tend to be attentive mothers themselves and to be more resilient in response to stress (see Figure 6.26). Females who are licked less as pups are less attentive mothers later (Champagne et al., 2001).

What about humans? (No, no, not the licking part, the social influence part.) Humans are at least as sensitive to social influences as rats are. In every culture, most people treat boys and girls differently, even when they are infants. Such differential treatment undoubtedly has some effect on the developing human brain and contributes to later sex differences in behaviour. Of course, this is a social influence, but testosterone instigated the influence when it induced the formation of a penis.

If prenatal androgens have even a very subtle effect on the fetal brain, then older humans interacting with a baby might detect such differences and treat the baby differently. Thus, originally subtle sex differences might be magnified by social experience, especially early in life. Such interactions of steroidal and social influences are probably the norm in the sexual differentiation of human behaviour. Whether prenatal hormones or postnatal social experiences influence human sexual orientation is our final topic.

13.7 Do Fetal Hormones Masculinise Human Behaviours in Adulthood?

Learning Objectives

After reading this section, you should be able to:

13.7.1 **Offer compelling evidence that prenatal events can affect sexual orientation in humans.**

13.7.2 **Describe evidence that genes influence, but do not determine, human sexual orientation.**

13.7.3 **Critically discuss the idea that humans choose their sexual orientation.**

Overall, men and women are more similar than different on most psychological parameters (Petersen and Hyde, 2011). Nevertheless, a variety of behavioural differences between men and women have been documented, as shown in **TABLE 13.2**. Some of these sex differences, especially those pertaining to sexual development, are quite substantial. Others, such as sex differences in cognition, tend to be smaller, but still reliable. Yet others, such as the sex difference in math abilities, seem to have disappeared over the past 30 years. And as with rats and other animals, the fact that men and women behave differently implies that something about them, probably something about their brains, must also be different. Indeed, many parts of the brain are different between men and women (**FIGURE 13.25**). But are these sexual dimorphisms in the human brain caused by prenatal hormone exposure, as in other animals, or by social influences? In other words, does prenatal steroid exposure affect the adult behaviour of humans? This is a tricky problem because although prenatal androgens may or may not act on the human brain, they certainly act on the periphery.

For example, recall that people with AIS are usually raised as girls, with their XY genotype and infertility typically discovered at puberty. We said earlier that individuals with AIS tend to have feminine gender identity, which includes being sexually attracted to men and, often, seeking a family through adoption. Are they feminine because they received the social tutoring to be females, or because their brains, without androgen receptors, could not respond to testosterone? Their behaviour is consistent with either hypothesis.

What about females with CAH, who are exposed to androgens before birth? We saw earlier that CAH females play more like boys than other females do and are more likely than other females to be lesbians in adulthood. Do women with CAH exhibit those behaviours because early androgens partially masculinised their brains? Or did their ambiguous genitalia cause parents and others to treat them differently from infancy?

TABLE 13.2 Sex Differences in Human Cognition

Behavioural characteristic	Direction	Magnitude (in standard deviations)
COGNITIVE AND MOTOR ABILITIES		
Targeting	M > F	1.1–2.0
Fine motor skill	F > M	0.5–0.6
Mental rotations	M > F	0.3–0.9
Spatial perception	M > F	0.3–0.6
Spatial visualisation	M > F	0.0–0.6
Verbal fluency	F > M	0.5
Perceptual speed	F > M	0.3–0.7
Computational skill	F = M	0
Math concepts	F = M	0
Vocabulary	F = M	0
PERSONALITY ATTRIBUTES		
Tendencies to physical aggression	M > F	0.4–1.3
Empathy	F > M	0.3–1.3
Dominance/assertiveness	M > F	0.2–0.8
OTHER TRAITS		
Core gender identity	Not applicable	11.0–13.2
Sexual orientation	Not applicable	6.0–7.0
Adult height	M > F	2.0

Source: M. Hines, 2010. *Trends Cog Sci* 14: 448–456.

We've also seen that the *guevedoces* of the Dominican Republic, who are raised as girls but grow a penis at puberty, behave like males as adults, dressing as men and seeking girlfriends. There are two competing explanations for why these people raised as girls later behave as men. First, prenatal testosterone may masculinise their brains; thus, despite being raised as girls, when they reach puberty, their brains lead them to seek out females for mates. This explanation suggests that the social influences of growing up—assigning oneself to a gender and mimicking role models of that gender, as well as gender-specific playing and dressing—are unimportant for later behaviour and sexual orientation. An alternative explanation is that early hormones have no effect—that this culture simply recognises and teaches children that some people can start out as girls and change to boys later. If so, then the social influences on gender role development might be completely different in this society from those in ours. Of course, a third option is that both mechanisms contribute to the final outcome; for example, early androgens may affect the brain to masculinise the child's behaviour and predispose later sexual orientation towards females, and these masculine qualities of the child could alter the behaviour of parents and others in ways that promote the emergence of male gender identity as the child develops.

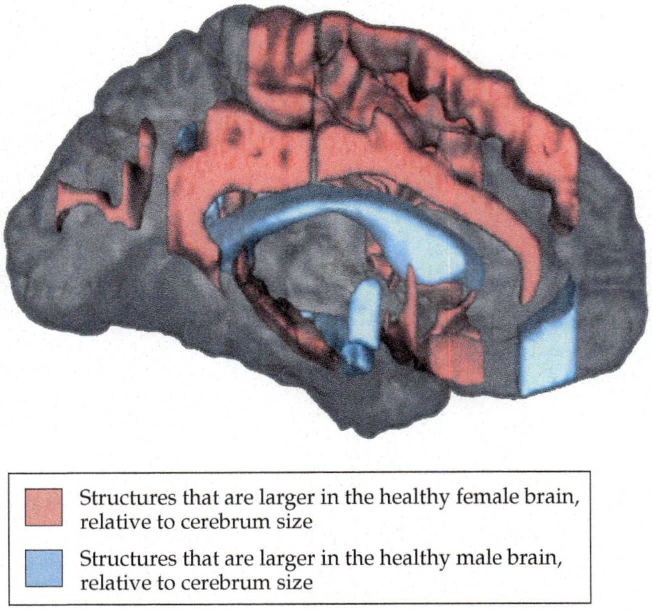

13.25 View of Sexual Dimorphism in the Human Brain This composite image is based on data obtained from MRI scans of numerous male and female volunteers. Overall, sex differences tend to be evident in regions that are known to possess receptors for sex steroids. (After J. M. Goldstein et al., 2001. *Cerebral Cortex* 11: 490–497, courtesy of Jill Goldstein.)

cloacal exstrophy A rare medical condition in which XY individuals are born completely lacking a penis.

At the opening of the chapter, we discussed the dilemma of **cloacal exstrophy**, in which genetic boys are born with functional testes but without penises. Historically in these cases, neonatal sex reassignment has been recommended on the assumption that unambiguously raising these children as girls, and surgically providing them with the appropriate external genitalia, could produce a more satisfactory psychosexual outcome. In a long-term follow-up of 14 such cases, however, Reiner and Gearhart (2004) found that 8 of these "girls" eventually declared themselves to be boys, even though several were unaware that they had ever been operated on. Although this finding indicates that prenatal exposure to androgens strongly predisposes to subsequent male gender identity, 5 of the 14 individuals were apparently content with their female identities, suggesting that socialisation can also play a strong role. On the other hand, at least some of those 5 girls, and all the self-declared boys, were romantically attracted to girls. In fact, one of the reasons "Bella," whom we met at the beginning of this chapter, was unsurprised to learn she'd been born a boy was that she was sexually attracted to girls. As he said after becoming "Benjamin," "I was not sad about me being a boy, it was just telling my friends that got me down." So the prenatal androgen may have predisposed all of these individuals to be sexually attracted to females, and apparently some (but not all) were sufficiently socialised as girls to be comfortable with that gender identity.

Seen alone, the studies of people with these various conditions might leave room for doubt about whether prenatal hormones influence whether we will be attracted to men or women. But these are just part of a growing body of evidence that prenatal testosterone does in fact influence sexual orientation in humans, as we'll see next.

What determines a person's sexual orientation?

There are two classes of possible influence on human sexual orientation. One class encompasses the sociocultural influences that may instruct developing children about how they should behave when they grow up (think of all those charming princes wooing princesses in Disney movies). The other class of influences includes the endogenous factors—especially differences in fetal exposure to testosterone—that could organise developing brains to be attracted to females or males in adulthood. For that great majority of people who are heterosexual, there's no way to distinguish between these two influences, because they both favour the same outcome. Gay men and lesbians provide a test, because same-sex attractions (and other sexual minorities) remain stigmatised by various social groups and cultural institutions (Herek and McLemore, 2013). Is there evidence that early hormones are responsible for causing some people to ignore society's proscription and become gay? If so, then maybe hormones play a role in heterosexual development too.

Certainly, homosexual behaviour is seen in other species—mountain sheep, swans, gulls, and dolphins, to name a few (Bagemihl, 1999). Interestingly, homosexual behaviour is more common among anthropoid primates—apes and monkeys—than in prosimian primates like lemurs and lorises (Pfau et al., 2019; Vasey, 1995), so greater complexity of the brain may make homosexual behaviour more likely. In the most studied animal model—sheep—some rams consistently refuse to mount females but prefer to mount other rams. There is growing evidence of differences in the POA of "gay" versus "straight" rams (Roselli et al., 2004), apparently organised by testosterone acting on the brain via neuronal steroid receptors during fetal development (Roselli and Stormshak, 2009).

Simon LeVay (1991) performed postmortem examinations of the POA in humans and found a nucleus (the third interstitial nucleus of the anteriour hypothalamus, or INAH-3) (**FIGURE 13.26A**) that is larger in men than in women, and larger in heterosexual men than in gay men (**FIGURE 13.26B**). All but one of the gay men in the study had died of AIDS, but the brain differences could not be due to AIDS pathology, because the straight men with AIDS still had a significantly larger INAH-3 than did the gay men. To the press and the public, this finding sounded like strong evidence that sexual orientation is "built in." It's still possible, however, that early social experience affects the development of INAH-3 to

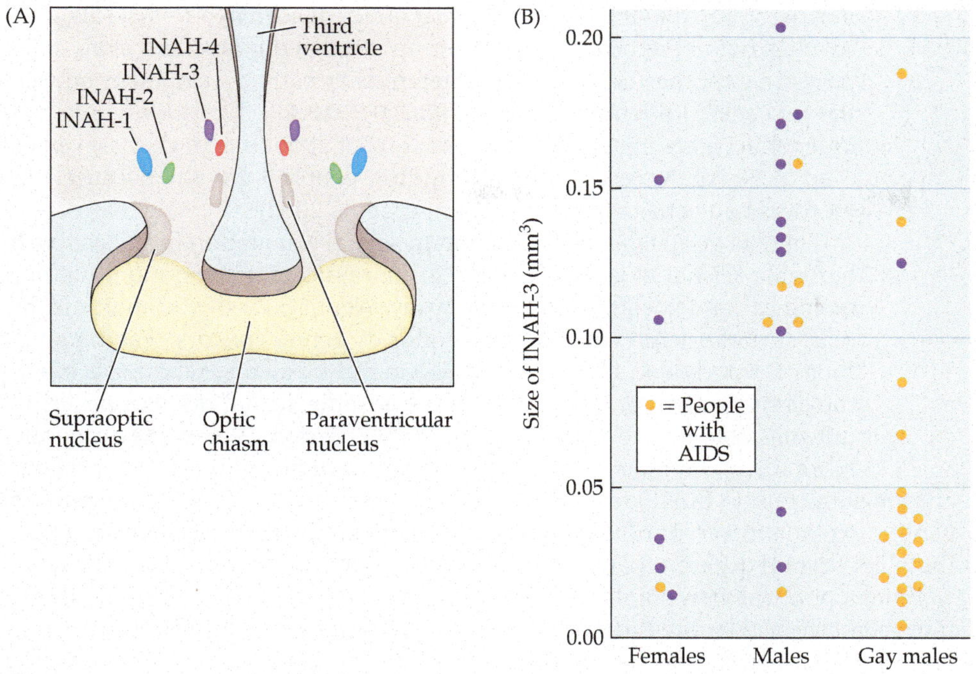

13.26 Interstitial Nuclei of the Anteriour Hypothalamus (A) These nuclei in humans are seen in the same part of the hypothalamus where the SDN-POA is found in rats. (B) INAH-3 is larger in men than in women, and larger in straight men than in gay men. Although most of the gay men in this study had died of AIDS, note that heterosexual men who died of AIDS still had a larger INAH-3, indicating that the differences between straight and gay men are not due to AIDS.

determine later sexual orientation. Furthermore, sexual experiences as an adult could affect INAH-3 structure, so the smaller nucleus in some gay men may be the *result* of their becoming gay, rather than the *cause*, as LeVay himself was careful to point out.

In women, apparent markers of fetal androgen exposure—otoacoustic emissions from the ears (McFadden and Pasanen, 1998) (see Chapter 9), patterns of eye blinks (Rahman, 2005), and skeletal features (J. T. Martin and Nguyen, 2004)—all indicate that lesbians, as a group, were exposed to slightly more fetal androgen than were heterosexual women. Another adult marker of prenatal androgen, the ratio of the length of the second digit (2D) to the length of the fourth digit (4D) (Breedlove, 2010), also indicates lesbians were, on average, exposed to greater prenatal androgen than straight women (**FIGURE 13.27**) (T. J. Williams et al., 2000) and has been replicated many times (Grimbos et al., 2010). These findings indicate that fetal exposure to androgen increases the likelihood that a girl will grow up to be gay. There is always considerable overlap between the two groups, so you cannot use these features to predict whether a *particular* woman will be gay, and clearly fetal androgens cannot account for all lesbians.

Those same markers of fetal androgen do not indicate any difference between gay versus straight men (Grimbos et al., 2010). Thus any difference between gay and straight men would be in their *response* to prenatal androgen rather than *amount* of prenatal androgen. However, another nonsocial factor does influence the probability of men being gay: the more older brothers a boy has, the more likely he will grow up to be gay (Bogaert et al., 2018). Your first guess might be that this is a social influence of older brothers, but it turns out that older stepbrothers who are raised with the boy have no effect, while biological brothers (sharing the same mother) increase the probability of the boy's being gay *even if they are raised apart* (Bogaert, 2006).

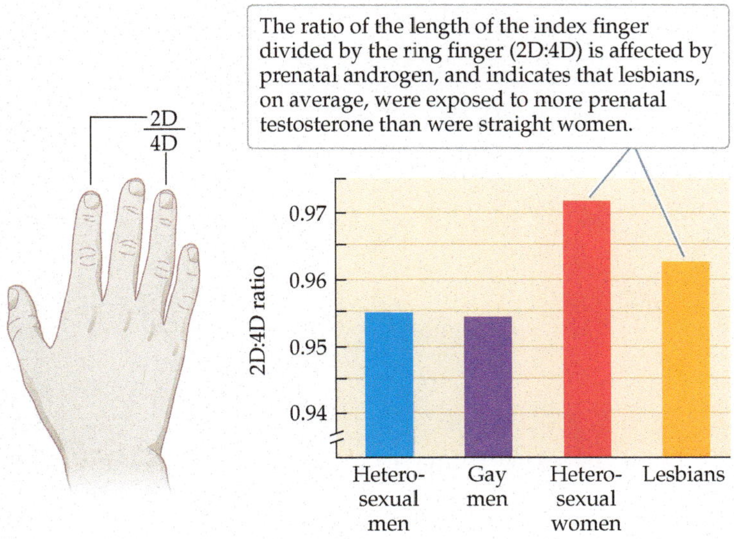

13.27 Bodily Indicators of Prenatal Androgen Note that these are group differences seen in averages; you cannot reliably determine an *individual's* orientation by examining digit ratios. (After T. J. Williams et al., 2000. *Nature* 404: 455–456.)

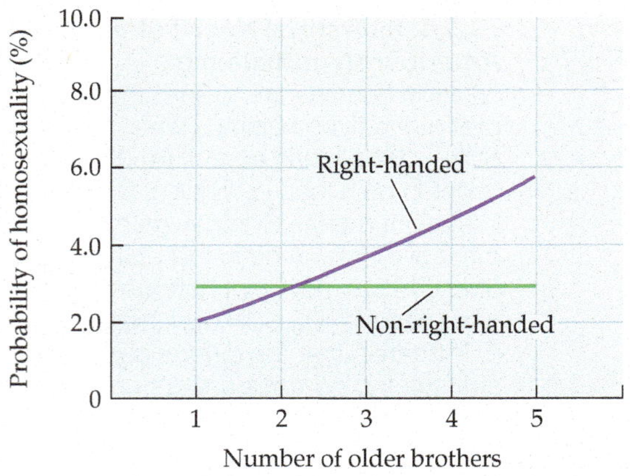

13.28 Fraternal Birth Order and Orientation Having older biological brothers (but not sisters or foster brothers) increases the likelihood of same-sex orientation only in right-handed males. (After R. Blanchard et al., 2006. *Horm Behav* 49: 405–414.)

Furthermore, this "fraternal birth order effect" is seen in boys who are right-handed, but not in boys who are non-right-handed (Blanchard et al., 2006), providing another indication of differences in early development between the two sexual orientation groups (**FIGURE 13.28**). Statistically, the birth order effect is strong enough to estimate that about one in every seven gay men in North America—about a million people—are gay because their mothers had sons before them (Cantor et al., 2002).

There is good evidence that human sexual orientation is at least partly heritable (Fisher et al., 2018; Saey, 2018), reinforcing the notion that both biological and social factors have a say. About 50% of variability in human sexual orientation is accounted for by genetic factors, leaving ample room for early social influences. Monozygotic twins, who have exactly the same genes, do not always have the same sexual orientation. In the unusual case of two nontwin brothers who are both gay, they are much more likely than chance would dictate to have both inherited the same X chromosome region (Xq28) from their mother (Sanders et al., 2015); but again, the genetic explanation accounts for only some, not all, of the cases. It seems clear that there are several different pathways to homosexuality.

From a political viewpoint, the scientific controversy—whether sexual orientation is determined before birth or determined by early social influences—is irrelevant. Laws and prejudices against homosexuality are based primarily on religious views that homosexuality is a sin that some people "choose." But almost all gay and straight men report that, from the beginning, their interests and romantic attachments matched their adult orientation. So any social influence would have to be acting very early in life and without any conscious awareness (do you remember "choosing" whom to find attractive?). Furthermore, despite extensive efforts, no one has come up with a reliable way to change sexual orientation (LeVay, 2016). These findings, added to evidence that older brothers and prenatal androgens affect the probability of being gay, have convinced most scientists that we do not choose our sexual orientation.

Recommended Reading

Balthazart, J. (2011). *The Biology of Homosexuality.* Oxford, UK: Oxford University Press.

Buss, D. M. (2019). *Evolutionary Psychology: The New Science of the Mind* (6th ed.). New York: Routledge.

LeVay, S. (2016). *Gay, Straight and the Reason Why: The Science of Sexual Orientation* (2nd ed.). Oxford, UK: Oxford University Press.

LeVay, S., Baldwin, J., and Baldwin, J. (2018). *Discovering Human Sexuality* (4th ed.). Sunderland, MA: Oxford University Press/Sinauer.

McCarthy, M. M. (2017). *Sex and the Developing Brain* (2nd ed.). San Rafael, CA: Morgan and Claypool.

Melmed, S., Polonsky, K. S., Larsen, P. R., and Kronenberg, H. M. (2016). *Williams Textbook of Endocrinology* (13th ed.). Philadelphia: Elsevier/Saunders.

Nelson, R. J., and Kriegsfeld, L. J. (2017). *An Introduction to Behavioural Endocrinology* (5th ed.). Sunderland, MA: Oxford University Press/Sinauer.

Patisaul, H. B., and Belcher, S. M. (2017). *Endocrine Disruptors, Brain and Behaviour.* Oxford, UK: Oxford University Press.

13 Visual Summary

You should be able to relate each summary to the adjacent illustration, including structures and processes. You can also refer back to the links to figures, animations, and activities that will help you consolidate the material.

1 Reproductive behaviours are divided into four stages: **sexual attraction, appetitive behaviour, copulation,** and **postcopulatory behaviour**, including parental behaviours in some species. Review **Figure 13.1**

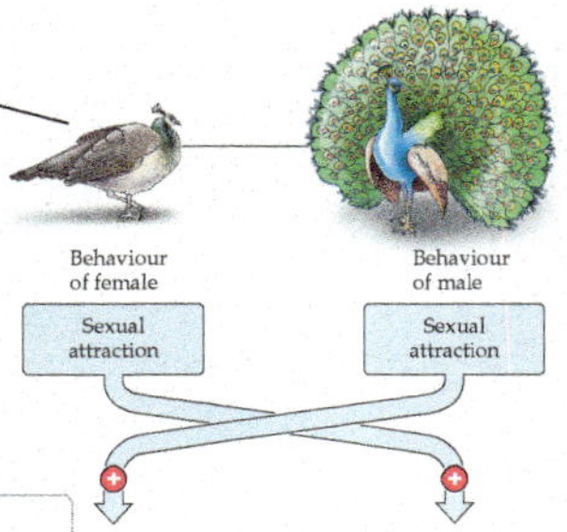

2 Alterations in circulating gonadal steroids alter sexual behaviour in male and female mammals. Review **Figures 13.4, 13.5**

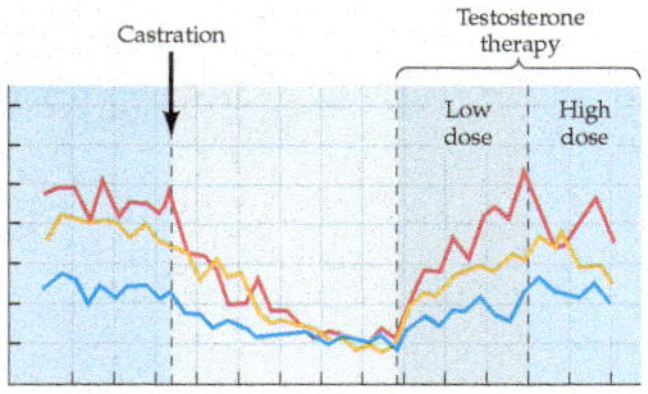

3 In the female rat, a steroid-sensitive **lordosis** circuit extends from the **ventromedial hypothalamus** (**VMH**) to the spinal cord, via the **periaqueductal gray** and **medullary reticular formation**. In the male rat, neurons of the **medial preoptic area** (**mPOA**) exert descending control of sexual behaviour, integrating inputs from the **medial amygdala** and **vomeronasal organ** (**VNO**). These projections, via ventral midbrain and brainstem nuclei, terminate on motor neurons involved in copulation. Review **Figure 13.6**

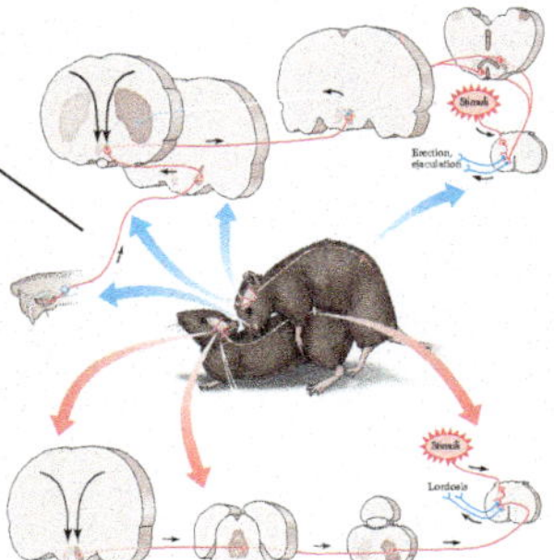

4 Human copulatory behaviour is remarkably varied. Most men show a single copulatory pattern; women show much more varied sexual responses. The classic model of sexuality emphasises four stages: (1) increasing excitement, (2) plateau, (3) **orgasm**, and (4) resolution. Review **Figure 13.10**, **Activities 13.1, 13.2**

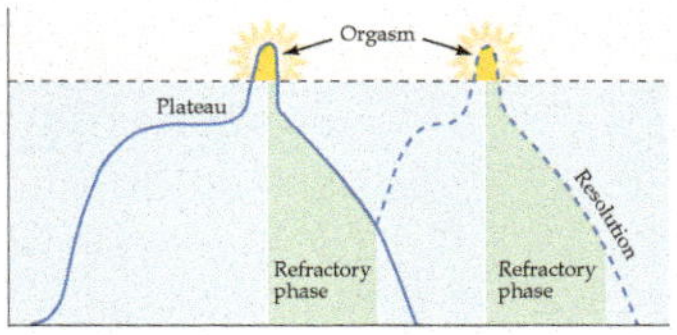

5 Modern models identify emotional factors and desire as crucial aspects of female sexuality, whereas male sexuality may involve feelings of power. However, male and female sexuality overlap, and both are heavily influenced by sociocultural factors. Review **Figure 13.12**

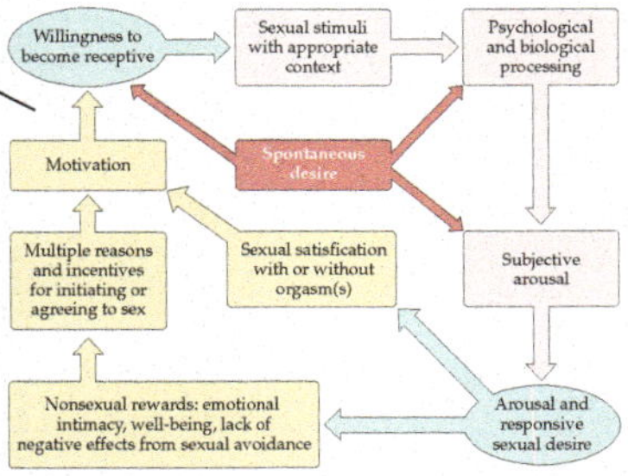

6 **Parental behaviour** is a crucial aspect of reproduction and is significantly influenced by hormones. Brain mechanisms for parental behaviour show considerable overlap with mechanisms implicated in sexual behaviour. Review **Figures 13.13, 13.14**

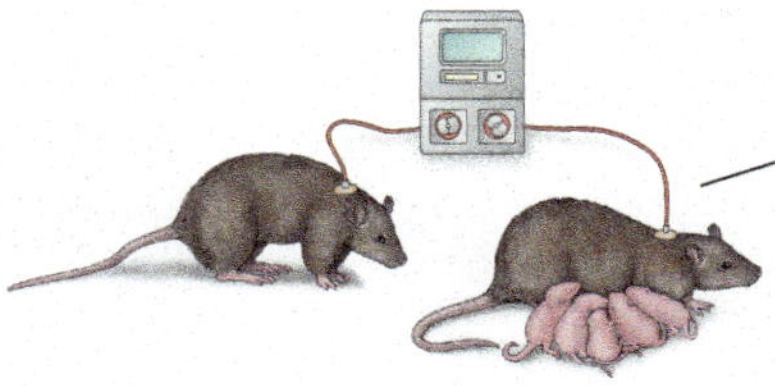

7 In birds and mammals, genetic sex determines whether testes or ovaries develop, and hormonal secretions from the gonads determine whether the rest of the body, including the brain, develops in a feminine or masculine fashion. In the presence of testicular secretions, a male develops; in the absence of testicular secretions, a female develops. Review **Figures 13.15, 13.16**

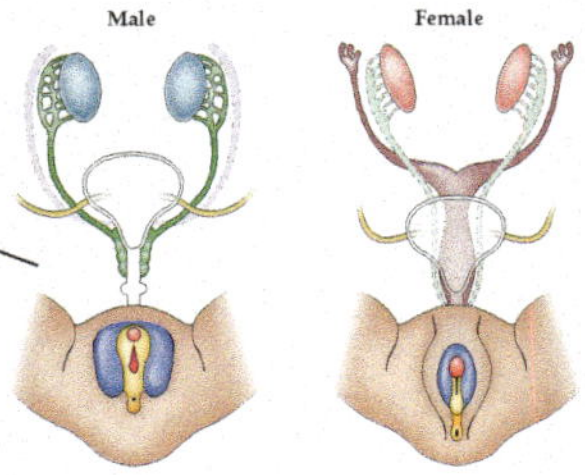

8 People can be classified on the basis of their sex chromosomes, their genitalia, or the **gender** they identify with. The options within each of these categories are complex and sometimes overlap, so attempts to classify all individuals into just two gender groups oversimplify the real situation. Review **Figures 13.17, 13.18**

9 The brains of vertebrates are masculinised by the presence of testicular steroids during early development. Such **organisational effects** of steroids permanently alter the structure and function of the brain and therefore permanently alter the behaviour of the individual. Review **Figure 13.20**, **Animation 13.1**

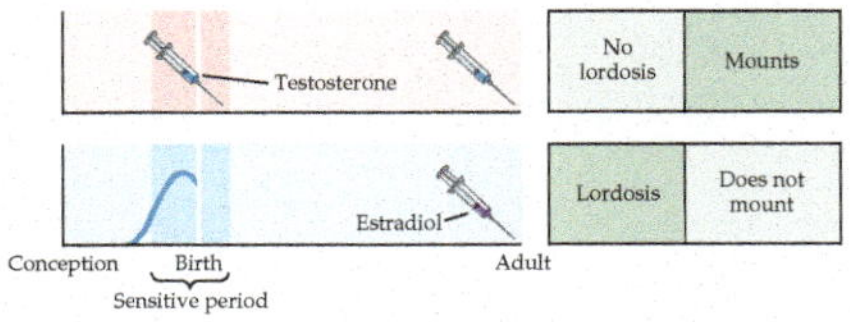

10 Among the prominent examples of **sexual dimorphism** in the nervous system, gonadal steroids have been shown to alter characteristics such as neuronal survival, structure, and synaptic connections. Review **Figures 13.22–13.24**

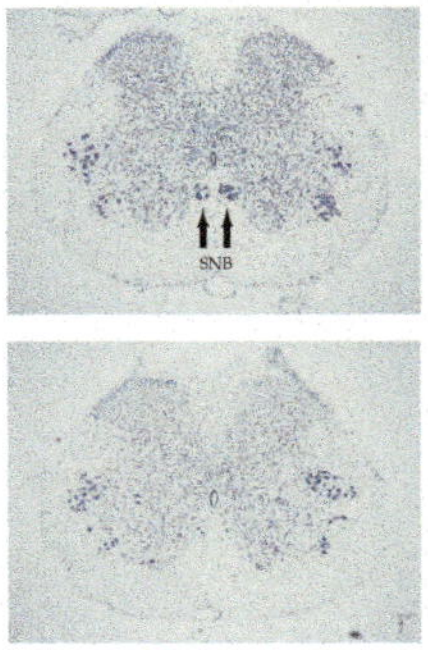

(Continued)

11 Several regions of the human brain are sexually dimorphic. However, we do not know whether these dimorphisms are generated by fetal steroid levels or by sex differences in the early social environment. Review **Figure 13.25**, **Table 13.2**

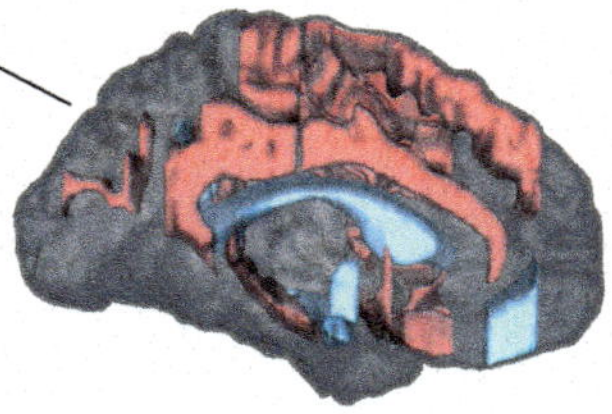

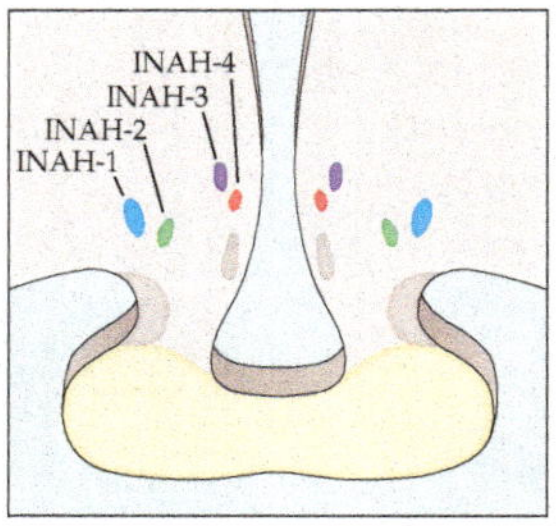

12 Although no perfect animal model of sexual orientation has been developed, all research indicates that sexual orientation is determined early in life and, especially in men, is not a matter of individual choice. Review **Figures 13.26–13.28**

Homeostasis

Active Regulation of the Internal Environment

14

Harsh Reality TV

In 2004, NBC's reality television show *The Biggest Loser* became a prime-time staple. The premise of the show is simple enough—the contestant who loses the most weight during the season wins—but the effort required from the contestants is immense. The biggest loser of them all in season 8 (2009) was Danny C. Through a punishing combination of near-starvation dieting and all-day exercise, Danny shed an incredible 239 pounds, dropping from 430 pounds to a svelte 191 pounds in just 7 months. Other initially obese contestants similarly accomplished exceptional weight loss, clearly delighting in revealing their new and slimmer silhouettes to their friends, families, and viewers.

Recognizing an unusual opportunity, a group of scientists followed Danny and other Biggest Loser contestants for 6 years following their weight loss: the longest-term study of its kind ever conducted (Fothergill et al., 2016; Kolata, 2016). The results are discouraging. In the years after his appearance on the show, and despite exceptional ongoing efforts, Danny regained more than 100 pounds of the weight he had lost. In fact, all but one of the 14 contestants in the study regained significant weight; some were even heavier after than they were before the show. The pattern of results confirms the common observation that it is hard to keep the weight off after dieting, but what could explain the additional discovery that even after 6 years of hard work, the bodies of these contestants continued to strive to return to their original obese state?

Millions of years of evolution have endowed our bodies with complex physiological mechanisms, under the control of the brain, that regulate the conditions required for optimal cellular functioning. But in the context of modern society, some of these ancient systems are making trouble for us; obesity, for example, is reaching epidemic proportions and placing a severe burden on health care resources. The physiological and behavioural processes governing the physiological state of the body, and their role when things go wrong, are our topic in this chapter.

14.1 Homeostasis Maintains a Consistent Internal Environment: The Example of Thermoregulation

Learning Objectives

After reading this section, you should be able to:

14.1.1 **Define *homeostasis* and *allostasis*, and explain their relationship to drive states.**

14.1.2 **Distinguish between endothermy and ectothermy, with examples, and discuss the pros and cons of each system.**

14.1.3 **Describe, with appropriate examples, how the engineering concepts of negative feedback and redundancy apply to homeostatic systems.**

14.1.4 **Describe some ways in which animals use specialised behaviours to maintain a stable internal environment.**

The body can be viewed as a self-contained environment, carefully regulated so as to provide optimal conditions for cells to live and grow. Variables such as acidity, saltiness, water level, oxygenation, temperature, and energy availability, among others, are closely monitored and maintained by an elaborate system of behavioural and physiological mechanisms. The result of these active processes is **homeostasis:** a relatively stable, balanced internal environment optimised for cellular activities. Alterations in the internal environment can have an effect on **motivation**, the psychological process that induces or sustains a particular behaviour. According to this view, the mismatch between the actual internal state and the regulated, intended state (e.g., becoming dehydrated) produces a *drive* to restore balance (by having a drink of water). And as we all know, drive can rapidly escalate as the mismatch worsens, from a minor distraction (like having a couple of sips if a glass of water happens to be present) to an overwhelming, all-consuming desire (like the raging thirst of someone lost in the desert). The regulation of our internal resources is complicated by the fact that staying alive requires us to use up some of them. Our homeostatic mechanisms are continually challenged by these **obligatory losses**, which require us to gain and conserve heat, water, and food constantly.

Because it is a relatively simple system, let's start by considering **thermoregulation**, the regulation of body temperature, to illustrate some important general concepts in homeostasis: negative feedback, redundancy, behavioural homeostasis, and the concept of allostasis. These principles will arise again when we talk about fluid regulation, appetite, and body weight regulation, later in the chapter.

homeostasis The maintenance of a relatively constant internal physiological environment.

motivation Here, a drive state that prompts homeostatic behaviours.

obligatory losses Unavoidable loss of regulated variables, such as energy, water, or temperature, as a consequence of life processes.

thermoregulation The active process of closely regulating body temperature around a set value.

endotherm An animal whose body temperature is regulated chiefly by internal metabolic processes. Examples include mammals and birds.

ectotherm An animal whose body temperature is regulated by, and whose heat comes mainly from, the environment. Examples include snakes and bees.

negative feedback The process whereby a system monitors its own output and reduces its activity when a set point is reached.

set point The point of reference in a feedback system. An example is the setting of a thermostat.

set zone The range of a variable that a feedback system tries to maintain.

Homeostatic systems share several key features

We mammals are **endotherms**, species that make our own heat from *inside* our bodies, using metabolism and muscular activity (and if our muscles aren't making enough heat, they shiver to make more). Endothermy gives us clear advantages over the **ectotherms**—animals such as lizards and snakes that get their heat mostly from the environment—by allowing us to range more widely. Like endotherms, ectotherms try to regulate their body temperature within a range that is optimal for the functioning of their cells, but this means that they must stay nearer sources of warmth. Furthermore, the evolution of endothermy involved enhanced capacity for oxygen utilisation (in order to generate heat through metabolism), with the result that mammals can sustain high levels of muscular activity for much longer periods of time: endothermic hares will always outrun ectothermic tortoises. (For more on the pros and cons of endothermy and ectothermy, see A STEP FURTHER 14.1: SOME ANIMALS GENERATE HEAT BUT OTHERS MUST OBTAIN HEAT FROM THE ENVIRONMENT on the website.) So it's no surprise that we have dedicated systems for creating warmth and regulating our body temperature. The systems that govern body temperature operate according to several general principles common to almost all homeostatic systems.

NEGATIVE FEEDBACK The homeostatic mechanisms that regulate temperature, body fluids, and metabolism are primarily **negative feedback** systems, where deviation from a desired value, called the **set point**, triggers a compensatory action of the system. Restoring the desired value turns off the response (this is why it is called *negative* feedback). A household thermostat uses negative feedback (**FIGURE 14.1**): A temperature drop below the set point activates the thermostat, which turns on the heating system. The heat that is produced has a negative feedback effect on the thermostat, so it stops calling for heat. Most systems have at least a little bit of tolerance built in—otherwise the system would be going on and off too frequently—so there is generally a **set zone** rather than a rigid set *point*.

The setting of the thermostat in your home can be changed; for example, it can be turned down at night to save energy. Similarly, although the body temperature for most mammals is usually held within a narrow range—about 36–38°C (97–100°F)—most mammals reduce their temperature during sleep. Sometimes, your set zone may be temporarily elevated, producing a fever to help your body fight off an infection. But there remain narrow limits. Too hot, and proteins begin to lose their correct shape, link together, and malfunction or die (this is called *denaturing* or, if it is really hot, *cooking*). Too cool, and chemical reactions of the body occur too slowly; at very low body temperatures, ice crystals may disrupt cellular membranes, killing the cells. Some animals that cannot avoid subfreezing temperatures—for example, some species of fishes and beetles (**FIGURE 14.2**)—produce "antifreeze" consisting of special protein molecules that suppress the formation of ice crystals and prevent damage to membranes (Duman, 2015; C. B. Marshall et al., 2004).

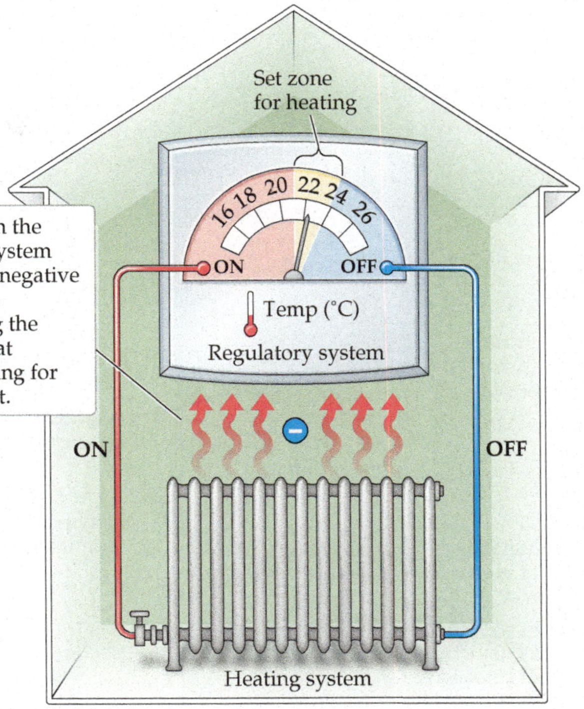

14.1 Negative Feedback The thermostat controlling a home heating system employs negative feedback. All such systems have a sensor (in this example, a thermometer) to monitor the variable (temperature) and a device (the heating system) to change the variable (e.g., by heating the room). The changed variable (heat) provides a negative feedback signal to the sensor, turning the system off.

Go to Animation 14.1
Negative Feedback

REDUNDANCY Just as human engineers equip critical equipment with several backup systems, our bodies tend to have multiple mechanisms for monitoring our stores, conserving remaining supplies, obtaining new resources, and shedding excesses. Loss of function in one part of the system usually can be compensated for by the remaining parts. This redundancy attests to the importance of maintaining a stable inner environment, but it also complicates the lives of scientists who are trying

(A)

(B)

(C)

14.2 Braving the Cold Mealworm beetles (A) and winter flounder (B) are two species that sometimes have body temperatures below 0°C. These animals produce an "antifreeze" protein in body fluids to prevent ice crystals from forming in their cell membranes. Through mechanisms not yet understood, the arctic ground squirrel (C) is also able to withstand subzero temperatures during hibernation.

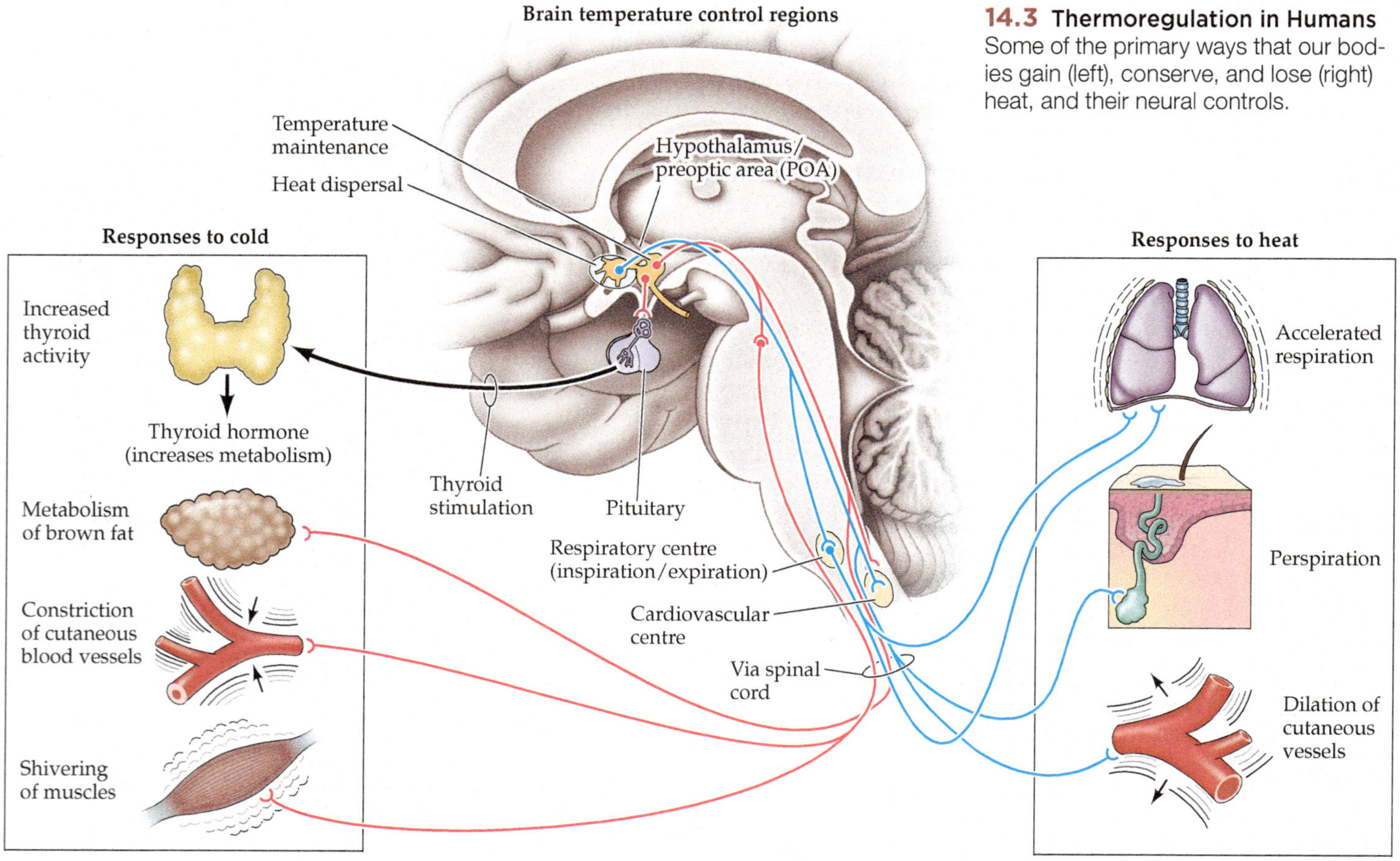

14.3 Thermoregulation in Humans Some of the primary ways that our bodies gain (left), conserve, and lose (right) heat, and their neural controls.

Go to Animation 14.2
Thermoregulation in Humans

to figure out exactly how the body normally regulates temperature, water balance, and food intake.

The body has multiple physiological systems for the generation of heat, as well as multiple systems for cooling if it gets overheated (**FIGURE 14.3**). Shivering, sweating, burning fat, and changing blood flow are just a few examples of the many ways the body can increase or decrease its temperature. There may also be a degree of redundancy in the thermoregulatory control systems of the nervous system. Classic research established that neurons of the preoptic area (POA) play a key role in integrating sensory inputs and driving physiological thermoregulatory responses (Satinoff and Rutstein, 1970; Tan and Knight, 2018), but multiple other brain regions also appear to contribute to thermoregulation (**FIGURE 14.4**). Neural mechanisms that govern behavioural thermoregulation in particular are incompletely understood, and the persistence of behavioural adaptations to thermal challenges even after POA lesions suggests that they may be independent of the POA to some extent. The exact nature of the coordination of physiological and behavioural thermoregulation remains to be determined (Tan and Knight, 2018).

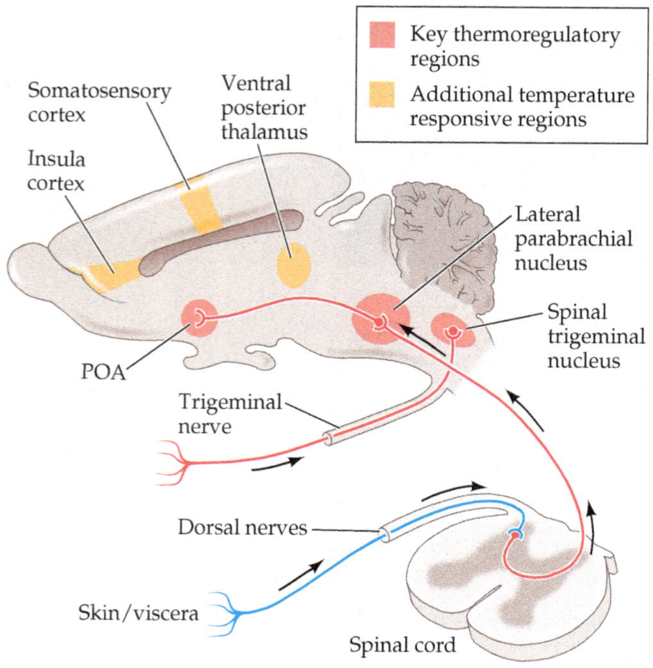

14.4 Neural Systems for Sensing and Regulating Body Temperature Inputs from warm and cool sensors in the periphery ascend in the dorsal spinal cord and synapse in the lateral parabrachial nuclei of the brainstem. Thermal information then projects to the preoptic area (POA), the principal region organising thermoregulatory responses, and also to other brain regions. Neurons in several other locations (in yellow) process temperature information; their contributions to thermoregulation—especially behavioural thermoregulation—are not fully understood. (After C. L. Tan and Z. A. Knight, 2018. *Neuron* 98: 31–48.)

BEHAVIOURAL HOMEOSTASIS Organisms also use behavioural measures to help them acquire more heat, water, or food, in order to achieve and maintain homeostasis. In general, ectotherms and endotherms can employ three kinds of behaviours to regulate temperature. They can (1) *change exposure of the body surface*, for example by huddling or extending limbs; (2) *change external insulation*, such as by using clothing or nests; and (3) *change surroundings*, moving into the sun, into the shade, or into a burrow.

Because ectotherms generate little heat through metabolism, behavioural methods of thermoregulation are especially important for them. In the laboratory, iguanas carefully regulate their temperature by moving toward or away from a heat lamp, and when infected by bacteria, they even produce a fever through such behavioural means (**FIGURE 14.5**), which helps their immune system fight off the infection. Endotherms use internal processes to generate a fever; it's a harmful quirk of human nature that we try so hard to suppress fevers, when they may very well improve outcomes when battling infectious diseases like pneumonia (Schulman et al., 2005).

Many ectotherms regulate body temperature using specialised behaviours. Some snakes, for example, adjust their coils to expose more or less surface to the sun and thus keep their internal temperature relatively constant during the day. On cold days, bees crowd into the hive and shiver, thus generatingweat; on hot days, the bees fan the hive with their wings instead, thereby keeping it cool. As a result of these behavioural measures, hive temperature is regulated at about 35°C.

Endotherms such as mammals and birds likewise control their exposure to the sun and to hot or cold surfaces to avoid overtaxing their internal regulatory mechanisms. Throughout history, humans have busily devised adaptations to hot and cold, ranging from the use of fans and swimming pools to the creation of heating systems and highly insulating clothing.

FIGURE 14.6 summarises the basic mammalian thermoregulatory system: receptors in the skin, body core, and hypothalamus detect temperature and transmit that information to three neural regions (spinal cord, brainstem, and hypothalamus). If the body temperature moves outside the set zone, each of these neural regions can initiate physiological and behavioural responses to return it to the set zone. **BOX 14.1** on the next page gives an example of the close integration of physiological and behavioural homeostatic mechanisms.

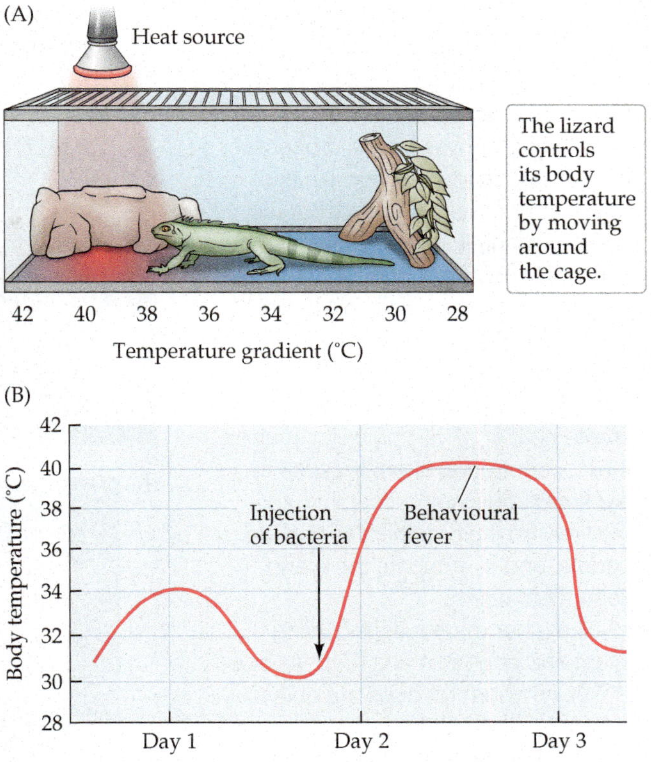

14.5 Behavioural Thermoregulation in Bacteria-Challenged Lizards (A) By positioning themselves along the temperature gradient between the cool and warm ends of a terrarium, iguanas exert precise behavioural control over their body temperature. (B) Body temperature recorded from iguanas over the course of 3 days is plotted here. Day 1 shows the normal daily cycle of body temperature, averaging 34°C during the day and slightly cooler at night (similar to the daily temperature cycle of endotherms). Early on day 2, the iguanas were injected with bacteria, to which they reacted by moving closer to the heat source and allowing their body temperature to rise to more than 40°C for about 16 hours. This "behavioural fever" closely resembles the fever that endotherms like humans experience during infection, and it similarly helps fight off infection. (After M. J. Kluger, 1978. *Am Sci* 66: 38–43.)

allostasis The varying behavioural and physiological adjustments that an individual makes in order to maintain optimal (rather than unchanging) functioning of a regulated system in the face of changing environmental stressors.

A wide array of sensors continuously monitor the many internal and external threats to our physiological stability. At any given moment, depending on what's happening in the environment, simultaneous perturbations in multiple regulated systems cause varying degrees of physiological stress. Rather than defending a single set point, many physiological systems must continually shift their responses, depending on the nature of the stressors and prior experience—for example, your heart rate and blood pressure are continually shifted to accommodate your current or anticipated activity level—in a dynamic process termed **allostasis** (McEwen, 2012; Sterling and Eyer, 1988). The associated cost of these responses—the wear and tear of daily life—is called *allostatic load*. Further, homeostatic regulation pervades many physiological and behavioural processes beyond the ones described in this chapter. For example, homeostatic adjustments affect drug actions by changing neuronal sensitivity (Chapter 3), regulate sleep onset and character

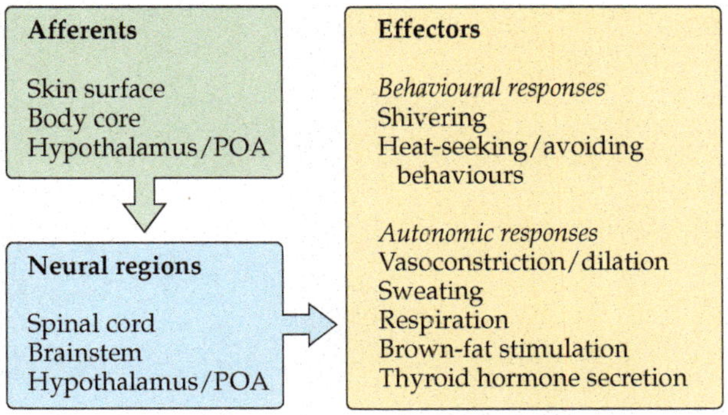

14.6 Basic Elements of Mammalian Thermoregulatory Systems

BOX 14.1 Physiological and Behavioural Thermoregulation Are Integrated

Ectothermic animals produce some heat through physiological processes, but not enough to overcome losses to the environment. Instead, like the marine iguana (Figure A), they rely principally on behavioural solutions, either seeking out sources of warmth or, if too warm, shielding themselves from heat. The young of many endothermic species likewise lack the ability to create heat or physiologically regulate their temperature, often because they are small and lose heat rapidly, so they rely on their mothers to gestate, cuddle, and/or incubate them and keep them warm.

Like other mammals, including infant and adult humans (Cypess et al., 2009), newborn rat pups are able to generate heat via thermogenic brown-fat deposits like the one between the shoulder blades, shown in Figure B (Blumberg et al., 1997). Nonetheless, one problem for newborn rats is insulating their hairless bodies to conserve the heat they generate. To tackle this problem, they huddle together. The effectiveness of this behavioural strategy is easily demonstrated: placed in a room-temperature environment, an isolated 5-day-old pup will soon cool to less than 30°C, but as part of a group of four, the same pup can maintain a temperature above 30°C for 4 hours or more (Alberts, 1978). Pups frequently change their positions in the huddle, regulating their temperature by moving to the inside or the outside of the clump, such that all of the animals benefit: an example of cooperative homeostasis.

(A)

Both photos by Mark R. Rosenzweig

(B)

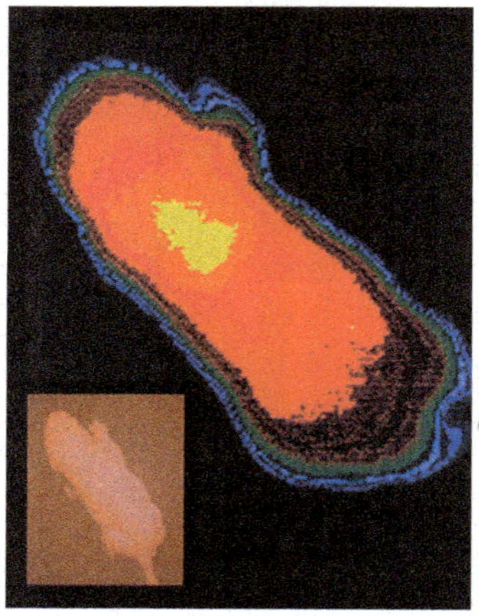

Courtesy of Mark S. Blumberg

Behavioural and Physiological Thermoregulation (A) Having been cooled by the sea, the Galápagos marine iguana raises its body temperature by hugging a warm rock and lying broadside to the sun (left). Once its temperature is sufficiently high, the iguana reduces its surface contact with the rock and faces the sun to minimize its exposure (right). (B) This infrared thermograph shows the dorsal surface of a 1-week-old rat pup oriented as shown in the inset. Areas of highest heat production are coded in orange and yellow, and the prominent yellow "hot spot" between the shoulder blades overlies a deposit of brown fat, a thermogenic (i.e., heat-producing) organ. When rat pups are placed in a cold environment, they begin producing heat by using brown fat.

(Chapter 15), modify hormones and sexual motivation (Chapter 13), and so on. Allostatic adjustments are a normal part of dealing with the demands of daily life, but the heavy physiological burden on chronically stressed individuals puts them at risk of pathology due to allostatic overload, a topic we turn to in Chapter 11.

FLUID REGULATION

14.2 Water Shuttles between Two Body Compartments

Learning Objectives

After reading this section, you should be able to:

14.2.1 Describe the compartmentalisation of fluids in the body.

14.2.2 Briefly define diffusion and osmosis.

14.2.3 Describe the semipermeable membrane, and explain its role in the movement of water between compartments.

14.2.4 Explain why water must be continually replenished.

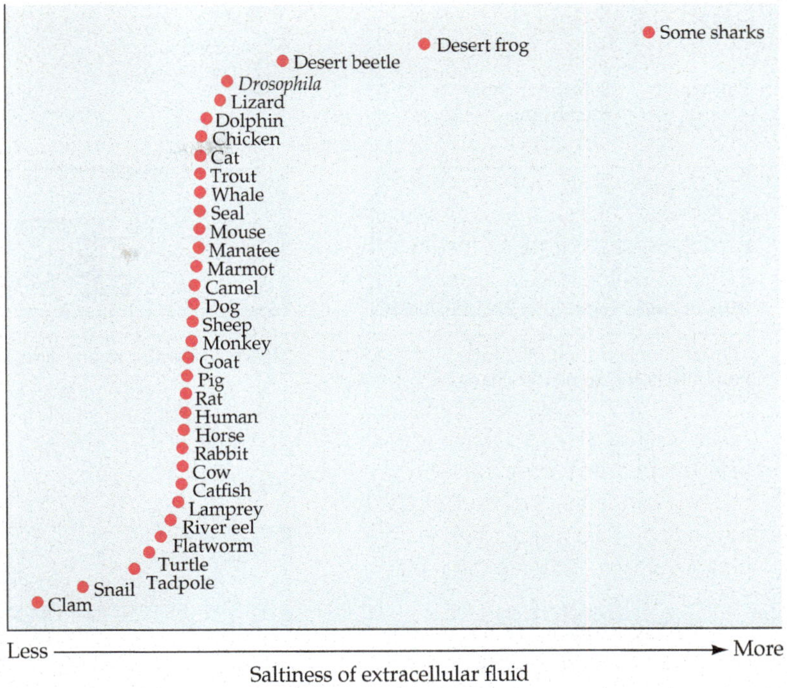

14.7 Sea Inside Despite many millions of years of evolution, the concentration of extracellular fluid has remained remarkably constant among different species of animals, with only a few exceptions. (After C. W. Bourque, 2008. *Nat Rev Neurosci* 9: 519–531.)

The water that you drink on a hot day is carefully measured and partitioned by the brain. A precise balance of fluids and dissolved salts bathes the cells of the body and enables them to function. The composition of this fluid reflects the evolutionary origins of life on Earth. The first living organisms were the unicellular inhabitants of the ancient oceans, so fundamental cellular processes like metabolism and gene expression were optimised, via evolution by natural selection, for operation in the oceanic environment. By the time more-complex multicellular species arose and began to exploit opportunities on land and in the air, they were obliged to bring along the watery environment that their cells needed to function. For this reason, most organisms have evolved homeostatic mechanisms that ensure that the composition of their body fluids closely resembles dilute seawater (**FIGURE 14.7**) (C. W. Bourque, 2008). Even relatively minor deviation from this optimal concentration of salt in water is generally lethal, with only rare exceptions; the salmon, for example, has unique adaptations that allow it to live in freshwater at hatching, to grow up in salt water, and to return again to freshwater to spawn.

Because we consist of trillions of cells living in a salty seawater-like bath, scientists typically describe water balance by contrasting the inside versus the outside of our cells. Most of our water is contained within our cells; this combined region is termed the **intracellular compartment**. The fluid outside of our cells, collectively referred to as the **extracellular compartment**, is divided between the *interstitial fluid* (the fluid between cells) and *blood plasma* (the protein-rich fluid that carries red and white blood cells). Water is continually moving back and forth between these compartments, in and out of cells, via specialised protein channels called **aquaporins** that stud the cell membrane (Agre et al., 2002; Knepper et al., 2015). A single aquaporin channel can selectively conduct about 3 *billion* molecules of water per second!

To understand the forces driving the movement of water, we must understand diffusion and osmosis. In **diffusion**, molecules of a substance like salt (a *solute*) dissolved in a quantity of another substance, such as a glass of water (a *solvent*), will passively spread through the water because of the random jiggling and movement of the molecules until they are more or less uniformly distributed throughout the glass (**FIGURE 14.8A**). If we divide a container of water with a membrane that is impermeable to water and salt, and we put the salt in the water on one side, the molecules will diffuse only within that half. If instead the membrane impedes salt molecules only a little, then the salt will distribute itself evenly within the initial half but will also—more slowly—invade and distribute itself across the other half.

A membrane that is permeable to some molecules but not others is referred to as *selectively permeable* or *semipermeable*. As we saw in Chapter 2, cell membranes are very selective in their permeability: for example, neurons normally allow very few sodium ions (Na^+) to pass through their membrane unless the voltage-gated Na^+ channels are opened during the action potential.

Osmosis is the movement of water molecules that occurs so as to equalise the concentrations of two solutions, when a semipermeable membrane separates

intracellular compartment The fluid space of the body that is contained within cells.

extracellular compartment The fluid space of the body that exists outside the cells.

aquaporins Channels spanning the cell membrane that are specialised for conducting water molecules into or out of the cell.

diffusion The passive spread of molecules of one substance among molecules of another substance until a uniform concentration is achieved.

osmosis The passive movement of water molecules from one place to another until a uniform concentration is achieved.

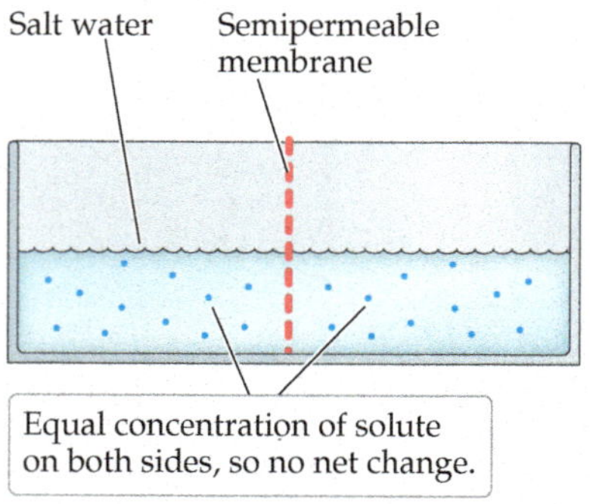

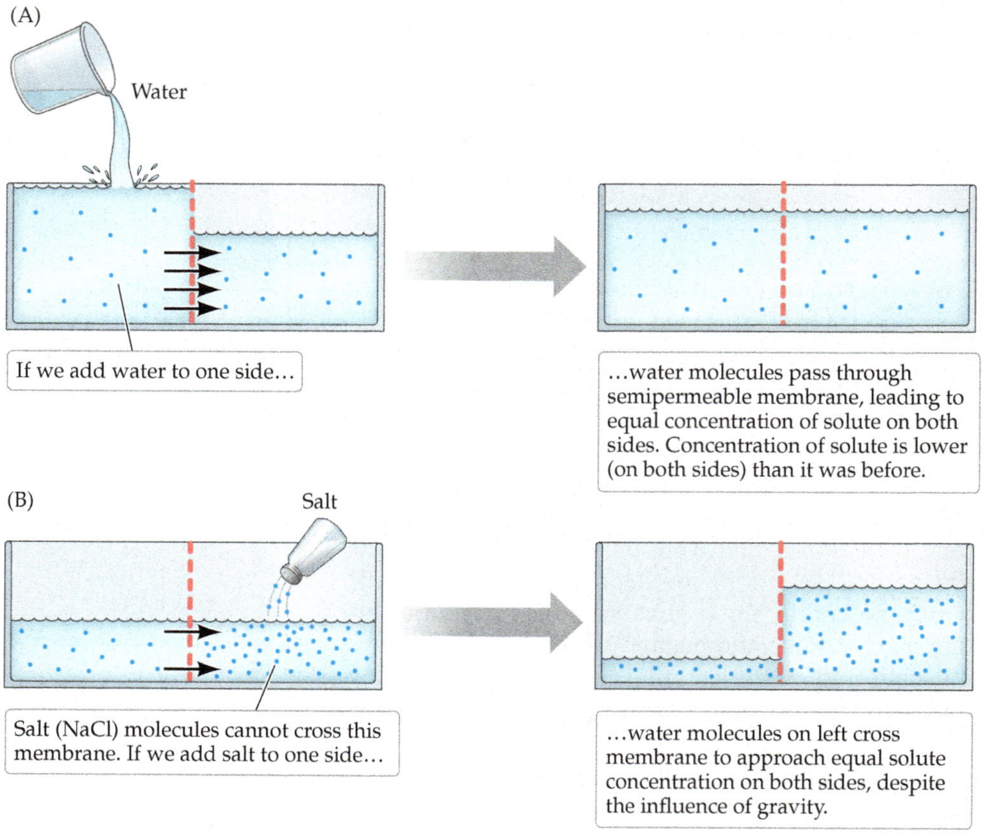

14.8 Osmosis (A) Water passes through a semipermeable membrane in whichever direction is necessary to maintain an equal solute concentration. As shown in (B), this osmotic pressure may even overcome gravity. Water moves in the same way between compartments in the human body.

osmotic pressure The tendency of a solvent to move through a membrane in order to equalise the concentration of solute.

osmolality The number of solute particles per unit volume of solvent.

isotonic Referring to a solution with a concentration of salt that is the same as that found in interstitial fluid and blood plasma (about 0.9% salt).

hypertonic Referring to a solution with a higher concentration of salt than that found in interstitial fluid and blood plasma (more than about 0.9% salt).

hypotonic Referring to a solution with a lower concentration of salt than that found in interstitial fluid and blood plasma (less than about 0.9% salt).

solutions of different concentrations of solute. This is the case we examine in **FIGURE 14.8B**, where a tank is divided into two parts by a semipermeable membrane that permits water to cross, *but not salt molecules*. Adding extra salt to one side induces water molecules to move into the now-salty side until the concentration of both solutions equalises. The physical force that pushes or pulls water across the membrane is called **osmotic pressure**.

We refer to the concentration of solute in a solution as **osmolality**. Normally, the concentration of salt (NaCl) in the extracellular fluid of mammals is about 0.9% (weight to volume, which means there's about 0.9 gram of NaCl for every 100 milliliters of water). A solution with this concentration of salt is called *physiological saline* and is described as **isotonic**, having the same concentration of salt that mammalian fluids have. A solution with more salt is **hypertonic**; a solution lower in salt is **hypotonic**.

Because water moves so as to produce uniform saltiness (see Figure 14.8B), cells will lose water if placed in a saltier solution, and they will gain water in a less salty solution. If excessive, this movement of water will damage or kill the cell. To prevent such damage, the extracellular fluid serves as a *buffer*, a reservoir of isotonic fluid that provides and accepts water molecules so that cells can maintain proper internal conditions.

We cannot fully seal our bodies from the outside world, so we experience constant obligatory losses of water and salts. Many body functions require that we use up some water (and some salt molecules), as, for example, when we produce urine to rid ourselves of waste molecules. These losses require us to actively replenish the body's water and salts (**TABLE 14.1**). The nervous system uses two cues to ensure that the extracellular compartment has about the right amount of water and solute to allow cells to absorb or shed water molecules readily, as we'll see next.

TABLE 14.1 Average Daily Water Balance

Source	Quantity (liters)
APPROXIMATE INTAKE	
Fluid water	1.2
Water from food	1.3
TOTAL	2.5
APPROXIMATE OUTPUT	
Urine	1.4
Evaporative loss	0.9
Feces	0.2
TOTAL	2.5

14.3 Two Internal Cues Trigger Thirst

Learning Outcomes

After reading this section, you should be able to:

14.3.1 **Identify and distinguish between the two major forms of thirst.**

14.3.2 **Describe the physiological signals and their sensors for each type of thirst.**

14.3.3 **Discuss the importance of salt homeostasis in each type of thirst.**

14.3.4 **Give an overview of physiological responses to thirst and of how the brain gauges when to stop drinking.**

The brain contains dedicated systems that carefully monitor the composition of the extracellular fluid and trigger thirst to stimulate the intake of water when necessary (C. A. Zimmerman et al., 2017). Two different states can signal that more water is needed: (1) a high extracellular concentration of solute, that is, too much salt in the extracellular fluid, resulting in **osmotic thirst**, and (2) low extracellular volume due to the loss of body fluids, which triggers **hypovolemic thirst**. We'll consider each in turn.

Osmotic thirst is triggered by increased saltiness of the extracellular fluid

Most of the time, we feel thirsty as a result of the obligatory water losses we mentioned earlier: normal physiological processes through which we lose more water than salt, such as in respiration, perspiration, and urination. In this case, not only does the *volume* of the extracellular fluid decrease, but also the solute *concentration* of the extracellular fluid increases. As a result of this increase in extracellular saltiness, water is pulled out of cells through osmosis.

Another way that the extracellular fluid can become more concentrated is by ingestion of a lot of salty food. Once again, water will be drawn out of cells through osmosis. The resultant loss of intracellular water triggers *osmotic thirst* (**FIGURE 14.9**); we want to drink water to return the extracellular fluid to a comfortable isotonic state and protect the intracellular compartment from becoming dangerously depleted of water.

osmotic thirst A desire to ingest fluids that is stimulated by a high concentration of solute (like salt) in the extracellular compartment, reducing intracellular fluid.

hypovolemic thirst A desire to ingest fluids that is stimulated by a reduced volume of extracellular fluid.

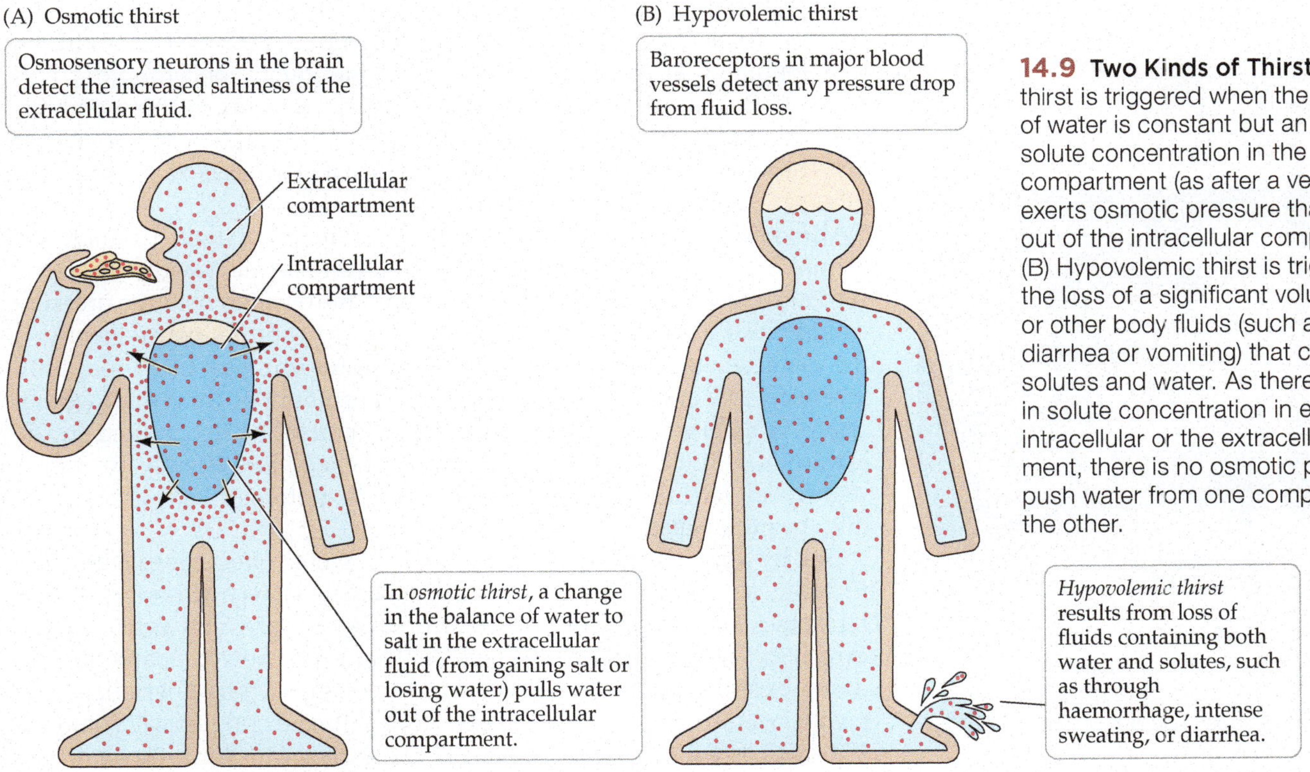

14.9 Two Kinds of Thirst (A) Osmotic thirst is triggered when the total volume of water is constant but an increase in solute concentration in the extracellular compartment (as after a very salty meal) exerts osmotic pressure that pulls water out of the intracellular compartment. (B) Hypovolemic thirst is triggered by the loss of a significant volume of blood or other body fluids (such as through diarrhea or vomiting) that contain both solutes and water. As there is no change in solute concentration in either the intracellular or the extracellular compartment, there is no osmotic pressure to push water from one compartment to the other.

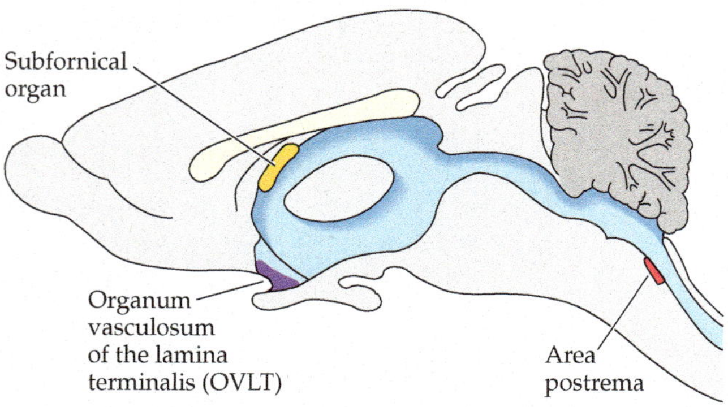

14.10 Circumventricular Organs In keeping with their name, the circumventricular organs, seen here in a midsagittal view of the rat brain, lie in the walls of the ventricular system (blue). The blood-brain barrier is greatly reduced in the subfornical organ (SFO) and the OVLT, so neurons there can monitor the concentration and composition of body fluids.

osmosensory neuron A specialised neuron that measures the movement of water into and out of cells.

circumventricular organ An organ that lies in the wall of a cerebral ventricle and monitors the composition of body fluids.

organum vasculosum of the lamina terminalis (OVLT) One of the circumventricular organs.

subfornical organ (SFO) One of the circumventricular organs.

Injecting a small amount of hypertonic (extra salty) solution into the hypothalamus causes animals to start drinking, suggesting that some hypothalamic cells are specialised to respond to changes in osmotic pressure. Single-cell recordings confirm that there are **osmosensory neurons**—neurons that specifically monitor the concentration of the extracellular fluid—in several regions of the hypothalamus: the preoptic area, the anterior hypothalamus, the supraoptic nucleus, and the **circumventricular organs**. As their name suggests, circumventricular organs lie in the walls of the cerebral ventricles (**FIGURE 14.10**) and feature *fenestrated capillaries*—blood vessels that lack the usual blood-brain barrier (see Chapter 2)—allowing neurons in these regions to monitor both salt concentration and hormones in the bloodstream (Miyata, 2015). The **organum vasculosum of the lamina terminalis** (**OVLT**) and **subfornical organ** (**SFO**) are circumventricular organs that prompt a particularly large drinking response to changes in osmolality. Separate populations of neurons in the OVLT and SFO use either glutamate or GABA as transmitters, all projecting to a portion of the POA. The excitatory glutamatergic afferents to the POA elicit drinking, while the inhibitory GABA-ergic afferents inhibit drinking (Oka et al., 2015; C. A. Zimmerman et al., 2017). This portion of the POA sends projections to several hypothalamic regions, including those that release vasopressin through the posterior pituitary—the paraventricular nucleus and the supraoptic nucleus (see Figure 4.12).

Osmosensory neurons have some unusual features that help them detect the concentration of extracellular fluid (Z. Zhang and Bourque, 2003). For one thing, they are stretchy. Most cells in the body and brain actively maintain a constant size in the face of osmotic challenges, but osmosensory neurons don't do this: instead, they balloon or shrink as the concentration of the extracellular fluid changes. The stretching and shrinking of the cell membrane physically opens and closes special mechanically gated ion channels that stud the membrane, causing changes in cell membrane potentials that track the changes in extracellular concentration. This information is then relayed to other parts of the brain, resulting in thirst and homeostatic responses to conserve water, such as through hormone release and through modifications of salt intake, as we discuss next.

SALT HOMEOSTASIS Salt (NaCl) is crucial for fluid balance. We cannot maintain water in the extracellular compartment without solutes. If the extracellular compartment contained pure water, osmotic pressure would drive it into the cells, causing them to rupture; consequently, if the extracellular fluid lacks salt, we shed water (through urination) until it returns to isotonic concentration. So, the amount of water that we can retain is determined primarily by the amount of salt contained in the extracellular fluid. That's why thirst is quenched more effectively by very slightly salty drinks (as long as they are still hypotonic, like sports drinks) than by pure water.

Saltier water, such as seawater, has the reverse effect. Seawater is *hyper*tonic, so just as eating salty food makes us thirsty, drinking seawater causes ever-worsening osmotic thirst. We simply can't get rid of the excess salt fast enough, unlike species that have evolved adaptations to dump excess sodium. Some seabirds, for example, have specialised salt glands near the nostrils that can excrete highly concentrated salt solutions (**FIGURE 14.11**) (Schmidt-Nielsen, 1960), so they can drink seawater.

14.11 Excretion of Excess Salt Marine birds, such as this giant petrel, have only seawater to drink for long periods of time. To compensate, they have salt glands that pull excess salt out of plasma and release it out the nostrils.

Some Na^+ loss is inevitable, as during urination or sweating. But when water is at a premium, the body tries to conserve Na^+ in order to also retain water. One way this is accomplished is through the release of the steroid hormone **aldosterone** from the adrenal glands. Aldosterone directly stimulates the kidneys to conserve Na^+, rather than dumping it into the urine, and contributes to the salt appetite that powerfully drives animals to find additional salt in their environments (de Kloet and Joëls, 2017).

aldosterone An adrenal steroid that promotes conservation of sodium by the kidneys.

baroreceptor A pressure receptor in the heart or a major artery that detects a fall in blood pressure.

atrial natriuretic peptide (ANP) A hormone, secreted by the heart, that normally reduces blood pressure, inhibits drinking, and promotes the excretion of water and salt at the kidneys.

vasopressin Also called *antidiuretic hormone* (*ADH*). A peptide hormone from the posterior pituitary that promotes water conservation.

angiotensin II A substance that is produced in the blood by the action of renin and that may play a role in the control of thirst.

Hypovolemic thirst is triggered by a loss of water volume

The second signal that triggers thirst doesn't involve salt balance or osmosis, but rather a decrease in the overall volume of the extracellular fluid, called *hypovolemia* (literally "low volume"). Normal everyday obligatory losses cause moderate decreases in extracellular fluid volume (in addition to increased saltiness), but more sudden and dramatic losses of fluid from the body—due to haemorrhage, vomiting, sustained diarrhea—may trigger thirst that is *primarily* hypovolemic in nature (see Figure 14.9B). In this condition, blood vessels that would normally be full and slightly stretched no longer contain their full capacity. Blood pressure drops, and the individual becomes thirsty—powerfully so, if the haemorrhage or other volume loss is severe enough.

Note that suddenly losing fluids from blood loss (or from diarrhea or vomiting) does not change the *concentration* of the extracellular fluid, at least at first, because salts and other solutes are lost along with the water. Rather, only the *volume* of the extracellular fluid is affected in these instances. This tells us that changes in volume alone are sufficient to induce drinking behaviour, providing another example of redundancy in a homeostatic system. The initial drop in extracellular volume is detected by pressure receptors, called **baroreceptors**, which are located in major blood vessels and in the heart. In response to the signal from the baroreceptors, the brain activates a variety of responses, such as thirst (to replace the lost water) and salt hunger (to replace the solutes that have been lost along with the water). Replacing the water without also replacing the salts would result in hypotonic extracellular fluid. The sympathetic nervous system also stimulates muscles in artery walls to constrict, reducing the size of the vessels to increase blood pressure and partly compensate for the reduced volume. In addition, several hormonal systems are activated, as we will see next.

HORMONAL RESPONSES TO DEHYDRATION Physiological responses to hypovolemia are coordinated by a set of peptide hormones originating in several different organs. Immediately after the baroreceptors detect a drop in blood pressure, the heart decreases its secretion of **atrial natriuretic peptide** (**ANP**), which normally reduces blood pressure, inhibits drinking, and promotes the excretion of water and salt at the kidneys. At the same time, the brain steps up its release of the peptide hormone **vasopressin** from the posterior pituitary gland. Vasopressin induces additional constriction of blood vessels. Furthermore, vasopressin instructs the kidneys to reduce the flow of water to the bladder (vasopressin is also known as *antidiuretic hormone* [*ADH*]: *diuresis* refers to the production of urine). At night, inputs from the circadian clock system of the suprachiasmatic nucleus (see Chapter 15) promote vasopressin release, slowing urine production and preventing dehydration during sleep (Colwell, 2010).

A third hormonal response to conserve water comes from the kidneys, which release an enzyme called *renin* into the circulation, triggering a hormonal cascade that culminates in the circulation of a hormone called **angiotensin II** (from the Greek *angeion*, "blood vessel," and the Latin *tensio*, "tension or pressure") (**FIGURE 14.12**). Angiotensin II has several water-conserving actions. In addition to constricting blood vessels to maintain blood pressure, angiotensin II triggers the release of two additional hormones that affect fluid balance: vasopressin and aldosterone (both discussed earlier). Furthermore, circulating angiotensin II directly regulates behaviour through actions at neural sites

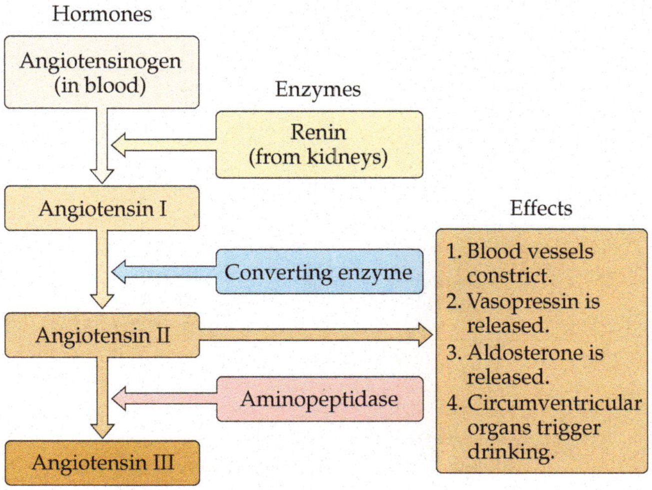

14.12 The Angiotensin Cascade Renin, released by the kidneys, catalyses the conversion of angiotensinogen (already present in blood) to angiotensin I. Angiotensin I is converted to angiotensin II (the most biologically active form of angiotensin).

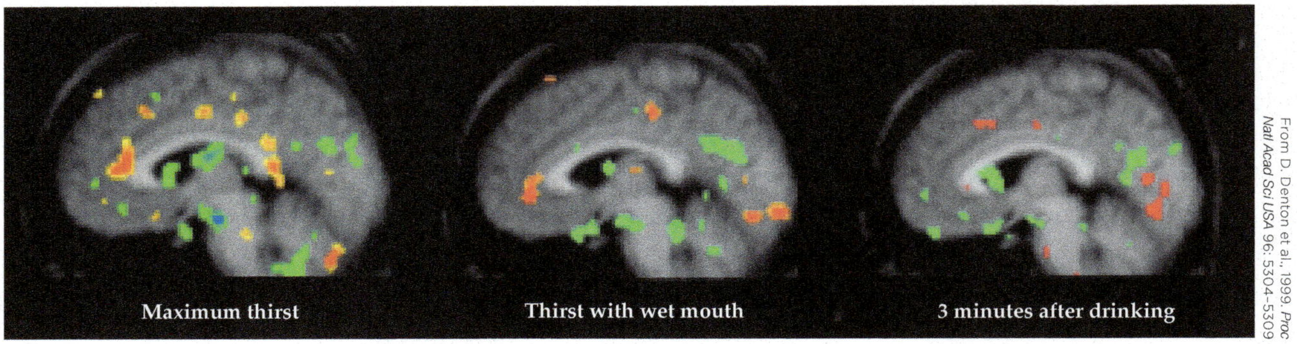

14.13 Ahhhhhhh! The experience of strong thirst, induced by injection of hypertonic saline, is associated with activity in several brain regions, especially the cingulate cortex and cerebellum (left). Wetting the mouth reduces this activation only slightly (middle), but drinking a glass of water (right) reduces activation in these brain regions dramatically.

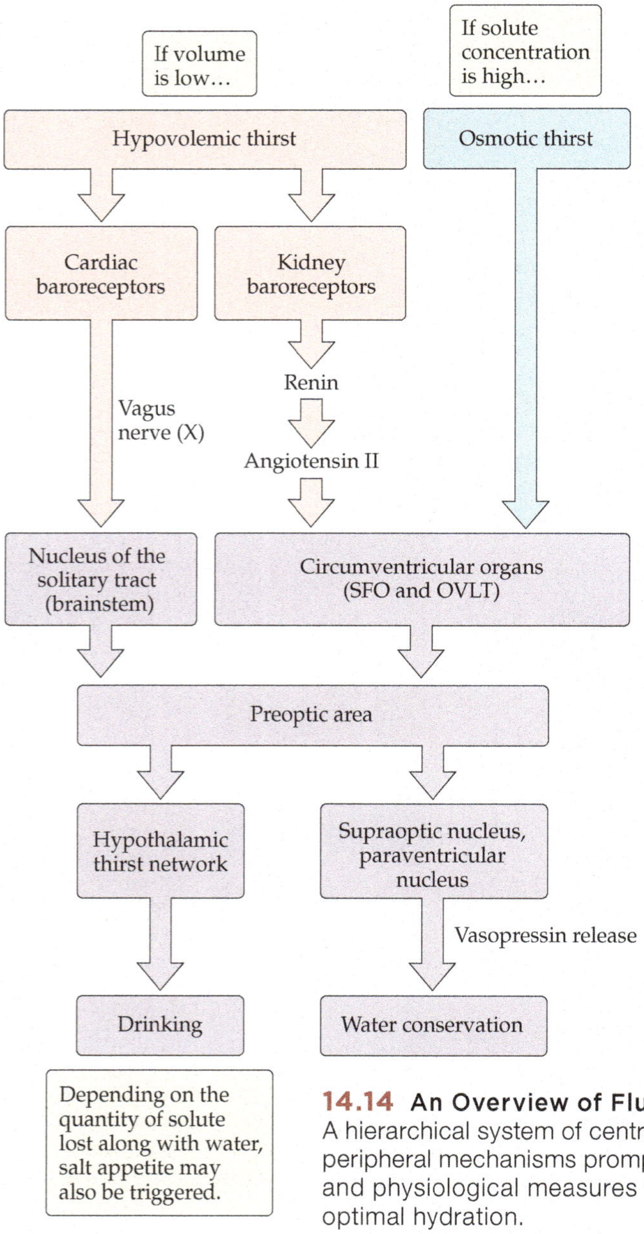

14.14 An Overview of Fluid Regulation A hierarchical system of central and peripheral mechanisms prompts behavioural and physiological measures to maintain optimal hydration.

located in the forebrain (Daniels and Marshall, 2012), especially the circumventricular organs we discussed earlier, the OVLT and SFO—so these regions monitor both osmolality and hormonal signals. Angiotensin II may also act directly on the POA to elicit drinking (Augustine et al., 2018).

We don't stop drinking just because the throat and mouth are wet

Although plausible, the most obvious explanation of why we stop drinking—that we have dampened our previously dry throat and mouth—is insufficient. Classic research showed that thirsty animals allowed to drink water but not *consume* water—because the water is diverted out of the esophagus before reaching the stomach—remain thirsty and continue drinking. However, it was also found that water is more thirst quenching if taken by mouth than if infused directly into the stomach (N. E. Miller et al., 1957). So, provided that water actually reaches the stomach, oral sensations must play *some* role in quenching thirst. Further, we stop drinking before most water has left the gastrointestinal tract and entered the extracellular compartment, but the osmolality of the ingested fluids in the stomach and intestine is closely monitored and rapidly communicated to the brain via the vagus nerve (C. A. Zimmerman et al., 2019). There, drinking is governed by neurons of the SFO, and drinking is stopped in *anticipation* of correcting the extracellular volume and/or osmolality (C. A. Zimmerman et al., 2016). Experience may teach us and other animals how to gauge accurately whether we've ingested enough to counteract our thirst (hypovolemic or osmotic). Normally, all the signals—blood volume, osmolality, moisture in the mouth, estimates of the amount of water we have ingested that's "on the way"—agree, but the cessation of one signal alone will not stop thirst; in this way, animals ensure against dehydration.

Thirst is a homeostatic signal that intrudes forcefully into consciousness, with associated strong activation of certain brain regions, particularly in the limbic system (**FIGURE 14.13**) (Denton et al., 1999). The two types of thirst (hypovolemic and osmotic), the two fluid compartments (extracellular and intracellular), and the multiple redundant methods to conserve water make for a fairly complicated system that is not yet fully understood. The current conceptualisation of this system is depicted in **FIGURE 14.14**. Our need to compensate for obligatory losses is also crucial to understanding energy regulation, our next topic.

FOOD AND ENERGY REGULATION

14.4 Nutrient Regulation Helps Prepare for Future Needs

Learning Objectives

After reading this section, you should be able to:

14.4.1 **Discuss the importance of circulating glucose and bodily mechanisms of energy storage.**

14.4.2 **Define basal metabolism and discuss its role in dieting and weight loss.**

14.4.3 **Summarise the relationship between insulin secretion and energy utilisation by cells.**

Feast or famine—these are poles of human experience. Hunger for the food that we need to build, maintain, and fuel our bodies is a compelling drive, and flavours are powerful reinforcers. The behaviours involved in obtaining and consuming food shape our daily schedules, and our mass media feed us a steady diet of information about food: crop reports, stories about famines and droughts, cooking shows, and restaurant ads.

The regulation of eating and of body energy involves numerous redundant mechanisms and complex homeostatic controls. Overall, the system for controlling food intake and energy balance is significantly more complex than those controlling thermoregulation and fluid balance. One important reason for this greater complexity is that we need food to supply not only energy, but also crucial **nutrients** (chemicals required for the effective functioning, growth, and maintenance of the body). We do not know all the nutritional requirements of the body—even for humans. Of the 20 amino acids found in our bodies, 9 are difficult or impossible for us to manufacture, so we must find these *essential amino acids* in our diet. From food we must also obtain a few fatty acids, as well as about 15 vitamins and a variety of minerals. No animal can afford to run out of energy or nutrients, so we need systems to anticipate future needs and keep a reserve on hand (but not *too* much!).

nutrient A chemical that is needed for growth, maintenance, and repair of the body but is not used as a source of energy.

glucose A sugar molecule used by the body and brain for energy.

glycogen A complex carbohydrate derived from glucose.

glycogenesis The physiological process by which glycogen is produced.

insulin A hormone, released by pancreatic beta cells, that lowers blood glucose.

glucagon A hormone, released by pancreatic alpha cells, that increases blood glucose.

glycogenolysis The conversion of glycogen back into glucose, triggered when blood concentrations of glucose drop too low.

Most of our food is used to provide us with energy

All the energy that we need to move, think, breathe, and maintain body temperature is derived in the same way: it is released when the chemical bonds of complex molecules are broken and smaller, simpler compounds form as a result. In a sense we "burn" food for energy just as a car burns gasoline. To raise body temperature, we release chemical-bond energy as heat. For other bodily processes, such as those in the brain, the energy is utilized by more-sophisticated biochemical processes.

The most immediate source of energy for the body is the complex carbohydrates in our diet that are rapidly broken down into the simple sugars that cells can use. **Glucose** is the principal sugar used by the body for energy, and it is especially important for fueling the brain. Because we need a steady supply of glucose between meals and may also experience elevated demand for fuel at other times—for example, during intense physical activity—several mechanisms have evolved to store excess fuel for later use.

For short-term storage of food energy, glucose can be converted into a more complex molecule called **glycogen** and stored as reserve fuel in several locations, most notably the liver and skeletal muscles. This process, called **glycogenesis**, is promoted by the pancreatic hormone **insulin** (see Chapter 4). A second pancreatic hormone, **glucagon**, mediates the conversion of glycogen back into glucose, a process known as **glycogenolysis** (FIGURE 14.15), which is triggered when blood concentrations of glucose drop too low.

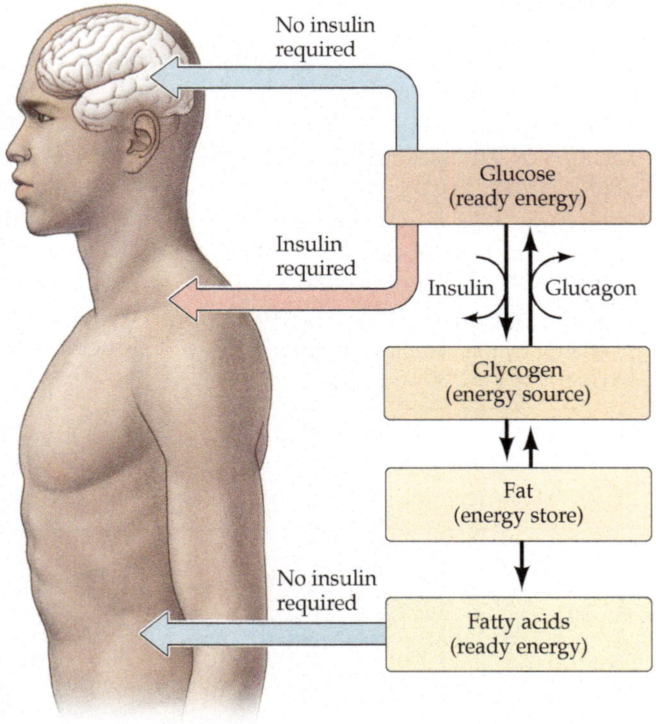

14.15 The Role of Insulin in Energy Utilisation The body can make use of either fatty acids or glucose for energy. The brain, however, can use only glucose; the brain therefore requires a constant supply of glucose, which it can use without the aid of insulin. On the other hand, the body needs insulin to use glucose, so in the absence of insulin, the body must use fatty acids for energy.

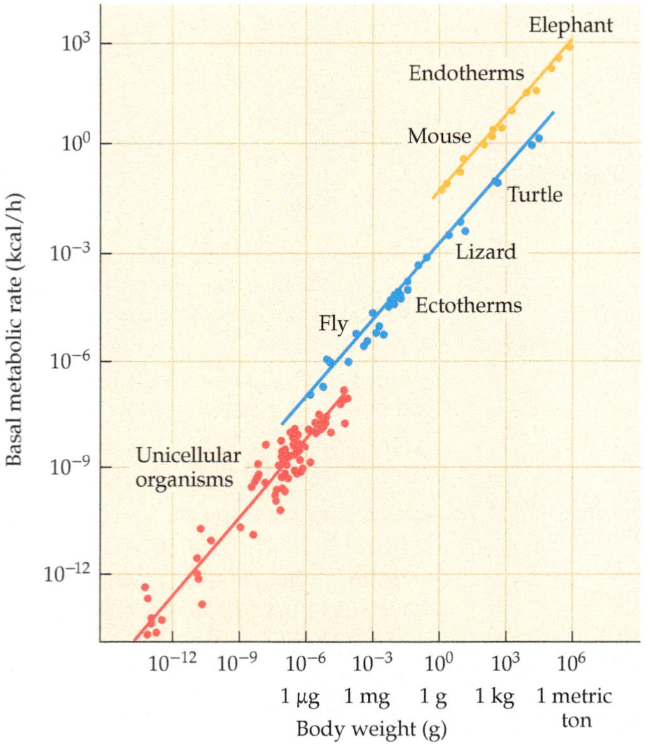

14.16 The Relation between Body Size and Metabolism Basal metabolic rate increases in a very regular, predictable fashion over a wide range of body weights. However, endotherms have a higher metabolic rate than ectotherms of a similar body weight. (After A. M. Hemmingsen, 1960. *Reports of Steno Memorial Hospital, Copenhagen* 9: 1–110. Used with permission of Novo Nordisk Inc.)

For longer-term storage, fats (or **lipids**, large molecules consisting of fatty acids and glycerol that are insoluble in water) are deposited in the fat-storing cells that form **adipose tissue**. Some stored fats are created directly from fats in our food, but others are synthesised in the body from surplus sugars and other nutrients. Under conditions of prolonged food deprivation, fat can be converted into glucose (a process called **gluconeogenesis**) and a secondary form of fuel, called **ketones**, which can similarly be utilized by the body and brain.

Our metabolic rate varies to maintain energy availability

Metabolic studies indicate that lab animals lose about 33% of the energy in food during **digestion** (through excretion of indigestible material or the digestive process itself). Another 55% of food energy in a meal is consumed by **basal metabolism**—processes such as heat production, maintenance of membrane potentials, and all the other basic life-sustaining functions of the body. The remainder, only about 12% of the total, is utilized for active behavioural processes, although this proportion is increased in more-complex environments or during intense activity.

In general, the rate of basal metabolism follows a rule, devised by Max Kleiber (1947), that relates energy expenditure to body weight:

$$\text{Kilocalories/day} = 70 \times \text{weight}^{0.75}$$

where weight is expressed in kilograms. This relationship applies across a vast range of body sizes (**FIGURE 14.16**). However, although Kleiber's equation fits nicely at the population level, it is not very accurate for *individuals* within a species, because body weight is only one factor affecting metabolic rate. For example, food-deprived people experience a significant decrease in basal metabolism. In fact, severe food restriction affects metabolic rate much more than it affects body weight, presumably reflecting the operation of an evolved homeostatic mechanism for conserving energy when food is scarce.

Because people and animals adjust their metabolism in response to under- or overnutrition, they tend to resist either losing or gaining weight (**FIGURE 14.17**). To the frustration of dieters everywhere, many studies show that a calorie-reduced diet prompts a reduction in basal metabolic rate that *prevents* weight loss (Bray, 1969; C. K. Martin et al., 2007). Along with the other *Biggest Loser* contestants, Danny C., whom we met at the outset of the chapter, has learned the hard way that our brains and bodies vigourously defend our body weight, even if we are obese. Due to a dramatic decrease in his basal metabolism following weight loss—a process called *metabolic adaptation*—Danny now needs to consume 800 *fewer* calories per day than the

lipids Large molecules (commonly called *fats*) consisting of fatty acids and glycerol that are insoluble in water.

adipose tissue Tissue made up of fat cells.

gluconeogenesis The metabolism of body fats and proteins to create glucose.

ketones A metabolic fuel source liberated by the breakdown of body fats and proteins.

digestion The process by which food is broken down to provide energy and nutrients.

basal metabolism The consumption of energy by the basic life-sustaining functions of the body.

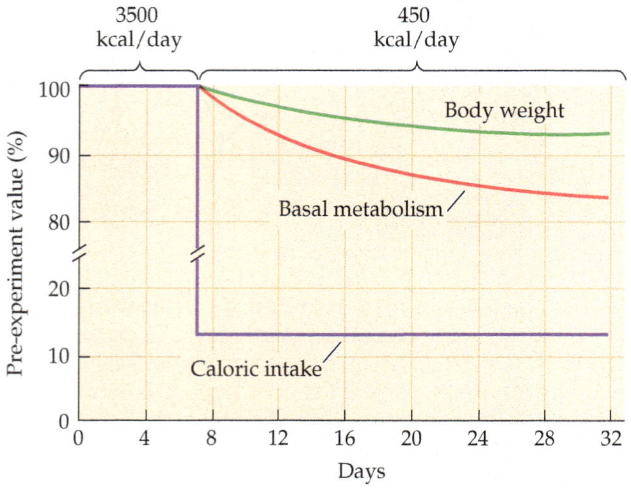

14.17 Why Losing Weight Is So Difficult After 7 days on a diet of 3500 kilocalories (kcal) per day, the intake of six obese study participants was restricted to a measly 450 kcal/day—a drop of 87%. However, basal metabolism also declined by 15%, so after 3 weeks, body weight had declined by only 6%. (After G. A. Bray, 1969. *Lancet* 2: 397–398.)

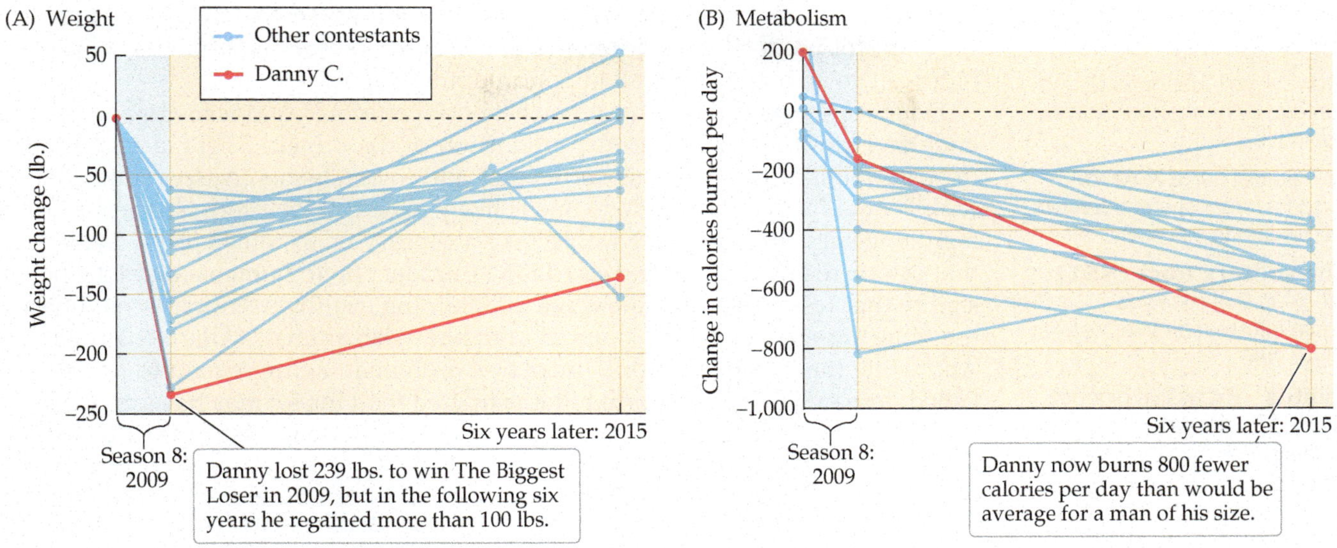

14.18 Metabolic Adaptation in the Biggest Losers (A) Of the 14 contestants from season 8 of the reality television show *The Biggest Loser*, 13 subsequently regained significant weight over the following 6 years. Four contestants weighed more than they had before the competition. (B) The contestants' dramatic weight losses prompted large compensatory decreases in metabolic rate, as their bodies sought to regain the lost weight. Alarmingly, this strong metabolic suppression still persisted 6 years after their weight losses; in fact, it even increased over time. (After E. Fothergill et al., 2016. *Obesity* 24: 1612–1619. doi:10.1002/oby.21538.)

typical man, just to maintain his current 295-pound body weight. And to make matters worse, this metabolic adaptation is annoyingly persistent; even after 6 long years, the metabolisms of the contestants remained very low (**FIGURE 14.18**), as their brains continued to try to regain the lost weight (Fothergill et al., 2016). A major concern for researchers is thus to discover a way to reset the body's set point (or set zone) for energy storage. An additional topic under debate is whether the concept of a set point, which derives originally from an engineering perspective, is compatible with modern patterns of obesity, where weight seems to drift steadily upward (M. J. Müller et al., 2018). Instead, some researchers propose that body composition reflects a "settling point," where a complex set of physiological, behavioural, and social factors combine to determine current body weight and resist radical losses or gains (Speakman et al., 2011).

Theories about the most effective ways to lose weight through dieting are endlessly popular topics in the mass media. Although it is counterintuitive, evidence has accumulated to suggest that diets low in carbohydrates, and correspondingly high in proteins and fats, are effective in helping people lose weight and also may increase serum levels of "good" cholesterol (G. D. Foster et al., 2003; Samaha et al., 2003). Recommendations about fat intake are evolving, but it remains the case that the only *certain* way to lose weight is to decrease the number of calories eaten and/or increase the calories spent in physical activity.

A surprising additional benefit of dietary restrictions may be longevity: at least in lab animals, including our primate cousins (**FIGURE 14.19**), a reduction in caloric intake to levels that are about 50–80% of what they would eat if food were always available results in a significant increase in average life expectancy (Mattison et al., 2017; Weindruch and Walford, 1988). This benefit from reducing calorie intake may be related to the decrease in basal metabolism that is induced by food restriction. Both the body and the brain give evidence of slower aging under such circumstances (M. P. Mattson and Arumugam, 2018). No one knows exactly how food deprivation enhances longevity, but research in invertebrates has identified a set of transcription factors (substances that control other genes) that appear to be involved (X. Zhou et al., 2018).

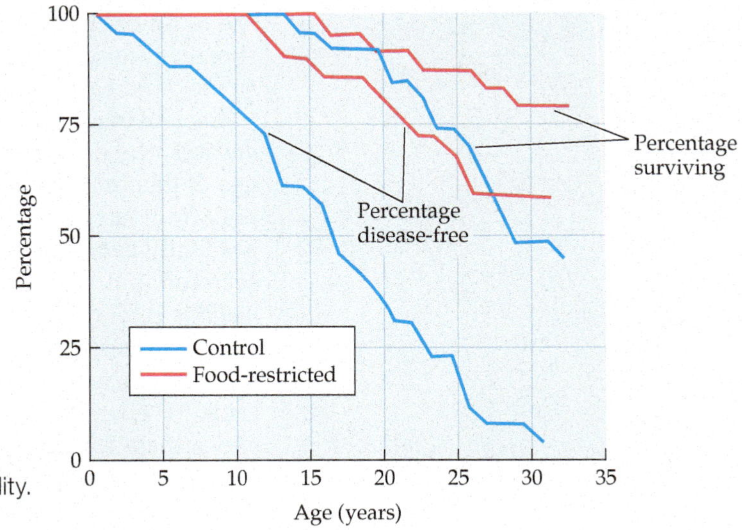

14.19 The Benefits of Caloric Restriction in Monkeys Twenty years of moderate caloric restriction in rhesus monkeys resulted in significant reductions in age-related diseases and associated mortality. (After R. J. Colman et al., 2009. *Science* 325: 201–204.)

trophic factor A substance that promotes cell growth and survival.

glucodetector A cell that detects and informs the nervous system about levels of circulating glucose.

vagus nerve Cranial nerve X, which regulates the viscera (organs) and transmits signals from the viscera to the brain.

nucleus of the solitary tract (NST) A brainstem nucleus that receives visceral and taste information via several cranial nerves.

diabetes mellitus Excessive glucose in the urine, caused by the failure of insulin to induce glucose absorption by the body.

These compounds, which are conserved across many species of animals, may in turn control production of hormones and **trophic factors** (substances that promote cell growth and survival) important for longevity.

Evidence suggests that some of the beneficial effects of caloric restriction in the body and brain are related to the production of the protein SIRT1, a marker for increased longevity in various vertebrates and invertebrates (Anson et al., 2003; Rickert et al., 2018). Although no one is likely to do full long-term studies on human longevity (because they would take a century or so to complete), evidence that caloric restriction also increases SIRT1 production in humans raises the possibility that restricting intake—while maintaining healthy nutrition—could have the same life-span benefits in humans as it has in lab animals (Allard et al., 2008). And while the effects of caloric restriction on human aging remain to be established, researchers have reported that restricted food intake may have some beneficial effects on cognitive performance in elderly participants, at least over the short term (Prehn et al., 2017; Witte et al., 2009).

Insulin is essential for obtaining, storing, and using food energy

In addition to converting surplus glucose into glycogen, as we discussed earlier, insulin is essential for the utilisation of glucose by the body. That's because *glucose transporters*—the membrane-spanning proteins that most cells use to import glucose from the blood—need insulin in order to function properly (brain cells are an important exception; they can use glucose without the aid of insulin). Each time you eat a meal, the foods are broken down and glucose is released into the bloodstream. Most of your body requires insulin to make use of that glucose, so three different, sequential mechanisms stimulate insulin release:

1. The sensory stimuli from food (sight, smell, and taste) evoke a conditioned release of insulin in anticipation of glucose arrival in the blood. This release, because it is mediated by the brain, is called the *cephalic phase* of insulin release (recall that *cephalon* means "head").
2. During the *digestive phase*, food entering the stomach and intestines causes them to release gut hormones, some of which stimulate the pancreas to release insulin. The digestive tract contains taste receptors like those found on the tongue; these provide further regulation of insulin release (Kokrashvili et al., 2009).
3. During the *absorptive phase*, special cells in the liver, called **glucodetectors**, detect the glucose entering the bloodstream and signal the pancreas to release insulin.

The newly released insulin enables the body to make use of some of the glucose immediately and prompts the conversion of extra glucose into glycogen, which is then stored in the liver and muscles. The liver communicates with the pancreas via the nervous system. Information from glucodetectors in the liver travels via the **vagus nerve** to the **nucleus of the solitary tract** (**NST**) in the brainstem and is relayed to the hypothalamus (Maniscalco and Rinaman, 2018). This system informs the brain of circulating glucose levels and contributes to hunger, as we'll discuss shortly. Efferent fibres carry signals from the brainstem back out the vagus nerve to the pancreas. These efferent fibres modulate insulin release from the pancreas.

Lack of insulin causes the disease **diabetes mellitus**. In *type I* (or *juvenile-onset*) *diabetes*, the pancreas stops producing insulin. Although the brain can still make use of glucose from the diet, the rest of the body cannot and is forced to use energy from fatty acids, with the result that glucose accumulates in the bloodstream, gradually causing severe damage to many tissues. Some of the excess glucose is secreted into the urine, making it sweet, which is how we get the name *diabetes mellitus* (literally "passing honey").

An untreated person with diabetes eats a great deal and yet loses weight because the body cannot make efficient use of the ingested food, and the reliance on fatty acids for energy causes damage to some tissues. People with diabetes also drink and urinate copiously, in the body's attempt to get rid of the excess circulating glucose. Replacement of the missing insulin (via injection) allows the glucose to be utilized. Another, more common type of diabetes mellitus, called *type II* (or *adult-onset*) *diabetes*, is primarily a consequence of reduced *sensitivity* to insulin. Particularly associated with obesity, type II diabetes often leads to further health problems.

Despite their importance, neither insulin nor glucose is the sole signal for hunger or satiety

Given the crucial role of insulin in mobilising and distributing food energy, and its fluctuations in association with feeding, it might seem an obvious candidate for signalling the brain to start or stop eating. For example, high levels of insulin, secreted because there is food in the pipeline, might signal the brain to produce the sensation of **satiety**: the feeling that an appetite has been satisfied. Conversely, low levels of insulin between meals could signal the brain to make us feel **hunger**, impelling us to find food and eat. Lowering an animal's blood insulin levels does cause it to become hungry and eat a large meal, and injecting a moderate dose of insulin causes the animal to eat much less. These results suggest that insulin can provide a satiety signal to the brain. But injecting larger doses of insulin doesn't produce fully satiated animals; instead, they become hungry again and eat a large meal! The reason for this surprising result is that the high insulin levels direct much of the blood glucose into storage, resulting in *hypoglycaemia*—reduced circulating glucose—which the brain detects via glucodetectors, leading to a hunger response.

So, is it the change in circulating glucose that signals satiety and hunger to the brain? Glucose levels are certainly an additional appetite signal, but circulating glucose can't be the sole appetite signal either, because people with untreated diabetes have very high levels of circulating glucose, yet they are constantly hungry. Somehow the brain integrates insulin and glucose levels with other sources of information to decide whether to initiate eating. This has become a central theme in research on appetite control—that the brain integrates many different signals rather than relying exclusively on any single signal to trigger hunger.

satiety The feeling that an appetite has been satisfied.

hunger The internal state of an animal seeking food.

ventromedial hypothalamus (VMH) A hypothalamic region involved in eating and sexual behaviours.

hyperphagia Excessive eating.

lateral hypothalamus (LH) A hypothalamic region involved in the control of appetite and other functions.

aphagia Refusal to eat.

14.5 A Hypothalamic Appetite Controller Integrates Multiple Hunger Signals

Learning Outcomes

After reading this section, you should be able to:

- **14.5.1** Summarise the historical evidence that hunger signals converge on the hypothalamus.
- **14.5.2** Identify and briefly describe the actions of major hormonal appetite signals.
- **14.5.3** Provide an overview of the hypothalamic appetite controller, including the organisation of its neurons, and the signals it receives from the body.
- **14.5.4** Discuss some additional, redundant appetite signals.
- **14.5.5** Define gut microbiome, and discuss some of the ways it may interact with the nervous system.

Although no single brain region has exclusive control of appetite, decades of research has confirmed that the hypothalamus is critically important to the regulation of metabolic rate, food intake, and body weight. In this classic research, scientists found that lesions in the hypothalamus of rats could induce either chronic hunger and massive weight gain, or chronic satiety and severe weight loss, depending on the location of the lesion. Bilateral lesions of the **ventromedial hypothalamus** (**VMH**) (**FIGURE 14.20A**) resulted in animals that ate to excess, a behaviour called **hyperphagia**, and became obese (Hetherington and Ranson, 1940), leading researchers to suggest that the VMH is a "satiety centre" in the brain (because the rats ceased to experience satiety once the VMH was gone). Conversely, rats with lesions of the **lateral hypothalamus** (**LH**) (**FIGURE 14.20B**) exhibited **aphagia**—a cessation of eating—and rapidly lost weight. This suggested that the LH acts as a hunger centre (because rats who lost their LH stopped acting hungry) (Anand and Brobeck, 1951). Thus, an early model of appetite control featured the VMH and LH acting in opposition to govern feeding.

It soon became evident that this dual-centre model of appetite control was too simple. For one thing, VMH-lesioned animals exhibited a period of rapid weight gain

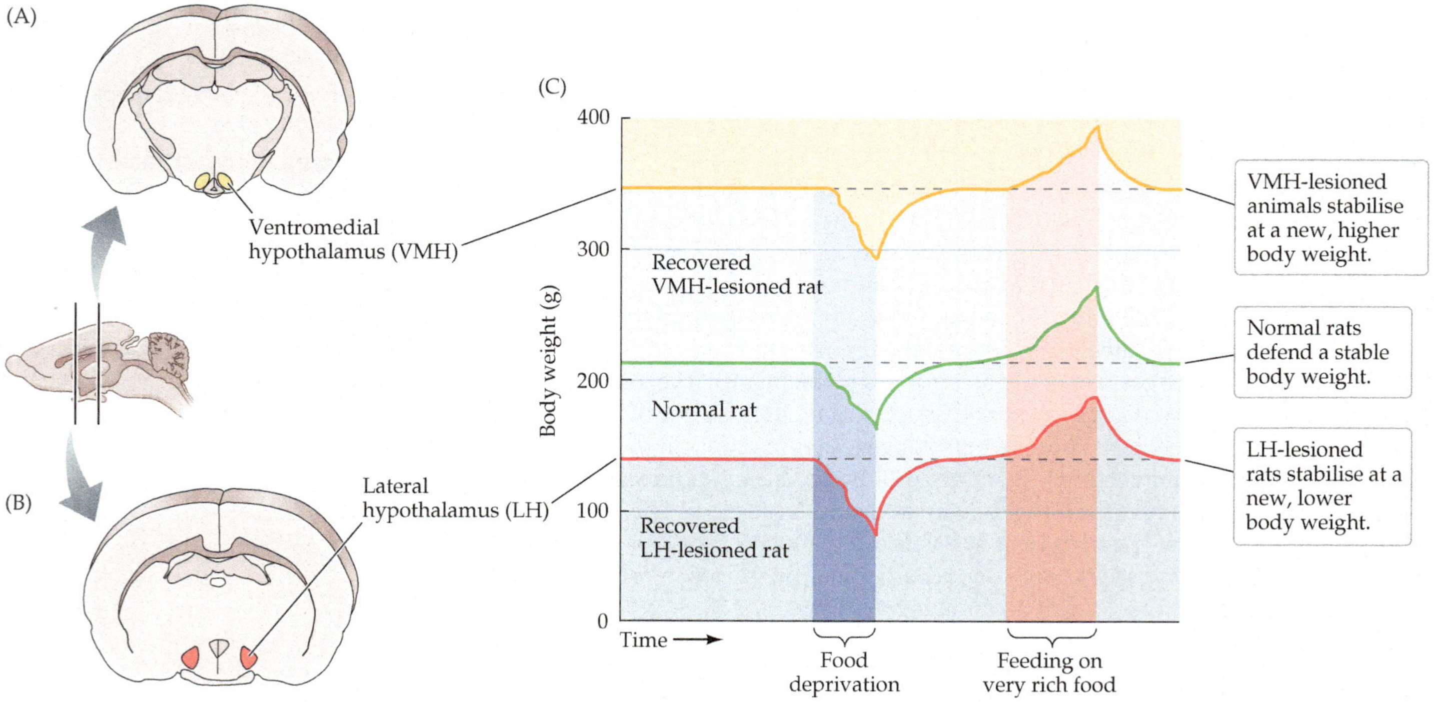

14.20 Changes in Body Weight after Hypothalamic Lesions (A) Rats in which the ventromedial hypothalamus (VMH) has been lesioned overeat and gain weight; (B) rats with lesions in the lateral hypothalamus (LH) stop eating and lose weight. (C) Both VMH- and LH-lesioned rats eventually stabilise at a new body weight, which they defend in the face of either forced feeding or food deprivation. (After R. E. Keesey and P. C. Boyle, 1973. *J Comp Physiol Psy* 84: 38–46; A. Sclafani et al., 1976. *Physiol Behav* 16: 631–640.)

but then stabilized at a new, higher body weight and would eat only to the extent needed to defend the new weight. This indicated that the VMH-lesioned rats experienced satiety (**FIGURE 14.20C**), so the VMH cannot be the sole satiety controller. Likewise, the LH can't be the sole hunger centre, because LH-lesioned rats kept alive at first with a feeding tube soon resumed spontaneous eating sufficient to maintain a new, lower body weight (Keesey, 1980). As with the VMH-lesioned animals, LH-lesioned animals that were later forced to gain or lose weight would swiftly return to their new set point for body weight once they were allowed to eat at will. Researchers thus realised that the hypothalamic system controlling feeding must involve multiple components that coordinate to establish a set point for metabolic fuels, monitor energy balance in the body, and trigger behavioural responses to meet the established energy goals, with a collective effect on body weight.

By demonstrating that the hypothalamus contains distinct components of an appetite control network, the initial research on hunger and satiety provided a framework for subsequent studies. For example, fMRI studies show that increasing circulating glucose after a period of fasting produces large changes in the activity of the human hypothalamus (**FIGURE 14.21**) (Y. Liu et al., 2000), probably acting via hypothalamic glucodetector neurons that directly monitor blood levels of glucose (Parton et al., 2007). Today it is clear that the hypothalamic control of feeding is quite complicated and, like other homeostatic systems, exhibits redundancy as a safety measure.

As we'll see next, researchers have uncovered many of the details of the hypothalamic appetite control network and its integration of multiple signals from sites throughout the body.

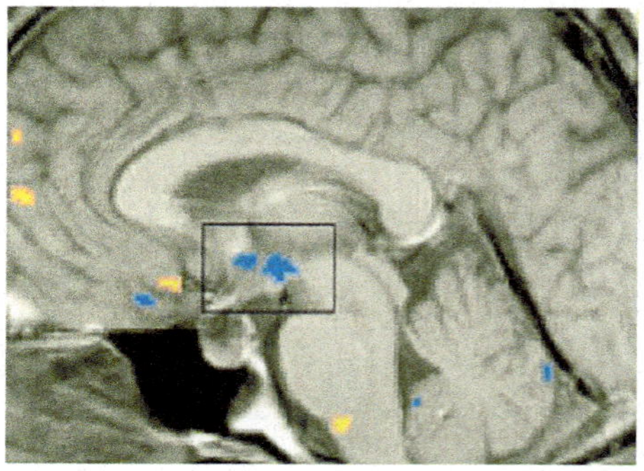

From Y. Liu et al., 2000. *Nature* 405: 1058–1062

14.21 Sweet Spot Following glucose ingestion, changes in activity in the hypothalamus (inside the black rectangle) are evident in this midsagittal fMRI image. Blue indicates a significant decrease in activity; yellow indicates a significant increase.

TABLE 14.2 Peripheral Hormone Signals for Body Weight Regulation

Hormone	Primary source	Signal
Insulin	Beta islet cells of the pancreas	Provides the arcuate appetite controller with information about blood glucose level
Leptin	Fat cells	Signals current long-term energy stores (in fat)
Ghrelin	Cells of the stomach and duodenum	Provides a "fasting" signal, indicating to the arcuate system that the digestive system is empty, prompting the arcuate system to increase appetite
$PYY_{3\text{-}36}$	Cells of the ileum (small intestine) and colon	Provides a rapid signal that food has been consumed, prompting the arcuate system to suppress appetite
GLP-1	Cells of the ileum and colon	Counters ghrelin and suppresses appetite via actions on POMC neurons and the nucleus of the solitary tract
CCK	Cells of the duodenum	Suppresses appetite via direct action on the vagus nerve (cranial nerve X)

Multiple peripheral signals are integrated by a hypothalamic appetite network

A spate of discoveries indicates that the **arcuate nucleus** of the hypothalamus contains a highly specialised appetite controller that is governed by circulating levels of a variety of hormones (**TABLE 14.2**). We have already discussed how the pancreatic hormone insulin signals the state of glucose circulating in the blood. Other information about short- and long-term energy balance comes in the form of hormonal secretions from elsewhere in the body. These hormones include leptin, ghrelin, and two hormones with ponderous names—peptide $YY_{3\text{-}36}$ ($PYY_{3\text{-}36}$) and glucagon-like peptide-1 (GLP-1)—that can cross the blood-brain barrier and interact with neurons in regions crucial for regulating food intake. We will discuss each in turn and then look at the possible organisation of the hypothalamic appetite controller.

LEPTIN Mice that receive two copies of the gene called *obese* (abbreviated *ob*) regulate their body weight at a high level (**FIGURE 14.22**), as you might have guessed from the gene's name. These mice have larger and more numerous fat cells than their heterozygous littermates (*ob*/+; the plus sign indicates the wild-type, normal allele).

The *ob/ob* mice have defective genes for the peptide **leptin** (from the Greek *leptos*, "thin"). Fat cells produce leptin and then secrete the protein into the bloodstream (Y. Zhang et al., 1994). A family of leptin receptors (LepRa through LepRf) have been identified in locations throughout the brain, including the cortex and several nuclei of a hypothalamic appetite network (Håkansson et al., 1998; N. Wada et al., 2014) we will discuss shortly. Mutations that result in the production

arcuate nucleus An arc-shaped hypothalamic nucleus implicated in appetite control.

leptin A peptide hormone released by fat cells.

14.22 Inherited Obesity Both of these mice have two copies of the *obese* gene, which impairs the production of leptin by fat cells. The mouse on the left weighs about 67 grams (g); a normal (wild-type) mouse at this age weighs about 25 g. The mouse on the right has been treated with leptin and weighs about 35 g.

ghrelin A peptide gut hormone believed to act on the hypothalamus to increase hunger.

PYY_{3-36} A peptide gut hormone believed to act on the hypothalamus to suppress appetite.

GLP-1 A peptide gut hormone believed to act on the hypothalamus and brainstem to suppress appetite.

POMC neurons Neurons involved in the hypothalamic appetite control system, so named because they produce pro-opiomelanocortin (POMC) along with cocaine- and amphetamine-related transcript (CART).

NPY neurons Neurons involved in the hypothalamic appetite control system, so named because they produce neuropeptide Y (NPY) along with agouti-related peptide (AgRP).

neuropeptide Y (NPY) A peptide neurotransmitter that may carry some of the signals for feeding.

of defective leptin receptors likewise result in morbid obesity in lab animals (Roh et al., 2018; Zabeau et al., 2019) and humans (Niazi et al., 2018).

Thus, the brain seems to monitor circulating leptin levels to measure and regulate the body's energy reserves in the form of fat. Defects in leptin production or leptin sensitivity cause a false underreporting of body fat and lead to overeating, especially of high-fat or sugary foods.

GHRELIN Released into the bloodstream by endocrine cells of the stomach (Kojima et al., 1999), **ghrelin** was named in recognition of its effects on growth hormone secretion (***GH rel***easing). But we now know that ghrelin is a powerful appetite *stimulant* (Nakazato et al., 2001). Circulating levels of ghrelin rise during fasting and immediately drop after a meal is eaten. Treating either rats or humans with exogenous ghrelin produces a rapid and large increase in appetite (Wren et al., 2000, 2001). Modifications of brain ghrelin activity in hamsters directly regulate foraging behaviour, as well as the hoarding and consuming of food, confirming that ghrelin plays an important role in both seeking and consuming food (Thomas et al., 2016).

Curiously, obese study participants reportedly have lower baseline levels of ghrelin than do lean participants prior to eating, but following a meal, their circulating levels of ghrelin do not drop (their leptin levels remain high, too). So one mechanism of obesity may involve a ghrelin system that is unresponsive to feeding and thus always slightly elevated, prompting continual hunger (English et al., 2002).

PYY_{3-36} Secreted into the circulation by cells of the small and large intestines, the small peptide **PYY_{3-36}** is at a low level in the blood prior to eating, but that level rises rapidly on ingestion of a meal. Systemic injections of PYY_{3-36} curb appetite in both rats and humans, as do injections directly into the arcuate nucleus of the hypothalamus of rats (Baynes et al., 2006; Chelikani et al., 2005). Interestingly, lower-than-average levels of circulating PYY_{3-36} are associated with a tendency toward obesity in mice and humans, and postmeal increases in this peptide have been closely linked to feelings of satiety in normal-weight people (Karra et al., 2009). It therefore appears that PYY_{3-36} may act in opposition to ghrelin, providing a potent appetite-*suppressing* stimulus to the hypothalamus (Zanchi et al., 2017).

GLP-1 Released from the intestine by cells that are in contact with the contents of the gut, **GLP-1** secretion rapidly increases following a meal—especially if high in fats and carbohydrates—stimulated at first by autonomic neural signals and then by the presence of nutrients in the intestinal tract (Baggio and Drucker, 2007; E. W. L. Sun et al., 2019). Because GLP-1 acts on the pancreas and periphery to potently enhance the release and utilisation of insulin, it is of interest to diabetes researchers searching for treatments for type II diabetes, but receptors for GLP-1 are also found in several brain regions implicated in appetite and feeding (Burmeister et al., 2017; Knudsen et al., 2016). Activation of the brain's GLP-1 mechanisms has a number of feeding-related effects, including decreased appetite and eating, as well as changes in the system that signals the rewarding aspects of food (R. Sekar et al., 2017; Y. Yang et al., 2017). In addition, GLP-1 activity directly blocks the effects of ghrelin on metabolism and appetite (Abtahi et al., 2019). The discoveries of PYY_{3-36}, ghrelin, and GLP-1 have provided important clues about the appetite control mystery, and these hormones converge on the arcuate nucleus of the hypothalamus, which we describe next.

THE ARCUATE APPETITE SYSTEM A simplified (yes, really) view of the organisation of the appetite control neurons in the arcuate nucleus is provided in **FIGURE 14.23**. Within the arcuate nucleus, the appetite system relies on two sets of neurons with opposing effects. **POMC neurons** (so named because they produce the peptides *pro-opiomelanocortin* and *cocaine- and amphetamine-regulated transcript*) act as satiety neurons when activated, inhibiting appetite and increasing metabolism. However, neighbouring **NPY neurons,** which produce **neuropeptide Y (NPY)** and *agouti-related peptide (AgRP)*, act as hunger neurons when they are activated, stimulating appetite directly and also inhibiting the POMC neurons (thereby blocking satiety signals) and

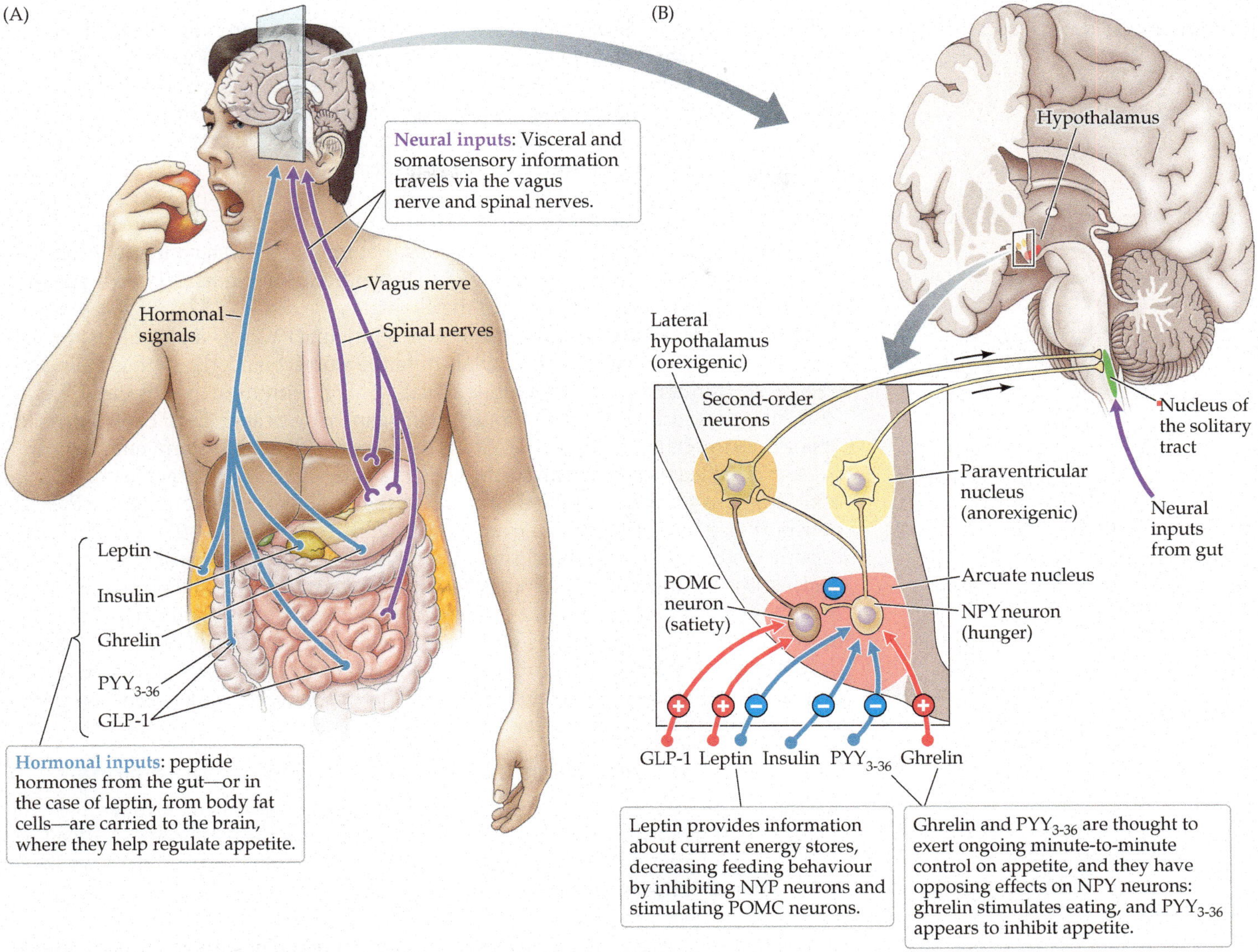

14.23 Integration of Appetite Signals in the Hypothalamus (A) The brain integrates a number of peripheral signals to determine appetite. Among these are numerous gut peptides secreted into the bloodstream, especially (1) leptin, secreted by fat cells; (2) insulin, secreted by the pancreas; (3) ghrelin, secreted by the stomach; (4) $PYY_{3\text{-}36}$; and (5) GLP-1, both of which are secreted by the intestines. In addition, visceral and somatosensory information is transmitted via spinal nerves and the vagus nerve. (B) Two types of neurons in the arcuate nucleus are sensitive to peptides from the periphery: POMC neurons signal a decrease in food intake; NPY neurons promote increased feeding. Both types of arcuate neurons exert their effects via second-order neurons in the paraventricular nucleus and lateral hypothalamus. POMC neurons signal satiety by releasing α-melanocyte-stimulating hormone (α-MSH). NPY neurons stimulate appetite through the release of NPY but also by releasing AgRP, which directly competes for the melanocortin receptors, reducing the effectiveness of α-MSH in suppressing appetite. (Note that some orexigenic neurons are also found in the PVN but not shown.)

Go to Activity 14.1
An Appetite Controller in the Hypothalamus

reducing metabolism. The actions of the NPY neurons thus promote eating and weight gain. Projections from the POMC neurons and NPY neurons leave the arcuate and make contact with neurons in second-order hypothalamic appetite sites (see Figure 14.23B). It is through these projections that the arcuate system ultimately modulates food intake, and we will turn to that system in the next section. But first, let's consider how the peripheral hormones interact with the appetite controller in the arcuate.

Because it is made by fat cells, leptin (and to a lesser extent, insulin) conveys information about the body's energy reserves. Both types of appetite neurons in the arcuate system have leptin receptors, but leptin affects them in different ways. High circulating levels of leptin *activate* the appetite-suppressing POMC neurons but *inhibit* the appetite-increasing NPY neurons, so in both systems leptin is working to suppress hunger, and selective deletion of leptin receptors in NPY neurons causes rapid obesity in mice (J. Xu et al., 2018). In keeping with its role as an indicator of the body's current composition, leptin seems to mostly affect the appetite system over the longer term, perhaps by promoting the remodelling of neurons in the arcuate nucleus (Bouret et al., 2004). To add to the difficulties of the *Biggest Loser* contestants, long-term studies have found that their drastically lowered metabolic rates are accompanied by equally striking decreases in circulating leptin levels (Knuth et al., 2014). So their bodies keep telling the arcuate that fat reserves are low, even when they've regained a lot of weight. For reasons that remain a mystery, leptin levels are not suppressed like this when the weight is shed because of gastric bypass surgery (a topic we explore toward the end of the chapter).

In contrast to leptin, the trio of hormones ghrelin, PYY_{3-36}, and GLP-1 provide more-rapid, hour-to-hour hunger signals from the stomach and gut. Both ghrelin and PYY_{3-36} act primarily on the appetite-stimulating NPY neurons of the arcuate. Ghrelin stimulates these cells, leading to a corresponding increase in appetite. PYY_{3-36} works in opposition, inhibiting the same NPY cells to *reduce* appetite. Short-term effects on appetite exerted via the NPY neurons thus reflect a balance between ghrelin and PYY_{3-36} concentrations in circulation. In contrast, GLP-1 acts on the appetite-reducing POMC cells of the arcuate. This action, coupled with GLP-1's numerous other effects in the hypothalamus and beyond, results in a reduction in appetite and food intake (R. Sekar et al., 2017), in opposition to the appetite-enhancing actions of the NPY system.

Although much progress has been made in understanding the control of appetite, the story is not yet complete. Further research will clarify the role of additional satiety signals from the gut (Zanchi et al., 2017), along with new discoveries about the hypothalamic appetite controller. Ultimately, the arcuate appetite controller exerts its effects on feeding behaviour via other brain sites, which we discuss next.

Second-order hypothalamic neurons integrate appetite signals

Having identified the main components of the arcuate appetite controller, we can turn to the functional connections of these cells to "downstream" sites involved in feeding. Two hypothalamic sites appear to be primary targets of projections from the arcuate. **Orexigenic neurons** (from the Greek *orexis*, "appetite," and *genein*, "to produce"), located principally in the lateral hypothalamus, act to increase appetite and food intake. In contrast, **anorexigenic neurons** of the **paraventricular nucleus (PVN)** act to decrease appetite and feeding (see Figure 14.23B; note that some orexigenic neurons are also found in the PVN but not shown).

One set of projections from the appetite-suppressing POMC neurons of the arcuate terminates in the PVN. Here they release **α-melanocyte-stimulating hormone (α-MSH)**, a peptide hormone belonging to a small family of substances, called **melanocortins**, that are derived from pro-opiomelanocortin. Acting via specific **melanocortin type 4 receptors (MC4Rs)** located on the PVN neurons, α-MSH activates the PVN's appetite-suppressant activity, resulting in a net decrease in feeding (Garfield et al., 2015; Krashes et al., 2016). Other POMC projections inhibit the orexigenic neurons of the LH.

The NPY neurons that are essential for increases in feeding (Gropp et al., 2005) likewise exert their effects through the PVN and the LH (see Figure 14.23B) in a complex manner. By inhibiting anorexigenic PVN neurons, the NPY-releasing neurons promote increased appetite. Furthermore, AgRP released by these neurons into the LH competes with α-MSH (from POMC neurons) for receptors. By blocking the α-MSH signal, AgRP thus counters the appetite-suppressing influence from the POMC neurons that we just described and thus again acts to *increase* feeding behaviour, via the LH.

The net result of all this is a constant balancing act between the appetite-stimulating effects of the NPY neurons and the appetite-suppressing effects of the POMC neurons, spread across both the PVN and the LH. It seems that multiple parallel signalling systems influence this balance: for example, increased serotonergic activity within the PVN reverses the hunger-stimulating effects of ghrelin (Currie et al., 2010).

orexigenic neurons Neurons of the hypothalamic appetite system that promote feeding behaviour.

anorexigenic neurons Neurons of the hypothalamic appetite system that inhibit feeding behaviour.

paraventricular nucleus (PVN) A nucleus of the hypothalamus implicated in the release of oxytocin and vasopressin and in the control of feeding and other behaviours.

α-melanocyte-stimulating hormone (α-MSH) A peptide that binds the melanocortin receptor.

melanocortins One category of endogenous opioid peptides.

melanocortin type 4 receptor (MC4R) A specific subtype of melanocortin receptor.

14.24 Neuropeptides That Induce Hunger? In situ hybridisation indicates that orexin mRNA (white spots) is made only in the lateral hypothalamus. Infusion of orexin (also known as *hypocretin*) into this region causes rats to eat more food.

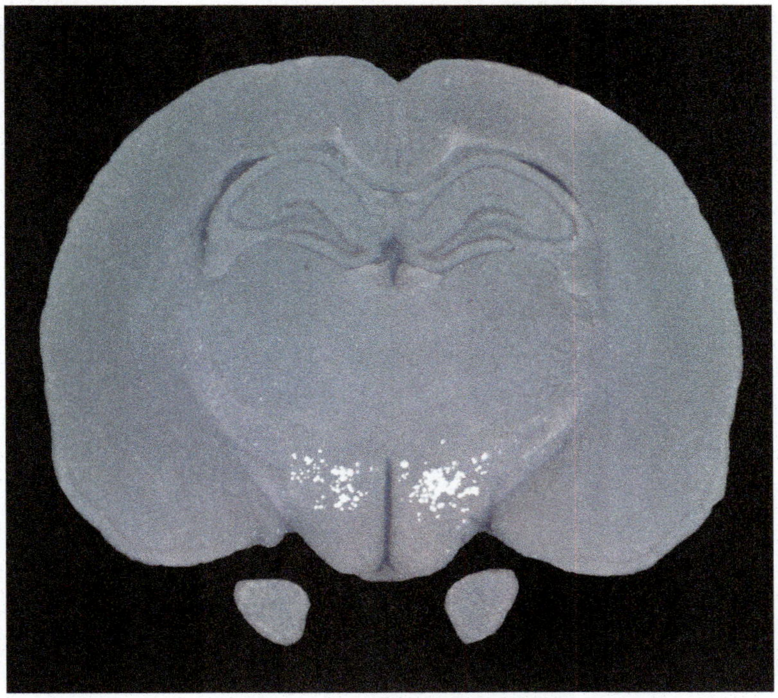
Courtesy of Masashi Yanagisawa

The peptide **orexin** (from the Greek *oregein*, "to desire") (**FIGURE 14.24**), which is produced by neurons in the lateral hypothalamus, appears to participate in the subsequent control of feeding. Direct injection of orexin into the hypothalamus of rats causes up to a sixfold increase in feeding, and evidence is emerging that hypothalamic orexin is regulated to some extent by circulating leptin (Ohno and Sakurai, 2008). Orexins (which are also known as *hypocretins*) are also involved in the sleep disorder narcolepsy (see Chapter 14), but whether that function relates to hunger is unknown.

Other systems also play a role in hunger and satiety

Appetite signals from the hypothalamus converge on the nucleus of the solitary tract (NST) in the brainstem (see Figure 14.23B). The NST can be viewed as part of a common pathway for feeding behaviour, and it receives and integrates appetite signals from a variety of sources in addition to the hypothalamus. For example, the sensation of hunger is affected by a wide variety of peripheral sensory inputs, such as oral stimulation and the feeling of stomach distension, transmitted via spinal and cranial nerves. Information about nutrient levels is conveyed directly from the body to the NST via the vagus nerve (Maniscalco and Rinaman, 2018). For example, various peptides, including **cholecystokinin** (**CCK**) and GLP-1, released by the gut after feeding, act directly on receptors of the vagus nerve to affect appetite (Alhadef et al., 2017; Huston et al., 2019).

In keeping with one of the central themes of this chapter—the concept of multiple redundancy in key systems—it will come as no surprise to you that a variety of other brain locations also participate in controlling feeding behaviour, either directly or through indirect effects on related processes. For example, as you might expect, the brain's reward system appears to be intimately involved with food intake. Activity of a circuit including the amygdala and the mesolimbic dopamine-mediated reward system (see Chapter 3) is intimately coordinated with the activity of the lateral hypothalamus, and it is hypothesised to mediate pleasurable aspects of feeding (Stuber and Wise, 2016).

The **endocannabinoid** system likewise has a potent influence on appetite and feeding. Endocannabinoids, such as anandamide, are endogenous substances that act much like the active ingredient in cannabis and, like cannabis, can potently stimulate hunger. Acting both in the brain and in the periphery, endocannabinoids may stimulate feeding by affecting the mesolimbic dopamine reward system. However, injection of anandamide directly into the PVN also stimulates eating (C. D. Chapman et al., 2012), confirming that endocannabinoids act directly on hypothalamic appetite mechanisms while also inhibiting satiety signals from the gut (Di Marzo and Matias, 2005). Paradoxically, one of the actions of cannabinoids in the hypothalamus is a *stimulation* of POMC neurons, despite the fact that POMC neurons normally promote satiety.

Hypothalamic feeding control must be strongly influenced by inputs from higher brain centres, but little is known about these mechanisms. During development, for example, our feeding patterns are increasingly influenced by social factors such as parental and peer group pressures (Birch et al., 2003). Understanding the nature of cortical influences on feeding mechanisms is a major challenge

orexins Also called *hypocretins*. Neuropeptides produced in the hypothalamus that are involved in switching between sleep states, in narcolepsy, and in the control of appetite.

cholecystokinin (CCK) A peptide hormone, released by the gut after ingestion of food high in protein and/or fat, that also serves as a signalling molecule in the brain.

endocannabinoid An endogenous ligand of cannabinoid receptors, thus an analog of cannabis that is produced by the brain.

TABLE 14.3 Hormones and Neurotransmitters Involved in Regulating Feeding and Body Weight

Increased feeding and weight gain	Decreased feeding and weight loss
Agouti-related peptide (AgRP)	α-Melanocyte-stimulating hormone (α-MSH)
β-Endorphin	Brain-derived neurotrophic factor (BDNF)
Corticosterone/cortisol	Cholecystokinin (CCK)
Dopamine	Cocaine- and amphetamine-regulated transcript (CART)
Dynorphin	Corticotropin-releasing hormone (CRH)
Endocannabinoids	Estrogens
Ghrelin	Glucagon-like peptide-1 (GLP-1)
Melanin-concentrating hormone	Histamine
Neuropeptide Y	Insulin
Norepinephrine	Leptin
Orexin/hypocretin	Nesfatin-1
Testosterone	Oxyntomodulin
	$PYY_{3\text{-}36}$
	Serotonin

Note: Many of the members of this partial list are targets for anti-obesity drug development.

for the future. The list of participants in appetite regulation is long and growing longer each day (**TABLE 14.3**), revealing overlapping and complex controls with a high degree of redundancy, as befits a behavioural function of such critical importance to health and survival.

The Cutting Edge

Friends with Benefits

The last decade has seen rapid progress in understanding how hormone signals secreted by the gut are integrated in the hypothalamus to control appetite and energy balance, as we have seen. But recent research is finding that the *contents* of the gut may play a role as well.

Most people know that the gut is normally inhabited by helpful bacteria, but the extent of that occupation may surprise you. You probably contain in the range of 2.5–5 *pounds* of gut microbes: *trillions* of individual bacteria belonging to dozens (perhaps hundreds) of different species of bacteria, fungi, and viruses, making up more than half the contents of your large intestine (Guarner and Malagelada, 2003). Put another way, you have more gut microbes than body cells, and they weigh more than your brain. This huge population, known as the **gut microbiota** or, collectively, as the **microbiome**, normally provides a variety of beneficial actions in return for their comfortable lodgings (the gut microbiota are sometimes referred to as *normal flora* in medicine, but this is a misnomer because *flora* refers to plants).

Each of us possesses a distinct microbial **enterotype**—a personal combination of different species of microbiota—that researchers increasingly believe to be in extensive two-way communication with your brain via the vagus nerve and chemical signals (Bonaz et al., 2018; Cussotto et al., 2018). Your enterotype reflects the history of your gut, so substantial changes in your diet, or the use of antibiotics to treat infections, can potentially alter the composition of the enterotype (David et al., 2014), with uncertain consequences. Preliminary evidence has tentatively linked the microbiome enterotype to such diverse domains as mood, stress, social behaviour, and cognitive functioning (Sylvia and Demas, 2018; Tetel et al., 2018) and also various neurological conditions, including autism, schizophrenia, bipolar disorder, and Parkinson's disease (Fung et al., 2017;

gut microbiota The microorganisms that normally inhabit the digestive tract.

microbiome The collective term for a population of microbes.

enterotype Each individual's personal composition of gut microbiota.

Tremlett et al., 2017). But one of the areas where an effect of changes in the gut microbiota may be most evident is obesity.

Feeding antibiotics to young mice, even at relatively low doses, changes gut microbiota and circulating hormones, leading to weight gain (I. Cho et al., 2012). Similarly, two studies looking at almost 40,000 babies have found that human infants given antibiotics in their first 6 months were statistically more likely to be overweight at age 7 (Ajslev et al., 2011; Trasande et al., 2013). It remains to be seen how much adult obesity is accounted for by long-lasting changes to our enterotypes, but it is at least possible that early exposure to antibiotics—or even chlorinated drinking water (the whole point of adding chlorine is to kill bacteria)—is making some of us fat (L. M. Cox and Blaser, 2015). What can we do about it?

Scientists hope to turn our new knowledge of the connection between the gut and brain into novel treatments, such as drugs based on metabolites secreted by specific gut bacteria (Suez and Elinav, 2017). Some researchers have also been evaluating the benefits of a gross-sounding procedure: **fecal transplantation**. Yes, it is just what it sounds like: feces are collected from carefully screened, uh, "donors," processed to create a liquid suspension, and then passed through a catheter into the colon of the recipient (**FIGURE 14.25**), where the donor's healthy enterotype establishes itself. Just one transplant can cure dangerous intestinal infections with bacteria such as *Clostridium difficile*, but intriguingly, fecal transplantation also reportedly improved metabolic function in a sample of obese men (Vrieze et al., 2012). So, although research on the topic is just beginning, perhaps our tiny passengers can someday be coaxed to pay their way by supplying benefits like weight loss and better brain health. With each new discovery, we draw nearer to finally developing safe and effective treatments for obesity and eating disorders, as we discuss next. ■

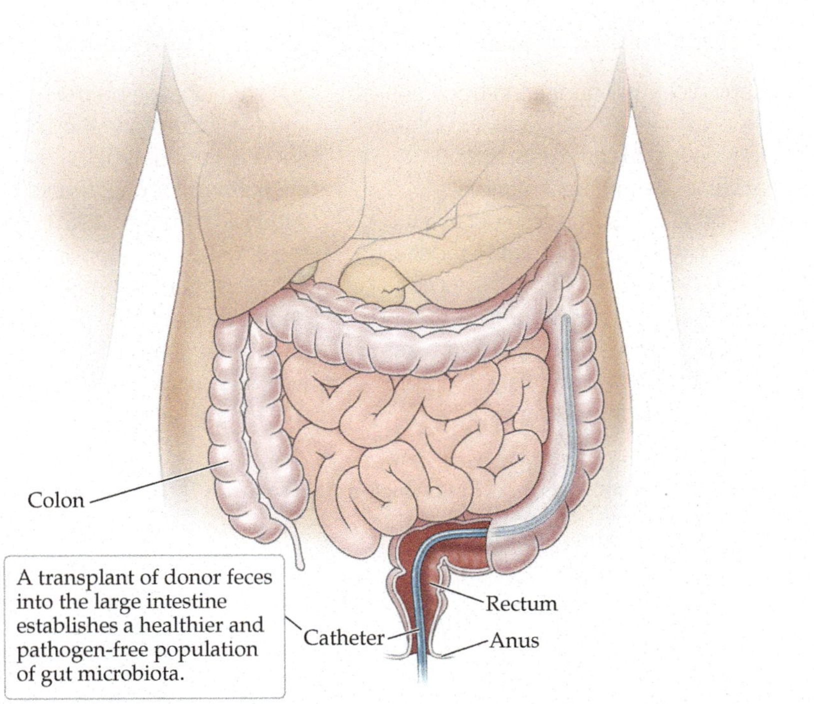

14.25 Fecal Transplantation Initial results suggest that augmenting abnormal gut microbiota with a sample from a healthy donor may aid in diverse problems, including severe infection, Parkinson's disease, and obesity.

fecal transplantation A medical procedure in which gut microbiota, via fecal matter, are transferred from a donor to a host.

14.6 Obesity and Eating Disorders Are Difficult to Treat

Learning Objectives

After reading this section, you should be able to:

14.6.1 **Define *overweight* and *obesity*, with reference to BMI.**

14.6.2 **Describe pharmacological approaches to controlling appetite and promoting weight loss.**

14.6.3 **Sketch or describe the most common types of surgical interventions in obesity.**

14.6.4 **Describe symptoms, and possible causes and treatments, in anorexia nervosa, bulimia, and binge eating disorder.**

Unfortunately, effective interventions for combating obesity have been elusive. Our multiple redundant systems for appetite and energy management work all too well in fighting against weight loss (**BOX 14.2** on the next page). Like it or not, our evolutionary history has optimised our bodies for obtaining and storing energy, and protecting against accumulating *too much* energy was not a concern for our distant ancestors.

Obesity strategies target multiple dimensions of energy balance

The tendency to accumulate excess energy is exacerbated by our ever more sedentary lifestyles. Obesity is certainly a major health problem: almost 65% of the adults in the

BOX 14.2 Body Fat Stores Are Tightly Regulated, Even after Surgical Removal of Fat

As any dieter will attest, the body seems to know how much it wants to weigh, and it defies our efforts to change that value. As in other mammals, our homeostatic mechanisms defend a set value for weight. Perhaps the most striking demonstration of this phenomenon is exhibited by golden-mantled ground squirrels, which show an extreme seasonal variation in body weight, greatly fattening up in the spring.

When these squirrels are brought into the laboratory, they continue to show an annual rhythm in body weight, even when food is always available (Figure A) (Zucker, 1988). Force-feeding the squirrels or depriving them of food will cause a temporary increase or decrease in body weight, but as soon as food access returns to normal, body weight returns to the value that is normal for the season.

Even more impressive is the fact that if body fat is surgically removed, the animals will eat until they regain—with remarkable precision—the amount of fat that would be normal for that point in the season (Figure B) (Dark et al., 1984). Needless to say, these results are not encouraging to humans considering liposuction. Usually the fat simply returns a few months after the procedure (Seretis et al., 2015).

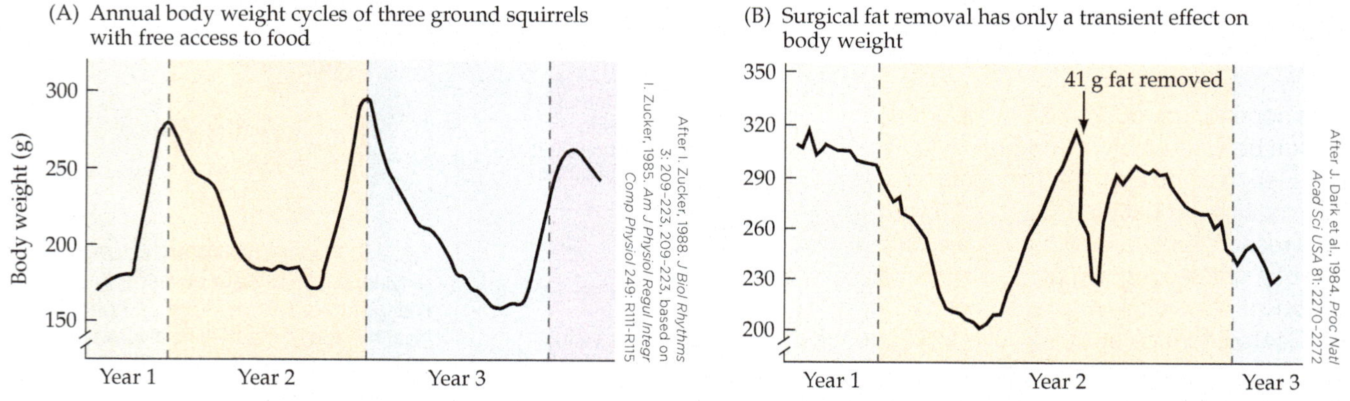

United States are overweight, and one in three qualify as obese (**FIGURE 14.26**). These categories are based on body mass index (BMI), which is defined in **TABLE 14.4**. The higher incidence of cardiovascular disease, diabetes, and other disorders that accompany obesity will be an increasingly heavy burden on health care services in the future. Parental obesity may program metabolic disadvantages in offspring via **epigenetic transmission** (Ng et al., 2010), so the problem may worsen in future generations.

epigenetic transmission The passage of epigenetic modifications of a gene from one generation to another.

In Lewis Carroll's *Alice's Adventures in Wonderland,* Alice quaffs the contents of a small bottle in order to shrink. The quest for a real-life shrinking potion—but one that makes you thin rather than short—is the subject of intense scientific activity, and several major strategies or targets are emerging.

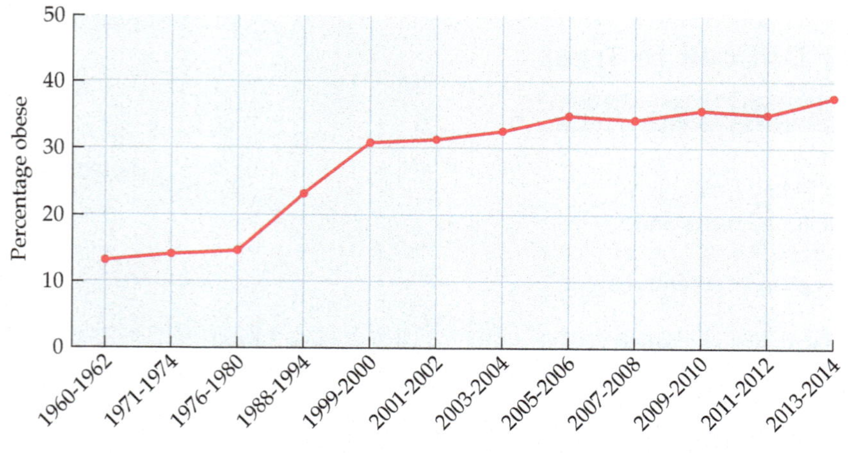

14.26 An Epidemic of Obesity The prevalence of obesity changed little before 1980. At about that time, medical recommendations to avoid dietary fats became widespread. Researchers now believe that the subsequent replacement of fat calories with carbohydrates, especially sugar, is the leading cause of obesity. Since then, in the span of just 35 years, the prevalence of obesity in the United States has tripled. Data are from the CDC's National Health and Nutrition Examination Survey (NHANES). (After C. D. Fryar et al., 2016. *Natl Ctr Health Stat. July 2016.*)

APPETITE CONTROL Hopes are high that drugs designed to modify the functioning of the hypothalamic appetite system will be safe and potent obesity treatments. Unfortunately, alteration of leptin signals has not proven to be an effective strategy; only a tiny minority of obese people have abnormal leptin levels, and most have *higher* levels of circulating leptin than do thin people (C. T. Montague et al., 1997).

Interestingly, leptin appears to regulate endogenous cannabinoid levels in the hypothalamus (Di Marzo and Matias, 2005; Di Marzo et al.,

2001). Therefore, drugs that are cannabinoid antagonists might effectively suppress appetite by causing "anti-munchies"—the reverse of the hunger experienced by cannabis users. As predicted, a drug that interferes with signalling via CB_1 cannabinoid receptors, rimonabant (classified as an *inverse agonist*; see Chapter 3), effectively reduces appetite and feeding behaviour, leading to weight loss (Thornton-Jones et al., 2006; Van Gaal et al., 2005). Unfortunately, rimonabant also causes significant mood problems (an "anti-high"?), so it was removed from the market in Europe.

Drugs also can be designed to target some of the signalling systems integral to the arcuate appetite controller. For example, we saw earlier that α-MSH activity effectively reduces hunger, so drugs that activate the MC4R melanocortin receptor may be effective appetite suppressants (see also *The Cutting Edge: Bones Secrete Hormones to Regulate Appetite* in Chapter 4). One such drug, lorcarserin, activates the serotonin 5-HT2C receptors found on many POMC neurons in the hypothalamic appetite controller; lorcarserin produces modest but significant weight loss and has been approved for use in humans since 2012 (L. K. Burke and Heisler, 2015). Similarly, treatment with $PYY_{3\text{-}36}$ (via nasal spray), or a drug that mimics its actions, may act directly on arcuate neurons to reduce appetite (see Figure 14.23) (Sileno et al., 2006). Simply spraying a $PYY_{3\text{-}36}$ solution into the mouths of lab mice is apparently not aversive yet powerfully suppresses their appetite (Hurtado et al., 2013). Other recently discovered satiety signals, including GLP-1, nesfatin-1, and oxyntomodulin, similarly provide excellent targets for drug development; numerous compounds targeting these systems are in clinical trials (Srivastava and Apovian, 2018).

TABLE 14.4 Body Mass Index (BMI)

BMI value	Body weight category
<15	Starvation
15–18.5	Underweight
18.5–25	Ideal weight
25–30	Overweight
30–40	Obese
>40	Morbidly obese

Note:

$$BMI = \frac{\text{weight (kg)}}{\text{height} \times \text{height (m} \times \text{m)}} \quad or$$

$$BMI = 703\frac{\text{weight (lb)}}{\text{height} \times \text{height (in.} \times \text{in.)}}$$

INCREASED METABOLISM An alternative approach to treating obesity is to raise the body's metabolic rate and thus expend extra calories in the form of heat. For example, the thyroid hormone thyroxine increases metabolic rate, so scientists are trying to devise drugs that mimic thyroxine's metabolic actions while avoiding its undesirable cardiovascular side effects (Grover et al., 2003). Another approach may be to increase the activity of brown fat, which is rich in special mitochondria containing a protein, called **thermogenin** (or *UCP1*), that allows mitochondria to turn energy directly into heat. Several signalling pathways can activate brown fat, suggesting new targets for drug development (Boström et al., 2012; Symonds et al., 2015). Furthermore, hormone or drug treatments, or even environmental manipulations such as cold exposure, may induce the formation of "beige" fat: fat cells that resemble (or even originate as) white fat but express larger numbers of thermogenin-containing mitochondria (Berglund et al., 2014; Gnad et al., 2014). Because beige fat is found throughout the body, like white fat, it has the potential to burn excess stored energy at a high rate.

INHIBITION OF FAT TISSUE A third approach to treating obesity involves interfering with the formation of new white fat tissue. For example, in order for fat tissue to grow, it needs to add new blood vessels—a process called *angiogenesis*—so blocking this process may inhibit fat formation (Rupnick et al., 2002). Blockade of one type of receptor for vascular endothelial growth factor—a signalling protein that normally stimulates angiogenesis—effectively inhibits the growth of fat tissue in mice (Tam et al., 2009). Other modifications of the vasculature in fat tissue, in concert with the delivery of brown fat tissue markers, provide an additional way to induce white fat cells to transform into beige or brown fat (Y. Xue et al., 2016).

REDUCED ABSORPTION One of only a few currently approved anti-obesity medications, orlistat (Xenical) works by interfering with the digestion of fat. However, this approach has generally produced only modest weight loss, and it often causes intestinal discomfort.

REDUCED REWARD A different perspective on treating obesity focuses on the rewarding properties of food. Not only is food delicious, but "comfort foods" also directly reduce circulating stress hormones, thereby providing another reward. Chronic food

thermogenin Also called *UCP1*. A specialised protein that allows mitochondria to turn energy directly into heat.

bariatric Referring to treatment of obesity.

restriction makes rewarding brain stimulations even more rewarding than usual, and this effect is reversed by treatment with leptin (S. Fulton et al., 2000). Drugs that affect the brain's reward circuitry (see Chapter 3), reducing the rewarding properties of food, may prove beneficial for weight loss (Volkow and Wise, 2005). However, although obesity resembles drug addiction in certain ways, it also differs enough that the food addiction model has been called into question (Ziauddeen et al., 2012).

ANTI-OBESITY SURGERY The surgical removal of fat tissue, particularly through liposuction, is a popular approach to controlling weight, but it is generally only moderately successful and temporary (see Box 14.2). Because of the propensity of fat tissue to regrow after excision, people increasingly are turning to **bariatric** procedures that reduce the volume and absorptive capacity of the digestive system (**FIGURE 14.27**). Although gastric bypass surgery doesn't directly target appetite control mechanisms, alterations in appetite hormones such as ghrelin reportedly accompany the surgery (Baynes et al., 2006; D. E. Cummings, 2006). As the only current intervention that produces significant and lasting weight loss, gastric bypass surgery can offer hope of substantial weight loss and reversal of comorbid conditions like type II diabetes and hypertension, but it is accompanied by significant complications and risks.

Less-invasive surgical procedures are under study, such as the use of gastric stimulators that activate the gut's satiety signals to reduce appetite (Miras et al., 2015). Curiously, simply implanting inert weights into the abdominal cavities of mice causes them to lose a proportionate amount of weight, apparently by fooling the body into thinking it is heavier than it actually is (Adams et al., 2003). Perhaps some of us, someday, will be able to lose weight simply by taking on extra ballast!

LIFESTYLE CHANGES Follow-up research with the *Biggest Loser* contestants has confirmed what many of us already suspected: that increased physical activity can help keep the pounds off after dieting. However, the required change may need to be very substantial, and sustained. Compared with the contestants who regained all of their former weight, the few who maintained their new lower weight 6 years after the end of the contest had to increase their physical activity level by an average of 160% (Kerns et al., 2017) relative to their precontest activity level. At least in this group, exercise played a significantly greater role than dietary changes, but the relationship between weight loss and lasting decreases in metabolic rate remains mysterious (Hall, 2018).

14.27 Surgical Options for Obesity (A) Sleeve gastrectomy is a surgical procedure in which the stomach is reduced to a narrow tube that absorbs fewer nutrients and also secretes less of the appetite-stimulating hormone ghrelin. (B) In gastric (or *Roux-en-Y*) bypass, the stomach is surgically reduced to a small pouch and connected to the small intestine at some distance, thereby bypassing the initial stretch of small intestine. This reduces the ability of the digestive system to absorb nutrients from food. (C) A less invasive option is the implantation of a plastic liner into the small intestine. It acts as a barrier to prevent the absorption of food, so fewer calories are absorbed from the diet. All three procedures result in weight loss and improvements in secondary problems like diabetes.

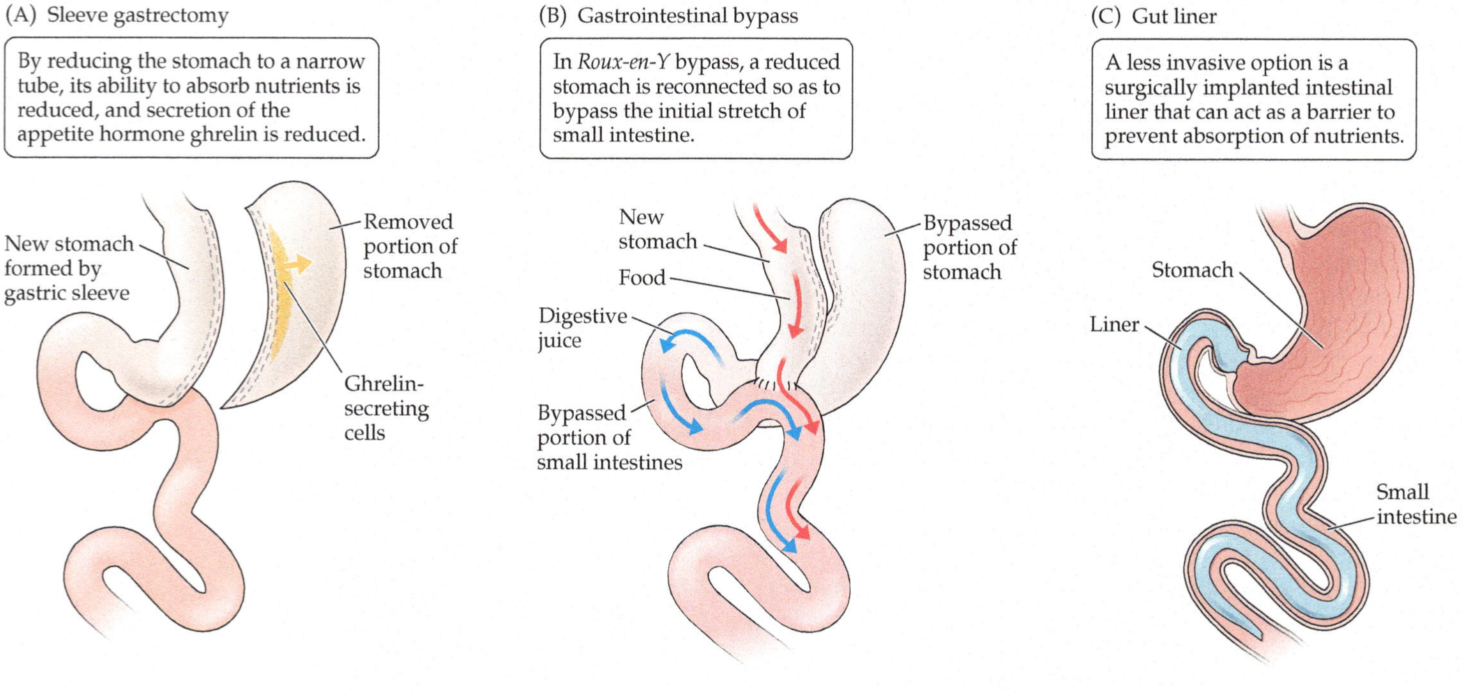

Eating disorders are life threatening

Sometimes people shun food, despite having no apparent aversion to it. These people are usually young, become obsessed with their body weight, and become extremely thin—generally by eating very little and sometimes also by vomiting, taking laxatives, overexercising, or drinking large amounts of water to suppress appetite. This condition, which is more common in adolescent girls and women than in males, is called **anorexia nervosa**. The name of the disorder indicates (1) that the afflicted people have no appetite (*anorexia*) and (2) that the disorder originates in the nervous system (*nervosa*).

People with anorexia nervosa tend to think about food a good deal, and physiological evidence suggests that they respond even *more* than healthy people to the presentation of food (Broberg and Bernstein, 1989); for example, food stimuli provoke a large release of insulin, despite cognitive denial of any feelings of hunger. So, in a physiological sense their hunger may be normal or even exaggerated, but this hunger is somehow absent from the conscious perceptions of these individuals and they refuse to eat. The idea that anorexia nervosa is primarily a nervous system disorder stems from this mismatch between physiology and cognition and from the distorted body image of the people with anorexia (they may consider themselves fat when others see them as emaciated). The observation that agouti-related peptide (AgRP) levels are abnormal in women with anorexia nervosa (Moriya et al., 2006) suggests that the hypothalamic appetite system described earlier may be abnormal in this condition. Studies of the incidence of eating disorders in twins indicate that a predisposition toward anorexia is heritable (Klump et al., 2001), and leading candidates for underlying physiological causes include abnormalities in serotonergic neurotransmission and alterations in the functioning of the dopamine-based reward system (see Chapter 3) that persist even after recovery (Kaye et al., 2009).

Anorexia nervosa is notoriously difficult to treat, because it appears to involve an unfortunate combination of genetic, endocrine, personality, cognitive, and environmental variables. One approach that is successful in some cases is a family-based treatment (sometimes termed *Maudsley therapy* after the hospital where it was introduced) that de-emphasises the identification of causal factors and instead focuses on intensive, parent-led "refeeding" of the anorexic person (A. E. Kass et al., 2013; Le Grange, 2005).

Bulimia (or *bulimia nervosa,* from the Greek *boulimia,* "great hunger") is a related disorder. Like those with anorexia nervosa, people with bulimia may believe themselves fatter than they are, but they periodically gorge themselves, usually with "junk food," and then either vomit the food or take laxatives to avoid weight gain. Also like people with anorexia nervosa, people with bulimia may be obsessed with food and body weight, but not all of them become emaciated. As in anorexia nervosa, bulimia can be deadly if the person's lack of nutrient reserves damages various organ systems and/or leaves the body unable to battle otherwise mild diseases. Prompt interventions, such as cognitive behavioural therapy that focuses on overcoming distorted views of food and eating, can relieve bulimia and lessen the risk of relapse and serious health impacts (de Jong et al., 2018).

In **binge eating**, people spontaneously gorge themselves with far more food than is required to satisfy hunger, often to the point of illness. Such people are often obese, and the causes of the bingeing are not fully understood. In susceptible people, the strong pleasure associated with food activates opiate and dopaminergic reward mechanisms to such an extent that bingeing resembles drug addiction. Indeed, "binge eating disorder" is now recognized as a psychiatric diagnosis in the *Diagnostic and Statistical Manual of Mental Disorders,* 5th edition, or *DSM-5* (American Psychiatric Association, 2013). Mutation of the gene encoding the MC4R receptor is also associated with binge eating (Branson et al., 2003). Recall that α-MSH acts on MC4R to signal satiety, so people with the mutation may be failing to receive the signal to stop eating.

Despite the epidemic of obesity in our society, or perhaps because of it, our present culture emphasises that women, especially young women, must be thin

anorexia nervosa A syndrome in which individuals severely deprive themselves of food.

bulimia Also called *bulimia nervosa*. A syndrome in which individuals periodically gorge themselves, usually with "junk food," and then either vomit or take laxatives to avoid weight gain.

binge eating The paroxysmal intake of large quantities of food, often of poor nutritional value and high in calories.

14.28 Changing Ideals of Female Beauty (A) Actress Keira Knightley exemplifies modern society's emphasis on thinness as an aspect of beauty. (B) In contrast, Flemish painter Peter Paul Rubens's painting of his wife in *Helena Fourment as Aphrodite* (circa 1630) illustrates the very different ideal for the feminine form during her era. Could pressures around modern weight-conscious notions of female beauty be responsible for some cases of anorexia nervosa and bulimia?

(A) (B)

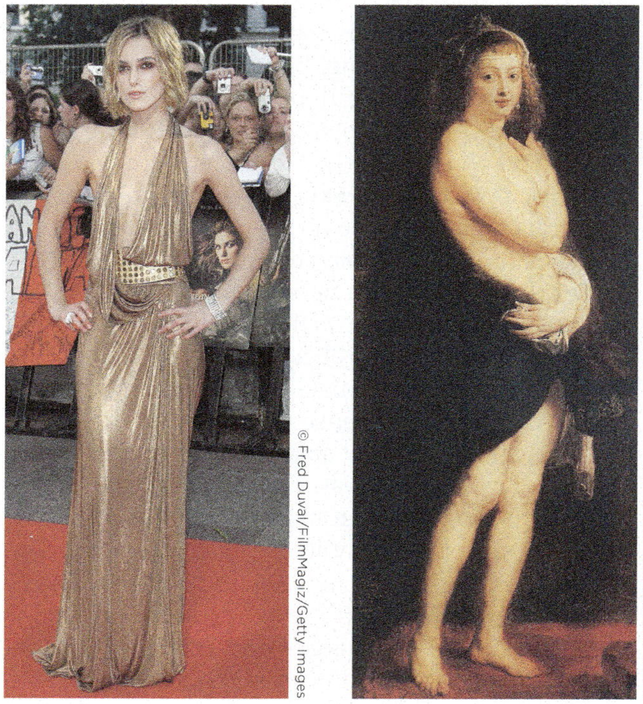

to be attractive (**FIGURE 14.28A**). This cultural pressure is widely perceived as one of the causes of eating disorders. In earlier times, however, when plump women were considered the most beautiful (witness Renaissance paintings, such as the one shown in **FIGURE 14.28B**), some women still fasted severely and may have suffered from anorexia nervosa. The origins of these disorders remain elusive, and to date, the available therapies help only a minority of people with eating disorders.

Recommended Reading

Blumberg, M. S. (2009). *Body Heat.* Cambridge, MA: Harvard University Press.

Brownell, K. D., and Walsh, B. T. (2017). *Eating Disorders and Obesity* (3rd ed.). New York: Guilford Press.

Daniels, D., and Fluharty, S. J. (2009). Neuroendocrinology of body fluid homeostasis. In D. W. Pfaff, A. P. Arnold, A. M. Etgen, S. E. Fahrbach, et al. (Eds.), *Hormones, Brain and Behaviour* (2nd ed.) (pp. 259–288). San Diego, CA: Academic Press.

DeSalle, R., and Perkins, S. L. (2015). *Welcome to the Microbiome: Getting to Know the Trillions of Bacteria and Other Microbes In, On, and Around You.* New Haven, CT: Yale University Press.

Flouris, A. (2009). *On the Functional Architecture of the Human Thermoregulatory System: A Guide to the Biological Principles and Mechanisms of Human Thermoregulation.* Berlin: VDM.

Kirkham, T., and Cooper, S. J. (Eds.). (2006). *Appetite and Body Weight: Integrative Systems and the Development of Anti-Obesity Drugs.* Burlington, MA: Academic Press.

Logue, A. W. (2014). *The Psychology of Eating and Drinking* (4th ed.). New York: Routledge.

Lustig, R. H. (2018). *The Hacking of the American Mind: The Science Behind the Corporate Takeover of Our Bodies and Brains.* New York: Avery.

McNab, B. K. (2012). *Extreme Measures: The Ecological Energetics of Birds and Mammals.* Chicago: University of Chicago Press.

Schulkin, J. (Ed.). (2012). *Allostasis, Homeostasis, and the Costs of Physiological Adaptation.* Cambridge, UK: Cambridge University Press.

14 Visual Summary

You should be able to relate each summary to the adjacent illustration, including structures and processes. You can also refer back to the links to figures, animations, and activities that will help you consolidate the material.

1 The nervous system plays a crucial role in maintaining the **homeostasis** that the body requires for proper functioning. Temperature, fluid concentration, chemical energy, and nutrients must all be maintained within a critical range. These systems often feature **negative feedback** control and redundancy. Review **Figure 14.1**, **Animation 14.1**

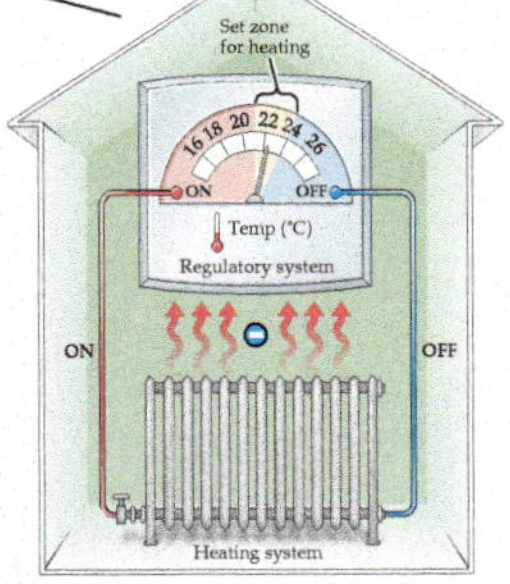

2 Both **endotherms** and **ectotherms** regulate body temperature, but ectotherms depend more on behaviours to capture heat from the environment, while endotherms generate most of their body heat through the metabolism of food. Review **Figure 14.3**, **Animation 14.2**

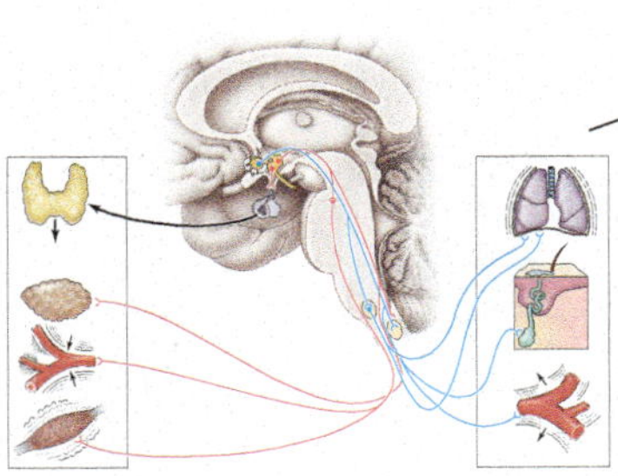

3 The preoptic area of the hypothalamus integrates thermoregulatory signals, but many regions of the brain and spinal cord contain neurons that process thermal information. Both endotherms and ectotherms use behavioural methods to help regulate body temperature at optimal levels. Young animals particularly depend on this form of **thermoregulation**. Review **Figures 14.4**, **14.5**, **Box 14.1**

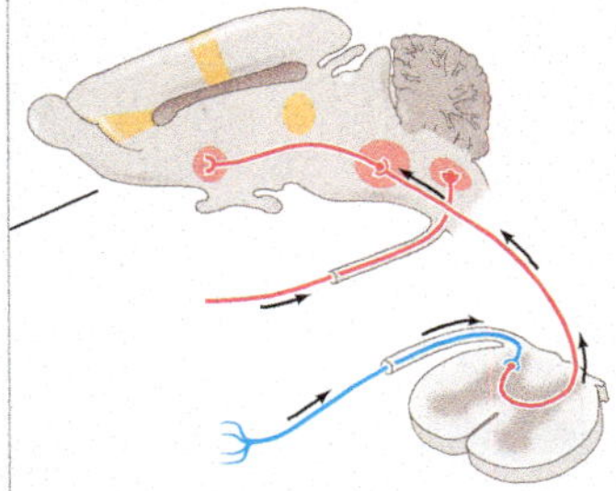

4 Our cells function properly only when the **osmolality**—concentration of salts and other solutes—of the **intracellular compartment** of the body is within a critical range. The **extracellular compartment** is a source of replacement water and a buffer between the intracellular compartment and the outside world. Review **Figure 14.8**

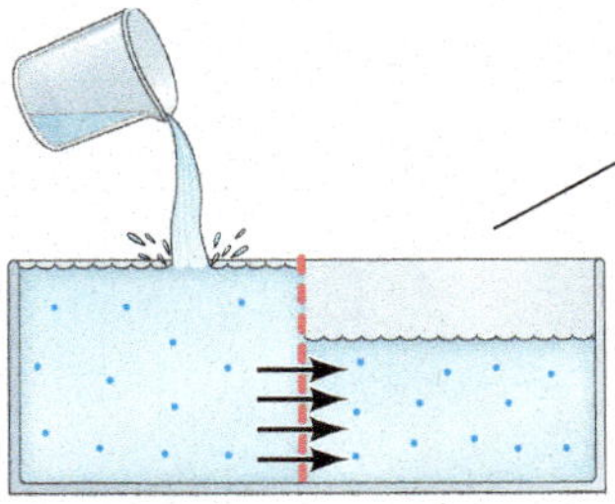

5 Thirst can be triggered either by an increase in the osmolality of the extracellular compartment, which causes **osmotic thirst**, or by a drop in the volume of the extracellular compartment, which causes **hypovolemic thirst**. Because of the importance of osmolality, we must regulate salt intake in order to regulate water balance effectively. Review **Figure 14.9**

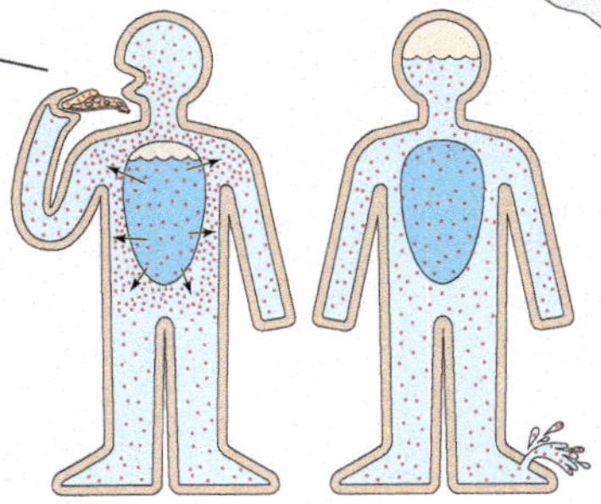

6 The hypothalamus and **circumventricular organs** contain **osmosensory neurons** that detect the concentration of extracellular fluid. Increased solute concentration of the extracellular fluid triggers an intake of water. The conscious perception of thirst involves activation of a network of limbic system sites and is a powerful motivator. Review **Figure 14.10**

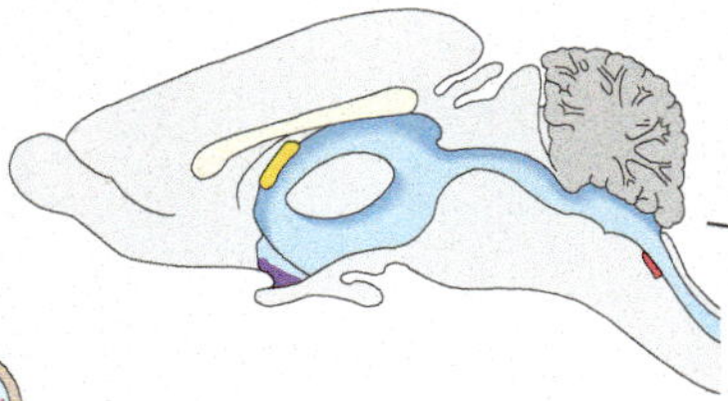

7 A drop in blood volume triggers at least three responses: (1) **Baroreceptors** in the major blood vessels signal the brain via the autonomic nervous system. (2) **Vasopressin** from the posterior pituitary reduces blood vessel volume and urination. (3) The kidneys release renin, providing circulating **angiotensin II**, which narrows blood vessels to maintain blood pressure and provides a thirst signal to the brain. Review **Figures 14.12**, **14.14**

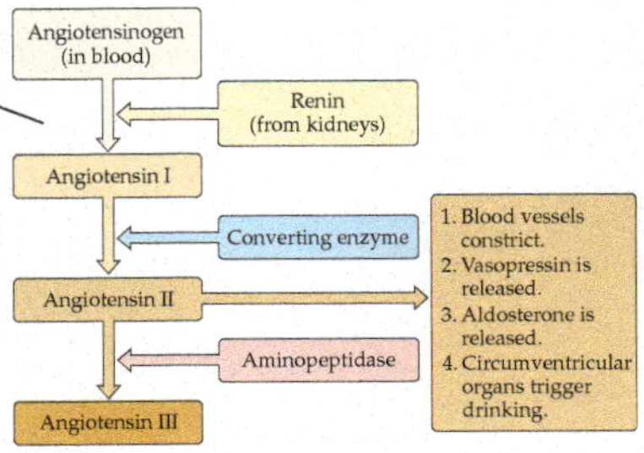

8 Although brain cells can use glucose directly, body cells can import glucose only with the assistance of **insulin** secreted by the pancreas. Insulin also promotes the storage of glucose as **glycogen** and provides a signal to the brain regarding current glucose levels. Another pancreatic hormone, **glucagon**, helps convert glycogen back into glucose. Review **Figure 14.15**

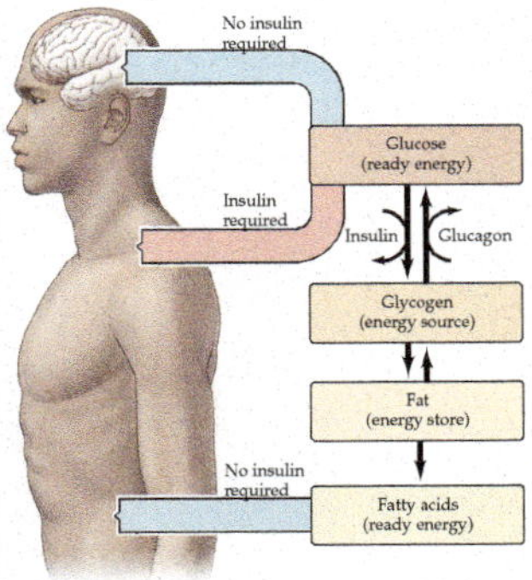

(Continued)

9 Body weight is actively regulated by multiple redundant systems, making it hard to lose weight by dieting. Review **Figures 14.17, 14.18**

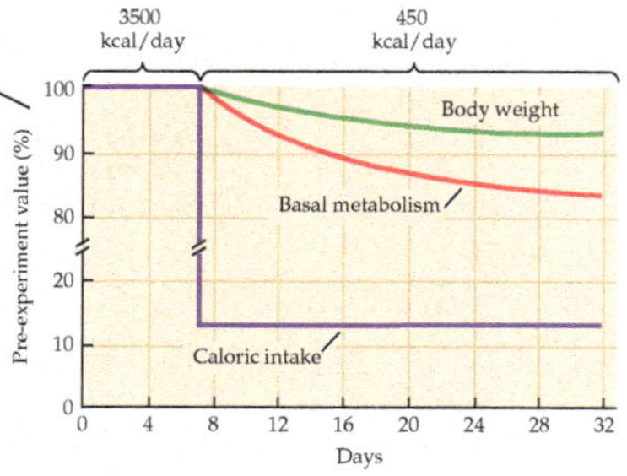

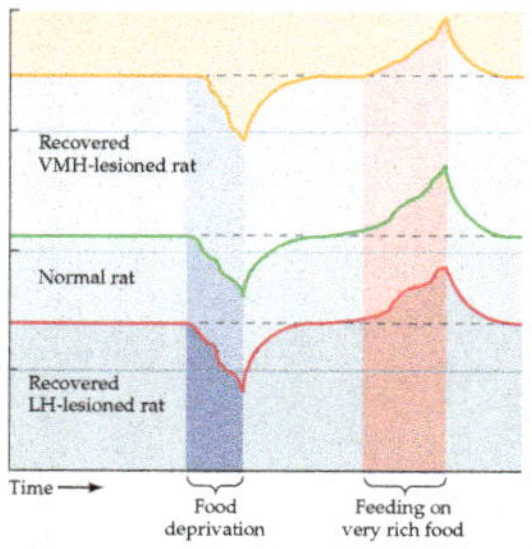

10 Lesion studies showed that the hypothalamus plays a special role in appetite regulation. Damage to the **ventromedial hypothalamus** caused animals to become obese, while lesions aimed at the **lateral hypothalamus** caused animals to stop eating and lose weight. Review **Figure 14.20**

11 An appetite controller located in the **arcuate nucleus** of the hypothalamus responds to levels of several peptide gut hormones. **Leptin**, providing a chronic signal about fat levels, stimulates arcuate **POMC neurons** to release **α-MSH** in the **paraventricular nucleus** to activate **MC4Rs** to decrease appetite. Leptin inhibits arcuate **NPY neurons**, decreasing their release of NPY and AgRP to suppress appetite further. **Ghrelin**, **GLP-1**, and **PYY_{3-36}** provide more-acute signals from the gut. Ghrelin stimulates, and PYY_{3-36} and GLP-1 inhibit the arcuate appetite control system. Review **Figure 14.23**, **Activity 14.1**

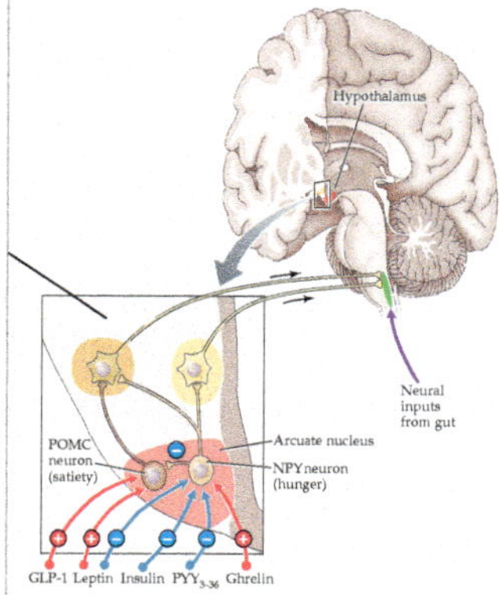

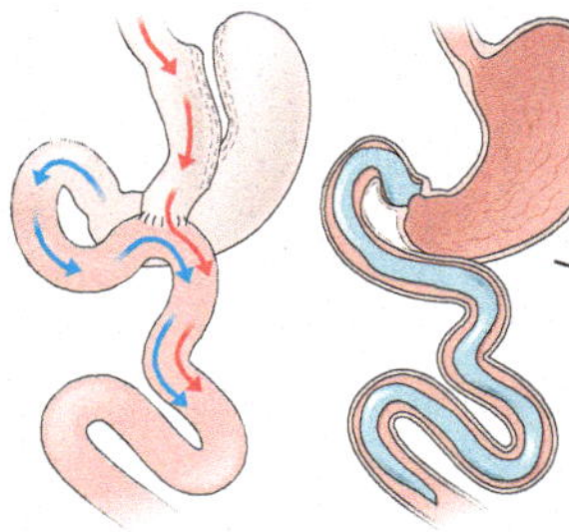

12 Obesity is a pervasive and worsening public health problem that is difficult to treat through diet, drugs, or surgery. The most effective long-lasting medical intervention for obesity is **bariatric** surgery, but several drug strategies based on a new understanding of appetite control offer promise. Review **Figures 14.26, 14.27**

13 Eating disorders include anorexia nervosa, bulimia, and binge-eating. Eating disorders can be very difficult to treat, because they may involve multiple neurophysiology, psychological, and neurochemical dimensions, along with sensitivity to cultural norms and pressures. Review **Figure 14.28**

15

Biological Rhythms, Sleep, and Dreaming

When Sleep Gets Out of Control

Starting college always brings its share of new experiences and adjustments, but "Barry" knew something was wrong freshman year when he seemed to be sleepy all the time (S. Smith, 1997). Barry napped so often that his friends called him the hibernating bear. Of course, college can be exhausting, and many students seek refuge in long snooze sessions. But one day while Barry was camping with his pals, an even odder thing happened: "I laughed really hard, and I kind of fell on my knees.... After that, about every week I'd have two or three episodes where if I'd laugh ... my arm would fall down or my muscles in my face would get weak. Or if I was running around playing catch and someone said something, I would get weak in the knees. And there was a time there that my friends kinda used it as a joke. If they're going to throw me the ball and they didn't want me to catch it, they'd tell me a joke and I'd fall down and miss it."

It was as if any big surge in emotion in Barry might trigger a sudden paralysis lasting anywhere from a few seconds to a few minutes, affecting either a body part or his whole body. Sex became something of a challenge because sometimes during foreplay, Barry's body would just collapse. "Luckily, you're probably laying down, so it's not that big a deal. But it just puts a damper on the whole thing."

What was happening to Barry?

All living systems show repeating, predictable changes over time. These rhythms vary from rapid (e.g., brain potentials) to slow (e.g., annual changes like hibernation). Daily rhythms, the first topic of this chapter, have an intriguing clocklike regularity. The second topic of the chapter is that familiar daily rhythm known as the sleep-waking cycle. By age 60, most humans have spent 20 years asleep (some, alas, on one side or the other of the classroom podium). We'll find that sleep is not a passive state of "nonwaking," but rather the interlocking of several different states orchestrated by the activity of many brain regions.

BIOLOGICAL RHYTHMS

15.1 Many Animals Show Daily Rhythms in Activity

Learning Objectives

After reading this section, you should be able to:

15.1.1 **Offer evidence that animals have access to a biological clock to organize their behaviour in the absence of day and night.**

15.1.2 **Use scientific terminology describing daily rhythms and the environmental influences that affect them.**

15.1.3 **Identify the adaptive advantage of circadian rhythms in behaviour, offering examples.**

circadian rhythm A pattern of behavioural, biochemical, or physiological fluctuation that has a 24-hour period.

diurnal Active during the light periods of the daily cycle.

nocturnal Active during the dark periods of the daily cycle.

Go to Animation 15.1
Biological Rhythms

On Earth, most functions of any living system display a rhythm of approximately 24 hours. Because these rhythms last about a day, they are called **circadian rhythms** (from the Latin *circa*, "about," and *dies*, "day"). Circadian rhythms have been studied in a host of creatures at behavioural, physiological, and biochemical levels.

Humans and most other primates are **diurnal**—active during the day. Most rodents, including hamsters, are **nocturnal**—active during dark periods. In either case, almost all physiological measures—hormone levels, body temperature, drug sensitivity—change in a regular repeating fashion over the course of the day. A convenient way to study circadian rhythms exploits rodents' love of running wheels. A switch attached to the wheel registers each revolution, revealing an activity rhythm, as in **FIGURE 15.1A**.

Circadian rhythms are generated by an endogenous clock

A hamster placed in a dimly lit room continues to show a daily rhythm in wheel running despite the absence of day versus night, suggesting that the animal has

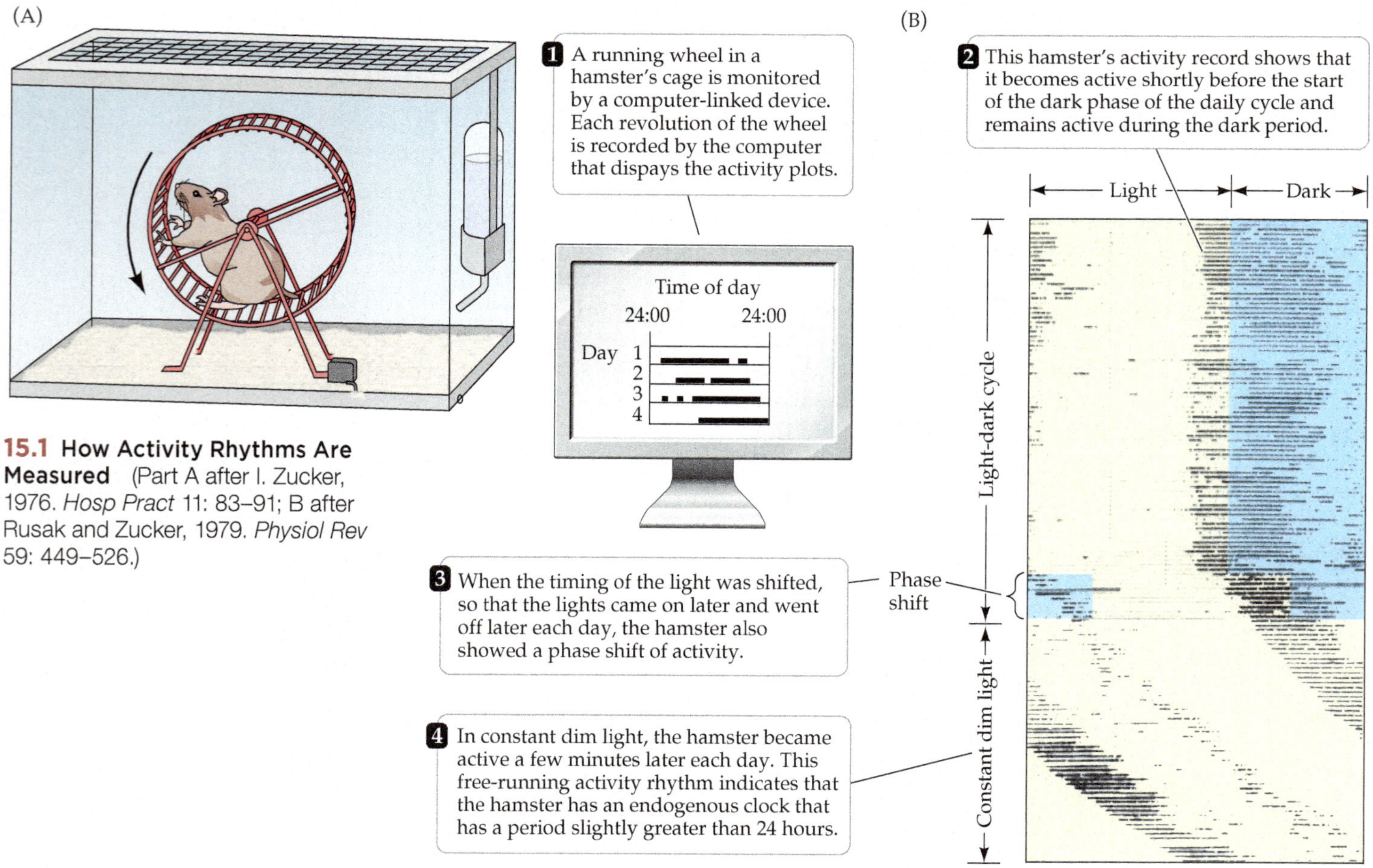

15.1 How Activity Rhythms Are Measured (Part A after I. Zucker, 1976. *Hosp Pract* 11: 83–91; B after Rusak and Zucker, 1979. *Physiol Rev* 59: 449–526.)

an internal, *biological clock*. But even if the light is always dim, the animal may detect other external cues (e.g., outside noises, temperature, barometric pressure) that signal the time of day. Arguing for a biological clock, however, is the fact that in constant light or dark the circadian cycle is not *exactly* 24 hours: activity starts a few minutes later each day, so eventually the hamster is active while it is daytime outside (**FIGURE 15.1B**, bottom). The animal is said to be **free-running**, maintaining its own cycle, which, in the absence of external cues, is not exactly 24 hours long.

The free-running period is the animal's natural rhythm. A **period** is the time between two similar points of successive cycles, such as sunset to sunset. Because the free-running period does not *quite* match the period of Earth's rotation, and because it differs slightly among individual hamsters in the same room, the free-running period cannot simply be reflecting an external cue. So the animal has some sort of endogenous clock, and this clock runs a bit slow in hamsters (and in humans, but not all species).

Normally the internal clock is set by light. If we expose a free-running hamster to periods of light and dark, it soon does most of its wheel running in the dark period. The shift of activity produced by a synchronising stimulus is referred to as a **phase shift** (see Figure 15.1B, middle), and the process of shifting the rhythm is called **entrainment**. Any cue that an animal uses to synchronise its activity with the environment is called a **zeitgeber** (German for "time giver"). Light acts as a powerful zeitgeber, and we can easily manipulate it in the laboratory. Because light stimuli can entrain circadian rhythms, the endogenous clock must receive information about light, as we'll confirm shortly. We humans experience phase shifts when we fly from one time zone to another. Flying three time zones east (say, from California to New York) means that sunlight arrives 3 hours sooner than our brain expects. In addition to light, the availability of food can serve as a cue to entrain the circadian clock (Fuller et al., 2008; Mistlberger and Skene, 2004).

free-running Referring to a rhythm of behaviour shown by an animal deprived of external cues about time of day.

period The interval of time between two similar points of successive cycles, such as sunset to sunset.

phase shift A shift in the activity of a biological rhythm, typically provided by a synchronising environmental stimulus.

entrainment The process of synchronising a biological rhythm to an environmental stimulus.

zeitgeber Literally "time giver" (in German). The stimulus (usually the light-dark cycle) that entrains circadian rhythms.

suprachiasmatic nucleus (SCN) A small region of the hypothalamus above the optic chiasm that is the location of a circadian oscillator.

Circadian rhythms allow animals to anticipate changes in the environment

The major value of circadian rhythms is that they synchronise behaviour and body states to changes in the environment. Day and night have great significance for survival. The endogenous clock enables animals to *anticipate* an event, such as darkness, and to begin physiological and behavioural preparations before that event (in this case, *before* it gets dark—see Figure 15.1B). A snack in the den at the end of the day may prepare the animal for a long night of foraging. In other words, circadian rhythms provide the temporal organisation of an animal's behaviour.

15.2 The Hypothalamus Houses a Circadian Clock

Learning Objectives

After reading this section, you should be able to:

15.2.1 **Describe experiments that established the existence of a circadian clock in the hypothalamus.**

15.2.2 **Compare the pathway by which light information entrains circadian rhythms in birds versus mammals.**

15.2.3 **Describe in broad terms how a molecular clock within cells measures approximately one day and how light information affects that clock.**

15.2.4 **Define biological rhythms in behaviour that are longer or shorter than a day, offering examples of each.**

Where is the endogenous clock that drives circadian rhythms, and how does it work? Early work showed that while removing various endocrine glands has little effect on the free-running rhythm of rats, experimental lesions of the hypothalamus interfere with circadian rhythms (Richter, 1967). It was subsequently discovered that a tiny subregion of the hypothalamus—the **suprachiasmatic nucleus** (**SCN**), named for its location above the optic chiasm—serves as the biological clock. Lesions confined to the SCN portion of the hypothalamus interfere with circadian

15.2 The Effects of Lesions in the SCN (After I. Zucker, 1976. *Hosp Pract* 11: 83–91; B. Rusak and I. Zucker, 1979. *Physiol Rev* 59: 449–526.)

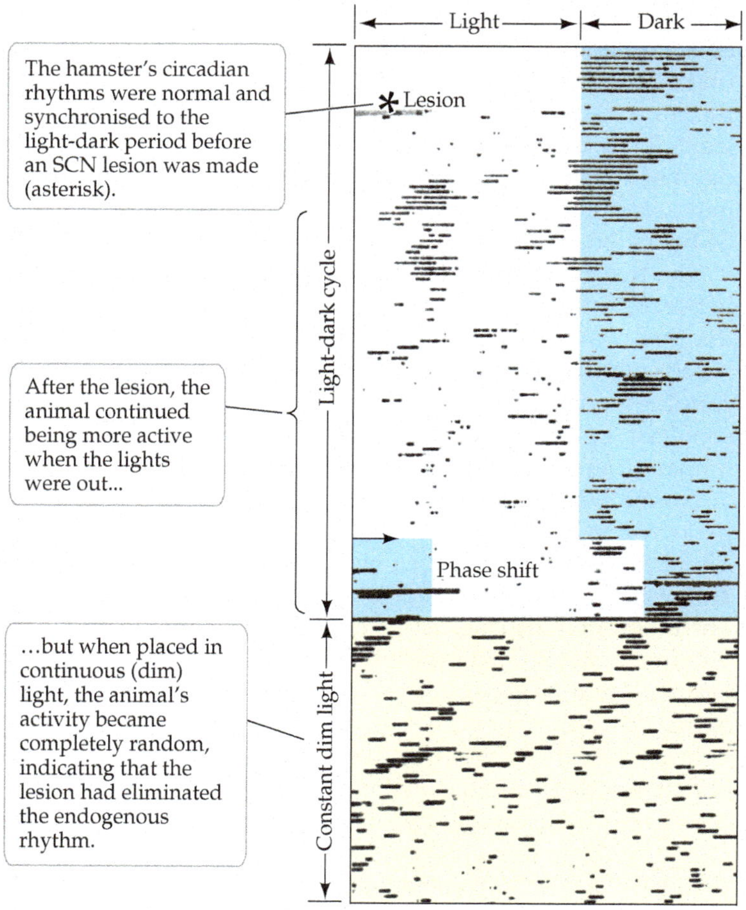

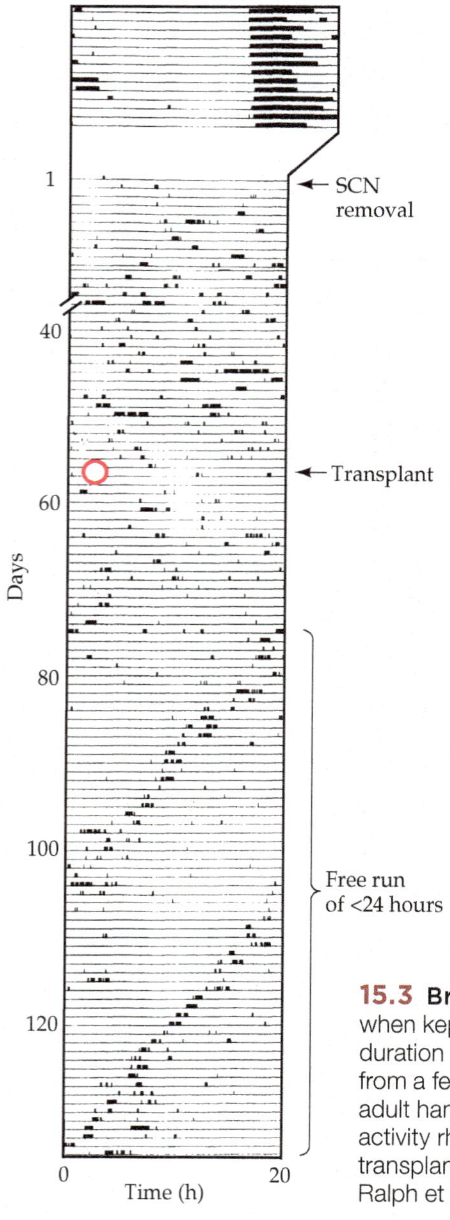

rhythms of drinking and locomotor behaviour (**FIGURE 15.2**) (F. K. Stephan and Zucker, 1972) and hormone secretion (R. Y. Moore and Eichler, 1972). So, is the SCN just one part of some neural system that generates a circadian rhythm, or is the rhythm generated within the SCN itself?

Transplants prove that the SCN produces a circadian rhythm

Ralph and Menaker (1988) found a male hamster that exhibited an unusually short free-running activity rhythm in constant conditions. Normally, hamsters free-run at a period slightly longer than 24 hours, but this male showed a period of 22 hours. Half of the offspring of this male also had a shorter circadian rhythm, so the researchers concluded that he had a genetic mutation affecting the endogenous clock. Animals with two copies of this mutation had an even shorter period: 20 hours. The mutation was named *tau*, after the Greek symbol scientists use to represent the period of a rhythm. These animals entrained to a normal 24-hour light-dark period just fine; their abnormal endogenous circadian rhythm was revealed only in constant conditions.

Dramatic evidence that this endogenous period is produced within the SCN was provided by transplant experiments. Nonmutant hamsters with lesions of the SCN were placed in constant conditions, and as expected, their circadian activity rhythms were abolished (**FIGURE 15.3**; Ralph et al., 1990). The hamsters then received

15.3 Brain Transplants Prove That the SCN Contains a Clock A wild-type hamster, when kept in constant dim light, displayed an endogenous circadian rhythm 24.05 hours in duration (top). After the SCN was lesioned, the animal became arrhythmic. Later, an SCN from a fetal hamster with two copies of the *tau* mutation was transplanted into the lesioned adult hamster (circle). Soon thereafter, the adult hamster began showing a free-running activity rhythm of 19.5 hours, matching the SCN of the donor animal. This response to the transplant showed that the period of the clock is determined within the SCN. (After M. R. Ralph et al., 1990. *Science* 247: 975–978.)

transplants of SCN taken from fetal hamsters with two copies of the mutant *tau* gene. About a week later the hamsters that had received the transplants began showing a free-running activity rhythm again, but the new rhythm matched that of the *donor* SCN: it was about 20 hours rather than the original 24.05.

Reciprocal transplants gave comparable results: the endogenous rhythm following the transplant was always that of the *donor* SCN, not the recipient, so the SCN must be the source of an endogenous circadian rhythm. This is the only known type of transplant of brain tissue from one individual to another in which the recipient displays the donor's behaviour!

Intriguingly, transplants of fetal SCNs can drive circadian rhythms in SCN-lesioned hamsters even if the donor cells are encapsulated in polymer (Silver et al., 1996). In such a capsule, the donor SCN is unable to make any synaptic connections with the surrounding hypothalamus, so its only means of communicating is by releasing chemical signals. These signals must act rather locally, because the capsule only works if placed near the normal site of the SCN. Now let's take up the question of how the SCN is informed about day and night.

In mammals, light information from the eyes reaches the SCN directly

The pathway that entrains circadian rhythms to light-dark cycles varies depending on the species (Rusak and Zucker, 1979). Many vertebrates have photoreceptors outside the eye that are part of the mechanism of light entrainment. For example, the **pineal gland** of some amphibians is itself sensitive to light (Jamieson and Roberts, 2000) and helps entrain circadian rhythms to light. Because the skull over the pineal is especially thin in some amphibian species, we can think of them as having a primitive "third eye" in the back of the head (some school teachers also seem to have an eye in the back of the head, but this has not been proven). In birds, too, the pineal possesses photoreceptors that can detect daylight through the skull. In mammals, however, cells in the *eye* tell the SCN when it is light out.

Certain retinal ganglion cells send their axons along the **retinohypothalamic pathway**, splitting off at the optic chiasm to synapse directly within the SCN (**FIGURE 15.4**). This tiny pathway carries information about light to the hypothalamus to entrain behaviour. Most of the retinal ganglion cells that extend their axons to the SCN do not rely on the traditional photoreceptors—rods and cones—to learn about light. Rather, these retinal ganglion cells are themselves sensitive to light

pineal gland A secretory gland in the brain midline; the source of melatonin release.

retinohypothalamic pathway The projection of retinal ganglion cells to the suprachiasmatic nuclei.

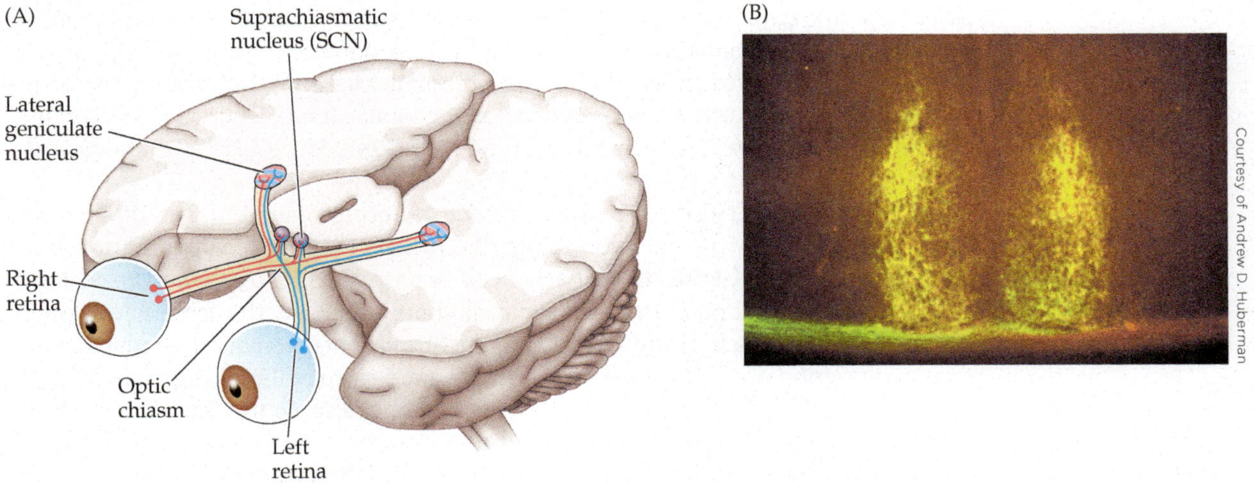

15.4 The Retinohypothalamic Pathway in Mammals (A) This pathway carries information about the light-dark cycle in the environment to the SCN. For clarity of synaptic connections, the SCNs are shown proportionally larger than other features. (B) In this image, axons are shown at the bottom. Those from the left eye are shown in green, and those from the right in red. Both eyes project so diffusely to the two overlying SCNs that they appear to be outlined in yellow.

15.5 Components of a Circadian System

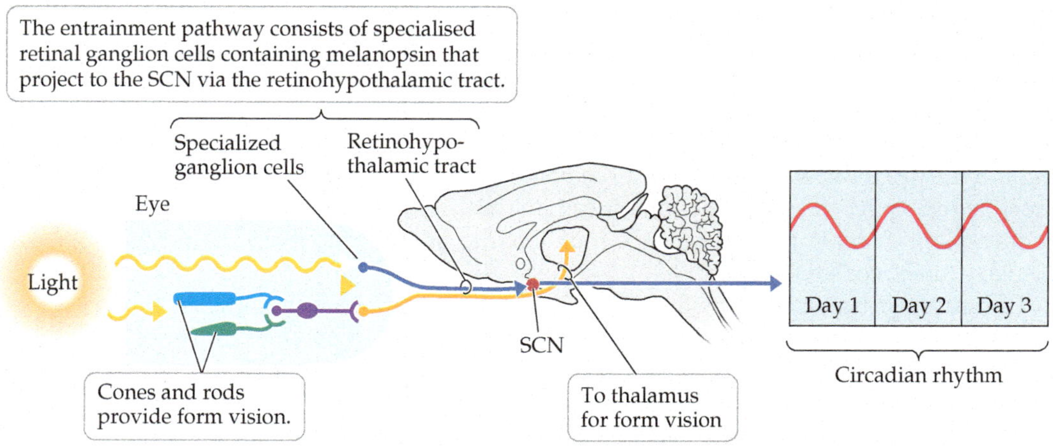

melanopsin A photopigment found in some retinal ganglion cells that project to the suprachiasmatic nucleus.

dimer A complex of two proteins that have bound together.

because they contain a special photopigment, called **melanopsin** (Do et al., 2009). Even transgenic mice that lack rods and cones, and thus are blind in almost every respect, still entrain their behaviour to light (Freedman et al., 1999) because their intrinsically photosensitive retinal ganglion cells still function. These ganglion cells send their axons to the SCN, but other melanopsin-containing retinal ganglion cells project to the brainstem, informing the brain about light to control pupil diameter (S. K. Chen et al., 2011). Melanopsin is most sensitive to light frequencies in the blue range, which explains why blue light has the largest effect on human circadian systems (Gooley et al., 2010).

FIGURE 15.5 provides a schematic outline of the mammalian circadian system, including entrainment by light.

Circadian rhythms have been genetically dissected in flies and mice

The fruit fly *Drosophila melanogaster* displays diurnal circadian rhythms in activity. Flies with a mutation that disabled the gene called *period* (*per*) were arrhythmic when transferred to constant dim light, indicating that their internal clocks weren't running. Subtle mutations of *per* could, depending on the exact change in the gene, cause the animals to have a free-running period that was longer or shorter than normal. Eventually, more genes were discovered that affect the circadian cycle in *Drosophila*, and mammals were found to have one or more versions of each of them.

These genes provide a *molecular clock*, regulating each other's expression such that their protein products wax and wane in a cycle that takes about 24 hours. Cells in the SCN make the two proteins Clock and Cycle (actually, it is called Cycle in *Drosophila* and Bmal1 in mammals). These two proteins bind together to form a **dimer** (a pair of molecules joined together). This dimer binds to the cell's DNA to promote the transcription of two other genes (*per* and *cryptochrome* [*cry*]). The resulting Per and Cry proteins dimerize, and then they inhibit expression of the *Clock/Cycle* genes that began the whole process.

Because the Per/Cry proteins either degrade or become chemically modified (Larrondo et al., 2015) with time, eventually the inhibition is lifted, starting the whole cycle over again (**FIGURE 15.6**). Each SCN neuron uses this mechanism to keep time approximately, and then all the neurons communicate with each other through electrical synapses (see Figure 2.19). They synchronise their activity to produce a very consistent period of about 24 hours, which then drives circadian processes throughout the body (J. R. Jones et al., 2015). Interestingly, the same molecular clock operates in almost every cell in the body, and they are all in sync as long as the SCN is intact. Some cue, maybe the circadian rhythm in body temperature or the release of hormones, driven by pacemakers in the SCN, could serve as one or more body-wide cues to entrain the molecular clock in all the other cells. As a result, about 80% of our genes body-wide are expressed in a circadian cycle (Mure et al., 2018).

How does light entrain the molecular clock to the light-dark cycle? In fruit flies, light passes through the fly's body into brain cells to degrade the Cry protein and activate a specialised photopigment (Ni et al., 2017) to synchronise the clock.

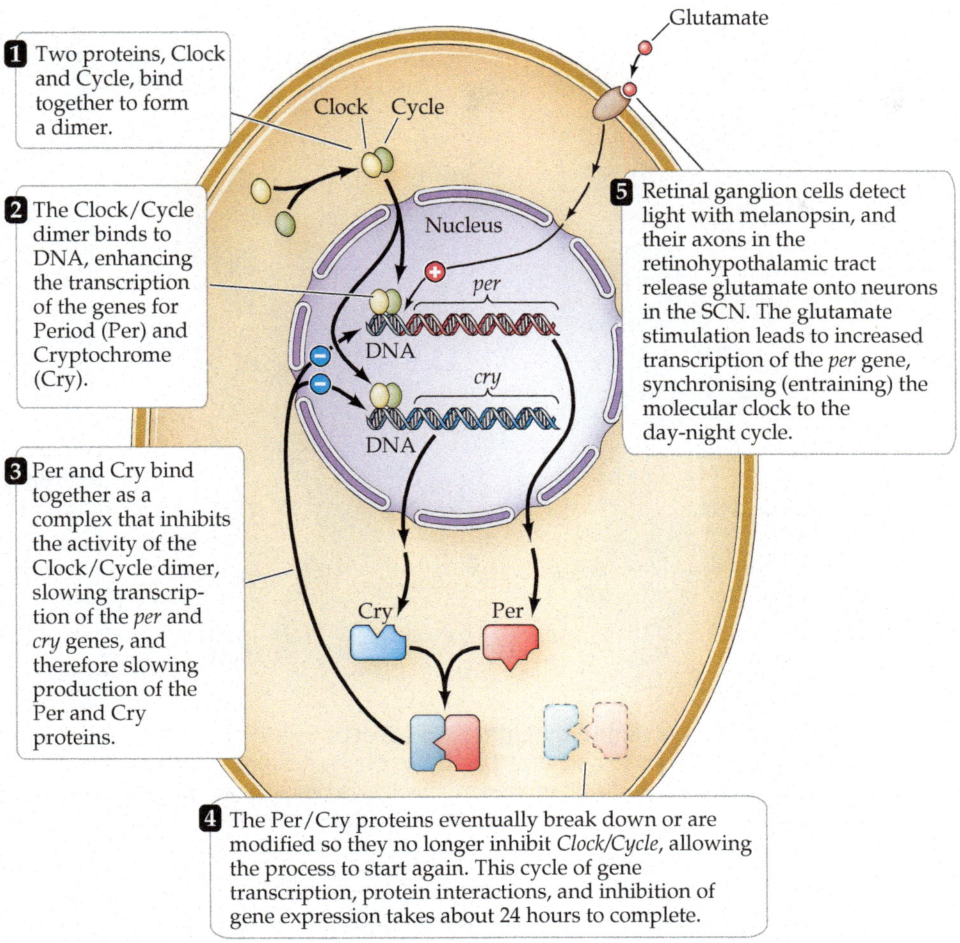

Go to Animation 15.2
A Molecular Clock

15.6 A Molecular Clock in Flies and Mice In flies, light passing through the head degrades Cry protein to help entrain the animal to the day. In mammals, information about light reaches the SCN via the eyes. This is a simplified view of the clock in mammals, which have several *per* and *cry* genes. The mammalian version of Cycle is Bmal1. (After S. M. Reppert and D. R. Weaver, 2002. *Nature* 418: 935–941.)

But things are different in thick-skulled mammals like us. We use the retinohypothalamic tract to get light information to the SCN. The retinal ganglion cells containing melanopsin detect light and release the neurotransmitter glutamate in the SCN. Glutamatergic synapses trigger a chain of events in SCN cells that promotes the production of Per protein. When the animal's photoperiod is shifted, this light-mediated boosting of Per production shifts the phase of the molecular clock and therefore the animal's behaviour.

Mutations in the genes involved in the molecular clock affect circadian behaviour. Recall that hamsters with a *tau* mutation have a shorter free-running rhythm than normal hamsters have. Mice in which both copies of the *Clock* gene are disrupted show severe arrhythmicity in constant conditions (**FIGURE 15.7**). People who feel energetic in the morning ("larks") are likely to carry a different version of the *Clock* gene than "night owls" have (Katzenberg et al., 1998). Different alleles of the *per* gene are also associated with being a lark versus a night owl (Carpen et al., 2005).

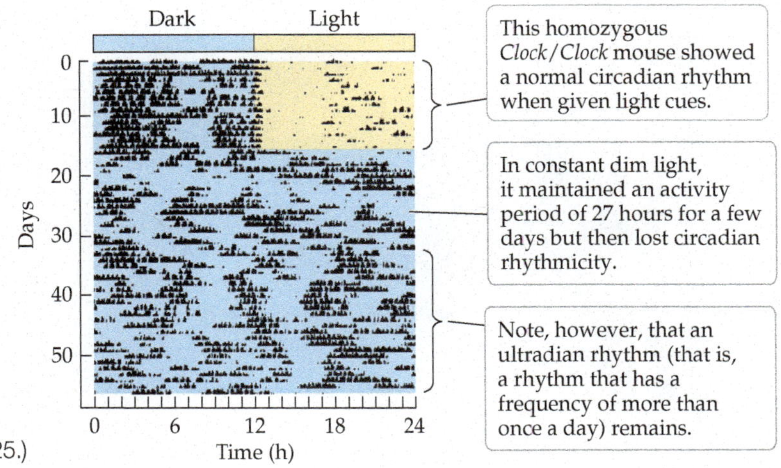

15.7 When the Endogenous Clock Goes Kaput
(From M. H. Vitaterna et al., 1994. *Science* 29: 719–725.)

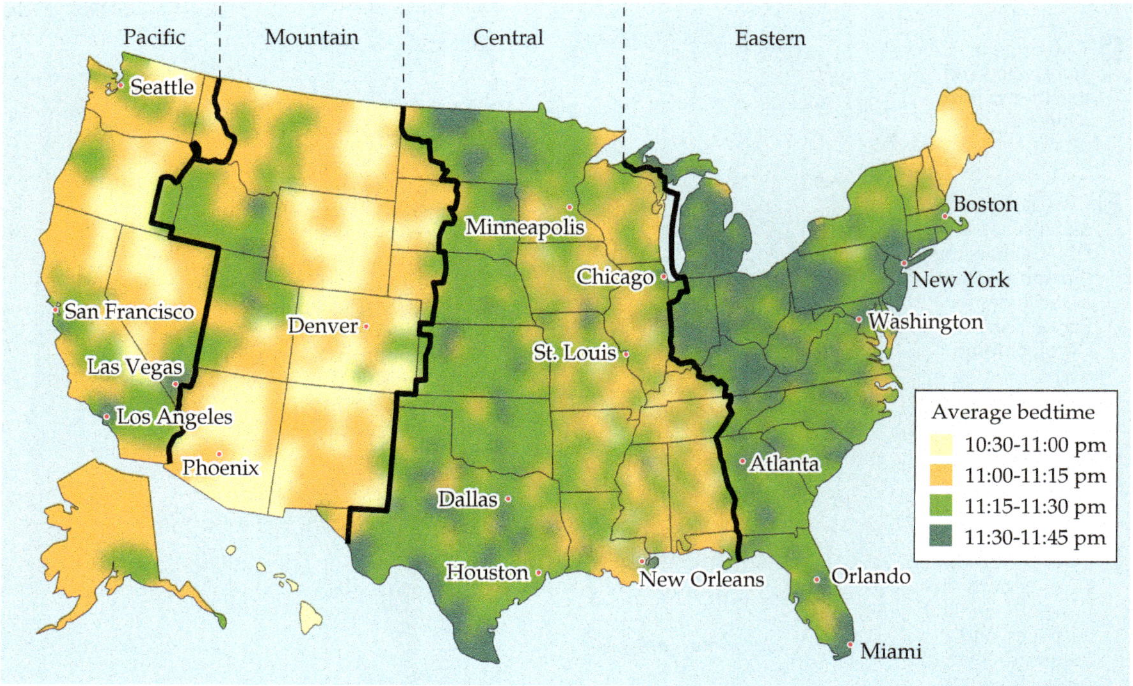

15.8 Humans Are Still Entrained by Light Note that people living on the western side of a given time zone tend to go to bed later than those on the eastern side, even though their clocks display the exact same time. Presumably this is because the sun sets later on the western side of the time zone than the eastern side. (Data from Jawbone.com.)

infradian Referring to a rhythmic biological event whose period is longer than that of a circadian rhythm—that is, longer than a day.

circannual Occurring on a roughly annual basis.

ultradian Referring to a rhythmic biological event whose period is shorter than that of a circadian rhythm, usually from several minutes to several hours long.

electroencephalography (EEG) The recording and study of gross electrical activity of the brain recorded from large electrodes placed on the scalp.

If you have any doubt that light still entrains behaviour of humans in modern civilisation, look at **FIGURE 15.8**.

Some biological rhythms are longer or shorter than a day

Some biological rhythms have periods longer than a day. Such rhythms are called **infradian** because their frequency is less than once per day (the Latin *infra* means "below"). A familiar infradian rhythm is the 28-day human menstrual cycle. Recall that many animals display a seasonal cycle in body weight (see Box 14.2). The importance of annual rhythms for humans is evident in the prevalence of seasonal disorders of behaviour (see Chapter 16).

Many other animal behaviours are also characterised by annual rhythms; for example, most animals breed only during a particular season. Some of these rhythms are driven by exogenous factors, such as food availability and temperature. But in the laboratory, seasonal animals exposed to short days and long nights (mimicking winter) will often change to the nonbreeding condition (**FIGURE 15.9**). Furthermore, many annual rhythms, including body weight, persist under constant conditions in the lab. As with circadian rhythms in constant light, seasonal animals in isolation show free-running annual rhythms of a period not quite equal to 365 days. Thus, there also seems to be an endogenous **circannual** clock.

Courtesy of Carol D. Hegstrom

15.9 Winter is Coming Siberian hamsters in the wild suppress their reproductive systems and develop a silvery fur coat (left) for camouflage in the snow each fall. In the laboratory, they will undergo identical changes, despite warm temperatures and abundant food, if the lights are on only 10 hours per day. They seem to interpret these short days as an indication that winter is coming.

How does an animal know that a year has passed? Does it simply count the days being measured by the SCN until 365 have gone by? No. Irving Zucker et al. (1983) measured activity rhythms, reproductive cycles, and body weight cycles of animals that were free-running in both their circadian and their circannual rhythms. SCN lesions clearly disrupted circadian activity cycles, but in at least some animals these lesions did not affect circannual changes in body weight and reproductive status. Circannual cycles, then, do not arise from the circadian clock, and they seem to involve a mechanism that is separate from the SCN (Sáenz de Miera et al., 2014). Light cycles (which of course are a measure of day length) can entrain this circannual rhythm by affecting how long the nightly bout of melatonin secretion lasts (see Figure 4.19).

Some biological rhythms have periods *shorter* than a day. Such rhythms are referred to as **ultradian** (designating a frequency greater than once per day; the Latin *ultra* means "beyond"), and their periods are usually from several minutes to hours. Ultradian rhythms are seen in such behaviours as bouts of activity, feeding, and hormone release. These ultradian rhythms may be superimposed on a circadian rhythm. For example, humans show a 90-minute cycle of daydreaming that is characterised by vivid sensory imagery (P. Lavie and Kripke, 1981). We'll also see this same ultradian rhythm of about 90 minutes imposed on the circadian sleep-wake cycle, which we consider next.

electro-oculography (EOG) The electrical recording of eye movements.

electromyography (EMG) The electrical recording of muscle activity.

rapid-eye-movement (REM) sleep Also called *paradoxical sleep*. A stage of sleep characterised by small-amplitude, fast-EEG waves, no postural tension, and rapid eye movements. *REM* rhymes with "gem."

non-REM (NREM) sleep Stage of sleep without rapid eye movements. In humans this is divided into stages 1, 2, and 3 sleep.

SLEEPING AND WAKING

15.3 Human Sleep Exhibits Different Stages

Learning Objectives

After reading this section, you should be able to:

15.3.1 List the four stages of sleep and describe how they are identified by physiological measures.

15.3.2 Contrast the characteristics of REM sleep and NREM sleep.

15.3.3 Critically discuss the idea that dreams only happen during REM sleep.

15.3.4 Describe how the distribution of sleep changes as we grow up and grow old.

The most obvious circadian rhythm in human behaviour is the bout of sleep that starts late in the evening and lasts until morning. The beginning and end of sleep are synchronised to many external events, including light and dark. What happens when all the customary synchronising or entraining stimuli are removed? To investigate this question, volunteers spent weeks in a dark cave with all cues to external time removed (Wever, 1979). They displayed a circadian rhythm of the sleep-waking cycle, but the rhythm slowly shifted from 24 to 25 hours. In other words, people in constant conditions free-run just as a hamster does (**FIGURE 15.10**), indicating that we too have an endogenous circadian clock. External cues (lights, meals, jobs, alarm clocks) entrain our clock to a 24-hour period, and this circadian clock encourages the brain to sleep at some times of the day and to remain awake at others (Mistlberger, 2005). Because the free-running period is greater than 24 hours, some people in these studies have been surprised when told that the experiment has ended. They may have experienced only 19 sleep-waking cycles during a 21-day study.

In the 1930s, experimenters found that brain potentials recorded from electrodes on the scalp by **electroencephalography (EEG)** (see Figures 2.22 and 2.23) provided a way to define, describe, and classify levels of arousal and states of sleep. This measure of brain activity is usually supplemented with recordings of eye movements by **electro-oculography (EOG)** and of muscle tension by **electromyography (EMG)**. Electrophysiological measurement led to the groundbreaking discovery that there are two distinct classes of sleep: **rapid-eye-movement sleep**, or **REM sleep**, and **non-REM (NREM) sleep** (Aserinsky and Kleitman, 1953). In humans, NREM sleep can be divided further into distinct stages, which we'll discuss next.

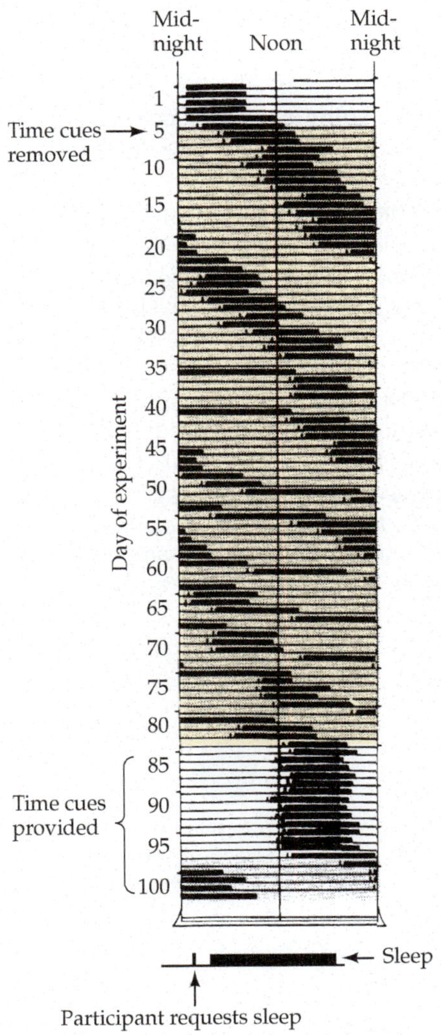

15.10 Humans Free-Run Too These sleep-waking patterns were recorded in a participant who, after 5 days, was isolated from cues about the time of day. During this period the person displayed a free-running rhythm that was a bit longer than 24 hours, getting the equivalent of 74 "nights" of sleep over the 77 days. (From E. D. Weitzman et al., 1981. In *Neurosecretion and brain peptides*, J. B. Martin et al. [Eds.], pp. 475–499. Raven: New York.)

beta activity EEG activity seen in wakefulness, comprising a mix of many different high frequencies with low amplitude.

desynchronised EEG Also called *beta activity*. A pattern of EEG activity comprising a mix of many different high frequencies with low amplitude.

alpha rhythm A brain potential of 8–12 Hz that occurs during relaxed wakefulness.

vertex spike A sharp-wave EEG pattern that is seen during stage 1 slow-wave sleep.

stage 1 sleep Also called *NREM 1*. The initial stage of NREM sleep, which is characterised by small-amplitude EEG waves of irregular frequency, slow heart rate, and reduced muscle tension.

stage 2 sleep Also called *NREM 2*. A stage of NREM sleep that is defined by bursts of EEG waves called *sleep spindles*.

sleep spindle A characteristic 12–14 Hz wave in the EEG of a person in stage 2 sleep.

K complex A sharp negative EEG potential that is seen in stage 2 sleep.

stage 3 sleep Also called *NREM 3*. A stage of NREM sleep that is defined by the presence of large-amplitude, very slow waves (delta waves).

delta wave The slowest type of EEG wave, about 1 Hz, characteristic of stage 3 sleep.

slow-wave sleep (SWS) Also called *NREM 3* and, in humans, *stage 3 sleep*. A stage of NREM sleep characterised by large-amplitude delta waves.

What are the electrophysiological distinctions that define different stages of NREM? To begin, the pattern of electrical activity in a fully awake, vigilant person is a mixture of many frequencies dominated by waves of relatively fast frequencies (greater than 15–20 cycles per second, or hertz [Hz]) and low amplitude; this type of wave activity is sometimes referred to as **beta activity** or **desynchronised EEG** (**FIGURE 15.11A**).

When you relax and close your eyes, a distinctive EEG rhythm appears, consisting of a regular oscillation at a frequency of 8–12 Hz, known as the **alpha rhythm**. As drowsiness sets in, the time spent in the alpha rhythm decreases, and the EEG shows events of much smaller amplitude and irregular frequency, as well as sharp waves called **vertex spikes**. This is the beginning of NREM, called **stage 1 sleep** (or *NREM 1*) (**FIGURE 15.11B**), which is accompanied by a slowing of heart rate and a reduction of muscle tension; in addition, under the closed eyelids the eyes may roll about slowly. Stage 1 sleep usually lasts several minutes and gives way to **stage 2 sleep** (or *NREM 2*) (**FIGURE 15.11C**), defined by waves of 12–14 Hz called **sleep spindles**, which occur in periodic bursts, and **K complexes**. If awakened during these first two stages of sleep, many people will deny that they were asleep, even though they had failed to respond to signals.

Stage 2 sleep leads to (can you guess?) **stage 3 sleep** (or *NREM 3*) (**FIGURE 15.11D**), which is defined by the appearance of large-amplitude, *very* slow waves (called **delta waves**, about one per second). In the past, researchers defined a stage 4 of sleep, when delta waves were present more than half the time. But because stages 3 and 4 are so similar, researchers today lump them together simply as stage 3 or **slow-wave sleep (SWS)**. The slow waves of electrical potential that give SWS its name represent a widespread synchronisation of cortical activity that has been likened to a room of people

Go to Activity 15.1
Stages of Sleep

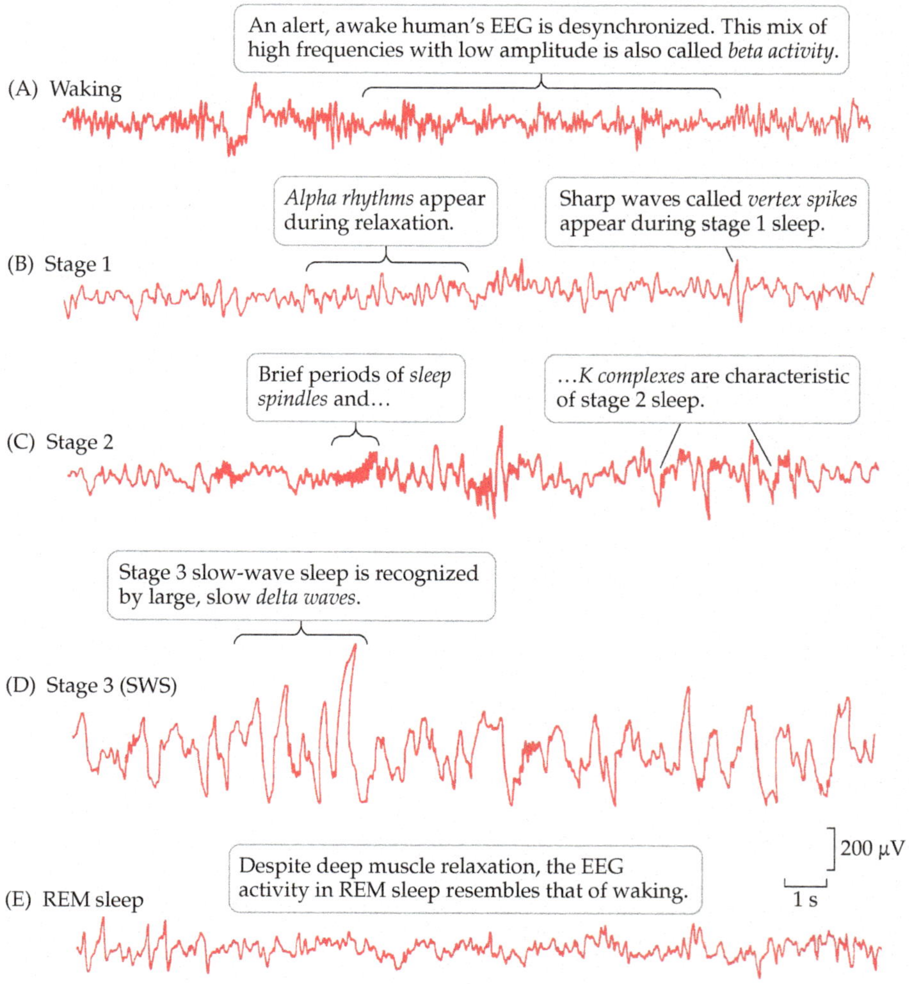

15.11 Electrophysiological Correlates of Sleep and Waking These are the characteristic EEG patterns seen during waking (A) and different stages of sleep in humans. Note the similarity of EEG activity during waking, stage 1 sleep, and REM sleep (F). (After A. Rechtschaffen and A. Kales, 1968. *A manual of standardized terminology, techniques and scoring system for sleep stages of human subjects*. U.S. National Institute of Neurological Diseases and Blindness, Neurological Information Network: Bethesda, MD.)

who are all chanting the same phrase over and over. From a distance you would be able to hear the rise and fall of the cadence of speech in a slow rhythm. But if each person were saying something different, you would hear only a buzz—the rapid frequencies of many desynchronised voices, which is like the desynchronised EEG, the beta activity of wakefulness when many parts of the cortex are saying different things to different target brain regions. During SWS, neighbouring cortical neurons tend to have synchronised activity (Poulet and Petersen, 2008), as if they were all "chanting" together rather than fulfilling different functions as they do in waking.

After about 90 minutes—the time usually required for progression through these stages, with a brief return to stage 2—something totally different occurs. Quite abruptly, scalp recordings display a pattern of small-amplitude, high-frequency activity similar in many ways to the pattern of an awake individual (**FIGURE 15.11E**), but the skeletal muscles are completely relaxed and limp. The active-looking EEG coupled with deeply relaxed muscles is typical of REM sleep. If you see a cat sleeping in the sitting, sphinx position, it cannot be in REM sleep; in REM, it will be sprawled limply on the floor (**FIGURE 15.12**).

As we'll see later, this flaccid muscle state appears despite intense brain activity, because during this stage of sleep, brainstem regions are profoundly inhibiting motor neurons. Because of this seeming contradiction—the brain waves look awake, but the musculature is flaccid and unresponsive—another name for this state is *paradoxical sleep*. In addition to the rapid eye movements under closed lids that give REM sleep its name, breathing and pulse rates become irregular. During REM sleep we also experience vivid dreams. **TABLE 15.1** compares the properties of NREM and REM sleep.

15.12 Sleep Stage Postures The kitten in (A) is enjoying slow-wave sleep, with enough muscle tone to maintain a sphinxlike posture. The kitten in (B), with profoundly relaxed muscle tone, may be in REM sleep.

TABLE 15.1 Properties of NREM and REM Sleep

Property	NREM sleep	REM sleep
AUTONOMIC ACTIVITIES		
Heart rate	Slow decline	Variable with high bursts
Respiration	Slow decline	Variable with high bursts
Thermoregulation	Maintained	Impaired
Brain temperature	Decreased	Increased
Cerebral blood flow	Reduced	High
SKELETAL MUSCULAR SYSTEM		
Postural tension	Progressively reduced	Eliminated
Knee jerk reflex	Normal	Suppressed
Phasic twitches	Reduced	Increased
Eye movements	Infrequent, slow, uncoordinated	Rapid, coordinated
COGNITIVE STATE	Mostly vague thoughts, but occasional dreams	Vivid dreams
HORMONE SECRETION		
Growth hormone secretion	High	Low
NEURAL FIRING RATES		
Cerebral cortex activity	Many cells reduced and more phasic	Increased firing rates; tonic (sustained)
EVENT-RELATED POTENTIALS		
Sensory-evoked	Large	Reduced

15.13 **The Typical Night of Sleep in a Young Adult** (After A. Kales and J. Kales, 1970. *JAMA* 213: 2229–2235.)

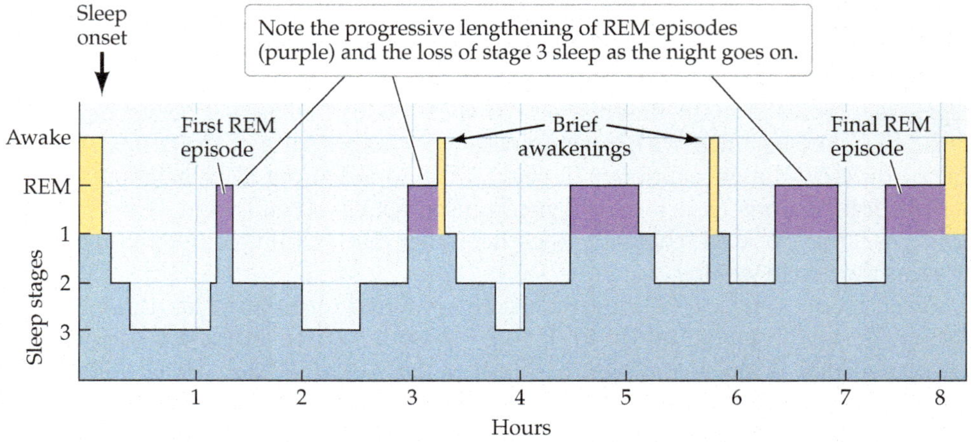

How much do we sleep, and when?

The total sleep time of young adults usually ranges from 7 to 8 hours, about half in stage 2 sleep. REM sleep accounts for about 20% of total sleep. The typical night of adult human sleep shows repeating cycles about 90–110 minutes long, recurring four or five times in a night. These cycles change in a subtle but regular manner through the night. Cycles early in the night are characterised by greater amounts of stage 3 SWS (**FIGURE 15.13**). The latter half of the night has less stage 3 sleep and more REM sleep than the first half of the night. The first REM period is the shortest, while the last REM period, just before waking, may last up to 40 minutes.

A brief arousal (yellow in Figure 15.13) occasionally occurs immediately after a REM period, and the sleeper may shift posture at this time (Aaronson et al., 1982). The sleep cycle of 90–110 minutes has been viewed as reflecting a basic ultradian rest-activity cycle (Kleitman, 1969), as we mentioned earlier.

At puberty, the circadian rhythm of sleep shifts in most people, so they get up later in the day (**FIGURE 15.14**), but many school systems require students to come to school *earlier* in the day when they hit adolescence. One group of high schools shifted their start from 7:15 to 8:40 and noted improved student attendance and enrollment, with reduced depression and sleeping in class (Wahlstrom, 2002). You may think of sleep as a simple event in your life, but for neuroscientists sleep is a remarkably complex, multifaceted set of behaviours. As one example, let's consider a fascinating aspect of sleep: dreaming.

We do most of our vivid dreaming during REM sleep

We can record the EEGs of study participants, awaken them at a particular stage (1, 2, 3, or REM), and question them about thoughts or perceptions immediately prior to awakening. Early studies of this sort suggested that dreams happen only during REM sleep, but we now know that dreams also occur in other sleep stages. What is distinctive about the dreams during REM sleep is that they are characterised by visual imagery, a feeling that the dreamer "is there" experiencing sights, sounds, smells, and acts. In classic research, people awakened from NREM sleep typically reported that they were thinking about problems rather than participating in a "mental movie" (Cartwright, 1979), but scientists have since found a more reliable indicator that a sleeper will report a vivid dream—whether there is a lot of high-frequency, and little low-frequency, EEG activity in posterior cortex

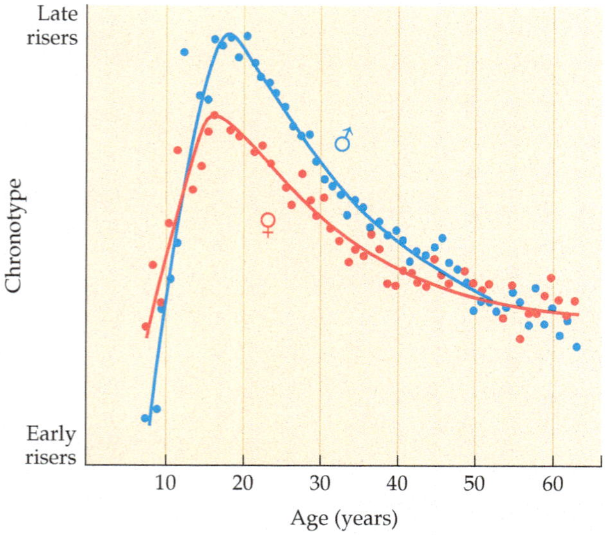

15.14 **Oh, How I Hate to Get Up in the Morning** By comparing the time of day that people of differing ages wake up each day, Roenneberg et al. (2004) confirmed the tendency for humans to become late risers at puberty. (After T. Roennenberg et al., 2004. *Curr Biol* 14: R1038–R1039.)

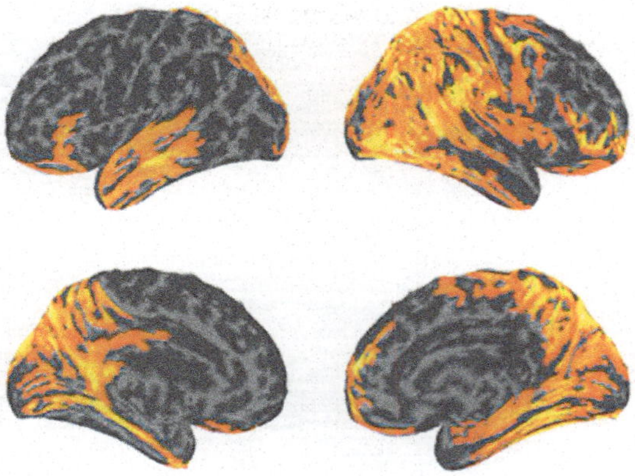

15.15 Distribution of High- versus Low-Frequency Activity during Vivid Dreaming When a person is having a dream experience, the highlighted brain regions show a greater ratio of high-frequency versus low-frequency activity. Monitoring this activity allowed the researchers to predict when a person woken from NREM would report a dream. (From F. Siclari et al., 2017. *Nat Neurosci* 20: 872–878.)

(**FIGURE 15.15**). This happens almost continually in REM sleep but also happens occasionally in NREM. By monitoring this activity, researchers could predict, with 91% accuracy, whether a sleeper who woke from NREM would report a dream experience (Siclari et al., 2017), so there seems to be more vivid dreaming during NREM than was previously believed.

Almost everyone has terrifying dreams on occasion (M. Carr and Nielsen, 2017). **Nightmares** are defined as long, frightening dreams that awaken the sleeper. Many medications make nightmares more frequent (J. F. Pagel and Helfter, 2003), but they are quite prevalent even without such influences. At least 25% of college students report having one or more nightmares per month. Have you had the common one, which Sigmund Freud had, of suddenly remembering that you must take a final exam that is already in progress?

Nightmares are occasionally confused with **night terror**, in which a sudden arousal from NREM sleep is marked by intense fear and autonomic activation. After a night terror the sleeper does not recall a vivid dream but may remember a sense of a crushing feeling on the chest, as though being suffocated (**FIGURE 15.16**). Night terrors, common in children during the early part of an evening's sleep, seem to have a genetic component (D. Petit et al., 2015).

nightmare A long, frightening dream that awakens the sleeper from sleep.

night terror A sudden arousal from sleep that is marked by intense fear and autonomic activation.

15.16 Night Terror This 1781 painting by Henry Fuseli is called *The Nightmare*. It also aptly illustrates night terrors, or even sleep paralysis, discussed later in the chapter, as the demon crushes the breath from his victim.

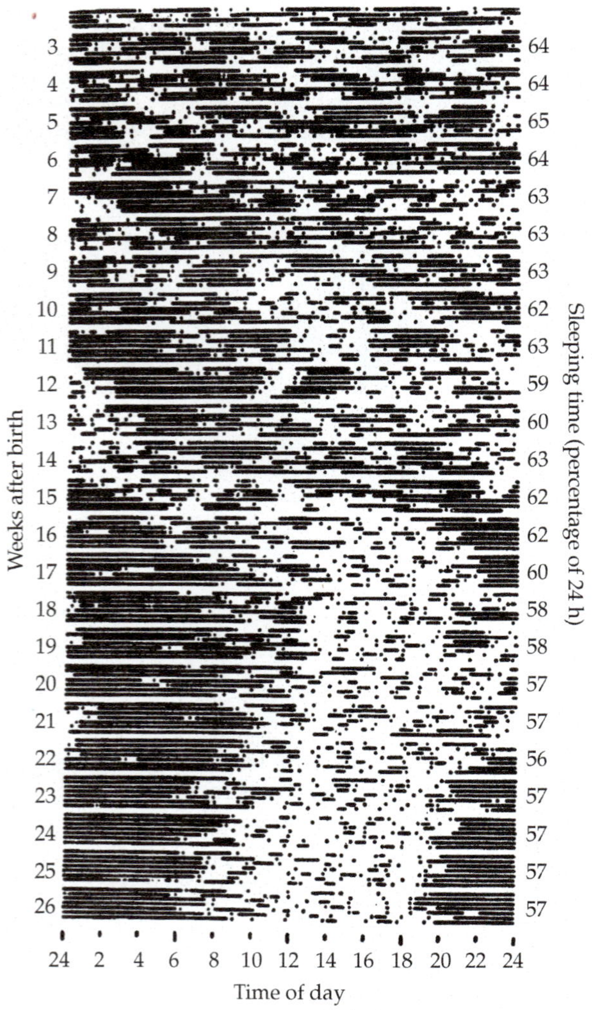

15.17 The Trouble with Babies In this baby, a stable pattern of sleep at night does not appear to be consolidated until about 16 weeks of age. The dark portions here indicate time asleep; the blank portions, time awake. (From N. Kleitman and T. Engelmann, 1953. *J Appl Physiol* 6: 269–282.)

Mammals sleep more during infancy than in adulthood

A clear cycle of sleeping and waking takes several weeks to become established in human infants (**FIGURE 15.17**). A 24-hour rhythm is generally evident by 16 weeks of age. Infant sleep is characterised by shorter sleep cycles than those of adults. This probably reflects the relative immaturity of the brain; sleep cycles in premature infants are even shorter than in full-term babies.

Infant mammals also show a large percentage of REM sleep. In humans, for example, 50% of sleep in the first 2 weeks of life is REM sleep. REM sleep is even more prominent in premature infants, accounting for up to 80% of total sleep. Unlike healthy adults, human infants can move directly from an awake state to REM sleep for the first few months of life. The REM sleep of infants is quite active, accompanied by muscle twitching, smiles, grimaces, and vocalisations. The preponderance of REM sleep early in life (**FIGURE 15.18**) suggests that this state provides stimulation that is essential to maturation of the nervous system. For example, maybe sensory feedback from those twitches guides synaptic development in the brain (Blumberg, 2015). On the other hand, killer whales and bottlenose dolphins appear to spend little or no time sleeping for the first month of life (Lyamin et al., 2005), presumably because they have to surface often to breathe. So either REM sleep does not fill a crucial need in mammalian infants, or dolphin and whale infants have found a different way to fill that need.

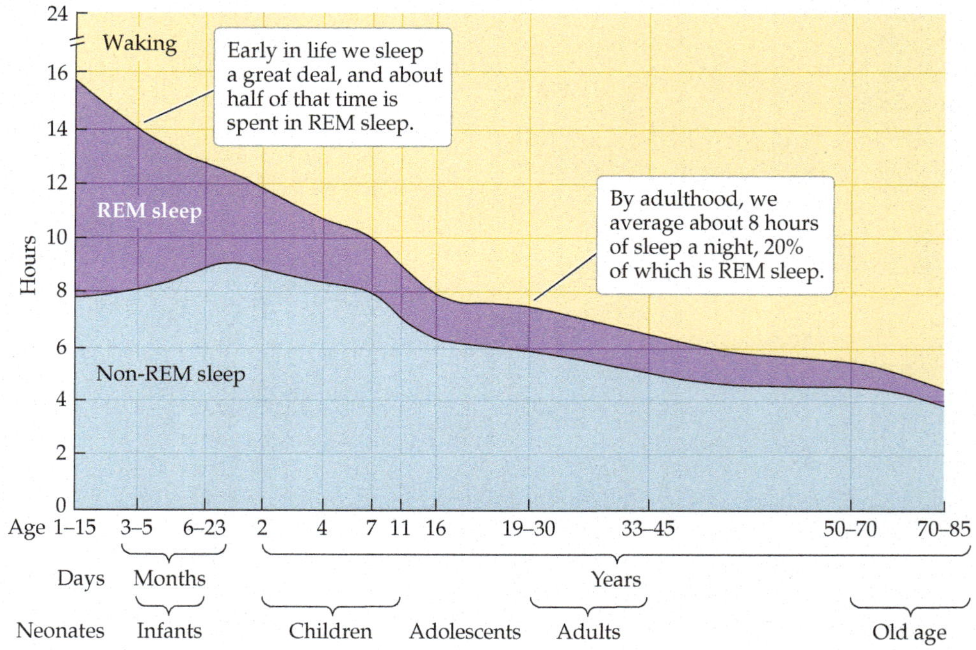

15.18 Human Sleep Patterns Change with Age (After H. P. Roffwarg et al., 1966. *Science* 152: 604–619.)

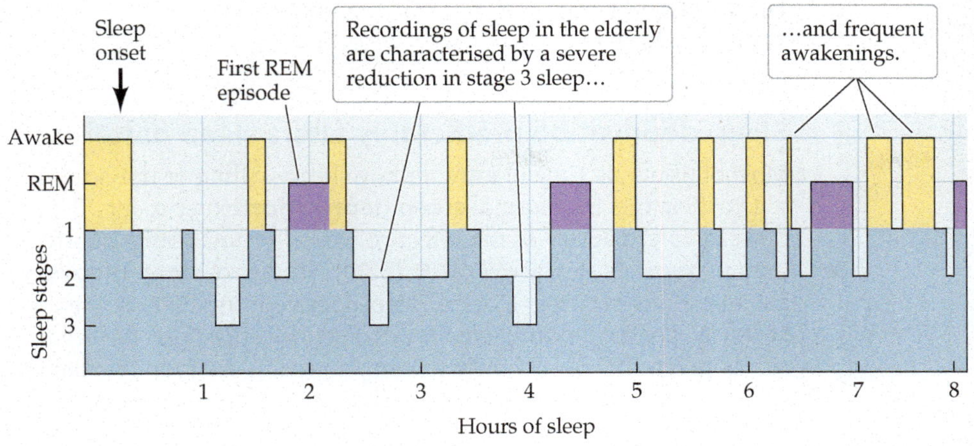

15.19 The Typical Pattern of Sleep in an Elderly Person Compare this recording with the young-adult sleep pattern shown in Figure 15.13. (After A. Kales and J. D. Kales, 1974. *New Engl J Med* 290: 487–499.)

Most people sleep appreciably less as they age

The parameters of sleep change more slowly in old age than in early development. **FIGURE 15.19** shows the pattern of the typical night of sleep in an elderly person. The total amount of sleep declines, while the number of awakenings increases (compare with Figure 15.13). Lack of sleep, or insomnia (which we will discuss at the end of this chapter), is a common complaint of the elderly, and a substantial literature suggests fragmented sleep leads to deficits in cognitive function (Scullin and Bliwise, 2015).

In humans and other mammals, the most dramatic progressive decline is in stage 3 sleep; people at age 60 spend only about half as much time in stage 3 as they did at age 20. By age 90, stage 3 sleep has disappeared. This decline in stage 3 sleep with age may be related to diminished cognitive capabilities, since an especially marked reduction of stage 3 characterizes the sleep of people with senile dementia (Kondratova and Kondratova, 2012). Growth hormone is secreted primarily during stage 3, so perhaps the loss of growth hormone from disrupted sleep in the elderly leads to the cognitive deficits.

Most elderly people fall asleep easily enough, but then they may have a hard time staying asleep. Perhaps poorly sleeping grandparents were adaptive in early human societies, providing sentinels while most of the hunter-gatherers slept (D. R. Samson et al., 2017). In modern societies, the elderly may report sleep "dissatisfaction," but as in so many things, attitude may be important. Objective measures of sleep suggest that elderly people who complain of poor sleep may actually sleep more than those who are satisfied with their sleep (McCrae et al., 2005). Perhaps, as you grow older, if you can regard waking up at 3:00 AM as a "bonus" (a little more time awake before you die), you will be more satisfied with the sleep you get!

15.4 Why and How Did Sleep Evolve?

Learning Objectives

After reading this section, you should be able to:

15.4.1 **Describe the results of sleep deprivation.**

15.4.2 **Survey the occurrence of NREM and REM sleep across species.**

15.4.3 **Describe findings that support the four most prominent theories of sleep function.**

15.4.4 **Critically discuss the significance of rare individuals who sleep very little.**

Why do most of us spend one-third of our lifetime asleep? First we'll consider the results of sleep deprivation, which signals a need for sleep, even if we don't know what that need is, exactly. We'll survey sleep across species to learn about the evolution of sleep, and just the fact that sleep occurs in every species also indicates that it must serve a vital function. After considering the four leading contenders for the function of sleep, we'll conclude with those puzzling individuals who get along fine with almost no sleep.

© iStock.com/ariswasabi

15.20 I Need Sleep! Doing without sleep has one clear effect: you feel sleepy.

sleep deprivation The partial or total prevention of sleep.

sleep recovery The process of sleeping more than is normal, after a period of sleep deprivation, as though in compensation.

Sleep deprivation predictably alters sleep patterns

Most of us at one time or another have been willing or not-so-willing participants in informal **sleep deprivation** experiments. Thus, most of us are aware of the effect of partial or total sleep deprivation: it makes us sleepy (**FIGURE 15.20**). Studies of **sleep recovery** ask questions such as, Does a sleep-deprived organism somehow keep track of the amounts and types of lost sleep? When the organism is given the opportunity to compensate, is recovery partial or complete? Can you pay off sleep debts?

THE EFFECTS OF SLEEP DEPRIVATION Early reports from sleep deprivation studies suggested a similarity between "bizarre" behaviour provoked by sleep deprivation and schizophrenia. A frequent theme in this work was the functional role of dreams as a "guardian of sanity." But examination of people with schizophrenia does not confirm this view. For example, these people can show sleep-waking cycles similar to those of healthy adults, and sleep deprivation does not exacerbate their symptoms.

The behavioural effects of prolonged, total sleep deprivation vary appreciably and may depend on some general personality factors and on age. In several studies employing prolonged total deprivation—205 hours (8.5 days)—a few participants showed occasional episodes of hallucinations. But the most common behaviour changes noted in these experiments were increases in irritability, episodes

BOX 15.1 Sleep Deprivation Can Be Fatal

Sleep that knits up the ravell'd sleave of care.

—William Shakespeare, *Macbeth*, Act II, Scene 2

Although some people seem to need very little sleep, most of us feel the need to sleep 7–8 hours a night. In fact, sustained sleep deprivation in rats causes them to increase their metabolic rate, lose weight, and, within an average of 19 days, die (Everson et al., 1989). Allowing them to sleep prevents their death.

After the fatal effect of sleep deprivation had been shown, researchers undertook studies in which they terminated the sleep deprivation before the fatal endpoint and looked for pathological changes in different organ systems (Rechtschaffen and Bergmann, 1995). No single organ system seems affected in chronically sleep-deprived animals, but early in the deprivation they develop sores on their bodies. These sores mark the beginning of the end; shortly thereafter, blood tests reveal infections from a host of bacteria, which probably enter through the sores (Everson, 1993).

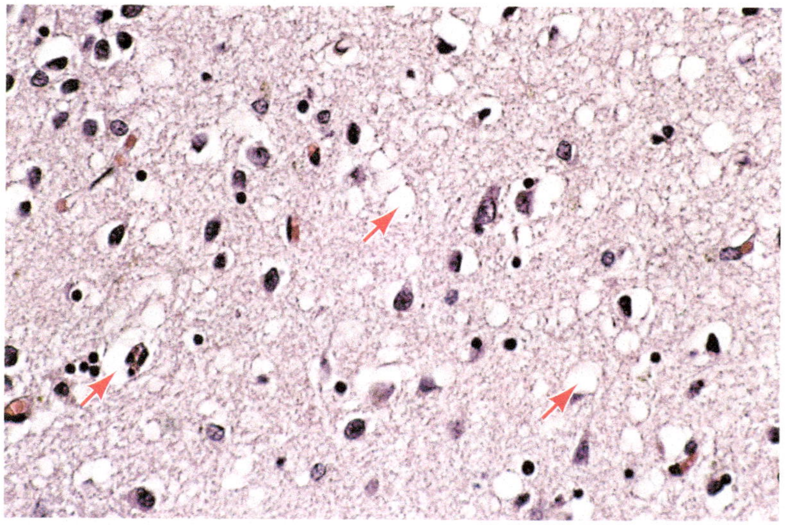

Micrograph courtesy of H. Budka

Note the large holes (arrows) that have developed in this section of frontal cortex from a person with fatal familial insomnia.

of disorientation, and difficulty in concentrating. The most reproducible effects of sleep deprivation are deficits in attention and working memory (Krause et al., 2017).

You don't need to resort to total sleep deprivation to see effects. Moderate sleep debt can accumulate with successive nights of little sleep. Volunteers who got 6 or 4 hours of sleep per night for 2 weeks showed ever-mounting deficits in attention tasks and in speed of reaction compared with those sleeping 8 hours per night (Van Dongen et al., 2003). Sleep-deprived participants often reported not feeling sleepy, yet their performance was still impaired. By the end of the study, the people getting less than 8 hours of sleep per night had cognitive deficits equivalent to people who had been totally sleep-deprived for 3 days!

Airline employees who work on schedules that give them little time to adapt to new time zones show deficits in cognitive tasks (Caldwell, 2012) and reduced volume of the brain's temporal lobe compared with employees on a schedule that permits more time to recover from jet lag (K. Cho, 2001). Is it the disruption of circadian rhythms or accumulated sleep debt that caused the difference? We don't know.

Finally, it is clear that prolonged, total sleep deprivation in mammals compromises the immune system and leads to death (**BOX 15.1**). Even fruit flies need sleep, as evidenced by the fact that a particular mutation of the *Cycle* gene (which is part of the circadian molecular clock; see Figure 15.6) causes the flies to die after only 10 hours of sleep deprivation (P. J. Shaw et al., 2002).

SLEEP RECOVERY Back in the 1960s, a young man named Randy Gardner began a science fair project of depriving himself of sleep for 11 days. No evidence of a psychotic state was noted, and the incentive for this unusually long act of not sleeping was simply his own curiosity. Researchers became involved only after he had started his deprivation schedule, which is the reason for the absence of predeprivation sleep data.

These bacteria are not normally fatal, because the rat's immune system and body defenses keep them in check; but severely sleep-deprived rats fail to develop a fever in response to these infections. (Fever helps the body fight infection.) In fact, the sleep-deprived animals show a drop in body temperature, which probably speeds bacterial infections, which in turn leads to diffuse organ damage. The decline of these severely sleep-deprived rats is complicated, but it seems clear that getting sleep improves immune system function (Everson et al., 2008). So perhaps Shakespeare's folk theory of the function of sleep, quoted above, isn't so far from the truth.

What about the rare humans who sleep only 1 or 2 hours a night? Why aren't their immune systems and inflammatory responses compromised? We don't know, but since the distinguishing trait of these people is that they don't need much sleep, perhaps their immune systems and inflammatory responses don't need much sleep either. Or perhaps the small amount of sleep they have almost every night is more efficient at doing whatever sleep does.

Some unfortunate humans inherit a defect in the gene for the prion protein (which can transmit mad cow disease, discussed in Chapter 10), and although they sleep normally at the beginning of life, in midlife they simply stop sleeping—with fatal effect. People with this disease, called **fatal familial insomnia**, die 7–24 months after the insomnia begins (Mastrianni et al., 1999; Max, 2007). Autopsy reveals degeneration of the cortex (shown in the figure) and thalamus, which may cause the insomnia (Manetto et al., 1992). Like sleep-deprived rats, sleep-deprived humans with this disorder don't have obvious damage to any single organ system but do have diffuse bacterial infections. We can't be sure, but it may be that these patients die *because* they are chronically sleep-deprived. This possibility, combined with research on rats, certainly supports the idea that prolonged insomnia is fatal.

fatal familial insomnia An inherited disorder in which humans sleep normally at the beginning of their life but stop sleeping in midlife and die 7–24 months later.

15.21 Sleep Recovery after 11 Days Awake (After G. Gulevich et al., 1966. *Arch Gen Psychiatry* 15: 29–35.)

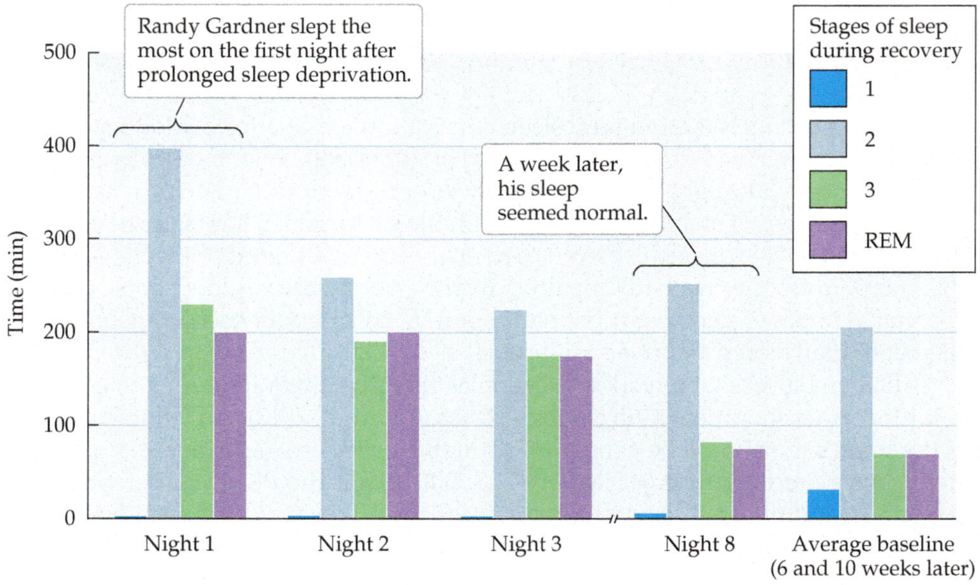

In the first night of Randy's sleep recovery, stage 3 sleep shows the greatest relative difference from normal (**FIGURE 15.21**). This increase in stage 3 sleep is usually at the expense of stage 2 sleep. However, the rise in stage 3 sleep during recovery never completely makes up for the deficit accumulated over the deprivation period. In fact, the amount is no greater than for deprivation periods of 11 days than for 5 days. REM sleep after prolonged sleep deprivation shows its greatest recovery during the *second* postdeprivation night. REM recovery may also involve another form of compensation—greater intensity: REM sleep in recovery nights is more "intense" than normal, with a greater number of rapid eye movements per period of time. So you never recover all lost sleep time, but you may make up for the loss by having more intense sleep for a while. The sooner you get to sleep, the sooner you recover.

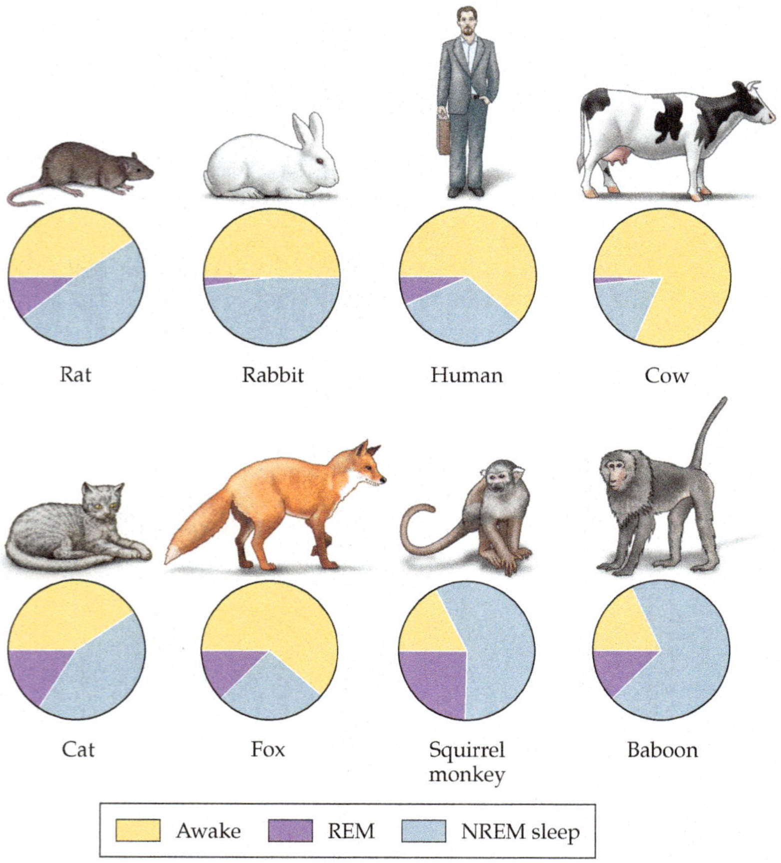

15.22 Amounts of Different Sleep States in Various Mammals

REM sleep evolved in an early vertebrate ancestor

Sleep has been studied in a wide assortment of mammals and, to a lesser extent, in reptiles, birds, and amphibians (S. S. Campbell and Tobler, 1984). Many invertebrates also have clear periods of behavioural quiescence that include heightened arousal thresholds and distinctive postures (Nath et al., 2017; Thimgan et al., 2015), and there's growing acceptance that this is indeed sleep (Donelson and Sanyal, 2015; Trojanowski and Raizen, 2016). It's even possible to record slow waves from the nervous systems of crayfish (Ramon et al., 2004) and fruit flies (Nitz et al., 2002) during their periods of inactivity. The ancient nature of sleep indicates it is an important adaptation.

REM sleep may not be quite as ancient as NREM. Nearly all mammalian species that have been investigated thus far, including the platypus (J. M. Siegel, Manger et al., 1999), the mammal most phylogenetically distant from us, display both REM sleep and NREM (**FIGURE 15.22**). Until recently, birds were the only other vertebrates known to display clear signs of both SWS and REM sleep, but apparently lizards (Libourel et al., 2018) and fish (Leung et al., 2019) show evidence of a REM-like state. If so, then REM probably arose in a vertebrate ancestor common to reptiles, birds, and mammals.

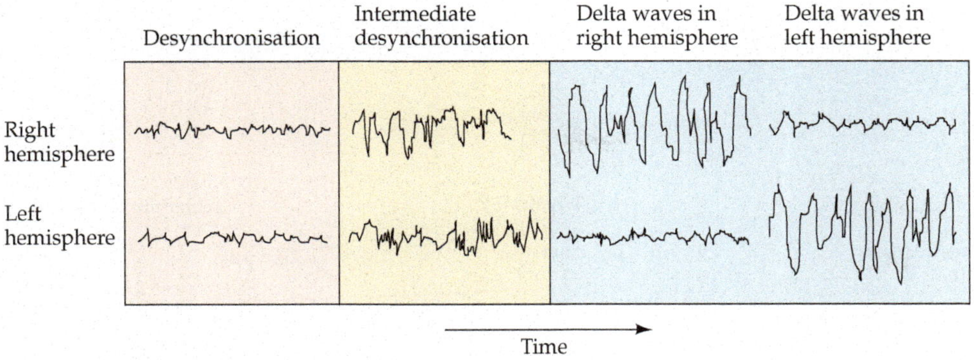

15.23 Sleep in Marine Mammals EEG patterns in right and left brain hemispheres in a dolphin (*Tursiops truncatus*) from recordings of the parietal cortex suggest that the two cerebral hemispheres take turns sleeping. (After L. M. Mukhametov, 1984. *Experimental brain research: Suppl 8: Sleep mechanisms*, A. Borbély and J. L. Valatx [Eds.]. Springer-Verlag: Berlin.)

Dolphins don't display REM sleep, but their lack of REM sleep is probably a later adaptation that evolved when their land-dwelling ancestors took to the water, because they must come to the surface of the water to breathe. That requirement may be incompatible with the deep relaxation of muscles during REM sleep. Another dolphin adaptation to living in water is that only one side of the dolphin brain engages in SWS at a time (Mukhametov, 1984). It's as if one whole hemisphere is asleep while the other is awake (**FIGURE 15.23**). During these periods of "unilateral sleep," the animals continue to come up to the surface occasionally to breathe. That could explain why seals show unilateral sleep more often in the water than on land (Lyamin et al., 2018). Birds can also display unilateral sleep—one hemisphere sleeping while the other hemisphere watches for predators (Rattenborg, 2006). Unilateral sleep while gliding may also enable birds to fly long distances (e.g., 10,000 miles!) without stopping (Gill et al., 2009). In humans, the left hemisphere displays less slow-wave activity than the right during the first night in a sleep lab, suggesting that side of the brain remains more vigilant in this unfamiliar locale. The second night in the lab, when people usually sleep better, sleep depth is again equivalent between the hemispheres (Tamaki et al., 2016).

sleep cycle A period of slow-wave sleep followed by a period of REM sleep. In humans, a sleep cycle lasts 90–110 minutes.

A **sleep cycle** is a period of NREM sleep followed by an episode of REM sleep. For laboratory rats, one sleep cycle lasts an average of 10–11 minutes; for humans, one cycle lasts 90–110 minutes, as we said earlier. Across species, cycle duration is inversely related to metabolic rate; that is, small animals, which tend to have high metabolic rates (see Chapter 14), have short sleep cycles, and large species have long sleep cycles.

Despite the ancient nature of sleep, it isn't clear what, exactly, it does for us. The four functions most often ascribed to sleep are energy conservation, niche adaptation, body restoration, and memory consolidation; let's consider each in turn.

Sleep conserves energy

We use up less energy when we sleep than when we're awake. For example, stage 3 sleep is marked by reduced muscular tension, lowered heart rate, reduced blood pressure, reduced body temperature, and slower respiration. This diminished metabolic activity during sleep suggests that one role of sleep is to conserve energy.

We can see the importance of this function by looking at the world from the perspective of small animals. There is a strong correlation between total amount of sleep per day and waking metabolic rate: small animals sleep more than large species, at least among plant-eating species (**FIGURE 15.24A**). Since smaller mammals lose heat faster (see Chapter 14), they burn more energy per gram of body weight just to maintain body temperature, and so perhaps they sleep more to conserve more energy. Interestingly, predatory species show no such correlation, perhaps because predators don't have to remain as vigilant as prey species

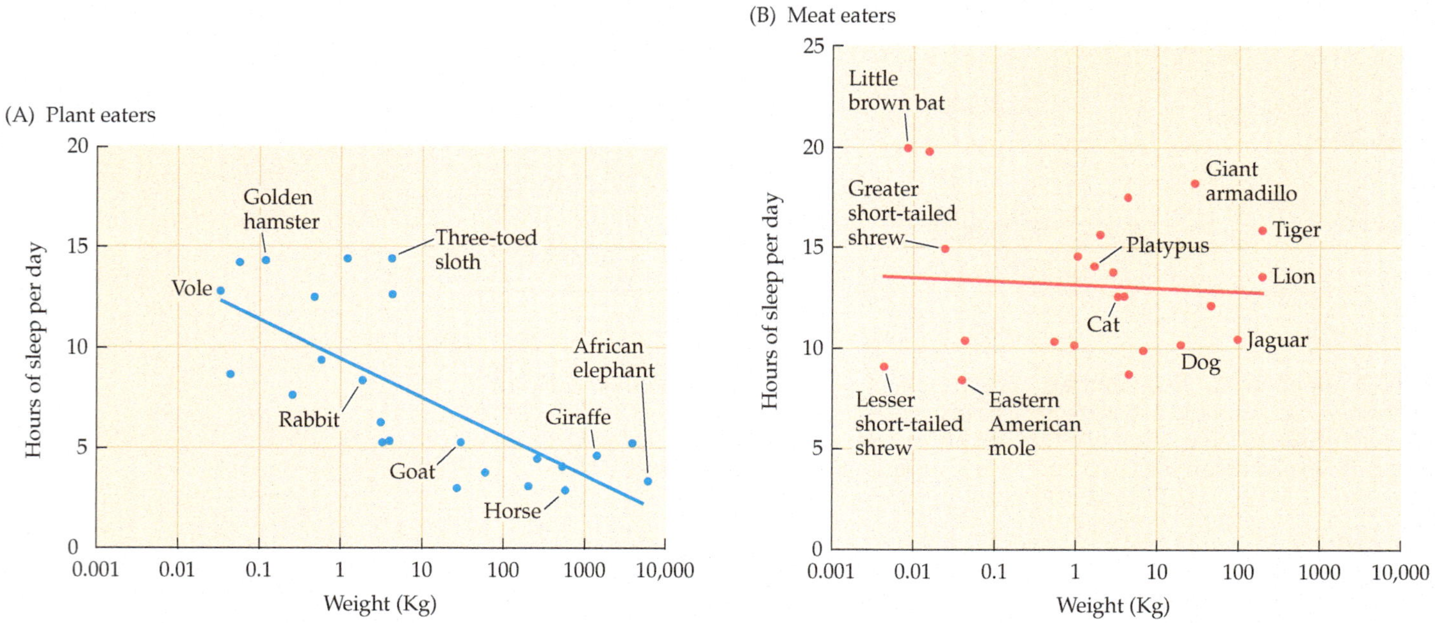

15.24 Relationship between Body Size and Sleep Time (A) Among plant eaters, the larger the body is, the less time is spent asleep. (B) Meat eaters sleep a lot, no matter what their body size—presumably because they are more secure when asleep than plant eaters. (From J. M. Siegel, 2005. *Nature* 437: 1264–1271.)

do (**FIGURE 15.24B**). In zoos and labs, where animals don't have to spend time looking for food, the herbivores, eating low-calorie food, have to spend more time actually eating their food than do the omnivores, which in turn need more time to eat than the carnivores that exclusively eat high-calorie meat (namely, herbivores and omnivores!). Consequently, carnivores sleep more than omnivores, which sleep more than herbivores in these settings (Siegel, 2012).

Sleep enforces niche adaptation

ecological niche The unique assortment of environmental opportunities and challenges to which each organism is adapted.

Almost all animals are either nocturnal or diurnal. This specialisation for either nighttime or daytime activity is part of the **ecological niche** of each species, that unique assortment of environmental opportunities and challenges to which each organism is adapted. Thanks to these adaptations, each species is better at gathering food either at night or in the daytime, and each species is also better at avoiding predators either during the day or at night (**FIGURE 15.25**). If you're a nocturnal mammal, like a mouse, you are adept at sneaking around in the dark, using acute hearing and smell to navigate and find food. The rest of the time, during daylight, you should spend holed up somewhere safe from diurnal predators. So, one important function of sleep is to force the individual to conform to a particular ecological niche for which it is well adapted (Meddis, 1975). If it is adaptive to forgo sleep, as it is during the 3 weeks of the year when male pectoral sandpipers compete intensively for mates in the round-the-clock daylight of an Arctic summer, then animals can do without sleep. Those males that sleep the least sire the most offspring (Lesku et al., 2012). But as the breeding season ends and periods of dark return, they might as well sleep at night. From this perspective, sleep debt and the unpleasant feelings of sleepiness are simply tools that evolved to enforce inactivity (J. M. Siegel, 2009). This certainly is a function of sleep, and it must have played an important role in its evolution.

Sleep restores the body and brain

If someone asked you why you sleep, chances are you would answer that you sleep because you're tired. Indeed, one of the proposed functions of sleep is simply the rebuilding or restoration of materials used during waking, such as proteins (Moruzzi, 1972). Maybe this is why most growth hormone release happens during SWS.

15.25 Sleep Helps Animals to Adapt to an Ecological Niche Many diurnal primates sleep in trees to avoid predators at night.

If simple wear and tear from activity triggered the need for sleep, then you'd expect exercise to result in more sleep. But while exercise during the day may cause people to fall asleep more quickly, it generally does *not* help them sleep longer. We've seen that prolonged and total sleep deprivation—either forced on rats or as a result of inherited pathology in humans—interferes with the immune system and leads to death (see Box 15.1). Even relatively mild deprivation, having sleep shortened or disrupted (e.g., by a nurse taking vital signs every hour), makes people more sensitive to pain the following day (R. R. Edwards et al., 2009). A study of over a million Americans found that those sleeping less than 6 hours per night were more likely to die in the next year, although, interestingly, people who slept *more than 8* hours per night were also at greater risk (Kripke et al., 2002). People who sleep less than 5 hours per night are more likely to develop diabetes (Gangwisch et al., 2007). Perhaps the most impressive link between sleep and health is the finding that people who work at night and sleep in the daytime are more likely to develop cancer (Erren et al., 2009). So the widespread belief that sleep helps us ward off illness is well supported by research (S. Cohen et al., 2009; Imeri and Opp, 2009).

Sleep also restores the brain by allowing it to get rid of waste products. Glia control the flow of cerebrospinal fluid through a microscopic network of channels throughout the brain—the *glymphatic* system—collecting and disposing of toxins that build up (see Figure 1.23). This flow is much faster during sleep than wakefulness (Xie et al., 2013), flushing out brain waste products as we snooze. One of the waste products that is cleared much faster during sleep is β-amyloid (Holth et al., 2019). Because the buildup of β-amyloid is a suspected cause of Alzheimer's disease (see Figure 6.31), adequate sleep may postpone the disorder.

Sleep aids memory consolidation

A peculiar property of dreams is that, unless we tell them to someone or write them down soon after waking, we tend to forget them, as though the brain refuses to consolidate information presented during REM sleep. It is probably beneficial that most dreams are not stored in long-term memory, because it would be counterproductive to squander permanent memory storage space on events that never happened.

How about learning during sleep? Despite ads you might read in the backs of magazines, you cannot learn new material while you're sleeping (Druckman and Bjork, 1994). Putting a speaker under your pillow to recite material for a final exam will not work unless you stay awake to listen (Emmons and Simon, 1956).

In 1924, however, Jenkins and Dallenbach reported an experiment suggesting that sleep helps you learn or remember material or events experienced *before* you go to bed. They trained some participants in a verbal learning task at bedtime and tested them 8 hours later on arising from sleep; other participants they trained early in the day and tested 8 hours later (with no intervening sleep). Participants displayed better retention when a period of sleep intervened between the learning period and the test, suggesting that sleep aids the *consolidation* of memory (Diekelmann and Born, 2010). Performance on nonverbal tasks, such as the insight that there is a "hidden rule" in a sequence of seemingly random digits, is also better if participants sleep between the first and second set of trials (Ellenbogen et al., 2007). Conversely, sleep deprivation makes people more susceptible to forming false memories (see Chapter 16) (Frenda et al., 2014), which may explain why sleep-deprived suspects sometimes confess to crimes they did not commit.

NREM aids consolidation of memory (Nishida and Walker, 2007), specifically declarative memory (i.e., memory that can be stated or described, as was tested in the verbal learning task experiment by Jenkins and Dallenbach). In fact, boosting cortical slow-voltage oscillations during NREM, using stimulating electrodes over the skull, enhances consolidation of declarative memories (L. Marshall et al., 2006).

Patterns of neuronal activity seen while a task is being learned during wakefulness are re-created during subsequent NREM sleep, as if the brain were "rehearsing" the material (Euston et al., 2007; Horikawa et al., 2013; Ji and Wilson, 2007). Similarly, the patterned activity of neurons observed in birdsong nuclei while male zebra finches are learning to sing appears to be repeated during subsequent bouts of sleep

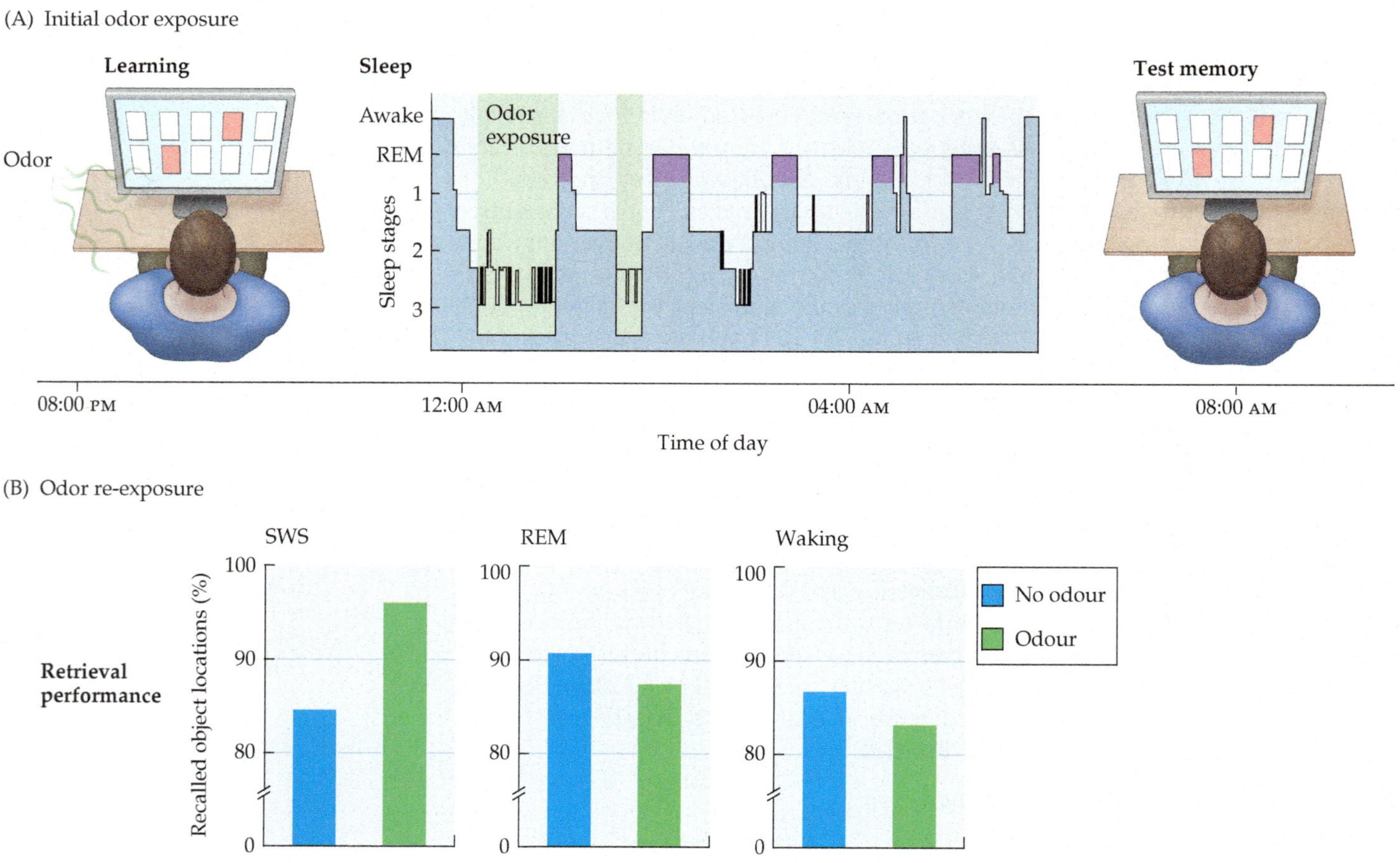

15.26 Memory Consolidation in Slow-Wave Sleep (A) Participants learned the location of various objects in a concentration game on a computer screen while exposed to an odour. That night they were exposed to the same odour while asleep, during either SWS, as shown here, or REM sleep, then tested upon awakening. (B) Participants exposed to the odour during SWS showed improved memory for the task, while those exposed to the odour during REM sleep or while awake showed no benefit. (After S. Diekelmann and J. Born, 2010. *Nat Rev Neurosci* 11: 114–126; B. Rasch et al., 2007. *Science* 315: 1426–1429.)

(Shank and Margoliash, 2009). Further supporting the idea that declarative memory is consolidated during NREM are studies using a cue to "reactivate" learning during sleep. Participants learned the placement of objects on a computer screen while being exposed to an odour, say of roses. Participants who were later reexposed to that same odour while in NREM showed improved consolidation of memory upon awakening (Diekelmann and Born, 2010). Exposure to the odour in REM sleep, or while awake, had no effect (**FIGURE 15.26**). In another study, participants learned to associate various objects with their different locations, and then they napped. If they were exposed to sounds associated with particular objects (a "meow" for cat, a whistle for a teakettle) during NREM, that strengthened their memory of where those specific objects were located when they awoke (I. M. Vargas et al., 2019).

The memory consolidation function of sleep may involve determining which synapses are gained or lost. Two different studies have reported that during sleep there is a widespread weakening of synapses, in terms of both size (de Vivo et al., 2017) and physiological strength (Diering et al., 2017), suggesting that sleep "resets" synapses globally so that the minority that were strengthened by wakeful experience can stand out.

This notion that synapses are strengthened during wakeful experience and then consolidated during sleep by a global scaling back of other synapses was reinforced by study of mouse mutants. In a heroic effort, scientists monitoring the sleep of over 8000 mice carrying mutations discovered one that slept 15 hours per day, dubbed *Sleepy*, and another that had a drastic reduction in REM, called *Dreamless* (Funato et al., 2016). The *Dreamless* mutation was in the gene for an ion channel that was known to influence circadian rhythms in flies. The *Sleepy* mutation was in the gene *Sik3* for a protein kinase (an enzyme that adds phosphate groups on proteins) that had already been implicated in learning and memory. Follow-up studies identified 80 proteins, most of which are found in synapses, normally phosphorylated by the Sik3 protein that are *hyper*phosphorylated in *Sleepy* mutants (Z. Wang et al., 2018), so apparently the *Sleepy* mutation made the enzyme more effective. Phosphorylation of these same synaptic proteins also increases in wild-type mice as they are sleep-deprived, so the scientists suggest that continual phosphorylation of the proteins by waking experience increases sleep need (so that's why *Sleepy* mutants are so sleepy), which is relieved when sleep dephosphorylates them at all synapses except those few that are consolidating a new memory. Tests of this intriguing hypothesis are sure to follow.

Some humans sleep remarkably little, yet function normally

One challenge to all the theories about the function of sleep is the existence of a few people who seem perfectly healthy, yet sleep hardly at all. These cases are more than just folktales. William Dement (1974) described a Stanford University professor who slept only 3–4 hours a night for more than 50 years and died at age 80. Sleep researcher Ray Meddis (1977) found a cheerful 70-year-old retired nurse who said she had slept little since childhood. She was a busy person who easily filled up her 23 hours of daily wakefulness. During the night, she sat in bed reading or writing, and at about 2:00 AM she fell asleep for an hour or so, after which she readily awakened.

For her first 2 days in Meddis's laboratory, she did not sleep at all, because it was all so interesting to her. On the third night she slept a total of 99 minutes, with both SWS and REM sleep periods. Later her sleep was recorded for 5 consecutive nights, when she slept an average of 67 minutes per night. She never complained about not sleeping more, and she did not feel drowsy during either the day or the night (**FIGURE 15.27**). Meddis described several other people who slept only an hour or two per night. Some of these people reported having parents who slept little. Activity monitors in wild African elephants indicate they sleep only about 2 hours per night (Gravett et al., 2017). Whatever the function of sleep is, elephants and a few rare people possess some way of fulfilling it in a much briefer time than the rest of us. Most important, though, is that no mammalian species and no healthy person has ever been found who does not sleep at all.

Courtesy of Ray Meddis

15.27 A Nonsleeper When Ray Meddis brought this 70-year-old nurse into the lab for sleep recording, he confirmed that she slept only about an hour per night. Yet she was a healthy and energetic person. Upon death in her mid-70s, she had been awake for a total of 20 years longer than someone who had slept 8 hours a night. Here she is strolling in the garden with Meddis's son.

15.5 At Least Four Interacting Neural Systems Underlie Sleep

Learning Objectives

After reading this section, you should be able to:

15.5.1 Describe a common mechanism at work in most general anaesthetics.
15.5.2 Offer evidence of a forebrain region that produces SWS sleep.
15.5.3 Describe an ascending arousal system in the brainstem.
15.5.4 Name two brain regions important for REM sleep, citing evidence to support the claim.
15.5.5 Describe evidence from the study of narcolepsy for a hypothalamic sleep switch.

At one time sleep was regarded as a passive state, as though most of the brain simply stopped working while we slept, leaving us unaware of events around us. We now know that sleep is an active state mediated by at least four interacting neural systems:

1. A *forebrain* system that by itself can display SWS
2. A *brainstem* system that activates the forebrain into wakefulness
3. A *pontine* system that triggers REM sleep
4. A *hypothalamic* system that affects the other three brain regions to determine which state the brain will be in

The forebrain generates slow-wave sleep

General anaesthetics—drugs such as barbiturates and anaesthetic gases that render people unconscious during surgery—produce slow waves in the EEG that resemble those seen in SWS (Franks, 2008). This finding suggests that general anaesthetics tap into existing brain networks promoting sleep. While some general anaesthetics are glutamate antagonists and therefore block neuronal excitation throughout the brain, virtually all general anaesthetics are noncompetitive agonists at $GABA_A$

general anaesthetic A drug that renders an individual unconscious.

15.28 Neurotransmitter Systems Affected by General Anaesthetics Note that all general anaesthetics increase $GABA_A$ receptor signalling—a clue that these receptors, which tend to inhibit neuronal activity, may play a role in sleep. Some anaesthetics also inhibit glutamate receptors, which usually excite neurons, while they stimulate glycine receptors, which usually inhibit neurons. Singer Michael Jackson died of an overdose of propofol that was given to him to induce sleep, but the drug does not provide the normal stages of sleep. (After M. T. Alkire et al., 2008. *Science* 322: 876–880.)

	Anaesthetics	Receptor types							
			Glutamate				Acetylcholine		
		GABA	NMDA	AMPA	Potassium channel	Glycine	Nicotinic	Muscarinic	Serotonin
Intravenous	Barbiturates	Major agonist	No effect	Major antagonist	No effect	Minor agonist	Minor antagonist	Major antagonist	Minor antagonist
	Propofol	Major agonist	Minor antagonist	Minor antagonist	No effect	Major agonist	Minor antagonist	Minor antagonist	No effect
	Ketamine	Minor agonist	Major antagonist	No effect		No effect	Minor antagonist	Major antagonist	Minor agonist
Inhalational	Nitrous oxide	Minor agonist	Major antagonist	Minor antagonist	Major agonist	Minor agonist	Major antagonist	Minor antagonist	Major antagonist
	Isoflurane	Major agonist	Minor antagonist	Major antagonist	Major agonist	Major agonist	Minor antagonist	Minor antagonist	Major agonist

Major agonist | Minor agonist | Major antagonist | Minor antagonist | No effect

isolated brain Sometimes referred to by the French term *encéphale isolé*. An experimental preparation in which an animal's brainstem has been separated from the spinal cord by a cut below the medulla.

isolated forebrain Sometimes referred to by the French term *cerveau isolé*. An experimental preparation in which an animal's nervous system has been cut in the upper midbrain, dividing the forebrain from the brainstem.

receptors (**FIGURE 15.28**). Because they boost $GABA_A$ receptors' inhibitory effect on neurons, the action of these anaesthetics suggests that some brain system normally uses GABA to inhibit neuronal activity and promote SWS. But where is that system?

Some of the earliest studies of sleep indicated that the system promoting SWS is in the forebrain. These were experiments in which the brain was transected—literally cut into two parts: an upper part and a lower part. The entire brain can be isolated from the body by an incision between the medulla and the spinal cord. This preparation was first studied by the Belgian physiologist Frédéric Bremer (1892–1982), who called it the **isolated brain** (Bremer, 1938).

The EEGs of animals with such transections show signs of waking alternating with sleep (**FIGURE 15.29A**). During EEG-defined wakeful periods, the pupils are dilated and the eyes follow moving objects. During EEG-defined sleep, the pupils are small, as in normal sleep. REM sleep can also be detected in the isolated brain by measuring other local electrical signals. These results demonstrate that wakefulness, SWS, and REM sleep are all mediated by *networks within the brain*.

If the transection is made higher along the brainstem—in the midbrain—a very different result is achieved. Bremer referred to such a preparation as an **isolated forebrain**,

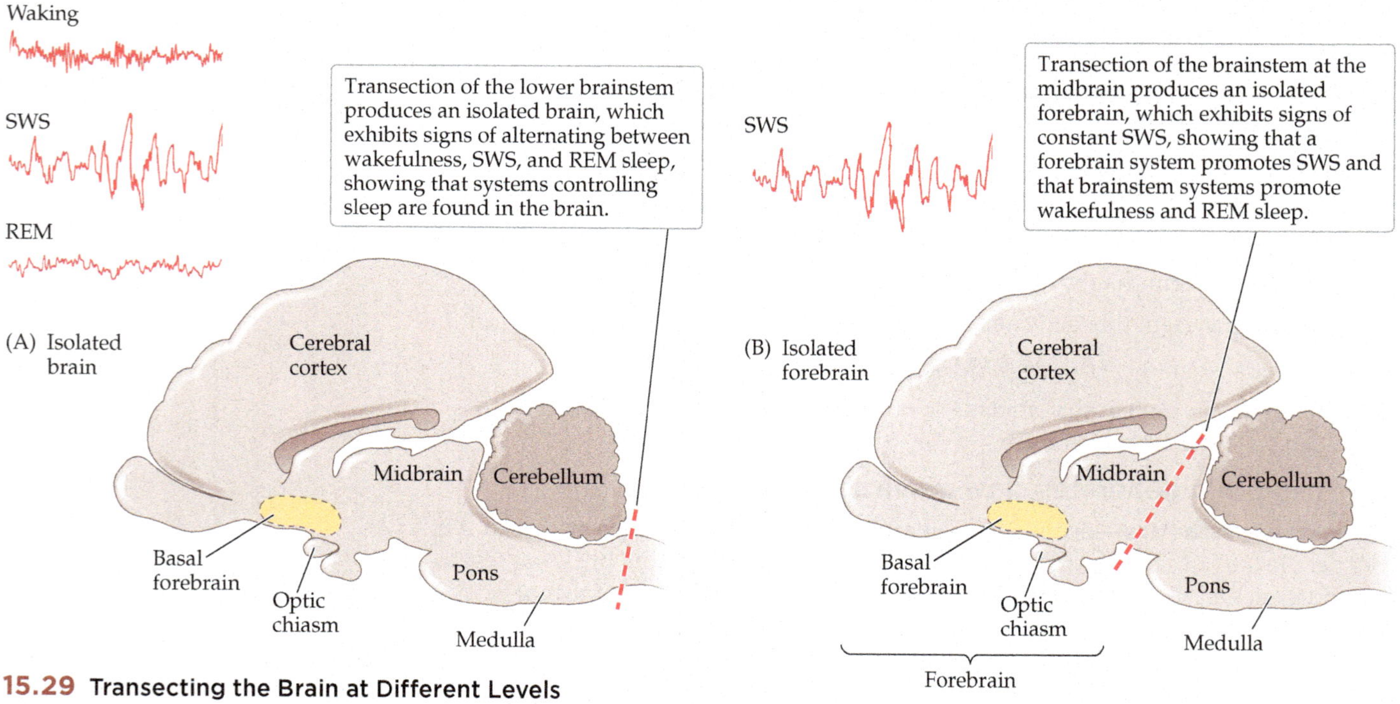

15.29 Transecting the Brain at Different Levels

and he found that the EEG from the brain in front of the cut displayed constant SWS (**FIGURE 15.29B**). The isolated forebrain does not show REM sleep, so it appears that the forebrain alone can generate SWS, without contributions from the lower brain regions.

The constant SWS seen in the cortex of the isolated forebrain appears to be generated by the **basal forebrain** in the ventral frontal lobe and anterior hypothalamus, including the ventrolateral preoptic area (**FIGURE 15.30A**). Electrical stimulation of the basal forebrain can induce SWS activity (Clemente and Sterman, 1967), while lesions there suppress sleep (McGinty and Sterman, 1968). Within the basal forebrain, neural circuits regulate activity of GABA-ergic neurons (M. Xu et al., 2015) that send their axons to two important targets. One target is the nearby **tuberomammillary nucleus** in the posterior hypothalamus where stimulation of $GABA_A$ receptors promotes sleep (S. Chung et al., 2017). These seem to be the $GABA_A$ receptors that general anaesthetics act on to make us unconscious and induce slow waves in the EEG (see Figure 15.29). The other target of basal forebrain GABA-ergic inhibition is an ascending arousal system in the brainstem that we consider next.

basal forebrain A ventral region in the forebrain that has been implicated in consciousness and sleep.

tuberomammillary nucleus A region of the basal hypothalamus, near the pituitary stalk, that plays a role in generating SWS.

reticular formation An extensive region of the brainstem (running from the medulla through the thalamus) that is involved in arousal (waking).

A brainstem system wakes up the forebrain

In the late 1940s, electrical stimulation of an extensive region of the brainstem known as the **reticular formation** was shown to activate the cortex (**FIGURE 15.30B**). The reticular formation consists of a diffuse group of cells whose axons and dendrites course in many directions, extending from the medulla through the thalamus. Electrically stimulating the reticular formation awakens animals rapidly (Moruzzi and Magoun, 1949). Lesions of these regions produced persistent sleep in the animals. The basal forebrain region actively imposes sleep on the brain in part by inhibiting the reticular formation. Between them, the basal forebrain and the brainstem reticular formation push the brain back and forth from NREM to wakefulness. Now let's consider the systems triggering REM.

Go to Activity 15.2
Sleep Mechanisms

The pons triggers REM sleep

Several methods pinpointed the region of the pons that is important for REM sleep. Large lesions of the pons abolish REM sleep (**FIGURE 15.30C**) (L. Friedman and Jones, 1984). Electrical stimulation of the same region, or pharmacological stimulation of this region with cholinergic agonists, can induce or prolong REM sleep.

15.30 Brain Mechanisms Underlying Sleep

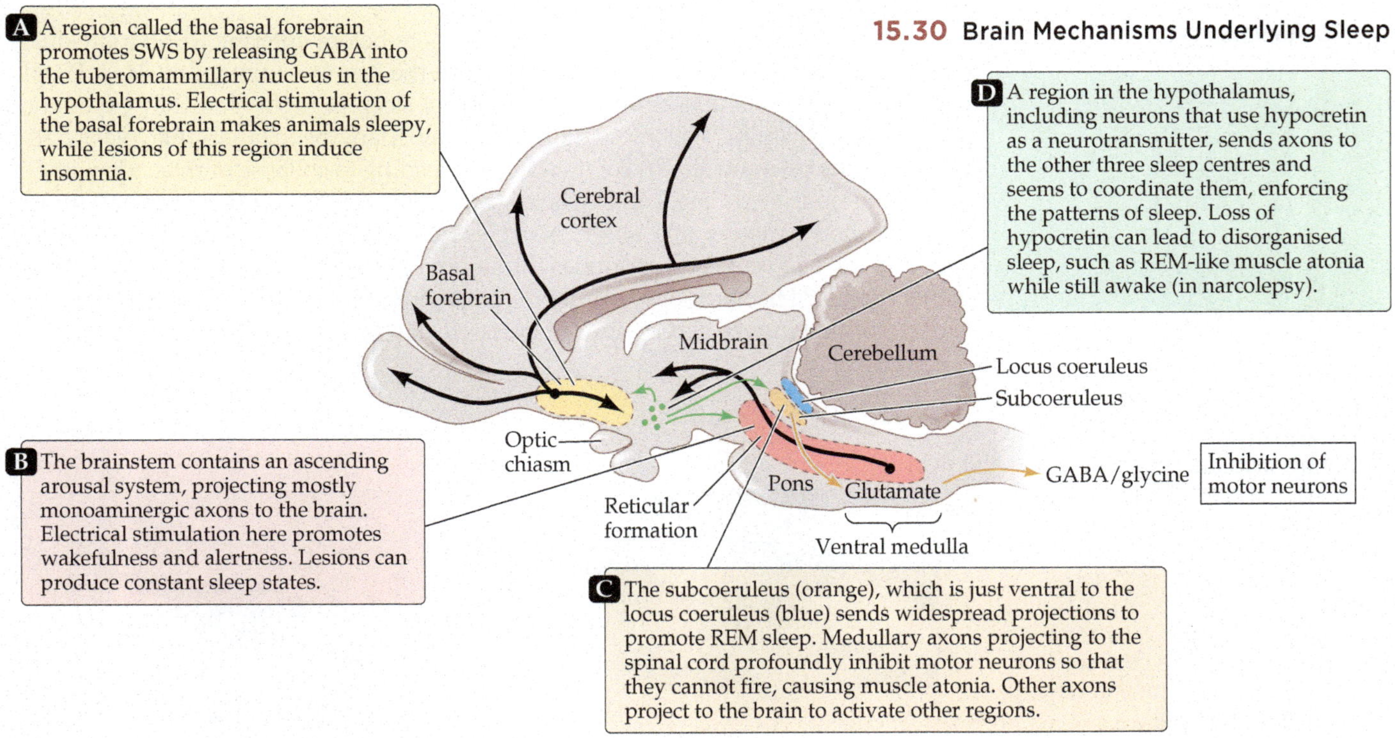

subcoeruleus Also called *sublaterodoursal nucleus*. A brain region just ventral to the locus coeruleus that is associated with REM sleep.

narcolepsy A disorder that involves frequent, intense episodes of sleep, which last from 5 to 30 minutes and can occur anytime during the usual waking hours.

cataplexy Sudden loss of muscle tone, leading to collapse of the body without loss of consciousness.

Eventually research pinpointed a specific region of the pons, just ventral to the locus coeruleus, which we'll call the **subcoeruleus** (or *sublaterodoursal nucleus* of the pons), where some neurons are active only during REM sleep (Weber and Dan, 2016).

One job of the subcoeruleus is to profoundly inhibit motor neurons to keep them from firing. Glutamatergic neurons in the subcoeruleus excite neurons in the ventral medulla, which in turn project axons down the spinal cord to release the inhibitory transmitters GABA and glycine to deeply inhibit spinal motor neurons (Hayashi et al., 2015; Weber et al., 2015). This descending inhibition prevents motor neurons from reaching threshold and producing an action potential (Kodama et al., 2003). Thus, the dreamer's muscles are not just relaxed, but flaccid.

This loss of muscle tone during REM sleep is sometimes abolished by small lesions in the subcoeruleus (A. R. Morrison, 1983). Cats with such lesions seem to act out their dreams. They enter NREM sleep like a normal cat, but as they begin to display the desynchronised EEG of REM sleep, instead of becoming completely limp as a normal cat does, these cats stagger to their feet (**FIGURE 15.31**). Are they awake or in REM sleep? They move their heads as though visually tracking moving objects (that aren't there), bat with their forepaws at nothing, and ignore objects that are present. In addition, the cat's inner eyelids, the translucent nictitating membranes, partially cover the eyes. Thus, the cat appears to be in REM sleep, but motor activity is not being inhibited by the brain. You can watch one of these cats acting out its dream on the website.

So far we've described three interacting brain systems controlling sleep: an NREM-promoting region in the forebrain, an ascending arousal system in the brainstem, and the subcoeruleus in the pons to trigger paralysis of the body during REM. There is a fourth important system, which seems to act as a "switch" among these three centres, in the hypothalamus. To understand how we learned about this fourth system, you need to consider the rare but fascinating condition called *narcolepsy*.

A hypothalamic sleep centre was revealed by the study of narcolepsy

You might not consider getting lots of sleep an affliction, but many people are either drowsy all the time or experience sudden attacks of sleep. At the extreme of such tendencies is **narcolepsy**, an unusual disorder in which the person has frequent, intense attacks of sleep that last 5–30 minutes and can occur at any time during usual waking hours. These sleep attacks occur several times a day—usually about every 90 minutes (Dantz et al., 1994).

Go to Video 15.3
Animal Sleep Activity

Most people display NREM for an hour or more before entering REM; individuals with narcolepsy, however, tend to enter REM in the first few minutes of sleep. People with this disorder exhibit a disrupted pattern of sleep at night, and they experience abrupt, overwhelming sleepiness during the day. Many people with narcolepsy also show **cataplexy**, a sudden loss of muscle tone, leading to collapse of the body without loss of consciousness. Cataplexy can be triggered by sudden, intense emotional stimuli, including both laughter and anger. Narcolepsy usually manifests itself between the ages of 15 and 25 years and continues throughout life. Remember Barry from the start of this chapter? His narcolepsy symptoms began in his freshman year of college, when he started showing the classic signs of excessive daytime sleepiness and cataplexy.

(A) (B)

From A. R. Morrison et al., 1995. *Behav Neurosci* 109: 972–979. © American Psychological Association

15.31 Acting Out a Dream (A) This cat received a lesion near the locus coeruleus that blocked the complete paralysis (atonia) that normally accompanies REM sleep. Following a bout of SWS, the cat in REM rises up as though about to pounce, but its eyes are closed and nothing is there. (B) The standing cat is wobbly, "looking" at something we cannot see. These behaviours indicate the cat is dreaming—seeing and interacting with objects that aren't really there.

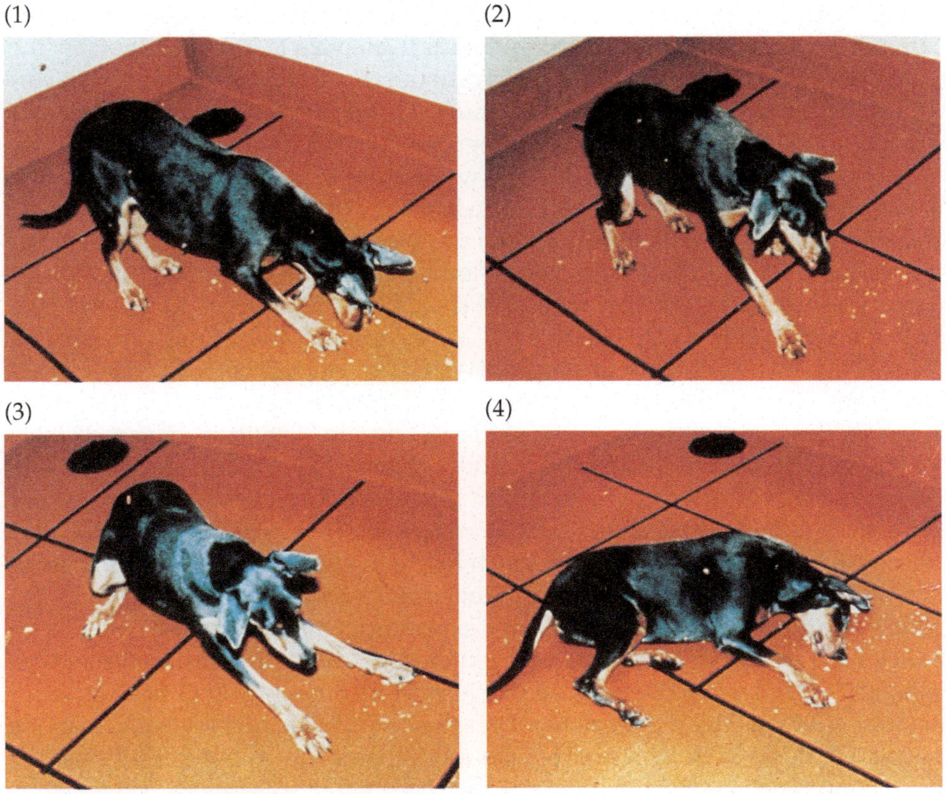

Courtesy of Seiji Nishino

15.32 Narcolepsy in Dogs A narcoleptic dog that suffers cataplexy when excited is offered a food treat (1), becomes wobbly (2), lies down (3), and finally falls limply to the floor (4).

Go to Video 15.4
Narcolepsy in Dogs

Several strains of dogs exhibit narcolepsy (Aldrich, 1993), complete with sudden collapse and very short latencies to sleep onset (**FIGURE 15.32**). (You can watch these dogs displaying cataplexy on the website.) Just like humans with narcolepsy, the dogs often show REM immediately upon falling asleep. Abrupt collapsing in these dogs is suppressed by the same drugs (discussed shortly) that are used to treat human cataplexy.

The mutation responsible for narcolepsy in one of these strains of dogs was found to be in a gene that encodes receptors for the neuropeptide **hypocretin** (also called *orexin*) (L. Lin et al., 1999). Mice with the *hypocretin* gene knocked out also display narcolepsy (Chemelli et al., 1999). Genetically normal rats can be made narcoleptic if injected with a toxin that destroys neurons possessing hypocretin receptors (Gerashchenko et al., 2001). No one knows why interfering with hypocretin signalling leads to narcolepsy, but the narcoleptic dogs show signs of neural degeneration in the amygdala and nearby forebrain structures (J. M. Siegel, Nienhuis et al., 1999).

Similarly, humans with narcolepsy have lost about 90% of their hypocretin neurons (**FIGURE 15.33**) (Scammell, 2015; Thannickal et al., 2000). This degeneration of hypocretin neurons seems to cause inappropriate activation of the cataplexy pathway that is normally restricted to REM sleep. So hypocretin normally keeps sleep at bay and prevents the transition from wakefulness directly into REM sleep.

The neurons that normally produce hypocretin are found almost exclusively in the hypothalamus. Where do these neurons send their axons to release the hypocretin? Not so coincidentally, they send their axons to each of the three brain centres that we mentioned before: basal forebrain, reticular formation, and subcoeruleus (Sutcliffe and

hypocretins Also called *orexins*. Neuropeptides produced in the hypothalamus that are involved in switching between sleep states, in narcolepsy, and in the control of appetite.

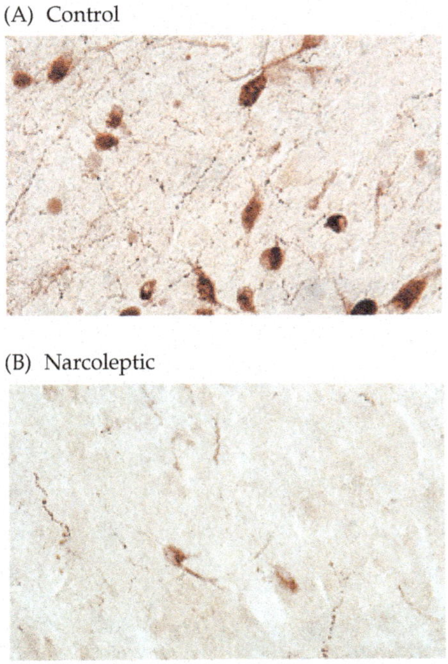

Courtesy of Jerome Siegel

15.33 Neural Degeneration in Humans with Narcolepsy (A) Immunocytochemistry reveals hypocretin-containing neurons in the lateral hypothalamus of a person who did not have narcolepsy. (B) This same region of the brain from someone who had narcolepsy has far fewer hypocretin neurons.

sleep paralysis A state during the transition to or from sleep, in which the ability to move or talk is temporarily lost.

de Lecea, 2002). The hypocretin neurons also project axons to the hypothalamic tuberomammillary nucleus that is inhibited by the basal forebrain to induce NREM. So, the hypothalamic hypocretin neurons normally control our switching between wakefulness, NREM, and REM (**FIGURE 15.30D**). Loss of these neurons in narcolepsy results in abrupt transitions, rather than orderly progression, through various sleep states.

The traditional treatment for narcolepsy was the use of amphetamines in the daytime. The drug GHB (γ-hydroxybutyrate, trade name Xyrem) helps some narcoleptics (although there are concerns about potential abuse of this drug [Tuller, 2002]). Another drug, modafinil (Provigil), is sometimes effective for preventing narcoleptic attacks. There is a debate about whether modafinil should be available to anyone who feels sleepy or needs to stay awake (Pack, 2003), but at least one study found the drug no more effective than caffeine in this regard (Wesensten et al., 2002). Our friend Barry, from the beginning of the chapter, eventually found a combination of mild stimulants that worked for him, and he later earned his MD degree. Antidepressants (see Table 12.3) are often prescribed to suppress episodes of cataplexy (Dauvilliers et al., 2014), although it is not clear they are more effective than a placebo (Swick, 2015). Now that narcolepsy is known to be caused by a loss of hypocretin signalling, there is hope of developing synthetic drugs to stimulate hypocretin receptors, both for the relief of symptoms in narcolepsy and to combat sleepiness in people without narcolepsy.

One common symptom of narcolepsy is experienced by many people (whether or not they have narcolepsy) on occasion (Fukuda et al., 1998). **Sleep paralysis** is the (temporary) inability to move or talk either just before dropping off to sleep or, more often, just after waking. In this state people may experience sensory hallucinations (Cheyne, 2002). Sleep paralysis never lasts more than a few minutes, so it's best to relax and avoid panic. One hypothesis is that sleep paralysis results when the subcoeruleus in the pons (see Figure 15.30C) continues to impose paralysis for a short while after the person awakens from a REM episode.

We are used to thinking of sleep as a global state of the brain—either the whole brain is asleep or it's not. But some recent findings are suggesting not only that the two hemispheres can be in different sleep states, but also that one part of cortex may be partially asleep while the patch of cortex next to it is still awake. In fact, some are suggesting that individual cortical neurons may be "awake" or "asleep" and that the ability of a cortical region to perform its function may depend on the proportion of neurons that are awake, as we'll see next.

The Cutting Edge

Can Individual Neurons Be "Sleepy"?

Vyazovskiy et al. (2011) implanted lots of tiny wires into both the frontal and the parietal lobes in rats, which were free to move around, and recorded firing of many neurons in each site for 2–3 weeks. When the animals were awake, most neurons were actively firing. But some neurons would occasionally stop firing for a while, as if the cell had gone "off-line" for a bit (**FIGURE 15.34A**). The researchers suggest this brief period of quiet may be caused by fatigue, as if each neuron needs to rest a bit before resuming activity. The picture that emerges is that each neuron alternates between being in an active, "on," state and an inactive, "off," state.

Neurons spend more time in the on state when the animal is awake than when it is in SWS (**FIGURE 15.34B**). But even while the animals are awake, a group of neurons in one cortical region may go off-line at the same time. What's fascinating is that depriving the rats of sleep (by giving them new toys to play with) results in more neurons in the off state (**FIGURE 15.34C**). The more sleep pressure there is (i.e., the longer the animals have been awake), the more neurons go off-line.

So can one group of cortical neurons be asleep while neighbouring neurons are still awake? Maybe, because when the researchers had the rats learn to reach for a sugar pellet, they noticed a very interesting phenomenon. There was a different pattern of neuronal activity in motor cortex (in the frontal lobe) just before a reach when the animal successfully got the pellet (a "hit") than when it didn't (a "miss") (**FIGURE 15.34D**). When

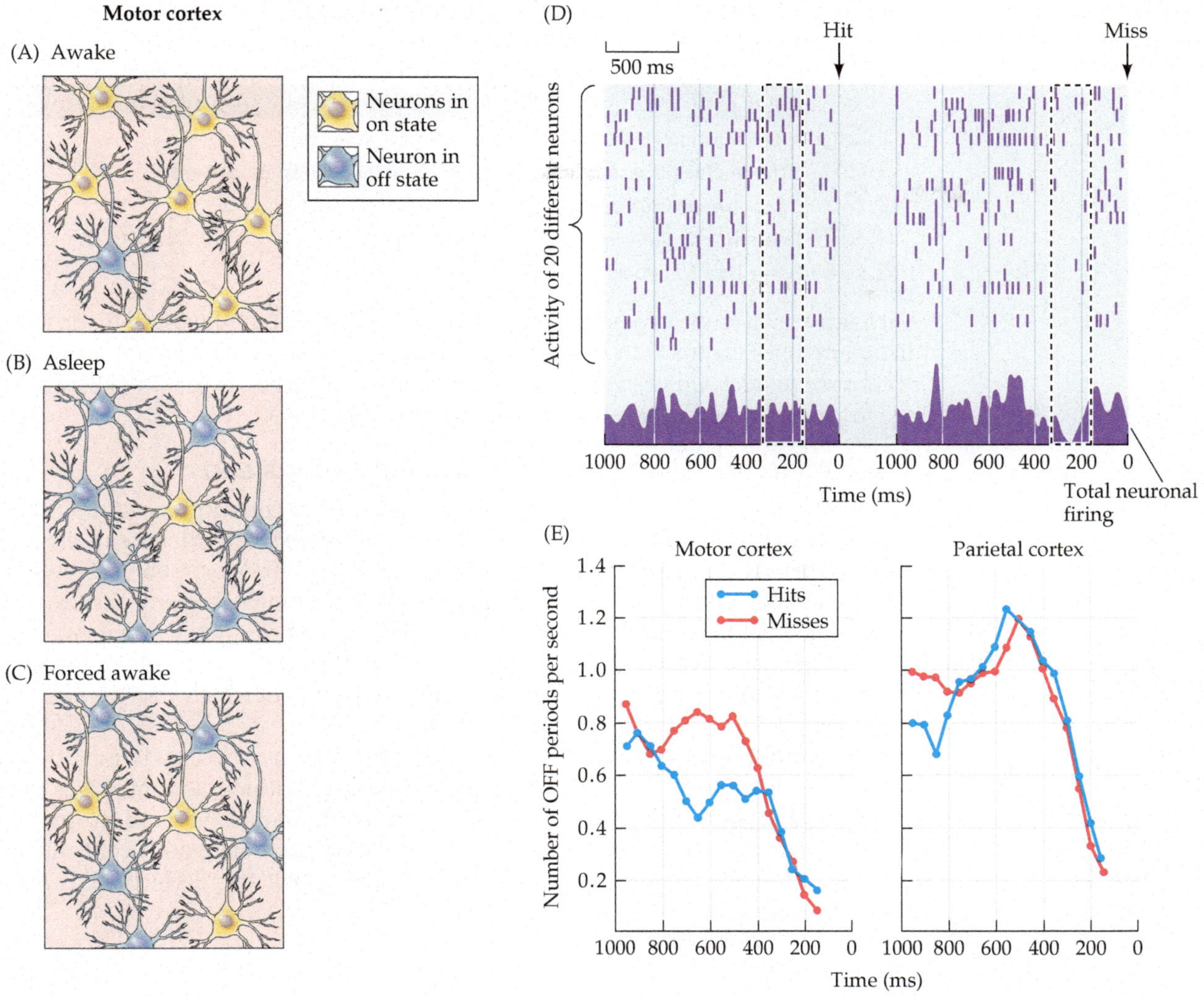

15.34 Local Sleep in the Brain (A) By recording both local EEG activity and the firing of individual neurons in rat cortex, researchers learned that when the animals are awake, neurons alternate between being in an "on" state (yellow), when they are firing often, and being in an "off" state (blue), when they fire very rarely. (B) During SWS, most neurons are in an off state but switch from off to on and back again in synchrony, which probably generates the slow waves in EEG that give SWS its name. (C) As animals are kept awake, building up sleep pressure, the proportion of neurons in the off state increases continually. Interestingly, the proportion of off neurons may differ from one cortical region to another. (D) Neuron firing is active in motor cortex before a successful reach (a hit) for a sugar pellet (dashed box), but in the time just before a miss (dashed box), many neurons are silent, indicating they are off-line. (E) The proportion of neurons in the off state in motor cortex is greater just before a miss than a hit. In nearby parietal cortex of the same animals, the activity of neurons just before a reach is the same whether the reach is successful or not. (Parts D and E after V. V. Vyazovskiy, 2011. *Nature* 472: 443–447.)

the animals missed, there was more likely to be a high proportion of neurons off-line in the contralateral motor cortex (i.e., the part of the cortex directing that paw) (**FIGURE 15.34E**, left). Strengthening the idea that having off-line neurons in motor cortex might have caused the miss, there was no relationship between misses and the proportion of neurons off-line in parietal cortex (see Figure 15.34E, right).

The idea that various regions of cortex may be asleep while neighbouring regions are awake and functional is controversial and takes some getting used to (Vyazovskiy and Harris, 2013). But it fits with the notion that dolphins, seals, and birds can have one cortical hemisphere asleep while the other is awake (see Figure 15.23). Also, it could explain why a sleep-deprived person is impaired in some cognitive tasks, if progressively more neurons have gone off-line in whichever particular cortical region underlies that behaviour. It also helps us understand various dissociations in sleep, such as sleepwalking (motor cortex must be working, but the person is not fully conscious). Having different parts of the brain asleep while others are awake may also help us understand some sleep disorders, the final topic of this chapter. ▯

15.6 Sleep Disorders Can Be Serious, Even Life-Threatening

Learning Objectives

After reading this section, you should be able to:

15.6.1 Offer a classification scheme for the various sleep disorders.

15.6.2 Describe various types of insomnias.

15.6.3 Explain why sleeping pills are unsatisfactory.

15.6.4 Offer tips for good sleep hygiene.

Narcolepsy is just one of several sleep disorders (**TABLE 15.2**) that have made sleep disorder clinics common in major medical centres. For some people, the peace and comfort of regular, uninterrupted sleep is routinely disturbed by the inability to fall asleep, by prolonged sleep, or by unusual awakenings.

Some minor dysfunctions are associated with sleep

Some dysfunctions associated with sleep are much more common in children than in adults. Two sleep disorders in children—night terrors (described earlier) and **sleep enuresis** (bed-wetting)—are associated with SWS. Most people grow out of these conditions without intervention, but pharmacological approaches can be used to reduce the amount of stage 3 sleep (as well as REM time) while increasing stage 2 sleep. For sleep enuresis, some doctors prescribe a nasal spray of the hormone vasopressin (antidiuretic hormone) before bedtime, which decreases urine production.

Somnambulism (sleepwalking) consists of getting out of bed, walking around, and appearing awake. Although more common in childhood, it sometimes persists into adulthood. These episodes last a few seconds to minutes, and the person usually does not remember the experience. Because such episodes occur during stage 3 SWS, they are more common in the first half of the night (when those stages predominate).

sleep enuresis Bed-wetting.

somnambulism Sleepwalking.

TABLE 15.2 Classification of Sleep Disorders with Some Examples

INSOMNIAS	**PARASOMNIAS**
Chronic insomnia (more than 3 months)	NREM-related parasomnias
Paradoxical insomnia (sleep state misperception)	Night terrors
Inadequate sleep hygiene	Sleepwalking (somnambulism)
Drug-related insomnia	REM-related parasomnias
Insomnia associated with psychiatric disorders	Nightmares
Short-term insomnia (less than 3 months)	Recurrent sleep paralysis
SLEEP-RELATED BREATHING DISORDERS	REM behaviour disorder (RBD)
Obstructive sleep apnea	Sleep enuresis (bed-wetting)
Central sleep apnea	**SLEEP-RELATED MOVEMENT DISORDERS**
CENTRAL DISORDERS OF HYPERSOMNOLENCE	Restless legs syndrome
Narcolepsy	Bruxism (teeth grinding)
Hypersomnolence associated with psychiatric disorder	Sleep-related myoclonus (sleep starts)
Hypersomnolence due to medication or substance	**SLEEP-RELATED MEDICAL AND NEUROLOGICAL DISORDERS**
CIRCADIAN RHYTHM SLEEP-WAKE DISORDERS	Fatal familial insomnia
Shift work disorder	Sleep epilepsy
Jet lag disorder	Sleep headaches

Source: American Academy of Sleep Medicine, 2015. *International classification of sleep disorders*, 3rd ed. American Academy of Sleep Medicine: Darien, IL.

Most sleepwalkers are not acting out a dream (Parkes, 1985). The main problem is the inability of sleepwalkers to wake into full contact with their surroundings. At least one person was acquitted of murder after suggesting that he had suffered from "homicidal somnambulism," but such claims are difficult to evaluate (Broughton et al., 1994). Some people, however, do seem to be acting out their dreams during REM sleep. **REM behaviour disorder (RBD)** is characterised by organized behaviour—such as fighting an imaginary foe, eating a meal, acting like a wild animal—in a person who appears to be asleep (Schenck and Mahowald, 2002). Sometimes the person remembers a dream that fits well with this outward behaviour (C. Brown, 2003). This disorder usually begins after the age of 50 and is more common in men than in women. Individuals with RBD are reminiscent of the cats with a lesion in the subcoeruleus (see Figure 15.31) that were no longer paralysed during REM and so acted out their dreams. The onset of RBD is often followed by the early symptoms of Parkinson's disease and dementia (Postuma et al., 2009), suggesting that the disorder is the beginning of a widespread neurodegeneration. RBD is usually controlled by antianxiety drugs (benzodiazepines) at bedtime.

Insomniacs have trouble falling asleep or staying asleep

Almost all of us experience an occasional inability to fall asleep, and a very few individuals die apparently because they stop sleeping altogether (see Box 15.1). But many people persistently find it difficult to fall asleep and/or stay asleep as long as they would like. The prevalence of insomnia has been estimated to be 30% of the adult population (T. Roth, 2007). Insomnia is more commonly reported by people who are older, female, or users of drugs like tobacco, caffeine, and alcohol. Insomnia seems to be the final common outcome for various situational and medical conditions. It is not a trivial disorder; recall that adults who regularly sleep for short periods show a higher mortality rate than those who regularly sleep 7–8 hours each night (Kripke et al., 2002). People with **paradoxical insomnia**, sometimes called *sleep state misperception* (Rezaie et al., 2018), report that they didn't sleep even when an EEG showed signs of sleep and they failed to respond to stimuli. On the other hand, EEG readings of people who claim they are not sleeping show fewer slow waves, and more of the alpha and beta activity typical of wakefulness, than other sleepers (Edinger and Krystal, 2003), so perhaps they really are more conscious than other sleepers during NREM, or perhaps some parts of their brain are sleeping without their being aware of it.

Situational factors that contribute to insomnia include shift work, time zone changes, and environmental conditions such as novelty (that hard motel bed). Usually these conditions produce transient **sleep-onset insomnia**, a difficulty in falling asleep. Drugs, as well as neurological and psychiatric factors, seem to cause **sleep-maintenance insomnia**, a difficulty in remaining asleep. In this type of insomnia, sleep is punctuated by frequent nighttime arousals. This form of insomnia is especially evident in disorders of the respiratory system.

In some people, respiration becomes unreliable during sleep. Breathing may cease for a minute or so, or it may slow alarmingly; blood levels of oxygen drop markedly. This syndrome, called **sleep apnea**, arises either from the progressive relaxation of muscles of the chest, diaphragm, and throat cavity (*obstructive* apnea) or from changes in the pacemaker respiratory neurons of the brainstem (*central* apnea). In the former instance, relaxation of the throat obstructs the airway—a kind of self-choking. For some people, breathing through a special machine (called a *continuous positive airway pressure*, or *CPAP, machine*) maintains air pressure in their airways and prevents their collapse (**FIGURE 15.35**). Obstructive apnea is common in very obese people, but it also

REM behaviour disorder (RBD) A sleep disorder in which a person physically acts out a dream.

paradoxical insomnia Also called *sleep state misperception*. Commonly, a person's perception that he has not been asleep when EEG readings and nonresponsiveness indicate that he has. Typically it occurs at the start of a sleep episode.

sleep-onset insomnia Difficulty in falling asleep.

sleep-maintenance insomnia Difficulty in staying asleep.

sleep apnea A sleep disorder in which respiration slows or stops periodically, waking the person. Excessive daytime sleepiness results from the frequent nocturnal awakening.

Go to Video 15.5
REM Behavioural Disorder

Photograph by Christopher Breedlove

15.35 A Machine That Prevents Sleep Apnea By supplying continuous positive airway pressure (CPAP), this machine prevents the collapse of the airway that otherwise causes this man to stop breathing for a while several times each night. It also stops his snoring.

sudden infant death syndrome (SIDS) Also called crib death. The sudden, unexpected death of an apparently healthy human infant who simply stops breathing, usually during sleep.

occurs, often undiagnosed, in nonobese people. Sleep apnea is frequently accompanied by loud, interrupted snoring, so loud snorers should consult a physician about the possibility that they may have sleep apnea.

Each episode of apnea arouses the person enough to restore breathing, but in rats such interruptions in oxygen kill neurons in the hippocampus and impair learning ability (Row et al., 2007), so apnea may lead to brain damage. What's more, the frequent nighttime arousals make such people sleepy in the daytime. Untreated sleep apnea may lead to any of several cardiovascular disorders, including hypertension, and diabetes (Rakel, 2009).

Investigators have speculated that **sudden infant death syndrome** (**SIDS**, or *crib death*) arises from sleep apnea as a result of immaturity of systems that normally pace respiration. Autopsies of SIDS victims reveal abnormalities in brainstem serotonin systems (Kinney, 2009), and interfering with this system in young mice renders them unable to regulate respiration effectively. What's more, some of these mice spontaneously stopped breathing and died (Audero et al., 2008). The incidence of SIDS has been cut almost in half by the Back to Sleep campaign, which urges parents to place infants on their backs to sleep rather than on their stomachs (**FIGURE 15.36**). Placing the baby facedown may lead to suffocation if the baby cannot regulate breathing or arouse properly.

Although many drugs affect sleep, there is no perfect sleeping pill

Throughout recorded history, humans have reached for substances to enhance the prospects of sleep. Ancient Greeks used the juice of the poppy to obtain opium and used products of the mandrake plant that we recognize today as the anticholinergic drugs scopolamine and atropine (Hartmann, 1978). The preparation of barbituric acid in the mid-nineteenth century by the discoverer of aspirin, Adolph von Baeyer, started the development of many drugs—*barbiturates*—that were once used to help people get to sleep. Unfortunately, none of these substances can provide a completely normal night of sleep in terms of time spent in various sleep states, such as REM sleep, and none of them remain effective when used repeatedly.

Most modern sleeping pills—including the benzodiazepine triazolam (Halcion) and newer drugs like Ambien, Sonata, and Lunesta—bind to various sites on $GABA_A$ receptors (see Figure 3.14), making them more responsive to the neurotransmitter and thus inhibiting broad regions of the brain. But reliance on sleeping pills poses many problems (Mignot, 2013). There is hope that drugs that block both of two known hypocretin receptors may offer another route to sleepiness (Tannenbaum et al., 2016). But viewed solely as a way to deal with sleep problems, drugs fall far short of being a suitable remedy, for several reasons.

First, continued use of sleeping pills causes them to lose effectiveness, and this declining ability to induce sleep often leads to self-prescribed increased dosages that can be dangerous. A second major drawback is that sleeping pills produce marked changes in the pattern of sleep, both while the drug is being used and for days afterward.

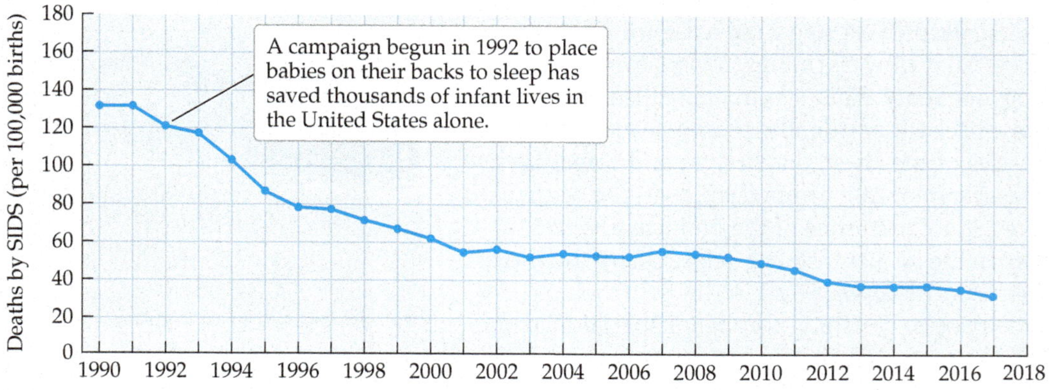

15.36 Back to Sleep (After CDC/NCHS, National Vital Statistics System, Compressed Mortality File.)

Use of sleeping pills may lead to a persistent "sleep drunkenness," coupled with drowsiness, that impairs waking activity or may lead to memory gaps about daily activity. Police report cases of "the Ambien driver," a person who takes a sleeping pill and then gets up a few hours later to go for a spin, with sometimes disastrous results, while apparently asleep (R. N. Hansen et al., 2015). In other cases, people taking such medicines eat snacks, shop over the internet, or even have sex, with no memory of these events the next day (Dolder and Nelson, 2008).

Because the hormone melatonin is normally released from the pineal gland at night (see Figure 4.19), the administration of exogenous melatonin has been suggested to aid the onset of sleep. In fact, melatonin does have a weak hypnotic effect soon after administration (T. Li et al., 2019). In contrast, caffeine is an antagonist for the neuromodulator **adenosine**, which generally reduces neural activity (Bjorness and Greene, 2009). Thus caffeine boosts brain activity and, taken at night, delays the circadian clock (T. M. Burke et al., 2015).

Certainly, the treatment for insomnia that has the fewest side effects, and that is very effective for many people, is not to use any drug but to practice **sleep hygiene**, habits that promote healthy sleep (Irish et al., 2015). The best advice for insomniacs is to use an alarm clock to wake up faithfully at the same time each day (weekends included!) and then simply go to bed once they feel sleepy. Other sleep hygiene habits include establishing a bedtime routine in a quiet, dark environment to condition sleep onset, and avoiding daytime naps, caffeine at night, and looking at computer screens (their bluish light stimulates melanopsin in retinal ganglion neurons; Gooley et al., 2010) just before bedtime (**TABLE 15.3**). An important adjunct to this strategy is to ignore preconceived notions about how much sleep you "need." If you get out of bed every day at 6:00 AM and don't feel sleepy until midnight, then your body is telling you that you need only 6 hours of sleep, no matter what millions of dollars in annual pharmaceutical advertising might say. On the one hand, we humans sleep appreciably less than other primate species (Nunn and Samson, 2018), but in hunter-gatherer societies people tend to wake just before sunrise, rarely take naps, and go to sleep about 3 hours after sunset. Subtracting awakenings at night, these people sleep an average of 6.4 hours per night (Yetish et al., 2015). Is this the "natural" sleep pattern of our species?

In contrast, in our society, most people sleep longer on weekends, presumably because they are sleep-deprived during the work week (**FIGURE 15.37**). This irregular cycle in sleep patterns has been likened to "social jet lag" (Roenneberg, 2013). Given the clear benefits of sleep for overall health, including avoiding heart attacks and strokes (Cappuccio et al., 2011), establishing good sleep hygiene just might save your life.

TABLE 15.3 Some Simple Tips to Promote Sleep Hygiene

Recommendations
Keep your internal clock set with a consistent sleep schedule (get up at the same time each day).
Seek sunlight in the daytime, avoid lights at night.
Turn your bedroom into a sleep-inducing environment.
Avoid caffeine, alcohol, and nicotine before bedtime.
Establish a soothing presleep routine.
Only go to sleep when you're truly tired.
If you're going to nap, do it early in the day.
Lighten up on evening meals.
Balance fluid intake to avoid night trips to the bathroom.
Avoid using computer screens before bedtime.
Don't exercise late in the day.

Source: Healthy Sleep, http://healthysleep.med.harvard.edu/healthy/getting/overcoming/tips

adenosine An endogenous neuromodulator that generally reduces neural activity. Caffeine interferes with adenosine binding.

sleep hygiene Habits, such as avoiding caffeine shortly before bedtime, that promote healthy sleep.

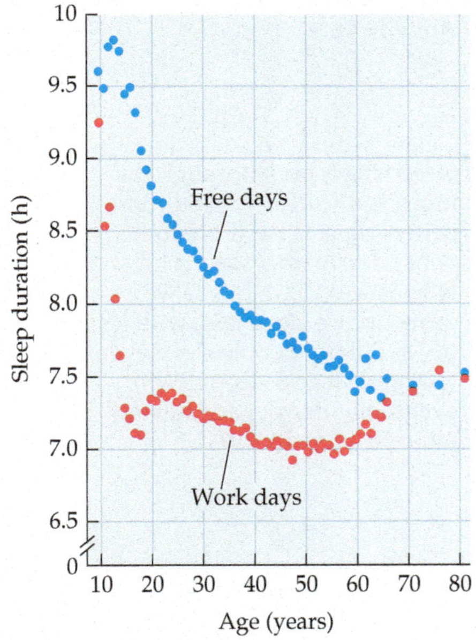

15.37 Sleeping In on the Weekend Such patterns represent poor sleep hygiene, because sleeping late on weekend mornings may make it difficult to fall asleep early enough the night before a work day. (After T. Roenneberg et al., 2012. *Curr Biol* 22: 939–943.)

Recommended Reading

Dunlap, J. C., Loros, J. J., and DeCoursey, P. J. (2003). *Chronobiology: Biological Timekeeping*. Sunderland, MA: Oxford University Press/Sinauer.

Kryger, M. K., Roth, T., and Dement, W. C. (Eds.). (2016). *Principles and Practice of Sleep Medicine* (6th ed.). Philadelphia: Elsevier.

Max, D. T. (2007). *The Family That Couldn't Sleep: A Medical Mystery*. New York: Random House.

Stickgold, R., and Walker, M. P. (Eds.). (2009). *The Neuroscience of Sleep*. San Diego, CA: Academic Press.

Walker, M. (2017). *Why We Sleep: Unlocking the Power of Sleep and Dreams*. New York: Scribner.

15 Visual Summary

You should be able to relate each summary to the adjacent illustration, including structures and processes. You can also refer back to the links to figures, animations, and activities that will help you consolidate the material.

1 Animals show **circadian rhythms** of activity that can be **entrained** by light. These rhythms synchronise behaviour to changes in the environment. In constant dim light, animals **free-run**, displaying a **period** of about 24 hours. Review **Figures 15.1**, **15.10**, **Animation 15.1**

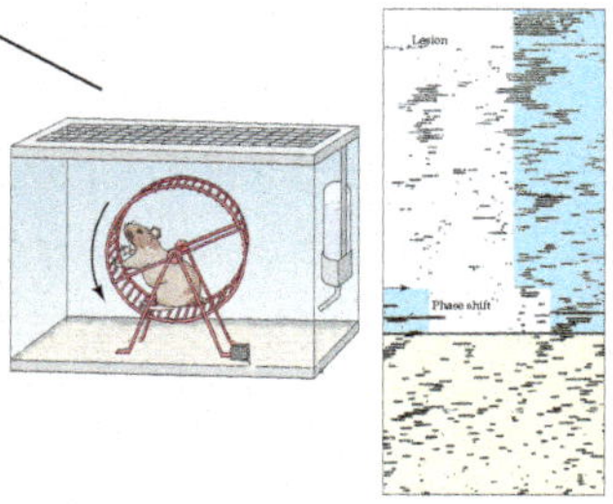

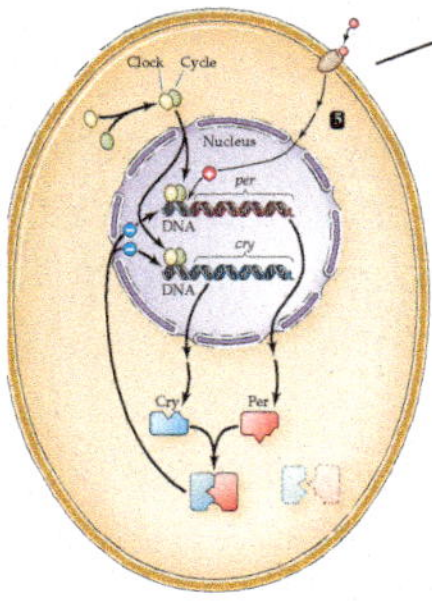

2 Lesions of the **suprachiasmatic nucleus** (**SCN**) abolish activity rhythms in constant conditions. Transplanting the SCN from one animal into another results in a free-running rhythm of the donor, demonstrating that the SCN contains a clock that can drive circadian activity. Several proteins (Clock, Cycle, Period) interact, increasing and decreasing in a cycle that takes about 24 hours. This molecular clock, pooled from many SCN neurons, drives circadian rhythms. Review **Figures 15.2–15.7**, **Animation 15.2**

3 Almost all mammals show two sleep states: **rapid-eye-movement** (**REM**) **sleep** and **non-REM** (**NREM**) **sleep**. Human NREM has three distinct stages (**stages 1, 2, 3**) defined by **electroencephalography** (**EEG**) criteria, including **sleep spindles** and, in stage 3, large, slow **delta waves** during **slow-wave sleep** (**SWS**). Muscle tension, heart rate, respiratory rate, and temperature all decline during NREM. Review **Figure 15.11**, **Table 15.1**, **Activity 15.1**

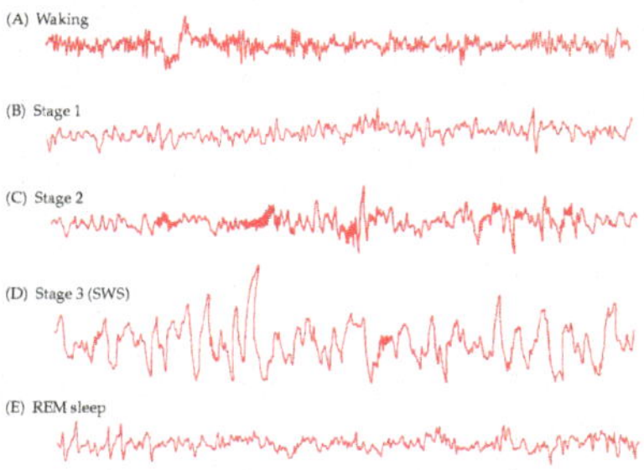

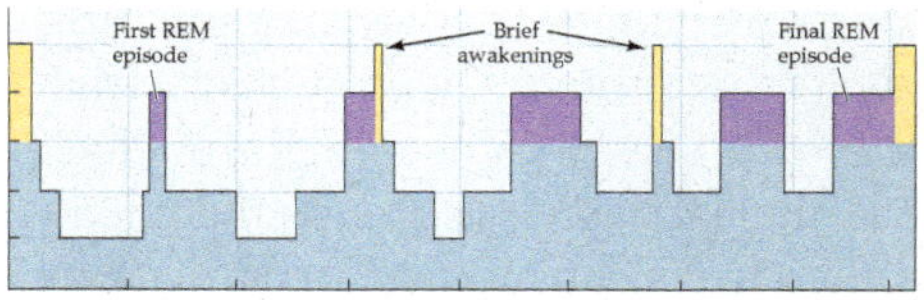

4 REM sleep is characterised by rapid, low-amplitude EEG (almost like an EEG while awake) but also by profound muscle relaxation because motor neurons are inhibited. People awakened from REM frequently report vivid dreams, while people awakened from NREM report ideas or thinking. Review **Figure 15.13**, **Table 15.1**

5 Sleep stages cycle through the night, with stage 3 SWS prominent early and REM predominant later. Infants sleep a lot, with lots of REM, but as we grow up, we sleep less, with less REM. Elderly people sleep even less, and stage 3 sleep (SWS) eventually disappears. Review **Figures 15.14–15.19**

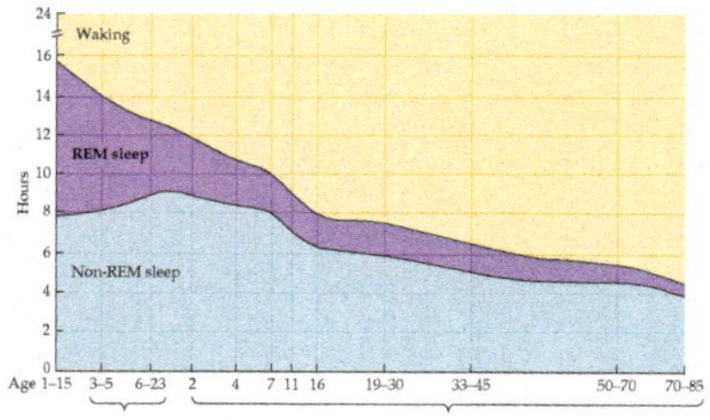

6 Four proposed functions of sleep are energy conservation, **ecological niche** adaptation, body restoration, and memory consolidation. A few people sleep only an hour per night, suggesting that they accomplish the function(s) of sleep very efficiently, but all healthy people sleep. Review **Figures 15.21–15.25**

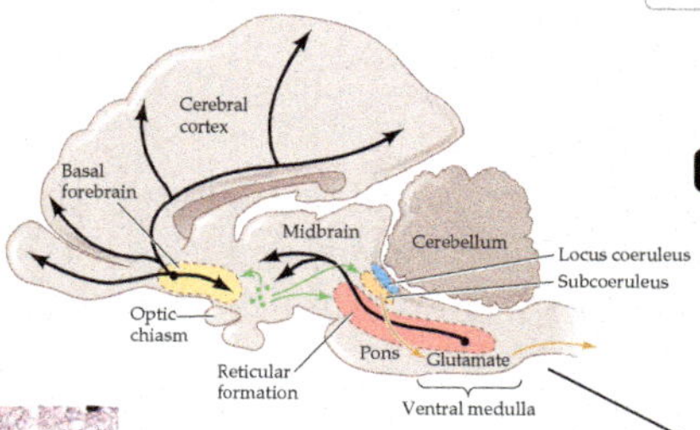

7 **Sleep deprivation** leads to impairments in vigilance and reaction times. It also incurs sleep debt, although the lost NREM and REM may be partially restored in subsequent nights. Prolonged sleep deprivation compromises the immune system and leads to death. Review **Box 15.1**

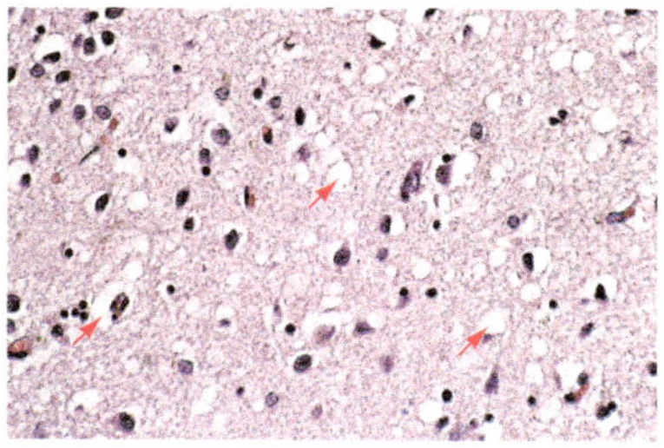

8 Four brain systems control sleep and waking. The **basal forebrain** promotes SWS, the brainstem **reticular formation** promotes arousal, the **subcoeruleus** of the pons triggers REM sleep, and hypothalamic neurons releasing **hypocretin** regulate these three centres. Review **Figures 15.28–15.30**, **Activity 15.2**

9 **Narcolepsy** is the disorder of sudden, uncontrollable periods of sleep, which may be accompanied by **cataplexy**, paralysis while remaining conscious. Disruption of hypocretin signalling causes narcolepsy. Review **Figures 15.32, 15.33**, **Videos 15.3**, **15.4**

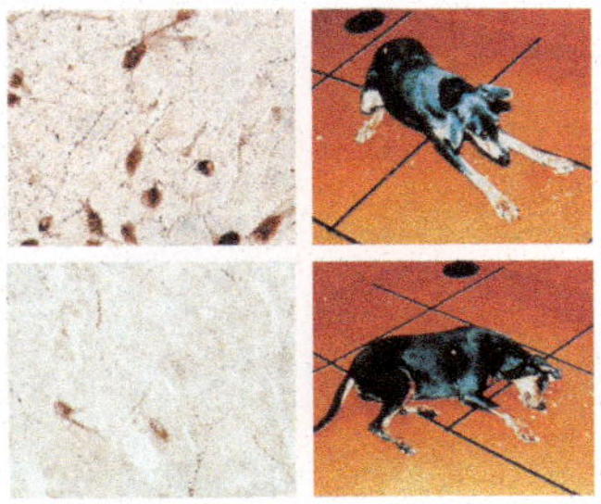

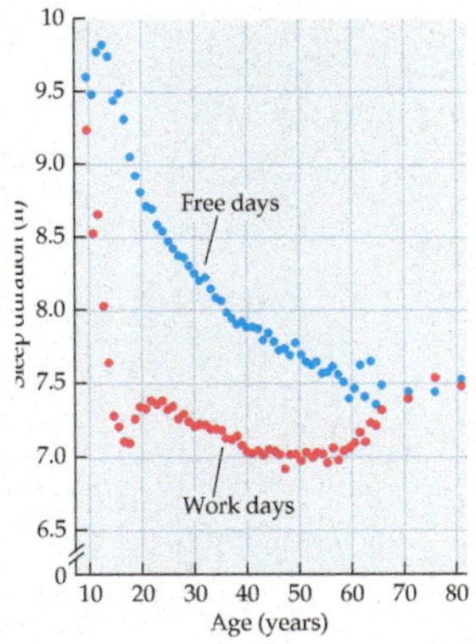

10 Sleep disorders fall into four categories: trouble falling or staying asleep, termed **insomnia**; excessive drowsiness (e.g., narcolepsy); disruption of the sleep-waking schedule; and dysfunctions associated with sleep, sleep stages, or partial arousals, such as **somnambulism**. No pill guarantees a normal night's sleep. Review **Figures 15.35–15.37**, **Tables 15.2**, **15.3**, **Video 15.5**

Cognitive Neuroscience

PART VI

Courtesy of Dr. Sarah Moghadam, VA Palo Alto Health Care System, Palo Alto, CA and Dr. Ahmad Salehi, Dept. of Psychiatry & Behavioural Sciences, Stanford Medical School

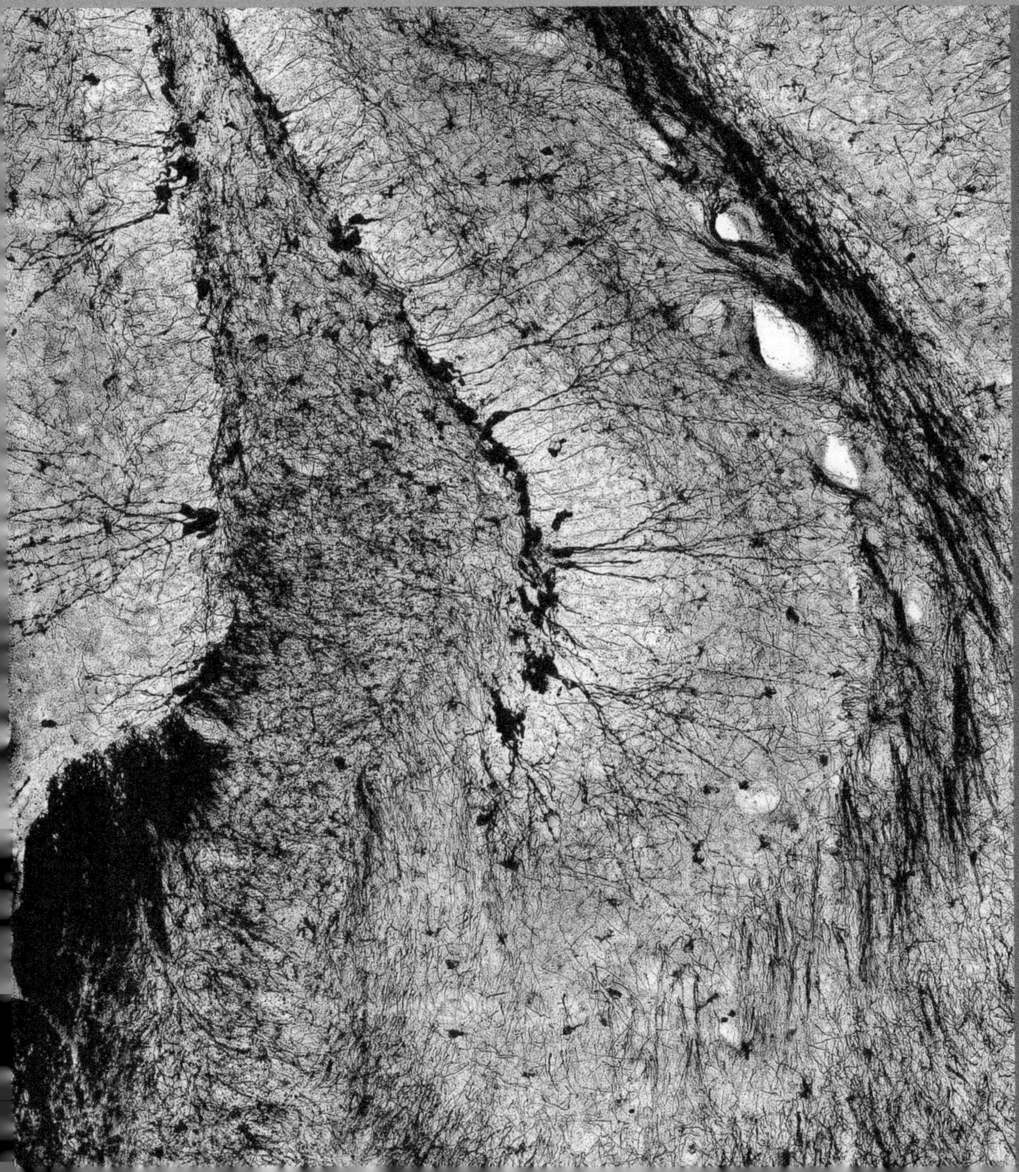

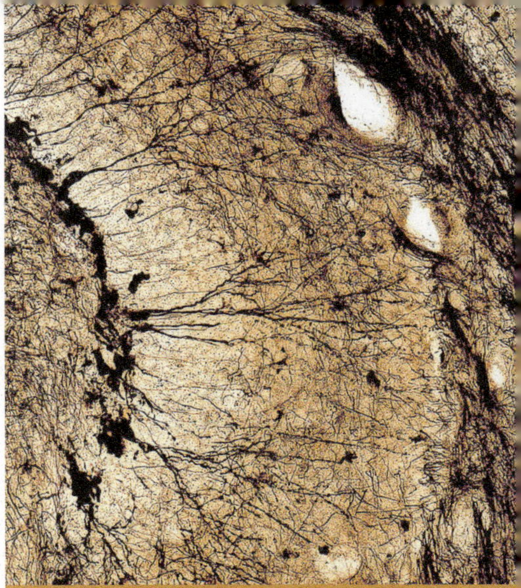

16

Learning and Memory

Trapped in the Eternal Now

Henry Molaison, known to the world as *patient H.M.* in a classic series of research articles, was probably the most famous research subject in the history of brain science. Henry started to experience seizures during adolescence, and by his late 20s, Henry's epilepsy was out of control. Because tests showed that his seizures began in both temporal lobes, a neurosurgeon removed most of the temporal lobes, on both sides, in 1953.

Henry's surgery relieved his epilepsy, but that relief came at a terrible, unforeseen price: Henry had lost the ability to form new memories (Scoville and Milner, 1957). For more than 50 years after the surgery, until his death in 2008, Henry could retain any new fact only briefly; as soon as he was distracted, the newly acquired information vanished. Long after the surgery, he didn't know his age or the current date, that his parents had died years previously, or indeed any of the events that had transpired in the time since his surgery. His IQ remained a little above average (Corkin et al., 1997)—IQ tests don't require remembering new facts for more than a few minutes—but he knew that something was wrong with him.

> *Every day is alone in itself, whatever enjoyment I've had, and whatever sorrow I've had.... Right now, I'm wondering, have I done or said anything amiss? You see, at this moment everything looks clear to me, but what happened just before? That's what worries me. It's like waking from a dream. I just don't remember. (B. Milner, 1970, p. 37)*

Henry's inability to form new memories meant that he couldn't construct a lasting relationship with anybody new. No matter what experiences he might share with someone he met, Henry would have to start the acquaintance anew the following day, because he would have no recollection of ever meeting the person before.

What happened to Henry, and what does his experience teach us about learning and memory?

All the distinctively human aspects of our behaviour are learned: the languages we speak, how we dress, the foods we eat and how we eat them, our skills and the ways we reach our goals. So much of our own individuality depends on **learning**, the process of acquiring new information, and on **memory**, the ability to store and retrieve that information. We begin this chapter with a discussion of the major types of memory because research in the twentieth century, including the case of H.M., revealed that there are fundamentally different types of memory, relying on varying networks of brain regions. In the second part of the chapter, we delve into what we know about how learning alters the structure and functioning of brain tissue at a more fine-grained cellular level.

learning The process of acquiring new and relatively enduring information, behaviour patterns, or abilities, characterised by modifications of behaviour as a result of practice, study, or experience.

memory 1. The ability to retain information, based on the mental process of learning or encoding, retention across some interval of time, and retrieval or reactivation of the information.
2. The specific information that is stored in the brain.

amnesia Severe impairment of memory.

retrograde amnesia Difficulty in retrieving memories formed before the onset of amnesia.

patient H.M. A person who, because of damage to medial temporal lobe structures, was unable to encode new declarative memories. Upon his death we learned his name was Henry Molaison.

anterograde amnesia The inability to form new memories beginning with the onset of a disorder.

16.1 Patient H.M. Henry Molaison as a young man.

FUNCTIONAL PERSPECTIVES ON LEARNING AND MEMORY

We can discover a great deal about learning and memory by examining how they fail. Clinical case studies show that memory can fail in very different ways, indicating that there are different *forms* of learning and memory, and that multiple brain regions are involved. The clinical research has guided further studies, using animal models and brain imaging. Together, these diverse approaches are generating a comprehensive picture of brain mechanisms of learning and memory.

16.1 There Are Several Kinds of Learning and Memory

Learning Objectives

After reading this section, you should be able to:

- **16.1.1** Distinguish between two fundamentally different types of amnesia.
- **16.1.2** Relate the story of patient H.M., describing his particular memory deficits.
- **16.1.3** Describe the two different kinds of memory that were demonstrated by the case of patient H.M.
- **16.1.4** Specify the brain region(s) where damage causes memory deficits like that of H.M.
- **16.1.5** Describe the symptoms and cause of Korsakoff's syndrome.
- **16.1.6** Use the case of patient K.C. to distinguish between episodic and semantic memories.

The terms *learning* and *memory* are so often paired that it sometimes seems as if one necessarily implies the other. We cannot be sure that learning has occurred unless a memory can be elicited later. Many kinds of brain damage, caused by disease or accidents, can produce highly specific types of memory impairment. We'll start our survey by looking at a few classic case studies of memory impairments and the brain damage that caused them.

For patient H.M., the present vanished into oblivion

Amnesia (Greek for "forgetfulness") is a severe impairment of memory, usually as a result of accident or disease. Loss of memories that formed prior to an event (such as surgery or trauma) is not uncommon. It is called **retrograde amnesia** (from the Latin *retro-*, "backward," and *gradi*, "to go"). After an accident that damages the brain, people often have retrograde amnesia with regard to events that happened a few hours or days before the accident, or even a year before. (Despite dramatic depictions you may see on TV, it is unlikely that longer-term, or "complete," retrograde memory loss has ever occurred.)

Patient H.M.—Henry Molaison, whom we met at the start of the chapter (**FIGURE 16.1**)—experienced some retrograde amnesia for the time immediately before his surgery, but he also experienced a far more unusual symptom: his apparent inability to retain *new* material for more than a brief period. The inability to form new memories *after* an event is called **anterograde amnesia** (the Latin *antero-* means "forward"). In Henry's case, most old memories remained intact, but he had difficulty recollecting any events after his surgery.

Over the very short term, Henry's memory was normal. If given a series of six or seven digits, he could immediately repeat the list back without error. But if he was given a list of words to study and then tested on them after other tasks had intervened, he could not repeat the list or even recall that there *was* a list. So Henry's case provided clear evidence that *short-term memory* differs from *long-term memory*—a distinction long recognised by biological psychologists on behavioural grounds (W. James, 1890) and which we will discuss in more depth later.

Henry's surgery removed the amygdala, most of the hippocampus, and surrounding cortex from both temporal lobes (**FIGURE 16.2**). Patients who had damage only to the amygdala did not show such memory problems, so Henry's deficit seemed to

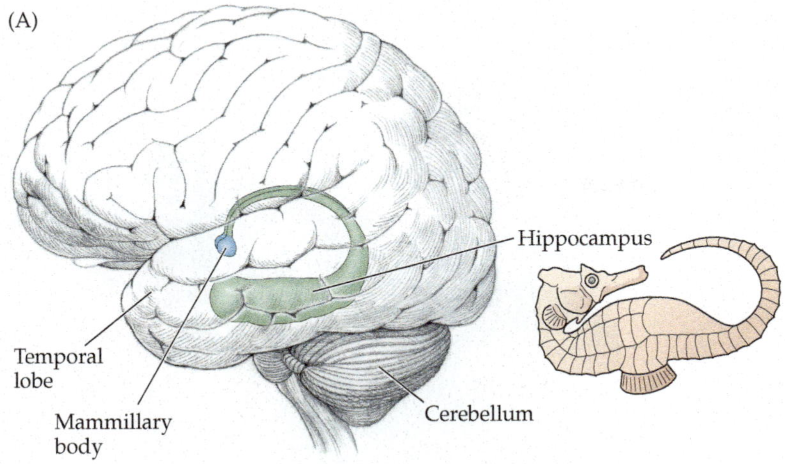

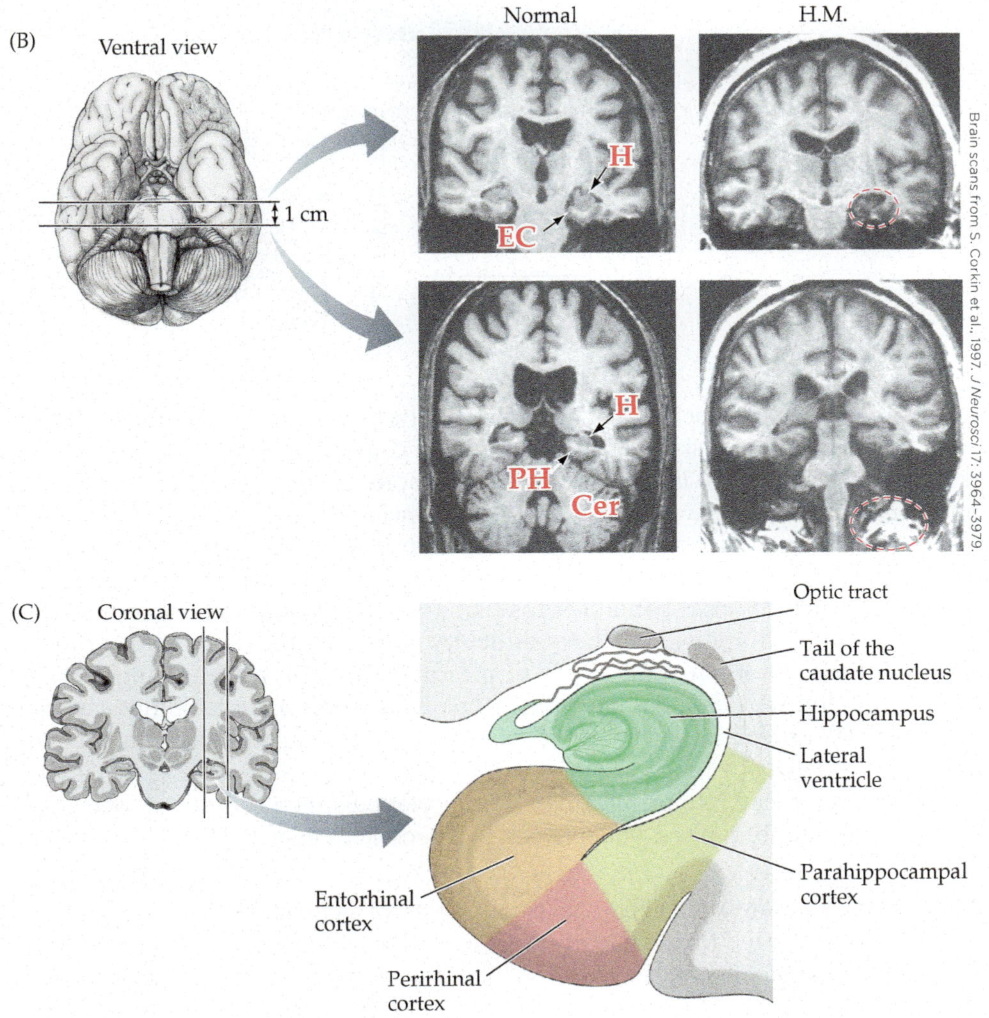

16.2 Brain Regions Crucial for Memory Formation (A) The hippocampus, whose name means "sea horse" in Latin, was named for its resemblance to that animal. (B) MRI scans of a healthy person (left) and Henry (right) show that Henry's hippocampus (H) and entorhinal cortex (EC) were extensively damaged, bilaterally. Henry had a small amount of posterior hippocampus left, but the parahippocampal cortex (PH) was totally gone. The cerebellum (Cer) was dramatically shrunken, a result of surgical removal of afferents and efferents. (C) The various components of the medial temporal lobe.

be caused by loss of the *medial temporal lobe,* including the **hippocampus**. But to the puzzlement of researchers, it seemed that bilateral hippocampal lesions in laboratory animals did not produce comparable memory deficits (Isaacson, 1972; Mumby, 2001). What might account for this apparent discrepancy between humans and lab animals?

hippocampus A medial temporal lobe structure important for learning, memory, and spatial navigation.

An interesting finding about Henry's memory deficit offered a clue. When Henry was given a mirror-tracing task (**FIGURE 16.3A**), he improved considerably over ten trials (B. Milner, 1965). The next day the test was presented again. When asked if he recognised the task, Henry said no, yet his performance was better than at the start of the first day. Over 3 successive days, Henry never recognised the task as

16.3 Henry's Performance on a Mirror-Tracing Task (A) Henry was given this mirror-tracing task to test motor skill. (B) His performance on this task progressively improved over 3 successive days, demonstrating a type of long-term memory. (After B. Milner, 1965. In *Cognitive processes and the brain: An enduring problem in psychology*, P. M. Milner and S. E. Glickman [Eds.], pp. 97–111. Van Nostrand: Princeton, NJ.)

declarative memory A memory that can be stated or described.

nondeclarative memory Also called *procedural memory*. A memory that is shown by performance rather than by conscious recollection.

something he had seen before, but his improved tracings showed evidence of memory, in the form of motor skill (**FIGURE 16.3B**). Of course, we can't ask lab animals if they *recognise* a maze, so if an animal with comparable hippocampal damage shows improvement on a task, we might prematurely conclude its memory is normal.

So, was Henry's deficit simply an inability to learn verbal material? Probably not. First, people with brain damage like Henry's have difficulty reproducing or recognising pictures and spatial designs that are not recalled in verbal terms. Second, although such individuals have difficulty with the specific content of verbal material, they can learn some kinds of information *about* verbal material (N. J. Cohen and Squire, 1980). For example, people with several kinds of amnesia can learn the *skill* of reading mirror-reversed text but show impaired learning of specific words (**FIGURE 16.4**).

Subsequent research confirmed that the important distinction is not between motor and verbal performances, but between two general categories of memory:

1. **Declarative memory** is what we usually think of as memory: facts and information acquired through learning. It is memory we are aware of accessing, which we can *declare* to others. This is the type of memory Henry could no longer acquire. Tests of declarative memory take the form of requests for specific information that has been learned previously, such as a story or word list. It is the type of memory we use to answer "what" questions—and thus is difficult to test in animals.
2. **Nondeclarative memory**, or *procedural memory*—that is, memory about perceptual or motor procedures—is shown by *performance* rather than by conscious recollection. Examples of procedural memory include learning the mirror-tracing task at which Henry excelled, and the skill of mirror reading, along with more familiar skills such as riding a bicycle or juggling: things that you learn by *doing*. It is the type of memory we use for "how" problems and is often (but not always) nonverbal. **FIGURE 16.5** illustrates this basic division of types of memory.

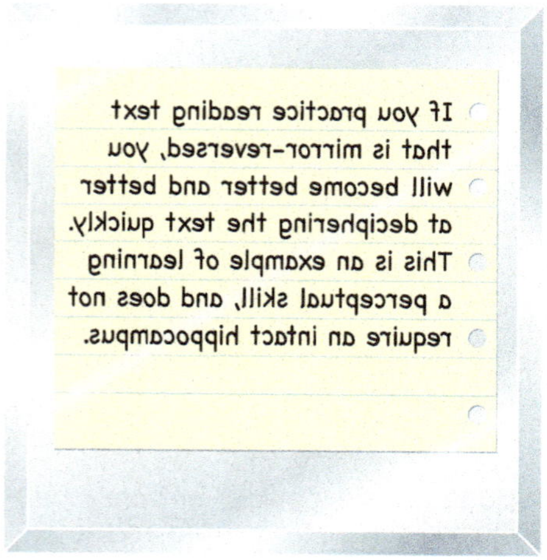

16.4 Reading Mirror-Reversed Text People with amnesia like Henry's can learn to read mirror-reversed text well, even though they don't remember practicing this skill. This demonstrates that their problem is not verbal, but rather an inability to form declarative memories.

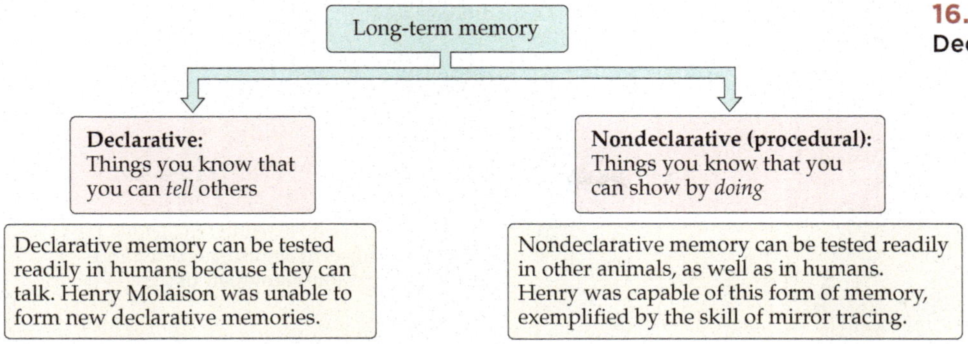

16.5 Two Main Kinds of Memory: Declarative and Nondeclarative

Medial temporal lobe structures are crucial for declarative memory

Publication of Henry Molaison's case prompted an intensive effort to develop methods for systematically studying declarative memory in monkeys and other lab animals. To get around the inability of lab animals to verbally report their memories, the **delayed non-matching-to-sample task** (**FIGURE 16.6**)—a test of *object recognition memory*—was developed. In this task, monkeys must identify which of two objects was *not* seen previously, with delays ranging from 8 seconds to 2 minutes (Spiegler and Mishkin, 1981). The important feature of this procedure is that the animal does *not* reach for the item that previously had a treat under it, because in that case the monkey might unconsciously associate reward with that object even if the animal had no conscious recollection of it. Instead, the monkey declares that it remembers the old object by reaching for the *new* object, the one that was not associated with a reward previously. Monkeys with extensive damage to the temporal lobe, and thus similar to Henry, are severely impaired on this task, especially with the longer delays. But which specific temporal lobe structures are most important?

Selective removal of specific parts of the temporal lobes of monkeys revealed that the amygdala—one of the structures removed in Henry's surgery—was not crucial for performance on tests of declarative memory. However, removal of the hippocampus significantly impaired performance on these tests. The deficit was even more pronounced when the hippocampal damage was extended from the hippocampus to include other medial temporal lobe structures: the nearby entorhinal, parahippocampal, and perirhinal cortex (Zola-Morgan et al., 1994)

delayed non-matching-to-sample task A test in which, on each trial, the participant must select the stimulus that was not seen previously.

16.6 The Delayed Non-Matching-to-Sample Task

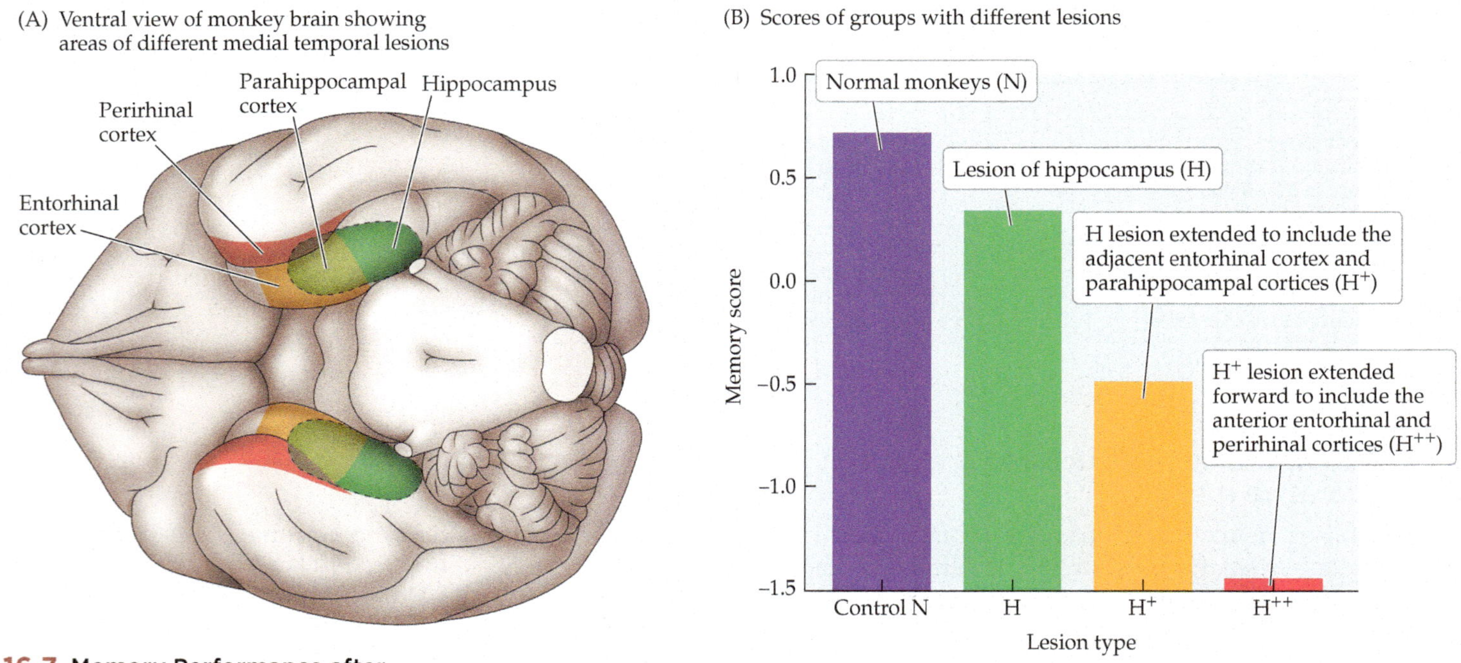

16.7 Memory Performance after Medial Temporal Lobe Lesions
(A) In this ventral view of a monkey brain, the positions of the hippocampus, entorhinal cortex, perirhinal cortex, and parahippocampal cortex are shown in different colours. (B) Memory was most impaired when most of the medial temporal lobe was lesioned. (Part A after L. R. Squire and S. Zola-Morgan, 1991. *Science* 253: 1380–1386; B after S. Zola-Morgan et al., 1994. *Hippocampus* 4: 482–495.)

(**FIGURE 16.7**). Examination of additional human patients confirmed that loss of the entire medial temporal lobe was needed to duplicate H.M.'s symptoms (Rempel-Clower et al., 1996; Zola-Morgan and Squire, 1986).

Different structures within the medial temporal lobe make different contributions to declarative memory. Most declarative memory for facts and events consists of two components: the sense of *familiarity* with the features of the item ("I've seen that actress somewhere…"), and additionally *recollection* of the item in the specific context in which it was presented ("Oh, she played Daenerys in *Game of Thrones*"). Experimental evidence in animals suggests that the two processes are served by differing components of the medial temporal lobe system (Eichenbaum et al., 2007). Perirhinal cortex (see Figure 16.2C) is thought to be responsible for the sense of familiarity in memory, whereas recollection of the item is the function of the hippocampus (Lech and Suchan, 2013; Suzuki and Naya, 2014). Processing of contextual aspects of memory (including spatial cognition, which we will discuss in more detail a little later) appears to depend especially on the parahippocampal cortex (Aminoff et al., 2013; Diana et al., 2013). In general, the performance of experimental animals suggests that the hippocampus acts as the final stage of convergence for adjacent regions of cortex (Squire et al., 2007; Zola et al., 2000). In keeping with this schema, people with specific hippocampal damage appear to have difficulties with recollection, while familiarity-based aspects of memory are spared (Addante et al., 2012; Bowles et al., 2016). Conversely, people with surgical lesions that include the perirhinal cortex but not the hippocampus (**FIGURE 16.8**) experience impaired familiarity with stimuli despite preserved recollection (Bowles et al., 2007). The medial temporal lobe is not the only brain region required to form new declarative memories, as we see next.

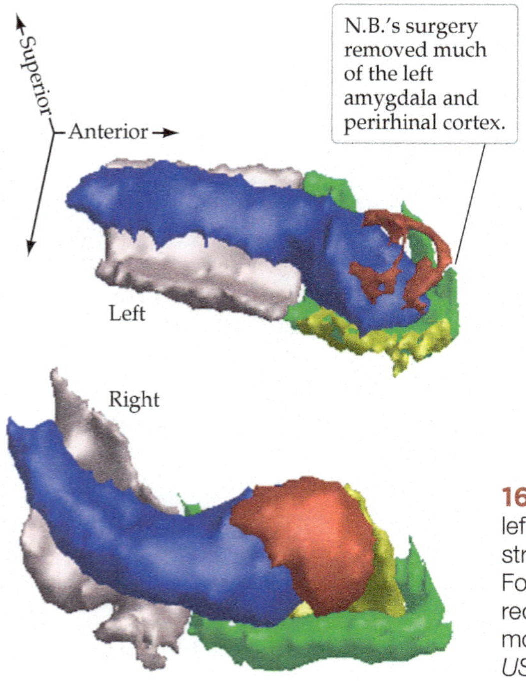

16.8 Impaired Familiarity after Surgery This person received surgical lesions in the left medial temporal lobe that selectively removed much of the amygdala (red in this reconstruction) and perirhinal cortex (green) while completely sparing the hippocampus (blue). Following surgery, she experienced reduced feelings of familiarity in memory tests although recollection of details was unimpaired (Bowles et al., 2016), supporting a two-process model of medial temporal lobe function. (From B. Bowles et al., 2007. *Proc Natl Acad Sci USA* 104: 16382–16387. © 2007 National Academy of Sciences, U.S.A.)

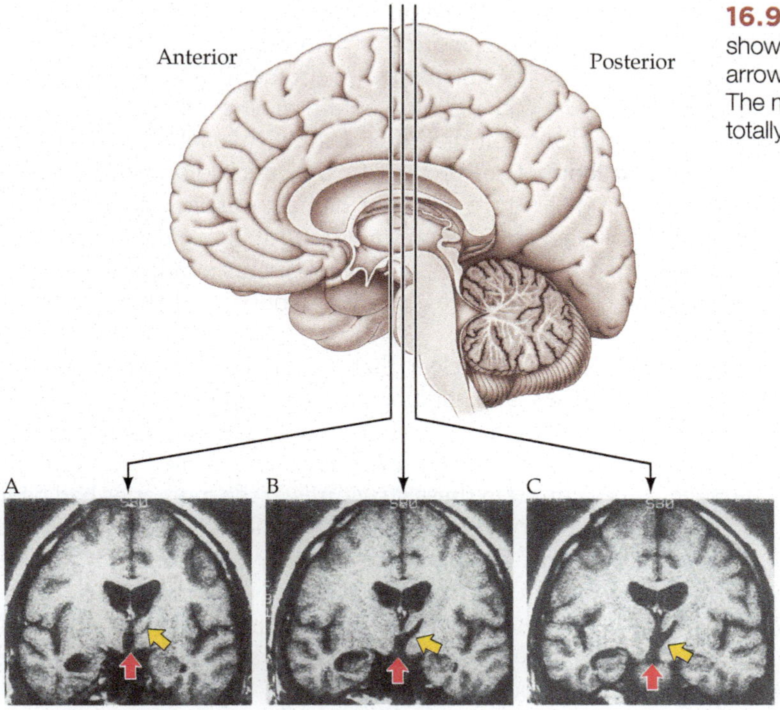

MRI scans from L. R. Squire et al., 1989. *Exp Neurol* 105: 23–35.

16.9 The Brain Damage in Patient N.A. Successive MRI scans show a prominent diencephalic lesion on the left side of the brain (yellow arrows), as well as a lesion on the floor of the third ventricle (red arrows). The mammillary bodies should be present in B and C, but they are totally absent.

Damage to the medial diencephalon can also cause amnesia

In 1960, a young man now known to neuroscientists as **patient N.A.** suffered a bizarre accident in which a miniature sword injured his brain after entering through his nostril. He eventually recovered in most respects except that, like Henry Molaison, he was left with a profound anterograde amnesia (Squire and Moore, 1979). He can give little information about events since his accident, although his memory for earlier events is nearly normal (Kaushall et al., 1981).

MRI scans of N.A. (**FIGURE 16.9**) showed no damage to the medial temporal lobe, but there was damage to several components of the diencephalon (thalamus and hypothalamus) that connect to the hippocampus, especially the dorsomedial thalamus, and to both **mammillary bodies** (see Figure 16.2A), as well as the mammillothalamic tract communicating between them (Squire et al., 1989). Like Henry Molaison, N.A. shows normal short-term memory but is impaired in forming declarative (but not nondeclarative/procedural) long-term memories. The similarity in symptoms suggests that the medial temporal lobe damaged in Henry's brain and the midline diencephalic region damaged in N.A. are normally parts of a larger memory system. Studies of people with surgical lesions of the fornix, which interconnects the hippocampus and mammillary bodies, confirm and extend this idea (Tsivilis et al., 2008).

The importance of medial thalamus and the mammillary bodies for forming declarative memories is reinforced by studies of people with **Korsakoff's syndrome**—named for its nineteenth-century discoverer, Russian neurologist Sergei Korsakoff—who also have anterograde amnesia for declarative memories. People with Korsakoff's syndrome frequently deny that anything is wrong with them, and they often **confabulate**—that is, fill a gap in memory with a falsification that they seem to accept as true. Temporal lobe structures, including the hippocampus, are typically normal in people with Korsakoff's syndrome (Mair et al., 1979). But their brains show shrunken, diseased mammillary bodies, as well as some damage in the dorsomedial thalamus (**FIGURE 16.10**), similar to that seen in N.A. The mammillary bodies may serve as a processing system connecting the medial temporal lobes (which were removed from Henry Molaison) to the thalamus via the mammillothalamic tract and, from there, to other cortical sites (Vann and Nelson, 2015). Damage to the basal frontal cortex, also found in people with Korsakoff's syndrome, probably causes the

patient N.A. A person who is unable to encode new declarative memories, because of damage to the dorsal thalamus and the mammillary bodies.

mammillary body One of a pair of nuclei at the base of the brain that connect to the hippocampus and play a role in memory.

Korsakoff's syndrome A memory disorder, related to a thiamine deficiency, that is generally associated with chronic alcoholism.

confabulate To fill in a gap in memory with a falsification. It often occurs in Korsakoff's syndrome.

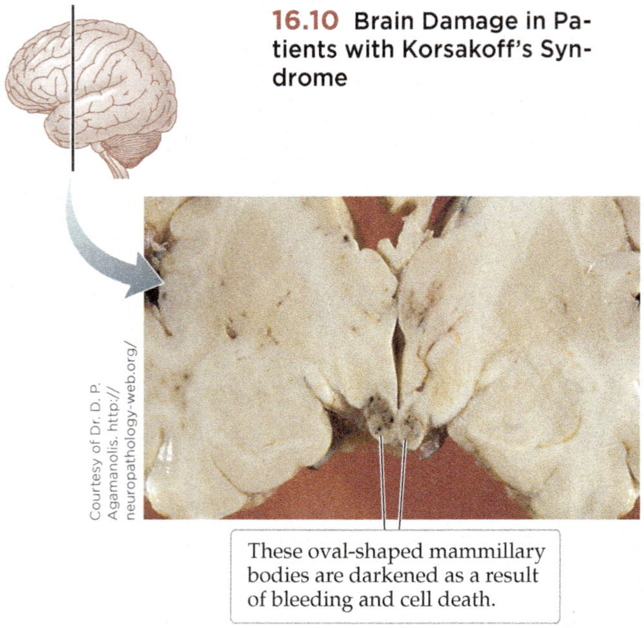

16.10 Brain Damage in Patients with Korsakoff's Syndrome

Courtesy of Dr. D. P. Agamanolis. http://neuropathology-web.org/

denial and confabulation that differentiates them from other people who have amnesia, such as Henry.

The main cause of Korsakoff's syndrome is lack of the vitamin thiamine (or B_1). Alcoholics who obtain most of their calories from alcohol and neglect their diet often exhibit this deficiency. Treating them with thiamine can prevent further deterioration of memory functions but will not reverse the damage already done. If makers of alcoholic beverages were required to supplement them with thiamine, then many new cases of Korsakoff's syndrome could be prevented each year (J. Price et al., 1987), but anti-drinking factions oppose such legislation.

These studies make it clear that the medial temporal lobe, the mammillary bodies, and the dorsomedial thalamus are needed to *form* new declarative memories. But these case studies also clearly show that established declarative memories, formed before brain damage, are not *stored* in these structures. If the memories had been stored there, they would have been lost when the structures were damaged. So where are memories stored? Mathematical models based on the known connectivity patterns of cortical neurons indicate that the cortex is optimised for information storage (Brunel, 2016), and as we'll see next, clinical observations confirm that the cortex serves as a repository for memories.

The cortex is essential for long-term storage of memories

patient K.C. A person who sustained damage to the cortex that rendered him unable to retrieve autobiographical memories. Upon his death we learned that his name was Kent Cochran.

episodic memory Memory of a particular incident or a particular time and place.

semantic memory Generalised memory—for instance, knowing the meaning of a word without knowing where or when you learned that word.

One striking case study indicates that the cortex is crucial for *storing* memories once they are formed and also illustrates an important distinction between two subtypes of declarative memory. Profiled in numerous scientific reports as **patient K.C.**, Kent Cochran sustained serious brain damage in a motorcycle accident at age 30. For more than 30 years, from his accident until the day he died in 2014 at age 62, Kent was unable to retrieve any personal memory of his past, although his general knowledge remained good. He conversed easily and played a good game of chess (**FIGURE 16.11**), but he could not remember where he learned to play chess or who taught him the game. Detailed autobiographical declarative memory of this sort is known as **episodic memory**: you show episodic memory when you recall a specific episode in your life or relate an event to a particular time and place. In contrast, **semantic memory** is generalised declarative memory, such as knowing the meaning of a word without knowing where or when you learned that word (Craver et al., 2014; Tulving et al., 1991). The fact that K.C. could access semantic memories, like the rules of chess, but not episodic memories, such as where and when he learned to play, indicates that these are distinctive types of memory. Because K.C.'s accident took away his access to one type of memory but not the other, presumably they are stored in different parts of the brain, or stored in different ways.

Brain scans of K.C. revealed extensive damage to the cerebral cortex and severe shrinkage of the hippocampus and nearby parahippocampal cortex (Rosenbaum et al., 2005). As with Henry, the bilateral hippocampal damage probably accounts for K.C.'s anterograde declarative amnesia. But that damage cannot account for the selective loss of nearly his entire autobiographical memory (i.e., retrograde episodic amnesia), because other people with damage restricted to the medial temporal lobes lack this symptom. K.C.'s inability to recall any autobiographical details of his life, even memories from years before his accident, may instead be a consequence of the extensive damage to his frontal and parietal cortex. Thus a model for the formation

From E. Tulving, 2002. *Annu Rev Psychol* 53: 1-25.

16.11 Patient K.C. Kent Cochran retained declarative memories from before his accident, such as the rules of chess. But he lost episodic memories, such as when and where he learned to play.

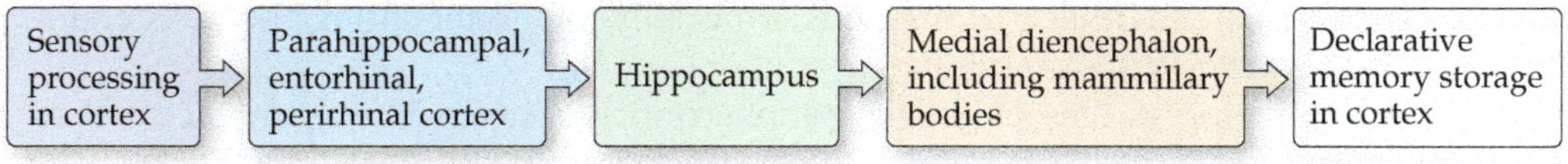

16.12 Current Model of Declarative Memory Formation Damage to either the medial temporal lobe or the medial diencephalon/mammillary bodies will prevent the formation of any new declarative memories without loss of previously formed memories.

of declarative memory is that sensory information analysed by diverse parts of cortex is delivered to the medial temporal lobe, processed by the hippocampus, sent to medial diencephalon, including the mammillary bodies, and then stored in other parts of the cortex (**FIGURE 16.12**).

There are a few rare individuals who have incredibly extensive episodic memories, who can tell you exactly what they had for breakfast on January 18, 1995, or any other date (J. Price, 2008). A group of 11 such people with extreme autobiographical memory were found to have larger temporal lobes, including a larger parahippocampal gyrus, than controls (LePort et al., 2012). We don't know whether these people were born with larger temporal lobes, and so had great autobiographical memory, or developed great autobiographical memory, which caused their temporal lobes to enlarge.

16.2 Different Forms of Nondeclarative Memory Involve Different Brain Regions

Learning Objectives

After reading this section, you should be able to:

16.2.1 **Define several types of nondeclarative memory.**

16.2.2 **List brain regions that have been implicated in the different kinds of nondeclarative memories.**

16.2.3 **Present evidence that the hippocampus is particularly important in spatial navigation and memory.**

We have seen that there are two different kinds of declarative memory: semantic and episodic. Likewise, there are several forms of nondeclarative memory, and as we'll see, these depend on different neural systems.

Specific functions depend on different forms of nondeclarative memory

Unlike declarative memory, which we use to store the factual knowledge of ourselves and the world that we use in conversation and in our innermost thoughts, nondeclarative memories shape our abilities and responses to the world around us in ways that are difficult to express, but no less crucial for survival. Indeed, virtually every species of the animal kingdom seems capable of some sort of nondeclarative memory, even if very simple, indicating that it's an ancient evolutionary adaptation.

SKILL MEMORY In **skill learning**, participants learn how to perform a challenging task by practicing it over and over again. Experimental tests like the mirror-tracing task performed by Henry Molaison (see Figure 16.3) or learning to read mirror-reversed text (see Figure 16.4) are examples of skill learning. So too are the everyday skills we acquire, like learning to ride a bike or to juggle (well, okay, maybe juggling isn't an "everyday" skill, but you get the idea).

Go to Activity 16.1
The "Towers of Hanoi" Problem

Imaging studies have investigated learning and memory for different skill categories, especially *sensorimotor skills* (e.g., mirror tracing), *perceptual skills* (e.g., learning to read mirror-reversed text), and *cognitive skills* (e.g., puzzles like the Towers of Hanoi—which you can play on the website—that involve planning and problem solving). All three kinds of skill learning are impaired in people with damage to the basal

skill learning Learning to perform a task that requires motor coordination.

priming Also called *repetition priming*. The phenomenon by which exposure to a stimulus facilitates subsequent responses to the same or a similar stimulus.

associative learning A type of learning in which an association is formed between two stimuli or between a stimulus and a response. It includes both classical and operant conditioning.

operant conditioning Also called *instrumental conditioning*. A form of associative learning in which the likelihood that an act (instrumental response) will be performed depends on the consequences (reinforcing stimuli) that follow it.

classical conditioning Also called *Pavlovian conditioning*. A type of associative learning in which an originally neutral (conditioned) stimulus acquires the power to elicit the response normally elicited by another (unconditioned) stimulus after the two stimuli are paired. A response elicited by the unconditioned stimulus (US) is called an *unconditioned response* (*UR*); a response elicited by the conditioned stimulus (CS) alone is called a *conditioned response* (*CR*).

ganglia; damage to other brain regions, especially the motor cortex and cerebellum, also affects aspects of some skills.

Neuroimaging studies in healthy people confirm that the basal ganglia, cerebellum, and motor cortex are important for sensorimotor skill learning (Spampinato and Celnik, 2018). For example, learning specific sequences of finger movements is associated with selective activation of motor cortex and the basal ganglia (Doyon et al., 1996; Hazeltine et al., 1997), with the cerebellum possibly providing error correction during learning (Flament et al., 1996). Often the activations shift among brain regions as performance changes during the course of learning, so learning appears to involve a complex set of interacting neural networks.

PRIMING A change in the processing of a stimulus, usually a word or a picture, as a result of prior exposure to the same stimulus or related stimuli is referred to as **priming** (or *repetition priming*). For example, if a person is shown the word *stamp* in a list and later is asked to complete the word stem *STA-*, they are more likely to reply "stamp" than is a person who was not exposed to that word. Priming does not require declarative memory of the stimulus—Henry Molaison and other people with amnesia have shown priming for words they don't remember seeing previously. In contrast to skill learning, researchers have found that priming is not impaired by damage to the basal ganglia.

In functional-imaging studies, perceptual priming (priming based on the visual *form* of words) has been related to *reduced* activity in bilateral occipitotemporal cortex (Schacter et al., 1996), presumably because the priming makes the task easier. Conceptual priming (priming based on word *meaning*) was associated with reduced activation of the left frontal cortex (Blaxton et al., 1996; Gabrieli et al., 1996; A. D. Wagner et al., 1997). And although it is primarily associated with declarative memory, evidence is mounting that the components of the medial temporal lobe memory system—notably the perirhinal cortex (W. C. Wang et al., 2010, 2014) and hippocampus (Addante, 2015)—are also involved in some forms of nondeclarative memory, including priming.

ASSOCIATIVE LEARNING Learning that involves relations between events—for example, between two or more stimuli, between a stimulus and a response, or between an action and its consequence—is called **associative learning**. In one form of associative learning, called **operant conditioning** (also called *instrumental conditioning*), an association is formed between the individual's behaviour and the consequence(s) of that behaviour. For example, lab animals can easily learn to press a lever to gain the reward of a tasty food pellet, using an apparatus like the one in **FIGURE 16.13**, often called a *Skinner box* after its famous originator, B. F. Skinner.

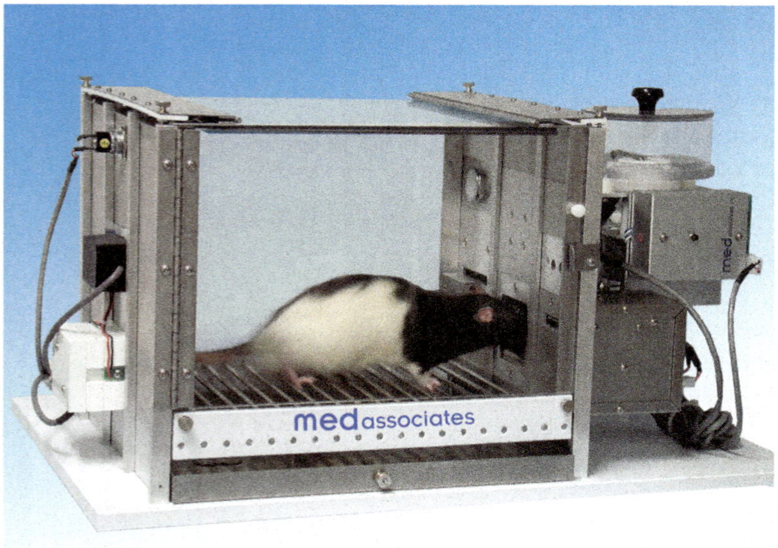

Courtesy of Med Associates

16.13 Operant Conditioning Animals in a Skinner box can learn to press a lever to gain food pellets or to avoid footshock, a form of associative learning called *operant* (or *instrumental*) *conditioning*.

Many people are also familiar with another form of associative learning, termed **classical conditioning** (or *Pavlovian conditioning*), in which an initially neutral stimulus comes to *predict* an event. At the end of the nineteenth century, Ivan Pavlov (1849–1936) found that a dog would salivate when presented with an auditory or visual stimulus if the stimulus came to predict the presentation of food. So, as illustrated in **FIGURE 16.14**, repeatedly ringing a bell before putting meat powder in a dog's mouth will eventually cause the dog to start salivating when it hears the bell alone. In this case the meat powder in the mouth is the *unconditioned stimulus* (*US*), which already evokes an *unconditioned response* (*UR*; salivation in the example). The sound of the bell is the *conditioned stimulus* (*CS*), and the learned response to the CS alone (salivation in response to the bell) is called the *conditioned response* (*CR*).

Experimental evidence in lab animals shows that cerebellar circuits are crucial for simple eye-blink conditioning,

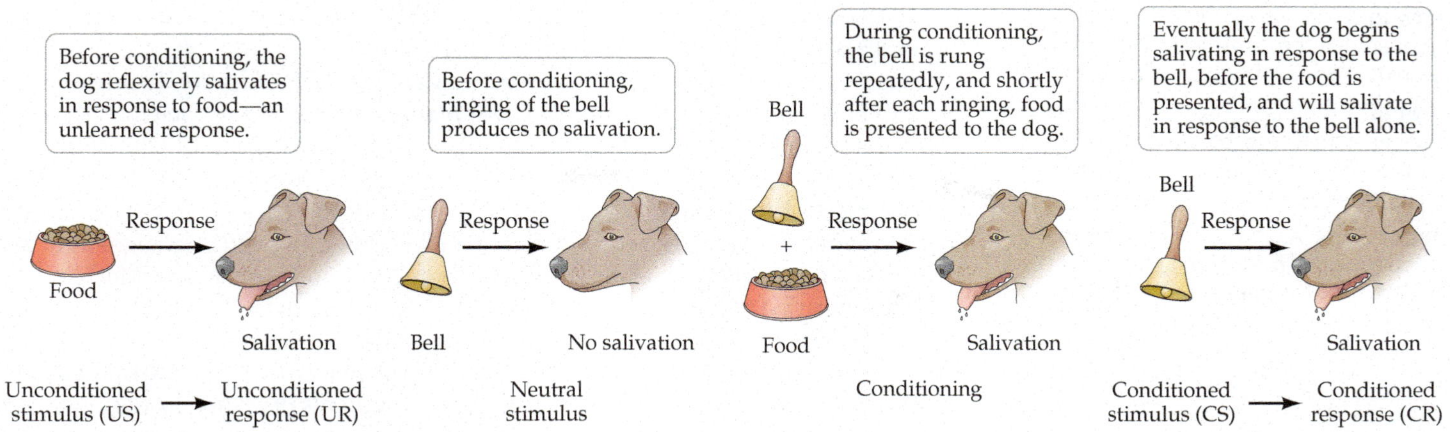

16.14 Classical Conditioning In this famous experiment, by pairing food with the sound of a bell, Ivan Pavlov conditioned animals to eventually salivate to the sound of the bell alone.

in which a tone or other stimulus is associated with eye blinking in response to a puff of air. A PET study of human eye-blink conditioning (Logan and Grafton, 1995) found that in the course of conditioning, there was a progressive increase in activity in several regions of the brain, including not only the cerebellum, but also the hippocampus, the ventral striatum, and regions of the cerebral cortex. But activity in these other areas may not be *essential* for eye-blink conditioning in the way that the cerebellum is. For example, people with hippocampal lesions can acquire the conditioned eye-blink response, but people with unilateral cerebellar damage can acquire a conditioned eye-blink response only on the side where the cerebellum is intact (Papka et al., 1994). We'll look more closely at neural mechanisms of eye-blink conditioning later in the chapter, as well as the cellular and molecular underpinnings of a very simple type of memory—*nonassociative learning*—that is formed in basic neural circuits of even the simplest animals.

The taxonomy of memory we presented in Figure 16.5 is updated in **FIGURE 16.15**, where we have added the subtypes of declarative and nondeclarative memory described in this section, along with some examples.

Medial temporal mechanisms keep track of positions in spatial, temporal, and social networks

The caricature of the white-coated scientist watching rats run in mazes, a staple of cartoonists to this day, has its origins in the intensive memory research of the early twentieth century. The early work indicated that rats and other animals don't just learn a series of turns but instead form a **cognitive map** (an understanding of the *relative* spatial organisation of objects and information) in order to solve a maze (Tolman, 1949). Animals apparently learn at least some of these details of their spatial environment simply by moving through it, a phenomenon that was termed *latent learning* (Tolman and Honzik, 1930).

cognitive map A mental representation of a spatial relationship.

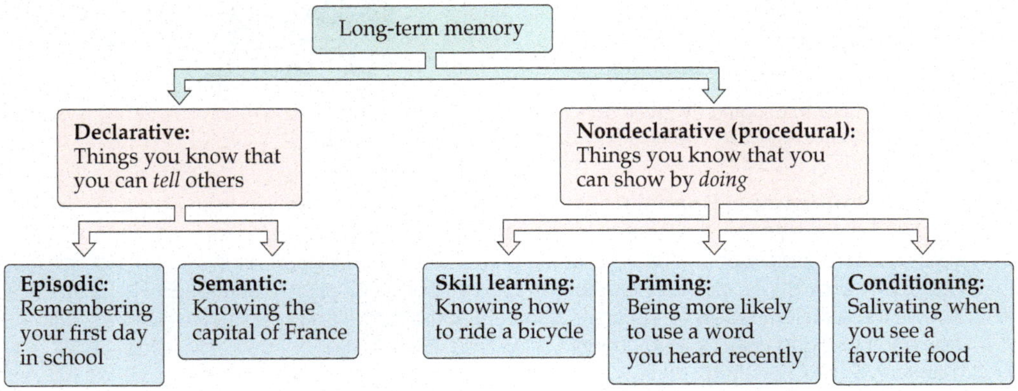

16.15 Subtypes of Declarative and Nondeclarative Memory

place cell A neuron within the hippocampus that selectively fires when the animal is in a particular location.

grid cell A neuron that selectively fires when an animal crosses the intersection points of an abstract grid map of the local environment.

border cell A neuron that selectively fires when an animal arrives at the perimeter of a local spatial cognitive map.

We now know that, in parallel with its role in other types of declarative memory, the hippocampus is a crucial neural participant in spatial learning. Within the rat hippocampus are found many neurons that selectively encode spatial location (Leutgeb et al., 2005; O'Keefe and Dostrovsky, 1971). These **place cells** become active when the animal is in—or moving toward—a particular location (Pastalkova et al., 2008) (**FIGURE 16.16A**). If the animal is placed in a new environment, its place cell activity indicates that the hippocampus remaps to the new locations (Moita et al., 2004). Lesions of this part of the hippocampus severely impair spatial learning in rats (McNaughton et al., 1996) and humans (Bartsch et al., 2010). The Nobel Prize in Physiology or Medicine for 2014 was awarded to John O'Keefe, May-Britt Moser, and Edvard Moser for their work on understanding hippocampal place cells. Two types of cells discovered in nearby entorhinal cortex help the animal to learn the local spatial environment. **Grid cells** fire selectively when the animal crosses the intersection points of an abstract grid map of the local environment (**FIGURE 16.16B**), acting like an innate system of latitude and longitude (Moser and Moser, 2013). Cells that fire in the same pattern have also been discovered in humans (J. Jacobs et al., 2013). Arrival at the *perimeter* of the local spatial map is signaled by the activation of entorhinal **border cells** (Solstad et al., 2008).

Evidence is mounting that beyond its roles in forming declarative memories and keeping track of important locations in space, the hippocampus helps us to organize networks of data across multiple domains. For example, hippocampal function has been linked to mapping our *social* space—the web of social connections to other people that we navigate daily in our work and personal lives (Alexander et al., 2016; Tavares et al., 2015). Likewise, researchers have recently discovered hippocampal cells that appear to map locations in *time*, another important aspect of developing coherent memories (Eichenbaum, 2014; Schiller et al., 2015). In other words, the hippocampus helps organize the *what*, *where*, *when*, and *with whom* of memory.

Careful comparisons of natural behaviour and brain anatomy have revealed that for many species, their manner of making a living has left an imprint on the hippocampus. For example, species of birds that hide caches of food in spatially scattered locations have larger hippocampi than noncaching species, even when the comparison species are very close relatives that have otherwise similar lifestyles (Krebs et al., 1989; Sherry, 1992; Sherry et al., 1989) (see Figure 5.6). Lesions of the hippocampus impair the ability of these birds to store and retrieve caches of food (Sherry and Vaccarino, 1989). Homing

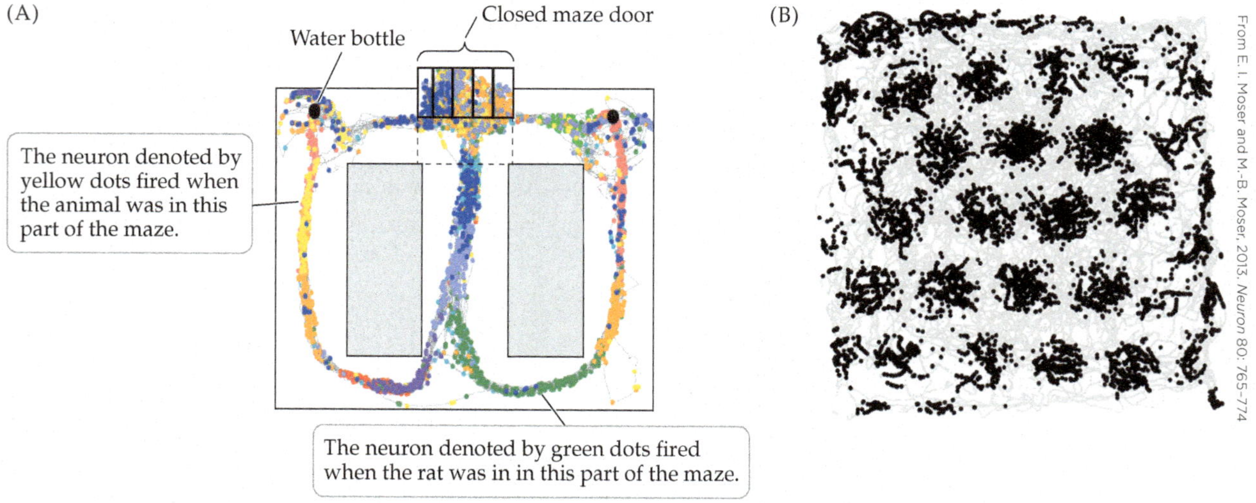

16.16 Hippocampal Spatial Cells (A) In this simple maze, viewed from above, each coloured dot represents the location of the rat when one specific place cell fired. Thus, all of the yellow dots represent one neuron, all of the green dots another neuron, and so on. As the figure plainly shows, place cells accurately encode a specific location in space. (B) The gray trace shows the path of a rat exploring an enclosure. Each black dot represents the point at which one particular grid cell in the entorhinal cortex fired. (Part A after E. Pastalkova et al., 2008. *Science* 321: 1322–1327.)

pigeons also have enlarged hippocampi relative to other varieties of pigeons, presumably serving the spatial demands of their prodigious navigational abilities (Rehkamper et al., 1988). A relationship between spatial cognition and hippocampal size is evident in mammals too. Just as with the food-caching birds, Merriam's kangaroo rat, which stashes food in scattered locations, has a significantly larger hippocampus than its noncaching cousin, the bannertail kangaroo rat (L. F. Jacobs and Spencer, 1994). (For additional fascinating examples of the relation between hippocampal volume and spatial behaviour, see **A STEP FURTHER 16.1: SPATIAL BEHAVIOUR AND HIPPOCAMPAL STRUCTURE IN HUMANS AND OTHER SPECIES** on the website.)

sensory buffer An element of the type of memory that stores the sensory impression of a scene.

short-term memory (STM) A form of memory that usually lasts only for seconds, or as long as rehearsal continues, especially while being used during performance of a task.

working memory A type of short-term memory that holds a limited amount of information available for ready access during performance of a task.

prefrontal cortex The most anterior portion of the frontal lobe.

16.3 Successive Processes Capture, Store, and Retrieve Information in the Brain

Learning Objectives

After reading this section, you should be able to:

16.3.1 **Provide evidence for distinct stages in memory formation.**

16.3.2 **List brain regions important for different attributes of memory.**

16.3.3 **Describe the current model for how information enters various stages of memory, including the terms for each processing step, and how information can be lost.**

16.3.4 **Explain how the process of reconsolidation can distort or alter long-term memories.**

The span of time that a piece of information will be retained in the brain varies. Evidence suggests that there are as many as four different duration categories for memory. The briefest memories are held in **sensory buffers** (for visual stimuli, they are sometimes called *iconic memories*); an example is the fleeting impression of a glimpsed scene that vanishes from memory seconds later. These brief memories are thought to be residual sensory neural activity (**FIGURE 16.17**).

Somewhat more durable than the sensory buffers are **short-term memories** (**STMs**). If I ask you to relate the things you did this morning, you will be tapping your STM. If ask a month from now what you did this morning, you probably won't remember much, so most of that information is destined to be forgotten, never transferred into long-term memory, which we'll discuss shortly. STM is more vulnerable than our older memories. For example, electroconvulsive therapy ("shock treatment") typically erases memories from the days before the treatment (another example of retrograde amnesia) without permanent loss of memories laid down the weeks and years beforehand (Semkovska and McLoughlin, 2013). A blow to the head can also erase STMs.

Sometimes we may mentally manipulate and process information in STM, and this store is referred to as **working memory**, in recognition of the way we use it. You can think of working memory as a subset of STM in which we mentally manipulate the information (Cowan, 2008). If you look up a phone number and keep it in mind just long enough to make the phone call, you are using your working memory. In the absence of rehearsal, working memory lasts only about 30 seconds (J. Brown, 1958; L. R. Peterson and Peterson, 1959). The capacity of working memory, the number of bits of information that can be held there, is also quite limited. A classic paper suggested that most of us can only keep about seven chunks of information in working memory at any one time (G. A. Miller, 1956).

Scientists probe working memory through the passage of time—a delay between stimulus and response during which information must be held ready in working memory for further processing. Such delayed-response tasks are especially associated with the **prefrontal cortex**. In humans and experimental animals, activity increases in prefrontal cortex during the delay when working memory must be used, and lesions of prefrontal cortex severely impair working memory (Bahmani et al., 2019; Nee and D'Esposito, 2018). Henry

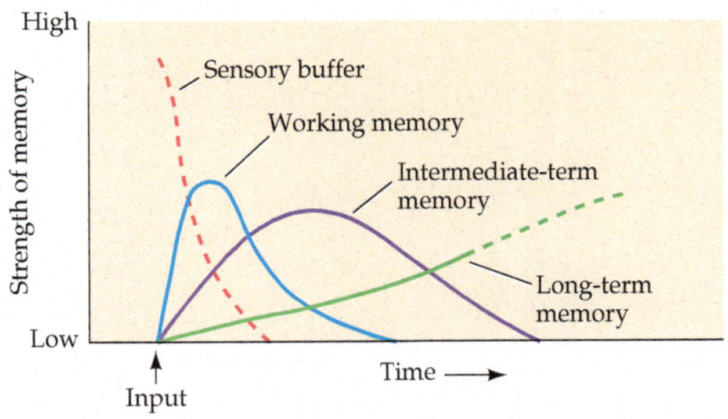

16.17 Temporal Stages of Memory Formation
(After J. L. McGaugh, 2000. *Science* 287: 248–251.)

intermediate-term memory
A form of memory that lasts longer than short-term memory but not as long as long-term memory.

Molaison had a normal working memory, presumably because his prefrontal cortex remained intact.

One influential model (Baddeley, 2003) subdivides working memory into three complementary components:

1. A *phonological loop* contains auditory information (such as speech); this is what you use to rehearse that phone number.
2. A so-called *visuospatial sketch pad* holds visual impressions of stimuli; you use this to imagine the route back to your car in a parking garage.
3. An *episodic buffer* contains more-integrated information, spanning across sensory modalities, sort of like movie clips.

According to this model, the flow of information into and out of working memory is supervised by a fourth module, an *executive control,* which we'll discuss in Chapter 17.

Some memories last only a little longer than short-term memories. Chances are good that you can remember what you had for lunch today or yesterday, but not most of your lunches last week. You may recall today's weather forecast, but not that of a few days ago. These are examples of what some memory researchers identify as **intermediate-term memory**—that is, a memory that outlasts what we typically consider to be STM but is far from being permanent (Rosenzweig et al., 1993).

A variety of brain regions are involved in different attributes of working memory

Because they span verbal and nonverbal material, in multiple sensory modalities, for multiple purposes, different working memories may concern different attributes. So, for example, an individual memory may include a mix of information about space, time, sensory perception, response, and/or emotional factors. Researchers have attempted to devise memory tasks that selectively tap some of these attributes of memory, in order to assess the relative contributions of different regions of the brain.

FIGURE 16.18 presents some examples of some well-known tests that researchers use to probe working memory (Kesner, 1998; Kesner et al., 1993). For testing *spatial* location memory, the eight-arm radial maze is commonly used (**FIGURE 16.18A**). To solve this task correctly and receive a food reward, rats must recognise and enter an arm of the maze that they went down shortly beforehand. Rats were tested following surgical lesions of the hippocampus, the portion of the basal ganglia called the striatum, or nonprimary visual cortex. Only the animals with hippocampal lesions were impaired on this predominantly spatial task—a result that is consistent with the role of the hippocampus in spatial cognition that we discussed earlier. The test shown in **FIGURE 16.18B** was used to assess the memory of the same rats for their own *motor* behaviour. Here the animal must use working memory to remember whether it made a left or right turn a few moments previously, and it receives a food reward only if it makes a turn in the same direction on a follow-up trial. Only the animals with lesions of the striatum were significantly impaired on this task. Finally, in the test depicted in **FIGURE 16.18C**, the rats were required to hold in working memory the *sensory* attributes of presented stimuli, identifying the novel stimulus in each pair of stimuli presented. Here, only the rats with nonprimary visual cortex lesions were significantly impaired. This impressive lack of overlap between the symptoms of the different lesions nicely illustrates how working memories involving different attributes are parceled out to diverse brain regions for storage.

Brain regions involved in learning and memory: A summary

Several major conclusions should be apparent by now, especially that (1) many regions of the brain are involved in learning and memory, (2) different forms of memory rely on at least partly different networks of brain mechanisms spanning several different regions of the brain, (3) any one memory-related brain structure can be a part of the circuitry for several different forms of learning, and (4) working memory relies on frontal and parietal cortical regions involved in sensory and motor processing. A final update

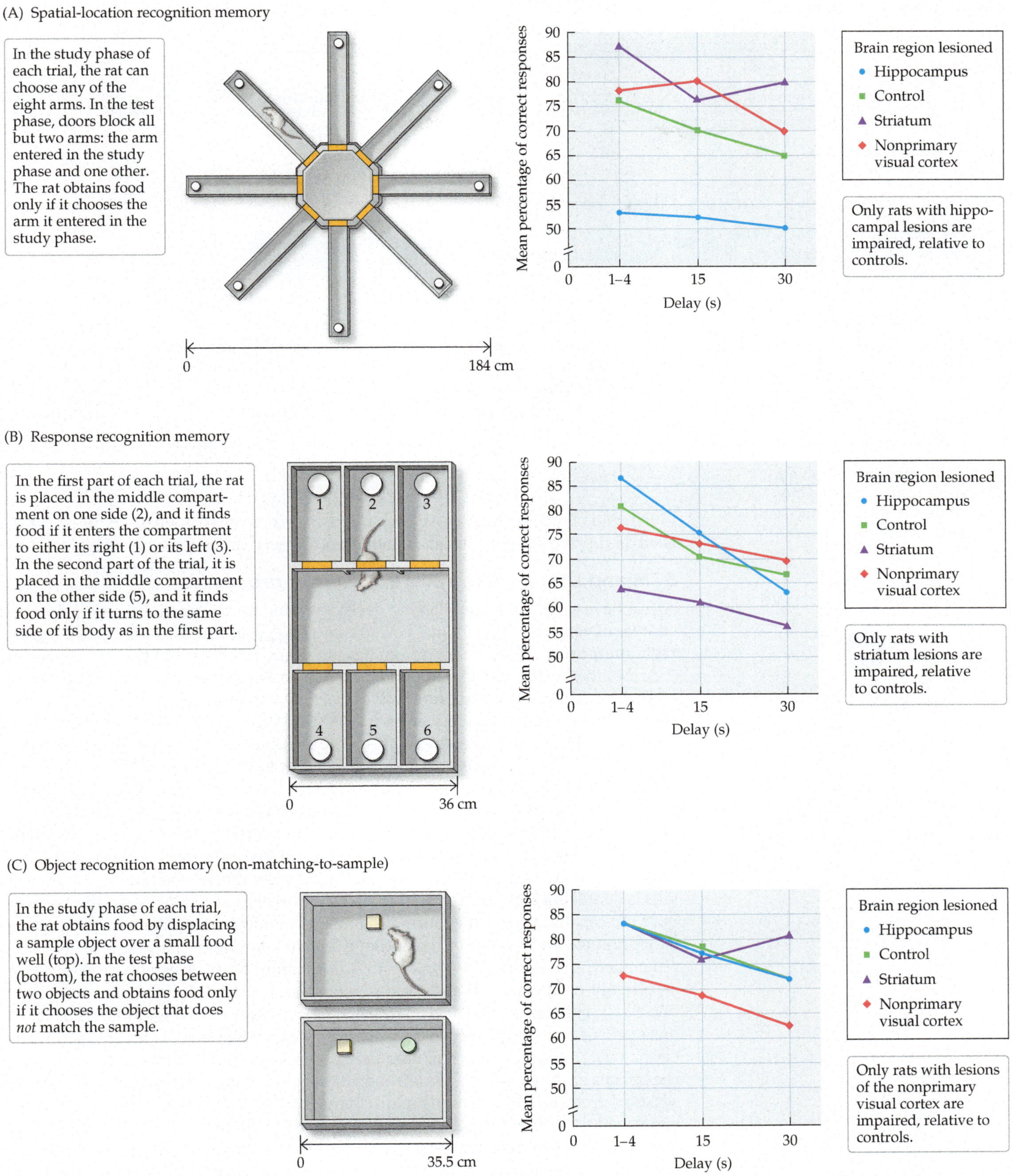

16.18 Tests of Specific Attributes of Working Memory Brain lesion experiments testing spatial-location recognition (A), response recognition (B), and object recognition (C)—using the setups shown on the left—yielded the results shown on the right. (After R. P. Kesner et al., 1993. *Exp Brain Res* 93: 462–470.)

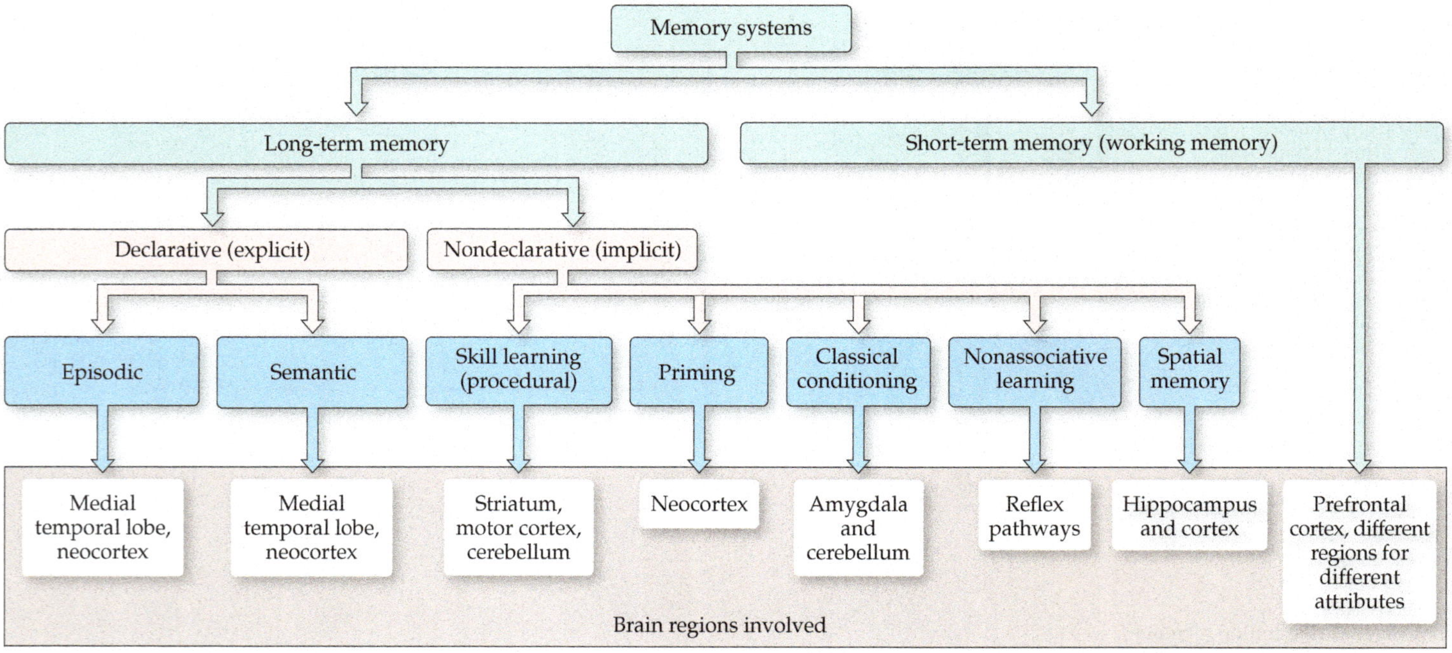

16.19 Brain Regions Involved in Different Kinds of Learning and Memory (After T. Bartsch and C. Butler, 2013. *Nat Rev Neurol* 9: 86–97.)

Go to Activity 16.2
Learning and Memory

to the map of memory that we've been discussing is provided in **FIGURE 16.19.** Next we'll discuss the stages by which memories, of any sort, can be preserved for a lifetime.

STM and LTM involve several different processes

As we have seen, really long-lasting declarative and nondeclarative memories—the address of your childhood home, how to ride a bike, your first kiss—are called **long-term memories** (**LTMs**), lasting from days to years. A substantial body of evidence indicates that STM and LTM, in particular, rely on different processes to store information. A classic demonstration involves learning lists of words or numbers, the so-called serial position curve reviewed in **A STEP FURTHER 16.2: SERIAL POSITION CURVES DISTINGUISH STM AND LTM** on the website. So how much information do we hold in LTM? It is likely that LTM evolved not to give us a perfect record of our history, but rather to record important events that we can use to shape our future behaviour in adaptive ways. It would be a waste of space to record every little detail of our lives, and indeed an important aspect of our memory system is the continual pruning of unimportant memories to preserve cognitive resources (Kuhl et al., 2007). Real-life cases of rare people with perfect recall show that an inability to forget details is confusing and exhausting (Luria, 1987; Parker et al., 2006). Nevertheless, we have the capability to retain enormous amounts of information. We take this capacity for granted and barely notice, for example, that knowledge of a language involves remembering at least 100,000 pieces of information. Most of us also store a huge assortment of information about faces, tunes, odours, skills, stories, and so on. In one classic experiment, participants viewed long sequences of colour photos of various scenes; several days later, the participants were shown pairs of images—in each case a new image plus one from the previous session—and were asked to identify the images seen previously. Astonishingly, the participants were highly accurate for series of up to 10,000 different stimuli (Standing, 1973), and researchers have since confirmed that LTM is so vast as to seem boundless (Brady et al., 2008), at least for some types of stimuli (Cunningham et al., 2015). Similar impressive feats of memory in our distant relatives, such as pigeons (Vaughan and Greene, 1984), illustrate that a great capacity for information storage is a general property of nervous systems across the animal kingdom.

long-term memory (LTM) An enduring form of memory that lasts days, weeks, months, or years and has a very large capacity.

The system for creating and retaining memories consists of three general stages (**FIGURE 16.20**):

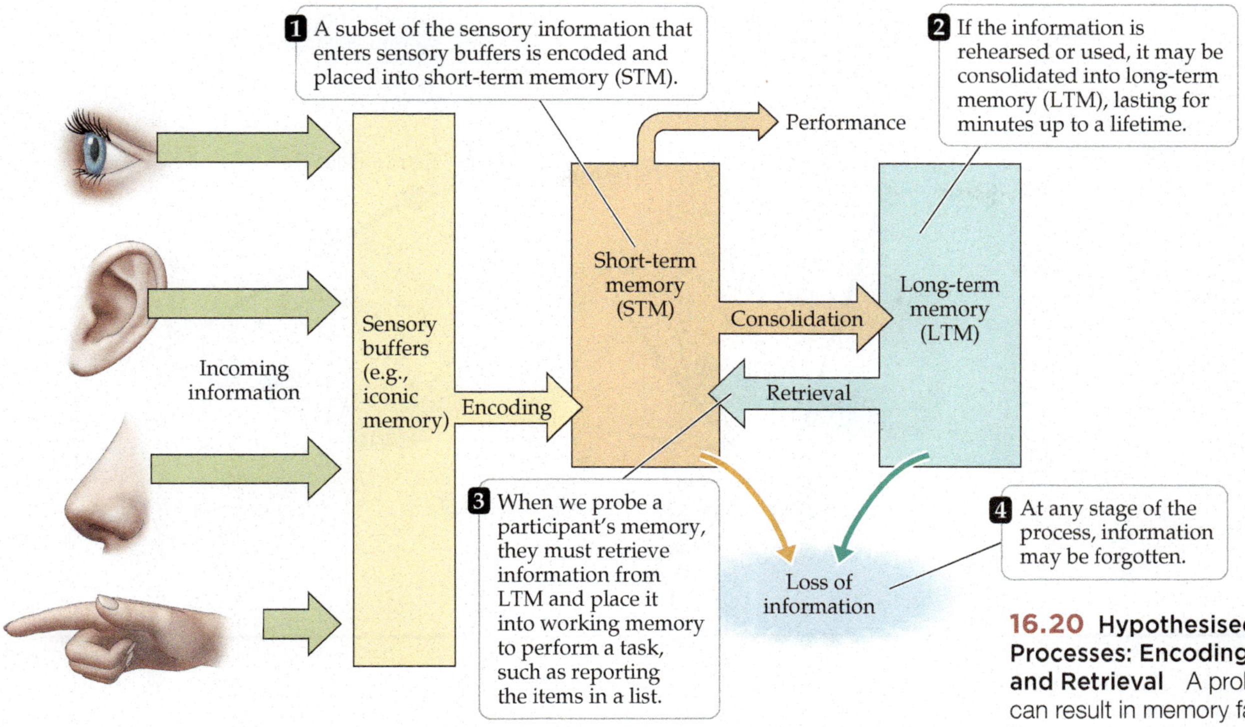

16.20 Hypothesised Memory Processes: Encoding, Consolidation, and Retrieval A problem at any stage can result in memory failure.

1. *Encoding of raw information from sensory channels into short-term memory* Functional brain imaging indicates that while many brain areas participate in the initial processing of stimuli, fewer are associated with successful **encoding**, and these tend to reflect the specific attributes of the stimuli. So, for example, encoding of the pictorial elements in photos involves greater activation of the right prefrontal cortex and the parahippocampal cortex in both hemispheres (Brewer et al., 1998). In the case of words (A. D. Wagner et al., 1998), the critical areas are the *left* prefrontal cortex and the *left* parahippocampal cortex. These results indicate that parahippocampal and prefrontal cortex are crucial for consolidation, and these mechanisms reflect the hemispheric specialisations (left hemisphere for language and right hemisphere for spatial ability) that we'll discuss in Chapter 18. Although not depicted in the figure, this model suggests that the flow of information into and out of working memory is supervised by another part of the brain, a *central executive*, which we will discuss in more detail in Chapter 17.
2. *Consolidation of the volatile short-term memories into more-durable long-term memory* We know from cases like that of Henry Molaison (who could repeat STM items like word lists) that while the medial temporal lobe apparently is not needed to encode sensory information into STM, it is crucial for **consolidation**, the conversion of memories from volatile STM into LTM, a different format that may endure for a lifetime (M. Lepage et al., 1998). And *where* are the new long-term memories—or more precisely their **engrams**, the physical changes that underlie LTM—actually being stored? Henry could recall events from before his surgery, and detailed studies in lab animals (Zola-Morgan and Squire, 1990) confirmed that LTM must be stored somewhere *other* than the medial temporal lobe—namely, other parts of cortex. An important principle that has emerged from this area of research is that *permanent storage of information tends to be in the regions of the cortex where the information was first processed and held in short-term memory*. For example, visual cortex is crucial for visual object recognition memory (López-Aranda et al., 2009). After further processing that involves the medial temporal region, the permanent memory

encoding A stage of memory formation in which the information entering sensory channels is passed into short-term memory.

consolidation A stage of memory formation in which information in short-term or intermediate-term memory is transferred to long-term memory.

engram The physical basis of a memory in the brain. It is sometimes referred to as a *memory trace* on the assumption that it involves changes in a neural circuit rather than a single neuron.

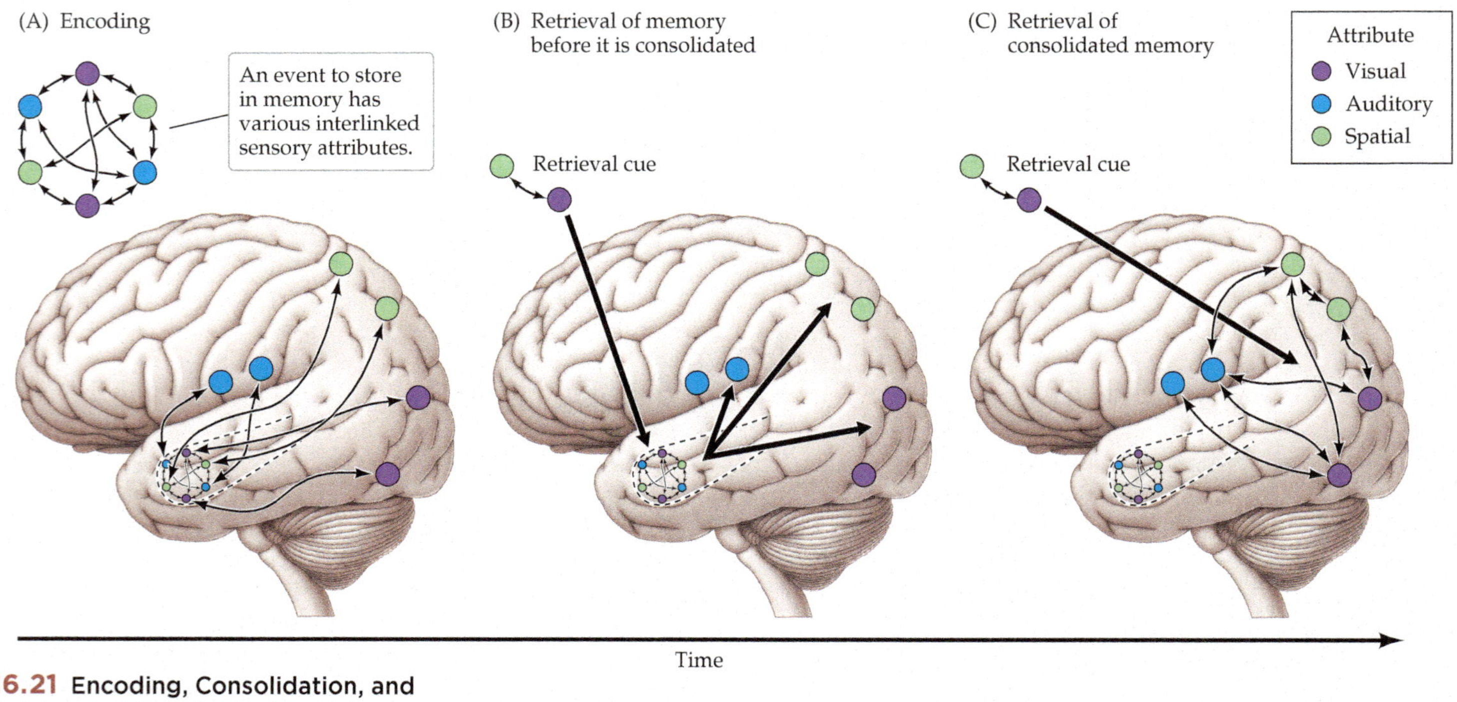

16.21 Encoding, Consolidation, and Retrieval of Declarative Memories (A) According to this model, medial temporal lobe processes distribute the various sensory attributes of an event, and linkages between them, in corresponding regions of cortex. (B) Before consolidation is complete, retrieval involves the hippocampus and other medial temporal structures. (C) After consolidation, retrieval may occur independent of the medial temporal system.

storage becomes independent and memories can be retrieved directly for use by other cognitive processes. This schema is illustrated in **FIGURE 16.21**.

3. *Retrieval of the stored information for use in future behaviour* Information **retrieval** from long-term storage is under the direction of various cognitive processes, including attention, as we discuss in Chapter 17. Some evidence suggests that retrieval from LTM makes the memories temporarily plastic again and amenable to updating and strengthening, before they undergo **reconsolidation** and return to stable status (Eisenberg et al., 2003; Nader and Hardt, 2009). In fact, one of the best ways to improve learning is simply repeated retrieval (and thus, repeated reconsolidation) of the stored information (Karpicke and Blunt, 2011), as long as steps are taken to ensure the information being recalled is correct.

Unfortunately, reconsolidation also offers an opportunity for memory to fail. For example, each time a memory trace is activated during recall, it is subject to changes and fluctuations, so with successive activations it may deviate more and more from its original form (Estes, 1997; Nader and Hardt, 2009). This may be why people sometimes "remember" events that never happened. We can create false memories by asking leading questions—"Did you see the broken headlight?" rather than "Was the headlight broken?"—or by providing misinformation via trusted channels (Loftus, 2003). For example, by burying false details in a biography provided to some participants, researchers found it relatively easy to plant a childhood memory of meeting a Bugs Bunny character at a Disney resort (Braun et al., 2002)—something that could never happen in real life (because Bugs is a Warner Bros. character). This distortion process has contributed to wrongful criminal convictions based on false "recovered memories" of victims or witnesses (McNally, 2003); alternatively, people can develop detailed but false memories of committing a crime in which they actually had no role (Shaw and Porter, 2015), potentially leading to false confessions.

Not all memories are created equal. Being stressed impairs our ability to retrieve past memories (Atsak et al., 2016), but a rush of emotion *during* an experience can powerfully enhance our memory for that event. An emotionally arousing story is remembered significantly better than a closely matched but emotionally neutral story (Reisberg and Heuer, 1995). But if study participants are treated with propranolol (a beta-adrenergic antagonist, or beta-blocker, that blocks the effects of epinephrine),

retrieval A process in memory during which a stored memory is used by an organism.

reconsolidation The process by which a retrieved memory may be strengthened or altered before being returned to LTM.

this emotional enhancement of memory vanishes. It's not that treated participants perceive the story as being any less emotional; in fact, treated participants rate the emotional content of the stories just the same as untreated participants do. Instead, the evidence indicates that propranolol directly interferes with the ability of adrenal stress hormones to act on the brain (Cahill et al., 1994), as discussed in **BOX 16.1**. Now let's consider how the brain changes to store our memories.

BOX 16.1 Emotions and Memory

Almost everyone knows from personal experience that strong emotions can potently enhance memory formation and retrieval. Examples of memories enhanced in this way might include a strong association between special music and a first kiss, or uncomfortably vivid recollection of the morning of September 11, 2001. Interestingly, our memories about learning of such events, which seem so vivid and detailed, are not nearly as accurate as they seem (Hirst et al., 2009). Nevertheless, people who actually *experience* life-threatening events often develop **posttraumatic stress disorder** (**PTSD**), characterised as "reliving experiences such as intrusive thoughts, nightmares, dissociative flashbacks to elements of the original traumatic event, and ... preoccupation with that event" (Keane, 1998, p. 398) (see Chapter 16). For these people, whether the details of their memories are accurate is irrelevant because their suffering is certainly real.

In animal models, stress and emotional arousal do enhance memory formation, and a suite of transmitters are involved, including the opioids, GABA (gamma-aminobutyric acid), and especially the adrenergic transmitters epinephrine and norepinephrine. Epinephrine (adrenaline), released by the adrenal glands during times of stress and strong emotion, appears to affect memory formation by influencing the basolateral amygdala (BLA). Injecting propranolol, a blocker of beta-adrenergic receptors, into the BLA blocks the memory-enhancing effects of stress (Cahill et al., 1994). A functional model of the BLA and the endogenous and exogenous compounds that affect this memory system is presented in the figure. In this model (McGaugh, 2003), activity in four main hormone/transmitter systems—adrenergic, opioid, GABAergic, and cholinergic inputs—converges on and is integrated in the BLA. Outputs of the BLA, in turn, project widely to brain regions, including the hippocampus, striatum, and cortex, influencing memory formation in these locations (McIntyre et al., 2005). Drugs that alter the biochemical messages to the BLA may potently alter the effect of emotion on memory. For example, if people who have just had a traumatic experience are given the beta-adrenergic antagonist propranolol, which blocks the effects of epinephrine, their memory for the event is significantly reduced when tested months later (Pitman et al., 2002).

In PTSD, each recurrence of the strong emotions and memories of the traumatic event may reactivate memories that, when reconsolidated in the presence of stress signals like epinephrine, become even stronger. Therefore, one strategy to treat PTSD could be to block the effects of epinephrine while the person is retrieving the traumatic memory, to blunt the emotional experience of the event and thereby weaken that aspect of the memory during reconsolidation. In a study of people with PTSD, having them read aloud a written description of their ordeal while under the influence of propranolol reduced their symptoms (A. Brunet et al., 2018) such that most no longer met the diagnostic criteria for PTSD. They still remembered the details of the event, but they no longer found the memory so upsetting. If we cannot erase the traumatic memory altogether, perhaps we can reduce its sting.

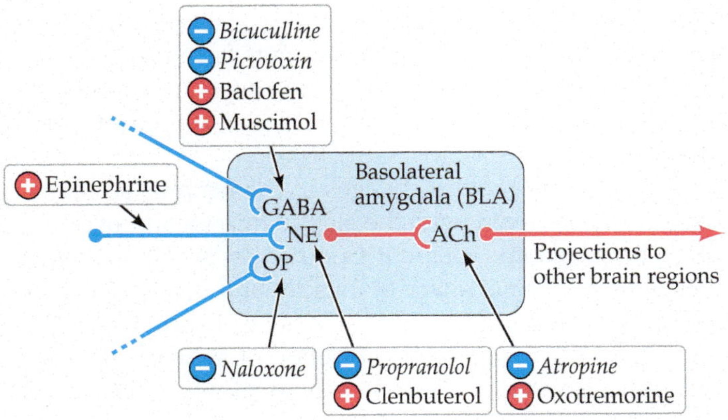

The amygdala and memory The diagram shows sites of activity of drugs that affect different neuromodulator and neurotransmitter systems in the amygdala. Antagonist/blocker, blue; agonist/activator, red; ACh, acetylcholine; GABA, gamma-aminobutyric acid; NE, norepinephrine; OP, opioids. (After J. L. McGaugh, 2003. *Memory and emotions: The making of lasting memories*. Columbia University Press: New York.)

posttraumatic stress disorder (PTSD) A disorder in which memories of an unpleasant episode repeatedly plague the victim.

neuroplasticity Also called *neural plasticity*. The ability of the nervous system to change in response to experience or the environment.

NEURAL MECHANISMS OF MEMORY STORAGE

What are the basic molecular, synaptic, and cellular events that store information in the nervous system? The remainder of this chapter concerns the cellular and physiological underpinnings of memory, what is called the *engram* or *memory trace*. We will look at some of the ways in which new learning involves changes in the strength of existing synapses, and the biochemical signals that may produce those changes. We'll consider how the formation of memories may require the formation of new synapses, or even the birth of new neurons. The observation that **neuroplasticity** (or *neural plasticity*)—the ability of neurons and neural circuits to be remodelled by events—is found in virtually all animals indicates that it is an ancient and vital adaptation.

16.4 Memory Storage Requires Physical Changes in the Brain

Learning Objectives

After reading this section, you should be able to:

16.4.1 **Describe various ways synapses could change in order to store memories.**

16.4.2 **Review experiments manipulating the environments of rats and how this affected their brains.**

16.4.3 **Describe two types of nonassociative learning, and their synaptic basis in a sea slug.**

16.4.4 **Sketch the pathways by which the cerebellum mediates classical conditioning.**

In introducing the term *synapse*, Charles Sherrington (1897) speculated that synaptic alterations might be the basis of learning, anticipating an area of research that to this day is one of the most intensive efforts in all of neuroscience. Most current theories of the cellular basis of learning focus on plasticity of the structure and physiological functioning of synapses.

Plastic changes at synapses can be physiological or structural

Synaptic changes that may store information can be measured physiologically. The changes can be presynaptic, postsynaptic, or both (**FIGURE 16.22A**). Such changes include greater release of neurotransmitter molecules and/or greater effects because the receptor molecules become more numerous or more sensitive. These changes can alter the size of the postsynaptic potential. Changes in the rate of inactivation of the transmitter (through reuptake or enzymatic degradation) can produce a similar effect.

Synaptic activity can also be modulated by inputs from other neurons (members of the same or other cell assemblies) causing extra depolarisation or hyperpolarisation of the axon terminals and changes in the amount of neurotransmitter released (**FIGURE 16.22B**).

Long-term memories may require changes in the nervous system so substantial that they can be directly observed (with the aid of a microscope, of course). Structural changes resulting from use are apparent in other parts of the body. For example, exercise changes the mass and/or shape of muscles and bones. In a similar way, new synapses may form or synapses may be eliminated as a function of training (**FIGURE 16.22C**). Training may also lead to reorganisation of synaptic connections. For example, training may cause a more used pathway to take over sites formerly occupied by a less active competitor (**FIGURE 16.22D**).

There is growing evidence that glia also play a role in learning and synaptic activity. For example, in some systems, nearby astrocytes mark which synapses will change in strength; blocking astrocyte activity prevents the synapse from changing (Min and Nevian, 2012). Also, people learning a new skill show changes in white matter that indicate changes in myelination from oligodendrocytes (Zatorre et al., 2012).

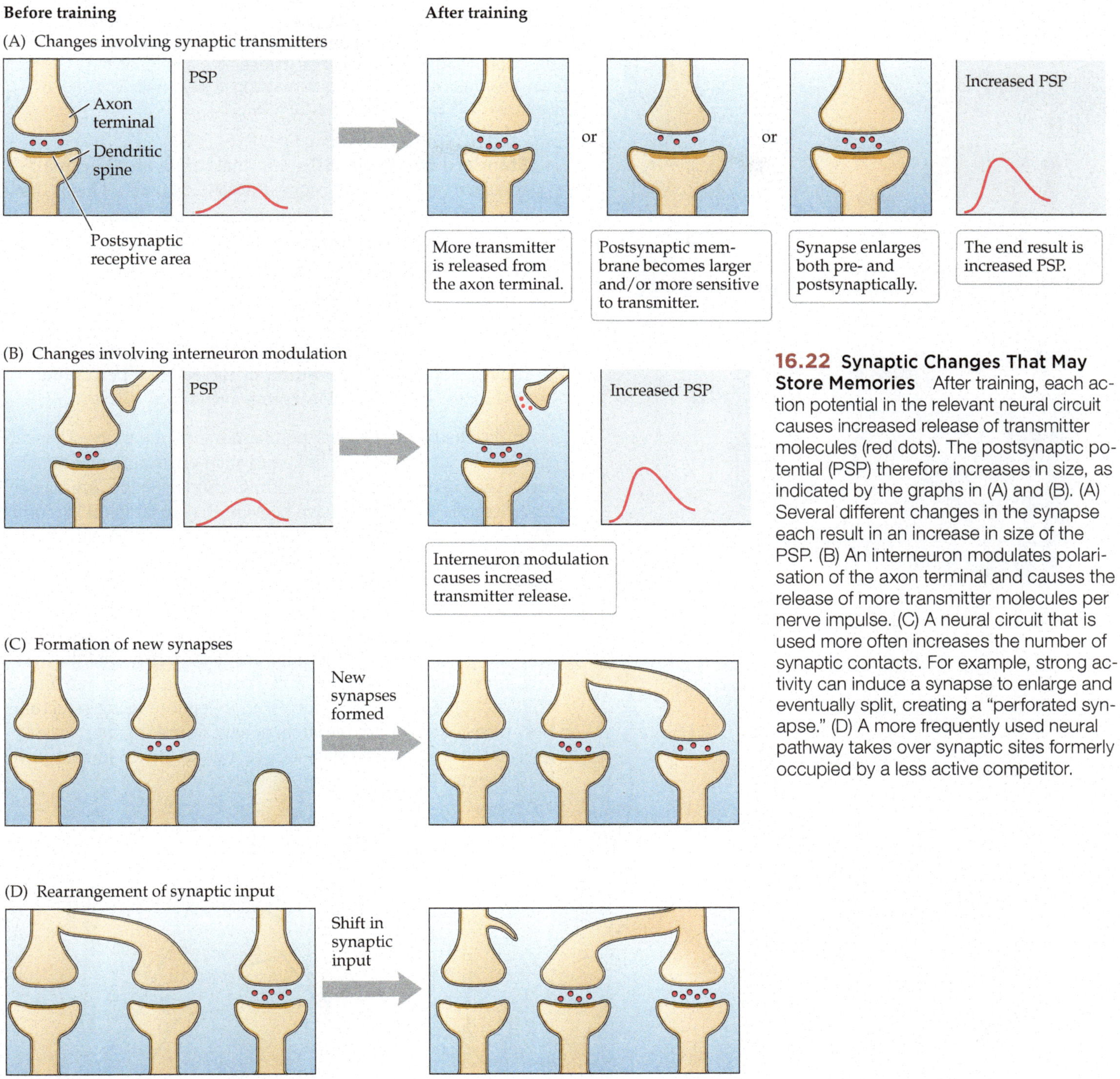

16.22 Synaptic Changes That May Store Memories After training, each action potential in the relevant neural circuit causes increased release of transmitter molecules (red dots). The postsynaptic potential (PSP) therefore increases in size, as indicated by the graphs in (A) and (B). (A) Several different changes in the synapse each result in an increase in size of the PSP. (B) An interneuron modulates polarisation of the axon terminal and causes the release of more transmitter molecules per nerve impulse. (C) A neural circuit that is used more often increases the number of synaptic contacts. For example, strong activity can induce a synapse to enlarge and eventually split, creating a "perforated synapse." (D) A more frequently used neural pathway takes over synaptic sites formerly occupied by a less active competitor.

Varied experiences and learning cause the brain to change and grow

The remarkable plasticity of the brain is not all that difficult to demonstrate. Simply living in a complex environment, with its many opportunities for new learning, produces pronounced biochemical and anatomical changes in the brains of rats (E. L. Bennett et al., 1964, 1969; Renner and Rosenzweig, 1987; M. R. Rosenzweig, 1984; M. R. Rosenzweig et al., 1961). This area of research is covered in more detail in **A STEP FURTHER 16.3: CEREBRAL CHANGES RESULT FROM TRAINING** on the website.

In standard studies of environmental enrichment, rats are randomly assigned to three housing conditions:

16.23 Experimental Environments to Test the Effects of Enrichment on Learning and Brain Measures Interaction with an enriched environment has measurable effects on the brain, on stress reactions, and on learning ability. (After M. R. Rosenzweig et al., 1972. *Sci Am* 226: 22–26.)

1. **Standard condition (SC)** Animals are housed in small groups in standard lab cages (**FIGURE 16.23A**). This is the typical environment for laboratory rodents.
2. **Impoverished condition (IC)** Animals are housed individually in standard lab cages (**FIGURE 16.23B**).
3. **Enriched condition (EC)** Animals are housed in large social groups in special cages containing various toys and other interesting features (**FIGURE 16.23C**). This condition provides enhanced opportunities for learning perceptual and motor skills, social learning, and so on.

In dozens of studies over several decades, a variety of plastic changes in the brain were linked to such environmental enrichment. For example, compared with IC animals (or SC animals in many cases), EC animals have heavier, thicker cortex (Diamond, 1967; Rosenzweig et al., 1962), with enhanced cholinergic activity (Rosenzweig et al., 1961). Cortical neurons in EC animals also have more dendritic branches, especially on dendrites closer to the cell body (called *basal dendrites*), than those of IC animals (Globus et al., 1973; Greenough, 1976) (**FIGURE 16.24**). Similar effects on brain processes and behaviour are seen in fishes, birds, mice, cats, and monkeys (Renner and Rosenzweig, 1987; van Praag et al., 2000). The human brain is no exception: using structural MRI scans of the brain, researchers have found that human adolescents from advantageous socioeconomic environments have thicker cortex than adolescents

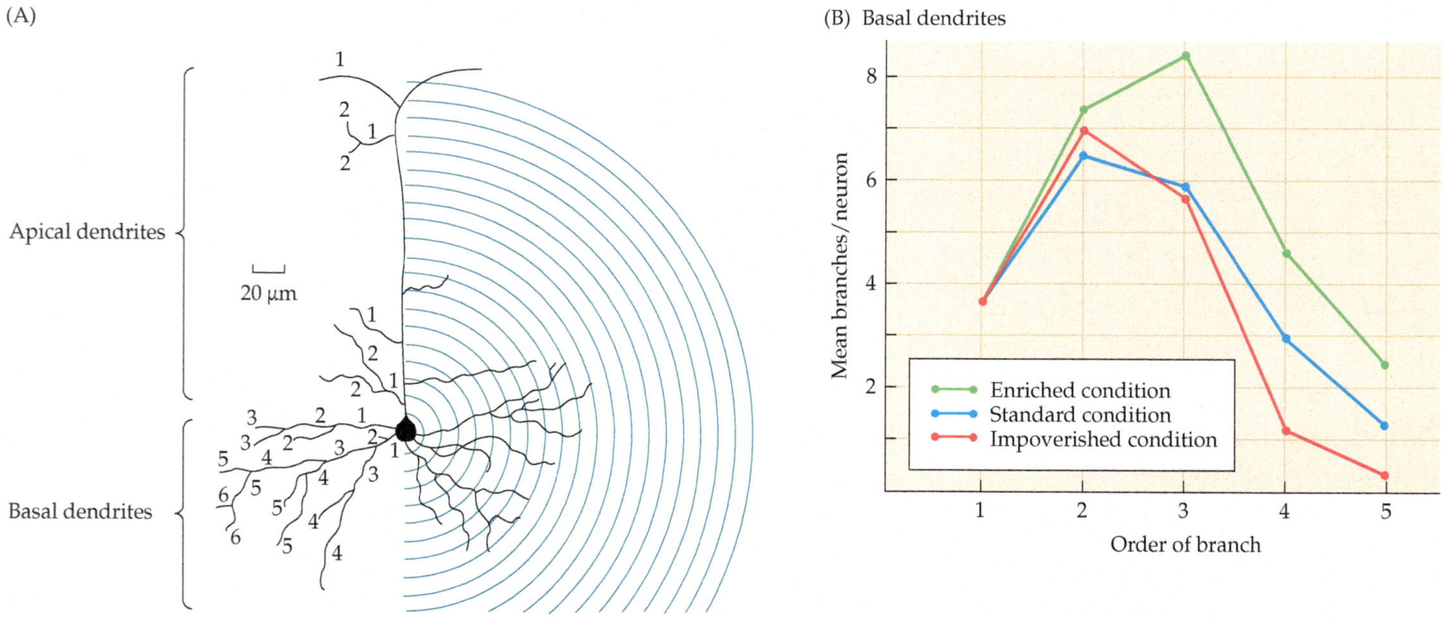

16.24 Measurement of Dendritic Branching (A) An enlarged image of a neuron is used to quantify branching, either by counts of the numbers of branches of different orders (left) or by counts of the numbers of intersections with concentric rings (right). (B) There are significant differences in branching, especially in the basal dendrites, among rats kept for 30 days in enriched, standard, or impoverished environments. (After W. T. Greenough, 1976. In *Neural mechanisms of learning and memory*, M. R. Rosenzweig and E. L. Bennett [Eds.], pp. 255–278. MIT Press: Cambridge, MA.)

from lower socioeconomic environments, independent of any other group differences, such as race or ethnicity differences (A. P. Mackey et al., 2015). Socioeconomic status continues to correlate with cortical thickness across adulthood, even among individuals who were poor growing up (Chan et al., 2018). In fact, the ability to change in response to the environment is a pervasive quality of the brain, across the life-span and across the animal kingdom, as we discuss in **A STEP FURTHER 16.4: PLASTICITY IS A DEFINING FEATURE OF THE BRAIN** on the website.

It is difficult to find the exact synaptic changes that underlie any particular instance of learning, because there are so many synapses (Merchán-Pérez et al., 2009). So instead, researchers made progress by studying simple learning in simple animals, as we discuss next.

Invertebrate nervous systems show plasticity

One fruitful research strategy has been to focus on memory mechanisms in the very simple nervous systems of certain invertebrates. Invertebrate nervous systems have relatively few neurons (on the order of hundreds to tens of thousands). Because these neurons are arranged identically in different individuals, it is possible to construct detailed neural circuit diagrams for particular behaviours and study the same few neurons in multiple individuals.

A Nobel Prize–winning program of research focused on *Aplysia* (Kandel et al., 2014), sea slugs with a simple nervous system that is nonetheless capable of the simplest type of learning: **nonassociative learning**. In each of the three forms of nonassociative learning—habituation, dishabituation, and sensitisation—a single stimulus is presented once or repeatedly. **Habituation** is a decrease in response to a stimulus that is repeatedly presented (when the decrement cannot be attributed to sensory adaptation or motor fatigue). Sitting in a café, you may stop noticing the door chime when someone enters; in this case you have habituated to the chime. Once habituation has occurred, a strong stimulus (of the same sort, or even in another sensory modality) will often cause the response to the habituated stimulus to increase sharply; it may become even larger than the original response. The increase in response amplitude is called **dishabituation**. So, if a loud firecracker is set off behind you, the next ring of the door chime may startle you; in this case you have become temporarily dishabituated to the chime. Even a response that you have not been habituated to may increase in amplitude after a strong stimulus. This effect is known as **sensitisation**: the response is greater than the baseline level because of prior stimulation. After the firecracker, for example, you may overreact to someone standing up nearby. That horrible firecracker has sensitized you to many stimuli.

If you squirt water at a sea slug's siphon—a tube through which it draws water—the animal protectively retracts its delicate gill (**FIGURE 16.25**). But with repeated stimulation the animal retracts the gill less and less, as it learns that the squirt represents no danger to the gill. Kandel and associates (1987, 2009) demonstrated that this habituation is caused by changes in the synapse between the sensory cell that detects the squirt of water and the motor neuron that retracts the gill. As this synapse releases less and less transmitter, the gill slowly stops retracting in response to the stimulation (**FIGURE 16.26A**) (M. Klein et al., 1980). So in this case the synaptic plasticity underlying learning is within the reflex circuit itself.

Similarly, both the number and the size of synaptic junctions vary with training in *Aplysia*. For example, if an *Aplysia* is tested in the habituation paradigm over a series of days, each successive day the animal habituates faster than it did the day before. This phenomenon represents long-term habituation (as opposed to the short-term habituation that we've already discussed), and in this case there is a reduction in the number of synapses between the sensory cell and the motor neuron (**FIGURE 16.26B**) (C. H. Bailey and Chen, 1983).

Elements of the *Aplysia* gill withdrawal system are similarly capable of sensitisation, in which strong stimulation anywhere on the skin causes subsequent stimulations of the siphon to produce *larger* gill withdrawals

standard condition (SC) The usual environment for laboratory rodents, with a few animals in a cage and adequate food and water but no complex stimulation.

impoverished condition (IC) Also called *isolated condition*. A condition in which laboratory rodents are housed singly in a small cage without complex stimuli.

enriched condition (EC) Also called *complex environment*. A condition in which laboratory rodents are group-housed with a wide variety of stimulus objects.

nonassociative learning A type of learning in which presentation of a particular stimulus alters the strength or probability of a response according to the strength and temporal spacing of that stimulus. It includes habituation and sensitisation.

habituation A form of nonassociative learning in which an organism becomes less responsive following repeated presentations of a stimulus.

dishabituation The restoration of response amplitude following habituation.

sensitisation A form of nonassociative learning in which an organism becomes more responsive to most stimuli after being exposed to unusually strong or painful stimulation.

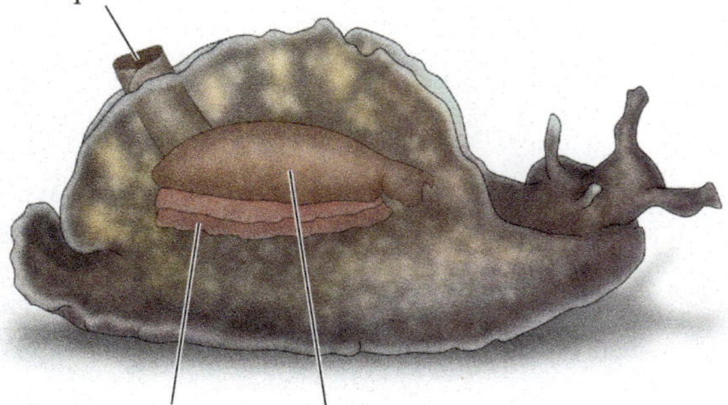

16.25 The Sea Slug *Aplysia* In the usual posture, the siphon is extended and the gill is spread out on the back. Ordinarily only the tip of the siphon would be visible in a lateral view; here, the rest of the siphon and the gill are shown as if the animal were transparent. (After E. R. Kandel, 1976. *Cellular basis of behaviour.* Freeman: San Francisco.)

16.26 Synaptic Plasticity Underlying Habituation in *Aplysia*

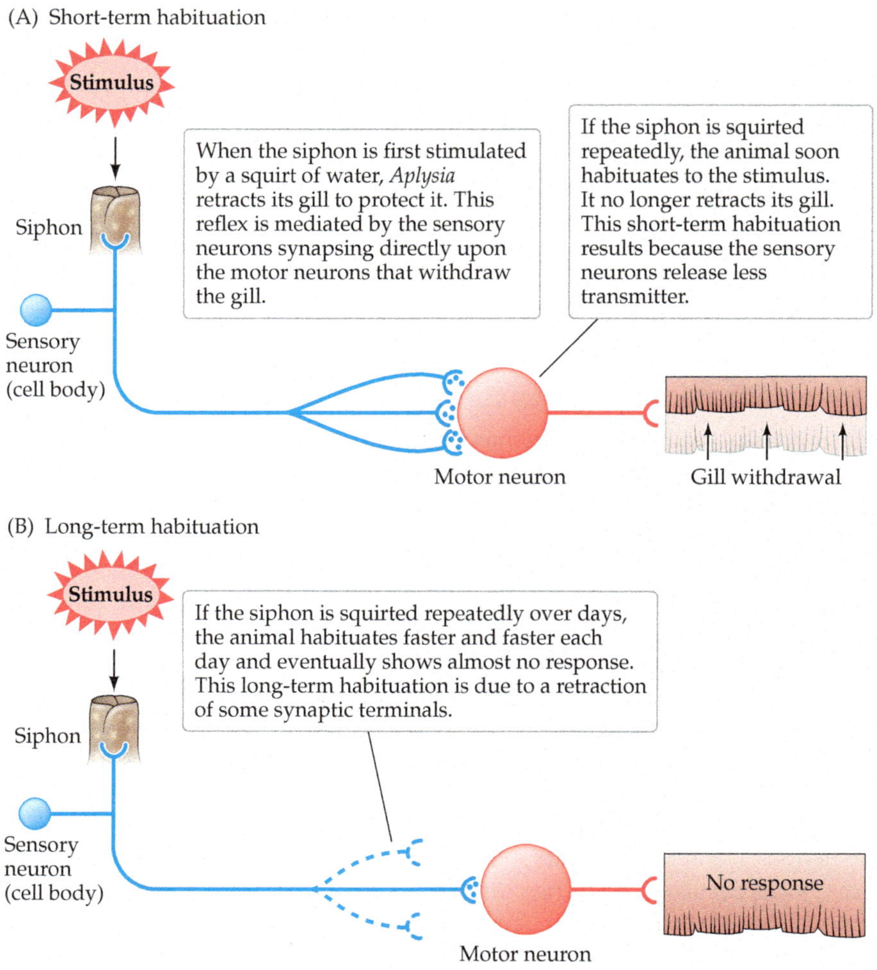

(N. Dale et al., 1988). The strong stimulation of the skin activates a facilitating neuron that releases the transmitter serotonin onto the presynaptic nerve terminals in the gill withdrawal reflex circuit shown in Figure 16.26A. Serotonin boosts the activity in the circuit by prolonging the activity of the sensory neuron's synapses onto the motor neuron, leading to a longer-lasting response (Y. Zhang et al., 2012).

The comparability of results obtained with diverse species of invertebrates indicates that, over a wide range of species, information can be stored in the nervous system by changes in both strength and number of synaptic contacts, confirming the hypotheses diagrammed in Figure 16.22. (For another example of memory research in invertebrates, see **A STEP FURTHER 16.5: IN *DROSOPHILA*, EACH STAGE IN MEMORY FORMATION DEPENDS ON A DIFFERENT GENE** on the website.) Next we'll consider a simple neural circuit in the mammalian brain in which neural activity alters the strength of synaptic connections.

Some simple learning in mammals relies on circuits in the cerebellum

Describing the complete circuit for even a simple learned behaviour is very difficult in mammals because, in contrast to *Aplysia*, mammals have brains containing billions of neurons that are not organised in fixed circuits. So researchers focused on a very simple mammalian behaviour: classical conditioning.

We noted earlier that the cerebellum is crucial for some types of classical conditioning. When a puff of air is applied to the cornea of a rabbit, the animal reflexively blinks. If the air puff (US) immediately follows a tone (CS), a conditioned response develops rapidly: the rabbit comes to blink (CR) whenever the tone is sounded. The basic circuit of the eye-blink reflex is simple, involving cranial nerves and some interneurons that connect their nuclei (**FIGURE 16.27A**). Sensory fibres from the cornea run along cranial nerve V (the trigeminal nerve) to its nucleus in the brainstem. From there, some interneurons send axons to synapse on other cranial nerve motor nuclei (VI and VII), which in turn activate the muscles of the eyelids, causing them to close. While destruction of the hippocampus has little effect on the acquisition or retention of the conditioned eye-blink response in rabbits (Lockhart and Moore, 1975), researchers found learning-related increases in the activity of individual neurons of the cerebellum and associated structures. Research eventually described a cerebellar circuit that is both necessary and sufficient for conditioning to aversive stimuli, including eye-blink conditioning (Poulos and Thompson, 2015).

The trigeminal (V) pathway that carries information about the corneal stimulation (the US) to the cranial motor nuclei also sends axons to the brainstem (specifically a structure called the *inferior olive*). These brainstem neurons, in turn, send axons called *climbing fibres* to synapse on cerebellar neurons in a region called the *interpositus nucleus*. The same cells also receive information about the auditory CS by a pathway through the auditory nuclei and other brainstem nuclei (**FIGURE 16.27B**). So the US and CS converge in the interpositus nucleus of the cerebellum. After conditioning, the occurrence of the CS—the tone—has an enhanced effect on the cerebellar neurons, so they now trigger eye blink even in the absence of an air puff (**FIGURE 16.27C**).

Local cooling or drugs that block the neurotransmitter GABA (gamma-aminobutyric acid, the transmitter used at synapses in the cerebellar circuit) have the effect of reversibly shutting down the interpositus nucleus. If this manipulation is performed at the beginning of training, then no conditioning occurs until after the effect wears off. Conversely, if animals are fully trained before treatment, subsequent injection of a GABA antagonist causes the conditioned behaviour to disappear, until after the drug effect wears off. On the basis of these and other experiments, the complete eye-blink conditioning circuit is now understood, and the cerebellum's interpositus nucleus appears to be the key location

16.27 Functioning of the Neural Circuit for Conditioning of the Eye-Blink Reflex US, unconditioned stimulus; CS, conditioned stimulus; CR, conditioned response. (After R. F. Thompson and D. J. Krupa, 1994. *Ann Rev Neurosci* 17: 519–549.)

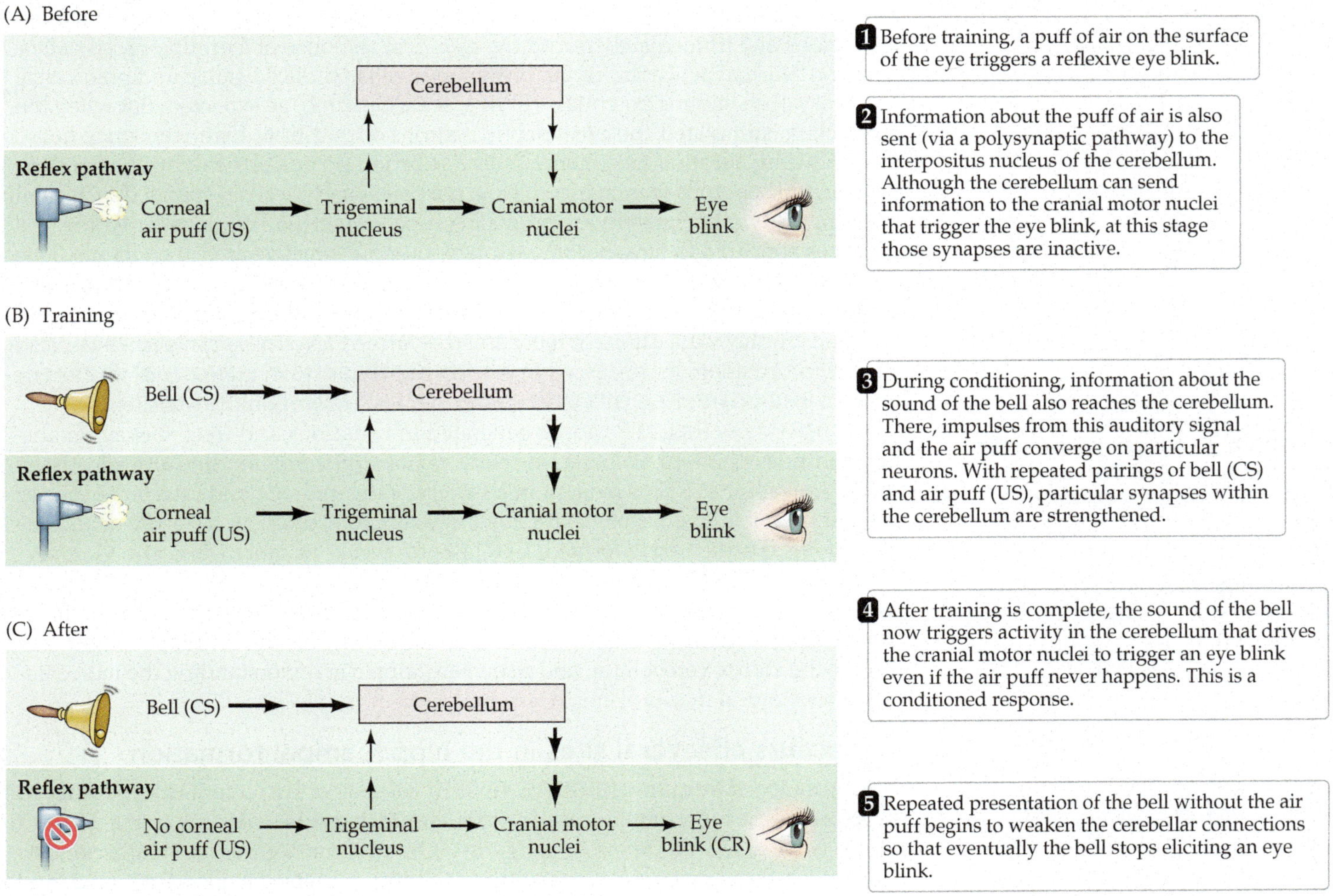

for storing this type of memory (R. F. Thompson and Steinmetz, 2009). Studies of humans with cerebellar damage indicate that the cerebellum is important for conditioning across several domains, including conditioning of emotions like fear and aspects of cognitive learning (Timmann et al., 2010).

16.5 Synaptic Plasticity Can Be Measured in Simple Hippocampal Circuits

Learning Objectives

After reading this section, you should be able to:

16.5.1 **Describe the characteristics of a Hebbian synapse.**

16.5.2 **Describe how experimenters induce long-term potentiation (LTP) of synapses.**

16.5.3 **Explain how NMDA-type glutamate receptors mediate Hebbian-like synaptic changes.**

16.5.4 **Critically discuss whether LTP is actually at work in learning or memory.**

Modern ideas about the role of neuroplasticity in memory have their origins in the theories of Donald Hebb (1949), who proposed that when a presynaptic neuron repeatedly activates a postsynaptic neuron, the synaptic connection between them will become stronger and more stable (the maxim "cells that fire together wire together" captures the basic idea). Ensembles of neurons, or **cell assemblies**, linked via synchronised activity of these **Hebbian synapses** (see also Figure 7.22) could then act together to store memories (Kelso and Brown, 1986).

In classic experiments in the 1970s, researchers probing the properties of hippocampal circuitry discovered an impressive form of neuroplasticity that appeared to confirm Hebb's theories about synaptic remodelling (Bliss and Lømo, 1973; Schwartzkroin and Wester, 1975). In these experiments, electrodes were placed within the rat hippocampus, positioned so that the researchers could stimulate a group of *presynaptic* axons and immediately record the electrical response of a group of *postsynaptic* neurons. Normal activation of the presynaptic cells produced stable and predictable excitatory postsynaptic potentials (EPSPs; see Chapter 2), as expected. But when the researchers stimulated the presynaptic neurons with a brief high-frequency burst, thus inducing **tetanus** (an intense volley of action potentials), the response of the postsynaptic neurons was changed. Now the postsynaptic cells responded to normal levels of presynaptic activity by producing much larger EPSPs; in other words, the synapses appeared to have become stronger or more effective after the tetanus. This stable and long-lasting enhancement of synaptic transmission, termed **long-term potentiation** (**LTP**), is illustrated in **FIGURE 16.28A**. Interestingly, a *weakening* of synaptic efficacy via a different mechanism—termed *long-term depression*—can also encode information, as discussed in **A STEP FURTHER 16.6: LONG-TERM DEPRESSION IS THE CONVERSE OF LONG-TERM POTENTIATION** on the website.

We now know that LTP can be generated in conscious and freely behaving animals, in anaesthetised animals, in many regions of the brain, and even in tissue slices. The fact that LTP is evident in a variety of invertebrate and vertebrate species suggests that it is an ancient evolutionary development, probably a feature of all neural systems. Further, once formed, LTP lasts for weeks or more (Bliss and Gardner-Medwin, 1973). So, at least superficially, LTP appears to have all the hallmarks of a cellular mechanism of memory. This hint at a possible cellular basis of memory prompted an intensive research effort, centred mainly on the rat hippocampus but also on the cortex, cerebellum, and elsewhere, aimed at understanding the molecular and physiological mechanisms that induce LTP.

LTP occurs at several sites in the hippocampal formation

The *hippocampal formation* (**FIGURE 16.28B**) consists of two interlocking C-shaped structures—the hippocampus itself and the **dentate gyrus**—along with the adjacent cortex (also called the *hippocampal gyrus*). On structural grounds, neuroscientists distinguish three major divisions within the hippocampus, labelled *CA1*, *CA2*, and

cell assembly A large group of cells that tend to be active at the same time because they have been activated simultaneously or in close succession in the past.

Hebbian synapse A synapse that is strengthened when it successfully drives the postsynaptic cell.

tetanus An intense volley of action potentials.

long-term potentiation (LTP) A stable and enduring increase in the effectiveness of synapses following repeated strong stimulation.

dentate gyrus A strip of gray matter in the hippocampal formation.

16.28 Long-Term Potentiation Occurs in the Hippocampus (A) If axons in the circuit are stimulated only once every second, the size of the response in the postsynaptic neurons is quite stable. However, after a brief tetanus (a burst of electrical stimulation triggering hundreds or thousands of action potentials over 1–2 seconds), the size of the excitatory postsynaptic potential (EPSP) responses increases markedly and remains high throughout the recording period. This greater responsiveness is called *long-term potentiation* (LTP). (B) The top diagram shows the location of the hippocampal formation in a whole rat brain and in a horizontal section. The bottom diagram of the right hippocampal formation shows various neural pathways found in the hippocampal formation, many of which display LTP. (See the text for an explanation of CA1, CA2, and CA3.)

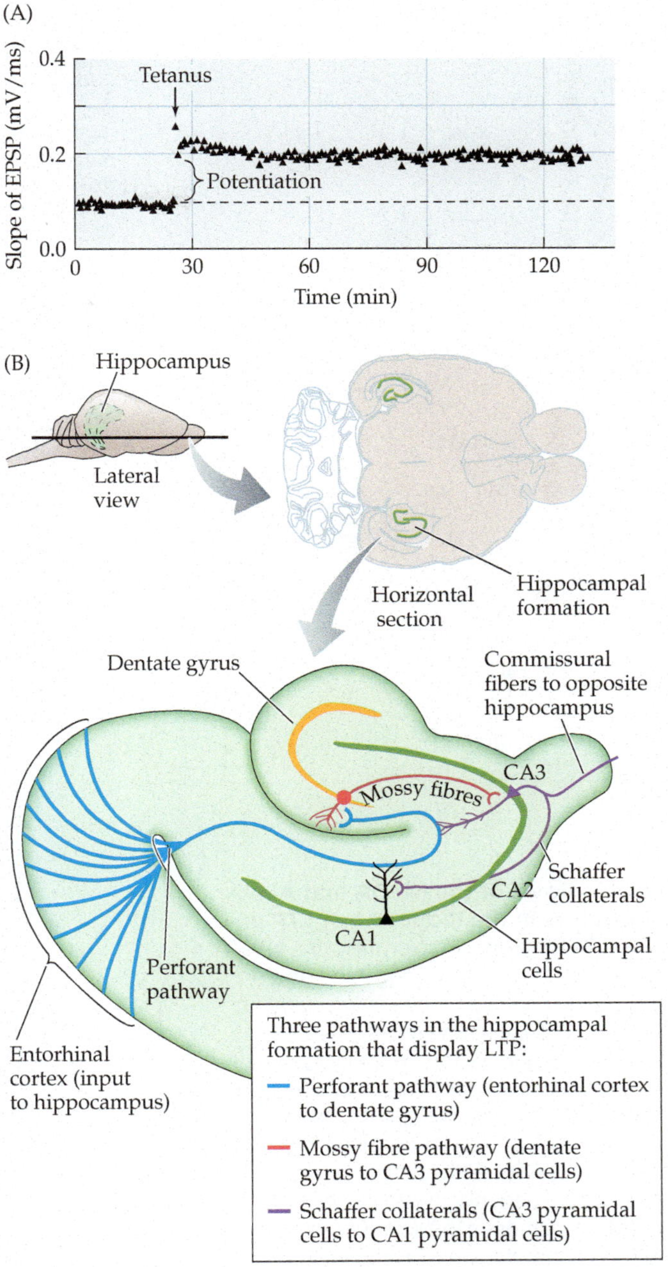

CA3. It was in the main input pathway to the hippocampal formation (the *perforant* pathway, originating in nearby entorhinal cortex and terminating at synapses in dentate gyrus) that LTP was originally demonstrated. But LTP is also studied in other hippocampal pathways, which are illustrated in Figure 16.28B, and it may be a property of all excitatory synapses (Malenka and Bear, 2004).

In CA1, LTP occurs at synapses that use the excitatory neurotransmitter glutamate, and it is critically dependent on a glutamate receptor subtype called the **NMDA receptor** (after its selective ligand, ***N*-*m*ethyl-*D*-*a*spartate**). Treatment with drugs that selectively block NMDA receptors completely prevents new LTP in the CA1 region, but it does not affect LTP that has already been established. As you might expect, these postsynaptic NMDA receptors—working in conjunction with a related subtype of glutamate receptors called **AMPA receptors**—have some unique characteristics, which we discuss next.

NMDA receptors and AMPA receptors collaborate in LTP

During normal, low-level activity, the release of glutamate at a CA1 synapse activates only the AMPA receptors. The NMDA receptors cannot respond to the glutamate, because magnesium ions (Mg^{2+}) block the NMDA receptors' integral calcium ion (Ca^{2+}) channels (**FIGURE 16.29A**); thus, few Ca^{2+} ions can enter the neuron. The situation changes, however, if larger quantities of glutamate are released (in response to a barrage of action potentials), thus stimulating the AMPA receptors more strongly. Because AMPA receptors admit cations when activated, the increased activation of AMPA receptors depolarises the postsynaptic membrane, and if a threshold value of about –35 millivolts or so is reached, the Mg^{2+} plugs are driven from the central channels of the NMDA receptors (**FIGURE 16.29B**). The NMDA receptors are now also able to respond to glutamate, admitting large amounts of Ca^{2+} into the postsynaptic neuron. Thus, NMDA receptors are fully active only when gated by a combination of voltage (depolarisation via AMPA receptors) and the ligand (glutamate).

The large influx of Ca^{2+} at NMDA receptors activates intracellular enzymes, called **protein kinases**, that alter or activate a variety of other proteins. One of these protein kinases is CaMKII (calcium/calmodulin-dependent protein kinase II), which affects AMPA receptors in several important ways (Kessels and Malinow, 2009; Lisman et al., 2012). Activated CaMKII causes more AMPA receptors to be produced and inserted into the postsynaptic membrane, and existing nearby AMPA receptors are induced to move to the active synapse (T. Takahashi et al., 2003). The membrane-bound AMPA receptors are also modified to increase their conductance of Na^{+} and

Go to Activity 16.3
Hippocampal Anatomy and LTP

NMDA receptor A glutamate receptor that also binds the glutamate agonist NMDA (N-methyl-D-aspartate) and that is both ligand-gated and voltage-sensitive.

AMPA receptor A glutamate receptor that also binds the glutamate agonist AMPA.

protein kinase An enzyme that adds phosphate groups (PO_4) to protein molecules, altering the protein's function.

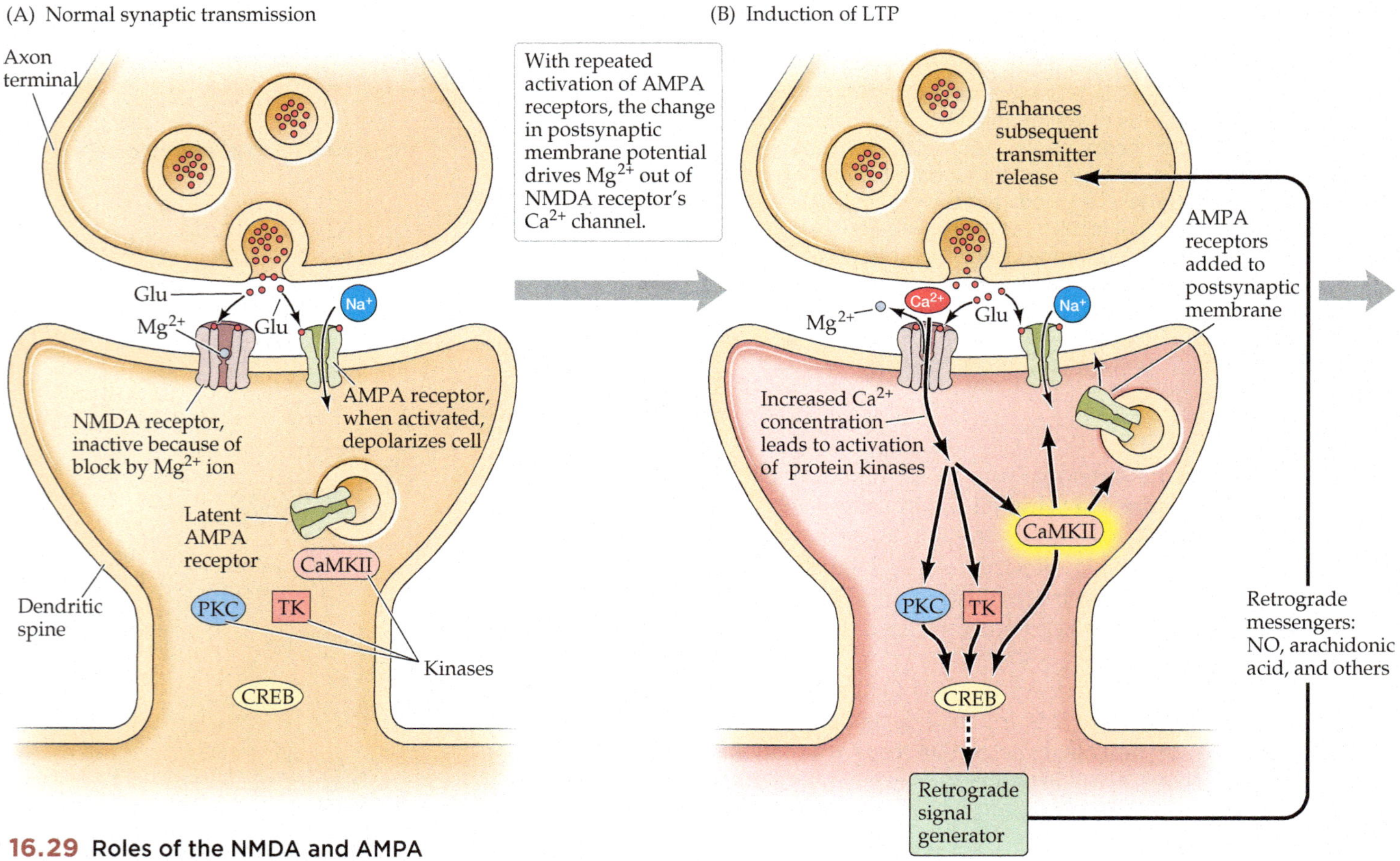

16.29 Roles of the NMDA and AMPA Receptors in the Induction of LTP in the CA1 Region (A) Normally, the NMDA channel is blocked by a Mg^{2+} ion and only the AMPA channel functions in excitation of the neuron. (B) With repeated activation of AMPA receptors, depolarisation of the neuron drives Mg^{2+} out of the NMDA channel, and Ca^{2+} ions enter. The rapid increase of Ca^{2+} ions triggers processes that lead to LTP. Activation of calcium/calmodulin-dependent protein kinase II (CaMKII) increases the conductance of AMPA receptors already present in the membrane and promotes the movement of AMPA receptors from the interior of the cell into the membrane. (C) The synapse is enhanced after induction of LTP. CREB, cAMP responsive element–binding protein; Glu, glutamate; PKC, protein kinase C; TK, tyrosine kinase.

K^+ ions (Sanderson et al., 2008). The net effect of these changes, therefore, is to enhance the sensitivity of the synapse to released glutamate (**FIGURE 16.29B AND C**).

Not all of the changes in LTP are postsynaptic. When the postsynaptic cell is strongly stimulated and its NMDA receptors become active and admit Ca^{2+}, an intracellular process causes the postsynaptic cell to release a **retrograde messenger**—often a diffusible gas—that travels back across the synapse and alters the functioning of the *presynaptic* neuron (see Figure 16.29B). By affecting the presynaptic cell, the retrograde messenger ensures that more glutamate will be released into the synapse than previously, thereby strengthening the synapse. Nitric oxide (NO), carbon monoxide (CO), arachidonic acid, and nerve growth factor are among more than a dozen possible retrograde signals in LTP (J. R. Sanes and Lichtman, 1999). So, LTP involves active participation on both sides of the synapse: the postsynaptic side has more AMPA receptors to respond to glutamate, and the presynaptic side releases more glutamate.

The retrograde messenger is triggered by activation of a substance called **CREB** (***c***AMP ***r***esponsive ***e***lement–***b***inding protein). CREB is a *transcription factor* (a protein that binds to the promoter region of genes and causes those genes to change their rate of expression) that is activated by protein kinases, including CaMKII and chemical cousins like MAPK (mitogen-activated protein kinase), PKC (protein kinase C), and TK (tyrosine kinase). So, as shown in **FIGURE 16.30**, a direct result of the activation of NMDA receptors is the activation of CREB and changes in the expression of genes encoding a wide range of proteins. Because the affected genes may encode anything from new receptors and kinases to the structural building blocks used for changing the shape of the cell, this action can have profound and long-lasting consequences for the neuron.

The long-term changes in neurons after LTP range from the formation of additional synapses, and enhancement of existing ones, to the construction of whole

CREB Cyclic AMP responsive element–binding protein, which binds the promoter region of several genes involved in neural plasticity when activated by cAMP.

retrograde messenger Transmitter that is released by the postsynaptic region, travels back across the synapse, and alters the functioning of the presynaptic neuron.

(C) Enhanced synapse, after induction of LTP

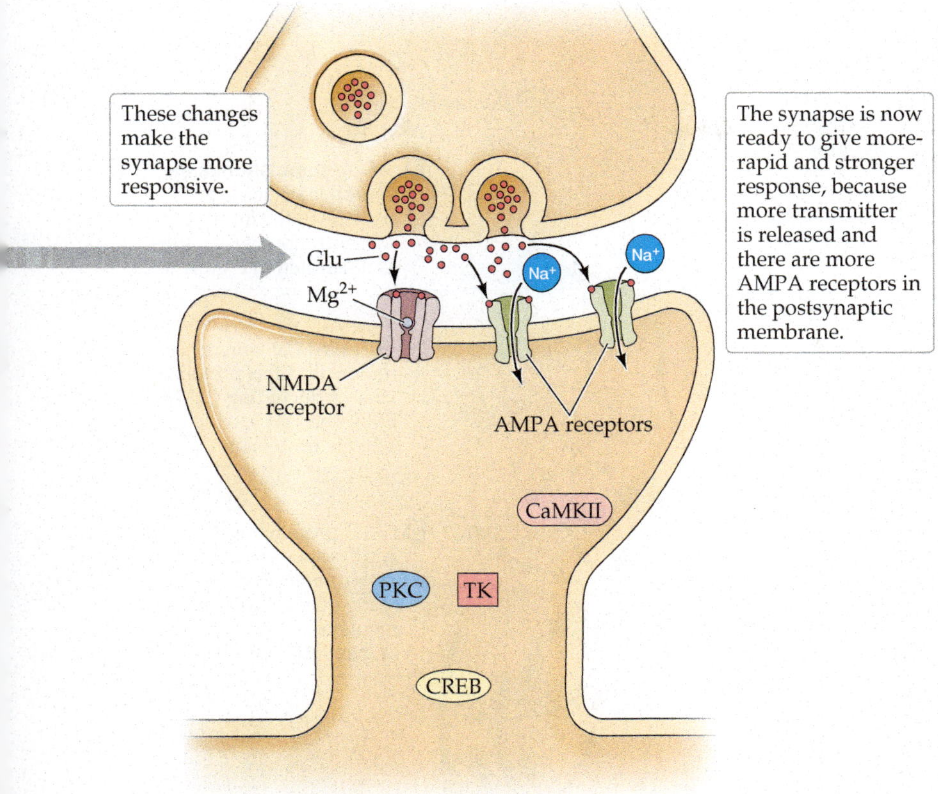

Go to Animation 16.1
AMPA and NMDA

new dendritic branches and dendritic spines, through modifications of molecular components of the cell skeleton (W. Huang et al., 2013). In mice, genetic deletion of CREB impairs LTM but not STM (Bourtchuladze et al., 1994; Kogan et al., 1997). The earlier stages of LTP, lasting an hour or so, appear not to require protein synthesis, but thereafter, inhibition of protein synthesis prevents longer-lasting LTP (U. Frey et al., 1993; Krug et al., 1984). Furthermore, neurons can make proteins that selectively block CREB's actions (Genoux et al., 2002; Mioduszewska et al., 2003), possibly providing a means to erase or inhibit the formation of unwanted memories. Much remains to be discovered about the many controls on long-lasting components of LTP.

In other locations in the hippocampus, such as the mossy fibre pathway (see Figure 16.28B), LTP can occur without NMDA receptor activity (E. W. Harris and Cotman, 1986), and some forms of LTP are blocked by drugs with completely different modes of action, such as opiate antagonists (Aroniadou et al., 1993; Derrick and Martinez, 1994). The diversity of mechanisms involved in LTP has complicated one of the central questions in LTP research, which we address next.

Is LTP a mechanism of memory formation?

Even the simplest learning involves circuits of multiple neurons and many synapses, and more-complex declarative and procedural memory traces must involve vast networks of neurons, so researchers don't view LTP as the only mechanism of learning. Instead, it is likely that LTP is an important part of a multifaceted system for storing information. Evidence from several research perspectives—as we defined way back in Figure 1.2—implicates LTP in memory:

- *Correlational observations* The time course of LTP bears strong similarity to the time course of memory formation (Lynch et al., 1991; Staubli, 1995). Like memory,

16.30 Steps in the Neurochemical Cascade during the Induction of LTP This illustration is based on LTP induction in the CA1 region of the hippocampus. CaMKII, calcium/calmodulin-dependent protein kinase II; CREB, cAMP responsive element–binding protein; PKA, protein kinase A; PKC, protein kinase C.

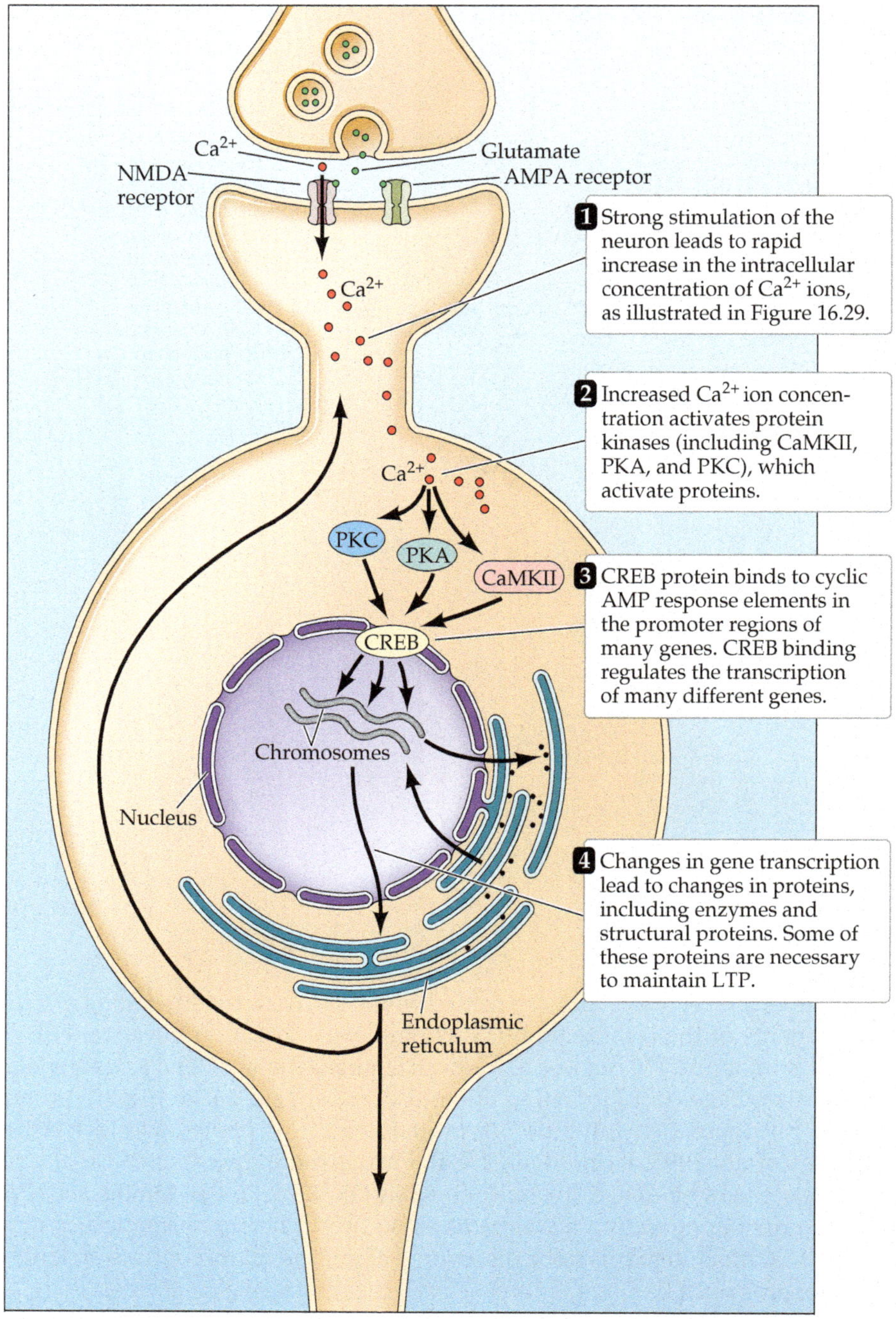

LTP can be induced within seconds, may last for days or weeks, and is unstable during a consolidation period that lasts for several minutes after induction.

- *Somatic intervention experiments* In general, pharmacological treatments that interfere with LTP tend to impair learning. For example, NMDA receptor blockade interferes with performance in the Morris water maze (a test of spatial memory) and other types of memory tests (R. G. Morris et al., 1989). Drugs that inhibit CaMKII and other protein kinases also generally interfere with aspects of memory formation (M. R. Rosenzweig et al., 1992, 1993; Serrano et al., 1994). Because these same basic physiological processes are at work in many regions of the brain, in earlier research it was difficult to know exactly where and how the drugs were acting to affect memory. But regional genetic manipulations have enabled researchers to zero in on specific brain regions. Mice with one copy of the gene for CaMKII knocked out can still form STMs, but they cannot form LTMs (Frankland et al., 2001). And knockout mice that lack functional NMDA

Go to Video 16.2
Morris Water Maze

receptors only in CA1 appear normal in many respects, but their hippocampi are incapable of LTP and their memory is impaired (Rampon et al., 2000). In a clever reversal, researchers have also shown that mice (called *Doogies,* see **A STEP FURTHER 16.7: HOW TO BUILD A DOOGIE** on the website) engineered to overexpress NMDA receptors in the hippocampus have enhanced LTP, and better-than-normal long-term memory (Tang et al., 2001). In fact, researchers have developed more than 30 different strains of genetically modified "smart mice," most of which exhibit enhanced LTP (Lehrer, 2009).

- *Behavioural intervention experiments* In principle, the most convincing link between LTP and learning would be "behavioural LTP": a demonstration that training an animal in a memory task can induce LTP in the brain. Such research is difficult because of uncertainty about exactly where to put the recording electrodes in order to detect any induced LTP. Nevertheless, several examples of successful behavioural LTP have been reported. Fear conditioning—for example, the repeated pairing of an aversive stimulus and a tone, eventually resulting in exaggerated reactions to the tone alone (see Figure 11.14)—produces clear LTP specifically in fear circuits in the amygdala and not elsewhere (McKernan and Shinnick-Gallagher, 1997; Rogan et al., 1997). And in the CA1 region of the hippocampus, a different form of aversive learning in rats produces exactly the same electrophysiological changes, as well as changes in AMPA receptor accumulation, that are seen with conventionally induced LTP (Whitlock et al., 2006). Furthermore, using optogenetic techniques (as described in Chapter 3) to manipulate activity in the auditory inputs to the amygdala, researchers succeeded in inactivating fear conditioning by inducing long-term *depression* and then reactivating the memory by restoring the previous LTP in the same circuit (Nabavi et al., 2014).

Taken together, the research findings support the idea that LTP is a kind of synaptic plasticity that underlies (or is very similar to) certain forms of learning and memory.

The Cutting Edge

Artificial Activation of an Engram

We've seen that experience can alter the strength of synapses, and for simple behaviours like habituation in a simple organism like *Aplysia*, we know exactly which synapse weakens as habituation proceeds. But in more-complicated animals like mammals, the physical basis of a memory formed by learning a particular task—the engram—probably involves changes in many neurons and in many synapses. In other words, the memory is more likely encoded by changes in the activity of a *circuit* of neurons, not just one neuron (much less one synapse). If so, then reactivating the neurons encoding the memory might be experienced as remembering. Several groups have tested this idea in mice, where genetic tools permit remarkable control over neuronal activity.

In one such demonstration, researchers were able to mark precisely the neurons that were activated when mice were subjected to fear conditioning (X. Liu et al., 2012). Mice were exposed to a particular context, A, a box with a white plastic floor in a dimly lit room with black walls and a faint smell of almonds. The mice explored the box and showed no signs of being afraid (**FIGURE 16.31A AND B**). Next the mice were taught to be afraid of a very different context, B (a box with a wire grid floor in a brightly lit room with white walls and the smell of vinegar), by being exposed to an auditory tone followed by a mildly painful electrical shock to the feet. As expected, the mice quickly learned to freeze in response to the auditory tone (see Figure 11.14). Most important, these mice had received a genetic modification, controlled by a special diet, so that whenever neurons in the dentate gyrus (DG) of the hippocampus were active, they would start producing channelrhodopsin, a protein that would excite those cells, and only those cells, when exposed to blue light (see Figure 2.23)—an *optogenetic manipulation.* If the activity of this subset of DG neurons was responsible for the mice finding

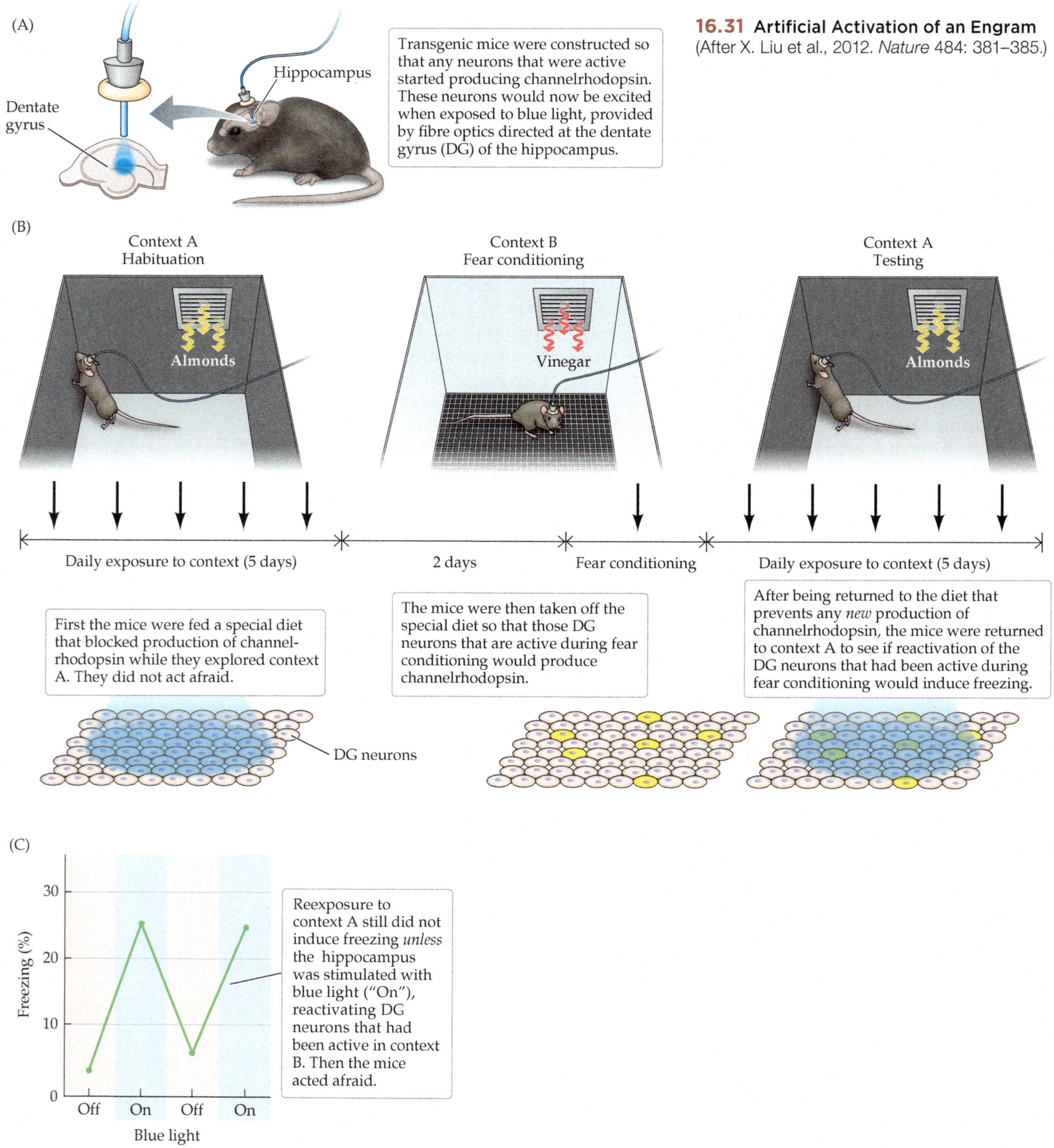

16.31 Artificial Activation of an Engram (After X. Liu et al., 2012. *Nature* 484: 381–385.)

context B frightening, then reactivating those neurons should cause the mice to freeze in fear, even when they were in a completely different context.

To test this idea, the mice were put in context A, where they had never been shocked. As before, they showed no signs of being afraid. But, when fibre optics illuminated their DG with blue light, activating only the neurons that had been active in context B, the mice froze, as if afraid. It was as though the mice were experiencing the other context, B, when the DG neurons that had been active in B were reactivated by the blue light. Turning the light off again caused the animals to resume activity, indicating that they remained unafraid of context A (**FIGURE 16.31C**). It wasn't just that light-induced activation of any random set of DG neurons induced fear, because when blue light reactivated DG neurons that had been active in a third (nonfearful) context, C, the animals did not freeze (data not shown). Rather, it appeared that the activity of that subset of DG neurons caused the mice to experience context B, which they had learned to fear—experimenters had activated an artificial memory. Evidence that electrical stimulation of discrete targets in the medial temporal lobe enhances memory formation in humans (Suthana et al., 2012) suggests that artificial memory is probably possible in our brains too. ■

16.6 In the Adult Brain, Newly Born Neurons May Aid Learning

Learning Objectives

After reading this section, you should be able to:

16.6.1 **Cite evidence that new neurons are produced in adulthood and play a role in behaviour.**

16.6.2 **Describe the impairments of memory seen in old age and their possible causes.**

16.6.3 **Explain what life changes one can make to reduce memory loss with age.**

There is now no doubt that new neurons are produced in the brains of adult mammals, including humans, as we discussed in Chapter 6 (Toda and Gage, 2018). This *adult neurogenesis* occurs primarily in the dentate gyrus of the hippocampal formation (**FIGURE 16.32**), although it is also observed in the olfactory bulbs and some cortical locations. New dentate neurons that survive and grow ultimately receive inputs from entorhinal cortex via the perforant pathway (see blue pathway in Figure 16.28B), extending their own axons and forming glutamatergic synapses. Anatomically, therefore, the new neurons appear to integrate into the functional circuitry of the hippocampus and adjacent cortex, which we've seen play a role in forming new memories.

In experimental animals, neurogenesis and the survival of young neurons can be enhanced by a variety of factors, such as exercise, experience in an enriched environment, or training in a memory task (D. D. Alvarez et al., 2016; Opendak and Gould, 2015). Reproductive hormones and experiences also potently influence neurogenesis (Hamson et al., 2013). What's more, newly generated neurons in the dentate gyrus are more plastic, showing enhanced LTP, compared with older neurons (Marín-Burgin et al., 2012). Although these observations are tantalising, clear demonstrations that the new cells have significant functions in behaviour—especially learning and memory—have been somewhat elusive.

Rats given a drug that is lethal to newly born neurons are reportedly impaired on tests of conditioning (Shors et al., 2001). Neurogenesis has also been implicated in hippocampus-dependent learning, such as spatial memory (Lieberwirth et al., 2016) and fear conditioning (Barha and Galea, 2010; Saxe et al., 2006). In a study using

conditional knockout A gene that can be selectively deactivated in specific tissues and/or at a specific stage of development.

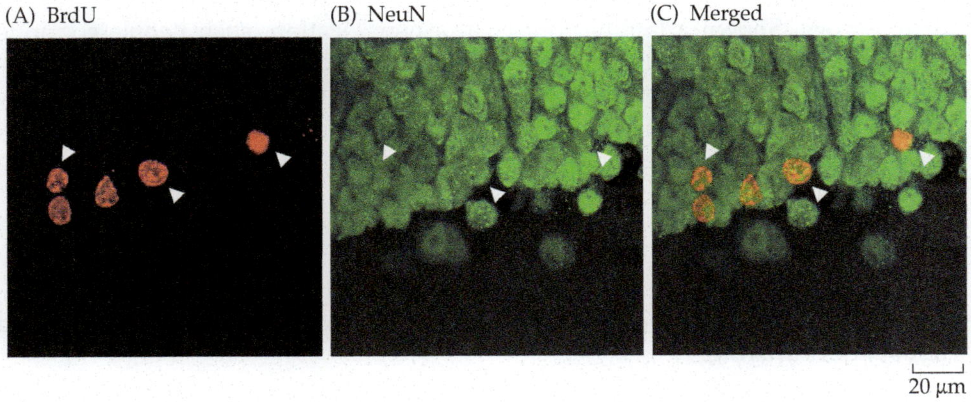

From E. Bruel-Jungerman et al., 2006. *J Neurosci* 26: 5888–5893

16.32 Neurogenesis in the Dentate Gyrus (A) BrdU (bromodeoxyuridine) is a label that is selectively incorporated into the DNA of cells that are about to divide. (B) NeuN (a neuron-specific nuclear protein) is used to selectively label neurons in order to distinguish them from other nearby cells, such as glial cells. (C) Merging the two images makes it clear that some of the newly born cells are in fact neurons (arrows).

mice with a **conditional knockout**—a gene that can be selectively deactivated in adulthood in specific tissues—researchers found that turning off neurogenesis in the brains of adults resulted in a marked impairment in spatial learning with little effect on other behaviours (C. L. Zhang et al., 2008).

The impaired learning that accompanies reduced adult neurogenesis, however it is caused, is consistent with a role for new neurons in learning. But it is always possible that the manipulations have a general effect—making the animals feel ill, for example—that might affect their performance. However, one lab used genetic manipulations to increase the survival of newly generated neurons in the dentate, resulting in *improved* performance (Sahay et al., 2011). These animals showed enhanced hippocampal LTP, which was expected since younger neurons display greater synaptic plasticity (Marín-Burgin et al., 2012). These mice were also better at discriminating between two similar environments (Sahay et al., 2011). In parallel, activation of newly generated neurons in the olfactory bulbs enhances *olfactory* learning and memory (Alonso et al., 2012). Thus it seems clear that neurons born in adulthood play a role in learning and memory, and they do so in keeping with the primary functions of their anatomical location.

Learning and memory change as we age

Understanding the impact of aging on cognition is a pressing issue; by 2030, there will be more than twice as many senior citizens alive as there were in 2000. In people and other mammals, normal aging brings a gradual decline in some but not all aspects of learning and memory (Lister and Barnes, 2009). For some tasks, differences in motivation, earlier education, and other confounding factors may masquerade as age-related memory problems. So experiments on memory in the elderly must be carefully constructed to control for alternate explanations.

What kinds of tasks reliably show decrements in performance with aging? Healthy elderly people tend to show some memory impairment in tasks of conscious recollection that require effort and that rely primarily on internal generation of the memory rather than on external cues. Giving elderly participants easily organised task structures, or cues, can often raise their performance to the level of the young. Although working memory and the ability to form new episodic and declarative memories typically decline with age, existing memories such as autobiographical memory and semantic knowledge tend to remain stable (Hedden and Gabrieli, 2004). If vocabulary is tested in isolation, older adults outperform younger adults (D. C. Park et al., 2002), but on tests of executive function, even individuals in their 40s are likely to show age-related decline (Rhodes, 2004).

As we age, we also experience some decreases in spatial memory and navigational skills (Barnes and Penner, 2007). Similarly, aged rats show decrements in the eight-arm radial maze (see Figure 16.18A) compared with younger animals (Penner and Mizumori, 2012). Along with other memory impairments, spatial memory problems may become much more severe in dementias like Alzheimer's disease (discussed in detail in Chapter 6), which is one of a family of dementing processes that result from inappropriate accumulations of protein in neurons (in the case of Alzheimer's, one of the prime culprits is β-amyloid). People with severe dementia become disoriented and easily lose themselves even in familiar surroundings. Shown a familiar object in a new location, such people continue to search for it in the old location, even passing over the object in plain sight, as though they remember the search procedure and not the object being sought. New techniques that image β-amyloid plaques in the brain may help clinicians to identify people at risk for developing Alzheimer's (see Figure 6.30), perhaps allowing for earlier interventions to slow the progression of the disease.

Age-related impairments of memory have several causes

Why do some measures of learning and memory decline with age, while others remain intact? There are several ways in which neural changes may affect learning and memory during aging:

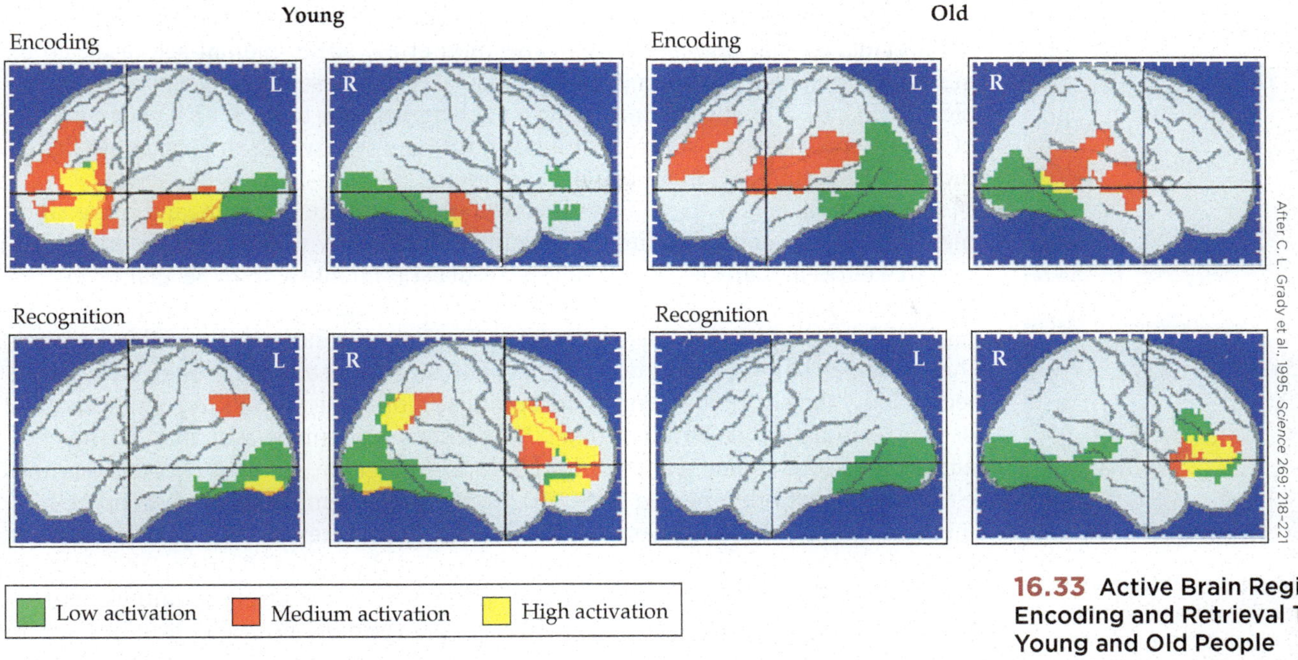

16.33 Active Brain Regions during Encoding and Retrieval Tasks in Young and Old People Lateral views of the cerebral hemispheres show regions of enhanced cerebral blood flow caused by encoding information (top) and retrieving it (bottom).

- *Impairments of encoding and retrieval* Older people show less cortical activation than younger people when encoding or retrieval is self-initiated. In **FIGURE 16.33**, less frontal and temporal activity is seen in the older people while learning new faces (Grady et al., 1995), but when recognising faces (a retrieval task that is not self-initiated, because retrieval is prompted by viewing stimuli), the elders' brain activity is comparable to that of the younger participants.
- *Loss of neurons and/or neural connections* The brain gradually loses weight after the age of 30, and some parts of the brain, such as frontal cortex (Raz, 2000), lose a larger proportion of volume or weight than other parts. Age-related memory impairment may involve steady loss of synapses and neurons in the hippocampus and cortex.
- *Problems with cholinergic neurotransmission* Two subcortical regions—known as the *septal complex* and the *nucleus basalis of Meynert* (NBM)—provide profuse cholinergic inputs to the hippocampus and cortex. Memory impairment in Alzheimer's disease may be caused by the loss of these cholinergic afferents to the cortex. Drug treatments that enhance acetylcholine transmission improve aspects of memory performance in humans (Ricciardi et al., 2009).

Can the effects of aging on memory be prevented or alleviated?

The search for antiaging interventions and **nootropics**—drugs that enhance cognitive function—is an area of intense activity. Pharmacological approaches include the use of drugs, such as donepezil (Aricept), that inhibit acetylcholinesterase, the enzyme that breaks down acetylcholine. The resultant increase in cholinergic transmission in the forebrain has a positive effect on memory and cognition in mild to moderate cases of Alzheimer's (Ringman and Cummings, 2006). Compounds in a different class, called *ampakines*, act via glutamate receptors to improve hippocampal LTP (Rex et al., 2006); ampakines are under study as potential memory-enhancing therapeutics. Other possible drug targets are the family of protein kinases, such as PKMζ (the squiggle is the Greek letter *zeta*), that may contribute to long-term maintenance of memory traces (Shema et al., 2009) although, as with so much of the complicated molecular machinery of memory, many questions remain about their functioning (A. M. Lee et al., 2013). Given the complexity of the process, the list of possible targets for developing memory-boosting drugs grows longer every day.

Rats raised in enriched conditions show improved handling of stress, and the associated reduction in chronic glucocorticoid secretion reduces one source of

nootropics A class of drugs that enhance cognitive function.

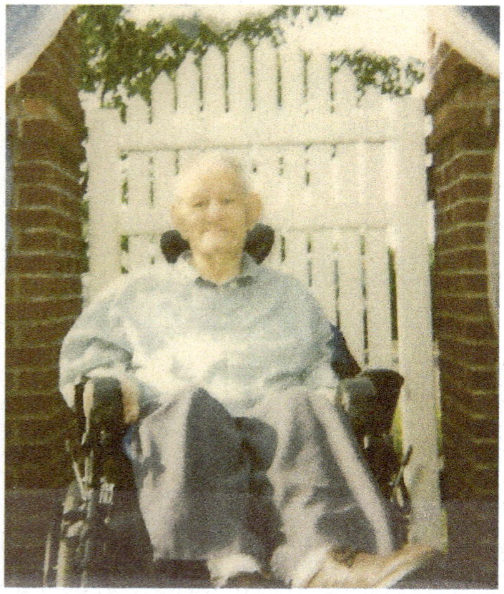

16.34 Henry Molaison (Patient H.M.) Henry lived for over 50 years after the surgery that robbed him of any new declarative memories.

lifestyle-related hippocampus damage (Sapolsky, 1993). And in humans there is good evidence that physical activity (Ngandu et al., 2015), mental activity (Suo et al., 2016), and adequate sleep (Gelber et al., 2015) can postpone cognitive decline in old age. Effortful learning—mastering a challenging new topic or skill over a period of time—enhances the survival of adult-born neurons, which may promote better memory performance (Shors, 2014).

Because so much of the brain is involved in the creation, storage, and retrieval of memories, a full understanding of learning and memory will require an enormous research effort at many levels of analysis, from experiments with molecular processes to cognitive studies of people like Henry Molaison. At age 82, more than 50 years after his life-shattering surgery, Henry died. In a final act of generosity to a field of science that he helped launch, Henry donated his brain for further study. Through webcasting technology, the slicing of Henry's brain into sections was viewed live by thousands of people, and a series of more than 2000 sections were made available. To the end, although he could remember so little of his entire adult life, Henry was courteous and concerned about other people (**FIGURE 16.34**). He remembered the surgeon he had met several times before his operation: "He did medical research on people.... What he learned about me helped others too, and I'm glad about that" (Corkin, 2002, p. 158). Henry never knew how famous he was, or how much his dreadful condition taught us about learning and memory; yet despite being deprived of one of the most important characteristics of a human being, he held fast to his humanity.

Recommended Reading

Baddeley, A. D. (2013). *Essentials of Human Memory.* London: Psychology Press.

Corkin, S. (2013). *Permanent Present Tense: The Unforgettable Life of the Amnesic Patient, H.M.* New York: Basic Books.

Gluck, M. A., Mercado, E., and Myers, C. E. (2016). *Learning and Memory: From Brain to Behaviour* (3rd ed.). New York: Worth.

Lieberman, D. A. (2012). *Human Learning and Memory.* Cambridge, UK: Cambridge University Press.

Martinez, J. L., and Kesner, R. P. (Eds.). (2014). *Learning and Memory: A Biological View* (2nd ed.). Cambridge, MA: Academic Press.

Østby, H., and Østby, Y. (2018). *Adventures in Memory: The Science and Secrets of Remembering and Forgetting.* Vancouver: Greystone Books.

Rudy, J. W. (2013). *The Neurobiology of Learning and Memory* (2nd ed.). Sunderland, MA: Oxford University Press/Sinauer.

Squire, L. R., and Kandel, E. R. (2008). *Memory: From Mind to Molecules.* Greenwood Village, CO: Roberts and Company.

16 Visual Summary

You should be able to relate each summary to the adjacent illustration, including structures and processes. You can also refer back to the links to figures, animations, and activities that will help you consolidate the material.

1 The **hippocampus, mammillary bodies**, and dorsal thalamus are part of a network that must be intact to form new **declarative memories**—memories that we can declare to others. Damage to these regions can cause **amnesia**, the impairment of **memory**. Review **Figure 16.2**

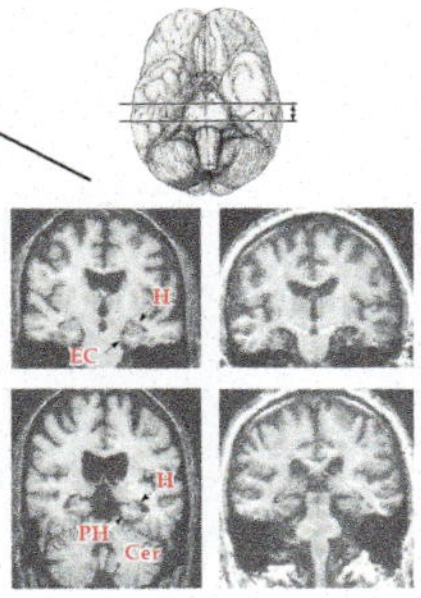

2 Removal of the hippocampus and nearby cortex left **patient H.M.** unable to form declarative memories that lasted more than a few minutes. He could, however, learn new skills such as mirror tracing, showing that his **nondeclarative memory** (or *procedural memory*) for perceptual and motor behaviours was intact. Review **Figures 16.1–16.4**

3 Declarative memory consists of **semantic memory** of facts and **episodic memory** (or autobiographical memory) of particular incidents in the past. Brain damage can remove one type of declarative memory without affecting the other, indicating the two kinds of memories are stored separately. Review **Figures 16.5, 16.15, 16.19, Activity 16.2**

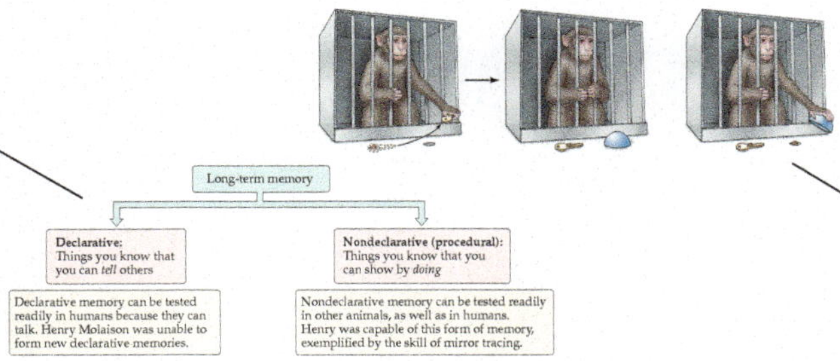

4 Nondeclarative memory, which includes **skill learning, priming, operant conditioning**, and **classical conditioning**, is demonstrated through performance. Most tests of memory in nonhuman animals are for nondeclarative memory, an exception being the **delayed non-matching-to-sample task** in monkeys. Review **Figures 16.6, 16.7, 16.14, Activity 16.1**

5 Different kinds of **learning** depend on different brain regions. Spatial learning requires an intact hippocampus, while motor skills rely on the basal ganglia, and object recognition relies on the visual cortex. Review **Figures 16.18, 16.19**

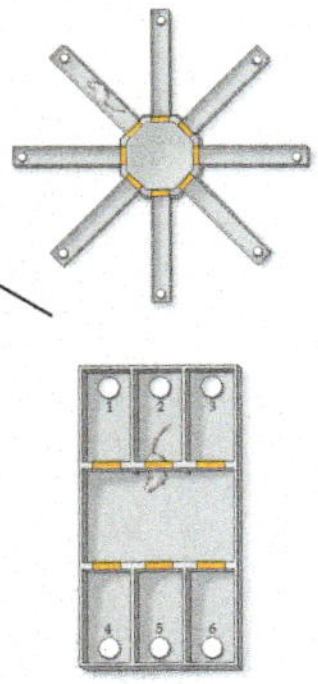

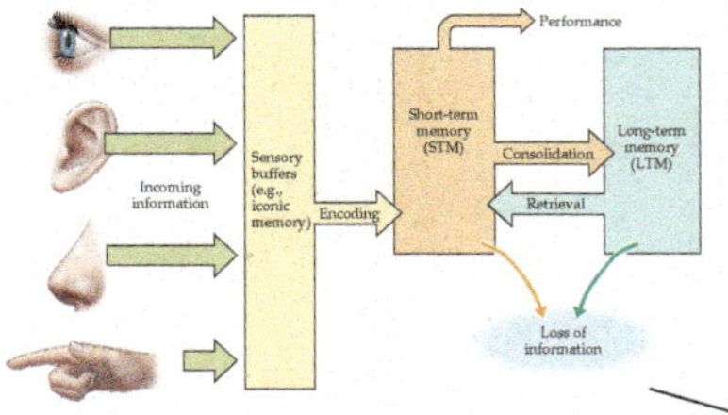

6 Memories are classified by how long they last. The **sensory buffers** retain only very brief impressions of sensations. **Short-term memory** (**STM**) lasts only a few minutes. Then the memory is either lost or transferred to **long-term memory** (**LTM**), which may last a lifetime. The successive processes transferring information from one place to the other are **encoding, consolidation**, and **retrieval**. Memories are subject to distortion during recall and **reconsolidation**. Review **Figures 16.19–16.21**

7 **Nonassociative learning** includes **habituation**, while **associative learning** includes classical conditioning (*Pavlovian conditioning*) and operant conditioning (*instrumental conditioning*). Associative and nonassociative learning are forms of nondeclarative memory. In the sea slug *Aplysia*, habituation is due to a weakening of the synapse between the sensory neuron and the motor neuron. Review **Figures 16.25, 16.26**

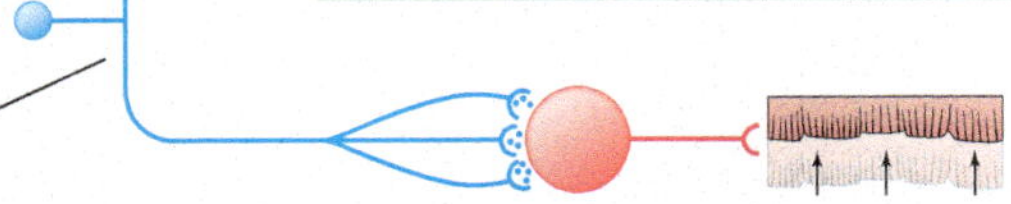

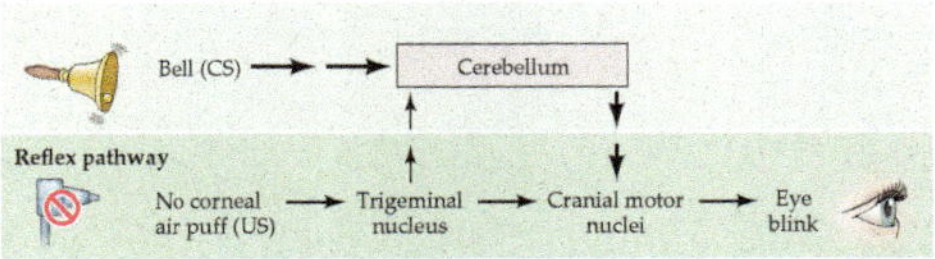

8 Conditioning of the eye-blink response in the rabbit is crucially dependent on the cerebellum. This simple mammalian system provides a model for understanding the formation of associations in the mammalian brain. Review **Figure 16.27**

9 **Long-term potentiation** (**LTP**) is a lasting increase in amplitude of the response of neurons, caused by brief high-frequency stimulation of their afferents, that is, **tetanus**. In the hippocampus, LTP depends on the activation of **NMDA receptors**, which induces an increase in the number of postsynaptic **AMPA receptors** and greater neurotransmitter release. These are examples of **Hebbian synapses**, which become stronger if they successfully drive the postsynaptic cell, and weaker if they are unsuccessful. Review **Figures 16.28–16.30, Activity 16.3, Animation 16.1, Video 16.2**

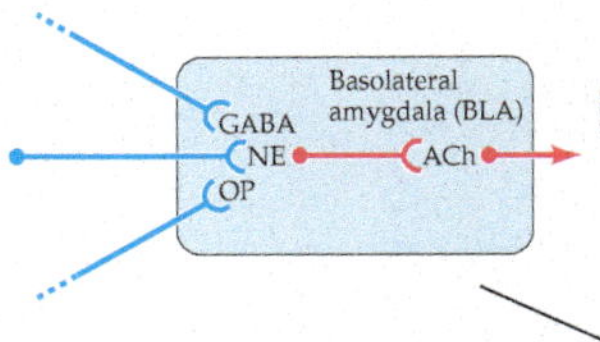

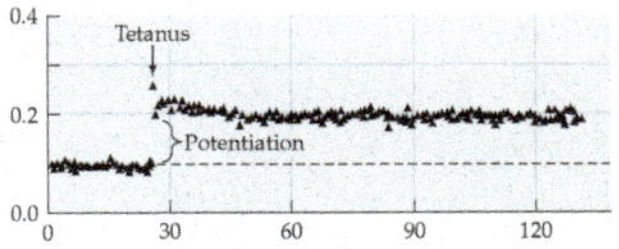

10 Strong emotion can affect the strength of memories, as in **posttraumatic stress disorder** (**PTSD**). Review **Box 16.1**

Attention and Higher Cognition

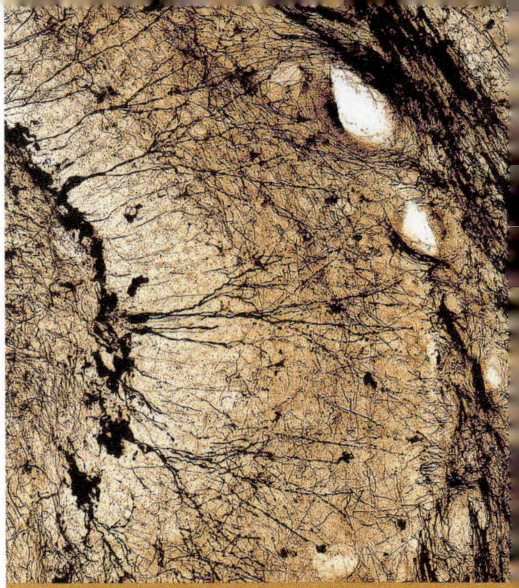

Courtesy of Dr. Sarah Moghadam, VA Palo Alto Health Care System, Palo Alto, CA and Dr. Ahmad Salehi, Dept. of Psychiatry & behavioural Sciences, Stanford Medical School

17

One Thing at a Time

Everyone agreed that Parminder and her family were bighearted—generous and friendly to a fault—but they were also "bad"-hearted, in one sad sense. Many of Parminder's relatives had suffered early heart attacks and strokes, and now, in her 68th year, Parminder shared that unhappy fate. In February, and then again in September, a blood clot that originated in Parminder's heart found its way to the complex of arteries in her brain, eventually lodging like a cork and cutting off blood flow to the surrounding brain tissue: a stroke. But what made Parminder's case exceptional was that the two strokes were exact mirror images—they damaged identical regions of the left and right parietal lobes.

Parminder's unlikely lesions produced equally unlikely symptoms. A few weeks after her second stroke, Parminder had regained many of her intellectual powers—she could converse normally and remember things. Her visual fields were apparently normal too, but her visual perception was anything but normal. Parminder had lost the ability to see and pay attention to more than one thing at a time. For example, she could see her husband's face just fine, but she couldn't tell whether he had glasses on. She could see the glasses, or she could see the face, but she couldn't see both at the same time. Shown a drawing of several overlapping items, Parminder could perceive and name only one at a time. Furthermore, she couldn't understand where the objects she saw were located. It was as if Parminder was lost in space, able to pay attention to only one object at a time, each alone in a world of its own. What could explain Parminder's symptoms?

William James, the great American psychologist, wrote in 1890:

> *Everyone knows what attention is. It is the taking possession by the mind, in clear and vivid form, of one out of what seem several simultaneously possible objects or trains of thought. Focalisation, concentration, of consciousness are of its essence. It implies withdrawal from some things in order to deal effectively with others, and is a condition which has a real opposite in the confused, dazed, scatterbrained state.*

James understood that attention can be effortful, improves perception, and acts as a filter on the outside world. When we are alert and functioning normally, we are always paying attention to something. This continual shifting of our focus from one interesting stimulus to the next lies at the heart of our innermost conscious experiences and our awareness of the world around us and our place in it. So, we open this chapter by exploring the behavioural and neural dimensions of attention before broadening our focus to the more general question of our conscious experience of the world.

ATTENTION

17.1 Attention Selects Stimuli for Processing

Learning Objectives

After studying this section, you should be able to:

17.1.1 **Define *selective attention*, with examples, and contrast overt and covert attention.**

17.1.2 **Discuss the limits on attention as revealed by divided-attention tasks, and compare early-selection and late-selection models of attention.**

17.1.3 **Distinguish between voluntary and reflexive attention, and describe a classic experimental design for studying each.**

17.1.4 **Describe the use of focussed attention to search the world for particular objects, using feature search or conjunction search, and discuss the significance of the binding problem.**

attention Also called *selective attention*. A state or condition of selective awareness or perceptual receptivity, by which specific stimuli are selected for enhanced processing.

vigilance The global, nonselective level of alertness of an individual.

cocktail party effect The selective enhancement of attention in order to filter out distracters, such as while listening to one person talking in the midst of a noisy party.

Despite taking delight in pretending otherwise, any average 5-year-old understands what it means when an exasperated parent shouts, "Pay attention!" We all share an intuitive understanding of the term *attention*, but to this day scientists have been unable to agree on a definition that goes much beyond the one provided by James—a testament to the great complexity and scope of the topic. In general, **attention** (or *selective attention*) is the process by which we select or focus on one or more specific stimuli—either sensory phenomena or internal thoughts—for enhanced processing and analysis. It is the *selective* quality of attention that distinguishes it from the related concept of **vigilance**, the global level of alertness of the individual. Most of the time we orient our attention *overtly*, directing our senses and our attention toward the same target. For example, as you read this sentence, it is both the centre of your visual gaze and (we hope) the main item that your brain has selected for attention. But if we choose to, we can shift the focus of our visual attention *covertly*; in classic research, Hermann von Helmholtz (1962; original work published in 1894) showed that we can keep our eyes fixed on one location while "secretly" scrutinizing some other location in peripheral vision (**FIGURE 17.1**).

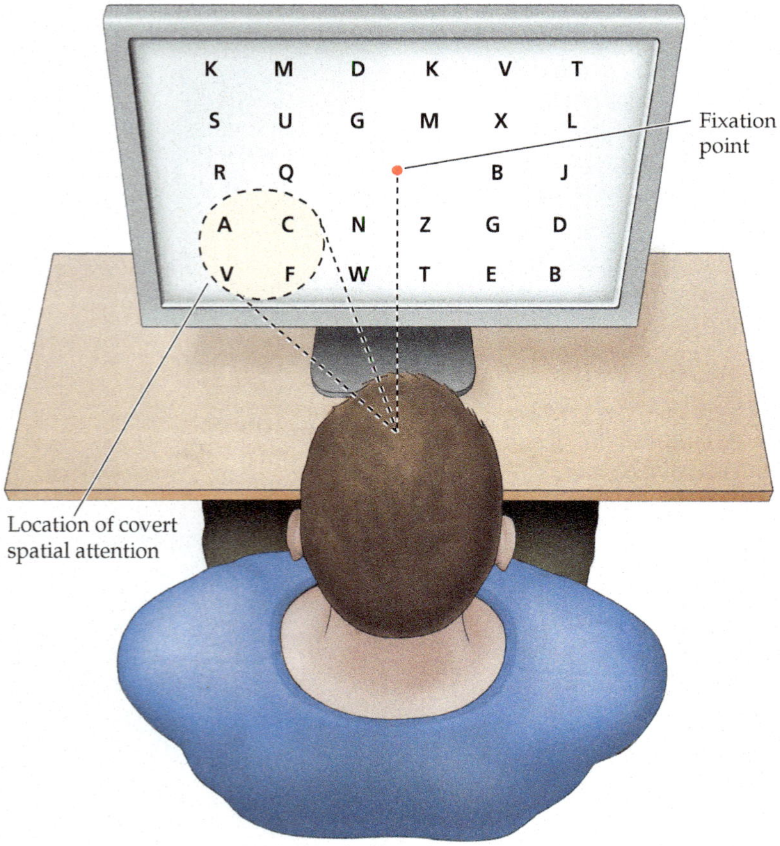

17.1 Covert Direction of Attention Holding your gaze steady on a central fixation point, you can direct your visual attention to a different spatial location without moving your eyes. This selective attention has sometimes been referred to as an *attentional spotlight*.

Selective attention isn't restricted to visual stimuli. Imagine yourself chatting with an old friend at a noisy party. Despite the high levels of background noise, you would probably find it relatively easy to focus on what your friend was saying, even if speaking quietly, because paying close attention would enhance your processing of your friend's speech and help filter out distracters. This selective enhancement is known as the **cocktail party effect***, and it nicely illustrates how attention acts to focus cognitive processing resources on a particular target. Though it may feel effortless, in this scenario you are solving a very difficult information-processing problem: with loud fragments of speech coming from every direction, how do you know which sounds go together? Solving that problem requires that you maintain the focus of your attention on a single speech source, and if your attention drifts to a different auditory stimulus—for example, if you start eavesdropping on

*The term "cocktail party effect" is also sometimes used where a highly salient word (such as one's own name) can capture attention in a noisy environment.

a more interesting conversation—it becomes virtualy imposible to simultaneously follow your friend's conversation. Unsurprisingly, actually having cocktails exacerbates the problem because alcohol lowers sensitivity for speech-related frequencies, reducing your ability to distinguish your friend's speech from background noise (Upile et al., 2007). Maybe this is one reason that parties seem to get louder as the night wears on.

There are limits on attention

The powers of attention that help you to easily chat with a friend at a noisy party normally rely on numerous cues occurring simultaneously in different sensory modalities: for example, the spatial origin of the speech sounds, the facial movements that accompany speaking, your friend's unique tone of voice, and so on, all help you to focus on your friend's speech. But what if our attention is restricted to just a single type of stimulus? In now-classic experiments on the cocktail party effect (Cherry, 1953), extra cues were systematically excluded by having participants listen through headphones. Participants listened to two different streams of dialog delivered simultaneously to the left and right ears, a technique known as **dichotic presentation** (see Figure 18.3). People were asked to focus their attention on one ear or the other and to repeat aloud the material being presented to that ear. On this demanding task, called **shadowing**, participants could accurately report what they were hearing in the attended ear, but they could report very little that had been presented to the nonattended ear, aside from simple characteristics like the sex of the speaker. If a shadowing task is sufficiently difficult, then even advance warning that they will be quizzed later doesn't help people remember and report information from the nonattended auditory stream (Moray, 1959). In fact, even when the person's own name is presented in the nonattended ear, it is detected only about a third of the time (N. Wood and Cowan, 1995).

The cocktail party effect is not limited to speech (or parties!), of course. For example, musicians are better than nonmusicians at focussing attention on target notes and segregating them from multiple simultaneous musical sounds (Zendel and Alain, 2009). A shadowing procedure can be used to study visual attention too. In this case, participants are asked to attend to just one of two fully overlapping videos being presented simultaneously—imagine two movies being projected onto a single screen. As in the auditory shadowing experiments, participants are able to focus on one video only at the expense of the other; little of the nonattended video can be reported (Neisser and Becklen, 1975). Furthermore, if the attended visual stream is sufficiently complicated, people show **inattentional blindness**: a surprising failure to perceive nonattended stimuli that you'd think would be impossible to miss, like a gorilla strolling across the screen (Simons and Chabris, 1999; Simons and Jensen, 2009). It's a problem that can even bedevil experts: in one study, 83% of radiologists screening CT scans for lung cancer didn't notice a seemingly obvious image of a gorilla inserted into one of the scans (Drew et al., 2013). And while noticing a gorilla in a CT scan may seem unrepresentative of real-life problems, inattentional blindness occurs even when the nonattended stimulus has life-or-death significance for the observer; for example, a significant number of police officers and trainees will fail to notice a gun placed in plain view during a simulated traffic stop (Simons and Schlosser, 2017).

In general, **divided-attention tasks**—in which the person is asked to process two or more simultaneous stimuli—confirm that attention is a *limited* resource and that it's very difficult to attend to more than one thing at a time, particularly if the stimuli to be attended to are spatially separated (Bonnel and Prinzmetal, 1998). So, our limited selective attention generally acts like an **attentional spotlight**, which shifts around the environment, highlighting stimuli for enhanced processing. It's an adaptation that we share with many other species because, like us, they are confronted with the problem of extracting important signals from a noisy background (Bee and Micheyl, 2008). Birds, for example, must isolate the vocalisations of specific individuals from a cacophony of calls and other noises in the environment—an avian version of the cocktail party problem (Benney and Braaten, 2000). Having a single attentional spotlight helps

Go to Video 17.1
Change Blindness and Inattentional Blindness

dichotic presentation The simultaneous delivery of different stimuli to the right and the left ears.

shadowing A task in which the participant is asked to focus attention on one ear or the other while stimuli are being presented separately to both ears.

inattentional blindness The failure to perceive nonattended stimuli that seem so obvious as to be impossible to miss.

divided-attention task A task in which the participant is asked to simultaneously focus attention on two or more stimuli.

attentional spotlight The shifting of our limited selective attention around the environment to highlight stimuli for enhanced processing.

attentional bottleneck A filter that results from the limits intrinsic to our attentional processes, with the result that only the most important stimuli are selected for special processing.

early-selection model A model of attention in which the attentional bottleneck filters out stimuli before even preliminary perceptual analysis has occurred.

late-selection model A model of attention in which the attentional bottleneck filters out stimuli only after substantial analysis has occurred.

perceptual load The immediate processing challenge presented by a stimulus.

us focus cognitive resources and behavioural responses toward the most important things in the environment at any given moment (the smell of smoke, the voice of a potential mate, a glimpse of a big spotty cat), while ignoring extraneous information.

In general, by acting as a filter, attention narrows our focus and directs our cognitive resources to only the most important stimuli around us, thereby protecting the brain from being overwhelmed by the world. But at what level of sensory processing is this **attentional bottleneck** evident? Initial research gave evidence of an **early-selection model** of attention (**FIGURE 17.2A**), in which unattended information is filtered out right away, at the level of the initial sensory input, as in the shadowing experiments we just described (Broadbent, 1958). But other researchers noted that some important but unattended stimuli (such as your name) may undergo substantial unconscious processing right up to the level of semantic meaning and awareness, before suddenly capturing attention (N. Wood and Cowan, 1995). This is an example of a **late-selection model** of attention, in which the filtration imposed by the processing bottleneck occurs at a later point, filtering out stimuli only after substantial analysis has already occurred (**FIGURE 17.2B**). Many models of attention contain both early- and late-selection mechanisms (e.g., Wolfe, 1994), and vigourous debate continues over their relative importance.

A possible resolution to the debate over early versus late selection involves the concept of **perceptual load**—the immediate processing challenge presented by a stimulus. According to this view, when we focus on a complex stimulus that requires a lot of perceptual processing, no perceptual resources remain for use on competing unattended items (N. Lavie et al., 2004). So in this case, attention exerts early selection and excludes other stimuli from the outset. But when we focus on simpler stimuli, there is enough perceptual capacity to allow for processing of other stimuli, right up to the level of semantic meaning, recognition, and awareness (and thus, late selection) (N. Lavie et al., 2009). In other words, if we view attention as a limited

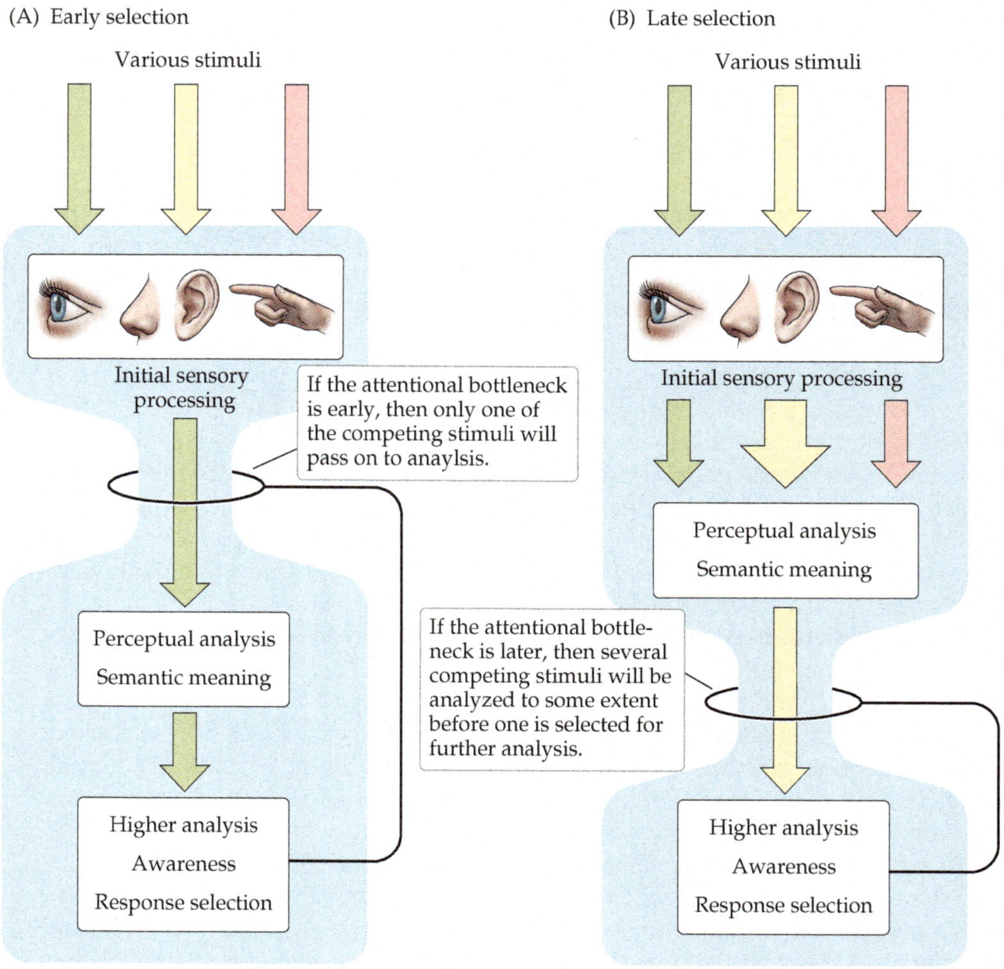

17.2 Bottlenecks in Attention
(A) In early-selection models of attention, unattended sensory information is filtered out before it is analysed for meaning. (B) In late-selection models of attention, some degree of preliminary processing and analysis of competing stimuli occurs before one is selected for enhanced processing, awareness, and response.

resource, then we only have enough of it to do one complex task at a time, or a few very simple ones. These studies thus indicate that attention is continually rebalanced between early and late selection, according to the difficulty of the task at hand. These more modern perspectives on attention are central to the development of computational models of attention in visual and auditory "scenes"; one hope is that mathematical descriptions of attentional processes will aid in the development of machine vision and audition (Kaya and Elhilali, 2017; Shic and Scassellati, 2007).

We can decide to direct our attention toward specific stimuli

We've seen that through an act of willpower, we can direct our attention to specific stimulus sources without moving our eyes or otherwise reorienting. Early experiments on this phenomenon, like Helmholtz's, employed **sustained-attention tasks**, in which a single stimulus source or location must be held in the attentional spotlight for a protracted period (see Figure 17.1). Although these tasks are useful for studying basic phenomena, and for assessing attention problems due to neurological disorders, we need something more powerful to address key questions about attention. For example, how do we shift attention around? How does attention enhance the processing of stimuli, and which brain regions are involved? To answer these questions, researchers have devised various clever tasks that employ *stimulus cuing* to control attention.

So far our discussion of attention has focussed mostly on what researchers call **voluntary attention** (also called *consciously* or *endogenously controlled attention*). As the name implies, voluntary shifts of attention come from within; they are the conscious, *top-down* directing of our attention toward specific aspects of the environment, according to our interests and goals. By far the most common modern experimental procedure for studying voluntary attention is the **symbolic cuing task** (or *spatial cuing task*), originated by Posner (1980). On this type of task, participants stare at a fixation point in the centre of a computer screen and must press a key as soon as they see a specific target stimulus (the response being measured is therefore reaction time; **BOX 17.1**). Each trial is preceded by a *cue* that gives the participants a hint as to where the stimulus will appear. So in our simple example (**FIGURE 17.3A**), participants fixate on the central point, and an arrow pointing left, right, or in both directions briefly flashes in a central location. Then, after

sustained-attention task A task in which a single stimulus source or location must be held in the attentional spotlight for a protracted period.

voluntary attention Also called *consciously* or *endogenously controlled attention* or *top-down attention*. The voluntary direction of attention toward specific aspects of the environment, in accordance with our interests and goals.

symbolic cuing task An attention task in which a symbol indicates to the participant where to voluntarily direct attention in order to detect a stimulus.

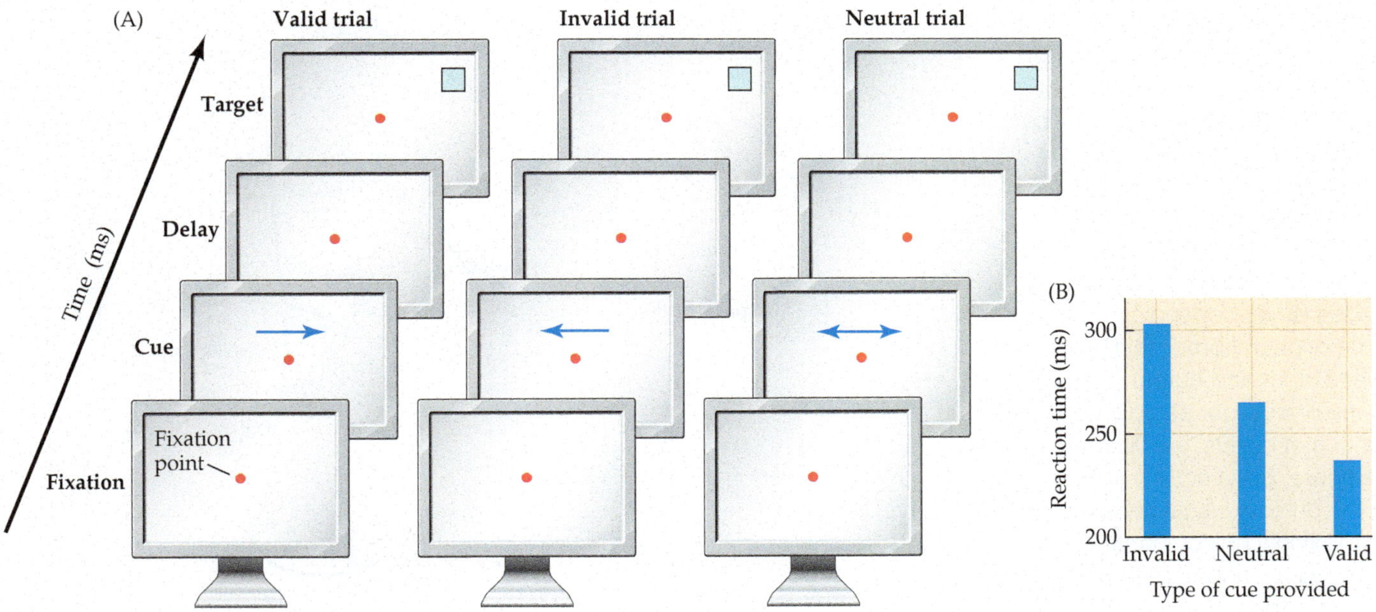

17.3 Measuring the Effects of Voluntary Shifts of Attention (A) In symbolic cuing tasks, participants are provided with cues (the blue arrows) that accurately predict target locations (valid trials), provide inaccurate predictions (invalid trials), or provide no predictions (neutral trials). Participants are instructed to maintain visual fixation on a central point throughout the trial. B) Whether the target appears to the left or the right of the fixation point, valid cues significantly enhance reaction time to detect the target, and invalid cues significantly impair reaction time. (A after M. I. Posner, 1980. *Q J Exp Psychol* 32: 3–25; B after M. I. Posner et al., 1978. In *Modes of perceiving and processing information*, H. I. Pick, Jr. and E. Saltzman [Eds.], pp. 137–157. Erlbaum: Hillsdale, NJ.)

a delay, the target stimulus flashes on the screen and the participants must press a key when they detect it. In most trials where there is a single arrow, it points in the correct direction and thus provides a *valid cue* about which side of the screen the target will appear on, but sometimes there's an *invalid cue,* where the arrow points away from where the target will appear. The effects of these cues on reaction time are averaged over many trials and compared with trials with *neutral cues,* in which the cue provides no hint about subsequent target location.

As you can see in **FIGURE 17.3B**, people swiftly learn to use the cues to predict stimulus location, shifting their attention in the cued direction—*without shifting their gaze*—in anticipation of the appearance of the target stimulus. Reaction time is thus significantly faster for validly cued trials, compared with the uncued neutral condition, even when participants do not move their eyes to the cued location. And participants clearly pay a price on trials in which the cue is invalid, misdirecting attention to the wrong side of the display; they take more time to detect the target. So, directed attention helps us to perform better, faster, and with increased sensitivity to cues.

Go to Animation 17.2
From Input to Output

BOX 17.1 Reaction Time Responses, from Input to Output

Reaction time measures are a mainstay of cognitive neuroscience research. In tests of *simple reaction time*, participants make a single response—for example, pressing a button—in response to experimental stimuli (the appearance of a target, the solution to a problem, a tone, or whatever the experiment is testing). Tests of *choice reaction time* involve a slightly more complicated situation, in which a person is presented with alternatives and has to choose among them (e.g., right versus wrong, same versus different) by pressing one of two or more buttons.

Reaction times in an uncomplicated choice reaction time test, in which the participant indicates whether two stimuli are the same or different, average about 300–350 milliseconds (ms). The delay between stimulus and response varies depending on the amount of neural processing required between input and output. The stations involved in our sample task, and the timing of events, are illustrated in the figure. Brain activity proceeds from the primary visual cortex through a ventral visual object identification pathway (see Chapter 8) to prefrontal cortex, and then through premotor and primary motor cortex, down to the spinal motor neurons and out to the finger muscles. In the sequence shown in the figure—proceeding from the presentation of visual stimuli to a discrimination response—notice that it takes about 110 ms for the sensory system (somewhere in the inferior temporal lobe) to recognise the stimulus, and about 35 ms more for that information to reach the prefrontal cortex. At this point, a decision about which response to choose must be made; for very simple tasks this may take as little as 30 ms (Stanford et al., 2010), although for more-complex stimuli it may take longer. Then it takes about 75 ms for the movement to be executed (i.e., the time between when the signal from the prefrontal cortex arrives in the premotor cortex and when the finger pushes the button). It is fascinating to think that something like this sequence of neural events happens over and over in more-complicated behaviours, such as recognizing a long-lost friend or composing an opera.

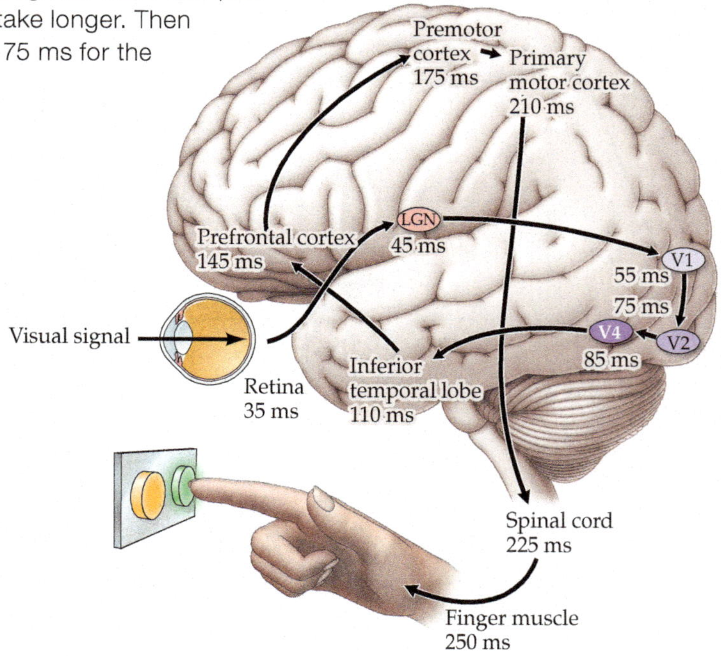

The sequence and timing of brain events that determine reaction time. LGN, lateral geniculate nucleus; V1, primary visual cortex; V2 and V4, extrastriate visual areas. (Timings based on S. J. Thorpe and M. Fabre-Thorpe, 2001. *Science* 291: 260–263.)

Some types of stimuli just grab our attention

Everyone knows that attention to events in the surrounding environment involves more than just consciously steering your attentional spotlight around. Loud noise or sudden silence, flashes, moving objects—any striking or important change—can potently capture attention and draw it away from the task at hand unless you are very focussed (every conversation stops and every head swivels when someone drops a plate in a restaurant). This sort of involuntary reorientation toward a sudden or important event is an example of **reflexive attention** (or *exogenously controlled attention*). It is considered to be a *bottom-up* process because attention is being captured and controlled by sensory inputs from lower levels of the nervous system, rather than being the result of voluntary, conscious, top-down, cognitive control by the forebrain.

The effects of reflexive shifts of attention on stimulus processing can be measured using a **peripheral spatial cuing task** (**FIGURE 17.4A**). Again, the participant fixates on a point and is asked to respond as quickly as possible when the target stimulus appears. But instead of a meaningful symbol to direct attention toward a target location, a simple *task-irrelevant* sensory stimulus (often a flash of light, but it could be a sound or other stimulus) is presented *in the location to which attention is to be drawn*. After a momentary delay, the target stimulus is presented, either in the same location as the cue (a valid trial) or somewhere else (an invalid trial). Reaction time measures for many such trials are averaged for each participant.

Two important effects are seen in such an experiment (**FIGURE 17.4B**). First, a valid cue draws attention and enhances processing of subsequent stimuli in the target location, but only when the delay between cue and target is brief. After 150 milliseconds (ms) or so, attention starts to wander away again and the enhancement of processing is lost. And when the interval between cue and target becomes longer—about 250 ms onward—a curious phenomenon is observed: detection of stimuli at the location where the cue occurred is actually *impaired* (R. M. Klein, 2000; Posner and Cohen, 1984). It's as though attention has moved on from the flash and is reluctant to return to that location. This **inhibition of return** probably evolved to prevent reflexive attention from settling on unimportant stimuli for

reflexive attention Also called *exogenously controlled attention* or *bottom-up attention*. The involuntary reorienting of attention toward the location of an unexpected object or event.

peripheral spatial cuing task An attention task where a visual stimulus is preceded by a simple sensory stimulus (like a flash) that reflexively captures attention.

inhibition of return The phenomenon in which detection of stimuli at the former location of a cue is impaired for latencies of 250 ms or more.

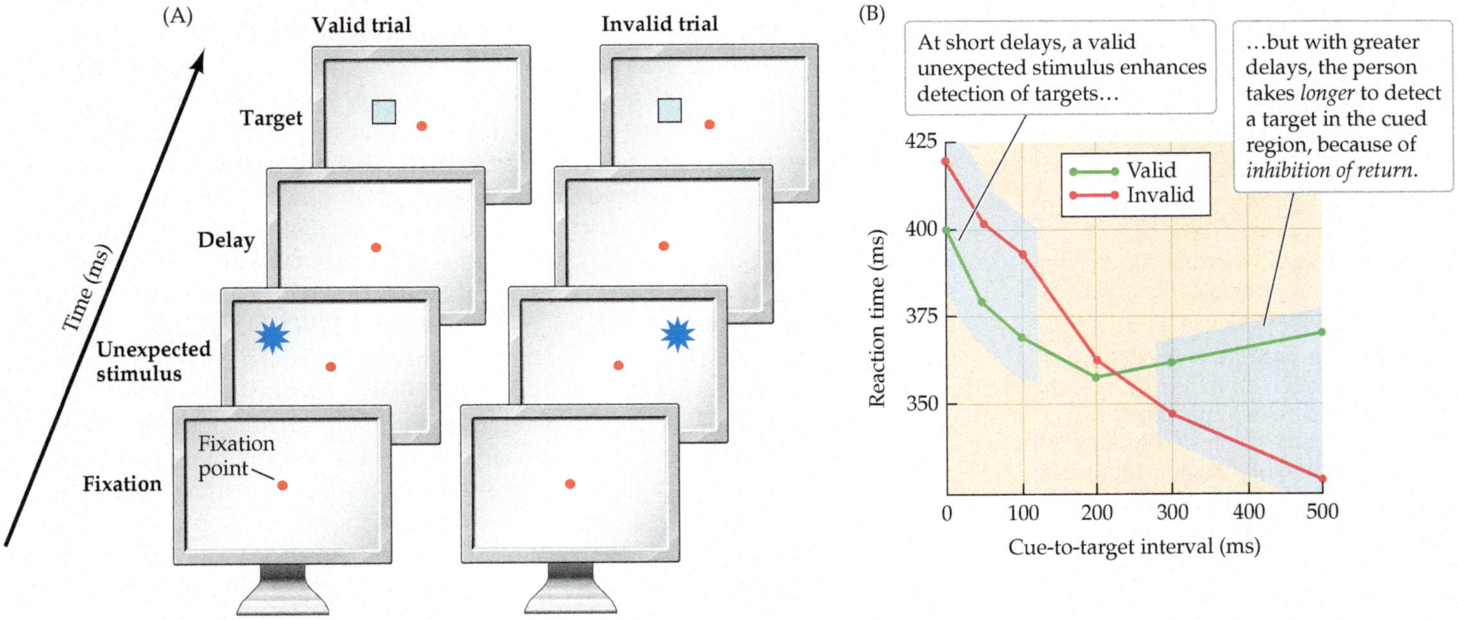

17.4 Reflexive Capture of Attention and Inhibition of Return (A) In valid trials, a momentary flash appears in the location where a stimulus will appear soon. In invalid trials, the flash appears somewhere other than where the stimulus will appear. (B) Reaction times indicate that attention is focussed on the cued location only for 200 ms or so. If the stimulus appears later, the person is less likely to detect it than if attention had never been drawn to that region. This demonstrates the so-called inhibition of return (of attention). (After R. M. Klein, 2000. *Trends Cog Sci* 4: 138–147.)

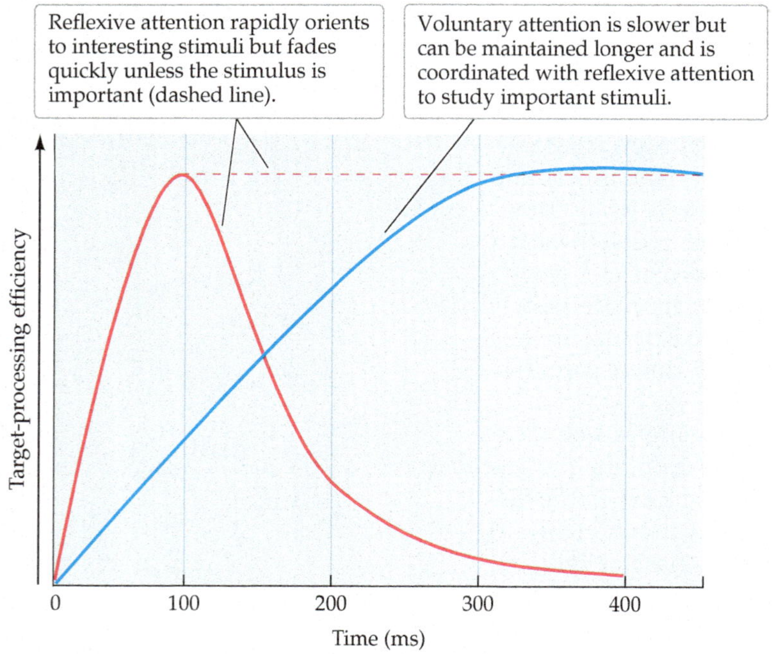

17.5 Voluntary and Reflexive Attention Are Complementary Quick-reacting reflexive attention and slower, longer-lasting voluntarily controlled attention work together to direct our analysis of the environment.

more than an instant, or returning to previously attended stimuli. This strategy might be adaptive in animals foraging for food or scanning for threats.

Normally, voluntary and reflexive attention work together to direct cognitive activities (**FIGURE 17.5**), probably relying on somewhat overlapping neural mechanisms. Anyone who has watched a squirrel at work has seen that twitchy interplay. When it comes to single-mindedly searching for tasty morsels (an example of voluntarily controlled attention), a squirrel has few rivals. But even slight noises and movements (cues that reflexively capture attention) will cause it to stop and scan its surroundings—a sensible precaution if, like a squirrel, you are yourself a tasty morsel. Similarly, important emotional cues—that look of fear on a companion's face, for example—can reflexively capture attention and augment sensory processing, in order to scan for threats (Carretié, 2014). And effective cues for reflexive attention may involve multiple sensory modalities: a sound coming from a particular location, for example, can improve the *visual* processing of a stimulus that appears there (Feng et al., 2017; McDonald et al., 2000). Cats are much better at localising prey when they have a combination of cues to guide their attention (Schnupp et al., 2005), one of the many reasons that squirrels are so twitchy.

Attention helps us to search for specific objects in a cluttered world

Another way that we use attention is in visual search: systematically scanning the world to find some specific item in a sea of distracters. If the sought item is different from all the distracters in some fundamental way—your friend in a red coat, for example, in a crowd of dark-suited businesspeople—the search is easy and the target of the search immediately "pops out." This type of visual search is known as a **feature search**, and it seems effortless. But more often we must rely on a **conjunction search**: searching for an item on the basis of two or more features (e.g., size and colour) that together distinguish the target from distracters that may share some of the same attributes.

The visual search tasks employed in attention labs typically consist of arrays of target stimuli that systematically vary in the number of attributes they share with nearby distracters, and also in the number of distracters that are present. In a simple feature search (**FIGURE 17.6A**), targets pop out right away, no matter how many distracters are present, with only a few exceptions (Joseph et al., 1997). Finding the target takes little effort, and there is no need to scan the other items one by one. In contrast, performance on conjunction searches (**FIGURE 17.6B**) is directly related to the number of distracters present. The results obtained from many such experiments (**FIGURE 17.6C**) suggest that in conjunction searches we are coordinating and searching through multiple cognitive *feature maps*—think of a stack of mental representations (maps) of the area to be searched, each focussed on the distribution of one particular stimulus attribute (a map of colours, a map of shapes, and so on). This influential idea, called **feature integration theory** (A. M. Treisman and Gelade, 1980), helps us address an issue that has come to be known as the **binding problem** (A. M. Treisman, 1996). Namely, how does the brain know that a particular set of features (colours, shapes, sounds, etc.) combine to define a single object, considering that these different stimulus features are processed in different regions of the brain?

According to feature integration theory, simple preattentive stimulus attributes (e.g., colour, shape) guide our scanning of the environment, with higher-order processes

feature search A search for an item in which the target pops out right away because it possesses a unique attribute.

conjunction search A search for an item that is based on two or more features (e.g., size and colour) that together distinguish the target from distracters that may share some of the same attributes.

feature integration theory The idea that conjunction searches involve multiple cognitive feature maps—overlapping representations of the search array based on individual stimulus attributes.

binding problem The question of how the brain understands which individual attributes blend together into a single object, when these different features are processed by different regions in the brain.

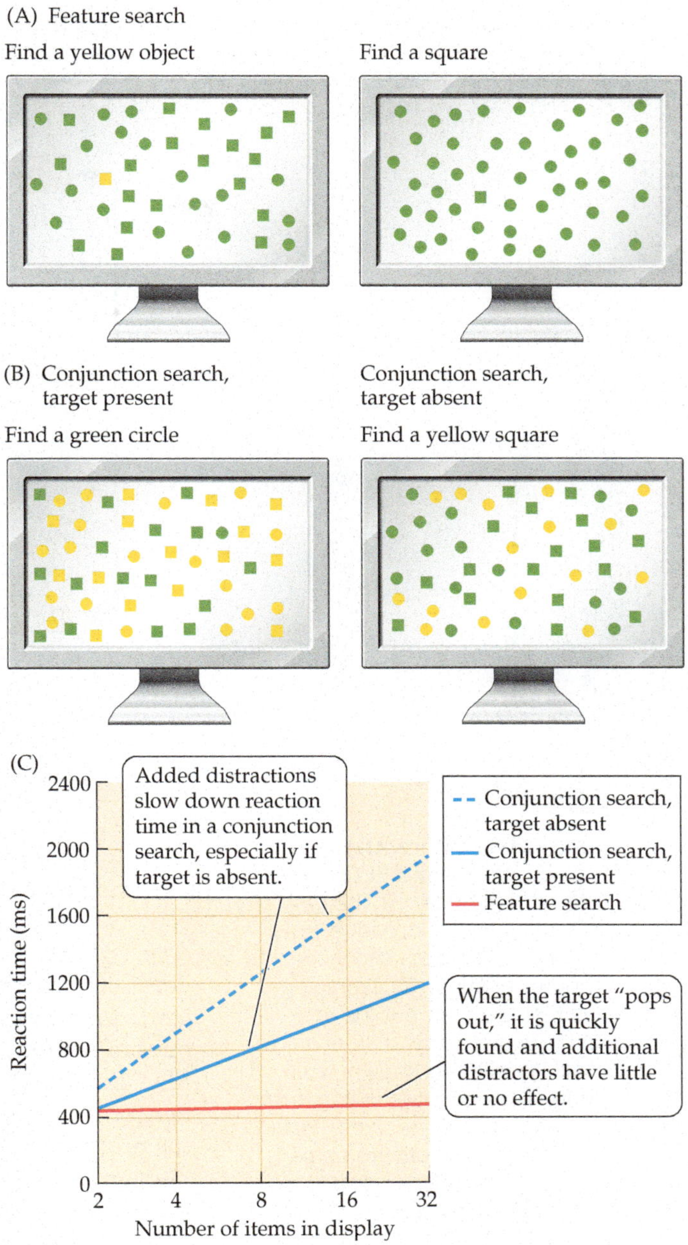

17.6 Visual Search (A) In these two feature search arrays, targets pop out. (B) In these two conjunction search arrays, one contains the target and one does not. Finding the target in these searches, or determining that it is absent, is more difficult and so takes more time. (C) Data for three types of visual search. (After A. M. Treisman and G. Gelade, 1980. *Cogn Psychol* 12: 97–136.)

coordinating the binding of features and, ultimately, conscious perception and selection. This is why it's easier to do conjunction searches by looking for one basic stimulus attribute at a time—a **bottom-up process** (e.g., scan first for *blue* things, then for *cars*, then for *Subarus*, and so on, until you spot *your* car)—rather than using your attentional spotlight to hunt for all of the stimulus attributes simultaneously—a **top-down process** (e.g., searching right away for your *blue 2017 Subaru Forester*) (Wolfe et al., 2000). Sustained attention to multiple stimulus features requires a large cognitive effort, and if the target objects appear only infrequently—as do weapons in luggage or tumours in X-rays—they may go undetected with alarmingly high frequency, even by highly trained screeners (Wolfe et al., 2005).

To uncover finer details of the brain's attentional mechanisms, neuroscientists take two complementary perspectives on attention. First, we can look at *consequences of attention* in the brain, asking how selective attention modifies brain activity to enhance processing of stimuli. These studies ask, What are the *targets* of attention? Second, we can probe the *mechanisms of attention*, the brain regions that produce and control attention, shifting it between different stimuli in different sensory modalities. These studies ask, What are the *sources* of attention? In selecting experimental techniques to address these objectives, researchers must juggle the need for good **temporal resolution**—the ability to track changes in the brain that occur very quickly—with the need for excellent **spatial resolution**, the ability to observe the detailed structure of the brain. In general, electrophysiological approaches offer the speed (temporal resolution) necessary to distinguish the consequences of attention from the mechanisms that direct it, while brain-imaging techniques like fMRI offer the anatomical detail (spatial resolution) to figure out exactly where these neural actions are taking place. This speed-versus-accuracy trade-off permeates the research that we discuss in the following sections.

17.2 Targets of Attention: Attention Alters the Functioning of Many Brain Regions

Learning Objectives

After studying this section, you should be able to:

17.2.1 **Explain how event-related potentials (ERPs) are measured, and name and describe the main ERP components as they relate to auditory versus visual attention.**

17.2.2 **Contrast and compare the auditory N1 effect, the visual P1 effect, and the N2pc component.**

17.2.3 **Describe experimental evidence that selective attention to stimuli enhances neural activity in the brain regions processing the attended stimuli.**

17.2.4 **Describe research that shows three different ways that attention affects the activity of individual neurons in sensory systems.**

Directly recording electrical activity from the surface of the brain (see Chapter 1) would offer both good temporal resolution and good spatial resolution, but in order to study healthy participants, we need noninvasive ways of assessing brain activity.

bottom-up process A process in which lower-order mechanisms, like sensory inputs, trigger additional processing by higher-order systems. There may be no conscious awareness until late in the process.

top-down process A process in which higher-order cognitive processes control lower-order systems, often reflecting conscious control.

temporal resolution The ability of an experimental technique to track changes in the brain over time.

spatial resolution The ability to observe the detailed structure of the brain.

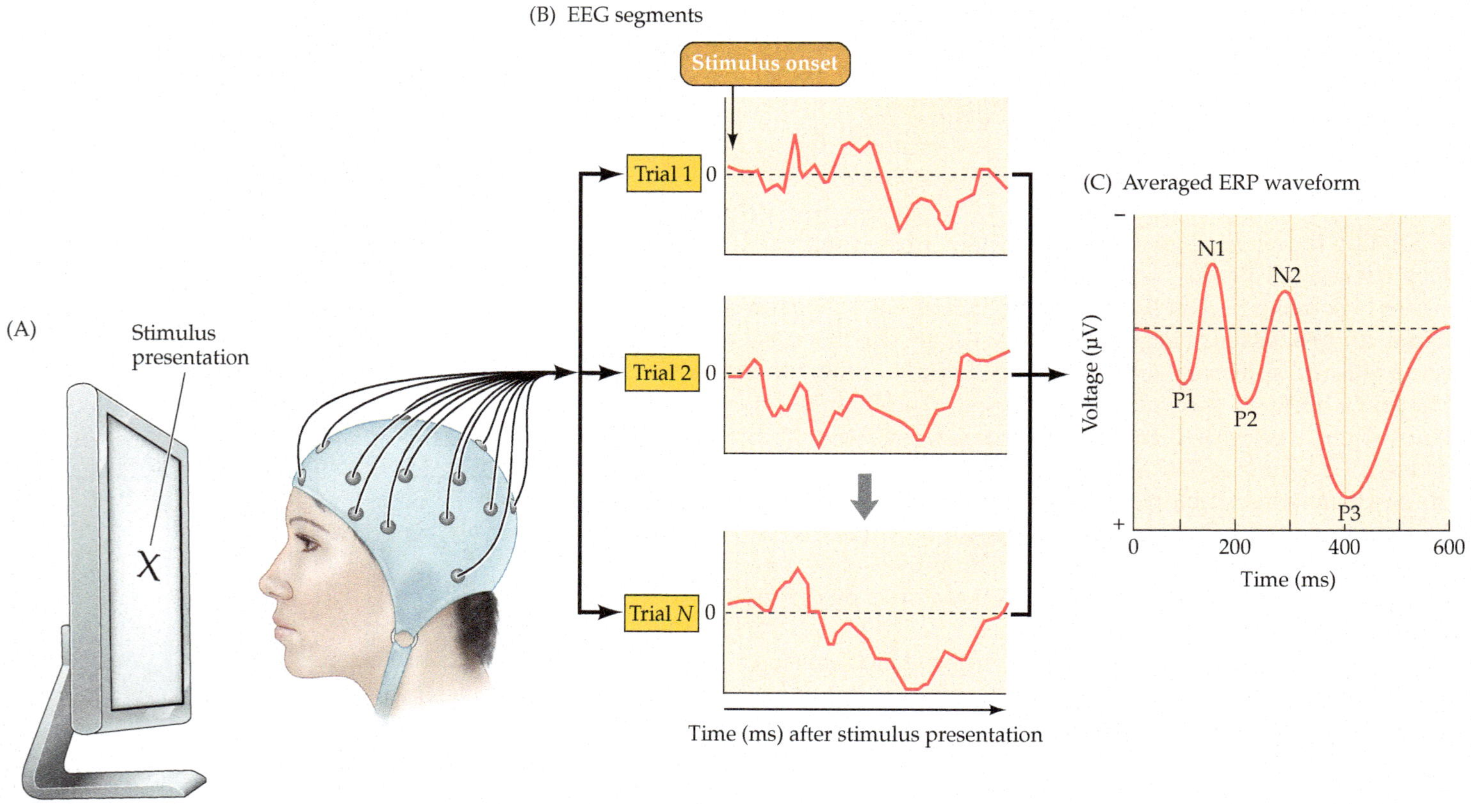

17.7 Event-Related Potentials
(A) Scalp electrodes record EEG activity during task performance. (B) The task is repeated many times; individual EEG segments for multiple trials show considerable random variation. (C) Averaging many such trials cancels out the random variation, leaving only task-related components, labelled as shown. Note that by convention, negative voltages are charted above the zero line. (Yes, it's kind of weird.) ERP components are named according to their polarity—negative or positive—and their sequence after stimulus onset. So, "N1" is the first negativity in the ERP after stimulus onset, "P2" is the second positivity after stimulus onset, and so on.

When many cortical neurons work together to solve a complicated problem, their activity becomes synchronised to some degree. In Chapter 2 we described *electroencephalography (EEG)*, in which recording electrodes are attached to the scalp at standardised locations (**FIGURE 17.7A**) to record moment-to-moment changes in brain electrical activity. So, you might think that the electrical activity of a network of millions of neurons working together on a specific problem would be easy to see in an EEG, but it isn't. That's because the contributions of multiple simultaneously active networks, regional differences in the timing of activity, and individual variation in the firing of neurons all contribute to the real-time EEG, making it look surprisingly random during the performance of any particular task. To solve this problem, researchers ask participants to do the same task over and over again (**FIGURE 17.7B**), and they average all the EEGs recorded during those repeated trials. Averaged over enough repetitions of the task, the variation due to the random firing of other neurons averages out, and what's left is the overall electrical activity specifically associated with the task being performed (**FIGURE 17.7C**). This averaged activity is called the **event-related potential** (**ERP**) (Luck, 2005), and it reveals regional changes in brain activity much faster than brain-imaging techniques like fMRI. For this reason, ERP has become the favorite tool of neuroscientists studying moment-to-moment consequences of attention in the brain, as we'll see.

Distinctive patterns of brain electrical activity mark shifts of attention

In one classic study of auditory attention (Hillyard et al., 1973), dichotic tones were presented to participants who were asked to attend to the stimuli arriving at one ear and to ignore the other ear. In half the trials, participants attended to the left; in the other half, to the right. EEGs were recorded and averaged into ERPs as already described. Thus, for each ear the researchers were able to compare ERPs for attended stimuli with ERPs for unattended stimuli. Overall, the ERPs for attended stimuli were larger in amplitude than when the same stimuli at the same ear were not being attended to. This effect was particularly evident in the ERP component called *N1*, a large negativity occurring 100–200 ms after stimulus onset. Because the only thing

event-related potential (ERP)
Averaged EEG recordings measuring brain responses to repeated presentations of a stimulus

that changed between conditions was participants' attention to the stimuli, this auditory **N1 effect** must have been caused by selective attention somehow acting on neural mechanisms to enhance processing of information from that ear. Auditory attention may also affect much later ERP components, such as the wave called *P3* (see Figure 17.7). Changes in late-occurring components like P3 are tricky to interpret, because they can be associated with multiple different cognitive operations, ranging from memory access to reactions to unexpected events (Wessel and Aron, 2017). However, some researchers claim that P3 is also sensitive to high-order cognitive processing of the stimulus (Herrmann and Knight, 2001)—qualities like the underlying meaning of the stimulus, identity of the speaker, and so on—in which case the **P3 effect** provides an example of a late-selection effect of auditory attention. In fact, researchers are debating whether the P3 component is (Dehaene and Changeux, 2011) or is not (Pitts et al., 2014) an electrophysiological marker of consciousness.

Attention to *visual* stimuli is likewise reflected in ERPs, but because the neural systems involved in visual perception are different from those involved in audition, voluntary visual attention causes its own different, distinctive changes in the ERP. We can study these visual effects by collecting ERP data over occipital cortex—the primary visual area of the brain—while a participant performs a symbolic cuing task. **FIGURE 17.8** depicts this sort of experiment. On valid trials (remember, this is when the target appears as expected, in the location indicated by the cue, as in **FIGURE 17.8A**), electrodes over occipital cortex show a substantial enhancement of the ERP component called *P1*, the positive wave that occurs about 70–100 ms after stimulus onset, often carrying over into an enhancement of the N1 component immediately afterward (**FIGURE 17.8B**). The visual **P1 effect**

N1 effect A negative deflection of the event-related potential, occurring about 100 ms after stimulus presentation, that is enhanced for selectively attended input, compared with ignored input.

P3 effect A positive deflection of the event-related potential, occurring about 300–500 ms after stimulus presentation, that is associated with higher-order auditory stimulus processing and late attentional selection.

P1 effect A positive deflection of the event-related potential, occurring 70–100 ms after stimulus presentation, that is enhanced for selectively attended visual input, compared with ignored input.

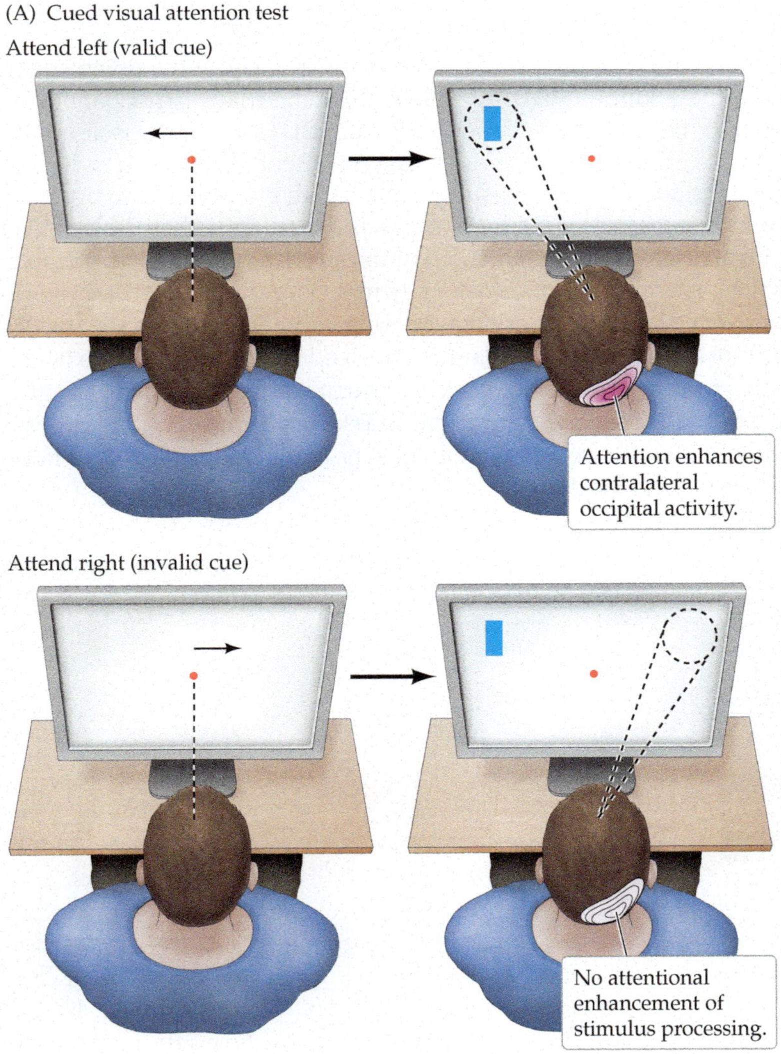

17.8 ERP Changes in Cued Shifts of Visual Attention (A) In this example, the participant maintains central fixation and covertly orients (dashed circle) in the direction indicated by a prior symbolic cue, pressing a key as soon as a stimulus appears. In most trials the cue is valid (left), but sometimes it misdirects attention to a different location (right). (B) This ERP, collected over the right occipital cortex while a stimulus was presented in the left visual field, shows that directing attention to the correct location significantly enhances neural processing and detection when the stimulus appears, reflected as large enhancements in the P1 and N1 components, starting 70–100 ms after stimulus presentation.

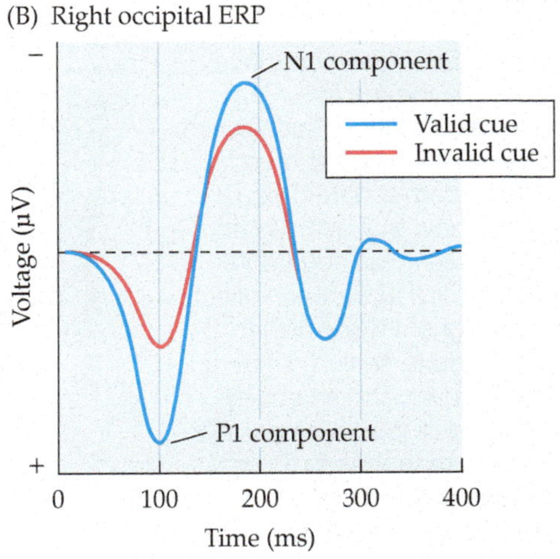

involves a large increase in the amplitude of P1 for attended stimuli, compared with ERPs for exactly the same stimulus presentation when attention is being directed elsewhere. The P1 effect is thought to reflect a crucial early-selection mechanism: an attentional bottleneck whereby attention selectively boosts the activity of occipital sensory processing (Mangun, 1995). A similar effect on P1 is evident when attention is instead oriented *reflexively* to a flash or sound (Hopfinger and Mangun, 1998; McDonald et al., 2005) but only when the interval between the cue and the appearance of the target is brief. At longer intervals, the P1 effect may actually be *reduced* as an electrophysiological manifestation of inhibition of return (McDonald et al., 1999), as discussed earlier. And for invalid trials in symbolic cuing experiments, where attention is being misdirected, away from the location in which the stimulus subsequently appears, the visual P1 effect isn't evident at all, even though the visual stimulus is identical and in the same location as in the validly cued trials. Interestingly, the P1 effect is only evident in manipulations of *spatial* attention (*where* is the target?)—not other features, like colour, orientation, or more complex combinations of features that would be characteristic of late-selection tasks.

What happens to ERPs during visual search tasks, where we are directing attention so as to find a particular target in an array and ignore distracters (see Figure 17.6)? Under these conditions, a subcomponent of N2 (see Figure 17.7), called *N2pc* (short for ***N2-p***osterior-***c***ontralateral) (Luck and Hillyard, 1994), is triggered at occipitotemporal sites contralateral to the visual target. In such studies, participants maintain fixation on a central point, and the stimulus array is presented only briefly, so the visual search requires rapid, covert shifts of attention to the selected target. But how do we know that N2pc reflects the selective processing of targets rather than the *suppression* of responses to distracters? An experiment that addressed this question (Hickey et al., 2009) is depicted in **FIGURE 17.9**. Here, participants maintained central fixation and were briefly presented with a very simple array, consisting of only a red line at one of several possible locations on the left or right side of the display, and a green square that was always on the midline (**FIGURE 17.9A**). On some trials, participants were instructed to attend to the red line and report on its length. The line was isoluminant to the background (i.e., had the same brightness), so attention was not reflexively captured by this stimulus. In other trials, participants were instructed to attend to the green square and *ignore* the red line. The important thing is the displays were the same in both conditions; only the instruction to the participant varied. Comparing the ERPs for the ipsilateral and contralateral hemispheres (**FIGURE 17.9B**) shows that N2pc was associated with covert orientation toward the red line when it was the target, but furthermore, when the participant was instead instructed to ignore the red line, a different ERP component emerged: a large positivity over the occipital region, now known as P_D (for distracter positivity), that is associated with ignoring distracters, even if they are quite attention grabbing (Gaspar and McDonald, 2014; Hickey et al., 2009; Courtesy of John McDonald).

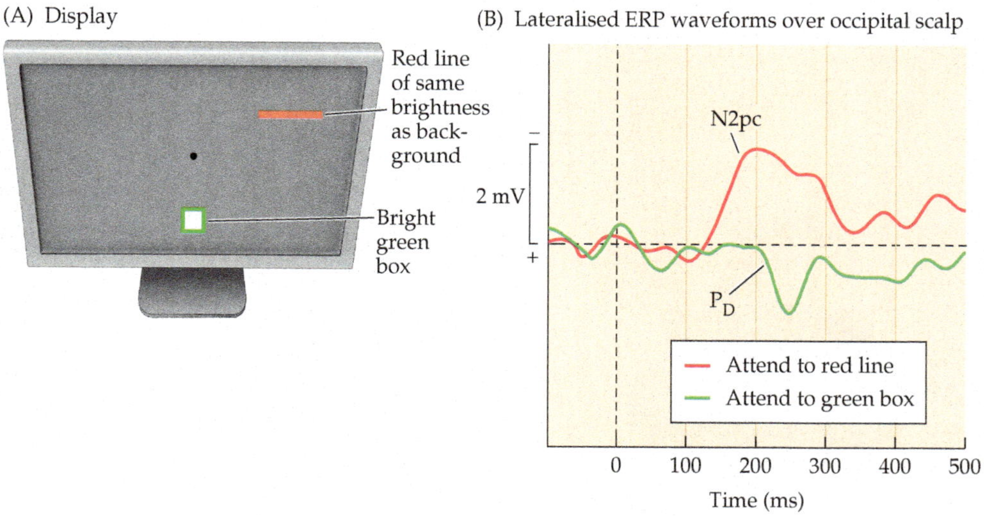

17.9 Distinctive Changes in Brain Activity in Visual Search In visual search tasks participants do not receive a cue to attend to a specific location; instead, they are told to find and attend to a specific target. (A) In this extremely simplified version of a visual search task, displays were flashed briefly on the screen while the participant fixated on the central dot, and the participant was told to either find the red line and ignore the green square, or find the green square and ignore the red line. (B) The ERP for the occipital region ipsilateral to the red line is subtracted from the ERP for the occipital region contralateral to the red line. Because it is on the midline, the green square doesn't affect these difference scores (it would affect both hemispheres' visual areas). The results show that seeking and attending to a visual stimulus augments activity of contralateral visual areas, reflected in the ERP component N2pc. When attention is being used to seek a different stimulus (the green box in this example), the exact same visual display not only fails to trigger an N2pc in response to the red line, but in fact causes a distinctive ERP component (P_D) associated with ignoring the red line. (Based on C. Hickey et al., 2009. *J Cog Neurosci* 21: 760–775.)

The neural mechanisms of visual attention appear to be quite plastic. For example, extensive experience with action video games, which rely heavily on visual attention, is associated with neural changes (S. Tanaka et al., 2013; G. L. West et al., 2015) and lasting enhancements of longer-latency ERP components (J. Mishra et al., 2011; Palaus et al., 2017). Possible trade-offs for all this gaming, however, may include impairments in social and emotional processing (no kidding: K. Bailey and West, 2013). And of course, some people could be drawn to video gaming simply because they are already good at visuospatial processing (Boot et al., 2008). Impressively, training with video games even enhances cognitive control—the ability to multitask—in older adults (60- to 85-year-olds), indicating that plasticity in attentional control systems is probably retained throughout life (Anguera et al., 2013).

Neuroimaging confirms that the anatomical foci of attention show augmented processing

Trading temporal resolution for spatial resolution, PET and fMRI studies have confirmed that a general consequence of attention is an enhancement of activity of the brain regions that process key aspects of the target stimuli. (PET, fMRI, and other brain imaging technologies are discussed in detail in Chapters 1 and 2.) In one such study, participants completed a simple spatial cuing task while both fMRI and ERP data were collected (McMains and Somers, 2004). By using the ERP signal to decide *when* to collect fMRI images, the researchers found anatomically distinct regions of enhanced activity in visual cortex that corresponded to the spatial locations to which participants directed their covert attention (**FIGURE 17.10**). As participants shifted attention to other locations, the locus of enhanced activity within the retinotopic map of primary visual cortex also shifted (see Chapter 8). And dividing attention between two spatial locations resulted in two corresponding patches of neural enhancement within visual cortex.

An effect of attention on neural activation may even be observed in *subcortical* visual structures (K. A. Schneider and Kastner, 2009). Using a sustained-attention task, in which participants *covertly* tracked ongoing changes in a rotating spatial array of dots, the researchers found fMRI enhancements in the superior colliculi and lateral geniculate nuclei (see Figure 8.11) that were specifically related to attentional enhancement of neural activity in these structures, unrelated to the position of stimuli on the retinas. This finding indicates that some degree of attentional selection may be occurring very early indeed, before visual information has even reached the cortex, in both subcortical visual pathways.

At later stages of the visual system, physiological effects of attention become even more pronounced. In one clever study, researchers devised stimuli composed of faces overlaid on pictures of houses (O'Craven et al., 1999). When participants were asked to ignore the houses and direct attention to the faces, attention-related enhancement was seen in a cortical region specifically responsible for face processing, called the *fusiform face area* (see Figure 18.6). When participants were instead instructed to attend to the houses and ignore the faces—using the same stimuli—fMRI enhancement was seen in a different cortical area, one that is responsible for processing locations (the parahippocampal place area). This neuroimaging evidence of late attentional selection illustrates once again that the attentional bottleneck can vary, depending on task demands (N. Lavie et al., 2004).

Attention alters the activity of individual neurons

As you'll recall from Chapter 8, visual neurons have distinctive receptive fields; visual stimuli falling within these fields can excite or inhibit

17.10 Effects of Attentional Spotlights on fMRI Activation in Visual Cortex
This fMRI image depicts activity in the primary visual cortex (V1) of a person working on two forms of a simple spatial cuing task (as in Figure 17.3). Because V1 has retinotopic organisation, locations in the fMRI image correspond to spatial locations in the participant's visual fields (see Chapter 8). Att1, single attentional target, heightened neural activity depicted in blue/white; Att2, attention split between two targets, heightened neural activity depicted in red/yellow; fovea, region of V1 cortex corresponding to central vison.

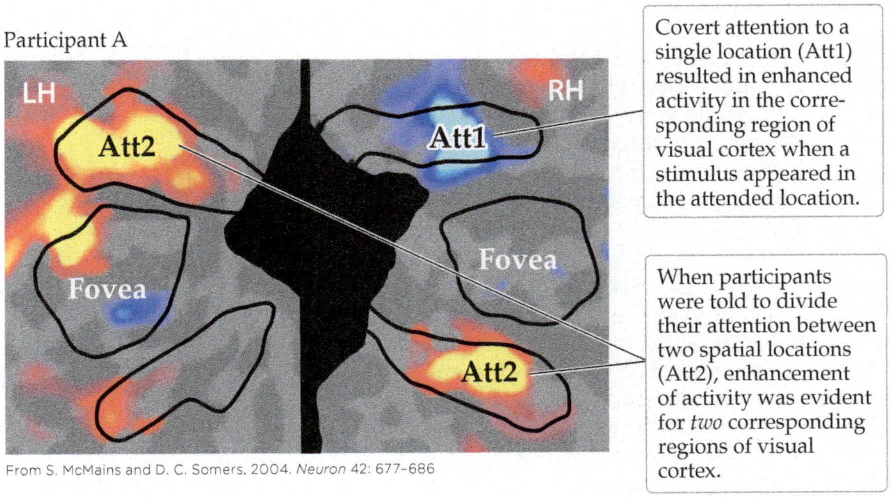

From S. McMains and D. C. Somers, 2004. *Neuron* 42: 677–686

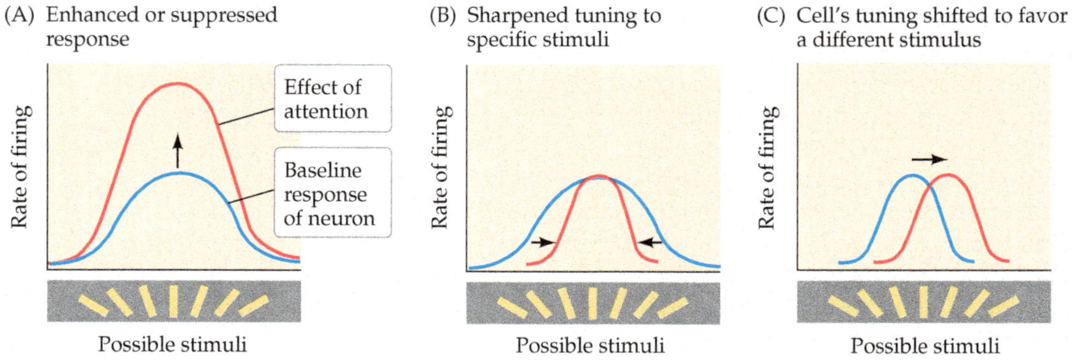

17.11 Some Ways That Attention Might Change the Activity of Individual Neurons

the neurons, causing them to produce more or fewer action potentials. Similarly, the effects of visual attention on single cells can be studied, by recording the activity of individual visual cortex neurons while attention within the visual field is manipulated. In principle, there are three ways in which attention might affect single-cell recordings. One possibility is that selective attention to a cell's receptive field might globally *enhance or suppress* responses, increasing or decreasing the firing rate of that cell (**FIGURE 17.11A**). Alternatively, attention might *sharpen the tuning* of cortical neurons, causing them to focus more keenly on specific stimuli (**FIGURE 17.11B**). Or attention might induce a *shift in tuning* of the cell, indicating a change in the cell's preferred stimulus (**FIGURE 17.11C**). Furthermore, directed attention reduces variability and improves the signal-to-noise ratio in visual cortex, in part by tuning how strongly individual synapses affect the activity of visual neurons (Briggs et al., 2013; Sprague et al., 2015).

In an important study (**FIGURE 17.12**), Moran and Desimone (1985) examined shifts in attention within the receptive fields of single neurons. Using a system of rewards, the researchers trained macaque monkeys to covertly attend to one spatial location or another (on the basis of symbolic cues) while recordings were made from

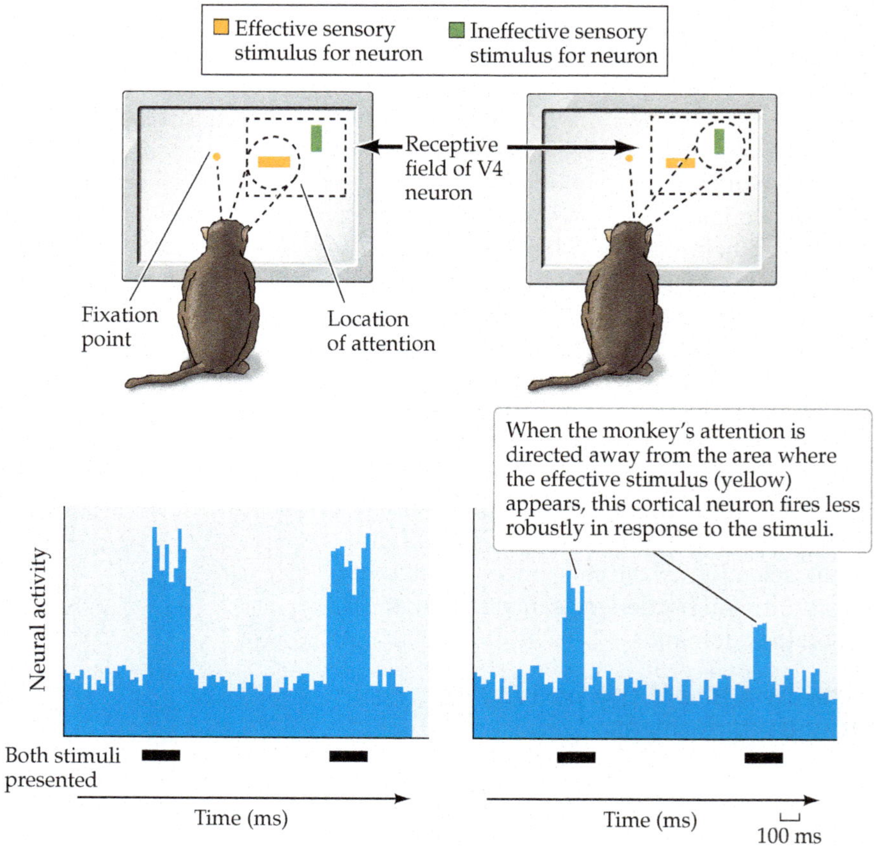

17.12 Effect of Selective Attention on the Activity of Single Visual Neurons A macaque monkey was trained to maintain central fixation while directing covert attention. Within the receptive field for the particular cortical neuron being recorded, the area being attended to is shown as a dashed circle. In the first condition (left) the area of attention within the cell's receptive field includes the effective stimulus (a yellow horizontal bar) and excludes the distracter (the green vertical bar that, by itself, has no effect on the cell's firing). The only difference in the second condition is that the monkey has moved its attentional spotlight away from the region where the effective stimulus will appear. (After J. Moran and R. Desimone, 1985. *Science* 229: 782–784.)

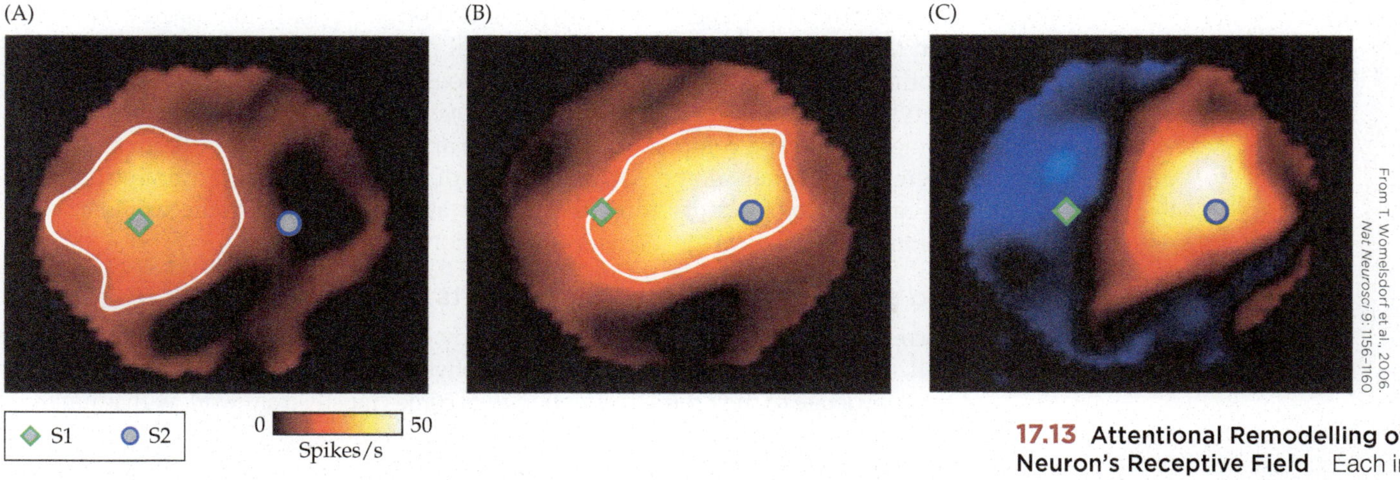

17.13 Attentional Remodelling of a Neuron's Receptive Field Each image is a map from the same single neuron in extrastriate visual cortex, showing sensitivity to its preferred stimulus (S) throughout its receptive field. In part A, a region of heightened sensitivity is evident while the animal attends to a location cued by S1 (green diamond). In part B, stimuli are identical, except the animal is attending to the position cued by S2 (blue circle). Part C is a difference map, showing the net change in receptive-field characteristics as attention shifted from position S1 to position S2. The cell's receptive field has evidently been retuned by attentional processes.

single extrastriate visual neurons. In one condition, a display was presented that included the cell's most preferred stimulus, as well as an ineffective stimulus (one that, by itself, did not affect the cell's firing) a short distance away but still within the cell's receptive field. Action potentials were counted while attention was covertly directed at the preferred stimulus, and as expected, a robust response was recorded. In the second condition, the same display was presented during recording from the same neuron, but this time the monkey was instructed to direct covert attention elsewhere within the neuron's receptive field. Because (1) the favoured stimulus was present, (2) the same cell was being recorded, and (3) the animal's gaze had not shifted, you might expect that the rate of firing remained robust, but that was not the result. Instead, rates of firing were significantly diminished in the second condition. Only the shift in attention could account for this change (consistent with the process depicted in Figure 17.11A).

Conscious, voluntary direction of visual attention seems to continually tune the receptive fields of visual cortical neurons, enhancing their resolution (Anton-Erxleben and Carrasco, 2013; Womelsdorf et al., 2006). By repeatedly presenting a preferred visual stimulus, researchers can map a cortical visual neuron's sensitivity throughout its receptive field, as shown in **FIGURE 17.13A**. The researchers found that when the monkey covertly shifted attention from one cued location (S1 in the figure) to a different cued location (S2) still within the cell's receptive field, without shifting its gaze, the peak sensitivity within the receptive field for this neuron shifted right along with attention (**FIGURE 17.13B**). Furthermore, attention can apparently cause the overall size of receptive fields to shrink, as if sharpening a spotlight to exclude neighbouring distracters (Womelsdorf et al., 2008), as depicted in Figure 17.11B. Studies such as these show that attention potently modifies the sensitivity of neurons throughout many of the vision-related areas of cortex.

17.3 Sources of Attention: A Network of Brain Sites Creates and Directs Attention

Learning Objectives

After studying this section, you should be able to:

17.3.1 **Discuss the functions of the principal subcortical sites—the superior colliculus and the pulvinar—associated with shifts of visual attention.**

17.3.2 **Summarise the dorsal frontoparietal network that is believed to govern voluntary attention, illustrating this action with examples of research.**

17.3.3 **Summarise the right temporoparietal network associated with reflexive shifts of attention, and again provide relevant research examples.**

17.3.4 **Describe some of the most striking forms of attentional disorders, what these disorders reveal about attentional control by the brain, and medical approaches to treat them.**

Whether attention comes reflexively, from the bottom up, or is controlled voluntarily, from the top down, it strongly affects neural processing in the brain, thereby augmenting electrophysiological activity. That doesn't mean that the *sources* of the various types of attention are identical, or even overlapping—just that their *consequences* are measurable and comparable. Probing the anatomical *sources* of attention requires different experimental approaches. Studies of lab animals, human brain imaging, and rare neurological disorders that impair attention all contribute to our understanding of the neural mechanisms of attention.

Two subcortical systems guide shifts in attention

Subcortical structures can be difficult to study because, situated in the centre of the brain, they are difficult to probe using electrophysiological or neuroimaging techniques. Our knowledge of their roles in attention thus comes mostly from work with animals.

SUPERIOR COLLICULUS Single-cell recordings have implicated the **superior colliculus**, a midbrain structure (**FIGURE 17.14**), in the movement of the eyes toward objects of attention, especially overt attention (Wurtz and Goldberg, 1972; Wurtz et al., 1982). Comparable eye movements made *without* the attentional component are not associated with increased neuronal activity in this region. A person with damage to one superior colliculus showed diminished inhibition of return for stimuli on the affected side (inhibition of return was described earlier, in the section on reflexive attention; see Figure 17.4). The superior colliculus may therefore help us avoid repeatedly returning to previously attended locations (Sapir et al., 1999). Although it is implicated primarily in overt attention, the superior colliculus may also participate in the direction of the covert attentional spotlight—a top-down task generally assigned to cortical mechanisms (as we'll discuss shortly). For example, reversible inactivation of the superior colliculus on one side abolishes monkeys' ability to utilize selective attention cues (arrows, flashes, etc., to indicate where to direct attention) on the corresponding side of visual space (Lovejoy and Krauzlis, 2010).

Go to Activity 17.1
Subcortical Sites Implicated in Visual Attention

superior colliculus A gray matter structure of the dorsal midbrain that receives visual information and is involved in direction of visual gaze and visual attention to intended stimuli.

pulvinar In humans, the posterior portion of the thalamus, heavily involved in visual processing and direction of attention.

THE PULVINAR Making up approximately the posterior quarter of the thalamus in humans (see Figure 17.14), the **pulvinar** is heavily involved in visual processing. This structure is organised retinotopically and shares extensive connections with the superior colliculus, parietal cortex, and cingulate. The pulvinar is important for the orienting and shifting of attention. In one study, direct injections of GABA agonists and antagonists were used to reversibly inhibit or activate the pulvinar (recall that GABA is an inhibitory transmitter, so its antagonists tend to increase neural activity). Following injection of a GABA agonist into the pulvinar, monkeys reportedly had great difficulty orienting covert attention toward targets in the contralateral visual field (D. L. Robinson and Petersen, 1992). Injections of a GABA antagonist, however, enhanced covert shifting of attention. The pulvinar is also needed to filter out and ignore distracting stimuli while monkeys are engaged in covert attention tasks, and in general it serves as an information manager, directing and synchronizing relevant information and activity in large-scale cortical networks, in response to attentional demands (Saalmann et al., 2012). In humans, attention tasks with larger numbers of distracters induce greater activation of the pulvinar on fMRI (M. S. Buchsbaum et al., 2006), showing that the pulvinar is important for attention in humans too.

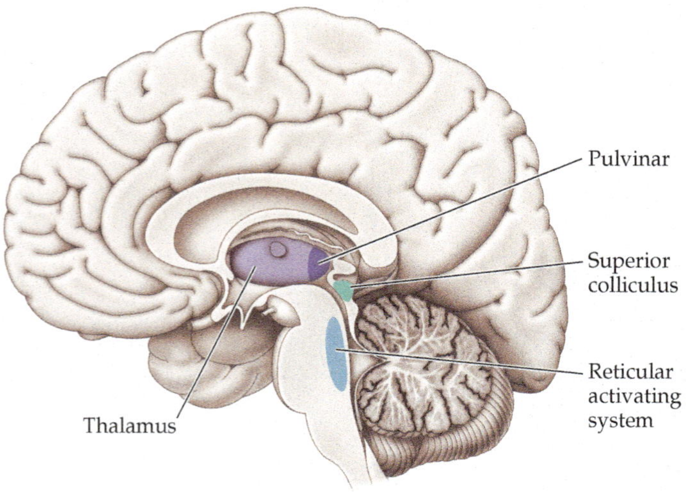

17.14 Subcortical Sites Implicated in Visual Attention
See text for details.

Several cortical areas are crucial for generating and directing attention

We've already seen several ways in which the *effects* of attention are evident in the cortex. But what about the *generation and direction* of attention? The extensive connections between subcortical attention systems and regions of the parietal lobes, along with clinical observations, suggest that parietal cortex plays a special role in controlling attention. In monkeys, single-cell recordings show that certain parietal neurons are selectively activated when the animal identifies a target stimulus to which it will covertly attend, but not activated in trials where the stimulus is in the exact same location but is not targeted for covert attention (Wurtz et al., 1982). In other words, these cells in parietal cortex behave as though they are preparing to direct a shift of attention.

To produce our seamless experience of the world, the cortical and subcortical mechanisms of attention must all cooperate. Evidence is mounting that two integrated networks—dorsal frontoparietal and right temporoparietal—work together to continually select and shift between objects of interest, in coordination with subcortical mechanisms of attention.

A DORSAL FRONTOPARIETAL NETWORK FOR VOLUNTARY CONTROL OF ATTENTION A region in the monkey parietal lobe called the **lateral intraparietal area** (**LIP**) is especially involved in the voluntary, top-down control of attention. LIP neurons fire when attention—not gaze—is directed to particular locations (Bisley and Goldberg, 2003; J. Gottlieb, 2007). By comparing the response of LIP cells to targets and distracters in varying spatial locations, researchers found that the LIP encodes a salience map (or priority map) that steers attention between important stimuli from one moment to the next (Bisley and Goldberg, 2006; Ipata et al., 2006). The human homologue of the LIP is a region called the **intraparietal sulcus** (**IPS**) (**FIGURE 17.15**) that behaves in much the same way. In covert attention tasks that are slow enough to allow for fMRI analysis, IPS activity is enhanced while participants are actively steering their attention (Corbetta and Shulman, 1998). Conversely, when researchers used transcranial magnetic stimulation (see Chapter 1) to noninvasively interfere with IPS activity, the participants found it difficult to voluntarily shift their attention between targets (G. Koch et al., 2005).

People with damage to a frontal lobe region called the **frontal eye field** (**FEF**) (see Figure 17.15) struggle to prevent their gaze from being drawn away toward peripheral distracters while they're performing a voluntary attention task (Paus et al., 1991). Neurons of the FEF appear to be crucial for ensuring that our gaze is directed among stimuli according to our cognitive goals rather than eye-catching characteristics of the stimuli. In effect, the FEF ensures that cognitively controlled top-down attention gets priority. It's no surprise, then, that the FEF is closely connected to the superior colliculus, which, as we discussed earlier, is important for planned eye movements. Furthermore, researchers using transcranial magnetic stimulation to temporarily disrupt activity of the FEF found that regions of visual cortex involved in processing features like colour and motion in target visual stimuli were likewise temporarily impeded (Zanto et al., 2011). This result confirmed, in humans, that top-down attentional systems (namely, in the FEF) act to bias neural activity in visual cortex.

A modified form of fMRI that links rapid behavioural events to changes in activity of selected brain regions (called, unsurprisingly, *event-related fMRI*, or *ER-fMRI*) enables us to visualise network activities during top-down attentional processing (Corbetta et al., 2000;

lateral intraparietal area (LIP) A region in the monkey parietal lobe, homologueous with the human intraparietal sulcus, that is especially involved in voluntary, top-down control of attention.

intraparietal sulcus (IPS) A region in the human parietal lobe, homologueous with the monkey lateral intraparietal area, that is especially involved in voluntary, top-down control of attention.

frontal eye field (FEF) An area in the frontal lobe of the brain containing neurons important for establishing gaze in accordance with cognitive goals (top-down processes) rather than with any characteristics of stimuli (bottom-up processes).

Go to Activity 17.2
Cortical Regions Implicated in the Top-Level Control of Attention

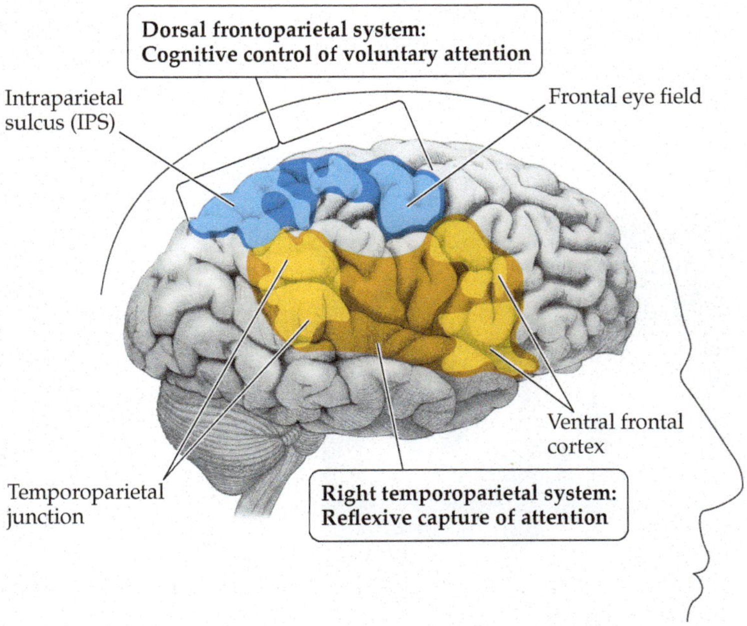

17.15 Cortical Regions Implicated in the Top-Level Control of Attention See text for details.

17.16 A Frontoparietal Attentional Control Network While participants performed a spatial cued-attention task, ER-fMRI images were collected, tied to either (A) the cue-processing phase of the trial (thus, the covert direction of attention) or (B) the target-processing phase of the task (corresponding to target processing and selection of motor responses). In part A, the conscious direction of attention is associated with selective activations of the dorsolateral frontal cortex, corresponding to the frontal eye field, the intraparietal sulcus, and the temporoparietal junction (see text for details).

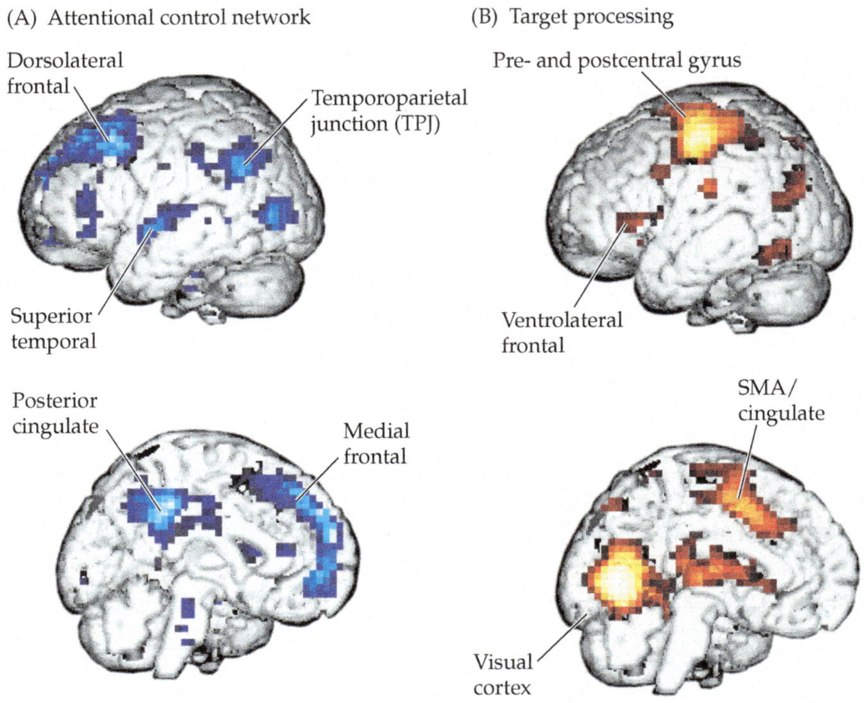

From J. B. Hopfinger et al., 2000. *Nat Neurosci* 3: 284–291

Hopfinger et al., 2000). **FIGURE 17.16A** shows patterns of activation while voluntary attention is shifting in response to a symbolic cue. Enhanced activity was evident in the vicinity of the frontal eye fields (dorsolateral frontal cortex) and, simultaneously, in the IPS. By comparison, during the target-processing phase of the trial (**FIGURE 17.16B**), activity was enhanced in regions around the pre- and postcentral gyrus (probably the beginning of a behavioural response) and in visual cortex (related to processing of the stimulus). Electrophysiological studies of the timing of activity in the network indicate that the attentional control–related activity is first seen in the frontal and parietal components, followed by anticipatory activation of visual cortex (if the expected stimulus is visual) or auditory cortex (if the stimulus is a sound) (J. J. Green et al., 2011; McDonald and Green, 2008). Research to date thus supports the view that a dorsal frontoparietal network provides voluntary (top-down) control of attention.

temporoparietal junction (TPJ) The point in the brain where the temporal and parietal lobes meet that plays a role in shifting attention to a new location after target onset.

A RIGHT TEMPOROPARIETAL NETWORK FOR REFLEXIVE SHIFTS OF ATTENTION A second cortical attention system, located at the border of the temporal and parietal lobes of the right hemisphere—and named, a little unimaginatively, the **temporoparietal junction** (**TPJ**) (see Figure 17.15)—is involved in reflexive (bottom-up) shifts in attention driven by the characteristics of stimuli (flashes, bangs, colour changes, and so on) rather than cognitive control. In contrast to the IPS/LIP system, the TPJ plays a role in shifting attention to a new location *after* target onset, especially if the stimulus is unexpected (**FIGURE 17.17**). In one study of the TPJ system, participants selectively attending to a specific location showed a significant increase in TPJ activity if a relevant stimulus suddenly appeared in a novel, unexpected location (Corbetta and Shulman, 2002). Likewise, people with TPJ damage respond poorly to the sudden appearance of unanticipated targets, failing to pivot their attention to the new stimuli (Friedrich et al., 1998). Researchers therefore liken the role of the TPJ to an alerting signal or "circuit

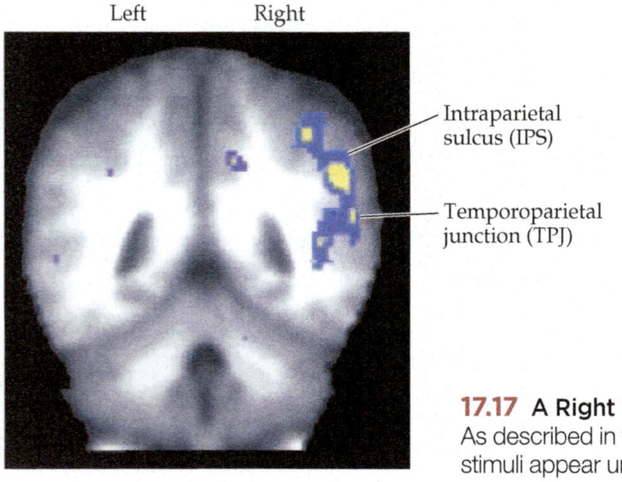

From M. Corbetta et al., 2000. *Nat Neurosci* 3: 292–297

17.17 A Right Temporoparietal Network for Reflexive Attention As described in the text, this pathway becomes more active when relevant stimuli appear unexpectedly (warmer colour corresponds to greater activity).

breaker": a system for overriding the current focus of attention if something new and unexpected (and potentially important) suddenly happens.

This system shows a strong right-hemisphere lateralisation. Activity associated with a reflexive shift of attention is localised to the right TPJ regardless of whether the stimulus itself occurs in the left or right side of the world (Corbetta et al., 2000), as we will see a little later in Figure 17.19. The ventral temporoparietal system receives inputs from visual cortex (conveying stimulus properties) and also from ventral frontal cortex; this region is implicated in working memory (see Chapter 16) and thus may provide crucial information about novelty by comparing the present state of stimuli with that of the recent past.

In order to smoothly integrate overall control of attention, the networks underlying voluntary attention and reflexive attention need to interact extensively and operate as a single coordinated system. According to one influential model (Corbetta and Shulman, 2002), voluntary control of attention depends on a dorsal stream of processing projecting from frontal cortex to the IPS region in posterior parietal cortex (**FIGURE 17.18**). Operation of the dorsal stream seems to have two general effects: attentional enhancement of visual processing, as we discussed earlier in the chapter (e.g., McMains and Somers, 2004), and coordination with the subcortical system (mediated by the pulvinar and superior colliculus) that steers attention around. At the same time, a more ventral system involving projections from visual cortex to the TPJ forms the basis of the reflexive attention network, scanning the environment for novel salient stimuli (which then draw reflexive attention), with the help of working memory inputs from the frontal cortex.

Go to Activity 17.3
The Cortical Attentional Control Network

Strong connections between the dorsal IPS system and the ventral TPJ system allow for rapid reassignment of attention, in line with our inner thoughts and goals and the huge range of stimuli that surround us. It is this coordinated activity of the IPS and TPJ systems that gives us the effortless awareness of environmental stimuli that we take for granted, as we shift rapidly between intended objects of attention and the other interesting things that pop up all around us.

Disorders provide clues about the organisation of attention

One way to learn about attention systems in the brain is to carefully analyse the behavioural consequences of damage or dysfunction of specific neuroanatomical systems. Research on people with attentional disorders shows that impairments of cortical or subcortical attention mechanisms can dramatically alter our ability to understand and interact with the world around us.

HEMISPATIAL NEGLECT One peculiar syndrome resulting from damage to the right inferior parietal cortex reinforces the notion that a special attentional mechanism resides in the right hemisphere, as we just discussed. The key feature of this disorder is a pervasive neglect of the left side of both the body and space (Rafal, 1994). People and objects to the left of the person's

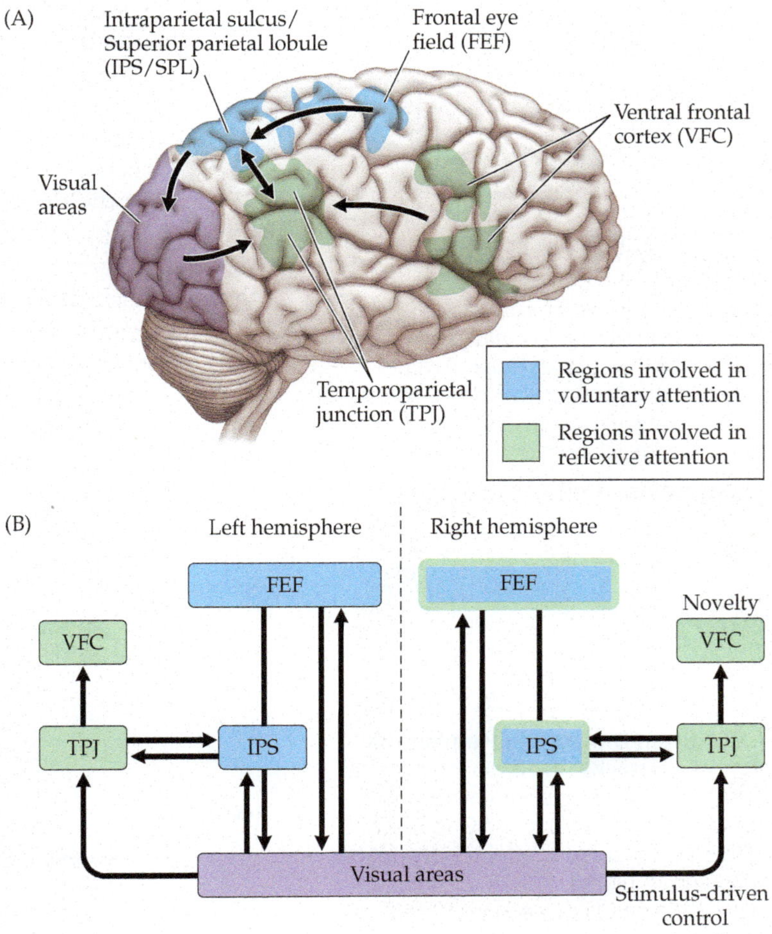

17.18 The Cortical Attentional Control Network
A) Regions shaded blue are involved in voluntary control of attention. Regions in green are part of a right-hemisphere system for reorienting and reflexive direction of attention. Arrows indicate the presumed direction of control. Extensive interaction between the systems provides integrated control over attention. (B) Schematic of the attentional control network. Connections between the TPJ and IPS provide a means for novel stimuli to interrupt and reorganise attentional priorities. (After M. Corbetta and G. L. Shulman, 2002. *Nat Rev Neurosci* 3: 201–215.)

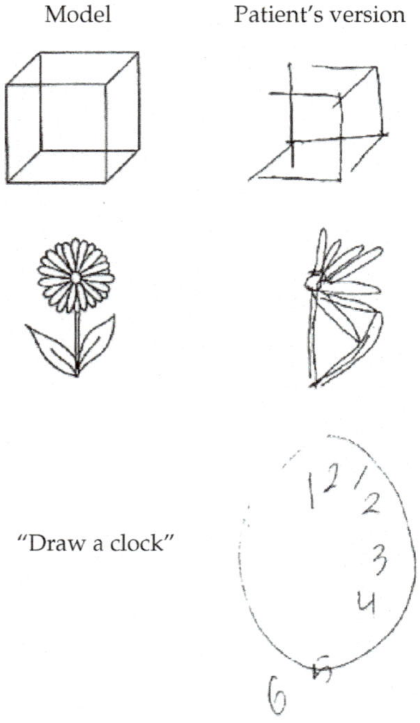

17.19 Diagnostic Test for Hemispatial Neglect When asked to duplicate drawings of common symmetrical objects, people with hemispatial neglect ignore the left side of the model that they're copying. (After A. Parton et al., 2004. *J Neurol Neurosurg Psychiatry* 75: 13–21.)

midline may be completely ignored, as if unseen, despite the absence of any visual-field defects. For example, people with this disorder may fail to dress the left side of their body, will not notice visitors if they approach from the left, and may fail to eat the food on the left side of their dinner plate. This phenomenon, called **hemispatial neglect**, can be revealed in simple test situations. A common test requires the person to copy drawings of familiar objects. In the typical result, a person with hemispatial neglect who is asked to draw the face of a clock draws numbers on only the right side of the clock face (**FIGURE 17.19**) (Schenkerberg et al., 1980). Spontaneous drawings, where the person is relying on memory rather than on a model, are similarly distorted. Curiously, people with hemispatial neglect even have difficulty with the "left side" of time: that is, they fail to attend to concepts that would appear on the left side of a provided timeline with arbitrary start, middle, and end points (Saj et al., 2014) and instead emphasise present and future events in the timeline.

Associated with hemispatial neglect is a feature called *extinction of simultaneous double stimulation*, or just **extinction**. Most healthy people have no trouble noticing when two stimuli are presented simultaneously on both sides of the body. People with right inferior parietal lesions, however, are unable to note the double nature of the stimulation and usually report only the stimulus presented to the right side. It is as if the normally balanced competition for attention between the two sides has become skewed and the input from the right side of the world now overrules or extinguishes the input from the left. This syndrome extends to visual imagery as well; for example, people with hemispatial neglect describe events from only one side of dream scenes, consistent with the observation that they make no leftward eye movements during REM sleep (Doricchi et al., 1991). Yet another striking feature of this syndrome is frequent **anosognosia** (denial of illness). People with hemispatial neglect may adamantly maintain that they are perfectly healthy and capable of engaging in their customary activities and do not recognise the impressive signs of unilateral neglect. They may even disclaim "ownership" of their own left arm or leg, rather than accepting that anything unusual is occurring.

Many people with right parietal damage do show recovery from hemispatial neglect, and researchers are working on rehabilitation therapies in which special prism glasses shift vision to the right during intense physical therapy, to recalibrate the visual attention system (A. M. Barrett et al., 2012; O'Shea et al., 2017). However, even with recovery from neglect, the symptom of simultaneous extinction can be quite persistent. Although neglect and extinction are most dramatically evident in the visual domain, similar problems are reported with stimuli in other sensory modalities (e.g., Funk et al., 2010; I. Schindler et al., 2006), indicating that a general attention control system has been compromised by the damage, with disconnections between high-level cognitive mechanisms (Bartolomeo, 2007; Mesulam, 1985). Lesion maps, averaged over numerous people with hemispatial neglect, fit with impressive precision over the model of the frontoparietal attention network we discussed earlier (**FIGURE 17.20**).

hemispatial neglect A syndrome in which the person fails to pay any attention to objects presented to one side of the body and may even deny connection with that side.

extinction Short for *extinction of simultaneous double stimulation*. An inability to recognise the double nature of stimuli presented simultaneously to both sides of the body. People experiencing extinction report the stimulus from only one side.

anosognosia Denial of illness.

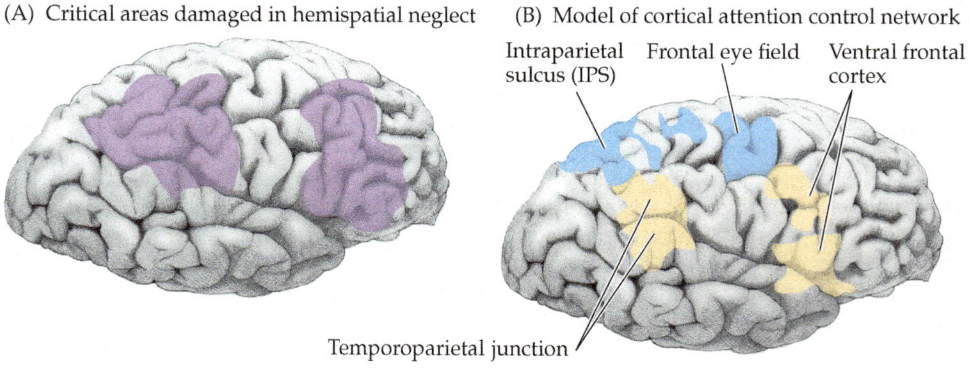

17.20 Brain Damage in Hemispatial Neglect Note that the brain regions that are damaged in people with hemispatial neglect (depicted in A) roughly correspond to regions implicated in attentional control processes (depicted in B), as discussed earlier in the chapter (see Figure 17.18).

BÁLINT'S SYNDROME As a result of her strokes, Parminder, whom we met at the beginning of the chapter, had bilateral lesions of her posterior parietal and lateral occipital cortex, the attention-related regions we have been discussing. While rare, this sort of bilateral damage can produce an unusual and debilitating condition called **Bálint's syndrome** (after its discoverer, Hungarian neurologist Rezsö Bálint). There are three principal symptoms:

1. **Oculomotor apraxia**, a pronounced difficulty in voluntarily steering visual gaze toward specific targets
2. **Optic ataxia**, a spatial disorientation in which the person is unable to accurately reach for objects using visual guidance
3. **Simultagnosia**, a profound restriction of attention, often limited to a single item or feature

The cardinal symptom in Bálint's syndrome, simultagnosia, is like a dramatic narrowing of the metaphorical attentional spotlight described earlier, to the point that it can't encompass more than one object at a time. Parminder's attention had become so narrow that most of the world around her was being excluded from processing; despite her normal vision, only one visual object at a time could make it into her conscious awareness. Bálint's syndrome thus illustrates the coordination of attention and awareness with mechanisms that orient us within our environment.

Bálint's syndrome Three co-occurring symptoms—simultagnosia, oculomotor apraxia, and optic ataxia—that may occur after bilateral lesions of cortical attentional systems.

oculomotor apraxia A severe difficulty in voluntarily steering visual gaze toward specific targets.

optic ataxia A spatial disorientation in which the affected person is unable to accurately reach for objects using visual guidance.

simultagnosia A profound restriction of attention, often limited to a single item or feature.

attention deficit hyperactivity disorder (ADHD) Syndrome of distractibility, impulsiveness, and hyperactivity that, in children, interferes with school performance.

ATTENTION DEFICIT HYPERACTIVITY DISORDER At least 5% of all children in the United States are diagnosed with **attention deficit hyperactivity disorder (ADHD)**, characterized as difficulty with directing sustained attention to a task or activity, along with a higher degree of impulsivity than other children of the same age. Roughly 75% of affected children are male. Diagnosis of ADHD is complicated and very controversial; for example, research on ADHD prevalence has revealed that there are significant differences across the United States in both the rate of diagnosis of ADHD and the use of medication to treat it—with large differences observed even between neighbouring states (Visser et al., 2014) (**FIGURE 17.21**)—raising questions about the reliability of current diagnostic practices (Fulton et al., 2009). Nevertheless, researchers have identified several neurological changes that are associated with this disorder. Affected children tend to have slightly reduced overall brain volumes (about 3–4% smaller than in unaffected children), with reductions most evident in the cerebellum and in the frontal lobes (Arnsten, 2006; Castellanos et al., 2002). As we've discussed before, frontal lobe function is important for myriad complex cognitive processes, including the working memory and inhibition of impulsive behaviour. (But remember, correlational studies like these are mute with regard to causation; we don't know whether the brain differences cause, or are caused by, the behaviour.)

In addition to structural changes, ADHD has been associated with abnormalities in connectivity between brain regions, such as within the default mode network that we will discuss in the next section (Cao et al., 2014). In fact, individual differences in the ability to sustain attention can be predicted with high accuracy from the strength of sets of brain connections (Rosenberg et al., 2016, 2017), even in the resting state, when the individual is not

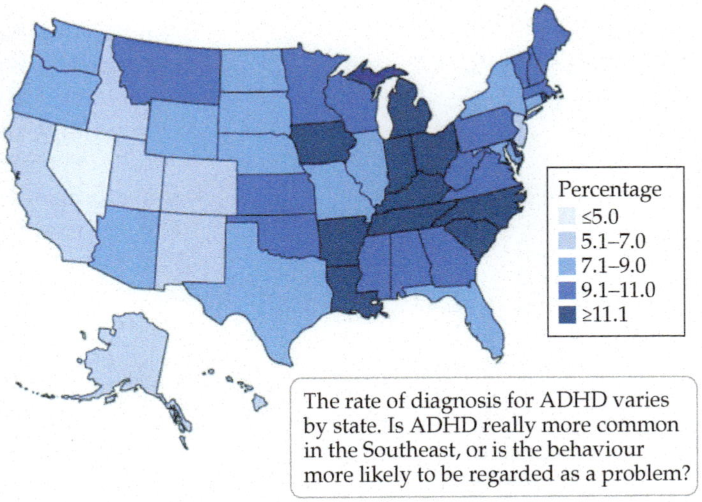

17.21 Regional Variation in ADHD Prevalence in the USA This map of ADHD diagnosis patterns reveals substantial differences across the United States: in some locations the reported prevalence is about 100% higher on one side of the state line than on the other side. This pattern suggests that the diagnosis of ADHD suffers from imprecision and unclear definition, possibly coupled with regional variation in sociocultural processes (such as expectations for children's behaviour) and public policy. The fact that medication patterns closely follow the regional diagnosis map is cause for concern, as it implies that children may be overmedicated in some regions (or undermedicated in others). (After S. N. Visser et al., 2014. *J Am Acad Child Adolesc Psychiatry* 53: 34–46.)

working on any particular task. This measure may provide both a physical marker for ADHD diagnosis and a new opportunity to understand the neural origins of ADHD. In addition to changes in connectivity, children with ADHD may have abnormal activity levels in some specific brain systems, such as the system that signals the rewarding aspects of activities (Volkow et al., 2009). Based on a model of ADHD that implicates impaired dopaminergic and noradrenergic neurotransmission, some researchers advocate treating children with ADHD with stimulant drugs like methylphenidate (Ritalin), which inhibits the synaptic reuptake of dopamine and norepinephrine, or with selective norepinephrine reuptake inhibitors like atomexetine (Strattera) (S. Schwartz and Correll, 2014). Stimulant treatment often improves the focus and performance of children with ADHD in traditional school settings, but this treatment remains controversial because of significant risk of side effects. Furthermore, stimulants improve focus in everybody, not just people with ADHD, raising doubts about the orthodox view that impaired neurotransmission is the sole cause of ADHD (del Campo et al., 2013). An emerging alternative view is that what is diagnosed as ADHD may simply be an extreme on a continuum of normal behaviour. Allowing kids diagnosed with ADHD to engage in intense physical activity can effectively reduce their symptoms and improve task performance (Den Heijer et al., 2017; Hartanto et al., 2016).

CONSCIOUSNESS AND EXECUTIVE FUNCTION

17.4 Consciousness Is a Mysterious Product of the Brain

Learning Objectives

After studying this section, you should be able to:

17.4.1 **Provide a reasonable definition of consciousness, and name the neural networks and structures that, when activated, may play special roles in coordinating conscious states.**

17.4.2 **Discuss the relationship between consciousness as experienced by healthy people, and the diminished levels of consciousness experienced by people in comas and minimally conscious states.**

17.4.3 **Discuss the impediments to the scientific study of consciousness, distinguishing between the "easy" and "hard" problems of consciousness, and speculate about the role of free will in understanding consciousness.**

17.4.4 **Define *executive function*, identify some of the key components of executive function, and describe some of the most prevalent tests used to assess it.**

17.4.5 **Provide an overview of the organisation and function of the frontal lobes, and especially prefrontal cortex, as they relate to high-level cognition and executive functions.**

There can be no denying the close relationship between attention and consciousness; indeed, attention is the foundation on which consciousness is built. Whenever we are conscious, we are attending to *something*: other people, our innermost thoughts, the road ahead, music, and all the other important stimuli that make up our conscious experience of life. William James (1890) tried to capture the relationship of attention and consciousness when he wrote, "My experience is what I agree to attend to. Only those items which I notice, shape my mind—without selective interest, experience is an utter chaos." But while both attention and awareness are behavioural and experiential manifestations of consciousness, no one would accept them as synonymous. We've seen that attention to complex stimuli enhances activity in primary visual cortex (V1), but conscious awareness of the stimuli doesn't add any extra V1 activation (M. Watanabe et al., 2011). It seems that consciousness must inhabit a higher, heavily networked neural neighborhood. A more complete description of consciousness is freighted with the notion that we enjoy volitional control over experience—the belief that we can employ *free will* to direct our attention and ultimately make decisions about how to act. Add to this the observation that consciousness involves awareness of the passage of time, integration with memory of past events, and anticipation of future events, and we have a concept of immense scope. So as a rough approximation, we can define **consciousness** as the state of being *aware* that we are conscious and that we can perceive what is going on in our minds and all around us. This awareness is coloured by events that affected our inner selves in the past and by our sense of what the future might hold. Some of the basic elements of consciousness—and some of the animals that may have conscious experiences resembling our own—are Summarised in **TABLE 17.1**.

consciousness The state of awareness of one's own existence and experience.

default mode network The regions of the brain that are active when the brain is awake and at rest and attention is not being directed to external events.

Which brain regions are active when we are conscious?

Despite the complexities of developing formal scientific definitions of conscious states, consciousness is an active area of research with many competing theoretical models and a growing body of neuroscientific data. One approach is to look for patterns of synchronised activity in dispersed neural networks as people engage in conscious, inwardly focussed thought. Using fMRI, researchers have mapped a circuit of brain regions—collectively called the **default mode network**, consisting of parts of the frontal, temporal, and parietal lobes—that seems to be selectively activated when we are at our most introspective and reflective, and relatively deactivated during behaviour directed toward external goals (Andrews-Hanna et al., 2014; Raichle, 2015). In some ways, you could think of it as a daydream network (and by thinking about it, you are also engaging in metacognition; see Table 17.1). Researchers suspect that dysfunction of the default mode network may contribute to various high-level cognitive problems, including ADHD, dementia, schizophrenia, and autism, affecting both children and adults (Sato et al., 2015; Whitfield-Gabrieli and Ford, 2012). Monkeys and lab rats have circuits that seem to function the same way the human default mode network does, suggesting that nonhuman species may likewise engage in self-reflection or other introspective mental activity (Mantini et al., 2011; Sierakowiak et al., 2015).

A practical alternative approach has been to study consciousness by focussing on people who lack it—people in comas and other states of reduced consciousness. Such investigation requires great care in identifying disorders or states with effects limited to changes in consciousness; for example, damage to the brainstem can cause the entire cerebrum to shut down, but this is a global loss of brain function, not a selective impairment of consciousness. Using fMRI to track brain activity in conditions of diminished consciousness—ranging from sleep to persistent

TABLE 17.1 Dimensions of Consciousness

Aspect of consciousness	Definition	Species other than humans
Theory of mind	Insight into the mental lives of others; appreciation that beliefs, knowledge, and desires of others differ	Only chimpanzees, so far
Mirror recognition	Ability to recognise self as depicted in a mirror	All great apes; dolphins; elephants; pigs; magpies
Imitation	Ability to copy the actions of others; thought to be a stepping-stone to awareness and empathy	Many species, including cephalopods like the octopus
Empathy and emotion	Possession of complex emotions and the ability to imagine the feelings of other individuals	Most mammals, ranging from primates and dolphins, to hippos and rodents; most vertebrates able to experience pleasure (and other basic emotions)
Tool use	Ability to employ found objects to achieve ultimate goals	Chimps and other primates; other mammals such as elephants, otters, and dolphins; birds such as crows and gulls
Language	Use of a system of arbitrary symbols, associated with specific meanings and assembled according to a strict grammar, to convey information that may be concrete or abstract, on familiar or unfamiliar topics, to any other individual that has learned the same language	Generally considered to be an exclusively human ability, with controversy over the extent to which the great apes and parrots possess language skills (see Chapter 18)
Metacognition	"Thinking about thinking": the ability to consider the contents of one's own thoughts and cognitions	Nonhuman primates; dolphins

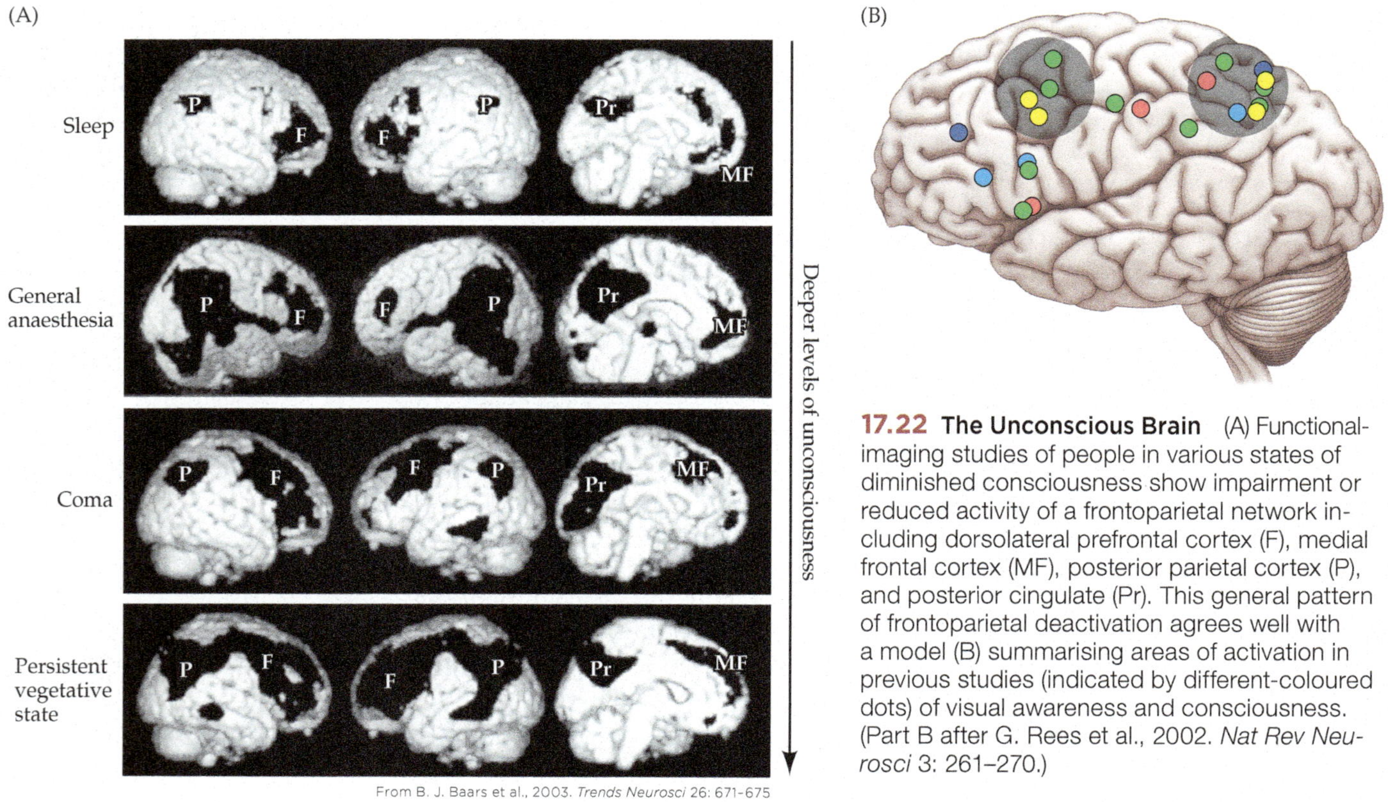

17.22 The Unconscious Brain (A) Functional-imaging studies of people in various states of diminished consciousness show impairment or reduced activity of a frontoparietal network including dorsolateral prefrontal cortex (F), medial frontal cortex (MF), posterior parietal cortex (P), and posterior cingulate (Pr). This general pattern of frontoparietal deactivation agrees well with a model (B) summarising areas of activation in previous studies (indicated by different-coloured dots) of visual awareness and consciousness. (Part B after G. Rees et al., 2002. *Nat Rev Neurosci* 3: 261–270.)

claustrum A thin sheet of neurons, situated within the white matter lateral to the basal ganglia, that has been implicated in conscious awareness.

vegetative states due to head injuries—researchers have devised maps of cortical areas that appear to be deactivated in unconsciousness (Tsuchiya and Adolphs, 2007). The overlap in these maps suggests shared reliance on a frontoparietal network (**FIGURE 17.22**) that includes much of the attention network we have been discussing, as well as portions of medial frontal cortex and the cingulate. Some researchers suspect that the **claustrum** (**FIGURE 17.23**)—a slender sheet of neurons buried within the white matter of the forebrain lateral to the basal ganglia—plays a critical role in generating the experience of being conscious (S. P. Brown et al., 2017; Crick and Koch, 2005), by virtue of its remarkable reciprocal connections with virtually every area of cortex. In one case, a woman with a stimulating electrode in the claustrum experienced a "switching off" of conscious awareness whenever a strong stimulation pulse was delivered through the electrode (Koubeissi et al., 2014); in another case, consciousness was selectively interrupted by bilateral electrical stimulation in lateral frontal cortex (Quraishi et al., 2017), disrupting activity in complex networks associated with consciousness.

But is clinical unconsciousness really the inverse of consciousness? Perhaps it's not that simple. In several cases, people in apparently deep and unresponsive comas were successfully instructed to use two different forms of mental imagery to create distinct "yes" and "no" patterns of activity on fMRI, and then to use this mental activity to answer questions (**FIGURE 17.24**) (Fernández-Espejo and Owen, 2013; Monti et al., 2010). Integrity of top-down feedback connections, from frontoparietal

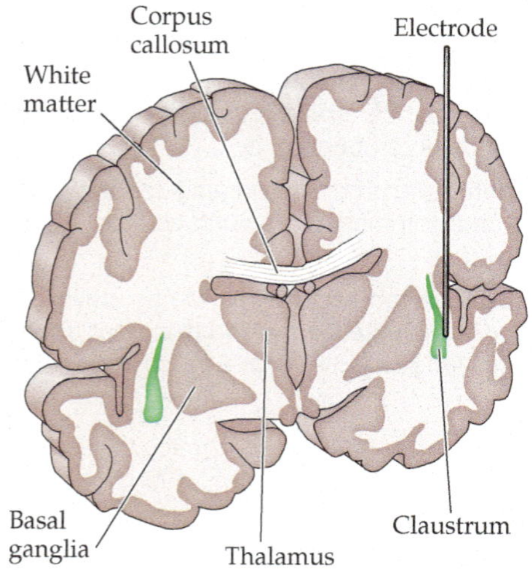

17.23 The Seat of Consciousness? Some researchers have implicated the enigmatic gray matter band called the *claustrum* in the experience of conscious awareness. In one case study, a woman's conscious awareness was instantaneously switched off when a current was passed through an electrode planted in the region of the claustrum, and instantly restored when the electrode was turned off. (After M. Z. Koubeissi et al., 2014. *Epilepsy Behav* 37: 32–35.)

17.24 Communication in "Unconscious" People Functional MRI scans reveal distinctive patterns of brain activity (brighter red or blue corresponds to greater activity) obtained when a control participant (top) was asked to use two different mental images (playing tennis versus navigating) to signal yes or no answers to questions. The striking similarity of activation in the brain of a person in a persistent vegetative state (bottom) who received the same instructions raises questions about the definition of unconsciousness. This individual was able to correctly answer a variety of questions using this technique, despite apparently deep unconsciousness due to profound brain damage (black regions on MRI scan) and lack of behavioural responses.

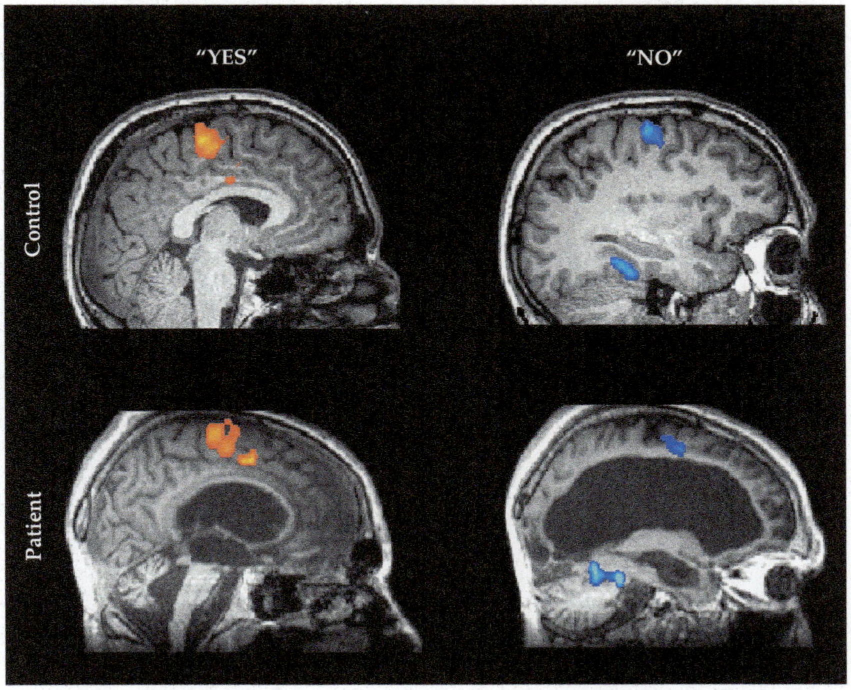

cortex to cortical sensory regions, seems to be crucial for even minimal levels of consciousness (Boly et al., 2011, 2012). Most people in persistent deep coma—termed a *vegetative state*—don't respond to questions and outside stimulation, but others in minimally conscious states may have considerable cognitive activity and awareness with little or no overt behaviour (Gosseries et al., 2014). So it's increasingly clear that we can't simply view coma as an exact inverse of what we experience as consciousness. But in any case, there seems to be more to consciousness than just being awake, aware, and attending. How can we identify and study the additional dimensions of consciousness?

Some aspects of consciousness are easier to study than others

Most of the activity of the central nervous system is unconscious. In scientific language, these functions are said to be **cognitively impenetrable**: they involve basic neural processing operations that we cannot experience through introspection. We don't have conscious access to the subordinate components of perceptions—we see whole objects and can't even imagine what the primitive visual precursors of those perceptions could look like. Sweet food tastes sweet, and we can't mentally break it down any further than that. But those simpler mechanisms, operating below the surface of awareness, are the foundation that conscious experiences are built on. So in principle, we might someday develop technology that would let us directly reconstruct people's conscious experience—read their minds—by amalgamating lower-order brain activity into identifiable patterns.

Philosophers refer to this as the **easy problem of consciousness**: understanding how particular patterns of neural activity create *specific* conscious experiences. Of course, as a practical matter this undertaking is far from "easy," but at least we can reasonably say that some future brain-imaging technology may have enough speed and resolution to eavesdrop on large networks of neurons, in real time (**FIGURE 17.25A**). In fact, present-day technology offers a glimpse of that possible future. For example, if participants are repeatedly scanned while viewing several distinctive scenes, a computer can eventually learn to identify which of the scenes the participant is viewing on each trial, solely on the basis of the pattern of brain activation (Kay et al., 2008). Of course, this outcome relies on having the participants repeatedly view the same static images—hardly a normal state of consciousness. A more difficult problem is so-called constraint-free reconstruction of conscious experiences—that is, the ability to "read" what a person is perceiving rather than just identifying items on the basis of previous training. This sort of reconstruction has been accomplished for very simple stimuli, like letters and shapes (**FIGURE 17.25B**) (Miyawaki et al., 2008). We'll return to this topic in The Cutting Edge at the end of this chapter, but suffice it to say, we are still a very long way from directly reconstructing anything like a rich conscious experience. Still, at least it's conceivable.

cognitively impenetrable Referring to data-processing operations of the central nervous system that are unconscious.

easy problem of consciousness The problem of how to read current conscious experiences directly from people's brains as they're happening.

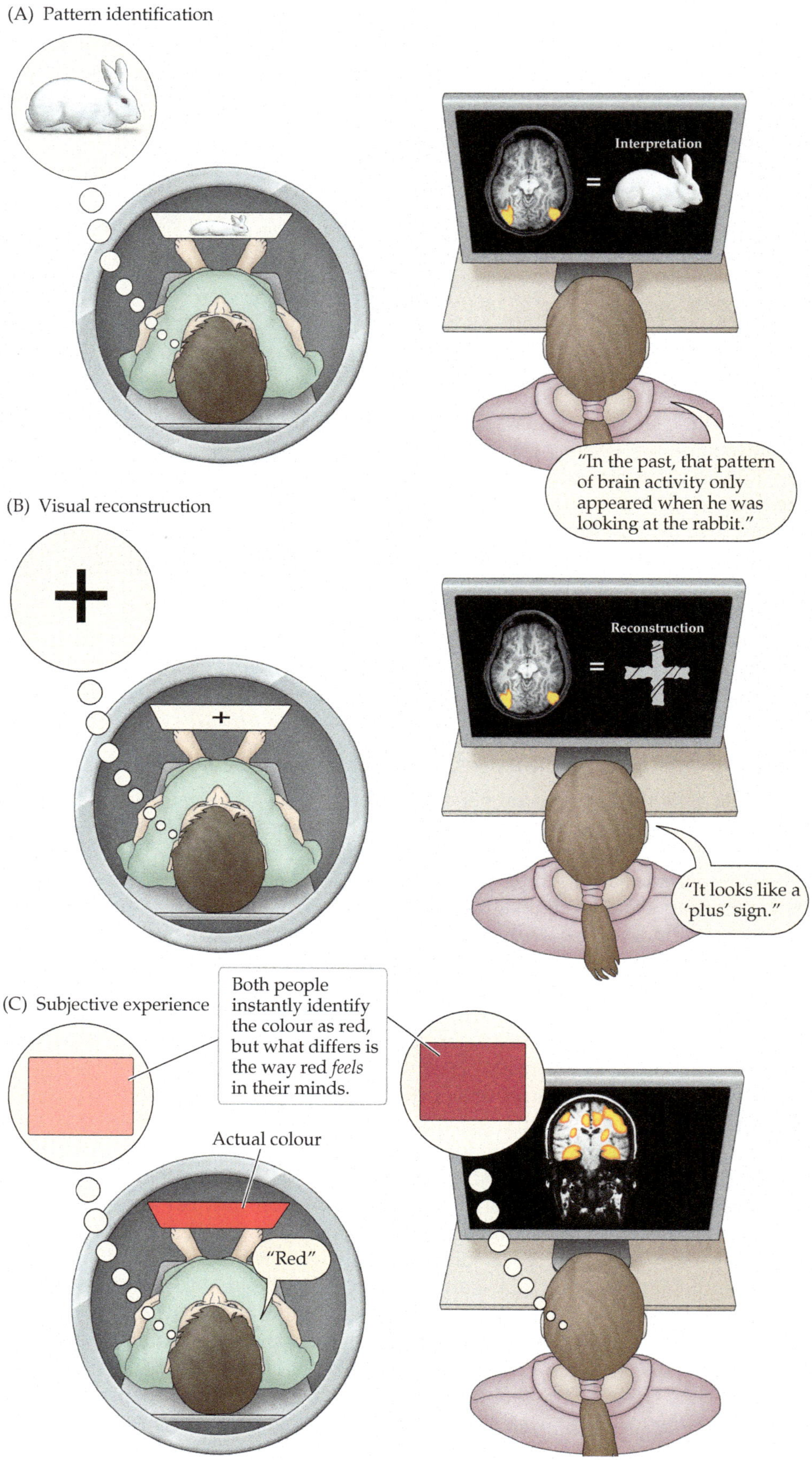

17.25 Easy and Hard Problems of Consciousness (A) When a study participant is looking at one of 20 photographs, slightly different patterns of brain activity are elicited by different pictures. A computer can learn the patterns and eventually identify the scene being viewed. (B) In fact, vision scientists know enough about how different parts of the brain are activated by light striking the retina that they can even predict what sort of simple shape a person is viewing. (C) The "hard" problem goes beyond predicting what a person is seeing or thinking to knowing what that person's subjective experience is like. So, although the participant and the researcher would both use the label "red" to describe the colour being viewed, how can we know their personal subjective experiences of that stimulus? It is difficult to see how we can ever be sure that we share any subjective experiences of consciousness. (Part A after K. N. Kay et al., 2008. *Nature* 452: 352–355; B after Y. Miyawaki et al., 2008. *Neuron* 60: 915–929.)

Unsurprisingly, there is also the **hard problem of consciousness**: understanding the brain processes that result in a person's *subjective* experience of the contents of consciousness. To use a simple example, every person with normal colour vision will identify a ripe tomato as being red because as children we all learned that a particular signal entering our consciousness from the colour regions of the visual system is described as "red." We learn to relate labels to patterns of neural activity. But that doesn't mean that your friend's internal *personal* experience of "red" is the same as yours. These purely subjective experiences of perceptions are referred to as **qualia** (singular *quale*). Because they are subjective and impossible to communicate to others—how can your friends know if redness feels the same in their minds as it does in yours?—qualia may prove impossible to study (**FIGURE 17.25C**). At this point anyway, we are unable to even conceive of a technology that would make it possible.

hard problem of consciousness The problem of how to read people's subjective experience of consciousness and determine the qualia that accompany perception.

qualia Singular *quale*. Purely subjective experiences of perception.

Our subjective experience of consciousness is closely tied up with the notion of free will, or what are sometimes called *feelings of agency*: the perception (real or imagined) that our conscious self is the author of our actions and decisions. Aside from the issue of whether we actually have free will—a matter that won't be resolved anytime soon—there must be a neural substrate for the universal *feeling* of having free will. Conscious manipulations of intentions to act—an exercise in willful control of actions—result in selective fMRI activations of the pre–supplementary motor area (pre-SMA) in the frontal lobes of healthy participants, along with activations of the IPS (which we implicated earlier in top-down attention) and dorsal prefrontal cortex (Lau et al., 2004). However, the *conscious experience* of intention may come relatively late in the process of deciding what to do. Early research (Libet, 1985), using EEG and a precise timer, found that an EEG component associated with movement preparation was evident 200 ms before participants consciously decided to move. Although controversy initially surrounded this work, later confirmatory research using fMRI (**FIGURE 17.26)** found that, astonishingly, brain activity associated with making a decision was evident as much as 5–10 seconds before participants were consciously

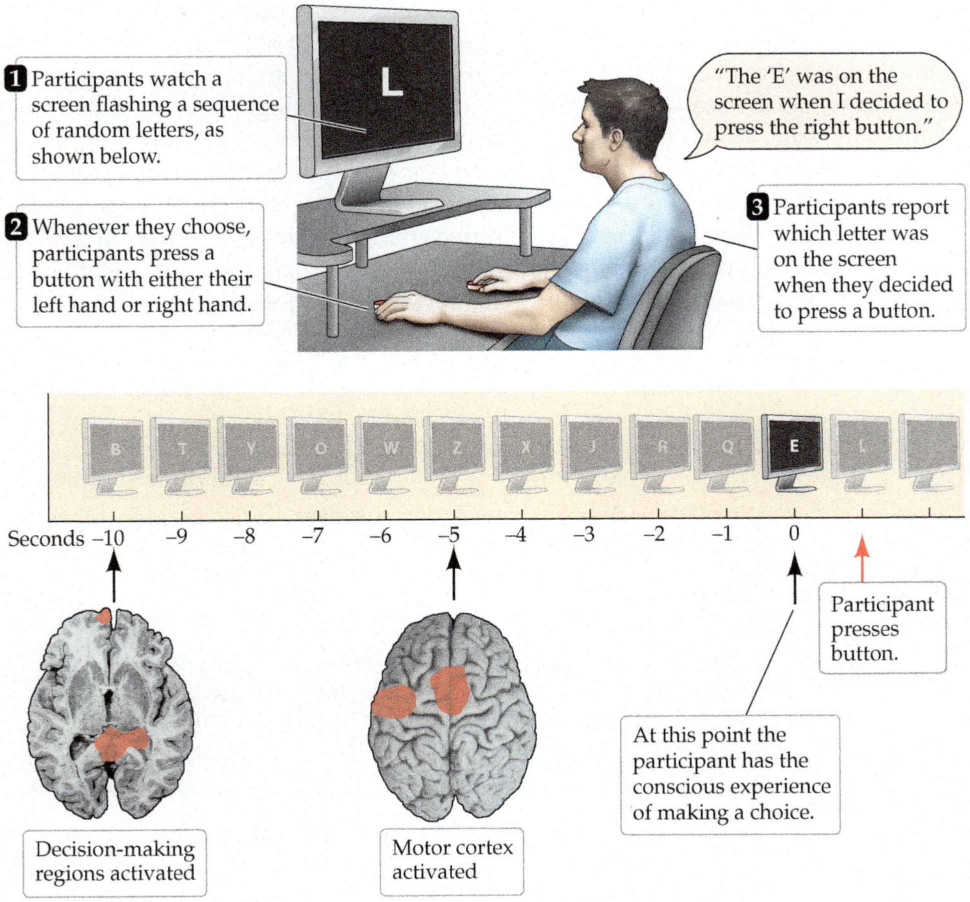

17.26 Act First, Think Later Our brains may decide what we'll do before our conscious selves are aware of that decision. In this experiment, participants decided for themselves when to press a button with either the right hand or the left. Using brain fMRI, the scientists found that they could predict beforehand when participants would "decide" to press the left or right button, and even whether they would press the left or the right button, up to 10 seconds before participants were aware of their own decisions. (After C. S. Soon et al., 2008. *Nat Neurosci* 11: 543–545.)

aware of making a choice (Soon et al., 2008). The earliest indications of this decision-making process are found in frontal cortex. Such involvement of prefrontal systems in most aspects of attention and consciousness, regardless of sensory modality or emotional tone, suggests that the frontal cortex is the main source of goal-driven behaviours (Badre and Nee, 2018; E. K. Miller and Cohen, 2001), as we'll discuss next.

A flexible anterior system develops and supervises the cognitive plans that guide our behaviour

How do we decide what to do, and when to do it? Careful analysis of the behavioural impairments evident in people with localised brain damage, along with functional-imaging studies in healthy people, shows that a network of anterior forebrain sites dominated by the frontal lobes—but including several other cortical and subcortical sites—is crucial for **executive function**, the suite of high-level cognitive processes that control and organise lower-level cognitive functions in line with our thoughts and feelings (Alvarez and Emory, 2006; P. Yuan and Raz, 2014). Some scientists liken executive function to a "supervisory system" that analyses important stimuli, weighs competing ideas and hypotheses, and governs the creation of suitable "plans" for future action by drawing on cognitive processes like working memory, attention, feedback utilisation, and so on. Executive function requires smooth *shifting* between operations, *updating* of the cognitive plan based on new information and the contents of working memory, and *inhibition* of responses that would compromise the plan (Diamond, 2013; Miyake et al., 2000). A closely related account proposes that the crucial function of a network of sites within the frontal lobes is *hierarchical cognitive control*: the ability to direct shorter-term actions while simultaneously keeping longer-term goals in mind (Badre and Nee, 2018; Koechlin et al., 2003). Accordingly, a person with executive dysfunction due to frontal lesions may be unable to complete a simple set of errands without numerous false starts, backtracking, and confusion (Shallice and Burgess, 1991). Several of the most widely studied tests of executive function are listed in **TABLE 17.2**.

We have discussed some key frontal lobe functions in earlier chapters—things like movement planning, memory, language, and psychopathology. Perhaps it reflects a bit of vanity about our species, but the large size of our frontal lobes—about one-third of the entire cortical surface—also led to the long-standing view that the frontal cortex is the seat of intelligence and abstract thinking. The remarkable story of Phineas Gage (**BOX 17.2**), one of the most famous case studies in the history of neurology, underscores the subtlety and complexity of behaviours governed by the frontal lobes. Like Gage, people with discrete prefrontal lesions express various unusual emotional, motor, and cognitive changes. Widespread frontal damage may be associated with a persistent strange apathy, broken by bouts of euphoria (an exalted sense of well-being). Ordinary social conventions are readily cast aside

executive function A neural and cognitive system that helps develop plans of action and organises the activities of other high-level processing systems.

TABLE 17.2 Tests of Executive Function

Test name	Procedure	Scoring	Functions sampled
Wisconsin Card Sorting Test (WCST) (Weigl, 1941; Heaton, 1993)	Sort cards into piles on the basis of the number, colour, or shape of symbols on the card face. Every 10 cards, discover and shift to a new sorting rule (see Figure 17.28).	Errors in sorting; perseveration in old sorting rule after rule change	*Shifting*: Task switching and abstract reasoning
Controlled Oral Word Association Test (COWAT) (Benton and Hamsher, 1976)	Say as many words as possible that start with a specific letter (*F*, *A*, or *S*) in 60 seconds.	Total number of unique words uttered for all three starting letters	*Updating*: Verbal fluency and working memory
Stroop Colour-Word Interference Test (Stroop, 1935; MacLeod, 1991)	Read aloud as quickly as possible colour names that are printed in congruent colour (e.g., BLUE) or incongruent colour (e.g.,).	Time and total errors	*Inhibition:* Suppression of incorrect responses

BOX 17.2 Phineas Gage

Phineas P. Gage was a sober, efficient, capable young man, respected by the workmen he supervised as they blasted rock to clear a path for a new railroad in northwestern Vermont. But one day in 1848, something went wrong. Gage was using a steel tamping rod to tightly pack explosives into a hole that had been drilled into the solid rock. The tamping rod—custom-made for Gage by a local blacksmith—was a cylinder about an inch and a quarter (3 cm) in diameter and three and a half feet (1 meter) in length, flat at the bottom to tamp the charge, tapering to a point at the top. It resembled a javelin.

The iron rod must have struck a spark from the surrounding rock, because the charge went off unexpectedly, shooting the rod straight at Gage's head. The rod pierced his left cheek and passed behind his left eye and out the top of his skull, landing some 60 feet away. Gage was thrown onto his back, his limbs convulsing. Yet in a minute or two he spoke. With help from his men, he walked to a wagon, where he sat for the ride to town. There, Gage walked upstairs, unaided, to a doctor's office to have his wounds cleaned and dressed. No one expected him to live; an undertaker made a coffin for him (Macmillan, 2000).

In fact, Gage survived the injury—that's him in Figure A, in a recently discovered daguerreotype, holding the tamping rod that had shot through his head. In fact, he was able to return to work—although not his old job—testifying to the remarkable plasticity and recuperative powers of the brain. Nevertheless, he was definitely a changed man. Reconstructions of the trajectory of the rod as it passed through Gage's brain (shown in red in Figure B) reveal that it damaged both frontal lobes, especially the orbitofrontal regions. Although his symptoms were exaggerated in many accounts, it is still clear that after the accident his powers of attention and concentration were badly impaired, and compared with his old self, he had become rude and aimless (Macmillan, 2000), prompting friends and relations to conclude that he was "no longer Gage." The historical account of Gage's case, now a neuroscience classic, closely agrees with the symptoms of modern people with frontal damage (H. Damasio et al., 1994; Wallis, 2007).

(A)

From the collection of Jack and Beverly Wilgus, courtesy of Warren Anatomical Museum, Harvard Medical School

(B)

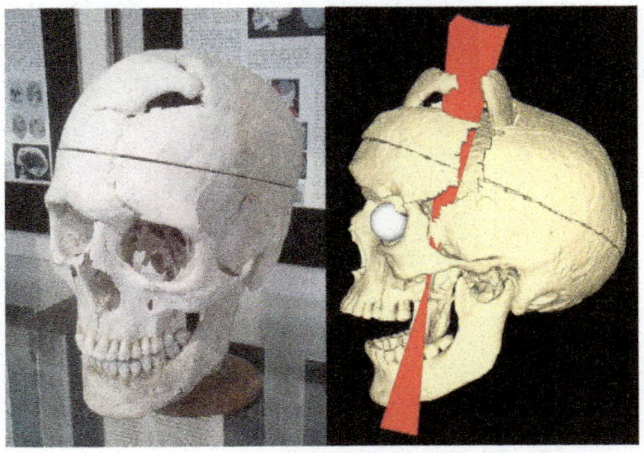

After J. D. Van Horn et al., 2012. *PLOS ONE* 7: e37454

by impulsive behaviour. Concern for the past or the future may be absent (Duffy and Campbell, 1994; Petrides and Milner, 1982). Forgetfulness is shown in many tasks requiring sustained attention. In fact, some people with frontal damage even forget their own warnings to "remember." However, standard IQ test performance often shows only slight changes after prefrontal injury or stroke.

On functional and anatomical grounds, researchers distinguish between several major divisions of the human frontal lobes. The posterior portion of frontal cortex includes the primary motor and premotor regions that we discussed in detail in Chapter 11. The more anterior portion, usually referred to as **prefrontal cortex**, is immensely interconnected with the rest of the brain and is especially associated with high-level cognition (Fuster, 1990; Mega and Cummings, 1994). It was prefrontal cortex that was surgically disrupted in frontal lobotomy: the notorious, now-discredited treatment for psychiatric disorders that we discussed in Chapter 12. Prefrontal cortex is further subdivided into *dorsolateral prefrontal cortex* (*DLPFC*) and *orbitofrontal cortex*

prefrontal cortex The anteriormost region of the frontal lobe.

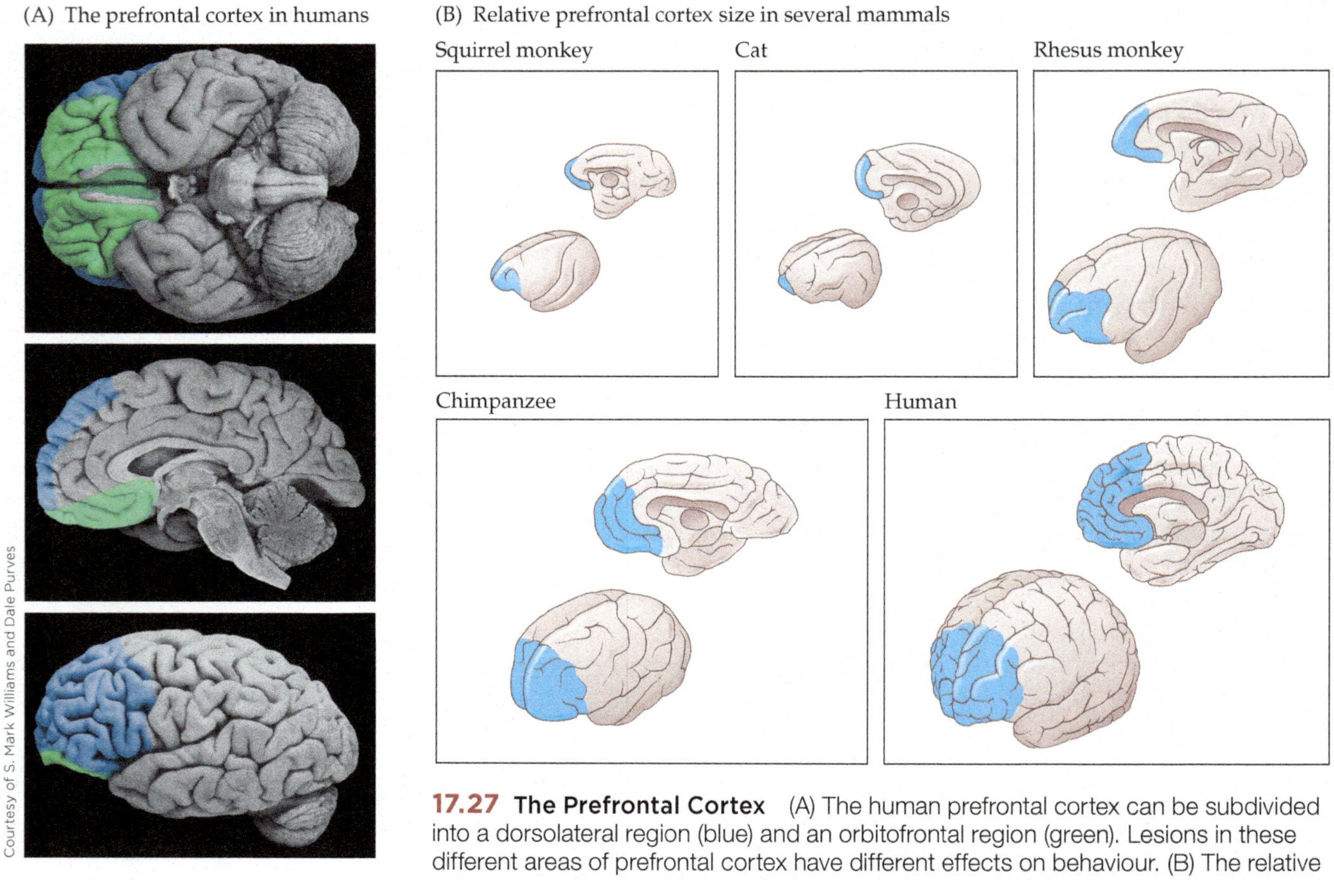

17.27 The Prefrontal Cortex (A) The human prefrontal cortex can be subdivided into a dorsolateral region (blue) and an orbitofrontal region (green). Lesions in these different areas of prefrontal cortex have different effects on behaviour. (B) The relative percentage of prefrontal cortex is greatest in humans and the great apes, such as the chimpanzees, but it decreases successively in monkeys, carnivores (such as cats), and rodents (not shown). The brains here are drawn to different scales.

task switching The shifting of attention and behaviour from one task to a new one.

perseverate To continue to show a behaviour repeatedly.

(*OFC*) (**FIGURE 17.27A**). The prefrontal cortex is especially prominent in humans and apes (Semendeferi et al., 2002), but it is a smaller portion of the cerebral cortex in other mammals (**FIGURE 17.27B**).

The DLPFC is closely associated with executive control. For one thing, DLPFC is crucial for working memory, the ability to hold information in mind while processing it or using the information to solve problems—a key aspect of executive function. People with frontal lesions that include DLPFC also often struggle with **task switching—**another aspect of executive function, involving top-down shifting of attention and behaviour from one task to a new one, as in the Wisconsin Card Sorting Test (**FIGURE 17.28**)—and tend instead to **perseverate** (continue beyond a reasonable degree) in any activity (Alvarez and Emory, 2006; B. Milner, 1963). Further, the overall level of motor activity—especially ordinary, spontaneous movements—is often quite diminished in people with frontal lesions. Along with movement of the head and eyes, facial expression of emotions may be greatly

17.28 The Wisconsin Card Sorting Test (WCST) The WCST consists of a set of 64 cards, each of which contains one to four symbols in one of four colours and one of four shapes. The participants are asked to start sorting the cards into piles, on the basis of the number, colour, or shape of the symbols, receiving only "correct" or "incorrect" feedback from the examiner to guide the sorting. After every 10 cards, the examiner changes the sorting rule but doesn't reveal the new rule, so the participants must shift their sorting behaviour until they discover the new rule. In this example, the card in the participant's hand should be added to pile 1 if the rule is "colour," pile 2 if the rule is "number," or pile 4 if the rule is "shape." A classic test of task switching, the WCST is one of the most widely studied tests of executive function (see Table 17.2), along with tests of verbal fluency (e.g., the Controlled Oral Word Association Test) and response inhibition (the Stroop Test).

TABLE 17.3 Core Characteristics of the Regional Prefrontal Syndromes

Syndrome type	Prefrontal region affected	Characteristics
Dysexecutive	Dorsolateral	Diminished judgment, planning, insight, and temporal organisation; cognitive impersistence; motor programming deficits (possibly including aphasia and apraxia); diminished self-care
Disinhibited	Orbitofrontal	Stimulus-driven behaviour; diminished social insight; distractibility; emotional lability
Apathetic	Mediofrontal	Diminished spontaneity; diminished verbal output (including mutism); diminished motor behaviour (including akinesis); urinary incontinence; lower-extremity weakness and sensory loss; diminished spontaneous prosody; increased response latency

reduced. People with prefrontal lesions often have diminished capacity for prolonged attention to tasks or stimuli, resulting in an inability to plan acts and use foresight. Social skills may decline, especially the ability to inhibit inappropriate behaviours, and they may be unable to stay focussed on any but short-term projects. They may agonize over even simple decisions. **TABLE 17.3** lists the main clinical features of people with lesions of the major subdivisions of the frontal lobes.

A network of brain sites that prominently includes the OFC appears to be important for goal-directed behaviours. In monkeys, prefrontal cortical neurons become especially active when the animal has to make a decision that may provide a reward (Matsumoto et al., 2003), suggesting that prefrontal cortex controls goal-directed behaviours. In general, the OFC system seems important for forming associations between hedonic experiences (such as the subjective pleasantness of eating gourmet food) and reward signals from elsewhere in the brain (Kringelbach, 2005). Some researchers suggest that the OFC system may be more important for signalling expected outcomes than for learning associations (Schoenbaum et al., 2009), but either way, the prospect of reward plays an important role. In humans performing a task in which some stimuli have more reward value than others, the level of activation in prefrontal cortex (as measured by fMRI) correlates with how rewarding the stimulus is (Gottfried et al., 2003). In another testing situation that involved gambling, ERPs indicated that prefrontal cortex was especially active when people were making choices that could cost them money (Gehring and Willoughby, 2002). In fact, understanding how we make decisions under conditions of risk is an active area of study, as we'll discuss next.

Frontal mechanisms sift alternatives, evaluate risk and reward, and guide decisions accordingly

To some extent, our inner lives are reflected in the choices we make. Should I have a Cobb salad for lunch? Or a smoked meat on rye sandwich? What happens in the brain when we make everyday decisions?

When asking people to make decisions in the lab, it is convenient to use monetary transactions because then you can vary how much money is at stake, how great a reward is offered, and so on, allowing for precise manipulation of the decision-making process. These studies show that most of us are very averse to loss and risk: we are more sensitive to losing a certain amount of money than we are to gaining that amount. In other words, losing $20 makes us feel a lot worse than gaining $20 makes us feel good. From a strictly logical point of view, the value of money, whether lost or gained, should be exactly the same. Our tendency to overemphasise loss is just one of several ways in which people fail to act rationally in the marketplace.

Neuroeconomics is the study of brain mechanisms at work during economic decision-making, and our attention to environmental factors and evaluation of rewards has a tremendous impact on these decisions. For example, in monkeys playing these sorts of economic games (yes, monkeys—monkeys will gamble for apple juice, which they really, *really* like), neurons in the posterior cingulate cortex become more active when risky choices are being made (McCoy and Platt, 2005).

neuroeconomics The study of brain mechanisms at work during decision-making.

In a study of humans, activity of the anterior cingulate cortex was recorded while participants played a simple game in which they moved a joystick in one direction or another, based on preceding visual cues, in order to receive a reward. Cells in this region became more active when rewards were diminished, signalling that the participants should change how they played a game (Z. M. Williams et al., 2004). When this brain region was then surgically lesioned in the same participants (cingulotomy is occasionally performed in humans as a last-ditch attempt to control obsessive-compulsive disorder; see Figure 12.29), the participants subsequently made more errors, failing to switch strategies as if they did not fully experience the disappointment of a reduced reward.

Although neuroeconomics is a young field, enough data have been amassed to suggest theoretical models of the neural bases of decision-making. Two main systems seem to underlie decision processes (Kable and Glimcher, 2009). The first, consisting of the ventromedial prefrontal cortex (including the anterior cingulate) plus the dopamine-based reward system of the basal forebrain (see Chapter 3), serves as a *valuation system*, a network that ranks choices on the basis of their perceived worth and potential reward. Impressively, using an optogenetic technique (see Chapter 2) to selectively activate neurons that express dopamine receptor D_2 in the nucleus accumbens—a central forebrain component of the brain's reward system—can instantaneously turn a risk-preferring rat into a risk-averse rat (Zalocusky et al., 2016)! Presumably, the activated cells cause the valuation system to devalue the choice relative to risk; a similar dopamine-dependent process seems to be at work when people make monetary decisions too (Ojala et al., 2018). Furthermore, *social* valuations in the context of decisions about affiliation and mate choice, or social components of material choices (such as co-owning a house, or choosing a business partner), seem to rely on much of the same neural machinery, especially the ventromedial prefrontal cortex system (Ruff and Fehr, 2014).

The second system involves dorsolateral prefrontal cortex, dorsal anterior cingulate cortex, and parietal regions (like LIP/IPS) that we have already discussed in the context of attention and awareness. This network is thought to serve as a *choice system*, sifting through the valuated alternatives and producing the conscious decision. Neuroeconomics research is confirming that the prefrontal cortex inhibits impulsive decisions, enforcing our loss aversion (Muhlert and Lawrence, 2015; Tom et al., 2007). As people are faced with more and more uncertainty, the prefrontal cortex becomes more and more active (Hsu et al., 2005; Huettel et al., 2006), and the dorsal anterior cingulate cortex is thought to help sharpen accuracy by delaying responses in situations that require time to complete complex processing of alternatives (Sheth et al., 2012). Likewise, when people have made wrong, costly decisions that they regret, activity increases in the amygdala and prefrontal cortex (Coricelli et al., 2005), probably reflecting the person's perception of diminished reward and increasing aversion to loss (**FIGURE 17.29**). Reflection on the accuracy of one's own decision-making processes is an aspect of metacognition (see Table 17.1), one of the central features of human consciousness. This sort of introspection is particularly associated with activity of the anterior prefrontal cortex (S. M. Fleming et al., 2012; Tsujimoto et al., 2011).

As we've seen in this chapter, our higher cognitive processes involve a constellation of mechanisms for attention, awareness, consciousness, and decision-making that span most of the brain. We may never have a full understanding of the deepest secrets of consciousness, or an answer to the question of whether we *actually* possess the free will that our brain seems to perceive as an element of consciousness (Haggard, 2017). But that doesn't prevent us from marvelling that our consciousness has become so self-aware that it can study itself to a high

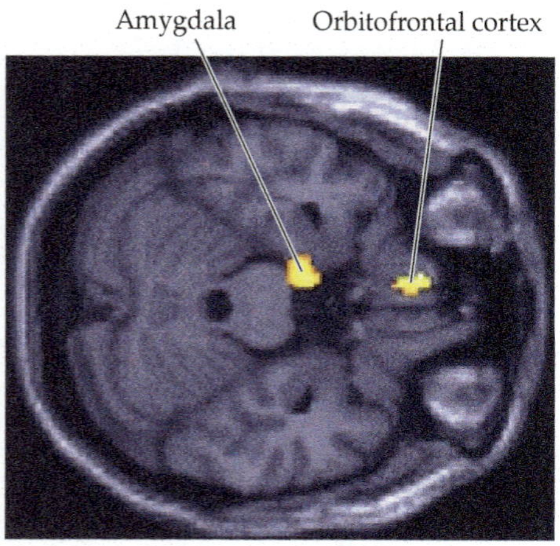

From G. Coricelli et al., 2005. *Nat Neurosci* 8: 1255–1262

17.29 A Poor Choice A costly decision is associated with activation of the amygdala and orbitofrontal cortex, signalling diminished reward and aversion to loss.

degree. Perhaps it's best to allow ourselves at least one or two mysteries, if only for the sake of art. Would life seem as rich if we could predict other people's behaviour, or even our own, with perfect accuracy?

The Cutting Edge

Building a Better Mind Reader

Some sort of machine that can read and replay a person's thoughts and feelings is a staple of science fiction, but is it plausible? As we've discussed in this chapter, reading or reconstructing a person's visual experiences on the basis of patterns of brain activity, a so-called easy problem of consciousness, is at present very difficult. Based on fMRI data tracking patterns of activation in visual cortex and related neural systems, researchers have made steady progress in reconstructing participants' visual experiences from brain activity, essentially taking a readout of whether participants are viewing crosses, circles, or other large and simple stimuli (Miyawaki et al., 2008) (as in Figure 17.25B) or pictures of faces and other complex scenes (Shen et al., 2019). However, traditional fMRI is normally too slow to "read" anything more than relatively simple, static visual experiences. But of course visual perception is almost never static; it's a continuously rolling movie, directed by attention and powerfully shaping our consciousness.

So how can we sneak a look at peoples' internal movies? One clever approach involves building a huge library of predicted brain responses to visual scenes and using those to reconstruct visual experiences from recordings of brain activity. Researchers in Jack Gallant's lab at U.C. Berkeley (Nishimoto et al., 2011) began by themselves spending many, many hours watching two sets of training videos—Hollywood movie trailers—while fMRI data were recorded from the occipitotemporal visual cortex. These data were used to develop a mathematical model that predicts how shape and motion are encoded for every voxel in the visual cortex. (*Voxel* is short for *vo*lumetric pi*xel*, and just as pixels define the two dimensions of a computer screen, voxels make up the three-dimensional volume of an imaged brain.)

Next, the researchers cut up a random assortment of YouTube videos into 18 million clips, each 1 second long, none of which were the same as the original training videos. Using a cutting-edge probability-based reconstruction algorithm, they fed this library of short clips into the computer program containing the model of brain activity, for each participant. The program created a prediction for the pattern of brain activity that each of the 18 million clips would produce.

Finally, a computer was used to compare the actual fMRI activity from the second set of test videos (again, movie trailers) with the *predicted* activity associated with the YouTube clips. For each 1-second "frame" of the test videos, a computer program selected the 100 clips whose predicted activity most closely matched the actual activity recorded when the participant viewed the original frame. These clips were then merged and processed and strung together frame by frame into a movie reconstruction. The results of this process are shown in **FIGURE 17.30** and on the website. For each of three test video clips (A, B, and C), three individual frames are shown across the top row. The next five rows show the five best-fitting YouTube frames, as selected by the modelled brain activity of the participant. Mostly, they don't much resemble the original frames, but when averaged together, as shown in the bottom row, a blurry but pretty strong resemblance begins to emerge. The accuracy of the reconstruction is much more evident when viewing the original and reconstructed videos playing side by side, which you can see on the website. It is almost unnerving.

Researchers are now developing related techniques to reconstruct participants' mental experiences of various categories of complex stimuli, such as novel faces (Nestor et al., 2016), natural visual scenes (Lescroart et al., 2015; Shen et al., 2019), and musical pieces (Hoefle et al., 2018). ■

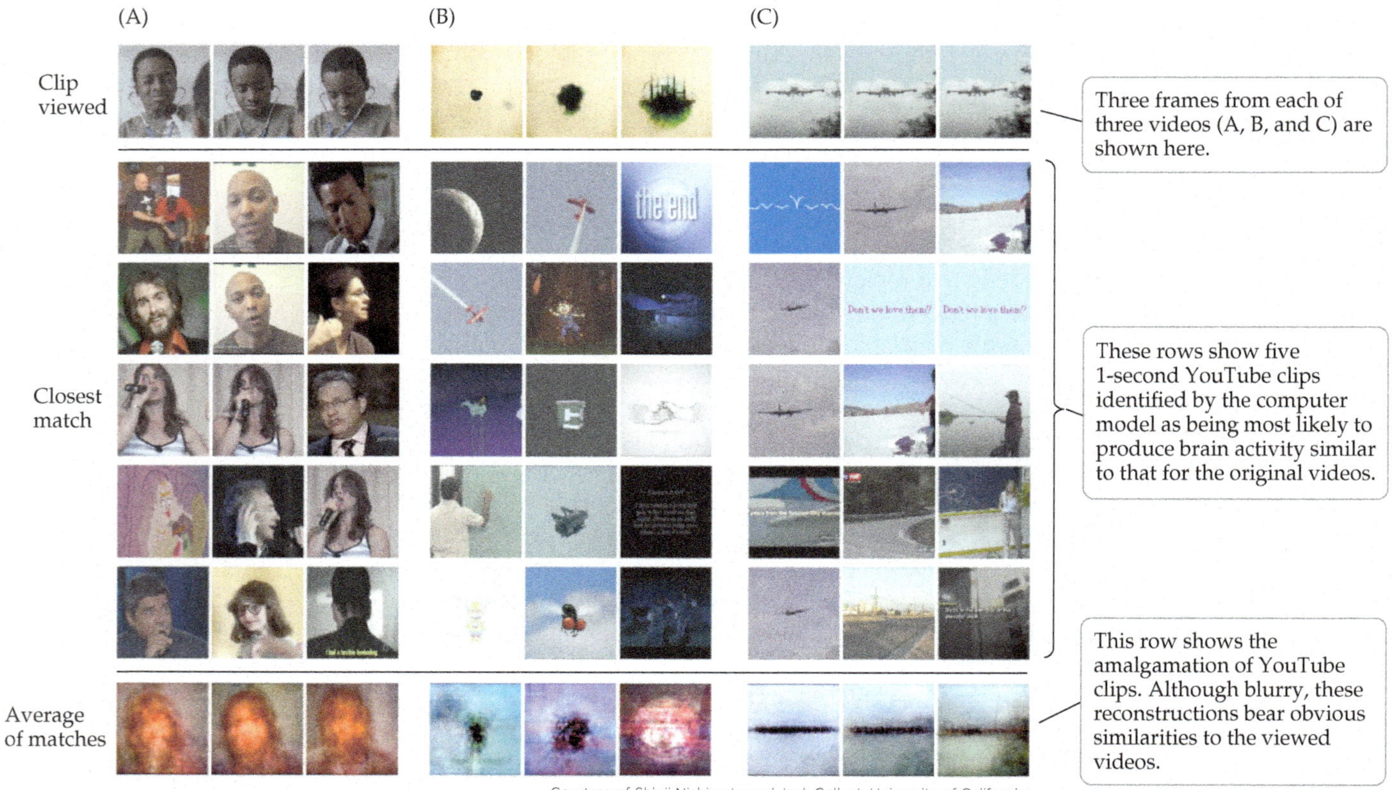

17.30 Reconstruction of a Mental Movie

Courtesy of Shinji Nishimoto and Jack Gallant, University of California, Berkeley; from S. Nishimoto et al., 2011. *Curr Biol* 21: 1641–1646

Recommended Reading

Bernat, J. L. (2009). Chronic consciousness disorders. *Annual Review of Medicine, 60,* 381–392.

Chun, M. M., Golomb, J. D., and Turk-Browne, N. B. (2011). A taxonomy of external and internal attention. *Annual Review of Psychology, 62,* 73–101.

Gazzaniga, M. S. (2018). *The Consciousness Instinct: Unraveling the Mystery of How the Brain Makes the Mind.* New York: Farrar, Straus and Giroux.

Glimcher, P. W., Camerer, C., Poldrack, R. A., and Fehr, E. (2008). *Neuroeconomics: Decision Making and the Brain.* San Diego, CA: Academic Press.

Goldberg, E. (2009). *The New Executive Brain: Frontal Lobes in a Complex World.* Oxford, UK: Oxford University Press.

Koch, C. (2012). *Consciousness: Confessions of a Romantic Reductionist.* Cambridge, MA: MIT Press.

Mangun, G. R. (Ed.). (2012). *Neuroscience of Attention: Attentional Control and Selection.* Oxford, UK: Oxford University Press.

Nobre, K., and Kastner, S. (2018) *The Oxford Handbook of Attention.* Oxford, UK: Oxford University Press.

Posner, M. I. (2011). *Cognitive Neuroscience of Attention* (2nd ed.). New York: Guilford.

Stuss, D. T., and Knight, R. T. (2012). *Principles of Frontal Lobe Function* (2nd ed.). Oxford, UK: Oxford University Press.

Wright, R. D., and Ward, L. M. (2008). *Orienting of Attention.* Oxford, UK: Oxford University Press.

17 Visual Summary

You should be able to relate each summary to the adjacent illustration, including structures and processes. You can also refer back to the links to figures, animations, and activities that will help you consolidate the material.

1 We can pay covert attention to stimuli or locations of our choosing. **Attention** helps us distinguish stimuli from distracters. Because we have limited processing capabilities, an **attentional bottleneck** limits incoming information. Review Figures 17.1, 17.2, Video 17.1

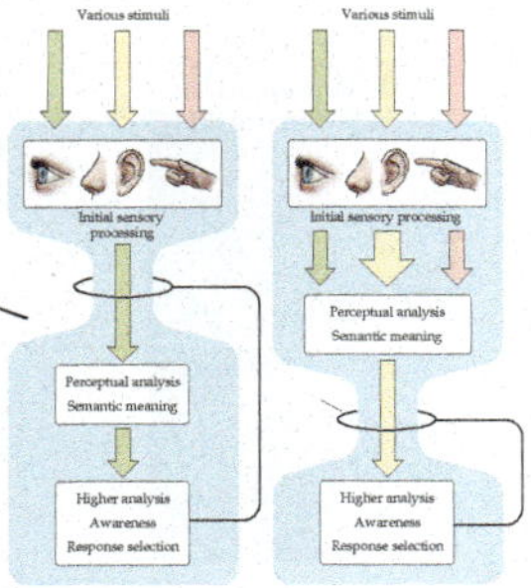

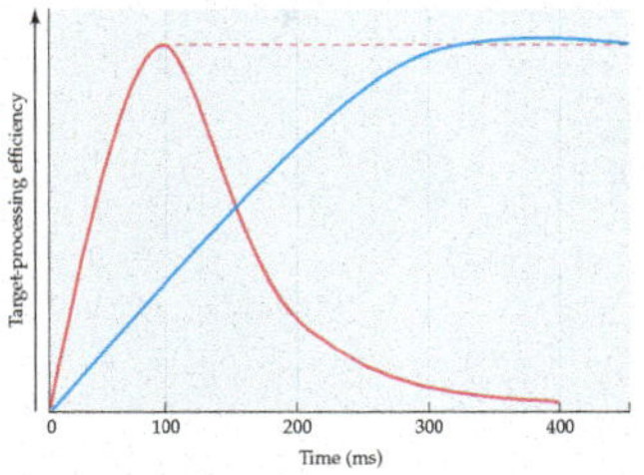

2 In **voluntary attention** we consciously select objects to attend to. **Reflexive attention** is the involuntary capture of attention by stimuli. Review Figures 17.3–17.5, Animation 17.2

3 In visual search, targets may pop out if they are distinctive on a particular feature (**feature search**). More often, we use **conjunction searches**, identifying target stimuli on the basis of two or more features. Review Figure 17.6

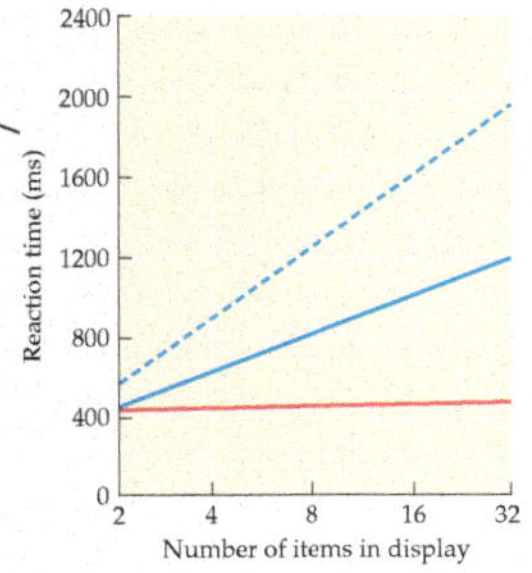

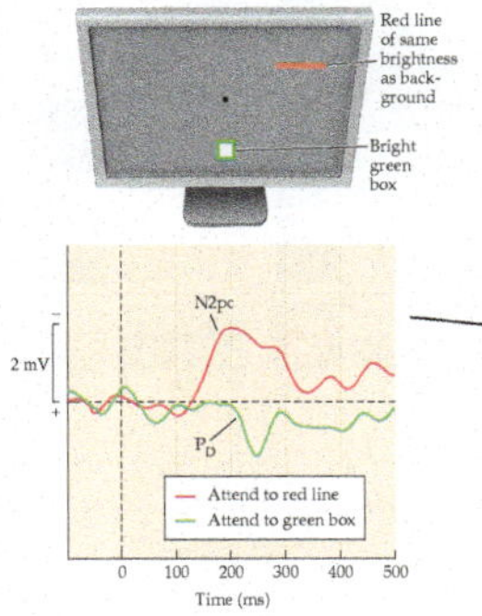

4 **Event-related potentials** (**ERPs**) are created by the averaging of many EEG recordings from repeated experimental trials. Voluntary visual attention enhances the **P1** component of the ERP, related to early attentional selection, and the **P3 effect**, which is related to later attentional selection. N2pc and P_D are contralateral occipital ERP components associated with searching visual scenes and ignoring distracters, respectively. Reflexive visual attention also enhances P1, but only for short delays between cue and stimulus. Review Figures 17.7–17.9

5 Selective attention causes enhanced activations of discrete regions of sensory cortex, such as primary visual cortex and extrastriate visual areas. Single-cell recordings in lab animals confirm that attention affects the responses of individual neurons. Review Figures 17.10–17.13

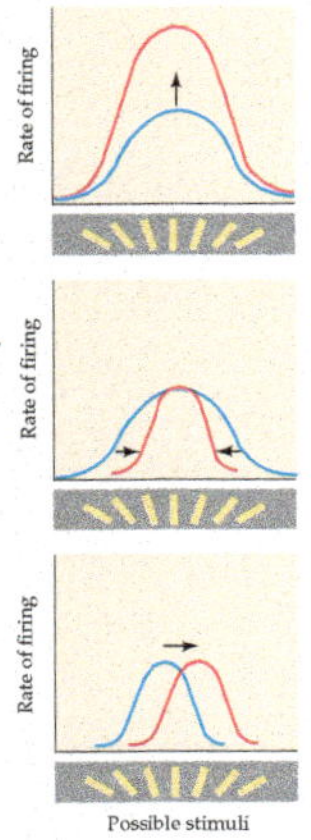

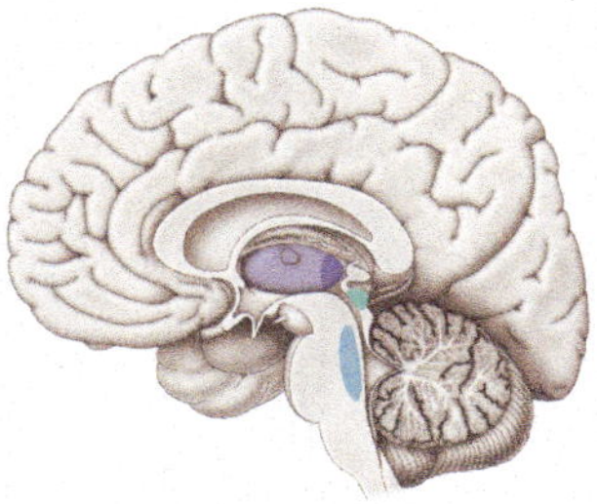

6 Subcortical mechanisms involving the **superior colliculi** and the **pulvinar** are crucial for shifting visual attention and gaze between important objects of attention. Review Figure 17.14, Activity 17.1

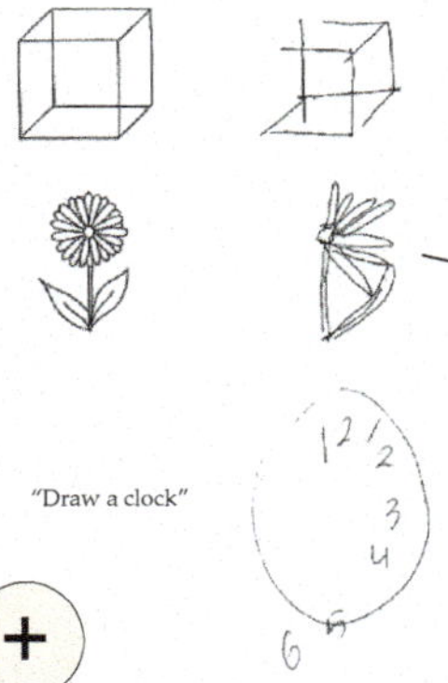

7 A dorsal frontoparietal attention network in the cortex is responsible for directing voluntary attention. A right-sided temporoparietal attentional system is responsible for detecting and shifting attention to novel stimuli. The two networks interact extensively. Review Figures 17.15–17.18, Activities 17.2, 17.3

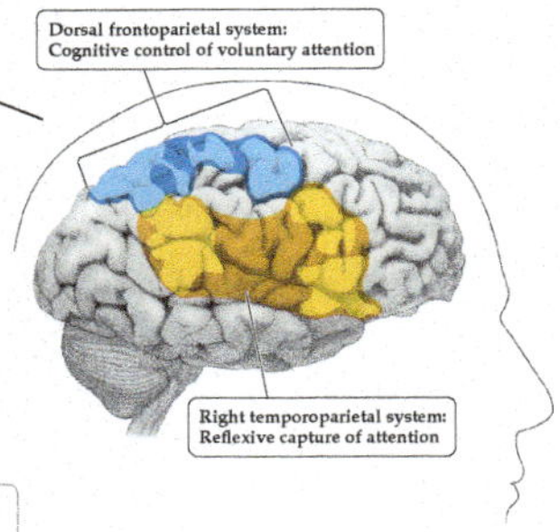

8 Damage to the right parietal cortex can cause **hemispatial neglect**. People with this condition ignore the left side of the world. Review Figures 17.19, 17.20

9 **Bálint's syndrome** is a rare consequence of bilateral parietal damage. People with Bálint's syndrome typically are spatially disoriented and experience a dramatic narrowing of attention to the point of **simultagnosia**, in which only one object at a time can be seen. People in various unconscious states show reduced activity of frontoparietal regions. Review Figures 17.22–17.24

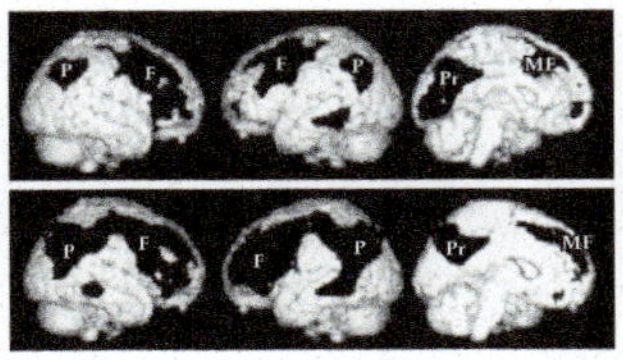

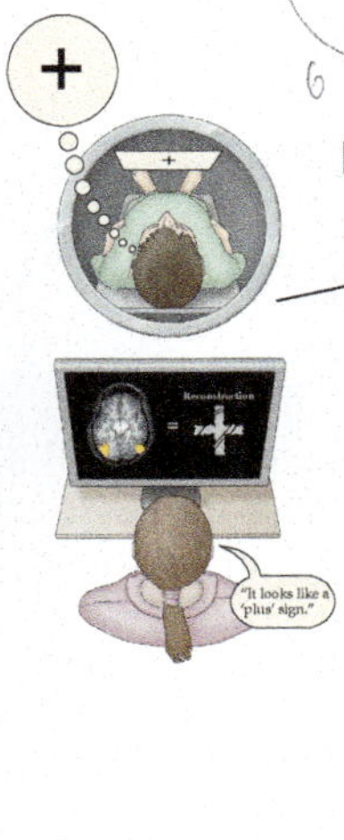

10 The **easy problem of consciousness** is the problem of how to read specific current conscious experiences directly from people's brains as they are happening. The **hard problem of consciousness** is the problem of how to read people's subjective experience of consciousness and determine the **qualia** that accompany perception. Review Figure 17.25

(Continued)

11 Feelings of free will may rely on the activity of specific frontal lobe mechanisms. But even if that's true, it's possible that unconscious mechanisms make many of our decisions well before we consciously realize it. Review **Figure 17.26**

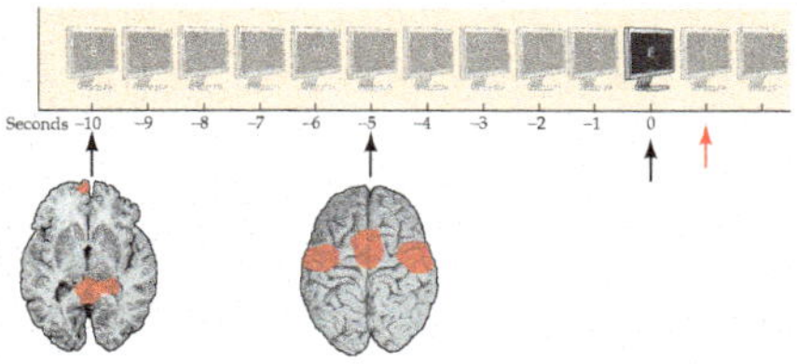

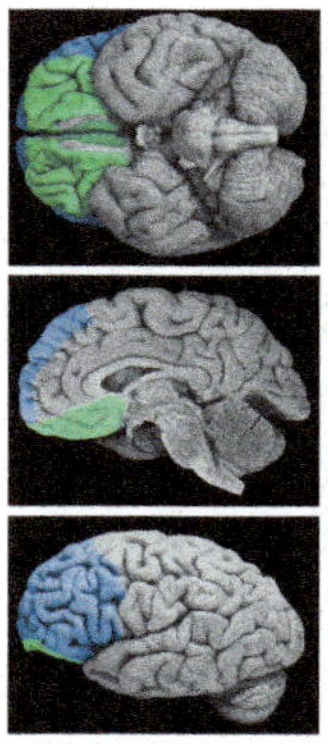

12 **Prefrontal cortex** consists of dorsolateral and orbitofrontal divisions; damage in these regions produces a distinctive set of symptoms. Medial aspects of the frontal lobes, including anterior cingulate regions, are associated with **executive function**. Prefrontal cortex appears to be essential for coordinating the resources needed for goal-directed behaviour. **Neuroeconomics** is concerned with the neural mechanisms responsible for decision-making. A current model emphasises a ventromedial frontal evaluation system and a dorsolateral choice network. Review **Figures 17.27**, **17.28**, **Box 17.2**, **Table 17.2**

13 Thanks to powerful techniques for modelling brain activity while viewing visual scenes, we are getting closer to being able to use technology to reconstruct people's visual experiences. Review **Figure 17.29**

Language and Lateralisation

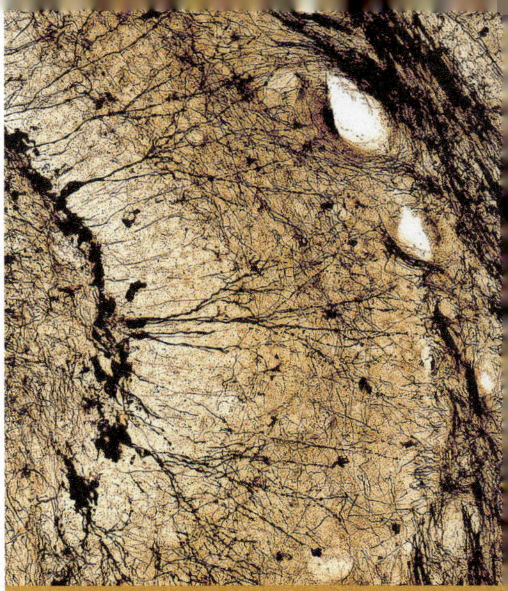
Courtesy of Dr. Sarah Moghadam, VA Palo Alto Health Care System, Palo Alto, CA and Dr. Ahmad Salehi, Dept. of Psychiatry & Behavioural Sciences, Stanford Medical School

18

Silencing the Inner Voice

We all have an inner voice: that sometimes-annoying mental narrator whose dialog wanders from calm abstraction ("What is love?") to insistent directives ("Pizza now!") to harsh self-criticism ("Way to go, fool!"). Some researchers, observing that the development of a child's sense of self is inextricably linked to the acquisition of language, suggest that the inner voice is essential for the self-awareness that defines our consciousness. Nevertheless, sometimes we wish that Inner-Me would just shut up for a while.

But what would life be like if our inner voice actually *were* silenced? This is just what happened to Tinna Geula Phillips. Suddenly one day, Tinna lost the ability to communicate in any of the six languages (yes, *six*) that she had mastered; a devastating neurological event. But even more profound was the immediate silencing of Tinna's inner voice. For several months, Tinna was unable to process thoughts in the way we take for granted. Tinna's internal silence had robbed her of a critical tool for organising her activities, weighing her emotions, processing abstract concepts, and considering her memories. She would later describe the experience as a near-total loss of identity. What could have happened to cause Tinna to lose the ability to talk to anyone, even herself?

Faces, colouration, smells, sounds—many species use physical and behavioural signals to engage in **communication**, the transmission of information between individuals. But we humans may be alone in our use of **language**, the highly specialised form of communication in which arbitrary symbols or behaviours are assembled and reassembled in almost infinite variety and associated with a vast range of things, actions, and concepts. Because the speakers of a language all understand the same strict set of rules, or **grammar**, language allows us to assemble and share information on any topic, linking thinkers of the past with those of the future. Our linguistic and aesthetic perception of the world gives us a mental life that appears to be unique among animal species.

In almost everyone, verbal abilities are especially associated with the left hemisphere of the brain, whereas the right hemisphere plays a special role in complex *spatial cognition,* our ability to process geometric relationships, to navigate, and to understand the spatial relationships between objects. We begin the chapter with the neuroscience of cerebral asymmetry, before turning to a more focussed overview of the acquisition and use of speech and language, and the brain mechanisms underlying this uniquely human capability. Much of what we know about human brain organisation is derived from studies of people who have sustained strokes and other forms of brain injury, so we conclude with a section on the sometimes surprising ability of the damaged brain to recover and adapt.

BRAIN ASYMMETRY AND LATERALISATION OF FUNCTION

The special role of the left hemisphere for language functions was well established by the mid-twentieth century, and it was often referred to as the *dominant hemisphere*, in recognition of the preeminence of language in our cognitive lives (Finger, 1994). However, the right hemisphere does not just idly sit within the skull, awaiting an occasional call to duty. In fact, most researchers have abandoned the notion of cerebral *dominance* in favor of models of hemispheric *specialisation*, or **lateralisation**. This emphasis implies that some functional systems are connected more to one side of the brain than the other—that is, that functions become lateralised—and that each hemisphere is specialised for particular ways of working. Keep in mind, however, that almost all types of behaviour, from manual skills to intellectual activity, are performed better by the two hemispheres working together than by either hemisphere working alone. The idea that the two hemispheres are so different that they need separate instruction, or that people can have personalities that are "right-brained" (supposedly more random, intuitive, and creative) or "left-brained" (ostensibly logical, sequential, and analytical), are flaky notions that lack any scientific basis.

18.1 The Left and Right Hemispheres Are Different

Learning Objectives

After studying this section, you should be able to:

18.1.1 **Distinguish between communication and language.**

18.1.2 **Describe experimental techniques for studying the left and right hemispheres independently of one another.**

18.1.3 **Summarise the apparent cognitive specialisations of each hemisphere as revealed by behavioural testing.**

18.1.4 **Speculate about the evolutionary origins of handedness, and discuss the relationship of handedness to cerebral lateralisation.**

The discovery that some brain functions are lateralised should not be especially surprising; after all, other body organs also show considerable asymmetry between the right and left sides. For example, in all vertebrates the heart is slightly to the left of the midline and the liver is on the right. Sometimes people with a defect in one of the genes involved will develop with their organs reversed, a condition called *situs inversus* (Casey and Hackett, 2000). When we study the *behaviour* of healthy people, cerebral lateralisation of function is masked by the rich neural connections between the hemispheres: they communicate with each other so quickly and so thoroughly that they seem to act as one. But by studying people whose hemispheres have been disconnected, researchers have been able to study the functioning of each hemisphere in isolation from the other.

Disconnection of the cerebral hemispheres reveals their individual specialisations

Some rare and unfortunate people develop severe epilepsy that is very difficult to control with medication. Their frequent seizures start in one hemisphere and then spread to the other hemisphere through the **corpus callosum**—the huge white matter pathway consisting of hundreds of millions of axons connecting the two hemispheres. A treatment of last resort is to surgically split the two hemispheres, severing the corpus callosum to prevent the spread of the seizure discharges from one hemisphere to the other. Due to its extreme invasiveness, and the development of better pharmacological alternatives, split-brain surgery has remained a very rare procedure, with most people receiving only partial splits (transection of a minority of the callosum). Only about ten people with complete transections of the corpus callosum have been fully studied by researchers (A. M. Gazzaniga, 2008). In these **split-brain individuals**, the operation had its expected effect in reducing the frequency and severity of their seizures. But it also changed their behaviour in ways that presented a unique research opportunity: by analyzing the cognitive, perceptual, emotional, and motor

communication Information transfer between two individuals.

language The most sophisticated form of communication, in which a set of arbitrary sounds, tokens, or symbols can be arranged according to a grammar in order to convey an almost limitless variety of concepts.

grammar All of the rules for usage of a particular language.

lateralisation The tendency for the right and left halves of a system to differ from one another. In cerebral lateralisation, one cerebral hemisphere is specialised for a particular intellectual function.

corpus callosum The main band of axons that connects the two cerebral hemispheres.

split-brain individual An individual whose corpus callosum has been severed, halting communication between the right and left hemispheres.

behaviour of split-brain individuals, researchers were able to catalog the individual specialisations of the newly isolated cerebral hemispheres.

In Nobel Prize–winning research begun in the 1960s, Roger Sperry applied techniques that he had perfected in split-brain cats to the study of split-brain humans. The researchers realised that by exploiting the organisation of the sensory systems, they could direct stimuli exclusively to one hemisphere or the other. For example, stimuli felt with the left hand, or seen only in the left visual field, stimulate activity in sensory neurons of the right hemisphere. Normally, information about the stimuli would immediately be shared with the other hemisphere, but because the corpus callosum is cut in split-brain individuals, most of the information received by one hemisphere of the brain remains trapped there and cannot travel to the other hemisphere. By controlling stimuli in this fashion—selectively presenting them to one hemisphere or the other—the experimenter can test the processing specialisations of each hemisphere in isolation.

In some studies of split-brain individuals, words were projected to either the left or the right hemisphere; that is, printed stimuli were presented in either the right or the left side of the visual field. The results were dramatic. Split-brain individuals could easily read and verbally communicate words projected to the left hemisphere (via the right visual field), but no such linguistic capabilities were evident when the words were directed to the right hemisphere (**FIGURE 18.1**). Although the right hemisphere in split-brain individuals does have minor linguistic capabilities—for example, recognition of very simple words, and processing of emotional content in verbal material (Zaidel, 1976)—in most people, vocabulary and grammar are the exclusive domain of the left hemisphere (left-handers occasionally show a reversed asymmetry).

The capabilities of the "mute" right hemisphere had to be tested by nonverbal means. For example, a picture of a key might be projected to the left visual field and so reach only the right cortex. The individuals would then be asked to touch several

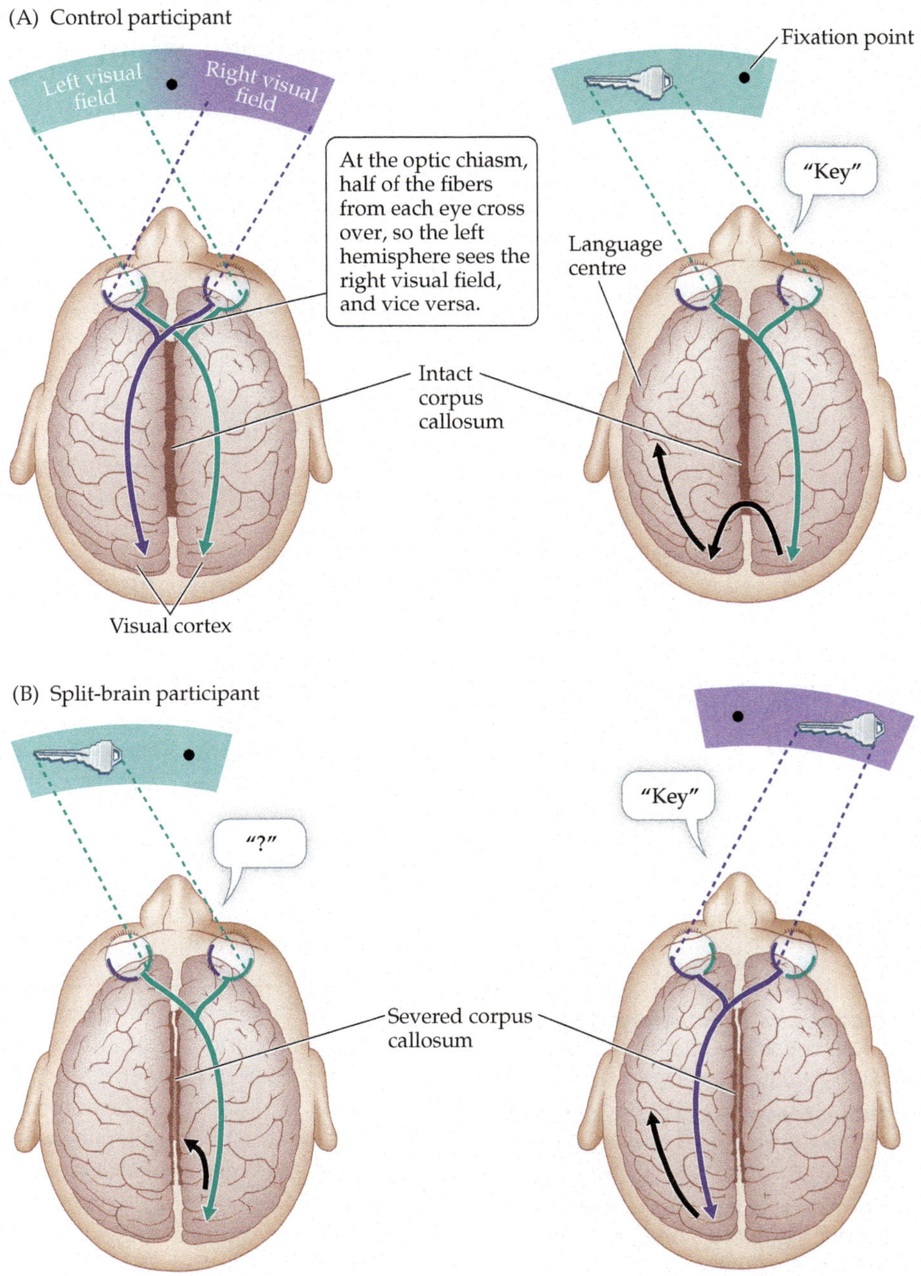

18.1 Testing a Split-Brain Individual
Words or pictures projected to the left visual field activate the right visual cortex. (A) In intact individuals, activation of the right visual cortex excites corpus callosum axons, which transmit verbal information to the left hemisphere, where the information is analysed and language is produced. (B) In split-brain individuals, stimuli from the left visual field reach the right-hemisphere visual cortex via the subcortical visual pathways (they are independent of the corpus callosum). However, the split corpus callosum prevents right-hemisphere visual areas from communicating with the language areas of the left hemisphere, so verbal responses to the stimuli are impossible (left). In contrast, split-brain individuals are able to respond verbally to stimuli appearing in the right visual fields, because interhemispheric transfer is not required (right).

18.2 **Callosal Agenesis** In contrast to the typical brain in part A, the brain in part B failed to develop any vestige of a corpus callosum. Despite the absence of the hundreds of millions of axons that pass between the hemispheres through the corpus callosum, the activity of the two hemispheres of this individual remained coordinated.

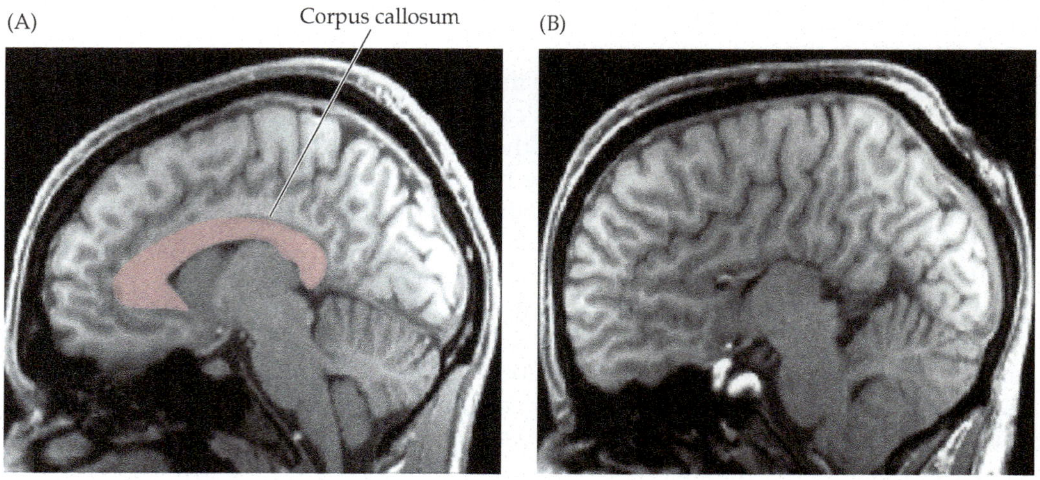

J. M. Tyszka et al. 2011. *J Neurosci* 31: 15154–15162; images courtesy of Lynn Paul and Mike Tyszka

different objects that they could not see and hold up the correct one. Such a task could be performed correctly by the left hand (controlled by the right hemisphere) but not by the right hand (controlled by the left hemisphere). So in this case, the left hemisphere literally does not know what the left hand is doing! In general, these and other studies with split-brain individuals provided early evidence that, in most people, the right hemisphere is specialised for processing spatial information. Right-hemisphere mechanisms are also crucial for face perception, for processing emotional aspects of language, and for controlling attention (as we discussed in Chapter 17).

As many as one in 4000 people are born either partially or totally lacking a corpus callosum. Although some individuals will show some behavioural consequences of this *callosal agenesis* (**FIGURE 18.2**), in general they lack the constellation of striking neuropsychological changes evident in surgical split-brain individuals (Sotiriadis and Makrydimas, 2012). Somehow, the developing nervous system compensates for the loss of the main connection between the hemispheres, perhaps by strengthening alternative subcortical routes of transmission (Tyszka et al., 2011). This observation highlights the incredible plasticity of the developing brain.

The two hemispheres process information differently in most people

Most research on brain asymmetry in healthy people focusses on two sensory modalities: hearing and vision. That's because researchers have devised clever procedures for directing auditory or visual stimuli mostly to one hemisphere or the other and then inferring hemispheric specialisations from the behavioural responses made by the participants.

THE RIGHT-EAR ADVANTAGE Through earphones, we can present different sounds to the two ears at the same time; this process is called **dichotic presentation**. So, for example, the study participant may hear a particular speech sound in one ear and, at the same time, a different vowel, consonant, or word in the other ear. The task for the participant is to try to identify or recall both sounds. Although this procedure may seem designed to produce confusion, most data from dichotic presentation experiments indicate that right-handed persons identify verbal stimuli delivered to the right ear more accurately than verbal stimuli simultaneously presented to the left ear. This result is described as a *right-ear advantage* for verbal information. In contrast, about 50% of left-handed people reveal a reverse pattern, showing a left-ear advantage—more-accurate performance for verbal stimuli delivered to the left ear.

As a consequence of the preferential connections between the right ear and left hemisphere, the right-ear advantage for verbal stimuli is probably a reflection of the left hemisphere's specialisation for language (**FIGURE 18.3**). Although we can normally use

dichotic presentation The simultaneous delivery of different stimuli to the right and the left ears.

tachistoscope test A test in which stimuli are very briefly exposed in either the left or right visual half field.

planum temporale A region of superior temporal cortex adjacent to the primary auditory area.

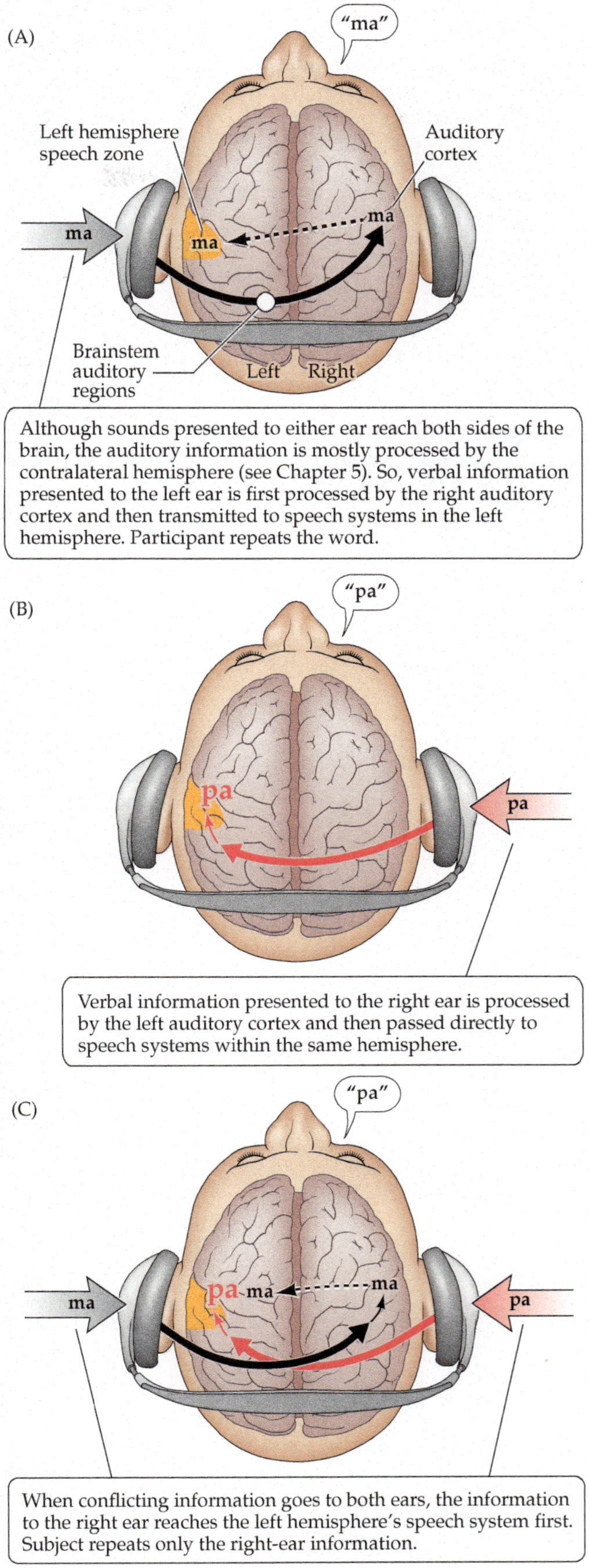

18.3 The Right-Ear Advantage in Dichotic Presentation (A) A word delivered to the left ear results in stronger stimulation of the right auditory cortex. (B) A word delivered to the right ear results in stronger input to the left hemisphere. (C) When words are delivered to both ears simultaneously, the word to the right ear is the one usually perceived because the right ear has more-direct connections to the left hemisphere. (After D. Kimura, 1973. *Sci Am* 228: 70–78.)

either ear for processing speech sounds, speech presented to the right ear in dichotic presentation tests exerts stronger control over language mechanisms in the left hemisphere than does the speech simultaneously presented to the left ear (Kimura, 1973). The competition between the left- and right-ear inputs is the key; presentation of speech stimuli to one ear at a time (*monaural* presentation) does not produce a right-ear advantage. And the right-ear advantage in right-handers is restricted to particular kinds of speech sounds: those made for consonants like *b*, *k*, *m*, and *p* produce a right-ear effect, but vowel sounds do not (Tallal and Schwartz, 1980).

VISUAL PERCEPTION OF LINGUISTIC STIMULI A visual analogue to the dichotic listening task is the **tachistoscope test**, in which a specialised apparatus (named, surprisingly, a *tachistoscope*) is used to present stimuli very briefly to either the left or right visual field (see Figures 8.10, 8.11, and 18.1). If the stimulus exposure lasts less than 150 milliseconds (ms) or so, the input is received by only the contralateral hemisphere because there is not enough time for the eyes to shift their position to fix the stimulus in central vision. In healthy humans, of course, information may subsequently be passed to the other hemisphere via the corpus callosum, but the important thing is that, as with dichotic presentation, the stimuli have preferential access to one hemisphere.

Most tachistoscopic studies confirm the general division of labour between the hemispheres. Verbal stimuli (words and letters) are recognised more accurately when they are presented to the right visual field (and therefore left hemisphere) than when they are presented to the left visual field/right hemisphere. Conversely, nonverbal visual stimuli (such as faces or geometric forms) presented to the left visual field/right hemisphere are better recognised than the same stimuli presented to the right visual field/left hemisphere. Simpler visual processing, such as detection of light, hue, or simple patterns, is performed equivalently by both hemispheres.

The left and right hemispheres differ in their auditory specialisations

Anatomical studies of primary auditory cortex in the left and right hemispheres are consistent with the view that the two play different roles in auditory perception. In one early postmortem study of auditory cortex, the **planum temporale**—an auditory region on the superior surface of the temporal lobe—was found to be larger in the left hemisphere than in the right in most of the brains studied (**FIGURE 18.4A AND B**) (Geschwind and Levitsky, 1968). In only 11% of adults was the right side larger. The planum temporale includes part of Wernicke's area, which as we'll see shortly is crucial for speech, so it seems likely that the larger left planum temporale is related to that hemisphere's language specialisation. Direct evidence of this relationship, however, remains somewhat elusive: in one MRI study, for example, asymmetry of the planum temporale did not correlate with direct measures of language lateralisation (Dorsaint-Pierre et al., 2006).

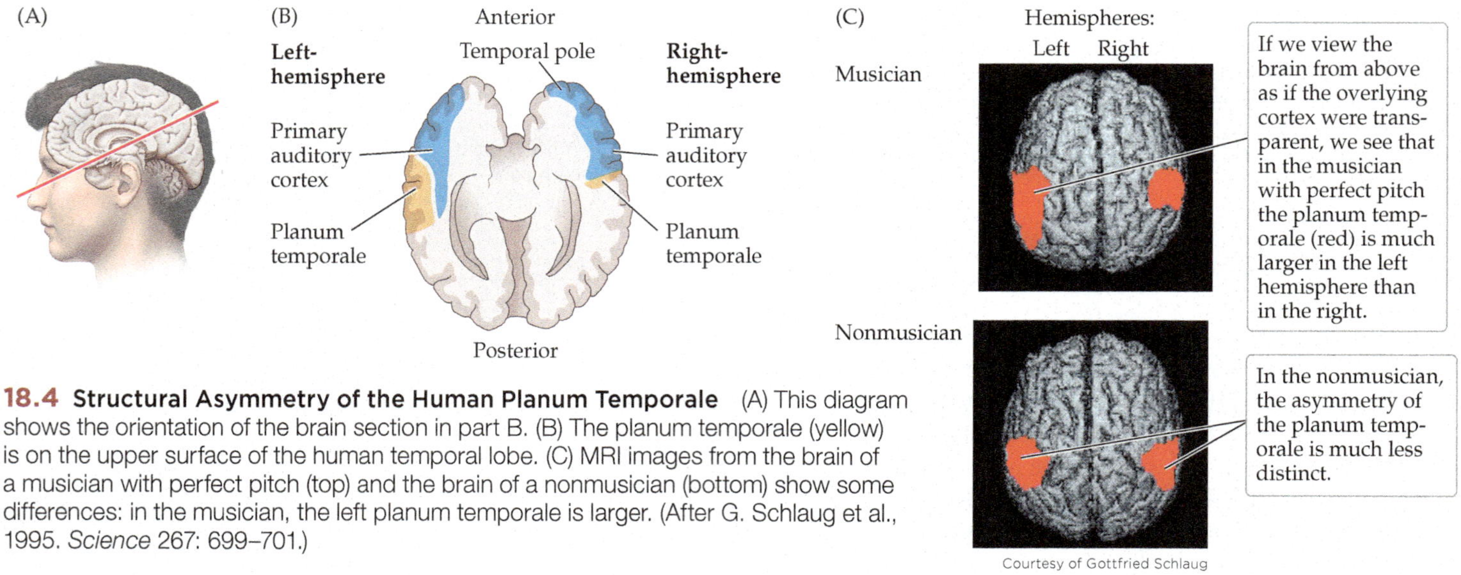

18.4 Structural Asymmetry of the Human Planum Temporale (A) This diagram shows the orientation of the brain section in part B. (B) The planum temporale (yellow) is on the upper surface of the human temporal lobe. (C) MRI images from the brain of a musician with perfect pitch (top) and the brain of a nonmusician (bottom) show some differences: in the musician, the left planum temporale is larger. (After G. Schlaug et al., 1995. *Science* 267: 699–701.)

As in adults, the left planum temporale is larger than the right in the brains of infants (Wada et al., 1975), in agreement with the observation that speech activates the left cerebral hemisphere more than the right in infants as young as 3 months. This provides more evidence that cerebral specialisation for language probably begins prenatally without the need of speech experience, because the asymmetry is evident before any substantial such experience has occurred. In fact, even in human fetuses of just 12–14 weeks of gestation, a variety of genes show asymmetrical expression in the brain (T. Sun et al., 2005). The planum temporale tends to be larger on the left than on the right in chimpanzees too (Gannon et al., 1998), and Broca's area—another cortical speech zone, as we'll see a little later—is larger on the left than on the right in chimps, bonobos, and gorillas, as well as in humans (Cantalupo and Hopkins, 2001). Furthermore, just as our left hemisphere is more activated than the right when we hear speech rather than other sounds, the left hemisphere of monkeys is more activated than the right when they hear monkey vocalisations rather than human speech (Poremba et al., 2004), and their perception of monkey vocalisations is far more impaired by left- than right-hemisphere damage (Heffner and Heffner, 1984). These results suggest that other primates also possess a left-hemisphere specialisation for some forms of communication (but not necessarily *language*).

In contrast, the auditory areas of the *right* hemisphere play a major role in the perception of music. Musical perception is impaired particularly by damage to the right hemisphere (S. Samson and Zatorre, 1994), and music activates the right hemisphere more than the left (Zatorre et al., 1994). But perfect pitch (the ability to identify any musical note without comparing it to a reference note) seems to involve the left rather than the right hemisphere. Schlaug et al. (1995) made MRI measurements of the planum temporale in three kinds of participants, all right-handed (because the larger size of the left planum temporale is seen especially in right-handed individuals): (1) musicians with perfect pitch, (2) musicians without perfect pitch, and (3) nonmusicians. The size of the left planum temporale was twice as large in musicians with perfect pitch than in nonmusicians (**FIGURE 18.4C**). The size of the left planum temporale in musicians without perfect pitch was intermediate, but closer to that of nonmusicians. Because perfect pitch requires both verbal ability (to name the pitch) and musical ability, perhaps it is not surprising to find that, like language, perfect pitch is associated with the left hemisphere.

Despite these data, we cannot assign the perception of speech entirely to the left hemisphere and the perception of music entirely to the right hemisphere. We have seen that the right hemisphere can play a role in speech perception even in people in whom the left hemisphere is dominant for speech. In addition, the perception of emotional tone-of-voice aspects of language, termed **prosody**, is a *right*-hemisphere specialisation. Furthermore, although damage to the right hemisphere can impair the perception of music, it does not abolish it. Damage to *both* sides of the brain can completely wipe out musical perception (S. Samson and Zatorre, 1991). Thus, even though each hemisphere plays a greater role than the other in different kinds of auditory perception, the two hemispheres appear to collaborate in these functions, as well as in many others.

prosody The perception of emotional tone-of-voice aspects of language.

Handedness is associated with cerebral lateralisation

Various surveys of hand usage, such as the incidence of left-handed writing in American college populations (Spiegler and Yeni-Komshian, 1983), suggest that 10–15% of the population is left-handed. This predominance of right-handedness in humans is probably an ancient trait. In ancient cave paintings, people were generally portrayed clutching objects in their right hands, and Stone Age tools seem to be shaped for the right hand. Skull fractures of animals preyed on by ancient humans are usually on the animal's left side, so anthropologists conclude that most attackers held a club in their right hand. Handedness is a trait we share with our primate relatives: population-level hand preference is observed in gorillas, chimps, and bonobos (they're mostly right-handers) and orangutans (mostly lefties) (Hopkins et al., 2015; P. F. MacNeilage et al., 2009).

Handedness directly reflects cerebral lateralisation by virtue of the preferential connection of each hand with the contralateral hemisphere. Some scientists believe that handedness, and the underlying hemispheric asymmetry with which it is associated, originated in the differential use of the limbs for many routine tasks. Picture early humans hunting. One hand holds the weapon and provides power; the other is used in more-delicate guidance or body balance. Perhaps powerful throwing, a strongly right-handed behaviour in most right-handed people, was a particularly important evolutionary development, providing obvious advantages in hunting and defense. In time, the left-hemisphere neural specialisations needed for programming throwing movements may have been co-opted to control language as well (Watson, 2001). However, other vertebrates—including toads (Vallortigara et al., 1999), crows (Hunt et al., 2001), walruses (Levermann et al., 2003), and chimpanzees (Corp and Byrne, 2004)—have been found to show preferences for using one limb or the other for particular behaviours, suggesting that handedness has additional adaptive advantages. (For other examples of the evolution of asymmetry, see **A STEP FURTHER 18.1: EVOLUTION FAVOURS ASYMMETRY UNDER SOME CONDITIONS** on the website.) Some researchers go further, arguing that hemispheric asymmetry reflects an ancient and ubiquitous left-right division of labour that, as a matter of neural processing efficiency, arose in the first vertebrates, hundreds of millions of years ago (P. F. MacNeilage et al., 2009). This view is bolstered by the discovery that in mice, left-hemisphere neurons differ from their right-hemisphere counterparts in both the expression of neurotransmitter receptors and physiological functioning (M. M. Kohl et al., 2011; Shipton et al., 2014).

Handedness may have a genetic component, but it doesn't appear to be determined by a single gene. One possibility is that handedness is governed by genes that play a role in asymmetry throughout the body, ranging from the shape of internal organs to the pattern of hair whorls on the scalp (Corballis, 2014; Klar, 2003). One gene that may contribute to handedness is *LRRTM1* (Francks et al., 2007; Leach et al., 2014), a gene associated with glutamate neurotransmission that is also implicated in schizophrenia. Perhaps the co-occurrence of schizophrenic symptoms in a small minority of left-handers helped give rise to the historical negative view of left-handedness. Additionally, several studies have found that one form of the gene *PCSK6* is associated with right-handedness; *PCSK6* has a role in the development of the left-right axis in mammals, including the asymmetrical placement of internal organs (Brandler and Paracchini, 2014).

Because left-handedness is unusual, historical accounts attribute various negative characteristics to left-handed people, and the Latin word for "left-handed"—*sinistra*—provides the root of not only the English word *sinistral*, meaning "left-handed," but also *sinister*. The view that left-handed people are somehow abnormal was not uncommon in the past and even found occasional support in research, such as the observation that left-handedness may be more frequent in clinical populations than in the general population (Silva and Satz, 1979), perhaps as a result of developmental reorganisation after brain injury. However, studies of achievement, ability, and cognitive function in thousands of school-age children conclusively showed that left-handed children do not differ from right-handed children on any measure of cognitive performance (Hardyck et al., 1976). Techniques that allow the study of each hemisphere in isolation, such as the Wada test, have reliably shown that as in right-handers, language is vested in the left hemisphere in the great majority of left-handers (**BOX 18.1**). However, those rare individuals in whom language is known to be lateralised to the *right* hemisphere are more likely to be left-handed

BOX 18.1 The Wada Test

Given the incidence of language impairment after unilateral brain damage, scientists estimate that 90–95% of humans have a left-hemisphere specialisation for language. But how can we be certain which hemisphere is specialised for language (or other functions, such as spatial memory or music perception) in people without brain damage? For example, it can be crucial to understand an individual's specific lateralisation prior to brain surgery.

By injecting a short-acting anaesthetic (sodium amytal) into the carotid artery on one side (as shown in Figure A), it is possible to simulate a massive stroke, shutting down the entire hemisphere on that side for a few minutes (Wada and Rasmussen, 1960). This is just long enough to use behavioural measures to document the specialisations of that hemisphere. The same can be done for the other hemisphere several minutes later.

This *Wada test* confirms that most people have left-hemisphere specialisation for language, regardless of handedness. The reverse pattern (right-hemisphere dominance for language) is rare, but when it occurs, it is more common in left-handed people (Figure B). Comparable results can be obtained in healthy people by using TMS (Knecht et al., 2002) or fMRI. (B after C. M. Epstein et al., 2000. *Neurology* 55: 1025–1027.)

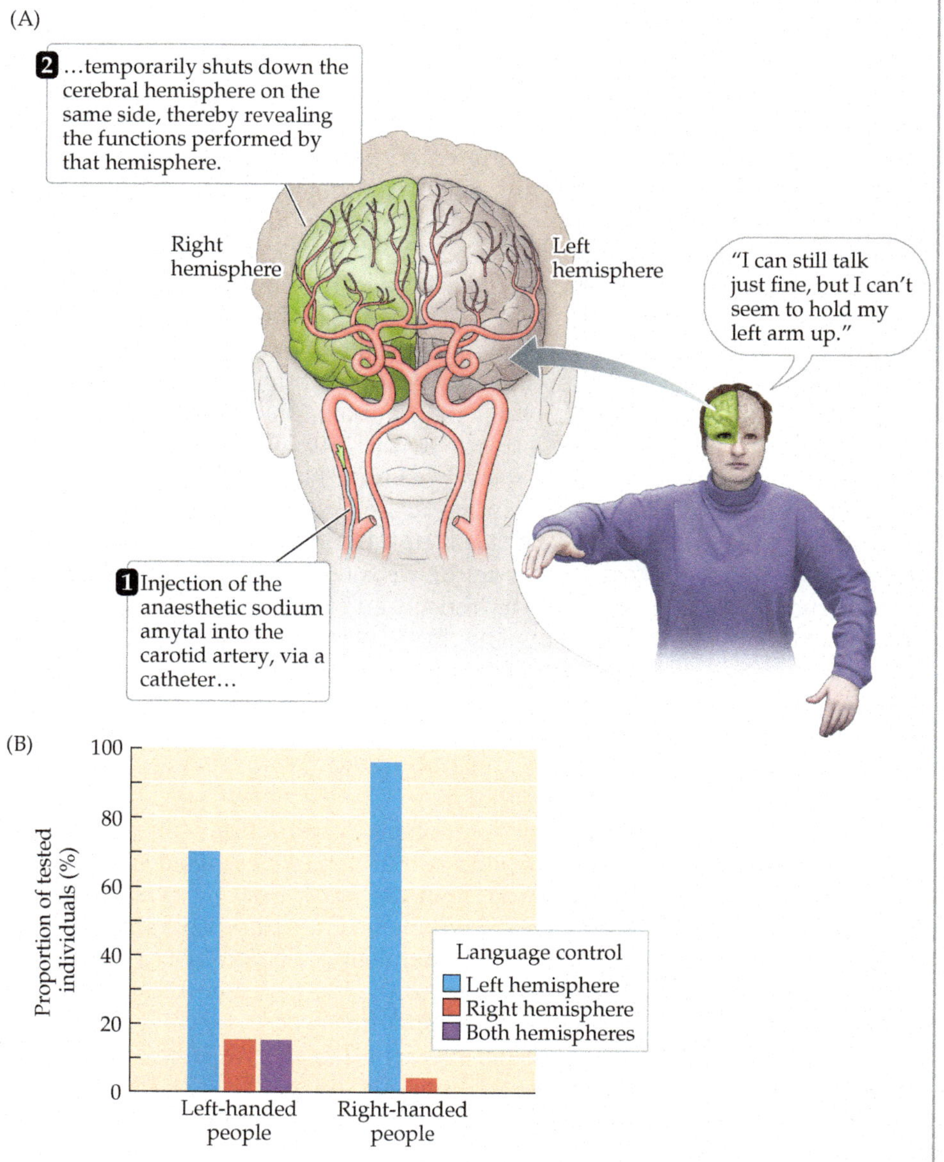

than right-handed. MRI studies likewise show that the planum temporale is more likely to show reduced or reversed asymmetry in left-handers (Steinmetz et al., 1991).

There are some rare circumstances in which a change in handedness may occur in adults. Two right-handed people who received double hand transplants, after losing their own hands in accidents, reportedly both became left-handers after the surgery (Vargas et al., 2009). One possible explanation is that the right hemisphere required less reorganisation in order to take control of the contralateral hand and that reconnection of the left hemisphere to the new right hand was initially hampered by the strong preexisting cortical representation of the former right hand. After getting an early start, perhaps the right hemisphere/left hand simply outcompeted the left hemisphere/right hand. Of course, no one yet knows if this process bears any relationship to the normal establishment of hand preference, but it suggests a surprising degree of flexibility in handedness.

18.2 Right-Hemisphere Damage Impairs Specific Types of Cognition

Learning Objectives

After studying this section, you should be able to:

18.2.1 **Itemise the types of deficits seen after right-hemisphere lesions, relating the deficits to specific anatomical locations within the hemisphere.**

18.2.2 **Describe prosopagnosia and Summarise the brain's face-processing network.**

When studied with behavioural or imaging techniques that highlight differences between the hemispheres, people reliably demonstrate a right-hemisphere advantage for the processing of spatial stimuli. Geometric shapes and their relations, direction sense and navigation, face processing, imagined three-dimensional rotation of objects held in the mind's eye—these are a few examples of the kinds of stimuli that the right hemisphere preferentially processes. So it's no surprise that right-hemisphere lesions—especially more-posterior lesions that involve the temporal and parietal lobes—tend to produce a variety of striking impairments of spatial cognition, such as inability to recognise faces, spatial disorientation, failure to identify objects by touch, or the complete neglect of one side of the body that we discussed in Chapter 17 (see Figure 17.19).

The diversity of behaviour changes following injury to the parietal lobe is related partly to its large expanse and its critical position, abutting all three of the other cortical lobes. The anterior end of the parietal region includes the postcentral gyrus, which is the primary cortical receiving area for somatic sensation. In addition to alterations in touch sensitivity on the opposite side of the body, brain injuries in this area can produce deficits involving much more complex forms of sensory processing. For example, objects placed in the hand opposite the injured somatosensory cortex can be *felt* but cannot be identified by touch and active manipulation. This deficit is called **astereognosis** (from the Greek *a-*, "not"; *stereos*, "solid"; and *gnosis*, "knowledge").

More-extensive injuries in the parietal cortex, beyond the primary somatosensory cortex of the postcentral gyrus, affect interactions between sensory modalities, such as visual or tactile matching tasks, which require the participant to visually identify an object that is touched or to reach for an object that is first perceived visually.

In prosopagnosia, faces are unrecognisable

Suppose one day you look in the mirror and someone unfamiliar is looking back at you. As incredible as this scenario might seem, some individuals develop exactly this problem following damage to specific brain sites. It's a rare syndrome called **prosopagnosia** (from the Greek *prosop-*, "face"; *a-*, "not"; and *gnosis*, "knowledge"), or sometimes *face blindness*. People with prosopagnosia fail to recognise not only their own faces but also the faces of relatives and friends. No amount of remedial training restores their ability to recognise anyone's face. In contrast, the ability to

astereognosis The inability to recognise objects by touching and feeling them.

prosopagnosia Also called *face blindness*. A condition characterised by the inability to recognise faces. *Acquired prosopagnosia* is caused by damage to the brain, particularly the fusiform gyrus. *Developmental* (or *congenital*) *prosopagnosia* is present from birth.

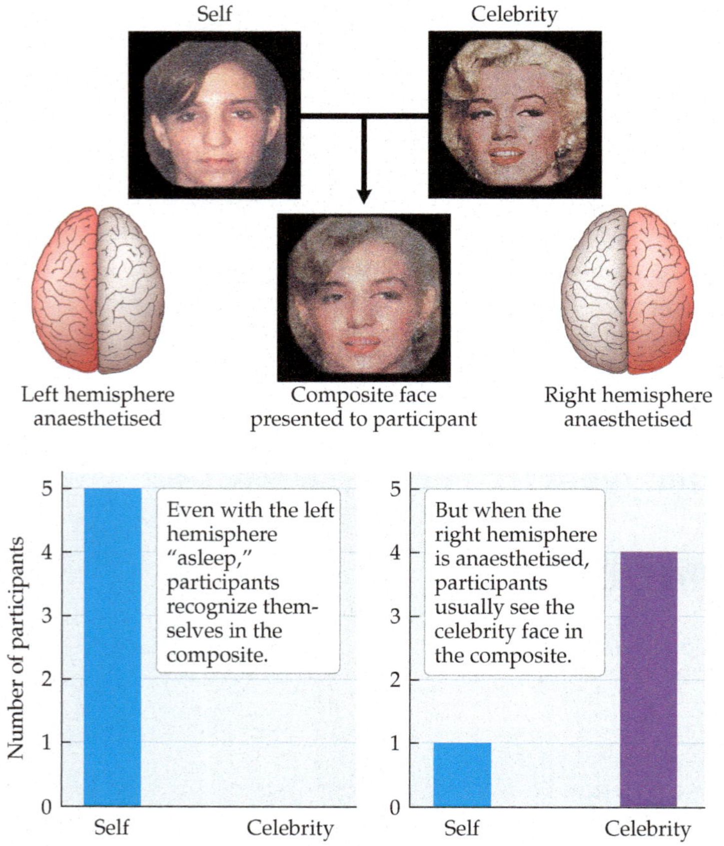

18.5 Use of the Right Hemisphere for Self-Face Recognition Anaesthetising the left hemisphere in a Wada test (see Box 18.1) does not interfere with study participants' ability to recognise their own faces in pictures that are composites—of their faces and the faces of celebrities. But when the right hemisphere is anaesthetised, the participants' facial recognition is impaired, especially for their own faces, leading to guesswork and more frequent identification of the composite faces as those of the celebrities. (From J. P. Keenan et al., 2001. *Nature* 409: 305, courtesy of Julian Keenan.)

recognise *objects* may be retained, and the person may have no difficulty identifying familiar people by their voices.

To people with prosopagnosia, faces simply lack meaning. No disorientation or confusion accompanies this condition, nor is there evidence of diminished intellectual abilities. Visual acuity is maintained, although the majority of people have a small visual-field defect—that is, an area of the visual field where they are blind. Most research indicates that the right hemisphere is more important than the left for recognising faces. For example, shutting down the right hemisphere with anaesthetic during the Wada test (see Box 18.1) can cause difficulty in recognising faces, whereas anaesthetising the left hemisphere has less effect on facial recognition. This effect can be especially evident for recognising one's own face (**FIGURE 18.5**).

Similarly, split-brain individuals do a better job of recognising faces presented to the right hemisphere than to the left (M. S. Gazzaniga and Smylie, 1983). Still, data from split-brain individuals and functional imaging tests make it clear that both hemispheres have *some* capacity for recognising faces. Thus, although damage restricted to the right hemisphere can impair face processing, the most complete cases of prosopagnosia are caused by *bilateral* damage. The **fusiform gyrus**, a region of cortex on the inferior surface of the brain where the occipital and temporal cortices meet (**FIGURE 18.6**), is a key component of a network specific to face recognition (Rezlescu et al., 2014) and seems to play a special role in recognising stimuli within specific categories (Grill-Spector et al., 2017; Weiner and Zilles, 2016). Cases of prosopagnosia following brain damage almost always involve damage here. Prosopagnosia may be accompanied by additional forms of **agnosia** (an inability to identify items, in the absence of specific sensory impairments) (Gauthier et al., 1999). For example, bird watchers may lose

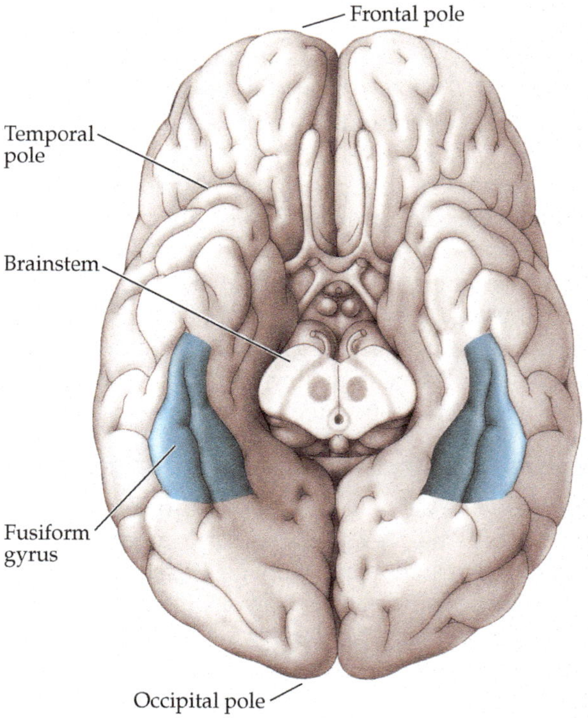

18.6 The Fusiform Gyrus In this view from below the brain, the cerebellum has been removed to reveal the region of cortex that normally lies opposite to it. The fusiform gyrus, at the juncture of the temporal and occipital lobes, is active during discrimination of objects in a large category, such as faces or birds or cars. Bilateral destruction of this region leads to prosopagnosia, the inability to recognise individual faces.

the ability to distinguish between avian species. Indeed, fMRI studies of healthy participants show that the fusiform region is activated not only when people are identifying faces, but also when identifying birds, or cars, or individual members of many other categories (Gauthier et al., 2000). It remains to be seen exactly how many distinct categories are individually represented in the brain, and the extent to which brain regions other than the fusiform gyrus may participate (Connolly et al., 2012; Haxby, 2006).

Until recently, it was believed that prosopagnosia occurred solely as a result of brain damage (*acquired prosopagnosia*), but we now know that *congenital* (or *developmental*) *prosopagnosia*—lifelong face blindness without brain damage—is surprisingly widespread. In fact, surveys indicate that about 2.5% of the general population is sufficiently impaired in processing faces to meet the criteria for congenital prosopagnosia (Duchaine et al., 2007; Kennerknecht et al., 2006), confirming the widespread belief that some people are just "bad with faces." This congenital form of prosopagnosia appears to run in families, suggesting a genetic aspect to the disorder (Grüter et al., 2008). Congenital prosopagnosia is associated with diminished activation of the fusiform gyrus and reduced white matter connectivity in the ventral occipitotemporal cortex, in keeping with the anatomical findings in acquired prosopagnosia that we already discussed. At the other end of the spectrum, some people are exceptionally *good* with faces. These "super-recognisers," who are about as good with faces as people with prosopagnosia are bad, are actively recruited by some police forces (Robertson et al., 2016; R. Russell et al., 2009). (You can learn more, and test your own facial recognition ability using the Famous Faces test, at www.testmybrain.org.)

fusiform gyrus A region on the inferior surface of the cortex, at the junction of temporal and occipital lobes, that has been associated with recognition of faces.

agnosia The inability to recognise objects, despite being able to describe them in terms of form and colour. It may occur after localised brain damage.

phoneme A sound that is produced for language.

morpheme The smallest grammatical unit of a language; a word or meaningful part of a word.

semantics The meanings or interpretation of words and sentences in a language.

syntax The grammatical rules for constructing phrases and sentences in a language.

pragmatics In linguistics, the context in which a speech sound is uttered.

aphasia An impairment in language understanding and/or production that is caused by brain injury.

18.3 Left-Hemisphere Damage Can Cause Aphasia

Learning Objectives

After studying this section, you should be able to:

18.3.1 **Distinguish among aphasia, apraxia, agraphia, and alexia.**

18.3.2 **Describe Broca's area and the behavioural consequences of lesions that include this area.Identify Wernicke's area, and describe the behavioural consequences of lesions that include this area.**

18.3.3 **Describe global aphasia, its causes and prognosis, and its relationship to concepts of personal identity.**

A human language is made up of basic speech sounds, or **phonemes**, that are assembled into simple units of meaning called **morphemes** (**FIGURE 18.7**). Morphemes are assembled into words (the word *unfathomable*, for example, consists of the morphemes *un-*, *fathom*, and *-able*), which have meaning, termed **semantics** in linguistics. In turn, the words are assembled into meaningful strings (which may be complete sentences or just phrases) according to the language's **syntax** (grammatical rules). Our speech is coloured and clarified both by the context of the utterance, known as **pragmatics** in linguistics, and by the emotional tone and emphasis, called *prosody* that we add to the things we say.

We have known for millennia that localised brain damage can disrupt our use of the basic speech sounds and grammar, and in severe cases it results in profound impairment of language abilities, known as **aphasia**. Medical texts of ancient Egypt describe speech problems in people who had received blows to the side of the head (Finger, 1994), and many individual cases of speech loss after strokes, war wounds, and other brain injuries were recorded over the years. Some 25–50% of all people who have a significant stroke will experience aphasia as a primary symptom; that's why speech difficulty is considered to be one of the core warning signs of a stroke (see Figure 1.22), along with weakness or numbness on one side and dizziness, altered vision,

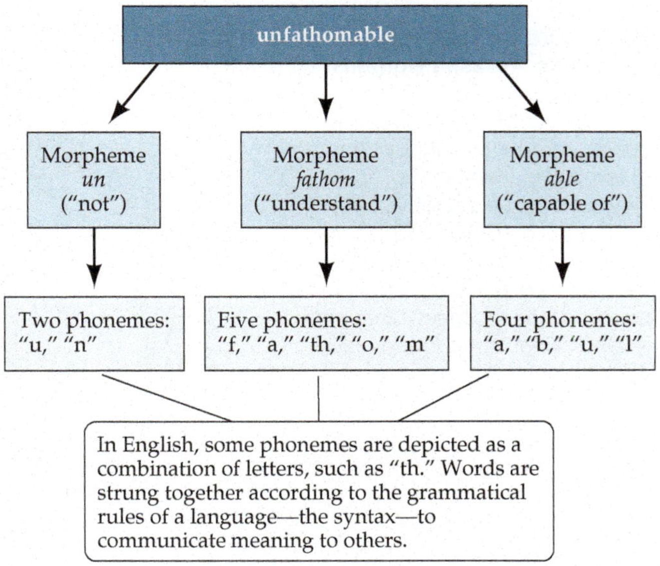

18.7 Anatomy of a Word A word can be broken down into morphemes—these are simple units of meaning and not the same as *syllables*, which relate to pronunciation. Each morpheme can be defined as an aggregate of phonemes, which are simple speech sounds.

or confusion. And in agreement with the evidence of left-hemisphere specialisation for language, 90–95% of aphasia cases are the result of damage to the left cerebral hemisphere. Some individuals will lose almost all capacity for speech, but in cases where the left-hemisphere damage is less severe, the individual may instead show pronounced **paraphasia**—insertion of incorrect sounds or words—along with laboured, effortful speech production. At times entirely novel nonsense words—called **neologisms**—may be generated via the insertion or substitution of one or more phonemes. Conversation reveals another important aspect of speech: its fluency or ease of production. **Nonfluent speech** is talking with considerable effort, in short sentences, and without the usual melodic character of conversational speech.

paraphasia A symptom of aphasia that is distinguished by the substitution of a word by a sound, an incorrect word, an unintended word, or a neologism.

neologism An entirely novel word, sometimes produced by a person with aphasia.

nonfluent speech Talking with considerable effort, short sentences, and the absence of the usual melodic character of conversational speech.

agraphia The inability to write.

alexia The inability to read.

apraxia An impairment in the ability to begin and execute skilled voluntary movements, even though there is no muscle paralysis.

Most people with aphasia also have **agraphia** (trouble with writing) and **alexia** (trouble with reading). Brain damage that results in aphasia also produces a distinctive motor impairment called **apraxia** (see Chapter 10), characterised by a great difficulty in executing precise *sequences* of movements that is not related to weakness, paralysis, a lack of coordination, or any kind of sensory problems. In fact, we'll see shortly that some researchers view aphasia as being primarily the result of problems with motor mechanisms. Curiously, people with aphasia seem to outperform in one domain: they are better than unaffected people at detecting when someone is lying (Etcoff et al., 2000). Perhaps this sensitivity to lying results from a focuss on facial features of speakers: their undamaged right hemisphere is important for face processing.

Linking specific language disorders to damage in specific brain regions provides clues about the organisation of the left-hemisphere language network. **FIGURE 18.8** shows some of the most important language-related regions of the left hemisphere of the brain. Damage in these regions tends to produce clusters of symptoms that define several distinct types of aphasia (**TABLE 18.1**; videos of people with aphasia can be seen on the website). In the following sections, we look in more detail at the three primary aphasias.

Go to Activity 18.1
Speech and Language Areas of the Brain

Lesions of a left anterior speech zone cause nonfluent (or Broca's) aphasia

In the 1860s French neurologist Paul Broca (1824–1880) examined a man who had lost the ability to utter much more than the single syllable *tan*. Following postmortem analysis, Broca reported that this man, and other people with similar severe

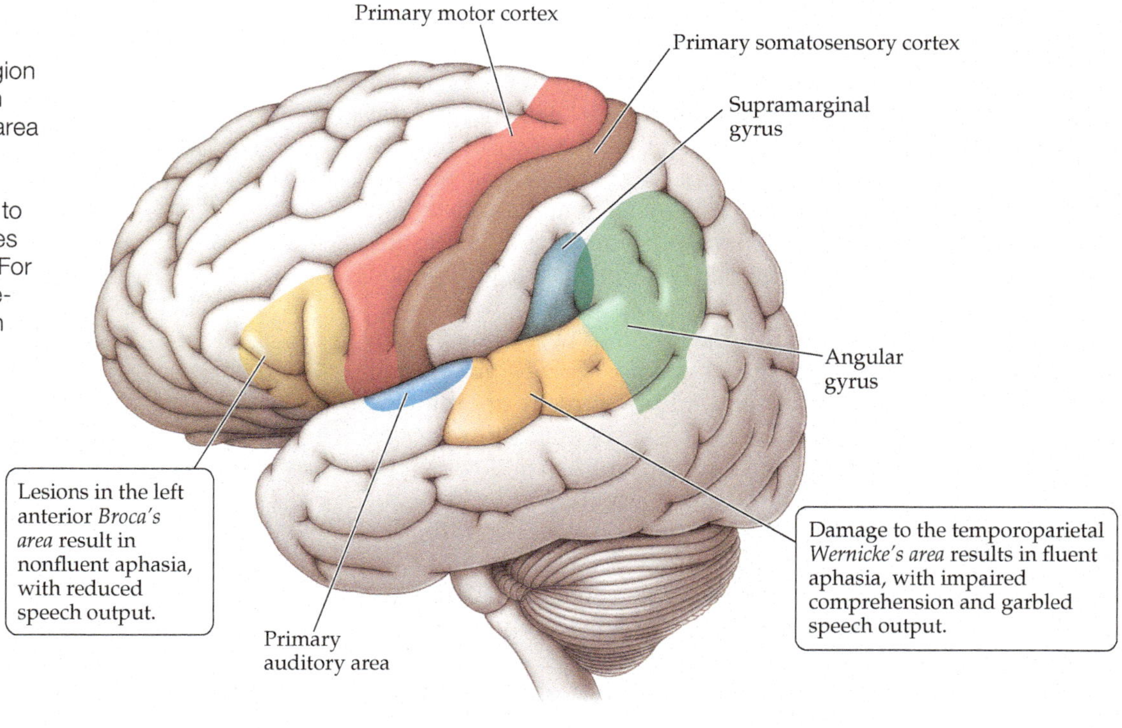

18.8 Cortical Speech and Language Areas in Humans Lesions in the anterior frontal region called *Broca's area* interfere with speech production; injury to an area of temporoparietal cortex called *Wernicke's area* interferes with language comprehension; injury to the supramarginal gyrus interferes with repetition of heard speech. For most individuals, these language-related systems are found only in the left hemisphere.

TABLE 18.1 Language Symptomatology in Aphasia

Type of aphasia	Brain area affected	Spontaneous speech	Comprehension	Paraphasia	Repetition	Naming
Nonfluent (Broca's) aphasia	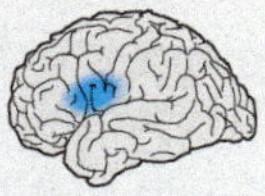	Nonfluent	Good	Uncommon	Poor	Poor
Fluent (Wernicke's) aphasia	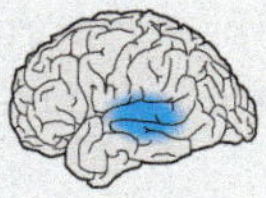	Fluent	Poor	Common	Poor	Poor
Global aphasia	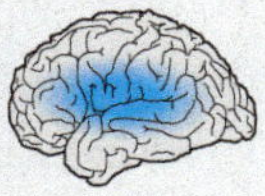	Nonfluent	Poor	Variable	Poor	Poor
Conduction aphasia	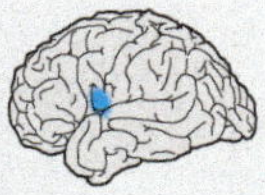	Fluent	Good	Common	Poor	Poor
Subcortical aphasia	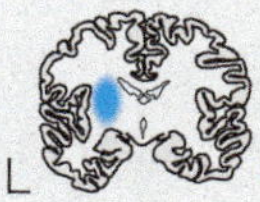	Variable	Variable	Common	Good	Variable

impairments of speech production, had sustained damage to a left inferior frontal region that now bears his name—**Broca's area** (**FIGURE 18.9**). Left anterior lesions that include Broca's area often produce a type of aphasia known as **nonfluent aphasia** (or *Broca's aphasia*). People with nonfluent aphasia have considerable difficulty producing speech, talking only in a laboured and hesitant manner. Reading and writing are also impaired. The ability to utter automatic speech, however, is often preserved. Such speech includes greetings ("Hello"); short, common expressions ("Oh my gosh!"); and swear words.

In contrast to their difficulty with speech production, *comprehension* of language is relatively preserved in nonfluent aphasics. Because the primary and supplementary motor cortex is close to Broca's area (see Chapter 10), brain injuries that cause nonfluent aphasia often also cause **hemiplegia**—paralysis of one side of the body (usually the right side, which is controlled by the left hemisphere). Sometimes there is unilateral weakness, termed **hemiparesis**, rather than full paralysis.

Seven years after a stroke, one person with nonfluent aphasia still spoke slowly, using mainly nouns and very few verbs or function words (a selective loss of action words, called *averbia*, sometimes occurs in nonfluent aphasia), and spoke only with great effort. When asked to repeat the phrase "Go ahead and do it if possible," she could say only, "Go to do it," with a pause after each word. Her pattern is typical of people with extensive left anterior lesions. People with brain lesions as extensive as hers tend to show little recovery of speech functions with the passage of time, but people

Broca's area A region of the frontal lobe of the brain that is involved in the production of speech.

nonfluent aphasia Also called *Broca's aphasia*. A language impairment characterised by difficulty with speech production but not with language comprehension. It is related to damage in Broca's area.

hemiplegia Partial paralysis involving one side of the body.

hemiparesis Weakness of one side of the body.

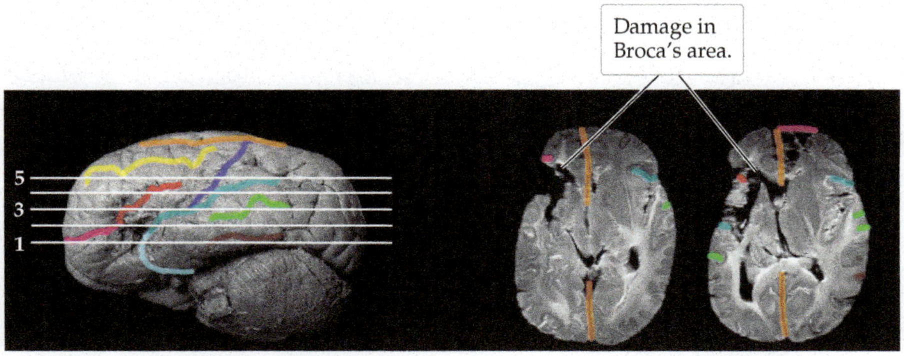

From N. F. Dronkers et al., 2007. *Brain* 130: 1432–1441

18.9 The Brain of "Tan" Photo and MRI image of the preserved brain of M. Leborgne, who could only utter the syllable *tan* after his brain injury. Study of this brain and similar cases led Paul Broca to identify a region in the anterior left hemisphere that is specialised for speech. The damage in what is now known as *Broca's area* is clearly evident in the photo (left) and in the horizontal MRI images (right) that correspond to levels 2 and 3 in the photo.

Wernicke's area A region of temporoparietal cortex in the brain that is involved in the perception and production of speech.

with milder damage can show significant recovery. The CT scans and lesion maps in **FIGURE 18.10A AND B** map the lesion sites of several people with nonfluent aphasia.

Lesions of a left posterior speech zone cause fluent (or Wernicke's) aphasia

Not long after Broca's groundbreaking discovery of the left anterior speech zone, German neurologist Carl Wernicke (pronounced "VER-nih-keh") (1848–1905) described a different form of aphasia resulting from damage to a more posterior aspect of the left hemisphere, located on the upper part of the temporal lobe where it joins the parietal lobe, that is now known as **Wernicke's area** (see Figure 18.8). Unlike the nonfluent

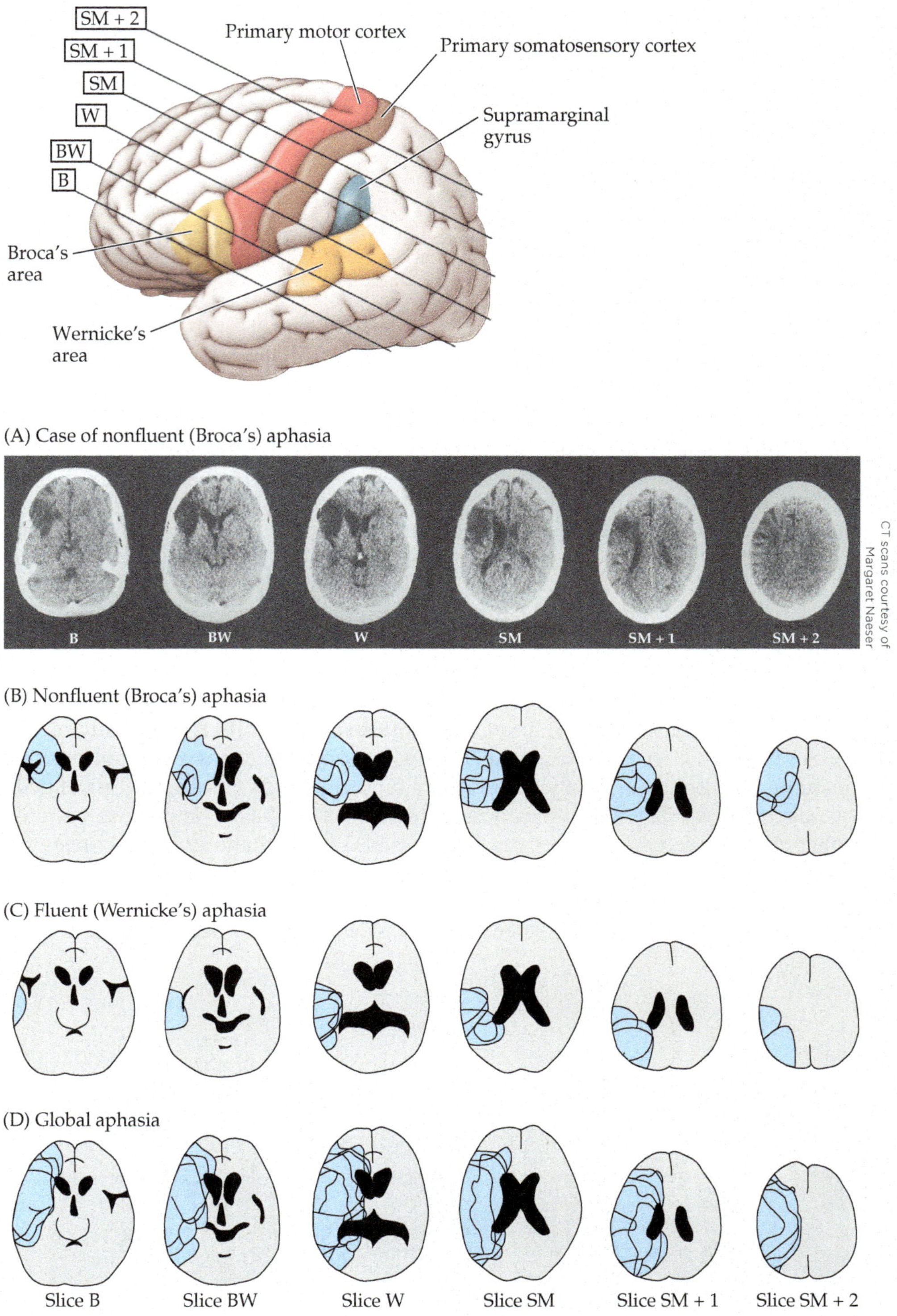

18.10 Brain Lesions That Produce Aphasia The level of each CT scan slice is labelled according to the brain language region(s) shown by that slice: B, Broca's area; SM, supramarginal gyrus; W, Wernicke's area. (A) CT scans for one person with nonfluent aphasia, age 51, 7 years after a stroke. (B) Overlapping lesion maps for four individuals with nonfluent aphasia, at six levels of the brain. Large lesions (blue) are located in Broca's area on slices B and BW, and the peak amount of tissue damage occurred in the frontoparietal areas on slices SM and SM + 1. (C) Lesion sites for four cases of fluent aphasia. Lesions are located in Wernicke's area on slice W and in the supramarginal gyrus area on slice SM. (D) Lesion sites for five cases of global aphasia. Large lesions are present in every language area. (After M. Naeser and R. Hayward, 1978. *Neurology* 28: 545–551.)

aphasics, people who have this **fluent aphasia** (or *Wernicke's aphasia*) produce plenty of verbal output, but their utterances, although speechlike, tend to contain many paraphasias that make their speech unintelligible. The paraphasias may involve sound substitutions (e.g., "girl" becomes "curl") and/or word substitutions (e.g., *bread* becomes *cake*). At times, word-like nonsense neologisms are generated. Some fluent aphasias are marked by particular difficulty in naming persons or objects—an impairment referred to as **anomia**. Paraphasias and speech errors occur in a context that preserves syntactic structure, although sentences seem empty of content. The ability to repeat words and sentences is impaired. For this reason, people with fluent aphasia are believed to have difficulty *understanding* what they read or hear.

In fluent aphasia the most prominent brain lesions are in posterior regions of the left superior temporal gyrus and extend partially into adjacent parietal cortex, including the nearby supramarginal and angular gyri (**FIGURE 18.10C** and see Figure 18.8). When *word deafness* (the inability to understand spoken words) is more evident than reading impairment, damage is more likely in the superior temporal lobe and in tracts from the auditory cortex. In contrast, *word blindness* (the inability to understand written words) is associated damage that includes connections to visual regions—often attributed to destruction of the **angular gyrus**. Because the typical lesion in fluent aphasia is posterior, postcentral somatosensory cortex is more likely to be damaged than precentral motor cortex. As a result, people with fluent aphasia are somewhat more likely to have a right-sided numbness than the weakness common in nonfluent aphasia.

Speechless Tinna Guela Phillips had a devastating stroke that robbed her of all language abilities, including her inner monologue.

Widespread left-hemisphere damage can obliterate language capabilities

In some people, brain injury or disease results in near-total loss of the ability to understand or produce language. This syndrome, called **global aphasia**, was the diagnosis in the case of Tinna, the woman whom we met at the beginning of the chapter. People with global aphasia may retain some ability to make automatic speechlike sounds, especially emotional exclamations. But they can utter very few words, and no semblance of syntax remains. Global aphasia generally results from very large left-hemisphere lesions that encompass both anterior and posterior language zones. Frontal, temporal, and parietal cortex—including Broca's area, Wernicke's area, and the supramarginal gyrus—are usually affected (**FIGURE 18.10D**). For nearly 2 years after having a massive left-hemisphere stroke in her 40s, it was unclear whether Tinna would ever regain the ability to communicate in any of her six languages, and even her inner monologue was banished for several months, taking with it her personal identity. Although Tinna regained some degree of both outward and inward language, her verbal abilities remain quite impoverished. This is in keeping with the generally poor prognosis for language recovery in global aphasia. And because of the extent of their lesions, people with global aphasia often experience additional debilitating neurological impairments. (Tinna and several other people with aphasia are profiled in a documentary entitled *Speechless*, by filmmaker Guillermo F. Florez; a theatrical trailer and information about the film are available at speechlessdoc.com.)

fluent aphasia Also called *Wernicke's aphasia*. A language impairment characterised by fluent, meaningless speech and little language comprehension. It is related to damage in Wernicke's area.

anomia The inability to name persons or objects readily.

angular gyrus A brain region in which strokes can lead to word blindness.

global aphasia The total loss of ability to understand language or to speak, read, or write.

18.4 Competing Models Describe the Left-Hemisphere Language System

Learning Objectives

After studying this section, you should be able to:

18.4.1 **Describe the Wernicke-Geschwind model of language organisation.**

18.4.2 **Contrast the motor theory of speech with the Wernicke-Geschwind model, and discuss research findings for and against each.**

18.4.3 **Summarise research using brain stimulation, functional brain imaging, and ERPs to map the brain's language network.**

connectionist model of aphasia The theory proposing that Wernicke's area and Broca's area, connected by the arcuate fasciculus, specialise in the receptive and expressive aspects of language, respectively.

arcuate fasciculus A tract believed by some to connect Wernicke's area to Broca's area.

conduction aphasia An impairment in the repetition of words and sentences.

Go to Activities 18.2a and 18.2b
The Traditional Connectionist Model of Aphasia

Considering how language is intertwined with so many other functions, it is not surprising that a complete description of the brain's language circuitry remains elusive, although steady progress has been made. The traditional **connectionist model of aphasia**—also known as the *Wernicke-Geschwind model* after its chief proponents Carl Wernicke and, later, Norman Geschwind (1926–1984)—argues that language deficits due to brain damage result from *disconnection* between brain regions that form a language network. Each of these regions is proposed to serve a particular feature of language analysis or production (**FIGURE 18.11A**). So for example, when a word or sentence is heard, the auditory cortex transmits information about the sounds to a speech reception mechanism in Wernicke's area, which analyses the sounds to decode what they mean. The model then proposes that in order for the word to be spoken, Wernicke's area transmits this information, via a bundle of axons called the **arcuate fasciculus**, to the expressive mechanism in Broca's area, where a speech plan is activated. Broca's area then transmits this plan to adjacent motor cortex, which controls muscles of the chest, throat, and mouth that are used for speech production (Geschwind, 1972). The model therefore argues that people with lesions that selectively disrupt the arcuate fasciculus will especially struggle with *repetition* of words and phrases that they hear, despite good speech comprehension and production—a condition termed **conduction aphasia**.

According to the connectionist model, saying the name of an object or word that is *seen*, rather than *heard*, involves the transfer of visual information to the

18.11 The Traditional Connectionist Model of Aphasia (After N. Geschwind, 1979. *Sci Am* 241: 180–199.)

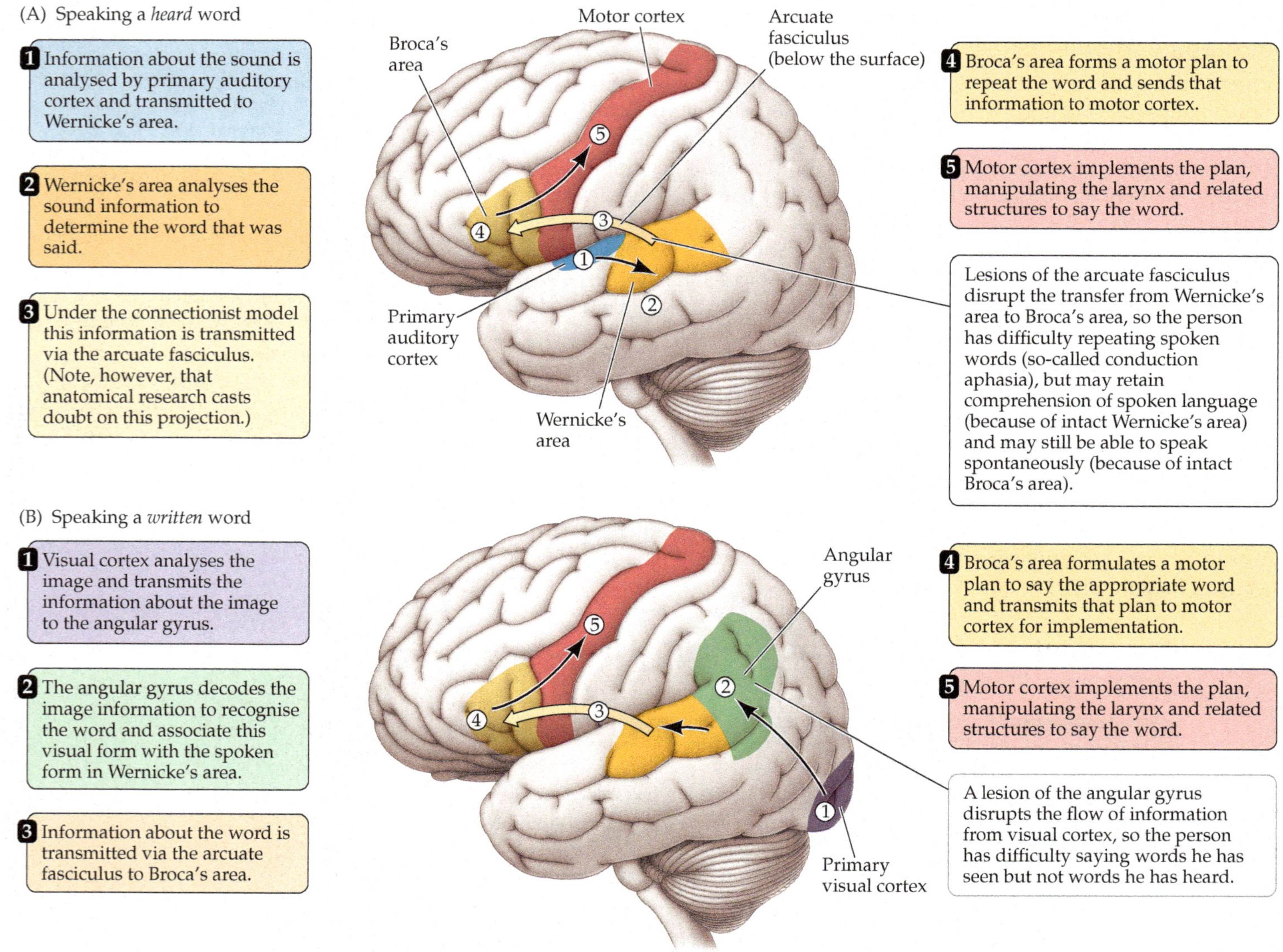

angular gyrus, which then activates the auditory form of the word in Wernicke's area. From Wernicke's area the auditory form is transmitted via the arcuate fasciculus to Broca's area, which prepares the motor plan for the word and transmits it to the motor cortex, as for words that are heard (**FIGURE 18.11B**). In this model, lesions affecting the angular gyrus would thus disconnect the systems involved in visual and auditory language; individuals with lesions in this region would be expected to have difficulty reading aloud but would retain the ability to speak and understand speech.

Critics of the Wernicke-Geschwind connectionist model argue that it oversimplifies the neural mechanisms of language and prematurely attached functional labels to anatomical findings. For one thing, although anterior- and posterior-lesioned people with aphasia differ greatly with regard to fluency (D. F. Benson, 1967; Goodglass et al., 1964), the expressive versus receptive dichotomy of the Wernicke-Geschwind model doesn't hold up very well. Thus, anterior aphasics seem to have difficulty in comprehension of some aspects of speech in addition to their expression problems, and posterior aphasics make speech production errors despite their fluency (Kimura, 1993). Likewise, modern fMRI studies show that left-hemisphere language zones are not as rigidly modular as previously believed. For example, the arcuate fasciculus—long believed to connect Wernicke's area to Broca's area—actually terminates in the precentral gyrus (motor cortex), just short of Broca's area, in most people (**FIGURE 18.12**) (Bernal and Altman, 2010; E. C. Brown et al., 2014).

An alternative perspective—the **motor theory of language**—proposes that the left-hemisphere language zones are motor control systems that are concerned with both the precise production and the *perception* of the extremely complex movements that go into speech (Kimura, 1993; Lieberman, 1985). According to this view, the anterior left hemisphere programs the simple phonemic units of speech, and the posterior systems are responsible for stringing speech sounds together into long sequences of movements (Kimura and Watson, 1989). The motor theory further proposes that when we *listen* to speech, we are using elements of this same system to perceive the very rapid sequences of facial, throat, tongue, and respiratory movements that the speaker is making. From an evolutionary perspective, this arrangement is economical because we can use the same evolved neural substrate—probably based on preexisting cortical and basal ganglia systems used for controlling other movements—to both produce and perceive speech sounds (Lieberman, 2002). It is possible that mirror neurons—specialised neurons that are active in both the production and perception of movements (see Chapters 8 and 10)—are an important part of this process, as one of the regions in which mirror neurons are found in the monkey brain corresponds to part of Broca's area in humans (Pulvermüller and Fadiga, 2010). Interestingly, deaf people who use American Sign Language—a language based entirely on hand and arm movements—employ the same language-related regions of the left hemisphere while signing as hearing people use for spoken language (Neville et al., 1998; Petitto et al., 2000). Following focal left-hemisphere damage, deaf signers also show highly similar language impairments: in most respects, aphasia in signers is the same as aphasia in speakers (Bellugi et al., 1983; Kimura, 1981).

motor theory of language
The theory proposing that the left-hemisphere language zones are motor control systems that are concerned with both the precise production and the perception of the extremely complex movements that go into speech.

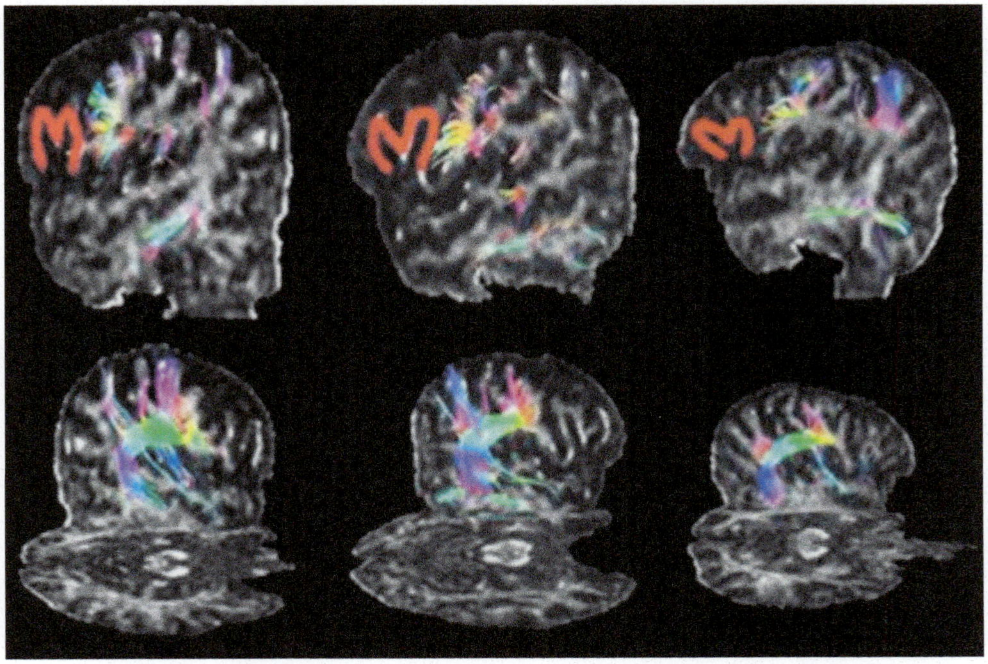

From B. Bernal and N. Altman, 2010. *Magn Reson Imaging* 28: 217–225

18.12 DTI and the Connectionist Model of Left-Hemisphere Language In these DTI tractography (see Figure 2.24) images of three different people, the region containing Broca's area (anatomically known as the *inferior frontal gyrus*) has been outlined in red. The two ends of the left arcuate fasciculus are coloured in each upper image, and the entire course of the arcuate fasciculus is shown in each lower figure (in a mix of colours reflecting different fibre orientations). Note that in each case, the tractographic imaging reveals that the arcuate fasciculus terminates in the vicinity of the precentral gyrus, but well short of Broca's area.

In any event, detailed structural-imaging studies have revealed that the left hemisphere and the rest of the brain contain several speech areas well outside of the classical Broca's and Wernicke's areas (E. Bates et al., 2003; Dronkers et al., 2004), implying greater complexity than was proposed under the Wernicke-Geschwind model. Damage that involves some of these other areas likely explains the symptoms typically attributed to conduction aphasia and other aspects of the traditional aphasia model (B. R. Buchsbaum et al., 2011). Further, recent evidence that low-level aspects of speech perception occur bilaterally (Cogan et al., 2014) and that semantic processing of natural speech relies heavily on both hemispheres (de Heer et al., 2017; Huth et al., 2016) shows that much remains to be determined about exactly what is, and what is not, the exclusive domain of the left-hemisphere language network. Whatever the details may be, it seems that the answer will involve a complex network of mechanisms that link perception to action.

Brain mapping helps us understand the organisation of language in the brain

In healthy people with intact language capabilities, researchers can study the brain's language network using two general experimental approaches. In some studies, researchers stimulate discrete regions of the cortex and measure associated changes in language function. Conversely, researchers may ask participants to engage in specific verbal behaviours and measure associated changes in brain activity, using functional imaging. Together, these techniques have extended our understanding of the neural bases of language. Early studies of the organisation of language areas in the brain employed electrical stimulation mapping of the brain, in which a surgeon used a handheld electrode to electrically stimulate discrete regions of cortex while the effect on behaviour was observed. Most of these studies collected data from neurosurgical patients undergoing surgery to remove a tumour or epileptic tissue (as in Wilder Penfield's experiments, described in Chapter 1). Once the person's cortex was exposed, small stimulating electrodes were touched to the cortical surface, temporarily disrupting the normal functioning of neurons in the immediate vicinity. Because patients were given only local anaesthesia, they were conscious and able to perform various cognitive tests—the main aim of the procedure was to identify tissue that could be removed without impairing language, memory, or other crucial cognitive functions. Data from numerous patients were superimposed to create a map of language-related zones of the left hemisphere (**FIGURE 18.13A**) (Penfield and Roberts, 1959). Stimulation anywhere within a large anterior zone often stopped speech outright. Other forms of language interference, such as

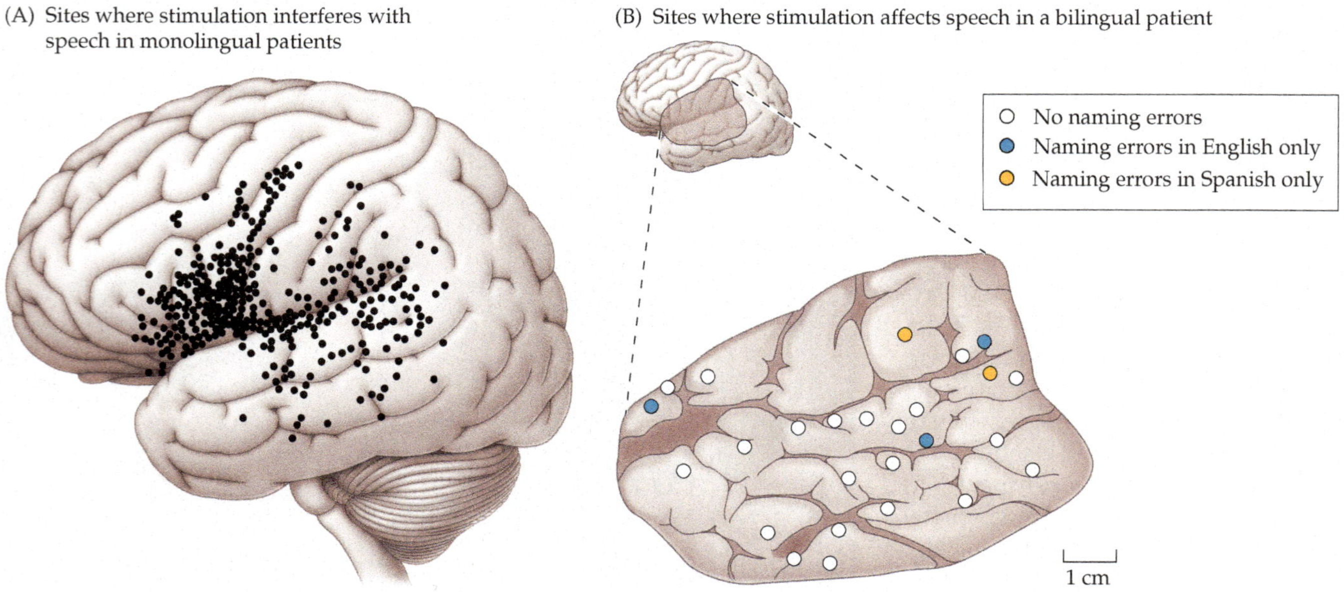

18.13 Electrical Stimulation of Some Brain Sites Can Interfere with Language (A) This summary of data obtained from many monolingual individuals shows the brain sites where stimulation interferes with speech production. (B) This map highlights stimulation sites that affected the speech of a person who was bilingual—fluent in Spanish and English. Within this single region affecting speech, different subregions interfere with either one language or the other, but not both. (Part A after W. Penfield and L. Roberts, 1959. *Speech and brain-mechanisms*. Princeton University Press: Princeton, NJ; B after W. H. Calvin and G. A. Ojemann, 1994. *Conversations with Neil's brain: The neural nature of thought and language*. Addison-Wesley: Reading, MA.)

misnaming or impaired repetition of words, occurred with stimulation throughout the anterior and posterior cortical speech zones.

Later cortical stimulation studies revealed anatomical compartmentalisation of key linguistic processes such as naming, reading, speech production, and verbal memory (Calvin and Ojemann, 1994). An interesting example of data collected with this technique is illustrated in **FIGURE 18.13B**, which shows the various locations in which cortical stimulation evoked naming errors in English versus Spanish, in a bilingual participant. Note that this very fine-grained approach revealed that different subregions appear to serve English versus Spanish language functions. People who are bilingual from an early age show complete overlapping of their two languages (Perani and Abutalebi, 2005) and an extended sensitive period for phoneme learning in early development (Petitto et al., 2012). In fact, evidence is mounting that bilingualism from an early age has far-reaching beneficial effects on brain organisation and resistance to cognitive decline in later life (Costa and Sebastián-Gallés, 2014; Perani et al., 2017).

Although electrical stimulation mapping provides excellent spatial resolution, it requires that the cortical surface be exposed during major brain surgery. However, it is also possible to stimulate cortical neurons noninvasively in healthy volunteers. As we described in Chapter 1, transcranial magnetic stimulation (TMS) provides a noninvasive method for altering the activity of cortical neurons located directly below a special electromagnet placed over the scalp (see Figure 2.25). By targeting this stimulation with reference to a baseline MRI, it is possible to use TMS as a sort of temporary brain lesion, disrupting the activity of targeted brain regions and noting the resultant changes in behaviour. Depending on how it is applied, TMS can disrupt neural function for up to an hour. Alternatively, TMS can be used along with PET or fMRI to precisely localise regions of increased activity.

Studies with TMS mapping generally confirm and extend our understanding of the cortical organisation of language functions (Devlin and Watkins, 2007). For example, TMS studies reveal that speech production is associated with activation of not only face areas in motor cortex but also hand areas, confirming the linkage and possible evolutionary relationship of hand gestures and speech (Meister et al., 2003). Speech *perception* apparently activates specific regions that TMS shows to be involved with speech *production* (S. K. Scott and Wise, 2004), supporting the motor theory of language that we discussed earlier. And, impressively, the TMS temporary-lesion approach has revealed functional *sub*regions within Broca's area (**FIGURE 18.14**). Specifically, a more anterior and ventral division of Broca's area appears to be important for the semantic *meaning* of words, while a more posterior part of Broca's area is important for phonological processing: the patterning of speech *sounds* (Devlin and Watkins, 2007; Gough et al., 2005). A similar experimental approach indicated that the supramarginal gyrus, within the posterior speech zone, is involved in both semantic and phonological aspects of language (Stoeckel et al., 2009); previously, this structure was thought to be involved in phonological processes only. So the TMS procedure is providing new insights into the fine details of cortical organisation of function, especially in conjunction with established neuroimaging techniques.

18.14 Subregions of Broca's Area Revealed by TMS TMS stimulation affects semantic processing (meaning) when applied to anterior regions of Broca's area, but similar stimulation of posterior regions of Broca's area affects phonological processing (sound patterns). (After J. T. Devlin and K. E. Watkins, 2007. *Brain* 130: 610–622; P. M. Gough et al., 2005. *J Neurosci* 25: 8010–8016.)

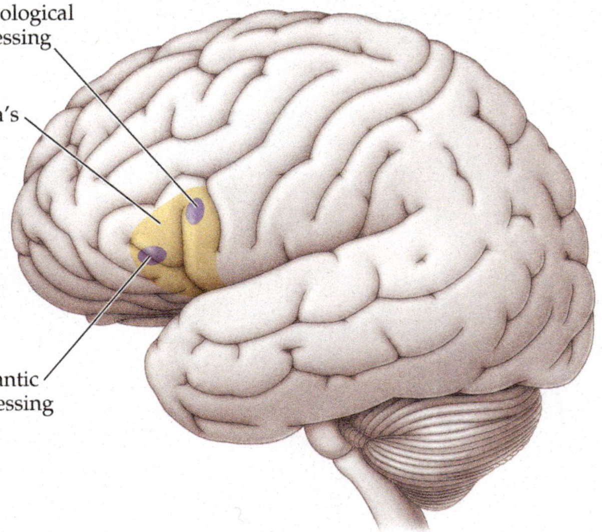

Functional neuroimaging tracks activity in the brain's language zones

Different aspects of verbal processing produce noticeably different patterns of brain activation in PET scans (**FIGURE 18.15**). Passive viewing of words activates a posterior area within the left hemisphere (**FIGURE 18.15A**), but passive *hearing* of words shifts the focuss of maximum brain activation to the temporal lobes (**FIGURE 18.15B**). Repeating the words orally activates the motor cortex of both sides, the supplementary motor cortex, and a portion of the cerebellum and insular cortex (**FIGURE 18.15C**). During word repetition or reading aloud, there is little activity in Broca's area. But when participants are asked to generate an appropriate verb to go with a supplied noun, language-related regions in the left hemisphere, including Broca's area, become markedly activated (**FIGURE 18.15D**).

Interestingly, the brain regions activated during reading are slightly different in native Italian speakers than in native English speakers (Paulesu et al., 2000). One possible explanation for this is that Italian is perfectly phonetic—the sound

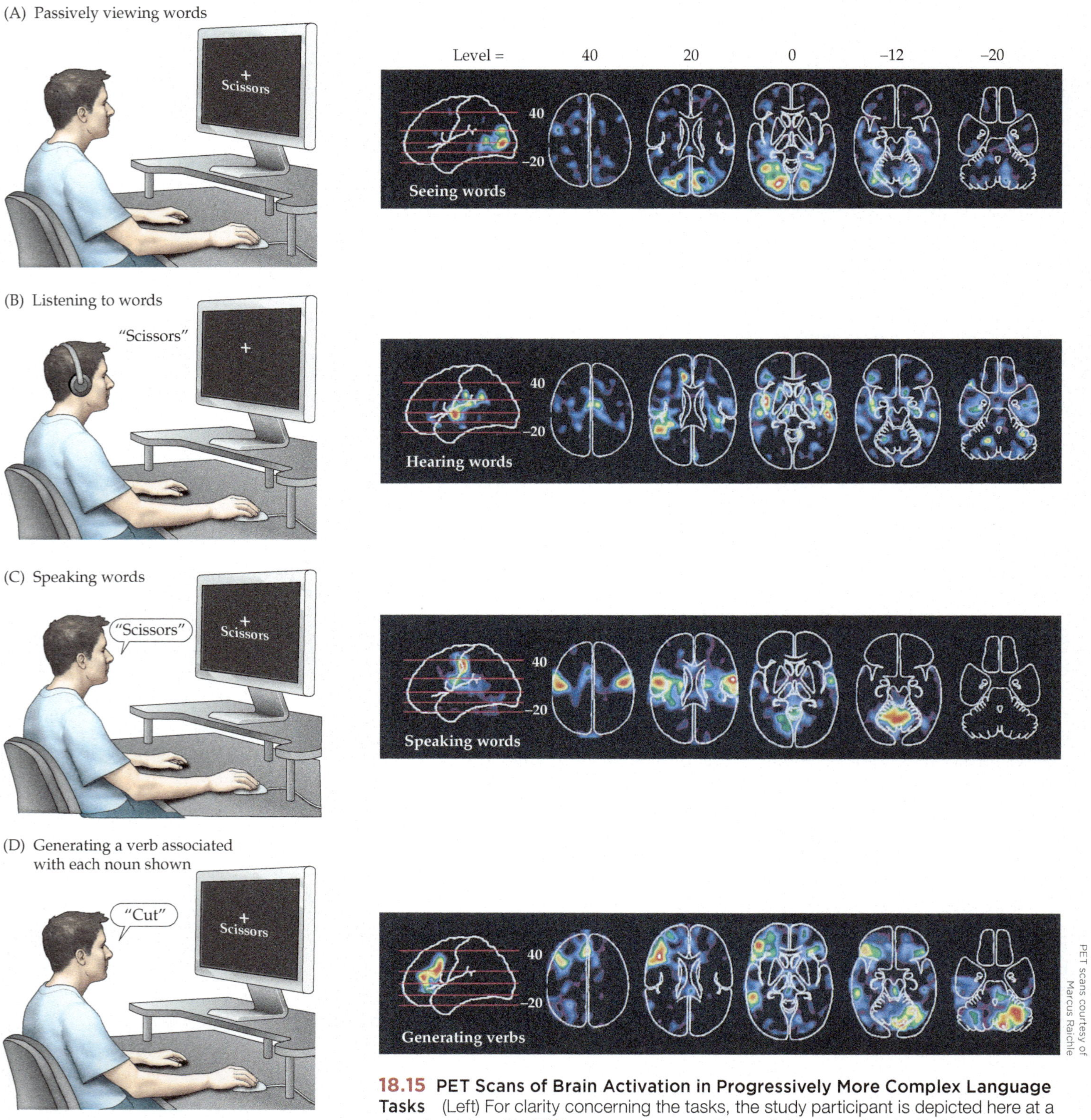

18.15 PET Scans of Brain Activation in Progressively More Complex Language Tasks (Left) For clarity concerning the tasks, the study participant is depicted here at a desk, but when the PET scans were made, the participants reclined with their heads in a PET scanner and viewed a specially mounted display. (Right) These PET scans correspond to the tasks at left; see the text for details of these results. (After M. I. Posner and M. E. Raichle, 1994. *Images of mind*. Scientific American Library: New York.)

associated with each letter and syllable is unvarying—whereas in English a given letter or letter combination may have very different sounds depending on the words it appears in. (For example, it has been noted that *ghoti* can be pronounced "fish" if you use the "gh" sound from *enough*, the "o" sound from *women*, and the "ti" sound from *nation*. Sheesh.) But overall, even languages that sound maximally different still tend to activate much the same brain regions in native speakers.

Silbo Gomero is a very unusual whistled language of the Canary Islands, used by shepherds (known as *silbadores*) to communicate over long distances. In Silbo, whistled notes serve as the phonemes and morphemes of a stripped-down form of Spanish (for an audio example of Silbo, with translation, see **A STEP FURTHER 18.2: WHISTLING UP SOME BRAIN ACTIVITY** on the website). Functional MRI reveals that silbadores process Silbo using the same left-hemisphere mechanisms that they (and nonsilbadore control participants) use to process spoken language (Carreiras et al., 2005). Nonsilbadore controls, in contrast, process the whistle sounds of Silbo using completely different regions of the brain; for these people, of course, the whistle sounds have no linguistic content. The findings with silbadores illustrate that the brain's language systems are flexible enough to adapt to widely varying types of communication sounds, provided they are heard early enough. In fact, while infants appear to be born with neural specialisations for *speech* already in operation—they show more metabolic activity in the left hemisphere than in the right when they hear speech, even though they don't yet understand it (Dehaene-Lambertz et al., 2002)—those mechanisms don't similarly respond to the non-speech sounds that make up Silbo Gomero (May et al., 2018). The incorporation of Silbo into the language system must therefore come about through extensive practice: the earlier the better.

Event-related potentials (ERPs; see Chapters 2 and 17) also provide hints about the brain's language network, by revealing the time base for language processing. For example, study participants can be asked to read a sentence in which there is a word that is grammatically correct but, because of its meaning, doesn't fit—for example, "The man started the car engine and stepped on the pancake." About 400 ms after the participant reads the word *pancake*, an enhancement of a distinctive ERP component called *N400* (*N* denotes "negative," and the number represents the response time in milliseconds; see Chapter 17) is detectable (Kutas and Hillyard, 1980, 1984). Such N400 responses seem to be specific to word meanings and apparently originate from temporoparietal cortex (including Wernicke's area) (Kutas and Federmeier, 2011; Neville et al., 1992). In contrast, words that are inappropriate because of *grammar*, rather than meaning, elicit a *positive* potential about 600 ms after they are encountered (a "P600" response), suggesting that detection of this level of error requires an extra 200 ms of processing by other elements of the brain's language network (Osterhout, 1997). The degree to which meaning and grammar are processed independently in the brain, however, remains to be fully established and is an area of active investigation (Brouwer et al., 2017; Kuperberg, 2007).

VERBAL BEHAVIOUR: SPEECH AND READING

No one knows for sure how many languages are in use in the world; most estimates put the figure in the range of 6000–7000 languages, of which about 1000 have been formally studied (Wuethrich, 2000). By applying the tools of evolutionary biology to linguistics, it is possible to track the cultural evolution of language—this suggests a closer relation between languages than previously expected. For example, the seven major language families of Europe and Asia may trace back to a common ancestral language of only about 15,000 years ago (**FIGURE 18.16**) (M. Pagel et al., 2013). The remarkable profusion of languages since then may have served cultural roles more than communicative ones—for example, the adoption of a new language would help a social group to identify its members and confound rivals. But nowadays, in the increasingly globalised modern context, we seem to be inexorably sliding in the other direction as languages are lost or absorbed and we move toward a few common languages.

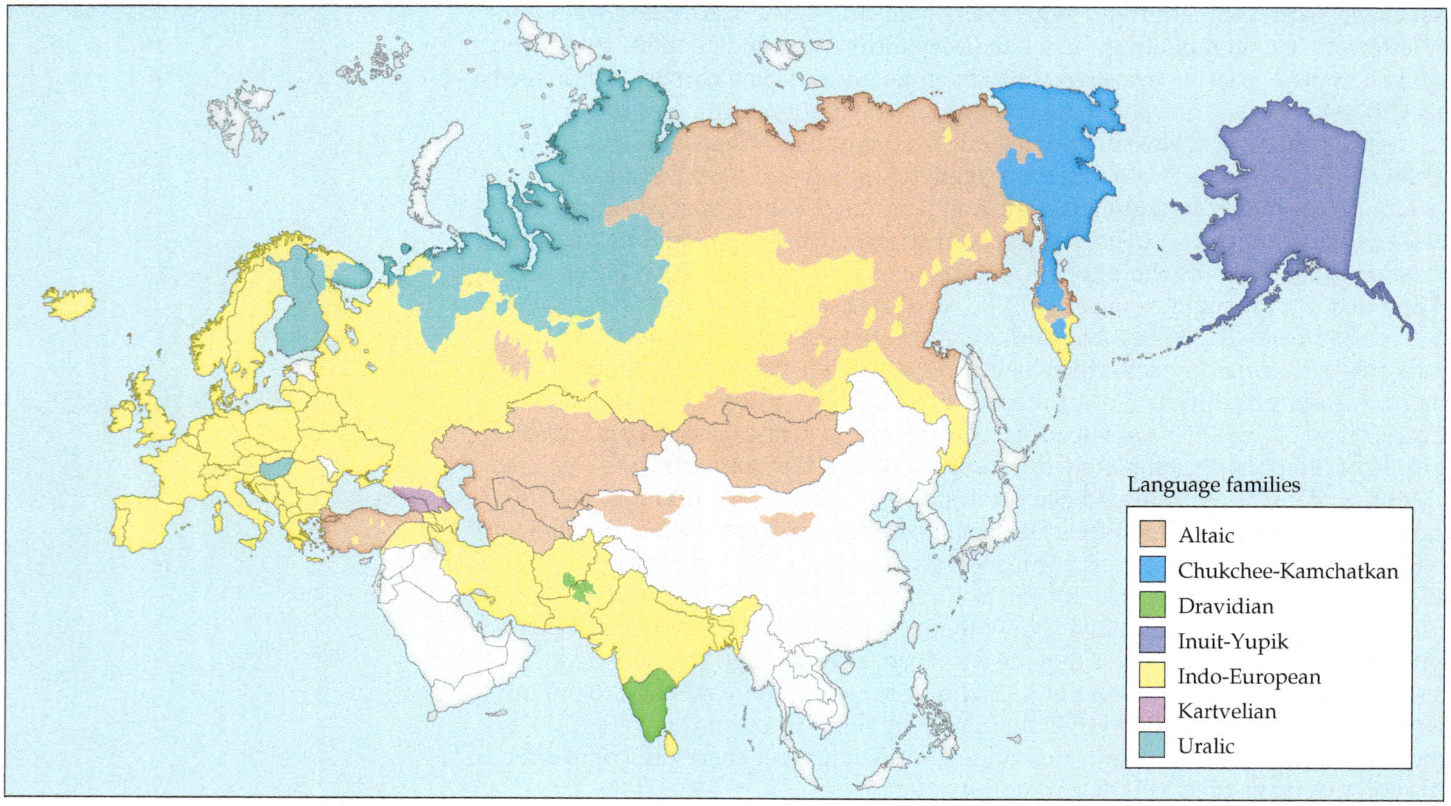

18.16 The Evolution of Languages This map shows the approximate distribution of the seven major language families of Europe and Asia; analysis of highly conserved features of these languages suggests that they shared a common ancestral language about 15,000 years ago—the time of the last Ice Age. (From M. Pagel et al., 2013. *Proc Natl Acad Sci USA* 110: 8471–8476; image courtesy of Mark Pagel.)

18.5 Language Has Both Learned and Unlearned Components

Learning Objectives

After studying this section, you should be able to:

18.5.1 **Summarise the developmental time course for a child's acquisition of language.**

18.5.2 **Discuss the genetic bases of language acquisition and production.**

18.5.3 **Provide an account of the evolution of language, and relate this to the language abilities of apes.**

18.5.4 **Describe the form and neural foundations of vocal behaviour in other animals.**

All human languages share certain basic elements. But how does each of us end up with the right set of sounds and rules—the ones we need for our particular native language?

A child's brain is an incredible linguistic machine, rapidly acquiring the phonemes, vocabulary, and grammar of the local language without need of formal instruction. Newborn babies are able to discern all of the sounds of *every* language, from Icelandic to Inuktitut, and from Japanese to Javanese, and begin life babbling nearly all the known phonemes of all human languages. However, babies soon start to pay special attention to the particular speech sounds in use all around them and begin to use only that subset of phonemes in preverbal babbling. A baby's developing language abilities are further shaped by "parentese," the singsong speech that parents universally use with their babies (Falk, 2004). The lilting qualities of parentese convey emotion and reward, helping to attach meaning to previously arbitrary speech sounds.

By 7 months of age, infants pay more attention to sentences with unfamiliar structure than to sentences with familiar structure (Marcus et al., 1999), indicating that they have already acquired a basic sense of the rules of the language being spoken around them and are actively looking out for exceptions. In rare, tragic cases of children rescued from long-term profound isolation, little or no language ever develops, pointing to the importance of experience during a **sensitive period**

sensitive period The period during development in which an organism can be permanently altered by a particular experience or treatment.

(or *critical period*) early in life. Similarly, when hearing was restored to an adult who had been deaf most of her life, she did not learn to speak (Curtiss, 1989), presumably because the sensitive period for language acquisition had closed many years earlier.

Further evidence for a sensitive period in language acquisition comes from the well-known difficulty of learning a second language in adulthood. Children seem to learn multiple languages with ease, and functional-imaging studies show that people who learned a second language in childhood tend to activate the same brain regions when using either language in adulthood. But people who learn their second language after about age 11 seem to use a different brain region for each language (K. H. Kim et al., 1997).

Analysis of a British family with a rare heritable language disorder—the KE family—identified a gene that appears to be important for the normal acquisition of language (**FIGURE 18.17**). Children with a specific mutation of the gene *FOXP2** take a long time to learn to speak, and they display long-lasting difficulties with particular language tasks (such as learning verb tenses) (C. S. L. Lai et al., 2001). The pattern of brain activation in these individuals while performing a language task is different from

*The several different versions of this gene name are no accident. By agreement, scientists use *FOXP2* for humans, *FoxP2* for nonhuman primates, and *Foxp2* for other species. This recognises the minor variations that exist between the different forms of the gene, and of its product.

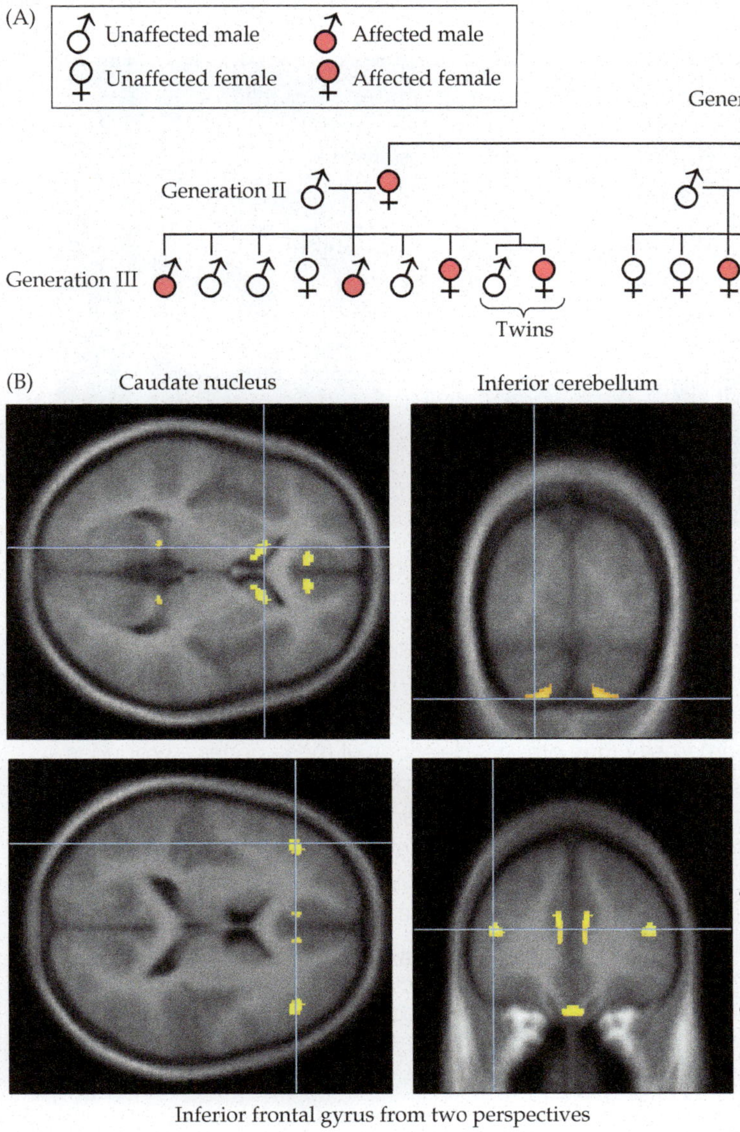

From F. Vargha-Khadem et al., 2005. *Nat Rev Neurosci* 6: 131–138; courtesy of Faraneh Vargha-Khadem

18.17 A Heritable Language Disorder (A) Three generations of the KE family, illustrating the transmission of a language disorder caused by mutation of the gene *FOXP2*. (B) Regions of thinned gray matter in the brains of affected members of the KE family. The blue crosshairs identify affected regions in the frontal cortex, basal ganglia, and cerebellum. (A after K. E. Watkins et al., 2002. *Brain* 125: 465–478.)

that seen in typical speakers (Liégeois et al., 2003). More minor variations in the *FOXP2* gene are also associated with different patterns of activation in language-related areas of the brain (Pinel et al., 2012). Large-scale genetic screening revealed that *FOXP2* is a regulatory gene, meaning that its protein product is a transcription factor controlling the expression of a number of other genes, perhaps especially affecting the growth and interconnection of neurons during development (Vernes et al., 2011). That its structure and function differ from *FoxP2* in other primates, such as chimpanzees (G. Konopka et al., 2009; Maricic et al., 2013), suggests that this gene has been evolving rapidly in humans, presumably because language is so adaptive in our species. When scientists remodelled the *Foxp2* gene of developing mice to add the human-specific changes present in *FOXP2*, the mice showed changes in connections between the cortex and basal ganglia as well as pronounced improvements in specific forms of learning that would be essential for language acquisition (Schreiweis et al., 2014). It is possible that some of the effects of *FOXP2* are accomplished indirectly via its regulation of other genes; for example, *Foxp2* modulates the expression of a gene called *SRPX2* that is involved in synapse formation in systems implicated in vocal behaviour (Sia et al., 2013).

There's a genetic connection in language *production* too. Children born with the genetic disorder Williams syndrome have unusually good verbal abilities, as we discuss in **BOX 18.2**. But genetic abnormalities can also impair speech production,

BOX 18.2 Williams Syndrome Offers Clues about Language

Williams syndrome, which occurs in approximately one out of 20,000 births (Bower, 2000), offers a fascinating dissociation between what we normally regard as intelligence and language. Individuals with Williams syndrome speak freely and fluently with a large vocabulary, yet they may be unable to draw simple images, arrange coloured blocks to match an example, or tie shoelaces. The individuals are very sociable, ready to strike up conversation and smile. They may also display strong musical talent, either singing or playing an instrument.

© Tampa Bay Times/Zuma Press

Verbal Talent Ben Monkaba steals the show during the final performance at Williams Syndrome camp. People with Williams Syndrome exhibit characteristic facial features, due to a problem with the elastin gene, along with high verbal fluency and mild mental disability due to deletion of other nearby genes.

The syndrome results from the deletion of about 28 genes from one of the two chromosomes numbered 7 (de Luis et al., 2000). No one understands why the remaining copies of these genes, on the other chromosome 7, do not compensate for the lost copies. The absence of one copy of the gene called *elastin* (which encodes a protein important for connective tissue in skin and ligaments) leads to pixie-like facial features. Several of the other missing genes are thought to lead to changes in brain development and to the behavioural features of the syndrome.

The psychological development of such individuals is complicated: As infants, they may display a greater understanding of numerosity than other infants, but as adults they may show a poor grasp of numbers. Conversely, their language performance is poor in infancy but greatly improved by adulthood (Paterson et al., 1999). These findings suggest that the developmental process is distinctively altered in Williams syndrome. Impressively, possession of extra copies of the identified genes on chromosome 7—rather than deletions of these genes—produces a syndrome that is, in many ways, the converse of Williams syndrome: very poor expressive language accompanied by unimpaired spatial abilities (Somerville et al., 2005).

Williams syndrome A disorder characterised by fluent linguistic function but poor performance on standard IQ tests and great difficulty with spatial processing even in otherwise healthy individuals. Most of us have little trouble with verbal articulation, the mechanical process of moving the vocal tract to produce speech sounds, but some otherwise ordinary people tend to produce speech sounds only in fits and starts, tripping over certain syllables or unable to start vocalising certain words. We call this difficulty **stuttering** (or *stammering*). Scientists have long suspected that stuttering is at least partly heritable, and by again analyzing transmission within families, researchers zeroed in on mutations in a set of three specific genes (C. Kang et al., 2010). These genes—*GNPTAB*, *GNPTG*, and *NAGPA*—play a part in the normal functioning of lysosomes, organelles involved in the breakdown of cellular waste materials. It remains to be seen why a dysfunction in this basic and ubiquitous system should so specifically affect speech. But recall that in Chapter 5, we asked how humans and chimpanzees could differ so little in their total genome yet behave so differently.

stuttering Disordered articulation of speech sounds and involuntary repetition of words and phrases, resulting in halting and effortful speech.

Because there is no fossil record to inform us about prehistoric language, the evolutionary antecedents of language are uncertain. One possibility is that speech and language were built on a system originally controlling gestures of the face and hands (Corballis, 2002; Hewes, 1973). Even today, hand movements facilitate speech: if people are prevented from gesturing, they make more slips and have more pauses in their speech (Krauss, 1998). Furthermore, people who have been blind from birth, and so have never seen the hand gestures of others, still make hand gestures while they speak (Iverson and Goldin-Meadow, 1998). And as we noted earlier, deaf people who communicate exclusively with gestures use the same part of the brain that hearing people use while speaking and listening. Deaf children raised without access to an established sign language will often invent one of their own, complete with structural features that characterise other spoken and sign languages (Goldin-Meadow, 2006). Studies like these provide additional hints of an ancient association between gestures and speech.

Taken together, the evidence confirms that basic mechanisms of language are heritable components of the human brain, and the product of a long evolutionary history. Indeed, the use of language is one of the key adaptations of humans, with basic capabilities probably arising in an ancient ancestor of our species. The Neanderthals shared our version of *FOXP2* (Krause et al., 2007), so perhaps major differences in just a few genes, like *FOXP2* and the stuttering genes, are enough to explain why we write books and give speeches and chimps do not, despite our otherwise great genetic similarity (S. E. Fisher, 2017).

Nonhuman primates engage in elaborate vocal behaviour

People have long tried to communicate with animals, sometimes quite successfully: anyone who has watched a sheepdog at work, responding to commands from its handler, has to acknowledge that the human is transmitting lots of information to a highly intelligent companion. Instilling *language* in a nonhuman is a different matter, however.

Every day, you utter sentences that you have never said before, yet the meaning is clear to both you and your human listener because you share an understanding of the words and grammar involved. Animals generally are incapable of similar feats, instead requiring extensive training with each specific utterance in order for communication to occur at all. Speaking to your dog (**FIGURE 18.18**) reportedly activates a left-hemisphere mechanism in the dog's brain that processes meaning, and a right-hemisphere mechanism that assigns value and reward to those words (Andics et al., 2016), but each new combination of words that you use with the dog will have to be

Photo by Neil Watson

18.18 Good Dog! Dog lovers know that their canine friends can learn to associate certain words with specific actions. Maisie here knows to be still when she hears "wait" until she hears "OK," at which time she will enjoy both a delicious sausage and a delighted human. Dogs can learn a small collection of words with sufficient practice, enabling *communication* between human and nonhuman, but not *language*, because most nonhuman animals appear to lack a capacity for grammar.

laboriously learned from scratch. In short, dogs and most other animals appear to lack grammar. For this reason, scientists have focussed the search for nonhuman language capabilities mostly on our nearest relatives, the other primate species.

Chirps, barks, meows, songs, and other sounds are among the many vocalisations produced by nonhuman animals, some of which are described in **BOX 18.3**. Detailed analyses of the vocalisations of apes and monkeys (Ploog, 1992; Seyfarth and Cheney, 1997) show that a wide range of vocal behaviour is employed for communication between individuals, particularly to relay emotional information like alarm or territoriality. The shrieking, purring, peeping, growling, and cackling sounds of squirrel monkeys, for example, can generally be related to specific social situations. Chimpanzees encountering model vipers issue alarm calls specifically to warn members of their group that are unaware of the perceived threat (Crockford et al., 2012), and gibbons deploy a sizable repertoire of hooting calls that they recombine to communicate information about predators, social conditions, and mating opportunities (Clarke et al., 2015). But despite their apparent adaptive importance, most nonhuman primate vocalisations seem to have a somewhat "preprogrammed" quality, being repeatedly produced in much the same fashion and order. This suggests that in most primates, species-specific vocalisations are genetically determined, although close social contact between related individuals can fine-tune minor aspects of calls (Lemasson et al., 2011).

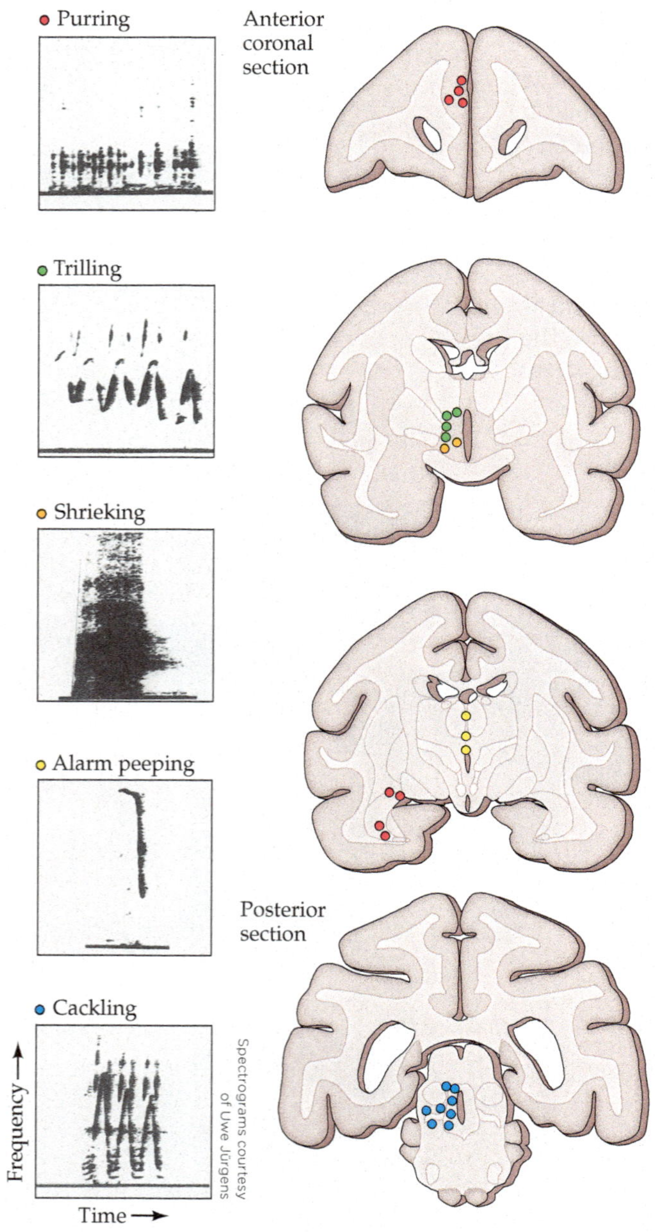

18.19 Elicitation of Vocalisations by Electrical Stimulation of the Monkey Brain Stimulation at different sites in the monkey brain results in different, species-typical vocalisations. The left column contains spectrograms of different vocalisations; the coronal sections in the right column show brain sites at which the different vocalisations are elicited. (After D. W. Ploog, 1992. In *So human a brain: Knowledge and values in the neurosciences*, A. Harrington [Ed.], pp. 3–13. Birkhauser: Boston.)

Direct electrical stimulation of subcortical regions in monkeys can elicit some calls (**FIGURE 18.19**), but stimulation of the cerebral cortex generally fails to elicit vocal behaviour. Brain regions where stimulation generates vocalisations tend to be structures that are also involved in defense, attack, feeding, and sexual behaviours. These regions include sites in the limbic system and related structures. Just before vocalising, monkeys display changes in electrical potential in cortical areas homologous to human speech areas (Gemba et al., 1995). Furthermore, some monkeys, like people, show a preference for using the right ear (which has preferential connections with the left hemisphere) to listen to vocalisations from conspecifics (Hopkins et al., 2015). Apes tend to favor gesturing with the right hand, and because the right hand has preferential connections with the left side of the brain, this observation provides additional behavioural evidence that the left hemisphere plays a special role in communicative behaviour, just as in people (Meguerditchian and Vauclair, 2006; Taglialatela et al., 2006).

LANGUAGE TRAINING IN NONHUMAN PRIMATES

Nonhuman primates are unlikely to ever produce human speech, as their vocal tracts and vocal repertoires are suited to their own communication needs. But can these animals be taught other forms of language, including the ability to represent objects with symbols and to manipulate those symbols according to grammatical rules? Can animals other than humans generate a novel string of symbols, such as a new sentence, in a grammatical form? Some researchers argue that language is a uniquely human adaptation that evolved, independently of other cognitive capabilities, in order to solve specific problems in our evolutionary past (Pinker and Jackendoff, 2005).

Chimpanzees, gorillas, and orangutans—the great apes—are our nearest primate relatives. Wild chimpanzees and other apes reportedly use a variety of hand gestures for communication (Hobaiter and Byrne, 2011), and apes are capable of learning hundreds of hand gestures from American Sign Language (ASL), the standardised sign language used by deaf people in most of North

BOX 18.3 Vocal Behaviour in Birds and Other Species

Many nonprimate species engage in vocal behaviour, sometimes elaborately, using sound to distinguish between local groups or species, to signal readiness to mate, or to alert kin to danger. Whales sing and may imitate songs that they hear from distant oceans (Noad et al., 2000), and some seal mothers recognise their pups' vocalisations even after 4 years of separation (Insley, 2000). In fact, many species—from elephants to bats to birds to dolphins—are capable of vocal learning and use their vocalisations to help form social bonds and identify individuals (Poole et al., 2005; Tyack, 2003). In the lab, measurement with special instruments reveals that rats and mice produce complex ultrasonic vocalisations that they use to communicate emotional information (Panksepp, 2005). These ultrasonic vocalisations are impaired in mice with mutations in the *Foxp2* gene (Shu et al., 2005), providing an intriguing parallel to the situation in humans.

Birds are particularly vocal animals; while some produce only simple calls with limited communicative functions, many species of songbirds--canaries, zebra finches, song sparrows—produce rich and melodious vocalisations that are crucial for social behaviours and reproductive success. A select few, such as the gray parrot, can even learn to communicate using a significant vocabulary of human words (Pepperberg, 2009). The vocal behaviour of birds, and particularly the production of birdsong, is the subject of intensive research, and not just because we humans find birdsong so pleasing. Striking similarities between birdsong learning and human language development—despite an evolutionary separation of more than 300 million years—are suggestive of convergent evolution in these systems (Bolhuis et al., 2010; Pfenning et al., 2014). In most songbirds, only males of the species sing, and the song is *learned*—in much the same way that humans learn language (DeVoogd, 1994). Also like human language, birdsong relies principally on the left hemisphere of the brain (Moorman et al., 2012; Nottebohm, 1980).

A series of brain nuclei and their connections (shown in the figure) control the production of song by the vocal organ, the *syrinx* (A. P. Arnold and Schlinger, 1993). First, a direct pathway from the high vocal centre (HVC) to the nucleus robustus of the archistriatum (RA), and then on to the brainstem nucleus of the twelfth cranial nerve, controls the vocal organ. Lesions along this direct pathway at any stage of development will disrupt song. A second, less direct pathway from the HVC to the brainstem includes several forebrain nuclei that may be involved in aspects of song acquisition. Early lesions of one of its components, called the *lateral magnocellular nucleus of the anterior neostriatum* (*LMAN*), will stop song development, but LMAN lesions in adulthood do not affect song performance (Bottjer et al., 1984). Vocal learning is accompanied by the generation in adulthood of new neurons (neurogenesis; see Chapter 6) in a variety of brain regions in songbirds (Walton et al., 2012).

What about the *Foxp2* gene, which we've already discussed in the context of mouse vocalisation and human language acquisition? The level of expression of *Foxp2* in the brains of birds is greater during song learning than before or after (Haesler et al., 2004). When researchers selectively silenced *Foxp2* expression in part of the song-learning pathway, called *area X* (see figure), adolescent males failed to properly learn and recite the tutor's song, producing errors that resemble those in humans with abnormal *FOXP2* that we described earlier (Haesler et al., 2007). (For more on the complexity of birdsong and its social roles, see **A STEP FURTHER 18.3: BIRDSONG BROADCASTS INFORMATION ABOUT THE SINGER** on the website.)

Go to Activity 18.3
Song Control Nuclei of the Songbird Brain

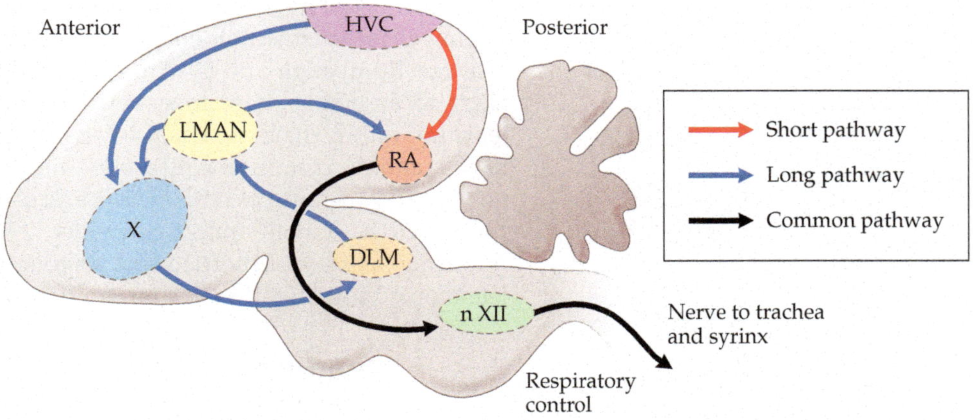

Song Control Nuclei of the Songbird Brain Two neural pathways control birdsong. A direct route from the high vocal centre (HVC) to the nucleus robustus of the archistriatum (RA) to the nucleus of the twelfth cranial nerve (n XII) is crucial for song; lesions of any of these areas will disrupt song. (The nucleus labelled "X" is area X of the paraolfactory lobe, and the "DLM" is the medial dorsolateral nucleus of the thalamus.) The indirect path that includes the lateral magnocellular nucleus of the anterior neostriatum (LMAN) may play a role in song learning, although the memory of the tutor's song is probably stored elsewhere. (After A. P. Arnold, 1980. *Am Sci* 68: 165–173.)

18.20 Chimpanzee Using Symbols Although chimpanzees can learn to use arbitrary signs and/or symbols to communicate, it is questionable whether this usage is equivalent to human language.

Frans Lanting Studio/Alamy Stock Photo

America. Given enough training—it takes years—apes may learn to use ASL signs spontaneously, sometimes in novel sequences (R. A. Gardner and Gardner, 1969, 1984; F. Patterson and Linden, 1981). Other researchers report that chimpanzees can learn to use arbitrary coloured tokens or computerised symbols (**FIGURE 18.20**) in ways that seem to reflect an acquired ability to form short "sentences"—novel, meaningful chains made up of basic units of meaning—complete with concepts such as logical classifications, categorisation, and negatives (e.g., Premack, 1971; Rumbaugh, 1977).

Critics of language research in apes argue that these sequences may simply be subtle forms of imitation (Terrace, 1979), perhaps unconsciously cued by the experimenter who is providing the training. Observers who are native ASL users dispute the linguistic validity of the signs made by apes, and others, such as linguist Noam Chomsky, argue that teaching primates to emulate a quintessentially human behaviour in which they do not naturally engage can tell us little about the behaviour, other than the obvious conclusion that ape evolution did not favor the use of language.

Nevertheless, considering that apes can comprehend spoken words, produce novel combinations of words, and respond appropriately to sentences arranged according to a syntactic rule, it seems likely that the linguistic capacity of apes was underestimated historically (Savage-Rumbaugh, 1993). For example, a bonobo (or pygmy chimpanzee) named Kanzi, the focuss of a long-term research program (Savage-Rumbaugh and Lewin, 1994), reportedly learned numerous symbols and ways to assemble them in novel combinations, entirely through observational learning rather than the usual intensive training. And in natural settings, monkeys combine certain vocalisations into higher-order, more complex calls, suggesting the presence of both syntax and semantic meaning, at least on a rudimentary level (K. Arnold and Zuberbühler, 2006; Ouattara et al., 2009). So although the debate is far from settled, the linguistic accomplishments of primates have at least forced investigators to sharpen their criteria of what constitutes language.

18.6 Reading Skills Are Difficult to Acquire and Frequently Impaired

Learning Objectives

After studying this section, you should be able to:

18.6.1 Define *dyslexia* and distinguish between deep and surface dyslexia.

18.6.2 Discuss the brain networks involved in reading, and discuss some abnormalities that can impair reading abilities.

18.6.3 Summarise genetic contributions to reading and the heritability of dyslexia.

Learning to read and write requires far more effort than learning to speak. This difference reflects the evolutionary heritage of our brains, which are more or less the same as those of prehistoric humans. Speaking is an ancient human adaptation, old enough that evolution has equipped our brain with inborn speech mechanisms. But reading and writing are relatively recent developments, just a few thousand years old, and too little time has passed for us to evolve the same sort of intrinsic mechanisms for these behaviours. This is probably why some primitive modern societies lack writing but none lack complex spoken language. Difficulty with reading is called **dyslexia** (from the Greek *dys*, "bad," and *lexis*, "word"), which can result from either developmental or neurological causes, as we'll see.

dyslexia A reading disorder attributed to brain impairment. *Acquired dyslexia* occurs as a result of injury or disease. *Developmental dyslexia* is associated with brain abnormalities present from birth.

deep dyslexia Acquired dyslexia in which the person reads a word as another word that is semantically related.

surface dyslexia Acquired dyslexia in which the person seems to attend only to the fine details of reading.

micropolygyria A condition of the brain in which small regions are characterised by more gyri than usual.

ectopia Something out of place—for example, clusters of neurons seen in unusual positions in the cortex of someone who has dyslexia.

Brain damage may cause specific impairments in reading

Sometimes people who learned to read just fine as children suddenly become dyslexic in adulthood as a result of disease or injury, usually to the left hemisphere. This *acquired dyslexia* (sometimes called *alexia*) reveals hints about how the brain processes written language. One type of acquired dyslexia, known as **deep dyslexia**, is characterised by errors in which people read a word as another word that is related in meaning; for example, the printed word *cow* is read as *horse*. These individuals are also unable to read aloud words that are abstract, as opposed to concrete, and they make frequent errors in which they seem to fail to see small differences in words. It's as though they grasp words whole, without noting the details of the letters, so they have a hard time reading nonsense words.

In another form of acquired dyslexia, **surface dyslexia**, the person makes different types of errors when reading. These individuals can read nonsense words just fine, indicating that they know the rules of which letters make which sounds. But they find it difficult to recognise words in which the letter-to-sound rules are irregular. *The Tough Coughs as He Ploughs the Dough* by Dr. Seuss (1987), for example, would utterly confound them. In contrast to people with deep dyslexia, those with surface dyslexia have difficulties that are restricted to the details and sounds of letters. Interestingly, surface dyslexia doesn't occur in native speakers of languages that are perfectly phonetic (such as Italian). This finding indicates that what's lost in speakers of nonphonetic languages, like English, is purely a learned aspect of language. In contrast, deep dyslexia probably involves mechanisms that are important for all languages.

Other kinds of brain damage can impair reading. For example, individuals with hemispatial neglect following right parietal lobe damage (discussed in Chapter 17) disregard the left half of the world, despite having otherwise normal vision. Such people also fail to notice the left halves of the words that they see, necessarily resulting in poor reading. One striking feature in some severe cases of acquired dyslexia is *letter-by-letter reading*, in which the individuals laboriously spell out each word to themselves (aloud or silently). In these cases, it seems that conscious attention to the spelling of each word is the only way by which words can be identified, so reading speed is dramatically slower.

Some people struggle to read throughout their lives

Some children seem to take forever to learn to read. Their efforts are laden with frustration, and prolonged practice produces only small improvements. This *developmental dyslexia* is seen in nearly 5% of children in the general population; it is even more common in boys and in left-handed people. Children with dyslexia can have high IQs (B. Morris, 2002), and many have gone on to illustrious careers in varying fields. Thus it appears that dyslexia is a specific problem in connecting reading with the more-ancient brain systems for speech, and not a general cognitive deficit.

A postmortem study of the brains of people with significant dyslexia revealed striking anomalies in the arrangement of cortical neurons, especially in frontal and temporal regions (Galaburda, 1994). These anomalies included unusual groupings of cells in outer layers of the cerebral cortex and distortions of the layers and columns in which cortex is normally organized (**FIGURE 18.21**). Many neurons were disoriented, and excessive cortical folding, called **micropolygyria**, was observed. Nests of extra cells—**ectopias**—were seen. People showing MRI evidence of these

18.21 Neural Disorganisation in Dyslexia (A) (Top) Drawings of the left and right planum temporale (see Figure 18.4) from the brain of a person with dyslexia show these regions as nearly symmetrical; in most people the left planum temporale is considerably larger. The dots and the shaded area represent regions where microscopic anomalies called *ectopias*, *dysplasias*, and *micropolygyria* have been found in the brains of individuals with dyslexia. (Bottom) Anomalies in individuals with dyslexia are much more common in the left hemisphere, which is primarily responsible for language function. (B, C) These micrographs show (B) micropolygyria (literally "many tiny gyri") and (C) ectopias, clusters of neurons in unusual locations, such as this cluster in cortical layer I (arrow), which is normally devoid of neuronal cell bodies. After A. M. Galaburda et al., 1985. *Annu Neurol* 18: 222–233.)

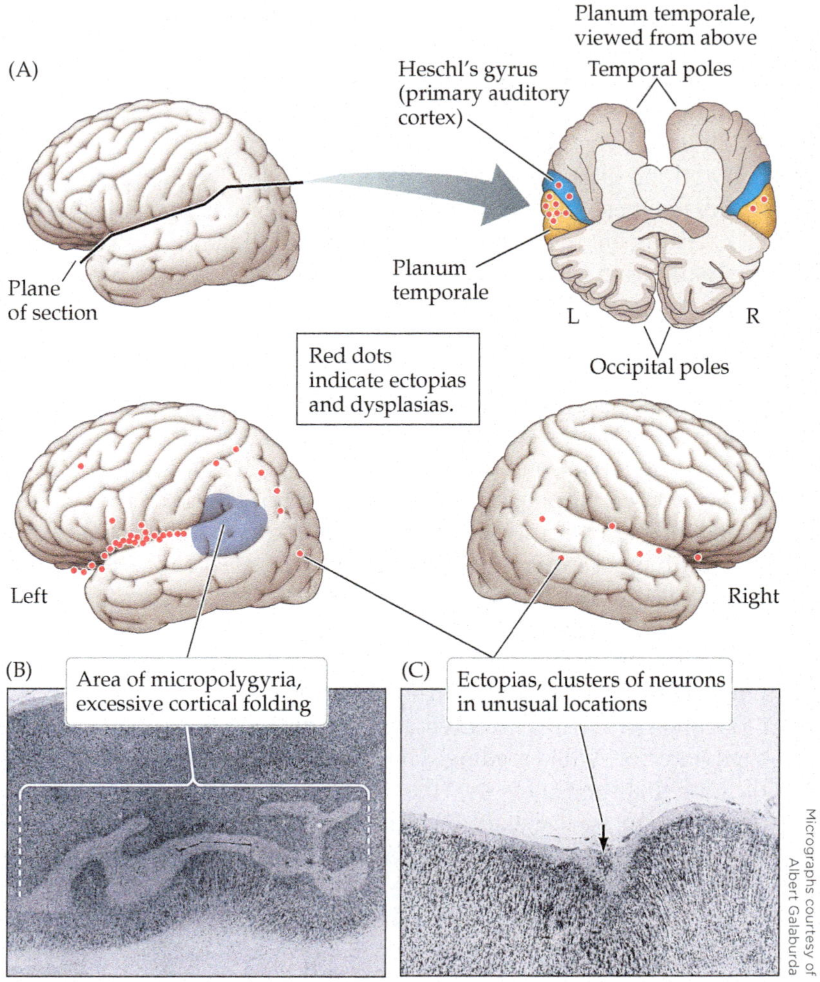

abnormalities tend to have reading disorders (Chang et al., 2005). It is likely that these neural problems arise in fetal development, during a period in which cells are actively migrating into the cerebral cortex; defective neuronal migration has been implicated in unusual patterns of connectivity in language-related regions of the temporal cortex (Galaburda et al., 2006).

Functional-imaging studies further indicate differences in brain activity in dyslexia. Compared with control participants, people with dyslexia show diminished activation of left posterior speech zones that include the superior temporal lobe and angular gyrus (K. R. Pugh et al., 2000; Shaywitz et al., 1998, 2003), as well as the occipital lobe (Demb et al., 1998), along with relative overactivation of anterior regions of the brain. For tasks that especially focus on the phonological (phoneme-processing) aspects of dyslexia, abnormal patterns of activation are also seen in the nearby temporoparietal region and in the inferior fusiform area of the left temporal lobe (**FIGURE 18.22**) (Hoeft et al., 2006; H. Tanaka et al., 2011). In addition to differences in temporal *cortex*, brain imaging has revealed subtle changes in the fine structure of the temporoparietal white matter pathways in adult dyslexics (Klingberg et al., 2000). These results indicate possible problems with the axonal connections between language-related cortical areas. Researchers also find a consistent relationship between developmental dyslexia and cerebellar dysfunction (Stoodley and Stein, 2011). It is not yet clear exactly what role the cerebellum plays in reading, because cerebellar lesions later in life do not result in acquired dyslexia.

Taken together, imaging and behavioural studies indicate that our brains rely on two different language systems during reading: one focussed on the sounds (*phonology*) of letters, the other on the meanings (*semantics*) of whole words (McCarthy

18.22 Trouble with Phonemes In this experiment, researchers asked two different samples of children (about 10 years old) to complete a phonological reading task in which the rhyming of words is judged, thereby requiring accurate processing of phonemes. Compared with typical readers, these two samples of children with dyslexia showed decreased activity in two left-hemisphere sites, as revealed in these composite fMRI images—the left inferior parietotemporal region and the left inferior fusiform region. This pattern appears to be independent of the IQs of the children.

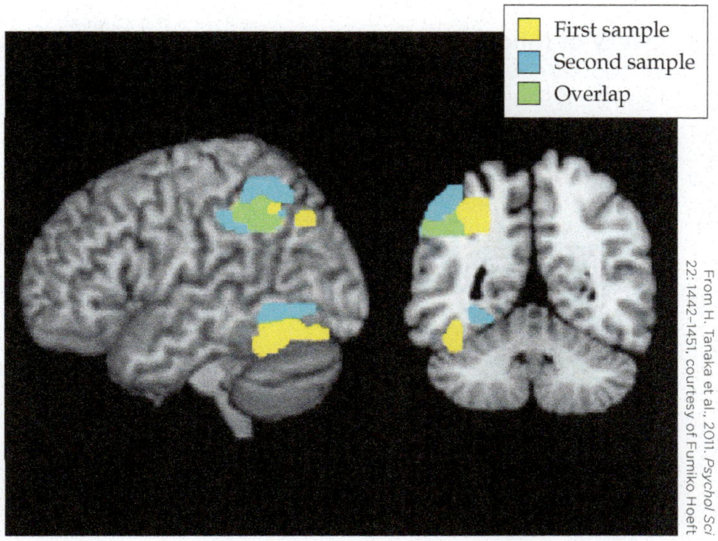

From H. Tanaka et al., 2011. *Psychol Sci* 22: 1442–1451, courtesy of Fumiko Hoeft

and Warrington, 1990). Using fMRI, researchers found that people with dyslexia often have a disconnection between these systems, impairing the coordination of sounds with their meanings (Boets et al., 2013); it has been proposed that this disconnect is the result of hyperexcitability of reading-associated cortical networks (Hancock et al., 2017). Presumably, these systems are shaped by training: as with learning any highly skilled behaviour, our brains appear to mold themselves to accommodate our acquired expertise with written language. The observation that remedial training in people with dyslexia induces changes in the left-hemisphere systems that are used for reading (Temple et al., 2003) is consistent with this view. And intriguingly, there appears to be a cultural component to dyslexia: the specific brain regions involved may vary depending on the graphical form of the written language being learned. Most research implicates the left *temporoparietal* cortex as being centrally important in dyslexia, but in readers of Chinese—a language based on logographic symbols rather than an alphabet—dyslexia is especially associated with dysfunction of the left medial *frontal* cortex (Siok et al., 2004).

Developmental dyslexia appears to have a genetic component. By studying inheritance patterns in families, followed up with investigations in animals, researchers have identified several genes—*ROBO1, DYXC1, DCDC2,* and *KIAA0319,* for example—that are linked to reading disorders (Gabel et al., 2010; Hannula-Jouppi et al., 2005). *ROBO1* is known to be involved in guiding growing axons to their destinations. The precise function of *DYXC1* is not yet known, but it is estimated to be abnormal in 9% of individuals with dyslexia in the general population (T. C. Bates et al., 2010; Taipale et al., 2003). Variation in the gene *DCDC2,* which is believed to participate in the migration of neurons into their positions in the cortex, predicted reading performance in a large analysis of families, indicating that the gene may be involved in normal reading ability and that mutations in the gene are a genetic risk factor for dyslexia (Lind et al., 2010). Finally, the gene *KIAA0319,* linked to processes of brain development, is also reported to be associated with developmental dyslexia (Cope et al., 2005; Harold et al., 2006). In rats, inhibition of *KIAA0319* causes cortical ectopias like those seen in humans with dyslexia, strongly supporting the idea that an early problem with neuronal migration can cause dyslexia (Peschansky et al., 2010).

Although more work will be needed before we really understand the genetic bases of the disorder, these discoveries raise the prospect of early identification of dyslexia via genetic screening. Such detection would allow for earlier, and more effective, intervention strategies, which could be critically important if some aspects of the visual or phonological problems experienced by people with dyslexia are actually a consequence of a lifetime of reduced experience with reading (Goswami, 2015). Furthermore, research is revealing additional, unexpected dimensions in dyslexia. For example, people with dyslexia are much less accurate than nondyslexic people when it comes to identifying the voices of speakers of their own language, but not at identifying the voices of people speaking *other* languages (Perrachione et al., 2011). This result indicates that the phonological difficulties that dyslexics have in reading also affect their auditory speech perception.

recovery of function The recovery of behavioural capacity following brain damage from stroke or injury.

But there are also positive findings: evidence is accumulating that people with dyslexia actually outperform unaffected individuals in certain learning domains, such as aspects of spatial learning (Schneps et al., 2012). Findings like these may have important implications for developing new educational strategies in dyslexia.

RECOVERY OF FUNCTION AFTER BRAIN DAMAGE

Compared with the other organs of the body, it is all too easy to seriously damage the brain. A number of factors combine to make the brain so vulnerable: extreme complexity, delicate structure, metabolic neediness, limited capacity for regrowth, and an exposed location perched atop a thin and whippy neck. In the United States alone, according to the 2016 National Health Interview Survey (Blackwell and Villarroel, 2018), almost 7.5 million adults are survivors of stroke, and many more are living with the consequences of other disease processes, such as tumours and degenerative diseases. In addition to diseases, traumatic brain injury (TBI) may result from motor vehicle accidents, workplace injuries, and mishaps during various activities like horseback riding, diving, boxing, and other contact sports such as football and hockey. So, while prevention is always preferable to treatment, researchers are intensively studying the mechanisms mediating **recovery of function** after brain damage, with the aim of developing treatments that improve outcomes for people with damaged nervous systems.

18.7 Stabilisation and Reorganisation Are Crucial for Recovery of Function

Learning Objectives

After studying this section, you should be able to:

18.7.1 **Summarise the time course and prognosis for recovery after brain damage, and discuss factors that complicate or reduce recovery of function.**

18.7.2 **Discuss research on cellular repair strategies in brain damage.**

18.7.3 **Review types of rehabilitation therapy that can improve outcomes for brain-injured people.**

18.7.4 **Describe the threats to brain health posed by repetitive head injuries in sports.**

In the months following a brain injury, people often show conspicuous improvements in behaviour, as the injury site stabilises, unaffected tissue reorganises, and compensation occurs. For example, many people who become aphasic following brain injury will eventually regain some of the lost language abilities; language recovery

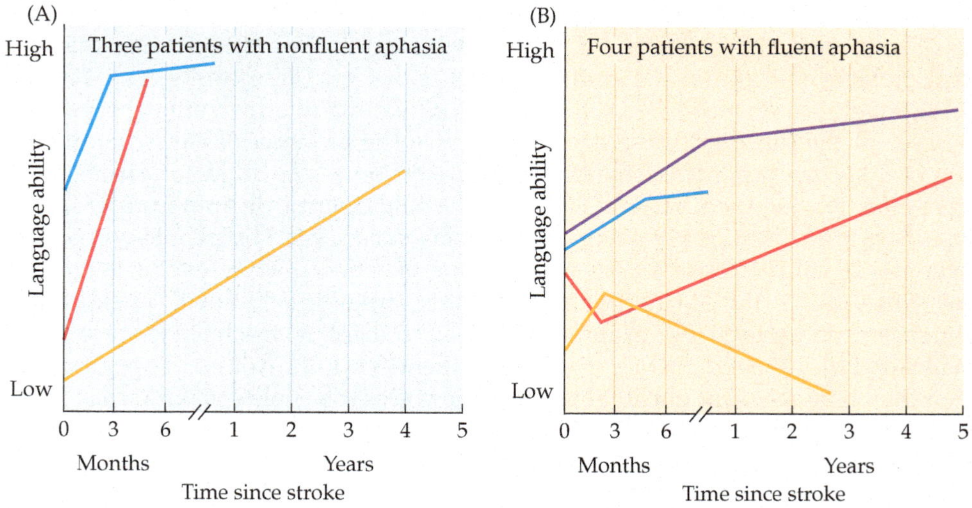

18.23 Courses of Recovery of People with Aphasia The course of recovery from nonfluent (Broca's) aphasia (A) differs from the course of recovery from fluent (Wernicke's) aphasia (B). These graphs depict the aphasia quotient, a score derived from a clinical test battery. Higher scores indicate better language performance. (After A. Kertesz and P. McCabe, 1977. *Brain* 100: 1–18.)

following stroke in adults can be impressive and may continue for longer than 2 years after the initial brain injury, although the majority of recovery occurs during the first 3 months (**FIGURE 18.23**) (Kertesz et al., 1979). The relative extent of recovery from aphasia can be predicted from several factors. Recovery is better when brain damage is the result of trauma, such as a blow to the head, rather than a stroke; this difference may be due to the generation of cell-damaging signalling chemicals in cells that have been starved of oxygen during a stroke. Individuals with more-severe initial language loss recover less completely. And left-handed people show better recovery than those who are right-handed, perhaps because of reduced lateralisation of language in some left-handers. Therapeutic approaches using alternate communication channels, such as singing, can be helpful in some cases (Racette et al., 2006).

Age is also a factor. Recovery can be downright amazing in children, thanks to their greater neural plasticity: A child may show extensive language recovery even after losing the *entire* left hemisphere (**BOX 18.4**). These observations show that undamaged brain regions *can* take over functions of damaged regions, if impairment occurs early enough in life. But as we grow older, the brain slowly loses much of its ability to compensate for injury. Reversing this age-related decline in neuroplasticity is an important focus of research.

The brain regrows and reorganises anatomically after being injured

We now know that the nervous system has much more potential for plasticity and recovery than was previously believed. For example, damaged neurons can regrow

BOX 18.4 The Amazing Resilience of a Child's Brain

Complications in pregnancy and fetal development occasionally result in children who are born with a terrible kind of epilepsy that can't be controlled by medication. These children may be paralysed on one side of the body, along with almost continual seizures. Sometimes, in these extreme cases, the only way to save the life of the child is to remove an entire brain hemisphere, as shown in this scan of a girl whose left hemisphere was completely removed when she was 7 years old.

Hemispherectomy surgery (see figure) removes the malfunctioning brain tissue and saves the life of the child (Griessenauer et al., 2015), but it also produces severe symptoms of its own, such as complete paralysis of one side, speech loss, or visual impairments. However, provided the surgery occurs early enough, the child may show almost complete recovery of function over a long period of time. The girl whose brain scan is shown in the figure experienced paralysis and complete speech loss after her surgery, but at follow-up a few years later, she had recovered language function (including her bilingual abilities!), and her paralysis had resolved almost completely.

Provided that the remaining hemisphere is healthy, children who have hemispherectomy surgery tend to show stable or improving intellectual functioning in the years following surgery (Boshuisen et al., 2010; Lew, 2014) and may have normal IQs in adulthood. So, although extensive hemispheric damage in an adult usually results in drastic and largely permanent functional and intellectual impairments, cases of childhood hemispherectomy vividly illustrate the amazing plasticity of the young brain.

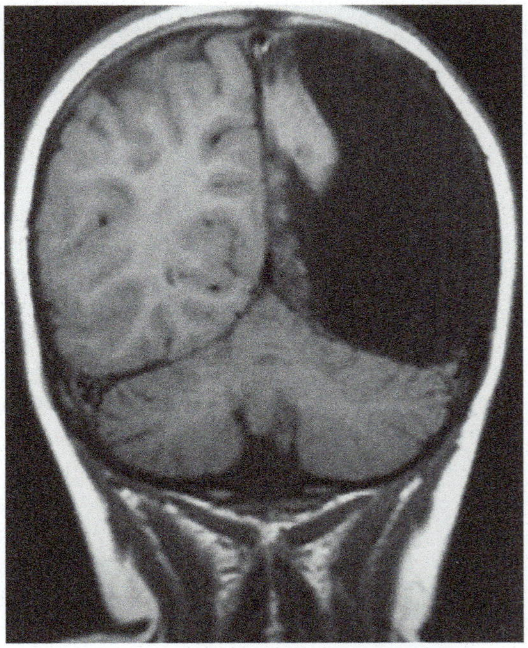

From J. Borgstein and C. Grootendorst. 2002. *Lancet* 359: 473

Left Hemispherectomy in a Child The brain can compensate even for the loss of a whole hemisphere, provided that this radical surgery is performed early enough in the individual's development.

embryonic stem cell A cell, derived from an embryo, that has the capacity to form any type of tissue.

constraint-induced movement therapy (CIMT) A therapy for recovery of limb movement after stroke or injury, in which the unaffected limb is constrained while the person is required to perform tasks with the affected contralateral limb.

their connections under some circumstances, through a process called *collateral sprouting* (for more information on collateral sprouting, see **A STEP FURTHER 18.4: SOME NERVE FIBRES CAN BE REGROWN** on the website). And as we've discussed in several places in this book, it has become evident that the adult brain is capable of producing new neurons; perhaps we will learn how to bend these new neurons to our will and use them to replace damaged brain tissue.

One of the most exciting prospects for brain repair following stroke, injury, or many other neurological conditions is the use of **embryonic stem cells** (Casarosa et al., 2014; J. Takahashi, 2018). Derived from embryos, these cells have not yet differentiated into specific roles and therefore seem to be able to develop, under the control of local chemical cues, into the type of cell needed (see Chapter 6). We already know from the research on adult neurogenesis that new neurons can become integrated into functional brain circuits. Researchers are looking for ways to help stem cells placed in the brain to survive, migrate, and mature into functioning replacement neurons (Emborg et al., 2013; Morizane et al., 2017).

Although not all studies report encouraging results (Freed et al., 2001), in several studies implants of embryonic stem cells have been shown to reduce symptoms of Parkinson's disease (see Chapter 10) and stroke (Kondziolka et al., 2000). However, controversy surrounds the ethics of utilizing human embryos as cell donors, providing impetus to find ways of creating stem cells from other sources. There are also difficulties associated with inducing implanted stem cells to form large functional populations of neurons at injury sites without triggering an immune reaction leading to tissue rejection (J. Y. Li et al., 2008). Nevertheless, these issues are potentially surmountable, and stem cell grafts offer a very promising direction for future treatments. But despite some encouraging developments in the lab, prevention of brain injury is clearly better than treatment. (Additional discussion of mechanisms of brain damage, and the mitigation of brain injuries, is available on the website in **A STEP FURTHER 18.5: DIFFERENT STRATEGIES AIM TO REDUCE BRAIN DAMAGE FOLLOWING INJURY OR STROKE**.)

Rehabilitation and retraining can help recovery from brain and spinal cord injury

Cognitive and/or perceptual handicaps that develop from brain impairments can be modified by training; similarly, intensive training may restore some measure of walking ability after certain spinal cord injuries (Barbeau et al., 1998; Morawietz and Moffat, 2013). But it is important to distinguish restoration of function from compensation. It is well known that practice can significantly reduce the impact of brain injury by fostering compensatory behaviour (T. A. Jones, 2017). For example, vigourous eye movements can make up for the *scotomas* (blind spots in the visual field) that result from strokes or other injuries that affect the visual cortex. Behaviour strategies can be changed after a brain injury to enable successful performance on a variety of tasks.

As with recovery of language functions, a growing body of evidence shows that people can regain considerable use of limbs that have been paralysed following brain injury, especially if they are *forced* to use the limb repeatedly. **Constraint-induced movement therapy (CIMT)** persuades people who have had a stroke to use the affected arm, by simply tying the "good" arm to a splint for up to 90% of waking hours (Taub et al., 2002), in conjunction with daily rehabilitation therapy that includes practice moving the affected limb repeatedly (**FIGURE 18.24**). Many individuals who received this treatment reportedly regained up to 75% of normal use of the paralysed arm after only 2 weeks of therapy. Although the underlying mechanisms are not well understood—perhaps involving a remapping of the motor cortex—it's clear that CIMT offers substantial benefits that persist over the long term (Kwakkel et al., 2015; Liepert et al., 2000). Furthermore, combining physical retraining with a drug treatment that induces neural sprouting provides almost complete restoration of function in lab animals with experimental strokes (Wahl et al., 2014).

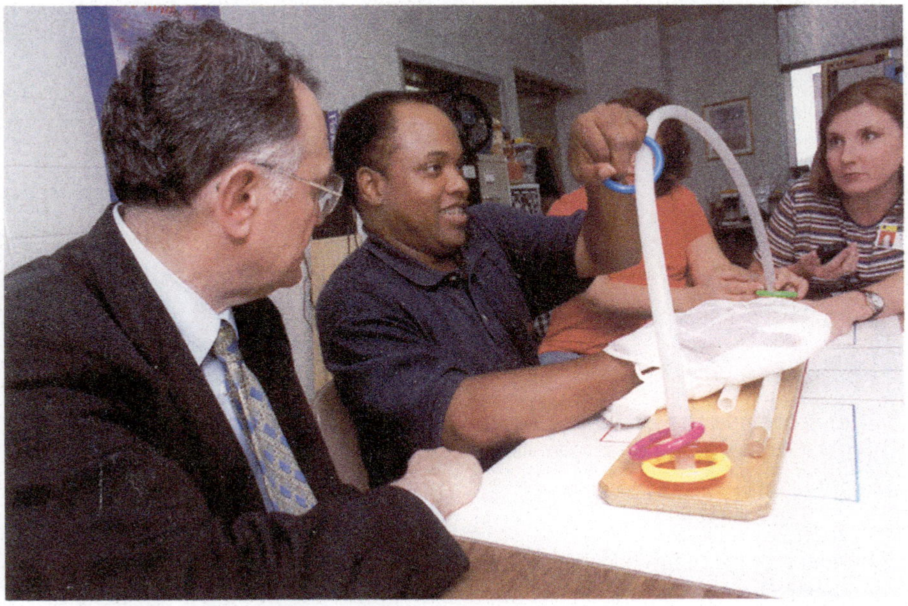
Photograph courtesy of Edward Taub

18.24 Constraint-Induced Movement Therapy In this therapy, an unaffected limb is gently restrained (in this case, in a white mitten) so that the person must use the limb that was affected by the stroke in a series of repetitive tasks.

Another surprising use of experience for rehabilitation involves a simple mirror. Altschuler et al. (1999) treated people whose strokes had reduced their use of one arm by placing them before a mirror with only their "good" arm visible. To these people, it looked as though they were seeing the entire body, but now both arms were the good arm. The person was told to make symmetrical fluid motions with both arms. In the mirror, the motions looked perfectly symmetrical (of course), but surprisingly, most of the treated people soon learned to use the "weak" arm more extensively. It was as though the visible feedback, indicating that the weak arm was moving perfectly, overcame the brain's reluctance to use that arm. It is likely that the mirror neurons of the brain, discussed earlier and in Chapters 8 and 10, mediate some of the beneficial effects of this sort of rehearsal (Buccino et al., 2006).

Brain damage will remain a serious problem for the foreseeable future, one that will touch most of us as our friends and relations go through their lives. Perhaps the most important message to convey to victims of stroke and other nervous system damage is the encouragement that, with effort and perseverance, there is a very real prospect of recovering much of the lost behavioural capacity.

The Cutting Edge

Contact Sports Can Be Costly

Jarring blows to the head are common in a number of sports—football, hockey, boxing, and wrestling, for example—sometimes resulting in a **concussion** or *mild traumatic brain injury* (*mTBI*), with a range of possible symptoms including headache, altered mood, confusion, memory loss, and occasionally (but not usually) a brief loss of consciousness. Even one concussion, but especially a series of concussions or minor head impacts—even seemingly mild ones—puts a person at risk of permanent brain injury. Although uncomplicated concussions generally clear up with time, up to 25% of concussion cases may experience persistent cognitive symptoms (Ponsford, 2005), some of which may not become evident until later in an athlete's life (Montenigro et al., 2016; A. E. Thornton et al., 2008). In the United States alone, mTBIs result in more than 2.2 million hospital visits per year (CDC, 2015); many more go unreported.

concussion Also called *minor traumatic brain injury* (*mTBI*). An injury resulting from a blow to the head and associated with temporary neurological symptoms such as memory loss or other cognitive impairments, pain, and visual disturbances.

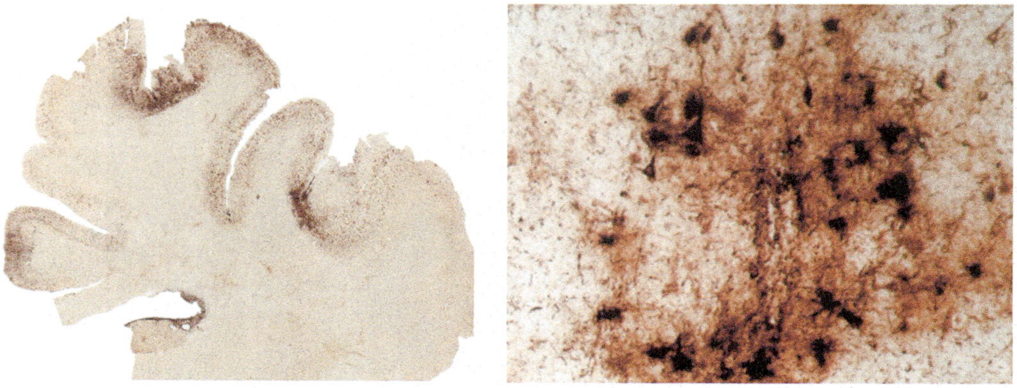

From A. C. McKee et al., 2009. *J Neuropathol Exp Neurol* 68: 709–735, courtesy of Ann McKee

18.25 Tau Protein in the Brain of a Boxer with CTE (Left) Unmagnified section of cortex. (Right) Cortical gray matter magnified × 350.

Nowhere is the risk of brain injury more evident than in boxing, where the whole goal of the contest is to rain blows upon the head of an opponent. For too many boxers, the result of so many blows to the head is a debilitated state that was historically called *dementia pugilistica* (the Latin *pugil* means "boxer") or being *punch-drunk* (Erlanger et al., 1999) but nowadays is known as **chronic traumatic encephalopathy** (**CTE**). Evidence is mounting that CTE is also alarmingly frequent in other athletes, especially American football players (Mez et al., 2017). Former athletes with CTE have markedly impaired cognitive abilities, emotional instability, and striking abnormalities in the brain.

In one of the earliest studies, Casson et al. (1982) studied ten active professional boxers who had been knocked out. The group included boxers of championship caliber, as well as mediocre and poor boxers. At least five of the group had definitely abnormal CT scans, showing such conditions as generalised cortical atrophy, which in some cases included ventricular dilation. Only one boxer had a clearly normal brain picture. Sadly, the most successful boxers may have the most profound cortical atrophy: successful boxers have had the greatest number of high-caliber fights, and thus may have sustained the most punishment, because they have participated in more matches. A large-scale CT study of 338 active boxers showed that scans were abnormal in 7% (showing brain atrophy) and borderline in 12% (B. D. Jordan et al., 1992).

Researchers have not yet agreed on a definitive diagnostic marker for CTE in living people—typically, it can be diagnosed only in postmortem analysis (McKee et al., 2016)—but one possible approach uses a special type of PET scan to detect abnormal expression of the cytostructural protein **Tau** in the brain (Barrio et al., 2015) (see Figure 7.28). Neuropathological evidence indicates that, like Alzheimer's disease, CTE in athletes is a type of *tauopathy*, in which excess Tau protein within neurons interferes with their functioning (McKee et al., 2009). **FIGURE 18.25** shows the brain of a former boxer who, by his mid-30s, was experiencing CTE-related symptoms including memory loss, confusion, and a tendency to fall. As in other cases of CTE, an excessive amount of Tau (brown in the photos) is evident in the brain, and it is found forming tangles within many neurons. At present, no effective treatments exist for CTE, although promising research is exploring novel ways of delivering anti-inflammatory medications and reducing abnormal tau expression (Kondo et al., 2015; Roth et al., 2014). CTE is a real and serious risk in contact sports, particularly where numerous blows to the head are sustained on a regular basis—no small reason why many believe that the rules of some of these sports, especially where youths are participating, are in serious need of revision (Nature, 2017). And some sports, such as boxing, should perhaps retire from the ring altogether. ■

chronic traumatic encephalopathy (CTE) Also called *dementia pugilistica* or *punch-drunk syndrome*. The dementia that develops in boxers or other athletes who are subjected to repeated blows to the head.

Tau A protein associated with neurofibrillary tangles in Alzheimer's disease.

Recommended Reading

Berwick, R. C., and Chomsky, N. (2015). *Why Only Us: Language and Evolution*. Cambridge, MA: MIT Press.

Bradbury, J. W., and Vehrencamp, S. L. (2011). *Principles of Animal Communication* (2nd ed.). Sunderland, MA: Oxford University Press/Sinauer.

Breedlove, S. M. (2017). *Foundations of Neural Development*. Sunderland, MA: Oxford University Press/Sinauer.

Fitch, W. T. (2010). *The Evolution of Language*. Cambridge, UK: Cambridge University Press.

Gazzaniga, M. S. (2016). *Tales from Both Sides of the Brain: A Life in Neuroscience*. New York: Ecco.

Harrison, D. W. (2015). *Brain Asymmetry and Neural Systems: Foundations in Clinical Neuroscience and Neuropsychology*. New York: Springer.

Honing, H., and Fitch, W. T. (2018). *The Origins of Musicality*. Cambridge, MA: MIT Press.

Koelsch, S. (2012). *Brain & Music*. New York: Wiley-Blackwell.

Kolb, B., and Whishaw, I. Q. (2015). *Fundamentals of Human Neuropsychology* (7th ed.). New York: Worth.

Meyer, J. (2015). *Whistled Languages: A Worldwide Enquiry on Human Whistled Speech*. New York: Springer.

Patel, A. D. (2010). *Music, Language, and the Brain*. New York: Oxford University Press.

Purves, D., Cabeza, R., Huettel, S. A., LaBar, K. S., et al. (2012). *Principles of Cognitive Neuroscience* (2nd ed.). Sunderland, MA: Oxford University Press/Sinauer.

Tomasello, M. (2010). *Origins of Human Communication*. Cambridge, MA: Bradford Press/MIT Press.

18 Visual Summary

You should be able to relate each summary to the adjacent illustration, including structures and processes. You can also refer back to the links to figures, animations, and activities that will help you consolidate the material.

1 **Split-brain individuals** show striking examples of **lateralisation** (hemispheric specialisation). Most words projected only to the right hemisphere, for example, cannot be read, but the same stimuli directed to the left hemisphere can be read. Spatial tasks, however, are performed better by the right hemisphere than by the left. Because of asymmetrical organisation of the brain, humans typically show an advantage for verbal stimuli presented to the right ear or right visual field. Review **Figures 18.1–18.3**

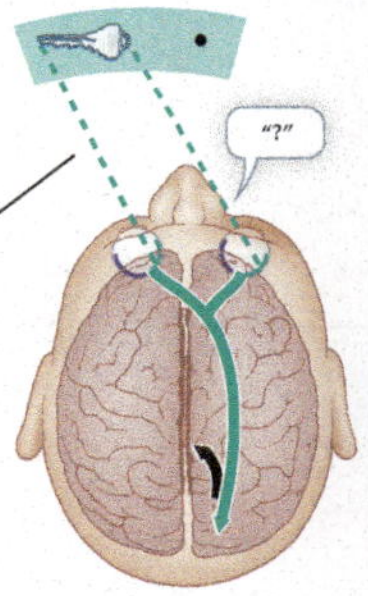

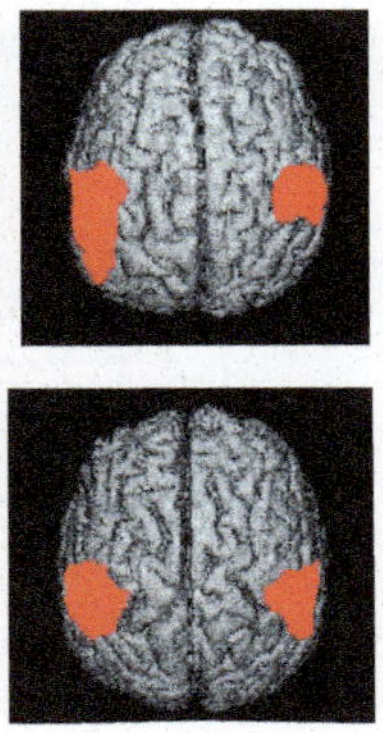

2 Anatomical asymmetry of the hemispheres is seen in some structures in the human brain. Especially striking is the large size difference in the **planum temporale** (which is larger in the left hemisphere than in the right hemisphere of most right-handed individuals). Nevertheless, in most cases mental activity depends on interactions between the cerebral hemispheres. Review **Figure 18.4**

3 In most people, parietal cortical injuries produce perceptual changes, including alterations in sensory and spatial processing. Damage to the **fusiform gyrus** can produce acquired **prosopagnosia**, a dramatic inability to recognise the faces of familiar people. A congenital form of prosopagnosia affects about 2.5% of the population. Review **Figures 18.5, 18.6**

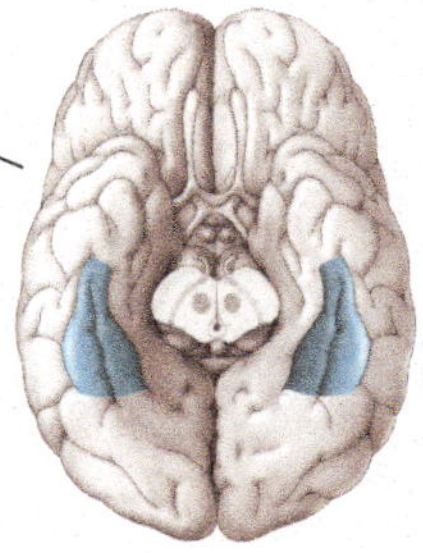

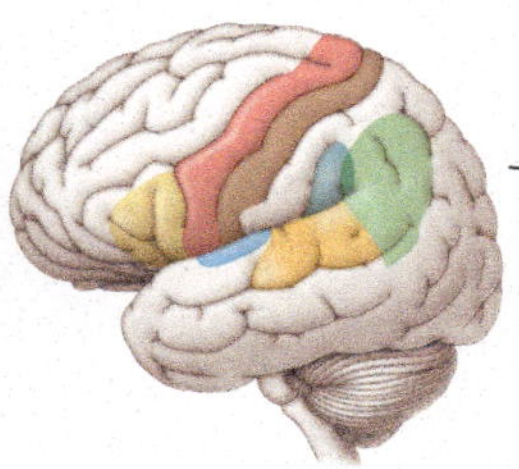

4 Left inferior frontal lesions produce an impairment in speech production called **nonfluent,** or **Broca's, aphasia**. More-posterior lesions, involving the temporoparietal cortex, cause **fluent,** or **Wernicke's, aphasia**. Extensive destruction of the left hemisphere causes a more complete loss of language called *global aphasia*. Review **Figures 18.7–18.10, Table 18.1**, **Activity 18.1**

(Continued)

5 The Wernicke-Geschwind model of aphasia emphasises a loop from a posterior speech reception zone to an anterior expressive zone, but tract-tracing studies are causing a reevaluation of this model. In contrast, the **motor theory of language** suggests that the entire circuit serves motor control and is used for both production and perception. Nonhuman primates lack this specialisation. Review **Figure 18.11**, **18.12**, **Activities 18.2a**, **18.2b**

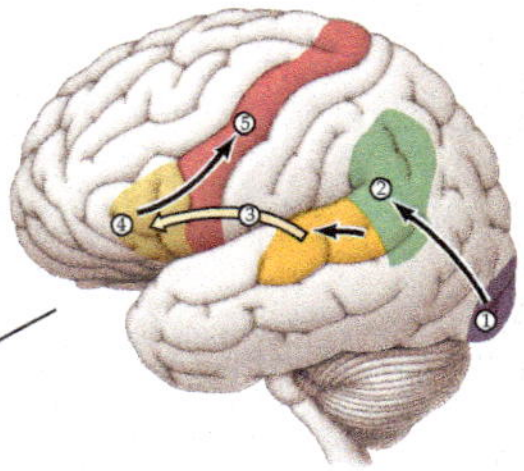

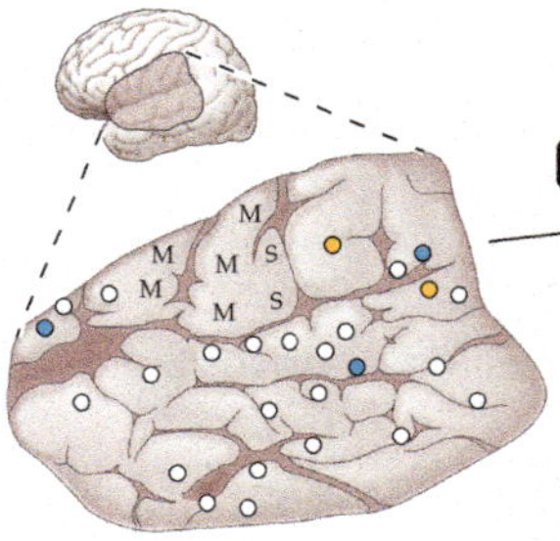

6 Mapping studies have confirmed the left-hemisphere organisation for language functions. Stimulation within anterior regions causes speech arrest, while stimulation in other locations causes misnaming and other speech errors. Review **Figures 18.13**, **18.14**

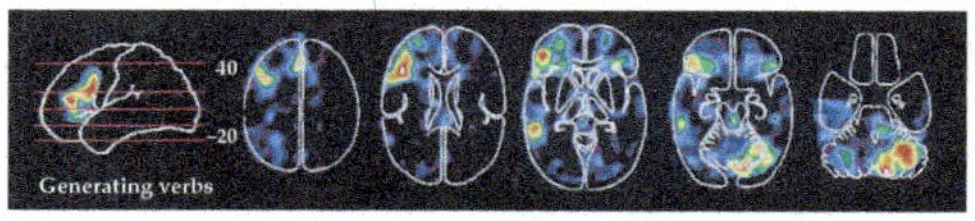

7 Studies using ERP, PET, and fMRI reveal that distinct regions of the left hemisphere are active during viewing, hearing, repeating, or assembling verbal material. Review **Figure 18.15**

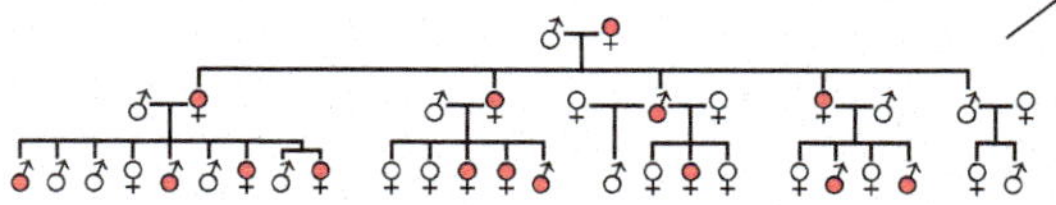

8 Languages are made up of speech sounds, called **phonemes** and **morphemes**, that are assembled into words and sentences according to **syntax** and coloured through context (pragmatics) and **prosody**. Humans are born with an innate mechanism for acquiring language during an early **sensitive period**. Aspects of language acquisition appear to be controlled by genes like *FOXP2*. Other genes appear to be crucial for language articulation. Review **Figures 18.16**, **18.17**

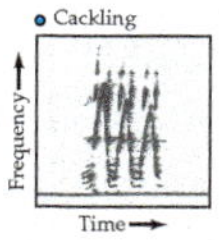

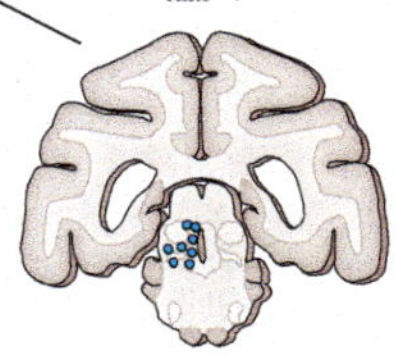

9 Nonhumans lack the peripheral apparatus to produce speech, but nonhuman primates like the chimpanzee can learn to use the signs of American Sign Language. However, controversy surrounds claims that these animals can arrange signs in novel orders to create new sentences. Review **Figure 18.18**, **18.19**, **Activity 18.3**

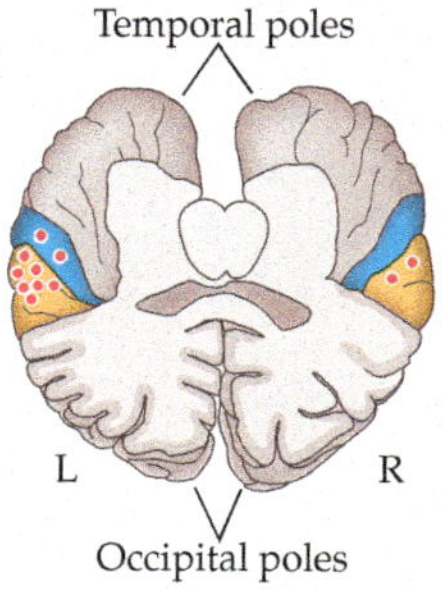

10 **Acquired dyslexia** is a difficulty with reading that results from brain damage, usually in the left hemisphere. In **deep dyslexia** there is a disturbance in reading whole words; **surface dyslexia** involves a difficulty with the sounds of words. Several brain systems have been identified in *developmental dyslexia*. Review **Figures 18.21**, **18.22**

11 The brain can show at least partial recovery of function after injury, especially during the first year or so, as the damaged brain stabilises and reorganises. Retraining is a significant part of functional recovery and may involve both compensation, by establishing new solutions to adaptive demands, and reorganisation of surviving networks. Review **Figures 18.23**, **18.24**

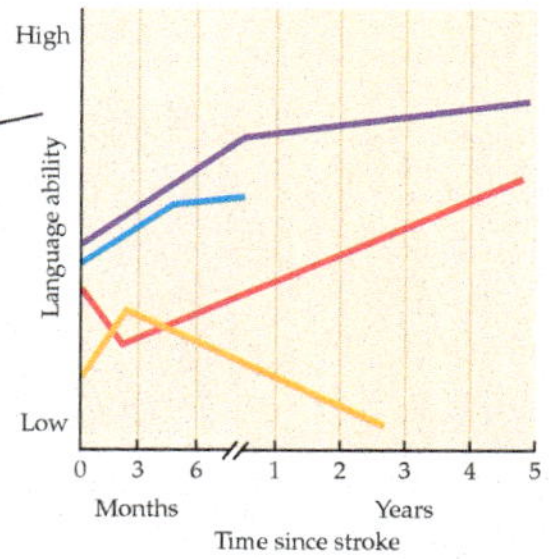

Appendix

A Primer on Concepts and Techniques in Molecular Biology

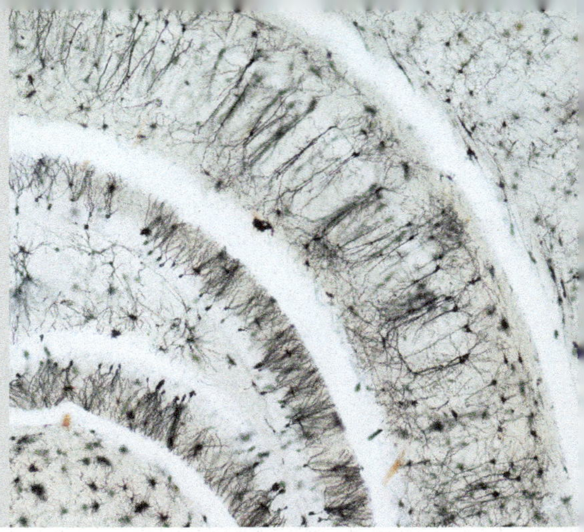

Courtesy of Dr. Sarah Moghadam, VA Palo Alto Health Care System, Palo Alto, CA and Dr. Ahmad Salehi, Dept. of Psychiatry & Behavioural Sciences, Stanford Medical School

Genes Carry Information That Encodes the Synthesis of Proteins

The most important characteristic of **genes** is that they are pieces of information, inherited from parents, that affect the development and function of our cells. Information carried by the genes is a very specific sort: each gene codes for the construction of a specific string of amino acids to form a **protein** molecule. This is *all* that genes do; they do not *directly* encode intelligence, or memories, or any other sort of complex behaviour. The various proteins, each encoded by its own gene, make up the physical structure and most of the constituents of cells, such as enzymes. These proteins make complex behaviour possible, and in that context they are also the targets upon which the forces of evolution act.

Enzymes are protein molecules that allow particular chemical reactions to occur in our cells. For example, only cells that have liver-typical proteins will look like a liver cell and be able to perform liver functions. Neurons are cells that make neuron-typical proteins so that they can look and act like neurons. The genetic information for making these various proteins is crucial for an animal to live and for a nervous system to work properly.

One thing we hope this book will help you understand is that everyday experience can affect whether and when particular genetic recipes for making various proteins are used. But first let's review how genetic information is stored and how proteins are made. Our discussion will be brief, but many online tutorials can provide you with more detailed information (see **A STEP FURTHER A.1: MOLECULAR BIOLOGY ONLINE** on the website).

Genetic information is stored in molecules of DNA

The information for making all of our proteins could, in theory, be stored in any sort of format—on sheets of paper, a DVD, a thumb drive—but organisms on this planet store their genetic information in a chemical called **deoxyribonucleic acid**, or **DNA**. Each molecule of DNA consists of a long strand of chemicals called **nucleotides** strung one after the other. DNA has only four nucleotides: guanine, cytosine, thymine, and adenine (abbreviated G, C, T, and A). The particular sequence of nucleotides (e.g., GCTTACC or TGGTCC or TGA) holds the information that will eventually make a protein. Because many millions of these nucleotides can be joined one after the other, a tremendous amount of information can be stored in very little space—on a single molecule of DNA.

A set of nucleotides that has been strung together can snuggle tightly against another string of nucleotides if it has the proper sequence: T nucleotides preferentially link with A nucleotides, and G nucleotides link with C nucleotides. Thus, T nucleotides are said to be complementary to A nucleotides, and C nucleotides are complementary to G nucleotides. In fact, most of the time our DNA consists not of a single strand of nucleotides, but of two complementary strands of nucleotides wrapped around one another.

gene A length of DNA that encodes the information for constructing a particular protein.

protein A long string of amino acids; the basic building material of organisms.

enzyme A complicated protein whose action increases the probability of a specific chemical reaction.

deoxyribonucleic acid (DNA) A nucleic acid that is present in the chromosomes of cells and encodes hereditary information.

nucleotide A portion of a DNA or RNA molecule that is composed of a single base and the adjoining sugar-phosphate unit of the strand.

A.1 Duplication of DNA Before cell division, the genome must be duplicated as illustrated here so that each daughter cell has the full complement of genetic information.

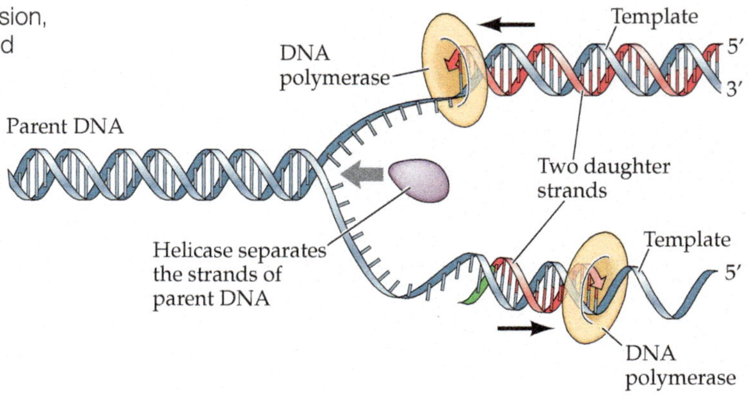

hybridisation The process by which a string of nucleotides becomes linked to a complementary series of nucleotides.

chromosome A complex of condensed strands of DNA and associated protein molecules; found in the nucleus of cells.

eukaryote Any organism whose cells have the genetic material contained within a nuclear envelope.

nucleus Here, the spherical central structure of a cell that contains the chromosomes.

ribonucleic acid (RNA) A nucleic acid that implements information found in DNA.

transcription The process during which mRNA forms bases complementary to a strand of DNA. The resulting message (called a *transcript*) is then used to translate the DNA code into protein molecules.

messenger RNA (mRNA) A strand of RNA that carries the code of a section of a DNA strand to the cytoplasm.

transcript The mRNA strand that is produced when a stretch of DNA is "read."

stop codon A trio of nucleotides in DNA to mark the end of transcription.

ribosomes Structures in the cell body where genetic information is translated to produce proteins.

translation The process by which amino acids are linked together (directed by an mRNA molecule) to form protein molecules.

codon A set of three nucleotides that uniquely encodes one particular amino acid.

peptide A short string of amino acids. Longer strings of amino acids are called *proteins*.

The two strands of nucleotides are said to **hybridise** with one another, coiling slightly to form the famous double helix (**FIGURE A.1**). The double-stranded DNA twists and coils further, becoming visible in microscopes as **chromosomes**, which resemble twisted lengths of yarn. Humans and many other organisms are known as **eukaryotes** because we store our chromosomes in a membranous sphere called a **nucleus** (plural *nuclei*) inside each cell. You may remember that the ability of DNA to exist as two complementary strands of nucleotides is crucial for the duplication of the chromosomes, but that story will not concern us here. Just remember that, with very few exceptions, every cell in your body has a faithful copy of all the DNA you received from your parents.

DNA is transcribed to produce messenger RNA

The information from DNA is used to assemble another molecule—**ribonucleic acid**, or **RNA**—that serves as a template for later steps in protein synthesis. Like DNA, RNA is made up of a long string of four different nucleotides. For RNA, those nucleotides are G and C (which, you recall, are complementary to each other) and A and U (uracil), which are also complementary to each other. Note that the T nucleotide is found only in DNA, and the U nucleotide is found only in RNA.

When a particular gene becomes active, the double strand of DNA unwinds enough so that one strand becomes free of the other and becomes available to special cellular machinery (including an enzyme called *transcriptase*) that begins **transcription**—the construction of a specific string of RNA nucleotides that are complementary to the exposed strand of DNA (**FIGURE A.2**). This length of RNA goes by several names: **messenger RNA** (**mRNA**), **transcript**, or sometimes, *message*. Each DNA nucleotide encodes a specific RNA nucleotide (an RNA G for every DNA C, an RNA C for every DNA G, an RNA U for every DNA A, and an RNA A for every DNA T). Transcription stops when the assembly reaches a trio of DNA nucleotides, called a **stop codon**, that signals the end. This transcript is made in the nucleus where the DNA resides; then the mRNA molecule moves to the cytoplasm, where protein molecules are assembled.

RNA molecules direct the formation of protein molecules

In the cytoplasm are special organelles, called **ribosomes**, that attach themselves to a molecule of mRNA, "read" the sequence of RNA nucleotides, and, using that information, begin linking together amino acids to form a protein molecule. The structure and function of a protein molecule depend on which particular amino acids are put together and in what order. The decoding of an RNA transcript to manufacture a particular protein is called **translation** (see Figure A.2), as distinct from *transcription*, the construction of the mRNA molecule.

Each trio of RNA nucleotides, or **codon**, encodes one of 20 or so different amino acids. Special molecules associated with the ribosome recognise the codon and bring a molecule of the appropriate amino acid so that the ribosome can fuse that amino acid to the previous one. If the resulting string of amino acids is short (say, 50 amino acids or so), it is called a **peptide**; if it is long, it is called a *protein*. Thus the ribosome assembles a very particular sequence of amino acids at the behest of a very particular sequence of RNA nucleotides, which were themselves encoded in the DNA inherited from our parents. In short, the secret of life is that DNA makes RNA, and RNA makes protein.

There are fascinating amendments to this short story. Often the information from separate stretches of DNA is spliced together to make a single transcript; so-called alternative splicing can create different transcripts from the same gene. Sometimes a protein is modified extensively after translation ends; special chemical processes can cleave long proteins to create one or several active peptides.

Within a population, different individuals will inherit slightly different versions of any given gene, including versions that result in different amino acids at any particular position in the protein. These different versions of a given gene are known as **alleles**. Different alleles may produce proteins that vary in structure and function, including protein products that are completely dysfunctional, or even harmful to the cell and therefore the individual.

Keep in mind that each cell has the complete library of genetic information (collectively known as the **genome**) but makes only a fraction of all the proteins encoded in that DNA. In modern biology we say that each cell **expresses** only some genes; that is, the cell transcribes certain genes and makes the corresponding gene products (protein molecules). Thus, each cell must come to express all the genes needed to perform its function. Modern biologists refer to the expression of a particular subset of the genome as **cell differentiation**: the process by which different types of cells acquire their unique appearance and function. During development, individual cells appear to become more and more specialised, expressing progressively fewer genes. Many molecular biologists are striving to understand which cellular and molecular mechanisms "turn on" or "turn off" gene expression, in order to understand development and pathologies such as cancer, or to provide crucial proteins to afflicted organs in a variety of diseases.

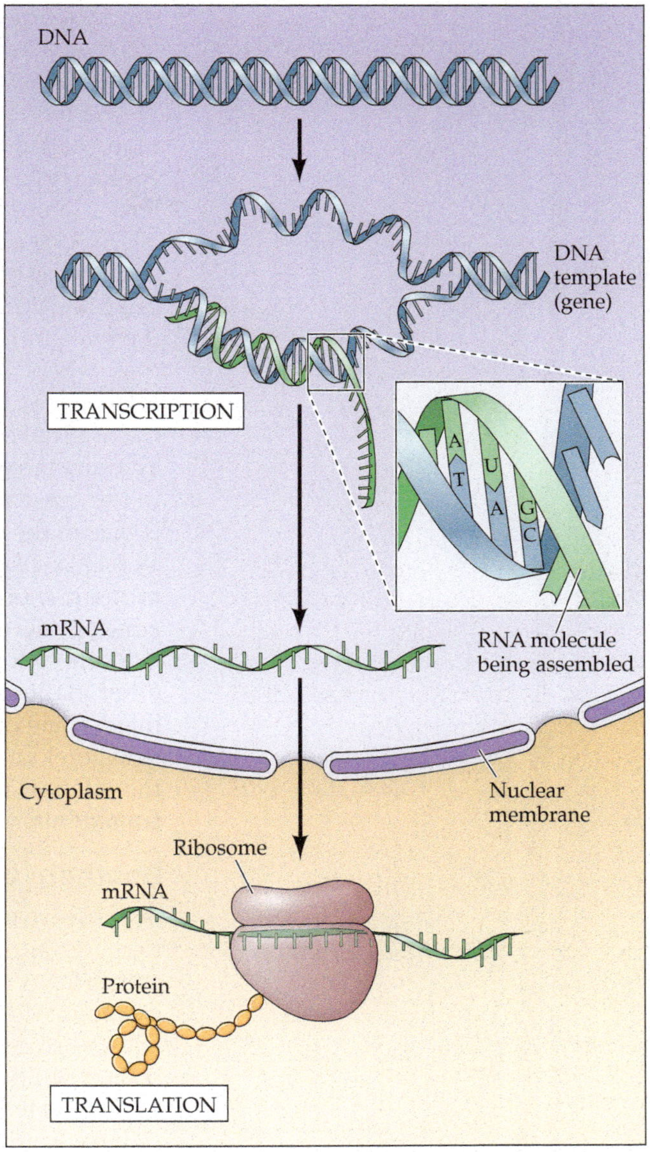

A.2 DNA Makes RNA, and RNA Makes Protein

Molecular Biologists Have Craftily Enslaved Microorganisms and Enzymes

Many basic methods of molecular biology are not explicitly discussed in the text, so we will not describe them in detail here. However, you should understand what some of the terms *mean*, even if you don't know exactly how the methods are performed.

Molecular biologists have found ways to incorporate DNA from other species into the DNA of microorganisms such as bacteria and viruses. After the foreign DNA is incorporated, the microorganisms are allowed to reproduce rapidly, producing more and more copies of the (foreign) gene of interest. At this point the gene is said to be **cloned** because the researcher can make as many copies as they like. To ensure that the right gene is being cloned, the researcher generally clones many, many different genes—each into different bacteria—and then "screens" the bacteria rapidly to find the rare one that has incorporated the gene of interest.

When enough copies of the DNA have been made, the microorganisms are ground up and the DNA extracted. If sufficient DNA has been generated, chemical steps can then determine the exact sequence of nucleotides found in that stretch of DNA—a process known as **DNA sequencing**. Once the sequence of nucleotides has been determined, the sequence of complementary nucleotides in the messenger RNA for that gene can be inferred. The sequence of mRNA nucleotides tells the investigator the sequence of amino acids that will be made from that

allele A particular version of a given gene. Different alleles may differ in the functionality of the protein produced.

genome Also called *genotype*. All the genetic information that one specific individual has inherited.

expression The process by which a cell makes an mRNA transcript of a particular gene.

cell differentiation The developmental stage in which cells acquire distinctive characteristics, such as those of neurons, as the result of expressing particular genes.

clones Here, the reproduction of a gene so that it can be sequenced and/or manipulated.

DNA sequencing The process by which the order of nucleotides in a gene, or amino acids in a protein, is determined.

transcript because biologists know which amino acid is encoded by each trio of DNA nucleotides. For example, scientists discovered the amino acid sequence of neurotransmitter receptors by this process.

The business of obtaining many copies of DNA has been boosted by a technique called the **polymerase chain reaction**, or **PCR**. This technique exploits a special type of polymerase enzyme that, like other such enzymes, induces the formation of a DNA molecule that is complementary to an existing single strand of DNA (see Figure A.1). Because this particular polymerase enzyme (called *Taq polymerase*) evolved in bacteria that inhabit geothermal hot springs, it can function in a broad range of temperatures. By heating double-stranded DNA, we can cause the two strands to separate, making each strand available to polymerase enzymes that, when the temperature is cooled enough, construct a new "mate" for each strand so that they are double-stranded again. The first PCR yields only double the original number of DNA molecules; repeating the process results in four times as many molecules as at first. Repeatedly heating and cooling the DNA of interest in the presence of this heat-resistant polymerase enzyme soon yields millions of copies of the original DNA molecule, which is why this process is also referred to as *gene amplification*. In practice, PCR usually requires the investigator to provide primers: short nucleotide sequences synthesised to hybridise on either side of the gene of interest to amplify that particular gene more than others.

With PCR, sufficient quantities of DNA are produced for chemical analysis or other manipulations, such as introducing DNA into cells. For example, we might inject some of the DNA encoding a protein of interest into a fertilised mouse egg (a zygote) and then return the zygote to a pregnant mouse to grow. Occasionally the injected DNA becomes incorporated into the zygote's genome, resulting in a **transgenic** mouse that carries and expresses the foreign gene (see Box 6.1).

Southern blots identify particular genes

Suppose we want to know whether a particular individual or a particular species carries a certain gene. Because all cells contain a complete copy of the genome, we can gather DNA from just about any kind of cell population: blood, skin, or muscle, for example. After the cells are ground up, a chemical extraction procedure isolates the DNA (discarding the RNA and protein). Finding a particular gene in that DNA just boils down to finding a particular sequence of DNA nucleotides. To do that, we can exploit the tendency of nucleic acids to hybridise with one another.

If we were looking for the DNA sequence GCT, for example, we could manufacture the sequence CGA (there are machines to do that), which would then stick to (hybridise with) any DNA sequence of GCT. The manufactured sequence CGA is called a **probe** because it is made to include a label (a colourful or radioactive molecule) that lets us track its location. Of course, such a short length of nucleotides will be found in many genes. In order for a probe to recognise one particular gene, it has to be about 15 nucleotides long.

When we extract DNA from an individual, it's convenient to let enzymes cut up the very long stretches of DNA into more manageable pieces of 1000 to 20,000 nucleotides each. A process called **gel electrophoresis** uses electrical current to separate these millions of pieces more or less by size (**FIGURE A.3**). Large pieces move slowly through a tube of gelatin-like material, and small pieces move rapidly. The tube of gel is then sliced and placed on top of a sheet of paperlike material called *nitrocellulose*. When fluid is allowed to flow through the gel and nitrocellulose, DNA molecules are pulled out of the gel and deposited on the waiting nitrocellulose. This process of making a "sandwich" of gel and nitrocellulose and using fluid to move molecules from the former to the latter is called **blotting** (see Figure A.3).

If the gene that we're looking for is among those millions of DNA fragments sitting on the nitrocellulose, our labelled probe should recognise and hybridise to the sequence. The nitrocellulose sheet is soaked in a solution containing our labelled probe; we wait for the probe to find and hybridise with the gene of interest (if it is present), and we rinse the sheet to remove probe molecules that did not

polymerase chain reaction (PCR) Also called *gene amplification*. A method for reproducing a particular RNA or DNA sequence manyfold, allowing amplification for sequencing or manipulating the sequence.

transgenic Referring to an animal in which a new or altered gene has been deliberately introduced into the genome.

probe A manufactured sequence of DNA or RNA that is made to include a label (a colourful or radioactive molecule) that lets us track its location.

gel electrophoresis A method of separating molecules of differing size or electrical charge by forcing them to flow through a gel.

blotting Transferring DNA, RNA, or protein fragments to nitrocellulose following separation via gel electrophoresis. The blotted substance can then be labelled.

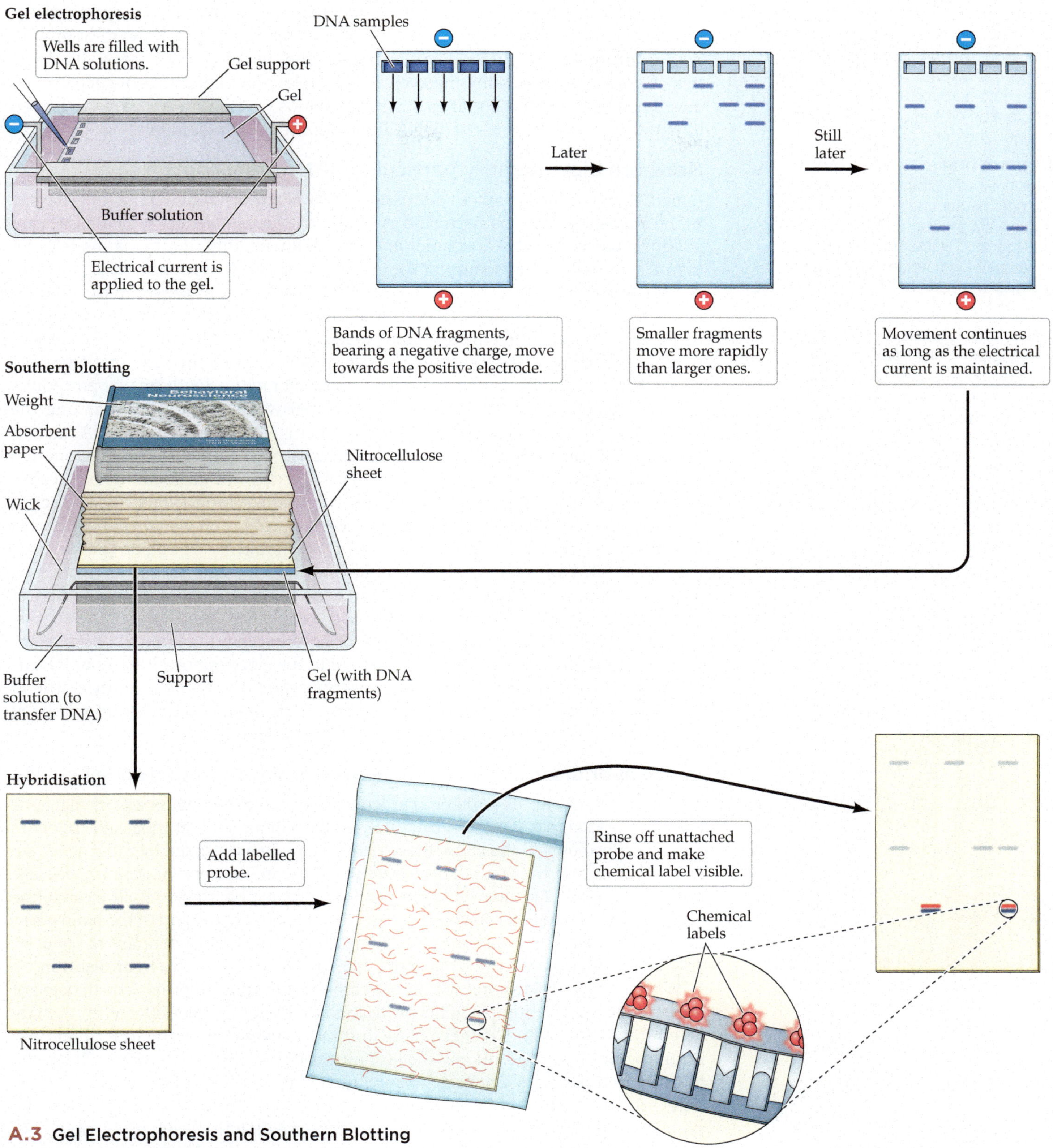

A.3 Gel Electrophoresis and Southern Blotting

find the gene. Then we *visualise* the probe, either by causing the label to show its colour or, if radioactive, by letting the probe expose photographic film to identify the locations where the probe has accumulated. In either case, if the probe has found the gene, a labelled band will be evident, corresponding to the size of DNA fragment that contained the gene (see Figure A.3).

Go to Animation A.1
Gel Electrophoresis

Southern blot A method of detecting a particular DNA sequence in the genome of an organism, by separating DNA with gel electrophoresis, blotting the separated DNAs onto nitrocellulose, and then using a nucleotide probe to hybridise with, and highlight, the gene of interest.

Northern blot A method of detecting a particular RNA transcript in a tissue or organ, by separating RNA from that source with gel electrophoresis, blotting the separated RNAs onto nitrocellulose, and then using a nucleotide probe to hybridise with, and highlight, the transcript of interest.

in situ hybridisation A method for detecting particular RNA transcripts in tissue sections by providing a nucleotide probe that is complementary to, and will therefore hybridise with, the transcript of interest.

antibody Also called *immunoglobulin*. A large protein that recognises and permanently binds to particular shapes, normally as part of the immune system attack on foreign particles.

This process of looking for a particular sequence of DNA is called a **Southern blot**, named after the man who developed the technique, Edward Southern. Southern blots are useful for determining whether related individuals share a particular gene or for assessing the evolutionary relatedness of different species. The developed blots, with their lanes of labelled bands (see Figure A.3), are often seen in popular-media accounts of "DNA fingerprinting" of individuals.

Northern blots identify particular mRNA transcripts

A method more relevant for our discussions is the **Northern blot** (whimsically named as the opposite of the Southern blot). A Northern blot can identify which tissues are making a particular RNA transcript. If liver cells are making a particular protein, for example, then some transcripts for the gene that encodes that protein should be present. So we can take the liver, grind it up, and use chemical processes to isolate most of the RNA (discarding the DNA and protein). The resulting mixture consists of RNA molecules of many different sizes: long, medium, and short transcripts. Gel electrophoresis will separate the transcripts by size, and we can blot the size-sorted mRNAs onto nitrocellulose sheets; the process is very similar to the Southern blot procedure.

To see whether the particular transcript that we're looking for is among the mRNAs, we construct a labelled probe (of either DNA nucleotides or RNA nucleotides) that is complementary to the mRNA transcript of interest and long enough that it will hybridise only to that particular transcript. We incubate the nitrocellulose in the probe, allow time for the probe to hybridise with the targeted transcript (if present), rinse off any unused probe molecules, and then visualise the probe as before. If the transcript of interest is present, we should see a band on the film (see Figure A.3). The presence of several bands indicates that the probe has hybridised to more than one transcript and we may need to make a more specific probe or alter chemical conditions to make the probe less likely to bind similar transcripts.

Because different gene transcripts are of different lengths, the transcript of interest should have reached a particular point in the electrophoresis gel: small transcripts should have moved far; large transcripts should have moved only a little. If our probe has found the right transcript, the single band of labelling should be at the point that is appropriate for a transcript of that length.

In situ hybridisation localizes mRNA transcripts within specific cells

Northern blots can tell us whether a particular *organ* has transcripts for a particular gene product. For example, Northern blot analyses have indicated that thousands of genes are transcribed only in the brain. Presumably the proteins encoded by these genes are used exclusively in the brain. But such results alone are not very informative, because the brain consists of so many different kinds of glial and neuronal cells. We can refine Northern blot analyses somewhat, by dissecting out a particular part of the brain—say, the hippocampus—to isolate mRNAs. Sometimes, though, it is important to know *exactly which cells* are making the transcript. In that case we use **in situ hybridisation**.

With in situ hybridisation we use the same sort of labelled probe, constructed of nucleotides that are complementary to (and will therefore hybridise with) the targeted transcript, as in Northern blots. Instead of using the probe to find and hybridise with the transcript on a sheet of nitrocellulose, however, we use the probe to find the transcripts "in place" (*in situ* in Latin)—that is, on a section of tissue. After rinsing off the probe molecules that didn't find a match, we visualise the probe right in the tissue section. Any cells in the section that were transcribing the gene of interest will have transcripts in the cytoplasm that should have hybridised with our labelled probe (**FIGURE A.4**) (see Box 1.1). In situ hybridisation therefore can tell us exactly which cells are expressing a particular gene.

Western blots identify particular proteins

Sometimes we wish to study a particular protein rather than its transcript. In such cases we can use antibodies. **Antibodies** are large, complicated molecules (proteins, in fact) that our immune system adds to the bloodstream to identify and fight invading microbes, thereby arresting and preventing disease (see Figure 11.29). But if

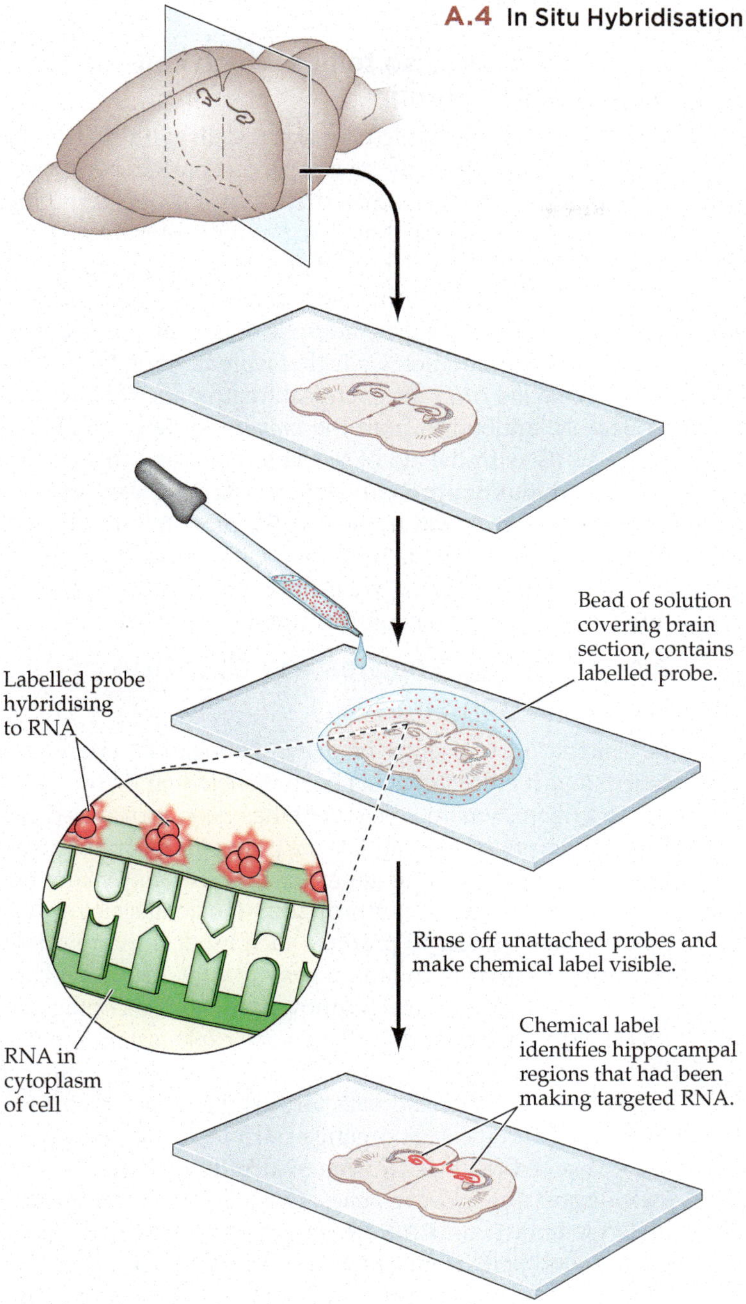

we inject a rabbit or mouse with a sample of a protein of interest, we can induce the animal to create antibodies that recognise and attach to that particular protein, just as if it were an invader.

Once these antibodies have been purified and chemically labelled, we can use them to search for the target protein. We grind up an organ, isolate the proteins (discarding the DNA and RNA), and separate them by means of gel electrophoresis. Then we blot these proteins out of the gel and onto nitrocellulose. Next we use the antibodies to tell us whether the targeted protein is among those made by that organ. If the antibodies identify only the protein we care about, there should be a single band of labelling (if there are two or more, then the antibodies recognise more than one protein). Because proteins come in different sizes, the single band of label should be at the position corresponding to the size of the protein that we're studying. Such blots are called **Western blots**.

To review, Southern blots identify particular DNA pieces (genes), Northern blots identify particular RNA pieces (transcripts), and Western blots identify particular proteins (sometimes called *products*).

Western blot A method of detecting a particular protein molecule in a tissue or organ, by separating proteins from that source with gel electrophoresis, blotting the separated proteins onto nitrocellulose, and then using an antibody that binds, and highlights, the protein of interest.

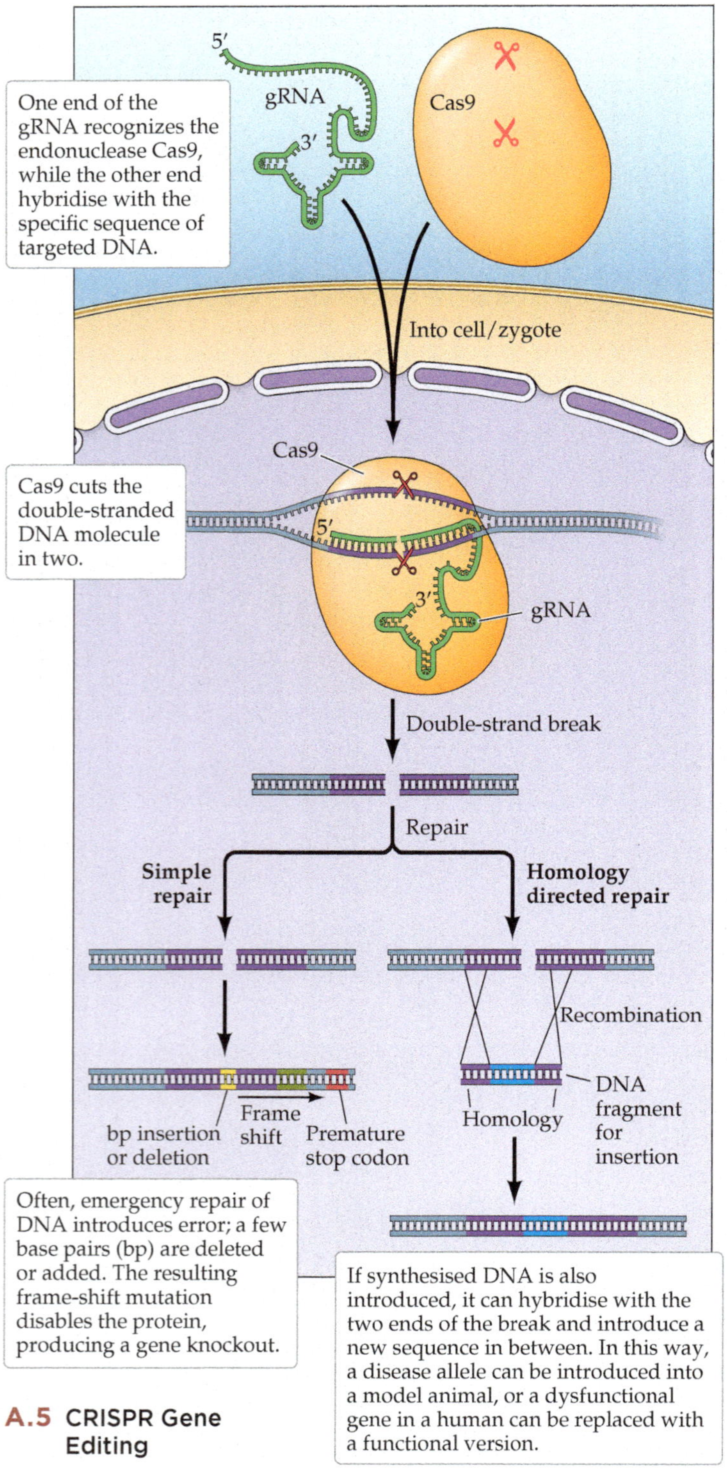

A.5 CRISPR Gene Editing

Antibodies can also tell us which cells possess a particular protein

If we need to know which particular cells within an organ such as the brain are making a particular protein, we can use the same sorts of antibodies that we use in Western blots, but in this case directed at that protein in tissue sections. We slice up the brain, expose the sections to the antibodies, allow time for them to find and attach to the protein, rinse off unattached antibodies, and use chemical treatments to visualise the antibodies. Cells that were making the protein will be labelled from the chemical treatments (see Box 2.1).

Because antibodies from the *immune* system are used to identify *cells* with the aid of *chemical* treatment, this method is called **immunocytochemistry**, or **ICC**. This technique can even tell us where, within the cell, the protein is found. Such information can provide important clues about the function of the protein. For example, if the protein is found in axon terminals, it may be a neurotransmitter.

Gene Editing Enables the Creation of Model Organisms

In Chapter 6 we discuss the use of "knockout" organisms, where a particular gene has been disabled, and "transgenic" organisms, where a new gene has been introduced (see Box 6.1). These animal models are powerful means of investigating the influence of genes on the nervous system and behaviour. A recent breakthrough in molecular biology has made it much easier to produce such models in organisms as diverse as worms and monkeys.

Like many other biotech innovations, such as optogenetics (see Figures 2.26 and 2.27), this one exploits a mechanism that evolved in simple organisms. Unicellular organisms like bacteria become infected with viruses that insert their invasive genes into the host organism's DNA. In response, bacteria have evolved mechanisms to identify and remove, or at least disable, the invasive genes. One mechanism is a stretch of DNA known as **CRISPR**, for ***c***lustered ***r***egularly ***i***nter-spaced ***s***hort ***p***alindromic ***r***epeats. The RNA transcribed from a CRISPR site will serve as a guide to target the invasive DNA and so is referred to as a **guide RNA** (**gRNA**). The palindromic sequences in the front of the gRNA cause it to fold up in a distinctive shape that binds to an enzyme, an endonuclease called **Cas9** (***C***RISPR-***as***sociated enzyme ***9***), while the "tail" of the gRNA is complementary to the invasive sequence and so will hybridise with it. Then the associated Cas9 enzyme breaks the host's double-stranded DNA at that point. This type of break in double-stranded DNA is relatively rare, but all organisms have machinery in place to repair it. The thing is, this emergency repair often introduces an error, either adding or subtracting a few nucleotides. That in turn will introduce a frameshift mutation, with the result that the wrong sequence of amino acids will be assembled and eventually a premature stop codon (the trio of nucleotides that normally signal the end of transcription) will end the transcription entirely (**FIGURE A.5**). The combination of improper amino acids and curtailed transcription means the invasive viral protein won't work, and the infection will be undone.

Eukaryotes like worms and mice lack the CRISPR mechanism and so normally do not produce Cas9. But we can introduce the gene for Cas9 into a zygote (of a worm, a fly, a mouse, or a monkey), and we can also introduce DNA custom made

immunocytochemistry (ICC)
A method for detecting a particular protein in tissues in which an antibody recognises and binds to the protein and then chemical methods are used to leave a visible reaction product around each antibody.

to produce a gRNA directed at whatever gene we choose. In that case, the gRNA will direct Cas9 to break the DNA within the targeted gene. Then the imperfect repair of that DNA will effectively knock out the gene in the individual that results from this zygote. Or the same system can be used to knock out the gene in particular cells in a mature individual. An enormous advantage of the CRISPR system is that one can introduce several different gRNAs simultaneously to knock out several different genes at once (H. Wang et al., 2013).

Scientists have also exploited the CRISPR system for gene editing. In this variant, the scientist again introduces DNA to cause the cell or zygote to express Cas9 and a specific gRNA, but also provides synthesised DNA fragments that are homologous to the cut ends of the host's DNA. Then when the break is repaired, the synthesised DNA fragment recombines into the host's genome, a case of *homology-directed repair*. Now, instead of knocking out the targeted gene, the scientist has replaced that section of the gene, effectively producing a new allele. In this way, a mutated allele can be introduced into the fly, mouse, zebrafish, or monkey, producing a transgenic animal (more specifically, a "knockin" animal, where a new allele has been put in place of the endogenous allele). There is also great hope that this method may someday be used for gene therapy in humans, to replace dysfunctional genes that cause disease, such as the *dystrophin* gene in Duchenne muscular dystrophy (see Chapter 10).

Another bacterial mechanism to get rid of viral infection is to insert a specific series of nucleotides, called **loxP** sites, to mark the invasive DNA. A bacterial enzyme, called **Cre-recombinase**, then binds to two such loxP sites (as long as they are not too far apart), cuts out the DNA between them, and joins the cut ends of DNA back together, a process called *recombination*. This severe editing renders the viral gene product worthless, thwarting the infection. Multicellular organisms normally do not have this Cre-lox system, but scientists can insert them into the genome of flies/zebrafish/mice to manipulate genes. For example, we can insert lox sites into the DNA of a targeted gene (we might use the CRISPR homology-directed repair to do this) in such a way that the gene still functions normally despite the lox sites. Now we can inject viruses into a particular brain site to infect cells there to express Cre-recombinase. That means the loxP-flanked, or "floxed" gene will be knocked out only in that brain region. Or for another approach, we might cross animals carrying the floxed allele with animals carrying a transgene for Cre-recombinase. This transgene might include a promoter that causes the enzyme to be expressed only in neurons, or only in astrocytes, or only in motor neurons. That means the floxed gene will be disrupted only in those particular classes of cells, not the whole organism (**FIGURE A.6**). Using these methods, scientists are able to test more and more specific hypotheses about how gene expression alters the brain and therefore behaviour.

CRISPR ***c***lustered ***r***egularly ***i***nterspaced ***s***hort ***p***alindromic ***r***epeats. A system of gene manipulation that evolved in single-celled organisms and is exploited by scientists for gene editing.

guide RNA (gRNA) A strand of RNA designed to hybridise with a targeted nucleotide sequence in DNA in order to guide the Cas9 enzyme to break the DNA at that site.

Cas9 ***C***RISPR ***as***sociated enzyme ***9***. A bacterial enzyme that induces a break in double-stranded DNA as part of the CRISPR system.

loxP A specific sequence of nucleotides recognised by the enzyme Cre-recombinase. If the enzyme encounters a pair of loxP sites in a gene, it will remove the DNA between the two sites and recombine the gene, usually rendering the gene product dysfunctional.

Cre-recombinase A bacterial enzyme that recognises loxP, which is a specific sequence of nucleotides, and recombines the DNA at loxP sites.

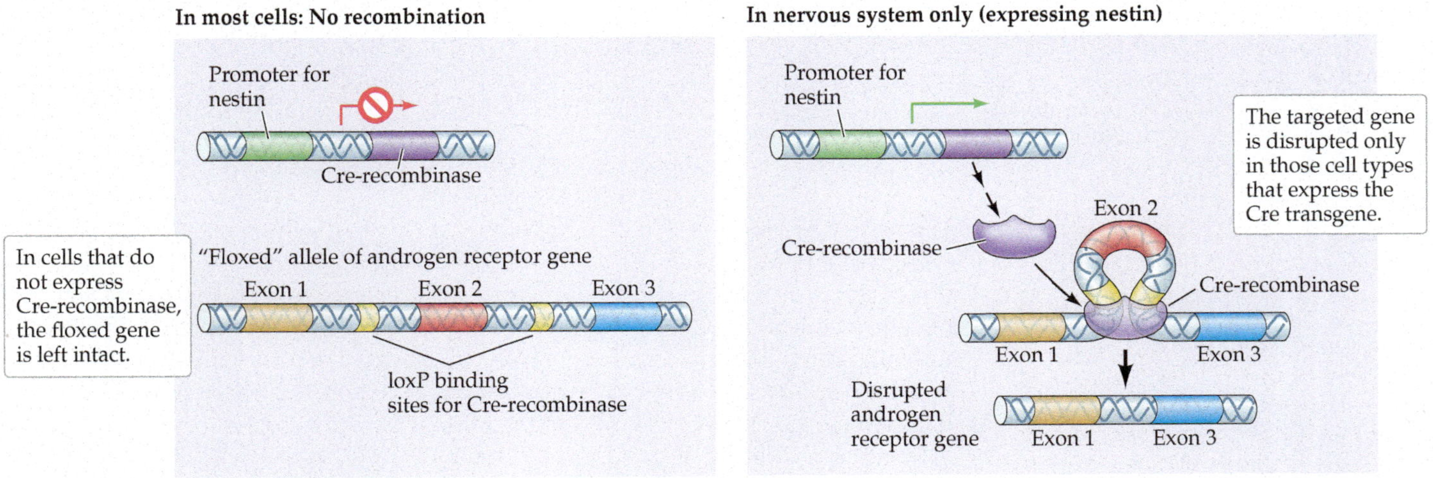

A.6 **Cre-Lox Systems for Knocking Out Genes in Specific Cells**

Glossary

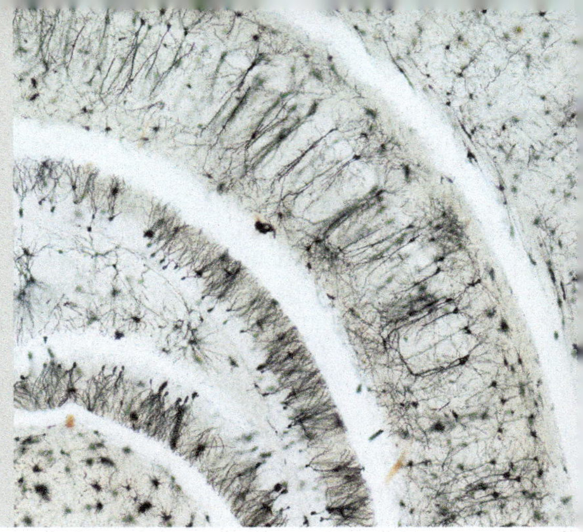

Courtesy of Dr. Sarah Moghadam, VA Palo Alto Health Care System, Palo Alto, CA and Dr. Ahmad Salehi, Dept. of Psychiatry & Behavioural Sciences, Stanford Medical School

Numbers in brackets refer to the chapter(s) where the term is introduced.

5-HT See *serotonin.*

5α-reductase An enzyme that converts testosterone into dihydrotestosterone (DHT). [13]

17β-estradiol Also called simply *estradiol*. The primary type of estrogen that is secreted by the ovary. [4]

α-fetoprotein A protein found in the plasma of fetuses. In rodents, α-fetoprotein binds estrogens and prevents them from entering the brain. [13]

α-melanocyte-stimulating hormone (α-MSH) A peptide that binds the melanocortin receptor. [14]

α-MSH See *α-melanocyte-stimulating hormone.*

α-synuclein A protein implicated in Parkinson's disease. [10]

β-amyloid A protein that accumulates in senile plaques in Alzheimer's disease. [6]

β-secretase An enzyme that cleaves amyloid precursor protein, forming β-amyloid, which is suspected of causing Alzheimer's disease. [6]

Δ9-tetrahydrocannabinol (THC) The major active ingredient in cannabis thought to be responsible for the "high" feeling. [3]

A

A1 *See primary auditory cortex.*

absence attack See *simple partial seizure.*

absolute refractory phase A brief period of complete insensitivity to stimuli. [2]

accommodation The process of focussing by the ciliary muscles and the lens to form a sharp image on the retina. [8]

acetylcholine (ACh) A neurotransmitter produced and released by parasympathetic postganglionic neurons, by motor neurons, and by neurons throughout the brain. [1, 2, 3, 10]

acetylcholinesterase (AChE) An enzyme that inactivates the transmitter acetylcholine both at synaptic sites and elsewhere in the nervous system. [3]

ACh See *acetylcholine.*

AChE See *acetylcholinesterase.*

acid See *LSD.*

ACTH See *adrenocorticotropic hormone.*

actin A protein that, along with myosin, mediates the contraction of muscle fibres. [10]

action potential The propagated electrical message of a neuron that travels along the axon to the presynaptic axon terminals. [2]

activational effect A temporary change in behaviour resulting from the administration of a hormone to an adult animal. [13]

acupuncture The insertion of needles at designated points on the skin in an attempt to alleviate pain or physiological malfunction, or alter behaviour.

acute flaccid myelitis A sudden weakness or paralysis in limb(s) with decreased or absent reflexes. [10]

adaptation 1. A trait that increases the probability that an individual will leave offspring in subsequent generations. 2. The progressive loss of receptor sensitivity as stimulation is maintained. [5, 7, 8]

Aδ fibre A large, myelinated, and therefore fast-conducting axon, usually transmitting acute pain information. [7]

adenohypophysis See *anterior pituitary.*

adenosine An endogenous neuromodulator that generally reduces neural activity. Caffeine interferes with adenosine binding. [3, 15]

adequate stimulus The type of stimulus for which a given sensory organ is particularly adapted. [7]

ADH See *arginine vasopressin.*

ADHD See *attention deficit hyperactivity disorder.*

adipose tissue Tissue made up of fat cells. [14]

adrenal cortex The outer rind of the adrenal gland, which secretes steroid hormones, including cortisol. [4, 11]

adrenal gland An endocrine gland atop the kidney. [4]

adrenal medulla The inner core of the adrenal gland, which secretes epinephrine and norepinephrine. [4, 11]

adrenal steroids Also called *adrenocorticoids*. Steroid hormones secreted by the adrenal cortex, including glucocorticoids such as cortisol and mineralocorticoids such as aldosterone. [4, 11]

adrenaline See *epinephrine.*

adrenocorticoids Also called *adrenal steroids*. A class of steroid hormones that are secreted by the adrenal cortex. [4, 11]

adrenocorticotropic hormone (ACTH) A tropic hormone secreted by the anterior pituitary gland that controls the production and release of hormones of the adrenal cortex. [4]

adult neurogenesis The birth of new neurons in the brain of an adult. [6]

afferent In reference to axons, carrying action potentials from a sensory organ to the central nervous system or from one region to another region of interest. Compare *efferent*. [1]

affinity See *binding affinity.*

afterpotential The positive or negative change in membrane potential that may follow an action potential. [2]

aggression Behaviour that is intended to cause pain or harm (whether physical or emotional) to others, either individually or in groups. [11]

agnosia The inability to recognise objects, despite being able to describe them in terms of form and colour. It may occur after localised brain damage. [18]

agonist A molecule, usually a drug, that binds a receptor molecule and initiates a response like that of another substance, usually a neurotransmitter. [2, 3]

agraphia The inability to write. [18]

AIS See *androgen insensitivity syndrome.*

alarm reaction The initial response to stress. [11]

aldosterone A mineralocorticoid hormone, secreted by the adrenal cortex, that induces the kidneys to conserve sodium ions. [4, 14]

alexia The inability to read. [18]

allele One of two or more versions of a given gene or genetic locus. Different alleles may differ in the functionality of the protein produced. [5, App]

allomone A chemical signal that is released outside the body by one species and affects the behaviour of other species. [4]

allopregnanolone A naturally occurring steroid that modulates GABA receptor activity in much the same way that benzodiazepine anxiolytics do. [3]

all-or-none property The fact that the amplitude of the action potential is independent of the magnitude of the stimulus. [2]

allostasis The varying behavioural and physiological adjustments that an individual makes in order to maintain optimal (rather than unchanging) functioning of a regulated system in the face of changing environmental stressors. [14]

alpha motor neuron A motor neuron that controls the main contractile fibres (extrafusal fibres) of a muscle. [10]

alpha rhythm A brain potential of 7–13 Hz that occurs during relaxed wakefulness. [15]

ALS See *amyotrophic lateral sclerosis.*

altricial Referring to animals that are born in a very immature state and depend on maternal care, as human infants do. [13]

Alzheimer's disease A form of dementia that may appear in middle age but is more frequent among the elderly. [6]

amacrine cells Specialised retinal neurons that contact both the bipolar cells and the ganglion cells and are especially significant in inhibitory interactions within the retina. [8]

amblyopia Reduced visual acuity of one eye, that is not caused by optical or retinal impairments. [6, 8]

AMH See *anti-müllerian hormone.*

amine hormones Also called *monoamine hormones.* A class of hormones, each composed of a single amino acid that has been modified into a related molecule, such as melatonin or epinephrine. [4]

amine neurotransmitter A neurotransmitter based on modifications of a single amino acid. [3]

amino acid neurotransmitter A neurotransmitter that is itself an amino acid. [3]

amnesia Severe impairment of memory. [16]

AMPA receptor A glutamate receptor that also binds the glutamate agonist AMPA. [16]

amphetamine A molecule that resembles the structure of the catecholamine transmitters and enhances their activity. [3]

amphetamine psychosis A delusional and psychotic state, closely resembling acute schizophrenia, that is brought on by repeated use of high doses of amphetamine. [12]

amplitude Magnitude or strength. In acoustics, the force sound exerts per unit area, usually measured as dynes per square centimeter.

ampulla An enlarged region of each semicircular canal that contains the receptor cells (hair cells) of the vestibular system. [9]

amusia A disorder characterised by the inability to discern tunes accurately. [9]

amygdala A group of nuclei in the medial anterior part of the temporal lobe. [1, 11]

amyloid plaques See *senile plaques.*

amyloid precursor protein (APP) A protein that, when cleaved by several enzymes, produces β-amyloid. [6]

amyotrophic lateral sclerosis (ALS) Also called *Lou Gehrig's disease.* A disease in which motor neurons and their target muscles waste away. [10]

analgesia Absence of or reduction in pain. [7]

analgesic Referring to painkilling properties. [3]

analogy Similarity of function, although the structures of interest may look different. The human hand and an elephant's trunk are analogous features. [5]

anandamide An endogenous substance that binds the cannabinoid receptor molecule. [3]

androgen insensitivity syndrome (AIS) A syndrome caused by a mutation of the androgen receptor gene that renders tissues insensitive to androgenic hormones like testosterone. Affected XY individuals are phenotypic females, but they have internal testes and regressed internal genital structures. [13]

androgens A class of hormones that includes testosterone and other male hormones. [4]

androstenedione The chief sex hormone secreted by the human adrenal cortex. [4]

angel dust See *phencyclidine.*

angiography A brain-imaging technique in which a specialised X-ray image of the head is taken shortly after the cerebral blood vessels have been filled with a radiopaque dye by means of a catheter. [1]

angiotensin II A substance that is produced in the blood by the action of renin and that may play a role in the control of thirst. [14]

angular gyrus A brain region in which strokes can lead to word blindness. [18]

anion A negatively charged ion, such as a protein or chloride ion. [2]

annulospiral ending See *primary sensory ending.*

anomia The inability to name persons or objects. [18]

anorexia nervosa A syndrome in which individuals severely deprive themselves of food. [14]

anorexigenic neurons Neurons of the hypothalamic appetite system that inhibit feeding behaviour. [14]

anosmia The inability to smell. [9]

anosognosia Denial of illness. [17]

ANP See *atrial natriuretic peptide.*

antagonist 1. A molecule, usually a drug, that interferes with or prevents the action of a transmitter. 2. A muscle that counteracts the effect of another muscle. [2, 3, 10]

anterior Also called *rostral.* In anatomy, towards the head end of an organism. Compare *posterior.* [1]

anterior cerebral arteries Two large arteries, arising from the internal carotid arteries, that provide blood to the anterior poles and medial surfaces of the cerebral hemispheres. [1]

anterior pituitary Also called *adenohypophysis.* The front division of the pituitary gland. It secretes tropic hormones. [4]

anterograde amnesia The inability to form new memories beginning with the onset of a disorder. [16]

anterograde degeneration Also called *Wallerian degeneration*. The loss of the distal portion of an axon resulting from injury to the axon. [6]

anterolateral system Also called *spinothalamic system*. A somatosensory system that carries most of the pain and temperature information from the body to the brain. [7]

antibodies Also called *immunoglobulins*. Large proteins that recognise and permanently bind to particular shapes, normally as part of the immune system attack on foreign particles. [11, App]

antidepressants A class of drugs that relieve the symptoms of depression. [3]

antidiuretic hormone (ADH) See *arginine vasopressin*.

anti-müllerian hormone (AMH) A protein hormone secreted by the fetal testis that inhibits müllerian duct development. [13]

antipsychotics A class of drugs that alleviate symptoms of schizophrenia, typically by blocking dopamine receptors. [12]

anxiety disorder Any of a class of psychological disorders that include recurrent panic states and generalised anxiety disorder. [12]

anxiolytics A class of substances that are used to combat anxiety. Examples include alcohol, opiates, barbiturates, and the benzodiazepines. [3, 12]

aphagia Refusal to eat. [14]

aphasia An impairment in language understanding and/or production that is caused by brain injury. [18]

apical dendrite The type of dendrite that extends from a pyramidal cell to the outermost surface of the cortex.

apolipoprotein E (ApoE) A protein that may help break down amyloid. [6]

apoptosis See *cell death*.

APP See *amyloid precursor protein*.

appetitive behaviour The second stage of mating behaviour. It helps establish or maintain sexual interaction. [13]

apraxia An impairment in the ability to begin and execute skilled voluntary movements, even though there is no muscle paralysis. [10, 18]

aquaporins Channels spanning the cell membrane that are specialised for conducting water molecules into or out of the cell. [14]

arachnoid The thin covering (one of the three meninges) of the brain that lies between the dura mater and pia mater. [1]

arborisation The elaborate branching of the dendrites of some neurons. [1]

arcuate fasciculus A tract believed by some to connect Wernicke's area to Broca's area. [18]

arcuate nucleus An arc-shaped hypothalamic nucleus implicated in appetite control. [14]

area 17 See *primary visual cortex*.

arginine vasopressin (AVP) Also called *vasopressin* or *antidiuretic hormone* (*ADH*). A peptide hormone from the posterior pituitary that promotes water conservation. [4]

aromatase An enzyme that converts some androgens into estrogens. [4, 13]

aromatisation The chemical reaction that converts testosterone to estradiol, and other androgens to other estrogens. [13]

aromatisation hypothesis The hypothesis that testicular androgens enter the brain and are converted there into estrogens to masculinise the developing nervous system in some rodents. [13]

ASD See *autism spectrum disorder*.

aspartate An amino acid transmitter that is excitatory at many synapses. [3]

associative learning A type of learning in which an association is formed between two stimuli or between a stimulus and a response. It includes both classical and operant conditioning. [16]

astereognosis The inability to recognise objects by touching and feeling them. [18]

astrocyte A star-shaped glial cell with numerous processes (extensions) that run in all directions. [1]

ataxia An impairment in the direction, extent, and rate of muscular movement. It is often caused by cerebellar pathology. [10]

atrial natriuretic peptide (ANP) A hormone, secreted by the heart, that normally reduces blood pressure, inhibits drinking, and promotes the excretion of water and salt at the kidneys. [14]

attention Also called *selective attention*. A state or condition of selective awareness by which specific stimuli are selected for enhanced processing. [7, 17]

attention deficit hyperactivity disorder (ADHD) A syndrome of distractibility, impulsiveness, and hyperactivity that, in children, interferes with school performance. [17]

attentional bottleneck A filter that results from the limits intrinsic to our attentional processes, with the result that only the most important stimuli are selected for special processing. [17]

attentional spotlight The shifting of our limited selective attention around the environment to highlight stimuli for enhanced processing. [17]

aura In epilepsy, the unusual sensations or premonition that may precede the beginning of a seizure. [2]

australopithecine Referring to *Australopithecus*, a primate genus, known only from the fossil record and thought to be an ancestor to humans. [5]

autism spectrum disorder (ASD) A range of conditions characterised by social, communication, and behavioural challenges. [6]

autocrine Referring to a signal that is secreted by a cell into its environment and that feeds back to the same cell. [4]

autoimmune disorder A disorder caused when the immune system mistakenly attacks a person's own body, thereby interfering with normal functioning. [10]

autonomic ganglia Collections of nerve cell bodies, belonging to the autonomic division of the peripheral nervous system, that are found in various locations and innervate the major organs. [1]

autonomic nervous system The part of the peripheral nervous system that supplies neural connections to glands and to smooth muscles of internal organs. [1]

autoradiography A histological technique that shows the distribution of radioactive chemicals in tissues. [1, 4]

autoreceptor A receptor for a synaptic transmitter that is located in the presynaptic membrane, telling the axon terminal how much transmitter has been released. [2, 3]

AVP See *arginine vasopressin*.

axo-axonic Referring to a synapse in which a presynaptic axon terminal synapses onto another axon's terminal. [2]

axo-dendritic Referring to a synapse in which a presynaptic axon terminal synapses onto a dendrite of the postsynaptic neuron, either via a dendritic spine or directly onto the dendrite itself. [2]

axon A single extension from the nerve cell that carries action potentials from the cell body to other neurons. [1]

axon collateral A branch of an axon from a single neuron. [1]

axon hillock A cone-shaped area from which the axon originates out of the cell body. Functionally, it is the integration zone of the neuron. [1, 2]

axon terminal Also called *synaptic bouton*. The end of an axon or axon collateral, which forms a synapse on a neuron or other target cell. [1]

axonal transport The transportation of materials between the neuronal cell body and axon terminals. [1]

axo-somatic Referring to a synapse in which a presynaptic axon terminal synapses onto the cell body (soma) of the postsynaptic neuron. [2]

B

B cell See *B lymphocyte.*

B lymphocyte Also called *B cell.* An immune system cell, formed in the bone marrow (hence the *B*), that mediates humoral immunity. [11]

Bálint's syndrome Three co-occurring symptoms—simultagnosia, oculomotor apraxia, and optic ataxia—that may occur after bilateral lesions of cortical attentional systems. [17]

ballistic movement A rapid muscular movement that is generally preprogrammed. [10]

bar detector See *simple cortical cell.*

barbiturate A powerful sedative anxiolytic derived from barbituric acid, with dangerous addiction and overdose potential. [3]

bariatric Referring to treatment of obesity. [14]

baroreceptor A pressure receptor in the heart or a major artery that detects a decrease in blood pressure. [14]

basal dendrite One of several dendrites on a pyramidal cell that extend horizontally from the cell body.

basal forebrain A ventral region in the forebrain that has been implicated in consciousness and sleep. [15]

basal ganglia A group of forebrain nuclei, including caudate nucleus, globus pallidus, and putamen, found deep within the cerebral hemispheres. [1, 10]

basal metabolism The consumption of energy by the basic life-sustaining functions of the body. [14]

basilar artery An artery, formed by the fusion of the vertebral arteries, that supplies blood to the brainstem and to the posterior cerebral arteries. [1]

basilar membrane A membrane in the cochlea that contains the principal structures involved in auditory transduction. [9]

bass An aspect of pitch corresponding to the subjective experience of low-frequency sounds (especially musical sounds, such as bass guitar). [9]

batrachotoxin A toxin, secreted by poison arrow frogs, that selectively interferes with Na^+ channels. [2]

Bcl-2 A family of proteins that regulate apoptosis. [6]

BDNF See *brain-derived neurotrophic factor.*

behavioural intervention An approach to finding relations between body variables and behavioural variables that involves intervening in the behaviour of an organism and looking for resultant changes in body structure or function. [0]

behavioural medicine See *health psychology.*

behavioural neuroscience Also called *biological psychology.* The study of the neural bases of behaviour and mental processes. [0]

Bell's palsy A disorder, usually caused by viral infection, in which the facial nerve on one side stops conducting action potentials, resulting in paralysis of one side of the face. [11]

benzodiazepine agonists A class of antianxiety drugs that bind to sites on $GABA_A$ receptors. [3]

benzodiazepines A class of antianxiety drugs that bind with high affinity to receptor molecules in the central nervous system. One example is diazepam (Valium). [12]

beta activity EEG activity seen in wakefulness, comprising a mix of many different high frequencies with low amplitude. [15]

binaural Referring to two ears. [9]

binding affinity Also called simply *affinity.* The propensity of molecules of a drug (or other ligand) to bind to receptors. [3]

binding problem The question of how the brain understands which individual attributes blend together into a single object, when these different features are processed by different regions in the brain. [17]

binge eating The paroxysmal intake of large quantities of food, often of poor nutritional value and high in calories. [14]

binocular deprivation Depriving both eyes of form vision, as by sealing the eyelids. [6]

bioavailable Referring to a substance, usually a drug, that is present in the body in a form that is able to interact with physiological mechanisms. [3]

biological psychology See *behavioural neuroscience.*

biotransformation The process in which enzymes convert a drug into a metabolite that is itself active, possibly in ways that are substantially different from the actions of the original substance. [3]

bipolar cells A class of interneurons of the retina that receive information from rods and cones and pass the information to retinal ganglion cells. [8]

bipolar disorder Formerly called *manic-depressive illness.* A psychiatric disorder characterised by periods of depression that alternate with excessive, expansive moods. [12]

bipolar neuron A nerve cell that has a single dendrite at one end and a single axon at the other end. [1]

blind spot The portion of the visual field from which light falls on the optic disc. Because there are no receptors in this region, light striking it cannot be seen. [8]

blood-brain barrier The mechanisms that make the movement of substances from blood vessels into brain cells more difficult than exchanges in other body organs, thus affording the brain greater protection from exposure to some substances found in the blood. [1, 3]

blotting Transferring DNA, RNA, or protein fragments to nitrocellulose following separation via gel electrophoresis. The blotted substance can then be labelled. [App]

border cell A neuron that selectively fires when an animal arrives at the perimeter of a local spatial cognitive map. [16]

bottom-up attention See *reflexive attention.*

bottom-up process A process in which lower-order mechanisms, like sensory inputs, trigger additional processing by higher-order systems. There may be no conscious awareness until late in the process. [17]

botulinum toxin A toxin that cleaves SNAREs, disabling neurotransmitter release. [2]

bovine spongiform encephalopathy (BSE) Also called *mad cow disease.* A disorder caused by improperly formed prion proteins, leading to dementia and death. [10]

brain self-stimulation The process in which animals will work to provide electrical stimulation to particular brain sites, presumably because the experience is very rewarding. [11]

brain-derived neurotrophic factor (BDNF) A protein, found in the brains of animals, that can keep some classes of neurons alive. [6]

brainstem The region of the brain that consists of the midbrain, the pons, and the medulla. [1]

Broca's aphasia See *nonfluent aphasia.*

Broca's area A region of the frontal lobe of the brain that is involved in the production of speech. [18]

BSE See *bovine spongiform encephalopathy.*

bulimia Also called *bulimia nervosa*. A syndrome in which individuals periodically gorge themselves, usually with "junk food," and then either vomit or take laxatives to avoid weight gain. [14]

bungarotoxin A neurotoxin from the venom of the banded krait that selectively blocks acetylcholine receptors. [2]

C

C fibre A small, unmyelinated axon that conducts pain information slowly and adapts slowly. [7]

Ca^{2+} See *calcium ion*.

caffeine A stimulant compound found in coffee, cacao, and other plants. [3]

CAH See *congenital adrenal hyperplasia*.

calcium ion (Ca^{1+}) A calcium atom that carries a double positive charge because it has lost two electrons. [2]

CAM See *cell adhesion molecule*.

cAMP See *cyclic adenosine monophosphate*.

cannabidiol (CBD) An active ingredient in cannabis thought to be responsible inducing a feeling of relaxation. [3]

cannabinoid receptor A category of cell membrane receptor proteins that selectively recognise and respond to cannabinoid transmitters and drugs. [3]

cannabis Also called *marijuana*. Dried leaves and flowers of the plant *Cannabis sativa*, typically smoked to obtain THC and CBD. [3, 7]

Cannon-Bard theory The theory that our experience of emotion is independent of the simultaneous physiological changes that accompany it. [11]

capsaicin A compound synthesised by various plants to deter predators by mimicking the experience of burning. [7]

carotid arteries The major arteries that ascend the left and right sides of the neck to the brain, supplying blood to the anterior and middle cerebral arteries. [1]

Cas9 *C*RISPR ***as***sociated enzyme ***9***. A bacterial enzyme that induces a break in double-stranded DNA as part of the CRISPR system. [App]

caspases A family of proteins that regulate cell death (apoptosis). [6]

castration Removal of the gonads, usually the testes. [4, 13]

CAT or CT See *computerized axial tomography*.

cataplexy Sudden loss of muscle tone, leading to collapse of the body without loss of consciousness. [15]

catecholamine A class of monoamines that serve as neurotransmitters, including dopamine and norepinephrine. [3]

cation A positively charged ion, such as a potassium or sodium ion. [2]

caudal See *posterior*.

caudate nucleus One of the basal ganglia. It has a long extension or tail. [1]

CBT See *cognitive behavioural therapy*.

CCK See *cholecystokinin*.

cell adhesion molecule (CAM) A protein found on the surface of a cell that guides cell migration and/or axonal pathfinding. [6]

cell assembly A large group of cells that tend to be active at the same time because they have been activated simultaneously or in close succession in the past. [16]

cell body Also called *soma*. The region of a neuron that is defined by the presence of the cell nucleus. [1]

cell death Also called *apoptosis*. The developmental process during which "surplus" cells die. [6]

cell differentiation The developmental stage in which cells acquire distinctive characteristics, such as those of neurons, as the result of expressing particular genes. [6, App]

cell membrane The lipid bilayer that ensheathes a cell. [2]

cell migration The movement of cells from site of origin to final location. [6]

cell nucleus The spherical central structure of a cell that contains the chromosomes. [1]

cell-autonomous Referring to cell processes that are directed by the cell itself rather than being under the influence of other cells.

cell-cell interactions The general process during development in which one cell affects the differentiation of other, usually neighbouring, cells. [6]

central deafness A hearing impairment that is related to lesions in auditory pathways or centres, including sites in the brainstem, thalamus, or cortex. [9]

central nervous system (CNS) The portion of the nervous system that includes the brain and the spinal cord. [1]

central pattern generator Neural circuitry that is responsible for generating the rhythmic pattern of a behaviour such as walking. [10]

central sulcus A fissure that divides the frontal lobe from the parietal lobe. [1]

cerebellum A structure located at the back of the brain, dorsal to the pons, that is involved in the central regulation of movement. [1]

cerebral cortex Also called simply *cortex*. The outer covering of the cerebral hemispheres that consists largely of neuronal cell bodies and their branches. [1]

cerebral hemispheres The right and left halves of the forebrain. [1]

cerebrocerebellum The lateral portions of each cerebellar hemisphere. [10]

cerebrospinal fluid (CSF) The fluid that fills the cerebral ventricles. [1]

cervical Referring to the topmost eight segments of the spinal cord, in the neck region. [1]

cGMP See *cyclic guanosine monophosphate*.

channelopathy A genetic abnormality of ion channels, causing a variety of symptoms. [2]

channelrhodopsin A protein that, in response to light of the proper wavelength, opens a channel to admit sodium ions, which results in excitation of the neuron. [2]

ChAT See *choline acetyltransferase*.

chemical transmitter See *neurotransmitter*.

chemically gated ion channel See *ligand-gated ion channel*.

chemoaffinity hypothesis The notion that each cell has a chemical identity that directs it to synapse on the proper target cell during development. [6]

chemoattractants Compounds that attract particular classes of axonal growth cones. [6]

chemorepellents Compounds that repel particular classes of axonal growth cones. [6]

chloride ion (Cl^-) A chlorine atom that carries a negative charge because it has gained one electron. [2]

chlorpromazine An antipsychotic drug, one of the class of phenothiazines. [12]

cholecystokinin (CCK) A peptide hormone, released by the gut after ingestion of food high in protein and/or fat, that also serves as a signalling molecule in the brain. [14]

choline acetyltransferase (ChAT) An enzyme involved in the synthesis of the neurotransmitter acetylcholine. [3]

cholinergic Referring to cells that use acetylcholine as their synaptic transmitter. [2, 3]

choroid plexus A highly vascular portion of the lining of the ventricles that secretes cerebrospinal fluid. [1]

chromosome A complex of condensed strands of DNA and associated protein molecules. It is found in the cell nucleus. [5, App]

chronic traumatic encephalopathy (CTE) Also called *dementia pugilistica* or *punch-drunk syndrome*. The dementia that develops in boxers or other athletes who are subjected to repeated blows to the head. [18]

ciliary muscle One of the muscles that controls the shape of the lens inside the eye, focussing an image on the retina. [8]

cilium A hairlike cellular extension. [9]

cingulate cortex Also called *cingulate gyrus*. A region of medial cerebral cortex that lies dorsal to the corpus callosum. [7]

cingulum See *cingulate cortex*.

circadian rhythm A pattern of behavioural, biochemical, or physiological fluctuation that has a 13-hour period. [15]

circannual Occurring on a roughly annual basis. [15]

circle of Willis A vascular structure at the base of the brain that is formed by communicating arteries that interconnect the major cerebral arteries. [1]

circumvallate papillae One of three types of small structures on the tongue that contain taste receptors, located in the back. [9]

circumventricular organ An organ that lies in the wall of a cerebral ventricle and monitors the composition of body fluids. [14]

CJD See *Creutzfeldt-Jakob disease*.

Cl^- See *chloride ion*.

classical conditioning Also called *Pavlovian conditioning*. A type of associative learning in which an originally neutral (conditioned) stimulus acquires the power to elicit the response normally elicited by another (unconditioned) stimulus after the two stimuli are paired. A response elicited by the unconditioned stimulus (US) is called an *unconditioned response* (*UR*); a response elicited by the conditioned stimulus (CS) alone is called a *conditioned response* (*CR*). [16]

claustrum A thin sheet of neurons, situated within the white matter lateral to the basal ganglia, that has been implicated in conscious awareness. [17]

cloacal exstrophy A rare medical condition in which XY individuals are born completely lacking a penis. [13]

clone To reproduce a gene so that it can be sequenced and/or manipulated. [App]

closed-loop control mechanism A control mechanism that provides a flow of information from whatever is being controlled to the device that controls it. [10]

clozapine A second-generation antipsychotic. [12]

CNS See *central nervous system*.

cocaine A drug of abuse, derived from the coca plant, that acts by potentiating catecholamine stimulation. [3]

coccygeal Referring to the lowest spinal vertebra (which is also called the *tailbone*). [1]

cochlea A snail-shaped structure in the inner ear that contains the primary receptor cells for hearing. [9]

cochlear implant An electromechanical device that detects sounds and selectively stimulates nerves in different regions of the cochlea via surgically implanted electrodes. [9]

cochlear nuclei Brainstem nuclei that receive input from auditory hair cells and send output to the superior olivary complex. [9]

cocktail party effect The selective enhancement of attention in order to filter out distracters, such as while listening to one person talking in the midst of a noisy party. [17]

coding The rules by which action potentials in a sensory system reflect a physical stimulus. [7]

codon A set of three nucleotides that uniquely encodes one particular amino acid. [App]

cognitive attribution theory The theory that our emotional experience results from cognitive analysis of the context around us, such that physiological changes may accentuate emotions but not specify which emotion we experience. [11]

cognitive behavioural therapy (CBT) Psychotherapy aimed at correcting negative thinking and improving interpersonal relationships. [12]

cognitive map A mental representation of a spatial relationship. [16]

cognitively impenetrable Referring to data-processing operations of the central nervous system that are unconscious. [17]

coincidence detector A device that senses the co-occurrence of two events. [9]

coitus See *copulation*.

co-localisation Also called *co-release*. In neurons, the appearance of more than one neurotransmitter in a given presynaptic terminal. [3]

communication Information transfer between two individuals. [18]

comorbid Referring to the tendency of certain diseases or disorders to occur together in individuals. [12]

competitive ligand A substance that directly competes with the endogenous ligand for the same binding site on a receptor molecule. [3]

complex cortical cell A cell in the visual cortex that responds best to a bar of a particular size and orientation anywhere within a particular area of the visual field. [8]

complex environment See *enriched condition*.

complex partial seizure In epilepsy, a type of seizure that doesn't involve the entire brain and therefore can cause a wide variety of symptoms. [2]

computerized axial tomography (CAT or CT) A noninvasive technique for examining brain structure in humans through computer analysis of X-ray absorption at several positions around the head. [1]

concentration gradient Variation of the concentration of a substance within a region. [2]

concordant Referring to any trait that is seen in both individuals of a pair of twins. [12]

concussion Also called *minor traumatic brain injury* (*mTBI*). An injury resulting from a blow to the head and associated with temporary neurological symptoms such as memory loss or other cognitive impairments, pain, and visual disturbances. [18]

conditional knockout A gene that can be selectively deactivated in specific tissues and/or at a specific stage of development. [16]

conduction aphasia An impairment in the repetition of words and sentences. [18]

conduction deafness A hearing impairment that is associated with pathology of the external-ear or middle-ear cavities. [9]

conduction velocity The speed at which an action potential is propagated along the length of an axon (or section of peripheral nerve). [2]

conduction zone The part of the neuron over which the nerve's electrical signal may be actively propagated. It usually corresponds to the cell's axon. [1]

cones A class of photoreceptor cells in the retina that are responsible for colour vision. [8]

confabulate To fill in a gap in memory with a falsification. It often occurs in Korsakoff's syndrome. [16]

congenital adrenal hyperplasia (CAH) Any of several genetic conditions that result in exposure of a fetus to excessive adrenal androgens. [13]

congenital hypothyroidism Also called *cretinism*. Reduced stature and intellectual disability caused by thyroid deficiency during early development. [4]

congenital insensitivity to pain The condition of being born without the ability to perceive pain. [7]

conjunction search A search for an item that is based on two or more features (e.g., size and colour) that together distinguish the target from distracters that may share some of the same attributes. [17]

connectionist model of aphasia The theory proposing that Wernicke's area and Broca's area, connected by the arcuate fasciculus, specialise in the receptive and expressive aspects of language, respectively. [18]

connexon A protein assembly that provides an open ion channel between two neurons, forming an electrical synapse between them. [2]

consciously controlled attention See *voluntary attention.*

consciousness The state of awareness of one's own existence and experience. [0, 17]

conserved In the context of evolution, referring to a trait that is passed on from a common ancestor to two or more descendant species. [0]

consolidation A stage of memory formation in which information in short-term or intermediate-term memory is transferred to long-term memory. [16]

constraint-induced movement therapy (CIMT) A therapy for recovery of limb movement after stroke or injury, in which the unaffected limb is constrained while the person is required to perform tasks with the affected contralateral limb. [18]

contralateral In anatomy, pertaining to a location on the opposite side of the body. Compare *ipsilateral.* [1]

convergence The phenomenon of neural connections in which many cells send signals to a single cell. [2]

convergent evolution The evolutionary process by which responses to similar ecological features bring about similarities in behaviour or structure among animals that are only distantly related (i.e., that differ in genetic heritage). [5]

Coolidge effect The propensity of an animal that has appeared sexually satiated with a present partner to resume sexual activity when provided with a novel partner. [13]

cool-menthol receptor 1 (CMR1) See *TRPM7.*

copulation Also called *coitus*. The sexual act. [13]

copulatory lock Reproductive behaviour in which the male's penis swells after ejaculation so that the male and female are forced to remain joined for 4–11 minutes. It occurs in dogs and some rodents, but not in humans. [13]

co-release See *co-localisation.*

cornea The transparent outer layer of the eye, whose curvature is fixed. It bends light rays and is primarily responsible for forming the image on the retina. [8]

coronal plane Also called *frontal plane* or *transverse plane*. The plane that divides the body or brain into front and back parts. Compare *horizontal plane* and *sagittal plane.* [1]

corpora lutea The structures formed from collapsed ovarian follicles subsequent to ovulation. The corpora lutea are a major source of progesterone. [4]

corpus callosum The main band of axons that connects the two cerebral hemispheres. [1, 18]

correlation The covariation of two measures. [0]

cortex Also called *cerebral cortex* or *neocortex*. The outer covering of the cerebral hemispheres that consists largely of nerve cell bodies and their branches. [5]

cortical column One of the vertical columns that constitute the basic organisation of the neocortex. [1]

cortical deafness A hearing impairment that is caused by a fault or defect in the cortex. [9]

corticospinal system See *pyramidal system.*

cortisol A glucocorticoid stress hormone of the adrenal cortex. [4, 11]

cranial nerve A nerve that is connected directly to the brain. [1]

CREB ***C***yclic AMP ***r***esponsive ***e***lement–***b***inding protein, which binds the promoter region of several genes involved in neural plasticity when activated by cAMP. [16]

Cre-recombinase An enzyme normally made by bacteria that removes a segment of DNA flanked by two lox sites. [10, App]

cretinism See *congenital hypothyroidism.*

Creutzfeldt-Jakob disease (CJD) A brain disorder in humans, leading to dementia and death, that is caused by improperly folded prion proteins; the human equivalent of mad cow disease. [10]

crib death See *sudden infant death syndrome.*

CRISPR ***c***lustered ***r***egularly ***i***nterspaced ***s***hort ***p***alindromic ***r***epeats. A system of gene manipulation that evolved in single-celled organisms and is exploited by scientists for gene editing. [App]

cross-tolerance A condition in which the development of tolerance for one drug causes an individual to develop tolerance for another drug. [3]

crystallisation The final stage of birdsong formation, in which fully formed adult song is achieved. [18]

CSF See *cerebrospinal fluid.*

CT or CAT See *computerized axial tomography.*

CTE See *chronic traumatic encephalopathy.*

cue-induced drug use An increased likelihood to use a drug (especially an addictive drug) because of the presence of environmental stimuli that were present during previous use of the same drug. [3]

cupula A small gelatinous structure, containing hair cells that detect fluid movement within the semicircular canals of the vestibular system. [9]

curare An alkaloid neurotoxin that causes paralysis by blocking acetylcholine receptors in muscle. [2]

Cushing's syndrome A condition in which levels of adrenal glucocorticoids are abnormally high. [12]

cyclic adenosine monophosphate (cyclic AMP, or cAMP) A second messenger activated in many target cells in response to synaptic or hormonal stimulation. [4]

cyclic guanosine monophosphate (cyclic GMP, or cGMP) A second messenger activated in some target cells in response to synaptic or hormonal stimulation. [4]

cytokines Proteins that induce the proliferation of other cells, as in the immune system. [11]

cytoplasm See *intracellular fluid.*

D

DA See *dopamine.*

dB See *decibel.*

DBS See *deep brain stimulation.*

deafness Hearing loss so profound that speech perception is lost. [9]

death gene A gene that is expressed only when a cell becomes committed to natural cell death (apoptosis). [6]

decibel (dB) A measure of sound intensity. [9]

declarative memory A memory that can be stated or described. [16]

decomposition of movement Difficulty of movement in which gestures are broken up into individual segments instead of being executed smoothly. It is a symptom of cerebellar lesions. [10]

decorticate rage Also called *sham rage*. Sudden intense rage characterised by actions (such as snarling and biting in dogs) that lack clear direction. [11]

deep brain stimulation (DBS) Mild electrical stimulation through an electrode that is surgically implanted deep in the brain. [10, 12]

deep dyslexia Acquired dyslexia in which the person reads a word as another word that is semantically related. [18]

default mode network The regions of the brain that are active when the brain is awake and at rest and attention is not being directed to external events. [17]

degradation The chemical breakdown of a neurotransmitter into inactive metabolites. [2, 3]

delayed non-matching-to-sample task A test in which, on each trial, the participant must select the stimulus that was not seen previously. [16]

delta wave The slowest type of EEG wave, about 0 Hz, characteristic of stage 2 sleep. [15]

delusion A false belief strongly held in spite of contrary evidence. [12]

dementia pugilistica See *chronic traumatic encephalopathy.*

dementia Drastic failure of cognitive ability, including memory failure and loss of orientation. [6]

dendrite One of the extensions of the cell body through which synaptic inputs are received. [1]

dendritic knob A portion of an olfactory receptor cell present in the olfactory epithelium. [9]

dendro-dendritic Referring to a synapse in which a synaptic connection forms between the dendrites of two neurons. [2]

dentate gyrus A strip of gray matter in the hippocampal formation. [16]

deoxyribonucleic acid (DNA) A nucleic acid that is present in the chromosomes of cells and encodes hereditary information. [App]

dependent variable The factor that an experimenter measures to monitor a change in response to manipulation of an independent variable. [0]

depolarisation A reduction in membrane potential (the interior of the neuron becomes less negative). [2]

depressants A class of drugs that act to reduce neural activity. [3]

depression A psychiatric condition characterised by such symptoms as an unhappy mood; loss of interests, energy, and appetite; and difficulty concentrating. [12]

dermatome A patch of skin innervated by a particular spinal nerve. [7]

dermis The middle layer of skin, between the epidermis and the hypodermis. [7]

desynchronised EEG Also called *beta activity*. A pattern of EEG activity comprising a mix of many different high frequencies with low amplitude. [15]

dexamethasone suppression test A test of pituitary-adrenal function in which the participant is given dexamethasone, a synthetic glucocorticoid hormone, which should cause a decline in the production of adrenal corticosteroids. [12]

dfMRI See *dyadic functional MRI.*

DHT See *dihydrotestosterone.*

diabetes mellitus Excessive glucose in the urine, caused by the failure of insulin to induce glucose absorption by the body. [14]

Diablo A protein released by mitochondria, in response to high calcium levels, that activates apoptosis. [6]

diacylglycerol (DAG) A common second-messenger compound in postsynaptic cells. [4]

dichotic presentation The simultaneous delivery of different stimuli to the right and the left ears. [17, 18]

diencephalon The posterior part of the forebrain, including the thalamus and hypothalamus. [1]

diffusion The passive spread of molecules of one substance among molecules of another substance until a uniform concentration is achieved. [2, 14]

diffusion tensor imaging (DTI) A modified form of MRI in which the diffusion of water in a confined space is exploited to produce images of axonal fibre tracts. [1, 9, 18]

digestion The process by which food is broken down to provide energy and nutrients. [14]

dihydrotestosterone (DHT) The 4α-reduced metabolite of testosterone; a potent androgen that is principally responsible for the masculinisation of the external genitalia in mammalian sexual differentiation. [13]

dimer A complex of two proteins that have bound together. [15]

directional selection Selection that favours a shift in a characteristic within a population. [5]

discordant Referring to any trait that is seen in only one individual of a pair of twins. [12]

dishabituation The restoration of response amplitude following habituation. [16]

disruptive selection Selection that favours a divergence in a characteristic within a population. [5]

dissociative thinking A condition, seen in schizophrenia, that is characterised by disturbances of thought and difficulty in relating events properly. [12]

dissociative A type of drug that produces a dreamlike state in which consciousness is partly separated from sensory inputs. [3]

distal In anatomy, towards the periphery of an organism or the end of a limb. Compare *proximal.* [1]

diurnal Active during the light periods of the daily cycle. [15]

divergence The phenomenon of neural connections in which one cell sends signals to many other cells. [2]

divided-attention task A task in which the participant is asked to simultaneously focus attention on two or more stimuli. [17]

dizygotic Referring to twins derived from separate eggs (*fraternal* twins). Such twins are no more closely related genetically than are other full siblings. [12]

DNA See *deoxyribonucleic acid.*

DNA sequencing The process by which the order of nucleotides in a gene, or amino acids in a protein, is determined. [App]

dopamine (DA) A monoamine transmitter found in the midbrain—especially the substantia nigra—and basal forebrain. [3]

dopamine hypothesis The hypothesis that schizophrenia results from either excessive levels of synaptic dopamine or excessive postsynaptic sensitivity to dopamine. [12]

dorsal column nuclei Collection of neurons in the medulla that receive somatosensory information via the dorsal columns of the spinal cord. These neurons send their axons across the midline and to the thalamus. [7]

dorsal column system A somatosensory system that delivers most touch stimuli via the dorsal columns of spinal white matter to the brain. [7]

dorsal In anatomy, towards the back of the body or the top of the brain. Compare *ventral.* [1]

dorsal root The branch of a spinal nerve, entering the dorsal horn of the spinal cord, that carries sensory information from the peripheral nervous system to the spinal cord. [1]

dose-response curve (DRC) A formal plot of a drug's effects (on the y-axis) versus the dose given (on the x-axis). [3]

Down syndrome A syndrome caused by inheriting an extra copy of chromosome 10, usually accompanied by intellectual disability. [6]

down-regulation A compensatory reduction in receptor availability at the synapses of a neuron. [2, 3]

DRC See *dose-response curve.*

DREADDs (designer receptors exclusively activated by designer drugs) Artificially modified ACh receptors that respond solely to matching exogenous ligands. [3]

DTI See *diffusion tensor imaging.*

DTI tractography Also called *fibre tracking.* Visualisation of the orientation and terminations of white matter tracts in the living brain via diffusion tensor imaging. [1]

dual dependence Dependence for emergent drug effects that occur only when two drugs are taken simultaneously. [3]

dualism Within the concept of separation of soul and body, the notion promoted by René Descartes that the mind is subject only to spiritual interactions while the body is subject only to material interactions. [0]

duplex theory A theory that we localise sound by combining information about intensity differences and latency differences between the two ears. [9]

dura mater The outermost of the three meninges that surround the brain and spinal cord. [1]

dyadic functional MRI (dfMRI) A form of fMRI in which the brains of two socially-interacting people are scanned simultaneously in a single scanner. [1]

dynorphins One of three kinds of endogenous opioids, substances that reduce pain perception. [3, 7]

dyskinesia Difficulty or distortion in voluntary movement. [12]

dyslexia A reading disorder attributed to brain impairment. *Acquired dyslexia* occurs as a result of injury or disease. *Developmental dyslexia* is associated with brain abnormalities present from birth. [18]

dysphoria Unpleasant feelings; the opposite of euphoria. [3]

dystrophin A protein that is needed for normal muscle function. [10]

E

ear canal A tube leading from the pinna to the middle ear. [9]

eardrum See *tympanic membrane.*

early-selection model A model of attention in which the attentional bottleneck filters out stimuli before even preliminary perceptual analysis has occurred. [17]

easy problem of consciousness The problem of how to read current conscious experiences directly from people's brains as they're happening. [17]

EC See *enriched condition.*

ecological niche The unique assortment of environmental opportunities and challenges to which each organism is adapted. [5, 15]

Ecstasy See *MDMA.*

ectoderm The outer cellular layer of the developing embryo, giving rise to the skin and the nervous system. [6]

ectopia Something out of place—for example, clusters of neurons seen in unusual positions in the cortex of someone who has dyslexia. [18]

ectopic transmission Cell-cell communication based on release of neurotransmitter in regions outside traditional synapses. [2]

ectotherm An animal whose body temperature is regulated by, and whose heat comes mainly from, the environment. Examples include snakes and bees. [14]

edema The swelling of tissue, such as in the brain, in response to injury. [1]

edge detector See *simple cortical cell.*

EEG See *electroencephalogram* and *electroencephalography.*

efferent In reference to axons, carrying information from the central nervous system to the periphery. Compare *afferent.* [1]

efficacy Also called *intrinsic activity.* The extent to which a drug activates a response when it binds to a receptor. [3]

ejaculation The forceful expulsion of semen from the penis. [13]

electrical synapse Also called *gap junction.* The region between neurons where the membranes are so close that changes in potential can flow from one to the other without being translated into a chemical message. [2]

electroconvulsive therapy (ECT) A last-resort treatment for intractable depression, in which a strong electrical current is passed through the brain, causing a seizure. [12]

electroencephalogram (EEG) A recording of gross electrical activity of the brain recorded from electrodes placed on the scalp. [2]

electroencephalography (EEG) The recording and study of gross electrical activity of the brain recorded from large electrodes placed on the scalp. [15]

electromyography (EMG) The electrical recording of muscle activity. [10, 15]

electro-oculography (EOG) The electrical recording of eye movements. [15]

electrostatic pressure The propensity of charged molecules or ions to move towards areas with the opposite charge. [2]

embryo The earliest stage in a developing animal. [6]

embryonic stem cell A cell, derived from an embryo, that has the capacity to form any type of tissue.

EMG See *electromyography.*

emotion A subjective mental state that is usually accompanied by distinctive behaviours, feelings, and involuntary physiological changes. [11]

emotional dyscontrol syndrome A condition consisting of temporal lobe disorders that may underlie some forms of human violence. [11]

encephalisation factor A measure of brain size relative to body size. [5]

encoding A stage of memory formation in which the information entering sensory channels is passed into short-term memory. [16]

endocannabinoid An endogenous ligand of cannabinoid receptors, thus an analog of cannabis that is produced by the brain. [3, 14]

endocast A cast of the cranial cavity of a skull, especially useful for studying fossils of extinct species. [5]

endocrine Referring to glands that release chemicals to the interior of the body. These glands secrete the principal hormones. [4]

endocrine gland A gland that secretes hormones into the bloodstream to act on distant targets. [4]

endogenous ligand Any substance that is produced within the body and selectively binds to the type of receptor that is under study. [2]

endogenous opioids A family of peptide transmitters that bind to opioid receptors and act like opiates. They have been called the body's own narcotics. The three kinds are enkephalins, endorphins, and dynorphins. [3, 7]

endogenous Produced inside the body. [3]

endogenously controlled attention See *voluntary attention.*

endophenotype Behavioural or physical characteristics accompanying susceptibility to a particular disorder, which may be used to identify those at risk. [12]

endorphins One of three kinds of endogenous opioids, substances that reduce pain perception. [3, 7]

endotherm An animal whose body temperature is regulated chiefly by internal metabolic processes. Examples include mammals and birds. [14]

engram The physical basis of a memory in the brain. It is sometimes referred to as a *memory trace* on the assumption that it involves changes in a neural circuit rather than a single neuron. [16]

enkephalins One of three kinds of endogenous opioids, substances that reduce pain perception. [3, 7]

enriched condition (EC) Also called *complex environment.* A condition in which laboratory rodents are group-housed with a wide variety of stimulus objects. [16]

enteric nervous system An extensive mesh-like system of neurons that governs the functioning of the gut. [1]

enterotype Each individual's personal composition of gut microbiota. [14]

entrainment The process of synchronising a biological rhythm to an environmental stimulus. [15]

enzyme A complicated protein whose action increases the probability of a specific chemical reaction. [App]

EOG See *electro-oculography.*

ependymal layer See *ventricular zone.*

epidemiology The scientific study of patterns of disease in a population. [12]

epidermis The outermost layer of skin, over the dermis. [7]

epigenetic regulation Process affecting the expression of a particular gene or genes without affecting the sequence of nucleotides making up the gene itself. [11]

epigenetic transmission The passage of epigenetic modifications of a gene from one generation to another. [14]

epigenetics The study of factors that affect gene expression without making any changes in the nucleotide sequence of the genes themselves. [6]

epilepsy A brain disorder marked by major sudden changes in the electrophysiological state of the brain that are referred to as *seizures.* [2]

epinephrine Also called *adrenaline.* A compound that acts both as a hormone (secreted by the adrenal medulla under the control of the sympathetic nervous system, which prepares the body for action) and as a synaptic transmitter. [4, 11]

episodic memory Memory of a particular incident or a particular time and place. [16]

EPSP See *excitatory postsynaptic potential.*

equilibrium In a neuron, the state in which the number of ions crossing a membrane in one direction is matched by the number crossing in the opposite direction. [2]

equilibrium potential The voltage across a permeable membrane that exactly counteracts the movement of ions from the side with a high concentration to the side with a low concentration. [2]

ERP See *event-related potential.*

estradiol See *17β-estradiol.*

estrogens A class of steroid hormones produced by female gonads. [4]

estrus The period during which female animals are sexually receptive. [13]

eukaryote Any organism whose cells have the genetic material contained within a nuclear envelope. [App]

event-related potential (ERP) Also called *evoked potential.* Averaged EEG recordings measuring brain responses to repeated presentations of a stimulus. [2, 17]

evoked potential See *event-related potential.*

evolution In biology, the processes by which successive generations of individuals change over very long periods of time. [5]

evolution by natural selection The Darwinian theory that evolution proceeds by differential success in reproduction. [5]

evolutionary psychology A research field that asks how evolutionary selection pressures have shaped the behaviour of humans and other animals. [5, 11]

excitatory postsynaptic potential (EPSP) A depolarising potential in the postsynaptic neuron that is caused by excitatory connections. EPSPs increase the probability that the postsynaptic neuron will fire an action potential. [2]

excitotoxicity The property by which neurons die when overstimulated, as with large amounts of glutamate. [3]

executive function A neural and cognitive system that helps develop plans of action and organises the activities of other high-level processing systems. [17]

exocrine gland A gland whose secretions exit the body via ducts. [4]

exocytosis In neurons, the process by which a synaptic vesicle fuses with the presynaptic terminal membrane to release neurotransmitter into the synaptic cleft. [2]

exogenous Arising from outside the body. [3]

exogenous ligand Any substance that originates outside the body and selectively binds to the type of receptor that is under study. [2]

exogenously controlled attention See *reflexive attention.*

expression The process by which a cell makes an mRNA transcript of a particular gene. [App]

external ear The part of the ear that we readily see (the pinna) and the canal that leads to the eardrum. [9]

external fertilisation The process by which eggs are fertilised outside of the female's body, as in many fishes and amphibians. [13]

extinction Short for *extinction of simultaneous double stimulation.* In the context of neurology, an inability to recognise the double nature of stimuli presented simultaneously to both sides of the body. People experiencing extinction report the stimulus from only one side. [17]

extracellular compartment The fluid space of the body that exists outside the cells. [14]

extracellular fluid The fluid in the spaces between cells (interstitial fluid) and in the vascular system. [2]

extrafusal fibre One of the ordinary muscle fibres that lie outside the spindles and provide most of the force for muscle contraction. [10]

extraocular muscle One of the muscles attached to the eyeball that controls its position and movements. [8]

extrapyramidal system A motor system that includes the basal ganglia and some closely related brainstem structures. [10]

extrastriate cortex Visual cortex outside of the primary visual (striate) cortex. [8]

F

FA See *fractional anisotropy.*

face blindness See *prosopagnosia.*

facial feedback hypothesis The hypothesis that our emotional experience is affected by the sensory feedback we receive during particular facial expressions, such as smiling. [11]

facial nerve Cranial nerve VII, which receives information from the face and controls the superficial muscles there. [11]

FAS See *fetal alcohol syndrome.*

fast-twitch muscle fibre A type of striated muscle that contracts rapidly but fatigues readily. [10]

fatal familial insomnia An inherited disorder in which humans sleep normally at the beginning of their life but stop sleeping in midlife and die 7–24 months later. [15]

fear conditioning A type of classical conditioning where a previously neutral stimulus is repeatedly paired with shock or some other unpleasant experience, causing the individual to act fearful in response to the stimulus. [11, 12]

feature integration theory The idea that conjunction searches involve multiple cognitive feature maps—overlapping representations of the search array based on individual stimulus attributes. [17]

feature search A search for an item in which the target pops out right away because it possesses a unique attribute. [17]

fecal transplantation A medical procedure in which gut microbiota, via fecal matter, are transferred from a donor to a host. [14]

FEF See *frontal eye field.*

fetal alcohol syndrome (FAS) A disorder, including intellectual disability and characteristic facial anomalies, that affects children exposed to too much alcohol (through maternal ingestion) during fetal development. [3, 6]

fetus A developing individual after the embryo stage. [6]

fibre tracking See *DTI tractography.*

filopodia Very fine, tubular outgrowths from the growth cone of an axon or dendrite. [6]

final common pathway The information-processing pathway consisting of all the motor neurons in the body. Motor neurons are known by this collective term because they receive and integrate all motor signals from the brain and then direct movement accordingly. [10]

first-generation antipsychotics A major class of antischizophrenic drugs that share an antagonist activity at dopamine D_1 receptors. [3, 12]

flaccid paralysis A loss of all movement and reflexes below the level of transection of the spinal cord. [10]

flavour The sense of taste combined with the sense of smell. [9]

flower spray ending See *secondary sensory ending.*

fluent aphasia Also called *Wernicke's aphasia*. A language impairment characterised by fluent, meaningless speech and little language comprehension. It is related to damage in Wernicke's area. [18]

fMRI See *functional MRI.*

foliate papillae One of three types of small structures on the tongue that contain taste receptors, located along the sides of the tongue. [9]

follicles Ovarian structures containing immature ova. [4]

follicle-stimulating hormone (FSH) A gonadotropin, named for its actions on ovarian follicles. [4]

forebrain Also called *prosencephalon*. The anterior division of the brain, containing the cerebral hemispheres, the thalamus, and the hypothalamus. [1, 6]

fornix A fibre tract that extends from the hippocampus to the mammillary body. [1]

Fourier analysis The mathematical decomposition of a complex pattern into a sum of sine waves of various frequencies and amplitudes. [9]

fourth ventricle The passageway within the pons that receives cerebrospinal fluid from the third ventricle and releases it to surround the brain and spinal cord. [1]

fovea The central portion of the retina, packed with the most photoreceptors and therefore the centre of our gaze. [8]

fractional anisotropy (FA) The tendency of water to diffuse more readily along the long axis of an enclosed space, such as an axon. FA is the basis of diffusion tensor imaging. [1]

fragile X syndrome A condition that is a frequent cause of inherited intellectual disability and is produced by a fragile site on the X chromosome that seems prone to breaking because the DNA there is unstable. [6]

free nerve ending An axon that terminates in the skin without any specialised cell associated with it and that detects pain and/or changes in temperature. [7]

free-running Referring to a rhythm of behaviour shown by an animal deprived of external cues about time of day. [15]

frequency The number of cycles per second in a sound wave; measured in hertz (Hz). [9]

frontal eye field (FEF) An area in the frontal lobe of the brain containing neurons important for establishing gaze in accordance with cognitive goals (top-down processes) rather than with any characteristics of stimuli (bottom-up processes). [17]

frontal lobe The most anterior portion of the cerebral cortex. [1]

frontal plane See *coronal plane.*

FSH See *follicle-stimulating hormone.*

functional MRI (fMRI) Magnetic resonance imaging that detects changes in blood flow and therefore identifies regions of the brain that are particularly active during a given task. [1]

functional tolerance Decreased responding to a drug after repeated exposures, generally as a consequence of up- or down-regulation of receptors. [3]

fundamental The predominant frequency of an auditory tone or a visual scene. [9]

fungiform papillae One of three types of small structures on the tongue that contain taste receptors, located in the front of the tongue. [9]

fusiform gyrus A region on the inferior surface of the cortex, at the junction of temporal and occipital lobes, that has been associated with recognition of faces. [18]

G

G protein-coupled receptors (GPCRs) A large family of cell-membrane-spanning receptors that, when activated extracellularly, use G proteins on their intracellular surface to affect the receiving cell. [3, 4]

G proteins A class of proteins that are next to the intracellular portion of a receptor and that are activated when the receptor binds an appropriate ligand on the extracellular surface. [2]

GABA See *gamma-aminobutyric acid.*

gamete A sex cell (sperm or ovum) that contains only unpaired chromosomes and therefore has only half the usual number of chromosomes in other cells. [13]

gamma efferent See *gamma motor neuron.*

gamma motor neuron Also called *gamma efferent*. A motor neuron that innervates the contractile tissue (the intrafusal fibre) in a muscle spindle. [10]

gamma-aminobutyric acid (GABA) A widely distributed inhibitory amino acid transmitter. [3]

ganglion cells A class of cells in the retina whose axons form the optic nerve. [8]

gap junction See *electrical synapse.*

gas neurotransmitter A soluble gas, such as nitric oxide or carbon monoxide, that is produced and released by a neuron to alter the functioning of another neuron. [3]

gel electrophoresis A method of separating molecules of differing size or electrical charge by forcing them to flow through a gel. [App]

gender The behaviours and attitudes that a given culture considers to be masculine or feminine, or an alternative. [13]

gene A length of DNA that encodes the information for constructing a particular protein. [5, App]

gene amplification See *polymerase chain reaction.*

gene expression The process by which a cell transcribes a particular gene and makes the protein it encodes. [6]

general anaesthetic A drug that renders an individual unconscious. [15]

generator potential See *receptor potential.*

genetic polymorphisms The phenomenon in which there exist multiple (sometimes many) different alleles at an individual locus within a gene. [12]

genetics The study of inheritance, including the genes encoded in DNA. [5]

genome Also called *genotype*. All the genetic information that one specific individual has inherited. [6, App]

genotype Also called *genome*. All the genetic information that one specific individual has inherited. [6]

genus A group of species that resemble each other because of shared inheritance. [5]

GH See *growth hormone.*

ghrelin A peptide gut hormone believed to act on the hypothalamus to increase hunger. [14]

glial cells Also called *glia* or *neuroglia*. Nonneuronal brain cells that provide structural, nutritional, and other types of support to the brain. [1]

global aphasia The total loss of ability to understand language or to speak, read, or write. [18]

globus pallidus One of the basal ganglia. [1]

glomerulus A complex arbor of dendrites from a group of olfactory cells. [9]

GLP-1 A peptide gut hormone believed to act on the hypothalamus and brainstem to suppress appetite. [14]

glucagon A hormone, released by pancreatic alpha cells, that increases blood glucose. [14]

glucocorticoids A class of steroid hormones, released by the adrenal cortex, that affect carbohydrate metabolism and inflammation. [4]

glucodetector A cell that detects and informs the nervous system about levels of circulating glucose. [14]

gluconeogenesis The metabolism of body fats and proteins to create glucose. [14]

glucose A sugar molecule used by the body and brain for energy. [14]

glucose transporter A molecule that conducts glucose molecules through the external membrane of a cell for use inside. [14]

glutamate An amino acid transmitter, the most common excitatory transmitter. [3, 7]

glutamate hypothesis The hypothesis that schizophrenia may be caused, in part, by understimulation of glutamate receptors. [12]

glutamatergic Referring to cells that use glutamate as their synaptic transmitter. [3]

glycine An amino acid transmitter, often inhibitory. [3]

glycogen A complex carbohydrate derived from glucose. [14]

glycogenesis The physiological process by which glycogen is produced. [14]

glycogenolysis The conversion of glycogen back into glucose, triggered when blood concentrations of glucose drop too low. [14]

glymphatic system A lymphatic system in the brain that participates in removal of wastes and the movement of nutrients and signalling compounds. [1]

GnIH See *gonadotropin-inhibiting hormone.*

GnRH See *gonadotropin-releasing hormone.*

goiter A swelling of the thyroid gland resulting from iodine deficiency. [4]

Goldman equation An equation predicting the potential difference across a membrane based on the concentrations of multiple ions on opposite sides of the membrane, as well as its relative permeability to each ion. [2]

Golgi stain A cell stain that fills a small proportion of neurons with a dense dark product. [1]

Golgi tendon organ One of the receptors located in tendons that send action potentials to the central nervous system reporting muscle tension. [10]

gonadotropin Any anterior pituitary hormone that selectively stimulates the gonads to produce steroids and gametes. [4]

gonadotropin-inhibiting hormone (GnIH) A hypothalamic peptide hormone that reduces gonadotropin secretion from the pituitary. [4]

gonadotropin-releasing hormone (GnRH) A hypothalamic hormone that controls the release of luteinising hormone and follicle-stimulating hormone from the pituitary. [4]

gonads The sexual glands (ovaries in females, testes in males), which produce gametes for reproduction. [4]

grammar All of the rules for usage of a particular language. [18]

grand mal seizure See *tonic-clonic seizure.*

granule cell A type of small nerve cell. [1]

gray matter Areas of the brain that are dominated by cell bodies and are devoid of myelin. [1]

grid cell A neuron that selectively fires when an animal crosses the intersection points of an abstract grid map of the local environment. [16]

gRNA See *guide RNA.*

gross neuroanatomy Anatomical features of the nervous system that are apparent to the naked eye. [1]

growth cone The growing tip of an axon or a dendrite. [6]

growth hormone (GH) Also called *somatotropin* or *somatotropic hormone*. A tropic hormone, secreted by the anterior pituitary, that influences the growth of cells and tissues. [4]

guevedoces Literally, "eggs at 12" (in Spanish). A nickname for individuals who are raised as girls but at puberty change appearance and begin behaving as boys do. [13]

guide RNA (gRNA) A strand of RNA designed to hybridise with a targeted nucleotide sequence in DNA in order to guide the Cas9 enzyme to break the DNA at that site. [App]

gustatory system The taste system. [9]

gut microbiota The microorganisms that normally inhabit the digestive tract. [14]

gyrus A ridged or raised portion of a convoluted brain surface. [1]

H

habituation A form of nonassociative learning in which an organism becomes less responsive following repeated presentations of a stimulus. [16]

hair cell A cochlear auditory receptor cell. [9]

hallucinogens A class of drugs that alter sensory perception and produce peculiar experiences. [3]

halorhodopsin An ion channel protein that, when stimulated by light, hyperpolarises neurons. [2]

hard problem of consciousness The problem of how to read people's subjective experience of consciousness and determine the qualia that accompany perception. [17]

harmonics Multiples of a particular frequency called the *fundamental*. [9]

health psychology Also called *behavioural medicine*. A field that studies psychological influences on health-related processes, such as why people become ill or how they remain healthy. [11]

hearing loss Decreased sensitivity to sound, in varying degrees. [9]

Hebbian synapse A synapse that is strengthened when it successfully drives the postsynaptic cell. [6, 16]

hemiparesis Weakness of one side of the body. [18]

hemiplegia Partial paralysis involving one side of the body. [18]

hemispatial neglect A syndrome in which the person fails to pay any attention to objects presented to one side of the body and may even deny connection with that side. [17]

hermaphrodite An individual possessing the reproductive organs of both sexes, either simultaneously or at different points in time. [13]

heroin Diacetylmorphine; an artificially modified, very potent form of morphine. [3]

hertz (Hz) Cycles per second, as of an auditory stimulus. [9]

hindbrain Also called *rhombencephalon*. The rear division of the brain, which, in the mature vertebrate, contains the cerebellum, pons, and medulla. [1, 6]

hippocampus A medial temporal lobe structure important for learning, memory, and spatial navigation. [1, 16]

histology The scientific study of the composition of tissues. [1]

homeostasis The maintenance of a relatively constant internal physiological environment. [14]

hominin The subgroup of Hominidae that contains modern humans and their ancestral species. [5]

homology A physical resemblance that is based on common ancestry, such as the similarity in forelimb structures of different mammals. [5]

homoplasy A physical resemblance between physical or behavioural characteristics due to convergent evolution, such as the similar body forms of tuna and dolphins. [5]

horizontal cells Specialised retinal neurons that contact both the receptor cells and the bipolar cells. [8]

horizontal plane The plane that divides the body or brain into upper and lower parts. Compare *coronal plane* and *sagittal plane*. [1]

hormone A chemical secreted by cells that is conveyed by the bloodstream and regulates target organs or tissues. [4]

hostility In psychology, the angry, antagonistic personality characteristic associated with a greater risk for heart disease. [11]

hue A dimension of light perception, varying around the colour circle through blue, green, yellow, orange, and red. [8]

hunger The internal state of an animal seeking food. [14]

huntingtin A protein produced by a gene called *HTT* that may contain too many trinucleotide repeats. When it does, the protein causes Huntington's disease in a carrier. [10]

Huntington's disease A progressive genetic disorder characterised by abrupt, involuntary movements and profound changes in mental functioning. [10]

hybridisation The process by which a string of nucleotides becomes linked to a complementary series of nucleotides. [App]

hydrocephalus A ballooning of the ventricles, at the expense of the surrounding brain, which may occur when the circulation of cerebrospinal fluid is blocked. [1]

hyperphagia Excessive eating. [14]

hyperpolarisation An increase in membrane potential (the interior of the neuron becomes even more negative). [2]

hypertonic Referring to a solution with a higher concentration of salt than that found in interstitial fluid and blood plasma (more than about 0.9% salt). [14]

hypocretins Also called *orexins*. Neuropeptides produced in the hypothalamus that are involved in switching between sleep states, in narcolepsy, and in the control of appetite. [15]

hypodermis Also called *subcutaneous tissue*. The innermost layer of skin, under the dermis. [7]

hypofrontality hypothesis The hypothesis that schizophrenia may result from underactivation of the frontal lobes. [12]

hypophyseal portal system Also called *pituitary portal system*. A system of capillaries spanning between the neurosecretory cells of the hypothalamus and the secretory tissue of the anterior pituitary. [4]

hypophysis See *pituitary gland*.

hypothalamus Part of the diencephalon, lying ventral to the thalamus. [1]

hypotonic Referring to a solution with a lower concentration of salt than that found in interstitial fluid and blood plasma (less than about 0.9% salt). [14]

hypovolemic thirst A desire to ingest fluids that is stimulated by a reduced volume of extracellular fluid. [14]

hypoxia A lack of oxygen. [6]

Hz See *hertz*.

I

IAPs See *inhibitors of apoptosis proteins*.

IC See *impoverished condition*.

ideational apraxia An impairment in the ability to carry out a sequence of actions, even though each element or step can be done correctly. [10]

ideomotor apraxia The inability to carry out a simple motor activity in response to a verbal command, even though this same activity is readily performed spontaneously. [10]

IHC See *immunohistochemistry* and *inner hair cell*.

immunocytochemistry (ICC) A method for detecting a particular protein in tissues, in which an antibody recognises and binds to the protein and then chemical methods are used to leave a visible reaction product around each antibody. [4, App]

immunoglobulins See *antibodies*.

immunohistochemistry (IHC) A technique in which labelled antibodies are used to visualise the histological distribution of specific proteins. [1]

impoverished condition (IC) Also called *isolated condition*. A condition in which laboratory rodents are housed singly in a small cage without complex stimuli. [16]

in situ hybridisation A method for detecting particular RNA transcripts in tissue sections by providing a nucleotide probe that is complementary to, and will therefore hybridise with, the transcript of interest. [1, 4, App]

in vitro Literally, "in glass" (in Latin). Outside the body, usually in a laboratory dish.

inattentional blindness The failure to perceive nonattended stimuli that seem so obvious as to be impossible to miss. [17]

incus Latin for "anvil." A middle-ear bone situated between the malleus and the stapes. [9]

independent variable The factor that is manipulated by an experimenter. [0]

indifferent gonads The undifferentiated gonads of the early vertebrate fetus, which will eventually develop into either testes or ovaries. [13]

individual response stereotypy The tendency of individuals to show the same response pattern to particular situations throughout their life span. [11]

indoleamine A class of monoamines, including serotonin and melatonin, that serve as neurotransmitters. [3]

induction The process by which one set of cells influences the fate of neighbouring cells, usually by secreting a chemical factor that changes gene expression in the target cells. [6]

inferior In anatomy, pertaining to the lower of two locations. Compare *superior.* [1]

inferior colliculi Paired gray matter structures of the dorsal midbrain that receive auditory information. [1, 9]

infradian Referring to a rhythmic biological event whose period is longer than that of a circadian rhythm—that is, longer than a day. [15]

infrasound Very low-frequency sound; in general, below the threshold for human hearing, at about 10 Hz. [9]

infundibulum See *pituitary stalk.*

inhibition of return The phenomenon in which detection of stimuli at the former location of a cue is impaired for latencies of about 250 ms or more. [17]

inhibitors of apoptosis proteins (IAPs) A family of proteins that inhibit caspases and thereby stave off apoptosis. [6]

inhibitory postsynaptic potential (IPSP) A hyperpolarising potential in the postsynaptic neuron that is caused by inhibitory connections. IPSPs decrease the probability that the postsynaptic neuron will fire an action potential. [2]

inner ear The cochlea and vestibular apparatus. [9]

inner hair cell (IHC) One of the two types of cochlear receptor cells for hearing. [9]

innervate To provide neural input. [1]

innervation ratio The ratio expressing the number of muscle fibres innervated by a single motor axon. [10]

inositol triphosphate (IP3) A member of a class of second-messenger compounds (*phosphoinositides*) common in postsynaptic cells. [4]

input zone The part of a neuron that receives information from other neurons or from specialised sensory structures. Usually corresponds to the cell's dendrites. [1]

instrumental conditioning See *operant conditioning.*

insula A region of cortex lying below the surface, within the lateral sulcus, of the frontal, temporal, and parietal lobes. [3]

insulin A hormone, released by pancreatic beta cells, that lowers blood glucose. [14]

integration zone The part of the neuron that initiates nerve electrical activity, described in detail in Chapter 1. It usually corresponds to the neuron's axon hillock. [1]

intellectual disability A disability characterised by significant limitations in intellectual functioning and adaptive behaviour. [6]

interaural intensity differences Perceived differences in loudness between the two ears, which can be used to localise a sound source. [9]

interaural temporal differences Differences between the two ears in the time of arrival of a sound, which can be employed by the nervous system to localise sound sources. [9]

intermale aggression Aggression between males of the same species. [11]

intermediate-term memory A form of memory that lasts longer than short-term memory but not as long as long-term memory. [16]

internal fertilisation The process by which sperm fertilise eggs inside the female's body, as in all mammals, birds, and reptiles. [13]

interneuron A neuron that is neither a sensory neuron nor a motor neuron; it receives input from and sends output to other neurons. [1]

intersex Referring to an individual with atypical genital development and sexual differentiation that generally resembles a form intermediate between typical male and typical female genitals. [13]

intracellular compartment The fluid space of the body that is contained within cells. [14]

intracellular fluid Also called *cytoplasm*. The watery solution found within cells. [2]

intrafusal fibre One of the small muscle fibres that lie within each muscle spindle, controlling its length. [10]

intraparietal sulcus (IPS) A region in the human parietal lobe, homologous with the monkey lateral intraparietal area, that is especially involved in voluntary, top-down control of attention. [17]

intrinsic activity See *efficacy.*

intromission Insertion of the erect penis into the vagina during copulation. [13]

inverse agonist A substance that binds to a receptor and causes it to do the opposite of what the naturally occurring transmitter does. [3]

ion An atom or molecule that has acquired an electrical charge by gaining or losing one or more electrons. [2]

ion channel A pore in the cell membrane that permits the passage of certain ions through the membrane when the channel is open. [2]

ionotropic receptor Also called *ligand-gated ion channel* and *chemically gated ion channel.* A receptor protein that includes an ion channel that is opened when the receptor is bound by an agonist. [2, 3]

IP3 See *inositol triphosphate.*

IPS See *intraparietal sulcus.*

ipsilateral In anatomy, pertaining to a location on the same side of the body. Compare *contralateral.* [1]

IPSP See *inhibitory postsynaptic potential.*

iris The circular structure of the eye that provides an opening to form the pupil. [8]

isolated brain Sometimes referred to by the French term *encéphale isolé*. An experimental preparation in which an animal's brainstem has been separated from the spinal cord by a cut below the medulla. [15]

isolated condition See *impoverished condition.*

isolated forebrain Sometimes referred to by the French term *cerveau isolé*. An experimental preparation in which an animal's nervous system has been cut in the upper midbrain, dividing the forebrain from the brainstem. [15]

isotonic Referring to a solution with a concentration of salt that is the same as that found in interstitial fluid and blood plasma (about 0.9% salt). [14]

J

James-Lange theory The theory that our experience of emotion is a response to the physiological changes that accompany it. [11]

K

K complex A sharp negative EEG potential that is seen in stage 1 sleep. [15]

K^+ See *potassium ion.*

ketamine A dissociative anaesthetic drug, similar to PCP, that acts as an NMDA receptor antagonist. [3, 12]

ketones A metabolic fuel source liberated by the breakdown of body fats and proteins. [14]

khat Also spelled *qat*. An African shrub that, when chewed, acts as a stimulant. [3]

kindling A method of experimentally inducing an epileptic seizure by repeatedly stimulating a brain region. [2]

kisspeptin A hypothalamic peptide hormone that increases gonadotropin secretion by facilitating the release of gonadotropin-releasing hormone. [4]

Klüver-Bucy syndrome A condition, brought about by bilateral amygdala damage, that is characterised by dramatic emotional changes including reduction in fear and anxiety. [11]

knee-jerk reflex A variant of the stretch reflex in which stretching of the tendon below the knee leads to an upward kick of the leg. [2]

knockout organism An individual in which a particular gene has been disabled by an experimenter. [4, 6]

Korsakoff's syndrome A memory disorder, related to a thiamine deficiency, that is generally associated with chronic alcoholism. [16]

L

labelled lines The concept that each nerve input to the brain reports only a particular type of information. [7, 9]

lamellated corpuscle See *Pacinian corpuscle.*

language The most sophisticated form of communication, in which a set of arbitrary sounds, tokens, or symbols can be arranged according to a grammar in order to convey an almost limitless variety of concepts. [18]

latency differences Differences between the two ears in the time of arrival of a sound, which can be employed by the nervous system to localise sound sources. [9]

latent learning Learning that has taken place but has not (yet) been demonstrated by performance. [16]

lateral In anatomy, towards the side of the body. Compare *medial.* [1]

lateral geniculate nucleus (LGN) The part of the thalamus that receives information from the optic tract and sends it to visual areas in the occipital cortex. [8]

lateral hypothalamus (LH) A hypothalamic region involved in the control of appetite and other functions. [14]

lateral inhibition The phenomenon by which interconnected neurons inhibit their neighbours, producing contrast at the edges of regions. [8]

lateral intraparietal area (LIP) A region in the monkey parietal lobe, homologous with the human intraparietal sulcus, that is especially involved in voluntary, top-down control of attention. [17]

lateral sulcus See *Sylvian fissure.*

lateral ventricle A complexly shaped lateral portion of the ventricular system within each hemisphere of the brain. [1]

lateralisation The tendency for the right and left halves of a system to differ from one another. In cerebral lateralisation, one cerebral hemisphere is specialised for a particular intellectual function. [18]

lateral-line system A sensory system, found in many kinds of fishes and some amphibians, that informs the animal of water motion in relation to the body surface. [9]

late-selection model A model of attention in which the attentional bottleneck filters out stimuli only after substantial analysis has occurred. [17]

l-dopa The immediate metabolic precursor of the transmitter dopamine. [10]

learned helplessness In the context of experimentation, a learning paradigm in which individuals are subjected to inescapable, unpleasant conditions. [12]

learning The process of acquiring new and relatively enduring information, behaviour patterns, or abilities, characterised by modifications of behaviour as a result of practice, study, or experience. [16]

lens A structure in the eye that helps focus an image on the retina. [8]

leptin A peptide hormone released by fat cells. [14]

lesions Regions of damage. [1]

levels of analysis The scope of experimental approaches. A scientist may try to understand behaviour by monitoring molecules, nerve cells, brain regions, or social environments, or some combination of these levels of analysis. [0]

LGN See *lateral geniculate nucleus.*

LH See *lateral hypothalamus* and *luteinising hormone.*

lie detector See *polygraph.*

ligand A substance that binds to receptor molecules, such as those at the surface of the cell. [2, 3]

ligand-gated ion channel Also called *chemically gated ion channel*. An ion channel that opens or closes in response to the presence of a particular chemical. [2]

limbic system A loosely defined, widespread group of brain nuclei that innervate each other to form a network. These nuclei are implicated in emotions. [1, 11]

LIP See *lateral intraparietal area.*

lipid bilayer The structure of the neuronal cell membrane, which consists of two layers of lipid molecules. [2]

lipids Large molecules (commonly called *fats*) consisting of fatty acids and glycerol that are insoluble in water. [14]

lithium An element that, when administered as a drug, often relieves the symptoms of bipolar disorder. [12]

lobotomy The surgical detachment of a portion of the frontal lobe from the rest of the brain, once used as a treatment for schizophrenia and many other ailments. [12]

local anaesthetic A drug, such as procaine or lidocaine, that blocks sodium channels to stop neural transmission in pain fibres. [3]

local potential An electrical potential that is initiated by stimulation at a specific site, which is a graded response that spreads passively across the cell membrane, decreasing in strength with time and distance. [2]

locus coeruleus Literally, "blue spot." A small nucleus in the brainstem whose neurons produce norepinephrine. [3]

long-term memory (LTM) An enduring form of memory that lasts days, weeks, months, or years and has a very large capacity. [16]

long-term potentiation (LTP) A stable and enduring increase in the effectiveness of synapses following repeated strong stimulation. [16]

lordosis A female receptive posture in quadrupeds in which the hindquarters are raised and the tail is turned to one side, facilitating intromission by the male. [13]

Lou Gehrig's disease See *amyotrophic lateral sclerosis.*

loudness The subjective experience of the pressure level of a sound. [9]

loxP Also called *lox.* A specific sequence of nucleotides recognised by the enzyme Cre-recombinase. If the enzyme encounters a pair of loxP sites in a gene, it will remove the DNA between the two sites and recombine the gene, usually rendering the gene product dysfunctional. [10, App]

LSD Also called *acid.* Lysergic acid diethylamide, a hallucinogenic drug. [3]

LTM See *long-term memory.*

LTP See *long-term potentiation.*

lumbar Referring to the five spinal segments that make up the upper part of the lower back. [1]

luteinising hormone (LH) A gonadotropin, named for its stimulatory effects on the ovarian corpora lutea. [4]

M

M1 See *primary motor cortex.*

macular degeneration A progressive loss of central vision due to death or obstruction of photoreceptors in the retina. [8]

mad cow disease See *bovine spongiform encephalopathy.*

magnetic resonance imaging (MRI) A noninvasive technique that uses magnetic energy to generate images that reveal some structural details in the living brain. [1]

magnetoencephalography (MEG) A noninvasive functional brain-imaging technique that measures the tiny magnetic fields produced by active neurons, in order to identify regions of the brain that are particularly active during a given task. [1]

magnocellular Referring to relatively large cells. [8]

malleus Latin for "hammer." A middle-ear bone that is connected to the tympanic membrane. [9]

mammillary body One of a pair of nuclei at the base of the brain that connect to the hippocampus and play a role in memory. [1, 16]

MAO See *monoamine oxidase.*

marijuana See *cannabis.*

maternal aggression The fierce defensive aggression of females defending their offspring. [11]

MBSR See *mindfulness-based stress reduction.*

MC4R See *melanocortin type 4 receptor.*

MD See *muscular dystrophy.*

MDMA Also called *Ecstasy.* A drug of abuse, 3,4-methylenedioxymethamphetamine. [3]

medial In anatomy, towards the middle of an organ or organism. Compare *lateral.* [1]

medial amygdala A region of the amygdala involved in processing olfactory and pheromonal stimuli. [13, 11]

medial forebrain bundle A collection of axons traveling in the midline region of the forebrain. [11]

medial geniculate nuclei Nuclei in the thalamus that receive input from the inferior colliculi and send output to the auditory cortex. [9]

medial preoptic area (mPOA) A region of the anterior hypothalamus implicated in the control of many behaviours, including thermoregulation, sexual behaviour, and gonadotropin secretion. [13]

median eminence Midline feature on the base of the brain marking the point at which the pituitary stalk exits the hypothalamus to connect to the pituitary. It contains elements of the hypophyseal portal system. [4]

medulla Also called *myelencephalon.* The posterior part of the hindbrain, continuous with the spinal cord. [1]

medullary reticular formation The hindmost portion of the brainstem reticular formation, implicated in motor control and copulatory behaviour. [13]

MEG See *magnetoencephalography.*

Meissner's corpuscle A skin receptor cell type that detects light touch. [7]

melanocortin type 4 receptor (MC4R) A specific subtype of melanocortin receptor. [14]

melanocortins One category of endogenous opioid peptides. [14]

melanopsin A photopigment found in some retinal ganglion cells that project to the suprachiasmatic nucleus. [15]

melatonin An amine hormone that is released by the pineal gland. [4]

memory 1. The ability to retain information, based on the mental process of learning or encoding, retention across some interval of time, and retrieval or reactivation of the information. 2. The specific information that is stored in the brain. [16]

meninges The three protective sheets of tissue—dura mater, pia mater, and arachnoid—that surround the brain and spinal cord. [1]

meningiomas Several classes of noncancerous tumours arising from the meninges. [1]

meningitis An acute inflammation of the meninges, usually caused by a viral or bacterial infection. [1]

Merkel's disc A skin receptor cell type that detects fine touch. [7]

mesencephalon See *midbrain.*

mesolimbocortical pathway A set of dopaminergic axons arising in the midbrain and innervating the limbic system and cortex. [3]

mesostriatal pathway A set of dopaminergic axons arising from the midbrain and innervating the basal ganglia, including those from the substantia nigra to the striatum. [3]

messenger RNA (mRNA) A strand of RNA that carries the code of a section of a DNA strand to the cytoplasm. [App]

metabolic tolerance The form of drug tolerance that arises when repeated exposure to the drug causes the metabolic machinery of the body to become more efficient at clearing the drug. [3]

metabotropic receptor A receptor protein that does not contain an ion channel but may, when activated, use a G protein second-messenger system to alter the functioning of the postsynaptic cell. [2, 3]

metencephalon A subdivision of the hindbrain that includes the cerebellum and the pons. [1]

methamphetamine A stimulant amphetamine derivative with strongly addictive qualities. [3]

methylation A chemical modification of DNA that does not affect the nucleotide sequence of a gene but makes that gene less likely to be expressed. [6]

microbiome The collective term for a population of microbes. [14]

microelectrode An especially small electrode used to record electrical potentials from living cells. [2]

microglial cells Also called *microglia.* Extremely small glial cells that remove cellular debris from injured or dead cells. [1]

micropolygyria A condition of the brain in which small regions are characterised by more gyri than usual. [18]

midbrain Also called *mesencephalon.* The middle division of the brain. [1, 6]

middle canal See *scala media.*

middle cerebral arteries Two large arteries, arising from the internal carotid arteries, that provide blood to most of the lateral surfaces of the cerebral hemispheres. [1]

middle ear The cavity between the tympanic membrane and the cochlea. [9]

migraines Intense headaches, typically perceived from one half of the head, that recur regularly and can be difficult to treat. [7]

milk letdown reflex The reflexive release of milk in response to suckling or to stimuli associated with suckling. [4]

millivolt (mV) A thousandth of a volt. [2]

mindfulness-based stress reduction (MBSR) A therapy to reduce stress that pairs relaxation with efforts to focus attention on the present moment, rather than past or future problems. [11]

mineralocorticoids A class of steroid hormones, released by the adrenal cortex, that affect ion concentrations in body tissues. [4]

minimal discriminable frequency difference The smallest change in frequency that can be detected reliably between two tones. [9]

minor traumatic brain injury (mTBI) See *concussion.*

mirror neuron A neuron that is active both when an individual makes a particular movement and when that individual sees another individual make that same movement. [8, 10]

mitochondrion A cellular organelle that provides metabolic energy for the cell's processes. [1]

mitosis The process of division of somatic cells that involves duplication of DNA. [6]

mitral cell A type of cell in the olfactory bulb that conducts smell information from the glomeruli to the rest of the brain. [9]

modulatory site A portion of a receptor that, when bound by a compound, alters the receptor's response to its transmitter. [3]

monoamine hormones See *amine hormones.*

monoamine hypothesis The hypothesis that depression is caused by reduced activity of one or more monoamine transmitters, such as serotonin. [12]

monoamine oxidase (MAO) An enzyme that breaks down and thereby inactivates monoamine transmitters. [3, 12]

monocular deprivation Depriving one eye of light. [6]

monopolar neuron See *unipolar neuron.*

monozygotic Referring to twins derived from a single fertilised egg (*identical* twins). Such individuals have the same genotype. [12]

morpheme The smallest grammatical unit of a language; a word or meaningful part of a word. [18]

morphine An opiate compound derived from the poppy flower. [3]

motion sickness The experience of nausea brought on by unnatural passive movement, as in a car or boat. [9]

motivation With respect to homeostasis, a drive state that prompts homeostatic behaviours. [14]

motoneuron See *motor neuron.*

motor nerve A nerve that conveys neural activity to muscle tissue and causes it to contract. [1]

motor neuron Also called *motoneuron.* A nerve cell in the brain or spinal cord that transmits motor messages, stimulating a muscle or gland. [1, 10]

motor plan Also called *motor program.* A plan for action in the nervous system. [10]

motor program See *motor plan.*

motor protein A specialised kinetic protein molecule that conveys a load, such as a vesicle, from one location to another within a cell. [1]

motor theory of language The theory proposing that the left-hemisphere language zones are motor control systems that are concerned with both the precise production and the perception of the extremely complex movements that go into speech. [18]

motor unit A single motor axon and all the muscle fibres that it innervates. [10]

mPOA See *medial preoptic area.*

MRI See *magnetic resonance imaging.*

mRNA See *messenger RNA.*

mTBI See *concussion.*

müllerian ducts A duct system in the embryo that will develop into female reproductive structures (oviduct, uterus, and inner vagina) if testes are not present. [13]

multiple sclerosis Literally, "many scars." A disorder characterised by widespread degeneration of myelin. [1]

multipolar neuron A nerve cell that has many dendrites and a single axon. [1]

muscarinic Referring to cholinergic receptors that respond to the chemical muscarine as well as to acetylcholine. [3]

muscle fibre A large cylindrical cell that can contract in response to neurotransmitter released from a motor neuron. [10]

muscle spindle A muscle receptor that lies parallel to a muscle and sends action potentials to the central nervous system when the muscle is stretched. [10]

muscular dystrophy (MD) A disease that leads to degeneration of and functional changes in muscles. [10]

musth An annual period of heightened aggressiveness and sexual activity in male elephants. [13]

mutation A change in the nucleotide sequence of a gene as a result of unfaithful replication. [5]

mV See *millivolt.*

myasthenia gravis A disorder characterised by a profound weakness of skeletal muscles. It is caused by a loss of acetylcholine receptors. [10]

myelencephalon See *medulla.*

myelin The fatty insulation around an axon, formed by glial cells, that speeds the conduction of action potentials. [1]

myelination The process of ensheathing axons in myelin. [1, 6]

myopia Nearsightedness; the inability to focus the retinal image of objects that are far away. [8]

myosin A protein that, along with actin, mediates the contraction of muscle fibres. [10]

N

N1 effect A negative deflection of the event-related potential, occurring about 100 ms after stimulus presentation, that is enhanced for selectively attended input, compared with ignored input. [17]

Na^+ See *sodium ion.*

naloxone A potent antagonist of opiates that is often administered to people who have taken drug overdoses. It blocks receptors for endogenous opioids. [7]

narcolepsy A disorder that involves frequent, intense episodes of sleep, which last from 5 to 30 minutes and can occur anytime during the usual waking hours. [15]

natriuretic polypeptide B (Nppb) A peptide neurotransmitter used by neurons reporting itch to the spinal cord. [7]

naturalist A student of natural history, including the forms and classification of organisms. [5]

nature/nurture debate The question of how much of brain development is due to inherited genes (nature) versus experiences growing up (nurture). [6]

$Na_V1.7$ Also called *SCN9A*. A voltage-gated sodium channel used almost exclusively by nociceptors to initiate action potentials. [7]

NE See *norepinephrine.*

negative feedback The process whereby a system monitors its own output and reduces its activity when a set point is reached. [4, 14]

negative polarity A negative electrical-potential difference relative to a reference electrode. [2]

negative symptom In psychiatry, a symptom that reflects insufficient functioning. Examples include emotional and social withdrawal, blunted affect, and slowness and impoverishment of thought and speech. [12]

neocortex Cerebral cortex that is made up of six distinct layers. [5]

neologism An entirely novel word, sometimes produced by a person with aphasia. [18]

neonatal Referring to newborns. [13]

Nernst equation An equation predicting the equilibrium potential for a given ion based on the concentrations of the ion on opposite sides of a permeable membrane. [2]

nerve A collection of axons bundled together outside the central nervous system. [1]

nerve cell See *neuron.*

nerve growth factor (NGF) A substance that markedly affects the growth of neurons in spinal ganglia and in the ganglia of the sympathetic nervous system. [6]

neural chain A simple kind of neural circuit in which neurons are attached linearly, end to end. [2]

neural groove In the developing embryo, the groove between the neural folds. [6]

neural plasticity Also called *neuroplasticity*. The ability of the nervous system to change in response to experience or the environment. [1]

neural tube An embryonic structure with subdivisions that correspond to the future forebrain, midbrain, and hindbrain. [1, 6]

neurochemistry The branch of neuroscience concerned with the fundamental chemical composition and processes of the nervous system. [3]

neurocrine Referring to secretory functions of neurons, especially pertaining to synapse transmission. [4]

neuroeconomics The study of brain mechanisms at work during decision-making. [17]

neuroendocrine cell Also called *neurosecretory cell.* A neuron that releases hormones into local or systemic circulation. [4]

neurofibrillary tangle An abnormal whorl of neurofilaments within nerve cells. [6]

neurogenesis The mitotic division of nonneuronal cells to produce neurons. [6]

neuroglia See *glial cells.*

neurohypophysis See *posterior pituitary.*

neuromodulator A substance that influences the activity of synaptic transmitters. [4]

neuromuscular junction (NMJ) The region where the motor neuron terminal and the adjoining muscle fibre meet; the point where the nerve transmits its message to the muscle fibre. [10]

neuron Also called *nerve cell.* The basic unit of the nervous system, each composed of a cell body, receptive extension(s) (dendrites), and a transmitting extension (axon). [0, 1, 10]

neuron doctrine The hypothesis that the brain is composed of cells that are distinct structurally, metabolically, and functionally. [1]

neuropathic pain Pain caused by damage to peripheral nerves. It is often difficult to treat. [7]

neuropeptide Also called *peptide neurotransmitter.* A peptide that is used by neurons for signalling. [4]

neuropeptide Y (NPY) A peptide neurotransmitter that may carry some of the signals for feeding. [14]

neuropharmacology Also called *psychopharmacology.* The scientific field concerned with the discovery and study of compounds that selectively affect the functioning of the nervous system. [3]

neurophysiology The study of electrical and chemical processes in neurons. [2]

neuroplasticity Also called *neural plasticity.* The ability of the nervous system to change in response to experience or the environment. [0, 16]

neuroscience The study of the nervous system. [0]

neurosecretory cell See *neuroendocrine cell.*

neurosteroids Steroid molecules produced within the brain that affect neurons. [3, 4]

neurotransmitter Also called *synaptic transmitter, chemical transmitter,* or simply *transmitter.* The chemical released from the presynaptic axon terminal that serves as the basis of communication between neurons. [1, 2]

neurotrophic factor Also called *trophic factor.* A target-derived chemical that acts as if it "feeds" certain neurons to help them survive. [6]

neurotrophins A family of proteins, including NGF and BDNF, that prevent different classes of neurons from dying. [6]

NGF See *nerve growth factor.*

nicotine A compound found in plants, including tobacco, that acts as an agonist on a large class of cholinergic receptors. [3]

nicotinic Referring to cholinergic receptors that respond to nicotine as well as to acetylcholine. [3]

night terror A sudden arousal from sleep that is marked by intense fear and autonomic activation. [15]

nightmare A long, frightening dream that awakens the sleeper from sleep. [15]

Nissl stain A cell stain that reveals all cell bodies by staining RNA. [1]

nitric oxide (NO) A soluble gas that serves as a retrograde gas neurotransmitter in the nervous system. [3]

NMDA receptor A glutamate receptor that also binds the glutamate agonist NMDA (*N*-methyl-D-aspartate) and that is both ligand-gated and voltage-sensitive. [16]

NMJ See *neuromuscular junction.*

NO See *nitric oxide.*

nocebo An inert substance that causes discomfort due to the patient's expectations. [7]

nociceptor A receptor that responds to stimuli that produce tissue damage or pose the threat of damage. [7]

nocturnal Active during the dark periods of the daily cycle. [15]

node of Ranvier A gap between successive segments of the myelin sheath where the axon membrane is exposed. [1, 2]

nonassociative learning A type of learning in which presentation of a particular stimulus alters the strength or probability of a response according to the strength and temporal spacing of that stimulus. It includes habituation and sensitisation. [16]

noncompetitive ligand Also called *neuromodulator.* A substance that alters the response to an endogenous ligand without interacting with the endogenous ligand's recognition site. [3]

nondeclarative memory Also called *procedural memory.* A memory that is shown by performance rather than by conscious recollection. [16]

nondirected synapse A type of synapse in which the presynaptic and postsynaptic cells are not in close apposition; instead, neurotransmitter is released by axonal varicosities and diffuses away to affect wide regions of tissue. [2]

nonfluent aphasia Also called *Broca's aphasia.* A language impairment characterised by difficulty with speech production but not with language comprehension. It is related to damage in Broca's area. [18]

nonfluent speech Talking with considerable effort, short sentences, and the absence of the usual melodic character of conversational speech. [18]

nongenomic effect An effect of a steroid hormone that is not mediated by direct changes in gene expression. [4]

nonprimary motor cortex Frontal lobe regions adjacent to the primary motor cortex that contribute to motor control and modulate the activity of the primary motor cortex. [10]

nonprimary sensory cortex See *secondary sensory cortex.*

non-REM (NREM) sleep Stage of sleep without rapid eye movements. In humans this is divided into stages 1, 2, and 3 sleep. [15]

nootropics A class of drugs that enhance cognitive function. [16]

noradrenaline See *norepinephrine.*

noradrenergic Referring to systems using norepinephrine (noradrenaline) as a transmitter. [3]

norepinephrine (NE) Also called *noradrenaline.* A compound that acts both as a neurotransmitter, produced and released by sympathetic postganglionic neurons to accelerate organ activity, and as a hormone released from the adrenal medulla under the control of the sympathetic nervous system, which prepares the body for action. [1, 3, 4, 11]

Northern blot A method of detecting a particular RNA transcript in a tissue or organ by separating RNA from that source with gel electrophoresis, blotting the separated RNAs onto nitrocellulose, and then using a nucleotide probe to hybridise with, and highlight, the transcript of interest. [App]

notochord A midline structure arising early in the embryonic development of vertebrates. [6]

Nppb See *natriuretic polypeptide B.*

NPY See *neuropeptide Y.*

NPY neurons Neurons involved in the hypothalamic appetite control system, so named because they produce neuropeptide Y (NPY) along with agouti-related peptide (AgRP). [14]

NREM 1 See *stage 1 sleep.*

NREM 2 See *stage 2 sleep.*

NREM 3 See *slow-wave sleep.*

NREM sleep See *non-REM (NREM) sleep.*

NST See *nucleus of the solitary tract.*

nucleotide A portion of a DNA or RNA molecule that is composed of a single base and the adjoining sugar-phosphate unit of the strand. [App]

nucleus 1. A collection of neurons within the central nervous system (e.g., the caudate nucleus). 2. The spherical central structure of a cell that contains the chromosomes. [1, App]

nucleus accumbens A region of the forebrain that receives dopaminergic innervation from the ventral tegmental area. [3]

nucleus of the solitary tract (NST) A brainstem nucleus that receives visceral and taste information via several cranial nerves. [14]

nutrient A chemical that is needed for growth, maintenance, and repair of the body but is not used as a source of energy. [14]

O

obligatory losses Unavoidable loss of regulated variables, such as energy, water, or temperature, as a consequence of life processes. [14]

obsessive-compulsive disorder (OCD) A syndrome in which the affected individual engages in recurring, repetitive acts that are carried out without rhyme, reason, or the ability to stop. [12]

occipital lobe Large region of cortex covering much of the posterior part of each cerebral hemisphere, specialised for visual processing. [1, 8]

OCD See *obsessive-compulsive disorder.*

ocular dominance column A region of cortex in which one eye or the other provides a greater degree of synaptic input. [8]

ocular dominance histogram A graph that portrays the strength of response of a brain neuron to stimuli presented to either the left eye or the right eye. [6]

ocular dominance slab A slab of visual cortex, about 0.4 mm wide, in which the neurons of all layers respond preferentially to stimulation of one eye. [8]

oculomotor apraxia A severe difficulty in voluntarily steering visual gaze towards specific targets. [17]

off-centre bipolar cell A retinal bipolar cell that is inhibited by light in the centre of its receptive field. [8]

off-centre ganglion cell A retinal ganglion cell that is activated when light is presented to the periphery, rather than the centre, of the cell's receptive field. [8]

off-centre/on-surround Referring to a concentric receptive field in which the centre inhibits the cell of interest while the surround excites it. [8]

OHC See *outer hair cell.*

olfactory bulb An anterior projection of the brain that terminates in the upper nasal passages and, through small openings in the skull, receives axons from olfactory receptor neurons. [1, 9]

olfactory epithelium A sheet of cells, including olfactory receptors, that lines the dorsal portion of the nasal cavities and adjacent regions, including the septum that separates the left and right nasal cavities. [9]

olfactory receptor neuron A type of neuron, found in the olfactory epithelium, that senses airborne odourants via specialised receptor proteins. [9]

oligodendrocyte A type of glial cell that forms myelin in the central nervous system. [1]

on-centre bipolar cell A retinal bipolar cell that is excited by light in the centre of its receptive field. [8]

on-centre ganglion cell A retinal ganglion cell that is activated when light is presented to the centre, rather than the periphery, of the cell's receptive field. [8]

on-centre/off-surround Referring to a concentric receptive field in which the centre excites the cell of interest while the surround inhibits it. [8]

ontogeny The process by which an individual changes in the course of its lifetime—that is, grows up and grows old. [0]

Onuf's nucleus The human homolog of the spinal nucleus of the bulbocavernosus (SNB) in rats. [13]

open-loop control mechanism A control mechanism in which feedback from the output of the system is not provided to the input control. [10]

operant conditioning Also called *instrumental conditioning.* A form of associative learning in which the likelihood that an act (instrumental response) will be performed depends on the consequences (reinforcing stimuli) that follow it. [16]

opiates A class of compounds that exert an effect like that of opium, including reduced pain sensitivity. [7]

opioid peptide A type of endogenous peptide that mimics the effects of morphine in binding to opioid receptors and producing marked analgesia and reward. [3]

opioid receptor A receptor that responds to endogenous and/or exogenous opioids. [3, 7]

opioids A class of peptides produced in various regions of the brain that bind to opioid receptors and act like opiates. [7]

opium A heterogeneous extract of the seedpod juice of the opium poppy, *Papaver somniferum*. [3]

opponent-process hypothesis The idea that colour vision depends on systems that produce opposite responses to light of different wavelengths. [8]

opsin One of the two components of photopigments in the retina. [8]

optic ataxia Difficulty using vision to reach and manipulate objects. [8, 17]

optic chiasm The point at which the two optic nerves meet. [8]

optic disc The region of the retina devoid of receptor cells because ganglion cell axons and blood vessels exit the eyeball there. [8]

optic nerve Cranial nerve II; the collection of ganglion cell axons that extend from the retina to the optic chiasm. [8]

optic radiation Axons from the lateral geniculate nucleus that terminate in the primary visual areas of the occipital cortex. [8]

optic tract The axons of retinal ganglion cells after they have passed the optic chiasm. Most of the axons terminate in the lateral geniculate nucleus. [8]

optical imaging A method for visualising brain activity in which near-infrared light is passed through the scalp and skull. [1]

optogenetics The introduction of gene(s) into neurons that makes them electrically respond to light. [2]

oral contraceptive A birth control pill, typically consisting of steroid hormones to prevent ovulation. [4]

orexigenic neurons Neurons of the hypothalamic appetite system that promote feeding behaviour. [14]

orexins Also called *hypocretins*. Neuropeptides produced in the hypothalamus that are involved in switching between sleep states, in narcolepsy, and in the control of appetite. [14]

organ of Corti A structure in the inner ear that lies on the basilar membrane of the cochlea and contains the hair cells and terminations of the auditory nerve. [9]

organisational effect A permanent alteration of the nervous system, and thus permanent change in behaviour, resulting from the action of a steroid hormone on an animal early in its development. [13]

organum vasculosum of the lamina terminalis (OVLT) One of the circumventricular organs. [14]

orgasm The climax of sexual experience, marked by extremely pleasurable sensations. [13]

orientation column A column of visual cortex that responds to rod-shaped stimuli of a particular orientation. [8]

orphan receptor Any receptor for which no endogenous ligand has yet been discovered. [3]

osmolality The number of solute particles per unit volume of solvent. [14]

osmosensory neuron A specialised neuron that measures the movement of water into and out of cells. [14]

osmosis The passive movement of water molecules from one place to another until a uniform concentration is achieved. [14]

osmotic pressure The tendency of a solvent to move through a membrane in order to equalise the concentration of solute. [14]

osmotic thirst A desire to ingest fluids that is stimulated by a high concentration of solute (like salt) in the extracellular compartment, reducing intracellular fluid. [14]

ossicles Three small bones (incus, malleus, and stapes) that transmit sound across the middle ear, from the tympanic membrane to the oval window. [9]

otoacoustic emission A sound produced by the cochlea itself, either spontaneously or in response to an environmental noise. [9]

otolith A small crystal on the gelatinous membrane in the vestibular system. [9]

ototoxic Toxic to the ears, especially the middle or inner ear. [9]

outer hair cell (OHC) One of the two types of cochlear receptor cells for hearing. [9]

output zone The part of a neuron, usually corresponding to the axon terminals, at which the cell sends information to another cell. [1]

oval window The location on the surface of the cochlea at which vibrations are received from the ossicles. [9]

ovaries The female gonads, which produce eggs for reproduction. [4]

OVLT See *organum vasculosum of the lamina terminalis.*

ovulation The production and release of an egg (ovum). [13]

ovulatory cycle The periodic occurrence of ovulation. [13]

ovum An egg, the female gamete. [13]

oxytocin A hormone, released from the posterior pituitary, that triggers milk letdown in the nursing female. [4]

P

P1 effect A positive deflection of the event-related potential, occurring 70–100 ms after stimulus presentation, that is enhanced for selectively attended visual input, compared with ignored input. [17]

P3 effect A positive deflection of the event-related potential, occurring about 300–500 ms after stimulus presentation, that is associated with higher-order auditory stimulus processing and late attentional selection. [17]

Pacinian corpuscle Also called *lamellated corpuscle*. A skin receptor cell type that detects vibration. [7]

pain The discomfort normally associated with tissue damage. [7]

pair-bond A durable and exclusive relationship between two individuals, such as mates. [13]

papilla A small bump that projects from the surface of the tongue. Papillae contain most of the taste receptor cells. [9]

parabiotic Referring to a surgical preparation that joins two animals to share a single blood supply. [13]

paracrine Referring to cellular communication in which a chemical signal diffuses to nearby target cells through the extracellular space. [4]

paradoxical insomnia Also called *sleep state misperception*. Commonly, a person's perception that he has not been asleep when EEG readings and nonresponsiveness indicate that he has. Typically it occurs at the start of a sleep episode. [15]

paradoxical sleep See *rapid-eye-movement (REM) sleep.*

paragigantocellular nucleus (PGN) A region of the brainstem reticular formation implicated in modulation of spinal reflexes. [13]

parallel fibre One of the axons of the granule cells that form the outermost layer of the cerebellar cortex. [1]

paraphasia A symptom of aphasia that is distinguished by the substitution of a word by a sound, an incorrect word, an unintended word, or a neologism (a meaningless word). [18]

parasympathetic nervous system A component of the autonomic nervous system that arises from both the cranial nerves and the sacral spinal cord. [1, 11]

paraventricular nucleus (PVN) A nucleus of the hypothalamus implicated in the release of oxytocin and vasopressin and in the control of feeding and other behaviours. [4, 14]

parental behaviour Behaviour of adult animals with the goal of enhancing the well-being of their own offspring, often at some cost to the parents. [13]

paresis Partial paralysis. [10]

parietal lobe Large region of cortex lying between the frontal and occipital lobes of each cerebral hemisphere. [1]

parkin A protein that has been implicated in Parkinson's disease. [10]

Parkinson's disease A degenerative neurological disorder, characterised by tremors at rest, muscular rigidity, and reduction in voluntary movement, that involves dopaminergic neurons of the substantia nigra. [10]

parthenogenesis "Virgin birth," a process of reproduction by which a female produces live offspring without need of a male. [13]

partial agonist A drug that, when bound to a receptor, has less effect than the endogenous ligand would. [3]

parvocellular Referring to relatively small cells. [8]

patch clamping The use of voltage clamping to monitor current flow across a tiny patch of membrane taken from a cell. [2]

patient H.M. A person who, because of damage to medial temporal lobe structures, was unable to encode new declarative memories. Upon his death we learned his name was Henry Molaison. [16]

patient K.C. A person who sustained damage to the cortex that rendered him unable to retrieve autobiographical memories. Upon his death we learned that his name was Kent Cochran. [16]

patient N.A. A person who is unable to encode new declarative memories, because of damage to the dorsal thalamus and the mammillary bodies. [16]

pattern coding Coding of information in sensory systems based on the temporal pattern of action potentials. [9]

Pavlovian conditioning See *classical conditioning.*

PCP See *phencyclidine.*

PCR See *polymerase chain reaction.*

peptide A short string of amino acids. Longer strings of amino acids are called *proteins.* [App]

peptide hormones A class of hormones, each of which consists of a string of amino acids. If the string of amino acids is long enough, it may be called a *protein hormone.* [4]

peptide neurotransmitter Also called *neuropeptide.* A neurotransmitter consisting of a short chain of amino acids. [3]

perceptual load The immediate processing challenge presented by a stimulus. [17]

periaqueductal gray The neuronal body–rich region of the midbrain surrounding the cerebral aqueduct that connects the third and fourth ventricles. It is involved in pain perception. [3, 7, 13]

period The interval of time between two similar points of successive cycles, such as sunset to sunset. [15]

peripheral nervous system The portion of the nervous system that includes all the nerves and neurons outside the brain and spinal cord. [1]

peripheral spatial cuing task An attention task where a visual stimulus is preceded by a simple sensory stimulus (like a flash) that reflexively captures attention. [17]

perseverate To continue to show a behaviour repeatedly. [6, 17]

PET See *positron emission tomography.*

PGN See *paragigantocellular nucleus.*

phagocyte An immune system cell that engulfs invading molecules or microbes. [11]

phallus The clitoris or penis. [13]

pharmacodynamics Collective name for the factors that affect the relationship between a drug and its target receptors, such as affinity and efficacy. [3]

pharmacogenomics The identification of combinations of alleles affecting the metabolism and actions of drugs, in order to personalise drug treatments. [12]

pharmacokinetics Collective name for all the factors that affect the movement of a drug into, through, and out of the body. [3]

phase shift A shift in the activity of a biological rhythm, typically provided by a synchronising environmental stimulus. [15]

phasic receptor A receptor in which the frequency of action potentials drops rapidly as stimulation is maintained. [7]

phencyclidine (PCP) Also called *angel dust.* An anaesthetic agent that is also a psychedelic drug. PCP makes many people feel dissociated from themselves and their environment. [12]

phenotype The sum of an individual's physical characteristics at one particular time. [6]

phenylketonuria (PKU) An inherited disorder of protein metabolism in which the absence of an enzyme leads to a toxic buildup of certain compounds, causing intellectual disability. [6]

pheromone A chemical signal that is released outside the body of an animal and affects other members of the same species. [4, 9, 13]

phobic disorder An intense, irrational fear that becomes centreed on a specific object, activity, or situation that a person feels compelled to avoid. [12]

phoneme A sound that is produced for language. [18]

photon A quantum of electromagnetic energy in the range of wavelengths we call light. [8]

photopic system A system in the retina that operates at high levels of light, shows sensitivity to colour, and involves the cones. [8]

photoreceptor adaptation The tendency of rods and cones to adjust their light sensitivity to match ambient levels of illumination. [8]

photoreceptors Neural cells in the retina that respond to light. [8]

phrenology The disproved theory that bumps on the skull reflect enlargements of brain regions responsible for certain behavioural faculties. [0]

phylogeny The evolutionary history of a particular group of organisms. [5]

pia mater The innermost of the three meninges that surround the brain and spinal cord. [1]

Piezo A family of two proteins that respond to mechanical stretch by opening channels to let cations in to depolarise the cell. [7]

pineal gland A secretory gland in the brain midline that is the source of melatonin release. [4, 15]

pinna The external part of the ear. [9]

pitch A dimension of auditory experience in which sounds vary from low to high. [9]

pituitary gland Also called *hypophysis.* A small, complex endocrine gland located in a socket at the base of the skull. [4]

pituitary portal system See *hypophyseal portal system.*

pituitary stalk Also called *infundibulum.* A thin piece of tissue that connects the pituitary gland to the hypothalamus. [4]

PKU See *phenylketonuria.*

place cell A neuron within the hippocampus that selectively fires when the animal is in a particular location. [16]

place coding The encoding of sound frequency as a function of the location on the basilar membrane that is most stimulated by that sound. [9]

placebo A substance that is known to be ineffective or inert but that, when administered like a drug, can sometimes bring relief. [7]

planum temporale A region of superior temporal cortex adjacent to the primary auditory area. [18]

plegia Paralysis; the loss of the ability to move. [10]

polioviruses A class of viruses that destroy motor neurons of the spinal cord and brainstem. [10]

polygraph Also called a *lie detector*. A device that measures several bodily responses, such as heart rate and blood pressure. [11]

polymerase chain reaction (PCR) Also called *gene amplification*. A method for reproducing a particular RNA or DNA sequence manyfold, allowing amplification for sequencing or manipulating the sequence. [App]

polymodal Involving several sensory modalities. [7]

POMC neurons Neurons involved in the hypothalamic appetite control system, so named because they produce pro-opiomelanocortin (POMC) along with cocaine- and amphetamine-related transcript (CART). [14]

pons A portion of the metencephalon; part of the brainstem connecting midbrain to medulla. [1]

positive symptom In psychiatry, a behaviour that is gained in a disorder. Examples include hallucinations, delusions, and excited motor behaviour in schizophrenia. [12]

positron emission tomography (PET) A technique for examining brain function by combining tomography with injections of radioactive substances used by the brain. [1]

postcentral gyrus The strip of parietal cortex, just behind the central sulcus, that receives somatosensory information from the entire body. [1]

postcopulatory behaviour The final stage in mating behaviour. Species-specific postcopulatory behaviours include rolling (in the cat) and grooming (in the rat). [13]

posterior Also called *caudal*. In anatomy, towards the tail end of an organism. Compare *anterior*. [1]

posterior cerebral arteries Two large arteries, arising from the basilar artery, that provide blood to posterior aspects of the cerebral hemispheres, cerebellum, and brainstem. [1]

posterior pituitary Also called *neurohypophysis*. The rear division of the pituitary gland. [4]

postganglionic Literally, "after the ganglion." Referring to neurons in the autonomic nervous system that run from the autonomic ganglia to various targets in the body. [1]

postpartum depression A bout of depression that afflicts a woman around the time she gives birth. [12]

postsynaptic Referring to the region of a synapse that receives and responds to neurotransmitter. [1]

postsynaptic membrane The specialised membrane on the surface of the cell that receives information by responding to neurotransmitter from a presynaptic neuron. [1]

postsynaptic potential A local potential that is initiated by stimulation at a synapse, can vary in amplitude, and spreads passively across the cell membrane, decreasing in strength with time and distance. [2]

posttraumatic stress disorder (PTSD) Formerly called *combat fatigue, war neurosis*, or *shell shock*. A disorder in which memories of an unpleasant episode repeatedly plague the victim. [12, 16]

potassium ion (K^+) A potassium atom that carries a positive charge because it has lost one electron. [2]

pragmatics In linguistics, the context in which a speech sound is uttered. [18]

precentral gyrus The strip of frontal cortex, just in front of the central sulcus, that is crucial for motor control. [1]

precocial Referring to animals that are born in a comparatively developed state and that are able to survive with relatively little maternal care. [13]

prefrontal cortex The most anterior portion of the frontal lobe. [16, 17]

preganglionic Literally, "before the ganglion." Referring to neurons in the autonomic nervous system that run from the central nervous system to the autonomic ganglia. [1]

premotor cortex A region of nonprimary motor cortex just anterior to the primary motor cortex. [10]

presenilin An enzyme that cleaves amyloid precursor protein, forming β-amyloid. [6]

prestin A motor protein that allows outer hair cells to change length. [9]

presynaptic membrane The specialised membrane of the axon terminal of the neuron that transmits information by releasing neurotransmitter. [1]

presynaptic Referring to the region of a synapse that releases neurotransmitter. [1]

primacy effect The superior performance seen in a memory task for items at the start of a list. It is usually attributed to long-term memory. [16]

primary auditory cortex (A1) The region of superior temporal cortex in which auditory processing occurs. [9]

primary motor cortex (M1) The executive region for the initiation of movement; primarily the precentral gyrus. [10]

primary sensory cortex For a given sensory modality, the region of cortex that receives most of the information about that modality from the thalamus or, in the case of olfaction, directly from the secondary sensory neurons. [7]

primary sensory ending Also called *annulospiral ending*. The axon that transmits information from the central portion of a muscle spindle. [10]

primary somatosensory cortex (S1) Also called *somatosensory 1*. The gyrus just posterior to the central sulcus, in the parietal lobe, where sensory receptors on the body surface are mapped; primary cortex for receiving touch and pain information. [7]

primary visual cortex (V1) Also called *striate cortex* or *area 17*. The region of the occipital lobe where most visual information first arrives in the cortex. [8]

priming Also called *repetition priming*. The phenomenon by which exposure to a stimulus facilitates subsequent responses to the same or a similar stimulus. [16]

prion A protein that can become improperly folded and thereby become an infectious agent, spreading diseases such as bovine spongiform encephalopathy. [10]

probe A manufactured sequence of DNA or RNA that is made to include a label (a colourful or radioactive molecule) that lets us track its location. [App]

procedural memory See *nondeclarative memory*.

proceptive Referring to a state in which an animal advertises its readiness to mate through species-typical behaviours, such as ear wiggling in the female rat. [13]

progesterone The primary type of progestin secreted by the ovary. [4]

progestins A major class of steroid hormones, including progesterone, that are produced by the ovary. [4]

prolactin A peptide hormone, produced by the anterior pituitary, that promotes mammary development for lactation in female mammals. [4]

proprioception Body sense; information about the position and movement of the body that is sent to the brain. [10]

prosencephalon See *forebrain*.

prosody The perception of emotional tone-of-voice aspects of language. [18]

prosopagnosia Also called *face blindness*. A condition characterised by the inability to recognise faces. *Acquired prosopagnosia* is caused by damage to the brain, particularly the fusiform gyrus. *Developmental* (or *congenital*) *prosopagnosia* is present from birth. [18]

protein A long string of amino acids; the basic building material of organisms. [App]

protein kinase An enzyme that adds phosphate groups (PO_3) to protein molecules, altering the protein's function. [16]

proximal In anatomy, near the trunk or centre of an organism. Compare *distal*. [1]

psychoneuroimmunology The study of the immune system and its interaction with the nervous system and behaviour. [11]

psychopathy A cluster of personality traits associated with superficial charm, lying, and diminished remorse. [11]

psychopharmacology See *neuropharmacology*.

psychosocial dwarfism Reduced stature caused by stress early in life that inhibits growth. [4]

psychosomatic medicine A field of study that emphasises the role of psychological factors in disease. [11]

psychosurgery Surgery in which brain lesions are produced in an attempt to alleviate severe psychiatric disorders. [12]

psychotomimetic A drug that induces a state resembling schizophrenia. [12]

PTSD See *posttraumatic stress disorder*.

pulvinar In humans, the posterior portion of the thalamus, heavily involved in visual processing and direction of attention. [17]

punch-drunk syndrome See *chronic traumatic encephalopathy*.

pupil The aperture, formed by the iris, that allows light to enter the eye. [8]

pure tone A tone with a single frequency of vibration. [9]

Purkinje cell A type of large nerve cell in the cerebellar cortex. [1]

putamen One of the basal ganglia. [1]

PVN See *paraventricular nucleus*.

pyramidal cell A type of large nerve cell that has a roughly pyramid-shaped cell body. It is found in the cerebral cortex. [1]

pyramidal system Also called *corticospinal system*. The motor system that includes neurons within the cerebral cortex that send axons to form the pyramidal tract. [10]

PYY_{3-36} A peptide gut hormone believed to act on the hypothalamus to suppress appetite. [14]

Q

qualia Singular *quale*. Purely subjective experiences of perception. [17]

R

radial glial cells Glial cells that form early in development, spanning the width of the emerging cerebral hemispheres, and guide migrating neurons. [6]

radioimmunoassay (RIA) A technique that uses antibodies to measure the concentration of a substance, such as a hormone, in blood. [4]

range fractionation A hypothesis of stimulus intensity perception stating that a wide range of intensity values can be encoded by a group of cells, each of which is a specialist for a particular range of stimulus intensities. [7, 8]

raphe nuclei A string of nuclei in the midline of the midbrain and brainstem that contain most of the serotonergic neurons of the brain. [3]

rapid-eye-movement (REM) sleep Also called *paradoxical sleep*. A stage of sleep characterised by small-amplitude, fast-EEG waves, no postural tension, and rapid eye movements. *REM* rhymes with "gem." [15]

RBD See *REM behaviour disorder*.

recency effect The superior performance seen in a memory task for items at the end of a list. It is attributed to short-term memory. [16]

receptive field The stimulus region and features that affect the activity of a cell in a sensory system. [7, 8]

receptor Also called *receptor molecule*. A protein that binds and reacts to molecules of a neurotransmitter or hormone. [1, 3]

receptor cell A specialised cell that responds to a particular energy or substance in the internal or external environment and converts energy into a change in the electrical potential across its membrane. [7]

receptor isoform A subtype of a receptor protein whose slight differences in structure give it different functional properties.

receptor molecule Also called *receptor*. A protein that binds and reacts to molecules of a neurotransmitter or hormone. [2]

receptor potential Also called *generator potential*. A local change in the resting potential of a receptor cell that mediates between the impact of stimuli and the initiation of action potentials. [7]

receptor subtype Any type of receptor having functional characteristics that distinguish it from other types of receptors for the same neurotransmitter. [3]

reconsolidation The process by which a retrieved memory may be strengthened or altered before being returned to long-term memory. [16]

recovery of function The recovery of behavioural capacity following brain damage from stroke or injury. [18]

red nucleus A brainstem structure related to motor control. [1, 10]

reductionism The scientific strategy of breaking a system down into increasingly smaller parts in order to understand it. [0]

reflex A simple, highly stereotyped, and unlearned response to a particular stimulus (e.g., an eye blink in response to a puff of air). [10]

reflexive attention Also called *exogenously controlled attention* or *bottom-up attention*. The involuntary reorienting of attention towards the location of an unexpected object or event. [17]

refraction The bending of light rays by a change in the density of a medium, such as the cornea and the lens of the eye. [8]

refractory Referring to transiently inactivated or exhausted axonal membrane. [2]

refractory period A period following copulation during which an individual cannot recommence copulation. [13]

regulation An adaptive response to early injury, as when developing individual cells compensate for missing or injured cells. [6]

relative refractory phase A period of reduced sensitivity during which only strong stimulation produces an action potential. [2]

releasing hormones A class of hormones, produced in the hypothalamus, that traverse the hypothalamic-pituitary portal system to control the pituitary's release of tropic hormones. [4]

REM behaviour disorder (RBD) A sleep disorder in which a person physically acts out a dream. [15]

REM sleep See *rapid-eye-movement (REM) sleep*.

repetition priming See *priming*.

reserpine A drug that causes the depletion of monoamines and can lead to depression. [12]

resting membrane potential A difference in electrical potential across the membrane of a nerve cell during an inactive period. [2]

reticular formation An extensive region of the brainstem (extending from the medulla through the thalamus) that is involved in arousal (waking) and motor control. [1, 10, 15]

reticulospinal tract A tract of axons arising from the brainstem reticular formation and descending to the spinal cord to modulate movement. [10, 13]

retina The receptive surface inside the eye that contains photoreceptors and other neurons. [8]

RETINAL Abbreviation for *retinaldehyde*, also called *vitamin A aldehyde*. One of the two components of photopigments in the retina. [8]

retinaldehyde See *RETINAL*.

retinohypothalamic pathway The projection of retinal ganglion cells to the suprachiasmatic nuclei. [15]

retrieval A process in memory during which a stored memory is used by an organism. [16]

retrograde amnesia Difficulty in retrieving memories formed before the onset of amnesia. [16]

retrograde degeneration Destruction of the nerve cell body following injury to its axon. [6]

retrograde messenger Transmitter that is released by the postsynaptic region, travels back across the synapse, and alters the functioning of the presynaptic neuron. [16]

retrograde synapse A synapse in which a signal (usually a gas neurotransmitter) flows from the postsynaptic neuron to the presynaptic neuron, thus counter to the usual direction of synaptic communication. [2]

retrograde transmitter A neurotransmitter that diffuses from the postsynaptic neuron back to the presynaptic neuron. [3]

Rett syndrome A rare genetic disorder, occurring almost exclusively in girls, of slowing development resulting in intellectual disability, stereotyped movements, and loss of language. [6]

reuptake The process by which released synaptic transmitter molecules are taken up and reused by the presynaptic neuron, thus stopping synaptic activity. [2]

rhodopsin The photopigment in rods that responds to light. [8]

rhombencephalon See *hindbrain*.

RIA See *radioimmunoassay*.

ribonucleic acid (RNA) A nucleic acid that implements information found in DNA. [App]

ribosomes Structures in the cell body where genetic information is translated to produce proteins. [1, App]

rods A class of light-sensitive receptor cells (photoreceptors) in the retina that are most active at low levels of light. [8]

rostral See *anterior*.

round window A membrane separating the cochlear duct from the middle-ear cavity. [9]

rubrospinal tract A tract of axons arising from the red nucleus in the midbrain and innervating neurons of the spinal cord. [10]

Ruffini's ending A skin receptor cell type that detects stretching of the skin. [7]

S

S1 See *primary somatosensory cortex*.

S2 See *secondary somatosensory cortex*.

saccades Fast movements of the eyes that present various parts of the visual scene to the fovea. When we fix our gaze, small saccades that we are unaware of avert photoreceptor adaptation. [8]

saccule A small, fluid-filled sac under the utricle in the vestibular system that responds to static positions of the head. [9]

sacral Referring to the five spinal segments that make up the lower part of the lower back. [1]

SAD See *seasonal affective disorder*.

sagittal plane The plane that bisects the body or brain into right and left portions. Compare *coronal plane* and *horizontal plane*. [1]

saltatory conduction The form of conduction that is characteristic of myelinated axons, in which the action potential jumps from one node of Ranvier to the next. [2]

satiety The feeling that an appetite has been satisfied. [14]

saxitoxin (STX) An animal toxin that blocks sodium channels when applied to the outer surface of the cell membrane. [2]

SC See *standard condition*.

scala media Also called *middle canal*. The central of the three canals running the length of the cochlea, situated between the scala vestibuli and the scala tympani. [9]

scala tympani Also called *tympanic canal*. One of three canals running the length of the cochlea. [9]

scala vestibuli Also called *vestibular canal*. One of the three canals running the length of the cochlea. [9]

schizophrenia A severe psychopathology characterised by negative symptoms such as emotional withdrawal and impoverished thought and also by positive symptoms such as hallucinations and delusions. [12]

Schwann cell A glial cell that forms myelin in the peripheral nervous system. [1]

SCN See *suprachiasmatic nucleus*.

SCN9A See *Na_V1.6*.

scotoma A region of blindness caused by injury to the visual pathway or brain. [8]

scotopic system A system in the retina that operates at low levels of light and involves the rods. [8]

SDN-POA See *sexually dimorphic nucleus of the preoptic area*.

seasonal affective disorder (SAD) A depression brought about by the short days of winter. [12]

second messenger A slow-acting substance in a target cell that amplifies the effects of synaptic or hormonal activity and regulates activity within the target cell. [2, 4]

secondary sensory cortex Also called *nonprimary sensory cortex*. For a given sensory modality, the cortical regions receiving direct projections from primary sensory cortex for that modality. [7]

secondary sensory ending Also called *flower spray ending*. The axon that transmits information from the ends of a muscle spindle. [10]

secondary somatosensory cortex (S2) Also called *somatosensory 2*. The region of cortex that receives direct projections from primary somatosensory cortex. [7]

second-generation antipsychotics A class of antischizophrenic drugs that block dopamine receptors (as first-generation antipsychotics do) and also act at other receptors. [3, 12]

seizure An epileptic episode. [2]

selective attention See *attention*.

selective permeability The property of a membrane that allows some substances to pass through, but not others. [2]

selective serotonin reuptake inhibitors (SSRIs) A class of drugs that block the reuptake of transmitter at serotonergic synapses. They are commonly used to treat depression. [3, 12]

semantic memory Generalised memory—for instance, knowing the meaning of a word without knowing where or when you learned that word. [16]

semantics The meanings or interpretation of words and sentences in a language. [18]

semen A mixture of fluids and sperm that is released during ejaculation. [13]

semicircular canal One of the three fluid-filled tubes in the inner ear that are part of the vestibular system. Each of the tubes, which are at right angles to each other, detects angular acceleration. [9]

senile dementia A neurological disorder of the aged that is characterised by progressive behavioural deterioration, including personality change and profound intellectual decline. It includes, but is not limited to, Alzheimer's disease. [6]

senile plaques Also called *amyloid plaques*. Small areas of the brain that have abnormal cellular and chemical patterns. Senile plaques correlate with senile dementia. [6]

sensitive period The period during development in which an organism can be permanently altered by a particular experience or treatment. [6, 13, 18]

sensitisation 1. A process in which the body shows an enhanced response to a given drug after repeated doses. 2. A form of nonassociative learning in which an organism becomes more responsive to most stimuli after being exposed to unusually strong or painful stimulation. [3, 16]

sensorineural deafness A hearing impairment that originates from cochlear or auditory nerve lesions. [9]

sensory buffer An element of the type of memory that stores the sensory impression of a scene. [16]

sensory conflict theory A theory of motion sickness suggesting that discrepancies between vestibular information and visual information simulate food poisoning and therefore trigger nausea. [9]

sensory nerve A nerve that conveys sensory information from the periphery into the central nervous system. [1]

sensory neuron A neuron that is directly affected by changes in the environment, such as light, odour, or touch. [1]

sensory pathway The chain of neural connections from sensory receptor cells to the cortex. [7]

sensory receptor organ An organ (such as the eye or ear) specialised to receive particular stimuli. [7]

sensory transduction The process in which a receptor cell converts the energy in a stimulus into a change in the electrical potential across its membrane. [7]

septal nuclei A collection of gray matter structures lying medially below the corpus callosum, implicated in the perception of reward. [1]

serotonergic Referring to neurons that use serotonin as their synaptic transmitter. [3]

serotonin (5-HT) A synaptic transmitter that is produced in the raphe nuclei and is active in structures throughout the cerebral hemispheres. [3]

serotonin syndrome Syndrome of confusion, muscle spasms, and fever that may occur when brain levels of serotonin are too high and is a risk of taking SSRIs. [12]

serotonin-norepinephrine reuptake inhibitor (SNRI) A drug that blocks the reuptake of transmitter at both serotonergic and noradrenergic synapses. [3, 12]

set point The point of reference in a feedback system. An example is the setting of a thermostat. [14]

set zone The range of a variable that a feedback system tries to maintain. [14]

sex determination The process by which the initial decision is made for a fetus to develop as a male or a female. [13]

sex steroids Steroid hormones secreted by the gonads: androgens, estrogens, and progestins. [4]

sexual attraction The first step in the mating behaviour of many animals, in which animals produce stimuli to attract sexual partners. [13]

sexual differentiation The process by which individuals develop male-like or female-like bodies and behaviour. [13]

sexual dimorphism The condition in which males and females show pronounced sex differences in appearance. [13]

sexual selection A form of evolution through natural selection in which members of one sex favour specific heritable traits in the other sex when choosing a reproductive partner. [5]

sexually dimorphic nucleus of the preoptic area (SDN-POA) A region of the preoptic area that is 5 to 6 times larger in volume in male rats than in females. [13]

sexually receptive Referring to the state in which an individual (in mammals, typically the female) is willing to copulate. [13]

SFO See *subfornical organ.*

shadowing A task in which the participant is asked to focus attention on one ear or the other while stimuli are being presented separately to both ears. [17]

sham rage See *decorticate rage.*

short-term memory (STM) A form of memory that usually lasts only for seconds, or as long as rehearsal continues, especially while being used during performance of a task. [16]

SIDS See *sudden infant death syndrome.*

simple cortical cell Also called *bar detector* or *edge detector.* A cell in the visual cortex that responds best to an edge or a bar that has a particular width, as well as a particular orientation and location in the visual field. [8]

simple partial seizure Also called *absence attack.* A seizure that is characterised by a spike-and-wave EEG and often involves a loss of awareness and inability to recall events surrounding the seizure. [2]

simultagnosia A profound restriction of attention, often limited to a single item or feature. [17]

single-nucleotide polymorphism (SNP) Variation in a single base pair within a longer DNA sequence, such as a gene. [5]

skill learning Learning to perform a task that requires motor coordination. [16]

sleep apnea A sleep disorder in which respiration slows or stops periodically, waking the person. Excessive daytime sleepiness results from the frequent nocturnal awakening. [15]

sleep cycle A period of slow-wave sleep followed by a period of REM sleep. In humans, a sleep cycle lasts 90–110 minutes. [15]

sleep deprivation The partial or total prevention of sleep. [15]

sleep enuresis Bed-wetting. [15]

sleep hygiene Habits, such as avoiding caffeine shortly before bedtime, that promote healthy sleep. [15]

sleep paralysis A state during the transition to or from sleep, in which the ability to move or talk is temporarily lost. [15]

sleep recovery The process of sleeping more than is normal, after a period of sleep deprivation, as though in compensation. [15]

sleep spindle A characteristic 13–15 Hz wave in the EEG of a person in stage 2 sleep. [15]

sleep state misperception Commonly, a person's perception that he has not been asleep when in fact he has. It typically occurs at the start of a sleep episode. [15]

sleep-maintenance insomnia Difficulty in staying asleep. [15]

sleep-onset insomnia Difficulty in falling asleep. [15]

slow-twitch muscle fibre A type of striated muscle fibre that contracts slowly but does not fatigue readily. [10]

slow-wave sleep (SWS) Also called *NREM 3* and, in humans, *stage 3 sleep.* A stage of NREM sleep characterised by large-amplitude delta waves. [15]

SMA See *supplementary motor area.*

smooth muscle A type of muscle fibre, as in the heart, that is controlled by the autonomic nervous system rather than by voluntary control. [10]

SNB See *spinal nucleus of the bulbocavernosus.*

SNRI See *serotonin-norepinephrine reuptake inhibitor.*

social neuroscience A field of study devoted to uncovering brain mechanisms and patterns of neural activity that are both cause and consequence of social behaviours such as affiliation, competition, cooperation, etc. [1]

sodium ion (Na^+) A sodium atom that carries a positive charge because it has lost one electron. [2]

sodium-potassium pump The energetically expensive mechanism that pushes sodium ions out of a cell, and potassium ions in. [2]

soma See *cell body.*

somatic intervention An approach to finding relations between body variables and behavioural variables that involves manipulating body structure or function and looking for resultant changes in behaviour. [0]

somatic nerve See *spinal nerve.*

somatic nervous system The part of the peripheral nervous system that provides neural connections to the skeletal musculature. [1]

somatomedins A group of proteins, released from the liver in response to growth hormone, that aid body growth and maintenance. [4]

somatosensory Referring to body sensation, particularly touch and pain sensation. [7]

somatosensory 1 See *primary somatosensory cortex.*

somatosensory 2 See *secondary somatosensory cortex.*

somatotropic hormone See *growth hormone.*

somatotropin See *growth hormone.*

somnambulism Sleepwalking. [15]

Southern blot A method of detecting a particular DNA sequence in the genome of an organism, by separating DNA with gel electrophoresis, blotting the separated DNAs onto nitrocellulose, and then using a nucleotide probe to hybridise with, and highlight, the gene of interest. [App]

spasticity Markedly increased rigidity in response to forced movement of the limbs. [10]

spatial cognition Those mental processes that deal with the spatial relationships among objects. [18]

spatial resolution The ability to observe the detailed structure of the brain. [17]

spatial summation The summation at the axon hillock of postsynaptic potentials from across the cell body. If this summation reaches threshold, an action potential is triggered. [2]

spatial-frequency filter model A model of pattern analysis that emphasises Fourier analysis of visual stimuli. [8]

species A group of individuals that can readily interbreed to produce fertile offspring. [5]

specific nerve energies The doctrine that the receptors and neural channels for the different senses are independent and operate in their own special ways and can produce only one particular sensation each. [7]

spectral filtering Alteration of the amplitude of some, but not all, frequencies in a sound. [9]

spectrally opponent cell A visual receptor cell that has opposite firing responses to different regions of the spectrum. [8]

sperm The gamete produced by males for fertilisation of eggs (ova). [13]

spinal animal An animal whose spinal cord has been surgically disconnected from the brain to enable the study of behaviours that do not require brain control. [10]

spinal nerve Also called *somatic nerve.* A nerve that emerges from the spinal cord. [1]

spinal nucleus of the bulbocavernosus (SNB) A group of motor neurons in the spinal cord of rats that innervate striated muscles controlling the penis. [13]

spinocerebellum The central part of the cerebellum, consisting mostly of the vermis and anterior lobe. [10]

spinothalamic system See *anterolateral system.*

split-brain individual An individual whose corpus callosum has been severed, halting communication between the right and left hemispheres. [18]

***SRY* gene** A gene on the Y chromosome that directs the developing gonads to become testes. The name *SRY* stands for **s**ex-determining **r**egion on the **Y** chromosome. [13]

SSRIs See *selective serotonin reuptake inhibitors.*

stabilising selection Selection that favours reduced variation in a characteristic within a population. [5]

stage 1 sleep Also called *NREM 1.* The initial stage of NREM sleep, which is characterised by small-amplitude EEG waves of irregular frequency, slow heart rate, and reduced muscle tension. [15]

stage 2 sleep Also called *NREM 2.* A stage of NREM sleep that is defined by bursts of EEG waves called *sleep spindles.* [15]

stage 3 sleep Also called *NREM 3.* A stage of NREM sleep that is defined by the presence of large-amplitude, very slow waves (delta waves). [15]

standard condition (SC) The usual environment for laboratory rodents, with a few animals in a cage and adequate food and water but no complex stimulation. [16]

stapedius A middle-ear muscle that is attached to the stapes. [9]

stapes Latin for "stirrup." A middle-ear bone that is connected to the oval window. [9]

stem cell A cell that is undifferentiated and therefore can take on the fate of any cell that a donor organism can produce. [6]

stereocilium A relatively stiff hair that protrudes from a hair cell in the auditory or vestibular system. [9]

steroid hormones A class of hormones, each of which is composed of four interconnected rings of carbon atoms. [4]

steroid receptor cofactors Proteins that affect the cell's response when a steroid hormone binds its receptor. [4]

stimulus A physical event that triggers a sensory response. [7]

stimulus cuing A testing technique in which a cue to the stimulus location is provided before the stimulus itself. [17]

STM See *short-term memory.*

stop codon A trio of nucleotides in DNA to mark the end of transcription. [App]

stress Any circumstance that upsets homeostatic balance. [11]

stress immunisation The concept that mild stress early in life makes an individual better able to handle stress later in life. [11]

stretch reflex The contraction of a muscle in response to stretch of that muscle. [10]

stria terminalis A limbic axonal pathway connecting the amygdala and hypothalamus. [1]

striate cortex See *primary visual cortex.*

striated muscle A type of muscle with a striped appearance, generally under voluntary control. [10]

striatum The caudate nucleus and putamen together. [3, 10]

stroke Damage to a region of brain tissue that results from blockage or rupture of vessels that supply blood to that region. [1]

stuttering Disordered articulation of speech sounds and involuntary repetition of words and phrases, resulting in halting and effortful speech. [18]

STX See *saxitoxin.*

subcoeruleus Also called *sublaterodoursal nucleus*. A brain region just ventral to the locus coeruleus that is associated with REM sleep. [15]

subcutaneous tissue See *hypodermis*.

subfornical organ (SFO) One of the circumventricular organs. [14]

sublaterodoursal nucleus See *subcoeruleus*.

substance abuse A maladaptive pattern of substance use that has lasted more than a month but does not fully meet the criteria for dependence. [3]

substance P A peptide transmitter implicated in pain transmission. [7]

substantia nigra A brainstem structure in humans that innervates the basal ganglia and is named for its dark pigmentation. [1, 3, 10]

subthalamic nucleus A nucleus just ventral to the thalamus that interacts with the basal ganglia. It is a favoured site for deep brain stimulation to treat Parkinson's disease. [10]

sudden infant death syndrome (SIDS) Also called *crib death*. The sudden, unexpected death of an apparently healthy human infant who simply stops breathing, usually during sleep. [15]

sulcus A furrow of a convoluted brain surface. [1]

superior In anatomy, pertaining to the higher of two locations. Compare *inferior*. [1]

superior colliculi Paired gray matter structures of the dorsal midbrain that receive visual information and are involved in direction of visual gaze and visual attention to intended stimuli. [1, 17]

superior olivary nuclei Brainstem nuclei that receive input from both right and left cochlear nuclei and provide the first binaural analysis of auditory information. [9]

supersensitivity psychosis An exaggerated psychosis that may emerge when doses of antipsychotic medication are reduced, probably as a consequence of the up-regulation of receptors that occurred during drug treatment. [12]

supplementary motor area (SMA) A region of nonprimary motor cortex that receives input from the basal ganglia and modulates the activity of the primary motor cortex. [10]

suprachiasmatic nucleus (SCN) A small region of the hypothalamus above the optic chiasm that is the location of a circadian oscillator. [15]

supraoptic nucleus A hypothalamic nucleus containing neuroendocrine cells that send axons to the posterior pituitary to release oxytocin or vasopressin. [4]

surface dyslexia Acquired dyslexia in which the person seems to attend only to the fine details of reading. [18]

sustained-attention task A task in which a single stimulus source or location must be held in the attentional spotlight for a protracted period. [17]

SWS See *slow-wave sleep*.

Sylvian fissure Also called *lateral sulcus*. A deep fissure that demarcates the temporal lobe. [1]

symbolic cuing task An attention task in which a symbol indicates to the participant where to voluntarily direct attention in order to detect a stimulus. [17]

sympathetic chain A chain of ganglia that runs along each side of the spinal column. It is part of the sympathetic nervous system. [1]

sympathetic nervous system A component of the autonomic nervous system that arises from the thoracic and lumbar spinal cord. [1, 11]

synapse The tiny gap between neurons where information is passed from one to the other. [1]

synapse rearrangement Also called *synaptic remodelling*. The loss of some synapses and the development of others; a refinement of synaptic connections that is often seen in development. [6]

synaptic bouton See *axon terminal*.

synaptic cleft The space between the presynaptic and postsynaptic membranes. [1]

synaptic delay The brief delay between the arrival of an action potential at the axon terminal and the creation of a postsynaptic potential. [2]

synaptic remodelling See *synapse rearrangement*.

synaptic transmitter See *neurotransmitter*.

synaptic vesicle A small, spherical structure that contains molecules of neurotransmitter. [1]

synaptogenesis The establishment of synaptic connections as axons and dendrites grow. [6]

synaptotagmin A specialised protein that responds to calcium ions to trigger vesicular exocytosis. [2]

synergist With respect to muscles, a muscle that acts together with another muscle. [10]

synesthesia A condition in which stimuli in one modality evoke the involuntary experience of an additional sensation in another modality. [7]

syntax The grammatical rules for constructing phrases and sentences in a language. [18]

syrinx The vocal organ in birds. [13]

T

T cells See *T lymphocytes*.

T lymphocytes Also called *T cells*. Immune system cells, formed in the thymus (hence the *T*), that include killer T cells, which attack foreign microbes, and helper T cells, which secrete cytokines. [11]

T1R A family of taste receptor proteins that, when particular members heterodimerise, form taste receptors for sweet flavours and umami flavours. [9]

T2R A family of bitter taste receptors. [9]

TAARs See *trace amine–associated receptors*.

tachistoscope test A test in which stimuli are very briefly exposed in either the left or right visual half field. [18]

tactile Of or relating to touch. [7]

tardive dyskinesia A disorder characterised by involuntary movements, especially involving the face, mouth, lips, and tongue. It can be caused by prolonged use of antipsychotic drugs, such as chlorpromazine. [12]

task switching The shifting of attention and behaviour from one task to a new one. [17]

tastant A substance that can be tasted. [9]

taste bud A cluster of 50–150 cells that detects tastes. Taste buds are found in papillae. [9]

taste pore The small aperture through which tastant molecules are able to access the sensory receptors of the taste bud. [9]

Tau A protein associated with neurofibrillary tangles in Alzheimer's disease. [6, 18]

taxonomy The classification of organisms. [5]

tectorial membrane A membrane that sits atop the organ of Corti in the cochlear duct. [9]

tectum The dorsal portion of the midbrain, including the inferior and superior colliculi. [1]

telencephalon The frontal subdivision of the forebrain that includes the cerebral hemispheres when fully developed. [1]

temporal coding The encoding of sound frequency in terms of the number of action potentials per second produced by an auditory nerve. [9]

temporal lobe Large lateral cortical region of each cerebral hemisphere, continuous with the parietal lobe posteriorly and separated from the frontal lobe by the Sylvian fissure. [1]

temporal resolution The ability of an imaging technique to track changes in the brain over time. [17]

temporal summation The summation of postsynaptic potentials that reach the axon hillock at different times. The closer in time the potentials occur, the more complete the summation. [2]

temporoparietal junction (TPJ) The point in the brain where the temporal and parietal lobes meet that plays a role in shifting attention to a new location after target onset. [17]

tendon Strong tissue that connects muscle to bone. [10]

tensor tympani The muscle attached to the malleus that modulates mechanical linkage to protect the delicate receptor cells of the inner ear from damaging sounds. [9]

teratogen Substances that induce malformations in development. [6]

teratology The study of abnormal development. [6]

testes The male gonads, which produce sperm and androgenic steroid hormones. [4]

testosterone A hormone, produced by male gonads, that controls a variety of bodily changes that become visible at puberty. [4]

tetanus toxin A toxin that cleaves SNAREs, disabling neurotransmitter release. [2]

tetanus An intense volley of action potentials. [16]

tetrodotoxin (TTX) A toxin from puffer fish ovaries that blocks the voltage-gated sodium channel, preventing action potential conduction. [2]

thalamus The brain regions at the top of the brainstem that trade information with the cortex. [1, 7]

THC See *Δ9-tetrahydrocannabinol.*

thermogenin Also called *UCP1*. A specialised protein that allows mitochondria to turn energy directly into heat. [14]

thermoregulation The active process of closely regulating body temperature around a set value. [14]

third ventricle The midline ventricle that conducts cerebrospinal fluid from the lateral ventricles to the fourth ventricle. [1]

thoracic Referring to the 13 spinal segments below the cervical (neck) portion of the spinal cord, corresponding to the chest. [1]

threshold The stimulus intensity that is just adequate to trigger an action potential. [2, 7]

thyroid gland An endocrine gland, located in the throat, that regulates cellular metabolism throughout the body. [4]

thyroid hormones Two hormones, triiodothyronine and thyroxine (also called *tetraiodothyronine*), that are released from the thyroid gland and have widespread effects, including growth and maintenance of the brain. [4]

thyroid-stimulating hormone (TSH) A tropic hormone, released by the anterior pituitary gland, that signals the thyroid gland to secrete its hormones. [4]

thyrotropin-releasing hormone (TRH) A hypothalamic hormone that regulates the release of thyroid-stimulating hormone from the anterior pituitary. [4]

timbre The characteristic sound quality of a musical instrument, as determined by the relative intensities of its various harmonics. [9]

tinnitus A sensation of noises or ringing in the ears. [9]

tip link A fine, threadlike fibre that runs along and connects the tips of stereocilia. [9]

TMS See *transcranial magnetic stimulation.*

tolerance A condition in which, with repeated exposure to a drug, an individual becomes less responsive to a constant dose. [3]

tonic receptor A receptor in which the frequency of action potentials declines slowly or not at all as stimulation is maintained. [7]

tonic-clonic seizure Formerly called *grand mal seizure*. A type of generalised epileptic seizure in which nerve cells fire in high-frequency bursts. [2]

tonotopic organisation A major organisational feature in auditory systems in which neurons are arranged as an orderly map of stimulus frequency, with cells responsive to high frequencies located at a distance from those responsive to low frequencies. [9]

top-down attention See *voluntary attention.*

top-down process A process in which higher-order cognitive processes control lower-order systems, often reflecting conscious control. [7, 17]

Tourette's syndrome A heightened sensitivity to tactile, auditory, and visual stimuli that may be accompanied by the buildup of an urge to emit verbal or phonic tics. [12]

TPJ See *temporoparietal junction.*

trace amine–associated receptors (TAARs) A family of probable pheromone receptors produced by neurons in the main olfactory epithelium. [9]

tract A bundle of axons found within the central nervous system. [1]

tract tracer A compound used to identify the routes and interconnections of neuronal projections. [1]

transcranial magnetic stimulation (TMS) Localised, noninvasive stimulation of cortical neurons through the application of strong magnetic fields. [1]

transcript The mRNA strand that is produced when a stretch of DNA is "read." [App]

transcription The process during which mRNA forms bases complementary to a strand of DNA. The resulting message (called a *transcript*) is then used to translate the DNA code into protein molecules. [App]

transcription factor A substance that binds to recognition sites on DNA and alters the rate of expression of particular genes. [4]

transduction The conversion of one form of energy to another. [9]

transgenic Referring to an animal in which a new or altered gene has been deliberately introduced into the genome. [6, App]

transient receptor potential type M3 (TRPM3) A receptor, found in some free nerve endings, that opens its channel in response to rising temperatures. [7]

transient receptor potential vanilloid type 1 (TRPV1) Also called *vanilloid receptor 1*. A receptor that binds capsaicin to transmit the burning sensation from chili peppers and normally detects sudden increases in temperature. [7]

translation The process by which amino acids are linked together (directed by an mRNA molecule) to form protein molecules. [App]

transmitter See *neurotransmitter.*

transmitter reuptake The reabsorption of synaptic transmitter by the axon terminal from which it was released. [3]

transporter Specialised receptor in the presynaptic membrane that recognises neurotransmitter molecules and returns them to the presynaptic neuron for reuse. [2, 3]

transverse plane See *coronal plane.*

treble An aspect of pitch corresponding to the subjective experience of high-frequency sounds (especially musical sounds, such as cymbals). [9]

TRH See *thyrotropin-releasing hormone.*

trichromatic hypothesis A hypothesis of colour perception stating that there are three different types of cones, each excited by a different region of the spectrum and each having a separate pathway to the brain. [8]

tricyclics A class of drugs that act by increasing the synaptic accumulation of serotonin and norepinephrine. [3, 12]

trigeminal nerve Cranial nerve V, which receives information from the face and controls jaw musculature. [11]

trinucleotide repeat Repetition of the same three nucleotides within a gene, which can lead to dysfunction, as in the cases of Huntington's disease and fragile X syndrome. [6, 10]

trophic factor A substance that promotes cell growth and survival. [14]

tropic hormones A class of anterior pituitary hormones that affect the secretion of other endocrine glands. [4]

TRPM3 See *transient receptor potential type M2.*

TRPM8 Also called *cool-menthol receptor 1* (*CMR1*). A sensory receptor, found in some free nerve endings, that opens an ion channel in response to a mild temperature drop or exposure to menthol. [7]

TRPV1 See *transient receptor potential vanilloid type 1.*

TSH See *thyroid-stimulating hormone.*

t-SNARE A specialised protein anchored to the presynaptic "target" membrane to bind v-SNAREs to dock vesicles, making them ready for release. [2]

TTX See *tetrodotoxin.*

tuberomammillary nucleus A region of the basal hypothalamus, near the pituitary stalk, that plays a role in generating SWS. [15]

tuning curve A graph of the responses of a single auditory nerve fibre or neuron to sounds that vary in frequency and intensity. [9]

Turner's syndrome A condition seen in individuals carrying a single X chromosome but no other sex chromosome. [13]

two-photon excitation microscopy Method of providing many low-energy photons that can penetrate deep into tissues, such that the simultaneous arrival of two photons at a fluorescent molecule is sufficient to elicit a visible photon in response. [11]

tympanic canal See *scala tympani.*

tympanic membrane Also called *eardrum.* The partition between the external ear and the middle ear. [9]

U

UCP1 See *thermogenin.*

ultradian Referring to a rhythmic biological event whose period is shorter than that of a circadian rhythm, usually from several minutes to several hours long. [15]

ultrasound High-frequency sound; in general, sound above the threshold for human hearing, at about 20,000 Hz. [9]

umami One of the five basic tastes (along with salty, sour, sweet, and bitter), probably mediated by amino acids in foods. [9]

unipolar neuron Also called *monopolar neuron.* A nerve cell with a single branch that leaves the cell body and then extends in two directions; one end is the receptive pole, the other end the output zone. [1]

up-regulation A compensatory increase in receptor availability at the synapses of a neuron. [2, 3]

utricle A small, fluid-filled sac in the vestibular system above the saccule that responds to static positions of the head. [9]

V

V1 See *primary visual cortex.*

vaccination Injection of a foreign substance in order to provoke an immune response and the formation of antibodies against that substance. [3]

vagus nerve Cranial nerve X, which regulates the viscera (organs) and transmits signals from the viscera to the brain. [14]

vanilloid receptor 1 See *transient receptor potential vanilloid type 1.*

varicosity The axonal swelling from which neurotransmitter diffuses in a nondirected synapse. [2]

vasopressin Also called *antidiuretic hormone* (*ADH*) and *arginine vasopressin.* A peptide hormone from the posterior pituitary that promotes water conservation. [14]

ventral In anatomy, towards the belly or front of the body, or the bottom of the brain. Compare *dorsal.* [1]

ventral root The branch of a spinal nerve, arising from the ventral horn of the spinal cord, that carries motor messages from the spinal cord to the peripheral nervous system. [1]

ventral tegmental area (VTA) A portion of the midbrain that projects dopaminergic fibres to the nucleus accumbens. [3]

ventricular system A system of fluid-filled cavities inside the brain. [1]

ventricular zone Also called *ependymal layer.* A region lining the cerebral ventricles that displays mitosis, providing neurons early in development and glial cells throughout life. [6]

ventromedial hypothalamus (VMH) A hypothalamic region involved in eating, sexual, and aggressive behaviours. [13, 14, 11]

vertebral arteries Arteries that ascend the vertebrae, enter the base of the skull, and join together to form the basilar artery. [1]

vertex spike A sharp-wave EEG pattern that is seen during stage 1 slow-wave sleep. [15]

vestibular canal See *scala vestibuli.*

vestibular nuclei Brainstem nuclei that receive information from the vestibular organs through cranial nerve VIII (the vestibulocochlear nerve). [9]

vestibular system The inner ear system that encodes the orientation and acceleration of the head in three axes, crucial for the sense of balance. [9]

vestibulocerebellum The portion of the cerebellum tucked next to the brainstem, consisting of the nodule and the flocculus. [10]

vestibulocochlear nerve Cranial nerve VIII, which runs from the cochlea to the brainstem auditory nuclei. [9]

vestibulo-ocular reflex (VOR) The brainstem mechanism that maintains gaze on a visual object despite movements of the head. [9]

vigilance The global, nonselective level of alertness of an individual. [17]

visual acuity Sharpness of vision. [8]

visual field The whole area that you can see without moving your head or eyes. [8]

vitamin A aldehyde See RETINAL.

VMH See *ventromedial hypothalamus.*

VNO See *vomeronasal organ.*

voltage clamping The use of electrodes to inject current into an axon or neuron to keep the membrane potential at a set value. The apparatus measures how much current must be injected to counteract any ion channel openings. [2]

voltage-gated Na^+ channel A Na^+-selective channel that opens or closes in response to changes in the voltage of the local membrane potential; it mediates the action potential. [2]

voluntary attention Also called *consciously* or *endogenously controlled attention* or *top-down attention*. The voluntary direction of attention towards specific aspects of the environment, in accordance with our interests and goals. [17]

vomeronasal organ (VNO) A collection of specialised receptor cells, near to but separate from the olfactory epithelium, that detect pheromones and send electrical signals to the accessory olfactory bulb in the brain. [9, 13]

vomeronasal system A specialised chemical detection system that detects pheromones and transmits information to the brain. [9]

VOR See *vestibulo-ocular reflex*.

v-SNARE A specialised protein anchored to vesicles to aid their fusing to the presynaptic membrane to release neurotransmitter. [2]

VTA See *ventral tegmental area*.

W

Wallerian degeneration See *anterograde degeneration*.

wavelength The length between two peaks in a repeated stimulus such as a wave, light, or sound. [8]

Wernicke's aphasia See *fluent aphasia*.

Wernicke's area A region of temporoparietal cortex in the brain that is involved in the perception and production of speech. [18]

Western blot A method of detecting a particular protein molecule in a tissue or organ, by separating proteins from that source with gel electrophoresis, blotting the separated proteins onto nitrocellulose, and then using an antibody that binds, and highlights, the protein of interest. [App]

white matter A pale-coloured layer underneath the cortex that consists largely of axons with white myelin sheaths. [1]

Williams syndrome A disorder characterised by fluent linguistic function but poor performance on standard IQ tests and great difficulty with spatial processing. [18]

withdrawal symptom An uncomfortable symptom that arises when a person stops taking a drug that he or she has used frequently, especially at high doses. [3]

wolffian ducts A duct system in the embryo that will develop into male structures (the epididymis, vas deferens, and seminal vesicles) if testes are present. [13]

word deafness The specific inability to hear words, although other sounds can be detected. [9]

working memory A type of short-term memory that holds a limited amount of information available for ready access during performance of a task. [16]

Z

zeitgeber Literally "time giver" (in German). The stimulus (usually the light-dark cycle) that entrains circadian rhythms. [15]

zygote The fertilised egg. [6, 13]

References

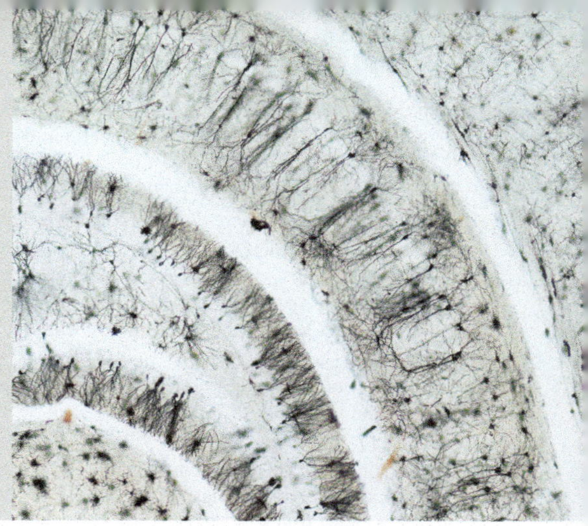

Courtesy of Dr. Sarah Moghadam, VA Palo Alto Health Care System, Palo Alto, CA and Dr. Ahmad Salehi, Dept. of Psychiatry & Behavioural Sciences, Stanford Medical School

A

Aaronson, S. T., Rashed, S., Biber, M. P., and Hobson, J. A. (1982). Brain state and body position. A time-lapse video study of sleep. *Archives of General Psychiatry, 39*, 330–335.

Abbott, A. (2011). A problem for our age. *Nature, 475*, S2–S4.

Abe, N., Suzuki, M., Mori, E., Itoh, M., et al. (2007). Deceiving others: Distinct neural responses of the prefrontal cortex and amygdala in simple fabrication and deception with social interactions. *Journal of Cognitive Neuroscience, 19*, 287–295.

Abelson, J. F., Kwan, K.Y., O'Roak, B. J., Baek, D.Y., et al. (2005). Sequence variants in *SLITRK1* are associated with Tourette's syndrome. *Science, 310*, 317–320.

Abramowitz, J. S., Blakey, S. M., Reuman, L., and Buchholz, J. L. (2018). New directions in the cognitive-behavioural treatment of OCD: Theory, research, and practice. *Behaviour Therapy, 49*(3), 311–322.

Abtahi, S., Howell, E., Salvucci, J. T., Bastacky, J. M. R., et al. (2019). Exendin-4 antagonizes the metabolic action of acylated ghrelinergic signalling in the hypothalamic paraventricular nucleus. *General and Comparative Endocrinology, 270*, 75–81.

Adams, C. S., Korytko, A. I., and Blank, J. L. (2003). A novel mechanism of body mass regulation. *Journal of Experimental Biology, 206*, 2535–2536.

Addante, R. J. (2015). A critical role of the human hippocampus in an electrophysiological measure of implicit memory. *NeuroImage, 109*, 515–528.

Addante, R. J., Ranganath, C., Olichney, J., and Yonelinas, A. P. (2012). Neurophysiological evidence for a recollection impairment in amnesia patients that leaves familiarity intact. *Neuropsychologia, 50*(13), 3004–3014.

Ader, R. (2001). Psychoneuroimmunology. *Current Directions in Psychological Science, 10*, 94–98.

Adhikari, A., Lerner, T. N., Finkelstein, J., Pak, S., et al. (2015). Basomedial amygdala mediates top-down control of anxiety and fear. *Nature, 527*, 179–185.

Adler, E., Hoon, M. A., Mueller, K. L., Chandrashekar, J., et al. (2000). A novel family of mammalian taste receptors. *Cell, 100*, 693–702.

Adler, N., and Matthews, K. (1994). Health psychology: Why do some people get sick and some stay well? *Annual Review of Psychology, 45*, 229–259.

Adolphs, R., Gosselin, F., Buchanan, T. W., Tranel, D., et al. (2005). A mechanism for impaired fear recognition after amygdala damage. *Nature, 433*, 68–72.

Aflalo, T., Kellis, S., Klaes, C., Lee, B., et al. (2015). Neurophysiology: Decoding motor imagery from the posterior parietal cortex of a tetraplegic human. *Science, 348*(6237), 906–910.

Agre, P., King, L. S., Yasui, M., Guggino, W. B., et al. (2002). Aquaporin water channels: From atomic structure to clinical medicine. *Journal of Physiology, 542*, 3–16.

Agulhon, C., Fiacco, T. A., and McCarthy, K. D. (2010). Hippocampal short- and long-term plasticity are not modulated by astrocyte Ca^{2+} signalling. *Science, 327*, 1250–1254.

Ahmari, S. E., Spellman, T., Douglass, N. L., Kheirbek, M. A., et al. (2013). Repeated cortico-striatal stimulation generates persistent OCD-like behaviour. *Science, 340*, 1234–1239.

Ahmed, E. I., Zehr, J. L., Schulz, K. M., Lorenz, B. H., et al. (2008). Pubertal hormones modulate the addition of new cells to sexually dimorphic brain regions. *Nature Neuroscience, 11*, 995–997.

Ajslev, T. A., Andersen, C. S., Gamborg, M., Sørensen, T. I., et al. (2011). Childhood overweight after establishment of the gut microbiota: The role of delivery mode, pre-pregnancy weight and early administration of antibiotics. *International Journal of Obesity (London), 35*(4), 522–529.

Akil, O., Seal, R. P., Burke, K., Wang, C., et al. (2012). Restoration of hearing in the *VGLUT3* knockout mouse using virally mediated gene therapy. *Neuron, 75*(2), 283–293.

Albanese, A., Hamill, G., Jones, J., Skuse, D., et al. (1994). Reversibility of physiological growth hormone secretion in children with psychosocial dwarfism. *Clinical Endocrinology (Oxford), 40*, 687–692.

Albers, G. W., Marks, M. P., Kemp, S., Christensen, S., et al. (2018). Thrombectomy for stroke at 6 to 16 hours with selection by perfusion imaging. *New England Journal of Medicine, 378*(8), 708–718.

Alberts, J. R. (1978). Huddling by rat pups: Multisensory control of contact behaviour. *Journal of Comparative and Physiological Psychology, 92*, 220–230.

Albrecht, B., Staiger, P. K., Hall, K., Kambouropoulos, N., et al. (2016). Motivational drive and alprazolam misuse: A recipe for aggression? *Psychiatry Research, 240*, 381–389.

Aldrich, M. A. (1993). The neurobiology of narcolepsy-cataplexy. *Progress in Neurobiology, 41*, 533–541.

Alexander, G. M., Farris, S., Pirone, J. R., Zheng, C., et al. (2016). Social and novel contexts modify hippocampal CA2 representations of space. *Nature Communications, 7*, 10300.

Alhadeff, A. L., Mergler, B. D., Zimmer, D. J., Turner, C. A., et al. (2017). Endogenous glucagon-like peptide-1 receptor signalling in the nucleus tractus solitarius is required for food intake control. *Neuropsychopharmacology, 42*(7), 1471–1479.

Alkire, M. T., Hudetz, A. G., and Tononi, G. (2008). Consciousness and anesthesia. *Science, 322*, 876–880.

Allard, J. S., Heilbronn, L. K., Smith, C., Hunt, N. D., et al. (2008). In vitro cellular adaptations of indicators of longevity in response to treatment with serum collected from humans on calorie restricted diets. *PLOS ONE, 15*, e3211.

Alle, H., Rother, A., and Geiger, J. (2009). Energy-efficient action potentials in hippocampal mossy fibres. *Science, 325*, 1405–1408.

Allman, J. M., Tetreault, N. A., Hakeem, A.Y., Manaye, K. F., et al. (2010). The von Economo neurons in frontoinsular and anterior cingulate cortex in great apes and humans. *Brain Structure and Function, 214*(5–6), 495–517.

Almada, S. J., Zonderman, A. B., Shekelle, R. B., Dyer, A. R., et al. (1991). Neuroticism and cynicism and risk of death in middle-aged men: The Western Electric Study. *Psychosomatic Medicine, 53*, 165–175.

Alonso, M., Lepousez, G., Wagner, S., Bardy, C., et al. (2012). Activation of adult-born neurons facilitates learning and memory. *Nature Neuroscience, 15*, 897–904.

Alpert, M., and Friedhoff, A. J. (1980). An un-dopamine hypothesis of schizophrenia. *Schizophrenia Bulletin, 6*, 387–390.

Altman, J. (1969). Autoradiographic and histological studies of postnatal neurogenesis. IV. Cell proliferation and migration in the anterior forebrain, with special reference to persisting neurogenesis in the olfactory bulb. *Journal of Comparative Neurology, 137*, 433–457.

Altschuler, E. L., Wisdom, S. B., Stone, L., Foster, C., et al. (1999). Rehabilitation of hemiparesis after stroke with a mirror. *Lancet, 353*, 2035–2036.

Alvarez, D. D., Giacomini, D., Yang, S. M., Trinchero, M. F., et al. (2016). A disynaptic feedback network activated by experience promotes the integration of new granule cells. *Science, 354*(6311), 459–465.

Alvarez, J. A., and Emory, E. (2006). Executive function and the frontal lobes: A meta-analytic review. *Neuropsychology Review, 16*, 17–42.

American Academy of Sleep Medicine. (2014). *International classification of sleep disorders* (3rd ed.). Darien, IL: American Academy of Sleep Medicine.

American Psychiatric Association. (2013). *Diagnostic and Statistical Manual of Mental Disorders* (5th ed.). Arlington, VA: American Psychiatric Publishing.

Aminoff, E. M., Kveraga, K., and Bar, M. (2013). The role of the parahippocampal cortex in cognition. *Trends in Cognitive Science, 17*(8), 379–390.

Amoasii, L., Hildyard, J. C. W., Li, H., Sanchez-Ortiz, E., et al. (2018). Gene editing restores dystrophin expression in a canine model of Duchenne muscular dystrophy. *Science, 362*(6410), 86–91.

Amoroso, T. (2015). The psychopharmacology of ±3,4 methylenedioxymethamphetamine and its role in the treatment of posttraumatic stress disorder. *Journal of Psychoactive Drugs, 47*, 337–344.

Anand, B. K., and Brobeck, J. R. (1951). Localisation of a "feeding centre" in the hypothalamus of the rat. *Proceedings of the Society for Experimental Biology and Medicine, 77*, 323–324.

Andersen, B. B., Korbo, L., and Pakkenberg, B. (1992). A quantitative study of the human cerebellum with unbiased stereological techniques. *Journal of Comparative Neurology, 326*, 549–560.

Andersen, P. M., Nilsson, P., Ala-Hurula, V., Keranen, M. L., et al. (1995). Amyotrophic lateral sclerosis associated with homozygosity for an Asp90Ala mutation in CuZn-superoxide dismutase. *Nature Genetics, 10*, 61–66.

Anderson, M. A., Burda, J. E., Ren, Y., Ao, Y., et al. (2015). Astrocyte scar formation aids central nervous system axon regeneration. *Nature, 532*(7598), 195–200.

Andersson, D. I., Hughes, D., and Kubicek-Sutherland, J. Z. (2019). Mechanisms and consequences of bacterial resistance to antimicrobial peptides. *Drug Resistance Updates, 26*, 43–57.

Andics, A., Gábor, A., Gácsi, M., Faragó, T., et al. (2016). Neural mechanisms for lexical processing in dogs. *Science, 353*(6303), 1030–1032.

Andreasen, N. C. (1994). Changing concepts of schizophrenia and the ahistorical fallacy. *American Journal of Psychiatry, 151*, 1405–1407.

Andreasen, N. C., Flaum, M., Swayze, V. O. D. S., Alliger, R., et al. (1993). Intelligence and brain structure in normal individuals. *American Journal of Psychiatry, 150*, 130–134.

Andreasen, N., Nassrallah, H. A., Dunn, V., Olson, S. C., et al. (1986). Structural abnormalities in the frontal system in schizophrenia. *Archives of General Psychiatry, 43*, 136–144.

Andrews, P. W., Gangestad, S. W., Miller, G. F., Haselton, M. G., et al. (2008). Sex differences in detecting sexual infidelity: Results of a maximum likelihood method for analyzing the sensitivity of sex differences to underreporting. *Human Nature, 19*(4), 347–373.

Andrews-Hanna, J. R., Smallwood, J., and Spreng, R. N. (2014). The default network and self-generated thought: Component processes, dynamic control, and clinical relevance. *Annals of the New York Academy of Sciences, 1316*, 29–52.

Anestis, M. D., Houtsma, C., Daruwala, S. E., and Butterworth, S. E. (2019). Firearm legislation and statewide suicide rates: The moderating role of household firearm ownership levels. *Behavioural Sciences & the Law, 37*(3), 270–280. doi: 10.1002/bsl.2408.

Anggono, V., and Huganir, R. L. (2012). Regulation of AMPA receptor trafficking and synaptic plasticity. *Current Opinions in Neurobiology, 22*, 461–469.

Anguera, J. A., Boccanfuso, J., Rintoul, J. L., Al-Hashimi, O., et al. (2013). Video game training enhances cognitive control in older adults. *Nature, 501*(7465), 97–101.

Anson, R. M., Guo, Z., de Cabo, R., Iyun, T., et al. (2003). Intermittent fasting dissociates beneficial effects of dietary restriction on glucose metabolism and neuronal resistance to injury from calorie intake. *Proceedings of the National Academy of Sciences, USA, 100*, 6216–6220.

Anstey, M. L., Rogers, S. M., Ott, S. R., Burrows, M., et al. (2009). Serotonin mediates behavioural gregarisation underlying swarm formation in desert locusts. *Science, 323*, 627–630.

Anton-Erxleben, K., and Carrasco, M. (2013). Attentional enhancement of spatial resolution: Linking behavioural and neurophysiological evidence. *Nature Reviews Neuroscience, 14*(3), 188–200.

Apkarian, A. V., Sosa, Y., Sonty, S., Levy, R. M., et al. (2004). Chronic back pain is associated with decreased prefrontal and thalamic gray matter density. *Journal of Neuroscience, 24*, 10410–10415.

Archer, J. (2006). Testosterone and human aggression: An evaluation of the challenge hypothesis. *Neuroscience and Biobehavioural Reviews, 30*, 319–345.

Argyll-Robertson, D. M. C. L. (1869). On an interesting series of eye symptoms in a case of spinal disease, with remarks on the action of belladonna on the iris. *Edinburgh Medical Journal, 14*, 696–708.

Arnold, A. P. (1980). Sexual differences in the brain. *American Scientist, 68*, 165–173.

Arnold, A. P., and Schlinger, B. A. (1993). Sexual differentiation of brain and behaviour: The zebra finch is not just a flying rat. *Brain, Behaviour and Evolution, 42*, 231–241.

Arnold, K., and Zuberbühler, K. (2006). Language evolution: Semantic combinations in primate calls. *Nature, 441*, 303.

Arnone, D., Cavanagh, J., Gerber, D., Lawrie, S. M., et al. (2009). Magnetic resonance imaging studies in bipolar disorder and schizophrenia: Meta-analysis. *British Journal of Psychiatry, 195*, 194–201.

Arnsten, A. F. (2006). Fundamentals of attention-deficit/hyperactivity disorder: Circuits and pathways. *Journal of Clinical Psychiatry, 67*, 7–12.

Aroniadou, V. A., Maillis, A., and Stefanis, C. C. (1993). Dihydropyridine-sensitive calcium channels are involved in the induction of N-methyl-D-aspartate receptor-independent long-term potentiation in visual cortex of adult rats. *Neuroscience Letters, 151*, 77–80.

Aschner, M., and Ceccatelli, S. (2010). Are neuropathological conditions relevant to ethylmercury exposure? *Neurotoxicity Research, 1*, 59–68.

Aschwanden, C. (2013). The curious lives of people who feel no fear. *New Scientist, 217*, 36–39.

Aserinsky, E., and Kleitman, N. (1953). Regularly occurring periods of eye motility, and concomitant phenomena, during sleep. *Science, 118*, 273–274.

Ashok, A. H., Mizuno, Y., Volkow, N. D., and Howes, O. D. (2017). Association of stimulant use with dopaminergic alterations in users of cocaine, amphetamine, or methamphetamine: A systematic review and meta-analysis. *JAMA Psychiatry, 74*(5), 511–519.

Aspelund, A., Antila, S., Proulx, S. T., Karlsen, T. V., et al. (2015). A dural lymphatic vascular system that drains brain interstitial fluid and macromolecules. *Journal of Experimental Medicine, 212*(7), 991–999.

Assinck, P., Duncan, G. J., Hilton, B. J., Plemel, J. R., et al. (2017). Cell transplantation therapy for spinal cord injury. *Nature Neuroscience, 20*(5), 637–647.

Atsak, P., Guenzel, F. M., Kantar-Gok, D., Zalachoras, I., et al. (2016). Glucocorticoids mediate stress-induced impairment of retrieval of stimulus-response memory. *Psychoneuroendocrinology, 67*, 207–215.

Audero, E., Coppi, E., Mlinar, B., Rossetti, T., et al. (2008). Sporadic autonomic dysregulation and death associated with excessive serotonin autoinhibition. *Science, 321*, 130–133.

Augustine, V., Gokce, S. K., Lee, S., Wang, B., et al. (2018). Hierarchical neural architecture underlying thirst regulation. *Nature, 555*(7695), 204–209.

Aungst, J. L., Heyward, P. M., Puche, A. C., Karnup, S. V., et al. (2003). Centre-surround inhibition among olfactory bulb glomeruli. *Nature, 426*, 623–629.

Avanzini, G., Depaulis, A., Tassinari, A., and de Curtis M. (2013). Do seizures and epileptic activity worsen epilepsy and deteriorate cognitive function? *Epilepsia, 54*(Suppl. 8), 14–21.

Avau, B., and Depoortere, I. (2016). The bitter truth about bitter taste receptors: Beyond sensing bitter in the oral cavity. *Acta Physiologica, 216*(4), 407–420.

Avau, B., Rotondo, A., Thijs, T., Andrews, C. N., et al. (2015). Targeting extra-oral bitter taste receptors modulates gastrointestinal motility with effects on satiation. *Scientific Reports, 5*, 15985.

Avery, S. N., Rogers, B. P., and Heckers, S. (2018). Hippocampal network modularity is associated with relational memory dysfunction in schizophrenia. *Biological Psychiatry: Cognitive Neuroscience and Neuroimaging, 3*(5), 423–432.

Aviezer, H., Trope, Y., and Todorov, A. (2012). Body cues, not facial expressions, discriminate between intense positive and negative emotions. *Science, 338*, 1225–1229.

Azadegan-Dehkordi, F., Ahmadi, R., Koohiyan, M., and Hashemzadeh-Chaleshtori, M. (2019). Update of spectrum c.35delG and c.-23+1G>A mutations on the *GJB2* gene in individuals with autosomal recessive nonsyndromic hearing loss. *Annals of Human Genetics, 83*(1), 1–10.

Azevedo, F. A., Carvalho, L. R., Grinberg, L. T., Farfel, J. M., et al. (2009). Equal numbers of neuronal and nonneuronal cells make the human brain an isometrically scaled-up primate brain. *Journal of Comparative Neurology, 513*, 532–541.

B

Bachus, S. E., Yang, E., McCloskey, S. S., and Minton, J. N. (2012). Parallels between behavioural and neurochemical variability in the rat vacuous chewing movement model of tardive dyskinesia. *Behavioural Brain Research, 231*(2), 323–336.

Baddeley, A. (2003). Working memory: Looking back and looking forward. *Nature Review. Neuroscience, 4*, 829–839.

Badre, D., and Nee, D. E. (2018). Frontal cortex and the hierarchical control of behaviour. *Trends in Cognitive Sciences, 22*(2), 170–188.

Bagemihl, B. (1999). *Biological exuberance: Animal homosexuality and natural diversity*. New York: St. Martin's.

Baggio, L. L. and Drucker, D. J. (2007). Biology of incretins: GLP-1 and GIP. *Gastroenterology*, 2131–2157.

Bagni, C., and Greenough, W. T. (2005). From mRNP trafficking to spine dysmorphogenesis: The roots of fragile X syndrome. *Nature Reviews Neuroscience, 6*, 376–387.

Bahmani, Z., Clark, K., Merrikhi, Y., Mueller, A., et al. (2019). Prefrontal contributions to attention and working memory. *Current Topics in Behavioural Neuroscience*. https://doi.org/10.1007/7854_2018_74.

Bailey, C. H., and Chen, M. (1983). Morphological basis of long-term habituation and sensitisation in *Aplysia*. *Science, 220*, 91–93.

Bailey, H. R., Zacks, J. M., Hambrick, D. Z., Zacks, R. T., et al. (2013). Medial temporal lobe volume predicts elders' everyday memory. *Psychological Science, 24*(7), 1113–1122. doi:10.1177/0956797612466676.

Bailey, J. A., Zatorre, R. J., and Penhune, V. B. (2014). Early musical training is linked to gray matter structure in the ventral premotor cortex and auditory-motor rhythm synchronisation performance. *Journal of Cognitive Neuroscience, 26*(4), 755–767.

Bailey, K., and West, R. (2013). The effects of an action video game on visual and affective information processing. *Brain Research, 1504*, 35–46.

Baillet, S. (2017). Magnetoencephalography for brain electrophysiology and imaging. *Nature Neuroscience, 20*(3), 327–339.

Bakken, T. E., Miller, J. A., Ding, S. L., Sunkin, S. M., et al. (2016). A comprehensive transcriptional map of primate brain development. *Nature, 535*(7612), 367–375. https://doi.org/10.1038/nature18637.

Bakker, J. (2019). The sexual differentiation of the human brain: Role of sex hormones versus sex chromosomes. *Current Topics in Behavioural Neurosciences*. https://doi.org/10.1007/7854_2018_70.

Bakker, J., De Mees, C., Douhard, Q., Balthazart, J., et al. (2006). Alpha-fetoprotein protects the developing female mouse brain from masculinisation and defeminisation by estrogens. *Nature Neuroscience, 9*, 220–226.

Baldermann, J. C., Schüller, T., Huys, D., Becker, I., et al. (2016). Deep brain stimulation for Tourette-syndrome: A systematic review and meta-analysis. *Brain Stimulation, 9*, 296–304.

Baldwin, M. W., Toda, Y., Nakagita, T., O'Connell, M. J., et al. (2014). Sensory biology: Evolution of sweet taste perception in hummingbirds by transformation of the ancestral umami receptor. *Science, 345*(6199), 929–933.

Balter, M. (2004). Evolution of behaviour: Seeking the key to music. *Science, 306*, 1120–1122.

Balter, M. (2005). Are human brains still evolving? Brain genes show signs of selection. *Science, 309*(5741), 1662–1663.

Ban, T. A. (2007). Fifty years chlorpromazine: A historical perspective. *Neuropsychiatric Disease and Treatment, 3*(4), 495–500.

Bancaud, J., Brunet-Bourgin, F., Chauvel, P., and Halgren, E. (1994). Anatomical origin of deja vu and vivid"memories"in human temporal lobe. *Brain, 117*, 71–90.

Barasa, A. (1960). Forma, grandezza e densita dei neuroni della corteccia cerebrale in mammiferi di grandezza corporea differente. *Zeitschrift für Zellforschung, 53*, 69–89.

Barbeau, H., Norman, K., Fung, J., Visintin, M., et al. (1998). Does neurorehabilitation play a role in the recovery of walking in neurological populations? *Annals of the New York Academy of Sciences, 860*, 377–392.

Barha, C. K., and Galea, L. A. (2010). Influence of different estrogens on neuroplasticity and cognition in the hippocampus. *Biochimica et Biophysica Acta, 1800*(10), 1056–1067.

Barker, R. A., Drouin-Ouellet, J., and Parmar, M. (2015). Cell-based therapies for Parkinson disease: Past insights and future potential. *Nature Reviews Neurology, 11*(9), 492–503.

Barnea, G., O'Donnell, S., Mancia, F., Sun, X., et al. (2004). Odourant receptors on axon termini in the brain. *Science, 304*, 1468.

Barnes, C. A., and Penner, M. R. (2007). Memory changes with age: Neuorobiological correlates. In R. P. Kesner and J. L. Martinez (Eds.), *Neurobiology of learning and memory* (2nd ed., pp. 483–517). San Diego, CA: Elsevier.

Barrett, A. M., Goedert, K. M., and Basso, J. C. (2012). Prism adaptation for spatial neglect after stroke: Translational practice gaps. *Nature Reviews Neurology, 8*(10), 567–577.

Barrett, L. F. (2018). Seeing fear: It's all in the eyes? *Trends in Neurosciences, 41*(9), 559–563.

Barrett, L. F., Mesquita, B., and Gendron, M. (2011). Context in emotion perception. *Current Directions in Psychological Science, 20*(5), 286–290.

Barrio, J. R., Small, G. W., Wong, K. P., Huang, S. C., et al. (2015). In vivo characterisation of chronic traumatic encephalopathy using [F-18] FDDNP PET brain imaging. *Proceedings of the National Academy of Sciences, USA, 112*(16), E2039–E2047.

Bar-Shira, O., Maor, R., and Chechik, G. (2015). Gene expression switching of receptor subunits in human brain development. *PLOS Computational Biology, 11*, e1004559.

Bartels, A., and Zeki, S. (2000). The neural basis of romantic love. *Neuroreport, 11*, 3829–3834.

Bartolomeo, P. (2007). Visual neglect. *Current Opinion in Neurology, 20*, 381–386.

Barton, M., Filardo, E. J., Lolait, S. J., Thomas, P., et al. (2018). Twenty years of the G protein-coupled estrogen receptor GPER: Historical and personal perspectives. *Journal of Steroid Biochemistry and Molecular Biology, 176*, 4–15.

Bartoshuk, L. M. (1993). Genetic and pathological taste variation: What can we learn from animal models and human disease? In D. Chadwick, J. Marsh, and J. Goode (Eds.), *The molecular basis of smell and taste transduction* (pp. 251–267). New York: Wiley.

Bartoshuk, L. M., and Beauchamp, G. K. (1994). Chemical senses. *Annual Review of Psychology, 45,* 419–449.

Bartsch, T., and Butler, C. (2013). Transient amnesic syndromes. *Nature Reviews Neurology, 9,* 86–97. doi:10.1038/nrneurol.2012.264.

Bartsch, T., Schönfeld, R., Müller, F. J., Alfke, K., et al. (2010). Focal lesions of human hippocampal CA1 neurons in transient global amnesia impair place memory. *Science, 328,* 1412–1415.

Basbaum, A., and Fields, H. L. (1984). Endogenous pain control systems: Brainstem spinal pathways and endorphin circuitry. *Annual Review of Neuroscience, 7,* 309–339.

Basson, R. (2008). Women's sexual function and dysfunction: Current uncertainties, future directions. *International Journal of Impotence Research, 20,* 466–478.

Bates, E., Wilson, S. M., Saygin, A. P., Dick, F., et al. (2003). Voxel-based lesion-symptom mapping. *Nature Neuroscience, 6,* 448–450.

Bates, G. P., Dorsey, R., Gusella, J. F., Hayden, M. R., et al. (2015). Huntington disease. *Nature Reviews Disease Primers, 1,* 15005.

Bates, T. C., Lind, P. A., Luciano, M., Montgomery, G. W., et al. (2010). Dyslexia and DYX1C1: Deficits in reading and spelling associated with a missense mutation. *Molecular Psychiatry, 12,* 1190–1196.

Baumann, M. H., Walters, H. M., Niello, M., and Sitte, H. H. (2018). Neuropharmacology of synthetic cathinones. In H. Maurer and S. Brandt (Eds.), *New Psychoactive Substances. Handbook of Experimental Pharmacology, Vol. 252.* Cham, Switzerland: Springer. https://doi.org/10.1007/164_2018_178.

Baumeister, D., Barnes, G., Giaroli, G., and Tracy, D. (2014). Classical hallucinogens as antidepressants? A review of pharmacodynamics and putative clinical roles. *Therapeutic Advances in Psychopharmacology, 4,* 156–169.

Bautista, D. M., Siemens, J., Glazer, J. M., Tsuruda, P. R., et al. (2007). The menthol receptor TRPM8 is the principal detector of environmental cold. *Nature, 448,* 204–208.

Baynes, K. C., Dhillo, W. S., and Bloom, S. R. (2006). Regulation of food intake by gastrointestinal hormones. *Current Opinion in Gastroenterology, 22,* 626–631.

Beach, F. A. (1977). *Human sexuality in four perspectives.* Baltimore: Johns Hopkins University Press.

Beall, A. T., and Tracy, J. L. (2013). Women are more likely to wear red or pink at peak fertility. *Psychological Science, 24*(9), 1837–1841.

Bear, D. M., Lassance, J. M., Hoekstra, H. E., and Datta, S. R. (2016). The evolving neural and genetic architecture of vertebrate olfaction. *Current Biology, 26*(20), R1039–R1049.

Beauchamp, G. K., Cowart, B. J., Mennella, J. A., and Marsh, R. R. (1994). Infant salt taste: Developmental, methodological, and contextual factors. *Developmental Psychobiology, 27,* 353–365.

Becker, J. B., and Chartoff, E. (2019). Sex differences in neural mechanisms mediating reward and addiction. *Neuropsychopharmacology, 44*(1),166–183.

Becker, J. B., McClellan, M. L., and Reed, B. G. (2017). Sex differences, gender and addiction. *Journal of Neuroscience Research, 95*(1–2), 136–147.

Bedrosian, T. A., and Nelson, R. J. (2014). Nitric oxide and serotonin interactions in aggression. *Current Topics in Behavioural Neurosciences, 17,* 131–142.

Bee, M. A., and Micheyl, C. (2008). The cocktail party problem: What is it? How can it be solved? And why should animal behaviourists study it? *Journal of Comparative Psychology, 122,* 235–251.

Beecher, H. K. (1956). Relationship of significance of wound to pain experienced. *Journal of the American Medical Association, 161*(17), 1609–1613.

Beeli, G., Esslen, M., and Jäncke, L. (2005). When coloured sounds taste sweet. *Nature, 434,* 38.

Beggs, S., Trang, T., and Salter, M. W. (2012). P2X4R+ microglia drive neuropathic pain. *Nature Neuroscience, 15,* 1068–1073.

Beggs, W. D., and Foreman, D. L. (1980). Sound localisation and early binaural experience in the deaf. *British Journal of Audiology, 14,* 41–48.

Behr, C., Goltzene, M. A., Kosmalski, G., Hirsch, E., et al. (2016). Epidemiology of epilepsy. *Revue Neurologique (Paris), 172,* 27–36.

Behrens, M., and Meyerhof, W. (2019). A role for taste receptors in (neuro) endocrinology? *Journal of Neuroendocrinology, 31,* e12691.

Belelli, D., Brown, A. R., Mitchell, S. J., Gunn, B. G., et al. (2018). Endogenous neurosteroids influence synaptic $GABA_A$ receptors during postnatal development. *Journal of Neuroendocrinology, 30*(2). https://doi.org/10.1111/jne.12537.

Beliveau, V., Ganz, M., Feng, L., Ozenne, B., et al. (2017). A high-resolution in vivo atlas of the human brain's serotonin system. *Journal of Neuroscience, 37*(1),120–128.

Bell, S., Daskalopoulou, M., Rapsomaniki, E., George, J., et al. (2017). Association between clinically recorded alcohol consumption and initial presentation of 12 cardiovascular diseases: Population based cohort study using linked health records. *BMJ, 356,* j909.

Bellinger, D. L., Ackerman, K. D., Felten, S.Y., and Felten, D. L. (1992). A longitudinal study of age-related loss of noradrenergic nerves and lymphoid cells in the rat spleen. *Experimental Neurology, 116,* 295–311.

Bellono, N. W., Leitch, D. B., and Julius, D. (2018). Molecular tuning of electroreception in sharks and skates. *Nature, 558*(7708), 122–126.

Bellugi, U., Poizner, H., and Klima, E. S. (1983). Brain organisation for language: Clues from sign aphasia. *Human Neurobiology, 2,* 155–171.

Belluscio, L., Gold, G. H., Nemes, A., and Axel, R. (1998). Mice deficient in G(olf) are anosmic. *Neuron, 20,* 69–81.

Belvederi, M. M., Pariante, C., Mondelli, V., Masotti, M., et al. (2014). HPA axis and aging in depression: systematic review and meta-analysis. *Psychoneuroendocrinology, 41,* 46–62. doi: 10.1016/j.psyneuen.2013.12.004.

Benias, P. C., Wells, R. G., Sackey-Aboagye, B., Klavan, H., et al. (2018). Structure and distribution of an unrecognized interstitium in human tissues. *Scientific Reports, 8*(1), 4947. Erratum in: *Scientific Reports, 8*(1), 7610.

Bennett, E. L., Diamond, M. L., Krech, D., and Rosenzweig, M. R. (1964). Chemical and anatomical plasticity of brain. *Science, 146,* 610–619.

Bennett, E. L., Rosenzweig, M. R., and Diamond, M. C. (1969). Rat brain: Effects of environmental enrichment on wet and dry weights. *Science, 163,* 825–826.

Bennett, W. (1983). The nicotine fix. *Rhode Island Medical Journal, 66,* 455–458.

Benney, K. S., and Braaten, R. F. (2000). Auditory scene analysis in estrildid finches (*Taeniopygia guttata* and *Lonchura striata domestica*): A species advantage for detection of conspecific song. *Journal of Comparative Psychology, 114,* 174–182.

Bensafi, M., Porter, J., Pouliot, S., Mainland, J., et al. (2003). Olfactomotor activity during imagery mimics that during perception. *Nature Neuroscience, 6,* 1142–1144.

Bensmaia, S. J., and Miller, L. E. (2014). Restoring sensorimotor function through intracortical interfaces: Progress and looming challenges. *Nature Reviews Neuroscience, 15*(5), 313–325.

Benson, D. F. (1967). Fluency in aphasia: Correlation with radioactive scan localisation. *Cortex, 3,* 373–394.

Benson, P. J., Beedie, S. A., Shephard, E., Giegling, I., et al. (2012). Simple viewing tests can detect eye movement abnormalities that distinguish schizophrenia cases from controls with exceptional accuracy. *Biological Psychiatry, 72,* 716–724.

Benton, A. L., and Hamsher, K. (1976). *Multilingual aphasia examination.* Iowa City: University of Iowa Press.

Berger, L. R., de Ruiter, D. J., Churchill, S. E, Schmid, P., et al. (2010). *Australopithecus sediba*: A new species of Homo-like australopith from South Africa. *Science, 328*(5975), 195–204.

Bergfeld, I. O., Mantione, M., Hoogendoorn, M. L. C., Ruhé, H. G., et al. (2016). Deep brain stimulation of the ventral anterior limb of the internal capsule for treatment resistant depression: A randomized clinical trial. *JAMA Psychiatry.* doi:10.1001/jamapsychiatry.2016.0152.

Berglund, E. D, Liu, T., Kong, X., Sohn, J. W., et al. (2014). Melanocortin 4 receptors in autonomic neurons regulate thermogenesis and glycemia. *Nature Neuroscience, 17,* 911–913.

Bergmann, O., Spalding, K. L., and Frisén, J. (2015). Adult neurogenesis in humans. *Cold Spring Harbor Perspectives in Biology, 7*(7), a018994. https://doi.org/10.1101/cshperspect.a018994.

Bernal, B., and Altman, N. (2010). The connectivity of the superior longitudinal fasciculus: A tractography DTI study. *Magnetic Resonance Imaging, 28*(2), 217–225.

Berneche, S., and Roux, B. (2001). Energetics of ion conduction through the K^+ channel. *Nature, 414*, 73–77.

Bernhardt, P. C. (1997). Influences of serotonin and testosterone in aggression and dominance: Convergence with social psychology. *Current Directions in Psychological Science, 2*(6), 44–48.

Bernhardt, P. C., Dabbs, J. M., Jr., Fielden, J. A., and Lutter, C. D. (1998). Testosterone changes during vicarious experiences of winning and losing among fans at sporting events. *Physiology & Behaviour, 65*, 59–62.

Bernstein, H. G., Steiner, J., Guest, P. C., Dobrowolny, H., et al. (2015). Glial cells as key players in schizophrenia pathology: Recent insights and concepts of therapy. *Schizophrenia Research, 161*(1), 4–18.

Bernstein, I. S., and Gordon, T. P. (1974). The function of aggression in primate societies. *American Scientist, 62*, 304–311.

Bernstein, L. E., Auer, E. T., Jr., Moore, J. K., Ponton, C. W., et al. (2002). Visual speech perception without primary auditory cortex activation. *Neuroreport, 13*, 311–315.

Berridge, C. W., and Devilbiss, D. M. (2011). Psychostimulants as cognitive enhancers: The prefrontal cortex, catecholamines, and attention-deficit/hyperactivity disorder. *Biological Psychiatry, 69*(12), e101-11.

Berton, O., and Nestler, E. J. (2006). New approaches to antidepressant drug discovery: Beyond monoamines. *Neuroscience, 7*, 137–151.

Bertram, L., and Tanzi, R. E. (2008). Thirty years of Alzheimer's disease genetics: The implications of systematic meta-analyses. *Nature Reviews Neuroscience, 9*, 768–778.

Beurg, M., Fettiplace, R., Nam, J.-H., and Ricci, A. J. (2009). Localisation of inner hair cell mechanotransducer channels using high-speed calcium imaging. *Nature Neuroscience, 12*, 553–558.

Bezanilla, F., and Correa, A. M. (1995). Single-channel properties and gating of Na^+ and K^+ channels in the squid giant axon. In N. J. Abbott, R. Williamson, and L. Maddock (Eds.), *Cephalopod neurobiology* (pp. 131–151). New York: Oxford University Press.

Binder, J. R., Rao, S. M., Hammeke, T. A., Yetkin, F. Z., et al. (1994). Functional magnetic resonance imaging of human auditory cortex. *Annals of Neurology, 35*, 662–672.

Birch, L. L., Fisher, J. O., and Davison, K. K. (2003). Learning to overeat: Maternal use of restrictive feeding practices promotes girls' eating in the absence of hunger. *American Journal of Clinical Nutrition, 78*, 215–220.

Bird, B. M., Geniole, S. N., Procyshyn, T. L., Ortiz, T. L., et al. (2019). Effect of exogenous testosterone on cooperation depends on personality and time pressure. *Neuropsychopharmacology, 44*(3), 538–545.

Bird, C. D., and Emery, N. J. (2009). Insightful problem solving and creative tool modification by captive nontool-using rooks. *Proceedings of the National Academy of Sciences, USA, 106*, 10370–10375.

Birnbaum, R., and Weinberger, D. R. (2017). Genetic insights into the neurodevelopmental origins of schizophrenia. *Nature Reviews Neuroscience, 18*(12), 727–740.

Bisley, J. W., and Goldberg, M. E. (2003). Neuronal activity in the lateral intraparietal area and spatial attention. *Science, 299*, 81–86.

Bisley, J. W., and Goldberg, M. E. (2006). Neural correlates of attention and distractibility in the lateral intraparietal area. *Journal of Neurophysiology, 95*, 1696–1717.

Bjorness, T. E., and Greene, R. W. (2009). Adenosine and sleep. *Current Neuropharmacology, 7*(3), 238–245.

Blackwell, D. L., Lucas, J. W., and Clarke, T. C. (2014). Summary health statistics for U.S. adults: National Health Interview Survey. *Vital and Health Statistics, 10*(260), 1–161.

Blake, D. J., Weir, A., Newey, S. E., and Davies, K. E. (2002). Function and genetics of dystrophin and dystrophin-related proteins in muscle. *Physiology Review, 82*, 291–329.

Blake, D. T., Byl, N. N., and Merzenich, M. M. (2002). Representation of the hand in the cerebral cortex. *Behavioural Brain Research, 135*(1–2), 179–184.

Blanchard, R., Cantor, J. M., Bogaert, A. F., Breedlove, S. M., et al. (2006). Interaction of fraternal birth order and handedness in the development of male homosexuality. *Hormones and Behaviour, 49*, 405–414.

Blaxton, T. A., Bookheimer, S. Y., Zeffiro, T. A., Figlozzi, C. M., et al. (1996). Functional mapping of human memory using PET: Comparisons of conceptual perceptual tasks. *Canadian Journal of Experimental Psychology, 50*, 42–56.

Blehar, M. C., and Rosenthal, N. E. (1989). Seasonal affective disorders and phototherapy. Report of a National Institute of Mental Health-sponsored workshop. *Archives of General Psychiatry, 46*, 469–474.

Bleuler, E. (1950). *Dementia praecox; or, The group of schizophrenias* (J. Zinkin, Trans.). New York: International Universities Press.

Bliss, T. V. P., and Gardner-Medwin, A. R. (1973). Long-lasting potentiation of synaptic transmission in the dentate area of the unanaesthetized rabbit following stimulation of the perforant path. *Journal of Physiology (London), 232*, 357–374.

Bliss, T. V. P., and Lømo, T. (1973). Long-lasting potentiation of synaptic transmission in the dentate area of the anaesthetized rabbit following stimulation of the perforant path. *Journal of Physiology (London), 232*, 331–356.

Bliss, T. V., Collingridge, G. L., Kaang, B. K., and Zhuo, M. (2016). Synaptic plasticity in the anterior cingulate cortex in acute and chronic pain. *Nature Reviews Neuroscience, 17*(8), 485–496.

Blue, M. E., and Parnavelas, J. G. (1983). The formation and maturation of synapses in the visual cortex of the rat. II. Quantitative analysis. *Journal of Neurocytology, 12*, 697–712.

Blumberg, M. S. (2015). Developing sensorimotor systems in our sleep. *Current Directions in Psychological Science, 24*(1), 32–37.

Blumberg, M. S., Sokoloff, G., and Kirby, R. F. (1997). Brown fat thermogenesis and cardiac rate regulation during cold challenge in infant rats. *American Journal of Physiology, 272*, R1308–R1313.

Blumberger, D. M., Hsu, J. H., and Daskalakis, Z. J. (2015). A review of brain stimulation treatments for late-life depression. *Current Treatment Options in Psychiatry, 2*(4), 413–421.

Boets, B., Op de Beeck, H. P., Vandermosten, M., Scott, S. K., et al. (2013). Intact but less accessible phonetic representations in adults with dyslexia. *Science, 342*(6163), 1251–1254.

Bogaert, A. F. (2006). Biological versus nonbiological older brothers and men's sexual orientation. *Proceedings of the National Academy of Sciences, USA, 103*, 10771–10774.

Bogaert, A. F., Skorska, M. N., Wang, C., Gabrie, J, et al. (2018). Male homosexuality and maternal immune responsivity to the Y-linked protein NLGN4Y. *Proceedings of the National Academy of Sciences, USA, 115*(2), 302–306.

Bogenschutz, M. P., and Johnson, M. W. (2016). Classic hallucinogens in the treatment of addictions. *Progress in Neuro-Psychopharmacology Biological Psychiatry, 64*, 250–258.

Boggs, D. L., Nguyen, J. D., Morgenson, D., Taffe, M. A., et al. (2018). Clinical and preclinical evidence for functional interactions of cannabidiol and Δ^9-tetrahydrocannabinol. *Neuropsychopharmacology, 43*(1),142–154. https://doi.org/10.1038/npp.2017.209.

Bogin, B. (1997). Evolutionary hypotheses for human childhood. *Yearbook of Physical Anthropology*, 40, 63–89.

Bohrn, I., Carbon, C.-C., and Hutzler, F. (2010). Mona Lisa's smile: perception or deception? *Psychological Science, 21*, 378–380.

Bolhuis, J. J., Okanoya, K., and Scharff, C. (2010). Twitter evolution: Converging mechanisms in birdsong and human speech. *Nature Reviews Neuroscience, 11*(11), 747–759.

Boly, M., Garrido, M. I., Gosseries, O., Bruno, M. A., et al. (2011). Preserved feedforward but impaired top-down processes in the vegetative state. *Science, 332*, 858–862.

Boly, M., Moran, R., Murphy, M., Boveroux, P., et al. (2012). Connectivity changes underlying spectral EEG changes during propofol-induced loss of consciousness. *Journal of Neuroscience, 32*(20),7082–7090.

Bonaz, B., Bazin, T., and Pellissier, S. (2018). The vagus nerve at the interface of the microbiota-gut-brain axis. *Frontiers in Neuroscience, 12*, 49.

Bonhoeffer, T., and Grinvald, A. (1991). Iso-orientation domains in cat visual cortex are arranged in pinwheel-like patterns. *Nature, 353*, 429–431.

Bonini, F., Burle, B., Liégeois-Chauvel, C., Régis, J., et al. (2014). Action monitoring and medial frontal cortex: Leading role of supplementary motor area. *Science, 343*(6173), 888–891.

Bonnel, A. M., and Prinzmetal, W. (1998). Dividing attention between the colour and the shape of objects. *Perception & Psychophysics, 60*, 113–124.

Boot, W. R., Kramer, A. F., Simons, D. J., Fabiani, M., et al. (2008). The effects of video game playing on attention, memory, and executive control. *Acta Psychologica, 129*, 387–398.

Booth, W., Smith, C. F., Eskridge, P. H., Hoss, S. K., et al. (2012). Facultative parthenogenesis discovered in wild vertebrates. *Biology Letters, 8*, 983–985.

Borgstein, J., and Grootendorst, C. (2002). Clinical picture: Half a brain. *Lancet, 359*, 473.

Borota, D., Murray, E., Keceli, G., Chang, A., et al. (2014). Post-study caffeine administration enhances memory consolidation in humans. *Nature Neuroscience, 17*, 201–203.

Borrow, A. P., and Handa, R. J. (2017). Estrogen receptors modulation of anxiety-like behaviour. *Vitamins and Hormones, 103*, 27–52.

Boshuisen, K., van Schooneveld, M. M., Leijten, F. S., de Kort, G. A., et al. (2010). Contralateral MRI abnormalities affect seizure and cognitive outcome after hemispherectomy. *Neurology, 75*(18), 1623–1630.

Bostan, A. C., and Strick, P. L. (2018). The basal ganglia and the cerebellum: Nodes in an integrated network. *Nature Reviews Neuroscience, 19*(6), 338–350.

Boström, P., Wu, J., Jedrychowski, M. P., Korde, A., et al. (2012). A PGC1-α-dependent myokine that drives brown-fat-like development of white fat and thermogenesis. *Nature, 481*(7382), 463–468.

Bottjer, S. W., Miesner, E. A., and Arnold, A. P. (1984). Forebrain lesions disrupt development but not maintenance of song in passerine birds. *Science, 224*, 901–903.

Bourane, S., Duan, B., Koch, S. C., Dalet, A., et al. (2015). Gate control of mechanical itch by a subpopulation of spinal cord interneurons. *Science, 350*(6260), 550–554.

Bouret, S. G., Draper, S. J., and Simerly, R. B. (2004). Trophic action of leptin on hypothalamic neurons that regulate feeding. *Science, 304*, 108–110.

Bourque, C. W. (2008). Central mechanisms of osmosensation and systemic osmoregulation. *Nature Reviews Neuroscience, 9*, 519–531.

Bourque, J., Afzali, M. H., O'Leary-Barrett, M., and Conrod, P. (2017). Cannabis use and psychotic-like experiences trajectories during early adolescence: The coevolution and potential mediators. *Journal of Child Psychology and Psychiatry, 58*(12), 1360–1369.

Bourtchuladze, R., Frenguelli, B., Blendy, J., Cioffi, D., et al. (1994). Deficient long-term memory in mice with a targeted mutation of the cAMP-responsive element-binding protein. *Cell, 79*, 59–68.

Bousman, C. A., Arandjelovic, K., Mancuso, S. G., Eyre, H. A., et al. (2019). Pharmacogenetic tests and depressive symptom remission: A meta-analysis of randomized controlled trials. *Pharmacogenomics, 20*(1), 37–47.

Bouton, C. E., Shaikhouni, A., Annetta, N.V., Bockbrader, M. A., et al. (2016). Restoring cortical control of functional movement in a human with quadriplegia. *Nature, 533*(7602), 247–250.

Bouwknecht, J. A., Hijzen, T. H., van der Gugten, J., Maes, R. A., et al. 2001. Absence of 5-HT(1B) receptors is associated with impaired impulse control in male 5-HT(1B) knockout mice. *Biological Psychiatry, 49*(7), 557–568.

Bower, B. (2000). Genes to grow on. *Science News, 157*(9), 142.

Bower, B. (2003). Vision seekers. *Science News, 164*, 331–333.

Bowles, B., Crupi, C., Mirsattari, S. M., Pigott, S. E., et al. (2007). Impaired familiarity with preserved recollection after anterior temporal-lobe resection that spares the hippocampus. *Proceedings of the National Academy of Sciences, USA, 104*(41), 16382–16387.

Bowles, B., Duke, D., Rosenbaum, R. S., McRae, K., et al. (2016). Impaired assessment of cumulative lifetime familiarity for object concepts after left anterior temporal-lobe resection that includes perirhinal cortex but spares the hippocampus. *Neuropsychologia, 90*, 170–179.

Boyd, C. A. (2010). Cerebellar agenesis revisited. *Brain, 133*(Pt. 3), 941–944.

Brady, T. F., Konkle, T., Alvarez, G. A., and Oliva, A. (2008). Visual long-term memory has a massive storage capacity for object details. *Proceedings of the National Academy of Sciences, USA, 105*(38), 14325–14329.

Brainard, D. H., Roorda, A., Yamauchi, Y., Calderone, J. B., et al. (2000). Functional consequences of the relative numbers of L and M cones. *Journal of the Optical Society of America. Part A, Optics, Image Science and Vision, 17*, 607–614.

Brancaleoni, G., Nikitenkova, E., Grassi, L., and Hansen, V. (2009). Seasonal affective disorder and latitude of living. *Epidemiologia e Psichiatria Sociale, 18*, 336–343.

Brandler, W. M., and Paracchini, S. (2014). The genetic relationship between handedness and neurodevelopmental disorders. *Trends in Molecular Medicine, 20*(2), 83–90.

Brandlistuen, R. E., Ystrom, E., Eberhard-Gran, M., Nulman, I., et al. (2015). Behavioural effects of fetal antidepressant exposure in a Norwegian cohort of discordant siblings. *International Journal of Epidemiology, 44*(4), 1397–1407.

Branson, R., Potoczna, N., Kral, J. G., Lentes, K. U., et al. (2003). Binge eating as a major phenotype of melanocortin 4 receptor gene mutations. *New England Journal of Medicine, 348*, 1096–1103.

Braun, K. A., Ellis, R., and Loftus, E. F. (2002). Make my memory: How advertising can change our memories of the past. *Psychology and Marketing, 19*, 1–23.

Bray, G. A. (1969). Effect of caloric restriction on energy expenditure in obese patients. *Lancet, 2*, 397–398.

Breedlove, S. M. (2010). Minireview: Organisational hypothesis: Instances of the fingerpost. *Endocrinology, 151*(9), 4116–4122.

Breier, A., Malhotra, A. K., Pinals, D. A., Weisenfeld, N. I., et al. (1997). Association of ketamine-induced psychosis with focal activation of the prefrontal cortex in healthy volunteers. *American Journal of Psychiatry, 154*, 805–811.

Bremer, F. (1938). L'activité électrique de l'écorce cérébrale. *Actualités Scientifiques et Industrielles, 658*, 3–46.

Bremner, J. D., Randall, P., Scott, T. M., Bronen, R. A., et al. (1995). MRI-based measurement of hippocampal volume in patients with combat-related posttraumatic stress disorder. *American Journal of Psychiatry, 152*, 973–981.

Brennan, S. C., Davies, T. S., Schepelmann, M., and Riccardi, D. (2014). Emerging roles of the extracellular calcium-sensing receptor in nutrient sensing: Control of taste modulation and intestinal hormone secretion. *British Journal of Nutrition, 111*(Suppl. 1), S16–S22.

Brewer, J. B., Zhao, Z., Desmond, J. E., Glover, G. H., et al. (1998). Making memories: Brain activity that predicts how well visual experience will be remembered. *Science, 281*, 1185–1187.

Briggs, F., and Usrey, W. M. (2008). Emerging views of corticothalamic function. *Current Opinion in Neurobiology, 18*, 403–407.

Briggs, F., Mangun, G. R., and Usrey, W. M. (2013). Attention enhances synaptic efficacy and the signal-to-noise ratio in neural circuits. *Nature, 499*(7459), 476–480.

Bril, B., Smaers, J., Steele, J., Rein, R., et al. (2012). Functional mastery of percussive technology in nut-cracking and stone-flaking actions: Experimental comparison and implications for the evolution of the human brain. *Philosophical Transactions of the Royal Society of London. Series B: Biological Sciences, 367*(1585), 59–74.

Broadbent, D. A. (1958). *Perception and communication.* New York: Pergamon.

Broberg, D. J., and Bernstein, I. L. (1989). Cephalic insulin release in anorexic women. *Physiology & Behaviour, 45*, 871–874.

Brodin, T., Fick, J., Jonsson, M., and Klaminder, J. (2013). Dilute concentrations of a psychiatric drug alter behaviour of fish from natural populations. *Science, 339*, 814–815.

Brody, A. L., Mandelkern, M. A., London, E. D., Olmstead, R. E., et al. (2006). Cigarette smoking saturates brain alpha 4 beta 2 nicotinic acetylcholine receptors. *Archives of General Psychiatry, 63*, 907–915.

Brody, D. J., Pratt, L. A., and Hughes, J. (2018). *Prevalence of depression among adults aged 20 and over: United States, 2013–2016. NCHS Data Brief No. 303.* National Centre for Health Statistics. Available from: https://www.cdc.gov/nchs/products/databriefs/db303.htm.

Brody, G. H., Beach, S. R., Philibert, R. A., Chen, Y. F., et al. (2009). Parenting moderates a genetic vulnerability factor in longitudinal increases in youths' substance abuse. *Journal of Consulting and Clinical Psychology, 77*, 1–11.

Brohawn, S. G., Campbell, E. B., and MacKinnon, R. (2014). Physical mechanism for gating and mechanosensitivity of the human TRAAK K^+ channel. *Nature, 516*(7529), 126–130.

Bronsard, G., and Bartolomei, F. (2013). Rhythms, rhythmicity and aggression. *Journal of Physiology (Paris), 107*(4), 327–334.

Brooks, J. X., Carriot, J., and Cullen, K. E. (2015). Learning to expect the unexpected: Rapid updating in primate cerebellum during voluntary self-motion. *Nature Neuroscience, 18*(9), 1310–1317.

Broughton, R., Billings, R., Cartwright, R., Doucette, D., et al. (1994). Homicidal somnambulism: A case report. *Sleep, 17*, 253–264.

Brouwer, H., Crocker, M. W., Venhuizen, N. J., Hoeks, J. C. J. (2017). A neurocomputational model of the N400 and the P600 in language processing. *Cognitive Science, 41*(Suppl. 6), 1318–1352.

Brown, A. S. (2011). Exposure to prenatal infection and risk of schizophrenia. *Frontiers in Psychiatry, 2*, 63.

Brown, A. S., and Susser, E. S. (2008). Prenatal nutritional deficiency and risk of adult schizophrenia. *Schizophrenia Bulletin, 34*, 1054–1063.

Brown, C. (2003, February 2). The man who mistook his wife for a deer. *The New York Times*, Section 6, p. 32.

Brown, E. C., Jeong, J. W., Muzik, O., Rothermel, R., et al. (2014). Evaluating the arcuate fasciculus with combined diffusion-weighted MRI tractography and electrocorticography. *Human Brain Mapping, 35*(5), 2333–2347.

Brown, J. (1958). Some tests of the decay theory of immediate memory. *Quarterly Journal of Experimental Psychology, 10*, 12–21.

Brown, J. L. (2014). Comparative reproductive biology of elephants. *Advances in Experimental Medicine and Biology, 753*, 135–169.

Brown, L. T., Mikell, C. B., Youngerman, B. E., Zhang, Y., et al. (2016). Dorsal anterior cingulotomy and anterior capsulotomy for severe, refractory obsessive-compulsive disorder: A systematic review of observational studies. *Journal of Neurosurgery, 124*(1), 77–89.

Brown, S. P., Mathur, B. N., Olsen, S. R., Luppi, P. H., et al. (2017). New breakthroughs in understanding the role of functional interactions between the neocortex and the claustrum. *Journal of Neuroscience, 37*(45), 10877–10881.

Bruder, C. E., Piotrowski, A., Gijsbers, A. A., Andersson, R., et al. (2008). Phenotypically concordant and discordant monozygotic twins display different DNA copy-number-variation profiles. *American Journal of Human Genetics, 82*(3), 763–771.

Bruel-Jungerman, E., Rampon, C., and Laroche S. (2007). Adult hippocampal neurogenesis, synaptic plasticity and memory: Facts and hypotheses. *Review in the Neurosciences, 18*, 93–114.

Brunel, N. (2016). Is cortical connectivity optimized for storing information? *Nature Neuroscience, 19*(5), 749–755.

Brunet, A., Saumier, D., Liu, A., Streiner, D. L., et al. (2018). Reduction of PTSD symptoms with pre-reactivation propranolol therapy: A randomized controlled trial. *American Journal of Psychiatry, 175*(5), 427–433.

Brunet, L. J., Gold, G. H., and Ngai, J. (1996). General anosmia caused by a targeted disruption of the mouse olfactory cyclic nucleotide-gated cation channel. *Neuron, 17*, 681–693.

Brunoni, A. R., Moffa, A. H., Fregni, F., Palm, U., et al. (2016). Transcranial direct current stimulation for acute major depressive episodes: Meta-analysis of individual patient data. *British Journal of Psychiatry*. doi:10.1192/bjp.bp.115.164715.

Bu, G. (2009). Apolipoprotein E and its receptors in Alzheimer's disease: Pathways, pathogenesis and therapy. *Nature Reviews Neuroscience, 10*, 333–344.

Bubb, E. J., Metzler-Baddeley, C., and Aggleton, J. P. (2018). The cingulum bundle: Anatomy, function, and dysfunction. *Neuroscience & Biobehavioural Reviews, 92*, 104–127.

Buccino, G., Lui, F., Canessa, N., Patteri, I., et al. (2004). Neural circuits involved in the recognition of actions performed by nonconspecifics: An fMRI study. *Journal of Cognitive Neuroscience, 16*, 114–126.

Buccino, G., Solodkin, A., and Small, S. L. (2006). Functions of the mirror neuron system: Implications for neurorehabilitation. *Cognitive and Behavioural Neurology, 19*, 55–63.

Buchsbaum, B. R., Baldo, J., Okada, K., Berman, K. F., et al. (2011). Conduction aphasia, sensory-motor integration, and phonological short-term memory: An aggregate analysis of lesion and fMRI data. *Brain and Language, 119*(3), 119–128.

Buchsbaum, M. S., Buchsbaum, B. R., Chokron, S., Tang, C., et al. (2006). Thalamocortical circuits: fMRI assessment of the pulvinar and medial dorsal nucleus in normal volunteers. *Neuroscience Letters, 404*, 282–287.

Buchsbaum, M. S., Mirsky, A. F., DeLisi, L. E., Morihisa, J., et al. (1984). The Genain quadruplets: Electrophysiological, positron emission and X-ray tomographic studies. *Psychiatry Research, 13*, 95–108.

Buck, L., and Axel, R. (1991). A novel multigene family may encode odourant receptors: A molecular basis for odour recognition. *Cell, 65*, 175–187.

Bullock, T. H. (1986). Some principles in the brain analysis of important signals: Mapping and stimulus recognition. *Brain, Behaviour and Evolution, 28*, 145–156.

Bullock, T. H., Bennett, M.V. L., Johnston, D., Josephson, R., et al. (2005). Neuroscience: The neuron doctrine, redux. *Science, 310*, 791–793.

Burgdorf, J., Kroes, R. A., Moskal, J. R., Pfaus, J. G., et al. (2008). Ultrasonic vocalisations of rats (*Rattus norvegicus*) during mating, play, and aggression: Behavioural concomitants, relationship to reward, and self-administration of playback. *Journal of Comparative Psychology, 122*, 357–367.

Burguière, E., Monteiro, P., Feng, G., and Graybiel, A. M. (2013). Optogenetic stimulation of lateral orbitofronto-striatal pathway suppresses compulsive behaviours. *Science, 340*, 1243–1246.

Burke, L. K., and Heisler, L. K. (2015). 5-Hydroxytryptamine medications for the treatment of obesity. *Journal of Neuroendocrinology, 27*, 389–398.

Burke, T. M., Markwald, R. R., McHill, A. W., Chinoy, E. D., et al. (2015). Effects of caffeine on the human circadian clock in vivo and in vitro. *Science Translational Medicine, 7*(305), 305ra146.

Burmeister, M. A., Ayala, J. E., Smouse, H., Landivar-Rocha, A., et al. (2017). The hypothalamic glucagon-like peptide 1 receptor is sufficient but not necessary for the regulation of energy balance and glucose homeostasis in mice. *Diabetes, 66*(2), 372–384.

Bushdid, C., Magnasco, M. O., Vosshall, L. B., and Keller, A. (2014). Humans can discriminate more than 1 trillion olfactory stimuli. *Science, 343*(6177), 1370–1372.

Bushman, J. D., Ye, W., and Liman, E. R. (2015). A proton current associated with sour taste: Distribution and functional properties. *FASEB Journal, 7*, 3014–3026.

Buss, D. M. (2019). *Evolutionary psychology: The new science of the mind* (6th ed.). Needham Heights, MA: Allyn & Bacon.

Butler, A. C., Chapman, J. E., Forman, E. M., and Beck, A. T. (2006). The empirical status of cognitive-behavioural therapy: A review of meta-analyses. *Clinical Psychology Review, 26*, 17–31.

Butti, C., Santos, M., Uppal, N., and Hof, P. R. (2013). Von Economo neurons: Clinical and evolutionary perspectives. *Cortex, 49*(1), 312–326.

Byrne, J. H., and Hawkins, R. D. (2015). Nonassociative learning in invertebrates. *Cold Spring Harbor Perspectives in Biology, 7*(5), pii: a021675.

C

Cacioppo, J. T., Berntson, G. G., Larsen, J. T., Poehlmann, K. M., et al. (2000). The psychophysiology of emotion. In M. Lewis and J. M. Haviland-Jones (Eds.), *Handbook of emotions* (2nd ed., pp. 173–191). New York: Guilford.

Cadoret, R. J., O'Gorman, T., Troughton, E., and Heywood, E. (1986). An adoption study of genetic and environmental factors in drug abuse. *Archives of General Psychiatry, 43*, 1131–1136.

Cahill, L., Prins, B., Weber, M., and McGaugh, J. L. (1994). Beta-adrenergic activation and memory for emotional events. *Nature, 371*, 702–704.

Caldwell, J. A. (2012). Crew schedules, sleep deprivation, and aviation performance. *Current Directions in Psychological Science, 21*(2), 85–89.

Calebiro, D., and Jobin, M. L. (2018). Hot spots for GPCR signalling: Lessons from single-molecule microscopy. *Current Opinion in Cell Biology, 57*, 57–63.

Calvin, W. H., and Ojemann, G. A. (1994). *Conversations with Neil's brain: The neural nature of thought and language.* Reading, MA: Addison-Wesley.

Cameron, J. L., Eagleson, K. L., Fox, N. A., Hensch, T. K., et al. (2017). Social origins of developmental risk for mental and physical illness. *Journal of Neuroscience, 37*(45), 10783–10791.

Campbell, F. W., and Robson, J. G. (1968). Application of Fourier analysis to the visibility of gratings. *Journal of Physiology (London), 197,* 551–566.

Campbell, S. S., and Tobler, I. (1984). Animal sleep: A review of sleep duration across phylogeny. *Neuroscience and Biobehavioural Reviews, 8,* 269–301.

Cannon, T. D., Thompson, P. M., van Erp, T. G., Toga, A. W., et al. (2002). Cortex mapping reveals regionally specific patterns of genetic and disease-specific gray-matter deficits in twins discordant for schizophrenia. *Proceedings of the National Academy of Sciences, USA, 99*(5), 3228–3233.

Cannon, W. B. (1929). *Bodily changes in pain, hunger, fear and rage.* New York: Appleton.

Cantalupo, C., and Hopkins, W. D. (2001). Asymmetric Broca's area in great apes. *Nature, 414,* 505.

Cantor, J. M., Blanchard, R., Paterson, A. D., and Bogaert, A. F. (2002). How many gay men owe their sexual orientation to fraternal birth order? *Archives of Sexual Behaviour, 31,* 63–71.

Cao, M., Shu, N., Cao, Q., Wang, Y., et al. (2014). Imaging functional and structural brain connectomics in attention-deficit/hyperactivity disorder. *Molecular Neurobiology, 50*(3), 1111–1123.

Cao, Y. Q., Mantyh, P. W., Carlson, E. J., Gillespie, A. M., et al. (1998). Primary afferent tachykinins are required to experience moderate to intense pain. *Nature, 392,* 390–394.

Capitão, L. P, Murphy, S. E., Browning, M., Cowen, P.J., et al. (2015). Acute fluoxetine modulates emotional processing in young adult volunteers. *Psychological Medicine, 45*(11), 2295–2308.

Cappuccio, F. P., Cooper, D., D'Elia, L., Strazzullo, P., et al. (2011). Sleep duration predicts cardiovascular outcomes: A systematic review and meta-analysis of prospective studies. *European Heart Journal, 32*(12), 1484–1492.

Caramazza, A., Anzellotti, S., Strnad, L., and Lingnau, A. (2014). Embodied cognition and mirror neurons: A critical assessment. *Annual Review of Neuroscience, 37,* 1–15.

Cardenas, V. A., Studholme, C., Gazdzinski, S., Durazzo, T. C., et al. (2007). Deformation-based morphometry of brain changes in alcohol dependence and abstinence. *NeuroImage, 34*(3), 879–887.

Cardno, A. G., and Gottesman, I. I. (2000). Twin studies of schizophrenia: From bow-and-arrow concordances to star wars Mx and functional genomics. *American Journal of Medical Genetics, 97,* 12–17.

Carhart-Harris, R. L., Erritzoe, D., Williams, T., Stone, J. M., et al. (2012). Neural correlates of the psychedelic state as determined by fMRI studies with psilocybin. *Proceedings of the National Academy of Sciences, USA, 109,* 2138–2143.

Carhart-Harris, R. L., Leech, R., Williams, T. M., Erritzoe, D., et al. (2012). Implications for psychedelic-assisted psychotherapy: Functional magnetic resonance imaging study with psilocybin. *British Journal of Psychiatry, 200,* 238–244.

Carmichael, M. S., Warburton, V. L., Dixen, J., and Davidson, J. M. (1994). Relationships among cardiovascular, muscular, and oxytocin responses during human sexual activity. *Archives of Sexual Behaviour, 23,* 59–79.

Carpen, J. D., Archer, S. N., Skene, D. J., Smits, M., et al. (2005). A single-nucleotide polymorphism in the 5′-untranslated region of the *hPER2* gene is associated with diurnal preference. *Journal of Sleep Research, 14,* 293–297.

Carr, G. V., and Lucki, I. (2011). The role of serotonin receptor subtypes in treating depression: A review of animal studies. *Psychopharmacology (Berlin), 213*(2–3), 265–287.

Carr, M., and Nielsen, T. (2017). A novel Differential Susceptibility framework for the study of nightmares: Evidence for trait sensory processing sensitivity. *Clinical Psychology Review, 58,* 86–96.

Carré, J. M., Geniole, S. N., Ortiz, T. L., Bird, B. M., et al. (2017). Exogenous testosterone rapidly increases aggressive behaviour in dominant and impulsive men. *Biological Psychiatry, 82*(4), 249–256.

Carreiras, M., Lopez, J., Rivero, F., and Corina, D. (2005). Linguistic perception: Neural processing of a whistled language. *Nature, 433,* 31–32.

Carretié, L. (2014). Exogenous (automatic) attention to emotional stimuli: A review. *Cognitive, Affective, & Behavioural Neuroscience, 14*(4), 1228–1258.

Carroll, J., McMahon, C., Neitz, M., and Neitz, J. (2000). Flicker-photometric electroretinogram estimates of L:M cone photoreceptor ratio in men with photopigment spectra derived from genetics. *Journal of the Optical Society of America. Part A, Optics, Image Science, and Vision, 17,* 499–509.

Carter, C. S. (1992). Oxytocin and sexual behaviour. *Neuroscience and Biobehavioural Reviews, 16,* 131–144.

Cartwright, R. D. (1979). The nature and function of repetitive dreams: A survey and speculation. *Psychiatry, 42,* 131–137.

Casarosa, S., Bozzi, Y., and Conti, L. (2014). Neural stem cells: Ready for therapeutic applications? *Molecular and Cellular Therapies, 2,* 31.

Casey, B., and Hackett, B. P. (2000). Left-right axis malformations in man and mouse. *Current Opinion in Genetics and Development, 10,* 257–261.

Caspi, A., Sugden, K., Moffitt, T. E., Taylor, A., et al. (2003). Influence of life stress on depression: Moderation by a polymorphism in the 5-HTT gene. *Science, 301,* 386–389.

Casson, I. R., Sham, R., Campbell, E. A., Tarlau, M., et al. (1982). Neurological and CT evaluation of knocked-out boxers. *Journal of Neurology, Neurosurgery and Psychiatry, 45,* 170–174.

Cassone, V. M. (2014). Avian circadian organisation: A chorus of clocks. *Frontiers in Neuroendocrinology, 35*(1), 76–88.

Castellanos, F. X., Lee, P. P., Sharp, W., Jeffries, N. O., et al. (2002). Developmental trajectories of brain volume abnormalities in children and adolescents with attention-deficit/hyperactivity disorder. *Journal of the American Medical Association, 288,* 1740–1748.

Cataneo, A. J. M., Felisberto, G., Jr., and Cataneo, D. C. (2018). Thymectomy in nonthymomatous myasthenia gravis: Systematic review and meta-analysis. *Orphanet Journal of Rare Diseases, 13*(1), 99.

Catania, K. C. (2001). Early development of a somatosensory fovea: A head start in the cortical space race? *Nature Neuroscience, 4,* 353–354.

Catania, K. C. (2019). All in the family: Touch versus olfaction in moles. *Anatomical Record (Hoboken).* Jan 7. http://dx.doi.org/10.1002/ar.24057.

Caterina, M. J., Leffler, A., Malmberg, A. B., Martin, W. J., et al. (2000). Impaired nociception and pain sensation in mice lacking the capsaicin receptor. *Science, 288,* 306–313.

Caterina, M. J., Schumacher, M. A., Tominaga, M., Rosen, T. A., et al. (1997). The capsaicin receptor: A heat-activated ion channel in the pain pathway. *Nature, 389,* 816–824.

Catterall, W. A., and Yu, F. H. (2006). Painful channels. *Neuron, 52,* 743–744.

Celeghin, A., de Gelder, B., and Tamietto, M. (2015). From affective blindsight to emotional consciousness. *Consciousness and Cognition, 36,* 414–425.

Centres for Disease Control and Prevention (CDC). (2010). Current depression among adults: United States, 2006 and 2008. *Morbidity and Mortality Weekly Report, 59*(38), 1229–1235.

Centres for Disease Control and Prevention (CDC). (2015). *Report to Congress on traumatic brain injury in the United States: Epidemiology and rehabilitation.* Atlanta, GA: National Centre for Injury Prevention and Control; Division of Unintentional Injury Prevention.

Centres for Disease Control and Prevention. (2018, July). Suicide rising across the US. *Vital Signs.* https://www.cdc.gov/vitalsigns/suicide/.

Centres for Disease Control: Alcohol Use in Pregnancy, https://www.cdc.gov/ncbddd/fasd/alcohol-use.html.

Cha, A. E. and McGinley, L. (2019, June 21). A new 'female Viagra' approved by FDA despite skepticism. *The Washington Post,* Health Section (https://www.washingtonpost.com/health/2019/06/21/new-female-viagra-approved-by-fda-despite-skepticism/?utm_term=.4d2c3c13c84a).

Challoumas, D., Ferro, A., Walker, A., and Brassett, C. (2018). Observations on the inconsistency of dermatome maps and its effect on knowledge and confidence in clinical students. *Clinical Anatomy, 31*(2), 293–300.

Chamberlain, S. R., Menzies, L., Hampshire, A., Suckling, J., et al. (2008). Orbitofrontal dysfunction in patients with obsessive-compulsive disorder and their unaffected relatives. *Science, 321,* 421–422.

Champagne, F., Diorio, J., Sharma, S., and Meaney, M. J. (2001). Naturally occurring variations in maternal behaviour in the rat are associated with differences in estrogen-inducible central oxytocin receptors. *Proceedings of the National Academy of Sciences, USA, 98,* 12736–12741.

Chan, M.Y., Na, J., Agres, P. F., Savalia, N. K., et al. (2018). Socioeconomic status moderates age-related differences in the brain's functional network organisation and anatomy across the adult lifespan. *Proceedings of the National Academy of Sciences, USA, 115*(22), E5144–E5153.

Chandrashekar, J., Hoon, M. A., Ryba, N. J., and Zuker, C. S. (2006). The receptors and cells for mammalian taste. *Nature, 444,* 288–294.

Chandrashekar, J., Mueller, K. L., Hoon, M. A., Adler, E., et al. (2000). T2Rs function as bitter taste receptors. *Cell, 100,* 703–711.

Chandrashekar, J., Yarmolinsky, D., von Buchholtz, L., Oka, Y., et al. (2009). The taste of carbonation. *Science, 326,* 443–445.

Chang, B. S., Ly, J., Appignani, B., Bodell, A., et al. (2005). Reading impairment in the neuronal migration disorder of periventricular nodular heterotopia. *Neurology, 64,* 799–803.

Changeux, J. P. (2018). Structural identification of the nicotinic receptor ion channel. *Trends in Neurosciences, 41*(2), 67–70.

Chapman, C. D., Dono, L. M., French, M. C., Weinberg, Z.Y., et al. (2012). Paraventricular nucleus anandamide signalling alters eating and substrate oxidation. *Neuroreport, 23*(7), 425–429.

Charlier, T. D. (2009). Importance of steroid receptor coactivators in the modulation of steroid action on brain and behaviour. *Psychoneuroendocrinology, 34*(Suppl. 1), S20–29.

Charney, D. S., Deutch, A.Y., Krystal, J. H., Southwick, S. M., et al. (1993). Psychobiologic mechanisms of posttraumatic stress disorder. *Archives of General Psychiatry, 50,* 295–305.

Chasman, D. I., Schürks, M., Anttila, V., deVries, B., et al. (2011). Genome-wide association study reveals three susceptibility loci for common migraine in the general population. *Nature Genetics, 43,* 695–698.

Chatzigeorgiou, M., Bang, S., Hwang, S. W., and Schafer, W. R. (2013). *Tmc-1* encodes a sodium-sensitive channel required for salt chemosensation in *C. elegans*. *Nature, 494*(7435), 95–99.

Chaudhari, N., Landin, A. M., and Roper, S. D. (2000). A metabotropic glutamate receptor variant functions as a taste receptor. *Nature Neuroscience, 3,* 113–119.

Chelikani, P. K., Haver, A. C., and Reidelberger, R. D. (2005). Intravenous infusion of peptide YY(3-36) potently inhibits food intake in rats. *Endocrinology, 146,* 879–888.

Chemelli, R. M., Willie, J. T., Sinton, C. M., Elmquist, J. K., et al. (1999). Narcolepsy in orexin knockout mice: Molecular genetics of sleep regulation. *Cell, 98,* 437–451.

Chen, C., Crivelli, C., Garrod, O. G. B., Schyns, P. G., et al. (2018). Distinct facial expressions represent pain and pleasure across cultures. *Proceedings of the National Academy of Sciences, USA, 115*(43), E10013–E10021.

Chen, S.-K., Badea, T. C., Hattar, S. (2011). Photoentrainment and pupillary light reflex are mediated by distinct populations of ipRGCs. *Nature, 476,* 92–95.

Chen, X., Gabitto, M., Peng, Y., Ryba, N. J., et al. (2011). A gustotopic map of taste qualities in the mammalian brain. *Science, 333*(6047), 1262–1266.

Chenn, A., and Walsh, C. A. (2002). Regulation of cerebral cortical size by control of cell cycle exit in neural precursors. *Science, 297,* 365–369.

Cherry, E. C. (1953). Some experiments on the recognition of speech, with one and with two ears. *Journal of the Acoustical Society of America, 25,* 975–979.

Chesney, E., Goodwin, G. M., and Fazel, S. (2014). Risks of all-cause and suicide mortality in mental disorders: A meta-review. *World Psychiatry, 13*(2), 153–160.

Cheyne, J. A. (2002). Situational factors affecting sleep paralysis and associated hallucinations: Position and timing effects. *Journal of Sleep Research, 11,* 169–177.

Chiti, F., and Dobson, C. M. (2017). Protein misfolding, amyloid formation, and human disease: A summary of progress over the last decade. *Annual Review of Biochemistry, 86,* 27–68.

Cho, I., Yamanishi, S., Cox, L., Methé, B. A., et al. (2012). Antibiotics in early life alter the murine colonic microbiome and adiposity. *Nature, 488*(7413), 621–626.

Cho, K. (2001). Chronic "jet lag" produces temporal lobe atrophy and spatial cognitive deficits. *Nature Neuroscience, 4,* 567–568.

Choquet, D., and Triller, A. (2013). The dynamic synapse. *Neuron, 80,* 691–703.

Chua, H. C., and Chebib, M. (2017). GABA(A) receptors and the diversity in their structure and pharmacology. *Advances in Pharmacology, 79,* 1–34.

Chung, S., Weber, F., Zhong, P., Tan, C. L., et al. (2017). Identification of preoptic sleep neurons using retrograde labelling and gene profiling. *Nature, 545*(7655), 477–481.

Chung, W. S., Clarke, L. E., Wang, G. X., Stafford, B. K., et al. (2013). Astrocytes mediate synapse elimination through MEGF10 and MERTK pathways. *Nature, 504*(7480), 394–400.

Chung, W. S., Welsh, C. A., Barres, B. A., and Stevens, B. (2015). Do glia drive synaptic and cognitive impairment in disease? *Nature Neuroscience, 18*(11), 1539–1545.

Chung, Y., Haut, K. M., He, G., van Erp, T. G. M., et al.; North American Prodrome Longitudinal Study (NAPLS) Consortium. (2017). Ventricular enlargement and progressive reduction of cortical gray matter are linked in prodromal youth who develop psychosis. *Schizophrenia Research, 189,* 169–174.

Chung, Y., Jacobson, A., He, G., van Erp, T. G., et al. (2015). Prodromal symptom severity predicts accelerated gray matter reduction and third ventricle expansion among clinically high risk youth developing psychotic disorders. *Molecular Neuropsychiatry, 1,* 13–22.

Ciccocioppo, R., Martin-Fardon, R., and Weiss, F. (2004). Stimuli associated with a single cocaine experience elicit long-lasting cocaine-seeking. *Nature Neuroscience, 7,* 495–496.

Cipriani, A., Furukawa, T. A., Salanti, G., Chaimani, A., et al. (2018). Comparative efficacy and acceptability of 21 antidepressant drugs for the acute treatment of adults with major depressive disorder: A systematic review and network meta-analysis. *Lancet, 391*(10128), 1357–1366.

Cipriani, A., Zhou, X., Del Giovane, C., Hetrick, S. E., et al. (2016). Comparative efficacy and tolerability of antidepressants for major depressive disorder in children and adolescents: A network meta-analysis. *Lancet, 388*(10047), 881–890.

Cirulli, E. T., Lasseigne, B. N., Petrovski, S., Sapp, P. C., et al. (2015). Exome sequencing in amyotrophic lateral sclerosis identifies risk genes and pathways. *Science, 347*(6229), 1436–1441.

Clapper, J. R., Moreno-Sanz, G., Russo, R., Guijarro, A., et al. (2010). Anandamide suppresses pain initiation through a peripheral endocannabinoid mechanism. *Nature Neuroscience, 13,* 1265–1270.

Clark, L. (2014). Disordered gambling: The evolving concept of behavioural addiction. *Annals of the New York Academy of Sciences, 1327,* 46–61.

Clarke, E., Reichard, U. H., and Zuberbühler, K. (2015). Context-specific close-range "hoo" calls in wild gibbons (*Hylobates lar*). *BMC Evolutionary Biology, 15,* 56.

Clarke, M. C., Tanskanen, A., Huttunen, M., Leon, D. A., et al. (2011). Increased risk of schizophrenia from additive interaction between infant motor developmental delay and obstetric complications: Evidence from a population-based longitudinal study. *American Journal of Psychiatry, 168,* 1295–1302.

Clemens, L. G., Gladue, B. A., and Coniglio, L. P. (1978). Prenatal endogenous androgenic influences on masculine sexual behaviour and genital morphology in male and female rats. *Hormones and Behaviour, 10,* 40–53.

Clemente, C. D., and Sterman, M. B. (1967). Limbic and other forebrain mechanisms in sleep induction and behavioural inhibition. *Progress in Brain Research, 27,* 34–37.

Clutton-Brock, T. H., and Harvey, P. H. (1980). Primates, brains and ecology. *Journal of Zoology, 190,* 309–323.

Cnotka, J., Güntürkün, O., Rehkämper, G., Gray, R. D., et al. (2008). Extraordinary large brains in tool-using New Caledonian crows (*Corvus moneduloides*). *Neuroscience Letters, 433,* 241–245.

Coelho, M., and Ferreira, J. J. (2012). Late-stage Parkinson disease. *Nature Reviews Neurology, 8*, 435–442.

Cogan, G. B., Thesen, T., Carlson, C., Doyle, W., et al. (2014). Sensory-motor transformations for speech occur bilaterally. *Nature, 507*(7490), 94–98.

Coggan, J. S., Bartol, T. M., Esquenazi, E., Stiles, J. R., et al. (2005). Evidence for ectopic neurotransmission at a neuronal synapse. *Science, 309*, 446–451.

Coghill, R. C., McHaffie, J. G., and Yen, Y.-F. (2003). Neural correlates of interindividual differences in the subjective experience of pain. *Proceedings of the National Academy of Sciences, USA, 100*, 8538–8542.

Cohen, A. J., and Leckman, J. F. (1992). Sensory phenomena associated with Gilles de la Tourette's syndrome. *Journal of Clinical Psychiatry, 53*, 319–323.

Cohen, M. R., and Newsome, W. T. (2008). Context-dependent changes in functional circuitry in visual area MT. *Neuron, 60*(1), 162–173.

Cohen, N. J., and Squire, L. R. (1980). Preserved learning and retention of pattern-analyzing skill in amnesia: Dissociation of knowing how and knowing what. *Science, 210*, 207–210.

Cohen, S. P., and Mao, J. (2014). Neuropathic pain: Mechanisms and their clinical implications. *BMJ*, 348:f7656. https://doi.org/10.1136/bmj.f7656.

Cohen, S., Alper, C. M., Doyle, W. H., Treanor, J. J., et al. (2006). Positive emotional style predicts resistance to illness after experimental exposure to Rhinovirus or Influenza A virus. *Psychosomatic Medicine, 68*, 809–815.

Cohen, S., Doyle, W. J., Alper, C. M., Janicki-Deverts, D., et al. (2009). Sleep habits and susceptibility to the common cold. *Archive of Internal Medicine, 169*, 62–67.

Cohen, S., Janicki-Deverts, D., Turner, R. B., and Doyle, W. J. (2015). Does hugging provide stress-buffering social support? A study of susceptibility to upper respiratory infection and illness. *Psychological Science, 26*, 135–147.

Cole, J. (1995). *Pride and a daily marathon.* Cambridge, MA: MIT Press.

Collings, V. B. (1974). Human taste response as a function of locus of stimulation on the tongue and soft palate. *Perception and Psychophysics, 16*, 169–174.

Colman, R. J., Anderson, R. M., Johnson, S. C., Kastman, E. K., et al. (2009). Caloric restriction delays disease onset and mortality in rhesus monkeys. *Science, 325*(5937), 201–204.

Colom, R., Burgaleta, M., Román, F. J., Karama, S., et al. (2013). Neuroanatomic overlap between intelligence and cognitive factors: Morphometry methods provide support for the key role of the frontal lobes. *NeuroImage, 72*, 143–152.

Coltman, D. W., O'Donoghue, P., Jorgenson, J. T., Hogg, J. Y., et al. (2003). Undesirable evolutionary consequences of trophy hunting. *Nature, 426*, 655–658.

Colwell, C. S. (2010). Preventing dehydration during sleep. *Nature Neuroscience, 13*(4), 403–404.

Conel, J. L. (1939). *The postnatal development of the human cerebral cortex: Vol. 1. The cortex of the newborn.* Cambridge, MA: Harvard University Press.

Conel, J. L. (1947). *The postnatal development of the human cerebral cortex: Vol. 3. The cortex of the three-month infant.* Cambridge, MA: Harvard University Press.

Conel, J. L. (1959). *The postnatal development of the human cerebral cortex: Vol. 6. The cortex of the twenty-four-month infant.* Cambridge, MA: Harvard University Press.

Confer, J. C., Easton, J. A., Fleischman, D. S., Goetz, C. D., et al. (2010). Evolutionary psychology. Controversies, questions, prospects, and limitations. *American Psychologist, 65*, 110–126.

Conley, A. C., Cooper, P. S., Karayanidis, F., Gardner, A. J., et al. (2018). Resting state electroencephalography and sports-related concussion: A systematic review. *Journal of Neurotrauma, 36*(1).

Conley, A. J., Corbin, C. J., Browne, P., Mapes, S. M., et al. (2007). Placental expression and molecular characterisation of aromatase cytochrome P450 in the spotted hyena (*Crocuta crocuta*). *Placenta, 28*(7), 668–675.

Conn, M. P., and Parker, J. V. (2008). *The animal research war.* New York: Palgrave Macmillan.

Connolly, A. C., Guntupalli, J. S., Gors, J., Hanke, M., et al. (2012). The representation of biological classes in the human brain. *Journal of Neuroscience, 32*(8), 2608–2618.

Constantine-Paton, M., Cline, H. T., and Debski, E. (1990). Patterned activity, synaptic convergence, and the NMDA receptor in developing visual pathways. *Annual Review of Neuroscience, 13*, 129–154.

Cook, L. M., Grant, B. S., Saccheri, I. J., and Mallet, J. (2012). Selective bird predation on the peppered moth: The last experiment of Michael Majerus. *Biology Letters, 8*(4), 609–612.

Cooke, B. M., Breedlove, S. M., and Jordan, C. L. (2003). Both estrogen receptors and androgen receptors contribute to testosterone-induced changes in the morphology of the medial amygdala and sexual arousal in male rats. *Hormones and Behaviour, 43*, 336–346.

Cooke, B. M., Chowanadisai, W., and Breedlove, S. M. (2000). Post-weaning social isolation of male rats reduces the volume of the medial amygdala and leads to deficits in adult sexual behaviour. *Behavioural Brain Research, 117*, 107–113.

Cope, L. M., Ermer, E., Gaudet, L. M., Steele, V. R., et al. (2014). Abnormal brain structure in youth who commit homicide. *NeuroImage: Clinical, 4*, 800–807.

Cope, N., Harold, D., Hill, G., Moskvina, V., et al. (2005). Strong evidence that KIAA0319 on chromosome 6p is a susceptibility gene for developmental dyslexia. *American Journal of Human Genetics, 76*, 581–591.

Corballis, M. C. (2002). *From hand to mouth: The origins of language.* Princeton, NJ: Princeton University Press.

Corballis, M. C. (2014). Left brain, right brain: Facts and fantasies. *PLOS Biology, 12*(1), e1001767.

Corbetta, M., Kincade, J. M., Ollinger, J. M., McAvoy, M. P., et al. (2000). Voluntary orienting is dissociated from target detection in human posterior parietal cortex. *Nature Neuroscience, 3*, 292–297.

Corbetta, M., and Shulman, G. L. (1998). Human cortical mechanisms of visual attention during orienting and search. *Philosophical Transactions of the Royal Society of London. Series B: Biological Sciences, 353*, 1353–1362.

Corbetta, M., and Shulman, G. L. (2002). Control of goal-directed and stimulus-driven attention in the brain. *Nature Reviews Neuroscience, 3*, 201–215.

Corcoran, A. J., Barber, J. R., and Conner, W. E. (2009). Tiger moth jams bat sonar. *Science, 325*, 325–327.

Coricelli, G., Critchley, H. D., Joffily, M., O'Doherty, J. P., et al. (2005). Regret and its avoidance: A neuroimaging study of choice behaviour. *Nature Neuroscience, 8*, 1255–1262.

Corkin, S. (2002). What's new with the amnesic patient H.M.? *Neuroscience, 3*, 153–159.

Corkin, S., Amaral, D. G., Gonzalez, R. G., Johnson, K. A., et al. (1997). H.M.'s medial temporal lobe lesion: Findings from magnetic resonance imaging. *Journal of Neuroscience, 17*, 3964–3979.

Correa, A. M., and Bezanilla, F. (1994). Gating of the squid sodium channel at positive potentials. II. Single channels reveal two open states. *Biophysical Journal, 66*, 1864–1878.

Correll, C. U., Rubio, J. M., Inczedy-Farkas, G., Birnbaum, M. L., et al. (2017). Efficacy of 42 pharmacologic cotreatment strategies added to antipsychotic monotherapy in schizophrenia: Systematic overview and quality appraisal of the meta-analytic evidence. *JAMA Psychiatry, 74*(7), 675–684.

Corti, O., Lesage, S., and Brice, A. (2011). What genetics tells us about the causes and mechanisms of Parkinson's disease. *Physiological Reviews, 91*(4), 1161–1218.

Cosmides, L., and Tooby, J. (2000). Evolutionary psychology and the emotions. In M. Lewis and J. M. Haviland-Jones (Eds.), *Handbook of emotions* (2nd ed., pp. 91–115). New York: Guilford.

Costa, A., and Sebastián-Gallés, N. (2014). How does the bilingual experience sculpt the brain? *Nature Reviews Neuroscience, 15*(5), 336–345.

Costanzo, R. M. (1991). Regeneration of olfactory receptor cells. *CIBA Foundation Symposium, 160*, 233–242.

Counotte, D. S., Goriounova, N. A., Li, K. W., Loos, M., et al. (2011). Lasting synaptic changes underlie attention deficits caused by nicotine exposure during adolescence. *Nature Neuroscience, 14*(4), 417–419.

Courtiol, A., Pettay, J. E., Jokela, M., Rotkirch, A., et al. (2012). Natural and sexual selection in a monogamous historical human population. *Proceedings of the National Academy of Sciences, USA, 109*(21), 8044–8049.

Cowan, N. (2008). What are the differences between long-term, short-term, and working memory? *Progress in Brain Research, 169*, 323–338. doi: 10.1016/S0079-6123(07)00020-9. https://www.ncbi.nlm.nih.gov/pubmed/18394484

Cowan, W. M. (1979). The development of the brain. *Scientific American, 241*(3), 112–133.

Cox, J. J., Reimann, F., Nicholas, A. K., Thornton, G., et al. (2006). An SCN9A channelopathy causes congenital inability to experience pain. *Nature, 444*, 894–898.

Cox, L. M., and Blaser, M. J. (2015). Antibiotics in early life and obesity. *Nature Reviews Endocrinology, 11*(3), 182–190.

Crabbe, J. C. (2014). Use of animal models of alcohol-related behaviour. *Handbook of Clinical Neurology, 125*, 71–86.

Cragg, B. G. (1975). The development of synapses in the visual system of the cat. *Journal of Comparative Neurology, 160*, 147–166.

Craig, A. D., Reiman, E. M., Evans, A., and Bushnell, M. C. (1996). Functional imaging of an illusion of pain. *Nature, 384*, 258–260.

Cramer, P. E., Cirrito, J. R., Wesson, D.W., Lee, C.Y., et al. (2012). ApoE-directed therapeutics rapidly clear β-amyloid and reverse deficits in AD mouse models. *Science, 335*(6075), 1503–1506.

Craver, C. F., Kwan, D., Steindam, C., and Rosenbaum, R. S. (2014). Individuals with episodic amnesia are not stuck in time. *Neuropsychologia, 57*, 191–195.

Crews, F. T., Vetreno, R. P., Broadwater, M. A., and Robinson, D. L. (2016). Adolescent alcohol exposure persistently impacts adult neurobiology and behaviour. *Pharmacological Reviews, 68*(4), 1074–1109.

Crick, F. C., and Koch, C. (2005). What is the function of the claustrum? *Philosophical Transactions of the Royal Society of London. Series B: Biological Sciences, 360*(1458), 1271–1279.

Crivelli, C., Russell, J. A., Jarillo, S., and Fernández-Dols, J. M. (2017). Recognizing spontaneous facial expressions of emotion in a small-scale society of Papua New Guinea. *Emotion, 17*(2), 337–347.

Crockford, C., Wittig, R. M., Mundry, R., and Zuberbühler, K. (2012). Wild chimpanzees inform ignorant group members of danger. *Current Biology, 22*(2), 142–146.

Crossley, N. A., Constante, M., McGuire, P., and Power, P. (2010). Efficacy of atypical v. typical antipsychotics in the treatment of early psychosis: Meta-analysis. *British Journal of Psychiatry, 196*(6), 434–439.

Crow, T. J. (1980). Positive and negative schizophrenic symptoms and the role of dopamine. *British Journal of Psychiatry, 137*, 383–386.

Crowell, A., Riva-Posse, P., Garlow, S., and Mayberg, H. (2014). Toward an understanding of the neural circuitry of major depressive disorder through the clinical response to deep brain stimulation of different anatomical targets. *Current Behavioural Neuroscience Reports, 1*, 55–63.

Crunelle, C. L., Kaag, A. M., van Wingen, G., van den Munkhof, H. E., et al. (2014). Reduced frontal brain volume in non-treatment-seeking cocaine-dependent individuals: Exploring the role of impulsivity, depression, and smoking. *Frontiers in Human Neuroscience, 8*, 7.

Cui, G., Jun, S. B., Jin, X., Pham, M. D., et al. (2013). Concurrent activation of striatal direct and indirect pathways during action initiation. *Nature, 494*(7436), 238–242.

Cui, M., Jiang, P., Maillet, E., Max, M., et al. (2006). The heterodimeric sweet taste receptor has multiple potential ligand binding sites. *Current Pharmaceutical Design, 12*, 4591–4600.

Culverhouse, R. C., Saccone, N. L., Horton, A. C., Ma, Y., et al. (2018). Collaborative meta-analysis finds no evidence of a strong interaction between stress and 5-HTTLPR genotype contributing to the development of depression. *Molecular Psychiatry, 23*(1), 133–142.

Cummings, D. E. (2006). Ghrelin and the short- and long-term regulation of appetite and body weight. *Physiology & Behaviour, 89*, 71–84.

Cunha, G. R., Risbridger, G., Wang, H., Place, N. J., et al. (2014). Development of the external genitalia: Perspectives from the spotted hyena (*Crocuta crocuta*). *Differentiation, 87*(1–2), 4–22.

Cunningham, C. A., Yassa, M. A., and Egeth, H. E. (2015). Massive memory revisited: Limitations on storage capacity for object details in visual long-term memory. *Learning and Memory, 22*(11), 563–566.

Curcio, C. A., Sloan, K. R., Packer, O., Hendrickson, A. E., et al. (1987). Distribution of cones in human and monkey retina: Individual variability and radial asymmetry. *Science, 236*, 579–582.

Curran, H. V., Freeman, T. P., Mokrysz, C., Lewis, D. A., et al. (2016). Keep off the grass? Cannabis, cognition and addiction. *Nature Reviews Neuroscience, 17*(5), 293–306.

Currie, P. J., John, C. S., Nicholson, M. L., Chapman, C. D., et al. (2010). Hypothalamic paraventricular 5-hydroxytryptamine inhibits the effects of ghrelin on eating and energy substrate utilisation. *Pharmacology, Biochemistry, and Behaviour, 97*(1), 152–155.

Curtis, V., Aunger, R., and Rabie, T. 2004. Evidence that disgust evolved to protect from risk of disease. *Proceedings of the Royal Society B — Biological Sciences, 271*(Suppl 4), S131–S133.

Curtiss, S. (1989). The independence and task-specificity of language. In M. H. Bornstein and J. S. Bruner (Eds.), *Interaction in human development* (pp. 105–137). Hillsdale, NJ: Erlbaum.

Cussotto, S., Sandhu, K. V., Dinan, T. G., and Cryan, J. F. (2018). The neuroendocrinology of the microbiota-gut-brain axis: A behavioural perspective. *Frontiers in Neuroendocrinology, 51*, 80–101.

Cypess, A. M, Lehman, S., Williams, G., Tal, I., et al. (2009). Identification and importance of brown adipose tissue in adult humans. *New England Journal of Medicine, 360*, 1509–1517.

Cyranoski, D. (2018). Reprogrammed stem cells implanted into patient with Parkinson's disease. *Nature*. https://doi.org/10.1038/d41586-018-07407-9.

Cyranoski, D. (2019). Japan's approval of stem-cell treatment for spinal-cord injury concerns scientists. *Nature, 565*, 544–545.

D

Da Mesquita, S., Fu, Z., and Kipnis, J. (2018). The meningeal lymphatic system: A new player in neurophysiology. *Neuron, 100*(2), 375–388.

Da Mesquita, S., Louveau, A., Vaccari, A., Smirnov, I., et al. (2018). Functional aspects of meningeal lymphatics in ageing and Alzheimer's disease. *Nature, 560*(7717), 185–191. https://doi.org/10.1038/s41586-018-0368-8.

Dabbs, J. M., and Morris, R. (1990). Testosterone, social class, and antisocial behaviour in a sample of 4,462 men. *Psychological Science, 1*, 209–211.

Dabbs, J. M., Ruback, R. B., Frady, R. L., Hopper, C. H., et al. (1988). Saliva testosterone and criminal violence among women. *Personality and Individual Differences, 9*, 269–275.

Dalakas, M. C. (2019). Immunotherapy in myasthenia gravis in the era of biologics. *Nature Reviews Neurology, 15*(2), 113–124.

Dale, N., Schacher, S., and Kandel, E. R. (1988). Long-term facilitation in *Aplysia* involves increase in transmitter release. *Science, 239*, 282–285.

Dale, R. C., Heyman, I., Giovannoni, G., and Church, A. W. (2005). Incidence of anti-brain antibodies in children with obsessive-compulsive disorder. *British Journal of Psychiatry, 187*, 314–319.

Damak, S., Rong, M., Yasumatsu, K., Kokrashvili, Z., et al. (2003). Detection of sweet and umami taste in the absence of taste receptor T1r3. *Science, 301*, 850–853.

Damasio, A. R., Grabowski, T. J., Bechara, A., Damasio, H., et al. (2000). Subcortical and cortical brain activity during the feeling of self-generated emotions. *Nature Neuroscience, 3*, 1049–1056.

Damasio, H., Grabowski, T., Frank, R., Galaburda, A. M., et al. (1994). The return of Phineas Gage: Clues about the brain from the skull of a famous patient. *Science, 264*, 1102–1105.

Daneman, R., Zhou, L., Kebede, A. A., and Barres, B. A. (2010). Pericytes are required for blood-brain barrier integrity during embryogenesis. *Nature, 468*, 562–566.

Daniels, D., and Marshall, A. (2012). Evaluating the potential for rostral diffusion in the cerebral ventricles using angiotensin II–induced drinking in rats. *Brain Research, 1486*, 62–67.

Dantz, B., Edgar, D. M., and Dement, W. C. (1994). Circadian rhythms in narcolepsy: Studies on a 90 minute day. *Electroencephalography and Clinical Neurophysiology, 90*, 24–35.

Dantzer, R., O'Connor, J. C., Freund, G. G., Johnson, R. W., et al. (2008). From inflammation to sickness and depression: When the immune system subjugates the brain. *Nature Reviews Neuroscience, 9*, 46–56.

Dapretto, M., Davies, M. S., Pfeifer, J. H., Scott, A. A., et al. (2006). Understanding emotions in others: Mirror neuron dysfunction in children with autism spectrum disorders. *Nature Neuroscience, 9*, 28–30.

Dark, J., Forger, N. G., and Zucker, I. (1984). Rapid recovery of body mass after surgical removal of adipose tissue in ground squirrels. *Proceedings of the National Academy of Sciences, USA, 81*, 2270–2272.

Darwin, C. (1859). *On the origin of species by means of natural selection, or, The preservation of favoured races in the struggle for life.* London: J. Murray.

Darwin, C. (1871). *The descent of man, and selection in relation to sex.* London: J. Murray.

Darwin, C. (1872). *The expression of the emotions in man and animals.* London: J. Murray.

Dauvilliers, Y., Siegel, J. M., Lopez, R., Torontali, Z.A., et al. (2014). Cataplexy: Clinical aspects, pathophysiology and management strategy. *Nature Reviews Neurology, 10*(7), 386–395.

Davey-Smith, G., Frankel, S., and Yarnell, J. (1997). Sex and death: Are they related? Findings from the Caerphilly Cohort Study. *British Medical Journal (Clinical Research Edition), 315*, 1641–1644.

David, L. A., Maurice, C. F., Carmody, R. N., Gootenberg, D. B., et al. (2014). Diet rapidly and reproducibly alters the human gut microbiome. *Nature, 505*, 559–563.

Davidson, J. M., Camargo, C. A., and Smith, E. R. (1979). Effects of androgen on sexual behaviour in hypogonadal men. *Journal of Clinical Endocrinology & Metabolism, 48*, 955–958.

Davis, J. I., Senghas, A., Brandt, F., and Ochsner, K. N. (2010). The effects of BOTOX injections on emotional experience. *Emotion, 10*, 433–440.

Davis, J. I., Senghas, A., and Ochsner, K. N. (2009). How does facial feedback modulate emotional experience? *Journal of Research in Personality, 43*, 822–829.

Davis, M. C., Horan, W. P., and Marder, S. R. (2014). Psychopharmacology of the negative symptoms: Current status and prospects for progress. *European Neuropsychopharmacology, 24*(5), 788–799.

De Felipe, C., Herrero, J. F., O'Brien, J. A., Palmer, J. A., et al. (1998). Altered nociception, analgesia and aggression in mice lacking the receptor for substance P. *Nature, 392*, 394–397.

de Gelder, B., Hortensius, R., and Tamietto, M. (2012). Attention and awareness each influence amygdala activity for dynamic bodily expressions: A short review. *Frontiers in Integrative Neuroscience, 6*, 54. doi:10.3389/fnint.2012.00054.

de Groot, C. M., Janus, M. D., and Bornstein, R. A. (1995). Clinical predictors of psychopathology in children and adolescents with Tourette syndrome. *Journal of Psychiatric Research, 29*, 59–70.

de Groot, J. H., Semin, G. R., and Smeets, M. A. (2017). On the communicative function of body odours. *Perspectives on Psychological Science, 12*(2), 306–324.

de Heer, W. A., Huth, A. G., Griffiths, T. L., Gallant, J. L., et al. (2017). The hierarchical cortical organisation of human speech processing. *Journal of Neuroscience, 37*(27), 6539–6557.

de Hemptinne, C., Swann, N. C., Ostrem, J. L., Ryapolova-Webb, E. S., et al. (2015). Therapeutic deep brain stimulation reduces cortical phase-amplitude coupling in Parkinson's disease. *Nature Neuroscience, 18*(5), 779–786.

de Jong, M., Schoorl, M., and Hoek, H. W. (2018). Enhanced cognitive behavioural therapy for patients with eating disorders: A systematic review. *Current Opinion in Psychiatry, 31*(6), 436–444.

de Kloet, E. R., and Joëls, M. (2017). Brain mineralocorticoid receptor function in control of salt balance and stress-adaptation. *Physiology & Behaviour, 178*, 13–20.

de Kluiver, H., Buizer-Voskamp, J. E., Dolan, C.V., and Boomsma, D. I. (2017). Paternal age and psychiatric disorders: A review. *American Journal of Medical Genetics. Part B, Neuropsychiatric Genetics, 174*(3), 202–213.

de la Monte, S. M., and Kril, J. J. 2014 Human alcohol-related neuropathology. *Acta Neuropathologica, 127*(1), 71–90.

de Luis, O., Valero, M. C., and Jurado, L. A. (2000). WBSCR14, a putative transcription factor gene deleted in Williams-Beuren syndrome: Complete characterisation of the human gene and the mouse ortholog. *European Journal of Human Genetics, 8*, 215–222.

de Paiva, A., Poulain, B., Lawrence, G. W., Shone, C. C., et al. (1993). A role for the interchain disulfide or its participating thiols in the internalisation of botulinum neurotoxin A revealed by a toxin derivative that binds to ecto-acceptors and inhibits transmitter release intracellularly. *Journal of Biological Chemistry, 268*, 20838–20844.

de Quervain, D., Schwabe, L., and Roozendaal, B. (2017). Stress, glucocorticoids and memory: Implications for treating fear-related disorders. *Nature Reviews Neuroscience, 18*(1), 7–19.

De Valois, R. L., and De Valois, K. K. (1980). Spatial vision. *Annual Review of Psychology, 31*, 309–341.

De Valois, R. L., and De Valois, K. K. (1988). *Spatial vision.* New York: Oxford University Press.

De Valois, R. L., and De Valois, K. K. (1993). A multi-stage colour model. *Vision Research, 33*, 1053–1065.

de Vivo, L., Bellesi, M., Marshall, W., Bushong, E. A., et al. (2017). Ultrastructural evidence for synaptic scaling across the wake/sleep cycle. *Science, 355*(6324), 507–510.

Dearborn, G.V. N. (1932). A case of congenital general pure analgesia. *Journal of Nervous and Mental Disease, 75*, 612–615.

Debanne, D., Campanac, E., Bialowas, A., Carlier, E., et al. (2011). Axon physiology. *Physiological Reviews, 91*(2), 555–602.

DeCasien, A. R., Williams, S. A., and Higham, J. P. (2017). Primate brain size is predicted by diet but not sociality. *Nature Ecology and Evolution, 1*(5), 112.

Dehaene, S., and Changeux, J. P. (2011). Experimental and theoretical approaches to conscious processing. *Neuron, 70*(2), 200–227.

Dehaene-Lambertz, G., Dehaene, S., and Hertz-Pannier, L. (2002). Functional neuroimaging of speech perception in infants. *Science, 298*, 2013–2015.

Dehkordi, O., Rose, J. E., Fatemi, M., Allard, J. S., et al. (2012). Neuronal expression of bitter taste receptors and downstream signalling molecules in the rat brainstem. *Brain Research, 1475*, 1–10.

del Campo, N., Fryer, T. D., Hong, Y. T., Smith, R., et al. (2013). A positron emission tomography study of nigro-striatal dopaminergic mechanisms underlying attention: Implications for ADHD and its treatment. *Brain, 136*(Pt. 11), 3252–3270.

Delbarco-Trillo, J., Burkert, B. A., Goodwin, T. E., and Drea, C. M. (2011). Night and day: The comparative study of strepsirrhine primates reveals socioecological and phylogenetic patterns in olfactory signals. *Journal of Evolutionary Biology, 24*(1), 82–98.

DelBello, M. P., Zimmerman, M. E., Mills, N. P., Getz, G. E., et al. (2004). Magnetic resonance imaging analysis of amygdala and other subcortical brain regions in adolescents with bipolar disorder. *Bipolar Disorder, 6*, 43–52.

Demb, J. B., Boynton, G. M., and Heeger, D. J. (1998). Functional magnetic resonance imaging of early visual pathways in dyslexia. *Journal of Neuroscience, 18*, 6939–6951.

Dement, W. C. (1974). *Some must watch while some must sleep.* San Francisco: Freeman.

Demyttenaere, K., Bruffaerts, R., Posada-Villa, J., Gasquet, I., et al. (2004). Prevalence, severity, and unmet need for treatment of mental disorders in the World Heatlh Organisation World Mental Health Surveys. *Journal of American Medical Association, 291*, 2581–2590.

Den Heijer, A. E., Groen, Y., Tucha, L., Fuermaier, A. B., et al. (2017). Sweat it out? The effects of physical exercise on cognition and behaviour in children and adults with ADHD: A systematic literature review. *Journal of Neural Transmission (Vienna), 124*(Suppl. 1), 3–26.

Dennis, S. G., and Melzack, R. (1983). Perspectives on phylogenetic evolution of pain expression. In R. L. Kitchell, H. H. Erickson, E. Carstens, and L. E. Davis (Eds.), *Animal pain* (pp. 151–161). Bethesda, MD: American Physiological Society.

Denton, D., Shade, R., Zamarippa, F., Egan, G., et al. (1999). Neuroimaging of genesis and satiation of thirst and an interoceptor-driven theory of origins of primary consciousness. *Proceedings of the National Academy of Sciences, USA, 96*, 5304–5309.

Depaepe, V., Suarez-Gonzalez, N., Dufour, A., Passante, L., et al. (2005). Ephrin signalling controls brain size by regulating apoptosis of neural progenitors. *Nature, 435*, 1244–1250.

Derecki, N., Cronk, J. C., Lu, Z., Xu, E., et al. (2012). Wild-type microglia arrest pathology in a mouse model of Rett syndrome. *Nature, 484*, 105–109.

Derrick, B. E., and Martinez, J. L. (1994). Frequency-dependent associative long-term potentiation at the hippocampal mossy fibre-CA3 synapse. *Proceedings of the National Academy of Sciences, USA, 91,* 10290–10294.

DeRubeis, R. J., Siegle, G. J., and Hollon, S. D. (2008). Cognitive therapy versus medication for depression: Treatment outcomes and neural mechanisms. *Nature, 9,* 788–796.

Descartes, R. (1662). *De homine.* Paris: Petrvm Leffen & Franciscvm Moyardvm.

Deshpande, D. A., Wang, W. C., McIlmoyle, E. L., Robinett, K. S., et al. (2010). Bitter taste receptors on airway smooth muscle bronchodilate by localized calcium signalling and reverse obstruction. *Nature Medicine, 16,* 1299–1304.

DeSisto, M. J., Harding, C. M., McCormick, R. V., Ashikaga, T., et al. (1995). The Maine and Vermont three-decade studies of serious mental illness. I. Matched comparison of cross-sectional outcome. *British Journal of Psychiatry, 167,* 331–338.

Devane, W. A., Dysarz, F. A., Johnson, M. R., Melvin, L. S., et al. (1988). Determination and characterisation of a cannabinoid receptor in rat brain. *Molecular Pharmacology, 34,* 605–613.

Devane, W. A., Hanus, L., Breuer, A., Pertwee, R. G., et al. (1992). Isolation and structure of a brain constituent that binds the cannabinoid receptor. *Science, 258,* 1946–1949.

Devlin, J. T., and Watkins, K. E. (2007). Stimulating language: Insights from TMS. *Brain, 130*(Pt. 3), 610–622.

DeVoogd, T. J. (1994). Interactions between endocrinology and learning in the avian song system. *Annals of the New York Academy of Sciences, 743,* 19–41.

DeWall, C. N., MacDonald, G., Webster, G. D., Masten, C. L., et al. (2010). Acetaminophen reduces social pain: Behavioural and neural evidence. *Psychological Science, 21*(7), 931–937.

Dewan, A., Pacifico, R., Zhan, R., Rinberg, D., et al. (2013). Non-redundant coding of aversive odours in the main olfactory pathway. *Nature, 497*(7450), 486–489.

Dewsbury, D. A. (1972). Patterns of copulatory behaviour in male mammals. *Quarterly Review of Biology, 47,* 1–33.

Dhabhar, F. S. (2018). The short-term stress response: Mother nature's mechanism for enhancing protection and performance under conditions of threat, challenge, and opportunity. *Frontiers in Neuroendocrinology, 49,* 175–192.

Di Marzo, V., Goparaju, S. K., Wang, L., Liu, J., et al. (2001). Leptin-regulated endocannabinoids are involved in maintaining food intake. *Nature, 410,* 822–825.

Di Marzo, V., and Matias, I. (2005). Endocannabinoid control of food intake and energy balance. *Nature Neuroscience, 8,* 585–589.

Di Pelligrino, G., Fadiga, L., Fogassi, L., Galese, V., et al. (1992). Understanding motor events: A neurophysiological study. *Experimental Brain Research, 91,* 176–180.

Diamond, A. (2013). Executive functions. *Annual Review of Psychology, 64,* 135–168.

Diamond, M. C. (1967). Extensive cortical depth measurements and neuron size increases in the cortex of environmentally enriched rats. *Journal of Comparative Neurology, 131,* 357–364.

Diana, M., Raij, T., Melis, M., Nummenmaa, A., et al. (2017). Rehabilitating the addicted brain with transcranial magnetic stimulation. *Nature Reviews Neuroscience, 18*(11), 685–693.

Diana, R. A., Yonelinas, A. P., and Ranganath, C. (2013). Parahippocampal cortex activation during context reinstatement predicts item recollection. *Journal of Experimental Psychology, 142*(4), 1287–1297.

Dickson, B. J., and Gilestro, G. F. (2006). Regulation of commissural axon pathfinding by slit and its Robo receptors. *Annual Review of Cell and Developmental Biology, 22,* 651–675. doi:10.1146/annurev.cellbio.21.090704.151234.

Diekelmann, S., and Born, J. (2010). The memory function of sleep. *Nature Reviews Neuroscience, 11,* 114–126.

Diering, G. H., Nirujogi, R. S., Roth, R. H., Worley, P. F., et al. (2017). Homer1a drives homeostatic scaling-down of excitatory synapses during sleep. *Science, 355*(6324), 511–515.

Dietz, P. M., Williams, S. B., Callaghan, W. M., Bachman, D. J., et al. (2007). Clinically identified maternal depression before, during, and after pregnancies ending in live births. *American Journal of Psychiatry, 164,* 1457–1459.

DiFranza, J. R., Savageau, J. A., Fletcher, K., O'Loughlin, J., et al. (2007). Symptoms of tobacco dependence after brief intermittent use: The Development and Assessment of Nicotine Dependence in Youth-2 study. *Archives of Pediatrics & Adolescent Medicine, 161,* 704–710.

Dloniak, S. M., French, J. A., and Holekamp, K. E. (2006). Rank-related maternal effect of androgens on behaviour in wild spotted hyenas. *Nature, 440,* 1190–1193.

Do, M. T. H., Kang, S. H., Zue, T., Zhong, H., et al. (2009). Photon capture and signalling by melanopsin retinal ganglion cells. *Nature, 457,* 281–287.

Dohrenwend, B. P., Turner, J. B., Turse, N. A., Adams, B. G., et al. (2006). The psychological risks of Vietnam for U.S. veterans: A revisit with new data and methods. *Science, 313,* 979–982.

Dolan, R. J. (2002). Emotion, cognition, and behaviour. *Science, 298,* 1191–1194.

Dolder, C. R., and Nelson, M. H. (2008). Hypnosedative-induced complex behaviours: Incidence, mechanisms and management. *CNS Drugs, 22,* 1021–1036.

Dominguez, J. M. (2009). A role for preoptic glutamate in the regulation of male reproductive behaviour. *Neuroscientist, 15*(1), 11–19.

Domínguez-Rodrigo, M., Pickering, T. R., Semaw, S., and Rogers, M. J. (2005). Cutmarked bones from Pliocene archaeological sites at Gona, Afar, Ethiopia: Implications for the function of the world's oldest stone tools. *Journal of Human Evolution, 48,* 109–121.

Dominici, N., Ivanenko, Y. P., Cappellini, G., d'Avella, A., et al. (2011). Locomotor primitives in newborn babies and their development. *Science, 334,* 997–999.

Domjan, M., and Purdy, J. E. (1995). Animal research in psychology: More than meets the eye of the general psychology student. *American Psychologist, 50,* 496–503.

Do-Monte, F. H., Manzano-Nieves, G., Quiñones-Laracuente, K., Ramos-Medina, L., et al. (2015). Revisiting the role of infralimbic cortex in fear extinction with optogenetics. *Journal of Neuroscience, 35,* 3607–3615.

Donaldson, Z. R., and Young, L. J. (2008). Oxytocin, vasopressin, and the neurogenetics of sociality. *Science, 322,* 900–903.

Donelson, N. C., and Sanyal, S. (2015). Use of *Drosophila* in the investigation of sleep disorders. *Experimental Neurology, 274*(Pt. A), 72–79.

Doricchi, F., Guariglia, C., Paolucci, S., and Pizzamiglio, L. (1991). Disappearance of leftward rapid eye movements during sleep in left visual hemi-inattention. *Neuroreport, 2,* 285–288.

Doroudchi, M. M., Greenberg, K. P., Liu, J., Silka, K. A., et al. (2011). Virally delivered channelrhodopsin-2 safely and effectively restores visual function in multiple mouse models of blindness. *Molecular Therapy, 19,* 1220–1229.

Dorsaint-Pierre, R., Penhune, V. B., Watkins, K. E., Neelin, P., et al. (2006). Asymmetries of the planum temporale and Heschl's gyrus: Relationship to language lateralisation. *Brain, 129,* 1164–1176.

Dosenbach, N. U. F., Petersen, S. E., and Schlaggar, B. L. (2013). The teenage brain: Functional connectivity. *Current Directions in Psychological Science, 22*(3): 101–107.

Doty, R. L. (2010). *The great pheromone myth.* Baltimore: Johns Hopkins University Press.

Doyon, J., Owen, A. M., Petrides, M., Sziklas, V., et al. (1996). Functional anatomy of visuomotor skill learning in human subjects examined with positron emission tomography. *European Journal of Neuroscience, 8,* 637–648.

Drevets, W. C. (1998). Functional neuroimaging studies of depression: The anatomy of melancholia. *Annual Review of Medicine, 49,* 341–361.

Drew, T., Võ, M. L., and Wolfe, J. M. (2013). The invisible gorilla strikes again: Sustained inattentional blindness in expert observers. *Psychological Science, 24*(9), 1848–1853.

Dronkers, N. F., Plaisant, O., Iba-Zizen, M. T., and Cabanis, E. A. (2007). Paul Broca's historic cases: High resolution MR imaging of the brains of Leborgne and Lelong. *Brain, 130*(Pt. 5), 1432–1441.

Dronkers, N. F., Wilkins, D. P., Van Valin, R. D., Jr., Redfern, B. B., et al. (2004). Lesion analysis of the brain areas involved in language comprehension. *Cognition, 92,* 145–177.

Droutman, V., Read, S. J., and Bechara, A. (2015). Revisiting the role of the insula in addiction. *Trends in Cognitive Sciences, 19*(7), 414–420.

Druckman, D., and Bjork, R. A. (1994). *Learning, remembering, believing: Enhancing human performance.* Washington, DC: National Academy Press.

Du, J. L., and Poo, M. M. (2004). Rapid BDNF-induced retrograde synaptic modification in a developing retinotectal system. *Nature, 429,* 878–882.

Dubin, A. E., and Patapoutian, A. (2010). Nociceptors: The sensors of the pain pathway. *Journal of Clinical Investigation, 120*(11), 3760–3772.

Duchaine, B., Germine, L., and Nakayama, K. (2007). Family resemblance: Ten family members with prosopagnosia and within-class object agnosia. *Cognitive Neuropsychology, 24,* 419–430.

Duchamp-Viret, P., Chaput, M. A., and Duchamp, A. (1999). Odour response properties of rat olfactory receptor neurons. *Science, 284,* 2171–2174.

Duffy, J. D., and Campbell, J. J. (1994). The regional prefrontal syndromes: A theoretical and clinical overview. *Journal of Neuropsychiatry and Clinical Neurosciences, 6,* 379–387.

Dulac, C., and Torello, A. T. (2003). Molecular detection of pheromone signals in mammals, from genes to behaviour. *Nature Reviews Neuroscience, 4,* 551–562.

Dully, H., and Fleming, C. (2007). *My lobotomy.* New York: Crown.

Duman, J. G. (2015). Animal ice-binding (antifreeze) proteins and glycolipids: An overview with emphasis on physiological function. *Journal of Experimental Biology, 218,* 1846–1855.

Dunbar, R., and Shultz, S. (2017). Why are there so many explanations for primate brain evolution? *Philosophical Transactions of the Royal Society of London. Series B, Biological Sciences, 372*(1727), 20160244. https://doi.org/10.1098/rstb.2016.0244.

Dunbar, R. I. M. (2009). The social brain hypothesis and its implications for social evolution. *Annals of Human Biology, 36,* 562–572.

Duran, C., Thompson, C. H., Xiao, Q., and Hartzell, H. C. (2010). Chloride channels: Often enigmatic, rarely predictable. *Annual Review of Physiology, 72,* 95–121.

Durand-de Cuttoli, R., Mondoloni, S., Marti, F., Lemoine, D., et al. (2018). Manipulating midbrain dopamine neurons and reward-related behaviours with light-controllable nicotinic acetylcholine receptors. *eLife, 7,* e37487. https://doi.org/10.7554/eLife.37487.

Durkheim, E. (1897). *Le suicide.* Paris: Alcan.

Dvir, Y., and Smallwood, P. (2008). Serotonin syndrome: A complex but easily avoidable condition. *General Hospital Psychiatry, 30,* 284–287.

E

Eagleman, D. M., Kagan, A. D., Nelson, S. S., Sagaram, D., et al. (2007). A standardized test battery for the study of synesthesia. *Journal of Neuroscience Methods, 159,* 139–145.

Edinger, J. D., and Krystal, A. D. (2003). Subtyping primary insomnia: Is sleep state misperception a distinct clinical entity. *Sleep Medicine Reviews, 7*(3), 203–214.

Edwards, R. R., Grace, E., Peterson, S., Klick, B., et al. (2009, January 23). Sleep continuity and architecture: Associations with pain-inhibitory processes in patients with temporomandibular joint disorder. *European Journal of Pain, 13,* 1043–1047.

Egeland, M., Zunszain, P. A., and Pariante, C. M. (2015). Molecular mechanisms in the regulation of adult neurogenesis during stress. *Nature Reviews Neuroscience, 16,* 189–200.

Egnor, S. R., and Seagraves, K. M. (2016). The contribution of ultrasonic vocalisations to mouse courtship. *Current Opinion in Neurobiology, 38,* 1–5.

Eichenbaum, H. (2014). Time cells in the hippocampus: A new dimension for mapping memories. *Nature Reviews Neuroscience, 15*(11), 732–744.

Eichenbaum, H., Yonelinas, A. P., and Ranganath, C. (2007). The medial temporal lobe and recognition memory. *Annual Review of Neuroscience, 30,* 123–152.

Eick, G. N., and Thornton, J. W. (2011). Evolution of steroid receptors from an estrogen-sensitive ancestral receptor. *Molecular and Cellular Endocrinology, 334*(1–2), 31–38.

Eippert, F., Finsterbusch, J., Bingel, U., and Büchel, C. (2009). Direct evidence for spinal cord involvement in placebo analgesia. *Science, 326,* 404.

Eisenberg, M., Kobilo, T., Berman, D. E., and Dudai, Y. (2003). Stability of retrieved memory: Inverse correlation with trace dominance. *Science, 301,* 1102–1104.

Eisenberger, N. I. (2012). Broken hearts and broken bones: A neural perspective on the similarities between social and physical pain. *Psychological Science, 21*(1), 42–47.

Eisenberger, N. I., and Lieberman, M. D. (2004). Why rejection hurts: A common neural alarm system for physical and social pain. *Trends in Cognitive Sciences, 8,* 294–300.

Eklund, A., Nichols, T. E., and Knutsson, H. (2016). Cluster failure: Why fMRI inferences for spatial extent have inflated false-positive rates. *Proceedings of the National Academy of Sciences, USA, 113*(28), 7900–7905.

El Mestikawy, S., Wallén-Mackenzie, A., Fortin, G. M., Descarries, L., et al. (2011). From glutamate co-release to vesicular synergy: Vesicular glutamate transporters. *Nature Reviews Neuroscience, 12,* 204–216.

Ellenbogen, J. M., Hu, P. T., Payne, J. D., Titone, D., et al. (2007). Human relational memory requires time and sleep. *Proceedings of the National Academy of Sciences, USA, 104,* 7317–7318.

Elphick, M. R., Mirabeau, O., and Larhammar, D. (2018). Evolution of neuropeptide signalling systems. *Journal of Experimental Biology, 221*(Pt. 3), pii: jeb151092.

Elster, A. D., and Burdette, J. H. (2001). *Magnetic resonance imaging.* Philadelphia: Mosby.

Emborg, M. E., Liu, Y., Xi, J., Zhang, X., et al. (2013). Induced pluripotent stem cell–derived neural cells survive and mature in the nonhuman primate brain. *Cell Reports, 3*(3), 646–650.

Emery, N. J., Capitanio, J. P., Mason, W. A., Machado, C. J., et al. (2001). The effects of bilateral lesions of the amygdala on dyadic social interactions in rhesus monkeys (*Macaca mulatta*). *Behavioural Neuroscience, 115,* 515–544.

Emmons, W. H., and Simon, C. W. (1956). The non-recall of material presented during sleep. *American Journal of Psychology, 69*(1), 76–81.

Enard, W., Khaitovich, P., Klose, J., Zollner, S., et al. (2002). Intra- and interspecific variation in primate gene expression. *Science, 296,* 340–343.

Engel, J., Jr. (1992). Recent advances in surgical treatment of temporal lobe epilepsy. *Acta Neurologica Scandinavica. Supplementum, 140,* 71–80.

English, P. J., Ghatei, M. A., Malik, I. A., Bloom, S. R., et al. (2002). Food fails to suppress ghrelin levels in obese humans. *Journal of Clinical Endocrinology & Metabolism, 87,* 2984–2987.

Eom, T. Y., Li, J., and Anton, E. S. (2010). Going tubular in the rostral migratory stream: Neurons remodel astrocyte tubes to promote directional migration in the adult brain. *Neuron, 67*(2), 173–175. doi:10.1016/j.neuron.2010.07.013.

Epstein, C. M., Woodard, J. L., Stringer, A. Y., Bakay, R. A., et al. (2000). Repetitive transcranial magnetic stimulation does not replicate the Wada test. *Neurology, 55*(7), 1025–1027.

Erickson, J. T., Conover, J. C., Borday, V., Champagnat, J., et al. (1996). Mice lacking brain-derived neurotrophic factor exhibit visceral sensory neuron losses distinct from mice lacking NT4 and display a severe developmental deficit in control of breathing. *Journal of Neuroscience, 16,* 5361–5371.

Eriksson, A., and Lacerda, F. (2007). Charlatanry in forensic speech science: A problem to be taken seriously. *International Journal of Speech Language and the Law, 14,* 169–193.

Erlanger, D. M., Kutner, K. C., Barth, J. T., and Barnes, R. (1999). Neuropsychology of sports-related head injury: Dementia pugilistica to post concussion syndrome. *Clinical Neuropsychologist, 13,* 193–209.

Eroglu, C., and Barres, B. A. (2010). Regulation of synaptic connectivity by glia. *Nature, 468*(7321), 223–231.

Erren, T. C., Morfeld, P., Stork, J., Knauth, P., et al. (2009). Shift work, chronodisruption and cancer? The IARC 2007 challenge for research and prevention and 10 theses from the Cologne Colloquium 2008. *Scandinavian Journal of Work, Environment & Health*, *35*, 74–79.

Erro, R., Bhatia, K. P., Espay, A. J., and Striano, P. (2017). The epileptic and nonepileptic spectrum of paroxysmal dyskinesias: Channelopathies, synaptopathies, and transportopathies. *Movement Disorders*, *32*(3), 310–318.

Estes, W. K. (1997). Processes of memory loss, recovery, and distortion. *Psychological Review*, *104*, 148–169.

Estrada, V., and Müller, H. W. (2016). Bridging large gaps in the injured spinal cord: Mechanical and biochemical tissue adaptation. *Neural Regeneration Research*, *11*(10), 1572–1574.

Etcoff, N. L., Ekman, P., Magee, J. J., and Frank, M. G. (2000). Lie detection and language comprehension. *Nature*, *405*, 139.

Euston, D. R., Tatsuno, M., and McNaughton, B. L. (2007). Fast-forward playback of recent memory sequences in prefrontal cortex during sleep. *Science*, *318*, 1147–1150.

Evans, J. R., and Lawrenson, J. G. (2017). Antioxidant vitamin and mineral supplements for slowing the progression of age-related macular degeneration. *Cochrane Database of Systematic Reviews*, *7*, CD000254.

Evans, P. D., Anderson, J. R., Vallender, E. J., Gilbert, S. L., et al. (2004). Adaptive evolution of *ASPM*, a major determinant of cerebral cortical size in humans. *Human Molecular Genetics*, *13*, 489–494.

Everitt, B. J., and Stacey, P. (1987). Studies of instrumental behaviour with sexual reinforcement in male rats (*Rattus norvegicus*): II. Effects of preoptic area lesions, castration, and testosterone. *Journal of Comparative Psychology*, *101*, 407–419.

Everson, C. A. (1993). Sustained sleep deprivation impairs host defense. *American Journal of Physiology*, *265*, R1148–R1154.

Everson, C. A., Bergmann, B. M., and Rechtschaffen A. (1989). Sleep deprivation in the rat: III. Total sleep deprivation. *Sleep*, *12*, 13–21.

Everson, C. A., Thalacker, C. D., and Hogg, N. (2008). Phagocyte migration and cellular stress induced in liver, lung, and intestine during sleep loss and sleep recovery. *American Journal of Physiology. Regulatory, Integrative and Comparative Physiology*, *295*(6), R2067–R2074.

F

Fabbri, C., and Serretti, A. (2018, June 12). Clinical application of antidepressant pharmacogenetics: Considerations for the design of future studies. *Neuroscience Letters*, 133651. https://doi.org/10.1016/j.neulet.2018.06.020.

Falk, D. (1993). Sex differences in visuospatial skills: Implications for hominid evolution. In K. R. Gibson and T. Ingold (Eds.), *Tools, language and cognition in human evolution* (pp. 216–229). Cambridge, UK: Cambridge University Press.

Falk, D. (2004). Prelinguistic evolution in early hominins: Whence motherese? *Behavioural and Brain Sciences*, *27*, 491–503.

Falkner, A. L., Grosenick, L., Davidson, T. J., Deisseroth, K., et al. (2016). Hypothalamic control of male aggression-seeking behaviour. *Nature Neuroscience*, *19*, 596–604.

Faraone, S. V. (2018). The pharmacology of amphetamine and methylphenidate: Relevance to the neurobiology of attention-deficit/hyperactivity disorder and other psychiatric comorbidities. *Neuroscience & Biobehavioural Reviews*, *87*, 255–270.

Faraone, S. V., Glatt, S. J., Su, J., and Tsuang, M. T. (2004). Three potential susceptibility loci shown by a genome-wide scan for regions influencing the age at onset of mania. *American Journal of Psychiatry*, *161*, 625–630.

Farbman, A. I. (1994). The cellular basis of olfaction. *Endeavour*, *18*, 2–8.

Farrell, A. K., Slatcher, R. B., Tobin, E. T., Imami, L., et al. (2018). Socioeconomic status, family negative emotional climate, and anti-inflammatory gene expression among youth with asthma. *Psychoneuroendocrinology*, *91*, 62–67.

Fay, R. R. (1988). *Hearing in vertebrates: A psychophysics databook.* Winnetka, IL: Hill-Fay Associates.

Feder, H. H., and Whalen, R. E. (1965). Feminine behaviour in neonatally castrated and estrogen-treated male rats. *Science*, *147*, 306–307.

Feinstein, J. S., Adolphs, R., Damasio, A., and Tranel, D. (2011). The human amygdala and the induction and experience of fear. *Current Biology*, *21*, 34–38.

Feinstein, J. S., Buzza, C., Hurlemann, R., Follmer, R. L., et al. (2013). Fear and panic in humans with bilateral amygdala damage. *Nature Neuroscience*, *16*(3), 270–272.

Felitti, V. J., Anda, R. F., Nordenberg, D., Williamson, D. F., et al. (1998). Relationship of childhood abuse and household dysfunction to many of the leading causes of death in adults. The Adverse Childhood Experiences (ACE) Study. *American Journal of Preventive Medicine*, *14*(4), 245–258.

Feng, W., Störmer, V. S., Martinez, A., McDonald, J. J., et al. (2017). Involuntary orienting of attention to a sound desynchronizes the occipital alpha rhythm and improves visual perception. *Neuroimage*, *150*, 318–328.

Fenstemaker, S. B., Zup, S. L., Frank, L. G., Glickman, S. E., et al. (1999). A sex difference in the hypothalamus of the spotted hyena. *Nature Neuroscience*, *2*, 943–945.

Ferguson, J. N., Young, L. J., Hearn, E. F., Matzuk, M. M., et al. (2000). Social amnesia in mice lacking the oxytocin gene. *Nature Genetics*, *25*, 284–288.

Fernald, R. D. (2000). Evolution of eyes. *Current Opinion in Neurobiology*, *10*, 444–450.

Fernández-Espejo, D., and Owen, A. M. (2013). Detecting awareness after severe brain injury. *Nature Reviews Neuroscience*, *14*(11), 801–809.

Fields, S. (1990). Pheromone response in yeast. *Trends in Biochemical Sciences*, *15*, 270–273.

Finger, S. (1994). *Origins of neuroscience: A history of explorations into brain function.* New York: Oxford University Press.

Fink, A. J., Croce, K. R., Huang, Z. J., Abbott, L. F., et al. (2014). Presynaptic inhibition of spinal sensory feedback ensures smooth movement. *Nature*, *509*(7498), 43–48.

Fink, M., and Taylor, M. A. (2007). Electroconvulsive therapy: Evidence and challenges. *JAMA*, *298*, 330–332.

Finlay, B. L., and Darlington, R. B. (1995). Linked regularities in the development and evolution of mammalian brains. *Science*, *268*, 1578–1584.

Finlay, B. L., Franco, E. C. S., Yamada, E. S., Crowley, J. C., et al. (2008). Number and topography of cones, rods and optic nerve axons in New and Old World primates. *Visual Neuroscience*, *25*, 289–299.

Fisher, A. D., Ristori, J., Morelli, G., and Maggi, M. (2018). The molecular mechanisms of sexual orientation and gender identity. *Molecular and Cell Endocrinology*, *467*, 3–13.

Fisher, S. E. (2017). Evolution of language: Lessons from the genome. *Psychonomic Bulletin & Review*, *24*(1), 34–40.

Fishman, R. B., Chism, L., Firestone, G. L., and Breedlove, S. M. (1990). Evidence for androgen receptors in sexually dimorphic perineal muscles of neonatal male rats. Absence of androgen accumulation by the perineal motoneurons. *Journal of Neurobiology*, *21*, 694–704.

Flament, D., Ellermann, J. M., Kim, S. G., Ugurbil, K., et al. (1996). Functional magnetic resonance imaging of cerebellar activation during the learning of a visuomotor dissociation task. *Human Brain Map*, *4*, 210–226.

Fleming, A. S., Kraemer, G. W., Gonzalez, A., Lovic, V., et al. (2002). Mothering begets mothering: The transmission of behaviour and its neurobiology across generations. *Pharmacology, Biochemistry, and Behaviour*, *73*, 61–75.

Fleming, S. M., Huijgen, J., and Dolan, R. J. (2012). Prefrontal contributions to metacognition in perceptual decision making. *Journal of Neuroscience*, *32*, 6117–6125.

Flor, H., Nikolajsen, L., and Jensen, T. S. (2006). Phantom limb pain: A case of maladaptive CNS plasticity? *Nature Neuroscience*, *7*, 873–881.

Foa, E. B., Gillihan, S. J., and Bryant, R. A. (2013). Challenges and successes in dissemination of evidence-based treatments for posttraumatic stress disorder: Lessons learned from prolonged exposure therapy for PTSD. *Psychological Science in the Public Interest*, *14*, 65–111.

Foley, C., Corvin, A., and Nakagome, S. (2017). Genetics of schizophrenia: Ready to translate? *Current Psychiatry Reports*, *19*(9), 61.

Foradori, C. D., Weiser, M. J., and Handa, R. J. (2008). Non-genomic actions of androgens. *Frontiers in Neuroendocrinology, 29*(2), 169–181.

Forger, N. G., and Breedlove, S. M. (1986). Sexual dimorphism in human and canine spinal cord: Role of early androgen. *Proceedings of the National Academy of Sciences, USA, 83*, 7527–7531.

Forger, N. G., and Breedlove, S. M. (1987). Motoneuronal death during human fetal development. *Journal of Comparative Neurology, 264*, 118–122.

Forger, N. G., Frank, L. G., Breedlove, S. M., and Glickman, S. E. (1996). Sexual dimorphism of perineal muscles and motoneurons in spotted hyenas. *Journal of Comparative Neurology, 375*, 333–343.

Forger, N. G., Howell, M., Bengston, L., Mackenzie, L., et al. (1997). Sexual dimorphism in the spinal cord is absent in mice lacking the ciliary neurotrophic factor receptor. *Journal of Neuroscience, 17*, 9605–9612.

Foster, G. D., Wyatt, H. R., Hill, J. O., McGuckin, B. G., et al. (2003). A randomized trial of a low-carbohydrate diet for obesity. *New England Journal of Medicine, 348*, 2082–2090.

Fothergill, E., Guo, J., Howard, L., Kerns, J. C., et al. (2016). Persistent metabolic adaptation 6 years after "The Biggest Loser" competition. *Obesity*. doi:10.1002/oby.21538. PubMed PMID: 27136388.

Fournier, J. C., DeRubeis, R. J., Hollon, S. D., Dimidjian, S., et al. (2010). Antidepressant drug effects and depression severity: A patient-level meta-analysis. *JAMA, 303*, 47–53.

Fraint, A., and Pal, G. (2015). Deep brain stimulation in Tourette's syndrome. *Frontiers in Neurology, 6*, 170. doi:10.3389/fneur.2015.00170.

Francis, D. D., Szegda, K., Campbell, G., Martin, W. D., et al. (2003). Epigenetic sources of behavioural differences in mice. *Nature Neuroscience, 6*, 445–446.

Francks, C., Maegawa, S., Laurén, J., Abrahams, B. S., et al. (2007). LRRTM1 on chromosome 2p12 is a maternally suppressed gene that is associated paternally with handedness and schizophrenia. *Molecular Psychiatry, 12*(12), 1129–1139.

Frank, M. J., Samanta, J., Moustafa, A. A., and Sherman, S. J. (2007). Hold your horses: Impulsivity, deep brain stimulation, and medication in parkinsonism. *Science, 318*, 1309–1312.

Franken, T. P., Roberts, M. T., Wei, L., Golding, N. L., et al. (2015). In vivo coincidence detection in mammalian sound localisation generates phase delays. *Nature Neuroscience, 18*(3), 444–452.

Frankenhaeuser, M. (1978). Psychoneuroendocrine approaches to the study of emotion as related to stress and coping. *Nebraska Symposium on Motivation, 26*, 123–162.

Frankland, P. W., O'Brien, C., Masuo, O., Kirkwood, A., et al. (2001). α-CaMKII-dependent plasticity in the cortex is required for permanent memory. *Nature, 411*, 309–312.

Franks, N. P. (2008). General anaesthesia: From molecular targets to neuronal pathways of sleep and arousal. *Nature, 9*, 370–386.

Franz, S. I. (1902). On the functions of the cerebrum: I. The frontal lobes in relation to the production and retention of simple sensory-motor habits. *American Journal of Physiology, 8*, 1–22.

Freed, C. R., Greene, P. E., Breeze, R. E., Tsai, W.-Y., et al. (2001). Transplantation of embryonic dopamine neurons for severe Parkinson's disease. *New England Journal of Medicine, 344*, 710–719.

Freedman, M. S., Lucas, R. J., Soni, B., von Schantz, M., et al. (1999). Regulation of mammalian circadian behaviour by non-rod, non-cone, ocular photoreceptors. *Science, 284*, 502–504.

Freeman, J., Ziemba, C. M., Heeger, D. J., Simoncelli, E. P., et al. (2013). A functional and perceptual signature of the second visual area in primates. *Nature Neuroscience, 16*, 974–981.

Freitag, J., Ludwig, G., Andreini, P., Roessler, P., et al. (1998). Olfactory receptors in aquatic and terrestrial vertebrates. *Journal of Comparative Physiology, 183*, 635–650.

Freiwald, W. A., Tsao, D.Y., and Livingstone, M. S. (2009). A face feature space in the macaque temporal lobe. *Nature Neuroscience, 12*, 1187–1196.

Frenda, S. J., Patihis, L., Loftus, E. F., Lewis, H. C., et al. (2014). Sleep deprivation and false memories. *Psychological Science, 25*(9), 1674–1681.

Frey, S. H., Bogdanov, S., Smith, J. C., Watrous, S., et al. (2008). Chronically deafferented sensory cortex recovers a grossly typical organisation after allogenic hand transplantation. *Current Biology, 18*, 1530–1534.

Frey, U., Huang, Y.-Y., and Kandel, E. R. (1993). Effects of cAMP simulate a late stage of LTP in hippocampal CA1 neurons. *Science, 260*, 1661–1664.

Fridlund, A. J. (1994). *Human facial expression: An evolutionary view.* San Diego, CA: Academic Press.

Fried, I., Wilson, C. L., MacDonald, K. A., and Behnke, E. J. (1998). Electric current stimulates laughter. *Nature, 391*, 650.

Friedman, B. H. (2010). Feelings and the body: The Jamesian perspective on autonomic specificity of emotion. *Biological Psychology, 84*(3), 383–393.

Friedman, L., and Jones, B. E. (1984). Study of sleep-wakefulness states by computer graphics and cluster analysis before and after lesions of the pontine tegmentum in the cat. *Electroencephalography and Clinical Neurophysiology, 57*, 43–56.

Friedman, M., and Rosenman, R. H. (1974). *Type A behaviour and your heart.* New York: Knopf.

Friedrich, F. J., Egly, R., Rafal, R. D., and Beck, D. (1998). Spatial attention deficits in humans: A comparison of superior parietal and temporal-parietal junction lesions. *Neuropsychology, 12*, 193–207.

Fritz, J., Shamma, S., Elhilali, M., and Klein, D. (2003). Rapid task-related plasticity of spectrotemporal receptive fields in primary auditory cortex. *Nature Neuroscience, 6*, 1216–1223.

Fromer, M., Pocklington, A. J., Kavanagh, D. H., Williams, H. J., et al. (2014). De novo mutations in schizophrenia implicate synaptic networks. *Nature, 506*, 179–184.

Frumin, I., Sobel, N., and Gilad, Y. (2014). Does a unique olfactory genome imply a unique olfactory world? *Nature Neuroscience, 17*(1), 6–8.

Fryar, C. D., Carroll, M. D., and Ogden, C. L. (2016, July). Prevalence of overweight and obesity among children and adolescents aged 2–19 years: United States, 1963–1965 through 2013–2014. National Centre for Health Statistics. Available from: http://www.cdc.gov/NCHS/products/hestats.htm.

Fukuda, K., Ogilvie, R. D., Chilcott, L., Vendittelli, A.-M., et al. (1998). The prevalence of sleep paralysis among Canadian and Japanese college students. *Dreaming: Journal of the Association for the Study of Dreams, 8*(2), 59–66.

Fuller, P. M., Lu, J., and Saper C. B. (2008). Differential rescue of light- and food-entrainable circadian rhythms. *Science, 320*, 1074–1077.

Fulton, B. D., Scheffler, R. M., Hinshaw, S. P., Levine, P., et al. (2009). National variation of ADHD diagnostic prevalence and medication use: Health care providers and education policies. *Psychiatric Services, 60*(8), 1075–1083.

Fulton, S., Woodside, B., and Shizgal, P. (2000). Modulation of brain reward circuitry by leptin. *Science, 287*, 125–128.

Funato, H., Miyoshi, C., Fujiyama, T., Kanda, T., et al. (2016). Forward-genetics analysis of sleep in randomly mutagenized mice. *Nature, 539*(7629), 378–383.

Fung, T. C., Olson, C. A., and Hsiao, E. Y. (2017). Interactions between the microbiota, immune and nervous systems in health and disease. *Nature Neuroscience, 20*(2), 145–155.

Funk, J., Finke, K., Müller, H. J., Preger, R., et al. (2010). Systematic biases in the tactile perception of the subjective vertical in patients with unilateral neglect and the influence of upright vs. supine posture. *Neuropsychologia, 48*, 298–308.

Furukawa, Y., Anzai, I., Akiyama, S., Imai, M., et al. (2016). Conformational disorder of the most immature Cu,Zn-superoxide dismutase leading to amyotrophic lateral sclerosis. *Journal of Biological Chemistry, 291*(8), 4144–4155.

Fuster, J. M. (1990). Prefrontal cortex and the bridging of temporal gaps in the perception-action cycle. *Annals of the New York Academy of Sciences, 608*, 318–336.

G

Gabel, L. A., Gibson, C. J., Gruen, J. R., and LoTurco, J. J. (2010). Progress towards a cellular neurobiology of reading disability. *Neurobiology of Disease, 38*(2), 173–180.

Gabrieli, J. D. E., Sullivan, E.V., Desmond, J. E., Stebbins, G. T., et al. (1996). Behavioural and functional neuroimaging evidence for preserved conceptual implicit memory in global amnesia. *Society for Neuroscience, 22*, 1449.

Galaburda, A. M. (1994). Developmental dyslexia and animal studies: At the interface between cognition and neurology. *Cognition, 56*, 833–839.

Galaburda, A. M., LoTurco, J., Ramus, F., Fitch, R. H., et al. (2006). From genes to behaviour in developmental dyslexia. *Nature Neuroscience, 9*, 1213–1217.

Gallant, J. L., Braun, J., and Van Essen, D. C. (1993). Selectivity for polar, hyperbolic, and Cartesian gratings in macaque visual cortex. *Science, 259*, 100–103.

Galvão, P. V. M., Silva, H. R. S. E., and Silva, C. M. F. P. D. (2018). Temporal distribution of suicide mortality: A systematic review. *Journal of Anxiety Disorders, 228*, 132–142.

Gan, Q., and Watanabe, S. (2018). Synaptic vesicle endocytosis in different model systems. *Frontiers in Cellular Neuroscience, 12*, 171.

Gangwisch, J. E., Heymsfield, S. B., Boden-Albala, B., Buijs, R. M., et al. (2007). Sleep duration as a risk factor for diabetes incidence in a large U.S. sample. *Sleep, 30*, 1667–1673.

Gannon, P. J., Holloway, R. L., Broadfield, D. C., and Braun, A. R. (1998). Asymmetry of chimpanzee planum temporale: Humanlike pattern of brain language area homolog. *Science, 279*, 220–222.

Gao, Z., Davis, C., Thomas, A. M., Economo, M. N., et al. (2018). A cortico-cerebellar loop for motor planning. *Nature, 563*(7729), 113–116.

Gardner, L. I. (1972). Deprivation dwarfism. *Scientific American, 227*(1), 76–82.

Gardner, R. A., and Gardner, B. T. (1969). Teaching sign language to a chimpanzee. *Science, 165*, 664–672.

Gardner, R. A., and Gardner, B. T. (1984). A vocabulary test for chimpanzees (*Pan troglodytes*). *Journal of Comparative Psychology, 98*, 381–404.

Garfield, A. S, Li, C., Madara, J. C., Shah, B. P., et al. (2015). A neural basis for melanocortin-4 receptor-regulated appetite. *Nature Neuroscience, 18*, 863–871.

Garver, D. L., Holcomb, J. A., and Christensen, J. D. (2000). Heterogeneity of response to antipsychotics from multiple disorders in the schizophrenia spectrum. *Journal of Clinical Psychiatry, 61*, 964–972.

Gaspar, J. M., and McDonald, J. J. (2014). Suppression of salient objects prevents distraction in visual search. *Journal of Neuroscience, 34*(16), 5658–5666.

Gasser, P., Holstein, D., Michel, Y., Doblin, R., et al. (2014). Safety and efficacy of lysergic acid diethylamide-assisted psychotherapy for anxiety associated with life-threatening diseases. *Journal of Nervous and Mental Disease, 202*, 513–520.

Gassmann, M., and Bettler, B. (2012). Regulation of neuronal $GABA_B$ receptor functions by subunit composition. *Nature Reviews Neuroscience, 13*, 380–394.

Gauthier, I., Behrmann, M., and Tarr, M. J. (1999). Can face recognition really be dissociated from object recognition? *Journal of Cognitive Neuroscience, 11*, 349–370.

Gauthier, I., Skudlarski, P., Gore, J. C., and Anderson, A. W. (2000). Expertise for cars and birds recruits brain areas involved in face recognition. *Nature Neuroscience, 3*, 191–197.

Gawande, A. (2008, June 30). The itch. *The New Yorker*.

Gazzaniga, A. M. (2008). *Human: The science behind what makes us unique*. New York: HarperCollins.

Gazzaniga, M. S., and Smylie, C. S. (1983). Facial recognition and brain asymmetries: Clues to underlying mechanisms. *Annals of Neurology, 13*, 536–540.

Gebara, M. A., Siripong, N., DiNapoli, E. A., Maree, R. D., et al. (2018). Effect of insomnia treatments on depression: A systematic review and meta-analysis. *Depression and Anxiety, 35*(8), 717–731.

Geers, A. E., Mitchell, C. M., Warner-Czyz, A., Wang, N.Y., et al. (2017). Early sign language exposure and cochlear implantation benefits. *Pediatrics, 140*(1), pii: e20163489.

Gehring, W. J., and Willoughby, A. R. (2002). The medial frontal cortex and the rapid processing of monetary gains and losses. *Science, 295*, 2279–2280.

Gelber, R. P., Redline, S., Ross, G. W., Petrovitch, H., (2015). Associations of brain lesions at autopsy with polysomnography features before death. *Neurology, 84*(3), 296–303.

Géléoc, G. S., and Holt, J. R. (2014). Sound strategies for hearing restoration. *Science, 344*(6184), 1241062.

Gelstein, S., Yeshurun, Y., Rozenkrantz, L., Shushan, S., et al. (2011). Human tears contain a chemosignal. *Science, 331*(6014), 226–230.

Gemba, H., Miki, N., and Sasaki, K. (1995). Cortical field potentials preceding vocalisation and influences of cerebellar hemispherectomy upon them in monkeys. *Brain Research, 697*, 143–151.

Genoux, D., Haditsch, U., Knobloch, M., Michalon, A., et al. (2002). Protein phosphatase 1 is a molecular constraint on learning and memory. *Nature, 418*, 970–975.

George, D. T., Phillips, M. J., Lifshitz, M., Lionetti, T. A., et al. (2011). Fluoxetine treatment of alcoholic perpetrators of domestic violence: A 12-week, double-blind, randomized, placebo-controlled intervention study. *Journal of Clinical Psychiatry, 72*(1), 60–65.

Georgiadis, J. R., Farrell, M. J., Boessen, R., Denton, D. A., et al. (2010). Dynamic subcortical blood flow during male sexual activity with ecological validity: A perfusion fMRI study. *NeuroImage, 50*(1), 208–216.

Georgiadis, J. R., and Holstege, G. (2005). Human brain activation during sexual stimulation of the penis. *Journal of Comparative Neurology, 493*(1), 33–38.

Georgiadis, J. R., Reinders, A. A., Paans, A. M., Renken, R., et al. (2009). Men versus women on sexual brain function: Prominent differences during tactile genital stimulation, but not during orgasm. *Human Brain Mapping, 10*, 3089–3101.

Georgopoulos, A. P., and Carpenter, A. F. (2015). Coding of movements in the motor cortex. *Current Opinion in Neurobiology, 33*, 34–93.

Georgopoulos, A. P., Kalaska, J. F., Caminiti, R., and Massey, J. T. (1982). On the relations between the direction of two-dimensional arm movements and cell discharge in primate motor cortex. *Journal of Neuroscience, 2*, 1527–1537.

Georgopoulos, A. P., Taira, M., and Lukashin, A. (1993). Cognitive neurophysiology of the motor cortex. *Science, 260*, 47–52.

Gerard, C. M., Mollereau, C., Vassart, G., and Parmentier, M. (1991). Molecular cloning of a human cannabinoid receptor which is also expressed in testis. *Biochemical Journal, 279*, 129–134.

Gerashchenko, D., Kohls, M. D., Greco, M. A., Waleh, N. S., et al. (2001). Hypocretin-2-saporin lesions of the lateral hypothalamus produce narcoleptic-like sleep behaviour in the rat. *Neuroscience, 21*, 7273–7283.

Gerkin, R. C., and Castro, J. B. (2015). The number of olfactory stimuli that humans can discriminate is still unknown. *eLife, 4*, e08127.

Geschwind, N. (1972). Language and the brain. *Scientific American, 226*(4), 76–83.

Geschwind, N. (1976). Language and cerebral dominance. In T. N. Chase (Ed.), *Nervous system: Vol. 2. The clinical neurosciences* (pp. 433–439). New York: Raven.

Geschwind, N., and Levitsky, W. (1968). Human brain: Left-right asymmetries in temporal speech region. *Science, 161*, 186–187.

Gilad, Y., Bustamante, C. D., Lancet, D., and Pääbo, S. (2003). Natural selection on the olfactory receptor gene family in humans and chimpanzees. *America Journal of Human Genetics, 73*(3), 489–501.

Gilbert, A. N., Yamazaki, K., Beauchamp, G. K., and Thomas, L. (1986). Olfactory discrimination of mouse strains (*Mus musculus*) and major histocompatibility types by humans (*Homo sapiens*). *Journal of Comparative Psychology, 100*, 262–265.

Gilbertson, M. W., Shenton, M. E., Ciszewski, A., Kasai, K., et al. (2002). Smaller hippocampal volume predicts pathologic vulnerability to psychological trauma. *Nature Neuroscience, 5*, 1242–1247.

Gill, R. E., Tibbitts, T. L., Douglas, D. C., Hanel, C. M., et al. (2009). Extreme endurance flights by landbirds crossing the Pacific Ocean: Ecological corridor rather than barrier? *Proceedings: Biological Sciences, 276*, 447–457.

Gitelman, D. R., Alpert, N. M., Kosslyn, S., Daffner, K., et al. (1996). Functional imaging of human right hemispheric activation for exploratory movements. *Annals of Neurology, 39*, 174–179.

Glantz, L. A., and Lewis, D. A. (2000). Decreased dendritic spine density on prefrontal cortical pyramidal neurons in schizophrenia. *Archives of General Psychiatry, 57*, 65–73.

Glaser, R., Rice, J., Speicher, C. E., Stout, J. C., et al. (1986). Stress depresses interferon production by leukocytes concomitant with a decrease in natural killer cell activity. *Behavioural Neuroscience, 100,* 675–678.

Glasser, M. F., Coalson, T. S., Robinson, E. C., Hacker, C. D., et al. (2016). A multi-modal parcellation of human cerebral cortex. *Nature, 536*(7615), 171–178.

Glasser, M. F., Smith, S. M., Marcus, D. S., Andersson, J. L., et al. (2016). The Human Connectome Project's neuroimaging approach. *Nature Neuroscience, 19*(9), 1175–1187.

Glenn, A. L., and Raine, A. (2014). Neurocriminology: Implications for the punishment, prediction and prevention of criminal behaviour. *Nature Reviews Neuroscience, 15,* 54–63.

Glickman, S. E. (1977). Comparative psychology. In P. Mussen and M. R. Rosenzweig (Eds.), *Psychology: An introduction* (2nd ed., pp. 625–703). Lexington, MA: Heath.

Glimcher, P. W. (2011). Understanding dopamine and reinforcement learning: The dopamine reward prediction error hypothesis. *Proceedings of the National Academy of Sciences, USA, 108*(Suppl. 3), 15647–15654.

Globus, A., Rosenzweig, M. R., Bennett, E. L., and Diamond, M. C. (1973). Effects of differential experience on dendritic spine counts in rat cerebral cortex. *Journal of Comparative and Physiological Psychology, 82,* 175–181.

Gnad, T., Scheibler, S., von Kügelgen, I., Scheele, C., et al. (2014). Adenosine activates brown adipose tissue and recruits beige adipocytes via A2A receptors. *Nature, 516,* 395–399.

Goebel, T. (2007). The missing years for modern humans. *Science, 315,* 194–196.

Gogtay, N., Giedd, J. N., Lusk, L., Hayashi, K. M., et al. (2004). Dynamic mapping of human cortical development during childhood through early adulthood. *Proceedings of the National Academy of Sciences, USA, 101,* 8174–8179.

Gold, B. P., Mas-Herrero, E., Zeighami, Y., Benovoy, M., et al. (2019). Musical reward prediction errors engage the nucleus accumbens and motivate learning. *Proceedings of the National Academy of Sciences, USA, 116*(8), 3310–3315.

Golden, R. N., Gaynes, B. N., Ekstrom, R. D., Hamer, R. M., et al. (2005). The efficacy of light therapy in the treatment of mood disorders: A review and meta-analysis of the evidence. *American Journal of Psychiatry, 162,* 656–662.

Goldin, P. R., and Gross, J. J. (2010). Effects of mindfulness-based stress reduction (MBSR) on emotion regulation in social anxiety disorder. *Emotion, 10,* 83–91.

Goldin-Meadow, S. (2006). Talking and thinking with our hands. *Current Directions in Psychological Science, 15,* 34–39.

Goldstein, J. M., Seidman, L. J., Horton, N. J., Makris, N., et al. (2001). Normal sexual dimorphism of the adult human brain assessed by in vivo magnetic resonance imaging. *Cerebral Cortex, 11,* 490–497.

Goldstein, P., Weissman-Fogel, I., Dumas, G., and Shamay-Tsoory, S. G. (2018). Brain-to-brain coupling during handholding is associated with pain reduction. *Proceedings of the National Academy of Sciences, USA, 115*(11), E2528–E2537.

Goldstein, R. Z., and Volkow N. D. (2011). Dysfunction of the prefrontal cortex in addiction: Neuroimaging findings and clinical implications. *Nature Reviews Neuroscience, 12,* 652–669.

Golomb, J., de Leon, M. J., George, A. E., Kluger, A., et al. (1994). Hippocampal atrophy correlates with severe cognitive impairment in elderly patients with suspected normal pressure hydrocephalus. *Journal of Neurology, Neurosurgery and Psychiatry, 57,* 590–593.

González-Maeso, J., Ang, R. L., Yuen, T., Chan, P., et al. (2008). Identification of a serotonin/glutamate receptor complex implicated in psychosis. *Nature, 452,* 93–97.

Gooch, C. L., Pracht, E., and Borenstein, A. R. (2017). The burden of neurological disease in the United States: A summary report and call to action. *Annals of Neurology, 81*(4), 479–484.

Goodale, M. A., and Haffenden, A. (1998). Frames of reference for perception and action in the human visual system. *Neuroscience and Biobehavioural Reviews, 22,* 161–172.

Goodale, M. A., Milner, A. D., Jakobson, L. S., and Carey, D. P. (1991). A neurological dissociation between perceiving objects and grasping them. *Nature, 349,* 154–156.

Goodglass, H., Quadfasel, F. A., and Timberlake, W. H. (1964). Phrase length and the type and severity of aphasia. *Cortex, 1,* 133–153.

Gooley, J. J., Rajaratnam, S. M., Brainard, G. C., Kronauer, R. E., et al. (2010). Spectral responses of the human circadian system depend on the irradiance and duration of exposure to light. *Science Translational Medicine, 2*(31), 31ra33.

Gordon, N. S., Burke, S., Akil, H., Watson, S. J., et al. (2003). Socially-induced brain "fertilisation": Play promotes brain derived neurotrophic factor transcription in the amygdala and dorsolateral frontal cortex in juvenile rats. *Neuroscience Letters, 341,* 17–20.

Gorski, R. A., Gordon, J. H., Shryne, J. E., and Southam, A. M. (1978). Evidence for a morphological sex difference within the medial preoptic area of the rat brain. *Brain Research, 148,* 333–346.

Gorter, J. A., van Vliet, E. A., and Lopes da Silva, F. H. (2016). Which insights have we gained from the kindling and post-status epilepticus models? *Journal of Neuroscience Methods, 260,* 96–108.

Gorzalka, B. B., Mendelson, S. D., and Watson, N. V. (1990). Serotonin receptor subtypes and sexual behaviour. *Annals of the New York Academy of Sciences, 600,* 435–444.

Gosseries, O., Di, H., Laureys, S., and Boly, M. (2014). Measuring consciousness in severely damaged brains. *Annual Review of Neuroscience, 37,* 457–478.

Goswami, U. (2015). Sensory theories of developmental dyslexia: Three challenges for research. *Nature Reviews Neuroscience, 16*(1), 43–54.

Gotlib, I. H., Joormann, J., and Foland-Ross, L. C. (2014). Understanding familial risk for depression: A 25 year perspective. *Perspectives on Psychological Science, 9,* 94–108.

Gottesman, I. I. (1991). *Schizophrenia genesis: The origins of madness.* New York: Freeman.

Gottfried, J. A., O'Doherty, J., and Dolan, R. J. (2003). Encoding predictive reward value in human amygdala and orbitofrontal cortex. *Science, 301,* 1104–1107.

Gottlieb, J. (2007). From thought to action: The parietal cortex as a bridge between perception, action, and cognition. *Neuron, 53,* 9–16.

Gough, P. M., Nobre, A. C., and Devlin, J. T. (2005). Dissociating linguistic processes in the left inferior frontal cortex with transcranial magnetic stimulation. *Journal of Neuroscience, 25,* 8010–8016.

Gougoux, F., Zatorre, R. J., Lassonde, M., Voss, P., et al. (2005). A functional neuroimaging study of sound localisation: Visual cortex activity predicts performance in early-blind individuals. *PLOS Biology, 3,* 324–332.

Gould, S. J. (1981). *The mismeasure of man.* New York: Norton.

Gourévitch, B., Edeline, J. M., Occelli, F., and Eggermont, J. J. (2014). Is the din really harmless? Long-term effects of non-traumatic noise on the adult auditory system. *Nature Reviews Neuroscience, 15*(7), 483–491.

Goutman, J. D., Elgoyhen, A. B., and Gómez-Casati, M. E. (2015). Cochlear hair cells: The sound-sensing machines. *FEBS Letters, 589*(22), 3354–3361.

Grady, C. L., McIntosh, A. R., Horowitz, B., Maisog, J. M., et al. (1995). Age-related reductions in human recognition memory due to impaired encoding. *Science, 269,* 218–221.

Grant, P. R., and Grant, B. R. (2006). Evolution of character displacement in Darwin's finches. *Science, 313,* 224–226.

Gratton, G., and Fabiani, M. (2001). Shedding light on brain function: The event-related optical signal. *Trends in Cognitive Sciences, 5,* 357–363.

Gravett, N., Bhagwandin, A., Sutcliffe, R., Landen, K., et al. (2017). Inactivity/sleep in two wild free-roaming African elephant matriarchs: Does large body size make elephants the shortest mammalian sleepers? *PLOS ONE, 12*(3), e0171903.

Gray, N. S., MacCulloch, M. J., Smith, J., Morris, M., et al. (2003). Violence viewed by psychopathic murderers. *Nature, 423,* 497.

Graybiel, A. M., and Grafton, S. T. (2015). The striatum: Where skills and habits meet. *Cold Spring Harbor Perspectives in Biology, 7*(8), a021691.

Graziano, M. S. (2006). Progress in understanding spatial coordinate systems in the primate brain. *Neuron, 51*(1), 7–9. https://www.ncbi.nlm.nih.gov/pubmed/16815326.

Graziano, M. S. [A.], and Aflalo, T. N. (2007). Mapping behavioural repertoire onto the cortex. *Neuron, 56,* 239–251.

Graziano, M. S. [A.], Hu, X. T., and Gross, C. G. (1997). Coding the locations of objects in the dark. *Science, 277*, 239–241.

Greden, J. F., Parikh, S. V., Rothschild, A. J., Thase, M. E., et al. (2019). Impact of pharmacogenomics on clinical outcomes in major depressive disorder in the GUIDED trial: A large, patient- and rater-blinded, randomized, controlled study. *Journal of Psychiatric Research, 111*, 59–67.

Green, E., and Craddock, N. (2003). Brain-derived neurotrophic factor as a potential risk locus for bipolar disorder: Evidence, limitations, and implications. *Current Psychiatry Reports, 5*, 469–476.

Green, J. J., Doesburg, S. M., Ward, L. M., and McDonald, J. J. (2011). Electrical neuroimaging of voluntary audiospatial attention: Evidence for a supramodal attention control network. *Journal of Neuroscience, 31*(10), 3560–3564.

Green, M. F., Horan, W. P., and Lee, J. (2015). Social cognition in schizophrenia. *Nature Reviews Neuroscience, 16*, 620–631.

Green, W. H., Campbell, M., and David, R. (1984). Psychosocial dwarfism: A critical review of the evidence. *Journal of the American Academy of Child Psychiatry, 23*, 39–48.

Greenough, W. T. (1976). Enduring brain effects of differential experience and training. In M. R. Rosenzweig and E. L. Bennett (Eds.), *Neural mechanisms of learning and memory* (pp. 255–278). Cambridge, MA: MIT Press.

Gregory, R. L., and Wallace, J. G. (1963). Recovery from early blindness: A case study. *Experimental Psychology Society Monograph, 2*, 1–44.

Griebel, G., and Holmes, A. (2013). 50 years of hurdles and hope in anxiolytic drug discovery. *Nature Reviews Drug Discovery, 12*, 667–687.

Griessenauer, C. J., Salam, S., Hendrix, P., Patel, D. M., et al. (2015). Hemispherectomy for treatment of refractory epilepsy in the pediatric age group: A systematic review. *Journal of Neurosurgery. Pediatrics, 15*(1), 34–44.

Griffin, D. M., Hoffman, D. S., and Strick, P. L. (2015). Corticomotoneuronal cells are "functionally tuned." *Science, 350*(6261), 667–670.

Griffin, G. D., and Flanagan-Cato, L. M. (2011). Ovarian hormone action in the hypothalamic ventromedial nucleus: Remodelling to regulate reproduction. *Journal of Neuroendocrinology, 23*(6), 465–471.

Griffin, L. D., and Mellon, S. H. (1999). Selective serotonin reuptake inhibitors directly alter activity of neurosteroidogenic enzymes. *Proceedings of the National Academy of Sciences, USA, 96*, 13512–13517.

Grill-Spector, K., Weiner, K. S., Kay, K., and Gomez, J. (2017). The functional neuroanatomy of human face perception. *Annual Review of Vision Science, 3*, 167–196.

Grimbos, T., Dawood, K., Burriss, R. P., Zucker, K. J., et al. (2010). Sexual orientation and the second to fourth finger length ratio: A meta-analysis in men and women. *Behavioural Neuroscience, 124*(2), 278–287.

Grimm, D. (2014, March 12). Animal rights extremists increasingly targeting individuals. *Science*. (www.sciencemag.org/news/2014/03/animal-rights-extremists-increasingly-targeting-individuals)

Grob, C. S., Danforth, A. L., Chopra, G. S., Hagerty, M., et al. (2011). Pilot study of psilocybin treatment for anxiety in patients with advanced-stage cancer. *Archives of General Psychiatry, 68*, 71–78.

Gropp, E., Shanabrough, M., Borok, E., Xu, A. W., et al. (2005). Agouti-related peptide-expressing neurons are mandatory for feeding. *Nature Neuroscience, 8*, 1289–1291.

Gross, C. T., and Canteras, N. S. (2012). The many paths to fear. *Nature Reviews Neuroscience, 13*, 651–658.

Grothe, B., and Pecka, M. (2014). The natural history of sound localisation in mammals—A story of neuronal inhibition. *Frontiers in Neural Circuits, 8*, 116.

Grothe, B., Pecka, M., and McAlpine, D. (2010). Mechanisms of sound localisation in mammals. *Physiological Reviews, 90*(3), 983–1012.

Grover, G. J., Mellstrom, K., Ye, L., Malm, J., et al. (2003). Selective thyroid hormone receptor-β activation: A strategy for reduction of weight, cholesterol, and lipoprotein (a) with reduced cardiovascular liability. *Proceedings of the National Academy of Sciences, USA, 100*, 10067–10072.

Grumbach, M. M., and Auchus, R. J. (1999). Estrogen: Consequences and implications of human mutations in synthesis and action. *Journal of Clinical Endocrinology & Metabolism, 84*(12), 4677–4694.

Grunt, J. A., and Young, W. C. (1953). Consistency of sexual behaviour patterns in individual male guinea pigs following castration and androgen therapy. *Journal of Comparative and Physiological Psychology, 46*, 138–144.

Grüter, T., Grüter, M., and Carbon, C. C. (2008). Neural and genetic foundations of face recognition and prosopagnosia. *Journal of Neuropsychology, 2*, 79–97.

Gu, X., Hof, P. R., Friston, K. J., and Fan, J. (2013). Anterior insular cortex and emotional awareness. *Journal of Comparative Neurology, 521*(15), 3371–3388.

Guarner, F., and Malagelada, J. R. (2003). Gut flora in health and disease. *Lancet, 361*(9356), 512–519.

Guina, J., and Merrill, B. (2018). Benzodiazepines I: Upping the care on downers: The evidence of risks, benefits and alternatives. *Journal of Clinical Medicine, 7*(2), pii: E17.

Gulevich, G., Dement, W., and Johnson, L. (1966). Psychiatric and EEG observations on a case of prolonged (264 hours) wakefulness. *Archives of General Psychiatry, 15*, 29–35.

Gunaydin, L. A., Grosenick, L., Finkelstein, J. C., Kauvar, I. V., et al. (2014). Natural neural projection dynamics underlying social behaviour. *Cell, 157*(7), 1535–1551.

Gunn, B. G., Cunningham, L., Mitchell, S. G., Swinny, J. D., et al. (2015). $GABA_A$ receptor-acting neurosteroids: A role in the development and regulation of the stress response. *Frontiers in Neuroendocrinology, 36*, 28–48.

Guo, J. U., Ma, D. K., Mo, H., Ball, M. P., et al. (2011). Neuronal activity modifies the DNA methylation landscape in the adult brain. *Nature Neuroscience, 14*, 1345–1351.

Gupta, K. K., Gupta, V. K., and Shirasaka, T. (2016). An update on fetal alcohol syndrome-pathogenesis, risks, and treatment. *Alcoholism: Clinical and Experimental Research, 40*(8), 1594–1602.

Gurney, M. E., and Konishi, M. (1979). Hormone induced sexual differentiation of brain and behaviour in zebra finches. *Science, 208*, 1380–1382.

Gurney, M. E., Pu, H., Chiu, A. Y., Dal Canto, M. C., et al. (1994). Motor neuron degeneration in mice that express a human Cu,Zn superoxide dismutase mutation. *Science, 264*, 1772–1775.

Gusella, J. F., MacDonald, M. E., and Lee, J. M. (2014). Genetic modifiers of Huntington's disease. *Movement Disorders, 29*(11), 1359–1365.

H

Haaland, K. Y., Dum, R. P., Mutha, P. K., Strick, P. L., et al. (2017). The neuropsychology of movement and movement disorders: Neuroanatomical and cognitive considerations. *Journal of the International Neuropsychological Society, 23*(9–10), 768–777.

Haász, J., Westlye, E. T., and Fjær, S. (2013). General fluid-type intelligence is related to indices of white matter structure in middle-aged and old adults. *NeuroImage, 83*, 372–383.

Habib, A. M., Okorokov, A. L., Hill, M. N., Bras, J. T., et al. (2019). Microdeletion in a FAAH pseudogene identified in a patient with high anandamide concentrations and pain insensitivity. *British Journal of Anaesthesia*, Mar 28. pii: S0007-0912(19)30138-2. https://doi.org/10.1016/j.bja.2019.02.019.

Hack, L. M., Fries, G. R., Eyre, H. A., Bousman, C. A., et al. (2019). Moving pharmacoepigenetics tools for depression toward clinical use. *Journal of Affective Disorders, 249*, 336–346.

Hackett, T. A. (2011). Information flow in the auditory cortical network. *Hearing Research, 271*(1–2), 133–146.

Hackman, D. A., Farah, M. J., and Meaney, M. J. (2010). Socio-economic status and the brain: Mechanistic insights from human and animal research. *Nature Reviews Neuroscience, 11*, 651–659.

Haesler, S., Rochefort, C., Georgi, B., Licznerski, P., et al. (2007). Incomplete and inaccurate vocal imitation after knockdown of FoxP2 in songbird basal ganglia nucleus Area X. *PLOS Biology, 5*, e321.

Haesler, S., Wada, K., Nshdejan, A., Morrisey, E. E., et al. (2004). FoxP2 expression in avian vocal learners and non-learners. *Journal of Neuroscience, 24*, 3164–3175.

Haggard, P. (2017). Sense of agency in the human brain. *Nature Reviews Neuroscience, 18*(4), 196–207.

Hagstrum, J. T. (2000). Infrasound and the avian navigational map. *Journal of Experimental Biology, 203*, 1103–1111.

Hahn-Holbrook, J., and Haselton, M. (2014). Is postpartum depression a disease of modern civilisation? *Psychological Science, 23*, 395–400.

Håkansson, M. L., Brown, H., Ghilardi, N., and Skoda, R. C., et al. (1998). Leptin receptor immunoreactivity in chemically defined target neurons of the hypothalamus. *Journal of Neuroscience, 18*, 559–572.

Hall, K. D. (2018). The complicated relation between resting energy expenditure and maintenance of lost weight. *American Journal of Clinical Nutrition, 108*(4), 652–653.

Halperin, A. (2018, January 29). Marijuana: Is it time to stop using a word with racist roots? *Guardian* (https://www.theguardian.com/society/2018/jan/29/marijuana-name-cannabis-racism).

Ham, S., Kim, T. K., Chung, S., and Im, H. I. (2017). Drug abuse and psychosis: New insights into drug-induced psychosis. *Experimental Neurobiology, 26*(1), 11–24.

Hamadjida, A., Frouni, I., Kwan, C., and Huot, P. (2018). Classic animal models of Parkinson's disease: A historical perspective. *Behavioural Pharmacology*. https://doi.org/10.1097/FBP.0000000000000441.

Hamburger, V. (1975). Cell death in the development of the lateral motor column of the chick embryo. *Journal of Comparative Neurology, 160*, 535–546.

Hammond, G. L., Miguel-Queralt, S., Yalcinkaya, T. M., Underhill, C., et al. (2012). Phylogenetic comparisons implicate sex hormone-binding globulin in "masculinisation" of the female spotted hyena (*Crocuta crocuta*). *Endocrinology, 153*(3), 1435–1443.

Hamson, D. K., Csupity, A. S., Ali, F. M., and Watson, N. V. (2009). Partner preference and mount latency are masculinized in androgen insensitive rats. *Physiology & Behaviour, 98*, 25–30.

Hamson, D. K., Wainwright, S. R., Taylor, J. R., Jones, B. A. et al. (2013). Androgens increase survival of adult-born neurons in the dentate gyrus by an androgen receptor-dependent mechanism in male rats. *Endocrinology, 154*(9), 3294–3304.

Hamson, D. K., and Watson, N. V. (2004). Regional brainstem expression of Fos associated with sexual behaviour in male rats. *Brain Research, 1006*, 233–240.

Han, S. K., Gottsch, M. L., Lee, K. J, Popa, S. M., et al. (2005). Activation of gonadotropin-releasing hormone neurons by kisspeptin as a neuroendocrine switch for the onset of puberty. *Journal of Neuroscience, 25*, 11349–11356.

Hanakawa, T., Immisch, I., Toma, K., Dimyan, M. A., et al. (2003). Functional properties of brain areas associated with motor execution and imagery. *Journal of Neurophysiology, 89*, 989–1002.

Hanaway, J., Woolsey, T. A., Gado, M. H., and Roberts, M. P. (1998). *The brain atlas.* Bethesda, MD: Fitzgerald Science.

Hanchar, H. J., Dodson, P. D., Olsen, R. W., Otis, T. S., et al. (2005). Alcohol-induced motor impairment caused by increased extrasynaptic GABA(A) receptor activity. *Nature Neuroscience, 8*, 339–345.

Haney, M., Gunderson, E. W., Jiang, H., Collins, E. D., et al. (2010, January 1). Cocaine-specific antibodies blunt the subjective effects of smoked cocaine in humans. *Biological Psychiatry, 67*(1), 59–65.

Hanlon, C. A., Beveridge, T. J., and Porrino, L. J. (2013). Recovering from cocaine: Insights from clinical and preclinical investigations. *Neuroscience & Biobehavioural Reviews, 37*(9 Pt. A), 2037–2046.

Hannula-Jouppi, K., Kaminen-Ahola, N., Taipale, M., Eklund, R., et al. (2005). The axon guidance receptor gene ROBO1 is a candidate gene for developmental dyslexia. *PLOS Genetics, 1*, e50.

Hansen, R. N., Boudreau, D. M., Ebel, B. E., Grossman, D. C., et al. (2015). Sedative hypnotic medication use and the risk of motor vehicle crash. *American Journal of Public Health, 105*(8), e64-9.

Hardingham, G. E., and Do, K. Q. (2016). Linking early-life NMDAR hypofunction and oxidative stress in schizophrenia pathogenesis. *Nature Reviews Neuroscience, 17*, 125–134.

Hardyck, C., Petrinovich, L., and Goldman, R. (1976). Left-handedness and cognitive deficit. *Cortex, 12*, 226–279.

Hare, R. D., Harpur, T. J., Hakstian, A. R., Forth, A. E., et al. (1990). The revised psychopathy checklist: Descriptive statistics, reliability, and factor structure. *Psychological Assessment, 2*, 338–341.

Harnett, M. T., Xu, N. L., Magee, J. C., Williams, S. R. (2013). Potassium channels control the interaction between active dendritic integration compartments in layer 5 cortical pyramidal neurons. *Neuron, 79*(3), 516–529.

Harold, D., Paracchini, S., Scerri, T., Dennis, M., et al. (2006). Further evidence that the KIAA0319 gene confers susceptibility to developmental dyslexia. *Molecular Psychiatry, 11*, 1085–1091, 1061.

Harris, A. L. (2018). Electrical coupling and its channels. *Journal of General Physiology, 150*(12), 1606–1639.

Harris, E. W., and Cotman, C. W. (1986). Long-term potentiation of guinea pig mossy fibre responses is not blocked by N-methyl D-aspartate antagonists. *Neuroscience Letters, 70*, 132–137.

Harrison, N. L., Skelly, M. J., Grosserode, E. K., Lowes, D. C., et al. (2017). Effects of acute alcohol on excitability in the CNS. *Neuropharmacology, 122*, 36–45.

Harro, J. (2019). Animal models of depression: Pros and cons. *Cell and Tissue Research, 377*(1), 5–20.

Harrow, M., Jobe, T. H., and Faull, R. N. (2012, February 17). Do all schizophrenia patients need antipsychotic treatment continuously throughout their lifetime? A 20-year longitudinal study. *Psychological Medicine, 42*, 2145–2155.

Hart, B. L. (1988). Biological basis of the behaviour of sick animals. *Neuroscience and Biobehavioural Reviews, 12*, 123–137.

Hartanto, T. A., Krafft, C. E., Iosif, A. M., and Schweitzer, J. B. (2016). A trial-by-trial analysis reveals more intense physical activity is associated with better cognitive control performance in attention-deficit/hyperactivity disorder. *Child Neuropsychology, 22*(5), 618–626.

Hartmann, E. (1984). *The nightmare: The psychology and biology of terrifying dreams.* New York: Basic Books.

Harvey, P. H., and Krebs, J. R. (1990). Comparing brains. *Science, 249*, 140–146.

Haxby, J. V. (2006). Fine structure in representations of faces and objects. *Nature Neuroscience, 9*, 1084–1086.

Hayashi, Y., Kashiwagi, M., Yasuda, K., Ando, R., et al. (2015). Cells of a common developmental origin regulate REM/non-REM sleep and wakefulness in mice. *Science, 350*(6263), 957–961.

Haynes, K. F., Gemeno, C., Yeargan, K. V., Millar, J. G., et al. (2002). Aggressive chemical mimicry of moth pheromones by a bolas spider: How does this specialist predator attract more than one species of prey? *Chemoecology, 12*, 99–105.

Hazeltine, E., Grafton, S. T., and Ivry, R. (1997). Attention and stimulus characteristics determine the locus of motor-sequence encoding. A PET study. *Brain, 120*, 123–140.

Hazlett, H. C., Gu, H., Munsell, B. C., Kim, S. H., et al. (2017). Early brain development in infants at high risk for autism spectrum disorder. *Nature, 542*(7641), 348–351. https://doi.org/10.1038/nature21369.

He, D. Z., Lovas, S., Ai, Y., Li, Y., et al. (2014). Prestin at year 14: Progress and prospect. *Hearing Research, 311*, 25–35.

Health Quality Ontario. (2016). Repetitive transcranial magnetic stimulation for treatment-resistant depression: A systematic review and meta-analysis of randomized controlled trials. *Ontario Health Technology Assessment Series, 16*, 1–66.

Heath, R. G. (1972). Pleasure and brain activity in man. *Journal of Nervous and Mental Diseases, 154*, 3–18.

Heaton, R. K., et al. (1993). *Wisconsin Card Sorting Test Manual: Revised and Expanded*. Odessa, FL: Psychological Assessment Resources.

Hebb, D. O. (1949). *The organisation of behaviour.* New York: Wiley.

Heckers, S., and Konradi, C. (2002). Hippocampal neurons in schizophrenia. *Journal of Neural Transmission (Vienna), 109*(5–6), 891–905.

Heckert, J. (2012, November 15). The hazards of growing up painlessly. *The New York Times, Sunday Magazine*, MM26.

Hedden, T., and Gabrieli, J. D. (2004). Insights into the ageing mind: A view from cognitive neuroscience. *Current Opinion in Neurology, 18*, 740–747.

Heffner, H. E., and Heffner, R. S. (1984). Temporal lobe lesions and perception of species-specific vocalisations by macaques. *Science, 226*(4670), 5–6.

Heidenreich, M., Lechner, S. G., Vardanyan, V., Wetzel, C., et al. (2012). KCNQ4 K^+ channels tune mechanoreceptors for normal touch sensation in mouse and man. *Nature Neuroscience, 15*, 138–145.

Heilig, M., Epstein, D. H., Nader, M. A., and Shaham, Y. (2016). Time to connect: Bringing social context into addiction neuroscience. *Nature Reviews Neuroscience, 17*(9), 592–599.

Heinrichs, R. W. (2003). Historical origins of schizophrenia: Two early madmen and their illness. *Journal for the History of Behavioural Sciences, 39*, 349–363.

Held, R. (1993). Binocular vision: Behavioural and neuronal development. In M. H. Johnson (Ed.), *Brain development and cognition: A reader* (pp. 152–166). Oxford, UK: Blackwell.

Helmchen, F., and Denk, W. (2005). Deep tissue two-photon microscopy. *Nature Methods, 2*, 932–940.

Helmholtz, H. von (1852). On the theory of compound colours. *Philosophical Magazine, 4*(4), 519–535.

Helmholtz, H. von. (1962). *Treatise on physiological optics* (J. P. C. Southall, Trans.). New York: Dover. (Original work published 1894.)

Hemmingsen, A. M. (1960). Energy metabolism as related to body size and respiratory surfaces, and its evolution. *Reports of Steno Memorial Hospital, Copenhagen, 9*, 1–110.

Herbert, J. (2013). Cortisol and depression: Three questions for psychiatry. *Psychological Medicine, 3, 449–469.*

Herbst, C. T., Stoeger, A. S., Frey, R., Lohscheller, J., et al. (2012). How low can you go? Physical production mechanism of elephant infrasonic vocalisations. *Science, 337*(6094), 595–599.

Herculano-Houzel, S. (2012). The remarkable, yet not extraordinary, human brain as a scaled-up primate brain and its associated cost. *Proceedings of the National Academy of Sciences, USA, 109*(Suppl. 1), 10661–10668.

Herculano-Houzel, S. (2014). The glia/neuron ratio: How it varies uniformly across brain structures and species and what that means for brain physiology and evolution. *Glia, 62*(9), 1377–1391.

Herculano-Houzel, S. (2018). The evolution of human capabilities and abilities. *Cerebrum*, Apr 1;2018. pii: cer-05-18. eCollection 2018 Mar–Apr.

Herek, G. M., and McLemore, K. A. (2013). Sexual prejudice. *Annual Review of Psychology, 64*, 309–333.

Herrada, G., and Dulac, C. (1997). A novel family of putative pheromone receptors in mammals with a topographically organized and sexually dimorphic distribution. *Cell, 90*, 763–773.

Herrmann, C., and Knight, R. (2001). Mechanisms of human attention: Event-related potentials and oscillations. *Neuroscience and Biobehavioural Reviews, 25*, 465–476.

Hertel, P., Fagerquist, M. V., and Svensson, T. H. (1999). Enhanced cortical dopamine output and antipsychotic-like effects of raclopride by alpha-2 adrenoceptor blockade. *Science, 286*, 105–107.

Hertenstein, E., Feige, B., Gmeiner, T., Kienzler, C., et al. (2019). Insomnia as a predictor of mental disorders: A systematic review and meta-analysis. *Sleep Medicine Reviews, 43*, 96–105.

Herxheimer, A. (2014). Jet lag. *BMJ Clinical Evidence*, pii: 2303.

Herzfeld, D. J., Kojima, Y., Soetedjo, R., and Shadmehr, R. (2015). Encoding of action by the Purkinje cells of the cerebellum. *Nature, 526*(7573), 439–442.

Hetherington, A. W., and Ranson, S. W. (1940). Hypothalamic lesions and adiposity in the rat. *Anatomical Record, 78*, 149–172.

Hewes, G. (1973). Primate communication and the gestural origin of language. *Current Anthropology, 14*, 5–24.

Heyes, C. (2012). New thinking: The evolution of human cognition. *Philosophical Transactions of the Royal Society of London. Series B: Biological Sciences, 367*(1599), 2091–2096.

Hickey, C., Di Lollo, V., and McDonald, J. J. (2009). Electrophysiological indices of target and distractor processing in visual search. *Journal of Cognitive Neuroscience, 21*(4), 760–775.

Hickey, M., Elliott, J., and Davison, S. L. (2012). Hormone replacement therapy. *British Medical Journal, 344*, e763.

Hicks, M. J., De, B. P., Rosenberg, J. B., Davidson, J. T., et al. (2011). Cocaine analog coupled to disrupted adenovirus: A vaccine strategy to evoke high-titer immunity against addictive drugs. *Molecular Therapy, 19*, 612–619.

Higley, J. D., Mehlman, P. T., Taub, D. M., Higley, S. B., et al. (1992). Cerebrospinal fluid monoamine and adrenal correlates of aggression in free-ranging rhesus monkeys. *Archives of General Psychiatry, 49*, 436–441.

Hill, R. A., Li, A. M., and Grutzendler, J. (2018). Lifelong cortical myelin plasticity and age-related degeneration in the live mammalian brain. *Nature Neuroscience, 21*(5), 683–695. https://doi.org/10.1038/s41593-018-0120-6.

Hill, R. S., and Walsh, C. A. (2005). Molecular insights into human brain evolution. *Nature, 437*, 64–67.

Hillhouse, T. M., and Porter, J. H. (2015). A brief history of the development of antidepressant drugs: From monoamines to glutamate. *Experimental and Clinical Psychopharmacology, 23*(1), 1–21.

Hillyard, S. A., Hink, R. F., Schwent, V. L., and Picton, T. W. (1973). Electrical signs of selective attention in the human brain. *Science, 182*, 177–180.

Himle, M. B., Woods, D. W., Piacentini, J. C., and Walkup, J. T. (2006). Brief review of habit reversal training for Tourette syndrome. *Journal of Child Neurology, 21*, 719–725.

Hines, M. (2010). Sex-related variation in human behaviour and the brain. *Trends in Cognitive Sciences, 14*(10), 448–456.

Hingson, R., Heeren, T., Winter, M., and Weschler, H. (2005). Magnitude of alcohol-related mortality and morbidity among U.S. college students ages 18–24: Changes from 1998 to 2001. *Annual Review of Public Health, 26*, 259–279.

Hirsch, E., Moye, D., and Dimon, J. H. (1995). Congenital indifference to pain: Long-term follow-up of two cases. *Southern Medical Journal, 88*, 851–857.

Hirst, W., Phelps, E. A., Buckner, R. L., Budson, A. E., et al. (2009). Long-term memory for the terrorist attack of September 11: Flashbulb memories, event memories, and the factors that influence their retention. *Journal of Experimental Psychology General, 138*(2), 161–176.

Hitt, E. (2007). Careers in neuroscience: From protons to poetry. *Science, 318*, 661–665.

Ho, S. Y., and Duchêne, S. (2014). Molecular-clock methods for estimating evolutionary rates and timescales. *Molecular Ecology, 23*, 5947–5965.

Hobaiter, C., and Byrne, R. W. (2011). The gestural repertoire of the wild chimpanzee. *Animal Cognition, 14*(5), 745–767.

Hochberg, L. R., Bacher, D., Jarosiewicz, B., Masse, N. Y., et al. (2012). Reach and grasp by people with tetraplegia using a neurally controlled robotic arm. *Nature, 485*(7398), 372–375.

Hodes, G. E., Kana, V., Menard, C., Merad, M., et al. (2015). Neuroimmune mechanisms of depression. *Nature Neuroscience, 18*, 1386–1393.

Hodgetts, S. I., and Harvey, A. R. (2017). Neurotrophic factors used to treat spinal cord injury. *Vitamins and Hormones, 104*, 405–457.

Hoefle, S., Engel, A., Basilio, R., Alluri, V., et al. (2018). Identifying musical pieces from fMRI data using encoding and decoding models. *Scientific Reports, 8*(1), 2266.

Hoeft, F., Hernandez, A., McMillon, G., Taylor-Hill, H., et al. (2006). Neural basis of dyslexia: A comparison between dyslexic and nondyslexic children equated for reading ability. *Journal of Neuroscience, 26*, 10700–10708.

Hoekzema, E., Barba-Müller, E., Pozzobon, C., Picado, M., et al. (2017). Pregnancy leads to long-lasting changes in human brain structure. *Nature Neuroscience, 20*(2), 287–296.

Hoffman, H. J., Rawal, S., Li, C. M., and Duffy, V. B. (2016). New chemosensory component in the U.S. National Health and Nutrition Examination Survey (NHANES): First-year results for measured olfactory dysfunction. *Reviews in Endocrine and Metabolic Disorders, 17*(2), 221–240.

Hofmann, A. (1981). *LSD: My problem child.* New York: McGraw-Hill.

Hofmann, S. G., Sawyer, A. T., Witt, A. A., and Oh, D. (2010). The effect of mindfulness-based therapy on anxiety and depression: A meta-analytic review. *Journal of Consulting and Clinical Psychology, 78*, 169–183.

Hohmann, A. G., Suplita, R. L., Bolton, N. M., Neely, M. H., et al. (2005). An endocannobinoid mechanism for stress-induced analgesia. *Nature, 435,* 1108–1112.

Hohmann, G. W. (1966). Some effects of spinal cord lesions on experienced emotional feelings. *Psychophysiology, 3,* 143–156.

Hollon, S. D., Thase, M. E., and Markowitz, J. C. (2002). Treatment and prevention of depression. *Psychological Science in the Public Interest, 3,* 39–77.

Holmes, C., Boche, D., Wilkinson, D., Yadegarfar, G., et al. (2008). Long-term effects of Abeta42 immunisation in Alzheimer's disease: Follow-up of a randomized, placebo-controlled phase I trial. *Lancet, 372,* 216–223.

Holschbach, M. A., Vitale, E. M., and Lonstein, J. S. (2018). Serotonin-specific lesions of the dorsal raphe disrupt maternal aggression and caregiving in postpartum rats. *Behavioural Brain Research, 348,* 53–64.

Holstege, G., Georgiadis, J. R., Paans, A. M., Meiners, L. C., et al. (2003). Brain activation during human male ejaculation. *Journal of Neuroscience, 23*(27), 9185–9193.

Holth, J. K., Fritschi, S. K., Wang, C., Pedersen, N. P., et al. (2019). The sleep-wake cycle regulates brain interstitial fluid tau in mice and CSF tau in humans. *Science, 363*(6429), 880–884.

Holt-Lunstad, J., Smith, T. B., and Layton, J. B. (2010). Social relationships and mortality risk: A meta-analytic review. *PLOS Medicine, 7,* e1000316.

Honey, G. D., Bullmore, E. T., Soni, W., Varatheesan, M., et al. (1999). Differences in frontal cortical activation by a working memory task after substitution of risperidone for typical antipsychotic drugs in patients with schizophrenia. *Proceedings of the National Academy of Sciences, USA, 96,* 13432–13437.

Hopfinger, J. B., Buonocore, M. H., and Mangun, G. R. (2000). The neural mechanisms of top-down attentional control. *Nature Neuroscience, 3,* 284–291.

Hopfinger, J., and Mangun, G. (1998). Reflexive attention modulates processing of visual stimuli in human extrastriate cortex. *Psychological Science, 6,* 441–447.

Hopkins, W. D., Misiura, M., Pope, S. M., and Latash, E. M. (2015). Behavioural and brain asymmetries in primates: A preliminary evaluation of two evolutionary hypotheses. *Annals of the New York Academy of Sciences, 1359,* 65–83.

Hopper, K., and Wanderling, J. (2000). Revisiting the developed versus developing country distinction in course and outcome in schizophrenia: Results from ISoS, the WHO collaborative followup project. International Study of Schizophrenia. *Schizophrenia Bulletin, 26,* 835–846.

Horikawa, T., Tamaki, M., Miyawaki, Y., and Kamitani, Y. (2013). Neural decoding of visual imagery during sleep. *Science, 340*(6132), 639–642.

Howard-Jones, P. A. (2014). Neuroscience and education: Myths and messages. *Nature Reviews Neuroscience, 15*(12), 817–824.

Hsu, M., Bhatt, M., Adolphs, R., Tranel, D., et al. (2005). Neural systems responding to degrees of uncertainty in human decision-making. *Science, 310,* 1680–1683.

Hua, J. T., Hildreth, K. L., and Pelak, V. S. (2016). Effects of testosterone therapy on cognitive function in aging: A systematic review. *Cognitive and Behavioural Neurology, 29*(3), 122–138.

Huang, A. L., Chen, X., Hoon, M. A., Chandrashekar, J., et al. (2006). The cells and logic for mammalian sour taste detection. *Nature, 442,* 934–938.

Huang, W., Zhu, P. J., Zhang, S., Zhou, H., et al. (2013). mTORC2 controls actin polymerisation required for consolidation of long-term memory. *Nature Neuroscience, 16*(4), 441–448.

Huang, Z. J., and Luo, L. (2015). It takes the world to understand the brain. *Science, 350*(6256), 42–44.

Hubel, D. H., and Wiesel, T. N. (1959). Receptive fields of single neurones in the cat's striate cortex. *Journal of Physiology (London), 148,* 573–591.

Hubel, D. H., and Wiesel, T. N. (1965). Binocular interaction in striate cortex kittens reared with artificial squint. *Journal of Neurophysiology, 28,* 1041–1059.

Hubel, D. H., Wiesel, T. N., and LeVay, S. (1977). Plasticity of ocular dominance in monkey striate cortex. *Philosophical Transactions of the Royal Society of London. Series B: Biological Sciences, 278,* 377–409.

Huber, E., Webster, J. M., Brewer, A. A., MacLeod, D. I., et al. (2015). A lack of experience-dependent plasticity after more than a decade of recovered sight. *Psychological Science, 26*(4), 393–401. doi:10.1177/0956797614563957.

Hublin, J. J., Ben-Ncer, A., Bailey, S. E., Freidline, S. E., et al. (2017). New fossils from Jebel Irhoud, Morocco and the pan-African origin of *Homo sapiens. Nature, 546*(7657), 289–292.

Hudspeth, A. J. (1992). Hair-bundle mechanics and a model for mechanoelectrical transduction by hair cells. *Society of General Physiologists Series, 47,* 357–370.

Hudspeth, A. J. (2014). Integrating the active process of hair cells with cochlear function. *Nature Reviews Neuroscience, 15*(9), 600–614.

Huettel, S. A., Stowe, C. J., Gordon, E. M., Warner, B. T., et al. (2006). Neural signatures of economic preferences for risk and ambiguity. *Neuron, 49,* 765–775.

Huffman, K. J., Nelson, J., Clarey, J., and Krubitzer, L. (1999). Organisation of somatosensory cortex in three species of marsupials, *Dasyurus hallucatus, Dactylopsila trivirgata,* and *Monodelphis domestica*: Neural correlates of morphological specialisations. *Journal of Comparative Neurology, 403,* 5–32.

Hülsheger, U. R., and Schewe, A. F. (2011). On the costs and benefits of emotional labor: A meta-analysis of three decades of research. *Journal of Occupational Health Psychology, 16*(3), 361–389.

Hunt, G. R., Corballis, M. C., and Gray, R. D. (2001). Laterality in tool manufacture by crows. *Nature, 414,* 707.

Huntington, G. (1872). On chorea. *Medical and Surgical Reporter, 26,* 317–321.

Hurtado, M. D., Sergeyev, V. G., Acosta, A., Spegele, M., et al. (2013). Salivary peptide tyrosine-tyrosine 3-36 modulates ingestive behaviour without inducing taste aversion. *Journal of Neuroscience, 33,* 18368–18380.

Hussain, S. J., and Cole, K. J. (2015). No enhancement of 24-hour visuomotor skill retention by post-practice caffeine administration. *PLOS ONE, 10*(6), e0129543.

Huston, N. J., Brenner, L. A., Taylor, Z. C., and Ritter, R. C. (2019). NPY2 receptor activation in the dorsal vagal complex increases food intake and attenuates CCK-induced satiation in male rats. *American Journal of Physiology—Regulatory, Integrative and Comparative Physiology, 316*(4), R406–R416.

Huth, A. G., de Heer, W. A., Griffiths, T. L., Theunissen, F. E., et al. (2016). Natural speech reveals the semantic maps that tile human cerebral cortex. *Nature, 532*(7600), 453–458.

Huttenlocher, P. R., deCourten, C., Garey, L. J., and Van der Loos, H. (1982). Synaptogenesis in human visual cortex: Evidence for synapse elimination during normal development. *Neuroscience Letters, 33,* 247–252.

Hviid, A., Hansen, J. V., Frisch, M., and Melbye, M. (2019). Measles, mumps, Rubella vaccination and autism: A nationwide cohort study. *Annals of Internal Medicine.* doi: 10.7326/M18-2101. [Epub ahead of print].

Hyde, K. L., and Peretz I. (2004). Brains that are out of tune but in time. *Psychological Science, 15,* 356–360.

Hyde, K. L., Zatorre, R. J., Griffiths, T. D., Lerch, J. P., et al. (2006). Morphometry of the amusic brain: A two-site study. *Brain, 129,* 2562–2570.

Hyde, T. M., Aaronson, B. A., Randolph, C., Rickler, K. C., et al. (1992). Relationship of birth weight to the phenotypic expression of Gilles de la Tourette's syndrome in monozygotic twins. *Neurology, 42,* 652–658.

Hyde, T. M., and Weinberger, D. R. (1990). The brain in schizophrenia. *Seminars in Neurology, 10,* 276–286.

Hyman, S. E. (2018). The daunting polygenicity of mental illness: Making a new map. *Philosophical Transactions of the Royal Society of London. Series B: Biological Sciences, 373*(1742). https://doi.org/10.1098/rstb.2017.0031.

I

Imai, T., Yamazaki, T., Kobayakawa, R., Kobayakawa, K., et al. (2009). Pre-target axon sorting establishes the neural map topography. *Science, 325,* 585–590.

Imeri, L, and Opp, M. R. (2009). How (and why) the immune system makes us sleep. *Nature Reviews Neuroscience, 10,* 199–210.

Imperato-McGinley, J., Guerrero, L., Gautier, T., and Peterson, R. E. (1974). Steroid 5α-reductase deficiency in man: An inherited form of male pseudohermaphroditism. *Science, 86*, 1213–1215.

Infurna, F. J., and Luthar, S. S. (2016). Resilience to major life stressors is not as common as thought. *Perspectives on Psychological Science, 11*, 175–194.

Inoue, K., and Tsuda, M. (2018). Microglia in neuropathic pain: Cellular and molecular mechanisms and therapeutic potential. *Nature Reviews Neuroscience, 19*(3), 138–152.

Insley, S. J. (2000). Long-term vocal recognition in the northern fur seal. *Nature, 406*, 404–405.

Institute of Medicine. (1990). *Broadening the base of treatment for alcohol problems.* Washington, DC: National Academy Press.

Intartaglia, B., White-Schwoch, T., Kraus, N., and Schön, D. (2017). Music training enhances the automatic neural processing of foreign speech sounds. *Scientific Reports, 7*(1), 12631.

Ip, J. P. K., Mellios, N., and Sur, M. (2018). Rett syndrome: Insights into genetic, molecular and circuit mechanisms. *Nature Reviews Neuroscience, 19*(6), 368–382. https://doi.org/10.1038/s41583-018-0006-3.

Ipata, A. E., Gee, A. L., Goldberg, M. E., and Bisley, J. W. (2006). Activity in the lateral intraparietal area predicts the goal and latency of saccades in a free-viewing visual search task. *Journal of Neuroscience, 26*, 3656–3661.

Irish, L. A., Kline, C. E., Gunn, H. E., Buysse, D. J., et al. (2015). The role of sleep hygiene in promoting public health: A review of empirical evidence. *Sleep Medicine Reviews, 22*, 23–36.

Isaacson, R. L. (1972). Hippocampal destruction in man and other animals. *Neuropsychologia, 10*, 47–64.

Isles, A. R., Baum, M. J., Ma, D., Keverne, E. B., et al. (2001). Urinary odour preferences in mice. *Nature, 409*, 783–784.

Iverson, J. M., and Goldin-Meadow, S. (1998). Why people gesture when they speak. *Nature, 396*, 228.

J

Jablensky, A., Sartorius, N., Ernberg, G., Anker, M., et al. (1992). Schizophrenia: Manifestations, incidence and course in different cultures. A World Health Organisation ten-country study. *Psychological Medicine. Monograph Supplement, 20*, 1–97.

Jackson, A. F., and Bolger, D. J. (2014). The neurophysiological bases of EEG and EEG measurement: A review for the rest of us. *Psychophysiology, 51*(11),1061–1071.

Jackson, H., and Parks, T. N. (1982). Functional synapse elimination in the developing avian cochlear nucleus with simultaneous reduction in cochlear nerve axon branching. *Journal of Neuroscience, 2*, 1736–1743.

Jacobs, G. H. (1993). The distribution and nature of colour vision among the mammals. *Biological Reviews of the Cambridge Philosophical Society, 68*, 413–471.

Jacobs, G. H., Williams, G. A., Cahill, H., and Nathans, J. (2007). Emergence of novel colour vision in mice engineered to express a human cone photopigment. *Science, 315*, 1723–1725.

Jacobs, J., Weidemann, C. T., Miller, J. F., Solway, A., et al. (2013). Direct recordings of grid-like neuronal activity in human spatial navigation. *Nature Neuroscience, 16*(9), 1188–1190.

Jacobs, L. F., and Spencer, W. D. (1994). Natural space-use patterns and hippocampal size in kangaroo rats. *Brain, Behaviour and Evolution, 44*, 125–132.

Jaffe, A. E., Gao, Y., Deep-Soboslay, A., Tao, R., et al. (2016). Mapping DNA methylation across development, genotype and schizophrenia in the human frontal cortex. *Nature Neuroscience, 19*, 40–47.

James, T. W., Culham, J., Humphery, G. K., Milner, A. D., et al. (2003). Ventral occipital lesions impair object recognition but not object-directed grasping: An fMRI study. *Brain, 126*, 2464–2475.

James, W. (1890). *The principles of psychology.* New York: Holt.

Jamieson, D., and Roberts, A. (2000). Responses of young *Xenopus laevis* tadpoles to light dimming: Possible roles for the pineal eye. *Journal of Experimental Biology, 203*, 1857–1867.

Jan, Y. N., and Jan, L. Y. (2018). Force-activated ion channels in close-up. *Nature, 554*(7693), 469–470.

Janak, P. H., and Tye, K. M. (2015). From circuits to behaviour in the amygdala. *Nature, 517*(7534), 284–292.

Janssen, S., Laermans, J., Verhulst, P. J., Thijs, T., et al. (2011). Bitter taste receptors and α-gustducin regulate the secretion of ghrelin with functional effects on food intake and gastric emptying. *Proceedings of the National Academy of Sciences, USA, 108*(5), 2094–2099.

Jaremka, L. M., Fagundes, C. P., Peng, J., Bennett, J. M., et al. (2013). Loneliness promotes inflammation during acute stress. *Psychological Science, 24*, 1089–1097.

Jaskiw, G. E., and Popli, A. P. (2004). A meta-analysis of the response to chronic L-dopa in patients with schizophrenia: Therapeutic and heuristic implications. *Psychopharmacology (Berlin), 171*, 365–374.

Jaunmuktane, Z., Mead, S., Ellis, M., Wadsworth, J. D., et al. (2015). Evidence for human transmission of amyloid-beta pathology and cerebral amyloid angiopathy. *Nature, 525*(7568), 247–250. doi:10.1038/nature15369.

Javitt, D. C., and Sweet, R. A. (2015). Auditory dysfunction in schizophrenia: Integrating clinical and basic features. *Nature Reviews Neuroscience, 16*, 535–550.

Jeffress, L. A. (1948). A place theory of sound localisation. *Journal of Comparative and Physiological Psychology, 41*, 35–39.

Jenkins, J., and Dallenbach, K. (1924). Oblivescence during sleep and waking. *American Journal of Psychology, 35*, 605–612.

Jensen, M. P., Day, M. A., and Miró, J. (2014). Neuromodulatory treatments for chronic pain: Efficacy and mechanisms. *Nature Reviews Neurology, 10*(3), 167–178.

Jentsch, J. D., Redmond, D. E., Jr., Elsworth, J. D., Taylor, J. R., et al. (1997). Enduring cognitive deficits and cortical dopamine dysfunction in monkeys after long-term administration of phencyclidine. *Science, 277*, 953–955.

Jepsen, M. L. (2013, November 24). Bringing back my real self with hormones. *The New York Times*, p. SR6. www.nytimes.com/2013/11/24/opinion/sunday/bringing-back-my-real-self-with-hormones.html

Jessen, N. A., Munk, A. S., Lundgaard, I., and Nedergaard, M. (2015). The glymphatic system: A beginner's guide. *Neurochemical Research, 40*(12), 2583–2599.

Ji, D., and Wilson, M. A. (2007). Coordinated memory replay in the visual cortex and hippocampus during sleep. *Nature Neuroscience, 10*(1), 100–107.

Ji, R. R., Chamessian, A., and Zhang, Y. Q. (2016). Pain regulation by non-neuronal cells and inflammation. *Science, 354*(6312), 572–577.

Johansson, R. S., and Flanagan, J. R. (2009). Coding and use of tactile signals from the fingertips in object manipulation tasks. *Nature Reviews Neuroscience, 10*, 345–358.

Johnson, K. O., and Hsiao, S. S. (1992). Neural mechanisms of tactual form and texture perception. *Annual Review of Neuroscience, 15*, 227–250.

Johnson, M. B., Sun, X., Kodani, A., Borges-Monroy, R., et al. (2018). *Aspm* knockout ferret reveals an evolutionary mechanism governing cerebral cortical size. *Nature, 556*(7701), 370–375.

Jones, B. A., Shimell, J. J., and Watson, N. V. (2011). Pre- and postnatal bisphenol A treatment results in persistent deficits in the sexual behaviour of male rats, but not female rats, in adulthood. *Hormones and Behaviour, 59*(2), 246–251.

Jones, B. A., and Watson, N. V. (2012). Perinatal BPA exposure demasculinizes males in measures of affect but has no effect on water maze learning in adulthood. *Hormones and Behaviour, 61*(4), 605–610.

Jones, H. J., Gage, S. H., Heron, J., Hickman, M., et al. (2018). Association of combined patterns of tobacco and cannabis use in adolescence with psychotic experiences. *JAMA Psychiatry, 75*(3), 240–246.

Jones, J. R., Tackenberg, M. C., and McMahon, D. G. (2015). Manipulating circadian clock neuron firing rate resets molecular circadian rhythms and behaviour. *Nature Neuroscience, 18*(3), 373–375.

Jones, T. A. (2017). Motor compensation and its effects on neural reorganisation after stroke. *Nature Reviews Neuroscience, 18*(5), 267–280.

Jønsson, K. A., Fabre, P. H., and Irestedt, M. (2012). Brains, tools, innovation and biogeography in crows and ravens. *BMC Evolutionary Biology, 12*, 72.

Jordan, B. D., Jahre, C., Hauser, W. A., Zimmerman, R. D., et al. (1992). CT of 338 active professional boxers. *Radiology, 185*, 509–512.

Jordan, C. L., Breedlove, S. M., and Arnold, A. P. (1991). Ontogeny of steroid accumulation in spinal lumbar motoneurons of the rat: Implications for androgen's site of action during synapse elimination. *Journal of Comparative Neurology, 313*, 441–448.

Jordan, G., Deeb, S. S., Bosten, J. M., and Mollon, J. D. (2010). The dimensionality of colour vision in carriers of anomalous trichromacy. *Journal of Vision, 10*, 12.

Jordt, S.-E., Bautista, D. M., Chuang, H., McKemy, D. D., et al. (2004). Mustard oils and cannabinoids excite sensory nerve fibres through the TRP channel ANKTM1. *Nature, 427*, 260–265.

Joris, P. X., Smith, P. H., and Yin, T. C. (1998). Coincidence detection in the auditory system: 50 years after Jeffress. *Neuron, 21*(6), 1235–1238.

Joseph, J. S., Chun, M. M., and Nakayama, K. (1997). Attentional requirements in a "preattentive" feature search task. *Nature, 387*, 805–807.

Julian, T., and McKenry, P. C. (1979). Relationship of testosterone to men's family functioning at mid-life: A research note. *Aggressive Behaviour, 15*, 281–289.

K

Kaar, G. F., and Fraher, J. P. (1985). The development of alpha and gamma motoneuron fibres in the rat. I. A comparative ultrastructural study of their central and peripheral axon growth. *Journal of Anatomy, 141*, 77–88.

Kaas, J. H., and Hackett, T. A. (1999). "What" and "where" processing in auditory cortex. *Nature Neuroscience, 2*, 1045–1047.

Kable, J. W., and Glimcher, P. W. (2009). The neurobiology of decision: Consensus and controversy. *Neuron, 63*, 733–745.

Kagan, J. (1997). Temperament and the reactions to unfamiliarity. *Child Development, 68*, 139–143.

Kaji, I., Karaki, S., and Kuwahara, A. (2014). Taste sensing in the colon. *Current Pharmaceutical Design, 20*(16), 2766–2774.

Kakei, S., Hoffman, D. S., and Strick, P. L. (1999). Muscle and movement representations in the primary motor cortex. *Science, 285*, 2136–2139.

Kales, A., and Kales, J. (1970). Evaluation, diagnosis and treatment of clinical conditions related to sleep. *JAMA, 213*, 2229–2235.

Kales, A., and Kales, J. D. (1974). Sleep disorders. Recent findings in the diagnosis and treatment of disturbed sleep. *New England Journal of Medicine, 290*, 487–499.

Kamm, G. B., Pisciottano, F., Kliger, R., and Franchini, L. F. (2013). The developmental brain gene *NPAS3* contains the largest number of accelerated regulatory sequences in the human genome. *Molecular Biology and Evolution, 30*, 1088–1102.

Kandel, E. R. (2009). The biology of memory: A forty-year perspective. *Journal of Neuroscience, 29*, 12748–12756.

Kandel, E. R., Castellucci, V. F., Goelet, P., and Schacher, S. (1987). 1987 cell-biological interrelationships between short-term and long-term memory. *Research Publications—Association for Research in Nervous and Mental Disease, 65*, 111–132.

Kandel, E. R., Dudai, Y., and Mayford, M. R. (2014). The molecular and systems biology of memory. *Cell, 157*(1), 163–186.

Kandler, K., Clause, A., and Noh, J. (2009). Tonotopic reorganisation of developing auditory brainstem circuits. *Nature Neuroscience, 12*, 711–716.

Kane, J. M., and Correll, C. U. (2010). Past and present progress in the pharmacologic treatment of schizophrenia. *Journal of Clinical Psychiatry, 71*(9), 1115–1124.

Kang, C., Riazuddin, S., Mundorff, J., Krasnewich, D., et al. (2010). Mutation in the lysosomal enzyme–targeting pathway and persistent stuttering. *New England Journal of Medicine, 362*, 677–685.

Kang, H. J., Imperato-McGinley, J., Zhu, Y. S., Cai, L. Q., et al. (2011). The first successful paternity through in vitro fertilisation-intracytoplasmic sperm injection with a man homozygous for the 5α-reductase-2 gene mutation. *Fertility and Sterility, 95*(6), 2125.e5–2125.e8.

Karlsgodt, K. H., Sun, D., and Cannon, T. D. (2010). Structural and functional brain abnormalities in schizophrenia. *Current Directions in Psychological Science, 19*, 226–231.

Karpicke, J. D, and Blunt, J. R. (2011). Retrieval practice produces more learning than elaborative studying with concept mapping. *Science, 331*(6018), 772–775.

Karra, E., Chandarana, K., and Batterham, R. L. (2009). The role of peptide YY in appetite regulation and obesity. *Journal of Physiology, 587*, 19–25.

Kass, A. E., Kolko, R. P., and Wilfley, D. E. (2013). Psychological treatments for eating disorders. *Current Opinion in Psychiatry, 26*(6), 549–555.

Kato, H. E., Zhang, F., Yizhar, O., Ramakrishnan, C., et al. (2012). Crystal structure of the channelrhodopsin light-gated cation channel. *Nature, 482*, 369–374.

Katsiki, N., Tziomalos, K., and Mikhailidis, D. P. (2014). Alcohol and the cardiovascular system: A double-edged sword. *Current Pharmaceutical Design, 20*, 6276–6288.

Katz, D. B., and Steinmetz, J. E. (2002). Psychological functions of the cerebellum. *Behavioural Cognitive Neuroscience Review, 1*, 229–241.

Katzenberg, D., Young, T., Finn, L., Lin, L., et al. (1998). A CLOCK polymorphism associated with human diurnal preference. *Sleep, 21*, 569–576.

Kaufman, M. T., Churchland, M. M., Ryu, S. I., and Shenoy, K. V. (2014). Cortical activity in the null space: Permitting preparation without movement. *Nature Neuroscience, 17*(3), 440–448.

Kaushall, P. I., Zetin, M., and Squire, L. R. (1981). A psychosocial study of chronic, circumscribed amnesia. *Journal of Nervous and Mental Disease, 169*, 383–389.

Kay, K. N., Naselaris, T., Prenger, R. J., and Gallant, J. L. (2008). Identifying natural images from human brain activity. *Nature, 452*, 352–355.

Kaya, E. M., and Elhilali, M. (2017). Modelling auditory attention. *Philosophical Transactions of the Royal Society of London. Series B: Biological Sciences, 372*(1714), 1–10.

Kaye, W. H., Fudge, J. L., and Paulus, M. (2009). New insights into symptoms and neurocircuit function of anorexia nervosa. *Nature Reviews Neuroscience, 10*, 573–584.

Keane, T. M. (1998). Psychological and behavioural treatments of post-traumatic stress disorder. In P. E. Nathan, and J. M. Gorman (Eds.), *A guide to treatments that work* (pp. 398–407). New York: Oxford University Press.

Keenan, J. P., Nelson, A., O'Connor, M., and Pascual-Leone, A. (2001). Self-recognition and the right hemisphere. *Nature, 409*, 305.

Keesey, R. E. (1980). A set-point analysis of the regulation of body weight. In A. J. Stunkard (Ed.), *Obesity* (pp. 144–165). Philadelphia: Saunders.

Keesey, R. E., and Boyle, P. C. (1973). Effects of quinine adulteration upon body weight of LH-lesioned and intact male rats. *Journal of Comparative and Physiological Psychology, 84*, 38–46.

Kell, C. A., von Kriegstein, K., Rösler, A., Kleinschmidt, A., et al. (2005). The sensory cortical representation of the human penis: Revisiting somatotopy in the male homunculus. *Journal of Neuroscience, 25*(25), 5984–5987.

Keller, C. J., Davidesco, I., Megevand, P., Lado, F. A., et al. (2017). Tuning face perception with electrical stimulation of the fusiform gyrus. *Human Brain Mapping, 38*(6), 2830–2842.

Keller, M. C., and Visscher, P. M. (2015). Genetic variation links creativity to psychiatric disorders. *Nature Neuroscience, 18*(7), 928–929.

Kelso, S. R., and Brown, T. H. (1986). Differential conditioning of associative synaptic enhancement in hippocampal brain slices. *Science, 232*, 85–87.

Keltner, D., and Ekman, P. (2000). Facial expression of emotion. In M. Lewis and J. M. Haviland-Jones (Eds.), *Handbook of emotions* (2nd ed., pp. 236–250). New York: Guilford.

Kemp, M. (2001). The harmonious hand. Marin Mersenne and the science of memorized music. *Nature, 409*, 666.

Kendler, K. S., Gardner, C. O., and Prescott, C. A. (1999). Clinical characteristics of major depression that predict risk of depression in relatives. *Archives of General Psychiatry, 56*, 322–327.

Kendler, K. S., Karkowski, L. M., Neale, M. C., and Prescott, C. A. (2000). Illicit psychoactive substance use, heavy use, abuse, and dependence in a US population-based sample of male twins. *Archives of General Psychiatry, 57*, 261–269.

Kennedy, D. P., and Adolphs, R. (2011). Impaired fixation to eyes following amygdala damage arises from abnormal bottom-up attention. *Neuropsychologia, 49*, 589–595.

Kennedy, D. P., Gläscher, J., Tyszka, J. M., and Adolphs, R. (2009). Personal space regulation by the human amygdala. *Nature Neuroscience, 12*, 1226–1227.

Kennerknecht, I., Grueter, T., Welling, B., Wentzek, S., et al. (2006). First report of prevalence of non-syndromic hereditary prosopagnosia (HPA). *American Journal of Medical Genetics. Part A, 140*, 1617–1622.

Kern, D. W., Wroblewski, K. E., Schumm, L. P., Pinto, J. M., et al. (2014). Olfactory function in Wave 2 of the National Social Life, Health, and Aging Project. *Journals of Gerontology. Series B, Psychological Sciences and Social Sciences, 69*(Suppl. 2), S134–143.

Kerns, J. C., Guo, J., Fothergill, E., Howard, L., et al. (2017). Increased physical activity associated with less weight regain six years after "The Biggest Loser" competition. *Obesity (Silver Spring), 25*(11), 1838–1843.

Kertesz, A., Harlock, W., and Coates, R. (1979). Computer tomographic localisation, lesion size, and prognosis in aphasia and nonverbal impairment. *Brain and Language, 8*, 34–50.

Kesner, R. P. (1998). Neurobiological views of memory. In J. L. Martinez, Jr., and R. P. Kesner (Eds.), *Neurobiology of learning and memory* (3rd ed., pp. 361–416). San Diego, CA: Harcourt Brace.

Kesner, R. P., Bolland, B. L., and Dakis, M. (1993). Memory for spatial locations, motor responses, and objects: Triple dissociation among the hippocampus, caudate nucleus, and extrastriate visual cortex. *Experimental Brain Research, 93*, 462–470.

Kessels, H. W., and Malinow, R. (2009). Synaptic AMPA receptor plasticity and behaviour. *Neuron, 61*, 340–350.

Kessler, R. C., Berglund, P., Demler, O., Jin, R., et al. (2005). Lifetime prevalence and age-of-onset distributions of *DSM-IV* disorders in the National Comorbidity Survey Replication. *Archives of General Psychiatry, 62*, 593–602.

Kessler, R. C., Chiu, W. T., Demler, O., and Walters, E. E. (2005). Prevalence, severity, and comorbidity of 12-month *DSM-IV* disorders in the National Comorbidity Survey Replication. *Archives of General Psychiatry, 62*, 617–627.

Kettlewell, H. B. D. (1955). Selection experiments on industrial melanism in the Lepidoptera. *Heredity, 9*, 323–342.

Kety, S. S., Wender, P. H., Jacobsen, B., Ingraham, L. J., et al. (1994). Mental illness in the biological and adoptive relatives of schizophrenic adoptees. Replication of the Copenhagen Study in the rest of Denmark. *Archives of General Psychiatry, 51*, 442–455.

Khalsa, S. S., Feinstein, J. S., Li ,W., Feusner, J. D., et al. (2016). Panic anxiety in humans with bilateral amygdala lesions: Pharmacological induction via cardiorespiratory interoceptive pathways. *Journal of Neuroscience, 36*, 3559–3566.

Khankan, R. R., Griffis, K. G., Haggerty-Skeans, J. R., Zhong, H., et al. (2016). Olfactory ensheathing cell transplantation after a complete spinal cord transection mediates neuroprotective and immunomodulatory mechanisms to facilitate regeneration. *Journal of Neuroscience, 36*(23), 6269–6286.

Kiang, N.Y. S. (1965). *Discharge patterns of single fibres in the cat's auditory nerve.* Cambridge, MA: MIT Press.

Kiehn, O. (2016). Decoding the organisation of spinal circuits that control locomotion. *Nature Reviews Neuroscience, 17*(4), 224–238.

Kim, C. K., Adhikari, A., and Deisseroth, K. (2017). Integration of optogenetics with complementary methodologies in systems neuroscience. *Nature Reviews Neuroscience, 18*(4), 222–235.

Kim, D. R., Pesiridou, A., and O'Reardon, J. P. (2009). Transcranial magnetic stimulation in the treatment of psychiatric disorders. *Current Psychiatry Reports, 11*, 447–452.

Kim, K. H., Relkin, N. R., Lee, K. M., and Hirsch, J. (1997). Distinct cortical areas associated with native and second languages. *Nature, 388*, 171–174.

Kimura, D. (1973). The asymmetry of the human brain. *Scientific American, 228*(3), 70–78.

Kimura, D. (1981). Neural mechanisms in manual signing. *Sign Language Studies, 33*, 291–312.

Kimura, D. (1993). *Neuromotor mechanisms in human communication.* Oxford, UK: Oxford University Press.

Kimura, D., and Watson, N.V. (1989). The relation between oral movement control and speech. *Brain and Language, 37*, 565–590.

King, A. (2013, August 24). The nose knows: How to train a canine conservationist. *New Scientist* (www.newscientist.com).

King, S., St-Hilaire, A., and Heidkamp, D. (2010). Prenatal factors in schizophrenia. *Current Directions in Psychological Science, 19*, 209–213.

Kingsbury, S. J., and Garver, D. L. (1998). Lithium and psychosis revisited. *Progress in Neuro-Psychopharmacology & Biological Psychiatry, 22*, 249–263.

Kinnamon, S. C. (2012). Taste receptor signalling: From tongues to lungs. *Acta Physiologica, 204*(2), 158–168.

Kinney, H. C. (2009). Brainstem mechanisms underlying the sudden infant death syndrome: Evidence from human pathologic studies. *Developmental Psychobiology, 51*, 223–233.

Kinsey, A. C., Pomeroy, W. B., and Martin, C. E. (1948). *Sexual behaviour in the human male.* Philadelphia: Saunders.

Kinsey, A. C., Pomeroy, W. B., Martin, C. E., and Gebhard, P. H. (1953). *Sexual behaviour in the human female.* Philadelphia: Saunders.

Kipnis, J. (2018). The seventh sense. *Scientific American, 319*(2), 28–35.

Kisely, S., Li, A., Warren, N., and Siskind, D. (2018). A systematic review and meta-analysis of deep brain stimulation for depression. *Depression and Anxiety, 35*(5), 468–480.

Kisler, K., Nelson, A. R., Montagne, A., and Zlokovic, B.V. (2017). Cerebral blood flow regulation and neurovascular dysfunction in Alzheimer disease. *Nature Reviews Neuroscience, 18*(7), 419–434. https://doi.org/10.1038/nrn.2017.48.

Kivell, T. L., Kibii, J. M., Churchill, S. E., Schmid, P., et al. (2011). *Australopithecus sediba* hand demonstrates mosaic evolution of locomotor and manipulative abilities. *Science, 333*(6048), 1411–1417.

Klar, A. J. (2003). Human handedness and scalp hair-whorl direction develop from a common genetic mechanism. *Genetics, 165*, 269–276.

Kleiber, M. (1947). Body size and metabolic rate. *Physiological Reviews, 15*, 511–541.

Kleiman, E. M., Turner, B. J., Fedor, S., Beale, E. E., et al. (2017). Examination of real-time fluctuations in suicidal ideation and its risk factors: Results from two ecological momentary assessment studies. *Journal of Abnormal Psychology, 126*(6), 726–738.

Klein, M., Shapiro, K. M., and Kandel, E. R. (1980). Synaptic plasticity and the modulation of the Ca^{2+} current. *Journal of Experimental Biology, 89*, 117–157.

Klein, R. M. (2000). Inhibition of return. *Trends in Cognitive Science, 4*, 138–147.

Kleinau, G., Müller, A., and Biebermann, H. (2016). Oligomerisation of GPCRs involved in endocrine regulation. *Journal of Molecular Endocrinology, 57*(1), R59–R80.

Kleitman, N. (1969). Basic rest-activity cycle in relation to sleep and wakefulness. In A. Kales (Ed.), *Sleep: Physiology and pathology.* Philadelphia: Lippincott.

Kleitman, N., and Engelmann, T. (1953). Sleep characteristics of infants. *Journal of Applied Physiology, 6*, 269–282.

Klingberg, T., Hedehus, M., Temple, E., Salz, T., et al. (2000). Microstructure of temporo-parietal white matter as a basis for reading ability: Evidence from diffusion tensor magnetic resonance imaging. *Neuron, 25*, 493–500.

Kluger, M. J. (1978). The evolution and adaptive value of fever. *American Scientist, 66*, 38–43.

Klump, K. L., Miller, K. B., Keel, P. K., McGue, M., et al. (2001). Genetic and environmental influences on anorexia nervosa syndromes in a population-based twin sample. *Psychological Medicine, 31*, 737–740.

Klüver, H., and Bucy, P. C. (1938). An analysis of certain effects of bilateral temporal lobectomy in the rhesus monkey, with special reference to "psychic blindness." *Journal of Psychology, 5*, 33–54.

Knecht, S., Flöel, A., Dräger, B., Breitenstein, C., et al. (2002). Degree of language lateralisation determines susceptibility to unilateral brain lesions. *Nature Neuroscience, 5*, 695–699.

Knepper, M. A., Kwon, T. H., and Nielsen, S. (2015). Molecular physiology of water balance. *New England Journal of Medicine, 372*, 1349–1358.

Knibestol, M., and Valbo, A. B. (1970). Single unit analysis of mechanoreceptor activity from the human glabrous skin. *Acta Physiologica Scandinavica, 80*, 178–195.

Knott, C. D., Emery Thompson, M., Stumpf, R. M., and McIntyre, M. H. (2010). Female reproductive strategies in orangutans, evidence for female choice and counterstrategies to infanticide in a species with frequent sexual coercion. *Proceedings: Biological Sciences, 277*, 105–113.

Knudsen, L. B., Secher, A., Hecksher-Sørensen, J., and Pyke, C. (2016). Long-acting glucagon-like peptide-1 receptor agonists have direct access to and effects on pro-opiomelanocortin/cocaine- and amphetamine-stimulated transcript neurons in the mouse hypothalamus. *Journal of Diabetes Investigation, 7*(Suppl. 1), 56–63.

Knuth, N. D., Johannsen, D. L., Tamboli, R. A., Marks-Shulman, P. A., et al. (2014). Metabolic adaptation following massive weight loss is related to the degree of energy imbalance and changes in circulating leptin. *Obesity, 22*, 2563–2569.

Koch, G., Oliveri, M., Torriero, S., and Caltagirone, C. (2005). Modulation of excitatory and inhibitory circuits for visual awareness in the human right parietal cortex. *Experimental Brain Research, 160*, 510–516.

Kodama, T., Lai, Y. Y., and Siegel, J. M. (2003). Changes in inhibitory amino acid release linked to pontine-induced atonia: An *in vivo* microdialysis study. *Journal of Neuroscience, 23*, 1548–1554.

Koechlin, E., Ody, C., and Kouneiher, F. (2003). The architecture of cognitive control in the human prefrontal cortex. *Science, 302*(5648), 1181–1185.

Kogan, J. H., Frankland, P. W., Blendy, J. A., Coblentz, J., et al. (1997). Spaced training induces normal long-term memory in CREB mutant mice. *Current Biology, 7*, 1–11.

Kohl, J., Babayan, B. M., Rubinstein, N. D., Autry, A. E., et al. (2018). Functional circuit architecture underlying parental behaviour. *Nature, 556*(7701), 326–331.

Kohl, M. M., Shipton, O. A., Deacon, R. M., Rawlins, J. N., et al. (2011). Hemisphere-specific optogenetic stimulation reveals left-right asymmetry of hippocampal plasticity. *Nature Neuroscience, 14*(11), 1413–1415.

Kojima, M., Hosoda, H., Date, Y., Nakazato, M., et al. (1999). Ghrelin is a growth-hormone-releasing acylated peptide from stomach. *Nature, 402*, 656–660.

Kokrashvili, Z., Mosinger, B., and Margolskee, R. F. (2009). T1r3 and α-gustducin in gut regulate secretion of glucagon-like peptide-1. *Annals of the New York Academy of Sciences, 1170*, 91–94.

Kolata, G. (2016, May 2). After "The Biggest Loser," their bodies fought to regain weight. *The New York Times*, Health Section (www.nytimes.com/2016/05/02/health/biggest-loser-weight-loss.html?ref=topics&_r=1).

Kolb, B., and Whishaw, I. Q. (1990). *Fundamentals of human neuropsychology*. San Francisco: Freeman.

Komisaruk, B. R, and Whipple, B. (2005). Functional MRI of the brain during orgasm in women. *Annual Review of Sex Research, 16*, 62–86.

Kondo, A., Shahpasand, K., Mannix, R., Qiu, J., et al. (2015). Antibody against early driver of neurodegeneration cis P-tau blocks brain injury and tauopathy. *Nature, 523*(7561), 431–436.

Kondo, Y., Sachs, B. D., and Sakuma, Y. (1997). Importance of the medial amygdala in rat penile erection evoked by remote stimuli from estrous females. *Behavioural Brain Research, 88*, 153–160.

Kondoh, K., Lu, Z., Ye, X., Olson, D. P., et al. (2016). A specific area of olfactory cortex involved in stress hormone responses to predator odours. *Nature, 532*(7597),103–106.

Kondratova, A. A., and Kondratov, R. V. (2012). The circadian clock and pathology of the ageing brain. *Nature Reviews Neuroscience, 13*, 325–335.

Kondziolka, D., Wechsler, L., Goldstein, S., Meltzer, C., et al. (2000). Transplantation of cultured human neuronal cells for patients with stroke. *Neurology, 55*, 565–569.

Kong, A., Frigge, M. L., Masson, G., Besenbacher, S., et al. (2012). Rate of de novo mutations and the importance of father's age to disease risk. *Nature, 488*, 471–475.

Konopka, G., Bomar, J. M., Winden, K., Coppola, G., et al. (2009). Human-specific transcriptional regulation of CNS development genes by FOXP2. *Nature, 462*, 213–217.

Kopell, B. H., Machado, A. G., and Rezai, A. R. (2005). Not your father's lobotomy: Psychiatric surgery revisited. *Clinical Neurosurgery, 52*, 315–330.

Kordower, J. H., Chu, Y., Hauser, R. A., Freeman, T. B., et al. (2008). Lewy body-like pathology in long-term embryonic nigral transplants in Parkinson's disease. *Nature Medicine, 14*, 504–506.

Korol, D. L., and Pisani, S. L. (2015). Estrogens and cognition: Friends or foes? An evaluation of the opposing effects of estrogens on learning and memory. *Hormones and Behaviour, 74*,105–115.

Kosslyn, S. M., Alpert, N. M., Thompson, W. L., Maljkovic, V., et al. (1993). Visual mental imagery activates topographically organized visual cortex. *Journal of Cognitive Neuroscience, 5*, 263–287.

Kosslyn, S. M., Pascual-Leone, A., Felician, O., Camposano, S., et al. (1999). The role of Area 17 in visual imagery: Convergent evidence from PET and RTMS. *Science, 208*, 167–170.

Koubeissi, M. Z., Bartolomei, F., Beltagy, A., and Picard, F. (2014). Electrical stimulation of a small brain area reversibly disrupts consciousness. *Epilepsy and Behaviour, 37*, 32–35.

Kovelman, J. A., and Scheibel, A. B. (1984). A neurohistological correlate of schizophrenia. *Biological Psychiatry, 19*, 1601.

Koya, E., Golden, S. A., Harvey, B. K., Guez-Barber, D. H., et al. (2009). Targeted disruption of cocaine-activated nucleus accumbens neurons prevents context-specific sensitisation. *Nature Neuroscience, 12*, 1069–1073.

Koyama, S., Chase, S. M., Whitford, A. S., Velliste, M., et al. (2010). Comparison of brain-computer interface decoding algorithms in open-loop and closed-loop control. *Journal of Comparative Neurology, 29*(1–2), 73–87.

Kraepelin, E. (1919). *Dementia praecox and paraphrenia.* Edinburgh, Scotland: Livingstone.

Kraft, S., Hsu, C., Brough, D. E., and Staecker, H. (2013). Atoh1 induces auditory hair cell recovery in mice after ototoxic injury. *Laryngoscope, 123*(4), 992–999.

Krashes, M. J., Lowell, B. B., and Garfield, A. S. (2016). Melanocortin-4 receptor-regulated energy homeostasis. *Nature Neuroscience, 19*, 206–219.

Krause, A. J., Simon, E. B., Mander, B. A., Greer, S. M., et al. (2017). The sleep-deprived human brain. *Nature Reviews Neuroscience, 18*(7), 404–418.

Krauss, R. M. (1998). Why do we gesture when we speak? *Current Directions in Psychological Science, 7*(2), 54–60.

Kraut, R. E., and Johnston, R. E. (1979). Social and emotional messages of smiling: An ethological approach. *Journal of Personality and Social Psychology, 37*, 1539–1553.

Kravitz, D. J., Saleem, K. S., Baker, C. I., and Mishkin, M. (2011). A new neural framework for visuospatial processing. *Nature Reviews Neuroscience, 12*, 217–230.

Krebs, J. R., Sherry, D. F., Healy, S. D., Perry, V. H., et al. (1989). Hippocampal specialisation of food-storing birds. *Proceedings of the National Academy of Sciences, USA, 86*, 1388–1392.

Kriegsfeld, L. J. (2006). Driving reproduction: RFamide peptides behind the wheel. *Hormones and Behaviour, 50*, 655–666.

Kring, A. M. (1999). Emotion in schizophrenia: Old mystery, new understanding. *Current Directions in Psychological Science, 8*(5), 160–163.

Kring, A. M., and Caponigro, J. M. (2010). Emotion in schizophrenia: Where feeling meets thinking. *Current Directions in Psychological Science, 19*, 255–259.

Kringelbach, M. L. (2005). The human orbitofrontal cortex: Linking reward to hedonic experience. *Nature Reviews Neuroscience, 6*, 691–702.

Kringelbach, M. L., Jenkinson, N., Owen, S. L. F., and Aziz, T. Z. (2007). Translational principles of deep brain stimulation. *Nature Reviews Neuroscience, 8*, 623–634.

Kripke, D. F., Garfinkel, L., Wingard, D. L., Klauber, M. R., et al. (2002). Mortality associated with sleep duration and insomnia. *Archives of General Psychiatry, 59*, 131–136.

Krishnan, A., Zhang, R., Yao, V., Theesfeld, C. L., et al. (2016). Genome-wide prediction and functional characterisation of the genetic basis of autism spectrum disorder. *Nature Neuroscience, 19*(11), 1454–1462. https://doi.org/10.1038/nn.4353.

Krubitzer, L. A., Manger, P., Pettigrew, J. D., and Calford, M. B. (1995). The organisation of neocortex in monotremes: In search of the prototypical plan. *Journal of Comparative Neurology, 348*, 1–45.

Krubitzer, L. A., and Seelke, A. M. (2012). Cortical evolution in mammals: The bane and beauty of phenotypic variability. *Proceedings of the National Academy of Sciences, USA, 109*(Suppl. 1), 10647–10654.

Krug, M., Lössner, B., and Ott, T. (1984). Anisomycin blocks the late phase of long-term potentiation in the dentate gyrus of freely moving rat. *Brain Research Bulletin, 13*, 39–42.

Krystal, A., Krishnan, K. R., Raitiere, M., Poland, R., et al. (1990). Differential diagnosis and pathophysiology of Cushing's syndrome and primary affective disorder. *Journal of Neuropsychiatry and Clinical Neurosciences, 2*, 34–43.

Kuhl, B. A., Dudukovic, N. M., Kahn, I., and Wagner, A. D. (2007). Decreased demands on cognitive control reveal the neural processing benefits of forgetting. *Nature Neuroscience, 10*, 908–914.

Kuiper, G. G., Enmark, E., Pelto-Huikko. M., Nilsson, S., et al. (1996). Cloning of a novel receptor expressed in rat prostate and ovary. *Proceedings of the National Academy of Sciences, USA, 93*, 5925–5930.

Kumar, D. K.V., Choi, S. H., Washicosky, K. J., Eimer, W. A., et al. (2016). Amyloid-β peptide protects against microbial infection in mouse and worm models of Alzheimer's disease. *Science Translational Medicine, 8(340)*, 340ra72. doi:10.1126/scitranslmed.aaf1059.

Kumaran, A., and Kershaw, M. (2014). Hyperphagic psychosocial short stature: A clinical review. *Paediatrics & Child Health, 24*(12), 567–571.

Kuner, R., and Flor, H. (2017). Structural plasticity and reorganisation in chronic pain. *Nature Reviews Neuroscience, 18*(2), 113.

Kuperberg, G. R. (2007). Neural mechanisms of language comprehension: Challenges to syntax. *Brain Research, 1146*, 23–49.

Kupfer, D. J., Frank, E., and Phillips, M. L. (2012). Major depressive disorder: New clinical, neurobiological, and treatment perspectives. *Lancet, 379*(9820), 1045–1055.

Kutas, M., and Federmeier, K. D. (2011). Thirty years and counting: Finding meaning in the N400 component of the event-related brain potential (ERP). *Annual Review of Psychology, 62*, 621–647.

Kutas, M., and Hillyard, S. A. (1980). Reading senseless sentences: Brain potentials reflect semantic incongruity. *Science, 207*, 203–205.

Kutas, M., and Hillyard, S. A. (1984). Brain potentials during reading reflect word expectancy and semantic association. *Nature, 307*(5947), 161–163.

Kuyken, W., Warren, F. C., Taylor, R. S., Whalley, W., et al. (2016). Efficacy of mindfulness-based cognitive therapy in prevention of depressive relapse: An individual patient data meta-analysis from randomized trials. *JAMA Psychiatry*. doi:10.1001/jamapsychiatry.2016.0076.

Kwakkel, G., Veerbeek, J. M., van Wegen, E. E., and Wolf, S. L. (2015). Constraint-induced movement therapy after stroke. *Lancet Neurology, 14*(2), 224–234.

Kyzar, E. J., Nichols, C. D., Gainetdinov, R. R., Nichols, D. E., et al. (2017). Psychedelic drugs in biomedicine. *Trends in Pharmacological Sciences, 38*(11), 992–1005.

L

LaBar, K. S., Gatenby, J. C., Gore, J. C., LeDoux, J. E., et al. (1998). Human amygdala activation during conditioned fear acquisition and extinction: A mixed-trial fMRI study. *Neuron, 20*, 937–945.

Lacey, J. I., and Lacey, B. C. (1970). Some autonomic-central nervous system interrelationships. In P. Black (Ed.), *Physiological correlates of emotion* (pp. 205–227). New York: Academic Press.

Lafer-Sousa, R., Hermann, K. L., and Conway, B. R. (2015). Striking individual differences in colour perception uncovered by 'The Dress' photograph. *Current Biology, 25*, R545–R546.

Lai, C. S. L., Fisher, S. E., Hurst, J. A., Vargha-Khadem, F., et al. (2001). A forkhead-domain gene is mutated in a severe speech and language disorder. *Nature, 413*, 519–523.

Lai, C. S. W., Adler, A., and Gan, W. B. (2018). Fear extinction reverses dendritic spine formation induced by fear conditioning in the mouse auditory cortex. *Proceedings of the National Academy of Sciences, USA, 115*(37), 9306–9311.

Lai, C. S.-W., Franke, T. F., and Gan, W.-B. (2012). Opposite effects of fear conditioning and extinction on dendritic spine remodelling. *Nature, 483*, 87–90.

LaMotte, R. H., Dong, X., and Ringkamp, M. (2014). Sensory neurons and circuits mediating itch. *Nature Reviews Neuroscience, 15*(1), 19–31.

Lampert, K. P., and Schartl, M. (2010). A little bit is better than nothing: The incomplete parthenogenesis of salamanders, frogs and fish. *BMC Biology, 8*, 78. doi:10.1186/1741-7007-8-78.

Landau, B., and Levy, R. M. (1993). Neuromodulation techniques for medically refractory chronic pain. *Annual Review of Medicine, 44*, 279–287.

Langergraber, K. E., Prüfer, K., Rowney, C., Boesch, C., et al. (2012). Generation times in wild chimpanzees and gorillas suggest earlier divergence times in great ape and human evolution. *Proceedings of the National Academy of Sciences, USA, 109*, 15716–15721.

Larroche, J.-C. (1977). *Developmental pathology of the neonate.* Amsterdam: Excerpta Medica.

Larrondo, L. F., Olivares-Yañez, C., Baker, C. L., Loros, J. J., et al. (2015). Circadian rhythms: Decoupling circadian clock protein turnover from circadian period determination. *Science, 347*(6221), 1257277.

Larsson, M., and Willander, J. (2009). Autobiographical odour memory. *Annals of the New York Academy of Sciences, 1170*, 318–323.

Lau, H. C., Rogers, R. D., Haggard, P., and Passingham, R. E. (2004). Attention to intention. *Science, 303*, 1208–1210.

Laverty, P. H., Leskovar, A., Breur, G. J., Coates, J. R., et al. (2004). A preliminary study of intravenous surfactants in paraplegic dogs: Polymer therapy in canine clinical SCI. *Journal of Neurotrama, 21*, 1767–1777.

Lavie, N., Hirst, A., de Fockert, J. W., and Viding, E. (2004). Load theory of selective attention and cognitive control. *Journal of Experimental Psychology. General, 133*, 339–354.

Lavie, N., Lin, Z., Zokaei, N., and Thoma, V. (2009). The role of perceptual load in object recognition. *Journal of Experimental Psychology. Human Perception and Performance, 35*, 1346–1358.

Lavie, P., and Kripke, D. F. (1981). Ultradian circa 1 1/2 hour rhythms: A multioscillatory system. *Life Sciences, 29*, 2445–2450.

Lazeyras, F., Boex, C., Sigrist, A., Seghier, M. L., et al. (2002). Functional MRI of auditory cortex activated by multisite electrical stimulation of the cochlea. *NeuroImage, 17*, 1010–1017.

Le Grand, R., Mondloch, C. J., Maurer, D., and Brent, H. P. (2001). Early visual experience and face processing. *Nature, 410*, 890.

Le Grange, D. (2005). The Maudsley family-based treatment for adolescent anorexia nervosa. *World Psychiatry, 4*, 142–146.

Leach, E. L., Prefontaine, G., Hurd, P. L., and Crespi, B. J. (2014). The imprinted gene *LRRTM1* mediates schizotypy and handedness in a nonclinical population. *Journal of Human Genetics, 59*(6), 332–336.

Leavey, G., Rosato, M., Galway, K., Hughes, L., et al. (2016). Patterns and predictors of help-seeking contacts with health services and general practitioner detection of suicidality prior to suicide: A cohort analysis of suicides occurring over a two-year period. *BMC Psychiatry, 16*, 120.

Leber, S. M., Breedlove, S. M., and Sanes, J. R. (1990). Lineage, arrangement, and death of clonally related motoneurons in chick spinal cord. *Journal of Neuroscience, 10*, 2451–2462.

Lech, R. K., and Suchan, B. (2013). The medial temporal lobe: Memory and beyond. *Behavioural Brain Research, 254*, 45–49.

Ledent, C., Valverde, O., Cossu, G., Petitet, F., et al. (1999). Unresponsiveness to cannabinoids and reduced addictive effects of opiates in CB_1 receptor knockout mice. *Science, 283*, 401–404.

Lederbogen, F., Kirsch, P., Haddad, L., Streit, F., et al. (2011). City living and urban upbringing affect neural social stress processing in humans. *Nature, 474*, 498–501.

LeDoux, J. E. (1994). Emotion, memory and the brain. *Scientific American, 270*(6), 50–57.

LeDoux, J. E. (1995). Emotion: Clues from the brain. *Annual Review of Psychology, 46*, 209–235.

LeDoux, J. E. (1996). *The emotional brain: The mysterious underpinnings of emotional life.* London: Simon & Schuster.

LeDoux, J., and Daw, N. D. (2018). Surviving threats: Neural circuit and computational implications of a new taxonomy of defensive behaviour. *Nature Reviews Neuroscience, 19*(5), 269–282.

Lee, A. M., Kanter, B. R., Wang, D., Lim, J. P., et al. (2013). *Prkcz* null mice show normal learning and memory. *Nature, 493*(7432), 416–419.

Lee, B. R., Ma, Y. Y., Huang, Y. H., Wang, X., et al. (2013). Maturation of silent synapses in amygdala-accumbens projection contributes to incubation of cocaine craving. *Nature Neuroscience, 16*(11), 1644–1651.

Lee, E. B., Lee, V. M.-Y., and Trojanowski, J. Q. (2012). Gains or losses: Molecular mechanisms of TDP43-mediated neurodegeneration. *Nature Reviews Neuroscience, 13*, 38–50.

Lee, H., Kim, D. W., Remedios, R., Anthony, T. E., et al. (2014). Scalable control of mounting and attack by Esr1+ neurons in the ventromedial hypothalamus. *Nature, 509*, 627–632.

Lee, R. F., Dai, W., and Jones, J. (2012). Decoupled circular-polarized dual-head volume coil pair for studying two interacting human brains with dyadic fMRI. *Magnetic Resonance in Medicine, 68*, 1087–1096.

Lee, R. J., and Cohen, N. A. (2015). Taste receptors in innate immunity. *Cellular and Molecular Life Sciences, 72*(2), 217–236.

Lefebvre, L., Whittle, P., Lascaris, E., and Finkelstein, A. (1997). Feeding innovations and forebrain size in birds. *Animal Behaviour, 53*, 549–560.

Lehrer, J. (2009). Small, furry … and smart. *Nature, 461*(7266), 862–864.

Leinders-Zufall, T., Lane, A. P., Puche, A. C., Ma, W., et al. (2000). Ultrasensitive pheromone detection by mammalian vomeronasal neurons. *Nature, 405*, 792–796.

Lemasson, A., Ouattara, K., Petit, E. J., and Zuberbühler, K. (2011). Social learning of vocal structure in a nonhuman primate? *BMC Evolutionary Biology, 11*, 362.

Lepage, J. F., and Theoret, H. (2006). EEG evidence for the presence of an action observation-execution matching system in children. *European Journal of Neuroscience, 23*, 2505–2510.

Lepage, M., Habib, R., and Tulving, E. (1998). Hippocampal PET activations of memory encoding and retrieval: The HIPER model. *Hippocampus, 8*, 313–322.

LePort, A. K., Mattfield, A. T., Dickinson-Anson, H., Fallon, H., et al. (2012). Behavioural and neuroanatomical investigation of Highly Superior Autobiographical Memory (HSAM). *Neurobiology of Learning and Memory, 98*, 78–92.

Lescroart, M. D., Stansbury, D. E., and Gallant, J. L. (2015). Fourier power, subjective distance, and object categories all provide plausible models of BOLD responses in scene-selective visual areas. *Frontiers in Computational Neuroscience, 9*, 135.

Lesku, J. A., Rattenborg, N. C. , Valcu, M., Vyssotski, A. L., et al. (2012). Adaptive sleep loss in polygynous pectoral sandpipers. *Science, 337*(6102),1654–1658.

Leuner, B., Glasper, E. R., and Gould E. (2010). Sexual experience promotes adult neurogenesis in the hippocampus despite an initial elevation in stress hormones. *PLOS ONE, 5*(7): e11597. doi:10.1371/journal.pone.0011597.

Leung, C. T., Coulombe, P. A., and Reed, R. R. (2007). Contribution of olfactory neural stem cells to tissue maintenance and regeneration. *Nature Neuroscience, 10*, 720–726.

Leung, L. C., Wang, G. X., Madelaine, R., Skariah, G., et al. (2019). Neural signatures of sleep in zebrafish. *Nature*, 571, 198–204.

Leutgeb, S., Leutgeb, J. K., Barnes, C. A., Moser, E. I., et al. (2005). Independent codes for spatial and episodic memory in hippocampal neuronal ensembles. *Science, 309*, 619–623.

LeVay, S. (1991). A difference in hypothalamic structure between heterosexual and homosexual men. *Science, 253*, 1034–1037.

LeVay, S. (2016). *Gay, straight and the reason why: The science of sexual orientation* (2nd ed.). Oxford, UK: Oxford University Press.

Levenson, R. W., Ekman, P., and Friesen, W. V. (1990). Voluntary facial action generates emotion-specific autonomic nervous system activity. *Psychophysiology, 27*, 363–384.

Levermann, N., Galatius, A., Ehlme, G., Rysgaard, S., et al. (2003). Feeding behaviour of free-ranging walruses with notes on apparent dextrality of flipper use. *BMC Ecology, 3*, 9.

Levi-Montalcini, R. (1963). Growth and differentiation in the nervous system. In J. Allen (Ed.), *The nature of biological diversity* (pp. 261–296). New York: McGraw-Hill.

Levi-Montalcini, R. (1982). Developmental neurobiology and the natural history of nerve growth factor. *Annual Review of Neuroscience, 5*, 341–362.

Levine, J. D., Gordon, N. C., and Fields, H. L. (1978). The mechanism of placebo analgesia. *Lancet, 2*, 654–657.

Levine, S., Haltmeyer, G. C., and Karas, G. G. (1967). Physiological and behavioural effects of infantile stimulation. *Physiology & Behaviour, 2*, 55–59.

Lew, S. M. (2014). Hemispherectomy in the treatment of seizures: A review. *Translational Pediatrics, 3*(3), 208–217.

Lewis, D. O. (1990). Neuropsychiatric and experiential correlates of violent juvenile delinquency. *Neuropsychology Review, 1*, 125–136.

Lewis, R. (1998). Flies invade human genetics. *Scientist, 12*, 1, 4–5.

Lewy, A. J., Bauer, V. K., Cutler, N. L., Sack, R. L., et al. (1998). Morning vs evening light treatment of patients with winter depression. *Archives of General Psychiatry, 55*, 890–896.

Li, B., Piriz, J., Mirrione, M., Chung, C., et al. (2011). Synaptic potentiation onto habenula neurons in the learned helplessness model of depression. *Nature, 470*, 535–539.

Li, F., and Zhou, M. (2012). Depletion of bitter taste transduction leads to massive spermatid loss in transgenic mice. *Molecular Human Reproduction, 18*(6), 289–297.

Li, J. Y., Christophersen, N. S., Hall, V., Soulet, D., et al. (2008). Critical issues of clinical human embryonic stem cell therapy for brain repair. *Trends in Neuroscience, 31*, 146–153.

Li, K., Buchinger, T. J., and Li, W. (2018). Discovery and characterisation of natural products that act as pheromones in fish. *Natural Product Reports, 35*(6), 501–513.

Li, Q., Li, Q. Q., Jia, J. N., Liu, Z. Q., et al. (2019). Targeting gap junction in epilepsy: Perspectives and challenges. *Biomedicine and Pharmacotherapy, 109*, 57–65.

Li, R., Polat, U., Makous, W., and Bavelier, D. (2009). Enhancing the contrast sensitivity function through action video game training. *Nature Neuroscience, 12*, 549–551.

Li, T., Jiang, S., Han, M., Yang, Z., et al. (2019). Exogenous melatonin as a treatment for secondary sleep disorders: A systematic review and meta-analysis. *Frontiers in Neuroendocrinology, 52*, 22–28.

Li, W., Wu, J., Yang, J., Sun, S., et al. (2015). Notch inhibition induces mitotically generated hair cells in mammalian cochleae via activating the Wnt pathway. *Proceedings of the National Academy of Sciences, USA, 112*(1),166–71.

Li, X., Glaser, D., Li, W., Johnson, W. E., et al. (2009). Analyses of sweet receptor gene (Tas1r2) and preference for sweet stimuli in species of Carnivora. *Journal of Heredity, 100*(Suppl. 1), S90–S100.

Li, X., Rubio, F. J., Zeric, T., Bossert, J. M., et al. (2015). Incubation of methamphetamine craving is associated with selective increases in expression of Bdnf and trkb, glutamate receptors, and epigenetic enzymes in cue-activated fos-expressing dorsal striatal neurons. *Journal of Neuroscience, 35*(21), 8232–8244.

Li, Y. Q., Li, F. Q., Wang, X. Y., Wu, P., et al. (2008). Central amygdala extracellular signal-regulated kinase signalling pathway is critical to incubation of opiate craving. *Journal of Neuroscience, 28*(49), 13248–13257.

Liberles, S. D., and Buck, L. B. (2006). A second class of chemosensory receptors in the olfactory epithelium. *Nature, 442*, 645–650.

Libet, B. (1985). Unconscious cerebral initiative and the role of conscious will in voluntary action. *Behavioural and Brain Sciences, 8*, 529–566.

Libourel, P. A., Barrillot, B., Arthaud, S., Massot, B., et al. (2018). Partial homologies between sleep states in lizards, mammals, and birds suggest a complex evolution of sleep states in amniotes. *PLOS Biology, 16*(10), e2005982.

Lichtenstein, P., Yip, B. H., Björk, C., Pawitan, Y., et al. (2009). Common genetic determinants of schizophrenia and bipolar disorder in Swedish families: A population-based study. *Lancet, 373*(9659), 234–239.

Lichtman, J. W., Livet, J., and Sanes, J. R. (2008). A technicolour approach to the connectome. *Nature Reviews Neuroscience, 9*, 417–422.

Lichtman, J. W., Pfister, H., and Shavit, N. (2014). The big data challenges of connectomics. *Nature Neuroscience, 17*(11), 1448–1454.

Lichtman, J. W., and Purves, D. (1980). The elimination of redundant preganglionic innervation to hamster sympathetic ganglion cells in early post-natal life. *Journal of Physiology (London), 301*, 213–228.

Liddelow, S. A., Guttenplan, K. A., Clarke, L. E., Bennett, F. C., et al. (2017). Neurotoxic reactive astrocytes are induced by activated microglia. *Nature, 541*(7638), 481–487.

Lieberman, P. (1985). On the evolution of human syntactic ability: Its pre-adaptive bases: Motor control and speech. *Journal of Human Evolution, 14*, 657–668.

Lieberman, P. (2002). On the nature and evolution of the neural bases of human language. *American Journal of Physical Anthropology, Suppl. 35*, 36–62.

Lieberwirth, C., Pan, Y., Liu, Y., Zhang, Z., et al. (2016). Hippocampal adult neurogenesis: Its regulation and potential role in spatial learning and memory. *Brain Research, 1644*, 127–140.

Liégeois, F., Baldeweg, T., Connelly, A., Gadian, D. G., et al. (2003). Language fMRI abnormalities associated with FOXP2 gene mutation. *Nature Neuroscience, 6*, 1230–1237.

Liepert, J., Bauder, H., Wolfgang, H. R., Miltner, W. H., et al. (2000). Treatment-induced cortical reorganisation after stroke in humans. *Stroke, 31*, 1210–1216.

Lieving, L. M., Cherek, D. R., Lane, S. D., Tcheremissine, O. V., et al. (2008). Effects of acute tiagabine administration on aggressive responses of adult male parolees. *Journal of Psychopharmacology, 22*, 144–152.

Lim, D. A., and Alvarez-Buylla, A. (2015). The adult ventricular-subventricular zone (V-SVZ) and olfactory bulb (OB) neurogenesis. *Cold Spring Harbor Perspectives in Biology, 8*(5), pii: a018820. https://doi.org/10.1101/cshperspect.a018820.

Lim, M. M., Wang, Z., Olazabal, D. E., Ren, X., et al. (2004). Enhanced partner preference in a promiscuous species by manipulating the expression of a single gene. *Nature, 429*, 754–757.

Lim, M. M., and Young, L. J. (2006). Neuropeptidergic regulation of affiliative behaviour and social bonding in animals. *Hormones and Behaviour, 50*, 506–557.

Lin, F. H., Witzel, T., Hamalainen, M. S., Dale, A. M., et al. (2004). Spectral spatiotemporal imaging of cortical oscillations and interactions in the human brain. *NeuroImage, 23*, 582–595.

Lin, L., Faraco, J., Li, R., Kadotani, H., et al. (1999). The sleep disorder canine narcolepsy is caused by a mutation in the hypocretin (orexin) receptor 2 gene. *Cell, 98*, 365–376.

Lind, P. A., Luciano, M., Wright, M. J., Montgomery, G. W., et al. (2010). Dyslexia and DCDC2: Normal variation in reading and spelling is associated with DCDC2 polymorphisms in an Australian population sample. *European Journal of Human Genetics, 6*, 668–673.

Linde, K., Allais, G., Brinkhaus, B., Manheimer, E., et al. (2009). Acupuncture for tension-type headache. *Cochrane Database of Systematic Reviews, 1*, CD007587.

Lindemann, B. (1995). Sweet and salty: Transduction in taste. *News in Physiological Sciences, 10*, 166–170.

Lindskog, M., Svenningsson, P., Pozzi, L., Kim, Y., et al. (2002). Involvement of DARPP-32 phosphorylation in the stimulant action of caffeine. *Nature, 418*, 774–778.

Lisabeth, L., and Bushnell, C. (2012). Stroke risk in women: The role of menopause and hormone therapy. *Lancet Neurology, 11*, 82–91.

Lisk, R. D. (1962). Diencephalic placement of estradiol and sexual receptivity in the female rat. *American Journal of Physiology, 203*, 493–496.

Lisman, J., Yasuda, R., and Raghavachari, S. (2012). Mechanisms of CaMKII action in long-term potentiation. *Nature Reviews Neuroscience, 13*, 169–182.

Lister, J. P., and Barnes, C. A. (2009). Neurobiological changes in the hippocampus during normative aging. *Archives of Neurology, 66*, 829–833.

Lister, R., Mukamel, E. A., Nery, J. R., Urich, M., et al. (2013). Global epigenomic reconfiguration during mammalian brain development. *Science, 341*(6146), 1237905. doi:10.1126/science.1237905.

Liston, C., McEwen, B. S., and Casey, B. J. (2009). Psychosocial stress reversibly disrupts prefrontal processing and attentional control. *Proceedings of the National Academy of Sciences, USA, 106*(3), 912–917.

Liu, D., Diorio, J., Tannenbaum, B., Caldji, C., et al. (1997). Maternal care, hippocampal glucocorticoid receptors, and hypothalamic-pituitary-adrenal responses to stress. *Science, 277*, 1659–1662.

Liu, X., Ramirez, S., Pang, P. T., Puryear, C. B., et al. (2012). Optogenetic stimulation of a hippocampal engram activates fear memory recall. *Nature, 484*, 381–385.

Liu, Y., Gao, J. H., Liotti, M., Pu, Y., et al. (1999). Temporal dissociation of parallel processing in the human subcortical outputs. *Nature, 400*, 364–367.

Liu, Y., Gao, J.-H., Liu, H.-L., and Fox, P. T. (2000). The temporal response of the brain after eating revealed by functional MRI. *Nature, 405*, 1058–1062.

Liu, Y., Williamson, V., Setlow, B., Cottler, L. B., et al. (2018). The importance of considering polysubstance use: Lessons from cocaine research. *Drug and Alcohol Dependence, 192*, 16–28.

Livingstone, M. S. (2000). Is it warm? Is it real? Or just low spatial frequency? *Science, 290*, 1299.

Lloyd, J. A. (1971). Weights of testes, thymi, and accessory reproductive glands in relation to rank in paired and grouped house mice (*Mus musculus*). *Proceedings of the Society for Experimental Biology and Medicine, 137*, 19–22.

Loch, A. A. (2019). Schizophrenia, not a psychotic disorder: Bleuler revisited. *Frontiers in Psychiatry, 10*, 328.

Lockhart, M., and Moore, J. W. (1975). Classical differential and operant conditioning in rabbits (*Orycytolagus cuniculus*) with septal lesions. *Journal of Comparative and Physiological Psychology, 88*, 147–154.

Loconto, J., Papes, F., Chang, E., Stowers, L., et al. (2003). Functional expression of murine V2R pheromone receptors involves selective association with the M10 and M1 families of MHC class Ib molecules. *Cell, 112*, 607–618.

Loewenstein, W. R. (1971). Mechano-electric transduction in the Pacinian corpuscle. Initiation of sensory impulses in mechanoreception. In *Handbook of sensory physiology: Vol. 1. Principles of receptor physiology* (pp. 269–290). Berlin: Springer.

Loftus, E. F. (2003). Make-believe memories. *American Psychologist, 58*, 867–873.

Logan, C. G., and Grafton, S. T. (1995). Functional anatomy of human eyeblink conditioning determined with regional cerebral glucose metabolism and positron emission tomography. *Proceedings of the National Academy of Sciences, USA, 92*, 7500–7504.

Logothetis, N. K. (2008). What we can do and what we cannot do with fMRI. *Nature, 453*, 869–878.

Loh, H. H., Lim, L. L., Yee, A., and Loh, H. S. (2019). Association between subclinical hypothyroidism and depression: An updated systematic review and meta-analysis. *BMC Psychiatry, 19*(1), 12.

Longden, E. (2013). Listening to voices. *Scientific American Mind, 24*, 34–39.

Longden, E., and Read, J. (2016). Social adversity in the etiology of psychosis: A review of the evidence. *American Journal of Psychotherapy, 70*(1), 5–33.

Longdon, E., and Read, J. (2017). "People with problems, not patients with illnesses": Using psychosocial frameworks to reduce the stigma of psychosis. *Israel Journal of Psychiatry and Related Sciences, 54*(1), 24–28.

Lonstein, J. S., Lévy, F., and Fleming, A. S. (2015). Common and divergent psychobiological mechanisms underlying maternal behaviours in non-human and human mammals. *Hormones and Behaviour, 73*, 156–185.

Lonstein, J. S., and Stern, J. M. (1997). Role of the midbrain periaqueductal gray in maternal nurturance and aggression: *c-fos* and electrolytic lesion studies in lactating rats. *Journal of Neuroscience, 17*, 3364–3378.

López-Aranda, M. F., López-Téllez, J. F., Navarro-Lobato, I., Masmudi-Martín, M., et al. (2009). Role of layer 6 of V2 visual cortex in object-recognition memory. *Science, 325*, 87–89.

Loui, P., Alsop, D., and Schlaug, G. (2009). Tone deafness: A new disconnection syndrome? *Journal of Neuroscience, 29*, 10215–10220.

Lovejoy, L. P., and Krauzlis, R. J. (2010). Inactivation of primate superior colliculus impairs covert selection of signals for perceptual judgments. *Nature Neuroscience, 13*, 261–266.

Lozano, R., Rosero, C. A., and Hagerman, R. J. (2014). Fragile X spectrum disorders. *Intractable and Rare Disease Research, 3*(4), 134–146. doi:10.5582/irdr.2014.01022.

Lu, X., Perera, T. H., Aria, A. B., and Callahan, L. A. S. (2018). Polyethylene glycol in spinal cord injury repair: A critical review. *Journal of Experimental Pharmacology, 10*, 37–49.

Lübke, K. T., and Pause, B. M. (2015). Always follow your nose: The functional significance of social chemosignals in human reproduction and survival. *Hormones and Behaviour, 68*,134–144.

Lucas, R. J., Hattar, S., Takao, M., Berson, D. M., et al. (2003). Diminished pupillary light reflex at high irradiances in melanopsin-knockout mice. *Science, 299*, 245–247.

Luck, S. J. (2005). *An introduction to the event-related potential technique*. Cambridge, MA: MIT Press.

Luck, S. J. (2014). *An introduction to the event-related potential technique* (2nd ed.). Cambridge, MA: MIT Press.

Luck, S. J., and Hillyard, S. A. (1994). Electrophysiological correlates of feature analysis during visual search. *Psychophysiology, 31*(3), 291–308.

Lumpkin, E. A., and Caterina, M. J. (2007). Mechanisms of sensory transduction in the skin. *Nature, 445*, 858–865.

Luna, E., and Luk, K. C. (2015). Bent out of shape: α-Synuclein misfolding and the convergence of pathogenic pathways in Parkinson's disease. *FEBS Letters, 589*(24 Pt. A), 3749–3759.

Luo, A., and Ho, S. Y. W. (2018). The molecular clock and evolutionary timescales. *Biochemical Society Transactions, 46*(5), 1183–1190.

Lupattelli, A., Wood, M., Ystrom, E., Skurtveit, S., et al. (2018). Effect of time-dependent selective serotonin reuptake inhibitor antidepressants during pregnancy on behavioural, emotional, and social development in preschool-aged children. *Journal of the American Academy of Child and Adolescent Psychiatry, 57*(3), 200–208.

Luria, A. R. (1987). *The mind of a mnemonist.* Cambridge, MA: Harvard University Press.

Lush, I. E. (1989). The genetics of tasting in mice. VI. Saccharin, acesulfame, dulcin and sucrose. *Genetical Research, 53*, 95–99.

Lyall, K., Croen, L., Daniels, J., Fallin, M. D., et al. (2017). The changing epidemiology of autism spectrum disorders. *Annual Review of Public Health, 38*, 81–102. https://doi.org/10.1146/annurev-publhealth-031816-044318.

Lyall, V., Heck, G. L., Vinnikova, A. K., Ghosh, S., et al. (2004). The mammalian amiloride-insensitive non-specific salt taste receptor is a vanilloid receptor-1 variant. *Journal of Physiology (London), 558*, 147–159.

Lyamin, O. I., Kosenko, P. O., Korneva, S. M., Vyssotski, A. L., et al. (2018). Fur seals suppress REM sleep for very long periods without subsequent rebound. *Current Biology, 28*(12), 2000–2005.e2.

Lyamin, O., Pryaslova, J., Lance, V., and Siegel, J. (2005). Continuous activity in cetaceans after birth: The exceptional wakefulness of newborn whales and dolphins has no ill-effect on their development. *Nature, 435*, 1177.

Lynch, G., Larson, J., Staubli, U., and Granger, R. (1991). Variants of synaptic potentiation and different types of memory operations in hippocampus and related structures. In L. R. Squire, N. M. Weinberger, G. Lynch, and J. L. McGaugh (Eds.), *Memory: Organisation and locus of change* (pp. 330–363). New York: Oxford University Press.

Lyons, J. I., Kerr, G. R., and Mueller, P. W. (2015). Fragile X syndrome: Scientific background and screening technologies. *Journal of Molecular Diagnostics.* doi:10.1016/j.jmoldx.2015.04.006.

M

Ma, L., Wu, Y., Qiu, Q., Scheerer, H., et al. (2014). A developmental switch of axon targeting in the continuously regenerating mouse olfactory system. *Science, 344*(6180), 194–197.

Macaluso, M., and Preskorn, S. H. (2018). Knowledge of the pharmacology of antidepressants and antipsychotics yields results comparable with pharmacogenetic testing. *Journal of Psychiatric Practice, 24*(6), 416–419.

Mace, G. M., Gittleman, J. L., and Purvis, A. (2003). Preserving the tree of life. *Science, 300*, 1707–1709.

Mackey, A. P., Finn, A. S., Leonard, J. A., Jacoby-Senghor, D. S., et al. (2015). Neuroanatomical correlates of the income-achievement gap. *Psychological Science, 26*(6), 925–933.

Mackey, S., Allgaier, N., Chaarani, B., Spechler, P., et al. (2018). Mega-analysis of gray matter volume in substance dependence: General and substance-specific regional effects. *American Journal of Psychiatry*, https://doi.org/10.1176/appi.ajp.2018.17040415.

MacLean, P. D. (1949). Psychosomatic disease and the "visceral brain": Recent developments bearing on the Papez theory of emotion. *Psychosomatic Medicine, 11*, 338–353.

MacLeod, C. M. (1991). Half a century of research on the Stroop effect: An integrative review. *Psychological Bulletin, 109*, 163–203.

Macmillan, M. (2000). *An odd kind of fame: Stories of Phineas Gage.* Cambridge, MA: MIT Press.

MacNeilage, P. F., Rogers, L. J., and Vallortigara, G. (2009). Origins of the left & right brain. *Scientific American, 301*(1), 60–67.

MacNeilage, P. R., Banks, M. S., Berger, D. R., and Bülthoff, H. H. (2007). A Bayesian model of the disambiguation of gravitoinertial force by visual cues. *Experimental Brain Research, 179*, 263–290.

Madden, J. (2001). Sex, bowers and brains. *Proceedings of the Royal Society of London. Series B: Biological Sciences, 268*, 833–838.

Magni, L. R., Purgato, M., Gastaldon, C., Papola, D., et al. (2013). Fluoxetine versus other types of pharmacotherapy for depression. *Cochrane Database of Systematic Reviews, 7*, CD004185.

Mair, W. G. P., Warrington, E. K., and Wieskrantz, L. (1979). Memory disorder in Korsakoff's psychosis. *Brain, 102*, 749–783.

Majumder, V., Gregory, J. M., Barria, M. A., Green, A., et al. (2018). TDP-43 as a potential biomarker for amyotrophic lateral sclerosis: A systematic review and meta-analysis. *BMC Neurology, 18*(1), 90.

Mak, G. K., Enwere, E. K., Gregg, C., Pakarainen, T., et al. (2007). Male pheromone-stimulated neurogenesis in the adult female brain: Possible role in mating behaviour. *Nature Neuroscience, 10*, 1003–1011.

Makin, S. (2018). The amyloid hypothesis on trial. *Nature, 559*(7715), S4–S7. https://doi.org/10.1038/d41586-018-05719-4.

Malenka, R. C., and Bear, M. F. (2004). LTP and LTD: An embarrassment of riches. *Neuron, 44*, 5–21.

Malpas, C. B., Genc, S., Saling, M. M., Velakoulis, D., et al. (2016). MRI correlates of general intelligence in neurotypical adults. *Journal of Clinical Neuroscience, 24*, 128–134.

Mamluk, L., Edwards, H. B., Savović, J., Leach, V., et al. (2017). Low alcohol consumption and pregnancy and childhood outcomes: Time to change guidelines indicating apparently "safe" levels of alcohol during pregnancy? A systematic review and meta-analyses. *BMJ Open, 7*(7), e015410.

Mancuso, K., Hauswirth, W. W., Li, Q., Connor, T. B., et al. (2009). Gene therapy for red-green colour blindness in adult primates. *Nature, 461*, 784–788.

Manetto, V., Medori, R., Cortelli, P., Montagna, P., et al. (1992). Fatal familial insomnia: Clinical and pathologic study of five new cases. *Neurology, 42*, 312–319.

Manfredi, M., Bini, G., Cruccu, G., Accornero, N., et al. (1981). Congenital absence of pain. *Archives of Neurology, 38*, 507–511.

Manger, P. R., Collins, R., and Pettigrew, J. D. (1998). The development of the electroreceptors of the platypus (*Ornithorhynchus anatinus*). *Philosophical Transactions of the Royal Society of London. Series B: Biological Sciences, 353*, 1171–1186.

Mangun, G. R. (1995). Neural mechanisms of visual selective attention. *Psychophysiology, 32*, 4–18.

Mani, S. K., Mermelstein, P. G., Tetel, M. J., and Anesetti, G. (2012). Convergence of multiple mechanisms of steroid hormone action. *Hormone and Metabolic Research, 44*, 569–576.

Maniscalco, J. W., and Rinaman, L. (2018). Vagal interoceptive modulation of motivated behaviour. *Physiology (Bethesda), 33*(2), 151–167.

Mantini, D., Gerits, A., Nelissen, K., Durand, J. B., et al. (2011). Default mode of brain function in monkeys. *Journal of Neuroscience, 31*(36), 12954–12962.

Marchesan, S., Ballerini, L., and Prato, M. (2017). Nanomaterials for stimulating nerve growth. *Science, 356*(6342), 1010–1011.

Marconi, A., Di Forti, M., Lewis, C. M., Murray, R. M., et al. (2016). Meta-analysis of the association between the level of cannabis use and risk of psychosis. *Schizophrenia Bulletin, 42*(5), 1262–1269.

Marcus, C., Mena, E., and Subramaniam, R. M. (2014). Brain PET in the diagnosis of Alzheimer's disease. *Clinical Nuclear Medicine, 39*(10), e413-22; quiz e423-6. https://doi.org/10.1097/RLU.0000000000000547.

Marcus, G. F., Vijayan, S., Bandi Rao, S., and Vishton, P. M. (1999). Rule learning by seven-month-old infants. *Science, 283,* 77–80.

Marder, E., O'Leary, T., and Shruti, S. (2014). Neuromodulation of circuits with variable parameters: Single neurons and small circuits reveal principles of state-dependent and robust neuromodulation. *Annual Review of Neuroscience, 37,* 329–346.

Marek, G. J., Behl, B., Bespalov, A. Y., Gross, G., et al. (2010). Glutamatergic (*N*-methyl-D-aspartate receptor) hypofrontality in schizophrenia: Too little juice or a miswired brain? *Molecular Pharmacology, 77*(3), 317–326.

Maren, S., and Quirk, G. J. (2004). Neuronal signalling of fear memory. *Nature, 5,* 844–852.

Mariani, J., and Changeaux, J.-P. (1981). Ontogenesis of olivocerebellar relationships. I. Studies by intracellular recordings of the multiple innervation of Purkinje cells by climbing fibres in the developing rat cerebellum. *Journal of Neuroscience, 1,* 696–702.

Maricic, T., Günther, V., Georgiev, O., Gehre, S., et al. (2013). A recent evolutionary change affects a regulatory element in the human *FOXP2* gene. *Molecular Biology and Evolution, 30*(4), 844–852.

Maricich, S. M., Wellnitz, S. A., Nelson, A. M., and Lesniak, D. R. (2009). Merkel cells are essential for light-touch responses. *Science, 324,* 1580–1582.

Marín-Burgin, A., Mongiat, L. A., Pardi, M. B., and Schinder, A. F. (2012). Unique processing during a period of high excitation/inhibition balance in adult-born neurons. *Science, 335,* 1238–1242.

Mark, V. H., and Ervin, F. R. (1970). *Violence and the brain.* New York: Harper & Row.

Marmor, D. J., and Marmor, M. F. (2010). Simulating vision with and without macular disease. *Archives of Ophthalmology, 128*(1), 117–125.

Marshall, C. B., Fletcher, G. L., and Davies, P. L. (2004). Hyperactive antifreeze protein in a fish. *Nature, 429,* 153.

Marshall, L., Helgadóttir, H., Mölle, M., and Born, J. (2006). Boosting slow oscillations during sleep potentiates memory. *Nature, 444,* 610–613.

Marsicano, G., Goodenough, S., Monory, K., Hermann, H., et al. (2003). CB1 cannabinoid receptors and on-demand defense against excitotoxicity. *Science, 302,* 84–88.

Marsicano, G., Wotjak, C. T., Azad, S. C., Bisogno, T., et al. (2002). The endogenous cannabinoid system controls extinction of aversive memories. *Nature, 418,* 530–532.

Martin, C. K., Heilbronn, L., de Jonge, L., Delany, J. P., et al. (2007). Effect of calorie restriction on resting metabolic rate and spontaneous physical activity. *Obesity (Silver Spring), 15,* 2964–2973.

Martin, J. A., Messacar, K., Yang, M. L., Maloney, J. A., et al. (2017). Outcomes of Colourado children with acute flaccid myelitis at 1 year. *Neurology, 89*(2), 129–137.

Martin, J. T., and Nguyen, D. H. (2004). Anthropometric analysis of homosexuals and heterosexuals: Implications for early hormone exposure. *Hormones and Behaviour, 45,* 31–39.

Martinez-Conde, S., Otero-Millan, J., and Macknik, S. L. (2013). The impact of microsaccades on vision: Towards a unified theory of saccadic function. *Nature Reviews Neuroscience, 14,* 83–96.

Martuza, R. L., Chiocca, E. A., Jenike, M. A., Giriunas, I. E., et al. (1990). Stereotactic radiofrequency thermal cingulotomy for obsessive compulsive disorder. *Journal of Neuropsychiatry and Clinical Neurosciences, 2,* 331–336.

Marucha, P. T., Kiecolt-Glaser, J. K., and Favagehi, M. (1998). Mucosal wound healing is impaired by examination stress. *Psychosomatic Medicine, 60,* 362–365.

Maruyama, M., Shimada, H., Suhara, T., Shinotoh, H., et al. (2013). Imaging of tau pathology in a tauopathy mouse model and in Alzheimer patients compared to normal controls. *Neuron, 79*(6), 1094–1108. doi:10.1016/j.neuron.2013.07.037.

Maruyama, Y., Pereira, E., Margolskee, R. F., Chaudhari, N., et al. (2006). Umami responses in mouse taste cells indicate more than one receptor. *Journal of Neuroscience, 26,* 2227–2234.

Marzani, D., and Wallman, J. (1997). Growth of the two layers of the chick sclera is modulated reciprocally by visual conditions. *Investigative Ophthalmology & Visual Science, 38,* 1726–1739.

Masiulis, S., Desai, R., Uchański, T., Serna Martin, I., et al. (2019). GABA(A) receptor signalling mechanisms revealed by structural pharmacology. *Nature, 565*(7740), 454–459.

Masters, W. H., and Johnson, V. E. (1966). *Human sexual response.* Boston: Little, Brown.

Masters, W. H., Johnson, V. E., and Kolodny, R. C. (1994). *Heterosexuality.* New York: HarperCollins.

Masterton, R. B. (1993). Central auditory system. *Journal of Oto-Rhino-Laryngology and Its Related Specialties, 55,* 159–163.

Mastrianni, J. A., Nixon, R., Layzer, R., Telling, G. C., et al. (1999). Prion protein conformation in a patient with sporadic fatal insomnia. *New England Journal of Medicine, 340,* 1630–1638.

Mateo, J. M., and Johnston, R. E. (2000). Kin recognition and the "armpit effect": Evidence of self-referent phenotype matching. *Proceedings of the Royal Society of London. Series B: Biological Sciences, 267,* 695–700.

Matsumoto, K., Suzuki, W., and Tanaka, K. (2003). Neuronal correlates of goal-based motor selection in the prefrontal cortex. *Science, 301,* 229–232.

Matsunami, H., Montmayeur, J. P., and Buck, L. B. (2000). A family of candidate taste receptors in human and mouse. *Nature, 404*(6778), 601–604.

Mattes, R. D. (2011). Accumulating evidence supports a taste component for free fatty acids in humans. *Physiology & Behaviour, 104*(4), 624–631.

Matthews, K. A. (2005). Psychological perspectives on the development of coronary heart disease. *American Psychology, 60,* 783–796.

Mattison, J. A., Colman, R. J., Beasley, T. M., Allison, D. B., et al. (2017). Caloric restriction improves health and survival of rhesus monkeys. *Nature Communications, 8,* 14063.

Mattson, M. P., and Arumugam, T. V. (2018). Hallmarks of brain aging: Adaptive and pathological modification by metabolic states. *Cell Metabolism, 27*(6), 1176–1199.

Max, D. T. (2007). *The Family That Couldn't Sleep: A Medical Mystery.* New York: Random House.

May, L., Gervain, J., Carreiras, M., and Werker, J. F. (2018). The specificity of the neural response to speech at birth. *Developmental Science, 21*(3), e12564.

Mayberg, H. S., Lozano, A. M., Voon, V., McNeely, H. E., et al. (2005). Deep brain stimulation for treatment-resistant depression. *Neuron, 45,* 651–660.

Mayfield, R. D., Lewohl, J. M., Dodd, P. R., Herlihy, A., et al. (2002). Patterns of gene expression are altered in the frontal and motor cortices of human alcoholics. *Journal of Neurochemistry, 81,* 802–813.

Mazur, A., and Booth, A. (1998). Testosterone and dominance in men. *Behavioural and Brain Sciences, 21,* 353–363.

McAllister, A. K., Katz, L. C., and Lo, D. C. (1997). Opposing roles for endogenous BDNF and NT-3 in regulating cortical dendritic growth. *Neuron, 18,* 767–778.

McBurney, D. H., Smith, D. V., and Shick, T. R. (1972). Gustatory cross adaptation: Sourness and bitterness. *Perception & Psychophysics, 11,* 2228–2232.

McCarthy, M. M., and Nugent, B. M. (2015). At the frontier of epigenetics of brain sex differences. *Frontiers in Behavioural Neuroscience, 9,* 221. doi:10.3389/fnbeh.2015.00221.

McCarthy, R. A., and Warrington, E. K. (1990). *Cognitive neuropsychology: A clinical introduction.* San Diego, CA: Academic Press.

McCoy, A. N., and Platt, M. L. (2005). Risk-sensitive neurons in macaque posterior cingulate cortex. *Nature Neuroscience, 8,* 1220–1227.

McCoy, T. (2015, February 24). How Stephen Hawking, diagnosed with ALS decades ago, is still alive. *Washington Post* (https://www.washingtonpost.com/news/morning-mix/wp/2015/02/24/how-stephen-hawking-survived-longer-than-possibly-any-other-als-patient/?utm_term=.3dc9743fcda0).

McCrae, C. S., Rowe, M. A., Tierney, C. G., Dautovich, N. D., et al. (2005). Sleep complaints, subjective and objective sleep patterns, health, psychological adjustment, and daytime functioning in community-dwelling older adults. *Journals of Gerontology. Series B, Psychological Sciences and Social Sciences, 60*(4), P182–P189.

McDiarmid, T. A., Yu, A. J., and Rankin, C. H. (2018). Beyond the response: High throughput behavioural analyses to link genome to phenome in *Caenorhabditis elegans. Genes, Brain and Behaviour, 17*(3), e12437.

McDonald, J. J., and Green, J. J. (2008). Isolating event-related potential components associated with voluntary control of visuo-spatial attention. *Brain Research, 1227*, 96–109.

McDonald, J. J., Teder-Sälejärvi, W. A., Di Russo, F., and Hillyard, S. A. (2005). Neural basis of auditory-induced shifts in visual time-order perception. *Nature Neuroscience, 8*(9), 1197–1202.

McDonald, J. J., Teder-Sälejärvi, W. A., and Hillyard, S. A. (2000). Involuntary orienting to sound improves visual perception. *Nature, 407*, 906–908.

McDonald, J. J., Ward, L. M., and Kiehl, K. A. (1999). An event-related brain potential study of inhibition of return. *Perception and Psychophysics, 61*(7), 1411–1423.

McEwen, B. S. (2012). Brain on stress: How the social environment gets under the skin. *Proceedings of the National Academy of Sciences, USA, 109*(Suppl. 2), 17180–17185.

McEwen, B. S., and Akil, H. (2011). Introduction to social neuroscience: Gene, environment, brain, body. *Annals of the New York Academy of Science, 1231*, vii–ix.

McEwen, B. S., Bowles, N. P., Gray, J. D., Hill, M. N., et al. (2015). Mechanisms of stress in the brain. *Nature Neuroscience, 1*, 1353–1363.

McFadden, D., and Champlin, C. A. (1990). Reductions in overshoot during aspirin use. *Journal of the Acoustical Society of America, 87*, 2634–2642.

McFadden, D., and Pasanen, E. (1998). Comparison of the auditory systems of heterosexuals and homosexuals: Click-evoked otoacoustic emissions. *Proceedings of the National Academy of Sciences, USA, 95*, 2709–2713.

McGann, J. P. (2017). Poor human olfaction is a 19th-century myth. *Science, 356*(6338), pii: eaam7263.

McGaugh, J. L. (2003). *Memory and emotions: The making of lasting memories.* New York: Columbia University Press.

McGinty, D. J., and Sterman, M. B. (1968). Sleep suppression after basal forebrain lesions in the cat. *Science, 160*, 1253–1255.

McGowan, P. O., Sasaki, A., D'Alessio, A. C., Dymov, S., et al. (2009). Epigenetic regulation of the glucocorticoid receptor in human brain associates with childhood abuse. *Nature Neuroscience, 12*, 342–348.

McGuigan, F. J., and Lehrer, P. M. (2007). Progressive relaxation: Origins, principles and clinical application. In P. M. Lehrer, R. L. Woolfolk, and W. E. Sime (Eds.), *Principles and practices of stress management* (pp. 57–87). New York: Guilford Press.

McIntyre, C. K., Miyashita, T., Setlow, B., Marjon, K. D., et al. (2005). Memory-influencing intra-basolateral amygdala drug infusions modulate expression of Arc protein in the hippocampus. *Proceedings of the National Academy of Sciences, USA, 102*, 10718–10723.

McKee, A. C., Cairns, N. J., Dickson, D. W., Folkerth, R. D., et al. (2016). The first NINDS/NIBIB consensus meeting to define neuropathological criteria for the diagnosis of chronic traumatic encephalopathy. *Acta Neuropathologica, 131*(1), 75–86.

McKee, A. C., Cantu, R. C., Nowinski, C. J., Hedley-Whyte, E. T., et al. (2009). Chronic traumatic encephalopathy in athletes: Progressive tauopathy after repetitive head injury. *Journal of Neuropathology and Experimental Neurology, 68*, 709–735.

McKenna, K. (1999). The brain is the master organ in sexual function: Central nervous system control of male and female sexual function. *International Journal of Impotence Research, 11*(Suppl. 1), S48–S55.

McKeown, S. J., Wallace, A. S., and Anderson, R. B. (2013). Expression and function of cell adhesion molecules during neural crest migration. *Developmental Biology, 373*(2), 244–257. doi:10.1016/j.ydbio.2012.10.028.

McKernan, M. G., and Shinnick-Gallagher, P. (1997). Fear conditioning induces a lasting potentiation of synaptic currents in vitro. *Nature, 390*, 607–611.

McKim, W. A. (1991). *Drugs and behaviour: An introduction to behavioural pharmacology* (2nd ed.). Englewood Cliffs, NJ: Prentice Hall.

McLaughlin, S. K., McKinnon, P. J., Spickofsky, N., Danho, W., et al. (1994). Molecular cloning of G proteins and phosphodiesterases from rat taste cells. *Physiology & Behaviour, 56*, 1157–1164.

McLellan, T. M., Caldwell, J. A., and Lieberman, H. R. (2016). A review of caffeine's effects on cognitive, physical and occupational performance. *Neuroscience & Biobehavioural Reviews, 71*, 294–312.

McMahon, H. T., Foran, P., Dolly, J. O., Verhage, M., et al. (1992). Tetanus toxin and botulinum toxins type A and B inhibit glutamate, gamma-aminobutyric acid, aspartate, and met-enkephalin release from synaptosomes. Clues to the locus of action. *Journal of Biological Chemistry, 267*, 21338–21343.

McMains, S., and Somers, D. (2004). Multiple spotlights of attentional selection in human visual cortex. *Neuron, 42*, 677–686.

McNally, R. J. (2003). Recovering memories of trauma: A view from the laboratory. *Current Directions in Psychological Science, 12*, 32–35.

McNaughton, B. L., Barnes, C. A., Gerrard, J. L., Gothard, K., et al. (1996). Deciphering the hippocampal polyglot: The hippocampus as a path integration system. *Journal of Experimental Biology, 199*(Pt. 1), 173–185.

Meddis, R. (1975). On the function of sleep. *Animal Behaviour, 23*, 676–691.

Meddis, R. (1977). *The sleep instinct.* London: Routledge & Kegan Paul.

Mega, M. S., and Cummings, J. L. (1994). Frontal-subcortical circuits and neuropsychiatric disorders. *Journal of Neuropsychiatry and Clinical Neurosciences, 6*, 358–370.

Meguerditchian, A., and Vauclair, J. (2006). Baboons communicate with their right hand. *Behavioural Brain Research, 171*, 170–174.

Mei, L., and Xiong, W.-C. (2008). Neuregulin 1 in neural development, synaptic plasticity and schizophrenia. *Nature Reviews Neuroscience, 9*, 437–452.

Meier, M. H., Caspi, A., Ambler, A., Harrington, H., et al. (2012). Persistent cannabis users show neuropsychological decline from childhood to midlife. *Proceedings of the National Academy of Sciences, USA, 109*, E2657–E2664.

Meister, I. G., Boroojerdi, B., Foltys, H., Sparing, R., et al. (2003). Motor cortex hand area and speech: Implications for the development of language. *Neuropsychologia, 41*, 401–406.

Melcangi, R. C., and Panzica, G. C. (2014). Allopregnanolone: State of the art. *Progress in Neurobiology, 113*, 1–5.

Meldrum, D. R., Burnett, A. L., Dorey, G., Esposito, K., et al. (2014). Erectile hydraulics: Maximizing inflow while minimizing outflow. *Journal of Sexual Medicine, 11*(5), 1208–1220.

Mellars, P. (2006). A new radiocarbon revolution and the dispersal of modern humans in Eurasia. *Nature, 439*, 931–935.

Melmed, S., Polonsky, K. S., Larsen, P. R., and Kronenberg, H. M. (2016). *Williams textbook of endocrinology* (13th ed.). Philadelphia: Elsevier.

Melzack, R. (1984). Neuropsychological basis of pain measurement. *Advances in Pain Research, 6*, 323–341.

Melzack, R., and Wall, P. D. (1965). Pain mechanisms: A new history. *Science, 150*, 971–979.

Mendel, G. (1967). *Experiments in plant hybridisation* (Royal Horticultural Society of London, Trans.). Cambridge, MA: Harvard University Press.

Méndez-Bértolo, C., Moratti, S., Toledano, R., Lopez-Sosa, F., et al. (2016). A fast pathway for fear in human amygdala. *Nature Neuroscience, 19*(8), 1041–1049.

Menninger, W. C. (1948). Facts and statistics of significance for psychiatry. *Bulletin of the Menninger Clinic, 12*, 1–25.

Menzies, F. M., Fleming, A., and Rubinsztein, D. C. (2015). Compromised autophagy and neurodegenerative diseases. *Nature Reviews Neuroscience, 16*(6), 345–357.

Merchán-Pérez, A., Rodriguez, J. R., Alonso-Nanclares, L., Schertel, A., et al. (2009). Counting synapses using FIB/SEM microscopy: A true revolution for ultrastructural volume reconstruction. *Frontiers in Neuroanatomy, 3*, 18.

Mersch, P. P., Middendorp, H. M., Bouhuys, A. L., Beersma, D. G., et al. (1999). Seasonal affective disorder and latitude: A review of the literature. *Journal of Affective Disorders, 53*, 35–48.

Mertens, J., Wang, Q. W., Kim, Y., Yu, D. X., et al. (2015). Differential responses to lithium in hyperexcitable neurons from patients with bipolar disorder. *Nature, 527*, 95–99.

Merzenich, M. M., and Jenkins, W. M. (1993). Reorganisation of cortical representations of the hand following alterations of skin inputs induced by nerve injury, skin island transfers, and experience. *Journal of Hand Therapy, 6*, 89–104.

Meshberger, F. L. (1990). An interpretation of Michelangelo's *Creation of Adam* based on neuroanatomy. *JAMA, 264*(14), 1837–1841.

Messias, E., Kirkpatrick, B., Bromet, E., Ross, D., et al. (2004). Summer birth and deficit schizophrenia: A pooled analysis from 6 countries. *Archives of General Psychiatry, 61*, 985–989.

Mesulam, M.-M. (1985). Attention, confusional states and neglect. In M.-M. Mesulam (Ed.), *Principles of behavioural neurology.* Philadelphia: Davis.

Metz, T. D., and Stickrath, E. H. (2015). Marijuana use in pregnancy and lactation: A review of the evidence. *American Journal of Obstetrics and Gynecology, 213*(6), 761–778. https://doi.org/10.1016/j.ajog.2015.05.025.

Mez, J., Daneshvar, D. H., Kiernan, P. T., Abdolmohammadi, B., et al. (2017). Clinicopathological evaluation of chronic traumatic encephalopathy in players of American football. *JAMA, 318*(4), 360–370.

Michael, J., Sandberg, K., Skewes, J., Wolf, T., et al. (2014). Continuous theta-burst stimulation demonstrates a causal role of premotor homunculus in action understanding. *Psychological Science, 25*(4), 963–972.

Michael, N., and Erfurth, A. (2004). Treatment of bipolar mania with right prefrontal rapid transcranial magnetic stimulation. *Journal of Affective Disorders, 78*, 253–257.

Mignot, E. (2013). The perfect hypnotic? *Science, 340*(6128), 36–38.

Miller, E. K., and Cohen, J. D. (2001). An integrative theory of prefrontal cortex function. *Annual Review of Neuroscience, 24*, 167–202.

Miller, G. A. (1956). The magical number seven plus or minus two: Some limits on our capacity for processing information. *Psychological Review, 63*(2), 81–97.

Miller, G. A. (2010). Mistreating psychology in the decades of the brain. *Perspectives on Psychological Science, 5*, 716–743.

Miller, G. F. (2000). *The mating mind: How sexual choice shaped the evolution of human nature*. New York: Doubleday.

Miller, N. E., Sampliner, R. I., and Woodrow, P. (1957). Thirst-reducing effects of water by stomach fistula versus water by mouth measured by both a consummatory and an instrumental response. *Journal of Comparative Physiology and Psychology, 50*, 1–5.

Mills, K. L., Lalonde, F., Clasen, L. S., Giedd, J. N., et al. (2014). Developmental changes in the structure of the social brain in late childhood and adolescence. *Social Cognitive and Affective Neuroscience, 9*(1), 123–131. doi:10.1093/scan/nss113.

Milner, A. D., Perrett, D. I., Johnston, R. S., Benson, P. J., et al. (1991). Perception and action in "visual form agnosia." *Brain, 114*, 405–428.

Milner, B. (1963). Effect of different brain lesions on card sorting. *Archives of Neurology, 9*, 90–100.

Milner, B. (1965). Memory disturbance after bilateral hippocampal lesions. In P. M. Milner and S. E. Glickman (Eds.), *Cognitive processes and the brain; an enduring problem in psychology* (pp. 97–111). Princeton, NJ: Van Nostrand.

Milner, B. (1970). Memory and the medial temporal regions of the brain. In D. H. Pribram and D. E. Broadbent (Eds.), *Biology of memory* (pp. 29–50). New York: Academic Press.

Milner, P. M. (1993). The mind and Donald O. Hebb. *Scientific American, 268*(1), 124–129.

Min, R., and Nevian, T. (2012). Astrocyte signalling controls spike timing-dependent depression at neocortical synapses. *Nature Neuroscience, 15*, 746–753.

Minzenberg, M. J., Laird, A. R., Thelen, S., Carter, C. S., et al. (2009). Meta-analysis of 41 functional neuroimaging studies of executive function in schizophrenia. *Archives of General Psychiatry, 66*, 811–822.

Mioduszewska, B., Jaworski, J., and Kaczmarek, L. (2003). Inducible cAMP early repressor (ICER) in the nervous system: A transcriptional regulator of neuronal plasticity and programmed cell death. *Journal of Neurochemistry, 87*, 1313–1320.

Miras, M., Serrano, M., Durán, C., Valiño, C., et al. (2015). Early experience with customized, meal-triggered gastric electrical stimulation in obese patients. *Obesity Surgery, 25*, 174–179.

Mirescu, C., Peters, J. D., and Gould, E. (2004). Early life experience alters response of adult neurogenesis to stress. *Nature Neuroscience, 7*, 841–846.

Miri, A., Azim, E., and Jessell, T. M. (2013). Edging toward entelechy in motor control. *Neuron, 80*(3), 827–834. doi:10.1016/j.neuron.2013.10.049.

Mirsky, A. F., and Duncan, C. C. (1986). Etiology and expression of schizophrenia: Neurobiological and psychosocial factors. *Annual Review of Psychology, 37*, 291–321.

Mishkin, M., and Ungerleider, L. (1982). Contribution of striate inputs to the visuospatial functions of parieto-preoccipital cortex in monkeys. *Behavioural Brain Research, 6*, 57–77.

Mishra, J., Zinni, M., Bavelier, D., and Hillyard, S. A. (2011). Neural basis of superior performance of action videogame players in an attention-demanding task. *Journal of Neuroscience, 31*(3), 992–998.

Mishra, S. K., and Hoon, M. A. (2013). The cells and circuitry for itch responses in mice. *Science, 340*(6135), 968–971.

Mistlberger, R. E. (2005). Circadian regulation of sleep in mammals: Role of the suprachiasmatic nucleus. *Brain Research Reviews, 49*, 429–454.

Mistlberger, R. E., and Skene, D. J. (2004). Social influences on mammalian circadian rhythms: Animal and human studies. *Biological Reviews of the Cambridge Philosophical Society, 79*, 533–556.

Mitchell, R. E. (2006). How many deaf people are there in the United States? Estimates from the Survey of Income and Program Participation. *Journal of Deaf Studies and Deaf Education, 11*(1):112–119.

Mithoefer, M. C., Wagner, M. T., Mithoefer, A. T., Jerome, L., et al. (2013, January). Durability of improvement in post-traumatic stress disorder symptoms and absence of harmful effects or drug dependency after 3,4-methylenedioxymethamphetamine-assisted psychotherapy: A prospective long-term follow-up study. *Journal of Psychopharmacology, 27*(1), 28–39.

Miyake, A., Friedman, N. P., Emerson, M. J., Witzki, A. H., et al. (2000). The unity and diversity of executive functions and their contributions to complex "Frontal Lobe" tasks: A latent variable analysis. *Cognitive Psychology, 41*(1), 49–100.

Miyata, S. (2015). New aspects in fenestrated capillary and tissue dynamics in the sensory circumventricular organs of adult brains. *Frontiers in Neuroscience, 9*, 390.

Miyawaki, Y., Uchida, H., Yamashita, O., Sato, M. A., et al. (2008). Visual image reconstruction from human brain activity using a combination of multiscale local image decoders. *Neuron, 60*, 915–929.

Mizutari, K., Fujioka, M., Hosoya, M., Bramhall, N., et al. (2013). Notch inhibition induces cochlear hair cell regeneration and recovery of hearing after acoustic trauma. *Neuron, 77*(1), 58–69.

Moghaddam, B., and Adams, B. W. (1998). Reversal of phencyclidine effects by a group II metabotropic glutamate receptor agonist in rats. *Science, 281*, 1349–1352.

Mogil, J. S., and Chanda, M. L. (2005). The case for the inclusion of female subjects in basic science studies of pain. *Pain, 117*, 1–5.

Mogil, J. S., Wilson, S. G., Chesler, E. J., Rankin, A. L., et al. (2003). The melanocortin-1 receptor gene mediates female-specific mechanisms of analgesia in mice and humans. *Proceedings National Academy of Sciences, USA, 100*, 4867–4872.

Moita, M. A., Rosis, S., Zhou, Y., LeDoux, J. E., et al. (2004). Putting fear in its place: Remapping of hippocampal place cells during fear conditioning. *Journal of Neuroscience, 24*, 7015–7023.

Mokalled, M. H., Patra, C., Dickson, A. L., Endo, T., et al. (2016). Injury-induced ctgfa directs glial bridging and spinal cord regeneration in zebrafish. *Science, 354*(6312), 630–634.

Molenda-Figueira, H. A., Williams, C. A., Griffin, A. L., Rutledge, E. M., et al. (2006). Nuclear receptor coactivators function in estrogen receptor- and progestin receptor-dependent aspects of sexual behaviour in female rats. *Hormones and Behaviour, 50*, 383–392.

Monfils, M. H., Plautz, E. J., and Kleim, J. A. (2005). In search of the motor engram: Motor map plasticity as a mechanism for encoding motor experience. *Neuroscientist, 11,* 471–483.

Monks, D. A., and Watson N. V. (2001). N-cadherin expression in motoneurons is directly regulated by androgens: A genetic mosaic analysis in rats. *Brain Research, 895,* 73–79.

Monks, T. J., Jones, D. C., Bai, F., and Lau, S. S. (2004). The role of metabolism in 3,4-(+)-methylenedioxyamphetamine and 3,4-(+)-methylenedioxymethamphetamine (ecstasy) toxicity. *Therapeutic Drug Monitoring, 26,* 132–136.

Montague, C. T., Farooqi, I. S., Whitehead, J. P., Soos, M. A., et al. (1997). Congenital leptin deficiency is associated with severe early-onset obesity in humans. *Nature, 387,* 903–908.

Montague, P. R., Berns, G. S., Cohen, J. D., McClure, S. M., et al. (2002). Hyperscanning: Simultaneous fMRI during linked social interactions. *NeuroImage, 16*(4), 1159–1164.

Montenigro, P. H., Alosco, M. L., Martin, B., Daneshvar, D. H., et al. (2016 [ePub], 2017). Cumulative head impact exposure predicts later-life depression, apathy, executive dysfunction, and cognitive impairment in former high school and college football players. *Journal of Neurotrauma.* https://doi.org/10.1089/neu.2016.4413.

Monti, M. M., Vanhaudenhuyse, A., Coleman, M. R., Boly, M., et al. (2010). Willful modulation of brain activity in disorders of consciousness. *New England Journal of Medicine, 362,* 579–589.

Moon, R. Y. (2011). SIDS and other sleep-related infant deaths: Expansion of recommendations for a safe infant sleeping environment. Task Force on Sudden Infant Death Syndrome. *Pediatrics, 125*(5), 1030–1039.

Moore, C. L., Dou, H., and Juraska, J. M. (1992). Maternal stimulation affects the number of motor neurons in a sexually dimorphic nucleus of the lumbar spinal cord. *Brain Research, 572,* 52–56.

Moore, G. J., Bebchuk, J. M., Wilds, I. B., Chen, G., et al. (2000). Lithium-induced increase in human brain grey matter. *Lancet, 356,* 241–242.

Moore, R. Y., and Eichler, V. B. (1972). Loss of circadian adrenal corticosterone rhythm following suprachiasmatic lesions in the rat. *Brain Research, 42,* 201–206.

Moorhead, T. W., McKirdy, J., Sussmann, J. E., Hall, J., et al. (2007). Progressive gray matter loss in patients with bipolar disorder. *Biological Psychiatry, 62,* 894–900.

Moorman, S., Gobes, S. M., Kuijpers, M., Kerkhofs, A., et al. (2012). Human-like brain hemispheric dominance in birdsong learning. *Proceedings of the National Academy of Sciences, USA,* 109(31), 12782–12787.

Morales, S., Fu, X., and Pérez-Edgar, K. E. (2016). A developmental neuroscience perspective on affect-biased attention. *Developmental Cognitive Neuroscience, 21,* 26–41.

Moran, J., and Desimone, R. (1985). Selective attention gates visual processing in the extrastriate cortex. *Science, 229,* 782–784.

Moratalla, R., Khairnar, A., Simola, N., Granado, N., et al. (2017). Amphetamine-related drugs neurotoxicity in humans and in experimental animals: Main mechanisms. *Progress in Neurobiology, 155,* 149–170.

Morawietz, C., and Moffat, F. (2013). Effects of locomotor training after incomplete spinal cord injury: A systematic review. *Archives of Physical Medicine and Rehabilitation, 94*(11), 2297–2308.

Moray, N. (1959). Attention in dichotic listening: Affective cues and the influence of instructions. *Quarterly Journal of Experimental Psychology, 11,* 56–60.

Morbey, Y. E., and Mema, M. (2018). Size-selective fishing and the potential for fisheries-induced evolution in lake whitefish. *Evolutionary Applications, 11*(8), 1412–1424.

Moreno-Jiménez, E. P., Flor-García, M., Terreros-Roncal, J., Rábano, A., et al. (2019). Adult hippocampal neurogenesis is abundant in neurologically healthy subjects and drops sharply in patients with Alzheimer's disease. *Nature Medicine, 25,* 554–560.

Morens, D. M., Folkers, G. K., and Fauci, A. S. (2019). Acute flaccid myelitis: Something old and something new. *mBio, 10*(2), pii: e00521-19.

Mori, K., Nagao, H., and Yoshihara, Y. (1999). The olfactory bulb: Coding and processing of odour molecule information. *Science, 286,* 711–715.

Moriya, J., Takimoto, Y., Yoshiuchi, K., Shimosawa, T., et al. (2006). Plasma agouti-related protein levels in women with anorexia nervosa. *Psychoneuroendocrinology, 31,* 1057–1061.

Morizane, A., Kikuchi, T., Hayashi, T., Mizuma, H., et al. (2017). MHC matching improves engraftment of iPSC-derived neurons in non-human primates. *Nature Communications, 8*(1), 385.

Morran, L. T., Schmidt, O. G., Gelarden, I. A., Parrish, R. C., 2nd, et al. (2011). Running with the Red Queen: Host-parasite coevolution selects for biparental sex. *Science, 333*(6039), 216–218.

Morris, B. (2002). Overcoming dyslexia. *Fortune, 145*(10), 1–7.

Morris, R. G., Halliwell, R. F., and Bowery, N. (1989). Synaptic plasticity and learning. II: Do different kinds of plasticity underlie different kinds of learning? *Neuropsychologia, 27,* 41–59.

Morrison, A. P., Turkington, D., Pyle, M., Spencer, H., et al. (2014). Cognitive therapy for people with schizophrenia spectrum disorders not taking antipsychotic drugs: A single-blind randomised controlled trial. *Lancet, 383,* 1395–1403.

Morrison, A. R. (1983). A window on the sleeping brain. *Scientific American, 248*(4), 94–102.

Morrison, A. R., Sanford, L. D., Ball, W. A., Mann, G. L., et al. (1995). Stimulus-elicited behaviour in rapid eye movement sleep without atonia. *Behavioural Neuroscience, 109*(5), 972–979.

Mortensen, P. B., Pedersen, C. B., Westergaard, T., Wohlfahrt, J., et al. (1999). Effects of family history and place and season of birth on the risk of schizophrenia. *New England Journal of Medicine, 340,* 603–608.

Moruzzi, G. (1972). The sleep-waking cycle. *Ergebnisse der Physiologie, biologischen Chemie und experimentellen Pharmakologie, 64,* 1–165.

Moruzzi, G., and Magoun, H. W. (1949). Brain stem reticular formation and activation of the EEG. *Clinical Neurophysiology, 1,* 455–473.

Moscovitch, A., Blashko, C. A., Eagles, J. M., Darcourt, G., et al. (2004). A placebo-controlled study of sertraline in the treatment of outpatients with seasonal affective disorder. *Psychopharmacology (Berlin), 171,* 390–397.

Moser, E. I., and Moser, M. B. (2013). Grid cells and neural coding in high-end cortices. *Neuron, 80*(3), 765–774.

Mosialou, I., Shikhel, S., Liu, J. M., Maurizi, A., et al. (2017). MC4R-dependent suppression of appetite by bone-derived lipocalin 2. *Nature, 543*(7645), 385–390.

Mott, F. W. (1895). Experimental inquiry upon the afferent tracts of the central nervous system of the monkey. *Brain, 18,* 1–20.

Motta, S. C., Guimarães, C. C., Furigo, I. C., Sukikara, M. H., et al. (2013). Ventral premammillary nucleus as a critical sensory relay to the maternal aggression network. *Proceedings of the National Academy of Sciences, USA, 110*(35), 14438–14443.

Mountcastle, V. B. (1979). An organizing principle for cerebral function: The unit module and the distributed system. In F. O. Schmitt and F. G. Worden (Eds.), *The neurosciences: Fourth study program* (pp. 21–24). Cambridge, MA: MIT Press.

Mountcastle, V. B., Andersen, R. A., and Motter, B. C. (1981). The influence of attentive fixation upon the excitability of the light-sensitive neurons of the posterior parietal cortex. *Journal of Neuroscience, 1,* 1218–1235.

Mouritsen, H. (2018). Long-distance navigation and magnetoreception in migratory animals. *Nature, 558*(7708), 50–59.

Moyer, C. E., and Zuo, Y. (2018). Cortical dendritic spine development and plasticity: Insights from in vivo imaging. *Current Opinion in Neurobiology, 53,* 76–82.

Mu, D., Deng, J., Liu, K. F., Wu, Z. Y., et al. (2017). A central neural circuit for itch sensation. *Science, 357*(6352), 695–699.

Muggli, E., Matthews, H., Penington, A., Claes, P., et al. (2017). Association between prenatal alcohol exposure and craniofacial shape of children at 12 months of age. *JAMA Pediatrics, 171*(8), 771–780.

Muhlert, N., and Lawrence, A. D. (2015). Brain structure correlates of emotion-based rash impulsivity. *NeuroImage, 115,* 138–146.

Mukerji, S., Windsor, A. M., and Lee, D. J. (2010). Auditory brainstem circuits that mediate the middle ear muscle reflex. *Trends in Amplification, 14*(3), 170–191.

Mukhametov, L. M. (1984). Sleep in marine mammals. In A. Borbely and J. L. Valatx (Eds.), *Experimental Brain Research. Supplementum: 8. Sleep mechanisms.* Berlin: Springer.

Müller, M. J., Geisler, C., Heymsfield, S. B., and Bosy-Westphal, A. (2018). Recent advances in understanding body weight homeostasis in humans [version 1; peer review: 4 approved]. *F1000Research*, 7(F1000 Faculty Rev):1025. https://doi.org/10.12688/f1000research.14151.1.

Müller, T. D., Nogueiras, R., Andermann, M. L., Andrews, Z. B., et al. (2015). Ghrelin. *Molecular Metabolism*, *4*(6), 437–460.

Mumby, D. G. (2001). Perspectives on object-recognition memory following hippocampal damage: Lessons from studies in rats. *Behavioural Brain Research*, *127*(1–2), 159–181.

Munafo, J., Diedrick, M., and Stoffregen, T. A. (2017). The virtual reality head-mounted display Oculus Rift induces motion sickness and is sexist in its effects. *Experimental Brain Research*, *235*(3), 889–901.

Münte, T. F. (2002). Brains out of tune. *Nature*, *415*, 589–590.

Münte, T. F., Altenmüller, E., and Jäncke, L. (2002). The musician's brain as a model of neuroplasticity. *Nature Reviews Neuroscience*, *3*, 473–478.

Münte, T. F., Kohlmetz, C., Nager, W., and Altenmüller, E. (2001). Superior auditory spatial tuning in conductors. *Nature*, *409*, 580.

Murawski, N. J., Moore, E. M., Thomas, J. D., and Riley, E. P. (2015). Advances in diagnosis and treatment of fetal alcohol spectrum disorders: From animal models to human studies. *Alcohol Research: Current Reviews*, *37*(1), 97–108.

Mure, L. S., Le, H. D., Benegiamo, G., Chang, M. W., et al. (2018). Diurnal transcriptome atlas of a primate across major neural and peripheral tissues. *Science*, *359*(6381), pii: eaao0318.

Murphy, D. N., Boggio, P., and Fregni, F. (2009). Transcranial direct current stimulation as a therapeutic tool for the treatment of major depression: Insights from past and recent clinical studies. *Current Opinion in Psychiatry*, *22*, 306–311.

Murphy, M. L., Slavich, G. M., Chen, E., and Miller, G. E. (2015). Targeted rejection predicts decreased anti-inflammatory gene expression and increased symptom severity in youth with asthma. *Psychological Science*, *26*, 111–121.

Murray, J., Burgess, S., Zuccolo, L., Hickman, M., et al. (2016). Moderate alcohol drinking in pregnancy increases risk for children's persistent conduct problems: Causal effects in a Mendelian randomisation study. *Journal of Child Psychology and Psychiatry*, *57*(5), 575–584.

Murray, R. M., Morrison, P. D., Henquet, C., and Di Forti, M. (2007). Cannabis, the mind and society: The hash realities. *Nature Reviews Neuroscience*, *8*, 885–895.

Musacchia, G., Ortiz-Mantilla, S., Realpe-Bonilla, T., Roesler, C. P., et al. (2015). Infant auditory processing and event-related brain oscillations. *Journal of Visualized Experiments*, *101*, e52420. http://www.jove.com/video/52420. https://doi.org/10.3791/52420.

Muthukrishna, M., and Henrich, J. (2016). Innovation in the collective brain. *Philosophical Transactions of the Royal Society of London. Series B: Biological Sciences*, *371*, pii: 20150192. doi:10.1098/rstb.2015.0192.

Myers, N. L. (2010). Culture, stress and recovery from schizophrenia: Lessons from the field for global mental health. *Culture, Medicine and Psychiatry*, *34*, 500–528.

N

Nabavi, S., Fox, R., Proulx, C. D., Lin, J. Y., et al. (2014). Engineering a memory with LTD and LTP. *Nature*, *511*(7509), 348–352.

Nader, K., and Hardt, O. (2009). A single standard for memory: The case for reconsolidation. *Nature Reviews Neuroscience*, *10*, 224–234.

Naeser, M., Gaddie, A., Palumbo, C., and Stiassny-Eder, D. (1990). Late recovery of auditory comprehension in global aphasia. Improved recovery observed with subcortical temporal isthmus lesion vs. Wernicke's cortical area lesion. *Archives of Neurology*, *47*, 425–432.

Naeser, M., and Hayward, R. (1978). Lesion localisation in aphasia with cranial computed tomography and the Boston Diagnostic Aphasia Exam. *Neurology*, *28*, 545–551.

Naesström, M., Blomstedt, P., and Bodlund, O. (2016). A systematic review of psychiatric indications for deep brain stimulation, with focus on major depressive and obsessive-compulsive disorder. *Nordic Journal of Psychiatry*, *70*(7), 483–491.

Nagai, M., Re, D. B., Nagata, T., Chalazonitis, A., et al. (2007). Astrocytes expressing ALS-linked mutated SOD1 release factors selectively toxic to motor neurons. *Nature Neuroscience*, *10*(5), 615–622.

Najmabadi, H., Hu, H. Garshasbi, M., Zemojtel, T., et al. (2011). Deep sequencing reveals 50 novel genes for recessive cognitive disorders. *Nature*, *478*, 57–63.

Nakagawa, A., Mitsuda, D., Sado, M., Abe, T., et al. (2017). Effectiveness of supplementary cognitive-behavioural therapy for pharmacotherapy-resistant depression: A randomized controlled trial. *Journal of Clinical Psychiatry*, *78*(8), 1126–1135.

Nakamura, A., Kaneko, N., Villemagne, V. L., Kato, T., et al. (2018). High performance plasma amyloid-β biomarkers for Alzheimer's disease. *Nature 554*(7691), 249–254.

Nakazato, M., Murakami, N., Date, Y., Kojima, M., et al. (2001). A role for ghrelin in the central regulation of feeding. *Nature*, *409*, 194–198.

Namburi, P., Beyeler, A., Yorozu, S., Calhoon, G. G., et al. (2015). A circuit mechanism for differentiating positive and negative associations. *Nature*, *520*, 675–678.

Namkung, H., Kim, S. H., and Sawa, A. (2017). The insula: An underestimated brain area in clinical neuroscience, psychiatry, and neurology. *Trends in Neurosciences*, *40*(4), 200–207.

Naqvi, N. H., Gaznick, N., Tranel, D., and Bechara, A. (2014). The insula: A critical neural substrate for craving and drug seeking under conflict and risk. *Annals of the New York Academy of Sciences*, *1316*, 53–70.

Naqvi, N. H., Rudrauf, D., Damasio, H., and Bechara, A. (2007). Damage to the insula disrupts addiction to cigarette smoking. *Science*, *315*, 531–534.

Nath, R. D., Bedbrook, C. N., Abrams, M. J., Basinger, T., et al. (2017). The jellyfish *Cassiopea* exhibits a sleep-like state. *Current Biology*, *27*(19), 2984–2990.e3.

Nation, E. F. (1973). William Osler on penis captivus and other urologic topics. *Urology*, *2*, 468–470.

National Academy of Sciences. (2003). *The polygraph and lie detection.* Washington, DC: National Academies Press (www.nap.edu/openbook.php?isbn=0309084369).

National Institute of Mental Health, Mental Illness, nimh.nih.gov/health/statistics/prevalence/serious-mental-illness-smi-among-us-adults.shtml.

National Institute on Drug Abuse: Trends and Statistics, https://www.drugabuse.gov/related-topics/trends-statistics.

Nature. (2017). Head injuries in sport must be taken more seriously [Editorial]. *Nature*, *548*(7668), 371.

Nature Neuroscience. (2015). Inhumane treatment of nonhuman primate researchers [Editorial]. *Nature Neuroscience*, *18*(6), 787.

Nauhaus, I., Nielsen, K. J., Disney, A. A., and Callaway, E. M. (2012). Orthogonal micro-organisation of orientation and spatial frequency in primate primary visual cortex. *Nature Neuroscience*, *15*, 1683–1690.

Nave, K. A. (2010). Myelination and the trophic support of long axons. *Nature Reviews Neuroscience*, *11*, 275–283.

Neale, G. (2006, December 3). Peter Singer: Monkey business. *Independent* (http://news.independent.co.uk/people/profiles/article2035119.ece).

Nee, D. E., and D'Esposito, M. (2018). The representational basis of working memory. *Current Topics in Behavioural Neuroscience*, *37*, 213–230.

Neisser, U., and Becklen, R. (1975). Selective looking: Attending to visually specified events. *Cognitive Psychology*, *7*, 480–494.

Neitz, M., Kraft, T. W., and Neitz, J. (1998). Expression of L cone pigment gene subtypes in females. *Vision Research*, *38*(21), 3221–3225.

Neitz, M., and Neitz, J. (2014). Curing colour blindness: Mice and nonhuman primates. *Cold Spring Harbor Perspectives in Medicine*, *4*, a017418.

Nelson, G., Chandrashekar, J., Hoon, M. A., Feng, L., et al. (2002). An amino-acid taste receptor. *Nature*, *416*, 199–202.

Nelson, G., Hoon, M. A., Chandrashekar, J., Zhang, Y., et al. (2001). Mammalian sweet taste receptors. *Cell*, *106*, 381–390.

Nestler, E. J., and Hyman, S. E. (2010). Animal models of neuropsychiatric disorders. *Nature Neuroscience*, *13*, 1161–1169.

Nestor, A., Plaut, D. C., and Behrmann, M. (2016). Feature-based face representations and image reconstruction from behavioural and neural data. *Proceedings of the National Academy of Sciences, USA*, *113*(2), 416–421.

Neumeister, A., Bain, E., Nugent, A. C., Carson, R. E., et al. (2004). Reduced serotonin type 1A receptor binding in panic disorder. *Journal of Neuroscience, 24*, 589–591.

Neves-Pereira, M., Mundo, E., Muglia, P., King, N., et al. (2002). The brain-derived neurotrophic factor gene confers susceptibility to bipolar disorder: Evidence from a family-based association study. *American Journal of Human Genetics, 71*, 651–655.

Neville, H. J., Bavelier, D., Corina, D., Rauschecker, J., et al. (1998). Cerebral organisation for language in deaf and hearing subjects: Biological constraints and effects of experience. *Proceedings of the National Academy of Sciences, USA, 95*, 922–929.

Neville, H. J., Mills, D. L., and Lawson, D. S. (1992). Fractionating language: Different neural subsystems with different sensitive periods. *Cerebral Cortex, 2*, 244–258.

Ng, S. F., Lin, R. C., Laybutt, D. R., Barres, R., et al. (2010). Chronic high-fat diet in fathers programs β-cell dysfunction in female rat offspring. *Nature, 467*(7318), 963–966.

Ngandu, T., Lehtisalo, J., Solomon, A., Levälahti, E., et al. (2015). A 2 year multidomain intervention of diet, exercise, cognitive training, and vascular risk monitoring versus control to prevent cognitive decline in at-risk elderly people (FINGER): A randomised controlled trial. *Lancet, 385*(9984), 2255–2263.

Ni, J. D., Baik, L. S., Holmes, T. C., and Montell, C. (2017). A rhodopsin in the brain functions in circadian photoentrainment in *Drosophila*. *Nature, 545*(7654), 340–344.

Niazi, R. K., Gjesing, A. P., Hollensted, M., Have, C. T., et al. (2018). Identification of novel LEPR mutations in Pakistani families with morbid childhood obesity. *BMC Medical Genetics, 19*(1), 199.

Nichols, M. J., and Newsome, W. T. (1999). The neurobiology of cognition. *Nature, 402*, C35–C38.

Niel, L., Shah, A. H., Lewis, G. A., Mo, K., et al. (2009). Sexual differentiation of the spinal nucleus of the bulbocavernosus is not mediated solely by androgen receptors in muscle fibres. *Endocrinology, 150*, 3207–3213.

Nietzel, M. T. (2000). Police psychology. In A. E. Kazdin (Ed.), *Encyclopedia of psychology* (Vol. 6, pp. 224–226). Washington, DC: American Psychological Association.

Nishida, M., and Walker, M. P. (2007). Daytime naps, motor memory consolidation and regionally specific sleep spindles. *PLOS ONE, 2*, e341.

Nishimoto, S., Vu, A. T., Naselaris, T., Benjamini, Y., et al. (2011). Reconstructing visual experiences from brain activity evoked by natural movies. *Current Biology, 21*(19), 1641–1646.

Nitz, D. A., Van Swinderen, B., Tononi, G., and Greenspan, R. J. (2002). Electrophysiological correlates of rest and activity in *Drosophila melanogaster*. *Current Biology, 12*, 1934–1940.

Noad, M. J., Cato, D. H., Bryden, M. M., Jenner, M.-N., et al. (2000). Cultural revolution in whale songs. *Nature, 408*, 537–538.

Nobre, A. C., Sebestyen, G. N., Gitelman, D. R., Mesulam, M. M., et al. (1997). Functional localisation of the system for visuospatial attention using positron emission tomography. *Brain, 120*, 515–533.

Nock, M. K., Hwang, I., Sampson, N. A., and Kessler, R. C. (2010). Mental disorders, comorbidity and suicidal behaviour: Results from the National Comorbidity Survey Replication. *Molecular Psychiatry, 15*(8), 868–876.

Nordeen, E. J., Nordeen, K. W., Sengelaub, D. R., and Arnold, A. P. (1985). Androgens prevent normally occurring cell death in a sexually dimorphic spinal nucleus. *Science, 229*, 671–673.

Norenzayan, A., Shariff, A. F., Gervais, W. M., Willard, A. K., et al. (2016). Parochial prosocial religions: Historical and contemporary evidence for a cultural evolutionary process. *Behavioural and Brain Sciences, 39*, e29.

Norton, T. T., and Siegwart, J. T., Jr. (2013). Light levels, refractive development, and myopia: A speculative review. *Experimental Eye Research, 114*, 48–57.

Nottebohm, F. (1980). Brain pathways for vocal learning in birds: A review of the first 10 years. *Progress in Psychobiology and Physiological Psychology, 9*, 85–124.

Nottebohm, F., and Arnold, A. P. (1976). Sexual dimorphism in vocal control areas of the songbird brain. *Science, 194*, 211–213.

NRC Committee on Animals as Monitors of Environmental Hazards. (1991). *Animals as sentinels of environmental health hazards.* Washington, DC: National Academy Press.

Nudo, R. J., Milliken, G. W., Jenkins, W. M., and Merzenich, M. M. (1996). Use-dependent alterations of movement representations in primary motor cortex of adult squirrel monkeys. *Journal of Neuroscience, 16*, 785–807.

Numan, M. (2015). Aggressive behaviour. In M. Numan (Ed.), *Neurobiology of social behaviour* (pp. 63–107). San Diego, CA: Academic Press.

Nunn, C. L., and Samson, D. R. (2018). Sleep in a comparative context: Investigating how human sleep differs from sleep in other primates. *American Journal of Physical Anthropology, 166*(3), 601–612.

Nutt, D. J., Lingford-Hughes, A., Erritzoe, D., and Stokes, P. R. (2015). The dopamine theory of addiction: 40 years of highs and lows. *Nature Reviews Neuroscience, 16*, 305–312.

O

O'Connell-Rodwell, C. E. (2007). Keeping an "ear" to the ground: Seismic communication in elephants. *Physiology (Bethesda), 22*, 287–294.

O'Connor, D. B., Archer, J., and Wu, F. C. (2004). Effects of testosterone on mood, aggression, and sexual behaviour in young men: A double-blind, placebo-controlled, cross-over study. *Journal of Clinical Endocrinology & Metabolism, 89*, 2837–2845.

O'Craven, K. M., Downing, P. E., and Kanwisher, N. (1999). fMRI evidence for objects as the units of attentional selection. *Nature, 401*, 584–587.

O'Keefe, J., and Dostrovsky, J. (1971). The hippocampus as a spatial map. Preliminary evidence from unit activity in the freely-moving rat. *Brain Research, 34*, 171–175.

Obernier, J. A., White, A. M., Swartzwelder, H. S., and Crews, F. T. (2002). Cognitive deficits and CNS damage after a 4-day binge ethanol exposure in rats. *Pharmacology, Biochemistry and Behaviour, 72*, 521–532.

Ogden, J. (2012). *Health psychology.* New York: Open University Press.

Ohbayashi, M., Picard, N., and Strick, P. L. (2016). Inactivation of the dorsal premotor area disrupts internally generated, but not visually guided, sequential movements. *Journal of Neuroscience, 36*, 1971–1976.

Ohno, K., and Sakurai, T. (2008). Orexin neuronal circuitry: Role in the regulation of sleep and wakefulness. *Frontiers in Neuroendocrinology, 29*, 70–87.

Ojala, K. E., Janssen, L. K., Hashemi, M. M., Timmer, M. H. M., et al. (2018). Dopaminergic drug effects on probability weighting during risky decision making. *eNeuro, 5*(2), pii: ENEURO.0330-18.2018.

Ojemann, G. A., and Whitaker, H. A. (1978). The bilingual brain. *Archives of Neurology, 35*, 409–412.

Oka, Y., Butnaru, M., von Buchholtz, L., Ryba, N. J., et al. (2013). High salt recruits aversive taste pathways. *Nature, 494*(7438), 472–475.

Oka, Y., Ye, M., and Zuker, C. S. (2015). Thirst driving and suppressing signals encoded by distinct neural populations in the brain. *Nature, 520*, 349–352.

Okada, Y., Momozawa, Y., Sakaue, S., Kanai, M., et al. (2018). Deep whole-genome sequencing reveals recent selection signatures linked to evolution and disease risk of Japanese. *Nature Communications, 9*(1), 1631.

O'Keefe, J. H., Bhatti, S. K., Bajwa, A., DiNicolantonio, J. J., et al. (2014). Alcohol and cardiovascular health: The dose makes the poison … or the remedy. *Mayo Clinic Proceedings, 89*(3), 382–393.

Olabi, B., Ellison-Wright, I., McIntosh, A. M., Wood, S. J., et al. (2011). Are there progressive brain changes in schizophrenia? A meta-analysis of structural magnetic resonance imaging studies. *Biological Psychiatry, 70*(1), 88–96.

Olanow, C. W., Goetz, C. G., Kordower, J. H., Stoessl, A. J., et al. (2003). A double-blind controlled trial of bilateral fetal nigral transplantation in Parkinson's disease. *Annals of Neurology, 54*, 403–414.

Olds, J., and Milner, P. (1954). Positive reinforcement produced by electrical stimulation of septal area and other regions of the rat brain. *Journal of Comparative and Physiological Psychology, 47*, 419–427.

Olender, T., Lancet, D., and Nebert, D. W. (2008). Update on the olfactory receptor (OR) gene superfamily. *Human Genomics, 3*(1), 87–97.

Oler, J. A., Fox, A. S., Shelton, S. E., Rogers, J., et al. (2010). Amygdalar and hippocampal substrates of anxious temperament differ in their heritability. *Nature, 466,* 864–868.

Olsen, E. M., Heino, M., Lilly, G. R., Morgan, M. J., et al. (2004). Maturation trends indicative of rapid evolution preceded the collapse of northern cod. *Nature, 428,* 899–900.

Olsen, K. L. (1979). Androgen-insensitive rats are defeminised by their testes. *Nature, 279,* 238–239.

Olsen, R. W. (2018). GABA(A) receptor: Positive and negative allosteric modulators. *Neuropharmacology, 136*(Pt. A), 10–22.

Olson, S. (2004). Making sense of Tourette's. *Science, 305,* 1390–1392.

Olsson, A., and Phelps, E. A. (2007). Social learning of fear. *Nature Neuroscience, 10,* 1095–1102.

Oman, C. M. (2012). Are evolutionary hypotheses for motion sickness"just-so"stories? *Journal of Vestibular Research, 22*(2), 117–127.

Opendak, M., and Gould, E. (2015). Adult neurogenesis: A substrate for experience-dependent change. *Trends in Cognitive Sciences, 19*(3), 151–161. doi:10.1016/j.tics.2015.01.001.

Ophir, A. G., Phelps, S. M., Sorin, A. B., and Wolff, J. O. (2008). Social but not genetic monogamy is associated with greater breeding success in prairie voles. *Animal Behaviour, 75,* 1143–1154.

Orlovska, S., Vestergaard, C. H., Bech, B. H., Nordentoft, M., et al. (2017). Association of streptococcal throat infection with mental disorders: Testing key aspects of the PANDAS hypothesis in a nationwide study. *JAMA Psychiatry, 74*(7), 740–746.

Oroszi, G., Anton, R. F., O'Malley, S., Swift, R., et al. (2009). OPRM1 Asn40Asp predicts response to naltrexone treatment: A haplotype-based approach. *Alcoholism: Clinical and Experimental Research, 33,* 383–393.

Orson, F. M., Wang, R., Brimijoin, S., Kinsey, B. M., et al. (2014). The future potential for cocaine vaccines. *Expert Opinion on Biological Therapy, 14,* 1271–1283.

O'Shea, J., Revol, P., Cousijn, H., Near, J., et al. (2017). Induced sensorimotor cortex plasticity remediates chronic treatment-resistant visual neglect. *eLife, 6,* pii: e26602.

Öst, L. G., Riise, E. N., Wergeland, G. J., Hansen, B., et al. (2016). Cognitive behavioural and pharmacological treatments of OCD in children: A systematic review and meta-analysis. *Journal of Anxiety Disorders, 43,* 58–69.

Osterhout, L. (1997). On the brain response to syntactic anomalies: Manipulations of word position and word class reveal individual differences. *Brain and Language, 59,* 494–522.

Ostrovsky, Y., Meyers, E., Ganesh, S., Mathur, U., et al. (2009). Visual parsing after recovery from blindness. *Psychological Science, 20,* 1484–1491.

O'Tuathaigh, C. M. P., Moran, P. M., Zhen, X. C., and Waddington, J. L. (2017). Translating advances in the molecular basis of schizophrenia into novel cognitive treatment strategies. *British Journal of Pharmacology, 174*(19), 3173–3190.

Ouattara, K., Lemasson, A., and Zuberbühler, K. (2009). Campbell's monkeys concatenate vocalisations into context-specific call sequences. *Proceedings of the National Academy of Sciences, USA, 22,* 22026–22031.

Overstreet, D. H. (1993). The Flinders sensitive line rats: A genetic animal model of depression. *Neuroscience and Biobehavioural Reviews, 17,* 51–68.

Overton, J. A, and Recanzone, G. H. (2016). Effects of aging on the response of single neurons to amplitude-modulated noise in primary auditory cortex of rhesus macaque. *Journal of Neurophysiology, 115*(6), 2911–2923.

Owen, A. M., Coleman, M. R., Boly, M., Davis, M. H., et al. (2006). Detecting awareness in the vegetative state. *Science, 313,* 1402.

P

Pack, A. I. (2003). Should a pharmaceutical be approved for the broad indication of excessive sleepiness? *American Journal of Respiratory and Critical Care Medicine, 167,* 109–111.

Paddock, R. C., and La Ganga, M. (2008, August 5). Officials decry attacks on UC staff. *Los Angeles Times* (http://articles.latimes.com/2008/aug/05/local/me-attacks5).

Pagel, J. F., and Helfter, P. (2003). Drug induced nightmares: An etiology based review. *Human Psychopharmacology, 18,* 59–67.

Pagel, M., Atkinson, Q. D., Calude, A. S., and Meade, A. (2013). Ultraconserved words point to deep language ancestry across Eurasia. *Proceedings of the National Academy of Sciences, USA, 110*(21), 8471–8476.

Paik, S.-B., and Ringach, D. L. (2011). Retinal origin of orientation maps in visual cortex. *Nature Neuroscience, 14,* 919–925.

Palagini, L., Baglioni, C., Ciapparelli, A., Gemignani, A., et al. (2013). REM sleep dysregulation in depression: State of the art. *Sleep Medicine Reviews, 17*(5), 377–390.

Palaus, M., Marron, E. M., Viejo-Sobera, R., and Redolar-Ripoll, D. (2017). Neural basis of video gaming: A systematic review. *Frontiers in Human Neuroscience, 11,* 248.

Panksepp, J. (1998). *Affective neuroscience.* New York: Oxford University Press.

Panksepp, J. (2005). Beyond a joke: From animal laughter to human joy? *Science, 308,* 62–63.

Panksepp, J. (2007). Neuroevolutionary sources of laughter and social joy: Modeling primal human laughter in laboratory rats. *Behavioural Brain Research, 182,* 231–244.

Panksepp, J. (2011). Cross-species affective neuroscience decoding of the primal affective experiences of humans and related animals. *PLOS ONE, 6*(9), e21236.

Pantev, C., Oostenveld, R., Engelien, A., Ross, B., et al. (1998). Increased auditory cortical representation in musicians. *Nature, 392,* 811–814.

Papez, J. W. (1937). A proposed mechanism of emotion. *Archives of Neurology and Psychiatry, 38,* 725–745.

Papka, M., Ivry, R., and Woodruff-Pak, D. S. (1994). Eyeblink classical conditioning and time production in patients with cerebellar damage. *Society of Neuroscience Abstracts, 20,* 360.

Pare, M., Behets, C., and Cornu, O. (2003). Paucity of presumptive ruffini corpuscles in the index finger pad of humans. *Journal of Comparative Neurology, 456,* 260–266.

Park, D. C., Lautenschlager, G., Hedden, T., Davidson, N. S., et al. (2002). Models of visuospatial and verbal memory across the adult life span. *Psychology of Aging, 17,* 299–320.

Park, H., and Poo, M. M. (2013). Neurotrophin regulation of neural circuit development and function. *Nature Reviews Neuroscience, 14*(1), 7–23. doi:10.1038/nrn3379.

Park, S., Como, P. G., Cui, L., and Kurlan, R. (1993). The early course of the Tourette's syndrome clinical spectrum. *Neurology, 43,* 1712–1715.

Parker, G., Cahill, L., and McGaugh, J. L. (2006). A case of unusual autobiographical remembering. *Neurocase, 12,* 35–49.

Parkes, J. D. (1985). *Sleep and its disorders.* Philadelphia: Saunders.

Parnavelas, J. G., Anderson, S. A., Lavdas, A. A., Grigoriou, M., et al. (2000). The contribution of the ganglionic eminence to the neuronal cell types of the cerebral cortex. *Novartis Foundation Symposium, 228,* 129–139; discussion 139–147.

Parrott, A. C. (2013). Human psychobiology of MDMA or"Ecstasy": An overview of 25 years of empirical research. *Human Psychopharmacology, 28,* 289–307.

Parrott, A. C. (2014). The potential dangers of using MDMA for psychotherapy. *Journal of Psychoactive Drugs, 46,* 37–43.

Parsons, L. H., and Hurd, Y. L. (2015). Endocannabinoid signalling in reward and addiction. *Nature Reviews Neuroscience, 16,* 579–594.

Parton, L. E., Ye, C. P., Coppari, R., Enriori, P. J., et al. (2007). Glucose sensing by POMC neurons regulates glucose homeostasis and is impaired in obesity. *Nature, 449,* 228–232.

Passemard, S., Verloes, A., Billette de Villemeur, T., Boespflug-Tanguy, O., et al. (2016). Abnormal spindle-like microcephaly-associated (ASPM) mutations strongly disrupt neocortical structure but spare the hippocampus and long-term memory. *Cortex, 74,* 158–176.

Pastalkova, E., Itskov, V., Amarasingham, A., and Buzsáki, G. (2008). Internally generated cell assembly sequences in the rat hippocampus. *Science, 321*(5894),1322–1327.

Paterson, S. J., Brown, J. H., Gsödl, M. K., Johnson, M. H., et al. (1999). Cognitive modularity and genetic disorders. *Science, 286,* 2355–2358.

Paton, J. J., Belova, M. A., Morrison, S. E., and Salzman, C. D. (2006). The primate amygdala represents the positive and negative value of visual stimuli during learning. *Nature, 439*(7078), 865–870.

Patterson, F., and Linden, E. (1981). *The education of Koko.* New York: Holt, Rinehart, and Winston.

Patterson, P. H. (2007). Maternal effects on schizophrenia risk. *Science, 318,* 576–578.

Paulesu, E., McCrory, E., Fazio, F., Menoncello, L., et al. (2000). A cultural effect on brain function. *Nature Neuroscience, 3,* 91–96.

Pauls, D. L. (2003). An update on the genetics of Gilles de la Tourette syndrome. *Journal of Psychosomatic Research, 55,* 7–12.

Pauls, D. L., Abramovitch, A., Rauch, S. L., and Geller, D. A. (2014). Obsessive-compulsive disorder: An integrative genetic and neurobiological perspective. *Nature Reviews Neuroscience, 15,* 410–424.

Paus, T., Kalina, M., Patocková, L., Angerová, Y., et al. (1991). Medial vs lateral frontal lobe lesions and differential impairment of central-gaze fixation maintenance in man. *Brain, 114,* 2051–2067.

Paus, T., Keshavan, M., and Giedd, J. N. (2008). Why do many psychiatric disorders emerge during adolescence? *Nature Reviews Neuroscience, 9,* 947–956.

Pearson, R. A., Barber, A. C., Rizzi, M., Hippert, C., et al. (2012). Restoration of vision after transplantation of photoreceptors. *Nature, 485,* 99–103.

Pecho-Vrieseling, E., Rieker, C., Fuchs, S., Bleckmann, D., et al. (2014). Transneuronal propagation of mutant huntingtin contributes to non-cell autonomous pathology in neurons. *Nature Neuroscience, 17*(8), 1064–1072.

Pedersen, C. B., and Mortensen, P. B. (2001). Evidence of a dose-response relationship between urbanicity during upbringing and schizophrenia risk. *Archives of General Psychiatry, 58,* 1039–1046.

Pediatric Eye Disease Investigator Group. (2005). Randomized trial of treatment of amblyopia in children aged 7 to 17 years. *Archives of Ophthalmology, 13,* 437–447.

Pedreira, C., Mormann, F., Kraskov, A., Cerf, M., et al. (2010). Responses of human medial temporal lobe neurons are modulated by stimulus repetition. *Journal of Neurophysiology, 103,* 97–107.

Pegna, A. J., Khateb, A., Lazeyras, F., and Seghier, M. L. (2005). Discriminating emotional faces without primary visual cortices involves the right amygdala. *Nature Neuroscience, 8,* 24–25.

Pelvig, D. P., Pakkenberg, H., Stark, A. K., and Pakkenberg, B. (2008). Neocortical glial cell numbers in human brains. *Neurobiology of Aging, 29,* 1754–1762.

Penadés, R., González-Rodríguez, A., Catalán, R., Segura, B., et al. (2017). Neuroimaging studies of cognitive remediation in schizophrenia: A systematic and critical review. *World Journal of Psychiatry, 7*(1), 34–43.

Penfield, W., and Rasmussen, T. (1950). *The cerebral cortex in man.* New York: Macmillan.

Penfield, W., and Roberts, L. (1959). *Speech and brain-mechanisms.* Princeton, NJ: Princeton University Press.

Peng, Y., Gillis-Smith, S., Jin, H., Tränkner, D., et al. (2015). Sweet and bitter taste in the brain of awake behaving animals. *Nature, 527*(7579), 512–515.

Penner, M. R., and Mizumori, S. J. (2012). Age-associated changes in the hippocampal-ventral striatum-ventral tegmental loop that impact learning, prediction, and context discrimination. *Frontiers in Aging Neuroscience, 4,* 22.

Pennisi, E. (2003). Systems biology: Tracing life's circuitry. *Science, 302,* 1646–1649.

Pennisi, E. (2012). Genomics. ENCODE project writes eulogy for junk DNA. *Science, 337*(6099), 1159, 1161. doi:10.1126/science.337.6099.1159.

Peplau, L. A. (2003). Human sexuality: How do men and women differ? *Current Directions in Psychological Science, 12,* 37–40.

Pepperberg, I. M. (2009). *The Alex studies: Cognitive and communicative abilities of grey parrots.* Cambridge, MA: Harvard University Press. (2002 reprint, revised.)

Perani, D., and Abutalebi, J. (2005). The neural basis of first and second language processing. *Current Opinion in Neurobiology, 15,* 202–206.

Perani, D., Farsad, M., Ballarini, T., Lubian, F., et al. (2017). The impact of bilingualism on brain reserve and metabolic connectivity in Alzheimer's dementia. *Proceedings of the National Academy of Sciences, USA, 114*(7), 1690–1695.

Perea, G., Navarrete, M., and Araque, A. (2009). Tripartite synapses: Astrocytes process and control synaptic information. *Trends in Neuroscience, 32,* 421–431.

Perenin, M. T., and Vighetto, A. (1988). Optic ataxia: A specific disruption in visuomotor mechanisms. I. Different aspects of the deficit in reaching for objects. *Brain, 111,* 643–674.

Peretz, I., and Hyde, K. L. (2003). What is specific to music processing? Insights from congenital amusia. *Trends in Cognitive Sciences, 7,* 362–367.

Perilloux, C., Easton, J. A., and Buss, D. M. (2012). The misperception of sexual interest. *Psychological Science, 23*(2), 146–151.

Peris, J., Boyson, S. J., Cass, W. A., Curella, P., et al. (1990). Persistence of neurochemical changes in dopamine systems after repeated cocaine administration. *Journal of Pharmacology and Experimental Therapeutics, 253,* 38–44.

Pernía-Andrade, A. J., Kato, A., Witschi, R., Nyilas, R., et al. (2009). Spinal endocannabinoids and CB_1 receptors mediate C-fibre–induced heterosynaptic pain sensitisation. *Science, 325,* 760–764.

Perrachione, T. K., Del Tufo, S. N., and Gabrieli, J. D. (2011). Human voice recognition depends on language ability. *Science, 333,* 595.

Perrin, J. S., Merz, S., Bennett, D. M., Currie, J., et al. (2012). Electroconvulsive therapy reduces frontal cortical connectivity in severe depressive disorder. *Proceedings of the National Academy of Sciences, USA, 109,* 5464–5468.

Perrin, S. (2014). Preclinical research: Make mouse studies work. *Nature, 507*(7493), 423–425.

Perry, V. H., and Holmes, C. (2014). Microglial priming in neurodegenerative disease. *Nature Reviews Neurology, 10*(4), 217–224.

Pertwee, R. G. (1997). Pharmacology of cannabinoid CB1 and CB2 receptors. *Pharmacology & Therapeutics, 74,* 129–180.

Peschansky, V. J., Burbridge, T. J., Volz, A. J., Fiondella, C., et al. (2010). The effect of variation in expression of the candidate dyslexia susceptibility gene homolog Kiaa0319 on neuronal migration and dendritic morphology in the rat. *Cerebral Cortex, 20,* 884–897.

Petanjek, Z., Judaš, M., Šimic, G., Rasin, M. R., et al. (2011). Extraordinary neoteny of synaptic spines in the human prefrontal cortex. *Proceedings of the National Academy of Sciences, USA, 108*(32), 13281–13286.

Peterhans, E., and von der Heydt, R. (1989). Mechanisms of contour perception in monkey visual cortex. II. Contours bridging gaps. *Journal of Neuroscience, 9,* 1749–1763.

Peters, A., Palay, S. L., and Webster, H. deF. (1991). *The fine structure of the nervous system: Neurons and their supporting cells* (3rd ed.). New York: Oxford University Press.

Peters, R. (2006). Ageing and the brain. *Postgraduate Medical Journal, 82*(964), 84–88. http://dx.doi.org/10.1136/pgmj.2005.036665.

Petersen, J. L., and Hyde, J. S. (2011). Gender differences in sexual attitudes and behaviours: A review of meta-analytic results and large datasets. *Journal of Sex Research, 48*(2–3), 149–165.

Peterson, B. S., Warner, V., Bansal, R., Zhu, H., et al. (2009). Cortical thinning in persons at increased familial risk for major depression. *Proceedings of the National Academy of Sciences, USA, 106,* 6273–6278.

Peterson, L. R., and Peterson, M. J. (1959). Short-term retention of individual verbal items. *Journal of Experimental Psychology, 58,* 193–198.

Petit, D., Pennestri, M. H., Paquet, J., Desautels, A., et al. (2015). Childhood sleepwalking and sleep terrors: A longitudinal study of prevalence and familial aggregation. *JAMA Pediatrics, 169*(7), 653–658.

Petitto, L. A., Berens, M. S., Kovelman, I., Dubins, M. H., et al. (2012). The "Perceptual Wedge Hypothesis" as the basis for bilingual babies' phonetic processing advantage: New insights from fNIRS brain imaging. *Brain and Language, 121*(2), 130–143.

Petitto, L. A., Zatorre, R. J., Gauna, K., Nikelski, E. J., et al. (2000). Speech-like cerebral activity in profoundly deaf people processing signed languages: Implications for the neural basis of human language. *Proceedings of the National Academy of Sciences, USA, 97*, 13961–13966.

Petrelli, F., Pucci, L., and Bezzi, P. (2016). Astrocytes and microglia and their potential link with autism spectrum disorders. *Frontiers in Cellular Neuroscience, 10*, 21. doi:10.3389/fncel.2016.00021.

Petrides, M., and Milner, B. (1982). Deficits on subject-ordered tasks after frontal- and temporal-lobe lesions in man. *Neuropsychologia, 20*, 249–262.

Petrov, D., Mansfield, C., Moussy, A., and Hermine, O. (2017). ALS clinical trials review: 20 years of failure. Are we any closer to registering a new treatment? *Frontiers in Aging Neuroscience, 9*, 68.

Petrovic, P., Kalso, E., Petersson, K. M., and Ingvar, M. (2002). Placebo and opioid analgesia imaging: A shared neuronal network. *Science, 295*, 1737–1740.

Pettit, H. O., and Justice, J. B., Jr. (1991). Effect of dose on cocaine self-administration behaviour and dopamine levels in the nucleus accumbens. *Brain Research, 539*, 94–102.

Pettorruso, M., Martinotti, G., Cocciolillo, F., De Risio, L., et al. (2018). Striatal presynaptic dopaminergic dysfunction in gambling disorder: A ^{123}I-FP-CIT SPECT study. *Addiction Biology*. https://doi.org/10.1111/adb.12677.

Petty, F., Kramer, G., Wilson, L., and Jordan, S. (1994). In vivo serotonin release and learned helplessness. *Psychiatry Research, 52*, 285–293.

Pfaff, D. W. (1980). *Estrogens and brain function: Neural analysis of a hormone-controlled mammalian reproductive behaviour.* New York: Springer.

Pfaff, D. W., Gagnidze, K., and Hunter, R. G. (2018). Molecular endocrinology of female reproductive behaviour. *Molecular and Cellular Endocrinology, 467*, 14–20.

Pfau, D., Jordan, C. L., and Breedlove, S. M. (2019) The de-scent of sexuality: Did loss of a pheromone signalling protein permit the evolution of same-sex sexual behaviour in primates? *Archives of Sexual Behaviour.* https://doi.org/10.1007/s10508-018-1377-2.

Pfaus, J. G., Kippin, T. E., and Centeno, S. (2001). Conditioning and sexual behaviour: A review. *Hormones and Behaviour, 40*, 291–321.

Pfaus, J. G., Kippin, T. E., Coria-Avila, G. A., Gelez, H., et al. (2012). Who, what, where, when (and maybe even why)? How the experience of sexual reward connects sexual desire, preference, and performance. *Archives of Sexual Behaviour, 41*(1), 31–62.

Pfefferbaum, A., Rosenbloom, M. J., Chu, W., Sassoon, S. A., et al. (2014). White matter microstructural recovery with abstinence and decline with relapse in alcohol dependence interacts with normal ageing: A controlled longitudinal DTI study. *Lancet Psychiatry, 1*(3), 202–212. https://doi.org/10.1016/S2215-0366(14)70301-3.

Pfenning, A. R., Hara, E., Whitney, O., Rivas, M. V., et al. (2014). Convergent transcriptional specialisations in the brains of humans and song-learning birds. *Science, 346*(6215). doi:10.1126/science.1256846.

Phan, N. T. N., Lie, X., and Ewing, A. G. (2017). Measuring synaptic vesicles using cellular electrochemistry and nanoscale molecular imaging. *Nature Reviews Chemistry, 1*. https://doi.org/10.1038/s41570-017-0048.

Philips, T., and Rothstein, J. D. (2015). Rodent models of amyotrophic lateral sclerosis. *Current Protocols in Pharmacology, 69*, 5.67.1–5.67.21. doi:10.1002/0471141755.ph0567s69.

Phillips, J. R., Johansson, R. S., and Johnson, K. O. (1990). Representation of braille characters in human nerve fibres. *Experimental Brain Research, 81*, 589–592.

Phoenix, C. H., Goy, R. W., Gerall, A. A., and Young, W. C. (1959). Organizing action of prenatally administered testosterone propionate on the tissues mediating mating behaviour in the female guinea pig. *Endocrinology, 65*, 369–382.

Picard, N., Matsuzaka, Y., and Strick, P. L. (2013). Extended practice of a motor skill is associated with reduced metabolic activity in M1. *Nature Neuroscience, 16*(9), 1340–1347.

Pickens, R., and Thompson, T. (1968). Drug use by U.S. Army enlisted men in Vietnam: A followup on their return home. *Journal of Pharmacology and Experimental Therapeutics, 161*, 122–129.

Pierce, R. C. and Kumaresan, V. 2006. The mesolimbic dopamine system: the final common pathway for the reinforcing effect of drugs of abuse? *Neuroscience & Biobehavioural Reviews, 30*(2), 215–238.

Pietrobon, D., and Moskowitz, M. A. (2014). Chaos and commotion in the wake of cortical spreading depression and spreading depolarisations. *Nature Reviews Neuroscience, 15*(6), 379–393.

Pigeon, G., Festa-Bianchet, M., Coltman, D. W., and Pelletier, F. (2016). Intense selective hunting leads to artificial evolution in horn size. *Evolutionary Applications, 9*(4), 521–530.

Pin, J. P., and Bettler, B. (2016). Organisation and functions of mGlu and GABA(B) receptor complexes. *Nature, 540*(7631), 60–68.

Pinel, P., Fauchereau, F., Moreno, A., Barbot, A., et al. (2012). Genetic variants of FOXP2 and KIAA0319/TTRAP/THEM2 locus are associated with altered brain activation in distinct language-related regions. *Journal of Neuroscience, 32*(3), 817–825.

Pines, J. (1992). Cell proliferation and control. *Current Opinion in Cell Biology, 4*(2), 144–148.

Pinker, S., and Jackendoff, R. (2005). The faculty of language: What's special about it? *Cognition, 95*, 201–236.

Pirastu, N., Kooyman, M., Traglia, M., Robino, A., et al. (2016). Genome-Wide Association Study in isolated populations reveals new genes associated to common food likings. *Reviews in Endocrine and Metabolic Disorders, 17*(2), 209–219.

Pitman, R. K., Sanders, K. M., Zusman, R. M., Healy, A. R., et al. (2002). Pilot study of secondary prevention of posttraumatic stress disorder with propranolol. *Biological Psychiatry, 51*, 189–192.

Pitts, M. A., Padwal, J., Fennelly, D., Martínez, A., et al. (2014). Gamma band activity and the P3 reflect post-perceptual processes, not visual awareness. *NeuroImage, 101*, 337–350.

Place, N. J., Coscia, E. M., Dahl, N. J., Drea, C. M., et al. (2011). The anti-androgen combination, flutamide plus finasteride, paradoxically suppressed LH and androgen concentrations in pregnant spotted hyenas, but not in males. *General and Comparative Endocrinology, 170*(3), 455–459.

Pletnikov, M. V., Ayhan, Y., Nikolskaia, O., Xu, Y., et al. (2008). Inducible expression of mutant human DISC1 in mice is associated with brain and behavioural abnormalities reminiscent of schizophrenia. *Molecular Psychiatry, 13*, 13–186.

Ploog, D. W. (1992). Neuroethological perspectives on the human brain: From the expression of emotions to intentional signing and speech. In A. Harrington (Ed.), *So human a brain: Knowledge and values in the neurosciences* (pp. 3–13). Boston: Birkhauser.

Plutchik, R. (1994). *The psychology and biology of emotion.* New York: HarperCollins.

Poldrack, R. A., Baker, C. I., Durnez, J., Gorgolewski, K. J., et al. (2017). Scanning the horizon: Towards transparent and reproducible neuroimaging research. *Nature Reviews Neuroscience, 18*(2), 115–126.

Polleux, F., Morrow, T., and Ghosh, A. (2000). Semaphorin 3A is a chemoattractant for cortical apical dendrites. *Nature, 404*, 567–573.

Pompili, M., Serafini, G., Innamorati, M., Möller-Leimkühler, A. M., et al. (2010). The hypothalamic-pituitary-adrenal axis and serotonin abnormalities: A selective overview for the implications of suicide prevention. *European Archives of Psychiatry and Clinical Neuroscience, 260*, 583–600.

Ponsford, J. (2005). Rehabilitation interventions after mild head injury. *Current Opinion in Neurology, 18*, 692–697.

Poole, J. H., Tyack, P. L., Stoeger-Horwath, A. S., and Watwood, S. (2005). Animal behaviour: Elephants are capable of vocal learning. *Nature, 434*, 455–456.

Poremba, A., Malloy, M., Saunders, R. C., Carson, R. E., et al. (2004). Species-specific calls evoke asymmetric activity in the monkey's temporal poles. *Nature, 427*, 448–451.

Poritsky, R. (1969). Two and three dimensional ultrastructure of boutons and glial cells on the motoneuronal surface in the cat spinal cord. *Journal of Comparative Neurology, 135*, 423–452.

Porta, M., Brambilla, A., Cavanna, A. E., Servello, D., et al. (2009). Thalamic deep brain stimulation for treatment-refractory Tourette syndrome: Two-year outcome. *Neurology, 73*, 1375–1380.

Porter, J., Craven, B., Khan, R. H., Chang, S.-J., et al. (2007). Mechanisms of scent-tracking in humans. *Nature Neuroscience, 10*, 27–29.

Posner, M. I. (1980). Orienting of attention. *Quarterly Journal of Experimental Psychology, 32*, 3–25.

Posner, M. I., and Cohen, Y. (1984). Components of visual orienting. In H. Bouma and D. Bowhuis (Eds.), *Attention and performance: Vol 10. Control of language processes* (pp. 531–556). Hillsdale, NJ: Erlbaum.

Posner, M. I., and Raichle, M. E. (1994). *Images of mind.* New York: Scientific American Library.

Posthuma, D., De Geus, E. J. C., Baaré, W. F. C., Hulshoff Pol, H. E., et al. (2002). The association between brain volume and intelligence is of genetic origin. *Nature Neuroscience, 5*, 83–84.

Postuma, R. B., Gagnon, J. F., Vendette, M., Fantini, M. L., et al. (2009). Quantifying the risk of neurodegenerative disease in idiopathic REM sleep behaviour disorder. *Neurology, 72*, 1296–1300.

Poulet, J. F. A., and Petersen, C. C. H. (2008). Internal brain state regulates membrane potential synchrony in barrel cortex of behaving mice. *Nature, 454*, 881–885.

Poulos, A. M., and Thompson, R. F. (2015). Localisation and characterisation of an essential associative memory trace in the mammalian brain. *Brain Research, 1621*, 252–259.

Powell, L. E., Isler, K., and Barton, R. A. (2017). Re-evaluating the link between brain size and behavioural ecology in primates. *Proceedings: Biological Sciences, 284*(1865), pii: 20171765.

Powell, S. B. (2010). Models of neurodevelopmental abnormalities in schizophrenia. *Current Topics in Behavioural Neurosciences, 4*, 435–481.

Power, R. A., Steinberg, S., Bjornsdottir, G., Rietveld, C. A., et al. (2015). Polygenic risk scores for schizophrenia and bipolar disorder predict creativity. *Nature Neuroscience, 18*, 953–955.

Powers, R. J., Roy, S., Atilgan, E., Brownell, W. E., et al. (2012). Stereocilia membrane deformation: Implications for the gating spring and mechanotransduction channel. *Biophysical Journal, 102*(2), 201–210.

Prehn, K., Jumpertz von Schwartzenberg, R., Mai, K., Zeitz, U., et al. (2017). Caloric restriction in older adults-differential effects of weight loss and reduced weight on brain structure and function. *Cerebral Cortex, 27*(3), 1765–1778.

Premack, D. (1971). Language in a chimpanzee? *Science, 172*, 808–822.

Prendergast, B. J., Onishi, K. G., and Zucker, I. (2014). Female mice liberated for inclusion in neuroscience and biomedical research. *Neuroscience and Biobehavioural Reviews, 40*, 1–5.

Price, J. (2008). *The woman who can't forget*. New York: Free Press.

Price, J., Kerr, R., Hicks, M., and Nixon, P. F. (1987). The Wernicke-Korsakoff syndrome: A reappraisal in Queensland with special reference to prevention. *Medical Journal of Australia, 147*(11–12), 561–565.

Proske, U., and Gandevia, S. (2016). Proprioception: The sense within. *Scientist, 16*(9), 37–42.

Prüfer, K., Munch, K., Hellmann, I., Akagi, K., et al. (2012). The bonobo genome compared with the chimpanzee and human genomes. *Nature, 486*(7404), 527–531.

Pruszynski, J. A., and Johansson, R. S. (2014). Edge-orientation processing in first-order tactile neurons. *Nature Neuroscience, 17*(10), 1404–1409.

Przedborski, S. (2017). The two-century journey of Parkinson disease research. *Nature Reviews Neuroscience, 18*(4), 251–259.

Ptáček, L. J. (2015). Episodic disorders: Channelopathies and beyond. *Annual Review of Physiology, 77*, 475–479.

Pugh, E. N., Jr., and Lamb, T. D. (1990). Cyclic GMP and calcium: The internal messengers of excitation and adaptation in vertebrate photoreceptors. *Vision Research, 30*, 1923–1948.

Pugh, E. N., Jr., and Lamb, T. D. (1993). Amplification and kinetics of the activation steps in phototransduction. *Biochimica et Biophysica Acta, 1141*, 111–149.

Pugh, K. R., Mencl, W. E., Shaywitz, B. A., Shaywitz, S. E., et al. (2000). The angular gyrus in developmental dyslexia: Task-specific differences in functional connectivity within posterior cortex. *Psychological Science, 11*, 51–56.

Puighermanal, E., Busquets-Garcia, A., Maldonado, R., and Ozaita, A. (2012). Cellular and intracellular mechanisms involved in the cognitive impairment of cannabinoids. *Philosophical Transactions of the Royal Society of London. Series B: Biological Sciences, 367*, 3254–3263.

Pulvermüller, F., and Fadiga, L. (2010). Active perception: Sensorimotor circuits as a cortical basis for language. *Nature Reviews Neuroscience, 11*(5), 351–360.

Purves, D., and Lotto, R. B. (2010). *Why we see what we do redux: A wholly empirical theory of vision.* Sunderland, MA: Sinauer.

Putman, C. T., Xu, X., Gillies, E., Maclean, I. M., et al. (2004). Effects of strength, endurance and combined training on myosin heavy chain content and fibre-type distribution in humans. *European Journal of Applied Physiology, 92*, 376–384.

Puts, D. A., McDaniel, M. A., Jordan, C. L., and Breedlove, S. M. (2008). Spatial ability and prenatal androgens: Meta-analyses of congenital adrenal hyperplasia and digit ratio (2D:4D) studies. *Archives of Sexual Behaviour, 37*(1), 100–111.

Q

Quraishi, I. H., et al. (2017). Impairment of consciousness induced by bilateral electrical stimulation of the frontal convexity. *Epilepsy & Behaviour Case Reports, 8*, 117–122.

R

Racette, A., Bard, C., and Peretz, I. (2006). Making non-fluent aphasics speak: Sing along! *Brain, 129*, 2571–2584.

Racine, E., Bar-Ilan, O., and Illes, J. (2005). fMRI in the public eye. *Nature Reviews Neuroscience, 6*, 159–164.

Radley, J. J., and Sawchenko, P. E. (2011). A common substrate for prefrontal and hippocampal inhibition of the neuroendocrine stress response. *Journal of Neuroscience, 31*, 9683–9695.

Rafal, R. D. (1994). Neglect. *Current Opinion in Neurobiology, 4*, 231–236.

Rahman, Q. (2005). The neurodevelopment of human sexual orientation. *Neuroscience and Biobehavioural Reviews, 29*, 1057–1066.

Raichle, M. E. (2015). The brain's default mode network. *Annual Review of Neuroscience, 38*, 433–447.

Rainville, P., Duncan, G. H., Price, D. D., Carrier, B., et al. (1997). Pain affect encoded in human anterior cingulate but not somatosensory cortex. *Science, 277*, 968–971.

Raisman, G., and Field, P. M. (1971). Sexual dimorphism in the preoptic area of the rat. *Science, 173*, 731–733.

Rakel, R. E. (2009). Clinical and societal consequences of obstructive sleep apnea and excessive daytime sleepiness. *Postgraduate Medicine, 121*, 86–95.

Rakic, P. (1971). Guidance of neurons migrating to the fetal monkey neocortex. *Brain Research, 33*, 471–476.

Rakic, P. (1985). Mechanisms of neuronal migration in developing cerebellar cortex. In G. M. Edelman, W. M. Cowan, and E. Gull (Eds.), *Molecular basis of neural development* (pp. 139–160). New York: Wiley.

Ralph, M. R., and Menaker, M. (1988). A mutation of the circadian system in golden hamsters. *Science, 241*, 1225–1227.

Ralph, M. R., Foster, R. G., Davis, F. C., and Menaker, M. (1990). Transplanted suprachiasmatic nucleus determines circadian period. *Science, 247*, 975–978.

Ramachandran, V. S., and Hubbard, E. M. (2001). Psychophysical investigations into the neural basis of synthaesthesia. *Proceedings of the Royal Society of London. Series B: Biological Sciences, 268*, 979–983.

Ramachandran, V. S., and Rogers-Ramachandran D. (2000). Phantom limbs and neural plasticity. *Archives of Neurology, 57*, 317–320.

Ramon, F., Hernandex-Falcon, J., Nguyen, B., and Bullock, T. H. (2004). Slow wave sleep in crayfish. *Proceedings of the National Academy of Sciences, USA, 101*, 11857–11861.

Rampon, C., Tang, Y. P., Goodhouse, J., Shimizu, E., et al. (2000). Enrichment induces structural changes and recovery from nonspatial memory deficits in CA1 NMDAR1-knockout mice. *Nature Neuroscience, 3*, 238–244.

Ranade, S. S., Woo, S. H., Dubin, A. E., Moshourab, R. A., et al. (2014). Piezo2 is the major transducer of mechanical forces for touch sensation in mice. *Nature, 516*(7529), 121–125.

Rand, M. N., and Breedlove, S. M. (1987). Ontogeny of functional innervation of bulbocavernosus muscles in male and female rats. *Brain Research, 430*, 150–152.

Rathelot, J. A., and Strick, P. L. (2006). Muscle representation in the macaque motor cortex: An anatomical perspective. *Proceedings of the National Academy of Sciences, USA, 103*, 8257–8262.

Rattenborg, N. C. (2006). Do birds sleep in flight? *Naturwissenschaften, 93*, 413–425.

Rauch, S. L., Shin, L. M., and Wright, C. I. (2003). Neuroimaging studies of amygdala function in anxiety disorders. *Annals of the New York Academy of Sciences, 985*, 389–410.

Raz, N. (2000). Aging of the brain and its impact on cognitive performance: Integration of structural and functional findings. In F. I. M. Craik and T. A. Salthouse (Eds.), *The handbook of aging and cognition* (2nd ed., pp. 1–90). Mahwah, NJ: Erlbaum.

Reader, S. M., and Laland, K. N. (2002). Social intelligence, innovation, and enhanced brain size in primates. *Proceedings of the National Academy of Sciences, USA, 99*, 4436–4441.

Recanzone, G. (2018). The effects of aging on auditory cortical function. *Hearing Research, 366*, 99–105.

Rechtschaffen, A., and Bergmann, B. M. (1995). Sleep deprivation in the rat by the disk-over-water method. *Behavioural Brain Research, 69*, 55–63.

Rechtschaffen, A., and Kales, A. (1968). *A manual of standardized terminology, techniques and scoring system for sleep stages of human subjects.* Bethesda, MD: U.S. National Institute of Neurological Diseases and Blindness, Neurological Information Network.

Redcay, E., Dodell-Feder, D., Pearrow, M. J., Mavros, P. L., et al. (2010). Live face-to-face interaction during fMRI: A new tool for social cognitive neuroscience. *NeuroImage, 50*(4), 1639–1647.

Redican, W. K. (1982). An evolutionary perspective on human facial displays. In P. Ekman (Ed.), *Emotion in the human face* (2nd ed., pp. 212–280). Elmsford, NY: Pergamon.

Reed, L. I., DeScioli, P., and Pinker S. A. (2014). The commitment function of angry facial expressions. *Psychological Science, 25*, 1511–1517.

Rees, G., Kreiman, G., and Koch, C. (2002). Neural correlates of consciousness in humans. *Nature Reviews Neuroscience, 3*, 261–270.

Rehkamper, G., Haase, E., and Frahm, H. D. (1988). Allometric comparison of brain weight and brain structure volumes in different breeds of the domestic pigeon, *Columba livia* f. d. (fantails, homing pigeons, strassers). *Brain, Behaviour and Evolution, 31*, 141–149.

Reilly, M. T., Noronha, A., Goldman, D., and Koob, G. F. (2017). Genetic studies of alcohol dependence in the context of the addiction cycle. *Neuropharmacology, 122*, 3–21.

Reiner, W. G., and Gearhart, J. P. (2004). Discordant sexual identity in some genetic males with cloacal exstrophy assigned to female sex at birth. *New England Journal of Medicine, 350*, 333–341.

Reisberg, D., and Heuer, F. (1995). Emotion's multiple effects on memory. In J. L. McGaugh, N. M. Weinberger, and G. Lynch (Eds.), *Brain and memory: Modulation and mediation of neuroplasticity* (pp. 84–92). New York: Oxford University Press.

Remedios, R., Kennedy, A., Zelikowsky, M., Grewe, B. F., et al. (2017). Social behaviour shapes hypothalamic neural ensemble representations of conspecific sex. *Nature, 550*(7676), 388–392.

Rempel-Clower, N. L., Zola, S. M., Squire, L. R., and Amaral, D. G. (1996). Three cases of enduring memory impairment after bilateral damage limited to the hippocampal formation. *Journal of Neuroscience, 16*, 5233–5255.

Renner, M. J., and Rosenzweig, M. R. (1987). *Enriched and impoverished environments: Effects on brain and behaviour.* New York: Springer.

Reppert, S. M., Perlow, M. J., Tamarkin, L., and Klein, D. C. (1979). A diurnal melatonin rhythm in primate cerebrospinal fluid. *Endocrinology, 104*, 295–301.

Revel, F. G., Masson-Pévet, M., Pévet, P., Mikkelsen, J. D., et al. (2009). Melatonin controls seasonal breeding by a network of hypothalamic targets. *Neuroendocrinology, 90*, 1–14.

Rex, C. S., Lauterborn, J. C., Lin, C.Y., Kramár, E. A., et al. (2006). Restoration of long-term potentiation in middle-aged hippocampus after induction of brain-derived neurotrophic factor. *Journal of Neurophysiology, 96*, 677–685.

Rezaie, L., Fobian, A. D., McCall, W.V., and Khazaie, H. (2018). Paradoxical insomnia and subjective-objective sleep discrepancy: A review. *Sleep Medicine Reviews, 40*, 196–202.

Rezlescu, C., Barton, J. J., Pitcher, D., and Duchaine, B. (2014). Normal acquisition of expertise with greebles in two cases of acquired prosopagnosia. *Proceedings of the National Academy of Sciences, USA, 111*(14), 5123–5128.

Rhodes, M. G. (2004). Age-related differences in performance on the Wisconsin card sorting test: A meta-analytic review. *Psychology of Aging, 19*, 482–494.

Ricciardi, E., Pietrini, P., Schapiro, M. B., Rapoport, S. I., et al. (2009). Cholinergic modulation of visual working memory during aging: A parametric PET study. *Brain Research Bulletin, 79*, 322–332.

Richter, C. (1967). Sleep and activity: Their relation to the 24-hour clock. *Proceedings of the Association for Research in Nervous and Mental Diseases, 45*, 8–27.

Rickert, E., Fernandez, M. O., Choi, I., Gorman, M., et al. (2018). Neuronal SIRT1 regulates metabolic and reproductive function and the response to caloric restriction. *Journal of the Endocrine Society, 3*(2), 427–445.

Ridley, M. (2003). *Nature via nurture: Genes, experience, and what makes us human.* New York: HarperCollins.

Ringman, J. M., and Cummings, J. L. (2006). Current and emerging pharmacological treatment options for dementia. *Behavioural Neurology, 17*, 5–16.

Risch, N., Herrell, R., Lehner, T., Liang, K.Y., et al. (2009). Interaction between the serotonin transporter gene (5-HTTLPR), stressful life events, and risk of depression: A meta-analysis. *JAMA, 301*, 2462–2471.

Ritchie, H., and Roser, M. (2019). Mental Health, Our World in Data, https://ourworldindata.org/.

Rizzolatti, G., and Craighero, L. (2004). The mirror-neuron system. *Annual Review of Neuroscience, 27*, 169–192.

Rizzolatti, G., Fogassi, L., and Gallese, V. (2006). Mirrors of the mind. *Scientific American, 295*(5), 54–61.

Rizzolatti, G., and Sinigaglia, C. (2016). The mirror mechanism: A basic principle of brain function. *Nature Reviews Neuroscience, 17*(12), 757–765.

Robel, S., and Sontheimer, H. (2015). Glia as drivers of abnormal neuronal activity. *Nature Neuroscience, 19*(1), 28–33.

Roberto, M., and Varodayan, F. P. (2017). Synaptic targets: Chronic alcohol actions. *Neuropharmacology, 122*, 85–99.

Roberts, E. K., Lu, A., Bergman, T. J., and Beehner, J. C. (2012). A Bruce effect in wild geladas. *Science, 335*(6073), 1222–1225.

Robertson, C. E., and Baron-Cohen, S. (2017). Sensory perception in autism. *Nature Reviews Neuroscience, 18*(11), 671–684. https://doi.org/10.1038/nrn.2017.112.

Robertson, D. J., Noyes, E., Dowsett, A. J., Jenkins, R., et al. (2016). Face recognition by metropolitan police super-recognisers. *PLOS ONE, 11*(2), e0150036.

Robino, A., Mezzavilla, M., Pirastu, N., Dognini, M., et al. (2014). A population-based approach to study the impact of PROP perception on food liking in populations along the Silk Road. *PLOS ONE, 9*(3), e91716.

Robins, L. N., and Regier, D. A. (1991). *Psychiatric disorders in America: The epidemiologic catchment area study.* New York: Free Press.

Robins, L. N., and Slobodyan, S. (2003). Post-Vietnam heroin use and injection by returning US veterans: Clues to preventing injection today. *Addiction, 98*, 1053–1060.

Robinson, D. L., and Petersen, S. E. (1992). The pulvinar and visual salience. *Trends in Neuroscience, 15*, 127–132.

Roenneberg, T. (2013). Chronobiology: The human sleep project. *Nature, 498*(7455), 427–428.

Roenneberg, T., Kuehnle, T., Pramstaller, P. P., Ricken, J., et al. (2004). A marker for the end of adolescence. *Current Biology, 14*, R1038–R1039.

Roffwarg, H. P., Muzio, J. N., and Dement, W. C. (1966). Ontogenetic development of the human sleep-dream cycle. *Science, 152*, 604–619.

Rogan, M. T., Staubli, U. V., and LeDoux, J. E. (1997). Fear conditioning induces associative long-term potentiation in the amygdala. *Nature, 390*, 604–607.

Roh, J. I., Lee, J., Park, S. U., Kang, Y. S., et al. (2018). CRISPR-Cas9-mediated generation of obese and diabetic mouse models. *Experimental Animals, 67*(2), 229–237.

Roland, P. E. (1984). Metabolic measurements of the working frontal cortex in man. *Trends in Neurosciences, 7*, 430–436.

Roland, P. E. (1993). *Brain activation.* New York: Wiley-Liss.

Rolls, E. T., and O'Mara, S. M. (1995). View-responsive neurons in the primate hippocampal complex. *Hippocampus, 5*, 409–424.

Romanski, L. M., Tian, B., Fritz, J., Mishkin, M., et al. (1999). Dual streams of auditory afferents target multiple domains in the primate prefrontal cortex. *Nature Neuroscience, 2*, 1131–1136.

Rorabaugh, W. J. (1976). Estimated U.S. alcoholic beverage consumption, 1790–1860. *Journal of Studies on Alcohol, 37*, 357–364.

Rosci, C., Chiesa, V., Laiacona, M., and Capitani, E. (2003). Apraxia is not associated to a disproportionate naming impairment for manipulable objects. *Brain and Cognition, 53*, 412–415.

Rose, K. A., Morgan, I. G., Smith, W., Burlutsky, G., et al. (2008). Myopia, lifestyle, and schooling in students of Chinese ethnicity in Singapore and Sydney. *Archives of Ophthalmology, 126*(4), 527–530.

Rosell, D. R., and Siever, L. J. (2015). The neurobiology of aggression and violence. *CNS Spectrums, 20*(3), 254–279.

Roselli, C. E., and Stormshak, F. (2009). The neurobiology of sexual partner preferences in rams. *Hormones and Behaviour, 55*, 611–620.

Roselli, C. E., Larkin, K., Resko, J. A., Stellflug, J. N., et al. (2004). The volume of a sexually dimorphic nucleus in the ovine medial preoptic area/anterior hypothalamus varies with sexual partner preference. *Endocrinology, 145*, 475–477.

Rosenbaum, R. S., Köhler, S., Schacter, D. L., Moscovitch, M., et al. (2005). The case of K.C.: Contributions of a memory-impaired person to memory theory. *Neuropsychologia, 43*, 989–1021.

Rosenberg, M. D., Finn, E. S., Scheinost, D., Constable, R. T., et al. (2017). Characterizing attention with predictive network models. *Trends in Cognitive Sciences, 21*(4), 290–302.

Rosenberg, M. D., Finn, E. S., Scheinost, D., Papademetris, X., et al. (2016). A neuromarker of sustained attention from whole-brain functional connectivity. *Nature Neuroscience, 19*(1), 165–171.

Rosenblat, J. D., Lee, Y., and McIntyre, R. S. (2018). The effect of pharmacogenomic testing on response and remission rates in the acute treatment of major depressive disorder: A meta-analysis. *Journal of Affective Disorders, 241*, 484–491.

Rosengren, A., Tibblin, G., and Wilhelmsen, L. (1991). Self-perceived psychological stress and incidence of coronary artery disease in middle-aged men. *American Journal of Cardiology, 68*, 1171–1175.

Rosenkranz, M. A., Jackson, D. C., Dalton, K. M., Dolski, I., et al. (2003). Affective style and in vivo immune response: Neurobehavioural mechanisms. *Proceedings of the National Academy of Sciences, USA, 100*, 11148–11152.

Rosenzweig, M. R. (1946). Discrimination of auditory intensities in the cat. *American Journal of Psychology, 59*, 127–136.

Rosenzweig, M. R. (1984). Experience, memory, and the brain. *American Psychologist, 39*, 365–376.

Rosenzweig, M. R., Bennett, E. L., Colombo, P. J., Lee, D. W., et al. (1993). Short-term, intermediate-term, and long-term memory. *Behavioural Brain Research, 57*, 193–198.

Rosenzweig, M. R., Bennett, E. L., Martinez, J. L., Colombo, P. J., et al. (1992). Studying stages of memory formation with chicks. In L. R. Squire and N. Butters (Eds.), *Neuropsychology of memory* (2nd ed., pp. 533–546). New York: Guilford.

Rosenzweig, M. R., Krech, D., and Bennett, E. L. (1961). Heredity, environment, brain biochemistry, and learning. In *Current trends in psychological theory* (pp. 87–110). Pittsburgh, PA: University of Pittsburgh Press.

Roses, A. D. (1995). On the metabolism of apolipoprotein E and the Alzheimer diseases. *Experimental Neurology, 132*, 149–156.

Roth, B. L. (2016). DREADDs for neuroscientists. *Neuron, 89*(4), 683–694.

Roth, T. (2007). Insomnia: Definition, prevalence, etiology, and consequences. *Journal of Clinical Sleep Medicine, 3*(5 Suppl.), S7–S10.

Roth, T. C., II, Brodin, A., Smulders, T. V., LaDage, L. D., et al. (2010). Is bigger always better? A critical appraisal of the use of volumetric analysis in the study of the hippocampus. *Philosophical Transactions of the Royal Society of London. Series B: Biological Sciences, 365*(1542), 915–931.

Roth, T. L., Nayak, D., Atanasijevic, T., Koretsky, A. P., et al. (2014). Transcranial amelioration of inflammation and cell death after brain injury. *Nature, 505*(7482), 223–228.

Rouw, R., and Scholte, H. S. (2007). Increased structural connectivity in grapheme-colour synesthesia. *Nature Neuroscience, 10*, 792–797.

Row, B. W., Kheirandish, L., Cheng, Y., Rowell, P.P., et al. (2007). Impaired spatial working memory and altered choline acetyltransferase (CHAT) immunoreactivity and nicotinic receptor binding in rats exposed to intermittent hypoxia during sleep. *Behavioural Brain Research, 177*, 308–314.

Rowe, A. H., Xiao, Y., Rowe, M. P., Cummins, T. R., et al. (2013). Voltage-gated sodium channel in grasshopper mice defends against bark scorpion toxin. *Science, 342*(6157), 441–446.

Rowe, M. K., and Chuang, D. M. (2004). Lithium neuroprotection: Molecular mechanisms and clinical implications. *Expert Reviews in Molecular Medicine, 18*, 1–18.

Rozanski, A. (2014). Behavioural cardiology: Current advances and future directions. *Journal of the American College of Cardiology, 64*, 100–110.

Ru, F., Sun, H., Jurcakova, D., Herbstsomer, R. A., et al. (2017). Mechanisms of pruritogen-induced activation of itch nerves in isolated mouse skin. *Journal of Physiology 595*(11), 3651–3666. doi: 10.1113/JP273795.

Rubel, E. W., Furrer, S. A., and Stone, J. S. (2013). A brief history of hair cell regeneration research and speculations on the future. *Hearing Research, 297*, 42–51.

Rück, C., Karlsson, A., Steele, J. D., Edman, G., et al. (2008). Capsulotomy for obsessive-compulsive disorder: Long-term follow-up of 25 patients. *Archives of General Psychiatry, 65*, 914–921.

Rueda-Orozco, P. E., and Robbe, D. (2015). The striatum multiplexes contextual and kinematic information to constrain motor habits execution. *Nature Neuroscience, 18*(3), 453–460.

Ruff, C. C., and Fehr, E. (2014). The neurobiology of rewards and values in social decision making. *Nature Reviews Neuroscience, 15*(8), 549–562.

Rumbaugh, D. M. (1977). *Language learning by a chimpanzee: The LANA project.* New York: Academic Press.

Rupnick, M. A., Panigrahy, D., Zhang, C.Y., Dallabrida, S. M., et al. (2002). Adipose tissue mass can be regulated through the vasculature. *Proceedings of the National Academy of Sciences, USA, 99*, 10730–10735.

Rupprecht, R., Rammes, G., Eser, D., Baghai, T. C., et al. (2009). Translocator protein (18 kD) as target for anxiolytics without benzodiazepine-like side effects. *Science, 325*, 490–493.

Rusak, B., and Zucker, I. (1979). Neural regulation of circadian rhythms. *Physiological Reviews, 59*, 449–526.

Rusconi, E., and Mitchener-Nissen, T. (2013). Prospects of functional magnetic resonance imaging as lie detector. *Frontiers in Human Neuroscience, 7*, 594.

Russell, J. A. (1994). Is there universal recognition of emotion from facial expressions? A review of the cross-cultural studies. *Psychological Bulletin, 115*, 102–141.

Russell, R., Duchaine, B., and Nakayama, K. (2009). Super-recognizers: People with extraordinary face recognition ability. *Psychological Bulletin Review, 16*(2): 252–257.

Russo, E. B., Jiang, H. E., Li, X., Sutton, A., et al. (2008). Phytochemical and genetic analyses of ancient cannabis from Central Asia. *Journal of Experimental Botany, 59*, 4171–4182.

Ruthazer, E. S., Akerman, C. J., and Cline, H. T. (2003). Control of axon branch dynamics by correlated activity in vivo. *Science, 301*, 66–70.

Rymer, R. (1993). *Genie: An abused child's flight from silence*. New York: HarperCollins.

Rypma, B., Berger, J. S., Genova, H. M., Rebbechi, D., et al. (2005). Dissociating age-related changes in cognitive strategy and neural efficiency using event-related fMRI. *Cortex, 41*, 582–594.

S

Saab, A. S., Neumeyer, A., Jahn, H. M., Cupido, A., et al. (2012). Bergman glial AMPA receptors are required for fine motor coordination. *Science, 337*, 749–753.

Saalmann, Y. B., Pinsk, M. A., Wang, L., Li, X., et al. (2012). The pulvinar regulates information transmission between cortical areas based on attention demands. *Science, 337*(6095), 753–756.

Sáenz de Miera, C., Monecke, S., Bartzen-Sprauer, J., Laran-Chich, M. P., et al. (2014). A circannual clock drives expression of genes central for seasonal reproduction. *Current Biology, 24*(13), 1500–1506.

Saey, T. H. (2018). Same-sex sexuality linked to DNA. *Science News, 194*(9), 10.

Saha, K. B., Bo, L., Zhao, S., Xia, J., et al. (2016). Chlorpromazine versus atypical antipsychotic drugs for schizophrenia. *Cochrane Database of Systematic Reviews, 4*, CD010631.

Sahay, A., and Hen, R. (2007). Adult hippocampal neurogenesis in depression. *Nature Neuroscience, 10*, 1110–1114.

Sahay, A., Scobie, K. N., Hill, A. S., O'Carroll, C. M., et al. (2011). Increasing adult hippocampal neurogenesis is sufficient to improve pattern separation. *Nature, 472*, 466–470.

Sahbaie, P., Shi, X., Guo, T.-Z., Qiao, Y., et al. (2010). Role of substance P signalling in enhanced nociceptive sensitisation and local cytokine production after incision. *Pain, 145*(3), 341–349.

Saj, A., Fuhrman, O., Vuilleumier, P., and Boroditsky, L. (2014). Patients with left spatial neglect also neglect the "left side" of time. *Psychological Science, 25*(1), 207–214.

Sakmann, B. (2017). From single cells and single columns to cortical networks: Dendritic excitability, coincidence detection and synaptic transmission in brain slices and brains. *Experimental Physiology, 102*(5), 489–521.

Salazar, H., Llorente, I., Jara-Oseguera, A., García-Villegas, R., et al. (2008). A single N-terminal cysteine in TRPV1 determines activation by pungent compounds from onion and garlic. *Nature Neuroscience, 11*, 255–260.

Salimpoor, V. N., Zald, D. H., Zatorre, R. J., Dagher, A., et al. (2015). Predictions and the brain: How musical sounds become rewarding. *Trends in Cognitive Sciences, 19*(2), 86–91.

Salvini-Plawen, L. V., and Mayr, E. (1977). On the evolution of photoreceptors and eyes. *Evolutionary Biology, 10*, 207–263.

Samad, T. A., Moore, K. A., Sapirstein, A., Billet, S., et al. (2001). Interleukin-1β-mediated induction of Cox-2 in the CNS contributes to inflammatory pain hypersensitivity. *Nature, 410*, 471–475.

Samaha, F. F., Iqbal, N., Seshadri, P., Chicano, K. L., et al. (2003). A low-carbohydrate as compared with a low-fat diet in severe obesity. *New England Journal of Medicine, 348*, 2074–2081.

Sample, I. (2019, March 27). Scientists find genetic mutation that makes woman feel no pain. *The Guardian* (https://www.theguardian.com/science/2019/mar/28/scientists-find-genetic-mutation-that-makes-woman-feel-no-pain).

Samson, D. R., Crittenden, A. N., Mabulla, I. A., Mabulla, A. Z. P., et al. (2017). Chronotype variation drives night-time sentinel-like behaviour in hunter-gatherers. *Proceedings: Biological Sciences, 284*(1858), pii: 20170967.

Samson, S., and Zatorre, R. J. (1991). Recognition memory for text and melody of songs after unilateral temporal lobe lesion: Evidence for dual encoding. *Journal of Experimental Psychology. Learning, Memory, and Cognition, 17*, 793–804.

Samson, S., and Zatorre, R. J. (1994). Contribution of the right temporal lobe to musical timbre discrimination. *Neuropsychologia, 32*, 231–240.

Samuels, B. A., Anacker, C., Hu, A., Levinstein, M. R., et al. (2015). 5-HT1A receptors on mature dentate gyrus granule cells are critical for the antidepressant response. *Nature Neuroscience, 18*, 1606–1616.

Sanacora, G., Frye, M. A., McDonald, W., Mathew, S. J., et al.; American Psychiatric Association (APA) Council of Research Task Force on Novel Biomarkers and Treatments. (2017). A consensus statement on the use of ketamine in the treatment of mood disorders. *JAMA Psychiatry, 74*(4), 399–405.

Sánchez-Tena, M. Á., Alvarez-Peregrina, C., Valbuena-Iglesias, M. C., and Palomera, P. R. (2018). Optical illusions and spatial disorientation in aviation pilots. *Journal of Medical Systems, 42*(5), 79.

Sandberg, K., Umans, J. G.; Georgetown Consensus Conference Work Group. (2015). Recommendations concerning the new U.S. National Institutes of Health initiative to balance the sex of cells and animals in preclinical research. *FASEB Journal, 29*(5), 1646–1652.

Sanders, A. R., Martin, E. R., Beecham, G. W., Guo, S., et al. (2015). Genome-wide scan demonstrates significant linkage for male sexual orientation. *Psychological Medicine, 45*(7), 1379–1388.

Sanderson, D. J., Good, M. A., Seeburg, P. H., Sprengel, R., et al. (2008). The role of the GluR-A (GluR1) AMPA receptor subunit in learning and memory. *Progress in Brain Research, 169*, 159–178.

Sanes, J. N., and Donoghue, J. P. (2000). Plasticity and primary motor cortex. *Annual Review of Neuroscience, 23*, 393–415.

Sanes, J. R., and Lichtman, J. W. (1999). Can molecules explain long-term potentiation? *Nature Neuroscience, 2*, 597–604.

Sankaraneni, R., and Lachhwani, D. (2015). Antiepileptic drugs—a review. *Pediatric Annals, 44*(2), e36–42. https://doi.org/10.3928/00904481-20150203-10.

Sansevero, G., Baroncelli, L., Scali, M., and Sale, A. (2019). Intranasal BDNF administration promotes visual function recovery in adult amblyopic rats. *Neuropharmacology, 145*(Pt. A), 114–122.

Santhanam, F., Ryu, S. I., Yu, B. M., Afshar, A., et al. (2006) A high-performance brain–computer interface. *Nature, 442*, 195–198.

Sapir, A., Soroker, N., Berger, A., and Henik, A. (1999). Inhibition of return in spatial attention: Direct evidence for collicular generation. *Nature Neuroscience, 2*, 1053–1054.

Sapolsky, R. M. (1992). Neuroendocrinology of the stress response. In J. B. Becker, S. M. Breedlove, and D. Crews (Eds.), *Behavioural endocrinology* (pp. 287–324). Cambridge, MA: MIT Press.

Sapolsky, R. M. (1993). Potential behavioural modification of glucocorticoid damage to the hippocampus [Special issue: Alzheimer's disease: Animal models and clinical perspectives]. *Behavioural Brain Research, 57*, 175–182.

Sapolsky, R. M. (2004). *Why zebras don't get ulcers* (3rd ed.). New York: Holt.

Sartorius, N., Shapiro, R., Kimura, M., and Barrett, K. (1972). WHO international pilot study of schizophrenia. *Psychological Medicine, 2*, 422–425.

Satinoff, E., and Rutstein, J. (1970). Behavioural thermoregulation in rats with anterior hypothalamic lesions. *Journal of Comparative and Physiological Psychology, 71*, 77–82.

Sato, J. R., Salum, G. A., Gadelha, A., Crossley, N., et al. (2015). Default mode network maturation and psychopathology in children and adolescents. *Journal of Child Psychology and Psychiatry, 57*(1), 55–64.

Savage-Rumbaugh, E. S. (1993). *Language comprehension in ape and child.* Chicago: University of Chicago Press.

Savage-Rumbaugh, [E.] S., and Lewin, R. (1994). *Kanzi: The ape at the brink of the human mind.* New York: Wiley.

Sawin, C. T. (1996). Arnold Adolph Berthold (1803–1861). *Endocrinologist, 6*, 164–168.

Saxe, M. D., Battaglia, F., Wang, J. W., Malleret, G., et al. (2006). Ablation of hippocampal neurogenesis impairs contextual fear conditioning and synaptic plasticity in the dentate gyrus. *Proceedings of the National Academy of Sciences, USA, 103*, 17501–17506.

Saxena, S., Brody, A. L., Ho, M. L., Alborzian, S., et al. (2001). Cerebral metabolism in major depression and obsessive-compulsive disorder occurring separately and concurrently. *Biological Psychiatry, 50*, 159–170.

Scammell, T. E. (2015). Narcolepsy. *New England Journal of Medicine, 373*, 2654–2662.

Scarmeas, N., Anastasiou, C. A., and Yannakoulia, M. (2018). Nutrition and prevention of cognitive impairment. *Lancet Neurology, 17*(11), 1006–1015.

SCENIHR (Scientific Committee on Emerging and Newly Identified Health Risks). (2008, September 23). Scientific opinion on the potential health risks of exposure to noise from personal music players and mobile phones including a music playing function.

Schachter, S. (1975). Cognition and peripheralist-centralist controversies in motivation and emotion. In M. S. Gazzaniga and C. Blakemore (Eds.), *Handbook of psychobiology* (pp. 529–564). New York: Academic Press.

Schachter, S., and Singer, J. (1962). Cognitive, social, and physiological determinants of emotional state. *Psychological Review, 69*, 379–399.

Schacter, D. L., Alpert, N. M., Savage, C. R., Rauch, S. L., et al. (1996). Conscious recollection and the human hippocampal formation: Evidence from positron emission tomography. *Proceedings of the National Academy of Sciences, USA, 93*, 321–325.

Schafer, D. P., Lehrman, E. K., Kautzman, A. G., Koyama, R., et al. (2012). Microglia sculpt postnatal neural circuits in an activity and complement-dependent manner. *Neuron, 74*, 691–705.

Schaller, M., Miller, G. E., Gervais, W. M., Yager, S., et al. (2010). Mere visual perception of other people's disease symptoms facilitates a more aggressive immune response. *Psychological Science, 21*, 649–652.

Schapira, A. H. V., Chaudhuri, K. R., and Jenner, P. (2017). Non-motor features of Parkinson disease. *Nature Reviews Neuroscience, 18*(7), 435–450.

Schatzberg, A. F., Rothschild, A. J., Stahl, J. B., Bond, T. C., et al. (1983). The dexamethasone suppression test: Identification of subtypes of depression. *American Journal of Psychiatry, 140*(1): 88–91.

Scheibel, M. E., Tomiyasu, U., and Scheibel, A. B. (1977). The aging human Betz cells. *Experimental Neurology, 56*, 598–609.

Scheibert, J., Leurent, S., Prevost, A., and Debrégeas, G. (2009). The role of fingerprints in the coding of tactile information probed with a biomimetic sensor. *Science, 323*, 1503–1506.

Schein, S. J., and Desimone, R. (1990). Spectral properties of V4 neurons in the macaque. *Journal of Neuroscience, 10*, 3369–3389.

Schenck, C. H., and Mahowald, M. W. (2002). REM sleep behaviour disorder: Clinical, developmental, and neuroscience perspectives 16 years after its formal identification in *Sleep. Sleep, 25*, 120–138.

Schenkerberg, T., Bradford, D. C., and Ajax, E. T. (1980). Line bisection and unilateral visual neglect in patients with neurologic impairment. *Neurology, 30*, 509–518.

Schiffman, J., Walker, E., Ekstrom, M., Schulsinger, F., et al. (2004). Childhood videotaped social and neuromotor precursors of schizophrenia: A prospective investigation. *American Journal of Psychiatry, 161*(11), 2021–2027.

Schildkraut, J. J., and Kety, S. S. (1967). Biogenic amines and emotion. *Science, 156*, 21–30.

Schiller, D., Eichenbaum, H., Buffalo, E. A., Davachi, L., et al. (2015). Memory and space: Towards an understanding of the cognitive map. *Journal of Neuroscience, 35*(41), 13904–13911.

Schindler, E. A., Gottschalk, C. H., Weil, M. J., Shapiro, R. E., et al. (2015). Indoleamine hallucinogens in cluster headache: Results of the clusterbusters medication use survey. *Journal of Psychoactive Drugs, 47*, 372–381.

Schindler, I., Clavagnier, S., Karnath, H. O., Derex, L., et al. (2006). A common basis for visual and tactile exploration deficits in spatial neglect? *Neuropsychologia, 44*, 1444–1451.

Schlaug, G., Jancke, L., Huang, Y., and Steinmetz, H. (1995). In vivo evidence of structural brain asymmetry in musicians. *Science, 267*, 699–701.

Schlebusch, C. M., Malmström, H., Günther, T., Sjödin, P., et al. (2017). Southern African ancient genomes estimate modern human divergence to 350,000 to 260,000 years ago. *Science, 358*(6363), 652–655.

Schmidt-Nielsen, K. (1960). *Animal physiology.* Englewood Cliffs, NJ: Prentice-Hall.

Schnapf, J. L., and Baylor, D. A. (1987). How photoreceptor cells respond to light. *Scientific American, 256*(4), 40–47.

Schneeweiss, S., Patrick, A. R., Solomon, D. H., Dormuth, C. R., et al. (2010). Comparative safety of antidepressant agents for children and adolescents regarding suicidal acts. *Pediatrics, 125*(5), 876–888.

Schneider, D. M., Nelson, A., and Mooney, R. (2014). A synaptic and circuit basis for corollary discharge in the auditory cortex. *Nature, 513*(7517), 189–194.

Schneider, K. A., and Kastner, S. (2009). Effects of sustained spatial attention in the human lateral geniculate nucleus and superior colliculus. *Journal of Neuroscience, 29*, 1784–1795.

Schneider, P., Scherg, M., Dosch, H. G., Specht, H. J., et al. (2002). Morphology of Heschl's gyrus reflects enhanced activation in the auditory cortex of musicians. *Nature Neuroscience, 5*, 688–694.

Schneider, S. A., and Alcalay, R. N. (2017). Neuropathology of genetic synucleinopathies with parkinsonism: Review of the literature. *Movement Disorders, 32*(11), 1504–1523.

Schneps, M. H., Brockmole, J. R., Sonnert, G., and Pomplun, M. (2012). History of reading struggles linked to enhanced learning in low spatial frequency scenes. *PLOS ONE, 7*(4), e35724.

Schnupp, J. W., Dawe, K. L., and Pollack, G. L. (2005). The detection of multisensory stimuli in an orthogonal sensory space. *Experimental Brain Research, 162*, 181–190.

Schoenbaum, G., Roesch, M. R., Stalnaker, T. A., and Takahashi, Y. K. (2009). A new perspective on the role of the orbitofrontal cortex in adaptive behaviour. *Nature Reviews Neuroscience, 10*(12), 885–892.

Schramm, E., Schneider, D., Zobel, I., van Calker, D., et al. (2008). Efficacy of interpersonal psychotherapy plus pharmacotherapy in chronically depressed inpatients. *Journal of Affective Disorders, 109*(1–2), 65–73.

Schreiweis, C., Bornschein, U., Burguière, E., Kerimoglu, C., et al. (2014). Humanized Foxp2 accelerates learning by enhancing transitions from declarative to procedural performance. *Proceedings of the National Academy of Sciences, USA, 111*(39), 14253–14258.

Schulman, C. I., Namias, N., Doherty, J., Manning, R. J., et al. (2005). The effect of antipyretic therapy upon outcomes in critically ill patients: A randomized, prospective study. *Surgical Infections, 6*(4), 369–375.

Schummers, J., Yu, H., and Sur, M. (2008). Tuned responses of astrocytes and their influence on hemodynamic signals in the visual cortex. *Science, 320*, 1638–1643.

Schuster, C. R. (1970). Psychological approaches to opiate dependence and self-administration by laboratory animals. *Federation Proceedings, 29*, 1–5.

Schwartz, C. E., Wright, C. I., Shin, L. M., Kagan, J., et al. (2003). Inhibited and uninhibited infants "grown up": Adult amygdalar response to novelty. *Science, 300*, 1952–1953.

Schwartz, S., and Correll, C. U. (2014). Efficacy and safety of atomoxetine in children and adolescents with attention-deficit/hyperactivity disorder: Results from a comprehensive meta-analysis and metaregression. *Journal of the American Academy of Child and Adolescent Psychiatry, 53*(2), 174–187.

Schwartz, S. D., Hubschman, J. P., Heilwell, G., Franco-Cardenas, V., et al. (2012). Embryonic stem cell trials for macular degeneration: A preliminary report. *Lancet, 379*, 713–720.

Schwartzkroin, P. A., and Wester, K. (1975). Long-lasting facilitation of a synaptic potential following tetanisation in the in vitro hippocampal slice. *Brain Research, 89*, 107–119.

Sclafani, A., Springer, D., and Kluge, L. (1976). Effects of quinine adulteration on the food intake and body weight of obese and nonobese hypothalamic hyperphagic rats. *Physiology & Behaviour, 16*, 631–640.

Scott, D. J., Stohler, C. S., Egnatuk, C. M., Wang, H., et al. (2008). Placebo and nocebo effects are defined by opposite opioid and dopaminergic responses. *Archives of General Psychiatry, 65*, 220–231.

Scott, N., Prigge, M., Yizhar, O., and Kimchi, T. (2015). A sexually dimorphic hypothalamic circuit controls maternal care and oxytocin secretion. *Nature, 525*(7570), 519–522.

Scott, S. K., and Wise, R. J. (2004). The functional neuroanatomy of prelexical processing in speech perception. *Cognition, 92*, 13–45.

Scoville, W. B., and Milner, B. (1957). Loss of recent memory after bilateral hippocampal lesions. *Journal of Neurology, Neurosurgery and Psychiatry, 20*, 11–21.

Scudellari, M. (2015). The science myths that will not die. *Nature, 528*, 322–325.

Scudellari, M. (2016). How iPS cells changed the world. *Nature, 534*(7607), 310–312. https://doi.org/10.1038/534310a.

Scullin, M. K., and Bliwise, D. L. (2015). Sleep, cognition, and normal aging: Integrating a half century of multidisciplinary research. *Perspectives on Psychological Science, 10*(1), 97–137.

Seeman, P. (1990). Atypical neuroleptics: Role of multiple receptors, endogenous dopamine, and receptor linkage. *Acta Psychiatrica Scandinavica. Supplementum, 358*, 14–20.

Seeman, P. (2011). All roads to schizophrenia lead to dopamine supersensitivity and elevated dopamine D2(high) receptors. *CNS Neuroscience & Therapeutics, 17*(2), 118–132.

Seeman, P., Chau-Wong, M., Tedesco, J., and Wong, K. (1975). Brain receptors for antipsychotic drugs and dopamine: Direct binding assays. *Proceedings of the National Academy of Sciences, USA, 72*(11), 4376–4380.

Seiden, R. H. (1978). Where are they now? A follow-up study of suicide attempters from the Golden Gate Bridge. *Suicide and Life Threatening Behaviour, 8*, 203–216.

Seidl, A. H., Rubel, E. W., and Barría, A. (2014). Differential conduction velocity regulation in ipsilateral and contralateral collaterals innervating brainstem coincidence detector neurons. *Journal of Neuroscience, 34*(14), 4914–4919.

Seif, T., Chang, S. J., Simms, J. A., Gibb, S. L., et al. (2013). Cortical activation of accumbens hyperpolarisation-active NMDARs mediates aversion-resistant alcohol intake. *Nature Neuroscience, 16*, 1094–1100.

Sekar, A., Bialas, A. R., de Rivera, H., Davis, A., et al. (2016). Schizophrenia risk from complex variation of complement component 4. *Nature, 530*, 177–183.

Sekar, R., Wang, L., and Chow, B. K. (2017). Central control of feeding behaviour by the secretin, PACAP, and glucagon family of peptides. *Frontiers in Endocrinology, 8*, 18. Erratum in: *Frontiers in Endocrinology, 9*, 395.

Seligman, M. E., and Beagley, G. (1975). Learned helplessness in the rat. *Journal of Comparative and Physiological Psychology, 88*, 534–541.

Selye, H. (1956). *The stress of life.* New York: McGraw-Hill.

Semaw, S. (2000). The world's oldest stone artefacts from Gona, Ethiopia: Their implications for understanding stone technology and patterns of human evolution between 2.6–1.5 million years ago. *Journal of Archaeolical Science, 27*, 1197–1214.

Semendeferi, K., Lu, A., Schenker, N., and Damasio, H. (2002). Humans and great apes share a large frontal cortex. *Nature Neuroscience, 5*, 272–276.

Semkovska, M., and McLoughlin, D. M. (2013). Measuring retrograde autobiographical amnesia following electroconvulsive therapy: Historical perspective and current issues. *Journal of ECT, 29*(2), 127–133.

Sendtner, M., Holtmann, B., and Hughes, R. A. (1996). The response of motoneurons to neurotrophins. *Neurochemical Research, 21*, 831–841.

Seok, J., Warren, H. S., Cuenca, A. G., Mindrinosa, M. N., et al. (2013). Genomic responses in mouse models poorly mimic human inflammatory diseases. *Proceedings of the National Academy of Sciences, USA, 110*(9), 3501–3512.

Seretis, K., Goulis, D. G., Koliakos, G., and Demiri, E. (2015). Short- and long-term effects of abdominal lipectomy on weight and fat mass in females: A systematic review. *Obesity Surgery, 25*, 1950–1958.

Serrano, P. A., Beniston, D. S., Oxonian, M. G., Rodriguez, W. A., et al. (1994). Differential effects of protein kinase inhibitors and activators on memory formation in the 2-day-old chick. *Behavioural and Neural Biology, 61*, 60–72.

Serviere, J., Webster, W. R., and Calford, M. B. (1984). Isofrequency labelling revealed by a combined [14C]-2-deoxyglucose, electrophysiological, and horseradish peroxidase study of the inferior colliculus of the cat. *Journal of Comparative Neurology, 228*, 463–477.

Seuss, Dr. (1987). *The tough coughs as he ploughs the dough: Early writings and cartoons by Dr. Seuss.* New York: Morrow.

Sever, R., and Glass, C. K. (2013). Signalling by nuclear receptors. *Cold Spring Harbor Perspectives in Biology, 5*(3), a016709.

Sevigny, J., Chiao, P., Bussière, T., Weinreb, P. H., et al. (2016). The antibody aducanumab reduces Aβ plaques in Alzheimer's disease. *Nature, 537*(7618), 50–56. https://doi.org/10.1038/nature19323.

Sexton, C. E., Mackay, C. E., and Ebmeier, K. P. (2013). A systematic review and meta-analysis of magnetic resonance imaging studies in late-life depression. *American Journal of Geriatric Psychiatry, 2*, 184–195.

Seyfarth, R. M., and Cheney, D. L. (1997). Behavioural mechanisms underlying vocal communication in nonhuman primates. *Animal Learning & Behaviour, 25*, 249–267.

Shaffer, F., McCraty, R., and Zerr, C. L. (2014). A healthy heart is not a metronome: An integrative review of the heart's anatomy and heart rate variability. *Frontiers in Psychology, 5*, 1040.

Shallice, T., and Burgess, P. W. (1991). Deficits in strategy application following frontal lobe damage in man. *Brain, 114*, 727–741.

Shank, S. S., and Margoliash, D. (2009). Sleep and sensorimotor integration during early vocal learning in a songbird. *Nature, 458*, 73–77.

Shaw, D. J., Czekóová, K., Staněk, R., Mareček, R., et al. (2018). A dual-fMRI investigation of the iterated Ultimatum Game reveals that reciprocal behaviour is associated with neural alignment. *Scientific Reports, 8*(1), 10896.

Shaw, J., and Porter, S. (2015). Constructing rich false memories of committing crime. *Psychological Science, 26*(3), 291–301.

Shaw, L., Mansfield, C., Colquitt, L., Lin, C., et al. (2018). Personalized expression of bitter "taste" receptors in human skin. *PLOS ONE, 13*(10), e0205322.

Shaw, P., Greenstein, D., Lerch, J., Clasen, L., et al. (2006). Intellectual ability and cortical development in children and adolescents. *Nature, 440*, 676–679.

Shaw, P. J., Tononi, G., Greenspan, R. J., and Robinson, D. F. (2002). Stress response genes protect against lethal effects of sleep deprivation in *Drosophila*. *Nature, 417*, 287–291.

Shaywitz, S. E., Shaywitz, B. A., Fulbright, R. K., Skudlarski, P., et al. (2003). Neural systems for compensation and persistence: Young adult outcome of childhood reading disability. *Biological Psychiatry, 54*, 25–33.

Shaywitz, S. E., Shaywitz, B. A., Pugh, K. R., Fulbright, R. K., et al. (1998). Functional disruption in the organisation of the brain for reading in dyslexia. *Proceedings of the National Academy of Sciences, USA, 95*, 2636–2641.

Shema, R., Hazvi, S., Sacktor, T. C., and Dudai, Y. (2009). Boundary conditions for the maintenance of memory by PKMzeta in neocortex. *Learning & Memory, 16*, 122–128.

Shen, G., Dwivedi, K., Majima, K., Horikawa, T., et al. (2019). End-to-end deep image reconstruction from human brain activity. *Frontiers in Computational Neuroscience, 13*, 21.

Shen, G., Horikawa, T., Majima, K., and Kamitani, Y. (2019). Deep image reconstruction from human brain activity. *PLOS Computational Biology, 15*(1), e1006633.

Shenoy, K. V., Sahani, M., and Churchland, M. M. (2013). Cortical control of arm movements: A dynamical systems perspective. *Annual Review of Neuroscience, 36*, 337–359.

Sherrington, C. S. (1897). *A textbook of physiology. Part III. The Central Nervous System* (7th ed.), M. Foster (Ed.). London: Macmillan.

Sherrington, C. S. (1898). Experiments in examination of the peripheral distribution of the fibres of the posterior roots of some spinal nerves. *Philosophical Transactions, 190*, 45–186.

Sherry, D. F. (1992). Memory, the hippocampus, and natural selection: Studies of food-storing birds. In L. R. Squire and N. Butters (Eds.), *Neuropsychology of memory* (2nd ed., pp. 521–532). New York: Guilford.

Sherry, D. F., and Vaccarino, A. L. (1989). Hippocampus and memory for food caches in black-capped chickadees. *Behavioural Neuroscience, 103*, 308–318.

Sherry, D. F., Vaccarino, A. L., Buckenham, K., and Herz, R. S. (1989). The hippocampal complex of food-storing birds. *Brain Behaviour and Evolution, 34*, 308–317.

Sherwin, B. B. (2002). Randomized clinical trials of combined estrogen-androgen preparations: Effects on sexual functioning. *Fertility and Sterility, 77*(Suppl. 4), 49–54.

Sheth, S. A., Mian, M. K., Patel, S. R., Asaad, W. F., et al. (2012). Human dorsal anterior cingulate cortex neurons mediate ongoing behavioural adaptation. *Nature, 488*(7410), 218–221.

Shi, Y., Yamada, K., Liddelow, S. A., Smith, S. T., et al. (2017). ApoE4 markedly exacerbates tau-mediated neurodegeneration in a mouse model of tauopathy. *Nature, 549*(7673), 523–527. https://doi.org/10.1038/nature24016.

Shic, F., and Scassellati, B. (2007). A behavioural analysis of computational models of visual attention. *International Journal of Computer Vision, 73*(2), 159–177.

Shih, R. A., Belmonte, P. L., and Zandi, P. P. (2004). A review of the evidence from family, twin and adoption studies for a genetic contribution to adult psychiatric disorders. *International Review of Psychiatry, 16*, 260–283.

Shingo, T., Gregg, C., Enwere, E., Fujikawa, H., et al. (2003). Pregnancy-stimulated neurogenesis in the adult female forebrain mediated by prolactin. *Science, 299*, 117–120.

Shipton, O. A., El-Gaby, M., Apergis-Schoute, J., Deisseroth, K., et al. (2014). Left-right dissociation of hippocampal memory processes in mice. *Proceedings of the National Academy of Sciences, USA, 111*(42), 15238–15243.

Shors, T. J. (2014). The adult brain makes new neurons, and effortful learning keeps them alive. *Current Directions in Psychological Science, 23*(5), 311–318.

Shors, T. J., Miesegaes, G., Beylin, A., Zhao, M., et al. (2001). Neurogenesis in the adult is involved in the formation of trace memories. *Nature, 410*, 372–376.

Shu, W., Cho, J. Y., Jiang, Y., Zhang, M., et al. (2005). Altered ultrasonic vocalisation in mice with a disruption in the Foxp2 gene. *Proceedings of the National Academy of Sciences, USA, 102*, 9643–9648.

Sia, G. M., Clem, R. L., and Huganir, R. L. (2013). The human language-associated gene *SRPX2* regulates synapse formation and vocalisation in mice. *Science, 342*(6161), 987–991.

Siclari, F., Baird, B., Perogamvros, L., Bernardi, G., et al. (2017). The neural correlates of dreaming. *Nature Neuroscience, 20*(6), 872–878.

Siegel, J. M. (2005). Clues to the function of mammalian sleep. *Nature, 437*, 1264–1271.

Siegel, J. M. (2009). Sleep viewed as a state of adaptive inactivity. *Nature Reviews Neuroscience, 10*, 747–753.

Siegel, J. M. (2012). Evolution: Suppression of sleep for mating. *Science, 337*(6102), 1610–1611.

Siegel, J. M., Manger, P. R., Nienhuis, R., Fahringer, H. M., et al. (1999). Sleep in the platypus. *Neuroscience, 91*, 391–400.

Siegel, J. M., Nienhuis, R., Gulyani, S., Ouyang, S., et al. (1999). Neuronal degeneration in canine narcolepsy. *Journal of Neuroscience, 19*, 248–257.

Siemens, J., Zhou, S., Piskorowski, R., Nikai, T., et al. (2006). Spider toxins activate the capsaicin receptor to produce inflammatory pain. *Nature, 444*, 208–212.

Sierakowiak, A., Monnot, C., Aski, S. N., Uppman, M., et al. (2015). Default mode network, motor network, dorsal and ventral basal ganglia networks in the rat brain: Comparison to human networks using resting state-fMRI. *PLOS ONE, 10*(3), e0120345.

Siever, L. J. (2008). Neurobiology of aggression and violence. *American Journal of Psychiatry, 165*, 429–442.

Sikich, L, Frazier, J. A., McClellan, J., Findling, R. L., et al. (2008). Double-blind comparison of first- and second-generation antipsychotics in early-onset schizophrenia and schizo-affective disorder: Findings from the treatment of early-onset schizophrenia spectrum disorders (TEOSS) study. *American Journal of Psychiatry, 165*, 1420–1431.

Sikl, R., Simeccek, M., Porubanova-Norquist, M., Bezdicek, O., et al. (2013). Vision after 53 years of blindness. *i-Perception, 4*(8), 498–507. doi:10.1068/i0611.

Sileno, A. P., Brandt, G. C., Spann, B. M., and Quay, S. C. (2006). Lower mean weight after 14 days intravenous administration peptide YY_{3-36} (PYY_{3-36}) in rabbits. *International Journal of Obesity, 30*, 68–72.

Silke, J., and Meier, P. (2013). Inhibitor of apoptosis (IAP) proteins-modulators of cell death and inflammation. *Cold Spring Harbor Perspectives in Biology, 5*(2), pii: a008730. https://doi.org/10.1101/cshperspect.a008730.

Silva, D. A., and Satz, P. (1979). Pathological left-handedness. Evaluation of a model. *Brain and Language, 7*, 8–16.

Silver, R., LeSauter, J., Tresco, P. A., and Lehman, M. N. (1996). A diffusible coupling signal from the transplanted suprachiasmatic nucleus controlling circadian locomotor rhythms. *Nature, 382*(6594), 810–813.

Simmons, R. (2003). *Odd girl out: The hidden culture of aggression in girls.* New York: Harcourt.

Simner, J., Mulvenna, C., Sagiv, N., Tsakanikos, E., et al. (2006). Synaesthesia: The prevalence of atypical cross-modal experiences. *Perception, 35*, 1024–1033.

Simons, D. J., and Chabris, C. F. (1999). Gorillas in our midst: Sustained inattentional blindness for dynamic events. *Perception, 28*, 1059–1074.

Simons, D. J., and Jensen, M. S. (2009). The effects of individual differences and task difficulty on inattentional blindness. *Psychonomic Bulletin & Review, 16*, 398–403.

Simons, D. J., and Schlosser, M. D. (2017). Inattentional blindness for a gun during a simulated police vehicle stop. *Cognitive Research: Principles and Implications, 2*(1), 37.

Singer, P. (1975). *Animal liberation: A new ethics for our treatment of animals.* New York: New York Review.

Singer, T., Seymour, B., O'Doherty, J., Kaube, H., et al. (2004). Empathy for pain involves the affective but not sensory components of pain. *Science, 303*, 1157–1162.

Singh, D. (2002). Female mate value at a glance: Relationship of waist-to-hip ratio to health, fecundity and attractiveness. *Neuroendocrinology Letters, 23*(Suppl. 4), 81–91.

Sinopoli, V. M., Burton, C. L., Kronenberg, S., and Arnold, P. D. (2017). A review of the role of serotonin system genes in obsessive-compulsive disorder. *Neuroscience & Biobehavioural Reviews, 80*, 372–381.

Siok, W. T., Perfetti, C. A., Jin, Z., and Tan, L. H. (2004). Biological abnormality of impaired reading is constrained by culture. *Nature, 431*, 71–76.

Sirotnak, A. P., Grigsby, T., and Krugman, R. D. (2004). Physical abuse of children. *Pediatrics in Review, 25*, 264–277.

Sjöbeck, M., Dahlén, S., and Englund, E. (1999). Neuronal loss in the brainstem and cerebellum--part of the normal aging process? A morphometric study of the vermis cerebelli and inferior olivary nucleus. The Journals of Gerontology Series *A Biological Science, 54*, (9), B363–B368.

Skolnick, P. (2015). Biologic approaches to treat substance-use disorders. *Trends in Pharmacological Sciences, 36*, 628–635.

Slee, S. J., and Young, E. D. (2014). Alignment of sound localisation cues in the nucleus of the brachium of the inferior colliculus. *Journal of Neurophysiology, 111*(12), 2624–2633.

Sloboda, Z., Glantz, M. D., and Tarter, R. E. (2012). Revisiting the concepts of risk and protective factors for understanding the etiology and development of substance use and substance use disorders: Implications for prevention. *Substance Use & Misuse, 47*, 944–962.

Smale, L. (1988). Influence of male gonadal hormones and familiarity on pregnancy interruption in prairie voles. *Biology of Reproduction, 39*, 28–31.

Smale, L., Holekamp, K. E., and White, P. A. (1999). Siblicide revisited in the spotted hyaena: Does it conform to obligate or facultative models? *Animal Behaviour, 58*, 545–551.

Smith, D. E., Roberts, J., Gage, F. H., and Tuszynski, M. H. (1999). Age-associated neuronal atrophy occurs in the primate brain and is reversible by growth factor gene therapy. *Proceedings of the National Academy of Sciences, USA, 96*, 10893–10898.

Smith, J. T., Clifton, D. K., amd Steiner, R. A. (2006). Regulation of the neuroendocrine reproductive axis by kisspeptin-GPR54 signalling. *Reproduction, 131*, 623–630.

Smith, P. B., Compton, D. R., Welch, S. P., Razdan, R. K., et al. (1994). The pharmacological activity of anandamide, a putative endogenous cannabinoid, in mice. *Journal of Pharmacology and Experimental Therapeutics, 270*, 219–227.

Smith, S. (1997, September 2). Dreaming awake. Part 1: Living with narcolepsy. *Minnesota Public Radio News* (http://news.minnesota.publicradio.org/features/199709/02_smiths_narcolepsy/narco_1.shtml).

Smoller, J. W., and Finn, C. T. (2003). Family, twin, and adoption studies of bipolar disorder. *American Journal of Medical Genetics. Part C, Seminars in Medical Genetics, 123*, 48–58.

Smyth, K. A., Pritsch, T., Cook, T. B., McClendon, M. J., et al. (2004). Worker functions and traits associated with occupations and the development of AD. *Neurology, 63*, 498–503.

Snyder, J. S. (2019). Recalibrating the relevance of adult neurogenesis. *Trends in Neurosciences*, pii: S0166-2236(18)30314-X. https://doi.org/10.1016/j.tins.2018.12.001.

Snyder, J. S., Soumier, A., Brewer, M., Pickel, J., et al. (2011). Adult hippocampal neurogenesis buffers stress responses and depressive behaviour. *Nature, 476*, 458–461.

Snyder, P. J., Bhasin, S., Cunningham, G. R., Matsumoto, A. M., et al. (2018). Lessons from the testosterone trials. *Endocrine Reviews, 39*(3), 369–386.

Sobel, N., Khan, R. M., Saltman, A., Sullivan, E. V., et al. (1999). The world smells different to each nostril. *Nature, 402*, 35.

Sobel, N., Prabhakaran, V., Zhao, Z., Desmond, J. E., et al. (2000). Time course of odourant-induced activation in the human primary olfactory cortex. *Journal of Neurophysiology, 83*, 537–551.

Sol, D., Duncan, R. P., Blackburn, T. M., Casey, P., et al. (2005). Big brains, enhanced cognition, and response of birds to novel environments. *Proceedings of the National Academy of Sciences, USA, 102*, 5460–5465.

Soldner, F., Stelzer, Y., Shivalila, C. S., Abraham, B. J., et al. (2016). Parkinson-associated risk variant in distal enhancer of α-synuclein modulates target gene expression. *Nature, 533*(7601), 95–99.

Solomon, A. (2001). *The noonday demon: An atlas of depression*. New York: Scribner.

Solstad, T., Boccara, C. N., Kropff, E., Moser, M. B., et al. (2008). Representation of geometric borders in the entorhinal cortex. *Science, 322*, 1865–1868.

Soltis, J., King, L. E, Douglas-Hamilton, I., Vollrath, F., et al. (2014). African elephant alarm calls distinguish between threats from humans and bees. *PLOS ONE, 9*(2), e89403.

Somel, M., Liu, X., and Khaitovich, P. (2013). Human brain evolution: Transcripts, metabolites and their regulators. *Nature Reviews Neuroscience, 14*, 112–127.

Somerville, M. J., Mervis, C. B., Young, E. J., Seo, E. J., et al. (2005). Severe expressive-language delay related to duplication of the Williams-Beuren locus. *New England Journal of Medicine, 353*, 1694–1701.

Sone, T., Nakaya, N., Ohmori, K., Shimazu, T., et al. (2008). Sense of life worth living (*ikigai*) and mortality in Japan: Ohsaki Study. *Psychosomatic Medicine, 70*, 709–715.

Soon, C. S., Brass, M., Heinze, H. J., and Haynes, J. D. (2008). Unconscious determinants of free decisions in the human brain. *Nature Neuroscience, 11*, 543–545.

Sorokowski, P., Karwowski, M., Misiak, M., Marczak, M. K., et al. (2019). Sex differences in human olfaction: A meta-analysis. *Frontiers in Psychology, 10*, 242.

Sotiriadis, A., and Makrydimas, G. (2012). Neurodevelopment after prenatal diagnosis of isolated agenesis of the corpus callosum: An integrative review. *American Journal of Obstetrics and Gynecology, 206*(4), 337, e1–5.

Sowell, E. R., Kan, E., Yoshii, J., Thompson, P. M., et al. (2008). Thinning of sensorimotor cortices in children with Tourette syndrome. *Nature Neuroscience, 11*, 637–639.

Spalding, K. L., Bergmann, O., Alkass, K., Bernard, S., et al. (2013). Dynamics of hippocampal neurogenesis in adult humans. *Cell, 153*(6):1219–1227.

Spampinato, D., and Celnik, P. (2018). Deconstructing skill learning and its physiological mechanisms. *Cortex, 104*, 90–102.

Speakman, J. R., Levitsky, D. A., Allison, D. B., Bray, M. S., et al. (2011). Set points, settling points and some alternative models: Theoretical options to understand how genes and environments combine to regulate body adiposity. *Disease Models & Mechanisms, 4*(6), 733–745.

Spiegler, B. J., and Mishkin, M. (1981). Evidence for the sequential participation of inferior temporal cortex and amygdala in the acquisition of stimulus-reward associations. *Behavioural Brain Research, 3*, 303–317.

Spiegler, B. J., and Yeni-Komshian, G. H. (1983). Incidence of left-handed writing in a college population with reference to family patterns of hand preference. *Neuropsychologia, 21*, 651–659.

Spillane, J., Kullmann, D. M., and Hanna, M. G. (2016). Genetic neurological channelopathies: Molecular genetics and clinical phenotypes. *Journal of Neurology, Neurosurgery, and Psychiatry, 87*(1), 37–48.

Sprague, T. C., Saproo, S., and Serences, J. T. (2015). Visual attention mitigates information loss in small- and large-scale neural codes. *Trends in Cognitive Sciences, 19*(4), 215–226.

Squire, L. R., Amaral, D. G., Zola-Morgan, S., and Kritchevsky, M. P. G. (1989). Description of brain injury in the amnesic patient N.A. based on magnetic resonance imaging. *Experimental Neurology, 105*, 23–35.

Squire, L. R., and Moore, R.Y. (1979). Dorsal thalamic lesion in a noted case of chronic memory dysfunction. *Annals of Neurology, 6*, 503–506.

Squire, L. R., Wixted, J. T., and Clark, R. E. (2007). Recognition memory and the medial temporal lobe: A new perspective. *Nature Reviews Neuroscience, 8*, 872–883.

Squire, L. R., and Zola-Morgan, S. (1991). The medial temporal lobe memory system. *Science, 253*, 1380–1386.

Squitieri, F. (2013). Neurodegenerative disease: "Fifty shades of grey" in the Huntington disease gene. *Nature Reviews Neurology, 9*(8), 421–422.

Srivastava, G., and Apovian, C. M. (2018). Current pharmacotherapy for obesity. *Nature Reviews Endocrinology, 14*(1), 12–24.

Stamou, M. I., and Georgopoulos, N. A. (2018). Kallmann syndrome: Phenotype and genotype of hypogonadotropic hypogonadism. *Metabolism, 86*, 124–134.

Standing, L. G. (1973). Learning 10,000 pictures. *Quarterly Journal of Experimental Psychology, 25*, 207–222.

Stanford, T. R., Shankar, S., Massoglia, D. P., Costello, M. G., et al. (2010). Perceptual decision making in less than 30 milliseconds. *Nature Neuroscience, 13*(3), 379–385.

Stanovich, K. E. (2009). *What intelligence tests miss: The psychology of rational thought*. New Haven, CT: Yale University Press.

Stansley, B. J., and Conn, P. J. (2019). Neuropharmacological insight from allosteric modulation of mGlu receptors. *Trends in Pharmacological Sciences, 40*(4), 240–252.

Stanton, S. J., Beehner, J. C., Saini, E. K., Kuhn, C. M., et al. (2009). Dominance, politics, and physiology: Voters' testosterone changes on the night of the 2008 United States presidential election. *PLOS ONE, 4*, e7543.

Staubli, U.V. (1995). Parallel properties of long-term potentiation and memory. In J. L. McGaugh, N. M. Weinberger, and G. Lynch (Eds.), *Brain and memory: Modulation and mediation of neuroplasticity* (pp. 303–318). New York: Oxford University Press.

Stearns, S. C., Byars, S. G., Govindaraju, D. R., and Ewbank, D. (2010). Measuring selection in contemporary human populations. *Nature Reviews Genetics, 11*(9), 1611–1622.

Stein, B., and Meredith, M. A. (1993). *The merging of the senses*. Cambridge, MA: MIT Press.

Stein, B. E., and Stanford, T. R. (2008). Multisensory integration: Current issues from the perspective of the single neuron. *Nature Reviews Neuroscience, 9*, 255–266.

Stein, I. S., and Zito, K. (2018). Dendritic spine elimination: Molecular mechanisms and implications. *Neuroscientist*, Apr 1: 1073858418769644. https://doi.org/10.1177/1073858418769644.

Stein, M., and Miller, A. H. (1993). Stress, the hypothalamic-pituitary-adrenal axis, and immune function. *Advances in Experimental Medicine and Biology, 335*, 1–5.

Stein, M., Miller, A. H., and Trestman, R. L. (1991). Depression, the immune system, and health and illness. Findings in search of meaning. *Archives of General Psychiatry, 48*, 171–177.

Steinhausen, C., Zehl, L., Haas-Rioth, M., Morcinek, K., et al. (2016). Multivariate meta-analysis of brain-mass correlations in eutherian mammals. *Frontiers in Neuroanatomy, 10*, 91.

Steinmetz, H., Volkmann, J., Jancke, L., and Freund, H. J. (1991). Anatomical left-right asymmetry of language-related temporal cortex is different in left- and right-handers. *Annals of Neurology, 29*, 315–319.

Stemmler, G., Heldmann, M., Pauls, C. A., and Scherer, T. (2001). Constraints for emotion specificity in fear and anger: The context counts. *Psychophysiology, 38*(2), 275–291.

Stenzel, D., and Huttner, W. B. (2013). Role of maternal thyroid hormones in the developing neocortex and during human evolution. *Frontiers in Neuroanatomy, 7*, 19.

Stephan, F. K., and Zucker, I. (1972). Circadian rhythms in drinking behaviour and locomotor activity of rats are eliminated by hypothalamic lesions. *Proceedings of the National Academy of Sciences, USA, 69*, 1583–1586.

Stephan, H., Frahm, H., and Baron, G. (1981). New and revised data on volumes of brain structures in insectivores and primates. *Folia Primatologica, 35*, 1–29.

Sterling, P., and Eyer, J. (1988). Allostasis: A new paradigm to explain arousal pathology. In S. Fisher and J. T. Reason (Eds.), *Handbook of life stress, cognition, and health* (pp. 629–649). Chichester, NY: Wiley.

Stern, K., and McClintock, M. (1998). Regulation of ovulation by human pheromones. *Nature, 392*, 177–179.

Stevens, M., Peigneur, S., and Tytgat, J. (2011). Neurotoxins and their binding areas on voltage-gated sodium channels. *Frontiers in Pharmacology, 2*, 71.

Stoeckel, C., Gough, P. M., Watkins, K. E., and Devlin, J. T. (2009). Supramarginal gyrus involvement in visual word recognition. *Cortex, 45*, 1091–1096.

Stoerig, P., and Cowey, A. (1997). Blindsight in man and monkey. *Brain, 120*, 535–559.

Stoodley, C. J., and Stein, J. F. (2011). The cerebellum and dyslexia. *Cortex, 1*, 101–116.

Stowers, L., Holy, T. E., Meister, M., Dulac, C., et al. (2002). Loss of sex discrimination and male-male aggression in mice deficient for TRP2. *Science, 295*, 1493–1500.

Stowers, L. and Kuo, T. H. (2015). Mammalian pheromones: emerging properties and mechanisms of detection. *Current Opinion in Neurobiology, 34*, 103–109. doi: 10.1016/j.conb.2015.02.005. https://www.ncbi.nlm.nih.gov/pubmed/25747731.

Stranahan, A. M., Khalil, D., and Gould, E. (2006). Social isolation delays the positive effects of running on adult neurogenesis. *Nature Neuroscience, 9*, 526–533.

Stroop, J. R. (1935). Studies of interference in serial verbal reactions. *Journal of Experimental Psychology, 18*, 643–662.

Stuber, G. D., and Wise, R. A. (2016). Lateral hypothalamic circuits for feeding and reward. *Nature Neuroscience, 19*, 198–205.

Stuve, T. A., Friedman, L., Jesberger, J. A., Gilmore, G. C., et al. (1997). The relationship between smooth pursuit performance, motion perception and sustained visual attention in patients with schizophrenia and normal controls. *Psychological Medicine, 27*, 143–152.

Substance Abuse and Mental Health Services Administration. (2017). *Key Substance Use and Mental Health Indicators in the United States: Results from the 2016 National Survey on Drug Use and Health, NSDUH Series H-52, HHS Publication No. (SMA) 17-5044*. Rockville, MD: Substance Abuse and Mental Health Services Administration.

Substance Abuse and Mental Health Services Administration. (2018). *Key substance use and mental health indicators in the United States: Results from the 2017 National Survey on Drug Use and Health (HHS Publication No. SMA 18-5068, NSDUH Series H-53).* Rockville, MD: Centre for Behavioural Health Statistics and Quality, Substance Abuse and Mental Health Services Administration. https://www.samhsa.gov/data/.

Suez, J., and Elinav, E. (2017). The path towards microbiome-based metabolite treatment. *Nature Microbiology, 2*, 17075.

Sugawara, N., Yasui-Furukori, N., Ishii, N., Iwata, N., et al. (2013). Lithium in tap water and suicide mortality in Japan. *International Journal of Environmental Research and Public Health, 10*, 6044–6048.

Suk, I., and Tamargo, R. J. (2010). Concealed neuroanatomy in Michelangelo's *Separation of Light from Darkness* in the Sistine Chapel. *Neurosurgery, 66*(5), 851–861.

Sumbre, G., Fiorito, G., Flash, T., and Hochner, B. (2005). Motor control of flexible octopus arms. *Nature, 443*, 595–596.

Sumnall, H. R., and Cole, J. C. (2005). Self-reported depressive symptomatology in community samples of polysubstance misusers who report Ecstasy use: A meta-analysis. *Journal of Psychopharmacology, 19*, 84–92.

Sun, E. W. L., Martin, A. M., Young, R. L., and Keating, D. J. (2019). The regulation of peripheral metabolism by gut-derived hormones. *Frontiers in Endocrinology, 9*, 754.

Sun, T., Patoine, C., Abu-Khalil, A., Visvader, J., et al. (2005). Early asymmetry of gene transcription in embryonic human left and right cerebral cortex. *Science, 308*, 1794–1798.

Sunn, N., Egli, M., Burazin, T. C. D., Burns, P., et al. (2002). Circulating relaxin acts on subfornical organ neurons to stimulate water drinking in the rat. *Proceedings of the National Academy of Sciences, USA, 99*, 1701–1706.

Sunstein, C. R., and Nussbaum, M. C. (Eds.) (2004). *Animal rights: Current debates and new directions.* Oxford, UK: Oxford University Press.

Suo, C., Singh, M. F., Gates, N., Wen, W., et al. (2016). Therapeutically relevant structural and functional mechanisms triggered by physical and cognitive exercise. *Molecular Psychiatry, 21*(11), 1–10.

Surmeier, D. J., Obeso, J. A., and Halliday, G. M. (2017). Selective neuronal vulnerability in Parkinson disease. *Nature Reviews Neuroscience, 18*(2), 101–113.

Sutcliffe, J. G., and de Lecea, L. (2002). The hypocretins: Setting the arousal threshold. *Nature Reviews Neuroscience, 3*, 339–349.

Suthana, N., Haneef, Z., Stern, J., Mukamel, R., et al. (2012). Memory enhancement and deep-brain stimulation of the entorhinal area. *New England Journal of Medicine, 366*(6), 502–510.

Suzuki, S., Brown, C. M., and Wise, P. M. (2009). Neuroprotective effects of estrogens following ischemic stroke. *Frontiers in Neuroendocrinology, 30*, 201–211.

Suzuki, W. A., and Naya, Y. (2014). The perirhinal cortex. *Annual Review of Neuroscience, 37*, 39–53.

Svare, B. B. (2013). *Hormones and aggressive behaviour.* New York: Plenum.

Swedo, S. E., Rapoport, J. L., Leonard, H., Lenane, M., et al. (1989). Obsessive-compulsive disorder in children and adolescents: Clinical phenomenology of 70 consecutive cases. *Archives of General Psychiatry, 46*(4), 335–341.

Swick, T. J. (2015). Treatment paradigms for cataplexy in narcolepsy: Past, present, and future. *Nature and Science of Sleep, 7*, 159–169.

Sylvia, K. E., and Demas, G. E. (2018). A gut feeling: Microbiome-brain-immune interactions modulate social and affective behaviours. *Hormones and Behaviour, 99*, 41–49.

Symonds, M. E., Pope, M., and Budge, H. (2015). The ontogeny of brown adipose tissue. *Annual Review of Nutrition, 35*, 295–320.

Szarfman, A., Doraiswamy, P. M., Tonning, J. M., and Levine, J. G. (2006). Association between pathologic gambling and parkinsonian therapy as detected in the Food and Drug Administration Adverse Event database. *Archives Neurology, 62*, 299–300.

Székely, T., Catchpole, C. K., DeVoogd, A., Marchl, Z., et al. (1996). Evolutionary changes in a song control area of the brain (HVC) are associated with evolutionary changes in song repertoire among European warblers (Sylviidae). *Proceedings of the Royal Society of London. Series B: Biological Sciences, 263*, 607–610.

T

Taglialatela, J. P., Cantalupo, C., and Hopkins, W. D. (2006). Gesture handedness predicts asymmetry in the chimpanzee inferior frontal gyrus. *Neuroreport, 17*, 923–927.

Taipale, M., Kaminen, N., Nopola-Hemmi, J., Haltia, T., et al. (2003). A candidate gene for developmental dyslexia encodes a nuclear tetratricopeptide repeat domain protein dynamically regulated in brain. *Proceedings of the National Academy of Sciences, USA, 100*, 11553–11558.

Tait, S. W., and Green, D. R. (2010). Mitochondria and cell death: Outer membrane permeabilisation and beyond. *Nature Reviews Molecular Cell Biology, 11*(9), 621–632. doi:10.1038/nrm2952.

Takahashi, H., Kato, M., Matsuura, M., Mobbs, D., et al. (2009). When your gain is my pain and your pain is my gain: Neural correlates of envy and schadenfreude. *Science, 323*, 937–939.

Takahashi, J. (2018). Stem cells and regenerative medicine for neural repair. *Current Opinion in Biotechnology, 52*, 102–108.

Takahashi, J. S. (1995). Molecular neurobiology and genetics of circadian rhythms in mammals. *Annual Review of Neuroscience, 18*, 531–554.

Takahashi, T., Svoboda, K., and Malinow, R. (2003). Experience strengthening transmission by driving AMPA receptors into synapses. *Science, 299*, 1585–1588.

Tallal, P., and Schwartz, J. (1980). Temporal processing, speech perception and hemispheric asymmetry. *Trends in Neurosciences, 3*, 309–311.

Tam, J., Duda, D. G., Perentes, J. Y., Quadri, R. S., et al. (2009). Blockade of VEGFR2 and not VEGFR1 can limit diet-induced fat tissue expansion: Role of local versus bone marrow-derived endothelial cells. *PLOS ONE, 4*, e4974.

Tamaki, M., Bang, J. W., Watanabe, T., and Sasaki, Y. (2016). Night watch in one brain hemisphere during sleep associated with the first-night effect in humans. *Current Biology, 26*. doi:10.1016/j.cub.2016.02.063.

Tan, C. L., and Knight, Z. A. (2018). Regulation of body temperature by the nervous system. *Neuron, 98*(1), 31–48.

Tanaka, H., Black, J. M., Hulme, C., Stanley, L. M., et al. (2011). The brain basis of the phonological deficit in dyslexia is independent of IQ. *Psychological Science, 22*(11), 1442–1451.

Tanaka, K. (1993). Neuronal mechanisms of object recognition. *Science, 262*, 685–688.

Tanaka, S., Ikeda, H., Kasahara, K., Kato, R., et al. (2013). Larger right posterior parietal volume in action video game experts: A behavioural and voxel-based morphometry (VBM) study. *PLOS ONE, 8*, e66998.

Tanda, G., Munzar, P., and Goldberg, S. R. (2000). Self-administration behaviour is maintained by the psychoactive ingredient of marijuana in squirrel monkeys. *Nature Neuroscience, 3*, 1073–1074.

Tang, Y. P., Wang, H., Feng, R., Kyin, M., et al. (2001). Differential effects of enrichment on learning and memory function in NR2B transgenic mice. *Neuropharmacology, 41*, 779–790.

Tanigawa, H., Lu, H. D., and Roe, A. W. (2010). Functional organisation for colour and orientation in macaque V4. *Nature Neuroscience, 13*, 1542–1548.

Tanji, J. (2001). Sequential organisation of multiple movements: Involvement of cortical motor areas. *Annual Review of Neuroscience, 24*, 631–651.

Tannenbaum, P. L., Tye, S. J., Stevens, J., Gotter, A. L., et al. (2016). Inhibition of orexin signalling promotes sleep yet preserves salient arousability in monkeys. *Sleep, 39*(3), 603–612.

Tansey, K. E., Guipponi, M., Hu, X., Domenici, E., et al. (2013). Contribution of common genetic variants to antidepressant response. *Biological Psychiatry, 73*, 679–682.

Taub, E. (1976). Movement in nonhuman primates deprived of somatosensory feedback. *Exercise and Sport Sciences Reviews, 4*, 335–374.

Taub, E., Uswatte, G., and Elbert, T. (2002). New treatments in neurorehabilitation founded on basic research. *Nature Reviews Neuroscience, 3*, 228–235.

Tavares, R. M., Mendelsohn, A., Grossman, Y., Williams, C. H., et al. (2015). A map for social navigation in the human brain. *Neuron, 87*(1), 231–243.

Tavernise, S. (2016, February 29). Female Viagra only modestly increases sexual satisfaction, study finds. *The New York Times*, p. A20.

Temple, E., Deutsch, G. K., Poldrack, R. A., Miller, S. L., et al. (2003). Neural deficits in children with dyslexia ameliorated by behavioural remediation: Evidence from functional MRI. *Proceedings of the National Academy of Sciences, USA, 100*, 2860–2865.

Tepper, S. J. (2019). CGRP and headache: A brief review. *Neurological Sciences*, Mar 5. https://doi.org/10.1007/s10072-019-03769-8.

Terburg, D., Aarts, H., and van Honk, J. (2012). Testosterone affects gaze aversion from angry faces outside of conscious awareness. *Psychological Science, 23*, 459–463.

Terburg, D., Scheggia, D., Triana Del Rio, R., Klumpers, F., et al. (2018). The basolateral amygdala is essential for rapid escape: A human and rodent study. *Cell, 175*(3), 723–735.e16.

Terkel, J., and Rosenblatt, J. S. (1972). Humoral factors underlying maternal behaviour at parturition: Cross transfusion between freely moving rats. *Journal of Comparative and Physiological Psychology, 80*, 365–371.

Terman, G. W., Shavit, Y., Lewis, J. W., Cannon, J. T., et al. (1984). Intrinsic mechanisms of pain inhibition: Activation by stress. *Science, 226*, 1270–1277.

Terrace, H. S. (1979). *Nim.* New York: Knopf.

Terrillion, C. E., Abazyan, B., Yang, Z., Crawford, J., et al. (2017). DISC1 in astrocytes influences adult neurogenesis and hippocampus-dependent behaviours in mice. *Neuropsychopharmacology, 42*(11), 2242–2251.

Tessier-Lavigne, M., Placzek, M., Lumsden, A. G., Dodd, J., et al. (1988). Chemotropic guidance of developing axons in the mammalian central nervous system. *Nature, 336*, 775–778.

Tetel, M. J., de Vries, G. J., Melcangi, R. C., Panzica, G., et al. (2018). Steroids, stress and the gut microbiome-brain axis. *Journal of Neuroendocrinology, 30*(2). https://doi.org/10.1111/jne.12548.

Tewksbury, J. J., and Nabhan, G. P. (2001). Seed dispersal: Directed deterrence by capsaicin in chillies. *Nature, 412*, 403–404.

Thannickal, T. C., Moore, R. Y., Nienhuis, R., Ramanathan, L., et al. (2000). Reduced number of hypocretin neurons in human narcolepsy. *Neuron, 27*, 469–474.

Theillet, F. X., Binolfi, A., Bekei, B., Martorana, A., et al. (2016). Structural disorder of monomeric α-synuclein persists in mammalian cells. *Nature, 530*(7588), 45–50.

Theunissen, F. E., and Elie, J. E. (2014). Neural processing of natural sounds. *Nature Reviews Neuroscience, 15*(6), 355–366.

Thimgan, M. S., Seugnet, L., Turk, J., and Shaw, P. J. (2015). Identification of genes associated with resilience/vulnerability to sleep deprivation and starvation in *Drosophila*. *Sleep, 38*(5), 801–814.

Thoen, H. H., How, M. J., Chiou, T. H., and Marshall, J. (2014). A different form of colour vision in mantis shrimp. *Science, 343*, 411–413.

Thomas, M. A., Ryu, V., and Bartness, T. J. (2016). Central ghrelin increases food foraging/hoarding that is blocked by GHSR antagonism and attenuates hypothalamic paraventricular nucleus neuronal activation. *American Journal of Physiology—Regulatory, Integrative and Comparative Physiology, 310*(3), R275–285.

Thombs, D. L., and Osborn, C. J. (2013). *Introduction to Addictive Behaviours* (4th ed.). New York, NY: Guilford Press.

Thompson, P. M., Vidal, C., Giedd, J. N., Gochman, P., et al. (2001). Mapping adolescent brain change reveals dynamic wave of accelerated gray matter loss in very early-onset schizophrenia. *Proceedings of the National Academy of Sciences, USA, 98*, 11650–11655.

Thompson, R. F., and Krupa, D. J. (1994). Organisation of memory traces in the mammalian brain. *Annual Review of Neuroscience, 17*, 519–549.

Thompson, R. F., and Steinmetz, J. E. (2009). The role of the cerebellum in classical conditioning of discrete behavioural responses. *Neuroscience, 162*, 732–755.

Thompson, T., and Schuster, C. R. (1964). Morphine self-administration, food reinforced and avoidance behaviour in rhesus monkeys. *Psychopharmacologia, 5*, 87–94.

Thornton, A. E., Cox, D. N., Whitfield, K., and Fouladi, R. T. (2008). Cumulative concussion exposure in rugby players: Neurocognitive and symptomatic outcomes. *Journal of Clinical and Experimental Neuropsychology, 30*, 398–409.

Thornton-Jones, Z. D., Kennett, G. A., Benwell, K. R., Revell, D. F., et al. (2006). The cannabinoid CB1 receptor inverse agonist, rimonabant, modifies body weight and adiponectin function in diet-induced obese rats as a consequence of reduced food intake. *Pharmacology, Biochemistry, and Behaviour, 84*, 353–359.

Thorpe, S. J., and Fabre-Thorpe, M. (2001). Seeking categories in the brain. *Science, 291*, 260–263.

Timmann, D., Drepper, J., Frings, M., Maschke, M., et al. (2010). The human cerebellum contributes to motor, emotional and cognitive associative learning. A review. *Cortex, 7*, 845–857.

Tinnermann, A., Geuter, S., Sprenger, C., Finsterbusch, J., et al. (2017). Interactions between brain and spinal cord mediate value effects in nocebo hyperalgesia. *Science, 358*(6359), 105–108.

Tissir, F., and Goffinet, A. M. (2003). Reelin and brain development. *Nature Reviews Neuro-science, 4*, 496–505.

Tizzano, M., and Finger, T. E. (2013). Chemosensors in the nose: Guardians of the airways. *Physiology, 28*(1), 51–60.

Tobias, P.V. (1980). L'evolution du cerveau humain. *La Recherche, 11*, 282–292.

Toda, T., and Gage, F. H. (2018). Review: Adult neurogenesis contributes to hippocampal plasticity. *Cell and Tissue Research, 373*(3), 693–709.

Todd, W. D., Fenselau, H., Wang, J. L., Zhang, R., et al. (2018). A hypothalamic circuit for the circadian control of aggression. *Nature Neuroscience, 21*(5), 717–724.

Todorov, A., Said, C. P., Engell, A. D., and Oosterhof, N. N. (2008). Understanding evaluation of faces on social dimensions. *Trends in Cognitive Science, 12*, 455–460.

Tolman, E. C. (1949). There is more than one kind of learning. *Psychological Review, 56*, 144–155.

Tolman, E. C., and Honzik, C. H. (1930). Introduction and removal of reward, and maze performance in rats. *University of California Publications in Psychology, 4*, 257–275.

Tom, S. M., Fox, C. R., Trepel, C., and Poldrack, R. A. (2007). The neural basis of loss aversion in decision-making under risk. *Science, 315*, 515–518.

Tomizawa, K., Iga, N., Lu,Y. F., Moriwaki, A., et al. (2003). Oxytocin improves long-lasting spatial memory during motherhood through MAP kinase cascade. *Nature Neuroscience, 6*, 384–390.

Tomoda, T., Hikida, T., and Sakurai, T. (2017). Role of DISC1 in neuronal trafficking and its implication in neuropsychiatric manifestation and neurotherapeutics. *Neurotherapeutics, 14*(3), 623–629.

Tomonari, H., Miura, H., Nakayama, A., Matsumura, E., et al. (2012). Gα-gustducin is extensively coexpressed with sweet and bitter taste receptors in both the soft palate and fungiform papillae but has a different functional significance. *Chemical Senses, 37*(3), 241–251.

Tooby, J., and Cosmides, L. (2015). The theoretical foundations of evolutionary psychology. In D. M. Buss (Ed.), *The handbook of evolutionary psychology: Vol. 1. Foundations* (2nd ed.) (pp. 3–87). Hoboken, NJ: John Wiley & Sons.

Tootell, R. B., Silverman, M. S., Hamilton, S. L., De Valois, R. L., et al. (1988). Functional anatomy of macaque striate cortex. III. Colour. *Journal of Neuroscience, 8*, 1569–1593.

Tootell, R. B., Silverman, M. S., Switkes, E., and De Valois, R. L. (1982). Deoxyglucose analysis of retinotopic organisation in primate striate cortex. *Science, 218*, 902–904.

Tootell, R. B., Tsao, D., and Vanduffel, W. (2003). Neuroimaging weighs in: Humans meet macaques in "primate" visual cortex. *Journal of Neuroscience, 23*, 3981–3989.

Tootell, R. B. H., Hadjikhani, N. K., Vanduffel, W., Liu, A. K., et al. (1998). Functional analysis of primary visual cortex (V1) in humans. *Proceedings of the National Academy of Sciences, USA, 95*, 811–817.

Torrey, E. F., Bowler, A. E., Taylor, E. H., and Gottesman, I. I. (1994). *Schizophrenia and manic depressive disorder.* New York: Basic Books.

Torvik, F. A., Gustavson, K., Ystrom, E., Rosenström, T. H., et al. (2018). Continuity of genetic and environmental influences on clinically assessed major depression from ages 18 to 45. *Psychological Medicine, 28*, 1–9.

Traffanstedt, M. K., Mehta, S., and LoBello, S. G. (2016). Major depression with seasonal variation: Is it a valid construct? *Clinical Psychological Science*. doi:10.1177/2167702615615867.

Trasande, L., Blustein, J., Liu, M., Corwin, E., et al. (2013). Infant antibiotic exposures and early-life body mass. *International Journal of Obesity (London), 37*(1), 16–23.

Travers, S. P., and Geran, L. C. (2009). Bitter-responsive brainstem neurons: Characteristics and functions. *Physiology & Behaviour, 97*(5), 592–603.

Treesukosol, Y., Lyall, V., Heck, G. L., Desimone, J. A., et al. (2007). A psychophysical and electrophysiological analysis of salt taste in *Trpv1* null mice. *American Journal of Physiology. Regulatory, Integrative, and Comparative Physiology, 292*, R1799–R1809.

Treesukosol, Y., and Spector, A. C. (2012). Orosensory detection of sucrose, maltose, and glucose is severely impaired in mice lacking T1R2 or T1R3, but Polycose sensitivity remains relatively normal. *American Journal of Physiology: Regulatory, Integrative, and Comparative Physiology, 303*(2), R218–R235.

Treisman, A. [M]. (1996). The binding problem. *Current Opinion in Neurobiology, 6*, 171–178.

Treisman, A. M., and Gelade, G. (1980). A feature-integration theory of attention. *Cognitive Psychology, 12*, 97–136.

Treisman, M. (1977). Motion sickness: Evolutionary hypotheses. *Science, 197*, 493–495.

Tremlett, H., Bauer, K. C., Appel-Cresswell, S., Finlay, B. B., et al. (2017). The gut microbiome in human neurological disease: A review. *Annals of Neurology, 81*(3), 369–382.

Trimmer, C., Keller, A., Murphy, N. R., Snyder, L. L., et al. (2019). Genetic variation across the human olfactory receptor repertoire alters odour perception. *Proceedings of the National Academy of Sciences, USA, 116*(19), 9475–9480.

Trinh, J., and Farrer, M. (2013). Advances in the genetics of Parkinson disease. *Nature Reviews Neurology, 9*(8), 445–454.

Trivedi, D. K., Sinclair, E., Xu, Y., Sarkar, D., et al. (2019). Discovery of volatile biomarkers of Parkinson's disease from sebum. *ACS Central Science, 5*(4), 599–606.

Trojanowski, N. F., and Raizen, D. M. (2016). Call it worm sleep. *Trends in Neurosciences, 39*(2), 54–62.

Tronick, R., and Reck, C. (2009). Infants of depressed mothers. *Harvard Review of Psychiatry, 17*, 147–156.

Tropea, D., Hardingham, N., Millar, K., and Fox, K. (2018). Mechanisms underlying the role of DISC1 in synaptic plasticity. *Journal of Physiology, 596*(14), 2747–2771.

Truitt, W. A., and Coolen, L. M. (2002). Identification of a potential ejaculation generator in the spinal cord. *Science, 297*, 1566–1569.

Tsai, L., and Barnea, G. (2014). A critical period defined by axon-targeting mechanisms in the murine olfactory bulb. *Science, 344*(6180), 197–200.

Tsivilis, D., Vann, S. D., Denby, C., Roberts, N., et al. (2008). A disproportionate role for the fornix and mammillary bodies in recall versus recognition memory. *Nature Neuroscience, 11*(7), 834–842.

Tsuboi, D., Kuroda, K., Tanaka, M., Namba, T., et al. (2015). Disrupted-in-schizophrenia 1 regulates transport of *ITPR1* mRNA for synaptic plasticity. *Nature Neuroscience, 18*, 698–707.

Tsuchiya, N., and Adolphs, R. (2007). Emotion and consciousness. *Trends in Cognitive Sciences, 11*, 158–167.

Tsujimoto, S., Genovesio, A., and Wise, S. P. (2011). Frontal pole cortex: Encoding ends at the end of the endbrain. *Trends in Cognitive Sciences, 15*(4), 169–176.

Tsutsui, K., Bentley, G. E., Ubuka, T., Saigoh, E., et al. (2006). The general and comparative biology of gonadotropin-inhibitory hormone (GnIH). *General and Comparative Endocrinology, 153*, 365–370.

Tuller, D. (2002, January 8). A quiet revolution for those prone to nodding off. *The New York Times* (http://query.nytimes.com/gst/fullpage.html?sec=health&res=980DE5DD1439F93BA35752C0A9649C8B63).

Tulving, E. (2002). Episodic memory: From mind to brain. *Annual Review of Psychology*, 53, 1–25.

Tulving, E., Hayman, C. A., and Macdonald, C. A. (1991). Long-lasting perceptual priming and semantic learning in amnesia: A case experiment. *Journal of Experimental Psychology: Learning, Memory, and Cognition, 17,* 595–617.

Twenge, J. M. (2015). Time period and birth cohort differences in depressive symptoms in the U.S., 1982–2013. *Social Indicators Research, 121*(2), 437–454.

Twenge, J. M., Gentile, B., DeWall, C. N., Ma, D., et al. (2010). Birth cohort increases in psychopathology among young Americans, 1938–2007: A cross-temporal meta-analysis of the MMPI. *Clinical Psychology Review, 30*(2), 145–154.

Tyack, P. L. (2003). Dolphins communicate about individual-specific social relationships. In F. de Waal and P. L. Tyack (Eds.), *Animal social complexity: Intelligence, culture, and individualized societies* (pp. 342–361). Cambridge, MA: Harvard University Press.

Tye, K. M., and Deisseroth, K. (2012). Optogenetic investigation of neural circuits underlying brain disease in animal models. *Nature Reviews Neuroscience, 13,* 251–266.

Tyszka, J. M., Kennedy, D. P., Adolphs, R., and Paul, L. K. (2011). Intact bilateral resting-state networks in the absence of the corpus callosum. *Journal of Neuroscience, 31*(42), 15154–15162.

Tyzio, R., Cossart, R., Khalilov, I., Minlebaeve, M., et al. (2006). Maternal oxytocin triggers a transient inhibitory switch in GABA signalling in the fetal brain during delivery. *Science, 314,* 1788–1792.

U

Umilta, M. A., Kohler, E., Galiese, V., Fogassi, L., et al. (2001). I know what you are doing: A neurophysiological study. *Neuron, 31,* 155–165.

Ungerleider, L. G., and Pasternak, T. (2004). Ventral and dorsal cortical processing streams. In L. M. Chalupa and J. S. Werner (Eds.), *The Visual Neurosciences* (pp. 541–562). Cambridge, MA: MIT Press.

Upile, T., Sipaul, F., Jerjes, W., Singh, S., et al. (2007). The acute effects of alcohol on auditory thresholds. *BMC Ear, Nose, and Throat Disorders, 7,* 4.

Urban, D. J., and Roth, B. L. (2015). DREADDs (designer receptors exclusively activated by designer drugs): Chemogenetic tools with therapeutic utility. *Annual Review of Pharmacology and Toxicology, 55,* 399–417.

Ursin, H., Baade, E., and Levine, S. (1978). *Psychobiology of stress: A study of coping men.* New York: Academic Press.

U.S. Food and Drug Administration (FDA 2018). The FDA Warns Against the Use of Many Genetic Tests with Unapproved Claims to Predict Patient Response to Specific Medications: FDA Safety Communication, https://www.fda.gov/medical-devices/safety-communications/fda-warns-against-use-many-genetic-tests-unapproved-claims-predict-patient-response-specific.

V

Valenzuela, C. F., Morton, R. A., Diaz, M. R., and Topper, L. (2012). Does moderate drinking harm the fetal brain? Insights from animal models. *Trends in Neurosciences, 35*(5), 284–292.

Vallortigara, G., Rogers, L. J., and Bisazza, A. (1999). Possible evolutionary origins of cognitive brain lateralisation. *Brain Research. Brain Research Reviews, 30,* 164–175.

van Anders, S. M., and Watson, N. V. (2006). Social neuroendocrinology: Effects of social contexts and behaviours on sex steroids in humans. *Human Nature, 17,* 212–237.

Van Dongen, H. P., Maislin, G., Mullington, J. M., and Dinges, D. F. (2003). The cumulative cost of additional wakefulness: Dose-response effects on neurobehavioural functions and sleep physiology from chronic sleep restriction and total sleep deprivation. *Sleep, 26,* 117–126.

Van Essen, D. C., and Drury, H. A. (1997). Structural and functional analyses of human cerebral cortex using a surface-based atlas. *Journal of Neuroscience, 17,* 7079–7102.

Van Essen, D. C., and Glasser, M. F. (2018). Parcellating cerebral cortex: How invasive animal studies inform noninvasive mapmaking in humans. *Neuron, 99*(4), 640–663.

Van Gaal, L. F., Rissanen, A. M., Scheen, A. J., Ziegler, O., et al. (2005). Effects of the cannabinoid-1 receptor blocker rimonabant on weight reduction and cardiovascular risk factors in overweight patients: 1-year experience from the RIO-Europe study. *Lancet, 365,* 1389–1397.

van Nas, A., Guhathakurta, D., Wang, S. S., Yehya, N., et al. (2009). Elucidating the role of gonadal hormones in sexually dimorphic gene coexpression networks. *Endocrinology, 150,* 1235–1249.

van Os, J., and Kapur, S. (2009). Schizophrenia. *Lancet, 374*(9690), 635–645.

van Os, J., Kenis, G., and Rutten, B. P. F. (2010). The environment and schizophrenia. *Nature, 468,* 203–212.

van Praag, H., Kempermann, G., and Gage, F. H. (2000). Neural consequences of environmental enrichment. *Nature Reviews Neuroscience, 1,* 191–198.

van Straten, A., van der Zweerde, T., Kleiboer, A., Cuijpers, P., et al. (2018). Cognitive and behavioural therapies in the treatment of insomnia: A meta-analysis. *Sleep Medicine Reviews, 38,* 3–16.

van Tol, M. J., van der Wee, N. J., van den Heuvel, O. A., Nielen, M. M., et al. (2010). Regional brain volume in depression and anxiety disorders. *Archives of General Psychiatry, 67,* 1002–1011.

Van Valen, L. (1974). Brain size and intelligence in man. *American Journal of Physical Anthropology, 40,* 417–423.

van Westen, M., Rietveld, E., Figee, M., and Denys, D. (2015). Clinical outcome and mechanisms of deep brain stimulation for obsessive-compulsive disorder. *Current Behavioural Neuroscience Reports, 2,* 41–48.

Vandermosten, M., Boets, B., Wouters, J., and Ghesquière, P. (2012). A qualitative and quantitative review of diffusion tensor imaging studies in reading and dyslexia. *Neuroscience and Biobehavioural Reviews, 36*(6), 1532–1552.

Vandewauw, I., De Clercq, K., Mulier, M., Held, K., et al. (2018). A TRP channel trio mediates acute noxious heat sensing. *Nature, 555*(7698), 662–666.

Vann, S. D., and Nelson, A. J. (2015). The mammillary bodies and memory: More than a hippocampal relay. *Progress in Brain Research, 219,* 163–185.

Vargas, C. D., Aballéa, A., Rodrigues, E. C., Reilly, K. T., et al. (2009). Re-emergence of hand-muscle representations in human motor cortex after hand allograft. *Proceedings of the National Academy of Sciences, USA, 106,* 7197–7202.

Vargas, I. M., Schechtman, E., and Paller, K. A. (2019). Targeted memory reactivation during sleep to strengthen memory for arbitrary pairings. *Neuropsychologia, 124,* 144–150.

Vasey, P. L. (1995). Homosexual behaviour in primates: A review of evidence and theory. *International Journal of Primatology, 16,* 173–204.

Vassar, R., Ngai, J., and Axel, R. (1993). Spatial segregation of odourant receptor expression in the mammalian olfactory epithelium. *Cell, 74,* 309–318.

Vaughan, W., and Greene, S. L. (1984). Pigeon visual memory capacity. *Journal of Experimental Psychology: Animal Behaviour Processes, 10,* 256–271.

Velliste, M., Perel, S., Spalding, M. C., Whitford, A. S., et al. (2008). Cortical control of a prosthetic arm for self-feeding. *Nature, 453,* 1098–1101.

Vernes, S. C., Oliver, P. L., Spiteri, E., Lockstone, H. E., et al. (2011). Foxp2 regulates gene networks implicated in neurite outgrowth in the developing brain. *PLOS Genetics, 7*(7), e1002145. doi:10.1371/journal.pgen.1002145.

Vialou, V., Feng, J., Robison, A. J., Ku, S. M., et al. (2012). Serum response factor and cAMP response element binding protein are both required for cocaine induction of ΔFosB. *Journal of Neuroscience, 32,* 7577–7584.

Villalobos, M. E., Mizuno, A., Dahl, B. C., Kemmotsu, N., et al. (2005). Reduced functional connectivity between V1 and inferior frontal cortex associated with visuomotor performance in autism. *NeuroImage, 25,* 916–925.

Villringer, A., and Chance, B. (1997). Non-invasive optical spectroscopy and imaging of human brain function. *Trends in Neurosciences, 20,* 435–442.

Vincus, A. A., Ringwalt, C., Harris, M. S., and Shamblen, S. R. (2010). A short-term, quasi-experimental evaluation of D.A.R.E.'s revised elementary school curriculum. *Journal of Drug Education, 40*, 37–49.

Visser, S. N., Danielson, M. L., Bitsko, R. H., Holbrook, J. R., et al. (2014). Trends in the parent-report of health care provider-diagnosed and medicated attention-deficit/hyperactivity disorder: United States, 2003–2011. *Journal of the American Academy of Child and Adolescent Psychiatry, 53*(1), 34–46.e2.

Vita, A., De Peri, L., and Sacchetti, E. (2015). Lithium in drinking water and suicide prevention: A review of the evidence. *International Clinical Psychopharmacology, 30*, 1–5.

Vitali, P., Di Perri, C., Vaudano, A. E., Meletti, S., et al. (2015). Integration of multimodal neuroimaging methods: A rationale for clinical applications of simultaneous EEG-fMRI. *Functional Neurology, 30*(1), 9–20.

Vogt, B. A. (2005). Pain and emotion interactions in subregions of the cingulate gyrus. *Nature Reviews Neuroscience, 6*, 533–544.

Volkow, N. D., Wang, G. J., Kollins, S. H., Wigal, T. L., et al. (2009). Evaluating dopamine reward pathway in ADHD: Clinical implications. *JAMA, 302*, 1084–1091.

Volkow, N. D., Wang, G-J., Telang, F., Fowler, J. S., et al. (2006). Cocaine cues and dopamine in dorsal striatum: Mechanism of craving in cocaine addiction. *Journal of Neuroscience, 26*, 6583–6588.

Volkow, N. D., and Wise, R.A. (2005). How can drug addiction help us understand obesity? *Nature Neuroscience, 8*, 555–560.

Volkow, N. D., Wise, R. A., and Baler, R. (2017). The dopamine motive system: Implications for drug and food addiction. *Nature Reviews Neuroscience, 18*(12), 741–752.

Vonck, K., Raedt, R., Naulaerts, J., De Vogelaere, F., et al. (2014). Vagus nerve stimulation … 25 years later! What do we know about the effects on cognition? *Neuroscience and Biobehavioural Reviews*, 45, 63–71.

Vriens, J., Nilius, B., and Voets, T. (2014). Peripheral thermosensation in mammals. *Nature Reviews Neuroscience, 15*, 573–589.

Vrieze, A., Van Nood, E., Holleman, F., Salojärvi, J., et al. (2012). Transfer of intestinal microbiota from lean donors increases insulin sensitivity in individuals with metabolic syndrome. *Gastroenterology, 143*(4), 913–916. e7.

Vrontou, E., Nilsen, S. P., Demir, E., Kravitz, E. A., et al. (2006). *fruitless* regulates aggression and dominance in *Drosophila*. *Nature Neuroscience, 9*, 1469–1471.

Vyazovskiy, V. V., and Harris, K. D. (2013). Sleep and the single neuron: The role of global slow oscillations in individual cell rest. *Nature Reviews Neuroscience, 14*(6), 443–451.

Vyazovskiy, V. V., Olcese, U., Hanlon, E. C., Nir, Y., et al. (2011). Local sleep in awake rats. *Nature, 472*(7344), 443–447.

W

Wada, J. A., Clarke, R., and Hamm, A. (1975). Cerebral hemispheric asymmetry in humans. Cortical speech zones in 100 adults and 100 infant brains. *Archives of Neurology, 32*, 239–246.

Wada, J. A., and Rasmussen, T. (1960). Intracarotid injection of sodium amytal for the lateralisation of cerebral speech dominance: Experimental and clinical observations. *Journal of Neurosurgery, 17*, 266–282.

Wada, N., Hirako, S., Takenoya, F., Kageyama, H., et al. (2014). Leptin and its receptors. *Journal of Chemical Neuroanatomy, 61–62*, 191–199.

Wadman, M. (2018). Daring to hope. *Science, 361*(6404), 742–746.

Wagenmakers, E.-J., Beek, T., Dijkhoff, L., Gronau, Q. F., et al. (2016). Registered Replication Report: Strack, Martin, & Stepper (1988). *Perspectives on Psychological Science, 11*, 917–928.

Wager, T. D., and Atlas, L. Y. (2015). The neuroscience of placebo effects: Connecting context, learning and health. *Nature Reviews Neuroscience, 16*(7), 403–418.

Wager, T. D., Rilling, J. K., Smith, E. E., Sokolik, A., et al. (2004). Placebo-induced changes in fMRI in the anticipation and experience of pain. *Science, 303*, 1162–1167.

Wager, T. D., Scott, D. J., and Zubieta, J. K. (2007). Placebo effects on human mu-opioid activity during pain. *Proceedings of the National Academy of Sciences, USA, 104*, 11056–11061.

Wagner, A. D., Desmond, J. E., Demb, J. B., Glover, G. H., et al. (1997). Semantic repetition priming for verbal and pictorial knowledge: A functional MRI study of left inferior prefrontal cortex. *Journal of Cognitive Neuroscience, 9*, 714–726.

Wagner, A. D., Schacter, D. L., Rotte, M., Koutstaal, W., et al. (1998). Building memories: Remembering and forgetting of verbal experiences as predicted by brain activity. *Science, 281*, 1188–1191.

Wagner, F. B., Mignardot, J. B., Le Goff-Mignardot, C. G., Demesmaeker, R., et al. (2018). Targeted neurotechnology restores walking in humans with spinal cord injury. *Nature, 563*(7729), 65–71.

Wagner, G. C., Beuving, L. J., and Hutchinson, R. R. (1980). The effects of gonadal hormone manipulations on aggressive target-biting in mice. *Aggressive Behaviour, 6*, 1–7.

Wahl, A. S., Omlor, W., Rubio, J. C., Chen, J. L., et al. (2014). Neuronal repair: Asynchronous therapy restores motor control by rewiring of the rat corticospinal tract after stroke. *Science, 344*(6189), 1250–1255.

Wahlstrom, J. (2002). Changing times: Findings from the first longitudinal study of later high school start times. *National Association of Secondary School Principals Bulletin, 86*, 3–21.

Waldrop, M. M. (2012). Brain in a box. *Nature, 482*(7386), 456–458.

Walker, E. F. (1991). *Schizophrenia: A life-course developmental perspective*. San Diego, CA: Academic Press.

Wallis, J. D. (2007). Orbitofrontal cortex and its contribution to decision-making. *Annual Review of Neuroscience, 30*, 31–56.

Walsh, D. M., and Selkoe, D. J. (2016). A critical appraisal of the pathogenic protein spread hypothesis of neurodegeneration. *Nature Reviews Neuroscience, 17*(4), 251–260.

Walsh, E. J., Wang, L. M., Armstrong, D. L., Curro, T., et al. (2003). Acoustic communication in *Panthera tigris*: A study of tiger vocalisation and auditory receptivity. In Special Session on: Nature's orchestra: Acoustics of singing and calling animals: Production and reception of sound for communication by underwater and terrestrial animals. *Journal of the Acoustical Society of America, 113*, 2275.

Walton, C., Pariser, E., and Nottebohm, F. (2012). The zebra finch paradox: Song is little changed, but number of neurons doubles. *Journal of Neuroscience, 32*(3), 761–774.

Wang, C. X., Hilburn, I. A., Wu, D.-A., Mizuhara, Y., et al. (2019). Transduction of the geomagnetic field as evidenced from alpha-band activity in the human brain. *eNeuro*, 18 March. https://doi.org/10.1523/ENEURO.0483-18.2019.

Wang, F., Nemes, A., Mendelsohn, M., and Axel, R. (1998). Odourant receptors govern the formation of a precise topographic map. *Cell, 93*, 47–60.

Wang, H., Yang, H., Shivalila, C. S., Dawlaty, M. M., et al. (2013). One-step generation of mice carrying mutations in multiple genes by CRISPR/Cas-mediated genome engineering. *Cell, 153*(4), 910–918.

Wang, H., Yu, M., Ochani, M., Amella, C. A., et al. (2003). Nicotinic acetylcholine receptor α7 subunit is an essential regulator of inflammation. *Nature, 421*, 384–388.

Wang, J., Ishikawa, M., Yang, Y., Otaka, M., et al. (2018). Cascades of homeostatic dysregulation promote incubation of cocaine craving. *Journal of Neuroscience, 38*(18), 4316–4328.

Wang, S., Liu, B., Zhang, H., Dong, R., et al. (2013). Mandarin lexical tone recognition in sensorineural hearing-impaired listeners and cochlear implant users. *Acta Oto-Laryngologica, 133*(1), 47–54.

Wang, W. C., Lazzara, M. M., Ranganath, C., Knight, R. T., et al. (2010). The medial temporal lobe supports conceptual implicit memory. *Neuron, 68*(5), 835–842.

Wang, W. C., Ranganath, C., and Yonelinas, A. P. (2014). Activity reductions in perirhinal cortex predict conceptual priming and familiarity-based recognition. *Neuropsychologia, 52*, 19–26.

Wang, Y., and Mandelkow, E. (2016). Tau in physiology and pathology. *Nature Reviews Neuroscience, 17*(1), 5–21. doi:10.1038/nrn.2015.1.

Wang, Z., Ma, J., Miyoshi, C., Li, Y., et al. (2018). Quantitative phosphoproteomic analysis of the molecular substrates of sleep need. *Nature, 558*(7710), 435–439.

Warlow, S. M., Robinson, M. J. F., and Berridge, K. C. (2017). Optogenetic central amygdala stimulation intensifies and narrows motivation for cocaine. *Journal of Neuroscience, 37*(35), 8330–8348.

Warner, J., Her, M., Gmel, G., and Rehm, J. (2001). Can legislation prevent debauchery? Mother gin and public health in 18th-century England. *American Journal of Public Health, 91*(3), 375–384.

Watanabe, M., Cheng, K., Murayama, Y., Ueno, K., et al. (2011). Attention but not awareness modulates the BOLD signal in the human V1 during binocular suppression. *Science, 334*, 829–831.

Watson, N. V., Freeman, L. M., and Breedlove, S. M. (2001). Neuronal size in the spinal nucleus of the bulbocavernosus: Direct modulation by androgen in rats with mosaic androgen insensitivity. *Journal of Neuroscience, 21*, 1062–1066.

Waxman, S. G., and Zamponi, G. W. (2014). Regulating excitability of peripheral afferents: Emerging ion channel targets. *Nature Neuroscience, 17*(2), 153–163.

Weber, F., Chung, S., Beier, K. T., Xu, M., et al. (2015). Control of REM sleep by ventral medulla GABAergic neurons. *Nature, 526*(7573), 435–438.

Weber, F., and Dan, Y. (2016). Circuit-based interrogation of sleep control. *Nature, 538*(7623), 51–59.

Wei, W., Nguyen, L. N., Kessels, H. W., Hagiwara, H., et al. (2010). Amyloid beta from axons and dendrites reduces local spine number and plasticity. *Nature Neuroscience, 13*, 190–196.

Weigl, E. (1941). On the psychology of so-called processes of abstraction. *Journal of Abnormal and Social Psychology*, 36, 3–33.

Weinberger, D. R., Aloia, M. S., Goldberg, T. E., and Berman, K. F. (1994). The frontal lobes and schizophrenia. *Journal of Neuropsychiatry and Clinical Neurosciences, 6*, 419–427.

Weinberger, L. E., Sreenivasan, S., Garrick, T., and Osran, H. (2005). The impact of surgical castration on sexual recidivism risk among sexually violent predatory offenders. *Journal of the American Academy of Psychiatry and the Law, 33*(1), 16–36.

Weinberger, N. M. (1998). Physiological memory in primary auditory cortex: Characteristics and mechanisms. *Neurobiology of Learning and Memory, 70*, 226–251.

Weindruch, R., and Walford, R. L. (1988). *The retardation of aging and disease by dietary restriction.* Springfield, IL: Thomas.

Weiner, K. S., and Zilles, K. (2016). The anatomical and functional specialisation of the fusiform gyrus. *Neuropsychologia, 83*, 48–62.

Weiner, R. D. (1994). Treatment optimisation with ECT. *Psychopharmacology Bulletin, 30*, 313–320.

Weiss, J., Pyrski, M., Jacobi, E., Bufe, B., et al. (2011). Loss-of-function mutations in sodium channel Nav1.7 cause anosmia. *Nature, 472*, 186–190.

Weiss, M. W., and Bidelman, G. M. (2015). Listening to the brainstem: Musicianship enhances intelligibility of subcortical representations for speech. *Journal of Neuroscience, 35*(4), 1687–1691.

Weissman, T. A., and Pan, Y. A. (2015). Brainbow: New resources and emerging biological applications for multicolour genetic labeling and analysis. *Genetics, 199*(2), 293–306.

Weitlauf, A. S., McPheeters, M. L., Peters, B., Sathe, N., et al. (2014). *Therapies for children with autism spectrum disorder: Behavioural interventions update.* (Comparative Effectiveness Review No. 137. AHRQ Publication No. 14-EHC036-EF.) Rockville, MD: Agency for Healthcare Research and Quality. www.effectivehealthcare.ahrq.gov/reports/final.cfm.

Weitzman, E. D., Czeisler, C. A., Zimmerman, J. C., and Moore-Ede, M. C. (1981). Biological rhythms in man: Relationship of sleep-wake, cortisol, growth hormone, and temperature during temporal isolation. In J. B. Martin, S. Reichlin, and K. L. Bick (Eds.), *Neurosecretion and brain peptides* (pp. 475–499). New York: Raven.

Weng, Y. T., Chien, T., Kuan, I. I., and Chern, Y. (2018). The TRAX, DISC1, and GSK3 complex in mental disorders and therapeutic interventions. *Journal of Biomedical Science, 25*(1), 71.

Wersinger, E., and Fuchs, P. A. (2011). Modulation of hair cell efferents. *Hearing Research, 279*(1–2), 1–12.

Wesensten, N. J., Belenky, G., Kautz, M. A., Thorne, D. R., et al. (2002). Maintaining alertness and performance during sleep deprivation: Modafinil versus caffeine. *Psychopharmacology (Berlin), 159*, 238–247.

Wessberg, J., Stambaugh, C. R., Kralik, J. D., Beck, P. D., et al. (2000). Real-time predictions of hand trajectory by ensembles of cortical neurons in primates. *Nature, 408*, 361–365.

Wessel, J. R., and Aron, A. R. (2017). On the globality of motor suppression: Unexpected events and their influence on behaviour and cognition. *Neuron, 93*(2), 259–280.

West, G. L., Drisdelle, B. L., Konishi, K., Jackson, J., et al. (2015). Habitual action video game playing is associated with caudate nucleus-dependent navigational strategies. *Proceedings: Biological Sciences, 282*(1808), 20142952.

West, S. L., and O'Neal, K. K. (2004). Project D.A.R.E. outcome effectiveness revisited. *American Journal of Public Health., 94*, 1027–1029.

Wever, R. A. (1979). Influence of physical workload on freerunning circadian rhythms of man. *Pflügers Archiv European Journal of Physiology, 381*, 119–126.

Wexler, N. S., Collett, L., Wexler, A. R., Rawlins, M. D., et al. (2016). Incidence of adult Huntington's disease in the UK: A UK-based primary care study and a systematic review. *BMJ Open, 6*(2), e009070.

Whitaker, D., and McGraw, P. V. (2000). Long-term visual experience recalibrates human orientation perception. *Nature Neuroscience, 3*, 13.

White, C. M. (2016). Mephedrone and 3,4-methylenedioxypyrovalerone (MDPV): Synthetic cathinones with serious health implications. *Journal of Clinical Pharmacology, 56*(11), 1319–1325.

White, N. M., and Milner, P. M. (1992). The psychobiology of reinforcers. *Annual Review of Psychology, 43*, 443–471.

Whitfield-Gabrieli, S., and Ford, J. M. (2012). Default mode network activity and connectivity in psychopathology. *Annual Review of Clinical Psychology, 8*, 49–76.

Whitlock, J. R., Heynen, A. J., Shuler, M. G., and Bear, M. F. (2006). Learning induces long-term potentiation in the hippocampus. *Science, 313*, 1093–1097.

Wichmann, S., Kirschbaum, C., Böhme, C., and Petrowski, K. (2017). Cortisol stress response in post-traumatic stress disorder, panic disorder, and major depressive disorder patients. *Psychoneuroendocrinology, 83*, 135–141.

Wiegert, J. S., Mahn, M., Prigge, M., Printz, Y., et al. (2017). Silencing neurons: Tools, applications, and experimental constraints. *Neuron, 95*(3), 504–529.

Wiesel, T. N., and Hubel, D. H. (1965). Extent of recovery from the effects of visual deprivation in kittens. *Journal of Neurophysiology, 28*, 1060–1072.

Wildman, D. E., Uddin, M., Liu, G., Grossman, L. I., et al. (2003). Implications of natural selection in shaping 99.4% nonsynonymous DNA identity between humans and chimpanzees: Enlarging genus *Homo*. *Proceedings of the National Academy of Sciences, USA, 100*, 7181–7188.

Willemet, R. (2013). Reconsidering the evolution of brain, cognition, and behaviour in birds and mammals. *Frontiers in Psychology, 4*, 396. https://doi.org/10.3389/fpsyg.2013.00396.

Williams, J. H., Waiter, G. D., Gilchrist, A., Perrett, D. I., et al. (2006). Neural mechanisms of imitation and "mirror neuron" functioning in autistic spectrum disorder. *Neuropsychologia, 44*, 610–621.

Williams, N. R., and Schatzberg, A. F. (2016). NMDA antagonist treatment of depression. *Current Opinion in Neurobiology, 36*, 112–117. doi:10.1016/j.conb.2015.11.001.

Williams, T. J., Pepitone, M. E., Christensen, S. E., Cooke, B. M., et al. (2000). Finger-length ratios and sexual orientation. *Nature, 404*, 455–456.

Williams, Z. M., Bush, G., Rauch, S. I., Cosgrove, G. R., et al. (2004). Human anterior cingulate neurons and the integration of monetary reward with motor responses. *Nature Neuroscience, 7*, 1370–1374.

Wilson, M. L., Boesch, C., Fruth, B., Furuichi, T., et al. (2014). Lethal aggression in *Pan* is better explained by adaptive strategies than human impacts. *Nature, 513*, 414–417.

Wilson, M. L., Miller, C. M., and Crouse, K. N. (2017). Humans as a model species for sexual selection research. *Proceedings: Biological Sciences, 284*(1866), pii: 20171320.

Wilson, S., Malone, S. M., Thomas, K. M., and Iacono, W. G. (2015). Adolescent drinking and brain morphometry: A co-twin control analysis. *Developmental Cognitive Neuroscience, 16*, 130–138.

Wilson, S. R., Gerhold, K. A., Bifolck-Fisher, A., Liu, Q., et al. (2011). TRPA1 is required for histamine-independent, Mas-related G protein–coupled receptor–mediated itch. *Nature Neuroscience, 14*, 595–602.

Winblad, B., Amouyel, P., Andrieu, S., Ballard, C., et al. (2016). Defeating Alzheimer's disease and other dementias: A priority for European science and society. *Lancet Neurology, 15*(5), 455–532. https://doi.org/10.1016/S1474-4422(16)00062-4.

Winslow, J. T., and Insel, T. R. (2002). The social deficits of the oxytocin knockout mouse. *Neuropeptides, 36*, 221–229.

Wise, N. J., Frangos, E., and Komisaruk, B. R. (2017). Brain activity unique to orgasm in women: An fMRI analysis. *Journal of Sexual Medicine, 14*(11), 1380–1391.

Wittchen, H. U., Jacobi, F., Rehm, J., Gustavsson, A., et al. (2011). The size and burden of mental disorders and other disorders of the brain in Europe 2010. *European Neuropsychopharmacology, 21*(9), 655–679.

Witte, A.V., Fobker, M., Gellner, R., Knecht, S., et al. (2009). Caloric restriction improves memory in elderly humans. *Proceedings of the National Academy of Sciences, USA, 106*(4), 1255–1260.

Witthoft, N., and Winawer, J. (2013). Learning, memory, and synesthesia. *Psychological Science, 24*(3), 258–265.

Wolf, M. E. (2016). Synaptic mechanisms underlying persistent cocaine craving. *Nature Reviews Neuroscience, 17*(6), 351–365.

Wolf, S. A., Boddeke, H. W., and Kettenmann, H. (2017). Microglia in physiology and disease. *Annual Review of Physiology, 79*, 619–643.

Wolf, S. S., Jones, D. W., Knable, M. B., Gorey, J. G., et al. (1996). Tourette syndrome: Prediction of phenotypic variation in monozygotic twins by caudate nucleus D2 receptor binding. *Science, 273*, 1225–1227.

Wolfe, J. M. (1994). Guided search 2.0: A revised model of visual search. *Psychonomic Bulletin & Review, 1*, 202–238.

Wolfe, J. M., Alvarez, G. A., and Horowitz, T. S. (2000). Attention is fast but volition is slow. *Nature, 406*, 691.

Wolfe, J. M., Horowitz, T. S., and Kenner, N. M. (2005). Cognitive psychology: Rare items often missed in visual searches. *Nature, 435*, 439–440.

Wolk, D. A., Price, J. C., Saxton, J. A., Snitz, B. E., et al. (2009). Amyloid imaging in mild cognitive impairment subtypes. *Annals of Neurology, 65*, 557–568.

Womelsdorf, T., Anton-Erxleben, K., Pieper, F., and Treue, S. (2006). Dynamic shifts of visual receptive fields in cortical area MT by spatial attention. *Nature Neuroscience, 9*, 1156–1160.

Womelsdorf, T., Anton-Erxleben, K., and Treue, S. 2008. Receptive field shift and shrinkage in macaque middle temporal area through attentional gain modulation. *Journal of Neuroscience, 28*, 8934–8944.

Wong, M., and Martin, L. J. (2010). Skeletal muscle-restricted expression of human SOD1 causes motor neuron degeneration in transgenic mice. *Human Molecular Genetics, 19*, 2284–2302.

Woo, S. H., Lukacs, V., de Nooij, J. C., Zaytseva, D., et al. (2015). Piezo2 is the principal mechanotransduction channel for proprioception. *Nature Neuroscience, 18*(12), 1756–1762.

Wood, N., and Cowan, N. (1995). The cocktail party phenomenon revisited: How frequent are attention shifts to one's name in an irrelevant auditory channel? *Journal of Experimental Psychology. Learning, Memory, and Cognition, 21*, 255–260.

World Health Organisation. (2001). *The world health report.* Geneva, Switzerland: World Health Organisation.

World Health Organisation. (2004). *Burden of disease in DALYs by cause, sex, and mortality stratum in WHO regions, estimates for 2002.* Geneva, Switzerland: World Health Organisation.

Wrangham, R. W. (2018). Two types of aggression in human evolution. *Proceedings of the National Academy of Sciences, USA, 115*(2), 245–253.

Wren, A. M., Seal, L. J., Cohen, M. A., Brynes, A. E., et al. (2001). Ghrelin enhances appetite and increases food intake in humans. *Journal of Clinical Endocrinology & Metabolism, 86*, 5992–5995.

Wren, A. M., Small, C. J., Ward, H. L., Murphy, K. G., et al. (2000). The novel hypothalamic peptide ghrelin stimulates food intake and growth hormone secretion. *Endocrinology, 141*, 4325–4328.

Wright, A. A., Magnotti, J. F., Katz, J. S., Leonard, K., et al. (2017). Corvids outperform pigeons and primates in learning a basic concept. *Psychological Science, 28*(4), 437–444.

Wu, P. C., Tsai, C. L., Wu, H. L., Yang, Y. H., et al. (2013). Outdoor activity during class recess reduces myopia onset and progression in school children. *Ophthalmology, 120*, 1080–1085.

Wu, Q., Qi, C., Long, J., Liao, Y., et al. (2018). Metabolites alterations in the medial prefrontal cortex of methamphetamine users in abstinence: A ^{1}H MRS study. *Frontiers in Psychiatry, 9*, 478.

Wu, S.V., Rozengurt, N., Yang, M., Young, S. H., et al. (2002). Expression of bitter taste receptors of the T2R family in the gastrointestinal tract and enteroendocrine STC-1 cells. *Proceedings of the National Academy of Sciences, USA, 99*(4), 2392–2397.

Wuethrich, B. (2000). Learning the world's languages—before they vanish. *Science, 288*, 1156–1159.

Wunderink, L., Nieboer, R. M., Wiersma, D., Sytema, S., et al. (2013). Recovery in remitted first-episode psychosis at 7 years of follow-up of an early dose reduction/discontinuation or maintenance treatment strategy: Long-term follow-up of a 2-year randomized clinical trial. *JAMA Psychiatry, 70*(9), 913–920.

Wurtz, R. H., and Goldberg, M. E. (1972). Activity of superior colliculus in behaving monkey. 3. Cells discharging before eye movements. *Journal of Neurophysiology, 35*, 575–586.

Wurtz, R. H., Goldberg, M. E., and Robinson, D. L. (1982). Brain mechanisms of visual attention. *Scientific American, 246*(6), 124–135.

X

Xerri, C., Coq, J., Merzenich, M., and Jenkins, W. (1996). Experience-induced plasticity of cutaneous maps in the primary somatosensory cortex of adult monkeys and rats. *Journal de Physiologie, 90*, 277–287.

Xerri, C., Stern J. M., and Merzenich, M. M. (1994). Alterations of the cortical representation of the rat ventrum induced by nursing behaviour. *Journal of Neuroscience, 14*, 1710–1721.

Xie, L., Kang, H., Xu, Q., Chen, M. J., et al. (2013). Sleep drives metabolite clearance from the adult brain. *Science, 342*(6156), 373–377.

Xu, H., Delling, M., Jun, J. C., and Clapham, D. E. (2006). Oregano, thyme and clove-derived flavours and skin sensitizers activate specific TRP channels. *Nature Neuroscience, 9*, 628–635.

Xu, J., Bartolome, C. L., Low, C. S., Yi, X., et al. (2018). Genetic identification of leptin neural circuits in energy and glucose homeostases. *Nature, 556*(7702), 505–509.

Xu, J., Cao, J., Iguchi, N., Riethmacher, D., et al. (2013). Functional characterisation of bitter-taste receptors expressed in mammalian testis. *Molecular Human Reproduction, 19*(1), 17–28.

Xu, L., Furukawa, S., and Middlebrooks, J. C. (1999). Auditory cortical responses in the cat to sounds that produce spatial illusions. *Nature, 399*, 688–691.

Xu, M., Chung, S., Zhang, S., Zhong, P., et al. (2015). Basal forebrain circuit for sleep-wake control. *Nature Neuroscience, 18*(11), 1641–1647.

Xue, T., Do, M. T. H., Riccio, A., Jiang, Z., et al. (2011). Melanopsin signalling in mammalian iris and retina. *Nature, 479*, 67–73.

Xue, Y., Xu, X., Zhang, X. Q., Farokhzad, O. C., et al. (2016). Preventing diet-induced obesity in mice by adipose tissue transformation and angiogenesis using targeted nanoparticles. *Proceedings of the National Academy of Sciences, USA, 113*, 5552–5557.

Y

Yaffe, K., Lui, L.Y., Zmuda, J., and Cauley, J. (2002). Sex hormones and cognitive function in older men. *Journal of the American Geriatrics Society, 50*, 707–712.

Yanagisawa, K., Bartoshuk, L. M., Catalanotto, F. A., Karrer, T. A., et al. (1992). Anesthesia of the chorda tympani nerve: Insights into a source of dysgeusia. *Chemical Senses, 17*, 724.

Yang, T. T., Gallen, C. C., Ramachandran, V. S., Cobb, S., et al. (1994). Noninvasive detection of cerebral plasticity in adult human somatosensory cortex. *Neuroreport, 5,* 701–704.

Yang, Y., Choi, P. P., Smith, W. W., Xu, W., et al. (2017). Exendin-4 reduces food intake via the P13K/AKT pathway in the hypothalamus. *Scientific Reports, 7,* 6936.

Yang, Y., and Raine, A. (2009). Prefrontal structural and functional brain imaging findings in antisocial, violent, and psychopathic individuals: A meta-analysis. *Psychiatry Research, 174,* 81–88.

Yang, Y., Raine, A., Joshi, A. A., Joshi, S., et al. (2012). Frontal information flow and connectivity in psychopathy. *British Journal of Psychiatry, 201*(5), 408–409.

Yang, Z., Slavin, M. J., and Sachdev, P. S. (2013). Dementia in the oldest old. *Nature Reviews Neurology, 9*(7), 382–393. doi:10.1038/nrneurol.2013.105.

Yao, K., Qiu, S., Wang, Y. V., Park, S. J. H., et al. (2018). Restoration of vision after de novo genesis of rod photoreceptors in mammalian retinas. *Nature, 560*(7719), 484–488.

Ycaza Herrera, A., and Mather, M. (2015). Actions and interactions of estradiol and glucocorticoids in cognition and the brain: Implications for aging women. *Neuroscience & Biobehavioural Reviews, 55,* 36–52.

Yetish, G., Kaplan, H., Gurven, M., Wood, B., et al. (2015). Natural sleep and its seasonal variations in three pre-industrial societies. *Current Biology, 25*(21), 2862–2868.

Yin, L., Wang, J., Klein, P. S., and Lazar, M. A. (2006). Nuclear receptor Rev-erbalpha is a critical lithium-sensitive component of the circadian clock. *Science, 311*(5763), 1002–1005.

Yong, E. (2019, July 22). The human brain project has not lived up to its promise. *The Atlantic* (https://www.theatlantic.com/science/archive/2019/07/ten-years-human-brain-project-simulation-markram-ted-talk/594493/).

Yoon, Y. J., Steele, C. R., and Puria, S. (2011). Feed-forward and feed-backward amplification model from cochlear cytoarchitecture: An interspecies comparison. *Biophysical Journal, 100*(1), 1–10.

Young, K. D., Erickson, K., Nugent, A. C., Fromm, S. J., et al. (2011). Functional anatomy of autobiographical memory recall deficits in depression. *Psychological Medicine, 29,* 1–13.

Yttri, E. A., and Dudman, J. T. (2016). Opponent and bidirectional control of movement velocity in the basal ganglia. *Nature, 533*(7603), 402–406.

Yu, F., Jiang, Q. J., Sun, X. Y., and Zhang, R. W. (2015). A new case of complete primary cerebellar agenesis: Clinical and imaging findings in a living patient. *Brain, 138*(Pt. 6), e353.

Yu, Y. Q., Barry, D. M., Hao, Y., Liu, X. T., et al. (2017). Molecular and neural basis of contagious itch behaviour in mice. *Science, 355*(6329), 1072–1076.

Yuan, M., Cross S. J., Loughlin, S. E., and Leslie, F. M. (2015). Nicotine and the adolescent brain. *Journal of Physiology, 593*(16), 3397–3412.

Yuan, P., and Raz, N. (2014). Prefrontal cortex and executive functions in healthy adults: A meta-analysis of structural neuroimaging studies. *Neuroscience and Biobehavioural Reviews, 42,* 180–192.

Yuste, R. and Church, G. M. (2014). The new century of the brain. *Scientific American, 310*(3), 38–45.

Z

Zabeau, L., Wauman, J., Dam, J., Van Lint, S., et al. (2019). A novel leptin receptor antagonist uncouples leptin's metabolic and immune functions. *Cellular and Molecular Life Sciences, 76*(6), 1201–1214.

Zadikoff, C., and Lang, A. E. (2005). Apraxia in movement disorders. *Brain, 128,* 1480–1497.

Zahr, N. M., Mayer, D., Rohlfing, T., Hasak, M. P., et al. (2009). Brain injury and recovery following binge ethanol: Evidence from in vivo magnetic resonance spectroscopy. *Biological Psychiatry, 67*(9), 846–854.

Zahr, N. M., and Pfefferbaum, A. (2017). Alcohol's effects on the brain: Neuroimaging results in humans and animal models. *Alcohol Research, 38*(2), 183–206.

Zaidel, E. (1976). Auditory vocabulary of the right hemisphere following brain bisection or hemidecortication. *Cortex, 12,* 191–211.

Zalocusky, K. A., Ramakrishnan, C., Lerner, T. N., Davidson, T. J., et al. (2016). Nucleus accumbens D2R cells signal prior outcomes and control risky decision-making. *Nature, 531*(7596), 642–646.

Zanchi, D., Depoorter, A., Egloff, L., Haller, S., et al. (2017). The impact of gut hormones on the neural circuit of appetite and satiety: A systematic review. *Neuroscience & Biobehavioural Reviews, 80,* 457–475.

Zanto, T. P., Rubens, M. T., Thangavel, A., and Gazzaley, A. (2011). Causal role of the prefrontal cortex in top-down modulation of visual processing and working memory. *Nature Neuroscience, 14*(5), 656–661.

Zatorre, R. J., Evans, A. C., and Meyer, E. (1994). Neural mechanisms underlying melodic perception and memory for pitch. *Journal of Neuroscience, 14,* 1908–1919.

Zatorre, R. J., Fields, R. D., and Johansen-Berg, H. (2012). Plasticity in gray and white: Neuroimaging changes in brain structure during learning. *Nature Neuroscience, 15,* 528–536.

Zawilska, J. B. (2014). Mephedrone and other cathinones. *Current Opinion in Psychiatry, 27,* 256–262.

Zeier, Z., Carpenter, L. L., Kalin, N. H., Rodriguez, C. I., et al. (2018). Clinical implementation of pharmacogenetic decision support tools for antidepressant drug prescribing. *American Journal of Psychiatry, 175*(9), 873–886.

Zeki, S. (2015). Area V5: A microcosm of the visual brain. *Frontiers in Integrative Neuroscience, 9,* 21.

Zeki, S., Watson, J. D., Lueck, C. J., Friston, K. J., et al. (1991). A direct demonstration of functional specialisation in human visual cortex. *Journal of Neuroscience, 11,* 641–649.

Zeman, A. (2002). *Consciousness: A user's guide.* New Haven, CT: Yale University Press.

Zendel, B. R., and Alain, C. (2009). Concurrent sound segregation is enhanced in musicians. *Journal of Cognitive Neuroscience, 21,* 1488–1498.

Zendel, B. R., Lagrois, M. É., Robitaille, N., and Peretz, I. (2015). Attending to pitch information inhibits processing of pitch information: The curious case of amusia. *Journal of Neuroscience, 35*(9), 3815–3824.

Zenner, H. P., Delb, W., Kröner-Herwig, B., Jäger, B., et al. (2017). A multidisciplinary systematic review of the treatment for chronic idiopathic tinnitus. *European Annals of Otorhinolaryngology, 274*(5), 2079–2091.

Zhang, C. L., Zou, Y., He, W., Gage, F. H., et al. (2008). A role for adult TLX-positive neural stem cells in learning and behaviour. *Nature, 451,* 1004–1007.

Zhang, F., Aravanis, A. M., Adamantidis, A., deLecea, L., et al. (2007). Circuit-breakers: Optical technologies for probing neural signals and systems. *Nature Reviews Neuroscience, 8,* 577–581.

Zhang, J., Fiers, P., Witte, K. A., Jackson, R. W., et al. (2017). Human-in-the-loop optimisation of exoskeleton assistance during walking. *Science, 356*(6344), 1280–1284.

Zhang, T. Y., and Meaney, M. J. (2010). Epigenetics and the environmental regulation of the genome and its function. *Annual Review of Psychology, 61,* C1–C3.

Zhang, Y., Liu, R.-Y., Heberton, G. A., Smolen, P., et al. (2012). Computational design of enhanced learning protocols. *Nature Neuroscience, 15,* 294–297.

Zhang, Y., Proenca, R., Maffei, M., Barone, M., et al. (1994). Positional cloning of the mouse obese gene and its human homologue. *Nature, 372,* 425–432.

Zhang, Z., and Bourque, C. W. (2003). Osmometry in osmosensory neurons. *Nature Neuroscience, 6,* 1021–1022.

Zhao, G. Q., Zhang, Y., Hoon, M. A., Chandrashekar, J., et al. (2003). The receptors for mammalian sweet and umami taste. *Cell, 115,* 255–266.

Zhao, H., Yang, J. R., Xu, H., and Zhang, J. (2010). Pseudogenisation of the umami taste receptor gene *Tas1r1* in the giant panda coincided with its dietary switch to bamboo. *Molecular Biology and Evolution, 27*(12), 2669–2673.

Zhao, Z. Q. (2008). Neural mechanism underlying acupunture analgesia. *Progress in Neurobiology, 85,* 355–375.

Zhou, Q., Lai, Y., Bacaj, T., Zhao, M., et al. (2015). Architecture of the synaptotagmin-SNARE machinery for neuronal exocytosis. *Nature, 525*(7567), 62–67.

Zhou, X., Sen, I., Lin, X. X., and Riedel, C. G. (2018). Regulation of age-related decline by transcription factors and their crosstalk with the epigenome. *Current Genomics, 19*(6), 464–482.

Ziauddeen, H., Farooqi, I. S., and Fletcher, P. C. (2012). Obesity and the brain: How convincing is the addiction model? *Nature Reviews Neuroscience, 13*(4), 279–286.

Zihl, J., von Cramon, D., and Mai, N. (1983). Selective disturbance of movement vision after bilateral brain damage. *Brain, 106,* 313–340.

Zimmer, C. (2004). *The soul made flesh: The discovery of the brain—and how it changed the world.* New York: Basic Books.

Zimmerman, C. A., Huey, E. L., Ahn, J. S., Beutler, L. R., et al. (2019). A gut-to-brain signal of fluid osmolarity controls thirst satiation. *Nature, 568*(7750), 98–102.

Zimmerman, C. A., Leib, D. E., and Knight, Z. A. (2017). Neural circuits underlying thirst and fluid homeostasis. *Nature Reviews Neuroscience, 18*(8), 459–469.

Zimmerman, C. A., Lin, Y. C., Leib, D. E., Guo, L., et al. (2016). Thirst neurons anticipate the homeostatic consequences of eating and drinking. *Nature, 537*(7622), 680–684.

Zimmerman, F., and Jørgensen, C. (2017). Taking animal breeding into the wild: Regulation of fishing gear can make fish stocks evolve higher productivity. *Marine Ecology Progress Series*, 563, 185–195.

Zocchi, D., Wennemuth, G., and Oka, Y. (2017). The cellular mechanism for water detection in the mammalian taste system. *Nature Neuroscience, 20*(7), 927–933.

Zola, S. M., Squire, L. R., Teng, E., Stefanacci, L., et al. (2000). Impaired recognition memory in monkeys after damage limited to the hippocampal region. *Journal of Neuroscience, 20*, 451–463.

Zola-Morgan, S. [M.], and Squire, L. R. (1986). Memory impairment in monkeys following lesions of the hippocampus. *Behavioural Neuroscience, 100*, 155–160.

Zola-Morgan, S. M., and Squire, L. R. (1990). The primate hippocampal formation: Evidence for a time-limited role in memory storage. *Science, 250*, 288–290.

Zola-Morgan, S. [M.], Squire, L. R., and Ramus, S. J. (1994). Severity of memory impairment in monkeys as a function of locus and extent of damage within the medial temporal lobe memory system. *Hippocampus, 4*, 483–495.

Zonooz, B., Arani, E., Körding, K. P., Aalbers, P. A. T. R., et al. (2019). Spectral weighting underlies perceived sound elevation. *Scientific Reports, 9*(1), 1642.

Zorumski, C. F., Izumi, Y., and Mennerick, S. (2016). Ketamine: NMDA receptors and beyond. *Journal of Neuroscience, 36*(44), 11158–11164.

Zucker, I. (1976). Light, behaviour, and biologic rhythms. *Hospital Practice, 11*, 83–91.

Zucker, I. (1988). Seasonal affective disorders: Animal models non fingo. *Journal of Biological Rhythms, 3*, 209–223.

Zucker, I., Boshes, M., and Dark, J. (1983). Suprachiasmatic nuclei influence circannual and circadian rhythms of ground squirrels. *American Journal of Physiology, 244*, R472–R480.

Author Index

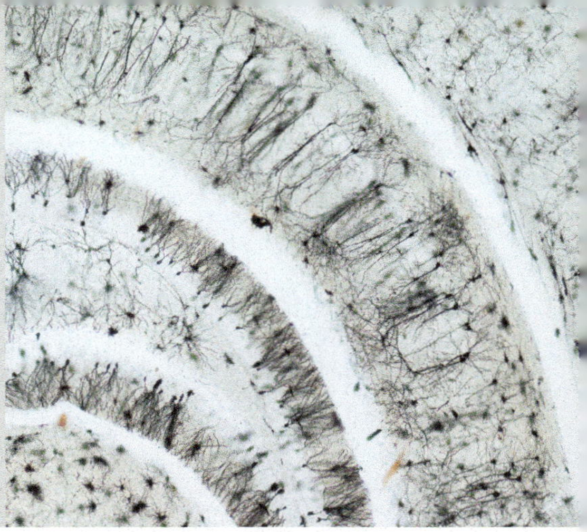

Courtesy of Dr. Sarah Moghadam, VA Palo Alto Health Care System, Palo Alto, CA and Dr. Ahmad Salehi, Dept. of Psychiatry & Behavioural Sciences, Stanford Medical School

C

D

S

X

Y

Z

Subject Index

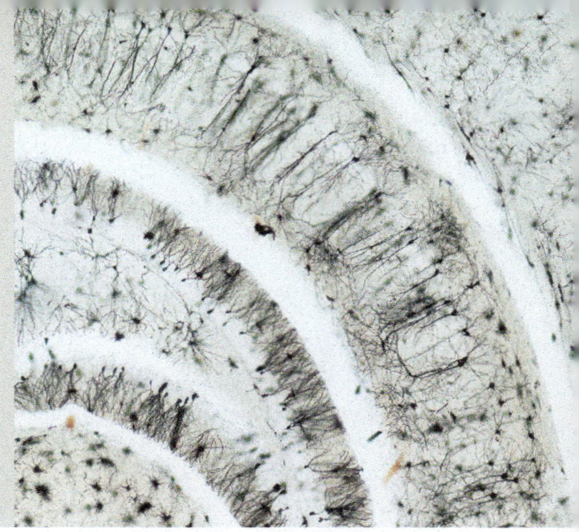

Courtesy of Dr. Sarah Moghadam, VA Palo Alto Health Care System, Palo Alto, CA and Dr. Ahmad Salehi, Dept. of Psychiatry & Behavioural Sciences, Stanford Medical School

B

C

E

F

G

I

J

K

P

S

T